CONTENTS:
1 v.
Computer optical disc (1)

The Physical Review

The First Hundred Years

The Physical Review

The First Hundred Years

A Selection of Seminal Papers
and Commentaries

Edited by

H. Henry Stroke

Department of Physics
New York University
New York, New York

American Institute of Physics — New York
The American Physical Society — Maryland

AIP Press
American Institute of Physics
500 Sunnyside Boulevard
Woodbury, NY 11797-2999

Library of Congress Cataloging-in-Publication Data
The physical review--the first hundred years : a selection of seminal papers and commentaries / edited by H. Henry Stroke.
p. cm.
Includes bibliographical references and indexes.
ISBN 1-56396-188-1
1. Physics. 2. Physical review. I. Stroke, H. Henry.
II. Physical review.
QC7.5.P48 1995 94-49713
530--dc20 CIP

10 9 8 7 6 5 4 3 2

The Physical Review—The First Hundred Years
is a selection of seminal papers
and commentaries highlighting the
developments in physics and their applications
presented in printed and electronic form.

The publication of this collection
is sponsored jointly by
The American Physical Society and the
American Institute of Physics
in celebration of the 100th anniversary of
The Physical Review.

CONTENTS

CONTENTS

FOREWORD

The American Institute of Physics (AIP) and The American Physical Society (APS) are proud to join in producing this centennial collection of noteworthy articles from *The Physical Review* and *Physical Review Letters* starting with the first issue of *The Physical Review* in 1893. Over the past hundred years, *The Physical Review* has been the place where many seminal ideas in physics first appeared and fired the imagination of physicists (as well as the broader scientific community) and often sparked entirely new fields of research. During this time, *The Physical Review* and *Physical Review Letters* became the most important physics journals in the world.

The origins in time and place of *The Physical Review* and The APS were intertwined and, about one-third of a century later, AIP was created to produce the journals of The APS and other physics societies. It is in the last two-thirds of the century that *The Physical Review* became the premier physics publication, its growth (in both pages and influence) reflecting the growth of the worldwide physics community.

As readers, authors, and referees (Ben Bederson points out in his Preface that we are all often one and the same), we have a long and abiding respect for *The Physical Review*. As undergraduates, we recall the pride in finding "Phys Rev" articles that we could begin to understand. "Phys Rev" articles became an intimate part of our education as graduate students; then, later, as with many physicists, *The Physical Review* contained most of the articles that were at the heart of our research careers. As we look over the articles reproduced here, the creative and interpretive wonder of our profession is laid out in a way that brings together the remarkable success of our collective endeavors and reflects the enduring significance of the research. We hope this compilation will serve to remind the scientific community how fundamental and pioneering these achievements have been in both intellectual and truly practical terms.

AIP and The APS look forward to continuing in the years ahead to serve *The Physical Review* and its creative authors, readers, referees, and editors as they document the advances of physics and the contributions of science to society.

Marc H. Brodsky
Executive Director, American Institute of Physics

Judy R. Franz
Executive Officer, The American Physical Society

PREFACE

This volume represents the first attempt by The American Physical Society to compose a collection from its distinguished heritage. Many physicists have agonized long hours in an effort to pare down the list of nominations for inclusion to a manageable size. Technology is coming to our rescue, at least in a limited way, by permitting expansion of the small number of articles that could be included in a book into a CD-ROM, which holds far more than the largest book of which we could possibly conceive.

The two people who played crucial roles in seeing this project to completion are H. Henry Stroke and Maria Taylor. Henry Stroke, professor of physics at New York University, who agreed in the first place to take on this project, shepherded it through all the stages that have culminated in what you are seeing here. Maria Taylor, Publisher of AIP Press at the American Institute of Physics, supplied much needed encouragement and continued interest that helped keep the project on track and moving forward. I also acknowledge the important contribution of Harry Lustig, Treasurer of The American Physical Society. Dr. Lustig was an early believer in this project. His support in supplying APS with resources, both moral as well as financial, was instrumental in seeing this project through to a successful completion.

We can all sympathize with Professor Stroke and his colleagues in undertaking this task. I am sure each of us has favorites that he or she might be chagrined to find omitted from this collection. I myself can single out just one example to illustrate the magnitude of the task. I am referring to the series of three articles by Lamb and Retherford published in *The Physical Review* in 1948 describing the now-renowned "Lamb shift" experiment. These articles not only described a masterpiece of scientific work—they also represented scientific *writing* of a very high order. One could read these articles almost as though they were describing a travel adventure. Lamb and Retherford took it upon themselves to convey to the reader the essence of scientific research—why the experiment was initiated in the first place, what unknown techniques had to be developed before the experiment could even have a chance of succeeding, what initial false steps turned into blind alleys—we do not see these described very much in today's scientific literature—how eventually the right paths were taken, then the triumph of observing the hoped-for signal, and finally the efforts to obtain as precise and as reliable numbers as possible. Along the way they pioneered techniques that later became virtual industries, e.g., the use of thermal dissociation to produce atomic hydrogen beams, the mention, almost in passing, of atomic recoil (in exciting the $n = 2$ states—a technique that I myself have fruitfully exploited in later years), the detection of metastables from cold surfaces, to name a few. To have included all three articles would have consumed no less than 5% of the full volume! Present day referees, abetted by us editors to encourage lean writing—a necessity in view of today's sheer volume of physics research—no longer favor such leisurely writing. There is a glimmer of hope on the horizon, however. When we fully enter the age of electronic publishing, where page counts no longer serve the constraint they do now, we may once again be able to permit scientific writing to expand to meet specific needs.

PREFACE

The editors, authors, readers, and referees (often all embodied in the same person) of *The Physical Review* (PR) and *Physical Review Letters* (PRL) in its various forms can look back with pride on the contributions our journals have made to the progress of science during this century of magnificent scientific achievement. I do not need to belabor the scientific content of this work. Not only does it speak for itself, but the several short essays prepared by our "Board of Selectors" eloquently describe their own perceptions; these present both a mosaic and a unified picture—as contradictory as this seems—of *The Physical Review/Physical Review Letters* past, and require no further elaboration by me. This content is, of course, what we are really celebrating. Our journals in the main are nothing more, or less, than the combined efforts of our past and present colleagues, Americans mainly in the early days but recently more and more representing international physics (well over 50% of our 1993–1994 submissions are from non-U. S. sources).

But there is more to the history of our journals than their scientific content. Its visionary editors, especially in the early days, were important in laying the groundwork for our journals' ultimate reputations, but they also themselves fostered American science by providing a friendly home for our domestic physicists. This project is not intended to be a history of *The Physical Review*; fortunately we do now have one, particularly, though not exclusively, concentrating on its early days at Cornell University. Professor Paul Hartman of Cornell has written a small volume called *A Memoir on The Physical Review—A History of the First Hundred Years*. This book has been published jointly by The American Physical Society and the American Institute of Physics, with the help of a generous grant from the Cornell University Physics Department.

As with so many other advances in modern science, it would have been virtually impossible to predict, from the vantage point of Cornell University of 1893, what has happened to our journals over the years. In 1994 PR/PRL produced about 97,000 printed pages, representing what has been estimated to constitute about one-fourth of all the physics literature published in the world. From a staff of one editor and a secretary (1893), we now have a full-time editorial professional staff of over 20 people at the editorial offices in Ridge, New York, about 15 "remote," i.e., part-time editors located mainly at universities throughout the country, assisted by another 100 or so office service people at Ridge and 20 more at Woodbury, New York, where The American Physical Society publishes *The Physical Review* through the American Institute of Physics. Apart from *Physical Review Letters*—a story in itself—there are now five Physical Reviews separately and independently edited. These are perhaps not very imaginatively called Physical Review A through E, but among them they encompass virtually all the subfields of physics. The subtitles of these journals are A: Atomic, Molecular, and Optical Physics; B: Condensed Matter; C: Nuclear Physics; D: Particles and Fields; and E: Statistical Physics, Plasmas, Fluids, and Related Many-Body Physics. It is worth emphasizing what is one of the unique features of our so-called "independently" edited journals. Because all full-time editors of all our journals are physically located at one site, an editorial coherence and consistency is obtained that would not have occurred otherwise. Editors of different journals can easily consult with each other, and lend their own expertise to each other as needed. This, we believe, is one of the factors that has helped earn our journals the reputations they now happily possess.

As is no surprise, the success of our journals brings with it important new challenges, and also serious concerns. This is not the place to bemoan the explosion of information that threatens to overwhelm us all (in fact, this project is itself an attempt, in part, to meet this challenge). Even so, some cautious words about the future of our journals would be appropriate here. Bearing in mind that this may continue to be read for some years to come, I do not propose to make any truly radical predictions, but rather to indicate what is by now quite obvious. First, while I do not venture to predict the future course of our *scientific* progress, it certainly appears that from the vantage point of 1995 looking forward, progress in physics shows absolutely no sign of slackening. As some subfields of physics reach "maturity"—sometimes used condescendingly to indicate areas that have been sufficiently plowed—others appear to crop up, often unpredictably, to offer vast openings of challenging new physics for fresh generations of physicists to explore happily. In my own area of atomic physics a number of such new developments have occurred very recently, e.g., atom trapping and atomic interferometry. These more than equal established areas in terms of the excitement and opportunity they offer for doing new physics. But even assuming that physics itself continues to thrive, what about our journals? Will *they* thrive? Will they even continue to exist 100 years from now? Knowing that there is no enforceable penalty for my being wrong, I can at least offer some modest predictions concerning the short-term future of our journals.

Clearly this future is closely linked to the electronic publishing and distribution revolutions. We are already witnessing the creation of electronic ("on-line") journals. Such a development offers the physicist and indeed all scientists new, and in some ways better, methods for the dissemination of physics—a fundamental goal of The American Physical Society. The cries of help from libraries due to escalating subscription costs, need for ever-increasing storage space, and continued explosion of information content, to list a few of the major problems that confront us, can be addressed in the nick of time by providing new methods beyond the printed page to deliver information to users. I am happy to say that The American Physical Society is deeply involved in exploring such fascinating new opportunities. It would be wonderful to again see the day when authors can prepare discursive articles, as did Lamb and Retherford, without requiring more library shelf space. In that new wonderful world of electronic publishing, authors could once again possess the luxury of discussing experimental excursions that failed and ideas that turned out to be wrong, and in the process make articles more fun to read and more informative as well. And isn't it a marvel of fate that so much of the technology that has led to the electronic publishing revolution stems directly from some of the science that first saw the light of day in the pages of our journals!

Benjamin Bederson
Editor-in-Chief, The American Physical Society
New York University, New York, NY

INTRODUCTION

When asked to undertake the editorship of a single volume of about 500–600 pages of reprints of papers that have been seminal in the development of science and technology drawn from the first 100 years of *The Physical Review* and *Physical Review Letters*, I must have had a moment of failing when I accepted: Two obvious papers that would certainly have to be included—those of Millikan, on the determination of the charge of the electron, and of Davisson and Germer, in which electron diffraction and hence its wave nature are reported, alone take up more than 70 pages! The project became a little more realistic when, after a discussion with Kurt Gottfried at Cornell University, the idea was proposed, and enthusiastically accepted by The American Physical Society (APS) and the American Institute of Physics (AIP), of supplementing the printed text of the most select (according to whatever criterion) with a CD-ROM containing all the papers chosen for the centenary volume: This was to be a forerunner of modern electronic publishing, now under contemplation for *The Physical Review* in its second century. Nonetheless, only salient pages of a number of the early, long papers could be included.

The early history of *The Physical Review* has been written by Paul Hartman[1] while a special issue of *Physics Today* (October 1993), *From Basic Research to High Technology*, was devoted to the centenary commemoration. The present work was to be a sampling of actual articles and letters. To make such a compilation was clearly a most ungrateful task. I was sure to incur the wrath of my physics colleagues in any case — for what is included and what is not included. This was an assignment without a unique solution. A simple procedure that some suggested was to use the Science Citation Index (SCI). I am grateful to SCI for providing me with a list of their most frequently cited articles. But we are well aware of the pitfalls of SCI, a particular one being that of papers so fundamental that they appear in the literature without citation. And then the SCI does not extend to the early days of *The Physical Review*. We did not adopt this approach.

So what was the criterion for the selection? Totally unscientific! First, save for very few exceptions (e.g., Supernova 1987), we chose to exclude the ten years prior to this publication. For obvious reasons we had to draw an arbitrary cutoff. This will leave for the next edition many exciting current developments. Length of articles played a part with some in placing them in the CD instead of the book, or leaving them out altogether.

H. Henry Stroke, a native of former Yugoslavia, is Professor of Physics at New York University and was chairman of the Physics Department from 1988–1991. He has been working on the interface of atomic and nuclear physics, studying the structure of the nucleus through electron-nuclear interactions. After earning a B.S. in Electrical Engineering at New Jersey Institute of Technology in 1949 and his Ph.D. at Massachusetts Institute of Technology in 1954, he continued research on radioactive atom spectroscopy at Princeton, MIT, NYU, and currently at ISOLDE (CERN). Work with laser spectroscopy took him to the laboratories Aimé Cotton and Kastler–Brossel in France, and, as a senior U.S. Humboldt Awardee, to the University of Munich and the Max Planck Institut für Quantenoptik in Garching. Recently he has been developing, with colleagues at the Institut de Physique Spatiale et Planétaire in Orsay and at CERN, low-temperature calorimetry techniques for particle spectroscopy and for neutrino mass and dark matter candidate searches.

I sent out more than 500 letters to physicists — young and old — at universities, industry, and national laboratories and to representatives of the divisions and sections of The APS, soliciting their input. The response was overwhelming — and in many cases obviously represented substantial and thoughtful effort. I express my thanks to them here. *APS News* also carried an announcement of the project, soliciting further input; this was less effective, but nonetheless useful. Well over 1000 entries were generated. The chosen selections are hopefully representative, certainly not complete, and, as I remarked earlier, do not constitute by any means a unique set.

Leonardo da Vinci is said to have been the last universal person, the last to make important contributions to the arts, science, and engineering. The century of *The Physical Review* has undoubtedly seen the last of a generation of universal physicists. I divided the entries into a dozen areas and asked colleagues with particular expertise in them to study the proposed papers in their fields, organize them, and write an introduction. There is inevitably a certain degree of overlap, and several articles fitting quite logically and cited in one section will be found in an equally legitimate different one. I am grateful to these colleagues for their immense contribution. Though clearly participating in the selection process, they were severely limited by the imposed space restriction; the approach of one (Paul Martin) is described in the condensed matter overview. I assured them that I will assume full responsibility in the face of all criticisms and that they will be held harmless! For persons cited in the text of the introductions, I adopted (at the suggestion of Martin Klein, Yale University) the uniform style of giving only the last names of persons cited in the chapter texts, unless there could be ambiguities, and include a list of their full names, as completely as available, at the back of the book.

Assembling such a compendium brings to light articles published in *The Physical Review* that one may not have expected to find there, but that were brought to our attention by respondents. Such a work is the 1937 paper of Čerenkov from Moscow on the first observation, following the theory of Frank and Tamm, of visible radiation from electrons that move in a medium at velocities greater than the speed of light in it (c/n, c—speed of light in vacuum, n—index of refraction). In the same year there was a posthumous publication of a young mathematical physicist, T. H. Gronwall, on the helium wave equation, work described by Dyson as an "unfinished symphony." Then there were casualties in papers caused by World War II, not just in *The Physical Review*. Abraham Pais recounts this in his history here, noting the sudden disappearance from *The Physical Review* of communications on the physics of nuclear fission. However, in 1940 there was still a letter in *The Physical Review* on spontaneous uranium fission, but from Leningrad, sent by cable on 14 June by Flerov and Petrjak and published two weeks later! Pais describes Flerov's later role in the USSR nuclear effort. We also still find in 1944 an article sent to *The Physical Review* from the USSR by Iwanenko and Pomeranchuk on the maximum energy obtainable with a betatron. Other casualties were more personal: In 1948 Hofstadter and Deutsch describe the scintillation counter as an important new instrument for nuclear spectroscopy. Credit is given to Kallmann, who invented this during the war in Germany but being part Jewish could not publish (or work). In fact, he used naphthalene, which he had in his home in the form of mothballs, to do the experiments (instead of sodium iodide, used more commonly afterwards). Then there were some papers that should have appeared in *The Physical Review* but did not. One such is the first realization of a laser by Maiman and co-workers: it appears that the initial reporting was rejected by the editor, Sam Goudsmit, because *The Physical Review* was "swamped by maser papers." As noted by Charles Townes, it then went to *Nature*. With several stages of editorial review, and possible appeal, and the existence today of

The APS Publications Oversight Committee to ensure due procedures, it is much less likely that this would be repeated today.

An examination, even if cursory, of the papers published since the inception of *The Physical Review* 100 years ago shows a constant interplay among theory, experiment, the invention of new techniques and instrumentation to advance the experiment, and unforeseen applications of the new techniques. It is also interesting to observe that many fundamental contributions came from a few forward-looking industrial research centers. Thus, as also described by others here, we have the perhaps forgotten examples of Steinmetz's work in 1896 on oscillating currents and of basic resonator studies by Blake and Fountain (1906), and the better-remembered work on thermionic emission by Child (1911) and Langmuir (1916), the x-ray tube of Coolidge (1913), quantitative x-ray analysis by A. W. Hull (1917), and in 1921 his work on the magnetron. We find other reminders, such as the mass spectrometer (Dempster, 1918), generation of millimeter waves (E. F. Nichols and Tear, 1923), electrometer tube (Du Bridge, 1931), surface ionization (J. B. Taylor and Langmuir, 1933), an electron microscope for biological samples (Marton, 1934), the work of Blodgett and Langmuir (1937) developing techniques for building up successive monolayers, the radiometer (Dicke, 1946). And then the work of Holstein (1947) on the transport of resonance radiation, so essential in our fluorescent lighting industry. In another domain, Migeotte reports in 1948 optical observation of methane in the Earth's atmosphere. On 1 September 1939, the day of the outbreak of World War II, the paper of N. Bohr and Wheeler on nuclear fission is published.

My dwelling here on the more applied contributions that have appeared in *The Physical Review* is certainly not to be weighed against all the fundamental work: It is obviously motivated by the current (but not unique) pressure on the physics community by government and society for *relevance*. This reminds me of the early 1950s when I was working at MIT with atomic beam magnetic resonance experiments on the hyperfine structure of cesium isotopes. A businessman cousin visited our laboratory and asked: "What good?" My reply as a young graduate student: "Pure scientific knowledge." Two years later, Zacharias (1954) built the first cesium atomic clock in our laboratory! In a recent monograph[2] Kleppner describes how far, 40 years later, a particular application of such quite innocuous fundamental physics study can reach. His illustration is the global positioning system, now used for sea and air, and perhaps in the future car, "navigation" — all relying on the precision atomic clocks. The potential for the industrial as well as mass markets is there. His plea is not to sacrifice basic research to direct it to *relevant* work, but for industry to "turn new technology into new products" and regain "imaginative design and manufacturing that is essential for being economically competitive." The 100 years of *The Physical Review* are a testimonial to the fact that the best way to achieve technological advances that ultimately benefit society is to support the foundations. It is perhaps appropriate to close this introduction with the words of Victor Weisskopf: "Science cannot develop unless it is pursued for the sake of pure knowledge and insight. It will not survive unless it is used intensely and wisely for the betterment of human conditions. Human existence depends upon two pillars: The urge to know more and compassion for others. The urge to know without compassion is inhuman; compassion without knowledge is ineffective." This has been the driving force for many around the world who have contributed to making *The Physical Review* what it has been in its first 100 years.

I would like to thank Maria Taylor and Taissa Kusma of the American Institute of Physics (AIP) for their invaluable help in producing this project. Their creativity with the book and the CD-ROM was indispensable. I thank Ben Bederson, Editor-in-Chief of *The Physical Review*, who initiated this endeavor and was a source of great support. I acknowledge the hospitality of the ISOLDE group at CERN, where I carried out much of this enterprise. There are a number of other people at AIP who were involved in putting the book together. I will mention those with whom I worked directly: Cynthia Blaut, Rose Ann Campise, Wendy LaGrego, Michele Matozzo, and Denise Weiss. I appreciate their cooperation and patience. R. Joseph Anderson and Tracy Keifer at the Niels Bohr Library, Center for History of Physics, cheerfully helped in my search for photographs. I am indebted to my wife, Norma, for her counsel on the artwork, and for her encouragement and understanding during the two and one half years which it has taken to bring this work to completion.

H. Henry Stroke
New York University, New York, NY

REFERENCES

1. P. Hartman, *A Memoir on The Physical Review—A History of The First Hundred Years* (AIP, New York, 1994).
2. D. Kleppner, *Golden Eggs from Science—Congress and the Future of Basic Science* (MIT, Cambridge, 1993) (unpublished); see also, Phys. Today **47**, 9 (January 1994).

ORGANIZATION OF THE COLLECTION

Over 1,000 articles from *The Physical Review* and *Physical Review Letters* are reproduced in this book and CD-ROM, arranged in 12 chapters according to different fields of physics. Each chapter has an introduction that gives a history and overview of the field, written by an expert in that field.

The book contains reproductions of 200 articles. Each chapter in the book begins with a chronological list of articles reproduced in the book and ends with a chronological list of those and additional papers contained on the CD-ROM. A complete author index at the end of the book gives the year of publication for all articles as well as the book page number for articles reproduced in the book.

The CD-ROM contains all the material in the book (except the photographs in the introductory material) as well as over 800 additional articles—a total of more than 7,500 pages. The information on the CD is organized for easy retrieval, display, and printing of individual works. It consists of a searchable database with authors, titles, years of publication, other bibliographic information, and the text of the new introductory material, as well as bit-mapped page images of the articles reproduced from *The Physical Review* and *Physical Review Letters*.

Individual articles and chapter introductions can be retrieved by searching on author names, title words, subject (chapter headings), or year of publication. The contents of the CD-ROM can also be browsed using an automatic page-turning (scrolling) feature. Articles can be displayed as well as printed.

System requirements and installation instructions for the CD-ROM are printed on the inside back cover of the book. Full instructions for using the CD are given in the "Help" section of the CD-ROM.

Chapter 1

THE PHYSICAL REVIEW THEN AND NOW

ABRAHAM PAIS

It might have been better had I entitled this overview "*The Physical Review*, Now and Then and Later." I shall in fact begin with the present situation, then go back some 50 years, and then further back, to the journal's beginnings. I shall conclude with brief thoughts about its future.[a,b]

Some years ago John Kenneth Galbraith wrote in an essay on his efforts at writing a history of economics:

"As one approaches the present, one is filled with a sense of hopelessness; in a year and possibly even in a month, there is now more economic comment in the supposedly serious literature than survives from the whole of the thousand years commonly denominated as the Middle Ages...anyone who claims to be familiar with it all is a confessing liar."[1] I believe that all physicists would subscribe to the same sentiments regarding their own professional literature. I do at any rate.

Let me remind you of some facts and figures.

In 1969 *The Physical Review* was split into four parts: A, B, C, and D, each appearing bimonthly, in two volumes per annum.[c] In 1994 their combined total number of pages was 69,412, and the total number of articles was 8704. For part D alone, dealing with the physics of particles and fields, my area of specialization, these numbers are 15,012 and 1,529, respectively.

For many years every issue of *The Physical Review* contained a number of letters to the Editor. Since 1958 these have been published in a separate journal, *Physical Review Letters*, which appeared in two volumes in 1994, containing 8456 pages with 1871 letters.

Edward L. Nichols (1854-1937), APS President. (Courtesy of Cornell University Archives, AIP Emilio Segrè Visual Archives.)

[a] *Address to The American Physical Society meeting in Washington, D.C., 13 April 1993.*

[b] *Work supported in part by the U. S. Department of Energy under Contract Grant Number DOE 91ER 40651, Task B.*

[c] *In 1993 a new part, E, made its first appearance.*

Physicist Abraham Pais (b. 1918) is a native of the Netherlands, where his ancestors settled sometime before 1700, having fled the Iberian Inquisition. In 1946 he came to the U.S. (citizen since 1954) first as Fellow, then as Professor at the Institute for Advanced Study in Princeton. In 1963 he joined Rockefeller University in New York, where he is now Detlev W. Bronk Emeritus Professor. Since 1978 he has devoted himself to the history of science. In 1983 his Einstein biography, *Subtle is the Lord...* (Oxford), won an American Book Award. Thereafter he published *Inward Bound, of Matter and Forces in the Physical World* (Oxford 1986); *Niels Bohr's Times* (Oxford 1991); and *Einstein Lived Here* (Oxford 1994). He received the Oppenheimer Prize in 1979, the Physica Prize of the Netherlands in 1992, and the Gemant Award in 1993. Also in that year he was named Officer of the Order of Oranje Nassau.

That is still not all. The above journals are published under the editorship of The American Physical Society. This society is one of the ten member societies of the American Institute of Physics, founded in 1931 "for the purpose of coordinating various societies whose interests are primarily in the field of physics and for the purpose of supporting their publications."[2] In 1994 the Institute published 16 journals, of which 8 are cover-to-cover translations of articles from the former Soviet Union. In addition, it publishes the general-interest magazines *Physics Today* and *Computers in Physics*. Furthermore, the member societies published through the Institute in that year another 21 journals. In addition, there were 28 member-Society journals published outside the American Institute of Physics; for example 12 by the American Geophysical Union.[d]

So far I have only quoted data on publications from the United States, my reason being that only for that country do I possess fairly complete information. I can add some significant numbers for the restricted domain of the physics of particles and fields. Detailed *worldwide* information on this subject is regularly published by central repositories for preprints, presently the common form of rapid dissemination. One of these, the Stanford Linear Accelerator Center (SLAC) preprint library, distributes a weekly bulletin listing documents received from all over the world, by title, author(s), and institution(s). The annual yield,[e] comprising articles, letters, and reviews, was 7471 in 1990. Thus at this time I would have to digest 20 publications per day, every day of the year, to claim familiarity with the whole body of only the high-energy physics literature.

[d] *I am much indebted to Dr. Andrzej Herczynski, formerly of the American Institute of Physics, who at my request collected the numbers cited in this paragraph.*

[e] *I owe this number to the SLAC librarians.*

Arthur G. Webster (1863–1923). (Courtesy of AIP Emilio Segrè Visual Archives.)

These numbers are meant to bring home my main message: There is not a soul on Earth who can read the deluge of physics publications in its entirety. As a result, it is sad but true that physics has irretrievably fallen apart from a cohesive to a fragmented discipline. I cannot give numbers for other areas of scholarly pursuit, but I am sure that elsewhere the development has been similar; *vide* Galbraith. It was not that long ago that people were complaining about two cultures. If we only had it that good today.

I have a nice spacious office, but even so have had, years ago, to stop subscribing to all journals except for the excellent quarterly *Reviews of Modern Physics*. Had I done otherwise, my shelf space would have been eaten up completely by the green monster, *The Physical Review*.

That is the situation today. One can only wonder, perhaps shudder, if one thinks ahead to what things will be like 100 years from now and beyond.

I next turn the clock back half a century.

I saw and read my first issue of *The Physical Review* in 1938. That happened in the office of George Uhlenbeck at the University of Utrecht in my native Holland, where I had begun graduate studies under his direction. The oldest *Physical Review* document I own is a photostat of the Bohr-Wheeler article of 1939 on nuclear fission.

Uhlenbeck was a highly gifted physicist. One of his remarkable traits was that he would read every issue of *The Physical Review* from cover to cover. Even then there may not have been many who did so, but at least it was still a manageable enterprise. In 1950, for example, four *Physical Review* volumes appeared with 447 articles on 3167 pages, not counting the letters to the Editor and the abstracts of papers presented at American Physical Society meetings, all still cozily published together with the full-fledged articles. So if you had the stomach for it, you could get by reading an average of 8½ pages per day.

My own first contribution to *The Physical Review* appeared in the November 1945 issue.[3] It was a letter to the Editor summarizing work on quantum electrodynamics I had done while living in hiding in wartime Holland. Boy, was I proud to see my name in print in the world's most prestigious physics journal! All through my career it has continued to be a source of satisfaction to have a paper accepted and published in *The Physical Review* and later also in *Physical Review Letters*. Between 1945 and 1981 I have published 80 papers in those two journals combined. That is about 1½ per year, a number that rises to a bit less than 3 if publications in other journals are included.

I do not mention these numbers to advertise my productivity but rather to make quite a different point. In the course of the past years, hundreds of postdoc applications have passed my desk. In more recent times it has not been rare to find attached a list of publications that includes ten or more items produced per annum. As all of us know, it is true that "publish or perish" is one of the

Fredrick Bedell (1868–1958). (Courtesy of Cornell University Archives, AIP Emilio Segrè Visual Archives.)

Ernest G. Merritt (1865–1948).
(Courtesy of Cornell University Archives, AIP Emilio Segrè Visual Archives.)

rules governing our careers. To this principle we are now obliged to add, I think, "publish too much and perish as well." There can be no doubt that our scientific journals are severely polluted these days.

I return to the year 1945 in order to recall the most important nonevent in the history of *The Physical Review*.

I spent World War II in Holland. During that time we were cut off from American journals. It was not the worst that happened to us then, yet we physicists very much missed the contact with American science.

On 16 June 1945, about a month after the cessation of hostilities in Europe, I noted in my diary: "*The Physical Review* issue of 1944 and first half of 1945 have arrived. Little news, we will catch up quickly." There was in fact just one paper that caused much excitement. It was written by Lars Onsager from Yale and reported the first exact calculation of a nontrivial phase transition, the case of the two-dimensional Ising model.[4]

I had been particularly eager for further news about nuclear fission—the great discovery of 1939 and the subject of my earliest theoretical calculations—and was astonished to find that not a single *Physical Review* publication dealt with that subject. It was entirely inconceivable that fission had not been researched further in the years 1941–1945. What was going on? Were the Americans pursuing the subject in secrecy? Were they making the bombs physicists had already been speculating about in 1939? I forgot about the whole business until I opened a Dutch daily of 7 August and saw the headline "The First Atomic Bomb on Japan," followed by the report that on 6 August at 8:15 local time, Hiroshima had been destroyed by what we later heard was a uranium-235 bomb. On 9 August I wrote in my diary: "Dammit, I was right... ."

It was not until 1990 that I learned from *Moscow News* (an English weekly published in Moscow) that as early as April 1942 one no less than Joseph Stalin had also been informed of *The Physical Review*'s wartime silence on fission.[5] That April Georgii Flerov (1913–1990), a well-known Soviet nuclear physicist, had written a letter to Stalin, made fully public in 1990 for the first time, addressed "Dear Iosif Vissarionovich," in which he pointed out that recent physics journals from the U.S., Britain, and Germany contained no reference whatsoever to fission. From this he had correctly inferred that scientists in those countries were up to something. He pleaded with Stalin to take an initiative "concerning the feasibility of the uranium problem." According to another Soviet expert, Flerov's letter may well have played a role in Soviet planning.[6]

One last word about this nonevent in physics publishing. The advent of World War II had made it imperative that further developments — on either side — be kept secret. Nevertheless, according to the famous Smyth report,[7] after the spring of 1939, "publication continued freely for about another year, although a few papers were withheld voluntarily by their authors." In the spring of 1940 a "Reference Committee" was organized by the U.S. National Research Council "to control publication policy in all fields of possible military interest... . This arrangement was very successful in preventing publication...[it] was a purely voluntary [procedure]; the scientists of the [U.S.] are to be congratulated on their complete cooperation." Thus it came about that the chemical separation of plutonium, achieved in 1941, was not published in the Letter columns of *The Physical Review* until 1946, when the following footnote was added: "This letter was received on [28 January 1941], but was voluntarily withheld until the end of the war... ."[8]

In mid-September 1946 I arrived in the United States, in time to give an invited paper at The American Physical Society meeting to be held in the building of the Engineering Societies in mid-town Manhattan starting on the 19th. I quote from the minutes of that meeting, which in those years still appeared in *The Physical Review:*

"This was a meeting of unprecedented character in more than one respect. It was confined to papers on three topics: cosmic-ray phenomena, theories of elementary particles, and the design and operation of accelerators of nuclear particles and electrons. Disparate as these three subjects may appear to be, the trend of physics is rapidly uniting them."[9]

It seems to me that this last phrase should be remembered as the opening salvo ushering in a new era at the frontiers of physics. Add to this another phrase that appeared one month later in the *Quarterly Progress Report of the Columbia University Radiation Laboratory*, where we find a proposal entitled "Microwave physics: experiment to determine the fine structure of the hydrogen atom (Lamb and Retherford)"[10]—and you see the curtain rising over the postwar physics scene, so new in style and content. Laboratories expand, theories start improving, new particles and new fields appear on the scene, *The Physical Review*'s volume

Henry D. Smyth (1898–1986). (Courtesy of *Physics Today* Collection, AIP Emilio Segrè Visual Archives.)

Niels Bohr (1885–1962) and **Ernest Rutherford (1871–1937).** (Courtesy of Niels Bohr Institute, AIP Emilio Segrè Visual Archives.)

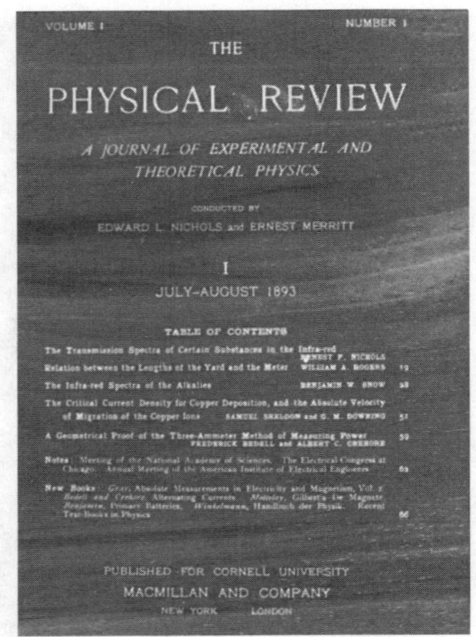

VOLUME I NUMBER I

THE

PHYSICAL REVIEW

A JOURNAL OF EXPERIMENTAL AND THEORETICAL PHYSICS

CONDUCTED BY

EDWARD L. NICHOLS and ERNEST MERRITT

I

JULY–AUGUST 1893

TABLE OF CONTENTS

PUBLISHED FOR CORNELL UNIVERSITY
MACMILLAN AND COMPANY
NEW YORK LONDON

Reproduction of the first cover of the ***Physical Review*** (1893).

explodes. I could, but shall not, talk for hours about what happened thereafter. Instead I shall turn the clock back another half century.

In the late 1970s, when I decided to try my hand at writing the history of physics, I had first to learn facts about the 19th century. I knew of course that branches of physics such as thermodynamics and electromagnetism date from that time, but it was new to me that the same is true for the terms "physicist" and "scientist," which even then did not at once receive a cordial reception. Michael Faraday wrote: "The Physicist is both to my mouth and ears so awkward that I think I shall never be able to use it." In the 1890s Lord Rayleigh wrote that he disliked the word scientist and had never used it himself.[11]

I also had to ferret out which American journals were available to physicists for publishing their results and thoughts. I found out that since the early 19th century they had, in order of first appearance, numerous options: the *American Journal of Science*, the *Journal of the Franklin Institute*, the *Proceedings of the American Academy of Arts and Sciences*, and the *Transactions of the Connecticut Academy of Arts and Sciences*. All those journals were of a general scientific nature, not exclusively devoted to physics.

That was the state of affairs when, 100 years ago, *The Physical Review* came into being as the first American journal for physics only.

In February 1893 E. L. Nichols (1854–1937), a physics professor at Cornell, received a letter from MacMillan and Company in New York,[f] in which was said: "We beg to say that we shall be very happy if arrangements can be made to publish the journal for the University."[12] *The Physical Review*, initially financed by the trustees of Cornell with an annual appropriation of $2400,[13] was ready

[f] *This company published the first 30 volumes of* The Physical Review.

John A. Wheeler (1911–) and **Hans G. Dehmelt (1922–).** (Photo by Ivan Massar. Courtesy of Alexandra Oleson, American Academy of Arts and Sciences.)

to roll. The first issue, dated July–August 1893, listed Nichols and Merritt (1865–1948) as editors; a third editor, Bedell (1868–1958), joined them soon thereafter. It was published under the auspices of Cornell and defined itself as "A journal of experimental and theoretical physics," a line that, along with Nichols's name, continued to grace the title page until the journal split into A—B—C—D. The first issue was 80 pages long. Volume 1 contained a mere 20 articles, spread over only 480 pages.

I found it most illuminating to read an early comment received by the editors: "The era of specialization is upon us."[14] Physics as a whole had become a specialty!

Next I should like to quote the first sentence of the first paper in the first issue of *The Physical Review*, written by E. F. Nichols (1869–1924): "Within a few years the study of obscure radiation has been greatly advanced by systematic inquiry into the laws of dispersion of infrared rays." Recall[15] that such studies in the infrared were crucial to Planck's discovery of the quantum theory in 1900. *The Physical Review* began in fact to appear just before the face of physics changed drastically due to the discoveries of x rays, radioactivity, the electron, and the quantum and relativity theories. It took some time, however, before all that novelty became reflected in the journal's pages.

Ernest F. Nichols (1869–1924). (Courtesy of AIP Emilio Segrè Visual Archives.)

More about the beginnings. *The Physical Review*, at first a bimonthly, became a monthly with Volume 5, in 1897, at a subscription rate of $5 per year. Until 1903 it arrived uncut and untrimmed on the reader's desk.[12] In those early years the journal, ever so much more humane than it is now, contained book reviews, obituaries of noted scientists, and announcements of meetings; it also ran advertisements. For some time after 1898, when publication of *Science Abstracts* started, copies of it were sent free to *Physical Review* subscribers.

Beginning with the first issue of 1903, *The Physical Review*'s title page carried these most important additional words: "Conducted with the cooperation of the American Physical Society" (APS).

The journal was already 6 years old when on 20 May 1899 at 10 a.m. some 40 physicists gathered in Fayerweather Hall at Columbia University to organize what came to be The APS. The meeting was called by Arthur Webster (1863–1923) of Clark University in Worcester, Massachusetts, who has been called the father of The APS.[16] It may be noted in passing that in 1904 Ernest Rutherford was elected to The APS Council.[17] For the first three years of its existence, The APS issued its own bulletins; thereafter these appeared in *The Physical Review*, for which, as of 1913, The APS assumed sole responsibility.[18] Thus came to an end the editorial leadership of Cornell and in particular of its three faculty members.

With no disrespect to others, the most important Chief Editor of *The Physical Review* in this century has been John Tate (1889–1950), who was in charge from 1926 until his untimely death. The initiatives he took were important and numerous. In

Felix Bloch (1905-1983), Robert F. Bacher (1905-), Owen Chamberlain (1920-), and **K. K. Darrow (1891-1982).** (Courtesy of *Physics Today* Collection, AIP Emilio Segrè Visual Archives.)

1929 he began semimonthly publication of *The Physical Review* and introduced its Letters to the Editor columns as well as the *Physical Review Supplements*, meant "to contain resumés, discussions of topics of unusual interest, and, in a broad sense, material that will give valuable aid to productive work in physics and yet that cannot appropriately be published in *The Physical Review*."[19] In 1930 the *Supplements* were renamed *Reviews of Modern Physics*. In 1931 Tate became the first editor of another new journal, *Physics*, which in 1937 was renamed *Journal of Applied Physics*.[g] In his obituary to Tate, Karl Darrow (1891–1982), another APS luminary,[h] suggested[20] that *The Physical Review* should be known as *Tate's Journal*, by analogy with *Poggendorf's Annalen* (now the *Annalen der Physik*).

It was also during the Tate years that *The Physical Review* became the world's leading physics journal, as two recollections will illustrate.

One is by Edward Condon (1902–1974), who was in Göttingen in 1927. In that year the physics department there ordered *The Physical Review* to arrive in one package containing a full year's issues. There was no urgency for doing otherwise.[22]

The other is by Heisenberg: "I certainly know that, I should say after 1930, I have read *The Physical Review* quite regularly. Then it started to become the leading journal, but not much before that time."[23]

To conclude, brief words about the future.

As time goes by, not only individuals but also libraries will be unable to cope with the amount of journal material that contin-

John T. Tate (1889–1950). (Courtesy of AIP Emilio Segrè Visual Archives.)

[g] *At the winter 1932 meeting of The APS, Tate reported that the finances of the Society's journals were now in the hands of the newly founded American Institute of Physics.*

[h] *Darrow was secretary of The APS from 1941 until 1967. He was nicknamed "Mr. American Physical Society." It has been written of him: "The manner in which he ran the APS will never be duplicated."*[21]

ues to amass. Likewise, the cost is rapidly becoming a serious limiting factor. As said earlier, in 1897 the subscription price for *The Physical Review* was $5 per year. By 1943 it had risen, but only to $15. In 1993 the full annual rate for the split journal is $1000.[1] It is indeed obvious and widely known that we are entering a period of transition during which the printed book must eventually cease to be the tool for partaking in the written word. Microfilm is only the beginning of what will substitute for books.

Should those developments make us weep? I think not. One may as well bewail the second law of thermodynamics. True, there is a loss involved for those who, like myself, are fond of holding the printed book in their hands, who love to indulge in browsing. That will have to go. Later generations will find new ways of indulging, however.

Most important at this time for keeping abreast of information is to be locked into the strongly developed modern oral tradition—the bane of future historians, who are forever locked out of this crucial flow of information.

It is true that one may at times deplore some lack of style that is now current, as, for example, in too-speedy publications (especially by theorists). It is true that one may at times deplore the role of too much publicity in this field; physics as front-page news is on the whole not a good thing. Yet none of this can detract from one fact: Good physics will always be recognized for what it is. Therein lies a common strength, a common enjoyment, and a common bond.

Sam Goudsmit (1902-1978). (Courtesy of AIP Emilio Segrè Visual Archives.)

REFERENCES

1. *New York Times* Book Review, October 1977.

2. K.T. Compton, *Phys. Today* **5**, 4 (February 1952).

3. A. Pais, *On the Theory of the Electron and the Nucleon*, Phys. Rev. **68**, 227–228 (1945).

4. L. Onsager, *Crystal Statistics*, Phys. Rev. **62**, 559(A) (1942); *Crystal Statistics I. A Two-Dimensional Model with an Order-Disorder Transition*, *ibid.* **65**, 117–149 (1944).

5. *Moscow News*, April 12, 1988.

6. Interview with Igor Golovin, in *Moscow News*, October 8, 1989. For more on the Soviet bombs, see A. Pais, *Phys. Today* **43**, 13 (August 1990); also the interview with Yuli Khariton, *The New York Times*, January 14, 1993.

7. H.D. Smyth, "A general account of the development of methods of using atomic energy for military purposes under the auspices of the United States Government 1940–1945" (U.S. Government Printing Office, Washington, D.C., 1945), pp. 26 and 27. For more details about the secrecy issue, see S. Weart, *Phys. Today* **29**, 23 (February 1976).

8. G.T. Seaborg, E.M. McMillan, J.W. Kennedy, and A.C. Wahl, *Radioactive Element 94 From Deuterons on Uranium*, Phys. Rev. **69**, 366–367 (1946); G.T. Seaborg, A.C. Wahl, and J.W. Kennedy, *ibid.* **69**, 367(L) (1946).

9. Phys. Rev. **70**, 784 (1946). The abstract of my talk, *Some General Aspects of the Self-Energy Problem*, appears on pp. 796–797 of that issue.

[1] *I thank Pat Odom of* The Physical Review *Editorial Office for providing these numbers.*

Karl T. Compton (1887–1954). (Courtesy of Karsh, AIP Emilio Segrè Visual Archives.)

10. W.E. Lamb, reprinted in *A Festschrift for I.I. Rabi*, edited by L. Motz, Trans. N. Y. Acad. Sci. Ser. II, **38**, 82 (1977).
11. S. Ross, *Scientist: The Story of a Word*, Ann. Sci. **18**, 65–85 (1962).
12. Quoted in P. Hartman, *A Memoir on The Physical Review—A History of the First Hundred Years* (AIP, New York, 1994), p. 25.
13. Reference 12, pp. 22 and 75.
14. Reference 12, p. 22.
15. See A. Pais, *Subtle is the Lord* (Oxford University, Oxford, 1982), Chap. 19.
16. For an eyewitness account of the meeting, see Frederick Bedell, *What Led to the Founding of the American Physical Society*, Phys. Rev. **75**, 1601–1604 (1949). For more on Webster, see Melba Phillips, *Arthur Gordon Webster, Founder of the APS*, Phys. Today **40**, 48–52 (June 1987). For the early years of The APS, see M. Phillips, *The American Physical Society: A Survey of Its First 50 Years*, Am. J. Phys. **58**, 219–230 (1990).
17. *Proceedings of the American Physical Society—Minutes of the Twenty-Second Meeting*, Phys. Rev. **18**, 116–117 (1904).
18. The statement of transfer is found in the 1 January 1913 issue, *The Physical Review—Announcement of the Transfer of the Review to the American Physical Society*, Phys. Rev. **1**, 1 (1913).
19. Phys. Rev. **33**, 276 (1929).
20. K.K. Darrow, *American Philosophical Society Yearbook 1950*, p. 325; for another obituary of Tate see A. Nier and J. H. Van Vleck, Biogr. Mem. Natl. Acad. Sci. **47**, 461 (1978).
21. W. W. Havens, *Karl K. Darrow*, Phys. Today **35**, 83–84 (November 1982).
22. E.U. Condon (private communication to F.J. Seitz, who in turn told me this story).
23. W. Heisenberg, interview by T.S. Kuhn, February 11, 1963, transcript in The Niels Bohr Library, College Park, MD.

Edward U. Condon (1902–1974), on right. (Courtesy of National Bureau of Standards, AIP Emilio Segrè Visual Archives.)

Chapter 2

ONE HUNDRED YEARS OF THE PHYSICAL REVIEW

VICTOR F. WEISSKOPF

It is a pleasure to participate at the centenary of such a venerable institution as *The Physical Review*. During this century it became the most important conveyor of modern physics. Long rows of the familiar green issues are found in the offices of any physicist here and abroad, not only on the shelves, but also on the desks, in constant use for work and information.

My own life span is of the same order of magnitude as that of *The Physical Review*. I began to be a physicist in 1926, one-third of a century after the journal's founding. Up to 1930 *The Physical Review* was not as highly valued in Europe as it should have been, partly because of European arrogance and partly because the important new physics of that time was done and published mostly in Europe. This was the birth of modern physics, under the impact of relativity and quantum mechanics. The former was the achievement of one man, the latter of a collaboration of a number of outstanding physicists. It is fitting to apply to that group a slightly altered statement by Winston Churchill praising the Royal Air Force: *Never have so few done so much in so short a time.*

Otto Stern (1888-1965) in Hamburg, Germany, 1927. (Courtesy of Niels Bohr Institute, AIP Emilio Segrè Visual Archives.)

Not all was done in Europe. Two of the most decisive experiments for quantum mechanics were executed and published in the U.S.: Compton's light scattering by electrons put the photon on the map, Davisson and Germer's electron interference experiments established the wave nature of electrons. Also, important papers by Tolman, Langmuir, and Oppenheimer appeared in *The Physical Review* during the 1920s. But there is no question: Up to 1930 the progress of modern physics, both experimental and theoretical, took place in Europe. Germany dominated the scene; the *Zeitschrift für Physik* was the leading journal. Up to 1930 American physics held a secondary position, derivative from Europe. Most of the American physicists who intended to play leading roles in the U.S. went to Europe for study.

All this changed dramatically around 1930. An internal reorientation took place in American physics, spurred by American

Victor F. Weisskopf is Professor of Physics Emeritus at Massachusetts Institute of Technology, Cambridge, and was head of the Physics Department from 1967 to 1973. His work includes the theory of quantum electrodynamics and nuclear physics. His awards and honors include the National Medal of Science, the Planck Medal, the Enrico Fermi Award, and the Wolf Prize. He received twenty-five honorary degrees from universities all over the world, and was director of CERN, the European Organization for Nuclear Research, Geneva from 1961 to 1966.

Richard C. Tolman (1881-1948). (Courtesy of AIP Emilio Segrè Visual Archives.)

personalities such as Condon, Kemble, Millikan, Oppenheimer, Rabi, Van Vleck, and others. After 1930 the U.S. changed from a derivative position to a central leading place in modern physics. This transformation was greatly helped but not primarily caused by the influx of German and Austrian refugees from the Nazis. German predominance ended because of Hitler's anti-intellectual and anti-Jewish drives. English became the international language in physics, *The Physical Review* became the leading journal replacing the *Zeitschrift für Physik*.

It was a most interesting period. In the "miracle year" of 1932, Chadwick discovered the neutron, C. D. Anderson and Neddermeyer discovered the positron, Fermi published his pioneering theory of the beta decay, and Urey discovered heavy water. True enough, the first and third achievements were accomplished in Europe, but the consequences were worked out in the U.S. For example, the physics of nuclear structure was intensely developed in America.

A personal remark: My first paper in *The Physical Review* was submitted in 1936, when I was still in Europe. It was on nuclear physics. I was in Copenhagen, where Niels Bohr invited a number of physicists such as Franck, Hevesy, Frisch, and myself, who could not find jobs because of Hitler's racial law. I thought my chance of getting a job in the U.S. would be enhanced if I published a paper in *The Physical Review* on a topic of special interest in the U.S. Bohr visited the U.S. and England every year to "sell his refugees." He succeeded in finding an instructorship for me in Rochester, New York. Maybe the paper in *The Physical Review* was of some help.

The sociology of physics before World War II was quite different from today. There were relatively few physicists. The yearly international conferences in Copenhagen dealing with everything that was new in physics had usually less than 60 participants. There was not much specialization, applications were rarely mentioned. Research teams were very small, financial support — the little that was necessary — came from universities or from foundations. There were few jobs available and they paid poorly; people were driven by idealism. It was a continuation of the intellectual and social tradition of the 19th century.

World War II changed all that. To the astonishment of government officials, physicists became successful engineers in military research and development (R&D) enterprises such as at the Radiation Laboratory, the Manhattan Project, the proximity fuse, etc. Scientists who were previously interested only in basic science designed and constructed the nuclear bomb, under the leadership of one of the most esoteric physicists, Oppenheimer, as well as the first nuclear reactor, under Fermi. Wigner designed reactors for plutonium production; Schwinger developed the theory of wave guides, essential for radar. This work was not published in *The Physical Review* because of secrecy.

After the war the public was under the impression that the physicists won the war. Of course, this was a vast exaggeration, but it is a fact that radar saved the United Kingdom and reduced the submarine threat to trans-Atlantic convoys, and that the atomic bomb led to an immediate end of the war with Japan. Physics and science in general earned a high reputation. This led to higher salaries and to generous financial support from government sources such as the Office of Naval Research (ONR), the National Science Foundation (NSF), created with the purpose of supporting basic research, and the Atomic Energy Commission, which supported research in nuclear and particle physics. The rationale for the support of basic science by government sources, irrespective of military and other applications, was twofold. First, the war experience engendered a strong belief that any basic science research will lead to useful applications; second, there was a desire to keep scientists happy and numerous since they might be needed again. The lavish support, without any regard as to the type of research, lasted for about a decade after the war; later on government sources became increasingly interested in more specific research directed at military or commercial applications. Still, basic science fared well until the 1970s.

The results of this support were truly amazing. The progress of natural science in the three decades after the war was outstanding. Science acquired a new face. It would be impossible in the frame of this survey to list all the significant advances. Many of the new results and discoveries were based upon the instrumental advances in electronics and nuclear physics due to war research. One of the most important new tools for all sciences was the computer. The development and improvement of this tool are perhaps the fastest that ever happened in technology. It brought about new methods of evaluation of experimental data, new ways to model and simulate natural processes. To quote Sam Schweber: "There are now three types of scientists: experimental, theoretical, and computational."

Percy W. Bridgman (1882–1961) and **Edwin C. Kemble (1889–1984).** (Photograph by J. Frenkel. Courtesy of AIP Emilio Segrè Visual Archives.)

George E. Uhlenbeck (1900-1988) and **Wolfgang Pauli (1900-1958)** at Lorentz Institute, Kamerlingh Onnes Laboratory, Leiden. (Courtesy of Uhlenbeck Collection, AIP Emilio Segrè Visual Archives.)

James Frank (1882–1964) and **Gustav Hertz (1887–1975).** (Courtesy of AIP Emilio Segrè Visual Archives.)

Most of the discoveries were made in the U.S. and published in *The Physical Review*. Our journal became more and more voluminous. The inflation of papers led to the following nasty joke: An extrapolation of the speed by which *The Physical Review* is filling the shelves leads to the conclusion that this speed will surpass light velocity in 2020. It is no contradiction to relativity because it contains no information!

The preponderance of the U.S. in the years 1946–1960 was overwhelming. The causes are obvious: Conditions in other regions after the war made intense scientific efforts difficult. Europe and East Asia had to be rebuilt. All the more we must admire certain pioneering efforts carried out mainly in England, Italy, and France, such as cosmic-ray research in England under Powell and in France under Leprince-Ringuet, and the important Italian meson absorption experiments by Conversi, Pancini, and Piccioni. Some of these results were published in *The Physical Review*. The situation was the reverse of the first quarter of the century: Europe's physics was mostly "derivative" and America's was "central." European and East Asian scientists had to spend some time in the U.S. in order to play a role at home.

Georg Von Hevesy (1885–1966). (Courtesy of Niels Bohr Institute, AIP Emilio Segrè Visual Archives.)

After 1960 the situation changed somewhat. European and East Asian scientists became more independent and were able to compete with their peers in the U.S. Large laboratories were founded abroad, such as CERN in Western Europe, Harwell in England, DESY in Germany, KEK in Japan, and the European Southern Sky Observatory in Chile. Older European and Japanese journals expanded and new ones were created, but *The Physical Review* remained dominant, especially with the separate publication of *Physical Review Letters*.

There was a definite change in the character and sociology of physics after the second world war. The large instrumentations that the generous support of physics made available needed large teams to exploit them. We find papers in *The Physical Review* with lists of names that fill almost a full page, especially in particle physics. A new type of physicist appeared, the team leader; he or she had to be an inventive physicist, a good organizer, a strong character to hold a hundred people together, and a politician to obtain financial support. This created a special type of person, quite different from the leaders in the past. The role of the subordinates in these monster teams created problems. How do they get sufficient credit for the work they perform?

Fortunately there are fields such as atomic and nuclear physics, quantum optics, and condensed matter physics, where small groups may still produce excellent physics.

The division of *The Physical Review* into five journals was certainly caused by the large number of papers, but it also testifies to a situation in today's physics that threatens the unity of our science. The articles in one division cannot be understood by the readers of another division. Even within one division (in particular in A and E), the papers are opaque to readers of a different subspecialty. This deplorable situation has two causes. One is increasing specialization: Physicists have no time and perhaps also no great interest to follow developments outside their specialty; there is too much going on in their own narrow fields. Rabi expressed it succinctly:[a]

> *"Science itself is badly in need of integration and unification. The tendency is more than the other way...Only the graduate student, poor beast of burden that he is, can be expected to know a little of each. As the number of physicists increases, each specialty becomes more self-sustaining and self-contained. Such Balkanization carries physics, and, indeed, every science further away from natural philosophy, which, intellectually, is the meaning and goal of science."*

There are journals that try to penetrate the specialization walls, such as *Reviews of Modern Physics*, *Science*, and *Scientific American*, but the articles in these journals increasingly have become surveys of recent research aimed again at the specialist.

But there is another, deeper reason. Physics is being divided into different levels that are increasingly decoupled. We may distinguish particle physics, including cosmology and astronomy, at the lowest level (no value judgment implied), nuclear physics at the next highest level, then atomic and molecular physics, then condensed matter, plasma physics, and biophysics. Each level has its "quasi-elementary" units, nucleons in parts of nuclear physics, nuclei and electrons in atomic and molecular physics, atoms or molecules in condensed matter, etc. Of course, at the lowest level, in particle physics, we believe we are dealing with the truly elementary units, until....

At the higher levels we may disregard the internal composition of the quasi-particles, since they are not excited at the characteristic lower energy exchanges; they behave as elementary

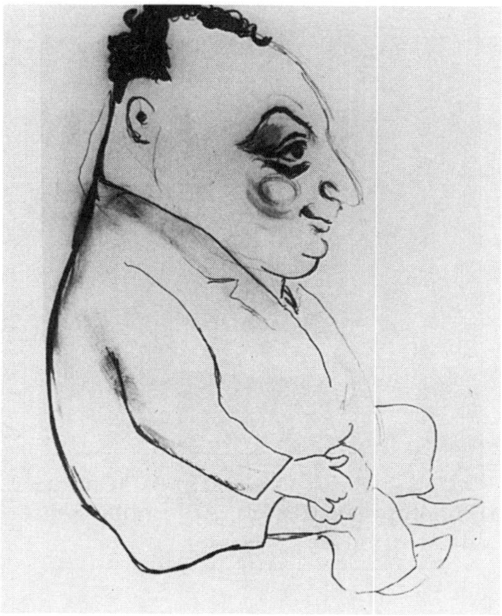

Wolfgang Pauli, caricature by Eugene Rabinowitsch, 1932. (Courtesy of Weisskopf Collection, AIP Emilio Segrè Visual Archives.)

[a] *I.I. Rabi, Science, The Center of Culture (World Publishing, New York, 1971), p. 92.*

Wolfgang Pauli (1900-1958). (Courtesy of Photopress, Zurich, AIP Emilio Segrè Visual Archives.)

units. "Effective" theories of interaction are devised, new concepts, laws, and properties emerge on the basis of the interactions between quasi-particles. These new concepts and laws are not in contradiction with the concepts and laws of lower levels, but they are not (yet?) derivable from them. This is why a "theory of everything," which could perhaps predict the elementary particles and their interactions, may remain a "theory of almost nothing," since at least 90% of all physics is outside the particle domain.

Thus the division of physics into decoupled levels is another reason for the separation into almost independent fields. The quark structure of nucleons plays no role whatsoever in higher levels, except in parts of nuclear physics. The same is true about the recent new insights into general relativity, the existence of black holes and the questions related to the gravitational waves, the big bang, and the 2.7° radiation filling the universe.

The phenomenon of decoupling among different fields of physics can also be interpreted as a symptom of the success of modern physics, encompassing increasingly wider fields of human experience. We do not need to fear for the future of physics. Of course there are problems and difficulties to overcome apart from the ones mentioned here, such as the reduced support of basic science compared to applied fields, and the rising costs of instrumentation in what is called "big science." However *The Physical Review* can look with confidence towards the future. It will remain the leading physics journal of the world for many decades to come; already today it publishes more papers from abroad than from the U. S., a truly international enterprise. Perhaps a Chinese or Japanese journal may take over in the far future.

Chapter 3

THE EARLY YEARS

MORTON HAMERMESH

Introduction

Papers Reprinted in Book

Theodore Lyman (1874-1954). (Courtesy of W. F. Meggers Collection, AIP Emilio Segrè Visual Archives.)

During the period 1893–1912, the development of physics in the U.S. was slow. E. F. Nichols and G. F. Hull made the first successful observation of radiation pressure (1901). Barnett studied electromagnetic induction (1912). O. W. Richardson investigated the energy distribution in thermionic emission (1911). The work of Coblentz in spectroscopy, of Child on electrical discharges, of Barus on atmospheric physics, and of Wood in optics also belongs to this period. (The great work of Michelson appeared in the *American Journal of Science* before the birth of *The Physical Re-*

Morton Hamermesh is now Emeritus Professor of Physics at the University of Minnesota. From 1948 to 1965 he worked at the Argonne National Laboratory and served as Associate Laboratory Director from 1963 to 1965. He has worked on nuclear theory, elementary particle theory, and accelerator design. He is the author of *Group Theory and Its Application to Physical Problems* (1962) and the translation of *Landau-Lifshitz Classical Theory of Fields* (1951).

Kenneth T. Bainbridge (1904-) with mass spectograph in his laboratory at Harvard University, 1938. (Courtesy of *Physics Today* Collection, AIP Emilio Segrè Visual Archives.)

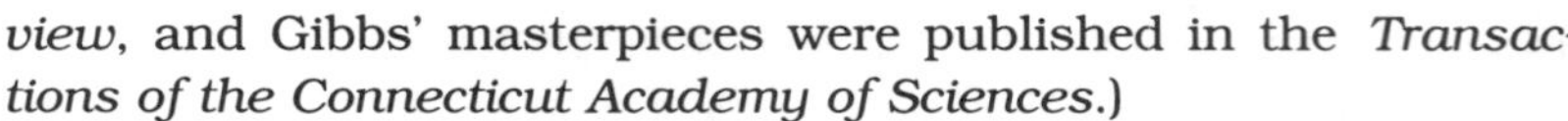

view, and Gibbs' masterpieces were published in the *Transactions of the Connecticut Academy of Sciences*.)

The period from 1912 to 1920 includes the early work of Bridgman on phase changes under pressure, which later led to the development of the whole field of studies at very high pressures. Birge did the first of his long series of works on determination of physical constants (1919). There is a paper of Ives, Coblentz, and Kingsbury on the mechanical equivalent of light (1915), the great work of Millikan on the charge of the electron (cf. the reprint in this compendium) and the 1914 paper of Lyman on short-wavelength spectral lines in hydrogen (also reprinted here).

William Duane (1872-1935). (Courtesy of AIP Emilio Segrè Visual Archives.)

Between 1920 and the early 1930s there was rapid growth in physics in the U.S. There is the work of Cady on piezoelectricity, the discovery of ferroelectricity by Valasek in 1920, and the paper of Condon in 1926 applying what we call the "Franck-Condon principle" to intensities in band systems of molecular spectra. There were advances in the design of particle accelerators, and studies of electron diffraction (cf. the reprint of Davisson and Germer). This was also the period of the discovery of modern quantum mechanics, which produced a whole school of theorists in the U.S. who applied the new theory to atomic, molecular, and solid state problems: Dennison, Kemble, Van Vleck, and Slater, whose influence grew into the next two decades. Eckart (1926) wrote his basic paper on operator calculus in quantum mechanics. This period also saw the development of mass spectrometry (by Dempster, Bainbridge, Nier, and others), which foreshadowed its future importance in planetary studies, space exploration, and technological applications. One of the stars of this period was Breit, whose work covered a wide range of subjects. His 1926 paper with Tuve (Phys. Rev. **28**, 554–575) on studies of the ionosphere used pulse techniques that foreshadowed the develop-

Arthur H. Compton (1892-1962) and **Francis Simon (1893-1956).** (Courtesy of AIP Emilio Segrè Visual Archives.)

ment of radar [see, also, Taylor and Hulbert, Phys. Rev. **27**, 189 (1926)]. His paper in 1929 (Phys. Rev. **34**, 553–573) on the effect of retardation on the interaction of a pair of electrons and his 1931 paper with Rabi on the Breit-Rabi equation are important works. Later in the 1930s we have the Breit-Wigner formula for nuclear resonance cross sections.

By the 1930s physics research in the U.S. was booming. The great influx of physicists from Europe began a torrent of immigration that would help make the United States a leading center of research and make *The Physical Review* the premier physics journal.

Harold C. Urey (1893-1981) at La Jolla, December 1958. (Courtesy of Argonne National Laboratory, AIP Emilio Segrè Visual Archives.)

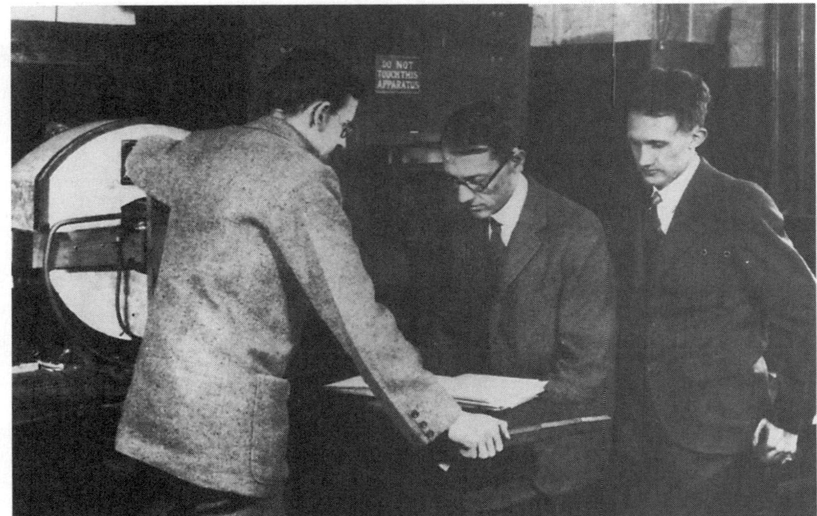

Clinton J. Davisson (1881-1959), Lester H. Germer (1896-1971), and **C. J. Calbick.** (Courtesy of Bell Laboratories, AIP Emilio Segrè Visual Archives.)

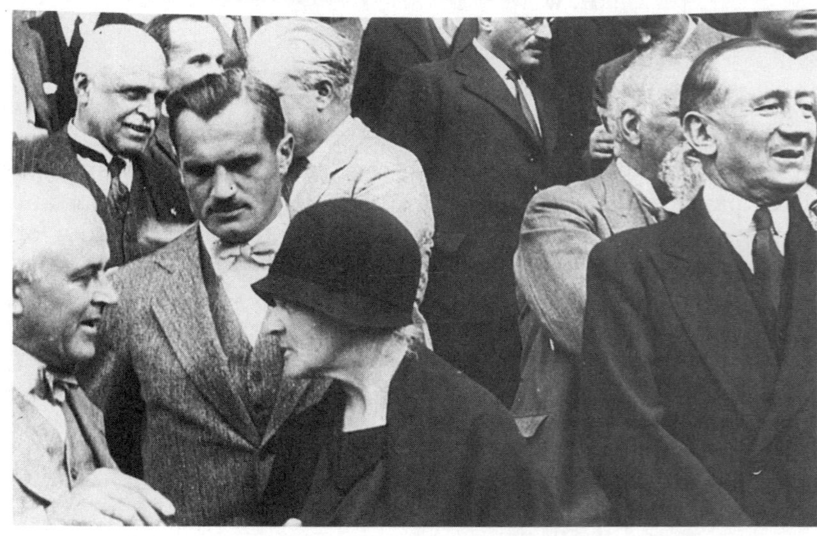

Robert A. Millikan (1868-1953), Arthur H. Compton (1892-1962), Marie Sklowdowska Curie (1867-1934), Guglielmo Marconi (1874-1937), and **Paul Ehrenfest (1880-1933)**, rear. (Courtesy of AIP Emilio Segrè Visual Archives.)

William D. Coolidge (1873-1975), Albert W. Hull (1890-1966), and **Irving Langmuir (1881-1951).** (Photograph by Francis Simon. Courtesy of AIP Emilio Segrè Visual Archives.)

Papers Reproduced on CD-ROM

E. F. Nichols. A study of the transmission spectra of certain substances in the infra-red, *Phys. Rev. 1st series* **I**, 1–3, 6, 7, 18 (1893)

E. F. Nichols and G. F. Hull. A preliminary communication on the pressure of heat and light radiation, *Phys. Rev. 1st series* **XIII**, 307–308, 317–320 (1901)

E. Rutherford and H. L. Cooke. A penetrating radiation from the Earth's surface, *Phys. Rev. 1st series* **XVI**, 183(A) (1903)

R. W. Wood and H. W. Springsteen. The magnetic rotation of sodium vapor, *Phys. Rev. 1st series* **XXI**, 41–51 (1905)

E. B. Rosa and N. E. Dorsey. The ratio of the electromagnetic and electrostatic units, *Phys. Rev. 1st series* **XXII**, 367–368(A) (1906)

G. B. Pegram and H. W. Webb. Heat developed in mass of thorium oxide, due to its radioactivity, *Phys. Rev. 1st series* **XXVI**, 410(A) (1908)

C. Barus and M. Barus. The grating interferometer, *Phys. Rev. 1st series* **XXXI**, 591–598 (1910)

C. D. Child. Discharge from hot CaO, *Phys. Rev. 1st series* **XXXII**, 492–511 (1911)

H. Fletcher. A verification of the theory of Brownian movements and a direct determination of the value of *NE* for gaseous ionization, *Phys. Rev. 1st series* **XXXIII**, 81–95, 106–110 (1911)

R. A. Millikan. On the elementary electrical charge and the Avogadro constant, *Phys. Rev.* **2**, 109–124, 133, 136–143 (1913)

W. H. Kadesch. The energy of photo–electrons from sodium and potassium as a function of the frequency of the incident light, *Phys. Rev.* **2**, 367–374 (1914)

T. Lyman. An extension of the spectrum in the extreme-violet, *Phys. Rev.* **3**, 504–505(A) (1914)

R. A. Millikan. A direct determination of "*h*", *Phys. Rev.* **4**, 73–75(A) (1914)

H. E. Ives, W. W. Coblentz, and E. F. Kingsbury. The mechanical equivalent of light, *Phys. Rev.* **5**, 269–274, 292–293 (1915)

H. E. Ives and O. Stuhlmann, Jr. The result of plotting the separation of homologous pairs against atomic numbers instead of atomic weights, *Phys. Rev.* **5**, 368–372 (1915)

W. Duane and F. L. Hunt. On x-ray wave-lengths, *Phys. Rev.* **6**, 166–171(A) (1915)

S. J. Barnett. Magnetization by rotation, *Phys. Rev.* **6**, 171–172(A) (1915)

A. W. Hull. A new method of x-ray crystal analysis, *Phys. Rev.* **10**, 661–664, 668–683 (1917)

R. W. Wood and F. L. Mohler. Resonance radiation of sodium vapor excited by one of the D lines, *Phys. Rev.* **11**, 70–80 (1918)

A. J. Dempster. A new method of positive ray analysis, *Phys. Rev.* **11**, 316–325 (1918)

H. J. van der Bijl. Theory of the thermionic amplifier, *Phys. Rev.* **12**, 171–180, 194–198 (1918)

J. Valasek. Piezoelectric and allied phenomena in Rochelle salt, *Phys. Rev.* **15**, 537–538(A) (1920)

W. D. Harkins and S. L. Madorsky. A graphical study of the stability relations of atom nuclei, *Phys. Rev.* **19**, 135–140, 142, 154–156 (1922)

A. H. Compton. A quantum theory of the scattering of x-rays by light elements, *Phys. Rev.* **21**, 483–502 (1923)

S. Dushman. Electron emission from metals as a function of temperature, *Phys. Rev.* **21**, 623–636 (1923)

A. H. Compton. The spectrum of scattered x-rays, *Phys. Rev.* **22**, 409–413 (1923)

A. H. Compton. A general quantum theory of the wave-length of scattered x-rays (Abstract), *Phys. Rev.* **24**, 168 (1924)

A. H. Compton and A. W. Simon. Directed quanta of scattered x-rays, *Phys. Rev.* **26**, 289–291, 299 (1925)

C. Eckart. Operator calculus and the solution of the equations of quantum dynamics, *Phys. Rev.* **28**, 711–726 (1926)

R. T. Birge. The most probable value of certain basic constants, *Phys. Rev.* **28**, 848(A) (1926)

J. B. Johnson. Thermal agitation of electricity in conductors, *Phys. Rev.* **29**, 367–368(A) (1927)

C. Davisson and L. H. Germer. Diffraction of electrons by a crystal of nickel, *Phys. Rev.* **30**, 705–740 (1927)

H. Nyquist. Thermal agitation of electric charge in conductors, *Phys. Rev.* **32**, 110–113 (1928)

R. W. Gurney and E. U. Condon. Quantum mechanics and radioactive disintegration, *Phys. Rev.* **33**, 127–140 (1929)

G. Breit. The effect of retardation on the interaction of two electrons, *Phys. Rev.* **34**, 553–573 (1929)

T. H. Johnson. Diffraction of hydrogen atoms, *Phys. Rev.* **37**, 847–861 (1931)

N. M. Gray and L. A. Wills. Note on the calculation of zero order eigenfunctions, *Phys. Rev.* **38**, 248–254 (1931)

H. C. Urey, F. G. Brickwedde, and G. M. Murphy. A hydrogen isotope of mass 2, *Phys. Rev.* **39**, 164–165(L) (1932)

H. C. Urey, F. G. Brickwedde, and G. M. Murphy. A Hydrogen isotope of mass 2 and its concentration, *Phys. Rev.* **40**, 1–15 (1932)

G. E. Uhlenbeck and L. Gropper. The equation of state of a non-ideal Einstein–Bose or Fermi–Dirac gas, *Phys. Rev.* **41**, 79–90 (1932)

K. T. Bainbridge. The isotopic weight of H^2, *Phys. Rev.* **42**, 1–2, 8–10 (1932)

K. T. Bainbridge. The equivalence of mass and energy, *Phys. Rev.* **44**, 123(L) (1933)

R. A. Beth. Direct detection of the angular momentum of light, *Phys. Rev.* **48**, 471(L) (1935)

E. Merritt. Edward Leamington Nichols, *Phys. Rev.* **53**, 1–2 (1938)

A. H. Spees and C. T. Zahn. The specific charge of the positron, *Phys. Rev.* **58**, 861–864 (1940)

Walter G. Cady (1874–1974). (Courtesy of AIP Emilio Segrè Visual Archives.)

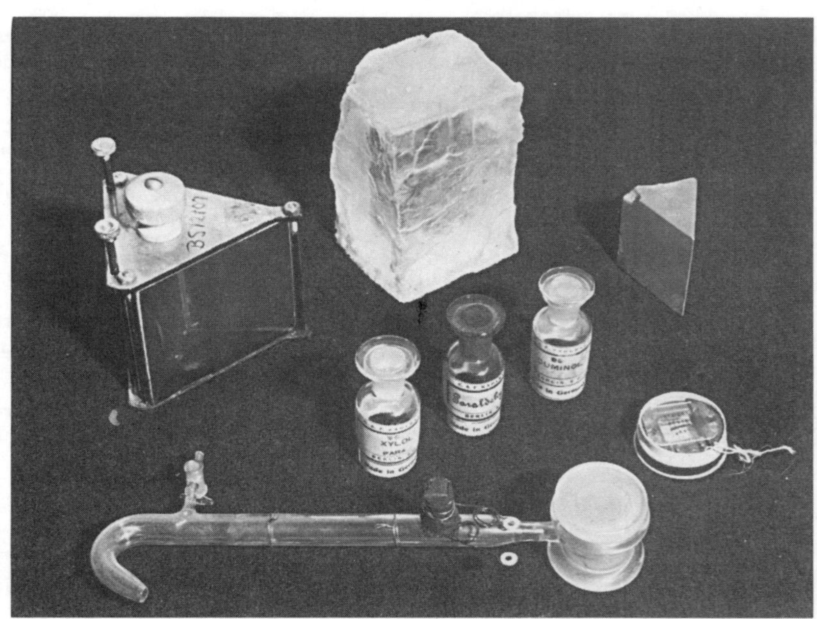

Natural NaCl prism and AgCl prism of Coblentz (1907), thermocouple and vacuum container. (Courtesy of AIP Emilio Segrè Visual Archives.)

ON THE ELEMENTARY ELECTRICAL CHARGE AND THE AVOGADRO CONSTANT.

By R. A. Millikan.

I. Introductory.

THE experiments herewith reported were undertaken with the view of introducing certain improvements into the oil-drop method[1] of determining e and N and thus obtaining a higher accuracy than had before been possible in the evaluation of these most fundamental constants.

In the original observations by this method such excellent agreement was found between the values of e derived from different measurements (l. c., p. 384) that it was evident that if appreciable errors existed they must be looked for in the constant factors entering into the final formula rather than in inaccuracies in the readings or irregularities in the behavior of the drops. Accordingly a systematic redetermination of all these constants was begun some three years ago. The relative importance of the various factors may be seen from the following review.

As is now well known the oil-drop method rested originally upon the assumption of Stokes's law and gave the charge e on a given drop through the equation

$$e_n = \frac{4}{3}\pi\left(\frac{9\eta}{2}\right)^{\frac{3}{2}}\left(\frac{1}{g(\sigma-\rho)}\right)^{\frac{1}{2}}\frac{(v_1+v_2)v_1^{\frac{1}{2}}}{F}, \qquad (1)$$

in which η is the coefficient of viscosity of air, σ the density of the oil, ρ that of the air, v_1 the speed of descent of the drop under gravity and v_2 its speed of ascent under the influence of an electric field of strength F.

The essential feature of the method consisted in repeatedly changing the charge on a given drop by the capture of ions from the air and in thus obtaining a series of charges with each drop. These charges showed a very exact multiple relationship under all circumstances—a fact which demonstrated very directly the atomic structure of the electric charge. If Stokes's law were correct the greatest common divisor of this series of charges should have been the absolute value of the elementary electrical charge. But the fact that this greatest common divisor failed to come out a constant when drops of different sizes were used showed that Stokes's

[1] R. A. Millikan, Phys. Rev., 32, pp. 349–397, 1911.

law breaks down when the diameter of a drop begins to approach the order of magnitude of the mean free path of a gas molecule. Consequently the following corrected form of Stokes's law for the speed of a drop falling under gravity was suggested.

$$v_1 = \frac{2}{9}\frac{ga^2(\sigma-\rho)}{\eta}\left\{1 + A\frac{l}{a}\right\}, \qquad (2)$$

in which a is the radius of the drop, l the mean free path of a gas molecule and A an undetermined constant. It is to be particularly emphasized that the term in the brackets was expressly set up merely as a first order correction term in *l/a and involved no theoretical assumptions of any sort;* further that *the constant A was empirically determined* through the use only of small values of l/a and that *the values of e and N obtained were therefore precisely as trustworthy as were the observations themselves.* This fact has been repeatedly overlooked in criticisms of the results of the oil-drop method.[1]

Calling then e_1 the greatest common divisor of all the various values of e_n found in a series of observations on a given drop there resulted from the combination of (1) and (2) the equation

$$e\left(1 + A\frac{l}{a}\right)^{\frac{3}{2}} = e_1, \qquad (3)$$

or

[1] Indeed M. Jules Roux (Compt. Rendu, 152, p. 1168, May, 1911) has attempted to correct my values of e and N by reducing some observations like mine which he made on droplets of sulphur, with the aid of a purely theoretical value of A which is actually approximately twice too large. The impossibility of the value of A which he assumes he would himself have discovered had he made observations on spheres of different sizes or at different pressures. Such observations whether made on solid spheres or on liquid spheres always yield a value of A about half of that assumed by Roux. Hence his value of e, viz., $e = 4.17 \times 10^{-10}$ rests on no sort of experimental foundation whatsoever. It rests rather on two erroneous assumptions, first the assumption of the correctness of the constants in Cunningham's theoretical equation (Proc. Roy. Soc., 83, p. 357; see also footnote 3, p. 380, Phys. Rev., Vol. 32)—constants which I shall presently show are in no case correct within the limits of experimental error even when inelastic impact is assumed, and second, the assumption that molecules make elastic impact against solid surfaces, an assumption which is completely incorrect as I had already proved by showing that the value of the "slip" term is the same for oil and air as for glass and air (Phys. Rev., Vol. 32, p. 382), which Knudsen also had proved experimentally to be erroneous (Knudsen, An. der Phys., 28, p. 75, 1909, and 35, p. 389) and which for theoretical reasons as well is plainly inadmissible, since were it correct Poiseuille's law could not hold for gases under any circumstances.

But even if Roux had assumed the correct value of A he would still have obtained results several per cent. too low, a fact which must be ascribed either to faulty experimental arrangements or to imperfect knowledge of the density of his sulphur spheres; *for solid spheres have been very carefully studied in the Ryerson Laboratory and are in fact found to yield results very close to those obtained with oil drops.* Solid spheres however are not nearly so well adapted to a precision measurement of e as are oil drops, since their density and sphericity are always matters of some uncertainty.

$$e = \frac{e_1}{\left(1 + A\frac{l}{a}\right)^{\frac{3}{2}}}. \qquad (4)$$

It was from this equation that e was obtained after A had been found by a graphical method which will be more fully considered presently.

The factors then which enter into the determination of e are: (1) The density factor, $\sigma - \rho$; (2) the electric field strength, F; (3) the viscosity of air, η; (4) the speeds, v_1 and v_2; (5) the drop radius, a; (6) the correction term constant, A.

Concerning the first two of these factors little need be said unless a question be raised as to whether the density of such minute oil drops might not be a function of the radius. Such a question is conclusively answered in the negative both by theory[1] and by the experiments reported in this paper.

Liquid rather than solid spheres were originally chosen because of the far greater certainty with which their density and sphericity could be known. Nevertheless I originally used liquids of widely different viscosities (light oil, glycerine, mercury) and obtained the same results with them all within the limits of error, thus showing experimentally that so far as this work was concerned, the drops all acted like rigid spheres. More complete proof of this conclusion is furnished both by the following observations and by other careful work on solid spheres soon to be reported in detail by Mr. J. Y. Lee.

The material used for the drops in the following experiments was the highest grade of clock-oil, the density of which, at 23° C., the temperature

[1] The pressure p_2 within an oil drop is given by

$$p_2 = k + \frac{\alpha}{2R}$$

where k is LaPlace's constant of internal pressure, α the constant of surface tension and R the radius. The difference $(p_2 - p_1)$ between the pressure within the oil drop and within the oil in bulk is then $\alpha/2R$. But the coefficient of compressibility of a liquid is defined by

$$\beta = \frac{v_2 - v_1}{v_1(p_2 - p_1)}.$$

Now β for oil of this sort never exceeds 70×10^{-6} megadynes per sq. cm. (see Landolt and Bornstein's tables), while α is about 35 dyne cm. R for the smallest drop used (Table XX.) is .00005 cm.; we have then

$$\frac{v_2 - v_1}{v_1} = \beta\frac{\alpha}{2R} = \frac{70 \times 10^{-12} \times 35}{.0001} = .000024.$$

The density of the smallest drop used is then 2 parts in 100,000 greater than that of the oil in bulk. The small drops could then only be appreciably denser than the larger ones if the oil were inhomogeneous and if the atomizing process selected the heavier constituents for the small drops. Such an assumption is negatived by the experimental results given in § 9.

at which the experiments were carried out, was found in two determinations made four months apart, to be .9199 with an error of not more than one part in 10,000.

The electric fields were produced by a 5,300-volt storage battery, the P.D. of which dropped on an average 5 or 10 volts during an observation of an hour's duration. The potential readings were taken, just before and just after a set of observations on a given drop, by dividing the bank into 6 parts and reading the P.D. of each part with a 900-volt Kelvin and White electrostatic voltmeter which showed remarkable constancy and could be read easily, in this part of the scale, with an accuracy of about 1 part in 2,000. This instrument was calibrated by comparison with a 750-volt Weston Laboratory Standard Voltmeter certified correct to 1/10 per cent. and actually found to have this accuracy by comparison with an instrument standardized at the Bureau of Standards in Washington. The readings of P.D. should therefore in no case contain an error of more than 1 part in 1,000. As a matter of fact 5,000 volt readings made with the aid of two different calibration curves of the K. & W. instrument made two years apart never differed by more than 1 or 2 parts in 5,000.

The value of F involves in addition to P.D. the distance between the plates, which was as before 16 mm. and correct to about .01 mm. (l. c., p. 351). Nothing more need be said concerning the first two of the above-mentioned factors. The last four however need especial consideration.

2. The Coefficient of Viscosity of Air.

This factor certainly introduces as large an element of uncertainty as inheres anywhere in the oil-drop method. Since it appears in equation (1) in the 3/2 power an uncertainty of 0.5 per cent. in η means an uncertainty of 0.75 per cent. in e. It was therefore of the utmost importance that η be determined with all possible accuracy. Accordingly two new determinations were begun three years ago in the Ryerson Laboratory, one by Mr. Lachlan Gilchrist and one by Mr. I. M. Rapp. Mr. Gilchrist, whose work has already been published,[1] used a constant deflection method (with concentric cylinders), which it was estimated (l. c.; p. 386) ought to reduce the uncertainty in η to 1 or 2 tenths of a per cent. The results have justified this estimate. Mr. Rapp used a form of the capillary tube method which it was thought was better adapted to an *absolute evaluation of* η than have been the capillary tube arrangements which have been commonly used heretofore.[2] Since Mr.

[1] Lachlan Gilchrist, Phys. Rev., 2d Ser., Vol. 1, p. 124.

[2] This investigation will shortly be published in full (Phys. Rev., 1913), hence only a bare statement will here be made of the results which are needed for the problem in hand.

Gilchrist completed his work at the University of Toronto, Canada, and Mr. Rapp made his computations and final reductions at Ursinus College, Pa., neither observer had any knowledge of the results obtained by the other. The two results agree within 1 part in 600. Mr. Rapp estimates his maximum uncertainty at 0.1 per cent., Mr. Gilchrist at 0.2 per cent. Mr. Rapp's work was done at 26° C. and gave $\eta_{26} = .00018375$. When this is reduced to 23° C., the temperature used in the following work, by means of formula (5)—a formula[1] which certainly can introduce no appreciable error for the range of temperature here used,—viz.,

$$\eta_t = 0.00018240 - 0.000000493(23 - t); \qquad (5)$$

there results

$$\eta_{23} = .00018227.$$

Mr. Gilchrist's work was done at 20.2° C. and gave $\eta_{20.2} = .0001812$. When this is reduced to 23° C. it yields

$$\eta_{23} = .00018257.$$

When this new work, by totally dissimilar methods, is compared with the best existing determinations by still other methods the agreement is exceedingly striking. Thus in 1905, Hogg[2] made at Harvard very careful observations on the damping of oscillating cylinders and obtained in three experiments at atmospheric pressure $\eta_{23} = 0.0001825$, $\eta_{15.6} = 0.0001790$ and $\eta_{18.6} = 0.0001795$. These last two reduced to 23° C., as above, are 0.0001826 and 0.0001817 respectively and the mean value of the three determinations is

$$\eta_{23} = 0.00018227.$$

Tomlinson's classical determination,[3] by far the most reliable of the nineteenth century, yielded when the damping was due primarily to "push" $\eta_{12.65° C.} = 0.00017746$; when it was due wholly to "drag" $\eta_{11.79° C} = 0.00017711$. These values reduced to 15° C., as above, are respectively 0.00017862 and 0.00017867. Hence we may take Tomlinson's direct determination as $\eta_{15} = 0.00017864$. This reduced to 23° C. by Tomlinson's own temperature formula (Holman's) yields $\eta_{23} = 0.00018242$. By the above formula it yields $\eta_{23} = 0.00018256$.

Grindley and Gibson using the tube method on so large a scale[4] (tube 1/8 inch in diameter and 108 feet long) as to largely eliminate the most

[1] See R. A. Millikan, Annalen der Physik, 1913, for a more extended discussion of this and other viscosity formulæ and measurements.

[2] J. L. Hogg, Proc. Amer. Acad., 40, 18, p. 611, 1905.

[3] Tomlinson, Phil. Trans., 177, p. 767, 1886.

[4] Grindley and Gibson, Proc. Roy. Soc., 80, p. 114, 1908.

fruitful sources of error in this method, namely, the smallness and ununiformity of the bore, obtained at room temperature the following results:[1] $\eta_{23.28° C.} = .00018347$, $\eta_{23.55° C.} = .00018241$, $\eta_{12.18° C.} = .00018257$, and $\eta_{15.4° C.} = .0001782$. These numbers, reduced to 23° C. as above, are respectively 18,245, 18,241, 18,201, and 18,195. The mean is 18,220. Grindley and Gibson's own formula, $\eta = .0001702\ \{1 + .00329t - .0000070t^2\}$, yields $\eta_{23} = .00018245$. We may take then Grindley and Gibson's direct determination as the mean of these two values, viz.: $\eta_{23} = .00018232$.

Collecting then the five most careful determinations of the viscosity of air which so far as I am able to discover have ever been made we obtain the following table.

TABLE I.

Air η_{23} = .00018227—Rapp. Capillary tube method. 1913.
Air η_{23} = .00018257—Gilchrist. Constant deflection method. 1913.
Air η_{23} = .00018227—Hogg. Damping of oscillating cylinder method. 1905.
Air η_{23} = .00018258—Tomlinson. Damping of pendular vibrations method. 1886.
Air η_{23} = .00018232—Grindley and Gibson. Flow through large pipe method. 1908.
Mean = .00018240

It will be seen, then, that every one of the five different methods which have been used for the absolute determination of η leads to a value which differs by less than 1 part in 1,000 from the above mean value η_{23} = .00018240. It is surely legitimate then to conclude that the absolute value of η for air is now known with an uncertainty of somewhat less than 1 part in 1,000.[2]

[1] These numbers represent the reduction to absolute C.G.S. units of all the observations which Grindley and Gibson made between 50° F. and 80° F.

[2] In obtaining the above mean I have chosen what, after careful study, I have considered to be the best determination by each of the five distinct methods. The transpiration method has been much more commonly used than have the others, and in general, the final result is in good agreement with other careful work by this method. Thus Rankine's final value (Proc. Roy. Soc., *A*, 83, p. 522, 1910) by a new modification of the capillary tube method, while probably not claiming an accuracy of more than .4 per cent., is, at 10.6° C., .0001767, a value which reduces to η_{23} = .0001828. Again Fisher's final formula (PHYS. REV., 28, p. 104, 1909) gives η_{23} = .00018218. Also Holman's much used formula (Phil. Mag., 21, p. 199, 1886, and Tomlinson, Phil. Trans., Vol. 177, part 2, p. 767, 1886) yields η_{23} = .00018237. In fact the only reliable work on η which I am able to find which is out of line with the value η_{23} = .00018240 is that by Breiterbach at Leipzig (Ann. der Phys., 5, p. 166, 1901) and that by Schultze (Ann. der Phys., 5, p. 157, 1901) and several other observers at the University of Halle who used Schultze's apparatus (Markowski, Ann. der Phys., 14, p. 742, 1904, and Tanzler, Verh. der D. Phys. Ges., 8, p. 222, 1906). None of these observers however were aiming at an absolute determination, but rather at the effects of temperature and the mixing of gases upon viscosity and their capillaries were too small (of the order .007 cm.) to make possible an absolute determination of high accuracy. Their agreement among themselves upon a value which is about 1.3 per cent. too high is partly accounted for by the fact that everal of them used the same tube. None of the m made any effort to eliminate the necessarily large error in the measurement of so small a bore (which appears in the result in the fourth power) by taking the mean of η from a considerable number of tubes.

A second question which might be raised in connection with η is as to whether the medium offers precisely the same resistance to the motion through it of a heavily charged drop as to that of an uncharged drop. This question has been carefully studied and definitely answered in the affirmative by the following work (cf. §§ 6 and 10).

3. The Speeds v_1 and v_2.

The accuracy previously attained in the measurement of the times of ascent and descent between fixed cross-hairs was altogether satisfactory, but the method which had to be employed for finding the magnifying power of the optical system, *i. e.*, for finding the actual distance of fall of the drop in centimeters, left something to be desired. This optical system was before a short-focus telescope of such depth of focus that it was quite impossible to obtain an accurate measure of the distance between the cross-hairs by simply bringing a standard scale into sharp focus immediately after focusing upon a drop. Accordingly, as stated in the original article, the standard scale was set up at the exact distance from the telescope of the pin-hole through which the drop entered the field. This distance could be measured with great accuracy but *the procedure assumed that the drop remained exactly at this distance throughout the whole of any observation, sometimes of several hours duration.* But if there were the slightest lack of parallelism between gravity and the lines of the electric field the drop would be obliged to drift slowly, and always in the same direction, away from this position, and a drift of 5 mm. was enough to introduce an error of 1 per cent. Such a drift could in no way be noticed by the observer if it took place in the line of sight; for the speeds of the drops were changing *very slowly* anyway because of evaporation, fall in the potential of the battery, etc., and a change in time due to such a drift would be completely masked by other causes of change. This source of uncertainty was well recognized at the time of the earlier observations and steps were taken at the beginning of the present work to eliminate it. It was in fact responsible for an error of nearly two per cent.

A new optical system was built, consisting of an achromatic objective of 28 mm. aperture and 12.5 cm. focal length and an eyepiece of 12 mm. focal length. The whole system was mounted in a support which could be moved bodily back and forth by means of a horizontal screw of ½ mm. pitch. In an observation the objective was 25 cm. distant from the drop, which was kept continually in sharp focus by advancing or withdrawing the whole telescope system. The depth of focus was so small that a motion of ½ mm. blurred badly the image of the drop. The eyepiece

was provided with a scale having 80 horizontal divisions and the distance between the extreme divisions of this scale (the distance of fall in the following experiments) could be regularly duplicated with an accuracy of at least 1 part in 1,000, by bringing a standard scale (Société Genevoise) into sharp focus. (The optical path when the scale was viewed was made exactly the same as when the drop was viewed.) The distance of fall, then, one of the most uncertain factors of the preceding determination, was now known with at least this degree of precision.

The accuracy of the *time* determinations can be judged from the data in Tables IV.–XIX. On account of the great convenience of a direct-reading instrument these time measurements were all made, not with a chronograph, as heretofore, but with a Hipp chronoscope which read to 0.002 second. This instrument was calibrated by comparison with the standard Ryerson Laboratory clock under precisely the same conditions as those under which it was used in the observations themselves and found to have an error between 0 and 0.2 per cent. depending upon the time interval measured. For the sake of enabling others to check all the computations herein contained if desired, as well as for the sake of showing what sort of consistency was attained in the measurement of time intervals there are given in Table II. the calibration readings for the 30 sec. interval and in Table III. the results of similar readings for all the intervals used.

Table II.

Chronoscope Readings for 30 Sec. Interval.	
29.962	29.990
29.988	29.958
29.986	29.920
29.930	29.972
29.964	29.976
30.002	30.006
29.940	29.979
29.998	30.018
29.930	29.926
29.967	29.972
Corr'n = +.1 per cent.	

Table III.

Clock Interval, Sec.	Chronoscope Interval.	Corr'n Applied, Per Cent.
6	6.0146	−0.26
10	10.0018	0.00
16	16.0080	0.00
20	19.9835	+0.07
30	29.9695	+0.10
40	39.9436	+0.14
60	59.9072	+0.16
114	113.795	+0.20
120	119.782	+0.20

The change in the per cent. correction with the time interval employed is due to the difference in the reaction times of the magnet and spring contact at make (beginning) and at break (end). All errors of this sort are obviously completely eliminated by making the calibration observations under precisely the same conditions as the observations on the drop. In Tables IV. to XIX. the recorded times are the uncorrected chronograph readings. The corrections are obtained by interpolation in the last column of Table III.

Under the head of possible uncertainties in the velocity determinations are to be mentioned also the effects of a distortion of the drop by the electric field. Such a distortion would increase the surface of the drop, and hence the speed imparted to it per dyne of *electric* force would not be the same as the speed imparted per dyne of gravitational force when the field was off and the drop had the spherical form. The following observations were made in such a way as to bring to light such an effect if it were of sufficient magnitude to exert any influence whatever upon the accuracy of the determination of e by this method (cf. §§ 6 and 10).

Similarly objection has been made to the oil-drop method on the ground that, on account of internal convection, fluid drops would not move through air with the same speed as solid drops of like diameter and mass. Such objection is theoretically unjustifiable in the case of oil drops of the sizes here considered.[1] Nevertheless the experimental demonstration of its invalidity is perhaps worth while and is therefore furnished below.

4. The Radius "a."

The radius of the drop enters only into the correction term (see equation 4) and so long as this is small need not be determined with a high degree of precision. It is most easily obtained by the following procedure which differs slightly from that originally employed (l. c., p. 379).

It will be seen that the equation (l. c., p. 353)

$$\frac{v_1}{v_2} = \frac{mg}{Fe - mg} \tag{6}$$

contains no assumption whatever save that a given body moves through a given medium with a speed which is proportional to the force acting upon it. Substitution in this equation of $m = \frac{4}{3}\pi a^3(\sigma - \rho)$ and the solution of the resulting equation for a gives

$$a = \sqrt[3]{\frac{3Fe}{4\pi g(\sigma - \rho)}\,\frac{v_1}{(v_1 + v_2)}}. \tag{7}$$

The substitution in this equation of an approximately correct value of e yields a with an error but one third as great as that contained in the assumed value of e. The radius of the drop can then be determined from (7) with a very high degree of precision as e becomes more and more accurately known. In the following work the value of e substituted in (7) to obtain a was 4.78×10^{-10} but the final value of e obtained would

[1] Hadamard, Compt. Rendus, 1911.

not have been appreciably different if the value substituted in (7) to obtain a had been 5 per cent. or 6 per cent. in error. The determination of a therefore introduces no perceptible error into the evaluation of e.

5. The Correction-Term Constant A.

This constant was before graphically determined (l. c., p. 379) by plotting the values of $e_1^{\frac{2}{3}}$ as ordinates and those of l/a as abscissæ and observing that if we let $x = l/a$, $y = e_1^{\frac{2}{3}}$ and $y_0 = e^{\frac{2}{3}}$ equation (3) may be written in the form

$$y_0(1 + Ax) = y \tag{8}$$

or

$$A = \frac{\frac{dy}{dx}}{y_0} = \frac{\text{slope}}{y\text{ intercept}}. \tag{9}$$

Now even if the slope were correctly determined by the former observations all of the above-mentioned sources of error would enter into the value of the intercept and hence would modify the value of A.

As a matter of fact however the accuracy with which the slope itself was determined could be much improved, for with the preceding arrangement it was necessary to make all the observations at atmospheric pressure and the only way of varying l/a was by varying a, *i. e.*, by using drops of different radii. But when a was very small the drops moved exceedingly slowly under gravity and the minutest of residual convection currents produced relatively large errors in the observed speeds, *i. e.*, in e_1. If for example the time of fall over a distance of 2 mm. is 20 minutes it obviously requires an extraordinary degree of stagnancy to prevent a drift in that time of say .2 mm. due to convection. But this would introduce an error of 10 per cent. into e_1. Furthermore with these slow drops Brownian movements introduce errors which can only be eliminated by taking a very large number of readings[1] and this is not in general feasible with such drops. It is quite impossible then by working at a single pressure to obtain from the graph mentioned above a line long enough (l. c., p. 379) to make the determination of its slope a matter of great precision. Accordingly in the new observations the variation of l/a was effected chiefly through the variation of l, *i. e.*, of the pressure p, rather than of a. This made possible not only the accurate evaluation of e, but also the solution of the interesting question as to the law of fall of a given drop through air at reduced pressures.

[1] Fletcher, Phys. Rev., 33, p. 92, 1911.

6. Method of Testing the Assumptions Involved in the Oil-Drop Method.

In order to make clear the method of treatment of the following observations a brief consideration of the assumptions underlying the oil-drop method must here be made. These assumptions may be stated thus:

1. The drag which the medium exerts upon a given drop is unaffected by its charge.

2. Under the conditions of observation the oil drops move through the medium essentially as would solid spheres. This assumption may be split into two parts and stated thus: Neither (2*a*) distortions due to the electric field nor (2*b*) internal convection within the drop modify appreciably the law of motion of an oil drop.[1]

3. The density of oil droplets is independent of their radius down to $a = .00005$ cm.

Of these assumptions (2*a*) is the one which needs the most careful experimental test.[2] It will be seen that it is contained in the fundamental equation of the method (see (7)) which may be written in the form

$$e_n = \frac{mg}{Fv_1}(v_1 + v_2). \qquad (10)$$

Or still more conveniently in the form

$$e_n = \frac{mgt_g}{F}\left(\frac{1}{t_g} + \frac{1}{t_F}\right), \qquad (11)$$

in which t_g and t_F are the respective time intervals required by the drop to fall under gravity and to rise under the field F the distance between the cross-hairs.

In order to see how the assumption under consideration can be tested let us write the corresponding equation after the same drop has caught n' additional units, namely,

$$e_{n+n'} = \frac{mgt_g}{F}\left(\frac{1}{t_g} + \frac{1}{t_F'}\right). \qquad (12)$$

The subtraction of (11) from (12) gives

$$e_{n'} = \frac{mgt_g}{F}\left(\frac{1}{t_F'} - \frac{1}{t_F}\right). \qquad (13)$$

[1] M. Brillouin has in addition suggested (see p. 149, La Théorie du Rayonnement et les Quanta) that the drops may be distorted by the molecular bombardment, but Einstein's reply (l. c., p. 150) to this suggestion is altogether unanswerable, and, in addition, such a distortion, if it existed, would make the value of e given by the oil-drop method *too small* instead of too large.

[2] Professor Lunn has however subjected it to a theoretical study and has in this way demonstrated its validity (Phys. Rev., XXXV., p. 227, 1912).

Now equations (11) and (12) show, since mgt_g/F remains constant, that as the drop changes charge the successive values of its charge are proportional to the successive values assumed by the quantity $(1/t_g + 1/t_F)$ and the elementary charge itself is obviously this same constant factor mgt_g/F multiplied by *the greatest common divisor* of all these successive values. It is to be observed too that since $1/t_g$ is in these experiments generally large compared to $1/t_F$ the value of this greatest common divisor, which will be denoted by $(1/t_g + 1/t_F)_0$, is determined primarily by the time of fall *under gravity*, and is but little affected by the time in the field. On the other hand equation (13) shows that the greatest common divisor of the various values of $(1/t_F' - 1/t_F)$, which will be designated by $(1/t_F' - 1/t_F)_0$, when multiplied by the same constant factor mgt_g/F, is also the elementary electrical charge. In other words $(1/t_g + 1/t_F)_0$ and $(1/t_F' - 1/t_F)_0$ are one and the same quantity, but while the first represents essentially a speed measurement when the field is off, the second represents a speed measurement in a powerful electric field. If then the assumption under consideration is correct we have two independent ways of obtaining the quantity which when multiplied by the constant factor mgt_g/F is the elementary electrical charge, but if on the other hand the distortion of the drop by the field modifies the law of motion of the oil drop through the medium then $(1/t_g + 1/t_F)_0$ and $(1/t_F' - 1/t_F)_0$ will not be the same. Now *a very careful experimental study of the relations of* $(1/t_g + 1/t_F)_0$ and $(1/t_F' - 1/t_F)_0$ *shows so perfect agreement that no effect of distortion in changing measurably the value of e can be admitted.*[1] (See Tables IV. to XIX.)

Turning next to assumption (1), this can be tested in three ways, all of which have been tried with negative results. First a drop containing from one up to six or seven elementary charges can be completely discharged and its time of fall under gravity when uncharged compared with its time when charged. Second, the multiple relationships shown in the successive charges carried by a given drop may be very carefully examined. They cannot hold exactly if when the drop is heavily charged it suffers a larger drag from the medium than when it is lightly charged. Third, when drops having widely different charges and different masses are

[1] It may be pointed out in passing that the above discussion brings to light a method of obtaining e which is independent of a viscosity measurement; for $(1/t_F' - 1/t_F)_0$ can be obtained for a body which is heavy enough to be weighed upon a micro-balance. Such a body would fall so rapidly that $1/t_g$ could not be measured, but it could be computed from the measurement of $1/t_F'$ and $1/t_F$ and the equation $(1/t_g + 1/t_F)_0 = (1/t_F' - 1/t_F)_0$. Either (12) or (13) could then be solved for e after m had been determined by direct weighing. A consideration of the sources of error in this method shows however that it cannot be made as accurate as the present method which involves the coefficient of viscosity of air.

brought to the same value of l/a by varying the pressure, the value of e_1 (which is proportional to $(v_1 + v_2)_0$), should come out smaller for the heavily than for the lightly charged drops. The following observations show that this is not the case.

The last criterion is also a test for (2*b*) for if internal convection modifies the speed of fall of a drop as Perrin wishes to assume that it may,[1] it must play a smaller and smaller rôle as the drop diminishes in size, hence varying l/a by diminishing a cannot be equivalent to varying l/a by increasing l. In other words the value of e_1 obtained from work on a large drop at a low pressure should be different from that obtained from work on a small drop at so high a pressure that l/a has the same value as for the large drop.

Finally if the density of a small drop is greater than that of a large one (see assumption 3) then, for a given value of l/a, the small drop will show a larger value of e_1 than the large one inasmuch as the computation of e_1 is based on a constant value of σ. *The fact, then, that for a given value of l/a the value of e_1 actually comes out independent of the radius or charge of the drops shows conclusively either that no one of these possible sources of error exists, or else that they neutralize one another so that for the purposes of this experiment they do not exist.* That they do not exist at all is shown by the independent theoretical and experimental tests mentioned above. This removes I think every criticism which has been suggested of the oil-drop method of determining e and N.

7. Summary of Improvements in Method.

In order to obtain the consistency shown in the following observations it was found necessary to take much more elaborate precautions to suppress convection currents in the air of the observing chamber than had at first been thought needful.

To recapitulate, then, the improvements which have been introduced into the oil-drop method, consist in (1) a redetermination of η; (2) an improved optical system; (3) an arrangement for observing speeds at all pressures; (4) the more perfect elimination of convection; (5) the experimental proof of the correctness of all the assumptions underlying the method, viz., (*a*) that a charge does not alter the drag of the medium on the charged body; (*b*) that the oil drops act essentially like solid spheres; (*c*) that the density of the oil drops is the same as the density of the oil in bulk.

[1] La Théorie du Rayonnement et les Quanta, p. 239—Rapports et Discussions de la Réunion tenue à Bruxelles, Novembre, 1911. Edited by Langevin and de Broglie. Gauthier-Villars.

8. The Experimental Arrangements.

The experimental arrangements are shown in Fig. 1. The brass vessel D was built for work at all pressures up to 15 atmospheres but since the present observations have to do only with pressures from 76 cm. down these were measured with a very carefully made mercury manometer m which at atmospheric pressure gave precisely the same reading as a

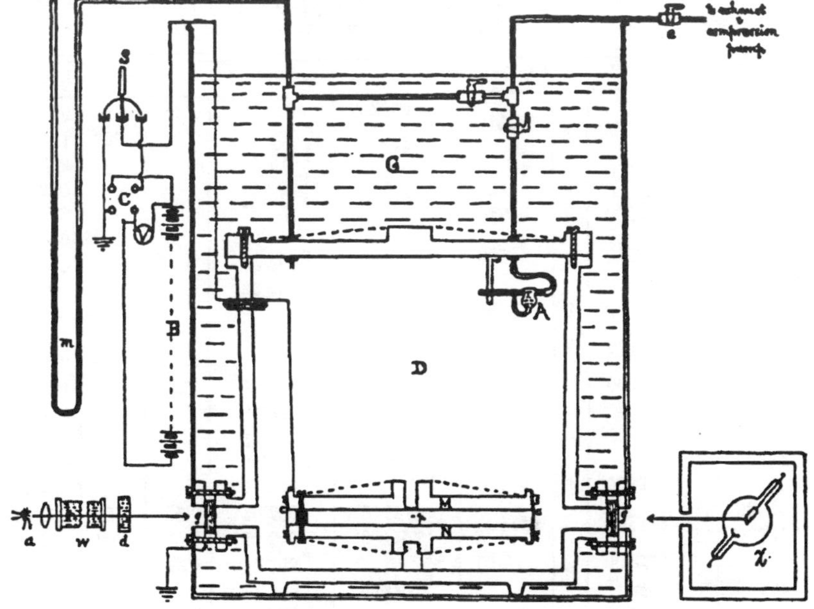

Fig. 1.

standard barometer. Complete stagnancy of the air between the condenser plates M and N was attained first by absorbing all of the heat rays from the arc A by means of a water cell w, 80 cm. long, and a cupric chloride cell[1] d, and second by immersing the whole vessel D in a constant temperature bath G of gas-engine oil (40 liters) which permitted, in general, fluctuations of not more than .02° C. during an observation. This constant temperature bath was found essential if such consistency of measurement as is shown below was to be obtained. A long search for causes of slight irregularity revealed nothing so important as this and after the bath was installed all of the irregularities vanished. The atomizer A was blown by means of a puff of carefully dried and dust-free air introduced through the cock e. The air about the drop p was ionized

[1] See Coblentz, Bulletin of the Bureau of Standards, Washington, D. C., Vol. 7, p. 660, 1911.

when desired by means of Rontgen rays from X which readily passed through the glass window g. To the three windows g (two only are shown) in the brass vessel D correspond, of course, three windows in the ebonite strip c which encircles the condenser plates M and N. Through the third of these windows, set at an angle of about 18° from the line Xpa and in the same horizontal plane, the oil drop is observed.

9. The Observations.

The record of a typical set of readings on a given drop is shown in Table IV. The first column, headed t_g, gives the successive readings on the time of descent under gravity. The fourth column, headed t_F, gives the successive times of ascent under the influence of the field F as measured on the Hipp chronoscope. These two columns contain all the data which is used in the computations. But in order to have a test of the stagnancy of the air a number of readings were also made with a stop-watch on the times of ascent through the first half and through the whole distance of ascent. These readings are found in the second and third columns, the times for the first half under the head $\frac{1}{2}t_F$, the times for the whole distance under the head t_F. It will be seen from these readings that there is no indication whatever of convection, since the readings for the one half distance have uniformly one half of the value of the readings for the whole distance, within the limits of error of a stop-watch measurement. This sort of a test was made on the majority of the drops, but since no further use is made of these stop-watch readings they will not be given in succeeding tables.

The fifth column, headed $1/t_F$, contains the reciprocals of the values in the fourth column after the correction found from Tables II. and III. has been applied. The sixth column contains the successive differences in the values of $1/t_F$ resulting from the capture of ions. The seventh column, headed n', contains the number of elementary units caught at each change, a number determined simply by observing by what number the quantity just before it in column 6 must be divided to obtain the constancy shown in the eighth column, which contains the successive determinations of $(1/t_F' - 1/t_F)_0$ (see § 6). Similarly the ninth column, headed n, gives the total number of units of charge on the drop, a number determined precisely as in the case of the numbers in the seventh column by observing by what numbers the successive values of $(1/t_g + 1/t_F)$ must be divided to obtain the constancy shown in the tenth column, which contains the successive values of $(1/t_g + 1/t_F)_0$. Since n' is always a small number and in some of the changes almost always has the value 1 or 2 its determination for any change is obviously never a matter of the slightest

uncertainty. On the other hand, n is often a large number, but with the aid of the known values of n' it can always be found with absolute certainty so long as it does not exceed say 100 or 150. It will be seen from the means at the bottom of the eighth and the tenth columns that in the case of this drop the two ways discussed in § 6 of obtaining the number which when multiplied by mgt_g/F is the elementary electrical charge yield absolutely identical results.

Table IV.

Drop No. 6.

t_g Sec.	$\frac{1}{2}t_F$ Sec.	t_F Sec.	t_F Sec.	$\frac{1}{t_F}$	$\left(\frac{1}{t_F'}-\frac{1}{t_F}\right)$	n'	$\frac{1}{n'}\left(\frac{1}{t_F'}-\frac{1}{t_F}\right)$	n	$\frac{1}{n}\left(\frac{1}{t_g}+\frac{1}{t_F}\right)$
11.848	39.9	80.2	80.708	.01236				18	.005366
11.890	11.2	22.4	22.366		.03234	6	.005390		
11.908			22.390	.04470				24	.005371
11.904	11.2	22.4	22.368		.03751	7	.005358		
11.882	70.6	140.4	140.565	.007192	.005348	1	.005348	17	.005375
11.906	39.9	79.6	79.600	.01254				18	.005374
11.838			34.748		.01616	3	.005387		
11.816			34.762	.02870				21	.005376
11.776			34.846						
11.840			29.286					22	.005379
11.904	14.6	29.3	29.236	.03414	.026872	5	.005375		
11.870	69.3	137.4	137.308	.007268	.021572	4	.005393	17	.005380
11.952	17.6	34.9	34.638	.02884				21	.005382
11.860					.01623	3	.005410		
11.846			22.104					24	.005386
11.912			22.268	.04507	.04307	8	.005384		
11.910			500.1	.002000				16	.005387
11.918			19.704		.04879	9	.005421	25	.005399
11.870			19.668	.05079	.03874	7	.005401		
11.888			77.630					18	.005390
11.894	38.9	77.6	77.806	.01285	.01079	2	.005395		
11.878	21.0	42.6	42.302	.02364				20	.005392
11.880					Means		.005386		.005384

Duration of exp. = 45 min.,
Plate distance = 16 mm.,
Fall distance = 10.21 mm.,
Initial volts = 5,088.8.
Final volts = 5,081.2.
Temperature = 22.82° C.,
Pressure = 75.62 cm.,
Oil density = .9199,
Air viscosity[1] = $1,824 \times 10^{-7}$,
Radius (a) = .000276 cm.,
l/a = .034,
Speed of fall = .08584 cm./sec.,
$e_1 = 4.991 \times 10^{-10}$.

[1] In the above and in all the following tables the computations were made on the basis of the assumption $\eta_{23} = 1,825 \times 10^{-7}$ instead of $\eta_{23} = 1,824 \times 10^{-7}$ (see § 2). The reduction to the latter value has been made only in the final value of e (see § 10).

10. Results and Discussion.

It will be seen at once from equation (4) that the value of e is simply the value of e_1 for which $l/a = 0$, so that if successive values of $e_1^{\frac{2}{3}}$ are plotted as abscissæ and of l/a as ordinates the intercept of the resulting curve on the $e_1^{\frac{2}{3}}$ axis is $e^{\frac{2}{3}}$. Furthermore if A is a constant then the curve in question is a straight line and A is the slope of this line divided by the

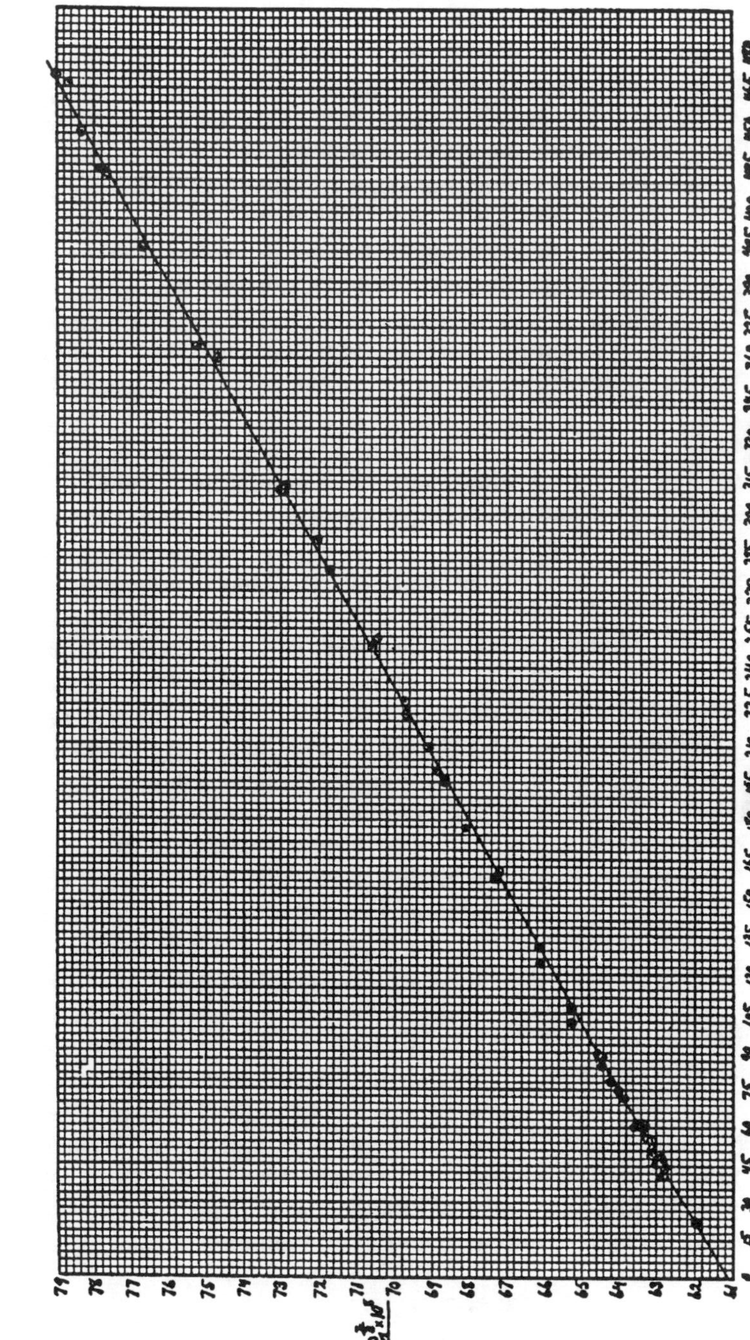

Fig. 2.

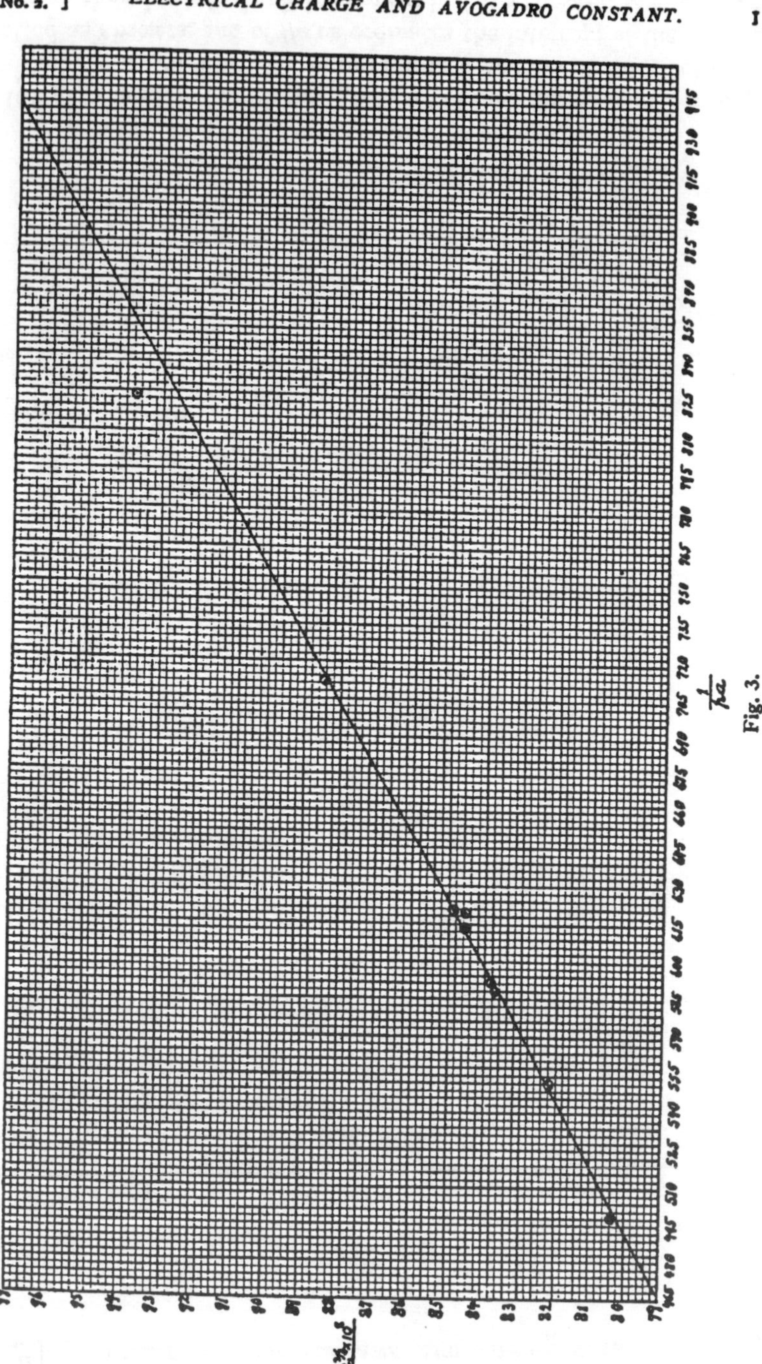

Fig. 3.

y intercept (see equation 9). In view of the uncertainty in l due to the fact that k in the equation $\eta = knm\bar{c}l$ has never been exactly evaluated, it was thought preferable to write the correction term to Stokes's law (see (2 and 3) in the form $(1 + b/pa)^{-1}$ instead of in the form $(1 + Al/a)^{-1}$ and then to plot $e_1^{2/3}$ against $1/pa$. Nevertheless in view of the greater ease of visualization of l/a all the values of this quantity corresponding to successive values of $1/pa$ are given in Table XX., k being taken, merely for the purposes of this computation, as .3502 (Boltzmann). Fig. 2 shows the graph obtained by plotting the values of $e_1^{2/3}$ against $1/pa$ for the first 51 drops of Table XX., and Fig. 3 shows the extension of this graph to twice as large values of $1/pa$ and $e_1^{2/3}$. It will be seen that there is not the slightest indication of a departure from a linear relation between $e_1^{2/3}$ and $1/pa$ up to the value $1/pa = 620.2$ which corresponds to a value of l/a of .4439 (see drop No. 58, Table XX.). Furthermore the scale used in the plotting is such that a point which is one division above or below the line in Fig. 2 represents in the mean an error of 2 in 700. *It will be seen from Figs. 2 and 3 that there is but one drop in the 58 whose departure from the line amounts to as much as 0.5 per cent. It is to be remarked, too, that this is not a selected group of drops but represents all of the drops experimented upon during 60 consecutive days*, during which time the apparatus was taken down several times and set up anew. It is certain then that an equation of the form (2) holds very accurately up to $l/a = .4$. The last drop in Fig. 3 seems to indicate the beginning of a departure from this linear relationship. Since such departure has no bearing upon the evaluation of e, discussion of it wil be postponed to another paper.

Attention may also be called to the completeness of the answers furnished by Figs. 2 and 3 to the questions raised in § 6. Thus drops No. 27 and 28 have practically identical values of $1/pa$ but while No. 28 carries, during part of the time, but 1 unit of charge (see Table XX.) drop No. 27 carries 29 times as much and it has about 7 times as large a diameter. Now if the small drop were denser than the large one (see assumption 3, § 6) or if the drag of the median upon the heavily charged drop were greater than its drag upon the one lightly charged (see assumption 1, § 6), then for both these reasons drop 27 would move more slowly relatively to drop 28 then would otherwise be the case and hence $e_1^{2/3}$ for 27 would fall below $e_1^{2/3}$ for drop 28. Instead of this the two $e_1^{2/3}$'s fall so nearly together that it is impossible to represent them on the present scale by two separate dots. Drops 52 and 56 furnish an even more striking confirmation of the same conclusion, for both drops have about the same value for l/a and both are exactly on the line though No. 56

carries at one time 68 times as heavy a charge as No. 52 and has three times as large a radius. In general the fact that Figs. 2 and 3 show no tendency whatever on the part of either the very small or the very large drops to fall above or below the line is experimental proof of the joint correctness of assumptions 1, 3, and 2*b* of § 6. The correctness of 2*a* was shown by the agreement throughout Tables IV. to XIX. between $1/n'(1/t_F' - 1/t_F)$ and $1/n(1/t_g + 1/t_F)$.

The values of $e^{2/3}$ and b obtained graphically from the y-intercept and the slope in Fig. 2 are $e^{2/3} = 61.13 \times 10^{-8}$ and $b = .0006254$, p being measured, for the purposes of Fig. 2 and of this computation in mm. of Hg at 23° C. and a being measured in cm. The value of A (equations 2 and 3) corresponding to this value of B is .874 instead of .817 as originally found. Cunningham's theory gives, in terms of the constants here used, $A = 788$.[1]

Instead however of taking the result of this graphical evaluation of $e^{2/3}$ it is more accurate to reduce each of the observations on $e_1^{2/3}$ to $e^{2/3}$ by means of the above value of B and the equation

$$e^{2/3}\left(1 + \frac{b}{pa}\right) = e_1^{2/3}. \tag{14}$$

The results of this reduction are contained in the last column of Table XX. These results illustrate very clearly the sort of consistency obtained in these observations. *The largest departure from the mean value found anywhere in the table amounts to 0.5 per cent., and "the probable error" of the final mean value computed in the usual way is 16 in 61,000.*

Instead however of using this final mean value as the most reliable evaluation of $e^{2/3}$ it was thought preferable to make a considerable number of observations at atmospheric pressure on drops small enough to make t_g determinable with great accuracy and yet large enough so that the whole correction term to Stokes's law amounted to but a few per cent., since in this case, even, though there might be a considerable error in the correction-term constant b, such error would influence the final value of e by an inappreciable amount. The first 23 drops of Table XX. represent such observations. It will be seen that they show slightly greater consistency than do the remaining drops in the table and that the correction-term reductions for these drops all lie between 1.3 per cent. (drop No. 1) and 5.6 per cent. (drop No. 23) so that even though b were in error by as much as 3 per cent. (its error is actually not more than .5 per cent.) $e^{2/3}$ would be influenced by that fact to the extent of but 0.1 per cent. The mean value of $e^{2/3}$ obtained from the first 23 drops is 61.12×10^{-8}, a

[1] PHYS. REV., 32, p. 380; also footnote.

number which differs by 1 part in 3,400 from the mean obtained from all the drops.

When correction is made for the fact that the numbers in Table XX. were obtained on the basis of the assumption $\eta_{23} = .0001825$, instead of $\eta_{23} = .0001824$ (see § 2) the final mean value of $e^{2/3}$ obtained from the first 23 drops is 61.085×10^{-8}. This corresponds to

$$e = 4.774 \times 10^{-10} \text{ electrostatic units.}$$

Since the value of the Faraday constant has now been fixed virtually by international agreement[1] at 9,650 absolute electromagnetic units and since this is the number N of molecules in a gram molecule times the elementary electrical charge, we have

$$N \times 4.774 \times 10^{-10} = 9{,}650 \times 2.9990 \times 10^{10};$$

$$\therefore \quad N = 6.062 \times 10^{23}.$$

Although the probable error in this number computed by the method of least squares from Table XX. is but one part in 3,000 it would be erroneous to infer that e and N are now known with that degree of precision, for there are four constant factors entering into all of the results in Table XX. and introducing uncertainties as follows. The coefficient of viscosity η which appears in the 3/2 power introduces into e and N a maximum possible uncertainty of 0.1 per cent. The distance between the condenser plates (16.00 mm.) is correct to .01 mm., and therefore, since it appears in the 1st power in e, introduces a maximum possible error of something less than 0.1 per cent. The voltmeter readings have a maximum possible error of rather less than 0.1 per cent., and carry this in the 1st power into e and N. The cross-hair distance which is uniformly duplicatable to one part in a thousand appears in the 3/2 power and introduces an uncertainty of no more than 0.1 per cent. The other factors introduce errors which are negligible in comparison. *The uncertainty in e and N is then that due to 4 continuous factors each of which introduces a maximum possible uncertainty of 0.1 per cent.* Following the usual procedure we may estimate the uncertainty in e and N as the square root of the sum of the squares of these four uncertainties, that is, as 2 parts in 1,000. We have then finally:

$$e = 4.774 \pm .009 \times 10^{-10}$$

and

$$N = 6.062 \pm .012 \times 10^{23}.$$

The difference between these numbers and those originally found by the oil-drop method, viz., $e = 4.891$ and $N = 5.992$ is due to the fact

[1] Atomic weight of silver 107.88. Electrochemical equivalent of silver 0.01118.

that this much more elaborate and prolonged study has had the effect of changing every one of the three factors η, A, and d (= cross-hair distance) in such a way as to lower e and to raise N. The chief change however has been due to the elimination of the faults of the original optical system.

11. Comparison with Other Measurements.

So far as I am aware, there is at present no determination of e or N by any other method which does not involve an uncertainty at least 15 times as great as that represented in the above measurements.

Thus the *radioactive method* yields in the hands of Regener[1] a count of the α particles which gives e with an uncertainty which he estimates at 3 per cent. This is as high a precision I think as has yet been claimed for any α particle count,[2] though Geiger and Rutherford's photographic registration[3] method will doubtless be able to improve it.

The *Brownian Movement* method yields results which fluctuate between Perrin's value[4] $e = 4.24 \times 10^{-10}$, and Fletcher's value,[5] 5.01×10^{-10}, with Svedberg's measurements[6] yielding the intermediate number 4.7×10^{-10}.

The *radiation method* of Planck[7] yields N as a product of $(c_2)^3$ and σ. The latest Reichsanstalt value of c_2 is 1.436[8] while Coblentz,[9] as the result of extraordinarily careful and prolonged measurements obtains 1.4456. The difference in these two values of $(c_2)^3$ is 2 per cent. Westphal[10] estimates his error in the measurement of σ at .5 per cent. though reliable observers differ in it by 5 per cent. or 6 per cent. We may take then 3 per cent. as the limit of accuracy thus far attained in measurements of e or N by other methods. *The mean results by each one of the three other methods fall well within this limit of the value found above by the oil-drop method.*

12. Computation of Other Fundamental Constants.

For the sake of comparison and reference, the following fundamental constants are recomputed on the basis of the above measurements:

[1] Regener, Sitz. Ber. d. k. Preuss. Acad., 37, p. 948, 1909.
[2] Rutherford and Geiger, Proc. Roy. Soc., 81, p. 155, 1908.
[3] Geiger and Rutherford, Phil. Mag., 24, p. 618, 1912.
[4] J. Perrin, C. R., 152, p. 1165, 1911.
[5] H. Fletcher, Phys. Rev., 33, p. 107, 1911.
[6] Svedberg, Arkiv f. Kemi, etc., utg. af K. Sv. Vetenšk. Akad., 2, 29, 1906. See also Svedberg, "Die Existenz der Molekule," p. 136. Leipzig, 1912.
[7] Planck, Vorlesungen über die Theorie der Wärmestrahlung, 2d edition, 1913, p. 166.
[8] See Planck, Vorles., p. 163.
[9] Coblentz, Journal of the Washington Academy of Sciences, Vol. 3, p. 178, April, 1913.
[10] Wm. H. Westphal, Verh. d. D. Phys. Ges., 13, p. 987, Dec., 1912.

1. The number n of molecules in 1 c.c. of an ideal gas at 0° 76 is given by

$$n = \frac{N}{V} = \frac{6.062 \times 10^{23}}{22,412} = 2.705 \times 10^{19}.$$

2. The mean kinetic energy of agitation E_0 of a molecule at 0° C. is given by

$$pV = \tfrac{1}{3}Nmu^2 = \tfrac{2}{3}NE_0 = RT,$$

$$\therefore\ E_0 = \frac{3}{2}\frac{p_0V_0}{N} = \frac{3 \times 1,013,700 \times 22,412}{2 \times 6.062 \times 10^{23}} = 5.621 \times 10^{-14} \text{ ergs}.$$

3. The constant ϵ of molecular energy defined by $E_0 = \epsilon T$ is given by

$$\epsilon = \frac{E_0}{T} = \frac{5.621 \times 10^{-14}}{273.11} = 2.058 \times 10^{-16}\,\frac{\text{ergs}}{\text{degrees}}.$$

4. The Boltzmann entropy constant k defined by $S = k \log W$ is given by[1]

$$k = \frac{R}{N} = \frac{p_0V_0}{TN} = \tfrac{2}{3}\epsilon = 1.372 \times 10^{-16}\,\frac{\text{ergs}}{\text{degrees}}.$$

All of these constants are known with precisely the accuracy attained in the measurement of e.

5. The Planck "Wirkungsquantum" h can probably be obtained considerably more accurately as follows than in any other way. From equation 292, p. 166, of the "Wärmestrahlung," we obtain[2]

$$h = \frac{k^{\frac{4}{3}}}{c}\left(\frac{48\pi\alpha}{a}\right)^{\frac{1}{3}} = \frac{(1.372 \times 10^{-16})^{\frac{4}{3}}}{2.999 \times 10^{10}} \cdot \left(\frac{48\pi 1.0823}{7.39 \times 10^{-15}}\right)^{\frac{1}{3}} = 6.620 \times 10^{-27}$$

which gives h with the same accuracy attainable in the measurement of $k^{\frac{4}{3}}/a$ in which a is the Stefan-Boltzmann constant. If Westphal's estimate of his error in the measurement of this constant is correct, viz., 0.5 per cent., it would introduce an uncertainty of but 0.2 per cent. into h. This is about that introduced by the above determination of $k^{\frac{4}{3}}$, hence the above value of h should not be in error by more than 0.4 per cent.

6. The constant c_2 of the Wien-Planck radiation law may also be computed with much precision from the above measurements. For also from equation 292 of the "Wärmestrahlung" we obtain

$$c_2 = \left(\frac{48\pi\alpha k}{a}\right)^{\frac{1}{3}} = \left(\frac{48\pi 1.0823\ 1.372 \times 10^{-16}}{7.39 \times 10^{-15}}\right)^{\frac{1}{3}} = 1.4470 \text{ cm. degrees}.$$

[1] See Planck's Vorles., p. 129.
[2] c = velocity of light, α = a numerical factor, and $a = 4\sigma/c$. Westphal's value of σ is 5.54×10^{-5} which corresponds to $a = 7.39 \times 10^{-15}$.

Since both k and a here appear in the 1/3 power, the error in c should be no more than 0.2 per cent., provided Westphal's error is no more than 0.5 per cent.

The difference between this and Coblentz's mean value, viz., *1.4456* is but 0.1 per cent. The agreement is then entirely satisfactory. A further independent check is found in the fact that Day and Sosman's location of the melting point of platinum at 1755° C.[1] is equivalent to a value of c_2 = *1.4475*.[2] On the other hand, the last Reichsanstalt value of c_2, viz., 1.437, is too low to fit well with these and Westphal's measurements. It fits perfectly however with a combination of the above value of e and Shakespear's[3] value of σ, viz., $\sigma = 5.67 \times 10^{-5}$.

13. Summary.

The results of this work may be summarized in the following table in which the numbers in the error column represent in the case of the first six numbers estimated limits of uncertainty rather than the so-called "probable errors" which would be much smaller. The last two constants however involve Westphal's measurements and estimates and Planck's equations as well as my own observations.

Table XXI.

Elementary electrical charge	$e = 4.774 \pm .009 \times 10^{-10}$
Number of molecules per gram molecule	$N = 6.062 \pm .012 \times 10^{23}$
Number of gas molecules per c.c. at 0° 76	$n = 2.705 \pm .005 \times 10^{19}$
Kinetic energy of a molecule at 0° C	$E_0 = 5.621 \pm .010 \times 10^{-14}$
Constant of molecular energy	$\epsilon = 2.058 \pm .004 \times 10^{-16}$
Constant of the entropy equation	$k = 1.372 \pm .002 \times 10^{-16}$
Elementary "Wirkungsquantum"	$h = 6.620 \pm .025 \times 10^{-27}$
Constant of the Wien displacement law	$c_2 = 1.4470 \pm 0030$

I take pleasure in acknowledging the able assistance of Mr. J. Yinbong Lee in making some of the above observations. Mr. Lee has also repeated with my apparatus the observations on oil at atmospheric pressure with results which are nearly as consistent as the above. Using my value of b he obtains, as a mean of measurements on 14 drops, a value of e which differs from the above by less than 1 part in 6,000, although its probable error computed as in the case of Table XX. is 1 part in 2,000.

Ryerson Physical Laboratory,
University of Chicago,
June 2, 1913.

[1] Amer. Jour. Sci., 30, p. 3, 1910.
[2] Coblentz, Journal of the Washington Academy of Sciences, Vol. 3, p. 13.
[3] G. A. Shakespear, Proc. Roy. Soc., 86, 180, 1911.

AN EXTENSION OF THE SPECTRUM IN THE EXTREME-VIOLET.[1]

BY THEODORE LYMAN.

THE researches of Schumann led him to extend the spectrum to the neighborhood of wave-length 1,250. His limiting wave-length was determined by the absorption of the fluorite which formed a necessary part of his apparatus. In 1904, I succeeded in pushing the limit to wave-length 1,030 by the use of a concave diffraction grating.

Recently I have renewed the attack on the problem with the result that I have succeeded in photographing the spectrum of hydrogen to wave-length 905. The extension is due not so much to any fundamental change in the nature of the apparatus as to an improvement in technique consequent on an experience of ten years.

It is a characteristic of the region investigated by Schumann between wave-lengths 1,850 and 1,250, that while hydrogen yields a rich secondary spectrum, with the possible exception of one line, no radiation has been discovered belonging to the primary spectrum. On the other hand, in the new region between the limit set by fluorite and wave-length 905, a disruptive discharge in hydrogen produces a primary spectrum of great interest made up of perhaps a dozen lines. These lines are always accompanied in pure hydrogen by members of the secondary spectrum but they may be obtained alone if helium containing a trace of hydrogen is employed.

Results obtained from vacuum tubes when a strong disruptive discharge is used, must always be interpreted with caution since the material torn from the

[1] Abstract of a paper presented at the Washington meeting of the Physical Society, April 24–25, 1914.

36

tube itself sometimes furnishes impurities. In the present case, it will be some time before the effect of such impurities can be estimated. However, it may be stated with some degree of certainty that the diffuse series predicted in this region by Ritz has been discovered. The first member at 1,216 is found to be greatly intensified by the disruptive discharge and the next line at 1,026 appears also, though very faintly. This diffuse series bears a simple relation to Balmer's formula. Following the same kind of argument, a sharp series corresponding to the Pickering series might be expected. The new region appears to yield two lines belonging to such a relation at positions demanded by calculation.

JEFFERSON PHYSICAL LABORATORY, HARVARD UNIVERSITY,
April 20, 1914.

Second Series May, 1923 Vol. 21, No. 5

THE PHYSICAL REVIEW

A QUANTUM THEORY OF THE SCATTERING OF X-RAYS BY LIGHT ELEMENTS

By Arthur H. Compton

Abstract

A quantum theory of the scattering of X-rays and γ-rays by light elements.—The hypothesis is suggested that when an X-ray quantum is scattered it spends all of its energy and momentum upon some particular electron. This electron in turn scatters the ray in some definite direction. The change in momentum of the X-ray quantum due to the change in its direction of propagation results in a recoil of the scattering electron. The energy in the scattered quantum is thus less than the energy in the primary quantum by the kinetic energy of recoil of the scattering electron. The corresponding *increase in the wave-length of the scattered beam* is $\lambda_\theta - \lambda_0 = (2h/mc)\sin^2\frac{1}{2}\theta = 0.0484\sin^2\frac{1}{2}\theta$, where h is the Planck constant, m is the mass of the scattering electron, c is the velocity of light, and θ is the angle between the incident and the scattered ray. Hence the increase is independent of the wave-length. *The distribution of the scattered radiation* is found, by an indirect and not quite rigid method, to be concentrated in the forward direction according to a definite law (Eq. 27). The total energy removed from the primary beam comes out less than that given by the classical Thomson theory in the ratio $1/(1+2\alpha)$, where $\alpha = h/mc\lambda_0 = 0.0242/\lambda_0$. Of this energy a fraction $(1+\alpha)/(1+2\alpha)$ reappears as scattered radiation, while the remainder is truly absorbed and transformed into kinetic energy of recoil of the scattering electrons. Hence, if σ_0 is the *scattering absorption coefficient* according to the classical theory, the coefficient according to this theory is $\sigma = \sigma_0/(1+2\alpha) = \sigma_s + \sigma_a$, where σ_s is the true scattering coefficient $[(1+\alpha)\sigma/(1+2\alpha)^2]$, and σ_a is the coefficient of absorption due to scattering $[\alpha\sigma/(1+2\alpha)^2]$. Unpublished experimental results are given which show that for graphite and the Mo–K radiation the scattered radiation is longer than the primary, the observed difference $(\lambda_{\pi/2} - \lambda_0 = .022)$ being close to the computed value .024. In the case of scattered γ-rays, the wave-length has been found to vary with θ in agreement with the theory, increasing from .022 A (primary) to .068 A ($\theta = 135°$). Also the velocity of secondary β-rays excited in light elements by γ-rays agrees with the suggestion that they are recoil electrons. As for the predicted variation of absorption with λ, Hewlett's results for carbon for wave-lengths below 0.5 A are in excellent agreement with this theory; also the predicted concentration in the forward direction is shown to be in agreement with the experimental results, both for X-rays and γ-rays. This remarkable *agreement between experiment and theory* indicates clearly that scattering is a quantum phenomenon and can be explained without introducing any new hypothesis as to the size of the electron or any new constants; also that a radiation quantum carries with it momentum as well as energy. The restriction to light elements is due to the assumption that the constraining forces acting on the scattering electrons are negligible, which is probably legitimate only for the lighter elements.

Spectrum of K-rays from Mo scattered by graphite, as compared with the spectrum of the primary rays, is given in Fig. 4, showing the change of wave-length.

Radiation from a moving isotropic radiator.—It is found that in a direction θ with the velocity, $I_\theta/I' = (1-\beta)^2/(1-\beta\cos\theta)^4 = (\nu_\theta/\nu')^4$. For the total radiation from a black body in motion to an observer at rest, $I/I' = (T/T')^4 = (\nu_m/\nu_m')^4$, where the primed quantities refer to the body at rest.

J. J. Thomson's classical theory of the scattering of X-rays, though supported by the early experiments of Barkla and others, has been found incapable of explaining many of the more recent experiments. This theory, based upon the usual electrodynamics, leads to the result that the energy scattered by an electron traversed by an X-ray beam of unit intensity is the same whatever may be the wave-length of the incident rays. Moreover, when the X-rays traverse a thin layer of matter, the intensity of the scattered radiation on the two sides of the layer should be the same. Experiments on the scattering of X-rays by light elements have shown that these predictions are correct when X-rays of moderate hardness are employed; but when very hard X-rays or γ-rays are employed, the scattered energy is found to be decidedly less than Thomson's theoretical value, and to be strongly concentrated on the emergent side of the scattering plate.

Several years ago the writer suggested that this reduced scattering of the very short wave-length X-rays might be the result of interference between the rays scattered by different parts of the electron, if the electron's diameter is comparable with the wave-length of the radiation. By assuming the proper radius for the electron, this hypothesis supplied a quantitative explanation of the scattering for any particular wave-length. But recent experiments have shown that the size of the electron which must thus be assumed increases with the wave-length of the X-rays employed,[1] and the conception of an electron whose size varies with the wave-length of the incident rays is difficult to defend.

Recently an even more serious difficulty with the classical theory of X-ray scattering has appeared. It has long been known that secondary γ-rays are softer than the primary rays which excite them, and recent experiments have shown that this is also true of X-rays. By a spectroscopic examination of the secondary X-rays from graphite, I have, indeed,

[1] A. H. Compton, Bull. Nat. Research Council, No. 20, p. 10 (Oct., 1922).

been able to show that only a small part, if any, of the secondary X-radiation is of the same wave-length as the primary.[1] While the energy of the secondary X-radiation is so nearly equal to that calculated from Thomson's classical theory that it is difficult to attribute it to anything other than true scattering,[2] these results show that if there is any scattering comparable in magnitude with that predicted by Thomson, it is of a greater wave-length than the primary X-rays.

Such a change in wave-length is directly counter to Thomson's theory of scattering, for this demands that the scattering electrons, radiating as they do because of their forced vibrations when traversed by a primary X-ray, shall give rise to radiation of exactly the same frequency as that of the radiation falling upon them. Nor does any modification of the theory such as the hypothesis of a large electron suggest a way out of the difficulty. This failure makes it appear improbable that a satisfactory explanation of the scattering of X-rays can be reached on the basis of the classical electrodynamics.

The Quantum Hypothesis of Scattering

According to the classical theory, each X-ray affects every electron in the matter traversed, and the scattering observed is that due to the combined effects of all the electrons. From the point of view of the quantum theory, we may suppose that any particular quantum of X-rays is not scattered by all the electrons in the radiator, but spends all of its energy upon some particular electron. This electron will in turn scatter the ray in some definite direction, at an angle with the incident beam. This bending of the path of the quantum of radiation results in a change in its momentum. As a consequence, the scattering electron will recoil with a momentum equal to the change in momentum of the X-ray. The energy in the scattered ray will be equal to that in the incident ray minus the kinetic energy of the recoil of the scattering electron; and since the scattered ray must be a complete quantum, the frequency will be reduced in the same ratio as is the energy. Thus on the quantum theory we should expect the wave-length of the scattered X-rays to be greater than that of the incident rays.

The effect of the momentum of the X-ray quantum is to set the

[1] In previous papers (Phil. Mag. 41, 749, 1921; Phys. Rev. 18, 96, 1921) I have defended the view that the softening of the secondary X-radiation was due to a considerable admixture of a form of fluorescent radiation. Gray (Phil. Mag. 26, 611, 1913; Frank. Inst. Journ., Nov., 1920, p. 643) and Florance (Phil. Mag. 27, 225, 1914) have considered that the evidence favored true scattering, and that the softening is in some way an accompaniment of the scattering process. The considerations brought forward in the present paper indicate that the latter view is the correct one.

[2] A. H. Compton, loc. cit., p. 16.

scattering electron in motion at an angle of less than 90° with the primary beam. But it is well known that the energy radiated by a moving body is greater in the direction of its motion. We should therefore expect, as is experimentally observed, that the intensity of the scattered radiation should be greater in the general direction of the primary X-rays than in the reverse direction.

The change in wave-length due to scattering.—Imagine, as in Fig. 1*A*,

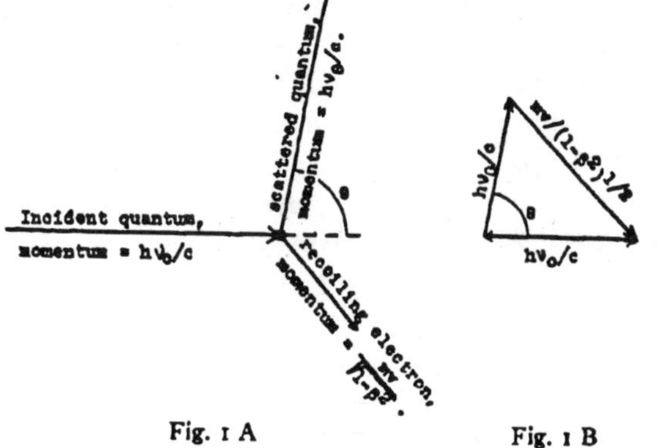

Fig. 1 A Fig. 1 B

that an X-ray quantum of frequency ν_0 is scattered by an electron of mass m. The momentum of the incident ray will be $h\nu_0/c$, where c is the velocity of light and h is Planck's constant, and that of the scattered ray is $h\nu_\theta/c$ at an angle θ with the initial momentum. The principle of the conservation of momentum accordingly demands that the momentum of recoil of the scattering electron shall equal the vector difference between the momenta of these two rays, as in Fig. 1*B*. The momentum of the electron, $m\beta c/\sqrt{1-\beta^2}$, is thus given by the relation

$$\left(\frac{m\beta c}{\sqrt{1-\beta^2}}\right)^2 = \left(\frac{h\nu_0}{c}\right)^2 + \left(\frac{h\nu_\theta}{c}\right)^2 + 2\frac{h\nu_0}{c}\cdot\frac{h\nu_\theta}{c}\cos\theta, \qquad (1)$$

where β is the ratio of the velocity of recoil of the electron to the velocity of light. But the energy $h\nu_\theta$ in the scattered quantum is equal to that of the incident quantum $h\nu_0$ less the kinetic energy of recoil of the scattering electron, *i.e.*,

$$h\nu_\theta = h\nu_0 - mc^2\left(\frac{1}{\sqrt{1-\beta^2}} - 1\right). \qquad (2)$$

We thus have two independent equations containing the two unknown quantities β and ν_θ. On solving the equations we find

$$\nu_\theta = \nu_0/(1 + 2\alpha\sin^2\tfrac{1}{2}\theta), \qquad (3)$$

where

$$\alpha = h\nu_0/mc^2 = h/mc\lambda_0. \tag{4}$$

Or in terms of wave-length instead of frequency,

$$\lambda_\theta = \lambda_0 + (2h/mc)\sin^2 \tfrac{1}{2}\theta. \tag{5}$$

It follows from Eq. (2) that $1/(1-\beta^2) = \{1 + \alpha[1 - (\nu_\theta/\nu_0)]\}^2$, or solving explicitly for β

$$\beta = 2\alpha \sin \tfrac{1}{2}\theta \, \frac{\sqrt{1 + (2\alpha + \alpha^2)\sin^2 \frac{1}{2}\theta}}{1 + 2(\alpha + \alpha^2)\sin^2 \frac{1}{2}\theta}. \tag{6}$$

Eq. (5) indicates an increase in wave-length due to the scattering process which varies from a few per cent in the case of ordinary X-rays to more than 200 per cent in the case of γ-rays scattered backward. At the same time the velocity of the recoil of the scattering electron, as calculated from Eq. (6), varies from zero when the ray is scattered directly forward to about 80 per cent of the speed of light when a γ-ray is scattered at a large angle.

It is of interest to notice that according to the classical theory, if an X-ray were scattered by an electron moving in the direction of propagation at a velocity $\beta' c$, the frequency of the ray scattered at an angle θ is given by the Doppler principle as

$$\nu_\theta = \nu_0 \Big/ \left(1 + \frac{2\beta'}{1-\beta'}\sin^2 \tfrac{1}{2}\theta\right). \tag{7}$$

It will be seen that this is of exactly the same form as Eq. (3), derived on the hypothesis of the recoil of the scattering electron. Indeed, if $\alpha = \beta'/(1-\beta')$ or $\beta' = \alpha/(1+\alpha)$, the two expressions become identical. It is clear, therefore, that so far as the effect on the wave-length is concerned, we may replace the recoiling electron by a scattering electron moving in the direction of the incident beam at a velocity such that

$$\bar{\beta} = \alpha/(1+\alpha). \tag{8}$$

We shall call $\bar{\beta}c$ the "effective velocity" of the scattering electrons.

Energy distribution from a moving, isotropic radiator.—In preparation for the investigation of the spatial distribution of the energy scattered by a recoiling electron, let us study the energy radiated from a moving, isotropic body. If an observer moving with the radiating body draws a sphere about it, the condition of isotropy means that the probability is equal for all directions of emission of each energy quantum. That is, the probability that a quantum will traverse the sphere between the angles θ' and $\theta' + d\theta'$ with the direction of motion is $\frac{1}{2}\sin\theta' d\theta'$. But

the surface which the moving observer considers a sphere (Fig. 2A) is

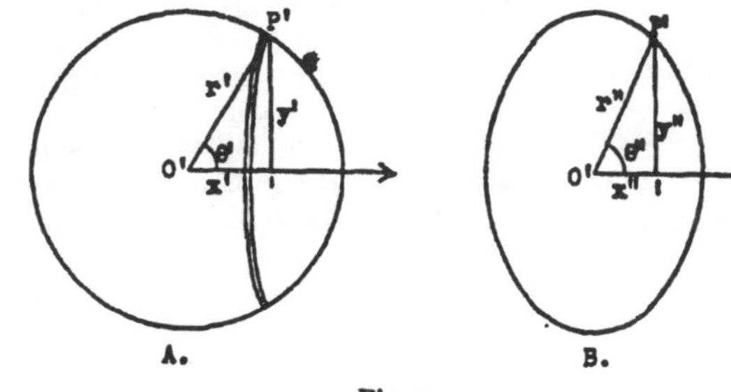

Fig. 2

considered by the stationary observer to be an oblate spheroid whose polar axis is reduced by the factor $\sqrt{1-\beta^2}$. Consequently a quantum of radiation which traverses the sphere at the angle θ', whose tangent is y'/x' (Fig. 2A), appears to the stationary observer to traverse the spheroid at an angle θ'' whose tangent is y''/x'' (Fig. 2B). Since $x' = x''/\sqrt{1-\beta^2}$ and $y' = y''$, we have

$$\tan\theta' = y'/x' = \sqrt{1-\beta^2}\, y''/x'' = \sqrt{1-\beta^2}\tan\theta'', \tag{9}$$

and

$$\sin\theta' = \frac{\sqrt{1-\beta^2}\tan\theta''}{\sqrt{1+(1-\beta^2)\tan^2\theta''}}. \tag{10}$$

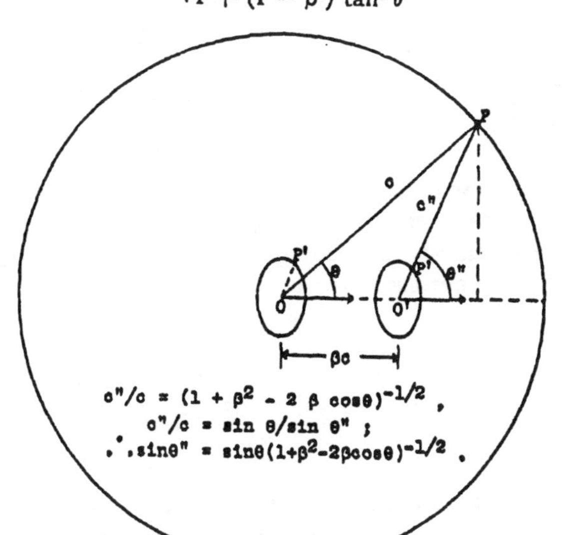

Fig. 3. The ray traversing the moving spheroid at P' at an angle θ'' reaches the stationary spherical surface drawn about O, at the point P, at an angle θ.

Imagine, as in Fig. 3, that a quantum is emitted at the instant $t = 0$, when the radiating body is at O. If it traverses the moving observer's sphere at an angle θ', it traverses the corresponding oblate spheroid, imagined by the stationary observer to be moving with the body, at an angle θ''. After 1 second, the quantum will have reached some point P on a sphere of radius c drawn about O, while the radiator will have moved a distance βc. The stationary observer at P therefore finds that the radiation is coming to him from the point O, at an angle θ with the direction of motion. That is, if the moving observer considers the quantum to be emitted at an angle θ' with the direction of motion, to the stationary observer the angle appears to be θ, where

$$\sin\theta/\sqrt{1 + \beta^2 - 2\beta\cos\theta} = \sin\theta'', \qquad (11)$$

and θ'' is given in terms of θ' by Eq. (10). It follows that

$$\sin\theta' = \sin\theta\frac{\sqrt{1-\beta^2}}{1-\beta\cos\theta}. \qquad (12)$$

On differentiating Eq. (12) we obtain

$$d\theta' = \frac{\sqrt{1-\beta^2}}{1-\beta\cos\theta}d\theta. \qquad (13)$$

The probability that a given quantum will appear to the stationary observer to be emitted between the angles θ and $\theta + d\theta$ is therefore

$$P_\theta d\theta = P_{\theta'}d\theta' = \tfrac{1}{2}\sin\theta' d\theta',$$

where the values of $\sin\theta'$ and $d\theta'$ are given by Eqs. (12) and (13). Substituting these values we find

$$P_\theta d\theta = \frac{1-\beta^2}{(1-\beta\cos\theta)^2}\cdot\tfrac{1}{2}\sin\theta d\theta. \qquad (14)$$

Suppose the moving observer notices that n' quanta are emitted per second. The stationary observer will estimate the rate of emission as

$$n'' = n'\sqrt{1-\beta^2},$$

quanta per second, because of the difference in rate of the moving and stationary clocks. Of these n'' quanta, the number which are emitted between angles θ and $\theta + d\theta$ is $dn'' = n''\cdot P_\theta d\theta$. But if dn'' per second are emitted at the angle θ, the number per second received by a stationary observer at this angle is $dn = dn''/(1 - \beta\cos\theta)$, since the radiator is approaching the observer at a velocity $\beta\cos\theta$. The energy of each quantum is, however, $h\nu_\theta$, where ν_θ is the frequency of the radiation as

received by the stationary observer.[1] Thus the intensity, or the energy per unit area per unit time, of the radiation received at an angle θ and a distance R is

$$I_\theta = \frac{h\nu_\theta\cdot dn}{2\pi R^2\sin\theta d\theta} = \frac{h\nu_\theta}{2\pi R^2\sin\theta d\theta}\frac{n'(1-\beta^2)^{3/2}}{(1-\beta\cos\theta)^3}\tfrac{1}{2}\sin\theta d\theta$$
$$= \frac{n'h\nu_\theta}{4\pi R^2}\frac{(1-\beta^2)^{3/2}}{(1-\beta\cos\theta)^3}. \qquad (15)$$

If the frequency of the oscillator emitting the radiation is measured by an observer moving with the radiator as ν', the stationary observer judges its frequency to be $\nu'' = \nu'\sqrt{1-\beta^2}$, and, in virtue of the Doppler effect, the frequency of the radiation received at an angle θ is

$$\nu_\theta = \nu''/(1-\beta\cos\theta) = \nu'[\sqrt{1-\beta^2}/(1-\beta\cos\theta)]. \qquad (16)$$

Substituting this value of ν_θ in Eq. (15) we find

$$I_\theta = \frac{n'h\nu'}{4\pi R^2}\frac{(1-\beta^2)^2}{(1-\beta\cos\theta)^4}. \qquad (17)$$

But the intensity of the radiation observed by the moving observer at a distance R from the source is $I' = n'h\nu'/4\pi R^2$. Thus,

$$I_\theta = I'[(1-\beta)^2/(1-\beta\cos\theta)^4] \qquad (18)$$

is the intensity of the radiation received at an angle θ with the direction of motion of an isotropic radiator, which moves with a velocity βc, and which would radiate with intensity I' if it were at rest.[2]

It is interesting to note, on comparing Eqs. (16) and (18), that

$$I_\theta/I' = (\nu_\theta/\nu')^4. \qquad (19)$$

[1] At first sight the assumption that the quantum which to the moving observer had energy $h\nu'$ will be $h\nu$ for the stationary observer seems inconsistent with the energy principle. When one considers, however, the work done by the moving body against the back-pressure of the radiation, it is found that the energy principle is satisfied. The conclusion reached by the present method of calculation is in exact accord with that which would be obtained according to Lorenz's equations, by considering the radiation to consist of electromagnetic waves.

[2] G. H. Livens gives for I_θ/I' the value $(1-\beta\cos\theta)^{-2}$ ("The Theory of Electricity," p. 600, 1918). At small velocities this value differs from the one here obtained by the factor $(1-\beta\cos\theta)^{-2}$. The difference is due to Livens' neglect of the concentration of the radiation in the small angles, as expressed by our Eq. (14). Cunningham ("The Principle of Relativity," p. 60, 1914) shows that if a plane wave is emitted by a radiator moving in the direction of propagation with a velocity βc, the intensity I received by a stationary observer is greater than the intensity I' estimated by the moving observer, in the ratio $(1-\beta^2)/(1-\beta)^2$, which is in accord with the value calculated according to the methods here employed.

The change in frequency given in Eq. (16) is that of the usual relativity theory. I have not noticed the publication of any result which is the equivalent of my formula (18) for the intensity of the radiation from a moving body.

This result may be obtained very simply for the total radiation from a black body, which is a special case of an isotropic radiator. For, suppose such a radiator is moving so that the frequency of maximum intensity which to a moving observer is ν_m' appears to the stationary observer to be ν_m. Then according to Wien's law, the apparent temperature T, as estimated by the stationary observer, is greater than the temperature T' for the moving observer by the ratio $T/T' = \nu_m/\nu_m'$. According to Stefan's law, however, the intensity of the total radiation from a black body is proportional to T^4; hence, if I and I' are the intensities of the radiation as measured by the stationary and the moving observers respectively,

$$I/I' = (T/T')^4 = (\nu_m/\nu_m')^4. \tag{20}$$

The agreement of this result with Eq. (19) may be taken as confirming the correctness of the latter expression.

The intensity of scattering from recoiling electrons.—We have seen that the change in frequency of the radiation scattered by the recoiling electrons is the same as if the radiation were scattered by electrons moving in the direction of propagation with an effective velocity $\bar{\beta} = \alpha/(1 + \alpha)$, where $\alpha = h/mc\lambda_0$. It seems obvious that since these two methods of calculation result in the same change in wave-length, they must also result in the same change in intensity of the scattered beam. This assumption is supported by the fact that we find, as in Eq. 19, that the change in intensity is in certain special cases a function only of the change in frequency. I have not, however, succeeded in showing rigidly that if two methods of scattering result in the same relative wave-lengths at different angles, they will also result in the same relative intensity at different angles. Nevertheless, we shall assume that this proposition is true, and shall proceed to calculate the relative intensity of the scattered beam at different angles on the hypothesis that the scattering electrons are moving in the direction of the primary beam with a velocity $\bar{\beta} = \alpha/(1 + \alpha)$. If our assumption is correct, the results of the calculation will apply also to the scattering by recoiling electrons.

To an observer moving with the scattering electron, the intensity of the scattering at an angle θ', according to the usual electrodynamics, should be proportional to $(1 + \cos^2\theta')$, if the primary beam is unpolarized. On the quantum theory, this means that the probability that a quantum will be emitted between the angles θ' and $\theta' + d\theta'$ is proportional to $(1 + \cos^2\theta')\cdot\sin\theta' d\theta'$, since $2\pi\sin\theta' d\theta'$ is the solid angle included between θ' and $\theta' + d\theta'$. This may be written $P_{\theta'} d\theta' = k(1 + \cos^2\theta')\sin\theta' d\theta'$.

The factor of proportionality k may be determined by performing the integration

$$\int_0^\pi P_{\theta'} d\theta' = k\int_0^\pi (1 + \cos^2\theta')\sin\theta' d\theta' = 1,$$

with the result that $k = 3/8$. Thus

$$P_{\theta'} d\theta' = (3/8)(1 + \cos^2\theta')\sin\theta' d\theta' \tag{21}$$

is the probability that a quantum will be emitted at the angle θ' as measured by an observer moving with the scattering electron.

To the stationary observer, however, the quantum ejected at an angle θ' appears to move at an angle θ with the direction of the primary beam, where $\sin\theta'$ and $d\theta'$ are as given in Eqs. (12) and (13). Substituting these values in Eq. (21), we find for the probability that a given quantum will be scattered between the angles θ and $\theta + d\theta$,

$$P_\theta d\theta = \tfrac{3}{8}\sin\theta d\theta\frac{(1 - \beta^2)\{(1 + \beta^2)(1 + \cos^2\theta) - 4\beta\cos\theta\}}{(1 - \beta\cos\theta)^4}. \tag{22}$$

Suppose the stationary observer notices that n quanta are scattered per second. In the case of the radiator emitting n'' quanta per second while approaching the observer, the n''th quantum was emitted when the radiator was nearer the observer, so that the interval between the receipt of the 1st and the n''th quantum was less than a second. That is, more quanta were received per second than were emitted in the same time. In the case of scattering, however, though we suppose that each scattering electron is moving forward, the nth quantum is scattered by an electron starting from the same position as the 1st quantum. Thus the number of quanta received per second is also n.

We have seen (Eq. 3) that the frequency of the quantum received at an angle θ is $\nu_\theta = \nu_0/(1 + 2\alpha\sin^2\tfrac{1}{2}\theta) = \nu_0/\{1 + \alpha(1 - \cos\theta)\}$, where ν_0, the frequency of the incident beam, is also the frequency of the ray scattered in the direction of the incident beam. The energy scattered per second at the angle θ is thus $nh\nu_\theta P_\theta d\theta$, and the intensity, or energy per second per unit area, of the ray scattered to a distance R is

$$I_\theta = \frac{nh\nu_\theta P_\theta d\theta}{2\pi R^2\sin\theta d\theta}$$

$$= \frac{nh}{2\pi R^2}\cdot\frac{\nu_0}{1+\alpha(1-\cos\theta)}\cdot\frac{3}{8}\cdot\frac{(1-\beta^2)\{(1+\beta^2)(1+\cos^2\theta)-4\beta\cos\theta\}}{(1-\beta\cos\theta)^4}.$$

Substituting for β its value $\alpha/(1 + \alpha)$, and reducing, this becomes

$$I = \frac{3nh\nu_0}{16\pi R}\frac{(1 + 2\alpha)\{1 + \cos^2\theta + 2\alpha(1 + \alpha)(1 - \cos\theta)^2\}}{(1 + \alpha - \alpha\cos\theta)^5}. \tag{23}$$

In the forward direction, where $\theta = 0$, the intensity of the scattered beam is thus

$$I_0 = \frac{3}{8\pi}\frac{nh\nu_0}{R^2}(1 + 2\alpha). \tag{24}$$

Hence

$$\frac{I_\theta}{I_0} = \frac{1}{2}\frac{1 + \cos^2\theta + 2\alpha(1 + \alpha)(1 - \cos\theta)^2}{\{1 + \alpha(1 - \cos\theta)\}^5}. \tag{25}$$

On the hypothesis of recoiling electrons, however, for a ray scattered directly forward, the velocity of recoil is zero (Eq. 6). Since in this case the scattering electron is at rest, the intensity of the scattered beam should be that calculated on the basis of the classical theory, namely,

$$I_0 = I(Ne^4/R^2m^2c^4), \tag{26}$$

where I is the intensity of the primary beam traversing the N electrons which are effective in scattering. On combining this result with Eq. (25), we find for the intensity of the X-rays scattered at an angle θ with the incident beam,

$$I = I\frac{Ne^4}{2R^2m^2c^4}\frac{1 + \cos^2\theta + 2\alpha(1 + \alpha)(1 - \cos\theta)^2}{\{1 + \alpha(1 - \cos\theta)\}^5}. \tag{27}$$

The calculation of the energy removed from the primary beam may now be made without difficulty. We have supposed that n quanta are scattered per second. But on comparing Eqs. (24) and (26), we find that

$$n = \frac{8\pi}{3}\frac{INe^4}{h\nu_0 m^2c^4(1 + 2\alpha)}.$$

The energy removed from the primary beam per second is $nh\nu_0$. If we define ***the scattering absorption coefficient*** as the fraction of the energy of the primary beam removed by the scattering process per unit length of path through the medium, it has the value

$$\sigma = \frac{nh\nu_0}{I} = \frac{8\pi}{3}\frac{Ne^4}{m^2c^4}\cdot\frac{1}{1 + 2\alpha} = \frac{\sigma_0}{1 + 2\alpha}, \tag{28}$$

where N is the number of scattering electrons per unit volume, and σ_0 is the scattering coefficient calculated on the basis of the classical theory.[1]

In order to determine the total energy truly scattered, we must integrate the scattered intensity over the surface of a sphere surrounding the scattering material, *i.e.*, $\epsilon_s = \int_0^\pi I_\theta \cdot 2\pi R^2 \sin\theta d\theta$. On substituting the value of I_θ from Eq. (27), and integrating, this becomes

$$\epsilon_s = \frac{8\pi}{3}\frac{INe^4}{m^2c^4}\frac{1 + \alpha}{(1 + 2\alpha)^2}.$$

[1] Cf. J. J. Thomson, "Conduction of Electricity through Gases," 2d ed., p. 325.

The *true scattering coefficient* is thus

$$\sigma_s = \frac{8\pi}{3}\frac{Ne^4}{m^2c^4}\frac{1 + \alpha}{(1 + 2\alpha)^2} = \sigma_0\frac{1 + \alpha}{(1 + 2\alpha)^2}. \tag{29}$$

It is clear that the difference between the total energy removed from the primary beam and that which reappears as scattered radiation is the energy of recoil of the scattering electrons. This difference represents, therefore, a type of true absorption resulting from the scattering process. The corresponding *coefficient of true absorption due to scattering* is

$$\sigma_a = \sigma - \sigma_s = \frac{8\pi}{3}\frac{Ne^4}{m^2c^4}\frac{\alpha}{(1 + 2\alpha)^2} = \sigma_0\frac{\alpha}{(1 + 2\alpha)^2}. \tag{30}$$

Experimental Test.

Let us now investigate the agreement of these various formulas with experiments on the change of wave-length due to scattering, and on the magnitude of the scattering of X-rays and γ-rays by light elements.

Wave-length of the scattered rays.—If in Eq. (5) we substitute the accepted values of h, m, and c, we obtain

$$\lambda_\theta = \lambda_0 + 0.0484 \sin^2 \tfrac{1}{2}\theta, \tag{31}$$

if λ is expressed in Angström units. It is perhaps surprising that the increase should be the same for all wave-lengths. Yet, as a result of an extensive experimental study of the change in wave-length on scattering, the writer has concluded that "over the range of primary rays from 0.7 to 0.025 A, the wave-length of the secondary X-rays at 90° with the incident beam is roughly 0.03 A greater than that of the primary beam which excites it."[1] Thus the experiments support the theory in showing a wave-length increase which seems independent of the incident wave-length, and which also is of the proper order of magnitude.

A quantitative test of the accuracy of Eq. (31) is possible in the case of the characteristic K-rays from molybdenum when scattered by graphite. In Fig. 4 is shown a spectrum of the X-rays scattered by graphite at right angles with the primary beam, when the graphite is traversed by X-rays from a molybdenum target.[2] The solid line represents the spectrum of these scattered rays, and is to be compared with the broken line, which represents the spectrum of the primary rays, using the same slits and crystal, and the same potential on the tube. The primary spectrum is, of course, plotted on a much smaller scale than

[1] A. H. Compton, Bull. N. R. C., No. 20, p. 17 (1922).

[2] It is hoped to publish soon a description of the experiments on which this figure is based.

the secondary. The zero point for the spectrum of both the primary and secondary X-rays was determined by finding the position of the first order lines on both sides of the zero point.

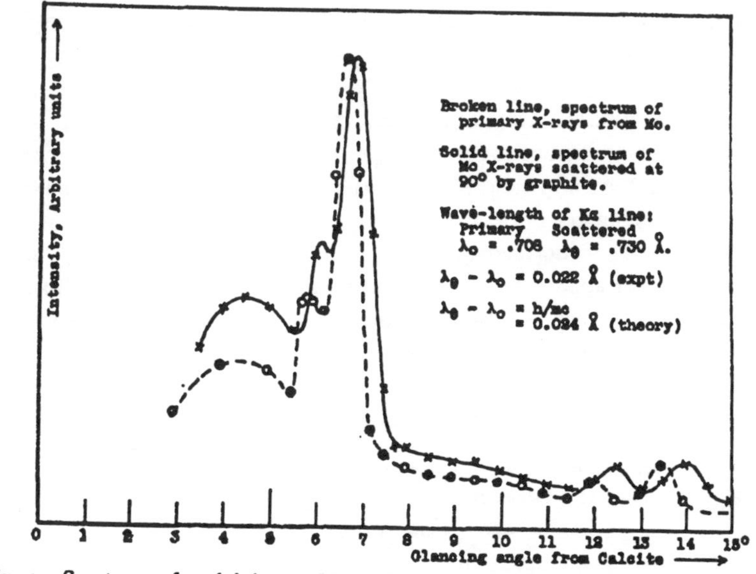

Fig. 4. Spectrum of molybdenum X-rays scattered by graphite, compared with the spectrum of the primary X-rays, showing an increase in wave-length on scattering.

It will be seen that the wave-length of the scattered rays is unquestionably greater than that of the primary rays which excite them. Thus the Kα line from molybdenum has a wave-length 0.708 A. The wave-length of this line in the scattered beam is found in these experiments, however, to be 0.730 A. That is,

$$\lambda_\theta - \lambda_0 = 0.022 \text{ A} \quad \text{(experiment)}.$$

But according to the present theory (Eq. 5),

$$\lambda_\theta - \lambda_0 = 0.0484 \sin^2 45^\circ = 0.024 \text{ A} \quad \text{(theory)},$$

which is a very satisfactory agreement.

The variation in wave-length of the scattered beam with the angle is illustrated in the case of γ-rays. The writer has measured[1] the mass absorption coefficient in lead of the rays scattered at different angles when various substances are traversed by the hard γ-rays from RaC. The mean results for iron, aluminium and paraffin are given in column 2 of Table I. This variation in absorption coefficient corresponds to a

[1] A. H. Compton, Phil. Mag. 41, 760 (1921).

difference in wave-length at the different angles. Using the value given by Hull and Rice for the mass absorption coefficient in lead for wave-length 0.122, 3.0, remembering[1] that the characteristic fluorescent absorption τ/ρ is proportional to λ^3, and estimating the part of the absorption due to scattering by the method described below, I find for the wave-lengths corresponding to these absorption coefficients the values given in the fourth column of Table I. That this extrapolation is very

TABLE I

Wave-length of Primary and Scattered γ-rays

	Angle	μ/ρ	τ/ρ	λ obs.	λ calc.
Primary..........	0°	.076	.017	0.022 A	(0.022 A)
Scattered.........	45°	.10	.042	.030	0.029
"	90°	.21	.123	.043	0.047
"	135°	.59	.502	.068	0.063

nearly correct is indicated by the fact that it gives for the primary beam a wave-length 0.022 A. This is in good accord with the writer's value 0.025 A, calculated from the scattering of γ-rays by lead at small angles,[2] and with Ellis' measurements from his β-ray spectra, showing lines of wave-length .045, .025, .021 and .020 A, with line .020 the strongest.[3] Taking $\lambda_0 = 0.022$ A, the wave-lengths at the other angles may be calculated from Eq. (31). The results, given in the last column of Table I., and shown graphically in Fig. 5, are in satisfactory accord with the measured values. There is thus good reason for believing that Eq. (5) represents accurately the wave-length of the X-rays and γ-rays scattered by light elements.

Velocity of recoil of the scattering electrons.—The electrons which recoil in the process of the scattering of ordinary X-rays have not been observed. This is probably because their number and velocity is usually small compared with the number and velocity of the photoelectrons ejected as a result of the characteristic fluorescent absorption. I have pointed out elsewhere,[4] however, that there is good reason for believing that most of the secondary β-rays excited in light elements by the action of γ-rays are such recoil electrons. According to Eq. (6), the velocity of these electrons should vary from 0, when the γ-ray is scattered forward, to $v_{max} = \beta_{max}c = 2c\alpha[(1+\alpha)/(1+2\alpha+2\alpha^2)]$, when the γ-ray quantum

[1] Cf. L. de Broglie, Jour. de Phys. et Rad. 3, 33 (1922); A. H. Compton, Bull. N. R. C., No. 20, p. 43 (1922).
[2] A. H. Compton, Phil. Mag. 41, 777 (1921).
[3] C. D. Ellis, Proc. Roy. Soc. A, 101, 6 (1922).
[4] A. H. Compton, Bull. N. R. C., No. 20, p. 27 (1922).

is scattered backward. If for the hard γ-rays from radium C, $\alpha = 1.09$, corresponding to $\lambda = 0.022$ A, we thus obtain $\beta_{max} = 0.82$. The effective velocity of the scattering electrons is, therefore (Eq. 8), $\bar{\beta} = 0.52$. These results are in accord with the fact that the average velocity of the

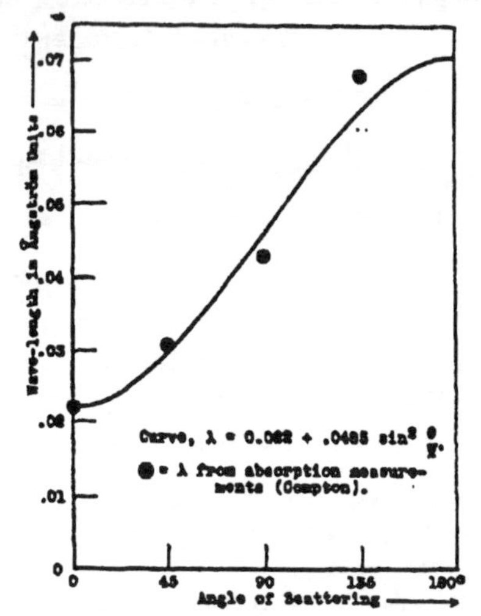

Fig. 5. The wave-length of scattered γ-rays at different angles with the primary beam, showing an increase at large angles similar to a Doppler effect.

β-rays excited by the γ-rays from radium is somewhat greater than half that of light.[1]

Absorption of X-rays due to scattering.—Valuable information concerning the magnitude of the scattering is given by the measurements of the absorption of X-rays due to scattering. Over a wide range of wave-lengths, the formula for the total mass absorption, $\mu/\rho = \kappa\lambda^3 + \sigma/\rho$, is found to hold, where μ is the linear absorption coefficient, ρ is the density, κ is a constant, and σ is the energy loss due to the scattering process. Usually the term $\kappa\lambda^3$, which represents the fluorescent absorption, is the more important; but when light elements and short wave-lengths are employed, the scattering process accounts for nearly all the energy loss. In this case, the constant κ can be determined by measurements on the longer wave-lengths, and the value of σ/ρ can then be estimated with considerable accuracy for the shorter wave-lengths from the observed values of μ/ρ.

Hewlett has measured the total absorption coefficient for carbon over a wide range of wave-lengths.[2] From his data for the longer wave-

[1] E. Rutherford, Radioactive Substances and their Radiations, p. 273.
[2] C. W. Hewlett, Phys. Rev. 17, 284 (1921).

lengths I estimate the value of κ to be 0.912, if λ is expressed in A. On subtracting the corresponding values of $\kappa\lambda^3$ from his observed values of μ/ρ, the values of σ/ρ represented by the crosses of Fig. 6 are obtained.

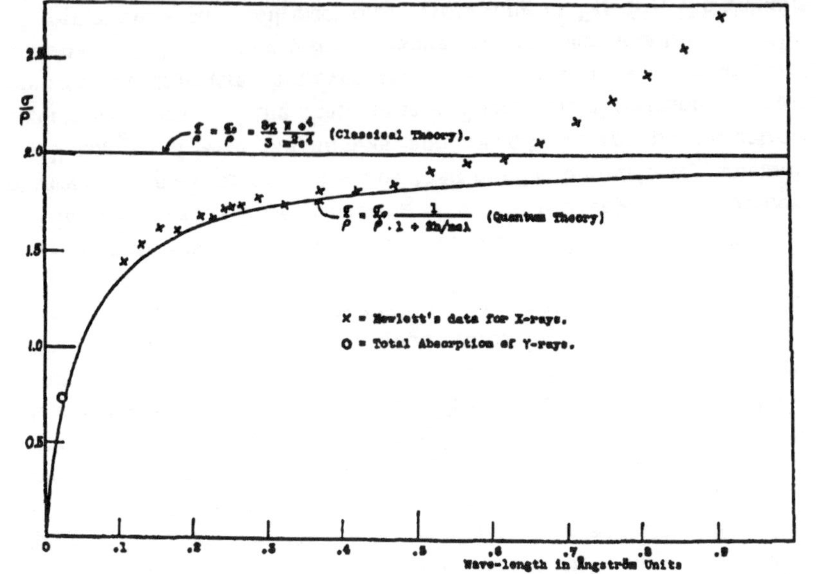

Fig. 6. The absorption in carbon due to scattering, for homogeneous X-rays.

The value of σ_0/ρ as calculated for carbon from Thomson's formula is shown by the horizontal line at $\sigma/\rho = 0.201$. The values of σ/ρ calculated from Eq. (28) are represented by the solid curve. The circle shows the experimental value of the total absorption of γ-rays by carbon, which on the present view is due wholly to the scattering process.

For wave-lengths less than 0.5 A, where the test is most significant, the agreement is perhaps within the experimental error. Experiments by Owen,[1] Crowther,[2] and Barkla and Ayers[3] show that at about 0.5 A the "excess scattering" begins to be appreciable, increasing rapidly in importance at the longer wave-lengths.[4] It is probably this effect which results in the increase of the scattering absorption above the theoretical value for the longer wave-lengths. Thus the experimental values of the absorption due to scattering seem to be in satisfactory accord with the present theory.

True absorption due to scattering has not been noticed in the case of

[1] E. A. Owen, Proc. Camb. Phil. Soc. 16, 165 (1911).
[2] J. A. Crowther, Proc. Roy. Soc. 86, 478 (1912).
[3] Barkla and Ayers, Phil. Mag. 21, 275 (1911).
[4] Cf. A. H. Compton, Washington University Studies, 8, 109 ff. (1921).

X-rays. In the case of hard γ-rays, however, Ishino has shown[1] that there is true absorption as well as scattering, and that for the lighter elements the true absorption is proportional to the atomic number. That is, this absorption is proportional to the number of electrons present, just as is the scattering. He gives for the true mass absorption coefficient of the hard γ-rays from RaC in both aluminium and iron the value 0.021. According to Eq. (30), the true mass absorption by aluminium should be 0.021 and by iron, 0.020, taking the effective wave-length of the rays to be 0.022 A. The difference between the theory and the experiments is less than the probable experimental error.

Ishino has also estimated the true mass scattering coefficients of the hard γ-rays from RaC by aluminium and iron to be 0.045 and 0.042 respectively.[2] These values are very far from the values 0.193 and 0.187 predicted by the classical theory. But taking $\lambda = 0.022$ A, as before, the corresponding values calculated from Eq. (29) are 0.040 and 0.038, which do not differ seriously from the experimental values.

It is well known that for soft X-rays scattered by light elements the total scattering is in accord with Thomson's formula. This is in agreement with the present theory, according to which the true scattering coefficient σ_s approaches Thomson's value σ_0 when $\alpha \equiv h/mc\lambda$ becomes small (Eq. 29).

The relative intensity of the X-rays scattered in different directions with the primary beam.—Our Eq. (27) predicts a concentration of the energy in the forward direction. A large number of experiments on the scattering of X-rays have shown that, except for the excess scattering at small angles, the ionization due to the scattered beam is symmetrical on the emergence and incidence sides of a scattering plate. The difference in intensity on the two sides according to Eq. (27) should, however, be noticeable. Thus if the wave-length is 0.7 A, which is probably about that used by Barkla and Ayers in their experiments on the scattering by carbon,[3] the ratio of the intensity of the rays scattered at 40° to that at 140° should be about 1.10. But their experimental ratio was 1.04, which differs from our theory by more than their probable experimental error.

It will be remembered, however, that our theory, and experiment also, indicates a difference in the wave-length of the X-rays scattered in different directions. The softer X-rays which are scattered backward are the more easily absorbed and, though of smaller intensity, may produce an

[1] M. Ishino, Phil. Mag. 33, 140 (1917).
[2] M. Ishino, loc. cit.
[3] Barkla and Ayers, loc. cit.

ionization equal to that of the beam scattered forward. Indeed, if α is small compared with unity, as is the case for ordinary X-rays, Eq. (27) may be written approximately $I_\theta/I_\theta' = (\lambda_0/\lambda_\theta)^3$, where I_θ' is the intensity of the beam scattered at the angle θ according to the classical theory. The part of the absorption which results in ionization is however proportional to λ^3. Hence if, as is usually the case, only a small part of the X-rays entering the ionization chamber is absorbed by the gas in the chamber, the ionization is also proportional to λ^3. Thus if i_θ represents the ionization due to the beam scattered at the angle θ, and if i_θ' is the corresponding ionization on the classical theory, we have $i_\theta/i_\theta' = (I_\theta/I_\theta')(\lambda_\theta/\lambda_0)^3 = 1$, or $i_\theta = i_\theta'$. That is, to a first approximation, the ionization should be the same as that on the classical theory, though the energy in the scattered beam is less. This conclusion is in good accord with the experiments which have been performed on the scattering of ordinary X-rays, if correction is made for the excess scattering which appears at small angles.

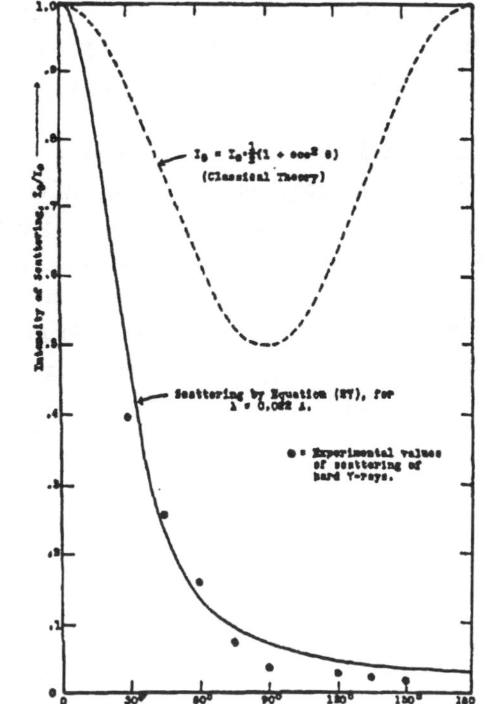

Fig. 7. Comparison of experimental and theoretical intensities of scattered γ-rays.

In the case of very short wave-lengths, however, the case is different. The writer has measured the γ-rays scattered at different angles by iron, using an ionization chamber so designed as to absorb the greater part of even the primary γ-ray beam.[1] It is not clear just how the ionization due to the γ-rays will vary with the wave-length under the conditions of the experiment, but it appears probable that the variation will not be great. If we suppose accordingly that the ionization measures the intensity of the scattered γ-ray beam, these data for the intensity are represented by the circles in Fig. 7. The experiments showed that the intensity at 90° was 0.074 times that predicted by the classical theory, or 0.037 I_0, where I_0 is the intensity of the scattering at the angle $\theta = 0$ as calculated on either the classical or the quantum theory. The absolute intensities of the scattered beam are accordingly plotted using I_0 as the unit. The solid curve shows the intensity in the same units, calculated according to Eq. (27). As before, the wave-length of the γ-rays is taken as 0.022 A. The beautiful agreement between the theoretical and the experimental values of the scattering is the more striking when one notices that there is not a single adjustable constant connecting the two sets of values.

Discussion

This remarkable agreement between our formulas and the experiments can leave but little doubt that the scattering of X-rays is a quantum phenomenon. The hypothesis of a large electron to explain these effects is accordingly superfluous, for all the experiments on X-ray scattering to which this hypothesis has been applied are now seen to be explicable from the point of view of the quantum theory without introducing any new hypotheses or constants. In addition, the present theory accounts satisfactorily for the change in wave-length due to scattering, which was left unaccounted for on the hypothesis of the large electron. From the standpoint of the scattering of X-rays and γ-rays, therefore, there is no longer any support for the hypothesis of an electron whose diameter is comparable with the wave-length of hard X-rays.

The present theory depends essentially upon the assumption that each electron which is effective in the scattering scatters a complete quantum. It involves also the hypothesis that the quanta of radiation are received from definite directions and are scattered in definite directions. The experimental support of the theory indicates very convincingly that a radiation quantum carries with it directed momentum as well as energy.

Emphasis has been laid upon the fact that in its present form the

[1] A. H. Compton, Phil. Mag. **41**, 758 (1921).

quantum theory of scattering applies only to light elements. The reason for this restriction is that we have tacitly assumed that there are no forces of constraint acting upon the scattering electrons. This assumption is probably legitimate in the case of the very light elements, but cannot be true for the heavy elements. For if the kinetic energy of recoil of an electron is less than the energy required to remove the electron from the atom, there is no chance for the electron to recoil in the manner we have supposed. The conditions of scattering in such a case remain to be investigated.

The manner in which interference occurs, as for example in the cases of excess scattering and X-ray reflection, is not yet clear. Perhaps if an electron is bound in the atom too firmly to recoil, the incident quantum of radiation may spread itself over a large number of electrons, distributing its energy and momentum among them, thus making interference possible. In any case, the problem of scattering is so closely allied with those of reflection and interference that a study of the problem may very possibly shed some light upon the difficult question of the relation between interference and the quantum theory.

Many of the ideas involved in this paper have been developed in discussion with Professor G. E. M. Jauncey of this department.

Washington University,
Saint Louis,
December 13, 1922

57. Thermal agitation of electricity in conductors. J. B. Johnson, Bell Telephone Laboratories, Inc.—Ordinary electric conductors are sources of random voltage fluctuations, as the result of thermal agitation of the electric charges in the conductor. The average effect of the fluctuations has been measured by means of a vacuum tube amplifier, where it manifests itself as a component of the phenomenon commonly called

"tube noise." A part of the "tube noise" arises in the first tube and other elements of the apparatus; the remainder in the input resistance, with a mean square voltage fluctuation $(V^2)_m$ which is proportional to the resistance R of that conductor. The ratio $(V^2)_m/R$, of the order of 10^{-18} watt at room temperature, is independent of the material and shape of the conductor, but is proportional to its absolute temperature. In the range of audio frequencies, at least, the noise contains all frequencies at equal amplitudes. The noise of an input resistance of only 5000 ohms may exceed that of the rest of the circuit, so that the limit of useful amplification is at times set by the thermal agitation of charges in the input resistance of the amplifier.

Second Series December, 1927 Vol. 30, No. 6

THE

PHYSICAL REVIEW

DIFFRACTION OF ELECTRONS BY A CRYSTAL OF NICKEL

By C. Davisson and L. H. Germer

Abstract

The intensity of scattering of a homogeneous beam of electrons of adjustable speed incident upon a single crystal of nickel has been measured as a function of direction. The crystal is cut parallel to a set of its {111}-planes and bombardment is at normal incidence. The distribution in latitude and azimuth has been determined for such scattered electrons as have lost little or none of their incident energy.

Electron beams resulting from diffraction by a nickel crystal.—Electrons of the above class are scattered in all directions at all speeds of bombardment, but at and near critical speeds sets of three or of six sharply defined beams of electrons issue from the crystal in its principal azimuths. Thirty such sets of beams have been observed for bombarding potentials below 370 volts. Six of these sets are due to scattering by adsorbed gas; they are not found when the crystal is thoroughly degassed. Of the twenty-four sets due to scattering by the gas-free crystal, twenty are associated with twenty sets of Laue beams that would issue from the crystal within the range of observation if the incident beam were a beam of heterogeneous x-rays, three that occur near grazing are accounted for as diffraction beams due to scattering from a single {111}-layer of nickel atoms, and one set of low intensity has not been accounted for. *Missing beams* number eight. These are beams whose occurrence is required by the correlations mentioned above, but which have not been found. The intensities expected for these beams are all low.

The spacing factor concerned in electron diffraction by a nickel crystal.—The electron beams associated with Laue beams do not coincide with these beams in position, but occur as if the crystal were contracted normally to its surface. The spacing factor describing this contraction varies from 0.7 for electrons of lowest speed to 0.9 for electrons whose speed corresponds to a potential difference of 370 volts.

Equivalent wave-lengths of the electron beams may be calculated from the diffraction data in the usual way. These turn out to be in acceptable agreement with the values of h/mv of the undulatory mechanics.

Diffraction beams due to adsorbed gas are observed except when the crystal has been thoroughly cleaned by heating. Six sets of beams of this class have been found; three of these appear only when the crystal is heavily coated with gas; the other three only when the amount of adsorbed gas is slight. The structure of the gas film giving rise to the latter beams has been deduced.

The investigation reported in this paper was begun as the result of an accident which occurred in this laboratory in April 1925. At that time we were continuing an investigation, first reported in 1921,[1] of the distribution-in-angle of electrons scattered by a target of ordinary (poly-

[1] Davisson & Kunsman, Science 64, 522, (1921).

crystalline) nickel. During the course of this work a liquid-air bottle exploded at a time when the target was at a high temperature; the experimental tube was broken, and the target heavily oxidized by the inrushing air. The oxide was eventually reduced and a layer of the target removed by vaporization, but only after prolonged heating at various high temperatures in hydrogen and in vacuum.

When the experiments were continued it was found that the distribution-in-angle of the scattered electrons had been completely changed. Specimen curves exhibiting this alteration are shown in Fig. 1. These curves are all for a bombarding potential of 75 volts. The electron beam is incident on the target from the right, and the intensities of scattering in different directions are proportional to the vectors from the point of bombardment to the curves. The upper curves (for different angles of incidence) are characteristic of the target prior to the accident. They are of the type

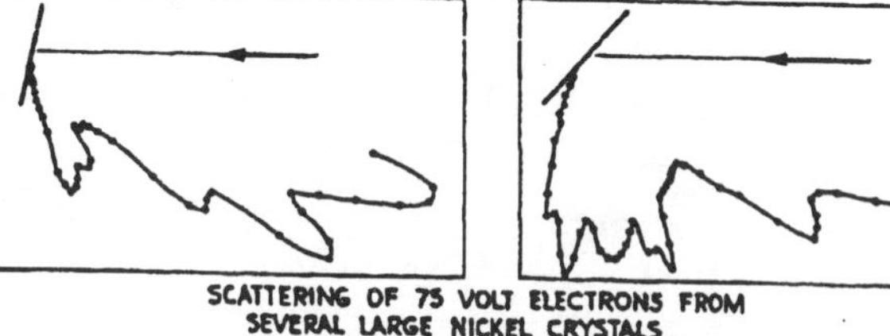

SCATTERING OF 75 VOLT ELECTRONS FROM
A BLOCK OF NICKEL (MANY SMALL CRYSTALS)

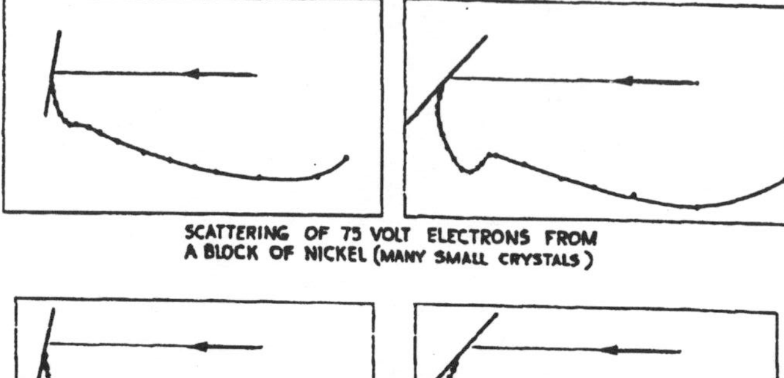

SCATTERING OF 75 VOLT ELECTRONS FROM
SEVERAL LARGE NICKEL CRYSTALS

Fig. 1. Scattering curves from nickel before and after crystal growth had occurred.

described in the note in "Science" in 1921, and are similar to curves that have been obtained for nickel in four or five other experiments. The lower curves—obtained after the accident—were the first of their sort to be observed. This marked alteration in the scattering pattern was traced to a re-crystallization of the target that occurred during the prolonged heating. Before the accident and in previous experiments we had been bombarding many small crystals, but in the tests subsequent to the accident we were bombarding only a few large ones. The actual number was of the order of ten.

It seemed probable from these results that the intensity of scattering from a single crystal would exhibit a marked dependence on crystal direction, and we set about at once preparing experiments for an investigation of this dependence. We must admit that the results obtained in these experiments have proved to be quite at variance with our expectations. It seemed to us likely that strong beams would be found issuing from the crystal along what may be termed its transparent directions—the directions in which the atoms in the lattice are arranged along the smallest number of lines per unit area. Strong beams are indeed found issuing from the crystal, but only when the speed of bombardment lies near one or another of a series of critical values, and then in directions quite unrelated to crystal transparency.

The most striking characteristic of these beams is a one to one correspondence, presently to be described, which the strongest of them bear to the Laue beams that would be found issuing from the same crystal if the incident beam were a beam of x-rays. Certain others appear to be analogues, not of Laue beams, but of optical diffraction beams from plane reflection gratings—the lines of these gratings being lines or rows of atoms in the surface of the crystal. Because of these similarities between the scattering of electrons by the crystal and the scattering of waves by three- and two-dimensional gratings a description of the occurrence and behavior of the electron diffraction beams in terms of the scattering of an equivalent wave radiation by the atoms of the crystal, and its subsequent interference, is not only possible, but most simple and natural. This involves the association of a wave-length with the incident electron beam, and this wave-length turns out to be in acceptable agreement with the value h/mv of the undulatory mechanics, Planck's action constant divided by the momentum of the electron.

That evidence for the wave nature of particle mechanics would be found in the reaction between a beam of electrons and a single crystal was predicted by Elsasser[2] two years ago—shortly after the appearance of L. de Broglie's original papers on wave mechanics. Elsasser believed, in fact, that evidence of this sort was already at hand in curves, published from these Laboratories,[3] showing the distribution-in-angle of electrons scattered by a target of polycrystalline platinum. We should like to agree with Elsasser in his interpretation of these curves, but are unable to do so. The maxima in the scattering curves for platinum are of the type of the single maximum in the curves for nickel shown in the upper half of Fig. 1, and are, we believe, unrelated to crystal structure.

Preliminary announcement of the main results contained in this paper was made in "Nature" for April 16, 1927. In the present article we give a more complete account of the experiments and additional data.

The Apparatus

The essential parts of the special apparatus, Fig. 2, used in these experiments are the "electron gun" G, the target T and the double Faraday box collector C. The electrons constituting the primary beam are emitted thermally from the tungsten ribbon F, and are projected from the gun into a field-free enclosure containing the target and collector; the outer walls of the gun, the target, the outer box of the collector and the box enclosing these parts are held always at the same potential. The beam of electrons

[2] W. Elsasser, Naturwiss. 13, 711 (1925).

[3] Davisson & Kunsman, Phys. Rev. 22, 242 (1923).

meets the target at normal incidence. High speed electrons scattered within the small solid angle defined by the collector opening enter the inner box of the collector, and from thence pass through a sensitive galvanometer. Electrons leaving the target with speeds appreciably less than the speed of the incident electrons are excluded from the collector by a retarding

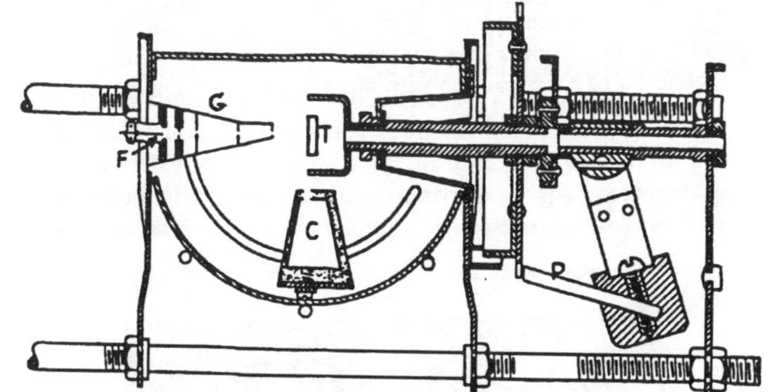

Fig. 2. Cross-sectional view of the experimental apparatus—glass bulb not shown.

potential between the inner and outer boxes. The angle between the axis of the incident beam and the line joining the bombarded area with the opening in the collector can be varied from 20 to 90 degrees. Also the target can be rotated about an axis that coincides with the axis of the incident beam. It is thus possible to measure the intensity of scattering in any direction in

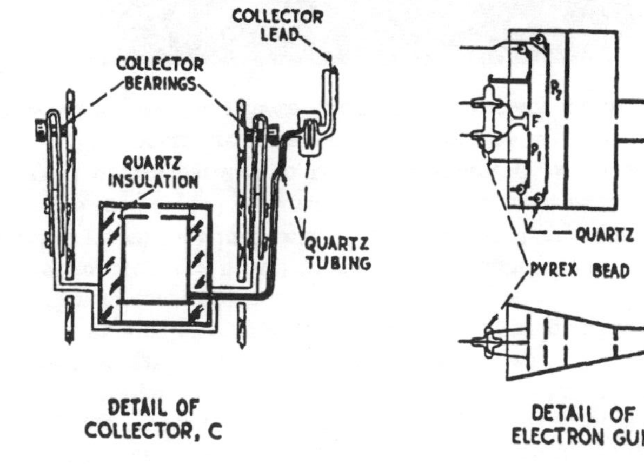

Fig. 3.

front of the target with the exception of directions lying within 20 degrees of the incident beam.

Details of the "electron gun" are shown in Fig. 3. The tungsten ribbon F lies in a rectangular opening in a nickel plate P_1. The purpose of this plate is to assist in concentrating the emission from the filament onto an opening in the parallel plate P_2. This is accomplished by making the potential of P_1 slightly more negative than that of the filament. The potential of P_2 relative to that of the filament is adjusted ordinarily to a rather high positive value.

The opening in P_2 is circular and slightly more than 1 mm in diameter. Some of the electrons passing through this opening continue on through apertures in a series of three plates that are at the same potential as the outer walls of the gun. It is the difference between this potential and that of the filament which determines the speed of the emergent beam. The first two of these apertures are 8 mm apart and are 1 mm in diameter; the diameter of the third is slightly greater. The geometry of these parts is such as to insure a well defined emergent beam relatively free from low speed secondary electrons. The gun was tested in a preliminary experiment, and was found to give a homogeneous beam. The distance from the end of the gun to the target is 7 mm.

The two parts of the collector (Fig. 3) are insulated from one another by blocks of clear quartz. The openings in the outer and inner boxes are circular,

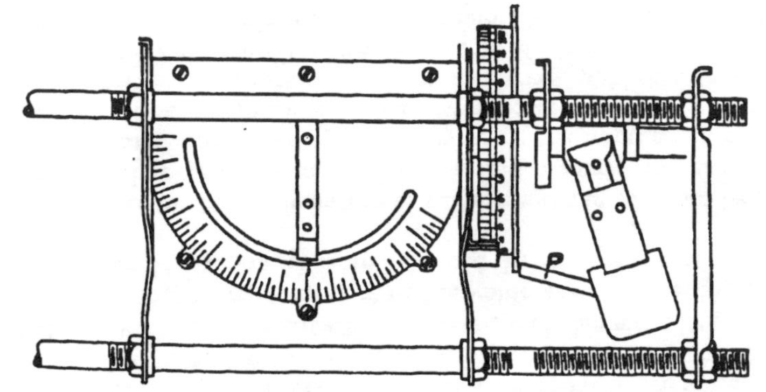

Fig. 4. Outside view of the experimental apparatus—glass bulb not shown—0.7 actual size.

their diameters being 1.0 mm and 2.0 mm respectively. These openings were made as near as possible to the side of the box adjacent to the gun in order to reduce to a minimum the unexplored region about the incident beam. The collector is suspended by arms from bearings outside the enclosing box (Fig. 4), and is free to rotate about a horizontal axis through the bombarded area and normal to the incident beam. The angular position of the collector is varied by rotating the whole tube, which is sealed from the pumps, about this axis. The lead to the inner box must be especially shielded from stray currents; it is enclosed in small quartz tubing from the point at which it leaves the outer box to the seal at which it leaves the tube. The distance from the bombarded area to the opening in the outer box is 11 mm.

The target is a block of nickel $8\times5\times3$ mm cut from a bar in which crystal growth had been induced by straining and annealing. The orientations of the largest crystals in the bar were determined by an examination of the optical reflections from crystal facets that had been developed by

etching. A cut was then made through one of the crystals approximately parallel to a set of its $\{111\}$-planes. One of the surfaces so exposed was polished, etched, examined and corrected, and became eventually the face of the target.

No particular care was taken in preparing the target to avoid straining or damaging the crystal. The cutting was done with a jeweler's saw; holes were drilled through the ends of the block, and nickel wires were passed through these to serve as supports. After this rather rough usage the target was heated in an auxiliary tube to near its melting point without its showing any indication of recrystallization.

The effect of etching a nickel crystal, either chemically or by vaporization, is to develop its surface into sets of facets parallel to its principal planes. Those parallel to the $\{111\}$-planes are developed most readily, but we have also observed others parallel to the $\{110\}$-planes.[4] Four sets of the predominant $\{111\}$-facets are, in general, exposed on a plane surface. If one of these is parallel to the general plane of the surface, as in the case of our

Fig. 5. Microphotograph of the nickel target.

target, the other sets have normals lying 20 degrees above the general plane of the surface and equally spaced in azimuth about the normal to the first set.

A microphotograph of the face of the target is shown in Fig. 5. The illumination is at normal incidence, and the large crystal shows white on account of the strong reflection from the $\{111\}$-facets lying in its surface. That these facets make up nearly the whole of the surface seems probable from the fact that in the visual examination of the crystal the reflections from the lateral facets were very weak. This conclusion may not, however, be stated without reservation, as the weakness of these reflections may indicate merely that some dimension of the individual facets is small compared with optical wave-lengths. The regions appearing black in the photograph are made up of crystals having no facets parallel to the surface. Those included in the large crystal and others adjacent to it are twinned with the main structure. The area selected for bombardment is shown enclosed in a circle.

[4] See also Potter and Sucksmith, Nature 119, 924 (1927), who found $\{100\}$-facets.

The target was mounted in a holder from which it was insulated, and the holder was fixed to the end of a hollow shaft mounted in bearings. A small tungsten filament mounted back of the target (not shown in Fig. 2) supplies electrons for heating the target by bombardment. Leads from the target and from this filament pass out through the hollow shaft and are connected through platinum brush contacts to other leads which are carried through seals in the tube.

The mechanism for rotating the target is shown in Figs. 2 and 4. When the tube is rotated counter-clockwise about the collector axis, to bring the collector into range in front of the target, the molybdenum plunger p (attached to a heavy pendulum) passes through an opening in a toothed wheel (attached to the shaft) and engages with a milled edge of a strip of molybdenum that is attached to the frame. The wheel and the target are then locked to the frame. When the tube is rotated clockwise until the main or longitudinal axis of the tube has passed slightly beyond the horizontal, the plunger disengages from the milled edge but still remains within the opening in the toothed wheel. The pendulum has a second degree of freedom (it revolves about a fixed hollow shaft coaxial with the shaft carrying the target) so that, by rotating the tube about its main axis, the pendulum and engaged wheel are rotated relative to the frame. The range of this rotation is only 20° or 30°, but by rotating the tube slightly further in the clockwise direction about the collector axis the plunger is disengaged from the wheel, and can be moved, by rotation again about the main axis, to a different opening in the toothed wheel. By these operations the target can be worked through any angle. Its azimuth is read from a scale ruled on the wheel. This scale and that for reading the position of the collector are shown in Fig. 4. The bearings throughout the tube, with the exception of one nickel on nickel bearing, are either molybdenum on molybdenum, or molybdenum on nickel.

Preparation of the Tube

The metal parts of the tube were preheated to 1000°C in a vacuum oven, and were then assembled and sealed into the bulb with the least possible delay. The bulb is of Pyrex and has sealed to it two auxiliary tubes, one containing cocoanut charcoal, and the other a misch metal vaporizer. This latter consists of a small pellet of misch metal attached to a molybdenum plate anode which may be bombarded from a nearby tungsten filament. The thermal contact between the pellet and the plate is reduced by the interposition of a narrow strip of molybdenum, so that the misch metal may be vaporized only by raising the plate to a very high temperature. The misch metal is vaporized when the pumping is nearly completed, and various of its constituents form solid compounds with the residual gas, thus improving and maintaining the vacuum.

During the pumping, which lasted several days, the tube itself and the tubing connecting it with the pumps were baked for hours at a time at 500°C, and the side tube containing charcoal was baked at an even higher temperature—about 550°C. This baking was alternated with heating by bombardment of such of the metal parts as could be reached from the filaments. The target in particular was heated several times to a temperature at which it vaporized freely. The tube was sealed from the pumps with the target at a high temperature, and the charcoal at 400 or 500°C and cooling. The pressure in the tube at the time was 2 or 3×10^{-6} mm of mercury. As soon as the tube containing charcoal had cooled sufficiently it was immersed in liquid air. No means were provided for measuring the pressure of the gas in the tube after sealing from the pumps, but from experience with similar tubes in which such measurements could be made we judge that its equilibrium value was 10^{-8} mm of mercury or less. The pumping equipment consisted of a three stage Gaede diffusion pump backed by a two stage oil pump.

The Crystal

It is important to have a clear picture of the arrangement of atoms presented to the incident beam by the crystal. The nickel crystal is of the face-centered cubic type. The {111}-plane is the plane of densest packing, and in this plane the atoms have a triangular arrangement. Looking directly downward onto a crystal cut to this plane (Fig. 6) one sees the atoms of the second plane below the centers of alternate triangles of the first plane, and the atoms of the third plane below the centers of the remaining triangles. The atoms of the fourth plane are below those of the first. The lines joining any second-layer atom with the three nearest first-layer atoms are {110}-directions in the crystal, and the lines joining it with the three next-nearest surface atoms are the orthogonal {100}-directions. It will be convenient to refer to the azimuths of these latter directions as {100}-azimuths. The azimuths of the {110}-directions are also those of the three lateral {111}-directions, already referred to, and we shall designate these as {111}-azimuths. We need also a designation for the azimuths that bisect the dihedral angles between adjacent members of the two sets already specified. There are six such azimuths and they will be referred to as {110}-azimuths.

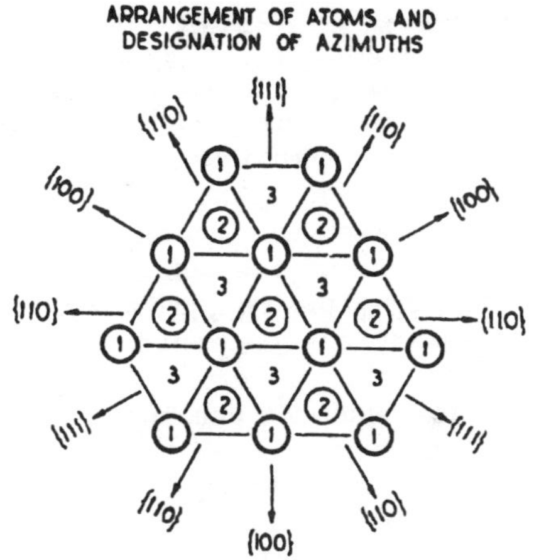

ARRANGEMENT OF ATOMS AND DESIGNATION OF AZIMUTHS

Fig. 6.

It follows from the trigonal symmetry of the crystal that if the intensity of scattering exhibits a dependence on azimuth as we pass from a {100}-azimuth to a next adjacent {111}-azimuth (60°), the same dependence must be exhibited in the reverse order as we continue on through 60° to the next following {100}-azimuth. Dependence upon azimuth must be an even function of period $2\pi/3$.

Distribution of Speeds among Scattered Electrons

The electrons leaving the target in any given direction appear always to have speeds that are distributed in one of two ways, depending upon whether the direction lies within or outside a diffraction beam. In the latter

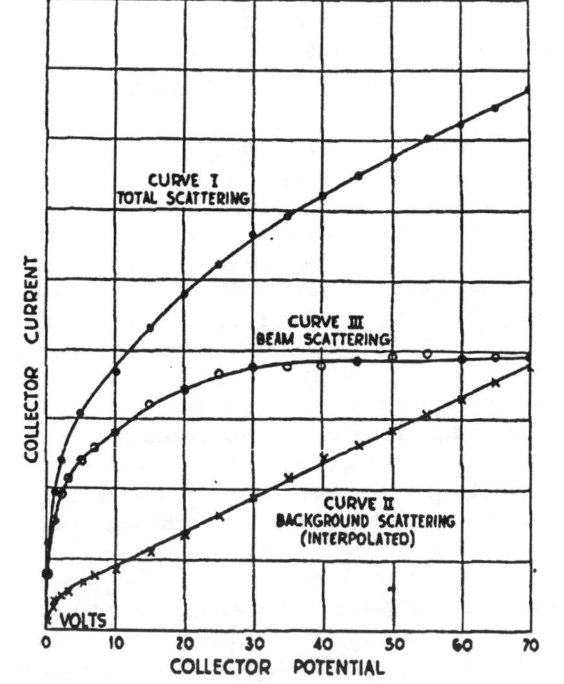

Fig. 7. Current to collector as a function of collector potential—bombarding potential 160 volts.

case—that of the electrons making up the "background" scattering—there is always a definite group having the speed of those in the incident beam. Below this speed there is in general a range over which the distribution in energy is very nearly uniform; and below this a range, ending with zero speed, in which the representation increases rapidly with decreasing energy.

These characteristics are inferred from the relation between collector current and the potential of the collector relative to that of the filament. The lower portion of a typical curve exhibiting this relation for the "background" scattering is shown in Fig. 7 (Curve II). The ordinate of a curve

of this sort is not, of course, an actual measure of the number of electrons entering the outer box with sufficient energy to reach the inner box. On account of the distortion of the field about the openings the probability that an electron entering the outer box with just sufficient energy to reach the walls of the inner box will actually do so is vanishingly small; the saturation current due to electrons of a given speed is attained only when the potential of the inner box is somewhat higher than that corresponding to their speed. The rounding off of the current-voltage curve at the top of the initial rise is due to some extent to this cause.

For this reason the group of high speed, or full speed, electrons is more nearly homogeneous than would be inferred from the current-voltage curve if no account were taken of this distortion. It seems probable, in fact, that the distortion accounts almost completely for the rounding off of the curve, and that the group is as nearly homogeneous as is permitted by the drop in potential along the filament and the initial speeds of the emitted electrons. This view is strongly supported by the observations of Becker,[5] Brown and Whiddington,[6] Sharman,[7] and Brinsmade[8] on the magnetic spectrum of electrons scattered by metals, and by similar curves obtained by Farnsworth.[9]

Within a diffraction-beam the distribution-in-speed is somewhat different. There is again a definite group of full speed electrons, but speeds just inferior to the maximum have much greater representation than among the background electrons. This is inferred from curve III in Fig. 7 which is representative of the current-voltage relation for electrons of this class. We shall return in a later section to a further consideration of the curves in this figure.

In studying the distribution in direction of the scattered electrons measurements have been confined, as nearly as possible, to the group of full speed electrons. The potential of the collector is set just high enough to admit all of this group. The ratio of collector to bombarding current is then of the order 10^{-4}, so that, by using bombarding currents of the order 10^{-6} ampere, collector currents are obtained that are easily measurable with a sensitive galvanometer. The total integrated current of full speed scattered electrons is from a tenth to a twentieth as great as the current of the incident beam.

Distribution of Directions among Full Speed Scattered Electrons

The current of full speed electrons entering the collector is proportional to the current incident upon the target and is otherwise a function of the bombarding potential and of the latitude and azimuth of the collector. Three simple types of measurement are thus possible in each of which two of the independent variables are held constant and the third is varied. When bombarding potential and azimuth are fixed and exploration is made in

[5] J. A. Becker, Phys. Rev. 23, 664 (1924).
[6] D. Brown and R. Whiddington, Nature 119, 427 (1927).
[7] C. F. Sharman, Proc. Camb. Phil. Soc. 23, 523 (1927).
[8] J. B. Brinsmade, Phys. Rev. 30, 494 (1927).
[9] H. E. Farnsworth, Phys. Rev. 25, 41 (1925).

latitude a dependence of current upon angle is observed which is generally of the form shown in Fig. 8; the current of scattered electrons is zero in the plane of the target and increases regularly to a highest value at the limit of observations—colatitude 20°. This type of dependence upon angle is essentially the same as is observed when the target is of ordinary nickel—made up of many small crystals.

When bombarding potential and latitude angle are fixed and exploration is made in azimuth a variation of collector current is always observed, and this exhibits always the three-fold symmetry required by the symmetry of the crystal. The curves show in general two sets of maxima—a set of three in the {111}-azimuths, and a set of three of different intensity in the {100}-azimuths. These crests and troughs in the azimuth curves are usually not pronounced.

In the third method of observation the position of the collector is fixed in one of the principal azimuths at one after another of a series of colatitude

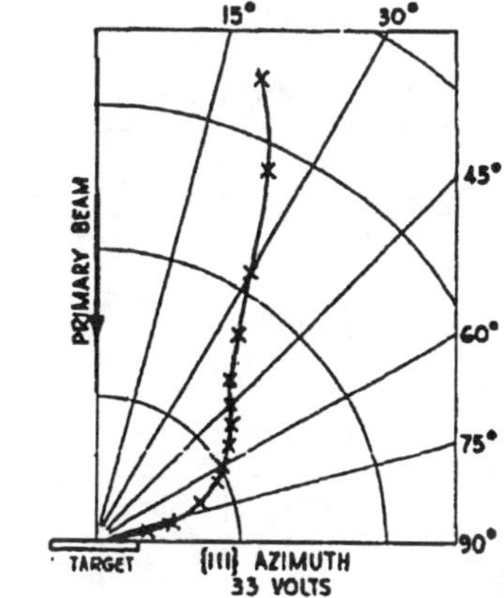

Fig. 8. Typical colatitude scattering curve for the single nickel crystal.

angles, and at each such setting the current to the collector is observed as a function of bombarding potential. It would be desirable in making observations of this sort to keep constant the current in the incident beam but, as there is no ready means of doing this, the current to the plate P_2 (Fig. 3) is kept constant instead. Beginning at colatitude 20° a series of such observations is made, over a predetermined voltage range, at 5° intervals to colatitude 80° or 85°. A portion of a set of curves constructed from such data is shown in Fig. 9.

The general trend of a single one of these curves is not significant as it is determined in part by variation with voltage of the bombarding current. The relative displacements among them, however, are significant as they indicate departures from the simple type of colatitude curve shown in Fig. 8. From the curves in Fig. 9 we see, for example, that the colatitude curves for bombarding potentials near 55 volts are characterized by exceptional intensities at colatitude angles near 50°. The data for constructing colatitude

curves for particular bombarding potentials are taken directly from such curves as those of Fig. 9, or the features in these latter curves are used as

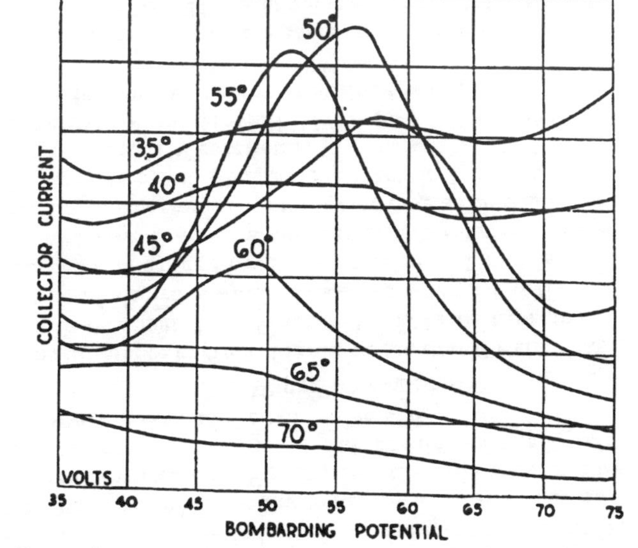

Fig. 9. Curves of collector current vs. bombarding potential—showing the development of the "54 volt beam." Azimuth {111}.

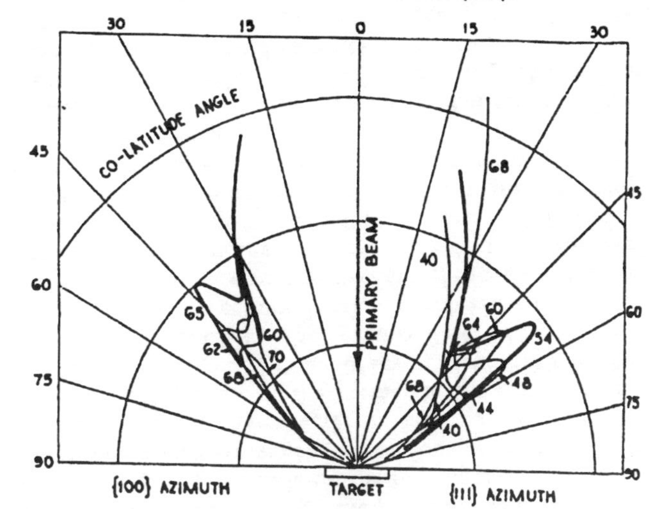

Fig. 10. Scattering curves showing the occurrence of the "54 volt" electron beam and the "65 volt" electron beam. (On each scattering curve is indicated the bombarding potential in volts.)

a guide to voltage-colatitude ranges requiring special study. This method has been employed in exploring the principal azimuths in the range from

15 to 350 volts, and two other azimuths, to be specified later, over the range from 200 to 300 volts. Every feature of the sort shown in Fig. 9 has been investigated.

The unusual and significant feature revealed by the curves of Fig. 9 is exhibited again in the set of colatitude curves on the right in Fig. 10. We see a slight hump at 60° in the colatitude curve for 40 volts, and observe that as the bombarding potential is increased this hump develops into a strong spur which reaches a maximum development at 54 volts in colatitude

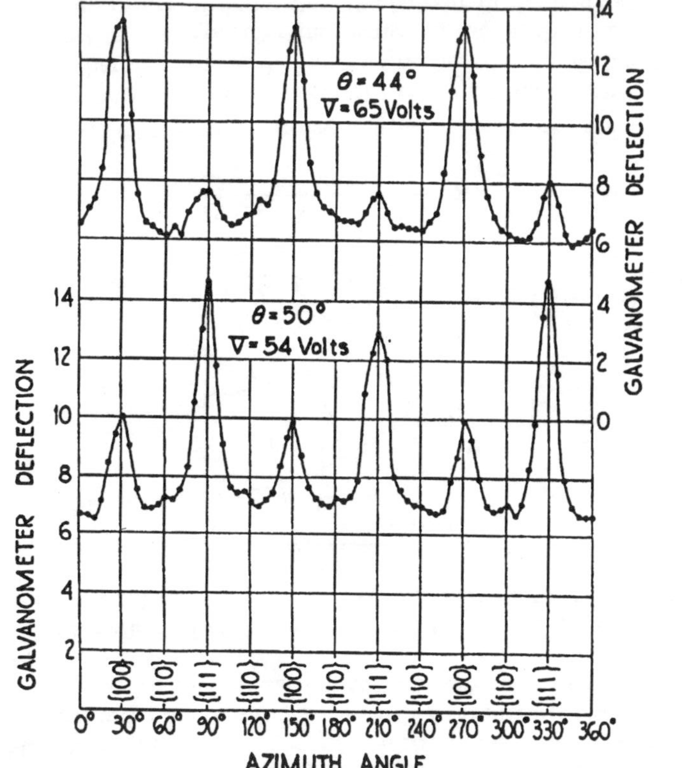

Fig. 11. Azimuth scattering curves through the "54-volt" electron beam and through the "65-volt" electron beam.

50°, then decreases in intensity and finally disappears at about 66 volts in colatitude 40°.

A section in azimuth through the center of this spur in its maximum development is shown in the lower curve of Fig. 11. The spur is sharp in azimuth as well as in latitude and is one of a set of three spurs as the symmetry of the crystal requires. The smaller peaks in the {100}-azimuth are sections of a similar set of spurs that attains its maximum development at 65 volts in colatitude 44°. A complete set of colatitude curves

for one of these spurs is shown on the left in Fig. 10. The upper curve in Fig. 11 is an azimuth curve through the peaks of these 65 volt spurs. The small peaks in the {111}-azimuth are the remnants of the "54 volt" spurs.

The colatitude angles at which the various spurs of a single set are strongest are found not to have exactly the same values. This is due apparently to imperfect alignment of the normal to the crystal planes and the axis of rotation of the target. In each of several sets that have been studied these angles are expressed by the formula $\theta=\theta_0+\Delta\theta\cos(\phi-\phi_0)$, where θ_0 is a constant for a given set and is taken to represent the colatitude angle at which all spurs in the set would be strongest if the alignment were perfect, and $\Delta\theta$ and ϕ_0 are constants that have the same values for all sets, 2° and 1° respectively. This is taken to mean that the axis of rotation is displaced about one degree from the normal to the crystal planes into azimuth 181°. The correction $2\cos(\phi-1°)$ degrees has been applied to all observed values of the colatitude angle θ.

The voltages at which the different spurs of a given set are strongest probably show a like variation. The differences are slight, however, and no attempt has been made to apply a voltage correction.

If we regard the spur as a feature superposed on the simple scattering curve the position of its maximum is falsified to some extent by the variation with angle of the background against which it appears. The method of correcting for this effect is indicated by the curves in Fig. 12. The end portions of the observed curve are joined by a curve of the known form of the simple relation (see e.g., Fig. 8), and the difference of these curves is plotted as the graph of the spur. The position of the maximum of this difference curve is taken as the true value of θ.

The Width of the Spur

From the difference curve in Fig. 12 we see that the spur has an apparent angular width of about 25°. What width is to be expected if the spur is due to a beam of electrons which is as sharply defined as the primary beam? This latter beam is defined by circular apertures 1 mm in diameter, and if we assume that the beam is a cylinder of this diameter, an equally sharp beam scattered at colatitude θ would extend over a colatitude arc of $(1\times\cos\theta)$ mm. The circular opening in the outer box is 1 mm in diameter, and its distance from the axis of rotation is 11 mm, so that the least possible value for the apparent colatitude width is $(1+\cos\theta)/11$ radians, or $5.2(1+\cos\theta)$ degrees. For the spur under consideration $\theta=50°$, and the calculated width is 8.5°.

The width in azimuth of the same spur is seen from the lower curve of Fig. 11 to be about 30°. The least value for this width is given by $5.2(2)/\sin\theta$ which for $\theta=50°$ amounts to 13.5°. Thus both in colatitude and in azimuth the observed width of the beam is more than double the least possible value.

It is to be expected, of course, that the observed values will be somewhat greater than those calculated, since it is hardly likely that the primary beam is as sharply defined as has been assumed; it is probably divergent as well as somewhat nonhomogeneous. There is no way, however, of investigating these matters. The most that can be said with certainty is that the spur is due to a beam of electrons whose definition is comparable in sharpness with that of the incident beam.

Distribution of Speeds among Electrons Constituting the Diffraction Beams

Assuming that the sharply defined beam is a distinct feature superposed upon the general background scattering, it is natural to inquire in what way

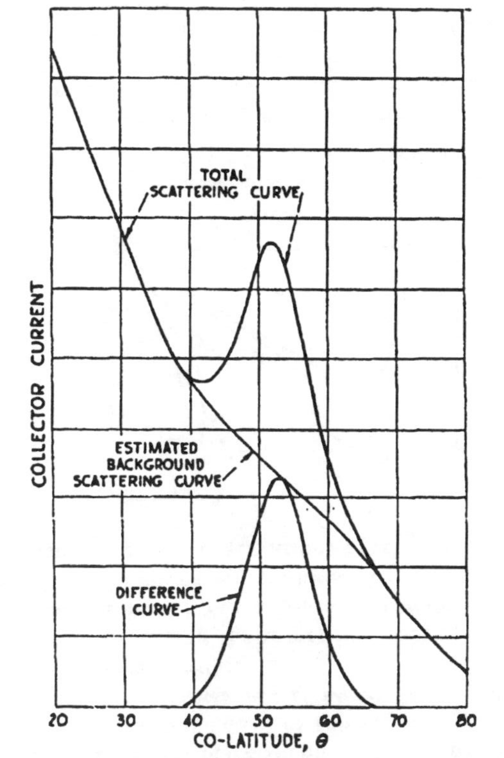

Fig. 12. Scattering curve of the "54 volt" beam, showing the method of determining the position of the maximum. (Angles not corrected for the tilt of the target).

the electrons constituting the whole of the superposed scattering are distributed in speed. Is the complete superposed scattering made up of full speed electrons only, or is it made up in part of electrons of lower speeds? This point has been investigated in several different ways. The fact seems to be that in addition to the group of full speed electrons observed in the distribution-in-direction measurements the complete superposed scattering includes also other electrons that have lost energy in various amounts up to about one quarter of their incident energy; it seems to include no electrons that have lost more than this amount.

These characteristics are inferred from the experimental results exhibited in Fig. 7. The upper curve was obtained with the collector in the center of a beam in the {100}-azimuth which is strongest at 160 volts. Two other curves (not shown) were then obtained with the position of the collector unaltered, but with the bombarding potential changed to 120 volts in one case, and to 200 volts in the other. The spur does not appear at either of these voltages, so that the current-voltage curves for these voltages are for "background" electrons only. These curves are not very different, so that a similar curve for 160 volts may be interpolated with considerable certainty. Curve II of Fig. 7 is this interpolated curve representing the current-voltage relation for "background" electrons only at 160 volts.

Curve III has been obtained by subtracting II from I, and consequently represents the current-voltage curve that would be obtained if the "background" scattering could be eliminated. We infer from the form of this curve that the electrons making up the whole of the superposed scattering are distributed in energy as described in the previous paragraph. The distribution is that which might be expected among the emergent electrons if a homogeneous beam were incident upon an extremely thin plate—a plate only one or two atoms in thickness.

Although the matters here considered require much more study than they have so far received, it is fairly clear that the superposed scattering is made up of beams of full speed and nearly full speed electrons that approximate the incident beam in sharpness, and that these beams appear only at and near certain critical bombarding potentials.

Positions and Voltages of Electron Beams

The work of investigating these beams and searching out new ones has progressed in several distinct stages. To begin with an exploration was made through the principal azimuths in the range 15 to 200 volts. Thirteen sets of beams were found, and these were described in our note to "Nature." The exploration was then extended to 350 volts, and eight additional sets were found. Up to this time the target had been heated last while the tube was still on the pumps.

After completing the exploration to 350 volts the target was strongly heated, and allowed to cool again to room temperature. The effect of this treatment was to increase generally the intensities of the beams without altering either the voltages at which they occurred or their positions. Three sets of beams only were exceptions to this rule, and these were the particular three sets which in our note in "Nature" we regarded as anomalous; the intensities of these were decreased.

These alterations in intensity resulted, we believe, from removal of gas from the surface of the target. In further tests it was found that immediately after bombardment, while the target is still hot, the beams are all weak; that they then increase in intensity as the target cools and that later, presumably as gas collects again on the surface, their intensities decrease. This final decrease in intensity was rather rapid after the first heat treatments, but after ten or a dozen heatings it was much slower—the intensities of the beams remaining for hours near their maximum values. This was the behavior of the normal beams. The three sets of anomalous beams were progressively weakened, and finally disappeared.

The alteration brought about in the "54 volt" beam by this degassing of the target is shown in Fig. 13. Curve *A* is reproduced from Fig. 10 and shows the beam in its maximum development in the earlier tests. Curve *B* is for the same beam from the cleaned target. The intensity has been increased between four and five fold, while the intensity of the background scattering has been decreased. For beams of higher voltage the increase in intensity was in general less marked, but even for beams above 300 volts intensities were at least doubled. *The ratio of full speed electrons scattered into any one of the most intense sets of beams to the total number scattered in all directions is about two-tenths.*

A further effect of cleaning the target has been to cause the appearance of certain new beams in the range below 200 volts. These beams are of peculiar interest. They are exceptionally sensitive to gas, and were entirely absent at the time the earlier observations were made.

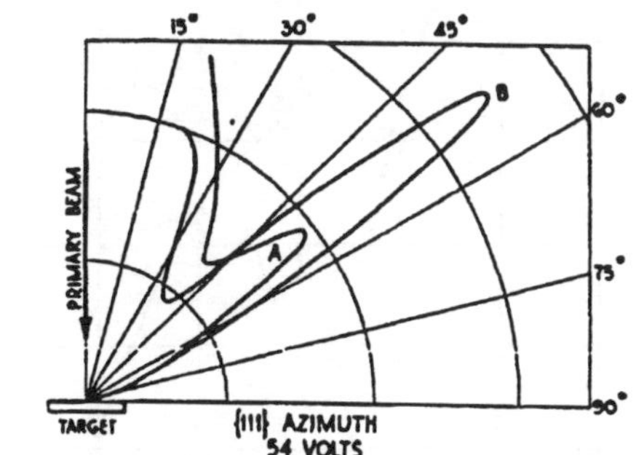

Fig. 13. The "54 volt" beam before and after heating the crystal by electron bombardment. *A*. Original condition as in Fig. 10. *B*. After heating the crystal.

We have found altogether thirty sets of beams, including those due to gas. Most of these are analogues of Laue beams, and the data for beams of this class are listed in Table I. The intensities in column 5 are estimated for a constant electron current bombarding the target, and for the target surface as free as possible from gas. In a few cases redetermined constants for the beams of Table I are somewhat different from those given in the note in "Nature." The beams not listed in Table I will be considered individually.

It would be possible to follow, with the beams listed in Table I, the procedure employed in our note in "Nature" (to point out a correspondence between the electron beams and the Laue beams that would issue from the same crystal if the incident beam were a beam of x-rays); then, with this suggestion of the wave nature of the phenomenon, to show that wave-lengths may be associated in a simple and natural way with the electron beams, and finally to compute these wave-lengths and show that they are in accord with the requirements of the undulatory mechanics. It is preferable, however, to start at once with the idea that a stream of electrons of speed v is in some way equivalent to a beam of radiation of wave-length h/mv, and to show to what extent the observations can be accounted for on this hypothesis. We assume that this radiation is scattered and absorbed by the atoms of the crystal and that, just as in the case of x-rays, strong diffraction beams result from coincidence in phase of all of the radiation scattered in some particular direction.

TABLE I

Space lattice electron beams

Azi.	Bombarding Potential V	Equivalent Wave-length $\lambda=(150/V)^{1/2}$	Colatitude θ	Beam Intensity (Arbitrary Scale)	Beam Int. / Background Int.
{111}	54 volts	1.67 A	50°	1.0	7.0
	106	1.19	28	0.4	1.4
	174	0.928	22	2.0	1.3
	181	0.910	55	0.7	1.0
	248	0.778	44	1.0	2.6
	258	0.762	<20	4.5	1.8
	343	0.661	34	3.0	1.5
	347	0.657	62	0.07	0.3
{100}	65	1.52	44	1.0	7.0
	126	1.09	28	2.0	3.8
	160	0.968	60	0.8	5.7
	190	0.889	20	2.0	1.3
	230	0.807	46	0.4	1.2
	292	0.716	<20	7.0	2.0
	310	0.695	70	0.15	0.8
	312	0.693	37	1.5	1.2
	370	0.636	57	0.15	0.4
{110}	143	1.024	56	0.2	0.9
	170	0.940	46	0.1	0.5
	188	0.893	43	0.3	1.0
	248	0.778	34	0.45	0.6

In considering the conditions under which such beams will occur it will be convenient to regard the crystal as built up of {111}-planes of atoms parallel to the principal facets, and to picture the radiation scattered by the crystal as made up of the contributions from all such planes. This viewpoint has a distinct advantage, in the present case, over regarding diffraction beams as built up of contributions regularly reflected from the Bragg atom-planes. The amplitude of the radiation proceeding in a given direction from the crystal is then to be regarded as the sum of the amplitudes (with due regard to phase) of the increments of radiation proceeding in the same direction from all such {111}-planes of atoms—or more precisely, the fractions of such increments that actually escape from the crystal.

If we imagine a system of Cartesian coordinates with its origin at the center of a surface atom, its positive z-axis extending outward from the facet, and its positive x-axis lying in one of the {110}-azimuths of the crystal, then atom centers occur at the points

$$x=(M+N+P)s/2$$
$$y=3^{1/2}(M-N+P/3)s/2$$
$$z=-2^{1/2}Ps/3$$

where $s(=a/2^{1/2})$ represents the least distance between atoms in a $\{111\}$-plane (the side of the elementary triangle), M and N are any integers, and P is zero or any positive integer.

If a beam of plane waves is incident upon any one of these atom planes along a line whose direction cosines are l_1, m_1, n_1, it may be shown that the radiation scattered by the individual atoms will be in phase along the directions

$$l_2=l_1+(p+r)\lambda/s$$
$$m_2=n_1+(p-r)\lambda/3^{1/2}s$$
$$n_2=\pm(1-l_2^2-m_2^2)^{1/2}$$

where λ represents the wave-length of the radiation, and p and r are any integers. If the waves meet the layer at normal incidence, as in our experiments, then $l_1=m_1=0$, $n_1=-1$, and

$$l_2=(p+r)\lambda/s$$
$$m_2=(p-r)\lambda/3^{1/2}s$$
$$n_2=\pm(1-l_2^2-m_2^2)^{1/2}$$

It may be shown that these are just the directions in which diffraction beams are to be expected if the plane of atoms is regarded as equivalent to a great number of line gratings. All of the atoms in the plane may be regarded as arranged on lines parallel to the line joining any two of them, and every such set of lines functions as a line grating. Diffraction beams due to each such grating occur in the plane normal to its lines, and satisfy the ordinary plane grating formula, $n\lambda=d\sin\theta$.

The grating constant d has its greatest value for the three plane gratings the lines of which are parallel to the sides of the elementary triangle. For these $d=d_1=3^{1/2}s/2$, and $n\lambda=(3^{1/2}s/2)\sin\theta$. The beams due to these gratings occur symmetrically in the $\{111\}$ and $\{100\}$-azimuths.

The longest wave-length that can give rise to a diffraction beam is found by setting $n=1$ and $\sin\theta=1$ in the grating formula in which d has its greatest value, i.e., it is the wave-length $3^{1/2}s/2$. When waves of this length are incident normally on the plane of atoms first order diffraction beams should appear at grazing emergence in the $\{111\}$ and $\{100\}$-azimuths—six beams in all. When the wave-length is decreased these beams should split into two sets of six beams each—one set moving upward toward the incident beam, and the other set moving downward. We are concerned with the upward moving set only. When the wave-length has been decreased to $3^{1/2}s/4$ second order beams should appear at grazing, and these should follow the course of the first order beams with still further decrease in the wave-length. In the meantime, however, six first order beams from the three gratings of second largest spacing $(d=d_2=s/2)$ should have appeared at grazing in the $\{110\}$-azimuths. And as the wave-length is shortened beams should appear in still other azimuths.

These are the diffraction beams to be expected from a single layer of atoms. In general, when the incident beam is being scattered simultaneously from a large number of such layers the diffraction beams from the different layers must emerge from the crystal out of phase with one another, and the amplitude of the resultant beam must be much smaller than that which a single layer should produce.

There are two conditions, however, under which the amplitude of a given set of beams may be as great as, or greater than, that due to a single layer. One of these is the well known case of the Laue beams in which the scattered waves contributed to a particular beam from the consecutive atom layers are in phase and reenforce one another by constructive interference; the other is the case in which the reduction in intensity of the radiation on passing through a single layer of atoms—due to scattering and absorption—is so great that no appreciable radiation emerges from the interior of the crystal to interfere with that scattered by the first layer. The resultant scattering in this case will be approximately that from a single layer of atoms. This condition will be most closely approached near grazing, since in this region the paths in the crystal over which radiation from the second and lower layers must escape are greatly lengthened. The new electron beams discovered below 200 volts after the crystal was thoroughly degassed appear to be of this latter type. We shall begin our discussion of the data with an examination of these "plane grating" beams.

The value of the spacing s for nickel is 2.48A, so that $d_1=3^{1/2}s/2=2.15$A, and first order beams should appear at grazing emergence in the $\{111\}$ and $\{100\}$-azimuths when $\lambda(=h/mv)$ has this value. Rewriting de Broglie's formula for λ in terms of the kinetic energy V of the electrons expressed in equivalent volts, we have:

$$\lambda(\text{in Angstrom units})=(150/V)^{1/2}$$

From this we calculate that the electron wave-length will have its critical value when $V=32.5$ volts.

The new beams in the $\{100\}$ and $\{111\}$-azimuths are shown in Figs. 14 and 15 in which current to the collector is plotted against bombarding potential for various colatitude angles. In both azimuths beams appear at grazing $(\theta=90°)$ at or very near the calculated voltage, and then move upward toward the incident beam as the voltage is increased. The low intensities of the beams near grazing are attributed to the general roughness of the surface and to absorption of the radiation scattered by the first layer by the atoms of the same layer. The final disappearance of these beams at about 30° above the surface is accounted for by interference between the radiation scattered by the first layer and that escaping from below. Over this range we expect wave-length and angular position of the beam to be related through the plane grating formula $\lambda=d_1\sin\theta$, or voltage and position through the equivalent relation $V^{1/2}\sin\theta=(32.5)^{1/2}=5.70$. That the observations are in accord with this requirement is shown by the values calculated from Figs. 14 and 15 and recorded in Table II.

TABLE II

Occurrence of "plane grating" electron beams.

θ	Azimuth —{100} V	$V^{1/2}\sin\theta$	Azimuth —{111} V	$V^{1/2}\sin\theta$	Azimuth —{110} V	$V^{1/2}\sin\theta$
85	32.0	5.64	32.5	5.68	97.5	9.83
80	33.0	5.66	34.0	5.75	100.0	9.85
75	35.0	5.72	35.0	5.72	103.5	9.83
70	36.0	5.64	36.5	5.68	108.0	9.77
65	38.5	5.63	35.0	5.37	112.5	9.62
60	42.5	5.65				

The difference in intensity between the beams shown in Figs. 14 and 15 is due apparently to a real dependence of intensity upon azimuth; the beams in the three {100}-azimuths are all more intense than the beams in the {111}-azimuths. We naturally try to account for this difference (which could not occur if the scattering were from a single atom layer) by supposing that although the extinction of the radiation in the metal is sufficiently great to leave first layer scattering predominant, it is not sufficiently great, even

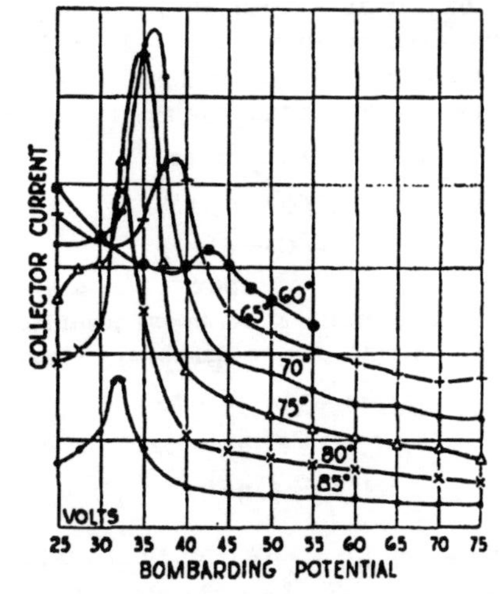

Fig. 14. Collector current vs. bombarding potential showing plane grating beams near grazing in {100}-azimuth.

near grazing, to suppress completely the escape of radiation from lower layers. The phase difference between first and second layer beams is not the same in the two azimuths, and as a consequence an intensity difference results. Whether the observed difference is in the sense to be expected will be considered later.

It was expected that second order beams corresponding to the ones just described would be found at grazing for $V=4\times32.5=130$ volts. These appear, however, to be entirely missing. We cannot account for this.

On the other hand the anticipated first order beams in the {110}-azimuths resulting from the atomic line gratings of second widest spacing are duly found. One of these beams is shown in Fig. 16. The grating constant is $s/2$,

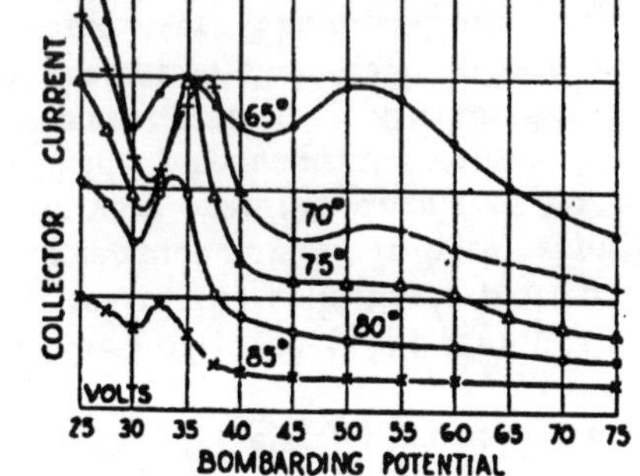

Fig. 15. Collector current vs. bombarding potential showing plane grating beams near grazing in {111}-azimuth.[10]

or $1/3^{1/2}$ as great as that of the former gratings, so that the beam should appear at grazing for $V=3\times32.5=97.5$ volts. The beam appears quite accurately at this voltage. It also conforms over the range through which

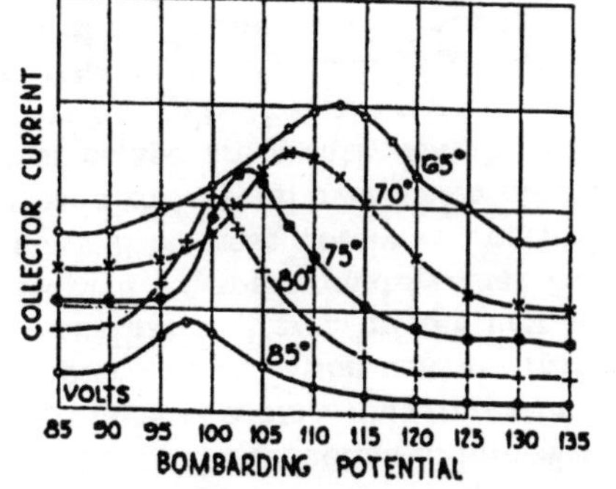

Fig. 16. Collector current vs. bombarding potential showing plane grating beams near grazing in {110}-azimuth.

it can be followed to its appropriate grating formula, $\lambda=d_2\sin\theta$, or $V^{1/2}\sin\theta=(97.5)^{1/2}=9.88$. Values of $V^{1/2}\sin\theta$ taken from Fig. 16 are given in the last section of Table II.

These beams, as has been mentioned, are extremely sensitive to the presence of gas on the surface of the target. They fall off in intensity as the

[10] In Fig. 15 the maxima near 50 volts in the curves for 65 and 70 degrees are due to the "54 volt" beam in early stages of its development. The curves in Fig. 15 should be compared with those in Fig. 9. The latter are for the target covered with gas; the "plane grating" beam is altogether lacking here, and the "54 volt" beam shows much more weakly than in the curves in Fig. 15. The rapid rise of the current in all colatitudes below 30 volts in Fig. 15 is not significant. It is due to a rapid increase of current in the incident beam which resulted, under the conditions of these measurements, from a focussing action within the electron gun.

gas collects, and disappear when the amount of gas on the surface is insufficient to affect more than slightly the intensity of beams occurring at higher voltages and smaller colatitude angles.

This behavior is consistent with the view already proposed that near grazing the intensity of scattering is determined almost entirely by the atoms in the topmost layer. When the crystal is clean this is a layer of regularly arranged nickel atoms which gives rise to the plane grating beams—when gas has collected on the surface it is a layer of gas atoms or molecules which may or may not give rise to diffraction effects of its own, but which, in either case, serves to absorb the radiation from the first layer of nickel atoms.

We thought that since the radiation in the grazing beams is scattered mostly by top layer atoms, the electrons in these beams would be more nearly homogeneous (made up more completely of full speed electrons) than those in beams at higher voltages and lower colatitude angles—beams more dependent for their intensities upon radiation scattered from atoms in lower layers. This appears, however, not to be the case. Current-voltage curves similar to those in Fig. 7 have been constructed for the grazing beams, and these indicate a distribution of speeds quite like that inferred from Curve III of Fig. 7.

We have looked also for the first order grazing beams from the plane gratings of third largest spacing, but have failed to find them. There should be twelve such beams (six in each of two new azimuths) appearing at 227 volts. It is to be expected, of course, that the higher voltage beams of this type will be weak on account of a less rapid extinction of the equivalent radiation in the metal. Here, as elsewhere, it appears a dependable procedure to infer characteristics of the equivalent radiation from known properties of electrons; if high speed electrons are less rapidly absorbed than low speed electrons, we may infer that short wave equivalent radiation is less rapidly absorbed than long.

We have looked (also without success) for other beams of this type grazing the inclined $\{111\}$-facets. The calculated positions of most of these beams fall so close to intense space lattice beams that it seems unlikely they could be found, even if present. On the other hand a careful exploration for one particular beam of this sort, the predicted position of which is not too close to any space lattice beam, was a complete failure.

Space Lattice Beams

The diffraction pattern for radiation of a given wave-length scattered by a single atom layer is made up, as has already been described, of the plane grating beams from all of the gratings that can be constructed from lines of atoms in the layer. The beams occur in planes normal to the lines of these gratings, and satisfy the plane grating formula $n\lambda = d \sin \theta$, where d represents the constant of the particular grating considered. This relation is represented graphically by the straight lines in Fig. 17, in which wave-length λ is plotted against $\sin \theta$, for beams occurring in the principal azimuths. For a given value of λ the positions of the beams in the various orders may be read off from these lines.

We consider now what diffraction effects are to be expected when the incident waves are scattered not by a single layer of atoms, but by a number of such layers piled one above the other. We see that if the various atom layers are similarly oriented, the lines in Fig. 17 represent the wave-length angle relation for the diffraction beams from each layer in the pile, and that, therefore, they must represent the same relation for the beams issuing from

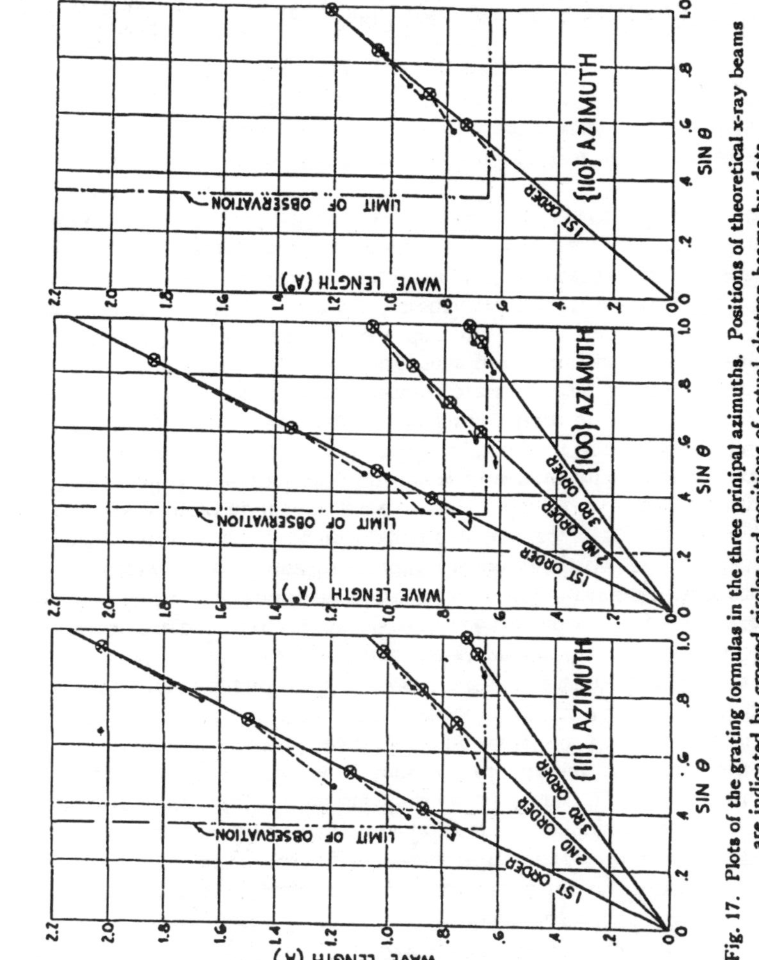

Fig. 17. Plots of the grating formulas in the three prinipal azimuths. Positions of theoretical x-ray beams are indicated by crossed circles and positions of actual electron beams by dots.

the pile of layers; the sum of the superposed beams is still the same set of beams. As far, therefore, as the positions of the beams are concerned the diffraction of waves by a pile of parallel and similarly oriented atom layers is the same as by a single layer.[11]

[11] It is assumed in this paragraph that the atom plane is large enough to be highly resolving and that we are not interested in discrepancies of the order of the width of its diffraction beams.

There is, however, a great difference in the intensities of the beams in the two cases. In the case of scattering from a single layer the intensity of a given beam would be found ordinarily not to change greatly as the wave-length and position of the beam are varied. In the case of scattering from a pile of such layers, however, the intensity of the emergent beam is greatly affected by interference among the beams from the individual layers. If a great number of layers contribute individual beams of about the same intensity, as is the case in x-ray scattering, the intensity of the resultant beam will ordinarily be quite low. When, however, the beams from the individual layers emerge from the crystal in phase the intensity of the resultant beam passes through a strong maximum. The wave-lengths for which this occurs depend upon the way in which the atom layers are piled up—that is, upon the separation between successive layers and upon the lateral shift from one layer to the next. If the wave-length of the radiation within the crystal is not the same as outside this also will have an effect upon the occurrence of intensity maxima.

The purpose of setting forth these rather elementary matters at such great length is to bring out clearly that the intense space lattice beams are always also plane grating beams. If the wave-length and position of such a beam is represented by a point in Fig. 17 the point must fall somewhere on one of the plane grating lines.

If the incident beam in the present experiments were a beam of heterogeneous x-rays an array of Laue diffraction beams would issue from the crystal. The wave-lengths and positions of all such beams in the three principal azimuths within the range of our observations ($\theta > 20°$, $\lambda > 0.63$A) are indicated in these figures by crossed cycles.

The plane grating lines for azimuths {111} and {100} are, of course, the same, but the occurrence of Laue beams in these azimuths is different. The wave-lengths that give rise to diffraction beams in the {111}-azimuth are not the same, except in such orders as are exactly divisible by three, as those that give rise to beams in the {100}-azimuth. And the same statement, *mutatis mutandis*, may be made, of course, in regard to the angles at which beams occur in the two azimuths. When the wave-lengths or angles of the beams of any order, not exactly divisible by three, are ordered according to magnitude the members belonging to one azimuth occur alternately with those belonging to the other. These characteristics result from the fact that to superpose the atoms of one layer upon those of an adjacent layer involves a displacement normal to the grating lines equal to $(2n+2/3)$ times the grating space—n representing any integer. If the odd fraction were one-half or zero, as in the {110}-azimuth, these differences would not occur.

In this same figure we have indicated by dots the wave-lengths and positions of all sets of electron beams listed in Table I. The first thing to be noted about the dots is that they fall generally along the basic plane grating lines, and to this extent satisfy the fundamental requirement that a space grating beam shall be first of all a plane grating beam. It will be noted also that they occur along these lines at about the same intervals as the crossed circles, and further that in the {111} and {100}-azimuths they exhibit the alternate occurrence characteristic of the crossed circles.

While the dots representing the sets of electron beams fall generally along the plane grating lines, one cannot fail to note that they actually fall off these lines—and systematically; they are above or to the left of the lines as one cares to view them. At the time of writing our note to "Nature" we believed that these departures could be accounted for by imperfections in the geometry of the apparatus. At present, with more data at our disposal, we are less certain that this is true. The fault of this explanation is that if the displacements of the dots from the lines result from imperfect alignment, angular displacement should be a function of angular position, and this condition seems not very well satisfied. The results of these investigations do not, however, rule out mechanical imperfection as the principal cause of these displacements, and for the present we shall assume it to be the only cause. We shall assume, that is, that the wave-lengths or voltages of the beams are correct but that the dots should be shifted to the right onto the lines. These shifts correspond to corrections in angle ranging from zero to about four degrees. We hope to remove the uncertainty here involved by measurements with a new tube now being constructed.

As the dots occur at about the same intervals along the lines as the crossed circles it is possible to associate each dot with a particular crossed circle—each set of electron beams, that is, with a particular set of x-ray beams, and consequently with a particular set of Bragg atom planes in the crystal. The associations which seem most natural are indicated by dotted lines in the figure. In the {110}-azimuth the association is imperfect as it appears that both the 0.893A and the 0.940A electron beams should be associated with the same x-ray beam. We have associated the stronger of these with the x-ray beam, and left the extremely weak electron beam at 0.940A without an apparent x-ray analogue. Also there is apparently a third order electron beam missing in the {111}-azimuth. We should expect this beam to be extremely weak.

We consider next the difference in position between each crossed circle and the location upon the line to which the associated dot is shifted as explained above. The positions of the lines are determined by the arrangement of atoms in a single plane and the scale factor of the structure, while the positions of the crossed circles are determined by the separation between successive planes and the lateral shift from one plane to the next. If the separation were decreased the crossed circles would all be moved downward along their lines; if it were increased they would be moved upward. Merely as a matter of description, therefore, we may say that a given dot when shifted occupies the position that its associated crossed circle would occupy if the separations between atom planes were decreased by a certain factor. This factor β may be calculated from the formula

$$\beta = \frac{\tan(\theta'/2)}{\tan(\theta_x/2)}$$

Here θ_x represents the colatitude angle of the Laue x-ray beam and θ' that of the associated electron beam after having been corrected to make the beam satisfy the grating formula exactly.

This factor β is found to be neither constant nor very regular in its behavior, but to increase generally with the speed of the electrons. The relation between the factor β and the bombarding potential is represented by the lower set of points in Fig. 18. There is a vague suggestion here that β approaches unity as a limiting value. If this is actually the case, it means, of course, that at sufficiently high voltages (short wave-lengths) there is no difference between the occurrence of x-ray and electron diffraction beams. This approach of the electron beam to the associated x-ray beam with

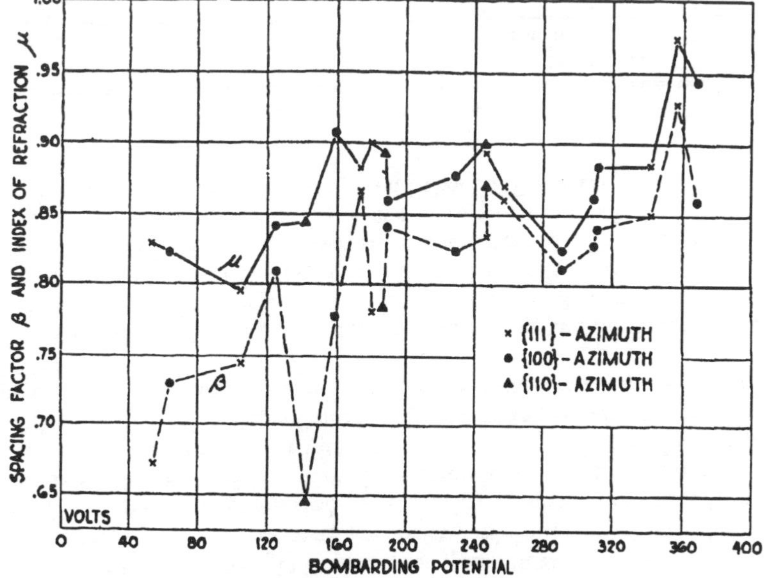

Fig. 18. Lower set of points—Spacing factor β vs. bombarding potential; Upper set of points—Index of Refraction μ vs. bombarding potential.

increasing voltage is shown very clearly by the series of three dots (Fig. 17) representing the first three orders of the beam whose first order is the 54 volt beam in the {111}-azimuth ($\lambda = 1.67$A). The first, second and third order beams occur at $\theta = 50°$, $55°$ and $62°$ respectively, while the corresponding x-ray beam occurs in each case at $\theta = 70°$ ($\sin\theta = 0.94$).

It has been suggested by Eckart[11] that this β factor may be interpreted as an index of refraction of the nickel crystal for the equivalent radiation. If the index of refraction of a crystal for radiation of a given wave-length is other than unity, diffraction will indeed occur as if the spacing between crystal planes were altered by a certain factor. This factor is not, however,

[11] Eckart, Proc. Nat. Acad. Sci. 13, 460 (1927).

equal to the index of refraction. For the case in which we are interested, that of normal incidence, and emergence from the same surface, the quantity appearing in the measurements as "spacing" factor β is given by

$$\mu = \frac{\beta}{\cos^2\psi + \beta^2 \sin^2\psi}$$

where ψ is the angle between the set of Bragg atom planes associated with the diffraction beam and the surface of the crystal. Values of μ calculated from this formula are represented by the upper set of points in Fig. 18. There is no greater regularity in the points representing μ than in those representing β.

There is the question, of course, as to whether the association of x-ray and electron beams indicated in Fig. 17 is the correct one. We might, for example, associate each dot in this figure with the next lower crossed circle instead of with the next higher, and there is at least one consideration favoring this procedure; the indices of refraction would then be slightly greater than unity for low speed electrons, and would approach unity for electrons of high speed. Such a dependence could be correlated with the increase in speed (decrease in wave-length) which an electron experiences on passing into a metal. In fact, we might naturally expect the index of refraction to be given by $\mu = [(V+\phi)/V]^{1/2}$, where ϕ represents the voltage equivalent of the work function of the metal.

The principal consideration opposed to this association is the absence of electron beams to be associated in the various orders with the x-ray beams of greatest wave-length. The absence of some of these might be accounted for by total reflection of the diffraction beam at the surface of the crystal. It seems unlikely, however, that the absence of analogues in three orders of the x-ray beams at $\theta = 70°$($\sin\theta = 0.94$) in the {111}-azimuth could be accounted for in this way. The absence of an electron beam to be associated with the first order x-ray beam of greatest wave-length in the {100}-azimuth would seem decisively against this alternative association were it not that the strong grazing beam in this azimuth (Fig. 14) may, in fact, be this analogue. This matter of association should, we think, be regarded as still an open question to be decided by further and more precise measurements. For the present, however, we shall adhere to the associations of Fig. 17.

Missing Beams

When electron and x-ray beams are associated as in Fig. 17 there is only one x-ray beam in the three principal azimuths for which an electron beam analogue predicted within the range of our observations has not been found. This is the third order beam in the {111}-azimuth already mentioned. There are, however, three Laue beams in range in two new azimuths, one of wave-length 0.78A at $\theta = 74°$ in the {331}-azimuth, and two of wave-lengths 0.81A and 0.76A at $\theta = 86°$ and $\theta = 68°$ in the {210}-azimuth. The first of these azimuths[12] lies 11° from the {110}-azimuth toward the nearest {111}-

[12] We have consistently designated each azimuth by the Miller indices of the densest plane of atoms the normal of which lies in the azimuth and in or above the surface of the crystal.

azimuth, and the second lies an equal interval on the opposite side of the $\{110\}$-azimuth. A thorough but fruitless search has been made for the electron beam companions of these three Laue beams.

The total number of x-ray beams for which electron beam companions are expected within the range of our observations is twenty-four, and for twenty of these the electron beam companions have been found. The four sets of missing electron beams are all beams whose intensities we should expect to be low. They are all short wave-length beams lying not far above the plane of the target.

Third Order Beams in the $\{111\}$ and $\{100\}$-Azimuths

The differences which have been noted between the occurrence of x-ray diffraction beams in the $\{111\}$ and $\{100\}$-azimuths do not extend, as has been mentioned, to orders that are exactly divisible by three; the wave-lengths and positions computed for the third order x-ray beams in one of these azimuths are identical with those computed for the third order beams in the other. Differences in intensity between corresponding beams might possibly result from a dependence of the scattering power of a single atom upon crystal direction, but, so far as we are aware, differences of this sort are not found in x-ray data.

It seemed reasonable to expect, therefore, that the third order electron beams in the $\{111\}$-azimuth would be found identical in voltage, position, and possibly also in intensity, with beams of the same order in the $\{100\}$-azimuth. This agreement, however, is not observed. Two pairs of beams are available for observation, and the beams of each pair have constants that are different. The data are given below in Table III.

Table III

Third order electron beams.

$\{111\}$-azimuth x-ray companion	θ	V	Int.	$\{100\}$-azimuth x-ray companion	θ	V	Int.
$\{55\bar{1}\}$	—	—	0.00	$\{711\}$	70°	310	0.15
$\{551\}$	60°	347	0.07	$\{822\}$	57	370	0.15

Resolving Power of the Crystal

It has been remarked (Fig. 10) that the space lattice beams are not sharply defined in voltage. The "54 volt" beam, for example, appears when the bombarding potential is about 40 volts and is still prominent in the curve for $V=64$ volts. In terms of equivalent wave-lengths the beam has appreciable intensity over a range $\Delta\lambda$ which is about one quarter as great as the equivalent wave-length of the beam at its maximum intensity. The corresponding ratios for beams of higher voltage are generally less, but even for beams for which V is greater than 300 volts values of $\Delta\lambda/\lambda$ are found as great as 0.07. We naturally conclude that as an optical instrument the crystal has low resolving power.

It might appear that this low resolving power is readily accounted for by the fewness of the atom planes contributing effectually to the scattering. In accounting for the "plane grating" beams we have already had occasion to assume that the extinction of the equivalent radiation in the metal is extremely rapid, and it may appear that no further assumption will be required to account for the low resolving power observed in the space lattice beams. It must be realized, however, that the resolving power of a space lattice is dependent not only upon the number of its atom planes, but also upon the number of grating lines in each plane, and that to ascribe the low resolving powers in the present case solely to the slight penetration of the radiation into the metal is to assume that the number of grating lines in each of the atom planes making up the lattice is relatively great—great compared with the number of effective planes. If the width of the lattice were the same as the width of the bombarded area this assumption would, of course, be amply justified. There are, however, at least two reasons for regarding the diffracting system not as a single lattice, but as a collection of very many extremely small lattices, similarly oriented but otherwise unrelated. A large part of the radiation proceeds, it would seem, from lattices not more than five or ten atom lines in width.

The first reason for ascribing this small width to the unit lattice is that on no other assumption have we been able to account for the great width of such peaks as those shown in Figs. 9, 14, 15 and 16. The width of these peaks is due in part, of course, to the width of the diffraction beam and to that of the collector opening, but these geometrical considerations are quite insufficient for the purpose. The curves referred to seem to show clearly that the resolving power of the topmost atom layer is quite low, and that consequently most of the radiation from this layer comes from independent gratings containing only a few atom lines each.

The second reason has to do with the relation between the voltage and the angular position of a space lattice beam in the various stages of its development. We have seen (Table II) that as a "plane grating" beam rises from the surface of the crystal the product $V^{1/2}\sin\theta$ remains constant. At one time we expected that the same relation would be found to obtain within a space lattice beam—that as the voltage was increased the beam would in all cases move upward to keep $V^{1/2}\sin\theta$ constant. This relation does in fact hold reasonably well for the "54 volt" beam. It holds less well, however, for the "65 volt" beam shown in Fig. 10, and fails entirely for most beams occurring at higher voltage. In many cases the total angular displacement during the growth and decay of the beam is not more than one-tenth that required by the plane grating formula. It may be shown, however, that the plane grating formula ($V^{1/2}\sin\theta=$constant) will describe the motion of the beam if the lattice is sufficiently wide, but that it will fail to do so otherwise. If the number of lines in the plane grating is not great compared with the number of atom planes, the plane grating relation will not obtain (except at the maximum of a beam—when the scattering from successive layers is exactly in phase) and departures from it will be in the direction of those that we have observed.

There are thus these two reasons for thinking that most of the scattered radiation proceeds from lattices not more than five or ten grating lines in width, and that the low resolving power of the plane gratings that make up the representative lattice is partly responsible for the low resolving power apparent in the space lattice beams.[14] It seems probable that the independent lattices are to be identified with the individual facets of the principal set—those parallel to the general plane of the target.

If the resolving power of the crystal were determined solely by the rate of extinction of the radiation in the metal, it would be possible to make use of the intensity-voltage relation within a given beam to calculate a coefficient of extinction of the metal for radiation of the wave-length of the beam, as suggested by Eckart (loc. cit.), or to estimate the number of atom planes effective in the scattering, as suggested by Patterson.[15] It may even be possible under the actual conditions of the scattering to evaluate these constants by making use also of the observed relation between the angular position of the beam and voltage. But the data which we have at present seem to us too inaccurate to make such calculations worth while. We hope to obtain data sufficiently precise for this purpose in the near future.

We have, however, calculated a rate of extinction from the data obtained for the "54 volt" beam with the surface of the target free from gas (not from the data of Fig. 10). As has already been mentioned, the product $V^{1/2}\sin\theta$ is very nearly constant in this beam, which means that its intensity-voltage relation is determined almost entirely by the rate of extinction. As the result of this calculation we have obtained the value 0.4 for the fraction by which the intensity of a beam of 54 volt equivalent radiation is reduced when such a beam passes normally through a single $\{111\}$-layer of nickel atoms. In terms of electrons this means presumably that when a 54 volt electron is incident normally upon such a layer the probability of its passing through without appreciable deflection or loss of energy is 0.6.[16]

Intensities of "Plane Grating" Beams

We return now to a further consideration of the intensity difference between the first order "plane grating" beams in the $\{100\}$ and $\{111\}$-azimuths (Figs. 14 and 15). If these were truly plane grating beams no difference of intensity should be observed. We have already pointed out, however, that it is only in the limit, as θ approaches $\pi/2$, that these beams are due to scattering from a single atom layer. They are actually space lattice beams, and their occurrence and behavior are described by the same mathematical formulas that describe the Laue type of beam. The intensity of a "plane grating" beam depends upon the efficiency with which a single

[14] It was shown in an earlier section that the apparent angular width of a space lattice beam (voltage constant) has more than double its least possible value as calculated from the dimensions of the apparatus. It seems probable from the present considerations that a considerable part of this additional width is to be ascribed to the low resolving power of the crystal.

[15] A. L. Patterson, Nature 120, 46, (1927).

[16] This question of resolving power has been considered recently by F. Zwicky (Proc. Nat. Acad. Sci. 13, 519 (1927)) in terms of the wave mechanics as formulated by Schroedinger and Born.

atom scatters the radiation incident upon it, and upon the rate of extinction of the radiation in the crystal. Such beams are not found when x-rays are scattered by a crystal because in this case both of these quantities are small.

The problem of finding which of the two beams under consideration should be the stronger resolves itself into finding in which azimuth the first order diffraction beams from successive atom layers are more nearly in phase near grazing emergence; the beam in this azimuth will be the more intense. These phase differences between beams from successive atom layers are readily deduced from the geometry of the lattice. Thus it may be shown that if the diffraction occurs as though the crystal were contracted normally to its surface to a fraction β of its normal spacing, this phase difference in the $\{111\}$-azimuth will be given by

$$\alpha_{111}=\frac{4\pi n}{3}\left[\beta 2^{1/2}\frac{1+\cos\theta}{\sin\theta}+1\right],$$

and in the $\{100\}$-azimuth by

$$\alpha_{100}=\frac{4\pi n}{3}\left[\beta 2^{1/2}\frac{1+\cos\theta}{\sin\theta}-1\right]$$

where n represents the order of the beam. Phase differences for $\theta=75°$ and for various values of β have been computed from these formulas, and are given in Table IV.

Table IV

First order phase differences for $\theta=75°$

Spacing Factor β	0.5	0.6	0.7	0.8
(Phase Diff)$_{111}$	$2\pi+1.8$	$2\pi+2.5$	$4\pi-3.0$	$4\pi-2.2$
(Phase Diff)$_{100}$	-0.3	0.4	1.2	2.0

From the results shown in Fig. 18 we expect that at 35 volts, the bombarding potential for these beams, the value of β will lie near 0.7. The phase difference in the $\{111\}$-azimuth is, therefore, about 3.0 radians and that in the $\{100\}$-azimuth about 1.2 radians. The beam in the $\{100\}$-azimuth should therefore be the stronger, and this is what is actually observed.

The Temperature Effect

It has been mentioned already that immediately after bombardment, while the target is still at a high temperature, the intensities of all beams, as far as observed, are low, and that as the target cools the intensities rise again to their normal values for room temperature. The behavior is illustrated by the curves in Fig. 19 in which the intensities of the beams at ($V=54$ volts, $\theta=50°$) and ($V=343$ volts, $\theta=34°$) in the $\{111\}$-azimuth are plotted against time. The temperature of the target at zero time was perhaps 1000°K. The data for these curves were taken during the same run—observations on one beam being alternated with those on the other.

This temperature effect has not yet been studied in detail. There seems no reason for doubting, however, that it is the analogue of the Debye temperature effect observed in the diffraction of x-rays. This view is supported by the evidence of the curves in Fig. 19 that the higher voltage (shorter wave-length) beam is the more sensitive to temperature.

Beams Due to Adsorbed Gas

The "plane grating" beams and space lattice beams for which data have been given in previous sections are the only beams observed when the target is, as we believe, free from gas. When the target is not free from gas still other beams appear. Three sets of these have already been mentioned as the ones referred to in our note to "Nature" as anomalous; one set occurs in each of the three principal azimuths, and all attain their maximum intensity at or close to $\theta=58°$, $V=110$ volts. These beams appear to depend for their existence upon the presence of a considerable amount of gas on the surface of the target; they were most intense before the first bombardment of the target, and were not found when the amount of adsorbed gas was known

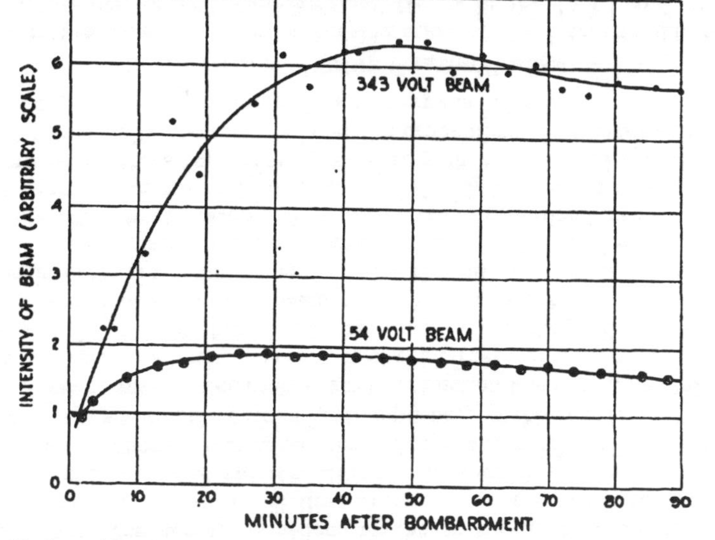

Fig. 19. Intensities of 54 volt beam and 343 volt beam vs. time, immediately after heating the target crystal.

to be slight. It should be possible to infer from the characteristics of these beams something about the arrangement of atoms to which they are due. It is clear, for example, that the arrangement is determined in part by the underlying nickel, since the beams occur in the principal azimuths of the nickel structure. We have been unable, however, to carry the deduction beyond this point.

With certain other beams due to gas we have been more successful. These constitute a family attaining greatest intensity when the quantity of adsorbed gas has a certain critical value. Various members of this family were observed first as beams of unusual behavior that appeared in places in which no beams were expected, and disappeared or changed in intensity for no apparent reason. Thus a beam was occasionally found in the {110}-azimuth at $\theta=75°$, $V=25$ volts. No such beam could result from scattering

by the nickel crystal unless our entire theory is fallacious; the first order grazing beam for this azimuth appears at $V=97.5$ volts, and no diffraction beam is possible for any lower voltage.

A correlation was eventually established between the intensity of this 25 volt beam and that of the normal grazing beam in the same azimuth. It was found that the 25 volt beam was strongest when the intensity of the grazing beam had been reduced by adsorbed gas to about one-fifth of its maximum value, and that when the grazing beam was at its maximum intensity the 25 volt beam was not to be found. Furthermore the 25 volt beam was not found when the surface was so contaminated that the normal grazing beam was absent; nor was it present when the 110 volt "anomalous" beams could be found.

It was noticed also that the voltage of this beam is almost exactly one quarter the voltage of the normal first order beam occupying the same position. The equivalent wave-length is thus twice that of the corresponding

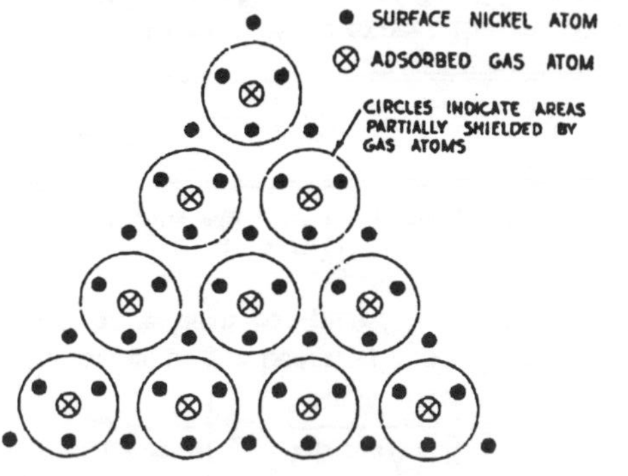

Fig. 20. Arrangement of gas atoms on the surface, and the topmost layer of nickel atoms.

first order beam, and the beam could be accounted for as radiation scattered by a layer of gas atoms of the same structure and orientation as the nickel atoms, but of twice the scale factor. Such a layer can be imagined built onto the surface layer of nickel atoms as indicated in Fig. 20. The first order diffraction beams from such a layer agree in wave-length and position with "one-half order beams" from the underlying lattice of nickel atoms, and as these latter have zero intensity the resultant beams should be those due to the single layer of gas atoms only. The beams should, therefore, move continuously upward as the voltage is increased without marked change in intensity. What is observed is not quite so simple; the beam can indeed be followed continuously from $\theta=85°$, $V=24$ volts up to $\theta=30°$, $V=98$ volts, but its intensity passes through broad maxima at $\theta=75°$, $V=25$ volts, and at $\theta=45°$, $V=50$ volts.

A possible explanation of this behavior may be pointed out. It will be noticed (Fig. 20) that the nickel atoms in the surface are of two sorts—those that are adjacent to gas atoms and those that are not. One-fourth of the nickel atoms are of this latter class, and they together form a layer of the same structure, orientation and scale factor as the gas atoms. If the two classes of nickel atoms scatter radiation in different amounts—if, for example, the gas atoms shield the adjacent nickel atoms more effectually than the non-adjacent—then, of course, the layer of nickel atoms as a whole will give rise to a differential diffraction beam capable of interference with that due to the gas layer, and broad maxima of the sort observed will result.

It was anticipated, of course, that beams of the same nature would be found in the $\{111\}$ and $\{100\}$-azimuths. These should appear at grazing at $V=32.5/4=8.1$ volts and should move upward with increasing voltage—reaching $\theta=20°$ at $V=70$ volts. Observations cannot be made for bombarding potentials as low as 8 volts. The beams have been picked up, however, in both azimuths at $\theta=60°$; $V=12$ volts, and have been followed upward to $\theta=25°$, $V=45$ volts. Broad maxima occur at $\theta=35°$, $V=25$ volts, and at $\theta=55°$, $V=13$ volts in the $\{100\}$-azimuth; and in the $\{111\}$-azimuth a broad maximum occurs at $\theta=55°$, $V=14$ volts, and another is indicated in the neighborhood of $\theta=25°$, $V=45$ volts. It should be possible from these data to calculate the separation, or at any rate the apparent separation, of the gas layer from the crystal—but this has not yet been attempted.

We have further observed that these beams cannot be made to appear when the temperature of the target is somewhat above that of the room; glowing the filament back of the target raises the temperature of the target sufficiently to eliminate them entirely, although under these conditions gas still collects on the target, and reduces the intensities of the grazing beams. The explanation of this behavior may be that the melting point of the two dimensional gas crystal is not far above room temperature. We have not yet observed whether the beams disappear sharply at a critical temperature.

Further Experiments with Gas

In a final series of experiments the effect was studied of introducing large amounts of gas into the tube. The liquid air was removed from the charcoal tube, and the behaviors of the "54 volt" beam and of one of the anomalous "110 volt" beams were observed as the charcoal was heated. The latter of these beams had not been observed for several weeks—not since its disappearance during the first heatings of the target.

The initial effect of increasing the gas pressure was to decrease the intensity of the "54 volt" beam. Its intensity was, however, greater than that represented by curve A in Fig. 13 until the charcoal temperature reached 350°C. At about this time the anomalous beam made its appearance. Maintaining the charcoal at 350°C the 54 volt beam decreased rapidly in intensity, vanishing entirely within a few minutes. In the meantime the intensity of the anomalous beam increased, reaching a maximum at about the time the 54 volt beam vanished, after which it too decreased and finally vanished. The gas pressure within the apparatus, as shown by ionization measurements, was perhaps 10^{-4} mm Hg at this time.

Heating was discontinued and liquid air was replaced on the charcoal. This caused the pressure to return to a very low value, but it did not bring

back either the 54 volt beam or the anomalous beam. Heating the target by electron bombardment did, however, bring back the 54 volt beam to about its maximum intensity (curve B, Fig. 13). At this time the anomalous beam was found to be absent. The 54 volt beam did not maintain its initial intensity but decreased rapidly, indicating that the vacuum condition of the tube as a whole had been greatly impaired by heating the charcoal. No further tests were made.

Summary of Electron Beams

Thirty sets of electron beams in all have been observed for bombarding potentials below 370 volts. Eleven of these occur in the $\{111\}$-azimuth, twelve in the $\{100\}$-azimuth and seven in the $\{110\}$-azimuth.

Twenty of the sets have been associated with twenty sets of Laue beams that would issue from the same crystal if the incident beam were x-rays.

Three sets are accounted for as "plane grating" beams which result from a preponderance of top layer scattering at angles near grazing emergence.

Six sets are attributed to scattering by adsorbed gas on the surface of the crystal, and the structure of the gas film giving rise to three of these beams has been inferred.

Some explanation has thus been given for twenty-nine of the thirty sets of beams. The remaining set occurs at $\theta=46°$ in the $\{110\}$-azimuth for $V=170$ volts, and is quite weak.

The explanations used in accounting for the observed beams require the occurrence of still other beams that have not been observed. The total of these missing beams is at least eight; four space lattice beams (one third order beam in the $\{111\}$-azimuth, one first order beam in the $\{210\}$-azimuth and two first order beams in the $\{331\}$-azimuth); and four "plane grating" beams (first order $\{210\}$ and $\{331\}$-azimuths, and second order $\{111\}$ and $\{100\}$-azimuths).

Discrepancies have also been noted between the characteristics of the third order space lattice beams in the $\{111\}$-azimuth and those of their companions in the $\{100\}$-azimuth.

The possibility of carrying through these investigations to their present stage has depended very largely upon the cooperation we have received from a number of our colleagues here in the laboratory. We are particularly indebted to Drs. H. D. Arnold and W. Wilson for the encouragement they have given us and for the benefit of their criticisms. We have had the benefit also, in technical matters, of discussions with Drs. L. W. McKeehan, K. K. Darrow and R. M. Bozorth.

We are indebted to Mr. H. T. Reeve for producing for us single nickel crystals of appropriate size—to Mr. C. J. Calbick for assistance in making the observations and for contributing not a little to their interpretation—and to Mr. G. E. Reitter for the great care with which he constructed the special apparatus and for his many contributions to its design.

Bell Telephone Laboratories, Inc.,
New York, N. Y.
August 27, 1927.

JULY, 1928 PHYSICAL REVIEW VOLUME 32

THERMAL AGITATION OF ELECTRIC CHARGE IN CONDUCTORS*

BY H. NYQUIST

ABSTRACT

The electromotive force due to thermal agitation in conductors is calculated by means of principles in thermodynamics and statistical mechanics. The results obtained agree with results obtained experimentally.

DR. J. B. JOHNSON[1] has reported the discovery and measurement of an electromotive force in conductors which is related in a simple manner to the temperature of the conductor and which is attributed by him to the thermal agitation of the carriers of electricity in the conductors. The work to be resported in the present paper was undertaken after Johnson's results were available to the writer and consists of a theoretical deduction of the electromotive force in question from thermodynamics and statistical mechanics.[2]

Consider two conductors each of resistance R and of the same uniform temperature T connected in the manner indicated in Fig. 1. The electromotive force due to thermal agitation in conductor I causes a current to be set up in the circuit whose value is obtained by dividing the electromotive force by $2R$. This current causes a heating or absorption of power in conductor II, the absorbed power being equal to the product of R and the square of the current. In other words power is transferred from conductor I to conductor II. In precisely the same manner it can be deduced that power is transferred from conductor II to conductor I. Now since the two conductors are at the same temperature it follows directly from the second law of thermodynamics that the power flowing in one direction is exactly equal to that flowing in the other direction. It will be noted that no assumption has been made as to the nature of the two conductors. One may be made of silver and the other of lead, or one may be metallic and the other electrolytic, etc.

Fig. 1.

It can be shown that this equilibrium condition holds not only for the total power exchanged by the conductors under the conditions assumed, but also for the power exchanged within any frequency. For, assume that this is not so and let A denote a frequency range in which conductor I delivers more power than it receives. Connect a non-dissipative network between the two conductors so designed as to interfere more with the transfer of energy

* A preliminary report of this work was presented before the Physical Society in February, 1927.

[1] See preceding paper.

[2] Cf. W. Schottky, Ann. d. Physik 57, 541 (1918).

in range A than in any other range, for example, a resonant circuit connected as indicated in Fig. 2 and resonant within the range A. Since there is equilibrium between the amounts of power transferred in the two directions before inserting the network, it follows that after the network is inserted more power would be transferred from conductor II to the conductor I than in the opposite direction. But since the conductors are at the same temperature, this would violate the second law of thermodynamics. We arrive, therefore, at the important conclusion that the electromotive force due to thermal agitation in conductors is a universal function of frequency, resistance and temperature and of these variables only.[3]

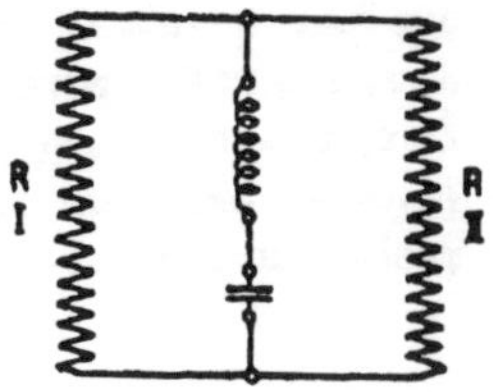

Fig. 2.

To determine the form of this function consider again two conductors each of resistance R connected as shown in Fig. 3 by means of a long non-dissipative transmission line, having an inductance L and a capacity C per unit length so chosen that $(L/C)^{1/2}=R$. In order to avoid radiation one conductor may be internal to the other. Under these conditions the lines has the characteristic impedance R, that is to say the impedance of any length of line when terminated at the far end in the impedance R presents the impedance R at the near end and consequently there is no reflection at either end of the line. Let the length of the line by l and the velocity of propagation

Fig. 3.

v. After thermal equilibrium has been established, let the absolute temperature of the system be T. There are then two trains of energy traversing the transmission line, one from left to right in the figure, being the power delivered by conductor I and absorbed by the conductor II, and another train in the reverse direction.

At any instant after equilibrium has been established, let the line be isolated from the conductors, say, by the application of short circuits at the two ends. Under these conditions there is complete reflection at the two ends and the energy which was on the line at the time of isolation remains trapped. Now, instead of describing the waves on the line as two trains traveling in opposite directions, it is permissible to describe the line as vibrating at its natural frequencies. Corresponding to the lowest frequency

[3] For a general treatment of the principle underlying the discussion of this paragraph reference is made to P. W. Bridgman, Phys. Rev. 31, 101 (1928).

the voltage wave has a node at each end and no intermediate nodes. The frequency corresponding to this mode of vibration is $v/2l$. The next higher natural frequency is $2v/2l$. For this mode of vibration there is a node at each end and one in the middle. Similarly there are natural frequencies $3v/2l$, $4v/2l$, etc. Consider a frequency range extending from ν cycles per second to $\nu+d\nu$ cycles per second, i.e., a frequency range of width $d\nu$. The number of modes of vibration, or degrees of freedom, lying within this range may be taken to be $2ld\nu/v$, provided l is taken sufficiently large to make this expression a great number. Under this condition it is permissible to speak of the average energy per degree of freedom as a definite quantity. To each degree of freedom there corresponds an energy equal to kT on the average, on the basis of the equipartition law, where k is the Boltzmann constant. Of this energy, one-half is magnetic and one-half is electric. The total energy of the vibrations within the frequency interval $d\nu$ is then seen to be $2lkTd\nu/v$. But since there is no reflection this is the energy within that frequency interval which was transferred from the two conductors to the line during the time of transit l/v. The average power, transferred from each conductor to the line within the frequency interval $d\nu$ during the time interval l/v is therefore $kTd\nu$.

It was pointed out above that the current in the circuit of Fig. 1 due to the electromotive force of either conductor is obtained by dividing the electromotive force by $2R$, and that the power transferred to the other conductor is obtained by multiplying the square of the current by R. If the square of the voltage within the interval $d\nu$ be denoted by $E^2d\nu$ we have, therefore

$$E^2d\nu = 4RkTd\nu \tag{1}$$

This is the expression for the thermal electromotive force in a conductor of pure resistance R and of temperature T. Let it next be required to find the corresponding expression for any network built up of impedance members of the common temperature T. Let the resistance R be connected, as shown in Fig. 4, to any such network having the impedance $R_\nu+iX_\nu$, where R_ν and X_ν may be any function of frequency. By reasoning entirely similar to that used above it is deduced that the power transferred from the conductor to the impedance network is equal to that transferred in the opposite direction. But the former is shown by simple circuit theory to be equal to

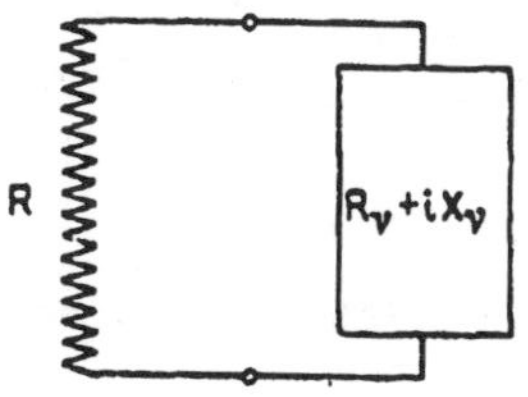

R

$R_\nu+iX_\nu$

Fig. 4.

$$E^2R_\nu d\nu/[(R+R_\nu)^2+X_\nu^2] \tag{2}$$

and the latter is similarly equal to

$$E_\nu^2Rd\nu/[(R+R_\nu)^2+X_\nu^2] \tag{3}$$

where $E_\nu^2d\nu$ is the square of the voltage within the frequency range $d\nu$. It follows that

$$E_\nu^2d\nu = 4R_\nu kTd\nu \tag{4}$$

for any network.

To put this relation in a form suitable for comparison with measurements let $Y(\omega)$ be the transfer admittance of any network from the member in which the electromotive force in question originates to a member in which the resulting current is measured. Let $\omega=2\pi\nu$ and let $R(\omega)=R_\nu$ be the resistance of the member in which the electromotive force is generated. We have then for the square of the measured current within the interval $d\nu$

$$I^2d\nu = E_\nu^2\,|\,Y(\omega)\,|^2d\nu = (2/\pi)kTR(\omega)\,|\,Y(\omega)\,|^2d\omega \tag{5}$$

Integrating from 0 to ∞

$$I^2 = (2/\pi)kT\int_0^\infty R(\omega)\,|\,Y(\omega)\,|^2d\omega \tag{6}$$

which is Eq. (1) in Johnson's paper.

It will be noted that such quantities as charge, number, and mass of the carriers of electricity do not appear explicitly in the formula for electromotive force These quantities influence R_ν, however, and, therefore, enter indirectly.

It is instructive to consider the equilibrium between the thermal agitation of the carriers of electricity in a conductor and the thermal agitation of molecules in a gas. Consider a semi-infinite tube filled with gas of temperature T and let the end be closed in a weightless inflexible piston forming the diaphragm of an ideal non-dissipative telephone receiver having no magnetic leakage. Such a receiver presents an electrical impedance which is a function of the mechanical impedance of the gas in the tube and which may be taken as R by choosing a suitable number of turns for the receiver element. Due to the bombardment of the diaphragm by the molecules in the gas, there will be an electromotive force at the terminals of the receiver. This electromotive force is, of course, in statistical equilibrium with that due to thermal agitation in a conductor of resistance R. It follows that it should be possible to calculate that electromotive force from the kinetic theory of gases, but this calculation would not be so direct as that given above, making use of a transmission line.

In what precedes the equipartition law has been assumed, assigning a total energy per degree of freedom of kT. If the energy per degree of freedom be taken

$$h\nu/(e^{h\nu/kT}-1) \tag{7}$$

where h is the Planck constant, the expression for the electromotive force in the interval $d\nu$ becomes

$$E_\nu^2d\nu = 4R_\nu h d\nu/(e^{h\nu/kT}-1). \tag{8}$$

Within the ranges of frequency and temperature where experimental information is available this expression is indistinguishable from that obtained from the equipartition law.

American Telephone and Telegraph Company,
April, 1928.

Second Series *February 1929* *Vol. 33, No. 2*

THE

PHYSICAL REVIEW

QUANTUM MECHANICS AND RADIOACTIVE DISINTEGRATION[1]

By R. W. Gurney and E. U. Condon

Abstract

Application of quantum mechanics to a simple model of the nucleus gives the phenomenon of radioactive disintegration. The statistical nature of the quantum mechanics gives directly disintegration as a chance phenomenon without any special hypothesis. §1 contains a presentation of those features of quantum mechanics which are here used and gives a simple calculation of the disintegration constant. §2 discusses the qualitative application of the model to the nucleus. §3 presents quantitative calculations amounting to a theoretical interpretation of the Geiger-Nuttall relation between the rate of disintegration and the energy of the emitted α-particle. In getting this relation one arrives at the rather remarkable conclusion that the law of force between emitted α-particle and the rest of the nucleus is substantially the same in all the atoms even where the decay rates stand in the ratio 10^{22}. §4 calls attention to the natural way in which the paradoxical results of Rutherford and Chadwick on the scattering of fast α-particles by uranium receive explanation with the model here used. §5 discusses certain limitations inherent in the methods employed.

The study of radioacitivity itself together with the application of it as a working source of high speed helium nuclei and electrons has played a fundamental role in the development of quantum physics. The scattering experiments of Rutherford and his associates gave the picture of the nuclear atom on which all of the success of modern atomic theory depends. Bohr's formulation of quantum postulates to be applied to such a model was a great step in the extension of knowledge of atomic structure and finally culminated in 1925 in the discovery by Heisenberg and by Schrödinger of a reformulation of mechanical laws which has subsequently proved extremely powerful in handling atomic structure problems. In this development of the last fifteen years little advance has been made on the problem of the structure of the nucleus.

It seems, however, that the new quantum mechanics has had sufficient success to justify the hope that it is competent to carry out an effective attack on the problem. The quantum mechanics has in it just those statistical elements which would seem appropriate to an explanation of the phenomenon

[1] An account of this work was first published in Nature for September 22, 1928. In a number of the Zeitschrift für Physik (51, 204, 1928) received here two weeks ago there appears a paper by Gamow who has arrived quite independently at the same basic idea as was presented in our letter and which is here treated in detail. Reports of this paper were also given at the Schenectady meeting of the National Academy of Sciences on November 20, 1928 and at the Minneapolis meeting of the American Physical Society on December 1, 1928.

of radioactive decay. This is the feature of the general problem with which we are concerned in this paper. We believe that the results provide at last an interpretation of nuclear disintegration which in its fundamental points is very close to the truth although it is necessarily quite incomplete.

The outstanding difficulties in the way of a good theoretical treatment of nuclear structure at present are mainly bound up with our lack of understanding of the quantum mechanics of the magnetism of the fundamental particles. This question has been much advanced this year by Dirac's extension of Pauli's theory of the spinning electron[2] but this remains essentially a theory of the behavior of one electron in an electromagnetic field. Not only is it apparently still unsatisfactory as such but this limitation must necessarily be disposed of in principle before the many body nuclear problem can be approached. And with that done there will remain the inevitable analytical difficulties.

Enough is known, however, to teach us that probably the magnetic interaction is not to be handled simply by an alteration of a potential energy function depending solely on the coordinates of the several interacting particles. This tends to detract from the value of arguments based simply on the use of quantum mechanics with the positional coordinates of the nuclear constituents. Nevertheless we shall restrict ourselves to the use of such methods in the discussion of the instability or capacity for spontaneous disintegration of a very much simplified nuclear model. The simplification to be made will consist in supposing that we can discuss the behavior of any one constituent by applying the quantum mechanics to it as a single body moving in a force field due to the rest of the nucleus.

The difference between quantum mechanics and classical mechanics which is here made responsible for the disintegration process is easily stated. In classical mechanics the orbit of a particle is entirely confined to those points in space at which its potential energy is less than its total energy. This is not true in quantum mechanics. Classically if a particle be moving in a basin of low potential energy and have not as much total energy as the maximum of potential energy surrounding the basin, it must *certainly* remain there for all time, unless it acquires the deficiency in energy somehow. But in quantum mechanics most statements of certainty are replaced by statements of probability. And the above statement must now be altered to read "· · · it may remain there for a long time but as time goes on the probability that it has escaped, even without change in its total energy, increases toward unity."

In §1 of this paper the detailed development of the argument leading to the conclusion of the preceding paragraph is given. In §2 we discuss its qualitative application to the nuclear disintegration problem. §3 is devoted to semi-quantitative estimates of the rates of decay.

1. Coupling of Motions of Equal Energy

Consider a particle of mass μ. It is sufficient to consider one degree of freedom; let the coordinate of the particle be x and let the forces be measured by the potential energy function $V(x)$.

[2] Dirac, Proc. Roy. Soc. A117, 610; A118, 351 (1928).

In classical mechanics the equations of motion possess the energy integral

$$p^2/2\mu+V(x)=W \tag{1}$$

(p = momentum) which, for values of x such that $W-V(x)<0$, can only be satisfied by p pure imaginary. Therefore, classically, one had the result that a particle could only be where $W-V(x)\geq 0$. An important consequence of this was that if there were several ranges of x for which $W-V(x)\geq 0$ separated by ranges where $W-V(x)<0$, then there were several different motions possible with the energy level W, each of which was wholly confined to one of these separate ranges. Thus in Fig. 1 for the energy level indicated there would be two distinct types of motion of the same energy W; one is a libration in the range I, and the other a libration in the range II.

These results are modified considerably by the new quantum mechanics. In the first place, Eq. (1) loses its validity and is replaced by an integral theorem, as Born[3] has shown, in which there is no longer a definite correlation between simultaneous value of position and momentum as (1) implies. The quantum mechanical form of (1) is, if $\psi(x)$ is Schroedinger's wave function

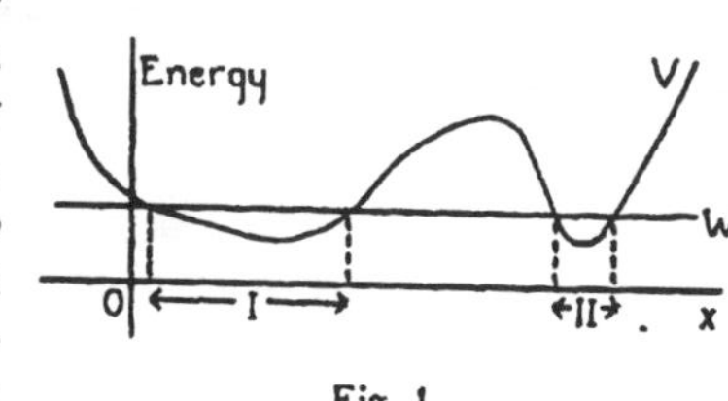

Fig. 1.

$$W=\int_{-\infty}^{+\infty}\left(\frac{h^2}{8\pi^2\mu}\frac{d\psi}{dx}\frac{d\bar{\psi}}{dx}+V(x)\psi\bar{\psi}\right)dx \tag{1a}$$

The lack of a precise correlation has been much emphasized by Heisenberg and by Bohr,[4] and is a general characteristic of quantum mechanics. From the new standpoint, one has to consider the behavior of Schrödinger's equation for the problem

$$\frac{d^2\psi}{dx^2}+\frac{8\pi^2\mu}{h^2}(W-V(x))\psi=0. \tag{2}$$

As is well known, in some problems there are solutions $\psi(W,x)$ for certain values of W which are finite and continuous everywhere. These are the "allowed" values of quantum theory. For the $\psi(W,x)$ which comes out of (2) as a by-product, Born has shown that its square may be satisfactorily interpreted as giving the probability that the particle lies between x and $x+dx$ when it is in the state of energy W. This is really the ground for requiring that ψ remain finite. For an energy level, such that $\psi(W,x)$, does not remain finite as $x\to\pm\infty$, the probability that it is not "at infinity" is vanishingly small, and therefore these states do not exist physically. Adopting the probability interpretation of $\psi(W,x)$ one has at once the result that there is a finite probability of being outside the range of the classical motion of that energy.

[3] Born, Zeits. f. Physik 38, 806 (1926).
[4] Heisenberg, Zeits. f. Physik 43, 172 (1927); Bohr, Nature, April 14, 1928.

A simple case is the lowest state of the harmonic oscillator, which has the energy $h\nu/2$. The $\psi(W,x)$ for this state is $e^{-1/2(x/a)^2}$ so $\psi^2=e^{-(x/a)^2}$ where a is the classical amplitude of motion associated with this energy. The probability of being outside the classical range is therefore

$$\frac{2\int_a^{\infty}e^{-(x/a)^2}dx}{\int_{-\infty}^{+\infty}e^{-(x/a)^2}dx}=0.157$$

or more than 15 percent.

When one studies the behavior of $\psi(W,x)$ from (2) for a $V(x)$ somewhat like the one in Fig. 1, he finds that, if the W is one for which ψ is finite everywhere, then ψ approaches zero very rapidly (exponential decrease) as $x\to\pm\infty$. In the neighborhood in which $W-V(x)$ is small, the function takes on appreciable values and has oscillatory character where $W-V(x)<0$, and non-oscillatory character elsewhere. Cases like that of Fig. 1 have been discussed by Hund[5] in connection with his studies of molecular spectra.

An important case is that in which the potential energy curve consists of a single "obstacle" or barrier as in Fig. 2, and the motion is one of insufficient energy, W, to clear the obstacle. In such cases there are *two* finite solutions $\psi_1(W,x)$, and $\psi_2(W,x)$ associated with each energy level, W, and so an arbitrary linear combination of them is also a solution of (2). Born has shown that there is always a combination of them which depends on x as $e^{+i\kappa x}$ and represents a pure left-to-right progressive wave motion as $x\to+\infty$. Such a solution for x large and negative can then be said to represent an incident left-to-right wave coming from the left side and a reflected wave which is not as strong as the incident wave. The interpretation is that the incident beam of particles is partly reflected and partly transmitted. In the range where $(W-V)<0$ the de Broglie wave-length h/p becomes imaginary, and so gives rise to an exponential behavior of ψ whose nearest analogue is, perhaps, in optics in the slight penetration of a refracted ray into a rarer medium even beyond the angle of total reflection where the refracted angle is imaginary. In this way, one can find the probability that a particle coming up from the left will get through the wall and escape to the right. The case illustrated in Fig. 3a for which

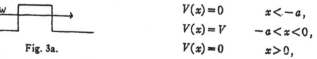

Fig. 2.

Fig. 3a.

$$V(x)=0 \qquad x<-a,$$
$$V(x)=V \qquad -a<x<0,$$
$$V(x)=0 \qquad x>0,$$

and for $0<W<V$, is a simple one with which to illustrate the nature of the calculation. For a given energy level, W, there are two ψ functions satisfying

[5] Hund, Zeits. f. Physik 40, 742 (1927); Wentzel, Zeits. f. Physik 38, 518 (1926).

the requirements of finiteness everywhere and of continuity for the ordinates and slopes at the discontinuities in $V(x)$.

These are readily found to be

$$\psi_1(W,x)=\begin{cases}\cosh\sigma_2 a\cdot\cos\sigma_1(x+a)-(\sigma_2/\sigma_1)\sinh\sigma_2 a\sin\sigma_1(x+a) & (x<-a)\\ \cosh\sigma_2 x & (-a<x<0)\\ \cos\sigma_1 x & (0<x)\end{cases}$$
$$\psi_2(W,x)=\begin{cases}-(\sigma_1/\sigma_2)\sinh\sigma_2 a\cdot\cos\sigma_1(x+a)+\cosh\sigma_2 a\sin\sigma_1(x+a) & (x<-a)\\ (\sigma_1/\sigma_2)\sinh\sigma_2 x & (-a<x<0)\\ \sin\sigma_1 x & (0<x)\end{cases}\quad(3)$$

where $\sigma_1=(2\pi/h)(2\mu W)^{1/2}$ and $\sigma_2=(2\pi/h)[2\mu(V-W)]^{1/2}$. To find the ψ function corresponding to a beam of particles incident from the left which is partly transmitted and partly reflected, one has to add these together in such a way that to the right of the obstacle there is only the pure left-to-right flow, i.e., one must take

$$\psi_1(W,x)+i\psi_2(W,x)$$

To the left of the obstacle, the ψ function represents the superposition of a left-to-right, or incident beam

$$\psi_{inc}=\left[\cosh\sigma_2 a-\frac{i}{2}\left(\frac{\sigma_1}{\sigma_2}-\frac{\sigma_2}{\sigma_1}\right)\cdot\sinh\sigma_2 a\right]e^{i\sigma_1(x+a)}$$

and a reflected beam

$$\psi_{ref}=\frac{i}{2}\left(\frac{\sigma_1}{\sigma_2}+\frac{\sigma_2}{\sigma_1}\right)\sinh\sigma_2 a e^{-i\sigma_1(x+a)}\qquad x<-a\quad(4)$$

The transmitted beam is simply

$$\psi_{tr}=e^{i\sigma_1 x}\qquad x>0\quad(5)$$

These expressions have, of course, the conservation property

$$(\psi\bar{\psi})_{inc}=(\psi\bar{\psi})_{ref}+(\psi\bar{\psi})_{tr}.$$

The probability that a particle coming up to the wall shall get through to the other side is simply $(\psi\bar{\psi})_{tr}\div(\psi\bar{\psi})_{inc}$ which for $e^{\sigma_2 a}\gg 1$ is clearly equal to

$$P_a(W)=16(W/V)(1-W/V)e^{-2\sigma_2 a}.\quad(6)$$

The controlling factor is the exponential term except when W/V is very near to 0 or 1.

For application to a theory of the pulling of electrons out of metals by electric fields Fowler and Nordheim[4] have derived the probability expression by similar methods for the curve of Fig. 3b, i.e.

$$V(x)=0\qquad x<0$$
$$V(x)=C-Fx\qquad x>0$$

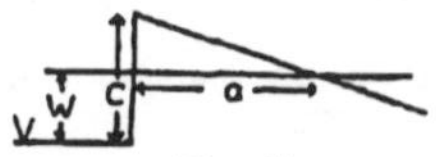

Fig. 3b.

[4] Fowler and Nordheim, Proc. Roy. Soc. A119, 1 (1928).

The probability that a particle of energy W get through the wall they find to be

$$P_b(W)=4[(W/C)(1-W/C)]^{1/2}\exp(-4k(C-W)^{3/2}/3F)$$

from their Eq. (18) p. 178. The term in the exponent can be written

$$4k(C-W)^{3/2}/3F=4\sigma_2/3a\quad(k^2=8\pi^2\mu/h^2)\quad(7)$$

to exhibit the similarity with the case of the square wall. Here a is the positive value of x for which $V(x)=W$, and σ_2 is defined as

$$\sigma_2=(2\pi/h)[2\mu(C-W)]^{1/2}.$$

The exponents in each of these cases can be written in the form

$$(4\pi/h)\int[2\mu(V-W)]^{1/2}dx$$

the integration extending across the barrier, the limits being the two places where $V(x)-W=0$.

Application of the method of approximate integration of Schrödinger's wave equation which was first used in quantum mechanics by Wentzel[5] indicates that such a result is quite general. The probability of getting through the wall at a single approach is governed essentially by the factor

$$\exp\left\{-(4\pi/h)\int[2\mu(V-W)]^{1/2}dx\right\}\quad(8)$$

being equal to it except for a factor of the order of magnitude of unity.

We have next to consider the case of a potential energy curve of the type shown in Fig. 4. According to classical mechanics there are two modes of motion associated with energy levels below the maximum such as W in the figure. One is a periodic motion in the range I while the other is an aperiodic motion in the range II. By the Bohr-Sommerfeld rule the periodic motions would give a discrete spectrum of allowed energy levels which would overlie the continuous spectrum associated with the aperiodic motions. On the quantum mechanics every energy level is allowed with the essential difference that *there are no energy levels with which two types of motion are associated.* With each energy level there is associated just one wave function $\psi(W,x)$ whose square gives the relative probability of being at different parts of the possible range of x. The $\psi(W,x)$ functions do show traces of the discreteness of the energy levels which the Bohr-Sommerfeld rule associates with the perodic motions in I, in an interesting way. The $\psi(W,x)$ for every W show sinusoidal oscillations as $x\to\infty$ and also oscillate in the range I. For most energies the amplitude of the oscillations in the range II is overwhelmingly large compared to that in range I, the ratio being of the order of $\exp\{(2\pi/h)\int[2\mu(V-W)]^{1/2}dx\}$ the integration extending across the barrier. This situation is just reversed however for little ranges of W values near those given by the old quantization rules. For these the amplitude in I is large compared to that in II in the same ratio. These then are the "allowed"

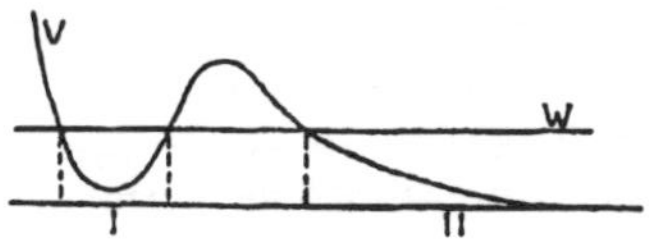

Fig. 4.

energy levels. It is not a stationary state for the particle to be in range I and remain there. But for certain energy levels there is an extraordinarily large probability of being in unit length of range I relative to unit length of range II.

We have to find the mean time which a particle remains in the range I before "leaking through" to the outer range II. This can be obtained from the following simple consideration. When the particle is at a place of x large and positive, $V(x)=0$ (Fig. 4) so the energy is all kinetic and the speed is therefore $(2W/\mu)^{1/2}$. The amount of time which the particle spends in unit length for x large is therefore $(\mu/2W)^{1/2}$. The time spent in a range of length a is therefore $a\ (\mu/2W)^{1/2}$. Now according to the wave-functions the probability of being in unit length of range I for one of the quasi-discrete energy values relative to the probability of being in unit length of range II is of the order $\exp\{(4\pi/h)\int[2\mu(V-W)]^{1/2}dx\}$. Therefore since the motion is aperiodic and the particle escaping from range I will in the mean only go through unit length of II once, the time T which must be spent in range I before getting through to range II is of the order of

$$T\sim a(\mu/2W)^{1/2}\exp\left\{(4\pi/h)\int[2\mu(V-W)]^{1/2}dx\right\}$$

where a is of the order of the breadth of range I.

Like all of the results of quantum mechanics this is to be interpreted as a probability result. So that if we start with a number of particles in the same allowed energy level in identical regions similar to range I, the number which leak out in time dt is governed by

$$dN=-N\lambda dt$$

which gives the usual exponential law of decay $N(t)=N_0e^{-\lambda t}$ where

$$\lambda=1/T. \tag{9}$$

The expression for T may be arrived at in a somewhat different way. One can think of the particle as executing its classical motion in range I, but as having at each approach to the barrier the probability of escaping to range II given by expression (8) above. The frequency of the periodic motion in I, which represents the number of approaches to the barrier in unit time, is of the order $a(\mu/2W)^{1/2}$ so the mean time of remaining in range I before excape comes out as the quotient of these two quantities as before. The reader will find it of interest to examine Oppenheimer's formula[7] for the pulling of electrons out of hydrogen atoms by an electric field. His formula for the mean time required for dissociation of the atom by a steady electric field splits naturally into a factor which is the classical frequency of motion in the Bohr orbit multiplied by an exponential probability factor of the type of expression (8) used in this paper.

2. Application to Radioactive Disintegration

After the exponential law in radioactive decay had been discovered in 1902, it soon became clear that the time of disintegration of an atom was

[7] Oppenheimer, Proc. Nat. Acad. Sci. 14, 363 (1928).

as independent of the previous history of the atom as it was of its physical condition. One could not for example suppose that an atom at its birth begins to lose energy by radiation and that its instability is the result of the drain of energy from the nucleus. On such a view it would be expected that the rate of decay would increase with the age of the atoms. When later it was observed that the number of atoms breaking up per second showed the fluctuations demanded by the laws of probability it became clear that the disintegrating depended solely on chance. This has been very puzzling so long as we have accepted a dynamics by which the behaviour of particles is definitely fixed by the conditions. We have had to consider the disintegration as due to the extraordinary conjunction of scores of independent events in the orbital motions of nuclear particles. Now, however, we throw the whole responsibility on to the laws of quantum mechanics, recognizing that the behaviour of particles everywhere is equally governed by probability.

From what was said in the preceding section it is clear that the property of the nucleus which we need to know in order to apply the theory is its potential energy curve; and this happens to be a property which we know fairly definitely. Outside a nucleus whose net charge is given by the atomic number we should expect to find a Coulomb inverse-square field of the appro-

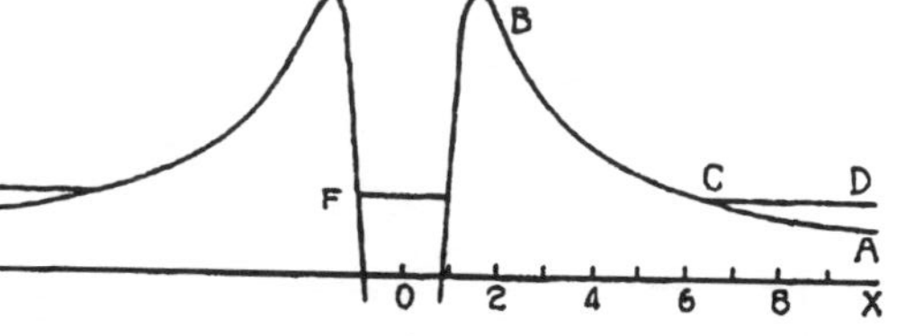

Fig. 5. The unit of abscissas is 10^{-12} cm. The horizontal line gives the energy of the α-particle emitted by uranium, 6.5×10^{-6} ergs.

priate strength. And it is well known that in experiments on the scattering of alpha particles from heavy nuclei the proper inverse-square field is found to extend through the whole accessible region. In Fig. 5 the curve AB is a plot of the potential energy of an alpha particle in this field against the distance from the centre of a nucleus of atomic number $Z=90$. To provide the attractive field which holds alpha particles in the nucleus it has long been recognized that the potential energy curve must turn over in the way shown in Fig. 5. And it has been shown, for example by Enskog,[8] that curves of this type may be obtained by giving the particle a magnetic moment.

In order to explain the ejection of a particle one has hitherto supposed that the praticle in the internal region received energy sufficient to raise it over the potential barrier. The suggestion that this energy was obtained by absorption of some ultra-penetrating radiation from outside never received wide acceptance. But it was necessary on classical mechanics to suppose that the emitted particle had received energy, if not from outside then from the other nuclear particles. Now the potential barrier which confines particles in the nucleus, i.e. the area under the curve in Fig. 5, is a region where

[8] Enskog, Zeits. f. Physik 45, 852 (1927).

the total energy would be less than the potential energy. And since the quantum mechanics endows particles with the new property of being able to penetrate such regions, this gives us at last a nucleus which can disintegrate without the absorption of energy.

We see that a mere qualitative application of the principles of quantum mechanics seems to account for the principal properties of radioactive atoms, most of which have been familiar for nearly thirty years. We have now to consider the question: How can nearly similar nuclei have periods of decay of anything from a small fraction of a second to over 10^9 years?. It has been shown above that in coupling the possible motions of a particle on either side of a potential barrier, the probability of transmission through the barrier is extremely sensitive to the area of the barrier; in fact the relation to it is exponential. In this way we shall show that we can obtain all rates of decay up to practical stability, and that from atoms whose potential curves are almost identical.

3. Quantitative Application

If the height of CD above OX in Fig. 5 gives the energy of the alpha-particle emitted, we have to consider the coupling of the motion along CD with the motion along EF inside the nucleus. We clearly do not have much choice in the area of the potential barrier we may take, since both the point C is fixed and the curve passing through C. For the purposes of numerical calculation we will compare radium A which has a period of 4.4 minutes (half-value period 3.05 minutes) with the extreme cases of uranium and

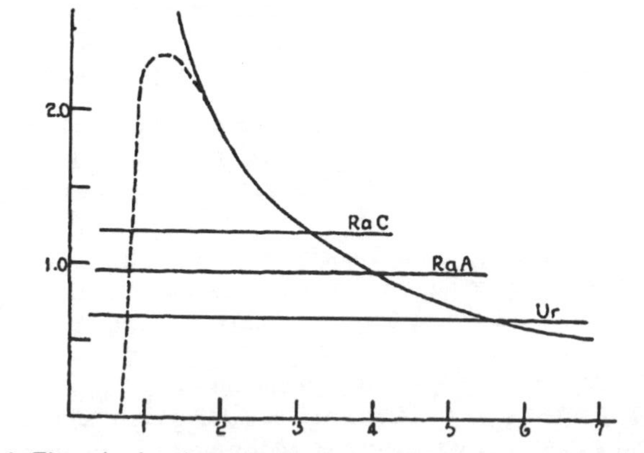

Fig. 6. The unit of ordinates is 10^{-6} ergs, and the unit of abscissas 10^{-12} cm.

radium C′, which have decay periods of about six thousand million years and a millionth of a second respectively.

An alpha-particle emitted by an element of atomic number Z escapes through the Coulomb field corresponding to (Z-2). Hence the potential energy for $Z=82$, which has been plotted in Fig. 6 is appropriate to radium A. The three horizontal lines in Fig. 6 give the energies of the alpha-particles

[9] Blackett, Proc. Roy. Soc. A107, 369 (1925).

from Ra A, Ra C′, and Uranium, which are 1.22×10^{-5}, 9.55×10^{-6}, and 6.5×10^{-6} ergs respectively. It is clear that the factor (V-W), which occurs in the expression (9), given above for the rate of decay, is simply the vertical distance between the horizontal line for W and the potential energy curve for the proper value of Z-2. In Fig. 7 is plotted a curve derived from Fig. 6 giving the value of $(4\pi/h)\ [2\mu(V-W)]^{1/2}$ for Ra A as a function of the radius. The upper curve is for uranium and the lower for Ra C′, derived from curves for the proper atomic numbers.

From these curves we can at once find how large a barrier we have to take in order to obtain any observed rate of decay. For the integral occurring in the exponent of expression (9) is merely the area that we will take under the curve in Fig. 7. Since the unit of abscissas taken is 10^{-12} cm and the unit of ordinates 10^{13} cm^{-1}, each of the squares in Fig. 7 has the dimensionless value 10; so that for an element whose potential barrier has an area of one

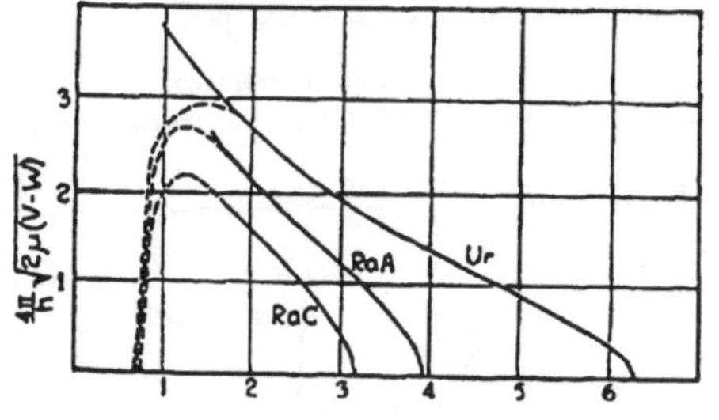

Fig. 7. Ordinates give the value of $(4\pi/h)[2\mu(V-W)]^{1/2}$ the unit being 10^{13} cm^{-1}. The unit of abscissas is 10^{-12} cm.

square on this diagram we should employ the factor e^{-10}. The broken line in Fig. 6 has been drawn so as to give for Ra A in Fig. 7 an area of approximately the value 53.7. For substituting this value in expression (9) together with $W=9.55\times10^{-6}$ and $a=10^{-12}$ we obtain the decay constant $8.45\times10^{20}\times e^{-53.7}=3.8\times10^{-3}$ sec.$^{-1}$, or the decay period $1/\lambda$ is 4.4 minutes in agreement with observation. In the expression for λ the precise value of the first factor is obviously unimportant, for if it were taken five times larger or smaller this would only alter the area of the required barrier by about 1 percent. The general size of the potential barrier in Fig. 6 that we have had to take seems to be a very reasonable one.

Now we reach an unexpected result. In drawing the areas for uranium and Ra C′ in Fig. 7 the continuous lines were predetermined, and the broken lines have been derived from the curve in Fig. 6, already used for Ra A. The values of the two areas are found to be 34.4 and 90, though the exact values depend on how the broken line is made to join the Coulomb potential curves. On substituting the values 34.4 and 90 in the expression for T we obtain for Ra C′ and uranium decay periods of the order of 10^{-6} sec. and 10^{10} years respectively, in agreement with observation.

It was already clear from Fig. 6 that we should obtain for all elements some qualitative agreement with the Geiger-Nuttall relation: the higher the energy of the alpha-particle the greater the rapidity of decay. But now we have found the unexpected result that the agreement is almost quantitative; that we do not have to choose a different potential energy inside the nucleus for each alpha-particle but having taken one potential curve for the whole series, it is the energy of the emitted alpha-particle which determines its own rate of decay. The mere fact that the velocity of the alpha-particle from Ra A, 1.69×10^9 cm per sec. is a little greater than the 1.4×10^9 cm per sec. of uranium, and a little less than the 1.92×10^9 cm per sec. of Ra C′, gives Ra A a decay period 10^{15} times as short as that of uranium and 10^8 times as long as that of Ra C′, in agreement with observation. Questions raised by this agreement with the factor of Geiger and Nuttall will be discussed in the last section of this paper. The radius 2×10^{-12} cm, at which we have taken the deviation from the inverse-square law, seems to be of the magnitude which our knowledge of the nucleus would lead us to expect.

Further we see at once why it is that no slow alpha-particles have been discovered. Although particles of ranges between 2.5 and 7 cm are plentifully distributed, no alpha-particles of energy less than 6.5×10^{-6} ergs have been found. But we now see from Figs. 6 and 7 that for particles of lower energy the area of the potential barrier increases very rapidly; so that for particles of range 2 cm or less the exponential factor would reduce the rate of decay of the element to a value at which its manifestation of radioactivity would be beyond the limits of detection.

Beta-ray disintegration.—It has been customary to assign the central core of the nucleus as the habitat of the nuclear electrons, with a potential energy curve of the type shown in Fig. 8. The outer slope AB again represents the Coulomb inverse-square field, as in Fig. 5. But since the charge of the electron is $-e$ instead of $+2e$ the potential energy is reversed in sign, and of half the magnitude of that in Fig. 5. There is nothing new in this assumed curve, although it looks somewhat artificial; this type of curve for the nuclear electron was obtained for example by Enskog in the paper referred to above.[8] What is new is the suggestion that an electron in the internal region again has a certain chance of penetrating the barrier, and of escaping at any time along CD with kinetic energy given by the height of CD above the axis.

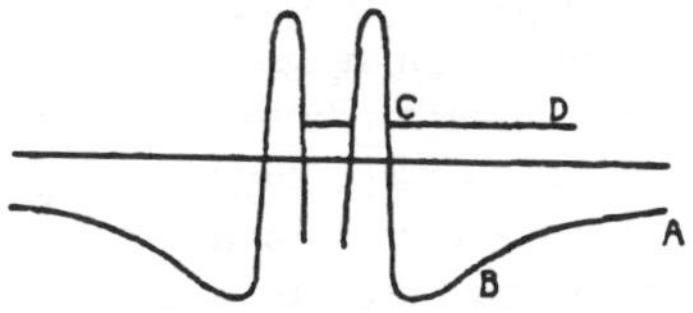

Fig. 8.

If we have alpha and beta-particles both with this chance of escaping from the nucleus, it might be thought that every radioactive element should be found to disintegrate part with expulsion of alpha-particles and part with beta-particles. But we would repeat that the chance of escape is extremely sensitive to the height to which the potential energy curve rises above the energy-level in question; and that if the size of this potential barrier be increased by a small factor the probability of escape may be decreased more than a million-fold. There seems then no reason why there should not be

the three types of disintegration: that in which the probability of escape is much greater for an alpha-particle than for an electron; that in which it is much greater for an electron than for an alpha-particle; and that in which the probabilities of escape are comparable. The last gives the branching type of disintegration as shown by Ra C, of which 99.97 percent emits beta-particles, and 0.03 percent alpha-particles. By taking this view of the disintegration process, we have raised the question: Does any radioactive element have a unique mode of disintegration, or does it merely appear unique in most cases because the secondary mode is a million times less frequent and escapes detection? The present discussion certainly favours the latter alternative. It need not surprise us then that so few cases of branching disintegration have so far been discovered, since it is unlikely (so far as we know) that the areas of the potential barriers will in many nuclei happen to have just that relative size which will give for alpha and beta-particles comparable probabilities of escape.

Artificial disintegration.—Blackett's cloud-chamber photographs[9] of artificial disintegration in nitrogen showed that the impinging alpha-particle was caught and retained by the nucleus. One is tempted to apply the present theory, using again the fact that the impinging alpha-particle may penetrate the barrier of potential, this time from the outside, instead of passing over the top as required by classical theory. But when we do this we are at once confronted by the fact that instead of approaching the barrier 10^{20} times per second, like a nuclear particle, our alpha-particles will only make one impact apiece. So it would seem that the capture of the alpha-particle could not be due to penetration. There is, however, another consideration; and that is that if the impinging alpha-particle have an energy very near that of an allowed but unoccupied nuclear energy-level, the chance of its penetrating the barrier at a single impact approaches unity. This property has already been referred to in section 1.

4. Experimental Evidence for the Penetration of Potential Barriers

The essential basis of the present theory is the assumed power of particles to pass through regions where their total energy would by classical mechanics be less than their total energy. For this property there is no direct experimental evidence in physics, although it follows from the laws of quantum mechanics. But in applying this to the nucleus we have found that we can actually *obtain* direct experimental evidence. Though on classical mechanics the passage of a particle through such a forbidden region was a manifest absurdity, it was found in 1925 by Rutherford and Chadwick[10] that that is exactly what the alpha-particles from uranium appear to do.

Consider the alpha-particle which the uranium nucleus emits during its disintegration. The alpha-particle will gain energy in escaping through the repulsive Coulomb field outside the nucleus. This energy is given on classical theory as $2Ze^2/r$. Even if the alpha-particle leaves its place in the nucleus

[10] Rutherford and Chadwick, Phil. Mag. 50, 889 (1925).

with no initial velocity, its energy cannot be less than this amount. The energy with which the alpha-particles leave the disintegrating Uranium atom is observed experimentally to be 6.5×10^{-6} ergs. On referring to Fig. 5, which was drawn for $Z=90$, we see that this energy corresponds to $r=6.3\times 10^{-12}$ cm and if any of the energy was initial energy and not acquired through falling through the repulsive field, the value of r would have to be greater than this value.

It was concluded that the inverse-square law of repulsive field could not possibly hold within this value of r. Consequently if we fire at the uranium nucleus an alpha-particle having slightly more energy than the 6.5 10^{-6} ergs, it should penetrate its structure to where the Coulomb law no longer holds; while still faster particles should penetrate, even when not fired directly at the nucleus. It was therefore disconcerting when, on examining the scattering of fast alpha-particles fired at uranium, Rutherford and Chadwick could find no indication of any departure from the inverse-square laws. The Coulomb field was found to hold inside the radius from which the uranium alpha-particle appeared to come. That is to say, the uranium alpha-particle appeared to emerge from a region where its kinetic energy was negative. To escape this conclusion Rutherford[11] supposed that the uranium alpha-particles before ejection are electrically neutral, having been neutralised by two electrons which they leave behind when they are ejected. This hypothesis succeeded in circumventing the paradox. But if we abandon classical mechanics, the paradox disappears, yielding us direct experimental evidence in favor of the phenomenon of quantum mechanics in which we are interested.

5. Discussion of Limitations

It must be clearly understood that although the Coulomb part of the potential curve outside the nucleus, represented by AB in Fig. 5 is necessarily common to all particles, the internal part is merely intended to represent the potential energy of a particular alpha-particle. And it must not be taken to represent a general central field common to many particles, such as we are so accustomed to in atomic structure. There is no reason why the internal field should be necessarily symmetrical about the center of the nucleus as drawn in Figs. 5 and 8. In fact, Rutherford[12] has suggested that the nucleus may have something analogous to a crystalline structure. If this caution is lost sight of, difficulties are encountered.

For the atom of each radioactive element contains within its nucleus not only the alpha-particle which it will itself emit, but also the alpha-particles destined to be emitted by its successors in the radioactive series. Now if the velocity of escape of the alpha-particles from each element were always less than that of those emitted by its predecessors, there would be no serious difficulty; for from an atom loaded with alpha-particles in various allowed energy levels, the particle in the highest level would have the

[11] Rutherford, Phil. Mag. 4, 580 (1927); Proc. Phys. Soc. 39, 370 (1927).
[12] Rutherford, Jour. Franklin Inst. 198, 743 (1924).

greatest probability of escape. This however is the opposite of what is observed; and we have to account for the subsequent emission of particles of higher energy than that emitted by the parent substance. We may do this by supposing either (a) that the alpha-particles of higher energy have in the parent element been confined by correspondingly high barriers; or (b) by supposing that the alpha-particles in the nucleus are not permanently in the high energy levels from which they emerge, but are temporarily raised up from lower levels. The latter seems to be a retrograde step, for the principle advantage of the present theory is that it has offered an escape from such processes.

If, however, we accept the former supposition (a), we see that the emission of one alpha-particle must profoundly modify the potential barrier which confines the alpha-particle destined to be emitted next. As we have shown, the Geiger-Nuttall relation seems to require that the barrier through which this alpha-particle emerges be approximately the same in all elements of the series. But until we know how this comes about, it seems inadvisable to discuss the Geiger-Nuttall relation in greater detail. In speaking of the energy of one particle in the nucleus, it must not be forgotten that we are making use of the simplification mentioned in the introduction: that of discussing one nuclear constituent alone.

Palmer Physical Laboratory,
Princeton University,
November 20, 1928.

Second Series *April 1, 1932* *Vol. 40, No. 1*

THE

PHYSICAL REVIEW

A HYDROGEN ISOTOPE OF MASS 2 AND ITS CONCENTRATION*

BY HAROLD C. UREY, F. G. BRICKWEDDE, AND G. M. MURPHY**

COLUMBIA UNIVERSITY AND THE BUREAU OF STANDARDS

(Received February 16, 1932)

ABSTRACT

In a recent paper Birge and Menzel pointed out that if hydrogen had an isotope with mass number two present to the extent of one part in 4500, it would explain the discrepancy which exists between the atomic weights of hydrogen as determined chemically and with the mass spectrograph, when reduced to the same standard. Systematic arrangements of atomic nuclei require the existence of isotopes of hydrogen H^2 and H^3 and helium He^5 to give them a completed appearance when they are extrapolated to the limit of nuclei with small proton and electron numbers. An isotope of hydrogen with mass number two has been found present to the extent of one part in about 4000 in ordinary hydrogen; no evidence for H^3 was obtained. The vapor pressures of pure crystals containing only a single species of the isotopic molecules H^1H^1, H^1H^2, H^1H^3 were calculated after postulating: (1) that the rotational and vibrational energies of the molecules are the same in the solid and gaseous states; (2) that in the Debye theory of the solid state, the Θ's are inversely proportional to the square roots of the molecular masses; (3) that the free energy of the gas is given by the free energy equation of an ideal monatomic gas; and (4) that there is a zero point lattice energy equal to $(9/8)R\Theta$ per mole. The calculated vapor pressures of the three isotopic molecules in equilibrium with their solids at the triple point for ordinary hydrogen are in the ratio $p_{11}:p_{12}:p_{13}=1:0.37:0.29$. The isotope was concentrated in three samples of gas by evaporating large quantities of liquid hydrogen and collecting the gas which evaporated from the last two or three cc. Sample I was collected from the end portion of six liters evaporated at atmospheric pressure and samples II and III from four liters, each, evaporated at a pressure only a few millimeters above the triple point.

These samples and ordinary hydrogen were investigated for the visible, atomic Balmer series spectra of H^1 and H^2 from a hydrogen discharge tube run in the condition favorable for the enhancement of the atomic spectrum and for the repression of the molecular spectrum, using the second order of a 21 foot grating with a dispersion of 1.31A per mm. When with ordinary hydrogen, the times of exposure required to just record the strong H^1 lines were increased 4000 times, very faint lines appeared at the calculated positions for the H^2 lines accompanying $H^1\beta$, $H^1\gamma$ and $H^1\delta$ on the short wave-length side and separated from them by between 1 and 2A. These lines do not agree in wave-length with any known molecular lines and they do not appear on the plates taken with the discharge tube operating under conditions favorable for the production of a strong molecular spectrum and the repression of the atomic spectrum. With ordinary hydrogen they were so weak that it was difficult to be sure that they were not irregular ghosts of the strongly overexposed atomic lines. Samples II and III evaporated near the triple point show these lines and another near $H^1\alpha$ greatly enhanced relative to the H^1 lines over those with ordinary hydrogen showing that these new lines are not ghosts, and that a considerable increase in the concentration of the isotope had been effected. With sample I, evaporated at the boiling point, no appreciable increase in concentration was detected. The new lines agree in wave-length with those calculated for an H^2 isotope.

The H^2 lines are broad as is to be expected for close unresolved doublets, but they are not as broad and diffuse as the H^1 lines, probably due to the smaller Doppler broadening. The $H^2\alpha$ line is resolved into a close doublet with a separation that agrees within the accuracy of the measurements with the observed separation for $H^1\alpha$.

Relative abundances were estimated by comparing the times required to just record photographically the corresponding H^1 and H^2 lines. The relative abundance of H^2 and H^1 in natural hydrogen is estimated to be about 1:4000 and in the concentrated samples about five times as great.

* Publication Approved by the Director of the Bureau of Standards of the U. S. Department of Commerce.

** H. C. Urey and G. M. Murphy, Columbia University, F. G. Brickwedde, Bureau of Standards.

THE possibility of the existence of isotopes of hydrogen has been discussed for a number of years. Older discussions involved Prout's hypothesis and dealt with the question as to whether hydrogen consisted of a mixture of isotopes, one having an atomic weight exactly one, and another or others with integral values, in such proportions as to give an average atomic weight of 1.008. The result of an exact determination in 1927 with the mass spectrograph by Aston[1] of the atomic weight of the hydrogen isotope of mass-number one not only proved that it is not integrally equal to unity but the agreement with the chemically determined value was so close that it was considered unlikely that hydrogen had more than the single isotope of mass-number one. The discovery of the oxygen isotopes by Giauque and Johnston[2] in 1929 showed that the chemical standard of atomic weights was not the same as that used by Aston and that agreement between the chemical determinations and Aston's values should not be expected. When the atomic weights of hydrogen as determined chemically and by the mass spectrograph are reduced to a common standard, the previous apparent agreement is destroyed and they differ. Birge and Menzel[3] showed that this discrepancy could be explained by the presence of an isotope of hydrogen of mass-number two, present to the extent of one part in 4500.

Quite independently of such a quantitative basis of prediction as is furnished by the agreement or disagreement of the atomic weights determined chemically and with the mass spectrograph, one may be led by other lines of reasoning to expect heavier isotopes of hydrogen and helium, as well, even though the atomic weights reduced to a common standard do agree, for it is only necessary to assume that they are so rare that they can not be detected

[1] F. W. Aston, Proc. Roy. Soc. (London) A115, 487 (1927).

[2] W. F. Giauque and H. L. Johnston, J. Am. Chem. Soc. 51, 1436 and 3528 (1929).

[3] R. T. Birge and D. H. Menzel, Phys. Rev. 37, 1669 (1931); F. Allison (J. Ind. Eng. Chem. 4, 9 (1932)) interprets two minima observed by his magneto-optical effect in water solutions of acids as due to two isotopes of hydrogen having mass numbers of 1 and 2. Whatever weight can be given to this method of detecting the *number* of isotopes of an element, certainly without some understanding of its dependence on mass the method gives no evidence in regard to the *masses* of the isotopes.

by atomic weight determinations within the limits of the experimental accuracy. The recent discoveries of rare isotopes emphasize that it may be impossible ever to disprove the existence of any nuclear species. Recent systematic arrangements of nuclear species[4] lead one to expect isotopes of hydrogen of masses 2 and 3 and an isotope of helium of mass 5. Beck leaves a place in his tables for H^3 and He^5. Johnston has question marks in his table for H^2, H^3, He^5 and Li^5. Urey makes no definite predictions but presents a proton-electron plot which shows the regularities very well, the three isotopes, H^2, H^3, and He^5 being required to give this plot a completed appearance.

Methods of Concentration

Birge and Menzel[3] remark that the discovery of a hydrogen isotope of higher mass-number by the methods of molecular spectra would be difficult though not impossible. The maximum abundance of an isotope of mass-number 2 which can be expected is that given by Birge and Menzel for if any isotope of higher mass number were present the abundance of the isotopes would all necessarily be less. It seemed essential to find some way of concentrating the heavier isotopes if they were to be detected by spectroscopic methods. Any of the various methods used for concentrating isotopes should be more effective in the case of these isotopes of hydrogen because of the large ratio of masses. Of these methods, that of fractional distillation should give the largest supply with the least effort. This method has been tried in a number of cases[5] but with little success except in the case of neon.[6]

The vapor pressures of the molecules H^1H^1, H^1H^2, H^1H^3 in equilibrium with their pure solids can be calculated if the following postulates are made: (1) the rotational and vibrational energies of the molecules are the same in the solid and gaseous states and thus need not be considered in the calculations of vapor pressures; (2) the free energy[7] of the solids can be calculated from the Debye theory of the solid state, assuming that the Θ's of the three solids are inversely proportional to the square roots of the molecular weights. (3) the free energy of the gas is given by the free energy equation of an ideal monatomic gas.

At equilibrium, the free energy of the gas is equal to the free energy of the solid, and since all the quantities may be evaluated, we may calculate the vapor pressures of the isotopic molecules. The free energy and entropy of hydrogen gas are given by the following expressions:

[4] H. L. Johnston, J. Am. Chem. Soc. 53, 2866 (1931); Harold C. Urey, J. Am. Chem. Soc. 53, 2872 (1931); Guido Beck, Z. Physik 47, 407 (1928); Henry A. Barton, Phys. Rev. 35, 408 (1930).

[5] F. A. Lindemann and F. W. Aston, Phil. Mag. (6) 37, 523 (1919); F. A. Lindemann, Phil. Mag. (6) 38, 173 (1919); H. G. Grimm, Zeits. f. phys. Chem. B2, 181 (1929); H. G. Grimm and L. Braun, Zeits. f. phys. Chem. B2, 200 (1929); P. Harteck and H. Striebel, Zeits. f. anorg. allgem. Chem. 194, 299 (1930).

[6] W. H. Keesom and H. van Dijk, Proc. Acad. Sci. Amsterdam 34, 42 (1931); H. van Dijk, Physica 11, 203 (1931).

[7] The "free energy" as used here refers to this term as defined by Lewis. See Lewis and Randall, Thermodynamics, McGraw-Hill, 1923, New York.

$$F_g = E_g + RT - TS_g \tag{1}$$

$$S_g = \frac{3}{2} R \ln M + \frac{5}{2} R \ln T - R \ln P + C + R \ln \cdot R \tag{2}$$

where M is the molecular weight, P is the pressure in atmospheres, R is the gas constant in cal. per mole per degree and C is the Sackur-Tetrode constant and equals -11.053 cal. per degree[8] and

$$E_g = \frac{3}{2} RT + \chi. \tag{3}$$

χ is the heat of vaporization at absolute zero from a hypothetical solid hydrogen without zero point energy, which for convenience is chosen as the standard reference energy state to which the internal energies of the solid and gaseous phases are referred. χ is assumed to be the same for the isotopic molecules. The differences between the values of the internal energy of the gas at the triple point of hydrogen (13.95°K) as calculated by Eq. (3) and by the more exact equations for a degenerate gas are negligibly small.

The free energy of solid hydrogen is given by:[9]

$$F_s = E' + T\Phi(M, T) + PV. \tag{4}$$

Because of the small volume of solid hydrogen the PV term may be neglected without serious error. The quantity E' is the zero point energy (Nullpunktsenergie) and must be included.[10] The function $\Phi(M, T)$ may be obtained from the Debye theory of specific heats.[9]

Solving these equations for $\ln P$ after equating (1) and (4) and dividing through by RT, we have:

$$\ln P = \frac{E'}{RT} + \frac{\Phi}{R} + \frac{3}{2} \ln M + \frac{5}{2} \ln T + \frac{C}{R} + \ln R - \frac{5}{2} - \frac{\chi}{RT}. \tag{5}$$

The only terms on the right of (5) which depend on the mass are the 1st, 2nd, and 3rd, since χ has been assumed to be the same for isotopic molecules. If we indicate the two molecules H^1H^1 and H^1H^2 by subscripts we have the ratio of their vapor pressures given by:

$$\ln P_{11}/P_{12} = \frac{1}{RT}(E'_{11} - E_{12}') + \frac{1}{R}(\Phi_{11} - \Phi_{12}) + \frac{3}{2} \ln M_{11}/M_{12}. \tag{6}$$

The quantity $\Phi(M,T)$ is a function of $h\nu/kT$, where ν is the characteristic frequency for the solid state; $h\nu/k$ for ordinary hydrogen as determined by Simon and Lange is 91.[11] Since the characteristic frequency ν is inversely proportional to the square root of the molecular weight, the argument of Φ may be determined for the isotopic molecules and the value of Φ taken from the

[8] R. T. Birge, Rev. Mod. Phys. 1, 65 (1929).

[9] Handbuch der Physik, Vol. X, p. 360–361, Julius Springer, 1926 Berlin.

[10] R. W. James, I. Waller and D. R. Hartree, Proc. Roy. Soc. A118, 334 (1928).

[11] Franz Simon and Fritz Lange, Zeits. f. Physik 15, 312 (1923).

tables.[12] The calculation of the ratio P_{11}/P_{12} is made for the temperature 13.95°K, the triple point for ordinary hydrogen. The ratio P_{11}/P_{13} is calculated in a similar way.

The numerical values for $h\nu/k$ and Φ are:

Molecule	M	$h\nu/k$	Φ
H^1H^1	2	91	− 0.1339
H^1H^2	3	74.29	− 0.2251
H^1H^3	4	64.36	− 0.3364

The value of the zero point energy is $9/8h\nu$ per molecule[13] and may be easily calculated for the isotopic molecules. The values of E' thus become $(9/8)R(h\nu)/k$; or $(9/8)R\Theta$. Substituting the numerical values in (6), we get:

$$P_{11}/P_{12} = 2.688, \quad P_{11}/P_{13} = 3.354.$$

If the calculation is carried through assuming that the zero point energy is zero, it is found that on this basis the *heavier* isotopic molecules should have the higher vapor pressures which is contrary to experience not only with the hydrogen isotopes but with all other isotopes.

This calculation of the ratios of the vapor pressures has been made for the solid state. A similar calculation cannot be made for the liquid state since the theory is inadequate. It seems reasonable to expect that differences between the vapor pressures of the isotopes should persist beyond the melting point and that a fractionation of the liquid solution should be possible.

The Rayleigh distillation formula integrated for ideal solutions is:

$$\left(\frac{1-N_0}{1-N}\right)^{\alpha/(1-\alpha)}\left(\frac{N}{N_0}\right)^{1/(1-\alpha)} = \frac{W_0}{W} \tag{7}$$

where N and N_0 are the mole fractions of the less volatile constituent left in the still and in the original sample respectively and W and W_0 are the moles of both constituents left in the still and in the original sample respectively, and α is the distribution coefficient equal to the ratio of the vapor pressure of the less volatile constituent to that of the more volatile constituents. If N_0 and N are small as compared to 1 as is the case for the distillation of these isotopes of hydrogen, this formula reduces to:

$$\left(\frac{N}{N_0}\right)^{1/(1-\alpha)} = \frac{W_0}{W}. \tag{8}$$

[12] Handbuch der Physik, Vol. X, p. 364–70, Julius Springer, 1926, Berlin.

[13] The mean zero point energy per degree of freedom is

$$\frac{\int_0^{\nu\max}\frac{h\nu}{2}\nu^2 d\nu}{\int_0^{\nu\max}\nu^2 d\nu} = \tfrac{3}{8}h\nu_{\max}$$

for $3N$ degrees of freedom, this gives $9/8\,Nh\nu$. Lindemann (Phil. Mag. 38, 173 (1919) showed that in order to make the calculated vapor pressures of the isotopes of lead at its boiling point equal to each other, as was found by experiment to be true, it was necessary to make the internal energy of isotopic, pure crystals at 0°K equal to $9/8\,h\nu$.

This formula has been used in estimating the increased concentrations expected.

If we assume that the mole fraction of H^2 is 1/4500 in the original hydrogen, that $\alpha = 1/2.688$ and that $W_0/W = 4000$, we secure about 4 mole percent as the value of N. Since we have not secured such high concentrations, we conclude that either the ratio of vapor pressures of the solids is quite different from those of the liquids at the same temperature, or that some of the assumptions made in regard to the solids are not sufficiently exact.[14] We have made this calculation in order to see whether the separation by fractionation was likely to be effective.

Preparation of the Concentrated Hydrogen Samples

Each of the different samples of hydrogen, which were later examined spectroscopically, was prepared from liquid hydrogen made by circulating about 400 cubic feet of free gas through a liquefier of the ordinary Hampson type in which, after precooling with liquid air boiling at reduced pressure, it was expanded from a pressure of about 2500 pounds per square inch to atmospheric pressure. As the liquid hydrogen was obtained it was collected in storage containers from which it was transferred to an unsilvered triple walled flask of about 1600 cm³ capacity in which the concentration of the isotope was effected. After filling the flask, the liquid hydrogen was allowed to evaporate until only about 1/3 or 1/4 remained, when the flask was refilled and the procedure was repeated until all the liquid had been transferred. The flask was connected by vacuum tight joints to the glass bulbs in which the hydrogen gas evaporating from the last two or three cubic centimeters of liquid was collected. These bulbs were connected to a Hyvac pump for exhaustion and flushing out previous to the collection of the final concentrate. Proper precautions were taken to prevent the entry of air into the system while the samples were being collected. This method of evaporation is somewhat less efficient than the method assumed in the calculation above and accounts, at least in part, for the lower efficiency observed.

Sample I was collected from the end portion of six liters of liquid hydrogen evaporated at atmospheric pressure, and samples II and III, each, from four liters evaporated at a pressure only a few millimeters above the triple point. The process of liquefaction could have had only a small effect in changing the relative concentrations of the isotopes since no appreciable increase in the concentration of the isotopic molecule H^1H^2 over that in ordinary hydrogen was detected for sample I obtained from six liters of liquid hydrogen evaporated at atmospheric pressure.

[14] Professor K. F. Herzfeld has called to our attention the possibility that the rotational states of the unsymmetrical molecules, H^1H^2 and H^1H^3, may not be the same as those of the gas, even though this is true for the symmetrical molecules. The center of mass of the unsymmetrical molecules does not coincide with the midpoint of the line of nuclei so that the rotation would take place in such a way that the H^1 atom would encounter the fields of force of other molecules to a greater extent thus changing the rotational levels in an unpredictable way.

Spectrum Analysis

It is possible to detect the hydrogen isotopes from the positions of the atomic lines, since the Balmer lines of any heavier isotopes will be displaced to the violet side of the H^1 Balmer lines. Assuming that the masses of the isotopic hydrogen nuclei of mass-numbers 2 and 3 are exactly twice and three times the mass of the proton, the calculated wave-lengths of the isotopic lines and the observed wave-lengths of the H^1 lines are:

	α	β	γ	δ
H^1	6562.793	4861.326	4340.467	4101.738
H^2	6561.000	4860.000	4339.282	4100.619
H^3	6560.400	4859.56$_6$	4338.882	4100.239.

The second order of a 21 foot grating with a dispersion of 1.3A per mm was used to analyze the spectrum from a Wood hydrogen discharge tube run in his so-called black stage.[15] This tube was 1 cm in diameter and was excited by a current of about 1 ampere at 3000 to 4000 volts; the radiation was sufficiently intense to record the $H^1\beta$ and $H^1\gamma$ lines in about 1 sec., though the lines were broad and unresolved under these conditions. By greatly decreasing the current and increasing the exposure time to about 16 sec., it was possible to resolve the $H^1\beta$ line into a doublet, but a simple calculation showed that the time of exposure necessary to record the isotope lines under conditions necessary to resolve them would be prohibitively long. We therefore worked with the high current density in order to decrease the exposure time.

The usual method of securing clean atomic hydrogen spectra by flowing moist hydrogen through the tube was not used, as the samples were limited in amount. They were not moistened by saturation with ordinary water since we did not wish to contaminate them with ordinary hydrogen from the water. The sample of hydrogen was contained in a glass bulb with two stop-cocks attached in series so that a small sample of hydrogen (about 2 cc) could be admitted to the discharge tube at one time. The stop-cock grease was a disadvantage since it was probably the source of the cyanogen bands in our tube which was troublesome when working with $H\delta$. The hydrogen gas was either, not moistened at all, in which case the molecular spectrum was rather strong, or, it was moistened by attaching near the electrodes small side tubes containing copper oxide or, by admitting oxygen gas in small amounts. The copper oxide in the side tubes was reduced by atomic hydrogen diffusing in from the discharge tube and water was formed. When oxygen was used, some of the oxygen bands and lines appeared which, however, caused no trouble. None of these methods of suppressing the molecular spectrum was as effective as the flowing stream of moist hydrogen gas and at times the molecular spectrum became intense in spite of all our efforts to keep the tube in a good black stage.

Before working on the evaporated samples of hydrogen, ordinary hydrogen was tried first in order to overcome any difficulties in the method of ex-

[15] R. W. Wood, Proc. Roy. Soc. (London) **97,** 455 (1920); **102,** 1 (1923); Phil. Mag. **42,** 729 (1921): **44,** 538 (1922).

citation. The sample of hydrogen evaporated at the boiling point (Sample I) was next investigated, but no isotopes present in the estimated concentrations could be found, though faint lines appeared at the calculated positions for H^2 lines. Returning then to ordinary hydrogen, these same lines were found with about the same intensity as in sample I. It was difficult to be certain that these lines were not irregular ghosts. All other lines near the Balmer lines could be accounted for as known molecular lines. Turning then to sample II, evaporated near the triple point, the H^2 lines were found greatly enhanced relative to the H^1 lines thus showing that an appreciable increased concentration of the H^2 isotope had been secured and that the lines could not be ghosts since their intensity varied relative to the known symmetrical ghosts. Sample III was investigated subsequently and found to have a higher concentration of H^2 than sample II.

The measurements on ordinary hydrogen will be discussed first. A great many plates were taken with ordinary hydrogen with the tube in the black stage and one with the tube in the white stage. (Copper oxide was blown into the discharge tube to produce an intense molecular spectrum.) In Table I we give the measurements made on plates (*34l, 35l*) showing the $H\beta$ and $H\gamma$ regions with the tube in the white stage, and measurements made on plates (*36l, 37l*) with the tube in the black stage. The times of exposure and currents through the tube were the same for all these plates. For comparison we give the wave-lengths given by Gale, Monk and Lee,[16] and by Finkelnburg[17] for the molecular spectrum in these regions and the calculated wave-lengths of the Balmer lines of H^2 and H^3. The positions of the H^1 lines were secured by taking the means of the positions of the symmetrical ghosts and all the lines were measured relative to the standard iron lines.

Table I.

		35*l*	37*l*	Gale, Monk and Lee[16]	Finkelnburg[17]
$H^1\beta$	4861.326	4861.322	4861.320	4861.328	
		4860.892	—	4860.806	
		4860.636	4860.633	—	4860.620
		4860.104	—	4860.108	4860.134
$H^2\beta$	4860.000	—	4859.975	—	—
$H^3\beta$	4859.566	—	—	—	—
		34*l*	36*l*		
$H^1\gamma$	4340.467	4340.465	4340.486	4340.470	4340.466
		4340.084	—	—	4340.154
		4339.847	4339.879	4339.817	4339.845
		4339.568	4339.599	4339.534	4339.538
$H^2\gamma$	4339.282	—	4339.318	—	—
$H^3\gamma$	4338.8$_{82}$	—	—	—	—

The discrepancies between our values and those of the other authors are rather large. In view of the fact that the molecular lines on our plates were so

[16] H. G. Gale, G. S. Monk and K. O. Lee, Astrophys. J. **67,** 89 (1928).

[17] W. Finkelnburg, Zeits. f. Physik **52,** 27 (1928).

weak that the measurements of their positions were very difficult, the agreement obtained was considered satisfactory. The $H^2\gamma$ line appeared as a slight irregularity on a microphotometer curve of the plate 34*l* but could not be measured with the comparator. The measurements on other plates taken of the atomic spectrum of ordinary hydrogen run very much the same, sometimes with other observed molecular lines on them. The average displacements of the H^2 lines from all plates taken with ordinary hydrogen are given in Table II.

The measurements of plates taken with the hydrogen of sample I under the same conditions as with ordinary hydrogen run very much the same as those for ordinary hydrogen. It was impossible by visual observation to be certain of any difference between the intensity of the H^2 lines on the plates

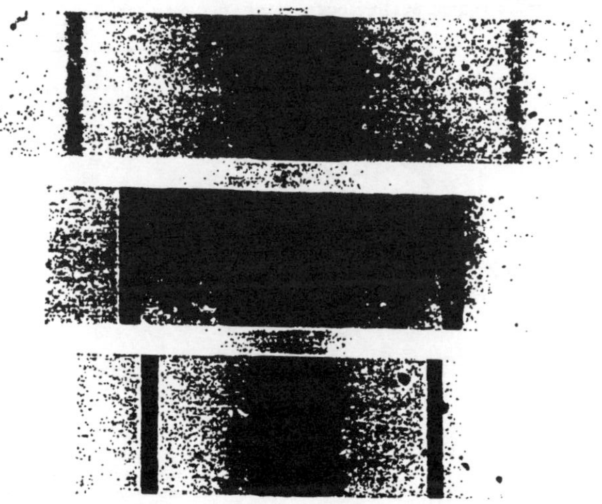

Fig. 1. Enlargement of the $H\alpha$, $H\beta$ and $H\gamma$ lines. The faint lines appearing on the high frequency side of the heavily over-exposed H^1 lines are the lines due to H^2. The symmetrical pair of lines in each case are ghosts.

for ordinary hydrogen and for sample I, although there were fewer molecular lines on sample I plates than on ordinary hydrogen plates. From this it was concluded that there was no appreciable increase in the concentration of the isotope H^2 in sample I evaporated from six liters of liquid hydrogen at atmospheric pressure over that in ordinary hydrogen and that at 20°K, the vapor pressures of the H^1H^1 and H^1H^2 isotopic molecules must be nearly, if not actually equal. The mean wave-length displacements of the H^2 lines from the H^1 lines on these plates are given in Table II. The agreement with the calculated displacements is better than in the case of ordinary hydrogen. This may indicate a greater ease of measurement due to an increased photographic density of the H^2 lines on sample I plates from which it might be concluded that there was a slight increase in the concentration of the heavier isotope.

When observations were made on samples II and III evaporated just above the triple point, the H^2 lines stood out so clearly from the background that there was no longer any possibility of confusing them with the molecular lines and no further measurements of the molecular lines were made. The measurement of the positions of the H^2 lines on these plates relative to the ghosts of the H^1 lines could be made with ease. The mean displacements listed in Table II for samples II and III are the most reliable ones obtained.

A mercury line falls at 4339.23, while the calculated wave-length of $H^2\gamma$

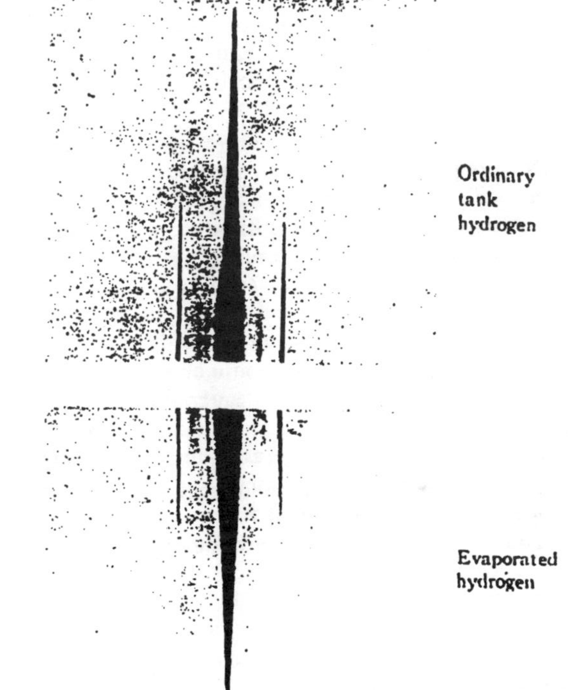

Fig. 2. The $H\beta$ lines for ordinary tank hydrogen, and sample II of the evaporated hydrogen. Although the intensity of the main line is about the same for both exposures, the $H^2\beta$ line is considerably more intense in the second case showing the increased concentration.

is 4339.282. Mercury got into our discharge tube due to various efforts we made to depress the molecular spectrum in the stationary gas. This occurred while we were working with samples II and III. On some plates this mercury line appears as a very faint black edge on the broad atomic line. In other cases, it was more intense and appeared as a very sharp line.

Fig. 1 shows enlarged prints of plates taken with sample II of the $H^1\alpha$, $H^1\beta$, and $H^1\gamma$ lines with the isotope lines appearing as faint companions on the high frequency side of the H^1 lines. The pair of symmetrical lines in each case are the ghosts. Fig. 2 shows the Hβ lines for ordinary tank hydrogen and for the evaporated hydrogen, sample II, the condition of the discharge and the time of exposure being approximately the same. The isotope H^2 line for sample II is considerably more intense than for ordinary hydrogen, showing that a considerable increase in the concentration of this isotope was affected by evaporation near the triple point. Similar plates have been obtained for Hγ. The $H^2\alpha$ line was obtained only with samples II and III.

TABLE II.[18]

	Hα	Hβ	Hγ	Hδ
Calcd. displacement	1.793	1.326	1.185	1.119
Obs.				
Ordinary hydrogen	—	1.346	1.206	1.145
Sample I	—	1.330	1.199	1.103
Samples II and III	1.791	1.313	1.176	1.088

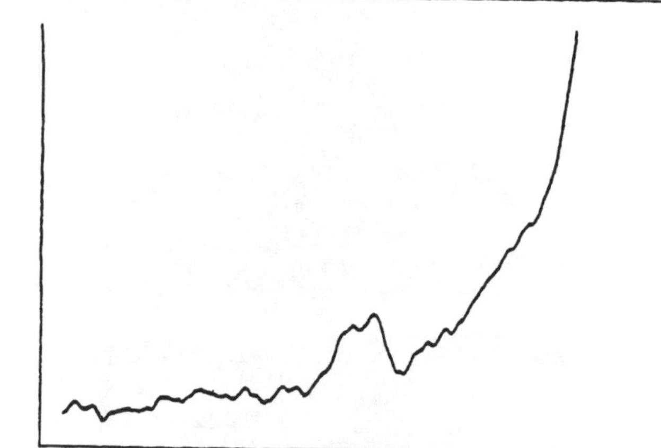

Fig. 3. Microphotometer curve of $H^2\alpha$ showing the doublet separation which is from 0.10 to 0.12A.

The lines of H^2 are broad as is to be expected if they consist of close unresolved doublets, but they are not so broad and diffuse as the lines of H^1 probably due to less Doppler broadening.[19] The $H^2\alpha$ line is just resolved into a close doublet on two plates. Visual settings on these lines with the comparator were difficult. One plate measured in this way gave a doublet separation of 0.16A. Microphotometer curves also show the doublet separation and the separation secured in this way is from 0.10 to 0.12A. These latter figures are the more reliable and agree with the observed separation of the $H^1\alpha$ line of 0.135A, being somewhat lower than the value for the well resolved line as is to be expected for partially resolved lines. Fig. 3 shows a microphotometer curve of this line. By itself it is not entirely convincing because of the irregularities due to grain size of the plate. That the resolution is real is proven by visual observation of the plate.

[18] In the letter to the Physical Review 39, 164 (1932) the wave-length displacement of the Hα line from sample II was given as 1.820. An error was made in the calculation of this displacement. The correct value for the one plate measured at that time is 1.778A. No measurements of plates with the mercury line near $H^2\gamma$ of more than slight intensity are included in this table. No plate of sample III showing the $H^2\gamma$ line is included for this reason.

[19] The abstract of this paper in the Bulletin of the American Physical Society states that the lines of H^2 have about the same breadth as the main lines. More careful study of our plates showed that the H^2 lines are distinctly narrower than the H^1 lines.

RELATIVE ABUNDANCE

When using ordinary hydrogen, the $H^2\beta$ line appears as a rather sharp line lying in a clear part of the plate between the region of halation from the main line and the main line itself. As the time of exposure is increased, the irradiated region and the region of halation build up the diffuse background of the plate so rapidly that the H^2 line does not become more distinct. However, in the case of sample II and sample III, the H^2 lines come out with a very much greater distinctness so that it is possible to secure these lines without bad halation from the main line. Thus there is no doubt that there has been a very distinct increase in concentration of the H^2 isotope relative to the H^1 isotope in the process of evaporation. It is difficult, however, to give an exact estimate of the relative abundance from the intensity of spectral lines which lie so close together with one so much more intense than the other. Moreover, a comparison of exposure times is not entirely satisfactory because we note that the H^2 lines are distinctly sharper than the H^1 lines so that if the same amounts of energy were emitted by the two varieties of atoms, the H^2 lines should appear to be the more intense, since this energy would fall in a narrower region on the plate. Comparison of the relative intensities of the ghosts of the H^1 lines and the H^2 lines meets with this same difficulty for the ghosts are distinctly more diffuse than the H^2 lines. The best that can be done, therefore, is to give rather rough estimates of the relative abundance judging from times of exposure.

In the case of ordinary hydrogen, it was found that when the discharge tube was running with such an intensity that the H^1 lines could be recorded within one second, that it required somewhat more than an hour to just detect the H^2 lines. It is, therefore, estimated that the relative abundance of the isotopes in ordinary hydrogen is about 1 in 4000 or less. We believe that the estimate of Birge and Menzel based on the atomic weights is consistent with our observations and that their estimate is probably the more reliable.

In the case of sample II, the $H^2\beta$ and $H^2\gamma$ lines could be photographed in ten minutes and the corresponding H^1 lines in one second. From this it is estimated that the relative abundance of H^2 in sample II was 1 in 600, but this, it is believed, is too high because in this case the discharge tube was running better than before and it should have been possible to photograph the H^1 lines in less than a second. Again, the intensities of the ghost lines produced by the grating used are about 1/200th of the intensities of the main lines and the H^2 lines have an intensity equal to about 1/4th of that of the ghost lines as determined by relative exposure times of 1 to 4. This gives a ratio of about 1 in 800. This is about the best estimate that we were able to make of the relative intensities in this sample. Sample III contains H^2 in larger amounts than sample II, perhaps as much as 1 part in 500 to 600.[10]

Fig. 4 shows microphotometer curves of the $H\beta$ lines from three samples of hydrogen. The plates were selected so that the densities of the ghost lines were as nearly alike as possible. Visual comparison of the plates shows that the variations in the densities of the ghost lines are such that the intensities

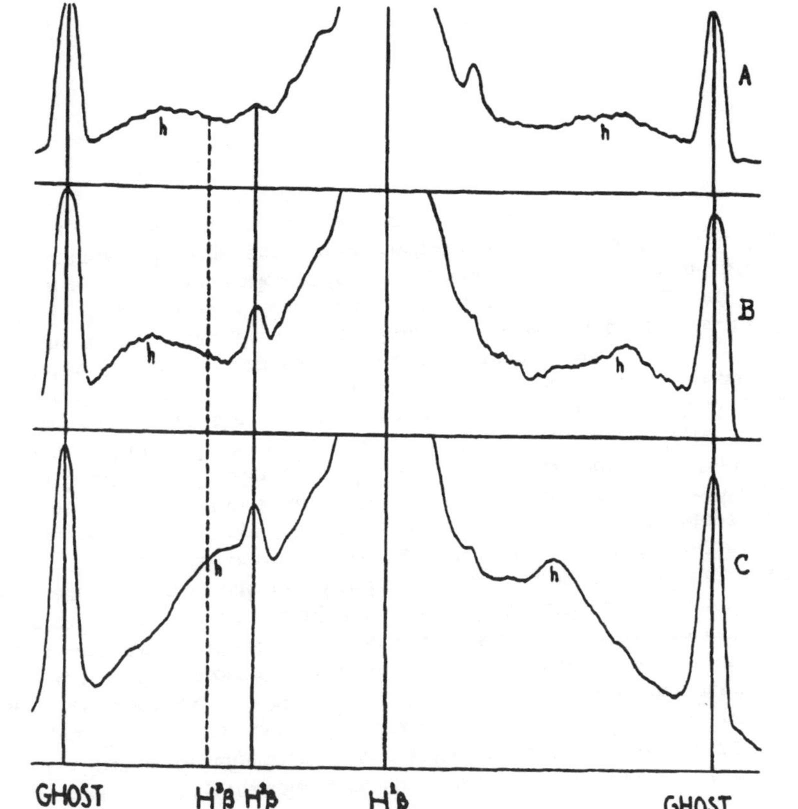

Fig. 4. Microphotometer curves of $H\beta$ for (A) ordinary hydrogen (B) Sample II (C) Sample III. The calculated position of $H^3\beta$ is indicated although there is no evidence for its existence from these curves. The *h*'s indicate regions of halation.

are in the order $A > B > C$. The heights of the microphotometric curves of the ghosts in Fig. 4 would not seem to substantiate this statement. The ghost

[10] Walker Bleakney (Bull. Am. Phys. Soc., Boston meeting) has found that the relative abundance in sample III is 1:1100 ±10 percent. This was determined after this paper was written. We have not revised our estimate since it was our best judgment based on our exposure times, but we believe his estimate to be better than ours since his method is the more reliable.

curves of B and C are higher than that of A because of a more continuous background in A due partly to the different distribution of the halation. The fourth order ghosts on the plates are not complicated in this way and visually have the intensity order $A > B > C$. The line to the right of the main H^1 line with an intensity in A greater than in either B or C is a molecular line. The increase in the intensity of the $H^2\beta$ line for samples II and III as recorded by curves B and C over that for ordinary hydrogen, Curve A, can easily be seen.

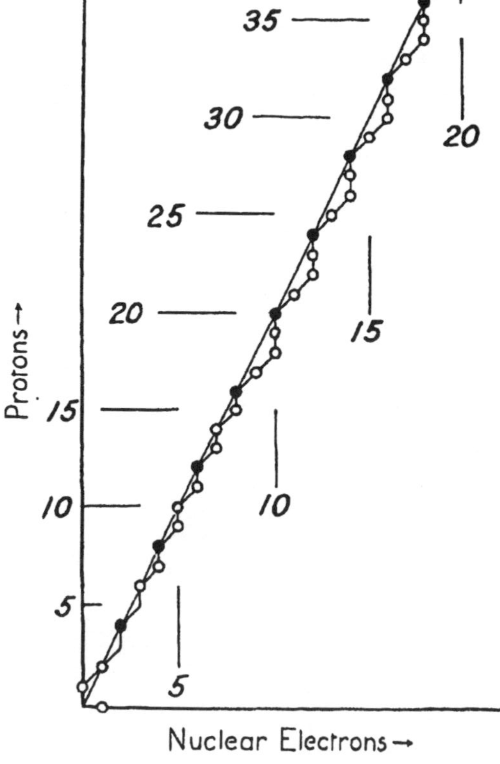

Fig. 5. The proton-electron plot of atomic nuclei.

This shows that the concentration of the H^2 isotope was markedly increased by evaporation at the triple point. The heights of the curves above the estimated continuous backgrounds are in the ratio $A:B:C = 4:16:17$, thus substantiating the estimates of increased concentration from exposure times.

A System of Atomic Nuclei

It is of interest to see how the H^2 nucleus fits into a system of atomic nuclei. Periodic systems have been proposed by several authors and are largely equivalent. The simple proton-electron plot shows regularities in a very good way and the accompanying figure (Fig. 5) shows the regularities up to A^{36}. The figure suggests that H^3 and He^5 should exist. No evidence for H^3 has as yet been found, but further concentration (see below) may yet show that this nuclear species exists. It should be possible to concentrate He^3 by the distillation of liquid helium, and this method may show that this nucleus also exists.

Other Methods for Concentrating the Heavier Isotopes of Hydrogen

It seems entirely feasible to construct a fractionating column that will greatly increase the efficiency of the distillation method for separating these isotopes. This method has the distinct advantage that it is capable of producing large samples. On the other hand, it requires rather large volumes of gas, so that after the isotope has been concentrated in small volumes by the fractional distillation and rectification of liquid hydrogen, further concentration may be better carried out using diffusion methods. Stern and Vollmer[21] used such a method in an attempt to find isotopes of hydrogen and oxygen, working on the hypothesis that the non-integral atomic weight of hydrogen might be due to a higher isotope. They report that a heavier hydrogen isotope is not present to the extent of 1 part in 100,000. Their negative result emphasizes the difficulties of diffusion methods which for success require carefully controlled conditions. Such an apparatus as has been described by Hertz[22] should be very effective for the separation of the hydrogen isotopes. Work is in progress on the construction of such an apparatus for further concentration beyond the state that we can reach with distillation methods.

The authors take pleasure in acknowledging their indebtedness to the Physics Department of Columbia University for the grating and other facilities used in this work and for its cooperation. We are also indebted to the Chemistry Department of New York University and, to Mr. R. L. Garman in particular, for the microphotometer curves.

[21] O. Stern and M. Vollmer, Ann. d. Physik (4) 59, 225 (1919).
[22] G. Hertz, Zeits. f. Physik 19, 35 (1923).

Chapter 4
ATOMIC PHYSICS
EDWARD GERJUOY

Introduction

Papers Reprinted in Book

Pieter Zeeman (1865-1943) in Holland, 1925. (Courtesy of Niels Bohr Institute, AIP Emilio Segrè Visual Archives.)

Edward Gerjuoy is Professor of Physics (Emeritus) at the University of Pittsburgh. He is a theoretical physicist whose research has spanned a number of areas, including nuclear physics, plasma physics, and acoustics, but the bulk of his publications have been in atomic and molecular collision theory. He is a Fellow of The American Physical Society and in the past has chaired several Society Committees. Presently he is a member of the Society's Panel on Public Affairs and Vice Chair of its Forum on Physics and Society.

ATOMIC PHYSICS

Robert W. Wood (1868-1955). (Courtesy of AIP Emilio Segrè Visual Archives.)

This overview surveys progress in atomic physics during the past century, with special reference to those *Physical Review* and *Physical Review Letters* publications (hereafter jointly denoted by the abbreviation "*PR*") that have been significant components of this progress. The domain of atomic physics is very broad, however; in particular, the subject matter of atomic physics includes subfields such as spectroscopy, fundamental quantum theory, laser physics, and nonrelativistic collisions (between electrons, atoms, and molecules, among other particles). Atomic physics also contributes importantly to, and receives important input from, many other broad areas of physical science: atmospheric physics, plasma physics, and chemical physics, for example. For these reasons the limited number of pages allotted to this overview prevents it from being a genuinely comprehensive survey of atomic physics; hopefully this overview nevertheless will convey some sense of the wide scope and great significance of the atomic physics publications that have been included in the CD-ROM accompanying the book. This overview cannot make specific reference to every atomic physics paper included in the CD-ROM, however; no inference whatsoever should be drawn from any such lack of mention. On the other hand, unless otherwise stated, every paper explicitly identified below has been included in the CD-ROM, and some have been included in the book as well.

EARLY THEORY. ATOMIC AND MOLECULAR SPECTRA

The foregoing understood, this overview begins with a discussion of the early (before about 1940) theoretical contributions to atomic physics that have been selected. Attainment of a more than rudimentary understanding of atomic physics necessarily had to await the advent of nonrelativistic quantum theory in its modern formulation, including the Schrödinger equation, the assignment of spin 1/2 to the electron, and the exclusion principle. Each of these crucial discoveries was published in European physics journals during the years 1925–1926. With very few exceptions, therefore, seminal atomic physics theory papers meriting inclusion in the CD-ROM did not appear in the pages of the *PR* before 1925. [One such exception is the contribution of Tolman (1924), showing that the value of the spontaneous transition probability from the upper level of a given spectral line, i.e., the Einstein A coefficient, could be inferred from line absorption measurements.] During the period 1925 through 1940, however, *PR* did publish a number of papers on atomic physics theory meriting inclusion in the CD-ROM or even in the book.

The 1926 European papers by Schrödinger showed that for purely Coulombic electron-proton interactions the Schrödinger equation yielded stationary state energy levels for atomic hydrogen that were identical with the energy levels obtained with Bohr's "old" quantum theory, the energy levels of which agreed very well with spectroscopic measurements. It also was known, however, that Bohr's theory had utterly failed to yield the energy levels of atomic helium, or of any other polyelectron atoms. After Schrödinger's triumph with atomic hydrogen, therefore, the obvious immediate task for atomic theorists was to determine whether the Schrödinger equation could correctly predict the

Phillip M. Morse (1903-1985). Woods Hole, 1979. (Courtesy of AIP Emilio Segrè Visual Archives.)

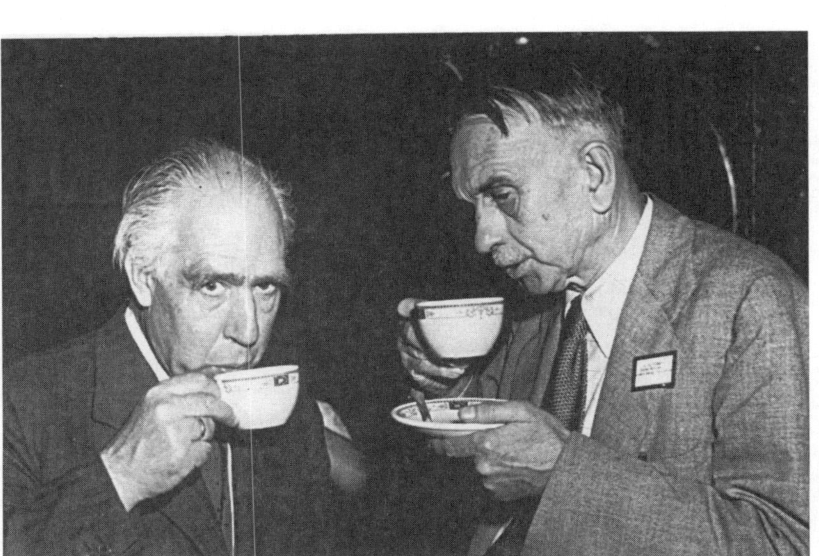

Niels Bohr (1885–1962) and **Richard C. Tolman (1881–1948).** (Courtesy of Princeton Photo Service, AIP Emilio Segrè Visual Archives.)

measured spectroscopic properties of polyelectron atoms, especially their spectral line frequencies, relative line intensities, and selection rules. But the Schrödinger equation cannot be solved analytically for any atomic system with more than one electron, even when all non-Coulombic (e.g., spin-orbit) interactions are neglected and the mass of the nucleus is set equal to infinity (so that the atomic center of mass always resides on the nucleus). Thus after 1926 the actual immediate task for atomic theorists was to develop methods for approximating the bound state wave functions and energy levels that exactly solve the Schrödinger equation for polyelectron atoms; moreover, these methods had to be usable with the limited computational facilities available at the time.

Contributions to this task began to be published in *PR* only a very few years after 1926. Indeed by the late 1920s and early 1930s *PR* already had published a number of significant advances in the theory of atomic and molecular spectra. Slater's papers probably were the most influential of these contributions; in particular, Slater [Phys. Rev. **34** (1929)] introduced the still widely employed "Slater determinants," which automatically satisfy the requirements of the exclusion principle, including spin. Other important theoretical calculations of atomic energy levels and/or energy level spacings include those of Breit and Wills (1933) and Van Vleck (1934), who applied his so-called "vector model" approach to molecules as well as atoms. The paper of Mulliken (1933) is one of a series that also made important contributions to molecular energy level calculations. Breit and Rabi (1931) derived the so-called Breit-Rabi equation for the splitting of atomic energy levels in an external magnetic field, a result that Rabi used in his first attempts to develop improved methods for measuring nuclear magnetic moments (see the discussion of MBMR in the following section).

John C. Slater (1900-1976) sitting, **Peter D. DeCicco (1939-1992)** and **George F. Koster (1927-)** standing. (Photograph by Ivan Massar Black Star, courtesy of The MIT Museum Collection.)

The molecular calculations described in the preceding paragraph were concerned primarily with the electronic components of the molecular energy when the nuclei of the atoms comprising the molecule lie in the vicinity of their normal equilibrium locations; other theoretical *PR* papers were concerned with the molecular rotational and vibrational energies, or with the shapes of interatomic potentials at internuclear separations far from equilibrium. In particular, Kronig and Rabi (1927) obtained analytic solutions to the Schrödinger equation for a rotating symmetric top (two equal principal moments of inertia), thereby yielding the exact rotational energies for a wide class of molecules; ammonia is an example. Morse (1929) introduced his famous "Morse potential" which, for a potential that very closely simulated actual internuclear interaction potentials, permitted exact analytic solutions of the Schrödinger equation for the nuclear vibrational energy levels of a diatomic molecule. Slater and Kirkwood (1931) considerably improved earlier calculations of the long-range van der Waals interactions between various pairs of atoms, e.g., two He atoms; their calculations later were further improved in the seminal Casimir and Polder paper (1948), which showed that taking the effects of retardation into account (which Slater and Kirkwood had not done) changes the long-range dependence of the van der Waals interaction from R^{-6} to R^{-7}.

The most far-reaching early advance in our understanding of molecular spectra, however, undoubtedly was that of Condon (1926), who introduced the so-called "Franck-Condon principle." In this paper Condon, following up on a proposal by Franck (not published in *PR*) for explaining experimental observations of molecular dissociation by photons, postulated that electronic transitions (and consequent photon emissions) between two different interatomic potential curves of a molecule necessarily occurred too rapidly for any internuclear distances to change appreciably. On this simple postulate alone, Condon was able to deduce many of the previously very puzzling features of observed molecular band spectra.

Since about 1940 calculating atomic and molecular energy levels has increasingly become the province of theoretical chemists rather than theoretical physicists. Nevertheless *PR* has continued to publish important papers on the theory of atomic and molecular spectra. Indeed, during the 1940s Racah published a series of papers, one of which [Racah, Phys. Rev. **61** (1942)] greatly advanced Slater's formulation of the theory of atomic spectra. Also, in 1958 and 1959 Pekeris—using the Rayleigh-Ritz variational minimum principle and taking advantage of the vast improvements in computing power since the 1930s (though his computers were pitiful compared to today's, of course)—computed the energies of the ground and first excited states of atomic He to a heretofore unheard-of accuracy of about ten significant figures, including relativistic and other corrections; his calculations, wherein he diagonalized 1000×1000 matrices, illustrate the growing practice of expanding the sought-for solutions of the

Giulio Racah (1909-1965).
(Courtesy of Igal Talmi.)

Joseph S. Levinger (1921-) and **Ugo Fano (1912-).** (Courtesy of *Physics Today* Collection, AIP Emilio Segrè Visual Archives.)

Schrödinger equation in function bases chosen less for their expected resemblance to the exact solutions than for their computational convenience.

EARLY EXPERIMENTS. MAGNETIC RESONANCE

Early (once again defined as before about 1940) progress in experimental atomic physics did not have to await the advent of nonrelativistic quantum theory in its modern formulation. Thus a number of important atomic physics experiments were published in *PR* before 1925. Many of the seminal early experiments, however, though falling under the atomic physics rubric—e.g., the discovery of the Compton effect [Compton (1923)] or the discovery of deuterium [Urey, Brickwedde, and Murphy (1932)]—are discussed elsewhere in this collection. Therefore the remainder of this overview will concentrate on those significant atomic physics experiments, whether before or after about 1940, that other overviews are unlikely to more than barely mention.

One early and very clever experimental paper is on the Lawrence and Beams (1928) measurement of the time delay for emission of an electron in the photoelectric effect; Lawrence and Beams inferred that this time delay was less than 3×10^{-9} s. Another experiment worth mentioning is that of Tate and Smith (1932), who passed electron beams through tubes containing gases (e.g., N_2, CO, and O_2) to measure the cross sections for ionization of the gases by electrons. Other than these the only early experimental papers published in *PR* that merit discussion are the marvelously ingenious and productive series of molecular beam measurements performed by Rabi and his co-workers, via what has become known as the "molecular beam magnetic resonance" (MBMR) method, which in this overview also denotes atomic beam measurements performed by basically the same magnetic resonance method.

H. B. G. Casimir (1909-). (Courtesy of *Physics Today* Collection, AIP Emilio Segrè Visual Archives.)

Rabi's point of departure was the so-called Stern-Gerlach type of molecular beam apparatus exemplified in the work of Estermann, O. C. Simpson, and Stern (1937). In this type of apparatus, a molecular beam traverses an inhomogeneous magnetic field, and the molecular magnetic moment is inferred from the transverse displacement of the beam. Rabi's initial modification of the type of apparatus Estermann used was the introduction of a second inhomogeneous magnetic field, whose gradient was opposite to the gradient of the first inhomogeneous field traversed by the beam [see Kellogg, Rabi, and Zacharias (1936)]; by adjusting the gradients and traversal lengths, the transverse deflections of the beam in the two inhomogeneous regions could be made to cancel, whereby the molecular magnetic moment could be inferred without errors stemming from uncertainties in the beam velocity distribution (which had plagued the Estermann experiment). Rabi further realized that this cancellation, once established, would be destroyed if the molecules in the beam then could be induced to make transitions to selected new magnetic sublevels while traversing the beam path between the two inhomogeneous regions: the original cancellation had taken advantage of the fact that the moments of the magnetic sublevels had the same value in those two regions. Rabi's MBMR method of accomplishing such transitions to new magnetic sublevels was to introduce into the space between the two inhomogeneous magnetic fields an additional, homogeneous, magnetic field, which also contained a tunable oscillating field. The sensitivity of the aforementioned destruction of cancellation to the oscillation frequency then permitted extremely accurate measurements of the energy spacing between the selected sublevels.

MBMR was first described in Rabi, Zacharias, Millman, and Kusch (1938); with this MBMR technique Rabi and his co-workers were able to measure accurately not only the nuclear magnetic moments of the proton, deuteron, and other nuclei, e.g., Li^6 and Li^7, but also the quadrupole moment of the deuteron [see Kellogg, Rabi, Ramsey, and Zacharias, Phys. Rev. **55** (1939)]. Other representative important papers by the Rabi group [e.g., Rabi, Millman, Kusch, and Zacharias (1939) placed in the Science and Technology chapter] have been included in the CD-ROM. It is noteworthy that Rabi's ability to extract these nuclear moments from his experimental data rested strongly on his grasp of the early advances in molecular spectroscopy theory discussed in the first section; without a grasp of that theory, especially of how level splittings in magnetic fields are related to the various spin and orbital angular momentum quantum numbers characterizing the levels, Rabi would not have known how to interpret the (sharply resonant) oscillating magnetic field frequencies at which—for any given molecular species in his beam—there was marked destruction of the transverse deflection cancellation. In addition, to infer the deuteron quadrupole moment from his measured D_2 beam resonance frequencies, Rabi had to rely on calculated values (from the then best available D_2 molecular wave functions) of

Jerrold R. Zacharias (1905–1986). (Courtesy of MIT, AIP Emilio Segrè Visual Archives.)

Willis E. Lamb (1913-). (Courtesy of *Newsweek*, AIP Emilio Segrè Visual Archives.)

the electric field gradients that the molecular electrons produce at the deuterons.

Rabi-type MBMR experiments, with some improvements, continued to be published in *The Physical Review* after 1940. Ramsey (1949) pointed out that the sharpness of the MBMR resonances, and thus the accuracy of the MBMR experiments, could be increased significantly by having the molecular beam traverse two distinct oscillating field regions (rather than merely one such region, as in the conventional Rabi MBMR apparatus). With this sort of refinement, the distinctions between magnetic resonance measurements on molecules in beams and on molecules in bulk matter [as in "nuclear magnetic resonance" (NMR), which has revolutionized the imaging of biological tissues *in vivo*], began to blur. NMR, stemming from Purcell, Torrey, and Pound (1946) and Bloch, Hansen, and Packard (1946), is outside the scope of this

Willis E. Lamb (1913-), Alfred Kastler (1902-1984), and **Hans Kopfermann (1895-1963).** (Courtesy of Speck, Heidelberg, AIP Emilio Segrè Visual Archives.)

overview. Ramsey (1952) presented a quite sophisticated theoretical treatment of the so-called "chemical shifts" seen in nuclear magnetic resonance frequencies, a phenomenon of importance in both MBMR and NMR; as Ramsey discusses, these chemical shifts result from (in effect) magnetic shielding, by its surrounding moving and spinning electrons, of the magnetically resonating particle being studied.

COLLISIONS. EXPERIMENT AND THEORY

A great many collision cross sections and other reaction coefficients are required for the full understanding of, for instance, fusion plasmas or the Earth's upper atmosphere. Starting well before 1925, therefore, and continuing to this day, atomic physics experimentalists the world over have been measuring such cross sections and reaction coefficients, for application as just indicated, and for the illumination such measurements can cast on the physical phenomena occurring in those collisions. The variety of "atomic" collisions that have been examined is almost limitless, including (and I am by no means being exhaustive) elastic scattering, excitation, de-excitation, and ionization of many species of atoms and molecules (in ground or excited states) by incident electrons, positrons, mesons, photons, neutral atoms, and ions (singly or multiply charged); electron capture by singly or multiply charged positive ions incident on atoms and molecules; dissociative recombination of molecular ions with electrons; etc. Of the many such experimental papers that have been published in *PR*, only comparatively few have been deemed sufficiently seminal to warrant inclusion in the CD-ROM.

An instructive 1959 experiment by Phelps illustrates the vast amount of information about collision rates that can be extracted, using modern techniques, from measurements on a gas (in this case Ne) in a container. Analysis of the measured optical absorption coefficients of various Ne lines enabled Phelps to infer, for a number of Ne excited states, diffusion coefficients, cross sections for collisional de-excitation by other Ne atoms, cross sections for collisional de-excitation by thermal electrons, and three-body collision rates. Dehmelt (1958) also used optical absorption measurements to infer the spin exchange cross section for collisions between thermal electrons and spin-oriented sodium atoms; the same experiment furnished an early direct measurement of the free-electron gyromagnetic ratio, determinations of which are further discussed in the next section. Knowledge of electron spin exchange collision cross sections in gases, notably NO, was the basis for the first reported evidence of positronium formation [Deutsch (1951)]. Madden and Codling (1963), using a chamber containing He, observed unusual (at the time) resonances in the He vacuum ultraviolet absorption spectrum that, in accordance with the theory of Fano (1961), represented photoionization of He($1s^2$) to various doubly excited, nominally discrete autoionizing levels [e.g., He(2s2p)] embedded in the He continuum energy spec-

Martin Deutsch (1917-) with angular correlation apparatus at Massachusetts Institute of Technology, 1959. (Courtesy of AIP Emilio Segrè Visual Archives.)

trum; this experiment apparently was the first use of synchrotron radiation to obtain results of interest in atomic physics.

Modern techniques also have made it possible to supplement collision experiments of the sort just described, involving gases in containers, with experiments employing crossed beams of the colliding particles (which may include photons). One of the first examples of a modern crossed-beam atomic collision experiment is the Fite and Brackmann (1958) measurement of the cross section for ionization of H(1s) by electrons. Gallagher and Cooke (1979) used laser pulses to excite a beam of Na atoms to high-lying, singly excited Rydberg states below the continuum, e.g., Na($1s^2 2s^2 2p^6 18p$), which levels then were ionized using a suitably delayed voltage pulse; these procedures enabled Gallagher and Cooke to determine the lifetimes of individual singly excited Na Rydberg states, the lifetimes of which were shown to be greatly reduced from their normally computed (zero temperature) values as a result of stimulated emission and absorption by the room-temperature blackbody radiation.

Predictions of collision cross sections of all sorts have been and still are the objectives of a large fraction of active atomic physics theorists. Comparisons of such predictions with experiment are used to test the theory and to reveal hitherto ignored physical effects (such as the above-described reduction of Na Rydberg state lifetimes by blackbody radiation). Reliable theoretical estimates of reaction rates also are needed because many collision rates important for the understanding of, e.g., plasma or atmospheric physics, cannot be measured in the laboratory. Since they generally involve solutions of the Schrödinger equation at energies in the continuum, however, calculations of collision rates of interest in atomic physics generally are much more difficult than the bound-state calculations discussed in the first section; for instance, calculations in the continuum cannot take advantage of the Rayleigh-Ritz minimum principle that has proved so useful for bound-state calculations (recall the Pekeris papers discussed above). Indeed, the difficulties with the continuum are so great that even today collision-rate calculations, though typically formulated to be consistent with the requirements of the exclusion principle, often assume the interactions are the dominant Coulombic-type only (in other words, often wholly ignore, e.g., the spin-orbit interactions that Slater's early bound-state calculations already were able to take into account).

Wade L. Fite (1925–). (Courtesy of General Atomic/General Dynamics, AIP Emilio Segrè Visual Archives.)

For these reasons, significant publications in atomic physics collision theory did not appear in *PR* until after World War II. Many of these publications contain results of general interest to collision theorists in many fields, not merely to atomic collision theorists. The most obvious illustration of such a paper is the derivation [Lippmann and Schwinger (1950)] of the famous Lippmann-Schwinger equation, which reformulates the Schrödinger equation as an integral equation; unfortunately, use of the equation does not wholly avoid boundary condition problems in collisions involving more than two particles [i.e., in collisions even as simple as electron-H(1s) scattering], as Foldy and Tobocman (1957) showed. Another very important paper of general interest for collision theory is that of Kohn (1948), who derived a variational principle that is usable for computing collision amplitudes; although this variational principle is not a minimum principle like the Rayleigh-Ritz, and therefore has no built-in mechanism for deciding when a new approximation is an improvement over a previous approximation (as is possible in bound-state calculations), nevertheless the Kohn variational principle has been very widely employed in atomic collision calculations, often quite successfully. Furthermore, minimal (or maximal) principles do exist for some collision problems, as Rosenberg, Spruch, and O'Malley (1960) showed.

On the other hand, many atomic physics collision theory papers, like the 1961 Fano paper mentioned above, deserve notice in this overview even though they do not have as much general interest for collision theory as the papers discussed in the preceding paragraph. O'Malley, Rosenberg, and Spruch (1962) showed that the well-known "effective range" expansion for the low-energy elastic scattering amplitude, much employed in the analysis of nuclear collisions (which typically involve short-range forces plus a purely Coulombic r^{-1} long-range repulsion between the colliding nuclei), must be modified for atomic collisions (wherein the dominant interactions between the colliding bodies, though Coulombic in origin and still long range, typically decrease more rapidly than r^{-1}, e.g., as r^{-4}). Gerjuoy and Baranger (1957) suggested that Breit-Wigner-type resonances [see Breit and Wigner (1936)], which have proved so useful for the unraveling of nuclear collision cross sections, should also be observable in atomic collisions; experiments, notably the crossed-beam experiment by Schulz (1963), have supported this suggestion. In 1953 Wannier, on the basis of a plausible, purely classical analysis (no quantum mechanics), predicted that near the ionization threshold the cross section for ionization of a neutral atom by incident electrons should be proportional to E^k, where E is the energy above threshold and $k = (-1 + \sqrt{91/3})/4 \cong 1.127$. The possibility of such a power law (which, it has been said, "should be unknown to science"), has spawned a great number of experimental and theoretical atomic physics papers; as yet, however, the range of validity (if any) of the Wannier threshold law has not been definitively ascertained.

THE LAMB SHIFT. TESTS OF QUANTUM ELECTRODYNAMICS

Although Pasternack (1938) already had suggested (from analysis of hydrogen Balmer line observations) that the $2^2S_{1/2}$ level of atomic hydrogen lies above the $2^2P_{1/2}$ level, the actual unequivocal demonstration of this fact by Lamb and Retherford (1947) made their paper one of the most important atomic physics experiments ever published. Unfortunately, because of space limitations, only

Walter Kohn (1923–). (Courtesy of AIP Emilio Segrè Visual Archives.)

Rabi's molecular beam apparatus, 1938. (Courtesy of Research Corporation, AIP Emilio Segrè Visual Archives.)

the abstract of the definitive article (1950) could be in the book; the full text of this remarkable paper is on the CD-ROM. Theorists very soon recognized that the source of this deviation from the predictions of the Dirac equation was the interaction of the hydrogenic electron with the radiation field [see Bethe (1947)]; very soon thereafter, theorists were able to compute the Lamb shift using quantum electrodynamics (QED) in essentially its present formulation.

QED as such is beyond the scope of this overview; this overview does survey a few atomic physics experiments, however, that (like the Lamb-Retherford experiment) test QED predictions. In particular, great efforts have been expended on attempts to measure with high precision the deviation of the electron's gyromagnetic ratio g from the exact value 2 implied by the Dirac equation; the existence of such a deviation was first postulated by Breit (1947), though not from QED considerations. To lowest order in the fine structure constant α, QED predicts the "anomaly" $a_e \equiv (g - 2)/2 = \alpha/2\pi \cong 0.001162$ [Schwinger (1948)]. The first direct measurement of g - 2, using techniques available in the 1940s, namely MBMR, yielded $a_e = 0.00119 \pm 0.00005$ [Kusch and Foley (1947) and (1948)]. In 1961 Schupp, Pidd, and Crane, using free electrons trapped for periods up to 300 μs in a magnetic "bottle" (not bound electrons, subject to Ramsey's chemical shifts, as Kusch and Foley were forced to employ), were able to improve the experimental accuracy to $a_e = 0.0011609 \pm 0.0000024$. Only some 15 years later Van Dyck, Schwinberg, and Dehmelt (1977), now able to observe a *single* electron that could be trapped for hundreds of seconds [see Wineland, Ekstrom, and Dehmelt (1973)], obtained the (then) almost unbelievably accurate result of $a_e = 0.001159652410 \pm 200 \times 10^{-12}$.

Other experiments that should be mentioned include Rich and Crane (1966), who, trapping positrons via the techniques of

Schupp, Pidd, and Crane, showed that to the accuracy of their experiment a_e for the positron equals a_e for the electron, a result required by charge, parity, time (CPT) reversal invariance; and Hänsch *et al.* (1975), who—with the aid of so-called "Doppler-free two-photon spectroscopy"—measured the Lamb shift of the hydrogen $1^2S_{1/2}$ ground state, i.e., the deviation of the hydrogen ground-state energy from the ground-state energy predicted by the Dirac equation (Lamb and Retherford had furnished information about the excited state $2^2S_{1/2} - 2^2P_{1/2}$ splitting only).

Hans G. Dehmelt (1922-). (Photograph by Joseph Freeman, University of Washington. Courtesy of AIP Emilio Segrè Visual Archives.)

MASERS AND LASERS. COHERENCE

The invention of the maser [Gordon, Zeiger, and Townes (1954) and (1955)] rested on an understanding of the theory discussed in the first section for the energy levels of NH_3 (the molecule employed by Gordon, Zeiger, and Townes). In its ground-state electronic configuration—the only configuration of present concern—NH_3 has a complex line spectrum at microwave frequencies, resulting from splitting of the ground-state electronic energy by a variety of interactions, including most importantly the quadrupole coupling between the nitrogen nucleus and the molecular electrons. Gordon, Zeiger, and Townes were able to prepare a beam of NH_3 molecules in the ground-state electronic configuration that had inverted populations (more molecules in higher energy levels than in lower). When the beam was sent through a tunable oscillating microwave cavity, the power level in the cavity increased sharply (i.e., was amplified) when the cavity was tuned to an NH_3 spectral line frequency, because a net emission was being resonantly induced from those inverted populations. Gordon, Zeiger, and Townes invented the word "maser," an acronym for "microwave amplification by stimulated emission of radiation," to denote this amplifying operation of their apparatus. When the beam current was sufficiently high, however, the spontaneous micro-

Dickinson W. Richards, Jr. (1895-1973), Polykarp Kusch (1911-1993), T. D. Lee (1926-), I. I. Rabi (1898-1988) and **André F. Cournand (1895-1988)** at Columbia University sending Lee off to receive his Nobel Prize, 1957. (Courtesy of AIP Emilio Segrè Visual Archives.)

Nicolaas Bloembergen (1920–) and **Norman Ramsey (1915–).** (Courtesy of Harvard University, AIP Emilio Segrè Visual Archives.)

wave emission from the molecules entering the cavity was large enough to compensate for the microwave energy losses from the cavity, and the apparatus functioned as a very narrow band microwave oscillator, not merely as an amplifier.

These maser properties greatly increased the practicality of a number of important experiments. The hydrogen maser [Kleppner *et al.* (1965)], which employs a hydrogen atomic beam sent into a tunable cavity centered at the hyperfine splitting frequency of H(1s), has especially desirable properties and has enabled highly accurate measurements of that frequency {1420405751.800 ± 0.028 Hz according to Crampton, Kleppner, and Ramsey [Phys. Rev. Lett. **11**, 338 (1963)]} and of its Stark shift. A hydrogen maser launched in a spacecraft has been used to verify general relativity-predicted frequency shifts [Vessot *et al.* (1980)].

Only three years after the invention of the maser, Schawlow and Townes (1958) published a design for an "optical maser" (not yet called the "laser"), which they believed would be able to maintain sustained oscillations at visible frequencies by taking advantage of externally induced inverted population levels in a "cavity" formed by a narrow tube with highly reflecting ends. The first to report an operating gas laser were Javan, Bennett, and Herriott (1961), who employed a mixture of He and Ne in a gaseous discharge. So much has been written about lasers, in the popular as well as scientific literature, that it appears pointless to devote any of this overview's limited space to further words about the laser's underlying physics, or to a summary of the myriad laser applications in science and technology. Semiclassical and more fully quantized theories of the laser have been given by Lamb (1964) and by Scully and Lamb (1967), respectively. An interesting scientific application of laser technology is that of Williams *et al.* (1976), who used lunar laser ranging to perform another test of general relativity. A number of "laser spectroscopy" schemes have been proposed for reducing the linewidths that complicate conventional spectroscopic observations. These schemes include Doppler-free two-photon spectroscopy, mentioned in an earlier section; "laser saturation spectroscopy," used by Hänsch *et al.* (1974) to obtain a precision value for the Rydberg (namely 109737.3143 ± 0.0010 cm^{-1}); and so-called "laser cooling" [Neuhauser *et al.* (1978)] to greatly reduce the velocities of the radiating atoms or molecules under study. Laser "trapping" of atoms and molecules was first envisioned by Ashkin (1970), who was able to laser trap micron-sized particles via radiation pressure alone.

Masers and lasers manifest their marvelous properties because they manage to induce many seemingly uncoupled atoms or molecules to radiate in concert, i.e., "coherently"; correspondingly, these devices generate coherent states of the radiation field. Dicke (1954) gave one of the first instructive discussions of coherent radiation. The credit for the first thorough development of the quantum theory of optical coherence is owed to Glauber [Phys. Rev. **130** (1963)], however [see also Glauber, Phys. Rev. **131**

(1963)]. Coherent states of the radiation field are related to so-called "squeezed states," wherein the fluctuation noise associated with one component of the radiation field is reduced at the expense of an increase in the fluctuation noise of a conjugate component [see Yuen (1976)]. Lasers have made it possible to observe other somewhat unintuitive consequences of radiation field coherence, including so-called "quantum beats," wherein photons of different frequencies appear to be interfering in a manner akin to acoustic beats. It is worth noting, however, that such quantum beats, between light waves originating from seemingly independent sources, were first demonstrated by Forrester, Gudmundsen, and P. O. Johnson (1955) before the invention of the laser, following a suggestion by Forrester, Parkins, and Gerjuoy (1947). In the Forrester apparatus, the light from a discharge wherein the Zeeman components had been separated by a magnetic field illuminated a photoelectric surface; the photoelectric current was passed through a microwave cavity, which manifested a sharp increase in output signal when (via variation of the magnetic field) the Zeeman splitting frequency was varied through the cavity resonant frequency. The Forrester experiment also was able to infer that the photoelectron-emission time delay was less than 10^{-10} s, an improvement on the Lawrence and Beams result discussed previously.

CONCLUSION. RECENT TRENDS

The atomic physics papers that have been discussed in this overview date predominantly from before 1970, for a number of reasons: Atomic physics is a long-established field; in any field of physics it is difficult to feel confident about the lasting merit of a paper that has been published too recently; and, the criteria for inclusion of a *PR* paper here rule out any that were published after 1983. Consequently, this overview, while hopefully meeting its objective of conveying a sense of the wide scope and great significance of the atomic physics papers that have been published in *The PR*, does not cover recent trends in atomic physics research. I will conclude, therefore, with a brief (certainly not comprehensive) survey of some recent atomic physics trends, especially those that seem likely to continue: (i) No let up in the bread-and-butter business of measuring and theoretically predicting collision cross sections and reaction coefficients of all sorts; the colliding particles are becoming more complicated, however (e.g., the collisions increasingly are involving heavier multiatom molecules), and chemists, both experimentalists and theorists, are increasingly performing such research. (ii) Increasing interest, theoretical and experimental, in the interactions of intense laser beams with atoms and molecules; the highly nonlinear couplings of the intense laser electromagnetic fields are difficult to treat theoretically and produce unexpected experimental results. (iii) Attempts to observe manifestations of chaos in quantum systems. (iv) Increasing interest in reactions of all kinds involving atoms in

Peter A. Franken (1928–). (Courtesy of AIP Gallery of Member Society Presidents, AIP Emilio Segrè Visual Archives.)

Chaim L. Pekeris (1908–1993). (Courtesy of *Physics Today* Collection, AIP Emilio Segrè Visual Archives.)

highly excited Rydberg states; such reactions test approximations (e.g., certain classical treatments), which are rarely examined in collisions involving atoms in lower states and also are thought to provide a vehicle for investigating quantum chaos. (v) Employment of polarized particles (atoms, electrons, etc.) in experimental investigations of collision rates; such experiments can yield otherwise unobtainable results and put greater strains on the theory. (vi) Use of high-energy particle accelerators to perform atomic physics experiments; by using merged-beam techniques, such experiments can even yield collision cross sections at very low energies in the center-of-mass system. (vii) Ever more accurate measurements of fundamental quantities, including some not mentioned in this overview (e.g., of the Lamb shift in He^+), taking advantage of laser cooling, single-atom trapping, etc.; such experiments should provide new tests not only of QED, but also of proposed theories violating parity conservation and time-reversal invariance. (viii) And finally, more frequent undertakings by atomic physics theorists of arduous "*a priori*" numerical computations that rely on the largest and fastest available computers and have little or no recourse to physically insightful approximations.

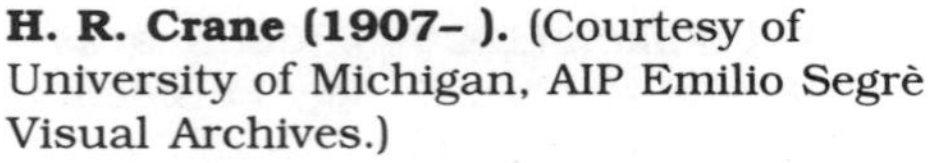

H. R. Crane (1907–). (Courtesy of University of Michigan, AIP Emilio Segrè Visual Archives.)

ATOMIC PHYSICS

Papers Reproduced on CD-ROM

P. S. Olmstead and K. T. Compton. Radiation potentials of atomic hydrogen, *Phys. Rev.* **22**, 559–565 (1923)

R. C. Tolman. Duration of molecules in upper quantum states, *Phys. Rev.* **23**, 693–709 (1924)

E. Condon. A theory of intensity distribution in band systems, *Phys. Rev.* **28**, 1182–1201 (1926)

R. de L. Kronig and I. I. Rabi. The symmetrical top in the undulatory mechanics, *Phys. Rev.* **29**, 262–269 (1927)

E. O. Lawrence and J. W. Beams. The element of time in the photoelectric effect, *Phys. Rev.* **32**, 478–485 (1928)

P. M. Morse. Diatomic molecules according to the wave mechanics. II. Vibrational levels, *Phys. Rev.* **34**, 57–64 (1929)

J. C. Slater. The theory of complex spectra, *Phys. Rev.* **34**, 1293–1322 (1929)

J. C. Slater. Atomic shielding constants, *Phys. Rev.* **36**, 57–64 (1930)

S. Goudsmit. Theory of hyperfine structure separations, *Phys. Rev.* **37**, 663–681 (1931)

J. C. Slater and J. G. Kirkwood. The van der Waals forces in gases, *Phys. Rev.* **37**, 682–697 (1931)

G. Breit and I. I. Rabi. Measurement of nuclear spin, *Phys. Rev.* **38**, 2082–2083(L) (1931)

J. T. Tate and P. T. Smith. The efficiencies of ionization and ionization potentials of various gases under electron impact, *Phys. Rev.* **39**, 270–277 (1932)

R. S. Mulliken. Electronic structures of polyatomic molecules and valence. II. General considerations, *Phys. Rev.* **41**, 49–71 (1932)

D. M. Dennison and G. E. Uhlenbeck. The two-minima problem and the ammonia molecule, *Phys. Rev.* **41**, 313–321 (1932)

R. S. Mulliken. Electronic structures of polyatomic molecules and valence. IV. Electronic states, quantum theory of the double bond, *Phys. Rev.* **43**, 279–302 (1933)

L. A. DuBridge. Theory of the energy distribution of photoelectrons, *Phys. Rev.* **43**, 727–741 (1933)

G. Breit and L. A. Wills. Hyperfine structure in intermediate coupling, *Phys. Rev.* **44**, 470–490 (1933)

E. U. Condon and C. W. Ufford. Relative multiplet transition probabilities from spectroscopic stability, *Phys. Rev.* **44**, 740–743 (1933)

J. H. Van Vleck. The Dirac vector model in complex spectra, *Phys. Rev.* **45**, 405–419 (1934)

G. M. Murphy and H. Johnston. The nuclear spin of deuterium, *Phys. Rev.* **45**, 550(L) (1934)

R. F. Bacher and S. Goudsmit. Atomic energy relations. I., *Phys. Rev.* **46**, 948–969 (1934)

J. M. B. Kellogg, I. I. Rabi, and J. R. Zacharias. The gyromagnetic properties of the hydrogens, *Phys. Rev.* **50**, 472–485 (1936)

I. I. Rabi. Space quantization in a gyrating magnetic field, *Phys. Rev.* **51**, 652–654 (1937)

I. Estermann, O. C. Simpson, and O. Stern. The magnetic moment of the proton, *Phys. Rev.* **52**, 535–545 (1937)

I. I. Rabi, J. R. Zacharias, S. Millman, and P. Kusch. A new method of measuring nuclear magnetic moment, *Phys. Rev.* **53**, 318(L) (1938)

S. Pasternack. Note on the fine structure of Hα and Dα, *Phys. Rev.* **54**, 1113(L) (1938)

J. M. B. Kellogg, I. I. Rabi, N. F. Ramsey, Jr., and J. R. Zacharias. An electrical quadrupole moment of the deuteron, *Phys. Rev.* **55**, 318–319(L) (1939)

R. P. Feynman. Forces in molecules, *Phys. Rev.* **56**, 340–343 (1939)

N. F. Ramsey, Jr. The rotational radiofrequency spectra of H_2, D_2, and HD in magnetic fields, *Phys. Rev.* **58**, 226–236 (1940)

G. Racah. Theory of complex spectra. I., *Phys. Rev.* **61**, 186–197 (1942)

G. Racah. Theory of complex spectra. II., *Phys. Rev.* **62**, 438–462 (1942)

G. Racah. Theory of complex spectra. III., *Phys. Rev.* **63**, 367–382 (1943)

W. E. Lamb, Jr. and R. C. Retherford. Fine structure of the hydrogen atom by a microwave method, *Phys. Rev.* **72**, 241–243 (1947)

A. T. Forrester, W. E. Parkins, and E. Gerjuoy. On the possibility of observing beat frequencies between lines in the visible spectrum, *Phys. Rev.* **72**, 728(L) (1947)

G. Breit. Does the electron have an intrinsic magnetic moment?, *Phys. Rev.* **72**, 984(L) (1947)

P. Kusch and H. M. Foley. Precision measurement of the ratio of the atomic '*g* values' in the $^2P_{3/2}$ and $^2P_{1/2}$ states of gallium, *Phys. Rev.* **72**, 1256–1257(L) (1947)

H. B. G. Casimir and D. Polder. The influence of retardation on the London–van der Waals forces, *Phys. Rev.* **73**, 360–372 (1948)

J. Schwinger. On quantum-electrodynamics and the magnetic moment of the electron, *Phys. Rev.* **73**, 416–417(L) (1948)

E. P. Wigner. On the behavior of cross sections near thresholds, *Phys. Rev.* **73**, 1002–1009 (1948)

P. Kusch and H. M. Foley. The magnetic moment of the electron, *Phys. Rev.* **74**, 250–263 (1948)

T. A. Welton. Some observable effects of the quantum-mechanical fluctuations of the electromagnetic field, *Phys. Rev.* **74**, 1157–1167 (1948)

W. Kohn. Variational methods in nuclear collision problems, *Phys. Rev.* **74**, 1763–1772 (1948)

J. B. French and V. F. Weisskopf. The electromagnetic shift of energy levels, *Phys. Rev.* **75**, 1240–1248 (1949)

ATOMIC PHYSICS

M. A. Biondi and S. C. Brown. Measurements of ambipolar diffusion in helium, *Phys. Rev.* **75**, 1700–1705 (1949)

J. H. Gardner and E. M. Purcell. A precise determination of the proton magnetic moment in Bohr magnetons, *Phys. Rev.* **76**, 1262–1263(L) (1949)

G. Racah. Theory of complex spectra. IV, *Phys. Rev.* **76**, 1352–1365 (1949)

M. A. Biondi and S. C. Brown. Measurement of electron–ion recombination, *Phys. Rev.* **76**, 1697–1700 (1949)

L. L. Foldy and S. A. Wouthuysen. On the Dirac theory of spin 1/2 particles and its non-relativistic limit, *Phys. Rev.* **78**, 29–36 (1950)

B. A. Lippmann and J. Schwinger. Variational principles for scattering processes. I., *Phys. Rev.* **79**, 469–480 (1950)

W. E. Lamb, Jr. and R. C. Retherford. Fine structure of the hydrogen atom. Part I, *Phys. Rev.* **79**, 549–572 (1950)

J. C. Slater. A simplification of the Hartree–Fock method, *Phys. Rev.* **81**, 385–390 (1951)

M. Deutsch. Evidence for the formation of positronium in gases, *Phys. Rev.* **82**, 455–456(L) (1951)

R. Sternheimer. On nuclear quadrupole moments, *Phys. Rev.* **84**, 244–253 (1951)

E. E. Salpeter and H. A. Bethe. A relativistic equation for bound-state problems, *Phys. Rev.* **84**, 1232–1242 (1951)

M. Deutsch and S. C. Brown. Zeeman effect and hyperfine splitting of positronium, *Phys. Rev.* **85**, 1047–1048(L) (1952)

J. Brossel and F. Bitter. A new "double resonance" method for investigating atomic energy levels. Application to Hg $^3P_1$, *Phys. Rev.* **86**, 308–316 (1952)

W. P. Allis and S. C. Brown. High frequency electrical breakdown of gases, *Phys. Rev.* **87**, 419–424 (1952)

M. Camac, A. D. McGuire, J. B. Platt, and H. J. Schulte. X-rays from mesic atoms, *Phys. Rev.* **88**, 134(L) (1952)

G. H. Wannier. The threshold law for single ionization of atoms or ions by electrons, *Phys. Rev.* **90**, 817–825 (1953)

W. B. Hawkins and R. H. Dicke. The polarization of sodium atoms, *Phys. Rev.* **91**, 1008–1009(L) (1953)

R. H. Dicke. Coherence in spontaneous radiation processes, *Phys. Rev.* **93**, 99–110 (1954)

J. P. Gordon, H. J. Zeiger, and C. H. Townes. Molecular microwave oscillator and new hyperfine structure in the microwave spectrum of NH_3, *Phys. Rev.* **95**, 282–284(L) (1954)

P.-O. Löwdin. Quantum theory of many-particle systems. I. Physical interpretations by means of density matrices, natural spin-orbitals, and convergence problems in the method of configurational interaction, *Phys. Rev.* **97**, 1474–1489 (1955)

E. P. Wigner. Lower limit for the energy derivative of the scattering phase shift, *Phys. Rev.* **98**, 145–147 (1955)

J. P. Gordon. Hyperfine structure in the inversion spectrum of $N^{14}H_3$ by a new high-resolution microwave spectrometer, *Phys. Rev.* **99**, 1253–1263 (1955)

S. H. Autler and C. H. Townes. Stark effect in rapidly varying fields, *Phys. Rev.* **100**, 703–722 (1955)

H. Ekstein. Theory of time-dependent scattering for multichannel processes, *Phys. Rev.* **101**, 880–890 (1956)

J. P. Wittke and R. H. Dicke. Redetermination of the hyperfine splitting in the ground state of atomic hydrogen, *Phys. Rev.* **103**, 620–631 (1956)

H. G. Dehmelt. Paramagnetic resonance reorientation of atoms and ions aligned by electron impact, *Phys. Rev.* **103**, 1125–1126(L) (1956)

P. Franken and S. Liebes, Jr. Magnetic moment of the proton in Bohr magnetons, *Phys. Rev.* **104**, 1197–1198(L) (1956)

L. L. Foldy and W. Tobocman. Application of formal scattering theory to many-body problems, *Phys. Rev.* **105**, 1099–1100 (1957)

E. Baranger and E. Gerjuoy. Helium excitation cross sections near threshold, *Phys. Rev.* **106**, 1182–1185 (1957)

H. G. Dehmelt. Spin resonance of free electrons polarized by exchange collisions, *Phys. Rev.* **109**, 381–385 (1958)

E. Gerjuoy. Outgoing boundary condition in rearrangement collisions, *Phys. Rev.* **109**, 1806–1814 (1958)

W. L. Fite and R. T. Brackmann. Collisions of electrons with hydrogen atoms. I. Ionization, *Phys. Rev.* **112**, 1141–1151 (1958)

C. L. Pekeris. Ground state of two-electron atoms, *Phys. Rev.* **112**, 1649–1658 (1958)

A. V. Phelps. Diffusion, de-excitation, and three-body collision coefficients for excited neon atoms, *Phys. Rev.* **114**, 1011–1125 (1959)

C. L. Pekeris. 1 ^1S and 2 ^3S states of helium, *Phys. Rev.* **115**, 1216–1221 (1959)

L. Rosenberg, L. Spruch, and T. F. O'Malley. Upper bounds on scattering lengths when composite bound states exist, *Phys. Rev.* **118**, 184–192 (1960)

F. T. Smith. Lifetime matrix in collision theory, *Phys. Rev.* **118**, 349–356 (1960)

A. V. Phelps and A. O. McCoubrey. Experimental verification of the "incoherent scattering" theory for the transport of resonance radiation, *Phys. Rev.* **118**, 1561–1565 (1960)

L. W. Anderson, F. M. Pipkin, and J. C. Baird, Jr. Hyperfine structure of hydrogen, deuterium, and tritium, *Phys. Rev.* **120**, 1279–1289 (1960)

A. A. Schupp, R. W. Pidd, and H. R. Crane. Measurement of the *g* factor of free, high-energy electrons, *Phys. Rev.* **121**, 1–17 (1961)

ATOMIC PHYSICS

U. Fano. Effects of configuration interaction on intensities and phase shifts, *Phys. Rev.* **124**, 1866–1878 (1961)

T. F. O'Malley, L. Rosenberg, and L. Spruch. Low-energy scattering of a charged particle by a neutral polarizable system, *Phys. Rev.* **125**, 1300–1310 (1962)

R. J. Glauber. The quantum theory of optical coherence, *Phys. Rev.* **130**, 2529–2539 (1963)

G. J. Schulz. Resonance in the elastic scattering of electrons in helium, *Phys. Rev. Lett.* **10**, 104–105 (1963)

R. P. Madden and K. Codling. New autoionizing atomic energy levels in He, Ne, and Ar, *Phys. Rev. Lett.* **10**, 516–518 (1963)

S. B. Crampton, D. Kleppner, and N. F. Ramsey. Hyperfine separation of ground-state atomic hydrogen, *Phys. Rev. Lett.* **11**, 338–340 (1963)

C. Lovelace. Practical theory of three-particle states. I. Nonrelativistic, *Phys. Rev.* **135**, B1225–B1249 (1964)

K. Codling and R. P. Madden. Optically observed inner shell electron excitation in neutral Kr and Xe, *Phys. Rev. Lett.* **12**, 106–108 (1964)

U. Fano and W. Lichten. Interpretation of Ar^{+}-Ar collisions at 50 KeV, *Phys. Rev. Lett.* **14**, 627–629 (1965)

A. Rich and H. R. Crane. Direct measurement of the g factor of the free positron, *Phys. Rev. Lett.* **17**, 271–275 (1966)

V. Franco. Diffraction theory of scattering by hydrogen atoms, *Phys. Rev. Lett.* **20**, 709–712 (1968)

R. K. Nesbet. Anomaly-free variational method for inelastic scattering, *Phys. Rev.* **179**, 60–70 (1969)

A. Ashkin. Acceleration and trapping of particles by radiation pressure, *Phys. Rev. Lett.* **24**, 156–159 (1970)

H. J. Andrä. Zero-field quantum beats subsequent to beam-foil excitation, *Phys. Rev. Lett.* **25**, 325–327 (1970)

R. Cohen, J. Lodenquai, and M. Ruderman. Atoms in superstrong magnetic fields, *Phys. Rev. Lett.* **25**, 467–469 (1970)

G. W. F. Drake. Theory of relativistic magnetic dipole transitions: Lifetime of the metastable $2\,^3S$ state of the heliumlike ions, *Phys. Rev. A* **3**, 908–915 (1971)

J. Macek and D. H. Jaecks. Theory of atomic photon-particle coincidence measurement, *Phys. Rev. A* **4**, 2288–2300 (1971)

G. Feinberg and J. Sucher. Calculation of the decay rate for $2\,^3S_1 \rightarrow 1\,^1S_0$ + one photon in helium, *Phys. Rev. Lett.* **26**, 681–684 (1971)

P. Lambropoulos. Spin-orbit coupling and photoelectron polarization in multiphoton ionization of atoms, *Phys. Rev. Lett.* **30**, 413–416 (1973)

D. Wineland, P. Eckstrom, and H. Dehmelt. Monoelectron oscillator, *Phys. Rev. Lett.* **31**, 1279–1282 (1973)

T. A. Patterson, H. Hotop, A. Kasdan, D. W. Norcross, and W. C. Lineberger. Resonances in alkali negative-ion photodetachment and electron affinities of the corresponding neutrals, *Phys. Rev. Lett.* **32**, 189–192 (1974)

T. W. Hänsch, M. H. Nayfeh, S. A. Lee, S. M. Curry, and I. S. Shahin. Precision measurement of the Rydberg constant by laser saturation spectroscopy of the Balmer α line in hydrogen and deuterium, *Phys. Rev. Lett.* **32**, 1336–1340 (1974)

S. B. Crampton and H. T. M. Wang. Duration of hydrogen-atom spin-exchange collisions, *Phys. Rev. A* **12**, 1305–1312 (1975)

P. F. Liao and J. E. Bjorkholm. Direct observation of atomic energy level shifts in two-photon absorption, *Phys. Rev. Lett.* **34**, 1–4 (1975)

T. W. Hänsch, S. A. Lee, R. Wallenstein, and C. Wieman. Doppler-free two-photon spectroscopy of hydrogen $1S$-$2S$, *Phys. Rev. Lett.* **34**, 307–309 (1975)

S. A. Lee, R. Wallenstein, and T. W. Hänsch. Hydrogen $1S$-$2S$ isotope shift and $1S$ Lamb shift measured by laser spectroscopy, *Phys. Rev. Lett.* **35**, 1262–1266 (1975)

H. P. Yuen. Two-photon coherent states of the radiation field, *Phys. Rev. A* **13**, 2226–2243 (1976)

E. M. Henley and L. Wilets. Parity nonconservation in Tl and Bi atoms, *Phys. Rev. A* **14**, 1411–1417 (1976)

E. J. Kelsey and L. Spruch. Retardation effects on high Rydberg states: A retarded R^{-5} polarization potential, *Phys. Rev. A* **18**, 15–25 (1976)

R. S. Van Dyck, Jr., P. B. Schwinberg, and H. G. Dehmelt. Precise measurements of axial, magnetron, cyclotron, and spin-cyclotron-beat frequencies on an isolated 1-meV electron, *Phys. Rev. Lett.* **38**, 310–314 (1977)

L. L. Lewis, J. H. Hollister, D. C. Soreide, E. G. Lindahl, and E. N. Fortson. Upper limit on parity-nonconserving optical rotation in atomic bismuth, *Phys. Rev. Lett.* **39**, 795–798 (1977)

P. E. G. Baird, M. W. S. M. Brimicombe, R. G. Hunt, G. J. Roberts, P. G. H. Sandars, and D. N. Stacey. Search for parity-nonconserving optical rotation in atomic bismuth, *Phys. Rev. Lett.* **39**, 798–801 (1977)

D. J. Wineland, R. E. Drullinger, and F. L. Walls. Radiation-pressure cooling of bound resonant absorbers, *Phys. Rev. Lett.* **40**, 1639–1642 (1978)

K. Bhadra, J. Callaway, and R. J. W. Henry. Electron-impact excitation of n=2 levels of helium at intermediate energies, *Phys. Rev. A* **19**, 1841–1851 (1979)

T. F. Gallagher and W. E. Cooke. Interactions of blackbody radiation with atoms, *Phys. Rev. Lett.* **42**, 835–839 (1979)

C. Wieman and T. W. Hänsch. Precision measurement of the $1S$ Lamb shift and of the $1S$-$2S$ isotope shift of hydrogen and deuterium, *Phys. Rev. A* **22**, 192–205 (1980)

W. D. Phillips and H. Metcalf. Laser deceleration of an atomic beam, *Phys. Rev. Lett.* **48**, 596–599 (1982)

J. V. Prodan, W. D. Phillips, and H. Metcalf. Laser production of a very slow monoenergetic atomic beam, *Phys. Rev. Lett.* **49**, 1149–1153 (1982)

A THEORY OF INTENSITY DISTRIBUTION IN BAND SYSTEMS*

BY EDWARD CONDON

ABSTRACT

A theory of the relative intensity of the various bands in a system of electronic bands is developed by an extension of an idea used by Franck in discussing the dissociation of molecules by light absorption. The theory predicts the existence of two especially favored values of the change in the vibrational quantum numbers, in accord with the empirical facts as discussed by Birge.

A means of calculating the intensity distribution from the known constants of the molecule is presented and shown to be in semi-quantitative agreement with the facts in the case of the following band systems: SiN, AlO, CO (fourth positive group of carbon), I_2 (absorption), O_2 (Schumann-Runge system), CN (violet system), CO (first negative group of carbon), N_2 (second positive group of nitrogen), and N_2 (first negative group of nitrogen).

In the case of I_2 there is a discrepancy, if Loomis' assignment of n'' values is used, which does not appear if Mecke's original assignment is used. It is suggested that at least some of the lower levels postulated by Mecke are real but that absorption from them always results in dissociation of the molecule and so they are not represented in the quantized absorption spectrum.

1. INTRODUCTION

THE problem of the explanation of relative or absolute intensities of spectral lines stands out today as one of the most important in the whole field of spectroscopy. As yet the problem has hardly received the necessary attention from experimental workers to make attempts at quantitative theoretical treatment profitable. The experimental difficulties of constructing controllable sources of excited materials seem to be yielding but slowly to the tremendous efforts toward their removal which are now being made. However, as is known, considerable progress has already been made in the discussion of relative intensities in the fine structure of lines (notably hydrogen Balmer lines in the Stark effect) and more recently in the relative intensities in multiplets in line spectra and of the various lines in the fine structure of individual bands.[1]

When the fine structure of lines or bands is under discussion the experimental problem is considerably easier, for in this case the difficulties of the unknown variation of photographic sensitivity with wave-

* A preliminary account of this paper was presented at the meeting of the Pacific Coast Section of the American Physical Society at Stanford University in March, 1926.

[1] Sommerfeld, Atombau und Spektrallinien. 4th ed. Chap. 5.

length do not make themselves felt appreciably. But when relative intensities are desired over long stretches of the spectrum, as between the members of a series of lines in atomic spectra or as in a system of bands in molecular spectra, the difficulties of photometric work present themselves in their full force. The result is that today there are few quantitative data in this field. In the case of the individual bands of a system,—the special subject of this article,—one has only the rough visual estimates of the observers; sometimes nothing better than a knowledge of the presence or absence of the band in question under very roughly described conditions of excitation and observation.

It seems worth while, however, to undertake an attempt at a theoretical correlation of the outstanding characteristics of the observed data with the quantum theory model for the emitter of the band spectra of diatomic molecules. The theory presented here is one of the relative likelihood of the various possible changes in the vibrational quantum number and makes no attempt to explain the distribution of molecules in the initial state. The distribution in the initial state, in cases approaching thermal equilibrium, should be governed by the Maxwell-Boltzmann distribution law. This is apparently the case *within the accuracy of the known data* for absorption spectra, where deviations from thermal equilibrium are least important.[2] The Maxwell-Boltzmann law seems also to hold remarkably well in some cases of violent electrical excitation in discharge tubes where one hardly expects to find thermal equilibrium. Perhaps this is a reflection of the fact that the Maxwell-Boltzmann law is one of a sort of maximum chaos, and it matters not whether the chaos results from disordered heat motions or irregular electrical conditions. To be more concrete, if n' is the vibration quantum number of the excited electronic state of the molecule and n'' that of the normal state of the molecule, then the theory presented here undertakes the discussion of what Birge has called an n' progression for the emission process or an n'' progression for the absorption process.

One expects the considerations of which use is made in the model discussed to be of an approximate nature only. They are based on certain kinematic relationships existing between the vibratory motions in the two electronic states concerned. As the simple mode of discussing the model is one which, without quantum theory, would yield fractional changes in the vibrational quantum numbers, the question arises as to the proper modification of the results of the continuous theory in order

[2] Chapter IV, Sec. 4, of National Research Council Report on Molecular Spectra (in press). The writer is indebted to Professor R. T. Birge for the opportunity to use the manuscript copy of this report.

to fit the discontinuous quantum theory. There is as yet no known way of treating this problem so the present results are necessarily approximate.

In view of the fact that the latest development in the disentanglement of the quantum enigma is in the direction of the substitution of another kinematics[3] for the description of mechanical systems whose sizes are of the order of Angstrom units it is thought best not to dwell too long on the attempt to formulate this treatment exactly. Rather it is recorded here as giving results in striking and partially satisfactory agreement with the facts although with the expectation that a better treatment must come through the application of the newer kinematics of Heisenberg, Born, and Jordan.

The modern quantum theory interpretation of electronic bands as presented in Sommerfeld's "Atombau"[4] and in the National Research Council Report on Molecular Spectra (in press) is adopted in this paper. The quantization of the motion of a diatomic molecule leads to energy levels associated with each electronic configuration which are specified by a rotational and a vibrational quantum number, m and n respectively, as given by the following formulas:

$$\frac{E}{hc}=A_n=B_n m^2=D_n m^4=F_n m^6= \cdots$$

$$A_n=\omega^0 n(1-\alpha n+\cdots)$$

$$B_n=B_0-\alpha n+\cdots$$

$$D_n=D_0+\cdots$$

All of the expressions are, in fact, infinite series but the terms indicated suffice for the analysis of most band systems. The derivatives of the potential energy curve at its equilibrium point are given in terms of the empirically determined, ω^0, x, B_0, α $\cdots$ by the formulas:[5]

$$V'(r_0)=0$$

$$V''(r_0)=c^2(2\pi\omega^0)^2\mu$$

$$V'''(r_0)=-\frac{6V''(r_0)\pi\sqrt{2B_0\mu c}}{\sqrt{h}}\left(\frac{\alpha\omega^0}{6B_0^2}+1\right)$$

$$V^{IV}(r_0)=\frac{2}{V''(r_0)}\left[\frac{5}{6}[V'''(r_0)]^2-8\frac{x}{ch\omega_0}[V''(r_0)]^3\right]$$

[3] Heisenberg, Zeits. f. Physik 33, 879 (1925); Born and Jordan, Zeits. f. Physik 34, 858 (1925); L. de Broglie, Ann. d. Physique 3, 22 (1925); Schrödinger, Ann. der Physik 79 459 (1926).

[4] Sommerfeld, Atombau, 4th ed., Chap. 9.

[5] The formulas are given by Born and Hückel, Phys. Zeits. 24, 1 (1925). The nomenclature used here is that of the Report on Molecular Spectra.

in which c is the velocity of light and μ is the harmonic mean of the masses of the two atoms in the molecule.

2. General Theory for Abrupt Structural Changes

The theory of transition probability which will now be developed is an outgrowth of a picture proposed by Franck for a mechanism for the dissociation of molecules by absorption of light.[6] Briefly Franck's picture is as follows:

In a "cold" gas the molecules are not vibrating and are in their lowest electronic level so that the nuclear motions are governed by $V_1(r)$. If now a light quantum is absorbed which is of sufficient energy to bring the molecule into an electronic excited state, it is natural to suppose that in thus changing the electronic energy of the molecule no other specific action is exerted by the light on the molecule. The absorption of light merely substitutes a new law of nuclear interaction, say $V_2(r)$ for the old one $V_1(r)$. But since this new one has a different equilibrium position, the atoms, at the instant after the absorption, will be away from equilibrium and so start to vibrate. Should it happen that

$$V_2(r_{01})>V_{2\max}$$

where $V_{2\max}$ is the maximum value of $V_2(r)$ in the range $r_{01}<r<\infty$ then the molecule will tend to execute oscillations of infinite amplitude, i.e., the molecule will dissociate. Thus Franck contemplates the photochemical dissociation of a molecule by a simple absorption of light.

It is natural to extend this point of view to the effect of an electron transition in a molecule either in absorption or emission and whether the molecule be vibrating in the initial state or not. The electron transition is supposed to happen in a negligibly short time as compared to the period of the nuclear vibrations. If the transition occurs at the instant when the separation of the atoms is r and the relative momentum is p_r then one supposes that the transition does not alter the instantaneous values of r and p_r but merely substitutes a new potential energy function, say $V_2(r)$, for the old one, say $V_1(r)$. The values of r and p_r at the instant of transition determine exactly a vibrational motion in the final state.

In general, of course, the vibrational motion so determined is not one of the states allowed by the quantum conditions. It becomes necessary, therefore, to suppose that the vibrational motion of the final state is not strictly governed by this principle but that it merely tends to take up the quantized vibration nearest to the one indicated by this principle.

[6] Franck, Trans. Faraday Society (1925).

Here a certain ambiguity appears which is akin to that which occurs in all other cases where attempts are made to reason accurately from properties of the orbits, other than their energies.

It is easy to give a general analytical formulation of the action governing the vibrational transition. Disregarding rotation and writing p for p_r, and q for r one has for the Hamiltonian functions of the initial and final states:

$$H_1=\frac{p^2}{2\mu}+V_1(q)$$

$$H_2=\frac{p^2}{2\mu}+V_2(q)$$

If the motion in either state is solved by the Hamilton-Jacobi method leading to the introduction of the action and angle variables, J_1w_1 for the initial state and J_2w_2 for the final state one finds relations of the form

$$\left.\begin{aligned}p&=p_1(J_1,w_1)\\ q&=q_1(J_1,w_1)\end{aligned}\right\}\quad\text{for the initial state,}$$

and similar relations for the final state. On the assumption underlying the work, if the electron transition takes place at an instant when the initial values of J and w are J_1,w_1 then the values of J and w after the transition, i.e., J_2 and w_2 will be given by the equations:

$$p_2(J_2,w_2)=p_1(J_1,w_1)$$

$$q_2(J_2,w_2)=q_1(J_1,w_1)$$

Or solving these equations for J_2 and w_2 one finds that:

$$J_2=J_2(J_1,w_1)$$

$$w_2=w_2(J_1,w_1)$$

so that the final motion is wholly determined by the initial motion. By the quantum conditions J_1 equals n_1h and J_2 equals n_2h where n_1 and n_2 are integers. It is thus seen that the solution for J_2 will not, in general, give an integral value. The phase of the final motion as given by w_2 plays no rôle in the spectroscopic theory. Moreover w_1 enters through a function which is periodic, with period 1 with respect to this variable. Thus the full range of possible values of J_2 corresponding to any J_1 is obtained by letting w_1 range from 0 to 1. From the general dynamical theory w_1 increases linearly with the time. Failing any other indications on this point, one may say that the electron transition is just as likely

to occur at one instant as at another, and thus there is a definite probability assigned to each range of J_2 to J_2+dJ_2.

It is clear that if all values of w_1 are equally probable then the values of J_2 corresponding to the small values of dJ_2/dw_1 will be strongly weighted. But $dJ_2/dw_1=0$ give the maxima and minima of the possible transitions so that these extremes are to be regarded as more probable than the other allowed jumps.

But the uncertainty of how to treat the fractional n_2 and the inaccuracy of the experimental data make it fruitless to attempt to consider such detailed questions as the accurate relative probabilities of the transitions. For a given J_1 there will be a maximum and minimum value of J_2 given by the theory. Moreover the most probable transitions, it will appear later, are to the extremes of J_2 if the times of electron transition be taken as equally probable. Therefore in discussing the relationship of the theory to empirical facts, only the extremes of J_2 for each J_1 have been computed.

Looking over the theory as just developed, it is to be emphasized that at least the extrema of J_2 for each J_1 are governed entirely by the energy functions $V_1(r)$ and $V_2(r)$. These in turn, as has been pointed out, are fixed by the positions of the bands in the spectrum. Thus there are no adjustable constants appearing in the theory so that *the vibrational transition probabilities are sharply correlated with the structure of the band system as regards position in the spectrum.*

The actual solution of the equations which present themselves in the case of the non-harmonic oscillator is prohibitively lengthy. In the following section the explicit formulas are developed for the case in which all derivatives of $V_2(r)$ and $V_1(r)$, higher than the second, vanish (harmonic oscillator). A graphical mode of treating the non-harmonic case is also presented.

3. Explicit Formulas for the Harmonic Oscillator

In the case in which $V_1(r)$ and $V_2(r)$ are parabolic the modes of vibration are simple harmonic in both initial and final states and the explicit development of the formulas is quite simpie. Thus let us assume:

$$V_1(q)=\tfrac{1}{2}K_1(q-q_{01})^2$$

$$V_2(q)=\tfrac{1}{2}K_2(q-q_{02})^2$$

The solution for p and q in terms of J and w for the harmonic oscillator is well-known.[7] One has for the final state:

[7] Born, Vorlesungen über Atommechanik, pp. 39, 57.

$$p_2(J_2, w_2) = \sqrt{2\mu\nu_2 J_2}\cos 2\pi w_2$$

$$q_2(J_2, w_2) = q_{02} + \frac{1}{2\pi}\sqrt{\frac{2J_2}{\mu\nu_2}}\sin 2\pi w_2$$

in which the frequency ν_2 is related to the mass μ and the spring constant K_2 by

$$\nu_2 = \frac{1}{2\pi}\sqrt{\frac{K_2}{\mu}}$$

Similarly for the initial state:

$$p_1(J_1 w_1) = \sqrt{2\mu\nu_1 J_1}\cos 2\pi w_1$$

$$q_1(J_1 w_1) = q_{01} + \frac{1}{2\pi}\sqrt{\frac{2J_1}{\mu\nu_1}}\sin 2\pi w_1$$

Equating the values of p and q at the instant of the electron transition and writing d for $q_{01} - q_{02}$:

$$\sqrt{2\mu\nu_2 J_2}\cos 2\pi w_2 = \sqrt{2\mu\nu_1 J_1}\cos 2\pi w_1$$

$$\frac{1}{2\pi}\sqrt{\frac{2J_2}{\mu\nu_2}}\sin 2\pi w_2 = d + \frac{1}{2\pi}\sqrt{\frac{2J_1}{\mu\nu_1}}\sin 2\pi w_1$$

Solving for J_2 in terms of J_1 and w_1 by elimination of w_2:

$$J_2 = 2\pi^2\mu\nu_2 d^2 + \frac{1}{2}\left(\frac{\nu_1}{\nu_2} + \frac{\nu_2}{\nu_1}\right)J_1 + 2d\sqrt{2\pi^2\mu\nu_2\frac{\nu_2}{\nu_1}}\sin 2\pi w_1 \cdot \sqrt{J_1}$$
$$+ \frac{1}{2}\left(\frac{\nu_1}{\nu_2} - \frac{\nu_2}{\nu_1}\right)\cos 4\pi w_1 \cdot J_1$$

The solution for w_2 in terms of J_1 and w_1 is readily obtained but it is not needed. Introducing the quantum integers n_1 and n_2 in place of J_1 and J_2:

$$n_2 = \frac{2\pi^2\mu\nu_2}{h}d^2 + \frac{1}{2}\left(\frac{\nu_1}{\nu_2} + \frac{\nu_2}{\nu_1}\right)n_1 + 2d\sqrt{\frac{2\pi^2\mu\nu_2}{h} \cdot \frac{\nu_2}{\nu_1}}\sin 2\pi w_1 \cdot \sqrt{n_1}$$
$$+ \frac{1}{2}\left(\frac{\nu_1}{\nu_2} - \frac{\nu_2}{\nu_1}\right)\cos 4\pi w_1 \cdot n_1$$

For brevity this will be written:

$$n_2 = A + Bn_1 + C\sqrt{n_1}\sin\theta + Dn_1\cos 2\theta,$$

defining the coefficients A, B, C, D by comparison with the preceding

equation. The extreme values of n_1 corresponding to any n_2 are found by the roots of $dn_1/d\theta = 0$ in the usual way. If $C/4D\sqrt{n_1}| > 1$

$$n_2 = A + Bn_1 \pm C\sqrt{n_1} \mp Dn_1$$

and if $|C/4D\sqrt{n_1}| < 1$ in addition to these also secondary maxima, given by $n_2 = (A - C^2/8D) \pm C^2/4D + (B+D)n_1$ the $\pm$ sign being taken the same as the sign of $C/4D$.

As has already been remarked, the extreme allowed transitions are the most probable ones and therefore these equations for the extrema give the bands which one expects to be strongest in a band system.

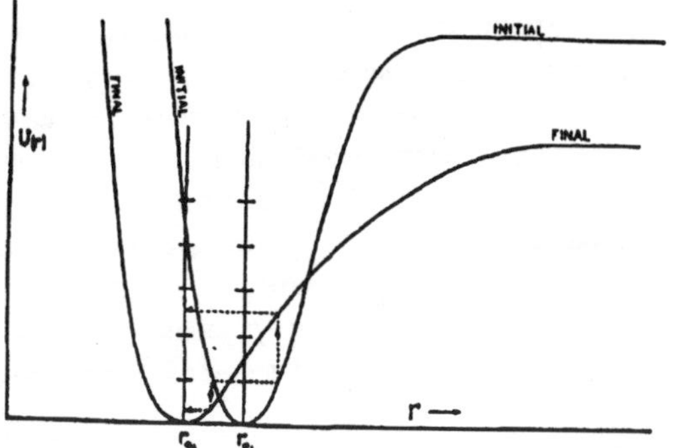

Fig. 1. Typical relation of the potential energy curves, illustrating graphical method of finding favored transitions.

The formulas for the non-harmonic oscillator can be obtained by successive approximations in a purely analytical way but it is easier to resort to an approximate graphical method. $V_1(r)$ and $V_2(r)$ are supposed to have been determined, in the neighborhood of r_{01} and r_{02} respectively from the analysis of the energy levels of the band system. Suppose these plotted on the same piece of paper. Also suppose the vibrational energy levels marked off as in Fig. 1. Then when the molecule vibrates in either electronic state with a known number of quanta it is easy to see what the amplitude of the motion is, for from the energy integral, $p = 0$ when

$$V(r) = W$$

if W is the energy of the vibratory motion. It is at the two extreme positions that the vibrator spends most of the time, i.e., the electronic transition is most likely to occur when the vibrator is in one of these extreme positions. At these extreme values of r for a vibratory motion given by $V_2(r)$, if $V_1(r)$ is suddenly made the law governing the motion then one sees directly from the figure what will be the amplitude and hence the energy of the new motion. Thus two most probable transitions are determined for each value of the quantum number in the initial state.

This graphical method, as will appear later, indicates how very sensitive the theory is to slight uncertainties in the functions, $V(r)$. In the next section are given the details of the application of the theory to all systems for which the necessary data are available.

4. Application to Known Systems

It will now be seen how far the model employed accounts for the main observed features of intensity distribution in band systems. These have

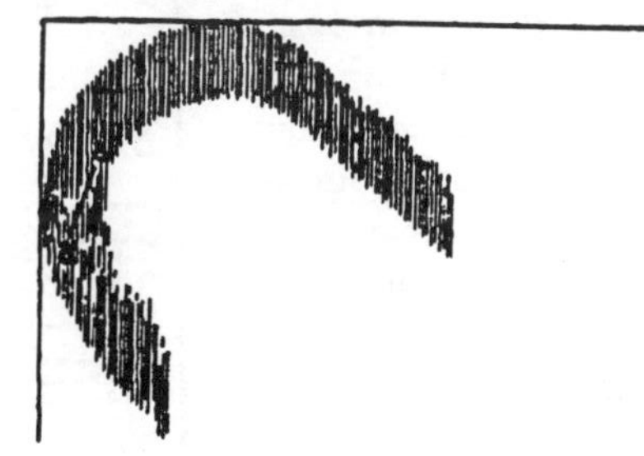

Fig. 2. Typical distribution of intensity in band system (n' ordinates, downward; n'', abscissas).

been discussed by Birge[8] who points out that it is a general rule that for each value of n' (the quantum number of the initial state for emission) there are two preferred values of $n'-n''$. On a double entry table as normally used, the n' being plotted downward as ordinates and the n'' to the left as abscissas the locus of the strong bands is a parabolic looking curve; i.e., the strong bands form a locus somewhat like the shaded part in Fig. 2. The size of this general locus with regard to the coordinate scale varies greatly from system to system, the branches being coincident for SiN, slightly separated for AlO, widely separated for CO while the scale for iodine is so great that only the part near the origin (a in the figure) is known. In this section the connection between these observed distributions and the theoretical predictions will be examined.

[8] Birge, Phys. Rev. **25**, 240 (1925), Abstract No. 23, also Chap. IV, Sec. 4, Report on molecular spectra.

The band systems to be treated are in order, SiN, silicon nitride; AlO, aluminum oxide; CO, 4th positive group of carbon; I_2, iodine; also O_2, Schumann-Runge system, CN, violet cyanogen; CO^+, first negative group of carbon, N_2, second positive group of nitrogen and N_2^+, first negative group of nitrogen. Of these nine systems, the first four are typical of the increasing scale of the intensity distribution and are discussed in greater detail. In CO and I_2 the deviations from the harmonic law of force are considerable. For these cases corrections for the deviations are discussed as fully as the data allow. For the remaining five systems the observed intensity data together with the theoretical curve calculated from an assumption of harmonic force law are presented as additional examples of the semi-quantitative correctness of the theory.

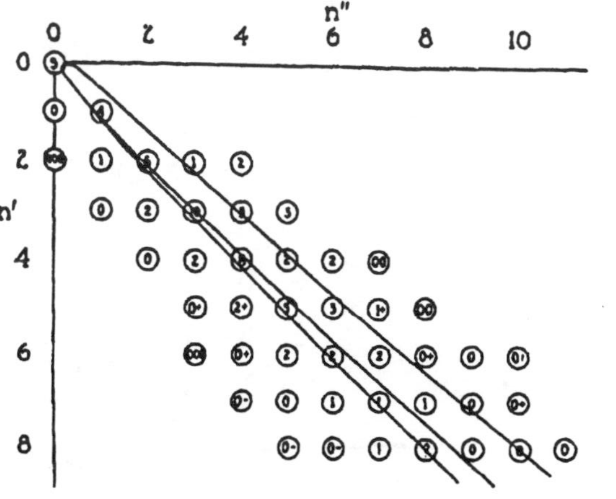

Fig. 3. The band system of silicon nitride.

In Table I are collected together all of the molecular constants which are needed for a discussion of the systems according to the harmonic force law. In every case the figures are taken from a large table of constants in Chapter IV of the National Research Council Report on Molecular Spectra. In the last four columns of the table are given the four constants A, B, C, D which occur in the formula for n'' in terms of n' as computed for the emission process. C.G.S. units are employed. A quantity with $'$ refers to the initial state and $''$ to the final state of the emission process. $\omega^{0\prime}$ is the ν_1 of the previous section expressed in cm^{-1} instead of sec^{-1}, similarly for $\omega^{0\prime\prime}$ and ν_2.

Turning now to the first system, one has available the estimated intensities by Mulliken.[9] These are plotted in Fig. 3. Here there is but a

[9] Mulliken, Phys. Rev. **26**, 319 (1925); Jevons, Proc. Roy. Soc. **89A**, 187 (1913–14).

TABLE I.—*Summary of molecular constants*

Carrier	μ $\times 10^{24}$	ν_e^* cm^{-1}	$\omega^{0'}$ cm^{-1}	$\omega^{0''}$ cm^{-1}	I_0' $\times 10^{40}$	I_0'' $\times 10^{40}$	r_0' $\times 10^8$	r_0'' $\times 10^8$	d $\times 10^8$	A	B	C	D	$(C/4D)^2$
SiN	15.417	24234.2	1016.3	1145.0	37.8	37.2	1.567	1.554	+0.013	0.0269*	1.007	+ 0.349	−0.119	0.537
AlO	16.60	20635.3	864.8	970.	46.01	43.38	1.665	1.617	+0.048	0.335	1.007	+ 1.23	−0.115	7.1
CO	11.317	64721.	1499.28	2147.74	17.3	14.9	1.24	1.17	+0.07	1.075	1.065	+ 2.485	−0.366	2.88
I_2	104.7	15598.3	126.52	213.76	1210.	787.	3.40	2.74	+0.66	88.0	1.140	+24.4	−0.548	124.
					1115.	720.5	3.622	2.622	+0.640	82.8		+23.62		
O_2	13.17	49359.3	708.	1565.	34.22	19.2	1.61	1.21	+0.40	29.8	1.331	+16.24	−0.879	21.4
CN	10.67	25799.8	2143.9	2055.64	14.140	14.647	1.151	1.172	−0.021	0.875	1.001	− 1.833	+0.043	113.
CO^+	11.317	45637.7	1704.42	2197.03	15.4	14.05	1.17	1.11	+0.06	0.809	1.033	+ 2.045	−0.256	3.98
N_2	11.558	29653.1	2018.66	1718.40	15.24	16.98	1.149	1.212	−0.063	0.712	1.013	− 1.56	+0.161	5.87
N_2^+	11.558	25565.9	2396.0	2187.4	13.35	14.41	1.075	1.117	−0.042	0.403	1.004	− 1.213	+0.092	10.9

* ν_e gives the frequency of the origin of the 0-0 band.

single ridge of strong bands. The figure shows the curve of most probable values of n'' as a function of n', calculated from the molecular constants according to the theory. As the theoretically double-branched curve is here scarcely an integer in width, it is natural that the double branch should have merged into one for the rough correspondence in question. This molecule is one which shows a doubling of the lower branch due to the cos 2θ term in the formula. The strong bands coincide exactly with this doubled line.

The next typical case is that of aluminum oxide. Here the experimental data both of Birge and also of Eriksson and Hulthén show clearly the double ridge of strong bands.[10] Here the curve computed from the molecular constants for the locus of the strong bands is open enough so that the branches should appear distinct, as in fact they are. The lower branch of the theoretical curve agrees precisely with the data, while the upper branch is slightly high. This distribution of intensity is by far the most common in the systems of bands which have thus far been analyzed on the quantum theory, as can be seen in Table I from inspection of the table of values of A, B, C, and D.

Extending the sequence of increasing scale for the distribution curve, the next example is the great system of the fourth positive group of carbon which is now known to be due to carbon mo-

[10] Birge has made estimates from unpublished spectrograms taken by him at the University of Wisconsin. Also Eriksson and Hulthén, Zeits. f. Physik 34, 775 (1925).

noxide.[11] In this one system, which extends from about 2650A to 1300A there are known more than 150 different bands. The great extent of this system over the frequency scale (40,000 cm^{-1}) naturally furnishes a severe test of any intensity theory. Unfortunately there are here no even semi-quantitative intensity data. In fact no one observer has ever photographed the entire system. Because of its great extent, however, the main features of the distribution are indicated by a diagram which shows simply which bands are observed. This accordingly has been done in Fig. 6. The full line gives the curve of the intensity distribution as computed by the simple theory based on the harmonic law of force.

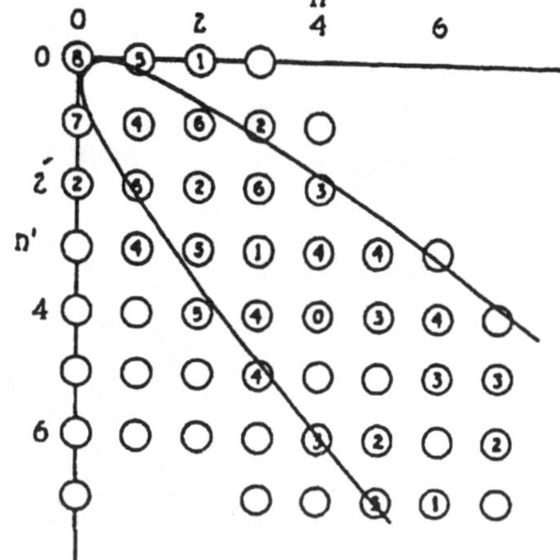

Fig. 4. The band system of AlO.

The analysis of the fine structure of the bands is so incomplete that it is only possible in a rough way to find the corrections due to deviations from the harmonic force law. In fact no fine structure work has been carried out on this system. Data on the initial state may be obtained since this, according to the analysis of Birge, is also the final state of the Angstrom CO bands. The moment of inertia in the final state is obtained from the infra-red CO rotation vibration bands. Besides the data given in Table I, the following figures on the 4th positive group are used in this work:

$$x'\omega^{0'} = 17.24 \qquad \alpha' = 0.023$$
$$x''\omega^{0''} = 12.703 \qquad \alpha'' = ?$$

By means of the formulas for the coefficients in the law of force already given, it is possible from these data to compute force law curves for both

[11] Birge, Nature 117, 229 (1926); Lowry, J.O.S.A. 8, 647 (1924).

states, but with some ambiguity and considerable uncertainty in the final state, for which no α value is known. Writing the expansion for the potential energy in the form

$$V=u_2(r-r_0)^2+u_3(r-r_0)^3+u_4(r-r_0)^4+\cdots$$

in which the unit of energy is spectroscopic wave numbers (cm^{-1}) while $(r-r_0)$ is expressed in Angstrom units one readily finds from the formulas previously given the following expressions for u_2 u_3 and u_4 in terms of the band structure constants:

$$u_2=\frac{10^{-16}V''(r_0)}{2!hc}=[0.95641-3](\omega^{0 2}\mu)$$

$$u_3=\frac{10^{-24}V'''(r_0)}{3!hc}=-[0.27924-1]u_2\sqrt{B_0\mu}\left(\frac{\alpha\omega^0}{6B_0^2}+1\right)$$

$$u_4=\frac{10^{-32}V^{IV}(r_0)}{4!hc}=\frac{1}{u_2}\left(\frac{5}{4}u_3^2-\frac{8}{3}u_2^3\frac{x}{\omega^0}\right)$$

In these expressions the unit of μ is 10^{-24} gm as in Table I, and the quantities in brackets are the logarithms of the coefficients of the corresponding quantities.

When the formulas are applied to the known data for the initial state in CO one obtains the following values:

$$u_2'=2.30\times10^5$$
$$u_3'=-6.00 \text{ ''}$$
$$u_4'=+8.84 \text{ ''}$$

For the final state a value of α is lacking. In computing the potential energy curve what seems like a reasonable value of α has been used, namely 0.02. This compares well with the known values for systems which have been analyzed. The coefficients found on this assumption are:

$$u_2''=4.718\times10^5$$
$$u_3''=-12.59 \text{ ''}$$
$$u_4''=+25.66 \text{ ''}$$

In Fig. 5 are drawn to scale the potential energy curves of the initial and final states in a form suitable for the application of the graphical method of predicting the intensity distribution. In Fig. 6 is given in the curve marked "corrected law" the theoretical position of the two ridges of intense bands as obtained by this method. It is clear that the corrected law improves the agreement for the lower branch while leaving almost unaltered the good fit of the other branch.

Passing now to the band system of iodine,[12] the largest known band system, it is essential to use the higher terms in the law of force since the change in the moment of inertia between the initial and the final states is very large.

The calculations make use of the following data in addition to that given in the table:

$$\alpha'=0.00015 \qquad x'\omega^{0 1'}=0.85$$
$$\alpha''=0.00011 \qquad x''\omega^{0 11'}=0.592$$

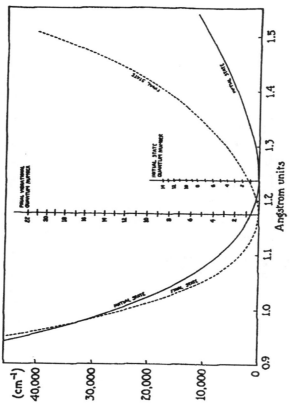

Fig. 5. The potential energy curves for the two electronic states involved in the fourth positive group of carbon (CO).

The resulting values of the coefficients in the potential energy function are:

	Initial	Final
u_2:	$+1.537\times10^4$	$+4.313\times10^4$
u_3:	-2.906 "	-6.002 "
u_4:	$+3.580$ "	$+4.014$ "

[12] The analysis is due to Mecke, Ann. d. Physik 71, 104 (1923), with Loomis' revision of the quantum assignment. The values of B_0^0 and α are due to Loomis (Chapter VI, Report on Molecular Spectra), that of α'', however, having been previously given by Kratzer and Sudholt, Zeits. f. Physik 33, 144 (1925). As Loomis points out, the values of α are quite uncertain. For new observations on resonance spectra, see Dymond, Zeits. f. Physik, 34, 553 (1925).

As in the case of CO, there are no intensity measurements available for this system of bands. The data are due to Mecke and give merely the presence or absence of the various bands. The system is thus plotted in Fig. 7. Here the data are for the *absorption* spectrum rather than for emission as in the examples previously given. To be consistent with the other examples, however, the terms "initial" and "final" in the foregoing set of constants have been chosen so as to refer to the *emission* process. In the absorption process, n'' becomes the initial quantum number and n' the final. Mecke has already commented on the peculiar intensity distribution in I_2, his Fig. 4 contrasting it with the more typical dis-

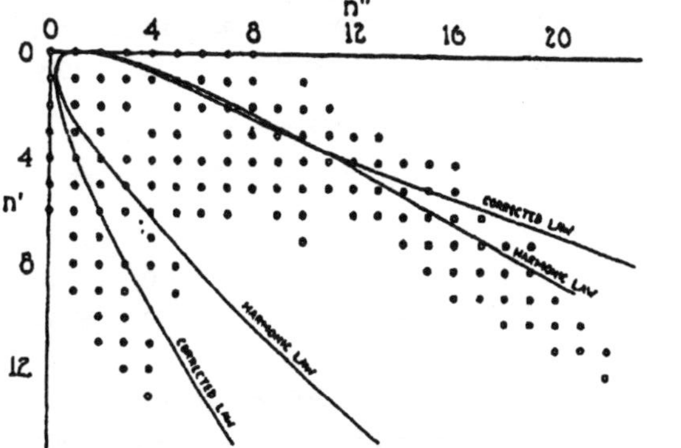

Fig. 6. The observed bands of the fourth positive group of carbon.

tribution of intensity of the violet CN-system. The distribution of intensity is, of course, profoundly modified by the distribution of the molecules among the initial states. This accounts for the great observed decrease in intensity of successive n' progressions, for increasing values of n''.

The curve in Fig. 7 is that which is obtained by the graphical process, using the potential energy curves given by the foregoing set of constants. Here evidently is a wide discrepancy between theory and experiment. The discrepancy may be removed by an alteration in the assignment of vibrational quantum numbers. Mecke's assignment of values of n'' is greater by four than the one, due to Loomis, on which the calculations have been based. Mecke's zero of n'' was chosen to include a place for four anti-Stokes terms in Wood's resonance spectrum. But if the theoretical intensity curve derived from Loomis' assignment is correct it is clear that a molecule in the zero-state initially would favor a transition to an energy state in excess of that required for the dissociation process.

This suggests that when the molecule absorbs light from the unexcited non-vibrating state it is always dissociated and no bands are obtained. It is therefore possible that lower vibrational states exist which contribute, by absorption from them, only to the continuous spectrum of iodine. The existence of such lower states is called for, of course, if there are really the so-called anti-Stokes terms in the resonance spectrum of Iodine. Loomis has, however, given an analysis of the resonance doublets which makes it appear that what were believed to be anti-Stokes terms in the $n'=26$ progression are really due to the excitation of some lines out of other n' progressions. Dymond has recorded observations of

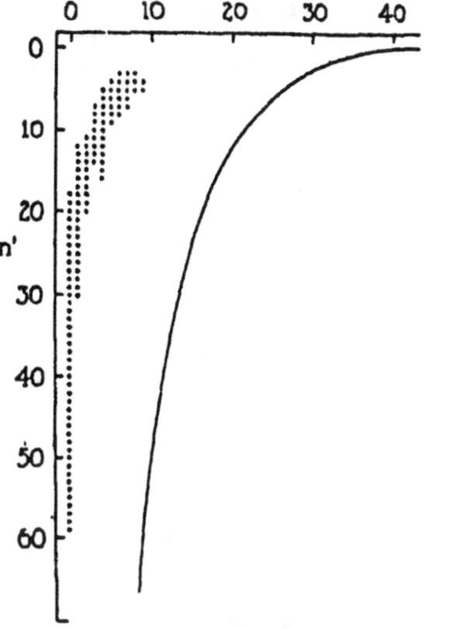

Fig. 7. The absorption bands of iodine (I_2).

anti-Stokes terms in resonance spectra excited by other lines than the green mercury line. It remains to be seen whether Loomis' analysis can be applied to them. It is clear that, since the theory here presented gives the same favored Δn values for absorption as for emission, transitions involving the lower levels in the unexcited state would not appear in the fluorescent spectrum for the same reason that they are not involved in the fully quantized absorption spectrum. Thus it is to be expected that anti-Stokes terms will not appear even though the necessary levels may exist in the molecule.

It is easy to see that a rather small change in the assignment of quantum numbers will make the theory agree with experiment. Full calculations were carried out for the case in which the n'' values were increased by 5 and the n' by 30. It was found that the discrepancy was about the same but in the other direction. Therefore a good fit could be obtained by some reassignment which makes smaller changes in n' and n''. It seems, however, that agreement on the intensities calls for reassignment of *both* n' and n'', since calculations show that good fits are not obtained if n' alone is altered. Iodine, then, may be tentatively regarded as in agreement with the theory.*

This completes the consideration of the four main sizes of the typical intensity distribution. Of the remaining five, the Schumann-Runge system of oxygen is much like that of iodine in that an unusually large change in the moment of inertia occurs during the electron transition. The other four are medium sized systems like that of AlO.

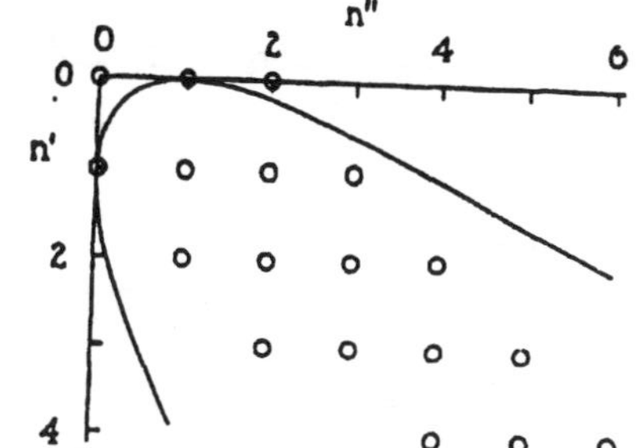

Fig. 8. The violet cyanogen system (CN).

The Schumann-Runge bands of oxygen are quite incompletely known.[13] This name is here given to the system whose $n''=0$ progression is known in absorption by the researches of Schumann and others and whose $n'=0$ progression was photographed in emission by Runge. The recognition of the fact that the Runge and Schumann bands are part of one system is due to Mulliken. The data on the absorption bands has recently been extended by the work of Leifson as interpreted by Birge. All that can be said here is that in absorption Leifson's strongest bands run from $n'=9$ to a limit at $n'=21$ with continuous spectrum beyond. Similarly the emission bands measured by Runge correspond to large values of Δn, the actual bands being $n'=0$, $n''=11$ to 17. Here is evidently a preference for large values of Δn which is quite in accord with the theory of this paper. The fine structure analysis of the bands thus

[13] Schumann, Smithsonian Contrib. **29**, no. 1413 (1903); Runge, Physica **1**, 254 (1921); Mulliken, Private communication to Professor Birge; Leifson, Astrophys. J. **63**, 73 (1926); Füchtbauer and Holm, Phys. Zeits. **26**, 345 (1925); Report on Molecular Spectra, Chap. IV, Section 7.

* See note on p. 1201.

far permits only the use of parabolic $V(r)$ curves, but these, when computed from the data of Table I show that in absorption the most probable transition is that beyond dissociation. The calculation shows that from the $n''=0$ state the molecule in the absorption process tends to take up over 1.2 volts of vibrational energy or 0.4 volts in excess of that needed for dissociation in the excited state for these bands. Similarly the curves indicate for the $n'=0$ progression of emission bands a large Δn, much larger in fact than that indicated directly by the Runge bands. Here the higher terms in the force law will undoubtedly operate to reduce the theoretical Δn, so that in this system there is as good agreement as can be hoped for in the present state of analysis of the bands.

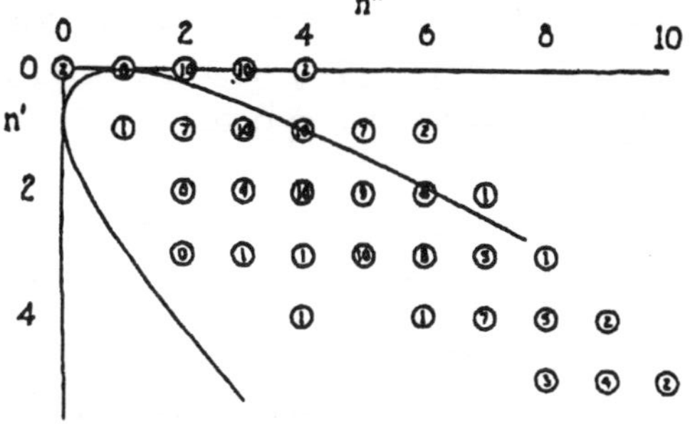

Fig. 9. The first negative group of carbon (CO^+).

The remaining four band systems to be considered present no new features of special interest. They are shown in Figs. 8, 9, 10, 11 in which the intensities are, when given, simply rough visual estimates from the plates. The theoretical curve in each case is based on the formula which assumes simple harmonic oscillators. The data for CN (Fig. 8), are from Heurlinger.[14]

The intensities given for CO^+ (Fig. 9) as well as the molecular constants used are based on the work of Blackburn and Johnson.[15] It is important to note that the theory predicts the existence of some strong emission bands in the far ultraviolet, along the lower branch of the theoretical curve. The single-branched intensity distribution the writer believes is entirely a consequence of the lack of complete experimental data. The intensity estimates for the second positive group of nitrogen (Fig. 10) are due to Birge, being made from inspection of spectrograms taken by E. P. Lewis.[8]

[14] Kayser's Handbuch, v. 5, p. 190.

[15] Johnson, Proc. Roy. Soc. **108A**, 343 (1925); Blackburn, Proc. Nat. Acad. **11**, 28 (1925).

The data on the first negative group of nitrogen (Fig. 11) are due to Merton and Pilley.[16]

In conclusion it seems permissible to claim that the theory outlined here gives a satisfactory semi-quantitative correlation between the

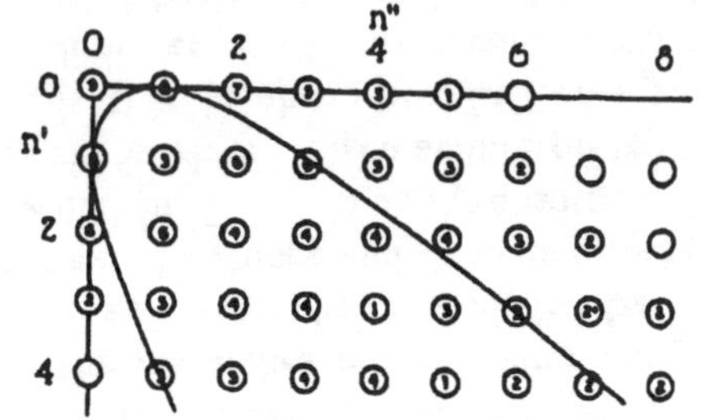

Fig. 10. The second positive group of nitrogen (N_2).

intensity distribution in band systems and the positional structure of the systems by means of a definite mechanical picture of the processes governing transition probabilities. The only previous attempt at a theoretical treatment of this topic is that of Lenz[17] who gave, by application of the correspondence principle to an over-simplified molecular

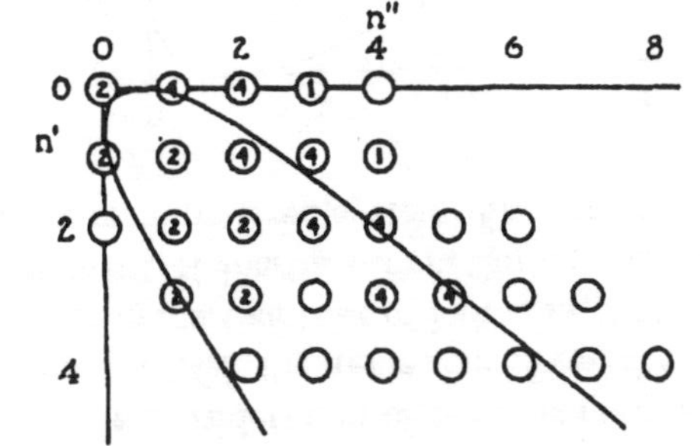

Fig. 11. The first negative group of nitrogen (N_2^+).

model, some general formulas which, however, were incapable of numerical application to specific cases. On the other hand, Lenz' theory does lead to an understanding of the many and seemingly irregular alternations of intensity in the $n'=26$ progression of the iodine emission spec-

[16] Merton and Pilley, Phil. Mag. **50**, 195 (1925).
[17] Lenz, Zeits. f. Phys. **25**, 299 (1924).

trum (Wood's resonance spectrum). This is a thing which is wholly unclear in terms of the theory given here.

It is a pleasure to acknowledge my indebtedness to Professor R. T. Birge who has generously given me the benefit of his knowledge of band spectra and their quantum interpretation. It is also appropriate again to call attention to the fact that my work is merely an extension of a leading thought on this subject by Professor J. Franck.

University of California,
Department of Physics,
July 27, 1926.

Note added to proof.—Since this was written I have learned of the experiments of H. Kuhn (Zeits. f. Physik **39**, 77 (1926)) which seem to show conclusively that the long series of absorption bands which have a convergence limit really do come from the lowest vibration level so that the values of n'' cannot be altered as suggested on page 1198.

Göttingen,
November 8, 1926.

DIATOMIC MOLECULES ACCORDING TO THE WAVE MECHANICS. II. VIBRATIONAL LEVELS

By Philip M. Morse
Palmer Physical Laboratory, Princeton University

(Received April 8, 1929)

Abstract

An exact solution is obtained for the Schroedinger equation representing the motions of the nuclei in a diatomic molecule, when the potential energy function is assumed to be of a form similar to those required by Heitler and London and others. The allowed vibrational energy levels are found to be given by the formula $E(n)=E_e+h\omega_0(n+1/2)-h\omega_0x(n+1/2)^2$, which is known to express the experimental values quite accurately. The empirical law relating the normal molecular separation r_0 and the classical vibration frequency ω_0 is shown to be $r_0^3\omega_0=K$ to within a probable error of 4 percent, where K is the same constant for all diatomic molecules and for all electronic levels. By means of this law, and by means of the solution above, the experimental data for many of the electronic levels of various molecules are analyzed and a table of constants is obtained from which the potential energy curves can be plotted. The changes in the above mentioned vibrational levels due to molecular rotation are found to agree with the Kratzer formula to the first approximation.

Introduction

The wave equation for the nuclear motion[1] of a diatomic mocluele of nuclear masses M_1 and M_2 and charges Z_1 and Z_2, is approximately

$$\nabla^2\psi+\frac{8\pi^2\mu}{h^2}[W-(c^2Z_1Z_2/r)+V_e(r)]\psi=0 \tag{1}$$

where $\mu=M_1M_2/(M_1+M_2)$, r is the distance between the nuclei, θ and ϕ the usual orientation angles measured from the center of gravity, and $V_e(r)$ the electronic energy calculated by considering the two nuclei as fixed in space a distance r apart.

The combination of the energy of repulsion and the electronic energy can be considered as a nuclear potential energy

$$E(r)=(c^2Z_1Z_2/r)-V_e(r).$$

The wave function Ψ can be considered as a product of three factors $\Psi=N\cdot\Phi(\phi)\cdot\Theta(\theta)\cdot R(r)/r$, where it can be shown that

$$\Phi=e^{ig\phi}$$

$$\Theta=\sin^g\theta\cdot P_j^g(\cos\theta)$$

where g and j are integers. The normalizing factor N is adjusted so that $\int\Psi\cdot\bar{\Psi}\,dv=1$.

When these functions have been substituted in the general equation an equation for R results,

[1] Born and Oppenheimer, Ann. d. Physik 84, 457 (1927).

$$\frac{d^2R}{dr^2}+\frac{j(j+1)R}{r^2}+\frac{8\pi^2\mu}{h^2}[W-E(r)]R=0. \tag{2}$$

Since we are primarily interested in vibrational levels, j is set equal to zero.

The function $E(r)$ is a complicated function of r and of the electronic quantum numbers, and is not known accurately for any molecule, so it is useless to try to substitute the actual expression for E in the equation. It is possible, however, to assume some functional form for E which gives curves of approximately the same shape as the actual curves, and whose constants can be carried through the calculations to the end and then adjusted to conform to the experimental data.

The forms for E which have been used are, among others, the two series[2] $a/r+b/r^2+c(r-r_0)^2+\cdots$, and $b'(r-r_0)^2+c'(r-r_0)^3+\cdots$. These series give a general equation for the allowed energy levels

$$W=-D+h\omega_0[(n+\tfrac{1}{2})-x(n+\tfrac{1}{2})^2+K_3(n+\tfrac{1}{2})^3+\cdots] \tag{3a}$$

where the constants ω_0, x, K, $\cdots$, are functions of a, b, c, etc., and so if W is known empirically, E can be determined.

There are several objections to such series forms for E. In the first place the effect of all the terms in $(r-r_0)$ to the power 3 or over (which terms are not always small) must be computed by perturbation methods, thus adding another approximation to a list already long. In the second place the series for E determined from the known values of ω_0, x, K_3, etc., does not converge for large values of r, and so the series is only applicable over a restricted range of r.

In the third place the experimental data show that constants $K_3\cdots$, are very much smaller than ω_0 or x, whereas the general series for E do not show that any such peculiarities should exist. In other words these series are too general.

We must search, then, for a form for E which will satisfy the following requirements: (1), It should come asymptotically to a finite value as $r\to\infty$; (2), It should have its only minimum point at $r=r_0$; (3), It should become infinity at $r=0$ (this need not be exactly true, however, the results are practically the same if E becomes very large at $r=0$); (4), It should exactly give the allowed energy levels as the finite polynomial.

$$W(n)=-D+h\omega_0[(n+\tfrac{1}{2})-x(n+\tfrac{1}{2})^2]. \tag{3}$$

The very small correction term coefficients $K_3\cdots$, can then be determined by perturbation methods with a reasonable expectation that these methods on such small quantities will give fairly accurate results.

A form will be chosen for E which satisfies requirements 1 and 2 exactly, and 3 approximately, and the problem will then be to show that the chosen form satisfies requirement 4.

A Solution of the Problem

The function which it is proposed to use here is the simple one

$$E(r)=De^{-2a(r-r_0)}-2De^{-a(r-r_0)}. \tag{4}$$

[2] E. Fues, Ann. d. Physik 80, 367 (1926).

This function has a minimum of $-D$ at $r=r_0$, comes asymptotically to zero at $r=\infty$ and in general gives curves of a very similar form to the few potential energy curves which have been calculated theoretically.[3,4,5] The only portion where it does not fit these curves is at $r=0$, where it should be infinity. But it will be seen that for the values of D and a used to fit the data $E(r)$ is between $100D$ and $10000D$ at $r=0$, a value so large that, as far as its effect on the energy levels and wave function goes, it is as good as infinity.

The frequency of classical small vibrations about r_0 is

$$\omega_0=(a/2\pi)(2D/\mu)^{1/2}. \tag{5}$$

If this form of E is substituted in Eq. (3), j set equal to zero, and the transformation $u=(r-r_0)$ made, then

$$\frac{d^2R}{du^2}+\frac{8\pi^2\mu}{h^2}[W-De^{-2au}+2De^{-au}]R=0. \tag{6}$$

The boundary conditions are now set that R must be finite, single valued and continuous in the range $-\infty\leqq u\leqq+\infty$. It will be found that for some allowed solutions in this case $r\Psi$ will not be zero at $r=0$. But in every case $r\Psi$ will be extremely small, and since the point $r=0$ is some distance outside the important interval where $W>E$ this discrepancy will not affect the values of the energy levels.

In other words, since we have admittedly not used the correct form for E, and since we presumably could not find the true solution for R anyway, we must content ourselves with a solution which deviates from the correct solution in a portion which has little effect on the values of the allowed energies, especially since this deviation is very small.

Make a second transformation, letting $y=e^{-au}$. Then

$$\frac{d^2R}{dy^2}+\frac{1}{y}\frac{dR}{dy}+\frac{8\pi^2\mu}{a^2h^2}\left[\frac{W}{y^2}+\frac{2D}{y}-D\right]R=0 \tag{7}$$

where now R must be finite, continuous and single valued over the range $\infty\geqq y\geqq 0$. Let $R=e^{-dy}\cdot(2dy)^{b/2}\cdot F(y)$. Then if

$$d=2\pi(2\mu D)^{1/2}/ah \tag{8}$$

$$W=-a^2h^2b^2/32\pi^2\mu \tag{9}$$

and if $z=2dy$, then the equation becomes

$$z(d^2F/dz^2)+(b+1-z)(dF/dz)+\left(\frac{8\pi^2\mu D}{a^2dh^2}-b/2-1/2\right)F=0. \tag{10}$$

The solution of this equation is a finite polynomial[6] if $(8\pi^2\mu D/a^2h^2d-b/2-1/2)=n$, an integer greater than zero. That is

$$b=4\pi(2\mu D)^{1/2}/ah-1-2n=k-1-2n \tag{11}$$

[3] O. Burrau, Klg. Danske. Vid. Selskab. 7, 14 (1927).
[4] Heitler and London, Zeits. f. Physik 44, 455 (1927).
[5] Morse and Stueckelberg, Phys. Rev. 33, 907 (1929).
[6] Schroedinger Ann. d. Physik 80, 483 (1926).

where $k\ (=4\pi(2\mu D)^{1/2}/ah)$ must be greater than unity to have a discrete energy spectrum. The positive values of b are the only ones for which R is finite over the range $0\leqq z\leqq\infty$. This means that n can have any integral value in the range $0\leqq 2n\leqq(k-1)$.

The solutions for F are the generalized Laguerre polynomials

$$F=L^b_{n+b}(z)=\frac{d^b}{dz^b}\left[e^z\cdot\frac{d^{n+b}}{dz^{n+b}}(z^{n+b}e^{-z})\right].$$

Usually both superscript and subscript-minus-superscript are taken to be integers, although the superscript need not be integral. Here b is a fraction, and fractional differentiation must be used to obtain the polynomial from the definition above. The polynomial can be obtained by the use of the recursion formula obtained directly from Eq. (10) without evoking the use of fractional differentiation, but it is felt desirable to use the general definition for the sake of uniformity. Such differentiation does not vitiate the general formulas of integration etc., which have been developed for these polynomials, as long as n is an integer.

The formulas needed for fractional differentiation are given here for convenience.

$$\frac{d^b(z^a)}{dz^b}=\frac{\Gamma(a+1)}{\Gamma(a-b+1)}z^{a-b}\ ;\quad \frac{d^b(e^{-z})}{dz^b}=e^{i\pi b}\cdot e^{-z}.$$

The application of this to the definition of F results in

$$L^b_{n+b}(z)=e^{i\pi(k-n-1)}\frac{\Gamma(k-n)}{n!}\left[x^n-(k-n-1)nx^{n-1}+\frac{(k-n-1)(k-n-2)n(n-1)x^{n-2}}{2!}\cdots\right]$$

The normalizing integral

$$\int_0^\infty z^{b-1}\cdot e^{-z}\cdot L^b_{n+b}(z)\cdot L^b_{m+b}(z)\cdot dz=N_{nm}$$

$$=\begin{cases}0 & (\text{if } n\neq m)\\[1ex] [\Gamma(k-n)]^2\displaystyle\sum_{s=0}^{n}\frac{\Gamma(k-2n+s-1)}{\Gamma(s-1)} & (\text{if } n=m)\end{cases} \tag{12}$$

so that the complete wave function R for the nuclear vibration is

$$(2da/N_{nn})^{1/2}\cdot e^{-de^{-a(r-r_0)}}\cdot[2de^{-a(r-r_0)}]^{(k-2n-1)/2}\cdot L^{k-2n-1}_{k-n-1}[2de^{-a(r-r_0)}]$$

The square of this, times the perturbing energy, is to be multiplied by dr as a "volume element" and the integral taken from zero to infinity to obtain the perturbation energy.

The allowed energy levels are obtained from Eqs. (9) and (11)

$$\begin{aligned}W(n)&=-a^2h^2(k-1-2n)^2/32\pi^2\mu\\&=-D+\frac{ah}{2\pi}(n+1/2)(2D/\mu)^{1/2}-a^2h^2(n+1/2)^2/8\pi^2\mu\\&=-D+h\omega_0(n+1/2)-(h^2\omega_0^2/4D)(n+1/2)^2\end{aligned} \tag{13}$$

from Eq. (5). As has been noted before, n takes on all integral values from zero to $(k-1)/2$. This is the first case noted of a Schroedinger equation giving a finite number of discrete energy levels.

Equation (13) is of the form of empirical Eq. (3), and it therefore is of the form which we set out to obtain. Since this equation expresses the empirical data so well in most cases, therefore the potential energy E, as given by Eq. (4), must have the same shape as the real potential energy throughout the range where Eq. (13) is a valid representation of the actual energy levels.

Thus if the lists of spectroscopically determined molecular constants give r_0 in Angstrom units and ω_0 and $\omega_0 x$ in wave-numbers, D is found in wave-numbers by the equation

$$D=\omega_0^2/4\omega_0 x \tag{15}$$

and the coefficient a is found by the relation

$$a=(8\pi^2 c\mu\omega_0 x/h)^{1/2}=0.2454(M\omega_0 x)^{1/2} \tag{16}$$

where $M=M_1M_2/(M_1+M_2)$, M_1 and M_2 being the atomic weights of the two nuclei in terms of oxygen 16.

With r_0 known and D and a determined, the potential energy curve corresponding to the data is given by Eq. (4), where E is in wave-numbers if r is in Angstrom units.

An Empirical Law for r_0

When the available lists of molecular constants were examined it was found that in many cases ω_0 and $\omega_0 x$ were known but r_0 was not known. Several writers[7] have made use of a relation $r_0^2\omega_0$=constant to obtain the unknown r_0's, but deviations from this relation are quite large.[8]

To find what law obtained, if any, between r_0 and ω_0, 21 cases were taken from Birge's table[9] where r_0 and ω_0 were both known. An equation, $\log \omega_0 - p \log r_0 = \log K$ was assumed, and the data were subjected to a least squares analysis. The most probable values of the constants were found to be $p=2.95$ and $K=2975$. If p be taken to be 3 then the equation becomes

$$r_0^3\omega_0=3000\text{A}^3/\text{cm} \tag{17}$$

to within a probable error of ± 120 and a maximum deviation of 420. This is about due to the probable error of the recorded values of r_0, for if these values had a probable error of 1.3 percent, then r_0^3 would be given to a probable error of 4 percent.

[7] Birge, Phys. Rev. 25, 240 (1925). Mecke, Zeits. f. Physik 32, 823 (1925).

[8] Since this paper has been sent to the editor, Professor Birge has kindly brought to the writer's notice the fact that the equation $r_0^2\omega_0=C_m$, where C_m has a different value for each molecule, is a better fit for some data than Eq. (17). In other words, the slope of the curve $\log r_0$ plotted against $\log \omega_0$ is nearer two for each individual molecule, but the slope of the band representing all molecules is nearer three, as given by Eq. (17). However, at least one value of r_0 must be known for a molecule before C_m can be known. Therefore Eq. (17) presents the only means of obtaining a value for r_0 for a molecule when no empirical value of r_0 for that molecule is available; and so it is useful, even if the probable error of the value so obtained were rather larger than the above least squares analysis would indicate.

[9] Mulliken, Phys. Rev. 32, 206 (1928) and Birge, Int. Crit. Tables, Vol. V, 411, (1929).

Table I.[10]

No.	Mol.	State	A wave-nos.	D wave-nos.	a	r_0 obs. A units	r_0 calc. A units	reference
1	BeO	$^1\Sigma$	41600	42300	2.12	1.33	1.27	(14)
2	BeO	$^1\Sigma$	72200	51700	1.76	1.36	1.30	(14)
3	BO	$^2\Sigma$	74200	75100	2.14	1.21	1.17	(9)
4	BO	$^2\Pi$	59800	36800	2.04	1.36	1.34	(9)
5	BO	$^2\Sigma$	82500	40000	1.99	1.31	1.33	(9)
6	AlO	—	33100	33600	2.06	1.62	1.46	(15)
7	AlO	—	70000	49800	1.51	1.66	1.52	(15)
8	C_2	$^3\Pi$	55900	56700	2.14	1.31	1.23	(9)
9	C_2	$^3\Pi$	49100	40600	2.65	1.27	1.19	(9)
10	CN	$^2\Sigma$	75600	76600	2.32	1.17	1.14	(9)
11	CN	$^2\Pi_i$	68700	55200	2.30	—	1.20	(9)
12	CN	$^2\Sigma$	78600	53900	2.88	1.15	1.12	(9)
13	CO	$^1\Sigma$	90500	91600	2.29	1.15	1.12	(9)
14	CO	$^3\Pi$	99200	51600	2.45	—	1.20	(9)
15	CO	$^3\Sigma$?	94900	38300	1.93	—	1.37	(9)
16	CO	$^1\Pi$	96700	32600	2.67	1.24	1.26	(9)
17	CO	$^1\Sigma$	108800	28800	4.55	1.12	1.12	(9)
18	CO	$^1\Pi$	145000	46300	2.86	—	1.16	(9)
19	CO^+	$^2\Sigma$	193100	79200	2.91	1.11	1.11	(9)
20	CO^+	$^2\Pi_i$	177300	42700	2.42	1.25	1.25	(9)
21	CO^+	$^2\Sigma$	189100	29600	3.17	1.17	1.21	(9)
22	F_2	$^1\Sigma$	33600	34200	2.39	1.4	1.37	(9)
23	F_2	$^1\Sigma$	24800	1420	3.21	—	2.11	(9)
24	H_2	$1^1\Sigma$	38000	40100	1.85	.76	.89	(11)
25	H_2	$2^1\Sigma$	116900	27600	.69	1.31	1.31	(16)
26	H_2	$2^3\Pi$	112000	19500	1.48	.97	1.08	(11)
27	H_2	$2^3\Sigma$	118200	24600	1.44	1.08	1.05	(11)
28	H_2	(C)	119200	21300	1.41	1.06	1.08	(11)
29	H_2	$3^3\Pi$	131500	21200	1.37	1.14	1.09	(11)
30	H_2	$3^1\Pi$	225000	114500	.61	—	1.08	(11)
31	H_2	$4^3\Pi$	136500	20400	1.38	1.14	1.10	(11)
32	H_2	$5^3\Pi$	140400	21800	1.32	1.17	1.10	(11)
33	H_2	$6^3\Pi$	143000	23000	1.28	1.17	1.11	(11)
34	H_2^+	—	143700	20600	1.36	1.06	1.11	(11)
35	I_2	—	19100	19200	1.50	2.66	2.41	(13)
36	I_2	—	20300	4770	1.80	3.01	2.86	(13)
37	N_2	$^1\Sigma$	94300	95500	2.56	—	1.09	(9)
38	N_2	$^3\Sigma$?	103200	37600	2.42	—	1.28	(9)
39	N_2	$^1\Pi$	119100	51000	2.41	—	1.21	(9)
40	N_2	$^3\Pi_n$	126000	51200	2.46	1.21	1.21	(9)
41	N_2	$^3\Pi_n$	150300	46100	3.31	1.15	1.14	(9)
42	N_2^+	$^2\Sigma$	202800	67000	2.62	1.12	1.11	(9)
43	N_2^+	$^2\Sigma$	224500	62900	3.10	1.08	1.08	(9)
44	NO	$^2\Pi_n$	61300	62200	2.55	1.15	1.16	(9)
45	NO	$^2\Sigma$	149200	106200	2.42	1.07	1.09	(9)
46	NO	$^2\Pi_n$	80300	35400	1.82	1.42	1.43	(9)
47	NO	—	102100	50000	3.49	—	1.09	(9)
48	O_2	$^3\Sigma$	53000	53800	2.34	1.21	1.24	(9)
49	O_2	$^1\Sigma$?	54500	42100	2.39	1.23	1.29	(9)
50	O_2	$^3\Sigma$	59000	10100	2.44	1.61	1.62	(9)
51	O_2^+	$^2\Pi$	164400	56100	2.82	—	1.16	(9)
52	O_2^+	$^2\Pi$	165400	14300	2.57	—	1.50	(9)
53	O_2^+	—	174800	23800	2.31	—	1.43	(9)
54	O_2^+	—	187700	19700	2.93	—	1.37	(9)
55	SiN	—	49400	50000	1.92	1.57	1.38	(12)
56	SiN	—	38200	14500	3.15	1.58	1.43	(12)

[10] It will be noticed that the molecules for which r_0 calc. differs markedly from r_0 obs. are the least symmetric molecules in the list [ie., $M_1M_2/(M_1+M_2)$ differs considerably from $(M_1+M_2)/4$]. Professor R. S. Mulliken has kindly suggested to the writer that perhaps the rule enunciated in Eq. (17) above only holds accurately for molecules where M_1 is about equal to M_2. Certainly the calculated values of r_0 for the perfectly symmetric molecules O_2, H_2 and N_2 give the most consistent check with the experimental values. To apply the rule to very unsymmetric molecules it may be necessary to introduce an "unsymmetry factor" of the type $[4M_1M_2/(M_1+M_2)^2]^{1/4}$ into the term $r_0^3\omega_0$. Curiously enough, those levels for which the above rule is not satisfactory are the ones whose vibrational levels fit Eq. (13) least satisfactorily.

Since six different neutral molecules and two different molecular ions were used in the analysis, it would appear that K is independent of molecular weight, of the electronic state, and of the net charge on the molecule. An independent check of this rule is discussed in the following paper.

Results

A table is given from which the potential energy curves can be plotted by means of the equation

$$E = A + De^{-2a(r-r_0)} - 2De^{-a(r-r_0)} \tag{18}$$

r_0 is given in Angstrom units and D, ω_0 and $\omega_0 x$ are given in wave-numbers. The values of r_0 in the column marked r_0 calc. were calculated from ω_0 by means of rule (17). The striking agreement between these values and the experimental values is to be noted. A is given so that all curves of neutral molecule and ion are reckoned from the lowest vibration level of the lowest electronic state of the neutral molecule. Figure 1 shows the levels for N_2 and N_2^+ plotted from these data.

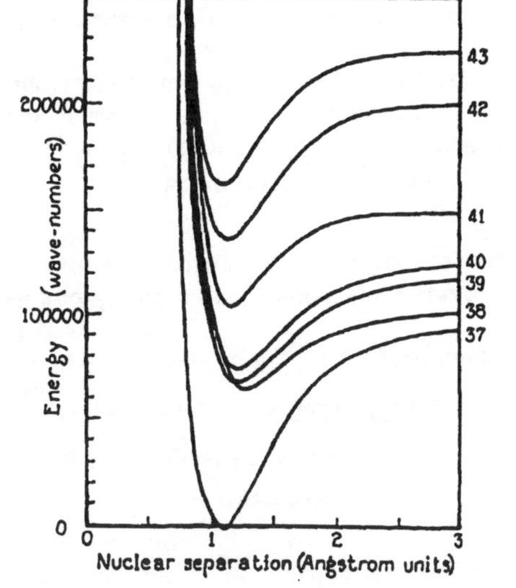

Fig. 1. Potential energy curves for nitrogen. Energy in wave-number and nuclear separation in Angstrom units. Numbers on curves refer to Table I.

Rotational Levels

When the rotational quantum number j is different from zero, the potential energy E in Eq. (4) is increased by an amount $E_j = j(j+1)h^2/8\pi^2\mu r_0^2$. Inasmuch as this increase only affects the wave function to an appreciable

[11] Birge, Proc. Nat. Acad. 14, 12 (1928).
[12] Jenkins and Laszlo, Proc. Roy. Soc. A122, 103 (1929).
[13] Birge, Molecular Spectra in Gases, N. R. C. Bull. 57, 230 (1927).
[14] Rosenthal and Jenkins, Phys. Rev. 33, 163 (1929).
[15] Pomeroy, Phys. Rev. 29, 57 (1927).
[16] Hyman and Birge, Nature 123, 277 (1929).

extent in the region near $r = r_0$ (where $W > E$), it can be expanded about this point,

$$E_j = \frac{h^2 j(j+1)}{8\pi^2 \mu r_0^2}\left[1 - 2\frac{(r-r_0)}{r_0} + 3\frac{(r-r_0)^2}{r_0^2} - \cdots\right].$$

In the range where E_j has any appreciable effect, $r - r_0$ is small compared to r_0, and since $h^2 j(j+1)/8\pi^2\mu r_0^2$, (which can be called R) is small compared to E for the usual values of j, this expansion can be added to the expansion for E, giving for the first two terms

$$E + E_j = -D + R - R^2/Da^2r_0^2 + a^2(D-R)(r - r_0 - R/r_0a^2D)^2$$

plus terms in higher powers of $(r - r_0)$.

These two terms can be considered as the first two terms of the expansion of

$$E + E_j = (D - R + R^2/Da^2r_0^2)e^{-2a(r-r_0-R/r_0a^2D)}$$
$$-2(D - R + R^2/Da^2r_0^2)e^{-a(r-r_0-R/r_0a^2D)} \tag{19}$$

indicating that to the first approximation D has decreased to $D - R + R^2/r_0^2a^2D$ and that r_0 has increased to $r_0 + R/r_0a^2D$. The resultant energy levels will be, to the first approximation

$$W(n,j) = -D + R + (ah/2\pi)(n+\tfrac{1}{2})\left(\frac{2D-2R}{\mu}\right)^{1/2}$$
$$-a^2h^2(n+\tfrac{1}{2})^2/8\pi^2\mu - R^2/Da^2r_0^2$$
$$= -D + h\omega_0(n+\tfrac{1}{2})[1 - h\omega_0(n+\tfrac{1}{2})/4D - h^2j(j+1)/16\pi^2D\mu r_0^2] \tag{20}$$
$$+(h^2j(j+1)/8\pi^2\mu r_0^2)[1 - h^2j(j+1)/16\pi^4\mu^2r_0^4\omega_0^2].$$

This agrees with the general Kratzer[17] formula to the first approximation. (i.e., as far as the above expansion is written).

The energy levels required by Eqs. (13) or (20) agree quite well with the experimentally determined levels for most molecules up to quite high values of n. In other words the $\Delta W(n)$ curve is a straight line for a considerable distance as n is increased. This indicates that the potential energy is effectively that given by Eq. (4) for a large range of r.

However there are some electronic states, usually the normal levels of the molecules, whose $\Delta W(n)$ curves are not straight lines, and therefore whose potential energy curves deviate somewhat from the form given by Eq. (4). This is not surprising, however, for the potential energies of initial states have usually much deeper minima than the rest, and would be expected to deviate most markedly from the standard form.

Such deviations from the straight line ΔW curve can be considered as due to an additional term in the potential energy of the form $E' = B/r + C/r^2 + \cdots$, where B and C etc., are very small compared to D. This perturbation can be dealt with in the same manner as E_j has been dealt with, and the values of $B, C, \cdots$, can be found by comparing the resulting formula for $\Delta W(n)$ with the data.

In most cases, however, the $\Delta W(n)$ curve deviates so little from the straight line that such a calculation is not necessary.

[17] Kratzer, Zeits. f. Physik 3, 289 (1920).

Second Series November 15, 1929 Vol. 34, No. 10

THE

PHYSICAL REVIEW

THE THEORY OF COMPLEX SPECTRA

By J. C. Slater
Jefferson Physical Laboratory, Harvard University

(Received June 7, 1929)

Abstract

Atomic multiplets are treated by wave mechanics, without using group theory. In part 1 Hund's scheme for multiplet classification is derived directly from theory. Part 2 is devoted to the computation of the energy distances between multiplets, and comparison of these distances with experiment in some typical examples. There is no treatment of the separations between the various terms of a multiplet, since that has been done elsewhere, but only between one multiplet and another. It is found that Hund's rule, that terms of large L and S values lie lowest, has no general significance; the present theory leads to the same results as the rule when it is obeyed experimentally, but many cases which were exceptions to that rule are in agreement with the theory. The method of calculation of multiplet distances is described in sufficient detail, with the necessary tables of coefficients, etc., so that further checks with experiment could easily be made.

THE theory of complex spectra is treated in this paper by the method of wave mechanics. The results of the calculation may be divided into two parts: first, the classification of the terms into multiplets; second, the energy values of these multiplets. The first part contains no new results of physical interest, for it leads precisely to Hund's[1] scheme of classification, and uses almost the identical steps that Hund uses. Its value lies in the fact that this well-known scheme is shown in an elementary way to follow directly from wave mechanics. The second part, however, is almost entirely new, and it leads to definite formulas for the intervals between the different multiplets in spectra, intervals which could previously be considered only very roughly from an empirical rule of Hund, which proves to have no general significance. These calculations, of course, give the intervals in terms of certain integrals, which we do not calculate, but merely estimate well enough to permit some comparison with experiment. The agreement is in general fairly good. A third part would be also included, dealing with the intervals within the multiplets, produced by the magnetic interaction, were

[1] F. Hund, Linienspektren und periodisches System der Elemente, Springer.

it not that Goudsmit[2] has already answered the question by a method closely analogous to that used here.

It will be noted that the objects of the present paper resemble closely those aimed at by Heisenberg, Wigner, Hund, Heitler, Weyl, and others,[3] who employ the methods of the group theory. That method is not used at all in the present calculation, and, in contrast, no mathematics but the simplest is required, until one actually comes to the computation of the integrals. This, it is believed, is in itself sufficient justification for paralleling to some extent work already done. The simplification is achieved largely by introducing the spin at the very beginning of the calculation, rather than later. Thus we need only consider antisymmetric wave functions, which can be treated very simply as determinants, and can avoid the other symmetry characters, with which the other papers have been mostly concerned. The process of building up wave functions of the proper symmetry by using determinants is not new; it is found in Dirac's earlier papers, and has been used, for example, by Waller and Hartree.[4] The spin function, however, is introduced into the determinant in the present paper in a new and more satisfactory way. The results which we obtain, concerning the diagonal term of the energy with respect to the antisymmetric wave functions, are of the sort found by the group theory, and also by the recent method of Dirac,[5] but it seems worth while to derive them in a simple fashion.

The essentially new results of the present paper, those relating to energy, in the second part, come from the fact that we consider the whole degeneracy, that coming from orbital as well as spin angular momentum, which most of the other papers have failed to do. The present method is essentially equivalent to the others in that it gives sums of energies, rather than individual energies. When it is applied to the case in which the degeneracy with respect to orbital angular momentum is not considered, it leads, as do the others, merely to the sum of all terms of a given multiplicity; the writer is indebted to Dr. Bloch for pointing out that in that case all the results of Heitler can be easily demonstrated by the present method. But we make the observation that, by considering the whole problem with all its degeneracy, the method of energy sums can be used much more effectively, so much so that in most important cases we can get the actual energies of the individual terms. In connection with this, it should be noted that in several cases the actual perturbation problems have been solved directly, rather than by the method of energy sums, obtaining results[6] that hold even when the magnetic energy is appreciable.

[2] S. Goudsmit, Phys. Rev. 31, 946 (1928).

[3] W. Heisenberg, Zeits. f. Physik 41, 239 (1927). E. Wigner, Zeits. f. Physik 40, 492, 883, etc. (1927). F. Hund, Zeits. f. Physik 43, 788 (1927). W. Heitler, Zeits. f. Physik 46, 47 (1928). H. Weyl, Gruppentheorie und Quantenmechanik, Hirzel, 1928.

[4] I. Waller and D. R. Hartree, Proc. Roy. Soc. A124, 119 (1929). The writer is indebted to Dr. Hartree for calling his attention to this paper.

[5] P. A. M. Dirac, Proc. Roy. Soc. A123, 714 (1929).

[6] W. V. Houston, Phys. Rev. 33, 297 (1929). J. A. Gaunt, Trans. Camb. Phil. Soc. (1929).

Part 1. The Classification of Multiplets

1. As a preface, we remind the reader of the theory of multiplets, as developed before wave mechanics. This theory considered the interactions of the various angular momenta of the parts of an atom, and treated the various energy levels resulting from different relative orientations of these parts. In a logical development of atomic structure it would follow directly after Bohr's theory of electron orbits; for Bohr went as far as was possible on the assumption that each electron moved in a field with spherical symmetry, so that its orientation with respect to the rest of the atom was a matter of no concern. We start then, as Bohr finished, with an atom composed of a number of electrons, each characterised by a total quantum number n, and an azimuthal quantum number l, the latter determining the angular momentum, and being often regarded as a vector (normal to the plane of the orbit, in Bohr's theory, but this is of no importance). The orientation of this vector in space was arbitrary, on account of the spherical symmetry. As an additional part of the formulation, we need to consider the électronic spin of Goudsmit and Uhlenbeck: each electron has an angular momentum (with consequent magnetic moment) entirely apart from the momentum l which it acquires by its motion. This additional angular momentum, denoted by s, (which like l, is often regarded as a vector), can likewise be oriented arbitrarily in space, without reference to the direction of l. Since the values which n, l, and s can take on, and the notation, are rather significant, we observe that l can equal 0, 1, 2, $\cdots$; n can be any integer greater than or equal to $(l+1)$; and s is always equal to 1/2. The values of n and l are denoted by symbols, as $1s$, $2s$, $\cdots$; $2p$, $3p\cdots$; $3d$, $4d$, $\cdots$; $4f$, $5f\cdots$; referring respectively to the cases where n and l are 10, 20, $\cdots$; 21, 31, $\cdots$; 32, 42, $\cdots$; etc. Then the structure of an entire atom in a given state can be given, as far as Bohr's theory is concerned, by giving the n and l of each electron, denoted as in the example: $(1s)^2\,(2s)^2\,(2p)^6\,(3s)^2$, meaning that there are two $(1s)$ electrons, two $(2s)$, six $(2p)$, two $(3s)$. Another state of the same atom (in this case magnesium) would be $(1s)^2\,(2s)^2\,(2p)^6\,(3s)\,(3p)$; and so on. These two states would be often abbreviated $3s$ and $3p$ respectively. The spin, having always the same value, need not be mentioned at this stage.

Having described the atom to the degree of accuracy considered by Bohr, we next must consider that the angular momentum vectors really are coupled together; there are differences of energy depending on orientation, and a consequent splitting up of each energy level into a number of different ones. In the commonest case, which alone we shall consider, the coupling can be considered in several stages. First, the l's of the various electrons group themselves into a vector sum L, which takes on only integral values, 0, 1, 2, $\cdots$, as the separate l's do. Each such arrangement gives a different energy, and a different term in the spectrum. The terms with L equal respectively to 0, 1, 2, 3, $\cdots$, are called S, P, D, F, $\cdots$ terms. Second, the spins of the electrons group themselves into a vector sum S, taking on the integral or half integral values (for any case, the orientation where all s's are

parallel is allowed; if it is integral, only integral S's are allowed, and if it is half integral, only half integers are allowed for S). The different values of S lead to terms of different multiplicities, as we shall see in a moment, and different energies; thus $S=0$ gives singlets, $S=1/2$ doublets, $S=2$ triplets, etc. Finally L and S are coupled together, their sum being called J, and being integral or half integral to correspond with S. Since J can vary from $|L+S|$ to $|L-S|$, there are (if $L>S$) $2S+1$ different terms so obtained. These terms are the various terms of a single multiplet, which is named singlet $(2S+1=1)$, doublet $(2S+1=2)$, etc., to agree with the number of terms. The energies involved in going from one value of L to another, or of S to another, are generally large; they are (as one sees from wave mechanics) electrostatic energies. On the other hand, the energy of orientation of L with respect to S is small, coming from magnetic interaction of the parts of the atom, so that the different levels of a multiplet (same L and S, different J) lie near enough together to be grouped together. If one neglects the magnetic interaction as a first approximation, the levels of a multiplet all have just the same energy. As to notation, we speak, for example, of ${}^3P_2$, meaning a triplet P term with $J=2$. And if we wish to indicate the state of the electrons which produce the multiplet terms, we write it as $(1s)^2\,(2s)^2\,(2p)^6\,(3s)\,(3p)\;{}^3P_2$, which would often be abbreviated $(3p)\;{}^3P_2$.

The process of coupling can also be described in the language of space quantization. If one has an angular momentum vector, as l, acted on by no torques, so that its direction is arbitrary, it is legitimate to choose any axis in space, and consider that the component of the vector along this direction can take up any one of a set of values, integrally spaced between the parallel and antiparallel orientations. Thus, if m_l is the component of l along this axis, m_l can equal l, $l-1$, $l-2$, $\cdots\;-l$. These $(2l+1)$ values are considered to denote $(2l+1)$ separate stationary states, all with the same energy. Now we follow by this method the process of coupling, beginning with the uncoupled vectors, and carrying it through to the case where the l's are coupled to give L, and the s's to give S, but the magnetic interactions are not considered. In the uncoupled case, each electron has its own m_l and m_s. As the coupling forces are introduced, torques appear which make the components of the individual l's and s's vary with time, so that the same space quantization is no longer possible. But the torques are internal; they cannot change the total angular momentum. The total sum of all the l's and the sum of all the s's, and their components along the axes, remain constant, and hence quantized: $\sum m_l$ and $\sum m_s$ are quantized even when the coupling has taken its full value. On the other hand, when the electrons are coupled, the atom consists of vectors L and S, each separately free, so that these can be space quantized, giving components M_L and M_S along the axis. We need now only identify the M_L and M_S for the coupled system with the $\sum m_l$ and $\sum m_s$ for the uncoupled one, to get the correspondence necessary to show what multiplets appear from any combination of electrons. The exact method is easily shown from an example.

For illustration, we take the coupling of two p electrons. Each has $l=1$; the first has principal quantum number n, the second n'. For each, we can have $m_l=1, 0, -1$; $m_s=1/2, -1/2$; that is, $2(2l+1)=6$ different sets (m_l, m_s), each representing a stationary state. With the combined atom, there are 36 combinations of one state of the first electron with one of the second. These are partly given in Table I. Two different notations are

Table I. *Two p electrons.*

First notation	Second notation	Σm_l	Σm_s	Notes
$(n\,1\,1\,\tfrac{1}{2})\ (n'\,1\,1\,\tfrac{1}{2})$	$\{(n\,1\,1)\ (n'\,1\,1)\}\ \{\ \}$	2	1	Exc
$(n\,1\,0\,\tfrac{1}{2})\ (n'\,1\,1\,\tfrac{1}{2})$	$\{(n\,1\,0)\ (n'\,1\,1)\}\ \{\ \}$	1	1	A
$(n\,1\,-1\,\tfrac{1}{2})\ (n'\,1\,1\,\tfrac{1}{2})$	$\{(n\,1\,-1)\ (n'\,1\,1)\}\ \{\ \}$	0	1	
$(n\,1\,1\,-\tfrac{1}{2})\ (n'\,1\,1\,\tfrac{1}{2})$	$\{(n'\,1\,1)\}\ \{(n\,1\,1)\}$	2	0	
$(n\,1\,0\,-\tfrac{1}{2})\ (n'\,1\,1\,\tfrac{1}{2})$	$\{(n'\,1\,1)\}\ \{(n\,1\,0)\}$	1	0	
$(n\,1\,-1\,-\tfrac{1}{2})\ (n'\,1\,1\,\tfrac{1}{2})$	$\{(n'\,1\,1)\}\ \{(n\,1\,-1)\}$	0	0	
$(n\,1\,1\,\tfrac{1}{2})\ (n'\,1\,0\,\tfrac{1}{2})$	$\{(n\,1\,1)\ (n'\,1\,0)\}\ \{\ \}$	1	1	A
$(n\,1\,0\,\tfrac{1}{2})\ (n'\,1\,0\,\tfrac{1}{2})$	$\{(n\,1\,0)\ (n'\,1\,0)\}\ \{\ \}$	0	1	Exc
$(n\,1\,-1\,\tfrac{1}{2})\ (n'\,1\,0\,\tfrac{1}{2})$	$\{(n\,1\,-1)\ (n'\,1\,0)\}\ \{\ \}$	−1	1	
etc.				

used in this table for comparison. First we simply give the four quantum numbers $(n\ l\ m_l\ m_s)$ for each of the electrons. For the other scheme of notation, which is often convenient in practice, we take advantage of the fact that each m_s is capable of only the two values $1/2$, $-1/2$: we set up two brackets, one containing the symbols $(n\ l\ m_l)$ of each electron with $m_s=1/2$, the second containing the symbols for those with $m_s=-1/2$. Such a bracket symbol, in which the arrangement of terms within a bracket is of no significance, is equivalent to a state of the first sort. After each symbol is given the value of $\sum m_l$ and of $\sum m_s$ for that state. Finally, after some of the terms, are notes which will be explained later.

The values of $\sum m_l$ and $\sum m_s$ are conveniently plotted by giving $\sum m_s$ as abscissa, $\sum m_l$ as ordinate, and putting a dot for each state. Thus in Fig. 1, a, we give a single p state; and in b, the points for the sum of the two (the numbers indicate the number of states with the same $\sum m_s$, $\sum m_l$). But now in Fig. 1, b, each point should also represent the M_L, M_S of one of the multiplets existing in the coupled atom. This can be brought about in but one way: by assuming the multiplets ${}^1S\,{}^1P\,{}^1D\,{}^3S\,{}^3P\,{}^3D$, whose separate representations are shown below (in Fig. 1, d—i), and whose patterns, if superposed, would just give Fig. 2b. Thus it must be that two p electrons produce these multiplets. But that is just what our other method gives; for two l's, each equal to unity, can add to give $L=0$, 1, 2, (S, P, D terms); while two s's, each 1/2, can give $S=0$, 1 (singlet, triplet). In every case, the two methods lead to the same result, just as here.

The scheme as described above will result, in an atom with many electrons, in an enormous variety of terms, comparatively few of which are realized in actual atoms. The principle limiting the number is the exclusion principle of Pauli. It cannot be stated in the language of vectors, but only in

terms of the space quantization just described. It is: no two electrons can have identical sets of numbers $(n\ l\ m_l\ m_s)$; further, two combinations of electrons which differ only in the interchange of the quantum numbers of two electrons are to be treated as identical. Stated in terms of the bracket method of writing the electronic combinations, this means that no two sets of quantum numbers $(n\ l\ m_l)$ within a bracket can be identical; and that the order of arrangement within a bracket is immaterial. By these restrictions, the number of terms is often enormously decreased. Many points on our diagrams are removed, and just the remaining ones must be fitted into multiplets, resulting in a much smaller number. For example, in our Table I, the two electrons have the same l; hence, if they have also the same total quantum number (that is, if $n=n'$), they cannot both have the same set of m_l and m_s. Thus the two terms marked *Exc* (and many other terms not

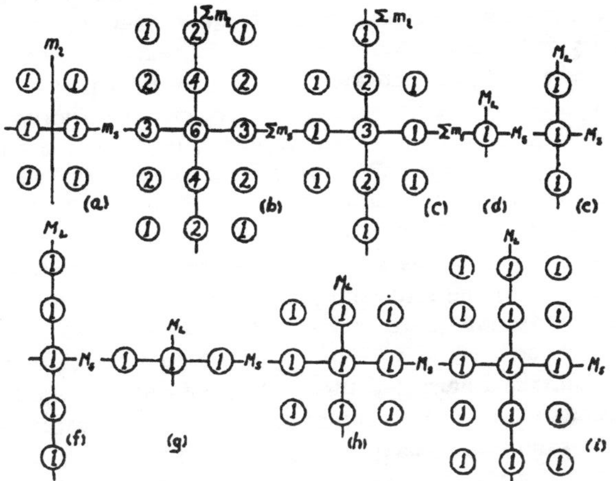

Fig. 1. (a) Single p electron; (b) Two p electrons; (c) Two equivalent p electrons; (d) ${}^1S$ (e) ${}^1P$; (f) ${}^1D$; (g) ${}^3S$; (h) ${}^3P$; (i) ${}^3D$.

included in the table) must be excluded; and the two terms marked A (and many more pairs) are to be considered identical. When the terms not allowed are removed, the new pattern in place of Fig. 1b proves to be Fig. 1c; and this is the superposition of ${}^3P\,{}^1S\,{}^1D$, which then are the multiplets allowed with two equivalent p electrons (that is, two electrons, each with $l=1$, and with the same n). As a general thing, the exclusion principle is active only when there are equivalent electrons; its many properties, as in limiting the number of electrons in a closed shell, are well known and need not be elaborated.

2. In quantum mechanics, we meet the problem of complex spectra as one step in the approximate solution of the wave equation for the atom. There are two fundamental principles which govern the structure of matter: quantum dynamics, and the principle of antisymmetry which shows itself

in the exclusion principle and the Fermi-Dirac statistics. The first states that only those energy levels are possible which are characteristic numbers of Schrödinger's equation for the system; the second further restricts the possible energies to those connected with characteristic functions antisymmetric in the electrons. Our whole object is to find such characteristic functions and numbers. And as before, the first step is to approximate by supposing that the separate electrons move in fields of force with spherical symmetry, so that there is no tendency for orientation of the various angular momenta. There are various ways of doing this; one scheme which gives good results is that of Hartree.[7] For simplicity in description we shall imagine that scheme modified slightly in one detail; according to it, each electron moves in a field of force slightly different from the others. We shall neglect the difference, assuming that all the electrons move in precisely the same field. And this field is to be so chosen as to give the best agreement with the correct values even without further corrections.

An electron moving in a central field of force, according to wave mechanics, is characterized by the same two quantum numbers n and l that we have previously described. The arbitrary direction of the angular momentum is, as before, most conveniently described by quantizing the component m_l in a fixed direction. Similarly the orientation of the spin is most conveniently given by specifying its component m_s in the same fixed direction. Thus each electron in an atom, in the approximation in which we can neglect the interactions of their rotations, is specified by the four quantum numbers $n\ l\ m_l\ m_s$, just as it was before. All relations of these numbers remain unchanged. But now, each electron has a wave function—a function of its coordinates (and, as we shall describe presently, of a coordinate representing its spin) depending on the numbers $n\ l\ m_l\ m_s$, which is a solution of Schrödinger's equation for a particle in a central field. We can denote the function for the ith electron by $u(n_i/x_i)$, where n_i stands for the four numbers $n_i l_i m_{li} m_{si}$, and x_i symbolizes the four coordinates (three of position, one of spin) of the ith electron. Now it is well known that the product of these functions, for all the electrons $(1 \cdots N)$ of the atom, gives a function which approximately satisfies Schrödinger's equation. That is, $u(n_1/x_1)u(n_2/x_2) \cdots u(n_N/x_N)$ is an approximate solution. But it is not antisymmetric in the electrons, so that it does not satisfy the exclusion principle. To build up an antisymmetric solution, we first note that we still have an approximate solution, connected with the same energy value, if we interchange any two x's, obtaining for example $u(n_1/x_2)\ u(n_2/x_1) \cdots u(n_N/x_N)$. We still have an approximation with the same energy if we make a linear combination of any such solutions. Then we can make the one possible combination which is antisymmetric, and it will both satisfy the exclusion principle, and will be an approximate solution of Schrödinger's equation. This combination is conveniently written as a determinant:

[7] D. R. Hartree, Proc. Camb. Phil. Soc. 24, 89 (1928). See also J. C. Slater, Phys. Rev. 32, 339 (1928) for discussion of Hartree's method and application of some of present results to it.

$$\begin{vmatrix} u(n_1/x_1) & u(n_1/x_2) & \cdots & u(n_1/x_N) \\ \cdot & \cdot & \cdots & \cdot \\ u(n_N/x_1) & u(n_N/x_2) & \cdots & u(n_N/x_N) \end{vmatrix}$$

It is obviously antisymmetric, for interchanging, say, x_1 and x_2 interchanges two columns of the determinant, which by a familiar property merely changes the sign. It can be shown that it is the only antisymmetric combination of these functions. And it leads at once to the familiar interpretation of the exclusion principle. For if two of the functions had the same quantum numbers (say $n_1 = n_2$, symbolizing equality of four quantum numbers), then the corresponding rows of the determinant would be identical (since they contain the functions $u(n_1/x_1) = u(n_2/x_1)$, $u(n_1/x_2) = u(n_2/x_2)$, etc.) and by another familiar rule, the determinant will vanish. Thus there is no solution corresponding to the case where two electrons have the same set of quantum numbers. Further, the determinant treats all electrons alike; hence we cannot count as separate two states which differ only by the interchange of the quantum numbers of two electrons. Our exclusion principle then coincides with the one previously described.

We now have, corresponding to each set of quantum numbers, or to each independent bracket expression of quantum numbers (as given in Table I), which is allowed by Pauli's principle, a single, antisymmetric function of the electrons, which is an approximate solution of Schrödinger's equation. Our next task is, beginning with this, to introduce the interaction between the various angular momenta, and to try to improve the agreement with Schrödinger's equation, without destroying the property of antisymmetry. We use essentially the method of perturbations. The fundamental result of this method is that, if we take the approximate but incorrect wave function, and compute the matrix of the real energy with respect to this, the diagonal terms of this matrix are good approximations to the actual energy values of the problem. The errors remaining are of the order of the square of the ratio of non-diagonal to diagonal terms. Since we can easily show in our case that the non-diagonal terms here are really small, this method will give a good approximation to the energy values. We are then to take the real energy operator (involving the interactions between electrons, rather than with fictitious central fields), find its matrix with respect to the wave functions already determined and take these diagonal terms as energy levels.

There is, however, one case in which our criterion for the accuracy of this approximation is not valid. This is the case where a number of terms lie close together. Then in the first place one can no longer say that the errors are as small as we have assumed; in the second place, since we are generally interested in the energy differences between the neighboring terms, we really demand a much greater accuracy than usual, to give this difference correctly. Thus this case—that of degeneracy—demands special treatment. We see that, in our case, we actually meet this difficulty, for the various wave functions with the same values of n's and l's, but different m_l's and m_s's, have very nearly the same energy, and yet the energy differences, giving the separations between multiplet levels, are just what we wish to find. Thus the problem of degeneracy is the essential part of the calculation for complex spectra. We may describe the situation this way: the non-diagonal energy terms connected with transitions between states with different n's and l's for the electrons are negligible; but those between states with the same n's and l's but different m_l's although they may be no greater

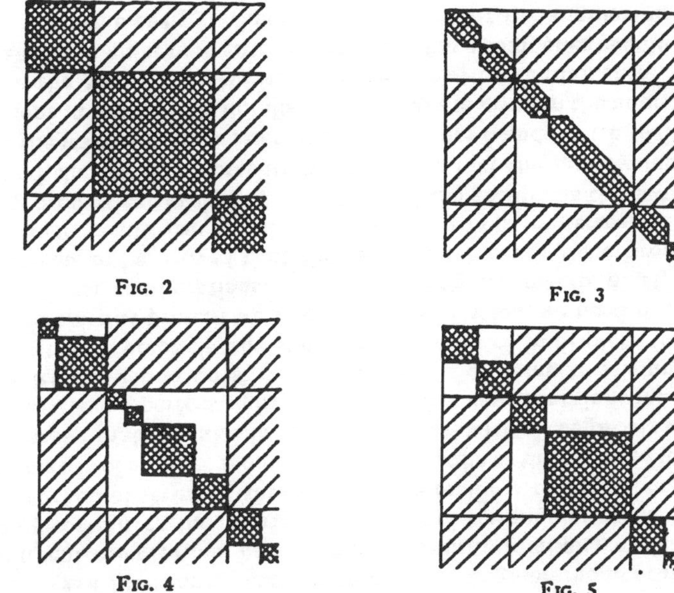

Figs. 2–5. Figures 2, 3, 4 represent the energy matrix schematically. In Fig. 2, it is with respect to the unperturbed wave functions. The terms in the double shaded squares represent transitions between two states with the same electron quantum numbers. These terms are all significant; those outside the squares are negligible. In Fig. 3 we have made the linear combinations to the perturbed wave functions, reducing each square to a diagonal matrix. Fig. 4 is with respect to unperturbed wave functions, as is Fig. 2; but we take account of the fact that there are no components between states of different $\sum m_l$, $\sum m_s$. In Fig. 4, energy levels are arranged according to the values of $\sum m_l$, $\sum m_s$. Fig. 5 (see Note 3) represents the matrix of angular momentum, in the final perturbed wave functions. Each square represents a separate multiplet. In Figs. 3 and 5 the terms are arranged according to multiplets.

numerically, are not negligible for our purpose. This situation can be described graphically. Suppose we make a scheme for the energy matrix, giving stationary states (the incorrect ones that we determine from the central field, denoted by the numbers $n_1 \cdots m_{sN}$) along the two sides of a square array, and putting in the matrix components. Then if we arrange together all terms with one set of n's and l's, then those with another set, and so on, we see that the matrix components in the double shaded regions

of Fig. 2 are all important (even the non-diagonal ones); but those in the singly shaded regions are negligible. The method of treatment is now this: we take a particular square of the matrix, as A, coming from transitions between different states with the same n's and l's. And, by linear combinations of the wave functions connected with these states, we reduce the matrix to a diagonal one. This can always be done for a finite matrix. Then we do the same thing with each such double-shaded square. The result after that is then a matrix of the form shown in Fig. 3. Here the only important non-diagonal terms have disappeared; so that we can take the diagonal terms as the approximate energy values of the real problem, with assurance that the errors are not large. We observe that, after this is done, there will still be just as many wave functions connected with each set of values of the n's and l's as before.

In our problem of complex spectra, there is a feature which greatly simplifies the calculation. For we shall prove[8] that the energy has no matrix components connected with transitions in which $\sum m_l$ or $\sum m_s$ change (if we neglect the magnetic energy, as we are doing for the present). To show the effect of this graphically, we can arrange the terms, first according to the n's and l's, as we have done; but under each such classification, we can arrange according to $\sum m_l$ and $\sum m_s$. Then the original energy matrix really has the simplified form shown in Fig. 4. And the process of removing the important non-diagonal terms reduces to separate linear combinations between the functions of each set having a given set of n's, l's, $\sum m_l$, and $\sum m_s$. Generally the number of functions actually present in one of these groups is very small; thus the problem by this method becomes very simple, and can be readily carried out. As a result of it, we see that the number of states with a given set of n's and l's, and a given $\sum m_l$ and $\sum m_s$, remains unchanged as we apply the interaction between angular momenta. And this is the essential point required in the classification of the terms.

3. The essential features of the perturbation theory as applied to complex atoms have been described; and we can now make connections with Hund's scheme for classifying the terms. Each unperturbed wave function can be described by its set of $(n\ l\ m_l\ m_s)$ for each electron (Pauli's principle being actually satisfied by our condition of antisymmetry). Then, just as before, we can make a table of all the possible terms with any set of n's and l's; we can find $\sum m_l$ and $\sum m_s$ for each of the terms; and we can make a diagram, as before, plotting $\sum m_l$ against $\sum m_s$, obtaining just the sort of point diagram that was shown in Fig. 1, b or c. The small square arrays of matrix components, in Fig. 4, are just the components between different terms represented by coincident points in Fig. 1, b and c. The components between terms lying at different points on the $\sum m_l$—$\sum m_s$ diagram are zero. Now we have just proved that, even when the interactions are considered, we still have the same diagram representing the number of states with each set of $\sum m_l$ and $\sum m_s$. To verify Hund's method completely, we need only show that we can correctly break up this diagram into a number of separate

[8] See Note 1 for the proof.

ones, as Fig. 1, d—i, each representing a separate multiplet. Obviously we can separate the diagram into rectangular arrays of points (that is, sets of wave functions with particular values of $\sum m_l$ and $\sum m_s$), but there are other conditions that must be satisfied to make these wave functions really represent a multiplet.

In looking for the conditions that a set of wave functions must satisfy to represent a multiplet, we first observe that, so long as we neglect magnetic energies, they must all have the same energy value. Now after we have made our linear combinations, we shall have perfectly definite wave functions, each with a perfectly definite energy value (the diagonal terms in Fig. 3). Without proof, it is not obvious that we can pick out sets of functions, with different $\sum m_l$ and $\sum m_s$, but all with the same energy. Even if we can (and we shall prove that it is possible),[9] the set of functions must satisfy further conditions. Somehow we must work in the fact that they all represent the same vector L, and the same S, but with different orientations in space. Now the wave functions corresponding to an angular momentum vector with different orientations have perfectly definite relations, which are expressed by computing the matrix components of the angular momentum. If, then, our group of terms of the same energy really is to represent a multiplet, the matrix of angular momentum must also have the required form. But we shall prove that this also is the case.[9] Thus, with these proofs given, we see that we have a perfect right to separate the wave functions into a group of multiplets, as Hund does. Since the method of doing this is unique, the result must give precisely the same classification of multiplets that Hund finds. And since at present we wish only classification, we need not actually make the linear combinations, and get the diagonal matrix of H, at all.

It remains to be proved that the terms, after making the proper linear combinations, really can be divided up into groups of terms, having the following properties: every wave function of the group has the same energy value; and the angular momentum matrix has just the same values that it would if the wave functions referred to the vectors L and S with their various orientations in space. The details of the proof are given in the notes[9]; but we can state the essential features. First we show, by fundamental methods, that the total angular momentum matrix can have components only between states of the same energy. Since we can show that the angular momentum matrix is not diagonal, this proves that there must be at least several states of each energy, in order that the angular momentum can have components between them. This breaks up the terms associated with a given set of n's and l's into sets each of the same energy. Next we show that each of these sets is just the sort we need for a multiplet. We do this by detailed consideration of the angular momentum matrix; the essential point being that the entire matrix can be determined uniquely from certain commutation relations between the x, y, z components of the angular momentum vector, and that these relations are just the same for the whole atom as for a single angular momentum vector, so that we must have just the same matrix. We then

[9] See Note 2.

have the complete proof that the wave functions of the atom really break up into sets each having all the properties of a multiplet; the $M_L—M_S$ diagram of such a multiplet is just as we assumed in the first section; and the result is that the method of finding what multiplets arise from any configuration of electrons, is precisely the method of Hund.

Part 2. Determination of the Energies

1. To find the energy values of the multiplets, the natural procedure is to solve the various problems of linear combination encountered above, find the correct wave functions, and compute the diagonal terms of the energy with respect to them. But we can find a simpler way, a way that even here will allow us to omit the actual calculation of the wave functions altogether. This is done by the use of the principle that the sum of a number of energy values is in some cases not changed by applying a perturbation, even if the individual values are. The principle is this: given a set of wave functions $u_1 \cdots u_i$; make a set of orthogonal linear combinations of them, $u_1' \cdots u_i'$. Consider the diagonal terms of a matrix, as the energy matrix, referred to the original wave functions: say $E_1 \cdots E_i$. Consider also the diagonal terms of the matrix of the same function, referred to the new wave functions, as $E_1' \cdots E_i'$. Then $E_1 + \cdots + E_i = E_1' + \cdots + E_i'$. In our case, we can use this as follows. We consider a set of terms, all with the same n's, l's, $\sum m_l$, $\sum m_s$ (that is, the terms connected with a small square in Fig. 4, or with coincident points in Fig. 2b and c). Before making our linear combinations, we can easily calculate the diagonal terms of the energy. We can add these terms for all the wave functions. Then the sum is equal to the sum of the corresponding energy values after making the linear combinations; that is, it is the sum of certain energy values that we wish to calculate. By proper use of this method, we can generally get the energies of all the multiplets.

To make the process clear, we shall illustrate by the case of two equivalent p electrons, shown in Fig. 2c. We compute the energy of the function $\sum m_s = 0, \sum m_l = 2$; that is, the energy of the configuration $\{(n11)\}\ \{(n11)\}$. But there is only one function connected with this point in the diagram. Hence the sum degenerates to one term. The process of making linear combinations is not necessary here; the corresponding square matrix array in Fig. 4 has but one row and column. Thus the energy we have found is also the energy of the term after interaction is considered; and comparison with (f) shows that it is the energy of the ^1D multiplet. Similarly the point $\sum m_s = 1, \sum m_l = 1$, gives the energy of ^3P. Again, $\sum m_s = 1, \sum m_l = 0$ or -1, and $\sum m_s = -1\ \sum m_l = 1, 0, -1$, should all give the energy of P; and in consequence of certain identities between the energy matrix components, these actually do give the same result as before. Also, the point $\sum m_s = 0$, $\sum m_l = 1$, has two wave functions, for the states $\{(n\,1\,0)\}\ \{(n1\,1)\}$ and $\{(n1\,1)\}\ \{(n1\,0)\}$. We calculate the energies of these terms, and add. And the result should, and does, come out to be the sum of the ^1D and ^3P energies. Finally, the point $\sum m_l = 0, \sum m_s = 0$ has three wave functions. We add their energies, and the result should be the sum of the energies of

^1D, ^3P, and ^1S. Since the first two of these energies are known, we need only subtract them from the sum to get the energy of ^1S. Thus the energies of all the multiplets are computed, without finding any perturbed wave functions at all. It is readily seen that extension of the same scheme will give complete information in all cases except where there is more than one multiplet of the same kind; for example, if a given configuration of electrons contains two ^2D multiplets, the method will give only the sum of the two terms. To get the individual terms, we must carry through the more elaborate scheme of taking linear combinations. But such cases are not found among the more important terms, so that in practical use the limitation is not important.

For actual use, the essential step is to find the energy values referred to the original wave functions; then it is simply a matter of adding and subtracting to get the energies of the multiplets. This is merely a problem of integration; for we know the original wave functions (they are the determinants mentioned above), and we can easily find the matrix components of energy with respect to them. The process is somewhat complicated, however, in many cases, and a careful arrangement of the arithmetic simplifies it greatly. To work out the actual method properly, we shall have to go more into detail than we have about the whole problem. This calculation will be made in sections 2 and 3; finally we give a number of examples of calculation of energy in section 4.

2. Our problem in the present section is this: to compute the diagonal term of the energy with respect to one of the approximate wave functions, written as a determinant, which we discussed earlier. If H represents the energy operator, and

$$u = \begin{vmatrix} u(n_1/x_1) & \cdots & u(n_1/x_N) \\ \cdot & \cdots & \cdot \\ u(n_N/x_1) & \cdots & u(n_N/x_N) \end{vmatrix}$$

is the wave function, we then have as the desired diagonal term

$$\frac{\int u^* H u dv}{\int u^* u dv}.$$

The functions u are determinants; that is, sums of the $N!$ terms formed from the product $u(n_1/x_1)\cdots u(n_N/x_N)$ by carrying out all permutations of the coordinates $x_1\cdots x_N$, even permutations having the coefficient $+1$, odd ones the coefficient -1. The product u^*u is then a double sum, and each of the integrals would be likewise; except that, in consequence of orthogonality relations, they really reduce to single sums, of which all the terms are identical.

We consider first the energy integral. The energy operator H consists, as we shall see, of three parts: first, a constant, which we need consider

123

no further; second, a sum of terms each depending on the coordinates of a single electron, and all the same in form, which we can write $H(x_1)+\cdots+H(x_N)$; third, a sum of terms (the Coulomb repulsions) each depending on the coordinates of two electrons, which we can write explicitly as Σ(all pairs) e^2/r, where r is the distance between the two. The whole integral then consists of a constant, and two triple sums, two of the summations being over the permutations found in u and u^*, the third over the terms in the energy operator. Consider a single term out of this triple sum, of the sort connected with the coordinates of just one electron. Then, if the permutations involved in u and u^* are different, at least two x's must have different quantum numbers in the product; for example, we may have

$$u^*(n'/x_k)u^*(n''/x_l)u(n'''/x_k)u(n''''/x_l),$$

where n' is different from n''', and n'' different from n''''. When we multiply this by a function of one electron, and integrate, there will always be at least one product (as $u^*(n'/x_k)u(n'''/x_k)$), which by orthogonality will integrate to zero. Thus the only terms which are not zero are those where we have the same permutation in u and in u^*; the double sum over permutations reduces to a single one. We are left, then, with

$$\Sigma(\text{permutations of } x_1\cdots x_N)\int u^*(n_1/x_1)\cdots u^*(n_N/x_N)$$
$$(H(x_1)+\cdots+H(x_N))u(n_1/x_1)\cdots u(n_N/x_N)dx_1\cdots dx_N.$$

A single integral from this sum, by normalization, immediately reduces to

$$\left[\int u^*(n_1/x_1)H(x_1)u(n_1/x_1)dx_1+\cdots+\int u^*(n_N/x_N)H(x_N)u(n_N/x_N)dx_N\right].$$

If we let these terms be $I(n_1),\cdots I(n_N)$, we see that the whole is simply the sum of the I's, counted $N!$ times; so that this part of the energy integral is $N!(I(n_1)+\cdots+I(n_N))$, where it will be observed that the sum of terms $H(x_i)$, each depending on the coordinates of one electron, has been changed into a sum of terms each depending on a single set of quantum numbers.

Next we take the terms e^2/r_{kl} in the energy. As before, we take a typical term of the triple summation. If it happens that the same permutation occurs both in u and u^*, then the integral will reduce to

$$\int u^*(n/x_k)u^*(n'/x_l)e^2/r_{kl}u(n/x_k)u(n'/x_l)dx_k dx_l,$$

which we will call $J(n;n')$. Each permutation will yield, from terms of this kind, the summation over all pairs $x_k x_l$; this reduces to the summation of J's over all pairs $n\ n'$; and the $N!$ permutations finally give, from terms of this sort, $N!\sum$(pairs of n's) $J(n;n')$. Next, it may be that the permutation in u^* differs from that in u by the interchange of two electron coordinates.

If in addition the term e^2/r_{kl} happens to refer to the particular coordinates $x_k x_l$ which have been interchanged, we have an integral which reduces to

$$\int u^*(n/x_l)u^*(n'/x_k)e^2/r_{kl}u(n/x_k)u(n'/x_l)dx_k dx_l,$$

which we call $K(n;n')$. Such terms all occur with negative sign; for, since the permutation in u^* differs by one interchange from that in u, one of the terms is always an even, the other an odd, permutation, resulting in a coefficient -1 in the double sum. Then by a similar argument to that above, these terms yield a contribution $-N!\sum$(pairs of n's) $K(n;n')$ to the energy integral. Finally, if the term of u^* differs from that of u by more than a single interchange, there will always be at least one coordinate whose wave functions will integrate to zero on account of orthogonality, so that no other terms exist.

Finally we need the normalization integral $\int u^* u\, dv$. Formally, we can find this as we did the first part of the energy integral, replacing $H(x_1)+\ldots+H(x_N)$ by unity. This results in replacing $\sum I(n)$ by unity; so that the integral is just $N!$. Thus division by the normalization factor simply removes the factor $N!$ which occurs in each term of the energy integral. We have then, for the diagonal term of the energy which we desire, constant$+\sum(n\text{'s})I(n)+\sum$(pairs of n's) $J(n;n')-\sum$(pairs of n's)$K(n;n')$.

3. Now we must consider the exact form of the wave function, to compute I, J and K. In connection with this, we must answer the question which no doubt will be felt at this point, as to the part which the spin plays in the calculations. We begin by considering the individual functions $u(n_i/x_i)$. Each such function is a solution of Schrödinger's equation for an electron in a central field. Let the potential of such a field be $U(xyz)$. Then Schrödinger's equation is

$$H_1 u=\left(-\frac{h^2}{8\pi^2 m}\left(\frac{\partial^2}{\partial x^2}+\frac{\partial^2}{\partial y^2}+\frac{\partial^2}{\partial z^2}\right)+U\right)u=\epsilon u.$$

Here u is a function of x, y, z, and certain coordinates representing the spin, depending parametrically on n, l, m_l, and m_s; ϵ is a function of the quantum numbers alone. To describe the spin, we proceed as Pauli[10] does: we use as a coordinate the component of the spin along our fixed axis, m_s (which thus appears both as quantum number and as coordinate). Thus we could write our equation

$$H_1 u(n\,l\,m_l\,m_s/x\,y\,z\,m_s)=\epsilon(n\,l\,m_l\,m_s)u(n\,l\,m_l\,m_s/x\,y\,z\,m_s).$$

Since H_1 is independent of the spin coordinate m_s, (neglecting magnetic interactions), we can separate variables, writing u as a product of a function of xyz, and a function of m_s:

$$u(n\,l\,m_l\,m_s/x\,y\,z\,m_s)=u(n\,l\,m_l/x\,y\,z)u(m_s/m_s).$$

[10] W. Pauli Jr., Zeits. f. Physik 43, 601 (1927).

The function $u(m_s/m_s)$ is to be interpreted as follows: m_s (the quantum number) can have two possible values, $\pm\frac{1}{2}$, and we have a different wave function for each quantum number. For example, we may have $u(\frac{1}{2}/m_s)$. But now assuming that the quantum number is $\frac{1}{2}$, we know that the wave function is different from zero only if the spin points along the positive axis, so that m_s (the coordinate) is $\frac{1}{2}$; for $m_s=-\frac{1}{2}$, there is no wave function. Thus $u(\frac{1}{2}/\frac{1}{2})=1$, $u(\frac{1}{2}/-\frac{1}{2})=0$. We can then write $u(\frac{1}{2}/m_s)=\delta(\frac{1}{2}/m_s)$. Similarly $u(-\frac{1}{2}/m_s)=\delta(-\frac{1}{2}/m_s)$; and we can write the whole symbolically as $u(m_s/m_s)=\delta(m_s/m_s)$. We now have

$$u(n\,l\,m_l\,m_s/x\,y\,z\,m_s)=u(n\,l\,m_l/x\,y\,z)\delta(m_s/m_s).$$

And for the first factor, the equation is

$$H_1 u(n\,l\,m_l/x\,y\,z)=\epsilon(n\,l)u(n\,l\,m_l/x\,y\,z),$$

where, since the energy does not depend on m_l or m_s, we have left out those quantum numbers in describing ϵ.

The solution of the central field problem is well known. We separate variables in spherical coordinates $r\,\theta\,\phi$. Then we have

$$u(n\,l\,m_l/x\,y\,z)=R(n\,l/r)\Theta(l\,m_l/\theta)\Phi(m_l/\phi)$$

where

$$\Theta(l\,m_l/\theta)=\left[\frac{(2l+1)(l-|m_l|)!}{(l+|m_l|)!}\right]^{1/2}P_l^{|m_l|}(\cos\theta)$$

where

$$P_l^{|m_l|}(\cos\theta)=\frac{1}{2^l l!}\sin^{|m_l|}\theta\frac{d^{|m_l|+l}(-\sin^2\theta)^l}{d(\cos\theta)^{|m_l|+l}}$$

and

$$\Phi(m_l/\phi)=e^{im_l\phi}/(2\pi)^{1/2}.$$

Thus we have

$$u(n\,l\,m_l\,m_s/x\,y\,z\,m_s)=R(n\,l/r)\Theta(l\,m_l/\theta)\Phi(m_l/\phi)\delta(m_s/m_s).$$

To proceed further, we must investigate the value of H operating on the product of u's. By definition,

$$H=-\frac{h^2}{8\pi^2 m}\sum_{i=1}^{N}\left(\frac{\partial^2}{\partial x_i^2}+\frac{\partial^2}{\partial y_i^2}+\frac{\partial^2}{\partial z_i^2}\right)+V,$$

where V is the potential energy, given by

$$V=\sum_{i=1}^{N}\left(-\frac{Ze^2}{r_i}\right)+\sum_{1}^{N}(i>j)\frac{e^2}{r_{ij}},$$

where r_i is the distance of the ith electron from the nucleus, r_{ij} the distance between the ith and jth electrons. Thus we have

$$Hu(n_1/x_1)\cdots u(n_N/x_N) = -\frac{h^2}{8\pi^2 m}\left(\frac{\partial^2 u(n_1/x_1)}{\partial x_1^2}+\frac{\partial^2 u(n_1/x_1)}{\partial y_1^2}+\frac{\partial^2 u(n_1/x_1)}{\partial z_1^2}\right)u(n_2/x_2)\cdots u(n_N/x_N)\cdots + Vu(n_1/x_1)\cdots u(n_N/x_N).$$

But by our assumptions,

$$\frac{-h^2}{8\pi^2 m}\left(\frac{\partial^2 u(n_1/x_1)}{\partial x_1^2}+\frac{\partial^2 u(n_1/x_1)}{\partial y_1^2}+\frac{\partial^2 u(n_1/x_1)}{\partial z_1^2}\right)=(\epsilon(n_1)-U(x_1))u(n_1/x_1),$$

etc. Thus we are left with

$$Hu(n_1/x_1)\cdots u(n_N/x_N)=\left[\sum_{i=1}^{N}(\epsilon(n_i)-U(x_i))+V\right]u(n_1/x_1)\cdots u(n_N/x_N).$$

This is of the form used above: Constant $+\sum H(x)+\sum e^2/r$, where constant $=\sum(n)\epsilon(n)$, and $H(x)=-U(x)-Ze^2/r$.

We are now ready to compute the integrals I, J, and K. We must first note that by integrating over the coordinates of one electron, we really mean integrating over the $dx\,dy\,dz$, and summing over the spins:

$$\int dv \equiv \sum_{-1/2}^{1/2}(m_s)\int dxdydz.$$

Thus, for example, we have the normalization and orthogonality of the individual wave functions:

$$\int u^*(n'/x)u(n''/x)$$

$$=\sum_{-1/2}^{1/2}(m_s)\delta(m_s'/m_s)\delta(m_s''/m_s)$$

$$\int u^*(n'l'm_l'/xyz)u(n''l''m_l''/xyz)dxdydz$$

$$=\delta(m_s'/m_s'')\delta(n'l'm_l'/n''l''m_l'').$$

This orthogonality is all that is needed in the proofs of the preceding paragraphs.

For the integral I, we have

$$I(nlm_lm_s)=\sum_{-1/2}^{1/2}(m_s)\delta(m_s/m_s)\delta(m_s/m_s)$$

$$\int u^*(nlm_l/xyz)H(xyz)u(nlm_l/xyz)dxdydz.$$

The summation over m_s merely reduces to the factor unity, independent of m_s. When we insert the value of $H(xyz)$, we note that the result is also

independent of m_l; for H is a function of r only, the functions of angles integrate to unity, and we are left with

$$I(nlm_lm_s)=I(nl)=-\int R^2(nl/r)(U(r)+Ze^2/r)4\pi r^2dr.$$

Next we are to find J and K. For J, we have

$$J(nlm_lm_s;n'l'm_l'm_s')$$

$$=\sum_{-1/2}^{1/2}(m_s)_k\sum_{-1/2}^{1/2}(m_s)_l\ \delta(m_s/(m_s)_k)\delta(m_s'/(m_s)_l)\delta(m_s/(m_s)_k)\delta(m_s'/(m_s)_l)$$

$$\int u^*(nlm_l/x_ky_kz_k)u^*(n'l'm_l'/x_ly_lz_l)e^2/r_{kl}u(nlm_l/x_ky_kz_k)u(n'l'm_l'/x_ly_lz_l)\,dx_kdy_kdz_kdx_ldy_ldz_l.$$

The summation again reduces to unity, independent of m_s and m_s'. Then our quantity reduces to the integral, which we can write

$$J(nlm_lm_s;n'l'm_l'm_s')=J(nlm_l;n'l'm_l').$$

Similarly we have

$$K(nlm_lm_s;n'l'm_l'm_s')$$

$$=\sum_{-1/2}^{1/2}(m_s)_k\sum_{-1/2}^{1/2}(m_s)_l\ \delta(m_s/(m_s)_l)\delta(m_s'/(m_s)_k)\delta(m_s/(m_s)_k)\delta(m_s'/(m_s)_l)$$

$$\int u^*(nlm_l/x_ly_lz_l)u^*(n'l'm_l'/x_ky_kz_k)e^2/r_{kl}u(nlm_l/x_ky_kz_k)u(n'l'm_l'/x_ly_lz_l)$$

$$dx_kdy_kdz_kdx_ldy_ldz_l.$$

The summation reduces to $\delta(m_s/m_s')$; so that we are left with

$$K(nlm_lm_s;n'l'm_l'm_s')=\delta(m_s/m_s')K(nlm_l;n'l'm_l'),$$

where the integral is symbolized by the last K. We thus observe that these exchange integrals only exist for electrons with spins parallel to each other.

We are now to compute

$$J(nlm_l;n'l'm_l')$$

and

$$K(nlm_l;n'l'm_l')$$

We recall the expression for u previously given. Also we use the familiar expansion

$$\frac{1}{r(xx')}=\sum(k,m)\frac{(k-|m|)!}{(k+|m|)!}\frac{r(a)^k}{r(b)^{k+1}}P_k^{|m|}(\cos\theta)P_k^{|m|}(\cos\theta')\exp(im(\phi-\phi'))$$

where $r(xx')$ is the distance between (xyz) and $(x'y'z')$, and where $r(a)$ is the smaller, $r(b)$ the greater, of r and r'. Forming the expression for J, we have integrals of the form $\int_0^{2\pi}\exp(im\phi)d\phi$ which vanish unless $m=0$.

The double sum over k and m thus reduces to a single one over k, m being always zero. Then we easily have

$$J(nlm_l;n'l'm_l') = \sum(k)a^k(lm_l;l'm_l')F^k(nl;n'l')$$

where

$$a^k(lm_l;l'm_l') = \frac{(2l+1)(l-|m_l|)!}{(l+|m_l|)!}\;\frac{(2l'+1)(l'-|m_l'|)!}{(l'+|m_l'|)!}$$
$$\int_0^\pi [P_l^{|m_l|}(\cos\theta)]^2 P_k^0(\cos\theta)\frac{\sin\theta}{2}d\theta$$
$$\int_0^\pi [P_{l'}^{|m_l'|}(\cos\theta')]^2 P_k^0(\cos\theta')\frac{\sin\theta'}{2}d\theta'$$

and

$$F^k(nl;n'l') = e^2(4\pi)^2\int_0^\infty\int_0^\infty R^2(nl/r)R^2(n'l'/r')\frac{r(a)^k}{r(b)^{k+1}}r^2r'^2drdr'.$$

One notes that the a's can be computed once for all, in terms of the spherical harmonics; the special properties of the atom in question appear only in the integrals F.

In a similar way we form K. Here the integrals over ϕ are of the form $\int_0^{2\pi}\exp\, i(m_l-m_l'+m)\phi\, d\phi$, vanishing unless $|m|=|m_l-m_l'|$. Thus we have

$$K(nlm_l;n'l'm_l') = \sum(k)b^k(lm_l;l'm_l')G^k(nl;n'l'),$$

where

$$b^k(lm_l;l'm_l') = \frac{(k-|m_l-m_l'|)!}{(k+|m_l-m_l'|)!}\;\frac{(2l+1)(l-|m_l|)!}{(l+|m_l|)!}\;\frac{(2l'+1)(l'-|m_l'|)!}{(l'+|m_l'|)!}$$
$$\left[\int_0^\pi P_l^{|m_l|}(\cos\theta)P_{l'}^{|m_l'|}(\cos\theta)P_k^{|m_l-m_l'|}(\cos\theta)\frac{\sin\theta}{2}d\theta\right]^2$$

and

$$G^k(nl;n'l') = e^2(4\pi)^2\int_0^\infty\int_0^\infty R(nl/r)R(n'l'/r)R(nl/r')R(n'l'/r')\frac{r(a)^k}{r(b)^{k+1}}r^2r'^2drdr'.$$

Using the values of the associated spherical harmonics one can compute the various a's and b's; although the writer has not succeeded in setting up closed formulas for them, since this would involve the integrals of products of three spherical harmonics, an unfamiliar form.[11] We give a table, including all the coefficients involved with s, p, d electrons (that is, $l, l' \leq 2$).

We have now obtained the diagonal term of energy which we desired: in section 2 we have found it in terms of certain integrals I, J, K, and in section 3 we have evaluated those integrals. Before passing to the examples, we should note one fact: that in finding the energy differences between multiplets, one needs only the integrals J and K, which do not depend

[11] See, however, J. A. Gaunt, l. c. Whether one has formulas or not, the table of values is certainly most convenient for computation.

TABLE OF $a^k(lm_l; l'm_l')$.
(Note: in cases with two ± signs, the two can be combined in any of the four possible ways).

Electrons	l	m_l	l'	m_l'	$k=0$	2	4
ss	0	0	0	0	1	0	0
sp	0	0	1	±1	1	0	0
	0	0	1	0	1	0	0
pp	1	±1	1	±1	1	1/25	0
	1	±1	1	0	1	−2/25	0
	1	0	1	0	1	4/25	0
sd	0	0	2	±2	1	0	0
	0	0	2	±1	1	0	0
	0	0	2	0	1	0	0
pd	1	±1	2	±2	1	2/35	0
	1	±1	2	±1	1	−1/35	0
	1	±1	2	0	1	−2/35	0
	1	0	2	±2	1	−4/35	0
	1	0	2	±1	1	2/35	0
	1	0	2	0	1	4/35	0
dd	2	±2	2	±2	1	4/49	1/441
	2	±2	2	±1	1	−2/49	− 4/441
	2	±2	2	0	1	−4/49	6/441
	2	±1	2	±1	1	1/49	16/441
	2	±1	2	0	1	2/49	−24/441
	2	0	2	0	1	4/49	36/441

TABLE OF $b^k(lm_l; l'm_l')$
(Note: in cases where there are two ± signs, the two upper, or the two lower, signs must be taken together).

Electrons	l	m_l	l'	m_l'	$k=0$	1	2	3	4
ss	0	0	0	0	1	0	0	0	0
sp	0	0	1	±1	0	1/3	0	0	0
	0	0	1	0	0	1/3	0	0	0
pp	1	±1	1	±1	1	0	1/25	0	0
	1	±1	1	0	0	0	3/25	0	0
	1	±1	1	∓1	0	0	6/25	0	0
	1	0	1	0	1	0	4/25	0	0
sd	0	0	2	±2	0	0	1/5	0	0
	0	0	2	±1	0	0	1/5	0	0
	0	0	2	0	0	0	1/5	0	0
pd	1	±1	2	±2	0	2/5	0	3/245	0
	1	±1	2	±1	0	1/5	0	9/245	0
	1	±1	2	0	0	1/15	0	18/245	0
	1	±1	2	∓1	0	0	0	30/245	0
	1	±1	2	∓2	0	0	0	9/245	0
	1	0	2	±2	0	0	0	15/245	0
	1	0	2	±1	0	1/5	0	24/245	0
	1	0	2	0	0	4/15	0	27/245	0
dd	2	±2	2	±2	1	0	4/49	0	1/441
	2	±2	2	±1	0	0	6/49	0	5/441
	2	±2	2	0	0	0	4/49	0	15/441
	2	±2	2	∓1	0	0	0	0	35/441
	2	±2	2	∓2	0	0	0	0	70/441
	2	±1	2	±1	1	0	1/49	0	16/441
	2	±1	2	0	0	0	1/49	0	30/441
	2	±1	2	∓1	0	0	6/49	0	40/441
	2	0	2	0	1	0	4/49	0	36/441

explicitly on the central field U at all. The integral I is needed only in finding the center of gravity of a multiplet system. The reason is that I depends only on n and l, and so is the same for all the various degenerate states with which we start our perturbation problem.

4 *Examples. One electron outside closed shells.* We need take but one case: a $3p$ electron outside completed K and L shells. Thus the scheme of electrons is $(1s)^2(2s)^2(2p)^6 3p$. We must now consider the various antisymmetric wave functions which are possible. By the method of symbolization mentioned above, in which we group in separate brackets the quantum numbers of electrons with parallel and antiparallel spins, we see that there are six wave functions, which we give below, together with the values of $\sum m_l$ and $\sum m_s$:

		Σm_l	Σm_s
$\{(100)(200)(211)(210)(21-1)(311)\}$	$\{(100)(200)(211)(210)(21-1)\}$	1	$\frac{1}{2}$
$\{(100)(200)(211)(210)(21-1)(310)\}$	$\{(100)(200)(211)(210)(21-1)\}$	0	$\frac{1}{2}$
$\{(100)(200)(211)(210)(21-1)(31-1)\}$	$\{(100)(200)(211)(210)(21-1)\}$	-1	$\frac{1}{2}$
$\{(100)(200)(211)(210)(21-1)\}$	$\{(100)(200)(211)(210)(21-1)(311)\}$	1	$-\frac{1}{2}$
$\{(100)(200)(211)(210)(21-1)\}$	$\{(100)(200)(211)(210)(21-1)(310)\}$	0	$-\frac{1}{2}$
$\{(100)(200)(211)(210)(21-1)\}$	$\{(100)(200)(211)(210)(21-1)(31-1)\}$	-1	$-\frac{1}{2}$

We note that the arrangement of $\sum m_l, \sum m_s$, is just that for a single multiplet ^2P. There are no cases in which more than one term has a given value of $\sum m_l, \sum m_s$; thus there is no need of applying the sum rule at all. The diagonal terms of the energy, computed with respect to these six wave functions, should give directly the energies of the six terms of the multiplet. But now we come back to our general principle; these six terms must have the same energy. We must actually compute the energies by our rules, and see that they are the same in each of the six functions. As has been mentioned before, we need only use the J and K terms. These terms, we recall, were $\sum(\text{pairs})J - \sum(\text{pairs with par. spins})K$. Now in these sums, many terms are the same for each of the six wave functions; all the terms, in fact, relating to pairs of electrons both in the closed shells. These terms can exert no influence on the multiplet separations, or anything of that sort. Thus we can leave them out, for our present purpose, as we left out the terms depending on the integrals I. The only terms we need retain are those in which our $3p$ electron is a member of the pairs. Thus for the first wave function we must compute the following:

$$2J(311;100)+2J(311;200)+2J(311;211)+2J(311;210)+2J(311;21-1)$$
$$-K(311;100)-K(311;200)-K(311;211)-K(311;210)-K(311;21-1).$$

For the second and third, we substitute respectively 310, 31-1 in place of 311; the fourth, fifth, and sixth evidently give the same three results already given. Let us now group these terms according to the shells that the $3p$'s interact with:

$$1s: 2J(311;100)-K(311;100)$$
$$2s: 2J(311;200)-K(311;200)$$
$$2p: 2J(311;211)+2J(311;210)+2J(311;21-1)-K(311;211)$$
$$-K(311;210)-K(311;21-1).$$

For the interaction with $1s$, we have

$$2J(311;100)=2\sum(k)a^k(11;00)F^k(31;10)=2F^0(31;10)$$
$$-K(311;100)=-\sum(k)b^k(11;00)G^k(31;10)=-\tfrac{1}{3}G^1(31;10).$$

Before going further, let us find the corresponding terms in the interaction of a 310 electron with the $1s$ shell:

$$2J(310;100)=2\sum(k)a^k(10;00)F^k(31;10)=2F^0(31;10)$$
$$-K(310;100)=-\sum(k)b^k(10;00)G^k(31;10)=-\tfrac{1}{3}G^1(31;10).$$

That is, the interaction integrals of an outer p electron with an s shell are the same whether the p electron has $m_l=1$ or 0 (or -1, as one immediately verifies). We shall show the same result to hold for the interaction with the $2p$ shell: for the (311) electron, the terms are

$$2J(311;211)+2J(311;210)+2J(311;21-1)-K(311;211)$$
$$-K(311;210)-K(311;21-1)$$
$$=2\sum(k)a^k(11;11)F^k(31;21)+2\sum(k)a^k(11;10)F^k(31;21)$$
$$+2\sum(k)a^k(11;1-1)F^k(31;21)-\sum(k)b^k(11;11)G^k(31;21)$$
$$-\sum(k)b^k(11;10)G^k(31;21)-\sum(k)b^k(11;1-1)G^k(31;21)$$
$$=2\sum(k)(a^k(11;11)+a^k(11;10)+a^k(11;1-1))F^k(31;21)$$
$$-\sum(k)(b^k(11;11)+b^k(11;10)+b^k(11;1-1))G^k(31;21)$$
$$=6F^0(31;21)-G^0(31;21)-\tfrac{2}{5}G^2(31;21).$$

Similarly for the (310) electron interacting with the $2p$ shell, we have

$$2\sum(k)(a^k(10;11)+a^k(10;10)+a^k(10;1-1))F^k(31;21)$$
$$-\sum(k)(b^k(10;11)+b^k(10;10)+b^k(10;1-1))G^k(31;21)$$
$$=6F^0(31;21)-G^0(31;21)-\tfrac{2}{5}G^2(31;21).$$

This agrees with the former value, showing that the interaction of either a 311 or a 310 (or, by a simple extension, a 31–1) electron with a completed p shell, give the same result, independent of m_l. Putting all these results together, all six levels of the ^2P multiplet give the same energy, by direct computation. Of course, this is merely a check of our general theorem that all the levels of any multiplet must have the same energy. The special properties of the a's and b's which lead to this result could be proved by use of that theorem.

Any configuration outside closed shells. We have just seen that the interaction energy of a single electron with a closed shell is independent of the m_l of the outer electron. We have proved this by direct computation for a p electron interacting with s or p shells, but we could extend the result to the general case, from our general theorem that all the terms of any multiplet have the same energy. But this has an important bearing on our general problem. For in any case an atom consists of a certain number of electrons outside closed shells, and the central closed shells. In the various unperturbed wave functions which we are to use, the outer electrons have different m_l's. In the energies of the terms, we are to compute all J's and K's connected with pairs of electrons in the atom. We have already seen that all pairs, both of which are in closed shells, will give identical contributions to each of the unperturbed terms. But now we can go further: all sets of pairs, in which one is an outer electron, the other one of the electrons of a closed shell, will, when summed over the electrons of the closed shell, give the same result for each unperturbed state. That is, as far as multiplet separations are concerned, closed shells exert no influence at all; they affect only the position of the whole set of multiplet terms. It is well known that the classification of the terms is independent of the existence of closed shells; this proves that the energy relations also depend only on the outer electrons.[13] For the rest of our examples, then, we shall consider only those electrons which are outside closed shells.

Two electrons outside closed shells, one in s state (helium, alkaline earths). Suppose we have the scheme (ns), $(n'p)$, for example. There are, in this case, the unperturbed wave functions symbolized by

Wave function	Σm_l	Σm_s
$\{(n\,0\,0)(n'\,1\,1)\}\ \{\ \}$	1	1
$\{(n\,0\,0)(n'\,1\,0)\}\ \{\ \}$	0	1
$\{(n\,0\,0)(n'\,1\,-1)\}\ \{\ \}$	−1	1
$\{(n\,0\,0)\}\ \{(n'\,1\,1)\}$	1	0
$\{(n\,0\,0)\}\ \{(n'\,1\,0)\}$	0	0
$\{(n\,0\,0)\}\ \{(n'\,1\,-1)\}$	−1	0
$\{(n'\,1\,1)\}\ \{(n\,0\,0)\}$	1	0
$\{(n'\,1\,0)\}\ \{(n\,0\,0)\}$	0	0
$\{(n'\,1\,-1)\}\ \{(n\,0\,0)\}$	−1	0
$\{\ \}\ \{(n\,0\,0)(n'\,1\,1)\}$	1	−1
$\{\ \}\ \{(n\,0\,0)(n'\,1\,0)\}$	0	−1
$\{\ \}\ \{(n\,0\,0)(n'\,1\,-1)\}$	−1	−1

By our general scheme of classification, we have a ^1P and ^3P term. The terms with $\sum m_l$, $\sum m_s$ equal respectively to (1 1), (0 1), (−1 1), (1 −1), (0 −1), (−1 −1), belong to the ^3P state. On the other hand, the remaining terms, as (1 0), are each degenerate. The sum of the energies of two such terms equals the sum of the ^1P and ^3P energy. Thus for example we have

$$^3P:\{(n\,0\,0)(n'\,1\,1)\}$$

$$^3P+{}^1P:\{(n\,0\,0)\}\{(n'\,1\,1)\}+\{(n'\,1\,1)\}\{(n\,0\,0)\}=2\{(n\,0\,0)\}\{(n'\,1\,1)\}.$$

That is, for the energies, we have

$$^3P:J(n\,0\,0;n'\,1\,1)-K(n\,0\,0;n'\,1\,1)$$

$$^3P+{}^1P=2J(n\,0\,0;n'\,1\,1).$$

Therefore $^1P=J(n\,0\,0;n'\,1\,1)+K(n\,0\,0;n'\,1\,1)$

The singlet and triplet are thus given by a definite value $\pm K(n\,0\,0;n'\,1\,1)$. This checks with Heisenberg's[14] calculation of this case; our integral K is readily seen to be the same exchange integral which he computes.

[13] See W. Heitler, Zeits. f. Physik 46, 70 (1928).
[14] W. Heisenberg, Zeits. f. Physik 39, 499 (1926).

Shell of equivalent p electrons. (Elements C, N, O, F, etc.) p^2. This is the case shown in Fig. 1c. There are 15 wave functions, of which we give those with $\sum m_l\geq 0, \sum m_s\geq 0$; for simplicity we omit the total quantum number from our descriptions, so that (1 0), for example, stands for (n 1 0). Then, correlating the $\sum m_l$, $\sum m_s$ with the multiplets represented (from the figure), we have

Multiplet	M_l	M_s	Wave function
^1D	2	0	$\{(1\,1)\}\ \{(1\,1)\}$
$^1D+{}^3P$	1 1	0 0	$\{(1\,0)\}\ \{(1\,1)\}$ $\{(1\,1)\}\ \{(1\,0)\}$
$^1D+{}^3P+{}^1S$	0 0 0	0 0 0	$\{(1\,1)\}\ \{(1\,-1)\}$ $\{(1\,0)\}\ \{(1\,0)\}$ $\{(1\,-1)\}\ \{(1\,1)\}$
^3P	1	1	$\{(1\,1)\ (1\,0)\}\ \{\ \}$
^3P	0	1	$\{(1\,1)\ (1\,-1)\}\ \{\ \}$

Now we have for the energies

$$^1D:J(n\,1\,1;n\,1\,1)=F^0(n\,1;n\,1)+\frac{1}{25}F^2(n\,1;n\,1)$$

$$^3P:J(n\,1\,1;n\,1\,0)-K(n\,1\,1;n\,1\,0)=F^0(n\,1;n\,1)-\frac{2}{25}F^2(n\,1;n\,1)-\frac{3}{25}G^2(n\,1;n\,1)$$

We note that, for equivalent electrons, the F's and G's of the same indices are equal. Hence

$$^3P=F^0(n\,1;n\,1)-\frac{5}{25}F^2(n\,1;n\,1).$$

We can check the same value from the other ^3P term ($\Sigma m_l=0$, $\Sigma m_s=1$) and from the two terms giving $^1D+{}^3P$. Finally we have

$$^1D+{}^3P+{}^1S:J(n\,1\,1;n\,1\,-1)+J(n\,1\,0;n\,1\,0)+J(n\,1\,-1;n\,1\,1)$$
$$=3F^0(n\,1;n\,1)+\frac{6}{25}F^2(n\,1;n\,1).$$

Therefore $^1S=F^0(n\,1;n\,1)+\frac{10}{25}F^2(n\,1;n\,1)$.

We note that the term $F^0(n\,1;n\,1)$ is common to all the levels; thus we can leave it out, as we have all the terms which do not affect the separation. We note from its definition that $F^2(n\,1;n\,1)$ is positive. Thus we see that of the three multiplets, the ^3P lies lowest, ^1D next, ^1S highest, in accordance with Hund's rule that the terms of largest L and S lie lowest. The separations are in a simple ratio: $^3P-{}^1D=6/25\ F^2(n\,1;n\,1)$, and $^1D-{}^1S=9/25\ F^2(n\,1;n\,1)$ so that the ratio is 2 to 3.

We should find an example in the lowest levels of C, $(1s)^2\,(2s)^2\,(2p)^2$; these levels, however, have not been observed, as far as the writer knows. For Si $(1s)^2\,(2s)^2\,(2p)^6\,(3s)^2\,(3p)^2$, however, the term values are[15] $^3P=65{,}615$,

[15] McLennan and Shaver, Roy. Soc. Canada 18, 1, (1924). A. Fowler, Proc. Roy. Soc. A123, 422 (1929).

$^1D=59{,}466$, $^1S=50{,}370$, giving $^3P-{}^1D=6149$, $^1D-{}^1S=9096$, the first giving $F^2(3\,1;3\,1)=25{,}602$, the second $F^2(3\,1;3\,1)=25{,}267$, in good agreement.

p^3. The method is so similar to that used for p^2 that the calculations need not be given in detail. The multiplets are $^4S\ {}^2D\ {}^2P$, the ^4S lying lowest and ^2P highest, again in agreement with Hund's rule. The separations are given by $^4S-{}^2D=9/25\ F^2(n\,1;n\,1)$ $^2D-{}^2P=6/25\ F^2(n\,1;n\,1)$. An example is found in the normal N spectrum, $(1s)^2\,(2s)^2\,(2p)^3$, observed by Compton and Boyce.[15] They find $^4S=117{,}345$, $^2D=98{,}143$, $^2P=88{,}537$, giving $^4S-{}^2D=19{,}200$, $^2D-{}^2P=9{,}600$. From the first, $F^2(2\,1;2\,1)=53{,}400$, and from the second, it is 40,000; a somewhat poorer agreement than before, probably on account of the tighter binding of the electrons.

p^4. Here the relations prove to be as in p^2 as regards energy, as well as in the arrangement of terms: $^3P-{}^1D=6/25\ F^2(n\,1;n\,1)$ $^1D-{}^1S=9/25\ F^2(n\,1;n\,1)$. An example is normal O, $(1s)^2\,(2s)^2\,(2p)^4$. Here it is estimated[16] that, counting terms up from ^3P as zero, $^1D=25{,}500$, and $^1S=65{,}000$, giving $^3P-{}^1D=25{,}500$, $^1D-{}^1S=39{,}500$. From the first, $F^2(2\,1;2\,1)=106{,}000$, and from the second 109,700, a very good agreement. It is interesting to note the increase in the integral from N to O on account of the tighter binding.

p^5 and p^6 yield each only one multiplet, so that they need not be considered.

Shell of equivalent d electrons (Iron group) d^2. The multiplets are $^3F\ {}^3P\ {}^1G\ {}^1D\ {}^1S$. When we work out the separations, however, it appears that they are not arranged in this order; the singlet terms are anomalous, disobeying Hund's rule, in that the ^1D lies lower than ^1G. Except for this, however, the arrangement is as we should expect, ^3F lying below ^3P, and being the lowest term of the combination. The separations are given by

$$^3F-{}^3P=(135/441)\ F^2(n\,2;n\,2)-(75/441)\ F^4(n\,2;n\,2)$$
$$^3F-{}^1G=(108/441)\ F^2(n\,2;n\,2)+(10/441)\ F^4(n\,2;n\,2)$$
$$^3F-{}^1D=(45/441)\ F^2(n\,2;n\,2)+(45/441)\ F^4(n\,2;n\,2)$$
$$^3F-{}^1S=(198/441)\ F^2(n\,2;n\,2)+(135/441)\ F^4(n\,2;n\,2)$$

The separations now depend on the two parameters $F^2(n\,2;n\,2)$ and $F^4(n\,2;n\,2)$, which bear no fixed relation to each other. Nevertheless we can estimate their relative magnitude. For by definition F^k is the integral of a certain function of r_1 and r_2, multiplied by r_a^k/r_b^{k+1} where r_b is the greater, r_a the less, of r_1 and r_2. Thus increasing k necessarily decreases the integrand, and hence the function F: $F^4(n\,2;n\,2)<F^2(n2\,;n\,2)$. Rough calculation indicates that the decrease is about to a half. If then we provisionally take F^4 to be half of F^2, we have the separations in the ratio

$$^3F-{}^3P:135-37=\ 98 \qquad {}^3F-{}^1D:\ 45+22=\ 67$$
$$^3F-{}^1G:108+\ 5=113 \qquad {}^3F-{}^1S:198+67=265.$$

The order of terms is thus expected to be $^3F\ {}^1D\ {}^3P\ {}^1G\ {}^1S$.

[15] K. T. Compton and J. C. Boyce, Phys. Rev. 33, 145 (1929).

[16] McLennan, McLeod, and Ruedy, Phil. Mag. Sept., 1928, p. 558. These values are estimated from the energy level diagram in that paper.

Experimentally we find an example in the normal spectrum of Ti,[17] $(1s)^2\,(2s)^2\,(2p)^6\,(3s)^2\,(3p)^6\,(4s)^2\,(3d)^2$. Here the order of terms is in fact just what our calculation predicts, the exception to Hund's rule being found experimentally. The observed separations are approximately

$$^3F-{}^3P=\ 8500 \qquad {}^3F-{}^1D=\ 7200$$
$$^3F-{}^1G=12100 \qquad {}^3F-{}^1S=15100.$$

If we assume that F^4 is really half of F^2, these give $F^2(n\,2;n\,2)$ equal respectively to 38,000, 47,000, 41,000, 72,000, in fair agreement except for the last one. This last results from the ^1S, which is not nearly so far above the rest of the terms as the theory would indicate—probably because the second order corrections for this term would be large, and would have the effect of depressing it. The agreement with observations can be somewhat, but not much, improved by taking a slightly different ratio of F^4 to F^2.

$d^3..d^8$. We shall give only the lowest terms for the rest of the d shell. The terms of highest multiplicity prove to lie in general lower than the others, as the rule would predict. Of these, we have for d^3: $^4F\ {}^4P$; d^4, ^5D; d^5, ^6S; d^6, ^5D; d^7, $^4F\ {}^4P$; d^8, $^3F\ {}^3P$. Evidently the only cases where there are significant separations in the multiplets of highest multiplicity are the $d^2\ d^3\ d^7\ d^8$ $F-P$ separations. When one calculates, one discovers the fact that all these are given by the same formula, $(135/441)\,F^2(n\,2;n\,2)-(75/441)\,F^4(n\,2;n\,2)$. This permits an interesting comparison with experiment: we can compare the observed separations for Ti (d^2), V(d^3), Co(d^7) and Ni (d^8). We should expect these to increase regularly with the number of d electrons. Experimentally this separation is about 8000 for Ti, 9500 for V, 14000 for Co, 15000 for Ni, indicating a fairly uniform increase of about 1000 to 1500 for the addition of one d electron.

Non-equivalent p's: Two p's: pp'. The multiplets are $^3D\ {}^3P\ {}^3S\ {}^1D\ {}^1P\ {}^1S$. The triplets are particularly interesting, and we give their separations. If the principal quantum numbers of the electrons are n and n', they are

$$^3D-{}^3P=-6/25F^2(n\,1;n'\,1)+2G^0(n\,1;n'\,1)-4/25G^2(n\,1;n'\,1)$$
$$^3D-{}^3S=9/25F^2(n\,1;n'\,1)-9/25G^2(n\,1;n'\,1).$$

The significant feature of these results is that they depend in such a complicated way on several integrals. That is, in a case like this— and it is the simplest set of multiplets that can be built up from non-equivalent electrons—we must not look for simple numerical relations between the separations, or even for a definite, fixed order for the terms. We may rather expect that, as we go from one element to another, the relative order of terms can change.

We find an example in N^+ $(1s)^2\,(2s)^2\,2p\,3p$.[18] Here the experimental separations are approximately $^3D-{}^3P=4000$ $^3D-{}^3S=2300$, disobeying Hund's rule, in that ^3S lies below ^3P. We can make from these figures a rough estimate of the integrals F and G. We note that G differs from F in having

[17] H. N. Russell, Astroph. Jour. 66, 283 (1927).

[18] A. Fowler, Proc. Roy. Soc. A107, 31 (1925).

the product $R(n/x_1)\, R(n'/x_1)$ in its integrand instead of $R^2(n/x_1)$. The first is less than the second—very much less if the orbits n and n' are of decidedly different size, for then either $R(n/x_1)$ or $R(n'/x_1)$ will be small through most of the range of x_1. Thus each G integral is small compared to the corresponding F. It is reasonable to suppose that G^0 is of the same order of magnitude as F^2. Further, G^2 will be smaller than G^0, by analogy with what we have already seen, and as a rough assumption we may take $G^2=\frac{1}{2}G^0$. Thus we have the two equations

$$4000=-(6/25)F^2+(48/25)G^0$$
$$2300=(9/25)F^2-(9/50)G^0,$$

giving $F^2(2\,1;3\,1)=7900$, $G^0(2\,1;3\,1)=3070$, $G^2(2\,1;3\,1)=1535$. These values are reasonable, but provide no definite check for the equations.

Five equivalent p's, one other p: p^5p'. Again we have ${}^3D\,{}^3P\,{}^3S\,{}^1D\,{}^1P\,{}^1S$, by combination of the p' with the ${}^2P$ of p^5. But now it proves that the triplet separations are quite different, and have a much simpler formula. We have

$${}^3D-{}^3P=(6/25)F^2(n\,1;n'\,1)$$
$${}^3D-{}^3S=-(9/25)F^2(n\,1;n'\,1)$$

depending on only one parameter (the others all cancel out, seemingly, although perhaps not really, by chance). Thus we can predict definitely the order of terms: ${}^3S$ lies lowest, then ${}^3D$, finally ${}^3P$, in direct contradiction to Hund's rule. We find an example, however, in Ne $(1s)^2\,(2s)^2\,(2p)^5\,3p$, Paschen's p terms,[19] and it definitely verifies the contradiction. Experimentally, Paschen's $2p_{10}$, which is the ${}^3S$, lies well below any others. On account of the wide multiplet separations it is impossible to show any good check of the ratio 2:3 for the ${}^3D-{}^3P$ and ${}^3D-{}^3S$ separations. If one takes centers of gravity of Paschen's terms, one finds

$${}^3D-{}^3P=570 \qquad {}^3D-{}^3S=-2023$$

the first giving $F^2(2\,1;\,3\,1)=2370$, the second 5620. The correct figure is probably between these. It is interesting to note the agreement as to order of magnitude of this with the same one, 7900, for N^+.

Notes

1. We wish to prove that the energy has no matrix components (taken with reference to the incorrect, approximate wave functions) connected with transitions in which $\sum m_l$ or $\sum m_s$ changes. The essential part of the proof is the demonstration that the energy operator H commutes with the operators connected with $\sum m_l$ or $\sum m_s$. For $\sum m_s$ the proof is obvious: $\sum m_s$ is a quantity depending only on the spins, H only on the coordinates, and operators depending on entirely independent quantities always commute. For $\sum m_l$, the essential point is that $\sum m_l$ is the operator connected with an infinitesimal rotation of space about the z axis, and on account of the fact that H is in-

[19] F. Paschen, Ann. d. Physik 60, 405 (1920).

dependent of orientation, such a rotation does not affect it, and so commutes with it. To be more specific, one can write the three components of angular momentum of an electron in operator form, using the operators for linear momentum. Put in spherical coordinates, the z component of angular momentum is represented by the operator $(h/2\pi i)\partial/\partial\phi$. Operating on a single electron wave function with the factor $e^{im_l\phi}$, this operator reduces to the multiplication by $m_lh/2\pi$, so that it has a diagonal matrix, and m_l measures the z component of angular momentum, in units $h/2\pi$. Then the operator connected with $\sum m_l$ is $-i\sum\partial/\partial\phi$, where the sum is over the ϕ's of the various electrons in the many-electron wave function; this operator is in the many electron problem again a diagonal matrix. This holds, we note, even with the incorrect, unperturbed wave functions. Except for the factor $-i$, this operator simply represents the change in the function it operates on, if all φ's are increased by the same amount; that is, if the whole electronic system is rotated rigidly. Then we have

$$\left(-i\sum\frac{\partial}{\partial\phi}\right)Hu=-i\sum\frac{\partial H}{\partial\phi}u-iH\sum\frac{\partial}{\partial\phi}u.$$

But on account of its spherical symmetry, $\Sigma\partial H/\partial\phi=0$. Hence

$$\left(-i\sum\frac{\partial}{\partial\phi}\right)Hu=H\left(-i\sum\frac{\partial}{\partial\phi}\right)u, \quad \text{or}$$
$$\left(\sum m_l\right)H-H\left(\sum m_l\right)=0,$$

showing that the operator $\sum m_l$ commutes with H.

Knowing that both $\sum m_l$ and $\sum m_s$ commute with H, and that both have diagonal matrices, the rest of the proof is simple. We merely write down the commutation laws in matrix form:

$$\sum(n'')\sum m_l(n'/n'')H(n''/n''')-H(n'/n'')\sum m_l(n''/n''')=0.$$

On account of the diagonal relation, this amounts to

$$\left(\sum m_l(n'/n')-\sum m_l(n'''/n''')\right)H(n'/n''')=0.$$

This cannot be satisfied unless either $\sum m_l(n'/n')=\sum m_l(n'''/n''')$ or unless $H(n'/n''')=0$; that is, H has matrix components different from zero only if $\sum m_l$ has the same values in initial and final states. The same proof holds for $\sum m_s$.

2. *Theorems regarding angular momentum.* It is assumed that we have wave functions which, although not exact, are the good approximations used in this paper; that is, linear combinations have been made so that the energy matrix has components different from zero only between states of decidedly different energy; it is diagonal as far as the states are concerned which come from one set of electron quantum numbers. We wish to show first that matrix components of angular momentum have non-negligible components only between states of the same energy (or diagonal term of the energy matrix). Suppose we let the operators connected with the three com-

ponents of orbital angular momentum be M_L^x, M_L^y, M_L^z, where M_L^z is the quantity usually denoted by M_L, and is equal to $\sum m_l$. Similarly for spin we have M_S^x, M_S^y, M_S^z. Now we have seen that M_L^z and M_S^z commute with H; and, since there is nothing peculiar about the z axis, we can equally well show that M_L^x, M_L^y, M_S^x, M_S^y commute with H. Thus for example $M_S^xH-H\,M_S^x=0$, or in matrix form

$$\sum(n'')M_S^x(n'/n'')H(n''/n''')-H(n'/n'')M_S^x(n''/n''')=0.$$

We can write this, for non-diagonal terms,

$$M_S^x(n'/n''')\left[H(n'''/n''')-H(n'/n')\right]$$
$$=-\sum(n''\neq n',n''')\left[M_S^x(n'/n'')H(n''/n''')-H(n'/n'')M_S^x(n''/n''')\right].$$

The non-diagonal terms of the energy, $H(n''/n''')$, etc., are by hypothesis different from zero only if n'' and n''' refer to states with different electron quantum numbers, and even so they are small of the first order. If the right side were precisely zero, we should have obviously the result that $M_S^x(n'/n''')$ was different from zero only if $H(n'''/n''')$ was equal to $H(n'/n')$ which is what we wished to show. As it is, this result is true only to the first order of small quantities: M_S^x can have other, small components, and $H(n'''/n''')$ can differ from $H(n'/n')$ by small quantities. Thus we should expect that our approximate wave functions for a multiplet would give only approximately equal energies. This is not true, however, when we actually calculate by our method; the energies are precisely equal. The reason can be easily seen. Let the states n' and n''' be both connected with the same electron quantum number. Then the matrix components on the left side of the equation above are all found in terms of integrals over wave functions of those particular electron quantum numbers. The matrices on the right, however, involve other states, n'', which must refer to different electron quantum numbers, so that the right side would involve different, independent integrals. The two sides could not be in general equal unless each was zero. Hence the result is: if the approximate wave functions are computed by our method, the angular momentum has components, between two states of the same electron quantum numbers, only if both states have the same energy. Its components between states of different electron quantum number are small of the first order. From now on, we can neglect the latter terms, as we did with the energy. Thus the matrix of the angular momentum is of the form shown in Fig. 5. This can be compared with Fig. 3, showing the matrix of the energy in the final wave functions; but Fig. 3 should now be numbered to show that the diagonal terms of energy within one of the small squares of Fig. 5 (that is, in a multiplet) are all equal.

We have shown that the wave functions connected with a set of n's and l's are divided into groups, each wave function of a group having the same energy value, and the angular momentum having components only between different wave functions of a group, not from one group to another. We wish now to show that the matrix components of angular momentum within such a group are really such as to indicate a vector of magnitude L, and

another S. To do this, we first note that M_L^x, M_L^y, M_L^z, M_S^x, M_S^y, M_S^z satisfy the commutation relations

$$M_L^yM_L^z-M_L^zM_L^y=-\frac{h}{2\pi i}M_L^x,\ \text{etc.},$$

$$M_S^yM_S^z-M_S^zM_S^y=-\frac{h}{2\pi i}M_S^x,\ \text{etc.},$$

$$M_L^x\,M_S^x-M_S^x\,M_L^x=0,\ \text{etc.}$$

We show these by proving them for the separate m_l's and m_s's of the separate electrons, and then combining by the relations $M_L^x=\sum m_l^x$, etc. For the separate electrons, the results for the m_l's are well known; for the m_s's, the results come directly from Pauli's theory of the electron spin. We also note; that M_L^z and M_S^z have diagonal matrices, with diagonal values M_L and M_S (in units of $h/2\pi$). Now it can be proved[20] that if we have a set of matrices related by the commutation rules $M_L^yM_L^z-M_L^z\,M_L^y=-(h/2\pi i)M_L^x$, etc.; if these have components only between a limited set of states and if M_L^z forms a diagonal matrix; then the whole matrix is uniquely determined, except for a phase constant; further, M_L^z must have integral, or half integral, characteristic values M_L, ranging from a value $-L$ to L; the states can be described by the values of M_L, so that can be taken as a quantum number. Then the matrix components prove to be given by

$$M_L^x(M_L/(M_L+1))M_L^x((M_L+1)/M_L)=\frac{1}{4}\left(\frac{h}{2\pi}\right)^2[L(L+1)-M_L(M_L+1)]\cdots$$
$$M_L^z(M_L/M_L)=M_L$$
$$[(M_L^x)^2+(M_L^y)^2+(M_L^z)^2](M_L/M_L')=\left(\frac{h}{2\pi}\right)^2L(L+1)\delta(M_L/M_L').$$

Such a set of matrix components is the unique description of a vector L, in its $(2L+1)$ possible orientations. Similar results apply to M_S.

In our case, these relations lead immediately to the following results. In the first place, the M_L^x's depend only on orbital coordinates, the M_S^x's only on spins. Thus the M_L^x's can have components only between two wave functions in which the spin appears in the same way (and which thus have the same M_S). Similarly M_S^x's can have components only between states of the same M_L. Therefore for each M_S there are a number of states of the multiplet, with different M_L's and having matrix components of the M_L^x's between them as given above. These different sets of states, on the other hand, can differ only in the spin function, so that the components of M_L^x's are the same for each set with each M_S. Similarly, for each M_L, there are a number of states, with different spin functions but with the same orbital functions, different M_S's, and components of M_S^x's just as above. The result is an array of terms which can be plotted in precisely the rectangular form used in the paper, and corresponding to the $(2L+1)$ orientations of the vector L, combined with the $(2S+1)$ orientations of S.

[20] Born, Heisenberg, and Jordan, Zeits. f. Physik 35, 557 (1926).

Measurement of Nuclear Spin

Hyperfine structures of spectral lines and alternating intensities of band spectra constitute at present the only available means of determining angular momenta of atomic nuclei. We wish to point out another means of finding nuclear spins. It is well known that the Stern Gerlach experiment allows one to determine the angular momentum of an electronic configuration. If the atom has an angular momentum $j(h/2\pi)$ there are $2j+1$ lines in the Stern Gerlach pattern for conditions where the velocity of the atomic beam is sharply defined. It is also obvious that if the inhomogeneous magnetic field used in the experiment is not too strong, the coupling of the electrons to the nucleus will not be destroyed. There will now be $(2j+1)$ $(2i+1)$ distinct states in a magnetic field, where $i(h/2\pi)$ is the angular momentum of the nucleus. It is possible, in some cases, to observe the pattern due to these states and to follow the transition to the strong field condition with $2j+1$ lines.

The number of Stern-Gerlach lines observed in a weak field, their positions, and the magnetic field strength necessary to bring about a partial transition to the strong field pattern will be seen to determine the value of the nuclear spin.

As an example consider an atom in a state

with inner quantum number $j=\frac{1}{2}$ and nuclear spin i. The energy of the atom in a magnetic field is

$$W=-\frac{\Delta W}{2(2i+1)}\pm\frac{\Delta W}{2}\left(1+\frac{2m}{i+\frac{1}{2}}x+x^2\right)^{1/2};$$

$$x=\frac{g\omega}{\Delta W},\quad \omega=\frac{eh}{4\pi mc}\quad H=\mu_0 H$$

where ΔW is the separation between the two hfs components in the absence of a magnetic field, g is the Landé g factor for the electronic configuration (e.g. 2 for ^2S terms), m is the magnetic quantum number and H is the magnetic field. The force due to an inhomogeneous magnetic field is

$$F=-\frac{dW}{dH}\frac{\partial H}{\partial y}$$

where dH/dy is the gradient of the magnetic field. Thus

$$F=\pm\frac{\frac{2m}{2i+1}+x}{2\left(1+\frac{4m}{2i+1}x+x^2\right)^{1/2}}\cdot g\mu_0\frac{\partial H}{\partial y}.$$

For

$$i=\tfrac{1}{2},\quad F=\pm\frac{x+m}{2(1+2mx+x^2)^{1/2}}\cdot g\mu_0\frac{\partial H}{\partial y}.$$

For

$$m=1,\ 0,\ 0,\ -1$$

$$F=\left(1,\frac{x}{(1+x^2)^{1/2}}-\frac{x}{(1+x^2)^{1/2}}-1\right)(g/2)\mu_0\frac{\partial H}{\partial y}.$$

In a weak field the Stern-Gerlach pattern should consist of three lines the central one having twice the intensity of the two deflected lines. In intermediate fields there are 4 lines and in strong fields (complete Paschen-Back effect for hfs) there are only two lines both of which are displaced. The weak field region may be said to correspond to $x<0.1$, the intermediate to $x=1$, and the strong to $x>3$. If ΔW measured in cm^{-1} is denoted by $\Delta\nu$ then H in gauss is

$$H=2.14\times 10^4\times(\Delta\nu)/g.$$

For the normal ^2S state of $Cs\Delta\nu=0.30\ \mathrm{cm}^{-1}$ and the low field region $x=0.1$ lies below 320 gauss while $x=3$ for 9.6×10^3 gauss. For $x=1/(8)^{1/2}=0.354$ the four lines will be equidistant. This corresponds to a field of $1.14\times 10^{+3}$ gauss.

The nuclear spins of Cs and Rb are at present being investigated with this method by one of us (I. I. R.).

G. Breit
I. I. Rabi

New York University,
Columbia University,
November 10, 1931.

A New Method of Measuring Nuclear Magnetic Moment*

It is the purpose of this note to describe an experiment in which nuclear magnetic moment is measured very directly. The method is capable of very high precision and extension to a large number and variety of nuclei.

Consider a beam of molecules, such as LiCl, traversing a magnetic field which is sufficiently strong to decouple completely the nuclear spins from one another and from the molecular rotation. If a small oscillating magnetic field is applied at right angles to a much larger constant field, a re-orientation of the nuclear spin and magnetic moment with respect to the constant field will occur when the frequency of the oscillating field is close to the Larmor frequency of precession of the particular angular momentum vector in question. This precession frequency is given by

$$\nu = \mu H/hi = g(i)\mu_0 H/h. \quad (1)$$

To apply these ideas a beam of molecules in a $^1\Sigma$ state (no electronic moment) is spread by an inhomogeneous magnetic field and refocused onto a detector by a subsequent field, somewhat as in the experiment of Kellogg, Rabi and Zacharias.[1] As in that experiment the re-orienting field is placed in the region between the two magnets. The homogeneous field is produced by an electromagnet capable of supplying uniform fields up to 6000 gauss in a gap 6 mm wide and 5 cm long. In the gap is placed a loop of wire in the form of a hairpin (with its axis parallel to the direction of the beam) which is connected to a source of current at radiofrequency to produce the oscillating field at right angles to the steady field. If a re-orientation of a spin occurs in this field, the subsequent conditions in the second deflecting field are no longer correct for refocusing, and the intensity at the detector goes down. The experimental procedure is to vary the homogeneous field for some given value of the frequency of the oscillating field until the resonance is observed by a drop in intensity at the detector and a subsequent recovery when the resonance value is passed.

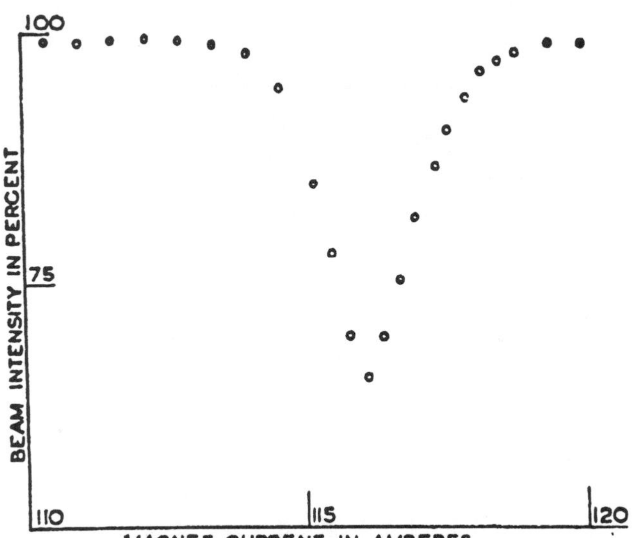

Fig. 1. Curve showing refocused beam intensity at various values of the homogeneous field. One ampere corresponds to about 18.4 gauss. The frequency of the oscillating field was held constant at 3.518×10^6 cycles per second.

The re-orientation process is more accurately described as one in which transitions occur between the various magnetic levels given by the quantum number m_i of the particular angular momentum vector in question. An exact solution for the transition probability was given by Rabi[2, 3] for the case where the variable field rotates rather than oscillates. However, it is more convenient experimentally to use an oscillating field, in which case the transition probability is approximately the same for *weak* oscillating fields *near* the resonance frequency, except that ϑ is replaced by $\vartheta/2$ in Eq. (13). With this replacement and with passage to the limit of weak oscillating fields, the formula becomes for the case of $i = \frac{1}{2}$

$$P(\tfrac{1}{2}, -\tfrac{1}{2}) = \frac{\vartheta^2}{(1-q)^2 + q\vartheta^2} \sin^2 \{\pi t r[(1-q)^2 + q\vartheta^2]^{\frac{1}{2}}\}, \quad (2)$$

where ϑ is $\frac{1}{2}$ the ratio of the oscillating field to the steady field, q is the ratio of the Larmor frequency of Eq. (1) to the frequency r of the oscillating field. The denominator of the expression is the familiar resonance denominator. The formula is generalized to any spin i by formula (17).[2]

In the theory of this experiment, t, in Eq. (2), is replaced by L/v, where L is the length of the oscillating region of the field, and v is the molecular velocity. $P(\frac{1}{2}, -\frac{1}{2})$ must then be averaged over the Maxwellian distribution of velocities. However, the first term is not affected by the velocity distribution if t is long enough for many oscillations to take place. The average value of the $\sin^2$ term over the velocity distribution is approximately $\frac{1}{2}$.

To produce deflections of the weakly magnetic molecules sufficient to make the apparatus sensitive to this effect, the beam is made 245 cm long; the first deflecting field is 52 cm in length and the second 100 cm.

We have tried this experiment with LiCl and observed the resonance peaks of Li and Cl. The effects are very striking and the resonances sharp (Fig. 1). A full account of this experiment, together with the values of the nuclear moments, will be published when the homogeneous field is recalibrated.

I. I. Rabi
J. R. Zacharias
S. Millman
P. Kusch

Hunter College (J. R. Z.),
Columbia University,
New York, N. Y.
January 31, 1938.

* Publication assisted by the Ernest Kempton Adams Fund for Physical Research of Columbia University.

[1] Kellogg, Rabi and Zacharias, Phys. Rev. 50, 472 (1936).
[2] Rabi, Phys. Rev. 51, 652 (1937).
[3] C. J. Gorter, Physica 9, 995 (1936). We are very much indebted to Dr. Gorter who, when visiting our laboratory in September 1937, drew our attention to his stimulating experiments in which he attempted to measure nuclear moments by observing the rise in temperature of solids placed in a constant magnetic field on which an oscillating field was superimposed. Dr. F. Bloch has independently worked out similar ideas but for another purpose (unpublished).

An Electrical Quadrupole Moment of the Deuteron*

The molecular beam magnetic resonance method[1] applied to HD molecules at liquid nitrogen temperature gives the magnetic moment of the proton and the deuteron. When applied to H_2 and D_2 molecules the method reveals the close groups of sharp resonance minima shown in Figs. 1 and 2. The resonance minima for H_2 agree in number and location with predictions based on the assumption of spin-spin magnetic interaction of the two protons $(\mu_1\cdot\mu_2/r^3-3(\mu_1\cdot r)(\mu_2\cdot r)/r^5)$ and a spin-orbit interaction of the protons with the rotation of the molecule $(2\mu_P\bar{H}I\cdot J)$. All symbols here have their usual significance except $\bar{H}$ which is the spin-orbit interaction constant. The only state of the molecule to be considered is the lowest rotational state of orthohydrogen: $J=1$ and total nuclear spin $I=1$. The nine energy levels which arise from these interactions and from the external magnetic field give six possible transitions for the nuclear spin because $\Delta M_N=\pm 1$. The pattern can be accounted for completely on the basis of the known value of the proton moment $(\mu_P=2.78)$ with the known value of the internuclear distance and a value for $\bar{H}$ of 27 gauss.

In the case of D_2 the deep central minimum arises from the states with $I=2$ and $J=0$. The six smaller peaks arise from the states with $J=1$ and $I=1$. The states with $J=2$ are not abundant enough at these low temperatures for observation. Since the internuclear distance in the D_2 molecule is the same as in H_2 and the mass is twice as great the spin-orbit interaction constant $\bar{H}$ is half as great. The deuteron magnetic moment $(\mu_D=0.853)$ is 0.307 times

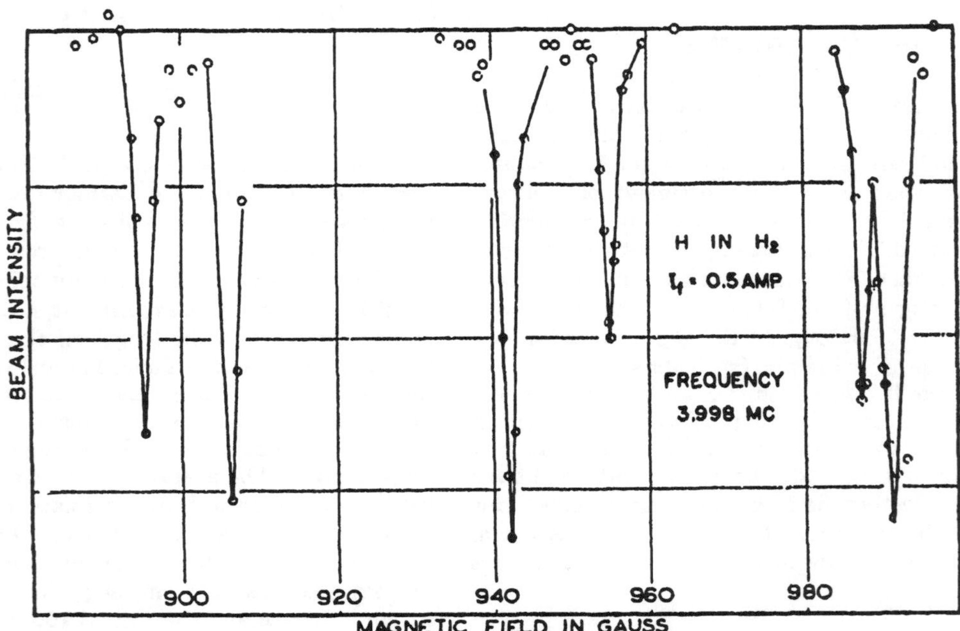

Fig. 1. Resonance minima for H in H_2.

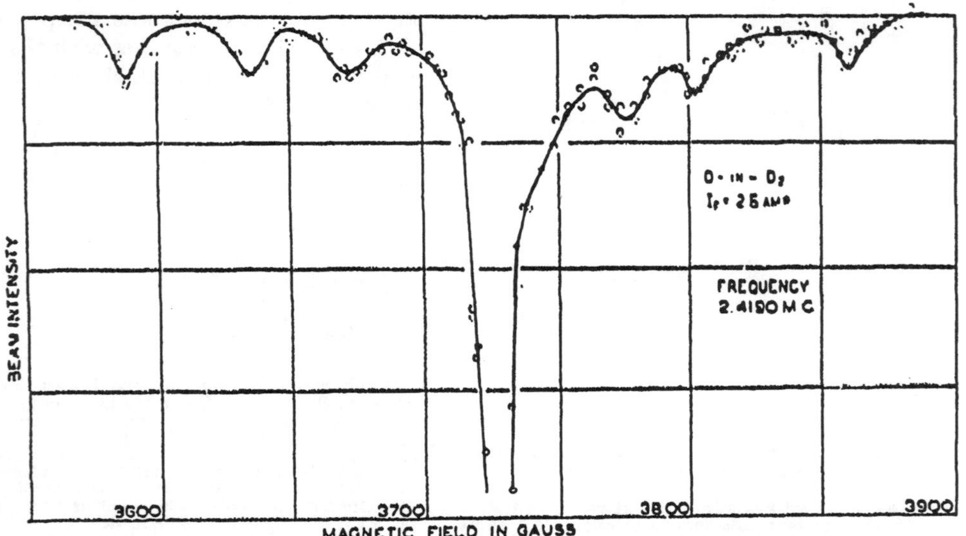

Fig. 2. Resonance minima for D in D_2.

that of the proton and therefore the magnetic spin-spin interaction is proportionately smaller. It was expected that the theory of the resonance minima for H_2 when applied to D_2 should give the locations of the minima from the constants given above. The displacements of the minima from the center should be much less than those of H_2. However experiment shows the displacements to be six times greater than the predicted values.

This effect can be accounted for by the presence of an electrical quadrupole moment in the deuteron. The interaction energy which gives rise to the large displacements is that of the nuclear quadrupole moment with the gradient of the molecular electric field. This form of interaction contributes to the nine energy levels exactly as the spin-spin and therefore appears as a larger spin-spin interaction.

To prove that the large displacements in D_2 are of nuclear origin rather than molecular, similar experiments were performed on the proton and the deuteron in the HD molecule. The group of resonance minima for H was narrow as expected and that for D had large displacements as in D_2. Furthermore, the experimentally evaluated spin-orbit interaction constant for D_2 is one-half as great as that for H_2 as predicted. We therefore believe that the apparent large spin-spin interaction is not magnetic, nor is it of molecular origin and must be a nuclear effect which behaves like a quadrupole moment.

To obtain the magnitude of this quadrupole moment one must know the molecular electric field. This value can be calculated from the various wave functions which have been suggested for the hydrogen molecule. The result of such a calculation by Dr. A. Nordsieck with Wang wave functions when combined with our data yields a quadrupole moment $Q=(3z^2-r^2)_{Av}$ of about 2×10^{-27} cm^2. The chief source of error lies in the inaccuracy of the wave functions.

The sign of the quadrupole moment may also be inferred from our measurements in two ways. Present indications are that it is positive, that is, the charge configuration is prolate along the spin axis. Full details of these experiments will be published later in this journal.

We are greatly indebted to Dr. Nordsieck for making available to us the results of his calculations and to Dr. Brickwedde of the National Bureau of Standards for preparation of the sample of pure HD used in the experiments. The experiments were supported in part by a grant from the Research Corporation.

J. M. B. KELLOGG
I. I. RABI
N. F. RAMSEY, JR.
J. R. ZACHARIAS

Columbia University,
Hunter College (JRZ),
New York, New York,
January 15, 1939.

* Publication assisted by the Ernest Kempton Adams Fund for Physical Research of Columbia University.

[1] J. M. B. Kellogg, I. I. Rabi, N. F. Ramsey and J. R. Zacharias, Bull. Am. Phys. Soc. Vol. 13, No. 7, Abs. 24 and 25.

FEBRUARY 1 AND 15, 1942 PHYSICAL REVIEW VOLUME 61

Theory of Complex Spectra. I

GIULIO RACAH
The Hebrew University, Jerusalem, Palestine

(Received November 14, 1941)

This paper gives a closed formula which entirely replaces for the two-electron spectra the previous lengthy calculations with the diagonal-sum method. Applications are also made to some configurations with three or more electrons and to the p^n configurations of the nuclei.

§1. INTRODUCTION

THE first-order perturbation energy for the terms of a given configuration was calculated at first by Slater.[1] In his classical paper he showed that the electrostatic interaction between two electrons depends on a very few integrals F^k and G^k, and he developed the diagonal-sum procedure for calculating the coefficients of these integrals; with this procedure he obtained numerical tables of coefficients for the two-electron configurations involving s, p, or d electrons. These tables were extended by several authors[2] to f electrons and to some configurations with three or more electrons.

But the diagonal-sum procedure has some deficiencies. Firstly, when two terms of a kind occur in a given configuration, this procedure will determine only the sum of their energies, and they can be separated only by other methods. Secondly, this method does not give general formulas, but only numerical tables; it is therefore impossible to make generalizations, and one must begin again for each new case with new and more complex calculations.[3]

It is the purpose of this paper to substitute for the numerical methods of Chapters VI and VII of TAS more general methods and more conformable to Chapter III of the same book.

---

[1] J. C. Slater, Phys. Rev. **34**, 1293 (1929).
[2] See E. U. Condon and G. H. Shortley, *Theory of Atomic Spectra* (Cambridge 1935), (which we shall denote by TAS), Chapters VI and VII, for definitions, notations and bibliographical indications.
[3] G. H. Shortley and B. Fried, Phys. Rev. **54**, 739 (1938).

§2. TWO-ELECTRON CONFIGURATIONS

If ω is the angle between the radii vectors of the two electrons, the coefficients f_k of F^k are[4] the eigenvalues of the matrix

$$(l_1l_2m_1m_2|P_k(\cos\omega)|l_1l_2m_1'm_2');\tag{1}$$

here P_k is the Legendre polynomial of the order k. The transformation which diagonalizes this matrix is $(l_1l_2LM|l_1l_2m_1m_2)$, and therefore

$$f_k(l_1l_2L)=\sum_{m_1m_2m_1'm_2'}(l_1l_2LM|l_1l_2m_1m_2)(l_1l_2m_1m_2|P_k(\cos\omega)|l_1l_2m_1'm_2')(l_1l_2m_1'm_2'|l_1l_2LM),\tag{2}$$

or

$$f_k(l_1l_2L)=(l_1l_2LM|P_k(\cos\omega)|l_1l_2LM).\tag{3}$$

In the same way, if $\pm g_k$ are the coefficients of G^k for the singlet and for the triplet terms, we have

$$g_k(l_1l_2L)=\sum_{m_1m_2m'_1m'_2}(l_1l_2LM|l_1l_2m_1m_2)(l_1l_2m_1m_2|P_k(\cos\omega)|l_2l_1m'_2m'_1)(l_1l_2m'_1m'_2|l_1l_2LM),\tag{4}$$

and in view of[5]

$$(l_1l_2m'_1m'_2|l_1l_2LM)=(-1)^{l_1+l_2-L}(l_2l_1m'_2m'_1|l_2l_1LM),\tag{5}$$

this becomes

$$g_k(l_1l_2L)=(-1)^{l_1+l_2-L}(l_1l_2LM|P_k(\cos\omega)|l_2l_1LM).\tag{6}$$

Slater calculated the matrix elements of P_k (cos ω) in the $l_1l_2m_1m_2$ scheme, and then obtained the eigenvalues of this operator by means of the diagonal-sum procedure; we will calculate the matrix elements of cos ω directly in the l_1l_2LM scheme by the method of Güttinger and Pauli,[6] and then calculate f_k and g_k with the ordinary methods of matrix calculations.

If $\mathbf{u}_i$ is the unit vector in the direction from the origin to the electron i, by comparing TAS 4^321 with TAS 9^311, we have

$$(l_i\vdots u_i\vdots l_i)=0,\quad (l_i\vdots u_i\vdots l_i-1)=(l_i-1\vdots u_i\vdots l_i)=\frac{1}{[(2l_i-1)(2l_i+1)]^{\frac{1}{2}}};\tag{7}$$

and since

$$\cos\omega=(\mathbf{u}_1\cdot\mathbf{u}_2),\tag{8}$$

introducing (7) in TAS 12^32 we find that the only non-vanishing elements of $(l_1l_2LM|\cos\omega|l_1'l_2'LM)$ are

$$\begin{aligned}
(l_1l_2LM|\cos\omega|l_1-1\ l_2-1\ LM)&=-\frac{[(l_1+l_2+L+1)(l_1+l_2+L)(l_1+l_2-L)(l_1+l_2-L-1)]^{\frac{1}{2}}}{2[(2l_1-1)(2l_1+1)(2l_2-1)(2l_2+1)]^{\frac{1}{2}}},\\
(l_1l_2LM|\cos\omega|l_1+1\ l_2-1\ LM)&=\frac{[(L+l_1-l_2+2)(L+l_1-l_2+1)(L+l_2-l_1)(L+l_2-l_1-1)]^{\frac{1}{2}}}{2[(2l_1+1)(2l_1+3)(2l_2-1)(2l_2+1)]^{\frac{1}{2}}},\\
(l_1l_2LM|\cos\omega|l_1-1\ l_2+1\ LM)&=\frac{[(L+l_1-l_2)(L+l_1-l_2-1)(L+l_2-l_1+2)(L+l_2-l_1+1)]^{\frac{1}{2}}}{2[(2l_1-1)(2l_1+1)(2l_2+1)(2l_2+3)]^{\frac{1}{2}}},\\
(l_1l_2LM|\cos\omega|l_1+1\ l_2+1\ LM)&=-\frac{[(l_1+l_2+L+3)(l_1+l_2+L+2)(l_1+l_2-L+2)(l_1+l_2-L+1)]^{\frac{1}{2}}}{2[(2l_1+1)(2l_1+3)(2l_2+1)(2l_2+3)]^{\frac{1}{2}}}.
\end{aligned}\tag{9}$$

From these formulas it is possible to calculate the matrix elements of P_k (cos ω) with the ordinary methods of matrix calculations; in order that these elements may have a value different from zero, k must satisfy the conditions

$$k+l_1+l'_1=2g_1,\quad k+l_2+l'_2=2g_2\tag{10}$$

[4] TAS $\S8^6$.
[5] TAS 14^3 7.
[6] Güttinger and Pauli, Zeits. f. Physik **67**, 743 (1931); TAS $\S10^3$ *et seq.*

(g_1 and g_2 are integers), and the so-called triangular conditions

$$|l_1-l'_1|\leqslant k\leqslant l_1+l'_1,\quad |l_2-l'_2|\leqslant k\leqslant l_2+l'_2; \tag{11}$$

if these conditions are satisfied, the final result is

$$(l_1l_2LM|P_k(\cos\omega)|l'_1l'_2LM)=\frac{(-1)^{g_1+g_2-k}(2g_1-2l_1)!(2g_2-2l'_2)!g_1!g_2!}{(g_1-k)!(g_1-l_1)!(g_1-l'_1)!(g_2-k)!(g_2-l_2)!(g_2-l'_2)!(2g_1+1)!(2g_2+1)!}$$
$$\left[\frac{(2l_1+1)(2l'_1+1)(2l_2+1)(2l'_2+1)(l_1+l_2+L+1)!(l'_1+l'_2+L+1)!\times(l_1+l_2-L)!(l'_1+l'_2-L)!(L+l_1-l_2)!(L+l'_2-l'_1)!}{(L+l'_1-l'_2)!(L+l_2-l_1)!}\right]^{\frac{1}{2}}$$
$$\cdot\sum_u(-1)^u\frac{(u+l'_1-l'_2)!(u+l_2-l_1)!(k+l_1+l'_2-u)!}{(u+L+1)!(u-L)!(l_1+l_2-u)!(l'_1+l'_2-u)!(k-l_1-l'_2+u)!}, \tag{12}$$

where, in the summation, u takes on all integral values consistent with the factorial notation, the factorial of a negative number being meaningless.

In order to demonstrate this formula, it suffices to verify that (12) reduces to $\delta(l_1l'_1)\delta(l_2l'_2)$ for $k=0$ and to (9) for $k=1$ and that, introducing (12) for $k=n-1$ and $k=n-2$ in the formula[7]

$$P_n(\cos\omega)=\frac{2n-1}{n}\cos\omega P_{n-1}(\cos\omega)-\frac{n-1}{n}P_{n-2}(\cos\omega) \tag{13}$$

written in matrix form, we obtain again (12) for $k=n$. These verifications are somewhat long, but they are not difficult and will be omitted for brevity.

It is remarkable that (12) has an unsymmetrical aspect: it is however possible (as is shown in the appendix) to transform this formula by means of algebraic identities and to replace it with[8]

$$(l_1l_2LM|P_k(\cos\omega)|l'_1l'_2LM)$$
$$=\frac{(-1)^{g_1+g_2-k-L}(l_1+l'_1-k-1)!!(k+l_1-l'_1-1)!!(k+l'_1-l_1-1)!!\times(l_2+l'_2-k-1)!!(k+l_2-l'_2-1)!!(k+l'_2-l_2-1)!!}{(k+l_1+l'_1+1)!!(k+l_2+l'_2+1)!!}$$
$$\cdot\left[\frac{(2l_1+1)(2l_2+1)(2l'_1+1)(2l'_2+1)(l_1+l_2-L)!(l'_1+l'_2-L)!\times(L+l_1-l_2)!(L+l'_1-l'_2)!(L+l_2-l_1)!(L+l'_2-l'_1)!}{(l_1+l_2+L+1)!(l'_1+l'_2+L+1)!}\right]^{\frac{1}{2}}$$
$$\cdot\sum_v(-1)^v\frac{(l_1+l_2+l'_1+l'_2+1-v)!}{(l_1+l_2-L-v)!(l'_1+l'_2-L-v)!(l_1+l'_1-k-v)!(l_2+l'_2-k-v)!v!\times(k+L-l_1-l'_2+v)!(k+L-l'_1-l_2+v)!}. \tag{12'}$$

Introducing (12′) in (3) and (6), and putting

$$v=l_1+l_2-L-w,$$

[7] Courant and Hilbert, *Methoden der Mathematischen Physik* (Springer, 1931), p. 73, Eq. (19).

[8] With the symbol $n!!$ we indicate the *semifactorial* of n, that is the product $1.3.5\ldots n$ if n is odd, and the product $2.4.6\ldots n$ if n is even.

we obtain

$$f_k(l_1l_2L)=\frac{(k-1)!!^4(2l_1-k-1)!!(2l_2-k-1)!!(2l_1+1)(2l_2+1)}{(2l_1+k+1)!!(2l_2+k+1)!!}$$
$$\times\sum_w(-1)^w\binom{l_1+l_2+L+1+w}{w}\binom{l_1+l_2-L}{w}\binom{L+l_1-l_2}{k-w}\binom{L+l_2-l_1}{k-w} \quad (14)$$

and

$$g_k(l_1l_2L)=(-1)^{l_1+l_2-L}\frac{(k+l_1-l_2-1)!!^2(k+l_2-l_1-1)!!^2(l_1+l_2-k-1)!!^2(2l_1+1)(2l_2+1)}{(l_1+l_2+k+1)!!^2}$$
$$\times\sum_w(-1)^w\binom{l_1+l_2+L+1+w}{w}\binom{l_1+l_2-L}{w}\binom{L+l_1-l_2}{k+l_1-l_2-w}\binom{L+l_2-l_1}{k+l_2-l_1-w} \quad (15)$$

The dependence on L of such formulas was already given by Kramers[9] for f_k and by Brinkman[10] for g_k by means of a group-theoretical procedure, but it was impossible, by such a general method, to give the first factor.

Putting

$$p=l_1(l_1+1)l_2(l_2+1),\quad s=l_1(l_1+1)+l_2(l_2+1),\quad \Lambda=L(L+1),\quad \lambda=\frac{\Lambda-s}{2}=(\mathbf{l}_1\cdot\mathbf{l}_2),\quad q=(l_1+l_2+1)^2, \quad (16)$$

we obtain, for the most common and important cases:

$$f_0=1,$$
$$f_2=\frac{6\lambda^2+3\lambda-2p}{(2l_1-1)(2l_1+3)(2l_2-1)(2l_2+3)}, \quad (17)$$
$$f_4=9\frac{70\lambda^4+350\lambda^3-10(6p-5s-39)\lambda^2-10(17p-6s-9)\lambda+3p(2p-4s-27)}{4(2l_1-3)(2l_1-1)(2l_1+3)(2l_1+5)(2l_2-3)(2l_2-1)(2l_2+3)(2l_2+5)},$$

and

$$g_{l_1-l_2}(l_1l_2L)=\frac{(-1)^{l_1+l_2-L}}{2^{2k}}\binom{2k}{k}\frac{\Lambda(\Lambda-2)(\Lambda-6)\cdots[\Lambda-(k-1)k]}{(2l_2+1)(2l_2+3)^2(2l_2+5)^2\cdots(2l_1-1)^2(2l_1+1)},$$
$$g_{l_1-l_2+2}(l_1l_2L)=\frac{(-1)^{l_1+l_2-L}}{2^{2k-1}}\binom{2k-2}{k-1}\Lambda(\Lambda-2)(\Lambda-6)\cdots[\Lambda-(k-3)(k-2)]. \quad (18)$$
$$\cdot\frac{k(2k-1)\Lambda^2-2(k-1)(2k-1)(q-k)\Lambda+(k-1)[(2k-3)q^2-(4k^2-6k-1)q+(k-1)k^2]}{(2l_2-1)^2(2l_2+1)(2l_2+3)^2(2l_2+5)^2\cdots(2l_1-1)^2(2l_1+1)(2l_1+3)^2}.$$

By means of these formulas all results of TAS and of Shortley and Fried[3] were checked, and a sole mistake was found: the coefficient of F_2 and of G_3 in the F terms of the configuration ff is not $+10$, as reported in TAS (p. 207), but -10, as given in the original paper of Condon and Shortley.[11]

§3. CONFIGURATIONS WITH THREE OR MORE ELECTRONS

The expression of the electrostatic interaction of two electrons as function of λ is not only important for a more rapid calculation of the two-electron terms: in the case of three or more electrons the methods of Chapter III of TAS give us the possibility of calculating the matrix elements of

[9] Kramers, Proc. Amst. Acad. 34, 965 (1931).
[10] Brinkman, Zeits. f. Physik 79, 753 (1932).
[11] Condon and Shortley, Phys. Rev. 37, 1030 (1931).

$\lambda_{ij}=(\mathbf{l}_i\cdot\mathbf{l}_j)$ in every complex case of vector coupling; and it is therefore possible to calculate the terms of more complex configurations, even if two or more terms of the same kind occur.

We will at first pay attention to some particular applications to atomic and nuclear spectra, that can be treated without matrix calculations; then we shall treat in detail the p^2p configuration, and give the results for the p^2l configuration. For other important configurations and for the cases of (jj) coupling the calculations are more complex, and a new general procedure for this purpose will therefore be developed in a later paper.

The first application was already made to the p^n configurations by Van Vleck,[12] who found empirically the formula of f_2 for p electrons, and expressed λ^2_{ij} as a linear function of λ_{ij} and of $(\mathbf{s}_i\mathbf{s}_j)$ by means of Dirac's vector model; but, as Van Vleck himself pointed out, such a procedure is not generally sufficient for d^n configurations.

This procedure suffices however for the terms of d^n with higher multiplicity. In these states all spins are parallel, and it follows therefore from the principle of antisymmetry that the possible values of each λ are 0 or -5 (F or P resultant); from this we have that

$$\lambda^2+5\lambda=0; \tag{19}$$

introducing this relation into the expressions

$$f_2(dd)=\frac{2\lambda^2+\lambda-24}{147},\quad f_4(dd)=\frac{35\lambda^4+175\lambda^3-585\lambda^2-2655\lambda-162}{7938}, \tag{20}$$

we obtain

$$f_2=-(3\lambda+8)/49,\quad f_4=(15\lambda-9)/441, \tag{21}$$

and therefore

$$E=\sum_{i<j}\left(F^0-\frac{3\lambda_{ij}+8}{49}F^2+\frac{15\lambda_{ij}-9}{441}F^4\right);$$

since for a d^n term

$$L(L+1)=6n+2\sum_{i<j}\lambda_{ij}, \tag{22}$$

we obtain that for all d^n configurations with $S=n/2$

$$E=\frac{n(n-1)}{2}F^0-\frac{\frac{3}{2}L(L+1)+n(4n-13)}{49}F^2+\frac{\frac{15}{2}L(L+1)-\frac{9}{2}n(n+9)}{441}F^4. \tag{23}$$

§4. THE NUCLEAR CONFIGURATIONS p^n

The calculation of the energy levels of a nuclear configuration p^n with symmetrical forces was made by Hund[13] with the diagonal-sum procedure. After very long calculations he obtained a numerical table for the energies of such configurations, and from this table he deduced empirical formulas for the interactions of Wigner and of Majorana. The direct calculation of such formulas is a remarkable application of the above developed methods.

Putting, as customary in order to avoid fractional coefficients,

$$F_0=F^0;\quad F_2=F^2/25, \tag{24}$$

we obtain from (17) that the normal (Wigner) interaction between two particles is

$$V_{ij}=F_0+(6\lambda_{ij}^2+3\lambda_{ij}-8)F_2; \tag{25}$$

[12] J. H. Van Vleck, Phys. Rev. 45, 412 (1934).
[13] Hund, Zeits. f. Physik 105, 202 (1937).

hence the Wigner interaction between all particles of the configuration is

$$V_W=\frac{n(n-1)}{2}F_0+\sum_{i<j}(6\lambda_{ij}^2+3\lambda_{ij}-8)F_2. \tag{26}$$

In order to calculate this sum, we cannot use Dirac's vector model, because the exclusion principle does not hold for a proton and a neutron; but we can observe that for p particles the operator

$$M_{ij}=\lambda_{ij}^2+\lambda_{ij}-1 \tag{27}$$

is Majorana's operator of position exchange, since it has the eigenvalue 1 for the symmetrical states S and D, and the eigenvalue -1 for the antisymmetrical state P: from (26) and (27) we obtain

$$V_W=\frac{n(n-1)}{2}F_0+\sum_{i<j}(6M_{ij}-3\lambda_{ij}-2)F_2. \tag{28}$$

The sum

$$\mathfrak{M}=\sum_{i<j}M_{ij} \tag{29}$$

is Hund's $(\alpha-\beta)$ and depends only from the symmetry character of the positional eigenfunction of the level: it is the difference between the number of symmetrically connected couples and the number of antisymmetrically connected couples. From (29) and from

$$\sum_{i<j}\lambda_{ij}=\tfrac{1}{2}L(L+1)-n, \tag{30}$$

we obtain

$$V_W=\frac{n(n-1)}{2}F_0+[6\mathfrak{M}-\tfrac{3}{2}L(L+1)-n(n-4)]F_2. \tag{31}$$

The Majorana interaction is

$$V_M=\sum_{i<j}M_{ij}V_{ij}=\mathfrak{M}F_0+\sum_{i<j}(\lambda_{ij}^2+\lambda_{ij}-1)(6\lambda_{ij}^2+3\lambda_{ij}-8)F_2 \tag{32}$$

and reduces to

$$V_M=\mathfrak{M}F_0+\sum_{i<j}(M_{ij}-3\lambda_{ij}+3)F_2 \tag{33}$$

in virtue of

$$\lambda_{ij}^3+2\lambda_{ij}^2-\lambda_{ij}-2=0 \tag{34}$$

(which expresses the fact that λ_{ij} has only the eigenvalues 1, -1, and -2) and of (27). In view of (29) and (30), (33) becomes

$$V_M=\mathfrak{M}F_0+[\mathfrak{M}-\tfrac{3}{2}L(L+1)+\tfrac{3}{2}n(n+1)]F_2. \tag{35}$$

Putting

$$F_0=A-\tfrac{2}{3}B \quad\text{and}\quad F_2=\tfrac{1}{3}B$$

in (31) and (35), we obtain Hund's formulas.

When two terms with the same L, R, S and $\mathfrak{M}$ occur, Hund could only calculate their sum, and said that his formulas hold for their mean values; since our method gives the energies of all terms separately, we can say that in the later case the first-order energies are the same, and that Hund's formulas hold for each term separately.

§5. THE CONFIGURATION p^3p

It follows from §2 that the interaction between two non-equivalent p electrons is

$$W_{pp}=F^0(np,n'p)+\frac{6\lambda^2+3\lambda-8}{25}F^2(np,n'p)\pm(-1)^L\left[G^0(np,n'p)+\frac{6\lambda^2+3\lambda-8}{25}G^2(np,n'p)\right], \tag{36}$$

where the upper sign holds when the spins are antiparallel, and the lower sign when the spins are parallel.

Following Dirac's vector model[14] we can substitute the operator

$$\frac{1+4(\mathbf{s}\cdot\mathbf{s}')}{2}=\frac{1+\mu}{2}. \tag{37}$$

for the double sign. For $(-1)^L$ we can substitute the operator (27).

Considering that λ satisfies the Eq. (34), we have that

$$W_{pp}=F^0(np,n'p)+\frac{6\lambda^2+3\lambda-8}{25}F^2(np,n'p)\\ -\frac{1+\mu}{2}\left[(\lambda^2+\lambda-1)G^0(np,n'p)+\frac{\lambda^2-2\lambda+2}{25}G^2(np,n'p)\right]. \tag{38}$$

Putting

$$F_0=F^0(np,np)+2F^0(np,n'p),\quad F_2=\tfrac{1}{25}F^2(np,np),\quad F'_2=\tfrac{1}{25}F^2(np,n'p),$$
$$G_0=G^0(np,n'p),\quad G_2=\tfrac{1}{25}G^2(np,n'p),$$

and marking by 1 and 2 the two np electrons, and by 3 the $n'p$ electron, we obtain for the electrostatic interaction of the $np^2n'p$ configuration:

$$E=F_0+(6\lambda_{12}^2+3\lambda_{12}-8)F_2+\sum_1^2{}_i(6\lambda_{i3}^2+3\lambda_{i3}-8)F'_2\\ -\sum_1^2{}_i\frac{1+\mu_{i3}}{2}[(\lambda_{i3}^2+\lambda_{i3}-1)G_0+(\lambda_{i3}^2-2\lambda_{i3}+2)G_2]. \tag{39}$$

The most convenient scheme for calculating this energy operator is the LS scheme of the parent ion np^2. In this scheme λ_{12} is diagonal, and the coefficient of F_2 is therefore the diagonal matrix

$$\|(\mathrm{L}|6\lambda_{12}^2+3\lambda_{12}-8|\mathrm{L}')\|=\begin{array}{c|ccc} & \mathrm{D} & \mathrm{P} & \mathrm{S}\\ \hline \mathrm{D} & 1 & 0 & 0\\ \mathrm{P} & 0 & -5 & 0\\ \mathrm{S} & 0 & 0 & 10\end{array} \tag{40}$$

In order to calculate the coefficients of F'_2, G_0 and G_2 we must at first calculate the matrices of λ_{i3}. From TAS 10^32 we have

$$\|(\mathrm{L}\vdots l_i\vdots\mathrm{L}')\|=\begin{array}{c|ccc} & \mathrm{D} & \mathrm{P} & \mathrm{S}\\ \hline \mathrm{D} & \frac{1}{2} & \pm\frac{1}{2\sqrt{3}} & 0\\ \mathrm{P} & \pm\frac{1}{2\sqrt{3}} & \frac{1}{2} & \pm\left(\frac{2}{3}\right)^{\frac{1}{2}}\\ \mathrm{S} & 0 & \pm\left(\frac{2}{3}\right)^{\frac{1}{2}} & \frac{1}{2}\end{array}, \tag{41}$$

[14] Dirac, Proc. Roy. Soc. A123, 714 (1929) and *Quantum Mechanics* (Oxford, 1930), Chapter 11.

the upper sign holding for $i=1$ and the lower for $i=2$; from this table and from TAS $12^3 2$ we obtain

$$\|(\mathrm{L}L|\lambda_{i3}|\mathrm{L}'L)\| = \begin{array}{c|ccc|} & \mathrm{D} & \mathrm{P} & \mathrm{S} \\ \hline \mathrm{D} & \dfrac{\Lambda-8}{4} & \mp\dfrac{[\Lambda(12-\Lambda)]^{\frac{1}{2}}}{4\sqrt{3}} & 0 \\ \mathrm{P} & \mp\dfrac{[\Lambda(12-\Lambda)]^{\frac{1}{2}}}{4\sqrt{3}} & \dfrac{\Lambda-4}{4} & \mp\dfrac{[\Lambda(6-\Lambda)]^{\frac{1}{2}}}{\sqrt{6}} \\ \mathrm{S} & 0 & \mp\dfrac{[\Lambda(6-\Lambda)]^{\frac{1}{2}}}{\sqrt{6}} & \dfrac{\Lambda-2}{4} \\ \hline \end{array}, \tag{42}$$

Λ having the meaning of (16); and this matrix gives us easily the coefficient of F'_2:

$$\|(\mathrm{L}L|\sum_{1}^{2}{}_i\,(6\lambda_{i3}^2+3\lambda_{i3}-8)|\mathrm{L}'L)\| = \begin{array}{c|ccc|} & \mathrm{D} & \mathrm{P} & \mathrm{S} \\ \hline \mathrm{D} & \dfrac{\Lambda^2-15\Lambda+40}{2} & 0 & \dfrac{\Lambda[(12-\Lambda)(6-\Lambda)]^{\frac{1}{2}}}{\sqrt{2}} \\ \mathrm{P} & 0 & \dfrac{-3\Lambda^2+21\Lambda-20}{2} & 0 \\ \mathrm{S} & \dfrac{\Lambda[(12-\Lambda)(6-\Lambda)]^{\frac{1}{2}}}{\sqrt{2}} & 0 & \dfrac{-5\Lambda^2+42\Lambda-64}{4} \\ \hline \end{array} \tag{43}$$

In order to calculate the coefficients of G_0 and G_2 we must also calculate the matrices of $\mu_{i3}=4(\mathbf{s}_i\cdot\mathbf{s}_3)$. From TAS $10^3 2$ and $12^3 2$ we obtain in the same way

$$\|(\mathrm{s}S|\mu_{i3}|\mathrm{s}'S)\| = \begin{array}{c|cc|} & \mathrm{s}=1 & \mathrm{s}=0 \\ \hline \mathrm{s}=1 & S(S+1)-\tfrac{11}{4} & \mp[[S(S+1)+\tfrac{1}{4}][\tfrac{15}{4}-S(S+1)]]^{\frac{1}{2}} \\ \mathrm{s}=0 & \mp[[S(S+1)+\tfrac{1}{4}][\tfrac{15}{4}-S(S+1)]]^{\frac{1}{2}} & S(S+1)-\tfrac{3}{4} \\ \hline \end{array} \tag{44}$$

and therefore

$$\left\|\left({}_{\mathrm{s}}\mathrm{L}\ {}^2L\left|\frac{1+\mu_{i3}}{2}\right|{}_{\mathrm{s}'}\mathrm{L}\ {}^2L\right)\right\| = \begin{array}{c|cc|} & {}^3\mathrm{L} & {}^1\mathrm{L} \\ \hline {}^3\mathrm{L} & \dfrac{1}{2} & \mp\dfrac{\sqrt{3}}{2} \\ {}^1\mathrm{L} & \mp\dfrac{\sqrt{3}}{2} & \dfrac{1}{2} \\ \hline \end{array} \tag{45}$$

and

$$\left({}^3\mathrm{L}\ {}^4L\left|\frac{1+\mu_{i3}}{2}\right|{}^3\mathrm{L}\ {}^4L\right)=1. \tag{45'}$$

TABLE I.

Atom	Configuration	$({}^4P-{}^4D)/({}^4D-{}^4S)$
Theory	$np^2n'p$	0.667
N I	$2p^23p$	0.521
N I	$2p^24p$	0.504
O II	$2p^23p$	0.520
S II	$3p^24p$	0.604

From (42) and (45) we get

$$\left\|\left({}^{\alpha}\mathrm{L}\;{}^2L\left|\sum_1^2{}_i\frac{1+\mu_{i3}}{2}(\lambda_{i3}^2+\lambda_{i3}-1)\right|{}^{\alpha'}\mathrm{L}'\;{}^2L\right)\right\|$$

$$= \begin{array}{c|ccc|} & {}^1\mathrm{D} & {}^3\mathrm{P} & {}^1\mathrm{S} \\ \hline {}^1\mathrm{D} & \dfrac{\Lambda^2-12\Lambda+24}{24} & \dfrac{(\Lambda-4)[\Lambda(12-\Lambda)]^{\frac{1}{2}}}{8} & \dfrac{\Lambda[(12-\Lambda)(6-\Lambda)]^{\frac{1}{2}}}{12\sqrt{2}} \\ {}^3\mathrm{P} & \dfrac{(\Lambda-4)[\Lambda(12-\Lambda)]^{\frac{1}{2}}}{8} & \dfrac{\Lambda^2-8\Lambda+8}{8} & \dfrac{(\Lambda-1)[\Lambda(6-\Lambda)]^{\frac{1}{2}}}{2\sqrt{2}} \\ {}^1\mathrm{S} & \dfrac{\Lambda[(12-\Lambda)(6-\Lambda)]^{\frac{1}{2}}}{12\sqrt{2}} & \dfrac{(\Lambda-1)[\Lambda(6-\Lambda)]^{\frac{1}{2}}}{2\sqrt{2}} & \dfrac{-5\Lambda^2+48\Lambda-60}{48} \\ \hline \end{array} \tag{46}$$

and

$$\left\|\left({}^{\alpha}\mathrm{L}\;{}^2L\left|\sum_1^2{}_i\frac{1+\mu_{i3}}{2}(\lambda_{i3}^2-2\lambda_{i3}+2)\right|{}^{\alpha'}\mathrm{L}'\;{}^2L\right)\right\|$$

$$= \begin{array}{c|ccc|} & {}^1\mathrm{D} & {}^3\mathrm{P} & {}^1\mathrm{S} \\ \hline {}^1\mathrm{D} & \dfrac{\Lambda^2-30\Lambda+240}{24} & \dfrac{(\Lambda-10)[\Lambda(12-\Lambda)]^{\frac{1}{2}}}{8} & \dfrac{\Lambda[(12-\Lambda)(6-\Lambda)]^{\frac{1}{2}}}{12\sqrt{2}} \\ {}^3\mathrm{P} & \dfrac{(\Lambda-10)[\Lambda(12-\Lambda)]^{\frac{1}{2}}}{8} & \dfrac{\Lambda^2-2\Lambda-40}{8} & \dfrac{(\Lambda-7)[\Lambda(6-\Lambda)]^{\frac{1}{2}}}{2\sqrt{2}} \\ {}^1\mathrm{S} & \dfrac{\Lambda[(12-\Lambda)(6-\Lambda)]^{\frac{1}{2}}}{12\sqrt{2}} & \dfrac{(\Lambda-7)[\Lambda(6-\Lambda)]^{\frac{1}{2}}}{2\sqrt{2}} & \dfrac{-5\Lambda^2+12\Lambda+156}{48} \\ \hline \end{array} \tag{47}$$

From (42) and (45′) we get

$$\left({}^3\mathrm{P}\;{}^4L\left|\sum_1^2{}_i\frac{1+\mu_{i3}}{2}(\lambda_{i3}^2+\lambda_{i3}-1)\right|{}^3\mathrm{P}\;{}^4L\right)=-\frac{\Lambda^2-8\Lambda+8}{4} \tag{46′}$$

and

$$\left({}^3\mathrm{P}\;{}^4L\left|\sum_1^2{}_i\frac{1+\mu_{i3}}{2}(\lambda_{i3}^2-2\lambda_{i3}+2)\right|{}^3\mathrm{P}\;{}^4L\right)=-\frac{\Lambda^2-2\Lambda-40}{4}. \tag{47′}$$

Introducing our results into (39) we obtain

$$\begin{aligned} {}^4D &= F_0-5F_2-F'_2-G_0-4G_2, \\ {}^4P &= F_0-5F_2+5F'_2-G_0-10G_2, \\ {}^4S &= F_0-5F_2-10F'_2+2G_0-10G_2. \end{aligned}$$

$$^2S = F_0 - 5F_2 - 10F'_2 - G_0 + 5G_2,$$

$$^2F = F_0 + F_2 + 2F'_2 - G_0 - G_2.$$

The energy matrix for the ^2D terms is

	^1D	^3P
^1D	$F_0+F_2-7F'_2+\frac{1}{2}G_0-4G_2$	$-\frac{3}{2}G_0+3G_2$
^3P	$-\frac{3}{2}G_0+3G_2$	$F_0-5F_2-F'_2+\frac{1}{2}G_0+2G_2$

and has the eigenvalues

$$^2D = F_0 - 2F_2 - 4F'_2 + \tfrac{1}{2}G_0 - G_2 \pm 3[(F_2 - F'_2 - G_2)^2 + \tfrac{1}{4}(G_0 - 2G_2)^2]^{\frac{1}{2}}.$$

The energy matrix for the ^2P terms is

	^1D	^3P	^1S
^1D	$F_0+F_2+7F'_2-\frac{1}{6}G_0-\frac{23}{3}G_2$	$(-G_0+2G_2)\sqrt{5}$	$(4F'_2-\frac{1}{3}G_0-\frac{1}{3}G_2)\sqrt{5}$
^3P	$(-G_0+2G_2)\sqrt{5}$	$F_0-5F_2+5F'_2+\frac{1}{2}G_0+5G_2$	$-G_0+5G_2$
^1S	$(4F'_2-\frac{1}{3}G_0-\frac{1}{3}G_2)\sqrt{5}$	$-G_0+5G_2$	$F_0+10F_2-\frac{1}{3}G_0-\frac{19}{3}G_2$

.

It follows from our results that the ratio $(^4P-{}^4D)/(^4D-{}^2S)$ has the theoretical value of $\frac{3}{2}$. The comparison with the experimental ratios[15] is given in Table I.

The deviations are of the same order as those of the np^3 configurations of the same elements.[16]

§6. THE CONFIGURATION p^2l

The terms of the configuration $np^2n'l$ can be calculated in the same way as those of $np^2n'p$. The only difference is that the coefficients of G^{l-1} and G^{l+2} in W_{pl} are polynomials of higher degree in λ, and must be reduced to the second degree by means of the equation

$$\lambda^3 + 2\lambda^2 - (l^2 + l - 1)\lambda - l(l+1) = 0, \tag{48}$$

which corresponds to the Eq. (34) of the pp case. This reduction cannot be carried out without specifying the value of l; it is possible however to avoid such direct reduction, by calculating at first the single values of $g_{l-1}(l1L)$ and $g_{l+1}(l1L)$ for the three possible values of L by means of (15), and then determining the polynomials of the second degree which assume these values for the values l, -1, and $-(l+1)$ of the variable λ. By this procedure we find that the electrostatic interaction between a p and a l electron can be expressed by means of the formula

$$W_{pl} = F^0(np, n'l) + [6\lambda^2 + 3\lambda - 4l(l+1)]F'_2 - \frac{1+\mu}{2}[(\lambda^2 + l\lambda - l)G_{l-1} + (\lambda^2 - (l+1)\lambda + l + 1)G_{l+1}]. \tag{49}$$

where[17]

$$F'_2 = \frac{F^2(np, n'l)}{5(2l-1)(2l+3)}, \quad G_{l-1} = \frac{3G^{l-1}(np, n'l)}{(2l+1)(2l-1)^2}, \quad G_{l+1} = \frac{3G^{l+1}(np, n'l)}{(2l+1)(2l+3)^2}. \tag{50}$$

From this point the calculations were carried out exactly in the same way as for p^2p, and give the following results

[15] See Bacher and Goudsmit, *Atomic Energy States* (McGraw-Hill, 1932).

[16] TAS, p. 198.

[17] Our definitions (50) differ in some cases by a factor 3 from the definitions of TAS, p. 177.

Quartets:

$$L=l+1:\quad F_0-5F_2-l(2l-1)F'_2-l(2l-1)G_{l-1}-2(l+1)G_{l+1},$$
$$L=l:\quad F_0-5F_2+(2l-1)(2l+3)F'_2-l(2l-1)G_{l-1}-(l+1)(2l+3)G_{l+1},$$
$$L=l-1:\quad F_0-5F_2-(l+1)(2l+3)F'_2+2lG_{l-1}-(l+1)(2l+3)G_{l+1}.$$

Doublets:

$$L=l+2:\quad F_0+F_2+2l(2l-1)F'_2-l(2l-1)G_{l-1}-G_{l+1},$$
$$L=l-2:\quad F_0+F_2+2(l+1)(2l+3)F'_2-G_{l-1}-(l+1)(2l+3)G_{l+1},$$
$$L=l+1:\quad F_0-2F_2-(l+3)(2l-1)F'_2+\tfrac{1}{2}(2l-1)G_{l-1}-G_{l+1}$$
$$\pm[[3F_2-3(2l-1)F'_2-\tfrac{1}{2}(l-1)(2l-1)G_{l-1}-(l+2)G_{l+1}]^2+3l(l+2)[\tfrac{1}{2}(2l-1)G_{l-1}-G_{l+1}]^2]^{\frac{1}{2}},$$
$$L=l-1:\quad F_0-2F_2-(l-2)(2l+3)F'_2-G_{l-1}-\tfrac{1}{2}(2l+3)G_{l+1}$$
$$\pm[[3F_2+3(2l+3)F'_2+(l-1)G_{l-1}-\tfrac{1}{2}(l+2)(2l+3)G_{l+1}]^2+3(l^2-1)[G_{l-1}+\tfrac{1}{2}(2l+3)G_{l+1}]^2]^{\frac{1}{2}}.$$

The energy matrix for the doublets with $L=l$ is of the third order and has the elements:

$$({}^1\mathrm{D}|E|{}^1\mathrm{D})=F_0+F_2-(2l-3)(2l+5)F'_2-\frac{2l^2-13l+12}{6}G_{l-1}-\frac{2l^2+17l+27}{6}G_{l+1},$$
$$({}^3\mathrm{P}|E|{}^3\mathrm{P})=F_0-5F_2+(2l-1)(2l+3)F'_2+\tfrac{1}{2}l(2l-1)G_{l-1}+\tfrac{1}{2}(l+1)(2l+3)G_{l+1},$$
$$({}^1\mathrm{S}|E|{}^1\mathrm{S})=F_0+10F_2-\tfrac{1}{3}l(2l-1)G_{l-1}-\tfrac{1}{3}(l+1)(2l+3)G_{l+1},$$
$$({}^1\mathrm{D}|E|{}^3\mathrm{P})=\tfrac{1}{2}[(2-l)G_{l-1}+(l+3)G_{l+1}][(2l-1)(2l+3)]^{\frac{1}{2}},$$
$$({}^3\mathrm{P}|E|{}^1\mathrm{S})=\tfrac{1}{2}[-(2l-1)G_{l-1}+(2l+3)G_{l+1}][2l(l+1)]^{\frac{1}{2}},$$
$$({}^1\mathrm{D}|E|{}^1\mathrm{S})=(2F'_2-\tfrac{1}{6}G_{l-1}-\tfrac{1}{6}G_{l+1})[2l(l+1)(2l-1)(2l+3)]^{\frac{1}{2}}.$$

APPENDIX

From the addition theorem for binomial coefficients

$$\sum_s \binom{x}{s}\binom{y}{z-s}=\binom{x+y}{z}, \tag{51}$$

putting $x=a-b,\ y=b,\ z=a-c$, we have

$$\frac{a!}{b!c!}=\sum_s \frac{(a-b)!(a-c)!}{(a-b-s)!(a-c-s)!(b+c-a+s)!s!}. \tag{52}$$

If y is negative, we can transform (51) by means of

$$\binom{y}{z-s}=(-1)^{z-s}\binom{z-s-y-1}{z-s} \tag{53}$$

and obtain

$$\sum_s (-1)^s\binom{x}{s}\binom{z-s-y-1}{z-s}=(-1)^z\binom{x+y}{z} \tag{54}$$

or

$$\sum_s (-1)^s\binom{x}{s}\binom{z-s-y-1}{z-s}=\binom{z-x-y-1}{z} \tag{54'}$$

putting $y=z-t-1$ we have from (54)

$$\sum_s (-1)^s \frac{(t-s)!}{s!(x-s)!(z-s)!} = (-1)^z \frac{(t-z)!(x+z-t-1)!}{x!z!(x-t-1)!} \quad \text{if} \quad x>t\geqslant z\geqslant 0, \tag{55}$$

and from (54′)

$$\sum_s (-1)^s \frac{(t-s)!}{s!(x-s)!(z-s)!} = \frac{(t-x)!(t-z)!}{x!z!(t-x-z)!} \quad \text{if} \quad t\geqslant x\geqslant 0,\ t\geqslant z\geqslant 0. \tag{55′}$$

Using repeatedly (52), and also (55) and (55′), we can transform the sum in (12) as follows:

$$\sum_u (-1)^u \frac{(u+l'_1-l'_2)!}{(u+L+1)!}\cdot\frac{(u+l_2-l_1)!}{(u-L)!(k-l_1-l'_2+u)!}\cdot\frac{(k+l_1+l'_2-u)!}{(l_1+l_2-u)!(l_1'+l'_2-u)!}$$

$$=\sum_{\alpha\beta u}(-1)^u\frac{(u+l'_1-l'_2)!}{(u+L+1)!}\cdot\frac{(L+l_2-l_1)!(l_2+l'_2-k)!}{(L+l_2-l_1-\alpha)!(l_2+l_2'-k-\alpha)!(k-l_2-l'_2-L+u+\alpha)!\alpha!}$$

$$\cdot\frac{(k+l'_2-l_2)!(k+l_1-l'_1)!}{(k+l'_2-l_2-\beta)!(k+l_1-l'_1-\beta)!(l'_1+l_2-k-u+\beta)!\beta!}$$

$$=\sum_{\alpha\beta}(-1)^{k-l_2-l'_2-L+\alpha}\frac{(L+l_2-l_1)!(l_2+l'_2-k)!(k+l'_2-l_2)!(k+l_1-l'_1)!}{(L+l_2-l_1-\alpha)!(l_2+l'_2-k-\alpha)!(k+l'_2-l_2-\beta)!(k+l_1-l'_1-\beta)!}$$

$$\cdot\frac{(l'_1+l_2+L-k-\alpha)!}{(l'_1+l_2+L+1-k+\beta)!}\cdot\frac{(\alpha+\beta)!}{(l'_1-l'_2-L+\alpha+\beta)!(L+l'_2-l'_1)!\alpha!\beta!}$$

$$=\sum_{\alpha\beta\gamma}(-1)^{k-l_2-l'_2-L+\alpha}\frac{(L+l_2-l_1)!(l_2+l'_2-k)!(k+l'_2-l_2)!(k+l_1-l'_1)!(l'_1+l_2+L-k-\alpha)!}{(L+l_2-l_1-\alpha)!(l_2+l'_2-k-\alpha)!(k+l'_2-l_2-\beta)!\times(k+l_1-l'_1-\beta)!(l'_1+l_2+L+1-k+\beta)!}$$

$$\cdot\frac{1}{(L+l'_2-l'_1-\gamma)!(\alpha-\gamma)!(l'_1-l'_2-L+\beta+\gamma)!\gamma!}$$

$$=\sum_\gamma(-1)^{k-l_2-l'_2-L+\gamma}\frac{(L+l_2-l_1)!(l_2+l'_2-k)!(k+l'_2-l_2)!(k+l_1-l'_1)!}{(L+l'_2-l'_1-\gamma)!\gamma!}$$

$$\cdot\frac{(l_1+l'_1-k)!(L+l'_1-l'_2)!}{(L+l_2-l_1-\gamma)!(l_2+l'_2-k-\gamma)!(l_1+l'_1-l_2-l'_2+\gamma)!}$$

$$\cdot\frac{(k+l_1+l'_1+1+\gamma)!}{(l'_1+l'_2+L+1)!(l_1+l_2+L+1)!(k+l_1-l'_2-L+\gamma)!(k+l'_1-l_2-L+\gamma)!}.$$

Putting $\gamma=l_2+l'_2-k-v$, this expression becomes

$$\sum_v(-1)^{v-L}\frac{(L+l_2-l_1)!(l_2+l'_2-k)!(k+l'_2-l_2)!(k+l_1-l'_1)!\times(l_1+l'_1-k)!(L+l'_1-l'_2)!(l_1+l_2+l'_1+l'_2+1-v)!}{(k+L-l'_1-l_2+v)!(l_2+l'_2-k-v)!(k+L-l_1-l'_2+v)!v!(l_1+l'_1-k-v)!\times(l_1+l_2-L-v)!(l'_1+l'_2-L-v)!(l'_1+l'_2+L+1)!(l_1+l_2+L+1)!};$$

introducing this result into (12) we obtain (12′).

Fine Structure of the Hydrogen Atom by a Microwave Method* **

WILLIS E. LAMB, JR. AND ROBERT C. RETHERFORD
Columbia Radiation Laboratory, Department of Physics, Columbia University, New York, New York

(Received June 18, 1947)

THE spectrum of the simplest atom, hydrogen, has a fine structure[1] which according to the Dirac wave equation for an electron moving in a Coulomb field is due to the combined effects of relativistic variation of mass with velocity and spin-orbit coupling. It has been considered one of the great triumphs of Dirac's theory that it gave the "right" fine structure of the energy levels. However, the experimental attempts to obtain a really detailed confirmation through a study of the Balmer lines have been frustrated by the large Doppler effect of the lines in comparison to the small splitting of the lower or $n=2$ states. The various spectroscopic workers have alternated between finding confirmation[2] of the theory and discrepancies[3] of as much as eight percent. More accurate information would clearly provide a delicate test of the form of the correct relativistic wave equation, as well as information on the possibility of line shifts due to coupling of the atom with the radiation field and clues to the nature of any non-Coulombic interaction between the elementary particles: electron and proton.

The calculated separation between the levels $2^2P_{1/2}$ and $2^2P_{3/2}$ is 0.365 cm^{-1} and corresponds to a wave-length of 2.74 cm. The great wartime advances in microwave techniques in the vicinity of three centimeters wave-length make possible the use of new physical tools for a study of the $n=2$ fine structure states of the hydrogen atom. A little consideration shows that it would be exceedingly difficult to detect the direct absorption of radiofrequency radiation by excited H atoms in a gas discharge because of their small population and the high background absorption due to electrons. Instead, we have found a method depending on a novel property of the $2^2S_{1/2}$ level. According to the Dirac theory, this state exactly coincides in energy with the $2^2P_{1/2}$ state which is the lower of the two P states. The S state in the absence of external electric fields is metastable. The radiative transition to the ground state $1^2S_{1/2}$ is forbidden by the selection rule $\Delta L=\pm 1$. Calculations of Breit and Teller[4] have shown that the most probable decay mechanism is double quantum emission with a lifetime of 1/7 second. This is to be contrasted with a lifetime of only 1.6×10^{-9} second for the non-metastable 2^2P states. The metastability is very much reduced in the presence of external electric fields[5] owing to Stark effect mixing of the S and P levels with resultant rapid decay of the combined state. If for any reason, the $2^2S_{1/2}$ level does not exactly coincide with the $2^2P_{1/2}$ level, the vulnerability of the state to external fields will be reduced. Such a removal of the accidental degeneracy may arise from any defect in the theory or may be brought about by the Zeeman splitting of the levels in an external magnetic field.

In brief, the experimental arrangement used is the following: Molecular hydrogen is thermally dissociated in a tungsten oven, and a jet of atoms emerges from a slit to be cross-bombarded by an electron stream. About one part in a hundred million of the atoms is thereby excited to the metastable $2^2S_{1/2}$ state. The metastable atoms (with a small recoil deflection) move on out of the bombardment region and are detected by the process of electron ejection from a metal target. The electron current is measured with an FP-54 electrometer tube and a sensitive galvanometer.

If the beam of metastable atoms is subjected to any perturbing fields which cause a transition to any of the 2^2P states, the atoms will decay while moving through a very small distance. As a result, the beam current will decrease, since the

* Publication assisted by the Ernest Kempton Adams Fund for Physical Research of Columbia University, New York.

** Work supported by the Signal Corps under contract number W 36-039 sc-32003.

[1] For a convenient account, see H. E. White, *Introduction to Atomic Spectra* (McGraw-Hill Book Company, New York, 1934), Chap. 8.

[2] J. W. Drinkwater, O. Richardson, and W. E. Williams, Proc. Roy. Soc. 174, 164 (1940).

[3] W. V. Houston, Phys. Rev. 51, 446 (1937); R. C. Williams, Phys. Rev. 54, 558 (1938); S. Pasternack, Phys. Rev. 54, 1113 (1938) has analyzed these results in terms of an upward shift of the S level by about 0.03 cm^{-1}.

[4] H. A. Bethe in *Handbuch der Physik*, Vol. 24/1, §43.

[5] G. Breit and E. Teller, Astrophys. J. 91, 215 (1940).

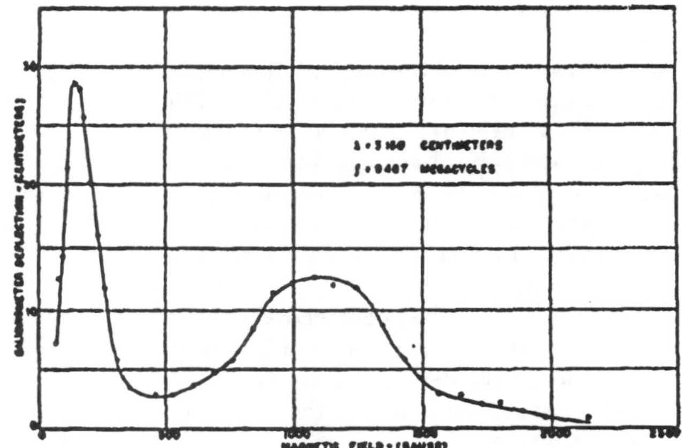

FIG. 1. A typical plot of galvanometer deflection due to interruption of the microwave radiation as a function of magnetic field. The magnetic field was calibrated with a flip coil and may be subject to some error which can be largely eliminated in a more refined apparatus. The width of the curves is probably due to the following causes: (1) the radiative line width of about 100 Mc/sec. of the ^2P states, (2) hyperfine splitting of the ^2S state which amounts to about 88 Mc/sec., (3) the use of an excessive intensity of radiation which gives increased absorption in the wings of the lines, and (4) inhomogeneity of the magnetic field. No transitions from the state $2^2S_{\frac{1}{2}}(m=-\frac{1}{2})$ have been observed, but atoms in this state may be quenched by stray electric fields because of the more nearly exact degeneracy with the Zeeman pattern of the ^2P states.

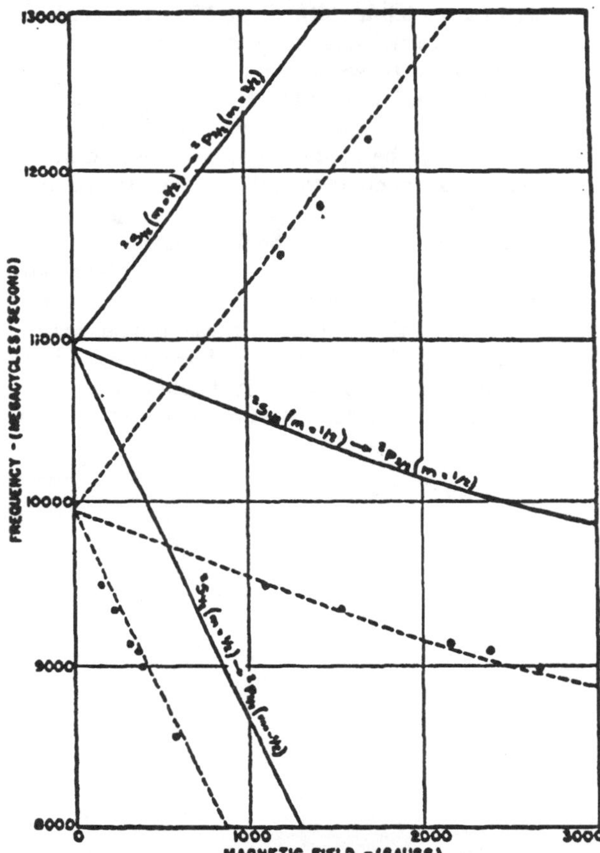

FIG. 2. Experimental values for resonance magnetic fields for various frequencies are shown by circles. The solid curves show three of the theoretically expected variations, and the broken curves are obtained by shifting these down by 1000 Mc/sec. This is done merely for the sake of comparison, and it is not implied that this would represent a "best fit." The plot covers only a small range of the frequency and magnetic field scale covered by our data, but a complete plot would not show up clearly on a small scale, and the shift indicated by the remainder of the data is quite compatible with a shift of 1000 Mc.

detector does not respond to atoms in the ground state. Such a transition may be induced by the application to the beam of a static electric field somewhere between source and detector. Transitions may also be induced by radiofrequency radiation for which $h\nu$ corresponds to the energy difference between one of the Zeeman components of $2^2S_{\frac{1}{2}}$ and any component of either $2^2P_{\frac{1}{2}}$ or $2^2P_{3/2}$. Such measurements provide a precise method for the location of the $2^2S_{\frac{1}{2}}$ state relative to the P states, as well as the distance between the latter states.

We have observed an electrometer current of the order of 10^{-14} ampere which must be ascribed to metastable hydrogen atoms. The strong quenching effect of static electric fields has been observed, and the voltage gradient necessary for this has a reasonable dependence on magnetic field strength.

We have also observed the decrease in the beam of metastable atoms caused by microwaves in the wave-length range 2.4 to 18.5 cm in various magnetic fields. In the measurements, the frequency of the r-f is fixed, and the change in the galvanometer current due to interruption of the r-f is determined as a function of magnetic field strength. A typical curve of quenching *versus* magnetic field is shown in Fig. 1. We have plotted in Fig. 2 the resonance magnetic fields for various frequencies in the vicinity of 10,000 Mc/sec. The theoretically calculated curves for the Zeeman effect are drawn as solid curves, while for comparison with the observed points, the calculated curves have been shifted downward by 1000 Mc/sec. (broken curves). The results indicate clearly that, contrary to theory but in essential agreement with Pasternack's hypothesis,[8] the $2^2S_{\frac{1}{2}}$ state is higher than the $2^2P_{\frac{1}{2}}$ by about 1000 Mc/sec. (0.033 cm^{-1} or about 9 percent of the spin relativity doublet separation. The lower frequency transitions $^2S_{\frac{1}{2}}(m=\frac{1}{2})\rightarrow$ $^2P_{\frac{1}{2}}(m=\pm\frac{1}{2})$ have also been observed and agree well with such a shift of the $^2S_{\frac{1}{2}}$ level. With the present precision, we have not yet detected any discrepancy between the Dirac theory and the doublet separation of the P levels. (According to most of the imaginable theoretical explanations of the shift, the doublet separation would not be affected as much as the relative location of the S and P states.) With proposed refinements in sensitivity, magnetic field homogeneity, and calibration, it is hoped to locate the S level with respect to each P level to an accuracy of at least ten Mc/sec. By addition of these frequencies and assumption of the theoretical formula $\Delta\nu = \frac{1}{16}\alpha^2 R$ for the doublet separation, it should be possible to measure the square of the fine structure constant times the Rydberg frequency to an accuracy of 0.1 percent.

By a slight extension of the method, it is hoped to determine the hyperfine structure of the $2^2S_{\frac{1}{2}}$ state. All of these measurements will be repeated for deuterium and other hydrogen-like atoms.

A paper giving a fuller account of the experimental and theoretical details of the method is being prepared, and this will contain later and more accurate data.

The experiments described here were discussed at the Conference on the Foundations of Quantum Mechanics held at Shelter Island on June 1–3, 1947 which was sponsored by the National Academy of Sciences.

Does the Electron Have an Intrinsic Magnetic Moment?

G. Breit
Yale University, New Haven, Connecticut
September 29, 1947

THE hyperfine structure of the ground term of H^1 and H^2 is greater[1] than expected from nuclear magnetic moments by, respectively, 0.26 and 0.31 percent. The difference between these values is less certain than the approximate value 0.28 percent and will be assumed to be insignificant.[2] If the electron had a small, Pauli-type, intrinsic magnetic moment[3] μ_e the observed and calculated values would differ.

The effect of $\mu_e\rho_3(\mathbf{H}\sigma)$ in the Hamiltonian (Dirac's notation) is to change the hfs interval factor to

$$A = \frac{2e\mu_N}{ij(j+1)} Im \int_0^\infty \left[-k + \frac{\mu_e p_0}{ek} \right] F^* G dr, \qquad (1)$$

where $k = l$, $-l-1$, respectively, for $j = l - \frac{1}{2}$, $l + \frac{1}{2}$. The functions F, G are, respectively, $-iF$, G of Roess.[4] Azimuthal, inner, and nuclear-spin quantum numbers are l, j, i. The magnetic field of nucleus at the electron is H, the nuclear magnetic moment is μ_N. The molecular-beam experiment gives $\mu_N/(\mu_0 - \mu_e)$, where $\mu_0 > 0$ is the Bohr magneton. The atomic-beam experiment determines,[5] according to Eq. (1), the quantity $\mu_N(\mu_0 + \mu_e/2)$. The theoretical ratio of the hfs to the molecular-beam value of μ_N contains, therefore, $1 - \mu_e/2\mu_0$ as a factor. To explain the observed discrepancy one needs $\mu_e/\mu_0 = -0.0056$, a small value which could have escaped detection. According to Eq. (1) the interval factors of s, $p_{1/2}$, $p_{3/2}$ terms contain μ_e in the factors $1 + \mu_e/2\mu_0$, $1 - \mu_e/2\mu_0$, $1 + \mu_e/4\mu_0$ apart from factor $1 - \mu_e/\mu_0$ which is needed in (1) if the apparent μ_N is substituted for the true value. In principle, ratios of interval factors for these terms could determine μ_e/μ_0.

One expects the following additional effects of μ_e: (a) A modification of the Landé g factor through factor $1 - 2\mu_e/\mu_0$ in $g - 1$. (b) The term $(-\mu_e)\rho_2(\mathbf{E}\sigma)$ caused by nuclear electric field contributes to the energy

$$\Delta E = -2\mu_e Im \int_0^\infty F^* G dr \cong \frac{2\mu_e(1+k)Z^4 Rch\alpha^2}{\mu_0(l+1)l(2l+1)n^3}, \qquad (2)$$

where n is the principal quantum number of a hydrogenic term, R is the Rydberg, and α is the fine structure constant. For $n = 2$, $Z = 1$ Eq. (2) gives

$$\Delta E = (-4, 4/3)(\mu_e/\mu_0)(Rch\alpha^2/16)$$

for $2s$, $2p_{1/2}$, respectively. The displacement of $2s$ with respect to $2p_{1/2}$ of hydrogen is $(-16/3)(\mu_e/\mu_0)(Rch\alpha^2/16)$, which is about 1/33 of the $2p_{3/2}$, $2s$ doublet separation for $\mu_e/\mu_0 = -0.0056$. The Lamb-Retherford $-2s + 2p_{1/2}$ separation is roughly 3 times the above value. It is doubtful that Bethe's[6] electrodynamic shift theory of the Lamb-Retherford effect[7] is as yet quantitative enough to exclude the possibility of about $\frac{1}{3}$ of the effect arising from another cause.

The presence of the Coulomb energy in p_0 in Eq. (1) makes the integral diverge for s terms. The integral converges, however, if the Coulomb energy is made finite at small distances. A cut-off of the integral at $r \sim e^2/mc^2$ makes the contribution of the Coulomb energy to the integral of the negligible order $\alpha^2 \log\alpha^{-2}$ of the term containing p_0. The quantity p_0 has accordingly been replaced by mc in the estimates.

It is not claimed that the electron has an intrinsic magnetic moment. Aesthetic objections could be raised against such a view. The only object of this note is to point out that the evidence considered above does not disprove a small μ_e of the order $\alpha\mu_0$.

If the discrepancy is due to an interaction between the electron and the nucleus of a local type, it is hard to see why it should have the same fractional value for the proton and the deuteron. In this case the effect would be practically confined to s terms, and one could, in principle, distinguish between it and the hypothesis of the intrinsic magnetic moment by comparing hfs interval factors for different spectroscopic terms.

[1] J. E. Nafe, E. B. Nelson, and I. I. Rabi, Phys. Rev. **71**, 914 (1947).
[2] Verbal communication from Professor I. I. Rabi. The writer is very grateful to Professors Rabi and Ramsey for this and other discussions of the subject.
[3] W. Pauli, *Handbuch der Physik* (Verlag Julius Springer, Berlin, 1933), Vol. 24/1, p. 211.
[4] C. G. Darwin, Proc. Roy. Soc. **A118**, 654 (1928); L. C. Roess, Phys. Rev. **37**, 532 (1931).
[5] G. Breit and F. W. Doermann, Phys. Rev. **36**, 1732 (1930), see p. 1737.
[6] H. A. Bethe, Phys. Rev. **72**, 339 (1947).
[7] Willis E. Lamb and Robert C. Retherford, Phys. Rev. **72**, 241 (1947).

Precision Measurement of the Ratio of the Atomic 'g Values' in the $^2P_{3/2}$ and $^2P_{1/2}$ States of Gallium*

P. KUSCH AND H. M. FOLEY
Columbia University, New York, New York
November 3, 1947

THE measurement of the frequencies associated with the Zeeman splittings of the energy levels in two different atomic states in a constant magnetic field permits a determination of the ratio of the g_J values of the atomic states. This determination involves only an accurate measurement of the frequencies, and does not require a knowledge of the value of the constant magnetic field.

Using the atomic beam magnetic resonance technique we have measured six lines in the Zeeman spectrum of the $^2P_{1/2}$ state, and five lines of the $^2P_{3/2}$ state of gallium at a field strength of 380 gauss. The spectrum is, of course, complicated by the level splittings produced by the nuclear magnetic moments and electric quadrupole moments. At the field strength employed in this experiment the nuclear energy level pattern is of an intermediate Paschen-Back character.

The procedure employed in the observations was to make a series of alternate measurements of the frequencies of the lines in the $^2P_{1/2}$ and $^2P_{3/2}$ states. In this way the effect of a drift in magnetic field was minimized.

Becker and Kusch[1] have recently determined with high precision the nuclear magnetic moment and electric quadrupole moment coupling coefficients as well as the nuclear g values of Ga^{69} and Ga^{71} in both states. Their determinations are independent of any knowledge of g_J in either state.

Their results are consistent with the less accurate values obtained by Renzetti[2] by the zero moment deflection method.

With the constants given by Becker and Kusch and with assumed values for g_J it is possible to determine the magnetic field from the observed frequencies. The field values obtained from either of the groups of lines in the $^2P_{1/2}$ or the $^2P_{3/2}$ states are constant within the experimental uncertainty. What is more, there is excellent consistency of the results obtained from the data on the two isotopes. The mean value of magnetic field determined from the $^2P_{1/2}$ lines is less than the mean field value for the $^2P_{3/2}$ lines by 0.65±0.02 gauss when the g_J values for the $^2P_{3/2}$ and $^2P_{1/2}$ states are taken to be 4/3 and 2/3, respectively. To remove this discrepancy we must assume for the ratio of the g_J values

$$\frac{g3/2}{g1/2}=2.00344\pm0.00012=2+\Delta.$$

If the electronic configuration in these states is accurately described by Russell-Saunders coupling the above discrepancy must be assigned to a change in the g value of the intrinsic moment of the electron or of the orbital moment from their accepted values. If the electron spin g value $g_S=2+\delta_S$ and the orbital g value $g_L=1+\delta_L$, then $\Delta=\frac{2}{3}\delta_S-3\delta_L$. Our present experiments, even assuming Russell-Saunders coupling, do not permit any evaluation of δ_S and δ_L. However, the discrepancy could be accounted for by taking $g_S=2.00229\pm0.00008$ and $g_L=1$, or alternatively $g_S=2$ and $g_L=0.99886\pm0.00004$.

The experiments reported in this note are of a preliminary nature. It is our intention to continue these studies with other atomic systems to clarify the role of the coupling in this phenomenon.

The authors are grateful to Professors G. Breit and I. I. Rabi for suggestions which stimulated these experiments.

* Publication assisted by the Ernest Kempton Adams Fund for Physical Research of Columbia University.

[1] To be published.

[2] N. A. Renzetti, Phys. Rev. 57, 753 (1940).

PHYSICAL REVIEW VOLUME 73, NUMBER 4 FEBRUARY 15, 1948

The Influence of Retardation on the London-van der Waals Forces

H. B. G. Casimir and D. Polder
Natuurkundig Laboratorium der N. V. Philips' Gloeilampenfabrieken, Eindhoven, Netherlands

(Received May 16, 1947)

The influence of retardation on the energy of interaction between two neutral atoms is investigated by means of quantum electrodynamics. As a preliminary step, Part I contains a discussion of the interaction between a neutral atom and a perfectly conducting plane, and it is found that the influence of retardation leads to a reduction of the interaction energy by a correction factor which decreases monotonically with increasing distance R. This factor is equal to unity for R small compared with the wave-lengths corresponding to the atomic frequencies, and is proportional to R^{-1} for distances large compared with these wave-lengths. In the latter case the total interaction energy is given by $-3\hbar c\alpha/8\pi R^4$, where α is the static polarizability of the atom. Although the problem of the interaction of two atoms discussed in Part II is much more difficult to handle mathematically, the results are very similar. Again the influence of retardation can be described by a monotonically decreasing correction factor which is equal to unity for small distances and proportional to R^{-1} for large distances. In the latter case the energy of interaction is found to be $-23\hbar c\alpha_1\alpha_2/4\pi R^7$.

PART I. GENERAL FEATURES OF THE PROBLEM

1. Introduction

THE problem treated in this paper, though apparently a somewhat academic exercise in quantum electrodynamics, arose directly from the work of Verwey and Overbeek[1] on the stability of colloidal systems. Starting from work of Hamaker, Verwey and Overbeek have in recent years developed a theory in which the attraction between colloidal particles is exclusively ascribed to London-van der Waals forces, the repulsion being accounted for by the interaction of electric double layers. In applying this

[1] E. J. W. Verwey, J. T. G. Overbeek, and K. van Nes, *Theory of the Stability of Lyophobic Colloids* (Elsevier Publishing Company, Inc., Amsterdam, in press); E. J. W. Verwey and J. T. G. Overbeek, Trans. Faraday Soc. (In press); E. J. W. Verwey, J. Phys. and Colloid Chem. 51, 631 (1947).

theory to suspensions of comparatively large particles, they found a discrepancy between their theory and the experimental results which could be removed only by assuming that at large distances the attractive force between two atoms decreases more rapidly than R^{-7}. Overbeek then pointed out that on the basis of the picture customarily used for visualizing London forces, an influence of retardation on the interaction is to be expected as soon as the distance between the particles becomes comparable to the wave-length corresponding to the atomic frequencies. Although this argument is suggestive, we have not succeeded in deriving an expression for the influence of retardation based on such a simple model, and we doubt very much whether a result can be obtained in that way. In this paper hardly any reference will be made to Overbeek's original considerations. Also, the application to the problems of colloid chemistry will not be touched upon but will be left for a future publication. We want, however, to emphasize our indebtedness to Overbeek's suggestion.

So far, problems of retardation have only occasionally been treated by means of quantum electrodynamics. There is, of course, the work of Møller[2] and its justification by Bethe and Fermi.[3] Also in the work of Breit[4] on the interaction of the electrons in the He atom, retardation is taken into account. In these cases, however, we have to deal with the influence of retardation on expressions containing the square of the electronic charge, which means that we can restrict ourselves to studying the interaction between electrons and the radiation field to a second approximation. In our case, which concerns the interaction between two neutral atoms, the approximation has to be pushed to the fourth order, as the usual expression for the London energy contains the fourth power of the electronic charge. We found, however, that what seemed to us the most essential features of the final result are already clearly revealed by a problem which can be treated by means of quite simple mathematics, involving only second-order perturbation theory, i.e., the interaction of a neutral atom with a perfectly conducting wall. According to classical ideas the energy should always be given by the interaction of the atomic dipole with its image, and retardation effects are to be expected when its distance from the wall becomes large. The result of a direct calculation by means of quantum electrodynamics, which will be given in Section 12, is not in disagreement with this notion. Yet the final result is rather unexpected. For short distances we find the usual value for the London energy between a neutral atom and a conducting wall, which is proportional to R^{-3}. With increasing R, however, the usual value must be multiplied by a monotonically decreasing factor, and for large values of R the London energy is found to be proportional to R^{-4} rather than to R^{-3}. It is remarkable that the asymptotic expression for large R contains Planck's constant and, in addition, the static polarizability of the atom as the only quantity characterizing the specific properties of the atom.

The calculations in Part II, dealing with the interaction of two neutral atoms, are much more complicated, but it is of interest to remark that here also the present-day formulation of quantum electrodynamics, if properly handled, is able to give an unambiguous result. For short distances the usual expression for the London energy, in this case being proportional to R^{-6}, is valid again, whereas for large distances the energy of interaction is proportional to R^{-7}. The asymptotic expression contains Planck's constant and the product of the polarizabilities of the two atoms.

2. Interaction of a Neutral Atom with a Perfectly Conducting Plane

Consider a region of space, defined by $0<x<L$, $0<y<L$, $0<z<L$, enclosed in a box with perfectly conducting walls. The eigenstates of the electromagnetic field in this box are described by solutions of Maxwell's equations satisfying the boundary condition that the tangential components of $\mathbf{E}$, the electric field, vanish at the walls. These solutions are easily found to be

$$\begin{aligned} E_x(\mathbf{k},\lambda) &= e_x(\mathbf{k},\lambda)\cos k_1x \sin k_2y \sin k_3z\cdot C_n, \\ E_y(\mathbf{k},\lambda) &= e_y(\mathbf{k},\lambda)\sin k_1x \cos k_2y \sin k_3z\cdot C_n, \\ E_z(\mathbf{k},\lambda) &= e_z(\mathbf{k},\lambda)\sin k_1x \sin k_2y \cos k_3z\cdot C_n, \end{aligned} \quad (1)$$

[2] C. Møller, Zeits. f. Physik 70, 786 (1931).
[3] H. Bethe und E. Fermi, Zeits. f. Physik 77, 296 (1932).
[4] G. Breit, Phys. Rev. 34, 353 (1929); Phys. Rev. 36, 383 (1930); Phys. Rev. 39, 616 (1932).

where $\mathbf{k}$ is a wave vector with components $k_i = n_i\pi/L$, with $n_i = 0, 1, 2, 3\cdots$, and $\mathbf{e}$ is a unit vector perpendicular to $\mathbf{k}$. To each vector $\mathbf{k}$ belong two vectors $\mathbf{e}$, corresponding to the two directions of polarization; they are indicated by the symbol λ ($\lambda=1, 2$). The normalization factor C_e is given by

$$C_e^2 = 16\pi\hbar c/kL^3. \tag{2}$$

In order to verify that this is the correct normalization we write for the vector potential of the electromagnetic field in the box:

$$\mathbf{A} = \sum_{k,\lambda} (A_{k\lambda}e^{-i\omega t} + A_{k\lambda}^{\dagger}e^{i\omega t})\mathbf{E}(\mathbf{k}, \lambda), \tag{3}$$

and determine the energy ϵ of the field:

$$\epsilon = \frac{1}{8\pi}\int (E^2+H^2)dv = \frac{1}{4\pi}\int E^2 dv$$

$$= \frac{1}{4\pi}\frac{L^3}{8}\sum_{k,\lambda}\frac{16\pi\hbar c}{kL^3}k^2(A_{k\lambda}^{\dagger}A_{k\lambda} + A_{k\lambda}A_{k\lambda}^{\dagger})$$

$$= \sum_{k,\lambda}\tfrac{1}{2}\hbar\omega(A_{k\lambda}^{\dagger}A_{k\lambda} + A_{k\lambda}A_{k\lambda}^{\dagger}). \tag{4}$$

In quantum electrodynamics A and $A^{\dagger}$ are operators satisfying:

$$A_{k\lambda}A_{l\mu}^{\dagger} - A_{l\mu}^{\dagger}A_{k\lambda} = \delta_{kl}\delta_{\mu\lambda}, \tag{5}$$

and the eigenvalues of $A_{k\lambda}^{\dagger}A_{k\lambda}$ are 0, 1, 2$\cdots$ In this way we have obtained the usual formulation of the quantization of the radiation field in an empty box.

We now consider the operator G of the interaction between a neutral atom and the radiation field:

$$G = \sum_j \left\{ -\frac{e}{mc}(\mathbf{p}_j\mathbf{A}) + \frac{e^2}{2mc^2}A^2 \right\}, \tag{6}$$

where the summation is over all electrons in the atom and $\mathbf{p}_j$ is the operator of the momentum of an electron. We determine the energy perturbation of the lowest level of the system consisting of the atom and the radiation field. Since A has no diagonal elements, there is no first-order perturbation proportional to e. Therefore we use second-order perturbation theory for the terms with e in G and first-order perturbation theory for the terms with e^2. In the course of the following calculation we shall determine the perturbation energies for the case in which the atom is situated at a very large distance from the walls of the box and for the case in which it is at a short distance from one of the walls. In both cases the result is given by a divergent series over the excited states of the atom, each term of which is a convergent sum over the excited states of the radiation field. The difference between the perturbation energies in both cases can be found without ambiguity and is finite; this difference will be interpreted as the energy of interaction between the atom and the wall so far as it is due to the interaction of the atom with the radiation field.

The second-order perturbation energy of the lowest level due to an operator H is given by

$$\Delta_2 E = -\sum_g \frac{|H_{0g}|^2}{E_g - E_0}. \tag{7}$$

In our system the excited states g are labeled by the index n for the states of the atom and by the indices k, λ for the states of the radiation field in which one light quantum is present. The energy difference between the level n, k, λ of our system and the lowest level is given by $\hbar c(k_n + k)$ and, therefore, the perturbation energy is

$$\Delta_2 E = -\frac{e^2}{m^2c^2}\sum_{n,k,\lambda}\frac{|(\mathbf{A}_{0;k\lambda}\mathbf{p}_{0;n})|^2}{\hbar c(k_n+k)} \tag{8}$$

(where $\mathbf{p} = \sum_j \mathbf{p}_j$), to which must be added the first-order perturbation due to terms with e^2; which according to the laws of matrix multiplication can be written as:

$$\Delta_1 E = \sum_j \frac{e^2}{2mc^2}\sum_{k,\lambda}\sum_{x,y,z}|A_{0;k\lambda}^x|^2. \tag{9}$$

In writing these formulae we have made one approximation: we have neglected the variation of the electromagnetic field inside the atom. It is well known that due to this approximation the contribution to the second-order perturbation energy from one excited level of the atom already becomes infinite since the integral over $\mathbf{k}$ does not converge for $|k|\to\infty$. Instead of taking these effects into account rigorously we shall introduce a factor $e^{-\gamma k}$, which makes the integral convergent, and put $\gamma=0$ in the final result. In reality it should be of the order of the radius of the atom, but this does not appreciably affect the results.

In order to obtain a simple expression for $\Delta_r E = \Delta_2 E + \Delta_1 E$, we make use of the sum rule

$$\delta_{xy} \sum_i \frac{e^2}{2mc^2} = \frac{e^2}{m^2c^2} \sum_n \frac{\frac{1}{2}(p_{0;\,n}{}^x p_{n;\,0}{}^y + p_{0;\,n}{}^y p_{n;\,0}{}^x)}{\hbar c k_n}, \quad (10)$$

and the relation $e p_{0;\,n} = -imck_n q_{0;\,n}$, where q is the operator of the total dipole moment. We find for the total perturbation energy

$$\Delta_r E = \frac{1}{\hbar c} \sum_{n,\,k,\,\lambda} \frac{k k_n}{k+k_n} |(\mathbf{A}_{0;\,k\lambda} \mathbf{q}_{0;\,n})|^2. \quad (11)$$

In order to simplify the problem we assume the zero state of the atom to be a state with angular momentum $J=0$, which means that matrix elements of $\mathbf{p}$ exist only between this state and the threefold degenerate states with $J=1$. The three wave functions belonging to a state with $J=1$ may be chosen without loss of generality in such a way that they have the same transformation properties under a rotation as x, y, and z. Then all cross products of the type $q_{0;\,n}{}^x q_{0;\,n}{}^y$ vanish and (11) can be written as:

$$\Delta_r E = \frac{1}{\hbar c} \sum_{n,\,k,\,\lambda} \frac{k k_n}{k+k_n} \sum_{x,\,y,\,z} |A_{0;\,k\lambda}{}^x|^2 |q_{0;\,n}{}^x|^2, \quad (12)$$

where n denotes the states with $J=1$. Substituting (3) and (1) in (12) we have

$$\Delta_r E = \frac{16\pi}{L^3} \sum_{n,\,k,\,\lambda} \frac{k_n}{k+k_n} \{ |q_{0;\,n}{}^x|^2 e_x{}^2(\mathbf{k},\lambda) \cos^2 k_1 x \sin^2 k_2 y \sin^2 k_3 z + |q_{0;\,n}{}^y|^2 e_y{}^2(\mathbf{k},\lambda) \sin^2 k_1 x \cos^2 k_2 y \sin^2 k_3 z + |q_{0;\,n}{}^z|^2 e_z{}^2(\mathbf{k},\lambda) \sin^2 k_1 x \sin^2 k_2 y \cos^2 k_3 z \}. \quad (13)$$

In order to carry out the summation over λ we use the relation

$$\sum_\lambda e_i{}^2(\mathbf{k},\lambda) = 1 - k_i{}^2/k^2. \quad (14)$$

We assume that the box is very large and therefore the summation over all values of $\mathbf{k}$ can be replaced by an integral. Since the integrand is an even function of k_i, the summation is equal to $L^3/8\pi^3$ times a threefold integral from $-\infty$ to $+\infty$ over k_1, k_2 and k_3. At the same time we introduce the convergence factor $e^{-\gamma k}$. We obtain

$$\Delta_r E = \frac{2}{\pi^2} \int\!\!\int\!\!\int_{-\infty}^{+\infty} dk_1 dk_2 dk_3 \sum_n \frac{k_n e^{-\gamma k}}{k+k_n} \left\{ |q_{0;\,n}{}^x|^2 \left(1 - \frac{k_1{}^2}{k^2}\right) \cos^2 k_1 x \sin^2 k_2 y \sin^2 k_3 z + |q_{0;\,n}{}^y|^2 \left(1 - \frac{k_2{}^2}{k^2}\right) \sin^2 k_1 x \cos^2 k_2 y \sin^2 k_3 z + |q_{0;\,n}{}^z|^2 \left(1 - \frac{k_3{}^2}{k^2}\right) \sin^2 k_1 x \sin^2 k_2 y \cos^2 k_3 z \right\}. \quad (15)$$

We assume that the distance of the atom from the walls $y=0$ and $z=0$ is very large so that the value of the integral does not change if we put $\sin^2 k_2 y$, $\cos^2 k_2 y$, $\sin^2 k_3 z$, and $\cos^2 k_3 z$ equal to $\frac{1}{2}$. When the distance from the wall $x=0$ is also very large, the same may be done with $\sin^2 k_1 x$ and $\cos^2 k_1 x$. For the difference between the perturbation energies in the case in which the atom is situated at a distance R from the wall $x=0$ and in the case in which the atom is at a large distance, we find, therefore:

$$\Delta_d E = \frac{1}{4\pi^2} \sum_n \int\!\!\int\!\!\int_{-\infty}^{+\infty} dk_1 dk_2 dk_3 \frac{k_n e^{-\gamma k}}{k+k_n} \left\{ |q_{0;\,n}{}^x|^2 \left(1 - \frac{k_1{}^2}{k^2}\right)(2\cos^2 k_1 R - 1) + |q_{0;\,n}{}^y|^2 \left(1 - \frac{k_2{}^2}{k^2}\right)(2\sin^2 k_1 R - 1) + |q_{0;\,n}{}^z|^2 \left(1 - \frac{k_3{}^2}{k^2}\right)(2\sin^2 k_1 R - 1) \right\}. \quad ($$

Introducing polar coordinates in the **k** space we obtain:

$$\Delta_d E = \sum_n \int_0^\infty -dk\left[\frac{e^{-\gamma k}}{2\pi i}\frac{k^2k_n}{(k+k_n)}\left\{|q_{0;\,n}{}^x|^2\frac{e^{2ikR}}{2kR}\left(\frac{2i}{2kR}-\frac{2}{4k^2R^2}\right)\right.\right.$$

$$\left.\left.+(|q_{0;\,n}{}^y|^2+|q_{0;\,n}{}^z|^2)\frac{e^{2ikR}}{2kR}\left(1+\frac{i}{2kR}-\frac{1}{4k^2R^2}\right)\right\}+\text{complex conjugate}\right]. \quad (17)$$

It is interesting to remark that the expression in { } suggests the existence of an interpretation of formula (17) on the basis of the correspondence principle. If multiplied by a factor k^3, the first term in the expression equals the energy of a complex dipole $q^x e^{-ikct}$ in the retarded electric field of a dipole $q^x e^{ikct}$ at a distance $2R$, and this second dipole might be interpreted as the "electrical image" of the complex conjugate of the first dipole, with regard to a perfectly conducting plane at a distance R from the first dipole. A similar interpretation can be given to the second term in { }, but we have not been able to find a general consideration, based on the correspondence idea, by means of which at least the form of Eq. (17) could be foretold. In this connection we should like to point out that in dealing with the behavior of an atom in an excited state we usually meet with a factor $1/(k-k_n)$ instead of the factor $1/(k+k_n)$ occurring here, and the result is mainly determined by the residue at $k=k_n$ so that one definite frequency is singled out and an interpretation in terms of oscillators with well-determined frequencies becomes possible.

We now proceed with the calculation of the energy of interaction between an atom and a conducting wall by taking into consideration the electrostatic interaction. The electrostatic energy between a dipole q^x at $x=R$ and a conducting wall at $x=0$ is

$$\epsilon_x{}^e = -(q^x)^2/8R^3. \quad (18)$$

For a dipole q^y or q^z the energy is:

$$\epsilon_{y,z}{}^e = -(q^{y,z})^2/16R^3. \quad (19)$$

The first-order perturbation energy of the lowest level of our system due to the electrostatic terms is:

$$\Delta_e E = -\left[\frac{2((q^x)^2)_{00}+((q^y)^2+(q^z)^2)_{00}}{16R^3}\right]$$

$$= -\frac{\sum_n(2|q_{0;\,n}{}^x|^2+|q_{0;\,n}{}^y|^2+|q_{0;\,n}{}^z|^2)}{16R^3}. \quad (20)$$

The total interaction energy between the atom and the wall is

$$\Delta_t E = \Delta_d E + \Delta_e E. \quad (21)$$

A closer examination of $\Delta_d E$ shows that the integrand in (17) remains finite at $k=0$, but that both the term within [ ] that is completely written out and its complex conjugate have a simple pole. We shall integrate each of these terms separately from ϵ to ∞ and let ϵ tend to zero afterwards. We now want to replace the integration along the real axis by the integration along path 1 in Fig. 1 for the first integral and along path 2 for the second. In the limit $\epsilon\to 0$, we easily find for the integration on the semicircle:

$$\frac{1}{16R^3}\sum_n(2|q_{0.\,n}{}^x|^2+|q_{0;\,n}{}^y|^2+|q_{0;\,n}{}^z|^2),$$

which exactly cancels the term $\Delta_e E$ in $\Delta_t E$. Introducing the variable $u=-ik$ for the integration from $i\epsilon$ to $i\infty$ and $u=ik$ for the integration from $-i\epsilon$ to $-i\infty$, we find

$$\Delta_t E = -\frac{1}{\pi}\sum_n\int_0^\infty \frac{k_n u^2 du}{u^2+k_n{}^2}\frac{e^{-2uR}}{2R}$$

$$\times\left\{2|q_{0;\,n}{}^x|^2\left(\frac{1}{2uR}+\frac{1}{4u^2R^2}\right)\right.$$

$$\left.+(|q_{0;\,n}{}^y|^2+|q_{0;\,n}{}^z|^2)\left(1+\frac{1}{2uR}+\frac{1}{4u^2R^2}\right)\right\}. \quad (22)$$

In the limit of very small distances R, it is

easily found that the formula reduces to

$$\Delta_t E(R\rightarrow 0)$$

$$= -\frac{1}{16R^3}\sum_n(2|q_{0;\,n}{}^x|^2+|q_{0;\,n}{}^y|^2+|q_{0;\,n}{}^z|^2), \quad (23)$$

being equal to the value of the London energy, derived by the elementary theory which takes into account the electrostatic interaction only. For very large R (R larger than all $\lambda_n = 2\pi/k_n$) it is immediately seen that (22) reduces to

$$\Delta_t E(R\rightarrow\infty)$$

$$= -\sum_n \frac{(|q_{0;\,n}{}^x|^2+|q_{0;\,n}{}^y|^2+|q_{0;\,n}{}^z|^2)}{4\pi k_n R^4}, \quad (24)$$

which can be written in terms of the static polarizability α of the atom:

$$\Delta_t E(R\rightarrow\infty) = -\frac{\hbar c}{8\pi R^4}(\alpha_x+\alpha_y+\alpha_z). \quad (25)$$

Because we wanted to see more clearly the role of the x, y, and z dipoles individually, we have not yet used the relation

$$\sum|q_{0;\,n}{}^x|^2=\sum|q_{0;\,n}{}^y|^2=\sum|q_{0;\,n}{}^z|^2=|q_{0;\,n}|^2, \quad (26)$$

where the summation extends over the three states with $J=1$, belonging to one degenerate level, which will be indicated from now by one symbol n. With the aid of this relation, Eq. (22) may be written as

$$\Delta_t E = -\frac{2}{\pi}\sum_n\int_0^\infty \frac{k_n u^3 du}{u^2+k_n^2}\frac{e^{-2uR}}{2R} \times|q_{0;\,n}|^2\left(1+\frac{2}{2uR}+\frac{2}{4u^2R^2}\right), \quad (27)$$

where each term of the sum over n represents the contribution of all three states with $J=1$ belonging to one degenerate level. In Fig. 2 we have given the result of a numerical calculation of the factor $\Delta_t E/\Delta_e E$ for the case in which only one excited level n (with $E_n = hc/\lambda_n$) gives a contribution to the London energy. It is seen that the value of the factor decreases monotonically with increasing R. It starts with the value 1 (for $R\rightarrow 0$), while for large R it is approximately equal to $3\lambda_n/2\pi^2 R$.

PART II. THE INTERACTION BETWEEN TWO ATOMS

1. Outline of Method

The energy of interaction between two neutral atoms will be determined by solving the following perturbation problem. The unperturbed states of the system, consisting of two atoms and the radiation field, will be assumed to be the states which are completely defined by the indication of the states of the two atoms and the state of the radiation field in empty space.

The perturbation operator, which is responsible for the interaction of the two atoms, contains the electrostatic interaction Q between the charged particles of the first atom with those of the second atom, the interaction G_A between the first atom and the radiation field, and the interaction G_B between the second atom and the radiation field.

With the aid of the perturbation operator G_A+G_B+Q we shall determine the energy perturbation; we have already remarked in Part I that the approximation has to be pushed to the fourth power of the electronic charge e. For this purpose we shall apply first-, second-, third-, and fourth-order perturbation theory. By the order of the perturbation theory we mean the degree in which the perturbation operator occurs in the expression for the energy perturbation. It does not indicate the power of the electronic charge occurring in this expression, as the perturbation operator contains terms with e as well as with e^2. The total result of the calculation will be divergent, but, as in Part I, we shall find a finite value for those terms that depend on the distance R between the two atoms; this value will be interpreted again as the total energy of interaction between the atoms.

In order to carry out the perturbation procedure we shall have to examine carefully the matrix elements occurring in the expression for the first-, second-, third-, and fourth-order perturbation in the energy. The examination does not give rise to special difficulties and, therefore, we shall only mention the results. Restricting outselves to those terms in the energy perturbation which depend on R and which are proportional to e^m, with $m\leqslant 4$, we find that in our problem there is no first-order perturbation in the energy. Further, matrix elements of Q occur only in the expression for the second-order perturbation in the energy, namely, as the products of two matrix elements of Q, and in the third-order perturbation in the energy, namely, as the products of one matrix element of Q, one matrix element of G_A, and one matrix element of G_B. All the other terms in the energy perturbations do not involve Q.

We have thus to deal with the following terms:

(a) Terms obtained by applying second-order perturbation theory with the electrostatic interaction, which itself is proportional to e^2. The result is proportional to e^4 and is equal to the usual expression for the London energy.

(b) Terms obtained by applying third-order perturbation theory, restricting ourselves to the terms involving Q. We shall carry out the perturbation procedure in a somewhat unconventional way by successively applying first-order perturbation theory for the electrostatic interaction and second-order perturbation theory for the interaction with the radiation field: We calculate to the first order of the perturbation (which is an approximation proportional to e^2) the wave functions of the two atoms coupled by electrostatic interaction, and we then proceed to calculate, in exactly the same way as in the first part of this paper, the second-order interaction energy of this compound system with the radiation field. Again the result is proportional to e^4.

(c) Terms obtained by determining the energy perturbation with the aid of the operator G_A+G_B, the electrostatic interaction now being omitted. Again we shall solve this perturbation problem in a somewhat unconventional way: We first calculate in the usual way the second-order interaction energy of the atom A with a radiation field. In this way we obtain also the interaction between the two atoms if for the vector-potential we do not use the matrix representations corresponding to the electromagnetic field in empty space, but the matrices corresponding to a regular solution for the system, consisting of atom B and the radiation field. It is obviously sufficient to know these matrix elements of the vector potential correctly to the order of approximation e^2. The vector potential satisfies Maxwell's equations also in quantum electrodynamics and the terms with e^2 in the vector potential can be derived, therefore, by means of classical formulae from the expression for the current involving first powers of e. This method is often referred to as Heisenberg's method.[5] The resulting energy perturbation is again proportional to e^4.

It may be remarked that the terms (b) and (c) can also be calculated by systematically writing down all the products of matrix elements occurring in the expressions for the second-, third-, and fourth-order perturbation in the energy. We have carried out this systematic calculation and found the result to be in agreement with the results derived in this paper.

In the following sections we shall first restate briefly the field theory and verify Heisenberg's method for the case of a single atom. This will also teach us in which way the singularities in the solution of the perturbation equations are to be avoided. In a subsequent section (Section 4) we calculate the terms mentioned under (a) and (b). The terms under (c) will be determined in Section 5, and the final result is discussed in Section 6.

Since the perturbation procedure involves a number of rather lengthy calculations, we do not want to go into the details but shall only mention the most important steps and give the results.

FIG. 1. Path of integration for the integral in (17).

2. The Radiation Field

We want to carry out the quantization of the radiation field by means of traveling waves, from which we demand periodicity in the x, y, and z directions with a period L. Using the

[5] W. Heisenberg, Ann. d. Physik (5) 9, 338 (1931).

notation of Section 12, we write for the vector potential

$$\mathbf{A}=\sum_{k,\lambda} cC_k\mathbf{e}(\mathbf{k},\lambda)\times\{A_{k,\lambda}e^{-i(\omega t-\mathbf{k}\mathbf{r})}+A_{k,\lambda}^{\dagger}e^{i(\omega t-\mathbf{k}\mathbf{r})}\},\quad(28)$$

where the values of the components of the wave vector $\mathbf{k}$ are restricted to $k_i=2\pi n_i/L$; with $n_i=0, \pm1, \pm2, \cdots$. For some details of the calculations of the following sections it is suitable to assume $\mathbf{e}(\mathbf{k},\lambda)=\mathbf{e}(-\mathbf{k},\lambda)$. The value of the normalization factor C_k,

$$C_k=(2\pi\hbar/\omega L^3)^{\frac{1}{2}},\quad(29)$$

can easily be verified by determining the energy of the field. In the following work we only need the matrix element of A or $A\dagger$ between the zero state of the field, $\psi(0, 0\cdots)$, and the state in which one light quantum is present, $\psi(0, 0\cdots1_{k\lambda}\cdots)$:

$$\int\psi^*(0,0\cdots)A_{k\lambda}\psi(0,0\cdot1_{k\lambda}\cdot)=1.\quad(30)$$

We shall always assume that L is very large so that a summation over all values of $\mathbf{k}$ can be replaced by $L^3/8\pi^3$ times an integration over all values of $\mathbf{k}$.

We want to remark that the vector potential $\mathbf{A}$ satisfies $\operatorname{div}\mathbf{A}=0$. For the interest of the next section we now write down the classical expression for the retarded vector potential satisfying the same condition and belonging to a periodic current $I=I_0\exp(ikct)$ inside an atom. Neglecting the dimensions of the atom as being small in comparison with R, it is easy to derive from Maxwell's equations that

$$A_z^{(\mathrm{ret})}=\frac{I_z}{c}\left[\frac{e^{-ikR}}{R}\left(-\frac{2}{ikR}+\frac{2}{k^2R^2}\right)-\frac{2}{k^2R^3}\right],$$

$$A_{x,y}^{(\mathrm{ret})}=\frac{I_{x,y}}{c}\left[\frac{e^{-ikR}}{R}\left(1+\frac{1}{ikR}-\frac{1}{k^2R^2}\right)+\frac{1}{k^2R^3}\right],\quad(31)$$

where for the sake of simplicity, we assumed $\mathbf{R}$ to point in the direction of the positive z axis. We shall use this formula also in Section 5.

3. Heisenberg's Method

The fundamental idea of Heisenberg's method is that even when the quantities of the electromagnetic field are considered as matrices, they satisfy Maxwell's equations. When we regard the elementary charge, e, as the perturbation parameter in the perturbation problem arising from the interaction, G, of the charged particles with the radiation field, it follows that, in order to determine the matrix elements of a field quantity to the order of approximation e^s, it is sufficient to know the matrix elements of the electric current to the order e^{s-1}.* In this section we shall give an illustration of Heisenberg's method by discussing the matrix element of the vector potential $(m;00\cdots|A|0;000\cdots)$ when one neutral atom is present in the radiation field. In the notation of the two states the first number indicates the level of the atom and the following numbers denote the number of quanta with different $(\mathbf{k}\lambda)$ in the radiation field. We shall first give a direct calculation of the matrix element in first approximation, and afterwards we shall verify that the result can be obtained with the aid of Maxwell's equations from the matrix element of the current in the zero approximation.

In the following calculation we assume that the atom is situated at $x=y=z=0$. The dimensions of the atom are assumed to be very small in comparison with the distance R at which we want to know the vector potential. We also neglect the variation of the electromagnetic field within the atom. The last approximation gives rise to divergencies of the type discussed in Section 12, but they can always be removed by introducing a factor $\exp(-\gamma|k|)$.

With the use of G as perturbation operator, we find for the first approximation of the zero state $\psi(0;00\cdots)$ of our system, consisting of the atom and the radiation field:

$$\psi'(0;0\cdots)=\psi(0;0\cdots)+\sum_{n,k,\lambda}\frac{eC_k}{m\hbar c}\cdot\frac{(\mathbf{e}(\mathbf{k},\lambda)\mathbf{p}_{n;0})^*}{k_n+k}\psi(n;0\cdots1_{k\lambda}\cdots),\quad(32)$$

* *Note added in proof.* This statement may be misleading. What is meant is, that since the current operator always contains one factor e explicitly, the matrix element multiplying this factor has only to be known to the order e^{s-1}.

and for the excited states we are interested in:

$$\psi'(n;00\cdots)=\psi(n;00\cdots)+\sum_{k,\lambda}\frac{eC_k}{m\hbar c}\frac{(\mathbf{e}(k,\lambda)\mathbf{p}_{0;n})}{k-k_n}\psi(0;0\cdots1_{k\lambda})+\cdots, \quad (33)$$

$$\psi'(0;0\cdots1_{k\lambda}\cdots)=\psi(0;0\cdots1_{k\lambda}\cdots)+\sum_n\frac{eC_k}{m\hbar c}\frac{(\mathbf{e}(\mathbf{k},\lambda)\mathbf{p}_{n;0})}{k_n-k}\psi(n;00\cdots)+\cdots. \quad (34)$$

In the last two formulae we have omitted a number of terms which we shall not need in the following calculations. The matrix element of the vector potential $\mathbf{A}$ in first approximation is now easily found with the aid of (32) and (33).

$$(m;0\cdots|\mathbf{A}|0;0\cdots)=\sum_{k,\lambda}\frac{eC_L^2}{m\hbar}\mathbf{e}(\mathbf{k},\lambda)(\mathbf{e}(\mathbf{k},\lambda)\mathbf{p}_{m;0})\times\left\{\frac{e^{ikr}}{k_m+k}-\frac{e^{-ikr}}{k_m-k}\right\}. \quad (35)$$

After a small calculation we find for the z component of $\mathbf{A}$ at the point $\mathbf{r}$, given by $x=y=0$, $z=R$:

$$(m;0\cdots|A_z|0;0\cdots)=-\frac{ep_{m;0}^{(z)}}{2\pi mc}\times\int_{-\infty}^{+\infty}\frac{dk}{k_m-k}\left\{\frac{e^{ikR}}{iR}\left(-\frac{2i}{kR}+\frac{2}{k^2R^2}\right)+\text{complex conjugate}\right\}\exp(-\gamma|k|), \quad (36)$$

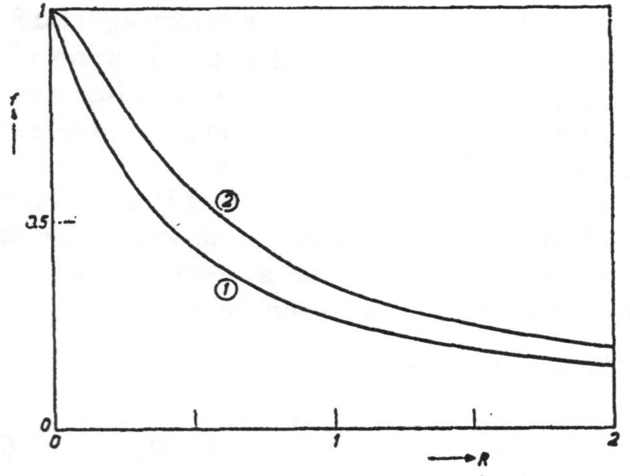

FIG. 2. Correction factor due to retardation for the contribution of one excited state to the usual London energy. (1) For the interaction between a neutral atom and a metallic wall. R is measured in units $\frac{1}{2}\lambda_n$. (2) For the interaction between two neutral atoms. R is measured in units λ_l.

FIG. 3. Path of integration for the integral in (36).

in which formula we have introduced again the factor $\exp(-\gamma|k|)$.

The integrand is regular at $k=0$, but has a simple pole at $k=k_m$. In order to see in which way this singularity must be avoided, we remark that finally we want to obtain a retarded expression for the vector potential, i.e., an expression in which the terms with $\exp(ikR)$ do not occur. (The time factor of our matrix element is $\exp(ik_mct)$, thus retarded expressions will contain a factor $\exp(-ik_mR)$.) For this purpose we write $k_m-i\zeta-k$ instead of k_m-k in the denominator, by means of which the singularity is now fixed at a small distance ζ below the real axis of the complex k plane. It will turn out that this procedure gives the desired retarded expression in the final result. For the integration from $-\infty$ to $+\infty$ we take the path illustrated in Fig. 3. We carry out the integration of the terms with $\exp(ikR)$ and $\exp(-ikR)$ separately. The integral of the terms first mentioned can be replaced by an integral over a closed contour with the aid of a large semicircle above the real axis, and since there are no singularities within this contour, the value of the integral is zero. The integral with the other terms can be replaced by an integral over a closed contour with the aid of a large semicircle below the real axis. Within this contour there are two singularities for the terms with $\exp(-ikR)$, one at $k=k_m-i\zeta$ and one at $k=0$. The residues at these points determine the value of the integral, which is now easily found to be

$$(m;0\cdots|A_z|0;0\cdots)=\frac{ep_{m;0}^z}{mc}\left[\frac{e^{-ik_mR}}{R}\left(-\frac{2}{ik_mR}+\frac{2}{k_m^2R^2}\right)-\frac{2}{k_m^2R^3}\right]. \quad (37)$$

With the foregoing calculation we have obtained the matrix elements of the vector potential in first approximation. The operation of the current is given by

$$I=\sum_i\frac{e}{m}\left(\mathbf{p}_i-\frac{e}{c}\mathbf{A}\right),$$

and, therefore, $ep_{m;0}{}^{i}/m$ are the corresponding matrix elements of the current in the zero approximation. It is seen, by comparing (37) with (31), that the matrix elements of the vector potential in first approximation could have been obtained immediately from the matrix elements of the current in the zero approximation.

In Section II2 we have explicitly restricted ourselves to retarded solutions of Maxwell's equations. The restriction to retarded solutions is also implied in the calculation of this section, namely, by the way in which the singularity at $k=k_m$ is avoided. It was found to be adequate to write $k_m-i\zeta$ instead of k_m in the denominator of (36), and as the complex conjugate of the function $\psi(m;00\cdots)$ occurred in the matrix element of the vector potential, we have to write $k_n+i\zeta$ instead of k_n in the denominator of (33) in order to stay in the domain of retarded expressions.

In the course of the following sections we shall also use Eq. (34), and we shall have to determine in which way singularities arising form the denominator in (34) must be avoided. Remarking that the perturbed eigenfunctions can be obtained from the unperturbed functions by means of a unitary transformation, we conclude that in (34) k_n must be replaced by $k_n-i\zeta$.

4. Perturbation Terms Involving Electrostatic Interaction

The electrostatic interaction between two neutral atoms A and B is given by

$$Q=\frac{\mathbf{q}_A\mathbf{q}_B}{R^3}-\frac{3(\mathbf{q}_A\mathbf{R})(\mathbf{q}_B\mathbf{R})}{R^5}, \tag{38}$$

when we neglect higher powers of the ratio between the atomic dimensions and $\mathbf{R}$. We shall assume at once that the vector $\mathbf{R}$, pointing from atom B to atom A, is in the direction of the positive z axis and that atom B is situated at $x=y=z=0$. Further, we shall assume, as in Section I2, that each of the two atoms has a state with $J=0$ as zero state. The second-order perturbation energy is now easily found to be

$$\Delta_0E=-\frac{1}{R^6}\sum_{l,m}\frac{(q_l{}^xq_m{}^x)^2+(q_l{}^yq_m{}^y)^2+4(q_l{}^zq_m{}^z)^2}{\hbar c(k_l+k_m)}. \tag{39}$$

In this notation the indices l and m denote the states with $J=1$ of the atoms A and B, respectively; $q_l{}^i$ is the matrix element of the total dipole moment between the zero state of atom A and the state l of this atom. The matrix elements are assumed to be real, which can be done without loss of generality. It is obvious that the symbols A and B can be omitted in this notation without giving rise to confusion. As in Section I2, we do not yet use relation (26). The expression (39), being the usual London energy between two neutral atoms, gives the terms mentioned under (a) in Section II1.

The calculation of the terms mentioned under (b) proceeds along the lines indicated in the outline of method. In the calculation we restrict ourselves to the terms which give a contribution proportional to the fourth power, or to a smaller power, of the electronic charge to the final result. The result will be divergent, but we determine the difference between the energy perturbation in the case of a distance R between the two atoms and the case of a very large distance; this difference will be finite.

The calculations are rather elaborate but do not give rise to special difficulties. Taking into consideration that for the zero states of the atoms $I=0$, we find finally:

$$\Delta_1E=+\frac{2}{R^6}\frac{(q_l{}^xq_m{}^x)^2+(q_l{}^yq_m{}^y)^2+4(q_l{}^zq_m{}^z)^2}{\hbar c(k_l+k_m)}-\frac{4}{\pi\hbar cR^4}\sum_{l,m}\int_0^\infty\frac{k_lk_mu^2du}{(k_l{}^2+u^2)(k_m{}^2+u^2)}$$
$$\times\left\{((q_l{}^xq_m{}^x)^2+(q_l{}^yq_m{}^y)^2)e^{-uR}\left(1+\frac{1}{uR}+\frac{1}{u^2R^2}\right)+4(q_l{}^zq_m{}^z)^2e^{-uR}\left(\frac{1}{uR}+\frac{1}{u^2R^2}\right)\right\}. \tag{40}$$

In the course of the calculations the two terms in (40) were obtained by means of the same procedure of complex integration as was applied to Eq. (17) in Section I2.

5. The Terms Not Involving the Electrostatic Interaction

So far we have carried out our perturbation procedures in an entirely symmetrical way with respect to the atoms A and B. In this section it is our aim to determine the terms mentioned under (c) in the outline of method. We have to solve a perturbation problem in which $G_A+G_B=G$ is the perturbation operator:

$$G=\sum_{i_A}\left\{-\frac{e}{mc}(\mathbf{p}_{i_A}\mathbf{A}_A)+\frac{e^2}{2mc^2}A_A{}^2\right\}+\sum_{i_B}\left\{-\frac{e}{mc}(\mathbf{p}_{i_B}\mathbf{A}_B)+\frac{e^2}{2mc^2}A_B{}^2\right\}. \quad (41)$$

Now we have to take recourse to an unsymmetrical attack on the situation. The reason is the following. We have already remarked several times that, in consequence of the form of the operator of the interaction between the charged particles and the radiation field, there is no first-order perturbation between an atom and the electromagnetic field that is proportional to e. The terms in G that are proportional to e only give rise to a second-order perturbation, while the terms with e^2 are responsible for the first-order perturbation. As we are interested only in terms with e^4 in the final expression for the energy perturbation, we now have the opportunity to solve our perturbation problem in two steps. First, we calculate the vector potential to the order of approximation e^2 in the system consisting of atom B+radiation field, and then proceed to determine the first- and second-order perturbation energy of the atom A in this perturbed electromagnetic field. This procedure is necessarily unsymmetric in the atoms A and B, but in the final stage of the calculation the asymmetry vanishes, as must be the case for any consistent treatment of our problem.

We label the states of the system (B+radiation field) with the index N. The energy perturbation of the atom A in the electromagnetic field of this system is now given by

$$\Delta_A E=\frac{1}{\hbar c}\sum_{l,N}\frac{k_N k_l}{k_N+k_l}\sum_{x,y,z}|A_{0;N}{}^x(A)|^2|q_{0;l}{}^x|^2. \quad (42)$$

Here we have at once combined the first-order and the second-order perturbation in the same way as was done in Section 12. The index l denotes the states with $J=1$ of the atom A.

We want to know $|A_{0;N}{}^x|^2$ to the order of approximation e^2. There are two possibilities: either the matrix element $A_{N;0}{}^i$ vanishes in zeroth approximation, i.e., terms not containing the factor e do not occur in the matrix element, or the matrix element does not vanish in zeroth approximation. In the first case, we only need to know the matrix element to the order of approximation e. Matrix elements of this type that do not vanish in this approximation, are found only if the state N is one of the states with wave functions $\psi'(m;0,0\cdots)$. This statement immediately follows from Heisenberg's method, discussed in Section 3. In that section the matrix element $(m;00|A_i|0;00\cdots)$ has already been calculated, so that we can at once write down the contribution to (42) from this special series of states N:

$$\Delta_a E=\frac{1}{\hbar c R^2}\sum_{l,m}\frac{k_m{}^3k_l}{k_m+k_l}(q_l{}^xq_m{}^x)^2\times\left\{\left(1-\frac{1}{k_m{}^2R^2}+\frac{2}{k_m{}^4R^4}\right)+\left[\frac{e^{ik_mR}}{k_m{}^2R^2}\left(1+\frac{i}{k_mR}-\frac{1}{k_m{}^2R^2}\right)+\text{complex conjugate}\right]\right\}. \quad (43)$$

In order to simplify the formula we have not written down terms proportional to $(q_l{}^yq_m{}^y)^2$ and $(q_l{}^zq_m{}^z)^2$. We shall omit these terms in all formulas of this section. In the second case, where the matrix element of the vector potential contains terms independent of e, we must know the vector potential to the order of approximation e^2. This case is realized only if the state N is of the type $\psi'(0;00\cdots1_{k\lambda}\cdots)$. By Heisenberg's method it is easily seen that the corresponding matrix element of the vector potential contains no terms proportional to e, so that we can write

$$(0;0\cdots1_{k\lambda}\cdots|A_i|0;0\cdots0\cdots)=A_{k\lambda;0}{}^{(0)i}+A_{k\lambda;0}{}^{(2)i}. \quad (44)$$

When we insert these matrix elements in (42), we shall have to deal only with the cross products:

$$A_{0;\,k\lambda}{}^{(0)i}A_{k\lambda;\,0}{}^{(2)i}+A_{0;\,k\lambda}{}^{(2)i}A_{k\lambda;\,0}{}^{(0)i},$$

the other products being either independent of R or proportional to a too high power of e.

The vector potential in second approximation is calculated by means of Heisenberg's method with the aid of the current in first approximation:

$$I_{k\lambda;\,0}{}^{i}=\int\psi'^{*}(0;0\cdots1_{k\lambda}\cdots)\times\left\{\sum_{jB}\frac{e}{m}\left(p_{jB}{}^{i}-\frac{e}{c}A_{B}{}^{i}\right)\right\}\psi'(0;0\cdots). \quad (45)$$

Here we use the wave functions given by (32) and (34), introducing at once the term $i\zeta$ in the denominator of (34). Collecting the terms with e^2 in (45) we find

$$I_{k\lambda;\,0}{}^{i}=\sum_{m}\frac{e^2C_k}{m^2\hbar c}|p_{m;\,0}{}^{i}|^2e_i(\mathbf{k}\lambda)\times\left(\frac{1}{k_m-k+i\zeta}+\frac{1}{k_m+k}\right)-\sum_{jB}\frac{e^2C_k}{m}e_i(\mathbf{k}\lambda), \quad (46)$$

and with the aid of (31) we find the matrix element $A_{k\lambda;0}{}^{(2)i}(A)$. The matrix element $A_{k\lambda;0}{}^{0i}$ is simply

$$A_{k\lambda;\,0}{}^{0i}(A)=cC_ke_i(\mathbf{k}\lambda)e^{-ik_zR}. \quad (47)$$

Applying the sum rule (10) to the last term in (46), we can now calculate the contribution to (42) arising from the states $\psi'(0;0\cdots1_{k\lambda}\cdots)$.

The calculation leads to a rather complicated integral over k, and the integrand requires a careful examination at the points $k=0$ and $k=k_m$ before the integral can be evaluated by means of complex integration. The way in which the singularity at $k=k_m$ must be avoided is prescribed unambiguously, however, by a term $i\zeta$ occurring in a denominator which originates from the corresponding denominator with the term $i\zeta$ in Eq. (46). Therefore, a straightforward evaluation of the integral is possible. We find finally:

$$\Delta_\beta E=\sum(q_l{}^xq_m{}^z)^2(B_1+B_2+B_3), \quad (48)$$

where

$$B_1=-\frac{1}{\hbar cR^2}\left[\frac{1}{k_mR^4}+\frac{k_m{}^2k_l}{k_m+k_l}\left(1-\frac{1}{k_m{}^2R^2}+\frac{1}{k_m{}^4R^4}\right)\right],$$

$$B_2=-\frac{1}{\hbar cR^4}e^{ik_mR}\left(1+\frac{i}{k_mR}-\frac{1}{k_m{}^2R^2}\right)\frac{k_mk_l}{k_m+k_l}+\text{complex conjugate},$$

$$B_3=\frac{4}{\pi\hbar c}\int_0^\infty\frac{k_lk_mu^2du}{(k_l{}^2+u^2)(k_m{}^2+u^2)}\times\frac{e^{-uR}}{R^4}\left(1+\frac{1}{uR}+\frac{1}{u^2R^2}\right)-\frac{2}{\pi\hbar c}\int_0^\infty\frac{k_lk_mu^4du}{(k_l{}^2+u^2)(k_m{}^2+u^2)}\times\frac{e^{-2uR}}{R^2}\left(1+\frac{1}{uR}+\frac{1}{u^2R^2}\right)^2.$$

6. Result and Discussion

We have now calculated all the terms that contribute to the energy of interaction Δ_LE of two neutral atoms in S states:

$$\Delta_LE=\Delta_qE+\Delta_3E+\Delta_\alpha E+\Delta_\beta E. \quad (49)$$

Fortunately a number of terms cancel, and the final result is comparatively simple. It may be remarked that this result has regained symmetry with respect to the atoms A and B. Adding at once the terms proportional to $(q_l{}^yq_m{}^y)^2$ and $(q_l{}^xq_m{}^x)^2$ we find:

$$\Delta_LE=-\frac{2}{\pi\hbar c}\sum_{l,m}\int_0^\infty\frac{k_lk_mu^4du}{(k_l{}^2+u^2)(k_m{}^2+u^2)}\frac{e^{-2uR}}{R^2}\times\left\{((q_l{}^xq_m{}^x)^2+(q_l{}^yq_m{}^y)^2)\left(1+\frac{1}{uR}+\frac{1}{u^2R^2}\right)^2+4(q_l{}^zq_m{}^z)^2\left(\frac{1}{uR}+\frac{1}{u^2R^2}\right)^2\right\}. \quad (50)$$

In the limit of a very small distance R, $R\ll\lambda_l=2\pi/k_l$, $R\ll\lambda_m$ it is easily found that the formula reduces to:

$$\Delta_LE(R\to0)=-\frac{1}{R^6}\sum_{l,m}\frac{(q_l{}^xq_m{}^x)^2+(q_l{}^yq_m{}^y)^2+4(q_l{}^zq_m{}^z)^2}{\hbar c(k_l+k_m)}, \quad (51)$$

being equal to the value of the London energy as derived by the elementary theory, which takes into account the electrostatic interaction only. For very large R (R larger than all λ_l and λ_m) it is found, after a short calculation, that (50) reduces to:

$$\Delta_L E(R\rightarrow\infty) = -\frac{1}{2\pi\hbar c R^7} \times\sum_{l,m}\frac{13[(q_l^x q_m^x)^2+(q_l^y q_m^y)^2]+20(q_l^z q_m^z)^2}{k_l k_m}, \quad (52)$$

which can be written in terms of the static polarizability of the atoms:

$$\Delta_L E(R\rightarrow\infty) = -\frac{\hbar c}{8\pi R^7}\{13(\alpha_x(A)\alpha_x(B) +\alpha_y(A)\alpha_y(B))+20\alpha_z(A)\alpha_z(B)\}. \quad (53)$$

So far we have not yet used the relations

$$\begin{aligned}\sum|q_l^x|^2=\sum|q_l^y|^2=\sum|q_l^z|^2=q_l^2,\\ \sum|q_m^x|^2=\sum|q_m^y|^2=\sum|q_m^z|^2=q_m^2,\end{aligned} \quad (54)$$

where the summation extends over the three states with $J=1$ belonging to one degenerate level, which will be indicated hereafter by one symbol l or m, respectively. With the aid of (54), Eq. (50) can now be written as:

$$\Delta_L E = -\frac{4}{\pi\hbar c}\sum_{l,m}\int_0^\infty \frac{k_l k_m u^4 du}{(k_l^2+u^2)(k_m^2+u^2)}(q_l q_m)^2 \times\frac{e^{-2uR}}{R^2}\left(1+\frac{2}{uR}+\frac{5}{u^2R^2}+\frac{6}{u^3R^3}+\frac{3}{u^4R^4}\right), \quad (55)$$

where each term of the sum over l, m represents the contribution from all three states with $J=1$ belonging to one degenerate level l together with all three states belonging to the level m.

Equation (53) may now be written

$$\Delta_L E(R\rightarrow\infty) = -\frac{23\hbar c}{4\pi R^7}\alpha(A)\alpha(B), \quad (56)$$

which is equal to

$$\Delta_L E(R\rightarrow\infty) = -\frac{23}{4\pi}\frac{\hbar c}{e^2}\cdot\frac{e^2}{R^7}\alpha(A)\alpha(B) \approx 251\frac{e^2}{R}\frac{\alpha(A)\alpha(B)}{R^6}. \quad (57)$$

In Fig. 2 we have given the result of a numerical calculation of the factor $\Delta_L E/\Delta_e E$ for the case in which the two atoms are identical and in which for both atoms only one excited state l (with $E_l=\hbar c/\lambda_l$) gives a contribution to the London energy. The factor decreases monotonically with increasing R. It starts with the value 1 (for $R\rightarrow 0$) while for large R it is approximately equal to $23\lambda_l/6\pi^3 R$.

The very simple form of Eq. (56) and the analogous formula (25) suggest that it might be possible to derive these expressions, perhaps apart from the numerical factors, by more elementary considerations. This would be desirable since it would also give a more physical background to our result, a result which in our opinion is rather remarkable. So far we have not been able to find such a simple argument.

On Quantum-Electrodynamics and the Magnetic Moment of the Electron

JULIAN SCHWINGER
Harvard University, Cambridge, Massachusetts
December 30, 1947

ATTEMPTS to evaluate radiative corrections to electron phenomena have heretofore been beset by divergence difficulties, attributable to self-energy and vacuum polarization effects. Electrodynamics unquestionably requires revision at ultra-relativistic energies, but is presumably accurate at moderate relativistic energies. It would be desirable, therefore, to isolate those aspects of the current theory that essentially involve high energies, and are subject to modification by a more satisfactory theory, from aspects that involve only moderate energies and are thus relatively trustworthy. This goal has been achieved by transforming the Hamiltonian of current hole theory electrodynamics to exhibit explicitly the logarithmically divergent self-energy of a free electron, which arises from the virtual emission and absorption of light quanta. The electromagnetic self-energy of a free electron can be ascribed to an electromagnetic mass, which must be added to the mechanical mass of the electron. Indeed, the only meaningful statements of the theory involve this combination of masses, which is the experimental mass of a free electron. It might appear, from this point of view, that the divergence of the electromagnetic mass is unobjectionable, since the individual contributions to the experimental mass are unobservable. However, the transformation of the Hamiltonian is based on the assumption of a weak interaction between matter and radiation, which requires that the electromagnetic mass be a small correction $(\sim(e^2/\hbar c)m_0)$ to the mechanical mass m_0.

The new Hamiltonian is superior to the original one in essentially three ways: it involves the experimental electron mass, rather than the unobservable mechanical mass; an electron now interacts with the radiation field only in the presence of an external field, that is, only an accelerated electron can emit or absorb a light quantum;* the interaction energy of an electron with an external field is now subject to a *finite* radiative correction. In connection with the last point, it is important to note that the inclusion of the electromagnetic mass with the mechanical mass does not avoid all divergences; the polarization of the vacuum produces a logarithmically divergent term proportional to the interaction energy of the electron in an external field. However, it has long been recognized that such a term is equivalent to altering the value of the electron charge by a constant factor, only the final value being properly identified with the experimental charge. Thus the interaction between matter and radiation produces a renormalization of the electron charge and mass, all divergences being contained in the renormalization factors.

The simplest example of a radiative correction is that for the energy of an electron in an external magnetic field. The detailed application of the theory shows that the radiative correction to the magnetic interaction energy corresponds to an additional magnetic moment associated with the electron spin, of magnitude $\delta\mu/\mu=(\frac{1}{2}\pi)e^2/\hbar c$ $=0.001162$. It is indeed gratifying that recently acquired experimental data confirm this prediction. Measurements on the hyperfine splitting of the ground states of atomic hydrogen and deuterium[1] have yielded values that are definitely larger than those to be expected from the directly measured nuclear moments and an electron moment of one Bohr magneton. These discrepancies can be accounted for by a small additional electron spin magnetic moment.[2] Recalling that the nuclear moments have been calibrated in terms of the electron moment, we find the additional moment necessary to account for the measured hydrogen and deuterium hyperfine structures to be $\delta\mu/\mu=0.00126$ ± 0.00019 and $\delta\mu/\mu=0.00131\pm 0.00025$, respectively. These values are not in disagreement with the theoretical prediction. More precise conformation is provided by measurement of the g values for the $^2S_{\frac{1}{2}}$, $^2P_{\frac{1}{2}}$, and $^2P_{3/2}$ states of sodium and gallium.[3] To account for these results, it is necessary to ascribe the following additional spin magnetic moment to the electron, $\delta\mu/\mu=0.00118\pm 0.00003$.

The radiative correction to the energy of an electron in a Coulomb field will produce a shift in the energy levels of hydrogen-like atoms, and modify the scattering of electrons in a Coulomb field. Such energy level displacements have recently been observed in the fine structures of hydrogen,[4] deuterium, and ionized helium.[5] The values yielded by our theory differ only slightly from those conjectured by Bethe[6] on the basis of a non-relativistic calculation, and are, thus, in good accord with experiment. Finally, the finite radiative correction to the elastic scattering of electrons by a Coulomb field provides a satisfactory termination to a subject that has been beset with much confusion.

A paper dealing with the details of this theory and its applications is in course of preparation.

* A classical non-relativistic theory of this type was discussed by H. A. Kramers at the Shelter Island Conference, held in June 1947 under the auspices of the National Academy of Sciences.

[1] J. E. Nafe, E. B. Nelson, and I. I. Rabi, Phys. Rev. 71, 914 (1947); D. E. Nagel, R. S. Julian, and J. R. Zacharias, Phys. Rev. 72, 971 (1947).

[2] G. Breit, Phys. Rev. 71, 984 (1947). However, Breit has not correctly drawn the consequences of his empirical hypothesis. The effects of a nuclear magnetic field and a constant magnetic field do not involve different combinations of μ and $\delta\mu$.

[3] P. Kusch and H. M. Foley, Phys. Rev. 72, 1256 (1947), and further unpublished work.

[4] W. E. Lamb, Jr. and R. C. Retherford, Phys. Rev. 72, 241 (1947).

[5] J. E. Mack and N. Austern, Phys. Rev. 72, 972 (1947).

[6] H. A. Bethe, Phys. Rev. 72, 339 (1947).

The Magnetic Moment of the Electron†

P. KUSCH AND H. M. FOLEY
Department of Physics, Columbia University, New York, New York
(Received April 19, 1948)

A comparison of the g_J values of Ga in the $^2P_{3/2}$ and $^2P_{1/2}$ states, In in the $^2P_{1/2}$ state, and Na in the $^2S_{1/2}$ state has been made by a measurement of the frequencies of lines in the *hfs* spectra in a constant magnetic field. The ratios of the g_J values depart from the values obtained on the basis of the assumption that the electron spin gyromagnetic ratio is 2 and that the orbital electron gyromagnetic ratio is 1. Except for small residual effects, the results can be described by the statement that $g_L=1$ and $g_S=2(1.00119\pm0.00005)$. The possibility that the observed effects may be explained by perturbations is precluded by the consistency of the result as obtained by various comparisons and also on the basis of theoretical considerations.

1. INTRODUCTION

ONE of the important conclusions derived from the relativistic Dirac theory of the electron is that the electron possesses an angular momentum of $\frac{1}{2}$ measured in units of $h/2\pi$ and that with this angular momentum is associated a magnetic moment of one Bohr magneton. This conclusion substantiates earlier conclusions based on an analysis of the experimental data on the anomalous Zeeman effect. Indeed, all relevant experimental data have been in substantial agreement with this conclusion.

A direct measurement of the electron moment can most easily be made by a measurement of the g value of an atomic energy state. Direct determinations of the g values of atomic states from measurements of the frequencies of Zeeman lines in known magnetic fields, as, for example, in the work of Kinsler and Houston,[1] have yielded no significant differences between the measured atomic gyromagnetic ratios and the values consequent from the Dirac theory.* Millman and Kusch[2] have measured the magnetic moments of various nuclei, in particular that of the proton, in terms of the magnetic moment of the electron, assumed to be one Bohr magneton. The magnetic moments so found agree with those dependent on a measurement of a magnetic field in terms of classical standards to within about 0.14 percent ±0.5 percent. This again indicates that the g value of the electron is 2, to within the stated precision. It seems certain that any discrepancy with the theoretical value will be small.

The growth of various techniques of microwave and r-f spectroscopy makes available a series of new tools for the investigation of the detailed structure of atomic spectra. These techniques make it possible to resolve extremely minute details of structure and to determine the relative positions of energy levels to a very high degree of precision. The recent experiments of Lamb and Retherford[3] on the fine structure of hydrogen indicate that the Dirac theory does not adequately describe the hydrogen atom and that all the detailed conclusions of the Dirac theory are, therefore, presumably suspect to some degree.

In the recent measurements of Nafe, Nelson, and Rabi[4] of the hyperfine spectrum of the ground states of atomic hydrogen and deuterium, deviations of the zero field level splittings from the values predicted from theory were found. The theoretical values depend on a knowledge of the nuclear magnetic moment of the nucleus (known only in terms of an assumed value of the electron moment) as well as on the assumption that the magnetic moment of the electron is one Bohr magneton. Breit[5] has suggested that the discrepancy may be removed by the assumption that the electron possesses an "intrinsic" mag-

† Publication assisted by the Ernest Kempton Adams Fund for physical research at Columbia University.

[1] L. E. Kinsler and W. V. Houston, Phys. Rev. 45, 104 (1934); 46, 533 (1934).

* See, however, a consideration of these measurements in Section 6.

[2] S. Millman and P. Kusch, Phys. Rev. 60, 91 (1941).

[3] W. E. Lamb, Jr. and R. C. Retherford, Phys. Rev. 72, 241 (1947).

[4] J. E. Nafe, E. B. Nelson, and I. I. Rabi, Phys. Rev. 71, 914 (1947). J. E. Nafe and E. B. Nelson, Phys. Rev. 73, 718 (1948).

[5] G. Breit, Phys. Rev. 72, 984 (1947).

netic moment, i.e., a magnetic moment over and above that deduced from the Dirac theory.

The present experiments were undertaken to utilize the power of the atomic beam magnetic resonance method to provide an accurate determination of the electron moment. Preliminary results of the present experiments were given in two Letters to the Editor.[6] Subsequent to the publication of preliminary results of our experiments, Schwinger[7] has published results of theoretical investigation which indicate that the magnetic moment of the electron is, indeed, to be modified as the result of the interaction of the electron with the radiation field.

A deviation of the magnetic moment of the electron from the accepted value of one Bohr magneton could be detected by a precise measurement of the magnetic moment of an atom in a state in which the coupling of the electron spin with the orbital angular momentum is sufficiently well known. An absolute measurement of the magnetic moment requires a measurement of the Zeeman splitting of the zero field energy level in a known magnetic field. At the present time it is difficult to produce magnetic fields which are accurately known in terms of absolute standards and of sufficient magnitude to be useful in atomic beam determinations of the Zeeman splittings of energy levels. However, the frequencies of lines in the Zeeman spectrum of atoms (that is, the differences between atomic energy levels) may be determined by use of readily available techniques to within one part in ten or twenty thousand, and where precision is limited by statistical errors, as, for example, those arising from the least count of an instrument, a considerable improvement in precision may be obtained by a suitable repetition of observations. From measurements of the frequencies of Zeeman lines in two atomic states arising in either the same or different atoms but in the same constant magnetic fields, it is possible to deduce the ratio of the values of the atomic gyromagnetic ratios of these two states. If the spin and orbit vectors are coupled in the same way in the two states, the measured ratio yields no information about the fundamental g values of the electron. If the spin orbit coupling in the two states is different, however, the electron spin g value may be determined in terms of the orbital g value, provided only that suitable information is available, either on experimental or theoretical grounds, as to the validity of the assumed coupling. The principal limitations on such an experimental determination of the electron spin g value are the accuracy in the determination of the line frequencies (limitations imposed by a frequency meter and by the line widths) and the stability and homogeneity of the magnetic field.

In actual practice the single atomic level described above is split into two or more h.f.s. levels because of the presence of the nuclear angular momentum. In such a case the interpretation of data on line frequencies becomes considerably more complicated.

2. THEORY OF THE EXPERIMENT

In this and the following section is developed the elementary theory on the basis of which the experimental data have actually been reduced. Corrections to the results of these sections are considered in Section 6.

The Hamiltonian of the interaction of the electrons in an atom with a constant applied magnetic field H may be written as:

$$\mathcal{H} = g_L\mu_0 L_z H_z + g_S\mu_0 S_z H_z, \qquad (1)$$

in which g_L and g_S are the orbital and electron spin gyromagnetic ratios, μ_0 is the Bohr magneton, and L_z and S_z are the operators of the z components of orbital and spin angular momentum. The diamagnetic energy is negligible at the fields employed in this experiment ($H<500$ gauss). For the calculation of matrix elements diagonal in the total atomic angular momentum J, i.e., neglecting atomic Paschen-Back effects, this Hamiltonian may be written:

$$\begin{aligned}\mathcal{H} &= g_L\mu_0 H_z(J|L_z|J)J_z + g_S\mu_0 H_z(J|S_z|J)J_z \\ &= g_L\mu_0 H_z \alpha_L J_z + g_S\mu_0 H_z \alpha_S J_z \\ &= g_J\mu_0 H_z J_z. \end{aligned} \qquad (2)$$

The constants α_L and α_S are determined by the electronic wave functions. For Russell-Saunders coupling:

$$\alpha_S = [J(J+1)+S(S+1)-L(L+1)]/2J(J+1), \qquad (3)$$

[6] P. Kusch and H. M. Foley, Phys. Rev. 72, 1256 (1947); H. M. Foley and P. Kusch, Phys. Rev. 73, 412 (1948).
[7] J. Schwinger, Phys. Rev. 73, 416 (1948).

TABLE I. The hyperfine structure separations and the nuclear g values of the atoms whose spectra have been observed.

	$(\Delta W/h)\times 10^{-6}$ sec.$^{-1}$	I	g_I
Na^{23}	1771.75	3/2	−0.0008039
Ga^{69}	2677.56	3/2	−0.0007239
Ga^{71}	3402.09	3/2	−0.0009218
In^{115}	11413	9/2	−0.000664

and

$$\alpha_L=[J(J+1)+L(L+1)-S(S+1)]/2J(J+1).$$

The ratio of the g_J values of two atomic states in the same or different atoms is:

$$g_{J1}/g_{J2}=[(g_L\alpha_{L1}+g_S\alpha_{S1})/(g_L\alpha_{L2}+g_S\alpha_{S2})], \quad (4)$$

in which it is assumed that the values of g_L and g_S are independent of the atomic state. If the fundamental gyromagnetic ratios differ from the conventional values by small amounts, then

$$g_S=2(1+\delta_S),$$
$$g_L=1+\delta_L,$$

and

$$g_{J1}/g_{J2}=[(2\alpha_{S1}+\alpha_{L1})/(2\alpha_{S2}+\alpha_{L2})]+2[(\alpha_{S1}\alpha_{L1}-\alpha_{L1}\alpha_{S2})/(2\alpha_{S2}+\alpha_{L2})^2]\times\{\delta_S-\delta_L\}. \quad (5)$$

Thus if the constants α_{S1}, α_{L1}, α_{S2}, α_{L2} are known from the state of coupling of the atomic levels, the quantity $\{\delta_S-\delta_L\}$ can be determined from the ratio of the atomic g_J values. Clearly, no experiment of this type can distinguish between an effect produced by a small change from the previously accepted values of the spin or the orbital gyromagnetic ratios.

The measurement of the g values of two different atomic energy states in an experiment in which the magnetic field is known in terms of classical standards would, in principle, yield the absolute values of g_S and of g_L separately. However, such a determination would depend on a precisely known value of e/m. Of the experimental methods which have been used to evaluate e/m, the determination from the Zeeman splitting of singlet states assumes that $g_L=1$, and the magnetic deflection results make use of the corresponding assumption of the Lorentz force law. Other determinations of e/m appear either to be of intrinsically low accuracy or to have a rather doubtful theoretical background. In the remainder of the present paper we shall adopt the convention $g_L=1$, and we shall express the experimentally determined quantity $(\delta_S-\delta_L)$ as δ_S.

If hyperfine interactions were absent or entirely negligible the ratio of the frequencies of the Zeeman lines of two atomic states would give directly the ratio of the atomic g_J values. By Zeeman lines we mean here lines resulting from transitions between the magnetic levels of a single atomic energy state. In the atomic states which were studied in the present experiments and at the values of the magnetic field at which lines were observed, the splitting of the energy levels into a hyperfine structure must be taken into account. The existence of this hyperfine structure complicates the analysis of the data from which the ratio of the g_J values is obtained; at the same time the possibility of observing a number of lines of different frequency resulting from hyperfine transitions within each atomic level gives a means of checking the self-consistency of data and improving the accuracy of the results.

3. DESCRIPTION OF THE ENERGY LEVELS

For the energy levels with $J=\frac{1}{2}$ the relevant part of the Hamiltonian is taken to be:

$$\mathcal{H}=a\mathbf{I}\cdot\mathbf{J}+g_I\mu_0I_zH_z+g_J\mu_0J_zH_z, \quad (6)$$

in which $\mathbf{I}$ is the nuclear spin operator and $\mathbf{J}$ is the operator of the total electronic angular momentum. The solutions of the secular equations for the energy eigenvalues are:[8]

$$W_{I\pm\frac{1}{2},m_F}=-[\Delta W/2(2I+1)]+g_I\mu_0H_zm_F\pm(\Delta W/2)[1+(4/(2I+1))m_Fx+x^2]^{\frac{1}{2}}. \quad (7)$$

The "weak field" quantum numbers $F=I\pm\frac{1}{2}$, m_F are used to designate the levels. The parameter, x, is defined by

$$x=(g_J-g_I)(\mu_0H/\Delta W), \quad (8)$$

and $\Delta W=2a$ is the zero field hyperfine separation.

The constants in the energy expression which are given in Table I have been determined with

[8] S. Millman, I. I. Rabi, and J. R. Zacharias, Phys. Rev. 53, 384 (1938).

precision, by Kusch, Millman, and Rabi[9,10] for sodium, by Becker and Kusch[11] for gallium, and by Hardy and Millman[12] for indium. The results of this paper demand some modification of the values of g_I as given in Table I. However, g_I enters into the expression for the energies of the states only as a correction term, and hence reasonable changes in g_I do not affect our results.[13]

For the $^2P_{3/2}$ level of gallium it is necessary to include in the energy expression a term corresponding to a nuclear electric quadrupole interaction. In this case the Hamiltonian becomes

$$\mathcal{H}=a\mathbf{I}\cdot\mathbf{J}+2b\mathbf{I}\cdot\mathbf{J}(2\mathbf{I}\cdot\mathbf{J}+1)+\mu_0H_z(g_JJ_z+g_II_z). \quad (9)$$

Terms due to moments of a higher order than the quadrupole moment are not included. The resulting secular determinant of sixteen rows and columns factors according to the total magnetic quantum number; thus there are two secular equations of the first, second, and third degrees, and one quartic equation. The expressions from which the energy levels may be determined are given below for the case in which $I=\frac{3}{2}$. The complexity of the expressions becomes much greater for $I>\frac{3}{2}$. To simplify the equations we have written:

$$x=(g_J-g_I)(\mu_0H/a), \quad (10a)$$

an expression reminiscent of that employed in describing the energy levels of the states for which $J=\frac{1}{2}$,

$$y=(g_J+g_I)(\mu_0H/a) \quad (10b)$$

and

$$r=b/a.$$

For

$$F=3,\ m_F=\pm 3,\ W/a=(9/4)(1+11r)\pm\tfrac{3}{2}y,$$
$$F=3,\ m_F=\pm 2,\ W/a=\tfrac{3}{4}(1+17r)\pm y+\tfrac{1}{2}[x^2+9(1+8r)^2]^{\frac{1}{2}},$$
$$F=2,\ m_F=\pm 2,\ W/a=\tfrac{3}{4}(1+17r)\pm y-\tfrac{1}{2}[x^2+9(1+8r)^2]^{\frac{1}{2}}. \quad (11)$$

The cubic equations may be simplified by use of the substitution

$$W/a=(W_0/a)+(\delta W/a), \quad (12)$$

where W_0/a is the weak field, first-order approximation to the energy of the level. For $F=3$, $m_F=\pm 1$

$$W_0/a=(9/4)(1+11r)\pm\tfrac{1}{2}y \quad (13a)$$

and

$$(\delta W/a)^3+8(1+3r)(\delta W/a)^2-(x^2-15-120r)(\delta W/a)-2x^2=0. \quad (13b)$$

The value of $\delta W/a$ is the same for $m_F=\pm 1$. For $F=2$, $m_F=\pm 1$,

$$W_0/a=-\tfrac{3}{4}(1-r)\pm\tfrac{1}{2}y \quad (14a)$$

and

$$(\delta W/a)^3-(1+48r)(\delta W/a)^2-(x^2-24r-576r^2+6)(\delta W/a)+(1+24r)x^2=0. \quad (14b)$$

For $F=1$, $m_F=\pm 1$,

$$W_0/a=-(11/4)(1-9r)\pm\tfrac{1}{2}y \quad (15a)$$

and

$$(\delta W/a)^3-(7-24r)(\delta W/a)^2-(x^2-10+120r)(\delta W/a)-3x^2=0. \quad (15b)$$

The quartic equation, whose roots determine the energies of the states for which $m_F=0$, is:

$$(W/a)^4+(5-99r)(W/a)^3+[(9/8)(2755r^2-266r-1)-(5/2)x^2](W/a)^2-[(9/16)(57123r^3-9487r^2-163r+47)+(9/4)(1-55r)x^2](W/a)+(4455/256)(1287r^4-1412r^3+114r^2+12r-1)-(135/32)\times(363r^2-26r-1)x^2+(9/16)x^4=0. \quad (16)$$

A graph showing the energies of the levels as a function of field has been given by BK. The interaction constants, a and b, have been deter-

---

[9] P. Kusch, S. Millman, and I. I. Rabi, Phys. Rev. 57, 765 (1940).

[10] S. Millman and P. Kusch, Phys. Rev. 58, 438 (1940).

[11] G. E. Becker and P. Kusch, Phys. Rev. 73, 584 (1948). (This reference is hereinafter referred to as BK.)

[12] T. C. Hardy and S. Millman, Phys. Rev. 61, 459 (1942).

[13] All the nuclear g values, with the exception of those of the gallium and indium isotopes, depend on the assumption that the g_J value for the $^2S_{\frac{1}{2}}$ state of Rb and Cs is 2. The g_I values given for the isotopes of gallium and indium depend on the assumption that the g_J value for the $^2P_{\frac{1}{2}}$ states of these atoms is $\frac{2}{3}$. Not only are these assumptions subject to correction, but they are inconsistent as well. To place the g_I values of Ga and In on the scale used for all other nuclei, a reduction of 0.25 percent is required. An arithmetical error made by Becker and Kusch just compensates for the required reduction of 0.25 percent in the case of Ga^{69}. The g_I values of Ga^{69}, Ga^{71}, In^{115} on the same scale as that used in all other cases are −0.0007239, −0.0009195, and −0.0006616. These changes have no effect at all on the results presented in this paper.

mined to be

for Ga^{69}:

$$a=190.790\times10^6 \text{ sec.}^{-1}, \quad b=2.6049\times10^6 \text{ sec.}^{-1};$$

and for Ga^{71}:

$$a=242.424\times10^6 \text{ sec.}^{-1}, \quad b=1.6416\times10^6 \text{ sec.}^{-1}.$$

It is to be noted that all expressions for the levels determine the energy of a level in the same units in which ΔW, a, and b are known. Experimentally, these quantities have been determined in sec.$^{-1}$ and observations of line frequencies, of course, lead to separations of energy levels expressed in terms of sec.$^{-1}$. The quantities x and y always contain the factor μ_0H/a (or $\mu_0H/\Delta W$) where a and ΔW are expressed in ergs. If we express all frequency measurements in terms of megacycles per second, then the quantity above becomes:

$$(\mu_0H/ha)\times10^{-6}=(eH/4\pi ma)\times10^{-6}=H'/a \quad (17)$$

or

$$H'=1.3998H.$$

In all subsequent discussion we use the quantity H' as a measure of the field.

It should be pointed out that in previous experiments with lithium and potassium in the $J=\frac{1}{2}$ states, observations made at very low fields and at high fields yielded self-consistent values of the interaction constants. This fact indicates that the Hamiltonian function given above is adequate to describe the hyperfine structure of alkali atoms in the $^2S_{\frac{1}{2}}$ state within experimental error. Thus it was unnecessary in the present experiment to carry out observations of all the possible transitions and over a range of values of the absolute magnetic field. For the $^2P_{3/2}$ state of gallium the self-consistency of the frequencies of four lines at a field strength of 380 gauss, under which conditions there is intermediate decoupling of the nuclear spin and the atomic angular momentum, was regarded as a confirmation of the adequacy of the Hamiltonian. Moreover, from a study of the h.f.s. pattern of gallium in the $^2P_{3/2}$ state at very low magnetic fields, BK conclude that no moments of a higher order than the quadrupole need to be included in the Hamiltonian to describe the energy levels.

For the accuracy required in these experiments, perturbation approximations to the energies of the levels were inadequate and, in all cases, accurate solutions to the secular equations were found.

In the present series of experiments the direction of the oscillating magnetic field was perpendicular to the constant magnetic field. Thus the allowed transitions are given, in weak field notation, by $\Delta F=\pm1, 0$, and $\Delta m_F=\pm1$. Under very strong magnetic field conditions the nuclear and atomic angular momenta are decoupled, and the transitions may be classified according to $\Delta m_J=\pm1$, $\Delta m_I=0$ or $\Delta m_J=0$, $\Delta m_I=\pm1$. Whether or not a particular field is "weak" or "strong" for a particular atomic state depends on the value of x (or y) which corresponds to that field. This, in turn, depends on the value of ΔW or a.

The decisions as to what magnetic field strength should be employed and which lines should be observed were made on the basis of the following considerations. Since the present experiments are directed toward the measurement of an atomic magnetic moment the lines to be observed should be selected to possess the greatest possible field sensitivity. Under very strong field conditions the field sensitivity of the transitions $\Delta m_I=\pm1$ is so small that these lines are not useful for the purpose of this experiment. The transitions $\Delta m_J=\pm1$ show adequate field sensitivity, but in most cases these lines occur at frequencies which are difficult to obtain experimentally. Under very weak field conditions the transitions $\Delta F=0$ are the most suitable for this experiment, as the frequencies of these lines are very nearly proportional to the field strength and are sufficiently field sensitive. Very weak fields ($H<100$ gauss) could not be employed in this experiment because of the residual inhomogeneity of the field.

The value of the magnetic field used in most of these experiments ($\sim$400 gauss) represents very weak field conditions for $^2P_{\frac{1}{2}}(In^{115})$, $^2P_{\frac{1}{2}}(Ga^{69})$; for $^2S_{\frac{1}{2}}(Na^{23})$ a considerable departure from weak field conditions appears, and for gallium in the $^2P_{3/2}$ state the field is very strong. Only a few lines in the spectrum of $^2P_{3/2}$(Ga) of frequency less than 10^9 sec.$^{-1}$ permit a satisfactory determination of g_J.

In practice the large field sensitivity of some

lines, particularly in the $^2S_{\frac{1}{2}}$(Na) spectrum, did not bring about a proportional increase in the precision of measurement of the field because of the lack of complete homogeneity in the constant field.

In principle it is possible, from the observed frequencies of lines in the spectra of atoms in two different states, to calculate directly the ratio g_{J1}/g_{J2}. However, such a procedure is extremely laborious. We have, instead, calculated the quantity H' for each observed line. If a discrepancy in the value of H' occurs for atoms in two different states and under conditions for which H' is known to be identical, the discrepancy may be removed by an adjustment of the g_J values. This is evident since in the expressions for the energies of the levels, H' always occurs in the product g_JH'.[14] Suppose the assumed values of g_J are g_{J1}^0 and g_{J2}^0 for two different states. The corresponding values of H' are H_1' and H_2'. $\Delta H' = H_1' - H_2'$. Then:

$$g_{J1}/g_{J2} = (g_{J1}^0/g_{J2}^0)[1+(\Delta H'/H')]. \quad (18)$$

4. APPARATUS AND PROCEDURE

The general procedures and instrumental requirements for the observation of lines in the radio frequency spectra of atoms have been discussed in a number of papers. The molecular beam apparatus used in the experiments described in this paper was originally designed as an apparatus for the study of the radio frequency spectra of molecules in very high magnetic fields. Accordingly, the deflecting fields are long and may be operated at high flux densities in order to deflect molecules with a total moment of the order of one nuclear magneton. The magnetic field in which transitions occur is 48 cm long and can operate at fields as high as twenty thousand gauss. Unfortunately, the field is not entirely homogeneous, and the use of a large fraction of the length of the transition field for the observation of extremely field sensitive atomic lines broadens the lines excessively. In all of the present experiments, the effective length of the transition field was reduced to 2 cm by the expedient of reducing the length of the r-f field through which the beam passes. The deflecting fields were operated at a very low level of flux density to permit refocussing of atoms whose moment is of the order of one Bohr magneton.

For the purpose of consecutive measurement of lines in the radio frequency spectra of two different atomic species, such as gallium and sodium, a special oven chamber was constructed in which an oven containing gallium and another containing sodium were mounted on a platform attached to a ground joint which could be rotated from the outside of the apparatus. This arrangement permitted rapid interchange and adjustment of the two ovens.

All frequency measurements were made on a General Radio heterodyne frequency meter, model 620A. The meter may be used to determine frequencies to about one part in twenty thousand, but the precision in reading may be somewhat greater or less than this, depending on the linearity of the scale at a particular reading, on the thermal stability of the meter and other factors. Because of the limit of accuracy in the location of the centers of the resonance curves imposed by their width, it was not worth while to use more accurate frequency measuring equipment. The uncertainties in frequency measurement imposed by the meter and by the line widths are statistical in character and the precision is improved by judicious repetition of observations. Readings on the wave meter are determined in terms of a crystal within the wave meter. The frequency of the crystal in the wave meter which we used differed from its nominal value by 0.0045 percent, as determined by comparison with signals from WWV. The correction is of no importance in the measurements of the frequencies of the lines of gallium, since BK used the identical meter in determining the constants of gallium. The correction has been applied to all cases, however.

All experimental data considered in the experiments discussed in this paper are measurements of the frequencies of spectral lines. No knowledge

[14] It is true that a term g_IH' also occurs. However, in the cases of Li^7, Na^{23}, Ga^{69}, Ga^{71}, and In^{115} the value of g_I has been determined in each case in terms of g_J of the ground state of the atom in which the nucleus occurs. The term g_IH' may then be written as $(g_I/g_J)g_JH'$ where the ratio g_I/g_J is a quantity independent of any assumption as to g_J. The only case in which g_IH' is an independent term is that of the $^2P_{3/2}$ state of gallium, where g_I is known in terms of $g_J(^2P_{\frac{1}{2}})$ and not in terms of $g_J(^2P_{3/2})$. However the term is so small that uncertainties in g_I have no appreciable effect on our results.

TABLE II. Observations of lines in the spectrum of the $^2P_{1/2}$ state of gallium.

Isotope	Line	Obs. $f\times10^{-6}$ sec.$^{-1}$	Calc. $f\times10^{-6}$ sec.$^{-1}$
71	$(2,\ 0)\leftrightarrow(2,\ -1)$	89.434	89.422
69	$(2,\ 0)\leftrightarrow(2,\ -1)$	89.867	89.870
71	$(1,\ 0)\leftrightarrow(1,\ -1)$	90.389	90.395
69	$(1,\ 0)\leftrightarrow(1,\ -1)$	90.637	90.634
71	$(2,\ -1)\leftrightarrow(2,\ -2)$	94.716	94.714
69	$(2,\ -1)\leftrightarrow(2,\ -2)$	96.848	96.848

of the magnetic field in which transitions occur is required, though it is necessary that transitions resulting in lines whose frequencies are to be compared, occur in the same magnetic field. In practice, a magnetic field is not entirely constant, and the rate of drift of the field depends upon a number of factors, of which the principal two are the condition of the storage cells which supply the exciting current to the magnet and the temperature stability of the d.c. circuit and the magnet. A drift is observed by noting a change in the frequency of a line with time and a very good correction for drift of field may be made by alternately measuring the frequencies of each of two lines to be compared and then, by suitable graphical or other methods, determining the frequencies of each of the lines at any arbitrary instant of time.

To reduce the time interval between observations on successive lines, and hence minimize the effects of a drifting field, all observations were made on the basis of a single frequency reading for each line. That is, no detailed plot was made of the reduction of beam intensity as a function of frequency; rather, the observer noted the frequency at which the maximum reduction of beam intensity occurred, the mean of the frequencies at which one-half the maximum reduction occurred or the mean frequency of any other pair of symmetrical points. On the basis of several of these observations of a single line,

TABLE III. The comparison of the apparent values of H' in the $^2P_{1/2}$ and $^2P_{3/2}$ states of gallium.

$I(^2P_{1/2})$ $f\times10^{-6}$ sec.$^{-1}$	$II(^2P_{1/2})$ H'	$III(^2P_{3/2})$ isotope	$IV(^2P_{3/2})$ transition	$V(^2P_{3/2})$ $f\times10^{-6}$ sec.$^{-1}$	$VI(^2P_{3/2})$ H'	VII $\Delta H'$
96.832	527.24	69	$(3,\ 0)\leftrightarrow(2,\ 1)$	492.33	528.27	1.03
96.751	526.84	71	$(3,\ -1)\leftrightarrow(3,\ -2)$	459.77	527.63	0.79
96.620	526.19	69	$(3,\ 0)\leftrightarrow(3,\ -1)$	443.92	527.08	0.89
					Mean	0.90

TABLE IV. The comparison of the apparent values of H' in the $^2P_{1/2}$ and $^2P_{3/2}$ states of gallium.

$I(^2P_{1/2})$ $f\times10^{-6}$ sec.$^{-1}$	$II(^2P_{1/2})$ H'	III isotope	$IV(^2P_{3/2})$ transition	$V(^2P_{3/2})$ $f\times10^{-6}$ sec.$^{-1}$	$VI(^2P_{3/2})$ H'	VII $\Delta H'$	VIII
90.909	528.92	69	$(3,\ 0)\leftrightarrow(3,\ -1)$	447.06	529.79	0.87	3
90.910	528.93	69	$(3,\ -1)\leftrightarrow(3,\ -2)$	486.90	529.85	0.92	3
90.913	528.94	71	$(3,\ -3)\leftrightarrow(3,\ -2)$	491.14	529.90	0.96	2
90.921	528.99	71	$(3,\ -1)\leftrightarrow(3,\ -2)$	463.23	529.91	0.92	1
					Mean	0.91	

the observer recorded a single reading on a wave meter.

5. EXPERIMENTAL RESULTS

a. The Determination of the Ratio of the g_J Value of Gallium in the $^2P_{3/2}$ State and in the $^2P_{1/2}$ State

To estimate the precision with which the frequency of a line may be determined, six lines in the spectrum of gallium in the $^2P_{1/2}$ state were measured in rapid succession. Preliminary observation had indicated that the field did not drift at a rapid rate just prior to the observations. Each line frequency is, therefore, the result of a single observation. By use of constants given by BK the frequencies of all lines were calculated, where the $(2,\ -1)\leftrightarrow(2,\ -2)$ line of Ga^{69} was used to determine the value of the magnetic field. The results are shown in Table II. The agreement between calculated and observed frequencies is very good. The results indicate, in general, that line frequencies are, indeed, measurable to within one part in twenty thousand. The constants given by BK and used in the calculations are consistent for the two isotopes of gallium and the relationships which determine the frequencies of the lines are valid to within the error in measurement of the line frequencies.

It should be pointed out, however, that all the lines recorded in Table II lie in the same frequency range and have very nearly the same frequency dependence on field. This means that inhomogeneities in the magnetic field, will affect all lines in a similar way and that similar agreement may not occur when two lines in a spectrum have a markedly different frequency dependence on field. In fact, for the case of Na^{23}, to be discussed later, a small systematic discrepancy occurs between the magnetic field as calculated from two lines with markedly different field dependence.

Three lines in the $^2P_{3/2}$ state of gallium were observed alternately with the $(2, -1)\leftrightarrow(2, -2)$ line of Ga^{69} in the $^2P_{1/2}$ state. The results are shown in Table III. Column I gives the frequency of the $(2, -1)\leftrightarrow(2, -2)$ line of Ga^{69} in the $^2P_{1/2}$ state in megacycles per second. Column II gives the value of H' as previously defined and calculated on the basis of the assumption that $g_J(^2P_{1/2}) = \frac{2}{3}$. Column III indicates the isotope of gallium in the spectrum of which the line tabulated in column IV has been measured. The lines in column IV are in the spectrum of the $^2P_{3/2}$ state. Column V lists the observed frequency in megacycles per second, and column VI the corresponding value of H' calculated on the basis of the assumption that $g_J(^2P_{3/2}) = 4/3$. Column VII lists the difference between the values of H' for the $^2P_{3/2}$ and the $^2P_{1/2}$ states. The data and results in Table III are to be considered as the result of an exploratory experiment. The rather large variation between the values of $\Delta H'$ can be ascribed to an insufficient number of successive observations of each line in the face of a drifting magnetic field and to insufficient care in locating the resonance minima. The rapid drift of field is apparent from an observation of the successive values of H' in columns II and VI. The deviation of the values of $\Delta H'$ from the mean can be explained by uncertainties, for the first line, of one part in ten thousand in frequency measurement, and for the second line of only slightly more than this.

In a repetition of the previous experiment, the frequencies of four lines in the spectrum of Ga in the $^2P_{3/2}$ state were measured alternately with the frequency of the $(1, 0)\leftrightarrow(1, -1)$ line of Ga^{69} in the $^2P_{1/2}$ state. The results are shown in Table IV. Column I gives the frequency of the $(1, 0)\leftrightarrow(1, -1)$ line of Ga^{69} in the $^2P_{1/2}$ state. The other columns are exactly as in Table III. Column VIII gives a weight assigned to each observation, based on an evaluation of experimental factors. It should be noted that the rate of drift of field for these observations is very much less than in the case of the data recorded in Table III. The particular lines in the spectrum of gallium in the $^2P_{3/2}$ state which have been measured are rather arbitrary; the lines themselves were not identified until rather elaborate reductions of data had been made. That is to say, observations were made at a field and within a frequency range known to contain lines with suitable frequency dependence on field. The assignment was subsequently made by use of the criterion that all lines in the spectrum of gallium in the $^2P_{3/2}$ state should determine substantially the same field. The assignment is unique. All the lines in the spectrum of gallium in the $^2P_{3/2}$ state which are recorded in the two tables have about the same frequency dependence on field with the exception of the first line in Table III, where the frequency dependence on field is about half as great as for the other lines. In all cases measurements have been made against a single line in the spectrum of gallium in the $^2P_{1/2}$ state, but auxiliary data were taken to make the assignment of the line definite.

TABLE V. Notation used in describing lines in the spectra of sodium, and of gallium in the $^2P_{1/2}$ state.

Line		Symbol
$(2, -2)\leftrightarrow(2, -1)$		Na*I*
$(1, 0)\leftrightarrow(1, -1)$		Na*II*
$(2, 0)\leftrightarrow(2, -1)$		Na*III*
$(1, 1)\leftrightarrow(1, 0)$		Na*IV*
$(2, 1)\leftrightarrow(2, 0)$		Na*V*
$(2, 2)\leftrightarrow(2, 1)$		Na*VI*
$(1, 0)\leftrightarrow(1, -1)$	69	Ga*I*
$(2, 0)\leftrightarrow(2, -1)$	69	Ga*II*

TABLE VI. The observed frequencies and apparent values of H' for several lines in the spectrum of sodium.

Time	Na*I*	Na*II*	Na*III*	Na*IV*	Na*V*	Na*VI*
3:10	418.10 (531.84)					
3:25						176.85 (531.77)
3:35	418.06 (531.80)					
3:45	418.00 (531.75)					
3:50	417.99 (531.74)					
3:55						176.83 (531.67)
4:15			260.51 (531.63)			
4:15			260.50 (531.60)			
4:30		261.34 (531.56)				
4:30		261.34 (531.56)				
4:35						176.81 (531.59)
4:40					206.77 (531.59)	
4:45				207.61 (531.53)		
4:50						176.78 (531.47)
5:00	417.72 (531.50)					

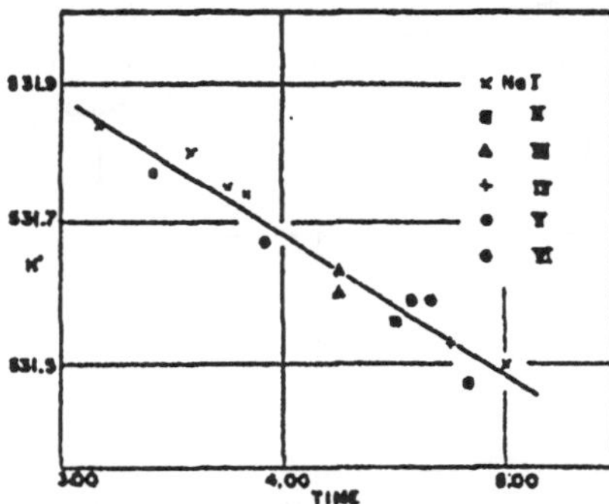

FIG. 1. Values of H' calculated from the observed frequencies of several lines in the spectrum of Na.

An uncertainty in the measurement of the frequency of the $(1, 0)\leftrightarrow(1, -1)$ line of Ga^{69} in the $^2P_{\frac{1}{2}}$ state of one part in twenty thousand gives rise to an uncertainty in the corresponding H' of 0.03. The same fractional uncertainty in the measurement of the frequencies of the lines in the spectrum of the $^2P_{3/2}$ state gives rise to an uncertainty in the corresponding value of H' of 0.02. Thus the maximum variation among the values of $\Delta H'$ can be ascribed entirely to the limitation of frequency determination. In our experimental arrangement, a better frequency meter could not be expected to increase the precision of the result to a significant extent, since the width of the resonance curves rendered a significant measurement of frequency to a higher precision impossible.

The variation among the values of $\Delta H'$ shown in Table IV are statistical in character. We conclude that for these observations $\Delta H' = 0.91 \pm 0.03$. The agreement of this value with the mean obtained from Table III indicates that the large scatter of values obtained in the first performance of the experiment is statistical.

On the basis of the assumption that $g_J(^2P_{3/2}) = 4/3$ and that $g_J(^2P_{\frac{1}{2}}) = \frac{2}{3}$, the ratio of the apparent field as calculated from lines in the spectrum of the $^2P_{3/2}$ state to that calculated from lines in the spectrum of the $^2P_{\frac{1}{2}}$ state is 1.00172 ± 0.00006. Evidently the field is independent of the mechanism of measurement. To satisfy this requirement we must make

$$g_J(^2P_{3/2}\text{Ga})/g_J(^2P_{\frac{1}{2}}\text{Ga}) = 2(1.00172 \pm 0.00006)$$

where it is assumed that the deviation of the g_J values from their nominal values is small.

b. The Determination of the Ratio of the g_J Value of Sodium in the $^2S_{\frac{1}{2}}$ State and That of Gallium in the $^2P_{\frac{1}{2}}$ State

Six lines of the spectrum of sodium and two lines of the spectrum of gallium were used in the determination of the ratio of the g_J values. To assist in the discussion of the observations, the notation given in Table V is used. All lines of sodium are those of the $^2S_{\frac{1}{2}}$ state and the lines of gallium are of the $^2P_{\frac{1}{2}}$ state.

As in the previous experiment, a test was made to determine the consistency of frequency measurements. In Table VI are given the observed frequencies of the six lines of sodium and the time at which each frequency was measured. Directly beneath each frequency in megacycles per second is given, in parentheses. the value of H' corresponding to the observed frequency and calculated on the basis of the assumption that $g_J(^2S_{\frac{1}{2}}) = 2$. The data of Table VI are plotted in Fig. 1, where the calculated values of H' are plotted against the time. A straight line is drawn through the points and the average deviation of the points from the mean line is less than 0.03 units of H'. The maximum deviation is 0.04.

TABLE VII. The observed frequencies and values of H' of several lines in the spectrum of sodium and gallium.

Time	Na*I*	Na*II*	Na*III*	Ga*I*	Ga*II*
6:30				91.093 (529.97)+0.00	
6:40					90.319 (529.93)−0.01
6:50				91.087 (529.93)+0.01	
7:05			260.32 (531.16)+0.00		
7:05		261.17 (531.14)−0.02			
7:15	417.33 (531.15)+0.02				
7:40				91.060 (529.78)−0.01	
7:40					90.295 (529.79)+0.00
7:58			260.27 (531.02)−0.01		
8:02		261.12 (531.01)−0.01			
8:14	417.17 (531.00)+0.01				
8:28				91.038 (539.65)−0.02	
8:35					90.275 (529.68)+0.02
8:45			260.21 (530.89)+0.00		
8:49		261.05 (530.85)−0.05			
9:10	417.04 (530.88)+0.03				
9:10	417.05 (530.89)+0.04				
9:26				91.014 (529.52)−0.01	
9:26					90.248 (529.52)−0.01

That is, the maximum deviation of the magnetic field as calculated from the separate lines is one part in thirteen thousand from the average as determined from all observations.

From the frequencies of the NaI and NaVI lines, at the same instant of time, we have found for $\Delta\nu$ of Na the value 1771.60×10^6 sec.$^{-1}$ A change in the assumed frequency of either line of one part in twenty thousand will alter the $\Delta\nu$ by about 0.25×10^6 sec.$^{-1}$. The agreement of the result with the value of 1771.75×10^6 sec.$^{-1}$ given by Millman and Kusch is therefore very good. It is to be noted that the value of the magnetic field employed in these experiments is poorly chosen for the determination of $\Delta\nu$ as compared to the fields used by Millman and Kusch. The widths of the sodium lines was roughly proportional to the rate of variation of line frequency with magnetic field and are presumably due to inhomogeneities in the magnetic field. The observed width of the NaI line was about 0.280 $\times10^6$ sec.$^{-1}$, that of the NaII and NaIII lines about 0.150×10^6 sec.$^{-1}$ and that of the NaVI line about 0.070×10^6 sec.$^{-1}$. The very good internal consistency noted above indicates that the measured center of the resonance curve is, indeed, the center of the line and that all line frequencies determine the same mean magnetic field.

Three lines in the sodium spectrum which have the greatest sensitivity to field were measured alternately with two lines in the spectrum of gallium. The data are given in Table VII. The time of observation is indicated in the first column, and in the other columns are given the observed line frequencies in megacycles per second. Directly below the observed frequency is recorded, in parentheses, the value of H' calculated on the basis of the assumption that $g_J(^2S_{\frac{1}{2}})=2$ and $g_J(^2P_{\frac{1}{2}})=\frac{2}{3}$. The quantity which appears directly after the value of H' is the difference between the observed H' and that calculated from a least squares solution of the data.

The data are graphically presented in Fig. 2. An independent least squares solution of the variation of H' with time for sodium and for gallium gives very nearly the same rate of drift of field in both cases, as expected. A combined least squares solution for both sets of lines, where the rate of drift of field is the same in both cases, yields the result

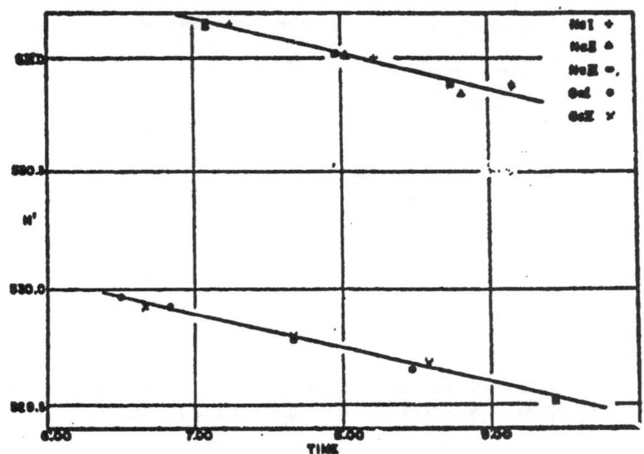

FIG. 2. Values of H' calculated from the observed frequencies of several lines in the spectra of Na and of Ga in the $^2P_{\frac{1}{2}}$ state.

$$\begin{aligned}&\text{for Na: } H'=531.319-0.002475t,\\&\text{for Ga: } H'=530.036-0.002475t,\end{aligned}\qquad(19)$$

where the time, t, is measured in minutes from 6:00. The difference between the observed and calculated values is indicated in the table. It is to be noted that the differences are all positive for the NaI line and negative or zero for the NaII and NaIII lines. The effect probably represents a slightly unsymmetrical broadening of the lines by the inhomogeneities of the field. The difference in the value of H' for Na and that for Ga is $531.319-530.036=1.283$. To this value we assign the arbitrary precision value of ±0.030, which is the sum of the mean deviation of the values of H' yielded by Eq. (19) from those calculated for the individually observed Na lines and of the corresponding quantity for Ga. This uncertainty is obviously somewhat greater than the probable error calculated from the internal consistency of the data and is given in view of the fact that a systematic error may occur.

On the basis of the assumption that $g_J(^2S_{\frac{1}{2}})=2$ and that $g_J(^2P_{\frac{1}{2}})=\frac{2}{3}$, the ratio of the apparent field as calculated from Na to that as calculated from Ga is 1.00242 ± 0.00006. Other data has given the value for the same quantity of 1.00247 ±0.00020. To satisfy the condition that the magnetic field is the same for a measurement of the frequencies of the gallium and sodium lines, we must set:

$$g_J(^2S_{\frac{1}{2}}\,\text{Na})/g_J(^2P_{\frac{1}{2}}\,\text{Ga})=3(1.00242\pm0.00006),$$

TABLE VIII. Notation used in identifying lines in the spectrum of indium in the $^2P_{\frac{1}{2}}$ state.

Line	Symbol
$(4, -4)\leftrightarrow(4, -3)$	In*I*
$(4, -3)\leftrightarrow(4, -2)$	In*II*
$(5, 5)\leftrightarrow(5, 4)$	In*III*
$(5, 4)\leftrightarrow(5, 3)$	In*IV*
$(5, 3)\leftrightarrow(5, 2)$	In*V*

where it is assumed that the deviation of the g_J values from their nominal values is small.

c. The Determination of the Ratio of the g_J Value of Na in the $^2S_{\frac{1}{2}}$ State and That of In in the $^2P_{\frac{1}{2}}$ State

The $\Delta\nu$ of In^{115} and its g_I have been determined by Hardy and Millman.[12] The low abundance of In^{113} together with the fact that its $\Delta\nu$ is almost identical to that of In^{115} makes it impossible to observe lines in the spectrum of In^{113} in our experiments. In fact, the frequencies of the In lines which we observe are predominantly due to the term $g_I\mu_0 H/h$, and higher order terms, involving $\Delta\nu$, contribute at most about 3 percent to the frequencies of the lines. Under these conditions the small variation of the $\Delta\nu$ between the two isotopes of indium would not permit the independent observation of lines in the spectrum of both isotopes in our apparatus.

For the purposes of the comparison of the g_J values, we have used three lines in the spectrum of sodium, previously identified as Na*I*, Na*II* and Na*III*. Five lines in the spectrum of indium have been used, identified as indicated in Table VIII. The data were observed in exactly the same way as in the case of the comparison of the g_J values of gallium and sodium and the arrangement of Table IX, wherein are recorded the observations on sodium and indium, is identical to the arrangement of Table VII.

It can be seen from the table that the rate of drift of magnetic field during the experiment was very small. In making a least squares reduction of the data, the observations on the Na*II* line

TABLE IX. The observed frequencies and values of H' for several lines in the spectrum of indium and sodium.

Time	In*I*	In*II*	In*III*	In*IV*	In*V*	Na*II*	Na*III*	Na*IV*
3:45	35.4104 (514.56) −0.03							
3:47		35.1936 (514.58) −0.02						
3:53			33.0891 (514.57) −0.03					
3:54				33.2855 (514.64) +0.04				
3:55					33.4769 (514.57) −0.04			
4:08						254.210 (515.90) +0.03		
4:35							254.989 (515.88) −0.01	
4:40						254.166 (515.92) +0.03		
5:00							254.980 (515.84) −0.07	
5:10	35.4194 (514.69) +0.02							
5:12		35.1982 (514.64) −0.03						
5:20			33.0973 (514.70) +0.02					
5:22				33.2909 (514.72) +0.04				
5:24					33.4855 (514.70) +0.02			
5:47						254.181 (515.96) +0.01		
5:52							255.00 (515.91) −0.05	
5:54						254.186 (515.97) +0.01		
6:02								203.323 (515.944) −0.02
6:16	35.4228 (514.74) +0.02							
6:30		35.2032 (514.72) −0.02						
6:40				33.2929 (514.76) −0.01				
6:45					33.4879 (514.74) −0.04			

have been weighted by two because of experimental factors which made it very much easier to locate this line than the other two lines in the spectrum of sodium. The data may be adequately represented by assuming a linear drift of field. If t is the time in minutes from 3:45, we can express the result

$$\begin{aligned} &\text{for In: } H' = 514.594 + 0.000861t; \\ &\text{for Na: } H' = 515.846 + 0.000861t. \end{aligned} \tag{20}$$

The difference in the value of H' for Na and In is 1.252. If all the sodium lines are weighted equally, this difference becomes 1.243. The first of these two values is believed to be the better one. The data are clearly inferior to those obtained in the case of gallium. To the value of 1.252 we assign the arbitrary precision value of ±0.050, obtained as in the previous experiment. This uncertainty is obviously somewhat greater than the probable error calculated from the internal consistency of the data.

On the basis of the assumption that $g_J(^2S_{\frac{1}{2}}) = 2$ and that $g_J(^2P_{\frac{1}{2}}) = \frac{2}{3}$, the ratio of the apparent field as calculated from Na to that as calculated from In is 1.00243±0.00010. To satisfy the condition that the magnetic field is the same for a measurement of the lines in each of the two spectra, we must set:

$$g_J(^2S_{\frac{1}{2}}\,\text{Na})/g_J(^2P_{\frac{1}{2}}\,\text{In}) = 3(1.00243 \pm 0.00010).$$

6. DISCUSSION OF RESULTS

The observed deviations of the ratios of the atomic g_J values from those calculated under the assumptions that pure LS coupling obtains and that the fundamental gyromagnetic ratios of spin and orbit have the values 2 and 1, respectively, may be attributed, in any particular case, to a deviation of the spin or orbit g values from their accepted values. It was shown in Section 2 that δ_S, defined by $g_S = 2(1+\delta_S)$, can be computed from the observed ratio of the g values of two atomic states of different spectral term classifications. This quantity has been evaluated from the results of the three relevant experiments. The results appear in Table X. The error assigned to δ_S in each case follows directly from the error assigned to the corresponding ratio of the g values. Thus from the experimental results, the difference between the value of δ_S which results from the $^2P_{3/2}(\text{Ga}) - {}^2P_{\frac{1}{2}}(\text{Ga})$ experiment and that which results from the $^2P_{\frac{1}{2}}(\text{Ga}) - {}^2S_{\frac{1}{2}}(\text{Na})$ experiment is probably real. It is, however, quite possible that this discrepancy may be accounted for by small deviations of the properties of the atomic systems from the simple description implicit in the theory underlying the calculations by means of which the experimental data has been reduced. It is to be emphasized that the numerical results are independent of the precise values of any of the fundamental atomic constants.

The agreement of the values of δ_S in Table X is very strong evidence in support of the hypothesis that the fundamental spin gyromagnetic ratio does, in fact, differ from the accepted integral value by an amount of very nearly the magnitude of the quantities given in Table X. An alternative explanation of these results in terms of perturbations (e.g., configuration interactions) of the individual atomic states would require that the perturbations on the g values of the four states should have just such magnitudes as to give the agreement noted above. Aside from the question of the possible magnitude of such perturbations (discussed below) it should be pointed out that three different atomic species and three states of different spectral classification are involved in these experiments. It is very unlikely that under these conditions there should occur the systematic effect shown in these results. Schwinger[7] has reported the results of a theoretical study of the interaction of an electron with the quantized electromagnetic field. By a canonical transformation of the Hamiltonian the infinite interaction terms are separated from the rest of the Hamiltonian and are shown to be properly included in the experimental mass of the electron. The evaluation of the remaining finite terms in the case of an applied magnetic field leads to an interaction energy term corresponding to a spin moment of the electron increased from the accepted value of one Bohr magneton by a factor $(1+\alpha/2\pi) = 1.00116$. This value is in very good agreement with our experimental results. The possible reasons for the small remaining discrepancies are discussed below.

The result of this experiment, that the electron spin g value differs from 2 by about 12 parts in ten thousand, may be compared with previous

TABLE X. Observed ratios of atomic g values and the corresponding values of δg.

Experimental ratio	δg
$g_J(^2P_{3/2}\text{Ga})/g_J(^2P_{1/2}\text{Ga}) = 2(1.00172 \pm 0.00006)$	0.00114 ± 0.00004
$g_J(^2S_{1/2}\text{Na})/g_J(^2P_{1/2}\text{Ga}) = 3(1.00242 \pm 0.00006)$	0.00121 ± 0.00003
$g_J(^2S_{1/2}\text{Na})/g_J(^2P_{1/2}\text{In}) = 3(1.00243 \pm 0.00010)$	0.00121 ± 0.00005

measurements of atomic g values in the Zeeman effect. In the classic work of Back[15] the uncertainty in the measured atomic g values is not less than one part in a thousand, and thus the effect would escape detection. In the somewhat more accurate work of Kinsler and Houston[1] on the determination of e/m from the Zeeman pattern in standard magnetic fields, singlet states were studied in almost all cases, so that no discrepancy corresponding to the effect described in the present paper could be found. However, in one experiment of this series, Kinsler[16] measured the g values of the two states $2p^5 3s\,^1P_1$(Ne) and $2p^5 3s\,^3P_1$(Ne). These two states arise from the same configuration and are the only states of this configuration with $J=1$. Application of the g sum rule to the experimental values yields 2.5017 ± 0.0016 for the sum of the atomic g values. In view of the assigned experimental uncertainty this value was considered by Kinsler to be in sufficiently good agreement with the value 2.5000 predicted on the basis of the accepted integral g values of the electron. Introduction of the spin moment g value reported in the present work gives a predicted value of the g sum of 2.5012.

There are several perturbations of the electronic states involved in the present experiments which in principle could bring about deviations of the atomic g values from the values given by the Russell-Saunders coupling formula. Estimates of the magnitude of these effects will be given in the following paragraphs.

In all of the atomic states employed in this experiment the configuration corresponds to a single valence electron outside of closed shells. Thus, the term classification is unique, and the question of the type of coupling does not arise for these states.

By far the largest configuration perturbation is the electrostatic interaction. If the excited configurations of the atomic system be described in first approximation by Russell-Saunders coupling it is well known that the electrostatic coupling will mix states only of the same values of total L and S. Thus the electrostatic coupling among pure LS states does not change the atomic g value in any approximation.

The effect on the g_J values of gallium and indium of the magnetic interactions with the excited configurations of the outer group of three electrons is small. The ordinary spin-orbit interaction with these configurations vanishes. A magnetic interaction with a ^4P state of magnitude 100 cm^{-1} would increase the $g_J(^2P_{1/2}\text{Ga})$ by about 1×10^{-5}. The effects in indium might be somewhat larger than in gallium. The $g_J(^2P_{3/2}\text{Ga})$ is affected by a still smaller amount than is the $g_J(^2P_{1/2}\text{Ga})$, as the g_J of all the $J=\frac{3}{2}$ terms of the interacting configurations differ very little from $g_J(^2P_{3/2})$. Spin-orbit interactions with excited states of the core will produce effects no larger than those described above, because of the high excitation of these levels.

It appears that an important perturbation of the g_J values of gallium or of indium would occur only if in the lower excited configurations, i.e., those of the outer three electrons in each case, there is a very considerable departure from Russell-Saunders coupling. In this case the electrostatic configuration mixing could bring about an appreciable alteration from the Russell-Saunders values of the g_J values of the states employed in the present experiment. The evidence as to the character of the coupling in the excited states of gallium and of indium is meager. In the $4s4p^2\,^4P$ term of gallium the interval rule appears to be obeyed accurately, while in the corresponding term $5s5p^2\,^4P$ of indium the fine structure interval ratio differs by 17 percent from the Landé rule. The relative positions of the terms arising from the $5s5p^2$ configuration of Sb*III* seem to be accounted for by LS coupling with the addition of electrostatic interaction with other configurations.[17]

The effect of configuration interactions on the g_J values of the ground states of the alkali metal

[15] E. Back and A. Landé, *Zeemaneffekt* (Verlag Julius Springer, Berlin, 1925). See also, R. Bacher and S. Goudsmit, *Atomic Energy States* (McGraw-Hill Book Company, Inc., New York, 1932).

[16] L. E. Kinsler, Phys. Rev. 46, 533 (1934).

[17] E. U. Condon and G. H. Shortley, *Theory of Atomic Spectra* (University Press, Cambridge, 1935).

atoms were investigated by Phillips,[18] who indicated that these effects are negligible.

Relativistic effects on the g_J values of a Dirac electron have been considered by Margenau[19] who showed that the atomic g_J value of a single electron is decreased by a factor $1-[(2J+1)/2(J+1)][\bar{T}/mc^2]$ for $J=L+\frac{1}{2}$ and $1-[(2J+1)/2J][\bar{T}/mc^2]$ for $J=L-\frac{1}{2}$. $\bar{T}$ is the mean kinetic energy. For the alkali metals the reduction in the g_J value is about 1 part in 10^5. For gallium and indium the effect is somewhat larger because of penetration of the core and because of the larger numerical factor in the relativistic expression for the $^2P_{\frac{1}{2}}$ state. When approximate account is taken of the penetration the g_J values of the $^2P_{\frac{1}{2}}$ states of gallium and indium are reduced by four parts in 10^5, and the $^2P_{3/2}$(Ga)g_J value is reduced by about two parts in 10^5.

The magnitude of the orientation dependence of the diamagnetic susceptibility, which in principle could affect the spectrum of the $^2P_{3/2}$(Ga) state, is negligible (less than one part in 10^5) at the fields employed in this experiment.

Incipient electronic Paschen-Back effect produces a perturbation in the line frequencies of the $^2P_{3/2}$(Ga) spectrum of about one part in 10^5.

The interaction of the valence electron spin with the diamagnetically induced moment of the core electrons reduces the g_J value of the atom. The calculation of this effect requires fairly accurate core and valence electron charge distributions. Such a calculation has been carried out for $^2S_{\frac{1}{2}}$(Na) with Hartree fields and a corresponding optical electron wave function. The decrease in the g_J value is less than one part in 10^5. It is believed that this effect is also negligible for the other atomic states.

The application of the corrections estimated in this section to the experimental values of the g_J ratios of Table X affects the calculated values of δ_S by about one percent. This magnitude is well within the experimental uncertainty and these corrections have not been applied in Table X.

We write $g_S=2(1.00119\pm0.00005)$ as the best value of the spin gyromagnetic ratio obtainable from the present experiments. The discrepancy between individual values of δ_S indicates the existence of small residual systematic effects and further analysis might indicate that g_S differs somewhat from the value given here.

The original motivation for these experiments came from the suggestion of G. Breit with regard to the intrinsic magnetic moment of the electron. The actual method of testing Breit's hypothesis by comparing the g values of atomic states was suggested to us by I. I. Rabi. We wish to express our thanks to both Dr. Breit and Dr. Rabi for many helpful discussions and to Miss Zelda Marblestone who performed a considerable fraction of the rather extensive calculations involved in this work.

[18] M. Phillips, Phys. Rev. 60, 100 (1941).
[19] H. Margenau, Phys. Rev. 57, 383 (1940).

PHYSICAL REVIEW VOLUME 79, NUMBER 3 AUGUST 1, 1950

Variational Principles for Scattering Processes. I

B. A. LIPPMANN
Nucleonics Division, Naval Research Laboratory, Washington, D. C.

AND

JULIAN SCHWINGER
Harvard University, Cambridge, Massachusetts

(Received April 10, 1950)

A systematic treatment is presented of the application of variational principles to the quantum theory of scattering.

Starting from the time-dependent theory, a pair of variational principles is provided for the approximate calculation of the unitary (collision) operator that describes the connection between the initial and final states of the system. An equivalent formulation of the theory is obtained by expressing the collision operator in terms of an Hermitian (reaction) operator; variational principles for the reaction operator follow. The time-independent theory, including variational principles for the operators now used to describe transitions, emerges from the time-dependent theory by restricting the discusson to stationary states. Specialization to the case of scattering by a central force field establishes the connection with the conventional phase shift analysis and results in a variational principle for the phase shift.

As an illustration, the results of Fermi and Breit on the scattering of slow neutrons by bound protons are deduced by variational methods.

I. INTRODUCTION

ALTHOUGH variational methods have long been applied to eigenvalue problems in many fields of physics, no systematic use had been made of variational procedures in connection with scattering processes until the period 1942–1946 when variational techniques, among others, were extensively employed in the solution of electromagnetic wave guide problems.[1] Variational formulations have also been devised for the treatment of neutron diffusion,[2] acoustical and optical diffraction,[3] and quantum-mechanical scattering problems.[4] Indeed,

[1] "Notes on Lectures by Julian Schwinger: Discontinuities in Waveguides," prepared by David S. Saxon, MIT Radiation Laboratory Report, February 1945.

[2] J. Schwinger, unpublished; R. E. Marshak, Phys. Rev. 71, 688 (1947).

[3] H. Levine and J. Schwinger, Phys. Rev. 74, 958 (1948); 75, 1423 (1949).

[4] J. Schwinger, "Lectures on Nuclear Physics," Harvard University, 1947; J. Schwinger, Phys. Rev. 72, 742 (1947); J. M. Blatt, Phys. Rev. 74, 92 (1948); W. Kohn, Phys. Rev. 74, 1763 (1948); J. M. Blatt and J. D. Jackson, Phys. Rev. 76, 18 (1949). Variational principles for scattering problems have also been independently developed by L. Hulthén, see Mott and Massey, *The Theory of Atomic Collisions* (Oxford University Press, London, 1949), 2nd ed., p. 128, and I. Tamm, J. Exp. Theor. Phys. USSR 18, 337 (1948); 19, 74 (1949).

such methods are applicable in any branch of physics where the fundamental equations can be derived from an extremum principle.

It is the purpose of this paper to describe the quantum mechanical time-dependent scattering theory and its variational reformulation. As a simple illustration of these methods, we consider the scattering of slow neutrons by protons bound in a molecule. This was first discussed by Fermi[5] in terms of an equivalent potential used in conjunction with the Born approximation. A more exact integral equation treatment was given by Breit,[6] with quite small ensuing corrections[7] to Fermi's theory. We shall show that the results of Fermi and Breit are easily derived from a variational treatment. Although one could consider, without difficulty, the scattering by any number of nuclei, the discussion will be restricted to the spin-dependent scattering by a single proton in an otherwise inert molecule of arbitrary mass. An extension to two protons, and in particular to the hydrogen molecule, is contained in an accompanying paper by one of us. Also included is an estimate of the error in the para-hydrogen scattering cross section calculated by Fermi's method.

II. TIME-DEPENDENT SCATTERING THEORY

We are concerned with the development in time of a system consisting of two interacting parts, which are such that the interaction energy approaches zero as the two parts are separated spatially. Correspondingly, the Hamiltonian is decomposed into the unperturbed Hamiltonian H_0, describing the two independent parts, and H_1, the energy of interaction. Since the problem is to describe the effect of H_1, it is convenient to remove the time dependence associated with H_0 from the Schrodinger equation

$$i\hbar[\partial\Psi'(t)/\partial t]=(H_0+H_1)\Psi'(t). \tag{1.1}$$

This is accomplished by the unitary transformation

$$\Psi'(t)=\exp(-iH_0t/\hbar)\Psi(t) \tag{1.2}$$

which yields

$$\begin{aligned} i\hbar[\partial\Psi(t)/\partial t]&=H_1(t)\Psi(t), \\ H_1(t)&=\exp(iH_0t/\hbar)H_1\exp(-iH_0t/\hbar). \end{aligned} \tag{1.3}$$

The initially non-interacting parts of the system are characterized by the state vector $\Psi(-\infty)$. On following the course of the interaction and the eventual separation of the two parts, we are led to the state vector $\Psi(+\infty)$, representing the final state of the system. This description can be made independent of the particular initial state by regarding the time development as the unfolding of a unitary transformation:

$$\Psi(t)=U_+(t)\Psi(-\infty),\quad U_+^+(t)U_+(t)=1. \tag{1.4}$$

[5] E. Fermi, Ricerca Scient. VII-II, 13 (1936).
[6] G. Breit, Phys. Rev. 71, 215 (1947).
[7] G. Breit and P. R. Zilsel, Phys. Rev. 71, 232 (1947); Breit, Zilsel, and Darling, Phys. Rev. 72, 576 (1947).

In particular,

$$\Psi(\infty)=S\Psi(-\infty),\quad S=U_+(\infty) \tag{1.5}$$

defines the collision operator, which generates the final state of the system from an arbitrary initial state. The operator $U_+(t)$ is to be obtained as the solution of the differential equation

$$i\hbar[\partial U_+(t)/\partial t]=H_1(t)U_+(t) \tag{1.6}$$

subject to the boundary condition

$$U_+(-\infty)=1. \tag{1.7}$$

It is also useful to introduce a unitary operator $U_-(t)$, which generates the state vector $\Psi(t)$ from the final state $\Psi(\infty)$,

$$\Psi(t)=U_-(t)\Psi(\infty)=U_-(t)S\Psi(-\infty). \tag{1.8}$$

Since the two operators are related by

$$U_+(t)=U_-(t)S \tag{1.9}$$

the operator $U_-(t)$ is evidently the solution of the equations

$$i\hbar[\partial U_-(t)/\partial t]=H_1(t)U_-(t),\quad U_-(\infty)=1. \tag{1.10}$$

Furthermore,

$$U_-(-\infty)=S^{-1} \tag{1.11}$$

which is the operator generating the initial state vector from the final state vector.

The differential equation for $U_+(t)$ can be replaced by the integral equation

$$\begin{aligned} U_+(t)&=1-(i/\hbar)\int_{-\infty}^{t}H_1(t')U_+(t')dt' \\ &=1-(i/\hbar)\int_{-\infty}^{\infty}\eta(t-t')H_1(t')U_+(t')dt' \end{aligned} \tag{1.12}$$

which incorporates the boundary condition (1.7). Here

$$\begin{aligned} \eta(t-t')&=1;\quad t>t'. \\ &=0;\quad t<t'. \end{aligned} \tag{1.13}$$

Similarly, $U_-(t)$ obeys the integral equation

$$\begin{aligned} U_-(t)&=1+(i/\hbar)\int_{t}^{\infty}H_1(t')U_-(t')dt' \\ &=1+(i\hbar/)\int_{-\infty}^{\infty}dt'H_1(t')U_-(t')\eta(t'-t). \end{aligned} \tag{1.14}$$

By considering the limit as $t\rightarrow\infty$ in (1.12) and $t\rightarrow-\infty$ in (1.14), we obtain

$$S=1-(i/\hbar)\int_{-\infty}^{\infty}H_1(t)U_+(t)dt \tag{1.15}$$

and

$$S^{-1}=1+(i/\hbar)\int_{-\infty}^{\infty}H_1(t)U_-(t)dt \qquad (1.16)$$

which are, of course, connected by (1.9).

The differential and integral equations characterizing $U_+(t)$ and $U_-(t)$ will now be replaced by equivalent variational principles from which the fundamental equations are obtained as conditions expressing the stationary property of a suitable expression. Furthermore, the stationary value of this quantity is just S, the collision operator. Hence the variational formulation of the problem also yields a practical means of approximate calculation, since errors in the construction of S will be minimized by employing a stationary expression.

We first consider

$$`S'=U_+(\infty)-\int_{-\infty}^{\infty}U_-^+(t)\left(\frac{\partial}{\partial t}+\frac{i}{\hbar}H_1(t)\right)U_+(t)dt, \qquad (1.17)$$

which is regarded as a function of the operator $U_+(t)$, subject only to the restriction (1.7), and of the Hermitian conjugate of the arbitrary operator $U_-(t)$. The change induced in $`S'$ by small, independent, variations of U_+ and U_- is

$$\delta`S'=(1-U_-(\infty))^+\delta U_+(\infty)-\int_{-\infty}^{\infty}\delta U_-^+(t)\left(\frac{\partial}{\partial t}+\frac{i}{\hbar}H_1(t)\right)U_+(t)dt+\int_{-\infty}^{\infty}\left[\left(\frac{\partial}{\partial t}+\frac{i}{\hbar}H_1(t)\right)U_-(t)\right]^+\delta U_+(t)dt. \qquad (1.18)$$

The requirement that $`S'$ be stationary with respect to arbitrary variations of $U_+$ and $U_-$, apart from the restriction (1.7), thus leads to the differential equations (1.6), (1.10) and the boundary condition (1.10) for $U_-(t)$. It is also evident from (1.17) that the stationary value of $`S'$ is the collision operator S, according to (1.5). A somewhat more symmetrical version of (1.17) is

$$S'=\tfrac{1}{2}(U_+(\infty)+U_-^+(-\infty))-\int_{-\infty}^{\infty}\left[\frac{1}{2}U_-^+(t)\frac{\partial U_+(t)}{\partial t}-\frac{1}{2}\frac{\partial U_-^+(t)}{\partial t}U_+(t)+\frac{i}{\hbar}U_-^+(t)H_1(t)U_+(t)\right]dt \qquad (1.19)$$

subject to the restrictions

$$U_+(-\infty)=U_-(\infty)=1. \qquad (1.20)$$

It is easily verified that $`S'$ is stationary with respect to variations of $U_+$ and $U_-$ about the solutions of the differential equations (1.6) and (1.10), subject to the boundary conditions (1.20), and that the stationary value of $`S'$ is S.

A variational basis for the integral equations (1.12) and (1.14) is provided by the expression

$$`S'=1-\frac{i}{\hbar}\int_{-\infty}^{\infty}[U_-^+(t)H_1(t)+H_1(t)U_+(t)]dt+\frac{i}{\hbar}\int_{-\infty}^{\infty}U_-^+(t)H_1(t)U_+(t)dt+\left(\frac{i}{\hbar}\right)^2\int_{-\infty}^{\infty}\int_{-\infty}^{\infty}U_-^+(t)H_1(t)\eta(t-t')\times H_1(t')U_+(t')dtdt'. \qquad (1.21)$$

Thus,

$$\delta`S'=\frac{i}{\hbar}\int_{-\infty}^{\infty}dt\delta U_-^+(t)H_1(t)\left[U_+(t)-1+\frac{i}{\hbar}\int_{-\infty}^{\infty}\eta(t-t')H_1(t')U_+(t')dt'\right]+\frac{i}{\hbar}\int_{-\infty}^{\infty}dt\left[U_-(t)-1-\frac{i}{\hbar}\int_{-\infty}^{\infty}dt'H_1(t')\times U_-(t')\eta(t'-t)\right]^+H_1(t)\delta U_+(t), \qquad (1.22)$$

which is indeed zero if U_+ and U_- satisfy their defining integral equations. It is also evident that the stationary value of $`S'$ is just the collision operator, in the form (1.15).

This variational principle differs from (1.17), or (1.19), in that no restrictions are imposed on U_+ and U_-, and that every integral contains the interaction operator H_1. The latter property implies that an adequate approximation to U_+ and U_- is required only during the actual process of interaction. Furthermore, the second type of variational principle will yield more accurate results than the first if the same approximate operators U_+ and U_- are employed. This is indicated by the results of inserting the simple but crude approximation

$$U_+(t)=U_-(t)=1 \qquad (1.23)$$

in (1.17) and (1.21). The former yields

$$S\simeq 1-(i/\hbar)\int_{-\infty}^{\infty}H_1(t)dt \qquad (1.24)$$

which is equivalent to the first Born approximation, while (1.21) gives

$$S\simeq 1-(i/\hbar)\int_{-\infty}^{\infty}H_1(t)dt+(i/\hbar)^2\int_{-\infty}^{\infty}\int_{-\infty}^{\infty}H_1(t)\times\eta(t-t')H_1(t')dtdt' \qquad (1.25)$$

the second Born approximation.

These approximate expressions for S illustrate a disadvantage of the variational principles thus far discussed; the unitary property is not guaranteed for an inexact S. It follows from (1.24), for example, that

$$S^+S\simeq 1+(1/\hbar^2)\left(\int_{-\infty}^{\infty} H_1(t)dt\right)^2. \quad (1.26)$$

A version of the theory that meets this objection is obtained on replacing the unitary operators $U_+(t)$ and $U_-(t)$ by

$$V(t)=U_+(t)2/(1+S)=U_-(t)2/(1+S^{-1}). \quad (1.27)$$

Note that

$$V(-\infty)=2/(1+S); \quad V(\infty)=2/(1+S^{-1})=2S/(1+S), \quad (1.28)$$

whence

$$\tfrac{1}{2}(V(\infty)+V(-\infty))=1 \quad (1.29)$$

and

$$V(\infty)=V^+(-\infty). \quad (1.30)$$

The property (1.29) leads us to write

$$V(\infty)=1-\tfrac{1}{2}iK; \quad V(-\infty)=1+\tfrac{1}{2}iK \quad (1.31)$$

while (1.30) supplies the information

$$K^+=K \quad (1.32)$$

the so-called reaction operator K is Hermitian. On remarking that

$$S=V(\infty)/V(-\infty) \quad (1.33)$$

we obtain

$$S=(1-\tfrac{1}{2}iK)/(1+\tfrac{1}{2}iK) \quad (1.34)$$

which represents the unitary S in terms of the Hermitian K. We shall now construct a variational principle for K in which the Hermitian property is assured.

Consider the operator 'K', defined by

$$\begin{aligned}'K' = &-\frac{i}{2}\int_{-\infty}^{\infty}\left(V^+(t)\frac{\partial V(t)}{\partial t}-\frac{\partial V^+(t)}{\partial t}V(t)\right)dt \\ &+\frac{1}{\hbar}\int_{-\infty}^{\infty}V^+(t)H_1(t)V(t)dt \\ &+\frac{i}{2}[(V(\infty)-V(-\infty)) \\ &-(V^+(\infty)-V^+(-\infty))] \end{aligned} \quad (1.35)$$

which is evidently Hermitian for arbitrary $V(t)$. The effect of a small variation in $V(t)$ and $V^+(t)$ is indicated by

$$\begin{aligned}\delta'K' = -\frac{1}{\hbar}\int_{-\infty}^{\infty}\Big[&\delta V^+(t)\left(i\hbar\frac{\partial}{\partial t}-H_1(t)\right)V(t) \\ &+\left[\left(i\hbar\frac{\partial}{\partial t}-H_1(t)\right)V(t)\right]^+\delta V(t)\Big]dt \\ &-\frac{i}{2}\Big[(V^+(\infty)-V^+(-\infty))\delta\left(\frac{V(\infty)+V(-\infty)}{2}\right) \\ &+\left(\frac{V^+(\infty)+V^+(-\infty)}{2}-1\right)\delta(V(\infty)-V(-\infty)) \\ &-\delta\left(\frac{V^+(\infty)+V^+(-\infty)}{2}\right)(V(\infty)-V(-\infty)) \\ &-\delta(V^+(\infty)-V^+(-\infty)) \\ &\times\left(\frac{V(\infty)+V(-\infty)}{2}-1\right)\Big]. \end{aligned} \quad (1.36)$$

If, therefore, $V(t)$ is restricted by the mixed boundary condition (1.29), 'K' is stationary with respect to variations about the solution of the differential equation

$$\left(i\hbar\frac{\partial}{\partial t}-H_1(t)\right)V(t)=0 \quad (1.37)$$

and the stationary value of 'K' equals K, according to (1.31) and (1.32).

The integral equation satisfied by $V(t)$ can be constructed from that obeyed by $U_+(t)$, or directly in the following manner. On integrating the differential equation (1.37) from $-\infty$ to t, and from ∞ to t, we obtain

$$\begin{aligned}V(t)&=V(-\infty)-\frac{i}{\hbar}\int_{-\infty}^{t}H_1(t')V(t')dt', \\ V(t)&=V(\infty)+\frac{i}{\hbar}\int_{t}^{\infty}H_1(t')V(t')dt'\end{aligned} \quad (1.38)$$

The addition of these equations yields, in consequence of the boundary condition (1.29),

$$V(t)=1-\frac{i}{2\hbar}\int_{-\infty}^{\infty}\epsilon(t-t')H_1(t')V(t')dt' \quad (1.39)$$

where

$$\begin{aligned}\epsilon(t-t')&=1; \quad t>t'. \\ &=-1; \quad t<t'.\end{aligned} \quad (1.40)$$

Conversely, the differential equation and boundary condition obeyed by $V(t)$ can be deduced from the integral equation. Note also that

$$K=i(V(\infty)-V(-\infty))=\frac{1}{\hbar}\int_{-\infty}^{\infty}H_1(t)V(t)dt. \quad (1.41)$$

A variational principal formulation of this integral equation is provided by the expression

$$
\begin{aligned}
{}^{\backprime}K' = & \frac{1}{\hbar}\int_{-\infty}^{\infty}(H_1(t)V(t)+V^+(t)H_1(t))dt \\
& -\frac{1}{\hbar}\int_{-\infty}^{\infty}V^+(t)H_1(t)V(t)dt \\
& -\frac{i}{2\hbar^2}\int_{-\infty}^{\infty}\int_{-\infty}^{\infty}V^+(t)H_1(t)\epsilon(t-t')H_1(t') \\
& \qquad\qquad \times V(t')dtdt' \qquad (1.42)
\end{aligned}
$$

which is obviously Hermitian for arbitrary $V(t)$. Now

$$
\begin{aligned}
\delta^{\backprime}K' = & -\frac{1}{\hbar}\int_{-\infty}^{\infty}\delta V^+(t)H_1(t)\Big[V(t)-1 \\
& +\frac{i}{2\hbar}\int_{-\infty}^{\infty}\epsilon(t-t')H_1(t')V(t')dt'\Big]dt \\
& -\frac{1}{\hbar}\int_{-\infty}^{\infty}\Big[V(t)-1+\frac{i}{2\hbar}\int_{-\infty}^{\infty}\epsilon(t-t')H_1(t') \\
& \qquad \times V(t')dt'\Big]+H_1(t)\delta V(t)dt \qquad (1.43)
\end{aligned}
$$

which is indeed zero if $V(t)$ satisfies the integral equation (1.39). Furthermore, the stationary value of 'K' is just (1.41), the correct reactor operator.

The abstract theory thus far developed can be made more explicit by introducing eigenfunctions, Φ_a, for the separated parts of the system, which will describe the initial and final states. Thus, since $S\Phi_a$ is the final state that emergies from the initial state Φ_a, the probability that the system will be found eventually in the particular state Φ_b, is

$$W_{ba}=|(\Phi_b, S\Phi_a)|^2=|S_{ba}|^2. \qquad (1.44)$$

It is slightly more convenient to deal with the operator

$$T=S-1, \qquad (1.45)$$

which generates the change in the state vector produced by the interaction process. The unitary property of S implies that

$$T^+T=-(T+T^+) \qquad (1.46)$$

and the probability that the system will be found in a particular final state differing from the initial one is

$$b\neq a;\quad W_{ba}=|T_{ba}|^2. \qquad (1.47)$$

Now, according to (1.15),

$$
\begin{aligned}
T_{ba} &= -(i/\hbar)\int_{-\infty}^{\infty}dt(\Phi_b, H_1(t)U_+(t)\Phi_a), \\
&= -(i/\hbar)\int_{-\infty}^{\infty}dt(\Phi_b, \exp(iH_0t/\hbar)H_1 \\
&\qquad \times\exp(-iH_0t/\hbar)U_+(t)\Phi_a). \qquad (1.48)
\end{aligned}
$$

It should be noted that Φ_b cannot be an exact eigenfunction of H_0, since a superposition of momentum states (wave packet) is required to produce the spatial localizability involved in the definite separation of the two parts of the system. An equivalent description is obtained, however, by introducing eigenfunctions of H_0,

$$H_0\Phi_b=E_b\Phi_b \qquad (1.49)$$

and simulating the cessation of interaction, arising from the separation of the component parts of the system, by an adiabatic decrease in the interaction strength as $t\rightarrow\pm\infty$. The latter can be represented by the factor $\exp(-\epsilon|t|/\hbar)$ where ϵ is arbitrarily small. Accordingly, (1.48) becomes

$$T_{ba}=-(i/\hbar)(\Phi_b, H_1\Psi_a^{(+)}(E_b)) \qquad (1.50)$$

where

$$
\begin{aligned}
\Psi_a^{(+)}(E) = & \int_{-\infty}^{\infty}dt\exp(i(E-H_0)t/\hbar) \\
& \times\exp(-\epsilon|t|/\hbar)U_+(t)\Phi_a. \qquad (1.51)
\end{aligned}
$$

Formula (1.16) for $S^{-1}-1=T^+$ leads, in a similar way, to

$$(T^+)_{ba}=(i/\hbar)(\Phi_b, H_1\Psi_a^{(-)}(E_b)) \qquad (1.52)$$

or equivalently,

$$T_{ab}=-(i/\hbar)(\Psi_a^{(-)}(E_b), H_1\Phi_b) \qquad (1.53)$$

in which

$$
\begin{aligned}
\Psi_a^{(-)}(E) = & \int_{-\infty}^{\infty}dt\exp(i(E-H_0)t/\hbar) \\
& \times\exp(-\epsilon|t|/\hbar)U_-(t)\Phi_a. \qquad (1.54)
\end{aligned}
$$

Determining equations for $\Psi_a^{(+)}(E)$ and $\Psi_a^{(-)}(E)$ can be obtained from (1.12) and (1.14), the integral equations for $U_+(t)$ and $U_-(t)$. Thus

$$
\begin{aligned}
\Psi_a^{(+)}(E) = & \int_{-\infty}^{\infty}dt\exp(i(E-E_a)t/\hbar)\exp(-\epsilon|t|/\hbar)\Phi_a \\
& -(i/\hbar)\int_0^{\infty}d\tau\exp(i(E-H_0)\tau/\hbar) \\
& \times\exp(-\epsilon\tau/\hbar)H_1\Psi_a^{(+)}(E) \qquad (1.55)
\end{aligned}
$$

and

$$
\begin{aligned}
\Psi_a^{(-)}(E) = & \int_{-\infty}^{\infty}dt\exp(i(E-E_a)t/\hbar)\exp(-\epsilon|t|/\hbar)\Phi_a \\
& +(i/\hbar)\int_0^{\infty}d\tau\exp(-i(E-H_0)\tau/\hbar) \\
& \times\exp(-\epsilon\tau/\hbar)H_1\Psi_a^{(-)}(E), \qquad (1.56)
\end{aligned}
$$

where $\tau=|t-t'|$. Now

$$\mp\frac{i}{\hbar}\int_0^\infty d\tau\exp(\pm i(E-H_0)\tau/\hbar)\exp(-\epsilon\tau/\hbar)$$

$$=\frac{1}{E\pm i\epsilon-H_0}=\frac{E-H_0}{(E-H_0)^2+\epsilon^2}\mp i\frac{\epsilon}{(E-H_0)^2+\epsilon^2}$$

$$=P\frac{1}{E-H_0}\mp i\pi\delta(E-H_0). \quad (1.57)$$

The last expression is a symbolic statement of the following integral properties possessed by the real and imaginary parts of (1.57) in the limit as $\epsilon\to 0$.

$$\lim_{\epsilon\to 0}\int_{-\infty}^{\infty}\frac{x}{x^2+\epsilon^2}f(x)dx=P\int_{-\infty}^{\infty}\frac{f(x)}{x}dx,$$
$$\lim_{\epsilon\to 0}\frac{1}{\pi}\int_{-\infty}^{\infty}\frac{\epsilon}{x^2+\epsilon^2}f(x)dx=f(0), \quad (1.58)$$

where P denotes the principal part of the integral and $f(x)$ is an arbitrary function. Therefore

$$\Psi_a^{(\pm)}(E)=2\pi\hbar\delta(E-E_a)\Phi_a+\frac{1}{E\pm i\epsilon-H_0}H_1\Psi_a^{(\pm)}(E) \quad (1.59)$$

and, on writing

$$\Psi_a^{(\pm)}(E)=2\pi\hbar\delta(E-E_a)\Psi_a^{(\pm)} \quad (1.60)$$

we obtain

$$\Psi_a^{(\pm)}=\Phi_a+\frac{1}{E_a\pm i\epsilon-H_0}H_1\Psi_a^{(\pm)}. \quad (1.61)$$

These equations provide a time-independent formulation of the scattering problem, in which the small positive or negative imaginary addition to the energy serves to select, automatically, outgoing or incoming scattered waves.

A matrix element of the operator T can now be expressed as

$$T_{ba}=-2\pi i\delta(E_a-E_b)\mathrm{T}_{ba} \quad (1.62)$$

where

$$\mathrm{T}_{ba}=(\Phi_b,H_1\Psi_a^{(+)})=(\Psi_b^{(-)},H_1\Phi_a) \quad (1.63)$$

are equivalent forms for an element of the association matrix T, which is defined only for states of equal energy. The resulting formula for the transition probability,

$$W_{ba}=4\pi^2[\delta(E_a-E_b)]^2|\mathrm{T}_{ba}|^2 \quad (1.64)$$

is to be interpreted by replacing one factor, $\delta(E_a-E_b)$, by its defining time integral

$$\delta(E_a-E_b)=\frac{1}{2\pi\hbar}\int_{-\infty}^{\infty}\exp(i/(E_a-E_b)t/\hbar)\times\exp(-\epsilon|t|/\hbar)dt;\quad \epsilon\to 0 \quad (1.65)$$

in which E_a-E_b must be placed equal to zero, in view of the second delta-function factor. The expression thus obtained

$$W_{ba}=\frac{2\pi}{\hbar}\delta(E_a-E_b)|\mathrm{T}_{ba}|^2\int_{-\infty}^{\infty}dt, \quad (1.66)$$

evidently describes the fact that transitions occur only between states of equal energy for the separated system, and with an intensity proportional to the total time of effective interaction. In the idealized limit $\epsilon\to 0$, the latter is infinitely large. However, we infer from (1.66) that the rate at which the transition probability increases is

$$w_{ba}=(2\pi/\hbar)\delta(E_a-E_b)|\mathrm{T}_{ba}|^2. \quad (1.67)$$

A somewhat more satisfactory derivation of this result follows from the evaluation of

$$w_{ba}=\frac{\partial}{\partial t}|(\Phi_b,U_+(t)\Phi_a)|^2 \quad (1.68)$$

which expresses the increase, per unit time, of the probability that the system, known to be initially in the state a, will be found at time t in the state b. Now

$$w_{ba}=\frac{i}{\hbar}(H_1(t)U_+(t)\Phi_a,\Phi_b)\times(\Phi_b,U_+(t)\Phi_a)+\text{complex conjugate}$$
$$=\frac{1}{\hbar^2}\int_{-\infty}^{t}dt'(\exp(i/(E_b-H_0)t/\hbar)U_+(t)\Phi_a,H_1\Phi_b)\times(\Phi_b,H_1\exp(i(E_b-H_0)t'/\hbar)\times U_+(t')\Phi_a)+\text{c.c.} \quad (1.69)$$

in which we have employed (1.12), and assumed that $b\neq a$. This can be simplified by noting that (1.51) and (1.60),

$$\int_{-\infty}^{\infty}dt\exp(i(E-H_0)t/\hbar)\exp(-\epsilon|t|/\hbar)U_+(t)\Phi_a=2\pi\hbar\delta(E-E_a)\Psi_a^{(+)} \quad (1.70)$$

imply that

$$\exp(-iH_0t/\hbar)U_+(t)\Phi_a=\exp(-iE_at/\hbar)\Psi_a^{(+)} \quad (1.71)$$

which is just the state vector, in the Schrödinger representation, of our idealized stationary state. Hence

$$w_{ba}=\frac{1}{\hbar^2}|\mathrm{T}_{ba}|^2\int_{-\infty}^{t}\exp(i(E_a-E_b)(t-t')/\hbar)dt'+\text{c.c.}$$
$$=\frac{2\pi}{\hbar}|\mathrm{T}_{ba}|^2\delta(E_a-E_b). \quad (1.72)$$

A simple expression for the total rate of transition from the initial state follows from the general property of the operator T contained in (1.46). On writing a matrix element of this operator relation and substituting (1.62), we obtain

$$4\pi^2 \sum_b \delta(E_a-E_b)\mathrm{T}_{ba}{}^*\delta(E_b-E_c)\mathrm{T}_{bc} = 2\pi i\delta(E_a-E_c)(\mathrm{T}_{ac}-\mathrm{T}_{ca}{}^*). \quad (1.73)$$

The factor $\delta(E_a-E_c)$ can be canceled and (1.73) then yields, for the special situation, $c=a$,

$$4\pi^2 \sum_b \delta(E_a-E_b)|\mathrm{T}_{ba}|^2 = -4\pi Im(\mathrm{T}_{aa}) \quad (1.74)$$

or

$$\sum_b w_{ba} = -(2/\hbar)\, Im(\mathrm{T}_{aa}). \quad (1.75)$$

The left side of this formula is not exactly the total rate of transition out of the state a, since $b=a$ is included in the summation. However, a single state makes no contribution to such a summation; a group of states is required. A relation of the type (75) is characteristic of a wave theory, in which the reduction in intensity of a plane wave passing through a scattering medium is accounted for by destructive interference between the original wave and the secondary waves scattered in the direction of propagation.

A variational formulation of Eq. (1.61) by means of a stationary expression for T_{ba} can be obtained from the variational principle (1.21). A matrix element of this operator equation reads

$$\begin{aligned}
{}^{\backprime}T'_{ba} = &-\frac{i}{\hbar}\int_{-\infty}^{\infty} dt[(\exp(i(E_a-H_0)t/\hbar)U_-(t)\Phi_b, H_1\Phi_a) \\
&+(\Phi_b, H_1\exp(i(E_b-H_0)t/\hbar)U_+(t)\Phi_a)] \\
&+\frac{i}{\hbar}\int_{-\infty}^{\infty} dt(\exp(-iH_0t/\hbar)U_-(t)\Phi_b, \\
&\qquad\times H_1\exp(-iH_0t/\hbar)U_+(t)\Phi_a) \\
&+\left(\frac{i}{\hbar}\right)^2\int_{-\infty}^{\infty} dt\int_{-\infty}^{t} dt'(\exp(-iH_0t/\hbar)U_-(t)\Phi_b, \\
&\times H_1\exp(-iH_0(t-t')/\hbar) \\
&\qquad\times H_1\exp(-iH_0t'/\hbar)U_+(t')\Phi_a)
\end{aligned} \quad (1.76)$$

in which the adiabatic reduction of H_1 for large $|t|$ has not been indicated explicitly. We now restrict ourselves to the class of stationary states, according to the assumption

$$\exp(-iH_0t/\hbar)U_\pm(t)\Phi_a = \exp(-iE_at/\hbar)\Psi_a{}^{(\pm)}. \quad (1.77)$$

The result of performing the time integrations is expressed by

$$\begin{aligned}
{}^{\backprime}\mathrm{T}'_{ba} = &(\Psi_b{}^{(-)}, H_1\Phi_a)+(\Phi_b, H_1\Psi_a{}^{(+)}) \\
&-(\Psi_b{}^{(-)}, H_1\Psi_a{}^{(+)}) \\
&+\left(\Psi_b{}^{(-)}, H_1\frac{1}{E+i\epsilon-H_0}H_1\Psi_a{}^{(+)}\right),
\end{aligned} \quad (1.78)$$

where E is the common energy of states a and b. We shall verify directly that (1.78) has the required properties. Thus

$$\begin{aligned}
\delta{}^{\backprime}\mathrm{T}'_{ba} = &\left(\delta\Psi_b{}^{(-)}, H_1\left(\Phi_a+\frac{1}{E+i\epsilon-H_0}H_1\Psi_a{}^{(+)}\right.\right. \\
&\left.\left.-\Psi_a{}^{(+)}\right)\right)+\left(\left(\Phi_b+\frac{1}{E-i\epsilon-H_0}H_1\Psi_b{}^{(-)}\right.\right. \\
&\left.\left.-\Psi_b{}^{(-)}\right), H_1\delta\Psi_a{}^{(+)}\right)
\end{aligned} \quad (1.79)$$

which is indeed zero for variations about the solutions of (1.61). Furthermore, it is a consequence of the latter equations that

$$\begin{aligned}
(\Psi_b{}^{(-)}, H_1\Psi_a{}^{(+)}) &- \left(\Psi_b{}^{(-)}, H_1\frac{1}{E+i\epsilon-H_0}H_1\Psi_a{}^{(+)}\right) \\
&= (\Psi_b{}^{(-)}, H_1\Phi_a) = (\Phi_b, H_1\Psi_a{}^{(+)})
\end{aligned} \quad (1.80)$$

so that the stationary value of ${}^{\backprime}\mathrm{T}'_{ba}$ is T_{ba}, according to (1.63).

A similar theory can be developed for the matrix elements of the operator K. It is easily shown that

$$K_{ba} = 2\pi\delta(E_a-E_b)\mathrm{K}_{ba}, \quad (1.81)$$

where

$$\mathrm{K}_{ba} = (\Phi_b, H_1\Psi_a{}^{(1)}) = (\Psi_b{}^{(1)}, H_1\Phi_a). \quad (1.82)$$

The time-independent state vector $\Psi_a{}^{(1)}$ describes a stationary state, according to the relation

$$\exp(-iH_0t/\hbar)V(t)\Phi_a = \exp(-iE_at/\hbar)\Psi_a{}^{(1)} \quad (1.83)$$

and obeys the equation

$$\Psi_a{}^{(1)} = \Phi_a + P\left(\frac{1}{E_a-H_0}\right)H_1\Psi_a{}^{(1)}. \quad (1.84)$$

A variational basis for (1.82) and (1.84) is provided by

$$\begin{aligned}
{}^{\backprime}K'_{ba} = {}^{\backprime}K'_{ab}{}^* = &(\Psi_b{}^{(1)}, H_1\Phi_a) \\
&+(\Phi_b, H_1\Psi_a{}^{(1)})-(\Psi_b{}^{(1)}, H_1\Psi_a{}^{(1)}) \\
&+\left(\Psi_b{}^{(1)}, H_1P\left(\frac{1}{E-H_0}\right)H_1\Psi_a{}^{(1)}\right).
\end{aligned} \quad (1.85)$$

The connection between the matrices $\mathbf{T}$ and $\mathbf{K}$ is obtained from

$$T = S-1 = -iK/(1+\tfrac{1}{2}iK) \quad (1.86)$$

on rewriting the latter as

$$T+\tfrac{1}{2}iKT = -iK. \quad (1.87)$$

Non-vanishing matrix elements of this operator relation are restricted to states of equal energy, according to

(1.62) and (1.81), whence

$$\mathbf{T}_{ba}+i\pi\sum_c \mathbf{K}_{bc}\delta(E_c-E)\mathbf{T}_{ca}=\mathbf{K}_{ba}, \quad (1.88)$$

where E is the common energy states a and b. An effective way to solve this equation is to construct the eigenfunctions of $\mathbf{K}$, which are defined by the eigenvalue equation

$$\sum_a \mathbf{K}_{ba}\delta(E_a-E)f_{aA}=\mathbf{K}_A f_{bA}. \quad (1.89)$$

Since $\mathbf{K}$ is an Hermitian matrix, the eigenvalues $\mathbf{K}_A$ are real, the eigenfunctions f_{aA} are orthogonal, and may be normalized according to

$$\sum_a f_{aA}^*\delta(E_a-E)f_{aB}=\delta_{AB}. \quad (1.90)$$

The matrix elements of $\mathbf{K}$ can be exhibited in terms of the eigenfunctions and eigenvalues of $\mathbf{K}$

$$\mathbf{K}_{ba}=\sum_A f_{bA}\mathbf{K}_A f_{aA}^*. \quad (1.91)$$

Equation (1.88) for $\mathbf{T}$ will then be satisfied by

$$\mathbf{T}_{ba}=\sum_A f_{bA}\mathbf{T}_A f_{aA}^*, \quad (1.92)$$

where

$$\mathbf{T}_A+i\pi\mathbf{K}_A\mathbf{T}_A=\mathbf{K}_A \quad (1.93)$$

or

$$\mathbf{T}_A=\mathbf{K}_A/(1+i\pi\mathbf{K}_A). \quad (1.94)$$

This is only to say that $\mathbf{T}$ is a function of $\mathbf{K}$ and therefore possesses the same eigenfunctions, while its eigenvalues are determined by those of $\mathbf{K}$. These eigenvalues can be conveniently expressed by introducing the real angles δ_A, according to

$$\mathbf{K}_A=-(1/\pi)\tan\delta_A \quad (1.95)$$

so that

$$\mathbf{T}_A=-(1/\pi)\sin\delta_A e^{i\delta_A}. \quad (1.96)$$

The resulting expression for the transition probability per unit time is

$$w_{ba}=(2/\pi\hbar)|\sum_A \sin\delta_A e^{i\delta_A}f_{bA}f_{aA}^*|^2\delta(E_a-E_b) \quad (1.97)$$

and the total probability per unit time for transitions from a particular state is given by

$$\sum_b w_{ba}=(2/\pi\hbar)\sum_A \sin^2\delta_A|f_{aA}|^2 \quad (1.98)$$

according to (1.97) or (1.75). Finally, the sum of the total transition probability per unit time over all initial states of the same energy is expressed by

$$\sum_{b,a} w_{ba}\delta(E_a-E)=(2/\pi\hbar)\sum_A \sin^2\delta_A. \quad (1.99)$$

These results are generalizations of familiar formulas obtained in the conventional phase shift analysis of the scattering of a particle by a central field of force. In the latter situation, the eigenfunctions of $\mathbf{K}$ are evident from symmetry considerations, namely the invariance of $\mathbf{K}_{ba}$ under a simultaneous rotation of $\mathbf{k}_a$ and $\mathbf{k}_b$, the propagation vectors that define the initial and final states. It may be inferred that the f_{aA} are spherical harmonics, considered as a function of the angles that define the direction of $\mathbf{k}_a$,

$$f_{aA}=CY_l^m(\mathbf{k}_a);\quad A\equiv l,m, \quad (1.100)$$

and that the eigenvalues of $\mathbf{K}$ depend only upon the order of the spherical harmonics, i.e., $\delta_A\equiv\delta_l$. The constant C is fixed by the normalization convention contained in (1.90), which now reads,

$$|C|^2\int Y_l^{m*}(\mathbf{k})Y_{l'}^{m'}(\mathbf{k})\rho d\Omega=\delta_{ll'}\delta_{mm'}. \quad (1.101)$$

Here $\rho d\Omega$ is the number of states, per unit energy range, associated with motion within the solid angle $d\Omega$. This occurs as a weight factor in a summation over states with equal energy, replacing the summation over all states as restricted by the factor $\delta(E_a-E)$. Explicitly,

$$\rho=\frac{p^2dp}{8\pi^3\hbar^3dE}=\frac{1}{8\pi^3\hbar}\frac{k^2}{v}, \quad (1.102)$$

if we consider a unit spatial volume. The second form in (1.102) expresses ρ in terms of the wave number k and v, the speed of the particle. With spherical harmonics that are normalized on a unit sphere (1.101) requires that

$$|C|^2=1/\rho=8\pi^3\hbar v/k^2. \quad (1.103)$$

We may now compute from (1.97) the probability, per unit time, that the particle is scattered from the direction of $\mathbf{k}_a$ into the solid angle $d\Omega$ around the direction of $\mathbf{k}_b$,

$$w=(2/\pi\hbar)|\sum_{l,m}\sin\delta_l e^{i\delta_l}|C|^2Y_l^m(\mathbf{k}_b)Y_l^{m*}(\mathbf{k}_a)|^2\rho d\Omega. \quad (1.104)$$

We then obtain the well-known expression of the differential cross section for scattering through an angle ϑ,

$$d\sigma(\vartheta)=(1/k^2)|\sum_l(2l+1)\sin\delta_l e^{i\delta_l}P_l(\cos\vartheta)|^2d\Omega \quad (1.105)$$

on dividing w by v, which measures the flux of incident particles, and employing the spherical harmonics addition theorem,

$$\sum_{m=-l}^{l} Y_l^m(\mathbf{k}_b)Y_l^{m*}(\mathbf{k}_a)=[(2l+1)/4\pi]P_l(\cos\vartheta), \quad (1.106)$$

where the Legendre polynomial $P_l(\cos\vartheta)$ is a function of ϑ, the angle between $\mathbf{k}_a$ and $\mathbf{k}_b$. The total scattering cross section is obtained from (1.98),

$$\begin{aligned}\sigma&=(2/\pi\hbar v)\sum_{l,m}\sin^2\delta_l|C|^2|Y_l^m(\mathbf{k}_a)|^2\\ &=(4\pi/k^2)\sum_l(2l+1)\sin^2\delta_l\end{aligned} \quad (1.107)$$

in consequence of

$$\sum_m |Y_l^m(\mathbf{k}_a)|^2 = (2l+1)/4\pi. \tag{1.108}$$

Since the total cross section is independent of the incidence direction, the same result follows immediately from (1.99).

We consider finally, the variational formulation of problems possessing the general character of the scattering by a central force field; namely, those in which the eigenfunctions of $\mathbf{K}$ are determined by symmetry considerations, and the basic question is to obtain the eigenvalues $\mathbf{K}_A$, or the phase angles δ_A. For this purpose, we notice that the inverse of (1.91) is

$$\sum_{b,a} f_{bB}^*\delta(E_b-E)\mathbf{K}_{ba}f_{aA}\delta(E_a-E) = \mathbf{K}_A\delta_{AB}. \tag{1.109}$$

On introducing the state vectors

$$\sum_a \Phi_a f_{aA}\delta(E_a-E) = \Phi_A, \tag{1.110}$$

and

$$\sum_a \Psi_a^{(1)} f_{aA}\delta(E_a-E) = \Psi_A^{(1)}, \tag{1.111}$$

the variational principle (1.85) becomes

$$\begin{aligned}-\frac{1}{\pi}\tan'\delta'_A\delta_{AB} &= (\Psi_B^{(1)}, H_1\Phi_A) \\ &\quad+(\Phi_B, H_1\Psi_A^{(1)})-(\Psi_B^{(1)}, H_1\Psi_A^{(1)}) \\ &\quad+\left(\Psi_B^{(1)}, H_1P\left(\frac{1}{E-H_0}\right)H_1\Psi_A^{(1)}\right).\end{aligned} \tag{1.112}$$

Note that Φ_A, or more exactly written $\Phi_{A,E}$, has the following orthogonality-normalization property:

$$\begin{aligned}(\Phi_{A,E}, \Phi_{B,E'}) &= \sum_a f_{aA}^*\delta(E_a-E)f_{AB}\delta(E_a-E') \\ &= \delta(E-E')\sum_a f_{aA}^*\delta(E_a-E)f_{aB} \\ &= \delta_{AB}\delta(E-E')\end{aligned} \tag{1.113}$$

and that the inverses of (1.110), (1.111) are

$$\Phi_a = \sum_A f_{aA}^*\Phi_A, \quad \Psi_a^{(1)} = \sum_A f_{aA}^*\Psi_A^{(1)} \tag{1.114}$$

which are expansions of these state vectors in eigenvectors of $\mathbf{K}$.

III. NEUTRON SCATTERING BY A BOUND PROTON

As an application of the variational methods discussed in the first section, we consider the scattering of slow neutrons by a proton bound in an otherwise inert molecule. If the momentum associated with the center of gravity of the whole system is assumed to be zero, the unperturbed Hamiltonian consists of two parts, one describing the relative motion of the neutron and the molecular center of gravity, the other being the Hamiltonian of the internal molecular motion,

$$H_0 = (\mathbf{p}_n^2/2\mu) + H_m. \tag{2.1}$$

Here

$$\mu = AM/(A+1) \tag{2.2}$$

is the reduced mass for relative motion of the neutron and molecule, while A is the molecular mass in units of M, the mass of the neutron. The perturbation is the neutron-proton interaction energy,

$$H_1 = V(\mathbf{r}_n - \mathbf{r}_p) \tag{2.3}$$

which also depends upon the spin operators of neutron and proton, σ_n and σ_p. The simplifying feature in this problem arises from the short range and large magnitude of the nuclear potential contrasted with the long range, weak molecular forces. The variational principle (1.78) requires a knowledge of the wave function representing the state vector only within the region of nuclear interaction, where the molecular force on the proton is negligible. Thus, the basic problem is the scattering of a neutron by a free proton, with essentially zero energy of relative motion. We therefore first consider some properties of the latter system.

The unperturbed Hamiltonian for a neutron and a free proton, in the system in which the center of gravity is at rest, is

$$\mathcal{H}_0 = \mathbf{p}^2/M, \tag{2.4}$$

where $\mathbf{p}$ is the relative momentum of the particles. If we temporarily omit the spin coordinates, the wave function φ, representing the unperturbed state vector Φ_a, is simply a constant in the limit of zero energy. This constant can be chosen as unity, corresponding to a unit spatial volume. The wave function representing the state vectors $\Psi_a^{(+)}$ and $\Psi_a^{(-)}$ will be denoted by $\psi(\mathbf{r})$. There is no distinction between outgoing and incoming waves in the limit of zero energy. Since the scattering is necessarily isotropic, $\mathbf{T}_{ba}$ is simply a constant, denoted by $\mathbf{t}$. According to (1.63), $\mathbf{t}$ is given by

$$\mathbf{t} = (\varphi, V\psi) = \int V(\mathbf{r})\psi(\mathbf{r})d\mathbf{r}, \tag{2.5}$$

where ψ obeys the integral equation (1.61).

$$\psi + (1/\mathcal{H}_0)V\psi = \varphi. \tag{2.6}$$

The connection between $\mathbf{t}$ and the S phase shift is obtained from (1.92) and (1.96),

$$\mathbf{t} = -|f|^2ka/\pi, \tag{2.7}$$

in which we have employed the zero energy limiting form,

$$\sin\delta \to ka; \quad k\to 0 \tag{2.8}$$

thereby introducing the scattering amplitude a. The constant f is fixed by the normalization condition (1.90),

$$|f|^2 4\pi\rho = 1 \tag{2.9}$$

where, (1.102),

$$4\pi\rho = k^2/2\pi^2\hbar v = kM/4\pi^2\hbar^2. \tag{2.10}$$

The second form of (2.10) follows from $\hbar k = \frac{1}{2}Mv$, the relation between the relative momentum and the relative velocity. Finally,

$$\mathbf{t} = -4\pi\hbar^2 a/M. \tag{2.11}$$

If the neutron-proton interaction operator is spin-dependent, $\mathbf{t}$ must be replaced by a matrix in the spin quantum numbers. The eigenfunctions of this spin matrix are those of the triplet and singlet states of resultant spin angular momentum. The associated eigenvalues of $\mathbf{t}$ are related to the triplet and singlet scattering amplitudes,

$$\mathbf{t}_{1,0}=-4\pi\hbar^2 a_{1,0}/M. \tag{2.12}$$

As in (1.92), the matrix $\mathbf{t}$ can be constructed as a linear combination of its eigenvalues, multiplied by coefficients which are the matrix elements of projection operators for the corresponding eigenvalues. The projection operators for the triplet and singlet states are well known to be

$$P_1=\tfrac{1}{4}(3+\sigma_n\cdot\sigma_p),\quad P_0=\tfrac{1}{4}(1-\sigma_n\cdot\sigma_p). \tag{2.13}$$

Hence, to include spin dependent interactions it is sufficient to regard $\mathbf{t}$ in (2.11) as a spin operator, with

$$a=a_1P_1+a_0P_0=\tfrac{1}{4}(3a_1+a_0)+\tfrac{1}{4}(a_1-a_0)\sigma_n\cdot\sigma_p. \tag{2.14}$$

We shall now perform an approximate but highly accurate evaluation of $\mathbf{T}_{ba}$, which describes the scattering of a neutron by a bound proton. For this purpose, (1.78) is written

$$\begin{aligned}{}'\mathbf{T}'_{ba}=&(\Psi_b^{(-)},V\Phi_a)+(\Phi_b,V\Psi_a^{(+)})\\&-(\Psi_b^{(-)},V\Psi_a^{(+)})-\left(\Psi_b^{(-)},V\frac{1}{\mathcal{H}_0}V\Psi_a^{(+)}\right)\\&+\left(\Psi_b^{(-)},V\left(\frac{1}{E+i\epsilon-H_0}+\frac{1}{\mathcal{H}_0}\right)V\Psi_a^{(+)}\right),\end{aligned} \tag{2.15}$$

In treating the spin dependent interactions, it is convenient to suppress spin functions and thus regard $\mathbf{T}_{ba}$ as a spin operator. The first approximation to be introduced concerns the wave function representing the state vector Φ_a, say $\Phi_a(\mathbf{r}_n,\mathbf{r})$. Here $\mathbf{r}_n$ is the neutron coordinate relative to the molecular center of gravity, while $\mathbf{r}$ symbolizes the set of internal molecular coordinates, including $\mathbf{r}_p$, the proton position vector relative to the molecular center of gravity. This wave function, describing the independent motion of the neutron and molecule, will have the form

$$\Phi_a(\mathbf{r}_n,\mathbf{r})=\exp(i\mathbf{k}_a\cdot\mathbf{r}_n)\cdot\chi_a(\mathbf{r}) \tag{2.16}$$

in which $\chi_a(\mathbf{r})$ is an internal molecular wave function. Now $\Phi_a(\mathbf{r}_n,\mathbf{r})$ in (2.15), only occurs multiplied by the short range nuclear potential $V(\mathbf{r}_n-\mathbf{r}_p)$. We shall therefore replace $\Phi_a(\mathbf{r}_n,\mathbf{r})$ by

$$\Phi_a(\mathbf{r}_p,\mathbf{r})=F_a(\mathbf{r}). \tag{2.17}$$

The error thereby incurred is of the order $(kr_0)^2$, where r_0 is a measure of the nuclear force range. Since the influence of molecular binding is only of interest for slow neutrons, $(kr_0)^2\gtrsim10^{-6}$, and we need not introduce a correction to compensate for this replacement.

A second approximation involves the last term of (2.15), which is small in comparison with the other terms, since molecular energies are negligible in comparison with the practically equal kinetic energies of neutron and proton during the nuclear interaction process. If we initially ignore the last term of (2.15) the latter reads

$$\begin{aligned}{}'\mathbf{T}'_{ba}=&(\Psi_b^{(-)},VF_a)+(F_b,V\Psi_a^{(+)})\\&-(\Psi_b^{(-)},V\Psi_a^{(+)})-\left(\Psi_b^{(-)},V\frac{1}{\mathcal{H}_0}V\Psi_a^{(+)}\right).\end{aligned} \tag{2.18}$$

The condition that $'\mathbf{T}'_{ba}$ be stationary is that $\Psi_a^{(\pm)}$ satisfy the relation

$$\Psi_a^{(\pm)}+(1/\mathcal{H}_0)V\Psi_a^{(\pm)}=F_a(\mathbf{r}). \tag{2.19}$$

On comparison with (2.6), it is evident that

$$\Psi_a^{(\pm)}=\psi(\mathbf{r}_n-\mathbf{r}_p)F_a(\mathbf{r}) \tag{2.20}$$

and the stationary value of $'\mathbf{T}'_{ba}$, an approximation to the correct $\mathbf{T}_{ba}$, is given by

$$\begin{aligned}\mathbf{T}_{ba}\simeq&(F_b,V\Psi_a^{(+)})=(F_b,V\psi F_a)\\&=\mathbf{t}\int F_b^*(\mathbf{r})F_a(\mathbf{r})d\mathbf{r}\end{aligned} \tag{2.21}$$

according to (2.5). This result,

$$\mathbf{T}_{ba}\simeq-\frac{4\pi\hbar^2}{M}a\int\exp[i(\mathbf{k}_a-\mathbf{k}_b)\cdot\mathbf{r}_p]\cdot\times\chi_b^*(\mathbf{r})\chi_a(\mathbf{r})d\mathbf{r} \tag{2.22}$$

is the Fermi approximation.

To include the last term in (2.15), we observe that it may be written, in terms of wave functions, as

$$\begin{aligned}&\int\Psi_b^{(-)*}(\mathbf{r}_n,\mathbf{r})V(\mathbf{r}_n-\mathbf{r}_p)\\&\times\left(\mathbf{r}_n,\mathbf{r}\left|\frac{1}{E+i\epsilon-H_0}+\frac{1}{\mathcal{H}_0}\right|\mathbf{r}_n',\mathbf{r}'\right)V(\mathbf{r}_n'-\mathbf{r}_p')\\&\times\Psi_a^{(+)}(\mathbf{r}_n',\mathbf{r}')d\mathbf{r}_nd\mathbf{r}d\mathbf{r}_n'd\mathbf{r}'.\end{aligned} \tag{2.23}$$

We shall again introduce an approximation which exploits the short range of V in comparison with molecular dimensions, namely, the replacement of (2.23) by

$$\begin{aligned}&\int\Psi_b^{(-)*}(\mathbf{r}_n,\mathbf{r})V(\mathbf{r}_n-\mathbf{r}_p)K^{(+)}(\mathbf{r},\mathbf{r}')V(\mathbf{r}_n'-\mathbf{r}_p')\\&\times\Psi_a^{(+)}(\mathbf{r}_n',\mathbf{r}')d\mathbf{r}_nd\mathbf{r}d\mathbf{r}_n'd\mathbf{r}',\end{aligned} \tag{2.24}$$

where

$$K^{(\pm)}(\mathbf{r},\mathbf{r}')=\left(\mathbf{r}_p,\mathbf{r}\left|\frac{1}{E\pm i\epsilon-H_0}+\frac{1}{\mathcal{H}_0}\right|\mathbf{r}_p',\mathbf{r}'\right). \tag{2.25}$$

The conditions that 'T'$_{ba}$ be stationary are then expressed by

$$\Psi_a^{(\pm)}+\frac{1}{\mathcal{H}_0}V\Psi_a^{(\pm)}=F_a(\mathbf{r})+\int K^{(\pm)}(\mathbf{r},\mathbf{r}')\times V(\mathbf{r}_n'-\mathbf{r}_p')\Psi_a^{(\pm)}(\mathbf{r}_n',\mathbf{r}')d\mathbf{r}_n'd\mathbf{r}' \quad (2.26)$$

which, in virtue of (2.5) and (2.6), imply that

$$\Psi_a^{(\pm)}=\psi(\mathbf{r}_n-\mathbf{r}_p)G_a^{(\pm)}(\mathbf{r}) \quad (2.27)$$

where $G_a^{(\pm)}(\mathbf{r})$ obeys the integral equation

$$G_a^{(\pm)}(\mathbf{r})-\mathrm{t}\int K^{(\pm)}(\mathbf{r},\mathbf{r}')G_a^{(\pm)}(\mathbf{r}')d\mathbf{r}'=F_a(\mathbf{r}). \quad (2.28)$$

This is a generalization of the integral equation obtained by Breit.

The stationary value of 'T'$_{ba}$ is given by

$$\mathrm{T}_{ba}\simeq(F_b, V\Psi_a^{(+)})=\mathrm{t}\int F_b^*(\mathbf{r})G_a^{(+)}(\mathbf{r})d\mathbf{r}. \quad (2.29)$$

The integral equation for $G_a^{(+)}(\mathbf{r})$ can be solved by successive substitutions,

$$G_a^{(+)}(\mathbf{r})=F_a(\mathbf{r})+\mathrm{t}\int K^{(+)}(\mathbf{r},\mathbf{r}')F_a(\mathbf{r}')d\mathbf{r}' +\mathrm{t}^2\int K^{(+)}(\mathbf{r},\mathbf{r}')K^{(+)}(\mathbf{r}',\mathbf{r}'')\times F_a(\mathbf{r}'')d\mathbf{r}'d\mathbf{r}''+\cdots, \quad (2.30)$$

which is evidently a power series expansion in a/l, where l is a characteristic molecular dimension. Since $a/l\sim10^{-3}$, the series converges rapidly and it is quite sufficient to retain only the first term beyond $F_a(\mathbf{r})$ to obtain an accurate estimate of the correction to Fermi's approximation. Therefore,

$$\mathrm{T}_{ba}\simeq\mathrm{t}\int F_b^*(\mathbf{r})F_a(\mathbf{r})d\mathbf{r}+\mathrm{t}^2\int F_b^*(\mathbf{r})K^{(+)}(\mathbf{r},\mathbf{r}')F_a(\mathbf{r}')\times d\mathbf{r}d\mathbf{r}'. \quad (2.31)$$

To construct $K^{(+)}(\mathbf{r},\mathbf{r}')$, we observe that

$$\left(\mathbf{r}_n,\mathbf{r}\left|\frac{1}{E+i\epsilon-H_0}\right|\mathbf{r}_n',\mathbf{r}'\right) =\sum_c \Phi_c(\mathbf{r}_n,\mathbf{r})\frac{1}{E+i\epsilon-E_a}\Phi_c^*(\mathbf{r}_n',\mathbf{r}')$$
$$=\sum_\gamma\int\frac{d\mathbf{k}}{(2\pi)^3}\exp(i\mathbf{k}\cdot\mathbf{r}_n)\cdot\chi_\gamma(\mathbf{r})\times\frac{1}{E+i\epsilon-(\hbar^2k^2/2\mu)-W_\gamma}\times\exp(-i\mathbf{k}\cdot\mathbf{r}_n')\chi_\gamma^*(\mathbf{r}'). \quad (2.32)$$

In the second version, the summation over the states of the system molecule plus free neutron is explicitly performed over the independent states of the molecule and of the neutron. For the evaluation of the corresponding matrix element of $1/\mathcal{H}_0$, it must be realized that the latter operator refers to the relative motion of neutron and proton only. Thus

$$\left(\mathbf{r}_n,\mathbf{r}\left|\frac{1}{\mathcal{H}_0}\right|\mathbf{r}_n',\mathbf{r}'\right)=\int\frac{d\mathbf{k}}{(2\pi)^3}\exp(i\mathbf{k}\cdot(\mathbf{r}_n-\mathbf{r}_p))\left(\frac{M}{\hbar^2k^2}\right)\times\exp(i\mathbf{k}\cdot(\mathbf{r}_n'-\mathbf{r}_p'))\times\delta\left(\frac{\mathbf{r}_n+\mathbf{r}_p}{2}-\frac{\mathbf{r}_n'+\mathbf{r}_p'}{2}\right)\delta(\mathbf{s}-\mathbf{s}'), \quad (2.33)$$

where $\mathbf{s}$ symbolizes the set of internal molecular coordinates, omitting $\mathbf{r}_p$. We are actually interested in (2.33) as $\mathbf{r}_n\rightarrow\mathbf{r}_p$ and $\mathbf{r}_n'\rightarrow\mathbf{r}_p'$. In this limit, $\delta(\mathbf{r}_n+\mathbf{r}_p/2-\mathbf{r}_n'+\mathbf{r}_p'/2)$ becomes $\delta(\mathbf{r}_p-\mathbf{r}_p')$ and we may employ the completeness relation for the molecular eigenfunctions,

$$\delta(\mathbf{r}_p-\mathbf{r}_p')\delta(\mathbf{s}-\mathbf{s}')=\delta(\mathbf{r}-\mathbf{r}')=\sum_\gamma\chi_\gamma(\mathbf{r})\chi_\gamma^*(\mathbf{r}'). \quad (2.34)$$

One can now combine (2.32) and (2.33) to form

$$K^{(+)}(\mathbf{r},\mathbf{r}')=\sum_\gamma\int\frac{d\mathbf{k}}{(2\pi)^3}\exp(i\mathbf{k}\cdot\mathbf{r}_p)\chi_\gamma(\mathbf{r})\times\left[\frac{1}{E+i\epsilon-(\hbar^2k^2/2\mu)-W_\gamma}+\frac{1}{(\hbar^2k^2/M)}\right]\times\exp(-i\mathbf{k}\cdot\mathbf{r}_p')\chi_\gamma^*(\mathbf{r}')$$
$$=\frac{M}{\hbar^2}\sum_\gamma\int\frac{d\mathbf{k}}{(2\pi)^3}\exp(i\mathbf{k}\cdot(\mathbf{r}_p-\mathbf{r}_p'))\times\left[\frac{2\mu/M}{k_\gamma^2+i\eta-k^2}+\frac{1}{k^2}\right]\chi_\gamma(\mathbf{r})\chi_\gamma^*(\mathbf{r}'). \quad (2.35)$$

Here

$$k_\gamma^2=(2\mu/\hbar^2)(E-W_\gamma) \quad (2.36)$$

and $\eta=(2\mu/\hbar^2)\epsilon$. The $\mathbf{k}$ integration in (2.35) involves the well-known integrals

$$\int\frac{d\mathbf{k}}{(2\pi)^3}\frac{\exp(i\mathbf{k}\cdot(\mathbf{r}_p-\mathbf{r}_p'))}{k^2-k_\gamma^2-i\eta}=\frac{\exp(ik_\gamma|\mathbf{r}_p-\mathbf{r}_p'|)}{4\pi|\mathbf{r}_p-\mathbf{r}_p'|} \quad (2.37)$$

and

$$\int\frac{d\mathbf{k}}{(2\pi)^3}\frac{\exp(i\mathbf{k}\cdot(\mathbf{r}_p-\mathbf{r}_p'))}{k^2}=\frac{1}{4\pi|\mathbf{r}_p-\mathbf{r}_p'|}, \quad (2.38)$$

where

$$k_\gamma=+\left(\frac{2\mu}{\hbar^2}(E-W_\gamma)\right)^{\frac{1}{2}};\quad W_\gamma<E$$
$$=+i\left(\frac{2\mu}{\hbar^2}(W_\gamma-E)\right)^{\frac{1}{2}};\quad W_\gamma>E \tag{2.39}$$

the propagating or attenuating nature of the spherical wave corresponding to whether or not the excitation of the molecular state γ is energetically possible. Finally, then

$$K^{(+)}(\mathbf{r},\mathbf{r}')=-\frac{M}{4\pi\hbar^2}\sum_\gamma \chi_\gamma(\mathbf{r})\chi_\gamma^*(\mathbf{r}')\times\frac{2\mu}{M}\frac{\exp(ik_\gamma|\mathbf{r}_p-\mathbf{r}_p'|)-1}{|\mathbf{r}_p-\mathbf{r}_p'|} \tag{2.40}$$

and

$$\mathrm{T}_{ba}\simeq-\frac{4\pi\hbar^2}{M}a\left[\int F_b^*(\mathbf{r})F_a(\mathbf{r})d\mathbf{r}+a\sum_\gamma\int F_b^*(\mathbf{r})\chi_\gamma(\mathbf{r})\chi_\gamma^*(\mathbf{r}')\times\frac{(2\mu/M)\exp(ik_\gamma|\mathbf{r}_p-\mathbf{r}_p'|)-1}{|\mathbf{r}_p-\mathbf{r}_p'|}F_a(\mathbf{r}')d\mathbf{r}d\mathbf{r}'\right]. \tag{2.41}$$

The ratio $2\mu/M$ ranges from unity, referring to a free proton, to 2, which applies to a proton bound in an infinitely heavy molecule. Our results for these situations are in agreement with those of Breit. In particular, for a free proton $k_\gamma=k$, since there is no internal molecular motion, and (2.41) reduces to

$$\mathrm{T}_{ba}=\mathrm{t}'=-\frac{4\pi\hbar^2}{M}a(1+ika). \tag{2.42}$$

This is simply the exact version of (2.7)

$$\mathrm{t}'=-\frac{1}{\pi}|f|^2\sin\delta e^{i\delta}=-\frac{4\pi\hbar^2}{M}\frac{1}{k}\frac{\tan\delta}{1-i\tan\delta} \tag{2.43}$$

with $\tan\delta$ replaced by ka, the low energy limiting form, but retaining the complex factor $1/(1-i\tan\delta)\simeq1+ika$. The latter has a negligible effect on transition probabilities in the energy range of interest, but is necessary to preserve the general conservation theorem (1.74). We shall, indeed, verify (1.74) for the more general expression (2.41). It is most evident from (2.31) and (2.32) that

$$-\frac{1}{\pi}Im\mathrm{T}_{aa}=\mathrm{t}^2\sum_e\int F_a^*(\mathbf{r})F_e(\mathbf{r})\delta(E-E_e)\times F_e^*(\mathbf{r}')F_a(\mathbf{r}')d\mathbf{r}d\mathbf{r}'$$
$$=\sum_e|\mathrm{T}_{ea}|^2\delta(E-E_e) \tag{2.44}$$

in which T_{ea} on the right side is computed from the Fermi approximation. This is in accordance with the approximate nature of (2.41).

THE PHYSICAL REVIEW

A journal of experimental and theoretical physics established by E. L. Nichols in 1893

SECOND SERIES, VOL. 79, No. 4 AUGUST 15, 1950

Fine Structure of the Hydrogen Atom.* Part I

WILLIS E. LAMB, JR., AND ROBERT C. RETHERFORD†
Columbia Radiation Laboratory, Columbia University, New York, New York

(Received April 10, 1950)

The fine structure of the hydrogen atom is studied by a microwave method. A beam of atoms in the metastable $2^2S_{1/2}$ state is produced by bombarding atomic hydrogen. The metastable atoms are detected when they fall on a metal surface and eject electrons. If the metastable atoms are subjected to radiofrequency power of the proper frequency, they undergo transitions to the non-metastable states $2^2P_{1/2}$ and $2^2P_{3/2}$ and decay to the ground state $1^2S_{1/2}$ in which they are not detected. In this way it is determined that contrary to the predictions of the Dirac theory, the $2^2S_{1/2}$ state does not have the same energy as the $2^2P_{1/2}$ state, but lies higher by an amount corresponding to a frequency of about 1000 Mc/sec. Within the accuracy of the measurements, the separation of the $2^2P_{1/2}$ and $2^2P_{3/2}$ levels is in agreement with the Dirac theory. No differences in either level shift or doublet separation were observed between hydrogen and deuterium. These results were obtained with the first working apparatus. Much more accurate measurements will be reported in subsequent papers as well as a detailed comparison with the quantum electrodynamic explanation of the level shift by Bethe.

Among the topics discussed in connection with this work are (1) spectroscopic observations of the H_α line, (2) early attempts to use microwaves to study the hydrogen fine structure, (3) existence of metastable hydrogen atoms, their properties and methods for their production and detection, (4) estimates of yield and r-f power requirements, (5) Zeeman and hyperfine structure effects, (6) quenching of metastable hydrogen atoms by electric and motional electric fields, (7) production of a polarized beam of metastable hydrogen atoms.

* Work supported jointly by the Signal Corps and ONR.

† Submitted by Robert C. Retherford in partial fulfillment of the requirements for the degree of Doctor of Philosophy in the Faculty of Pure Science, Columbia University.

Evidence for the Formation of Positronium in Gases*

MARTIN DEUTSCH
Laboratory for Nuclear Science and Engineering, and Department of Physics, Massachusetts Institute of Technology, Cambridge, Massachusetts
(Received March 13, 1951)

THE distribution of time delays between the emission of a nuclear gamma-ray from the decay of Na^{22} and the appearance of an annihilation quantum has been measured for positrons stopping in a large number of gases and gas mixtures, extending earlier measurements.[1] A complete interpretation of the results appears to be fairly complex and will be attempted in a more extensive communication. At this time we want to report on some definite proof of the abundant formation of positronium, the bound electron-positron system analogous to the hydrogen atom. The ground state of this atom is expected[2] to have a lifetime against two-quantum annihilation of about 10^{-10} sec if the spins are antiparallel and slightly over 10^{-7} sec if they are parallel,[3] decaying by three-quantum annihilation. Ore[4] has shown that, in general, there is no mechanism for the rapid destruction of the triplet state at low energies. One may expect, however, that in a gas containing molecules with an odd number of electrons, such as nitric oxide, the triplet state would be converted very rapidly to the singlet by an electron exchange. In NO this requires essentially no energy at room temperature, the energy difference for the two spin orientation being only about 13 millivolts. The cross section for this process may be very large, since it is a resonance effect involving the coulomb interaction rather than a magnetic conversion. (The latter is expected to be very slow.[4]) Thus, even a small admixture of NO should cause the very rapid annihilation of those positrons which would otherwise have decayed by three-quantum annihilation with a period of the order of 10^{-7} sec. Figure 1 shows this phenomenon. In nitrogen the number of delayed counts is reduced by a factor of three by the addition of 3 percent NO. These counts then appear instead in the "prompt" channel, not shown in Fig. 1. In freon the effect is even greater, hardly any delayed coincidences remaining upon addition of NO. The residue in N_2 is probably due to free positrons.

Oxygen has a similar but less pronounced action. In pure O_2 the number of delayed coincidences is about half as great as in N_2 (Fig. 1).

Confirmation of the above interpretation comes from the direct observation of the continuous gamma-ray spectrum due to the

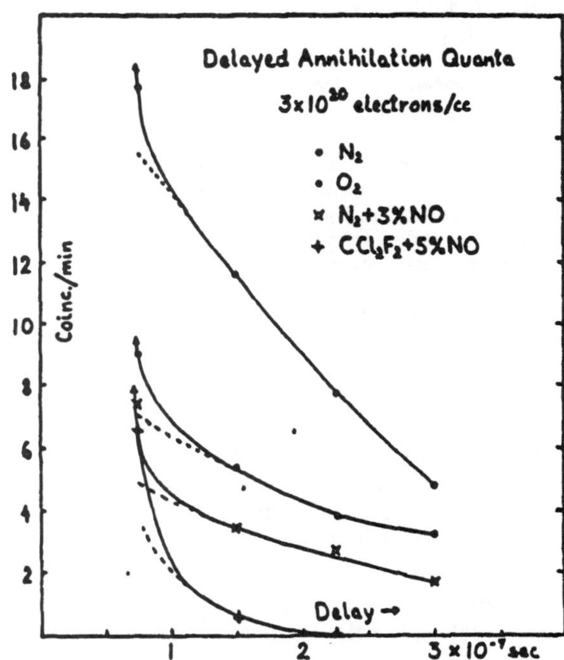

FIG. 1. Decay curves of positrons in several gases. The dotted lines are corrected for time resolution of the instrument.

three-quantum annihilation of triplet positronium in nitrogen. Figure 2 shows pulse height spectra obtained in a NaI scintillation spectrometer deliberately adjusted for low resolution to gain stability. It is seen that the photoelectron peak of the 510-kev radiation is markedly lower in pure N_2 while there is an increase of the number of lower energy electrons, compared with the spectrum from N_2+NO in which we expect almost pure two-

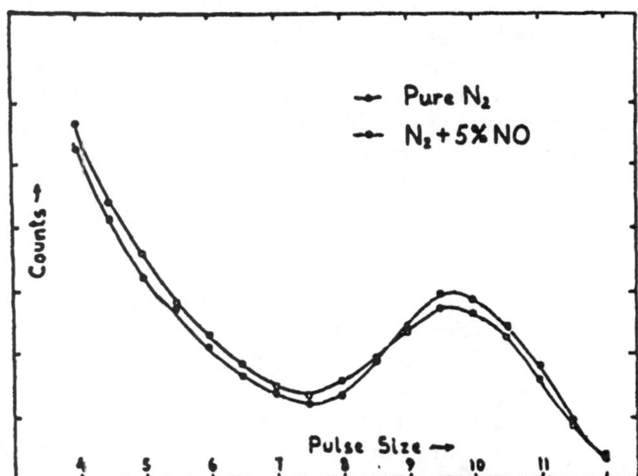

FIG. 2. Part of the scintillation spectrum of two-quantum (solid circles) and two-quantum plus three-quantum (open circles) annihilation. The uncertainty of individual points is indicated by the size of the circles.

quantum annihilation. The large number of small pulses present in both curves is due to scattered radiation from the walls of the gas vessel. The annihilation spectrum in pure O_2 resembles that in N_2+NO.

The slight shift of the photoelectron peak to lower energies in N_2 is due to the large number of quanta just below 510 kev from the three-quantum process[3] falling within the experimental width of the two-quantum line. A similar phenomenon may explain the low value of the annihilation energy obtained by DuMond.[5]

* Supported in part by the joint program of the AEC and ONR.
1 J. W. Shearer and M. Deutsch, Phys. Rev. 76, 462 (1949).
2 J. Pirenne, Arch. Sci. phys. et nat. 29, 257 (1947).
3 A. Ore and J. L. Powell, Phys. Rev. 75, 1696 (1949).
4 A. Ore, Yearbook 1949 (University of Bergen), No. 12.
5 J. W. M. DuMond, Phys. Rev. 81, 468 (1951).

Coherence in Spontaneous Radiation Processes

R. H. DICKE
Palmer Physical Laboratory, Princeton University, Princeton, New Jersey
(Received August 25, 1953)

By considering a radiating gas as a single quantum-mechanical system, energy levels corresponding to certain correlations between individual molecules are described. Spontaneous emission of radiation in a transition between two such levels leads to the emission of coherent radiation. The discussion is limited first to a gas of dimension small compared with a wavelength. Spontaneous radiation rates and natural line breadths are calculated. For a gas of large extent the effect of photon recoil momentum on coherence is calculated. The effect of a radiation pulse in exciting "super-radiant" states is discussed. The angular correlation between successive photons spontaneously emitted by a gas initially in thermal equilibrium is calculated.

IN the usual treatment of spontaneous radiation by a gas, the radiation process is calculated as though the separate molecules radiate independently of each other. To justify this assumption it might be argued that, as a result of the large distance between molecules and subsequent weak interactions, the probability of a given molecule emitting a photon should be independent of the states of other molecules. It is clear that this model is incapable of describing a coherent spontaneous radiation process since the radiation rate is proportional to the molecular concentration rather than to the square of the concentration. This simplified picture overlooks the fact that all the molecules are interacting with a common radiation field and hence cannot be treated as independent. The model is wrong in principle and many of the results obtained from it are incorrect.

A simple example will be used to illustrate the inadequacy of this description. Assume that a neutron is placed in a uniform magnetic field in the higher energy of the two spin states. In due course the neutron will spontaneously radiate a photon via a magnetic dipole transition and drop to the lower energy state. The probability of finding the neutron in its upper energy state falls exponentially to zero.[1,2]

If, now, a neutron in its ground state is placed near the first excited neutron (a distance small compared with a radiation wavelength but large compared with a particle wavelength and such that the dipole-dipole interaction is negligible), the radiation process would, according to the above hypothesis of independence, be unaffected. Actually, the radiation process would be strongly affected. The initial transition probability would be the same as before but the probability of finding an excited neutron would fall exponentially to one-half rather than to zero.

The justification for these assertions is the following: The initial state of the neutron system finds neutron 1 excited and neutron 2 unexcited. (It is assumed that the particles have nonoverlapping space functions, so that particle symmetry plays no role.) This initial state may be considered to be a superposition of the triplet and singlet states of the particles. The triplet state is capable of radiating to the ground state (triplet) but the singlet state will not couple with the triplet system. Consequently, only the triplet part is modified by the coupling with the field. After a long time there is still a probability of one-half that a photon has not been emitted. If, after a long period of time, no photon has been emitted, the neutrons are in a singlet state and it is impossible to predict which neutron is the excited one.

On the other hand, if the initial state of the two neutrons were triplet with $s=1$, $m_s=0$ namely a state with one excited neutron, a photon would be certain to be emitted and the transition probability would be just double that for a lone excited neutron. Thus, the presence of the unexcited neutron in this case doubles the radiation rate.

In recent years the excitation of correlated states of atomic radiating systems with the subsequent emission of spontaneous coherent radiation has become an important technique for nuclear magnetic resonance research.[3] The description usually given of this process is a classical one based on a spin system in a magnetic field. The purpose of this note is to generalize these results to any system of radiators with a magnetic or electric dipole transition and to see what effects, if any, result from a quantum mechanical treatment of the radiation process. Most of the previous work[4] was quite early and not concerned with the problems being considered here. In a subsequent article to be published in the *Review of Scientific Instruments* some of these results will be applied to the problem of instrumentation for microwave spectroscopy.

In this treatment the gas as a whole will be considered as a single quantum-mechanical system. The problem will be one of finding those energy states representing correlated motions in the system. The spontaneous emission of coherent radiation will accompany transitions between such levels. In the first problem to be considered the gas volumes will be assumed to have

[1] W. Heitler, *The Quantum Theory of Radiation* (Clarendon Press, Oxford, 1936), first edition, p. 112.
[2] E. P. Wigner and V. Weisskopf, Z. Physik 63, 54 (1930).
[3] E. L. Hahn, Phys. Rev. 77, 297 (1950); 80, 580 (1950).
[4] E.g., W. Pauli, *Handbuch der Physik* (Springer, Berlin, 1933), Vol. 24, Part I, p. 210; G. Wentzel, *Handbuch der Physik* (Springer, Berlin, 1933), Vol. 24, Part I, p. 758.

dimensions small compared with a radiation wavelength. This case, which is of particular importance for nuclear magnetic resonance experiments and some microwave spectroscopic applications, is treated first quantum mechanically and then semiclassically, the radiation process being treated classically. A classical model is also described. In the next case to be considered the gas is assumed to be of large extent. The effect of molecular motion on coherence and the effect on coherence of the recoil momentum accompanying the emission of a photon are discussed. Finally, the two principal methods of exciting coherent states by the absorption of photons from an intense radiation pulse or the emission of photons by the gas are discussed. Calculations of these two effects are made for the gas system initially in thermal equilibrium. The effect of photon emission on inducing coherence is discussed as a problem in the angular correlation of the emitted photons.

DIPOLE APPROXIMATION

The first problem to be considered is that of a gas confined to a container the dimensions of which are small compared with a wavelength. It is assumed that the walls of the container are transparent to the radiation field. In order to avoid difficulties arising from collision broadening it will be assumed that collisions do not affect the internal states of the molecules. It will be assumed that the transition under question takes place between two nondegenerate states of the molecule. The assumption of nondegeneracy is made in order to limit the scope of the problem to its bare essentials. It might be assumed that nondegenerate states are present as a result of a uniform static electric or magnetic field acting on the gas. Actually, for many of the questions being discussed it is not essential that the degeneracies be split. Also, it will be assumed that there is insufficient overlap in the wave functions of separate molecules to require that the wave functions be symmetrized.

Since it is assumed that internal coordinates of the individual molecules are unaffected by collisions and but two internal states are involved for each molecule, the wave function for the gas may be written conveniently in a representation diagonal in the center-of-mass coordinates and the internal energies of the molecules. The internal energy coordinate takes on only two values. Omitting for the moment the radiation field, the Hamiltonian for an n molecule gas can be written

$$H=H_0+E\sum_{j=1}^{n}R_{j3}, \tag{1}$$

where $E=\hbar\omega=$molecular excitation energy. Here H_0 acts on the center-of-mass coordinates and represents the translational and intermolecular interaction energies of the gas. ER_{j3} is the internal energy of the jth molecule and has eigenvalues $\pm\frac{1}{2}E$. H_0 and all the R_{j3} commute with each other. Consequently, energy eigenfunctions may be chosen to be simultaneous eigenfunctions of H_0, R_{13}, R_{23}, $\cdots$, R_{n3}.

Let a typical energy state be written as

$$\psi_{gm}=U_g(\mathbf{r}_1\cdots\mathbf{r}_n)[++-+\cdots]. \tag{2}$$

Here $\mathbf{r}_1\cdots\mathbf{r}_n$ designates the center-of-mass coordinates of the n molecules, and $+$ and $-$ symbols represent the internal energies of the various molecules. If the number of $+$ and $-$ symbols are denoted by n_+ and n_-, respectively, then m is defined as

$$m=\tfrac{1}{2}(n_+-n_-),\qquad n=n_++n_-=\text{number of gaseous molecules}. \tag{3}$$

If the energy of motion and mutual interaction of the molecules is denoted by E_g, then the total energy of the system is

$$E_{gm}=E_g+mE. \tag{4}$$

It is evident that the index m is integral or half-integral depending upon whether n is even or odd. Because of the various orders in which the $+$ and $-$ symbols can be arranged, the energy E_{gm} has a degeneracy

$$\frac{n!}{(\frac{1}{2}n+m)!(\frac{1}{2}n-m)!}. \tag{5}$$

This degeneracy has its origin in the internal coordinates only.

In addition, the wave function may have additional degeneracy from the center-of-mass coordinates. It should be noted in this connection that the degeneracy of the total wave function will depend upon whether or not the molecules are regarded as distinguishable or not.

If the molecules are indistinguishable, the symmetry of U_g will depend upon the symmetries of the wave function under interchanges of internal coordinates. For example, the states with all molecules excited are symmetric under an interchange of the internal coordinates of any two molecules. Consequently, for these states U_g must be symmetric for Bose molecules and antisymmetric for Fermi molecules. The limitations of symmetry are normally without physical significance as it is assumed that the gas is of such low density that the various molecules have nonoverlapping wave functions.

Of the Hamiltonian equation (1), H_0 operates on the center-of-mass coordinates only and gives

$$H_0U_g=E_gU_g, \tag{6}$$

whereas R_{j3} operates on the plus or minus symbol in the jth place corresponding to the internal energy of the jth molecule. Except for the factor $\frac{1}{2}$, it is analogous to one of the Pauli spin operators. As operators similar to the other two Pauli operators are also needed in this development, the properties of all three are listed here.

$$\begin{aligned} &\qquad\;\; j\\ &\qquad\;\;\downarrow\\ R_{j1}[\cdots&\pm\cdots]=\tfrac{1}{2}[\cdots\mp\cdots],\\ R_{j2}[\cdots&\pm\cdots]=\pm\tfrac{1}{2}i[\cdots\mp\cdots],\\ R_{j3}[\cdots&\pm\cdots]=\pm\tfrac{1}{2}[\cdots\pm\cdots]. \end{aligned} \tag{7}$$

It is also convenient to define the operators

$$R_k=\sum_{j=1}^{n} R_{jk}, \quad k=1, 2, 3, \tag{8}$$

and the operator

$$R^2=R_1^2+R_2^2+R_3^2. \tag{9}$$

In this notation the Hamiltonian becomes

$$H=H_0+ER_3, \tag{10}$$

and

$$R_3\psi_{gm}=m\psi_{gm}. \tag{11}$$

To complete the description of the dynamical system, there must be added to the Hamiltonian that of the radiation field and the interaction term between field and the molecular system.

For the purpose of definiteness the ineraction of a molecule with the electromagnetic field will be assumed to be electric dipole. The main results are actually independent of the type of coupling. The interaction energy of the jth molecule with the electromagnetic field can be written as

$$-\mathbf{A}(\mathbf{r}_j)\cdot\sum_{k=1}^{N-1}\frac{e_k}{m_k c}\mathbf{P}_k. \tag{12}$$

Here the configuration coordinates of the molecule are taken to be the center-of-mass coordinates and the coordinates relative to the center of mass of any $N-1$ of the N particles which constitute the jth molecule. e_k and m_k are the charge and mass of the kth particle, and $\mathbf{P}_k$ is the momentum conjugate to the position of the kth particle relative to the center of mass. The molecule is assumed electrically neutral.

Since $\mathbf{P}_k$ is an odd operator, it has only off-diagonal elements in a representation with internal energy diagonal. Hence the general form of Eq. (12) is

$$-\mathbf{A}(\mathbf{r}_j)\cdot(\mathbf{e}_1R_{j1}+\mathbf{e}_2R_{j2}). \tag{13}$$

$\mathbf{e}_1$ and $\mathbf{e}_2$ are constant real vectors the same for all molecules. The total interaction energy then becomes

$$H_1=-\sum_j \mathbf{A}(\mathbf{r}_j)\cdot(\mathbf{e}_1R_{j1}+\mathbf{e}_2R_{j2}). \tag{14}$$

Since the dimensions of the gas cell are small compared with a wavelength, the dependence of the vector potential on the center of mass of the molecules can be omitted and the interaction energy (12) becomes

$$H_1=-\mathbf{A}(0)\cdot(\mathbf{e}_1R_1+\mathbf{e}_2R_2). \tag{15}$$

Since the interaction term Eq. (15) does not contain the center-of-mass coordinates, the selection rule on the molecular motion quantum number g is $\Delta g=0$. Consequently there is no Doppler broadening of the transition frequency. This results solely from the small size of the gas container.[5]

The operators R_1, R_2, and R_3, apart from a factor of $\hbar$, obey the same commutation relations as the three components of angular momentum. Consequently, the interaction operator Eq. (15) obeys the selection rule $\Delta m=\pm1$. In general, it has nonvanishing matrix elements between a given state Eq. (2) and a large number of states with $\Delta m=\pm1$. In order to simplify the calculation of spontaneous radiation transitions, it is desirable that a set of stationary states be selected in such a way that the interaction term has matrix elements joining a given state with, at most, one state of higher and lower energy, respectively. Because of the very close analogy between this formalism and that of a system of particles of spin $\frac{1}{2}$, known results can be taken over from the spin formalism.

In a manner similar to an angular momentum formalism,[6] the operations H and R^2 commute; consequently, stationary states can be chosen to be eigenstates of R^2. These new states are linear combinations of the states of Eq. (2). The operator R^2 has eigenvalues $r(r+1)$. r is integral or half-integral and positive, such that

$$|m|\leq r\leq\tfrac{1}{2}n. \tag{16}$$

The eigenvalue r will be called the "cooperation number" of the gas. Denote the new eigenstates by

$$\psi_{gmr}. \tag{17}$$

Here

$$H\psi_{gmr}=(E_g+mE)\psi_{gmr}, \tag{18}$$

$$R^2\psi_{gmr}=r(r+1)\psi_{gmr}. \tag{19}$$

The degeneracy of the stationary states is not completely removed by introducing R^2. The state (g, m, r) has a degeneracy

$$\frac{n!(2r+1)}{(\frac{1}{2}n+r+1)!(\frac{1}{2}n-r)!}. \tag{20}$$

The complete set of eigenstates ψ_{gmr} may be specified in the following way: the largest value of m and r is

$$r=m=\tfrac{1}{2}n.$$

This state is nondegenerate in the internal coordinates and may be written as

$$\psi_{g,\frac{1}{2}n,\frac{1}{2}n}=U_g\cdot[++\cdots+]. \tag{20a}$$

All the states with this same value of $r=\frac{1}{2}n$, but with different values of m, are nondegenerate also and may be generated as[7]

$$\psi_{gmr}=[(R^2-R_3^2-R_3)^{-\frac{1}{2}}(R_1-iR_2)]^{r-m}\psi_{grr}. \tag{21}$$

The operator R_1-iR_2 reduces the m index by unity every time it is applied and the fractional power operator is to preserve the normalization of the wave function.[8] The fractional power operator is defined as having positive eigenvalues only.

[5] R. H. Dicke, Phys. Rev. 89, 472 (1953).

[6] E. U. Condon and G. H. Shortley, *The Theory of Atomic Spectra* (Cambridge University Press, Cambridge, 1935), pp. 45–49.

[7] See reference 6, p. 48, Eq. (3).

[8] See reference 6, p. 48.

The state $\psi_{g,\frac{1}{2}n-1,\frac{1}{2}n}$ is one of n states with this value of m. The remaining $n-1$ states should be chosen to be orthogonal to this state, orthogonal to each other, and normalized. Since these remaining $n-1$ states are not states of $r=\frac{1}{2}n$, they must be states of $r=\frac{1}{2}n-1$, the only other possibility. Again the complete set of states with this value of r can be generated using Eq. (21), where now $r=\frac{1}{2}n-1$, and the operator in Eq. (21) is applied to each of the $n-1$ orthogonal states of $r=m=\frac{1}{2}n-1$. This procedure can be repeated until all possible values of r are exhausted, in which case all the stationary states have been defined.

With this definition of the stationary states, the interaction energy operator has matrix elements joining a given state of the gas to but two other states. Aside from the factor involving the radiation field operator, the matrix elements of the interaction energy may be written[8]

$$(g,r,m|\mathbf{e}_1R_1+\mathbf{e}_2R_2|g,r,m\mp1)=\tfrac{1}{2}(\mathbf{e}_1\pm i\mathbf{e}_2)[(r\pm m)(r\mp m+1)]^{\frac{1}{2}}. \quad (23)$$

Transition probabilities will be proportional to the square of the matrix elements. In particular, the spontaneous radiation probabilities will be

$$I=I_0(r+m)(r-m+1). \quad (24)$$

Here, by setting $r=m=\frac{1}{2}$, it is evident that I_0 is the radiation rate of a gas composed of one molecule in its excited state. I_0 has the value[9]

$$I_0=\frac{4}{3}\frac{\omega^2}{c}\left|\left(\sum_k\frac{e_k\mathbf{P}_k}{m_kc}\right)_{+-}\right|^2=\frac{1}{3}\frac{\omega^2}{c}|\mathbf{e}_1-i\mathbf{e}_2|^2=\frac{1}{3}\frac{\omega^2}{c}(e_1^2+e_2^2). \quad (25)$$

If $m=r=\frac{1}{2}n$ (i.e., all n molecules excited),

$$I=nI_0. \quad (26)$$

Coherent radiation is emitted when r is large but $|m|$ small. For example, for even n let

$$r=\tfrac{1}{2}n,\quad m=0;\quad I=\tfrac{1}{2}n(\tfrac{1}{2}n+1)I_0. \quad (27)$$

This is the largest rate at which a gas with an even number of molecules can radiate spontaneously. It should be noted that for large n it is proportional to the square of the number of molecules.

Because of the fact that with the choice of stationary states given by Eq. (21) a given state couples with but one state of lower energy, this radiation rate [Eq. (27)], is an absolute maximum. Any superposition state will radiate at the rate

$$I=I_0\sum_{r,m}P_{r,m}(r+m)(r-m+1)=I_0\langle(R_1+iR_2)(R_1-iR_2)\rangle, \quad (28)$$

where $P_{r,m}$ is the probability of being in the state r, m.

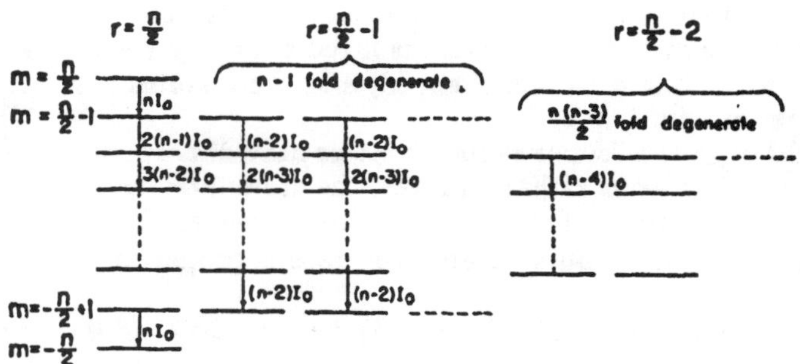

FIG. 1. Energy level diagram of an n-molecule gas, each molecule having 2 nondegenerate energy levels. Spontaneous radiation rates are indicated. $E_m=mE$.

There are no interference terms. Consequently, no superposition state can radiate more strongly than Eq. (27). An energy level diagram which shows the relative magnitudes of the various radiation probabilities is given in Fig. 1.

States with a low "cooperation number" are also highly correlated but in such a way as to have abnormally low radiation rates. For example, a gas in the state $r=m=0$ does not radiate at all. This state, which exists only for an even number of molecules, is analogous to a classical system of an even number of oscillators swinging in pairs oppositely phased.

The energy trapping which results from the internal scattering of photons by the gas appears naturally in the formalism. As an example, consider an initial state of the gas for which one definite molecule, and only this molecule, is excited. The gas at first radiates at the normal incoherent rate for a short time and thereafter fails to radiate. The probability of a photon's being emitted during the radiating period is $1/n$. These results follow from the fact that the assumed state is a linear superposition of the various states with $m=1-n/2$, and that $1/n$ is the probability of being in the state $r=\frac{1}{2}n$. The probability that the energy will be "trapped" is $(n-1)/n$. This is analogous to the radiation by a classical oscillator when $n-1$ similar unexcited oscillators are near. The solution of this classical problem shows that only $1/n$ of the excitation energy is radiated. The remainder appears in nonradiating normal modes of the system.

For want of a better term, a gas which is radiating strongly because of coherence will be called "super-radiant." There are two obvious ways in which a "super-radiant" state may be excited. First, if all the molecules be excited, the gas is in the state characterized by

$$r=m=\tfrac{1}{2}n. \quad (29)$$

As the system radiates it passes to states of lower m with r unchanged. This will take the system to the "super-radiant" region $m\sim0$.

Another way in which such a state can be excited is to start with the gas in its ground state,

$$r=-m=\tfrac{1}{2}n, \quad (30)$$

[8] Reference 1, p. 106.

and irradiate it with a pulse of radiation.[9a] If the pulse is sufficiently intense, the system is lifted to energy states with $m\sim 0$ but with r unchanged, and these states are "super-radiant."

Although the "super-radiant" states have abnormally large spontaneous radiation rates, the stimulated emission rate is normal. For example, with the system in the state m, r, the stimulated emission rate is proportional to

$$(r+m)(r-m+1)-(r+m+1)(r-m)=2m. \quad (30a)$$

With $m>0$ this is the normal incoherent stimulated emission rate. For $m<0$ this becomes the negative of the incoherent absorption rate.

As has been pointed out, the pulse technique for exciting "super-radiant" states is commonly used in nuclear magnetic resonance experiments. Here there is one important point that needs clarification, however. Instead of starting in the highly organized state given by Eq. (30) the pulse is applied to a system that is in thermal equilibrium at high temperatures. For example, if the system be a set of proton spins, the energy necessary to turn a spin over in the magnetic field may be about

$$E\sim 10^{-5}kT. \quad (31)$$

Under these conditions the two spin states of the proton are very nearly equally populated and it might be expected that thermal equilibrium would imply a badly disorganized system. The randomness in the initial state does not imply, however, complete randomness in m and r. For a gas with n, large states of low r have a high degeneracy. These states have a high statistical weight and are favored. However, Eq. (16) sets a lower bound on r for any m. The result is a relatively small range of values of m and r. For a system with n molecules in thermal equilibrium the mean square deviation from the mean of m is

$$n/4-\bar{m}^2/n. \quad (32)$$

Here $\bar{m}$ is the mean of m and is for high temperatures equal to

$$\bar{m}=-\tfrac{1}{4}nE/kT. \quad (33)$$

For a definite value of m the mean value of $r(r+1)$ is

$$m^2+\tfrac{1}{2}n, \quad (34)$$

and the mean square deviation is

$$\tfrac{1}{4}n^2-m^2. \quad (35)$$

The expression (32)–(35) may be easily derived using the density matrix formalism assuming the appropriate statistical ensemble.

It is hence clear that if

$$\bar{m}^2\gg n\gg 1, \quad (36)$$

the percentage deviation from the mean of m is small, that the percent deviation from the mean of $r(r+1)$ is small, and that the mean of $r(r+1)$ is approximately the smallest value compatible with the mean value of m. Thus, in the case of a gas system at high temperature, for sufficiently large n, values of m and r cluster to such an extent that the system may be considered as approximately in a state of definite $r=m=-nE/4kT$. If this gas is excited by a pulse of the proper intensity to excite states $m\sim 0$, the radiation rate after the pulse is approximately

$$I\cong I_0 r(r+1)\cong I_0 n^2(E/4kT)^2, \quad (37)$$

which is proportional to n^2 and hence coherent. A better calculation good for all temperatures gives the result [see Eq. (78) with $\theta=90°$]

$$I=\tfrac{1}{4}I_0 n(n-1)\tanh^2(E/2kT)+\tfrac{1}{2}nI_0. \quad (37a)$$

SEMICLASSICAL TREATMENT

For the spontaneous radiation from super-radiant states ($m\sim 0$) a semiclassical treatment is generally adequate. This method, which is a generalization of the well-known picture used in describing radiation from a nuclear spin system,[10] treats the molecular systems quantum mechanically but calculates the radiation process classically. In the following calculation the gas system will be assumed to be excited by a radiation pulse, which excites it from thermal equilibrium to a set of super-radiant states. To calculate the radiation rate, the expectation value of the electric dipole moment is treated as a classical dipole. When the gas contains a large number of molecules the dipole moment of the gas as a whole should be given by the sum of the expectation values of the individual dipole moments.

In thermal equilibrium the gas may be considered as having n_- molecules in the ground state and n_+ molecules in the excited state. A molecule which is initially in its ground state is assumed to be thrown into a superposition state of + and − by the radiation pulse. It is assumed that there is a unity probability ratio. The internal part of the wave function of the molecules after the pulse is given by

$$\psi_+=\frac{1}{\sqrt{2}}\left\{[+]\exp\left(-i\frac{\omega}{2}t\right)+[-]\exp i\left(\frac{\omega}{2}t+\delta\right)\right\}. \quad (38)$$

This is the most general form for ψ_+ apart from a possible multiplication phase factor. Here δ is a phase given by the phase of the exciting pulse. In a similar way a molecule in the excited state has its wave function converted to

$$\psi_-=\frac{1}{\sqrt{2}}\left\{[-]\exp i\frac{\omega}{2}t-[+]\exp\left(-i\frac{\omega}{2}t-i\delta\right)\right\}. \quad (39)$$

Instead of calculating the expectation value of the electric dipole moment it is more convenient to calculate the expectation value of the polarization current of the

[9a] See F. Bloch and I. I. Rabi, Revs. Modern Phys. 17, 237 (1945), for a discussion of the effect of a pulse on the analogous spin-½ system.

[10] F. Bloch, Phys. Rev. 70, 460 (1946).

jth molecule given by

$$\left(\sum_{k=1}^{N-1}\frac{e_k\mathbf{P}_k}{m_k}\right)=c\langle \mathbf{e}_1R_{j1}+\mathbf{e}_2R_{j2}\rangle$$

$$=\pm\tfrac{1}{2}c[\mathbf{e}_1\cos(\omega t+\delta)+\mathbf{e}_2\sin(\omega t+\delta)]. \quad (40)$$

The plus sign is obtained from the plus state, Eq. (38), and the negative sign from Eq. (39). Note the oscillating time dependence which results from the states being energy-superposition states. The polarization current for the gas as a whole is then

$$\mathbf{j}=(n_+-n_-)(c/2)[\mathbf{e}_1\cos(\omega t+\delta)+\mathbf{e}_2\sin(\omega t+\delta)]. \quad (41)$$

The radiation rate calculated classically is then[11]

$$I=\frac{2}{3}\frac{\omega^2}{c_3}|\mathbf{J}^2|=\frac{1}{12}\frac{\omega^2}{c}(n_+-n_-)^2(e_1^2+e_2^2). \quad (42)$$

In thermal equilibrium $n_+/n_-=\exp(-E/kT)$, from which

$$n_+-n_-=n\tanh(E/2kT). \quad (43)$$

Substituting into Eq. (42) gives the classical radiation rate

$$I=\frac{1}{12}\frac{\omega^2}{c}n^2(e_1^2+e_2^2)\tanh^2\left(\frac{E}{2kT}\right). \quad (44)$$

This may be compared with the quantum-mechanical result [Eq. (37a) and Eq. (25)]. For large n the two results are equal.

CLASSICAL MODEL

When the gas is in a state of definite "cooperation number" r which has a very large value, it is possible to represent it in its interaction with the electromagnetic field by a simple classical model. The energy-level spacing and the matrix elements joining adjacent levels are similar to those of a rotating top of large angular momentum and carrying an electric dipole moment. The details depend upon $\mathbf{e}_1$ and $\mathbf{e}_2$, which in turn depend on the nature of the original states. Let us consider a specific example. Assume that the radiators are atoms having a $^1P_1$ excited state and a $^1S_0$ ground state. Assume that the degeneracy of the excited state is split by a magnetic field in the z direction and that the $m_l=1$ excited level is being used. Under these conditions $\mathbf{e}_1$ and $\mathbf{e}_2$ are orthogonal to each other and the z axis, and the system has energy levels and interactions with the field identical with those of a spinning top having an electric dipole moment along its axis and precessing about the z axis as a result of an interaction with a static *electric* field in that direction. Consequently, since large quantum numbers are involved, to a good approximation the gas can be replaced by this classical model, which consists of a spinning top, in calculating both the interaction of the field on the gas and *vice versa.*

[11] Reference 1, p. 26.

RADIATION LINE BREADTH AND SHAPE

Under conditions for which the above "classical model" is valid, it is easy to calculate the natural line breadth and shape factor. This is of considerable importance in microwave spectroscopy. It has been customary to regard the natural line breadth as too small to be of any practical importance. However, as will be seen below, when coherence is properly taken into account the natural radiation breadth of the line may be far from negligible.

Using the above classical model, the angle between the spin axis and the z axis (the polar angle) will be designated as φ. In this approximation the quantum number m may be replaced by

$$m=r\cos\varphi, \quad (44a)$$

from which, using Eq. (24), the radiation rate becomes

$$I=I_0r^2\sin^2\varphi. \quad (44b)$$

Also, the internal energy of the gas is

$$mE=rE\cos\varphi. \quad (44c)$$

Balancing the radiation rate to the energy loss of the gas gives

$$\dot{\varphi}=(I_0r/E)\sin\varphi,$$

from which, assuming $\varphi=90°$ if $t=0$,

$$\sin\varphi=\operatorname{sech}(\alpha t),$$

where $\alpha=I_0r/E$. The radiated wave has the following form as a function of time:

$$A(t)=\begin{cases}e^{i\omega t}\sin\varphi, & t>0,\\ 0, & t<0,\end{cases}\qquad \hbar\omega=E.$$

The Fourier transform gives the line shape and has the value

$$\mathfrak{a}(\beta)=\left(\frac{\pi}{2}\right)^{\frac{1}{2}}\frac{1}{\alpha}\operatorname{sech}\left(\frac{\pi}{2}\frac{\beta-\omega}{\alpha}\right). \quad (44d)$$

It should be noted that this is not of the usual Lorentz form. The line width at half-intensity points is

$$\Delta\omega=1.12I_0r/E=1.12\gamma r. \quad (44e)$$

Here γ is the line width at half-intensity points for the radiation from isolated single molecules. Putting in the maximum value of r gives a line breadth of $\Delta\omega=1.12\gamma n/2$, which is generally very substantially larger than γ.

RADIATION FROM A GAS OF LARGE EXTENT

A classical system of simple harmonic oscillators distributed over a large region of space can be so phased relative to each other that coherent radiation is obtained in a particular direction. It might be expected also that the radiating gas under consideration would have energy levels such that spontaneous radiation occurs coherently in one direction.

It will be assumed that the gas occupies a region having dimensions generally larger than radiation wavelength but small compared with the reciprocal of the natural line width,

$$\Delta k = \Delta\omega / c.$$

It is necessary to turn again to the general expression for the interaction term in the Hamiltonian equation (13). The vector potential operator can be expanded in plane waves:

$$\mathbf{A}(\mathbf{r}) = \sum_{\mathbf{k}'}[\mathbf{v}_{\mathbf{k}'} \exp(i\mathbf{k}'\cdot\mathbf{r}) + \mathbf{v}_{\mathbf{k}'}^* \exp(-i\mathbf{k}'\cdot\mathbf{r})]. \quad (45)$$

$\mathbf{v}_{\mathbf{k}'}$ and its Hermitian adjoint $\mathbf{v}_{\mathbf{k}'}^*$ are photon destruction and creation operators, respectively. After substituting Eq. (45) into (13), the interaction term becomes

$$H_1 = -\tfrac{1}{2}\sum_{\mathbf{k}'} \mathbf{v}_{\mathbf{k}'}\cdot(\mathbf{e}_1 - i\mathbf{e}_2)\sum_{j=1}^{n} R_{j+} \exp(i\mathbf{k}'\cdot\mathbf{r}_j)$$

$$-\tfrac{1}{2}\sum_{\mathbf{k}'} \mathbf{v}_{\mathbf{k}'}^*\cdot(\mathbf{e}_1 + i\mathbf{e}_2)\sum_{j=1}^{n} R_{j-} \exp(-i\mathbf{k}'\cdot\mathbf{r}_j), \quad (46)$$

where $R_{j\pm} = R_{j1} \pm iR_{j2}$. In this expression, terms involving the product of the photon creation operator and the "excitation operator" R_{j+}, etc., have been dropped as these terms do not lead to first-order transitions for which energy is conserved. The form of Eq. (46) suggests defining the operators:

$$R_{\mathbf{k}1} = \sum_j (R_{j1} \cos\mathbf{k}\cdot\mathbf{r}_j - R_{j2} \sin\mathbf{k}\cdot\mathbf{r}_j),$$
$$R_{\mathbf{k}2} = \sum_j (R_{j1} \sin\mathbf{k}\cdot\mathbf{r}_j + R_{j2} \cos\mathbf{k}\cdot\mathbf{r}_j). \quad (47)$$

In terms of these operators the interaction energy becomes

$$H_1 = -\tfrac{1}{2}\sum_{\mathbf{k}'}(\mathbf{v}_{\mathbf{k}'}\cdot\mathbf{e}R_{\mathbf{k}'+} + \mathbf{v}_{\mathbf{k}'}^*\cdot\mathbf{e}^*R_{\mathbf{k}'-}), \quad (48)$$

where

$$R_{\mathbf{k}'\pm} = R_{\mathbf{k}'1} \pm iR_{\mathbf{k}'2} = \sum_{j=1}^{n} R_{j\pm} \exp(\pm i\mathbf{k}'\cdot\mathbf{r}_j),$$

$$\mathbf{e} = \mathbf{e}_1 - i\mathbf{e}_2.$$

For every direction of propagation $\mathbf{k}$ there are two orthogonal polarizations $\mathbf{v}_{\mathbf{k}}$ of $\mathbf{A}$. By a proper choice of polarization basis, the dot product of one of the basic polarizations with $\mathbf{e}$ can be assumed zero. This radiation oscillator is never excited and can be ignored. The orthogonal polarization is the one which couples with the gas. The polarization of emitted or absorbed radiation is uniquely given by the direction of propagation and need not be explicitly indicated.

The operators of Eq. (47), together with R_3, obey the angular momentum commutation relations. The operator

$$R_{\mathbf{k}}^2 = R_{\mathbf{k}1}^2 + R_{\mathbf{k}2}^2 + R_3^2 \quad (49)$$

commutes with the operators of Eq. (47) and with R_3. In Eq. (49) $\mathbf{k}$ is regarded as a fixed index. This operator does not commute with another one of the same type having a different index. Omitting for a moment the translational part of the wave function, wave functions may be so chosen as to be simultaneous eigenfunctions of the internal energy ER_3 and $R_{\mathbf{k}}^2$. They may be written as ψ_{mr} and are generated by an expression analogous to Eq. (21):

$$R_{\mathbf{k}}^2\psi_{mr} = r(r+1)\psi_{mr}, \quad ER_3\psi_{mr} = mE\psi_{mr}. \quad (50)$$

By analogy with the development leading to Eq. (24) it is clear that these states represent correlated states of the gas for which radiation emitted in the $\mathbf{k}$ direction is coherent. Thus, coherence is limited to a particular direction only, provided the initial state of the gas is given by a function of the same type as Eq. (50). The selection rules for the absorption or emission of a photon with momentum $\mathbf{k}$ are

$$\Delta r = 0, \quad \Delta m = \pm 1. \quad (51)$$

The spontaneous radiation rate in the direction $\mathbf{k}$ is given by Eq. (24), where I and I_0 are now to be interpreted as radiation rates per unit solid angle in the direction $\mathbf{k}$. This may be written as

$$I(\mathbf{k}) = I_0(\mathbf{k})[(r+m)(r-m+1)]. \quad (51a)$$

If a photon is emitted or absorbed having a momentum $\mathbf{k}' \neq \mathbf{k}$, the selection rules are

$$\Delta r = \pm 1, 0; \quad \Delta m = \pm 1. \quad (52)$$

To prove this, it may be noted that the commutation relations of the $2n$ operators

$$R_{j1}' = R_{j1} \cos(\mathbf{k}\cdot\mathbf{r}_j) - R_{j2} \sin(\mathbf{k}\cdot\mathbf{r}_j),$$
$$R_{j2}' = R_{j1} \sin(\mathbf{k}\cdot\mathbf{r}_j) + R_{j2} \cos(\mathbf{k}\cdot\mathbf{r}_j), \quad (53)$$

with those of Eq. (47) are of the same type as denoted by Condon and Shortley[12] as **T**. The selection rules satisfied by these operators are of the type given by Eq. (52).[13] The operators of Eq. (47), with $\mathbf{k} = \mathbf{k}'$, may be expressed as linear combinations of those of Eq. (53). Hence the operators of Eq. (47), with $\mathbf{k}$ replaced by $\mathbf{k}'$, satisfy the selection rules given by Eq. (52).

As was discussed previously in the dipole approximation, super-radiant states may be excited by irradiating the gas with radiation until states in the vicinity of $m=0$ are excited. In the present case the incident radiation is assumed to be plane with a propagation vector $\mathbf{k}$. After excitation the gas radiates coherently in the $\mathbf{k}$ direction. Because of the selection rules Eq. (52), radiation in directions other than $\mathbf{k}$ tends to destroy the coherence with respect to the direction $\mathbf{k}$ by causing transitions generally to states of lower r.

DOPPLER EFFECT

Because of the occurrence of the center-of-mass coordinates in the "cooperation" operator Eq. (49), it fails to commute with H_0 [Eq. (1)]; hence eigenstates of $R_{\mathbf{k}}^2$ are generally not stationary. This is equivalent to the fact that relative motion of classical oscillators will gradually destroy the coherence of the emitted radiation. If, on the other hand, a set of classical oscillators all move with the same velocity, the state of coherence

[12] Reference 6, p. 59.
[13] Reference 6, pp. 60–61.

is stationary. The corresponding question in the case of the quantum mechanical system is whether there exist simultaneous eigenstates of H and R_k^2 such that coherent radiation is emitted in a transition from one state to another. By starting with the state defined by

$$\psi_{srr}=(\exp i\mathbf{s}\cdot\sum_j \mathbf{r}_j)\cdot[+++\cdots+],\quad r=n/2, \tag{54}$$

and using the method leading to Eq. (21), there is obtained the set of states

$$\psi_{smr}=[(R_k^2-R_3^2-R_3)^{-\frac{1}{2}}(R_{k1}-iR_{k2})]^{r-m}\psi_{srr}. \tag{55}$$

If it is assumed that the gas is free, the functions Eq. (55) are simultaneous eigenfunctions of H and R_k^2. Consequently, the coherence in the $\mathbf{k}$ direction is stationary.

These states are analogous to the classical oscillators all moving with the same speed. Note one important difference, however; from Eq. (55) the momentum of an excited molecule is always

$$\mathbf{p}_+=\hbar\mathbf{s}, \tag{56}$$

whereas if a molecule is in its ground state the momentum, as given by Eq. (55), is

$$\mathbf{p}_-=\hbar(\mathbf{s}-\mathbf{k}), \tag{57}$$

the difference being the recoil momentum of the photon. Thus, the coherent states Eq. (55) are always a superposition of states such that the excited molecules have one momentum and the unexcited have another. Hence it is clear that the recoil momentum given to a molecule when it radiates in the $\mathbf{k}$ direction does not produce a molecular motion which destroys the coherence but rather is required to preserve the coherence.

The gain or loss in photon energy which has its origin in the Doppler effect is equal to the loss or gain in the kinetic energy of a radiator which results from the photon-induced recoil. Expressed as a fractional shift in photon frequency, this is

$$\frac{\Delta\omega}{\omega}=\frac{\hbar(\mathbf{S}-\frac{1}{2}\mathbf{k})\cdot\mathbf{k}}{Mck}. \tag{58}$$

Here M is the molecular mass. For energy states such that $|m|\ll n/2$, Eq. (58) can be written as

$$\frac{\Delta\omega}{\omega}=\frac{\mathbf{v}\cdot\mathbf{k}}{ck}. \tag{59}$$

Where $\mathbf{v}$ is the total momentum of the gas divided by its total mass. Equation (59) is the usual classical expression for the Doppler shift for a radiator moving with a velocity $\mathbf{v}$. Consequently, for the highly correlated states $|m|\sim 0$ the Doppler effect can be described in classical terms.

The stationary states Eq. (55) do not form a complete set. In particular, the final state, a photon being emitted or absorbed with a momentum not $\mathbf{k}$, is not one of these states. The set of stationary states may be made complete by adding all the other possible orthogonal plane wave states, each being characterized by a definite momentum and internal energy for each molecule. With this set of orthogonal states, matrix elements can be easily calculated for transitions from the states given by Eq. (55) to states in which photons appear having momenta not equal to $\mathbf{k}$. These matrix elements are found to have a magnitude characteristic of the incoherent radiation process. It should be noted that only for one magnitude of $\mathbf{k}$ as well as for direction are the matrix elements of a coherent transition obtained.

PULSE-INDUCED COHERENCE RADIATION

It will be assumed in this section that a gas initially in thermal equilibrium is illuminated for a short time by an intense radiation pulse. The intensity and angular dependence of the spontaneous radiation emitted after the pulse will be calculated. In order to avoid the difficulties associated with motional effects, the molecules will be assumed so massive that their center-of-mass coordinates can be represented by small stationary wave packets. The center-of-mass coordinates will be then treated as time-independent parameters in the equation. It is assumed that the intensity of the exciting radiation pulse is so great that the fields acting on the gas during the pulse can be considered as described classically. The spontaneous radiation rate after the exciting pulse will be calculated quantum mechanically.

Because the initial state of the gas is a mixed state describing thermodynamic equilibrium, it is convenient to use the density matrix formalism.[14] It will be assumed that one has an ensemble of gas systems statistically identical and that what one is calculating is certain ensemble averages.

For a pure state, Eq. (28) shows that the spontaneous radiation rate in the $\mathbf{k}'$ direction can be written as the expectation value

$$I(\mathbf{k}')=I_0(\mathbf{k}')\langle R_{k'+}R_{k'-}\rangle. \tag{60}$$

For a state which may be mixed or pure using the density matrix formalism this becomes the trace

$$I(\mathbf{k}')=I_0(\mathbf{k}')\,\mathrm{tr}R_{k'-}\rho R_{k'+}. \tag{62}$$

Here the density matrix is defined as the ensemble mean

$$\rho=[\psi\psi^*]_{\mathrm{Av}}. \tag{63}$$

In Eq. (63) the wave function ψ is interpreted as a column vector and the * is the Hermitian adjoint. The symbol $[\ ]_{\mathrm{Av}}$ signifies an ensemble mean.

Assume that the exciting radiation pulse is in the form of a plane wave in the $\mathbf{k}$ direction. The fields which act on the various molecules differ only in their arrival time. The Hamiltonian of the system can be written

$$H=\hbar\omega R_3-\sum_j \mathbf{A}_j(t)\cdot(\mathbf{e}_1R_{j1}+\mathbf{e}_2R_{j2}). \tag{64}$$

Here $\mathbf{A}_j(t)$ is a classical field quantity and

$$\mathbf{A}_j(t)=0,\quad \begin{matrix} t<t_j, \\ t>t_j+\tau \end{matrix} \tag{65}$$

[14] R. C. Tolman, *The Principles of Statistical Mechanics* (Clarendon Press, Oxford, 1938), p. 325.

where t_j is the arrival time of the radiation pulse at the jth molecule. Neglecting for the moment the interaction term, the time dependence of the wave function can be given by the unitary transformation

$$\psi(t)=\exp(-i\omega tR_3)\cdot\psi(0). \tag{66}$$

In general, the wave function after the interaction with the electromagnetic field can be obtained through a unitary transformation on the wave function prior to the pulse. The wave function of the gas after the radiation pulse has passed completely over the gas can be related to that before by

$$\psi'(t)=\exp(-i\omega tR_3)T\psi(0). \tag{67}$$

Here T is a unitary matrix which represents the effect of the pulse on the gas. To find the most general form of T it is convenient to consider the effect of the pulse on a particular molecule. Since this molecule has only two internal states of interest, its wave function can be regarded as a spinor in a pseudo "spin space." Then, apart from a multiplicative phase factor which has no physical significance, any unitary transformation can be represented as a rotation in "spin space." Any arbitrary rotation can be represented as a rotation about the No. 3 axis followed by a rotation about an axis perpendicular to No. 3. Except for the arrival time the radiation pulse is identical in its effect on each molecule of the gas. The operator T can be written then as the product

$$T=\exp[i\omega\textstyle\sum_j t_jR_{j3}]\cdot\prod_l \exp i\left[\frac{\theta}{2}(R_{l+}\alpha+R_{l-}\alpha^*)+\theta' R_{l3}\right]\cdot\exp[-i\omega\textstyle\sum_j t_jR_{j3}]. \tag{67a}$$

The first and second rotations are through angles of θ' and θ, respectively, and the phase of α determines the direction of the 2nd rotation axis. It is assumed that $|\alpha|=1$ and that the arrival time at the jth molecule is

$$t_j=(1/\omega)\mathbf{k}\cdot\mathbf{r}_j. \tag{67b}$$

Equation (67a) becomes Eq. (68) after making use of (67b):

$$T=\exp i\frac{\theta}{2}(R_{\mathbf{k}+}\alpha+R_{\mathbf{k}-}\alpha^*)\cdot\exp i\theta' R_3. \tag{68}$$

It should be noted that the effect of the different times of arrival of the pulse at the various molecules is contained in $\mathbf{k}\cdot\mathbf{r}_j$ which appears in $R_{\mathbf{k}\pm}$ in Eq. (68).

The reason for choosing this transformation to be a rotation about No. 3 followed by a perpendicular rotation is that the rotation about No. 3 is the same as a time displacement and has no effect since the initial state is assumed to be one of thermal equilibrium.

Assume that the initial density matrix can be written as

$$\rho_0=\frac{\exp(-ER_3/kT)}{\operatorname{tr}\exp(-ER_3/kT)}=2^{-n}\prod_j(1-\gamma R_{j3}), \tag{69}$$

$$\gamma=2\tanh(E/2kT).$$

The density matrix after the radiation pulse is

$$\rho(t)=\exp(-i\omega tR_3)\cdot T\rho_0T^{-1}\exp(i\omega tR_3). \tag{70}$$

The spontaneous radiation rate after the exciting pulse is given by Eq. (62) which becomes

$$I(\mathbf{k}')=I_0(\mathbf{k}')\operatorname{tr}T\rho_0T^{-1}R_{\mathbf{k}'+}R_{\mathbf{k}'-}, \tag{71}$$

since R_3 commutes with $R_{\mathbf{k}'+}R_{\mathbf{k}'-}$. The radiation rate is thus independent of the time after the exciting pulse. This is because the effect of the radiated field on the gas has been neglected. Equation (71) is to be interpreted as the radiation rate immediately after the exciting pulse. Since ρ_0 and R_3 commute, Eq. (71) can be written as

$$I(\mathbf{k}')=I_0(\mathbf{k}')\operatorname{tr}\exp[\tfrac{1}{2}i\theta(R_{\mathbf{k}+}\alpha+R_{\mathbf{k}-}\alpha^*)]\cdot\rho_0\cdot\exp[-\tfrac{1}{2}i\theta(R_{\mathbf{k}+}\alpha+R_{\mathbf{k}-}\alpha^*)]\cdot R_{\mathbf{k}'+}R_{\mathbf{k}'-}. \tag{72}$$

It is desirable to transform ρ_0 before evaluating the trace

$$\rho'=\exp[\tfrac{1}{2}i\theta(R_{\mathbf{k}+}\alpha+R_{\mathbf{k}-}\alpha^*)]\cdot\rho_0\exp[-\tfrac{1}{2}i\theta(R_{\mathbf{k}+}\alpha+R_{\mathbf{k}-}\alpha^*)]=2^{-n}\textstyle\prod_j(1-\gamma R_{j3}{}^\dagger); \tag{73}$$

where

$$R_{j3}{}^\dagger=R_{j3}\cos\theta-\tfrac{1}{2}i(R_{j+}'\alpha-R_{j-}'\alpha^*)\sin\theta. \tag{74}$$

The primed operators are obtained from Eq. (53) as

$$R_{j\pm}'=R_{j1}'\pm iR_{j2}'=R_{j\pm}\exp(\pm i\mathbf{k}\cdot\mathbf{r}_j). \tag{75}$$

The trace in Eq. (72) can now be evaluated to give

$$I(\mathbf{k}')=I_0(\mathbf{k}')\sum_{jl}\operatorname{tr}2^{-n}\prod_s(1-\gamma R_{s3}{}^\dagger)R_{j+}''R_{l-}''. \tag{76}$$

The double prime is Eq. (75) referred to the $\mathbf{k}'$ direction. To evaluate the trace the following relations are needed: For A_i and B_j functions of the R's of molecules i and j,

$$\begin{gathered}\operatorname{tr}A_iB_j=2^{-n}\operatorname{tr}A_i\operatorname{tr}B_j,\\ \operatorname{tr}R_{j3}=\operatorname{tr}R_{j\pm}=0,\quad \operatorname{tr}R_{j3}{}^2=2^{n-2},\\ \operatorname{tr}R_{j+}R_{j-}=\operatorname{tr}R_{j-}R_{j+}=2^{n-1}.\end{gathered} \tag{77}$$

The final result is

$$I(\mathbf{k}')=I_0(\mathbf{k}')\cdot\tfrac{1}{2}n[1-\cos\theta\cdot\tanh(E/2kT)+\tfrac{1}{2}\sin^2\theta\cdot\tanh^2(E/2kT)\cdot(n|[\exp i(\mathbf{k}-\mathbf{k}')\cdot\mathbf{r}]_{Av}|^2-1)]. \tag{78}$$

Here the symbol $[\ ]_{Av}$ signifies a mean over all the molecules of the gas. For the example considered in Eq. (37a) this mean is unity, and Eq. (37a) follows by integrating over all directions of the emitted radiation. Aside from the factor $I_0(\mathbf{k}')$, the directional dependence of the emitted radiation is given by this mean. This factor is identical with the distribution factor for radiation about a set of classical isotropic radiators which have been excited by a plane wave. Consequently, for a θ of 90° and $n\tanh^2(E/kT)$ large compared with unity, the angular distribution of radiation is just the classical one.

The physical significance of the angle θ is that $\sin^2\tfrac{1}{2}\theta$

is the probability of the pulse exciting a molecule in its ground state. Also, if the exciting pulse is a constant amplitude wave of frequency ω during the duration of the pulse, the angle θ is proportional to the product of pulse amplitude and duration.

If the radiating system consists of a set of particles of spin $\frac{1}{2}$ in a uniform magnetic field, the angle θ has a geometrical significance. The initial state of a particle will have spin parallel or antiparallel to the field. The radiofrequency pulse will change its state such that its spin axis will be tipped through an angle θ. Note that if $\theta=180°$ the populations of the + and − populations have been just interchanged, corresponding to a transition from a positive temperature T to the negative temperature $-T$.[15] $\theta=90°$ corresponds to the excitation of molecules to energy superposition states Eqs. (38) and (39) for which the gas is radiating coherently.

ANGULAR CORRELATION OF SUCCESSIVE PHOTONS

The system to be considered here is assumed to be initially in thermal equilibrium. It is allowed to radiate spontaneously. The angular correlation between successive photons is calculated. This correlation was implicit in some of the earlier development, for example in Eq. (51a). As an example, consider a gas composed of widely separated molecules, all excited. Assume that a photon is emitted in the $\mathbf{k}$ direction. The radiation rate for the second photon in this direction is by Eq. (51a).

$$I(\mathbf{k})=I_0(\mathbf{k})2(n-1). \tag{79}$$

This is twice the incoherent rate. It is not hard to show that for an intermolecular spacing large compared with a radiation wavelength the radiation rate averaged over all directions is the incoherent rate. Hence from Eq. (79) the radiation probability in the direction $\mathbf{k}$ has twice the probability averaged over all directions.

In the problem to be considered, the system will consist initially of the gas in thermal equilibrium having a temperature T (possibly negative) and a photonless field. The molecules will be assumed fixed in position and with intermolecular distances large compared with a radiation wavelength. Photons are observed to be emitted in the directions $\mathbf{k}_1, \mathbf{k}_2, \cdots, \mathbf{k}_{s-1}$ and only these photons are emitted. The problem is one of finding the radiation rate in the $\mathbf{k}_s$ direction for the next photon.

Stated more exactly, it is assumed that there is an ensemble of gaseous systems, each with its own external radiation field. Every member of the ensemble which is capable of radiating will eventually radiate a photon. Those members which radiate their first photon into a small solid angle in the direction $\mathbf{k}_1$, are selected to form a new ensemble. For this second ensemble the time zero is taken to be the time that a photon was detected for each member of the ensemble.

It is convenient to calculate correlations for the gas systems forming a microcanonical distribution having an energy per gas system of m_0E. The results for a canonical distribution with a temperature T can subsequently be determined as an average over the microcanonical distributions.

Since the initial state of the system is assumed photonless, it is sufficient to give the explicit dependence of the initial density matrix on the molecular coordinates. Except for normalization this can be written as a projection operator for states of molecular energy m_0E. A particularly useful form for this density matrix is

$$\rho_0=\frac{\sum_{q=1}^{n}\exp 2\pi i\frac{q}{n}(R_3-m_0)}{\operatorname{tr}\sum_{q=1}^{n}\exp 2\pi i\frac{q}{n}(R_3-m_0)}. \tag{80}$$

This is a convenient way to write the density matrix because of the relation

$$\exp\left(2\pi i\frac{q}{n}R_3\right)=\prod_j \exp\left(2\pi i\frac{q}{n}R_{j3}\right)=\prod_j\left[\cos\left(\pi\frac{q}{n}\right)+2iR_{j3}\sin\left(\pi\frac{q}{n}\right)\right]. \tag{81}$$

Here the product is over $j=1, \cdots, n$. To illustrate the importance of Eq. (81) the trace appearing in the denominator D of Eq. (80) will be calculated using the relations Eq. (77).

$$\begin{aligned}D&=\sum_{q=1}^{n}\exp\left(-2\pi i\frac{q}{n}m_0\right)\cdot\operatorname{tr}\prod_j\left[\cos\left(\pi\frac{q}{n}\right)+2iR_{j3}\sin\left(\pi\frac{q}{n}\right)\right]\\&=\sum_{q=1}^{n}2^n\exp\left(-2\pi i\frac{q}{n}m_0\right)\cdot\cos^n\left(\pi\frac{q}{n}\right)\\&=\frac{n!n}{(\frac{1}{2}n+m_0)!(\frac{1}{2}n-m_0)!},\quad |m_0|<\frac{n}{2}\\&=2n\quad\text{for } |m_0|=n/2.\end{aligned} \tag{82}$$

After one photon has been emitted and absorbed in the photon detector, the system is again photonless and its density matrix is (see Appendix 1)

$$\rho_1=(R_{\mathbf{k}_1-}\rho_0R_{\mathbf{k}_1+})/(\operatorname{tr}R_{\mathbf{k}_1-}\rho_0R_{\mathbf{k}_1+}). \tag{83}$$

After $s-1$ photons it is

$$\rho_{s-1}=\frac{R_{\mathbf{k}_{s-1}-}\cdots R_{\mathbf{k}_1-}\rho_0R_{\mathbf{k}_1+}\cdots R_{\mathbf{k}_{s-1}+}}{\operatorname{tr}R_{\mathbf{k}_{s-1}-}\cdots R_{\mathbf{k}_1-}\rho_0R_{\mathbf{k}_1+}\cdots R_{\mathbf{k}_{s-1}+}}. \tag{84}$$

The R's are defined in Eqs. (48) and (47) or (46). The radiation rate in the $\mathbf{k}_s$ direction immediately after the $s-1$ photon is from Eq. (62)

$$I(\mathbf{k}_s)=I_0(\mathbf{k}_s)\operatorname{tr}R_{\mathbf{k}_s-}\rho_{s-1}R_{\mathbf{k}_s+}. \tag{85}$$

Note that $s\leqslant\frac{1}{2}n+m_0$. For any l, $R_{l\pm}^2=0$. Consequently,

[15] E. M. Purcell and R. V. Pound, Phys. Rev. 81, 279 (1951).

the numerator of Eq. (84) can be written

$$\frac{1}{(s-1)!}\sum_{u,v\cdots=1}^{s-1}\sum_{u',v'\cdots=1}^{s-1}\sum_{j,l\cdots=1}^{n}$$
$$\times \exp i[(\mathbf{k}_u-\mathbf{k}_{u'})\cdot\mathbf{r}_j+(\mathbf{k}_v-\mathbf{k}_{v'})\cdot\mathbf{r}_l+\cdots]. \quad (86)$$
$$R_{j-}R_{l-}\cdots\rho_0\cdots R_{l+}R_{j+}.$$

Each of the above sums is over $s-1$ indices, including only terms for which all $s-1$ indices take on different values. The trace of the expression appears in the denominator of Eq. (84). In order to evaluate this trace it is necessary first to evaluate

$$\operatorname{tr}R_{j-}R_{l-}\cdots\rho_0\cdots R_{l+}R_{j+}=\operatorname{tr}\rho_0\cdots R_{l+}R_{l-}R_{j+}R_{j-}$$
$$=\operatorname{tr}\rho_0\cdots(\tfrac{1}{2}+R_{l3})(\tfrac{1}{2}+R_{j3}). \quad (87)$$

If Eqs. (80), (81), and (82) are substituted into Eq. (87), and use is made of Eq. (77) and the equality

$$\operatorname{tr}[\cos(\pi q/n)+2iR_{j3}\sin(\pi q/n)](\tfrac{1}{2}+R_{j3})$$
$$=2^{n-1}\exp(i\pi q/n), \quad (87a)$$

Eq. (87) becomes

$$=\frac{2^{n-s+1}}{D}\sum_{q=1}^{n}\exp\left[i\pi\frac{q}{n}(s-1-2m_0)\right]\cdot\cos^{n-s+1}\left(\pi\frac{q}{n}\right)$$
$$=\frac{(n-s+1)!(\tfrac{1}{2}n+m_0)!}{n!(\tfrac{1}{2}n+m_0-s+1)!},\quad |m_0|<\tfrac{1}{2}n \text{ or } |m_0|=\tfrac{1}{2}n,\ s=1$$
$$=\tfrac{1}{2},\quad |m_0|=\tfrac{1}{2}n,\ s>1. \quad (88)$$

Making use of Eq. (88) the denominator of Eq. (84) can be written as

$$=P_{s-1}\frac{(n-s+1)!(\tfrac{1}{2}n+m_0)!}{n!(\tfrac{1}{2}n+m_0-s+1)!},\quad |m_0|<\tfrac{1}{2}n \text{ or}$$
$$|m_0|=\tfrac{1}{2}n,\quad s=1$$
$$=\tfrac{1}{2}P_{s-1},\quad m_0=\tfrac{1}{2}n,\ s>1, \quad (89)$$

where

$$P_{s-1}=\frac{1}{(s-1)!}\sum_{u,v\cdots=1}^{s-1}\sum_{u',v'\cdots=1}^{s-1}\sum_{j,l\cdots=1}^{n}$$
$$\times\exp i[(\mathbf{k}_u-\mathbf{k}_{u'})\cdot\mathbf{r}_j+(\mathbf{k}_v-\mathbf{k}_{v'})\cdot\mathbf{r}_l+\cdots],\quad s>1$$
$$P_0=1. \quad (90)$$

Here, as before, each of the above sums is over $s-1$ indices, including only terms for which all $s-1$ indices take on different values. If Eq. (84) is substituted into Eq. (85), the numerator is Eq. (89) with s increased by one unit. Consequently, substituting Eq. (89) into Eq. (85),

$$I(\mathbf{k}_s)=I_0(\mathbf{k}_s)\frac{P_s(\tfrac{1}{2}n+m_0-s+1)}{P_{s-1}(n-s+1)}. \quad (91)$$

To restate the meaning of this equation, $I(\mathbf{k}_s)$ is the radiation probability per unit time per unit solid angle in the direction $\mathbf{k}_s$; $I_0(\mathbf{k}_s)$ is the corresponding radiation probability for a single isolated excited molecule. It has been assumed that the gas was initially in the energy state m_0E [see Eq. (3)] with a random distribution over the degeneracy of this state. The gas was observed to radiate photons $\mathbf{k}_1$, $\mathbf{k}_2$, $\cdots\mathbf{k}_{s-1}$ previously to $\mathbf{k}_s$. Equation (91) is the radiation rate immediately after the $\mathbf{k}_{s-1}$ photon was observed. As a check on the correctness of this expression, note that the incoherent rate is obtained if $s=1$. Also, for $m_0=\tfrac{1}{2}n$ and $\mathbf{k}_1=\mathbf{k}_2=\cdots=\mathbf{k}_s=\mathbf{k}$, the radiation rate Eq. (91) agrees with Eq. (51a).

It should be noted that Eq. (91) is independent of the ordering of the subscripts 1, $\cdots$, $s-1$. Consequently, the angular distribution of the s photon is dependent upon the direction of a previous photon but is independent of the previous photon's position in the sequence of prior photons.

For a gas which contains a large number of randomly positioned molecules and for which previous photons have either been emitted in the direction $\mathbf{k}_s$ or in quite different directions, the radiation rate [Eq. (91)] is approximately equal to the incoherent rate times the number of photons previously emitted in this direction plus one.

Perhaps the case of most physical interest is where $s=2$. In this case Eq. (91) becomes

$$I(\mathbf{k}_2)=I_0(\mathbf{k}_2)\frac{\tfrac{1}{2}n+m_0-1}{n-1}[n|[\exp i\Delta\mathbf{k}\cdot\mathbf{r}]_{\mathrm{Av}}|^2+n-2],$$
$$\Delta\mathbf{k}=\mathbf{k}_2-\mathbf{k}_1. \quad (92)$$

The symbol $[\ ]_{\mathrm{Av}}$ signifies an average over all the molecular positions.

In case of a gas system at a temperature T, Eq. (91) must be averaged over all possible values of m_0 to give

$$\bar{I}(\mathbf{k}_s)=I_0(\mathbf{k}_s)\frac{P_s\displaystyle\sum_{m_0=s-\frac{1}{2}n-1}^{\frac{1}{2}n}(\tfrac{1}{2}n+m_0+1-s)\frac{n!}{(\tfrac{1}{2}n+m_0)!(\tfrac{1}{2}n-m_0)!}\exp\left(-\frac{m_0E}{kT}\right)}{(n-s+1)P_{s-1}\displaystyle\sum_{m_0=s-\frac{1}{2}n-1}^{\frac{1}{2}n}\frac{n!\exp(-m_0E/kT)}{(\tfrac{1}{2}n+m_0)!(\tfrac{1}{2}n-m_0)!}}. \quad (93)$$

For $|E/kT|\ll1$ and $s\ll n$, Eq. (93) can be approximated by

$$\bar{I}(\mathbf{k}_s)=I_0(\mathbf{k}_s)\frac{(\tfrac{1}{2}n+\bar{m}_0+1-s)P_s}{(n-s+1)P_{s-1}}, \quad (94)$$

where

$$\bar{m}_0=-\tfrac{1}{4}nE/kT.$$

It is a pleasure to acknowledge the assistance of the author's colleague, Professor A. S. Wightman, who read the manuscript and made a number of helpful suggestions.

APPENDIX I

It is assumed that the system consists initially of a gas with an energy m_0E and a photonless radiation field. A photon and only one photon is observed to be emitted. The effect of the photon emission on the state of the system is required.

There are two separate effects to be considered. First there is the effect on the state of the system which has its origin in the interaction between the field and gas. Second there is the effect of the observation which determines that a photon and one photon only has been emitted, that this photon was emitted in the **k** direction, and that the photon was absorbed in the detector. The first part of the problem is solved using Schrödinger's equation. The Hamiltonian of the system is

$$H=\hbar\omega R_3+H_0+H',\quad H_0=\sum_{\mathbf{k}'}H_{\mathbf{k}'},$$
$$H'=-\tfrac{1}{2}\sum_{\mathbf{k}'}[\mathbf{v}_{\mathbf{k}'}\cdot\mathbf{e}R_{\mathbf{k}'+}+\mathbf{v}_{\mathbf{k}'}^*\cdot\mathbf{e}^*R_{\mathbf{k}'-}]. \tag{95}$$

Here $H_{\mathbf{k}'}$ is the energy of the $\mathbf{k}'$ radiation oscillator. Assume a pure state represented by a wave function ψ_0 at a time $t=0$. Assume that ψ_0 is an eigenstate of R_3 and is photonless. At some later time it is

$$\psi(t)=\exp(-iHt/\hbar)\psi_0=\left(1-\frac{i}{\hbar}Ht-\frac{H^2}{2\hbar^2}t^2+\cdots\right)\psi_0. \tag{96}$$

For the quadratic and higher powers of t each term will be a sum of products of H' and $(H_0+\hbar\omega R_3)$. However, the interaction term H' consists of sums of terms of the type

$$U_{\mathbf{k}'}=\mathbf{v}_{\mathbf{k}'}\cdot\mathbf{e}R_{\mathbf{k}'+} \tag{97}$$

and its Hermitian adjoint. The operator $U_{\mathbf{k}'}$ consists of the product of a photon annihilation operator and a gas excitation operator. It converts an eigenstate of R_3 and H_0 into another such or it gives zero. The most general term operating on ψ_0 in Eq. (96) is therefore a product of powers of $H_0+\hbar\omega R_3$ and terms of the type $U_{\mathbf{k}'}$ and $U_{\mathbf{k}'}^*$ taken in various orders. In each of these terms $H_0+\hbar\omega R_3$ always operates on an eigenfunction and consequently can be moved to the end of the product as a number, the eigenvalue. Consequently $\psi(t)$ becomes

$$\psi(t)=[1+\sum_{\mathbf{k}'}g_{\mathbf{k}'}(t)U_{\mathbf{k}'}^*+\sum_{\mathbf{k}'}h_{\mathbf{k}'}(t)U_{\mathbf{k}'}U_{\mathbf{k}'}^*+\sum_{\mathbf{k}'\mathbf{k}''}g_{\mathbf{k}'\mathbf{k}''}(t)U_{\mathbf{k}'}^*U_{\mathbf{k}''}^*+\cdots]\psi_0. \tag{98}$$

The g's and h's are numbers, functions of the time. It may be noted that since ψ_0 represents a photonless state, an annihilation operator for a given radiation oscillator $\mathbf{k}'$ appears only if preceded by the corresponding creation operator.

Assuming that at the time t a photon measurement is made which indicates the presence of photon **k** and no other photons, the wave function after the measurement is

$$\psi'=P_{\mathbf{k}}\psi, \tag{99}$$

where the operator $P_{\mathbf{k}}$ is a projection operator for the **k** photon state.

$$P_{\mathbf{k}}=\frac{H_{\mathbf{k}}}{\hbar\omega_{\mathbf{k}}}\prod_{\mathbf{k}'}'\left(\frac{\hbar\omega_{\mathbf{k}'}-H_{\mathbf{k}'}}{\hbar\omega_{\mathbf{k}'}}\right). \tag{100}$$

The product is over all $\mathbf{k}'\neq\mathbf{k}$. Two-photon excitation of one radiation oscillator has been neglected.

$$\psi'=[g_{\mathbf{k}}(t)U_{\mathbf{k}}^*+\sum_{\mathbf{k}'}H_{\mathbf{k}\mathbf{k}'}(t)U_{\mathbf{k}}^*U_{\mathbf{k}'}U_{\mathbf{k}'}^*+\sum_{\mathbf{k}}I_{\mathbf{k}\mathbf{k}'}(t)U_{\mathbf{k}'}U_{\mathbf{k}'}^*U_{\mathbf{k}}^*+\cdots]\psi_0. \tag{101}$$

In summing over the direction of $\mathbf{k}'$ in the second and third terms above, the expression

$$R_{\mathbf{k}'+}R_{\mathbf{k}'-}=\sum_{ab}\exp[i\mathbf{k}'\cdot(\mathbf{r}_a-\mathbf{r}_b)]\cdot R_{a+}R_{b-} \tag{102}$$

appears under the integral. By expanding the exponential in spherical harmonics it can be seen that for $a\neq b$ this integral vanishes, as it has been assumed that

$$\mathbf{k}'\cdot(\mathbf{r}_a-\mathbf{r}_b)\gg1\ \text{for}\ a\neq b.$$

It should be indicated that the angular dependence is not wholly in the exponential in Eq. (102) but exists in part in the square of the dot product of $\mathbf{e}$ and $\mathbf{v}_{\mathbf{k}'}$. However, this contribution to the angular dependence includes only spherical harmonics of finite degree in fact with $l<3$. As the only terms which need to be included in Eq. (102) are $a=b$, Eq. (102) becomes

$$R_{\mathbf{k}'+}R_{\mathbf{k}'-}=\tfrac{1}{2}+R_3+(\text{terms from } a\neq b). \tag{103}$$

Independent of its position in a series of products of U's the expression on the right side of Eq. (103) will operate on an eigenfunction and becomes an eigenvalue which can be removed as a number. In the higher-order terms in Eq. (101) $U_{\mathbf{k}'}$ and $U_{\mathbf{k}'}^*$ may not appear adjacent to each other, but if they do not, some other pair such as $U_{\mathbf{k}''}U_{\mathbf{k}''}^*$ will appear, and after removing this as an eigenvalue another such pair will occur, and eventually the $\mathbf{k}'$ pair will be adjacent. Consequently, to all orders in the expansion

$$\psi'=f(t)U_{\mathbf{k}}^*\psi_0, \tag{104}$$

where f is a function of the time of observation. As the photon detector also absorbs the photon, the wave function must be multiplied by the annihilation operator $\mathbf{e}\cdot\mathbf{v}_{\mathbf{k}}$. This gives, except for the time factor,

$$\psi''\sim R_{\mathbf{k}-}\psi_0, \tag{105}$$

which is another photonless state but with one quantum less energy.

If the initial density matrix ρ_0 contains only photonless states of the same energy m_0E, then from Eqs. (63) and (105) it is transformed to

$$\rho_1=R_{\mathbf{k}-}\rho_0R_{\mathbf{k}+}/\mathrm{tr}(R_{\mathbf{k}-}\rho_0R_{\mathbf{k}+}), \tag{106}$$

representing the photonless state of the ensemble of systems after the emission, detection, and absorption of photon described by **k**.

Molecular Microwave Oscillator and New Hyperfine Structure in the Microwave Spectrum of NH_3†

J. P. Gordon, H. J. Zeiger,* and C. H. Townes
Department of Physics, Columbia University, New York, New York
(Received May 5, 1954)

An experimental device, which can be used as a very high resolution microwave spectrometer, a microwave amplifier, or a very stable oscillator, has been built and operated. The device, as used on the ammonia inversion spectrum, depends on the emission of energy inside a high-Q cavity by a beam of ammonia molecules. Lines whose total width at half-maximum is six to eight kilocycles have been observed with the device operated as a spectrometer. As an oscillator, the apparatus promises to be a rather simple source of a very stable frequency.

A block diagram of the apparatus is shown in Fig. 1. A beam of ammonia molecules emerges from the source and enters a system of focusing electrodes. These electrodes establish a quadrupolar cylindrical electrostatic

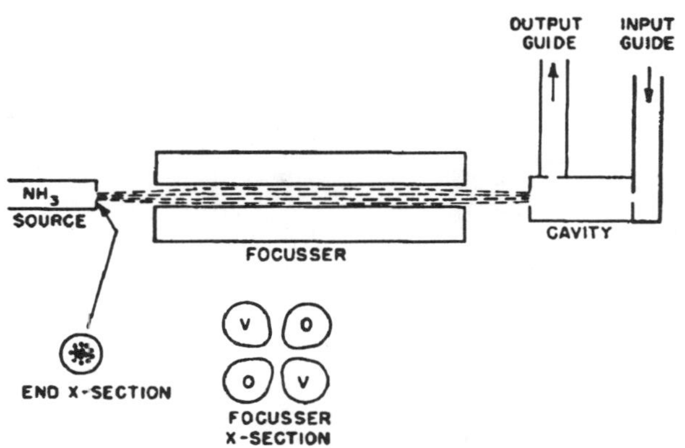

FIG. 1. Block diagram of the molecular beam spectrometer and oscillator.

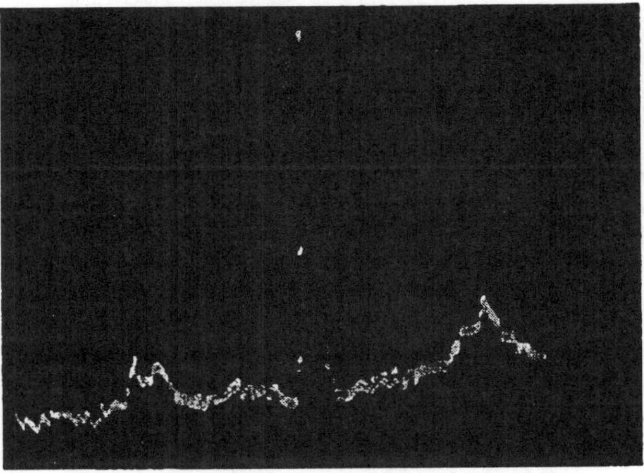

FIG. 2. A typical oscilloscope photograph of the NH_3, $J=K=3$ inversion line at 23 870 Mc/sec, showing the resolved magnetic satellites. Frequency increases to the left.

field whose axis is in the direction of the beam. Of the inversion levels, the upper states experience a radial inward (focusing) force, while the lower states see a radial outward force. The molecules arriving at the cavity are then virtually all in the upper states. Transitions are induced in the cavity, resulting in a change in the cavity power level when the beam of molecules is present. Power of varying frequency is transmitted through the cavity, and an emission line is seen when the klystron frequency goes through the molecular transition frequency.

If the power emitted from the beam is enough to maintain the field strength in the cavity at a sufficiently high level to induce transitions in the following beam, then self-sustained oscillations will result. Such oscillations have been produced. Although the power level has not yet been directly measured, it is estimated at about 10^{-8} watt. The frequency stability of the oscillation promises to compare favorably with that of other possible varieties of "atomic clocks."

Under conditions such that oscillations are not maintained, the device acts like an amplifier of microwave power near a molecular resonance. Such an amplifier may have a noise figure very near unity.

High resolution is obtained with the apparatus by utilizing the directivity of the molecules in the beam. A cylindrical copper cavity was used, operating in the *TE*011 mode. The molecules, which travel parallel to the axis of the cylinder, then see a field which varies in amplitude as $\sin(\pi x/L)$, where x varies from 0 to L. In particular, a molecule traveling with a velocity v sees a field varying with time as $\sin(\pi vt/L)\sin(\Omega t)$, where Ω is the frequency of the rf field in the cavity. A Fourier analysis of this field, which the molecule sees from $t=0$ to $t=L/v$, gives a frequency distribution whose amplitude drops to 0.707 of its maximum at points separated by a $\Delta\nu$ of $1.2v/L$. The cavity used was twelve centimeters long, and the most probable velocity of ammonia molecules in a beam at room temperature is 4×10^4 cm/sec. Since the transition probability is proportional to the square of the field amplitude, the resulting line should have a total width at half-maximum given by the above expression, which in the present case is 4 kc/sec. The observed line width of 6–8 kc/sec is close to this value.

The hyperfine structure of the ammonia inversion transitions for $J=K=2$ and $J=K=3$ has been examined, and previously unresolved structure due to the reorientation of the hydrogen spins has been observed. Figure 2 is a typical scope photograph of these new magnetic satellites on the 3,3 line. The observed spectra for the 3,3 line is shown in Fig. 3, which contains all the observed hyperfine structure components, including the quadrupole reorientation transitions of the nitrogen nucleus, which have been previously observed as single lines.

Within the resolution of the apparatus, the hyperfine structures of the upper and lower inversion levels are

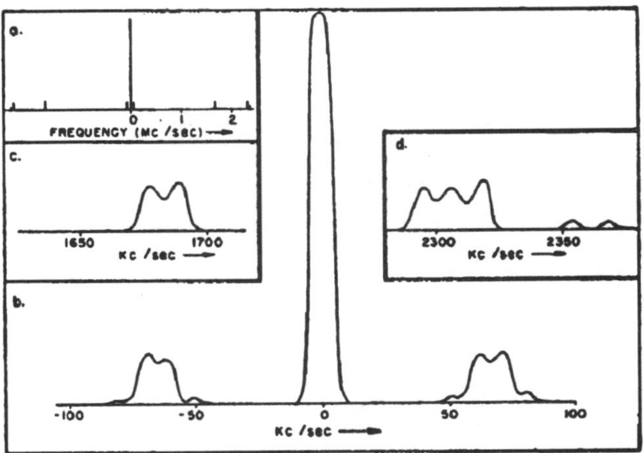

FIG. 3. The observed hyperfine spectrum of the 3,3 inversion line. (a) Complete spectrum, showing the spacings of the quadrupole satellites. (b) Main line with magnetic satellites. (c) Structure of the inner quadrupole satellites. (d) Structure of the outer quadrupole satellites. The quadrupole satellites on the low-frequency side of the main line are the mirror images of those shown, which are the ones on the high-frequency side.

identical, as evidenced by the fact that the main line is not split. Symmetry considerations require that the hydrogen spins be in a symmetric state under 120-degree rotations about the molecular axis. Thus for the 3,3 state, $I_H=3/2$, and one expects each of the quadrupole levels to be further split into four components by the interaction of the hydrogen magnetic moments with the various magnetic fields of the molecule. At the present writing, the finer details of the expected magnetic splittings have not been worked out.

This type of apparatus has considerable potentialities as a more general spectrometer. Since the effective dipole moments of molecules depend on their rotational state, some selection of rotational states could be effected by such a focuser. Similarly, a focuser using magnetic fields would allow spectroscopy of atoms. Sizable dipole moments are required for a strong focusing action, but within this limitation, the device may prove to have a fairly general applicability for the detection of transitions in the microwave region.

The authors would like to acknowledge the expert help of Mr. T. C. Wang during the latter stages of this experiment.

† Work supported jointly by the Signal Corps, the U. S. Office of Naval Research, and the Air Force.

* Carbide and Carbon post-doctoral Fellow in Physics, now at Project Lincoln, Massachusetts Institute of Technology, Cambridge, Massachusetts.

Spin Resonance of Free Electrons Polarized by Exchange Collisions*†

H. G. Dehmelt
Department of Physics, University of Washington, Seattle, Washington
(Received September 16, 1957)

An experiment is described in which thermal electrons, $t_e \approx 400°K$, become polarized in detectable numbers by undergoing exchange collisions with oriented sodium atoms during which the atom orientation is transferred to the electrons. The collisions establish interrelated equilibrium values for the atom and electron polarizations which depend upon the balance between the polarizing agency acting upon the atoms (optical pumping) and the disorienting relaxation effects acting both on atoms and electrons. When now the electrons are furthermore artificially disoriented by gyromagnetic spin resonance, an additional reduction of the atom polarization ensues which is detected by an optical monitoring technique, thereby allowing a determination of the free-electron spin g factor, g_e. Since it was experimentally convenient, at this stage only the ratio $g_J(\mathrm{Na})/g_e = 1.000026 \pm 0.00003$ was determined, showing no significant difference between g_e and $g_J(\mathrm{Na})$, the g factor of the $^2S_{\frac{1}{2}}$ sodium ground state. From the experimental strength and width of the electron disorientation signal a lower limit was obtained for the sodium exchange cross section with thermal electrons: $Q > 2.3 \times 10^{-14}$ cm^2. This may be compared with a theoretical exchange cross section, $Q = 2.3 \times 10^{-14}$ cm^2, which is derived under the assumption that the $3s^2\,{}^1S_0$ state of the Na$^-$ ion has essentially zero binding energy, thereby causing strong singlet scattering while the triplet scattering is negligible in comparison. Spin-orbit coupling during collisions of the electrons with the atoms of the inert argon buffer employed to slow down wall diffusion is discussed as the chief cause for the shortness of the observed free-electron spin relaxation time, $T_e \approx 6 \times 10^{-5}$ sec.

INTRODUCTION

CONSIDERABLE interest exists in experimental determinations of the free-electron spin magnetic moment μ_e in terms of the Bohr magneton μ_0 with accuracies high enough to provide further tests for the theoretical values,

$$\mu_e/\mu_0 = 1 + (\alpha/2\pi) + \text{higher terms},$$

obtained from quantum electrodynamics.[1–3] There are experimental values available[4,5] with an accuracy of about 10^{-6} for $g_J(H)/g_p$, the ratio of the g factors of the hydrogen ground state to that of the proton, which after a small relativistic bound state correction yield accurate g_e/g_p values that can be combined with other experimental data for[6] μ_p/μ_0 to obtain the desired ratio μ_e/μ_0. However, a *direct* experimental determination of the free-electron spin g factor in terms of g_p or $g_J(H)$ with an accuracy of 10^{-6} or better would be highly desirable. Various experimental schemes[7] have been proposed to accomplish this; however, no accuracies higher than 5×10^{-3} appear to have been reported so far. The present experiment was carried out on thermal electrons and an accuracy of 3×10^{-5} was achieved in preliminary measurements which also indicated that an increase in accuracy by one or two orders of magnitude should be possible.

PRINCIPLE OF EXPERIMENT AND APPARATUS

The electrons were polarized by allowing them to undergo exchange collisions with oriented sodium atoms in which the total spin component with respect to the axis of orientation, a magnetic field H_0, is conserved and the orientation of the atoms is transferred to the initially unpolarized electrons by exchange of the spin directions. Electrons and atoms, the latter polarized by optical pumping, were allowed to diffuse in an inert buffer, argon or helium at pressures of a few centimeters Hg. Since the electron-sodium collisions tend to equalize the polarization ratios of sodium atoms and electrons, the mere presence of free electrons reduced the sodium equilibrium polarization because the disorienting relaxation effects acting upon the electrons are passed on to the sodium atoms. In the same fashion, resonance disorientation of the electrons by a magnetic rf field of the proper frequency fulfilling the gyromagnetic resonance condition,

$$\nu_e = g_e \mu_0 H_0 / h,$$

caused a further decrease in the sodium orientation. This orientation decrease was detected by an optical-absorption monitoring technique, thereby allowing a determination of the free-electron spin g factor g_e. A typical experiment (see Fig. 1) employed a spherical

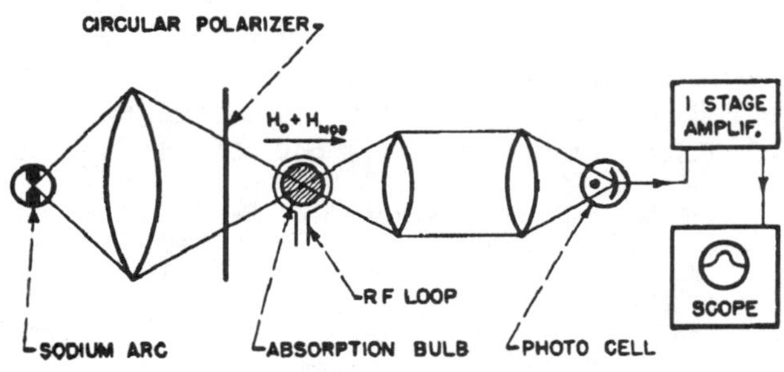

Fig. 1. Free-electron spin resonance apparatus.

* Supported by the U. S. Office of Ordnance Research.
† Early results of this work were reported at the 123rd meeting of the American Association for the Advancement of Science, December 26, 1956.

1 J. Schwinger, Phys. Rev. 73, 416 (1948).
2 R. Karplus and N. Kroll, Phys. Rev. 77, 536 (1950).
3 C. M. Sommerfield, Phys. Rev. 107, 328 (1957).
4 Koenig, Prodell, and Kusch, Phys. Rev. 88, 191 (1952).
5 R. Beringer and M. A. Heald, Phys. Rev. 95, 1474 (1954).
6 J. H. Gardner and E. M. Purcell, Phys. Rev. 76, 1262 (1949).
7 Reviewed by H. A. Tolhoek, Revs. Modern Phys. 28, 277 (1956).

200-cm^3 absorption bulb heated to around 140°C which contained 70 mm of 2.5×10^{18} atoms per cm^3 argon and sodium vapor of a density of about 8×10^9 per cm^3, corresponding roughly to a pressure of 10^{-7} mm. The sodium density N was estimated from the ratio of the transmitted to the indident light fluxes I_1/I_0 with the help of the formula[7a] for "line absorption"

$$1-(I_1/I_0)\approx1.4(\pi\ln2)^{\frac{1}{2}}r_0cNd/\Delta\nu_A,$$

which holds for small absorption and the source width equal to the absorber width $\Delta\nu_A$, r_0 being the classical electron radius and d the thickness of the absorbing layer. The value 3×10^9 sec^{-1} was taken for the pressure-broadened absorption line width $\Delta\nu_A$. For the purpose of ionization and generation of free electrons, the absorption bulb was placed between two capacitor plates (not shown in Fig. 1) to which 25-Mc/sec rf pulses of around 10^{-3} sec duration at a repetition rate of about 10 per second were applied synchronously with the sweep of the oscilloscope used for the observation of the disorientation signals and the sawtooth modulating field H_{mod}. Under these conditions, after each discharge pulse the electron temperature drops within 50 microseconds to the gas temperature. The electron density,[8,9] n, decays due to volume recombination approximately according to $1/n=(1/n_0)+\alpha t$, since wall diffusion of the electrons and ions can be neglected as long as the observation interval is short compared with the average ambipolar wall diffusion time $T_D=(1/D_a)(R/\pi)^2$ which, with $D_a(\text{argon})=91$ cm^2/sec at 1 mm pressure, is about 1 sec. Rough measurements of the rf conductivity σ of the decaying plasma at 25 Mc/sec indicated a decay of n from 3.2×10^8 cm^{-3} to 1.6×10^8 cm^{-3} during the 0.1 sec long usable portion of the observation interval, the connection[10] between n and σ being given by $n=(m/e^2)\omega_c\sigma$. The value 2×10^{10} sec^{-1} was assumed for the electron collision frequency ω_c in the argon buffer.

The optical system[11] now functioned in the following way: Light from the sodium arc was made circular polarized by a commercial polarizing plate and served to orient the sodium atoms in the bulb by optical pumping.[12] The transmitted light then was focussed upon a vacuum photocell connected to an oscilloscope through an amplifier. The amplified photocurrent was a measure of the sodium orientation P since the oriented atoms absorb less than unoriented ones. The axial magnetic field $H_0\approx21.4$ gauss was provided by a Helmholtz coil 30 inches in diameter. The modulation field was furnished by a separate ring coil. The rf loop was energized from a 62.08-Mc oscillator for the electron spin disorientation field and simultaneously from a second tunable oscillator in order to induce consecutively the four $\Delta m_F=\pm1$ transitions between the magnetic sublevels of the $F=2$ sodium hfs level at about 15.10, 15.36, 15.63, and 15.92 Mc which served to calibrate the magnetic field H_0.

INTERDEPENDENCE OF POLARIZATIONS

We now consider the factors relating the electron and sodium polarizations p and P. Of the N atoms, disregarding their nuclear spins, and the n electrons contained in one cubic centimeter, N_+ and n_+ have their spins up, N_- and n_- have them down. We define $P=(N_+-N_-)/N$; $p=(n_+-n_-)/n$. The cross section for exchange of the spin direction when oppositely oriented electrons and atoms meet will be denoted by Q. For simplicity the atoms are considered at rest, while at absolute temperature t_e the electrons are assumed to move with fixed speed $v=(3kt_e/m)^{\frac{1}{2}}$, their velocity distribution being neglected. Then the time variation of n_+ due to electron-sodium collisions alone is given by $\dot{n}_+=vQ[N_+n_--N_-n_+]$ which leads to $\dot{p}=f(P-p)$ for the electron polarization and $\dot{P}=F(p-P)$ for the atom polarization. Here the frequency of collision of an electron with sodium atoms, $f=vQN$, and that for a sodium atom to be hit by electrons, $F=vQn$, have been introduced. On the other hand, the atoms are continuously polarized by optical pumping and depolarized by relaxation effects of characteristic time T_a. In the absence of free electrons the time variation of P due to these processes is described[11] by

$$\dot{P}=cI_0(\bar{P}-P)-(1/T_a)P,$$

which can be rewritten

$$\dot{P}=(1/\tau)(P_I-P),\quad \tau=T_a/(cI_0T_a+1),$$

where $P_I=cI_0T_a\bar{P}/(cI_0T_a+1)$ is the equilibrium polarization corresponding to a finite light intensity I_0 and $\bar{P}$ is the saturation polarization obtainable with the optical pumping. The relaxation effects of characteristic time T_e acting on the electrons in the absence of sodium atoms would cause their polarization to decay according to $\dot{p}=-(1/T_e)p$. By combining all these contributions to $\dot{P}$ and $\dot{p}$ when the corresponding processes are simultaneously present, we get

$$\dot{p}=f(P-p)-(1/T_e)p,$$
$$\dot{P}=F(p-P)+(1/\tau)(P_I-P).$$

By setting $\dot{p}$ and $\dot{P}$ equal to zero, we now obtain the equilibrium polarizations $\mathbf{p}$ and $\mathbf{P}$,

$$\mathbf{p}=fT_e(fT_e+F\tau+1)^{-1}P_I,$$
$$\mathbf{P}=(fT_e+1)(fT_e+F\tau+1)^{-1}P_I.$$

The effect of resonance disorientation of the electrons

[7a] A. C. G. Mitchell and M. W. Zemansky, *Resonance Radiation and Excited Atoms* (Cambridge University Press, 1934), Chap. III.
[8] M. A. Biondi and S. C. Brown, Phys. Rev. 75, 1700 (1949).
[9] A. von Engel, *Ionized Gases* (Clarendon Press, Oxford, 1955).
[10] Cf. e.g., H. Belcher and T. M. Sugden, Proc. Roy. Soc. (London) A201, 480 (1950).
[11] Cf. H. G. Dehmelt, Phys. Rev. 105, 1487 (1957); 105, 1924 (1957).
[12] Cf., e.g., A. Kastler, J. Opt. Soc. Am. 47, 460 (1957).

by a magnetic rf field[13] is equivalent to shortening the electron relaxation time. It can be described by substituting for $1/T_e$ the modified quantity $1/T_e' = (1/T_e) + (1/T_{rf})$, where T_{rf} is the rf disorientation time which we define for exact resonance and which is connected with the rf field amplitude H_1 and the characteristic time T_2^* by $\omega_1^2 T_2^* T_{rf} = 1$. Here $\omega_1 = \pi g_e \mu_0 H_1/h$ would be the precession circular frequency in the field $H_1/2$. The time $T_2^* = 1/\pi\Delta\nu$ is a measure of the total experimental electron line width $\Delta\nu$ and, assuming small atom polarization, is given by

$$1/T_2^* \approx (1/T_e) + f + (1/T_2'),$$

while T_2' represents the contribution by the magnetic field inhomogeneity. As a consequence of the foregoing, the atom polarization is a function of the rf field H_1 acting upon the electrons, $\mathbf{P} = \mathbf{P}(H_1)$. For a given light intensity I_0 and therefore given τ, a measure of the optical signal resulting when the magnetic field $H_0 + H_{\text{mod}}$ is swept through the resonance value is provided by

$$S(H_1) = [\mathbf{P}(0) - \mathbf{P}(H_1)]/P_I,$$

the maximum possible signal being

$$S(\infty) = fT_e F\tau(F\tau+1)^{-1}(fT_e + F\tau + 1)^{-1}.$$

Radio-frequency saturation will become appreciable for $H_1 > H_1^*$, the latter quantity being defined by $S(H_1^*) = \frac{1}{2}S(\infty)$. For the corresponding critical disorientation time T_{rf}^*, one obtains

$$1/T_{rf}^* = f(F\tau+1)^{-1} + (1/T_e).$$

In order to use the experimentally observed signal $S_{\text{exp}} \approx 0.1$ to put a lower limit on Q, we note that

$$0.1 = S_{\text{exp}} < S(\infty) < fT_e F\tau = nNT_e \tau v^2 Q^2.$$

With the experimental values $v = 1.1\times10^7$ cm sec^{-1}, $N = 8\times10^9$ cm^{-3}, $n = 1.6\times10^8$ cm^{-3}, $\tau = 2\times10^{-2}$ sec, and $T_e = 6\times10^{-5}$ sec, we have

$$Q > 2.3\times10^{-14}\ \text{cm}^2.$$

Here T_e was obtained from the experimental line width data by assuming

$$\Delta\nu(\text{electron}) - 4\Delta\nu(\text{Na}) \approx 1/\pi T_e \approx 5.6\times10^3\ \text{sec}^{-1}.$$

EXCHANGE CROSS SECTION

The large observed exchange cross section Q can be understood as follows: There is evidence that the $3s^2\,{}^1S_0$ state of the Na$^-$ ion exists,[14,15] its binding energy W being close to zero. In this case the cross section for singlet s scattering, Q_- (only s-wave scattering need be considered at the low energies of interest here) can be expressed approximately for small energies E of the impinging electron by[16,17]

$$Q_- = 4\pi(\hbar^2/2m)(E + |W|)^{-1}.$$

This approaches the maximum possible cross section $4\pi\lambda^2$ for $|W| \ll E$. The cross section for triplet scattering, Q_+, should be much smaller than Q_- since no bound triplet level exists and the above resonance effect does not occur. Therefore, as an approximation we neglect Q_+. Under this assumption one finds for the exchange cross section, $Q = \frac{1}{4}Q_-$. For thermal electrons (400°K) and $|W| \ll E$, the exchange cross section‡ then assumes nearly its upper limit $\bar{Q}$,

$$\bar{Q} = (\pi\hbar^2/3mkt_e) = 2.3\times10^{-14}\ \text{cm}^2,$$

which in accordance with the earlier simplifying assumption of a fixed electron velocity has not been averaged over the electron energy distribution. In order to see that the exchange cross section Q is one-fourth as large as the singlet scattering cross section Q_-, we consider the asymptotic behavior of a mixed state Ψ which consists of equal parts of properly symmetrized singlet and triplet states,

$$\Psi = \tfrac{1}{2}[f(1)g(2) + f(2)g(1)][v_+(1)v_-(2) - v_+(2)v_-(1)] + \tfrac{1}{2}[f(1)g(2) - f(2)g(1)][v_+(1)v_-(2) + v_+(2)v_-(1)].$$

For the case of interest here, that the electron 1 in the free state f is initially at a large distance from the scattering atom in whose ground state g the electron 2 moves, $f(2)g(1)$ is nearly zero and can be neglected. The state then reduces to $\Psi = f(1)g(2)\,v_+(1)v_-(2)$, which corresponds to a definite situation where the electron 1 is free and has its spin up while the bound one 2 has its spin down; v_+ and v_- denote the spin functions with $m_s = +1$ and $m_s = -1$, respectively. If we now, as usual, represent $f(1)$ as a plane wave, Ψ will be associated with an electron stream of current density j of which $j/2$ will correspond to the singlet and $j/2$ to the triplet wave. Since only the singlet part is assumed to be scattered by the atom, the total scattered (singlet) current is given by $i_- = \frac{1}{2}jQ_-$. This current now consists of electrons 50% of which have exchanged their spins with the scattering atom. For the total current of spin-exchanged electrons, $i = \frac{1}{2}i_-$, we obtain therefore

$$i = \tfrac{1}{4}jQ_-, \quad \text{or} \quad Q = \tfrac{1}{4}Q_-.$$

ELECTRON SPIN RELAXATION

The main electron spin relaxation mechanism appears to be spin-orbit coupling during electron-argon collisions, which can be fairly accurately analyzed. First

[13] Cf., e.g., Bloembergen, Purcell, and Pound, Phys. Rev. **73**, 679 (1948).

[14] G. Glocker, Phys. Rev. **46**, 111 (1934).

[15] D. R. Hartree and W. Hartree, Proc. Cambridge Phil. Soc. **34**, 550 (1938).

[16] E. Wigner, Z. Physik **83**, 253 (1933).

[17] N. F. Mott and H. S. W. Massey, *Theory of Atomic Collisions* (Clarendon Press, Oxford, 1947), Chap. 2.

‡ For the exchange cross section Q to approach its maximum value, $\bar{Q} = \pi\lambda^2$, It would be sufficient that a level of one multiplicity, singlet or triplet, bound or virtual, lie much closer to zero than the free electron energy E, while the closest level of the other multiplicity is much further away than E.

we try to find the angle α through which a spin precesses during such a collision. Noting that simultaneously appreciable precession angles and scattering cross sections will occur only in p scattering and that the penetrating parts of the orbitals of a free, low-energy p electron around an argon atom and of a loosely bound p electron in a potassium atom should be very similar, we can calculate α from the doublet splitting $\delta\nu$ [cm^{-1}] and the classical period of revolution, $T=h^3(4\pi^2me^4)^{-1}n^3$, for potassium p orbitals of high principal quantum number n. We obtain

$$\alpha=2\pi c\sin\vartheta T\delta\nu=2.8\times10^{-5}\sin\vartheta n^3\delta\nu,$$

where ϑ is the angle which the spin direction makes with the resultant of spin and orbital angular momentum. Numerically, with $n^3\delta\nu=10^3$ cm^{-1} which for $n>12$ is practically constant, we find

$$\alpha=0.028\sin\vartheta.$$

The relaxation time T_e which is associated with the random walk steps α which the tip of the unit spin vector executes on the unit sphere and with the frequency of collisions with argon atoms, f_p, is given by[17a]

$$1/T_e=\tfrac{1}{2}f_p\langle\alpha^2\rangle_{Av}=\tfrac{1}{2}vq_pN_A\langle\alpha^2\rangle_{Av}.$$

With $\langle\sin^2\vartheta\rangle_{Av}\approx\frac{2}{3}$ and the experimental value $T_e=6\times10^{-5}$ sec and taking $N_A=2.5\times10^{18}$ cm^{-3}, $v=1.1\times10^7$ cm/sec, we find from this for q_p, the partial cross section for p scattering,

$$q_p=2.5\times10^{-18}\text{ cm}^2.$$

This value can be compared with a theoretical value extrapolated from Holtsmark's calculations.[18] Since in the limit of large de Broglie wavelength or small electron energy E the exact shape of the short-range scattering potential is immaterial, a square well may be substitured, for which it can be shown[17] that $q_p\propto E^2$ for $E\to0$. In this way one finds $q_p=1.03\times10^{-24}t_e^2$ cm^2, where t_e is the absolute electron temperature. Again for simplicity we have not taken an average over the energy distribution of the electrons, assuming instead a fixed energy $E=\frac{3}{2}kt_e$. With $t_e=400$°K we get numerically

$$q_p=1.65\times10^{-19}\text{ cm}^2.$$

The strong temperature dependence of q_p and therefore T_e is in agreement with the experimentally observed quenching of the electron signal by weak electric rf fields which heat up the electrons. It is likely that even the electric rf field associated with the H_1 field caused appreciable heating of the electrons since the electron signal got weaker and weaker with increasing frequency and at 120 Mc/sec no signal could be observed in the present apparatus. The assumption of some increase in t_e by this H_1 heating and proper averaging would make the theoretical and experimental q_p values more nearly equal also.

[17a] D. Pines and C. P. Slichter, Phys. Rev. 100, 1014 (1955).
[18] T. Holtsmark, Z. Physik 55, 437 (1929).

ELECTRON-SODIUM g-FACTOR RATIO

As the chief goal of the present experiment a preliminary determination of the free electron spin g factor in terms of the g factor of the $S_{\frac{1}{2}}$ Na ground state was carried out by comparing the free-electron precession frequency ν_e with the sum of the four $\Delta m_F=\pm1$ transition frequencies associated with the $F=2$ hfs level of the sodium atoms in the same sample and the same magnetic field. The sodium transitions were observed also by the optical method discussed earlier. During a run the free electron resonance was continuously displayed on the oscilloscope screen while the four Na resonances ν_1 to ν_4 were consecutively superimposed on the free-electron resonance. Typical resonances are shown in Fig. 2. The sodium frequency sum has the value

$$\textstyle\sum\nu=(g_J\mu_0+2\mu_I)H_0/h,$$

where g_J denotes the electronic g factor of the sodium ground state and μ_I is the magnetic moment of the sodium nucleus. By using the field-independent ratio, $(\nu_e-\sum\nu)/\nu_e$, we can now form with $I=\frac{3}{2}$

$$g_J/g_e=1-3(g_I/g_e)-(\nu_e-\textstyle\sum\nu)/\nu_e.$$

Employing the atomic-beam value[19] $g_J/g_I=-2487.8$,

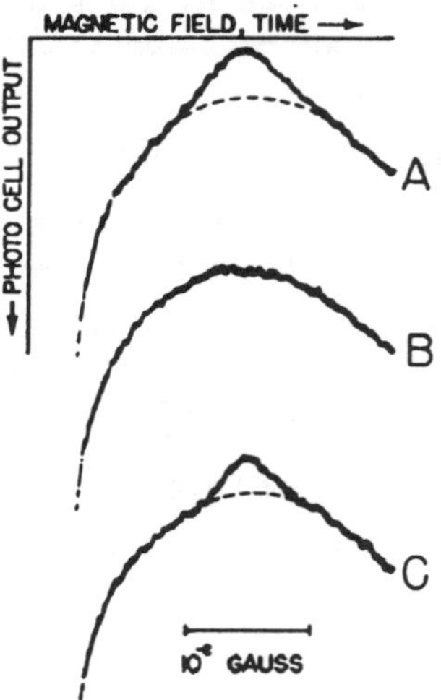

FIG. 2. Oscilloscope traces of electron and sodium resonances. The blank trace B was obtained by operating the equipment with the frequencies of the rf fields for electron and sodium resonances adjusted to off-resonance values. This trace served only to establish the dashed baselines which are shown in trace A, depicting the electron signal at about 62.1 Mc/sec and in trace C which displays one of the four sodium signals, namely that at about 15.9 Mc/sec. The signal peak intensities have been adjusted to about the same value and one easily notices the larger width of the electron resonance on a magnetic field scale. On a frequency scale the electron and sodium signal line widths turn out to be about 14.4 kc/sec and 2.2 kc/sec, respectively. The high light-intensity spikes at the beginning of the traces are the result of the ionizing discharge pulse at the beginning of each sweep cycle. The slow curvature of the remaining part of the baseline was either connected with afterglow effects in the decaying plasma or imperfections in the electronic equipment which were not further analyzed.

[19] P. Kusch and H. Taub, Phys. Rev. 75, 1477 (1948).

in place of g_e/g_I, we finally have

$$g_J/g_e = 1 + 1.2059\times 10^{-3} - (\nu_e - \sum \nu)/\nu_e.$$

With our preliminary experimental value for

$$(\nu_e - \sum \nu)/\nu_e = (118 \pm 3)\times 10^{-5},$$

we now obtain,

$$g_J/g_e = 1.000026 \pm 0.00003,$$

showing no difference in our limit of accuracy between the g factors of the free electron and the sodium ground state. Further experiments with the aim of improving the experimental accuracy and extending the method to much lower buffer gas pressures and eventually to near vacuum are in progress.

ACKNOWLEDGMENTS

The author wishes to express his appreciation for discussions with Dr. G. C. Wick of the Carnegie Institue of Technology and with his colleagues, especially Dr. R. Geballe and Dr. E. M. Henley. J. Jonson built the various absorption bulbs while F. Thoene designed the Helmholz coils and N. Pakinas helped with the electronic equipment.

PHYSICAL REVIEW VOLUME 130, NUMBER 6 15 JUNE 1963

The Quantum Theory of Optical Coherence*

ROY J. GLAUBER
Lyman Laboratory of Physics, Harvard University, Cambridge, Massachusetts
(Received 11 February 1963)

The concept of coherence which has conventionally been used in optics is found to be inadequate to the needs of recently opened areas of experiment. To provide a fuller discussion of coherence, a succession of correlation functions for the complex field strengths is defined. The nth order function expresses the correlation of values of the fields at $2n$ different points of space and time. Certain values of these functions are measurable by means of n-fold delayed coincidence detection of photons. A fully coherent field is defined as one whose correlation functions satisfy an infinite succession of stated conditions. Various orders of incomplete coherence are distinguished, according to the number of coherence conditions actually satisfied. It is noted that the fields historically described as coherent in optics have only first-order coherence. On the other hand, the existence, in principle, of fields coherent to all orders is shown both in quantum theory and classical theory. The methods used in these discussions apply to fields of arbitrary time dependence. It is shown, as a result, that coherence does not require monochromaticity. Coherent fields can be generated with arbitrary spectra.

I. INTRODUCTION

CORRELATION, it has long been recognized, plays a fundamental role in the concept of optical coherence. Techniques for both the generation and detection of various types of correlations in optical fields have advanced rapidly in recent years. The development of the optical maser, in particular, has led to the generation of fields with a range of correlation unprecedented at optical frequencies. The use of techniques of coincidence detection of photons[1,2] has, in the same period, shown the existence of unanticipated correlations in the arrival times of light quanta. The new approaches to optics, which such developments will allow us to explore, suggest the need for a fundamental discussion of the meaning of coherence.

The present paper, which is the first of a series on fundamental problems of optics, is devoted largely to defining the concept of coherence. We do this by constructing a sequence of correlation functions for the field vectors, and by discussing the consequences of certain assumptions about their properties. The definition of coherence which we reach differs from earlier ones in several significant ways. The most important difference, perhaps, is that complete coherence, as we define it, requires that the field correlation functions satisfy an infinite succession of coherence conditions. We are led then to distinguish among various orders of incomplete coherence, according to the number of conditions satisfied. The fields traditionally described as coherent in optics are shown to have only first-order coherence. The fields generated by the optical maser, on the other hand, may have a considerably higher order of coherence. A further difference between our approach and previous ones is that it is constructed to apply to fields of arbitrary time dependence, rather than just to those which are, on the average, stationary in time. We have also attempted to develop the discussion in a fully quantum theoretical way.

It would hardly seem that any justification is necessary for discussing the theory of light quanta in quantum theoretical terms. Yet, as we all know, the successes of classical theory in dealing with optical experiments have been so great that we feel no hesitation in introducing optics as a sophomore course. The quantum theory, in other words, has had only a fraction of the influence upon optics that optics has historically had upon quantum theory. The explanation, no doubt, lies in the fact that optical experiments to date have paid very little attention to individual photons. To the extent that observations in optics have been confined to the measurement of ordinary light intensities, it is not surprising that classical theory has offered simple and essentially correct insights.

Experiments such as those on quantum correlations suggest, on the other hand, the growing importance of studies of photon statistics. Such studies lie largely outside the grasp of classical theory. To observe that the quantum theory is fundamentally necessary to the treatment of these problems is not to say that the semiclassical approach always yields incorrect results. On the contrary, correct answers to certain classes of problems of photon statistics[3] may be found through adaptations of classical methods. There are, however, distinct virtues to knowing where such methods succeed and where they do not. For that reason, as well as for its intrinsic interest, we shall formulate the theory in quantum theoretical terms from the outset. Quite a few of our arguments can easily be paraphrased in classical terms. Several seem to be new in the context of classical theory.

We shall try to construct this paper so that it can be followed with little more than a knowledge of elementary quantum mechanics. Since its subject matter is, in the deepest sense, quantum electrodynamics, we begin with

* Supported in part by the U. S. Air Force Office of Scientific Research.

[1] R. Hanbury Brown and R. Q. Twiss, Nature 177, 27 (1956); Proc. Roy. Soc. (London) A242, 300 (1957); A243, 291 (1957).

[2] G. A. Rebka and R. V. Pound, Nature 180, 1035 (1957).

[3] E. M. Purcell, Nature 178, 1449 (1956).

a section which describes the few simple aspects of that subject which are referred to later.

II. ELEMENTS OF FIELD THEORY

The observable quantities of the electromagnetic field will be taken to be the electric and magnetic fields which are represented by a pair of Hermitian operators, $\mathbf{E}(\mathbf{r}t)$ and $\mathbf{B}(\mathbf{r}t)$. The state of the field will be described by means of a state vector, $|\rangle$, on which the fields operate from the left, or by means of its adjoint, $\langle|$, on which they operate from the right. Since we shall use the Heisenberg representation, the choice of a fixed state vector specifies the properties of the field at all times. The theory is constructed, by whatever formal means, so that in a vacuum the field operators $\mathbf{E}(\mathbf{r}t)$ and $\mathbf{B}(\mathbf{r}t)$ satisfy the Maxwell equations

$$\begin{aligned} \nabla\cdot\mathbf{E}&=0,\\ \nabla\times\mathbf{E}&=-\frac{1}{c}\frac{\partial\mathbf{B}}{\partial t},\\ \nabla\times\mathbf{B}&=\frac{1}{c}\frac{\partial\mathbf{E}}{\partial t},\\ \nabla\cdot\mathbf{B}&=0. \end{aligned} \tag{2.1}$$

We omit the source terms in the equations since, for the present, we are more interested in the fields themselves than the explicit way in which they are generated or detected. It follows from the Maxwell equations that the electric field operator obeys the wave equation

$$\left(\nabla^2-\frac{1}{c^2}\frac{\partial^2}{\partial t^2}\right)\mathbf{E}(\mathbf{r}t)=0, \tag{2.2}$$

and the magnetic field operator does likewise.

One of the essential respects in which quantum field theory differs from classical theory is that two values of the field operators taken at different space-time points do not, in general, commute with one another. The components of the electric field, which is the only field we shall discuss at length, obey a commutation relation of the general form

$$[E_\mu(\mathbf{r}t),E_\nu(\mathbf{r}'t')]=D_{\mu\nu}(\mathbf{r}-\mathbf{r}',t-t'). \tag{2.3}$$

That the tensor function $D_{\mu\nu}$ has as arguments the coordinate differences $\mathbf{r}-\mathbf{r}'$ and $t-t'$ follows from the invariance of the theory under translations in space and time. We shall not need any further details of the function $D_{\mu\nu}$, but may mention that it vanishes when the four-vector $(\mathbf{r}-\mathbf{r}', t-t')$ lies outside the light cone, i.e., for $(\mathbf{r}-\mathbf{r}')^2>c^2(t-t')^2$. The vanishing of the commutator, for points with spacelike separations, corresponds to the fact that measurements of the stated field components at such points can be carried out to arbitrary accuracy. Such accuracy is attainable since no disturbances can propagate through the field rapidly enough to reach one point from the other.

An important element of the discussion in this paper will be the separation of the electric field operator $\mathbf{E}(\mathbf{r}t)$ into its positive and negative frequency parts. The separation is most easily accomplished when the time dependence of the operator is represented by a Fourier integral. If, for example, the field operator has a representation

$$\mathbf{E}(\mathbf{r}t)=\int_{-\infty}^{\infty}\mathfrak{e}(\omega,\mathbf{r})e^{-i\omega t}d\omega, \tag{2.4}$$

where the Hermitian property is secured by the relation $\mathfrak{e}(-\omega,\mathbf{r})=\mathfrak{e}^\dagger(\omega,\mathbf{r})$, then we define the positive frequency part of $\mathbf{E}$ as

$$\mathbf{E}^{(+)}(\mathbf{r}t)=\int_{0}^{\infty}\mathfrak{e}(\omega,\mathbf{r})e^{-i\omega t}d\omega, \tag{2.5}$$

and the negative frequency part as

$$\mathbf{E}^{(-)}(\mathbf{r}t)=\int_{-\infty}^{0}\mathfrak{e}(\omega,\mathbf{r})e^{-i\omega t}d\omega, \tag{2.6}$$

$$=\int_{0}^{\infty}\mathfrak{e}^\dagger(\omega,\mathbf{r})e^{-i\omega t}d\omega. \tag{2.7}$$

It is evident from these definitions that the field is the sum of its positive and negative frequency parts,

$$\mathbf{E}(\mathbf{r}t)=\mathbf{E}^{(+)}(\mathbf{r}t)+\mathbf{E}^{(-)}(\mathbf{r}t). \tag{2.8}$$

The two parts, regarded separately, are not Hermitian operators; the fields they represent are intrinsically complex, and mutually adjoint,

$$\mathbf{E}^{(-)}(\mathbf{r}t)=\mathbf{E}^{(+)\dagger}(\mathbf{r}t). \tag{2.9}$$

In the absence of a Fourier integral representation of $\mathbf{E}(\mathbf{r}t)$, the positive and negative frequency parts of the field may be defined more formally as the limits of the integrals,

$$\mathbf{E}^{(+)}(\mathbf{r}t)=\lim_{\eta\to+0}\frac{1}{2\pi i}\int_{-\infty}^{\infty}\frac{\mathbf{E}(\mathbf{r},t-\tau)}{\tau-i\eta}d\tau, \tag{2.10}$$

$$\mathbf{E}^{(-)}(\mathbf{r}t)=-\lim_{\eta\to+0}\frac{1}{2\pi i}\int_{-\infty}^{\infty}\frac{\mathbf{E}(\mathbf{r},t-\tau)}{\tau+i\eta}d\tau. \tag{2.11}$$

It follows from the intrinsically different time dependences of $\mathbf{E}^{(+)}(\mathbf{r}t)$ and $\mathbf{E}^{(-)}(\mathbf{r}t)$ that they act to change the state of the field in altogether different ways, one associated with photon absorption, the other with photon emission. In particular, the positive frequency part, $\mathbf{E}^{(+)}(\mathbf{r}t)$, may be shown[4] to be a photon annihilation operator. Applied to an n-photon state it produces an $(n-1)$-photon state. Further applications of $\mathbf{E}^{(+)}(\mathbf{r}t)$

[4] See, for example, P. A. M. Dirac, *The Principles of Quantum Mechanics* (Oxford University Press, New York, 1947), 3rd ed., pp. 239–242.

reduce the number of photons present still further, but the regression must end with the state in which the field is empty of all photons. It is part of the definition of this state, which we represent as $|\text{vac}\rangle$, that

$$\mathbf{E}^{(+)}(\mathbf{r}t)|\text{vac}\rangle=0. \tag{2.12}$$

The adjoint relation is

$$\langle\text{vac}|\mathbf{E}^{(-)}(\mathbf{r}t)=0. \tag{2.13}$$

Since the operator $\mathbf{E}^{(+)}(\mathbf{r}t)$, annihilates photons, its Hermitian adjoint, $\mathbf{E}^{(-)}(\mathbf{r}t)$, must create them; applied to an n-photon state it produces an $(n+1)$-photon state. In particular, the state

$$\mathbf{E}^{(-)}(\mathbf{r}t)|\text{vac}\rangle$$

is a one-photon state.

It has become customary, in discussions of classical theory, to regard the electric field $\mathbf{E}(\mathbf{r}t)$ as the quantity one measures experimentally, and to think of the complex fields $\mathbf{E}^{(\pm)}(\mathbf{r}t)$ as convenient, but fictitious, mathematical constructions. Such an attitude can only be held be held in the classical domain, where quantum phenomena play no essential role. The frequency ω of a classical field must be so low that the quantum energy $\hbar\omega$ is negligible. In such a case, we can not tell whether a classical test charge emits or absorbs quanta. In measuring a classical field strength, $\mathbf{E}(\mathbf{r}t)$, we implicitly sum the effects of photon absorption and emission which are described individually by the fields $\mathbf{E}^{(+)}(\mathbf{r}t)$ and $\mathbf{E}^{(-)}(\mathbf{r}t)$.

Where quantum phenomena are important the situation is usually quite different. Experiments which detect photons ordinarily do so by absorbing them in one or another way. The use of any absorption process, such as photoionization, means in effect that the field we are measuring is the one associated with photon annihilation, the complex field $\mathbf{E}^{(+)}(\mathbf{r}t)$. We need not discuss the details of the photoabsorption process to find the appropriate matrix element of the field operator. If the field makes a transition from the initial state $|i\rangle$ to a final state $|f\rangle$ in which one photon, polarized in the μ direction, has been absorbed, the matrix element takes the form

$$\langle f|E_\mu^{(+)}(\mathbf{r}t)|i\rangle. \tag{2.14}$$

We shall define an ideal photon detector as a system of negligible size (e.g., of atomic or subatomic dimensions) which has a frequency-independent photoabsorption probability. The advantage of imagining such a detector, as we shall show more explicitly in a later paper, is that the rate at which it records photons is proportional to the sum over all final states $|f\rangle$ of the squared absolute values of the matrix elements (2.14). In other words, the probability per unit time that a photon be absorbed by an ideal detector at point $\mathbf{r}$ at time t is proportional to

$$\begin{aligned}\sum_f|\langle f|E_\mu^{(+)}(\mathbf{r}t)|i\rangle|^2 &=\sum_f\langle i|E_\mu^{(-)}(\mathbf{r}t)|f\rangle\langle f|E_\mu^{(+)}(\mathbf{r}t)|i\rangle\\ &=\langle i|E_\mu^{(-)}(\mathbf{r}t)E_\mu^{(+)}(\mathbf{r}t)|i\rangle.\end{aligned} \tag{2.15}$$

We may verify immediately from (2.12) that the rate at which photons are detected in the empty, or vacuum, state vanishes.

The photodetector we have described is the quantum-mechanical analog of what, in classical experiments, has been called a square-law detector. It is important to bear in mind that such a detector for quanta measures the average value of the product $E_\mu^{(-)}E_\mu^{(+)}$, and not that of the square of the real field $E_\mu(\mathbf{r}t)$. Indeed, it is easily seen from the foregoing work that the average value of $E_\mu^2(\mathbf{r}t)$ does not vanish in the vacuum state;

$$\langle\text{vac}|E_\mu^2(\mathbf{r}t)|\text{vac}\rangle>0.$$

The electric field in the vacuum undergoes zero-point oscillations which, in the correctly formulated theory, have nothing to do with the detection of photons.

Recording photon intensities with a single detector does not exhaust the measurements we can make upon the field, though it does characterize, in principle, virtually all the classic experiments of optics. A second type of measurement we may make consists of the use of two detectors situated at different points $\mathbf{r}$ and $\mathbf{r}'$ to detect photon coincidences or, more generally, delayed coincidences. The field matrix element for such transitions takes the form

$$\langle f|E_\mu^{(+)}(\mathbf{r}'t')E_\mu^{(+)}(\mathbf{r}t)|i\rangle, \tag{2.16}$$

if both photons are required to be polarized along the μ axis. The total rate at which such transitions occur is proportional to

$$\begin{aligned}\sum_f|\langle f|E_\mu^{(+)}(\mathbf{r}'t')E_\mu^{(+)}(\mathbf{r}t)|i\rangle|^2\\ =\langle i|E_\mu^{(-)}(\mathbf{r}t)E_\mu^{(-)}(\mathbf{r}'t')E_\mu^{(+)}(\mathbf{r}'t')E_\mu^{(+)}(\mathbf{r}t)|i\rangle.\end{aligned} \tag{2.17}$$

Such a total rate is to be interpreted as a probability per unit (time)^2 that one photon is recorded at $\mathbf{r}$ at time t and another at $\mathbf{r}'$ at time t'. Photon correlation experiments of essentially the type we are describing were performed by Hanbury Brown and Twiss[1] in 1955 and have, subsequently, been performed by others.[2]

Whatever may be the practical difficulties of more elaborate experiments, we may at least imagine the possibility of detecting n-fold delayed coincidences of photons for arbitrary n. The total rate per unit (time)^n for such coincidences will be proportional to

$$\langle i|E_\mu^{(-)}(\mathbf{r}_1t_1)\cdots E_\mu^{(-)}(\mathbf{r}_nt_n)E_\mu^{(+)}(\mathbf{r}_nt_n)\cdots E_\mu^{(+)}(\mathbf{r}_1t_1)|i\rangle,\quad n=1,2,3,\cdots. \tag{2.18}$$

The entire succession of such expectation values, therefore, possesses a simple physical interpretation.

In closing this survey we add a note on the commutation rules obeyed by the fields $\mathbf{E}^{(+)}$ and $\mathbf{E}^{(-)}$. It is easy to find these rules from the relation (2.3) for the real field $\mathbf{E}$, by decomposing its dependence on the two variables t and t' into positive and negative frequency parts. If the function $D_{\mu\nu}$ has the Fourier transform

$$D_{\mu\nu}(\mathbf{r}-\mathbf{r}', t-t') = \int_{-\infty}^{\infty} \mathfrak{D}_{\mu\nu}(\omega, \mathbf{r}-\mathbf{r}')e^{-i\omega(t-t')}d\omega, \quad (2.19)$$

we see immediately that the commutator (2.3) has no part which is of positive frequency in both its t and t' dependences. Neither does it have any part of negative frequency in both its time dependences. It follows that all values of the field $\mathbf{E}^{(+)}(\mathbf{r}t)$ commute with one another. and so too do those of $\mathbf{E}^{(-)}(\mathbf{r}t)$, i.e., we have

$$[E_\mu^{(+)}(\mathbf{r}t), E_\nu^{(+)}(\mathbf{r}'t')] = 0, \quad (2.20)$$

$$[E_\mu^{(-)}(\mathbf{r}t), E_\nu^{(-)}(\mathbf{r}'t')] = 0, \quad (2.21)$$

for all points $\mathbf{r}t$ and $\mathbf{r}'t'$, and all μ and ν. Products of the $\mathbf{E}^{(+)}$ operators or products of the $\mathbf{E}^{(-)}$ operators such as occur in (2.18) may, therefore, be freely rearranged, but the operators $\mathbf{E}^{(+)}$ and $\mathbf{E}^{(-)}$ do not, in general, commute.

III. FIELD CORRELATIONS

The electromagnetic field may be regarded as a dynamical system with an infinite number of degrees of freedom. Our knowledge of the condition of such a system is virtually never so complete or so precise in practice as to justify the use of a particular quantum state $|\rangle$ in its description. In the most accurate preparation of the state of a field which we can actually accomplish some parameters, usually an indefinitely large number of them, must be regarded as random variables. Since there is no possibility in practice of controlling these parameters, we can only hope ultimately to compare with experiment quantities which are averages over the distributions of the unknown parameters.

Our actual knowledge of the state of the field is specified fully by means of a density operator ρ which is constructed as an average, over the uncontrollable parameters, of an expression bilinear in the state vector. If $|\rangle$ is a precisely defined state of the field corresponding to a particular set of random parameters, the density operator is defined as the averaged outer product of state vectors

$$\rho = \{|\rangle\langle|\}_{\text{av}}. \quad (3.1)$$

The weightings to be used in the averaging are the ones which best describe the actual preparation of the fields. It is clear from the definition that ρ is Hermitian, $\rho^\dagger = \rho$.

The average of an observable $\mathcal{O}$ in the quantum state $|\rangle$ is the expectation value, $\langle|\mathcal{O}|\rangle$. It is the average of this quantity over the randomly prepared states which we compare with experiment. The average taken in this twofold sense may be written as

$$\{\langle|\mathcal{O}|\rangle\}_{\text{av}} = \text{tr}\{\rho\mathcal{O}\}, \quad (3.2)$$

where the symbol tr stands for the trace, or the sum of the diagonal matrix elements. Since we require the average of the unit operator to be one, we must have $\text{tr}\rho = 1$. These considerations show that the average counting rate of an ideal photodetector, which is proportional to (2.15) in a completely specified quantum state of the field, is more generally proportional to

$$\text{tr}\{\rho E_\mu^{(-)}(\mathbf{r}t)E_\mu^{(+)}(\mathbf{r}t)\} \quad (3.3)$$

when the state is less completely specified.

It is convenient at this point, as a simplification of notation, to confine our attention to a single vector component of the electric field. We suppose, for the present, that all of our detectors are fitted with polarizers and record only photons polarized parallel to an arbitrary unit vector $\mathbf{e}$. (If $\mathbf{e}$ is chosen as a complex unit vector, $\mathbf{e}^*\cdot\mathbf{e} = 1$, the photons detected may have arbitrary elliptical polarization.) We then introduce the symbols $E^{(+)}$ and $E^{(-)}$ for the projections of the complex fields in the direction $\mathbf{e}$ and $\mathbf{e}^*$,

$$E^{(+)}(\mathbf{r}t) = \mathbf{e}^*\cdot\mathbf{E}^{(+)}(\mathbf{r}t) \quad (3.4)$$

$$E^{(-)}(\mathbf{r}t) = \mathbf{e}\cdot\mathbf{E}^{(-)}(\mathbf{r}t). \quad (3.5)$$

We resume a fully general treatment of photon polarizations in Sec. V.

The field average (3.3) which determines the counting rate of an ideal photodetector is a particular form of a more general type of expression whose properties are of considerable interest. In the more general expression, the fields $E^{(-)}$ and $E^{(+)}$ are evaluated at different space-time points. Statistical averages of the latter type furnish a measure of the correlations of the complex fields at separated positions and times. We shall define such a correlation function, $G^{(1)}$, for the $\mathbf{e}$ components of the complex fields as

$$G^{(1)}(\mathbf{r}t,\mathbf{r}'t') = \text{tr}\{\rho E^{(-)}(\mathbf{r}t)E^{(+)}(\mathbf{r}'t')\}. \quad (3.6)$$

Only the values of this function at $\mathbf{r}=\mathbf{r}'$ and $t=t'$ are needed to predict the counting rate of an ideal photodetector. However, other values of the function become necessary, quite generally, when we use as detectors less ideal systems such as actual photo-ionizable atoms. In actual photodetectors the absorption of photons can not be localized too closely, either in space or in time. Atomic photo ionization rates must be written, in general, as double integrals, over a microscopic range, of all the variables in $G^{(1)}(\mathbf{r}t,\mathbf{r}'t')$. Our interest in the function $G^{(1)}$ extends to widely spaced values of its variables as well. That field correlations may extend over considerable intervals of distance and time is essential to the idea of coherence, which we shall shortly discuss.

As we have noted earlier, our interest in averages of the field operators extends beyond quadratic ones. Just as we generalized the expression for the photon detection rate to define $G^{(1)}$, we may generalize the expression

(2.17) for the photon coincidence rate and thereby define a second-order correlation function,

$$G^{(2)}(\mathbf{r}_1t_1\mathbf{r}_2t_2,\mathbf{r}_3t_3\mathbf{r}_4t_4) = \mathrm{tr}\{\rho E^{(-)}(\mathbf{r}_1t_1)E^{(-)}(\mathbf{r}_2t_2)E^{(+)}(\mathbf{r}_3t_3)E^{(+)}(\mathbf{r}_4t_4)\}. \quad (3.7)$$

This too is a function whose values, even at widely separated arguments, interest us.

In view of the possibility of discussing n-photon coincidence experiments for arbitrary n it is natural to define an infinite succession of correlation functions $G^{(n)}$. It is convenient in writing these to abbreviate a set of coordinates $(\mathbf{r}_j,t_j)$ by a single symbol, x_j. We then define the nth-order correlation function as

$$G^{(n)}(x_1\cdots x_n,x_{n+1}\cdots x_{2n}) = \mathrm{tr}\{\rho E^{(-)}(x_1\cdots E^{(-)}x_n)E^{(+)}(x_{n+1})\cdots E^{(+)}(x_{2n})\}. \quad (3.8)$$

The correlation functions have a number of simple properties. It is easily verified that interchanging the arguments in $G^{(1)}$ leads to the complex conjugate function

$$G^{(1)}(\mathbf{r}'t',\mathbf{r}t)=\{G^{(1)}(\mathbf{r}t,\mathbf{r}'t')\}^*. \quad (3.9)$$

The same type of relation holds for all of the higher order functions

$$G^{(n)}(x_{2n}\cdots x_1)=\{G^{(n)}(x_1\cdots x_{2n})\}^*. \quad (3.10)$$

Furthermore, the commutation relations (2.20) and (2.21) show us that $G^{(n)}$ is unchanged by any permutation of its arguments $(x_1\cdots x_n)$, or its arguments $(x_{n+1}\cdots x_{2n})$. The fact that the complex fields $E^{(\pm)}$ individually satisfy the wave equation (2.2) leads to another useful property of the $G^{(n)}$. The nth-order function satisfies $2n$ different wave equations, one for each of its arguments x_j, $(j=1,\cdots,2n)$.

A large number of inequalities satisfied by the functions $G^{(n)}$ may be derived from the positive definite character of the density operator ρ. Derivations of several classes of these are presented in the Appendix. We confine ourselves, in this section, to mentioning some of the simpler and more useful inequalities, those which are linear or quadratic in the correlation functions. It is clear from (3.10) that all of the functions $G^{(n)}(x_1\cdots x_n,x_n\cdots x_1)$ are real. The linear inequalities assert that these functions are positive definite as well. We have then, in particular for $n=1$, the self-evident relation

$$G^{(1)}(x_1,x_1)\geq 0, \quad (3.11)$$

and for arbitrary n

$$G^{(n)}(x_1\cdots x_n,x_n\cdots x_1)\geq 0. \quad (3.12)$$

These relations simply affirm that the average photon intensity of a field and the average coincidence counting rates are all intrinsically positive.

The simplest of the quadratic inequalities takes the form

$$G^{(1)}(x_1,x_1)G^{(1)}(x_2,x_2)\geq |G^{(1)}(x_1,x_2)|^2. \quad (3.13)$$

Higher order inequalities of this type are given by

$$G^{(n)}(x_1\cdots x_n,x_n\cdots x_1)G^{(n)}(x_{n+1}\cdots x_{2n},x_{2n}\cdots x_{n+1}) \geq |G^{(n)}(x_1\cdots x_n,x_{n+1}\cdots x_{2n})|^2, \quad (3.14)$$

which holds for arbitrary n. Different forms of these relations are obtained by permuting or equating coordinates. Various other inequalities are proved in the Appendix along with those noted.

It is interesting to note that when the number of quanta present in the field is bounded, the sequence of functions $G^{(n)}$ terminates. If the density operator restricts the number of photons present to be smaller than or equal to some value M, the properties of $E^{(\pm)}$ as annihilation and creation operators show that $G^{(n)}=0$ for $n>M$.

Classical correlation functions bearing some analogy to $G^{(1)}$ have received a great deal of discussion in recent years, mainly in connection with the theory of noise in radio waves. A detailed application of the classical correlation theory to optics has been made by Wolf.[5] At the core of Wolf's analysis is a single correlation function Γ, defined as an average over an infinite time span of the product of two fields, evaluated at times separated by a fixed interval. The procedure of time averaging restricts the application of such an approach to the treatment of field distributions which are statistically stationary in time.

If we were to restrict the character of our density operator ρ to describe only stationary field distributions (e.g., by choosing ρ to commute with the field Hamiltonian) our function $G^{(1)}(\mathbf{r}t,\mathbf{r}'t')$ would depend only on the difference of the two times, $t-t'$. In that case the function $G^{(1)}$ would, in the classical limit (strong, low-frequency fields), agree numerically[6] with Wolf's function Γ. It should be clear, however, that the concepts of correlation and ultimately of coherence are quite useful in the discussion of nonstationary field distributions. The correlation functions $G^{(n)}$ which we have defined are ensemble averages rather than time averages and hence remain well-defined in fields of arbitrary time dependence.

IV. COHERENCE

The term "coherence" has had long if somewhat varied use in areas of physics concerned with the electromagnetic field. In physical optics the term is used to denote a tendency of two values of the field at distantly separated points or at greatly separated times to take on correlated values. When optical means are used to superpose the fields at such points (e.g., as in Young's two-slit experiment) intensity fringes result. The possibility of producing such fringes in hypothetical superposition experiments epitomizes the optical definition of

[5] M. Born and E. Wolf, *Principles of Optics* (Pergamon Press, Inc., London, 1959), Chap. X. An extensive bibliography is given there.

[6] This is true provided Wolf's "disturbance" field V behaves ergodically and is identified with $E^{(+)}$.

coherence. The definition has remained a satisfactorily explicit one only as long as optical experiments were confined to measuring field intensities, or more generally quantities quadratic in the field strengths. We have already noted that the photon correlation experiment of Hanbury Brown and Twiss,[1] performed in 1955, is of an altogether new type and measures the average of a quartic expression.[7] The study of quantities of fourth and higher powers in the field strengths is the basis of all work in the recently developed area of nonlinear optics. It appears safe to assume that the number of such experiments will increase in the future, and that the concept of coherence should be extended to apply to them.

Another pressing reason for sharpening the meaning of coherence is provided by the recent development of the optical maser. The maser produces light beams of narrow spectral bandwidth which are characterized by field correlations extending over quite long ranges. Such light is inevitably described as coherent, but the sense in which the term is used has not been made adequately clear. If the sense is simply the optical one then, as we shall see, it may scarcely do justice to the potentialities of the device. The optical definition does not at all distinguish among the many ways in which fields may vary while remaining equally correlated at all pairs of points. That much greater regularities may exist in the field variations of a maser beam than are required by the optical definition of coherence may be seen by comparing the maser beam with the carrier wave of a radio transmitter. The latter type of wave ideally possesses a stability of amplitude which optically coherent fields need not have.[8] Furthermore, the field values of such a wave possess correlations of a much more detailed sort than the optical definition requires. These are properties best expressed in terms of the higher order correlation functions $G^{(n)}$, for $n>1$.

To discuss coherence in quantitative terms it is convenient to introduce normalized forms of the correlation functions. Corresponding to the first-order function $G^{(1)}$ we define

$$g^{(1)}(\mathbf{r}t,\mathbf{r}'t')=\frac{G^{(1)}(\mathbf{r}t,\mathbf{r}'t')}{\{G^{(1)}(\mathbf{r}t,\mathbf{r}t)G^{(1)}(\mathbf{r}'t',\mathbf{r}'t')\}^{1/2}}. \tag{4.1}$$

It is immediately seen from (3.13) that $g^{(1)}$ obeys the inequality

$$|g^{(1)}(\mathbf{r}t,\mathbf{r}'t')|\leq 1. \tag{4.2}$$

For $\mathbf{r}=\mathbf{r}'$, $t=t'$ we have, of course, $g^{(1)}\equiv 1$.

The normalized forms of the higher order correlation functions are defined as

$$g^{(n)}(x_1\cdots x_{2n})=G^{(n)}(x_1\cdots x_{2n})/\prod_{j=1}^{2n}\{G^{(1)}(x_j,x_j)\}^{1/2}. \tag{4.3}$$

These functions, for $n>1$, are not, in general, restricted in absolute value as is $g^{(1)}$.

We shall try in this paper to give the concept of coherence as precise a definition as is both realizable in physical terms, and useful as well.[9] We, therefore, begin by stating an infinite sequence of conditions on the functions $g^{(n)}$ which are to be satisfied by a fully coherent field. These necessary conditions for coherence are that the normalized correlation functions all have unit absolute magnitude,

$$|g^{(n)}(x_1\cdots x_{2n})|=1,\quad n=1,2\cdots. \tag{4.4}$$

That there exist at least some states which meet these conditions at all points of space and time is immediately clear from the example of a classical plane wave, $E^{(+)}\sim\exp[i(\mathbf{k}\cdot\mathbf{r}-\omega t)]$. We shall presently show that the class of coherent fields is vastly larger than that of individual plane waves.

The conditions (4.4) on the functions $g^{(n)}$ are stated only as necessary ones and need not be construed as defining coherence completely. We shall shortly, in fact, sharpen the definition somewhat further. It is worth noting at this point, however, that not all of the fields which have been described as "coherent" in the past meet the set of conditions (4.4) even approximately. There may be some virtue, therefore, in constructing a hierarchy of orders of coherence to discuss fields which do not have that property in its fullest sense. We shall state as a condition necessary for first-order coherence that $|g^{(1)}(\mathbf{r}t,\mathbf{r}'t')|=1$. More generally, for a field to be characterized by nth order coherence we shall require $|g^{(j)}|=1$ for $j\leq n$. For fields which occur in practice, one can not expect relations such as these to hold exactly for all points in space and time. We shall, therefore, often employ the term nth order coherence more loosely to mean that the first n coherence conditions are fairly accurately satisfied over appreciable intervals of the variables surrounding all points $x_1=x_2=\cdots=x_{2n}$.

The definition of coherence which has been used to date in all studies of physical optics corresponds only to first-order coherence. The most coherent fields which have been generated by optical means prior to the development of the maser, in fact, lack second and higher order coherence. On the other hand, the optical maser, functioning with ideal stability, may produce fields which are coherent to all orders.

The various orders of coherence may, in principle, be distinguished fairly directly in experimental terms. The inequality (3.12), which states that the n-fold coincidence counting rate is positive, requires that $g^{(n)}(x_1\cdots x_n,x_n\cdots x_1)$ be positive. If the field in question possesses nth-order coherence, it must, therefore, have

$$g^{(j)}(x_1\cdots x_j,x_j\cdots x_1)=1, \tag{4.5}$$

[7] R. J. Glauber, Phys. Rev. Letters 10, 84 (1963). The particular field referred to as incoherent in that note may have first-order coherence if it is monochromatic, but not second- or higher order coherence.

[8] This point has been noted with particular clarity by M. J. E. Golay, Proc. IRE 49, 959 (1961); also 50, 223 (1962).

[9] A brief account of this work was presented by R. J. Glauber, in Proceedings of the Third International Conference on Quantum Electronics, Paris, France, 1963 (to be published).

for $j \leq n$. It follows from the definitions of the $g^{(j)}$ that the corresponding values of the correlation functions $G^{(j)}$ factorize, i.e.,

$$G^{(j)}(x_1 \cdots x_j, x_j \cdots x_1) = \prod_{i=1}^{j} G^{(1)}(x_i, x_i), \tag{4.6}$$

for $j \leq n$. These relations mean, in observational terms, that the rate at which j-fold delayed coincidences are detected by our ideal photon counters, reduces to a product of the detection rates of the individual counters.[7] In photon coincidence experiments of multiplicity up to and including n, the photon counts registered by the individual counters may then be regarded as statistically independent events. No tendency of photon counts to be statistically correlated will be evident in j-fold coincidence experiments for $j \leq n$.

The experiments of Hanbury Brown and Twiss[1] were designed to detect correlations in the fluctuating outputs of two photomultipliers. These detectors were placed in fields made coherent with one another (in the optical sense) through the use of monochromatic, pinhole illumination and a semitransparent mirror. The photocurrents of the two detectors were observed to show a positive correlation for small delay times, rather than independent fluctuations. A similar experiment has been performed by Rebka and Pound,[2] using coincidence counting equipment. Their experiment, performed with a more monochromatic beam and better geometrical definition, shows an explicit correlation in the counting probabilities of the two detectors. These observations verify that light beams from ordinary sources such as discharge tubes, when made optimally coherent in the first-order sense, still lack second-order coherence.

The coherence conditions (4.4) can also be stated as a requirement that the functions $|G^{(n)}(x_1 \cdots x_{2n})|$ factorize into a product of $2n$ functions of the same form, each dependent on a single space-time variable,

$$|G^{(n)}(x_1 \cdots x_{2n})| = \prod_{j=1}^{2n} \{G^{(1)}(x_j, x_j)\}^{1/2}. \tag{4.7}$$

This statement of the necessary conditions for coherence suggests that it may be convenient to give a stronger definition to coherence by regarding it as a factorization property of the correlation functions,

Let us suppose that there exists a function $\mathcal{E}(x)$, independent of n, such that the correlation functions for all n may be expressed as the products

$$G^{(n)}(x_1 \cdots x_n, x_{n+1} \cdots x_{2n}) = \mathcal{E}^*(x_1) \cdots \mathcal{E}^*(x_n) \mathcal{E}(x_{n+1}) \cdots \mathcal{E}(x_{2n}). \tag{4.8}$$

It is immediately clear that these functions satisfy the conditions (4.4) and (4.7). To show that fields with such correlations exist we need only refer again to the case of a classical plane wave. In fact, any classical field of predetermined (i.e., nonrandom) behavior has correlation functions which fall into this form, and such fields are at times called coherent in communication theory. We shall, therefore, adopt the *factorization* conditions (4.8) as the definition of a coherent field and turn next to the question of how they may be satisfied in the quantum domain.

If it were possible for the field to be in an eigenstate of the operators $E^{(+)}$ and $E^{(-)}$, the correlation functions for such states would factorize immediately to the desired form. The operators $E^{(+)}(\mathbf{r}t)$ and $E^{(-)}(\mathbf{r}'t')$ do not commute, however, so no state can be an eigenstate of both in the usual sense. Not only are these operators non-Hermitian, but the failure of each to commute with its adjoint shows that $E^{(+)}$ and $E^{(-)}$ are non-normal as well. Operators of this type can not, as a rule, be diagonalized at all, but may nonetheless have eigenstates. In general, we must distinguish between their left and right eigenstates; the two types need not occur in mutually adjoint pairs. The operator $E^{(+)}(\mathbf{r}t)$, in particular, has no left eigenstates, but does have right eigenstates[10] corresponding to complex eigenvalues for the field, which are functions of position and time. We shall suppose that $|\,\rangle$ is a right eigenstate of $E^{(+)}$ and that the equation it satisfies takes the form

$$E^{(+)}(\mathbf{r}t)|\,\rangle = \mathcal{E}(\mathbf{r}t)|\,\rangle, \tag{4.9}$$

in which the function $\mathcal{E}(\mathbf{r}t)$ is to be interpreted as the complex eigenvalue. The Hermitian adjoint of this relation shows us that the conjugate state, $\langle\,|$, is a left eigenstate of $E^{(-)}(\mathbf{r}t)$,

$$\langle\,|E^{(-)}(\mathbf{r}t) = \langle\,|\mathcal{E}^*(\mathbf{r}t). \tag{4.10}$$

The density operator for such states is simply the projection operator, $\rho = |\,\rangle\langle\,|$. It follows immediately from these relations that the correlation functions $G^{(n)}$ all factorize into the form of Eq. (4.8). In other words, the state of the field defined by Eqs. (4.9) or (4.10) meets our definition precisely and is fully coherent. We shall discuss the properties of such states[11] at length in the paper to follow. For the present it may suffice to say that we can find an eigenstate $|\,\rangle$ which corresponds to the choice, as an eigenvalue, of any function $\mathcal{E}(\mathbf{r}t)$ which satisfies certain conditions. One condition, which is clear from Eq. (4.9), is that $\mathcal{E}(\mathbf{r}t)$ must satisfy the wave equation. The other, which corresponds to the positive frequency character of $E^{(+)}$, is that $\mathcal{E}(\mathbf{r}t)$, when regarded as a function of a complex time variable, be analytic in the lower half-plane. The eigenstates which correspond to different fields $\mathcal{E}(\mathbf{r}t)$ are not mutually orthogonal, but nontheless form a natural basis for the discussion of photon detection problems. We have introduced them

[10] States of the harmonic oscillator which have an analogous property were introduced in a slightly different but related connection by E. Schrödinger, Naturwiss. 14, 664 (1926). The electromagnetic field, as is well known, may be treated as an assembly of oscillators.

[11] Some of the properties of these states have already been noted in references 7 and 9.

here only to demonstrate the possibility of satisfying the coherence conditions in quantum theory. Such quantum states do not exhaust the possibility of describing coherent fields. Statistical mixtures, for example, of the states for which the eigenvalues $\mathcal{E}(rt)$ differ by constant phase factors satisfy the coherence conditions equally well.

The fields which have been described as most coherent in optical contexts have tended to be those of the narrowest spectral bandwidth. If coherent fields in optics have necessarily been chosen as monochromatic ones, it is because that has been virtually the only means of securing appreciably correlated fields from intrinsically chaotic sources. For this reason, perhaps, there has been a natural tendency to associate the concept of coherence with monochromaticity. The association was, in fact, made an implicitly rigid one by earlier discussions[5] of optical (i.e., first-order) coherence which were applicable only to statistically stationary fields. By extending the definition of coherence to nonstationary fields we see that it places no constraint on the frequency spectrum. Coherent fields exist corresponding to eigenvalues $\mathcal{E}(rt)$ with arbitrary spectra. The coherence conditions restrict randomness of the fields rather than their bandwidth.

Having defined full coherence by means of the factorization conditions (4.8), we may now use them in defining the various orders of coherence. We shall speak of mth-order coherent fields when the conditions (4.8) are satisfied for $n\leq m$, a definition which accords with our earlier conditions on $|g^{(j)}|$.

Photon correlation experiments have shown the importance of distinguishing between the first two orders of coherence. At the other end of the scale, we have shown that there exist, in principle at least, states which are fully coherent. We are entitled to ask, therefore, whether the intermediate orders of coherence will also be useful classifications. In the absence of any experimental information, we can only guess that they may be useful, though perhaps not in the sharp sense in which we we have defined them. One may easily imagine the possibility that, for light sources such as the maser, the correlation functions $G^{(n)}$ show gradually increasing departure from the factored forms (4.8) as n increases, even when the variables $x_1\cdots x_{2n}$ are not too widely separated. In such contexts the order of coherence can only be defined approximately.[12] Something of the same approximate character must be present in all applications of the definitions we have given. The field correlations we have discussed can extend over great intervals of distance and time, though never infinite ones in practice. Coherence conditions, such as $|g^{(n)}|=1$, can only be met within a finite range of relative values of the coordinates $x_1\cdots x_{2n}$. It is only within such ranges, and therefore as an approximation, that we can speak of coherence at all.

[12] The characterization we have given the nth-order coherent fields is, in principle, an accurately realizable one, however. States with such properties may be constructed in a variety of ways. The factorization conditions can be met for $j\leq n$, for example, by suitably chosen states in which the number of photons present may take on any value up to n. The correlation functions of order $j>n$ then vanish, as we have noted earlier. The vanishing of these correlation functions for states with bounded numbers of quanta shows, incidentally, that no bound can be placed on the photon number in a fully coherent field.

V. COHERENCE AND POLARIZATION

We have to this point, in the interest of simplicity, dealt only with the projections of the fields along a single (possibly complex) unit vector $\mathbf{e}$. To take fuller account of the vector nature of the fields we must define tensor rather than scalar correlation functions. The first-order function is taken to be

$$G_{\mu\nu}^{(1)}(x,x')=\mathrm{tr}\{\rho E_\mu^{(-)}(x)E_\nu^{(+)}(x')\},\qquad(5.1)$$

in which the indices μ and ν label Cartesian components. This function satisfies the symmetry relation

$$G_{\nu\mu}^{(1)}(x',x)=\{G_{\mu\nu}^{(1)}(x,x')\}^*,\qquad(5.2)$$

and is shown in the appendix to obey the inequalities,

$$G_{\mu\mu}^{(1)}(x,x)\geq 0\qquad(5.3)$$

and

$$G_{\mu\mu}^{(1)}(x,x)G_{\nu\nu}^{(1)}(x',x')\geq|G_{\mu\nu}^{(1)}(x,x')|^2.\qquad(5.4)$$

The photon intensities which can be detected at the space-time point x are found from $G_{\mu\nu}^{(1)}(x,x')$ for $x'=x$. We shall abbreviate this 3×3 matrix as $\mathbf{G}^{(1)}(x)$, and use it as the basis of a brief discussion of polarization correlations in three dimensions, a subject which seems to have received little attention in comparison to plane polarizations. The symmetry relation (5.2) for $x'=x$ shows that the intensity matrix $\mathbf{G}^{(1)}(x)$ is Hermitian; an argument given in the Appendix shows it to be a positive definite matrix as well. It follows that $\mathbf{G}^{(1)}(x)$ has positive real eigenvalues, $\lambda_p(x)$, ($p=1,2,3$), which correspond to a set of (generally, complex) eigenvectors. The eigenvectors, which we write as $\mathbf{e}^{(p)}$ satisfy

$$\mathbf{G}^{(1)}(x)\cdot\mathbf{e}^{(p)*}=\lambda_p\mathbf{e}^{(p)*},$$
$$\mathbf{e}^{(p)}\cdot\mathbf{G}^{(1)}(x)=\lambda_p\mathbf{e}^{(p)}.\qquad(5.5)$$

If the three eigenvalues $\lambda_p(x)$ are all different, *it is clear* that the three eigenvectors must be orthogonal; if not they may be chosen so. If the eigenvectors are normalized to obey the relations

$$\mathbf{e}^{(p)}\cdot\mathbf{e}^{(q)*}=\delta_{pq},\qquad(5.6)$$

their components form the unitary matrix which diagonalizes $\mathbf{G}^{(1)}(x)$. The eigenvectors, or equivalently the unitary matrix, are determined by a set of eight independent real parameters.

A tensor product, such as

$$\mathbf{e}^{(p)}\cdot\mathbf{G}^{(1)}(x)\cdot\mathbf{e}^{(q)*}=\lambda_p\delta_{pq},\qquad(5.7)$$

expresses the correlation, at the point x, of the field components in the $\mathbf{e}^{(p)}$ and $\mathbf{e}^{(q)}$ directions. It is clear, then, that there always exist a set of three (complex) orthogonal polarization vectors such that the field components in these directions are statistically uncorrelated. The eigenvalues λ_p correspond to the intensities for these polarizations. For quantitative discussions of polarization it is convenient to define the normalized intensities $I_p=\lambda_p/\sum_q\lambda_q$, $(p=1,2,3)$, which sum to unity, $\sum_p I_p=1$. When the normalized intensities are all equal to $\frac{1}{3}$ we have the case of an isotropic field, as in a hohlraum filled with thermal radiation.

The triad of eigenvectors at a point in an arbitrary field depends, in general, on time as well as position. If the density operator, ρ, represents a stationary ensemble, however, the triad becomes fixed. A particular example which has been studied in minute detail in optics is that of a beam of plane waves.[5,13] In that case, since the fields are transverse, one of the eigenvectors may be chosen as the beam direction and obviously corresponds to the eigenvalue zero. The net polarization of the beam is usually defined as the magnitude of the difference of the normalized intensities, $|I_1-I_2|$, which correspond to the remaining two eigenvalues.

We next define the higher order correlation functions as

$$G^{(n)}_{\mu_1\cdots\mu_{2n}}(x_1\cdots x_n,x_{n+1}\cdots x_{2n})=\mathrm{tr}\{\rho E_{\mu_1}{}^{(-)}(x_1)\cdots \times E_{\mu_n}{}^{(-)}(x_n)E_{\mu_{n+1}}{}^{(+)}(x_{n+1})\cdots E_{\mu_{2n}}{}^{(+)}(x_{2n})\}. \quad (5.8)$$

These functions are unchanged by simultaneous permutations of the coordinates $(x_1\cdots x_n)$ and the indices $(\mu_1\cdots\mu_n)$; they are likewise invariant under permutations of the $(x_{n+1}\cdots x_{2n})$ and $(\mu_{n+1}\cdots\mu_{2n})$. They satisfy the symmetry relation

$$G^{(n)}_{\mu_{2n}\cdots\mu_1}(x_{2n}\cdots x_1)=\{G^{(n)}_{\mu_1\cdots\mu_{2n}}(x_1\cdots x_{2n})\}^* \quad (5.9)$$

and are shown, in the Appendix, to obey the inequalities

$$G^{(n)}_{\mu_1\cdots\mu_n\mu_n\cdots\mu_1}(x_1\cdots x_n,x_n\cdots x_1)\geq 0 \quad (5.10)$$

and

$$G^{(n)}_{\mu_1\cdots\mu_n\mu_n\cdots\mu_1}(x_1\cdots x_n,x_n\cdots x_1) \times G^{(n)}_{\mu_{n+1}\cdots\mu_{2n}\mu_{2n}\cdots\mu_{n+1}}(x_{n+1}\cdots x_{2n},x_{2n}\cdots x_{n+1}) \geq |G^{(n)}_{\mu_1\cdots\mu_n\mu_{n+1}\cdots\mu_{2n}}(x_1\cdots x_n,x_{n+1}\cdots x_{2n})|^2. \quad (5.11)$$

As in our earlier discussion of coherence, it is convenient to make use of the normalized correlation functions

$$g^{(n)}_{\mu_1\cdots\mu_{2n}}(x_1\cdots x_{2n}) =G^{(n)}_{\mu_1\cdots\mu_{2n}}(x_1\cdots x_{2n})/\prod_{j=1}^{2n}\{G^{(1)}_{\mu_j\mu_j}(x_j,x_j)\}^{1/2}. \quad (5.12)$$

The necessary conditions for full coherence are

$$|g^{(n)}_{\mu_1\cdots\mu_{2n}}(x_1\cdots x_{2n})|=1, \quad (5.13)$$

which must hold for all components $\mu_1\cdots\mu_{2n}$, as well as all n. It is clear, however, that these conditions do not constitute an adequate definition of coherence, since they are not, in general, invariant under rotations of the coordinate axes. We therefore turn once again to a definition of coherence as a factorization property of the correlation functions.

We define full coherence to hold when the set of correlation functions $G^{(n)}$ may be expressed as products of the components of a vector field $\mathcal{E}_\mu(x)$, $(\mu=1,2,3)$, i.e.,

$$G^{(n)}_{\mu_1\cdots\mu_{2n}}(x_1\cdots x_n,x_{n+1}\cdots x_{2n}) =\prod_{j=1}^{n}\mathcal{E}^*_{\mu_j}(x_j)\prod_{l=n+1}^{2n}\mathcal{E}_{\mu_l}(x_l), \quad (5.14)$$

where it is understood that the vector field $\mathcal{E}_\mu(x)$ is independent of n. It is immediately clear, from the transformation properties of the definition, that a field coherent in one coordinate frame is equally coherent in any rotated frame. Furthermore, all of the normalized correlation functions $g^{(n)}$, which follow from the definition, satisfy the conditions (5.13).

The coherence conditions (5.14) imply that the field is fully polarized in the direction of the vector $\boldsymbol{\mathcal{E}}(x)$ at each point x. The formal way of seeing this is to note that the intensity matrix $G_{\mu\nu}{}^{(1)}(x,x)$, which we discussed earlier in general terms, reduces for a coherent field to,

$$G_{\mu\nu}{}^{(1)}(x,x)=\mathcal{E}_\mu{}^*(x)\mathcal{E}_\nu(x). \quad (5.15)$$

Such a matrix represents an unnormalized projection operator for the direction of $\boldsymbol{\mathcal{E}}(x)$. It obviously has, as an eigenvector in the sense of Eq. (5.5), the vector $\mathcal{E}_\mu(x)$ itself. The corresponding eigenvalue is the full intensity $\sum_\mu|\mathcal{E}_\mu(x)|^2$. The two remaining eigenvalues, which correspond to orthogonal directions, clearly vanish.

It is interesting to note that for coherent fields many of the inequalities stated earlier, e.g., (3.13), (3.14), (5.4), (5.11), reduce to statements of equality. This reduction holds quite generally, as is shown in the Appendix, for those inequalities of quadratic and higher degree in the correlation functions.

The arguments by which we exhibit fields satisfying the coherence conditions, are essentially unchanged from the previous section. In particular, as we shall discuss in the next paper, there exist states which are simultaneously right eigenstates of all three components of $E_\mu{}^{(+)}(\mathbf{r}t)$ and correspond to a set of three complex eigenvalues $\mathcal{E}_\mu(\mathbf{r}t)$. Such states satisfy the coherence conditions (5.14) precisely.

If we have chosen to discuss only the correlations of the electric field in this paper, it is because that field plays the dominant role in all detection mechanisms for photons of lower frequency than x rays. It is not difficult to construct correlation functions which involve the magnetic field as well as the electric field, and perhaps these too will someday prove useful. One method is to use the relativistic field tensor, $F_{\mu\nu}$, in precisely the way we have used the field E_μ. The field tensor may be written as a 4×4 antisymmetric matrix, made up of the

[13] Most of these studies have been confined to stationary, quasimonochromatic beams. See, for example, G. B. Parrent, Jr., and P. Roman, Nuovo Cimento 15, 370 (1960).

components of both the electric and magnetic fields. The nth-order correlation function for the complex components of those fields would have $4n$ four-valued indices. Coherence may then be defined as a requirement that the correlation functions all be separable into the the products of 4×4 antisymmetric fields, just as Eq. (5.14) requires a separation into products of three-vector field components. The advantage of such a definition is to make it clear that coherence is a relativistically invariant concept; that a field which is coherent in any one Lorentz frame is coherent in any other. Fields which are coherent in this relativistic sense are automatically coherent in the more limited senses we have described earlier.

ACKNOWLEDGMENTS

The author is grateful to the Research Laboratory of the American Optical Company and its director, Dr. S. M. MacNeille, for partial support of this work.

APPENDIX

In this section we derive a number of inequalities obeyed by the correlation functions defined in the paper. Fundamentally, these relations are all consequences of a single inequality

$$\operatorname{tr}\{\rho A^\dagger A\}\geq 0, \tag{A1}$$

which holds for arbitrary choice of the operator A. To prove this inequality, we note that the density operator ρ is Hermitian and can always be diagonalized, i.e., we can find a set of basis states such that the matrix representation of ρ is

$$\langle k|\rho|l\rangle=\delta_{kl}p_k. \tag{A2}$$

The numbers p_k may be interpreted as probabilities associated with the states $|k\rangle$. They are, therefore, non-negative, $p_k\geq 0$; which is to say that ρ is a positive definite operator. The normalization condition on the density operator, $\operatorname{tr}\rho=\sum p_k=1$, shows that not all the p_k vanish. The trace (A1) may be reduced, in the representation defined by (A2), to the form

$$\operatorname{tr}\{\rho A^\dagger A\}=\textstyle\sum_k p_k\langle k|A^\dagger A|k\rangle. \tag{A3}$$

The diagonal matrix elements on the right of (A3) are all non-negative since they may be expressed as a sum of squared absolute values,

$$\begin{aligned}\langle k|A^\dagger A|k\rangle&=\textstyle\sum_l\langle k|A^\dagger|l\rangle\langle l|A|k\rangle\\&=\textstyle\sum_l|\langle l|A|k\rangle|^2.\end{aligned} \tag{A4}$$

This statement completes the proof of (A1), since the trace is invariant under unitary transformations of the basis states.

The trace which occurs in the inequality (A1) has the same basic structure as all of the correlation functions $G^{(n)}$. Various inequalities relating the correlation functions follow, more or less directly, from different choices of the operator A. If, for example, we choose A to be $E^{(+)}(x)$, as defined by (3.4), we find the inequality (3.11),

$$G^{(1)}(x,x)\geq 0. \tag{A5}$$

If we choose A to be the n-fold product $E^{(+)}(x_1)\cdots E^{(+)}(x_n)$ we find the inequality (3.12),

$$G^{(n)}(x_1\cdots x_n,x_n\cdots x_1)\geq 0. \tag{A6}$$

The proofs are no different if the components of the three-dimensional field are used in place of $E^{(+)}$, i.e., if a component index μ_j is associated with each coordinate x_j. Hence, we have also derived (5.3) and (5.10).

The remaining inequalities are of second and higher degree in the correlation functions. Those obeyed by the first-order function, $G^{(1)}$, may be found as follows: We choose at random a set of m space-time points $x_1\cdots x_m$, and consider as the operator A,

$$A=\sum_{j=1}^{m}\lambda_jE^{(+)}(x_j), \tag{A7}$$

where the superposition coefficients $\lambda_1\cdots\lambda_m$ are an arbitrary set of complex numbers. When we substitute (A7) into the basic inequality, (A1), we find

$$\sum_{i,j}\lambda_i^*\lambda_jG^{(1)}(x_i,x_j)\geq 0. \tag{A8}$$

In other words, the set of correlation functions $G^{(1)}(x_i,x_j)$ for $i,\ j=1,\ \cdots,\ m$ forms the matrix of coefficients of a positive definite quadratic form. It follows, in particular, that the determinant of the matrix is non-negative,

$$\det[G^{(1)}(x_i,x_j)]\geq 0\quad i,\ j=1,\ \cdots,\ m. \tag{A9}$$

For $m=1$ this inequality is simply (A5). For $m=2$ it becomes the one noted in the text as (3.13),

$$G^{(1)}(x_1,x_1)G^{(1)}(x_2,x_2)\geq|G^{(1)}(x_1,x_2)|^2. \tag{A10}$$

For larger values of m the inequalities are perhaps best left in the form (A9). When tensor components are introduced, we have only to replace the coordinate x_j in the proofs by the combination of x_j and a tensor index μ_j. The relation (5.4) thereby follows from the form of (A10). If, in particular for $m=3$, we choose the three coordinates to be the same and the tensor indices all different, i.e., we choose

$$A=\sum_{\nu=1}^{3}\lambda_\nu E_\nu^{(+)}(x), \tag{A11}$$

we find that the 3×3 matrix $G_{\mu\nu}^{(1)}(x,x)$ is positive definite, a property used in the text in the discussion of polarizations.

Since the succession of inequalities which follows from (A1) is endless, we only mention the quadratic ones for the higher order functions. To find these, we choose a set of $2n$ coordinates at random and let A be any operator of the form

$$A=\lambda_1 E^{(+)}(x_1)\cdots E^{(+)}(x_n)+\lambda_2 E^{(+)}(x_{n+1})\cdots E^{(+)}(x_{2n}). \quad \text{(A12)}$$

The positive definiteness of the quadratic form which results from substituting this expression in (A1) shows that the inequality (3.14) must hold. When vector indices are attached to the operators $E^{(+)}$, the same proof leads to (5.11).

We have noted in the text that, for the particular case of coherent fields, the inequalities of second degree in the correlation functions reduce to equalities. The reason for the reduction lies in the way the correlation functions factorize. The factorization causes all of the second and higher order determinants involved in the statement of positive definiteness conditions [e.g., (A9)] to vanish.

ACCELERATION AND TRAPPING OF PARTICLES BY RADIATION PRESSURE

A. Ashkin
Bell Telephone Laboratories, Holmdel, New Jersey 07733
(Received 3 December 1969)

Micron-sized particles have been accelerated and trapped in stable optical potential wells using only the force of radiation pressure from a continuous laser. It is hypothesized that similar accelerations and trapping are possible with atoms and molecules using laser light tuned to specific optical transitions. The implications for isotope separation and other applications of physical interest are discussed.

This Letter reports the first observation of acceleration of freely suspended particles by the forces of radiation pressure from cw visible laser light. The experiments, performed on micron-sized particles in liquids and gas, have yielded new insights into the nature of radiation pressure and have led to the discovery of stable optical potential wells in which particles were trapped by radiation pressure alone. The ideas can be extended to atoms and molecules where one can predict that radiation pressure from tunable lasers will selectively accelerate, trap, or separate the atoms or molecules of gases because of their large effective cross sections at specific resonances. The author's interest in radiation pressure from lasers stems from a realization of the large magnitude of the force, and the observation that it could be utilized in a way which avoids disturbing thermal effects. For instance a power $P=1$ W of cw argon laser light at $\lambda=0.5145$ μm focused on a lossless dielectric sphere of radius $r=\lambda$ and density = 1 gm/cc gives a radiation pressure force $F_{rad}=2qP/c=6.6\times10^{-5}$ dyn, where q, the fraction of light effectively reflected back, is assumed to be of order 0.1. The acceleration = 1.2×10^{8} cm/sec^2 $\cong 10^5$ times the acceleration of gravity.

Historically,[1,2] the main problem in studying radiation pressure in the laboratory has been the obscuring effects of thermal forces. These are caused by temperature gradients in the medium surrounding an object and, in general, are termed radiometric forces.[3] When the gradients are caused by light, and the entire particle moves, the effect is called photophoresis.[3,4] These forces are usually orders of magnitude larger than radiation pressure. Even with lasers, photophoresis usually completely obscures radiation pressure.[5] In our work, radiometric effects were avoided by suspending relatively transparent particles in relatively transparent media. We operated free of thermal effects at 10^3 times the power densities of Ref. 5.

The first experiment used transparent latex spheres[6] of 0.59-, 1.31-, and 2.68-μm diam freely suspended in water. A TEM$_{00}$-mode beam of an argon laser of radius $w_0=6.2$ μm and $\lambda=0.5145$ μm was focused horizontally through a glass cell 120 μm thick and manipulated to focus on single particles. See Fig. 1(a). Results were observed with a microscope. If a beam with milliwatts of power hits a 2.68-μm sphere off center, the sphere is simultaneously drawn in to the beam axis and accelerated in the direction of the light. It moves with a limiting velocity of microns per second until it hits the front surface of the glass cell where it remains trapped in the beam. If the beam is blocked, the sphere wanders off by Brownian motion. Similar effects occur with the other sphere sizes but more power is required for comparable velocities. When mixed, one can accelerate 2.68-μm spheres and leave 0.585-μm spheres behind. The particle velocities and the trapping on the beam axis can

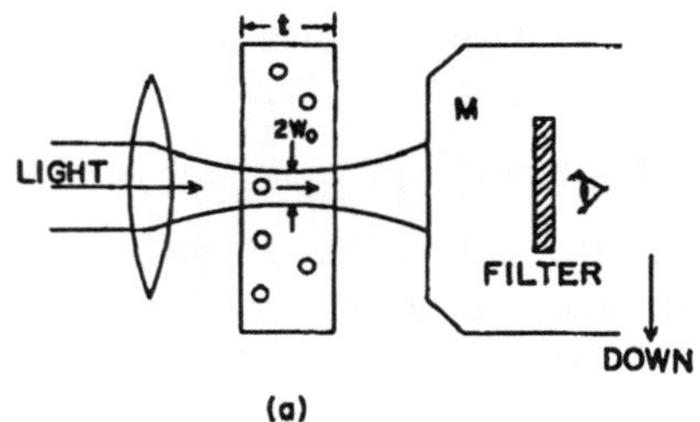

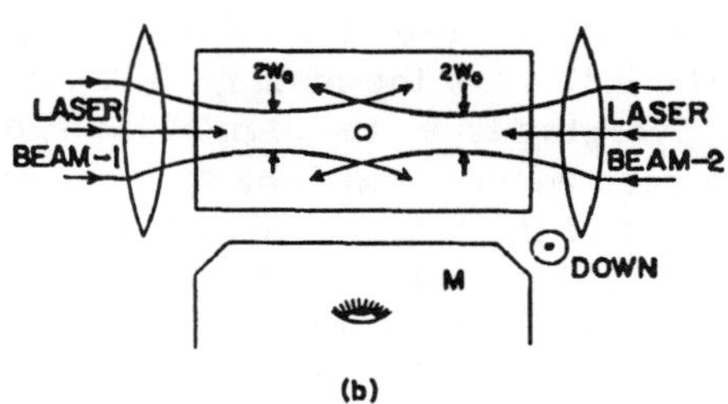

FIG. 1. (a) Geometry of glass cell, $t=120$ μm, for observing micron particle motions in a focused laser beam with a microscope M. (b) The trapping of a high-index particle in a stable optical well. Note position of the TEM$_{00}$-mode beam waists.

be understood as follows (see Fig. 2): The sphere of high index $n_H = 1.58$ is situated off the beam axis, in water of lower index $n_L = 1.33$. Consider a typical pair of rays symmetrically situated about the sphere axis B. The stronger ray (a) undergoes Fresnel reflection and refraction (called a deflection here) at the input and output faces. These result in radiation pressure forces F_R^i, F_R^o (the input and output reflection forces), and F_D^i, F_D^o (the input and output deflection forces), directed as shown. Although the magnitudes of the forces vary considerably with angle Φ, qualitatively the results are alike for all Φ. The radial (r) components of F_D^i, F_D^o are much larger than F_R^i and F_R^o (by ~10 at $\Phi = 25°$). All forces give accelerations in the $+z$ direction. F_R^i and F_R^o cancel radially to first order. F_D^i and F_D^o add radially in the $-r$ direction, thus the net radial force for the stronger ray is inward toward higher light intensity. Similarly the symmetrical weak ray (b) gives a net force along $+z$ and a net outward but weaker radial force. Thus the sphere as a whole is accelerated inward and forward as observed. To compute the z component of the force for a sphere on axis, one integrates the perpendicular (s) and parallel (p) components of the plane-polarized beam over the sphere. This yields an effective $q = 0.062$. This geometric optic result (neglecting diffraction) is identical with the asymptotic limit of a wave analysis by Debye[2] for an incident plane wave. He finds $q = 0.06$. From the force we get the limiting velocity v in a viscous medium using Stokes's law. For $r \ll w_0$,

$$v = 2qPr/3c\pi w_0^2 \eta, \tag{1}$$

where η is the viscosity. For $P = 19$ mW, $w_0 = 6.2$ μm, and a sphere of $r = 1.34$ μm in water ($\eta = 1\times10^{-2}$ P), one computes $v = 29$ μm/sec. We measured $v = 26 \pm 5$ μm/sec which is good agreement. In the above, the sphere acts as a focusing lens. If one reverses the relative magnitudes of the indices of the media, the sphere becomes a diverging lens, the sign of the radial deflection forces reverse, and the sphere should be pushed out of the beam. This prediction was checked experimentally in an extreme case of a low-index sphere in a high-index medium, namely an air bubble. Bubbles, of ~8-μm diam, were generated by shaking a high-viscosity medium consisting of an 80% by weight mixture of glycerol in water. It was found that the bubbles were always pushed out of the light beam as they were accelerated along, as expected. In the same

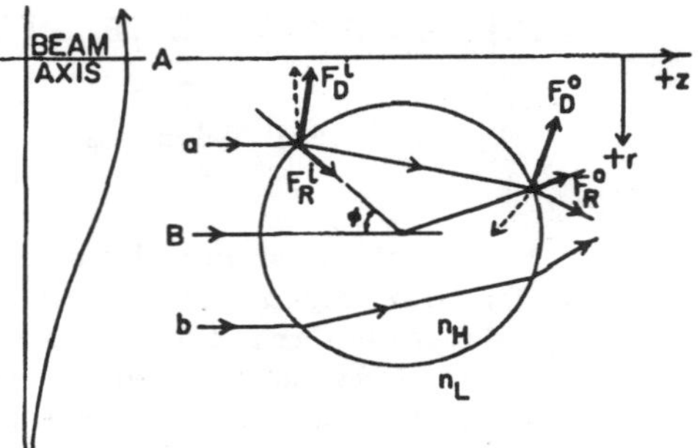

FIG. 2. A dielectric sphere situated off the axis A of a TEM_{00}-mode beam and a pair of symmetric rays a and b. The forces due to a are shown for $n_H > n_L$. The sphere moves toward $+z$ and $-r$.

medium of $n = 1.44$, the 2.68-μm spheres of $n = 1.58$ were still focusing. At higher powers the bubbles are expected to deform. This would result in a deformation contribution to the radiation pressure force as postulated by Askaryon.[7] Our observation of the attraction of high-index spheres into regions of high light intensity is related to the deformation of a liquid surface postulated by Kats and Kantorovich.[8]

The experimentally observed radial inward force on the high-index spheres suggest a means of constructing a true optical potential well or "optical bottle" based on radiation pressure alone. If one has two opposing equal TEM_{00} Gaussian beams with beam waists located as shown in Fig. 1(b), then a sphere of high index will be in stable equilibrium at the symmetry point as shown (i.e., any displacement gives a restoring force). Such trapping was observed experimentally in an open cell filled with 2.68-μm spheres in water as sketched in Fig. 1(b). Here the entire beam is viewed at once. Particles are observed by their brilliant scattered light. With 128 mW in only one of the beams, a maximum particle velocity of ~220 μm/sec was observed as particles traversed the entire near field. The calculated velocity is 195 μm/sec. For trapping, the two opposing beams were introduced. Particles that drift near either beam are drawn in, accelerated to the stable equilibrium point, and stop. To check for stability one can interrupt one beam for a moment. This causes the particle to accelerate rapidly in the remaining beam. When the opposing beam is turned on again the particle returns to its equilibrium point, only more slowly since it is now acted on by the differential force. Interrupting the other beam reverses the behavior. In other experiments, ~5-μm-diam water droplets from an atomizer were accelerated in air with a single beam. At 50 mW, velocities ~0.25 cm/sec were observed. Such motions could be seen with the naked eye. The behavior of the droplets was in qualitative agreement with expectation.

In our experiments it is clear that we have discriminated against radiometric forces. These forces push more strongly on hot surfaces and would push high-index spheres and bubbles out of the beam; whereas our high-index spheres were drawn into the beam. Even the observed direction of acceleration along the beam axis is the opposite of the radiometric prediction. A moderately absorbing focusing sphere concentrates more heat on the downstream side of both the ball and the medium and should move upstream into the light (negative photophoresis).[9] For water drops in air we can invoke the well-confirmed formula of Hettner[10] and compute the temperature gradient needed across a 5-μm droplet to account radiometrically for the observed velocity of 0.25 cm/sec. From Stokes's formula, $F=2.1\times10^{-7}$ dyn. Hettner's formula then requires a gradient of 0.5°C across the droplet. No such gradients are possible with the 50 mW used. For water and glycerol the gradients are also very low.

The extension to vacuum of the present experiments on particle trapping in potential wells would be of interest since then any motions are frictionless. Uniform angular acceleration of trapped particles based on optical absorption of circular polarized light or use of birefringent particles is possible. Only destruction by mechanical failure should limit the rotational speed. In vacuum, particles will heat until they are cooled by thermal radiation or vaporize. With the minimum power needed for levitation, micron spheres will assume temperatures of hundreds to thousands of degrees depending on the loss. The ability to heat in vacuum without contaminating containing vessels is of interest. Acceleration of neutral spheres to velocities ~10^6-10^7 cm/sec is readily possible using powers that avoid vaporization. In this regard one could attempt to observe and use the resonances in radiation pressure predicted by Debye[2] for spheres with specific radii. The separation of micron- or submicron-sized particles by radiation pressure based on radius as demonstrated experimentally could also be useful [see Eq. (1)].

Finally, the extension of the ideas of radiation pressure from laser beams to atoms and molecules opens new possibilities. In general, atoms and molecules are quite transparent. However, if one uses light tuned to a particular transition, the interaction cross section can be much larger than geometric. For example, an atom of sodium has $\pi r^2=1.1\times10^{-15}$ cm^2 whereas, from the absorption coefficient,[11] the cross section σ_T at temperature T for the D_2 resonance line at $\lambda=0.5890$ μm is $\sigma_T=1.6\times10^{-9}$ cm$^2=0.5\lambda^2$ for $T<40$°K (region of negligible Doppler broadening). The absorption and isotropic reradiation by spontaneous emission of resonance radiation striking an atom results in an average driving force or pressure in the direction of the incident light. We shall attempt to show that radiation pressure from a laser beam on resonance can work as an actual optical gas pump and operate against significant gas pressures. Figure 3(a) shows a schematic version of such a pump. Imagine two chambers initially filled with sodium vapor, for example. A transparent pump tube of radius w_0 is uniformly filled with laser light tuned to the D_2 line of Na from the left. Let the total optical power P and the pressure p_0 be low enough to neglect light depletion and absorption saturation. Most atoms are in the ground state. The average force on an atom is $P\sigma_T/c\pi w_0^2$ and is constant along the pump. Call x_{cr} the critical distance. It is the distance traveled by an atom in losing its average kinetic energy $\frac{1}{2}mv_{av}^2$. That is, $Fx_{cr}=\frac{1}{2}mv_{av}^2\cong kT$. The variation of pressure in a gas

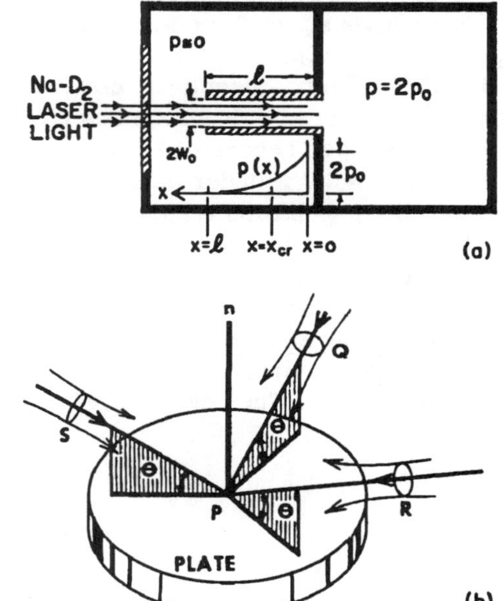

FIG. 3. (a) Schematic optical gas pump and graph of Na pressure $p(x)$. (b) Geometry of gas confinement about point P of a plane surface.

with a constant force is exponential at equilibrium. Thus

$$p(x) = p_0 e^{-Fx/kT} = p_0 e^{-x/x_{cr}}, \quad (2)$$

$$x_{cr} = \pi w_0^2 ckT/P\sigma_T. \quad (3)$$

Next, consider higher power. Saturation sets in. Population equalization occurs between upper and lower levels for those atoms of the Doppler-broadened line of width $\Delta\nu_D$, within the natural width $\Delta\nu_n$ of line center. A "hole" is burned in the absorption line and the power penetrates more deeply into the gas. But there is a net absorption, even when saturated, due to the ever-present spontaneous emission from the upper energy level. The average force per atom also saturates and is constant along the tube. Its value is $(h/\tau_n\lambda)(\Delta\nu_n/\Delta\nu_D)$, where τ_n is the upper level natural lifetime.[12] Lastly, we consider the effect of collision broadening due to a buffer gas on the force per atom. With collision one replaces $1/\tau_n$ by $(1/\tau_n + 1/\tau_L)$ and $\Delta\nu_n$ by $(\Delta\nu_n + \Delta\nu_L)$ in the average saturated force, where $\Delta\nu_L = \frac{1}{2}\pi\tau_L$ is the Lorentz width. This enhances the force greatly. Then

$$x_{cr} = \frac{kT\lambda}{h}\left(\frac{\tau_n\tau_L}{\tau_n + \tau_L}\right)\left(\frac{\Delta\nu_D}{\Delta\nu_n + \Delta\nu_L}\right). \quad (4)$$

As an example, consider Na vapor at $p_0 = 10^{-3}$ Torr ($n_0 = 3.4\times10^{13}$ atoms/cc and $T = 510°K$), buffered by helium at 30 Torr. Take a tube of $l = 20$ cm with diameter $2w_0 = 10^{-2}$ cm. For $\tau_n = 1.48\times10^{-8}$ sec, $\Delta\nu_D = 155\Delta\nu_n$ (at $T = 510°K$), and $\Delta\nu_L \cong 30\Delta\nu_n$,[13] one finds $x_{cr} = 1.5$ cm and $l = 20$ cm $= 13.3x_{cr}$. Thus $p(l) = 2p_0 e^{-13.3} = 2\times10^{-3}\times1.7\times10^{-6} = 3.4\times10^{-9}$ Torr. Essentially complete separation has occurred. This requires a total number of photons per second of $2\pi w_0^2 x_{cr} n_0/(1/\tau_n + 1/\tau_L) \cong 1.7\times10^{19} \cong 6$ W. Under saturated conditions there is little radiation trapping of the scattered light. Almost all the incident energy leaves the gas without generating heat. The technique applies for any combination of gases. Even different isotopes of the same atom or molecule could be separated by virtue of the isotope shift of the resonance lines. The possibilities for forming atomic or molecular beams with specific energy states and for studying chemical reaction kinetics are clear. The possibility of obtaining significant population inversions by resonant gas pumping remains to be evaluated. One can also show that gas can be optically trapped at the surface of a transparent plate. For example [see Fig. 3(b)], three equal TEM_{00}-mode beams with waists at points Q, R, and S directed equilaterally at point P, at some angle θ, result in a restoring force for displacements of an atom about P. Gas trapped about P could serve as a windowless gas target in many experimental situations. The perfection of accurately controlled frequency-tunable lasers is crucial for this work.

It is a pleasure to acknowledge stimulating conversations with many colleagues; in particular, J. G. Bergman, E. P. Ippen, J. E. Bjorkholm, J. P. Gordon, R. Kompfner, and P. A. Wolff. I thank J. M. Dziedzic for making his equipment and skill available.

---

[1]E. F. Nichols and G. F. Hull, Phys. Rev. 17, 26, 91 (1903).

[2]P. Debye, Ann. Physik 30, 57 (1909).

[3]N. A. Fuchs, The Mechanics of Aerosols (The Macmillan Company, New York, 1964).

[4]F. Ehrenhaft and E. Reeger, Compt. Rend 232, 1922 (1951).

[5]A. D. May, E. G. Rawson, and E. H. Hara, J Appl. Phys. 38, 5290 (1967); E. G. Rawson and E. H. May, Appl. Phys. Letters 8, 93 (1966).

[6]Available from the Dow Chemical Company.

[7]G. A. Askar'yan, Zh. Eksperim. i Teor. Fiz.—Pis'ma Redakt. 9, 404 (1969) [translation: JETP Letters 9, 241 (1969)].

[8]A. V. Kats and V. M. Kantorovich, Zh. Eksperim. i Teor. Fiz.—Pis'ma Redakt. 9, 192 (1969) [translation: JETP Letters 9, 112 (1969)].

[9]See Ref. 3, p. 60.

[10]G. Z. Hettner, Physics 37, 179 (1926); Ref. 3, p. 57.

[11]A. C. G. Mitchell and M. W. Zemansky, Resonance Radiation and Excited Atoms (Cambridge University Press, New York, 1969), p. 100.

[12]J. P. Gordon notes that power broadening occurs at still higher powers. This increases the hole width and the average force $\sim\sqrt{P}$.

[13]See Ref. 11, p. 166.

Doppler-Free Two-Photon Spectroscopy of Hydrogen 1S-2S*

T. W. Hänsch,† S. A. Lee, R. Wallenstein,‡ and C. Wieman§
Department of Physics, Stanford University, Stanford, California 94305
(Received 23 December 1974)

We have observed the 1S-2S transition in atomic hydrogen and deuterium by Doppler-free two-photon spectroscopy, using a frequency-doubled pulsed dye laser at 2430 Å. Simultaneous recording of the absorption spectrum of the Balmer-β line at 4860 Å, using the fundamental dye-laser output, allowed us to precisely compare the energy intervals 1S-2S and 2S,P-4S,P,D and to determine the Lamb shift of the 1S ground state to be 8.3±0.3 GHz (D) and 8.6±0.8 GHz (H).

We have observed transitions from the 1S ground state of atomic hydrogen and deuterium to the metastable 2S state, using Doppler-free two-photon spectroscopy.[1-3] The atoms are excited by absorption of two photons of wavelength 2430 Å, provided by a frequency-doubled pulsed dye laser, and the excitation is monitored by observing the subsequent collision-induced 2P-1S fluorescence at the L_α wavelength 1215 Å. Linewidths smaller than 2% of the Doppler width were achieved with two counter-propagating light beams, whose Doppler shifts cancel. The fundamental dye-laser wavelength at resonance 4860 Å coincides with the visible Balmer-β line, and simultaneous recording of the absorption profile of this line permits a precise comparison of the energy intervals 1S-2S and 2S,P-4S,P,D. From our first preliminary measurements we have determined the Lamb shift of the 1S ground state to be 8.3±0.3 GHz (D) and 8.6±0.8 GHz (H), in good agreement with theory. The only previous measurement of the Lamb shift of the 1S state of deuterium, 7.9±1.1 GHz, has been reported by Herzberg,[4] who used a difficult absolute-wavelength measurement of the L_α line. The hydrogen-1S Lamb shift has never been measured before.

Numerous authors[2,3,5] have pointed out that it would be very desirable to observe the 1S-2S transition in hydrogen by Doppler-free two-photon spectroscopy. The $\frac{1}{7}$-sec lifetime of the 2S state promises ultimately an extremely narrow resonance width. The resolution obtained in the present experiment is already better than that achieved in our recent study of the Balmer-α line by saturation spectroscopy,[6] and the implications for a future even more precise measurement of the Rydberg constant are obvious.

We utilized a dye-laser system, consisting of a pressure-tuned dye-laser oscillator with optional confocal-filter interferometer[7] and two subsequent dye-laser amplifier stages, pumped by the same 1-MW nitrogen laser (Molectron UV 1000) at 15 pulses/sec. This laser generates 10-nsec-long pulses of 30–50-kW peak power at 4860 Å with a bandwidth of about 120 MHz (1–2 GHz without confocal filter). A 1-cm-long crystal of lithium formate monohydrate (Lasermetrics) generates the second harmonic with a peak power of about 600 W. A detailed description of this laser system will be published elsewhere.

The ground-state hydrogen atoms are produced by dissociation of H_2 or D_2 gas in a Wood-type discharge tube (1 m long, 8 mm diam, typically 0.1–0.5 Torr, 15 mA). The atoms are carried by gas flow and diffusion through a folded transfer tube about 25 cm in length into the Pyrex observation chamber (Fig. 1). This chamber has two side arms with quartz Brewster windows to transmit the uv laser light and a MgF_2 (originally LiF) window for the observation of the emitted L_α photons. A thin coating of syrupy phosphoric acid is applied to all Pyrex walls to reduce the catalytic recombination of the atoms.

The uv laser light is focused into the chamber

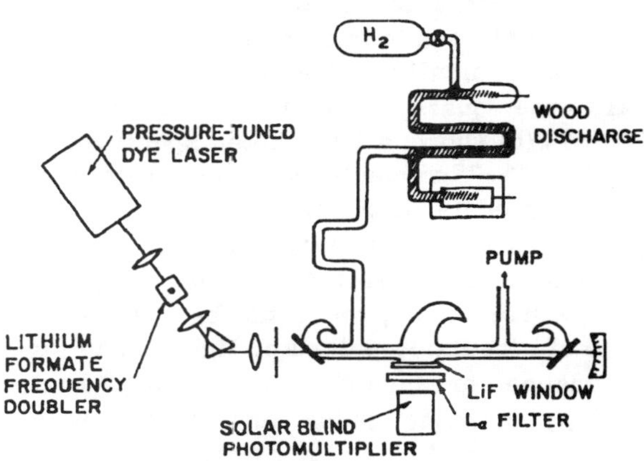

FIG. 1. Experimental setup.

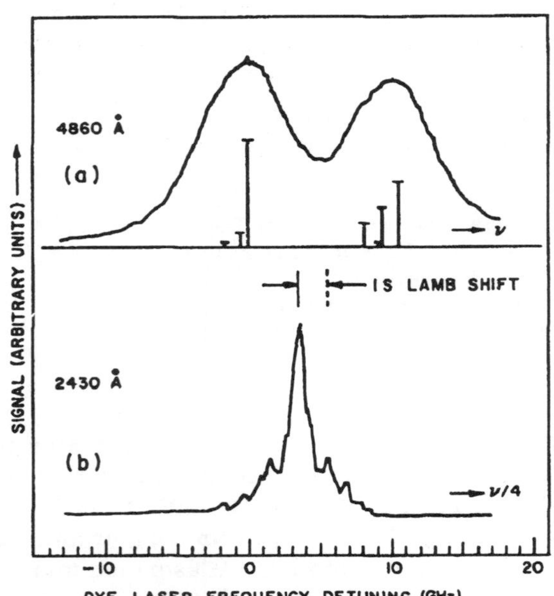

FIG. 2. (a) Absorption profile of the deuterium Balmer-β line with theoretical fine structure; (b) simultaneously recorded two-photon resonance of deuterium $1S$-$2S$.

to a spot size of about 0.5 mm diam. The transmitted beam is refocused into the cell by a spherical mirror to provide a standing-wave field. The separation between illuminated region and MgF_2 window is kept small (1–2 mm) to reduce the loss of L_α photons due to resonance trapping.

The transmitted L_α photons are detected by a solar-blind photomultiplier (EMR 541 J). An interference filter with 6% transmission at 1215 Å reduces the off-resonance background signal to less than one registered photon per several hundred laser pulses. The multiplier output is processed by a gated integrator (Molectron LSDS) with an effective gate opening time of 200 μsec and is electronically divided by a signal proportional to the square of the uv laser intensity.

Figure 2(b) shows a two-photon spectrum of deuterium $1S$-$2S$ recorded with moderate resolution (no confocal-filter interferometer). The low Doppler-broadened pedestal is caused by two-photon excitation by each of the linearly polarized uv beams individually and could be eliminated by the use of circularly polarized light.[3] The signal at resonance corresponds to about 10–20 registered L_α photons per pulse, and remains within the same order of magnitude when the H_2 or D_2 gas is diluted by He up to a ratio of 1000:1. The expected decrease in the number of excited metastable atoms is apparently largely compensated over a wide range by a concomitant reduction of the loss of L_α photons due to resonance trapping and quenching. Despite the lack of a near-resonant intermediate state, the two-photon fluorescence is comfortably strong, and it should easily be possible to observe the signal at a considerably lower total gas pressure (10^{-4} Torr), where pressure broadening and shifts would become unimportant if the $2S$-$2P$ transitions were induced by an applied rf field.

Figure 2(a) shows the absorption spectrum of the deuterium Balmer-β line which was simultaneously recorded by sending a small fraction of the blue dye-laser light through a 15-cm-long center section of the positive column of a Wood-discharge tube (0.2 Torr D_2, 25 mA). The positions of the indicated theoretical fine-structure components were located by a computer fit of the line profile. We have measured the separation of the $1S$-$2S$ resonance from the strongest component ($2P_{3/2}$-$4D_{5/2}$) in the Balmer-β spectrum to be 3.38 ± 0.08 GHz for deuterium and 3.3 ± 0.2 GHz for hydrogen (in terms of the blue-light frequency). The corresponding theoretical separations[8] are 3.420 and 3.422 GHz, respectively. These separations would be larger if the $1S$ state were not raised above its Dirac value by the Lamb shift (theoretically 8.172 GHz for D and 8.149 GHz for H) and our measurement can be interpreted as a determination of the ground-state Lamb shift. A considerable improvement in accuracy can be expected when a high-resolution saturation spectrum[6] of the Balmer-β line is used for the comparison.

A two-photon spectrum of hydrogen $1S$-$2S$ with the laser operating in its high-resolution mode is shown in Fig. 3 (scan time about 2 min). The linewidth is limited by the laser bandwidth of about 120 MHz (in the blue). The spectrum reveals two hyperfine components, separated by the difference of the hyperfine splittings of lower and upper states, as expected from the selection rule $\Delta F = 0$.[2]

It is not difficult to compare the observed signal strength with theoretical estimates, using Eq. (7) of Ref. 2, derived for steady-state conditions. In the present experiment the atoms are excited by light pulses whose time duration τ is short compared to the inverse linewidth Γ_e of the two-photon transition, and which have a near–Fourier transform-limited bandwidth $\Delta\omega \approx \tau^{-1}$. One can show with the help of time-dependent perturbation theory[9] that Eq. (7) of Ref. 2 in this case still

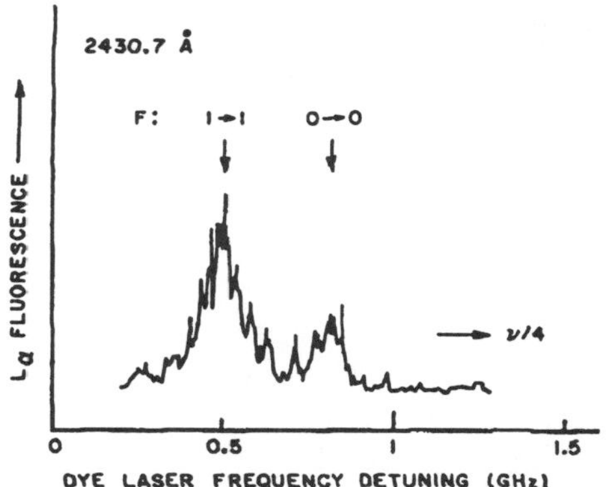

FIG. 3. High-resolution two-photon spectrum of hydrogen 1S-2S with resolved hyperfine splitting.

correctly predicts the effective two-photon transition rate if one replaces the linewidth Γ_e by the laser bandwidth $\Delta\omega$. We have numerically evaluated this expression in a calculation similar to that of Ref. 5, but with inclusion of the continuum in the summation over intermediate states, and obtain a two-photon transition rate per atom of $\Gamma \approx 7\times10^{-4}I^2/\Delta\omega$, where the light intensity I is measured in W/cm^2, the laser bandwidth $\Delta\omega$ in MHz, and Γ in sec^{-1}.

For a comparison with the experiment we consider a H_2 partial pressure of 2×10^{-4} Torr and 10% dissociation, i.e., a density of 1.4×10^{13} 1S atoms/cm^3, so that the loss by resonance trapping is not important. Our estimate then predicts about 3×10^5 excited metastable atoms per pulse over a 1-cm path length. With a detection solid angle of $0.02\times4\pi$ sr, a filter transmission of 6%, and a multiplier quantum efficiency of 10%, we expect about thirty registered photons per laser pulse, in reasonable agreement with the present observations.

We have also calculated the shift of the 1S-2S two-photon resonance frequency caused by the intense uv radiation (ac Stark effect). By numerically evaluating Eq. (1) of Liao and Bjorkholm,[10] we estimate an intensity shift of about 5.5 Hz/(W/cm^2) or less than 2 MHz under the present operating conditions. The light intensity and hence the shift can, in principle, be reduced without loss of resonance signal, by decreasing the laser bandwidth and increasing the interaction time with the atoms. A hundredfold improvement in resolution should be obtainable with the present pulsed dye-laser system, if the hydrogen cell is placed inside a narrow-band confocal-filter interferometer.

We are grateful to Professor T. Fairchild for lending us a superb L_α interference filter of his design. And we thank Professor N. Fortson for stimulating discussions and Professor A. Schawlow for his continuous stimulating interest in this research.

---

*Work supported by the National Science Foundation under Grant No. MPS74-14786A01, by the U. S. Office of Naval Research under Contract No. ONR-0071, and by a Grant from the National Bureau of Standards.

†Alfred P. Sloan Fellow 1973–1975.

‡NATO Postdoctoral Fellow.

§Hertz Foundation Predoctoral Fellow.

[1]L. S. Vasilenko, V. P. Chebotaev, and A. V. Shishaev, Pis'ma Zh. Eksp. Teor. Fiz. 12, 161 (1970) [JETP Lett. 12, 113 (1970)].

[2]B. Cagnac, G. Grynberg, and F. Biraben, J. Phys. (Paris) 34, 845 (1973).

[3]F. Biraben, B. Cagnac, and G. Grynberg, Phys. Rev. Lett. 32, 643 (1974); M. D. Levenson and N. Bloembergen, Phys. Rev. Lett. 32, 645 (1974); T. W. Hänsch, K. C. Harvey, G. Meisel, and A. L. Schawlow, Opt. Commun. 11, 50 (1974).

[4]G. Herzberg, Proc. Roy. Soc., Ser. A 234, 516 (1956).

[5]E. V. Baklanov and V. P. Chebotaev, Opt. Commun. 12, 312 (1974).

[6]T. W. Hänsch, M. H. Nayfeh, S. A. Lee, S. M. Curry, and I. S. Shahin, Phys. Rev. Lett. 32, 1336 (1974).

[7]R. Wallenstein and T. W. Hänsch, Appl. Opt. 13, 1625 (1974).

[8]J. D. Garcia and J. E. Mack, J. Opt. Soc. Amer. 55, 654 (1965).

[9]A. Gold, in *Quantum Optics, Proceedings of the International School of Physics "Enrico Fermi," Course 42*, edited by R. J. Glauber (Academic, New York, 1969).

[10]P. F. Liao and J. E. Bjorkholm, Phys. Rev. Lett. 34, 1 (1975).

Precise Measurements of Axial, Magnetron, Cyclotron, and Spin-Cyclotron-Beat Frequencies on an Isolated 1-meV Electron

R. S. Van Dyck, Jr., P. B. Schwinberg, and H. G. Dehmelt
Department of Physics, University of Washington, Seattle, Washington 98195
(Received 5 November 1976)

A sensitive frequency-shift technique is employed to monitor the magnetic quantum state of a single electron stored in a compensated Penning trap. The electron sees a weak parabolic magnetic pseudopotential in addition to the electric well, which causes the axial oscillation frequency to have a slight magnetic quantum state dependence. Transitions at both the spin-cyclotron-beat (anomaly) frequency and the cyclotron frequency have been measured in the same environment to yield a magnetic anomaly $a_e = (1\,159\,652\,410 \pm 200) \times 10^{-12}$.

The first measurement of the electron-spin magnetic moment on *free* (relativistic) electrons was carried out in 1953 by Crane and co-workers[1] giving $|a_e| \leq 5\times 10^{-3}$ for the "anomaly" defined as $a_e \equiv (\nu_s - \nu_c)/\nu_c$. Here, ν_s and ν_c denote spin and cyclotron frequencies, respectively, in the non-relativistic limit. This study led to the famous University of Michigan "$g-2$" experiments,[1] finally yielding $a_e = (1\,159\,656\,700 \pm 3500)\times 10^{-12}$ which was previously the most accurate experimental value. However, the first definite value,[1] $a_e = (1116 \pm 40)\times 10^{-6}$, was obtained in 1958 at the

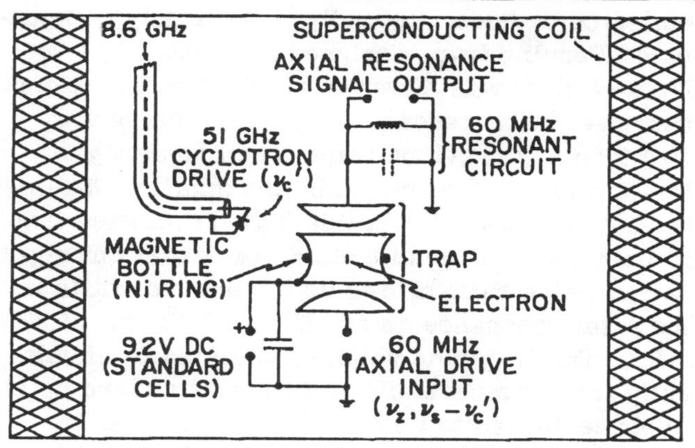

FIG. 1. Geonium spectroscopy experiment (schematic). This apparatus allows the measurement of the cyclotron frequency, ν_c', and the spin-cyclotron-beat (or anomaly) frequency, $\nu_a' = \nu_s - \nu_c'$, on a single electron stored in a Penning trap at $\simeq 4$°K ambient. Detection is via Rabi-Landau level-dependent shifts in the continuously monitored axial resonant frequency, ν_z, induced by a weak magnetic bottle.

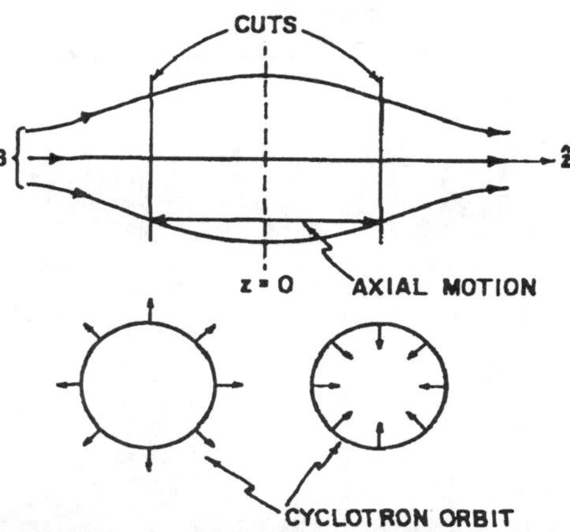

FIG. 2. Mechanism for inducing spin flips by the electron's axial motion in the magnetic bottle. From the electron's frame of reference, a magnetic field is seen rotating at ν_c', but modulated by the axial motion at ν_a', yielding sidebands at $\nu_c' \pm \nu_a'$ with $\nu_s = \nu_c' + \nu_a'$.

University of Washington. None of the above workers simultaneously measured ν_c; hence in 1969, Gräff and co-workers[1,2] first measured both $\nu_s - \nu_c$ and ν_c on electrons in a Penning trap, obtaining $a_e = (1\,159\,660 \pm 300) \times 10^{-9}$. Sharper cyclotron resonances[2,3] had been reported earlier by Walls.

Our experiment, proposed[4] and briefly reported upon earlier,[5] is the culmination of work initiated at the University of Washington several years ago by one of the authors after the realization that for a Penning trap the small shift $-\delta e$ in ν_c due to its electric field is constant throughout its volume and measurable.[2] The experiment (see Fig. 1) is carried out on a *single* electron, confined in such a trap at ~ 4°K, whose axial oscillatory resonance at frequency $\nu_z \simeq 60$ MHz is easily observed.[6] This system may profitably be looked upon as a synthetic ultraheavy metastable atom, which we chose to call "geonium" because ultimately the electron is bound to the earth via the trap structure and the magnet. The geonium energy levels[7] are given by

$$h^{-1}E_{mnkq} = m\nu_s + (n+\tfrac{1}{2})\nu_c' + (k+\tfrac{1}{2})\nu_z - (q+\tfrac{1}{2})\nu_m, \qquad (1)$$

where $\nu_c' \equiv \nu_c - \delta_e$ is the shifted cyclotron frequency with[8] $2\delta_e\nu_c' = \nu_z^2$. For an ideal axially symmetric trap the magnetron frequency ν_m equals δ_e. Spin flips and excitation of the cyclotron resonance are detected by making ν_z slightly dependent on spin and cyclotron quantum states, $m = \pm\frac{1}{2}$, $n = 0, 1, 2, 3, \ldots$. By use of weak magnetic bottle[9] the axial magnetic field, B_z, becomes $B_0 + B_2 z^2$, where $B_0 \simeq 18.3$ kG and $B_2 \simeq 120$ G/cm^2 or 300 G/cm^2 for two different applied bottles. The resulting pseudopotential $-\langle\mu_z\rangle B_z$ for a magnetic moment $\langle\mu_z\rangle \simeq (2n + 2m + 1)\mu_B$ will perturb the large electrostatic potential energy to yield a frequency shift, $\delta\nu_z(m,n) \simeq (m + n + \frac{1}{2})\delta$ with $\delta = 1.0$ or 2.5 Hz depending on the bottle used. The cyclotron resonance at 51 GHz is easily measured via excitation to $n \gg kT/h\nu_c' \simeq 1.5$ by a very modest microwave drive. However, the observation of spin flips is normally obscured by the rapid and random variation of n due to thermal background radiation since the spontaneous emission lifetime of the $n = 1$ level is $\simeq 1$ sec and $n \leq 4$ levels are frequently populated. It is nevertheless possible to determine m from the minimum values $\delta\nu_z = 0$ or $+\delta$ by following $\delta\nu_z\{m, n(t)\}$ for $\simeq 30$ sec because spin relaxation is extremely slow. In order to minimize the effect of magnetic field fluctuations, spin flips are induced by an auxiliary forced axial oscillation whose amplitude is kept *smaller than thermal* (4°K) through the magnetic bottle field (see Fig. 2) at $\nu_a' \equiv \nu_s - \nu_c'$, chosen near ν_z. Under the assumption of a thermally excited cyclotron orbit circling, for simplicity, about the trap axis at $z < 0$, the outward pointing radial components of the bottle field, B_r, appear to the electron as a field rotating at ν_c'. The forced axial oscillation at ν_a' now periodically carries the electron into the $z > 0$ region where the direction of B_r is reversed. This causes the electron to see a field rotating not at ν_c' but at ν_c'

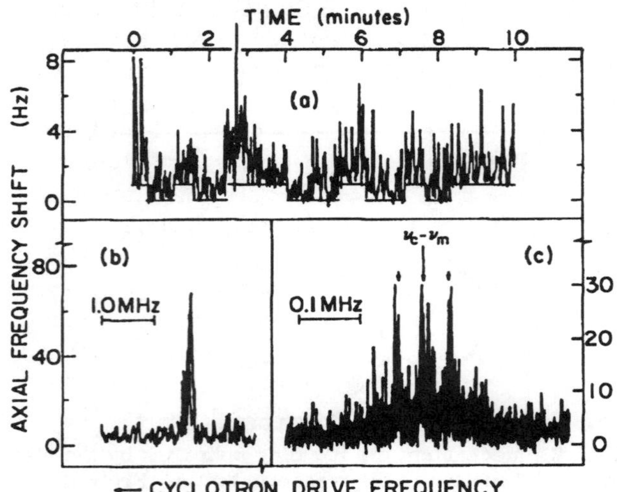

FIG. 3. Observed axial frequency shift, $\delta\nu_z(m,n)$, for a locked electron. (a) Spin flips are signaled in the 1.0-Hz bottle by ± 1.0-Hz changes in the $n=0$ plateau when ν_a' is applied continuously. The spikes reflect thermal cyclotron excitation. (b) Externally excited cyclotron resonances for the 2.5-Hz bottle with large locking drive and (c) the same for the 1.0-Hz bottle with weak locking drive. For (c), $\nu_c - \nu_m = 51\,073\,965$ kHz.

$\pm\nu_a'$ instead. As $\nu_c' + \nu_a'$ equals ν_s, this component induces spin flips. Thus, upon relating ν_s and ν_c to the experimental data by means of Eq. (1) we can obtain[8] $a_e\,(\text{expt}) = (\nu_a' - \delta_e)/(\nu_c' + \delta_e)$. The magnetron frequency ν_m may be indirectly measured because the electron also absorbs radiation at $\nu_z + \nu_m$. The resulting agreement of the measured ν_m and calculated δ_e values to within one part in 10^4 or 3 Hz constitutes experimental proof that the dc trapping potential is effectively axially symmetric as assumed in the derivation of Eq. (1). The $\nu_z + \nu_m$ absorption has been vital in the present apparatus in continually radially pushing[10] the electron into the saddle point. Here the ν_m-sideband spectrum of the ν_c' resonance, Fig. 3(c), indicates that indeed the magnetron radius has been reduced to $\lesssim 50\ \mu$m. This assures that the electron sees the same magnetic field throughout an entire run and helps stabilize the axial frequency.

The apparatus, as shown schematically in Fig. 1, makes use of two important prior experimental accomplishments. First, Wineland, Ekstrom, and Dehmelt[6] had shown that one single electron could be isolated in a Penning trap and that its forced axial harmonic motion could be detected. The trapped electron was obtained by ionizing a residual gas atom inside the trap with a 1-keV electron beam (< 50 nA) emitted from a field-emission cathode. Secondly, a specially compensated Penning trap, designed and tested by Van Dyck *et al.*,[11] was shown to yield a hundred times narrower axial resonant linewidth (one part in 10^7) than any previous trap. The frequency of this motion is determined by the applied ring-to-endcap potential. When this high-resolution trap is immersed in liquid helium, the excellent signal-to-noise ratio allows us to observe shifts in the axial resonance as small as 0.5 Hz out of 60 MHz. The trap's dc voltage, obtained predominantly from temperature-stabilized standard cells, is phase-locked to a frequency synthesizer by transforming the driven signal, available at the upper cap, into an additive correction voltage which reflects any frequency shift, $\delta\nu_z(t)$. The weak magnetic bottle obtained from a fine Ni wire wound concentrically around the ring electrode is shown in Fig. 1 for the case of $\delta = 1.0$ Hz. A supplementary large iron ring placed outside the vacuum envelope reinforces the Ni ring for the case of $\delta = 2.5$ Hz. For the excitation of the cyclotron resonance, it was convenient to mount a small Schottky diode outside the vacuum envelope, but near the gap between cap and ring. The diode, which acts simultaneously as a frequency multiplier and antenna, is driven from a stabilized $\nu_c'/6$ microwave source through a small coaxial lead.

The above apparatus was now used to measure ν_c' and ν_a' as simultaneously as feasible. (Since it is kept constant, ν_z is known throughout the experiment.) For $B_z = 18.3$ kG and $T = 4°$K, the electron spends about half the time in the $n=0$ level. Thus, the $\delta\nu_z(t)$ signal will exhibit a very well defined $n=0$ plateau with a series of positive spikes, as shown in Fig. 3(a) (corresponding to thermally excited cyclotron levels). The cyclotron resonance can be detected through the growth in the positive spikes upon irradiating the electron with the microwave antenna [see Fig. 3(b)]. In this way, the resonance has been measured in the limit of a very small axial drive, as shown in Fig. 3(c). This operation is necessary because of the B_2z^2 term in the magnetic field; the square of the small thermal and large drive amplitudes yields a cross term which effectively enhances the field jitter, thus producing broad resonances. Likewise, the average of the term yields a shift which is ≤ 0.1 ppm for weak locking drives (when $\delta = 1.0$ Hz). In addition, there are also the distinctive motional magnetron sidebands[2] [see Fig. 3(c)] which we attribute to a gradient associated with the twisted ends of the Ni wire producing the mag-

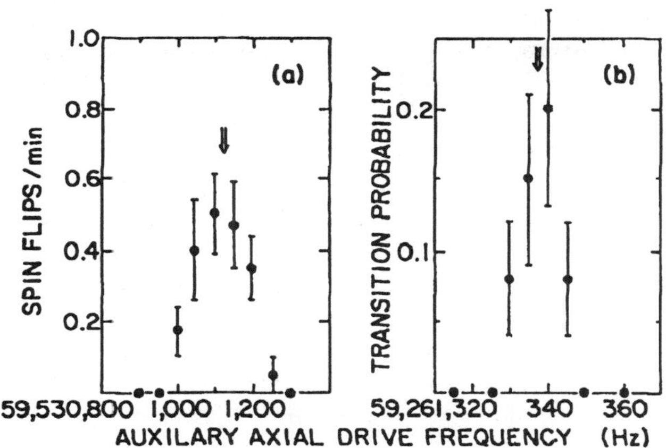

FIG. 4. Anomaly frequency resonances. (a) Data obtained for $\delta = 2.5$ Hz with continuous detection by counting $n = 0$ plateau changes (spin flips) in a fixed time interval (~ 20 min); (b) data obtained for $\delta = 1.0$ Hz with alternating detection by counting $n = 0$ plateau changes in ~ 20 alternating, 1-min excitation/detection time intervals. The error bars indicate counting statistics.

netic bottle. A recent method of alternately pulsing on the microwave and axial drive fields has been shown to reduce the line broadening to 0.1 ppm and to make any shift undetectable. As evident in Fig. 3(a), spin flips have been produced by the previously described mechanism, adjusted to create a rotating magnetic field of ~0.2 μG. During a typical anomaly run, there will be both a rf drive at the electron's axial resonant frequency, ν_z, used to lock the electron, and an off-resonance auxiliary axial rf drive near ν_a', used to induce spin flips. If both are on continuously, again the resonance is broad, as shown in Fig. 4(a). However, if the two drives are alternated in order to eliminate the enhanced magnetic field jitter, sharp resonances like that shown in Fig. 4(b) can be obtained.

The anomaly is computed using the measured ν_c' and ν_a' values and δ_e calculated from Eq. (1). The alternating method in the 1.0-Hz bottle typically yields an accuracy per run of ± 0.25 ppm using a linewidth as the limit of our uncertainty. [For a sample run, we had $\nu_c' = 51\,072\,915\,(10)$ kHz, $\nu_a' = 59\,261\,337.5\,(4.5)$ Hz, and $\nu_z = 59\,336\text{-}170.14\,(10)$ Hz.] The present average of eight runs to date using the 1.0-Hz bottle and the alternating method yields

$$a_e(\text{expt}) = (1\,159\,652\,410 \pm 200) \times 10^{-12},$$

where the error is the sum of the standard deviation from the average (± 100) and the estimated maximum systematic error in ν_c' (± 100). This accuracy is comparable to that obtained by quantum electrodynamic calculations. For three separate numerical determinations of the sixth-order terms in the anomaly,[12] $\{a_e(\text{expt}) - a_e(\text{theory})\}$ is 300 ± 650, − 70 ± 310, and − 960 ± 530 (all in units of 10^{-12}). Each of the theoretical errors includes ± 245 (in quadrature) contributed by the uncertainty of the fine-structure constant.[13] Our empirical dimensionless ratio $(\nu_s/\nu_c) = 1.001\,159\,652\,410\,(200)$ is the most precisely known characteristic parameter of an elementary particle! It may be compared with[14,15] $(\nu_s/\nu_c)_{\mu+} = 1.001\,165\,895\,(27)$ and $(\nu_s/\nu_c)_p = 2.792\,845\,600\,(1100)$, where it has been more difficult to develop adequate theoretical models. Future operation of our experiment at 60 kG promises a factor of 3 reduction in bottle broadening $\propto B_2/B_0$. In addition, with averaging and line splitting, accuracies of parts in 10^8 for the anomaly will become feasible with relativistic corrections still too small (order 10^{-9}) to be measured!

We thank the National Science Foundation for support.

---

[1]Descriptions of early work are given in the review article of A. Rich and J. Wesley, Rev. Mod. Phys. 44, 250 (1972).

[2]H. G. Dehmelt, in *Atomic Masses and Fundamental Constants 5* (Plenum, New York, 1976), p. 499.

[3]H. Dehmelt and F. Walls, Phys. Rev. Lett. 21, 127 (1968); preliminary anomaly data by F. Walls and T. Stein, Phys. Rev. Lett. 31, 975 (1973).

[4]H. Dehmelt and P. Ekstrom, Bull. Am. Phys. Soc. 18, 727 (1973); H. Dehmelt, P. Ekstrom, D. Wineland, and R. Van Dyck, Bull. Am. Phys. Soc. 19, 572 (1974).

[5]R. Van Dyck, P. Ekstrom, and H. Dehmelt, Bull. Am. Phys. Soc. 21, 827 (1976), and Nature (London) 262, 776 (1976).

[6]D. Wineland, P. Ekstrom, and H. Dehmelt, Phys. Rev. Lett. 31, 1279 (1973).

[7]A. A. Sokolov and Yu. G. Pavlenko, Opt. Spectrosc. 22, 1 (1967).

[8]This result holds well within the accuracy needed in this experiment even if the electric quadrupole and magnetic fields are misaligned to a much greater extent than is possible in our apparatus. (L. S. Brown, private communication.)

[9]Our scheme owes something to Stern and Gerlach and also to a proposal by F. Bloch, Physica (Utrecht) 19, 821 (1953), unrealized so far: L. Knight (1965), see Refs. 1, 2.

[10]To be discusses more fully in a later article; also see H. Dehmelt, Nature (London) 262, 777 (1976).

[11]R. S. Van Dyck, D. J. Wineland, P. A. Ekstrom, and H. G. Dehmelt, Appl. Phys. Lett. 28, 446 (1976).

[12]M. Levine and R. Roskies, Phys. Rev. D 14, 2191 (1976); M. Levine and J. Wright, Phys. Rev. D 8, 3171

(1973); R. Carroll, Phys. Rev. D 12, 2344 (1975); P. Cvitanovic and T. Kinoshita, Phys. Rev. D 10, 4007 (1974).

[13]P. T. Olsen and E. R. Williams, cited by Dehmelt, Ref. 2, p. 538.

[14]J. Bailey *et al.*, Phys. Lett. 55B, 420 (1975).

[15]Least-squares adjusted; see E. R. Cohen and B. N. Taylor, J. Phys. Chem. Ref. Data 2, 663 (1973).

Chapter 5
NUCLEAR PHYSICS
HERMAN FESHBACH

Introduction

Papers Reprinted in Book

Ernest O. Lawrence (1901–1958). (Courtesy of Lawrence Radiation Laborarory, AIP Emilio Segrè Visual Archives.)

Herman Feshbach has been a member of the Physics Department at Massachusetts Institute of Technology since 1941. He was the Head of the Department at MIT from 1973 to 1983, Institute Professor from 1983 to 1987, and Professor Emeritus from 1987 to the present. He also served as President of The American Physical Society in 1980. His areas of research include nuclear reactions, the optical model, doorway states, and multistop nuclear reactions. He is the author of *Methods of Theoretical Physics* (with P. M. Morse), *Nuclear Structure* (with A. de Shalit), and *Nuclear Reactions.* His awards include the APS Bonner Prize and the National Medal of Science.

Melba Phillips (1907-). (Courtesy of AIP Emilio Segrè Visual Archives.)

It is often remarked that nuclear physics began in 1932. This is an oversimplification, as natural radioactivity had been thoroughly explored, atomic masses had been measured, and Rutherford had observed a nuclear reaction before 1932. In 1931 Van de Graaff had invented his electrostatic generator, and Lawrence and Livingston the cyclotron, while Cockcroft and Walton had achieved higher voltages by means of a voltage-doubling scheme in 1930, and in 1932 they disintegrated lithium using 400 keV protons. On the theoretical side, the use of quantum mechanics for α-particle decay by Gamow and Condon and Gurney (1929) was very revealing. But a paradox remained. The Bose statistics of ^{14}N could not be explained if it consisted of protons and electrons, as Ehrenfest and Oppenheimer (1931) had shown that such a system would obey Fermi-Dirac statistics. The discovery of the neutron by Chadwick in 1932 resolved this issue. In the same year, deuterium was isolated by Urey providing a simple nucleus, the deuteron, for study.

The next years prior to World War II were spent in mapping out the properties of the nuclear system. Using proton-proton scattering, White (1936), and independently Tuve, Heydenburg, and Hafstad (1936), demonstrated experimentally the existence of a short-range nuclear force. The analysis was carried out by Breit, Condon, and Present (1936). Charge independence of nuclear forces had been proposed earlier by Young (1935) and was confirmed in this analysis. Cassen and Condon (1936) formalized this result using isospin, an internal symmetry that considered the neutron and proton as different states of the nucleon. This was the first introduction of an internal symmetry into physics.

Later, the discovery of the quadrupole moment of the deuteron by Kellogg *et al.* (1939) revealed the presence of tensor forces. Their impact was analyzed by Rarita and Schwinger (1941).

The use of neutrons, e.g. Ra-Be sources, led to the discovery of artificial radioactivity. A big step in the understanding of β decay was made by Fermi in his theory of 1934, making use of Pauli's suggestion that β decay involved a new particle, the neutrino. Neutron-induced reactions led to the observation of neutron resonances whose analysis was provided by the Breit-Wigner formula (1937). A statistical approach by Weisskopf (1937) led to his evaporation model. Neutrons were also used to produce new nuclei, such as the transuranics by Seaborg and his colleagues, as well as the lighter element technetium, filling an empty spot in the periodic table. And of course a major experimental discovery was that of fission, which was explicated by N. Bohr and Wheeler (1939).

Spins, magnetic moments, and electromagnetic transition probabilities yield enough information to make an attempt of the theory of light nuclei possible. In 1937 Feenberg and Wigner developed such a theory for nuclei up to ^{40}Ca using the two-body forces popular at that time, which did not include spin-orbit or tensor forces as these were yet to be discovered. In 1937 Wigner made use of the group theory SU(4), a symmetry that such forces would satisfy. A similar theory was developed by Hund. Saturation as indicated by the semiempirical mass law of von Weizsäcker remained a problem.

Applications of nuclear physics to industrial, medical, and other sciences began to be developed before World War II. Perhaps the most exciting development was Bethe's (1939), in which the source of energy production in stars was traced to nuclear reactions.

This fruitful period, summarized up to 1936 by Bethe, Bacher, and Livingston in their *Reviews of Modern Physics* articles, was then interrupted by World War II.

The period following the war to the present witnessed a spectacular flowering of the subject. Research directions indicated by pre-war studies were pursued, new phenomena were uncovered, and a relatively complete theoretical framework was developed. These became possible because of the expansion in experimental capability. Beams of electrons, photons, pions, kaons, neutrons, protons, deuterons, helium and ^3H, and heavier nuclei ("heavy ions") of good quality over a wide range in energy, which could be polarized where appropriate, are available. Concurrently, the sensitivity of detectors increased enormously in this period. What has been learned? Some salient results follow.

Franz N. D. Kurie (1907-1954). (Courtesy of U.S. Navy, AIP Emilio Segrè Visual Archives.)

Nuclear Forces

Charge independence was confirmed as well as the possibility of exchange reactions $n + p \rightarrow p + n$. At long distances the nucleon-nucleon potential is given by one pion exchange. At small interparticle distances the nucleon-nucleon wave function is very small. This may be simulated by a hard core [Jastrow (1951)]. Semiempirical nucleon-nucleon potentials fitting the two-particle data have been developed. The low-energy nucleon-nucleon data were parameterized by the effective range theory [Schwinger

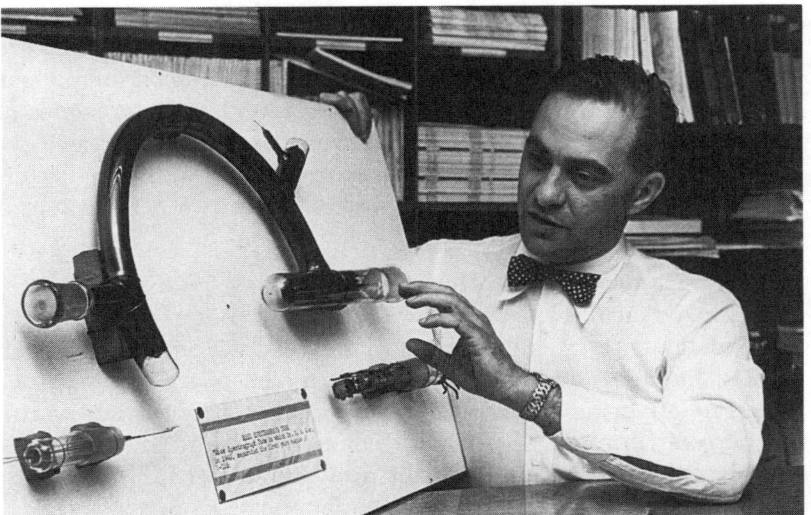

Alfred O. C. Nier (1911-1994). (Courtesy of University of Minnesota, AIP Emilio Segrè Visual Archives.)

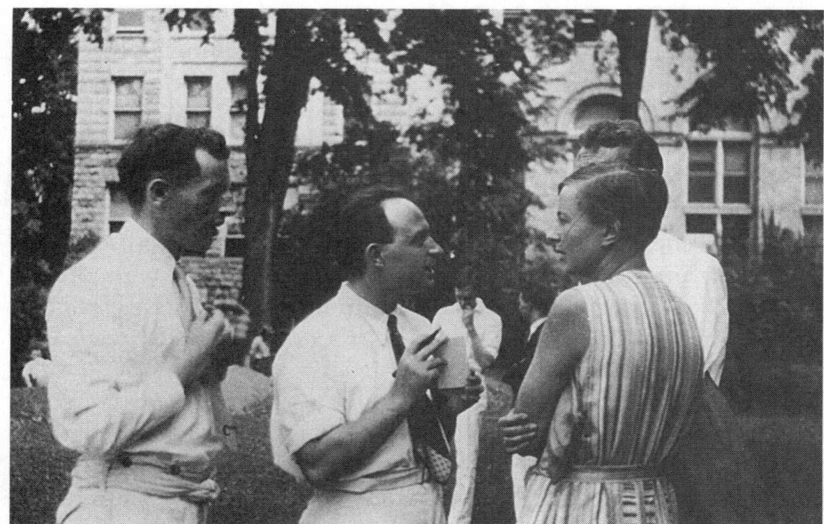

Enrico Fermi (1901-1954) (second from left) and **Maria Goeppert-Mayer (1906-1972).** (Courtesy of Goudsmit Collection, AIP Emilio Segrè Visual Archives.)

(1947), Bethe (1949), and Blatt and Jackson (1949)], which extracted the relevant quantities measured by experiment. At higher energies phase shift analysis was used [Breit *et al.* (1960)]. This became possible only when polarization parameters [Wouters (1951)] in addition to angular distributions were measured.

The Nucleon

Excited states of the nucleon exist. One formed by the pion-nucleon interaction is called the delta (Δ). The formation of the Δ and the resultant Δ-hole state govern the pion-nucleus scattering. Information on the behavior of the Δ in the nuclear medium can be obtained from such experiments.

Electron Scattering

By analyzing electron-nuclear scattering, the distribution of charge and current in the nucleus can be obtained. Hofstadter (1953) was awarded a Nobel Prize for such experiments. The results are compared with theoretical predictions, a comparison that will be commented on below. There were surprises that were understood when compared with a sophisticated theory of nuclear structure. Scattering of electrons by ^{3}He revealed the presence of pion exchange currents flowing between the nucleons. These experiments are relatively easy to analyze because of the very well-known relatively weak electromagnetic electron-nucleon interaction.

Neutron-Nuclear Interaction

Improved facilities provided the resolution that permitted mapping of low-energy, neutron-induced resonances [Garg *et al.* (1971)] for many nuclei. A statistical model of the spacing between neighboring resonance was developed, which leads to the Wigner distribution. This is now considered by many as an example of quantum chaos. Porter and Thomas (1956) devised a statistical distribution of the widths for the resonances.

Nuclear Structure

A remarkable and most important set of discoveries pointed to the existence of comparatively simple modes of motion of a complex system, the nucleus consisting of strongly interacting nucleons. As it appears now, such modes of motion are present for any complex system consisting of a finite number of particles, such as molecules, nuclei, and recently, metallic clusters. These are present even though the interparticle forces are very different.

Shell Model

Mayer (1948) and Jensen (1949) showed that one could understand the magic numbers—mass numbers for which the binding is larger than neighboring nuclei—if one assumed that each nucleon moves independently in a mean field, which is an average of the action of the other nucleons. The Pauli principle plays an important role. A relatively strong spin-orbit component not included in earlier prewar calculations is needed to obtain the magic numbers for the heavier nuclei. Much evidence, such as the existence of isomeric nuclei [M. Goldhaber and Sunyar (1951)], and the orbital angular momentum of the nucleon transferred to a nucleus in a (d,p) stripping reaction, was obtained to both correlate the existence of the shells and delineate their properties. It was an astounding development for which a Nobel Prize was awarded.

Unified Model

Stimulated by the existence of large quadrupole moments and rotational level structure, A. Bohr and Mottelson produced another surprise, namely rotational and vibrating nuclei. Later, they, together with Pines (1958), identified "superconducting" nuclei. Bohr and Mottelson, together with their colleagues, especially Nilsson, were able to relate the properties of these nuclei to the single-particle orbits in a deformed mean field—that is, a mean

J. H. D. Jensen (1907-) at 1934 Copenhagen Conference, Bohr Institute. (Courtesy of Weisskopf Collection, AIP Emilio Segrè Visual Archives.)

field that was not spherical. Qualitatively, one must realize that the mean field is dynamic; it can be deformed, and it can vibrate.

Giant Resonances

Soon after the war's end, a broad resonance induced by the gamma-ray interaction with nuclei was discovered [Baldwin and Klaiber (1947)]. It was shown to be dipole in nature. Later, other collective modes induced by electromagnetic means, such as the giant quadrupole resonance, the monopole resonance, and so on, were found to exist. The dipole resonance was explained as an oscillation of the protons against the neutrons [M. Goldhaber and E. Teller (1948)].

Another type of resonance is the isobar analog resonance, which is the expression of the isospin symmetry broken by the Coulomb interaction and the neutron-proton mass difference [J. D. Anderson and Wong (1961)]. Spin resonances and the Gamow-Teller resonance, which involves oscillation in both the charge and spin, were observed. These are collective vibratory degrees of freedom that exist in complex nuclei although strong internucleon forces are acting.

Shape resonances were first seen in the $^{12}C + ^{12}C$ reaction [Bromley *et al.* (1960)]. Later research found similar resonances in the collision of 4n nuclei. These systems are referred to as "nuclear molecules." An equivalent description considers them to be superdeformed states of the compound system. In the case of $^{12}C + ^{12}C$, the resonances are states of ^{24}Mg. Structure of this kind was also found in fissioning nuclei [Fubini, Blons, Michaudon, and Paya (1968)].

These resonating states are not exact solutions of the nuclear Schrödinger equation. When the energy resolution is improved sufficiently, the resonance disappears, being replaced by a rapidly fluctuating, chaotic cross section. More on this later.

Maurice Goldhaber (1911-). (Courtesy of *Physics Today* Collection, AIP Emilio Segrè Visual Archives.)

Heavy Ions

The collision of two nuclei yields a rich set of phenomena [Breit *et al.* (1952)]. Only a few will be mentioned here. The first is the production of very high spin states approaching spins for which the spinning nucleus is no longer stable. The recently discovered superdeformed nuclei were produced in such a reaction. Because of the relatively strong Coulomb forces, the cross section for the Coulomb excitation of nuclei becomes very probable. Finally, the transfer of a substantial number of nucleons to one of the interacting nuclei produces nuclei relatively far from the valley of stability and more generally new nuclei. A striking phenomenon, referred to as deep inelastic scattering [Kaufmann and Wolfgang (1959)] should be mentioned. For this collision type, nearly all of the initial kinetic energy is transformed into internal energy of the colliding nuclei. Theoretical developments include the time-dependent Hartree-Fock method [Koonin *et al.* (1977)], in which the nuclear wave functions are taken to be time-dependent Slater determinants. A second makes use of the Uehling and Uhlenbeck version of the Lorentz-Boltzmann equation, while a third, the

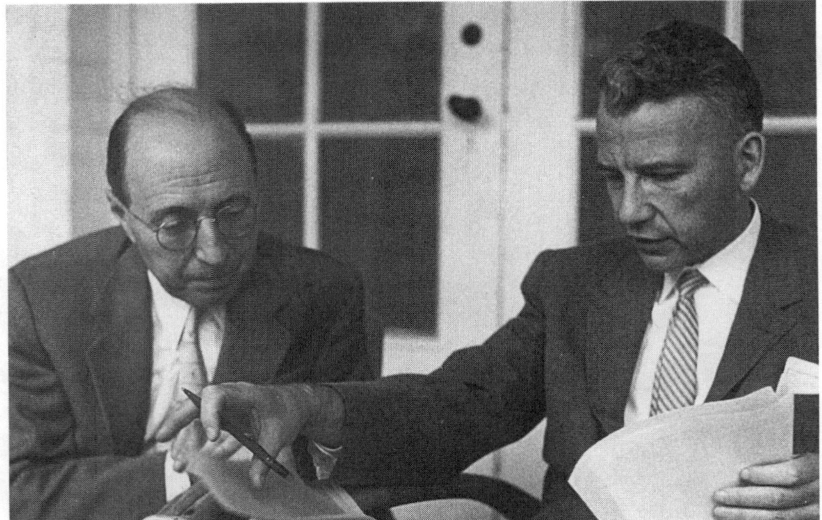

E. P. Wigner (1902-1995) and **H. H. Barschall (1915-)** in April 1960. (Photograph by Bob Davis. Courtesy of AIP Emilio Segrè Visual Archives.)

cascade model [Chen *et al.* (1968)], follows the path of each nucleon through the nucleus numerically, using the Monte Carlo procedure to select their trajectories.

Most recently, a new field has been opened—the collision of relativistic heavy ions, which will permit the study of high-temperature–high-density nuclear matter. Perhaps a quark-gluon plasma [Bjorken (1983)] will be formed.

Pion and Kaon Interaction with Nuclei

For the pion interaction we select the study of the propagation of the Δ, the nucleon isobar, within the nuclear medium. Charge exchange reactions in which $\pi^- \rightarrow \pi^0$, or double charge exchange $\pi^- \rightarrow \pi^+$, study fundamental properties of the nucleus. An interesting consequence of kaon interaction with nuclei is the production of *hypernuclei*—that is, nuclei in which one of the particles is a Λ.

Nuclear Many-Body Theory

The explanation of these striking features of the complex nuclear system is a focus of research in theoretical nuclear physics. The treatment of the strong short-range interaction was advanced by Bethe and Goldstone and incorporated into many-body theory in the Brueckner Hartree-Fock method [Brueckner *et al.* (1955)]. Negele (1970), using many-body theory, calculated the properties of the ground states of nuclei beginning with an empirical nucleon-nucleon force. A very good agreement with the results obtained from electron scattering was obtained.

Nuclear Reactions

Prewar experiment and theory were dominated by the compound nucleus concept. Characteristically, the cross sections for compound nuclear reactions vary rapidly with energy, with angular distributions that are symmetric about 90°. The analysis of such experiments, in which there are many resonances that may

Rudolf E. Peierls (1907–1995), Gerald E. Brown (1926-), and **Victor F. Weisskopf (1908–).** (Courtesy of Rudolf Peierls Collection, AIP Emilio Segrè Visual Archives.)

overlap, makes use of the Wigner R matrix theory [Wigner and Eisenbud (1947)].

However, in the postwar period, reactions whose angular distributions were sharply peaked in the forward direction and whose cross sections varied slowly with energy were observed. The phenomenon first appeared in particle transfer reactions such as (d,p) and later in inelastic nucleon scattering. These are referred to as direct reactions [S. T. Butler (1950)]. Regularities in energy with a variety of targets of the total neutron cross section led to the formulation of the optical model [Feshbach, Porter, and Weisskopf (1954)]. In this model, the mean field was employed to obtain the elastic scattering with the difference that the potential was complex. There are two origins of the imaginary component. One is the inelastic lead to an absorption in the elastic channels. Second, an energy average is implied so that only those components whose delay time is of the order of or less than $\hbar/\delta E$, where δE is the scale of the energy average, contribute to the optical model wave function. An additional absorption is thus provided by the compound nuclear states whose lifetime is large. The optical model describes the prompt reactions.

The behavior of the direct reactions can also be interpreted as prompt reactions, thus unifying them with the optical model.

Energy scales lying between the δE of the optical model and that corresponding to the lifetime of the compound nuclear state will also reveal structure. This was evidenced by two phenomena. One is very direct. The width of the giant resonances is much larger than compound nuclear widths, yet much smaller than the optical model δE. Second, the statistical Weisskopf theory of nuclear reactions (to be described below) was found to be wanting as the projectile energy is increased. The explanation exploits the possibility that the final system is formed before there has been enough time for the formation of the compound nucleus. It is therefore referred to as a precompound reaction. By using different energy averaging scales, one can isolate corresponding time delays—that is, interaction times. The longest of these corresponds to the optical model, while the next smaller interaction time leads to the giant resonances. These are referred to as doorway states, since the system must pass through them before a more complex state is formed. These remarks are substantiated by the fact that when the energy resolution is made sufficiently small, the giant resonance dissolves into a series of fluctuations in energy, which correspond to long time delays.

In the course of these developments, another mode of nuclear reactions was discovered. This is called the multistep direct reaction, in which the final state is achieved by a series of one or more direct reactions [Bonetti *et al.* (1982)].

The statistical theory of reactions asserts that the nuclear wave functions are so complex that one can assume that the phases of the transition matrix to differing channels are random. Applications of this concept led in 1937 to the evaporation theory [Weisskopf (1937)] of nuclear reactions, in which the concept of temperature for nuclei was introduced. This was later extended to take account of the effects of angular momentum and parity conservation [Hauser and Feshbach (1952)]. Both of these developments made use of the Bohr theory of the compound nucleus. To extend beyond that constraint, statistical theories of multistep processes have been developed.

The irregular fluctuating behavior of nuclear cross sections obtained in good resolution experiments, referred to as Ericson fluctuations, has been analyzed in terms of distribution functions giving the probability that a given cross section will appear. These fluctuations are chaotic in nature and have been seen with other small systems.

Herman Feshbach (1917-). (Courtesy of Donna Coveney/MIT.)

Weak Interactions

During the late 1930s and during the postwar period, β-decay transitions were extensively used in nuclear spectroscopy, help-

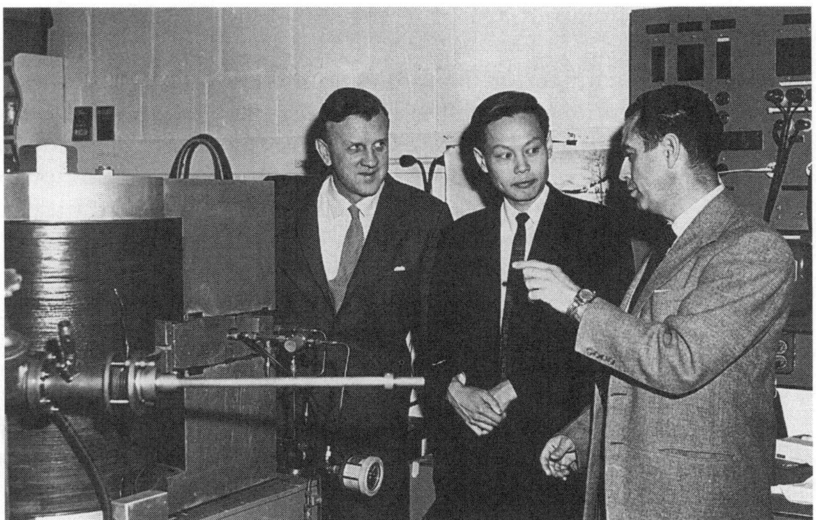

Emil J. Konopinski (1911-1990), C.-N. Yang (1922-), and **Lawrence M. Langer (1913-).** (Courtesy of AIP Emilio Segrè Visual Archives.)

C.-S. Wu (1913-). (Courtesy of *Physics Today* Collection, AIP Emilio Segrè Visual Archives.)

ing to determine the properties of nuclear levels. This was accompanied by the inclusion of Gamow-Teller (1936) transitions as well as consideration of both forbidden and allowed transitions.

This experience with β decays proved to be very useful for experiments in which parity conservation was tested. The crucial experiment of Wu *et al.* (1957) studied the decay of ^{60}Co, demonstrating parity nonconservation. The helicity of the neutrino was measured in an experiment of M. Goldhaber *et al.* (1958) using ^{152}Sm. Neutrino-induced reactions were observed by Reines and Cowan (1953) using $CdCl_2$. The flux of solar neutrinos was determined by Davis *et al.* (1968) using the absorption of neutrinos by ^{37}Cl to form ^{37}A. One sees in these examples the usefulness of the knowledge of nuclear properties in the study of the fundamental weak interactions. Experiments now in progress will test the properties of the standard model using once again nuclear properties to isolate various components predicted by that model.

CONCLUSIONS

In these pages I have attempted to describe the progress of nuclear physics in the 50 years up to 1983. The discussion is idiosyncratic. Other writers might have highlighted different aspects, although I am confident that there would be a large overlap. Of course, many developments have been omitted. But it is striking how very much has been accomplished. Speaking broadly, we have discovered the characteristic structures and reactions of systems consisting of relatively small numbers of particles. This has important applications to other systems consisting of a small particle number, such as molecules, metallic clusters, small conductors, and nucleons. Many aspects remain to be elucidated, and more discoveries will be made. Nowadays, nuclear physicists have turned their attention to the standard interaction and to quantum chromodynamics (QCD). In particular, the study of the nonperturbative QCD is a focus of interest, and the possibility of a new form of matter formed in the collision of relativistic heavy ions remains.

The close connection of nuclear physics with astrophysics should be mentioned. Briefly, the energy production [Bethe and Critchfield (1938)] and therefore the history of stars depends on various features of nuclear structure and reactions. The problems associated with supernovae will require nuclear input, e.g., the equation of state of nuclear matter, before they will be resolved. And the possibility of astronomical systems involving strange particles as well as nucleons will be informed by the study of hypernuclei.

Applications of nuclear physics continued to be developed in the postwar period. They are numerous, too numerous to be listed here. But they have been of importance not only to astrophysicists, but also to a number of other fields of science, industry, energy production, and accelerator development.

NUCLEAR PHYSICS

Papers Reproduced on CD-ROM

P. Ehrenfest and J. R. Oppenheimer. Note on the statistics of nuclei, *Phys. Rev.* **37**, 333–338 (1931)

R. J. Van De Graaff. A 1,500,000 volt electrostatic generator, *Phys. Rev.* **38**, 1919–1920(A) (1931)

E. O. Lawrence and M. S. Livingston. The production of high speed light ions without the use of high voltages, *Phys. Rev.* **40**, 19–35 (1932)

L. H. Thomas. The interaction between a neutron and a proton and the structure of H^3, *Phys. Rev.* **47**, 903–909 (1935)

E. J. Konopinski and G. E. Uhlenbeck. On the Fermi theory of β-radioactivity, *Phys. Rev.* **48**, 7–12 (1935)

J. R. Oppenheimer and M. Phillips. Note on the transmutation function for deuterons, *Phys. Rev.* **48**, 500–502 (1935)

L. A. Young. Note on the interaction of nuclear particles, *Phys. Rev.* **48**, 913–915 (1935)

M. G. White. Scattering of high energy protons in hydrogen, *Phys. Rev.* **49**, 309–316 (1936)

G. Breit and E. Wigner. Capture of slow neutrons, *Phys. Rev.* **49**, 519–531 (1936)

W. A. Fowler, L. A. Delsasso, and C. C. Lauritsen. Radioactive elements of low atomic number, *Phys. Rev.* **49**, 561–574 (1936)

E. Wigner and G. Breit. Capture of slow neutrons, *Phys. Rev.* **49**, 642(A) (1936)

G. Gamow and E. Teller. Selection rules for the β-disintegration, *Phys. Rev.* **49**, 895–899 (1936)

M. A. Tuve, N. P. Heydenburg, and L. R. Hafstad. The scattering of protons by protons, *Phys. Rev.* **50**, 806–825 (1936)

G. Breit, E. U. Condon, and R. D. Present. Theory of scattering of protons by protons, *Phys. Rev.* **50**, 825–845 (1936)

B. Cassen and E. U. Condon. On nuclear forces, *Phys. Rev.* **50**, 846–849 (1936)

E. Amaldi and E. Fermi. On the absorption and the diffusion of slow neutrons, *Phys. Rev.* **50**, 899–928 (1936)

E. Feenberg and E. Wigner. On the structure of the nuclei between helium and oxygen, *Phys. Rev.* **51**, 95–106 (1937)

E. Wigner. On the consequences of the symmetry of the nuclear Hamiltonian on the spectroscopy of nuclei, *Phys. Rev.* **51**, 106–119 (1937)

J. G. Hoffman, M. S. Livingston, and H. A. Bethe. Some direct evidence on the magnetic moment of the neutron, *Phys. Rev.* **51**, 214–215 (1937)

L. W. Alvarez. Nuclear *K* electron capture, *Phys. Rev.* **52**, 134–135(L) (1937)

J. Schwinger and E. Teller. The scattering of neutrons by ortho- and parahydrogen, *Phys. Rev.* **52**, 286–295 (1937)

V. Weisskopf. Statistics and nuclear reactions, *Phys. Rev.* **52**, 295–303 (1937)

A. J. F. Siegert. Note on the interaction between nuclei and electromagnetic radiation, *Phys. Rev.* **52**, 787–789 (1937)

J. Schwinger. On the spin of the neutron, *Phys. Rev.* **52**, 1250(L) (1937)

J. R. Oppenheimer and R. Serber. Note on boron plus proton reactions, *Phys. Rev.* **53**, 636–638 (1938)

V. F. Weisskopf. Excitation of nuclei by bombardment with charged particles, *Phys. Rev.* **53**, 1018(L) (1938)

H. A. Bethe and C. L. Critchfield. The formation of deuterons by proton combination, *Phys. Rev.* **54**, 248–254 (1938)

L. W. Alvarez. The capture of orbital electrons by nuclei, *Phys. Rev.* **54**, 486–497 (1938)

H. A. Bethe. Energy production in stars, *Phys. Rev.* **55**, 103(L) (1939)

A. O. Nier. The isotopic constitution of uranium and the half-lives of the uranium isotopes. I., *Phys. Rev.* **55**, 150–153 (1939)

N. Bohr. Resonance in uranium and thorium disintegrations and the phenomenon of nuclear fission, *Phys. Rev.* **55**, 418–419(L) (1939)

H. A. Bethe. Energy production in stars, *Phys. Rev.* **55**, 434–456 (1939)

L. Szilard and W. H. Zinn. Instantaneous emission of fast neutrons in the interaction of slow neutrons with uranium, *Phys. Rev.* **55**, 799–800(L) (1939)

E. T. Booth, J. R. Dunning, and F. G. Slack. Delayed neutron emission from uranium, *Phys. Rev.* **55**, 876(L) (1939)

N. Bohr and J. A. Wheeler. The mechanism of nuclear fission, *Phys. Rev.* **56**, 426–450 (1939)

W. H. Zinn and L. Szilard. Emission of neutrons by uranium, *Phys. Rev.* **56**, 619–624 (1939)

L. W. Alvarez and F. Bloch. A quantitative determination of the neutron moment in absolute nuclear magnetons, *Phys. Rev.* **57**, 111–122 (1940)

A. O. Nier, E. T. Booth, J. R. Dunning, and A. V. Grosse. Nuclear fission of separated uranium isotopes, *Phys. Rev.* **57**, 546(L) (1940)

A. O. Nier, E. T. Booth, J. R. Dunning, and A. V. Grosse. Further experiments on fission of separated uranium isotopes, *Phys. Rev.* **57**, 748(L) (1940)

Flerov and Petrjak. Spontaneous fission of uranium, *Phys. Rev.* **58**, 89(L) (1940)

D. R. Hamilton. On directional correlation of successive quanta, *Phys. Rev.* **58**, 122–131 (1940)

W. Rarita and J. Schwinger. On the neutron–proton interaction, *Phys. Rev.* **59**, 436–452 (1941)

NUCLEAR PHYSICS

R. E. Marshak. Forbidden transitions in β-decay and orbital electron capture and spins of nuclei, *Phys. Rev.* **61**, 431–449 (1942)

G. T. Seaborg, E. M. McMillan, J. W. Kennedy, and A. C. Wahl. Radioactive element 94 from deuterons on uranium, *Phys. Rev.* **69**, 366–367(L) (1946)

J. W. Kennedy, G. T. Seaborg, E. Segrè, and A. C. Wahl. Properties of 94(239), *Phys. Rev.* **70**, 555–556 (1946)

E. P. Wigner. Resonance reactions, *Phys. Rev.* **70**, 606–618 (1946)

G. C. Baldwin and G. S. Klaiber. Photo-fission in heavy elements, *Phys. Rev.* **71**, 3–10 (1947)

E. P. Wigner and L. Eisenbud. Higher angular momenta and long range interaction in resonance reactions, *Phys. Rev.* **72**, 29–41 (1947)

J. Schwinger. A variational principle for scattering problems, *Phys. Rev.* **72**, 742(A) (1947)

R. Serber. The production of high energy neutrons by stripping, *Phys. Rev.* **72**, 1008–1016 (1947)

R. Serber. Nuclear reactions at high energies, *Phys. Rev.* **72**, 1114–1115 (1947)

M. Deutsch. High efficiency, high speed scintillation counters for beta- and gamma-rays, *Phys. Rev.* **73**, 1240(A) (1948)

M. Goldhaber and G. Scharff-Goldhaber. Identification of beta-rays with atomic electrons, *Phys. Rev.* **73**, 1472–1473(L) (1948)

M. G. Mayer. On closed shells in nuclei, *Phys. Rev.* **74**, 235–239 (1948)

C. N. Yang. On the angular distribution in nuclear reactions and coincidence measurements, *Phys. Rev.* **74**, 764–772 (1948)

M. Goldhaber and E. Teller. On nuclear dipole vibrations, *Phys. Rev.* **74**, 1046–1049 (1948)

O. Haxel, J. H. D. Jensen, and H. E. Suess. On the "magic numbers" in nuclear structure, *Phys. Rev.* **75**, 1766(L) (1949)

J. M. Blatt and J. D. Jackson. On the interpretation of neutron–proton scattering data by the Schwinger variational method, *Phys. Rev.* **76**, 18–37 (1949)

H. A. Bethe. Theory of the effective range in nuclear scattering, *Phys. Rev.* **76**, 38–50 (1949)

C. H. Townes, H. M. Foley, and W. Low. Nuclear quadrupole moments and nuclear shell structure, *Phys. Rev.* **76**, 1415–1416(L) (1949)

J. S. Levinger and H. A. Bethe. Dipole transitions in the nuclear photo-effect, *Phys. Rev.* **78**, 115–129 (1950)

A. H. Snell, F. Pleasonton, and R. V. McCord. Radioactive decay of the neutron, *Phys. Rev.* **78**, 310–311(L) (1950)

J. M. Robson. Radioactive decay of the neutron, *Phys. Rev.* **78**, 311–312(L) (1950)

J. Rainwater. Nuclear energy level argument for a spheroidal nuclear model, *Phys. Rev.* **79**, 432–434 (1950)

S. T. Butler. On angular distributions from (d, p) and (d, n) nuclear reactions, *Phys. Rev.* **80**, 1095–1096(L) (1950)

J. M. Blatt and L. C. Biedenharn. The angular dependence of scattering reaction cross sections, *Phys. Rev.* **82**, 123(L) (1951)

O. Chamberlain, E. Segrè, and C. Wiegand. Experiments on proton–proton scattering from 120 to 345 Mev, *Phys. Rev.* **83**, 923–932 (1951)

L. F. Wouters. Detection of the azimuthal polarization of the n–p interaction at 150 Mev, *Phys. Rev.* **84**, 1069–1070(L) (1951)

H. A. Bethe and S. T. Butler. A proposed test of the nuclear shell model, *Phys. Rev.* **85**, 1045–1046(L) (1952)

H. H. Barschall. Regularities in the total cross sections for fast neutrons, *Phys. Rev.* **86**, 431(L) (1952)

G. Breit, M. H. Hull, Jr., and R. L. Gluckstern. Possibilities of heavy ion bombardment in nuclear studies, *Phys. Rev.* **87**, 74–80 (1952)

W. Hauser and H. Feshbach. The inelastic scattering of neutrons, *Phys. Rev.* **87**, 366–373 (1952)

D. W. Miller, R. K. Adair, C. K. Bockelman, and S. E. Darden. Total cross sections of heavy nuclei for fast neutrons, *Phys. Rev.* **88**, 83–90 (1952)

D. L. Hill and J. A. Wheeler. Nuclear constitution and the interpretation of fission phenomena, *Phys. Rev.* **89**, 1102–1145 (1953)

G. Scharff-Goldhaber. Excited states of even-even nuclei, *Phys. Rev.* **90**, 587–602 (no tables) (1953)

K. Alder and A. Winther. The theory of Coulomb excitation of nuclei, *Phys. Rev.* **91**, 1578(L) (1953)

V. L. Fitch and J. Rainwater. Studies of x-rays from mu-mesonic atoms, *Phys. Rev.* **92**, 789–800 (1953)

R. Hofstadter, H. R. Fechter, and J. A. McIntyre. High-energy electron scattering and nuclear structure determinations, *Phys. Rev.* **92**, 978–987 (1953)

A. de-Shalit and M. Goldhaber. Mixed configurations in nuclei, *Phys. Rev.* **92**, 1211–1218 (1953)

P. C. Gugelot. Level densities of nuclei from the inelastic scattering of 18-Mev protons, *Phys. Rev.* **93**, 425–433 (1954)

C. L. McClelland, H. Mark, and C. Goodman. Electric excitation of low-lying levels in separated Wolfram isotopes, *Phys. Rev.* **93**, 904–905(L) (1954)

K. A. Brueckner, C. A. Levinson, and H. M. Mahmoud. Two-body forces and nuclear saturation. I. Central forces, *Phys. Rev.* **95**, 217–228 (1954)

H. Feshbach, C. E. Porter, and V. F. Weisskopf. Model for nuclear reactions with neutrons, *Phys. Rev.* **96**, 448–464 (1954)

G. Scharff-Goldhaber and J. Weneser. System of even-even nuclei, *Phys. Rev.* **98**, 212–214(L) (1955)

A. M. Lane, R. G. Thomas, and E. P. Wigner. Giant resonance interpretation of the nucleon–nucleus interaction, *Phys. Rev.* **98**, 693–701 (1955)

NUCLEAR PHYSICS

K. A. Brueckner, R. J. Eden, and N. C. Francis. High-energy reactions and the evidence for correlations in the nuclear ground-state wave function, *Phys. Rev.* **98**, 1445–1455 (1955)

A. Ghiorso, B. G. Harvey, G. R. Choppin, S. G. Thompson, and G. T. Seaborg. New element mendelevium, atomic number 101, *Phys. Rev.* **98**, 1518–1519(L) (1955)

C. E. Porter and R. G. Thomas. Fluctuations of nuclear reaction widths, *Phys. Rev.* **104**, 483–491 (1956)

O. Chamberlain, E. Segrè, R. D. Tripp, C. Wiegand, and T. Ypsilantis. Experiments with 315-Mev polarized protons: proton–proton and proton–neutron scattering, *Phys. Rev.* **105**, 288–301 (1957)

C. S. Wu, E. Ambler, R. W. Hayward, D. D. Hoppes, and R. P. Hudson. Experimental test of parity conservation in beta decay, *Phys. Rev.* **105**, 1413–1414(L) (1957)

J. J. Griffin and J. A. Wheeler. Collective motions in nuclei by the method of generator coordinates, *Phys. Rev.* **108**, 311–327 (1957)

A. Bohr, B. R. Mottelson, and D. Pines. Possible analogy between the excitation spectra of nuclei and those of the superconducting metallic state, *Phys. Rev.* **110**, 936–938 (1958)

D. M. Chase, L. Wilets, and A. R. Edmonds. Rotational-optical model for scattering of neutrons, *Phys. Rev.* **110**, 1080–1092 (1958)

G. E. Brown and M. Bolsterli. Dipole state in nuclei, *Phys. Rev. Lett.* **3**, 472–476 (1959)

G. Breit, M. H. Hull, Jr., K. E. Lassila, and K. D. Pyatt, Jr. Phase-parameter representation of proton–proton scattering from 9.7 to 345 Mev, *Phys. Rev.* **120**, 2227–2249 (1960)

D. A. Bromley, J. A. Kuehner, and E. Almqvist. Resonant elastic scattering of C^{12} by carbon, *Phys. Rev. Lett.* **4**, 365–367 (1960)

V. W. Hughes, D. W. McColm, K. Ziock, and R. Prepost. Formation of muonium and observation of its Larmor precession, *Phys. Rev. Lett.* **5**, 63–65 (1960)

T. Ericson. Fluctuations of nuclear cross sections in the "continuum" region, *Phys. Rev. Lett.* **5**, 430–431 (1960)

A. de-Shalit. Core excitations in nondeformed, odd-A, nuclei, *Phys. Rev.* **122**, 1530–1536 (1961)

J. D. Anderson and C. Wong. Evidence for charge independence in medium weight nuclei, *Phys. Rev. Lett.* **7**, 250–252 (1961)

G. A. Peterson and W. C. Barber. Deuteron magnetic dipole disintegration by 180° electron scattering, *Phys. Rev.* **128**, 812–820 (1962)

A. M. Lane. New term in the nuclear optical potential: Implications for (p,n) mirror state reactions, *Phys. Rev. Lett.* **8**, 171–172 (1962)

H. L. Anderson, C. S. Johnson, and E. P. Hincks. μ-mesonic x-ray energies and nuclear radii for fourteen elements from $Z=12$ to 50, *Phys. Rev.* **130**, 2468–2480 (1963)

A. Klein and A. K. Kerman. Collective motion in finite many-particle systems. II., *Phys. Rev.* **138**, B1323–B1332 (1965)

K. Chen, Z. Fraenkel, G. Friedlander, J. R. Grover, J. M. Miller, and Y. Shimamoto. Vegas: A Monte Carlo simulation of intranuclear cascades, *Phys. Rev.* **166**, 949–967 (1968)

D. Bodansky, D. D. Clayton, and W. A. Fowler. Nucleosynthesis during silicon burning, *Phys. Rev. Lett.* **20**, 161–164 (1968)

R. Davis, Jr., D. S. Harmer, and K. C. Hoffman. Search for neutrinos from the Sun, *Phys. Rev. Lett.* **20**, 1205–1209 (1968)

A. Fubini, J. Blons, A. Michaudon, and D. Paya. Short-range intermediate structure observed in the ^{237}Np neutron subthreshold fission cross section, *Phys. Rev. Lett.* **20**, 1373–1375 (1968)

J. W. Negele. Structure of finite nuclei in the local-density approximation, *Phys. Rev. C* **1**, 1260–1321 (1970)

J. C. Vanderleeden and F. Boehm. Experiments on parity nonconservation in nuclear forces. I. Gamma transitions in Ta^{181} and Lu^{175}, *Phys. Rev. C* **2**, 748–760 (1970)

J. B. Garg, J. Rainwater, and W. W. Havens, Jr. Neutron resonance spectroscopy. VII. Ti, Fe, and Ni, *Phys. Rev. C* **3**, 2447–2463 (1971)

A. Arima and F. Iachello. Collective nuclear states as representations of a SU(6) group, *Phys. Rev. Lett.* **35**, 1069–1072 (1975)

S. E. Koonin, K. T. R. Davies, V. Maruhn-Rezwani, H. Feldmeier, S. J. Krieger, and J. W. Negele. Time-dependent Hartree–Fock calculations for ^{16}O+^{16}O and ^{40}Ca+^{40}Ca reactions, *Phys. Rev. C* **15**, 1359–1374 (1977)

M. Forte, B. R. Heckel, N. F. Ramsey, K. Green, G. L. Greene, J. Byrne, and J. M. Pendlebury. First measurement of parity-nonconserving neutron-spin rotation: The tin isotopes, *Phys. Rev. Lett.* **45**, 2088–2092 (1980)

R. Bonetti, L. Colli Milazzo, I. Doda, and P. E. Hodgson. Analyzing powers in the ^{58}Ni$(\vec{p},p')$ reaction calculated with the statistical multistep direct emission theory, *Phys. Rev. C* **26**, 2417–2423 (1982)

P. J. Twin, B. M. Nyakó, A. H. Nelson, J. Simpson, M. A. Bentley, H. W. Cranmer-Gordon, P. D. Forsyth, D. Howe, A. R. Mokhtar, J. D. Morrison, J. F. Sharpey-Schafer, and G. Sletten. Observation of a discrete-line superdeformed band up to $60\hbar$ in ^{152}Dy, *Phys. Rev. Lett.* **57**, 811–814 (1986)

10. A 1,500,000 volt electrostatic generator. ROBERT J. VAN DE GRAAFF, *National Research Fellow, Princeton University.*—The application of extremely high potentials to discharge tubes affords a powerful means for the investigation of the atomic nucleus and other fundamental problems. The electrostatic generator here described was developed to supply suitable potentials for such investigations. In recent preliminary trials, spark-gap measurements showed a potential of approximately 1,500,000 volts, the only apparent limit being brush discharge from the whole surface of the 24-inch spherical electrodes. The generator has the basic advantage of supplying a direct steady potential, thus eliminating certain difficulties inherent in the applica-

tion of non-steady high potentials. The machine is simple, inexpensive, and portable. An ordinary lamp socket furnishes the only power needed. The apparatus is composed of two identical units, generating opposite potentials. The high potential electrode of each unit consists of a 24-inch hollow copper sphere mounted upon a 7 foot upright Pyrex rod. Each sphere is charged by a silk belt running between a pulley in its interior and a grounded motor driven pulley at the base of the rod. The ascending surface of the belt is charged near the lower pulley by a brush discharge, maintained by a 10,000 volt transformer kenotron set, and is subsequently discharged by points inside the sphere.

APRIL 1, 1932 PHYSICAL REVIEW VOLUME 40

THE PRODUCTION OF HIGH SPEED LIGHT IONS WITHOUT THE USE OF HIGH VOLTAGES

BY ERNEST O. LAWRENCE AND M. STANLEY LIVINGSTON
UNIVERSITY OF CALIFORNIA

(Received February 20, 1932)

ABSTRACT

The study of the nucleus would be greatly facilitated by the development of sources of high speed ions, particularly protons and helium ions, having kinetic energies in excess of 1,000,000 volt-electrons; for it appears that such swiftly moving particles are best suited to the task of nuclear excitation. The straightforward method of accelerating ions through the requisite differences of potential presents great experimental difficulties associated with the high electric fields necessarily involved. The present paper reports the development of a method that avoids these difficulties by means of the multiple acceleration of ions to high speeds without the use of high voltages. The method is as follows: Semi-circular hollow plates, not unlike duants of an electrometer, are mounted with their diametral edges adjacent, in a vacuum and in a uniform magnetic field that is normal to the plane of the plates. High frequency oscillations are applied to the plate electrodes producing an oscillating electric field over the diametral region between them. As a result during one half cycle the electric field accelerates ions, formed in the diametral region, into the interior of one of the electrodes, where they are bent around on circular paths by the magnetic field and eventually emerge again into the region between the electrodes. The magnetic field is adjusted so that the time required for traversal of a semi-circular path within the electrodes equals a half period of the oscillations. In consequence, when the ions return to the region between the electrodes, the electric field will have reversed direction, and the ions thus receive second increments of velocity on passing into the other electrode. Because the path radii within the electrodes are proportional to the velocities of the ions, the time required for a traversal of a semi-circular path is independent of their velocities. Hence if the ions take exactly one half cycle on their first semi-circles, they do likewise on all succeeding ones and therefore spiral around in resonance with the oscillating field until they reach the periphery of the apparatus. Their final kinetic energies are as many times greater than that corresponding to the voltage applied to the electrodes as the number of times they have crossed from one electrode to the other. This method is primarily designed for the acceleration of light ions and in the present experiments particular attention has been given to the production of high speed protons because of their presumably unique utility for experimental investigations of the atomic nucleus. Using a magnet with pole faces 11 inches in diameter, a current of 10^{-9} ampere of 1,220,000 volt-protons has been produced in a tube to which the maximum applied voltage was only 4000 volts. There are two features of the developed experimental method which have contributed largely to its success. First there is the focussing action of the electric and magnetic fields which prevents serious loss of ions as they are accelerated. In consequence of this, the magnitudes of the high speed ion currents obtainable in this indirect manner are comparable with those conceivably obtainable by direct high voltage methods. Moreover, the focussing action results in the generation of very narrow beams of ions—less than 1 mm cross-sectional diameter—which are ideal for experimental studies of collision processes. Of hardly less importance is the second feature of the method which is the simple and highly effective means for the correction of the magnetic field along the paths of the ions. This makes it possible, indeed easy, to operate the tube effectively

with a very high amplification factor (i.e., ratio of final equivalent voltage of accelerated ions to applied voltage). In consequence, this method in its present stage of development constitutes a highly reliable and experimentally convenient source of high speed ions requiring relatively modest laboratory equipment. Moreover, the present experiments indicate that this indirect method of multiple acceleration now makes practicable the production in the laboratory of protons having kinetic energies in excess of 10,000,000 volt-electrons. With this in mind, a magnet having pole faces 114 cm in diameter is being installed in our laboratory.

INTRODUCTION

THE classical experiments of Rutherford and his associates[1] and Pose[2] on artificial disintegration, and of Bothe and Becker[3] on excitation of nuclear radiation, substantiate the view that the nucleus is susceptible to the same general methods of investigation that have been so successful in revealing the extra-nuclear properties of the atom. Especially do the results of their work point to the great fruitfulness of studies of nuclear transitions excited artificially in the laboratory. The development of methods of nuclear excitation on an extensive scale is thus a problem of great interest; its solution is probably the key to a new world of phenomena, the world of the nucleus.

But it is as difficult as it is interesting, for the nucleus resists such experimental attacks with a formidable wall of high binding energies. Nuclear energy levels are widely separated and, in consequence, processes of nuclear excitation involve enormous amounts of energy—millions of volt-electrons.

It is therefore of interest to inquire as to the most promising modes of nuclear excitation. Two general methods present themselves; excitation by absorption of radiation (gamma radiation), and excitation by intimate nuclear collisions of high speed particles.

Of the first it may be said that recent experimental studies [4,5] of the absorption of gamma radiation in matter show, for the heavier elements, variations with atomic number that indicate a quite appreciable nuclear effect. This suggests that nuclear excitation by absorption of radiation is perhaps a not infrequent process, and therefore that the development of an intense artificial source of gamma radiation of various wave-lengths would be of considerable value for nuclear studies. In our laboratory, as elsewhere, this being attempted.

But the collision method appears to be even more promising, in consequence of the researches of Rutherford and others cited above. Their pioneer investigations must always be regarded as really great experimental achievements, for they established definite and important information about nuclear processes of great rarity excited by exceedingly weak beams of bombarding particles—alpha-particles from radioactive sources. Moreover, and this is the point to be emphasized here, their work has shown strikingly the

[1] See Chapter 10 of Radiations from Radioactive Substances by Rutherford, Chadwick and Ellis.

[2] H. Pose, Zeits. f. Physik **64**, 1 (1930).

[3] W. Bothe and H. Becker, Zeits. f. Physik **66**, 1289 (1930).

[4] G. Beck, Naturwiss. **18**, 896 (1930).

[5] C. Y. Chao, Phys. Rev. **36**, 1519 (1930).

great fruitfulness of the kinetic collision method and the importance of the development of intense artificial sources of alpha-particles. Of course it cannot be inferred from their experiments that alpha-particles are the most effective nuclear projectiles: the question naturally arises whether lighter or heavier particles of given kinetic energy would be more effective in bringing about nuclear transitions.

A beginning has been made on the theoretical study of the nucleus and a partial answer to this question has been obtained. Gurney and Condon[6] and Gamow[7] have independently applied the ideas of the wave mechanics to radioactivity with considerable success. Gamow[8] has further considered along the same lines the penetration into the nucleus of swiftly moving charged particles (with excitation of nuclear transitions in mind) and has concluded that, for a given kinetic energy, the lighter the particle the greater is the probability that it will penetrate the nuclear potential wall. This result is not unconnected with the smaller momentum and consequent longer wave-length of the ligher particles; for it is well-known that transmission of matter waves through potential barriers becomes greater with increasing wave-lengths.

If the probability of nuclear excitation by a charged particle were mainly dependent on its ability to penetrate the nuclear potential wall, electrons would be the most effective. However, there is considerable evidence that nuclear excitation by electrons is negligible. It suffices to mention here the current view that the average density of the extra-nuclear electrons is quite great in the region of the nucleus, i.e., that the nucleus is quite transparent to electrons; in other words, there are no available stable energy levels for them.

On the other hand, there is evidence that there are definite nuclear levels for protons as well as alpha-particles;[9] indeed, there is some justification for the view that the general principles of the quantum mechanics are applicable in the nucleus to protons and alpha particles. It is not possible at the present time to estimate the relative excitation probabilities of the protons and alpha particles that succeed in penetrating the nucleus. However, it does seem likely that the greater penetrability of the proton* is an advantage outweighing any differences in their excitation characteristics. Protons thus appear to be most suited to the task of nuclear excitation.

Though at present the relative efficacy of protons and alpha-particles cannot be established with much certainty, it does seem safe to conclude at least that the most efficacious nuclear projectiles will prove to be swiftly moving ions, probably of low atomic number. In consequence it is important to develop methods of accelerating ions to speeds much greater than have heretofore been produced in the laboratory.

[6] Gurney and Condon, Phys. Rev. **33**, 127 (1929).
[7] Gamow, Zeits. f. Physik **51**, 204 (1928).
[8] Gamow, Zeits. f. Physik **52**, 514 (1929).
[9] J. Chadwick, J. E. R. Constable, E. C. Pollard, Proc. Roy. Soc. A130, 463 (1930).
* According to Gamow's theory a one million volt-proton has as great a penetrating power as a sixteen million volt alpha-particle.

The importance of this is generally recognized and several laboratories are developing techniques of the production and the application to vacuum tubes of high voltages for the generation of high speed electrons and ions. Highly significant progress in this direction has been made by Coolidge,[10] Lauritsen,[11] Tuve, Breit, Hafstad, Dahl,[12] Brasch and Lange,[13] Cockroft and Walton,[14] Van de Graaff[15] and others, who have developed several distinct techniques which have been applied to voltages of the order of magnitude of one million.

These methods involving the direct utilization of high voltages are subject to certain practical limitations. The experimental difficulties go up rapidly with increasing voltage; there are the difficulties of corona and insulation and also there is the problem of design of suitable high voltage vacuum tubes.

Because of these difficulties we have thought it desirable to develop methods for the acceleration of charged particles that do not require the use of high voltages. Our objective is two fold: first, to make the production of particles having kinetic energies of the order of magnitude of one million volt-electrons a matter that can be carried through with quite modest laboratory equipment and with an experimental convenience that, it is hoped, will lead to a widespread attack on this highly important domain of physical phenomena; and second, to make practicable the production of particles having kinetic energies in excess of those producible by direct high voltage methods—perhaps in the range of 10,000,000 volt-electrons and above.

A method for the multiple acceleration of ions to high speeds, primarily designed for heavy ions, has recently been described in this journal.[16] The present paper is a report of the development of a method for the multiple acceleration of light ions.[17] Particular attention has been given to the acceleration of protons because of their apparent unique utility in nuclear studies. In the present work relatively large currents of 1,220,000 volt-protons have been generated and there is foreshadowed in the not distant future the production of 10,000,000 volt-protons.

The Experimental Method

In the method for the multiple acceleration of ions to high speeds, recently described,[16] the ions travel through a series of metal tubes in synchronism with an applied oscillating electric potential. It is so arranged that as an

[10] W. D. Collidge, Am. Inst. E. Eng. **47**, 212 (1928).
[11] C. C. Lauritsen and R. D. Bennett, Phys. Rev. **32**, 850 (1928).
[12] M. A. Tuve, G. Breit, L. R. Hafstad and O. Dahl, Phys. Rev. **35**, 66 (1930); M. A. Tuve, L. R. Hafstad, O. Dahl, Phys. Rev. **39**, 384, (1932).
[13] A. Brasch and J. Lange, Zeits. f. Physik **70**, 10 (1931).
[14] J. J. Cockroft and E. T. S. Walton, Proc. Roy. Soc. **A129**, 477 (1930).
[15] R. S. Van de Graaff, Schenectady Meeting American Physical Society, 1931.
[16] D. H. Sloan and E. O. Lawrence, Phys. Rev. **38**, 2021 (1931).
[17] This method was first described before the September, 1930, meeting of the National Academy of Sciences (Lawrence and Edlefsen, Science **72**, 376–377 (1930)). Later before the American Physical Society (Lawrence and Livingston, Phys. Rev. **37**, 1707, (1931)) results of a preliminary study of the practicability of the method were given. Further work was reported in a Letter to the Editor of the Physical Review (Lawrence and Livingston, Phys. Rev. **38**, 834 (1931).

ion travels from the interior of one tube to the interior of the next there is always an accelerating field, and the final velocity of the ion on emergence from the system corresponds approximately to a voltage as many times greater than the applied voltage between adjacent tubes as there are tubes. The method is most conveniently used for the acceleration of heavy ions; for light ions travel faster and hence require longer systems of tubes for any given frequency of applied oscillations.

The present experimental method makes use of the same principle of repeated acceleration of the ions by a similar sort of resonance with an oscillating electric field, but has overcome the difficulty of the cumbersomely long accelerating system by causing, with the aid of a magnetic field, the ions to circulate back and forth from the interior of one electrode to the interior of another.

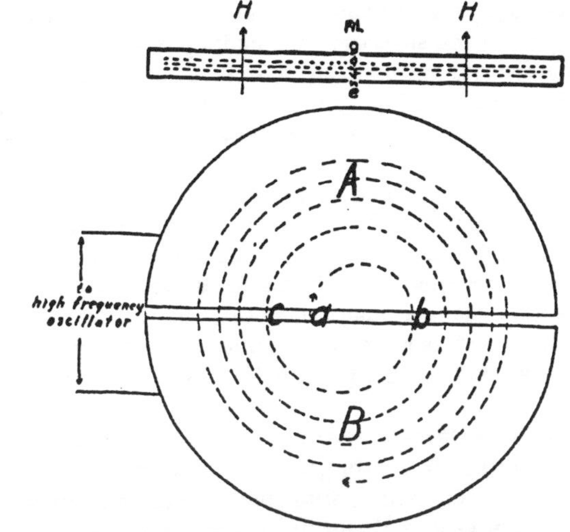

Fig. 1. Diagram of experimental method for multiple acceleration of ions.

This may be seen most readily by an outline of the experimental arrangement (Fig. 1). Two electrodes A, B in the form of semi-circular hollow plates are mounted in a vacuum tube in coplanar fashion with their diametral edges adjacent. By placing the system between the poles of a magnet, a magnetic field is introduced that is normal to the plane of the plates. High frequency electric oscillations are applied to the plates so that there results an oscillating electric field in the diametral region between them.

With this arrangement it is evident that, if at one moment there is an ion in the region between the electrodes, and electrode A is negative with respect to electrode B, then the ion will be accelerated to the interior of the former. Within the electrode the ion traverses a circular path because of the magnetic field, and ultimately emerges again between the electrodes; this is indicated in the diagram by the arc a .. b. If the time consumed by the ion in making the semi-circular path is equal to the half period of the electric oscillations, the electric field will have reversed and the ion will receive a second acceleration, passing into the interior of electrode B with a higher velocity. Again it travels on a semi-circular path (b .. c), but this time the radius of curvature is greater because of the greater velocity. For all velocities (neglecting variation of mass with velocity) the radius of the path is proportional to the velocity, so that the time required for traversal of a semi-circular path is independent of the ion's velocity. Therefore, if the ion travels its first half circle in a half cycle of the oscillations, it will do likewise on all succeeding paths. Hence it will circulate around on ever widening semi-circles from the interior of one electrode to the interior of the other, gaining an increment of energy on each crossing of the diametral region that corresponds to the momentary potential difference between the electrodes. Thus, if, as was done in the present experiments, high frequency oscillations having peak values of 4000 volts are applied to the electrodes, and protons are caused to spiral around in this way 150 times, they will receive 300 increments of energy, acquiring thereby a speed corresponding to 1,200,000 volts.

It is well to recapitulate these remarks in quantitative fashion. Along the circular paths within the electrodes the centrifugal force of an ion is balanced by the magnetic force on it, i.e., in customary notation,

$$\frac{mv^2}{r} = \frac{Hev}{c}\,. \qquad (1)$$

It follows that the time for traversal of a semi-circular path is

$$t = \frac{\pi r}{v} = \frac{\pi mc}{He} \qquad (2)$$

which is independent of the radius r of the path and the velocity v of the ion. The particle of mass m and charge e thus may be caused to travel in phase with the oscillating electric field by suitable adjustment of the magnetic field H: the relation between the wave-length λ of the oscillations and the corresponding synchronizing magnetic field H is in consequence

$$\lambda = \frac{2\pi mc^2}{He}\,. \qquad (3)$$

Thus for protons and a magnetic field of 10,000 gauss the corresponding wave-length is 19.4 meters; for heavier particles the proper wave-length is proportionately longer.*

It is easily shown also that the energy V in volt-electrons of the charged particles arriving at the periphery of the apparatus on a circle of radius r is

* It should be mentioned that, for a given wave-length, the ions resonate with the oscillations when magnetic fields of 1/3, 1/5, etc., of that given by Eq. (3) are used. Such types of resonance were observed in the earlier experimental studies. In the present experiments, however, the high speed ions resulting from the primary type of resonance only were able to pass through the slit system to the collector, because of the high deflecting voltages used.

$$V = 150 \frac{H^2 r^2}{c^2} \frac{e}{m} . \tag{4}$$

Thus, the theoretical maximum producible energy varies as the square of the radius and the square of the magnetic field.

Experimental Arrangement

The experimental arrangement is shown diagrammatically in some detail in Fig. 2. Fig. 3 is a photograph of the brass vacuum tube with cover removed showing the filament, the accelerating electrode, the deflecting plates and slit system, the probe in front of the first slit mounted on a ground joint and the Faraday collector behind the last slit. An external view of the apparatus is shown in Fig. 4. Here the tube is shown between the magnet pole faces, connected with the oscillator, the vacuum system and hydrogen generator. This gives a good general idea of the modest extent of the equipment involved for the generation of protons having energies somewhat in excess of 1,000,000 volt-electrons. The control panel and electrometer, being on the other side, are not shown in the picture. The description of the apparatus follows.

The accelerating system. Though there are obvious advantages in applying the high frequency potentials with respect to ground to both accelerating electrodes, in the present experiments it was found convenient to apply the high frequency voltage to only one of the electrodes, as indicated in Fig. 2. This electrode was a semi-circular hollow brass plate 24 cm in diameter and 1 cm thick. The sides of the hollow plate were of thin brass so that the interior of the plate had approximately these dimensions. It was mounted on a water-cooled copper re-entrant tube which in turn passed through a copper to glass seal. The electrode insulated in this way was mounted in an evacuated brass box having internal dimensions 2.6 cm by 28.6 cm by 28.6 cm, there being thus a lateral clearance between the electrode and walls of the brass chamber of 8 mm.

The brass box itself constituted the other electrode of the accelerating system. Across the mid-section of the brass chamber parallel to the diametral edge of the electrode *A* was placed a brass dividing wall *S* with slits of the same dimensions as the opening of the nearby electrode. This arrangement gave rise to the same type of oscillating electric fields as would have been produced had there been used two insulated semi-circular electrodes with their diametral edges adjacent and parallel.

The source of ions. An ideal source of ions is one that delivers to the diametral region between the electrodes large quantities of ions with low components of velocity normal to the plane of the accelerators. This requirement has most conveniently been met in the present experiments merely by having a filament placed above the diametral region from which a stream of electrons pass down along the magnetic lines of force, generating ions of gases in the tube. The ions so formed are pulled out sideways by the oscillating electric field. The electrons are not drawn out because of their very small radii of curvature in the magnetic field. Thus, the beam of electrons is col-

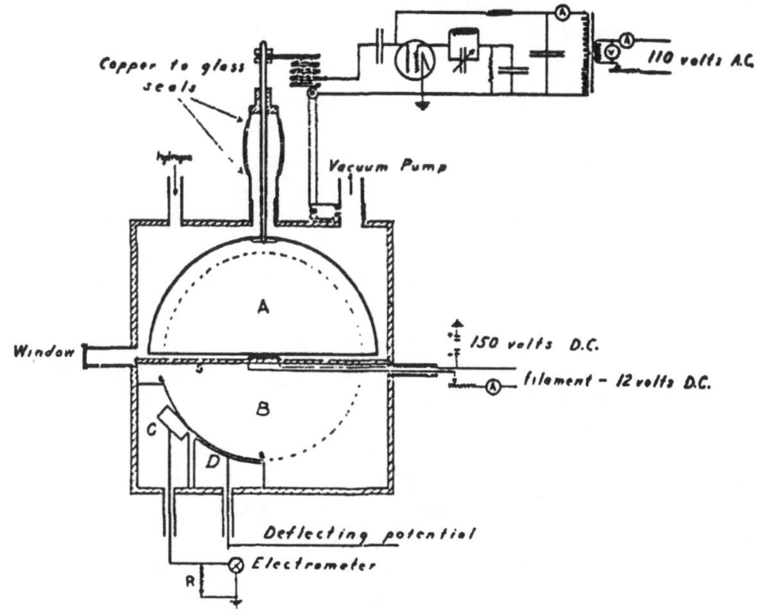

Copper to glass seals
Hydrogen
Vacuum Pump
110 volts A.C.
Window
A
B
S
C
D
150 volts D.C.
filament – 12 volts D.C.
Deflecting potential
Electrometer
R

Fig. 2. Diagram of apparatus for the multiple acceleration of ions.

Fig. 3. Tube for the multiple acceleration of light ions—with cover removed.

limated and the ions are formed with negligible initial velocities right in the region where they are wanted. The oscillating electric field immediately draws them out and takes them on their spiral paths to the periphery. This arrangement is diagrammatically shown in the upper part of Fig. 1.

Fig. 4. External view of apparatus for generation of 1,220,000 volt protons.

The magnetic field. This experimental method requires a highly uniform magnetic field normal to the plane of the accelerating system. For example, if the ions are to circulate around 100 times, thereby gaining energy corresponding to 200 times the applied voltage, it is necessary that the magnetic field be uniform to a fraction of one percent. A general consideration of the matter leads one to the conclusion that, if possible, the magnetic field should be constant to about 0.1 percent from the center outward. Though this presumably

difficult requirement has been met easily by an empirical method of field correction, the magnet used in the present experiments has pole faces machined as accurately as could be done conveniently. Its design was quite similar to that of Curtis.[15] The pole faces were 11 inches in diameter and the gap separation was $1\frac{1}{2}$ inches. Armco iron was used throughout the magnetic circuit. The magnetomotive force was provided by two coils of number 14 double cotton covered wire of 2,000 turns each. No water cooling was incorporated, for the magnet was not intended for high fields. In practice the magnet would give a field of 14,000 gauss for considerable periods without overheating. The pole faces were made parallel to about 0.2 percent and so it was to be expected that the magnetic field produced would be highly uniform. Exploration with a bismuth spiral confirmed this expectation, since it failed to show an appreciable variation of the magnetic field in the region between the poles, excepting within an inch of the periphery.

The collector system. In planning a suitable arrangement for collecting the high speed ions at the periphery of the apparatus, it was clearly desirable to devise something that would collect the high speed ions only and which would also measure their speeds. One might regard it as legitimate to suppose that the magnetic field itself and the distance of the collector from the center of the system would determine the speeds of the ions collected. This would be true provided there were no scattering and reflection of ions. To eliminate these extraneous effects a set of 1 mm slits was arranged on a circle $a \ldots a$, as shown in Fig. 2, of radius about 12 percent greater than the circle, indicated by the dotted line in the figure, having its center at the center of the tube and a radius of 11.5 cm. The two circles were tangent at the first slit as shown. The ions on arrival at the first slit would be traveling presumably on circles approximately like the dotted line, and hence would not be able to pass through the second and third slits to the Faraday collector C. Electrostatic deflecting plates D, separated by 2 mm, were placed between the first two slits, making possible the application of electrostatic fields to increase the radius of curvature of the paths of the high speed ions sufficiently to allow them to enter the collector. By applying suitable high potentials to the deflecting system in this way, only correspondingly high speed ions were registered.

The collector currents were measured by an electrometer shunted with a suitable high resistance leak.

The oscillator. The high frequency oscillations applied to the electrode were supplied by a 20 kilowatt Federal Telegraph water-cooled power tube in a "tuned plate tuned grid" circuit, for which the diagram of Fig. 2 is self-explanatory.

The Focussing Actions

When one considers the circulation of the ions around many times as they are accelerated to high speeds in this way, one wonders whether in practice an appreciable fraction of those starting out can ever be made to

[15] L. F. Curtis, Jour. Op. Soc. Am. 13, 73 (1926).

arrive at the periphery and to pass through a set of slits perhaps 1 mm wide and 1 cm long. The paths of the ions in the course of their acceleration would be several meters, and, because of the unavoidable spreading effects of space charge, thermal velocities and contact electromotive forces, as well as inhomogeneities of the applied fields, it would appear that the effective solid angle of the peripheral slit for the ions starting out would be exceedingly small.

Fortunately, however, this does not turn out to be the case. The electric and magnetic fields have been so arranged that they provide extremely strong focussing actions on the spiraling ions, which keep them circulating close to the median plane of the accelerating system.

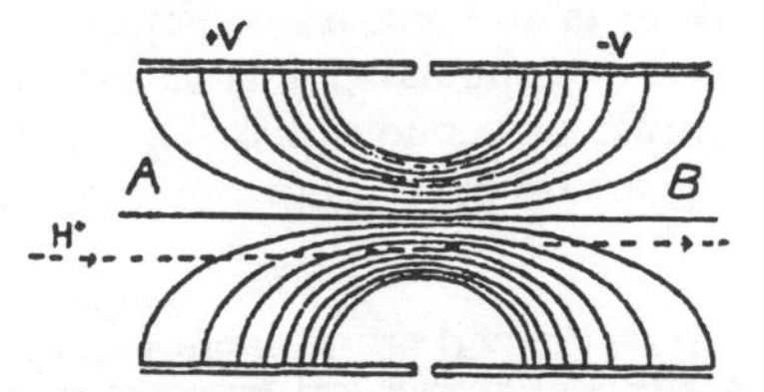

Fig. 5. Diagram indicating the focussing action of the electric field between the accelerating electrodes.

Fig. 5 shows the focussing action of the electric fields. There is depicted a cross-section of the diametral region between the accelerating electrodes with the nature of the field indicated by lines of force. There is shown also a dotted line which represents qualitatively the path of an ion as it passes from the interior of one electrode to the interior of the other. It is seen that, since it is off the median plane in electrode A, on crossing to B it receives an inward displacement towards the median plane. This is because of the existence of the curvature of the field, which over certain regions has an appreciable component normal to the plane, as indicated. If the velocity of the ion is very

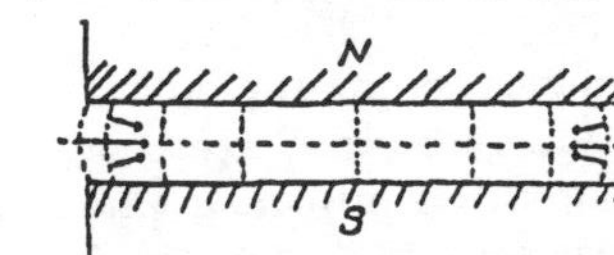

Fig. 6. Diagram indicating focussing action of magnetic field.

high in comparison to the increment of velocity gained in going from plate A to plate B, its displacement towards the center will be relatively small and, to the first approximation, it may be described as due to the ion having been accelerated inward on the first half of its path across and accelerated outward by an equal amount during the remainder of its journey, the net result being a displacement of the ion towards the center without acquiring a net transverse component of velocity. In general, however, the outward acceleration during the second half will not quite compensate the inward acceleration of the first, resulting in a gain of an inward component of velocity as well as an inward displacement. In any event, as the ion spirals around it will migrate back and forth across the median plane and will not be lost to the walls of the tube.

The magnetic field also has a focussing action. Fig. 6 shows diagrammatically the form of the field produced by the magnet. In the central region of the pole faces the magnetic field is quite uniform and normal to the plane of the faces; but out near the periphery the field has a curvature. Ions traveling on circles near the periphery experience thereby magnetic forces, indicated by the arrows. If the circular path is on the median plane then the magnetic force is towards the center in that plane. If the ion is traveling in a circle off the median plane, then there is a component of magnetic force that accelerates it towards the median plane, thereby giving effectively a focussing action.

We have experimentally examined these two focussing actions, using a probe in front of the first slit of the collector system that could be moved up and down across the beam by means of a ground joint (see Fig. 3). It was

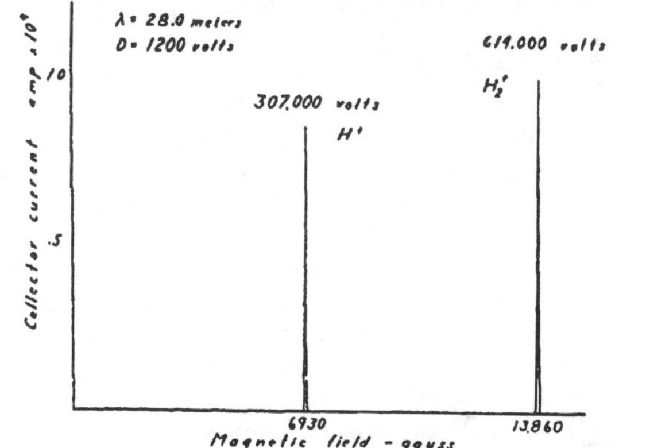

Fig. 7. Ion current to Faraday collector as a function of the magnetic field with oscillations of 28 meters wave-length applied to the accelerating electrodes.

found that the focussing actions were so powerful that *the beam of high speed ions had a width of less than one millimeter*. Such a narrow beam of ions of course is ideal for many experimental studies.

As a further test of the focussing action of the two fields, the median plane of the accelerating system was lowered 3 mm with respect to the plane of symmetry of the magnetic field. It was found that the high speed ion beam at the periphery traveled in a plane that was between the planes of symmetry of the two fields showing that both focussing actions were operative and at the periphery were of the same order of magnitude.

Experimental Results

As a typical example there is shown in Fig. 7 a plot of the ion current to the Faraday collector as a function of the magnetic field for applied oscillations of wave-length 28 meters and with hydrogen in the tube. It is seen that there are only two narrow ranges of magnetic field strength over which ion currents are observed; both correspond exactly to expectations, the one at

6930 gauss involving the resonance of protons, the other, hydrogen molecule ions.

For each wave-length used, *the magnetic field giving the greatest current to the collector agreed precisely with the theoretically expected value.* This is illustrated in Fig. 8 where the curves represent the theoretical hyperbolic relations between wave-length and magnetic field (Eq. 3) for protons and hydrogen molecule ions, and the circles represent the experimental observations. The magnetic fields were measured with a bismuth spiral and the oscillation wave-lengths were determined with a General Radio wavemeter. No effort was made to obtain considerable precision in these measurements, and in consequence their accuracy was hardly greater than 1 percent.

The variation with applied high frequency voltage of the widths of the resonance peaks agreed also with theoretical expectations. It was found that as the voltage was reduced the peaks became sharper, and indeed, with voltages such that the ions were required to spiral around fifty and more times to reach

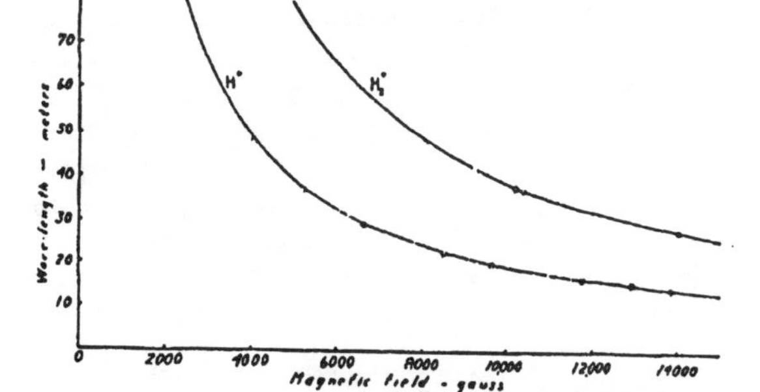

Fig. 8. Magnetic fields producing resonance of ions with oscillations of various wave-lengths: the curves are the theoretical relations (Eq. (3)) for H^+ and H_2^+ ions and the circles are the experimental observations.

the periphery, the ion currents diminished practically to zero when the magnetic field was changed a few tenths of one percent from the optimum value. This sharpness of resonance is understandable when it is remembered that the time required for an ion to execute one of its semi-circular paths is inversely proportional to the magnetic field. If, for example, the magnetic field were one percent greater or less than the resonance value, the ions would find themselves completely out of phase with the oscillations after having made fifty revolutions in the tube. In Fig. 7 the peaks exhibit an appreciable width, and indeed they extend over a one percent range of magnetic field. In most of the experiments, however, the ions circulated around many more times resulting in peaks of such restricted breadth as scarcely to be discernible in a diagram of this sort.

It is of course evident that the upper limit to the number of times the ions will circulate is determined by the degree of uniformity of the average value of the magnetic field along the spiral paths. Indeed, it would seem difficult to construct a magnet with pole faces giving fields of sufficient uniformity to

allow more than 100 accelerations of the ions. But happily there is a very simple empirical way of correcting for the lack of uniformity of the field, that makes possible a surprisingly large voltage amplification. This is accomplished by insertion of thin sheets of iron between the tube and the magnet; either in the central region or out towards the periphery, as may be needed. If the magnetic field is, on the average, slightly less out towards the periphery so that the ions lag in phase more and more with respect to the oscillations as they spiral around, they may be brought back into step again by the insertion near the periphery of a strip of iron of suitable width, thickness and extension. If, on the other hand, the ions tend to get ahead in phase in this region, an effective correction can be made by inserting a suitable iron sheet in the central region.

It should be emphasized in this connection that the requirement is not that the magnetic field has to be uniform everywhere to the extent indicated above; small deviations from uniformity are allowable provided that the average value of the magnetic field over the paths of the ions is such that they traverse successive revolutions in equal intervals of time. Thus, small magnetic field adjustments can be accomplished by increasing or decreasing the field over small portions of successive circular paths of the ions. In the present experiments the most satisfactory adjustment was made by the insertion of a sheet of iron 0.025 cm thick having a shape much like an exclamation point extending radially with the thick end 8 cm wide in the central region and the narrow end 3 cms wide at the periphery. Insertion of this correcting "shim" *increased the amplification factor* (that is, the ratio of the equivalent voltage of the ions arriving at the collector to the maximum high frequency voltage applied to the tube) *from about 75 to about 300.* These figures are of necessity somewhat rough estimates, because no means were conveniently at hand to measure the high frequency voltages applied to the tube. Our estimates are based solely on sparking distances in air, and hence it is not unlikely that the voltage amplifications were even greater.

The greatest voltage amplification was obtained when generating the highest speed ions, 1,220,000 volt-protons. In all our work we have found the experimental method to be increasingly effective in this regard, as in others, as we go to higher voltages.

For example, the optimum pressure of hydrogen in the tube has been found to increase from less than 10^{-4} mm of Hg when generating 200,000 volt-protons to more than 10^{-3} mm when producing 1,000,000 volt-protons. By the optimum pressure is meant the pressure that gives the largest current to the collector for a given electron emission from the filament. The reason for this is, of course, connected with the fact that the effective mean free path of the spiralling particles increases with voltage.

Examples of the observed variation with voltage on the deflecting plates of the ion currents to the collector are shown in Fig. 9. Each curve is for a particular resonance condition; curve *A*, for example, was obtained when protons resonated with 37.5 meter oscillations in a magnetic field of 5180 gauss, thereby theoretically resulting in the arrival of 172,000 volt-protons at the first slit of the collector system. The wave-lengths used and the theoretically expected equivalent voltages of the ions generated in each instance is indicated in the figure. It is seen that, the higher the equivalent voltage of the ions, the higher was the required deflecting voltage to obtain the maximum ion currents to the collector. Indeed, within the experimental error, the optimum deflecting voltage was proportional to the theoretical kinetic energies of the ions (calculated from Eq. (4)) and was quite independent of the magnitude of the high frequency voltage applied to the accelerating electrode. *These observations constitute incontrovertible evidence that the ions arriving at the collector actually had the high speeds theoretically expected. The observed absolute magnitudes of the deflecting voltages also agreed with theoretical calculations* within the experimental uncertainty of the paths of the ions before entering the deflecting system. Because of the considerable width of the ion source (the filament was 2.5 cm long) the effective center of

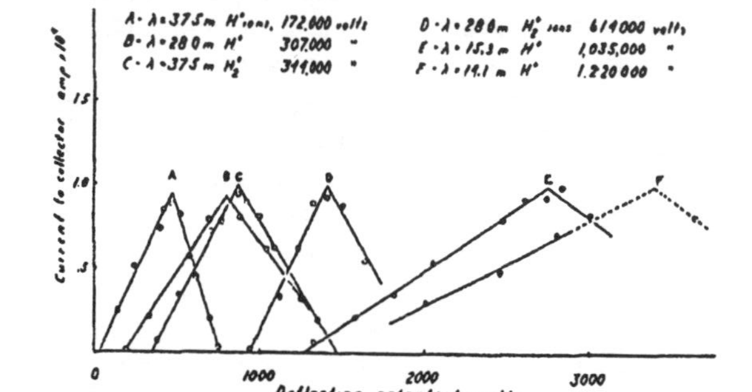

Fig. 9. Ion currents to the Faraday collector as a function of the voltage applied to the deflecting plates. The optimum deflecting voltages are seen to be proportional to the theoretically calculated kinetic energies of the ions (indicated in the figure in volts), thus proving that the ions arriving at the collector actually have the theoretically expected high speeds.

the circular paths of the ions at the periphery was quite broad. This fact together with the slit widths accounted for the absolute range of deflecting voltages over which ion currents reached the collector.

Discussion

The present experiments have accomplished one of the objectives set forth in the introduction, namely, the development of a convenient method for the production of protons having kinetic energies of the order of magnitude of 1,000,000 volt-electrons. It is well to emphasize two particular features that have contributed more than anything else to the effectiveness of the method: the *focussing actions of the electric and magnetic fields*, and the *simple means of empirically correcting the magnetic field* by the introduction of suitable iron strips. The former has solved the practical problem of generation of intense high speed ion beams of restricted cross-section so much desired in studies of collision processes. The latter has eliminated the problem of uniformity of magnetic field, making possible voltage amplifications of more than 300. This in turn has practically eliminated any difficulties associated

with generation and application to the accelerating electrodes of required high frequency voltages. In consequence, we have here a source of high speed light ions that is readily constructed and assembled in a relatively small laboratory space out of quite modest laboratory equipment. The beam of ions so produced has valuable characteristics of convenience and flexibility for many experimental investigations; there are obvious advantages of a steady beam of high speed ions of but one millimeter diameter generated in an apparatus on an ordinary laboratory table. Moreover, the apparatus evolved in the present work is in no respects capricious, but functions always in a satisfactorily predictable fashion. This is illustrated by the fact that the accelerating tube can be taken apart and reassembled, and then within a few hours after re-evacuation steady beams of 1,200,000 protons can always be obtained.

But it is perhaps of even more interest to inquire as to the practical limitations of the method; to see what extensions and developments are foreshadowed by the present experiments.

Of primary importance is the probable experimental limitation on the producible proton energies. The practical limit is set by the size of the electromagnet available; for the final equivalent voltage of the ions at the periphery is proportional to the square of the magnetic field strength and to the square of the radius of the path. For protons, it is not feasible to use magnetic fields much greater than employed in the present work (about 14,000 gauss) because of the difficulties of application of suitably higher frequency oscillations—that is to say, it is not desirable to go much below 14 meters wave-length. However, it is entirely practicable to use a much larger magnet than that employed in the present experiments. At the present time a magnet having pole faces 114 cm in diameter is being installed in our laboratory. As will be seen from Eq. (4), a magnetic field of 14,000 gauss over such a large region *makes possible the production of* 25,000,000 *volt-protons*.

Of course, it may be argued that there are other difficulties which preclude ever reaching such a range of energies. For example, there is the question of whether it is possible to obtain such a great amplification factor that the high frequency voltages necessarily applied to the accelerating electrodes are low enough to be realizable in practice. In the present experiments an amplification of 300 was obtained with no great effort, and it would seem that with more careful correction of the field this amplification could be considerably increased at higher voltages. In the higher range of speeds the variation of mass with velocity begins to be appreciable, but presents no difficulty as it can be allowed for by suitable alteration of the magnetic field in the same empirical manner as is done to correct its otherwise lack of uniformity.

Assuming then a voltage amplification of 500, the production of 25,000,000 volt-protons would require 50,000 volts at a wave-length of 14 meters applied across the accelerators; thus, 25,000 volts on each accelerator with respect to ground. *It does appear entirely feasible to do this*, although to be sure a considerable amount of power would have to be supplied because of the capacity of the system.

Of similar interest is the matter of maximum obtainable beam intensities. In the present experiments no efforts have been made to obtain high intensities and the collector currents have usually been of the order of magnitude of 10^{-9} amp. Using the present method of generation of the ions, there are two factors that can be drawn upon to increase the yield of high speed ions—the electron emission and the pressure of hydrogen in the tube. The electron emission can easily be increased from 10 to 100 times over that used in the present experiments. The effective free paths of the protons increase with voltage so that, as was found to be the case, the maximum usable pressure of hydrogen is governed by the setting in of a high frequency discharge in the tubes due to the voltage on the accelerators. This appears to occur at a pressure greater than 10^{-3} mm of Hg; the reason the critical pressure is so high is probably to be associated with the quenching action of the magnetic field. These considerations make it seem reasonable to expect that, using the present ion source, *high speed ion currents of as much as 0.1 microampere can readily be obtained.*

At all events, it seems that the focussing of the spiralling ions is so effective that a quite considerable portion of those starting out arrive at the collector and that the beam intensity is determined largely by the source. *This method of multiple acceleration is capable of yields of the same order of magnitude as would conceivably result from the direct application of high voltages.*

For a given experimental arrangement the energy of the ions arriving at the collector varies inversely as their masses and directly as their charges. Thus, the large magnet mentioned above makes possible the production of 12,500,000 volt hydrogen molecule ions and doubly charged helium ions (alpha-particles) as well as 25,000,000 volt-protons. Moreover, generating the theoretically maximum value of ion energies becomes much easier with increasing atomic weight because the wave-length of the applied high frequency oscillations increases in a like ratio. For example, using a magnetic field of 14,000 gauss over a region 114 cm in diameter, 2,800,000 volt nitrogen ions could be generated by applying 123 meter oscillations. Broadly speaking, then, the apparatus is well adapted to the production of ions of all the elements up to atomic weight 25 having kinetic energies in excess of 1,000,000 volt-electrons.

We wish to express our gratitude and thanks to the Committee-on-Grants-in-Aid of the National Research Council, the Federal Telegraph Company through the courtesy of Dr. Leonard F. Fuller, Vice-President, the Research Corporation, and the Chemical Foundation for their generous assistance which has made these experiments possible.

Capture of Slow Neutrons

G. BREIT AND E. WIGNER, *Institute for Advanced Study and Princeton University*

(Received February 15, 1936)

Current theories of the large cross sections of slow neutrons are contradicted by frequent absence of strong scattering in good absorbers as well as the existence of resonance bands. These facts can be accounted for by supposing that in addition to the usual effect there exist transitions to virtual excitation states of the nucleus in which not only the captured neutron but, in addition to this, one of the particles of the original nucleus is in an excited state. Radiation damping due to the emission of γ-rays broadens the resonance and reduces scattering in comparison with absorption by a large factor. Interaction with the nucleus is most probable through the s part of the incident wave. The higher the resonance region, the smaller will be the absorption. For a resonance region at 50 volts the cross section at resonance may be as high as 10^{-19} cm^2 and 0.5×10^{-20} cm^2 at thermal energy. The estimated probability of having a nuclear level in the low energy region is sufficiently high to make the explanation reasonable. Temperature effects and absorption of filtered radiation point to the existence of bands which fit in with the present theory.

1. INTRODUCTION

BETHE,[1] Fermi,[2] Perrin and Elsasser,[3] Beck and Horsley[4] gave theories of the anomalously large cross sections of nuclei for the capture of slow neutrons. These theories are essentially alike and explain the anomalously large capture cross sections as a sort of resonance of the s states of the incident particle. Resonance is usually helpful in causing a large scattering as well as a large probability of capture and it has been shown [H. B. Eq. (35)] that large scattering is to be expected by nuclei showing anomalously large capture at thermal energies. This consequence of the current theories is apparently in contradiction with experiment, there being no evidence of a large scattering in good absorbers. It also follows from current theories that with very few exceptions the capture cross section should vary inversely as the velocity of the slow neutrons. Experiments on selective absorption recently performed[5] indicate that there are absorption bands characteristic of different nuclei and it appears from the experiments of Szilard[6] that these bands have fairly well-defined edges. It has been pointed out by Van Vleck[7] that it is hard and probably impossible to reconcile the difference in internal phase required by the Bethe-Fermi theory with reasonable pictures of the structure of the nucleus. The combined evidence of experimental results and theoretical expectation is thus against a literal acceptance of the current theories and it is our purpose to outline an extension which is capable of explaining the above facts by a mechanism similar to that used for the inverse of the Auger effect by Polanyi and Wigner.[8]

It will be supposed that there exist quasi-stationary (virtual) energy levels of the system nucleus+neutron which happen to fall in the region of thermal energies as well as somewhat above that region. The incident neutron will be supposed to pass from its incident state into the quasi-stationary level. The excited system formed by the nucleus and neutron will then jump into a lower level through the emission of γ-radiation or perhaps at times in some other fashion. The presence of the quasi-stationary level, Q, will also affect scattering because the neutron can be returned to its free condition during the mean life of Q. If the probability of γ-ray emission from Q were negligible there would be in fact strong scattering at the resonance, the scattering cross section being then

[1] H. A. Bethe, Phys. Rev. **47**, 747 (1935). We refer to this paper as H. B. in the text.

[2] E. Amaldi, O. d'Agostino, E. Fermi, B. Pontecorvo, F. Rasetti, E. Segrè, Proc. Roy. Soc. **A149**, 522 (1935).

[3] Perrin and Elsasser, Comptes rendus **200**, 450 (1935).

[4] Beck and Horsley, Phys. Rev. **47**, 510 (1935).

[5] Moon and Tillman, Nature **135**, 904 (1935); Bjerge and Westcott, Proc. Roy. Soc. **A150**, 709 (1935); Arsimovitch, Kourtschatow, Miccovskii and Palibin, Comptes rendus **200**, 2159 (1935); Ridenour and Yost, Phys. Rev. **48**, 383 (1935); Pontecorvo, Ricerca scientifica **6–7**, 145 (1935).

[6] L. Szilard, Nature **136**, 950 (1935).

[7] J. H. Van Vleck, Phys. Rev. **48**, 367 (1935).

[8] O. K. Rice, Phys. Rev. **33**, 748 (1929); **35**, 1551 (1930); **38**, 1943 (1931); J. Chem. Phys. **1**, 375 (1933). A similar process was used by M. Polanyi and E. Wigner, Zeits. f. Physik **33**, 429 (1925).

of the order of the square of the wave-length. Estimates of order of magnitude show that it is reasonable to assign 12 volts to the "half-value breadth" of Q due to radiation damping and that the "half-value breadth" due to passing back into the free state is about one-fortieth of the above amount. This means that when the system passes into the state Q it radiates practically immediately and the neutron has no time to be rescattered. It will, in fact, be seen from the calculations that follow that the ratio of scattering to absorption is essentially the ratio of the corresponding half value breadths. The hardness of the emitted γ-rays is of primary importance for the small ratio of scattering to absorption because it makes the probability of γ-ray emission sufficiently high. Inasmuch as the interesting phenomena occur for low energies we may suppose that in most cases the coupling of the incident state occurs through its s state, i.e., in virtue of head on collisions. It will be seen, however, that the possibility of obtaining observable effects by means of p states is not excluded even though it is less probable and leads to smaller cross sections. Calculation shows that with resonances of the type considered here one may obtain appreciable probability of capture at energies of the order of 1000 volts. It is possible to have at such energies cross sections of roughly 10^{-22} cm^2 with a half-value breadth of about 20 volts. It is therefore not necessary to ascribe all large cross sections to neutrons of thermal velocities and the probability of finding a quasi-stationary level in a suitable region is not so small as to make the process improbable.

We are presenting below the theory of capture on this basis in some detail not because we believe it to be a final theory but because further development may be helped by having the preparatory structure well cemented.

2. Theory of Damping

The process of absorption from the continuum into a quasi-stationary level and a subsequent reemission of a photon is related to the phenomena of predissociation discussed by O. K. Rice[8] who made the first application of quantum mechanics to this type of process since Dirac's first approach.[9] It is essential for us to consider two continua and in this respect the present problem is more general. It resembles closely the problem of absorption of light from a level a to a level c which is strongly damped by radiation in jumps to a third level b. The absorption from a to c corresponds to the transition of the neutron into the quasi-stationary level and the jumps from c to b correspond to the emission of γ-rays in a transition to a more stable level of the nucleus. The absorption probabilities can be obtained by using the principle of detailed balance from the solution which represents emission[10] from the level c to the levels a, b or else by a direct application of the theory of absorption.[11] The usual theory as developed for either process is not accurate enough to represent the effect of the variation of matrix elements with velocity which is essential for our purpose, inasmuch as it is responsible for the existence of two regions of large absorption. The usual type of calculation will now be generalized so as to take the variation into account.

(a) Calculation of the absorption and scattering process

Let a_s denote the probability amplitudes of states in which the neutron is free and in a state s. Similarly let b_r stand for the probability amplitude of a state in which the neutron is captured and there is a photon r emitted and let c be the probability amplitude of the quasi-stationary state having energy $h\nu$. The states r, s are here considered to be discrete but very closely spaced in energy. The average spacing of the levels r, s are written $\Delta E_r = h\Delta\nu_r$, $\Delta E_s = h\Delta\nu_s$ so that the number of levels s per unit energy range is $1/\Delta E_s$. The matrix element of the interaction energy responsible for transitions from a_s to c, c to b_r will be written, respectively,

$$M_s = hA_s, \quad M_r = hB_r. \tag{1}$$

The damping constants for c due, respectively, to the possibility of emitting a_s or b_r are then[12]

[9] P. A. M. Dirac, Zeits. f. Physik 44, 594 (1927).
[10] V. Weisskopf and E. Wigner, Zeits. f. Physik 63, 54 (1930).
[11] V. Weisskopf, Ann. d. Physik 9, 23 (1931).
[12] G. Breit, Rev. Mod. Phys. 5, 91, 104, 117 (1933).

$$(4\pi\tau_a)^{-1}=\Gamma_s=[\pi\overline{|A_s|^2}/\Delta\nu_s]_{\nu_s=\nu_0};\qquad (2)$$

$$(4\pi\tau_b)^{-1}=\Gamma_r=[\pi\overline{|B_r|^2}/\Delta\nu_r]_{\nu_r=\nu_0};\ \Gamma=\Gamma_s+\Gamma_r,$$

where τ_a, τ_b are respective mean lives of c due to emission of a_s and b_r. The quantities Γ represent one-half of the "half-value breadth" measured in frequency. In discussing line emission and absorption [10, 11] the directional averages of $|A_s|^2$ and $|B_r|^2$ can be taken for any energy within the breadth of the line because the line can be usually considered to be sharp. In the present case it will be necessary to distinguish among directional averages of $|A_s|^2$ for different energies.

The states s will be thought of as plane waves modified by a central field due to the nucleus and satisfying boundary conditions at the surface of a fundamental cube of volume V. The equations satisfied by a_s, b_r, c are

$$\left(\frac{d}{2\pi i dt}+\nu_s\right)a_s=A_sc;\quad \left(\frac{d}{2\pi i dt}+\nu_r\right)b_r=B_rc,\qquad (3)$$

$$\left(\frac{d}{2\pi i dt}+\nu\right)c=\sum A_s^*a_s+\sum B_r^*b_r.$$

In these equations the influence of only one quasi-stationary level is taken into account and for this reason they are not quite accurate. They are sufficiently good for the present purpose because it will be supposed that different quasi-stationary levels do not fall closely together. At $t=0$ it will be supposed that

$$a_s=\delta_{ss_0},\quad b_r=0,\quad c=0,\quad (t=0).\qquad (4)$$

An approximate solution of (3) satisfying this initial condition can be obtained by forming a linear combination of

$$c=e^{-2\pi i(\nu-i\Gamma')t};$$

$$a_s=A_s[e^{-2\pi i(\nu-i\Gamma')t}-e^{-2\pi i\nu_st}]/(\nu_s-\nu+i\Gamma'),\qquad (5)$$

$$b_r=B_r[e^{-2\pi i(\nu-i\Gamma')t}-e^{-2\pi i\nu_rt}]/(\nu_r-\nu+i\Gamma'),$$

with

$$\Gamma'=[\pi\overline{|A_s|^2}/\Delta\nu_s+\pi\overline{|B_r|^2}/\Delta\nu_r]_{\text{resonance region}},\qquad (5')$$

and

$$a_{s_0}=e^{-2\pi i(\nu_0-i\gamma)t};\quad c=A_{s_0}^*e^{-2\pi i(\nu_0-i\gamma)t}/(\nu-\nu_0-i\Gamma),$$

$$a_s=A_sA_{s_0}^*[e^{-2\pi i(\nu_0-i\gamma)t}-e^{-2\pi i\nu_st}]/(\nu_s-\nu_0+i\gamma)(\nu-\nu_0-i\Gamma),\qquad (6)$$

$$b_r=B_rA_{s_0}^*[e^{-2\pi i(\nu_0-i\gamma)t}-e^{-2\pi i\nu_rt}]/(\nu_r-\nu_0+i\gamma)(\nu-\nu_0-i\Gamma);$$

$$\Gamma=[\pi\overline{|A_s|^2}/\Delta\nu_s+\pi\overline{|B_r|^2}/\Delta\nu_r]_{\nu_s=\nu_0}.$$

In Eq. (6) $s\neq s_0$. The quantities γ and $\nu_0-\nu_{s_0}$ are small compared with Γ; they will go to zero with increasing volume. From (3), one finds for them the equation:

$$(\nu_{s_0}-\nu_0+i\gamma)(\nu-\nu_0-i\Gamma)=|A_{s_0}|^2,\qquad (7)$$

so that

$$\gamma=|A_{s_0}|^2\Gamma/[(\nu-\nu_0)^2+\Gamma^2];$$
$$\nu_0=\nu_{s_0}+(\nu_0-\nu)(\gamma/\Gamma).\qquad (8)$$

In obtaining Eq. (7) the approximations

$$\sum_s{}'|A_s|^2\frac{1-e^{2\pi i(\nu_0-\nu_s-i\gamma)t}}{\nu_s-\nu_0+i\gamma}=\pi i\overline{|A_s|^2}/\Delta\nu_s\qquad (9)$$

are made. These correspond to replacing the sums by integrals and extending the range of integration from $\nu_s=-\infty$ to $\nu_s=+\infty$ and similarly for ν_r. In addition it is supposed that $\overline{|A_s|^2}$, $\overline{|B_r|^2}$ vary so slowly through the region in which the integrand is large that they may be taken outside the integral sign. These approximations are, therefore, valid only if the contributions to the sums (9a), (9b) are localized in a sharp maximum. Such a maximum exists for $\nu_s\cong\nu_0$ because: (1) γ vanishes as the fundamental volume is increased and therefore one may consider $\gamma t\ll 1$ and (2) for any $\nu_s-\nu_0$ it is possible to choose t sufficiently large to make $|\nu_s-\nu_0|t\gg 1$. For such times the most important part of the integrand oscillates rapidly with ν_s. However for $|\nu_s-\nu_0|\sim\gamma$, the values of t which satisfy $\gamma t\ll 1$

are always such that $|\nu_s-\nu_0|t \ll 1$. The integrand is thus not oscillatory for $\nu_s=\nu_0\pm\gamma$ and the values of $\overline{|A_s|^2}$, $\overline{|B_r|^2}$ on the right side of (9) are to be understood as corresponding to $\nu_s=\nu_0$ with an uncertainty of the order γ. It can be verified by calculation that the contribution to (9) due to a finite region at a distance $|\nu_s-\nu_0|\gg\gamma$ contributes imaginary quantities decreasing exponentially with $2\pi|\nu_s-\nu_0|t$ and real quantities which contribute to a frequency shift[9] of ν. For the present this shift will be neglected. Eqs. (6) are thus approximate solutions which become increasingly better as t increases, provided $\gamma t \ll 1$. In our application Γ is mostly due to the radiation damping Γ_r. The directional averages of $|B_r|^2$ vary smoothly since the energy of the γ-ray is of the order of several million volts and is large compared to Γ.

The quantity Γ' which enters (5) is not determined accurately by the present method because $\overline{|A_s|^2}$ which enters in this case is some sort of average over the resonance width. This complication causes no trouble because: (a) for times $t\gg 1/4\pi\Gamma'$ the rates of emission of states a_s, b_r are, respectively, $4\pi\gamma\Gamma_s/\Gamma$, $4\pi\gamma\Gamma_r/\Gamma$ and depend[13] only on Γ and not on Γ'; (b) the largeness of Γ_r in comparison with Γ_s makes $|\Gamma'-\Gamma|\ll\Gamma$. Thus Γ' is of importance only in determining the initial transients but not the steady rate of absorption. This can be expected from the fact that the solutions (6) represent a condition in which s_0 is absorbed at the rate $4\pi\gamma$. The addition of the "emission solution" (5) is only needed to enforce the condition $c=0$ at $t=0$; it modifies the emission of states b_r, a_s during times comparable with the mean life of the nucleus but leaves them unchanged over longer periods very similarly to the way in which analogous transient conditions are of no importance in the absorption of monochromatic radiation by classical vibrating systems.

The total cross section σ which corresponds to the disappearance of the incident states s_0 is given by

$$\sigma=4\pi\gamma V/v, \tag{10}$$

where v is the neutron velocity because the modified plane waves denoted by s were normalized in the volume V and thus represent states of density $1/V$.

The number of possible plane waves in V per unit frequency range is

$$1/\Delta\nu_s=4\pi V/v\Lambda^2, \tag{11}$$

where Λ is the de Broglie wave-length. From (2), (10), (11) we have

$$\sigma=\gamma\Lambda^2/\Delta\nu_s=\frac{\Lambda^2}{\pi}S\frac{\Gamma_s\Gamma}{(\nu-\nu_0)^2+\Gamma^2}. \tag{12}$$

Here the statistical factor S takes account of the fact that the state s_0 may be more or less effective in its coupling to the quasi-stationary level than the average modified plane wave in the same energy region. If the quasi-stationary level has an orbital angular momentum $L\hbar$ and if there is no spin orbit interaction then $|A_{s_0}|^2=(2L+1)\overline{|A_s|^2}$ because coupling to c can take place only through $1/(2L+1)$ of the total number of states. Thus.

$$S=2L+1 \tag{13}$$

in these special circumstances. For s terms $S=1$. The total cross section

$$\sigma=\sigma_c+\sigma_s,$$

where σ_s is the cross section due to scattering and σ_c is the cross section due to capture.. We have

$$\sigma_c=\frac{\Lambda^2}{\pi}S\frac{\Gamma_s\Gamma_r}{(\nu-\nu_0)^2+\Gamma^2};\quad \sigma_s=\frac{\Lambda^2}{\pi}S\frac{\Gamma_s^2}{(\nu-\nu_0)^2+\Gamma^2}. \tag{14}$$

The above value of σ_s corresponds to the value $\Sigma|a_s|^2$ and does not take into account the fact that there is scattering in the abscence of the quasi-stationary level. If this is strong one must correct σ_s for interference of the states s with the spherical wave present in s_0. In the applications made below the scattering effect due to either cause will be small and the correction need not be considered. According to (14) the extra scattering can be expected to be of the order Γ_s/Γ_r times the capture and is quite small for small Γ_s.

It should be noted that the order of magnitude of σ_c at resonance is changed by taking into account the radiation damping. If this were

[13] Appendix I.

neglected and if one were to calculate simply by using Einstein's emission probability for the stationary states of matter then one would obtain an incorrect value,

$$\sigma_c' = \frac{\Lambda^2}{\pi} S \frac{\Gamma_s \Gamma_r}{(\nu-\nu_0)^2+\Gamma_s^2}. \qquad (14')$$

For resonance $\sigma_c/\sigma_c' = \Gamma_s^2/\Gamma^2$ and approximately $\int \sigma_c dE / \int \sigma_c' dE$ is Γ_s/Γ_r. No paradox is involved here because it is not legitimate to apply Einstein's emission probability formula to levels separated by less than their breadth due to radiation damping. Eq. (14′) gives too high values to the cross section. If $\nu - \nu_0 \gg \Gamma$ there is no difference between σ_c' and σ_c. For sufficiently large values of $\nu - \nu_0$ the discussion which led to Eqs. (13), (14) will break down because Dirac's frequency shift[9] is neglected in these formulas. A more complete formal discussion including the frequency shift is given in Appendix I. The calculation shows that one should change the frequency of the quasi-stationary level ν by

$$\nu \rightarrow \nu - \int \frac{|A_s|^2}{\Delta\nu_s} \frac{d\nu_s}{\nu_s - \nu_0} - \int \frac{|B_r|^2}{\Delta\nu_r} \frac{d\nu_r}{\nu_r - \nu_0} \qquad (15)$$

where the integrations are extended over the complete range of states s, r and where the principal values of the integrals are to be taken. The last part of (15) represents the frequency shift due to electromagnetic radiation and can be incorporated in ν as a constant because ν_0 need be varied only in a range small in comparison with the frequency of the γ-ray. It is dangerous to take this shift into account on account of the well known inconsistency of quantum electrodynamics. The second term on the right side of Eq. (15) is due to interactions between free neutron states and the quasi-stationary state. It is physically correct and it is necessary in order to bring about agreement between (14) and calculations away from resonance by means of the Einstein emission probabilities. The shift is large in the applications. Nevertheless changes in it are small in the relatively small range of values which need be considered and its effect is therefore primarily that of displacing the resonance frequency by a constant amount.

(b) Resonance of one-body systems

The above discussion cannot be applied directly to cases in which resonance consists simply in a sharp increase of the wave function of one neutron to a maximum inside the nucleus because there is no intermediate state c under such conditions in the same sense as in the previous section. For low velocity neutrons such resonance can be sharp for states with $L \geqq 1$. Formally one could try to apply the discussion already given by starting with wave functions which are solutions of the wave equation for an infinitely high barrier somewhat outside the nucleus. The difference between the actual height of the barrier and ∞ can be then treated as a perturbation essentially responsible for the matrix elements hA_s. Such a procedure leads apparently to correct results which can be verified by other methods. It is troublesome to justify it completely because the region where the infinite barrier must be erected should be such that the wave functions within are small for all energies. It is preferable to use a more direct calculation for such a case. We consider a plane wave of neutrons incident on the nucleus. Resonance takes place to the wave functions of angular momentum $L\hbar$. We surround the nucleus by a large perfectly reflecting sphere of radius R and we calculate the rate at which states of angular momentum $L\hbar$ disappear by radiation. There is no essential restriction on the possibility of forming wave packets out of the plane waves if we admit only those states L which satisfy the boundary conditions on the sphere. The radius will be made finally infinitely large and the spacing between the levels infinitely small. This provides the necessary flexibility for the formation of the wave packets.

The spacing between successive possible neutron levels is given by

$$\Delta\nu = v/2R. \qquad (16)$$

The radial function will be expressed as F/r where F will be by definition a sine wave with unit amplitude at a large distance from the nucleus. The normalized wave function is then $Y_L(F/r)(2/R)^{\frac{1}{2}}$ where Y_L is a spherical harmonic normalized so as to have $\int |Y_L|^2 d\Omega = 1$. The wave function for the bound state will be written

$$Y_{L\pm 1} f/r; \quad \int_0^\infty f^2 dr = 1. \qquad (17)$$

The damping constant which corresponds to the emission of radiation from the state F is obtained by using the formula for Einstein's emission probability and is

$$\gamma_R = (C/R) \left| \int_0^\infty F f r dr \right|^2, \qquad (18)$$

where

$$C = \frac{32\pi^3 e'^2 \nu^3}{3hc^3} \frac{L+\frac{1}{2}\pm\frac{1}{2}}{2L+1}, \qquad (18')$$

the upper sign applying to jumps $L \rightarrow L+1$ and $e' \sim e/2$ is the effective charge of the neutron nucleus system. As $R \rightarrow \infty$, both $\Delta\nu$ and γ_R decrease towards zero but their ratio remains constant. The cross section due to capture computed directly from the emission probability is

$$\sigma_C' = (2L+1)\Lambda^2 \gamma_R/\Delta\nu. \qquad (19)$$

If this expression approaches Λ^2 then $\gamma_R/\Delta\nu$ becomes comparable with unity and the levels are close enough together to make Eq. (19) meaningless. It is then necessary to take into account the mutual influence of neighboring levels. This can be done by means of the damping matrix.[14] The successive states of angular momentum $L\hbar$ will be denoted by indices j, l and their probability amplitudes by a_j. These satisfy

$$\left(\frac{d}{2\pi i dt}+\nu_j\right)a_j=i\Sigma\gamma^{jl}a_l, \tag{20}$$

where

$$\gamma^{jl}=CJ_j^*J_l/R;\quad J_j=\int F_j f r dr. \tag{20'}$$

In our case only states with the same magnetic quantum number can interact so that a complete specification of the states is obtained through their energy. Solutions of (20) in which all quantities vary as $\exp\{-2\pi i(\nu_0-i\gamma)t\}$ correspond as closely as possible to the notion of a stationary state decaying under influence of radiation damping. From (20) one obtains

$$(\nu_j-\nu_0+i\gamma)a_j=i\Sigma\gamma^{jl}a_l. \tag{20''}$$

These equations with the complex eigenwert $\nu_0-i\gamma$ can be reduced making use of the fact that γ^{jl} is a matrix of rank 1. Thus eliminating the a_l one finds

$$1=\frac{iC}{R}\Sigma\frac{|J_j|^2}{\nu_j-\nu_0+i\gamma} \tag{21}$$

for the secular equation which determines ν_0 and γ. This equation will be solved approximately for the case of sharp resonance. The resonance will be supposed to take place at an energy $h\nu_F$ and to have a "half-value breadth" $2h\Gamma_F$. Close to resonance

$$|J_j|^2=\frac{\Gamma_F^2|I|^2}{(\nu_j-\nu_F)^2+\Gamma_F^2}, \tag{22}$$

where $|I|^2$ is the maximum value of $|J|^2$. This approximation will usually apply only in a region of a few Γ_F. The value of Γ_F can be estimated using [15]

$$4\pi\Gamma_F=v_r/\int\bar{G}^2dr, \tag{23}$$

where $\bar{G}$ is F for resonance, v_r is the velocity at resonance, and the integration is to be carried through the range of large values of $\bar{G}$. The quantity Γ_F is analogous to Γ_s of section (a). The state represented by $\bar{G}$ is analogous to the quasi-stationary state of section (a). In order to bring out the analogy we introduce a damping constant similar to the previous Γ_r

$$\Gamma_R=C|I|^2/2\int\bar{G}^2dr=2\pi C|I|^2\Gamma_F/v_r, \tag{23'}$$

which is the damping constant of the state represented by $\bar{G}$ when that state is normalized within the nucleus and its immediate vicinity. Substituting (23') into (21), replacing the sum by an integral everywhere except in the vicinity of ν_0 and performing the summation in that region on the assumption that the $\Delta\nu$ can be considered as equal to each other in that region gives

$$1=i\left\{a\cot\left[\frac{\pi(\nu_{j0}-\nu_0+i\gamma)}{\Delta\nu_{j0}}\right]+ia\right.$$
$$\left.+\int_0^\infty\frac{v_r|J_j|^2\Gamma_R d\nu_j}{\pi v_j|I|^2\Gamma_F(\nu_j-\nu_0+i\gamma)}\right\} \tag{24}$$

$$a=\left[\frac{v_r|J_j|^2\Gamma_R}{v_j|I|^2\Gamma_F}\right]_{\nu_0},$$

where the integral must be extended over all ν_j and the region around ν_0 is integrated over the real axis. The quantity ν_{j0} is any one of the ν_j located so close to ν_0 that the variation in $\Delta\nu$ in between can be neglected. In the approximation of Eq. (22) the integration over the resonance region Γ_F leads to an equation which to within a sufficient approximation reduces to

$$1+itTh=(Th+it)(a+ib);\quad b=\frac{v_{j0}|I|^2q}{v_r|J_{j0}|^2(1+q^2)} \tag{25}$$

with

$$\nu_0-\nu_F=q\Gamma_F;\quad Th=\tanh\frac{\pi\gamma}{\Delta\nu_{j0}};\quad t=\tan\frac{\pi(\nu_0-\nu_{j0})}{\Delta\nu_{j0}}. \tag{25'}$$

By eliminating t

$$Th+1/Th=a+1/a+b^2/a, \tag{25''}$$

which has the approximate solution

$$1/Th=a+1/a+b^2/a. \tag{26}$$

For values of ν_0 which lie in the region where Eq. (22) applies and where $\Delta\nu_{j0}\sim\Delta\nu_r$ we have approximately

$$\frac{\pi\gamma}{\Delta\nu}=\frac{\Gamma_R\Gamma_F}{\Gamma_R^2+\Gamma_F^2(1+q^2)}, \tag{26'}$$

where it is supposed that $\Gamma_R\gg\Gamma_F$. If, however, $\Gamma_F\gg\Gamma_R$ then

$$\frac{\pi\gamma}{\Delta\nu}=\frac{\Gamma_R}{\Gamma_F(1+q^2)}, \tag{26''}$$

which is equivalent to using the γ_R of Eq. (18), (19); in this case one may compute using emission probabilities. If one is so far away from resonance that $b^2/a<a$, $1/a$ Eq. (25'') gives

$$\gamma=\gamma^{j_0j_0}, \tag{26'''}$$

provided the right side is $\ll 1$. Here again the simple emission point of view applies. For $\Gamma_R\gg\Gamma_F$ all regions are approximated by

$$\frac{\pi\gamma}{\Delta\nu_{j0}}=\frac{v_r|J_{j0}|^2\Gamma_R\Gamma_F(1+q^2)}{v_{j0}|I|^2[\Gamma_R^2+\Gamma_F^2(1+q^2)]} \tag{27}$$

The treatment of scattering by means of the damping matrix is somewhat involved and will not be reproduced here. The phase shift due to $\nu_0-\nu_{j0}$ when added to the phase shift already present in F_{j0} gives the phase shift required. The scattering is diminished by Γ_R in much the same way as it was diminished by it in section (a). By comparing (27) with (19)

$$\sigma_C=(2L+1)\frac{\Lambda^2 v_r|J_{j0}|^2\Gamma_R\Gamma_F(1+q^2)}{\pi v_{j0}|I|^2[\Gamma_R^2+\Gamma_F^2(1+q^2)]}, \tag{28}$$

which is similar to Eq. (14), close to resonance. The factors $|J|^2/|I|^2$ and v_r/v_{j0} take into account the deviations from the dependence of $|J|^2$ on ν given by (22). In (14) this is

[14] G. Breit, Rev. Mod. Phys. 5, 117 (1933); G. Breit and I. S. Lowen, Phys. Rev. 46, 590 (1934).

[15] G. Breit and F. L. Yost, Phys. Rev. 48, 203 (1935). See also Eq. (32').

analogous to the dependence of Γ_r on ν_{r0} combined with Dirac's frequency shift.

(c) Sharpness of resonance for single-body problem

The upper limit of integration in Eq. (23) has been left indefinite. By Green's theorem

$$\frac{d}{dr}\left[F_1\frac{dF_2}{dr}-F_2\frac{dF_1}{dr}\right]+\frac{2\mu}{\hbar^2}(E_2-E_1)F_1F_2=0,$$

where F_1, F_2 correspond to energies E_1, E_2 and need not be regular at $r=0$. Hence [16]

$$\frac{\partial}{\partial r}\left[F^2\frac{\partial}{\partial E}\frac{\partial F}{F\partial r}\right]+\frac{2\mu}{\hbar^2}F^2=0. \tag{29}$$

In this section let F_i be the function inside the nucleus, and let F stand for the regular solution of the wave equation for $r\times$radial function on the absence of the nuclear field. The normalization of F is such as to make it a sine wave $\sin(kr+\varphi)$ of unit amplitude at ∞. Similarly G is defined as satisfying the same differential equation as F but it is to be 90° out of phase with F at ∞ i.e. $\cos(kr+\varphi)$. The regular solution of the differential equation in the presence of the nuclear field, normalized in the same way as F and G, will be called $\bar{F}$. At the nuclear radius r_0

$$\bar{F}^2=G^2\bigg/\left\{\left[G^2\left(\frac{G'}{G}-\frac{F_i'}{F_i}\right)\right]^2+\left[FG\left(\frac{F'}{F}-\frac{F_i'}{F_i}\right)\right]^2\right\}. \tag{30}$$

Here the accent stands for differentiation with respect to kr. At resonance $F_i'/F_i=G'/G$ and the second term in the curly bracket is then 1, while the first term is zero. As E changes to either side of the resonance value E_r the first term may become 1 for $E=E_r\pm\Delta E$ where ΔE is properly chosen. The half-value breadth is then $2\Delta E$ and $\Delta E=h\Gamma_F$. The value of ΔE can be estimated by

$$\Delta E\frac{\partial}{\partial E}\left[G^2\left(\frac{G'}{G}-\frac{F_i'}{F_i}\right)\right]_{r_0}=1. \tag{31}$$

Using Eq. (29) and calculation the $\partial/\partial E$ for $E=E_r$ one obtains a result which can be expressed in terms of integrals up to R where R is any value of r which is greater than r_0. The function which is G for $r>r_0$ and $F_i(G/F_i)_{r_0}$ for $r<r_0$ is continuous at r_0 and at resonance its derivative with respect to r is also continuous. The function will be called $\bar{G}$ for $0<r<\infty$. We have then [17]

$$\frac{E}{\Delta E}=k\int_0^R\bar{G}^2dr+\left[\frac{G^2E\partial}{k\partial E}\frac{\partial G}{G\partial r}\right]_{r=k};\quad \Delta E=h\Gamma_F. \tag{32}$$

The right side of this result is independent of R and is finite. The term outisde the integral should be included in Eq. (23) changing

$$\int G^2dr\to\int_0^R\bar{G}^2dr+\left[\frac{G^2E\partial}{k^2\partial E}\frac{\partial G}{G\partial r}\right]_{r=k}. \tag{32'}$$

Eq. (32) has a well-defined meaning only if resonance is sharp. Otherwise the $\partial/\partial E$ entering in Eq. (31) cannot be supposed to be sufficiently constant through the half-breadth $2h\Gamma_F$. It cannot be expected to hold for the broad S resonance discussed by Bethe.

(d) Capture by p states

For a potential well of constant depth

$$F_i=\sin z/z-\cos z;\qquad F=\sin\rho/\rho-\cos\rho;$$

$$G=\cos\rho/\rho+\sin\rho, \tag{33}$$

where

$$z=Kr;\quad \rho=kr;\quad K=\mu v_i/\hbar;\quad k=\mu v/\hbar \tag{33'}$$

v_i, v being, respectively, the velocities inside and outside the nucleus. The resonance condition is

$$z\sin z/[\sin z/z-\cos z]=\rho\cos\rho/[\cos\rho/\rho+\sin\rho].$$

For slow neutrons $\rho\ll 1$ the right side is $\ll\rho^2$ and therefore very small. The first resonance point is obtained for $z=\pi-\epsilon$, $\epsilon\sim\rho^2/\pi$. It will suffice to take $z=\pi$. By substituting into Eq. (32) it follows that

$$\Delta E/E=2\rho/3=4\pi r_0/3\Lambda. \tag{33''}$$

For $E=(1/40)$ volt, $\Lambda=1.8\times10^{-8}$ cm, $h\Gamma_F=5.8\times10^{-6}$ volt. For 3-MEV γ-rays a reasonable value of $h\Gamma_R$ is 5.8 volts. The cross section at resonance is by Eq. (28) $3\Lambda^2\Gamma_F/\pi\Gamma_R=300\times10^{-24}$ cm². Since scattering is of the order Γ_F/Γ_R times capture the scattering cross section is small. According to Eq. (33'') the cross section at resonance for p terms with $\Gamma_R\gg\Gamma_F$ can be expected to vary as v and $h\Gamma_F$ as v^3. The range in which p terms can be expected to give large capture cross sections and small scattering is therefore roughly from 1/40 volt to 1 volt. At higher velocities $h\Gamma_F$ is likely to be higher than $h\Gamma_R$. In the absence of an apparent reason for nuclear p levels to fall in this narrow velocity range, an explanation in terms of p terms although possible is improbable on account of the small range of neutron velocities required.

3. Capture Through s Wave

(a)

This section will contain the calculation of the A_s used in 2a. It is supposed that the system

---

[16] J. A. Wheeler. We are indebted to Dr. Wheeler for communicating to us other applications of this relation.

[17] Cf. Eq. (22) reference 15. In calculations with Coulombian fields it is sometimes convenient to transform Eq. (32) of the text into

$$\frac{E}{\Delta E}=G^2\left[\frac{k}{F_i^2}\int_0^{r_0}F_i^2dr-\frac{k}{F^2}\int^{r_0}F^2dr-\frac{EG^2\partial}{kr\partial E}\left(\frac{kr}{FG}\right)\right]$$

all quantities outside the integrals being taken for $r=r_0$.

"nucleus+neutron" can be treated in first approximation by means of an effective central field acting on the neutron. The difference between the Hamiltonian of the system and the Hamiltonian corresponding to the central field will be called H'. On account of this difference there exist transitions from the s wave of the incident state to quasi-stationary excited states of the "nucleus +neutron" system. Normalizing the s waves within a sphere of radius R the wave function inside the nucleus is

$$C \sin Kr/r; \qquad C^{-2} = [1 + (U/E)\cos^2 Kr_0] 2\pi R,$$

where $$K^2/k^2 = (U+E)/E$$

and U is the depth of the potential hole. The interaction energy H' involves besides r also internal coordinates x. The wave function of the whole system in the incident state may be written $C\psi_0(x)\sin Kr/r$ and in the quasi-stationary state $\psi_Q(r, x)$. The matrix element M_s of Eq. (1) is then

$$M_s = \int \psi_Q(r, x) H' C \sin Kr \psi_0(x) dv/r, \tag{34}$$

where dv is the volume element of the whole system. The state Q is by definition such that the integral of $|\psi_Q|^2$ through nuclear dimensions is unity. The order of magnitude of M_s is therefore

$$M_s = C\bar{H} r_0^{\frac{1}{2}}, \tag{34'}$$

where $\bar{H}$ is an average of H' through the nucleus and may have reasonably a value of 0.5 MEV. It cannot be specified further without detailed calculation which would probably be unsatisfactory in the present state of nuclear theory. Since $\Delta E = hv/2R$,

$$h\Gamma_s \cong \frac{\bar{H}^2 r_0}{2\Lambda U \cos^2 Kr_0}. \tag{34''}$$

According to Eq. (14)

$$\sigma_c = \frac{\Lambda r_0}{2\pi} \frac{\bar{H}}{U\cos^2 Kr_0} \frac{\bar{H} h\Gamma_r}{h^2(\Gamma_r+\Gamma_s)^2 + (E-E_r)^2}, \tag{35}$$

where E_r is the value of E for resonance. According to this formula there are two maxima for σ_c, one for $E = E_r$ and one for $E \doteq 0$. The expected cross sections are given in Table I to about ten percent accuracy. The numbers correspond to $\Lambda(kT) = 1.8 \times 10^{-8}$ cm; $\Lambda(1 \text{ volt}) = 2.9 \times 10^{-9}$ cm; $r_0 = 3 \times 10^{-13}$ cm; $U\cos^2 Kr_0 = 10^7$ volts. For $E_r = 1/40$, 1 volt the table shows large cross sections at thermal energies and above. The condition is similar to Bethe's except for a relatively sharper resonance determined by $h\Gamma_r$. For 50 volts one sees the development of two maxima one at resonance and one at thermal energies. For $E_r = 1000$ and 10,000 volts the maximum at thermal energies decreases as E_r^{-2} and the maximum at resonance roughly as $E_r^{-\frac{1}{2}}$. For such high values of E_r scattering has a chance of becoming comparable with absorption or even greater than the absorption at resonance. In the thermal energy region the $1/v$ law is obeyed for high values of E_r; for low E_r the maximum at E_r interferes with the $1/v$ law and the region of its validity is displaced below thermal energies.

TABLE I. *Calculated cross sections for neutron capture.*

Position of Resonance (volts)	$\bar{H}$ (MEV)	$h\Gamma_r$ (volts)	$h\Gamma_s$ at resonance (volts)	$10^{24}\sigma$ (cm²) Resonance	Thermal Energies
1/40	0.1	10	0.01	90 000	
	.1	1	.01	900 000	
1	.1	10	.05	14 000	
	.1	1	.05	140 000	
50	.1	10	.37	2 000	3500
	.5	10	9	13 400	80000
	.1	1	0.37	11 000	350
	.5	1	9	4 800	9000
1000	.1	10	1.6	320	9
	.5	10	40	420	200
10000	.1	10	5	60	0.09
	.5	10	125	18	2

In Table I only the effect of a quasi-stationary level at E_r is considered. In addition there may be effects of other levels as well as radiation jumps of the kind considered by Bethe and Fermi which do not depend on the existence of virtual levels. It is thus probable that in most cases there is a region with a $1/v$ dependence although it may be at times masked by a resonance region.

(b) Dirac's frequency shift

In the above estimates the effect of Dirac's frequency shift was neglected. This is given by

$$(h\Delta\nu)_D=\int_0^\infty \frac{h\Gamma_s}{\pi}\frac{dE}{E-E_0}$$

$$=\frac{\bar{H}^2 r_0}{\pi\Lambda_0 x_0}\int_0^\infty \frac{x^2dx}{(x^2-x_0^2)(x^2+a^2)},$$

where $x=E^{\frac{1}{2}}$, $a^2=U\cos^2 Kr_0$ and the value of $h\Gamma_s$ was substituted by means of Eq. (34''). Here the subscript 0 refers to the neutron energy E_0 and the principal value of the integral is understood. Evaluating the expression

$$(h\Delta\nu)_D=\frac{\bar{H}^2 r_0 U^{\frac{1}{2}}\cos Kr_0}{2\Lambda_0(E_0+U\cos^2 Kr_0)E_0^{\frac{1}{2}}}. \quad (36)$$

The shift is seen to be of the order of 3000 times $h\Gamma_s$ for $E_0=1$ volt. The shift is nearly independent of the velocity. In the approximation of Eq. (36)

$$\frac{d(h\Delta\nu)_D}{dE_0}=\frac{h\Gamma_s}{E_0}\frac{E_0^{\frac{1}{2}}}{U^{\frac{1}{2}}|\cos Kr_0|}, \quad (36')$$

which shows that the variation in the shift is small and of the order of $2\times10^{-5}(E-E_r)$ for $\bar{H}=0.1$ MEV.

4. Discussion

(a) Absence of scattering

According to Dunning, Pegram, Fink and Mitchell[18] the elastic scattering of slow neutrons by Cd is less than one percent of the number captured. According to A. C. G. Mitchell and E. J. Murphy[19] scattering as detected by silver is about the same as absorption for Fe, Pb, Cu, Zn, Sn while for Hg scattering is about 1/80 of the absorption. In the later communication of Mitchell and Murphy[19] it is also found that Ag, Hg, Cd are poor scatterers of slow neutrons detected by silver. It is interesting that Ag shows small scattering in these experiments because the detection took place by means of silver and that Hg and Cd show small scattering because they have large absorption cross sections.[18] The observation of scattering by a material having large absorption is difficult because the neutrons entering the material are absorbed before they can be scattered and it is possible that to some extent the failure to observe scattering in good absorbers is due to this cause. The absence of observed scattering in the region of strong absorption is therefore not a surprise, particularly in view of the relatively small numbers of neutrons available for experimentation. It seems more significant, however, that strong absorbers do not show, so far, strong scattering in any velocity region because, according to the Fermi-Bethe theory, the scattering cross section should be large in a wide range of energies. The experimental evidence says little about the ratio of scattering to absorption near resonance. It indicates that this ratio is less than 1/10 in most cases. It is impossible, therefore, to ascertain definitely the ratio Γ_s/Γ_r until more detailed experimental data are available. According to Table I the condition $\Gamma_s/\Gamma_r<1/10$ can be satisfied in many ways up to velocities of over 1000 ev.

(b) Magnitude of interaction with internal states and probability of internal state in required region

In Table I arbitrary assignments of values of Γ_s, Γ_r were made. It will be noted that at low neutron velocities the desired large capture cross sections are easily obtained through relatively wide bands having a half-value breadth $2\Gamma_r$. Keeping Γ_r fixed one can decrease the interaction energy $\bar{H}$ to 10,000 ev for $h\Gamma_r=1$ volt, $E_r=1$ volt and still have a cross section of 1000×10^{-24} cm^2 in an energy range up to 2 volts. In some cases relatively weak radiative transitions will come into consideration leading to smaller Γ_r. For such transitions $\bar{H}$ need not be as large as 10,000 ev in order to have cross sections of 1000×10^{-24} cm^2 in the resonance region. For the large energies involved in nuclear structure it is reasonable to expect interaction energies of the order of 10,000 volts between practically any pair of levels not isolated by a selection rule and interaction energies of the order 100,000 volts between a great many levels.

There are about ten elements among 72 observed that show cross sections of more than

[18] J. R. Dunning, G. B. Pegram, G. A. Fink and D. P. Mitchell, Phys. Rev. 48, 265 (1935).

[19] A. C. G. Mitchell and E. J. Murphy, Phys. Rev. 48, 653 (1935). Cf. also Bull. Am. Phys. Soc. 11, paper 27, Feb. 4, 1936.

500×10^{-24} cm^2. Allowing for the fact that there are more isotopes than elements it appears fair to say that the chance of such an anomalous cross section is about 1/20. One can try to account for these solely by the low velocity regions which exist for any resonance level, thus probably overestimating the necessary number of levels. In order that $\sigma_c > 500\times10^{-24}$ cm^2 at 1/40 volt for a nucleus having $r_0 = 10^{-12}$ cm and $h\Gamma_r = 10$ volts the resonance region must be not farther than at $|E_r| = \bar{H}/420$ from thermal energies by Eq. (35). We do not wish $h\Gamma_s$ at thermal energies to be greater than 0.1 volt so as not to have too much scattering and therefore $\bar{H}$ should be below 2×10^5 ev at the higher E_r. Thus E_r should be kept below about 460 volts in order to give the large capture cross sections for $E=1/40$ volt together with small scattering. A level below ionization will also be effective in producing an increased absorption. The observed number of large absorptions corresponds in this way to one level in 900 volts for 1/20 of nuclei or one level every 18,000 volts for a single nucleus. In addition some cross sections will be caused by direct resonance. Just how many is uncertain but it is clear that such effects exist in Cd, Ag, Au, Rh, In.

The average spacing between the γ ray levels of Th C'' as given in Gamow's book is about 100,000 volts and this is apparently the order of magnitude usual for γ-ray levels of radioactive nuclei. There appears to be no reason why the energy levels found through the analysis of γ-ray spectra should include all the nuclear levels and there may be as many as one level in 20,000 of a kind that may be responsible for coupling to incident neutrons. It should be remembered here that some of the levels may be active even though the coupling is weak so that more possibilities are likely to matter than for the γ-rays of radioactive nuclei.

For a complicated configuration of particles it seems reasonable to consider a total number of 100 possible levels per configuration because protons and neutrons can be combined separately to give different states. On this basis we deal with an average spacing between configurations of about 2 MEV which is not excessively small. It is, of course, impossible to prove anything definitely without calculating the levels; this appears to be premature at present on account of uncertainties in nuclear theories.

(c) Existence of two maxima

According to the calculation given above it is expected that there will be in general two maxima one of which should be at resonance and another at $v=0$. According to the experiments of Rasetti, Segrè, Fink, Dunning and Pegram[20] the $1/v$ law is not obeyed by Cd but is obeyed by Ag. Cadmium has therefore a resonance region close to thermal velocities. In the classification of Fermi and Amaldi[21] this region must be affected by the C group since absorption measurements by the Li ionization chamber which was used in these experiments agree for most elements with the measurements of Fermi and Amaldi on the C group.[22] The verification of the $1/v$ law for Ag by the rotating wheel indicates that in Ag the resonance band is located above thermal energies. This conclusion is in agreement with the smallness of the temperature effect for the A neutrons detected by silver which was recently established by Rasetti and Fink.[23] Since Rh behaves similarly to Ag in these temperature experiments Rh also has a resonance region above thermal velocities. Fermi and Amaldi have evidence that D neutrons, which affect Rh, are different from A neutrons which affect Ag. It is very probable that both of these groups lie above the thermal region and they may reasonably cover a range of 30 volts inasmuch as the B group overlaps weakly with both A and D.

According to Szilard[6] In shows strong selective effects outside the C group and according to Fermi and Amaldi[21] the same period of In (54 min.) detects the D group. The number of neutrons in the groups is presumably in the ratios $C/80 = B/20 = D/15 = A/1$. One could try to conclude that the order of increasing energies is C, B, D, A on the assumption that the number of neutrons increases towards low energies. Such a conclusion is dangerous because little is known about the velocity distribution, because within

[20] F. Rasetti, E. Segrè, G. Fink, J. R. Dunning and G. B. Pegram, Phys. Rev. 49, 103 (1936).

[21] E. Amaldi and E. Fermi, Ricerca scientifica 2, 9 (1936); E. Fermi and E. Amaldi, Recerca scientifica 2, 1 (1936).

[22] Unpublished results of F. Rasetti. We are very grateful to Professor Rasetti for informing us of these results.

[23] F. Rasetti and George A. Fink, Bull. Am. Phys. Soc. 11, Paper 28, Feb. 4, 1936.

each group there may be several bands at different velocities, and also because the number of expected neutrons in a group should depend on its width. Temperature effects show that practically all captures increase as the energy is lowered. The effects are strongest[24] for Cu, V are smaller for Ag, Dy weaker for Rh and weakest for I. The absorption coefficient for C neutrons is, however, larger for Rh than for Ag indicating that the smaller temperature effect in Rh is due to a relatively greater importance in it of a band above thermal energies. All temperature effects agree in indicating the presence of a region in which the $1/v$ law is followed approximately but again no definite conclusion about the order of bands is possible. The low temperature effect in I would tend to indicate that its absorption region is high and detection-absorption experiments on I and Br tend to indicate that their bands are isolated from the others discussed here; perhaps these isotopes have resonance bands at higher energies. A new band was recently discovered in Au by Frisch, Hevesy and McKay[25] which represents strong absorption on a weaker background. The large number of selective effects observed makes the present explanation reasonable and the existence of a region of low energies in which the absorption decreases with energy is seen to fit in well with expectation.

(d) Other possibilities

One may consider weak long range forces as a possible explanation of the same phenomenon. Potentials of the order of neutron energies in a region comparable with the neutron wave-length would produce strong effects on absorption and scattering. For thermal energies the wave-length is of atomic dimensions and one would therefore expect the binding energy of deuterium compounds to be different from that of hydrogen compounds by an amount comparable to 1/40 volt if such potentials were present. Such energy differences do not exist. It would be possible to devise potentials which fall off sufficiently rapidly with distance to make the interaction potential negligible for chemical binding and which would cover a total region appreciably larger than the nucleus. Such hypotheses seem improbable without additional argument. Besides special relations between the phase integrals through the nuclear interior and the part of the range of force outside the nucleus would have to be set up in order to make absorption large and scattering small. It is improbable that the large number of bands could be accounted for by any single particle picture.

Forces between electrons and neutrons even though they may exist are not likely to have much to do with the bands Thus it has been shown by Condon[26] that electron neutron interactions would give rise to scattering cross sections varying roughly as the square of the atomic number Z on the assumption that the electron-neutron forces alone are responsible for the scattering. Forces inside the nucleus must also be supposed to contribute to the phase shifts responsible for scattering. Since these forces also vary with Z one could obtain a more complicated dependence of the scattering cross section by suitably adjusting the nucleus-neutron and electron-neutron potentials. On such a picture one could try to account for sharp resonances by making the electron neutron interaction repulsive. However, Condon's calculation shows that isotope shifts would be also produced by these interactions. It is improbable that the isotope shift is due solely to neutron-electron interaction because the deviation from the inverse square law inside the nucleus due to smearing out of protons produces a considerably larger effect than the observed shift. But it would also be unreasonable to try to combine the proton and neutron effects in the nucleus so as to have each large but their difference small. It is therefore probable that the electron-neutron interaction is not much larger than that which corresponds to the observed isotope shift. Since the density of the Fermi-Thomas distribution varies for small r as r^{-1} the effective potential acting on the neutron will become high for small r. However, calculation shows that it is not high in a wide enough region to account for sharp resonances if the limitation due to the isotope shift is considered.

[24] P. B. Moon and R. R. Tillman, Proc. Roy. Soc. **A153**, 476 (1936).

[25] O. R. Frisch, G. Hevesy and H. A. C. McKay, Nature **137**, 149 (1936).

[26] E. U. Condon, in press. We are indebted to Professor Condon for showing us his manuscript before publication.

Bombardment of light nuclei with charged particles has also shown the existence of resonances. Thus there are resonances[27] for the emission of γ-rays in proton bombardment of Li, C, F and similarly there are the well known resonances in disintegrations produced by α particles. Experimental methods have not been very suitable so far for the detection of resonance regions on account of the scarcity of monochromatic sources and the necessity of using thin films. In Li protons are apparently able to produce γ-rays in two ways; by resonance at 450 kv and by another process at higher energies. In fluorine there are several peaks. In carbon there was an indication of the main resonance peak being double. It appears possible that many more levels will be detected inasmuch as neutron experiments indicate a high density of levels. Calculations on the radiative capture of carbon under proton bombardment[15] lead to a higher yield than is observed by a factor of several thousand. In these calculations the capture was supposed to occur by a jump from the p state of the incident wave to an s state of the N^{13} nucleus. The calculated half-value breadth due to proton escape from the quasi-stationary p level was of the order $h\Gamma_P \sim 10{,}000$ ev and thus much larger than the width due to radiation damping. The yield in thick targets under these conditions is nearly independent of the special value of $h\Gamma_P$. It is clear from the formulas given here for neutron capture that one can decrease the theoretically expected yield either by ascribing the capture to a transition having a small probability of radiation (small Γ_r) such as would correspond to quadruple or other forbidden transitions or else by using an intermediate state of excitation of the nucleus with a small transition probability to the incident state of the proton (small Γ_s). In the latter case this transition probability would have to be made so small as to have $\Gamma_s < \Gamma_r$ and the observed width of resonance would have to be ascribed to experimental effects. If $\Gamma_s < \Gamma_r$ the thick target yield depends on Γ_s and is proportional to it for small Γ_s. The apparent disagreement between theory and experiment previously found for carbon is thus not alarming from the many-body point of view and supports the belief that excitation states of the nucleus have often to do with the simultaneous excitation of more than one particle.

The excitation states responsible for the neutron absorption bands make it possible for a fast neutron to lose energy by inelastic impact with the nucleus. Estimates show that the cross sections for such processes are likely to be small when energy losses are high. The cross section is estimated to be

$$\frac{\Lambda_1}{4\pi\Lambda_2}\,\frac{\bar{\Pi}^2 r_0^2}{U^2\cos^4 Kr_0},$$

where Λ_1, Λ_2 are, respectively, neutron wavelengths in the incident and final states. For large energy losses $\Lambda_2 \gg \Lambda_1$ and only a small effect need be expected. The excitation levels responsible for neutron capture will give small values Λ_1/Λ_2. Excitation levels located lower are more favorable and probably the excitation of Pb to about 1.5 MEV has to do with such a possibility.[28]

We are very grateful to Professors R. Ladenburg and F. Rasetti for interesting discussions of the experimental material.

Appendix I

Variation of damping constant with energy and Dirac's frequency shift

Eq. (6) of the text lead to [cf. Eqs. (126′) to (129′) of reference in footnote (12)]

$$\begin{aligned}(\nu_{s0}-\nu_0+i\gamma)(\nu-\nu_0+i\gamma)=|A_{s0}|^2 \\ +(\nu_{s0}-\nu_0+i\gamma)\Big[\Sigma'|A_s|^2\frac{1-e^{2\pi i(\nu_0-\nu_s-i\gamma)t}}{\nu_s-\nu_0+i\gamma} \\ +\Sigma|B_r|^2\frac{1-e^{2\pi i(\nu_0-\nu_r-i\gamma)t}}{\nu_r-\nu_0+i\gamma}\Big]\end{aligned} \quad (38)$$

which determines Γ by comparison with (7) $[\Gamma \gg \gamma]$.

It is by no means natural that this equation can be satisfied because the right side depends on t. If the A_s as well as the B_r were all essentially equal and if γ were great in comparison with the frequency differences of consecutive levels the sums could be transformed into integrals in the well-known way[10–13] so that (9) as well as (6) would follow. We shall attempt here a more exact procedure.

Consider the Σ_s' in the square brackets. It is natural to divide the range of ν_s into two parts: one for which $|\nu_s-\nu_0|>a\gg\gamma$ and one for which $|\nu_s-\nu_0|\leqq a$. Since

[27] L. R. Hafstad and M. A. Tuve, Phys. Rev. 48, 306 (1935); P. Savel, Comptes rendus 198, 1404 (1934), Ann. de physique 4, 88 (1935).

[28] J. Chadwick and M. Goldhaber, Proc. Roy. Soc. A151, 479 (1935).

$\overline{|A_s|^2}/\Delta\nu_s$ changes slowly this quantity will be replaced by a constant in $|\nu_s-\nu_0|\leq a$ and its value may be taken to be that at ν_0 for the evaluation of the contribution of this region. We have then to consider

$$(\overline{|A_s|^2})_{\nu_0}\sum_{\nu_0-a}^{\nu_0+a}{}' \frac{1-e^{2\pi i(\nu_s-\nu_0-i\gamma)t}}{\nu_s-\nu_0+i\gamma}. \tag{39a}$$

An exact evaluation of this sum is not simple because γ and $\Delta\nu$ are of the same order of magnitude and the replacement of (39a) by an integral is somewhat objectionable. This point has never been completely cleared up and we have only qualitative arguments in favor of the correctness of the replacement of (39a) by an integral. For $\nu_s-\nu_0$ of the order of a few γ such a replacement is indeed meaningless but fortunately this region is not vital for $t\ll 1/\gamma$ since the numerator of (39a) is then small. For larger $|\nu_s-\nu_0|$ the terms of (39a) vary more smoothly and finally they become rapidly oscillating for $|\nu_s-\nu_0|t\gg 1$ which can be satisfied simultaneously with $t\ll 1/\gamma$ provided $a\gg\gamma$. The smallness of γ is thus not as serious as might appear from the fact that $\gamma/\Delta\nu\sim 1$. It should also be observed that the treatment of Rice[8] using real eigenwerte for a single one-dimensional continuum is in agreement with replacing (39a) by an integral. The result of doing so is given by (9).

In addition one has the contribution of $|\nu_s-\nu_0|>a$. This integral can be treated neglecting γ because it is of interest to evaluate γ only to quantities of order $\gamma/(\nu-\nu_0)$ and because the discussion is supposed to apply only to $\gamma t\ll 1$. This integration gives

$$\left(\int_0^{\nu_0-a}+\int_{\nu_0+a}^{\infty}\right)\frac{\overline{|A_s|^2}}{\Delta\nu_s}\frac{d\nu_s}{\nu_s-\nu_0} \tag{39b}$$

which means that the principal value of the $\int$ is understood. Similarly one obtains a contribution due to $|B_r|^2$. These two integrals give the Dirac frequency shift which is included in Eq. (15).

As stated in the text the difference between Γ' and Γ does not affect the absorption for $t\gg 1/\Gamma$. Thus for the initial condition given by Eq. (4)

$$|b_r|^2=\frac{|B_r|^2|A_{s0}|^2}{(\nu-\nu_0)^2+\Gamma^2}\left\{\frac{1+e^{-4\pi\gamma t}-2e^{-2\pi\gamma t}\cos 2\pi(\nu_r-\nu_0)t}{(\nu_r-\nu_0)^2+\gamma^2}+\frac{1+e^{-4\pi\Gamma' t}-2e^{-2\pi\Gamma' t}\cos 2\pi(\nu_r-\nu_0)t}{(\nu_r-\nu)^2+\Gamma'^2}+\text{cross product term}\right\}.$$

Only the first fraction in the curly brackets contributes to the steady increase of $\Sigma|b_r|^2$ in times $\gg 1/\Gamma$. Its contribution is

$$\frac{\pi|A_{s0}|^2\overline{|B_r|^2}}{\gamma[(\nu-\nu_0)^2+\Gamma^2]\Delta\nu_r}(1-e^{-4\pi\gamma t}).$$

The last factor is for practical purposes $4\pi\gamma t$. The second and third terms in the curly bracket give terms $\exp(-4\pi\Gamma' t)$, $\exp(-2\pi\Gamma' t)$ and constants. The first two kinds die off and the last kind represents the effect of transients which do not matter in the long run, so that for times not too large as compared with $1/\gamma$ and yet great as compared with $1/\Gamma$ one may consider the rates of change of $\Sigma|b_r|^2$ and of $\Sigma'|a_s|^2$ to be $4\pi\gamma\Gamma_r/\Gamma$ and $4\pi\gamma\Gamma_s/\Gamma$. These are the results used in the text.

Energy Production in Stars

In several recent papers,[1–3] the present author has been quoted for investigations on the nuclear reactions responsible for the energy production in stars. As the publication of this work which was carried out last spring has been unduly delayed, it seems worth while to publish a short account of the principal results.

The most important source of stellar energy appears to be the reaction cycle:

$$\begin{aligned} &C^{12}+H^1=N^{13}\ (a),\ N^{13}=C^{13}+\epsilon^+\ (b)\\ &C^{13}+H^1=N^{14}\ (c)\\ &N^{14}+H^1=O^{15}\ (d),\ O^{15}=N^{15}+\epsilon^+\ (e)\\ &N^{15}+H^1=C^{12}+He^4\ (f). \end{aligned} \qquad (1)$$

In this cycle, four protons are combined into one α-particle (plus two positrons which will be annihilated by two electrons). The carbon and nitrogen isotopes serve as catalysts for this combination. There are no alternative reactions between protons and the nuclei $C^{12}C^{13}N^{14}$; with N^{15}, there is the alternative process

$$N^{15}+H^1=O^{16},$$

but this radiative capture may be expected to be about 10,000 times less probable than the particle reaction (f). Thus practically no carbon and nitrogen will be consumed and the energy production will continue until all protons in the star are used up. At the present rate of energy production, the hydrogen content of the sun (35 percent by weight[4]) would suffice for 3.5×10^{10} years.

The reaction cycle (1) is preferred before all other nuclear reactions. Any element *lighter* than carbon, when reacting with protons, is destroyed permanently and will not be replaced. E.g., Be^9 would react in the following way:

$$\begin{aligned} &Be^9+H^1=Li^6+He^4\\ &Li^6+H^1=Be^7\\ &Be^7+\epsilon^-=Li^7\\ &Li^7+H^1=2He^4. \end{aligned}$$

Therefore, even if the star contained an appreciable amount of Li, Be or B when it was first formed, these elements would have been consumed in the early history of the star. This agrees with the extremely low abundance of these elements (if any) in the present stars. These considerations apply also to the heavy hydrogen isotopes H^2 and H^3.

The only abundant and very light elements are H^1 and He^4. Of these, He^4 will not react with protons at all because Li^5 is unstable, and the reaction between two protons, while possible, is rather slow[5] and will therefore be much less important[6] in ordinary stars than the cycle (1).

Elements heavier than nitrogen may be left out of consideration entirely because they will react more slowly with protons than carbon and nitrogen, even at temperatures much higher than those prevailing in stars. For the same reason, reactions between α-particles and other nuclei are of no importance.

To test the theory, we have calculated (Table I) the energy production in the sun for several nuclear reactions, making the following assumptions:

(1) The temperature at the center of the sun is 2×10^7 degrees. This value follows from the integration of the

TABLE I. *Energy production in the sun for several nuclear reactions.*

Reaction	Average Energy Production ϵ (erg/g sec.)
$H^1+H^1=H^2+\epsilon^++f.$*	0.2
$H^2+H^1=He^3$	3×10^{16}
$Li^7+H^1=2He^4$	4×10^{14}
$B^{10}+H^1=C^{11}+f.$	3×10^{8}
$B^{11}+H^1=3He^4$	10^{10}
$N^{14}+H^1=O^{15}+f.$	3
$O^{16}+H^1=F^{17}+f.$	10^{-4}

* "$+f.$" means that the energy production in the reactions following the one listed, is included. E.g. the figure for the $N^{14}+H^1$ includes the complete chain (1).

Eddington equations with any reasonable "star model."[4] The "point source model" with a convective core which is a very good approximation to reality gives 2.03×10^7 degrees.[7] The same calculation gives 50.2 for the density at the center of the sun. The central temperature is probably correct to within 10 percent.

(2) The concentration of hydrogen is assumed to be 35 percent by weight, that of the other reacting element 10 percent. In the reaction chain (1), the concentration of N^{14} was assumed to be 10 percent.

(3) The ratio of the average energy production to the production at the center was calculated[7] from the temperature-density dependence of the nuclear reaction and the temperature-density distribution in the star.

It is evident from Table I that only the nitrogen reaction gives agreement with the observed energy production of 2 ergs/g sec. All the reactions with lighter elements would give energy productions which are too large by many orders of magnitude if they were abundant enough, whereas the next heavier element, O^{16}, already gives more than 10,000 times too small a value. In view of the extremely strong dependence on the atomic number, the agreement of the nitrogen-carbon cycle with observation is excellent.

The nitrogen-carbon reactions also explain correctly the dependence of mass on luminosity for main sequence stars. In this connection, the strong dependence of the reaction rate on temperature ($\sim T^{18}$) is important, because massive stars have much greater luminosities with only slightly higher central temperatures (e.g., Y Cygni has $T=3.2\times10^7$ and $\epsilon=1200$ ergs/g sec.).

With the assumed reaction chain, there will be no appreciable change in the abundance of elements heavier than helium during the evolution of the star but only a transmutation of hydrogen into helium. This result which is more general than the reaction chain (1) is in contrast to the commonly accepted "Aufbauhypothese."

A detailed account of these investigations will be published soon.

H. A. BETHE

Cornell University,
Ithaca, New York,
December 15, 1938.

[1] C. F. v. Weizsaecker, Physik. Zeits. 39, 639 (1938).
[2] J. Oppenheimer and R. Serber, Phys. Rev. 54, 540 (1938).
[3] G. Gamow, Phys. Rev. (in print).
[4] B. Strömgren, Ergebn. d. Exak. Naturwiss. 16 (1937).
[5] H. Bethe and C. Critchfield, Phys. Rev. 54, 248 (1938).
[6] Only for very cool stars (red dwarfs) the H+H reaction may become important.
[7] The author is indebted to Mr. Marshak for these calculations.

The Mechanism of Nuclear Fission

NIELS BOHR
University of Copenhagen, Copenhagen, Denmark, and The Institute for Advanced Study, Princeton, New Jersey

AND

JOHN ARCHIBALD WHEELER
Princeton University, Princeton, New Jersey

(Received June 28, 1939)

On the basis of the liquid drop model of atomic nuclei, an account is given of the mechanism of nuclear fission. In particular, conclusions are drawn regarding the variation from nucleus to nucleus of the critical energy required for fission, and regarding the dependence of fission cross section for a given nucleus on energy of the exciting agency. A detailed discussion of the observations is presented on the basis of the theoretical considerations. Theory and experiment fit together in a reasonable way to give a satisfactory picture of nuclear fission.

INTRODUCTION

THE discovery by Fermi and his collaborators that neutrons can be captured by heavy nuclei to form new radioactive isotopes led especially in the case of uranium to the interesting finding of nuclei of higher mass and charge number than hitherto known. The pursuit of these investigations, particularly through the work of Meitner, Hahn, and Strassmann as well as Curie and Savitch, brought to light a number of unsuspected and startling results and finally led Hahn and Strassmann[1] to the discovery that from uranium elements of much smaller atomic weight and charge are also formed.

The new type of nuclear reaction thus discovered was given the name "fission" by Meitner and Frisch,[2] who on the basis of the liquid drop model of nuclei emphasized the analogy of the process concerned with the division of a fluid sphere into two smaller droplets as the result of a deformation caused by an external disturbance. In this connection they also drew attention to the fact that just for the heaviest nuclei the mutual repulsion of the electrical charges will to a large extent annul the effect of the short range nuclear forces, analogous to that of surface tension, in opposing a change of shape of the nucleus. To produce a critical deformation will therefore require only a comparatively small energy, and by the subsequent division of the nucleus a very large amount of energy will be set free.

Just the enormous energy release in the fission process has, as is well known, made it possible to observe these processes directly, partly by the great ionizing power of the nuclear fragments, first observed by Frisch[3] and shortly afterwards independently by a number of others, partly by the penetrating power of these fragments which allows in the most efficient way the separation from the uranium of the new nuclei formed by the fission.[4] These products are above all characterized by their specific beta-ray activities which allow their chemical and spectrographic identification. In addition, however, it has been found that the fission process is accompanied by an emission of neutrons, some of which seem to be directly associated with the fission, others associated with the subsequent beta-ray transformations of the nuclear fragments.

In accordance with the general picture of nuclear reactions developed in the course of the last few years, we must assume that any nuclear transformation initiated by collisions or irradiation takes place in two steps, of which the first is the formation of a highly excited compound nucleus with a comparatively long lifetime, while

[1] O. Hahn and F. Strassmann, Naturwiss. **27**, 11 (1939); see, also, P. Abelson, Phys. Rev. **55**, 418 (1939).

[2] L. Meitner and O. R. Frisch, Nature **143**, 239 (1939).

[3] O. R. Frisch, Nature **143**, 276 (1939); G. K. Green and Luis W. Alvarez, Phys. Rev. **55**, 417 (1939); R. D. Fowler and R. W. Dodson, Phys. Rev. **55**, 418 (1939); R. B. Roberts, R. C. Meyer and L. R. Hafstad, Phys. Rev. **55**, 417 (1939); W. Jentschke and F. Prankl, Naturwiss. **27**, 134 (1939); H. L. Anderson, E. T. Booth, J. R. Dunning, E. Fermi, G. N. Glasoe and F. G. Slack, Phys. Rev. **55**, 511 (1939).

[4] F. Joliot, Comptes rendus **208**, 341 (1939); L. Meitner and O. R. Frisch, Nature **143**, 471 (1939); H. L. Anderson, E. T. Booth, J. R. Dunning, E. Fermi, G. N. Glasoe and F. G. Slack, Phys. Rev. **55**, 511 (1939).

the second consists in the disintegration of this compound nucleus or its transition to a less excited state by the emission of radiation. For a heavy nucleus the disintegrative processes of the compound system which compete with the emission of radiation are the escape of a neutron and, according to the new discovery, the fission of the nucleus. While the first process demands the concentration on one particle at the nuclear surface of a large part of the excitation energy of the compound system which was initially distributed much as is thermal energy in a body of many degrees of freedom, the second process requires the transformation of a part of this energy into potential energy of a deformation of the nucleus sufficient to lead to division.[5]

Such a competition between the fission process and the neutron escape and capture processes seems in fact to be exhibited in a striking manner by the way in which the cross section for fission of thorium and uranium varies with the energy of the impinging neutrons. The remarkable difference observed by Meitner, Hahn, and Strassmann between the effects in these two elements seems also readily explained on such lines by the presence in uranium of several stable isotopes, a considerable part of the fission phenomena being reasonably attributable to the rare isotope U^{235} which, for a given neutron energy, will lead to a compound nucleus of higher excitation energy and smaller stability than that formed from the abundant uranium isotope.[6]

In the present article there is developed a more detailed treatment of the mechanism of the fission process and accompanying effects, based on the comparison between the nucleus and a liquid drop. The critical deformation energy is brought into connection with the potential energy of the drop in a state of unstable equilibrium, and is estimated in its dependence on nuclear charge and mass. Exactly how the excitation energy originally given to the nucleus is gradually exchanged among the various degrees of freedom and leads eventually to a critical deformation proves to be a question which needs not be discussed in order to determine the fission probability. In fact, simple statistical considerations lead to an approximate expression for the fission reaction rate which depends only on the critical energy of deformation and the properties of nuclear energy level distributions. The general theory presented appears to fit together well with the observations and to give a satisfactory description of the fission phenomenon.

For a first orientation as well as for the later considerations, we estimate quantitatively in Section I by means of the available evidence the energy which can be released by the division of a heavy nucleus in various ways, and in particular examine not only the energy released in the fission process itself, but also the energy required for subsequent neutron escape from the fragments and the energy available for beta-ray emission from these fragments.

In Section II the problem of the nuclear deformation is studied more closely from the point of view of the comparison between the nucleus and a liquid droplet in order to make an estimate of the energy required for different nuclei to realize the critical deformation necessary for fission.

In Section III the statistical mechanics of the fission process is considered in more detail, and an approximate estimate made of the fission probability. This is compared with the probability of radiation and of neutron escape. A discussion is then given on the basis of the theory for the variation with energy of the fission cross section.

In Section IV the preceding considerations are applied to an analysis of the observations of the cross sections for the fission of uranium and thorium by neutrons of various velocities. In particular it is shown how the comparison with the theory developed in Section III leads to values for the critical energies of fission for thorium and the various isotopes of uranium which are in good accord with the considerations of Section II.

In Section V the problem of the statistical distribution in size of the nuclear fragments arising from fission is considered, and also the questions of the excitation of these fragments and the origin of the secondary neutrons.

Finally, we consider in Section VI the fission effects to be expected for other elements than thorium and uranium at sufficiently high neutron velocities as well as the effect to be anticipated in

[5] N. Bohr, Nature **143**, 330 (1939).
[6] N. Bohr, Phys. Rev. **55**, 418 (1939).

thorium and uranium under deuteron and proton impact and radiative excitation.

I. Energy Released by Nuclear Division

The total energy released by the division of a nucleus into smaller parts is given by

$$\Delta E = (M_0 - \Sigma M_i)c^2, \tag{1}$$

where M_0 and M_i are the masses of the original and product nuclei at rest and unexcited. We have available no observations on the masses of nuclei with the abnormal charge to mass ratio formed for example by the division of such a heavy nucleus as uranium into two nearly equal parts. The difference between the mass of such a fragment and the corresponding stable nucleus of the same mass number may, however, if we look apart for the moment from fluctuations in energy due to odd-even alternations and the finer details of nuclear binding, be reasonably assumed, according to an argument of Gamow, to be representable in the form

$$M(Z, A) - M(Z_A, A) = \tfrac{1}{2}B_A(Z - Z_A)^2, \tag{2}$$

where Z is the charge number of the fragment and Z_A is a quantity which in general will not be an integer. For the mass numbers $A = 100$ to 140 this quantity Z_A is given by the dotted line in Fig. 8, and in a similar way it may be determined for lighter and heavier mass numbers.

B_A is a quantity which cannot as yet be determined directly from experiment but may be estimated in the following manner. Thus we may assume that the energies of nuclei with a given mass A will vary with the charge Z approximately according to the formula

TABLE I. *Values of the quantities which appear in Eqs.* (6) *and* (7), *estimated for various values of the nuclear mass number A. Both* B_A and δ_A *are in Mev.*

A	Z_A	B_A	δ_A	A	Z_A	B_A	δ_A
50	23.0	$3._5$	$2._8$	150	62.5	1.2	$1._5$
60	27.5	$3._2$	$2._8$	160	65.4	1.1	$1._3$
70	31.2	$2._6$	$2._7$	170	69.1	1.1	$1._2$
80	35.0	$2._3$	$2._7$	180	72.9	1.0	$1._2$
90	39.4	$2._0$	$2._7$	190	76.4	1.0	$1._1$
100	44.0	$2._0$	$2._6$	200	80.0	0.9_5	$1._1$
110	47.7	1.7	$2._4$	210	83.5	0.9_2	$1._1$
120	50.8	1.5	$2._1$	220	87.0	0.8_8	$1._1$
130	53.9	1.3	$1._9$	230	$90._6$	0.8_6	$1._0$
140	58.0	1.2	$1._8$	240	$93._9$	0.8_3	$1._0$

$$M(Z, A) = C_A + \tfrac{1}{2}B_A'(Z - \tfrac{1}{2}A)^2 + (Z - \tfrac{1}{2}A)(M_p - M_n) + 3Z^2e^2/5r_0A^{\frac{1}{3}}. \tag{3}$$

Here the second term gives the comparative masses of the various isobars neglecting the influence of the difference $M_p - M_n$ of the proton and neutron mass included in the third term and of the pure electrostatic energy given by the fourth term. In the latter term the usual assumption is made that the effective radius of the nucleus is equal to $r_0A^{\frac{1}{3}}$, with r_0 estimated as 1.48×10^{-13} from the theory of alpha-ray disintegration. Identifying the relative mass values given by expressions (2) and (3), we find

$$B_A' = (M_p - M_n + 6Z_Ae^2/5r_0A^{\frac{1}{3}})/(\tfrac{1}{2}A - Z_A) \tag{4}$$

and

$$B_A = B_A' + 6e^2/5r_0A^{\frac{1}{3}} = (M_p - M_n + 3A^{\frac{2}{3}}e^2/5r_0)/(\tfrac{1}{2}A - Z_A). \tag{5}$$

The values of B_A obtained for various nuclei from this last relation are listed in Table I.

On the basis just discussed, we shall be able to estimate the mass of the nucleus (Z, A) with the help of the packing fraction of the known nuclei. Thus we may write

$$M(Z, A) = A(1 + f_A) + \tfrac{1}{2}B_A(Z - Z_A)^2 - \tfrac{1}{2}\delta_A \left\{\begin{array}{ll} +0 & A \text{ odd} \\ & A \text{ even, } Z \text{ even} \\ +\tfrac{1}{2}\delta_A & A \text{ even, } Z \text{ odd} \end{array}\right\}, \tag{6}$$

where f_A is to be taken as the average value of the packing fraction over a small region of atomic weights and the last term allows for the typical differences in binding energy among nuclei according to the odd and even character of their neutron and proton numbers. In using Dempster's measurements of packing fractions we must recognize that the average value of the second term in (6) is included in such measurements.[7] This correction, however, is, as may be read from Fig. 8, practically compensated by the influence of the third term, owing to the fact that the great majority of nuclei studied in the mass spectrograph are of even-even character.

From (6) we find the energy release involved in electron emission or absorption by a nucleus unstable with respect to a beta-ray

[7] A. J. Dempster, Phys. Rev. 53, 869 (1938).

transformation:

$$E_\beta = B_A\{|Z_A - Z| - \tfrac{1}{2}\} - \delta_A \left\{\begin{array}{ll} +0 & A \text{ odd} \\ & A \text{ even, } Z \text{ even} \\ +\delta_A & A \text{ even, } Z \text{ odd} \end{array}\right\}. \quad (7)$$

This result gives us the possibility of estimating δ_A by an examination of the stability of isobars of even nuclei. In fact, if an even-even nucleus is stable or unstable, then δ_A is, respectively, greater or less than $B_A\{|Z_A - A| - \tfrac{1}{2}\}$. For nuclei of medium atomic weight this condition brackets δ_A very closely; for the region of very high mass numbers, on the other hand, we can estimate δ_A directly from the difference in energy release of the successive beta-ray transformations

$$UX_I \rightarrow (UX_{II},\ UZ) \rightarrow U_{II},$$
$$MsTh_I \rightarrow MsTh_{II} \rightarrow RaTh,\ RaD \rightarrow RaE \rightarrow RaF.$$

The estimated values of δ_A are collected in Table I.

Applying the available measurements on nuclear masses supplemented by the above considerations, we obtain typical estimates as shown in Table II for the energy release on division of a nucleus into two approximately equal parts.[8]

Below mass number $A \sim 100$ nuclei are energetically stable with respect to division; above this limit energetic instability sets in with respect

TABLE II. *Estimates for the energy release on division of typical nuclei into two fragments are given in the third column. In the fourth is the estimated value of the total additional energy release associated with the subsequent beta-ray transformations. Energies are in Mev.*

ORIGINAL	TWO PRODUCTS	DIVISION	SUBSEQUENT
$_{28}Ni^{61}$	$_{14}Si^{30,\,31}$	−11	2
$_{50}Sn^{117}$	$_{25}Mn^{58,\,59}$	10	12
$_{68}Er^{167}$	$_{34}Se^{83,\,84}$	94	13
$_{82}Pb^{206}$	$_{41}Nb^{103,\,103}$	120	32
$_{92}U^{239}$	$_{46}Pd^{119,\,120}$	200	31

to division into two nearly equal fragments, essentially because the decrease in electrostatic

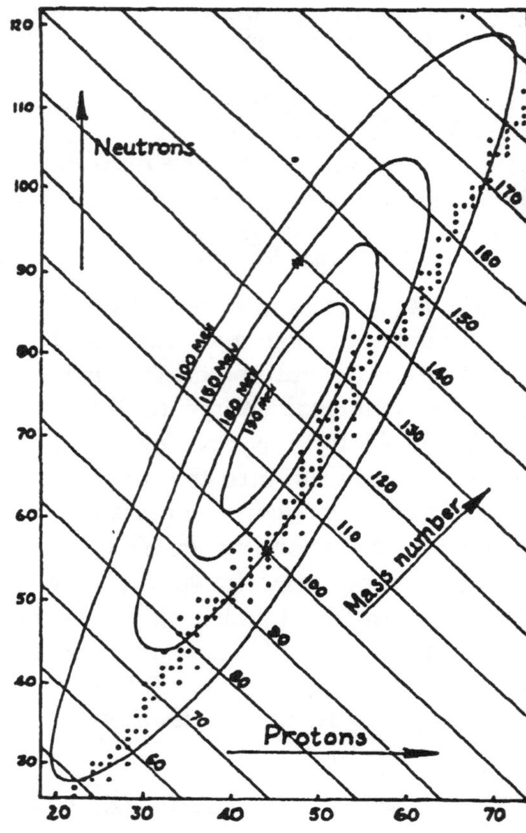

FIG. 1. The difference in energy between the nucleus $_{92}U^{239}$ in its normal state and the possible fragment nuclei $_{44}Ru^{100}$ and $_{48}Cd^{139}$ (indicated by the crosses in the figure) is estimated to be 150 Mev as shown by the corresponding contour line. In a similar way the estimated energy release for division of U^{239} into other possible fragments can be read from the figure. The region in the chart associated with the greatest energy release is seen to be at a distance from the region of the stable nuclei (dots in the figure) corresponding to the emission of from three to five beta-rays.

energy associated with the separation overcompensates the desaturation of short range forces consequent on the greater exposed nuclear surface. The energy evolved on division of the nucleus U^{239} into two fragments of any given charge and mass numbers is shown in Fig. 1. It is seen that there is a large range of atomic masses for which the energy liberated reaches nearly the maximum attainable value 200 Mev; but that for a given size of one fragment there is only a small range of charge numbers which correspond to an energy release at all near the maximum value. Thus the fragments formed by division of uranium in the *energetically* most favorable way lie in a narrow band in Fig. 1, separated from the region of the stable nuclei by an amount which corresponds to the change in nuclear charge

[8] Even if there is no question of actual fission processes by which nuclei break up into more than two comparable parts, it may be of interest to point out that such divisions in many cases would be accompanied by the release of energy. Thus nuclei of mass number greater than $A = 110$ are unstable with respect to division into three nearly equal parts. For uranium the corresponding total energy liberation will be $\sim$210 Mev, and thus is even somewhat greater than the release on division into two parts. The energy evolution on division of U^{239} into four comparable parts will, however, be about 150 Mev, and already division into as many as 15 comparable parts will be endothermic.

associated with the emission of three to six beta-particles.

The amount of energy released in the beta-ray transformations following the creation of the fragment nuclei may be estimated from Eq. (7), using the constants in Table I. Approximate values obtained in this way for the energy liberation in typical chains of beta-disintegrations are shown on the arrows in Fig. 8.

The magnitude of the energy available for beta-ray emission from typical fragment nuclei does not stand in conflict with the stability of these nuclei with respect to spontaneous neutron emission, as one sees at once from the fact that the energy change associated with an increase of the nuclear charge by one unit is given by the *difference* between binding energy of a proton and of a neutron, plus the neutron-proton mass difference. A direct estimate from Eq. (6) of the binding energy of a neutron in typical nuclear fragments lying in the band of greatest energy release (Fig. 1) gives the results summarized in the last column of Table III. The comparison of the figures in this table shows that the neutron binding is in certain cases considerably smaller than the energy which can be released by beta-ray transformation. This fact offers a reasonable explanation as we shall see in Section V for the delayed neutron emission accompanying the fission process.

II. Nuclear Stability with Respect to Deformations

According to the liquid drop model of atomic nuclei, the excitation energy of a nucleus must be

TABLE III. *Estimated values of energy release in beta-ray transformations and energy of neutron binding in final nucleus, in typical cases; also estimates of the neutron binding in the dividing nucleus. Values in Mev.*

Beta-Transition		Release	Binding
$_{40}Zr^{99}$	$_{41}Nb^{99}$	6.3	8.2
$_{41}Nb^{100}$	$_{42}Mo^{100}$	7.8	8.6
$_{46}Pd^{125}$	$_{47}Ag^{125}$	7.8	6.7
$_{47}Ag^{125}$	$_{48}Cd^{125}$	6.5	5.0
$_{49}In^{130}$	$_{50}Sn^{130}$	7.6	7.1
$_{52}Te^{140}$	$_{53}I^{140}$	5.0	3.5
$_{53}I^{140}$	$_{54}Xe^{140}$	7.4	5.9
Compound Nucleus			
	$_{92}U^{235}$		5.4
	$_{92}U^{236}$		6.4
	$_{92}U^{239}$		5.2
	$_{90}Th^{233}$		5.2
	$_{91}Pa^{232}$		6.4

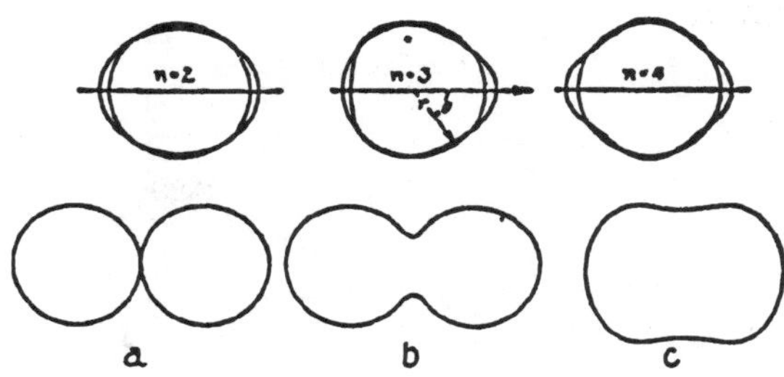

FIG. 2. Small deformations of a liquid drop of the type $\delta r(\theta) = \alpha_n P_n(\cos\theta)$ (upper portion of the figure) lead to characteristic oscillations of the fluid about the spherical form of stable equilibrium, even when the fluid has a uniform electrical charge. If the charge reaches the critical value (10×surface tension×volume)$^{\frac{1}{2}}$, however, the spherical form becomes unstable with respect to even infinitesimal deformations of the type $n=2$. For a slightly smaller charge, on the other hand, a finite deformation (c) will be required to lead to a configuration of *unstable equilibrium*, and with smaller and smaller charge densities the critical form gradually goes over (c, b, a) into that of two uncharged spheres an infinitesimal distance from each other (a).

expected to give rise to modes of motion of the nuclear matter similar to the oscillations of a fluid sphere under the influence of surface tension.[9] For heavy nuclei the high nuclear charge will, however, give rise to an effect which will to a large extent counteract the restoring force due to the short range attractions responsible for the surface tension of nuclear matter. This effect, the importance of which for the fission phenomenon was stressed by Frisch and Meitner, will be more closely considered in this section, where we shall investigate the stability of a nucleus for small deformations of various types[10] as well as for such large deformations that division may actually be expected to occur.

Consider a small arbitrary deformation of the liquid drop with which we compare the nucleus such that the distance from the center to an arbitrary point on the surface with colatitude θ is changed (see Fig. 2) from its original value R

[9] N. Bohr, Nature 137, 344 and 351 (1936); N. Bohr and F. Kalckar, Kgl. Danske Vid. Selskab., Math. Phys. Medd. 14, No. 10 (1937).

[10] After the formulae given below were derived, expressions for the potential energy associated with spheroidal deformations of nuclei were published by E. Feenberg (Phys. Rev. 55, 504 (1939)) and F. Weizsäcker (Naturwiss. 27, 133 (1939)). Further, Professor Frenkel in Leningrad has kindly sent us in manuscript a copy of a more comprehensive paper on various aspects of the fission problem, to appear in the U.S.S.R. "Annales Physicae," which contains a deduction of Eq. (9) below for nuclear stability against arbitrary small deformations, as well as some remarks, similar to those made below (Eq. (14)) about the shape of a drop corresponding to unstable equilibrium. A short abstract of this paper has since appeared in Phys. Rev. 55, 987 (1939).

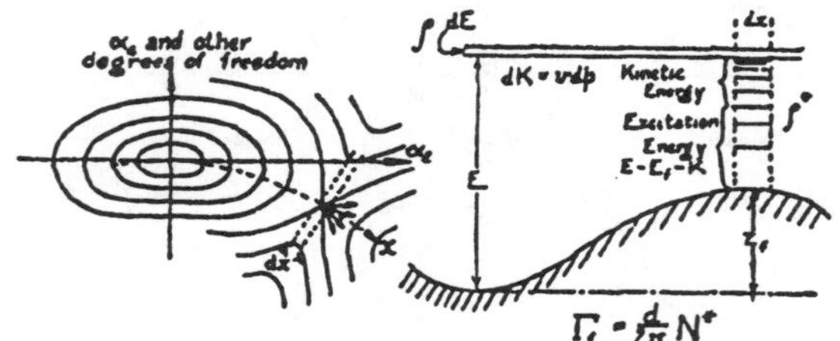

FIG. 3. The potential energy associated with any arbitrary deformation of the nuclear form may be plotted as a function of the parameters which specify the deformation, thus giving a contour surface which is represented schematically in the left-hand portion of the figure. The pass or saddle point corresponds to the critical deformation of unstable equilibrium. To the extent to which we may use classical terms, the course of the fission process may be symbolized by a ball lying in the hollow at the origin of coordinates (spherical form) which receives an impulse (neutron capture) which sets it to executing a complicated Lissajous figure of oscillation about equilibrium. If its energy is sufficient, it will in the course of time happen to move in the proper direction to pass over the saddle point (after which fission will occur), unless it loses its energy (radiation or neutron re-emission). At the right is a cross section taken through the fission barrier, illustrating the calculation in the text of the probability per unit time of fission occurring.

to the value

$$r(\theta)=R[1+\alpha_0+\alpha_2P_2(\cos\theta)+\alpha_3P_3(\cos\theta)+\cdots], \quad (8)$$

where the α_n are small quantities. Then a straightforward calculation shows that the surface energy plus the electrostatic energy of the comparison drop has increased to the value

$$\begin{aligned}E_{S+E}=&4\pi(r_0A^{\frac{1}{3}})^2O[1+2\alpha_2^2/5+5\alpha_3^2/7+\cdots\\&+(n-1)(n+2)\alpha_n^2/2(2n+1)+\cdots]\\&+3(Ze)^2/5r_0A^{\frac{1}{3}}[1-\alpha_2^2/5-10\alpha_3^2/49-\cdots\\&-5(n-1)\alpha_n^2/(2n+1)^2-\cdots],\end{aligned} \quad (9)$$

where we have assumed that the drop is composed of an incompressible fluid of volume $(4\pi/3)R^3=(4\pi/3)r_0^3A$, uniformly electrified to a charge Ze, and possessing a surface tension O. Examination of the coefficient of α_2^2 in the above expression for the distortion energy, namely,

$$4\pi r_0^2OA^{\frac{2}{3}}(2/5)\{1-(Z^2/A)\times[e^2/10(4\pi/3)r_0^3O]\} \quad (10)$$

makes it clear that with increasing value of the ratio Z^2/A we come finally to a limiting value

$$(Z^2/A)_{\text{limiting}}=10(4\pi/3)r_0^3O/e^2, \quad (11)$$

beyond which the nucleus is no longer stable with respect to deformations of the simplest type. The actual value of the numerical factors can be calculated with the help of the semi-empirical formula given by Bethe for the respective contributions to nuclear binding energies due to electrostatic and long range forces, the influence of the latter being divided into volume and surface effects. A revision of the constants in Bethe's formula has been carried through by Feenberg[11] in such a way as to obtain the best agreement with the mass defects of Dempster; he finds

$$r_0 \doteqdot 1.4\times10^{-13}\text{ cm}, \quad 4\pi r_0^2O\doteqdot 14\text{ Mev}. \quad (12)$$

From these values a limit for the ratio Z^2/A is obtained which is 17 percent greater than the ratio $(92)^2/238$ characterizing U^{238}. Thus we can conclude that nuclei such as those of uranium and thorium are indeed near the limit of stability set by the exact compensation of the effects of electrostatic and short range forces. On the other hand, we cannot rely on the precise value of the limit given by these semi-empirical and indirect determinations of the ratio of surface energy to electrostatic energy, and we shall investigate below a method of obtaining the ratio in question from a study of the fission phenomenon itself.

Although nuclei for which the quantity Z^2/A is slightly less than the limiting value (11) are stable with respect to small arbitrary deformations, a larger deformation will give the long range repulsions more advantage over the short range attractions responsible for the surface tension, and it will therefore be possible for the nucleus, when suitably deformed, to divide spontaneously. Particularly important will be that critical deformation for which the nucleus is just on the verge of division. The drop will then possess a shape corresponding to unstable equilibrium: the work required to produce any infinitesimal displacement from this equilibrium configuration vanishes in the first order. To examine this point in more detail, let us consider the surface obtained by plotting the potential energy of an arbitrary distortion as a function of the parameters which specify its form and magnitude. Then we have to recognize the fact that the

[11] E. Feenberg, Phys. Rev. 55, 504 (1939).

potential barrier hindering division is to be compared with a pass or saddle point leading between two potential valleys on this surface. The energy relations are shown schematically in Fig. 3, where of course we are able to represent only two of the great number of parameters which are required to describe the shape of the system. The deformation parameters corresponding to the saddle point give us the critical form of the drop, and the potential energy required for this distortion we will term the critical energy for fission, E_f. If we consider a continuous change in the shape of the drop, leading from the original sphere to two spheres of half the size at infinite separation, then the critical energy in which we are interested is the lowest value which we can at all obtain, by suitable choice of this sequence of shapes, for the energy required to lead from the one configuration to the other.

Simple dimensional arguments show that the critical deformation energy for the droplet corresponding to a nucleus of given charge and mass number can be written as the product of the surface energy by a dimensionless function of the charge mass ratio:

$$E_f = 4\pi r_0^2 O A^{2/3} f\{(Z^2/A)/(Z^2/A)_{\text{limiting}}\}. \quad (13)$$

We can determine E_f if we know the shape of the nucleus in the critical state; this will be given by solution of the well-known equation for the form of a surface in equilibrium under the action of a surface tension O and volume forces described by a potential φ:

$$\kappa O + \varphi = \text{constant}, \quad (14)$$

where κ is the total normal curvature of the surface. Because of the great mathematical difficulties of treating large deformations, we are however able to calculate the critical surface and the dimensionless function f in (13) only for certain special values of the argument, as follows: (1) if the volume potential in (14) vanishes altogether, we see from (14) that the surface of unstable equilibrium has constant curvature; we have in fact to deal with a division of the fluid into spheres. Thus, when there are no electrostatic forces at all to aid the fission, the critical energy for division into two equal fragments will just equal the total work done against surface tension in the separation process, i.e.,

$$E_f = 2 \cdot 4\pi r_0^2 O (A/2)^{2/3} - 4\pi r_0^2 O A^{2/3}. \quad (15)$$

From this it follows that

$$f(0) = 2^{1/3} - 1 = 0.260. \quad (16)$$

(2) If the charge on the droplet is not zero, but is still very small, the critical shape will differ little from that of two spheres in contact. There will in fact exist only a narrow neck of fluid connecting the two portions of the figure, the radius of which, r_n, will be such as to bring about equilibrium; to a first approximation

$$2\pi r_n O = (Ze/2)^2/(2r_0(A/2)^{1/3})^2 \quad (17)$$

or

$$r_n/r_0 A^{1/3} = 0.66\left(\frac{Z^2}{A}\right) \Big/ \left(\frac{Z^2}{A}\right)_{\text{limiting}} \quad (18)$$

To calculate the critical energy to the first order in Z^2/A, we can omit the influence of the neck as producing only a second-order change in the energy. Thus we need only compare the sum of surface and electrostatic energy for the original nucleus with the corresponding energy for two spherical nuclei of half the size in contact with each other. We find

$$\begin{aligned} E_f = {} & 2 \cdot 4\pi r_0^2 O (A/2)^{2/3} - 4\pi r_0^2 O A^{2/3} \\ & + 2 \cdot 3(Ze/2)^2/5r_0(A/2)^{1/3} \\ & + (Ze/2)^2/2r_0(A/2)^{1/3} - 3(Ze)^2/5r_0 A^{1/3}, \end{aligned} \quad (19)$$

from which

$$E_f/4\pi r_0^2 O A^{2/3} \equiv f(x) = 0.260 - 0.215x, \quad (20)$$

provided

$$x = \left(\frac{Z^2}{A}\right) \Big/ \left(\frac{Z^2}{A}\right)_{\text{limiting}} = (\text{charge})^2/\text{surface tension} \times \text{volume} \times 10 \quad (21)$$

is a small quantity. (3) In the case of greatest actual interest, when Z^2/A is very close to the critical value, only a small deformation from a spherical form will be required to reach the critical state. According to Eq. (9), the potential energy required for an infinitesimal distortion will increase as the square of the amplitude, and

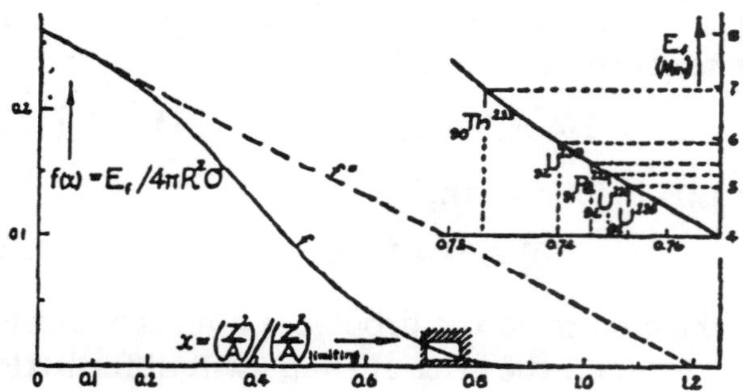

FIG. 4. The energy E_f required to produce a critical deformation leading to fission is divided by the surface energy $4\pi R^2O$ to obtain a dimensionless function of the quantity $x=(\text{charge})^2/(10\times\text{volume}\times\text{surface tension})$. The behavior of the function $f(x)$ is calculated in the text for $x=0$ and $x=1$, and a smooth curve is drawn here to connect these values. The curve $f^*(x)$ determines for comparison the energy required to deform the nucleus into two spheres in contact with each other. Over the cross-hatched region of the curve of interest for the heaviest nuclei the surface energy changes but little. Taking for it a value of 530 Mev, we obtain the energy scale in the upper part of the figure. In Section IV we estimate from the observations a value $E_f \sim 6$ Mev for U^{239}. Using the figure we thus find $(Z^2/A)_{\text{limiting}}=47.8$ and can estimate the fission barriers for other nuclei, as shown.

will moreover have the smallest possible value for a displacement of the form $P_2(\cos\theta)$. To find the deformation for which the potential energy has reached a maximum and is about to decrease, we have to carry out a more accurate calculation. We obtain for the distortion energy, accurate to the fourth order in α_2, the expression

$$\begin{aligned}\Delta E_{S+E}=&4\pi r_0^2OA^{2/3}[2\alpha_2^2/5+116\alpha_2^3/105\\&+101\alpha_2^4/35+2\alpha_2^2\alpha_4/35+\alpha_4^2]\\&-3(Ze)^2/5r_0A^{1/3}[\alpha_2^2/5+64\alpha_2^3/105\\&+58\alpha_2^4/35+8\alpha_2^2\alpha_4/35+5\alpha_4^2/27], \quad (22)\end{aligned}$$

in which it will be noted that we have had to include the terms in α_4^2 because of the coupling which sets in between the second and fourth modes of motion for appreciable amplitudes. Thus, on minimizing the potential energy with respect to α_4, we find

$$\alpha_4=-(243/595)\alpha_2^2 \quad (23)$$

in accordance with the fact that as the critical form becomes more elongated with decreasing Z^2/A, it must also develop a concavity about its equatorial belt such as to lead continuously with variation of the nuclear charge to the dumbbell shaped figure discussed in the preceding paragraph.

With the help of (23) we obtain the deformation energy as a function of α_2 alone. By a straightforward calculation we then find its maximum value as a function of α_2, thus determining the energy required to produce a distortion on the verge of leading to fission:

$$\begin{aligned}E_f/4\pi r_0^2OA^{2/3}=f(x)=&98(1-x)^3/135\\&-11368(1-x)^4/34425+\cdots \quad (24)\end{aligned}$$

for values of Z^2/A near the instability limit.

Interpolating in a reasonable way between the two limiting values which we have obtained for the critical energy for fission, we obtain the curve of Fig. 4 for f as a function of the ratio of the square of the charge number of the nucleus to its mass number. The upper part of the figure shows the interesting portion of the curve in enlargement and with a scale of energy values at the right based on the surface tension estimate of Eq. (12) and a nuclear mass of $A=235$. The slight variation of the factor $4\pi r_0^2OA^{2/3}$ among the various thorium and uranium isotopes may be neglected in comparison with the changes of the factor $f(x)$.

In Section IV we estimate from the observations that the critical fission energy for U^{239} is not far from 6 Mev. According to Fig. 4, this corresponds to a value of $x=0.74$, from which we conclude that $(Z^2/A)_{\text{limiting}}=(92)^2/239\times0.74=47.8$. This result enables us to estimate the critical energies for other isotopes, as indicated in the figure. It is seen that protactinium would be particularly interesting as a subject for fission experiments.

As a by product, we are also able from Eq. (12) to compute the nuclear radius in terms of the surface energy of the nucleus; assuming Feenberg's value of 14 Mev for $4\pi r_0^2O$, we obtain $r_0=1.47\times10^{-13}$ cm, which gives a satisfactory and quite independent check on Feenberg's determination of the nuclear radius from the packing fraction curve.

So far the considerations are purely classical, and any actual state of motion must of course be described in terms of quantum-mechanical concepts. The possibility of applying classical pictures to a certain extent will depend on the smallness of the ratio between the zero point amplitudes for oscillations of the type discussed above and the nuclear radius. A simple calculation gives for the square of the ratio in question the result

$$\langle \alpha_n^2 \rangle_{\text{Av; zero point}} = A^{-7/6} \times \{(\hbar^2/12M_p r_0^2)/4\pi r_0^2 O\}^{\frac{1}{2}} n^{\frac{1}{2}} (2n+1)^{\frac{1}{2}} \times \{(n-1)(n+2)(2n+1) - 20(n-1)x\}^{-\frac{1}{2}}. \quad (25)$$

Since $\{(\hbar^2/12M_p r_0^2)/4\pi r_0^2 O\}^{\frac{1}{2}} \doteqdot \frac{1}{3}$, this ratio is indeed a small quantity, and it follows that deformations of magnitudes comparable with nuclear dimensions can be described approximately classically by suitable wave packets built up from quantum states. In particular we may describe the critical deformations which lead to fission in an approximately classical way. This follows from a comparison of the critical energy $E_f \sim 6$ Mev required, as we shall see in Section IV, to account for the observations on uranium, with the zero point energy

$$\tfrac{1}{2}\hbar\omega_2 = A^{-\frac{1}{2}} \{4\pi r_0^2 O \cdot 2(1-x)\hbar^2/3M_p r_0^2\}^{\frac{1}{2}} \sim 0.4 \text{ Mev} \quad (26)$$

of the simplest mode of capillary oscillation, from which it is apparent that the amplitude in question is considerably larger than the zero point disturbance:

$$\langle \alpha_2^2 \rangle_{\text{Av}} / \langle \alpha_2^2 \rangle_{\text{Av; zero point}} \approx E_f / \tfrac{1}{2}\hbar\omega_2 \sim 15. \quad (27)$$

The drop with which we compare the nucleus will also in the critical state be capable of executing small oscillations about the shape of unstable equilibrium. If we study the distribution in frequency of these characteristic oscillations, we must expect for high frequencies to find a spectrum qualitatively not very different from that of the normal modes of oscillation about the form of stable equilibrium. The oscillations in question will be represented symbolically in Fig. 3 by motion of the representative point of the system in configuration space normal to the direction leading to fission. The distribution of the available energy of the system between such modes of motion and the mode of motion leading to fission will be determining for the probability of fission if the system is near the critical state. The statistical mechanics of this problem is considered in Section III. Here we would only like to point out that the fission process is from a practical point of view a nearly irreversible process. In fact if we imagine the fragment nuclei resulting from a fission to be reflected without loss of energy and to run directly towards each other, the electrostatic repulsion between the two nuclei will ordinarily prevent them from coming into contact. Thus, relative to the original nucleus, the energy of two spherical nuclei of half the size is given by Eq. (19) and corresponds to the values $f^*(x)$ shown by the dashed line in Fig. 4. To compare this with the energy required for the original fission process (smooth curve for $f(x)$ in the figure), we note that the surface energy $4\pi r_0^2 O A^{\frac{2}{3}}$ is for the heaviest nuclei of the order of 500 Mev. We thus have to deal with a difference of $\sim 0.05 \times 500$ Mev $= 25$ Mev between the energy available when a heavy nucleus is just able to undergo fission, and the energy required to bring into contact two spherical fragments. There will of course be appreciable tidal forces exerted when the two fragments are brought together, and a simple estimate shows that this will lower the energy discrepancy just mentioned by something of the order of 10 Mev, which is not enough to alter our conclusions. That there is no paradox involved, however, follows from the fact that the fission process actually takes place for a configuration in which the sum of surface and electrostatic energy has a considerably smaller value than that corresponding to two rigid spheres in contact, or even two tidally distorted globes; namely, by arranging that in the division process the surface surrounding the original nucleus shall not tear until the mutual electrostatic energy of the two nascent nuclei has been brought down to a value essentially smaller than that corresponding to separated spheres, then there will be available enough electrostatic energy to provide the work required to tear the surface, which will of course have increased in total value to something more than that appropriate to two spheres. Thus it is clear that the two fragments formed by the division process will possess internal energy of excitation. Consequently, if we wish to reverse the fission process, we must take care that the fragments come together again sufficiently distorted, and indeed with the distortions so oriented, that contact can be made between projections on the two surfaces and the surface tension start drawing them together while the electrostatic repulsion between the effective electrical centers of gravity of the two parts is still not excessive. The probability that two atomic nuclei in any actual encounter will be suitably excited and possess the proper phase relations so that union will be possible to form a compound system will be extremely small. Such union processes, converse to fission, can be expected to occur for unexcited nuclei only when we have available much more kinetic energy than is released in the fission processes with which we are concerned.

The above considerations on the fission process, based on a comparison between the properties of a nucleus and those of a liquid drop, should be supplemented by remarking that the distortion which leads to fission, although associated with a greater effective mass and lower quantum frequency, and hence more nearly approaching the possibilities of a classical description than any of the higher order oscillation frequencies of the nucleus, will still be characterized by certain specific quantum-mechanical properties. Thus there will be an essential ambiguity in the definition of the critical fission energy of the order of magnitude of the zero point energy, $\hbar\omega_2/2$, which however as we have seen above is only a relatively small quantity. More important from the point of view of nuclear stability will be the possibility of quantum-mechanical tunnel effects, which will make it possible for a nucleus to divide even in its ground state by passage through a portion of configuration space where classically the kinetic energy is negative.

An accurate estimate for the stability of a heavy nucleus against fission in its ground state will, of course, involve a very complicated mathematical problem. In natural extension of the well-known theory of α-decay, we should in principle determine the probability per unit time of a fission process, λ_f, by the formula

$$\lambda_f(=\Gamma_f/\hbar)=5(\omega_f/2\pi)$$
$$\times\exp -2\int_{P_1}^{P_2}\{2(V-E)\sum_i m_i(dx_i/d\alpha)^2\}^{\frac{1}{2}}d\alpha/\hbar. \quad (28)$$

The factor 5 represents the degree of degeneracy of the oscillation leading to instability. The quantum of energy characterizing this vibration is, according to (26), $\hbar\omega\sim0.8$ Mev. The integral in the exponent leads in the case of a single particle to the Gamow penetration factor. Similarly, in the present problem, the integral is extended in configuration space from the point P_1 of stable equilibrium over the fission saddle point S (as indicated by the dotted line in Fig. 3) and down on a path of steepest descent to the point P_2 where the classical value of the kinetic energy, $E-V$, is again zero. Along this path we may write the coordinate x_i of each elementary particle m_i in terms of a certain parameter α. Since the integral is invariant with respect to how the parameter is chosen, we may for convenience take α to represent the distance between the centers of gravity of the nascent nuclei. To make an accurate calculation on the basis of the liquid-drop model for the integral in (28) would be quite complicated, and we shall therefore estimate the result by assuming each elementary particle to move a distance $\frac{1}{2}\alpha$ in a straight line either to the right or the left according as it is associated with the one or the other nascent nucleus. Moreover, we shall take $V-E$ to be of the order of the fission energy E_f. Thus we obtain for the exponent in (28) approximately

$$(2ME_f)^{\frac{1}{2}}\alpha/\hbar. \quad (29)$$

With $M=239\times1.66\times10^{-24}$, $E_f\sim6$ Mev$=10^{-5}$ erg, and the distance of separation intermediate between the diameter of the nucleus and its radius, say of the order $\sim1.3\times10^{-12}$ cm, we thus find a mean lifetime against fission in the ground state equal to

$$1/\lambda_f\sim10^{-21}\exp[(2\times4\times10^{-22}\times10^{-5})^{\frac{1}{2}}1.3$$
$$\times10^{-12}/10^{-27}]\sim10^{30}\text{ sec.}\sim10^{22}\text{ years.} \quad (30)$$

It will be seen that the lifetime thus estimated is not only enormously large compared with the time interval of the order 10^{-15} sec. involved in the actual fission processes initiated by neutron impacts, but that this is even large compared with the lifetime of uranium and thorium for α-ray decay. This remarkable stability of heavy nuclei against fission is as seen due to the large masses involved, a point which was already indicated in the cited article of Meitner and Frisch, where just the essential characteristics of the fission effect were stressed.

III. Break-up of the Compound System as a Monomolecular Reaction

To determine the fission probability, we consider a microcanonical ensemble of nuclei, all having excitation energies between E and $E+dE$. The number of nuclei will be chosen to be exactly equal to the number $\rho(E)dE$ of levels in this energy interval, so that there is one nucleus in each state. The number of nuclei which divide per unit time will then be $\rho(E)dE\Gamma_f/\hbar$, according to our definition of Γ_f. This number will be equal to the number of nuclei in the transition state which pass outward over the fission barrier per unit time.[11a] In a unit distance measured in the direction of fission there will be $(dp/h)\rho^*(E-E_f-K)dE$ quantum states of the microcanonical ensemble for which the momentum and kinetic energy associated with the fission distortion have values in the intervals dp and $dK=vdp$, respectively. Here ρ^* is the density of those levels of the compound nucleus in the transition state which arise from excitation of all degrees of freedom other than the fission itself. At the initial time we have one nucleus in each of the quantum states in question, and consequently the number of fissions per unit time will be

$$dE\int v(dp/h)\rho^*(E-E_f-K)=dEN^*/h, \quad (31)$$

where N^* is the number of levels in the transition state available with the given excitation. Comparing with our original expression for this number, we have

$$\Gamma_f=N^*/2\pi\rho(E)=(d/2\pi)N^* \quad (32)$$

for the fission width expressed in terms of the level density or the level spacing d of the compound nucleus.

The derivation just given for the level width will only be valid if N^* is sufficiently large compared to unity; that is, if the fission width is comparable with or greater than the level spacing. This corresponds to the conditions under which a correspondence principle treatment of the fission distortion becomes possible. On the other hand, when the excitation exceeds by only a little the critical energy, or falls below E_f, specific quantum-mechanical tunnel effects will begin to become of importance. The fission probability will of course fall off very rapidly with decreasing excitation energy at this point, the mathematical expression for the reaction rate eventually going over into the penetration formula of Eq. (28); this, as we have seen above, gives a negligible fission probability for uranium.

The probability of neutron re-emission, so important in limiting the fission yield for high excitation energies, has been estimated from statistical arguments by various authors, especially Weisskopf.[12] The result can be derived in a very simple form by considering the microcanonical ensemble introduced above. Only a few changes are necessary with respect to the reasoning used for the fission process. The transition state will be a spherical shell of unit thickness just outside the nuclear surface $4\pi R^2$; the critical energy is the neutron binding energy, E_n; and the density ρ^{**} of excitation levels in the transition state is given by the spectrum of the residual nucleus. The number of quantum states in the microcanonical ensemble which lie in the transition region and for which the neutron momentum lies in the range p to $p+dp$ and in the solid angle $d\Omega$ will be

$$(4\pi R^2\cdot p^2dpd\Omega/h^3)\rho^*(E-E_n-K)dE. \quad (33)$$

We multiply this by the normal velocity $v\cos\theta=(dK/dp)\cos\theta$ and integrate, obtaining

$$dE(4\pi R^2\cdot 2\pi m/h^3)\int\rho^*(E-E_n-K)KdK \quad (34)$$

for the number of neutron emission processes occurring per unit time. This is to be identified with $\rho(E)dE(\Gamma_n/\hbar)$. Therefore we have for the probability of neutron emission, expressed in energy units, the result

$$\Gamma_n=(1/2\pi\rho)(2mR^2/\hbar^2)\int\rho^{**}(E-E_n-K)KdK$$
$$=(d/2\pi)(A^{\frac{2}{3}}/K')\sum_i K_i \quad (35)$$

in complete analogy to the expression

$$\Gamma_f=(d/2\pi)\sum_i 1 \quad (36)$$

[11a] For a general discussion of the ideas involved in the concept of a transition state, reference is made to an article by E. Wigner, Trans. Faraday Soc. 34, part 1, 29 (1938).

[12] V. Weisskopf, Phys. Rev. 52, 295 (1937).

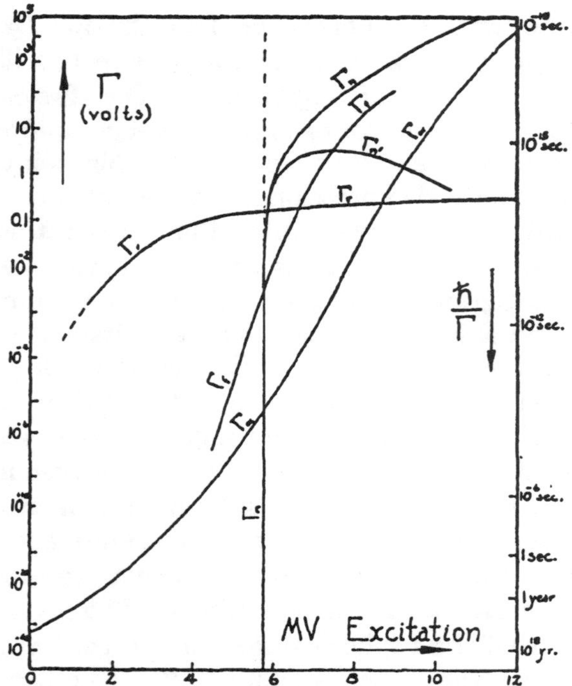

FIG. 5. Schematic diagram of the partial transition probabilities (multiplied by $\hbar$ and expressed in energy units) and their reciprocals (dimensions of a mean lifetime) for various excitation energies of a typical heavy nucleus. Γ_r, Γ_f, and Γ_α refer to radiation, fission, and alpha-particle emission, while $\Gamma_{n'}$ and Γ_n determine, respectively, the probability of a neutron emission leaving the residual nucleus in its ground state or in any state. The latter quantities are of course zero if the excitation is less than the neutron binding, which is taken here to be about 6 Mev.

for the fission width. Just as the summation in the latter equation goes over all those levels of the nucleus in the transition state which are available with the given excitation, so the sum in the former is taken over all available states of the residual nucleus, K_i denoting the corresponding kinetic energy $E-E_n-E_i$ which will be left for the neutron. K' represents, except for a factor, the zero point kinetic energy of an elementary particle in the nucleus; it is given by $A^{\frac{1}{3}}\hbar^2/2mR^2$ and will be 9.3 Mev if the nuclear radius is $A^{\frac{1}{3}}1.48\times10^{-13}$ cm.

No specification was made as to the angular momentum of the nucleus in the derivation of (35) and (36). Thus the expressions in question give us averages of the level widths over states of the compound system corresponding to many different values of the rotational quantum number J, while actually capture of a neutron of one- or two-Mev energy by a normal nucleus will give rise only to a restricted range of values of J. This point is of little importance in general, as the widths will not depend much on J, and therefore in the following considerations we shall apply the above estimates of Γ_f and Γ_n as they stand. In particular, d will represent the average spacing of levels of a given angular momentum. If, however, we wish to determine the partial width $\Gamma_{n'}$ giving the probability that the compound nucleus will break up leaving the residual nucleus in its ground state and giving the neutron its full kinetic energy, we shall not be justified in simply selecting out the corresponding term in the sum in (35) and identifying it with $\Gamma_{n'}$. In fact, a more detailed calculation along the above lines, specifying the angular momentum of the microcanonical ensemble as well as its energy, leads to the expression

$$\Sigma(2J+1)\Gamma_{n'}{}^{J} = (2s+1)(2i+1)(d/2\pi)(R^2/\lambda\!\!\!^{-2}) \quad (37)$$

for the partial neutron width, where the sum goes over those values of J which are realized when a nucleus of spin i is bombarded by a neutron of the given energy possessing spin $s=\frac{1}{2}$.

The smallness of the neutron mass in comparison with the reduced mass of two separating nascent nuclei will mean that we shall have in the former case to go to excitation energies much higher relative to the barrier than in the latter case before the condition is fulfilled for the application of the transition state method. In fact, only when the kinetic energy of the emerging particle is considerably greater than 1 Mev does the reduced wave-length $\lambda\!\!\!^{-}=\lambda/2\pi$ of the neutron become essentially smaller than the nuclear radius, allowing the use of the concepts of velocity and direction of the neutron emerging from the nuclear surface.

The absolute yield of the various processes initiated by neutron bombardment will depend upon the probability of absorption of the neutron to form a compound nucleus; this will be proportional to the converse probability $\Gamma_{n'}/\hbar$ of a neutron emission process which leaves the residual neutron emission process which leaves the residual nucleus in its ground state. $\Gamma_{n'}$ will vary as the neutron velocity itself for low neutron energies; according to the available information about nuclei of medium atomic weight, the width in volts is approximately 10^{-3} times the square root of the neutron energy in volts.[13] As the neutron energy increases from thermal values to 100 kev, we have to expect then an increase of $\Gamma_{n'}$ from something of the order of 10^{-4} ev to 0.1 or 1 ev. For high neutron energies we can use Eq. (37), according to which $\Gamma_{n'}$ will increase as the neutron energy itself, except as compensated by the decrease in level spacing as higher excitations are attained. As an order of magnitude estimate, we can take the level spacing in U to decrease from 100 kev for the lowest levels to 20 ev at 6 Mev (capture of thermal neutrons) to $\frac{1}{3}$ ev for $2\frac{1}{2}$-Mev neutrons. With $d=\frac{1}{3}$ ev we obtain $\Gamma_{n'}=(1/2\pi\times 5)(239^{\frac{2}{3}}/10)2\frac{1}{2}\div\frac{1}{2}$ ev for neutrons from the D+D reaction. The partial neutron width will not exceed for any energy a value of this order of magnitude, since the decrease in level spacing will be the dominating factor at higher energies.

The compound nucleus once formed, the outcome of the competition between the possibilities of fission, neutron emission, and radiation, will be determined by the relative magnitudes of Γ_f, Γ_n, and the corresponding radiation width Γ_r. From our knowledge of nuclei comparable with thorium and uranium we can conclude that the radiation width Γ_r will not exceed something of the order of 1 ev, and moreover that it will be nearly constant for the range of excitation energies which results from neutron absorption (see Fig. 5). The fission width will be extremely small for excitation energies below the critical energy E_f, but above this point Γ_f will become appreciable, soon exceeding the radiation width and rising almost exponentially for higher energies. Therefore, if the critical energy E_f required for fission is comparable with or greater than the excitation consequent on neutron capture, we have to expect that radiation will be more likely than fission; but if the barrier height is somewhat lower than the value of the neutron binding, and in any case if we irradiate with sufficiently energetic neutrons, radiative capture will always be less probable than division. As the speed of the bombarding neutrons is increased, we shall not expect an indefinite rise in the fission yield, however, for the output will be governed by the competition in the compound system between the possibilities of fission and of neutron emission. The width Γ_n which gives the probability of the latter process will for energies less than something of the order of 100 kev be equal to $\Gamma_{n'}$, the partial width for emissions leaving the residual nucleus in the ground state, since excitation of the product nucleus will be energetically impossible. For higher neutron energies, however, the number of available levels in the residual nucleus will rise rapidly, and Γ_n will be much larger than $\Gamma_{n'}$, increasing almost exponentially with energy.

In the energy region where the levels of the compound nucleus are well separated, the cross sections governing the yield of the various processes considered above can be obtained by direct application of the dispersion theory of Breit and Wigner.[14] In the case of resonance, where the energy E of the incident neutron is close to a special value E_0 characterizing an isolated level of the compound system, we shall have

$$\sigma_f=\pi\lambda\!\!\!^{-2}\frac{2J+1}{(2s+1)(2i+1)}\frac{\Gamma_{n'}\Gamma_f}{(E-E_0)^2+(\Gamma/2)^2} \tag{38}$$

and

$$\sigma_r=\pi\lambda\!\!\!^{-2}\frac{2J+1}{(2s+1)(2i+1)}\frac{\Gamma_{n'}\Gamma_r}{(E-E_0)^2+(\Gamma/2)^2}. \tag{39}$$

for the fission and radiation cross sections. Here $\lambda\!\!\!^{-}=\hbar/p=\hbar/(2mE)^{\frac{1}{2}}$ is the neutron wave-length divided by 2π, i and J are the rotational quantum numbers of the original and the compound nucleus, $s=\frac{1}{2}$, and $\Gamma=\Gamma_n+\Gamma_r+\Gamma_f$ is the total width of the resonance level at half-maximum.

In the energy region where the compound nucleus has many levels whose spacing, d, is comparable with or smaller than the total width, the dispersion theory cannot be directly applied due to the phase relations between the contributions of the different levels. A closer discussion[15] shows, however, that in cases like fission and radiative capture, the cross section will be obtained by summing many terms of the form (38) or (39). If the neutron wave-length is large compared with nuclear dimensions, only those states of the compound nucleus will contribute to the

[13] H. A. Bethe, Rev. Mod. Phys. 9, 150 (1937).

[14] G. Breit and E. Wigner, Phys. Rev. 49, 519 (1936). Cf. also H. Bethe and G. Placzek, Phys. Rev. 51, 450 (1937)

[15] N. Bohr, R. Peierls and G. Placzek, Nature (in press).

sum which can be realized by capture of a neutron of zero angular momentum, and we shall obtain

$$\sigma_f = \pi\lambda^2\Gamma_{n'}(\Gamma_f/\Gamma)(2\pi/d)\times\begin{cases}1 \text{ if } i=0\\ \frac{1}{2} \text{ if } i>0\end{cases}, \quad (40)$$

$$\sigma_r = \pi\lambda^2\Gamma_{n'}(\Gamma_r/\Gamma)(2\pi/d)\times\begin{cases}1 \text{ if } i=0\\ \frac{1}{2} \text{ if } i>0\end{cases}. \quad (41)$$

On the other hand, if λ becomes essentially smaller than R, the nuclear radius (case of neutron energy over a million volts), the summation will give

$$\sigma_f = \frac{\pi\lambda^2\sum(2J+1)\Gamma_{n'}}{(2s+1)(2i+1)}(\Gamma_f/\Gamma)(2\pi/d) = \pi R^2\Gamma_f/\Gamma, \quad (42)$$

$$\sigma_r = \pi R^2\Gamma_r/\Gamma. \quad (43)$$

The simple form of the result, which follows by use of the equation (37) derived above for $\Gamma_{n'}$, is of course an immediate consequence of the fact that the cross section for any given process for fast neutrons is given by the projected area of the nucleus times the ratio of the probability per unit time that the compound system react in the given way to the total probability of all reactions. Of course for extremely high bombarding energies it will no longer be possible to draw any simple distinction between neutron emission and fission; evaporation will go on simultaneously with the division process itself; and in general we shall have to expect then the production of numerous fragments of widely assorted sizes as the final result of the reaction.

IV. Discussion of the Observations

A. The resonance capture process

Meitner, Hahn, and Strassmann[16] observed that neutrons of some volts energy produced in uranium a beta-ray activity of 23 min. half-life whose chemistry is that of uranium itself. Moreover, neutrons of such energy gave no noticeable yield of the complex of periods which is produced in uranium by irradiation with either thermal or fast neutrons, and which is now known to arise from the beta-instability of the fragments arising from fission processes. The origin of the activity in question therefore had to be attributed to the ordinary type of radiative capture observed in other nuclei; like such processes it has a resonance character. The effective energy E_0 of the resonance level or levels was determined by comparing the absorption in boron of the neutrons producing the activity and of neutrons of thermal energy:

$$E_0 = (\pi kT/4)[\mu_{\text{thermal}}(B)/\mu_{\text{res}}(B)]^2 = 25\pm10 \text{ ev}. \quad (44)$$

The absorption coefficient in uranium itself for the activating neutrons was found to be 3 cm²/g, corresponding to an effective cross section of $3 \text{ cm}^2/\text{g}\times238\times1.66\times10^{-24} \text{ g}=1.2\times10^{-21} \text{ cm}^2$. If we attribute the absorption to a single resonance level with no appreciable Doppler broadening, the cross section at exact resonance will be twice this amount, or 2.4×10^{-21} cm²; if on the other hand the true width Γ should be small compared with the Doppler broadening

$$\Delta = 2(E_0kT/238)^{\frac{1}{2}} = 0.12 \text{ ev},$$

we should have for the true cross section at exact resonance $2.7\times10^{-21}\Delta/\Gamma$, which would be even greater.[17] If the activity is actually due to several comparable resonance levels, we will clearly obtain the same result for the cross section of each at exact resonance.

According to Nier[18] the abundances of U^{235} and U^{234} relative to U^{238} are 1/139 and 1/17,000; therefore, if the resonance absorption is due to either of the latter, the cross section at resonance will have to be at least $139\times2.4\times10^{-21}$ cm² or 3.3×10^{-19} cm². However, as Meitner, Hahn and Strassmann pointed out, this is excluded (cf. Eq. (39)) because it would be greater in order of magnitude than the square of the neutron wavelength. In fact, $\pi\lambda^2$ is only 25×10^{-21} cm² for 25-volt neutrons. Therefore we have to attribute the capture to $U^{238}\rightarrow U^{239}$, a process in which the spin changes from $i=0$ to $J=\frac{1}{2}$. We apply the

[16] L. Meitner, O. Hahn and F. Strassmann, Zeits. f. Physik 106, 249 (1937).

[17] We are using the treatment of Doppler broadening given by H. Bethe and G. Placzek, Phys. Rev. 51, 450 (1937).

[18] A. O. Nier, Phys. Rev. 55, 150 (1939).

resonance formula (39) and obtain

$$25\times 10^{-21}\times 4\Gamma_{n'}\Gamma_r/\Gamma^2 = 2.7\times 10^{-21}(\Delta/\Gamma) \text{ or } 2.4\times 10^{-21} \quad (45)$$

according as the level width $\Gamma=\Gamma_{n'}+\Gamma_r$ is or is not small compared with the Doppler broadening. In any case, we know[13] from experience with other nuclei for comparable neutron energies that $\Gamma_{n'}\ll\Gamma_r$; this condition makes the solution of (45) unique. We obtain $\Gamma_{n'}=\Gamma_r/40$ if the total width is greater than $\Delta=0.12$ ev; and if the total width is smaller than Δ we find $\Gamma_{n'}=0.003$ ev. Thus in neither case is the neutron width less than 0.003 ev. Comparison with observations on elements of medium atomic weight would lead us to expect a neutron width of $0.001\times(25)^{\frac{1}{2}}=0.005$ ev; and undoubtedly $\Gamma_{n'}$ can be no greater than this for uranium, in view of the small level spacing, or equivalently, in view of the small probability that enough energy be concentrated on a single particle in such a big nucleus to enable it to escape. We therefore conclude that $\Gamma_{n'}$ for 25-volt neutrons is approximately 0.003 ev.

Our result implies that the radiation width for the U^{239} resonance level cannot exceed $\sim$0.12 ev; it may be less, but not much less, first, because values as great as a volt or more have been observed for Γ_r in nuclei of medium atomic weight, and second, because values of a millivolt or more are observed in the transitions between individual levels of the radioactive elements, and for the excitation with which we are concerned the number of available lower levels is great and the corresponding radiation frequencies are higher.[13] A reasonable estimate of Γ_r would be 0.1 ev; of course direct measurement of the activation yield due to neutrons continuously distributed in energy near the resonance level would give a definite value for the radiation width.

The above considerations on the capture of neutrons to form U^{239} are expressed for simplicity as if there were a single resonance level, but the results are altered only slightly if several levels give absorption. However, the contribution of the resonance effect to the radiative capture cross section for *thermal* neutrons does depend essentially on the number of levels as well as their strength. On this basis Anderson and Fermi have been able to show that the radiative capture of slow neutrons cannot be due to the tail at low energies of only a single level.[19] In fact, if it were, we should have for the cross section from (39)

$$\sigma_r(\text{thermal})=\pi\lambda_{th}{}^2\Gamma_{n'}(\text{thermal})\Gamma_r/E_0{}^2; \quad (46)$$

since $\Gamma_{n'}$ is proportional to neutron velocity, we should obtain at the effective thermal energy $\pi kT/4=0.028$ ev.

$$\sigma_r(\text{thermal})\sim 23\times 10^{-18}\times 0.003(0.028/25)^{\frac{1}{2}}0.1/(25)^2 \sim 0.4\times 10^{-24}\text{ cm}^2. \quad (47)$$

Anderson and Fermi however obtain for this cross section by direct measurement 1.2×10^{-24} cm².

The conclusion that the resonance absorption at the effective energy of 25 ev is actually due to more than one level gives the possibility of an order of magnitude estimate of the spacing between energy levels in U^{239} if for simplicity we assume random phase relations between their individual contributions. Taking into consideration the factor between the observations and the result (47) of the one level formula, and recalling that levels below thermal energies as well as above contribute to the absorption, we arrive at a level spacing of the order of $d=20$ ev as a reasonable figure at the excitation in question.

B. Fission produced by thermal neutrons

According to Meitner, Hahn and Strassmann[20] and other observers, irradiation of uranium by thermal neutrons actually gives a large number of radioactive periods which arise from fission fragments. By direct measurement the fission cross section for thermal neutrons is found to be between 2 and 3×10^{-24} cm² (averaged over the actual mixture of isotopes), that is, about twice the cross section for radiative capture. No appreciable part of this effect can come from the isotope U^{239}, however, because the observations on the $\sim$25-volt resonance capture of neutrons by this nucleus gave only the 23-minute activity; the inability of Meitner, Hahn, and Strassmann to find for neutrons of this energy any appreciable yield of the complex of periods

[19] H. L. Anderson and E. Fermi, Phys. Rev. 55, 1106 (1939).

[20] L. Meitner, O. Hahn and F. Strassmann, Zeits. f. Physik 106, 249 (1937).

now known to follow fission indicates that for slow neutrons in general the fission probability for this nucleus is certainly no greater than 1/10 of the radiation probability. Consequently, from comparison of (38) and (39), the fission cross section for this isotope cannot exceed something of the order $\sigma_f(\text{thermal})=(1/10)\sigma_r(\text{thermal})=0.1\times10^{-24}$ cm². From reasoning of this nature, as was pointed out in an earlier paper by Bohr, we have to attribute practically all of the fission observed with thermal neutrons to one of the rarer isotopes of uranium.[21] If we assign it to the compound nucleus U^{235}, we shall have 17,000 $\times2.5\times10^{-24}$ or 4×10^{-20} cm² for σ_f(thermal); if we attribute the division to U^{236}, σ_f will be between 3 and 4×10^{-22} cm².

We have to expect that the radiation width and the neutron width for slow neutrons will differ in no essential way between the various uranium isotopes. Therefore we will assume $\Gamma_{n'}(\text{thermal})=0.003(0.028/25)^{\frac{1}{2}}=10^{-4}$ ev. The fission width, however, depends strongly on the barrier height; this is in turn a sensitive function of nuclear charge and mass numbers, as indicated in Fig. 4, and decreases strongly with decreasing isotopic weight. Thus it is reasonable that one of the lighter isotopes should be responsible for the fission.

Let us investigate first the possibility that the division produced by thermal neutrons is due to the compound nucleus U^{235}. If the level spacing d for this nucleus is essentially greater than the level width, the cross section will be due principally to one level ($J=\frac{1}{2}$ arising from $i=0$), and we shall have from

$$\sigma_f=\pi\lambda\!\!\!^{-2}\frac{2J+1}{(2s+1)(2i+1)}\frac{\Gamma_{n'}\Gamma_f}{(E-E_0)^2+(\Gamma/2)^2} \tag{38}$$

the equation

$$\Gamma_f/[E_0^2+\Gamma^2/4]=4\times10^{-20}/23\times10^{-18}\times10^{-4}=17(\text{ev})^{-1}. \tag{48}$$

Since $\Gamma>\Gamma_f$, this condition can be put as an inequality,

$$E_0^2<(\Gamma/4)(4/17)-\Gamma) \tag{49}$$

from which it follows first, that $\Gamma\leq4/17$ ev, and second, that $|E_0|<2/17$ ev. Thus the level would have to be very narrow and very close to thermal energies. But in this case the fission cross section would have to fall off very rapidly with increasing neutron energy; since $\lambda\!\!\!^{-}\propto1/v$, $E\propto v^2$, $\Gamma_{n'}\propto v$, we should have according to (38) $\sigma_f\propto1/v^5$ for neutron energies greater than about half a volt. This behavior is quite inconsistent with the finding of the Columbia group that the fission cross section for cadmium resonance neutrons (~0.15 ev) and for the neutrons absorbed in boron (mean energy of several volts) stand to each other inversely in the ratio of the corresponding neutron velocities $(1/v)$.[22] Therefore, if the fission is to be attributed to U^{235}, we must assume that the level width is greater than the level spacing (many levels effective); but as the level spacing itself will certainly exceed the radiative width, we will then have a situation in which the total width will be essentially equal to Γ_f. Consequently we can write the cross section (40) for overlapping levels in the form

$$\sigma_f=\pi\lambda\!\!\!^{-2}\Gamma_{n'}2\pi/d. \tag{50}$$

From this we find a level spacing

$$d=23\times10^{-18}\times10^{-4}\times2\pi/4\times10^{-20}=0.4\text{ ev}$$

which is unreasonably small: According to the estimates of Table III, the nuclear excitations consequent on the capture of slow neutrons to form U^{235} and U^{239} are approximately 5.4 Mev and 5.2 Mev, respectively; moreover, the two nuclei have the same odd-even properties and should therefore possess similar level distributions. From the difference ΔE between the excitation energies in the two cases we can therefore obtain the ratio of the corresponding level spacings from the expression exp $(\Delta E/T)$. Here T is the nuclear temperature, a low estimate for which is 0.5 Mev, giving a factor of exp 0.6=2. From our conclusion in IV-A that the order of magnitude of the level spacing in U^{239} is 20 ev, we would expect then in U^{235} a spacing of the order of 10 ev. Therefore the result of Eq. (51) makes it seem quite unlikely that the fission observed for the thermal neutrons can be due to the rarest uranium isotope; we consequently attribute it almost entirely to the reaction $U^{235}+n_{th}\rightarrow U^{236}\rightarrow$fission.

[21] N. Bohr, Phys. Rev. 55, 418 (1939).

[22] Anderson, Booth, Dunning, Fermi, Glasoe and Slack, reference 4.

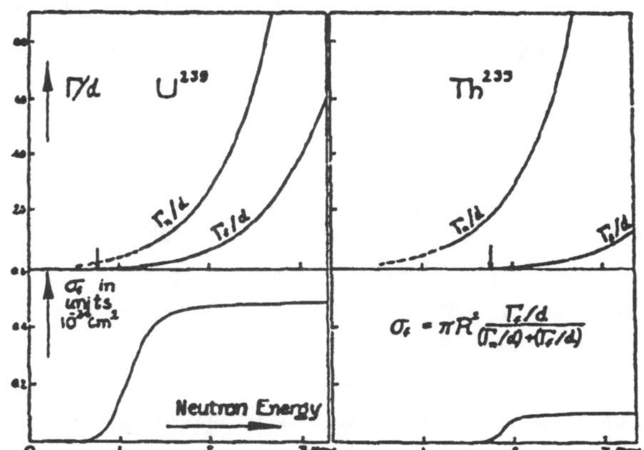

FIG. 6. Γ_n/d and Γ_f/d are the ratios of the neutron emission and fission probabilities (taken per unit of time and multiplied by $\hbar$) to the average level spacing in the compound nucleus at the given excitation. These ratios will vary with energy in nearly the same way for all heavy nuclei, except that the entire fission curve must be shifted to the left or right according as the critical fission energy E_f is less than or greater than the neutron binding E_n. The cross section for fission produced by fast neutrons depends on the ratio of the values in the two curves, and is given on the left for $E_f - E_n = (\frac{3}{4})$ Mev and on the right for $E_f - E_n = 1\frac{3}{4}$ Mev, corresponding closely to the cases of U^{239} and Th^{233}, respectively.

We have two possibilities to account for the cross section $\sigma_f(\text{thermal}) \sim 3.5 \times 10^{-22}$ presented by the isotope U^{235} for formation of the compound nucleus U^{236}, according as the level width is smaller than or comparable with the level spacing. In the first case we shall have to attribute most of the fission to an isolated level, and by the reasoning which was employed previously, we conclude that for this level

$$\Gamma_f/[E_0^2+\Gamma^2/4] = [(2s+1)(2i+1)/(2J+1)]0.15(\text{ev})^{-1} = R. \quad (52)$$

If the spin of U^{235} is $\frac{3}{2}$ or greater, the right-hand side of (52) will be approximately 0.30 $(\text{ev})^{-1}$; but if i is as low as $\frac{1}{2}$, the right side will be either 0.6 or 0.2 $(\text{ev})^{-1}$. The resulting upper limits on the resonance energy and level width may be summarized as follows:

	$i \geq \frac{3}{2}$	$i=\frac{1}{2}, J=0$	$i=\frac{1}{2}, J=1$	
$\Gamma < 4/R =$	13	7	20 ev	(53)
$\lvert E_0 \rvert < 1/R =$	3	1.7	5 ev.	

On the other hand, the indications[22] for low neutron energies of a $1/v$ variation of fission cross section with velocity lead us as in the discussion of the rarer uranium isotope to the conclusion that either E_0 or $\Gamma/2$ or both are greater than several electron volts. This allows us to obtain from (52) a lower limit also to Γ_f:

$$\Gamma_f = R[E_0^2+\Gamma^2/4] > 10 \text{ to } 400 \text{ ev}. \quad (54)$$

In the present case, the various conditions are not inconsistent with each other, and it is therefore possible to attribute the fission to the effect of a single resonance level.

We can go further, however, by estimating the level spacing for the compound nucleus U^{236} According to the values of Table III, the excitation following the neutron capture is considerably greater than in the case U^{239}, and we should therefore expect a rather smaller level spacing than the value ~ 20 ev estimated in the latter case. On the other hand, it is known that for similar energies the level density is lower in even even than odd even nuclei. Thus the level spacing in U^{236} may still be as great as 20 ev, but it is undoubtedly no greater. From (54) we conclude then that we have probably to do with a case of overlapping resonance levels rather than a single absorption line, although the latter possibility is not entirely excluded by the observations available.

In the case of overlapping levels we shall have from Eq. (40)

$$\sigma_f = (\pi \lambda\!\!\!^{-2}/2)\Gamma_{n'}(2\pi/d) \quad (55)$$

or consequently a level spacing

$$d = (23\times10^{-18}/2)\times10^{-4} \times 2\pi/3.5\times10^{-22} = 20 \text{ ev}; \quad (56)$$

and as we are attributing to the levels an unresolved structure, the fission width must be at least 10 ev. These values for level spacing and fission width give a reasonable account of the fission produced by slow neutrons.

C. Fission by fast neutrons

The discussion on the basis of theory of the fission produced by fast neutrons is simplified first by the fact that the probability of radiation can be neglected in comparison with the probabilities of fission and neutron escape and second by the circumstance that the neutron wavelength $/2\pi$ is small in comparison with the nuclear radius ($R \sim 9\times10^{-13}$ cm) and we are in the region of continuous level distribution. Thus the fission cross section will be given by

$$\sigma_f = \pi R^2 \Gamma_f/\Gamma \sim 2.4\times10^{-24}\Gamma_f/(\Gamma_f+\Gamma_n). \quad (57)$$

or, in terms of the ratio of widths to level spacing,

$$\sigma_f \sim 2.4\times 10^{-24}(\Gamma_f/d)/[(\Gamma_f/d)+(\Gamma_n/d)]. \quad (58)$$

According to the results of Section III,

$$\Gamma_n/d=(1/2\pi)(A^{\frac{1}{3}}/10 \text{ Mev})\sum_i K_i \quad (59)$$

and

$$\Gamma_f/d=(1/2\pi)N^*. \quad (60)$$

In using Eq. (58) it is therefore seen that we do not have to know the level spacing d of the compound nucleus, but only that of the residual nucleus (Eq. (59)) and the number N^* of available levels of the dividing nucleus in the transition state (Eq. 60).

Considered as a function of energy, the ratio of fission width to level spacing will be extremely small for excitations less than the critical fission energy; with increase of the excitation above this value Eq. (60) will quickly become valid, and we shall have to anticipate a rapid rise in the ratio in question. If the spacing of levels in the transition state can be compared with that of the lower states of an ordinary heavy nucleus ($\sim$50 to 100 kev) we shall expect a value of $N^*=10$ to 20 for an energy 1 Mev above the fission barrier; but in any case the value of Γ_f/d will rise almost linearly with the available energy over a range of the order of a million volts, when the rise will become noticeably more rapid owing to the decrease to be expected at such excitations in the level spacing of the nucleus in the transition state. The associated behavior of Γ_f/d is illustrated in curves in Fig. 6. It should be remarked that the specific quantum-mechanical effects which set in at and below the critical fission energy may even show their influence to a certain extent above this energy and produce slight oscillations in the beginning of the Γ_f/d curve, allowing possibly a direct determination of N^*. How the ratio Γ_n/d will vary with energy is more accurately predictable than the ratio just considered. Denoting by K the neutron energy, we have for the number of levels which can be excited in the residual (=original) nucleus a figure of from $K/0.05$ Mev to $K/0.1$ Mev, and for the average kinetic energy of the inelastically scattered neutron $\sim K/2$, so that the sum K_i in (59) is easily evaluated, giving us, if we express K in Mev,

$$\Gamma_n/d \backsim 3 \text{ to } 6 \text{ times } K^2. \quad (61)$$

This formula provides as a matter of fact however only a rough first orientation, since for energies below $K=1$ Mev it is not justified to apply the evaporation formula (a transition occurring until for slow neutrons Γ_n/d is proportional to velocity) and for energies above 1 Mev we have to take into account the gradual decrease which occurs in level spacing in the residual nucleus, and which has the effect of increasing the right-hand side of (61). An attempt has been made to estimate this increase in drawing Fig. 6.

The two ratios involved in the fast neutron fission cross section (58) will vary with energy in the same way for all the heaviest nuclei; the only difference from nucleus to nucleus will occur in the critical fission energy, which will have the effect of shifting one curve with respect to another as shown in the two portions of Fig. 6. Thus we can deduce the characteristic differences between nuclei to be expected in the variation with energy of the fast neutron cross section.

Meitner, Hahn, and Strassmann observed that fast neutrons as well as thermal ones produce in uranium the complex of activities which arise as a result of nuclear fission, and Ladenburg, Kanner, Barschall, and van Voorhis have made a direct measurement of the fission cross section for 2.5 Mev neutrons, obtaining 0.5×10^{-24} cm^2 ($\pm$25 percent).[23] Since the contribution to this cross section due to the U^{235} isotope cannot exceed $\pi R^2/139 \sim 0.02\times 10^{-24}$ cm^2, the effect must be attributed to the compound nucleus U^{239}. For this nucleus however as we have seen from the slow neutron observations the fission probability is negligible at low energies. Therefore we have to conclude that the variation with energy of the corresponding cross section resembles in its general features Fig. 6a. In this connection we have the further observation of Ladenburg *et al.* that the cross section changes little between 2 Mev and 3 Mev.[23] This points to a value of the critical fission energy for U^{239} definitely less

[23] R. Ladenburg, M. H. Kanner, H. H. Barschall and C. C. van Voorhis, Phys. Rev. 56, 168 (1939).

than 2 Mev in excess of the neutron binding. Unpublished results of the Washington group[24] give $\sigma_f = 0.003\times10^{-24}$ at 0.6 Mev and 0.012×10^{-24} cm² at 1 Mev. With the Princeton observations[23] we have enough information to say that the critical energy for U^{239} is not far from $\frac{3}{4}$ Mev in excess of the neutron binding (~5.2 Mev from Table III):

$$E_f(U^{239}) \sim 6 \text{ Mev}. \quad (62)$$

A second conclusion we can draw from the absolute cross section of Ladenburg *et al.* is that the ratio of (Γ_f/d) to (Γ_n/d) as indicated in the figure is substantially correct; this confirms our presumption that the energy level spacing in the transition state of the dividing nucleus is not different in order of magnitude from that of the low levels in the normal nucleus.

The fission cross section of Th^{232} for neutrons of 2 to 3 Mev energy has also been measured by the Princeton group; they find $\sigma_f = 0.1\times10^{-24}$ cm² in this energy range. On the basis of the considerations illustrated in Fig. 6 we are led in this case to a fission barrier $1\frac{3}{4}$ Mev greater than the neutron binding; hence, using Table III,

$$E_f(Th^{233}) \sim 7 \text{ Mev}. \quad (63)$$

A check on the consistency of the values obtained for the fission barriers is furnished by the possibility pointed out in Section II and Fig. 4 of obtaining the critical energy for all nuclei once we know it for one nucleus. Taking $E_f(U^{239}) = 6$ Mev as standard, we obtain $E_f(Th^{233}) = 7$ Mev, in good accord with (63).

As in the preceding paragraph we deduce from Fig. 4 $E_f(U^{236}) = 5\frac{1}{4}$ Mev, $E_f(U^{235}) = 5$ Mev. Both values are *less* than the corresponding neutron binding energies estimated in Table III, $E_n(U^{236}) = 6.4$ Mev, $E_n(U^{235}) = 5.4$ Mev. From the values of $E_n - E_d$ we conclude along the lines of Fig. 6 that for thermal neutrons Γ_f/d is, respectively, ~5 and ~1 for the two isotopes. Thus it appears that in both cases the level distribution will be continuous. We can estimate the as yet entirely unmeasured fission cross section of the lightest uranium isotope for the thermal neutrons from

$$\sigma_f = \pi\lambda^2\Gamma_n \cdot 2\pi/d. \quad (64)$$

[24] Reported by M. Tuve at the Princeton meeting of the American Physical Society, June 23, 1939.

d will not be much different from what it is for the similar compound nucleus U^{239}, say of the order of 20 ev. Thus

$$\sigma_f(\text{thermal}, U^{235}) \sim 23\times10^{-18}\times10^{-4}\times2\pi/20 \sim 500 \text{ to } 1000\times10^{-24} \text{ cm}^2, \quad (65)$$

which is of course practically the same figure which holds for the next heaviest compound nucleus.

The various values estimated for fission barriers and fission and neutron widths are summarized in Fig. 7. The level spacing f for past neutrons has been estimated from its value for slow neutrons and the fact that nuclear level densities appear to increase, according to Weisskopf, approximately exponentially as $2(E/a)^{\frac{1}{2}}$, where a is a quantity related to the spacing of the lowest nuclear levels and roughly 0.1 Mev in magnitude.[25] The relative values of Γ_n, Γ_f and d for fast neutrons in Fig. 7, being obtained less indirectly, will be more reliable than their absolute values.

V. Neutrons, Delayed and Otherwise

Roberts, Meyer and Wang[26] have reported the emission of neutrons following a few seconds after the end of neutron bombardment of a

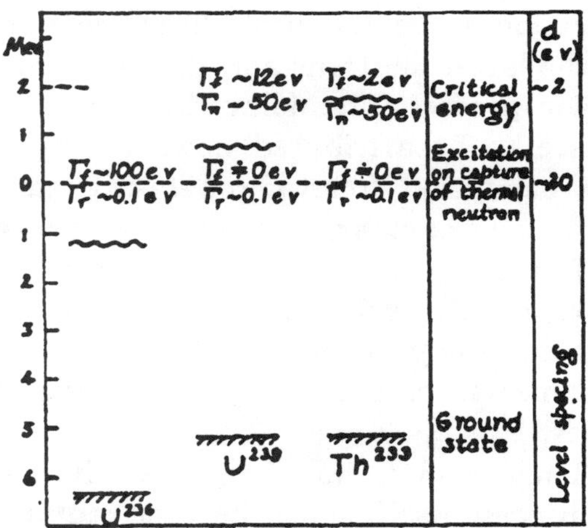

Fig. 7. Summary for comparative purposes of the estimated fission energies, neutron binding energies, level spacings, and neutron and fission widths for the three nuclei to which the observations refer. For fast neutrons the values of Γ_f, Γ_n, and d are less reliable than their ratios. The values in the top line refer to a neutron energy of 2 Mev in each case.

[25] V. Weisskopf, Phys. Rev. 52, 295 (1937).

[26] R. B. Roberts, R. C. Meyer and P. Wang, Phys. Rev. 55, 510 (1939).

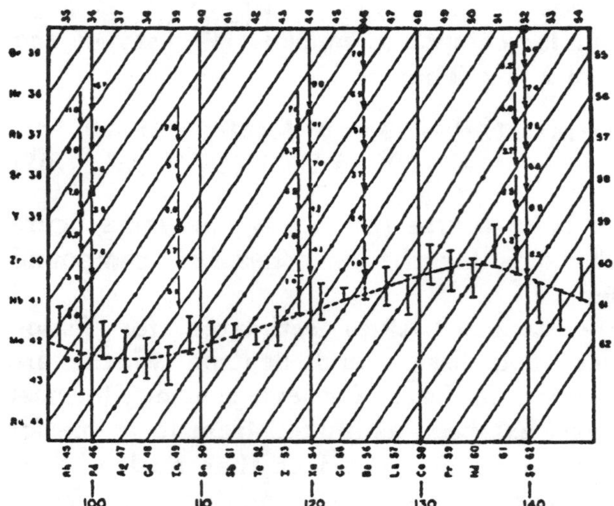

FIG. 8. Beta-decay of fission fragments leading to stable nuclei. Stable nuclei are represented by the small circles; thus the nucleus $_{50}Sn^{120}$ lies just under the arrow marked 4.1; the number indicates the estimated energy release in Mev (see Section I) in the beta-transformation of the preceding nucleus $_{49}In^{120}$. Characteristic differences are noted between nuclei of odd and even mass numbers in the energy of successive transformations, an aid in assigning activities to mass numbers. The dotted line has been drawn, as has been proposed by Gamow, in such a way as to lie within the indicated limits of nuclei of odd mass number; its use is described in Section I.

thorium or uranium target. Other observers have discovered the presence of additional neutrons following within an extremely short interval after the fission process.[27] We shall return later to the question as to the possible connection between the latter neutrons and the mechanism of the fission process. The delayed neutrons themselves are to be attributed however to a high nuclear excitation following beta-ray emission from a fission fragment, for the following reasons:

(1) The delayed neutrons are found only in association with nuclear fission, as is seen from the fact that the yields for both processes depend in the same way on the energy of the bombarding neutrons.

(2) They cannot, however, arise during the fission process itself, since the time required for the division is certainly less than 10^{-12} sec., according to the observations of Feather.[27a]

(3) Moreover, an excitation of a fission fragment in the course of the fission process to an energy sufficient for the subsequent evaporation of a neutron cannot be responsible for the *delayed* neutrons, since even by radiation alone such an excitation will disappear in a time of the order of 10^{-13} to 10^{-15} sec.

(4) The possibility that gamma-rays associated with the beta-ray transformations following fission might produce any appreciable number of photoneutrons in the source has been excluded by an experiment reported by Roberts, Hafstad, Meyer and Wang.[28]

(5) The energy release on beta-transformation is however in a number of cases sufficiently great to excite the product nucleus to a point where it can send out a neutron, as has been already pointed out in connection with the estimates in Table III. Typical values for the release are shown on the arrows in Fig. 8. The product nucleus will moreover have of the order of 10^4 to 10^5 levels to which beta-transformations can lead in this way, so that it will also be overwhelmingly probable that the product nucleus shall be highly excited.

We therefore conclude that the delayed emission of neutrons indeed arises as a result of nuclear excitation following the beta-decay of the nuclear fragments.

The actual probability of the occurrence of a nuclear excitation sufficient to make possible neutron emission will depend upon the comparative values of the matrix elements for the beta-ray transformation from the ground state of the original nucleus to the various excited states of the product nucleus. The simplest assumption we can make is that the matrix elements in question do not show any systematic variation with the energy of the final state. Then, according to the Fermi theory of beta-decay, the probability of a given beta-ray transition will be approximately proportional to the fifth power of the energy release.[29] If there are $\rho(E)dE$ excitation levels of the product nucleus in the range E to $E+dE$, it will follow from our assumptions that the probability of an excitation in the same energy interval will be given by

$$w(E)dE = \text{constant } (E_0-E)^5\rho(E)dE, \quad (66)$$

[27] H. L. Anderson, E. Fermi and H. B. Hanstein, Phys. Rev. 55, 797 (1939); L. Szilard and W. H. Zinn, Phys. Rev. 55, 799 (1939); H. von Halban, Jr., F. Joliot and L. Kowarski, Nature 143, 680 (1939).

[27a] N. Feather, Nature 143, 597 (1939).

[28] R. B. Roberts, L. R. Hafstad, R. C. Meyer and P. Wang, Phys. Rev. 55, 664 (1939).

[29] L. W. Nordheim and F. L. Yost, Phys. Rev. 51, 942 (1937).

where E_0 is the total available energy. According to (66) the probability $w(E)$ of a transition to the excited levels in a unit energy range at E reaches its maximum value for the energy $E = E_{max}$ given by

$$E_{max} = E_0 - 5/(d \ln \rho/dE)_{E_{max}} = E_0 - 5T, \quad (67)$$

where T is the temperature (in energy units) to which the product nucleus must be heated to have on the average the excitation energy E_{max}. Thus the most probable energy release on beta-transformation may be said to be five times the temperature of the product nucleus. According to our general information about the nuclei in question, an excitation of 4 Mev will correspond to a temperature of the order of 0.6 Mev. Therefore, on the basis of our assumptions, to realize an average excitation of 4 Mev by beta-transformation we shall require a total energy release of the order of $4+5\times0.6=7$ Mev.

The spacing of the lowest nuclear levels is of the order of 100 kev for elements of medium atomic weight, decreases to something of the order of 10 ev for excitations of the order of 8 Mev, and can, according to considerations of Weisskopf, be represented in terms of a nuclear level density varying approximately exponentially as the square root of the excitation energy.[23] Using such an expression for $\rho(E)$ in Eq. (66), we obtain the curve shown in Fig. 9 for the distribution function $w(E)$ giving the probability that an excitation E will result from the beta-decay of a typical fission fragment. It is seen that there will be appreciable probability for neutron emission if the neutron binding is somewhat less than the total energy available for the beta-ray transformation. We can of course draw only general conclusions because of the uncertainty in our original assumption that the matrix elements for the various possible transitions show no systematic trend with energy. Still, it is clear that the above considerations provide us with a reasonable qualitative account of the observation of Booth, Dunning and Slack that there is a chance of the order of 1 in 60 that a nuclear fission will result in the delayed emission of a neutron.[30]

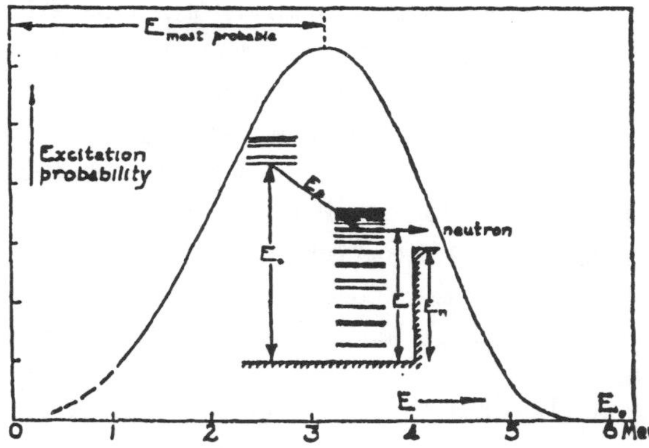

FIG. 9. The distribution in excitation of the product nuclei following beta-decay of fission fragments is estimated on the assumption of comparable matrix elements for the transformations to all excited levels. With sufficient available energy E_0 and a small enough neutron binding E_n it is seen that there will be an appreciable number of delayed neutrons. The quantity plotted is probability per unit range of excitation energy.

Another consequence of the high probability of transitions to excited levels will be to give a beta-ray spectrum which is the superposition of a very large number of elementary spectra. According to Bethe, Hoyle and Peierls, the observations on the beta-ray spectra of light elements point to the Fermi distribution in energy in the elementary spectra.[31] Adopting this result, and using the assumption of equal matrix elements discussed above, we obtain the curve of Fig. 10 for the qualitative type of intensity distribution to be expected for the electrons emitted in the beta-decay of a typical fission fragment. As is seen from the curve, we have to expect that the great majority of electrons will have energies much smaller in value than the actual transformation energy which is available. This is in accord with the failure of various observers to find any appreciable number of very high energy electrons following fission.[32]

The half-life for emission of a beta-ray of 8 Mev energy in an elementary transition will be something of the order of 1 to 1/10 sec., according to the empirical relation between lifetime and energy given by the first Sargent curve. Since we have to deal in the case of the nuclear fragments with transitions to 10^4 or 10^5 excited levels, we should therefore at first sight expect an extremely short lifetime with respect to electron emission. However, the existence of a

[30] E. T. Booth, J. R. Dunning and F. G. Slack, Phys. Rev. 55, 876 (1939).

[31] H. A. Bethe, F. Hoyle and R. Peierls, Nature 143, 200 (1939).

[32] H. H. Barschall, W. T. Harris, M. H. Kanner and L. A. Turner, Phys. Rev. 55, 989 (1939).

sum rule for the matrix elements of the transitions in question has as a consequence that the individual matrix elements will actually be very much smaller than those involved in beta-ray transitions from which the Sargent curve is deduced. Consequently, there seems to be no difficulty in principle in understanding lifetimes of the order of seconds such as have been reported for typical beta-decay processes of the fission fragments.

In addition to the delayed neutrons discussed above there have been observed neutrons following within a very short time (within a time of the order of at most a second) after fission.[27] The corresponding yield has been reported as from two to three neutrons per fission.[33] To account for so many neutrons by the above considered mechanism of nuclear excitation following beta-ray transitions would require us to revise drastically the comparative estimates of beta-transformation energies and neutron binding made in Section I. As the estimates in question were based on indirect though simple arguments, it is in fact possible that they give misleading results. If however they are reasonably correct, we shall have to conclude that the neutrons arise either from the compound nucleus at the moment of fission or by evaporation from the fragments as a result of excitation imparted to them as they separate. In the latter case the time required for neutron emission will be 10^{-13} sec. or less (see Fig. 5). The time required to bring to rest a fragment with 100 Mev kinetic energy, on the other hand, will be at least the time required for a particle with average velocity 10^9 cm/sec. to traverse a distance of the order of 10^{-3} cm. Therefore the neutron will be evaporated before the fragment has lost much of its translational energy. The kinetic energy per particle in the fragment being about 1 Mev, a neutron evaporated in nearly the forward direction will thus have an energy which is certainly greater than 1 Mev, as has been emphasized by Szilard.[34] The observations so far published neither prove nor disprove the possibility of such an evaporation following fission.

[33] Anderson, Fermi and Hanstein, reference 27. Szilard and Zinn, reference 27. H. von Halban, Jr., F. Joliot and L. Kowarski, Nature 143 680 (1939).

[34] Discussions, Washington meeting of American Physical Society, April 28, 1939.

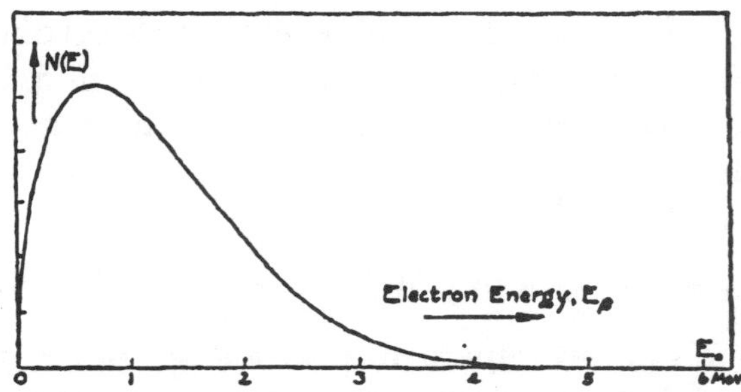

Fig. 10. The superposition of the beta-ray spectra corresponding to all the elementary transformations indicated in Fig. 9 gives a composite spectrum of a general type similar to that shown here, which is based on the assumption of comparable matrix elements and simple Fermi distributions for all transitions. The dependent variable is number of electrons per unit energy range.

We consider briefly the third possibility that the neutrons in question are produced during the fission process itself. In this connection attention may be called to observations on the manner in which a fluid mass of unstable form divides into two smaller masses of greater stability; it is found that tiny droplets are generally formed in the space where the original enveloping surface was torn apart. Although a detailed dynamical account of the division process will be even more complicated for a nucleus than for a fluid mass, the liquid drop model of the nucleus suggests that it is not unreasonable to expect at the moment of fission a production of neutrons from the nucleus analogous to the creation of the droplets from the fluid.

The statistical distribution in size of the fission fragments, like the possible production of neutrons at the moment of division, is essentially a problem of the dynamics of the fission process, rather than of the statistical mechanics of the critical state considered in Section II. Only after the deformation of the nucleus has exceeded the critical value, in fact, will there occur that rapid conversion of potential energy of distortion into energy of internal excitation and kinetic energy of separation which leads to the actual process of division.

For a classical liquid drop the course of the reaction in question will be completely determined by specifying the position and velocity in configuration space of the representative point of the system at the instant when it passes over the potential barrier in the direction of fission. If the energy of the original system is only infinitesimally greater than the critical energy, the representative point of the system must cross the barrier very near the saddle point and with a very small velocity. Still, the wide range of directions available for the velocity vector in this multidimensional space, as suggested schematically in Fig. 3, indicates that production of a considerable variety of fragment sizes may be expected even at energies very close to the threshold for the division process. When the excitation energy increases above the critical fission energy, however, it follows from the statistical arguments in Section III that the representative point of the system will in general pass over the fission barrier at some distance from the saddle point. With general displacements of the representative point along the ridge of the barrier away from the saddle point there are associated asymmetrical deformations from the critical form, and we therefore have to anticipate a somewhat larger difference in size of the fission fragments as more energy is made available to the nucleus in the transition state. Moreover, as an influence of the finer details of nuclear binding, it will also be expected that the relative probability of observing fission fragments of odd mass number will be less when we have to do with the division of a compound nucleus of even charge and mass than one with even charge and odd mass.[35]

VI. Fission Produced by Deuterons and Protons and by Irradiation

Regardless of what excitation process is used, it is clear that an appreciable yield of nuclear fissions will be obtained provided that the excitation energy is well above the critical energy for fission and that the probability of division of the compound nucleus is comparable with the probability of other processes leading to the break up of the system. Neutron escape being the most important process competing with fission, the latter condition will be satisfied if the fission energy does not much exceed the neutron binding, which is in fact the case, as we have seen, for the heaviest nuclei. Thus we have to expect for these nuclei that not only neutrons but also sufficiently energetic deuterons, protons, and gamma-rays will give rise to observable fission.

A. Fission produced by deuteron and proton bombardment

Oppenheimer and Phillips have pointed out that nuclei of high charge react with deuterons of not too great energy by a mechanism of polarization and dissociation of the neutron-proton binding in the field of the nucleus, the neutron being absorbed and the proton repulsed.[36] The excitation energy E of the newly formed nucleus is given by the kinetic energy E_d of the deuteron diminished by its dissociation energy I and the kinetic energy K of the lost proton, all increased by the binding energy E_n of the neutron in the product nucleus:

$$E=E_d-I-K+E_n. \qquad (68)$$

The kinetic energy of the proton cannot exceed E_d+E_n-I, nor on the other hand will it fall below the potential energy which the proton will have in the Coulomb field at the greatest possible distance from the nucleus consistent with the deuteron reaction taking place with appreciable probability. This distance and the corresponding kinetic energy K_{min} have been calculated by Bethe.[37] For very low values of the bombarding energy E_D, he finds $K_{min}\sim 1$ Mev; when E_d rises to equality with the dissociation energy $I=2.2$ Mev he obtains $K_{min}\sim E_d$; and even when the bombarding potential reaches a value corresponding to the height of the electrostatic barrier, K_{min} still continues to be of order E_d, although beyond this point increase of E_d produces no further rise in K_{min}. Since the barrier height for single charged particles will be of the order of 10 Mev for the heaviest nuclei, we can therefore assume $K_{min}\sim E_d$ for the ordinarily employed values of the deuteron bombarding energy. We conclude that the excitation energy of the product nucleus will have only a very small probability of exceeding the value

$$E_{max}\sim E_n-I. \qquad (69)$$

Since this figure is considerably less than the

[35] S. Flugge and G. v. Droste also have raised the question of the possible influence of finer details of nuclear binding on the statistical distribution in size of the fission fragments, Zeits. f. physik. Chemie B42 274 (1939).

[36] R. Oppenheimer and M. Phillips, Phys. Rev. 48, 500 (1935).

[37] H. A. Bethe, Phys. Rev. 53, 39 (1938).

estimated values of the fission barriers in thorium and uranium, we have to expect that Oppenheimer-Phillips processes of the type discussed will be followed in general by radiation rather than fission, unless the kinetic energy of the deuteron is greater than 10 Mev.

We must still consider, particularly when the energy of the deuteron approaches 10 Mev, the possibility of processes in which the deuteron as a whole is captured, leading to the formation of a compound nucleus with excitation of the order of

$$E_d+2E_n-I\sim E_d+10\text{ Mev.} \quad (70)$$

There will then ensue a competition between the possibilities of fission and neutron emission, the outcome of which will be determined by the comparative values of Γ_f and Γ_n (proton emission being negligible because of the height of the electrostatic barrier). The increase of charge associated with the deuteron capture will of course lower the critical energy of fission and increase the probability of fission relative to neutron evaporation compared to what its value would be for the original nucleus at the same excitation. If after the deuteron capture the evaporation of a neutron actually takes place, the fission barrier will again be decreased relative to the binding energy of a neutron. Since the kinetic energy of the evaporated neutron will be only of the order of thermal energies (≈ 1 Mev), the product nucleus has still an excitation of the order of E_d+3 Mev. Thus, if we are dealing with the capture of 6-Mev deuterons by uranium, we have a good possibility of obtaining fission at either one of two distinct stages of the ensuing nuclear reaction.

The cross section for fission in the double reaction just considered can be estimated by multiplying the corresponding fission cross section (42) for neutrons by a factor allowing for the effect of the electrostatic repulsion of the nucleus in hindering the capture of a deuteron:

$$\sigma_f\sim\pi R^2e^{-P}\{\Gamma_f(E')/\Gamma(E') + [\Gamma_n(E')/\Gamma(E')][\Gamma_f(E'')/\Gamma(E'')]\}. \quad (71)$$

Here P is the new Gamow penetration exponent for a deuteron of energy E and velocity v:[38]

$$P=(4Ze^2/\hbar v)\{\arccos x^{\frac{1}{2}}-x^{\frac{1}{2}}(1-x)^{\frac{1}{2}}\}, \quad (72)$$

[38] H. A. Bethe, Rev. Mod. Phys. **9**, 163 (1937).

with $x=(ER/Ze^2)$. πR^2 is the projected area of the nucleus. E' is the excitation of the compound nucleus, and E'' the average excitation of the residual nucleus formed by neutron emission. For deuteron bombardment of U^{238} at 6 Mev we estimate a fission cross section of the order of

$$\pi(9\times10^{-13})^2\exp(-12.9)\sim10^{-29}\text{ cm}^2 \quad (73)$$

if we make the reasonable assumption that the probability of fission following capture is of the order of magnitude unity. Observations are not yet available for comparison with our estimate.

Protons will be more efficient than deuterons for the same bombarding energy, since from (72) P will be smaller by the factor $2^{\frac{1}{2}}$ for the lighter particles. Thus for 6-Mev protons we estimate a cross section for production of fission in uranium of the order

$$\pi(9\times10^{-13})^2\exp(-12.9/2^{\frac{1}{2}})(\Gamma_f/\Gamma)\sim10^{-28}\text{ cm}^2,$$

which should be observable.

B. Photo-fission

According to the dispersion theory of nuclear reactions, the cross section presented by a nucleus for fission by a gamma-ray of wavelength $2\pi\lambda\!\!\!^{-}$ and energy $E=\hbar\omega$ will be given by

$$\sigma_f=\pi\lambda\!\!\!^{-2}(2J+1)/2(2i+1)\frac{\Gamma_{r'}\Gamma_f}{(E-E_0)^2+(\Gamma/2)^2} \quad (74)$$

if we have to do with an isolated absorption line of natural frequency E_0/h. Here $\Gamma_{r'}/\hbar$ is the probability per unit time that the nucleus in the excited state will lose its entire excitation by emission of a single gamma-ray.

The situation of most interest, however, is that in which the excitation provided by the incident radiation is sufficient to carry the nucleus into the region of overlapping levels. On summing (74) over many levels, with average level spacing d, we obtain

$$\sigma_f=\pi\lambda\!\!\!^{-2}[(2J_m+1)/2(2i+1)](2\pi/d)\Gamma_{r'}\Gamma_f/\Gamma. \quad (75)$$

Without entering into a detailed discussion of the orders of magnitude of the various quantities involved in (75), we can form an estimate of the cross section for photo-fission by comparison with the yields of photoneutrons reported by various observers. The ratio of the cross sections in question will be just Γ_f/Γ_n, so that

$$\sigma_f = (\Gamma_f/\Gamma_n)\sigma_n. \tag{76}$$

The observed values of σ_n for 12 to 17 Mev gamma-rays are $\sim 10^{-26}$ cm^2 for heavy elements.[39] In view of the comparative values of Γ_f and Γ_n arrived at in Section IV, it will therefore be reasonable to expect values of the order of 10^{-27} cm^2 for photo-fission of U^{238}, and 10^{-28} cm^2 for division of Th^{232}. Actually no radiative fission was found by Roberts, Meyer and Hafstad using the gamma-rays from 3 microamperes of 1-Mev protons bombarding either lithium or fluorine.[40] The former target gives the greater yield, about 7 quanta per 10^{10} protons, or 8×10^5 quanta/min. altogether. Under the most favorable circumstances, all these gamma-rays would have passed through that thickness, $\sim$6 mg/cm^2, of a sheet of uranium from which the fission particles are able to emerge. Even then, adopting the cross section we have estimated, we should expect an effect of

$$8\times 10^5\times 10^{-27}\times 6\times 10^{-3}\times 6.06 \times 10^{23}/238 \sim 1 \text{ count/80 min}; \tag{77}$$

which is too small to have been observed. Consequently, we have as yet no test of the estimated theoretical cross section.

Conclusion

The detailed account which we can give on the basis of the liquid drop model of the nucleus, not only for the possibility of fission, but also for the dependence of fission cross section on energy and the variation of the critical energy from nucleus to nucleus, appears to be verified in its major features by the comparison carried out above between the predictions and observations. In the present stage of nuclear theory we are not able to predict accurately such detailed quantities as the nuclear level density and the ratio in the nucleus between surface energy and electrostatic energy; but if one is content to make approximate estimates for them on the basis of the observations, as we have done above, then the other details fit together in a reasonable way to give a satisfactory picture of the mechanism of nuclear fission.

---

[39] W. Bothe and W. Gentner, Zeits. f. Physik 112, 45 (1939).

[40] R. B. Roberts, R. C. Meyer and L. R. Hafstad, Phys. Rev. 55, 417 (1939).

PHYSICAL REVIEW VOLUME 70, NUMBERS 9 AND 10 NOVEMBER 1 AND 15, 1946

Resonance Reactions

EUGENE P. WIGNER
Princeton University, Princeton, New Jersey
(Received August 1, 1946)

The considerations of a previous note are extended to include the possibility of several resonance levels. It is shown, in the case of resonance scattering, that R, which is the tangent of the phase shift divided by the wave number, is the sum of the reciprocals of linear functions of the energy (Eq. (12)), each term corresponding to one resonance level. All the coefficients in this expression for R are real, energy independent constants. As a result, it appears most natural to write the cross section (Eq. (12a)) as the square of a ratio of two expressions which are themselves fractional expressions of the energy. It is possible to write the cross section also as the square of a single fractional expression of the energy (Eq. (13)). However, the coefficients of the fractional expression are then slowly varying functions of the energy and are not real but subject to other, more involved limitations. The results are quite similar if, in addition to scattering, a reaction is possible also. The cross sections can be represented, most naturally, by means of the squares of the elements of a matrix (Eq. (33)). However, this matrix is the quotient of two matrices which involve the matrix $\mathfrak{R}$ and only $\mathfrak{R}$ is a simple function (Eq. (33a)) of the energy. The cross sections can be evaluated in a simple closed form only if either there is only one pair of reaction products possible (in addition to the reacting pair) (Eqs. (35), (35a)), or if there are only two resonances present. It is shown, however, that the cross sections can be represented also in the usual form (Eq. (42)) but the "constants" of this form are not strictly independent of energy and not real any more but subject to more involved restrictions. It turns out that the cross section becomes zero between consecutive resonances if only elastic scattering is possible. The elastic scattering cross section does not become zero in general for any value of the energy if a nuclear reaction or inelastic scattering is also possible. If the collision can yield, instead of the colliding pair of particles, only another pair, the cross section for the production of this pair will become zero between successive resonances if the product of certain real quantities has the same sign for both resonances. If the collision may result in any of three or more pairs of particles (e.g., $H^2+H^2\rightarrow H^1+H^3$, or He^3+neutron, or $He^4+\gamma$) no cross section will vanish, in general, for any value of the energy. The considerations of the present paper are restricted to the case in which the relative angular momenta of the reaction products as well as of the reacting particles vanishes.

INTRODUCTION

It was attempted, in a previous article,[1] to give a derivation of the resonance formula using a minimum of arbitrary assumptions. It was found useful to consider the wave function in that part of the configuration space in which all particles of the two colliding nuclei are close together. This part of the wave function may be called internal wave function. A reasonable generalization of the usual resonance formulae could be obtained by assuming that the internal wave function is, within the resonance region, in first approximation independent of energy and also the same, no matter whether the system was formed by the collision of the two particles appearing on the left side or of the two particles appearing on the right side of the equation describing the nuclear reaction (thus, e.g., that the internal wave function is the same no matter whether the compound state is obtained by the collision of Li^6 with H^2, or of Li^7 with H^1 or of He^4 with He^4). On the basis of this assumption, which is supposed to be valid in the first approximation, the second approximation could be calculated and gave the formulae for scattering and collision cross sections in question. An *a posteriori* justification of the procedure could be obtained by estimating the third approximation. This was found not to affect the results obtained in the second approximation to any appreciable extent if the constants appearing in the second approximation have values similar to those which seem to be, on the present experimental evidence, the usual ones and if the inaccuracy of the wave function in the inside of the internal region is not much greater than on the surface of it. This last assumption restricts the validity of the formulae obtained to a region which is limited by the resonance levels nearest to the one considered.

The present note will deal with the case of several resonance levels with a view, in particular, to obtaining the behavior of the cross sections, etc. in the region between two resonance

[1] E. P. Wigner, Phys. Rev. 70, 15, 1946. This paper will be referred to as "previous note."

energies.[2] The greater generality of the conditions, from the physical point of view, will be reflected, mathematically, in the assumption that the internal wave function is, in the first approximation, a linear combination, with energy dependent coefficients, of several energy independent wave functions, corresponding to the several resonance levels which play a role. In order to simplify the analysis, it will be further assumed that there are energy values, the so-called resonance levels, for which the phase shift is just π, i.e., for which the outgoing wave has just opposite sign to what would be the outgoing wave if there was no nuclear interaction present between the colliding particles. This assumption will be formulated more sharply later for the case in which the compound state can be obtained from more than one pair of nuclear particles, i.e., for the case in which a nuclear reaction is possible. This assumption was not made in the note referred to above and inasmuch as the final formulae show the existence of an energy value as postulated here, the assumption has been proved in the case of a single resonance level.

RESONANCE SCATTERING

We now proceed with the consideration of the case in which the compound nucleus can disintegrate only in one way which is then the same way in which it was formed. No reaction can take place in this case and the only phenomenon which occurs is that of scattering. We shall assume that the relative angular momentum of the colliding particles is zero. The wave function in the peripheral region of the configuration space (outside the internal region) is then

$$\varphi_1=(4\pi)^{-\frac{1}{2}}r^{-1}u_1^{-1}[\exp(-ik_1r) - U_1\exp(ik_1r)]\psi(i). \quad (1)$$

The same convention is being used here as before: φ_1 is the wave function of the stationary state with energy E_1 and unit flux of the incoming wave, k_1 is the wave vector $v_1=u_1^2=\hbar k_1/M$ the relative velocity of the particles for energy E_1. The $\psi(i)$ is the normalized, real wave function of the internal coordinates of the colliding particles, r is their distance. U_1 is the quantity to be determined; it gives by the equation $U_1=\exp(2i\delta_1)$ the phase shift and hence the scattering cross section at energy E_1. Equation (1) is valid in the peripheral region, i.e., if $r>a$. In the internal region, we define

$$\varphi_1=\Psi_1 \quad (r<a). \quad (2)$$

Integration of the equation $\varphi_2^*H\varphi_1-\varphi_1(H\varphi_2)^*=(E_1-E_2)\varphi_1\varphi_2^*$ (where H is the Hamilton operator) over the internal region gives by means of Green's theorem

$$-\frac{\hbar^2ik_1}{2Mu_1u_2}(\exp(ik_2a)-U_2^*\exp(-ik_2a))(-\exp(-ik_1a)-U_1\exp(ik_1a))$$
$$+\frac{\hbar^2ik_2}{2Mu_1u_2}(\exp(ik_2a)+U_2^*\exp(-ik_2a))(\exp(-ik_1a)-U_1\exp(ik_1a))=(E_1-E_2)\int\Psi_2^*\Psi_1. \quad (3)$$

The last integral, as all integrals of Ψ, has to be extended over the internal region. It was pointed out in the previous note that (3) can be derived also by considering the material balance in the internal region for the wave function

$$\varphi_1\exp(-iE_1t/\hbar)+\varphi_2\exp(-iE_2t/\hbar),$$

which is a non-stationary solution of Schrodinger's equation.

One can eliminate the exponentials from (3) by the substitution

$$\bar{U}=Ue^{2ika};\quad \bar{\Psi}=\Psi e^{ika}, \quad (4)$$

after which (3) becomes

$$k_1(1-\bar{U}_2^*)(1+\bar{U}_1)+k_2(1+\bar{U}_2^*)(1-\bar{U}_1)$$
$$=-2iMu_1u_2(E_1-E_2)\hbar^{-2}\int\Psi_2^*\Psi_1. \quad (5)$$

This is quite analogous to Eq. (8) of the preceding note.[1] The reality condition is obtained by noting that the conjugate imaginary of (1) and

[2] P. L. Kapur and R. Peierls, Proc. Roy. Soc. A166, 277 (1938); H. A. Bethe and G. Placzek, Phys. Rev. 51, 450 (1937); F. Kalckar, J. R. Oppenheimer, and R. Serber, Phys. Rev. 52, 273 (1937); H. A. Bethe, Rev. Mod. Phys. 9, 71 (1937) (pages 101–117); G. Breit, Phys. Rev. 58, 1068 (1940); 69, 472 (1946).

(2) must be, apart from a constant, identical with (1) and (2). The constant is easily seen to be $-U_1^*$. This gives

$$U_1^*U_1=|U_1|^2=1;\quad \Psi_1^*=-U_1^*\Psi_1. \tag{6}$$

Both equations are valid, of course, not only for the quantities with the index 1 but for any other index, corresponding to any other value of the energy, as well. The first shows that, in a stationary state, the intensities of incoming and outgoing waves are equal. Both equations hold equally for the unbarred quantities for which they are written and for the barred quantities of (4).

It was assumed in the previous note that all $\Psi_1, \Psi_2, \cdots$, etc. are multiples of one definite Ψ. This assumption will be generalized now to the assumption that the Ψ are linear combinations, with energy dependent coefficients, of a set of energy independent functions. Evidently, there is a certain arbitrariness in the choice of these functions. It will be assumed that these functions $\Psi_\lambda, \Psi_\mu, \cdots$, etc., are the internal wave functions for energy the values $E_\lambda, E_\mu, \cdots$ for which U becomes -1, i.e., for the various centers of resonance levels. We have, hence,

$$\Psi_1=\alpha_{1\lambda}\Psi_\lambda+\alpha_{1\mu}\Psi_\mu+\cdots; \tag{7}$$

$$U_\lambda=U_\mu=\cdots=-1;\quad \alpha_{\lambda\mu}=\delta_{\lambda\mu}. \tag{7a}$$

Because of (7a) and (6), the Ψ_λ are real. The assumptions embodied in (7), (7a) do not follow from simple physical postulates. A comparison with the results of the previous note will show, however, that they are reasonable generalizations of the results obtained there. It is also believed that they could be derived, by a little more algebra, in a way similar to that given there, from the same postulates, and that they could be verified afterwards in an entirely similar fashion to that used there.

Applying (5) for $E_1=E_\lambda$, $E_2=E_\mu$ and neglecting the difference between barred and unbarred quantities, the left side of (5) vanishes because of (7a) and one has

$$\int \Psi_\mu^*\Psi_\lambda=\delta_{\mu\lambda}c_\lambda, \tag{8}$$

i.e., that the internal wave functions of the various resonance levels are orthogonal. It then follows that, in general,

$$\int \Psi_2^*\Psi_1=\alpha_{2\lambda}^*\alpha_{1\lambda}c_\lambda+\alpha_{2\mu}^*\alpha_{1\mu}c_\mu+\cdots. \tag{9}$$

Applying now (5) for $E_2=E_\lambda$ gives

$$2k_1(1+U_1)=-2iMu_\lambda u_1(E_1-E_\lambda)h^{-2}\alpha_{1\lambda}c_\lambda$$

or

$$\alpha_{1\lambda}=ih^2k_1(1+U_1)/Mu_\lambda u_1(E_1-E_\lambda)c_\lambda. \tag{10}$$

Similar equations hold, of course, for $\alpha_{1\mu}$, etc. so that all the α are determined by (10) in terms of the U.

Dividing now (5) by $k_1k_2(1+U_1)(1+U_2^*)$ gives because of (9) and (10)

$$\frac{1}{k_2}\frac{1-U_2^*}{1+U_2^*}+\frac{1}{k_1}\frac{1-U_1}{1+U_1}=-\frac{2ih^2(E_1-E_2)}{M}\left\{\frac{1}{v_\lambda c_\lambda(E_1-E_\lambda)(E_2-E_\lambda)}+\frac{1}{v_\mu c_\mu(E_1-E_\mu)(E_2-E_\mu)}+\cdots\right\}$$

$$=2ih\left\{\frac{1}{k_\lambda c_\lambda}\left(\frac{1}{E_1-E_\lambda}-\frac{1}{E_2-E_\lambda}\right)+\frac{1}{k_\mu c_\mu}\left(\frac{1}{E_1-E_\mu}-\frac{1}{E_2-E_\mu}\right)+\cdots\right\}. \tag{11}$$

It then follows that the R_1 defined by (12) becomes

$$\frac{i}{k_1}\frac{1-U_1}{1+U_1}=R_1$$

$$=\frac{2h}{k_\lambda c_\lambda(E_\lambda-E_1)}+\frac{2h}{k_\mu c_\mu(E_\mu-E_1)}+\cdots+R_\infty, \tag{12}$$

where R_∞ is independent of energy. The real quantities R defined in (12) are M/h times the quantities S defined in the previous note; (12) itself is the generalization of (14a) given there. The c_λ, c_μ, etc., are, because of (8), all positive real quantities, they correspond to $(2/b)^2$ of the previous note. The notation R has been chosen for the quantity appearing in (12) because it has the dimension of a length and because its square is closely related by (13) to the scattering cross section. However, both R and also the

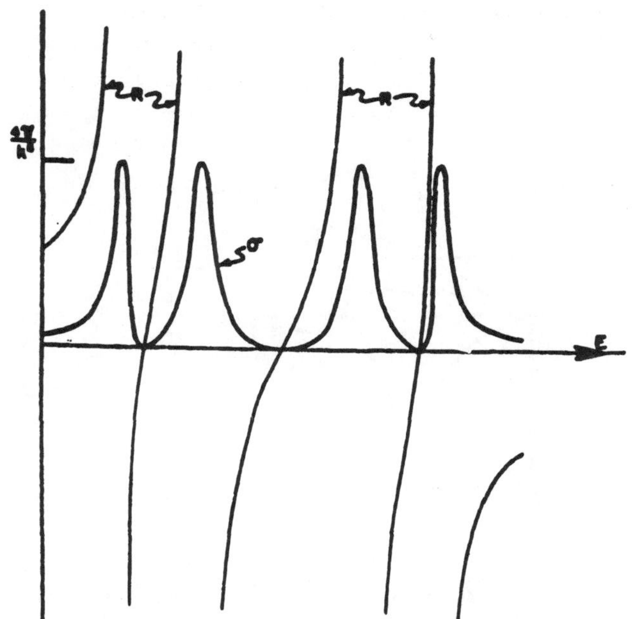

FIG. 1. Resonance scattering. R, i.e., tangent of phase shift divided by the wave number, in arbitrary units and scattering cross section σ in units of maximum possible cross section $4\pi/k^2$ (schematic).

constant R_∞ can be negative as well as positive. From $U=e^{2i\delta}$ and the formula for the cross section in terms of the phase shift δ one obtains for the scattering cross section

$$\sigma(E)=\frac{4\pi}{k^2}\sin^2\delta=\frac{4\pi R^2}{1+k^2R^2}. \qquad (12a)$$

This shows that the cross section is $4\pi R^2$ as long as $kR\ll 1$. It assumes its maximum possible value of $4\pi/k^2$ if $R=\infty$.

Figure 1 shows the general trend of R as function of energy and also σ as function of energy. It shows, in the instance represented, the familiar pattern of sharp lines. As long as the lines are as well separated as in Fig. 1, the width of the line at E_λ from half-maximum to half-maximum is given by

$$\Gamma_\lambda=4\hbar/c_\lambda. \qquad (12b)$$

Equation (12a) can also be written in the form

$$\sigma(E)=4\pi|S|^2;\quad S=R/(1-ikR). \qquad (13)$$

S is the quotient of two rational functions of E, i.e., itself a rational function. Since the degree of its numerator is not higher than the degree of its denominator, it can be written in the form

$$S=S_0+\sum_\nu\frac{s_\nu}{F_\nu-E}. \qquad (13a)$$

The sum of (13a) contains as many terms as are resonance levels, i.e., as there are terms in the sum of (12). If R_∞ of (12) vanishes, S_0 will vanish also. Although (13), (13a) are, mathematically, completely equivalent to (12) and (12a) and although they give a simpler expression for the cross section than (12) and (12a) do, the latter seem to me preferable for three reasons. First, the symbols s and F which occur in (13a) are in general complex while those in (12) are all real. Second, there are no restrictions on the constants E_λ, c_λ, except that they are real and the latter ones positive. The s_ν and F_ν of (13c) satisfy complicated equations (not to be given here) which express the fact that $R=S(1+ikS)^{-1}$ is real. Finally, the E_λ, c_λ are strictly independent of energy while s_ν and F_ν depend on the energy through k. They are, of course, slowly varying functions of the energy and their energy dependence is really significant only for very low k, e.g., for neutrons near the thermal region.

A rather good approximation for S is given by

$$S_{\text{appr}}=\sum_\lambda\frac{\frac{1}{2}k\Gamma_\lambda/k_\lambda}{E_\lambda-\frac{1}{2}ik\Gamma_\lambda/k_\lambda-E}, \qquad (14)$$

which is an expression often given in the literature.[3] Equation (14) is accurate as long as the energy E is not in the neighborhood of more than one resonance level E_λ, neighborhood meaning a distance of the order Γ_λ. It is always accurate if the energy actually coincides with one of the resonance levels and a good approximation for all E if the resonance levels are all distinct, i.e., their distance is greater than their width. As a result, (14) is, under ordinary conditions, a very good approximation for every E. One

[3] Cf. e.g., H. A. Bethe and G. Placzek, reference 2. The development leading to (13), (13a) is very similar to sections 4 and 5 of G. Breit's last paper (cf. reference 2). Breit already pointed out the limitations of this representation of the resonance formula as they manifest themselves in his model. In particular, he noted that although, in first approximation, the absolute value of our ks_ν is oppositely equal to the imaginary part of F_ν (cf. (14)), this does not hold rigorously and that, in general, the absolute value of ks_ν is greater than the imaginary part of F_ν. This is an inequality which follows from the equations to which the s_ν and F_ν are subject.

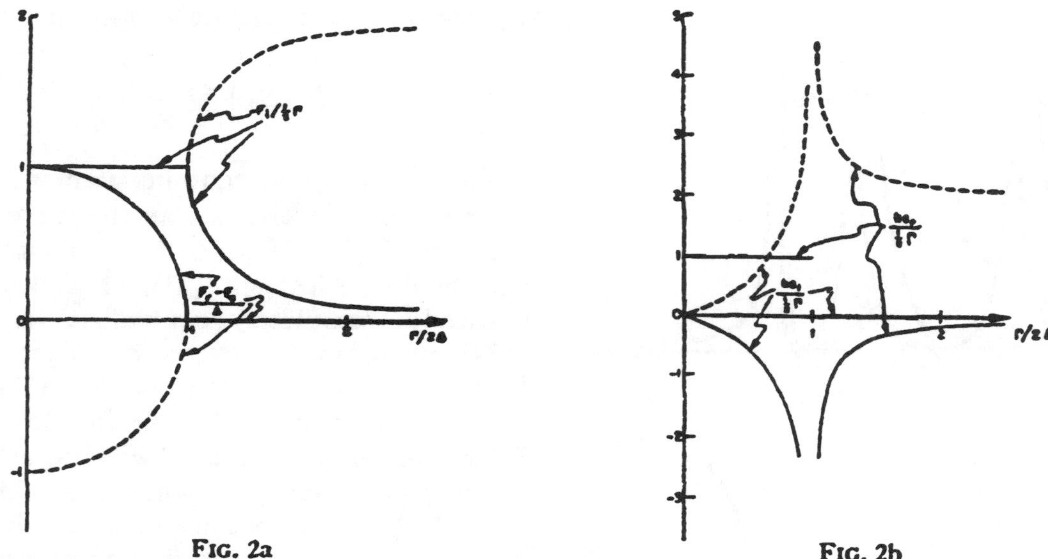

FIG. 2a FIG. 2b

FIG. 2. Relation between two descriptions of resonance scattering. Two resonances at $E_0-\Delta$ and $E_0+\Delta$, each of width Γ, give apparent resonances at F. The real parts F_r of the two F are given in Fig. 2a in units of Δ (one of them by a broken line), the negative imaginary parts in units of $\frac{1}{2}\Gamma$. The real parts of the two F are the same for $\Gamma>2\Delta$, the imaginary parts for $\Gamma<2\Delta$. Figure 2b gives the real and imaginary parts s_r and s_i of the strengths of the two apparent resonances in units $\frac{1}{2}\Gamma$.

sees this also by calculating

$$S-S_{appr}=(1-ikR)^{-1}\times\sum_{\lambda\neq\mu}\frac{\frac{1}{2}k\Gamma_\lambda/k_\lambda}{E_\lambda-\frac{1}{2}ik\Gamma_\lambda/k_\lambda-E}\,\frac{\frac{1}{2}k\Gamma_\mu/k_\mu}{E_\mu-E}. \tag{14a}$$

Because of the $\lambda\neq\mu$ condition, one of the factors after the summation sign is always small if all the resonances are distinct. The second factor can become very large if $E=E_\mu$ but is, in this case, compensated by the factor before the summation sign. In spite of all this, (14) can become grossly inaccurate, e.g., between two levels the width of which is comparable with their distance. In this case (13) with S_{appr} substituted for S can even become larger than the maximum possible width $4\pi/k^2$. The reason is, of course, that the approximate values of s_ν and F_ν (i.e., $\frac{1}{2}k\Gamma_\nu/k_\nu$ and $E_\nu-\frac{1}{2}ik\Gamma_\nu/k_\nu$) which were adopted in (14) do not satisfy the conditions to which the actual s_ν and F_ν are subject.

Figures 2a and 2b illustrate the behavior of the quantities s_ν and F_ν in the particularly simple case of only two resonance levels. These are assumed to be at $E_0-\Delta$ and $E_0+\Delta$ and to have both the same width Γ. The energy E_0 is supposed to be so high as compared with Δ or Γ that the variation of k with energy can be neglected. Figure 2a gives the real part and the negative imaginary part of the two $F-E_0$, in units of Δ and $\frac{1}{2}\Gamma$, respectively, both as function of $\Gamma/2\Delta$. The former is ±1 in the approximation in which (14) is valid, the latter 1. One sees that the imaginary part of both F is indeed $-\frac{1}{2}\Gamma$ as long as $\Gamma<2\Delta$ but that this does not hold if $\Gamma>2\Delta$. In the latter case, both F are purely imaginary, the real parts of both having gone to zero as Γ approached Δ. The s are shown in Fig. 2b. They are complex for $\Gamma<2\Delta$ but real for $\Gamma>2\Delta$. The absolute value of ks is always greater than the imaginary part of the corresponding F, a fact which has been recognized already by Breit.[3] It is worth while to note that, in spite of this, the absolute value of kS never exceeds 1. Inasmuch as we believe that the Γ and E_λ are the physically significant quantities, these figures illustrate the somewhat artificial nature of the representation (13), (13a).

RESONANCE REACTIONS

The treatment of the case in which, in addition to elastic scattering a real nuclear reaction (or, at least, inelastic scattering) is possible, differs from the above one in that the peripheral wave function has, instead of (1), the form

$$\varphi_{1f}=\sum_l (4\pi)^{-\frac{1}{2}}r_l^{-1}u_{1l}^{-1}[\delta_{fl}\exp(-ik_{1l}r_l) - U_{1fl}\exp(ik_{1l}r_l)]\psi(i_l). \tag{15}$$

In this as in the following formulae, the indices $j, l, \cdots$ etc. denote pairs of nuclei which can react with each other or appear as products of the reaction. The $\psi(i_l)$ is the product of the real normalized wave functions of the pair l. (E.g., j can denote the pair Li^6+H^2; l the pair Li^7+H^1; m the pair He^4+He^4 etc.) The r_l is the relative distance of the nuclei forming the pair l, the relative velocity of this pair for the energy value E_1 of the total system is denoted by $v_{1l}=u_{1l}^2=\hbar k_{1l}/M_l$, the M_l being the relative mass of the pair l. In (15), because of the δ_{jl}, only nuclei of the pair j approach each other; the U_{1jl} are the elements of the collision matrix. The square $|U_{1jl}|^2$ gives the number of pairs l formed if one pair of the kind j collides.

Equation (15) is valid in the peripheral region. In the internal region, i.e., in the part of the configuration space in which all particles are close together, we have instead of (2)

$$\varphi_{1j}=\Psi_{1j}. \tag{15a}$$

In order to calculate the second approximation we need an equation analogous to (3). This can be obtained again by integration of the equation

$$\varphi_{2j}{}^*H\varphi_{1l}-\varphi_{1l}(H\varphi_{2j})^*=(E_1-E_2)\varphi_{1l}\varphi_{2j}{}^*$$

over the internal region. This yields by Green's theorem

$$-\sum_m \frac{\hbar^2 ik_{1m}}{2M_m u_{1m}u_{2m}}[\delta_{jm}\exp(ik_{2m}a)-U_{2jm}{}^*\exp(-ik_{2m}a)][-\delta_{lm}\exp(-ik_{1m}a)-U_{1lm}\exp(ik_{1m}a)]$$

$$+\sum_m \frac{\hbar^2 ik_{2m}}{2M_m u_{1m}u_{2m}}[\delta_{jm}\exp(ik_{2m}a)+U_{2jm}{}^*\exp(-ik_{2m}a)][\delta_{lm}\exp(-ik_{1m}a)-U_{1lm}\exp(ik_{1m}a)]$$

$$=(E_1-E_2)\int\Psi_{2j}{}^*\Psi_{1l}. \tag{16}$$

In order to eliminate the exponentials, one can again introduce barred quantities

$$\bar{\Psi}_l=\Psi_l\exp(ik_l a);\quad \bar{U}_{lm}=U_{lm}\exp[i(k_m+k_l)a]. \tag{17}$$

With this substitution, and after multiplication with $\exp[i(k_{1l}-k_{2j})a]$, (16) goes over into

$$i\hbar\sum_m(\delta_{lm}+\bar{U}_{1lm})u_{1m}u_{2m}{}^{-1}(\delta_{jm}-\bar{U}_{2jm}{}^*)+i\hbar\sum_m(\delta_{lm}-\bar{U}_{1lm})u_{1m}{}^{-1}u_{2m}(\delta_{jm}+\bar{U}_{2jm}{}^*)$$

$$=2(E_1-E_2)\int\bar{\Psi}_{2j}{}^*\bar{\Psi}_{1l}. \tag{18}$$

This is analogous to Eq. (22) of the previous note. The bars will be left off again forthwith.

We shall now make the assumption mentioned at the end of the Introduction more precise. It will be assumed that there are "resonance energies" $E_\lambda, E_\mu, \cdots$, etc. for which the matrices $\mathfrak{U}_\lambda=\|U_{\lambda lm}\|$ have a characteristic value -1. The corresponding normalized characteristic vectors will be denoted by $\mathfrak{B}_\lambda$ with components $\beta_{\lambda m}$ so that $\mathfrak{U}_\lambda\mathfrak{B}_\lambda=-\mathfrak{B}_\lambda$, or, more in detail,

$$\sum_m U_{\lambda lm}\beta_{\lambda m}=-\beta_{\lambda l}. \tag{19}$$

This amounts to the assumption that, at the resonance energies, there are superpositions of incident waves

$$\sum_l \beta_{\lambda l}(4\pi)^{-\frac{1}{2}}r_l{}^{-1}u_{\lambda l}{}^{-1}\exp(-ik_{\lambda l}r_l)\psi(i_l), \tag{20a}$$

for which the outgoing wave has, apart from the sign, the same value which it would have if there was no nuclear reaction or scattering, but that the sign of the outgoing wave is opposite. This means that the outgoing wave for the superposition of incoming waves given by (20a) is

$$\sum_l \beta_{\lambda l}(4\pi)^{-\frac{1}{2}}r_l{}^{-1}u_{\lambda l}{}^{-1}\exp(ik_{\lambda l}r_l)\psi(i_l). \tag{20b}$$

The internal wave function for the incoming wave (20a) (and the outgoing wave (20b)) will be denoted by Ψ_λ and it will further be assumed that these are the $\Psi_\lambda, \Psi_\mu, \cdots$ which permit one to express, in first approximation, the internal wave functions linearly for all energies and all incident waves:

$$\Psi_{1j}=\alpha_{1\lambda}{}^j\Psi_\lambda+\alpha_{1\mu}{}^j\Psi_\mu+\cdots. \tag{21}$$

The coefficients α depend both on the energy (E_1 in (21)) and also on the pair of particles (j in (21)) the collision of which is represented by the wave function Ψ_{1j} of (15a). The fact that the internal wave function is Ψ_λ for energy E_λ if the incoming wave is given by (20a) means that

$$\Psi_\lambda = \sum_j \beta_{\lambda j} \Psi_{\lambda j}$$
$$= \sum_j \beta_{\lambda j}(\alpha_{\lambda\lambda}{}^j \Psi_\lambda + \alpha_{\lambda\mu}{}^j \Psi_\mu + \cdots). \quad (22)$$

This is equivalent with

$$\sum_j \beta_{\lambda j} \alpha_{\lambda\mu}{}^j = \delta_{\lambda\mu}, \quad (23)$$

which is the analogon of the second equation of (7a). Since the Ψ_λ, Ψ_μ, $\cdots$, etc. can be considered as solutions of the same characteristic value problem with the same homogeneous boundary conditions embodied in (20a), (20b), they are mutually orthogonal

$$\int \Psi_\mu{}^* \Psi_\lambda = \delta_{\mu\lambda} c_\lambda; \quad c_\lambda = c_\lambda{}^* > 0, \quad (24)$$

which is the analogon of (8). Equation (24) follows also from the more general equation (18) if one inserts λ for 1, μ for 2 and multiplies it with $\beta_{\lambda l}\beta_{\mu j}{}^*$, sums over l and j and considers (19), the symmetric nature of $\mathfrak{U}$ and the first part of (22).

No assumption analogous to the one contained in (21) was made in the previous note and it is, in fact, believed that this assumption follows from the other more general assumptions made in the Introduction. This was explicitly demonstrated in the previous note for the case of a single resonance level: the energy E_λ was denoted there by E_0 (there was only one E_λ), the corresponding vector $\mathfrak{b}_\lambda$ was denoted there by $\mathfrak{u}_0\mathfrak{b}$ (with components $u_{0l}\beta_l$), and all internal wave functions were the same apart from constant factors. The existence of both the E_0 and the vector $\mathfrak{b}$ was derived there while it is assumed here because it simplifies the analysis to such a great extent.

Since (20b) is the conjugate complex of (20a), (the vector $\mathfrak{b}$ will be shown in (26) to be real) the peripheral wave function is real and the same must hold then for Ψ_λ. This also follows from the more general equations which are the generalizations of (6).

$$\sum_l U_{1jl}{}^* U_{1lm} = \delta_{jm}; \quad -\sum_l U_{1jl}{}^* \Psi_{1l} = \Psi_{1j}{}^* \quad (25)$$

One can easily convince oneself that (25) holds both for the unbarred and also for the barred quantities defined in (17). The first equation of (25) reads, in matrix notation $\mathfrak{U}_1{}^*\mathfrak{U}_1 = 1$ and is, because of the unitary nature of $\mathfrak{U}$, equivalent with the statement that $\mathfrak{U}$ is symmetric. Both (25) and (16) or (18) are rigorous, i.e., do not involve any approximations. However, (21) has only approximate validity.

Introducing now (21) into (18), the right side of this goes over into

$$2(E_1 - E_2) \sum_\lambda \alpha_{2\lambda}{}^{j*} \alpha_{1\lambda}{}^l c_\lambda. \quad (18')$$

We can define the matrix product of two vectors $\mathbf{x}$ and $\mathbf{y}$ as the matrix $\mathbf{x}\times\mathbf{y}$ the j, l element of which is $(\mathbf{x}\times\mathbf{y})_{jl} = x_j y_l$; (18) then becomes in the matrix-vector notation which has been used also in the previous note

$$i\hbar(1+\mathfrak{U}_1)\mathfrak{u}_1\mathfrak{u}_2{}^{-1}(1-\mathfrak{U}_2{}^\dagger)$$
$$+ i\hbar(1-\mathfrak{U}_1)\mathfrak{u}_1{}^{-1}\mathfrak{u}_2(1+\mathfrak{U}_2{}^\dagger)$$
$$= 2(E_1 - E_2) \sum_\lambda c_\lambda(\alpha_{1\lambda} \times \alpha_{2\lambda}{}^*). \quad (18a)$$

All matrices (and vectors) have as many dimensions as there are ways of disintegration for the compound state, the indices 1, 2 give the energy values E_1, E_2 to which the quantities refer, $\mathfrak{u}_1$, $\mathfrak{u}_2$ are diagonal matrices with diagonal elements u_{1l}, u_{2l}, respectively, $\alpha_{1\lambda}$ is a vector with components $\alpha_{1\lambda}{}^l$, etc. Equation (19) then becomes

$$\mathfrak{U}_\lambda \mathfrak{b}_\lambda = -\mathfrak{b}_\lambda, \quad (19a)$$

while (23) goes over into

$$\mathfrak{b}_\lambda \cdot \alpha_{\lambda\mu} = \delta_{\lambda\mu}. \quad (23a)$$

It may be worth while to remark that because of (19a) and the symmetric nature of $\mathfrak{U}_\lambda$, the vectors $\mathfrak{b}_\lambda$ can be chosen to be real and that they can be assumed to be normalized

$$\mathfrak{b}_\lambda = \mathfrak{b}_\lambda{}^*; \quad \mathfrak{b}_\lambda \cdot \mathfrak{b}_\lambda = \sum_l \beta_{\lambda l}{}^2 = 1. \quad (26)$$

One can substitute E_μ for E_2 in (18a) and apply it to the vector $\mathfrak{b}_\mu{}^*$, i.e., substitute E_μ for E_2 in (18), multiply it with $\beta_{\mu j}{}^*$, and sum over j. Because of the symmetry of $\mathfrak{U}_\mu$ and (19), the second term on the left side drops out, the first simplifies considerably and one has

$$2i\hbar(1+\mathfrak{U}_1)\mathfrak{u}_1\mathfrak{u}_\mu^{-1}\beta_\mu^* = 2(E_1-E_\mu)\sum_\lambda c_\lambda(\alpha_{1\lambda}\times\alpha_{\mu\lambda}^*)\beta_\mu^* = 2(E_1-E_\mu)c_\mu\alpha_{1\mu}. \quad (27)$$

The last part follows from (23) or (23a) and the general equation

$$(\mathbf{x}\times\mathbf{y})\mathbf{z}=(\mathbf{y}\cdot\mathbf{z})\mathbf{x}. \quad \text{(I)}$$

From (27), the vector $\alpha_{1\mu}$ becomes

$$\alpha_{1\mu}=\frac{i\hbar}{c_\mu(E_1-E_\mu)}(1+\mathfrak{U}_1)\mathfrak{u}_1\mathfrak{u}_\mu^{-1}\beta_\mu. \quad (27a)$$

As was pointed out before, β_μ can be assumed to be real. Inserting (27a) into (18a) one obtains for the right side of the latter

$$\sum_\lambda \frac{2(E_1-E_2)\hbar^2}{c_\lambda(E_1-E_\lambda)(E_2-E_\lambda)}(1+\mathfrak{U}_1)\mathfrak{u}_1 \times(\mathfrak{u}_\lambda^{-1}\beta_\lambda\times\mathfrak{u}_\lambda^{-1}\beta_\lambda)\mathfrak{u}_2(1+\mathfrak{U}_2^\dagger) \quad (28)$$

because of the general equation

$$(\mathfrak{a}\mathbf{x}\times\mathfrak{b}\mathbf{y})=\mathfrak{a}(\mathbf{x}\times\mathbf{y})\mathfrak{b}' \quad \text{(II)}$$

valid for all matrices $\mathfrak{a}$, $\mathfrak{b}$ ($\mathfrak{b}'$ being the transposed of $\mathfrak{b}$) and all vectors $\mathbf{x}$, $\mathbf{y}$. Thus (18a) becomes after multiplication with $\mathfrak{u}_1^{-1}(1+\mathfrak{U}_1)^{-1}$ from the left, $(1+\mathfrak{U}_2^\dagger)^{-1}\mathfrak{u}_2^{-1}$ from the right and division by $i\hbar$

$$\mathfrak{u}_2^{-1}(1-\mathfrak{U}_2^\dagger)(1+\mathfrak{U}_2^\dagger)^{-1}\mathfrak{u}_2^{-1}+\mathfrak{u}_1^{-1}(1+\mathfrak{U}_1)^{-1}(1-\mathfrak{U}_1)\mathfrak{u}_1^{-1} -\sum_\lambda\left[\frac{2i\hbar}{c_\lambda(E_1-E_\lambda)}-\frac{2i\hbar}{c_\lambda(E_2-E_\lambda)}\right]\mathfrak{u}_\lambda^{-1}(\beta_\lambda\times\beta_\lambda)\mathfrak{u}_\lambda^{-1}=0. \quad (29)$$

It then follows that the terms which depend on E_1 (and similarly the terms which depend on E_2) give an energy independent matrix. One can clearly replace in (29) all the diagonal matrices $\mathfrak{u}^{-1}=\mathfrak{v}^{-\frac{1}{2}}$ by the diagonal matrices $\mathfrak{q}^{-1}=\mathfrak{k}^{-\frac{1}{2}}$ and obtain for the matrix $\mathfrak{R}_1$

$$\mathfrak{R}_1=i\mathfrak{q}_1^{-1}(1+\mathfrak{U}_1)^{-1}(1-\mathfrak{U}_1)\mathfrak{q}_1^{-1} =\mathfrak{R}_\infty+\sum_\lambda\frac{2\hbar}{c_\lambda(E_\lambda-E_1)}\mathfrak{q}_\lambda^{-1}(\beta_\lambda\times\beta_\lambda)\mathfrak{q}_\lambda^{-1}. \quad (30)$$

This is the analogon of the final result (12) for scattering and represents the general solution of our problem just as (36), (36a) of the previous note represented the general solution in the case of a single resonance level. $\mathfrak{R}_\infty$ is an arbitrary real symmetric matrix, the E_λ are arbitrary real energy values, the c_λ are positive real numbers the β_λ real vectors and they can be assumed to be normalized, or, if not, $c_\lambda=1$ can be assumed.

The matrix $\mathfrak{U}_1-1$ can be expressed in terms of $\mathfrak{R}_1$ as

$$\mathfrak{U}_1-1=2i\mathfrak{q}_1\mathfrak{R}_1\mathfrak{q}_1/(1-i\mathfrak{q}_1\mathfrak{R}_1\mathfrak{q}_1) \quad (31)$$

and the cross section for the transformation of pair j into pair l is

$$\sigma_{jl}(E_1)=\frac{\pi}{k_{1j}^2}|(\mathfrak{U}_1-1)_{jl}|^2 =\frac{4\pi}{k_{1j}^2}\left|\left(\frac{\mathfrak{q}_1\mathfrak{R}_1\mathfrak{q}_1}{1-i\mathfrak{q}_1\mathfrak{R}_1\mathfrak{q}_1}\right)_{jl}\right|^2. \quad (31a)$$

Although both this expression and that for $\mathfrak{R}_1$ (30) is quite simple, it does not seem possible to express in general the $\sigma_{jl}(E)$ in closed form in terms of the E, k_j, k_l, the energy independent quantities E_λ, β_λ, c_λ, and $\mathfrak{R}_\infty$.

HIGHER SPINS

It is easy to generalize (30), (31a) for the case that the nuclei of the pair j, say, have spins j and j', i.e., angular momenta $j\hbar$ and $j'\hbar$, respectively. If our assumption remains valid, that only those states need to be taken into account in which the angular momentum of the motion of the pair around their common center of mass vanishes, one can consider the collision system formed by the pair j as a mixture of states for which the total angular momentum has one of the values

$$J=|j-j'|,\ |j-j'|+1,\ \cdots,\ j+j'-1,\ j+j'. \quad (32)$$

This total angular momentum consists of the vector sum of the spins of the particles making up the pair j since, according to assumption, the relative motion of the particles has no angular momentum. From the wave functions with a definite J only such compound states Ψ can be reached which have the same J. As a consequence, one must consider as many sets of resonance levels as are J in (32) and define for each set an $\mathfrak{R}^J$ according to (30) in which, then, the summation is to be extended only over the compound states with spin J. The total cross section then becomes the weighted sum of expressions (31a), corresponding to the components of the mixture (32)

$$\sigma_{jl}(E)=\frac{4\pi k_l}{k_j}\sum_{J=|j-j'|}^{j+j'} g_J\left|\left(\frac{\mathfrak{R}^J}{1-iq\mathfrak{R}^J q}\right)_{jl}\right|^2,$$

$$g_J=\frac{2J+1}{(2j+1)(2j'+1)}. \qquad (33)$$

The k_j, q, $\mathfrak{R}^J$ all refer[4] to the energy E in (33). It is, probably, unnecessary to remark that (30) can be written in a somewhat simpler form

$$\mathfrak{R}^J(E)=\mathfrak{R}_\infty{}^J+\sum_\lambda{}^J\frac{\gamma_\lambda\times\gamma_\lambda}{E_\lambda-E}, \qquad (33a)$$

where the J on the summation sign indicates that the summation is to be extended only over those λ for which the angular momentum of the compound state Ψ_λ is $J\hbar$. The real vectors γ_λ are, as are the β_λ and the $\mathfrak{R}_\infty{}^J$, independent of energy

$$\gamma_\lambda=(2\hbar/c_\lambda)^{\frac{1}{2}}q_\lambda{}^{-1}\beta_\lambda,$$
$$\gamma_{\lambda j}=(2\hbar/c_\lambda)^{\frac{1}{2}}k_{\lambda j}{}^{-\frac{1}{2}}\beta_{\lambda j}. \qquad (33b)$$

They are, of course, not normalized.

The $\mathfrak{R}_\infty{}^J$ of (33a) could be, off hand, arbitrary real symmetric matrices. They describe that part of the reaction or scattering which would be present even in the absence of resonance levels. It has been argued, in the previous note, that these $\mathfrak{R}_\infty$ can be considered to be diagonal matrices as long as we can speak of pure resonance reactions. Even if this should not be the case, two remarks will be applicable to the matrix elements $R_{\infty jl}$. First, since they correspond to non-resonance reactions, their order of magnitude will not be greater than the nuclear radius. As a result, their product with $q_jq_l=(k_jk_l)^{\frac{1}{2}}$ will be small compared to 1 except if the energy of both pairs j and l is quite high. Second, $R_{\infty jl}(k_jk_l)^{\frac{1}{2}}$ will be particularly small if either the pair j, or the pair l, consists of a light quantum and a nucleus. If both j and l are such pairs, the process for which $R_{\infty jl}(k_jk_l)^{\frac{1}{2}}$ is responsible is essentially a Compton effect on a nucleus or a similar straight scattering term. Even if only one pair, say j, contains a light quantum, $q_jR_{\infty jl}q_l$ will be a high order correction term which is, e.g., consistently neglected in the usual treatments of light absorption or emission. It therefore appears to be justified to neglect $R_{\infty jl}$ if either of the pairs j or l contains a light quantum.

Since every symmetric matrix can be written as a sum of matrices $(\gamma_\rho\times\gamma_\rho)$, it is possible, formally, to set the $\mathfrak{R}_\infty{}^J$ equal to zero if one admits a few terms in the sum of (33a) for which E_λ is infinite but the γ_λ also infinite in such a way that the corresponding terms give a finite contribution to (33a), replacing $\mathfrak{R}_\infty{}^J$. This is particularly tempting in the case of the (n, γ) reaction. For these reactions, all pairs $l, \cdots$, etc. contain a light quantum (in addition to the normal or excited state of the product nucleus), except one, the initial state j. Because of the remarks made before, all the matrix elements of $\mathfrak{R}_\infty{}^J$ will vanish except the jj element and $\mathfrak{R}_\infty{}^J$ can be replaced by a single term $(\gamma_\infty\times\gamma_\infty)$ with a vector γ_∞ all the components of which vanish, except the j component. Although the elimination of $\mathfrak{R}_\infty{}^J$, which has just been described, is a very formal one, it will be adopted later because it simplifies the formulae at least from a formal point of view. We shall write then

$$\mathfrak{R}^J(E)=\sum_\lambda{}^J\frac{\gamma_\lambda\times\gamma_\lambda}{E_\lambda-E}. \qquad (33c)$$

However, when using (33c) we must keep in

[4] Strictly speaking, the matrix in (33) is undefined because $\mathfrak{R}$ and $q\mathfrak{R}q$ do not commute in general. A comparison with (32) shows that the proper definition is

$$\frac{\mathfrak{R}}{1-iq\mathfrak{R}q}=q^{-1}\frac{q\mathfrak{R}q}{1-iq\mathfrak{R}q}q^{-1}. \qquad (33')$$

The fraction on the right side is completely defined since it is a function of a single matrix $q\mathfrak{R}q$. The right side of (33′) will be meant always when the left side is written for brevity.

mind that it has a few (or, in the case of an (n, γ) reaction, one) finite terms with $E_\lambda = \infty$.

EVALUATION OF (33)

Although the preceding formulae give, in principle, a complete solution of our problem, they are unsuited not only to practical calculations but even for obtaining a qualitative picture of the variation of the cross sections. The difficulty in using (33) (or (31a) and (30)) consists in the evaluating of the elements of the matrix $\mathfrak{q}\mathfrak{R}\mathfrak{q}(1-i\mathfrak{q}\mathfrak{R}\mathfrak{q})^{-1}$ if only the matrix elements of $\mathfrak{R}$ are given.

This difficulty can be easily overcome if there are only two states involved, i.e., if only one pair of reaction products is possible. On the whole, this is an exceptional case as even the (n, γ) reaction has several possible end products corresponding to the different excited states of the product nucleus. Examples in question may be however, some (n, γ) reactions of light elements, and, perhaps, reactions of the kind $Li^7 + n = Be^7 + H^1$.

If there is, in addition to the original pair of nuclei, only one pair of reaction products possible, the matrix $\mathfrak{q}\mathfrak{R}\mathfrak{q}$ becomes two-dimensional and its reciprocal can be found easily. One obtains for

$$\frac{\mathfrak{q}\mathfrak{R}\mathfrak{q}}{1-i\mathfrak{q}\mathfrak{R}\mathfrak{q}} = (1-D-i(S_{jj}+S_{ll}))^{-1} \times \begin{Vmatrix} S_{jj}-iD & S_{jl} \\ S_{jl} & S_{ll}-iD \end{Vmatrix}, \quad (34)$$

in which

$$S_{jj} = k_j\left(R_{jj\infty} + \sum_\lambda \frac{\gamma_{\lambda j}^2}{E_\lambda - E}\right),$$

$$S_{jl} = (k_j k_l)^{\frac{1}{2}}\left(R_{jl\infty} + \sum_\lambda \frac{\gamma_{\lambda j}\gamma_{\lambda l}}{E_\lambda - E}\right) \quad (34a)$$

$$S_{ll} = k_l\left(R_{ll\infty} + \sum_\lambda \frac{\gamma_{\lambda l}^2}{E_\lambda - E}\right),$$

$$D = S_{jj}S_{ll} - S_{jl}^2.$$

It is assumed that all compound states have the same J. As a result of (34), the cross section for scattering becomes

$$\sigma_{jj} = \frac{4\pi}{k_j^2} \frac{S_{jj}^2 + D}{(1-D)^2 + (S_{jj}+S_{ll})^2} \quad (35)$$

while the reaction cross section is

$$\sigma_{jl} = \frac{4\pi}{k_j^2} \frac{S_{jl}^2}{(1-D)^2 + (S_{jj}+S_{ll})^2}. \quad (35a)$$

The remarkable feature of this last expression is that it goes through zero at least between any two consecutive resonances for which $\gamma_{\lambda j}\gamma_{\lambda l}$ has the same sign since, evidently, S_{jl} goes through zero between two such points. It is similar in this respect, to the scattering cross section of Fig. 1 which applies if no reaction is possible. However, the scattering cross section given by (35), which applies if a reaction is possible in addition to the scattering, does not exhibit this feature any more, since it can vanish only if both S_{jj} and D vanish for the same E. Even the reaction cross section does not become zero between consecutive maxima if either more than one pair of reaction products are possible or if there are compound states with more than one value of J. The former case will be investigated below. In the latter case the cross section is, according to (33), a weighted average of expressions of the form (35a). This could vanish only if the expression (35a) would vanish for the same E for every J. Such an occurrence has "zero probability."

If several pairs of reaction products are possible, the matrices $\mathfrak{R}$ and $\mathfrak{q}\mathfrak{R}\mathfrak{q}$ have more than two dimensions and the reciprocal of $1-i\mathfrak{q}\mathfrak{R}\mathfrak{q}$ becomes an involved expression. It seems worth while, in that case, to adopt the convention of (33c) to eliminate the constant matrix $\mathfrak{R}_\infty^J$ which occurs in (33a). The J will again be omitted in the following transformation.

Because of (II), one can write for the denominator of the expression in (33)

$$1 - i\mathfrak{q}\mathfrak{R}\mathfrak{q} = 1 - i\sum_\lambda \frac{(\mathfrak{q}\gamma_\lambda \times \mathfrak{q}\gamma_\lambda)}{E_\lambda - E}. \quad (36)$$

We shall try to write for

$$(1-i\mathfrak{q}\mathfrak{R}\mathfrak{q})^{-1} = 1 + i\sum_{\mu\nu} A_{\mu\nu}(\mathfrak{q}\gamma_\mu \times \mathfrak{q}\gamma_\nu). \quad (36a)$$

The $A_{\mu\nu}$ are, of course, functions of the energy. The product of the right sides of (36a) and (36) must give 1. It can be evaluated by means of the general equation

$$(\mathfrak{x}\times\mathfrak{y})(\mathfrak{z}\times\mathfrak{w}) = (\mathfrak{y}\cdot\mathfrak{z})(\mathfrak{x}\times\mathfrak{w}). \quad (III)$$

One obtains for the $A_{\mu\nu}$ the equations

$$A_{\mu\lambda}-\frac{\frac{1}{2}i}{E_\lambda-E}\sum_\nu A_{\mu\nu}\Gamma_{\nu\lambda}=\frac{\delta_{\mu\lambda}}{E_\lambda-E}, \quad (37)$$

where the $\Gamma_{\nu\lambda}$ are the scalar products

$$\tfrac{1}{2}\Gamma_{\nu\lambda}=(\mathfrak{q}\gamma_\nu\cdot\mathfrak{q}\gamma_\lambda)=\sum_j k_j\gamma_{\nu j}\gamma_{\lambda j}. \quad (37a)$$

The Γ with two indices are, in contrast to the Γ of (13a), functions of the energy because of the k which enters (37a). They are, however, slowly varying functions of the energy. The $A_{\mu\nu}$ will turn out to be rapidly varying functions of E.

Both the $\Gamma_{\nu\lambda}$ and the $A_{\mu\lambda}$ are symmetric in their two indices: $\Gamma_{\nu\lambda}=\Gamma_{\lambda\nu}$; $A_{\mu\lambda}=A_{\lambda\mu}$. As a result of (36a), (37), and (III), the matrix in (33) can also be written as[4]

$$\frac{\Re}{1-i\mathfrak{q}\Re\mathfrak{q}}=\sum_{\mu\lambda} A_{\mu\lambda}(\gamma_\mu\times\gamma_\lambda). \quad (38)$$

This is how far the transformation of (33) can be carried easily without making approximations.

At this point, the possibility of a transformation similar to that given by (13) should be mentioned. Equation (37) is evidently a matrix equation although the rows and columns of the matrices occurring in it do not refer to the different reaction products as in the case of the matrices $\mathfrak{q}$, $\Re$, etc., but refer to the different resonance levels λ, μ, ν, etc. After multiplication with $E_\lambda-E$, (37) can be written as

$$\mathfrak{A}(\mathcal{E}-E1)-\tfrac{1}{2}i\mathfrak{A}\mathcal{G}=1. \quad (39)$$

In this, $\mathcal{E}$ is the diagonal matrix with the diagonal elements E_λ, E_μ, etc., $\mathfrak{A}$ and $\mathcal{G}$ are symmetric matrices with elements $A_{\mu\nu}$ and $\Gamma_{\mu\nu}$. It follows from (39) that

$$\mathfrak{A}=(\mathcal{E}-\tfrac{1}{2}i\mathcal{G}-E1)^{-1}. \quad (39a)$$

If the symmetric matrix $\mathcal{E}-\frac{1}{2}i\mathcal{G}$ has no double characteristic values—which would be, after all, an exceptional case—it will be possible to bring it into the diagonal form $\mathcal{F}$

$$\mathcal{F}=\mathcal{T}(\mathcal{E}-\tfrac{1}{2}i\mathcal{G})\mathcal{T}^{-1}. \quad (39b)$$

Because of the symmetry of $\mathcal{E}-\frac{1}{2}i\mathcal{G}$, one can assume that $\mathcal{T}$ is a complex orthogonal matrix

$$\mathcal{T}^{-1}=\mathcal{T}' \quad (39c)$$

One has because of (39b)

$$\mathcal{E}-\tfrac{1}{2}i\mathcal{G}-E1=\mathcal{T}^{-1}(\mathcal{F}-E1)\mathcal{T} \quad (40)$$

and

$$\mathfrak{A}=\mathcal{T}'(\mathcal{F}-E1)^{-1}\mathcal{T}. \quad (40a)$$

The diagonal elements F_ν of $\mathcal{F}$, as well as the elements $T_{\mu\nu}$ of $\mathcal{T}$ will still be slowly varying functions of the energy. However, the elements of $\mathfrak{A}$ will be

$$A_{\mu\lambda}=\sum_\nu T_{\nu\mu}(F_\nu-E)^{-1}T_{\nu\lambda} \quad (40b)$$

at least in the neighborhood of the real part of F, rapidly varying functions of E.

Because of (40b), (38) becomes

$$\frac{\Re}{1-i\mathfrak{q}\Re\mathfrak{q}}=\sum_{\mu\lambda}\sum_\nu\frac{T_{\nu\mu}T_{\nu\lambda}}{F_\nu-E}(\gamma_\mu\times\gamma_\lambda)$$
$$=\sum_\nu\frac{(\mathbf{s}_\nu\times\mathbf{s}_\nu)}{F_\nu-E}, \quad (41)$$

wherein the vector $\mathbf{s}_\nu$ is defined by

$$s_{\nu l}=\sum_\lambda T_{\nu\lambda}\gamma_{\lambda l}. \quad (41a)$$

Finally, (33) gives for the cross section

$$\sigma_{jl}(E)=\frac{4\pi k_l}{k_j}\sum_J g_J\left|\sum_\nu{}^J\frac{s_{\nu j}s_{\nu l}}{F_\nu-E}\right|. \quad (42)$$

There is, of course, a separate set of F_ν and $s_{\nu j}$ for every value of J and the sum of (42) should be taken in this sense. Equation (42) is the analogon of (13), (13a) and is subject to the same kind of limitations as are those equations: the quantities $s_{\nu l}$, $s_{\nu j}$, F_ν are in general complex: they are subject to rather complicated equations which express the fact that the matrices $\mathcal{E}$ and $\mathcal{G}$ are real; they are functions of the energy although slowly varying functions. Except for these limitations (42) is a very concise form of the resonance formula.[5] It should be remembered, however, that the expressions (34), (35) give the cross sections in terms of the $R_{jj\infty}$, $R_{jl\infty}$, $R_{ll\infty}$, E_λ, γ_λ, etc. which are all real, subject to no equations and independent of energy. Of course, they apply only in the case of the simplest re-

[5] Again, Breit's work anticipates many of the results derived here. Although Breit's paper deals with a rather special model, a comparison of his formulae with ours shows that the model used by him already exhibits practically all the features which prevail in the general case.

actions in which only one pair of reaction products is possible.

In order to derive an expression for the cross sections which is not subject to the limitations to which (42) is subject, one will try to avoid the doubtful diagonalization (39b) and try to solve (37) for the $A_{\mu\lambda}$ more directly. If several reaction products are possible, one will obtain a manageable expression only if one resorts to approximations, and it will be assumed that the energy differences are larger than the widths Γ. It will then turn out that the $A_{\mu\mu}$ are, in general, considerably larger than the $A_{\mu\lambda}$ with $\mu\neq\lambda$. One can, therefore, obtain a first approximation for $A_{\mu\mu}$ by using the equation $\lambda=\mu$ of (37) and neglecting the $A_{\mu\nu}$ terms with $\mu\neq\nu$. This gives

$$A_{\mu\mu}=\frac{1}{E_\mu-E-\frac{1}{2}i\Gamma_{\mu\mu}}. \tag{43}$$

One then obtains for $A_{\mu\lambda}$ from (37) by keeping, from the sum, only the terms $\nu=\lambda$ and $\nu=\mu$ and using (43) for the latter

$$A_{\mu\lambda}=\frac{\frac{1}{2}i\Gamma_{\mu\lambda}}{(E_\mu-E-\frac{1}{2}i\Gamma_{\mu\mu})(E_\lambda-E-\frac{1}{2}i\Gamma_{\lambda\lambda})}, \quad (\mu\neq\lambda). \tag{43a}$$

This can be considered to be small as compared with (43) because the energy differences are, in general, larger than the Γ. For this reason, the terms with $\lambda\neq\mu$ may be neglected in first approximation in (38) and one obtains with (33)

$$\sigma_{jl}(E)=\frac{4\pi k_l}{k_j}\sum_J g_J\left|\sum_\mu{}^J\frac{\gamma_{\mu j}\gamma_{\mu l}}{E_\mu-E-\frac{1}{2}i\Gamma_{\mu\mu}}\right|^2. \tag{44}$$

This equation actually has the form of (42) and constitutes an approximation to it in the same sense as (14) is an approximation to (13c). One sees the analogy between (44) and the customary one level formula, perhaps most easily if one denotes

$$(q_j\gamma_{\mu j})^2=k_j\gamma_{\mu j}^2=\tfrac{1}{2}\Gamma_{\mu j}, \tag{44a}$$

so that, because of (37a), $\sum_j\Gamma_{\mu j}=\Gamma_{\mu\mu}$. One then sees that $\Gamma_{\mu\mu}$ is what is usually denoted by Γ_μ and that the matrix elements, which are the $q_j\gamma_{\mu j}$, are real. This shows that $s_{\nu j}$ of (42) are also approximately real, i.e., their complex phases are small. However, if the Γ are of the same order of magnitude as the energy differences between resonances, (44) may become grossly inaccurate, just as (14) was inaccurate under similar conditions.

In the next approximation, it is necessary to take the $A_{\mu\mu}$ with $\mu\neq\lambda$ into account. Using (43a) for these, the matrix needed for (33) becomes

$$\frac{\mathfrak{R}}{1-i\mathfrak{q}\mathfrak{R}\mathfrak{q}}=\sum_\mu\frac{(\gamma_\mu\times\gamma_\mu)}{E_\mu-E-\frac{1}{2}i\Gamma_{\mu\mu}}+\sum_{\lambda\neq\mu}\frac{\frac{1}{2}i\Gamma_{\mu\lambda}(\gamma_\mu\times\gamma_\lambda)}{(E_\mu-E-\frac{1}{2}i\Gamma_{\mu\mu})(E_\lambda-E-\frac{1}{2}i\Gamma_{\lambda\lambda})} \tag{45}$$

In this approximation, it is still easy to write (45) in the form (41). The F_μ are the same as in (44)

$$F_\mu=E_\mu-\tfrac{1}{2}i\Gamma_{\mu\mu} \tag{45a}$$

and the s_μ becomes

$$\mathbf{s}_\mu=\gamma_\mu+\sum_{\lambda\neq\mu}\frac{\frac{1}{2}i\Gamma_{\mu\lambda}}{F_\lambda-F_\mu}\gamma_\lambda. \tag{45b}$$

One sees that neither the F nor the $\mathbf{s}$ are entirely independent of the energy. The $\mathbf{s}$ are not real any more as they were in first approximation. However, in the present approximation, i.e., up to terms of the order $\Gamma/(E_\lambda-E_\mu)$, the imaginary part of F_ν is still equal to the square of the length of $q\gamma_\nu$.

The above is intended to show that if one uses equations of the form (42) without taking the restrictions on the F and $\mathbf{s}$ into consideration, one automatically restricts oneself to the case in which the level width is small compared with the spacing of the levels. Under such conditions, the matrix elements are at least approximately real. As one approaches the case in which the Γ are of the same order as the spacing of the levels, the $\mathbf{s}$ become complex in general. However, there remain relations which replace the reality condition which is valid for $\Gamma\ll E_\lambda-E_\mu$. It does not appear justifiable to make statistical statements on the cross sections without taking these conditions into account. Of course, (33) always remains valid, but it does not seem to be easy to make statistical statements on the basis thereof without resorting to somewhat crude assumptions.

One can calculate the $A_{\mu\mu}$ in the following approximation by taking into account all terms in the Eq. (37) with $\lambda=\mu$ but using for $A_{\mu\nu}$ the approximation (43a). This gives

$$(E_\mu-E-\tfrac{1}{2}i\Gamma_{\mu\mu})A_{\mu\mu}=1-\tfrac{1}{4}\sum_{\nu\neq\mu}\frac{\Gamma_{\nu\mu}{}^2}{(E_\mu-E-\frac{1}{2}i\Gamma_{\mu\mu})(E_\nu-E-\frac{1}{2}i\Gamma_{\nu\nu})}.$$

Since the last term herein is a correction term one may write for the right side $1/(1+\frac{1}{4}\sum\cdots)$ where $\sum\cdots$ is the sum on the right side. This gives then

$$A_{\mu\mu}=\frac{1}{E_\mu-E-\frac{1}{2}i\Gamma_{\mu\mu}+\frac{1}{4}\sum_\nu{}'\,\Gamma_{\mu\nu}{}^2/(E_\nu-E-\frac{1}{2}i\Gamma_{\nu\nu})}. \tag{46}$$

The term with $\nu=\mu$ has to be omitted in the summation over ν in the denominator. In general, (46) is still approximate. However, one can easily convince oneself that (46) is actually accurate if there is, in addition to μ, only one more resonance level. The approximation corresponding to (46) for $A_{\mu\lambda}$ with $\mu\neq\lambda$ is

$$A_{\mu\lambda}=\frac{\frac{1}{2}i\Gamma_{\mu\lambda}{}^2}{(E_\mu-E-\frac{1}{2}i\Gamma_{\mu\mu})(E_\lambda-E-\frac{1}{2}i\Gamma_{\lambda\lambda})\left(\Gamma_{\mu\lambda}-\frac{1}{2}i\sum_\nu{}''\dfrac{\Gamma_{\nu\lambda}\Gamma_{\nu\mu}}{E_\nu-E-\frac{1}{2}i\Gamma_{\nu\nu}}\right)+\frac{1}{4}\Gamma_{\mu\lambda}{}^3}. \tag{46a}$$

In the summation over ν in the denominator the terms $\nu=\mu$ and $\nu=\lambda$ have to be omitted so that the sum over ν vanishes if there are only two resonance levels, μ and λ. Under this condition (46a) is accurate and differs only by the last term in the denominator from (43a).

Evidently, the last equations are too complicated for being of great practical value. This can hardly to be expected otherwise since our result is quite general except that we assumed only short range forces and considered only collisions in which the angular momentum of the motion of the colliding particles vanishes. Hence (33) should be able to represent the variation of the various cross sections with energy under almost arbitrary conditions. In most cases, if special conditions prevail which are different from those assumed above, it may be easiest to refer back to (33). One property of the cross sections that is of fairly great generality can be, however, easily checked by means of the above formulae, that is that the reaction cross section goes through zero between successive maxima, at least if $\gamma_{\lambda i}\gamma_{\lambda j}$ has the same sign for both, if only two pairs of products are possible but that this does not occur if several products can be formed.

THE PHYSICAL REVIEW

A journal of experimental and theoretical physics established by E. L. Nichols in 1893

SECOND SERIES, VOL. 74, NO. 3 — AUGUST 1, 1948

On Closed Shells in Nuclei*

MARIA G. MAYER
Argonne National Laboratory and Institute for Nuclear Studies, University of Chicago, Chicago, Illinois
(Received April 16, 1948)

Experimental facts are summarized to show that nuclei with 20, 50, 82, or 126 neutrons or protons are particularly stable.

It has been suggested in the past that special numbers of neutrons or protons in the nucleus form a particularly stable configuration.[1] The complete evidence for this has never been summarized, nor is it generally recognized how convincing this evidence is. That twenty neutrons or protons (Ca^{40}) form a closed shell is predicted by the Hartree model. A number of calculations support this fact.[2] These considerations will not be repeated here. In this paper, the experimental facts indicating a particular stability of shells of 50 and 82 protons and of 50, 82, and 126 neutrons will be listed.

I. ISOTOPIC ABUNDANCES

The discussion in this section will be mostly confined to the heavy elements, which for this purpose may be defined as those with atomic number greater than 33; selenium would be the first "heavy" element. For these elements, the isotopic abundances show a number of striking regularities which are violated in very few cases.

(a) For elements with even Z, the relative abundance of a single isotope is not greater than 60 percent. This becomes more pronounced with increasing Z; for $Z>40$, relative abundances greater than 35 percent are not encountered. The exceptions to this rule are given in Table I.

(b) The isotopic abundances are not symmetrically distributed around the center, but the light, neutron-poor isotopes have low abundances. The concentration of the lightest isotope is, as a rule, less than 2 percent. The exceptions to this rule are listed in Table II.

It is seen that the violations of these two regularities occur practically only at neutron numbers 50 and 82. Only the case of ruthenium in Table II, which is not a very pronounced exception, does not fall into one of these groups.

The case of samarium, where the lightest isotope has an isotopic abundance of 3 percent, is only a bare violation of the rule and may not seem striking. However, what is extraordinary, the next heavier even isotope of samarium, Sm^{146} with 84 neutrons, which one would expect to find in greater concentration, does not exist at all.

II. NUMBER OF ISOTONES

Figures 1 and 2 reproduce the parts of the table by Segrè in the region of nuclei with 50

---

* This document is based on work performed under Contract Number W-31-109-eng-38 for the Atomic Energy Commission at the Argonne National Laboratory. Submitted for declassification on February 13th, 1948.

[1] W. Elsasser, J. de phys. et rad. 5, 625 (1934).

[2] E. Wigner, Phys. Rev. 51, 947 (1937); W. H. Barkas, Phys. Rev. 55, 691 (1939).

TABLE I. Even nuclei with $Z>32$ with isotopic abundance greater than 60 percent.

Element	Abundance in percent	Number of neutrons
Sr^{88}	82	50
Ba^{138}	71.66	82
Ce^{140}	90	82

TABLE II. Lightest isotopes of elements with even $Z>32$ with isotopic abundance greater than 2 percent.

Element	Abundance in percent	Number of neutrons
Zr^{90}	48	50
Mo^{92}	15.5	50
Ru^{96}	5	52
Nd^{142}	25.95	82
Sm^{144}	3	82

and 82 neutrons, respectively. For 82 neutrons, there exist seven stable nuclei, which, for convenience, shall be called isotones. For neutron number 50 there exist six naturally occurring isotones, of which one, Rb^{87}, is β-active, however, with a lifetime of 10^{11} years and a maximum β-energy of 0.25 Mev. The average number of isotones for odd neutrons number is somewhat less than one; the same number for even N varies as a rule between three and four. The greatest number of isotones, attained only once in the periodic table, is seven for neutron number 82; six isotones are encountered once only, and for neutron number 50. Five isotones are found five times, namely, for $N=20$, 28, 58, 74, and 78. The frequency of $N=28$ is probably due to the stability of Ca^{48}, with 20 protons, that of $N=74$ to the stability of Sn^{124}, with 50 protons. As few as two isotones for even N are found only three times for heavy nuclei, namely, for neutron numbers 84, 86, and 120.

III. THE SLOPE OF THE CENTER AND THE EDGES OF THE STABILITY CURVE

In the case of neutron number $N=82$ two isotones of odd Z are found, La and Pr. The same is the case for $N=50$, where the unstable but long-lived Rb^{87} and Y^{89} differ only in proton number. Only one other case where nuclei of different odd Z have the same number of neutrons is encountered in the periodic table, namely, that of Cl^{37} (abundance 24.6 percent) and K^{39} (abundance 93.3 percent); this is the case of 20 neutrons. The case of 82 neutrons is most pronounced, since the La and Pr isotopes in question have isotopic abundances of 100 percent.

As Fig. 2 shows, the isotones Nd^{142} and Sm^{144} are both the lightest isotopes of their respective elements. Here, the limit of the stability for neutron-poor isotopes stays at constant neutron number. Exactly the same is true for $N=50$ (Fig. 1). This situation does not occur anywhere else in the periodic table.

The limit of stability for neutron-rich isotopes also stays at constant neutron number for $N=50$ and $N=82$, namely, the pairs of isotones, Kr^{86}, Sr^{88} and Xe^{136}, Ba^{138} are the heaviest isotopes of their elements. Such a case is encountered once more in the periodic table: Ca^{48} and Ti^{50} are the heaviest isotopes of their respective elements and have the same neutron number $N=28$.

IV. THE CASE OF 20 AND 50 PROTONS

Ca, with 20 protons, has five isotopes, which is not too unusual for this region of the periodic table. The difference in mass number between its heaviest and lightest isotope is eight mass numbers, which is quite outstanding, since this difference does not exceed four for elements in this neighborhood.

Sn, $Z=50$, has without exception the greatest number of isotopes of any element, namely, 10. Its heaviest and lightest nuclei differ by 12 neutrons. Such a spread of isotopes is encountered in only one other case, namely, at Xe, where it may be attributed to the stability of Xe^{136} with 82 neutrons.

Incidentally, the next largest difference, 10, in mass number between heaviest and lightest isotope of an element, is encountered once only, in samarium, and may be attributed to the unusual stability of Sm^{144} with 82 neutrons.

V. THE CASE OF 82 PROTONS AND 126 NEUTRONS

Lead, $Z=82$, is the end of all radioactive chains. It has only four stable isotopes, of which the heaviest one, Pb^{208}, has 126 neutrons.

Evidence for the stability of 82 protons and 126 neutrons can be obtained from the energies of radioactive decay. If, for constant value of the charge of the resultant nucleus the energies of α-decay are plotted against the neutron num-

										Gd^{152} 0.2		Gd^{154} 1.5
										Eu^{151} 49		Eu^{153} 50.9
				Sm^{144} 3			Sm^{147} 17	Sm^{148} 14	Sm^{149} 15	Sm^{150} 5		Sm^{152} 26
				Nd^{142} 25.95	Nd^{143} 13	Nd^{144} 22.6	Nd^{145} 9.2	Nd^{146} 16.5		Nd^{148} 6.8		Nd^{150} 5.95
				Pr^{141} 100								
Ce^{136} <1		Ce^{138} <1		Ce^{140} 90		Ce^{142} 10						
				La^{139} 100								
Ba^{134} 2.43	Ba^{135} 6.59	Ba^{136} 7.81	Ba^{137} 11.32	Ba^{138} 71.66								
Cs^{133} 100												
Xe^{132} 26.96		Xe^{134} 10.54		Xe^{136} 8.95								
Te^{130} 33.1												
$N=78$	79	80	81	82	83	84	85	86	87	88	89	90

Fig. 1.

ber of the resultant nucleus, a sharp dip in energy is encountered when N drops below 126, indicating a larger binding energy for the 126th neutron. From these considerations, Elsasser[1] estimates the discontinuity in neutron binding energy at 126 neutrons to be 2.2 Mev or larger, the discontinuity in proton binding energy at $Z=82$ to be 1.6 Mev. These relations have been studied in detail by A. Berthelot.[2a]

VI. ABSOLUTE ABUNDANCE

Absolute abundances are notoriously uncertain. The best estimates are probably contained in the book by Goldschmidt.[3] For the light elements, the abundances vary erratically; for heavy elements, from about Se on, they remain roughly constant. In the region of heavy elements, the following abundance peaks are apparent. At Zr (50 neutrons), at Sn (50 protons); at Ba (82 neutrons), at W and at lead (82 protons or 126 neutrons). In Goldschmidt's plot of abundance against neutron number, page 127, the Zr and Ba peaks are seen to be at neutron number 50 and 82 and become much more pronounced and narrow, whereas the peak at Sn, $Z=50$, as well as the peak at W, become much broader than in the plot against Z.

Most trustworthy among absolute abundances is probably the relative abundance of the rare earths, since these are not likely to have been appreciably fractionated in the earth's crust. The case of 82 neutrons falls just on the edge of this region. According to Goldschmidt's data on the abundance of rare earths in eruptive rocks (which are probably more reliable chemical analyses than the abundance in meteorites), the

[2a] A. Berthelot, J. de phys. et rad. (8) 3, 17, 52 (1942).
[3] V. M. Goldschmidt, *Geochemische Verteilungsgesetze der Elemente* (Norske Videnskaps Akademi, Oslo, 1938), Fig. 1, page 117, or Fig. 2, page 118.

						Ru^{96} 5		Ru^{98} 2.2	Ru^{99} 12	Ru^{100} 14	Ru^{101} 22	Ru^{102} 30
				Mo^{92} 15.5		Mo^{94} 8.7	Mo^{95} 16.3	Mo^{96} 16.8	Mo^{97} 8.7	Mo^{98} 25.4		Mo^{100} 8.6
						Cb^{93} 100						
				Zr^{90} 48	Zr^{91} 11.5	Zr^{92} 22		Zr^{94} 17		Zr^{96} 1.5		
				Y^{89} 100								
Sr^{84} 0.56		Sr^{86} 9.86	Sr^{87} 7.02	Sr^{88} 82.56								
		Rb^{85} 72.3		Rb^{87} 27.7								
Kr^{82} 11.53	Kr^{83} 11.53	Kr^{84} 57.1		Kr^{86} 17.47								
Br^{81} 49.4												
Se^{80} 48.0		Se^{82} 9.3										
$N=46$	47	48	49	50	51	52	53	54	55	56	57	58

Fig. 2.

abundances of rare earths heavier than samarium are reasonably constant, except that the elements with even Z are about 5.7 times as abundant as those with odd Z. Of the lighter rare earths, however, praesodymium ($N=82$) is about 8 times and lanthanum ($N=82$) about 27 times as abundant as the average of the odd rare earth with greater Z. Nd, with a 26 percent isotopic composition of isotopes with $N=82$, is about five times as abundant; cerium, with 90 percent composition of isotopes with $N=82$, is about twelve times as abundant as the average of the heavier even rare earths. In the composition of meteors the differences are not quite as striking, but still very pronounced.

VII. DELAYED NEUTRON EMITTERS

If 50 or 82 neutrons form a closed shell, and the 51st and 83rd neutrons have less than average binding energy, one would expect especially low binding energies for the last neutron in Kr^{87} and Xe^{137}, which have 51 and 83 neutrons, respectively, and the smallest charge compatible with a stable nucleus with 50 or 82 neutrons, respectively. It so happens that the only two delayed neutron emitters identified are these two nuclei.[4]

The fission products Br^{87} ($N=52$), as well as I^{137} ($N=84$), have not enough energy to evaporate a neutron, and undergo β-decay; in the resultant nuclei, Kr^{87} and Xe^{137}, the binding energy of the last neutron is small enough to allow neutron evaporation.

VIII. ABSORPTION CROSS SECTIONS[5]

The neutron absorption cross sections for nuclei containing 50, 82, or 126 neutrons seem all to be unusually low. This is seen very clearly in the measurements of Griffiths[6] with Ra γ-Be neutrons, and those of Mescheryakov[7] with neutrons from $a(d,d)$ reaction. These measurements extend from mass number 51 to 209. In general,

[4] A. H. Snell, Y. S. Levinger, E. D. Meiners, Jr., M. B. Sampson, and R. G. Wilkinson, Phys. Rev. 72, 545 (1947).

[5] The author is indebted to Dr. Katherine Way, who pointed out the connection of the closed shells with neutron absorption cross sections.

[6] J. H. E. Griffiths, Proc. Roy. Soc. 170, 513 (1939).

[7] C. R. Mescheryakov, C. R. U.S.S.R. 48, 555 (1945).

the cross sections increase with increasing mass number. Griffiths investigates, of the nuclei in question, yttrium (89) with 50 neutrons and lanthanum and praesodymium with 82 neutrons. The activation cross section for yttrium is the smallest he observes for any element; it is about 20 to 30 times smaller than the cross sections in that region of mass number. There is a very pronounced dip of cross sections for lanthanum and praesodymium; the cross section of Pr^{141} is about one-seventh of the average of this region, and that of La^{139} is still smaller by a factor 3. Mescheryakov investigates, among others, La, Pr, barium (138), and bismuth (209). He finds a similar dip at La and Pr, and finds that the cross section of Ba^{138} with 82 neutrons is even lower, namely, less than 0.03 of that of lanthanum. The cross section of bismuth with 126 neutrons is even smaller. The only other unusually small cross section which Griffiths finds is that of thallium (122 or 124 neutrons), which is about the same as that for praesodymium.

Recent experiments by Hughes[8] with fission neutrons have shown exceptionally low neutron absorption cross sections for Pb^{208}, Bi^{209} (126 neutrons) and for Ba^{138} (82 neutrons).

IX. ASYMMETRIC FISSION

It is somewhat tempting to associate the existence of the closed shells of 50 and 82 neutrons with the dissymmetry of masses encountered in the fission process. U^{235} contains $143=82+50+11$ neutrons. It appears that the probable fissions are such that one fragment has at least 82, one other at least 50, neutrons.

X. THEORETICAL ESTIMATE OF THE DISCONTINUITY IN BINDING ENERGIES

It is possible to make an estimate of the change in neutron binding energy at, for instance, 82 neutrons. There exists the semi-empirical formula for the mass of an atom[9] with mass number A and charge Z.

$$M_{A,Z}=A-0.00081Z-0.00611A+0.014A^{2/3}+0.083(A/2-Z)^2A^{-1}+0.000627Z^2A^{-1/3}+\delta; \quad (1)$$

with $\delta=0$ for A odd, $\delta=-0.036A^{-3/4}$ for A even, Z even, $\delta=+0.036A^{-3/4}$ for A odd, Z odd. For odd A, this formula permits the calculation of the value of Z for which the energy is a minimum. For Z less than 50 and for neutron numbers greater than 82, the calculated curve is in good agreement with the position of, for instance, the nuclei of odd Z. Between $Z=50$ and $N=82$, however, the experimental values of Z seem to be below the theoretical curve. The disagreement can be explained by a definite shift of the stability line at 82 neutrons. This shift of the stability line can be explained by a change in binding energy of about 2 Mev. Also, according to the formula (1), xenon (136), with 82 neutrons, should be unstable by about 2 Mev, whereas it is undoubtedly stable; Sm^{144} should be unstable against K-capture by 0.6 Mev, whereas Ba^{140}, with 84 neutrons, which is unstable, would be just stable according to formula (1).

Whereas these calculations are undoubtedly very uncertain, they may serve as an estimate of the order of magnitude of the discontinuity in the binding energies. Since the average neutron binding energy in this region of the periodic table is about 6 Mev, the discontinuities represent only a variation of the order of 30 percent. This situation is very different from that encountered at the closed shells of electrons in atoms where the ionization energy varies by several hundred percent. Nevertheless, the effect of closed shells in the nuclei seems very pronounced.

[8] D. J. Hughes, private communication.

[9] N. Bohr and J. A. Wheeler, Phys. Rev. 56, 426 (1939); G. B. von Albada, Astrophys. J. 105, 393 (1947).

On the "Magic Numbers" in Nuclear Structure

OTTO HAXEL
Max Planck Institut, Göttingen
J. HANS D. JENSEN
Institut f. theor. Physik, Heidelberg
AND
HANS E. SUESS
Inst. f. phys. Chemie, Hamburg
April 18, 1949

A SIMPLE explanation of the "magic numbers" 14, 28, 50, 82, 126 follows at once from the oscillator model of the nucleus,[1] if one assumes that the spin-orbit coupling in the Yukawa field theory of nuclear forces leads to a strong splitting of a term with angular momentum l into two distinct terms $j=l\pm\frac{1}{2}$.

If, as a first approximation, one describes the field potential of the nucleons already present, acting on the last one added, as that due to an isotropic oscillator, then the energy levels are characterized by a single quantum number $r=r_1+r_2+r_3$, where r_1, r_2, r_3 are the quantum numbers of the oscillator in 3 orthogonal directions. Table I, column 2 shows the multiplicity of a term with a given value of r, column 3 the sum of all multiplicities up to and including r. Isotropic anharmonicity of the potential field leads to a splitting of each r-term according to the orbital angular momenta l (l even when r is odd, and vice versa), as in Table I, column 4. Finally, spin-orbit coupling leads to the l-term splitting into $j=l\pm\frac{1}{2}$, columns 5 and 6, whose multiplicities are listed in column 7.

The "magic numbers" (column 8) follow at once on the assumption of a particularly marked splitting of the term with the highest angular momentum, resulting in a "closed shell

TABLE I. Classification of nuclear states.

1 Oscillator-quantum number r	2 Multiplicity	3 Sum of all multiplicities	4 Orbital momentum l	5 Total angular momentum j	6 l_j-symbol	7 Multiplicities	8 Magic numbers
1	2	2	0	1/2	$s_{1/2}$	2	
2			1	3/2	$p_{3/2}$	4	
	6	8		1/2	$p_{1/2}$	2	
3			2	5/2	$d_{5/2}$	6	14
				3/2	$d_{3/2}$	4	
	12	20	0	1/2	$s_{1/2}$	2	
4			3	7/2	$f_{7/2}$	8	28
				5/2	$f_{5/2}$	6	
			1	3/2	$p_{3/2}$	4	
	20	40		1/2	$p_{1/2}$	2	
5			4	9/2	$g_{9/2}$	10	50
				7/2	$g_{7/2}$	8	
			2	5/2	$d_{5/2}$	6	
				3/2	$d_{3/2}$	4	
	30	70	0	1/2	$s_{1/2}$	2	
6			5	11/2	$h_{11/2}$	12	82
				9/2	$h_{9/2}$	10	
			3	7/2	$f_{7/2}$	8	
				5/2	$f_{5/2}$	6	
			1	3/2	$p_{3/2}$	4	
	42	112		1/2	$p_{1/2}$	2	
7			6	13/2	$i_{13/2}$	14	126
				11/2	$i_{11/2}$	12	
			4	9/2	$g_{9/2}$	10	

structure" for each completed r-group, together with the highest j-term of the next succeeding r-group. This classification of states is in good agreement with the spins and magnetic moments of the nuclei with odd mass number, so far as they are known at present. The anharmonic oscillator model seems to us preferable to the potential well model,[2] since the range of the nuclear forces is not notably smaller than the nuclear radius.

A more detailed account will appear in three communications to Naturwissenschaften.[3]

[1] See, e.g., H. A. Bethe and R. Bacher, Rev. Mod. Phys. 8, 82 (1937), pars. 32–34.
[2] Which anyhow does not lead to a very different term-sequence compared with that of an anharmonic oscillator, see reference 1.
[3] (a) Haxel, Jensen, and Suess, Naturwiss. (in press). (b) Suess, Haxel, and Jensen, Naturwiss. (in press). (c) Jensen, Suess, and Haxel, Naturwiss. (in press).

High-Energy Electron Scattering and Nuclear Structure Determinations*,†,‡

R. Hofstadter, H. R. Fechter, and J. A. McIntyre
Department of Physics and W. W. Hansen Laboratories, Stanford University, Stanford, California
(Received June 26, 1953)

Electrons of energies 125 and 150 Mev are deflected from the Stanford linear accelerator and brought to a focused spot of dimensions 3 mm×15 mm at a distance of 9 feet from a double magnet deflecting system. The focus is placed at the center of a brass-scattering chamber of diameter 20 inches. Thin foils are inserted in the chamber and elastically-scattered electrons from these foils pass through thin aluminum windows into the vacuum chamber of a double focusing analyzing magnet of the inhomogeneous field type. The energy resolution of the magnet has been about 1.5 percent in these experiments. This resolution is enough to separate clearly hydrogen or deuterium elastic peaks from carbon peaks in the same scattering target. The energy loss in the foils is readily measurable. In the case of light nuclei, e.g., H, D, Be, C, the shift of the peak of the elastic curve as a function of scattering angle indicates the recoil of the struck nucleus. Relative angular distributions are measured for Be, Ta, Au, and Pb. It is possible to interpret these data in terms of a variable charge density within the nucleus.

I. INTRODUCTION

In a recent publication[1] it was suggested that the gold nucleus does not have a sharp boundary. In this paper it is proposed to amplify this statement and to present data on other nuclei which tend to bear out this conclusion.

Guth[2] first pointed out that the finite size of the nucleus should produce large deviations from the expected scattering resulting from a point charge whenever the electron wavelength is of the order of nuclear dimensions. In principle, such deviations might be wholly or partially ascribed to departures from the Coulomb law of electric interaction at very small distances. In the following discussions it will be assumed that the Coulombian interaction holds at small distances and that the departures, if any, from point charge scattering are assignable to the finite dimensions of the nuclear charge distribution.

Other authors have subsequently considered the problem of the finite size of nuclei in relation to scattering experiments.[3-8] Parzen, Smith, and Schiff have dealt with energies of the order of 100 Mev and higher, while the other authors have been concerned mainly with lower energies. Parzen has made an exact calculation at a high energy (100 Mev). Unfortunately, a numerical error crept into Parzen's work and his published scattering curve cannot be considered reliable.[9] For nuclei having a uniform or spherical shell distribution of charge all Born approximation calculations predict maxima and minima in the angular distribution. These features are essentially diffraction phenomena and are similar to the observations in electron diffraction studies of atoms. The first Born approximation[2,4,7] for these models produces the first of a set of zeros in the angular intensity pattern at those angles where

$$qa \cong 3.5, \tag{1}$$

$$q = (4\pi/\lambda)\sin(\theta/2), \tag{2}$$

in which λ is the wavelength of the incident electrons and "a" is the rms value of the nuclear radius, calculated with the charge as the weighting factor. Yennie *et al.*[10] have shown that for a uniform model the exact calculation provides a result in which the first diffraction minimum is practically washed out and the second and third less pronounced than in the Born approximation but approximately in the same angular positions. Experimental indications of a deviation from point charge scattering have been found by Lyman, Hanson, and Scott[11] at an electron energy of 15.7 Mev. $\lambdabar = \lambda/2\pi$ is, in this case, of the order of 1.25×10^{-12} cm and, if one uses a conventional radius for gold ($R=8.1\times10^{-13}$ cm, $a=6.3\times10^{-13}$ cm), qa is found to be close to or less than unity. Thus, no maxima or minima are to be expected in the experiment of Lyman *et al.*, although marked deviations in the angular distribution were expected and, in fact, found. The experimental data proved to be consistent with a uniformly charged model of the nucleus in which

$$R = r_0 A^{\frac{1}{3}}, \tag{3}$$

where r_0 was 1.45×10^{-13} cm. In gold a twenty percent smaller radius gave a very slightly improved fit of the

* This work was initiated and aided at all stages by a grant from the Research Corporation. It was supported by the joint program of the U. S. Office of Naval Research and the U. S. Atomic Energy Commission. In the latter stages of the work, support has been received from the Office of Scientific Research, Air Research, and Development Command.

† This material was presented in part at the April 29, 30–May 1, 1953 Meeting of the American Physical Society in Washington, D. C.

‡ Portions of the theoretical interpretation were revised in proof.

[1] Hofstadter, Fechter, and McIntyre, Phys. Rev. **91**, 422 (1953).

[2] E. Guth, Wiener Anzeiger Akad. Wissenschaften No. 24, 299 (1934).

[3] G. Parzen, Phys. Rev. **80**, 261 (1950); **80**, 355 (1950).

[4] L. R. B. Elton, Proc. Phys. Soc. (London) **A63**, 1115 (1950); **65**, 481 (1952); Phys. Rev. **79**, 412 (1950).

[5] H. Feshbach, Phys. Rev. **84**, 1206 (1951); **88**, 295 (1952).

[6] L. K. Acheson, Phys. Rev. **82**, 488 (1951).

[7] J. H. Smith, Ph.D. thesis, Cornell University, 1951 (unpublished).

[8] L. I. Schiff, following paper, Phys. Rev. **93**, 988 (1953).

[9] G. Parzen (private communication). The error was first found by Elizabeth Baranger (private communication).

[10] Yennie, Wilson, and Ravenhall have recently recalculated the exact scattering at high energies (private communication).

[11] Lyman, Hanson, and Scott, Phys. Rev. **84**, 626 (1951).

data. Lyman *et al.* have noted that the smaller radius in gold might indicate a more dense packing of protons in the interior of the nucleus. A suggestion of this type had been made previously by Born and Yang.[12]

The experiments to be described below were intended to search for possible clear-cut signs of nuclear finite dimensions and by this means to find nuclear radii and charge distributions.

II. APPARATUS

Only the main features of the experimental equipment will be described at this time. Figure 1 shows the principal experimental arrangement. A monoenergetic group of electrons is deflected by a system of two magnets from the main beam of the Stanford linear accelerator.[13] The first magnet bends and disperses the beam and the second magnet bends in a reverse direction and refocuses the spread-out beam. A slit at position S determines the width of the accepted energy band, in this case about 3 percent. The initial width of the beam entering the first magnet is determined by the collimator C, in this case a $\frac{1}{4}$-inch cylindrical hole in a uranium block 1.0-inch long. A well focused beam emerges from the second magnet and closes to a small spot 9 feet from the second magnet. The position of the spot is accurately given by simple first-order calculation of trajectories. The size of the spot is approximately 1-mm high and 3-mm wide for a $\frac{1}{16}$-inch collimator and 3 mm×15 mm for a $\frac{1}{4}$-inch collimator. The beam has extremely little divergence because of the 9.0-foot focal distance and small slit S used (0.75). As many as 2×10^8 electrons per pulse have been measured in the focused beam, sixty pulses occurring per second, each lasting about 0.5 microsecond. The beam is quite free of any gamma rays produced at the collimator and slit because of the double deflection. The beam stopper B prevents gamma rays produced at C from traveling down the accelerator tube and producing unwanted background in the experimental area.

The focused beam is directed towards a scattering target, usually a thin foil of any one of various materials. The scattering target is placed at the center of a brass scattering chamber of diameter twenty inches and twelve inches high. The scattering chamber is built in the form of a large bell-jar which can be detached readily from the main base plate. The whole deflection system and scattering chamber are evacuated to high vacuum. The base of the scattering chamber contains electrical lead-in connections and provisions for mounting and internally moving one or more scintillation counters on a large ring gear whose position is controllable remotely. A thin aluminum window (0.006 in.) stretches around the chamber from about $-150°$ to $150°$ over a vertical distance of three inches. Scattered electrons from the target emerge from the chamber

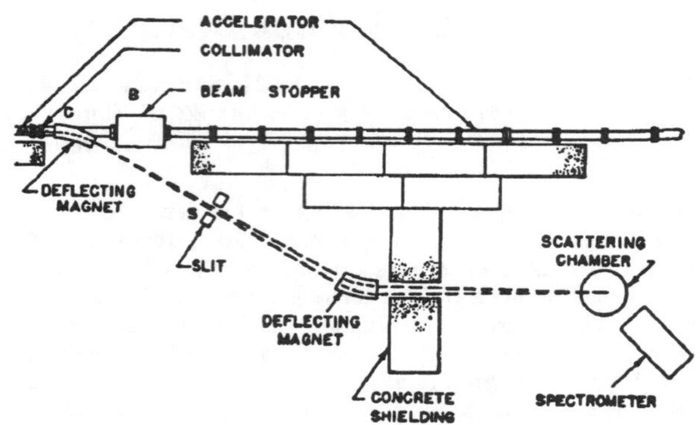

FIG. 1. The experimental arrangement of the electron-scattering system.

through the aluminum window and pass through about one inch of air before reaching the entrance port of the vacuum chamber of the analyzing magnet M. The C-shaped vacuum chamber lies between the shaped pole pieces of the inhomogeneous field magnet. The entrance and exit ports of the vacuum chamber are fitted with 3-mil aluminum windows. The magnet M is similar to the design of Snyder *et al.*,[14] weighs about two and a half tons and has a mean radius of curvature of 16 inches. It is located on the movable platform of a modified twin 40-mm anti-aircraft gun mount kindly lent us by the U. S. Navy, with the cooperation of the U. S. Office of Naval Research.[15] The magnet is of the double focusing variety and employs an inhomogeneous field of index $n=\frac{1}{2}$. Fields up to 12 or 13 kilogauss have been obtained and maintained over long periods of time without excessive heating. The field is usually maintained constant by manual regulation of the current.

A rotatable foil-holder that can be remotely controlled is now being installed in the scattering chamber. The experiments to be described were carried out with a substitute foil-holder which could be rotated and raised or lowered. The raising or lowering operation was remote, but the rotation was not. The holder could accommodate two foils at a time.

After the scattered electrons are analyzed in the magnet they leave the vacuum chamber and are collimated by a $\frac{1}{2}$-in. cylindrical hole in a lead block $2\frac{1}{2}$ inches long. Behind this collimator a conical Čerenkov counter, four inches long, detects the electrons admitted by the collimator. The Čerenkov counter is made of highly polished lucite and is one inch at the narrow input end and 1.5 in. at the output end where the lucite is coupled via heavy silicone oil to a 6292 Dumont

[12] M. Born and L. M. Yang, Nature **166**, 399 (1950).

[13] Among those mainly responsible for this system are Dr. J. A. McIntyre and Dr. W. K. H. Panofsky.

[14] Snyder, Rubin, Fowler, and Lauritsen, Rev. Sci. Instr. **21**, 852 (1950); C. W. Li, Ph.D. thesis, California Institute of Technology, 1951 (unpublished).

[15] The gun mount was modified and machined most capably by the machine shop of the Mare Island Shipyard of the U. S. Navy. We are very grateful to the officers and civilians of the yard for the excellent job done. The U. S. Office of Naval Research very kindly helped us make the necessary arrangements with the shipyard.

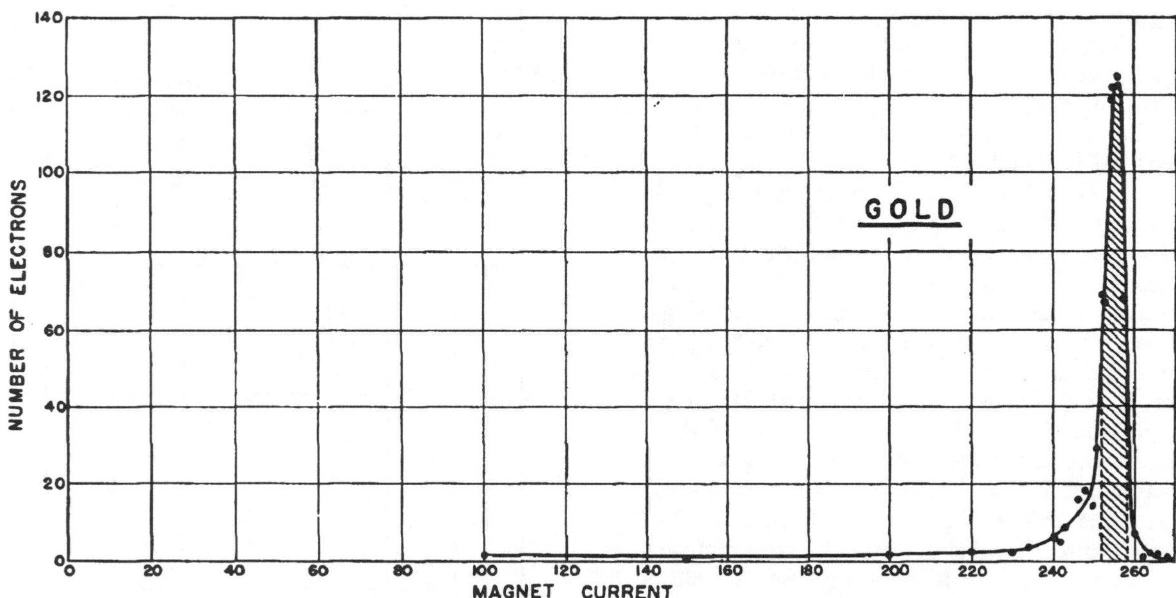

FIG. 2. An elastic-scattering curve in gold taken at 125 Mev, at a scattering angle of 35°, and with a 2-mil foil set at 45° with respect to the beam. The shaded portion of the peak shows the fraction of electrons counted in an individual peak setting. The abscissa is proportional to electron energy.

Photomultiplier. The pulses from the photomultiplier are amplified in an Elmore model 501 amplifier and fed to a gated scaler (gate 12 microseconds long) and also to an oscilloscope viewed by a monitor photomultiplier which has been made to act as a counter.[16] Both counters are gated by the main trigger signal of the linear accelerator. The biases against which both counters operate are adjusted so that the two agree on number of counts. Good plateaus in counting rates are thus obtained. An effort is made to count not more than one count a second so that pileup and loss of counts may be avoided. A large lead shield surrounds the Čerenkov counter and photomultiplier and has greatly helped in avoiding background troubles.

The main beam passing through the scattering target is monitored by a helium-filled ionization chamber designed by W. C. Barber of this laboratory. Dr. Barber has kindly calibrated and tested the chamber with 25-Mev electrons and has verified that it does not saturate under beam intensities up to 2×10^8 electrons per pulse, where a pulse lasts about 0.3 microsecond, and where the beam is about 2 cm^2 or larger on entering the chamber. Under the conditions of the experiments here reported the ion chamber does not saturate. The output of the chamber is brought to a charge integrator of a conventional type.

The deflecting magnets are presently stabilized by an electronic regulator to better than one part in a thousand. The analyzing magnet is manually controlled by an operator who reads the voltage across a shunt in series with the current. A Rubicon potentiometer is used to read the shunt voltage. With careful control the analyzing magnet current may be maintained constant between limits of ±0.1 percent during a run of 10 minutes duration.

The magnet has been calibrated by using the known energy of the Am^{241} α particles and by measurement of the magnetic field at the center of the magnet.

The angular position of the magnet on the gun mount stand is controlled remotely and measured by a combination of high and low speed selsyn indicators. The error in determining position is less than 0.1°.

III. EXPERIMENTAL PROCEDURE

Because of the temporary nature of the target holder the scattering foils were not rotated during the angular runs. This procedure has the advantage that one always uses the same region of the scattering foil provided that the beam spot does not shift during a run. When setting the analyzing magnet from one side of zero to the other, it is of course necessary to rotate the target foil and this was done. Also when points beyond 90° were examined, the foil was rotated to a new fixed position. A typical setting of the foil for a run between 35° and 90° was with foil plane at 45° with respect to the beam. The lineup of the beam was carefully carried out at the beginning of each run by a photographic method.

At a given angular position of the analyzing magnet an "elastic curve" can be obtained by measuring the total number of counts per unit integrated beam for various settings of the magnetic field. The magnetic field settings are assumed to be proportional to the magnet currents and the former are also proportional to the electronic momenta. For these high energies, the currents are therefore also proportional to the electronic energies. Typical elastic curves so obtained are shown in

[16] R. Hofstadter and J. A. McIntyre, Rev. Sci. Instr. 21, 52 (1950).

Figs. 2, 3, 4, and 5. The abscissa on these curves is proportional to the magnet current, and in all cases except Fig. 2 the zero of abscissa is far off to the left of the ordinate axis. The curve of magnetic field against magnet current shows little saturation in the region here investigated[14] (about 100 amperes).

The elastic curves show typical bremsstrahlung tails on the low-energy side and sharper cutoffs on the high-energy side, as expected. Figure 2 shows the appearance of an elastic curve for gold and indicates also the fraction of electrons collected in individual peak settings. Figure 3 shows a typical displacement between the peak of the main beam and that of the transmitted beam at zero degrees, the difference being the result of the energy loss in the gold foil (2-mils thick). Figures 3 and 4 show that the elastic curves have the same appearance at angular settings of 0°, 35°, and 70° and presumably, therefore, at other angles. Figure 5 and Figs. 1 and 2 of our earlier communication[1] show that the elastic curves of hydrogen and deuterium in polyethylene as well as beryllium shift with the angular position. This shift has been explained[1] by the recoil of the struck nucleus (H, D, Be, C). With improved resolution this recoil shift will permit scattering measurements to be made on unseparated isotopes in the same foil and also with the elements of compounds. In the case of Ta, Au, and Pb the recoil shift is too small to be observed at the present time. In those cases in which the elastic curves have the same appearance at all angles, it is sufficient to measure the total elastic yield at a given angle by an average of two measurements taken at positions on either side of the peak, for example, at abscissas 326 and 330 in Fig. 4 for tantalum. All angular distributions have been measured in this way. Measurements in light elements, for example, Be or C, must be taken at the positions of the shifted peaks. Figure 6 shows how the Be elastic curves shift with angular position. In fact, in Be the elastic curve changes in appearance as well as shifts with change in angular position. The peculiar curve in Be at 90° requires further investigation. It is possible that inelastic scattering may be observed in such light elements at large angles.

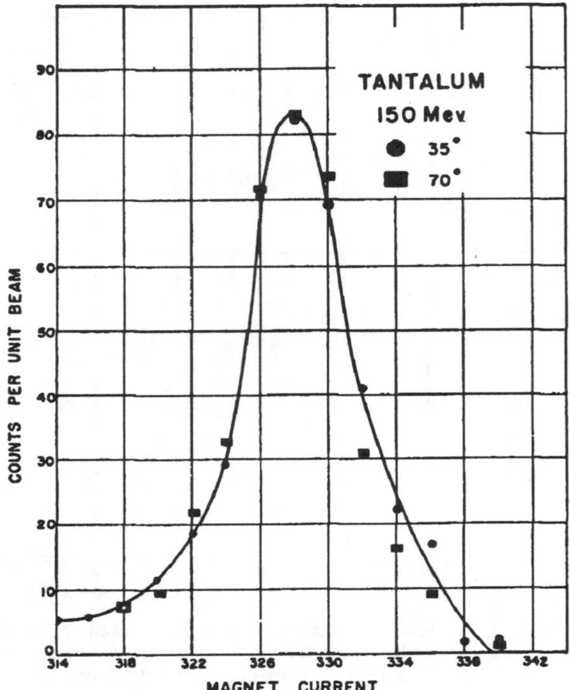

Fig. 4. Elastic-scattering curves in tantalum at 150 Mev. The data at 35° and 70° are essentially identical when normalized to the same peak value. The foil was 2.6 mils thick and was placed at 30° with respect to the direction of the beam.

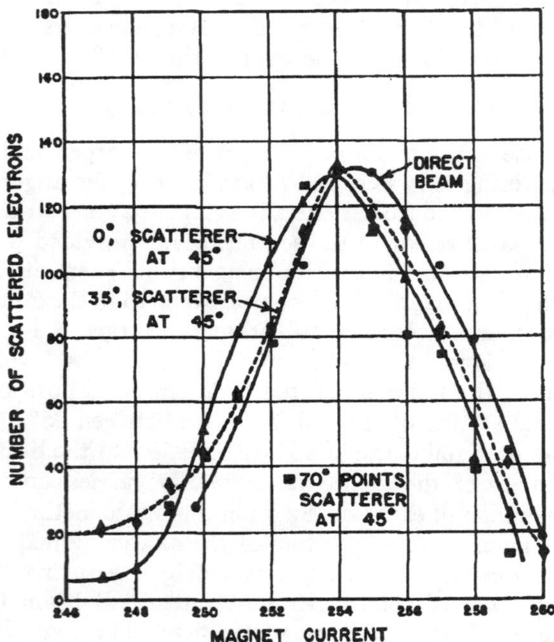

Fig. 3. Elastic-scattering curves for a 2-mil gold foil at 125 Mev. The foil was placed at 45° with respect to the beam. The abscissa is proportional to electron energy. Some of the curves were observed with a CsBr(Tl) detector by a dc method and have higher residual backgrounds (directly and indirectly the result of neutrons) than the normal Čerenkov detector curves.

The shift of the peak in Be is the result not only of the recoil energy but also of the energy loss of the incoming and outgoing electrons in the target foil. At a scattering angle of 45°, and a target setting of 45° with respect to the beam, the average energy loss in the 50-mil beryllium target is 0.49 Mev, and the recoil shift is 0.46 Mev. At 70° the corresponding figures are 0.51 Mev and 1.04 Mev, and at 90° they are 0.58 Mev and 1.58 Mev. The expected shifts relative to 35° are therefore 0.60 Mev for 70° and 1.21 Mev for 90°. The observed relative shifts are 0.60 ± 0.20 Mev and 1.20 ± 0.20 Mev and are in good agreement with the calculations. Similarly, good agreement has been obtained for the hydrogen and deuterium shifts. In fact the H, D shifts relative to carbon or some other standard may be used to measure the energy of the incident beam. With the present resolution and the accompanying experimental error in measuring the shifts, the calculated beam energy is of the order of 140 ± 20 Mev while the

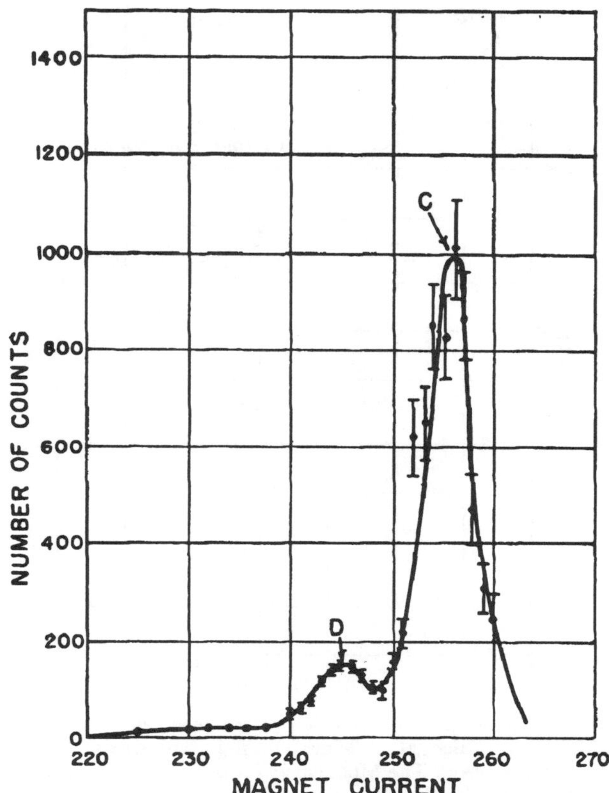

FIG. 5. An elastic curve at 125 Mev, showing the deuterium and carbon peaks observed in deuterated polyethylene at a scattering angle of 65°.

value obtained by calibration with alpha particles gives a value of 125±5 Mev. Obviously, the method can be refined.

A further check on the internal consistency of the scattering data is obtained by comparison of the ratio of the areas under the carbon and hydrogen or deuterium peaks in Fig. 1 of this paper and also in Fig. 2 of reference 1. The ratio in Fig. 5 is $\sigma_C/\sigma_D = 14\pm5$ and

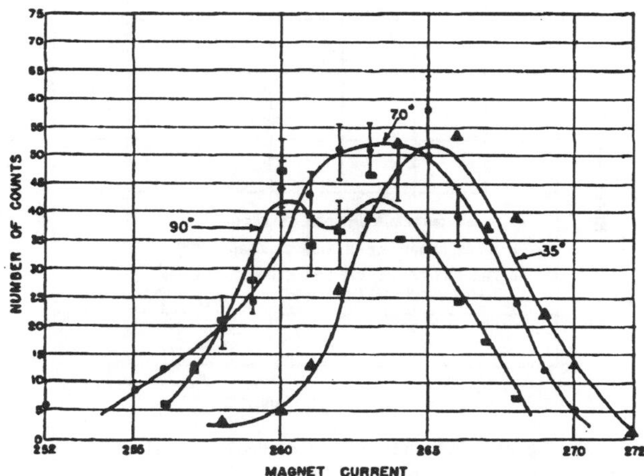

FIG. 6. Elastic curves in Be at 125 Mev at 35°, 70°, and 90°, observed with a 50-mil scattering target. The elastic peak shifts to lower energies at larger angles of scattering and also broadens as it shifts.

for Fig. 2 of reference 1 the ratio $\sigma_C/\sigma_H = 20\pm5$, where σ_C, σ_H, σ_D, are the respective cross sections. The accuracy of these measurements is not high because of the carbon bremsstrahlung background which must be subtracted from the H or D peaks. An average of the two results gives 17 which is close to the ratio, $Z_C^2/2(Z_{H,D})^2 = 18$, of the scattering from carbon and hydrogen or deuterium in polyethylene. Again, with better resolution this measurement of relative areas can be greatly improved, and, in fact, this method suggests itself for measuring the hydrogen and deuterium scattering cross sections relative to carbon as a standard.

To check the over-all behavior of the scattering measurements, two tests have been made: (1) the scattering

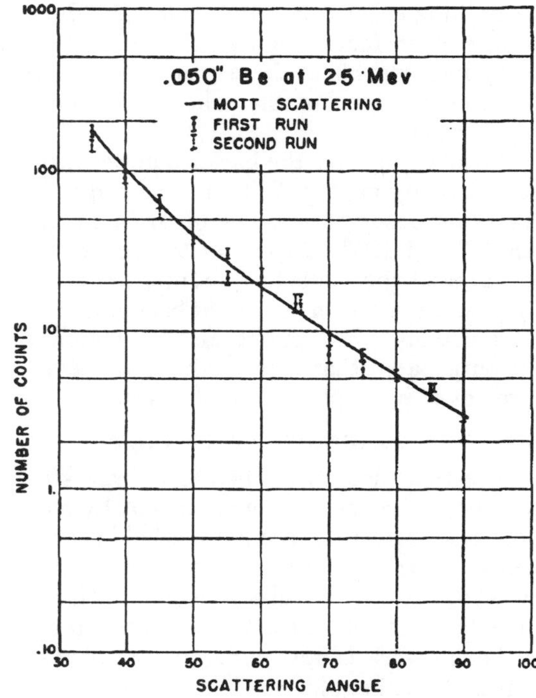

FIG. 7. The angular distribution of electrons scattered from a Be target, 50 mils thick, at 25 Mev. The target plane was at 45° with respect to the beam. The Mott curve for a point charge is shown. Arbitrary normalization is made at 35°.

of Be has been examined in the angular range 35°–90° at 25±2 Mev, and (2) the scattering from a gold foil of $\frac{1}{4}$-mil thickness has been studied in the same angular range at 25 Mev. In case (1), because of the relatively low energy, the Be nucleus may be approximated by a point charge and the exact calculations of Feshbach[5] and Parzen[3] for a point charge may be compared with the experimental data. In case (2) the observed angular distribution may be compared with the data of Lyman *et al.* at 15.7 Mev.

Figure 7 shows the Be data at 25 Mev using a 50-mil foil. The data are in excellent agreement with the accurate calculations and also with the Mott formula, which for low Z's is very close to the accurate formulas.

Figure 8 shows the data for gold found at 25 Mev

with a ¼-mil foil. The observed curve is seen to lie above the Mott curve, just as in the work of Lyman *et al.*[11] Our data show slightly less rise relative to the Mott curve as compared with the data of Lyman *et al.*, but this is to be expected, to some extent, because of the higher energy and therefore the shorter wavelength relative to nuclear size. On the whole, the agreement is better than might have been anticipated, since our windows were not designed for energies as low as 25 Mev. It is gratifying that the results for Be and Au at 25 Mev are so well in accord with expectation.

To check whether some systematic error might be prejudicing results on one side of zero preferentially the data for tantalum at 150 Mev have been observed on both sides of zero. Figure 9 shows the data and shows there is nothing special on either side. The data agree very well, and, in fact, the only point seriously off is at 80° on the left side where there was a steel supporting post blocking out part of the solid angle seen by the magnet chamber.

It is also pleasing that the background counting rate has been virtually absent. Without the target in place we have never observed a background count at angles between 35° and 120°. If some of the concrete shields and lead around the Čerenkov counter are removed it is easily possible however to obtain background counts. These are usually small pulses in the photomultiplier and it is anticipated that with further study even these can be removed by applying a higher bias.

IV. CORRECTIONS TO DATA

The corrections usually applied to the raw scattering data have been excellently summarized by Lyman *et al.*[11] Many of these corrections are not needed in the present work, because the data to be presented are entirely of a relative kind. A precise absolute calibration of the scattering data has not been attempted up to the present time, although work is now under way to obtain absolute cross sections in gold.

Since we shall be concerned only with relative data, there is probably no need to take into account such small corrections as the radiative type of Schwinger[17] which were calculated with Born approximation and depend very little on angle.

In order to avoid a variable loss of counts due to pileup, we have made an attempt to count approximately at the same rate at all angles so that a dead-time correction, if present, will be the same for all positions. Actually the dead time correction is negligible.

The largest corrections for multiple scattering are encountered in the case of the gold foils. For most measurements a foil 2 mils thick was used. In one run, to get more intensity at angles larger than 120° a four-mil foil was used. For the 2-mil foil the multiple scattering correction amounts to −0.3 percent at 35°, to −0.1 percent at 90°, and to −0.1 percent at 140°.

[17] J. Schwinger, Phys. Rev. **75**, 898 (1949).

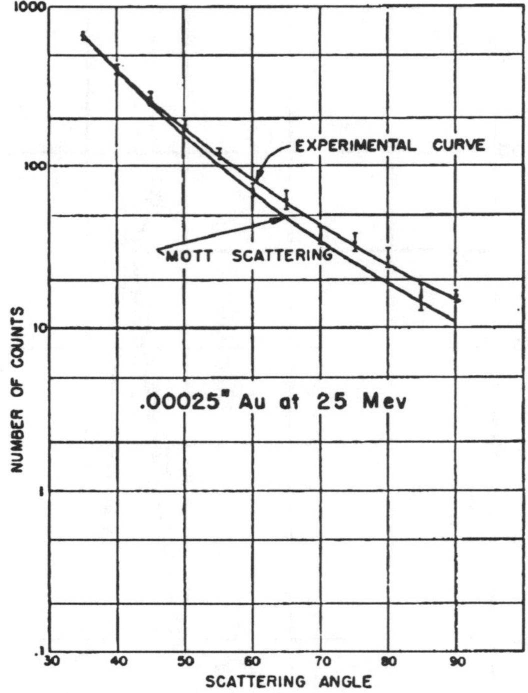

FIG. 8. The angular distribution of scattered electrons from a gold foil, ¼-mil thick, at 25 Mev. The foil plane was at 45° with respect to the beam. The Mott curve for a point charge is shown. Arbitrary normalization is made at 35°.

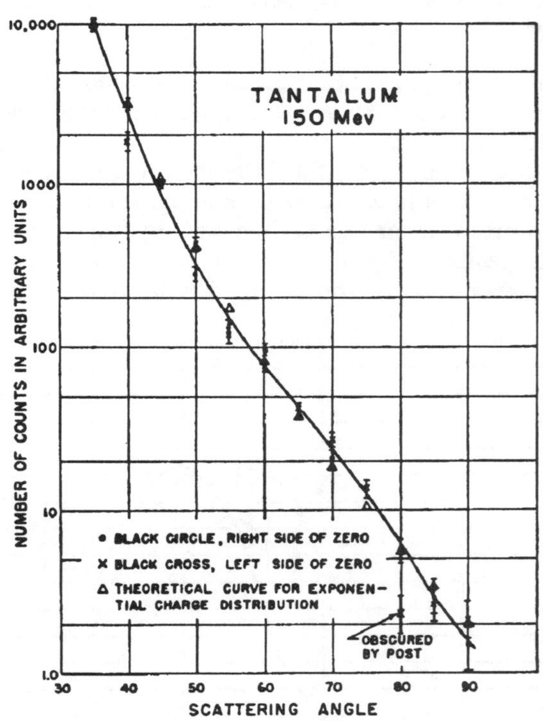

FIG. 9. The angular distribution of scattered electrons from a tantalum foil, 2.6-mil thick, at 150 Mev. The foil plane was at 30° relative to the beam direction. Curves on the left and right of zero are shown. A theoretical curve based on an exponential charge distribution is shown as well as the Feshbach point charge curve. All curves are normalized arbitrarily at 35°.

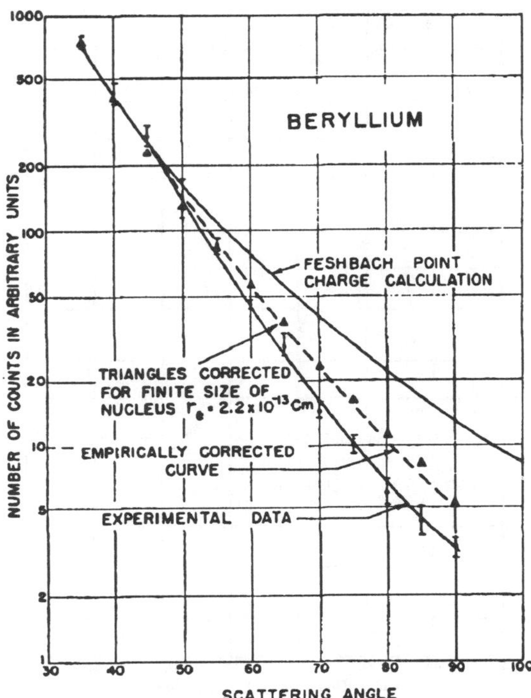

FIG. 10. The angular distribution of scattered electrons from a beryllium foil, 50-mils thick, at 125 Mev. The experimental curve has been corrected empirically for the broadening observed in the elastic curves at larger scattering angles. (See Fig. 6.) The dashed curve is the corrected curve. A theoretical curve based on the first Born approximation for an exponential charge distribution is shown. Also shown is the point charge calculation of Feshbach. Arbitrary normalization of all curves is made at 35°.

Thus, multiple scattering corrections in the target foil are unnecessary for the accuracy involved in this work. The multiple scattering in the aluminum windows is of the order of 0.4° and can be neglected since it is 4 times as small as the rms scattering angle in the gold foil. The beryllium 50-mil foil has an rms scattering angle of 0.6° which is negligible also.

The errors resulting from double (large-angle) scattering are estimated to be 0.15 percent at 90° for two mils of gold at 125 Mev and 0.01 percent at 150° under the same conditions. For 50 mils of Be at 125 Mev the errors are 1.5 percent at 150° and 0.04 percent at 90°. Hence, all double scattering corrections are ignored since they are very small effects.

The geometrical corrections for the aperture can be estimated from the effective aperture which is approximately 0.8 square inch at twelve inches. A calculation similar to that of Lyman *et al.* leads to corrections of a few tenths of a percent, which are thus negligible for our purposes.

The angular resolution of our scattering results depends on the size of the beam spot on the target foil and on the effective aperture of the entrance port of the analyzing magnet. Each of these contributions is about the same at the present time and each contributes about 2°, fairly independently of angles between 35° and 140°, for a target foil setting of 45°. Hence, our angular acceptance width is about ±4°, or a total of 8°. Structure in the scattering curves within such small angular ranges would not be resolved in our experiments. On the other hand, such fine structure is not expected.

Radiative straggling and electron-electron straggling affect the shape of the elastic-scattering peaks. Since in all the cases we have studied, with the exception of Be, the elastic profile is the same at all angles, no relative corrections for these effects need to be made. As a matter of fact, the same argument applies to the Schwinger correction.

With the exception of Be, all corrections are extremely small and will be ignored. In the case of Be (Fig. 6) the elastic profile changes as a function of angle, because of the combination of the recoil effect and the energy loss straggling in the target. Both effects are appreciable for Be. The correction has been taken into account empirically by measuring the areas under the elastic curves taken at various angles. At 90° the area is approximately 1.5 times the area at 35° when both curves are normalized to the same peak values. Hence, a correction of 50 percent is applied to the counting rate at 90°. At 35° the correction is zero, and a smooth curve has been drawn in Fig. 10 (the dashed line) to represent the corrected data at intermediate angles. Since the cross section varies rather violently with angle, the largest correction of 50 percent produces only a mild effect.

V. RESULTS

The relative angular distributions have been measured in Be, Au, and Pb at 125 Mev and in Ta at 150 Mev. In addition, as mentioned previously, check runs

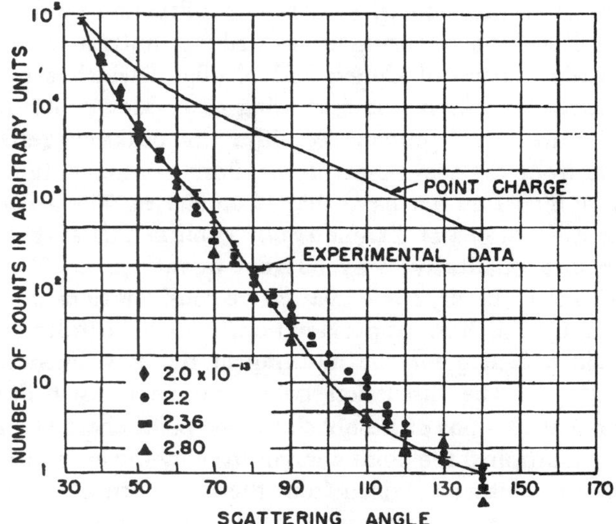

FIG. 11. The angular distribution of electrons scattered from a 2-mil gold foil at 125 Mev. The point charge calculation of Feshbach is indicated. Theoretical points based on the first Born approximation for exponential charge distributions are shown. Values of $\alpha = 2.0$, 2.2, 2.36, 2.8×10^{-13} cm are chosen to demonstrate the sensitivity of the angular distribution to change of radius. All curves are normalized arbitrarily at 35°.

were made on Be and Au at 25 Mev. The experimental curves are shown in Figs. 9, 10, 11, 12. The limits of errors indicated in the figures are entirely of statistical origin. In all cases except gold, the background counting rates were zero so that the present data represent the actual numbers taken in a run. In only one case, namely Au beyond 120°, was a small background observed in one of the counters. The other counter showed no background. The small background makes the data beyond 120° slightly less reliable than the other data.

Near 100° on the right side of zero it has not been possible to obtain data because of the presence of a steel post which is used as a support for the upper half of the scattering chamber. There is a similar post near 80° on the left-hand side of zero. In future experiments it will be possible to move the posts to other positions while taking data near these two angles.

In occasional runs we have noticed that after long periods of time the data do not check exactly. For example, after a couple of hours of running a point at 70° may change by 30 percent or so. Invariably the neighboring points will be down by the same factor. Thus, a slow drift in some part of the counting system or magnetic field appears at irregular times. The explanation of this effect is being sought. However, we do not feel that this effect is significant because the relative counting rates agree with those first obtained to better than 30 percent.

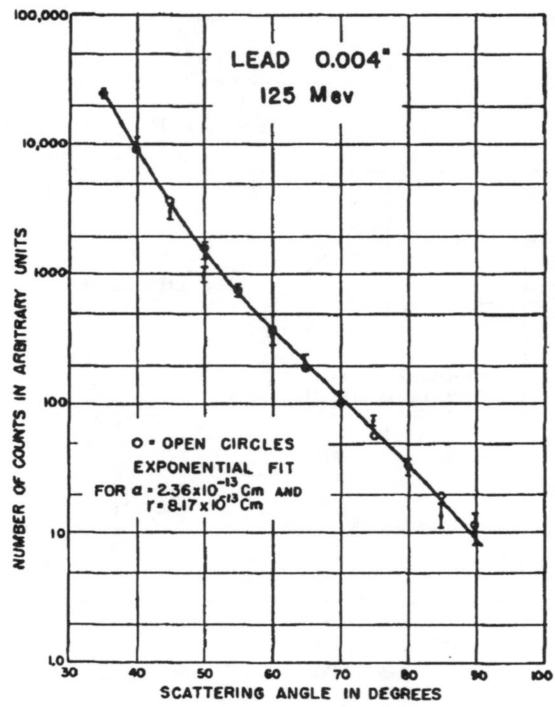

FIG. 12. The angular distribution of electrons scattered from a 4-mil lead foil at 125 Mev. A theoretical curve based on the first Born approximation for an exponential charge distribution is shown. Arbitrary normalization is made at 35°.

VI. ANALYSIS OF THE DATA

The experimental distributions lie far below the point charge calculations of Feshbach[5] which are shown typically in Fig. 11. Within the resolution of the experimental data (points were taken every 5 degrees apart) there is no pronounced evidence of diffraction minima or maxima. These curves are in striking contrast to the beautiful curves of Cohen and Neidigh[18] which show diffraction peaks in the scattering of 22-Mev protons.

The absence of pronounced diffraction peaks suggests that, from the viewpoint of the Born approximation, heavy nuclei do not have sharp boundaries. The accurate calculations of Yennie *et al.*[10] confirm this suggestion in a qualitative way but not quantitatively. We shall sketch briefly some qualitative considerations provided by the Born approximation, merely in order to obtain a feeling for the meaning of the experimental results. In the discussion below we have used the exact calculations of Feshbach for the point charge and have multiplied the point charge curve by the appropriate form factors obtained from the first Born approximation[2,4] for various assumed charge-density distributions.

Among the types of charge-density (ρ) distributions tried[19] were the following:

(A) exponential

$$\rho=\rho_0 e^{-r/a}; \tag{4}$$

(B) "half-uniform and half-Gaussian"

$$\rho=\rho_1,\quad 0\leqslant r\leqslant c;\quad \rho=\rho_1\exp[-\tfrac{1}{2}(r-c)^2/d^2],\quad r\geqslant c; \tag{5}$$

(C) gaussian

$$\rho=\rho_2\exp(-\tfrac{1}{2}r^2/r_0^2); \tag{6}$$

(D) uniform

$$\rho=\rho_3,\quad 0\leqslant r\leqslant c;\quad \rho=0,\quad r\geqslant c. \tag{7}$$

The ρ_0, ρ_1, ρ_2, and ρ_3 are all constants with the dimensions of charge density. For these charge distributions the root-mean-square radii, calculated with the charge as the weighting factor are, respectively:

$$\text{(A)}\quad r_e=3.46a, \tag{8}$$

$$\text{(B)}\quad r_{ug}=2.31c\quad \text{for}\quad d=c, \tag{9}$$

$$\text{(C)}\quad r_g=1.732r_0, \tag{10}$$

$$\text{(D)}\quad r_u=0.775c. \tag{11}$$

In the following, we discuss the data obtained for the various elements.

[18] B. L. Cohen and R. V. Neidigh, Phys. Rev. (to be published). We wish to thank Dr. A. M. Weinberg and Dr. B. L. Cohen for informing us of these results. See also J. W. Burkig and B. T. Wright, Phys. Rev. 82, 451 (1951).

[19] See also L. I. Schiff (reference 8) for other charge distributions and relevant remarks. A slightly different approach from ours has been used in Schiff's paper.

Gold

The "best fit" with the experimental data was obtained for the exponential charge distribution with $\alpha=2.3\pm0.3\times10^{-13}$ cm and $r_e=7.95\times10^{-13}$ cm. The experimental data and the exponential fit are shown in Fig. 11. The half-uniform and half-Gaussian model provides a "best fit" which is somewhat inferior to the exponential fit and requires $c=2.46\times10^{-13}$ cm and $r_{ug}=5.7\times10^{-13}$ cm. This angular distribution is a poor fit at angles greater than 110° where it drops off too rapidly. The pure Gaussian model (C) is also not as good as the exponential but gives a best fit for $r_0=2.48\times10^{-13}$ cm with $r_g=4.3\times10^{-13}$ cm. As in the last case the pure Gaussian drops off too rapidly beyond 115°. These rms radii are to be compared with the usual rms value of about 6.3×10^{-13} cm for gold and are of the correct order of magnitude.

The Born approximation is clearly quite poor for a uniform distribution of charge because of its true zeros. Therefore, no attempt will be made to use the Born approximation for this model. On the other hand, the exact calculations of Yennie *et al.*,[10] carried out both for a uniform charge distribution and an exponential charge distribution, show that the exponential fits better because of the absence of diffraction structure, which does occur for the uniform distribution. In either event, the most important conclusion is that the exact theoretical curves reflect the very steep falloff of cross section with angle shown by the experiments. The large departures from a point charge are, therefore, satisfactorily demonstrated in theory, as well as experiment. The rather extreme exponential model, when treated exactly, appears to fit the experimental data (35°–120°) quite well, but it is by no means ruled out that other less violent models with, e.g. Gaussian tails, might not do equally well. To see the qualitative effect of a small rounding off of the edge of a uniform charge distribution, a uniform model and Gaussian tail with parameter $d=0.2c$ in Eq. (5) has been tried in Born approximation but does not remove the zeros. This model simply moves the zeros out a little towards larger angles. A uniform model with a small exponential tail has a similar behavior, as shown by Smith.[7]

Lead

The lead foil used in the experiments was an isotopic mixture in the natural proportions and was 4-mils thick. The experimental curve for 125 Mev is given in Fig. 12. It may be seen that the points are fitted quite well by a theoretical Born curve based on an exponential model. The theoretical curve is hardly distinguishable from the experimental curve. The radius obtained from the best exponential fit is $r_e=8.17\times10^{-13}$ cm or $\alpha=2.36\times10^{-13}$ cm. This value is slightly larger than the radius for gold in the ratio 1.03.

Tantalum

The experimental curve for 2.6 mils of tantalum at 150 Mev is given in Fig. 9. By fitting with an exponential charge distribution corresponding to $\alpha=2.80\pm0.3\times10^{-13}$ cm or $r_e=9.7\times10^{-13}$ cm, an excellent reproduction of the data is obtained. Again, as in lead, the theoretical Born points are scarcely distinguishable from the experimental ones. While the tantalum should have a radius slightly smaller than lead or gold according to the rule expressed by Eq. (3), the scattering data indicate a larger radius. Relative to lead the rms radius is 1.18 times as large. This result may possibly have some connection with the extremely large quadrupole moment of Ta.

Beryllium

Beryllium has been studied at 125 Mev. The results obtained with a 50-mil scattering target are shown in Fig. 10. The curve which has been empirically corrected for the change in elastic profile as a function of angle is shown as a dashed line in the figure. The point charge curve of Feshbach is shown in open circles. In the case of beryllium the Born approximation should be quite valid. The triangles represent the point charge curve as modified by the "best fit" form factor corresponding to an exponential charge distribution with $r_e=2.2\times10^{-13}$ cm or $\alpha=6.36\times10^{-14}$ cm. This radius is quite a bit smaller than the radii measured for heavy nuclei and is in good agreement with what might be expected from Eq. (3). In the case of such a small nucleus as Be, the form factor can be chosen quite arbitrarily to correspond either to the exponential, Gaussian, or uniform model, since all give essentially the same behavior. For example, the best fit for a uniformly charged model gives $r_u=1.90\times10^{-13}$ cm and for a Gaussian model $r_g=1.96\times10^{-13}$ cm.

VII. CONCLUSIONS

With the angular resolution attained in these experiments, the absence of diffraction maxima and minima in the observed scattering of 125- and 150-Mev electrons from Ta, Au, and Pb suggests tentatively the concept that these nuclei have charge distributions tapering off gradually from near the center to the outside. This conclusion is not definite at this time because both theory and experiment are in preliminary stages.

The differences between the nuclear diffraction patterns observed in the scattering of 22-Mev protons[18] and the electron scattering results reported here may perhaps reflect the facts that nuclei interact with protons through short-range forces (also to a lesser extent through Coulomb forces) and are not transparent to protons of 22-Mev energy, while nuclei interact with electrons through long-range Coulomb forces and are transparent to electrons. Hence, the elastically-scattered protons interact effectively only with the outer edges of the nucleus giving the impression of a sharp boundary. Electrons interact with the entire nuclear volume.

It might be wondered whether the peaked charge distribution suggested here could influence arguments concerning saturation of nuclear forces. A simple electrostatic calculation was carried out for the exponential charge distribution with $\alpha = 2.3\times10^{-13}$ cm for gold. The result obtained indicates that the Coulomb energy is changed by not more than a factor of two relative to that of the uniform distribution. Hence, on this score there will be no serious change regarding nuclear saturation.

It is recognized that the charge distribution in heavy nuclei tentatively suggested by this work differs rather seriously from the uniform model generally proposed. For this reason we are attempting to improve the accuracy of the experiments by increasing the angular resolution, energy resolution and stability of all parts of the apparatus.

ACKNOWLEDGMENTS

We wish to thank Mrs. P. Hanson, Mr. G. Masek, and assistant crew members for their kind and efficient help in running the accelerator. Miss E. Wiener and Mr. J. Fregeau have kindly assisted in taking some of the data. V. Prosper, E. Wright, B. Chambers, B. Stuart, F. Renga, and P. Abreu deserve our special thanks for the skill they have shown in building and designing various parts of our scattering apparatus. Mr. R. H. Helm was of great assistance in designing the magnet support and in constructing the spectrometer. We appreciate the help of Drs. W. A. Fowler and W. Whaling of California Institute of Technology, who provided us with the drawings and advice needed in constructing the analyzing magnet, and Dr. I. Perlman of the University of California Radiation Laboratory, for lending us an alpha-particle calibration source. We are grateful to the Office of Naval Research which gave us early support in this project and particularly to Dr. Urner Liddel, Mr. F. Niemann, and Lt. M. S. Jones who helped us obtain the gun mount for this work. We wish to thank Dr. L. I. Schiff and Dr. W. E. Lamb, Jr., who have given us valuable advice in theoretical matters. Dr. E. Guth has read our work critically and has our special thanks for his comments. Drs. D. R. Yennie and D. G. Ravenhall and Mr. R. N. Wilson are to be thanked most cordially for their kindness in permitting us to use their results before publication. We acknowledge with thanks the recent support of the Office of Scientific Research of the Air Research and Development Command. Finally we are grateful for a grant received from the Research Corporation which enabled this project to obtain its start.

Model for Nuclear Reactions with Neutrons*

H. FESHBACH, C. E. PORTER,† AND V. F. WEISSKOPF
Department of Physics and Laboratory for Nuclear Science, Massachusetts Institute of Technology, Cambridge, Massachusetts

(Received June 28, 1954)

A simple model is proposed for the description of the scattering and the compound nucleus formation by nucleons impinging upon complex nuclei. It is shown that, by making appropriate averages over resonances, an average problem can be defined which is referred to as the "gross-structure" problem. Solution of this problem permits the calculation of the average total cross section, the cross section for the formation of the compound nucleus, and the part of the elastic-scattering cross section which does not involve formation of the compound nucleus. Unambiguous definitions are given for the latter cross sections.

The model describing these properties consists in replacing the nucleus by a one-body potential which acts upon the incident nucleon. This potential $V=V_0+iV_1$ is complex; the real part represents the average potential in the nucleus; the imaginary part causes an absorption which describes the formation of the compound nucleus. As a first approximation a potential is used whose real part V_0 is a rectangular potential well and whose imaginary part is a constant fraction of the real part $V_1=\zeta V_0$.

This model is used to reproduce the total cross sections for neutrons, the angular dependence of the elastic scattering, and the cross section for the formation of the compound nucleus. It is shown that the average properties of neutron resonances, in particular the ratio of the neutron width to the level spacing, are connected with the gross-structure problem and can be predicted by this model.

The observed neutron total cross sections can be very well reproduced in the energy region between zero and 3 Mev with a well depth of 42 Mev, a factor ζ of 0.03, and a nuclear radius of $R=1.45\times10^{-13}A^{\frac{1}{3}}$ cm. The angular dependence of the scattering cross section at 1 Mev is fairly well reproduced by the same model. The theoretical and experimental values for the ratios of neutron width to level distance at low energies and the reaction cross sections at 1 Mev do not agree too well but they show a qualitative similarity.

I. INTRODUCTION

THIS paper deals with the interaction of nuclear particles with complex nuclei in nuclear reactions. A model is proposed for the description of the energy exchange between the incoming particle and the target nucleus. The considerations are restricted to neutron reactions with incident energies between 0 and 20 Mev.

One usually describes the interaction of nuclear particles with complex nuclei by means of the concept of a compound nucleus which is formed after the nucleon has entered the nucleus. Before the striking success of the nuclear shell model was known, it was generally assumed that the quantum state formed by the particle entering the nucleus is one in which the motions of all particles are intimately coupled. We will refer to this assumption as the "strong-coupling model." These ideas led to certain general qualitative conclusions in regard to the cross sections for nuclear reactions. Several authors[1] have attempted in previous papers to derive approximate expressions for the cross sections of nuclear reactions with a minimum of special assumptions in addition to the main assumption of the validity of the strong-coupling model. We summarize the main results of these qualitative considerations.

1. Particle widths.—The particle widths of nuclear resonances with respect to particle emission are related in a general way to the average spacing D of the levels of the compound nucleus. For example, the width for the emission of neutrons with zero-orbital angular momentum is given approximately by

$$\Gamma_n \approx (2/\pi)(k/K)D, \tag{1.1}$$

and the widths for the emission of other particles are equal to the above expression multiplied by the penetration factor of the potential barrier. Here k is the wave number of the incoming particle, and K is the wave number in the interior of the nucleus; K is of the order of 10^{13} cm^{-1}.

2. Potential scattering.—The elastic scattering arises from a superposition of a resonance amplitude and a slowly varying potential scattering amplitude. The former is important only in the immediate vicinity of the resonance; the latter is equal to the scattering amplitude of an impenetrable sphere of a radius approximately equal to the nuclear radius.

3. Neutron total cross section.—The neutron total cross section averaged over resonances is equal to the total cross section of a spherical potential well whose depth is such as to give rise to an internal wave number $K\sim10^{13}$ cm^{-1} and which possesses an absorption for the incoming waves such that the waves are absorbed inside within distances of the order K^{-1}. These conditions were expressed approximately by Feshbach and Weisskopf[2] in the form of a boundary condition on the incoming wave function u/r at the nuclear boundary:

$$du/dr=-iKu.$$

Formula (1.1) and the other consequences of the strong-coupling model have been found correct as to the order of magnitude. However, as a consequence of point (3), the neutron total cross sections when averaged

* This work was supported in part by the U. S. Office of Naval Research and the U. S. Atomic Energy Commission.

† Now at Brookhaven National Laboratory, Upton, New York.

[1] H. Bethe, Phys. Rev. 57, 1125 (1940); Feshbach, Peaslee, and Weisskopf, Phys. Rev. 71, 145 (1947); H. Feshbach and V. F. Weisskopf, Phys. Rev. 76, 1550 (1949); E. P. Wigner, Phys. Rev. 73, 1002 (1948); E. P. Wigner, Am. J. Phys. 17, 99 (1949).

[2] H. Feshbach and V. F. Weisskopf, Phys. Rev. 76, 1550 (1949).

over resonances should all be smooth functions of the energy which decrease monotonically with increasing energy and whose form is rather similar for all atomic numbers A. Also the dependence on A at constant energy should show a continuous, slowly increasing trend with increasing A. The measurements of neutron total cross sections by the Wisconsin group and by others have clearly demonstrated that this is not so. The neutron total cross sections exhibit typical deviations from the predictions of the strong-coupling model. (See Fig. 3.) The shape of the energy dependence of the neutron cross sections changes significantly over the range of A; however, this change is not random but gradual. Nuclei with small differences in A show almost the same behavior. One concludes, therefore, that these characteristic shapes do not depend on detailed features of nuclear structure but on some general properties which vary slowly with A, say, the nuclear radius.

The success of the shell model has cast some doubt upon the fundamental assumptions of the strong-coupling model. Does the particle necessarily form a "compound state" after entering the nucleus? The shell structure furnishes much evidence that a nucleon can move freely within the nucleus without apparently changing the quantum state of the target nucleus. This is a consequence of observations made at the ground state and at low excitation energies, and it is questionable whether this apparent absence of interaction between one nucleon and the rest is valid also at those excitation energies (~8 Mev) which are created in nuclear reactions with neutrons of a few Mev. Furthermore, there is some reason to believe that at higher energies, 15 Mev and up, the interaction between the entering nucleon and the target is appreciable, since the reaction cross sections at those energies have been found[3] to be equal to the geometrical cross sections $[\pi(R+\lambda)^2]$. Hence, for such energies it happens rarely that a neutron enters the nucleus and leaves it again without sharing its energy with the rest.

It seemed worth while, therefore, to investigate the consequences of a reduced interaction between the nucleons for the theory of nuclear reactions in the energy region of a few Mev. This reduced interaction will manifest itself in the following way: The incident nucleon can penetrate into the nucleus and move within the boundaries of the nucleus *without* forming a compound state. Hence, in this case, the target nucleus acts upon the incoming nucleon as a potential well. The actual formation of a compound state occurs only with a probability smaller than unity, once the particle has entered the nucleus. It has a finite chance of leaving the nucleus without having formed a state in which it has exchanged energy or momentum with the rest of the nucleus. The formation of the compound state then would have the aspect of an absorption. Hence, the effect of the nucleus upon the incident particle could be described as the effect of a potential well with absorption, where the absorption coefficient within the well would be an adjustable parameter. It is obvious that this description represents an oversimplification which naturally cannot reproduce all features of nuclear reactions. Specifically, it will not reproduce any resonance phenomena which are connected with the many possible quantum states of the compound system. We therefore expect that this model will at best describe only the features of nuclear reactions after averaging over the resonances of the compound nucleus.

The formulation of this attempt to construct a simple model for nuclear reactions requires a study of the definitions of the various cross sections; in particular, the meaning of the cross section for the formation of a compound nucleus must be clarified.

We introduce the following cross sections: σ_t, the total cross section, which can be split into

$$\sigma_t=\sigma_{el}+\sigma_r,$$

where σ_{el} is the elastic scattering cross section, and σ_r is the "reaction cross section." The former is defined as the cross section for scattering without change of the quantum state of the nucleus. The particle leaves by the same channel by which it has entered. The elastic scattering has an angular dependence which we express by the differential cross section $d\sigma_{el}/d\Omega$,

$$\sigma_{el}=\int\frac{d\sigma_{el}}{d\Omega}(\theta)d\Omega.$$

The reaction cross section includes all processes in which the residual nucleus is different from or in a state different from that of the target nucleus. These are all processes whose exit channels differ from the entrance channel. It will be practical later on to subdivide the elastic cross section into two parts:

$$\sigma_{el}=\sigma_{se}+\sigma_{ce}.$$

We call the second part, σ_{ce}, the "compound elastic" cross section. It is the part of the elastic scattering which comes from the formation of the compound nucleus and the subsequent emission of the incident particle into the entrance channel. The first part we call "shape elastic" cross section; this is the part of the elastic scattering which occurs without the formation of a compound. The exact definition of this split will be given in Sec. II. We note that such definitions will be possible only for the average cross sections, averaged over an energy interval containing many resonances, if such resonances are present.

On the basis of the compound nucleus assumption, we consider all actual reactions to occur after compound formation. Hence, we introduce a cross section σ_c of compound nucleus formation:

$$\sigma_c=\sigma_{ce}+\sigma_r,$$

[3] Phillips, Davis, and Graves, Phys. Rev. 88, 600 (1952).

and obtain, naturally,

$$\sigma_t = \sigma_{se} + \sigma_c.$$

The nuclear model which we propose here is expected to predict only the cross sections σ_{se} and σ_c. It considers only the conditions in the entrance channel, that is, in that part of the phase space in which the target nucleus is in its initial state. Hence, the compound nucleus formation is considered as an *absorption* of the incident beam, although part of it, namely σ_{ce}, leads to an elastic scattering process. The model consists in describing these conditions by means of a one-particle problem. The nucleus is replaced by a complex potential,

$$V = V_0 + iV_1, \tag{1.2}$$

acting upon the incoming neutron. The scattering which the neutron suffers in (1.2) should reproduce the shape elastic scattering σ_{se}; and the absorption which is caused by the imaginary part V_1 should reproduce the compound nucleus formation.

It is probable that the potential functions in (1.2) vary somewhat with the incident energy. For example, one might expect an increase of the imaginary part with increasing energy. If an approximate description of the facts is possible by means of a potential (1.2), the shape of the potential will be indicative of the type of nuclear interaction which a neutron suffers in the nucleus. The real part V_0 would describe the average potential energy of the neutron within the nucleus, and its shape would give indications as to the form of the potential "well" inside the nucleus. It is similar to the potential encountered in the shell model of the nucleus, although we do not pretend that an incident neutron of several Mev is faced with exactly the same potential which acts upon the nucleus in the ground state. The imaginary part V_1 would indicate the strength and location of the processes that lead to an energy exchange between the incoming neutron and the target nucleus.

We expect the potential V to depend in a simple way upon the mass number A. Its dependence on r should be similar for all nuclei. The simplest choice would be a square-well potential:

$$V_0 = -U \quad \text{for} \quad r < R,$$
$$V_0 = 0 \quad \text{for} \quad r > R,$$
$$V_1 = \zeta V_0.$$

In general, we might express it in the form $V = V(r/R)$, $R = r_0 A^{\frac{1}{3}}$. However, there might be a region near $r = R$ in which the features depend on r and not on (r/R); the thickness of that part of the potential which represents the surface might be independent of the radius.

With a given $V(r)$ and its dependence on A, it is possible to calculate the cross sections σ_t, σ_{se}, and σ_c, each as functions of energy and mass number, and also the angular dependence of the scattering. The next section contains the definitions of the cross sections involved, and the following sections describe the technique of calculating the cross sections and their comparison with experimental material.

II. THEORY OF AVERAGE CROSS SECTIONS

All nuclear cross sections exhibit strong fluctuations with energy which are generally referred to as resonances, especially in the lower part of our energy range. As the energy increases, the width of the resonances increases too; and, for not too light nuclei, the width becomes comparable or larger than the level distance at energies above a few Mev. Hence, we find the cross sections at higher energies to be smooth functions of energy with little fluctuation. We will refer to the lower-energy region as the "resonance region" and the upper as the "continuum region."

The behavior of the cross sections in the resonance region does not lend itself to a description by a simple one-particle potential (1.2) because of the rapid fluctuations with energy. However, the averages of the cross sections taken over an interval I, which includes many resonances, will be shown to be the cross sections belonging to a new scattering problem with slowly varying phases, which we will call the "gross-structure" problem. In this problem it is possible to define cross sections for the formation of a compound nucleus which also includes the compound elastic scattering. *It is this gross-structure problem and not the actual rapidly varying cross sections which we intend to describe by means of a one-particle problem with the potential (1.2).*

We bombard a nucleus X with particles a and consider the total cross section σ_t, the elastic cross section σ_{el}, and the reaction cross section σ_r. $\sigma_t = \sigma_{el} + \sigma_r$. Each of these cross sections will be subdivided into their parts coming from different angular momenta l, e.g.,

$$\sigma_t = \sum_l \sigma_t^{(l)}. \tag{2.1}$$

These cross sections can be expressed in terms of the amplitudes of the wave which describes the situation in the entrance channel. We consider the subwave u_l/r in the entrance channel with the orbital angular momentum l (r is the channel coordinate), and we write the wave in the form for

$r \to \infty$,

$$\varphi_l \to \text{const}[\exp(-i(kr - \tfrac{1}{2}l\pi)) - \eta_l \exp(+i(kr - \tfrac{1}{2}l\pi))]. \tag{2.2}$$

The complex reflection factor η_l is connected with the complex phase shift φ_l by $\eta_l = \exp(2i\varphi_l)$, and the cross sections are given by the well-known expressions for the elastic cross section:

$$\sigma_{el}^{(l)} = \pi\lambda\!\!\!^{-2}(2l+1)|1-\eta_l|^2, \tag{2.3}$$

and for the reaction cross section,

$$\sigma_r^{(l)}=\pi\lambda\!\!\!^{-2}(2l+1)(1-|\eta_l|^2), \tag{2.3a}$$

where $\lambda\!\!\!^{-}$ is the wavelength of the incoming particle divided by 2π.

The reflection factor η_l is a complicated function of the energy of the incoming particle. It exhibits rapid fluctuations coming from the numerous close-spaced resonances of the compound nucleus. We will make the assumption that one can average over these fluctuations; that is, we assume that the average reflection factor,

$$\bar{\eta}_l(\epsilon)=\frac{1}{I}\int_{\epsilon-I/2}^{\epsilon+I/2}\eta_l(\epsilon')d\epsilon', \tag{2.4}$$

is a smooth function of ϵ if the interval I contains many close-spaced resonances. We also define average cross sections in the same way, and we can write

$$\begin{aligned}\bar{\sigma}_{el}^{(l)}&=\pi\lambda\!\!\!^{-2}(2l+1)\overline{|1-\eta_l|^2},\\ \bar{\sigma}_r^{(l)}&=\pi\lambda\!\!\!^{-2}(2l+1)(1-\overline{|\eta_l|^2}),\end{aligned} \tag{2.5}$$

where the bar over an expression signifies its average over the interval I. It is also assumed that I is much smaller than the energy ϵ such that slowly varying functions of ϵ, like $\lambda\!\!\!^{-2}$, need not be averaged.

One can easily verify the following relations:

$$\bar{\sigma}_{el}^{(l)}=\pi\lambda\!\!\!^{-2}(2l+1)\{|1-\bar{\eta}_l|^2-|\bar{\eta}_l|^2+\overline{|\eta_l|^2}\}, \tag{2.6}$$

and especially

$$\bar{\sigma}_t^{(l)}=\pi\lambda\!\!\!^{-2}(2l+1)\{|1-\bar{\eta}_l|^2+1-|\bar{\eta}_l|^2\}. \tag{2.7}$$

Hence, the average total cross section depends only upon the average reflection factor (2.4). [This follows directly from the fact that the total cross section is a *linear* function of the real part of the phase η_l.]

We now divide the average elastic cross section into two parts, the "shape elastic" cross section $\sigma_{se}^{(l)}$ and the "compound elastic" cross section[4] $\sigma_{ce}^{(l)}$, by writing

$$\begin{aligned}\sigma_{se}^{(l)}&=\pi\lambda\!\!\!^{-2}(2l+1)|1-\bar{\eta}_l|^2,\\ \sigma_{ce}^{(l)}&=\pi\lambda\!\!\!^{-2}(2l+1)\{\overline{|\eta_l|^2}-|\bar{\eta}_l|^2\}.\end{aligned} \tag{2.8}$$

Furthermore, we combine $\sigma_{ce}^{(l)}$ and $\bar{\sigma}_r^{(l)}$ into a new cross section $\sigma_c^{(l)}$, which we call the cross section for the formation of the compound nucleus

$$\sigma_c^{(l)}=\sigma_{ce}^{(l)}+\bar{\sigma}_r^{(l)}=\pi\lambda\!\!\!^{-2}(2l+1)\{1-|\bar{\eta}_l|^2\}. \tag{2.9}$$

We can see from (2.3) and (2.3a) that $\sigma_{se}^{(l)}$ and $\sigma_c^{(l)}$ have just the form of a scattering and a reaction cross section of a new and different problem, whose phase is the slowly varying function $\bar{\eta}_l$. In other words, by replacing η_l with $\bar{\eta}_l$, we obtain a new problem, which we have called the "gross-structure problem." The elastic scattering cross section σ_{se} of *this* problem is only part of the actual scattering; it is the "shape elastic" scattering. The other part, the "compound elastic," appears incorporated into the absorption or reaction cross section σ_c of the gross-structure problem together with the actual reaction cross section.

One is therefore led to consider the "compound elastic" scattering as that part which comes from the formation of the compound nucleus and its subsequent decay into the entrance channel, hence its incorporation into σ_c. After the averaging, σ_{ce} appears as part of the absorption from the incoming beam, which corresponds to the idea that the formation of the compound nucleus can be considered as an absorption whatever happens afterwards, re-emission or not.

It is the gross structure problem which we intend to reproduce by the interaction of the incident particle with the potential (1.2). The resulting scattering cross section should represent the shape elastic scattering, and the resulting absorption cross section should represent the compound formation. The latter contains the part σ_{ce} of the actual scattering.

When the energy is high enough above the resonance region that the continuum region is reached, the cross sections and phases are no longer rapidly varying functions of energy. Then the gross-structure problem is equal to the actual one and $\bar{\eta}_l=\eta_l$. It follows from (2.9) that $\sigma_{ce}=0$. One also can see this from an application of the compound nucleus assumption to the continuum region. The overlap of the resonances can be interpreted as a consequence of the fact that the probability Γ_α^s of the decay of the compound nucleus in the state s into the entrance channel α is much smaller than the probability of the decay into other channels. This follows from the well-known relation that any channel width Γ_α^s cannot be larger than $D/2\pi$ (D is the distance between resonances of the same J value). Hence, if the total width is much larger than D, the contribution to Γ from decays other than the one through α must be overwhelming. In the continuum region, therefore, the cross section for the formation of the compound nucleus is identical to the average reaction cross section $\bar{\sigma}_r$, and σ_{ce} is negligible.

We now illustrate the averaging process described above by using cross sections as given by the Breit-Wigner formula. We consider a nucleus with resonances at the energies ϵ_s, and we restrict our considerations to neutrons with $l=0$. We also restrict the discussion to low energies so that the following two magnitudes are small: One is kR and the other is Γ/D, with R the nuclear radius, and Γ and D the average values of the total width of and the distance between neutron resonances.

[4] This terminology will become obvious later on. B. T. Feld [*Experimental Nuclear Physics*, edited by E. Segrè (John Wiley and Sons, Inc., New York, 1953), Vol. 2] calls σ_{ce} the "capture elastic" cross section.

We have derived in the Appendix exact and approximate expressions for the scattering amplitude η_0 and for the cross sections in this energy region. For the present purposes, we will use the following form (see A.14b), which is valid in a region D_s including a resonance ϵ_s as indicated:

$$\eta_0 = e^{-2ikR'(\epsilon)}\left(1 - \frac{i\Gamma_\alpha^s}{\epsilon - \epsilon_s + i\Gamma^s/2}\right) + \eta_0^*,$$

$$\eta_0^* = e^{-2ikR'(\epsilon)}\left(\frac{\Gamma_\alpha^s}{\epsilon - \epsilon_s + i\Gamma^s/2} G_1 + iG_2 + G_3\right), \quad (2.10)$$

for

$$\epsilon_s + \epsilon_{s-1} < 2\epsilon < \epsilon_{s+1} + \epsilon_s.$$

Here R' is a length and a slowly varying function of the energy. (A function is slowly varying if it changes value appreciably only over intervals large compared to D.) The length R' is of the order of magnitude of nuclear dimensions. It plays the role of a scattering length and takes on both positive and negative values. The quantities Γ_α^s and Γ^s are the partial width and the total width, respectively. The terms G_1, G_2, and G_3 are real functions of ϵ of the following order of magnitude:

$$G_1 \sim \Gamma/D,\ G_2 \sim (\Gamma_\alpha/D)[(\Gamma/D) + kR],\ G_3 \sim \Gamma_\alpha/D, \quad (2.11)$$

where the omission of the superscript signifies the average value of the magnitude in the interval I.

The first term in η_0 incorporates the contribution from the resonance level ϵ_s, whereas η_0^* contains the contribution from the other resonances; the first term in η_0^* represents interference effects between the resonance ϵ_s and other resonances. It will appear later that η_0^* contributes negligibly to the average of η_0.

The cross sections in the immediate neighborhood of the resonance ($|\epsilon - \epsilon_s| \ll D_s$) follow from (2.5) and (2.10) by neglecting η_0^*, since, in that region, they contribute terms much smaller than the others.

$$\sigma_r^{(0)} = \pi\lambda\!\!\!^-{}^2 \frac{\Gamma_\alpha^s(\Gamma^s - \Gamma_\alpha^s)}{(\epsilon - \epsilon_s)^2 + (\Gamma^s/2)^2},$$

$$\sigma_{el}^{(0)} = \pi\lambda\!\!\!^-{}^2 \left|(e^{2ikR'} - 1) + \frac{i\Gamma_\alpha^s}{\epsilon - \epsilon_s + i\Gamma^s/2}\right|^2, \quad (2.12)$$

$$|\epsilon - \epsilon_s| \ll D.$$

The reaction cross section is just the sum over β of the one-level Breit-Wigner cross sections

$$\sigma_{\alpha\beta}^{(0)} = \frac{\Gamma_\alpha^s \Gamma_\beta^s}{(\epsilon - \epsilon_s)^2 + (\Gamma^s/2)^2} \pi\lambda\!\!\!^-{}^2 \quad (2.13)$$

for the reaction leading from the entrance channel α to an exit channel β. (Γ_β^s is the partial width of decay into the channel β.)

The elastic cross section contains a "potential" scattering amplitude

$$P = e^{2ikR'(\epsilon)} - 1,$$

which corresponds to a scattering at a hard sphere of a radius R', where R' is not identical to but only of the order of magnitude[6] of the nuclear radius and is a slowly varying function of the energy.

We now determine the average value of the scattering amplitude η_0 over the resonances in the interval I:

$$\bar{\eta}_0 = \frac{1}{I}\int_I \eta_0 d\epsilon = \left\langle \frac{1}{D_s}\int_{D_s} \eta_0(\epsilon) d\epsilon \right\rangle_I,$$

where the symbol $\langle\ \ \rangle_I$ signifies an average taken over all resonances within I. The random position of resonances allows us to write

$$\bar{\eta}_0 = \frac{1}{D}\int_{\epsilon_s - D/2}^{\epsilon_s + D/2} \eta_0(\epsilon) d\epsilon, \quad (2.14)$$

where D is the average level distance within the interval I; typical average values of Γ^s and Γ_α^s should be used in the expression (2.10) for η_0.

Evaluation of (2.14) gives

$$\bar{\eta}_0 = e^{-2ikR'}[1 - (\pi\Gamma_\alpha/D)], \quad (2.15)$$

when all magnitudes of the order $(\Gamma_\alpha\Gamma)/D^2$ or $(\Gamma_\alpha/D)kR'$ or smaller are neglected. It is seen in each interval D_s that the main contribution to the average comes from the main resonance. The contribution of neighboring resonances which are expressed by η_0^* in (2.10) contribute only to expressions which are smaller than (2.14) by a factor of the order Γ/D or kR.

We now use (2.14) for the calculation of the "shape elastic" scattering and get, with the help of (2.8),

$$\sigma_{se}^{(0)} = \pi\lambda\!\!\!^-{}^2 |(e^{2ikR'} - 1) + \pi\Gamma_\alpha/D|^2. \quad (2.16)$$

For small kR', this becomes

$$\sigma_{se}^{(0)} = 4\pi R'^2[1 + (\pi\Gamma_\alpha/2kR'D)^2]. \quad (2.17)$$

The magnitude $[\pi\Gamma_\alpha/2(kR'D)]^2$ is usually rather small. [It is of the order of 10^{-2}; see, for example, the estimate in Blatt and Weisskopf,[5] Chap. VIII, Eq. (7.14).] Hence, $\sigma_{se}^{(0)}$ is very nearly equal to $4\pi R'^2$ for $kR' \ll 1$.

We get the cross section for the formation of the compound nucleus according to (2.9),

$$\sigma_c^{(0)} = 2\pi^2\lambda\!\!\!^-{}^2(\Gamma_\alpha/D)(1 - \tfrac{1}{2}\pi\Gamma_\alpha/D). \quad (2.18)$$

[6] The appearance of the length R' is a consequence of our general treatment of the nuclear resonance in the Appendix. In the special derivation of the Breit-Wigner formula, as given in Feshbach, Peaslee, and Weisskopf (reference 1) or J. M. Blatt and V. F. Weisskopf, *Theoretical Nuclear Physics* (John Wiley and Sons, Inc., New York, 1952), assumptions are made which make R' constant and equal to the nuclear radius R. It is shown in this paper that these assumptions probably are valid only in special cases as in the case of strong coupling, for example.

The average total cross section then becomes

$$\sigma_t^{(0)}=\sigma_{se}^{(0)}+\sigma_c^{(0)}=4\pi R'^2+2\pi^2\lambda^2\Gamma_\alpha/D. \quad (2.19)$$

The second term in this expression is proportional to $(1/v)$.

It is interesting to compare σ_c with the average of σ_r, which, according to (2.12), is

$$\bar{\sigma}_r^{(0)}=2\pi^2\lambda^2(\Gamma_\alpha/D)(\Gamma-\Gamma_\alpha)/\Gamma. \quad (2.20)$$

Hence, the difference between the two, the "compound elastic" scattering, is [neglecting the small factor $\pi\Gamma_\alpha/(2D)$]

$$\sigma_{ce}^{(l)}=2\pi^2\lambda^2\Gamma_n^2/D\Gamma. \quad (2.21)$$

This is just the average of that part of the elastic scattering (2.12) which corresponds to the resonance amplitude only, namely, of

$$\sigma=\pi\lambda^2\frac{(\Gamma_\alpha^s)^2}{(\epsilon-\epsilon_s)^2+(\Gamma^s/2)^2}.$$

It is the cross section which one would get for the re-emission into the entrance channel from the Breit-Wigner expression (2.13).

It is significant that expressions (2.15), (2.18), and (2.19) do not contain the total width Γ but only the channel width Γ_α. The "gross" properties (total, shape elastic, and compound nucleus formation cross sections) are independent of the nature of the other exit channels. They would remain unchanged, for example, if the exit channels $\beta\neq\alpha$ were closed. It would only increase σ_{ce} at the expense of $\bar{\sigma}_r$, as seen in (2.20) and (2.21). This is connected with the fact that a change of Γ with constant Γ_α changes only the width of the resonance, but not its area.

At the energies considered here, the cross section for the formation of the compound nucleus contains only magnitudes (Γ_α and D), which can be determined by studying the neutron resonances. Hence, investigations of slow neutron resonances are useful to check the theoretical predictions of σ_c at low energy. The "shape-elastic" scattering, on the other hand, in this energy region is almost entirely given by $4\pi R'^2$ and is therefore essentially independent of the neutron resonance values. Apart from the small correction $\pi^3\lambda^2(\Gamma_\alpha/D)^2$, it is equal to the potential scattering as shown in (2.12) and, therefore, can be measured also by studying the cross sections near and between resonances.

III. POTENTIAL-WELL MODEL

In this section we shall employ a potential-well model to determine the gross-structure cross sections. We have adopted for the purposes of a preliminary survey the simplest type of potential well:

$$V=-V_0(1+i\zeta), \quad r<R, \qquad V=0, \quad r>R, \quad (3.1)$$

where V_0 and ζ are constants and R is the nuclear radius. The use of the complex potential is necessary to obtain nonzero values for the cross section for the formation of the compound nucleus. A similar model in which $\zeta\sim 1$ was employed by Bethe.[6] Fernbach, Serber, and Taylor[7] have used the same model in order to describe nuclear scattering at very high energies. A model in which $\zeta=0$ was used by Ford and Bohm[8] in discussing zero-energy cross sections. It is essential that the crudeness of this model be emphasized. We have, for example, omitted any spin-orbit terms which play an important role in the shell model, but which we expect will not affect the over-all qualitative features which we seek here. The constants in (3.1) may well turn out to be energy dependent. We particularly expect this for ζ, since we know that $\bar{\sigma}_r$ is large at high energies, while the success of the shell model indicates that ζ should be zero for the ground states of nuclei.

We give some of the details of the calculations with potential (3.1). For each l we calculate the value of the logarithmic derivative,

$$f_l=R(u_l'/u_l)_{r=R}. \quad (3.2)$$

The average reflection factor $\bar{\eta}_l$ is then

$$\bar{\eta}_l=e^{-2i\delta_l}\left(1-\frac{2s_l}{M_l+iN_l}\right), \quad (3.3)$$

where

$$\delta_l=\tan^{-1}(-j_l(x)/n_l(x)), \quad (3.4a)$$

$$\Delta_l+is_l=1+xh_l'(x)/h_l(x), \quad (3.4b)$$

$$M_l=s_l-\mathrm{Im}f_l, \quad N_l=-\Delta_l+\mathrm{Re}f_l. \quad (3.4c)$$

The functions j_l, n_l, and h_l are the spherical Bessel, Neumann, and Hankel functions, respectively, while x is, as usual, kR.[9] $h_l'(x)$ is the derivative of $h_l(x)$ with respect to x; Δ_l and s_l are both real magnitudes and are defined as the real and imaginary part of the expression on the right of (3.4b).

For potential (3.1), f_l may be written down directly

$$f_l=1+Xj_l'(X)/j_l(X), \quad (3.5)$$

where

$$X^2=x^2+X_0^2(1+i\zeta), \quad X_0^2=(2m/\hbar^2)V_0R^2.$$

This is, however, not the most convenient form for determining the real (Re) and imaginary (Im) parts of f_l. We have instead employed recurrence relations for these quantities based on recurrence relations for j_l.

[6] H. Bethe, Phys. Rev. 57, 1125 (1940).
[7] Fernbach, Serber, and Taylor, Phys. Rev. 75, 1352 (1949).
[8] K. W. Ford and D. Bohm, Phys. Rev. 79, 745 (1950).
[9] This follows the notation of Morse, Lowan, Feshbach, and Lax, U. S. Navy Department of Research and Inventions Report No. 62.1R, 1945 (unpublished).

For $l=0$, we get

$$f_0=X\cot X,$$
$$\mathrm{Re}f_0=\frac{X_1\sin 2X_1+X_2\sinh 2X_2}{\cosh 2X_2-\cos 2X_1}, \tag{3.6}$$
$$\mathrm{Im}f_0=\frac{X_2\sin 2X_1-X_1\sinh 2X_2}{\cosh 2X_2-\cos 2X_1},$$

where $X=X_1+iX_2$. The recurrence relations which follow from

$$f_l=\frac{X^2}{l-f_{l-1}}-l$$

are

$$\mathrm{Re}f=\frac{(X_1^2-X_2^2)(l-\mathrm{Re}f_{l-1})-2X_1X_2\,\mathrm{Im}f_{l-1}}{(l-\mathrm{Re}f_{l-1})^2+(\mathrm{Im}f_{l-1})^2}-l, \tag{3.7}$$

$$\mathrm{Im}f_l=\frac{(X_1^2-X_2^2)\,\mathrm{Im}f_{l-1}+2X_1X_2(l-\mathrm{Re}f_{l-1})}{(l-\mathrm{Re}f_{l-1})^2+(\mathrm{Im}f_{l-1})^2}. \tag{3.8}$$

The asymptotic expression for f_l,

$$f_l\xrightarrow[X\to\infty]{}X\cot(X-\tfrac{1}{2}l\pi), \tag{3.9}$$

unfortunately cannot be generally employed. The fractional error in (3.9) is $l(l+1)/(X\sin 2X)$, from which we learn that (3.9) is not sufficiently accurate for $l\geqslant 2$, while for $l=1$ it will fail for small X or for $X=n\pi$.

The total cross section, as well as the cross section for the formation of the compound nucleus, may be easily obtained

$$\frac{\bar{\sigma}_t^{(l)}}{\pi R^2}=\frac{4}{x^2}(2l+1)\left[\sin^2\delta_l+s_l\frac{M_l\cos 2\delta_l-N_l\sin 2\delta_l}{M_l^2+N_l^2}\right],$$

$$\frac{\sigma_c^{(l)}}{\pi R^2}=\frac{4}{x^2}(2l+1)s_l\left[\frac{-\mathrm{Im}f_l}{M_l^2+N_l^2}\right]\equiv\frac{(2l+1)T_l}{x^2}, \tag{3.10}$$

$$\bar{\sigma}_t=\sum\nolimits_l\bar{\sigma}_t^{(l)},\quad \sigma_c=\sum\nolimits_l\sigma_c^{(l)},$$

where the T_l may be interpreted as penetrabilities.

These cross sections will have characteristic large-scale resonances, which are present in the experimental data. In the $l=0$ case, these resonances occur when

$$X_1=(X_0^2+x^2)^{\frac{1}{2}}=(n+\tfrac{1}{2})\pi+\frac{X_0^2\zeta^2}{2(2n+1)\pi},$$

where n is an integer and where we have assumed that $\zeta X_0^2/n\pi\ll 1$. The width of the large-scale resonance is $2x\hbar^2/mR^2$, which in the experimental range is of the order of Mev. For a given energy, the $l=0$ cross section will give maxima as a function of R. The width of these maxima against changes in R is approximately $(2xR/X^2)$, independent of R.

The angular distribution for shape elastic scattering is

$$\frac{d\sigma_{se}}{d\Omega}=\frac{\lambda\!\!\!^{-2}}{4}\Big|\sum_l(2l+1)(1-\bar{\eta}_l)P_l(\cos\theta)\Big|^2.$$

Therefore

$$\frac{1}{R^2}\frac{d\sigma_{se}}{d\Omega}=(\mathrm{Re}\textstyle\sum)^2+(\mathrm{Im}\textstyle\sum)^2, \tag{3.11}$$

where

$$\mathrm{Im}\textstyle\sum=-\frac{x}{4}\sum_l\frac{\sigma_t^{(l)}}{\pi R^2}P_l(\cos\theta),$$
$$\mathrm{Re}\textstyle\sum=-\frac{1}{2x}\sum_l(2l+1)\left[\sin 2\delta_l-2s_l\frac{M_l\sin 2\delta_l+N_l\cos 2\delta_l}{M_l^2+N_l^2}\right]. \tag{3.12}$$

Before we can compare the theory with the experimental data on angular distributions, it is necessary to add the compound elastic scattering. From our general qualitative ideas, we may break the process up into the formation of the compound nucleus and the re-emission of the incident particle into a particular l state, which will naturally have a very definite associated angular distribution. The result is particularly simple in the case of a target nucleus of spin zero and an incident particle of spin zero, since here the angular momentum of the incident particle cannot change in an elastic scattering process. We may therefore write

$$\frac{d\sigma_{ce}}{d\Omega}=\sum\sigma_c^{(l)}|Y_{l0}|^2w_l, \tag{3.13}$$

where Y_{l0} are the normalized spherical harmonics and w_l is the probability that the compound nucleus formed by the absorption of a particle of angular momentum l will decay by emission of the same particle without loss of energy or change in angular momentum.

This simple result cannot be applied to the neutron case because of the possibility of spin changes of the neutron and re-orientation of the spin of the target nucleus without any change in the energy of either the neutron or the target nucleus. The formalism which needs to be used here has been worked out by Hauser and Feshbach[10] and by Wolfenstein.[11] The target nucleus and neutron system is now characterized by the spin of the target nucleus I, its z component m, and spin of the neutron i, the channel spin $s(\mathbf{s}=\mathbf{i}+\mathbf{I})$, the angular momentum of the incident neutron l, and its z component which is zero, and of course the parity of the system. The compound nucleus will have a total angular momentum J, z component m, and will decay into a residual nucleus of spin I' and a particle of spin i'. These form a final channel spin $s'(\mathbf{s}'=\mathbf{i}'+\mathbf{I}')$, z component $m-m'$. The system will have an angular momentum l', z component m'.

[10] W. Hauser and H. Feshbach, Phys. Rev. 87, 366 (1952).
[11] L. Wolfenstein, Phys. Rev. 82, 690 (1951).

To obtain the desired cross section, we must now introduce the assumptions of the statistical nuclear theory. We assume that, upon appropriate averaging, the various J levels do not interfere, that there is no residual interference between the various l's which can form the given compound state J, or between the various l's into which it can decay. We then break up the process of compound elastic scattering into the cross section for the formation of the compound nucleus in state J, with incident particles of angular momentum l, multiplied by the probability that it will decay by emission of a particle of angular momentum l', leaving the residual nucleus in the ground state with spin I. The formation process is given first by the cross section for the formation of the compound nucleus which, because of our simple assumption (3.1), depends only on l and is $\sigma_c^{(l)}$. This must be multiplied by the probability of forming the system with angular momentum J, using incident particles of angular momentum l. Again, because of the absence of any spin-dependent forces, this is simply the square of the Clebsch-Gordan coefficient $|(ls0m|lsJm)|^2$. On the emission side, we will need the probability of forming J with particles of angular momentum l'. This is given by $|(l's'm'm-m'|l's'Jm)|^2$. We need the relative probability of decay with emission of l' particles leaving the nucleus in the ground state which we will denote by $w(l')\leqslant 1$. The limitations of the relative probability of different kinds of emission arising from angular momentum conservation are contained in the Clebsch-Gordan coefficients. The function w contains all the other dependence. Because of our assumption of spin-independent forces in Eq. (3.1), it will depend only on l' and the parity of the system. The angular distribution of the emitted particles is $|Y_{l',m'}|^2$. Combining these results, we have

$$\frac{d\sigma_{ce}}{d\Omega}=\frac{1}{(2i+1)(2I+1)}\sum \sigma_c^{(l)}|(ls0m|lsJm)|^2 \times|(l's'm'm-m'|l's'Jm)|^2 w(l')|Y_{l'm'}|^2. \quad (3.14)$$

The indicated sums are over m, m', s, s', l, l', and J. The spin factor in front arises from the average over initial spin states, which involves the sum over m and m'.

By employing methods due to Racah[12] and discussed by Blatt and Biedenharn,[13] the sums over m and m' may be performed yielding

$$\frac{d\sigma_{ce}}{d\Omega}=\frac{1}{4\pi(2I+1)(2i+1)}\sum\frac{\sigma_c^{(l)}w(l')}{2l+1} \times Z(lJlJ;sL)Z(l'Jl'J;s'L)P_L(\cos\theta), \quad (3.15)$$

where the Z factors are defined by Biedenharn, Blatt, and Rose[14] and for which tables[15] are available. The sums are over J, l, l', and L. Only even L will occur. This result is given in reference 10.[16] We have not introduced a specific notation to describe the role of parity, so it should be understood that parity is conserved both in the formation and in the decay of the compound nucleus.

The total compound elastic cross section may be easily evaluated from (3.15) and gives the expected result

$$\sigma_{ce}=\sum\frac{2J+1}{(2I+1)(2i+1)(2l+1)}\sigma_c^{(l)}w(l').$$

Expression (3.15) simplifies considerably in two special cases (a) $I\gg 1$ and (b) $I=0$. In case (a), it follows from the sum rule (see reference 10),

$$\sum_{s'l'}(2s'+1)|Z(lJlJ;sL)Z(l'Jl'J;s'L)|P_L = (2J+1)^2(2l+1), \quad (3.16)$$

that $d\sigma_{ce}/d\Omega$ is approximately independent of angle. We note that if $I\gg 1$, the factor $2s'+1$ is approximately a constant, the error being of the order of $(1/I)$. In case (b) we note that $s=s'=\frac{1}{2}$ and that $l=l'$ because of parity conservation. We therefore find for this case (placing $i=\frac{1}{2}$)

$$\frac{d\sigma_{ce}}{d\Omega}=\sum_{l,L}\frac{\sigma_c^{(l)}w(l)}{4\pi(4l+2)}[Z^2(l,l+\tfrac{1}{2},l,l+\tfrac{1}{2};\tfrac{1}{2},L) +Z^2(l,l-\tfrac{1}{2},l,l-\tfrac{1}{2};\tfrac{1}{2},L)]P_L. \quad (3.17)$$

The factors w which may be computed as outlined in reference 10 depend on the details of the levels of the residual nucleus. There it is shown that

$$w(l')=T_{l'}(E)/\sum_{pqs}T_p(E_q'). \quad (3.18)$$

The quantity $w(l')$ lies between 0 and 1. The values of T_p are calculated in reference 10 under the assumption of strong coupling. The ideas underlying the present theory would change these factors to those given by Eq. (3.10). Since the compound elastic scattering is not very large compared to the shape elastic, we have only determined the upper limit for σ_{ce}, which is given by putting $w(l')=1$. We expect σ_{ce} to be near this upper limit at energies for which there is little inelastic scattering or capture, and to be near zero when inelastic scattering or other nuclear reactions are appreciable.

[12] G. Racah, Phys. Rev. 61, 186 (1942); 62, 438 (1942).

[13] J. M. Blatt and L. C. Biedenharn, Revs. Modern Phys. 24 258 (1952).

[14] Biedenharn, Blatt, and Rose, Revs. Modern Phys. 24, 249 (1952).

[15] L. C. Biedenharn, Oak Ridge National Laboratory Report ORNL-1501, May 28, 1953 (unpublished).

[16] It is worth while noting that Eq. (3.15) may be derived from the general analysis of Blatt and Biedenharn by combining the definitions of average cross section as given in Sec. II and the statistical assumptions. The chief elements of the latter are (1) nonoverlap of resonances and (2) random phases for the scattering matrix so that, upon averaging over possible ways of forming the compound nucleus, interference terms average to zero.

IV. ISOLATED RESONANCES

We should like to establish a correspondence between the parameters describing a single compound nucleus resonance and the parameters which describe the average potential (3.1). This is most easily done for the low-energy case. We evaluate the cross sections for very low energy on the basis of the potential (3.1) and by comparing them with the expressions for the average cross sections which were derived in Sec. II in terms of the resonance parameters. The only two resonance parameters entering here are the ratio Γ_α/D of neutron width[17] to level distance and the radius R' of the potential scattering.

We start with the evaluation of the results from (3.1). The only contribution comes from $l=0$ and we get from (3.3)

$$\bar{\eta}_0=e^{-2ix}\frac{f_0+ix}{f_0-ix}=e^{-2ix(1-\alpha)}, \tag{4.1}$$

where α is a complex number:

$$\alpha=(1/x)\tan^{-1}(x/f_0)\cong 1/f_0,$$

and f_0 is given by the expressions (3.6).

This should be compared with (2.14) in order to express the two relevant magnitudes R' and Γ_α/D in terms of X_1 and X_2. Equating (2.14) and (4.1) gives in the limit of $k\rightarrow 0$, a limit which also implies $\Gamma_\alpha/D\rightarrow 0$:

$$R'=R(1-\alpha_1),\quad (\pi/2x)(\Gamma_\alpha/D)=\alpha_2,$$

where α_1 and α_2 are the real and imaginary parts of α.

From (3.6) we can easily obtain the following relations:

$$\alpha=f_0^{-1}=\alpha_1+i\alpha_2,$$

$$\alpha_1=\frac{1}{|X|^2}\frac{X_2B-X_1A\sin 2X_1}{B^2+2A\cos^2X_1},$$

$$\alpha_2=\frac{1}{|X|^2}\frac{X_1B-X_2A\sin 2X_1}{B^2+2A\cos^2X_1},$$

with $A=1/(2\cosh^2X_2)$, $B=\tanh X_2$.

We now distinguish two limiting cases: the cases of strong and weak coupling. In the first case the absorption is so strong that the neutron is completely absorbed in a distance of a nuclear radius within nuclear matter: $\exp(-X_2)\ll 1$. In the case of weak coupling we assume $X_2\ll 1$.

Hence we get, for strong coupling: $A\rightarrow 0$, $B\rightarrow 1$, and

$$\alpha_1=X_2/|X|^2=1/X_2',\quad \alpha_2=X_1/|X|^2=1/X_1',$$

and $R'=R(1-1/X_2')$, $\Gamma_\alpha/D=2x/\pi X_1'$. The length R' is almost equal to R since $(X_2')^{-1}$ is a small magnitude. The expression for Γ_α/D is the same as that used by Feshbach, Peaslee, and Weisskopf with the only exception that X_1' replaces X_1. The former magnitude is somewhat larger than X_1.[18]

Strong coupling therefore leads essentially to the same results as Feshbach, Peaslee, and Weisskopf: The potential scattering length is roughly equal to R and $\Gamma_\alpha/D=2(x/\pi X_1')$.

In the weak coupling approximation we get

$$A=\tfrac{1}{2},\quad B=X_2\ll 1,\quad |X|^2=X_1^2,$$

and hence

$$R'=R\left(1-\frac{1}{2X_1}\frac{\sin 2X_1}{X_2^2+\cos^2X_1}\right),$$

$$\frac{\Gamma_\alpha}{D}=\frac{2x}{\pi X_1}\beta,\quad \beta=X_2\frac{1-(1/2X_1)\sin 2X_1}{X_2^2+\cos^2X_1}.$$

Here R' and β are functions of X_1, and hence of R, with a characteristic resonance denominator. The shapes of these functions are reminiscent of optical dispersion and absorption curves, respectively. The maximum in β occurs when $X_0\cong(n+\frac{1}{2})\pi$ (n integer), the value of β being $2/[(n+\frac{1}{2})\pi\zeta]$ and the width of the peak at half-maximum $(n+\frac{1}{2})\pi\zeta$. The minimum value of β is about $(n+\frac{1}{2})\pi\zeta/2$. Figure 1 shows both magnitudes plotted as a function of X_0 for a value of $\zeta=0.03$.

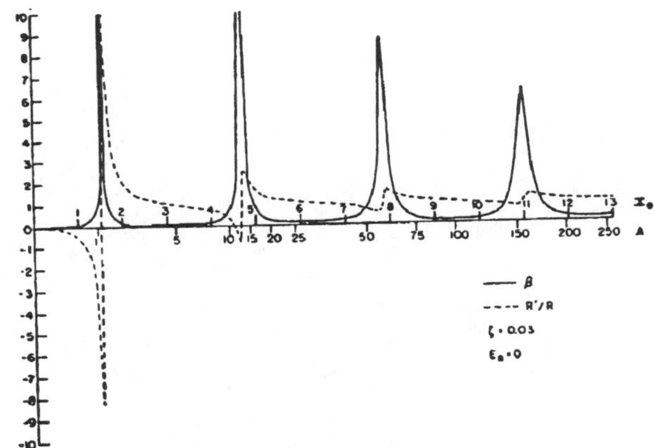

FIG. 1. Potential scattering length R' and the ratio Γ_α/D of the neutron width to the level distance at low energy as a function of $X_0=K_0R$ for $\zeta=0.03$. R' is plotted in units of R and Γ_α/D is given in the form of the parameter $\beta=(\pi/2)(V_0/\epsilon)^{\frac{1}{2}}(\Gamma_\alpha/D)$, where ϵ is the energy of the neutron. The atomic-weight scale corresponding to X_0 is shown also for a potential-well depth $V_0=42$ Mev and radii $R=1.45\times10^{-13}A^{\frac{1}{3}}$ cm.

[17] From here on we use the symbol Γ_α for "neutron width" since the entrance channel α is a *neutron* channel in all cases which we treat in this paper.

[18] It is plausible to assume that, in the case of strong coupling, the imaginary part of the potential is of the same order as the real part. An imaginary part that is much larger than the real one would imply that the absorption takes place over distances small compared to the wavelength in the interior. Hence X_1' is about twice as large as X_0 which leads to a Γ_α/D half as large as in Feshbach, Peaslee, and Weisskopf. This strong coupling result is somewhat more consistent than the result in Feshbach, Peaslee, and Weisskopf. In the latter paper the boundary condition was chosen such that the wave inside the nucleus is a sine wave $\sin(Kr-\delta)$, an assumption that is contrary to the idea of strong compound nucleus formation. In fact, a wave $\exp(+\Omega r)\sin(Kr-\delta)$ with $\Omega\sim K$ would be more consistent and does lead to the same result as the one above.

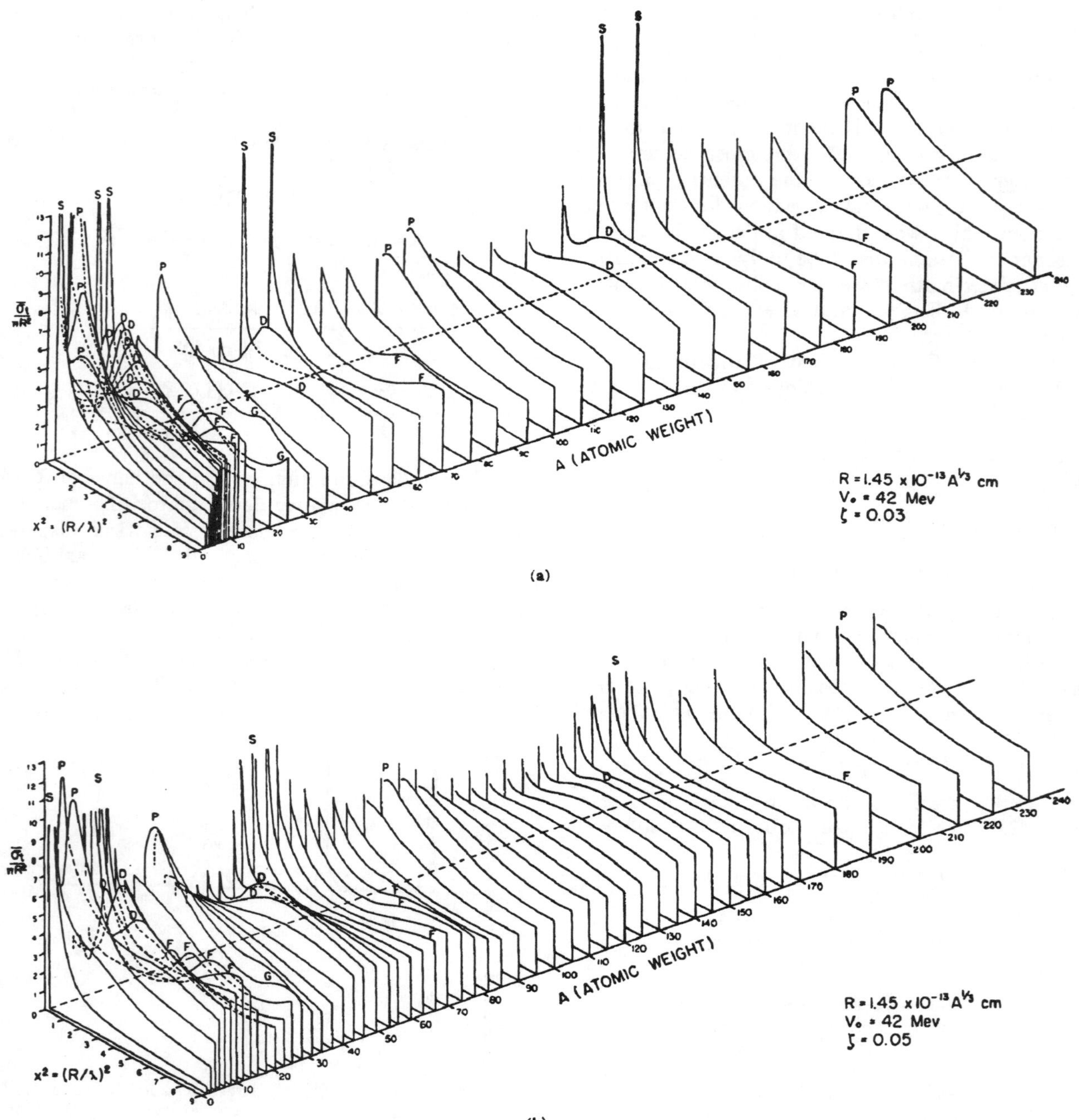

FIG. 2. (a) Calculated neutron total cross sections as a function of energy and mass number, for a well depth $V_0=42$ Mev, radius $R=1.45\times10^{-13}A^{\frac{1}{3}}$, $\zeta=0.03$. The energy ϵ is expressed in terms of

$$x^2=[A^{\frac{2}{3}}\cdot A/10(A+1)]\epsilon,$$

where ϵ is in Mev. (b) The same for $\zeta=0.05$.

V. COMPARISON WITH EXPERIMENTAL RESULTS

Figure 2a shows a profile presentation of the calculations of the neutron total cross sections on the basis of the potential (3.1) with a depth $V_0=42$ Mev and a radius $R=1.45\times10^{-13}A^{\frac{1}{3}}$ cm. The constant ζ is assumed to be 0.03 which corresponds to an absorption coefficient of $\kappa=4.2\times10^{11}$ cm^{-1} in nuclear matter for neutrons of zero energy in free space. This means that the intensity of a beam of slow neutrons is reduced in nuclear matter to $1/e$ at a distance of $\kappa^{-1}=2.4\times10^{-12}$ cm. The cross sections are plotted as a function of the energy in units of $x^2=(R/\lambdabar)^2$ and of the atomic weight. The letters denoting the maxima indicate the character of the resonance causing the maximum. Figure 3 contains a profile presentation of the observed cross sections plotted against the same coordinates.

The experimental curves in Fig. 3 are averages over resonances. For higher A and small level distance this

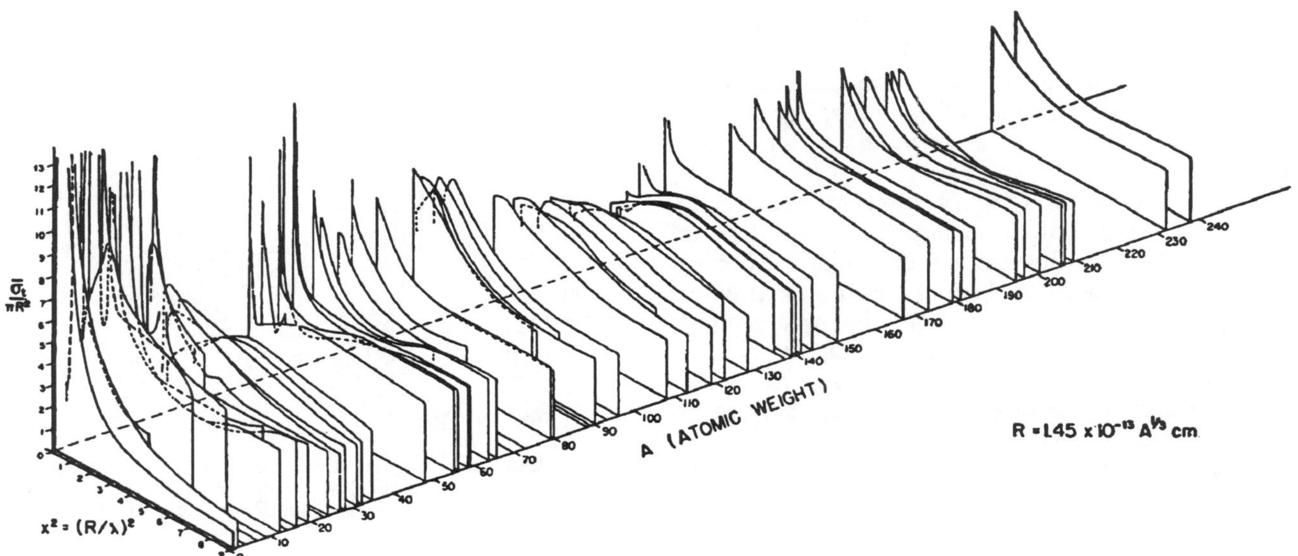

FIG. 3. Observed neutron total cross sections as a function of energy and mass number. The energy is expressed in terms of x^2 as in Fig. 2, (a) and (b).

average was done by the measuring apparatus itself, for lower A the averaging was done in the drawing. The theoretical and the experimental curves do not extend to zero energy. They are broken off at an energy of about 50 kev. As is well known, the curves should go approximately as $\epsilon^{-\frac{1}{2}}$ at very low energy. The experimental curves are compiled from measurements by many workers.[19–25]

The comparison of these two figures shows that the theory can account for a number of striking features of the experimental results. In particular, the theory reproduces the drop of the cross sections at low energies in the regions $A \sim 40$ and $100 < A < 140$.

It also reproduces the large cross sections at low energy in the regions $A \sim 60$, $A \sim 90$, and $A \sim 150$. The large values at $A \sim 90$ are ascribed to a P resonance; whereas the other two regions are supposed to contain S resonances. P resonances are expected to fall off towards low energies; whereas the S resonances merge directly with the $(1/v)$ rise. The observed energy dependence indicates the P-resonance behavior in the region $A \sim 90$ and shows typical S-resonance behavior at $A \sim 60$ and 150. There is an indication of P-resonance at low energies for $A \sim 30$ as the theory predicts.

The theory also reproduces the type of maxima (D maxima) which are found for energies corresponding to $x^2 \sim 3$ in the regions $A \sim 40$, and $A \sim 140$. It seems that the predicted F-wave maximum near $A \sim 200$ is also observed. It is remarkable that one finds reasonably good agreement in the shape of the curves even at very low atomic numbers: $A < 20$.[26]

We are using here a different depth of the potential than in the calculations published previously by the same authors.[27] The previous calculations were based upon a well depth of only $V_0 = 19$ Mev. The change to $V_0 = 42$ Mev was suggested by Adair[28] and improves the agreement considerably. At the time of the first calculation only measurements for $A > 60$ were used. The similarity between the theoretical results for $V_0 = 42$ Mev and $V_0 = 19$ Mev for $A > 60$ can be explained as follows:

S-wave maxima at low energy occur if $RK_0 = r_0 A^{\frac{1}{3}} K_0 \cong (n+\frac{1}{2})\pi$ and P maxima if $r_0 A^{\frac{1}{3}} K_0 \cong n\pi$, where $K_0 = (2mV_0/\hbar^2)^{\frac{1}{2}}$. For $V_0 = 19$ Mev and $r_0 = 1.45 \times 10^{-13}$ cm, one gets therefore S maxima at $A \sim 38$ and 170, and P maxima at $A \sim 11$ and 90. For $V_0 = 42$ and the same r_0 one gets S maxima at $A \sim 11$, 55, and 150; P maxima at $A \sim 27$, 90, and 216. Hence, the P maximum near 90 and the S maximum near 160 are reproduced by both potential depths. The behavior of the curves in the neighborhood of these maxima also must be similar, in particular, the depression at low energy for values of A just below an S maximum. However, the experimental data for nuclei below $A = 60$ definitely indicate another S maximum near 55 and a strong low-energy depression for $A \sim 40$ as predicted by $V_0 = 42$. These features are

[19] H. H. Barschall, Phys. Rev. **86**, 431 (1952).
[20] Miller, Adair, Bockelman, and Darden, Phys. Rev. **88**, 83 (1952).
[21] Walt, Becker, Okazaki, and Fields, Phys. Rev. **89**, 1271 (1953).
[22] Okazaki, Darden, and Walton, Phys. Rev. **93**, 461 (1954).
[23] N. Nereson and S. Darden, Phys. Rev. **89**, 775 (1953); Phys. Rev. **94**, 1678 (1954); and unpublished data on Li and B (private communication).
[24] C. F. Cook and T. W. Bonner, Phys. Rev. **94**, 651 (1954); McCrary, Taylor, and Bonner, unpublished data on Li (private communication).
[25] *Neutron Cross Sections*, U. S. Atomic Energy Commission Report AECU-2040 (Technical Information Division, Department of Commerce, Washington, D. C., 1952), and three supplements (unpublished).
[26] See also C. E. Porter, Bull. Am. Phys. Soc. **29**, No. 5, 25 (1954).
[27] Feshbach, Porter, and Weisskopf, Phys. Rev. **90**, 166 (1953).
[28] R. K. Adair, Phys. Rev. **94**, 737 (1954).

not reproduced by the theoretical curves for $V_0=19$. We therefore believe that $V_0=42$ Mev yields a better model. It should be noted that the agreement is not very sensitive to a change of potential V_0 with a corresponding change of r_0 such that $V_0^{\frac{1}{2}}r_0$ stays constant.

The shapes of the total cross section curves are quite sensitive to the value of the absorption constant ζ. An increase of ζ flattens the maxima and minima. Strong fluctuations in the calculated curves occur only at lower energies for values of $x<1.5$. This is below 2 Mev at $A\sim60$ and below 0.6 Mev at $A\sim200$. At higher energies the contributions of the numerous angular momenta prevent the appearance of any pronounced maxima or minima. Therefore the determination of ζ by fitting the calculated curves to the experimental ones only gives the value of ζ for relatively low energies. We cannot exclude a change of ζ at energies of, say, more than 1 Mev or fluctuations in ζ from one value of A to another although below 1 Mev it seems that ζ cannot vary much as a function of A. In the low-energy region the determination is quite accurate. A change of ζ to 0.05 or to 0.02 would give rise to a worse agreement with experiments. Figure 2(b) shows the total cross sections for $\zeta=0.05$, and it is obvious that the maxima and minima are not as pronounced as in the experimental data.[29]

We now turn to the calculations of the angular distribution of the elastic scattering. Figure 4 shows the experimental results at 1 Mev as measured by Walt and Barschall.[30] The most characteristic features are the flat distributions around $A\sim60$, a very strong forward peaking and a rise at backward angles at $A\sim140$, and the appearance of a second maximum at 90° around $A\sim180$. The calculation of the angular dependence (Fig. 5) is not unambiguous since the amount of compound elastic scattering is difficult to determine. Furthermore, the angular dependence of the compound elastic scattering depends upon the spin of the target nucleus. We therefore have shown in Fig. 5 the calculated angular distribution of the shape elastic scattering only. In Fig. 6 the compound elastic scattering is added in full which would correspond to the case in which the compound state decays exclusively via the entrance channel. The target spin was assumed to be zero. The actual $d\sigma_{el}/d\Omega$ must lie somewhere between Fig. 5 and Fig. 6. For nuclei with strong inelastic scattering, Fig. 5 should be the better approximation.

It is seen from Figs. 5 and 6 that some of the main features are again reproduced by the theory. The flatness of the distribution around $A\sim60$ comes from the fact that the P contribution is very weak in this region and, at small angles, of opposite phase to the S scattering. This occurs always at values KR somewhat below a P resonance. The second maximum at 90° at high mass numbers is not too well reproduced. The

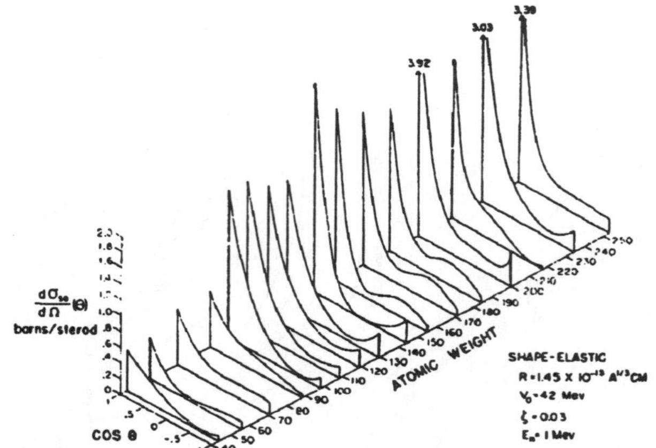

FIG. 5. Calculated angular distribution of the elastic neutron scattering (shape elastic only) as a function of $\cos\theta$ and the mass number A for a well $V_0=42$ Mev, $R=1.45A^{\frac{1}{3}}\times10^{-13}$ cm, and $\zeta=0.03$.

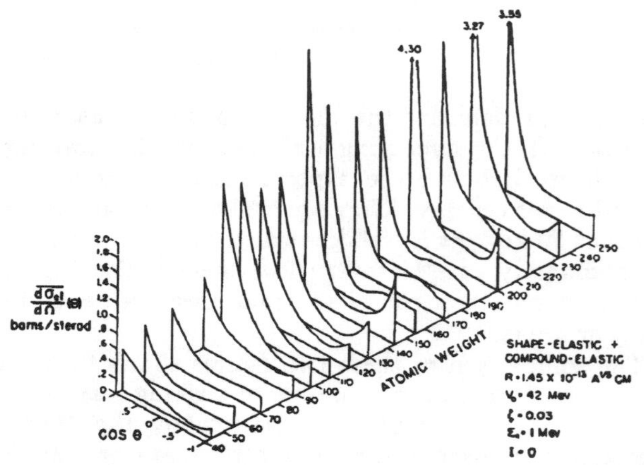

FIG. 6. Calculated angular distribution of the elastic neutron scattering (shape elastic plus maximum compound elastic) as a function of $\cos\theta$ and A for a well $V_0=42$ Mev, $R=1.45A^{\frac{1}{3}}\times10^{-13}$ cm, and $\zeta=0.03$.

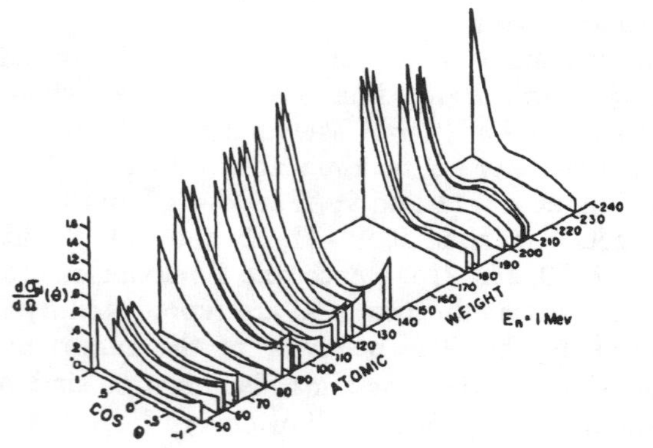

FIG. 4. Observed angular distribution (in barns/sterad) of the elastic 1-Mev neutron scattering as a function of $\cos\theta$ and the mass number A as measured by Walt and Barschall.

[29] The disagreement is worse for high values of A. This might be an indication of a slight decrease of ζ with the mass number. If the absorption were concentrated in a surface layer of given thickness, one would expect a similar effect [see M. H. Johnson and E. Teller, Phys. Rev. 93, 357 (1954)].

[30] M. Walt and H. H. Barschall, Phys. Rev. 93, 1062 (1954).

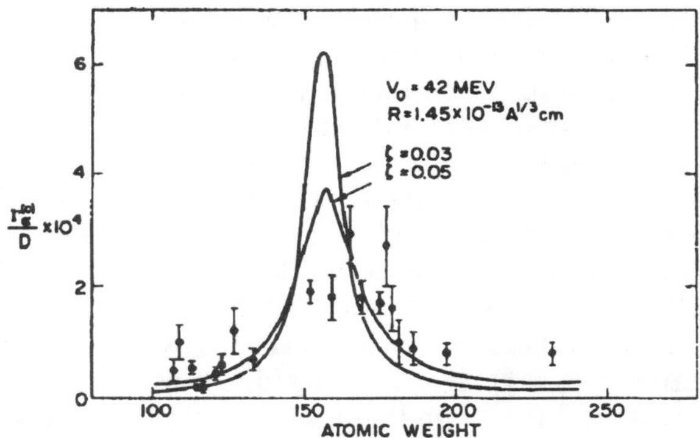

FIG. 7. Ratio $\Gamma_\alpha^{(0)}/D$ of neutron width to level distance for low energies as a function of A. Here $\Gamma_\alpha^{(0)}=\Gamma_\alpha(\epsilon^{(0)}/\epsilon)^{\frac{1}{2}}$ is the "reduced" width, and $\epsilon^{(0)}$ is taken to be 1 ev. The curves represent the calculated values for $\zeta=0.03$ and 0.05. The points represent the observed values and the limits of error.

theory shows it only between $A=150$ and $A=200$. The angular dependence above $A=200$ does not seem to agree too well with the experiment. The angular distributions are not very sensitive to the choice of constants. The results with $V_0=19$ Mev are not very different from the ones shown here.

The agreement with experiments is also less satisfactory for the cross section σ_c for the formation of the compound nucleus. It is difficult to measure σ_c directly since it includes the compound elastic scattering besides the reaction cross section, and the former cannot easily be separated from the shape elastic scattering. At very low energies, however, the formation of the compound nucleus can be measured by studying the individual resonances (see Sec. III). The relevant magnitude is the ratio Γ_α/D of the neutron width to the level distance, averaged over a number of neighboring resonances. The theoretical values Γ_α/D expected on the basis of $V_0=42$ Mev are shown in Fig. 7 together with a compilation[31] of the measurements[32–41] of Γ_α/D. Only recently has it been possible to measure the neutron widths of several resonances in one isotope, so that the average Γ_α and the level distance can be determined to some degree of reliability. It is seen that the expected maximum of Γ_α/D at $A\sim155$ is noticeable, but it is not as strong as the theory predicts for the same value of ζ which gives the best fit for the total cross sections ($\zeta=0.03$). Also the values off peak are somewhat larger than predicted.

The fact that the resonance at $A\sim155$ is not as strong as expected might be connected with the large deviations from sphericity which are ascribed to the nuclei in this region.[42] If the shape of the potential well is ellipsoidal, one would expect results which roughly represent averages over the spherical results taken over radii which lie between the smallest and the largest axis. This would give rise to a flattening of the maxima and a rise of the wings in the theoretical curves of Fig. 7.[43]

Although no direct measurement of the formation of the compound nucleus is possible, the measurements of inelastic cross sections σ_{in} or reaction cross sections σ_r can be used to compare with the theoretical predictions of σ_c. Evidently σ_{in} and σ_r must be smaller than σ_c. The difference $\sigma_c-\bar{\sigma}_r$ is the compound elastic cross section which is expected to be rather small if inelastic scattering or other reactions are strong enough to compete for the decay of the compound state.

Walt and Barschall have determined inelastic scattering cross sections σ_{in} at 1 Mev by subtracting the elastic scattering from the total scattering. The values of σ_{in} should be less than or equal to the theoretical values of σ_c.

Figure 8 shows a comparison between the observed inelastic cross sections and the calculated σ_c at 1 Mev as functions of A. The observed values are of the expected order of magnitude, but they do not agree with the theoretical curve. The absence of the maximum at $A=50$ might be explained by the fact that the compound elastic scattering is relatively high for these nuclei. The same fact explains the low value of the inelastic cross section in lead and bismuth. However, the expected maxima at $A=90$ and 150 seem to occur at higher values of A. We have no explanation for these discrepancies.

There are many measurements of inelastic cross sections at somewhat higher energies. They all indicate that the values are not too far from $\pi(R+\lambda\!\!\!^{-})^2$, which is

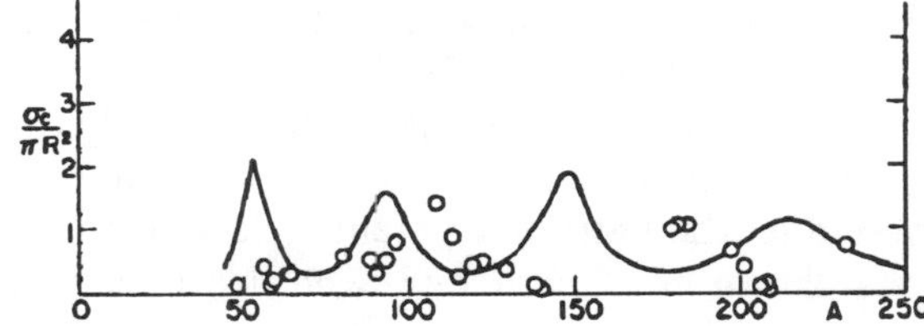

FIG. 8. The calculated cross section σ_c for compound nucleus formation at $E_n=1$ Mev and the observed reaction cross section at 1 Mev as determined by Walt and Barschall. R is taken to be $1.45\times10^{-13}A^{\frac{1}{3}}$ cm. In the calculations the parameters V_0 and ζ were taken to be 42 Mev and 0.03, respectively.

[31] R. S. Carter *et al.*, Phys. Rev. (to be published).
[32] F. G. P. Seidl, Hughes, Palvesky, Levin, Kato, and Sjöstrand, Phys. Rev. 95, 476 (1954); and private communication.
[33] R. S. Carter and J. A. Harvey, Phys. Rev. 95, 645(A) (1954).
[34] Foote, Landon, and Sailor, Phys. Rev. 92, 656 (1953).
[35] Sailor, Landon, and Foote, Phys. Rev. 93, 1292 (1954).
[36] Pilcher, Carter, and Stolovy, Phys. Rev. 95, 645(A) (1954); and private communication.
[37] Hughes, Kato, and Levin, Phys. Rev. 92, 1094 (1953); and private communication.
[38] R. L. Christensen, Phys. Rev. 92, 1509 (1953).
[39] Melkonian, Havens, and Rainwater, Phys. Rev. 92, 702 (1953).
[40] L. Bollinger, unpublished data on Sb (private communication). We wish to thank Dr. Bollinger for making his results available in advance of publication.
[41] V. E. Pilcher and R. S. Carter (private communication).

[42] A. Bohr and B. R. Mottelson, Kgl. Danske Videnskab. Selenskab Mat. fys. Medd. 27, 16 (1953).
[43] This thought was suggested to us by A. Bohr and B. R. Mottelson.

the value one would expect if the neutron wave were totally absorbed in contrast to our findings of $\zeta \sim 0.03$. Especially the measurements at 14 Mev[3] indicate this fact. On the basis of this evidence one would conclude that the value of ζ is strongly energy dependent and reaches a value ($\zeta \gtrsim 0.12$) corresponding to almost total absorption in a medium-sized nucleus certainly at 14 Mev but most likely already at energies as low as 4.5 Mev. The latter conclusion is based upon measurements of inelastic cross sections by Lonsjo, Taylor, and Bonner.[44] In this connection it is interesting that the calculations of Morrison, Muirhead, and Rosser[45] also give a very strong increase with energy of the absorption of nucleons in nuclear matter just in the region which corresponds to incident neutrons of 1 Mev. These calculations are based on the Goldberger method[46] of the scattering of free particles with the application of the Pauli principle. The effect of the exclusion principle alone causes a sharp drop of the energy exchange with decreasing energy.[47]

It is apparent that our model is much less successful in reproducing the strength of compound nucleus formation than in reproducing the total and elastic scattering. It gives too much variation with A of Γ_α/D at low energies and probably too little compound nucleus production at 1 Mev and higher, although it is possible to explain the discrepancies at higher energy by assuming that ζ increases with energy above 1 Mev.

The discrepancies may come from two possible sources: (A) The potential $V(r)$ as given by (3.1) may not be the shape best fitted for the model. (B) The attempt of this paper, the description of the gross behavior of a nucleus by a complex one-particle potential, may be unsuccessful. In connection with (A) it must be noted that the potential (3.1) necessarily is an oversimplified version, since it is physically impossible that the potential well actually has a discontinuity in the form of a sudden jump at $r=R$. It might be that a rounding-off of the corners of the potential well will improve the agreement with experiments.

The smoothing of the edges of the square-well potential was of significance for the interpretation of the elastic proton scattering with heavy nuclei. This scattering has been measured with protons of an energy of about 18 Mev by Gugelot,[48] Burkig and Wright,[49] and by Cohen and Neidigh.[50] The results cannot be interpreted on the basis of a potential (3.1) with sharp edge, as shown by Chase and Rohrlich.[51] However, Woods and Saxon[52] have shown recently that a rounding-off even within the small interval of 0.5×10^{-13} cm changes the results considerably and brings them into much better agreement with the experiments.

It is possible, therefore, that the smoothing of the discontinuity of V at $r=R$ would also improve the agreement of theory and experiment in respect to compound nucleus formation. It would decrease the reflection of the neutron wave at the nuclear surface and hence increase the cross section σ_c when all other constants (V_0,R,ζ) are unchanged. It remains to be seen whether the rounding-off of the potential well improves the agreement with respect to σ_c and Γ_α/D and with respect to the angular distribution of the elastic cross section, without destroying the agreement of the total and elastic cross sections.

Calculations are under way to investigate these possibilities.

It must be pointed out that one should never expect any exact agreement between the predictions based upon a model of this type and the observed cross sections. The very nature of this attempt to describe a complicated many-body problem by a simple one-body potential implies that the model can only contain the main features of the situation. Apart from this general limitation it should be kept in mind that we have used here a potential which has a particularly simple dependence on the radius and on the mass number. We have assumed the same radial dependence for the real and imaginary part which is very probably too strict an assumption. We have neglected spin-dependent forces as observed by Adair and co-workers,[53] and we have excluded any special features connected with the shell structure.

The purpose of the proposed approach is to connect some characteristic salient features of the nuclear cross sections with simple nuclear properties rather than to construct a theory which will produce the exact quantitative details of the observations.

ACKNOWLEDGMENTS

We are greatly indebted to many experimental physicists who in the course of the last few years have discussed with us their results before publication and have helped us adjust our models to the latest values. In particular, we owe special gratitude to H. H. Barschall, M. Walt, and collaborators at Wisconsin; R. K. Adair, J. A. Harvey, D. J. Hughes at Brookhaven; N. Nereson of Los Alamos; and T. W. Bonner of the Rice Institute for their invaluable help and advice and readiness to furnish their newest results.

We also wish to express our gratitude to those who have participated in the important, but tedious, work of computing. Our thanks are offered to Fern Abrams,

[44] Lonsjo, Taylor, and Bonner (private communication). We are grateful to the authors for showing us their results before publication.
[45] Morrison, Muirhead, and Rosser, Phil. Mag. 44, 1326 (1953).
[46] M. Goldberger, Phys. Rev. 74, 1269 (1948).
[47] V. F. Weisshopf, Science 113, 101 (1951).
[48] P. C. Gugelot, Phys. Rev. 87, 525 (1952).
[49] J. W. Burkig and B. T. Wright, Phys. Rev. 82, 451 (1951).
[50] B. L. Cohen and R. V. Neidigh, Phys. Rev. 93, 282 (1954).
[51] D. M. Chase and F. Rohrlich, Phys. Rev. 94, 81 (1954).
[52] R. D. Woods and D. S. Saxon (private communication). We are grateful to the authors for showing us their results before publication.
[53] Darden, Field, and Adair, Phys. Rev. 93, 931 (1953).

Harvey Amster, Betty Campbell, Elgie Ginsburgh, Dr. A. Glassgold, Mida Karakashian, Barbara Levine, Edith Moss, Hannah Paul, Evelyn Walker, and Hannah Wasserman at the Massachusetts Institute of Technology; Elaine Scheer at Nuclear Development Associates, Inc.; and Jane Levin and Phyllis Levy at Brookhaven National Laboratory.

One of us (C.E.P.) wishes to acknowledge the helpful cooperation he received at Brookhaven National Laboratory during the final stage of this work prior to publication.

APPENDIX

The Scattering Amplitude at Low Energies

The scattering amplitude η is the diagonal element $S_{\alpha\alpha}$ of the scattering matrix, where the index α refers to the entrance channel. The matrix S is given by the following expression [see Blatt and Weisskopf, formula (X, 4.11)]:

$$S_{\alpha\beta}=\exp[-i(k_\alpha+k_\beta)R]S_{\alpha\beta}'. \tag{A.1}$$

Here R is the nuclear radius, and

$$S'=(1+i\mathcal{R}')/(1-i\mathcal{R}'), \tag{A.1a}$$

where $\mathcal{R}'$ is connected with the derivative matrix $\mathcal{R}$ by

$$\mathcal{R}_{\alpha\beta}'=(k_\alpha k_\beta)^{\frac{1}{2}}\mathcal{R}_{\alpha\beta},$$

and k_α, k_β are the channel wave numbers at the energy E. The matrix $\mathcal{R}$ is defined on page 545 of Blatt and Weisskopf. It can be expressed in the following form [see Sec. X (4.22)]:

$$\mathcal{R}_{\alpha\beta}=\sum_s \frac{y_{s\alpha}y_{s\beta}}{E_s-E}, \tag{A.2}$$

where E_s are the resonance energies and the $y_{s\alpha}$ are magnitudes which are connected with the channel widths Γ_α^s (partial widths for the decay via the channel α):

$$\Gamma_\alpha^s=2k_\alpha y_{s\alpha}^2. \tag{A.3}$$

Each $y_{s\alpha}$ is real but its sign might be positive or negative. We make the reasonable assumption that the signs are distributed at random.

In what follows we will assume that we work in an energy region for which, first, $k_\alpha R\ll 1$, and, second, $\Gamma^s\ll D_s$, where D_s is the interval

$$D_s=\tfrac{1}{2}(E_{s+1}-E_{s-1}),$$

which includes the resonance E_s from mid-point to mid-point. We also assume that the values Γ_α^s and D_s have the same order of magnitude for all resonances in an energy interval I which includes many resonances but which is small compared to energy intervals occurring in single-particle problems (say $I\sim 10$ kev for heavy nuclei).

Let us surround each resonance E_s by an energy interval D_s from $\frac{1}{2}(E_{s-1}+E_s)$ to $\frac{1}{2}(E_s+E_{s+1})$. In this interval we can write the matrix $\mathcal{R}_{\alpha\beta}'$ in the form

$$\mathcal{R}_{\alpha\beta}'=\frac{y_{s\alpha}'y_{s\beta}'}{E_s-E}+g_{\alpha\beta}, \tag{A.4}$$

where $y_{s\alpha}'=(k_\alpha)^{\frac{1}{2}}y_{s\alpha}$. The matrix $g_{\alpha\beta}$ has no singularities in D_s and is given by

$$g_{\alpha\beta}=\sum_{t\neq s}\frac{y_{t\alpha}'y_{t\beta}'}{E_t-E}. \tag{A.4a}$$

We now estimate the order of magnitude of $g_{\alpha\beta}$. Here it is important to take into account that the signs of the $y_{t\alpha}'$ are distributed at random over the different resonances t and the different channels α.

First we split $g_{\alpha\beta}$ into the contributions of neighboring and far-off levels:

$$g_{\alpha\beta}=g_{\alpha\beta}'+r_{\alpha\beta},$$
$$g_{\alpha\beta}'=\sum_{t\neq s}{}'\frac{y_{t\alpha}'y_{t\beta}'}{E_t-E}, \tag{A.4b}$$
$$r_{\alpha\beta}=\sum_{t\neq s}{}''\frac{y_{t\alpha}'y_{t\beta}'}{E_t-E},$$

where the prime on the summation sign means that the sum should be extended only over resonances within the interval I, and the double prime means extension over the resonances outside I. Because of the random signs of $y_{t\alpha}'$, the terms in the sums (A.4b) have random signs for $\alpha\neq\beta$ and only the immediate-neighbor resonances contribute appreciably to $g_{\alpha\beta}$; on the same grounds $r_{\alpha\beta}$ can be neglected. We, then, obtain the estimate:

$$|g_{\alpha\beta}'|\cong|g_{\alpha\beta}|\sim(\Gamma_\alpha\Gamma_\beta)^{\frac{1}{2}}/D,\quad r_{\alpha\beta}\approx 0,\quad \alpha\neq\beta. \tag{A.5}$$

We understand by Γ_α (without superscript) the average value of Γ_α^s in the interval I. For $\alpha=\beta$, all terms have the same sign for $E_t>E$ or $E_t<E$. Since there are roughly an equal number of levels above and below E_s in the interval I, we get the following estimate:

$$g_{\alpha\alpha}'\sim\Gamma_\alpha/D. \tag{A.5a}$$

The order of magnitude of the contribution $r_{\alpha\alpha}$ from the faraway levels is quite undetermined. However, the scattering cross section between resonances turns out to be $4\pi R^2(1+r_{\alpha\alpha}/x)^2$. Experimentally, we know that this cross section is of the order of $4\pi R^2$, and hence we conclude $r_{\alpha\alpha}\sim k_\alpha R\equiv x$.

It follows from (A.5) and (A.5a) that the $g_{\alpha\beta}$ are all small compared to unity in the energy region considered, and we proceed to expand (A.1) in powers of g. For this purpose we introduce the *factorable* matrix

$$T_{\alpha\beta}=-y_{s\alpha}'y_{s\beta}'/\Delta,\quad \Delta=E-E_s. \tag{A.6}$$

The following relation holds:

$$T^n = B^{n-1}T, \quad n\geqslant 1, \qquad B=\sum_\beta T_{\beta\beta}. \tag{A.7}$$

The quantity B is a number and it is connected with the total width. $\Gamma^s=\sum_\beta \Gamma_\beta^s$ of the resonance s:

$$B=-\tfrac{1}{2}\Gamma^s/\Delta. \tag{A.8}$$

Hence, if a matrix is a function of T which can be expressed as a power series with *powers larger than zero*

$$A(T)=\sum_{n=1}^{\infty} a_n T^n,$$

we get the relation

$$A_{\alpha\beta}=B^{-1}A(B)T_{\alpha\beta}. \tag{A.9}$$

We may now expand (A.1), noting that $\mathcal{R}'=T+g$:

$$S'=-1+\frac{2}{1-i(T+g)}$$
$$=-1+\frac{2}{1-iT}+\frac{1}{1-iT}(2ig)\frac{1}{1-iT}$$
$$-\frac{2}{1-iT}g\frac{1}{1-iT}g\frac{1}{1-iT}+\cdots.$$

From (A.9) we have $1/(1-iT)=1+T/(1-iB)$, so that

$$S'=1+\frac{2iT}{1-iB}+2ig-2g^2-2\frac{(gT+Tg)}{1-iB}+S^*,$$

$$S^*=-2i\frac{TgT}{(1-iB)^2}-2i\frac{Tg^2+gTg+g^2T}{(1-iB)}$$
$$-2\frac{TgTg+Tg^2T+gTgT}{(1-iB)^2}+2i\frac{TgTgT}{(1-iB)^3}. \tag{A.10}$$

The scattering amplitude η_0 is the diagonal element $S_{\alpha\alpha}$. We first note the cross sections which would follow from (A.10) when we put $g=0$: $(x=k_\alpha R)$

$$\eta_0=e^{-2ix}\left(1+\frac{2iT_{\alpha\alpha}}{1-iB}\right)=e^{-2ix}\left(1-\frac{i\Gamma_\alpha^s}{\Delta+i\Gamma^s/2}\right), \tag{A.11}$$

and hence we get the well-known expressions:

$$\sigma_{el}^{(0)}=\pi\lambda\!\!\!^{-2}\left|2x+\frac{\Gamma_\alpha^s}{\Delta+i\Gamma^s/2}\right|^2,$$
$$\sigma_r^{(0)}=\pi\lambda\!\!\!^{-2}\frac{\Gamma_\alpha^s(\Gamma^s-\Gamma_\alpha^s)}{\Delta^2+(\Gamma^s/2)^2}. \tag{A.12}$$

We now proceed to neglect all terms in the expansion (A.10) which would give rise to terms in σ_{el} and σ_r of the order of $\delta\sigma_r$ and $\delta\sigma_{el}$, respectively, with

$$\delta\sigma_r=\sigma_c^{(0)}f, \quad \delta\sigma_{el}=\sigma_c^{(0)}f, \tag{A.13a}$$

where $\sigma_c^{(0)}=\pi\lambda\!\!\!^{-2}\Gamma_\alpha\Gamma/[\Delta^2+(\Gamma^2/4)]$ and f is a small number,

$$f\sim x \quad\text{or}\quad f\sim\Gamma/D. \tag{A.13b}$$

This means we will neglect cross sections which are small by a factor Γ/D or x, compared to $\sigma_c^{(0)}$, which can be regarded as the one-level value of the cross section for the formation of the compound nucleus.

It will be shown below that all terms of (A.10) contained in S^* give rise to corrections in the cross sections of the order (A.13) or smaller. Hence, within the accuracy (A.13), we can write S' in the form:

$$S'=1+\frac{2iT}{1-iB}+2ig-2g^2-\frac{2(gT+Tg)}{1-iB}. \tag{A.14}$$

To prove this point, we first examine the effect of a small addition $\Delta\eta'$ to $\eta'=S_{\alpha\alpha}'$ on the cross sections: We get

$$\Delta\sigma_{el}=\pi\lambda\!\!\!^{-2}2\,\mathrm{Re}[(e^{2ix}-\eta')\Delta\eta'^*]\lesssim 4\pi\lambda\!\!\!^{-2}|\Delta\eta'|,$$
$$\Delta\sigma_r=\pi\lambda\!\!\!^{-2}2\,\mathrm{Re}[\eta'^*\Delta\eta']\leqslant 2\pi\lambda\!\!\!^{-2}|\Delta\eta'|, \tag{A.15}$$

where $\mathrm{Re}(a)$ is the real part of a. The former relation follows from the fact that $|e^{2ix}-\eta'|\leqslant 2$ and the latter relation from the fact that $|\eta'|\leqq 1$. As an example, we discuss the omission of the first term in the expression S^* in (A.10). We find according to (A.5) and (A.5a)

$$|\Delta\eta'|\sim|(TgT)_{\alpha\alpha}/(1-iB)^2|\approx\frac{\Gamma_\alpha\Gamma}{\Delta^2+(\Gamma^s/2)^2}[(\Gamma/D)+x],$$

and hence the contributions to the cross section of this term are negligible according to (A.15) and (A.13). Similar considerations show that the other neglected terms in (A.10) contribute the same or less to the cross sections.

We now single out the diagonal element of S' because of its significance for the scattering amplitude. We can write $S_{\alpha\alpha}'$ from (A.14) and (A.13b) in the following form:

$$S_{\alpha\alpha}'\cong\exp(2ir_{\alpha\alpha})\left[\left(1+\frac{2iT_{\alpha\alpha}}{1-iB}\right)+2ig_{\alpha\alpha}'-2(g'^2)_{\alpha\alpha}-\frac{4(g'T)_{\alpha\alpha}}{1-iB}\right].$$

This expression differs from the diagonal element of (A.14) by terms which are of the order S^* and therefore negligible, as, for example, $r_{\alpha\alpha}(g'T)_{\alpha\alpha}/(1-iB)$. We can write it in the form

$$S_{\alpha\alpha}'=\exp(2ir_{\alpha\alpha})\times\left[1+\frac{2iT_{\alpha\alpha}}{1-iB}+\frac{T_{\alpha\alpha}}{1-iB}G_1+iG_2+G_3\right], \quad (A.14a)$$

or, according to (A.1)

$$S_{\alpha\alpha}=\eta_0=\exp(-2ik_\alpha R')\times\left[1+\frac{2iT_{\alpha\alpha}}{1-iB}+\frac{T_{\alpha\alpha}}{1-iB}G_1+iG_2+G_3\right], \quad (A.14b)$$

where

$$R'=R-r_{\alpha\alpha}/k_\alpha. \quad (A.16)$$

$$G_1=4\sum_{t\neq s}{}'\sum_{\beta}\frac{y_{t\alpha}'y_{t\beta}'(y_{s\beta}'/y_{s\alpha}')}{E_t-E},$$
$$G_2=2g_{\alpha\alpha}', \quad (A.17)$$
$$G_3=-2(g'^2)_{\alpha\alpha}.$$

This is the form which is used in the text. The orders of magnitude of these real functions are given by (2.11).

The following simplification can be used if one calculates the scattering cross section σ_{el} and the transfer cross sections $\sigma_{\alpha\beta}$ (cross section of the reaction $\alpha\rightarrow\beta$):

$$\sigma_{\alpha\beta}^{(0)}=\pi\lambda\!\!\!^{-2}|S_{\alpha\beta}'|^2 \quad (A.18)$$

within the limits of accuracy (A.13). It turns out that the last two terms in (A.14) give rise to nonnegligible contributions only to σ_r when expression (2.3a) is used. In expression (2.3) for σ_{el} and in expression (A.18) for $\sigma_{\alpha\beta}$, the two last terms of (A.14) give rise to contributions which can be neglected according to (A.13).[54] Hence for the calculation of $\sigma_{\alpha\beta}^{(0)}$ and $\sigma_{el}^{(0)}$ we may use the shorter form

$$S'=1+[2iT/(1-iB)]+2ig,$$

or

$$S_{\alpha\beta}'\simeq\delta_{\alpha\beta}\exp(2ir_{\alpha\alpha})+2i\sum_r{}'\frac{y_{r\alpha}'y_{r\beta}'}{E-E_r+i\Gamma^r/2}, \quad (A.19)$$

where the sum is extended over all resonances within the interval I. Actually the imaginary part $+i\Gamma^r/2$ in the denominator of (A.16) should be found only in the term $r=s$, but the addition in the other terms leads only to errors smaller than (A.13). We then get for the cross section $\sigma_{\alpha\beta}$

$$\sigma_{\alpha\beta}=\pi\lambda\!\!\!^{-2}|S_{\alpha\beta}'|^2=4\pi\lambda\!\!\!^{-2}\left|\sum_r{}'\frac{y_{r\alpha}'y_{r\beta}'}{E-E_r+i\Gamma^r/2}\right|^2, \quad (A.20)$$

and for the scattering cross section,

$$\sigma_{el}=\pi\lambda\!\!\!^{-2}|e^{2ix}-S_{\alpha\alpha}'|^2,$$
$$=\pi\lambda\!\!\!^{-2}\left|\exp(2ik_\alpha R')-1+\sum_r\frac{i\Gamma_\alpha^r}{E-E_r+i\Gamma^r/2}\right|^2. \quad (A.21)$$

According to (A.19), the value of $\sigma_{\alpha\beta}$ goes to zero between two resonances E_s and E_{s+1} if the sign of $y_{s\alpha}'y_{s\beta}'$ is the same for both resonances. If the sign is opposite, no zero occurs. Note that this statement is good only to the accuracy (A.13). At the zero of (A.17) the actual cross section might still be of the order (A.13).

We note in (A.21) that the potential scattering amplitude $\exp(2ikR')-1$ corresponds to the scattering by an impenetrable sphere of radius R' as given by (A.16). The quantity R' itself is a function of the energy, which is slowly varying and changes only over intervals much larger than D.

The forms (A.15) and (A.16) correspond to the Breit-Wigner formulas used in the literature before the more exact investigations by Wigner and Eisenbud.[55] The amplitudes contain characteristic sums over the contributions of the different resonances with the imaginary contribution $i\Gamma^r/2$ in the denominator. It has been pointed out repeatedly that the forms (A.17) and (A.18) are not exactly correct. We have shown, however, that they are valid within the errors given by (A.13).

[54] At first sight this seems puzzling since $\sigma_r=\sum_\beta\sigma_{\alpha\beta}$. It must be remembered that Eq. (2.3a) uses the diagonal element of S; whereas (A.18) uses off-diagonal elements. The connection between these elements is established by the unitary nature of S: $1-|S_{\alpha\alpha}|^2=\sum_{\beta\neq\alpha}(S_{\alpha\beta})^2$. In order to insure the validity of this equation up to the order g^2, one must include the last two terms of (A.14) in $S_{\alpha\alpha}$, but it is not necessary to include them in $S_{\alpha\beta}'$.

[55] E. P. Wigner and L. Eisenbud, Phys. Rev. 72, 29 (1947).

PHYSICAL REVIEW VOLUME 98, NUMBER 5 JUNE 1, 1955

High-Energy Reactions and the Evidence for Correlations in the Nuclear Ground-State Wave Function*

K. A. BRUECKNER, R. J. EDEN,† AND N. C. FRANCIS
Indiana University, Bloomington, Indiana
(Received January 13, 1955)

High-energy nuclear reactions which depend strongly on nucleon position correlations in the nuclear ground state are analyzed and shown to give evidence for the existence of marked correlation effects. The following high-energy experiments are considered: nuclear photoeffect, meson absorption in nuclei, deuteron pickup, proton-proton scattering in a nucleus, and meson production in proton-nucleus collisions. The corresponding cross sections depend on a nucleon momentum distribution which can be represented at high energies by a single function giving reasonable agreement with all the experiments considered. This momentum distribution differs substantially from that for the shell model of the nucleus and thus provides strong evidence for correlation in the nuclear ground-state wave function.

The transformation methods developed in previous papers are used to provide a unified theory of the above five processes. The momentum distribution predicted by this theory is estimated by two methods each of which gives close agreement with the experimentally determined function in the relevant energy ranges.

I. INTRODUCTION

IN the last few years a considerable body of evidence has been accumulated which provides information about the ground state of nuclei. This evidence comes primarily from quite different types of experiment and contains, as we shall show, upon first examination apparent contradictions in the information given about the ground state. One type of evidence, that perhaps is best known, comes from the study of ground and low excited states of nuclei and is encompassed in the very successful shell model theory which has been useful and accurate in predictions and understanding of nuclear properties. We shall not attempt to summarize this evidence on the theory here; we only comment that the central feature of the shell model is the assumption that nucleons move in the independent particle states of a uniform potential. The success of the shell model as usually formulated is very intimately connected with this assumption since the existence of long mean free paths and independent particle motion are reflections of the absence of two-body interactions and of the absence of correlations in the ground-state wave function.

The second body of evidence which has direct bearing on the nuclear ground state comes from high-energy experiments. It is the purpose of this paper to summarize this evidence and show how it may be reconciled with the knowledge of nuclear structure derived from low-energy experiments. We consider the following reactions: deuteron pickup, meson capture, high-energy photonuclear effect, high-energy proton-nucleus collisions, and meson production in high-energy proton-nucleus collisions. These high-energy reactions are all similar in that they provide in effect a method of observation with great resolving power since they allow us to probe nuclear structure with particles of wavelength less than the typical nucleon spacing in a nucleus. Consequently we can expect to resolve details of the structure which are not accessible to us if we restrict ourselves to observations at low energy with particles of large wavelength. As we shall see, the information we obtain from the high-energy experiments is in contradiction with the shell model as usually formulated and requires a change in the interpretation of the low-energy nuclear phenomena and their relation to the ground-state wave function.

This new interpretation has been described in

* Supported in part by a grant from the National Science Foundation and the Office of Naval Research and the U. S. Atomic Energy Commission.

† Smithson Research Fellow of the Royal Society, on leave of absence from Clare College, Cambridge, England.

previous papers[1] and is further discussed in Sec. III of the present paper with particular reference to the high-energy experimental information which is described in Sec. II.

In the first part of Sec. III we develop the formalism required for relating the nuclear ground-state wave function to the shell model wave function. In the second part of Sec. II this formalism is utilized to give a unified development of the theories of the nuclear photoeffect, meson capture, and deuteron pickup. We find that the cross sections for these processes depend on a momentum distribution function in the same way as has previously been stated, for example by Chew and Goldberger,[2] provided that the energies involved are sufficiently high. However, our momentum distribution function differs in principle from that of Chew and Goldberger because of our use of a different form of the impulse approximation which appears to have more general validity (this method was first used by Heidmann[3]). Finally in Sec. III we note the cross section formulas for high energy, for proton-proton scattering in nuclei and meson production by protons bombarding nuclei, with particular reference to their dependence on a momentum distribution function.

In Sec. IV we apply the theory to the various experiments. Since our momentum distribution appears in the formulas for cross sections in the same way as that used by previous authors, we are able to make use of earlier analyses of experiments to determine the momentum distribution from experiment. It is found that these experiments can be fitted to reasonable accuracy by a suitable Gaussian momentum distribution in the energy range 50 to 100 Mev. We compare this experimental distribution with our theory in two ways, one of these involves the use of a Hulthén wave function for two nucleons, and the second is partly phenomenological in that we insert an experimental value for the two nucleon scattering matrix. It is found that both these methods give a momentum distribution in agreement with the experimental one in the relevant energy range.

Finally in Sec. V we summarize our conclusions from the evidence presented in this paper.

II. EXPERIMENTAL EVIDENCE

A. Deuteron Pickup

This process is the ejection by fast neutrons of fast deuterons from nuclei[4]; the original theory was developed by Chew, and Goldberger[2] and later modified by Heidmann.[3] The phenomenon occurs in the following manner: a fast neutron (90 Mev, for example) in passing through a nucleus occasionally encounters a proton with such a momentum that the relative momentum of the neutron and proton can be accommodated in the deuteron wave function. When this occurs, it is possible for the neutron to "pickup" the proton and emerge as a deuteron. It is apparent that the probability of this process is a sensitive function of the momentum distribution of the proton in the ground-state wave function; consequently the empirical observations can be used to deduce properties of the wave function. The theory as it has been developed is only a Born approximation; the calculations of Chew and Goldberger and of Heidmann are somewhat different in form, but both agree on a simple dependence on the ground-state wave function and probably can be used to draw qualitative conclusions. The result given by Chew and Goldberger is that the cross section depends most strongly on a factor

$$N(\mathbf{k})=\left|\int\psi_I(\mathbf{r})e^{i\mathbf{k}\cdot\mathbf{r}}d\mathbf{r}\right|^2, \tag{1}$$

where $\psi_I(\mathbf{r})$ is the initial wave function of the picked up proton. Their analysis showed that the experiments were consistent with

$$N(\mathbf{k})=\frac{\alpha}{\pi^2}\frac{1}{(\alpha^2+\mathbf{k}^2)^2}, \tag{2}$$

with $\alpha^2/2m=(18)$ Mev. This distribution departs very markedly from a Fermi gas or from an independent-particle function in that a much stronger admixture of high-momentum components is predicted. Thus the pickup process is evidence for a strongly correlated wave function.

B. Meson Capture from Low Bohr Orbits

This process was first observed by Panofsky[5] and has since then been extensively studied. The theoretical analysis[6] shows that it is possible, from a study of the capture of π mesons at rest in light elements, particularly in hydrogen, deuterium, and carbon, to derive some striking information about nuclear structure. The interesting observed feature of the meson capture is that while the reactions in deuterium,

$$\pi^-+\mathrm{D}\rightarrow 2n,\quad \pi^-+\mathrm{D}\rightarrow 2n+\gamma,$$

occur with a ratio of about 2:1; in carbon, the ratio of the cross sections for the two processes

$$\pi^-+\mathrm{C}\rightarrow \mathrm{star},\quad \pi^-+\mathrm{C}\rightarrow \mathrm{star}+\gamma$$

has changed to a number greater than 65. It is theoretically expected that the γ emission, since it occurs

[1] Brueckner, Levinson, and Mahmoud, Phys. Rev. 95, 217 (1954); K. A. Brueckner, Phys. Rev. 96, 508 (1954) and Phys. Rev. 97, 1353 (1955); K. A. Brueckner and C. A. Levinson, Phys. Rev. 97, 1344 (1955); R. J. Eden and N. C. Francis, Phys. Rev. 97, 1366 (1955).

[2] G. F. Chew and M. L. Goldberger, Phys. Rev. 77, 470 (1950).

[3] J. Heidmann Phys. Rev. 80, 171 (1950).

[4] The experimental evidence is that obtained by J. Hadley and H. F. York, Phys. Rev. 80, 345 (1950); K. A. Brueckner and W. Powell, Phys. Rev. 75, 1274 (1949).

[5] Panofsky, Aamodt, and Hadley, Phys. Rev. 81, 565 (1951).

[6] Brueckner, Serber, and Watson, Phys. Rev. 84, 258 (1951), referred to as BSW in the text.

essentially as a one nucleon process

$$\pi^- + p \rightarrow n + \gamma$$

will be only weakly affected by the other nucleons. It is possible therefore to reach conclusions on the relative transition rates for the nonradiative capture in deuterium and carbon (or in other light and medium heavy nuclei).

The theory of the capture next is used to show that the nonradiative capture rate is a sensitive measure of the probability of finding two closely associated nucleons in the ground state of the nucleus, since the capture involves large momentum transfers and consequently can take place only through the cooperative effect of at least two nucleons. Qualitatively it is obvious that if in heavy nuclei the nucleons were randomly distributed, the probabilities of finding two nucleons close together is small, and the meson capture rate would be small, in contract to the effect observed. The result of a quantitative analysis of the various phenomena affecting the radiative and nonradiative capture showed that the nucleons are indeed highly correlated.

According to BSW, the transition rate depends on a factor $P(z_M)$ which is the probability of finding two nucleons with a separation $z_M \approx \hbar/\Delta p$, where Δp is the momentum transferred to the two nucleons. For an uncorrelated nucleus, $P(z_M)$ is just the nuclear density $1/v$, where v is the total volume $(4/3)\pi r_0^3 A$. Thus it is useful to define a correlation factor f by the equation

$$P(z_M) = f\cdot[(4/3)\pi r_0^3 A]^{-1}.$$

The analysis of BSW showed that $f \approx 35$ and thus that the wave function for the nuclear ground state must depart very markedly from that for an uncorrelated system. We will return to an alternative formulation of this result later in Sec. IV.

C. Photonuclear Effect at High Energy

The cross section for the photoejection of high-energy (50–200 Mev) protons from nuclei has been known for some time to be much larger than would be given by an independent-particle model. The presence of a large high-energy component of fast protons indicates quite unambiguously the existence of high-momentum components in the ground-state wave function. The process has been analyzed by Levinthal and Silverman[7] and by Levinger,[8] all in the dipole approximation. The former authors show that the observed cross section is in good agreement with the Chew-Goldberger[2] momentum distribution for the ground-state wave function. The analysis of Levinger[8] uses a somewhat different method of calculation to which we will return later, but he also concludes that the experiment shows the marked departure at high-momentum values of the wave function from that for an uncorrelated system.

[7] C. Levinthal and A. Silverman, Phys. Rev. 82, 822 (1951).
[8] J. S. Levinger, Phys. Rev. 84, 43 (1951).

D. Proton-Nucleus Scattering and Meson Production in Proton-Nucleus Collision

These processes,

$$p + \text{nucleus} \rightarrow \text{nucleus} + p',$$
$$p + \text{nucleus} \rightarrow \text{nucleus} + \pi,$$

can be used to give quite quantitative indications of the departures of the ground-state nuclear wave function from that of an independent particle model. In these cases a contribution to the cross section comes from each component of the nucleus ground-state momentum distribution. Thus if the cross section for the elementary process involving free nucleons is known, then the observed cross section is a function only of the momentum distribution. The analyses of Henley[9] and of Wolff[10] show that the momentum distribution is determined with fair accuracy, and is fitted well by a Gaussian

$$|\psi(\mathbf{p})|^2 = \frac{1}{\alpha^3\pi^{\frac{3}{2}}}\exp(-\mathbf{p}^2/\alpha^2), \tag{3}$$

where α^2 is such that the mean kinetic energy is 19.3 Mev. This distribution does not depart as strongly from an uncorrelated wave function as does the wave function used by Chew and Goldberger[2]; the difference, however, is still very marked, as is shown in Fig. 1.

E. Summary of Experimental Results

All the experiments discussed in this section are similar in that phenomena are observed which depend strongly on strong correlation effects in the coordinate space wave function or, equivalently, on an appreciable admixture of high momentum components in the momentum space wave function. Thus it is evident that the nuclear ground-state wave function cannot describe nucleons moving as independent particles, and reinterpretation of the independent particle model is necessary. In the next section we shall show how this can be done in the general case, using the techniques and concepts we have developed for treating the nuclear ground state. We shall also make detailed applications to specific cases of particular interest.

III. GENERAL THEORY

A. General Description of Method

The processes considered in the previous section have an important common feature; an appreciable contribution to the cross sections comes from an initial nuclear ground state in which at least one nucleon has a high momentum. We make the physical assumption that this high momentum is the result of strong inter-

[9] E. M. Henley, Phys. Rev. 85, 204 (1952); this paper lists the relevant experimental references.
[10] P. A. Wolff, Phys. Rev. 87, 434 (1952); see also the experimental results given by Cladis, Hess, and Mayer, Phys. Rev. 87, 425 (1952).

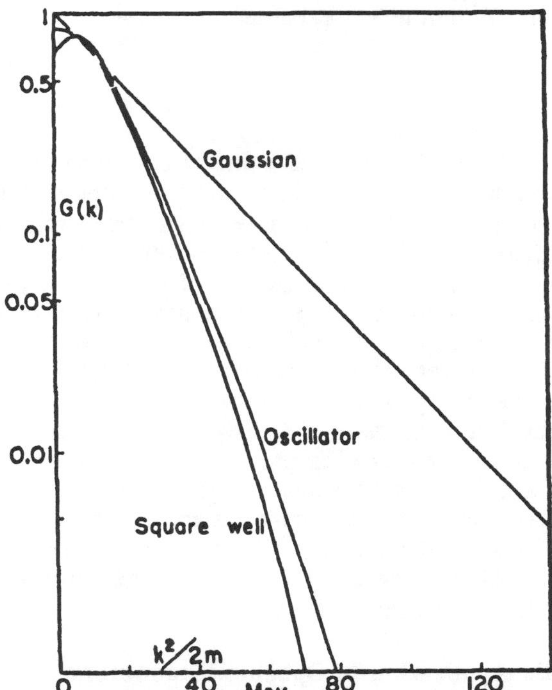

FIG. 1. Momentum distribution $G(k)$ of 8 neutrons and 8 protons in the independent-particle states of a square well with infinite walls and of a harmonic oscillator well. For comparison the Gaussian distribution of Eq. (3) is also given.

actions between a pair of nucleons; indeed the experiments[11] seem to verify this assumption. Consequently the ejection of a fast nucleon by the photoprocess, by meson capture or as a member of the deuteron in deuteron pickup, will usually be associated with the ejection of another fast nucleon which was originally paired with the directly ejected particle. Thus the process corresponds to ejection of a fast pair of nucleons from the ground state, with the residual nucleus only weakly excited. This assumption (or a stronger assumption) is explicit or implicit in the theory of all the high-energy processes we have considered.

It is at this point that the usefulness of the high-energy processes in the study of nuclear structure is particularly apparent. As we will see in the following development of the theory, if the ground-state function is weakly correlated as for a Fermi gas or an independent particle model, then the matrix elements will vanish in the former case or be very small in the latter case. Since the predominant low-momentum components in the wave function make very little or no contribution to the matrix elements, the importance of the high momentum components is greatly enhanced and hence it is possible to get detailed information about this aspect of the wave function. Before proceeding to the theory of these processes, we shall first make some brief remarks on the nature of the ground-state wave function and in particular on our interpretation of the shell model and its reconciliation with the simultaneous success of the shell model and with the high-energy phenomena which interest us.

It is well known from both experiment and theory that the nucleon-nucleon interactions are strong and short ranged. Consequently if the same forces act when nucleons are immersed in a many-body medium, one will very naturally expect to observe under appropriate experimental conditions very appreciable correlations in the nuclear wave function. On the other hand, the success of the shell model has often been assumed to indicate that the two-body forces in nuclear matter are in fact much weaker and long-ranged and can lead in an excellent approximation to a uniform Hartree field acting on the nucleons. The origin of this effect might be, for example, a strongly nonlinear behavior of the meson fields so that a very large damping effect modifies and smooths out the forces in nuclear matter. This effect can arise from many-body forces or from a nonlinearity in the meson field equations. In either case the effective potential felt by one nucleon would not have the rapidly varying spatial dependence which would result if the two-body forces remained strong, and a uniform potential would be a good approximation. A direct consequence would then be that the nuclear wave function would be weakly correlated, in disagreement with the high-energy experiments we are analyzing. It is also perhaps worth commenting here that the theoretical expectations for the character of the two-body and many-body interactions also strongly suggest that the strong two-body forces are still effective in a medium of nuclear density.

Thus we must adopt a picture of strong two-body nuclear potentials and modify our views of the shell model. The concepts and techniques which we have developed are discussed in detail in other papers[1]; the essential points may be summarized in the following way. We require that the shell model (or independent particle) wave function be a description not of nuclear motion but of a "collective particle" motion, the actual nucleon wave function being generated from the "shell model" wave function by a transformation. This transformation has (among other effects) the effect of introducing correlations and hence high momentum components into the wave function. Under certain conditions the behavior of the shell model "particles" is very nucleon-like, but this approximate identification is not generally valid. We see, in fact, a complete breakdown of the approximate description in the region of strong correlations or high momenta, where the departure of the simple shell model states from actual nuclear states becomes particularly marked. Stated in other terms, a consequence of our description of the nucleus is that the departure of shell model states from nuclear states is not very appreciable if observations of the state are

[11] Byfield, Kessler, and Lederman, Phys. Rev. 86, 17 (1952); see also reference 16. The effect of correlations involving more than two nucleons will become apparent only in the high-energy tail of the spectrum of the ejected nucleons, this will not influence appreciably the total cross sections of the processes we consider.

made which depend only on averages over space or time intervals which are large compared with characteristic correlation distances or fluctuation frequencies in the state. In the other extreme, observations at high frequencies or short wavelengths readily detect the departure of the ground state from that for an uncorrelated system.

We next shall summarize the necessary formalism which we need in the following discussions. For further details of notation and explanation the reader should refer to previous papers.[1] The nuclear ground-state wave function $\Psi_0(A)$ is in general a complicated function of the coordinates of the nucleons and contains marked correlations as a result of the strength of the two body interactions. It is related to the "shell-model" or "independent particle" wave function $\Phi_0(A)$ by a transformation function or "model operator" F, i.e.,

$$\Psi_0(A)=F\Phi_0(A). \tag{4}$$

Since $\Phi_0(A)$ is a weakly correlated function (a degenerate Fermi gas for the lowest energy state), F has the effect of introducing correlations into the wave function. Clearly, therefore, explicit knowledge of the transformation F and thus of the wave function $\Psi_0(A)$ is necessary in our problem, depending as it does on the correlations in the wave function. This is in marked contrast to determination of the ground-state energy, for example, where only the transformed Hamiltonian (the independent particle Hamiltonian) need be known and there are no departures of the particle (not nucleon) motion from independent particle motion.

The explicit form of the transformation, which has been used successfully in the considerations of other ground-state properties, is given by the following set of coupled equations:

$$F=1+\frac{1}{e}\sum_{ij} I_{ij}F_{ij}, \tag{5}$$

$$F_{ij}=1+\frac{1}{e}\sum_{lm\neq ij} I_{lm}F_{lm}, \tag{6}$$

where the "energy denominator" e is

$$e=E_0-\sum_i T_i-V_c, \tag{7}$$

and the energy eigenvalue E_0 is determined by

$$(E_0-\sum T_i-V_c)\Phi_0(A)=0. \tag{8}$$

The quantities I_{ij} and V_c are simply related to the two-body scattering operators t_{ij} which are defined by the equations:

$$t_{ij}=v_{ij}+v_{ij}(1/e)t_{ij}, \tag{9}$$

where v_{ij} is the potential between nucleons i and j. The operators I_{ij} are those parts of the t_{ij} which are nondiagonal with respect to the nuclear states; finally, the the uniform potential V_c is defined by

$$V_c=\tfrac{1}{2}\sum_{ij} t_{cij}, \tag{10}$$

where t_{cij} is the diagonal part of t_{ij}.

Some features of these results and of the transformation F are easily seen. The incoherent or nondiagonal operators I_{ij} cause transitions from the uncorrelated independent-particle state $\Phi_0(A)$; the effect is closely analogous to an inelastic scattering of particles a pair at a time out of the Fermi gas to excited states. Consequently the departures of the wave function $\Psi_0(A)$ from $\Phi_0(A)$ are very closely related to the details of the inelastic scattering of nucleon by nucleon and thus to the strength and range of the two-body potentials.

We shall in the following parts of this section make explicit applications of these principles to the high-energy phenomena we have summarized in the previous section. Where the development follows the work done by other authors, we shall abbreviate the discussion where this can be conveniently done.

B. Application to Specific Phenomena

(1) *Nuclear Photoeffect*

The cross section for the production of high-energy protons by γ radiation of a nucleus in dipole approximation is obtained from the matrix element H_{0f} given by

$$H_{0f}=\left(\Psi_f(A),\ \sum_{i=1}^{Z} e\mathbf{r}_i\cdot\mathbf{A}\Psi_0(A)\right), \tag{11}$$

where $e\mathbf{r}_i$ is the charge moment for the ith proton. We approximate to the final state wave function by making use of our physical assumption that two of the nucleons have high momenta so the wave function is approximately separable, i.e., we assume

$$\Psi_f(A)=\Psi_f(1,2)\Psi_f(A-2). \tag{12}$$

We suppose that nucleon "1" is a proton; the associated nucleon "2" must then be a neutron since the contribution from two protons is zero in the dipole approximation. The assumption of a product wave function is equivalent to the neglect of corrections due to antisymmetrization between the high-momentum and low-momentum particles, but $\Psi_f(1,2)$ and $\Psi_f(A-2)$ will be separately antisymmetrized. With a final wave function of this form we can neglect contributions to H_{0f} from terms not involving the position $\mathbf{r}_1$ associated with proton 1.

We may use the method of partial closure to eliminate the wave function $\Psi_f(A-2)$ from the cross section provided that the rest of the matrix element depends only weakly on the energies of excitation E_f of the residual nucleus. This will be the case if the energies of the ejected particles are high; it is also possible to correct approximately for the error made in the closure sum by altering the energy conservation law to include a mean excitation energy $\bar{E}_f$; i.e., we can set

$$E_0+E_\gamma=k_1^2/2m+k_2^2/2m+\bar{E}_f, \quad (13)$$

where $\bar{E}_f$ can either be estimated or regarded as a phenomenological parameter. Making this approximation, we find

$$\sum_f |H_{0f}|^2=ZN\int d\mathbf{r}_3\cdots\int d\mathbf{r}_A \times\left|\int\int\Psi_f^*(1,2)e\mathbf{r}_1\cdot\mathbf{A}\Psi_0(1,2,\cdots,A)d\mathbf{r}_1d\mathbf{r}_2\right|^2, \quad (14)$$

the factor of NZ coming from the number of ways of choosing the protons (1) and neutrons (2). This result also neglects interference terms which is consistent with our assumptions about the separability of the final state wave function. In this matrix element the final wave function can be determined with considerable accuracy since it is probably safe to assume that this is simply given by the solution to the two-body problem at high energy neglecting the effects of the nuclear medium. Thus only the initial wave function is unknown and its specifications largely determines the matrix element and the cross section.

It might be pointed out at this point that in the assumptions we have just made in treating the final states, the possible attenuation by nuclear collision of the outgoing proton wave have been neglected. One consequence of this assumption is that the calculated cross section will depend on the total number of protons in the nucleus, i.e., that it is a volume effect. The best evidence for this assumption comes from the photoeffect itself[8] and also from the process of meson capture in nuclei as studied by Byfield, Kessler, and Lederman.[11] We shall in this section and also in discussing the remaining phenomena neglect the possibility of absorption of the outgoing particles and as a consequence overestimate the cross sections by a factor depending on the process considered. The appropriate correction factor can usually be estimated without difficulty.

The wave function $\Psi_0(A)$ defined by Eqs. (4) and (5) to (10) is extremely difficult to give explicitly because of its very complicated dependence on the incoherent operators I_{ij}. If, however, we are willing to assume that the necessary correlation is largely a result of two-body interactions, then we can approximate to the function F by expanding it to first order in I_{ij}. Consequently we cannot expect fully quantitative knowledge of the wave function in the important high-momentum region, although the error introduced is probably not enough to alter the order of magnitude of the result. In this approximation, we take

$$\Psi_0(1,2,\cdots,A)=\left(1+\frac{1}{e}\sum t_{ij}\right)\Phi_0(1,2,\cdots,A). \quad (15)$$

We have replaced I_{ij} by t_{ij} in F since here only off-diagonal matrix elements contribute to the cross section. The wave function $\Phi_0(1,2,\cdots,A)$ can be written (treating neutrons and protons separately):

$$\frac{1}{\sqrt{(ZN)}}\sum_{lm}(\phi_l(1)\phi_m(2))\Phi_0^{lm}(3,\cdots,A) \quad (16)$$

with $\Phi_0^{lm}(3,\cdots,A)$ normalized to unity. Only the terms involving $\mathbf{r}_1$ and $\mathbf{r}_2$ will make any important contribution to the cross section, hence after integrating over the variables $\mathbf{r}_3,\cdots,\mathbf{r}_A$ we get

$$\sum_f|H_{0f}|^2=ZN\sum_{lm}\left|\frac{1}{\sqrt{(ZN)}}\int\int\Psi_f^*(1,2)e\mathbf{r}_1\cdot\mathbf{A}F_{12}\phi_l(1)\phi_m(2)d\mathbf{r}_1d\mathbf{r}_2\right|^2, \quad (17)$$

where l is summed over proton and m over neutron states, and

$$F_{12}=1+(1/e)t_{12}. \quad (18)$$

We define a quantity $\chi_{lm}'(1,2)$ by the equation

$$\chi_{lm}'(1,2)=F_{12}\phi_l(1)\phi_m(2) \quad (19)$$

(the prime distinguishes this from a symmetrized form used later). Then $\chi_{lm}'(1,2)$ will give the high-momentum components for the nucleons 1 and 2 in the nucleus. It is important to note that our approximation method is based on the fact that only high-momentum components will give large contributions to the matrix element so that it has been possible to neglect terms in Eq. (15) which do not refer to particles 1 and 2. For these reasons it would be misleading to regard $\chi_{lm}'(1,2)$ as a kind of two-body wave function in the nucleus.[12] We get now

$$\sum_f|H_{0f}|^2=\sum_{lm}\left|\int\int\Psi_f^*(1,2)e\mathbf{r}_1\cdot\mathbf{A}\chi_{lm}'(1,2)d\mathbf{r}_1d\mathbf{r}_2\right|^2. \quad (20)$$

This can be compared (see Levinger[8]) with the deuteron photoeffect, for which

$$|H_{0f}^D|^2=\left|\int\int\Psi_f^*(1,2)e\mathbf{r}_1\cdot\mathbf{A}\Psi_D(1,2)d\mathbf{r}_1d\mathbf{r}_2\right|^2, \quad (21)$$

where $\Psi_D(1,2)$ denotes the deuteron wave function.

Thus a comparison of the experimentally determined matrix elements for the photodisintegration of the nucleus (with fast proton ejection) and of the deuteron can be used to give us rather directly a relation between the deuteron and ground-state wave functions. This technique is particularly useful if we do not wish (or are unable) to give explicitly the final two-particle wave function $\Psi_f^*(1,2)$ since this enters very nearly as a

[12] A function analogous to our $\chi_{lm}^{(1,2)}$ is introduced on physical grounds by Levinger (reference 8) following a method of Heidmann (reference 3).

common factor into both matrix elements [Eqs. (20) and (21)].

(2) *Meson Capture*

The operator describing the absorption of a meson by a nucleon can be written[6,13]

$$a\phi(1)+b\sigma_1\cdot\nabla\phi(1), \tag{22}$$

where $\mathbf{r}_1$ denotes the coordinate of the nucleon which absorbs the meson, and $\phi(1)$ denotes the meson field at $\mathbf{r}_1$. If we assume the meson to be in an S state, $\phi(1)$ will be sufficiently slowly varying over the nucleus that we can replace it by some mean value ϕ_0, and similarly replace $\nabla\phi(1)$ by a mean gradient written $\nabla\phi_0$.

The matrix element for meson absorption in the nucleus is given by

$$H_{0f}=\sum_i(\Psi_f(A),\{a\phi_0+b\sigma_i\cdot\nabla\phi_0\}\Psi_0(A)). \tag{23}$$

We assume as with the photoeffect that two nucleons in the final state have high momenta and hence their wave function is separable from that for the residual (A-2) nucleons. Then

$$\Psi_f(A)=\Psi_f(1,2)\Psi_f(A-2). \tag{24}$$

Write θ_0 for the operator $a\phi_0+b\sigma_1\cdot\nabla\phi_0$, and let

$$\Psi_f(1,2)=F_{12}\Phi_f(1,2), \tag{25}$$

where $\Phi_f(1,2)$ denotes a plane wave state for the two outgoing nucleons. Then, using partial closure to eliminate $\Psi_f(A-2)$ and transforming $\Psi_0(A)$, we get for the nucleon pair (1,2):

$$\sum_f|H_{0f}|^2=\int d\mathbf{r}_3\cdots\int d\mathbf{r}_A\left|\int\int\Phi_f^*(1,2)(F_{12}{}^\dagger\theta_0)\times F_{12}\Phi_0(1,2,\cdots,A)d\mathbf{r}_1d\mathbf{r}_2\right|^2. \tag{26}$$

Hence

$$\sum_f|H_{0f}|^2=\frac{2}{A^2}\sum_{lm}\left|\int\int\Phi_f^*(1,2)(F_{12}{}^\dagger\theta_0)\chi_{lm}(1,2)d\mathbf{r}_1d\mathbf{r}_2\right|^2, \tag{27}$$

where the $\chi_{lm}(1,2)$ is properly antisymmetrized, i.e.,

$$\chi_{lm}(1,2)=(1/\sqrt{2})F_{12}\{\phi_l(1)\phi_m(2)-\phi_m(1)\phi_l(2)\}. \tag{19a}$$

It is useful to compare this with the corresponding process for meson absorption in the deteron for which

$$|H_{0f}{}^D|^2=\left|\int\Phi_f^*(1,2)(F_{12}{}^\dagger\theta_0)\Psi_D(1,2)d\mathbf{r}_1d\mathbf{r}_2\right|^2. \tag{28}$$

One result is that it is possible, if we wish to interpret the operator $(F_{12}{}^\dagger\theta_0)$ as a phenomonological meson absorption operator, to use the experiment to deduce relationships between the nuclear wave function and the deuteron wave function $\Psi_D(1,2)$. This method was used by BSW[6] in their analysis.

(3) *Deuteron Pickup*[14]

The Born approximation matrix element for this process is

$$H_{0f}=(\Psi_f(0,1,2)\Psi_f(A-2),V(\mathbf{r}_0-\mathbf{r}_1)e^{i\mathbf{k}_0\cdot\mathbf{r}_0}\Psi_0(A)). \tag{29}$$

Then following the same techniques as those used in the previous paragraphs, we can bring this to the form

$$\sum_f|H_{0f}|^2=\int d\mathbf{r}_3\cdots\int d\mathbf{r}_A\left|\int\Psi_f^*(0,1,2)V(\mathbf{r}_0-\mathbf{r}_1)\times e^{i\mathbf{k}_0\cdot\mathbf{r}_0}F_{12}\Phi_0(1,\cdots,A)d\mathbf{r}_0d\mathbf{r}_1d\mathbf{r}_2\right|^2. \tag{30}$$

If we assume

$$\Psi_f(0,1,2)=e^{i\mathbf{K}(\mathbf{r}_0+\mathbf{r}_1)/2}\phi(\mathbf{r}_0-\mathbf{r}_1)e^{i\mathbf{k}_2\cdot\mathbf{r}_2}, \tag{31}$$

where

$$\Psi_0(0,1)=e^{i\mathbf{K}(\mathbf{r}_0+\mathbf{r}_1)/2}\phi(\mathbf{r}_0-\mathbf{r}_1) \tag{32}$$

is the deuteron wave function, we obtain

$$\sum_f|H_{0f}|^2=\frac{2}{A^2}\sum_{lm}\left|\int e^{-i\mathbf{k}_1\cdot\mathbf{r}_1}e^{-i\mathbf{k}_2\cdot\mathbf{r}_2}\chi_{lm}(1,2)d\mathbf{r}_1d\mathbf{r}_2\right|^2\times\left|\int\phi^*(\mathbf{r})V(\mathbf{r})e^{i(\mathbf{k}-\mathbf{K}/2)\cdot\mathbf{r}}d\mathbf{r}\right|^2. \tag{33}$$

The first factor corresponds to the momentum distribution for nucleons 1 and 2 in the nucleus. In our notation nucleon 1 is picked up and goes off as part of the deuteron $\Psi_0(0,1)$ and nucleon 2 is a recoil nucleon which we have represented by a plane wave. The second factor can be evaluated by substituting an explicit form for the relative coordinate part $\phi(\mathbf{r})$ of the deuteron wave function and using the Schrödinger equation to rewrite $V(\mathbf{r})$ in terms of the binding energy and momenta.

(4) *Proton-Proton Scattering in Nuclei and Meson Production by Protons Bombarding Nuclei*

For a high-energy incident proton ejecting a second proton from a nucleus it is a good approximation to suppose that the incident proton collides only with the one nuclear particle.[10] The same approximation can be made in considering the production of mesons by a proton colliding with a nucleus.[9] It is worth noting that both these processes are possible if the second proton (nucleon) was not bound in the nucleus but was free. This has the effect that for a limited range of the momenta of the outgoing particles a large contribution to the cross section will come from considering the

[13] See, for example, K. A. Brueckner and K. M. Watson, Phys. Rev. 83, 1 (1951).

[14] The theory in this paragraph is similar to that of Heidmann (reference 3) except (1) we derive the correlations in the ground state wave function by transformation theory which allows for the effects of the nuclear medium whereas Heidmann obtains them as physical grounds, and (2) we use partial closure to eliminate the final state.

nucleus to be a Fermi gas, i.e., from taking the unit term in Eq. (5) which gives the transformation of the a Fermi gas, i.e., from taking the unit term in Eq. (5) which gives the transformation of the Fermi gas wave function to the actual nuclear wave function. However, if one looks instead at momentum regions which require a large momentum value for the proton initially in the nucleus, the principal contribution to these cross sections will come from matrix elements obtained by methods analogous to those considered in the previous sections. These involve the quantity $\chi_{lm}(1,2)$ and lead to a cross section which depends on a known factor (or one which can be found by comparison with experiment) and the momentum distribution. The analysis of p-p scattering in the nucleus has been given by Wolff,[10] and that of meson production in the nucleus by Henley.[9] We will not repeat these calculations using our methods as they would involve a lengthy analysis of approximations which are made by physical arguments. However, for comparison we note the formula for the relevant cross section given by Wolff and by Henley.

For proton-proton scattering in the nucleus, Wolff[10] obtains the cross section for an incoming proton of momentum p and final nucleons of momentum q and s:

$$\frac{d^2\sigma}{d_A dq}=\frac{4\pi M^2}{\hbar^4 p}\sum|a|^2\int\frac{d\mathbf{k}}{(2\pi)^3}N(\mathbf{k})\times\delta\left(p^2-q^2-(\mathbf{p}+\mathbf{k}-\mathbf{q})^2-\frac{2M}{\hbar^2}B_{if}\right)\cdot\frac{q^2}{(2\pi)^3},\quad(34)$$

where $\mathbf{k}=(\mathbf{q}+\mathbf{s}-\mathbf{k})$ and $N(k)$ is the momentum distribution for a nucleon in the nucleus. The detailed form of $N(k)$ as predicted by our method will be discussed in the next section.

The cross section for the creation of a π meson of momentum q is given by Henley[9]:

$$\frac{d^2\sigma_A{}^0(\pi^+)}{dTd\Omega_q}=\frac{1}{v_0}\int\left[Z\frac{d^2\sigma(p-p)}{dTd\Omega_q}v_H+(A-Z)\frac{d^2\sigma(p-n)}{dTd\Omega_q}v_R\right]\rho(\mathbf{k})d\mathbf{k},\quad(35)$$

where $\rho(\mathbf{k})$ is the momentum distribution for the struck nucleon in the nucleus, and $\sigma(p-p)$, $\sigma(p-n)$ denote the cross sections for collisions of free nucleons; v_0 and v_r represent the relative velocities of the target nucleus and nucleon with respect to the incident nucleon; T is the meson kinetic energy.

IV. APPLICATION OF THE THEORY TO EXPERIMENT

A. Dependence of the Theoretical Cross Sections on a Momentum Distribution in the Nucleus

In discussing the high-energy phenomena, we have shown that in each case the matrix element depends most strongly on the ground-state wave function through a term

$$f(\mathbf{k}_1,\mathbf{k}_2)=\frac{2}{A^2}\sum_{lm}\left|\int(2\pi]^{-3}e^{-i\mathbf{k}_1\cdot\mathbf{r}_1}e^{-i\mathbf{k}_2\cdot\mathbf{r}_2}\times\chi_{lm}(\mathbf{r}_1\mathbf{r}_2)d\mathbf{r}_1d\mathbf{r}_2\right|^2,\quad(36)$$

where k_1 and k_2 are the final momenta of the two fast ejected nucleons and χ_{lm} is, in the region of high momenta, an approximation to the initial wave function of the two nucleons. This result can be brought to a simpler form if we introduce relative and center-of-mass coordinates and further assume that

$$\chi_{lm}(\mathbf{r}_1,\mathbf{r}_2)=v^{-\frac{1}{2}}\chi_{lm}(\mathbf{r}_1-\mathbf{r}_2)e^{i(\mathbf{k}_l+\mathbf{k}_m)\cdot(\mathbf{r}_1+\mathbf{r}_2)/2}.\quad(37)$$

This is equivalent to the reasonable assumption that the center-of-mass moves with the momenta characteristic of the Fermi gas and that the departures from a Fermi gas occur only in the dependence on the relative coordinate. Making this approximation we find

$$f(\mathbf{k}_1,\mathbf{k}_2)=\frac{2}{A^2}\sum_{lm}\left|\int(2\pi)^{-\frac{3}{2}}e^{-i(\mathbf{k}_1-\mathbf{k}_2)\cdot\mathbf{r}/2}\chi_{lm}(\mathbf{r})d\mathbf{r}\right|^2\times\delta(\mathbf{k}_1+\mathbf{k}_2-\mathbf{k}_l-\mathbf{k}_m),\quad(38)$$

where we have replaced the Kronecker delta function giving total-momentum conservation by the Dirac delta function, using the relation

$$(\delta_{\mathbf{k}_1+\mathbf{k}_2,\mathbf{k}_l+\mathbf{k}_m})^2=[(2\pi)^{\frac{3}{2}}v^{-\frac{1}{2}}]^2\delta(\mathbf{k}_1+\mathbf{k}_2-\mathbf{k}_l-\mathbf{k}_m),\quad(39)$$

a further simplification is possible if we assume (as we have previously) that k_1, $k_2 \gg k_l$, k_m and consequently that the momentum conservation gives approximately $\mathbf{k}_1=-\mathbf{k}_2$. In this approximation we can also relax the restrictions arising from momentum conservation and sum freely over l,m. This is related to the approximation made in using partial closure and as there we can introduce a correction by altering the energy conservation law to take account of a mean excitation of the residual nucleus.

We thus are led (dropping the delta function) to the final result

$$g(\mathbf{k}_1)=f(\mathbf{k}_1,-\mathbf{k}_1)\approx\frac{2}{A^2}\sum_{lm}\left|\int\frac{e^{-i\mathbf{k}_1\cdot\mathbf{r}}}{(2\pi)^{\frac{3}{2}}}\chi_{lm}(\mathbf{r})d\mathbf{r}\right|^2.\quad(40)$$

The summation over the $A^2/2$ pairs of states lm is equivalent (with the factor $2/A^2$) to averaging over a Fermi gas. This result is now closely analogous to the results used by Chew-Goldberger, Levinthal-Silverman, Henley, and Wolff, who have all deduced a dependence of the cross section on a factor which they call the Fourier transform of the wave function of the *single* particle involved; i.e., they introduce a function $N(\mathbf{k})$

$$N(\mathbf{k})=\left|\int \frac{e^{-i\mathbf{k}\cdot\mathbf{r}}}{(2\pi)^{\frac{3}{2}}}\phi(\mathbf{r})d\mathbf{r}\right|^2, \quad (41)$$

where $\phi(\mathbf{r})$ is assumed to be the particle wave function. Our result differs however in that $\chi_{lm}(\mathbf{r})$ is a function depending on the relative coordinate for a pair of nucleons and in that an average is to be carried out over the initial state of the two ejected nucleons.

B. Determination of the Function $g(\mathbf{k})$ from Experiment

We shall not attempt to make explicit calculations for the various phenomena; instead we shall make use of the empirical determinations of the function $N(\mathbf{k})$ which have been made. In the case of deuteron pickup,[15] proton-proton scattering in nuclei,[10] and meson production by protons as nuclei,[9] it has been shown that a function

$$N(\mathbf{k})=\exp(-k^2/\alpha^2)(\alpha^2\pi^{\frac{1}{2}})^{-1}, \quad (42)$$

with $\alpha^2/2m\approx 14$ Mev gives a satisfactory fit to the experiments. We have also made an analysis of the photoproton data similar to that made by Levinthal and Silverman[7] using this Gaussian distribution. The results are shown in Fig. 2 which indicates that the shape of the predicted distribution agrees well with experiment. The magnitude of the predicted cross sections is considerably larger than the results of Levinthal and Silverman[7] and a factor of 10 has been introduced to renormalize the predicted cross section. The discrepancy is not so large, however, if comparison is made with the results of Keck[16] and Walker[17] which are a factor of 5 to 7 larger. Consequently we conclude that the Gaussian momentum distribution which is identical with that used in the other analyses also approximately fits the photoproton data, at least as well as the accuracy of the treatment would lead one to expect.

The remaining process of meson capture can also be easily brought into a form which allows comparison with the assumed Gaussian distribution. This is most simply done if we make use of the result of BSW[6] which shows that the ratio

$$\left|\int e^{-i\mathbf{p}\cdot\mathbf{r}}\phi(\mathbf{r})d\mathbf{r}\right|^2 \Big/ \left|\int e^{-i\mathbf{p}\cdot\mathbf{r}}\psi_D(\mathbf{r})d\mathbf{r}\right|^2 \geqslant 2.39, \quad (43)$$

where $\psi_D(\mathbf{r})$ is the deuteron wave function and p is the momentum carried off by one nucleon $[p\sim(M\mu)^{\frac{1}{2}}]$. Using for ψ_D a Hulthén wave function with standard

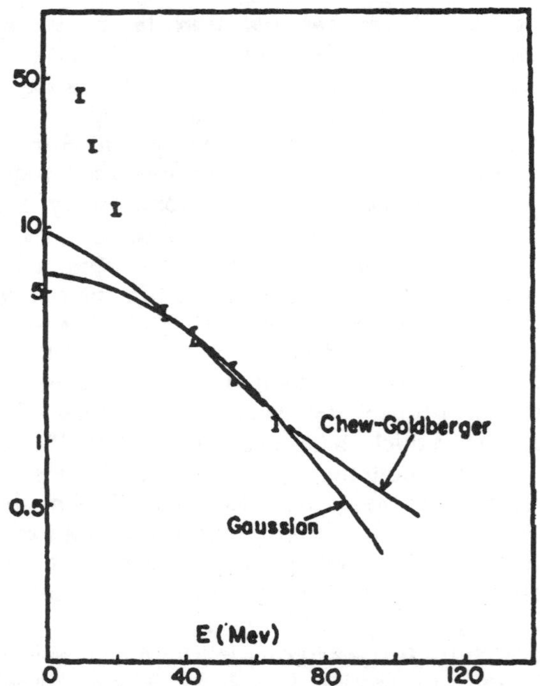

FIG. 2. Theoretical predictions for the photoproton process based on a Chew-Goldberger and on the Gaussian distribution of Eq. (3). The experimental data are taken from the paper of Levinthal and Silverman (see reference 7). The theoretical results are normalized to the experiments at 41.5 Mev by a factor of 0.75 for the Chew-Goldberger and 0.15 for the Gaussian.

parameters, and taking for $\phi(\mathbf{r})$ a normalized Gaussian $N_\beta \exp(-\beta^2 r^2/2)$ it is possible to determine a value for $\beta^2/2M$ which agrees with the empirically determined ratio. The result is that $\beta^2/2M\geqslant 14.4$ Mev which agrees very well with the results of the other experiments.

We therefore can well represent the results of all of the experiments by the same momentum distribution. Returning to our definition of $g(k)$ [Eq. (40)], we thus find that if our theory predicts correctly the result,

$$\frac{2}{A^2}\sum_{lm}\left|\int \frac{e^{-i\mathbf{p}\cdot\mathbf{r}}}{(2\pi)^{\frac{3}{2}}}\chi_{lm}(\mathbf{r})d\mathbf{r}\right|^2 = N_\alpha \exp(-p^2/\alpha^2), \quad (44)$$

for high values of the momentum p (corresponding to energy per particle of 75 to 150 Mev), then we can also expect to find good agreement with the high-energy experiments.

C. Calculation of Momentum Distribution

In this section we shall attempt to determine the function $g(k)$ defined in Eq. (40). As we shall show, there are two rather different methods which lead to essentially the same result. Let us consider the function $\chi_{lm}(\mathbf{r})$, which is the wave function for the relative coordinate $\mathbf{r}$. We first approximate this in a manner similar to that used by Levinger and by Heidmann,

[15] The use of a Gaussian distribution to fit deuteron pickup has been tested by Henley (see reference 9).
[16] J. Keck, Phys. Rev. 85, 410 (1952).
[17] D. Walker, Phys. Rev. 81, 634 (1951).

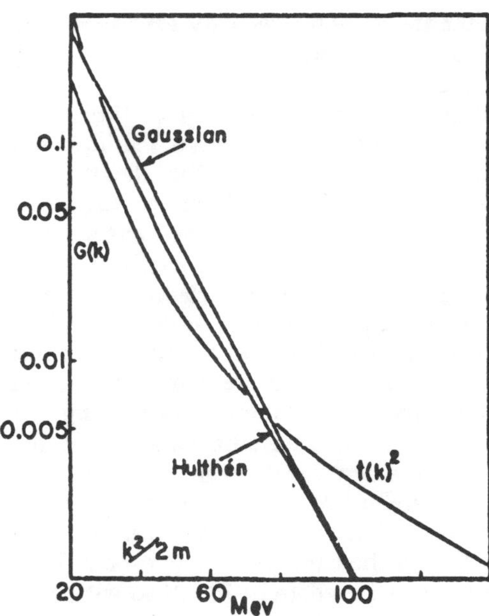

FIG. 3. Calculated momentum distributions for particles in the nucleus compared with the empirically derived Gaussian of Eq. (3). The results of the two methods discussed in the text are given, labelled Hulthén for the result of Eq. (48) and t^2 for the result of Eq. (52).

i.e., we write for the S-wave part of the function

$$\chi_{lm}(\mathbf{r})=\frac{1}{v^{\frac{1}{2}}}\left[\frac{\sin(kr+\delta)}{\sin\delta}-f(r)\right]\Big/\left[r(\alpha^2+k_{lm}{}^2)^{\frac{1}{2}}\right],\quad(45)$$

which is correctly normalized in the nuclear volume v. $1/\alpha$ is the scattering length; $f(r)$ is a function which is equal to one at the origin and depends in detail on the explicit choice of the potential. At small distances this can also be written as

$$\chi_{lm}(\mathbf{r})=\frac{1}{v^{\frac{1}{2}}}\left[e^{-\alpha r}-f(r)\right]/\left[r(\alpha^2+k_{lm}{}^2)^{\frac{1}{2}}\right].\quad(46)$$

For our simple example we take for $f(r)$ the result appropriate to a Hulthén function

$$f(r)=e^{-\beta r}.\quad(47)$$

We use the Hulthén function for the following reasons: (i) the Hulthén function gives a cross section which is of approximately the correct magnitude at the high energies we consider, (ii) the major contribution to the cross section from the Hulthén potential comes from S-wave scattering, (iii) a repulsive core potential would require calculations for at least one higher partial wave since the S-wave is anomalously small in the energy region of interest. In addition it would lead to a result which would be very sensitive to the potential shape in the region of the core boundary where little is known about its detailed form.

Inserting the wave function of Eq. (46) into Eq. (44), we find

$$g(\mathbf{p})=\frac{2}{A^2}\sum_{lm}\frac{(4\pi)^2}{(2\pi)^3}\frac{A}{v}\left[\frac{(\beta^2-\alpha^2)}{(p^2+\beta^2)(p^2+\alpha^2)}\right]^2(\alpha^2+k_{lm}{}^2)^{-1}.\quad(48)$$

We approximate to the average over the Fermi gas by using Levinger's result $\langle(\alpha^2+k_{lm}{}^2)^{-1}\rangle_{Av}\approx 4/k_f{}^2$, where k_f is the maximum momentum in the Fermi gas. This result is given in Fig. 3 together with the Gaussian.

The second method which we use makes use to a greater extent of knowledge of the scattering cross sections. We go back to the original definition of $\chi_{lm}(r)$ and write

$$g(\mathbf{p})=\frac{A}{v}\frac{2}{A^2}\sum_{lm}\frac{1}{(2\pi)^3}\left|\int e^{-i\mathbf{p}\cdot\mathbf{r}}\left(r\left|1+\frac{1}{e}t\right|r'\right)\times\phi_{lm}(r')dr'dr\right|^2.\quad(49)$$

Using a relative momentum·plane wave function for ϕ_{lm}, performing the integrations, and making use of the fact that the energy denominator "e" is simply $E-p^2/M$ if we neglect interactions of the final particles, we find

$$g(\mathbf{p})=\frac{A}{v}\frac{2}{A^2}\sum_{lm}\frac{1}{(2\pi)^3}\left[(\mathbf{p}|t|\mathbf{k}_{lm})^2/(E-p^2/M)\right]^2.\quad(50)$$

This can be evaluated in a very simple way if we assume that $(\mathbf{p}|t|\mathbf{k}_{lm})$ depends only on the momentum difference (as it would in Born approximation for example) and determine it experimentally. Making this assumption and making use of the relation between t and the scattering cross section,

$$t^2(\mathbf{p}-\mathbf{k}_{lm})=\left(\frac{4\pi}{M}\right)^2\frac{d\sigma(\mathbf{p}-\mathbf{k}_{lm})}{d\Omega},\quad(51)$$

we find

$$g(\mathbf{p})=\frac{A}{v}\frac{1}{(2\pi)^3}\left(\frac{4\pi}{p^2}\right)^2\frac{d\sigma(\mathbf{p})}{d\Omega},\quad(52)$$

where we have made use of $p^2/M\gg E$ and $k_{lm}\ll p$ to carry out the summation. To evaluate this result, we take the average differential scattering cross section at 90 degrees in the center-of-mass system for neutron-proton and proton-proton collisions, choosing an energy which gives the correct momentum transfer. The result is given in Fig. 3.

It is apparent that either of the methods we have used gives essentially the same results and that the agreement with the empirically derived Gaussian distribution is satisfactory over the energy range of primary interest. Therefore we can, as emphasized in the first part of this section, conclude that these methods will give correct order-of-magnitude predictions for the high-energy phenomena we have described.

V. CONCLUSIONS

We have analyzed evidence derived from a variety of high-energy experiments which has bearing on the problem of nuclear structure. This evidence is particularly significant since it is for these (or similar) processes that the possible departure of the nuclear ground-state wave function from an independent-particle wave function is most apparent. The result predicted uniformly by the group of quite diverse experiments which we have examined is that the nuclear ground-state wave function must have a very marked admixture of high-momentum components and hence must depart quite appreciably from an independent-particle-model wave function. Consequently it follows that the usual assumptions of the shell-model theory of the nucleus, that the particles move independently in a uniform potential, cannot be other than very approximately correct.

To investigate quantitatively the general conclusion drawn from experiment, we have made use of methods recently developed by us in other studies of the nucleus. These methods lead to a nuclear model which appears to agree well with many general details of the structure of the nucleus and also with the detailed properties predicted by the shell model, but does not assume the existence of the independent nucleon motion. An essential assumption of this method is that the nuclear forces acting between nucleons in dense nuclear matter are still very nearly the same as those acting between free nucleons and hence very strong and short ranged. An immediate consequence of this assumption is that the presence of marked correlation effects in the nuclear wave function is to be expected. Conversely, the experimental observation of such correlation effects implies directly that the strong two-body forces have not been "damped out" by nonlinear effects or cancelled by many-body effects.

In applying this nuclear theory to the study of the high-energy phenomena, we have made several simplifying assumptions to bring the theory into easily manageable form. The most important assumption made is that in the very-high-momentum region the effects which interest us are primarily due to the intimate association of pairs of nucleons. In this form our methods are analogous to those used by Heidmann and by Levinger although their origin and interpretation is rather different. The final expressions for the momentum distribution have been evaluated by making use of two quite different approximations which give results in reasonable agreement with each other and with the momentum distribution derived from experiment. It is to be emphasized that the theoretical predictions are very sensitive to the choice of the two-body interaction, which we have assumed to be identical with that acting on free particles. Thus we have shown not only that appreciable correlation effects are present but also that the presence of other nucleons in the dense nuclear matter cannot appreciably modify the two-body interactions.

In conclusion we would like to remark that although we have in this paper emphasized the departure of the nuclear wave function from an independent nuclear wave function, the great importance of the departure is manifested only in the specific high-energy processes (or similar cases) we have considered. As we have pointed out in other papers, the effects are not nearly so pronounced in many low-energy phenomena where the independent particles of the shell model can be identified more closely with nucleons.

Experimental Test of Parity Conservation in Beta Decay*

C. S. Wu, *Columbia University, New York, New York*

AND

E. Ambler, R. W. Hayward, D. D. Hoppes, and R. P. Hudson, *National Bureau of Standards, Washington, D. C.*

(Received January 15, 1957)

In a recent paper[1] on the question of parity in weak interactions, Lee and Yang critically surveyed the experimental information concerning this question and reached the conclusion that there is no existing evidence either to support or to refute parity conservation in weak interactions. They proposed a number of experiments on beta decays and hyperon and meson decays which would provide the necessary evidence for parity conservation or nonconservation. In beta decay, one could measure the angular distribution of the electrons coming from beta decays of polarized nuclei. If an asymmetry in the distribution between θ and $180°-\theta$ (where θ is the angle between the orientation of the parent nuclei and the momentum of the electrons) is observed, it provides unequivocal proof that parity is not conserved in beta decay. This asymmetry effect has been observed in the case of oriented Co^{60}.

It has been known for some time that Co^{60} nuclei can be polarized by the Rose-Gorter method in cerium magnesium (cobalt) nitrate, and the degree of polarization detected by measuring the anisotropy of the succeeding gamma rays.[2] To apply this technique to the present problem, two major difficulties had to be overcome. The beta-particle counter should be placed *inside* the demagnetization cryostat, and the radioactive nuclei must be located in a *thin surface* layer and polarized. The schematic diagram of the cryostat is shown in Fig. 1.

To detect beta particles, a thin anthracene crystal $\frac{3}{8}$ in. in diameter$\times\frac{1}{16}$ in. thick is located inside the vacuum chamber about 2 cm above the Co^{60} source. The scintillations are transmitted through a glass window and a Lucite light pipe 4 feet long to a photomultiplier (6292) which is located at the top of the cryostat. The Lucite head is machined to a logarithmic spiral shape for maximum light collection. Under this condition, the Cs^{137} conversion line (624 kev) still retains a resolution of 17%. The stability of the beta counter was carefully checked for any magnetic or temperature effects and none were found. To measure the amount of polarization of Co^{60}, two additional NaI gamma scintillation counters were installed, one in the equatorial plane and one near the polar position. The observed gamma-ray anisotropy was used as a measure of polarization, and, effectively, temperature. The bulk susceptibility was also monitored but this is of secondary significance due to surface heating effects, and the gamma-ray anisotropy alone provides a reliable measure of nuclear polarization. Specimens were made by taking good single crystals of cerium magnesium nitrate and growing on the upper surface only an additional crystalline layer containing Co^{60}. One might point out here that since the allowed beta decay of Co^{60} involves a change of spin of

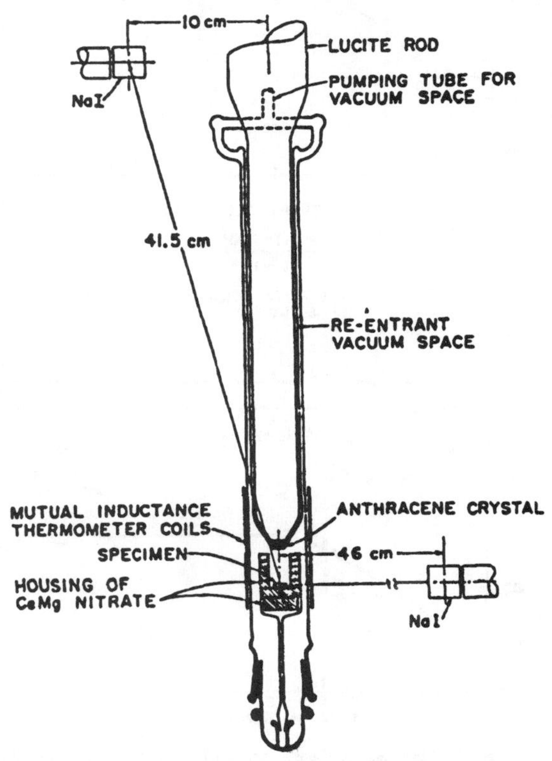

Fig. 1. Schematic drawing of the lower part of the cryostat.

one unit and no change of parity, it can be given only by the Gamow-Teller interaction. This is almost imperative for this experiment. The thickness of the radioactive layer used was about 0.002 inch and contained a few microcuries of activity. Upon demagnetization, the magnet is opened and a vertical solenoid is raised around the lower part of the cryostat. The whole process takes about 20 sec. The beta and gamma counting is then started. The beta pulses are analyzed on a 10-channel pulse-height analyzer with a counting interval of 1 minute, and a recording interval of about 40 seconds. The two gamma counters are biased to accept only the pulses from the photopeaks in order to discriminate against pulses from Compton scattering.

A large beta asymmetry was observed. In Fig. 2 we have plotted the gamma anisotropy and beta asymmetry *vs* time for polarizing field pointing up and pointing down. The time for disappearance of the beta asymmetry coincides well with that of gamma anisotropy. The warm-up time is generally about 6 minutes, and the warm counting rates are independent of the field direction. The observed beta asymmetry does not change sign with reversal of the direction of the demagnetization field, indicating that it is not caused by remanent magnetization in the sample.

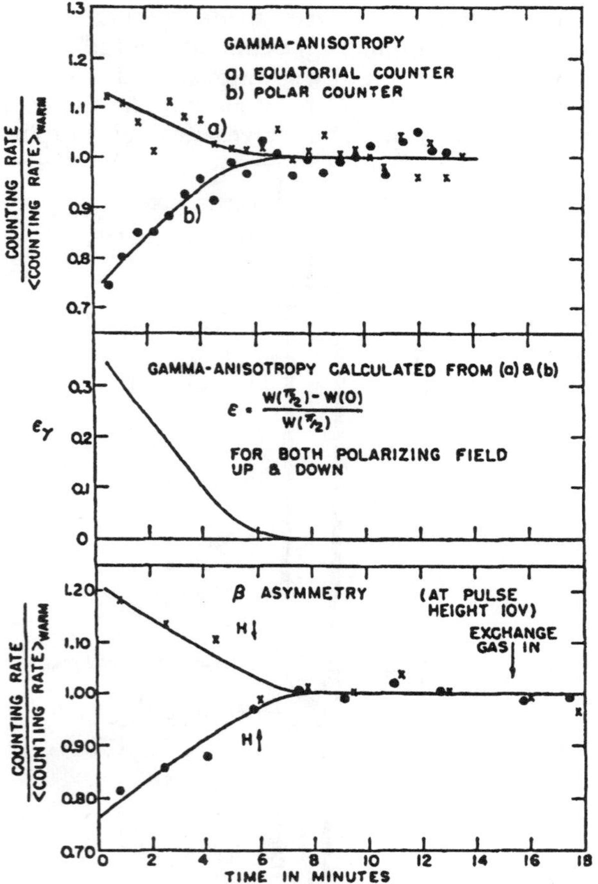

Fig. 2. Gamma anisotropy and beta asymmetry for polarizing field pointing up and pointing down.

The sign of the asymmetry coefficient, α, is negative, that is, the emission of beta particles is more favored in the direction opposite to that of the nuclear spin. This naturally implies that the sign for C_T and C_T' (parity conserved and parity not conserved) must be opposite. The exact evaluation of α is difficult because of the many effects involved. The lower limit of α can be estimated roughly, however, from the observed value of asymmetry corrected for backscattering. At velocity $v/c\approx 0.6$, the value of α is about 0.4. The value of $\langle I_z\rangle/I$ can be calculated from the observed anisotropy of the gamma radiation to be about 0.6. These two quantities give the lower limit of the asymmetry parameter $\beta(\alpha=\beta\langle I_z\rangle/I)$ approximately equal to 0.7. In order to evaluate α accurately, many supplementary experiments must be carried out to determine the various correction factors. It is estimated here only to show the large asymmetry effect. According to Lee and Yang[3] the present experiment indicates not only that conservation of parity is violated but also that invariance under charge conjugation is violated.[4] Furthermore, the invariance under time reversal can also be decided from the momentum dependence of the asymmetry parameter β. This effect will be studied later.

The double nitrate cooling salt has a highly anisotropic g value. If the symmetry axis of a crystal is not set parallel to the polarizing field, a small magnetic field will be produced perpendicular to the latter. To check whether the beta asymmetry could be caused by such a magnetic field distortion, we allowed a drop of $CoCl_2$ solution to dry on a thin plastic disk and cemented the disk to the bottom of the same housing. In this way the cobalt nuclei should not be cooled sufficiently to produce an appreciable nuclear polarization, whereas the housing will behave as before. The large beta asymmetry was not observed. Furthermore, to investigate possible internal magnetic effects on the paths of the electrons as they find their way to the surface of the crystal, we prepared another source by rubbing $CoCl_2$ solution on the surface of the cooling salt until a reasonable amount of the crystal was dissolved. We then allowed the solution to dry. No beta asymmetry was observed with this specimen.

More rigorous experimental checks are being initiated, but in view of the important implications of these observations, we report them now in the hope that they may stimulate and encourage further experimental investigations on the parity question in either beta or hyperon and meson decays.

The inspiring discussions held with Professor T. D. Lee and Professor C. N. Yang by one of us (C. S. Wu) are gratefully acknowledged.

* Work partially supported by the U. S. Atomic Energy Commission.

[1] T. D. Lee and C. N. Yang, Phys. Rev. 104, 254 (1956).

[2] Ambler, Grace, Halban, Kurti, Durand, and Johnson, Phil. Mag. 44, 216 (1953).

[3] Lee, Oehme, and Yang, Phys. Rev. (to be published).

FLUCTUATIONS OF NUCLEAR CROSS SECTIONS IN THE "CONTINUUM" REGION*

Torleif Ericson†
Lawrence Radiation Laboratory, University of California, Berkeley, California
(Received August 29, 1960; revised manuscript received October 14, 1960)

The assumption that the matrix elements of the compound nucleus are randomly distributed with respect to phase and magnitude has in recent years been very successful for the understanding of the fluctuations of the partial widths and the distribution of the level spacings of slow-neutron resonances.[1-3]

The purpose of this Letter is to examine the nontrivial consequences of the statistical assumption in the so-called "continuum" region, in which a large number of compound states overlap owing to the short lifetime of the compound nucleus. It is generally concluded that the excitation of a large number of intermediate compound states automatically implies that cross sections are smooth functions of the energy and that angular distributions are symmetrical around 90° to the beam direction. It will be shown that the proper statistical prediction for experiments performed with good energy resolution in the incident beam is that (1) partial cross sections and angular distributions fluctuate even though a large number of intermediate states are excited, (2) the formation and decay of the compound nucleus are independent only on the average, and (3) the fluctuations can be used to determine the lifetime of the compound nucleus in the "continuum" region.

Consider the compound-nucleus reaction proceeding from the initial state $|\alpha\rangle-$, i.e., the target nucleus and the incident wave—to the final state $|\alpha'\rangle$, a particular state of the final nucleus and the corresponding emitted wave. We assume the experiment to be performed with infinitely good energy resolution in the incident beam. The scattering matrix $S_{\alpha\alpha'}$ can be divided into two parts, one, S_α, leading into the compound nucleus, and one, $S_{\alpha'}$, leading out of the compound nucleus. The intermediate compound states $|i\rangle$ of energy E_i are excited with probability amplitudes $f(E,E_i)$ which are approximately of the form

$$f(E,E_i)\propto\frac{1}{(E-E_i)+i\Gamma/2}. \tag{1}$$

The "width" Γ depends in principle on the intermediate state $|i\rangle$. At high excitation this dependence will be weak, because Γ will be a sum of a large number of partial widths, of which no single one dominates. This situation should be realized at some 3-4 Mev above neutron binding energy in heavy nuclei, but is not very well realized at lower excitation. As we discuss the case of high excitation, we will neglect the fluctuations in Γ and take it to be a constant. The "width" Γ is related to the lifetime τ of the compound nucleus by $\Gamma=\hbar/\tau$ by the uncertainty principle. Equation (1) expresses that the compound states within a region of the order of Γ are excited simultaneously and must be treated coherently; we therefore call Γ the coherence energy. The reaction cross section $\sigma_{\alpha\alpha'}(E)$ can be written

$$\sigma_{\alpha\alpha'}(E) \propto |\sum_i \langle\alpha|S_\alpha|i\rangle f(E,E_i)\langle i|S_{\alpha'}|\alpha'\rangle|^2. \quad (2)$$

The matrix elements $\langle\alpha|S_\alpha|i\rangle$ and $\langle i|S_{\alpha'}|\alpha'\rangle$, and consequently their product, have random phases. If the coherence energy Γ encompasses a large number of intermediate states $|i\rangle$, the sum in Eq. (2) consists of a large number of terms of random phases and becomes a random number, the real and imaginary part of which have Gaussian distributions. Therefore, if transitions to different final states are compared under otherwise identical conditions, the transition-matrix elements are independently random, because the factors $\langle i|S_{\alpha'}|\alpha'\rangle$ are uncorrelated and have uncorrelated phases. The partial cross sections $\sigma_{\alpha\alpha'}(E)$ therefore fluctuate essentially like neutron widths in spite of the large number of compound states excited.

Furthermore, intermediate states $|i\rangle$ are excited essentially only within the coherence energy Γ. If the incident energy is changed by an amount much larger than Γ, the intermediate states are entirely different. The matrix element, thus the cross section, is different from its previous value, in general. On the other hand, if the energy change is small compared with Γ, essentially the same states are excited; the matrix element and cross section are practically unchanged. As a measure of the region of energy over which the correlation in the cross section is important, we evaluate the mean value over E of $\sigma_{\alpha\alpha'}(E)\sigma_{\alpha\alpha'}(E+\epsilon)$ from Eq. (2):

$$\langle\sigma_{\alpha\alpha'}(E)\sigma_{\alpha\alpha'}(E+\epsilon)\rangle_{\text{av}} = \langle\sigma_{\alpha\alpha'}(E)\rangle_{\text{av}}^2\left[1+\frac{1}{1+(\epsilon/\Gamma)^2}\right]. \quad (3)$$

Therefore we conclude that a partial reaction cross section will fluctuate as a function of incident energy with a typical period of the order of the coherence energy Γ. Because Γ is directly related to the compound-nucleus lifetime, these fluctuations, in principle, provide a means for measuring the extremely short lifetime of highly excited nuclei.

We point out that the usual statement of independence of formation and decay of the compound nucleus is not valid in this type of experiment: The fluctuations do not necessarily occur at the same excitation energy, if the same compound nucleus is formed by different means. In the "continuum" region this independence is a consequence of averaging over many residual states, or over an energy interval much larger than the coherence energy.

A more exact discussion should include angular momentum. The main results remain unchanged. For the integrated partial cross sections, states of opposite parity contribute independently, slightly reducing the amplitude of the fluctuations. The partial angular distribution and polarization fluctuate as functions of the incident energy over energy regions of the order of Γ. Only after an average has been performed are the ordinary statements, of symmetry about 90° to the beam direction and no polarization, valid.

The total reaction cross section is usually a sum of a large number of fluctuating partial cross sections $\sigma_{\alpha\alpha'}$ and it will therefore usually have fluctuations of small amplitude only. When only few exit channels are available, the fluctuations are appreciable and occur also in the total cross section. Fluctuations in the total cross section have been observed by Cranberg[4] with high-resolution neutrons on Ti and Fe. At 2.5-Mev incident neutron energy and 2-kev resolution the total cross section exhibits a fine structure which, for Fe, has a half-width of 5 kev, corresponding to an approximate lifetime of 10^{-19} sec. The expected spacing of levels of spin 1/2 is indicated to be of the order of a few kev by slow-neutron resonances; the spacings of all excited states are considerably smaller.

---

*This work was done under the auspices of the U. S. Atomic Energy Commission.

†Present address: CERN, Geneva, Switzerland, on leave from the Institute of Theoretical Physics, Lund, Sweden.

[1]C. E. Porter and R. G. Thomas, Phys. Rev. 104, 483 (1956).

[2]E. P. Wigner, in Proceedings of the Conference on Neutron Physics by Time-of-Flight, Gatlinburg, Tennessee, 1956 [Oak Ridge National Laboratory Report ORNL-2309 (unpublished)].

[3]L. Landau and Y. Smorodinsky, lectures on the Theory of the Atomic Nucleus, Moscow, 1956 (unpublished).

[4]Dr. L. Cranberg (private communication). See also L. Cranberg et al., in Proceedings of the Conference on Neutron Physics by Time-of-Flight, Gatlinburg, Tennessee, 1956 [Oak Ridge National Laboratory Report ORNL-2309 (unpublished)], p. 148.

EVIDENCE FOR CHARGE INDEPENDENCE IN MEDIUM WEIGHT NUCLEI*

J. D. Anderson and C. Wong
Lawrence Radiation Laboratory, University of California, Livermore, California
(Received July 26, 1961)

The importance of isotopic spin considerations ($\Delta T = 0$) in "mirror nuclei" (p,n) reactions has been pointed out by Bloom et al.[1] If one goes to nonmirror nuclei such as V^{51}, Cr^{51}, the (p,n) reaction is assumed to go as follows: The incoming proton reacts with an $f_{7/2}$ neutron, exchanges its charge, and is emitted as a neutron. In the initial state there are eight $f_{7/2}$ neutrons available and three $f_{7/2}$ protons. Since, by definition of isotopic spin state (charge independence), all the nuclear interactions within the initial and final nucleus are the same, the Q for the (p,n) reaction between V^{51} ($T = \frac{5}{2}$) and its analog state in Cr^{51} is the Coulomb energy difference. It is the purpose of this Letter to point out that direct-reaction neutrons from the (p,n) reaction on medium weight nuclei do indeed leave the residual nucleus in a state which is the analog of the ground state of the bombarded nucleus.

It was pointed out previously[2] that the energy and angular distribution of the continuum neutrons from the $V^{51}(p,n)Cr^{51}$ reaction are in agreement with predictions of the statistical model of the compound nucleus for proton bombarding energies up to 8 Mev. Using time-of-flight techniques[2,3] to measure the neutron spectra, direct-reaction neutrons are observed for proton bombarding energies above 10 Mev. These neutrons do not come from groups leaving Cr^{51} in low-lying states as might be expected from a simple Born approximation calculation in which one considers only the radial overlap of the wave functions describing the initial and final states. These neutrons are observed to come from a level (or levels) in Cr^{51} at 6.5 Mev.

Neutron spectra at $\theta_L = 23°$ have been measured for proton energies between 9 and 13 Mev using a self-supporting 8-mg/cm² vanadium foil and a 10-meter flight path. The neutron spectra for three incident proton energies are shown plotted in Fig. 1. The compound statistical model predicts that if a sufficient number of nuclear levels are involved in both the compound nucleus and the residual nucleus, the energy distribution of neutrons emitted is given by

$$P(E)dE = KE\omega(E_{\text{excit}})\sigma_c(E)dE,$$

where $\omega(E_{\text{excit}})$ is the level density of the residual nucleus at its excitation energy, which is determined by the incident proton energy, the Q value of the reaction, and the emitted neutron energy; $\sigma_c(E)$ is the reaction cross section for the inverse reaction between the excited residual nucleus and a neutron of energy E; and the constant K is a function of the incident proton energy. Assuming that $\sigma_c(E)$ is a slowly varying function of neutron energy, the energy dependence of the level density of the residual nucleus is proportional to $P(E)/E$. The relative neutron spectra transformed to the center-of mass system and divided by the energy of the emitted neutron in the c.m. system are plotted as a function of the excitation of the residual nucleus in Fig. 2. From Fig. 2 it is clear that $P(E)/E$ (allowing for the change in energy resolution as a function of neutron energy) is only a function of the excitation of the residual nucleus up to $E_{\text{excit}} = 6.5$ Mev, indicating the assumption of compound nucleus formation in this region is valid. However, a fluctu-

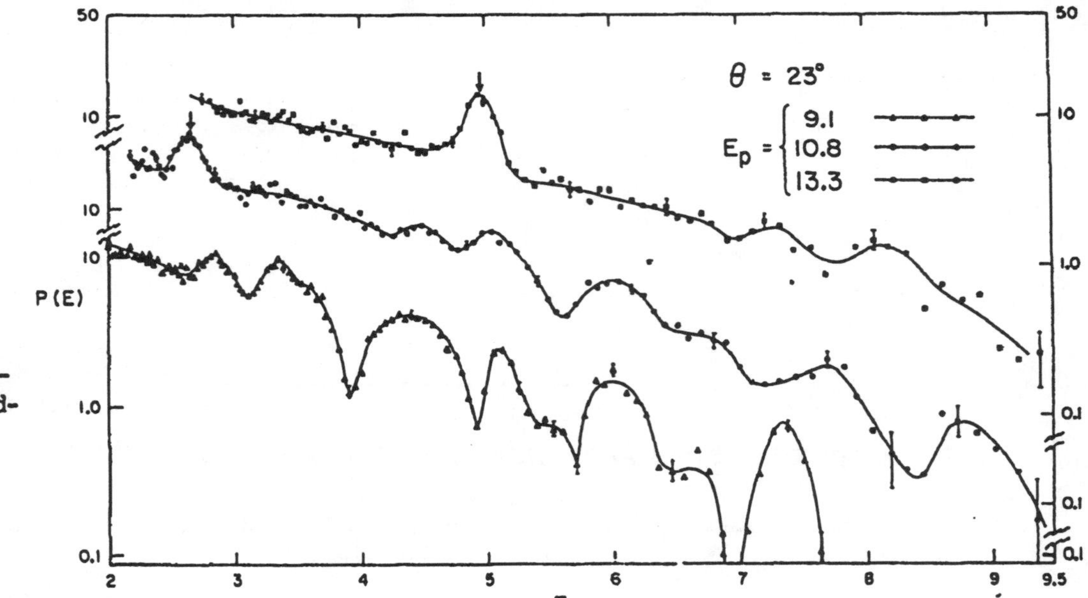

FIG. 1. Neutron spectra from proton bombardment of V^{51}.

ation in level density of a factor of 2 to 3 at $E_{\text{excit}} = 6.5$ Mev with a hundred levels being involved is incompatible with this level (or levels) being populated by a compound nucleus. As one increases the proton bombarding energy the continuum neutron cross section ($E_{\text{excit}} = 6.5$ Mev) decreases from 17 mb/sr Mev at $E_p = 10.8$ Mev to 4.5 mb/sr Mev at $E_p = 13.3$ Mev, while the cross section for the "level" remains at 2 mb/sr. The decrease in cross section for the continuum neutrons is in excellent agreement with our previous measurement of the Cr^{51} level density[2,4] and is to be expected on the basis of a compound nucleus while the direct-reaction cross section is expected to vary only slowly with energy which is borne out by the present experimental data. Thus it is concluded that the level at 6.5 Mev in Cr^{51} is excited via a direct reaction.

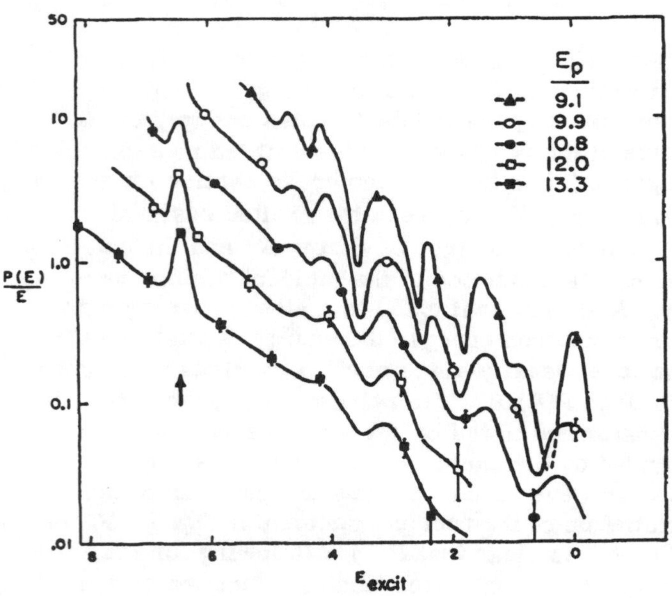

FIG. 2. The neutron spectra transformed to the center-of-mass system and divided by the energy of the emitted neutron are plotted as a function of the excitation of the residual nucleus Cr^{51}. The arrow denotes the position of the analog state.

If one calculates the V^{51}-Cr^{51} Coulomb energy difference using the results of Swamy and Green,[5] one obtains $\Delta E_c = 8.2$ Mev for $R = 1.25 A^{1/3}$ f. This is in excellent agreement with our measured value of 8.0 ± 0.2 Mev. [This is the (p,n) ground state $Q(-1.54$ Mev) plus the residual excitation in Cr^{51} (6.5 Mev).] Our conclusion is that the level in Cr^{51} at 6.5-Mev excitation is the analog of the ground state of V^{51} and is therefore a $T = \frac{5}{2}$ state.

Additional evidence for analog states is found in the $Fe^{56}(p,n)Co^{56}$ and $Co^{59}(p,n)Ni^{59}$ neutron spectra. Our Q values for the (p,n) reaction to the analog states are again in excellent agreement (see Table I) with the Coulomb energy differences calculated from reference 5.

Further experiments involving additional target nuclei and measurements of the angular distributions of the direct-reaction group are currently under way.

It is a pleasure to acknowledge the assistance of

Table I. Q values for the (p,n) reactions to the analog state.

	Q (measured) (Mev)	ΔE_C (calculated) (Mev)
$V^{51}(p,n)Cr^{51}$ [a]	8.0 ±0.2	8.2
$Fe^{56}(p,n)Co^{56}$ [a]	8.9 ±0.2	9.0
$Co^{59}(p,n)Ni^{59}$ [a]	9.1 ±0.2	9.2

[a]Denotes isotopic spin analog state.

J. W. McClure and B. D. Walker in obtaining the data and of S. D. Bloom in its interpretation.

*This work was done under the auspices of the U. S. Atomic Energy Commission.

[1]S. D. Bloom, N. K. Glendenning, and S. A. Moszkowski, Phys. Rev. Letters 3, 98 (1959).

[2]R. D. Albert, J. D. Anderson, and C. Wong, Phys Rev. 120, 2149 (1960).

[3]C. Wong, J. D. Anderson, C. C. Gardner, J. W. McClure, and M. P. Nakada, Phys. Rev. 116, 164 (1959).

[4]J. D. Anderson, C. Wong, J. W. McClure, and B. D. Walker (to be published).

[5]N. Swamy and A. Green, Phys. Rev. 119, 1719 (1958).

Chapter 6
STATISTICAL PHYSICS
JOEL L. LEBOWITZ

Introduction

Papers Reprinted in Book

Lars Onsager (1903-1976). (Courtesy of Joel L. Lebowitz, Rutgers University.)

Joel L. Lebowitz is the George William Hill Professor of Mathematics and Physics and Director of the Center for Mathematical Sciences Research at Rutgers University. He came to the United States in 1946 and was a student of Melba Phillips, Peter Bergmann, and Lars Onsager. His work has been mainly in statistical mechanics with an outsider's interest in foundational questions. His awards include the A. Cressy Morrison Award in Natural Sciences from the New York Academy of Sciences and the Boltzmann Medal from the IUPAP Committee on Thermodynamics and Statistical Physics.

STATISTICAL PHYSICS

Amedeo Avogadro (1776–1856). (Courtesy of N.B.S. Archives, AIP Emilio Segrè Visual Archives.)

The Physical Review is a relative latecomer to the world physics scene. Although it started almost 120 years after the founding of the United States, its clientele was rather small in the beginning. It was not until the 1930s, when scientists (and others) with "wrong" ancestors and beliefs became hunted species in certain parts of Europe, that physics and other sciences began their explosive stage of growth in the U.S. With this growth came the flowering of *The Physical Review*. By the 1940s *The Physical Review* was the leading physics journal in the world—a position it and its offspring, *Physical Review Letters*, maintain to this day. Dyson[1] describes Vol. 73 of *The Physical Review*, which contains the issues from January to June 1948, his first year in the United States, as follows: "Almost every paper in it is interesting, and many of them are memorable. In 1948 issues of the journal were thin enough to be read from cover to cover. Many of us did just that... . The paper that impressed me most in 1948 and still impresses me today is entitled 'Relaxation Effects in Nuclear Magnetic Resonance Absorption,' a monumental piece of work, 34 pages long, by Bloembergen, Purcell, and Pound... . In the same volume of *The Physical Review* are many other wonderful papers on the most diverse subjects: Alpher, Bethe, and Gamow on the origin of chemical elements; Wataghin on the formation of chemical elements inside stars; E. Teller on the change of physical con-

Edward Teller (1908–) and **Enrico Fermi (1901–1954).** (Courtesy of AIP Emilio Segrè Visual Archives.)

stants; Lewis, Oppenheimer, and Wouthuysen on the multiple production of mesons; Foley and Kusch on the experimental discovery of the anomalous magnetic moment of the electron; Schwinger on the theoretical explanation of the anomalous moment; and Dirac on the quantum theory of localizable dynamical systems. This was a vintage year for historic papers and I mention these seven just to give the flavor of what physicists were doing in the early post-war years."

Dyson's description illustrates the difficulty, nay impossibility, of selecting a seminal sample from among 90 years of *The Physical Review* papers. It should therefore be clear that this extremely space-constrained collection is not a scholarly compendium of either the "most important," "most original," "most clever," "most memorable," or "most anything" papers that appeared in *The Physical Review* or *Physical Review Letters* prior to 1984. It is rather a very small sample of such papers collected from responses to the solicitation letter sent out by H. Henry Stroke to a small sample of physicists, from conversations with a small sample of friends, from references in a small sample of books or reviews I happened to have on my shelves, from some very time-limited sample visits to the library, and finally from inherently faulty personal recollections, judgments, and prejudices. In particular, I believe that mathematically rigorous results are, in appropriate situations, not just "decorations on the cake"; they are essential for getting the physics right. Given the above, I apologize in advance for all sins of omission[a] and commission and only hope

[a] *In many cases some of the most basic papers in a given area were omitted just because of their length. For example, Onsager's solution of the Ising model is represented in the book by an abstract. In other cases a whole area had to be excluded, e.g., Kac potentials, one-dimensional quantum systems. Many of these can be found in the recent excellent compilation with comments by Daniel Mattis.*[2]

Lars Onsager (1903-1976), N. Meyer, Nico van Kampen (1921-), and Julian H. Gibbs (1924-1983). (Courtesy of *Physics Today* Collection, AIP Emilio Segrè Visual Archives.)

that the selection gives the reader a feeling of how exciting good science can be.

Having made my disclaimer, I take this opportunity to give a brief, highly personal, selective overview of developments in nonequilibrium and equilibrium statistical mechanics in order to put the works contained here in some context.

NONEQUILIBRIUM

I am a bit younger and much less precocious than Dyson, so although I came to the United States a year or two before him, I did not start reading *The Physical Review* until 1953, when I was a graduate student of Peter Bergmann's at Syracuse University. It was also in that year that I first met Lars Onsager, whose 1931 paper "Reciprocal Relations in Irreversible Processes II" is the earliest one in this section of the book. (Note that this article was submitted on 9 November and published in the 15 December issue of that year.)

Onsager's paper is the second of two works (the first one was too long for the book and is included in the CD) devoted to the deep underlying symmetry properties of the matrix connecting thermodynamic fluxes and forces. While particular examples of such symmetries had been known for a long time, it was Onsager who first understood that they are a general feature of transport in systems close to equilibrium. These symmetries are a direct consequence of the reversibility of the laws governing the time evolution of the microscopic constituents of macroscopic matter and the additional (mild) hypothesis that the time evolution of a macroscopic system following its preparation in a macrostate out of equilibrium is the same as that following a spontaneous fluctuation from equilibrium to that macrostate. The "Onsager relations" regarding the symmetry of the transport coefficients $L_{ij} = L_{ji}$ are, next to the second law of thermodynamics, the only known results about nonequilibrium phenomena that have a generality resembling that of equilibrium thermodynamics. They are in fact related to Boltzmann's probabilistic formulation of the second law connecting macroscopic entropy with microscopic phase space volume. The connection between microscopic and macroscopic evolutions and the role that the large number of microscopic degrees of freedom play in determining universal features of macroscopic behavior was further developed by Einstein in his beautiful quantitative theory of Brownian motion and in his relation between equilibrium fluctuations and macroscopic entropy. The works of Boltzmann and Einstein are the direct forerunners of Onsager's work.

The fluctuation-dissipation theorem of Callen and Welton[3] extends the Boltzmann-Einstein-Onsager connection between spontaneous fluctuations in equilibrium systems and the macroscopic response of the system when not in equilibrium. By considering the response to an externally imposed perturbation of the Hamiltonian describing the time evolution of the ensemble density in the Gibbs formalism of statistical mechanics, Callen and Welton were led to a general expression for (some) frequency-dependent transport coefficients in terms of the Fourier transform of the time evolution of equilibrium fluctuations. The same circle of ideas leads also to the Green-Kubo[4] formulas for transport coefficients and to the Onsager-Machlup[5] formulation of probabilities for fluctuations in the time evolution of macroscopic systems.

The above general results apply to linear transport in systems close to equilibrium. Their extension to systems far from equilibrium is still a very active research area. We should not be surprised, however, if it turns out that such general formulations may not be appropriate for some of the most interesting phenomena in such systems, including those most relevant to us, e.g., us, in our universe. These phenomena may not be capturable by general formulas; to quote Dyson again, "God is in the details." Still, there have been major advances in our understanding of the potentialities inherent in nonlinear dynamical systems. These include the discovery of solitons in integrable systems represented here by the paper of Zabusky and Kruskal.[6] Another discovery, or more properly a belated realization by the physics (as opposed to the mathematics) community, was that most dynamical systems, even those with only a few degrees of freedom, are *very far* from integrable—rather, the trajectories describing their time evolution are subject to sensitive dependence on initial conditions, parameters, perturbations; in short, they exhibit deterministic chaos.[7]

To put the discussion of chaos into this centennial framework, one should at least mention the seminal work of Poincaré. It was Poincaré who first realized the deep difference between linear and nonlinear evolutions with the inherent tendency for complex, apparently chaotic, behavior of the latter. Following Poincaré, the torch of nonlinearity was mainly kept alive in what is now the Former Soviet Union (FSU). In the FSU, following a tradition es-

Ludwig E. Boltzmann (1844–1906). "Boltzmann at the Blackboard" cartoon by Dr. K. Prizbram. (Courtesy of University of Vienna, AIP Emilio Segrè Visual Archives.)

J. Willard Gibbs (1839–1903). (Courtesy of Burndy Library, AIP Emilio Segrè Visual Archives.)

Yakov B. Sinai (1935–), Marta Aizenman, and **Joel L. Lebowitz (1930-).** (Courtesy of Joel L. Lebowitz, Rutgers University.)

tablished by Lyapunov, the work of mathematicians such as Kolmogorov, Bogolyubov, Rokhlin, and members of their schools, helped develop many of the fundamental ideas of the modern analytical theory of nonlinear dynamical systems. The works of Arnold, Chirikov, and Sinai from the FSU and of Birkhoff, Smale, Siegel, and Moser from the West have been particularly relevant for our understanding of deterministic chaos. (I never heard of Poincaré's work in any of my courses and first learned of it from Prigogine and Ford.)

The analytic insights of Poincaré were subsequently visualized clearly through the use of that modern (it occurred in my lifetime) invention, the computer, by Hénon and Heiles, reprinted in Ref. 7. It was the computer too through which Alder and Wainright[8] discovered the first example of the now ubiquitous "long time tails" of correlations present in many-body dynamical systems. Before the advent of computer simulations, there was a general belief (if my memory serves me right) that correlations typically showed exponential decay in time—the power law decay of conserved densities satisfying diffusion equations being truly exceptional. One of the first uses of the computer in statistical mechanics was the computation by Fermi, Pasta, and Ulam[9] (their famous Los Alamos report was never published in any journal) of the time evolution of a chain of 32 oscillators nonlinearly coupled via a cubic and/or quartic Hamiltonian. They found to their surprise that there was no equipartition of energy among the different (linear) normal modes on the time scale of their computation.[b] This was later understood to be a consequence of the celebrated Kolmogorov-Arnold-Moser (KAM) theorem.[10] Another computer

[b] *Note that even smooth trajectories can give rise to ergodic behavior of projections, as in the Lissajous figure generated by two incommensurate frequencies. For systems with many degrees of freedom, any nonlinearity was expected to produce ergodicity or at least equipartition.*

Berni J. Alder (1925–), Mary Ann Marrsigh, and **Tom Wainwright.** (Courtesy of AIP Emilio Segrè Visual Archives.)

first from Los Alamos was the development of Monte Carlo methods by Metropolis, A. W. Rosenbluth, M. N. Rosenbluth, A. H. Teller, and E. Teller.[11] (The story has it that this was "cooked up" in the course of a dinner.)

Some important items missing from *The Physical Review* and therefore from this collection are the papers by Lorenz,[12] Ruelle and Takens,[13] and Feigenbaum.[14] Lorenz was the first to make clear the connection between the chaotic behavior of a dynamical system with a small number of degrees of freedom and the onset of turbulence in a driven fluid. Ruelle and Takens then showed that the onset of turbulence need not occur through an infinite sequence of bifurcations, as proposed by Landau, but can result directly from the interaction of three or more modes. The work of Ahlers and of Gollub and Swinney[15] beautifully confirms this. (I recall the excitement of going with Ruelle to Swinney's lab to watch the experiment.) Feigenbaum, aided by a very modest computer, in what is for me still one of the best examples of the good uses of computers in science, discovered the existence of a simple "universal" (but far from obvious) feature in a large class of dynamical systems: period doubling leading to chaos. One of the first experimental papers to confirm this scenario clearly was by Libchaber and Maurer.[16] Pomeau and Manneville[17] found still another route to the onset of turbulence, that of intermittency—a route also confirmed by experiment.

The situation is much less satisfactory with regard to fully developed turbulence, perhaps *the* central problem of contemporary classical statistical physics. While the past 100 years have certainly seen important theoretical work on this subject, including some deep insights by L. Richardson, Kolmogorov, Onsager, Kraichnan, and others (mostly published in other journals), a full understanding remains for the sesquicentennial celebration of *The Physical Review*. The celebrated Kolmogorov scaling law is represented in the book by a later independent short announcement by Onsager.[18] (Apparently Onsager had this result for some years, but the war delayed its publication.[19])

Mitchell Feigenbaum (1944–), on left. (Courtesy of Joel L. Lebowitz, Rutgers University.)

EQUILIBRIUM

Let me turn now to the simpler (but far from simple) equilibrium systems. Here too our story in this collection begins with Onsager. In an abstract of less than 70 words (which was in keeping with his generally sparing and precise utterances), Onsager announced that he had obtained in closed form the partition function of the ferromagnetic Ising model in two dimensions and found that "For an infinite crystal, the specific heat near $T = T_c$ is proportional to $-log|T - T_c|$." This astonishing result can be said to mark the beginning of the modern era of equilibrium statistical mechanics. The full 1944 paper describing this work is on the CD-ROM. Its precursor, the 1941 paper by Kramers and Wannier,[20] in which the exact value of T_c is found by duality, is also there.[a]

Jerry P. Gollub (1944–). (Courtesy of Jerry P. Gollub, Haverford College.)

T. D. Lee (1926–) and **C.-N. Yang (1922–).** (Courtesy of Alan W. Richards, AIP Emilio Segrè Visual Archives.)

The Ising model was the first many-body system with nontrivial interactions solved exactly, and even after so many years and so many alternative derivations it is still an astonishing intellectual feat. I can still remember from my first year of graduate school the admiration in Mark Kac's voice when speaking of Onsager's solution in a colloquium at Syracuse University in 1953. He described there his work with Berlin,[21] on the solution of the much simpler spherical model—"but the solution of the Ising model was a superheroic effort resisting all assaults until Onsager threw at it his heavy machinery and the problem collapsed under its weight." The formula for the spontaneous magnetization of the Ising model was announced by Onsager at a conference at Cornell in 1948 and repeated in Florence in 1949, but no hint of its origin was given.[19] This was supplied by Yang in 1952.[22]

Also in the same year Lee and Yang[23] proved a remarkable theorem about the zeros of the partition function of the ferromagnetic Ising model in an external magnetic field h: They must all be on the unit circle in the complex fugacity plane z [in suitable units $z = exp(\beta h)$]. The Lee and Yang paper was the companion paper to one by Yang and Lee[24] [which pointed out the relationship between phase transitions and zeros (singularities) in the (logarithm of the) grand-canonical partition function]. Real singularities such as phase transitions occur only when the size of the system becomes truly macroscopic (formally in the infinite volume or thermodynamic limit), when these zeros can approach the real positive z axis and give rise to nonanalyticities in the thermodynamic functions. (The whole subject of existence of the thermodynamic limit and the equivalence of ensembles in that limit is missing from this collection—the original papers by van Hove, van Kampen, Ruelle, Fisher, Dyson, and Lenard and others did not appear in *The Physical Review*; see Refs. 25 and 26.)

The Lee-Yang theorem about the location of the zeros implies that the free energy of the Ising model with ferromagnetic pair interactions, considered as a function of the temperature T and

Harry L. Swinney (1939–). (Photograph by Austin Holiday. Courtesy of Harry L. Swinney, University of Texas, Austin.)

magnetic field h, is an analytic function of h for $h \neq 0$. This is true in any dimension and extends also to continuous classical and even quantum spins with Ising-type symmetry, as long as there are only ferromagnetic pair interactions—see Asano.[27] Combining the Lee-Yang theorem with the Griffiths' (and other) inequalities, one can prove, cf. Ref. 25, that the magnetization is a real analytic function of h and T whenever $h \neq 0$, or $T > T_c$. For $T < T_c$ the dependence of the magnetization on the field is discontinuous at $h = 0$, and the Ising system then has exactly two "pure phases," with spontaneous magnetization $\pm m^*(T)$. There are many other beautiful exact results known and worth knowing about the Ising model, but space limitations have permitted only a very few examples on the CD; see Refs. 25 and 28 for many more.

The main reason for the continued interest in the two-dimensional ferromagnetic Ising model is that it is the paradigm of a spontaneous breaking of symmetry with an order parameter $m^*(T)$, whose emergence (for $h = 0$) at $T < T_c$ produces long-range correlations in the system. There are similar easily defined order parameters for the Ising antiferromagnet and the Heisenberg magnet, where the magnetization has a continuous symmetry, as well as for other phase transitions in which there is a classically describable symmetry breaking. The critical exponents and even the existence of a phase transition depend strongly of course on symmetry and dimension, as shown in particular in the work of Hohenberg and of Mermin and Wagner.[29]

Oliver Penrose (1929–) with his daughter, Rebecca, in 1964. (Photograph by Jean Penrose. Courtesy of Ann Benedeuce.)

One thing that was unclear in the early 1950s, however, was what plays the role of the order parameter or what quantity develops long-range correlations in the purely quantum-mechanical normal-to-superfluid transition in He^4 or in the superconducting transition. The answer came in the work of Penrose and Onsager,[30] who noted that the asymptotic value of the one-particle density matrix $\rho_1(\mathbf{r}\text{-}\mathbf{r}')$ can, for interacting bosons, be interpreted as the square of the macroscopic wave function of the condensate: a generalization to interacting systems of the Bose condensate in an ideal gas. Yang[31] extended this idea to the superconducting transition in interacting fermions, where it shows up as "off-diagonal long-range order" in the two-particle density matrix. (The seminal works of Landau and of Bogolyubov should also be noted here.)

Following Onsager's 1942 work there was a 25-year hiatus in obtaining exact solutions for nontrivial interacting systems. The next major breakthrough came when Lieb obtained the exact asymptotic number of configurations of a six-vertex model on the two-dimensional square lattice[32]—a two-dimensional version of a model introduced by Pauling to explain the residual entropy of ice. In this model the oxygen atoms sit at the vertices of the square lattice and there is one hydrogen atom on each bond, which can be in one of two positions (closer to one or the other of the two oxygens connected by the bond). There is a constraint, however, which forbids any oxygen to have more than two close hydrogens, and this makes the calculation of the number of configura-

Linus C. Pauling (1901–1994). (Courtesy of California Institute of Technology, AIP Emilio Segrè Visual Arhcives.)

Gregory H. Wannier (1911–). (Courtesy of *Physics Today* Collection, AIP Emilio Segrè Visual Archives.)

tions very nontrivial. The fact that the final answer found by Lieb for the square lattice, using a Bethe-Ansatz trick, has the simple form of (3/2)log(4/3) for the entropy per oxygen atom, is to me still a mystery. (I repeat my offer of 1967 of a bottle of champagne for a simple derivation.) Lieb's results, which include exact solutions of several other related models as well as of some one-dimensional quantum models,[2,28] were soon extended by Baxter.[33] While "doing his sums" on a boat returning from the U.S. to Australia, he obtained an exact solution of the eight-vertex model in which the constraint on the position of the hydrogen atoms is relaxed a bit, but there is an energy cost for different configurations. This work was the progenitor of several other exact solutions by Baxter and led in a natural way to the discovery of a whole class of solvable two-dimensional lattice models, including some with a continuously varying critical index; see Ref. 28 for a review. This presented a challenge to the prevailing orthodoxy of universality, which held that critical indices depend only on the spatial dimension and symmetry of the order parameter. The discrepancy was resolved by Kadanoff and Wegner.[34]

The story of critical-point phenomena and their universality is one of the continuing sagas of statistical mechanics spawned in part by Onsager's solution of the Ising model.[c] Prior to that it was generally accepted, despite some clear experimental evidence to the contrary, that the behavior in the vicinity of a critical point is at least qualitatively correctly described by mean field theory. These theories—van der Waals for fluids, Bragg-Williams for magnets, and Landau theory in general—give "superuniversal" clas-

[c] *It is worth noting here that the only continuum system with short-range interactions for which one can rigorously prove the existence of a phase transition is the Widom-Rowlinson model (Ruelle, Ref. 35). The situation is different for systems with long-range "Kac potentials," for which a modified form of mean field theory can be proven in a suitable limit, but that is a different story, not included in this collection; see Ref. 25(a).*

Leon C. P. Van Hove (1924-1990) and **Cecil F. Powell (1903-1969).** (Courtesy of CERN, AIP Emilio Segrè Visual Archives.)

Arthur Jaffe (1937–), Leo Kadanoff (1937–), and **David Ruelle (1935–).** (Courtesy of Joel L. Lebowitz, Rutgers University.)

sical critical exponents: They are the same in all dimensions and for all symmetries. The two-dimensional Ising model, however, has exponents very different from the classical ones and also from those found experimentally for three-dimensional systems. On the other hand, these exponents agree with experiments on highly anisotropic layered spin systems. Furthermore, Onsager's solution showed that while the critical temperature of the two-dimensional Ising model depended on the ratio of the coupling strengths in the x and y directions, the critical exponents did not. This suggests strongly the kind of universality mentioned earlier. I remember Fisher as the leading proponent of this view.[36] It was he and others such as Domb who, by cleverly combining results from exact solutions, series expansions, thermodynamics, statistical mechanical inequalities, and experiments, started to bring order to this new universality; see the very nice collection of articles in Ref. 37. [The series expansion method was pioneered by the King's College group in London led by Domb; see Ref. 37. It was greatly aided by the Padé technique; see Baker[38] and Ref. 25(a) for the analysis of these series.]

It was clear by the mid-1960s from both theory and experiment that the universality of critical phenomena cried out for a universal formulation and explanation.[36] This came with the scaling theory of Widom[39] and the Kadanoff block transformation.[40] It culminated in Wilson's renormalization group[d] (RG) formulation of critical phenomena.[41] The RG was made into a practical tool for computing critical exponents by the ε expansion about four di-

Kenneth G. Wilson (1936–). (Courtesy of AIP Meggers Gallery of Nobel Laureates, AIP Emilio Segrè Visual Archives.)

[d] *It is hard to give a precise definition of the RG. As noted by Benfatto and Gallavotti in their introduction to lecture notes on the subject, "The notion of renormalization group is not well defined. One usually means a theory in which scale invariance ideas are involved and appear technically as invariance or covariance properties, of various quantities, with respect to a noninvertible transformation of coordinates. The noninvertibility is an essential feature. It is supposed to permit reducing effectively the difficulty of the problem."*[42]

Paul C. Martin (1931–) and **Henry Ehrenreich (1928–).** (Courtesy of Martha Mooney, Harvard University.)

mensions developed by Wilson and Fisher.[43] Computations were further facilitated by the development of finite-size scaling methods and by the systematic use of field theory techniques (see Fisher and Barber[44] and Ref. 25). Universality was given a deeper meaning (at least for two-dimensional systems) through the use of conformal invariance by Polyakov[45] and others in the FSU. This led to a "full classification" of two-dimensional critical points by Friedan, Qiu, and Shenker.[46]

The Wilson RG had antecedents in field theory, Gell-Mann and Low,[47] and in the application of those field theory ideas to statistical mechanics by Di Castro and Jona-Lasinio.[48] Following Wilson's work it has spread and had enormous influence on almost all fields of science. It provides a method for quantitative analysis of the "essential" features of a large class of nonlinear phenomena exhibiting self-similar structures. This includes not only scale invariant critical systems, where fluctuations are "infinite" on the microscopic spatial and temporal scale, but also fractals, dynamical systems exhibiting Feigenbaum period doubling, KAM theory, singular behavior in nonlinear partial differential equations, and "chaos." Even where not directly applicable, the RG often provides a paradigm for the analysis of complex phenomena. The paper by Halperin, Hohenberg, and Ma[49] brings the powerful machinery of the RG to bear on nonequilibrium phenomena near the critical point. For a recent review, see Refs. 42 and 50. On the cautionary side one should remember that there are still some serious open problems concerning the nature of the RG transformation of Hamiltonians for statistical mechanical systems, i.e., for critical phenomena. A lot of mathematical work remains to be done to make it into a well-defined theory of phase transitions.[51]

I will end this brief personal overview by noting that this chapter also contains Onsager's original abstract about the nematic phase for elongated molecules and the tobacco mosaic virus and the paper by Halperin and Nelson[52] on the hexatic phase. These papers, as well as the beautiful experimental papers on critical phenomena and low temperature fluids in this collection, deserve an introduction of their own and much more space than they get here. I am not able to discuss them or any of the other goodies, included and omitted, for lack of space, time, and competence. In any case, what is included is just the tip of the iceberg. There are many, many more where these came from, and the reader is urged to settle down for a good read with any randomly picked volume of *The Physical Review* or *Physical Review Letters*. She or he will be amply rewarded for the effort.

REFERENCES

1. F. Dyson, *George Green and Physics*, Phys. World **6**, 33–38 (August 1993).
2. *The Many-Body Problem: An Encyclopedia of Exactly Solved Models in One Dimension*, edited by D. C. Mattis (World Scientific, Singapore, 1993).
3. H. B. Callen and T. E. Welton, *Irreversibility and Generalized Noise*, Phys. Rev. **83**, 34–40 (1951).

4. M. S. Green, *Markoff Random Processes and the Statistical Mechanics of Time-Dependent Phenomena. II. Irreversible Processes in Fluids*, J. Chem. Phys. **22**, 398–413 (1954); R. Kubo, *Statistical-Mechanical Theory of Irreversible Processes. I. General Theory and Simple Applications to Magnetic and Conduction Problems*, J. Phys. Soc. Jpn. **12**, 570–586 (1957). For a historical perspective see H. Nakano, *Linear Response Theory: A Historical Perspective*, Int. J. Mod. Phys. B **7**, 2397–2467 (1992).

5. L. Onsager and S. Machlup, *Fluctuations and Irreversible Processes*, Phys. Rev. **91**, 1505–1512 (1953); *Fluctuations and Irreversible Processes. II. Systems with Kinetic Energy*, *ibid.* **91**, 1512–1515 (1953).

6. N. Zabusky and M. Kruskal, *Interaction of "Solitons" in a Collisionless Plasma and the Recurrence of Initial States*, Phys. Rev. Lett. **15**, 240–243 (1965).

7. For a good overview, see introduction and articles in *Universality in Chaos*, edited by P. Cvitanović (Hilger, London, 1984). For an excellent nontechnical exposition see D. Ruelle, *Chance and Chaos* (Princeton University, Princeton, 1991).

8. B. Alder and T. Wainright, *Velocity Autocorrelations for Hard Spheres*, Phys. Rev. Lett. **18**, 988–990 (1967); *Decay of the Velocity Autocorrelation Function*, Phys. Rev. A **1**, 18–21 (1970).

9. E. Fermi, J. Pasta, and S. Ulam, *Collected Papers of E. Fermi* (University of Chicago Press, Chicago, 1965), Vol. II, p. 978. For a historical perspective see J. Ford, *The Fermi-Pasta-Ulam Problem: Paradox Turns Discovery*, Phys. Rep. **213**, 272–310 (1992).

10. For a good exposition, see G. Gallavotti, *Quasi-Integrable Mechanical Systems*, edited by K. Osterwalder and R. Stora (Elsevier, Amsterdam, 1986), pp. 539–624.

11. N. Metropolis, A. W. Rosenbluth, M. N. Rosenbluth, A. H. Teller, and E. Teller, *Equation of State Calculations by Fast Computing Machines*, J. Chem. Phys. **21**, 1087–1092 (1953).

12. E. N. Lorenz, *Deterministic Nonperiodic Flow*, J. Atm. Sci. **20**, 130–141 (1963).

13. D. Ruelle and F. Takens, *On the Nature of Turbulence*, Commun. Math. Phys. **20**, 167–192 (1971).

14. M. J. Feigenbaum, *Quantitative Universality for a Class of Nonlinear Transformations*, J. Stat. Phys. **19**, 25–52 (1978); *The Universal Metric Properties of Nonlinear Transformations*, *ibid.* **21**, 669–706 (1979).

15. G. Ahlers, *Low-Temperature Studies of the Rayleigh-Bénard Instability and Turbulence*, Phys. Rev. Lett. **33**, 1185–1188 (1974); J. Gollub and H. Swinney, *Onset of Turbulence in a Rotating Fluid*, *ibid.* **35**, 927–930 (1975).

Bertrand I. Halperin (1941–) and **David R. Nelson (1951–).** (Courtesy of Martha Mooney, Harvard University.)

Nicholas C. Metropolis (1915–), Yoshio Shimamoto (1924–), Philip M. Morse (1903–1985), James N. Snyder (1923–), and **Donald O. Smith (1925–).** (Courtesy of AIP Emilio Segrè Visual Archives.)

General Chapman and **Hendrik Kramers (1894–1952).** (Courtesy of Leo Rosenthal - Pix, Inc., AIP Emilio Segrè Visual Archives.)

16. A. Libchaber and J. Maurer, *Une Expérience de Rayleigh-Bénard de Géométrie Réduite: Multiplication, Accrochage et Démultiplication de Fréquences*, J. Phys. (Paris) Colloq. **C3**, 51–56 (1980).

17. Y. Pomeau and P. Maneville, *Intermittent Transition to Turbulence in Dissipative Dynamical Systems*, Commun. Math. Phys. **74**, 189–197 (1980).

18. L. Onsager, *The Distribution of Energy in Turbulence*, Phys. Rev. **68**, 286(A) (1945).

19. For an informal and informative description of Onsager's work, see H. C. Longuet-Higgins and M. E. Fisher, in *Biographical Memoirs*, National Academy of Sciences **60**, 183–432 (1991). It and other historical material will appear in a forthcoming Onsager memorial issue of J. Stat. Phys. (January 1995).

20. H. A. Kramers and G. H. Wannier, *Statistics of the Two-Dimensional Ferromagnet. Part I*, Phys. Rev. **60**, 252–276 (1941).

21. T. Berlin and M. Kac, *The Spherical Model of a Ferromagnet*, Phys. Rev. **86**, 821–835 (1951).

22. C. N. Yang, *The Spontaneous Magnetization of a Two-Dimensional Ising Model*, Phys. Rev. **85**, 808–816 (1952).

23. T. D. Lee and C. N. Yang, *Statistical Theory of Equations of State and Phase Transitions. II. Lattice Gas and Ising Model*, Phys. Rev. **87**, 410–419 (1952).

24. C. N. Yang and T. D. Lee, *Statistical Theory of Equations of State and Phase Transitions. I. Theory of Condensation*, Phys. Rev. **87**, 404–409 (1952).

25. (a) G. A. Baker, *Quantitative Theory of Critical Phenomena* (Academic, New York, 1990); (b) B. Simon, *The Statistical Mechanics of Lattice Gases* (Princeton University, Princeton, 1993).

26. J. Lebowitz and E. Lieb, *Existence of Thermodynamics for Real Matter with Coulomb Forces*, Phys. Rev. Lett. **22**, 631–634 (1969); E. Lieb and W. Thirring, *Bound for the Kinetic Energy of Fermions Which Proves the Stability of Matter*, *ibid.* **35**, 687–689 (1975); E. Lieb, *The Stability of Matter*, Rev. Mod. Phys. **48**, 553–569 (1976).

27. T. Asano, *Lee-Yang Theorem and the Griffiths Inequality for the Anisotropic Heisenberg Ferromagnet*, Phys. Rev. Lett. **24**, 1409–1411 (1970); see also E. Lieb and A. Sokal, *A General Lee-Yang Theorem for One-Component and Multicomponent Ferromagnets*, Commun. Math Phys. **80**, 153–179 (1981).

28. R. Baxter, *Exactly Solved Models in Statistical Mechanics* (Academic, New York, 1982).

29. P. Hohenberg, *Existence of Long-Range Order in One and Two Dimensions*, Phys. Rev. **158**, 383–386 (1967); N. Mermin and H. Wagner, *Absence of Ferromagnetism or Antiferromagnetism in One- or Two-Dimensional Isotropic Heisenberg Models*, Phys. Rev. Lett. **17**, 1133–1136 (1966); *ibid.* **17**, 1307(E) (1966).

30. O. Penrose and L. Onsager, *Bose-Einstein Condensation in Liquid Helium*, Phys. Rev. **104**, 576–584 (1956).

31. C. N. Yang, *Concept of Off-Diagonal Long-Range Order and the Quantum Phase of Liquid He and of Superconductors*, Rev. Mod. Phys. **34**, 694–704 (1962).

32. E. Lieb, *Exact Solution of the Problem of the Entropy of Two-Dimensional Ice*, Phys. Rev. Lett. **18**, 692–694 (1967).

33. R. Baxter, *Eight-Vertex Model in Lattice Statistics*, Phys. Rev. Lett. **26**, 832–833 (1971).

34. L. P. Kadanoff and F. Wegner, *Some Critical Properties of the Eight-Vertex Model*, Phys. Rev. B **4**, 3989–3993 (1971).

35. D. Ruelle, *Existence of a Phase Transition in a Continuous Classical System*, Phys. Rev. Lett. **27**, 1040–1041 (1971).

36. See M. E. Fisher, *Correlation Functions and the Critical Region of Simple Fluids*, J. Math. Phys. **5**, 944–962 (1964).

37. *Cooperative Phenomena Near a Phase Transition; A Biography with Selected Readings* edited by H. E. Stanley (MIT, Cambridge, 1973); C. Domb and other articles, in *Phase Transitions and Critical Phenomena*, edited by C. Domb and M. S. Green (Academic, New York, 1974), Vol. 3.

38. G. A. Baker, Jr., *Application of the Padé Approximation Method to the Investigation of Some Magnetic Properties of the Ising Model*, Phys. Rev. **124**, 768–774 (1961).

39. B. Widom, *Equation of State in the Neighborhood of the Critical Point*, J. Chem. Phys. **43**, 3898–3905 (1965).

40. L. P. Kadanoff, *Scaling Laws for Ising Models Near T_c*, Physics **2**, 263–272 (1966).

41. K. Wilson, *Renormalization Group and Critical Phenomena. I. Renormalization Group and the Kadanoff Scaling Picture*, Phys. Rev. B **4**, 3174–3183 (1971); *Renormalization Group and Critical Phenomena. II. Phase-Space Cell Analysis of Critical Behavior*, *ibid.* **4**, 3184–3205 (1971); *Feynman Graph Expansion for Critical Exponents*, Phys. Rev. Lett. **28**, 548–551 (1972).

42. G. Benfatto and G. Gallavotti, *Renormalization Group*, Lecture Notes, Rome (1993) (Princeton University Press, 1995, to be published).

43. K. Wilson and M. E. Fisher, *Critical Exponents in 3.99 Dimensions*, Phys. Rev. Lett. **28**, 240–243 (1972).

44. M. E. Fisher and M. Barber, *Scaling Theory for Finite-Size Effects in the Critical Region*, Phys. Rev. Lett. **28,** 1516–1519 (1972).

45. A. M. Polyakov, *Quantum Geometry of Bosonic Strings*, Phys. Lett. **103B**, 207–210 (1981); *Quantum Theory of Fermionic Strings*, *ibid.* **103B**, 211–213 (1981).

46. D. Friedan, Z. Qiu, and S. Shenker, *Conformal Invariance, Unitarity, and Critical Exponents in Two Dimensions*, Phys. Rev. Lett. **52**, 1575–1578 (1984).

47. M. Gell-Mann and F. Low, *Quantum Electrodynamics at Small Distances*, Phys. Rev. **95**, 1300–1312 (1954).

48. C. Di Castro and G. Jona-Lasinio, *On The Microscopic Foundation of Scaling Laws*, Phys. Lett. A **29**, 322–323 (1969).

49. B. Halperin, P. Hohenberg, and S. Ma, *Calculations of Dynamic Critical Properties Using Wilson's Expansion Methods*, Phys. Rev. Lett. **29**, 1548–1551 (1972).

50. N. Goldenfeld, *Lectures on Phase Transitions and the Renormalization Group*, Frontiers in Physics Series (Addison Wesley, Reading, MA, 1992), Vol. 85.

51. R. B. Griffiths and P. A. Pearce, *Position-Space Renormalization-Group Transformations: Some Proofs and Some Problems*, Phys. Rev. Lett. **41**, 917–920 (1978); A. C. D. van Enter, R. Fernández, and A. D. Sokal, *Regularity Properties and Pathologies of Position-Space Renormalization-Group Transformations: Scope and Limitations of Gibbsian Theory*, J. Stat. Phys. **72**, 879–1167 (1993); F. Martinelli and E. Olivieri, *Some Remarks on Pathologies of Renormalization-Group Transformations for the Ising Model*, *ibid.* **72**, 1169–1177 (1993).

52. B. Halperin and D. Nelson, *Theory of Two-Dimensional Melting*, Phys. Rev. Lett. **41**, 121–124 (1978).

Fritz London (1900-1957), circa 1954. (Courtesy of *Physics Today* Collection, AIP Emilio Segrè Visual Archives.)

Rodney J. Baxter (1940–), John F. Nagle (1939–), Elliot H. Lieb (1932–), Joel L. Lebowitz (1930–), Harold Friedman (1923–), George Stell (1933–), and **Michael E. Fisher (1931–)** at the Lars Onsager Symposium, Trondheim, Norway (1993). (Courtesy of Joel L. Lebowitz, Rutgers University.)

Papers Reproduced on CD-ROM

I. Langmuir and K. H. Kingdon. Thermionic phenomena due to alkali vapors. Part II, *Phys. Rev.* **21**, 381(A) (1923)

G. E. Uhlenbeck and L. S. Ornstein. On the theory of the Brownian motion, *Phys. Rev.* **36**, 823–841 (1930)

L. Onsager. Reciprocal relations in irreversible processes. I., *Phys. Rev.* **37**, 405–426 (1931)

L. Onsager. Reciprocal relations in irreversible processes. II., *Phys. Rev.* **38**, 2265–2279 (1931)

F. London. On the Bose–Einstein condensation, *Phys. Rev.* **54**, 947–954 (1938)

E. C. Kemble. The quantum-mechanical basis of statistical mechanics, *Phys. Rev.* **56**, 1146–1164 (1939)

H. A. Kramers and G. H. Wannier. Statistics of the two-dimensional ferromagnet. Part I, *Phys. Rev.* **60**, 252–276 (1941)

L. Onsager. Anisotropic solutions of colloids, *Phys. Rev.* **62**, 558(A) (1942)

L. Onsager. Crystal statistics, *Phys. Rev.* **62**, 559(A) (1942)

J. Ashkin and E. Teller. Statistics of two-dimensional lattices with four components, *Phys. Rev.* **64**, 178–184 (1943)

L. Onsager. Crystal statistics. I. A two-dimensional model with an order–disorder transition, *Phys. Rev.* **65**, 117–149 (1944)

L. Onsager. The distribution of energy in turbulence, *Phys. Rev.* **68**, 286(A) (1945)

C. T. Lane, H. A. Fairbank, and W. M. Fairbank. Second sound in liquid helium II, *Phys. Rev.* **71**, 600–605 (1947)

F. London. On the problem of the molecular theory of superconductivity, *Phys. Rev.* **74**, 562–573 (1948)

E. M. Purcell and R. V. Pound. A nuclear spin system at negative temperature, *Phys. Rev.* **81**, 279–280(L) (1951)

R. Kikuchi. A theory of cooperative phenomena, *Phys. Rev.* **81**, 988–1003 (1951)

H. B. Callen and T. A. Welton. Irreversibility and generalized noise, *Phys. Rev.* **83**, 34–40 (1951)

C. N. Yang. The spontaneous magnetization of a two-dimensional Ising model, *Phys. Rev.* **85**, 808–816 (1952)

T. H. Berlin and M. Kac. The spherical model of a ferromagnet, *Phys. Rev.* **86**, 821–835 (1952)

C. N. Yang and T. D. Lee. Statistical theory of equations of state and phase transitions. I. Theory of condensation, *Phys. Rev.* **87**, 404–409 (1952)

T. D. Lee and C. N. Yang. Statistical theory of equations of state and phase transitions. II. Lattice gas and Ising model, *Phys. Rev.* **87**, 410–419 (1952)

R. B. Potts. Spontaneous magnetization of a triangular Ising lattice, *Phys. Rev.* **88**, 352 (1952)

L. Van Hove. The occurrence of singularities in the elastic frequency distribution of a crystal, *Phys. Rev.* **89**, 1189–1193 (1953)

R. P. Feynman. Atomic theory of liquid helium near absolute zero, *Phys. Rev.* **91**, 1301–1308 (1953)

L. Onsager and S. Machlup. Fluctuations and irreversible processes, *Phys. Rev.* **91**, 1505–1512 (1953)

D. Bohm and D. Pines. A collective description of electron interactions: III. Coulomb interactions in a degenerate electron gas, *Phys. Rev.* **92**, 609–625 (1953)

D. Pines. A collective description of electron interactions: IV. Electron interaction in metals, *Phys. Rev.* **92**, 626–636 (1953)

G. Placzek and L. Van Hove. Crystal dynamics and inelastic scattering of neutrons, *Phys. Rev.* **93**, 1207–1214 (1954)

R. P. Feynman. Atomic theory of the two-fluid model of liquid helium, *Phys. Rev.* **94**, 262–277 (1954)

L. Van Hove. Correlations in space and time and Born approximation scattering in systems of interacting particles, *Phys. Rev.* **95**, 249–262 (1954)

M. Lax. Generalized theory of mobility, *Phys. Rev.* **100**, 1808(A) (1955)

F. J. Dyson. General theory of spin-wave interactions, *Phys. Rev.* **102**, 1217–1230 (1956)

O. Penrose and L. Onsager. Bose–Einstein condensation and liquid helium, *Phys. Rev.* **104**, 576–584 (1956)

M. Gell-Mann and K. A. Brueckner. Correlation energy of an electron gas at high density, *Phys. Rev.* **106**, 364–368 (1957)

E. T. Jaynes. Information theory and statistical mechanics, *Phys. Rev.* **106**, 620–630 (1957)

H. Palevsky, K. Otnes, K. E. Larsson, R. Pauli, and R. Stedman. Excitation of rotons in helium II by cold neutrons, *Phys. Rev.* **108**, 1346–1347(L) (1957)

J. K. Percus and G. J. Yevick. Analysis of classical statistical mechanics by means of collective coordinates, *Phys. Rev.* **110**, 1–13 (1958)

J. L. Yarnell, G. P. Arnold, P. J. Bendt, and E. C. Kerr. Excitations in liquid helium: Neutron scattering measurements, *Phys. Rev.* **113**, 1379–1386 (1959)

P. C. Martin and J. Schwinger. Theory of many-particle systems. I., *Phys. Rev.* **115**, 1342–1373 (1959)

V. Bargmann, L. Michel, and V. L. Telegdi. Precession of the polarization of particles moving in a homogeneous electromagnetic field, *Phys. Rev. Lett.* **2**, 435–436 (1959)

R. J. Elliott, B. R. Heap, D. J. Morgan, and G. S. Rushbrooke. Equivalence of the critical concentrations in the Ising and Heisenberg models of ferromagnetism, *Phys. Rev. Lett.* **5**, 366–367 (1960)

V. A. Vyssotsky, S. B. Gordon, H. L. Frisch, and J. M. Hammersley. Critical percolation probabilities (bond problem), *Phys. Rev.* **123**, 1566–1567 (1961)

STATISTICAL PHYSICS

G. A. Baker, Jr. Application of the Padé approximant method to the investigation of some magnetic properties of the Ising model, *Phys. Rev.* **124**, 768–774 (1961)

R. Zwanzig. Memory effects in irreversible thermodynamics, *Phys. Rev.* **124**, 983–992 (1961)

N. Byers and C. N. Yang. Theoretical considerations concerning quantized magnetic flux in superconducting cylinders, *Phys. Rev. Lett.* **7**, 46–49 (1961)

L. Onsager. Magnetic flux through a superconducting ring, *Phys. Rev. Lett.* **7**, 50 (1961)

E. Lieb and D. Mattis. Theory of ferromagnetism and the ordering of electronic energy levels, *Phys. Rev.* **125**, 164–172 (1962)

B. J. Alder and T. E. Wainwright. Phase transition in elastic disks, *Phys. Rev.* **127**, 359–361 (1962)

M. H. Kalos. Monte Carlo calculations of the ground state of three- and four-body nuclei, *Phys. Rev.* **128**, 1791–1795 (1962)

P. Heller and G. B. Benedek. Nuclear magnetic resonance in MnF_2 near the critical point, *Phys. Rev. Lett.* **8**, 428–432 (1962)

M. S. Wertheim. Exact solution of the Percus–Yevick integral equation for hard spheres, *Phys. Rev. Lett.* **10**, 321–323 (1963)

J. L. Lebowitz. Exact solution of generalized Percus–Yevick equation for a mixture of hard spheres, *Phys. Rev.* **133**, A895–A899 (1964)

A. Rahman. Triplet correlations in liquids, *Phys. Rev. Lett.* **12**, 575–577 (1964)

W. L. McMillan. Ground state of liquid He^4, *Phys. Rev.* **138**, A442–A451 (1965)

N. J. Zabusky and M. D. Kruskal. Interaction of "solitons" in a collisionless plasma and the recurrence of initial states, *Phys. Rev. Lett.* **15**, 240–243 (1965)

F. P. Buff, R. A. Lovett, and F. H. Stillinger, Jr. Interfacial density profile for fluids in the critical region, *Phys. Rev. Lett.* **15**, 621–623 (1965)

J. L. Lebowitz and J. K. Percus. Mean spherical model for lattice gases with extended hard cores and continuum fluids, *Phys. Rev.* **144**, 251–258 (1966)

K. Kawasaki. Correlation-function approach to the transport coefficients near the critical point. I., *Phys. Rev.* **150**, 291–306 (1966)

J. R. Clow and J. D. Reppy. Temperature dependence of the superfluid density in He II near T_λ^*, *Phys. Rev. Lett.* **16**, 887–889 (1966)

W. R. Abel, A. C. Anderson, and J. C. Wheatley. Propagation of zero sound in liquid He^3 at low temperatures, *Phys. Rev. Lett.* **17**, 74–78 (1966)

H. E. Stanley and T. A. Kaplan. Possibility of a phase transition for the two-dimensional Heisenberg model, *Phys. Rev. Lett.* **17**, 913–915 (1966)

N. D. Mermin and H. Wagner. Absence of ferromagnetism or antiferromagnetism in one- or two-dimensional isotropic Heisenberg models, *Phys. Rev. Lett.* **17**, 1133–1136, 1307(E) (1966)

R. B. Griffiths. Thermodynamic functions for fluids and ferromagnets near the critical point, *Phys. Rev.* **158**, 176–187 (1967)

P. C. Hohenberg. Existence of long-range order in one and two dimensions, *Phys. Rev.* **158**, 383–386 (1967)

L. Verlet. Computer "experiments" on classical fluids. I. Thermodynamical properties of Lennard–Jones molecules, *Phys. Rev.* **159**, 98–103 (1967)

E. H. Lieb. Exact solution of the problem of the entropy of two-dimensional ice, *Phys. Rev. Lett.* **18**, 692–694 (1967)

R. A. Ferrell, N. Menyhárd, H. Schmidt, F. Schwabl, and P. Szépfalusy. Dispersion in second sound and anomalous heat conduction at the lambda point of liquid helium, *Phys. Rev. Lett.* **18**, 891–894 (1967)

E. H. Graf, D. M. Lee, and J. D. Reppy. Phase separation and the superfluid transition in liquid He^3-He^4 mixtures, *Phys. Rev. Lett.* **19**, 417–419 (1967)

J. S. Langer and M. E. Fisher. Intrinsic critical velocity of a superfluid, *Phys. Rev. Lett.* **19**, 560–563 (1967)

B. I. Halperin and P. C. Hohenberg. Generalization of scaling laws to dynamical properties of a system near its critical point, *Phys. Rev. Lett.* **19**, 700–703 (1967)

L. P. Kadanoff and J. Swift. Transport coefficients near the liquid–gas critical point, *Phys. Rev.* **166**, 89–101 (1968)

H. E. Stanley. Spherical model as the limit of infinite spin dimensionality, *Phys. Rev.* **176**, 718–722 (1968)

E. H. Lieb and F. Y. Wu. Absence of Mott transition in an exact solution of the short-range, one-band model in one dimension, *Phys. Rev. Lett.* **20**, 1445–1448 (1968)

C. J. Pearce, J. A. Lipa, and M. J. Buckingham. Velocity of second sound near the λ point of helium, *Phys. Rev. Lett.* **20**, 1471–1473 (1968)

B. M. McCoy and T. T. Wu. Random impurities as the cause of smooth specific heats near the critical temperature, *Phys. Rev. Lett.* **21**, 549–551 (1968)

G. Ahlers. Thermal conductivity of He I near the superfluid transition, *Phys. Rev. Lett.* **21**, 1159–1162 (1968)

B. I. Halperin and P. C. Hohenberg. Scaling laws for dynamic critical phenomena, *Phys. Rev.* **177**, 952–971 (1969)

J. L. Lebowitz and E. H. Lieb. Existence of thermodynamics for real matter with Coulomb forces, *Phys. Rev. Lett.* **22**, 631–634 (1969)

R. J. Birgeneau, H. J. Guggenheim, and G. Shirane. Neutron scattering from K_2NiF_4: A two-dimensional Heisenberg antiferromagnet, *Phys. Rev. Lett.* **22**, 720–723 (1969)

J. Ginibre. Simple proof and generalization of Griffiths' second inequality, *Phys. Rev. Lett.* **23**, 828–830 (1969)

B. J. Alder and T. E. Wainwright. Decay of the velocity autocorrelation function, *Phys. Rev. A* **1**, 18–21 (1970)

STATISTICAL PHYSICS

P. Berge, P. Calmetter, C. Laj, M. Tournarie, and B. Volochine. Dynamics of concentration fluctuations in a binary mixture in the hydrodynamical and nonhydrodynamical regimes, *Phys. Rev. Lett.* **24**, 1223–1225, 1468(E) (1970)

T. Asano. Lee–Yang theorem and the Griffiths inequality for the anisotropic Heisenberg ferromagnet, *Phys. Rev. Lett.* **24**, 1409–1411 (1970)

J. R. Dorfman and E. G. D. Cohen. Velocity correlation functions in two and three dimensions, *Phys. Rev. Lett.* **25**, 1257–1260 (1970)

M. Blume, V. J. Emery, and R. B. Griffiths. Ising model for the λ transition and phase separation in He^3-He^4 mixtures, *Phys. Rev. A* **4**, 1071–1077 (1971)

K. G. Wilson. Renormalization group and critical phenomena. I. Renormalization group and the Kadanoff scaling picture, *Phys. Rev. B* **4**, 3174–3183 (1971)

K. G. Wilson. Renormalization group and critical phenomena. II. Phase-space cell analysis of critical behavior, *Phys. Rev. B* **4**, 3184–3205 (1971)

L. P. Kadanoff and F. J. Wegner. Some critical properties of the eight-vertex model, *Phys. Rev. B* **4**, 3989–3993 (1971)

J. V. Sengers and P. H. Keyes. Scaling of the thermal conductivity near the gas–liquid critical point, *Phys. Rev. Lett.* **26**, 70–73 (1971)

R. J. Birgeneau, R. Dingle, M. T. Hutchings, G. Shirane, and S. L. Holt. Spin correlations in a one-dimensional Heisenberg antiferromagnet, *Phys. Rev. Lett.* **26**, 718–721 (1971)

R. J. Baxter. Eight-vertex model in lattice statistics, *Phys. Rev. Lett.* **26**, 832–833 (1971)

D. Ruelle. Existence of a phase transition in a continuous classical system, *Phys. Rev. Lett.* **27**, 1040–1041 (1971)

K. G. Wilson and M. E. Fisher. Critical exponents in 3.99 dimensions, *Phys. Rev. Lett.* **28**, 240–243 (1972)

K. G. Wilson. Feynman-graph expansion for critical exponents, *Phys. Rev. Lett.* **28**, 548–551 (1972)

D. D. Osheroff, R. C. Richardson, and D. M. Lee. Evidence for a new phase of solid He^3, *Phys. Rev. Lett.* **28**, 885–888 (1972)

M. E. Fisher and M. N. Barber. Scaling theory for finite-size effects in the critical region, *Phys. Rev. Lett.* **28**, 1516–1519 (1972)

E. K. Riedel and F. J. Wegner. Tricritical exponents and scaling fields, *Phys. Rev. Lett.* **29**, 349–352 (1972)

E. Brézin, D. J. Wallace, and K. G. Wilson. Feynman-graph expansion for the equation of state near the critical point (Ising-like case), *Phys. Rev. Lett.* **29**, 591–594 (1972)

G. Ahlers and D. S. Greywall. Second-sound velocity and superfluid density near the tricritical point in He^3-He^4 mixtures, *Phys. Rev. Lett.* **29**, 849–852 (1972)

D. D. Osheroff, W. J. Gully, R. C. Richardson, and D. M. Lee. New magnetic phenomena in liquid He^3 below 3 mK, *Phys. Rev. Lett.* **29**, 920–923 (1972)

B. I. Halperin, P. C. Hohenberg, and S-k. Ma. Calculation of dynamic critical properties using Wilson's expansion methods, *Phys. Rev. Lett.* **29**, 1548–1551 (1972)

P. C. Martin, E. D. Siggia, and H. A. Rose. Statistical dynamics of classical systems, *Phys. Rev. A* **8**, 423–437 (1973)

M. J. Ablowitz, D. J. Kaup, A. C. Newell, and H. Segur. Nonlinear-evolution equations of physical significance, *Phys. Rev. Lett.* **31**, 125–127 (1973)

J. D. Weeks, G. H. Gilmer, and H. J. Leamy. Structural transition in the Ising-model interface, *Phys. Rev. Lett.* **31**, 549–551 (1973)

R. J. Baxter and F. Y. Wu. Exact solution of an Ising model with three-spin interactions on a triangular lattice, *Phys. Rev. Lett.* **31**, 1294–1297 (1973)

Th. Niemeijer and J. M. J. van Leeuwen. Wilson theory for spin systems on a triangular lattice, *Phys. Rev. Lett.* **31**, 1411–1414 (1973)

R. Graham. Generalized thermodynamic potential for the convection instability, *Phys. Rev. Lett.* **31**, 1479–1482 (1973)

M. Hénon. Integrals of the Toda lattice, *Phys. Rev. B* **9**, 1921–1923 (1974)

G. Ahlers. Low-temperature studies of the Rayleigh–Bénard instability and turbulence, *Phys. Rev. Lett.* **33**, 1185–1188 (1974)

E. H. Lieb and W. E. Thirring. Bound for the kinetic energy of fermions which proves the stability of matter, *Phys. Rev. Lett.* **35**, 687–689 (1975)

J. P. Gollub and H. L. Swinney. Onset of turbulence in a rotating fluid, *Phys. Rev. Lett.* **35**, 927–930 (1975)

Y. Imry and S-k. Ma. Random-field instability of the ordered state of continuous symmetry, *Phys. Rev. Lett.* **35**, 1399–1401 (1975)

D. Sherrington and S. Kirkpatrick. Solvable model of a spin-glass, *Phys. Rev. Lett.* **35**, 1792–1796 (1975)

J. Fröhlich, B. Simon, and T. Spencer. Phase transitions and continuous symmetry breaking, *Phys. Rev. Lett.* **36**, 804–806 (1976)

J. Swift and P. C. Hohenberg. Hydrodynamic fluctuations at the convective instability, *Phys. Rev. A* **15**, 319–328 (1977)

D. Forster, D. R. Nelson, and M. J. Stephen. Large-distance and long-time properties of a randomly stirred fluid, *Phys. Rev. A* **16**, 732–749 (1977)

H. van Beijeren. Exactly solvable model for the roughening transition of a crystal surface, *Phys. Rev. Lett.* **38**, 993–996 (1977)

D. J. Thouless. Maximum metallic resistance in thin wires, *Phys. Rev. Lett.* **39**, 1167–1169 (1977)

D. R. Nelson and J. M. Kosterlitz. Universal jump in the superfluid density of two-dimensional superfluids, *Phys. Rev. Lett.* **39**, 1201–1205 (1977)

STATISTICAL PHYSICS

I. Rudnick. Critical surface density of the superfluid component in ^{4}He films, *Phys. Rev. Lett.* **40**, 1454–1455 (1978)

D. J. Bishop and J. D. Reppy. Study of the superfluid transition in two-dimensional ^{4}He films, *Phys. Rev. Lett.* **40**, 1727–1730 (1978)

B. I. Halperin and D. R. Nelson. Theory of two-dimensional melting, *Phys. Rev. Lett.* **41**, 121–124 (1978)

R. B. Griffiths and P. A. Pearce. Position-space renormalization-group transformations: Some proofs and some problems, *Phys. Rev. Lett.* **41**, 917–920 (1978)

V. L. Pokrovsky and A. L. Talapov. Ground state, spectrum, and phase diagram of two-dimensional incommensurate crystals, *Phys. Rev. Lett.* **42**, 65–67 (1979)

M. Aizenman. Instability of phase coexistence and translation invariance in two dimensions, *Phys. Rev. Lett.* **43**, 407–409 (1979)

D. J. Wallace and R. K. P. Zia. Euclidean group as a dynamical symmetry of surface fluctuations: The planar interface and critical behavior, *Phys. Rev. Lett.* **43**, 808–812 (1979)

J. S. Langer. Eutectic solidification and marginal stability, *Phys. Rev. Lett.* **44**, 1023–1026 (1980)

D. B. Abraham. Solvable model with a roughening transition for a planar Ising ferromagnet, *Phys. Rev. Lett.* **44**, 1165–1168 (1980)

M. E. Fisher and W. Selke. Infinitely many commensurate phases in a simple Ising model, *Phys. Rev. Lett.* **44**, 1502–1505 (1980)

B. Derrida. Random-energy model: Limit of a family of disordered models, *Phys. Rev. Lett.* **45**, 79–82 (1980)

N. Andrei. Diagonalization of the Kondo Hamiltonian, *Phys. Rev. Lett.* **45**, 379–382 (1980)

N. H. Packard, J. P. Crutchfield, J. D. Farmer, and R. S. Shaw. Geometry from a time series, *Phys. Rev. Lett.* **45**, 712–716 (1980)

J. E. Avron, L. S. Balfour, C. G. Kuper, J. Landau, S. G. Lipson, and L. S. Schulman. Roughening transition in the ^{4}He solid–superfluid interface, *Phys. Rev. Lett.* **45**, 814–817 (1980)

Y. Gefen, B. B. Mandelbrot, and A. Aharony. Critical phenomena on fractal lattices, *Phys. Rev. Lett.* **45**, 855–858 (1980)

B. Shraiman, C. E. Wayne, and P. C. Martin. Scaling theory for noisy period-doubling transitions to chaos, *Phys. Rev. Lett.* **46**, 935–939 (1981)

J. Fröhlich and T. Spencer. Kosterlitz–Thouless transition in the two-dimensional plane rotator and Coulomb gas, *Phys. Rev. Lett.* **46**, 1006–1009 (1981)

M. Aizenman. Proof of the triviality of φ_d^4 field theory and some mean-field features of Ising models for $d > 4$, *Phys. Rev. Lett.* **47**, 1–4 (1981)

E. D. Siggia and A. Zippelius. Pattern selection in Rayleigh–Bénard convection near threshold, *Phys. Rev. Lett.* **47**, 835–838 (1981)

H. Sompolinsky. Time-dependent order parameters in spin-glasses, *Phys. Rev. Lett.* **47**, 935–938 (1981)

T. A. Witten, Jr. and L. M. Sander. Diffusion-limited aggregation, a kinetic critical phenomenon, *Phys. Rev. Lett.* **47**, 1400–1403 (1981)

L. P. Kadanoff. Scaling for a critical Kolmogorov-Arnold-Moser trajectory, *Phys. Rev. Lett.* **47**, 1641–1643 (1981)

S. Fishman, D. R. Grempel, and R. E. Prange. Chaos, quantum recurrences, and Anderson localization, *Phys. Rev. Lett.* **49**, 509–512 (1982)

S. Katz, J. L. Lebowitz, and H. Spohn. Phase transitions in stationary nonequilibrium states of model lattice systems, *Phys. Rev. B* **28**, 1655–1658 (1983)

P. Grassberger and I. Procaccia. Characterization of strange attractors, *Phys. Rev. Lett.* **50**, 346–349 (1983)

G. Parisi. Order parameter for spin-glasses, *Phys. Rev. Lett.* **50**, 1946–1948 (1983)

A. Brandstäter, J. Swift, H. L. Swinney, A. Wolf, J. D. Farmer, E. Jen, and P. J. Crutchfield. Low-dimensional chaos in a hydrodynamic system, *Phys. Rev. Lett.* **51**, 1442–1445, 1814(E) (1983)

Pierre C. Hohenberg (1934–), Michael C. Cross (1952–), and **Guenter Ahlers (1934–).** (Courtesy of Guenter Ahlers, University of California, Santa Barbara.)

DECEMBER 15, 1931 PHYSICAL REVIEW VOLUME 38

RECIPROCAL RELATIONS IN IRREVERSIBLE PROCESSES. II.

By Lars Onsager
Department of Chemistry, Brown University

(Received November 9, 1931)

Abstract

A general reciprocal relation, applicable to transport processes such as the conduction of heat and electricity, and diffusion, is derived from the assumption of microscopic reversibility. In the derivation, certain average products of fluctuations are considered. As a consequence of the general relation $S = k \log W$ between entropy and probability, different (coupled) irreversible processes must be compared in terms of entropy changes. If the displacement from thermodynamic equilibrium is described by a set of variables $\alpha_1, \cdots, \alpha_n$, and the relations between the rates $\dot{\alpha}_1, \cdots, \dot{\alpha}_n$ and the "forces" $\partial S/d\alpha_1, \cdots, \partial S/d\alpha_n$ are linear, there exists a quadratic dissipation-function,

$$2\Phi(\dot{\alpha}, \dot{\alpha}) \equiv \Sigma\rho_j\dot{\alpha}_i\dot{\alpha}_i = dS/dt = \dot{S}(\alpha, \dot{\alpha}) \equiv \Sigma(\partial S/d\alpha_j)\dot{\alpha}_j$$

(denoting definition by $\equiv$). The symmetry conditions demanded by microscopic reversibility are equivalent to the variation-principle

$$\dot{S}(\alpha, \dot{\alpha}) - \Phi(\dot{\alpha}, \dot{\alpha}) = \text{maximum},$$

which determines $\dot{\alpha}_1, \cdots, \dot{\alpha}_n$ for prescribed $\alpha_1, \cdots, \alpha_n$. The dissipation-function has a statistical significance similar to that of the entropy. External magnetic fields, and also Coriolis forces, destroy the symmetry in past and future; reciprocal relations involving reversal of the field are formulated.

I. Introduction

In a previous communication[1] a reciprocal theorem for heat conduction in an anisotropic medium was derived from the principle of microscopic reversibility, applied to fluctuations. In the following we shall derive reciprocal relations for irreversible processes in general, particularly transport processes: conduction of electricity and heat, and diffusion.

As before we shall assume that the average regression of fluctuations will obey the same laws as the corresponding macroscopic irreversible processes. In (I) we considered fluctuations in "aged" systems, i.e., systems which have been left isolated for a length of time that is normally sufficient to secure thermodynamic equilibrium. For dealing with the conduction of heat we naturally considered the fluctuations of the distribution of heat, and we studied the behavior of the quantities $\alpha_1, \alpha_2, \alpha_3$ = the total displacements of heat in the directions x_1, x_2 and x_3, respectively.

We brought the laws of irreversible processes into the theory of fluctuations by studying averages

$$\overline{\alpha_1(t)\alpha_2(t+\tau)} = \lim_{t''\to\infty} \frac{1}{t''-t'} \int_{t=t'}^{t=t''} \alpha_1(t)\alpha_2(t+\tau)dt \tag{1.1}$$

[1] L. Onsager, Phys. Rev. 37, 405 (1931). Cited in the following as (I).

of the values of two fluctuating quantities α_1 and α_2 (in this particular case displacements of heat in two perpendicular directions) observed with a prescribed time interval τ. The condition of microscopic reversibility we applied in the form:

$$\overline{\alpha_1(t)\alpha_2(t+\tau)} = \overline{\alpha_2(t)\alpha_1(t+\tau)}\,. \tag{1.2}$$

The calculation of averages of the type (1.1) involves several steps. First of all, something must be known about the distribution of values of the fluctuating quantities $\alpha_1, \alpha_2 \cdots$; we shall employ standard methods for calculating the average products:

$$\overline{\alpha_1^2},\ \overline{\alpha_1\alpha_2}, \cdots \tag{1.3}$$

with but a slight variation. In addition, we must know the average changes (of quantities $\alpha_2, \alpha_3 \cdots$) which accompany a given deviation α_1' of a quantity α_1 from its normal value $\bar{\alpha}_1(=0)$. On this basis a certain *initial state* of an irreversible process is associated with the displacement $\alpha_1 = \alpha_1'$; the average regression towards the normal state will obey the ordinary macroscopic laws governing such processes.

The average regression is described by the functions:

$$\overline{\alpha_i}(\tau, \alpha_j')$$

defined as the average of the quantity α_i, taken over all cases, (picked at random for an aged system), in which, τ seconds earlier, the quantity α_j *had* the value α_j'. Whenever these functions are linear in α_j', which will be the common case when reasonable variables α are chosen, the knowledge of the averages (1.3) suffices for evaluating (1.1).

In (I), §4, we derived reciprocal relations for heat conduction in an anisotropic body, showing that the conducting properties of the most general (triclinic) crystal can be represented by a *symmetrical tensor* (ellipsoid). In that particular case it was not necessary to calculate the averages (1.3), nor to determine completely the state associated with a displacement $\alpha_1 = \alpha_1'$, because the necessary information could be derived from considerations of symmetry. Even so, these considerations were based on the proposition that, in regard to the probability for a given distribution of energy, the different volume elements of a *homogeneous* crystal would be equivalent, so that the anisotropy of the crystal could be neglected in this particular connection. This proposition involves the fundamental principles of statistical mechanics although it obviates a part of the general mathematical apparatus.

We shall review the derivation briefly. If J_1, J_2, J_3 denote the components of the heat flow along the coordinate axes x_1, x_2, x_3, respectively, and T the absolute temperature, the phenomenological laws of heat conduction in a triclinic crystal take the general form

$$\begin{aligned} J_1 &= L_{11}X_1 + L_{12}X_2 + L_{13}X_3 \\ J_2 &= L_{21}X_1 + L_{22}X_2 + L_{23}X_3 \\ J_3 &= L_{31}X_1 + L_{32}X_2 + L_{33}X_3 \end{aligned} \tag{1.4}$$

where X_1, X_2, X_3 are the components of the "force" on the heat flow

$$X_1 = -(1/T)\partial T/dx_1;\quad X_2 = -(1/T)\partial T/dx_2;\quad X_3 = -(1/T)\partial T/dx_3 \tag{1.5}$$

(Carnot). In order to derive the reciprocal relations

$$L_{12} = L_{21};\quad L_{23} = L_{32};\quad L_{31} = L_{13}$$

we considered the fluctuations of the moments

$$\alpha_1 = \int \epsilon x_1 dV \qquad \alpha_2 = \int \epsilon x_2 dV \tag{1.6}$$

of the distribution of energy, $\epsilon = \epsilon(x_1, x_2, x_3)$. When we chose the external boundary of the crystal spherical, center at the origin, all questions pertaining to the *instantaneous* distribution of energy have spherical symmetry (cf. above). Thus

$$\overline{\alpha_1} = \overline{\alpha_2} = 0;\quad \overline{\alpha_1^2} = \overline{\alpha_2^2};\quad \overline{\alpha_1\alpha_2} = 0, \tag{1.7}$$

and with a displacement of energy there is associated a temperature gradient in the same direction:

$$\overline{\partial T/dx_1}^{\alpha_1',\alpha_2'} = -T\overline{X}_1(\alpha_1', \alpha_2') = CT\alpha_1' \qquad -T\overline{X}_2(\alpha_1', \alpha_2') = CT\alpha_2' \tag{1.8}$$

where C is for our immediate purposes a mere constant. (Certain rather trivial considerations are needed to justify our assumption of a *linear* relation between displacement (α_1', α_2') and gradient $(T\overline{X}_1, T\overline{X}_2)$. When we take that much for granted, the more special form (1.8) follows from the spherical symmetry.)

The gradient (1.8) determines the heat flow J according to (1.4), and the rate of change $\dot{\alpha}$ of the displacement α is the same as the total flow

$$\dot{\alpha}_1(\alpha_1', \alpha_2', \alpha_3') = d\alpha_1/dt = \int J_1 dV = -CV(L_{11}\alpha_1' + L_{12}\alpha_2' + L_{13}\alpha_3')$$
$$\dot{\alpha}_2(\alpha_1', \alpha_2', \alpha_3') = -CV(L_{21}\alpha_1' + L_{22}\alpha_2' + L_{23}\alpha_3'). \tag{1.9}$$

Then in a short interval of time Δt

$$\overline{\alpha_2}(\Delta t, \alpha_1') = \overline{\alpha_2}(0, \alpha_1') + \dot{\alpha}_2(\alpha_1')\Delta t = 0 - L_{21}CV\alpha_1'\Delta t,$$

where $\bar{\alpha}_2(0, \alpha_1')$ vanishes by symmetry. From this, obviously

$$\overline{\alpha_1(t)\alpha_2(t+\Delta t)} = \overline{\alpha_1'\overline{\alpha_2}(\Delta t, \alpha_1')} = -L_{21}CV\overline{\alpha_1^2}\Delta t$$

and, by analogy

$$\overline{\alpha_2(t)\alpha_1(t+\Delta t)} = -L_{12}CV\overline{\alpha_2^2}\Delta t.$$

Since, by (1.7): $\overline{\alpha_1^2} = \overline{\alpha_2^2}$, the requirement:

$$\overline{\alpha_1(t)\alpha_2(t+\Delta t)} = \overline{\alpha_2(t)\alpha_1(t+\Delta t)}$$

for microscopic reversibility imposes the condition:

$$L_{12} = L_{21}.$$

The more general case of simultaneous transport of heat, electricity and matter (diffusion) in isotropic or anisotropic media leads us to consider the fluctuations of a set of variables $\alpha_1, \alpha_2, \cdots$, measuring displacements of heat, electricity and matter, eventually in different directions. We shall have to actually evaluate the averages (1.3); symmetry considerations can yield the necessary information only in a few cases, as above. The calculation of (1.3) involves directly Boltzmann's fundamental relation between entropy S and probability W:

$$S = k \log W + \text{constant}$$

The exceedingly general character of this relation is the reason that the *rates* of irreversible processes, not solely the ultimate equilibrium, are subject to reciprocal laws, in which *different processes have to be compared in terms of the entropy changes involved.*

Sometimes, of course, it may be more convenient to employ other thermodynamic potentials, particularly the free energy, the main reason being that conditions of *mechanical* equilibrium (pressure, elastic) frequently enter into the laws of irreversible processes, and that the description in terms of energy involves more familiar derived functions (thermodynamic potential, electromotive force, electrical resistance). Above, we purposely considered the "force" on heat, although the concept was not necessary for dealing with the problem in hand, just to demonstrate how the concept of "force" could be extended beyond the familiar.

Fundamentally, however, the *entropy* is the simplest among the thermodynamic potentials, and it is the only one that will serve our purposes in all cases. In our example, where displacements $\alpha_1, \alpha_2, \alpha_3$ of heat are considered, the state of the system being determined by these displacements, so that

$$S = \sigma(\alpha_1, \alpha_2, \alpha_3),$$

the temperature gradients are essentially the same as $\partial S/d\alpha_1$ etc., or rather

$$\partial\sigma/d\alpha_r = \partial(1/T)/dx_r.$$

In order to see this, we recall the fundamental thermodynamic relation

$$\delta S = (1/T)(\delta E - \delta A) - (\mu/T)\delta m,$$

where E = energy; A = work; m = amount of substance; μ = Gibbs' thermodynamic potential. The amount of heat added to a volume element is $\delta Q = \delta E - \delta A$. Now if there is a uniform gradient of temperature (or $1/T$ in the r-direction, then an amount of heat δQ transported a distance Δx_r changes the entropy by the amount

$$\delta S = \delta Q \cdot \Delta(1/T) = \delta Q \cdot \Delta x_r \cdot \partial(1/T)/dx_r = \delta\alpha_r \cdot \partial(1/T)/dx_r,$$

whereby $\delta\alpha_r$ measures a displacement of heat (in units cm × cal.).

Similarly, where a displacement α of matter (in the x direction) is considered,

$$\partial\sigma/d\alpha = -\partial(\mu/T)/dx,$$

and if α is a displacement of electricity, then

$$\partial\sigma/d\alpha = X/T,$$

where X is the intensity of the electric field.

According to the empirical laws of transport processes the flow J of matter, heat or electricity is proportional to the gradient of the corresponding specific potential, that is

$$J \sim -\text{grad } T$$

for heat conduction and

$$J \sim X; \; J \sim -\text{grad } \mu,$$

respectively, for electrical conduction (Ohm's law) and isothermal diffusion (alternative form of Fick's law). In the following we shall write these empirical relations in the general form

$$d\alpha_r/dt = \dot{\alpha}_r \sim \partial S/d\alpha_r,$$

where the rate of displacement $\dot{\alpha}$ is essentially the same as the flow J, (by definition), except for a volume factor. Whenever different transport processes interfere with each other the simple proportionality is replaced by a system of linear relations

$$\dot{\alpha}_r = G_{r1}\partial\sigma/d\alpha_1 + \cdots + G_{rn}\partial\sigma/d\alpha_n, \quad (r = 1, \cdots, n), \qquad (1.11)$$

where again $S = \sigma(\alpha_1, \cdots \alpha_n)$. In taking for granted the linear form (1.11) we are still making use of empirical laws, mostly familiar, which are understood and expected from simple and equally familiar kinetic considerations of very wide scope. Examples have been enumerated in (I), §§ 1–2, and need not be repeated here.

Our object is to show that the condition (1.2) for microscopic reversibility leads to the general reciprocal relation

$$G_{rs} = G_{sr}. \qquad (1.12)$$

2. General Theory of Fluctuations

It was shown by L. Boltzmann that a mechanical theory of molecules requires a statistical interpretation of the second law of thermodynamics. Thermodynamic equilibrium is explained as a statistical equilibrium of elementary processes, and Boltzmann gave a direct relation between the entropy S and the "thermodynamic probability" W of a thermodynamic state:

$$S = k \log W + \text{const.}, \qquad (2.1)$$

where k is the gas constant per molecule (1.371×10^{-16} erg/degree). The apparent rest associated with thermodynamic equilibrium is explained by the

smallness of the factor k. According to (2.1), under circumstances which normally lead to such equilibrium, the probability

$$e^{\Delta S/k}$$

for a deviation involving an entropy change ΔS (necessarily negative), is appreciable only when ΔS is (at the outmost) of the order of magnitude of k. The *fluctuations* permitted by this restriction can be observed only in very favorable cases, for example the opalescence of liquids near the critical point[2] and Brownian motion of small particles in liquids,[3] or of a mirror in delicate elastic suspension.[4]

The premises and the consequences of Boltzmann's principle (2.1) have been discussed by A. Einstein[5] to an extent which will be practically sufficient for our purposes. It is essential that a thermodynamic equilibrium state, specified in terms of energy and external parameters (volume etc.), is incompletely specified from a molecular point of view; the quantity W measures the number (or extent) of different possibilities for realizing a given thermodynamic state. In order to calculate W one needs a complete (molecular) theory of the system in hand: If one assumes that molecules obey the laws of classical mechanics, then W equals an extension in phase-space, while on the basis of quantum theory W equals the number of stationary states corresponding to the prescribed energy. However, as Einstein has pointed out, the calculation of fluctuations according to (2.1) is *independent of all special assumptions regarding the laws which may govern elementary processes* (we must of course assume that these laws do permit statistical equilibrium of some kind).

We shall have to make certain *general assumptions* about *aged systems*, i.e., systems which have been left isolated for a length of time that is normally sufficient to secure thermodynamic equilibrium. We expect that such a system will in the course of time pass through *all* the (thermodynamic) states $\Gamma^1, \Gamma^2, \cdots \Gamma^l$ that are compatible with the conditions of isolation, whereby the energy, the values of external parameters (volume, etc.) and the numbers of indestructible elementary particles (atoms, molecules) are prescribed.[6] In the course of a long time t the system will spend a total t_r of time intervals in the state Γ^r; we expect that $t_1, t_2 \cdots t_l$ will be proportional to the regions $W_1, W_2 \cdots W_l$. This statement contains an assumption even if $W_1, W_2 \cdots$ are considered as unknown, namely that $t_1/t, t_2/t \cdots$ will be fully determined by the nature of the system, (and the conditions of isolation), independently of the initial state. Granted this assumption, we may define $W_1, W_2 \cdots$ as pro-

[2] M. v. Smoluchowski, Ann. d. Physik [4], 25, 205 (1909). Theory accompanied by a general discussion of fluctuations is given by A. Einstein, Ann. d. Physik [4], 33, 1275 (1910).

[3] A. Einstein, Ann. d. Physik [4], 17, 549 (1905). M. v. Smoluchowski, Ann. d. Physik [4], 21, 756 (1906).

[4] P. Zeeman and O. Houdyk, Proc. Acad. Amsterdam 28, 52 (1925). W. Gerlach, Naturwiss. 15, 15 (1927). G. E. Uhlenbeck and S. Goudsmit, Phys. Rev. 34, 145 (1929).

[5] Einstein, reference 2.

[6] In a discussion of the fundamental questions involved, W. Schottky introduces the term "resistent groups." Ann. d. Physik [4], 68, 481 (1922).

portional t_1/t, $t_2/t \cdots$, without reference to any detailed theory of the system. (On the basis of the underlying picture, every W is still a large integer, equal to the number of "microscopic" states contained in a given thermodynamic state. Here, however, we are only interested in the ratios of $W_1, W_2, \cdots$). The various assumptions involved in this application of Boltzmann's principle may be summarized in the formula:

$$S_r = k \log (t_r/t) + \text{const.} \tag{2.2}$$

where S_r is the entropy of the state Γ^r; we shall refer to t_r/t as the *probability* for this state.

A thermodynamic state $\Gamma^r = \Gamma(\alpha_1^r, \cdots \alpha_n^r)$ may be defined in terms of given values $\alpha_1^r, \cdots \alpha_n^r$ for such variables $\alpha_1, \cdots \alpha_n$ as can be measured by ordinary means. From a statistical point of view we must allow latitudes $\Delta\alpha_1, \cdots \Delta\alpha_n$ in this specification (the probability for a region of no extension equals zero). We have to introduce a distribution-function

$$f(\alpha_1, \cdots, \alpha_n)$$

and the probability for the state Γ^r becomes equal to the integral of $f(\alpha_1, \cdots \alpha_n)$ over the region

$$\alpha_1^{(r)} < \alpha_1 < \alpha_1^{(r)} + \Delta\alpha_1$$

$$\cdots\cdots\cdots$$

$$\alpha_n^{(r)} < \alpha_n < \alpha_n^{(r)} + \Delta\alpha_n.$$

Then (2.2) takes the form

$$S_r = k \log f(\alpha_1^{(r)}, \cdots \alpha_n^{(r)}) + k \log (\Delta\alpha_1 \cdots \Delta\alpha_n) + \text{const.} \tag{2.3}$$

Our only direction for the appropriate choice of latitudes $\Delta\alpha$ is that they ought to be taken of the same order of magnitude as the common fluctuations of the quantities α in the state Γ^r. This convention takes care of all important cases; because thermodynamic measurements of entropy are possible only for equilibrium states.[7] A more accurate specification of $\Delta\alpha_1, \cdots \Delta\alpha_n$ is unnecessary because, say, doubling each $\Delta\alpha$ will change the right side of (2.3) only by the amount $nk \log 2$, where $k = 1.371 \times 10^{-16}$ erg/degree, and an entropy difference of this magnitude is far too small to affect any measurement. Actually, where reasonable variables $\alpha_1, \cdots \alpha_n$ are chosen, the order of magnitude of the product $\Delta\alpha_1, \Delta\alpha_2 \cdots \Delta\alpha_n$ varies so little that the contribution $k \log (\Delta\alpha_1' \cdots \Delta\alpha_n'/\Delta\alpha_1'' \cdots \Delta\alpha_n'')$ to the entropy difference between two thermodynamic states is entirely negligible in comparison with $k \log(f(\alpha_1', \cdots \alpha_n')/f(\alpha_1'', \cdots \alpha_n''))$; it is the factor $f(\alpha_1, \cdots \alpha_n)$ that causes the tremendous difference in the probabilities of different thermodynamic states, and is responsible for measurable entropy differences. Thus, as long as we restrict ourselves to cases where S_r is a measured entropy, we may neglect the variability of the term $k \log (\Delta\alpha_1 \cdots \Delta\alpha_n)$ on the right side of (2.3) and write

[7] We allow ourselves, following the custom of thermodynamics, to consider states that may be approximated in some way by equilibrium states. Cf. Schottky, reference 6.

$$S_r = k \log f(\alpha_1^{(r)}, \cdots \alpha_n^{(r)}) + \text{const.} \tag{2.4}$$

So far we have assumed that the variables $\alpha_1, \cdots \alpha_n$ define the state of the system in hand *completely* from the thermodynamic (phenomenological) point of view. As pointed out by Einstein[8] the relation (2.4) remains valid in cases where this definition is incomplete. We merely have to adopt the convention that among all the states which fulfill the given specifications, we select the one with the greatest entropy. This theorem again depends on the probabilities for different states being of different order of magnitude, so that the given set of values of $\alpha_1, \cdots \alpha_n$ will be realized by the chosen state much more frequently than by all other states taken together. We shall summarize these results for subsequent applications. The greatest entropy allowed by a given set $\alpha_1', \cdots \alpha_n'$ of values of the variables $\alpha_1, \cdots \alpha_n$ we denote by

$$\sigma_{1\cdots n}(\alpha_1', \cdots \alpha_n').$$

The corresponding (thermodynamic) state we denote by

$$\Gamma'_{1\cdots n} = \Gamma_{1\cdots n}(\alpha_1', \cdots \alpha_n').$$

Then

$$\sigma_{1\cdots n}(\alpha_1', \cdots \alpha_n') = k \log f(\alpha_1', \cdots \alpha_n') + \text{const.} \tag{2.5}$$

gives the probability for finding the variables $\alpha_1, \cdots \alpha_n$ with a given set of values $\alpha_1', \cdots \alpha_n'$. Practically every time when $\alpha_1, \cdots \alpha_n$ assume this set of values, the system will be in the state $\Gamma_{1\cdots n}(\alpha_1', \cdots \alpha_n')$.

We shall calculate at once certain averages which will be needed for the subsequent derivations. We denote by $f_p(\alpha_p)$ the distribution-function for the variable α_p. We have according to (2.5):

$$k \log f_p(\alpha_p) = \sigma_p(\alpha_p) + \text{const.}, \tag{2.6}$$

where $\sigma_p(\alpha_p')$ is the greatest entropy possible when $\alpha_p = \alpha_p'$, realized by the state $\Gamma_p(\alpha_p')$. The function f_p is determined by (2.6) together with the condition

$$\int_{-\infty}^{\infty} f_p(\alpha_p) d\alpha_p = 1.$$

We obtain from (2.6) by differentiation

$$k df_p/d\alpha_p = f_p(\alpha_p) d\sigma_p/d\alpha_p,$$

assuming that the differential quotient exists. We shall also assume that the entropy $\sigma_p(\alpha_p)$ attains its maximum for a finite value α_p^0 of α_p, corresponding to the equilibrium state Γ^0, and that $(\alpha_p - \alpha_p^0) f_p(\alpha_p)$ approaches zero for large values of $|\alpha_p - \alpha_p^0|$.[9] Then it is easy to calculate the average

$$\overline{(\alpha_p - \alpha_p^0) d\sigma_p/d\alpha_p} = \int_{-\infty}^{\infty} (\alpha_p - \alpha_p^0)(d\sigma_p/d\alpha_p) f_p(\alpha_p) d\alpha_p$$

$$= k \int_{-\infty}^{\infty} (\alpha_p - \alpha_p^0)(df_p/d\alpha_p) d\alpha_p.$$

[8] Einstein, reference 2.

[9] Otherwise this function must have an infinite number of maxima. All the conditions stated are fulfilled whenever α_p is a resonable thermodynamic variable.

Integration by parts yields

$$k\left|(\alpha_p - \alpha_p{}^0)f_p(\alpha_p)\right|_{-\infty}^{\infty} - k\int_{-\infty}^{\infty} f_p(\alpha_p)d\alpha_p.$$

The first term vanishes, and the second equals $-k$; thus

$$\overline{(\alpha_p - \alpha_p{}^0)d\sigma_p/d\alpha_p} = -k. \tag{2.7}$$

In the same manner, and under the same assumptions we find

$$\overline{(\alpha_p - \alpha_p{}^0)\partial\sigma_{1\ldots n}/d\alpha_p}$$

$$= \int_{-\infty}^{\infty}\cdots\int(\alpha_p - \alpha_p{}^0)(\partial\sigma/d\alpha_p)f_{1\ldots n}d\alpha_1\cdots d\alpha_n = -k, \tag{2.8a}$$

and

$$\overline{(\alpha_q - \alpha_q{}^0)\partial\sigma_{1\ldots n}/d\alpha_p} = 0, \quad (p \neq q). \tag{2.8b}$$

In the following we shall find it convenient to apply the simple formula (2.7) directly. However, it seems desirable to show the connection with the ordinary formulas for the averages of the products $(\alpha_p-\alpha_p{}^0)(\alpha_q-\alpha_q{}^0)$. We must assume that the entropy $\sigma_{1\ldots n}$ can be expressed by a multiple power series, and that the abridged Taylor development

$$\sigma_{1\ldots n}(\alpha_1, \cdots \alpha_n) = S_0 + \tfrac{1}{2}\sum_{p,q=1}^{n}\eta_{pq}(\alpha_p - \alpha_p{}^0)(\alpha_q - \alpha_q{}^0), \tag{2.9}$$

where

$$\eta_{pq} = \eta_{qp} = \left[\partial^2\sigma_{1\ldots n}/d\alpha_p d\alpha_q\right]_{\alpha_1=\alpha_1{}^0;\ \cdots;\ \alpha_n=\alpha_n{}^0}, \tag{2.10}$$

will suffice in the entire region of values of $\alpha_1, \cdots \alpha_n$, for which the contribution to any of the averages (2.8 a, b) is at all appreciable. (Since $f\sim\exp(\sigma/k)$, the maximum of $f_{1\ldots n}$ is very sharp). Then we may substitute in (2.8 a, b):

$$\partial\sigma_{1\ldots n}/d\alpha_p = \sum_{p=1}^{n}\eta_{pq}(\alpha_q - \alpha_q{}^0), \tag{2.11}$$

and we obtain a system of linear equations

$$\sum_{r=1}^{n}\eta_{pr}\overline{(\alpha_r - \alpha_r{}^0)(\alpha_q - \alpha_q{}^0)} = -k\delta_{pq} = \begin{cases} -k, & (p = q)\\ 0, & (p \neq q),\end{cases} \tag{2.12}$$

from which the mean squares and products of fluctuations may be computed.

3. The Regression of Fluctuations

We are accustomed to observe that the course of an irreversible process taking place in an isolated system is entirely determined by the initial thermodynamic state according to definite laws, such as the laws for conduction of heat. On the basis of a statistical interpretation of the second law of thermodynamics, no process can be completely predetermined by an initial thermodynamic state; because such a state is itself incompletely defined (from a

molecular point of view; cf. §2). However, we can understand a predetermination with practical certainty, within limits of the order of magnitude of ordinary fluctuations, whereby much greater deviations will be very rare. From this statistical point of view we may still interpret the predictions of irreversible processes from empirical laws as valid for averages taken over a large number of similar cases, which in this connection means cases of irreversible processes starting from the same initial *thermodynamic* state.

Strictly speaking, this rule does not specify uniquely the more refined "microscopic" (molecular) interpretation of laws derived from relatively crude "macroscopic" observations. It makes a considerable difference whether we take an average of the type $\bar{\alpha}=(\alpha'+\alpha'')/2$ or one of the type $\bar{\alpha}=[(\alpha'^5+\alpha''^5)/2]^{1/5}$. However, in all important concrete cases the natural answer to this question will be obvious. For example, if α is a total displacement of heat, itself the sum of many local displacements whose changes depend on local conditions, there may be no doubt that the straight average $\bar{\alpha}=(\alpha'+\alpha'')/2$ is correct.

Now we are able to solve the problem of predicting the *average regression* of fluctuations: Suppose that we start out with a certain isolated system, and watch the fluctuations of the variables $\alpha_1, \cdots \alpha_n$ for a great length of time. Whenever the values of $\alpha_1, \cdots \alpha_n$ happen to be (simultaneously) $\alpha_1', \cdots \alpha_n'$, we make a record of the values which these variables (and perhaps other quantities $\alpha_{n+1}, \cdots \alpha_{n+p}$) assume τ *seconds later*. The averages of such records we denote by

$$\overline{\alpha_1}(\tau, \alpha_1', \cdots \alpha_n'), \cdots \overline{\alpha_{n+p}}(\tau, \alpha_1', \cdots \alpha_n').$$

We know that almost every time when $\alpha_1=\alpha_1'; \cdots; \alpha_n=\alpha_n'$, the system will be in the (phenomenological) state $\Gamma'_{1\ldots n}=\Gamma(\alpha_1', \cdots \alpha_n')$, and the average course of an irreversible process following that state, described by the functions

$$\overline{\alpha_1}(\tau, \Gamma'_{1\ldots n}), \cdots \overline{\alpha_{n+p}}(\tau, \Gamma'_{1\ldots n}),$$

we know from macroscopic experiments. These functions may be considered as properties (in an extended sense) of the state $\Gamma'_{1\ldots n}$. The "normal" (common) properties of states corresponding to prescribed values $\alpha_1', \cdots \alpha_n'$ of the fluctuating quantities $\alpha_1, \cdots \alpha_n$ are certainly those of the state $\Gamma_1' \cdots {}_n$. The question whether we are allowed to interchange "normal" and average properties must be decided from the consideration of individual cases, as outlined above.

Assuming that the variables $\alpha_1, \cdots \alpha_n$ are suitable in this regard, we have:

$$\overline{\alpha_i}(\tau, \alpha_1', \cdots \alpha_n') = \overline{\alpha_i}(\tau, \Gamma'_{1\ldots n}), \quad (i = 1, \cdots, n + p), \tag{3.1}$$

as a general rule for predicting the average regression of fluctuations from the laws of irreversible processes.

4. Reciprocal Relations

For a discussion of the requirements of microscopic reversibility the averages

$$A_{ji}(\tau) = \overline{\alpha_j(t)\alpha_i(t+\tau)} = \overline{\alpha_j'\overline{\alpha_i}(\tau, \alpha_j')} \qquad (4.1)$$

afford a convenient point of attack. The quantities $A_{ij}(\tau)$ may also be defined as time averages (1.1)

$$A_{ji}(\tau) = \lim_{t''\to\infty} \frac{1}{t''-t'} \int_{t=t'}^{t=t''} \alpha_j(t)\alpha_i(t+\tau)dt.$$

In the following it will be convenient to assume that the variables $\alpha_1, \cdots \alpha_n$ measure deviations from the thermodynamic equilibrium, whereby their averages $\bar{\alpha}_1, \cdots \bar{\alpha}_n$ for this state (and also the "normal" values α_j^0) vanish:

$$\overline{\alpha_i} = \alpha_i^0 = 0, \quad (i = 1, \cdots n). \qquad (4.2)$$

The assumption of microscopic reversibility requires that, if α and β be two quantities which depend only on the configuration of molecules and atoms, *the event $\alpha = \alpha'$, followed τ seconds later by $\beta = \beta'$, will occur just as often as the event $\beta = \beta'$, followed τ seconds later by $\alpha = \alpha'$*. The same will be true if α and β depend on the velocities of elementary particles in such a manner that they are not changed when the velocities are reversed, for example, when α depends on the distribution of energy in a system.[10] If α_j and α_i are two such "reversible" variables of a reversible system, then obviously

$$A_{ji}(\tau) = \overline{\alpha_j(t)\alpha_i(t+\tau)} = \overline{\alpha_i(t)\alpha_j(t+\tau)} = A_{ij}(\tau). \qquad (4.3)$$

We shall consider cases where the course of an irreversible process starting from any state of the type $\Gamma_{1\ldots n}$ can be described by a set of linear differential equations of the form (1.11):

$$\frac{d\overline{\alpha_i}}{dt} = \dot{\alpha}_i = \sum_{r=1}^{n} G_{ir} \frac{\partial\sigma_{1\ldots n}(\alpha_1, \cdots \alpha_n)}{d\alpha_r}, \quad (i = 1, \cdots n). \qquad (4.4)$$

According to (3.1) we have

$$\overline{\alpha_i}(\tau, \alpha_j') = \overline{\alpha_i}(\tau, \Gamma_j'),$$

where Γ'_j is the state of maximum entropy for a given value α_j' of the variable α_j. Mathematically this state is characterized by the relations

$$\begin{aligned} \alpha_j &= \alpha_j' \\ \partial\sigma_{1\ldots n}/d\alpha_r &= 0, \quad (r \neq j), \end{aligned} \qquad (4.5)$$

and we have for the set of values of $\alpha_1, \cdots \alpha_n$ determined by these conditions:

$$\partial\sigma_{1\ldots n}/d\alpha_j = \left[d\sigma_j/d\alpha_j\right]_{\alpha_j=\alpha_j'}. \qquad (4.6)$$

[10] Cf. (I) p. 418.

From (4.4) we have in a short interval of time Δt

$$\overline{\alpha_i}(\Delta t, \alpha_j') = \overline{\alpha_i}(0, \alpha_j') + \dot{\alpha}_i\Delta t = \overline{\alpha_i}(0, \alpha_j') + \sum_{r=1}^{n} G_{ir}(\partial\sigma_{1\ldots n}/d\alpha_r)\Delta t.$$

or substituting (4.5) and (4.6):

$$\overline{\alpha_i}(\Delta t, \alpha_j') = \overline{\alpha_i}(0, \alpha_j') + G_{ij}(d\sigma_j(\alpha_j')/d\alpha_j')\Delta t.$$

Calculating according to (4.1) the average $A_{ji}(\Delta t)$ we obtain

$$A_{ji}(\Delta t) = \overline{\alpha_j(t)\alpha_i(t+\Delta t)} = \overline{\alpha_j'\overline{\alpha_i}(0, \alpha_j')} + G_{ij}\Delta t\overline{\alpha_j d\sigma_j/d\alpha_j},$$

or, observing (2.7) and the convention (4.2):

$$A_{ji}(\Delta t) = A_{ji}(0) - k\Delta t G_{ij}. \qquad (4.7)$$

Similarly, of course

$$A_{ij}(\Delta t) = A_{ij}(0) - k\Delta t G_{ji}.$$

Applying the condition (4.3) for microscopic reversibility we find

$$G_{ij} = G_{ji}, \qquad (4.8)$$

as announced at the end of §1. The importance of considering fluctuations for the derivation of this result is apparent from the occurrence of Boltzmann's constant k in (4.7).

5. The Principle of the Least Dissipation of Energy

The symmetry relation (4.8) contains the important reciprocal relations in transport processes. An alternative form of (4.8) is convenient for many applications, and commands considerable intrinsic interest. The description (4.4) of a set of simultaneous irreversible processes may be rewritten in the form

$$\frac{\partial\sigma_{1\ldots n}(\alpha_1, \cdots \alpha_n)}{d\alpha_i} = \sum_{r=1}^{n} \rho_{ir}\dot{\alpha}_r, \quad (i = 1, \cdots n), \qquad (5.1)$$

where, according to the equations

$$\sum_{r=1}^{n} \rho_{ir}G_{rj} = \sum_{r=1}^{n} G_{ir}\rho_{rj} = \delta_{ij} = \begin{cases} 1, & (i = j) \\ 0, & (i \neq j), \end{cases} \qquad (5.2)$$

the coefficients (ρ_{ij}) form the inverse matrix of (G_{ij}), which enter into (4.4). The symmetry relations (4.8) may be replaced by the equivalent

$$\rho_{ij} = \rho_{ji}, \quad (i = 1, \cdots n). \qquad (5.3)$$

We introduce the *dissipation-function*

$$\Phi(\dot{\alpha}, \dot{\alpha}) \equiv \tfrac{1}{2} \sum_{i,j} \rho_{ij}\dot{\alpha}_i\dot{\alpha}_j, \qquad (5.4)$$

and incorporate the symmetry relations (5.3) into the description of irreversible processes by writing

$$\partial\sigma_{1\ldots n}(\alpha_1, \cdots \alpha_n)/d\alpha_i = \partial\Phi(\dot{\alpha}, \dot{\alpha})/d\dot{\alpha}_i, \qquad (5.5)$$

in place of (5.1). Further, if we define a function

$$\dot{S}(\alpha, \dot{\alpha}) \equiv \sum_{r=1}^{n} (\partial\sigma_{1\ldots n}/d\alpha_r)\dot{\alpha}_r, \tag{5.6}$$

representing the rate of increase of the entropy, we can formulate a *variation-principle*, namely

$$\delta[\dot{S}(\alpha, \dot{\alpha}) - \Phi(\dot{\alpha}, \dot{\alpha})] = 0. \tag{5.7}$$

Our convention is that only the velocities $\dot{\alpha}_1, \cdots \dot{\alpha}_n$ should be varied, thus

$$\delta[\dot{S}(\alpha, \dot{\alpha}) - \Phi(\dot{\alpha}, \dot{\alpha})] = \sum_{i=1}^{n} \left(\frac{\partial\sigma_{1\ldots n}}{d\alpha_i} - \frac{\partial\Phi}{d\dot{\alpha}_i} \right) \delta\dot{\alpha}_i = 0,$$

according to (5.5). The variation principle (5.7), which we shall call the *principle of the least dissipation of energy*, for reasons mentioned in (I), §6, provides convenient means for transforming the reciprocal relations (5.3), or (4.8), to cases where the conventional description of irreversible processes involves an infinite number of variables, for example the temperatures in all parts of a space. The dissipation-function equals half the rate of production of entropy

$$2\Phi(\dot{\alpha}, \dot{\alpha}) = \dot{S}(\alpha, \dot{\alpha}), \tag{5.8}$$

because of (5.5), (5.6) and (5.4), which may be written

$$\Phi(\dot{\alpha}, \dot{\alpha}) = \tfrac{1}{2} \sum_{i,j} \rho_{ij}\dot{\alpha}_i\dot{\alpha}_j = \tfrac{1}{2} \sum_{i} \dot{\alpha}_i \partial\Phi/d\dot{\alpha}_i.$$

It is evident from (5.8) that $\Phi(\dot{\alpha}, \dot{\alpha})$ must be essentially positive (definite or semidefinite), because the second law of thermodynamics demands $\dot{S} \geqq 0$. Therefore the extremum given by (5.7) is always a maximum

$$\dot{S}(\alpha, \dot{\alpha}) - \Phi(\dot{\alpha}, \dot{\alpha}) = \text{maximum}. \tag{5.9}$$

Applications of this principle will be given in a later publication; in (I), §§4–5, a special result was derived by a direct method.

It is worth pointing out that in the dissipation-function has a direct statistical significance. A detailed discussion would be out of place in this article, where a compact presentation of important theorems is intended, but we may state without derivation the result, which is an extension of Boltzmann's principle (2.1). The equilibrium condition of thermodynamics

$$S = \text{maximum}$$

characterizes the most probable state, and the probability W for a state $\Gamma(\alpha_1, \cdots \alpha_n)$ is given by Boltzmann's principle

$$k \log W(\alpha_1, \cdots \alpha_n) = S(\alpha_1, \cdots \alpha_n) + \text{const.};$$

for the precise interpretation of this theorem we must refer to the discussion in §2. In a similar manner, Eq. (5.9) describes the most probable course of an

irreversible process. It is also possible to show, under assumptions approximately equivalent to those that are necessary for deriving, (simultaneously), (5.5) and (2.12), that the probability

$$W(\Gamma', \Delta t, \Gamma'') = W(\alpha_1', \cdots \alpha_n', \Delta t, \alpha_1'', \cdots \alpha_n'')$$

for the states $\Gamma' = \Gamma(\alpha_1', \cdots \alpha_n')$ and $\Gamma'' = \Gamma(\alpha_1'', \cdots \alpha_n'')$ occurring at the times t' and $t'' = t' + \Delta t$, respectively, is given by the formula

$$k \log W(\Gamma', \Delta t, \Gamma'') = S' + S'' - \frac{\Phi(\Delta\alpha, \Delta\alpha)}{\Delta t} + \text{const.}, \tag{5.10}$$

where $S' = S(\alpha_1', \cdots \alpha_n')$, $S'' = S(\alpha_1'', \cdots \alpha_n'')$, and

$$\Phi(\Delta\alpha, \Delta\alpha) = \tfrac{1}{2} \sum_{i,j} \rho_{ij}(\alpha_i'' - \alpha_i')(\alpha_j'' - \alpha_j').$$

(Needless to say, we assume that we are dealing with an aged system.)

6. Nonreversible Systems

As mentioned in (I), §7, we know from our macroscopic experience certain conservative dynamical systems which do not exhibit dynamical reversibility, namely systems where external magnetic fields are acting, and systems whose motion is described relatively to a rotating frame of coordinates, the rotation being equivalent to a field of Coriolis forces. In such cases, where the macroscopic laws of motion are non-reversible, the microscopic motion cannot be reversible.

In dealing with cases of this kind it is advantageous to consider the intensities of magnetic and Coriolis fields as variable external parameters of the system in hand. Then macroscopic dynamical systems subject to external magnetic and Coriolis forces have the following symmetry with regard to reversal of the time: If $[q] = [Q(t - t_0)]$ is a possible motion (succession of configurations $[q]$) of a system left to itself in a magnetic (or Coriolis) field of intensity Θ, then the reverse succession of configurations $[q] = [Q(t_0 - t)]$ is a possible motion of the same system when placed in a field of intensity $-\Theta$. Further, let α and β be two functions of the state (and of the parameters) of the system in hand, such that their values are not changed when all the velocities in the system are reversed, (simultaneously with Θ). Then, when we consider the fluctuations in an aged system, as in §4, the succession of events $\alpha = \alpha'$, $\beta = \beta'$, with an intervening lapse of time τ, will occur in a system placed in a field of intensity $+\Theta$, just as often as the succession of events $\beta = \beta'$, $\alpha = \alpha'$, (with a time-interval τ), will occur in a system placed in a field of intensity $-\Theta$.

If we may apply this symmetry condition to the motion of elementary particles, and α_i, α_j are two "reversible" dynamical variables, the averages (4.1) will be functions of the time τ and the field intensity Θ:

$$A_{ji}(\Theta, \tau),$$

and we have the symmetry condition

$$\Lambda_{ji}(\Theta, \tau) = \Lambda_{ji}(-\Theta, -\tau) = \Lambda_{ij}(-\Theta, \tau). \tag{6.1}$$

Supposing that a certain irreversible process can be described in the form (4.4), the coefficients G_{ji} being functions of Θ,

$$\frac{d\alpha_i}{dt} = \dot{\alpha}_i = \sum_{j=1}^{n} G_{ij}(\Theta) \frac{\partial \sigma_{1\ldots n}(\alpha_1, \cdots \alpha_n)}{d\alpha_j}, \quad (i = 1, \cdots n), \tag{6.2}$$

we can derive (4.7) as before

$$\Lambda_{ji}(\Theta, \Delta t) = \Lambda_{ji}(\Theta, 0) - k\Delta t G_{ij}(\Theta)$$
$$\Lambda_{ij}(-\Theta, \Delta t) = \Lambda_{ij}(-\Theta, 0) - k\Delta t G_{ji}(-\Theta),$$

and upon applying the symmetry condition (6.1) we find

$$G_{ij}(\Theta) = G_{ji}(-\Theta). \tag{6.3}$$

This theorem contains a reciprocal relation between the Nernst effect and the Ettingshausen effect, which has been derived previously by P. W. Bridgman[11] and by H. A. Lorentz[12] on a quasi-thermodynamic basis.

[11] P. W. Bridgman, Phys. Rev. 24, 644 (1924); Fourth Solvay Congress ("Conductibilité Electrique des Metaux"), 352 (1924).

[12] H. A. Lorentz, Fourth Solvay Congress, 354 (1924).

Abstracts of Contributed Papers

1. Anisotropic Solutions of Colloids. LARS ONSAGER, *Yale University.*—The solutions of certain colloids comprised of highly asymmetrical particles—plates or rods—are known to form anisotropic phases at remarkably low concentrations. For tobacco mosaic virus (rods), isotropic solutions containing 2–3 percent virus are in equilibrium with anisotropic phases containing 3–4.5 percent, respectively, according to the amount of electrolyte present. This phenomenon can be explained as a result of repulsive forces by the observation that the mutual co-volume of two swarms of parallel rods (or plates) is roughly proportional to the sine of the angle between their orientation, and larger than the volume of the particles by a factor which is proportional to the asymmetry. The case of rods is particularly simple in that the virial coefficients of order higher than 2 in Mayer's expansion are small, and a quantitative theory is possible. The computed ratio of concentrations at equilibrium is 1.34. The predicted osmotic pressure of the anisotropic phase is nearly proportional to the concentration, in fact, slightly greater than $3cRT/V$.

6. **Crystal Statistics.** LARS ONSAGER, *Yale University.*—The partition function for the Ising model of a two-dimensional "ferromagnetic" has been evaluated in closed form. The results of Kramers and Wannier concerning the "Curie point" T_c have been confirmed, including their conjecture that the maximum of the specific heat varies linearly with the logarithm of the size of the crystal. For an infinite crystal, the specific heat near $T=T_c$ is proportional to $-\log|(T-T_c)|$.

23. The Distribution of Energy in Turbulence. LARS ONSAGER, *Yale University, New Haven, Connecticut.*—The dissipation of energy by turbulence is regarded as primarily a "violet catastrophe." The velocity field of a liquid has an infinite set of Fourier components, whose mutual modulation redistributes the energy among more and more components which belong to ever increasing wave-numbers. In actual liquids this subdivision of energy is intercepted by the action of viscosity, which destroys the energy more rapidly the greater the wave number. However, various experiments indicate that the viscosity has a negligible effect on the primary process; hence one may inquire about the laws of turbulent dissipation in an ideal fluid. The modulation of a given Fourier component of the motion is mostly due to those others which belong to wave numbers of comparable magnitude. Some important applications of this principle are known; but it has not been pointed out before that the subdivision of the energy must be a stepwise process, such that an n-fold increase of the wave number is reached by a number of steps of the order $\log n$. For such a cascade mechanism that part of the energy density which is associated with large wave numbers should depend on the total volume rate of dissipation Q only. Then dimensional considerations require that the energy per component of wave number k equal (universal factor) $Q^{2/3}k^{-11/3}$. The corresponding correlation-function for the velocities at two points r apart has the form $R(r)=1-(\text{const.})\, r^{2/3}$.

PHYSICAL REVIEW VOLUME 83, NUMBER 1 JULY 1, 1951

Irreversibility and Generalized Noise*

HERBERT B. CALLEN AND THEODORE A. WELTON†
Randal Morgan Laboratory of Physics, University of Pennsylvania, Philadelphia, Pennsylvania

(Received January 11, 1951)

A relation is obtained between the generalized resistance and the fluctuations of the generalized forces in linear dissipative systems. This relation forms the extension of the Nyquist relation for the voltage fluctuations in electrical impedances. The general formalism is illustrated by applications to several particular types of systems, including Brownian motion, electric field fluctuations in the vacuum, and pressure fluctuations in a gas.

I. INTRODUCTION

THE parameters which characterize a thermodynamic system in equilibrium do not generally have precise values, but undergo spontaneous fluctuations. These thermodynamic parameters are of two classes: the "extensive" parameters,[1] such as the volume or the mole numbers, and the "intensive" parameters[2] or "generalized forces," such as the pressure or chemical potentials.

An equation relating particularly to the fluctuations in voltage (a "generalized force") in linear electrical systems was derived many years ago by Nyquist,[3] and such voltage fluctuations are generally referred to as Nyquist or Johnson "noise." The voltage fluctuations are related, not to the standard thermodynamic parameters of the system, but to the electrical resistance. The Nyquist relation is thus of a form unique in physics, correlating a property of a system in *equilibrium* (i.e., the voltage fluctuations) with a parameter which characterizes an irreversible process (i.e., the electrical resistance). The equation, furthermore, gives not only the mean square fluctuating voltage, but provides, in addition, the frequency spectrum of the fluctuations. The proof of the relation is based on an ingenious union of the second law of thermodynamics and a direct calculation of the fluctuations in a particular simple system (an ideal transmission line).

It has frequently been conjectured that the Nyquist relation can be extended to a general class of dissipative systems other than merely electrical systems. Yet, to our knowledge, no proof has been given of such a generalization, nor have any criteria been developed for the type of system or the character of the "forces" to which the generalized Nyquist relation may be applied. The development of such a proof and of such criteria is the purpose of this paper (Secs. II, III, and IV). The general theorem thus establishes a relation between the "impedance" in a general linear dissipative system and the fluctuations of appropriate generalized "forces."

Several illustrative applications are made of the general theorem. The viscous drag of a fluid on a moving body is shown to imply a fluctuating force, and application of the general theorem immediately yields the fundamental result of the theory of Brownian motion. The existence of a radiation impedance for the electromagnetic radiation from an oscillating charge is shown to imply a fluctuating electric field in the vacuum, and application of the general theorem yields the Planck radiation law. Finally, the existence of an acoustic radiation impedance of a gaseous medium is shown to imply pressure fluctuations, which may be related to the thermodynamic properties of the gas. The theorem thus correlates a number of known effects under one general principle and is able to predict a class of new relations.

In the final section of the paper, we discuss an intuitive interpretation of the principles underlying the theorem.

It is felt that the relationship between equilibrium fluctuations and irreversibility which is here developed provides a method for a general approach to a theory of irreversibility, using statistical ensemble methods. We are currently investigating such an approach.

II. THE DISSIPATION

A system may be said to be dissipative if it is capable of absorbing energy when subjected to a time-periodic perturbation (as an electrical resistor absorbs energy from an impressed periodic voltage). The system may be said to be linear if the power dissipation is quadratic in the magnitude of the perturbation. For a linear dissipative system, an impedance may be defined, and the proportionality constant between the power and the square of the perturbation amplitude is simply related to the impedance [in the electrical case, Power $=(\text{voltage})^2\cdot R/|Z|^2$].

In the present section we treat the applied perturbation by the usual quantum mechanical perturbation

* This work was supported in part by the ONR.

† Now at Oak Ridge National Laboratory, Oak Ridge, Tennessee.

[1] For the theory of fluctuations of extensive parameters see Fowler, *Statistical Mechanics* (Cambridge University Press, London, 1936), second edition; or Tolman, *Principles of Statistical Mechanics* (Oxford University Press, London, 1938). A recent development of the theory is given by M. J. Klein and L. Tisza, Phys. Rev. 76, 1861 (1949).

[2] A statistical mechanical theory of fluctuations of intensive parameters will be given in a subsequent paper by R. F. Greene and H. B. Callen.

[3] H. Nyquist, Phys. Rev. 32, 110 (1928). A very neat derivation and an interesting discussion is given by J. C. Slater, Radiation Laboratory Report; "Report on Noise and the Reception of Pulses," February 3, 1941, unpublished.

methods and thus relate the power dissipation to certain matrix elements of the perturbation operator. We thereby show that for small perturbations, a system with densely distributed energy levels is dissipative and linear, and we obtain certain pertinent information relative to the impedance function.

Let the hamiltonian of the system in the absence of the perturbation be H_0, a function of the coordinates $q_1 \cdots q_K \cdots$ and momenta $p_1 \cdots p_K \cdots$ of the system. In the presence of the perturbation, the hamiltonian is

$$H = H_0(\cdots q_K \cdots p_K \cdots) + VQ(\cdots q_K \cdots p_K \cdots). \quad (2.1)$$

where Q is a function of the coordinates and momenta, and V is a function of time which measures the instantaneous magnitude of the perturbation.

Again invoking the electrical case as a clarifying example, we may have V as the impressed voltage and $Q = \sum e_i x_i / L$, where e_i is the charge on the ith particle, x_i is its distance from one end of the resistor, and L is the total length of the resistor.

If the applied perturbation varies sinusoidally with time, we have

$$V = V_0 \sin\omega t. \quad (2.2)$$

We may now employ standard time-dependent perturbation theory to compute the power dissipation. Let $\psi_1, \psi_2 \cdots \psi_n \cdots$ be the set of eigenfunctions of the unperturbed hamiltonian H_0, so that

$$H_0\psi_n = E_n\psi_n, \quad (2.3)$$

and let the true wave function be ψ. Expanding ψ in terms of the ψ_n,

$$\psi = \sum_n a_n(t)\psi_n, \quad (2.4)$$

and substituting into the Schroedinger equation for ψ,

$$H_0\psi + V_0 \sin\omega t Q\psi = i\hbar\partial\psi/\partial t, \quad (2.5)$$

one obtains a set of first-order equations for the coefficients $a_n(t)$, which may readily be integrated. If the energy levels of the system are densely distributed, one thus finds that the total induced transition probability of a system initially in the state ψ_n is

$$\tfrac{1}{2}\pi V_0^2\hbar^{-1}\{|\langle E_n+\hbar\omega|Q|E_n\rangle|^2\rho(E_n+\hbar\omega) + |\langle E_n-\hbar\omega|Q|E_n\rangle|^2\rho(E_n-\hbar\omega)\}, \quad (2.6)$$

where the symbol $\langle E_n+\hbar\omega|Q|E_n\rangle$ indicates the matrix element of the operator corresponding to Q between the state with eigenvalue $E_n+\hbar\omega$ and the state with eigenvalue E_n. The symbol $\rho(E)$ indicates the density-in-energy of the quantum states in the neighborhood of E, so that the number of states between E and $E+\delta E$ is $\rho(E)\delta E$.

Each transition from the state ψ_n to the state with eigenvalue $E_n+\hbar\omega$ is accompanied by the absorption of energy $\hbar\omega$, and each transition from ψ_n to the state with eigenvalue $E_n-\hbar\omega$ is accompanied by the emission of energy $\hbar\omega$. Thus the rate of absorption of energy by a system initially in the nth state is

$$\tfrac{1}{2}\pi V_0^2\omega\{|\langle E_n+\hbar\omega|Q|E_n\rangle|^2\rho(E_n+\hbar\omega) - |\langle E_n-\hbar\omega|Q|E_n\rangle|^2\rho(E_n-\hbar\omega)\}. \quad (2.7)$$

To predict the behavior of a real thermodynamic system, we must average over-all initial states, weighting each according to the Boltzmann factor $\exp(-E_n/kT)$. Let the weighting factor be $f(E_n)$, so that

$$f(E_n+\hbar\omega)/f(E_n) = f(E_n)/f(E_n-\hbar\omega) = \exp(-\hbar\omega/kT). \quad (2.8)$$

The power dissipation is, then,

$$\text{Power} = \tfrac{1}{2}\pi V_0^2\omega\sum_n\{|\langle E_n+\hbar\omega|Q|E_n\rangle|^2\rho(E_n+\hbar\omega) - |\langle E_n-\hbar\omega|Q|E_n\rangle|^2\rho(E_n-\hbar\omega)\}f(E_n). \quad (2.9)$$

The summation over n may be replaced by an integration over energy

$$\sum_n(\ ) \rightarrow \int_0^\infty (\ )\rho(E)dE, \quad (2.10)$$

whence

$$\text{Power} = \tfrac{1}{2}\pi V_0^2\omega\int_0^\infty \rho(E)f(E) \times\{|\langle E+\hbar\omega|Q|E\rangle|^2\rho(E+\hbar\omega) - |\langle E-\hbar\omega|Q|E\rangle|^2\rho(E-\hbar\omega)\}dE. \quad (2.11)$$

We thus find that a small periodic perturbation applied to a system, the eigenstates of which are densely distributed in energy, leads to a power dissipation quadratic in the perturbation. For such a linear system it is possible to define an impedance $Z(\omega)$, the ratio of the force V to the response $\dot{Q}$, where all quantities are now assumed to be written in standard complex notation,

$$V = Z(\omega)\dot{Q}. \quad (2.12)$$

The instantaneous power is $V\dot{Q}R/|Z|$, and the average power is

$$\text{Power} = \tfrac{1}{2}V_0^2R(\omega)/|Z(\omega)|^2, \quad (2.13)$$

where $R(\omega)$, the resistance, is the real part of $Z(\omega)$.

If the applied perturbation is not sinusoidal, but some general function of time $V(t)$, and if $v(\omega)$ and $\dot{q}(\omega)$ are the fourier transforms of $V(t)$ and $\dot{Q}(t)$, the impedance is defined in terms of the fourier transforms:

$$v(\omega) = Z(\omega)\dot{q}(\omega). \quad (2.14)$$

In this notation we then obtain, for the general linear dissipative system,

$$R/|Z|^2 = \pi\omega\int_0^\infty \rho(E)f(E)\{|\langle E+\hbar\omega|Q|E\rangle|^2\rho(E+\hbar\omega) - |\langle E-\hbar\omega|Q|E\rangle|^2\rho(E-\hbar\omega)\}dE. \quad (2.15)$$

III. THE FLUCTUATION

We have, in the previous section, considered a system to which is applied a force V, eliciting a response $\dot{Q}$. We now consider the system to be left in thermal equilibrium, with no applied force. We may expect, even in this isolated condition, that the system will exhibit a spontaneously fluctuating $\dot{Q}$, which may be associated with a spontaneously fluctuating force. We shall see, in this section, that such a spontaneously fluctuating force does in fact exist, and we shall find its magnitude.

Let $\langle V^2\rangle$ be the mean square value of the spontaneously fluctuating force, and let $\langle \dot{Q}^2\rangle$ be the mean square value of the spontaneously fluctuating $\dot{Q}$. Although we shall be primarily interested in $\langle V^2\rangle$, we shall find it convenient to compute $\langle \dot{Q}^2\rangle$ and to obtain $\langle V^2\rangle$ from Eq. (2.14).

Consider that the system is known to be in the nth eigenstate. The hermitian property of H_0 causes the expectation value of $\dot{Q}$, $\langle E_n|\dot{Q}|E_n\rangle$, to vanish. The mean square fluctuation of $\dot{Q}$ is therefore given by the expectation value of $\dot{Q}^2$ or $\langle E_n|\dot{Q}^2|E_n\rangle$. Then

$$\begin{aligned}\langle E_n|\dot{Q}^2|E_n\rangle &= \textstyle\sum_m\langle E_n|\dot{Q}|E_m\rangle\langle E_m|\dot{Q}|E_n\rangle\\ &= \hbar^{-2}\textstyle\sum_m\langle E_n|H_0Q-QH_0|E_m\rangle\\ &\qquad\times\langle E_m|H_0Q-QH_0|E_n\rangle\\ &= \hbar^{-2}\textstyle\sum_m(E_n-E_m)^2|\langle E_m|Q|E_n\rangle|^2. \end{aligned}\tag{3.1}$$

Introducing a frequency ω by

$$\hbar\omega = |E_n-E_m|, \tag{3.2}$$

the summation over m may be replaced by two integrals over ω (one for $E_n<E_m$ and one for $E_n>E_m$):

$$\begin{aligned}\langle E_n|\dot{Q}^2|E_n\rangle &= \hbar^{-2}\int_0^\infty(\hbar\omega)^2|\langle E_n+\hbar\omega|Q|E_n\rangle|^2\\ &\quad\times\rho(E_n+\hbar\omega)\hbar d\omega+\hbar^{-2}\int_0^\infty(\hbar\omega)^2\\ &\quad\times|\langle E_n-\hbar\omega|Q|E_n\rangle|^2\rho(E_n-\hbar\omega)\hbar d\omega.\\ &= \int_0^\infty\hbar\omega^2\{|\langle E_n+\hbar\omega|Q|E_n\rangle|^2\rho(E_n+\hbar\omega)\\ &\quad+|\langle E_n-\hbar\omega|Q|E_n\rangle|^2\rho(E_n-\hbar\omega)\}d\omega. \end{aligned}\tag{3.3}$$

The fluctuation actually observed in a real thermodynamic system is obtained by multiplying the fluctuation in the nth state by the weighting factor $f(E_n)$, and summing

$$\begin{aligned}\langle \dot{Q}^2\rangle &= \sum_n f(E_n)\int_0^\infty\hbar\omega^2\{|\langle E_n+\hbar\omega|Q|E_n\rangle|^2\rho(E_n+\hbar\omega)\\ &\quad+|\langle E_n-\hbar\omega|Q|E_n\rangle|^2\rho(E_n-\hbar\omega)\}d\omega. \end{aligned}\tag{3.4}$$

As in Eq. (2.10), the summation over n may be replaced by an integration over the energy spectrum if we introduce the density factor $\rho(E)$. Thus we finally obtain

$$\begin{aligned}\langle \dot{Q}^2\rangle &= \int_0^\infty\hbar\omega^2\bigg[\int_0^\infty\rho(E)f(E)\{|\langle E+\hbar\omega|Q|E\rangle|^2\rho(E+\hbar\omega)\\ &\quad+|\langle E-\hbar\omega|Q|E\rangle|^2\rho(E-\hbar\omega)\}dE\bigg]d\omega, \end{aligned}\tag{3.5}$$

or, utilizing the definition (2.14) of the impedance,

$$\begin{aligned}\langle V^2\rangle &= \int_0^\infty|Z|^2\hbar\omega^2\bigg[\int_0^\infty\rho(E)f(E)\\ &\quad\times\{|\langle E+\hbar\omega|Q|E\rangle|^2\rho(E+\hbar\omega)\\ &\quad+|\langle E-\hbar\omega|Q|E\rangle|^2\rho(E-\hbar\omega)\}dE\bigg]d\omega. \end{aligned}\tag{3.6}$$

IV. THE GENERALIZED NYQUIST RELATION

In the two previous sections we have computed $R/|Z|^2$ and $\langle V^2\rangle$. These quantities involve the constructs

$$\begin{aligned}&\int_0^\infty\rho(E)f(E)\{|\langle E+\hbar\omega|Q|E\rangle|^2\rho(E+\hbar\omega)\\ &\qquad\pm|\langle E-\hbar\omega|Q|E\rangle|^2\rho(E-\hbar\omega)\}dE, \end{aligned}\tag{4.1}$$

the negative sign being associated with $R/|Z|^2$ and the positive sign with $\langle V^2\rangle$. We shall now see that the two values of (4.1) are simply related, and thus establish the desired relation between $\langle V^2\rangle$ and $R(\omega)$.

Consider first the value of (4.1) corresponding to the negative sign, which we denote by $C(-)$.

$$\begin{aligned}C(-) &= \int_0^\infty f(E)|\langle E+\hbar\omega|Q|E\rangle|^2\rho(E+\hbar\omega)\rho(E)dE\\ &\quad-\int_0^\infty f(E)|\langle E-\hbar\omega|Q|E\rangle|^2\rho(E)\rho(E-\hbar\omega)dE. \end{aligned}\tag{4.2}$$

In the second integral we note that $\langle E-\hbar\omega|Q|E\rangle$ vanishes for $E<\hbar\omega$, and making the transformation $E\rightarrow E+\hbar\omega$ in the integration variable, we obtain

$$\begin{aligned}C(-) &= \int_0^\infty|\langle E+\hbar\omega|Q|E\rangle|^2\rho(E+\hbar\omega)\rho(E)f(E)\\ &\quad\times\{1-f(E+\hbar\omega)/f(E)\}dE. \end{aligned}\tag{4.3}$$

By Eq. (2.8) this becomes

$$\begin{aligned}C(-) &= \{1-\exp(-\hbar\omega/kT)\}\int_0^\infty|\langle E+\hbar\omega|Q|E\rangle|^2\\ &\quad\times\rho(E+\hbar\omega)\rho(E)f(E)dE. \end{aligned}\tag{4.4}$$

If $C(+)$ denotes the value of (4.1) corresponding to the positive sign, we obtain, in an identical fashion,

$$C(+)=\{1+\exp(-\hbar\omega/kT)\}\int_0^\infty |\langle E+\hbar\omega|Q|E\rangle|^2 \times\rho(E+\hbar\omega)\rho(E)f(E)dE. \quad (4.5)$$

With these alternative expressions for (4.1), we can write, from Eq. (2.15),

$$R(\omega)/|Z(\omega)|^2=\pi\omega\{1-\exp(-\hbar\omega/kT)\} \times\int_0^\infty |\langle E+\hbar\omega|Q|E\rangle|^2\rho(E+\hbar\omega)\rho(E)f(E)dE, \quad (4.6)$$

and from Eq. (3.6),

$$\langle V^2\rangle=\int_0^\infty |Z|^2\hbar\omega^2\{1+\exp(-\hbar\omega/kT)\} \times\int_0^\infty |\langle E+\hbar\omega|Q|E\rangle|^2\rho(E+\hbar\omega)\rho(E)f(E)dEd\omega. \quad (4.7)$$

Comparison of these equations yields directly our fundamental theorem:

$$\langle V^2\rangle=(2/\pi)\int_0^\infty R(\omega)E(\omega, T)d\omega, \quad (4.8)$$

where

$$E(\omega, T)=\tfrac{1}{2}\hbar\omega+\hbar\omega[\exp(\hbar\omega/kT)-1]^{-1}. \quad (4.9)$$

It may be recognized that $E(\omega, T)$ is, formally, the expression for the mean energy at the temperature T of an oscillator of natural frequency ω.

At high temperatures, $E(\omega, T)$ takes its equipartition value

$$E(\omega, T)\simeq kT, \quad (kT\gg\hbar\omega) \quad (4.10)$$

and the generalized Nyquist relation takes its most familiar form

$$\langle V^2\rangle\simeq(2/\pi)kT\int R(\omega)d\omega. \quad (4.11)$$

To reiterate then, a system with a generalized resistance $R(\omega)$ exhibits, in equilibrium, a fluctuating force given by Eq. (4.8) or, at high temperature, by Eq. (4.11).

We shall now consider a few specific applications of this theorem. The application to the electrical case is obvious, the general Eq. (4.8) being identical with the Nyquist relation if the force V is interpreted as the voltage. The content of the general theorem is, however, clarified by considering certain less trivial applications.

V. APPLICATION TO BROWNIAN MOTION

The fundamental result of the theory of the Brownian motion of a small particle immersed in a fluid is that the particle moves in response to a randomly fluctuating force $F(t)$ (with components F_x, F_y, F_z) such that

$$\langle F_x^2\rangle=(2/\pi)kT\eta\int d\omega. \quad (5.1)$$

Here η is a frictional constant, so defined that the viscous drag on a particle moving with velocity v is

$$\text{Frictional force}=-\eta v. \quad (5.2)$$

(If, in particular, the particle is spherical, η is known by Stokes' law as $6\pi\cdot$(viscosity)$\cdot$(radius).)

It is interesting to recall briefly the rather complicated and circuitous chain of reasoning by which the above result is obtained. One first makes the *assumption* that the particle moves in response to a randomly fluctuating force which has a *constant*, but unknown, spectral density. (The spectral density is, in actuality, not constant, and Eq. (5.1) is not valid at high frequencies.) By application of the theory of stochastic processes, one is then able to predict the distribution functions for either the displacement or the velocity of the particle.[4] The distribution function for displacement yields the diffusion constant, which in turn may be related by the Einstein relation[5] to the frictional constant η, thus evaluating the spectral density.[6] Alternatively, the distribution function for velocity yields the energy, which is known by the equipartition theorem and which therefore evaluates the spectral density, yielding Eq. (5.1).

We now apply our general formalism to the Brownian motion. We assume the existence of a viscous force as given by Eq. (5.2). The system of a particle in a fluid, the particle being acted on by an external force, is then dissipative and linear. The real part of the impedance is simply η (the inertial mass of the particle giving a pure reactance of $m\omega$). We conclude immediately, in accordance with Eq. (4.8), that a particle in a fluid is acted upon by a spontaneously fluctuating force for which

$$\langle F_x^2\rangle=(2/\pi)\eta\int_0^\infty E(\omega, T)d\omega. \quad (5.3)$$

For high temperatures or low frequencies, $(\hbar\omega\ll kT)$; this reduces to Eq. (5.1).

VI. ELECTRIC DIPOLE RADIATION RESISTANCE AND ELECTRIC FIELD FLUCTUATIONS IN THE VACUUM

An oscillating electric charge radiates energy, leading to a radiation resistance. We shall see that this radiation resistance implies a fluctuating electric field as given by the Planck radiation law.

[4] See M. C. Wang and G. E. Uhlenbeck, Revs. Modern Phys. 17, 323 (1945); and J. L. Doob, Ann. Math. 43, 351 (1942).

[5] See A. Einstein, *Investigations on the Theory of the Brownian Movement* (Dutton and Company, New York); or A. Einstein, Ann. Physik 17, 549 (1905).

[6] A similar analysis has been applied to the flow of heat by L. S. Ornstein and J. M. W. Milatz, Physica 6, 1139 (1939).

Consider a dipole, of charge e, displacement x, and dipole moment $p=ex$. Let one charge be fixed and let the other oscillate so that

$$P=P_0 \sin\omega t. \tag{6.1}$$

It is well known that the electric dipole radiation leads to a dissipative force[7]

$$F_d=-\tfrac{2}{3}e^2c^{-3}d^2v/dt^2, \tag{6.2}$$

where v is the velocity of the moving charge. The equation of motion of this charge is

$$mdv/dt+m\omega_0^2x+F_d=F, \tag{6.3}$$

where F is the applied force, and ω_0 is the natural frequency associated with the intra-dipole binding force. Inserting (6.1) in (6.3) we get

$$F=mP_0e^{-1}(\omega_0^2-\omega^2)\sin\omega t+\tfrac{2}{3}e\omega^3c^{-3}P_0\cos\omega t. \tag{6.4}$$

One may note that the average rate of radiation of energy $\langle Fv\rangle$ is

$$\langle Fv\rangle=\tfrac{1}{2}(\tfrac{2}{3}e\omega^3c^{-3}P_0)(\omega P_0e^{-1})=\tfrac{1}{3}\omega^4c^{-3}P_0^2. \tag{6.5}$$

The real part of the impedance is obtained by taking the ratio of the in-phase component of F to v. Thus

$$R(\omega)=(\tfrac{2}{3}e\omega^3c^{-3}P_0)/(\omega P_0e^{-1})=\tfrac{2}{3}e^2c^{-3}\omega^2. \tag{6.6}$$

According to our general theorem, we now deduce that there exists a randomly fluctuating force $e\mathcal{E}_x$ on the charge, and hence a randomly fluctuating electric field $\mathcal{E}_x$, such that

$$\langle e^2\mathcal{E}_x^2\rangle=(2/\pi)\int_0^\infty E(\omega,T)\tfrac{2}{3}e^2c^{-3}\omega^2d\omega,$$

or

$$\langle\mathcal{E}_x^2\rangle=(4/3)\pi^{-1}c^{-3}\times\int_0^\infty\{\tfrac{1}{2}\hbar\omega+\hbar\omega[\exp(\hbar\omega/kT)-1]^{-1}\}\omega^2d\omega. \tag{6.7}$$

This conclusion can be put into a more familiar form by utilizing the fact that the energy density in an isotropic radiation field is simply

$$\text{Energy density}=\langle\mathcal{E}^2\rangle/4\pi=3\langle\mathcal{E}_x^2\rangle/4\pi \tag{6.8}$$

whence

$$\text{Energy density}=\pi^{-2}c^{-3}\int_0^\infty\{\tfrac{1}{2}\hbar\omega+\hbar\omega[\exp(\hbar\omega/kT)-1]^{-1}\}\omega^2d\omega. \tag{6.9}$$

The first term in this equation gives the familiar infinite "zero-point" contribution, and the second term gives the Planck radiation law.[8]

[7] W. Heitler, *The Quantum Theory of Radiation* (Oxford University Press, London, 1936).

[8] The interaction of free electron and radiation field has been discussed from a somewhat different point of view by W. Pauli, Z. Physik 18, 272 (1923); A. Einstein and P. Ehrenfest, Z. Physik 19, 301 (1923).

VII. ACOUSTIC RADIATION RESISTANCE AND PRESSURE FLUCTUATIONS IN A GAS

We now consider the acoustic radiation from a small oscillating sphere in a gaseous medium. This radiation leads to a radiation impedance which, in accordance with our general theorem, implies a fluctuating pressure in the gas.

The wave equation for the propogation of pressure waves in the gas is

$$\nabla^2P=c^{-2}\partial^2P/\partial t^2, \tag{7.1}$$

where c is the velocity of sound in the gas. Let the radius of the sphere be a, and let

$$a=a_0+e^{-i\omega t}\delta a \tag{7.2}$$

so that the sphere expands and contracts sinusoidally. The boundary condition to be satisfied by the pressure waves at $r=a_0$ is

$$\rho\partial^2a/\partial t^2=-\partial P/\partial r \quad\text{at}\quad r=a_0, \tag{7.3}$$

where ρ is the equilibrium value of the density. The solution of these equations is readily found to be

$$P=r^{-1}P_0\exp(iKr-i\omega t), \tag{7.4}$$

where

$$K=\omega/c \tag{7.5}$$

and

$$P_0=-\rho\omega^2a_0^2\delta a[1+iKa_0]\cdot[1+(Ka_0)^2]^{-1}\times\exp(-iKa_0). \tag{7.6}$$

Thus, the compressive force acting on the surface of the sphere is

$$F=4\pi a_0P_0\exp(iKa_0-i\omega t), \tag{7.7}$$

and defining the radiation impedance as the ratio of complex force to complex velocity, we find

$$\begin{aligned}Z&=F/[-i\omega\exp(-i\omega t)\delta a]\\ Z&=[4\pi a_0^2\rho c(Ka_0)^2-i4\pi a_0^2\rho cKa_0]/[1+(Ka_0)^2].\end{aligned} \tag{7.8}$$

The generalized Nyquist relation now states that a sphere immersed in a gas will experience a fluctuating compressive force, such that

$$\langle F^2\rangle=(2/\pi)\int E(\omega,T)4\pi a_0^2\rho c(\omega a_0/c)^2\times[1+(\omega a_0/c)^2]^{-1}d\omega. \tag{7.9}$$

The fluctuating pressure is the compressive force per unit area on a vanishingly small sphere.

$$\langle P^2\rangle=\lim_{a_0\to0}\langle F^2\rangle/(4\pi a_0^2)^2, \tag{7.10}$$

or

$$\langle P^2\rangle=\tfrac{1}{2}\pi^{-2}c^{-1}\rho\int E(\omega,T)\omega^2d\omega. \tag{7.11}$$

This result may be checked by a direct computation paralleling the standard derivation of the Planck radiation law for the electromagnetic modes in a vacuum. The number of acoustic modes with frequency between ω and $\omega+d\omega$ is $\frac{1}{2}\pi^{-2}c^{-3}\omega^2 d\omega$, and the acoustic energy density is

$$\text{Energy density} = \int E(\omega, T)\tfrac{1}{2}\pi^{-2}c^{-3}\omega^2 d\omega. \quad (7.12)$$

Employing the relation that the acoustic energy density is proportional to the mean square excess pressure

$$\text{Energy density} = \rho^{-1}c^{-2}\langle P^2\rangle, \quad (7.13)$$

we again obtain Eq. (7.11).

It is interesting to compare the above result with the pressure fluctuations at a boundary of the gas. The proximity to the boundary, and the shape of the boundary, may be expected to influence the radiation impedance and hence the pressure fluctuations. We consider the pressure fluctuations immediately contiguous to a plane rigid boundary, and we shall find that for this simple case, the mean square pressure fluctuation is just twice that in the volume of the gas.

Consider a plane wall bounding a semi-infinite region containing the gas. If the wall contains a circular piston of radius a_0, the radiation resistance is[9]

$$R = \pi a_0^2 \rho c[1 - ca_0^{-1}\omega^{-1}J_1(2\omega a_0/c)], \quad (7.14)$$

where J_1 indicates the first order bessel function. The fluctuating force acting on a circular area in a plane boundary is therefore

$$\langle F^2\rangle = (2/\pi)\int E(\omega, T)\pi a_0^2 \rho c \times[1 - ca_0^{-1}\omega^{-1}J_1(2\omega a_0/c)]d\omega, \quad (7.15)$$

and the fluctuating pressure is

$$\langle P^2\rangle = \lim_{a_0\to 0}\langle F^2\rangle/(\pi a_0^2)^2 \quad (7.16)$$

or

$$\langle P^2\rangle = \rho\pi^{-2}c^{-1}\int E(\omega, T)\omega^2 d\omega. \quad (7.17)$$

Thus the mean square fluctuating wall pressure, as given by (7.17), is just twice the mean square fluctuating volume pressure, as given by (7.11). This factor of two clearly arises from the fact that the pressure waves in the gas must have velocity nodes at the wall. Fluctuations in the neighborhood of the wall may be found by treating the radiation from an oscillating sphere near a reflecting boundary.

Finally, it will be noted that the above equations for pressure fluctuations involve the velocity of sound in the gas, which is not a usual thermodynamic parameter. This quantity may, however, be expressed in terms of standard thermodynamic quantities. Thus for frequencies which are sufficiently high that the compressions in the acoustic waves may be considered to be adiabatic, we have[9]

$$c^2 = C_P C_V^{-1}\rho^{-1}\mathcal{K}_T^{-1}, \quad (7.18)$$

where C_P and C_V are the specific heats at constant pressure and volume, ρ is the density, and $\mathcal{K}_T$ is the isothermal compressibility. For these frequencies, the pressure fluctuations in the volume of the gas are thus given by

$$\langle P^2\rangle = \tfrac{1}{2}\pi^{-2}\rho^2\mathcal{K}_T C_V C_P^{-1}\int E(\omega, T)\omega^2 d\omega. \quad (7.19)$$

VIII. CONCLUSION

The generalized Nyquist relation establishes a quantitative correlation between dissipation, as described by the resistance, and certain fluctuations. It seems to be possible to give an intuitive interpretation of such a connection.

A dissipative process may be conveniently considered to involve the interaction of two systems, which we characterize as the "source system" and the "dissipative system." The dissipative system, explicitly considered in Secs. II and III, is necessarily a system with densely distributed energy levels and is capable of absorbing energy when acted upon by a periodic force. The source system is the system which provides this periodic force and which delivers energy to the dissipative system.

Assume the source system to be first isolated from the dissipative system and to be given some internal energy. If the source system is a simple dynamical system, its subsequent dynamics will be periodic (as, for instance, the oscillations of a pendulum or of a polyatomic molecule). The system may be thought of as possessing a sort of internal coherence. If, now, the source system is allowed to act on the dissipative system, this internal coherence is destroyed, the periodic motion vanishes and the energy is sapped away, and the source system is left at last with only the random disordered energy ($\simeq kT$) characteristic of thermal equilibrium. This loss of coherence within the source system may be thought of as being *caused* by the random fluctuations generated by the dissipative system and acting on the source system. The dissipation thus appears as the macroscopic manifestation of the disordering effect of the Nyquist fluctuations and, as such, is necessarily quantitatively correlated with the fluctuations.

An analogy which is perhaps useful is provided by the historical development of the theory of spontaneous radiation from excited atoms. After the initial development of quantum mechanics, it was found impossible to compute the spontaneous transition probabilities for an isolated excited atom, and this dissipative process appeared to be outside the existing structure of dynamics. With the development of quantum electrodynamics, however, the dissipation could be computed,

[9] P. M. Morse, *Vibration and Sound* (McGraw-Hill Book Company, Inc., New York, 1936).

and it was found that the "spontaneous" transitions could be consistently considered to be induced by the random fluctuations of the electromagnetic field in the vacuum. In this case, of course, the excited atom plays the role of the source system, and the "vacuum" plays the role of the dissipative system.

It would thus appear that a reasonable approach to the development of a theory of linear irreversible processes is through the development of the theory of fluctuations in equilibrium systems. Certain results in this connection will be given in subsequent papers by Richard F. Greene and one of the authors (H.B.C.).

PHYSICAL REVIEW VOLUME 87, NUMBER 3 AUGUST 1, 1952

Statistical Theory of Equations of State and Phase Transitions. I. Theory of Condensation

C. N. YANG AND T. D. LEE
Institute for Advanced Study, Princeton, New Jersey
(Received March 31, 1952)

A theory of equations of state and phase transitions is developed that describes the condensed as well as the gas phases and the transition regions. The thermodynamic properties of an infinite sample are studied rigorously and Mayer's theory is re-examined.

I. INTRODUCTION

THIS and a subsequent paper will be concerned with the problem of a statistical theory of equations of state and phase transitions. This problem has always interested physicists both from the practical viewpoint of seeking for a workable theory of properties of matter (such as a theory of liquids) and also from the more academic viewpoint of understanding the occurrence of the discontinuities associated with phase transitions in the thermodynamic functions.

The work reported in this paper is quite general and fairly abstract. We are returning in a subsequent paper to the illustration and application of the methods here outlined. In order to present the work of this present paper in its proper perspective, it may be helpful if we outline briefly the history of our own thinking on the subject.

About a year ago one of us was able to make progress[1] with the problem of the spontaneous magnetization of the Ising model, taking advantage of some special properties of this problem when treated by the Onsager-Kaufman method.[2] We then noted that the solution there obtained was also the solution of another, physically quite different, but formally identical, problem. This is the problem of a lattice gas with attractive interaction between nearest neighbors. We were thus able to follow in detail the behavior of such a lattice gas, which in many ways should reveal the features of an actual gas. In particular, we were able to study and characterize the condensation phenonenon, and to identify the liquid, gas, and transition regions in the $p-v$ diagram. The isotherms thus obtained are flat in the transition region and rise very rapidly with increasing density in the liquid phase. At this point, we were led to compare the specific solution with the well-known work[3] of Mayer on the theory of condensation of gases. In particular we were led to inquire as to why, in Mayer's theory, the isotherms stay flat beyond the condensation point and do not give the equation of state for the liquid phase. It soon became apparent that this difference lay, not in the difference of the models, but in the inadequacy of Mayer's method for dealing with a condensed phase. This led to a study of the analytical behavior of the grand partition function of an assembly of interacting atoms, and we were able, as in the special case mentioned above, to identify and characterize quite generally the condensation phenomena. These general conclusions will be presented in the present paper.

The problem is approached by allowing the fugacity to take on complex values. Although only real values of the fugacity are of any physical interest, the analytical behavior of the thermodynamic functions can only be completely revealed by going into the complex plane, whereby one is able to obtain a description of the condensed phases as well as the gas phase and the transition regions. This approach is of a very general nature and can be applied to other problems of phase transitions such as ferromagnetism, order disorder transition, etc. It will be emphasized that also this approach can lead to practical approximation methods for the description of systems undergoing transitions. These points will be discussed in paper II.

The physical conclusions of this paper derive from some mathematical results which we shall state in the form of two theorems. Due to the nature of the problem (which involves a double limiting process) it is imperative to have mathematical rigor preserved throughout. The proofs are necessarily of a mathematical nature and will be given in the appendix.

II. INTERACTION

We consider a monatomic gas with the interaction

$$U=\sum u(r_{ij}), \tag{1}$$

where r_{ij} is the distance between the ith and jth atoms. The following assumptions are made about the nature of these interactions:

(1) The atoms have a finite impenetrable core of diameter a, so that $u(r)=+\infty$ for $r\leqq a$.

(2) The interaction has a finite range b so that

$$u(r)=0 \text{ for } r\geqq b.$$

(3) $u(r)$ is nowhere minus infinity.

The theory can be easily generalized to include many body forces and forces with a weak long tail such as

[1] C. N. Yang, Phys. Rev. 85, 808 (1952).

[2] L. Onsager, Phys. Rev. 65, 117 (1944); B. Kaufman, Phys. Rev. 76, 1232 (1949).

[3] J. E. Mayer, J. Chem. Phys. 5, 67 (1937); J. E. Mayer and Ph. G. Ackermann, J. Chem. Phys. 5, 74 (1937); J. E. Mayer and S. F. Harrison, J. Chem. Phys. 6, 87, 101 (1938); B. Kahn and G. E. Uhlenbeck, Physica 5, 399 (1938). M. Born and K. Fuchs, Proc. Roy. Soc. (London) A166, 391 (1938).

van der Waals' force. But for clarity we shall first treat only interactions with the properties enumerated above.

Consider a box of volume V kept at a constant temperature T. If it is allowed to exchange atoms with a reservoir at a given chemical potential μ per atom, the relative probability of having N atoms in the box is

$$Q_N y^N/N!,$$

where

$$Q_N=\int\cdots\int_V d\tau_1\cdots d\tau_N \exp(-U/kT) \qquad (2)$$

is the configurational part of the partition function for N atoms and

$$y=(2\pi mkT/h^2)^{\frac{3}{2}}\exp(\mu/kT). \qquad (3)$$

The quantities m, k, and h have the usual meanings,

The grand partition function of the gas in the volume V is

$$\mathfrak{Q}_V=\sum_{N=0}^{M}\frac{Q_N}{N!}y^N, \qquad (4)$$

where M is the maximum number of atoms that can be crammed into V.

III. THE LIMIT OF INFINITE VOLUME

The average pressure and the average density of such a gas in V are calculable in terms of $\mathfrak{Q}_V$ by the standard treatment of statistical mechanics, and are evidently dependent on V. In thermodynamics, however, one is only interested in an infinite sample and the thermodynamic functions are limits of these average quantities as $V\rightarrow\infty$. The pressure p and density ρ are accordingly given by

$$\frac{p}{kT}=\operatorname{Lim}_{V\rightarrow\infty}\frac{1}{V}\log\mathfrak{Q}_V, \qquad (5)$$

$$\rho=\operatorname{Lim}_{V\rightarrow\infty}\frac{\partial}{\partial\log y}\frac{1}{V}\log\mathfrak{Q}_V. \qquad (6)$$

The question of whether these limits do exist is usually not discussed.[4] It is, however, generally believed that in the gas phase such limits do exist and that (5) and (6) give the correct equation of state. At the point of condensation and in the liquid phase the situation has been extremely unclear. As a matter of fact doubts have been raised[5] as to whether the equation of state of both the liquid and the gas phase can be obtained from the same interaction (1) through the considerations of statistical mechanics.

We shall try to resolve these problems and prove that (5) and (6) do give a complete description of the equation of state of both the gas and condensed phases. In fact in paper II we shall give a concrete example in which it is seen how the same partition function describes both phases, and in which the two-phase-equilibrium region is exactly known.

We first state the following:

Theorem 1.—(Proved in Appendix I.) For all positive real values of y, $V^{-1}\log\mathfrak{Q}_V$ approaches, as $V\rightarrow\infty$, a limit which is independent of the shape of V. Furthermore, this limit is a continuous, monotonically increasing function of y.

The assumption is made, of course, that the shape of V is not so queer that its surface area increases faster than $V^{\frac{2}{3}}$.

One might be tempted to conclude from the independence of the limit on the shape of V that the system under consideration exists only in fluid phases (i.e., gas and liquid) with no elastic resistance against shearing strain. It is to be emphasized that this is not the case. The independence of the limit on the shape of V is not due to the lack of elastic resistance against shearing strain, but rather due to the fact that for an infinite sample changing the shape of V does not produce a strain in the interior which might serve to differentiate between a fluid and a solid. This is so because the strain at the boundary only penetrates to a finite depth and is inconsequential for an infinite sample.

To study the limit of $(\partial/\partial\log y)V^{-1}\log\mathfrak{Q}_V$ we notice that $\mathfrak{Q}_V$ is a polynomial in y of finite degree M. This is a direct consequence of the assumed impenetrable core of the atoms. It is therefore possible to factorize $\mathfrak{Q}_V$ and write

$$\mathfrak{Q}_V=\prod_{i=1}^{M}\left(1-\frac{y}{y_i}\right), \qquad (7)$$

where $y_1,\cdots y_M$ are the roots of the algebraic equation

$$\mathfrak{Q}_V(y)=0. \qquad (8)$$

Evidently none of these roots can be real and positive, since all the coefficients in the polynomial $\mathfrak{Q}_V$ are positive.

As V increases these roots move about in the complex y plane and their number M increases (essentially) linearly with V. Their distribution in the limit $V\rightarrow\infty$ gives the complete analytic behavior of the thermodynamic functions in the y plane. In fact one can prove the following:

Theorem 2.—(Proved in Appendix II.) If in the complex y plane a region R containing a segment of the positive real axis is always free of roots, then in this region as $V\rightarrow\infty$ all the quantities:

$$\frac{1}{V}\log\mathfrak{Q}_V,\quad \left(\frac{\partial}{\partial\log y}\right)\frac{1}{V}\log\mathfrak{Q}_V,\quad \left(\frac{\partial}{\partial\log y}\right)^2\frac{1}{V}\log\mathfrak{Q}_V\cdots,$$

[4] The behavior of the partition function Q_N as the volume approaches infinite was discussed by L. von Hove, Physica 15, 951 (1949), where it is proved that $N^{-1}\log Q_N$ approaches a limit as the volume approaches infinity at constant density. His proof is similar to our proof of theorem 1.

[5] There was apparently some discussion on this point at the International Conference held in Amsterdam, 26 November 1937. The doubts can perhaps be formulated in the form of the question: "How can the gas molecules know when they have to coagulate to form a liquid or solid?" See p. 391 of reference 3 (Born and Fuchs).

approach limits which are analytic with respect to y. Furthermore the operations $(\partial/\partial \log y)$ and $\text{Lim}_{V\to\infty}$ commute in R so that, e.g.,

$$\underset{V\to\infty}{\text{Lim}}\frac{\partial}{\partial \log y}\frac{1}{V}\log\mathcal{Q}_V=\frac{\partial}{\partial \log y}\underset{V\to\infty}{\text{Lim}}\frac{1}{V}\log\mathcal{Q}_V. \quad (9)$$

This gives, together with (5) and (6),

$$\rho=(\partial/\partial \log y)(p/kT). \quad (10)$$

IV. PHASE TRANSITIONS

The quantity $(\partial/\partial \log y)V^{-1}\log\mathcal{Q}_V$ does not, however, always approach a limit ρ for all values of y. Physically this must evidently be as the density of the system does not assume a single value at the point of condensation. It is clear therefore that the problem of phase transition is intrinsically related to the form of the regions R described in theorem 2. We discuss the following cases:

(1) The roots of $\mathcal{Q}_V(y)=0$ do not close in onto the positive real axis of y as $V\to\infty$, or more exactly, there exists a region R which contains the whole positive real axis and is free of roots.

In this case from the two theorems one concludes that the pressure and density of the system are analytic functions of y (along the positive real axis). They are related by Eq. (10). Furthermore, p is an increasing function of y. We shall show in Appendix III that ρ is also an increasing function of y. Consequently in the $p-\rho$ diagram, on the isotherm p increases analytically as the specific volume v decreases. The system under consideration is thus a single phase system (see Fig. 1).

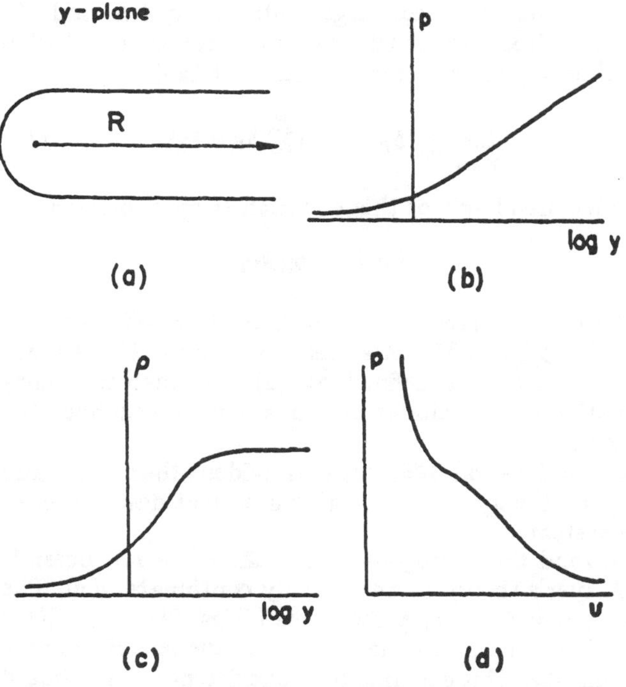

Fig. 1. Analytical behavior at a given temperature of thermodynamic functions for a single phase system. The quantity y is defined by Eq. (3) in the text. The region R is free of roots of Eq. (8). Notice that the density ρ of (c) is proportional to the slope of the p—log$y$ curve in (b).

(2) The roots of $\mathcal{Q}_V(y)=0$ do close in onto the real axis as $V\to\infty$, say at the points $y=t_1, t_2$; and regions R_1, R_2, and R_3 free of roots enclose, respectively, the three segments of the positive real axis as in Fig. 2(a).

By the same reasoning as in the previous case one concludes, within any one of these three segments, that the system exists in a single phase, that p and ρ are analytic and increasing functions of y, that ρ is $(kT)^{-1}(\partial p/\partial \log y)$, and that on the isotherm p increases analytically as v decreases.

At the points $y=t_1, t_2$ the pressure p is continuous (by theorem 1), but its derivative ρ has in general a discontinuity. By Appendix III one shows easily that ρ increases across the discontinuity. The functions p and ρ are schematically plotted in Figs. 2(b) and 2(c) which together give the isotherm in Fig. 2(d).

As the temperature varies the points t_1 and t_2 will in general move along the y axis. If at a certain temperature T_c the roots cease to close in onto one of the points, say t_1, then T_c is the critical temperature for the transition phase 1$\leftrightarrow$ phase 2. If, on the other hand, t_1 and t_2 merge together at a particular temperature T_0, we would then have a triple point at that temperature.

It may be remarked that at $y=t_1$ or t_2 the density ρ may in some cases be continuous (although its derivative will in general be discontinuous). At the critical temperature this will happen, but not at neighboring temperatures. If, however, this happens over an extended temperature range, one would have a transition of second (or higher) order.

It is clear therefore that phase transitions of the system occur only at the points on the positive real y axis onto which the roots of $\mathcal{Q}(V)=0$ close in as $V\to\infty$. For other values of the fugacity y a single phase system obtains.

As mentioned before, the theory can be easily generalized to include many body forces and forces with a weak long tail. In fact, the generalization does not lead to any alterations of the conclusions reached above.

Generalization can also be made to other kinds of phase transitions such as order-disorder phenomena and ferromagnetism, as will be discussed in paper II. The study of the equations of state and phase transitions can thus be reduced to the investigation of the distribution of roots of the grand partition function. In many cases, as will be seen in paper II, such distributions turn out to have some surprisingly simple regularities.

V. COMPARISON WITH MAYER'S THEORY

We first notice that by expanding in powers of y one obtains from (7)

$$\frac{1}{V}\log\mathcal{Q}_V=\sum_{l=1}^{\infty}b_l(V)y^l, \quad (11)$$

where

$$b_l(V) = \frac{-1}{lV} \sum_{j=1}^{M} \left(\frac{1}{y_j}\right)^l. \tag{12}$$

Combining (11) and (3) we have

$$\frac{Q_N}{N!} = \text{coefficient of } y^N \text{ in } \exp[V \sum_{l=1}^{\infty} b_l y^l]. \tag{13}$$

Comparison of this equation with Mayer's theory shows that the b_l's defined by (12) are identical with the reducible cluster integrals defined[6] by Mayer. It is interesting to notice that these reducible cluster integrals are, according to (12), closely related to the moments of the roots y_j of Eq. (8). It should be emphasized that in both (12) and in Mayer's definition the b_l's are functions of the volume V. It is evident from Mayer's definition that they approach definite limits $b_l(\infty)$ as $V \to \infty$.

In Mayer's theory the cluster integrals b_l are replaced from the very beginning by their limiting values $b_l(\infty)$. He then considers the series

$$\chi(y) = \sum_{1}^{\infty} b_l(\infty) y^l \tag{14}$$

and its analytical continuation along the positive real axis. If one calls the first singularity of $\chi(y)$ along the positive real axis t_1, one shows in Mayer's theory that

(1) for densities ρ less than

$$\rho_1 = \lim_{y \to t_1 -} y\chi'(y), \tag{15}$$

the system exists in a single phase;

(2) for $\rho \geqq \rho_1$, the pressure p (at a given temperature) becomes independent of the density. Consequently, one identifies the density ρ_1 as the density of the gas at condensation.

An essential difficulty of Mayer's theory is that it does not admit of the existence of a liquid phase with finite density, since the isotherm remains horizontal for all specific volume less than ρ_1^{-1}. This is clearly due to the replacement of the volume dependent b_l's by their limiting values. The question is therefore often raised[7] as to exactly at what point on the isotherm Mayer's theory breaks down.

In the present theory by retaining the volume dependence of the partition function $\mathfrak{Q}_V$ we do not encounter these difficulties. To clarify the relationship with Mayer's theory, we refer back to Fig. 2(a) and draw a small circle C within R_1 with the center at the origin. The series

$$\sum_{l=1}^{\infty} b_l(V) y^l$$

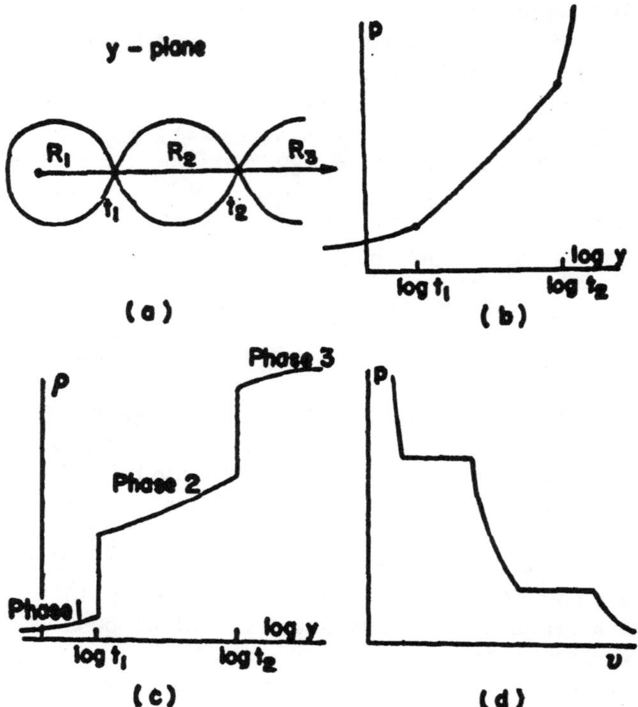

Fig. 2. Analytical behavior at a given temperature of thermodynamic functions for a system that undergoes two phase transitions. The transitions occur at t_1 and t_2 which are the points at which the roots of Eq. (8) close in onto the positive real y axis. The regions R_1, R_2, and R_3 are free of roots. The three phases 1, 2, and 3 are indicated in (c). The horizontal parts of (d) represent two-phase-equilibrium regions.

is easily shown[8] to converge uniformly in the circle C. By a well-known mathematical theorem on double limiting processes one concludes that in C

$$\lim_{V\to\infty} \sum_{1}^{\infty} b_l(V) y^l = \sum_{1}^{\infty} b_l(\infty) y^l. \tag{16}$$

The left-hand side of this equation is by definition

$$\lim_{V\to\infty} V^{-1} \log \mathfrak{Q}_V,$$

and the right-hand side $\chi(y)$. Therefore within C the function $\chi(y)$ in Mayer's theory is indeed $(kT)^{-1}$ times the pressure p as defined by (5). By analytical continuation one concludes that this holds throughout the region R_1.

In the interval $0 \leqq y < t_1$ it is evident that $\rho < \rho_1$ and Mayer's theory is seen to give a correct description of the system.

Beyond the point $y = t_1$ (i.e., $y \geqq t_1$) it is not possible in Mayer's theory to analytically continuate $\chi(y)$. The $p - \log y$ and $\rho - \log y$ diagrams [Figs. 2(b) and 2(c)] therefore exist in his theory only to the left of the first singularity. This explains the nonexistence of the liquid phase in Mayer's theory.

---

[6] See for example Eq. (13.5) in J. E. Mayer and Mayer, *Statistical Mechanics* (John Wiley and Sons, Inc., New York, 1946), p. 280.

[7] See, for example, reference 3 (Kahn and Uhlenbeck), p. 415.

[8] All roots y_j have absolute values larger than the radius r of the circle C. By Eq. (12) we have $|b_l| \leqq (M/V) l^{-1} r^{-l}$. But M/V is bounded. Hence the statement.

We thus remark:

(1) Throughout the gas phase (i.e., $\rho<\rho_1$) Mayer's theory gives correct results.

(2) For $\rho\geqq\rho_1$ Mayer's conclusion that the $p-v$ diagram becomes horizontal is, as already mentioned, incorrect for high densities due to the existence of the liquid phase. It is not even justified for densities immediately above ρ_1, as for transitions of high order the isotherm does not even have any horizontal part at all.

We are indebted to Professor J. R. Oppenheimer for criticism and comment.

APPENDIX I

To prove theorem 1 we first establish the following:

Lemma 1.—Let V and W be two cubes of linear dimensions L and $L+(b/2)$, respectively. Keeping b fixed one has as $L\rightarrow\infty$,

$$\mathrm{Lim}\,W^{-1}(\log\mathfrak{Q}_W-\log\mathfrak{Q}_V)=0. \tag{17}$$

Proof.—Put V completely inside W and write $\mathfrak{Q}_W$ as the sum of contributions $A_0, A_1\cdots$ from configurations with zero, one . . . atoms outside of V. Now (a) since the interaction has finite range, each atom interacts with at most a finite and definite number of other atoms. Also (b) the available volume for the first atom outside of V is $\Delta=W-V$. If the volume of the impenetrable core of an atom is α, the available volume for the second atom outside of V is less than $\Delta-\alpha$, the third, $\Delta-2\alpha$, etc. Combining (a) and (b) one concludes that

$$A_m\leqq\beta^m[\Delta(\Delta-\alpha)(\Delta-2\alpha)\cdots(\Delta-m\alpha+\alpha)/m!]\mathfrak{Q}_V, \tag{18}$$

where β is a constant. (This inequality is obtained by comparing the contributions to A_m and $\mathfrak{Q}_V$ of a distribution of atoms with, say, N atoms inside of V and m atoms outside.) Adding all A's one obtains

$$\mathfrak{Q}_W\leqq\mathfrak{Q}_V(1+\beta\alpha)^{\Delta/\alpha}.$$

But $\Delta\sim L^2$, and clearly

$$\mathfrak{Q}_V\leqq\mathfrak{Q}_W.$$

Hence the lemma.

Lemma 2.—Let W_i be a cube of linear dimension 2^iL and $\mathfrak{Q}_i$ an abbreviation for $\mathfrak{Q}_{W_i}$. Then

$$\lim_{i\rightarrow\infty} W_i^{-1}\log\mathfrak{Q}_i=K \text{ exists.}$$

Proof.—Consider W_j to be built up from 8^{j-i} smaller cubes $W_i(j>i)$. Evidently the number of atoms interacting across the boundaries of the small cubes is at most proportional to the area of such boundaries and hence is less than $8^j2^{-i}\gamma$, (γ=constant). Should one neglect these interactions, $\mathfrak{Q}_j$ would become $\mathfrak{Q}_i$ raised to the power 8^{j-i}. The inclusion of these terms violates this identity by not more than a constant factor β raised to the power 8^j2^{-i}, i.e.,

$$\log\mathfrak{Q}_j\leqq8^{j-i}\log\mathfrak{Q}_i+8^j2^{-i}\gamma\log\beta. \tag{19}$$

Next draw within each small cube W_i a concentric cube V_i with linear dimension $2^iL-(b/2)$. Since b is the range of interaction, clearly atoms in different V_i's do not interact. Hence,

$$8^{j-i}\log\mathfrak{Q}_{V_i}\leqq\log\mathfrak{Q}_j. \tag{20}$$

Equations (19) and (20) give

$$W_i^{-1}\log\mathfrak{Q}_{V_i}\leqq W_j^{-1}\log\mathfrak{Q}_j\leqq W_i^{-1}\log\mathfrak{Q}_i+2^{-i}\gamma L^{-3}\log\beta. \tag{21}$$

The last term approaches zero as $i\rightarrow\infty$. Also by Lemma 1, $W_i^{-1}(\log\mathfrak{Q}_i-\log\mathfrak{Q}_{V_i})\rightarrow0$. Thus as $j>i\rightarrow\infty$ $\cdot W_j^{-1}\log\mathfrak{Q}_j-W_i^{-1}\log\mathfrak{Q}_i\rightarrow0$. Hence the lemma.

Proof of the theorem.—Given any $\epsilon>0$ there exists by Lemma 2 a large enough box W such that

$$|K-W^{-1}\log\mathfrak{Q}_W|<\epsilon.$$

In fact by the same reasoning as used in proving that lemma, one easily sees that this can also be made true of any box Ω which can by adding partitions be divided into cubes of size W:

$$|K-\Omega^{-1}\log\mathfrak{Q}_\Omega|<\epsilon.$$

Now consider a volume V of arbitrary shape. For sufficiently large V one can build two Ω-type boxes Ω_1 and Ω_2 such that Ω_1 is contained in V and Ω_2 contains V and that

$$|(\Omega_1/\Omega_2)-1|<\epsilon.$$

Using

$$\mathfrak{Q}_{\Omega_1}\leqq\mathfrak{Q}_V\leqq\mathfrak{Q}_{\Omega_2},$$

one proves easily that $V^{-1}\log\mathfrak{Q}_V$ also approaches K.

That this limit monotonically increases with y follows from the same property of $V^{-1}\log\mathfrak{Q}_V$. That it is also continuous follows from the observation that $(\partial/\partial\log y)V^{-1}\log\mathfrak{Q}_V$ has a finite and definite upper bound (equal to the density of closest packing).

APPENDIX II

We first prove the following:

Lemma 3.—Consider the series

$$\sum_{l=0}^{\infty} b_l(V)z^l=S_V(z),$$

where

$$|b_l(V)|\leqq A\sigma^{-l},$$

A and σ being positive constants. For all real z between $-\sigma$ and $+\sigma$ assume $\mathrm{Lim}S_V(z)$ to exist as $V\rightarrow\infty$. Then (a) $\mathrm{Lim}b_l(V)$ as $V\rightarrow\infty$ exists and will be denoted by $b_l(\infty)$. (b) $S_V(z)$ approaches the limit

$$\sum_{l=0}^{\infty} b_l(\infty)z^l, \tag{22}$$

as $V\rightarrow\infty$ for all $|z|<\sigma$. The series (22) is convergent for all $|z|<\sigma$.

Proof.—(a) Evidently $b_0(\infty)$ exists and is equal to $\text{Lim}S_V(0)$ as $V\rightarrow\infty$. To prove the existence of $\text{Lim}b_1(V)$: Given any real ϵ between 0 and $\sigma/2$ consider the convergence of $S_V(\epsilon)$ and $b_0(V)$ as $V\rightarrow\infty$. There exists a volume V_0 such that for any volumes V and W greater than V_0 one has

$$|S_V(\epsilon)-S_W(\epsilon)|\le\epsilon^2, \tag{23}$$

$$|b_0(V)-b_0(W)|<\epsilon^2. \tag{24}$$

But

$$|\sum_{l=2}^{\infty} b_l(V)\epsilon^l|\leqq\frac{A\epsilon^2}{1-\epsilon\sigma^{-1}}\leqq 2A\epsilon^2. \tag{25}$$

The same is true if one replaces V by W. Using

$$\epsilon b_1(V)=S_V(\epsilon)-b_0(V)-\sum_{l=2}^{\infty} b_l(V)\epsilon_l,$$

one proves easily with the aid of (23), (24), and (25) that

$$\epsilon|b_1(V)-b_1(W)|<(2+4A)\epsilon^2.$$

Hence $\text{Lim}b_1(V)$ exists as $V\rightarrow\infty$. Similar proof holds for the other b_l's.

(b) The series $\sum b_l(V)z^l$ evidently converges uniformly in z for $|z|<\sigma$. The lemma follows from a well-known theorem on double limits.

Proof of theorem 2.—Consider first a circle C, lying inside R, with its center at the point $y=\eta$ along the positive real axis. We shall first prove the theorem inside this circle. Making the displacement $z=y-\eta$ we express (7) in the form

$$\mathfrak{Q}_V=\prod_{i=1}^{M}\left(1-\frac{z}{z_i}\right)\left(\frac{z_i}{y_i}\right), \tag{26}$$

where $z_i=y_i-\eta$ are the roots of $\mathfrak{Q}_V$. Expanding $V^{-1}\log\mathfrak{Q}_V$ in powers of z one obtains

$$\frac{1}{V}\log\mathfrak{Q}_V=\sum_{l=0}^{\infty}[b_l(V)z^l, \tag{27}$$

where

$$b_l(V)=\frac{-1}{Vl}\sum_{i=1}^{M}\left(\frac{1}{z_i}\right)^l \text{ for } l\geqq 1, \tag{28}$$

and

$$b_0(V)=\frac{1}{V}\sum_{i=1}^{M}\log\frac{z_i}{y_i}. \tag{29}$$

If σ is the radius of C, since C is free of roots we have $|z_i|\geqq\sigma$. Hence by (28)

$$|b_l(V)|\leqq(M/V)l^{-1}\sigma^{-l} \text{ for } l\geqq 1.$$

But M/V is bounded; hence we can use Lemma 3 and the theorem is proved in C.

By similar arguments we can extent the theorem into a circle C' lying inside R with its center inside C. One can easily prove the theorem in the whole region R by repeating this process.

APPENDIX III

To prove that ρ is an increasing function of y it is only necessary to show for any finite V the inequality

$$\frac{d^2}{(d\log y)^2}\log\mathfrak{Q}_V>0.$$

Now $\mathfrak{Q}_V$ is a polynomial in y with positive coefficients. Regarding the various terms of $\mathfrak{Q}_V$ as relative probabilities we have obviously

$$\frac{d}{d\log y}\log\mathfrak{Q}_V=\langle N\rangle,$$

where $\langle\ \rangle$ means "average." Also

$$\frac{d^2}{(d\log y)^2}\log\mathfrak{Q}_V=\langle N^2\rangle-\langle N\rangle^2=\langle(\Delta N)^2\rangle,$$

which is always positive. Here ΔN is the deviation of N from the average

$$\Delta N=N-\langle N\rangle.$$

Bose-Einstein Condensation and Liquid Helium

OLIVER PENROSE* AND LARS ONSAGER
Sterling Chemistry Laboratory, Yale University, New Haven, Connecticut

(Received July 30, 1956)

The mathematical description of B.E. (Bose-Einstein) condensation is generalized so as to be applicable to a system of interacting particles. B.E. condensation is said to be present whenever the largest eigenvalue of the one-particle reduced density matrix is an extensive rather than an intensive quantity. Some transformations facilitating the practical use of this definition are given.

An argument based on first principles is given, indicating that liquid helium II in equilibrium shows B.E. condensation. For absolute zero, the argument is based on properties of the ground-state wave function derived from the assumption that there is no "long-range configurational order." A crude estimate indicates that roughly 8% of the atoms are "condensed" (note that the fraction of condensed particles need not be identified with ρ_s/ρ). Conversely, it is shown why one would not expect B.E. condensation in a solid. For finite temperatures Feynman's theory of the lambda-transition is applied: Feynman's approximations are shown to imply that our criterion of B.E. condensation is satisfied below the lambda-transition but not above it.

1. INTRODUCTION

THE analogy between liquid He^4 and an ideal Bose-Einstein gas was first recognized by London.[1,2] He suggested that the lambda-transition in liquid helium could be understood as the analog for a liquid of the transition[3,4] which occurs in an ideal B.E. (Bose-Einstein) gas at low temperatures. The fact[5] that no lambda-transition has been found in He^3 supports London's viewpoint. Further support comes from recent theoretical work[6–8] which shows in more detail how a system of interacting particles can exhibit a transition corresponding to the ideal-gas transition.

Tisza[2,9] showed that the analogy between liquid He^4 and an ideal B.E. gas is also useful in understanding the transport properties of He II. Below its transition temperature a B.E. gas in equilibrium has a characteristic property: a finite fraction of the particles occupy the lowest energy level. Tisza reasoned that the presence of these "condensed" particles would make necessary a special two-fluid hydrodynamical description for such a gas. His idea that this two-fluid description applies also to He II has been strikingly verified by many experiments.[2,10,11]

In theoretical treatments where the forces between helium atoms are taken into account, the ideal-gas analogy takes on forms differing widely from one treatment to another. For example, Matsubara[6] and Feynman[7] account for the lambda-transition by writing the partition function for liquid helium in a form similar to the corresponding expression for an ideal gas. In Bogolyubov's theory[12–14] of the superfluidity of a system of weakly repelling B.E. particles, it is the distribution of the momenta of the particles which resembles that of an ideal gas. Yet another form for the analogy has been suggested by Penrose[15]; this work will be discussed in more detail in Sec. 4 below. A further complication is that the excitation theory of superfluidity[16,17] is apparently independent of the ideal-gas analogy (though Bogolyubov's work[12] suggests that there actually is a connection).

The object of the present paper is, first, to unify the varied forms of the ideal-gas analogy mentioned above by showing how they are all closely related to a single *criterion for B.E. condensation*, applicable in either a liquid or a gas, and, secondly, to give an argument based on first principles indicating that this criterion actually is satisfied in He II. The relation between B.E. condensation and the excitation theory of superfluidity will be discussed in a later paper.

2. PRELIMINARY DEFINITIONS

We make the usual approximation of representing liquid He^4 by a system of N interacting spinless B.E. particles, each of mass m, with position and momentum vectors $\mathbf{q}_1\cdots\mathbf{q}_N$ and $\mathbf{p}_1\cdots\mathbf{p}_N$, respectively. The Hamiltonian is taken to be

$$H=\sum_j p_j^2/2m+\sum_{i<j} U_{ij}. \tag{1}$$

Here U_{ij} stands for $U(|\mathbf{q}_i-\mathbf{q}_j|)$, where $U(r)$ is the interaction energy of two He^4 atoms separated by a

* Present address: Imperial College, London, England.

[1] F. London, Nature 141, 643 (1938); Phys. Rev. 54, 947 (1938).

[2] F. London, *Superfluids* (John Wiley and Sons, Inc., New York, 1954), Vol. 2, especially pp. 40–58, 199–201.

[3] A. Einstein, Sitzber. preuss. Akad. Wiss. physik-math. Kl. 1924, 261; *ibid.* 1925, 3, 18.

[4] For bibliography, see P. T. Landsberg, Proc. Cambridge Phil. Soc. 50, 65 (1954).

[5] E. F. Hammel in *Progress in Low-Temperature Physics*, edited by C. J. Gorter (North Holland Publishing Company, Amsterdam, 1955), Vol. 1, pp. 78–107.

[6] T. Matsubara, Progr. Theoret. Phys. Japan 6, 714 (1951).

[7] R. P. Feynman, Phys. Rev. 91, 1291 (1953).

[8] G. V. Chester, Phys. Rev. 100, 455 (1955).

[9] L. Tisza, Nature 141, 913 (1938).

[10] V. Peshkov, J. Phys. U.S.S.R. 8, 381 (1944); 10, 389 (1946).

[11] E. Andronikashvili, J. Phys. U.S.S.R. 10, 201 (1946).

[12] N. N. Bogolyubov, J. Phys. U.S.S.R. 11, 23 (1947).

[13] N. N. Bogolyubov and D. N. Zubarev, Zhur. Eksptl. i Teort. Fiz. 28, 129 (1955); English translation in Soviet Phys. 1, 83 (1955).

[14] D. N. Zubarev, Zhur. Eksptl. i Teort. Fiz. 29, 881 1955

[15] O. Penrose, Phil. Mag. 42, 1373 (1951).

[16] L. D. Landau, J. Phys. U.S.S.R. 5, 71 (1941).

[17] R. P. Feynman, Phys. Rev. 94, 262 (1954).

distance r, and $|\mathbf{q}_i-\mathbf{q}_j|$ means the length of the vector $\mathbf{q}_i-\mathbf{q}_j$ (except when the artifice of periodic boundary conditions is used, in which case $|\mathbf{q}_i-\mathbf{q}_j|$ means the length of the shortest vector congruent to $\mathbf{q}_i-\mathbf{q}_j$). Many-body interactions are omitted from (1), but including them would make no essential difference. The interaction between the He^4 atoms and those of the container is also omitted from (1); these could be included, but it is simpler to represent the container by a closed geometrical surface, considering only configurations for which all particles are within or on this surface, and imposing a suitable boundary condition on the wave function when any particle is on the surface. This boundary condition must be chosen to make H Hermitian. We denote the volume inside the container by V, and integrations over V by $\int_V\cdots d^3\mathbf{x}$ or $\int\cdots d^3\mathbf{x}$ or $\int\cdots d\mathbf{x}$.

As always in statistical mechanics, we are concerned here with very large values of N. Therefore we can often neglect quantities (for example $N^{-\frac{1}{3}}$) which are small when N is very large. A relation holding approximately by virtue of N being very large will be written in one of the forms $A\cong B$ or $A=B+o(1)$. These mean respectively that A/B is approximately 1 and that $A-B$ is negligible compared with 1, when N is large enough. We shall also use the notation $A=e^{O(1)}$ to mean that positive upper and lower bounds are known for A, but that a relation of the form $A\cong$const has not been established. Evidently $A\cong\text{const}>0$ implies $A=e^{O(1)}$, but the example $A=2+\sin N$ shows that the converse does not hold in general. We shall use the phrase "A is finite" to mean $A=e^{O(1)}$.

We can give more precise meanings to the symbols $\cong$, etc., by considering not a single system but an infinite sequence of systems with different values of N. The boundary conditions for the different members of the sequence should be the same, and should be specified on boundary surfaces of the same shape but of sizes such that N/V is independent of N. Then, if the quantities A, B, etc., are defined for each member of the sequence, $A\cong B$ means $\lim_{N\to\infty}(A/B)=1$, $A=B+o(1)$ means $\lim_{N\to\infty}(A-B)=0$, and $A=e^{O(1)}$ means that positive constants a_1, a_2, and N_1 exist such that $N>N_1$ implies that $a_1<A<a_2$.

We shall use Dirac's notation[18] for matrix elements and for eigenvalues of operators.

3. A GENERALIZED CRITERION OF B.E. CONDENSATION

It is characteristic of an ideal B.E. gas in equilibrium below its transition temperature that a finite fraction of the particles occupies the lowest single-particle energy level. Using the notation of Sec. 2, we can therefore give the following criterion of B.E. condensation[4] for an ideal gas in equilibrium:

$$\begin{aligned}\langle n_0\rangle_{Av}/N&=e^{O(1)}\leftrightarrow\text{B.E. condensation,}^{19}\\ \langle n_0\rangle_{Av}/N&=o(1)\leftrightarrow\text{no B.E. condensation,}\end{aligned}\tag{2}$$

where $\langle n_0\rangle_{Av}$ is the average number of particles in the lowest single-particle level and the sign $\leftrightarrow$ denotes logical equivalence. This criterion has meaning for noninteracting particles only, because single-particle energy levels are not defined for interacting particles.

To generalize the criterion (2), we rewrite it in a form which has meaning even when there are interactions. This can be done by using von Neumann's statistical operator,[20] σ, whose position representative[21] $\langle q_1'\cdots q_N'|\sigma|q_1''\cdots q_N''\rangle$ is known as the density matrix.[22] We define a reduced statistical operator, σ_1, as follows[23,24]:

$$\sigma_1=N\,\mathrm{tr}_{2\cdots N}(\sigma),\tag{3}$$

where $\mathrm{tr}_{2\ldots N}(\sigma)$ means the trace of σ taken with respect to particles $2\cdots N$ but not particle 1. For an ideal gas in equilibrium, the eigenstates of σ_1 are the single-particle stationary states, and the corresponding eigen values are the average numbers of particles in these stationary states.[25] Consequently, (2) may be rewritten as follows:

$$\begin{aligned}n_M/N&=e^{O(1)}\leftrightarrow\text{B.E. condensation,}\\ n_M/N&=o(1)\leftrightarrow\text{no B.E. condensation,}\end{aligned}\tag{4}$$

where n_M denotes the largest eigenvalue of σ_1. In this form the criterion has meaning for interacting as well as for noninteracting particles, since σ_1 is defined in either case; thus (4) provides a suitable generalization of the ideal-gas criterion (2).

According to our criterion (4), B.E. condensation cannot occur in a Fermi system, because[26] the exclusion principle implies that $0\leqslant n_M\leqslant 1$. For Bose systems, however, the only general restriction on n_M is $0\leqslant n_M\leqslant N$, a consequence of the identity $\mathrm{tr}(\sigma_1)=N$ and the fact that σ_1 is positive semidefinite.

The application of (4) is most direct when the system satisfies periodic boundary conditions and is spatially uniform (a homogeneous phase in the thermodynamic sense). For, in this case, the reduced density matrix $\langle\mathbf{q}'|\sigma_1|\mathbf{q}''\rangle$ is a function of $\mathbf{q}'-\mathbf{q}''$ only, and specifying this function is equivalent to specifying the single-particle momentum distribution. In fact, the momentum

[18] P. A. M. Dirac, *The Principles of Quantum Mechanics* (Oxford University Press, London, 1947).

[19] A possible alternative to this equation is $\langle n_0\rangle_{Av}/N\cong\text{const}>0$, but the weaker form used in (2) is easier to apply.

[20] J. von Neumann, *Mathematical Foundations of Quantum Mechanics* (Princeton University Press, Princeton, 1955), Chap. 4.

[21] P. A. M. Dirac, reference 18, Chap. 3.

[22] P. A. M. Dirac, reference, pp. 130–135.

[23] K. Husimi, Proc. Phys. Math. Soc. Japan 22, 264 (1940).

[24] J. de Boer, Repts. Progr. Phys. 12, 313–316 (1949).

[25] K. Husimi, reference 23, Eq. (10.6).

[26] See, for example, P-O. Löwdin, Phys. Rev. 97, 1474 (1955).

representative[21] of σ_1, given by

$$\langle \mathbf{p}'|\sigma_1|\mathbf{p}''\rangle \equiv \int\int \langle \mathbf{p}'|\mathbf{q}'\rangle d\mathbf{q}'\langle \mathbf{q}'|\sigma_1|\mathbf{q}''\rangle d\mathbf{q}''\langle \mathbf{q}''|\mathbf{p}''\rangle$$

$$= V^{-1}\int\int \exp[i(\mathbf{p}''\cdot\mathbf{q}''-\mathbf{p}'\cdot\mathbf{q}')/\hbar] \times\langle \mathbf{q}'|\sigma_1|\mathbf{q}''\rangle d\mathbf{q}'d\mathbf{q}''$$

is a diagonal matrix, so that n_M is the largest diagonal element of this matrix. Now, since

$$\langle \mathbf{p}_1'\cdots\mathbf{p}_N'|\sigma|\mathbf{p}_1'\cdots\mathbf{p}_N'\rangle$$

is the probability distribution in the (discrete) momentum space of N particles,

$$\langle \mathbf{p}_1'|\sigma_1|\mathbf{p}_1'\rangle = N\sum_{\mathbf{p}_2'}\cdots\sum_{\mathbf{p}_N'}\langle \mathbf{p}_1'\cdots\mathbf{p}_N'|\sigma|\mathbf{p}_1'\cdots p_N'\rangle$$

must be the average number of particles with momentum $\mathbf{p}_1'$. [To confirm this interpretation, note that[14]

$$\langle\sum_j f(\mathbf{p}_j)\rangle_{\mathrm{av}} = \sum_{\mathbf{p}'}\langle \mathbf{p}'|\sigma_1|\mathbf{p}'\rangle f(\mathbf{p}')$$

for arbitrary $f(\mathbf{p})$.] Therefore, according to (4), B.E. condensation is present for a spatially uniform system with periodic boundary conditions whenever a finite fraction of the particles have identical momenta. The work of Bogolybov[12] shows that this form of the criterion is satisfied in a system of weakly interacting B.E. particles at very low temperatures.

4. ALTERNATIVE FORMS OF THE CRITERION

When the system is not spatially uniform, it is more difficult to diagonalize σ_1, and some transformations of the criterion (4) are useful. The simplest of these depends on the following inequality:

$$n_M^2 \leqslant \sum_\alpha n_\alpha^2 \leqslant n_M \sum_\alpha n_\alpha = n_M N, \tag{5}$$

where the n_α's are the eigenvalues of σ_1; the fact that $\sum_\alpha n_\alpha = \mathrm{tr}(\sigma_1) = N$ follows from (3). We define

$$A_2 \equiv N^{-2}\int_V\int_V |\langle \mathbf{q}'|\sigma_1|\mathbf{q}''\rangle|^2 d^3\mathbf{q}'d^3\mathbf{q}''. \tag{6}$$

It is clear that $A_2 = N^{-2}\,\mathrm{tr}(\sigma_1^2) = N^{-2}\sum n_\alpha^2$, and hence, by (5), that $(n_M/N)^2 \leqslant A_2 \leqslant n_M/N$. It follows that

$$A_2 = e^{O(1)} \leftrightarrow n_M/N = e^{O(1)}, \text{ while } A_2 = o(1) \leftrightarrow n_M/N = o(1).$$

The following criterion is therefore equivalent to (4):

$$\begin{aligned} A_2 &= e^{O(1)} \leftrightarrow \text{B.E. condensation,} \\ A_2 &= o(1) \leftrightarrow \text{no B.E. condensation.} \end{aligned} \tag{7}$$

Another form of the criterion depends on inequalities satisfied by

$$A_1 \equiv (NV)^{-1}\int_V\int |\langle \mathbf{q}'|\sigma_1|\mathbf{q}''\rangle| d^3\mathbf{q}'d^3\mathbf{q}''. \tag{8}$$

Unlike A_2, this quantity has no simple interpretation in terms of the eigenvalues of σ_1/N, but it is easier to use. An upper bound for A_1 comes from the fact that $(A_1N/V)^2$, the square of the mean value of the function $|\langle \mathbf{q}'|\sigma_1|\mathbf{q}''\rangle|$, cannot exceed $A_2(N/V)^2$, the mean value of the square of this function; therefore we have

$$A_1^2 \leqslant A_2. \tag{9}$$

To find a lower bound for A_1, we use the fact that σ_1 is positive semidefinite (this follows intuitively from the probability interpretation of the eigenvalues of σ_1/N; alternatively it can be proved rigorously from (3) and the fact that σ is positive semidefinite). Since σ_1 has no negative eigenvalues, its square root is Hermitian. Applying the Schwartz inequality to the two state vectors $\sigma_1^{\frac{1}{2}}|\mathbf{q}'\rangle$ and $\sigma_1^{\frac{1}{2}}|\mathbf{q}''\rangle$, we obtain

$$|\langle \mathbf{q}'|\sigma_1|\mathbf{q}''\rangle| \leqslant [\langle \mathbf{q}'|\sigma_1|\mathbf{q}'\rangle\langle \mathbf{q}''|\sigma_1|\mathbf{q}''\rangle]^{\frac{1}{2}} \leqslant \alpha N/V, \tag{10}$$

where $\alpha N/V$ is any upper bound of $\langle \mathbf{q}'|\sigma_1|\mathbf{q}'\rangle$. Combining (6), (8), and (10), we find

$$A_2 \leqslant \alpha A_1. \tag{11}$$

For any physical system, α can be chosen independent of N; for $\langle \mathbf{q}'|\sigma_1|\mathbf{q}'\rangle$ is the average number density at the point $\mathbf{q}'$ and cannot become indefinitely large. For example, in a liquid at thermal equilibrium, $\langle \mathbf{q}'|\sigma_1|\mathbf{q}'\rangle$ is approximately N/V except near the boundary, so that α can be chosen just greater than 1. Treating α as finite, and combining (7) with (9) and (11), we obtain

$$\begin{aligned} A_1 &= e^{O(1)} \leftrightarrow \text{B.E. condensation,} \\ A_1 &= o(1) \leftrightarrow \text{no B.E. condensation.} \end{aligned} \tag{12}$$

A third form of the criterion is valuable when the reduced density matrix $\langle \mathbf{q}'|\sigma_1|\mathbf{q}''\rangle$ has the asymptotic form $\Psi(\mathbf{q}')\Psi^*(\mathbf{q}'')$ for large $|\mathbf{q}'-\mathbf{q}''|$ (Ψ^* is the complex conjugate of Ψ). Some consequences of assuming this asymptotic relation for He II were discussed by Penrose.[15] Here we formulate the assumption as follows:

$$|\langle \mathbf{q}'|\sigma_1|\mathbf{q}''\rangle - \Psi(\mathbf{q}')\Psi^*(\mathbf{q}'')| \leqslant (N/V)\gamma(|\mathbf{q}'-\mathbf{q}''|), \tag{13}$$

where the (non-negative) function $\gamma(r)$ is independent of N and satisfies

$$\lim_{r\to\infty}\gamma(r) = 0. \tag{14}$$

To use (13), we need the following lemma:

$$\Gamma(\mathbf{x}) \equiv V^{-1}\int \gamma(|\mathbf{x}'-\mathbf{x}|)d^3\mathbf{x}' = o(1). \tag{15}$$

Proof.—Let ϵ be an arbitrary positive number. Then, by (14), there exists a number R (depending on ϵ) such that

$$0 \leqslant \gamma(r) < \tfrac{1}{2}\epsilon \quad \text{if} \quad r > R. \tag{16}$$

We also have

$$0 \leqslant \gamma(r) < \gamma_M + \tfrac{1}{2}\epsilon \quad \text{if} \quad 0 \leqslant r \leqslant R,$$

where γ_M is the maximum of the function $\gamma(r)$. Using (16) in (15) we obtain $0 \leqslant \Gamma(\mathbf{x}) < \frac{1}{2}\epsilon + V^{-1}\gamma_M V_R(\mathbf{x})$, where $V_R(\mathbf{x})$ is the volume of $\mathbf{x}'$-space for which $|\mathbf{x}'-\mathbf{x}| \leqslant R$. Since $V_R(\mathbf{x}) \leqslant 4\pi R^3/3$, we can ensure that $0 \leqslant \Gamma < \epsilon$ by choosing $V > 8\pi\gamma_M R^3/3\epsilon$. By the definition of a limit, this implies $\lim_{V\to\infty}\Gamma = 0$; that is, (15) is true.

We can now obtain a criterion of B.E. condensation, using the relation

$$(NV)^{-1}\int\int |\langle \mathbf{q}'|\sigma_1|\mathbf{q}''\rangle - \Psi(\mathbf{q}')\Psi^*(\mathbf{q}'')|d\mathbf{q}'d\mathbf{q}'' = o(1), \tag{17}$$

which follows from (13) and (15). Combining this with (8) with the help of the elementary inequality $-|u-v| \leqslant |u|-|v| \leqslant |u-v|$ gives

$$A_1 = (NV)^{-1}\left[\int |\Psi(\mathbf{x})|d^3\mathbf{x}\right]^2 + o(1). \tag{18}$$

Using (12), we obtain the criterion, valid whenever (13) holds:

$$V^{-1}\int |\Psi|d^3\mathbf{x} = e^{O(1)} \leftrightarrow \text{B.E. condensation},$$
$$V^{-1}\int |\Psi|d^3\mathbf{x} = o(1) \leftrightarrow \text{no B.E. condensation}. \tag{19}$$

The function Ψ has a simple interpretation when B.E. condensation is present: we can show that $\Psi(\mathbf{x})$ is a good approximation to the eigenfunction of the matrix $\langle \mathbf{q}'|\sigma_1|\mathbf{q}''\rangle$ corresponding to the eigenvalue n_M, and also that its normalization is

$$n_\Psi \equiv \int |\Psi(\mathbf{x})|^2 d^3\mathbf{x} \cong n_M. \tag{20}$$

We note first that all eigenvalues of the matrix

$$N^{-1}\langle \mathbf{q}'|\tau|\mathbf{q}''\rangle \equiv N^{-1}[\langle \mathbf{q}'|\sigma_1|\mathbf{q}''\rangle - \Psi(\mathbf{q}')\Psi^*(\mathbf{q}'')]$$

are $o(1)$, since, by (17), (8), and (12), a system whose reduced density matrix was τ would not show B.E. condensation. It follows that

$$f\{\varphi(\mathbf{x})\} \equiv N^{-1}\int\int \varphi^*(\mathbf{q}')\langle \mathbf{q}'|\sigma_1|\mathbf{q}''\rangle\varphi(\mathbf{q}'')d\mathbf{q}'d\mathbf{q}'' = |(\varphi,\Psi)|^2/N + o(1), \tag{21}$$

where φ is an arbitrary normalized function and

$$(\varphi,\Psi) \equiv \int \varphi^*(\mathbf{x})\Psi(\mathbf{x})d^3\mathbf{x}.$$

The arbitrary function $\varphi(\mathbf{x})$ in (21) can be written in the form $\varphi(\mathbf{x}) = n_\Psi^{-\frac{1}{2}}[a\Psi(\mathbf{x}) + b\Phi(\mathbf{x})]$, where Φ is chosen to make $(\Psi,\Phi)=0$ and $(\Phi,\Phi)=n_\Psi$, and where (since φ is normalized) $|a|^2+|b|^2=1$. Inserting this expression for φ into (21) gives

$$f\{\varphi\} = |a|^2 n_\Psi/N + o(1) = (1-|b|^2)n_\Psi/N + o(1).$$

Now, the maximum value of $f\{\varphi\}$ is n_M/N, and it is attained when φ equals φ_M, the normalized eigenfunction of $\langle \mathbf{q}'|\sigma_1|\mathbf{q}''\rangle$ corresponding to the eigenvalue n_M. The last expression for $f\{\varphi\}$ shows that this maximum is $n_\Psi/N + o(1)$ and is attained with $|b| = o(1)$. It follows that $n_M \cong n_\Psi$ in agreement with (20), and also that

$$\int |a n_\Psi^{-\frac{1}{2}}\Psi(\mathbf{x}) - \varphi_M(\mathbf{x})|^2 d^3\mathbf{x} = |b|^2 = o(1). \tag{22}$$

This equation tells us that $\Psi(\mathbf{x})$ is to a good approximation proportional to $\varphi_M(\mathbf{x})$. In view of these results, we may call Ψ the *wave function of the condensed particles*, and n_Ψ/N the *fraction of condensed particles*.

5. GROUND STATE OF A B.E. FLUID

In this section we derive some general properties of the ground-state wave function. These will be needed in Sec. 6.

Let us define the ground-state wave function $\psi(\mathbf{x}_1\cdots\mathbf{x}_N)$ to be the real symmetric function which minimizes the expression

$$\int\cdots\int[\hbar^2\sum_i(\nabla_i\psi)^2/2m + \sum_{i<j} U_{ij}\psi^2]d\mathbf{x}_1\cdots d\mathbf{x}_N, \tag{23}$$

while at the same time satisfying the boundary conditions and the normalization condition. The Euler equation of this variation problem shows that ψ satisfies Schrödinger's equation for the Hamiltonian (1). Now, the function $|\psi|$ also conforms to the above definition, and so it too satisfies Schrödinger's equation. The first derivative of $|\psi|$ must therefore be continuous wherever the potential energy is finite. This is possible only if ψ does not change sign. We may therefore take ψ to be non-negative. Suppose now that ψ_1 and ψ_2 are two different non-negative functions conforming to the above definition. Then, since Schrödinger's equation is linear, $\psi_1 - \psi_2$ also conforms to the definition, and (by the result just proved) does not change sign; but this contradicts the original assumption that both ψ_1 and ψ_2 are normalized. Hence the above definition yields a unique, non-negative function[27] ψ.

For a fluid phase, we can obtain further information if we assume [28] that there is *no long-range configurational*

[27] These properties of the ground-state wave function are fairly well known [see, for example, R. P. Feynman, Phys. Rev. 91, 1301 (1953)], but the authors have seen no proof in the literature. A proof for the special case $N=1$ (to which no symmetry requirements apply) is given by R. Courant and D. Hilbert, *Methoden der Mathematischen Physik* (Verlag Julius Springer, Berlin, 1931), Vol. 1, Chap. 6, Secs. 6, 7.

[28] A similar principle is often used in the classical theory of liquids—for example by J. E. Mayer and E. W. Montroll, J. Chem. Phys. 9, 2 (1941). Its use here amounts to assuming that

order. By this we mean that there is a finite "range of order" R with the following property: for any two concentric spheres S_1 and S_2 with radii R_1 and R_1+R, respectively, the relative probabilities of the various possible configurations of particles inside S_1 are approximately[29] independent of the situation outside S_2. By the "situation" outside S_2, we mean here the number of particles ouside S_2, their positions, and the position of the part of the boundary surface outside S_2.

This assumption implies that, if the configuration of the particles inside S_1 is altered while everything else remains the same, then the probability density in configuration space changes by a factor approximately independent of the situation outside S_2. Hence, if the point $\mathbf{x}_i$ is inside S_1, then $\nabla_i \log\psi$ is approximately independent of the situation outside S_2. This is true for any choice of S_1 provided S_1 encloses $\mathbf{x}_i$, and in particular it is true when R_1 is vanishingly small. Therefore, $\nabla_i \log\psi$ is independent of the situation outside a sphere $S(\mathbf{x}_i)$ with center $\mathbf{x}_i$ and radius R.

Setting $i=1$ in this result and integrating shows that ψ can be written in the form

$$\psi(\mathbf{x}_1\cdots\mathbf{x}_N)=\theta(\mathbf{x}_2\cdots\mathbf{x}_N)\chi(\mathbf{x}_1;\mathbf{x}_2\cdots\mathbf{x}_N), \tag{24}$$

where the functions θ and χ are symmetric in $\mathbf{x}_2\cdots\mathbf{x}_N$, and χ is approximately[30] independent of the situation outside $S(\mathbf{x}_1)$.

The function θ in (24) has a simple physical meaning. To find this, we write the Schrödinger equation satisfied by ψ in the form

$$-(\hbar^2/2m)\sum_j[\nabla_j{}^2\log\psi+(\nabla_j\log\psi)^2]+\sum_{i<j}U_{ij}=\text{const.} \tag{25}$$

Taking the gradient with respect to x_i ($i\neq 1$) and substituting from (24), we obtain after some rearrangement

$$\begin{aligned}\nabla_i\{&-(\hbar^2/2m)\textstyle\sum_j{}'[\nabla_j{}^2\log\theta+(\nabla_j\log\theta)^2]+\sum_j{}''U_{ij}\}\\ &=(\hbar^2/2m)\textstyle\sum_j[\nabla_j{}^2+2(\nabla_j\log\theta)\cdot\nabla_j]\nabla_i\log\chi\\ &\quad+(\hbar^2/m)\textstyle\sum_j(\nabla_j\log\chi)\cdot\nabla_j\nabla_i\log\psi-\nabla_i U_{1i},\end{aligned} \tag{26}$$

where $\sum_j{}'$ means a sum with the $j=1$ term omitted, and $\sum_j{}''$ means a sum with the $j=1$ and $j=i$ terms omitted. Since the left member of (26) does not contain $\mathbf{x}_1$, the right member must be independent of $\mathbf{x}_1$. To evaluate the right member, we may therefore choose $\mathbf{x}_1$ to make $|\mathbf{x}_i-\mathbf{x}_1|>2R$. The properties of χ then imply that $\nabla_i\log\chi\simeq 0$, so that the first sum vanishes approximately; they also imply that the summand in the second sum is negligible unless $|\mathbf{x}_j-\mathbf{x}_1|\leqslant R$. The argument preceding (24) shows, however, that $\nabla_i\log\psi$ is independent of $\mathbf{x}_j$ unless $|\mathbf{x}_j-\mathbf{x}_i|\leqslant R$. Since $|\mathbf{x}_j-\mathbf{x}_1|$ and $|\mathbf{x}_j-\mathbf{x}_i|$ cannot both be less than R (because $|\mathbf{x}_1-\mathbf{x}_i|>2R$), the summand in the second sum is always negligible. The term $\nabla_i U_{1i}$ also vanishes, because the interaction has a short range. Thus the entire right member of (26) vanishes approximately.[31] The expression in curly brackets is therefore approximately independent of $\mathbf{x}_i$ for $i=2\cdots N$. This means that $\theta(\mathbf{x}_2\cdots\mathbf{x}_N)$ approximately satisfies an equation, analogous to (25), which is equivalent to Schrödinger's equation for a system of $N-1$ particles. Since θ is non-negative, it must therefore have the form

$$\theta(\mathbf{x}_2\cdots\mathbf{x}_N)\simeq c\vartheta(\mathbf{x}_2\cdots\mathbf{x}_N), \tag{27}$$

where c is a constant and ϑ is the normalized ground-state wave function for $N-1$ particles.

A simple illustration of (24) is provided by a type of approximation to $\psi(\mathbf{x}_1\cdots\mathbf{x}_N)$ used by various authors[32,33,13]:

$$\psi(\mathbf{x}_1\cdots\mathbf{x}_N)\propto[\prod_j\mu(\mathbf{x}_j)]\prod_{i<j}\omega(|\mathbf{x}_i-\mathbf{x}_j|), \tag{28}$$

where $\omega(r)\to 1$ when r is large. In this approximation, (24) can be satisfied by taking

$$\theta(\mathbf{x}_2\cdots\mathbf{x}_N)\propto[\prod_i{}'\mu(\mathbf{x}_i)]\prod_{i<j}{}'\omega(|\mathbf{x}_i-\mathbf{x}_j|), \tag{29}$$

$$\chi(\mathbf{x}_1;\mathbf{x}_2\cdots\mathbf{x}_N)=\mu(\mathbf{x}_1)\prod_i{}'\omega(|\mathbf{x}_i-\mathbf{x}_1|), \tag{30}$$

where Π' means a product with all $i=1$f actors omitted. It is clear that (29) is consistent with (27), and that, if R is large enough, χ as defined in (30) is approximately independent of the positions of the particles outside $S(\mathbf{x}_1)$.

6. LIQUID HELIUM-4 AT ABSOLUTE ZERO

At absolute zero, the density matrix is given by

$$\langle\mathbf{q}_1'\cdots\mathbf{q}_N'|\sigma|\mathbf{q}_1''\cdots\mathbf{q}_N''\rangle=\psi(\mathbf{q}_1'\cdots\mathbf{q}_N')\psi(\mathbf{q}_1''\cdots\mathbf{q}_N''),$$

since the ground-state wave function ψ is real and normalized. The reduced density matrix is therefore

$$\langle\mathbf{q}'|\sigma_1|\mathbf{q}''\rangle=N\int\cdots\int\psi(\mathbf{q}',\mathfrak{x})\psi(\mathbf{q}'',\mathfrak{x})d\mathfrak{x}, \tag{31}$$

the probability density in configuration space is qualitatively similar to the corresponding probability density for a classical liquid. The importance of this principle for the ground state of a quantum liquid was noted by A. Bijl, Physica 7, 869 (1940).

[29] The meaning of the word "approximately" is purposely left vague, since it would complicate the discussion too much to attempt a rigorous formulation. As we see it, a rigorous formulation would have to depend on a limit operation $R\to\infty$: that is, it would assume that the approximation of statistical independence could be made arbitrarily good by choosing R large enough.

[30] If the theory were formulated more rigorously (see reference 29), the corresponding property of χ might be $\nabla_j\chi(\mathbf{x}_1;\mathbf{x}_2\cdots\mathbf{x}_N)\leqslant K(|\mathbf{x}_j-\mathbf{x}_1|)$ where $K(r)\to 0$ in a suitable way as $r\to\infty$.

[31] Only a rigorous treatment can completely justify the implicit assumption that the sum of N negligible terms is itself negligible. The present methods can, however, be used to show that the contribution of a given j value to the sums in the right member of (26) is negligible compared with its contribution to the sums in the left (with a finite number of exceptions, for which $|\mathbf{x}_j-\mathbf{x}_1|\leqslant R$ and the contributions on both sides are negligible).

[32] A. Bijl, reference 28.

[33] R. B. Dingle, Phil. Mag. 40, 573 (1949).

where $\mathfrak{x}$ and $d\mathfrak{x}$ are abbreviations for $\mathbf{x}_2\cdots\mathbf{x}_N$ and $d^3\mathbf{x}_2\cdots d^3\mathbf{x}_N$, respectively. For a preliminary discussion of (31), we use a crude approximation to ψ suggested by Feynman[7]:

$$\psi(\mathbf{x}_1\cdots\mathbf{x}_N)\simeq(\Omega_N)^{-\frac{1}{2}}F_N(\mathbf{x}_1\cdots\mathbf{x}_N), \tag{32}$$

where Ω_N is a normalizing constant, and $F_N(\mathbf{x}_1\cdots\mathbf{x}_N)$ by definition takes the value 1 whenever $\mathbf{x}_1\cdots\mathbf{x}_N$ is a possible configuration for the centers of N hard spheres of diameter d and the value 0 for all other configurations. Here $d\simeq 2.6$A is the diameter of a He^4 atom. The approximation amounts to using (28) with $\mu=1$ and with $\omega(r)=0$ for $r<d$ and $\omega=1$ for $r\geqslant d$.

The normalization integral corresponding to (32) shows that $\Omega_N/N!$ is the configurational partition function for a classical system of N noninteracting hard spheres. Moreover, the integral in (31) is now closely related to the pair distribution function for $N+1$ hard spheres, defined as follows[24]:

$$n_2(\mathbf{q}',\mathbf{q}'')\equiv(N+1)N\int\cdots\int F_{N+1}(\mathbf{q}',\mathbf{q}'',\mathfrak{x})d\mathfrak{x}/\Omega_{N+1}. \tag{33}$$

Under the approximation (32), the integrand in (31) is $1/\Omega_N$ times that in (33) when $|\mathbf{q}'-\mathbf{q}''|\geqslant d$, so that

$$\langle\mathbf{q}'|\sigma_1|\mathbf{q}''\rangle=z^{-1}n_2(\mathbf{q}',\mathbf{q}'')\quad\text{if}\quad|\mathbf{q}'-\mathbf{q}''|\geqslant d. \tag{34}$$

Here $z\equiv(N+1)\Omega_N/\Omega_{N+1}$ is the activity of the hard-sphere system. The physical meaning of n_2 shows that, for large $|\mathbf{q}'-\mathbf{q}''|$, n_2 tends to $(N/V)^2$. Hence (13) can be satisfied by taking $\Psi\cong\text{const}\cong z^{-\frac{1}{2}}N/V$ (except, possibly, near the boundary). Hence, by (19), B.E. condensation is present; moreover, by (20) and the discussion following (20), the fraction of condensed particles is

$$n_M/N\cong n_\Psi/N\cong\Psi^2V/N\cong N/Vz. \tag{35}$$

The right member of (35) can be calculated from the virial series for hard spheres.[34] Taking the density of He II to be 0.28 times the density at closest packing, we obtain the result 0.08. Thus, Feynman's approximation (32) implies that B.E. condensation is present in He II at absolute zero and that the fraction of condensed particles is about 8%.

The above discussion makes it plausible that a treatment based on the true wave function will also indicate the presence of B. E. condensation. To supply such a treatment, we first substitute from (24) and (27) into (31). This yields

$$\langle\mathbf{q}'|\sigma_1|\mathbf{q}''\rangle=c^2N\langle\chi(\mathbf{q}';\mathfrak{x})\chi(\mathbf{q}'';\mathfrak{x})\rangle_\vartheta, \tag{36}$$

where, for any function $f(\mathfrak{x})$

$$\langle f\rangle_\vartheta\equiv\langle f(\mathfrak{x})\rangle_\vartheta\equiv\int\cdots\int f(\mathfrak{x})\vartheta^2(\mathfrak{x})d\mathfrak{x} \tag{37}$$

is the expectation value of $f(\mathfrak{x})$ in the ground state of a liquid of $N-1$ particles whose configuration is $\mathfrak{x}\equiv\mathbf{x}_2\cdots\mathbf{x}_N$.

In studying (36) it will be convenient to look on $\mathbf{q}'$ and $\mathbf{q}''$ as parameters and to treat $\chi'\equiv\chi(\mathbf{q}';\mathfrak{x})$ and $\chi''\equiv\chi(\mathbf{q}'';\mathfrak{x})$ as variables depending on the configuration $\mathfrak{x}$ of a liquid of $N-1$ particles. The correlation coefficient[35] of χ' and χ'' is defined by

$$\rho(\mathbf{q}',\mathbf{q}'')\equiv\frac{\langle\chi'\chi''\rangle_\vartheta-\langle\chi'\rangle_\vartheta\langle\chi''\rangle_\vartheta}{[\langle\chi'^2\rangle_\vartheta-\langle\chi'\rangle_\vartheta^2]^{\frac{1}{2}}[\langle\chi''^2\rangle_\vartheta-\langle\chi''\rangle_\vartheta^2]^{\frac{1}{2}}}. \tag{38}$$

Now, it was shown in Sec. 5 that χ' is independent of the "situation" outside $S(\mathbf{q}')$ and that χ'' is independent of the situation outside $S(\mathbf{q}'')$. By applying the principle of no long-range configurational order, given in Sec. 5, to the ground state of a liquid of $N-1$ particles, with the sphere S_1 chosen large enough to enclose both $S(\mathbf{q}')$ and $S(\mathbf{q}'')$, we find that $\rho(\mathbf{q}',\mathbf{q}'')$ is independent of V for large enough V. By applying the same principle with S_1 this time taken to coincide with $S(\mathbf{q}')$, we find that χ' and χ'' are approximately statistically independent if $S(\mathbf{q}'')$ is entirely outside S_2; that is, $\rho(\mathbf{q}',\mathbf{q}'')$ approximately vanishes if $|\mathbf{q}'-\mathbf{q}''|>3R$.

We can now show that (13) holds, with

$$\Psi(\mathbf{q}')=cN^{\frac{1}{2}}\langle\chi'\rangle_\vartheta. \tag{39}$$

For, substituting (36) and (39) into the left member of (13) gives $c^2N[\langle\chi'\chi''\rangle_\vartheta-\langle\chi'\rangle_\vartheta\langle\chi''\rangle_\vartheta]$, which, by (38), is less than $c^2N\rho(\mathbf{q}',\mathbf{q}'')[\langle\chi'^2\rangle_\vartheta\langle\chi''^2\rangle_\vartheta]^{\frac{1}{2}}$. Setting $\mathbf{q}'=\mathbf{q}''$ in (36) shows that this last expression equals $\rho(\mathbf{q}',\mathbf{q}'')\times[\langle\mathbf{q}'|\sigma_1|\mathbf{q}'\rangle\langle\mathbf{q}''|\sigma_1|\mathbf{q}''\rangle]^{\frac{1}{2}}$. Therefore, by (10) and the properties of $\rho(\mathbf{q}',\mathbf{q}'')$ given above, (13) can be satisfied by making $\gamma(|\mathbf{q}'-\mathbf{q}''|)\geqslant\alpha\rho(\mathbf{q}',\mathbf{q}'')$ for every $\mathbf{q}'$ and $\mathbf{q}''$.

If the distance from $\mathbf{q}'$ to the boundary exceeds $2R$, then $\langle\chi'\rangle_\vartheta$ and $\langle\chi'^2\rangle_\vartheta$ are (approximately) positive constants independent of N and $\mathbf{q}'$. For we may take the sphere S_1 defined in Sec. 5 to be $S(\mathbf{q}')$; then the relative probabilities for the various configurations of particles inside $S(\mathbf{q}')$—on which alone χ' depends—are independent of N and the relative positions of S_2 and the boundary. It follows that $V^{-1}\int_V\langle\chi'\rangle_\vartheta d^3\mathbf{q}'\cong\text{const}>0$ and also, by (36), that $c^2N\cong\text{const}>0$ since $\langle\mathbf{q}'|\sigma_1|\mathbf{q}'\rangle\cong N/V$ if $\mathbf{q}'$ is far from the boundary. Applying the criterion (19) to the Ψ defined in (39), we conclude that B.E. condensation is present in liquid He^4 at absolute zero.

The above discussion would not lead one to expect B.E. condensation in a solid, because the assumption of no long-range configurational order is valid for a fluid phase only. In fact, it can be argued that a solid does *not* show B.E. condensation, at least for $T=0$°K. We assume that a solid at $T=0$°K is a perfect crystal

[34] M. N. Roxenbluth and A. W. Rosenbluth, J. Chem. Phys. **22**, 881 (1954).

[35] H. Cramer, *Mathematical Methods of Statistics* (Princeton University Press, Princeton, 1946), p. 277.

—i.e., that there exists a set of lattice sites such that ψ is small unless one particle is near each lattice site.[36] In the expression (31) for the reduced density matrix, therefore, the integrand will be appreciable only if every one of the points $\mathbf{x}_2\cdots\mathbf{x}_N$ is near a separate lattice site, while both $\mathbf{q}'$ and $\mathbf{q}''$ are near the remaining site. When $|\mathbf{q}'-\mathbf{q}''|$ is large, this last condition cannot be fulfilled, so that $\langle\mathbf{q}'|\sigma_1|\mathbf{q}''\rangle$ will tend to 0 for large $|\mathbf{q}'-\mathbf{q}''|$. This indicates that the function Ψ of (13) will be 0, so that, by (20), there is no B.E. condensation in a solid at absolute zero.

Our result that B.E. condensation occurs in liquid He^4 at $T=0°\mathrm{K}$ must now be extended to nonzero temperatures. (The need for such an extension is illustrated by the example of a two-dimensional B.E. gas, which[4] shows B.E. condensation at $T=0°\mathrm{K}$ but not for $T\neq 0°\mathrm{K}$.) This will be done in the next section.

7. B.E. CONDENSATION AND THE LAMBDA-TRANSITION

Feynman,[7] and also Matsubara,[6] have studied the lambda-transition in liquid helium by expressing the partition function in the form

$$Z=\sum_{\{m_l\}}\left[\prod_l (m_l!\,l^{m_l})\right]^{-1}\mathrm{tr}(Pe^{-\beta H}),\qquad(40)$$

where the sum is over all partitions of the number N (that is, over all sets $\{m_l\}$ of non-negative integers satisfying $\sum_l m_l=N$), P is any permutation containing m_l cycles[37] of length l ($l=1\cdots N$), and $\beta\equiv 1/kT$ with $k\equiv$Boltzmann's constant. Evaluating (40) with the help of approximations for $\mathrm{tr}(Pe^{-\beta H})$, they showed how it could exhibit a transition, which they identified with the lambda-transition. In the present section, we shall show that Feynman's approximations also imply that the criterion (4) of B.E. condensation is satisfied for He II in equilibrium.

The statistical operator for thermal equilibrium is

$$\sigma=(N!Z)^{-1}\sum_P Pe^{-\beta H},\qquad(41)$$

where the sum is over all permutations P of the N particles. Feynman's path integral[7] for the density matrix shows that the position representative of (41) is non-negative. Therefore the corresponding reduced density matrix, calculated according to (3), is also non-negative, so that the quantity defined in (8) is now given by

$$A_1=(N!ZV)^{-1}\sum_P\int\int d\mathbf{q}'_1 d\mathbf{q}_1''\langle\mathbf{q}_1'|\mathrm{tr}_{2\ldots N}Pe^{-\beta H}|\mathbf{q}_1''\rangle.\qquad(42)$$

All permutations corresponding to a given partition $\{m_l\}$ and also having particle No. 1 in a cycle of given length L contribute equal terms to the above sum, since a suitable relabeling of the particles $2\cdots N$ will turn any one such term into any other. Collecting together, for each $\{m_l\}$ and L, the $(L/N)N!/\prod_l(m_l!l^{m_l})$ equal terms, we can write A_1 as a sum over L and $\{m_l\}$, obtaining

$$A_1=N^{-1}\sum_L L\langle m_L A_{1,L}\{m_l\}\rangle.\qquad(43)$$

Here we have defined, for any function $f\equiv f\{m_l\}$ depending on the set of numbers $\{m_l\}$, a quantity

$$\langle f\rangle\equiv Z^{-1}\sum_{\{m_l\}}\left[\prod_l(m_l!l^{m_l})\right]^{-1}f\{m_l\}\,\mathrm{tr}(Pe^{-\beta H}),\qquad(44)$$

where P is any permutation corresponding to the partition $\{m_l\}$. We have also defined

$$A_{1,L}\{m_l\}\equiv\frac{\int\int d\mathbf{q}_1'd\mathbf{q}_1''\langle\mathbf{q}_1'|\mathrm{tr}_{2\ldots N}Pe^{-\beta H}|\mathbf{q}_1''\rangle}{V\,\mathrm{tr}(Pe^{-\beta H})},\qquad(45)$$

where P is any permutation which corresponds to the partition $\{m_l\}$ and also has particle No. 1 in a cycle of length L.

To use (43), we introduce two approximations due to Feynman.[7,38] The first is

$$\langle\mathbf{q}_1'\cdots\mathbf{q}_N'|Pe^{-\beta H}|\mathbf{q}_1''\cdots\mathbf{q}_N''\rangle$$
$$\simeq K\lambda^{-3N}\phi(\mathbf{q}_1'\cdots\mathbf{q}_N')\phi(\mathbf{q}_1''\cdots\mathbf{q}_N'')$$
$$\times\exp\left[-(\pi/\lambda^2)\sum_j(\mathbf{q}_j'-\mathbf{q}_{Pj}'')^2\right],\qquad(46)$$

where K is a constant, λ means $h(2\pi m'kT)^{-\frac{1}{2}}$ with m' an effective mass, and $\phi(\mathbf{x}_1\cdots\mathbf{x}_N)$ is a normalized non-negative symmetric function which reduces to the ground-state wave function when $T\to 0°\mathrm{K}$. (We deviate slightly from Feynman's usage: he does not take ϕ to be normalized.) Feynman's other approximation is used in evaluating integrals over configuration space involving (46); it is to replace the factor containing ϕ by its value averaged over the region of integration and to replace each factor $\exp[-\pi(\mathbf{x}_i-\mathbf{x}_j)^2/\lambda^2]$ by

$$G(\mathbf{x}_i-\mathbf{x}_j)\equiv p(|\mathbf{x}_i-\mathbf{x}_j|)\exp[-\pi(\mathbf{x}_i-\mathbf{x}_j)^2/\lambda^2],\qquad(47)$$

where $p(0)\equiv 1$ and $p(r)$ for $r>0$ is the radial distribution function, tending to 1 as $r\to\infty$.

Using these approximations in (45), we obtain

$$A_{1,L}\{m_L\}\simeq A_{1,\infty}\delta_L/f_L,\qquad(48)$$

[36] For equilibrium at a temperature $T\neq 0°\mathrm{K}$, a few atoms will be in interstitial positions far from their proper lattice sites. The fraction of interstitial atoms will be $e^{-W/kT}$, where W is the energy required to excite one atom from a lattice to an interstitial site. This fraction tends to 0 as T tends to $0°\mathrm{K}$.

[37] For the definition of a cycle, see R. P. Feynman, reference 7, or H. Margenau and G. M. Murphy, *The Mathematics of Physics and Chemistry* (D. Van Nostrand Company, Inc., New York, 1943), p. 538.

[38] For a critical discussion of these approximations, see G. V. Chester, Phys. Rev. 93, 1412 (1954).

where (with $\mathfrak{x}$ standing, as before, for $\mathbf{x}_2\cdots\mathbf{x}_N$)

$$A_{1,\infty}\equiv V^{-1}\int\cdots\int\phi(\mathbf{q}',\mathfrak{x})\phi(\mathbf{q}'',\mathfrak{x})d\mathbf{q}'d\mathbf{q}''d\mathfrak{x}, \tag{49}$$

$$\delta_L\equiv V^{-1}\int\cdots\int\prod_{j=1}^{L}G(\mathbf{x}_{j+1}-\mathbf{x}_j)d\mathbf{x}_1\cdots d\mathbf{x}_{L+1}, \tag{50}$$

$$f_L\equiv\int\cdots\int G(\mathbf{x}_1-\mathbf{x}_L)\prod_{j=2}^{L}G(\mathbf{x}_j-\mathbf{x}_{j-1})d\mathbf{x}_1\cdots d\mathbf{x}_L. \tag{51}$$

[For $L=1$ we interpret (51) as $f_1\equiv V$.]

To find the order of magnitude of $A_{1,\infty}$, we replace ϕ in (49) by the ground-state wave function ψ (since ϕ is qualitatively similar to ψ and both are normalized). Then, by (31) and (8), $A_{1,\infty}$ roughly equals the value of A_1 for $T=0°K$; this is finite, by (12) and the result of Sec. 6. Feynman[7] has suggested using the approximation (32) for ϕ as well as for ψ; this leads, by (18) and (35), to the rough estimate $A_{1,\infty}\simeq 0.08$.

We estimate δ_L by replacing the integrals over $d\mathbf{x}_2\cdots d\mathbf{x}_{L+1}$ in (50) by the corresponding infinite integrals. This gives $\delta_L\simeq\delta_1{}^L$, which, when combined with (48) and (43), yields

$$A_1/A_{1,\infty}\simeq N^{-1}\sum_L L\langle m_L\rangle\delta_1{}^L/f_L. \tag{52}$$

To study (52), we note that Feynman's approximations (46) and (47) also imply[7] $\mathrm{tr}(Pe^{-\beta H})\simeq K\lambda^{-3N}\times\prod_l(f_l{}^{m_l})$. Substituting this into (40) and (44) yields

$$\langle Lm_L/f_L\rangle=Q_{N-L}/Q_N, \tag{53}$$

where

$$Q_M\equiv\sum_{\{m_l\}}\prod_l(f_l/l)^{m_l}/m_l!, \tag{54}$$

the sum being over all partitions of the arbitrary integer M. Equation (54) is just Mayer's expression[39] for the configurational partition function of an imperfect gas of M particles with cluster integrals $b_l=f_l/lV$. Therefore if $z\equiv Q_{N-1}/Q_N\cong\mathrm{const}$ is the activity of this imperfect gas when it contains N particles, the approximation $Q_{N-L}/Q_N\cong z^L$ will hold provided $L\ll N$.

Using this approximation with (53) and (52), we obtain

$$A_1/A_{1,\infty}\simeq N^{-1}\sum_{L=1}^{N}(z\delta_1)^L\simeq N^{-1}(1-z\delta_1)^{-1}=o(1) \tag{55}$$

provided that $z\delta_1\cong\mathrm{const}<1$. Feynman's work[7] shows that this condition holds above the transition temperature; therefore, since $A_{1,\infty}=e^{O(1)}$ and (12) holds, there is no B.E. condensation in HeI.

This argument fails below the transition temperature, where $z\delta_1\cong 1$. To study this case, we combine (52) with the identity $1=N^{-1}\sum_L L\langle m_L\rangle$ [which follows from (40) and (44)] and use (53); this gives

$$1-A_1/A_{1,\infty}\simeq N^{-1}\sum_L(f_L-\delta_1{}^L)Q_{N-L}/Q_N. \tag{56}$$

Feynman[7] estimates that, unless L is a small integer,

$$f_L\simeq(L^{-\frac{5}{2}}V\Delta+1)\delta_1{}^L, \tag{57}$$

where

$$\Delta\equiv\left[3\delta_1\Big/8\pi^2\int_0^\infty G(r)r^4dr\right]^{\frac{3}{2}}.$$

Therefore, although the approximation $Q_{N-L}/Q_N\simeq z^L$ is no longer legitimate in (52), it is still legitimate in (56), according to (57), a convergent series results even though $z\delta_1\cong 1$:

$$1-A_1/A_{1,\infty}\cong\mathrm{const}+N^{-1}\sum_{L=1}^{\infty}V\Delta L^{-\frac{5}{2}}\cong\mathrm{const}, \tag{58}$$

where the first "const" takes care of the error due to the failure of (57) for small L. Feynman's work[7] shows that the right-hand side of (58) is less than 1 below the transition. Hence $A_1/A_{1,\infty}\cong\mathrm{const}>0$, and, by (12), B.E. condensation does occur in He II.

The deductions we have made from Feynman's approximations can be paraphrased as follows: the quantity $\langle A_{1,L}\{m_l\}\rangle$ is very small if $L\ll(V\Delta)^{\frac{2}{5}}$ (where Δ is finite), and equals the finite quantity $A_{1,\infty}$ if $L\gg(V\Delta)^{\frac{2}{5}}$. Hence, by (43), $A_1/A_{1,\infty}$ equals the contribution of large L values to the sum $N^{-1}\sum L\langle m_L\rangle$; that is, it equals the fraction of particles in large cycles. Above the lambda-transition this fraction is negligible, so that, by (12), there is no B.E. condensation; below the transition this fraction is finite, so that B.E. condensation is present.

8. DISCUSSION

Equation (4) provides a mathematical definition of B.E. condensation, applicable for a system of interacting particles as well as for an ideal gas. Physically, the definition means that B.E. condensation is present whenever a finite fraction—n_M/N— of the particles occupies one single-particle quantum state, φ_M. The definitions of n_M and φ_M are given in Sec. 3 and Sec. 4, respectively. Even for an ideal gas, our definition is more general than the usual one, since here φ_M is not necessarily the lowest single-particle energy level. The close relation between our definition of B.E. condensation and London's suggested[1,2] "condensation in momentum space" is illustrated in the last paragraph of Sec. 3 above, where it is shown that under suitable conditions φ_M actually is an eigenstate of momentum.

The reasoning of Secs. 5, 6, and 7 indicates that liquid helium II satisfies our criterion of B.E. condensation. For $T=0°$K the only physical assumption used is that a quantum liquid—as distinct from a solid—lacks long-range configurational order (though the mathematical treatment of this assumption is not yet

[39] J. E. Mayer and M. G. Mayer, *Statistical Mechanics* (John Wiley and Sons, Inc., New York, 1940), pp. 277–282.

completely rigorous). For $T>0°K$, some fairly crude approximations, taken from Feynman's theory of the lambda-transition, have to be introduced. This part of the theory is therefore open to improvement—possibly in the form of a more rigorous proof that Feynman's implied criterion for B.E. condensation [the importance of long cycles in the sum (40) for the partition function] is equivalent to our criterion (4) at thermal equilibrium. Despite these imperfections, however, our analysis would appear to strengthen materially the case put forward previously by London[1,2] and Tisza[9] for the importance of B.E. condensation in the theory of liquid helium.

We have not considered here how B.E. condensation is related to superfluidity and to the excitation theory[16,17] of liquid helium. This will be done in another paper, where some of the results already obtained by Bogolyubov[12] for weakly repelling B.E. particles will be extended[40] to the case of interacting He^4 atoms.

9. ACKNOWLEDGMENTS

The authors are indebted to the National Science Foundation and to the United States Educational Commission in the United Kingdom for financial support. They would also like to thank Dr. G. V. Chester and Dr. D. W. Sciama for helpful discussions.

[40] A brief account of this work was given at the National Science Foundation Conference on Low-Temperature Physics and Chemistry, Baton Rouge, Louisiana, December, 1955 (unpublished).

Excitation of Rotons in Helium II by Cold Neutrons

H. Palevsky,* K. Otnes, K. E. Larsson, R. Pauli, and R. Stedman
Aktiebolaget Atomenergi, Stockholm, Sweden
(Received October 14, 1957)

THE theories of helium II developed by Landau[1] and Feynman[2] conclude that in helium below the λ point there exist relatively simple periodic motions of the fluid, *viz.*, phonons and rotons, having mean free paths long compared to their wavelengths. Although in the past few years neutron experiments have clearly demonstrated the existence of phonons in solids, neutron experiments to data[3] have failed to demonstrate any marked difference in helium measurements below and above the λ point. Earlier this year Cohen and Feynman[4] pointed out that on the basis of the Landau-Feynman theories, there should be a measurable difference in the energy distribution of scattered neutrons from He I as compared to He II. This letter is to report the preliminary results of such a measurement.

A neutron beam from the Stockholm Reactor[5] was filtered through 8 in. of polycrystalline Be, resulting in a neutron spectrum characterized by a sharp rise in intensity at 3.96 A, and varying approximately as $1/\lambda^5$ for wavelengths greater than 3.96 A (see Fig. 1). With this incident spectrum, the energy spectrum of the neutrons scattered at 90° from the incoming direction, by a 10-cm sample of liquid helium, at $T=1.2$ and 4.2°K, was measured. The energy was inferred from the measured time-of-flight of the neutrons using the Stockholm slow chopper and associated electronic equipment.[6]

Figure 1 shows the incident spectrum as measured by elastically scattering the incoming beam through 90° by means of a thin vanadium sample. (The scattering amplitude of vanadium is nearly totally incoherent.) The ordinate in the figure represents the counting rate in a 30-μsec time interval at the time indicated on the abscissa. The sharp rise in intensity at 3.96 A is the

Bragg "cutoff" in Be. The position of the cutoff corresponds to twice the separation of the most widely spaced planes in the crystal [(01·0) in Be]. For wavelengths longer than the Bragg cutoff, coherent elastic scattering can take place only in the forward direction, and therefore Be becomes transparent for neutrons of longer wavelengths. The background counting rate on the same scale as Fig. 1 is approximately 500 counts. Figure 2 shows the results with liquid helium acting as the scatterer. The dashed line gives the position of the Be cutoff given in Fig. 1. The light solid line represents the measured spectrum with helium at 1.4°K. At this temperature about 93% of the helium is in the superfluid state. The scattered spectrum exhibits the same shape as the incoming spectrum; however; the position of the cutoff is displaced some 305 μsec towards greater time-of-flight (lower energy). The sharp rise of the cutoff together with the displacement in energy represents the excitation of a single roton in helium, of momentum $K=2.14\pm0.07$ A^{-1} ($K=2\pi/\lambda$) and energy $E=10.7\pm0.5$°K. From an analysis of the limiting

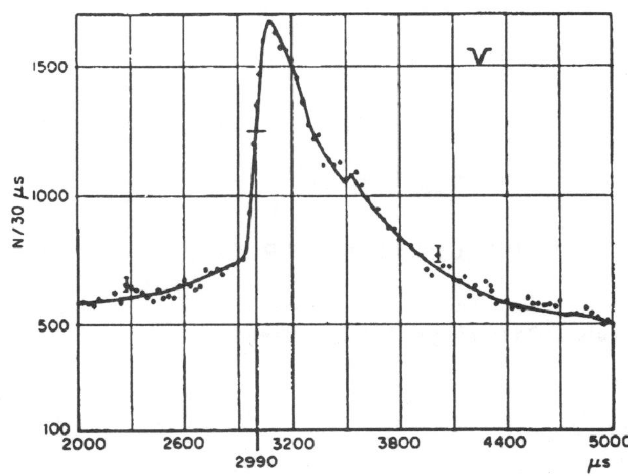

FIG. 1. The incident neutron spectrum as determined by scattering from vanadium. The ordinate gives the number of counts accumulated in a 30-μsec interval (channel) at the time indicated on the abscissa. The neutron wavelength is directly proportional to time-of-flight, and 2990 μsec represent 3.96 A.

resolution of the apparatus it is possible to estimate that the lower limit for the mean free path of a 2.14 A^{-1} roton at $T=1.4$°K is $\geq 2\times10^{-6}$ cm. This figure should be compared to an upper limit of $\leq 2.5\times10^{-6}$ cm as calculated from the Landau-Khalatnikov theory.[7] The heavy solid line shows the spectrum of scattered neutrons for the liquid at 4.2°K at which temperature the concentration of superfluid is zero. The spectrum loses the sharp cutoff completely, indicating that there are many-collision processes present. Qualitatively the spectrum has the shape one would expect from a normal fluid.

It is possible by the filter technique to trace out the dispersion curve in the region of the roton excitations, either by using filters with different Bragg cutoffs or by changing the angle between the incident and scattered neutrons. By using a BeO filter, a point on the dispersion curve at $K=1.80\pm0.06$ has been determined to have an energy of 10.0 ± 0.7°K. Measurements are now under way to trace out the region between $K=1.80$ and 2.14 A^{-1} by changing the angular settings, and from these measurements an effective mass for the roton excitations will be determined. A complete description of the experimental details will be published in a separate paper.

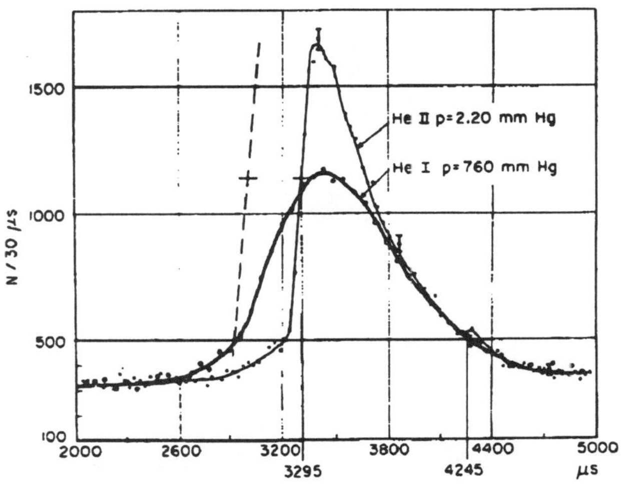

FIG. 2. The ordinate and abscissa same as Fig. 1. The light solid line represents the scattered neutrons from He at 1.4°K; the heavy solid line from He at 4.2°K. The dashed line gives the position of the Be cutoff as shown in Fig. 1.

We wish to thank Professor J. O. Linde and Mr. K. Svensson of the KTH in Stockholm for their aid with the cryogenics of the experiment. The time-of-flight electronics was designed by Mr. J. Björkman at AB Atomenergi. Professor J. Yvon, Commissariat à l'Energie Atomique, France, kindly lent us the BeO for the filter. One of us (H.P.) is indebted to Dr. M. Cohen and Professor R. P. Feynman for many enlightening discussions concerning the nature of helium II. Finally one of us (H.P.) is indebted to Dr. S. Eklund and the Swedish Atomic Energy Company for his very enjoyable stay in Sweden.

* Guest scientist on leave from Brookhaven National Laboratory, Upton, New York.

[1] L. Landau, J. Phys. (U.S.S.R.) **5**, 71 (1941); **11**, 91 (1947).
[2] R. P. Feynman, Phys. Rev. **91**, 1291, 1301 (1953); **94**, 262 (1954).
[3] D. G. Henshaw and D. G. Hurst, Phys. Rev. **91**, 1922 (1953); P. Egelstaff and H. London, Jener Report No. 7, *Proceedings of Kjeller Conference on Heavy Water Reactors*, 1953 (Joint Establishment for Nuclear Energy Research, Kjeller, 1953); Sommers, Dash, and Goldstein, Phys. Rev. **97**, 855 (1955).
[4] M. Cohen and R. P. Feynman, Phys. Rev. **107**, 13 (1957).
[5] S. Eklund, J. Nuclear Energy **1**, 93 (1954).
[6] Larsson, Stedman, and Palevsky, J. Nuclear Energy (to be published).
[7] L. Landau and I. Khalatnikov, J. Exptl. Theoret. Phys. (U.S.S.R.) **19**, 637, 709 (1949).

NUCLEAR MAGNETIC RESONANCE IN MnF_2 NEAR THE CRITICAL POINT*

P. Heller†
Gordon McKay Laboratory, Harvard University, Cambridge, Massachusetts

and

G. B. Benedek
Department of Physics, Massachusetts Institute of Technology, Cambridge, Massachusetts

(Received May 8, 1962)

The F^{19} nuclear resonance in MnF_2, first observed by Shulman and Jaccarino,[1,2] has been studied in detail in the very interesting temperature region around the paramagnetic-antiferromagnetic critical temperature $T_N = 67.4°K$. Our experimental system permits temperature measurement and control to within one millidegree for as long as six hours, any temperature variation across the sample being much less than a millidegree. Temperatures were measured with a platinum resistance thermometer calibrated very carefully using the vapor pressure-temperature curve for nitrogen.[3] An inductive coupling arrangement used with a very high frequency spectrometer,[4] plus a standard Pound-Knight-Watkins spectrometer, made possible nuclear magnetic resonance observations at frequencies from 23 to 110 Mc/sec without making any changes inside the low-temperature apparatus. Nuclear resonances of width up to 3 Mc/sec could be detected with this system.

We have determined the temperature dependence of the F^{19} resonance frequency ν_{19} in zero applied field in the antiferromagnetic state between T_N and 52°K. Our data start at $T = T_N - 0.005K°$ at which temperature ν_{19} is about 8 Mc/sec. This frequency corresponds to a reduction of the sublattice magnetization to 5% of the magnetization at absolute zero. For the first few degrees below the Néel point, the data very accurately fit the expression

$$\frac{M(T)}{M(0)} = \frac{\nu_{19}(T)}{\nu_{19}(0)} = A(T_N - T)^R, \quad (1)$$

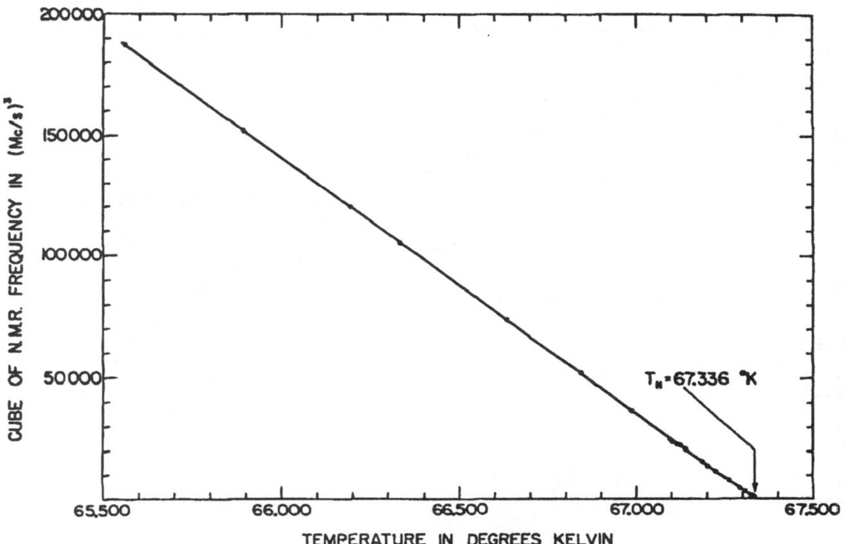

FIG. 1. Temperature dependence of the cube of the F^{19} nuclear resonance frequency for the first 1.8 degrees below T_N. The points lie on the straight line shown to within the experimental uncertainty of about 5 millidegrees.

where $R = 0.335 \pm 0.01$, $A = (0.295 \pm 0.001)(\mathrm{K}°)^{-R}$, and $\nu_{19}(0) = 159.978$ Mc/sec. The accuracy with which Eq. (1) fits the data can be seen in Fig. 1 which plots $\nu_{19}^3(T)$ versus the temperature for the first 1.8 degrees below T_N. Extrapolation over the last 5 millidegrees enables us to determine that

$$T_N = (67.336 \pm 0.003)°\mathrm{K}. \qquad (2)$$

Since Eq. (1) does not have the correct asymptotic behavior as $T \to 0$, it cannot be expected to hold far below T_N. The experimental results depart from the values given by Eq. (1) by 0.02K°, 0.10K°, and 0.5K° at 61°K, 58.5°K, and 53°K, respectively. The experimental data for the range $52°\mathrm{K} < T < T_N$ are plotted in Fig. 2. The cube root law that we observe shows that the sublattice magnetization rises much faster than the law $M \propto (T_N - T)^{1/2}$ predicted by the molecular field theory just below T_N. The molecular field theory prediction is shown together with our data for $\nu_{19}(T)$ in Fig. 2. The precision of the nuclear resonance measurements, as illustrated in Figs. 1 and 2 and Eq. (1), demonstrates the power of this method[5,6] in determining the sublattice magnetization in the theoretically difficult, but important, critical region.

By observing the shift of the nuclear resonances on applying a magnetic field H_0 along the C (antiferromagnetic) axis, we were able to determine the susceptibility $\chi_{\parallel}$ for each of the two Mn^{++} sublattices separately near T_N. The susceptibilities of the up and down sublattices were found to be equal and their sum is in good agreement with the ordinary macroscopic susceptibility measurements.[7]

In the antiferromagnetic state the F^{19} linewidths are strongly anisotropic, depending on the direction of the vector sum $\vec{H}_{\mathrm{nucl}}$ of the applied field and the field produced by the Mn^{++} spins at the F^{19} nucleus. The lines are broadest for $\vec{H}_{\mathrm{nucl}}$ along C and narrowest for $\vec{H}_{\mathrm{nucl}}$ along A. (See Fig. 3.) For $\vec{H}_{\mathrm{nucl}}$ along C, the linewidth is given to within 20% by

$$\delta\nu_C = 490\ [\mathrm{kc/sec}\,(\mathrm{K}°)^{1/2}](T_N - T)^{-1/2};$$

$$0.02\mathrm{K}° < T_N - T < 15\mathrm{K}°. \qquad (3)$$

In this 15-degree temperature range the linewidth increases from 100 kc/sec to 3000 kc/sec. In the paramagnetic state the line shows an extraordinarily rapid broadening as T_N is approached. The broadening is anisotropic, being more marked at a given temperature with the applied field along C than along A. (See Fig. 4.) For $0.04\mathrm{K}° < T - T_N < 10\mathrm{K}°$ the linewidths are given to within 15% by

$$H_0 \text{ along } C\text{:}\quad \delta\nu_C = 95\ \mathrm{kc/sec}\,[1 + 0.33\mathrm{K}°/(T - T_N)],$$

$$H_0 \text{ along } A\text{:}\quad \delta\nu_A = 85\ \mathrm{kc/sec}\,[1 + 0.12\mathrm{K}°/(T - T_N)]. \qquad (4)$$

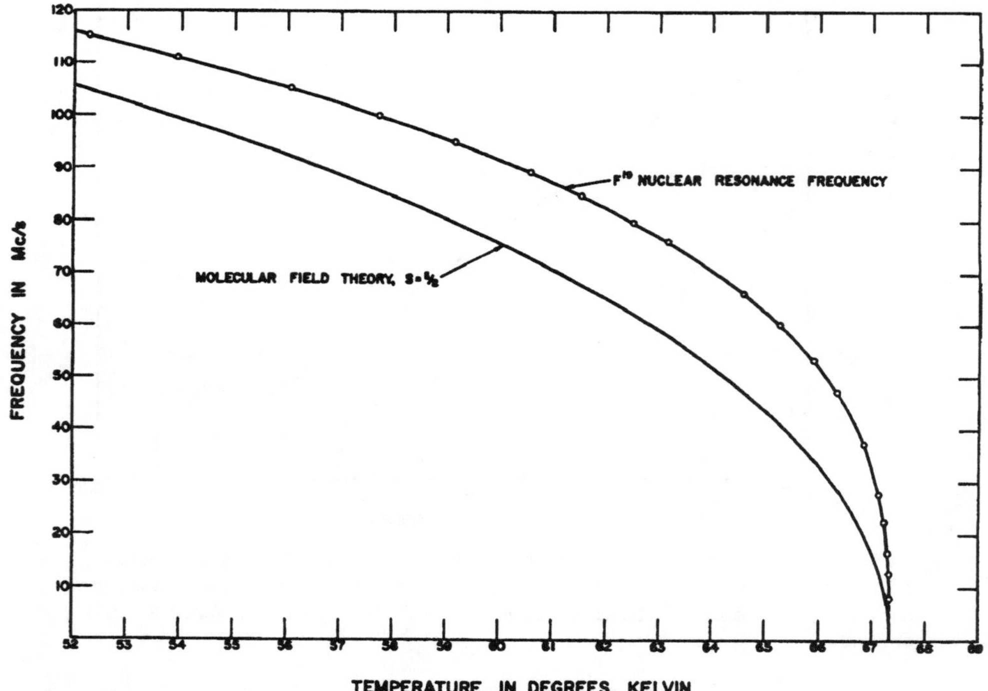

FIG. 2. Temperature dependence of the F^{19} resonance frequency in the range 52°K $< T < T_N$. The circles greatly exaggerate the size of the experimental uncertainties. The lower curve is computed using the molecular field theory.

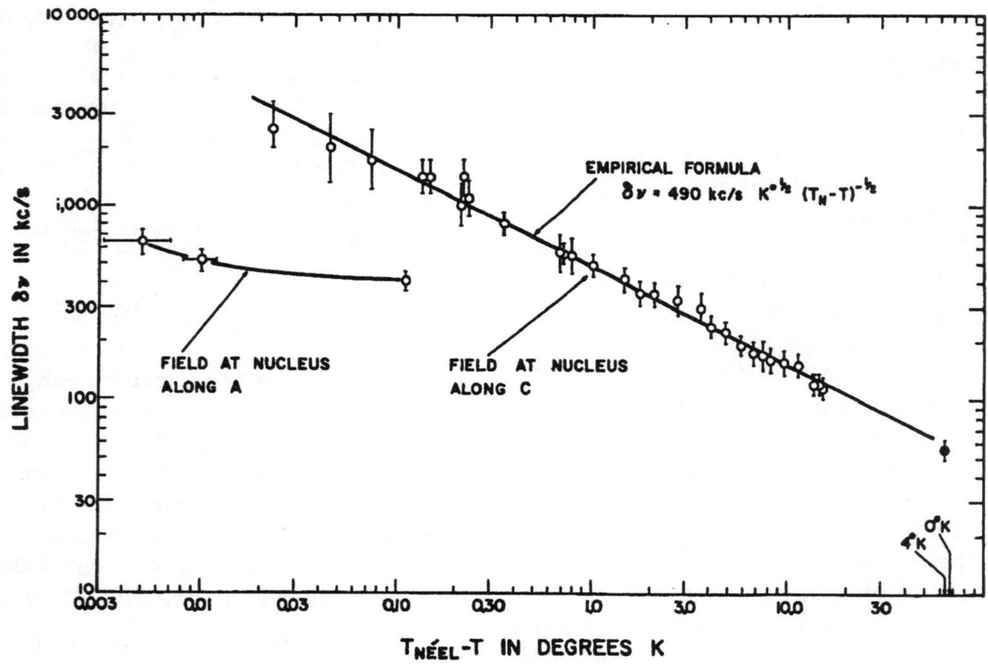

FIG. 3. Temperature dependence of the F^{19} linewidths in the antiferromagnetic state. The single point at 4°K is from reference 2.

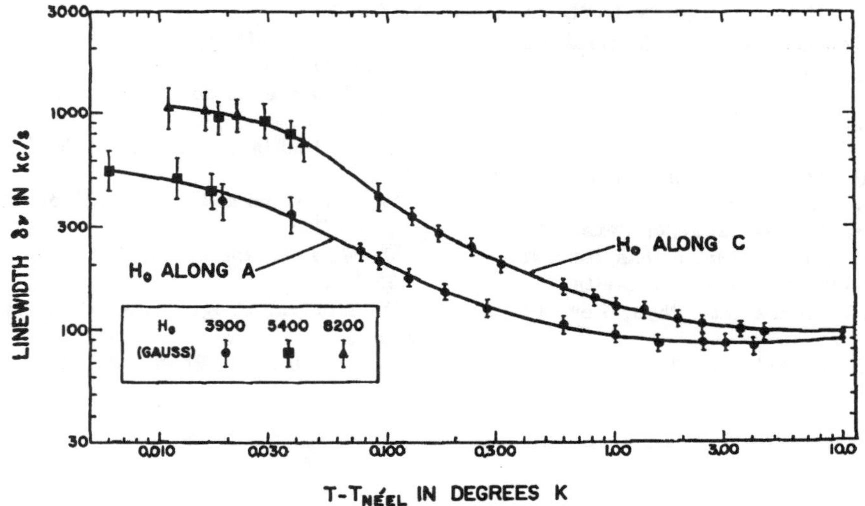

FIG. 4. Temperature dependence of the F^{19} linewidths in the paramagnetic state just above the Néel point.

Note that in the last degree above T_N the line broadens by about a factor of ten. It has been proposed[1,8] that the linewidth of the F^{19} resonance is determined by exchange narrowing of the hyperfine interaction between the F^{19} nucleus and the nearest neighbor Mn^{++} spins. If this is the case, the present measurements provide rather direct experimental information on the temperature dependence of the time-space correlation function for the manganese spins in the critical region.

The rapid fade-out of the line which accompanies this broadening provides an extremely sensitive method for measuring changes in the Néel temperature. In particular we have measured the depression of the Néel temperature due to an applied field. In Table I we list the results obtained, together with the predicted Néel point shifts on the molecular field model. We have also measured the effect of hydrostatic pressure on T_N. Our result,

$$dT_N/dP = 303 \pm 3 \text{ millidegrees per 1000 kg/cm}^2, \tag{5}$$

Table I. Effect on T_N of raising applied field H_0 from 5.4 kG to 8.25 kG.

Field direction	ΔT_N in millidegrees Expt.	Mol. field theory
A axis	-2.8 ± 1	-0.87
C axis	-7.5 ± 1	-2.6

agrees well with the estimate of Benedek and Kushida[4] obtained by a quite different method. The pressure dependence of T_N shows that the Néel point is itself a function of the temperature because of the effect of thermal expansion. Using our pressure data and the thermal expansion measurements of Gibbons,[9] we have estimated the importance of this effect in fitting the $\nu_{19}(T)$ data to Eq. (1) with $T_N = T_N(T)$. Our estimate shows that, with this correction, the coefficient *A* in (1) should be about one percent less than stated above. The effect on the exponent *R* is negligible.

Just below T_N we observe, to our surprise, lines corresponding to nuclei in both the antiferromagnetic and paramagnetic phases, the line at the paramagnetic location fading rapidly with decreasing temperature and disappearing entirely at about 25 millidegrees below T_N. At 5 millidegrees below T_N the areas under the two lines are roughly equal. We do not understand this effect. One obvious possibility is that it is due to an inhomogeneity in the sample's Néel temperature caused, perhaps, by internal strains. Such a mechanism would broaden the line below T_N. However, the actual linewidths for $\vec{H}_{\text{nucl}}$ along *A* are several times larger than possible for such a mechanism. Furthermore, the linewidths for $\vec{H}_{\text{nucl}}$ along *C* fall off more slowly as the temperature is reduced than can be predicted on this basis.

We are greatly indebted to Dr. V. Jaccarino and Dr. H. Guggenheim of the Bell Telephone Laboratories for providing us with oriented single crystals of MnF_2. It is a pleasure to thank Dr. D. Gill, Dr. J. Jeener, and Dr. G. Seidel for many helpful discussions.

---

*Research supported by the U. S. Joint Services and Advanced Research Projects Agency.

†Raytheon Predoctoral Fellow 1959-60; Texaco Predoctoral Fellow 1960-61; now at Massachusetts Institute of Technology, Cambridge, Massachusetts.

[1]R. G. Shulman and V. Jaccarino, Phys. Rev. 108, 1219 (1957).

[2]V. Jaccarino and R. G. Shulman, Phys. Rev. 107, 1196 (1957).

[3]G. T. Armstrong, J. Research Natl. Bur. Standards 53, 263 (1954).

[4]G. B. Benedek and T. Kushida, Phys. Rev. 118, 46 (1960).

[5]N. J. Poulis and G. E. G. Hardeman, Physica 18, 391 (1952).

[6]V. Jaccarino and L. R. Walker, J. phys. radium 20, 341 (1959).

[7]J. W. Stout and M. Griffel, J. Chem. Phys. 18, 1455 (1950).

[8]T. Moriya, Progr. Theoret. Phys. (Kyoto) 16, 641 (1956).

[9]D. F. Gibbons, Phys. Rev. 115, 1194 (1959).

INTERACTION OF "SOLITONS" IN A COLLISIONLESS PLASMA AND THE RECURRENCE OF INITIAL STATES

N. J. Zabusky

Bell Telephone Laboratories, Whippany, New Jersey

and

M. D. Kruskal

Princeton University Plasma Physics Laboratory, Princeton, New Jersey

(Received 3 May 1965)

We have observed unusual nonlinear interactions among "solitary-wave pulses" propagating in nonlinear dispersive media. These phenomena were observed in the numerical solutions of the Korteweg-deVries equation

$$u_t + uu_x + \delta^2 u_{xxx} = 0. \tag{1}$$

This equation can be used to describe the one-dimensional, long-time asymptotic behavior of small, but finite amplitude: shallow-water waves,[1,2] collisionless-plasma magnetohydrodynamic waves,[2] and long waves in anharmonic crystals.[3,4] Furthermore, the interaction and "focusing" in space-time of the solitary-wave pulses allows us to give a phenomenological description (some aspects of which we can already explain analytically) of the near recurrence to the initial state in numerical calculations for a discretized weakly-nonlinear string made by Fermi, Pasta, and Ulam (FPU).[4,5]

Spatially periodic numerical solutions of the Korteweg-deVries equation were obtained with a scheme that conserves momentum and almost conserves energy.[6] For a variety of initial conditions normalized to an amplitude of 1.0 and for small δ^2, the computational phenomena observed can be described in terms of three time intervals. (I) Initially, the first two terms of Eq. (1) dominate and the classical overtaking phenomenon occurs; that is, u steepens in regions where it has a negative slope. (II) Second, after u has steepened sufficiently, the third term becomes important and serves to prevent the formation of a discontinuity. Instead, oscillations of small wavelength (of order δ) develop on the left of the front. The amplitudes of the oscillations grow and finally each oscillation achieves an almost steady amplitude (which increases linearly from left to right) and has a shape almost identical to that of an individual solitary-wave solution of (1). (III) Finally, each such "solitary-wave pulse" or "soliton" begins to move uniformly at a rate (relative to the background value of u from which the pulse rises) which is linearly proportional to its amplitude. Thus, the solitons spread apart. Because of the periodicity, two or more solitons eventually overlap spatially and interact nonlinearly. Shortly after the interaction, they reappear virtually unaffected in size or shape. In other words, solitons "pass through" one another without losing their identity. Here we have a nonlinear physical process in which interacting localized pulses do not scatter irreversibly.

It is desirable to elaborate the concept of the soliton, for it plays such an important role in explaining the observed phenomena. We seek stationary solutions of (1) in a frame moving with velocity c. We substitute

$$u = U(x - ct)$$

into (1) and obtain a third-order nonlinear ordinary differential equation for u. This has periodic solutions representing wave trains, but to explain the concept of a soliton we are interested in a solution which is asymptotically constant at infinity ($u = u_\infty$ at $x = \pm\infty$). The result[7] of such a calculation is

$$u = u_\infty + (u_0 - u_\infty)\,\mathrm{sech}^2[(x - x_0)/\Delta], \tag{2}$$

where u_0, u_∞, and x_0 are arbitrary constants and

$$\Delta = \delta[(u_0 - u_\infty)/12]^{-1/2}, \tag{3}$$

and

$$c = u_\infty + (u_0 - u_\infty)/3. \tag{4}$$

Thus, the larger the pulse amplitude and the smaller δ, the narrower is the pulse. The surprising thing is that these pulses, which are strict solutions only when completely isolated, can exist in close proximity and interact without losing their form or identity (except mo-

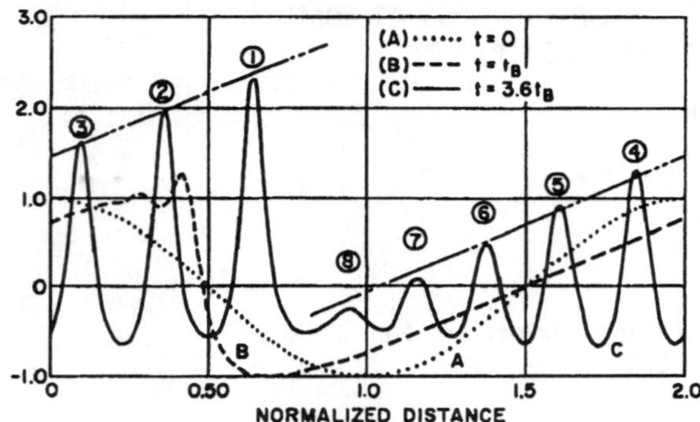

FIG. 1. The temporal development of the wave form $u(x)$.

mentarily while they "overlap" substantially).

The numerical calculations in which these phenomena were observed were made starting with $\delta = 0.022$ and the periodic initial condition

$$u|_0 = \cos\pi x. \tag{5}$$

Thus, initially, $\max|\delta^2 u_{xxx}|/\max|uu_x| = 0.004$ so the third term can be neglected and we are dealing with the equation $u_t + uu_x = 0$. Its solution is given by the implicit relation

$$u = \cos\pi(x - ut), \tag{6}$$

and we find that u tends to become discontinuous at $x = \frac{1}{2}$ and $t = T_B = 1/\pi$, the breakdown time. Figure 1, curve A, gives the initial condition, and curve B shows the function at T_B. The slight oscillatory structure for $x < \frac{1}{2}$ is due to the third derivative which we have neglected in arriving at the approximate solution (6). Curve C at $t = 3.6T_B$ shows a train of solitons (numbered 1-8), which have developed from the oscillations. A so-far unexplained property of these solutions is the linear variation of the amplitude of the largest pulses. Table I gives the amplitudes of the pulses, their observed and calculated widths [Eq. (3)], and their observed and calculated velocities [Eq. (4)]. We note that the calculated and observed widths and velocities of the first seven solitons are in very good agreement.

Figure 2 gives the space-time trajectories of the solitons. The vertical axis is normalized in terms of the recurrence time T_R ($T_R = 30.4T_B$ for this computation), which is the time it would take all the solitons to overlap or "focus" at a common spatial point. The diagram at the right of Fig. 2 shows the amplitude of soliton no. 1 (horizontally) versus time (vertically). The observed velocity of each soliton given in the table is calculated as the slope of the straight line drawn tangent to its trajectory over the time interval $0.0975T_R$ to $0.133T_R$.

When solitons of very different amplitude approach, their trajectories deviate from straight lines (accelerate) as they "pass through" one another. During the overlap time interval the

Table I. Soliton properties—observed[a] and calculated values ($\delta = 0.022, t = 3.6T_B$).

Pulse no.	Amplitudes (observed) $-u_\infty$	$u_0 - u_\infty$	Width (Δ) observed	calculated	Velocity (c) observed	calculated
1	325	1739	0.0455	0.0456	227	254
2	401	1597	0.0475	0.0476	110	131
3	491	1485	0.0492	0.0493	0	4
4	544	1318	0.0522	0.0516	−99	−105
5	574	1115	0.0567	0.0568	−169	−202
6	584	885	0.0636	0.0639	−273	−289
7	558	610	0.0769	0.0767	−361	−354
8	453	302	0.099	0.109	−443	−353

[a]The observed quantities (excluding c) are obtained from the numerical values of u and u_{xx} at $t = 3.6T_B$ (Fig. 1, curve C). u_0 is the observed maximum of each pulse; $u_0 - u_\infty$ (and therefore u_∞) and Δ_{obs} are obtained from the minimum value of u_{xx}:

$$\Delta_{\text{obs}} = (24\delta^2 |\min u_{xx}|_{x=x_0}|)^{1/4},$$

$$u_0 - u_\infty = 12\delta^2/\Delta_{\text{obs}}^2.$$

The observed values of c are obtained by measuring slopes on Fig. 2 at $t = 3.5T_B = 0.115T_R$, as described below. The calculated values of Δ are obtained from Eq. (3) with $\delta = 0.022$ and $(u_0 - u_\infty)$ obtained from column 3. The calculated values of c are obtained from (4) with u_∞ and $(u_0 - u_\infty)$ obtained from columns 2 and 3.

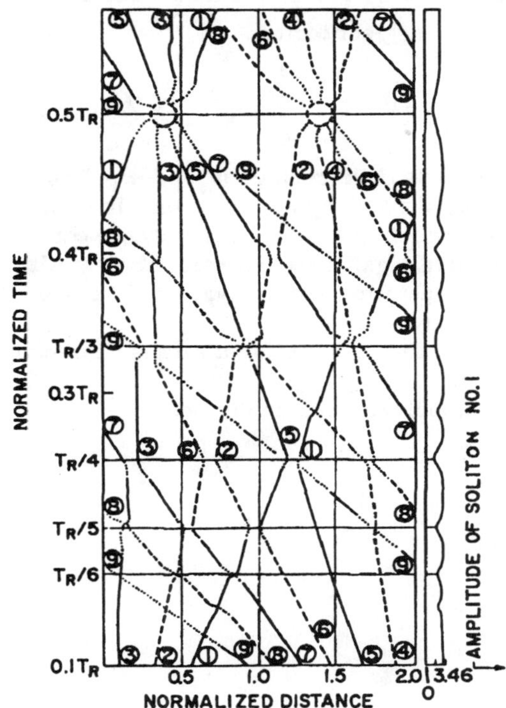

FIG. 2. Soliton trajectories on a space-time diagram beginning at $t = 0.1T_R = 3.04\,T_B$. The diagram at the right shows the variation of the amplitude of soliton no. 1 as a function of time.

joint amplitude of the interacting solitons decreases (in contradistinction to what would happen if two pulses overlapped linearly). This is evident at $t = T_R/6$ where solitons 1 and 7, 2 and 8, and 3 and 9 overlap. When the amplitudes of approaching solitons are comparable they seem to exchange amplitudes and therefore velocities. They do not have to approach very close to one another for this "transition" to occur. This is evident at $t = T_R/4$ where solitons 1 and 5, 2 and 6, 3 and 7, and 4 and 8 overlap (and similarly at $t = T_R/3$, etc.). In general, when solitons approach we have replaced the solid lines (odd-numbered solitons) and dashed lines (even-numbered solitons) by dots, as it is not yet clear how to describe what happens during a close interaction. At $t = 0.5T_R$ all the odd solitons overlap at $x = 0.385$ and all the even ones at $x = 1.385$, and because of the nonlinear interaction one cannot follow the crests, and so the regions are circled. The waveform of u which results is mostly composed of, and has the form of, the second harmonic of the initial waveform.

In conclusion, we should emphasize that at T_R all the solitons arrive almost in the same phase and almost reconstruct the initial state through the nonlinear interaction. This process proceeds onwards, and at $2T_R$ one again has a "near recurrence" which is not as good as the first recurrence. Tuck,[8] at the Los Alamos Scientific Laboratories, observed this phenomenon as well as eventual "superrecurrences" in calculations for a similar problem. We can understand these phenomena in terms of soliton interactions. For $t > T_R$ the successive focusings get poorer due to solitons arriving more and more out of phase with each other and then eventually gets better again when their phase relationship changes. Furthermore, because the solitons are remarkably stable entities, preserving their identity through numerous interactions, one would expect this system to exhibit thermalization (complete energy sharing among the corresponding linear normal modes) only after extremely long times, if ever.

The authors would like to thank G. S. Deem for assistance in programming and reducing the numerical data.

---

[1]J. J. Stoker, Water Waves (Interscience Publishers, Inc., New York, 1957). See Secs. 10.9 through 10.11 and Eq. (10.11.20).

[2]C. S. Gardner and G. K. Morikawa, Courant Institute of Mathematical Sciences Report No. NYO 9082, 1 May 1960 (unpublished).

[3]N. J. Zabusky, in Proceedings of the Conference on Mathematical Models in the Physical Sciences, edited by Stefan Drobot (Prentice-Hall, Inc., New York, 1963), p. 99.

[4]M. D. Kruskal, "Progress on the Fermi, Pasta, Ulam Problem" (to be published).

[5]E. Fermi, J. R. Pasta, and S. Ulam, Los Alamos Scientific Laboratory Report No. LA-1940, May 1955 (unpublished). See reference 3 for a review of the problem.

[6]We restrict ourselves to solutions of (1) periodic in x with period 2 so that we need only consider the interval $0 \leq x < 2$ with periodic (cyclic) boundary conditions. For numerical purposes we replaced (1) with

$$u_i^{j+1} = u_i^{j-1} - \tfrac{1}{3}(k/h)(u_{i+1}^{\,j} + u_i^{\,j} + u_{i-1}^{\,j})(u_{i+1}^{\,j} - u_{i-1}^{\,j}) - (\delta^2 k/h^3)(u_{i+2}^{\,j} - 2u_{i+1}^{\,j} + 2u_{i-1}^{\,j} - u_{i-2}^{\,j}),$$

$$i = 0,\ 1, \cdots, 2N-1,$$

where a rectangular mesh has been used with temporal and spatial intervals of k and $h = 1/N$, respectively. That is, the function $u(x,t)$ is approximated by $u_i^{\,j} = u(ih, jk)$. In performing the calculations we used periodic (cyclic) boundary conditions $u_i{}^j = u_{i+2N}{}^j$. The momentum

$$\sum_{i=0}^{2N-1} u_i^j$$

is identically conserved, for if one sums both sides of the equation with respect to i, then the quantities multiplied by k vanish identically. The energy

$$\sum_{i=0}^{2N-1} \tfrac{1}{2}(u_i^j)^2$$

is almost conserved, that is, the above quantity is invariant if we neglect terms

$$(k^2/6) \sum_{i=0}^{2N-1} u_i(u_i)_{ttt} + O(k^4).$$

This is evident if we replace $(u_i^{j+1} - u_i^{j-1})/2k$ by $\partial u_i/\partial t$, multiply through by u_i, and sum. In practice, those runs which are numerically stable conserve the energy to five significant figures. The details of numerical computation and analysis will be published in the near future.

[7]See reference 2, p. 18 for a similar calculation with $u_\infty = 0$.

[8]J. Tuck, private communication.

ABSENCE OF FERROMAGNETISM OR ANTIFERROMAGNETISM IN ONE- OR TWO-DIMENSIONAL ISOTROPIC HEISENBERG MODELS*

N. D. Mermin† and H. Wagner‡
Laboratory of Atomic and Solid State Physics, Cornell University, Ithaca, New York
(Received 17 October 1966)

It is rigorously proved that at any nonzero temperature, a one- or two-dimensional isotropic spin-S Heisenberg model with finite-range exchange interaction can be neither ferromagnetic nor antiferromagnetic. The method of proof is capable of excluding a variety of types of ordering in one and two dimensions.

The number of exact results on the presence or absence of phase transitions in systems with realistically short-range interactions is small: Van Hove has proved that there are no phase transitions in a one-dimensional classical gas with hard-core and finite-range interactions, and Griffiths has proved that the Ising model is ferromagnetic in more than one dimension.[1] More recently Hohenberg[2] has shown that an inequality due to Bogoliubov[3] can be used to exclude conventional superfluid or superconducting ordering in one or two dimensions. We have found that a similar application of the Bogoliubov inequality leads to a surprisingly elementary but rigorous argument that the one- and two-dimensional isotropic Heisenberg models with interactions of finite range can be neither ferromagnetic nor antiferromagnetic at nonzero temperature. These conclusions have long been suspected, being suggested by calculations of the elementary excitations from the ordered state,[4] as well as by considerations of the energetics of domain walls.[5] In view of the present degree of activity in critical-point studies, it nevertheless seems worth recording that these very plausible results can be proved rigorously.[6]

We prove that there can be no spontaneous magnetization or sublattice magnetization in an isotropic Heisenberg model with finite-range interactions at temperature T, by showing that for sufficiently small fields h,

$$|s_z| < \frac{\text{Const.}}{T^{1/2}} \frac{1}{|\ln|h||^{1/2}} \text{ (2 dimensions)}, \quad (1)$$

$$|s_z|h| < \frac{\text{Const.}}{T^{2/3}} |h|^{1/3} \text{ (1 dimension)}, \quad (2)$$

where s_z can be taken as either the infinite-volume limit of the magnetization per particle in a uniform field h, or the infinite-volume limit of the difference of the two sublattice magnetizations per particle, in a field of magnitude h and opposite sign on the two sublattices.

The proof exploits Bogoliubov's inequality,[8]

$$\tfrac{1}{2}\langle\{A,A^\dagger\}\rangle\langle[[C,H],C^\dagger]\rangle \geq k_B T|\langle[C,A]\rangle|^2, \quad (3)$$

where H is the Hamiltonian and $\langle X\rangle = \mathrm{Tr}Xe^{-\beta H}/\mathrm{Tr}e^{-\beta H}$ with $\beta = 1/k_BT$. The following elementary proof of the inequality is designed to make it clear that (3) is valid provided only that the operators A and C are such that the ensemble averages on the left-hand side of (3) exist. Define

$$(A,B) = \sum_{ij}{}' \langle i|A|j\rangle^* \langle i|B|j\rangle \frac{W_i - W_j}{E_j - E_i},$$

where the sum is over all pairs from a complete set of energy eigenstates, excluding pairs with the same energy; $W_i = e^{-\beta E_i}/\mathrm{Tr}e^{-\beta H}$. Note first that

$$0 < (W_i - W_j)/(E_j - E_i) < \tfrac{1}{2}\beta(W_i + W_j),$$

and, therefore,

$$(A,A) \leq \tfrac{1}{2}\beta\langle\{A,A^\dagger\}\rangle. \quad (4)$$

It is also easily verified that (A,B) satisfies all the definitions of an inner product necessary to establish the Schwartz inequality

$$(A,A)(B,B) \geq |(A,B)|^2. \quad (5)$$

Finally, if one chooses $B = [C^\dagger, H]$, then

$$(A,B) = \langle[C^\dagger, A^\dagger]\rangle,$$
$$(B,B) = \langle[C^\dagger,[H,C]]\rangle. \quad (6)$$

Equation (3) follows from (4)-(6).

To apply the inequality to the Heisenberg model, let

$$H = -\sum_{\vec{R}\vec{R}'} J(\vec{R}-\vec{R}')\vec{S}(\vec{R})\cdot\vec{S}(\vec{R}') - h\sum_{\vec{R}} S_z(\vec{R})e^{-i\vec{K}\cdot\vec{R}}. \quad (7)$$

To rule out ferromagnetism we will take $\vec{K}$ to be 0, and to exclude antiferromagnetism, we chose it such that $e^{i\vec{K}\cdot\vec{R}} = 1$ when $\vec{R}$ connects sites in the same sublattice, and -1 when it connects sites in different sublattices; $\vec{R}$ and $\vec{R}'$ run over the sites of a Bravais lattice with the usual periodic boundary condition to insure the presence of just N spins; $J(0) = 0$, $J(-\vec{R}) = J(\vec{R})$, and J is of finite range.[7] Define

$$\vec{S}(\vec{k}) = \sum_{\vec{R}} e^{-i\vec{k}\cdot\vec{R}}\vec{S}(\vec{R}), \quad \vec{S}(\vec{R}) = (1/N)\sum_{\vec{k}} e^{i\vec{k}\cdot\vec{R}}\vec{S}(\vec{k}),$$

$$J(\vec{k}) = \sum_{\vec{R}} e^{-i\vec{k}\cdot\vec{R}} J(\vec{R}), \quad J(\vec{R}) = (1/N)\sum_{\vec{k}} e^{i\vec{k}\cdot\vec{R}} J(\vec{k})$$

(where sums over $\vec{k}$ are always restricted to the first Brillouin zone).

If we take $C = S_+(\vec{k})$, $A = S_-(-\vec{k}-\vec{K})$, then (3) gives

$$\tfrac{1}{2}\langle\{S_+(\vec{k}+\vec{K}), S_-(-\vec{k}-\vec{K})\}\rangle \geq N k_B T s_z \Big\{(1/N)\sum_{\vec{k}'}[J(\vec{k}) - J(\vec{k}'-\vec{k})]$$
$$\times\langle S_z(-\vec{k}')S_z(\vec{k}') + \tfrac{1}{4}\{S_+(\vec{k}'), S_-(-\vec{k}')\}\rangle + (N/2)hs_z\Big\}^{-1}, \quad (8)$$

where $s_z = (1/N)\sum_{\vec{R}}\langle S_z(\vec{R})e^{i\vec{K}\cdot\vec{R}}\rangle$. The denominator on the right-hand side of (8) is positive, being of the form (B,B), and is therefore less than

$$(1/N)|\sum_{\vec{R}}J(\vec{R})(1-e^{i\vec{k}\cdot\vec{R}})\sum_{\vec{k}'}e^{-i\vec{k}'\cdot\vec{R}}\langle S_z(-\vec{k}')S_z(\vec{k})+\tfrac{1}{4}\{S_+(\vec{k}'),S_-(-\vec{k}')\}\rangle|$$

$$+\tfrac{1}{2}N|hs_z|\langle N\sum_{\vec{R}}|J(\vec{R})|(1-\cos\vec{k}\cdot\vec{R})S(S+1)+\tfrac{1}{2}N|hs_z|$$

$$\langle\tfrac{1}{2}N[\sum_{\vec{R}}R^2|J(\vec{R})|S(S+1)k^2+|hs_z|]. \tag{9}$$

We have used the fact that

$$\sum_{\vec{k}'}\langle S_i(\vec{k}')S_i(-\vec{k}')\rangle=N\sum_{\vec{R}}\langle S_i(\vec{R})S_i(\vec{R})\rangle=N^2\langle S_i(\vec{R}_0)S_i(\vec{R}_0)\rangle \tag{10}$$

independent of $\vec{R}_0$. Replacing the denominator by this upper bound and summing both sides of (8) over $\vec{k}$, we may conclude that[8]

$$S(S+1)>2k_BTs_z^2(1/N)\sum_{\vec{k}}[S(S+1)\sum_{\vec{R}}R^2|J(\vec{R})|k^2+|hs_z|]^{-1} \tag{11}$$

In the infinite-volume limit (11) becomes

$$S(S+1)>\frac{2k_BTs_z^2}{\rho}\int_{\text{first zone}}\frac{d\vec{k}}{(2\pi)^d}[S(S+1)\sum_{\vec{R}}R^2|J(\vec{R})|k^2+|hs_z|]^{-1} \tag{12}$$

where $1/\rho$ is the volume per spin and d is the number of dimensions. This inequality is strengthened if we integrate only over a sphere contained in the first Brillouin zone, so if k_0 is the distance of the nearest Bragg plane from the origin in $\vec{k}$ space, then

$$s_z^2<\frac{2\pi\rho S(S+1)}{k_0^2}\frac{\omega}{kT}\frac{1}{\ln(1+\omega/|hs_z|)} \tag{13}$$

(2 dimensions),

$$|s_z|^3<|h|\omega\left(\frac{S(S+1)}{2kT\tan^{-1}[(\omega/|hs_z|)^{-1/2}]}\right)^2 \tag{14}$$

(1 dimension),

$$\omega=\sum_{\vec{k}}S(S+1)k_0^2R^2|J(\vec{R})|.$$

In the limit of small h these reduce to (1) and (2).

The following additional points are of some interest:

(1) If the coupling is anisotropic the argument is inconclusive, but if $J_y=J_z\neq J_x$, then the same conclusions are reached for aligning fields in the z direction.

(2) Our inequality rules out only spontaneous magnetization or sublattice magnetization. It does not exclude the possibility of other kinds of phase transitions. For example, a state with $s_z=0$ but $(\partial s_z/\partial h)_T=\infty$ as $h\to 0$ is not inconsistent with (1) or (2).

(3) A very similar argument[9] rules out the existence of long-range crystalline ordering in one or two dimensions, without making the harmonic approximation.

(4) Since our conclusions hold whatever the magnitude of S, one would expect them to apply to classical spin systems. We can, in fact, prove them directly by purely classical arguments in such cases.

This study was stimulated by our hearing of P. C. Hohenberg's argument in the superfluid case. We have also had very useful conversations with G. V. Chester, M. E. Fisher, and J. Langer.

*Work supported by the National Science Foundation under Contract No. GP-5517.

†Alfred P. Sloan Foundation Fellow.

‡Permanent address: Max-Planck Institute für Physik, München, Germany.

[1]L. Van Hove, Physica 16, 137 (1950); R. B. Griffiths, Phys. Rev. 136, A437 (1964). See also R. Peierls, Proc. Cambridge Phil. Soc. 32, 477 (1936).

[2]P. C. Hohenberg, to be published.

[3]N. N. Bogoliubov, Physik. Abhandl. Sowjetunion 6, 1, 113, 229 (1962). See also H. Wagner, Z. Physik 195, 273 (1966).

[4]F. Bloch, Z. Physik 61, 206 (1930). Bloch discusses only the ferromagnetic case, but in spite of some suggestions to the contrary [see D. Mattis, The Theory of Magnetism (Harper and Row Publisher, Inc., New York, 1965), p. 244], his analysis leads to similar conclusions in the antiferromagnetic case.

[5]See C. H. Herring and C. Kittel, Phys. Rev. 81, 869 (1951), footnote on p. 873, and also G. Wannier, Elements of Solid State Theory (Cambridge University Press, Cambridge, England, 1959), pp. 111-113.

[6]For example, the doubts recently raised by H. E. Stanley and T. A. Kaplan, Phys. Rev. Letters 17, 913 (1966), on the validity of Wannier's conclusions can now be laid to rest, although their alternative suggestion of a transition to a state without spontaneous magnetization is not inconsistent with our theorem.

[7]Actually it is enough that $\sum_{\vec{R}} R^2|J(\vec{R})|$ converge.

[8]We have not hesitated to sacrifice better bounds for simpler expressions [e.g., $\langle S_x^2(\vec{R}_0)+S_y^2(\vec{R}_0)\rangle < S(S+1)$], but such crudities affect only the constants in (1) and (2) and not the dependence on h and T.

[9]N. D. Mermin and H. Wagner, to be published.

ERRATUM

ABSENCE OF FERROMAGNETISM OR ANTIFERROMAGNETISM IN ONE- OR TWO-DIMENSIONAL ISOTROPIC HEISENBERG MODELS. N. D. Mermin and H. Wagner [Phys. Rev. Letters 17, 1133 (1966)].

The following typographical errors should be noted:

Equation (2), left-hand side should be $|s_z|$ (instead of $|s_z|h|$).

Equation (8), first line, s_z should be s_z^2.

Equation (8), first line, following the summation should be $[J(\vec{k}')-J(\vec{k}'-\vec{k})]$ (instead of $[J(\vec{k})-J(\vec{k}'-\vec{k})]$).

Equation (9), first line, $S_z(-\vec{k}')S_z(\vec{k})$ should be $S_z(-\vec{k}')S_z(\vec{k}')$.

Equation (9), second and third lines, both left braces (⟨) should be less-than signs (<).

Twelfth line from the bottom of p. 1135, left column, the summation should be over $\vec{R}$, not $\vec{k}$.

Equation (14), the argument of $\tan^{-1}$ should be $[\omega/|hs_z|]^{1/2}$ (instead of $[\omega/|hs_z|)^{-1/2}$).

EXACT SOLUTION OF THE PROBLEM OF THE ENTROPY OF TWO-DIMENSIONAL ICE

Elliott H. Lieb*
Department of Physics, Northeastern University, Boston, Massachusetts
(Received 16 February 1967)

The entropy of two-dimensional ice has been found by the transfer-matrix method. Entropy $= Mk \ln W$, with $M =$ No. of molecules and $W = (\frac{4}{3})^{3/2}$.

At low temperatures ice has a residual entropy caused, presumably, by an indeterminacy of the crystal structure. The oxygen atoms constitute a periodic crystal lattice that is hydrogen bonded. The hydrogen atoms are not at the centers of the bonds, however, so that there are two possible states for each bond corresponding to the two positions of the hydrogen atom relative to the bond midpoint. Nevertheless, not all bond configurations are allowed, for there is a constraint called the "ice condition" such that for the four bonds emanating from each oxygen atom, exactly two of the bonds must have the hydrogen atoms close to the oxygen atom.

This problem has received a good deal of theoretical and numerical attention,[1-6] and the best numerical estimates of the entropy are in excellent agreement with experiment.[6] While the problem has also attracted the attention of mathematicians, no exact analytic solution of the problem has heretofore been obtained.

We have succeeded in solving the two-dimensional version of the problem which may be formulated as follows: Let the vertices of a square $N\times N$ net (as in the Ising model) represent the oxygen atoms, and on each bond draw an arrow (up or down for vertical bonds and left or right for horizontal bonds). The "ice condition" is that there must be precisely two arrows into each vertex. If $M=N^2$, then for large M the number of arrangements will be W^M, where W is to be calculated. The entropy is $Mk\ln W$. If we ignore the "ice condition," then obviously $W=4$.

The best numerical estimate[6] for W in two dimensions was $W=1.540\pm0.001$. Our exact result is

$$W=(\tfrac{4}{3})^{3/2}=1.539\,600\,7.$$

The calculation uses the well-known transfer-matrix formalism which we briefly outline here. A configuration of the lattice consists of N rows of N vertical arrows alternating with N rows of N horizontal arrows. Let φ^1 represent a definite configuration of the first "vertical" row. There are obviously 2^N choices for φ^1. Likewise, let φ^j be the configuration of the jth "vertical" row. If φ and φ' are the configurations of two successive "vertical" rows, let $A(\varphi,\varphi')$ be the number of ways of placing arrows on the intervening "horizontal" row such that the ice condition is satisfied at every vertex of that "horizontal" row. Thus, A is a 2^N square matrix whose entries are integers, and Z, the total number of ways of correctly placing arrows on the lattice, is then $Z=\mathrm{Tr}A^N$ (assuming the lattice to be wrapped on a torus). As usual, $Z=\lambda^N$, where λ is the largest eigenvalue of A.

In general, a state φ' differs from φ by the replacement of certain "up" arrows by "down" arrows, namely, a +− exchange, or the reverse, which is a −+ exchange. A little reflection yields the following matrix elements for $A(\varphi,\varphi')$: (i) $A(\varphi,\varphi)=2$; (ii) for $\varphi\neq\varphi'$, $A(\varphi,\varphi')=1$ if there is a +− exchange between every pair of −+ exchanges, and vice versa; and (iii) $A(\varphi,\varphi')=0$, otherwise. If we regard a state φ as a state of N spin-$\frac{1}{2}$ particles on a line, then the above rule is equivalent to $A=A_L+A_R$, where

$$A_L=1+\sum_{i<j}S_i^-S_j^+ +\sum_{i<j<k<l}S_i^-S_j^+S_k^-S_l^+ +\cdots+(S_1^-S_2^+\cdots S_{N-1}^-S_N^+)$$

(assuming N is even) and $A_R=A_L^+$. Since $S^z=\sum_1^N S_i^z$ is a constant, we must decide which S^z subspace has the largest eigenvalue. It can be shown, as expected, that $S^z=0$ has the largest.

If ψ is an eigenvector of A, let $f(x_1,\cdots,x_n)$ be the amplitude in ψ of the state with up arrows (spins) at the sites $x_1<x_2<\cdots<x_n$ (we are interested in $n=\frac{1}{2}N$). Further reflection shows that f satisfies

$$\lambda f(x_1,\cdots,x_n) =\sum_{y_1=1}^{x_1}\sum_{y_2=x_1}^{x_2}\cdots\sum_{y_n=x_{n-1}}^{x_n} f(y_1,\cdots,y_n) +\sum_{y_1=x_1}^{x_2}\sum_{y_2=x_2}^{x_3}\cdots\sum_{y_n=x_n}^{N} f(y_1,\cdots,y_n). \quad (1)$$

On the right-hand side of (1) it is to be understood that f is replaced by zero if any $y_i=y_{i+1}$ (e.g., $y_2=y_1=x_1$).

We make the following Ansatz for f: Let $\{k\}=k_1,\cdots,k_n$ be a set of distinct numbers and let

$$f(x_1,\cdots,x_n)=\sum_P{}^! A(P)\exp\{i\sum_{i=1}^n k_{P(i)}x_i\}, \quad (2)$$

where the sum is on $n!$ permutations and $A(P)$ is some set of $n!$ coefficients. Now, if we insert a given plane wave $\exp\{i\sum k_ix_i\}$ into the first sum in (1), we get (assuming no $k_i=0$)

$$\{\Pi_j[1-\exp(ik_j)]^{-1}\}[e^{ik_1}-e^{ik_1(x_1+1)}][e^{ik_2x_1}-e^{ik_2(x_2+1)}]\cdots\{\exp[ik_nx_{n-1}]-\exp[ik_n(x_n+1)]\}. \quad (3)$$

Expanding the above product gives 2^n terms. One of these is proportional to the same plane wave we started with and is desirable. All the others are unwanted because one or more x_i's fail to appear.

The situation is saved, however, because we are obliged to subtract from (3) those terms in (1) for which $y_i = y_{i+1} = x_i$, and these have the same character as the unwanted terms. A similar situation obtains for the second sum in (1).

By choosing the $A(P)$ correctly we can eliminate all the unwanted terms. The rule (which can be proved by induction) is this: If P and Q are two permutations which differ only in the jth and $(j+1)$th position, then

$$A(P) = A(Q)B(k, q), \tag{4}$$

where $k_{P(j)} = k_{Q(j+1)} = k$ and $k_{P(j+1)} = k_{Q(j)} = q$, and where

$$B(k, q) = -[1 + T(k)T(q) - T(k)][1 + T(k)T(q) - T(q)]^{-1}$$

with $T(k) = \exp(ik)$. Finally, periodicity comes in through the n conditions

$$\exp(ik_i n) = \prod_{j\neq i} B(k_i, k_j). \tag{5}$$

The eigenvalue, λ, is the coefficient of the wanted term from A_L and A_R, namely,

$$\lambda = \{\prod_{j=1}^{n} (1-\exp ik_j)^{-1}\}\{1 + \exp(i\sum_1^n k_j)\}. \tag{6}$$

It will be recognized that our wave function as defined by (4) and (5) is exactly the same as that for the anisotropic one-dimensional Heisenberg model[7]:

$$H = -\sum_{i=1}^{N} S_i^x S_{i+1}^x + S_i^y S_{i+1}^y + \tfrac{1}{2} S_i^z S_{i+1}^z. \tag{7}$$

Our eigenvalue (6) is different, however. For n even, the solution to (5) is such that no $k = 0$ and thus our previous analysis is correct. Furthermore, the maximum eigenstate of the transfer matrix and the ground state of (7) are identical because both are characterized by the fact that $f(x_1, \cdots, x_n) > 0$. For this state,

$$\sum_{j=1}^{n} k_j = 0.$$

As $N \to \infty$ one introduces a density function $\rho(k)$ for the k's. This function satisfies an integral equation which, fortunately, can be solved exactly for $S^z = 0$. All the details are in Ref. 7 where it is shown that in terms of a new variable α defined by $e^{ik} = (e^{i\mu} - e^{\alpha})(e^{i\mu+\alpha} - 1)^{-1}$ (with $\cos\mu = -\frac{1}{2}$), the density function is given by $R(\alpha) = [4\mu \cosh(\pi\alpha/2\mu)]^{-1}$.

Thus, we have

$$N^{-1} 2 \ln\lambda = -\int dk\, \rho(k) \ln(2 - 2\cos k),$$

$$= -\frac{1}{2\pi}\int_{-\infty}^{\infty} \frac{3d\alpha}{4\cosh(3\alpha/4)} \ln\left(1 - \frac{3}{1 + 2\cosh\alpha}\right),$$

$$= 3\ln(\tfrac{4}{3}).$$

Therefore, $W = \lambda^{1/N} = (\frac{4}{3})^{3/2}$.

I should like to thank Professor S. Sherman and Dr. J. Nagle for introducing me to the problem.

---

*Work supported by National Science Foundation Grant No. GP-6851.

[1]L. Onsager and M. Dupuis, Rendiconti della Scuola Internazionale di Fisica (Enrico Fermi), X Corso, "Termodinamica dei Processi Irreversibili" (Società Italiana di Fisica, Bologna, 1960), p. 294.

[2]L. Pauling, J. Am. Chem. Soc. 57, 2680 (1935). For a more recent review see L. Pauling, The Nature of the Chemical Bond and the Structure of Molecules and Crystals; An Introduction to Modern Structural Chemistry (Cornell University Press, Ithaca, New York, 1960), 3rd ed. See also L. K. Runnels, Sci. Am. 215, 118 (1966).

[3]W. F. Giauque and J. W. Stout, J. Am. Chem. Soc. 58, 1144 (1936).

[4]H. Takahasi, Proc. Phys.-Math. Soc. Japan 23, 1069 (1941).

[5]E. A. DiMarzio and F. H. Stillinger, Jr., J. Chem. Phys. 40, 1577 (1964).

[6]J. F. Nagle, J. Math. Phys. 7, 1484 (1966).

[7]C. N. Yang and C. P. Yang, Phys. Rev. 150, 321, 327 (1966).

EXISTENCE OF THERMODYNAMICS FOR REAL MATTER WITH COULOMB FORCES

J. L. Lebowitz*
Belfer Graduate School of Science, Yeshiva University, New York, New York 10033

and

Elliott H. Lieb†
Department of Mathematics, Massachusetts Institute of Technology, Cambridge, Massachusetts 02139
(Received 3 February 1969)

It is shown that a system made up of nuclei and electrons, the constituents of ordinary matter, has a well-defined statistical-mechanically computed free energy per unit volume in the thermodynamic (bulk) limit. This proves that statistical mechanics, as developed by Gibbs, really leads to a proper thermodynamics for macroscopic systems.

In this note we wish to report the solution to a classic problem lying at the foundations of statistical mechanics.

Ever since the daring hypothesis of Gibbs and others that the equilibrium properties of matter could be completely described in terms of a phase-space average, or partition function, $Z = \mathrm{Tr}e^{-\beta H}$, it was realized that there were grave difficulties in justifying this assumption in terms of basic microscopic dynamics and that such delicate matters as the ergodic conjecture stood in the way. These questions have still not been satisfactorily resolved, but more recently still another problem about Z began to receive attention: Assuming the validity of the partition function, is it true that the resulting properties of matter will be extensive and otherwise the same as those postulated in the science of thermodynamics? In particular, does the thermodynamic, or bulk, limit exist for the free energy derived from the partition function, and if so, does it have the appropriate convexity, i.e., stability properties?

To be precise, if N_j are an unbounded, increasing sequence of particle numbers, and Ω_j a sequence of reasonable domains (or boxes) of volume V_j such that $N_j/V_j \to \text{constant} = \rho$, does the free energy per unit volume

$$f_j = -kT(V_j)^{-1}\ln Z(\beta, N_j, \Omega_j) \tag{1}$$

approach a limit [called $f(\beta,\rho)$] as $j\to\infty$, and is this limit independent of the particular sequence and shape of the domains? If so, is f convex in the density ρ and concave in the temperature β^{-1}? Convexity is the same as thermodynamic stability (non-negative compressibility and specific heat).

Various authors have evolved a technique for proving the above,[1,2] but always with one severe drawback. It had to be assumed that the interparticle potentials were short range (in a manner to be described precisely later), thereby excluding the Coulomb potential which is the true potential relevant for real matter. In this note we will indicate the lines along which a proof for Coulomb forces can be and has been constructed. The proof itself, which is quite long, will be given elsewhere.[3] We will also list here some additional results for charged systems that go beyond the existence and convexity of the limiting free energy.

To begin with, a sine qua non for thermodynamics is the stability criterion on the N-body Hamiltonian $H=E_K+V$. It is that there exists a constant $B \geq 0$ such that for all N,

$$V(r_1,\cdots,r_N) > -BN \quad \text{(classical mechanics)}, \tag{2}$$

$$E_0 > -BN \quad \text{(quantum mechanics)}, \tag{3}$$

where E_0 is the ground-state energy in infinite space. (Classical stability implies quantum-mechanical stability, but not conversely.) Heuristically, stability insures against collapse. From the mathematical point of view, it provides a lower bound to f_j in (1). We wish to emphasize that stability of the Hamiltonian (H stability), while necessary, is insufficient for assuring the existence of thermodynamics. For example, it is trivial to prove H stability for charged particles all of one sign, and it is equally obvious that the thermodynamic limit does not exist in this case.

It is not too difficult to prove classical and thus also quantum-mechanical H stability for a wide variety of short-range potentials or for charged particles having a hard core.[2,4] But real charged particles require quantum mechanics and the recent proof of H stability by Dyson and Lenard[5] is as difficult as it is elegant. They show that stability will hold for any set of charges and masses provided that the negative particles and/or the positive ones are fermions.

The second requirement in the canonical proofs[1] is that the potential be tempered, which is to say that there exist a fixed r_0 and constants $C \geq 0$ and $\epsilon > 0$ such that if two groups of N_a and N_b particles are separated by a distance $r > r_0$, their interparticle energy is bounded by

$$V(N_a \oplus N_b) - V(N_a) - V(N_b) \leq C r^{-(3+\epsilon)} N_a N_b. \tag{4}$$

Tempering is roughly the antithesis of stability because the requirements that the forces are not too repulsive at infinity insures against "explosion." Coulomb forces are obviously not tempered and for this reason the canonical proofs have to be altered. Our proof, however, is valid for a mixture of Coulomb and tempered potentials and this will always be understood in the theorems below. It is not altogether useless to include tempered potentials along with the true Coulomb potentials because one might wish to consider model systems in which ionized molecules are the elementary particles.

Prior to explaining how to overcome the lack of tempering we list the main theorems we are able to prove. These are true classically as well as quantum mechanically. But first three definitions are needed:

(D1) We consider s species of particles with charges e_i, particle numbers $N^{(i)}$, and densities $\rho^{(i)}$. In the following N and ρ are a shorthand notation for s-fold multiplets of numbers. The conditions for H stability (see above) are assumed to hold.

(D2) A neutral system is one for which $\sum_1^s N^{(i)} \times e_i = 0$, alternatively $\sum_1^s \rho^{(i)} e_i = 0$.

(D3) The ordinary s-species grand canonical partition function is

$$\sum_{N^{(s)}=0}^{\infty} \cdots \sum_{N^{(1)}=0}^{\infty} \prod_1^s z_i^{N^{(i)}} Z(N,\Omega). \tag{5}$$

The neutral grand canonical partition function is the same as (5) except that only neutral systems enter the sum.

The theorems are the following:

(T1) The canonical, thermodynamic limiting free energy per unit volume $f(\beta,\rho)$ exists for a neutral system and is independent of the shape of the domain for reasonable domains. Furthermore, $f(\beta,\rho^{(1)},\rho^{(2)},\cdots)$ is concave in β^{-1} and jointly convex in the s variables $(\rho^{(1)},\cdots,\rho^{(s)})$.

(T2) The thermodynamic limiting microcanonical[6] entropy per unit volume exists for a neutral system and is a concave function of the energy per unit volume. It is also independent of domain shape for reasonable shapes and it is equal to the entropy computed from the canonical free energy.

(T3) The thermodynamic limiting free energy per unit volume exists for both the ordinary and the neutral grand canonical ensembles and are independent of domain shape for reasonable domains. Moreover, they are equal to each other and to the neutral canonical free energy per unit volume.

Theorem 3 states that systems which are not charge neutral make a vanishingly small contribution to the grand canonical free energy. While this is quite reasonable physically, it does raise an interesting point about nonuniform convergence because the ordinary and neutral partition functions are definitely not equal if we switch off the charge before passing to the thermodynamic limit, whereas they are equal if the limits are taken in the reverse order.

An interesting question is how much can charge neutrality be nonconserved before the free energy per unit volume deviates appreciably from its neutral value? The answer is in theorem 4.

(T4) Consider the canonical free energy with a surplus (i.e., imbalance) of charge Q and take the thermodynamic limit in either of three ways: (a) $QV^{-2/3}\to 0$; (b) $QV^{-2/3}\to\infty$; (c) $QV^{-2/3}\to$ const. In case (a) the limit is the same as for the neutral system while in case (b) the limit does not exist, i.e., $f\to\infty$. In case (c) the free energy approaches a limit equal to the neutral-system free energy plus the energy of a surface layer of charge Q as given by elementary electrostatics.

We turn now to a sketch of the method of proof and will restrict ourselves here to the neutral canonical ensemble. As usual, one first proves the existence of the limit for a standard sequence of domains. The limit for an arbitrary domain is then easily arrived at by packing that domain with the standard ones. The basic inequality that is needed is that if a domain Ω containing N particles is partitioned into D domains $\Omega_1, \Omega_2, \cdots, \Omega_D$ containing $N_1, N_2, \cdots, N_D$ particles, respectively, and if the interdomain interaction be neglected, then

$$Z(N,\Omega) \geq \prod_1^D Z(N_i, \Omega_i). \tag{6}$$

If Ω is partitioned into subdomains, as above, plus "corridors" of thickness $>r_0$ which are devoid of particles, one can use (4) to obtain a useful bound on the tempered part of the omitted interdomain interaction energy. We will refer to these energies as surface terms.

The normal choice[1] for the standard domains are cubes C_j containing N_j particles, with C_{j+1} being composed of eight copies of C_j together with corridors, and with $N_{j+1}=8N_j$. Neglecting surface terms one would have from (6) and (1)

$$f_{j+1} \leq f_j. \tag{7}$$

Since f_j is bounded below by H stability, (7) implies the existence of a limit. To justify neglect of the surface terms one makes the corridors increase in thickness with increasing j; although V_j^C, the corridor volume, approaches ∞ one makes $V_j^C/V_j\to 0$ in order that the limiting density not vanish. The positive ϵ of (4) allows one to accomplish these desiderata.

Obviously, such a strategy will fail with Coulomb forces, but fortunately there is another way to bound the interdomain energy. The essential point is that it is not necessary to bound this energy for all possible states of the systems in the subdomains; it is only necessary to bound the "average" interaction between domains, which is much easier. This is expressed mathematically by using the Peierls-Bogoliubov inequality[7] to show that

$$Z(N,\Omega) \geq e^{-\beta U} \prod_1^D Z(N_i, \Omega_i), \tag{8}$$

where U is the average interdomain energy in an ensemble where each domain is independent. U consists of a Coulomb part, U_C, and a tempered part, U_t, which can be readily bounded.[1]

We now make the observation, which is one of the crucial steps in our proof, that independently of charge symmetry U_C will vanish if the subdomains are spheres and are overall neutral. The rotation invariance of the Hamiltonian will produce a spherically symmetric charge distribution in each sphere and, as Newton[8] observed, two such spheres would then interact as though their total charges (which are zero) were concentrated at their centers.

With this in mind we choose spheres for our standard domains. Sphere S_j will have radius $R_j = p^j$ with p an integer. The price we pay for using spheres instead of cubes is that a given one, S_k, cannot be packed arbitrarily full with spheres S_{k-1} only. We prove, however, that it can be packed arbitrarily closely (as $k\to\infty$) if we use all the previous spheres $S_{k-1}, S_{k-2}, \cdots S_0$. Indeed for the sequence of integers $n_1, n_2, \cdots, n_j = (p-1)^{j-1}p^{2j}$ we can show that we can simultaneously pack n_j spheres S_{k-j} into S_k for $1\leq j\leq k$. The fractional volume of S_k occupied by the S_{k-j} spheres is $\varphi_j = p^{-3j}n_j$, and from (8) we then have

$$f_k \leq \varphi_1 f_{k-1} + \varphi_2 f_{k-2} + \cdots + \varphi_k f_0, \tag{9}$$

and

$$\sum_1^\infty \varphi_j = 1. \tag{10}$$

[Note that the inequality (6) is correct as it stands for pure Coulomb forces because U_C in (8) is identically zero. If short-range potentials are included there will also be surface terms, as in the cube construction, but these present only a technical complication that can be handled in the same manner as before.[1]] While Eq. (9) is more complicated than (7), it is readily proven explicitly that f_k approaches a limit as $k \to \infty$. [Indeed, it follows from the theory of the renewal equation[9] that (9) will have a limit if $\sum_1^\infty j\varphi_j < \infty$.]

The possibility of packing spheres this way is provided by the following geometrical theorem which plays the key role in our analysis. We state it without proof, but we do so in d dimensions generally and use the following notation: σ_d = volume of a unit d-dimensional sphere $= \frac{4}{3}\pi$ in three dimensions and $\alpha_d = (2^d - 1)2d^{\frac{1}{2}}$.

(T5) Let $p \geq \alpha_d + 2^d \sigma_d^{-1}$ be a positive integer. For all positive integers j, define radii $r_j = p^{-j}$ and integers $n_j = (p-1)^{j-1} p^{j(d-1)}$. Then it is possible to place simultaneously $\bigcup_j (n_j$ spheres of radius $r_j)$ into a unit d-dimensional sphere so that none of them overlap.

The minimum value of p required by the theorem in three dimensions is 27.

Many of the ideas presented here had their genesis at the Symposium on Exact Results in Statistical Mechanics at Irvine, California, in 1968, and we should like to thank our colleagues for their encouragement and stimulation: M. E. Fisher, R. Griffiths, O. Lanford, M. Mayer, D. Ruelle, and especially A. Lenard.

---

*Work supported by Air Force Office of Scientific Research, U. S. Air Force under Grant No. AFOSR 68-1416.

†Work supported by National Science Foundation Grant No. GP-9414.

[1]These developments are clearly expounded in M. E. Fisher, Arch. Ratl. Mech. Anal. 17, 377 (1964); D. Ruelle, Statistical Mechanics (W. A. Benjamin, Inc., New York, 1969). For a synopsis, see also J. L. Lebowitz, Ann. Rev. Phys. Chem. 19, 389 (1968).

[2]R. B. Griffiths, Phys. Rev. 176, 655 (1968), and footnote 6a in A. Lenard and F. J. Dyson [J. Math. Phys. 9, 698 (1968)]; O. Penrose, in Statistical Mechanics, Foundations and Applications, edited by T. Bak (W. A. Benjamin, Inc., New York, 1967), p. 98.

[3]E. H. Lieb and J. L. Lebowitz, "The Constitution of Matter," to be published.

[4]L. Onsager, J. Phys. Chem. 43, 189 (1939); M. E. Fisher and D. Ruelle, J. Math. Phys. 7, 260 (1966).

[5]F. J. Dyson and A. Lenard, J. Math. Phys. 8, 423 (1967); A. Lenard and F. J. Dyson, J. Math. Phys. 9, 698 (1968); F. J. Dyson, J. Math. Phys. 8, 1538 (1967).

[6]R. B. Griffiths, J. Math. Phys. 6, 1447 (1965).

[7]K. Symanzik, J. Math. Phys. 6, 1155 (1965).

[8]I. Newton, in Mathematical Principles, translated by A. Motte, revised by F. Cajori (University of California Press, Berkeley, Calif., 1934), Book 1, p. 193, propositions 71, 76.

[9]W. Feller, An Introduction to Probability Theory and Its Applications (J. Wiley & Sons, New York, 1957), 2nd ed. Vol. 1, p. 290.

PHYSICAL REVIEW A VOLUME 1, NUMBER 1 JANUARY 1970

Decay of the Velocity Autocorrelation Function*

B. J. Alder and T. E. Wainwright
Lawrence Radiation Laboratory, University of California, Livermore, California 94550
(Received 10 July 1969)

Molecular-dynamic studies of the behavior of the diffusion coefficient after a long time s have shown that the velocity autocorrelation function decays as s^{-1} for hard disks and as $s^{-3/2}$ for hard spheres, at least at intermediate fluid densities. A hydrodynamic similarity solution of the decay in velocity of an initially moving volume element in an otherwise stationary compressible viscous fluid agrees with a decay of $(\eta s)^{-d/2}$, where η is the viscosity and d is the dimensionality of the system. The slow decay, which would lead to a divergent diffusion coefficient in two dimensions, is caused by a vortex flow pattern which has been quantitatively compared for the hydrodynamic and molecular-dynamic calculations.

A previous study[1] of the diffusion coefficient has shown that the velocity autocorrelation function has a long positive tail, indicating a surprising persistence of velocities. Subsequently,[2] the collective nature of this persistence was established by the observation that the value of the diffusion coefficient depends strongly on the number of particles, particularly in two dimensions where the results did not seem to converge as larger systems were investigated. Finally, by studying the velocity correlation between a molecule and its neighborhood, a vortex flow pattern was found on a microscopic scale which could qualitatively explain the tail. Since the persistence of the vortex flow is long compared to the mean collision time, it is natural to ask whether a hydrodynamic model could calculate such vortex motion, and hence, the behavior of the velocity autocorrelation function for long times. This paper addresses itself to that question.

In such a hydrodynamic model, a fluid is imagined to be at rest except that a small volume element is given an initial velocity. A compression wave develops in front of this region and a rarefaction wave to the rear. When the sound waves have separated, the residual flow is in the form of a double vortex in two dimensions, or a vortex ring in three dimensions. At late times, the circulatory flow approaches that of an incompressible fluid, and hence the velocity decays solely due to the influence of the shear viscosity. It should be emphasized at the outset that this hydrodynamic model differs conceptually from the Stokes-Einstein model which also relates the diffusion coefficient to the viscosity. In that model, a sphere representing a molecule is assumed to slow down adiabatically in a viscous fluid; that is, the retarding force at each instant of time is assumed to be the steady-state value, which is proportional to the velocity, so that the velocity decays exponentially. In the model described here, on the other hand, a transient solution of the Navier-Stokes equation is carried out to find the long-time behavior of the initially moving volume element.

The initially moving volume element is made equal in size to the average volume per molecule and is given a velocity comparable to the root-mean-square molecular velocity. The subsequent motion of the fluid is then calculated by direct numerical integration of the Navier-Stokes equation.[3] Both Eulerian and Lagrangian formulations have been used successfully. A comparison of the flow pattern between the hydrodynamic and molecular-dynamic calculation at a fairly late time is given in Fig. 1. The nearly quantitative agreement obtained lends credence to the applicability of the model. The values for the viscosity η and $y = Pv/NkT - 1$ used in the hydrodynamic calculation were obtained from molecular-dynamic calculations at the same density. The comparison in Fig. 1 was made at a fairly late time, so that the flow pattern had approached the hydrodynamic regime, but not so late that there was any interference from the sound waves coming over the periodic border of the finite system, nor so late that the velocity had decayed to such a small value as to prevent an accurate determination. A correction of $1/N-1$ has been added to the velocity autocorrelation function as calculated by molecular dynamics. This is because whenever a given molecule has a velocity v, the average velocity of the other molecules is $-v/N-1$ in a system of N molecules where momentum is conserved.

Figure 2 illustrates, in the case of hard spheres, the agreement of the autocorrelation function $\rho(s)$ from the molecular-dynamic calculation with that from the hydrodynamic calculation. The latter is simply the velocity at the center of the flow pattern divided by the initial velocity. The two disagree at short times, as might be expected. The molecular-dynamic velocity autocorrelation function shows an initial exponential decay lasting for a few mean collision times s. However, at times great-

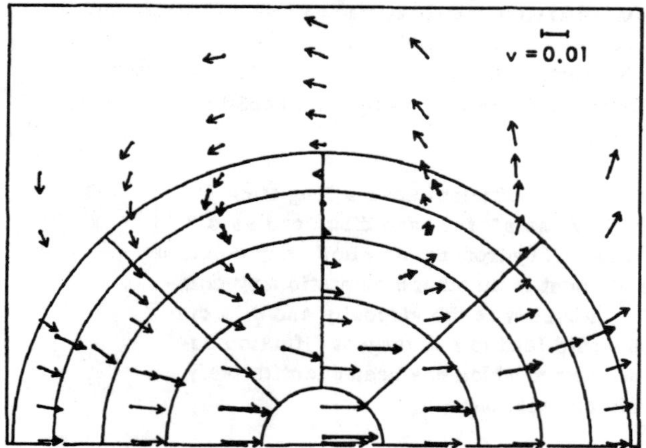

FIG. 1. Statistically averaged velocity field around a central disk from molecular dynamics (heavy arrows) compared to that given by the hydrodynamic model (light arrows). Because of symmetry only half the plane is shown. The scale of distance is indicated by the size of the central disk as shown by the smallest half-circle. The sizes of the other four concentric circles have been determined so as to include roughly six neighboring particles each. These semicircles have been partitioned further into four parts, as indicated by the lines, so as to have a measure of direction relative to the velocity vector of the central particle at zero time. The size of the arrows indicates the magnitude of the velocity (the scale of velocity is indicated as 0.01 of the initial velocity in the upper right-hand corner) and the direction of the arrow is determined by the parallel and perpendicular components of the velocity (relative to that of the central particle initially) averaged over all the particles in that section at a particular time. The arrow is hence drawn at the center of the section. A correction of $1/N-1$ has been added to the parallel component. The comparison is made at 9.9 collision times where the molecular-dynamic and hydrodynamic velocity autocorrelations begin to nearly agree, as seen on the graph by the velocity vectors of the central particle. (See also Fig. 3.) In the molecular-dynamics run, 224 hard disks were used at an area relative to close packing of 2. For the hydrodynamic run, the conditions are given in Table I.

er than about 10 mean collision times, both calculations show a decay like $s^{-3/2}$.

Figure 3 illustrates the same agreement at various densities in the case of hard disks where the decay is like s^{-1}. Figure 3 shows furthermore that the $1/N-1$ correction brings into agreement the velocity autocorrelation functions calculated in molecular-dynamic systems of various sizes and that the long-time behavior of the hydrodynamic solution is independent of the initial velocity. It was also found that the long-time hydrodynamic solution does not depend upon the bulk viscosity. Heat conductivity was not included in the hydrodynamic calculation. The late-time kinks seen in Fig. 3 in the velocity autocorrelation functions calculated for 504 particle systems are caused by the arrival of sound waves from the periodic images. The arrival time of these interferences can be predicted by the hydrodynamical model.

A simple analysis of the hydrodynamical model

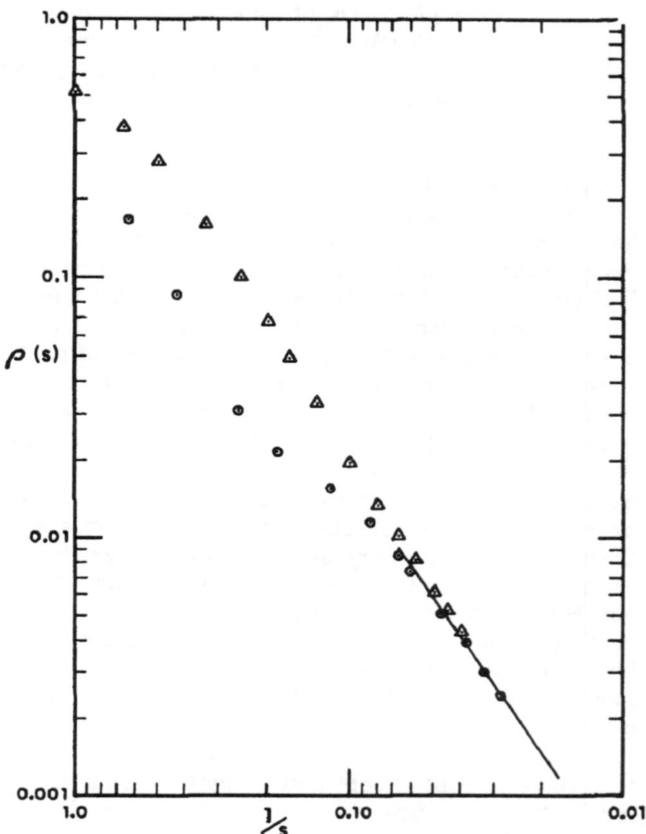

FIG. 2. Comparison of the velocity autocorrelation function $\rho(s)$ as a function of time (in terms of mean collision times s) between the hydrodynamic model (circles) and a 500-hard-sphere molecular-dynamic calculation (triangles) at a volume relative to close packing of 3 on a log-log plot. The straight line is drawn with a slope corresponding to $s^{-3/2}$. To the molecular dynamics $\rho(s)$ a correction of $1/N-1$ has been added. Furthermore, the function has only been graphed up to the time where serious interference between neighboring periodically repeated systems is indicated. In the hydrodynamic calculation the viscosity predicted by the Enskog theory has been used while the molecular-dynamic calculations indicate a 2% larger value. A value of pv/NkT of 3.03 was employed, and the initial velocity of the fluid volume element was normalized to unity for comparison purposes. If the initially moving cylindrical region is made to have the same volume as that corresponding to the volume per particle in the molecular system, a distance and hence a time scale can be obtained to make the above comparison.

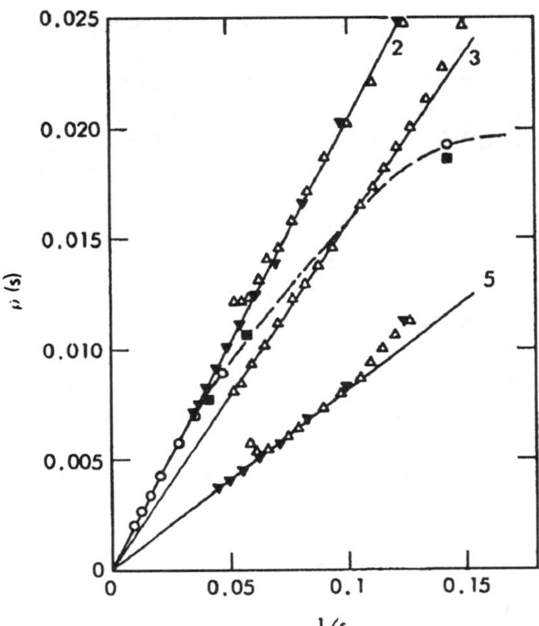

FIG. 3. The decay of the velocity autocorrelation function at large times for hard disks at three densities: $A/A_0 = 2, 3$, and 5. The closed and open triangles refer to molecular-dynamic runs of 986 and 504 particles, respectively. A $1/N-1$ correction to the molecular-dynamic results has been applied. At A/A_0 of 2 and 5 the 504-particle results include the initial deviations due to the interference of neighboring cells at the boundary while all other results have not been plotted beyond the point where serious interference is indicated. The dashed line represents the results of a hydrodynamic run at A/A_0 of 2 (see Table I for conditions) in which the initially moving square area element was given two different velocities, the root-mean-square molecular velocity (squares) and that $\frac{1}{12}$th as large (circles).

shows that a similarity solution exists for the circulatory flow at late times. The linear dimensions of the flow pattern increase at $(\nu s)^{1/2}$ and, since total momentum is conserved, the velocity decays as $(\nu s)^{-d/2}$, where ν is the kinematic viscosity (η divided by the density) and d is the dimensionality of the system. This result verifies the observed behavior.

Table I lists the values of the decay constants α, found by molecular dynamics in two dimensions, where $\rho(s) = \alpha s^{-1}$. Since the hydrodynamic model predicts $\rho(s)$ is proportional to $(\nu s)^{-1}$, $\alpha_H = \alpha\eta s/\eta_0 s_0 = \alpha\eta/\eta_0 y$ should be a constant; the collision rate being proportional to y. The zero subscript indicates the low-density Boltzmann values for the reference system. The value of α_H is π^{-1}, as accurately as it can be determined from the numerical hydrodynamic calculations at a number of different densities and also according to the analytic asymptotic solution. This solution does, however, depend on the empirically supported assumption that the two sound waves carry off $\frac{1}{2}$ of the original momentum, independent of all parameters involved in the calculation, the remaining half being involved in the vortex flow.

The slight remaining density dependence of α_H and its small disagreement with the hydrodynamic solution can be ascribed to the unrealistic nature of the hydrodynamical model. The hydrodynamic flow will not carry the molecule appreciably away from the center of the vortex pattern; but, in fact, in an actual system a molecule has a density-dependent probability of diffusing away from the center. At low densities, particularly, intermolecular diffusion can carry molecules away from the center to a distance comparable with the size of the vortex pattern. To account approximately for the diffusive motion, the vortex flow pattern has to be sampled over a spreading Gaussian distribution representing the probability that the molecule has moved away from the center. This argument leads to the following correction factor F:

$$F = \frac{\int \exp(-r^2 D_0/D_E s)\exp(-r^2\eta_0/\eta s) r\,dr}{\int \exp(-r^2 D_0/D_E s) r\,dr}$$

$$= \frac{\eta/\eta_0}{D_E/D_0 + \eta/\eta_0} \, .$$

The Enskog value of the diffusion coefficient D_E is used because it is intended to describe only the diffusion of the molecules among its neighbors and not the collective motion of the neighborhood for which the hydrodynamic model is used.

A comparison of the last two columns of Table I shows that this correction factor F accounts for the density dependence of α_H to within a few per-cent, that is, within the accuracy of the determina-

TABLE I. Values of the decay coefficient α.

A/A_0	α	y	η/η_0	D_E/D_0	$\alpha_H = \alpha\eta/\eta_0 y$	$\alpha_H\pi$	F
2	0.206	2.42	3.39	0.375	0.29	0.91	0.90
3	0.157	1.08	1.66	0.560	0.24	0.76	0.75
5	0.082	0.50	1.29	0.725	0.21	0.66	0.64

tion of α and η/η_0. The above argument leads to the prediction that in the low-density limit $F = \frac{1}{2}$. Thus, the velocity autocorrelation function in two dimensions decays as s^{-1} at any finite density, leading to a divergent diffusion coefficient at any nonzero density. This result is in contradiction to previous theories on the density expansion of the diffusion coefficient away from the low-density limit. The study of the late-time autocorrelation function at very low densities by molecular dynamics is unfortunately very difficult since the system must be so large that a molecule undergoes many collisions before a sound wave travels across the size of the system.

The hydrodynamic model, as discussed so far, cannot reverse the velocity of the region initially in motion, and thus cannot reproduce the negative part of the velocity autocorrelation found at high densities. This deficiency can be remedied at least qualitatively by the inclusion of visco-elastic forces in the Navier-Stokes equations. These forces can be obtained from the autocorrelations of the elements of the stress tensor as calculated by molecular dynamics. A trial calculation at $A/A_0 = 1.4$ has shown that negative autocorrelation functions can be obtained in this way, but that at very late times, in agreement with molecular dynamics, the function becomes again positive and decays like s^{-1}.

We wish to thank E. D. Giroux and J. A. Viecelli for invaluable help with the hydrodynamic calculations and M. A. Mansigh similarly with the molecular-dynamic calculations.

---

*Work performed under the auspices of the U.S. Atomic Energy Commission.

[1]B. J. Alder and T. E. Wainwright, Phys. Rev. Letters 18, 988 (1967).

[2]B. J. Alder and T. E. Wainwright, J. Phys. Soc. Japan, Suppl. 26, 267 (1968).

[3]M. L. Wilkins, in Methods in Computational Physics (Academic Press Inc., New York, 1964), Vol. 3, p. 211

PHYSICAL REVIEW B VOLUME 4, NUMBER 9 1 NOVEMBER 1971

Renormalization Group and Critical Phenomena. I. Renormalization Group and the Kadanoff Scaling Picture*

Kenneth G. Wilson
Laboratory of Nuclear Studies, Cornell University, Ithaca, New York 14850
(Received 2 June 1971)

The Kadanoff theory of scaling near the critical point for an Ising ferromagnet is cast in differential form. The resulting differential equations are an example of the differential equations of the renormalization group. It is shown that the Widom-Kadanoff scaling laws arise naturally from these differential equations if the coefficients in the equations are analytic at the critical point. A generalization of the Kadanoff scaling picture involving an "irrelevant" variable is considered; in this case the scaling laws result from the renormalization-group equations only if the solution of the equations goes asymptotically to a fixed point.

The problem of critical behavior in ferromagnets (and other systems) has long been a puzzle.[1] Consider the Ising model of a ferromagnet; the partition function is

$$Z(K,h)=\sum_{\{s\}}\exp\Big(K\sum_{\vec n}\sum_{\vec\imath}s_{\vec n}\,s_{\vec n+\vec\imath}+h\sum_{\vec n}s_{\vec n}\Big), \qquad (1)$$

where $K=-J/kT$, J is a coupling constant, $s_{\vec n}$ is the spin at lattice site $\vec n$, $\sum_{\vec\imath}$ is a sum over nearest-neighbor sites, and h is a magnetic field variable. The spin $s_{\vec n}$ is restricted to be ±1; $\sum_{\{s\}}$ means a sum over all possible configurations of the spins. T is the temperature, and k is Boltzmann's constant. The partition function is a sum of exponentials each of which is analytic in K and h. Therefore one would expect the partition function itself to be analytic in K and h. In fact, however, the partition function is singular for $K=k_c$ and $h=0$, where K_c is the critical value of K. To be precise, the singularity occurs only in the infinite-volume limit, in which case one calculates the free-energy density

$$F(K,h)=\lim_{V\to\infty}\frac{1}{V}\ln Z(K,h), \qquad (2)$$

where V is the volume of the system.

Because of the infinite-volume limit there is no formal contradiction in the result that $F(K, h)$ is singular at $K=K_c$. The problem is that the methods one has for calculating sums such as (1) do not easily lead to singular behavior, so it is extremely difficult to get a good understanding of critical singularities working directly with the sum of Eq. (1). What one would like to do is to transform the problem of calculating $F(K, h)$ into a form where it is natural for F to have singularities at $K=K_c$, and where one may hope the nature of the singularity will be more easily seen than from Eq. (1).

In this paper it will be suggested that an appropriate reformulation is in terms of a group, namely, the renormalization group.[2] It has already been suggested that the renormalization group is important for understanding critical phenomena.[3] The function of this paper is to explain what the renormalization group is and what the assumptions are that make it useful.

The renormalization group is a nonlinear transformation group of the kind that occurs in classical mechanics. The equations of motion of a classical system with time-independent potentials define transformations on phase space which form a group. The finite transformations of the group are the transformations induced by a finite translation in time; the infinitesimal transformation is defined by the equations of motion themselves. It will be shown how a translation group can arise in the analysis of critical behavior. This group is called the renormalization group for historical reasons (the connection with renormalization will be explained at the conclusion of paper II[4]). The infinitesimal transformation of the renormalization group is analogous to an equation of motion, and we shall use the language of differential equations rather than the language of group theory in the remainder of this paper.

The advantage of a reformulation of Eq. (1) in terms of the differential equations of the renormalization group is that it allows the singularities of the critical point to occur naturally. Before setting up these differential equations we shall show with a simple classical example how singularities can be generated from an equation of motion. Consider the equation

$$\frac{dx}{dt}=-\frac{dV}{dx}(x)\,, \tag{3}$$

where $V(x)$ is the function shown in Fig. 1. One can think of this equation as describing the motion of a ball rolling on a hill with height given by $V(x)$. Equation (3) is not strictly speaking the equation of motion for said ball, but qualitatively the solution of this equation is similar to the solution of the second-order equation one should write down (this is true in particular if there are frictional forces which prevent the ball from rolling back and forth in the valleys near x_A and x_B of Fig. 1). Let x_C be the location of the maximum of $V(x)$, i.e., the top of the hill. If the ball is released at any point $x>x_C$ (on the left of x_C) then the ball rolls down to the point x_A and stops. If it is released to the right of x_C it rolls to x_B and stops. This means the final position of the ball is a discontinuous function of its initial location. To be precise, let the position of the ball at time t be $x(t, x_0)$ where x_0 is the initial location (at time 0). Then the function $x(\infty, x_0)$ is a discontinuous function of x_0. There is nothing mysterious about this discontinuity, it is just that a small change in the initial condition, from x_0 slightly less than x_C to x_0 slightly greater than x_C, can be amplified by the passage of time until for very large t the difference in position is the difference in x_B-x_A. With an infinite length of time available one can get an infinite amount of amplification, thus leading to a discontinuity in $x(\infty, x_0)$ as a function of x_0 whereas $x(t, x_0)$ for finite t is continuous. It is assumed here that the potential $V(x)$ is analytic in x, as indicated by Fig. 1, so the discontinuity in $x(\infty, x_0)$ at $x_0=x_C$ cannot be blamed on any singularity in $V(x)$ itself.

The basic proposal of this paper is that the singularities at the critical point of a ferromagnet can be understood as arising from the $t=\infty$ limit of the solution of a differential equation. In order to develop an understanding of how one relates critical behavior to a differential equation, we shall set up Kadanoff's scaling picture[5] in differential form. Kadanoff's original hypothesis which led to the Widom-Kadanoff scaling laws[1] was that near the critical point one could imagine blocks of spins acting as a unit, i.e., all spins in a block would be up or down simultaneously. Kadanoff then argued that one could treat all spins in a block as a single effective spin, and one could write an effective Hamiltonian in the Ising form for these effective spins. He then showed how these assumptions

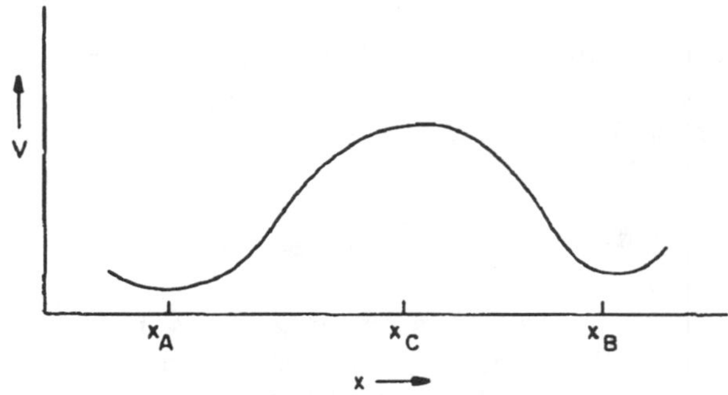

FIG. 1. Potential $V(x)$ with minima at x_A and x_B and a maximum at x_C.

lead to scaling laws.

The idea that blocks of spins act as a unit near the critical temperature does not stand up to close examination; in fact only very near zero temperature ($K \gg K_c$) is it true.[5] The reason for discussing the Kadanoff hypothesis is that it leads to a very simple set of differential equations. By studying these differential equations, one builds an understanding of how critical singularities can emerge from a set of equations of motion. In Paper II we will discuss a generalization of the differential equations which can be more realistic than the Kadanoff block hypothesis. The intuitive picture of how the critical singularities arise from the differential equations will not be changed by the generalization, although a wider range of critical singularities is possible with the generalized equations.

In short the Kadanoff block picture, although absurd, will be the basis for generalizations which are not absurd, and it is helpful to understand the Kadanoff picture in differential form before studying these generalizations.

Imagine an infinite cubic lattice divided into cubic blocks L lattice sites on a side. Each block contains L^3 lattice sites. The total spin in a block is the sum of the L^3 spins in the block. According to the Kadanoff block picture, all the spins in the block are aligned together, so the total spin has only two values, $+L^3$ or $-L^3$. The blocks of spins themselves define a lattice of spins, but the lattice spacing of the blocks is L times the spacing of individual spins. Introduce a spin variable $s'_{\vec{m}}$ associated with block $\vec{m}$. Normalize $s'_{\vec{m}}$ so $s'_{\vec{m}} = \pm 1$, i.e., $s'_{\vec{m}}$ is $L^{-3} \times$ (total spin in block $\vec{m}$). The interactions between blocks involve only nearest-neighbor blocks, and the magnetic field couples to each block separately. This suggests that the interaction energy of the blocks can be expressed in Ising form [Eq. (1)] except that one must substitute new constants K_L and h_L for the original parameters K and h.

Kadanoff proposes in particular that the total free energy of the original Ising model is the same as the free energy of the blocks calculated using the block parameters. In practice this equivalence is expressed in terms of the free-energy density rather than the total free energy. Let $F(K, h)$ be the free energy per lattice site of the original Ising Hamiltonian of Eq. (1). The free energy per block of the block Hamiltonian is simply $F(K_L, h_L)$. If the total free energy is the same for both, then[6]

$$F(K, h) = L^{-3} F(K_L, h_L) . \tag{4}$$

In the Kadanoff picture one can also compute the correlation length using the block Hamiltonian. Let $\xi(K, h)$ be the correlation length for the original Hamiltonian in units of the lattice spacing. Then $\xi(K_L, h_L)$ is the correlation length of the block Hamiltonian, in units of the block spacing. For the two to agree, one must have

$$\xi(K, h) = L \xi(K_L, h_L). \tag{5}$$

The Kadanoff picture is, in summary, that there exists effective coupling parameters K_L and h_L such that Eqs. (4) and (5) hold, for any L. Kadanoff also requires that correlation functions for large distances be calculable through the block Hamiltonian, but this will not be assumed here. Kadanoff's picture requires that L be an integer, but we shall assume that L can be a continuous variable, in order to be able to write differential equations in L. Kadanoff restricts L to be much less than $\xi(K, h)$; we shall allow L to be arbitrary.

The differential equations of the renormalization group will be, in the Kadanoff picture, equations for K_L and h_L. So far nothing has been said about how to compute K_L and h_L. Kadanoff proposed definite forms for the dependence of K_L and h_L on L, namely,

$$K_L = K_c - \epsilon L^y , \tag{6}$$

$$h_L = h L^x , \tag{7}$$

where[7]

$$\epsilon = K_c - K \tag{8}$$

and Eqs. (6) and (7) are valid only for $L \ll \xi(K, h)$. Here we shall first derive differential equations for K_L and h_L and show later that the solution of the differential equations has Kadanoff's form.

To obtain the general form of the differential equations for K_L and h_L, we note the following. The constants K_{2L} and h_{2L} are functions of K_L and h_L but not of L separately. The change from K_L and h_L to K_{2L} and h_{2L} is equivalent to making new blocks of size $2L$ out of old blocks of size L. Each new block is a cube containing eight old blocks. But in writing an effective Hamiltonian with constants K_L and h_L, one has substituted a lattice for the old blocks; having made this substitution the Hamiltonian does not know what the size L of the old blocks was. Regardless of the value of L, the change to $2L$ is simply a matter of combining eight lattice sites to make the new block, so K_{2L} and h_{2L} must be the same function of K_L and h_L for any L.

This continues to be true if one goes from L to $3L$, L to $4L$, etc. Generalizing to the continuous case, we assume this is true also for going from L to $(1+\delta)L$, for small δ. This means $\delta L\, dK_L/dL$ can depend on K_L and h_L but not L separately:

$$\frac{dK_L}{dL} = \frac{1}{L} u(K_L, h_L{}^2). \tag{9}$$

We expect u to depend only on h_L^2 owing to the symmetry of the Ising Hamiltonian for[8] $h_L \to -h_L$. The analogous equation for h_L is

$$\frac{dh_L}{dL} = L^{-1} h_L v(K_L, h_L^2) . \tag{10}$$

Equations (9) and (10) are the renormalization-group equations suggested by the Kadanoff block picture. Because of the questionable validity of this picture one would expect the differential equations to be equally questionable. Actually this is not so; there is another way to derive the differential equations which involves only minimal assumptions, such that the differential equations become essentially a tautology. Namely, let us *define* K_L and h_L to be the solutions of Eqs. (4) and (5). That is, we assume that one has found the exact solution of the Ising model, as a function of K and h, and regard Eqs. (4) and (5) as giving implicit definitions of K_L and h_L for any L. Assume that these equations have a unique solution for any L. The differential equations are now trivial to obtain: Differentiate both Eqs. (4) and (5) with respect to L. One gets

$$0 = L^{-3}\left(-\frac{3}{L}F(K_L, h_L) + \frac{dK_L}{dL}\frac{\partial F(K_L, h_L)}{\partial K_L} + \frac{dh_L}{dL}\frac{\partial F(K_L, h_L)}{\partial h_L}\right), \tag{11}$$

$$0 = L\left(\frac{\xi(K_L, h_L)}{L} + \frac{dK_L}{dL}\frac{\partial \xi(K_L, h_L)}{\partial K_L} + \frac{dh_L}{dL}\frac{\partial \xi(K_L, h_L)}{\partial h_L}\right). \tag{12}$$

These equations can be solved for dK_L/dL and dh_L/dL, to give Eqs. (9) and (10), with

$$u(K_L, h_L^2) = \left(3F(K_L, h_L)\frac{\partial \xi}{\partial h_L}(K_L, h_L) + \xi(K_L, h_L)\frac{\partial F}{\partial h_L}(K_L, h_L)\right) \times\left(\frac{\partial F}{\partial K_L}(K_L, h_L)\frac{\partial \xi}{\partial h_L}(K_L, h_L) - \frac{\partial \xi}{\partial K_L}(K_L, h_L)\frac{\partial F}{\partial h_L}(K_L, h_L)\right)^{-1} \tag{13}$$

and a similar formula for $v(K_L, h_L^2)$.[9] Note that $u(K_L, h_L^2)$ does not depend on L (except through K_L and h_L), as expected. This result means that one has an explicit formula for the function $u(K, h^2)$ in terms of $F(K, h)$ and $\xi(K, h)$ and their derivatives.

The differential equations thus obtained may not be very interesting. The reason is this. The idea of converting the Ising problem into a group was that the singularities at the critical point should result from solving the group equations. This makes sense only if the differential equations are themselves free of singularities at K_c. Referring back to the analogy of a ball on the hill, the function $V(x)$ can be analytic in x and still have $x(\infty, x_0)$ be discontinuous in x_0. Analogously we would like $u(K, h^2)$ and $v(K, h^2)$ to be analytic at the critical point. The trouble with Eq. (13) is that it expresses $u(K, h^2)$ in terms of $F(K, h)$ and $\xi(K, h)$, both of which are singular at the critical point, implying that u is singular also.[10] Most previous formulations of the renormalization group used equations such as Eq. (13) to define the functions u and v which appear in the renormalization-group equations. This has been the cause of much confusion about the purpose of the renormalization group.

The Kadanoff block hypothesis suggests that $u(K, h^2)$ and $v(K, h^2)$ will indeed be analytic at the critical point. In the block picture one should be able to construct K_L and h_L just by adding up interactions of individual spins within a block or across the boundary between two blocks; it is hard to see how this simple addition over a finite region can lead to singular expressions for K_L and h_L as a function of K and h, if L is fixed. More generally this process should give K_{nL} and h_{nL} as analytic functions of K_L and h_L for fixed n; specializing to $n=1+\delta$, the functions u and v should be analytic. However, in the spirit of the Kadanoff approach one does not try to get specific forms for $u(K, h^2)$ and $v(K, h^2)$ because this would require that one take literally the idea that all spins within a block act as a unit. An explicit realization of the renormalization-group equations will be presented in Paper II.[4]

Another feature of the renormalization-group equations suggested by the Kadanoff block picture is the following: In classical physics it is usually easier to write down the equation of motion, such as Eq. (3), than to write down the solution of the equation. Typically the potential $V(x)$ is a simple function, or easily approximated by a simple function. The solution of the equation can be much more complicated, especially if one has coupled differential equations to solve. The Kadanoff block hypothesis suggests that the renormalization-group differential equations will also be simpler than their solution. This is because the smaller L is, the fewer the interactions that have to be summed to give K_L and h_L. It is not immediately obvious that $L=1+\delta$ is easier to compute than $L=2$, but this is not the point. The point is that $L=2$, 3, or 4 is much easier to compute than $L=10\,000$; one expects $L=1+\delta$ also to be easier than $L=10\,000$.

If it is true that the renormalization-group equation is easier to write down than its solution, then it is natural to try to solve the Ising model by first obtaining the renormalization-group equations and then trying to integrate them. Specializing to the problem of critical behavior, what one thinks of doing is integrating the equations until L is of order ξ. When L is of order ξ, $\xi(K_L, h_L)$ is of order 1 [by Eq. (5)], which means K_L and h_L must be well away from their critical values; then it is easy to compute $\xi(K_L, h_L)$ and $F(K_L, h_L)$ by other means and reconstruct $F(K, h)$ and $\xi(K, h)$ from Eqs. (4) and (5). This integration procedure can be carried out for K and h near their critical values, and if $u(K, h)$ and $v(K, h)$ are analytic at the critical point, one obtains immediately the Widom-Kadanoff scaling laws.

To make the calculation precise, let us proceed as follows. Imagine that one is at a temperature slightly above T_c. This means K is slightly smaller than K_c. As L increases K_L must decrease, so as to go away from K_c. This ensures that $\xi(K_L, h_L)$ decreases as L increases, as required by Eq. (5). Pick a value for K_L well away from K_c, say $K_c/2$; let us integrate the renormalization-group equations until a value of L is reached for which $K_L = K_c/2$, then stop and compute $F(K, h)$ and $\xi(K, h)$ from Eqs. (4) and (5).

If K and h have exactly the critical values K_c and 0, respectively, then one must have $K_L = K_c$ and $h_L = 0$ for all L. The reason for this is that $\xi(K_c, 0)$ is infinite; therefore $\xi(K_L, h_L)$ must be infinite for all L. For $\xi(K_L, h_L)$ to be infinite, K_L and h_L must have the critical values. Hence $K_L \equiv K_c$ and $h_L \equiv 0$ must be a solution of the renormalization group equations, which is true only if

$$u(K_c, 0) = 0 . \tag{14}$$

Comparing this result to the classical analog of the ball on the hill, the point $K = K_c$ is analogous to one of the points of equilibrium for the ball ($x = x_A$, or x_B, or x_C).

Now let K and h be near the critical values K_c and 0. For small values of L, namely for $L \ll \xi(K, h)$, the effective correlation length $\xi(K_L, h_L)$ will be large and K_L and h_L will also be near the critical values. For this range of L one can use a linearized form of the renormalization group equations to compute K_L and h_L. The linearized equations for K_L and h_L are

$$\frac{dK_L}{dL} \simeq \frac{1}{L}(K_L - K_c)y , \tag{15}$$

$$\frac{dh_L}{dL} \simeq \frac{1}{L} h_L x , \tag{16}$$

where x and y are constants:

$$y = \frac{\partial u}{\partial K}(K_c, 0) , \tag{17}$$

$$x = v(K_c, 0) . \tag{18}$$

In writing these equations we have assumed that u and v are differentiable at the critical point; this is how one uses in practice the analyticity predicted from the Kadanoff block picture. Using the linearized equations of the renormalization group is analogous to replacing $V(x)$ by a quadratic form when x is near x_C.

The solutions of Eqs. (15) and (16) are the formulas (6) and (7) proposed by Kadanoff. Assume that this approximation is valid until $K_L = K_c/2$. Then one can solve for the value of L giving $K_L = K_c/2$:

$$K_c/2 = \epsilon L^y \tag{19}$$

so

$$L = (K_c/2\epsilon)^{1/y} , \tag{20}$$

which is a scaling law for L as a function of ϵ. For this value of L, h_L is

$$h_L = h(K_c/2\epsilon)^{x/y} . \tag{21}$$

Hence one can compute $F(K, h)$ and $\xi(K, h)$ to be

$$F(K, h) = (K_c/2\epsilon)^{-3/y} F[K_c/2, h(K_c/2\epsilon)^{x/y}] , \tag{22}$$

$$\xi(K, h) = (K_c/2\epsilon)^{1/y} \xi[K_c/2, h(K_c/2\epsilon)^{x/y}] . \tag{23}$$

These formulas are scaling laws; the functions $F(K, h)$ and $\xi(K, h)$ depending on two variables have been reduced to explicit powers of ϵ (i.e., $T - T_c$) multiplying functions depending only on the single variable $h\epsilon^{-x/y}$. For consequences of these laws see Ref. 1.

To be accurate we should have integrated the nonlinear equations once $K_L - K_c$ became large. What can one say about the solution K_L and h_L in this case? One can say the following: Let

$$K_L = \phi(L, h_0) , \tag{24}$$

$$h_L = \psi(L, h_0) \tag{25}$$

be the solution of the exact equations (9) and (10) over the range $0 < L < \infty$ satisfying the boundary conditions

$$\phi(1, h_0) = K_c/2 , \tag{26}$$

$$\psi(1, h_0) = h_0 , \tag{27}$$

that is, boundary conditions well away from the critical point. For $L \ll 1$ this solution should be near the critical point. Hence for $L \ll 1$, K_L and h_L should be solutions of the linearized equations, giving

$$\phi(L, h_0) \approx \phi(h_0)L^y + K_c , \tag{28}$$

$$\psi(L, h_0) \approx \psi(h_0)L^x \tag{29}$$

for $L \ll 1$ where $\phi(h_0)$ and $\psi(h_0)$ are constants depending on h_0. Now the exact renormalization-group equations have translational symmetry when written in terms of $\ln L$, just as the classical equation (3) does. Hence the functions

$$K_L = \phi(aL, h_0) , \tag{30}$$

$$h_L = \psi(aL, h_0) \tag{31}$$

are solutions of the renormalization-group equations where a is any constant. For $aL \ll 1$ this solution reduces to

$$K_L = \phi(h_0)(aL)^y + K_c, \tag{32}$$

$$h_L = \psi(h_0)(aL)^x. \tag{33}$$

This solution can be matched to any given initial condition for K and h; if K and h are near their critical values then K and h can be matched to the asymptotic forms (32) and (33), giving

$$\epsilon = a^y \phi(h_0), \tag{34}$$

$$h = a^x \psi(h_0). \tag{35}$$

This means that

$$h\epsilon^{-x/y} = \psi(h_0)/[\phi(h_0)]^{1/y}, \tag{36}$$

which is an implicit equation for h_0; h_0 depends only on the single variable $h\epsilon^{-x/y}$. Then, from Eq. (34), we have

$$a = [\epsilon/\phi(h_0)]^{1/y}. \tag{37}$$

But now the value of L for which $K_L = K_c/2$ is $L = a^{-1}$, from the boundary condition on ϕ [Eq. (26)]; and for this value of L, $h_L = h_0$ [Eq. (27)]. Using Eqs. (4) and (5) one again gets scaling laws for F and ξ, for example

$$\xi(K, h) = [\epsilon/\phi(h_0)]^{-1/y} \xi(K_c/2, h_0), \tag{38}$$

where h_0 depends only on $h\epsilon^{-x/y}$. This scaling law is the same as (23) except for a different functional dependence on the scaling variable $h\epsilon^{-x/y}$.

These results can be related to the classical analogue of the ball on a hill. Consider only the case of no magnetic field ($h=0$) in which case $h_L = 0$ for all L and one has left a single dependent variable K_L. With $h=0$, h_0 is zero also; $\phi(0)$ is a constant ϕ_0 and the formula for $\xi(K, 0)$ is

$$\xi(K, 0) = (\epsilon/\phi_0)^{1/y}\xi_0, \tag{39}$$

where ξ_0 is $\xi(\frac{1}{2}K_c, 0)$.

The point $K = K_c$ is analogous to the top of the hill ($x = x_C$). This is because if K is different from K_c then $K_L - K_c$ increases as L increases, and this is analogous to the ball rolling away from the top of the hill. In other words the point $K = K_c$ is analogous to a point of unstable equilibrium in classical mechanics. What is the analogue of the correlation length $\xi(K, 0)$? From Eq. (5), $\xi(K, 0)$ is proportional to L where L is chosen to make $K_L = K_c/2$. This is analogous to computing the time $t_{1/2}$ for the ball to roll half-way down the hill. As the initial location x_0 of the ball approaches the top of the hill, this time $t_{1/2}$ increases, becoming infinite as $x_0 \to x_C$. The scaling law for ξ is obtained by computing the dependence of L on $K - K_c$ for K near K_c. This is analogous to finding the dependence of $t_{1/2}$ on $x_0 - x_C$. Most of the time $t_{1/2}$ is spent near the top of the hill, and so the dependence of $t_{1/2}$ on $x_0 - x_C$ can be determined to a good approximation from linearized equations about the unstable equilibrium point x_C.

This completes the discussion of the simplest renormalization-group equations following from the Kadanoff block picture. One sees from this how the idea of differential equations with analytic coefficients leads to singularities at the critical point satisfying the Widom-Kadanoff scaling laws. The critical point corresponds to an unstable equilibrium point of the differential equation, and the singularity at the critical point is a consequence of the infinite time required to move away from a point of unstable equilibrium.

The problem with the simple renormalization-group equations discussed earlier is that there is at present no hope of showing that the functions $u(K, h^2)$ and $v(K, h^2)$ are analytic at the critical point. Without the analyticity, the renormalization-group equations become a tautology, as explained earlier. Instead of trying to prove that u and v are analytic, one can try to generalize the renormalization-group equations in the hope that analyticity will be easier to establish for the generalization. The generalizations which the author has been able to construct are rather complicated, involving an infinite number of L-dependent coupling constants. To prepare for these generalizations it is worth discussing the nature of the renormalization-group equations when they involve one additional coupling constant q_L. The initial variable q will be an "irrelevant" variable in Kadanoff's language.[11] It might be the coefficient of a second-nearest-neighbor coupling, for example. The variable q will prove to be irrelevant only in certain respects that will be explained later.

Imagine that the renormalization-group equations involve three L-dependent variables K_L, q_L, and h_L. We think of K_L and q_L as coefficients of first- and second-nearest-neighbor couplings which are even to the exchange $s'_m \to -s'_m$ while h_L still multiplies the spins s'_m themselves which are odd to this exchange. Hence one expects the form of the renormalization-group equations to be

$$\frac{dK_L}{dL} = \frac{1}{L} u(K_L, q_L, h_L^2), \tag{40}$$

$$\frac{dq_L}{dL} = \frac{1}{L} w(K_L, q_L, h_L^2), \tag{41}$$

$$\frac{dh_L}{dL} = \frac{1}{L} h_L v(K_L, q_L, h_L^2). \tag{42}$$

The initial values of K_L, q_L, and h_L (for $L=1$) are denoted K, q, and h. We assume the same rules for computing the free energy and the correlation length as before, namely,

$$F(K, q, h) = L^{-3} F(K_L, q_L, h_L), \tag{43}$$

$$\xi(K, q, h) = L\, \xi\, (K_L, q_L, h_L). \tag{44}$$

We now expect there will be a line of critical points: for each value of q there should be a critical value $K_c(q)$ for K. If the initial values of K, q, and h lie on the critical line ($K = K_c(q)$, $h = 0$) then the solution K_L, q_L, h_L must lie on the critical line for all L, in order that $\xi(K_L, q_L, h_L)$ be infinite for all L. Hence, if $K = K_c(q)$ and $h = 0$ one can write $K_L = K_C(q_L)$ and $h_L = 0$; the equation for q_L becomes

$$\frac{dq_L}{dL} = \frac{1}{L} w_c(q_L), \tag{45}$$

where

$$w_c(q_L) = w(K_c(q_L), q_L, 0). \tag{46}$$

This is a one-dimensional equation of motion, which is directly analogous to the classical equation of motion for the ball on a hill discussed earlier. In the limit $L \to \infty$ there is the possibility that q_L approaches an equilibrium point of the differential equation. This is not the only possibility; it is also possible that $q_L \to \infty$ as $L \to \infty$. We shall study only the case that q_L approaches as equilibrium point, this being the case that is easiest to study. This means we will not give a completely general discussion of the solution of the renormalization-group equations. In fact a general discussion is very complicated and becomes hopeless when the equations are generalized further.

Let the equilibrium point approached by q_L be q_c, and let $K_c(q_c)$ be K_c. If one chooses $q = q_c$ and $K = K_c$, then $q_L \equiv q_c$ for all L, $K \equiv K_c$ for all L so one has an equilibrium point of the full renormalization-group equations.

Let us study the solutions of the renormalization-group equations in linearized form about the equilibrium point K_c, q_c. The linearized equations are

$$L\frac{dK_L}{dL} = y_{11}(K_L - K_c) + y_{12}(q_L - q_c), \tag{47}$$

$$L\frac{dq_L}{dL} = y_{21}(K_L - K_c) + y_{22}(q_L - q_c), \tag{48}$$

$$L\frac{dh_L}{dL} = x h_L, \tag{49}$$

where

$$y_{11} = \frac{\partial u}{\partial K}(K_c, q_c, 0), \tag{50}$$

$$y_{12} = \frac{\partial u}{\partial q}(K_c, q_c, 0), \tag{51}$$

$$y_{21} = \frac{\partial w}{\partial K}(K_c, q_c, 0), \tag{52}$$

$$y_{22} = \frac{\partial w}{\partial q}(K_c, q_c, 0), \tag{53}$$

and

$$x = v(K_c, q_c, 0). \tag{54}$$

There will, in general, be two linearly independent solutions of the first two equations behaving as a power of L, say[12]

$$K_L - K_c = L^y, \; q_L - q_c = r_y L^y \tag{55}$$

and

$$K_L - K_c = L^z, \quad q_L - q_c = r_z L^z. \tag{56}$$

One of these solutions must be decreasing as $L \to \infty$, since if the initial value of K is $K_c(q)$ then $K_L \to K_c$ and $q_L \to q_c$ as $L \to \infty$, and this is not possible if both y and z are positive. Let z therefore be the negative exponent. The exponent y must be positive so that K_L goes away from K_c if $K \neq K_c(q)$. What this means is that the point K_c, q_c must be analogous to a saddle point in classical mechanics which is stable from one direction but unstable in the orthogonal direction.

One can set up a classical analog to the coupled equations for K_L and q_L. Consider only the case $h_L \equiv 0$, for simplicity. A classical analog is the pair of equations

$$\frac{dK}{dt}(t) = -\frac{\partial V}{\partial K}(K, q), \tag{57}$$

$$\frac{dq}{dt}(t) = -\frac{\partial V}{\partial q}(K, q). \tag{58}$$

This is not as general as Eqs. (40) and (41), since it is not true in general that two functions u and w can be written as the gradient of a potential. But we can illustrate the idea that K_c, q_c define a saddle point with this example. Think of Eqs. (57) and (58) as describing the motion of a ball on a two-dimensional terrain, with $V(K, q)$ being the elevation at the point (K, q). It is convenient to illustrate $V(K, q)$ by a contour map, as in Fig. 2. Figure 2 shows a depression at the point P_A (the origin); P_B is the top of a hill and P_C a saddle point. There is a ridge going down the hill to the saddle and a gully from the saddle to the bottom of the depression. q_0 is an arbitrary value of q. If one starts the ball at coordinates q_0 and a small value of K, it will simply roll to the bottom of the depression. If one increases the initial value of K, the ball will still roll to P_A, until one starts at the point P_D exactly on the ridge. In this case the ball rolls to the saddle point P_C and stops. If one starts just short of P_D, the ball rolls almost to P_C and then goes down the gully to P_A. The ball moves very slowly when it is near P_C and a large time elapses before it moves an appreciable distance down the gully.

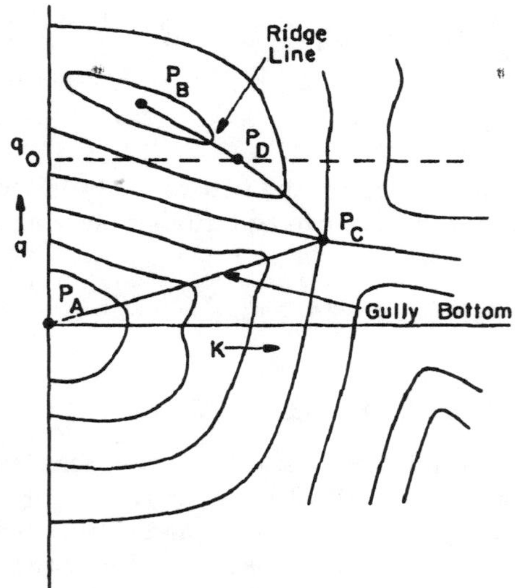

FIG. 2. Potential $V(K,q)$ plotted by contours of constant V. The minimum of V is at the origin (P_A); the maximum is at P_B and P_C is a saddle point. The ridge line is the trajectory going to P_C; the gully line is the trajectory from P_C to P_A. Unmarked lines are contours. The dashed line marks an arbitrary value (q_0) for q.

The ridge line is analogous to the critical line $K=K_c(q)$ and the saddle point P_c is analogous to the point K_c, q_c. Note that the initial location can be anywhere along the ridge line, and still the ball will roll down to the saddle point P_C, or if one starts just short of the ridge line, the path of the ball will just miss P_C and go down the gully line, independently of where along the ridge line one started. It is in this sense that q is an irrelevant variable. To be fairer, there is one relevant and one irrelevant variable in the pair (K,q) but whether K or q is called the irrelevant variable does not matter. If one looks at the functions K_L, q_L starting from an initial point close to the critical line, then as L increases (K_L, q_L) approaches the saddle point (K_c, q_c). By $L=5$ or 10 one should be close to K_c, q_c (unless $z \simeq 0$ so the rate of approach is slow). Then K_L and q_L change very little until L becomes of order of the correlation length $\xi(K,q)$, at which point K_L and q_L move away from the saddle point. In the classical analog, the large L part of the curve K_L, q_L is the gully line, and this is independent of the irrelevant variable. Also the value of L at which $K_L = K_c/2$ should be essentially independent of where along the ridge line one starts; how large L is is determined by how far from the ridge line one starts.

Returning to the solution of the linearized equations, the general solution is a linear combination of the three simple power solutions, namely,

$$K_L = K_c - \epsilon L^y + \eta L^z, \tag{59}$$

$$q_L = q_c - \epsilon r_y L^y + \eta r_z L^z, \tag{60}$$

$$h_L = hL^x, \tag{61}$$

where ϵ and η depend on the initial values K and q:

$$\epsilon = \frac{r_z(K-K_c)-(q-q_c)}{r_y - r_z}, \tag{62}$$

$$\eta = \frac{r_y(K-K_c)-(q-q_c)}{r_y - r_z}. \tag{63}$$

One is on the critical line if $\epsilon = h = 0$ and $\eta \neq 0$, for then $K_L \to K_c$, $q_L \to q_c$ when $L \to \infty$. One is near the critical line away from K_c and q_c if $\epsilon \ll \eta$ and h is small. Then for $L \sim 1$, the η term dominates, and since z is negative, K_L and q_L approach K_c and q_c (i.e., the ball is rolling down the ridge to the saddle point). For large enough L the ϵ term dominates and K_L and q_L move away from K_c and q_c. The η term continues to decrease and can be neglected. Since η measures the initial location along the critical line, independence of η is independence of the initial location on the critical line. One can now compute the value of L for which $K_L = K_c/2$, giving

$$K_c/2 \simeq \epsilon L^y \tag{64}$$

or

$$L = (K_c/2\epsilon)^{1/y} \tag{65}$$

as before. For this value of L,

$$q_L = q_c - r_y(K_c/2), \tag{66}$$

$$h_L = h(K_c/2\epsilon)^{x/y}. \tag{67}$$

The important property of these formulas is that q_L is a constant independent of h, ϵ, or η, while h_L again depends only on the ratio $h\epsilon^{-x/y}$. So from Eqs. (43) and (44) one gets the same form of scaling laws for $F(K,h,q)$ and $\xi(K,h,q)$ as was obtained earlier when there was no irrelevant variable q. Since $K=K_c(q)$ for $\epsilon=0$, it follows from Eq. (62) that ϵ is proportional to $K-K_c(q)$, which is proportional to $T-T_c(q)$. Hence no matter what q is, ϵ is the customary temperature variable $T-T_c$ apart from a meaningless normalization factor.

So far nothing has been said about initial conditions below T_c, i.e., $K > K_c(q)$. This is easily handled by the same analysis except that when K_L moves away from K_c it increases instead of decreasing. Hence one cannot compute the value of L for which $K_L = K_c/2$. Instead one can find the value of L for which $K_L = 2K_c$, say. In the classical anlaog, when $K > K_c(q)$ the ball rolls down the opposite side of the hill from the gully, and one measures the time required to reach the line $K=2K_c$ instead of the time to reach $K_c/2$. It must be noted, however, that for $K > K_c(q)$ and $h=0$ one is on the boundary between the two low-temperature phases (the two phases corresponding to positive or negative magnetization). As L increases K_L moves away from the critical value $K_c(q)$, but h_L is still zero so one is still on the boundary between the two phases. So for the renormaliztion group to be useful below T_c one must be able to calculate the free energy and correlation length by other methods on the phase boundary well below T_c.

It was important in the analysis that z be negative. If z were positive one would have a point like P_B unstable in all directions instead of a saddle point. The only way to reach P_B is to sit on it to start with which means fixing both K and q. In general a fixed point with both y and z as well as x positive is still a critical point but one which requires fixing three thermodynamic quantities instead of two.[13] An exceptional circumstance arises when $z=0$; the analysis of this case is complicated but still feasible. The principal results are that the critical point may or may not require fixing a third thermodynamic variable, and there can be logarithmic violations of the scaling laws.

The above calculations used the linearized form of the renormalization-group equations, but the conclusions will not be changed by using the nonlinear equations.

To summarize the results of this analysis, one starts with values of K and q near the critical line $K=K_c(q)$. The functions K_L and q_L have an initial transient behavior in which K_L and q_L adjust to the critical values K_c and q_c. In other words, while the initial Ising Hamiltonian can have an arbitrary fraction of second-nearest-neighbor coupling, by the time one has gone to a block Hamiltonian with reasonably large block size the couplings have become essentially fixed at K_c and q_c. The Hamiltonians for larger L are therefore independent of the initial fraction of second-nearest-neighbor coupling, and the critical behavior of the theory is likewise independent of the initial fraction.

In fact the parameters of the critical behavior such as the exponents x and y are determined by the renormalization-group differential equation, rather than the initial Hamiltonian. If one knows the form of the functions u, v, and w, one can calculate the exponents x and y by finding the saddle point K_c, q_c of the differential equation and then solving the linearized equations about the saddle point. In principle one then has two choices for how to find the exponents x and y, one choice being as just described, the other choice being to compute the partition function directly and extract the critical exponents. However there is a difference in the two approaches in that the first choice (working from the differential equation) is a well-posed problem, while solving for the partition function is not. Solving for a saddle point of the differential equation is a well-posed problem in the sense that small modifications of the functions u, v, and w make only small changes in K_c, q_c, and x and y; in other words one can get approximate values of K_c and q_c by requiring that u and w only vanish approximately. In contrast, to get the singular part of the free energy requires a calculation to infinite precision since the singular part approaches zero as $K\to K_c(q)$ and one must know how it approaches zero; the singular term exists underneath a finite nonsingular term which is why infinite precision is required.

Clearly, if the renormalization-group picture of critical behavior is correct, it is important to derive the differential equation of the renormalization group and try to find the critical saddle point of the equation. Extrapolating from the analysis of this paper, it is clear that the renormalization-group equations can involve any number of coupling constants, not just two or three; one will still get the Widom-Kadanoff scaling laws provided the solution of the equations approaches a saddle point when one is at the critical temperature. If there are n coupling constants instead of three, the only change will be that there will be $n-2$ linearly independent initial transient solutions of the linearized equations, instead of one, and hence $n-2$ irrelevant variables. However the equations with n variables are complicated, if n is large, and it is not at all certain that critical behavior would be caused by a saddle point. The solutions of the renormalization-group equations might instead approach a limit cycle[14] or go off to infinity or go into irregular oscillations (ergodic or turbulent?) as $L\to\infty$. So while generalized renormalization-group equations are capable of reproducing the Widom-Kadanoff theory, they are also capable of producing more challenging types of behavior.

I have benefitted from many discussions with Professor M. Fisher, Professor B. Widom, Professor L. Kadanoff, Professor D. Jasnow, Professor M. Wortis, and many others in the field of critical phenomena.

---

*Supported by the National Science Foundation.

[1]For reviews of the theory of critical phenomena, see M. Fisher, Rept. Progr. Phys. 30, 731 (1967); L. P. Kadanoff *et al.*, Rev. Mod. Phys. 39, 395 (1967).

[2]The original references on the renormalization group are M. Gell-Mann and F. E. Low, Phys. Rev. 95, 1300 (1954); E. C. G. Stueckelberg and A. Petermann, Helv. Phys. Acta 26, 499 (1953); N. N. Bogoliubov and D. V. Shirkov, *Introduction to the Theory of Quantized Fields* (Interscience, New York, 1959), Chap. VIII. For a recent review see K. Wilson, Phys. Rev. D 3, 1818 (1971).

[3]C. DiCastro and G. Jona-Lasinio, Phys. Letters 29A, 322 (1969). Renormalization-group ideas are also used in the program of Migdal and others: See A. A. Migdal,

Zh. Eksperim. i Teor. Fiz. 59, 1015 (1970) [Sov. Phys. JETP 32, 552 (1971)], and references cited therein.

[4]K. Wilson, following paper, Phys. Rev. B 4, 3184 (1971), referred to as II.

[5]L. P. Kadanoff, Physics 2, 263 (1966); and L. P. Kadanoff *et al.* (Ref. 1). Kadanoff assumed that a block of spins of size L would have a magnetization $\pm L^{d-\psi}$ with ψ unknown. In this paper there is no need to use this precise form of Kadanoff's hypothesis, so for simplicity ψ will be set equal to zero. The choice $\psi=0$ is false, as explained here; but even with ψ an undetermined parameter no one has been able to justify Kadanoff's idea that a block of spins acts as a unit with only two possible values for its magnetization.

[6]This scaling law can actually be true only for the singular part of F. The reason is that at the critical point one has $K_L = K_c$ and $h_L = 0$ for all L (see below) and therefore $F(K_c, 0) = L^{-3}F(K_c, 0)$, which is true only if $F(K_c, 0)$ is zero or ∞. In practice the singular part of F is zero at the critical point (it is only the derivatives of F which diverge at the critical point).

[7]I have chosen ϵ to be $K_c - K$ instead of $K - K_c$ so that ϵ is positive for temperatures above T_c.

[8]The original Ising interaction is invariant to the transformation $K \to K$, $h \to -h$, and $s_{\vec{n}} \to -s_{\vec{n}}$ for all $\vec{n}$. This means that the free energy F cannot depend on the sign of h, so it could be written as a function of h^2. One would expect this symmetry to be apparent in the effective Ising model; namely, one would expect that $K_L \to K_L$, $h_L \to -h_L$, and $s_{\vec{m}}^L \to -s_{\vec{m}}^L$ under this transformation. Hence K_L should depend only on h^2 but h_L should change sign when h changes sign.

[9]The function $u(K_L, h_L^2)$ depends only on h_L^2 (i.e., is even in h_L) because $F(K_L, h_L)$ and $\xi(K_L, h_L)$ are even functions of h_L.

[10]In quantum field theory, where the renormalization group first arose, there is a corresponding analyticity problem; namely, the renormalized amplitudes of quantum electrodynamics diverge when the electron mass goes to zero yet it is assumed that the function (called ψ in Gell-Mann and Low, Ref. 2) appearing in the renormalization-group equations has a finite limit for zero electron mass. The ψ function was defined by Gell-Mann and Low in terms of renormalized amplitudes, so the assumption seems questionable. But Gell-Mann and Low showed that ψ does have a zero-mass limit through fourth order in perturbation theory for which ψ can be calculated explicitly. For further discussion see Gell-Mann and Low (Ref. 2) or Wilson (Ref. 2). This analyticity problem is ignored in Bogoliubov and Shirkov (Ref. 2).

[11]L. P. Kadanoff (private communication). In the anisotropic Heisenberg model of a ferromagnet an irrelevant variable could be the ratio of the $\sigma_x \cdot \sigma_x$ nearest-neighbor coupling to the $\sigma_z \cdot \sigma_z$ nearest-neighbor coupling. (See, however, Ref. 13.)

[12]In special cases equations like Eqs. (47) and (48) may have solutions behaving as L^y and $L^y \ln L$ instead of L^y and L^z. This cannot occur here (it is shown below that $z \neq y$).

[13]An example where this is relevant is in the theory of the Heisenberg ferromagnet, where the fixed point associated with the isotropic ferromagnet is expected to be unstable with respect to adding anisotropic interactions to the Hamiltonian as well as to changing the temperature or magnetic field. See D. Jasnow and M. Wortis, Phys. Rev. 176, 739 (1968); E. Riedel and F. Wegner, Z. Physik 225, 195 (1969), Phys. Rev. Letters 24, 730 (1970); 24, 930(E) (1970).

[14]For an introduction to the theory of limit cycles, see N. Minorsky, *Nonlinear Oscillations* (Van Nostrand, Princeton, N. J., 1962), Chap. 3. See also K. Wilson, in Ref. 2.

PHYSICAL REVIEW B VOLUME 4, NUMBER 11 1 DECEMBER 1971

Some Critical Properties of the Eight-Vertex Model*

Leo P. Kadanoff and Franz J. Wegner†
Department of Physics, Brown University, Providence, Rhode Island 02912

(Received 16 July 1971)

The eight-vertex model solved by Baxter is shown to be equivalent to two Ising models with nearest-neighbor coupling interacting with one another via a four-spin coupling term. The critical properties of the model in the weak-coupling limit are in agreement with the scaling hypothesis. In this limit where $\alpha \to 0$, the critical indices obey $\gamma/\gamma_0 = \beta/\beta_0 = \nu/\nu_0 = 1 - \frac{1}{2}\alpha$, $\delta/\delta_0 = \eta/\eta_0 = 1$, with the subscripts zero denoting the index values for the ordinary two-dimensional Ising model.

In a recent publication, Baxter[1] has found the free energy for the eight-vertex problem and shown that α is a continuous function of the interaction constants. This continuous variation of a critical index contradicts the hypothesis of smoothness[2] or universality[3] often postulated for near-critical problems.

One way of seeing the source of this behavior is to rephrase the eight-vertex problem as an Ising model.[4] Imagine a spin placed at the interstitial points of the lattice as in Fig. 1. An arrow to the right (or upward) corresponds to the case in which the adjacent spins are parallel; a leftward or downward arrow makes the adjacent spins antiparallel. Then, the four combinatorical factors a, b, c, and d corresponding to the vertices shown can all be represented by a factor in the partition function

$$A e^{K^- \sigma_1 \sigma_4 + K^+ \sigma_2 \sigma_3 + \lambda \sigma_1 \sigma_2 \sigma_3 \sigma_4},$$

and we obtain the complete partition function

$$\sum_{\{\sigma_r = \pm 1\}} \prod_{j,k} A \exp(K^+ \sigma_{j,k} \sigma_{j+1,k+1} + K^- \sigma_{j+1,k} \sigma_{j,k+1} + \lambda \sigma_{j,k} \sigma_{j+1,k+1} \sigma_{j+1,k} \sigma_{j,k+1}), \quad (1)$$

in which next-nearest-neighbor spins are coupled by interaction constants $K^{\pm}$ depending upon the direction of the diagonal. The factor λ couples all four spins. The precise connection is that

$$a = A e^{K^+ + K^- + \lambda},$$
$$b = A e^{-(K^+ + K^-) + \lambda},$$
$$c = A e^{K^+ - K^- - \lambda},$$
$$d = A e^{-(K^+ - K^-) - \lambda}. \quad (2)$$

The constant A does not, of course, enter into the critical properties.

The Baxter solution shows that this Ising-type problem has a very new kind of singularity at the critical point, namely, one in which the singularity in the specific heat as $\epsilon \sim (b + c + d - a)/a$ goes to zero is of the form $\epsilon^{-\alpha}$ with α being a function of the parameters, namely,[5]

$$\sin \frac{\pi \alpha}{4(1 - \frac{1}{2}\alpha)} = \tanh 2\lambda . \quad (3)$$

This result seems at first to contradict the smoothness hypothesis[2] which suggests that critical indices should not change their value unless there is a symmetry change. However, this eight-vertex model certainly has a different set of symmetries than the usual two-dimensional Ising model. Notice that at $\lambda = 0$, the lattice with $j + k =$ (even integer) does not interact with the lattice with $j + k =$ (odd integer). Even at $\lambda \neq 0$ for $T > T_c$, i.e., $\epsilon > 0$, the spins on these two sublattices are uncorrelated. Therefore, the Ising form of the eight-vertex model can be viewed as having two lattices with "independent" ferromagnetic transitions which occur at exactly the same temperature. The coupling between these two lattices is of the form

$$\lambda \textstyle\sum_r u_r ; \quad u_r = \mathcal{S}_r^{(1)} \mathcal{S}_r^{(2)} , \quad (4)$$

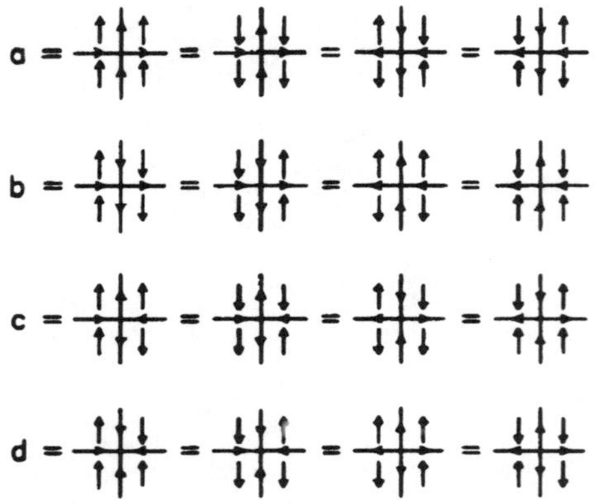

FIG. 1. The correspondence between the Ising-spin configurations, the eight-vertex configurations, and the Boltzmann factors a, b, c, d.

where $\mathcal{E}_r^{(1)}$ and $\mathcal{E}_r^{(2)}$ are the energy densities on the two sublattices. This kind of coupling leaves the spontaneous magnetizations on the two lattices free to point in either the same or in opposite directions. Since this two-sublattice symmetry is very different from that of the usual Ising model, it is not surprising that the critical indices of the Baxter solution are, in general, different from those of the Onsager solution.

A second unexpected feature of the solution is that α varies continuously with λ. The scaling hypothesis usually rules out this idea as is shown in the discussion of Ref. 3. However, there is one special case in which the scaling idea does permit the continuous variation of critical indices—if there is a term in the Hamiltonian of the form of $\lambda\sum_r \tilde{u}_r$ and $\tilde{u}_r$ scales as $1/r^d$ (d denotes the dimension of the lattice).

To see why this particular scaling is so significant, recall the definition of scaling: In the critical region, the phase transition is supposed to be described by fluctuating local quantities, e.g., the magnetization density and the energy density, which we can write as $O_\alpha(r)$. The α distinguishes among different quantities. Let O be a product of n different quantities of this type,

$$O=\prod_{i=1}^{n} O_{\alpha_i}(r_i)\,, \tag{5a}$$

and take each pair of operators in the product to be separated by a distance $|\vec{r}_i-\vec{r}_j|$ of the order of magnitude of R, with R much greater than a lattice constant and much smaller than a coherence length. Then the statement $O_\alpha(r)$ scales as $1/r^{x_\alpha}$ means precisely[6] that

$$\langle O\rangle_{K,\lambda}\sim\frac{1}{R^x}\,,\quad x=\sum_{j=1}^{n} x_{\alpha_j} \tag{5b}$$

for $n\geq 2$.

Here the x_α's are critical indices which describe the behavior of the fluctuating variables. For the ordinary two-dimensional Ising model, the magnetization has an index $x_\sigma=\frac{1}{8}$, and the energy density $x_{\mathcal{E}}=1$.

If these indices vary with λ, then the derivative of $\langle O\rangle_{K,\lambda}$ contains a term like $R^{-x}\ln R$, in particular,

$$\frac{\partial\langle O\rangle_{K,\lambda}}{\partial\lambda}=-\langle O\rangle_{K,\lambda}\sum_{j=1}^{n}\left(\frac{\partial x_{\alpha_j}}{\partial\lambda}\right)\ln R+\ldots\,, \tag{6}$$

where the $\cdots$ represents terms which are not logarithmic in R. Therefore, these logarithmic terms are signals of continuously varying critical indices.

To see how this logarithm can arise, notice that if λ is conjugate to u_r, which contains a term $\tilde{u}_r$, then

$$\frac{\partial}{\partial\lambda}\langle O\rangle_{K,\lambda}=\sum_r\langle O\tilde{u}_r\rangle_{K,\lambda}+\cdots. \tag{7}$$

According to the operator algebra concept, when a product of two operators which are relatively close to one another appears inside a correlation function, their product may be approximately replaced according to

$$O_\alpha(r_1)O_\beta(r_2)=\sum_\gamma A_{\alpha\beta,\gamma}(\vec{r}_1-\vec{r}_2)\,O_\gamma(\tfrac{1}{2}(r_1+r_2))\,, \tag{8}$$

where, according to scaling,

$$A_{\alpha\beta,\gamma}=\frac{a_{\alpha\beta,\gamma}[(\vec{r}_1-\vec{r}_2)/(|r_1-r_2|)]}{|r_1-r_2|^{x_\alpha+x_\beta-x_\gamma}} \tag{9}$$

for separations $|r_1-r_2|$ large in comparison to the lattice constant. In the particular case in which O_β is $\tilde{u}$, which scales as $1/r^d$, then the product in (8) contains a term of the form

$$O_\alpha(r_1)u(r)=\frac{a_\alpha O_\alpha(\frac{1}{2}(r_1+r))}{|r_1-r|^d}+\cdots\,, \tag{10}$$

when O_α and $\tilde{u}$ are scalars under rotation. Here a_α is, of course, the particular coefficient which appears in the reduction formula (9) when $\alpha=\gamma$ and $O_\beta=\tilde{u}$.

As a result, the sum in (7) contains a succession of terms

$$\frac{\partial}{\partial\lambda}\langle O\rangle_{K,\lambda}=\sum_{j=1}^{n}\;\sum_{|r-r_j|\ll R}\;\frac{a_j}{|r-r_j|^d}\langle O\rangle_{K,\lambda}+\cdots \tag{11}$$

which corresponds to $|r-r_j|$ being much smaller than the average separation distance $|r_i-r_j|\sim R$. Here the $+\cdots$ include all terms in which all separations are at least of order R. The logarithms then appear in $\sum_r$. In two dimensions one obtains

$$\sum_{r<R}1/r^2\approx 2\pi\ln(R/a_0)\,. \tag{12}$$

When Eqs. (11) and (12) are combined, a set of logarithms appears in the derivative. A comparison with Eq. (6) then shows that

$$\frac{\partial x_\alpha}{\partial\lambda}=-2\pi a_\alpha\,. \tag{13}$$

We apply this result to the model solved by Baxter

for the particular case $\lambda=0$. At $\lambda=0$, the operator

$$\bar{u}_r=\delta\mathcal{E}_r^{(1)}\delta\mathcal{E}_r^{(2)} \qquad (14)$$

does indeed scale as $1/r^2$ if $\delta\mathcal{E}_r^{(1)}$ and $\delta\mathcal{E}_r^{(2)}$ are the deviations of the energy on the two sublattices from their critical values. To see this, calculate $\langle\bar{u}_r\bar{u}_0\rangle$ at $\lambda=0$ and at the critical point. At $\lambda=0$, the two sublattices are independent so that

$$\langle\bar{u}_r\bar{u}_0\rangle=\langle\delta\mathcal{E}_r^{(1)}\delta\mathcal{E}_0^{(1)}\rangle^2 . \qquad (15)$$

However, the statement that $x_\mathcal{E}=1$ at $\lambda=0$ implies that at criticality for large r

$$\langle\delta\mathcal{E}_r^{(1)}\delta\mathcal{E}_0^{(1)}\rangle=q/2\pi r^2 . \qquad (16)$$

From Ref. 7 we obtain $q=4/\pi$. (Note that nearest neighbors in the sublattices are separated by $\sqrt{2}$.) The correlation function on the left-hand side of (15) is then $(q/2\pi r^2)^2$ and, consequently, $\bar{u}_r$ scales as $1/r^4$.

Because $\bar{u}_r$ has this special value of the scaling index, the critical phenomena theory indicates that the critical indices can vary continuously in λ. Conversely, if there is no operator with index d, then there can be no continuous variation of the critical indices.

To find the first variation in the critical index for the energy, calculate

$$\delta\mathcal{E}_{r_1}^{(1)}\bar{u}(r_2)\approx\delta\mathcal{E}_{r_1}^{(1)}\delta\mathcal{E}_{r_2}^{(1)}\delta\mathcal{E}_{r_2}^{(2)} . \qquad (17)$$

According to Eq. (17), as r_1 approaches r_2 at $\lambda=0$, the product of the energy fluctuations on lattice (1) can be replaced by a constant divided by $|r_1-r_2|^2$. In particular,

$$\delta\mathcal{E}_{r_1}^{(1)}\bar{u}(r_2)\approx\frac{q}{2\pi|r_1-r_2|^2}\delta\mathcal{E}_{r_2}^{(2)} . \qquad (18)$$

Note, however, that this result is not of the form (10), needed to reach Eq. (13). To achieve the form (10), we consider the combinations

$$\delta\mathcal{E}_r^{\pm}=\delta\mathcal{E}_r^{(1)}\pm\delta\mathcal{E}_r^{(2)} , \qquad (19)$$

which have a simple symmetry under the interchange of the two lattices. Equations (18) and (19) give

$$\delta\mathcal{E}_{r_1}^{\pm}\bar{u}(r_2)=\pm\frac{q}{2\pi|r_1-r_2|^2}\delta\mathcal{E}_{r_2}^{\pm}$$

when r_1 and r_2 are relatively close together compared to all other distances but $|r_1-r_2|$ is large compared to a lattice constant. It now follows that $\mathcal{E}_r^{\pm}$ scales as $1/r^{x_\pm}$ with Eq. (10) giving

$$\frac{dx_\pm}{d\lambda}=\mp q .$$

Since the scaling index is 1 at $\lambda=0$, we find that for small λ

$$x_\pm=1\mp\lambda q . \qquad (20a)$$

A similar argument applied to $\sigma_r^{(1)}$ indicates that at $\lambda=0$, $dx_\sigma/d\lambda=0$ so that to first order in λ

$$x_\sigma=\tfrac{1}{8} . \qquad (20b)$$

To derive this result, notice that for r_1 close to r_2,

$$\sigma_{r_1}^{(1)}u_{r_2}=\sigma_{r_1}^{(1)}\delta\mathcal{E}_{r_2}^{(1)}\delta\mathcal{E}_{r_2}^{(2)}$$

contains no term which is like $\sigma_r^{(1)}$ since this expression contains a reference to fluctuations on lattice 2. Hence the coefficient a in Eq. (13) vanishes. It follows that $\eta=2x_\sigma$ does not change to first order in λ.

From these results and scaling theory, we can predict all critical indices to first order in λ. For example, the deviation of energy from criticality contains a singular term of the form

$$\delta\mathcal{E}\sim\xi^{-x_+} ,$$

where ξ is the correlation length, since x_+ is the index which goes with the energy. Also the free energy contains a singluar term like

$$\delta F\sim\xi^{-2} .$$

Since $\xi\sim\epsilon^{-\nu}$, $\delta F\sim\epsilon^{2-\alpha}$, and $\delta\mathcal{E}\sim\epsilon^{1-\alpha}$, we find

$$(2-x_+)\nu=1$$

or

$$\nu=1-q\lambda , \qquad (21a)$$

$$\alpha=2\lambda q . \qquad (21b)$$

Equation (21b) is in agreement with Baxter's result, Eq. (3). Similarly, scaling theory implies that

$$\langle\sigma\rangle\sim\xi^{-x_\sigma}=(-\epsilon)^{\nu x_\sigma}$$

on the coexistence curve. Thus, $\beta=\nu x_\sigma$ yields

$$\beta=\tfrac{1}{8}(1-q\lambda) \qquad (21c)$$

to first order in λ. Thus, we find all the critical indices by using the assumption that scaling holds.

To check this assumption we use first-order perturbation theory. There is a term $\lambda\sum_r\mathcal{E}_r^{(1)}\mathcal{E}_r^{(2)}$ in $-\beta H$. This term may be written as

$$\lambda(\langle\mathcal{E}_r^{(1)}\rangle\langle\mathcal{E}_r^{(2)}\rangle+\Delta\mathcal{E}_r^{(1)}\langle\mathcal{E}_r^{(2)}\rangle$$
$$+\Delta\mathcal{E}_r^{(2)}\langle\mathcal{E}_r^{(1)}\rangle+\Delta\mathcal{E}_r^{(1)}\Delta\mathcal{E}_r^{(2)}) . \qquad (22)$$

For simplicity, set $K^+=K^-=K$. Now consider any expectation value $\langle O\rangle_{K,\lambda}$ where O is a product of terms $O^{(1)}O^{(2)}$ with $O^{(1)}$ containing spins on the first sublattice and $O^{(2)}$ containing spins on the second sublattice. To zeroth order, we may write

$$\langle O\rangle_{K,\lambda}=\langle O^{(1)}\rangle(\epsilon^0)\langle O^{(2)}\rangle(\epsilon^0)+\text{order }\lambda . \qquad (23)$$

Here $\epsilon^0=K_c^0-K$, with K_c^0 being the critical value of K at $\lambda=0$.

In first-order perturbation theory,

$$\frac{d}{d\lambda}\langle O\rangle_{K,\lambda}=\langle\mathcal{E}^{(2)}\rangle\langle O^{(2)}\rangle(\epsilon^0)\sum_r\langle\Delta\mathcal{E}_r^{(1)}O^{(1)}\rangle_{K,\lambda=0}$$

$$+\langle \mathcal{E}^{(1)}\rangle\langle O^{(1)}\rangle(\epsilon^0)\sum_r\langle\Delta\mathcal{E}_r^{(2)}O^{(2)}\rangle_{K,\lambda=0}$$

$$+\sum_r\langle O^{(1)}\Delta\mathcal{E}_r^{(1)}\rangle_{K,\lambda=0}\langle O^{(2)}\Delta\mathcal{E}_r^{(2)}\rangle_{K,\lambda=0}\,. \quad (24)$$

Since $\mathcal{E}$ is conjugate to K, the first two terms in (24) generate derivatives with respect to K of $\langle O^{(1)}\rangle_{K,\lambda=0}$ and $\langle O^{(2)}\rangle_{K,\lambda=0}$. To first order in λ, we may replace Eq. (24) by

$$\langle O^{(1)}O^{(2)}\rangle_{K,\lambda}\approx\langle O^{(1)}\rangle(\epsilon^*)\langle O^{(2)}\rangle(\epsilon^*)$$

$$-\lambda\sum_r\langle O^{(1)}\Delta\mathcal{E}_r^{(1)}\rangle\langle O^{(2)}\Delta\mathcal{E}_r^{(2)}\rangle\,, \quad (25)$$

with

$$\epsilon^*=\epsilon^0-\lambda\langle\mathcal{E}\rangle_{\lambda=0}\,. \quad (26)$$

Equation (26) gives a renormalized $T-T_c$. Near T_c, $\langle\mathcal{E}\rangle$ is given by

$$\langle\mathcal{E}\rangle=\tfrac{1}{2}\sqrt{2}-p\epsilon^0-q\epsilon^0\ln|\epsilon^0|\,, \quad (27)$$

with p being a new constant and q being the same as the constant defined by Eq. (16). [Eq. (27) can be obtained, e.g., from Eq. (97) of Ref. 8. Note that $2K_c(0)=\ln\operatorname{ctg}(\frac{1}{8}\pi)$.] To first order in λ, we can write

$$\epsilon^*=(\epsilon^0-\lambda\tfrac{1}{2}\sqrt{2}-p\lambda\epsilon^0)(1-\lambda q\ln|\epsilon^0|)$$

$$=\epsilon(1-\lambda q\ln|\epsilon|)=\epsilon e^{-\lambda q\ln|\epsilon|}=\epsilon|\epsilon|^{-\lambda q}\,, \quad (28)$$

with

$$\epsilon=\epsilon^0-\lambda\tfrac{1}{2}\sqrt{2}-p\lambda\epsilon^0 \quad (29)$$

being essentially $K_c(\lambda)-K$. The shift in K_c given by Eq. (29) checks directly against the value given by Baxter's solution.

Equations (25) and (28) may now be used to evaluate critical indices directly. When $O=\sigma_r^{(1)}$ and $T<T_c$, we find, to first order in λ,

$$\langle\sigma^{(1)}\rangle_{K,\lambda}=\pm(-\epsilon^*)^{\beta_0}=\pm(-\epsilon)^{\beta_0(1-q\lambda)}\,,$$

where β_0 is the magnetization index for the Onsager solution. This direct solution then recovers the scaling result (21c). Similarly, the two-spin correlation functions which have the form

$$\langle\sigma_0^{(1)}\sigma_r^{(1)}\rangle_{K,\lambda=0}=\frac{1}{r^{1/4}}f(r/\xi_0)$$

when $\xi_0\sim|\epsilon^0|^{-1}$ become, to first order,

$$\langle\sigma_0^{(1)}\sigma_r^{(1)}\rangle_{K,\lambda}=\frac{1}{r^{1/4}}f(r/\xi)+\text{order }\lambda^2 \quad (30)$$

when $\xi\sim|\epsilon|^{-1+\lambda q}$. It follows that the two-spin correlation function has a scaling form to first order in λ, and that the coherence length index is correctly given by Eq. (21a). Also, an integration of Eq. (30) over all r gives

$$\gamma=\tfrac{7}{4}(1-q\lambda)\,, \quad (31)$$

as would be predicted by scaling.

In fact, we can see that scaling holds quite generally to first order in λ. Consider the behavior of a product of n spins on the first sublattice at positions $r_1^{(1)},\ldots,r_n^{(1)}$ and m spins on the second at $r_1^{(2)},\ldots,r_m^{(2)}$. If they are far enough separated so that scaling holds at $\lambda=0$, then

$$\langle\sigma_{r_1}\cdots\sigma_{r_n}\rangle_{K,\lambda=0}=|\epsilon^0|^{n\beta}f_n(\{r_i/\xi_0\}) \quad (32a)$$

while the correlation with an energy density at R takes the form

$$\langle\sigma_{r_1}\cdots\sigma_{r_n}\delta\mathcal{E}_R\rangle_{K,\lambda=0}=\frac{|\epsilon^0|^{n\beta}}{\xi_0}f_{n,1}\left(\left\{\frac{r_i}{\xi_0}\right\},\frac{R}{\xi_0}\right). \quad (32b)$$

Then, Eq. (25) implies that to first order in λ,

$$\langle\sigma_{r_1^1}\cdots\sigma_{r_n^1}\sigma_{r_1^2}\cdots\sigma_{r_m^2}\rangle$$

$$=|\epsilon|^{(n+m)\beta}\left[f_n\left(\left\{\frac{r_i^{(1)}}{\xi}\right\}\right)f_m\left(\left\{\frac{r_i^{(2)}}{\xi}\right\}\right)\right.$$

$$\left.+\lambda\int\frac{d^2R}{\xi^2}f_{n,1}\left(\left\{\frac{r_i^{(1)}}{\xi}\right\},\frac{R}{\xi}\right)f_{m,1}\left(\left\{\frac{r_i^{(2)}}{\xi}\right\},\frac{R}{\xi}\right)\right], \quad (33)$$

which indicates that for large separations among all the spins their correlation functions obey all the scaling laws, at least to first order in λ.

It now follows that thermodynamic derivatives with respect to magnetic fields inserted in $-\beta H$ as $h_1\sum_r\sigma_r^{(1)}+h_2\sum_r\sigma_r^{(2)}$ must obey scaling in the form

$$\frac{\partial^n}{\partial h_1^n}\frac{\partial^m}{\partial h_2^m}\ln Z\Big|_{h_1=h_2=0}\sim|\epsilon|^{(n+m)\beta-[(n+m)-1]2\nu}\,. \quad (34)$$

In this way, we see that the free energy in the presence of magnetic fields contains a scaling term of the form

$$\epsilon^{2\nu}g\left(\frac{h_1}{\epsilon^\Delta},\frac{h_2}{\epsilon^\Delta}\right), \quad (35)$$

with $\Delta=2\nu-\beta$, and that this scaling term dominates all derivatives with respect to magnetic fields. Therefore the critical exponent $\delta=\Delta/\beta$ does not change to first order in λ.

To check the critical index α in first-order perturbation theory, we consider the free-energy density

$$-\beta f_\lambda=-\beta f_0+\lambda\langle\mathcal{E}^{(1)}\rangle\langle\mathcal{E}^{(2)}\rangle\,. \quad (36)$$

From Eqs. (27) and (29) we find that the nonanalytic contributions are

$$-\beta f_{\text{sing}}=-q\epsilon^2\ln|\epsilon|+\lambda q^2\epsilon^2(\ln|\epsilon|)^2\,, \quad (37)$$

which in first order in λ can be written

$$(|\epsilon|^{2-2q\lambda}-\epsilon^2)/2\lambda\,,$$

which again checks against Eq. (21b).

From Eq. (20a) we obtain the ratio $(2-x_+)/(2-x_-)=1+2\lambda q$ to first order in λ. To check this ratio we consider the model with an interaction constant

K_1 for the two-spin interactions in the first sublattice and an interaction constant K_2 in the second sublattice. In first order in λ we obtain similarly to Eq. (25)

$$\langle O^{(1)} O^{(2)} \rangle_{K_1 K_2 \lambda} \approx \langle O^{(1)} \rangle (\epsilon_1^*) \langle O^{(2)} \rangle (\epsilon_2^*) + \lambda \sum_r \langle O^{(1)} \Delta \mathcal{E}_r^{(1)} \rangle \langle O^{(2)} \Delta \mathcal{E}_r^{(2)} \rangle , \quad (38)$$

with

$$\epsilon_1^* = \epsilon_1^0 - \lambda \langle \mathcal{E}_2 \rangle_{\lambda=0} . \quad (39)$$

In first order in λ we can write

$$\epsilon_1^* = \epsilon_1 - \lambda q \epsilon_2 \ln |\epsilon_2| , \quad (40)$$

$$\epsilon_1 = \epsilon_1^0 - \lambda \tfrac{1}{2} \sqrt{2} + p \lambda \epsilon_2^0 , \quad (41)$$

and similar equations hold for ϵ_2 and ϵ_2^*. Therefore the critical line $\epsilon_1 = 0$ for $\lambda = 0$ moves to $\epsilon_1^* = 0$ at λ, that is,

$$\epsilon_1 = \lambda q \epsilon_2 \ln |\epsilon_2| \quad (42a)$$

in first-order perturbation theory. This equation as well as

$$\epsilon_2 = \lambda q \epsilon_1 \ln |\epsilon_1| \quad (42b)$$

can be written, in first order in λ,

$$|\epsilon_1 + \epsilon_2| = |\epsilon_1 - \epsilon_2|^{1+2\lambda q} . \quad (43)$$

Since $\epsilon_1 + \epsilon_2$ and $\epsilon_1 - \epsilon_2$ are conjugate to the energy densities $\mathcal{E}_\pm$, the exponent in Eq. (43) is expected to be $(2 - x_+)/(2 - x_-)$, in agreement with Eq. (20a).

APPENDIX

When shown the results of this paper, Wilson drew our attention to a similar problem in field theory which was studied by Wilson,[9] Callen,[10] and Symanzik.[11] In field theory the operator Φ^4 corresponds to the operator $\bar{u}_r$. In the free-field limit Φ^4 has the critical index (dimension) d, but in first-order perturbation theory its critical index changes. This leads to a breakdown of scaling.

According to Baxter's solution, the critical index α changes continuously with λ, Eq. (3). Therefore we expect $\bar{u}_r$ to scale like $1/r^2$ for any λ. Wilson and Fisher urged us to show this in first-order perturbation theory.

To see this, note that for $r_1 \neq r$ the operator

$$O_\alpha(r_1) \bar{u}(r) = \bar{u}(r_1) \bar{u}(r)$$

is even under the Kramers-Wannier (KW) transformation of one sublattice only. (Under the KW transformation of sublattice 1 $\delta\mathcal{E}_r^{(1)}$ is odd and $\delta\mathcal{E}_r^{(2)}$ is even.) Since $\bar{u}(r)$ is odd under this transformation, a_α vanishes. Thus, according to Eq. (13), the critical exponent x_u does not change in first order in λ.

We are indebted to Dr. K. Wilson for his comments.

---

*Work supported in part by the National Science Foundation.

†On leave from the Institut für Festkörperforschung of the Kernforschungsanlage Jülich, Germany.

[1]R. J. Baxter, Phys. Rev. Letters 26, 832 (1971).

[2]M. E. Fisher, Phys. Rev. Letters 16, 11 (1966); R. B. Griffiths, *ibid.* 24, 1479 (1970).

[3]L. P. Kadanoff, in Newport Conference on Phase Transitions, 1970 (unpublished); and in *Proceedings of the Enrico Fermi Summer School of Physics, Varenna*, 1970, edited by M. S. Green (Academic, New York, to be published).

[4]The relation between the eight-vertex model and the Ising model was independently given by R. Y. Wu, Phys. Rev. B (to be published).

[5]For positive interaction constants K^A and small λ the phase transition takes place for $a = b + c + d$. To obtain $\epsilon \sim (b + c + d - a)/a$ and Eq. (3) we use the symmetry property $Z(a, b, c, d) = Z(c, d, a, b)$, Eq. (11) of C. Fan and F. Y. Wu, Phys. Rev. B 2, 723 (1970).

[6]L. P. Kadanoff, Phys. Rev. Letters 23, 1430 (1969).

[7]R. Hecht, Phys. Rev. 158, 557 (1967); T. Stephenson, J. Math. Phys. 7, 1123 (1966).

[8]C. Domb, Advan. Phys. 9, 149 (1960).

[9]K. Wilson, Phys. Rev. D 2, 1478 (1970).

[10]C. Callen, Phys. Rev. D 2, 1541 (1970).

[11]K. Symanzik, Commun. Math. Phys. 18, 227 (1970).

Eight-Vertex Model in Lattice Statistics

R. J. Baxter
Research School of Physical Sciences, The Australian National University, Canberra, A.C.T. 2600, Australia
(Received 25 February 1971)

The solution of the zero-field "eight-vertex" model is presented. This model includes the square lattice Ising, dimer, ice, F, and KDP models as special cases. It is found that in general the free energy has a branch-point singularity at a phase transition, with a continuously variable exponent.

It has been pointed out[1] that many of the previously solved two-dimensional lattice models, notably the Ising and "ice"-type models, can be regarded as special cases of a more general model. Adopting the arrow terminology used by Lieb,[3] we can formulate this model as follows: Place arrows on the bonds of a square, N-by-N lattice and allow only those configurations with an even number of arrows pointing into each vertex. Then there are eight possible different configurations of arrows at each vertex (hence our name for the model), as shown in Fig. 1. Next we assign energies $\epsilon_1, \cdots, \epsilon_8$ to these vertex configurations and the problem is to evaluate the partition function

$$Z = \sum \exp(-\beta \sum_{j=1}^{8} N_j \epsilon_j), \tag{1}$$

where the summation is over all allowed configurations of arrows on the lattice, and N_j is the number of vertices of type j.

Let $\omega_j = \exp(-\beta\epsilon_j)$, and suppose the model to be unchanged by reversal of all arrows (in ferroelectric terminology this implies that there are no electric fields). Then we can write

$$\omega_1 = \omega_2 = a, \quad \omega_3 = \omega_4 = b,$$
$$\omega_5 = \omega_6 = c, \quad \omega_7 = \omega_8 = d. \tag{2}$$

We have succeeded in solving this model for arbitrary values of a, b, c, and d, and outline here the main results. The details of the derivation will be published elsewhere.[3]

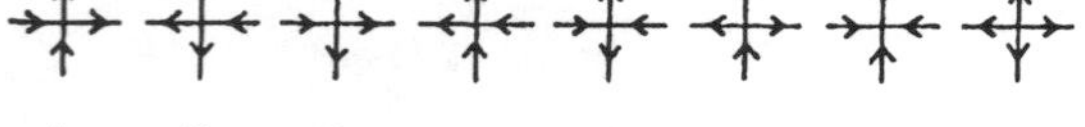

1 2 3 4 5 6 7 8

FIG. 1. The eight arrow configurations allowed at a vertex.

The method of attack on the problem was guided by some recent results[4] for an inhomogeneous system satisfying the "ice condition" ($d = 0$), in which we observed that the Bethe *Ansatz* approach worked provided that the transfer matrices of any two rows commuted. This led to an algebraic identity which later reflection has shown to be the diagonal representation of an identity between the transfer matrices and another matrix function Q.

Applying these ideas to the $d \neq 0$ situation, suppose that a, b, c, and d can vary from row to row. We find that the transfer matrices for any two rows commute provided that

$$a:b:c:d = \mathrm{sn}(\eta - v) : \mathrm{sn}(\eta + v) : \mathrm{sn}(2\eta) : -k\,\mathrm{sn}(2\eta)\,\mathrm{sn}(\eta - v)\,\mathrm{sn}(\eta + v), \tag{3}$$

where $\mathrm{sn}(u)$ is the usual elliptic function[5] of modulus k; k and η are fixed, but v can vary from row to row. Thus we can write the transfer matrix of a typical row as $T(v)$, where v is a variable parameter.

We now look for a matrix identity of the type mentioned above and find that there exists a matrix function $Q(v)$ (in general nonsingular) such that[6]

$$\zeta(v)T(v)Q(v) = \varphi(v-\eta)Q(v+2\eta) + \varphi(v+\eta)Q(v-2\eta), \tag{4}$$

where

$$\zeta(v) = [c^{-1}H(2\eta)\Theta(v-\eta)\Theta(v+\eta)]^N, \tag{5}$$

$$\varphi(v) = [\Theta(0)H(v)\Theta(v)]^N, \tag{6}$$

$H(u)$ and $\Theta(u)$ being the elliptic theta functions[5] of modulus k. Both $T(v)$ and $Q(v)$ are 2^N-by-2^N matrices, and $Q(v)$ commutes with any matrix $T(u)$ or $Q(u)$. Thus there exists a representation (independent of v) in which the matrices in (4) are diagonal, and we can look at one such diagonal element.

From the quasiperiodic properties of $Q(v)$ we can show that it must be possible to write each of its diagonal elements in the form

$$Q(v) = \prod_{r=1}^{\frac{1}{2}N} H(v-v_r)\Theta(v-v_r) \tag{7}$$

(provided that N is even). Setting $v=v_1, \cdots, v_{N/2}$ in (4), the left-hand side vanishes and we get $N/2$ equations for $v_1, \cdots, v_{N/2}$. These can in principle be solved and the corresponding diagonal element (eigenvalue) of $T(v)$ can be calculated from (4). (These equations are analogous to the equations for the wave numbers in the Bethe *Ansatz.*) The partition function for a large lattice can then be calculated in the usual way from the maximum eigenvalue.

Let K and K' be the complete elliptic integrals[5] of the first kind of moduli k and k'. Define

$$\tau = \pi K'/2K, \quad \eta = iK\lambda/\pi, \quad v = iK\alpha/\pi. \tag{8}$$

Then we find that the free energy per vertex, f, is given by

$$-\beta f = \lim_{N\to\infty} N^{-1}\ln Z = -\beta\epsilon_5 + 2\sum_{n=1}^{\infty} \frac{\sinh^2[(\tau-\lambda)n][\cosh(n\lambda)-\cosh(n\alpha)]}{n\sinh(2n\tau)\cosh(n\lambda)}, \tag{9}$$

provided that k, λ, and α are real and

$$0<k<1, \quad |\alpha|<\lambda<\tau. \tag{10}$$

For given values of a, b, c, and d (constant throughout the lattice), k, η, and v (and hence τ, λ, and α) are to be calculated from (3). The restrictions (10) are equivalent to $a>0$, $b>0$, $d>0$, and $c>a+b+d$. This set of values of a, b, c, and d we call the principal domian, and we see that it corresponds to a generalization of the ordered F-model state. Fortunately we can map any set of values a, b, c, d into the principal domain (or its boundaries—as f is continuous these present no problems) by using the symmetry relations (9)-(12) of Ref. 1 (writing a, b, c, d for u_1, u_2, u_3, u_4).

Inside the principal domain f is an analytic function of a, b, c, and d. Thus as the temperature T is varied a phase transition can occur only when a, b, c, or d (or their appropriately mapped values) cross a boundary of the principal domain, and in general this will correspond to just one of a, b, d, or $c-a-b-d$ becoming zero. If two or more become zero simultaneously, we have a more complicated situation—the F and KDP models are in this category—which is not discussed here.

If the mapped values of a, b, or d become zero, we find that f is the same analytic function of T on both sides of the boundary value, so there is no phase transition. However, if $c-a-b-d$ has a simple zero at some value T_c of T, we find that f can be written as the sum of an analytic and a singular function of T. The analytic part is the same on both sides of T_c, while near T_c the singular part is proportional either to

$$\cot(\pi^2/2\mu)|T-T_c|^{\pi/\mu}, \tag{11a}$$

or, if $\pi/2\mu = m$ integer, to

$$2\pi^{-1}(T-T_c)^{2m}\ln|T-T_c|, \tag{11b}$$

where $0<\mu<\pi$ and

$$\cos\mu = (ab-cd)/(ab+cd). \tag{12}$$

The constant of proportionality which multiplies (11) is the same on both sides of T_c.

We conclude that if a given eight-vertex model has a phase transition, then in general the free energy has a branch-point singularity. Note that the exponent π/μ of this singularity can range continuously from one to "infinity." All previously solved cases are in a sense deceptive in that they correspond to very special values of the exponent. For instance, the Ising model and the "free-fermion" model[1] correspond to $\mu=\pi/2$, and we can think of the F model as a limiting case in which $\mu\to 0$, i.e., the singularity becomes of infinitely high order.

The author is indebted to Professor Lieb for stimulating his interest in the above problem.

[1]C. Fan and F. Y. Wu, Phys. Rev. B 2, 723 (1970).

[2]E. H. Lieb, Phys. Rev. 162, 162 (1967), and Phys. Rev. Lett. 18, 1046 (1967), and 19, 108 (1967).

[3]R. J. Baxter, to be published.

[4]R. J. Baxter, Stud. Appl. Math. 50, 51 (1971).

[5]I. S. Gradshteyn and I. M. Ryzhik, *Tables of Integrals, Series, and Products* (Academic, New York, 1965), pp. 904-925.

[6]To derive (4) we have negated a. As Z is unaffected by this transformation, this has no effect on the subsequent equations.

Critical Exponents in 3.99 Dimensions*

Kenneth G. Wilson and Michael E. Fisher
Laboratory of Nuclear Studies and Baker Laboratory, Cornell University, Ithaca, New York 14850
(Received 11 October 1971)

Critical exponents are calculated for dimension $d=4-\epsilon$ with ϵ small, using renormalization-group techniques. To order ϵ the exponent γ is $1+\frac{1}{6}\epsilon$ for an Ising-like model and $1+\frac{1}{5}\epsilon$ for an XY model.

A generalized Ising model is solved here for dimension $d=4-\epsilon$ with ϵ small. Critical exponents[1] are obtained to order ϵ or ϵ^2. For $d>4$ the exponents are mean-field exponents[1] independent of ϵ; below $d=4$ the exponents vary continuously with ϵ. For example, the susceptibility exponent γ is $1+\frac{1}{6}\epsilon$ to order ϵ for $\epsilon>0$, and 1 exactly for $\epsilon<0$. The definitions for nonintegral d are trivial for the calculations reported here but may be more difficult for exact calculations to higher orders in ϵ. The exponents will be calculated using a recursion formula derived elsewhere[2] which represents critical behavior approximately in three dimensions but turns out to be exact to order ϵ (see the end of this paper). Exponents will also be obtained for the classical planar Heisenberg model (XY model) and a modified form of Baxter's eight-vertex model.[3]

The background for the recursion formula is as follows.[4] Let $s_{\vec{r}}$ be the spin at site $\vec{r}$, and let its range be $-\infty < s_{\vec{r}} < \infty$. Let $\sigma_{\vec{k}}$ be the Fourier-transform variable $\sum_{\vec{r}} \exp(i\vec{k}\cdot\vec{r}) s_{\vec{r}}$. Define a "block-spin" variable $s_l(\vec{x})$ to be

$$s_l(\vec{x}) = 2^{l(d/2-1)} \int_l \sigma_{\vec{k}} \exp[-i\vec{k}\cdot(2^l\vec{x})] d^d k, \qquad (1)$$

where $\int_l$ means the integration is restricted to the range $|\vec{k}| > 2^{-l}$. The variable $s_l(\vec{x})$ is, very roughly, the sum over all spins $s_{\vec{r}}$ in a block of length 2^l surrounding the point $2^l\vec{x}$.[5] There is an effective interaction (in Kadanoff's sense[6]) of Landau-Ginsberg form for the block-spin variable[7]:

$$\mathcal{H}_l = -\int [\tfrac{1}{2}\nabla s_l(\vec{x})\cdot\nabla s_l(\vec{x}) + Q_l(s_l(\vec{x}))] d^3x. \qquad (2)$$

For $l=0$ this specifies the interaction of interest. The function Q_l is obtained from the recursion formulas

$$Q_{l+1}(y) = -2^d \ln[I_l(2^{1-d/2}y)/I_l(0)], \qquad (3)$$

$$I_l(z) = \int_{-\infty}^{\infty} dy \exp[-y^2 - \tfrac{1}{2}Q_l(y+z) - \tfrac{1}{2}Q_l(-y+z)]. \qquad (4)$$

The initial function $Q_0(y)$ may be chosen as

$$Q_0(y) = r_0 y^2 + u_0 y^4, \qquad (5)$$

where the constant r_0 is varied to locate the critical point of the model, and the y^4 term is present so that the model is not the Gaussian model.[8] One must choose $u_0 \geq 0$ to avoid a divergent integral in (4). The effective interaction $\mathcal{H}_l$ determines the spin-spin correlation function through

$$\hat{G}(\vec{k}) \sim (k^2)^{-1} I_{l(k)}^{-1}(0) \int_{-\infty}^{\infty} dy\, y^2 \exp[-y^2 - Q_{l(k)}(y)], \qquad (6)$$

where $l(k) \sim -\log_2(ka)$ and a is the lattice spacing. (This formula is only an order-of-magnitude estimate.[2]) To derive these results, the partition function Z was first defined as a functional integral over all $\sigma_{\vec{k}}$ ($|\vec{k}| \lesssim 1$) of the initial Boltzmann factor $\exp(\mathcal{H}_0)$.[9] The recursion formula was obtained by performing the functional integral over $\sigma_{\vec{k}}$ for a factor-of-2 range of $|\vec{k}|$; qualitative approximations were made to ensure the simple form (2) for the effective interaction $\mathcal{H}_l$.[2,10]

General considerations (confirmed by a numerical study[2]) show the following. At the critical point $r_0 = r_c$ (with u_0 fixed), the function $Q_l(y)$ normally approaches a limit $Q_c(y)$ for $l \to \infty$.[11] The function $Q_c(y)$ is a "fixed point" of the recursion formula, namely an l-independent solution. The fixed point is unstable to changes in r_0; for $r_0 \simeq r_c$ and reasonably large l, $Q_l(y)$ has the form

$$Q_l(y) \simeq Q_c(y) + (r_0 - r_c)\lambda^l R_c(y), \qquad (7)$$

where λ is a constant. The critical exponents[1] ν and γ are given by $2\nu = \gamma = 2(\ln 2)/\ln\lambda$; the approximations made in deriving the recursion formula enforce $\eta = 0$. For $d=3$ the numerical work gave $2\nu = 1.217$.

If $u_0 = 0$, the solution of the recursion formula approaches a different fixed point. For this "Gaussian" fixed point,[2] $Q_l(y)$ has the form (7) for any r_0, but now $Q_c(y) \equiv 0$, $R_c(y) = y^2$, $r_c = 0$, $\lambda = 4$, and $2\nu = \gamma = 1$.

There is a common expectation that critical exponents become the mean-field exponents at $d = 4$.[12] This suggests that the nontrivial fixed point coincides with the Gaussian fixed point for $d=4$ and can be calculated analytically for $d \simeq 4$. This is correct; the calculation is remarkably easy to perform and will be summarized here.

Let ϵ be small; let the initial constants r_0 and u_0 be of order ϵ. Then by induction in l one finds that $Q_l(y)$ has the form

$$Q_l(y) = r_l y^2 + u_l y^4 + O(\epsilon^3), \qquad (8)$$

where r_l and u_l are of order ϵ. The recursion formulas for r_l and u_l are[13]

$$r_{l+1} = 4[r_l + 3u_l(1+r_l)^{-1} - 9u_l^2] + O(\epsilon^3), \qquad (9)$$

$$u_{l+1} = (1 + \epsilon \ln 2)u_l - 9u_l^2 + O(\epsilon^3). \qquad (10)$$

A fixed point is a solution $r_l = r$, $u_l = u$ of the recursion formulas independent of l. It is evident from Eq. (10) that there are two fixed points: the Gaussian fixed point $r = u = 0$, and a nontrivial fixed point

$$u = u^* = \tfrac{1}{9}\epsilon \ln 2 + O(\epsilon^2), \qquad (11)$$

$$r = r^* = -\tfrac{4}{9}\epsilon \ln 2 + O(\epsilon^2). \qquad (12)$$

Clearly the two fixed points coincide for $d=4$ and u^* is small for ϵ small.

Arbitrary initial values r_0 and u_0 will not correspond to either fixed point. Therefore the two fixed points compete to determine the asymptotic behavior for $l \to \infty$ of $Q_l(y)$ for a given initial value of u_0. In consequence there are *domains* of initial values associated with each fixed point. In the present case it is easy to determine the domains of u_0 from Eq. (10). The value of r_0 must be fixed at its critical value [$r_c = -4u_0 + O(\epsilon^2)$] once u_0 is given. One sees from Table I that the nontrivial fixed point wins the competition for $\epsilon > 0$ (unless $u_0 = 0$), and the Gaussian fixed point wins for $\epsilon < 0$. To obtain γ and ν for $d < 4$ one

TABLE I. Domains of initial values for the two fixed points of the generalized Ising model. Only $u_0 \lesssim \epsilon$ is considered; "unphysical" means negative values of u_0.

Dimension	Fixed point Gaussian	Nontrivial
$d<4$	$u_0=0$	$0<u_0$
$d>4$	$0\leq u_0$	unphysical

must determine the constant λ for the nontrivial fixed point. This one does by linearizing the recursion formulas for small departures from the fixed point and looking for solutions of the form $\Delta r_l = A\lambda^l$, $\Delta u_l = B\lambda^l$, where A and B are independent of l and proportional to $(r_0 - r_c)$. One gets an eigenvalue equation for λ; to order ϵ this equation is

$$\lambda\begin{bmatrix}A\\B\end{bmatrix}=\begin{bmatrix}4-12u & 12\\ 0 & 1+\epsilon\ln 2-18u\end{bmatrix}\begin{bmatrix}A\\B\end{bmatrix}. \tag{13}$$

The largest eigenvalue λ determines the unstable solution of Eq. (7) and is used in the calculation of ν and γ. For $\epsilon<0$ one gets the Gaussian exponents. For $\epsilon>0$ one gets $\lambda=4-12u^*$ giving γ $1+\frac{1}{6}\epsilon$. A more accurate calculation[14] gives

$$2\nu=\gamma=1+\tfrac{1}{6}\epsilon+\epsilon^2(\tfrac{1}{36}+\tfrac{1}{54}\ln 2),\quad \epsilon>0. \tag{14}$$

For $\epsilon=1$ this differs from the numerical result 1.217 by only 0.010. These results are plotted in Fig. 1.

The above analysis is easily extended to (classical) models where the spin $s_{\vec{r}}$ has n components $s_{\vec{r}j}$, as in the Heisenberg model. The recursion formulas are still Eqs. (3) and (4) except that y and z are vectors $\vec{y}$ and $\vec{z}$, $\int dy$ is replaced by $\int d^n y$, and $y^2=\vec{y}\cdot\vec{y}$. Consider the following initial form for $Q_0(y)$, for $n=2$:

$$Q_0(y)=r_0(y_1^2+y_2^2)+u_0(y_1^4+y_2^4)+g_0 y_1^2 y_2^2. \tag{15}$$

For $g_0=0$ one has two independent Ising-like models. For $g_0=2u_0$ the model has the rotational symmetry of the XY model. For $g_0=6u_0$ the model turns out again to be two independent Ising-like models if one uses the variables $x_1=(y_1+y_2)/\sqrt{2}$ and $x_2=(y_1-y_2)/\sqrt{2}$. For other values of g_0 the model involves two Ising-like models with $g_0 y_1^2 \times y_2^2$, providing, in the language of Kadanoff and Wegner,[15] an energy-energy–type coupling of the two: The model resembles the reformulation[15] of Baxter's eight-vertex model.[3]

The critical behavior has been computed to order ϵ for these models. The essential recursion

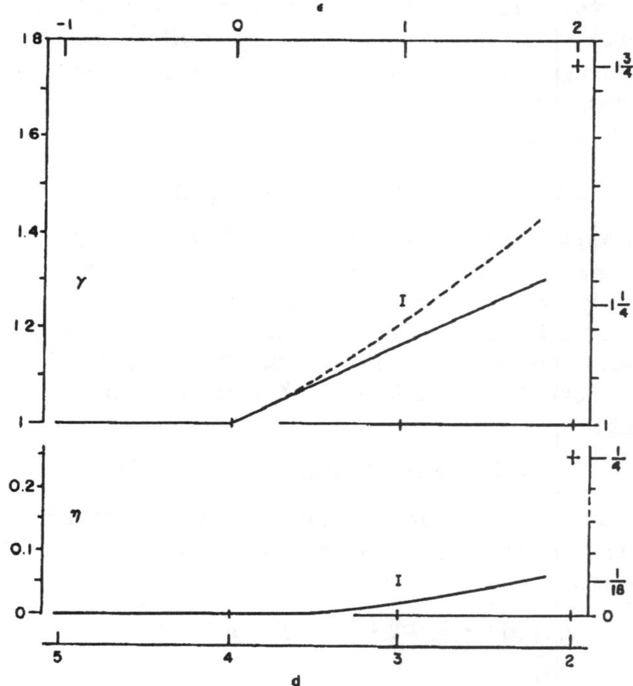

FIG. 1. Plot of the susceptibility exponent γ and the critical-point correlation exponent η versus dimension d to leading order in $\epsilon=4-d$. The dashed curve represents the truncated expansion (14). The special values indicated by "+" and "I" are for the standard spin-$\frac{1}{2}$ Ising models in two and three dimensions.

formulas are

$$u_{l+1}=(1+\epsilon\ln 2)u_l-9u_l^2-\tfrac{1}{4}g_l^2, \tag{16}$$

$$g_{l+1}=(1+\epsilon\ln 2)g_l-6u_l g_l-2g_l^2. \tag{17}$$

There are four fixed points:

$u=g=0$ (Gaussian),

$u=\frac{1}{9}\epsilon\ln 2,\quad g=0$ (Ising-like),

$u=\frac{1}{18}\epsilon\ln 2,\quad g=6\lambda$ (Ising-like),

$u=\frac{1}{10}\epsilon\ln 2,\quad g=2\lambda$ (XY-like).

The XY-like fixed point is the most stable one for $\epsilon>0$ and gives the critical behavior for any initial condition with g_0 in the range $0<g_0<6u_0$. The critical exponents for this regime are $2\nu=\gamma=1+\frac{1}{5}\epsilon$. The Ising-like roots are less stable and give the critical behavior only for $g_0=0$ or $g_0=6u_0$. The Gaussian root is the least stable root giving the critical behavior only for $u_0=g_0=0$. The ranges $g_0<0$ and $g_0>6u_0$ are anomalous; with this type of initial condition, iteration of the recursion formula gave values of g_l which increase without limit in magnitude. This takes one outside the range of validity of the recursion formulas (16) and (17); we do not know the critical behavior for this range of g_0.

To check these calculations an *exact* calculation was performed to determine γ to order ϵ, and η to order ϵ^2. The effective interaction $\mathcal{K}_l$ was allowed to be the general form

$$\mathcal{K}_l = -\iint d^3x\, d^3y\, u_{2l}(\vec{x}-\vec{y})s_l(\vec{x})s_l(\vec{y}) - \int d^3x_1 \cdots \int d^3x_4\, u_{4l}(\vec{x}_1, \cdots, \vec{x}_4)s_l(\vec{x}_1)\cdots s_l(\vec{x}_4) - \cdots. \quad (18)$$

Exact recursion formulas for u_{2l}, u_{4l}, etc. were obtained as power series in the non-Gaussian terms, by integrating exactly the functional integral over $\sigma_{\vec{k}}$ with $|\vec{k}|$ restricted to the range $b^{-l} < |\vec{k}| < b \cdot b^{-l}$, where b was left arbitrary. To obtain a fixed point it was necessary to use a more general scale factor $b^{l(d-2+\eta)/2}$ in Eq. (1). By induction it was shown that u_{4l} is of order ϵ, u_{6l} of order ϵ^2, u_{8l} of order ϵ^3, etc. A nontrivial fixed point was found for $\epsilon > 0$ with exponents $2\nu = \gamma = 1 + \frac{1}{6}\epsilon$ to order ϵ and $\eta = \frac{1}{54}\epsilon^2$ to order ϵ^2 (as plotted in Fig. 1).[16] Thus the result for 2ν and γ from the approximate recursion formulas is exact to order ϵ, but the result $\eta = 0$ is incorrect in order ϵ^2. The calculations for the Heisenberg and modified Baxter models are exact to order ϵ.

The exact results obtained here for d near 4 complement the exact solutions of two-dimensional models. Qualitative results concerning the competition between different fixed points and corresponding sets of exponents are probably true more generally. The analysis described here is a simple and powerful method for obtaining such results.

---

*Work supported in part by the National Science Foundation and the Advanced Research Projects Agency through the Materials Science Center at Cornell University.

[1]For general background on critical phenomena and definitions of the critical exponents, see M. E. Fisher, Rep. Progr. Phys. 30, 731 (1967); L. P. Kadanoff *et al.*, Rev. Mod. Phys. 39, 395 (1967).

[2]K. G. Wilson, Phys. Rev. B 4, 3174, 3184 (1971).

[3]R. J. Baxter, Phys. Rev. Lett. 26, 832 (1971).

[4]See Wilson, Ref. 2, for all details.

[5]The definition of $S_l(\vec{x})$ involves several scale factors: $2^{l(d/2-1)}$, 2^l, and a factor (not shown) depending on kT. These factors are introduced so that $s_l(\vec{x})$ is of order 1 (at the critical point) and so that $\vec{x}$ is measured in units of the block spacing.

[6]L. P. Kadanoff, Physics 2, 263 (1966).

[7]The function $\exp[-Q_l(s)]$ is, very roughly, the probability distribution for the total (normalized) spin s of a block.

[8]T. H. Berlin and M. Kac, Phys. Rev. 86, 821 (1952).

[9]The temperature has been absorbed into constants like r_0; see Ref. 2.

[10]For a brief exposition of these approximations see K. G. Wilson, Cornell University Report No. CLNS-142 (to be published). For a detailed analysis of the case $d=4$, also using renormalization-group methods, see A. I. Larkin and D. E. Khamel'nitskii, Zh. Eksp. Teor. Fiz. 56, 2087 (1969) [Sov. Phys. JETP 29, 1123 (1969)], Appendix 2.

[11]The function $Q_c(y)$ is plotted in Fig. 4 of Ref. 2.

[12]See, for example, E. Helfand and J. S. Langer, Phys. Rev. 160, 437 (1967), or Ref. 2.

[13]These equations are Eqs. (4.29) and (4.30) of Ref. 2.

[14]For this calculation one must add a $w_l y^6$ term to Eq. (8); w_l is of order ϵ^3.

[15]L. P. Kadanoff and F. J. Wegner, Phys. Rev. B 4, 3989 (1971). See also F. Y. Wu, Phys. Rev. B 4, 2312 (1971).

[16]This is a very small value for η; see, in this connection, A. A. Migdal, Zh. Eksp. Teor. Fiz. 59, 1015 (1970) [Sov. Phys. JETP 32, 552 (1971)].

Evidence for a New Phase of Solid He^3 †

D. D. Osheroff, R. C. Richardson, and D. M. Lee
Laboratory of Atomic and Solid State Physics, Cornell University, Ithaca, New York 14850
(Received 10 February 1972)

Measurements of the melting pressure of a sample of He^3 containing less than 40-ppm He^4 impurities, self-cooled to below 2 mK in a Pomeranchuk compression cell, indicate the existence of a new phase in solid He^3 below 2.7 mK of a fundamentally different nature than the anticipated antiferromagnetically ordered state. At lower temperatures, evidence of possibly a further transition is observed. We discuss these pressure measurements and supporting temperature measurements.

On the basis of measured values of the solid-He^3 spin exchange energy J, defined by $\mathcal{H}_{ex}=-2J \times \sum_{i<j} I_i \cdot I_j$, it has been assumed that near 2.0 mK solid He^3 would order antiferromagnetically by a second-order phase transition.[1] In this Letter we present evidence that at 2.7 mK solid He^3 undergoes a phase transition of a nature fundamentally different from that which had been expected, and that the ordered state is most probably not the simple antiferromagnetic one assumed. The refrigeration device, pressure transducer, and thermometry employed in our measurements are described, the evidence is presented, and a brief discussion follows.

The method of compressional cooling of He^3 to obtain temperatures as low as 2 mK is by now well established.[2-4] The present apparatus, shown in Fig. 1, employs a pressure amplifier which consists of a set of beryllium-copper bellows connected by a rigid piston. The pressure amplifier enables a moderate He^4 pressure (< 10 atm) in the upper chamber, generated externally, to compress and solidify the He^3 in the lower chamber. Although sufficient volume changes can be generated to solidify the entire 12-cm^3 He^3 sample, seldom was over 40% solid ever formed in the experiments to be discussed.

The apparatus was attached directly to the mixing chamber of a dilution refrigerator for precooling and thermal isolation.[5] Above about 5 mK the compression process was highly revers-

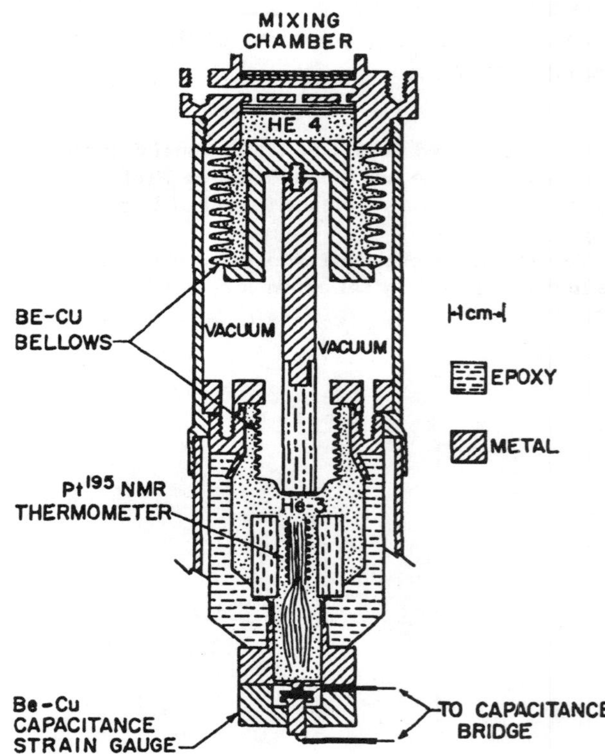

FIG. 1. Pomeranchuk cooling and pressure-measuring apparatus.

ible in the present device, and no heating associated with the motion of the bellows has been observed. Rates of compression (and decompression) producing solidification as rapidly as 2×10^{-2} mole/min were easily obtained. A typical rate of solidification, 1×10^{-3} mole/min, produced a cooling of about 0.25 mK/min.

The beryllium-copper capacitative pressure transducer[6] was calibrated at 1 K against a quartz secondary pressure standard. No hysteresis in the gauge greater than 2×10^{-3} atm, the reproducibility of the secondary standard, was observed during the calibration. The gauge reproducibility was better than 1×10^{-4} atm, and its resolution was better than 5×10^{-6} atm.

The thermometer shown in Fig. 1 consisted of 0.83 g of 0.001-in.-diam insulated reference-grade platinum wire obtained from the Sigmund Cohn Company. Temperatures were obtained by measuring the nuclear susceptibility of the platinum with a cw NMR spectrometer. A comparison of this thermometer with a similar one made of copper showed agreement better than 1% between 40 and 4 mK. Heating due to the applied rf signal caused less than a 2% temperature error at 3 mK, and the thermal relaxation time between the thermometer and the He^3 bath was measured to be shorter than 3 min at 3 mK in a 1.5-kOe magnetic field.

The thermometer was calibrated using the He^3 melting curve[7] as measured by Johnson *et al.*,[8] shifted downward in pressure by 0.24 psi to account for differences between pressure calibrations. From 30 to 3.5 mK, the melting curve obtained using the described apparatus agreed to better than 1% with the shifted curve of Johnson *et al.*[8] Below about 3.5 mK the thermometer indicated temperatures above those given by the melting curve. A lowest measured temperature of 2.7 mK was obtained repeatedly with the thermometer shown in Fig. 1, and also with one other platinum and one copper thermometer in various experiments during the past year. It is assumed on the basis of the pressure measurements that much lower temperatures were generated in the cell, and that the behavior of the thermometers near 3 mK may be associated with changes in the properties of the solid He^3 in contact with the thermometers.

Figure 2 shows the time evolution of pressure in the cell during a compression in zero (<10 Oe) magnetic field, and serves to illustrate, as accurately as possible, the phenomena we observe. From $t=0$ to point D, the rate of change of the cell volume with time (rate of compression) was constant and very uniform. The features of interest labeled A, B, and C are discussed below.

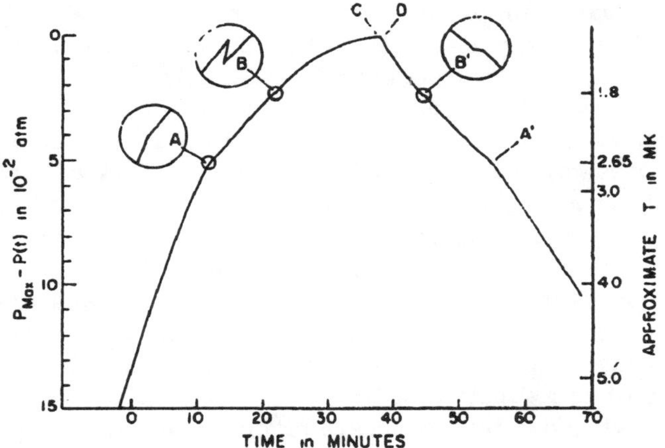

FIG. 2. Time evolution of the pressure in the Pomeranchuk cell during compression and subsequent decompression.

At point A, at 33.9053 atm, we observed an abrupt change in the slope of the pressure-versus-time (pressurization) curve. This pressure corresponds to a temperature of 2.65 mK on the shifted melting curve of Johnson *et al.*,[8] extrapolated from 2.8 to 2.65 mK. The change in slope, by roughly a factor of 1.8, occurred within a pressure interval of less than 3×10^{-4} atm. The change was highly reproducible both upon warming and cooling, provided that the time interval over which the measurement was made was less than about 15 min. Because the solid He^3 specific heat is immense below 3 mK, and its thermal conductivity poor, the bulk solid is thought never to reach thermal equilibrium with the liquid in these experiments. By working within a sufficiently short time period, the effects caused by the thermal relaxation of the solid are minimized.

At B, at 33.9279 atm, 2.26×10^{-2} atm above $P(A)$, a sudden drop in pressure of about 3×10^{-4} atm was observed upon cooling, accompanied by no change in slope of the pressurization curve such as had been observed at A. By varying the rate of pressurization from 6×10^{-4} atm/min to 1×10^{-2} atm/min hysteresis of about 3×10^{-3} atm in B was observed. The effect observed at B', upon warming through decompression, showed no such hysteresis, and was basically different in nature from the effect at B, as is shown by the inset. B' always occurred at least 3×10^{-4} atm lower in pressure than B.

At C, 33.9575 atm, 5.22×10^{-2} atm above $P(A)$,

a maximum pressure was reached in the cell. This pressure, P_{max}, is thought to be very near to the maximum in the ^{3}He melting curve. Not only was P_{max} reproducible over a period of weeks, but it was independent of the rate of compression for sufficiently high rates. Further, the time evolution of the cell pressure upon warming from P_{max} suggested that the melting curve might be nearly horizontal in the vicinity of P_{max}.

At D in Fig. 2, the motion of the piston was reversed, and the features A' and B' already discussed were traced out by the pressure in the cell as a function of time.

All these features were highly reproducible, independent of the fraction solid in the cell, the position of the bellows, and the temperature of the dilution refrigerator mixing chamber. Even after the cell had been altered to accept a new thermometer, no changes in A, B, or C were observed.

The magnetic field dependences of A, B, and C have been studied in fields up to 13.4 kOe. Both A and C shifted to lower pressures, with $P(A) = -6.76\times10^{-3}H^2$, and $P(C) = -2.08\times10^{-2}H^2$ with P in atm and H in kOe. The hysteresis between B and B' became quite large, and the pressure drop at B increased in magnetic fields. The drop in pressure associated with B did not occur in fields above about 4 kOe, presumably having been shifted to a pressure above P_{max}.

Because the He^3 melting curve is itself depressed in magnetic fields, the temperature shift associated with $P(A)$ is uncertain. Most likely, $T(A)$ is not shifted strongly in the fields applied in these experiments. In the course of this work, the magnetic field dependence of the melting curve was measured, and found to be in good agreement with calculations based on the high-temperature Heisenberg expansion of Baker *et al.*[10] in fields up to 13.4 kOe, down to 5 mK. This result is in contradiction to the measurements of Johnson, Rapp, and Wheatley[11] which suggested a much stronger field dependence than theory predicts.

Although the platinum thermometer failed to follow the melting curve below 3.5 mK, direct evidence of the thermal nature of the transition at A in Fig. 2 was obtained using it. When the compression was stopped at a pressure near P_{max}, and the melting pressure was allowed to drop slowly ($\dot{T} \simeq 7$ μK/min) as the bulk solid relaxed thermally, the results shown in Fig. 3 were obtained. The upper curve, representing the cell pressure versus time, shows only a slight dimple

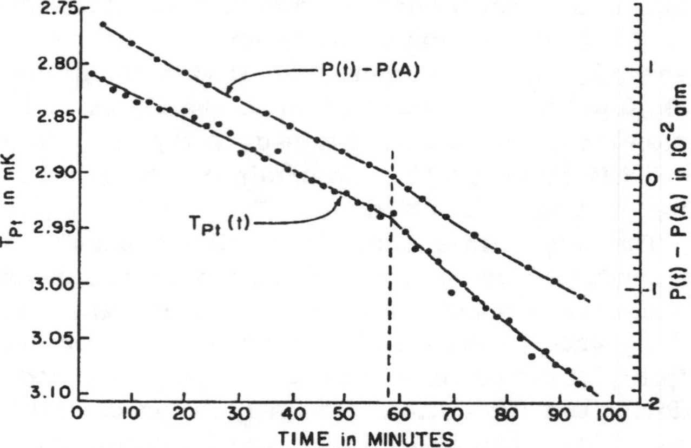

FIG. 3. Plots of pressure versus time and platinum temperature versus time in the neighborhood of the anomaly occuring at 2.65 mK.

at $P(A)$, but in the vicinity of this dimple, the characteristic change in slope by a factor of 1.8 is still easily measured. The lower curve, representing the temperature of the platinum thermometer during the same time interval, shows an abrupt change in slope at precisely the time that the cell pressure reached $P(A)$. This break in the warming curve of the platinum thermometer provides strong evidence that the phenomena observed with the strain gauge are related to the thermal properties of the helium under study.

To interpret the results we have obtained, we have had to make some basic assumptions. The most important of these, already discussed above, is that in the rapid compression-decompression experiments, the dominant thermal reservoir being cooled by the solidification process is that of the liquid He^3, and not the solid. The very lengthy solid thermal relaxation times we have measured during compression experiments, coupled with the independence of the phenomena observed on the rate or sense of the compression, and the extremely sharp nature of the transition observed at A, all support this hypothesis. If one also assumes that no large changes in the molar volume of the solid occur at A, the discontinuity in dP/dt at A in Fig. 2 must be caused by a discontinuity in dP/dT.

A microscopic description of the solid which exists below 2.7 mK cannot be obtained from these measurements. On the other hand, several surprising features of the new solid phase can be inferred from the data by means of the Clausius-Clapeyron equation,

$$\left(\frac{dP}{dT}\right)_{melt} = \frac{S_{sol} - S_{liq}}{V_{sol} - V_{liq}}.$$

First, since the liquid entropy is negligible at 2.7 mK, a sudden change in the slope of the melting curve, dP/dT, corresponds to a large change in the solid entropy. The estimated decrease in entropy is from $S \sim 0.6R$ above the transition to $\sim 0.45R$ below the transition. In rapid compression experiments in zero magnetic field, the narrow pressure interval over which the change takes place, less than 3×10^{-4} atm, corresponds to a temperature interval of approximately 10 μK. It is therefore quite likely that the effect corresponds to a first-order phase transition.

The second observation is that the pressure change we measure from 2.7 to 0 mK, $P(C)-P(A) = 5.22\times10^{-2}$ atm, is too large to be consistent with a simple antiferromagnetic transition at A. In fact, the pressure change we observe is twice as large as that calculated by Johnson *et al.*[8] on the basis of an antiferromagnetic transition. In order to obtain sufficient pressure change from 2.7 to 0 mK through integration of the Clausius-Clapeyron equation,

$$\Delta P = \int \frac{(S_{\rm sol} - S_{\rm liq})\,dT}{V_{\rm sol} - V_{\rm liq}} \approx \int \frac{S_{\rm sol}}{\Delta V}\,dT,$$

to agree with the value presented above, one is forced to hold the solid entropy nearly constant over a broad temperature region below the 2.7-mK transition temperature. This possible behavior of the solid entropy is, in fact, also suggested by the nearly constant slope of $P(t)$ between A and B in Fig. 2. We know of no physical system which furnishes a precedent for the entropy behavior we postulate here.

Finally, in the experiment of Sites *et al.*[4] and in subsequent work, we have measured values of the nuclear magnetic susceptibility of solid He^3 which are too large to be consistent with an antiferromagnetic transition at 2.7 mK. Perhaps the transition at 2.7 mK corresponds to a crystallographic phase change, induced by a complex coupling between the spin and phonon systems. The new crystallographic phase might have quite different magnetic properties from the high-temperature phase. The sharp break in the warming curve of the platinum thermometer shown in Fig. 3 seems to support this hypothesis. The thermometer can only measure the temperature of the phonon bath surrounding it. In the disordered solid, the temperature of the phonon bath is most strongly influenced by the temperature at the solid liquid interface. Once the liquid is below 2.7 mK, the phonon system becomes locked by spin interactions to the temperature of the interface between the two solid phases. The fact that the limiting temperature measured by the platinum thermometer was never below 2.7 mK suggests that the solid phase interface never reached the platinum wires.

We wish to thank Linton Corruccini for his assistance in the construction of the epoxy compression cell, and William Halperin for assistance in the design of the strain gauge. We would like to acknowledge many helpful conversations about these experiments with Professor John Reppy, Professor Neil Ashcroft, Professor Michael Fisher, Professor Robert Guyer, Professor James Krumhansl, and Professor John Wilkins.

---

†Work supported in part by the National Science Foundation under Grant No. GP-24179, and by the Advanced Research Projects Agency through the Cornell Materials Science Center.

[1]See, for example, the summary given by R. A. Guyer, R. C. Richardson, and L. I. Zane, Rev. Mod. Phys. 43, 532 (1971).

[2]R. T. Johnson, O. G. Symko, and J. C. Wheatley, Phys. Rev. Lett. 23, 1017 (1969).

[3]R. T. Johnson and J. C. Wheatley, J. Low Temp. Phys. 2, 423 (1970).

[4]J. R. Sites, D. D. Osheroff, R. C. Richardson, and D. M. Lee, Phys. Rev. Lett. 23, 836 (1969).

[5]For a detailed description of an early model of the cryostat used in these experiments, see J. R. Sites, Ph.D. thesis, Cornell University (unpublished).

[6]G. C. Straty and E. D. Adams, Rev. Sci. Instrum. 40, 1393 (1969).

[7]For a discussion of this technique, see R. A. Scribner and E. D. Adams, in the Fifth Symposium on Temperature, 1971, U. S. National Bureau of Standards (U. S. GPO, Washington, D. C., to be published).

[8]R. T. Johnson, O. V. Lounasmaa, R. Rosenbaum, O. G. Symko, and J. C. Wheatley, J. Low Temp. Phys. 2, 403 (1970).

[9]L. Goldstein, Phys. Rev. 171, 194 (1968).

[10]G. A. Baker, H. E. Gilbert, J. Eve, and G. S. Rushbrooke, Phys. Rev. 164, 800 (1967).

[11]R. T. Johnson, R. E. Rapp, and J. C. Wheatley, to be published. A preliminary report of this work was given by R. T. Johnson, in the Quantum Crystals Conference at Banff, Canada, September 1971 (unpublished).

Calculation of Dynamic Critical Properties Using Wilson's Expansion Methods

B. I. Halperin and P. C. Hohenberg
Bell Laboratories, Murray Hill, New Jersey 07974

and

Shang-keng Ma*
Department of Physics, University of California, San Diego, La Jolla, California 92037
(Received 10 October 1972)

The dynamic critical behavior of a continuum analog of the kinetic Ising model is studied using a generalization of Wilson's expansion methods. Results are found which disagree with the mode-mode coupling approach and the conventional (Van Hove) theory.

A major advance in our understanding of the properties of a system near its critical point has resulted from the discovery of methods for obtaining static critical exponents as power series expansions in $\epsilon = 4-d$,[1] and in $1/n$.[2] (d is the dimensionality of the system, and n the dimensionality of the order parameter.) We have extended these techniques to calculate the exponent for the time-dependent critical behavior of a simple system—the time-dependent Ginzburg-Landau (TDGL) model. This is a continuum generalization of the kinetic Ising model introduced by Glauber,[3] in which time dependence is introduced via weak coupling to an infinite heat reservoir at each lattice site. In the case where the order parameter is not conserved, we find a characteristic frequency $\omega_k \sim k^{2+c\eta}$, with $c \geq 0$ in all cases. This disagrees with the conventional (Van Hove[4]) prediction ($c=-1$), and results inferred from mode-mode coupling theories.[5] When the order parameter is conserved, on the other hand, we find $\omega_k \sim k^{4-\eta}$, in agreement with these theories.

The TDGL model is described by the equations[6]

$$\frac{\partial s_\alpha}{\partial t} = -\left(\frac{\Gamma}{k_BT}\right)\frac{\delta H}{\delta s_\alpha} + \eta_\alpha, \tag{1}$$

$$\frac{1}{k_BT}\frac{\delta H}{\delta s_\alpha} = (r_0 - \nabla^2 + 4u_0\sum_{\alpha'} s_{\alpha'}{}^2)s_\alpha - h. \tag{2}$$

Here $s_\alpha(\vec{x}, t)$ is the α component of the order parameter at position $\vec{x}$ and time t, $\vec{x}$ is a point in a d-dimensional space, and α varies from 1 to n; the Hamiltonian H is the one employed by Wilson,[1] except for the inclusion of an infinitesimal *position*- and *time-dependent* magnetic field $k_BTh(\vec{x}, t)$; and η_α is a Langevin noise source with mean zero, and correlation function

$$\langle \eta_\alpha(\vec{x}, t)\eta_{\alpha'}(\vec{x}', t')\rangle = 2\Gamma\delta(\vec{x}-\vec{x}')\delta(t-t')\delta_{\alpha\alpha'}. \tag{3}$$

It is understood that only fluctuations with wave vectors smaller than a cutoff Λ of order unity are to be included in the above, but frequencies may range from $-\infty$ to ∞. If Γ is proportional to ∇^2, then the integral over space of s_α is *conserved*, and $\partial s_\alpha/\partial t$ is described by a diffusion equation for $T > T_c$. We shall primarily discuss the case where Γ is a constant, so that the value of the $k=0$ component of s_α is *not conserved*, i.e., it relaxes at a finite frequency above T_c. In either case, the static properties are the same, and all thermodynamic properties and equal-time correlation functions are identical to those calculated by Wilson.[1]

We shall study the linear response function $\chi(k, \omega)$, which relates the expectation value of s_α to the time-dependent field h, for small values of the wave vector $\vec{k}$ and frequency ω. For $T = T_c$, according to the dynamic scaling hypothesis,[7] we expect to have

$$\chi^{-1}(k, \omega) = k^{2-\eta}f(\omega/\omega_k), \tag{4}$$

$$\omega_k \equiv \Gamma k^z, \tag{5}$$

where z is an unknown exponent, and f is a dimensionless function. Note that χ is equal to the static correlation function $g(k) \sim k^{\eta-2}$ when $\omega = 0$, so that f must approach a finite constant when its argument approaches zero. The exponent η has been calculated by Wilson[1] to be $\eta = \epsilon^2(n+2)/2(n+8)^2 + O(\epsilon^3)$. For $2 < d < 4$, Ma[2] has found

$$\eta = 4n^{-1}[(4/d)-1]\sin\pi(\tfrac{1}{2}d-1)[\pi(\tfrac{1}{2}d-1)B(\tfrac{1}{2}d-1, \tfrac{1}{2}d-1)]^{-1} + O(n^{-2}),$$

where $B(u, v)$ is the beta function.

It is convenient to write $\chi(k, \omega)$ in the form

$$\chi(k, \omega)^{-1} = G_0^{-1}(k, \omega) + \Sigma(k, \omega), \tag{6}$$

where G_0^{-1} is the value of χ^{-1} when $u_0 = 0$, namely

$$G_0^{-1} = -i\omega/\Gamma + r_0 + k^2. \tag{7}$$

A perturbation expansion for $\Sigma(k, \omega)$ can be developed, whose general structure is described in Ref. 6 for the TDGL model of a superconductor ($n = 2$). A typical diagram of order u_0^2 for the self-energy is shown in Fig. 1(a). The value of this diagram is

$$\Sigma_{(2)}(k, \omega) = 24(n+2)(4u_0)^2 \int \frac{d^d k_1}{(2\pi)^d} \frac{d^d k_2}{(2\pi)^d} \frac{d\omega_1}{2\pi} \frac{d\omega_2}{2\pi} \times G_0(\vec{k} - \vec{k}_1 - \vec{k}_2, \omega - \omega_1 - \omega_2) \frac{1}{\omega_1} \operatorname{Im} G_0(k_1, \omega_1) \frac{1}{\omega_2} \operatorname{Im} G_0(k_2, \omega_2). \tag{8}$$

Equation (8) in fact gives the entire contribution to Σ of order u_0^2, other than a term independent of k and ω, which contributes to a renormalization of T_c, but does not affect the correlation functions at T_c.

Now let us examine (4) and (8) near four dimensions. In the static case, Wilson[1] has argued that the asymptotic form of $g(k)$ at $T = T_c$ and $k \to 0$ can be obtained directly from the perturbation series for g provided one chooses $u_0 = 2\pi^2\epsilon/(n+8)$ (correct to lowest order in ϵ), and systematically expands all diagrams in powers of ϵ. We assert that the same procedure will yield the correct form for the dynamic correlation function.[8] To apply the Wilson prescription, we first note that when $r_0 = u_0 = 0$, which is the correct limit for the critical point in four dimensions, χ^{-1} is given by (7), and the exponent z of (5) has the value 2. If we take the limit where $k = 0$, while ω/Γ is small but finite, we expect *in general* that χ^{-1} will have a finite value. According to dynamic scaling [Eq. (4)], χ^{-1} will then have the form

$$\chi^{-1} = c_1 \omega^{1-\lambda}, \tag{9}$$

where $1 - \lambda = (2 - \eta)/z$. Furthermore in the limit $\epsilon \to 0$, we should have $\lambda \to 0$, and $c_1 \to i\Gamma^{-1}$.

In order to compute $\chi^{-1}(0, \omega)$ to order ϵ^2, we need only evaluate (8) at $d = 4$, with $k = 0$ and $r_0 = 0$. The leading term for small ω is proportional to $i\omega \ln\omega$. The coefficient of this term may be evaluated, and we find

$$\chi(0, \omega)^{-1} \sim -i(\omega/\Gamma)[1 - u_0^2 b_1 \ln\omega], \tag{10}$$

where $b_1 = 3(n+2)\ln(\frac{4}{3})/8\pi^4$. This is compatible with (9) and the expressions for η and u_0 if and only if $\lambda = b_1 u_0^2$, or

$$z = 2 + c\eta, \tag{11}$$

with $c = 6\ln(\frac{4}{3}) - 1$, to order ϵ^2. We note that c is independent of n in this case.

In order to check the dynamic scaling assumption (4) to order ϵ^2, we consider Eq. (8) in the limit $\omega \to 0$, $k \to 0$, with $\omega/\Gamma k^2$ finite. We find

$$\chi^{-1}(k, \omega) = k^2[1 - u_0^2 b_2 \ln k] - i(\omega/\Gamma)[1 - u_0^2 b_3 \ln k] + u_0^2 k^2 \Phi(\omega/\Gamma k^2), \tag{12}$$

where $b_2 = (n+2)/8\pi^4$, $b_3 = 2b_1$, and Φ is a regular function of its argument, in the sense that all derivatives exist at $\omega/\Gamma k^2 \to 0$. Using Wilson's expression for η and Eq. (11), it is now easy to check that Eq. (12) is consistent with the dynamic scaling assumption (4) to order ϵ^2, with the function f identified as

$$f(x) = 1 - ix + \epsilon^2[2\pi^2/(n+8)]^2\Phi(x). \tag{13}$$

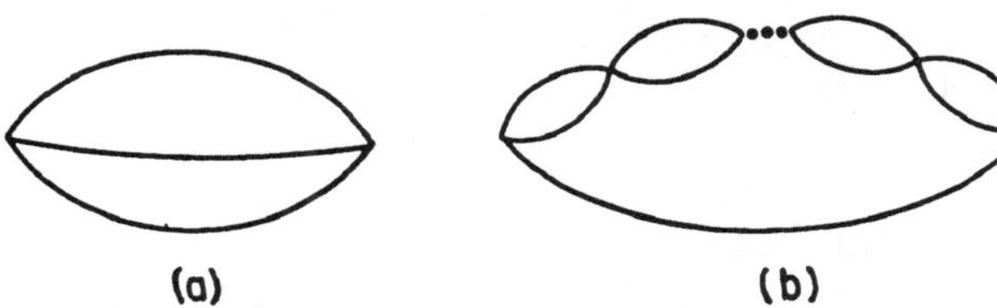

FIG. 1. (a) Diagram contributing to the self-energy to order ϵ^2. (b) General form of the self-energy to order $1/n$.

In the limit $\omega/\Gamma k^2 \to \infty$, of course, $\Phi(\omega/\Gamma k^2)$ diverges as $ib_1(\omega/\Gamma k^2)\ln(\omega/\Gamma k^2)$, in order to make contact with (9) and (10).

Let us now turn to the expansion in $1/n$. As in the static case,[2] in order to evaluate χ^{-1} to first order in $1/n$, we must sum the diagrams shown in Fig. 1(b). This yields, for $T = T_c$,

$$\mathrm{Im}\Sigma(k,\omega) = \left(\frac{2\omega}{\Gamma n}\right)\int_{-\infty}^{\infty}\frac{d\nu}{2\pi\nu}\int\frac{d^dp}{(2\pi)^d}\frac{I(p,\nu)}{[\Gamma^{-2}(\omega-\nu)^2+(\vec{k}-\vec{p})^4]}, \tag{14}$$

$$I(p,\nu) = 2\,\mathrm{Im}\left\{\int\frac{d^dq}{(2\pi)^d}q^{-2}(\vec{p}+\vec{q})^{-2}\left[\frac{q^2+(\vec{p}+\vec{q})^2}{q^2+(\vec{p}+\vec{q})^2-i\nu\Gamma^{-1}}\right]\right\}^{-1}. \tag{15}$$

Taking $k=0$, and ω small but finite, we again find $\mathrm{Im}\Sigma$ to be proportional to $\omega\ln\omega$; comparing with Eqs. (9), (4), and (5), we find Eq. (11), with

$$c = \left(\frac{4}{4-d}\right)\left\{\frac{dB(\frac{1}{2}d-1,\frac{1}{2}d-1)}{8\int_0^{1/2}dx[x(2-x)]^{d/2-2}}-1\right\}, \tag{16}$$

where we have used Ma's expression[2] for η, given in the equation preceding (6). It may be shown from Eq. (16) that for $d=4$, $c=6\ln(\frac{4}{3})-1$, in accordance with our previous result, while for $d=3$, $c=\frac{1}{2}$, and for $d=2$, $c=0$.

The case where the order parameter is *conserved* is similar to that considered above. Equations (1)–(8) are essentially unchanged, except that Γ is everywhere replaced by $\lambda_0 k^2$, where λ_0 is the "unrenormalized transport coefficient" for s. When we evaluate the self-energy in Eq. (8), however, for the case where $\omega \to 0$ and $k \to 0$, with $\omega/\lambda_0 k^2$ small but finite, we find no term of order $(\omega/\lambda_0 k^2)\ln\omega$. Thus there is no correction to the conventional form, $\chi^{-1}(0,\omega) \sim i\omega/\lambda_0 k^2$. We therefore have $\omega_k \sim \lambda_0 k^{4-\eta}$, as predicted by the conventional theory. Similarly, if we consider (8) in the case where $\omega/\lambda_0 k^4$ is a finite constant, we find

$$\Sigma(k,\omega) \sim -\eta k^2\ln k + b_4 k^2\psi(\omega/\lambda_0 k^4), \tag{17}$$

where b_4 is a constant of order ϵ^2, and ψ is a dimensionless regular function of $\omega/\lambda_0 k^4$. Again, this result shows that the conventional theory is correct in this case, and that the dynamic scaling assumption is verified to order ϵ^2.

Let us comment on the relevance of our calculations to other systems. For static properties the universality hypothesis leads one to believe that the exponents of the "Ginzburg-Landau" model[1] will be the same as those for other systems with the same symmetry and dimensionality. For dynamic critical phenomena, we again expect to have a certain degree of universality, but the classes of systems having the same exponents clearly must be smaller than for static properties alone. For example, the two systems considered here (order parameter conserved or not conserved) have identical static properties, but have different dynamic exponents. On the other hand, renormalization-group arguments[8] certainly support the idea that for $n=1$, our continuum TDGL models should have the same critical exponents as their discrete kinetic Ising model counterparts, independent of the details of the lattice, or of the coupling to the reservoirs, etc.

The TDGL models differ obviously from some of the more commonly considered models, such as the isotropic Heisenberg ferromagnet or antiferromagnet, in that they have no propagating hydrodynamic modes for the order parameter, even for $T < T_c$. The TDGL models are more analogous to anisotropic Heisenberg models, where there are also no propagating hydrodynamic modes. Examples include the uniaxial ferromagnet, where the order parameter is conserved, and models where the order parameter is not conserved, such as the uniaxial *anti*ferromagnet or the anisotropic ferromagnet without a symmetry axis. The TDGL models differ even from these, however, in that the assumption of an infinite heat reservoir at every site eliminates any effects that energy conservation might have on the dynamic critical properties. Within the spirit of mode-mode coupling theory,[5] one might ask whether the absence of a low-frequency thermal diffusion mode in the TDGL models will affect their dynamic critical exponents. In contrast to the TDGL and kinetic Ising models, the isotropic Heisenberg ferromagnet or antiferromagnet does not seem to approach a simple Gaussian fixed point for their dynamic properties as $d \to 4$ or $n \to \infty$, and the methods of the present paper are not directly applicable. For example, by studying the Heisenberg ferromagnet in the limit of long-range forces, or by using dynamic scaling, one finds that the exponent z does not reach its conventional value ($z=4$) in four dimensions.[9]

The frequencies ω_k found in the present paper

are always equal to or smaller than those predicted by the conventional theory. This is consistent with a rigorous theorem that has been proved[10] for purely dissipative models of the present type, and may be contrasted with cases such as the Heisenberg ferromagnet or antiferromagnet where the characteristic frequencies, according to dynamic scaling[7] or the mode-mode theories,[5] are always larger than the conventional predictions. A characteristic frequency slower than the conventional theory has previously been obtained for the two-dimensional kinetic Ising model by Yahata and Suzuki,[11] based on high-temperature series expansions. Specifically, these authors found a relaxation frequency $\omega \sim \kappa^2$ for the $k=0$ component of the magnetization above T_c(i.e., $z=2$), whereas the conventional theory predicts $z=\frac{7}{4}$ for this case. We note that the result $z=2$ ($c=0$) for two dimensions is precisely the one obtained in Eq. (16) to leading order in $1/n$. Recently, however, Schneider, Stoll, and Binder[12] have studied the same model using a Monte Carlo molecular-dynamics approach, and find results which agree with the conventional theory for the relaxation of the uniform mode.

*Alfred P. Sloan Foundation Fellow.

[1]K. G. Wilson and M. E. Fisher, Phys. Rev. Lett. 28, 240 (1972); K. G. Wilson, Phys. Rev. Lett. 28, 548 (1972).

[2]R. Abe, Progr. Theor. Phys. (to be published); M. Suzuki, to be published; K. G. Wilson, private communication; S. Ma, Phys. Rev. Lett. 29, 1311 (1972), and to be published; R. A. Ferrell and D. J. Scalapino, Phys. Rev. Lett. 29, 413 (1972).

[3]R. Glauber, J. Math. Phys. (N. Y.) 4, 294 (1963).

[4]L. Van Hove, Phys. Rev. 93, 1374 (1954).

[5]K. Kawasaki, Progr. Theor. Phys. 40, 706 (1968); G. E. Laramore and L. P. Kadanoff, Phys. Rev. 187, 619 (1969); E. Riedel and F. Wegner, Phys. Lett. 32A, 273 (1970).

[6]See, for instance, J. R. Tucker and B. I. Halperin, Phys. Rev. B 3, 3768 (1971), and references therein.

[7]R. A. Ferrell *et al.*, Phys. Rev. Lett. 18, 891 (1967); B. I. Halperin and P. C. Hohenberg, Phys. Rev. 177, 952 (1969).

[8]The justification for this procedure in terms of Wilson's renormalization group ideas is similar to one that can be given for the static case. In particular, we have studied the infinitesimal generator of the renormalization group for the dynamic system in the vicinity of the Gaussian fixed point [$u_0 = r_0 = 0$ in Eq. (2)], and we find that in four dimensions there is only one nontrivial zero eigenvalue, just as in the static case. The corresponding low-lying eigenvalue can be prevented from "contaminating" the ϵ expansion of the exponents by properly choosing a single parameter in the Hamiltonian, specifically by choosing u_0 according to Wilson's prescription. Details will be published elsewhere.

[9]P. C. Hohenberg, M. De Leener, and P. Résibois, to be published.

[10]K. Kawasaki, Phys. Rev. 148, 375 (1966).

[11]H. Yahata and M. Suzuki, J. Phys. Soc. Jap. 27, 1421 (1969).

[12]T. Schneider, E. Stoll, and K. Binder, Phys. Rev. Lett. 29, 1080 (1972).

Low-Temperature Studies of the Rayleigh-Bénard Instability and Turbulence

Guenter Ahlers
Bell Laboratories, Murray Hill, New Jersey 07974
(Received 20 August 1974)

Low-temperature techniques were applied to the study of a hydrodynamic instability. High-precision results for the Nusselt number N as a function of the Rayleigh number R for liquid and gaseous helium revealed no singularities in $N(R)$, except at the convective threshold R_c. For $R > 2.19R_c$, a new turbulent state was found and characterized by measuring the frequency spectrum and the amplitude of $N(R)$.

I wish to report on a number of quantitative experimental results relating to heat transport by thermal convection at low temperatures in liquid and gaseous He^4. The measurements were made on a horizontal layer of the fluid heated from below. They thus pertain to the Rayleigh-Bénard instability,[1] a particularly simple case of a hydrodynamic instability which has caused considerable interest[2-6] among physicists recently. The present work exploits some of the experimental advantages of low-temperature techniques[7] which permit thermal measurements of very high resolution and great accuracy. In addition to providing accurate measurements of the onset of convection and of the heat transport by the fluid under a wide range of conditions, the experiments reveal a transition to, and provide a quantitative description of, a new turbulent state. The properties of this state are described rather well by a theory developed recently by McLaughlin and Martin.[6]

The apparatus has been described in detail else·

where.[7] Measurements were made on samples contained in the part of the apparatus referred to as the "probe" in Ref. 7. These samples had cylindrical symmetry, with a height h of 0.088 $\pm$ 0.002 cm and a diameter D of 0.927 $\pm$ 0.002 cm. Their top and bottom boundaries were provided by isothermal copper plates having thermal relaxation times of 10^{-3} sec. The walls consisted of 0.013-cm-thick stainless steel. All heat-conductivity measurements were corrected for wall conduction. One of the advantages of the low-temperature environment is the virtual absence of any other parallel heat-transport mechanisms.

The effective thermal conductivity λ_{eff} was determined by imposing a time-independent heat current Q, and measuring the temperature increase ΔT of the bottom plate while holding constant the temperature of the top plate. One can express λ_{eff} in terms of the Nusselt number $N = \lambda_{eff}/\lambda$, where λ is the thermal conductivity of the fluid at rest. It is expected[1] that N is a function of only two independent dimensionless parameters, the Rayleigh number

$$R = g\alpha_P \Delta T h^3/\nu\kappa, \tag{1}$$

and the Prandtl number $\sigma = \nu/\kappa$.[1] Here g is the gravitational acceleration, α_P the isobaric thermal expansion coefficient, ν the kinematic viscosity, and κ the thermal diffusivity. In the present experiments, I could obtain $0.6 \lesssim \sigma \lesssim 1.4$. The results exhibited no dependence upon σ.

It is expected[1] that $\lambda_{eff} = \lambda$ for all R less than some critical value R_c because in that case the fluid remains at rest. For $R > R_c$, $\lambda_{eff} > \lambda$ because there is a contribution to the heat transport from fluid flow. Experimental values of N are shown as a function of $r \equiv R/R_c$ in Fig. 1. Although the region $r < 1$ is not adequately represented in the figure, the temperature resolution of about 10^{-7} K made it possible to measure N with a precision of 0.1% or better for $r \gtrsim 0.01$. For all $r \lesssim 0.93$, N was equal to unity within experimental error.

Much of the theory of hydrodynamic stability employs an approximation due to Boussinesq,[1] in which it is assumed that ΔT is small in the sense that all the parameters which enter into R may be regarded as constant. In order to describe the effect of departures from the Boussinesq approximation, I define a parameter $B \equiv \delta R/\langle R\rangle$. Here $\delta R = |R_b - R_t|$, where R_b and R_t correspond to the fluid at the hot and cold boundaries, respectively, and $\langle R\rangle$ is an average based upon the mean values of the properties of the fluid. By

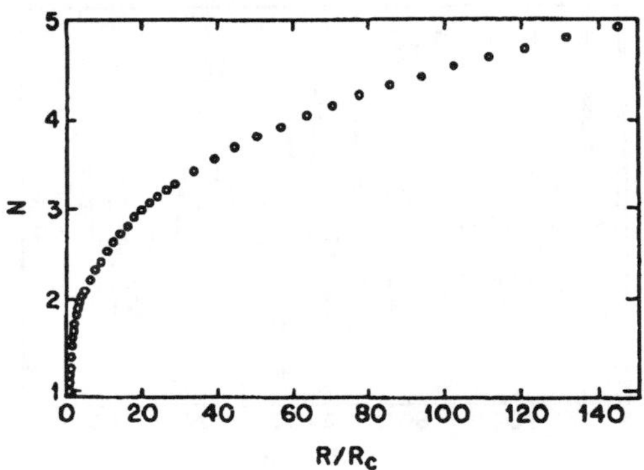

FIG. 1. The Nusselt number N as a function of the reduced Rayleigh number $r = R/R_c$.

varying the sample pressure and temperature, I could investigate the range $0.01 \lesssim B \lesssim 0.66$, and I found that $\langle R\rangle_c$ was constant within a random scatter of 1%. Systematic errors in $\langle R\rangle_c$ due to systematic errors in ν and h were large, yielding $\langle R\rangle_c = 1840 \pm 150$, consistent with the theoretical value.[1]

For the range $1.07 \leqslant r \leqslant 2.5$, the results for N in a particular sample ($B = 0.016$, $\sigma = 1.17$) could be represented within a random scatter of 0.1% by $N - 1 = f(\epsilon)$, with

$$f(\epsilon) = 1.034\epsilon + 0.981\epsilon^3 - 0.866\epsilon^5, \tag{2}$$

where $\epsilon \equiv 1 - R_c/R$. Results for all other samples differed from Eq. (2) by no more than possible systematic experimental errors, which for large B tended to be as large as 2% because of the large Q and ΔT involved in the measurement. Equation (2) can be compared with measurements for fluids with $\sigma \gtrsim 450$ by Pallas,[8] whose sample also had cylindrical symmetry.[9] His values are higher than Eq. (2) by about 9% of $N - 1$. Although the reason for the difference is not known, the large difference in σ between the two investigations should be noted. Equation (2) is consistent with the theoretical prediction that $N - 1 \sim \epsilon$,[1] but inconsistent with the analysis of recent light-scattering measurements[5] which would correspond to $N - 1 \sim \epsilon^{1.2}$.

Although one would expect Eq. (2) to be valid for $\epsilon \gtrsim 10^{-5}$,[3] I find a rather large range $0.95 \lesssim r \lesssim 1.05$ over which the data for N are "rounded." The contribution δN, equal to $N - f(\epsilon)$ for $r > 1$ and to $N - 1$ for $r \leqslant 1$, can be represented by

$$\delta N = 0.025 \exp[-(\epsilon/0.0554)^2]. \tag{3}$$

Within 0.1% of N this "rounding" is independent of B. It is possible, however, that this effect is

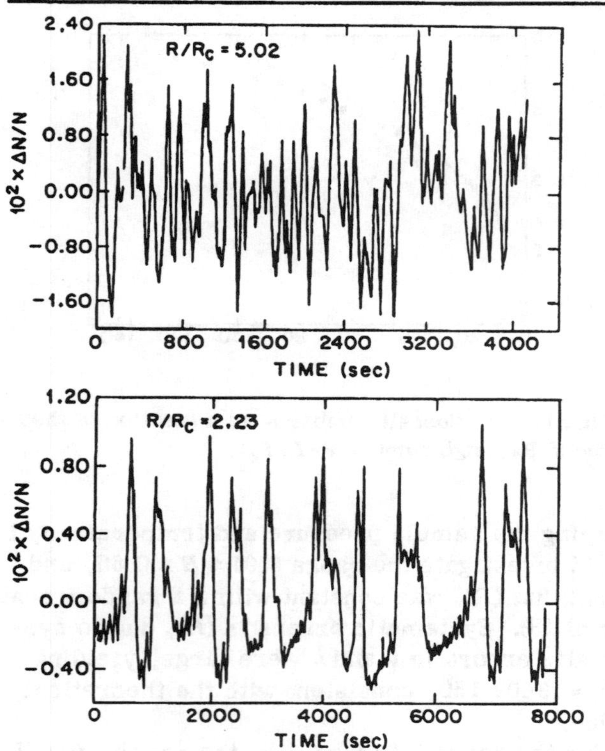

FIG. 2. Time dependence of N for two values of R/R_c.

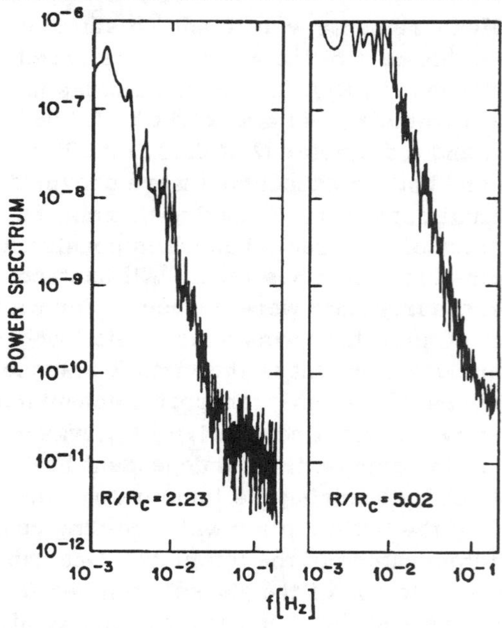

FIG. 3. The power spectrum [Fourier transform of $N(t)$] for two values of R/R_c.

peculiar to my particular samples, caused for instance by a slight variation of h in the horizontal plane.

For $30 \lesssim r \lesssim 150$, the data can be represented by

$$N-1=0.77(r-1)^{0.334}. \tag{4}$$

The exponent in Eq. (4) is somewhat higher than most of those reported previously[10]; but perhaps my value of r was not sufficiently large to determine the value pertinent to the large-r limit.

There have been persistent reports[11-13] of heat-flux transitions which manifest themselves as singularities in $N(R)$. I have examined the data by plotting the deviations from Eqs. (2) and (4) on high-resolution graphs. Within a precision of 0.1%, the data can be represented by a function with a continuous derivative. Thus, there is no evidence for discrete heat-flux transitions over the range $1<r\lesssim 150$, although with the resolution of the present experiment I should easily have seen singularities of the type reported by others.[11-13]

At $r=r_T\cong 2$, a transition occurred for all the samples from a region where N at constant heat current Q was independent of time to a region where N was time dependent. I shall refer to the time-dependent state as being turbulent. The possibility of observing the time dependence without inserting disturbing probes into the fluid is another of the advantages of the low-temperature experiment. The heat capacity of the end plates was negligible compared to that of the fluid, and the temperature of the hot plate readily followed fluctuations in the liquid. The high thermal conductivity of copper assured that the fluctuations were averaged in the horizontal plane. I investigated the turbulent state in detail at 4.515 K and at a pressure of 2.38 bar ($B=0.01$, $\nu=2.9\times10^{-4}$ cm^2/sec, $\kappa=3.36\times10^{-4}$ cm^2/sec). I found $r_T=2.19\pm0.03$. For $r>r_T$, I used the time average of N in Fig. 1 and for Eqs. (2) and (4). The fractional deviations from the mean values are shown in Fig. 2 for two values of r. Amplitudes were typically of the order of 0.5% of N, and fluctuations became more rapid with increasing r. For all $r<r_T$, there was no measurable time dependence of N although the experimental noise was about a factor of 20 smaller than the amplitude of $N(t)$ for r just above r_T. Since the time dependence was not periodic, I examined it in more detail by calculating the Fourier transform of $N(t)$. Each transform was based on approximately 10^3 data points, with sampling rates between 0.4 sec^{-1} at small r and 4 sec^{-1} at large r. Two such power spectra are shown in Fig. 3. In or-

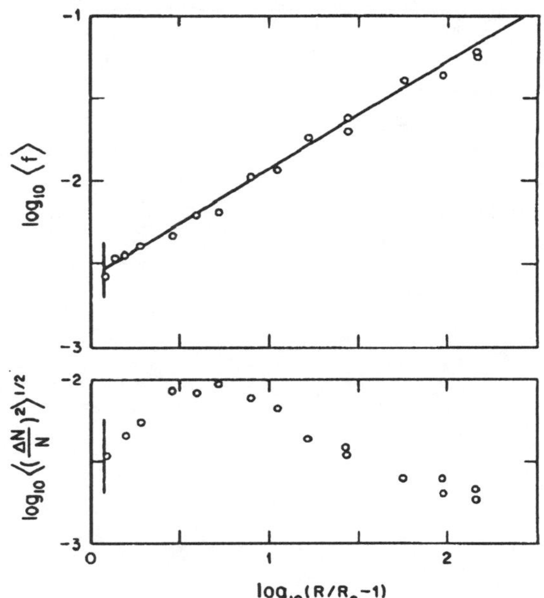

FIG. 4. The first moments $\langle f\rangle$ in hertz of the power spectra, and the rms amplitudes of $N(t)$, as a function of R/R_c-1 on logarithmic scales. To the left of the vertical lines in the figures, there was no time dependence.

der to characterize them by a single frequency, I calculated their first moments $\langle f\rangle$ which are shown in Fig. 4. They could be described by

$$\langle f\rangle = 2.7\times10^{-3}(r-1)^{0.65}\text{ Hz}. \quad (5)$$

Equation (5) extrapolates to $\langle f\rangle=0$ for $r=1$; but the observed time dependence ceased discontinuously at r_T where $\langle f\rangle = 3.0\times10^{-3}$ Hz. The amplitude of N, also shown in Fig. 4, is not readily described by a simple function; but it is also discontinuous at r_T and nonzero only for $r>r_T$.

The results for helium should be comparable to those obtained at higher temperatures for air, because air has the very similar Prandtl number $\sigma\cong0.7$. For air, singularities in $N(R)$ have been reported, and on the basis of local temperature measurements in the fluid these singularities have been associated with transitions to turbulent states.[12,13] I have seen no singularities in $N(R)$, and the values of $\langle f\rangle$ in Fig. 4, when reduced by the vertical viscous diffusion time h^2/ν, tend to be an order of magnitude smaller than qualitative frequency estimates for air based on local temperature measurements.[12] The smallest r at which a singularity in $N(t)$ has been reported for air is 3.3,[12] which is a factor of 1.5 larger than r_T in the present experiment. There seems to be no convincing evidence that the previous observations pertain to the same state of the system as the present measurements. My observations are in rather good agreement with the theoretical results of McLaughlin and Martin.[6] These authors obtain a transition to a turbulent state for $r_T=1.6$, with an amplitude of about 1.5% of N for $r>r_T$. They find a mean frequency which is within a factor of 2 or 3 of the experimental value, and very little change in the slope of $N(R)$ at r_T.

I am grateful to P. A. Fleury for calling my attention to this problem, to J. E. Graebner for the use of his automatic data acquisition system and for the calculations of the Fourier transforms, and to P. C. Hohenberg for numerous stimulating discussions.

---

[1]See, for instance, S. Chandrasekhar, *Hydrodynamic and Hydromagnetic Stability* (Clarendon Press, Oxford, England, 1961).

[2]V. M. Zaitsev and M. I. Shliomis, Zh. Eksp. Teor. Fiz. 59, 1583 (1970) [Sov. Phys. JETP 32, 866 (1971)].

[3]R. Graham, Phys. Rev. Lett. 31, 1479 (1973); W. A. Smith, Phys. Rev. Lett. 32, 1164 (1974).

[4]H. Haken, Phys. Lett. 46A, 193 (1973).

[5]P. Berge and M. Dubois, Phys. Rev. Lett. 32, 1041 (1974).

[6]J. B. McLaughlin and P. C. Martin, following Letter [Phys. Rev. Lett. 33, 1189 (1974)].

[7]G. Ahlers, Phys. Rev. A 3, 696 (1971), and 8, 530 (1973).

[8]S. G. Pallas, Ph.D. thesis, University of Texas, Austin, Texas, 1972 (unpublished).

[9]For a review, see E. L. Koschmieder, in *Advances in Chemical Physics*, edited by I. Prigogine and S. A. Rice (Wiley, New York, 1974), Vol. 26, p. 177.

[10]T. Y. Chu and R. J. Goldstein, J. Fluid Mech. 60, 141 (1973), and references therein.

[11]W. V. R. Malkus, Proc. Roy. Soc., Ser. A 225, 196 (1954).

[12]R. Krishnamurti, J. Fluid Mech. 60, 285 (1973), and references therein.

[13]W. Brown, J. Fluid Mech. 60, 539 (1973), and references therein.

Onset of Turbulence in a Rotating Fluid*

J. P. Gollub†‡ and Harry L. Swinney
Physics Department, City College of the City University of New York, New York, New York 10031
(Received 17 July 1975)

Light-scattering measurements of the time-dependent local radial velocity in a rotating fluid reveal three distinct transitions as the Reynolds number is increased, each of which adds a new frequency to the velocity spectrum. At a higher, sharply defined Reynolds number all discrete spectral peaks suddenly disappear. Our observations disagree with the Landau picture of the onset of turbulence, but are perhaps consistent with proposals of Ruelle and Takens.

Thirty years ago, Landau proposed[1] that the turbulent state of a fluid results from a large number of discrete transitions or bifurcations, each of which causes the velocity field to oscillate with a different frequency f_i, until for sufficiently large i the motion appears chaotic, although the time correlation functions $C(\tau)$ of the velocity field do not strictly go to zero as $\tau \rightarrow \infty$. The Landau picture has been presumed applicable to a large class of systems, including the rotating fluid that we have studied. Systems in a second class (which we will not mention further) exhibit inverted bifurcations, where the transition to turbulence is hysteretic, and usually no periodic regime precedes the onset of chaotic behavior.

The Landau picture has been challenged by Ruelle and Takens,[2] who propose on the basis of abstract mathematical arguments that the motion should be aperiodic with exponentially damped correlation functions after three or four bifurcations to time-dependent states. Recently McLaughlin and Martin[3] have performed numerical calculations on a truncated set of equations applicable to Rayleigh-Bénard convection, and they found a sharp transition to aperiodic behavior following a periodic regime, in qualitative agreement with the arguments of Ruelle and Takens.

A great variety of periodic and chaotic states have been observed in past experiments on rotating[4] and convecting[5,6] systems. These experiments have not examined the onset of aperiodicity in sufficient detail to distinguish between what we term the Landau and Ruelle-Takens pictures. In contrast, Ahlers[7] has recently observed and characterized a sharp transition to aperiodic behavior in sensitive heat-flux measurements on convecting liquid helium; however, the periodic states which presumably precede the transition were not observed.

We present here the first detailed measurements of a *local* property that shows a sequence of periodic regimes followed by a sharp and reversible transition to an aperiodic state, as defined by the vanishing of all discrete spectral peaks (or equivalently, the decay of the time correlation functions). Specifically, we have studied the radial velocity in a fluid rotating between concentric cylinders. The observed behavior clearly contradicts the Landau model of the onset of turbulence.

In our experiments the fluid (water) was confined between an inner rotating stainless-steel cylinder of radius $r_1 = 2.224$ cm and a stationary precision-bore glass tube of inner radius $r_2 = 2.540$ cm. The gap d was uniform to within 1% over its entire length. The fluid height in the cell was 6.25 cm and the cell temperature was 27.5 ± 0.1°C. The ten-cycle average of the rotation period T was constant to within 0.3%.

The local radial velocity V_r was observed by an optical heterodyne technique using an optical arrangement described elsewhere.[8] The scattering volume was located at the center of the gap between the cylinders, and its largest dimension was 150 μm, about 0.05 of the gap. Thus the observations are essentially local measurements, and no significant spatial averaging is involved. The time-dependent frequency of the photocurrent oscillation, which is proportional to $V_r(t)$, was

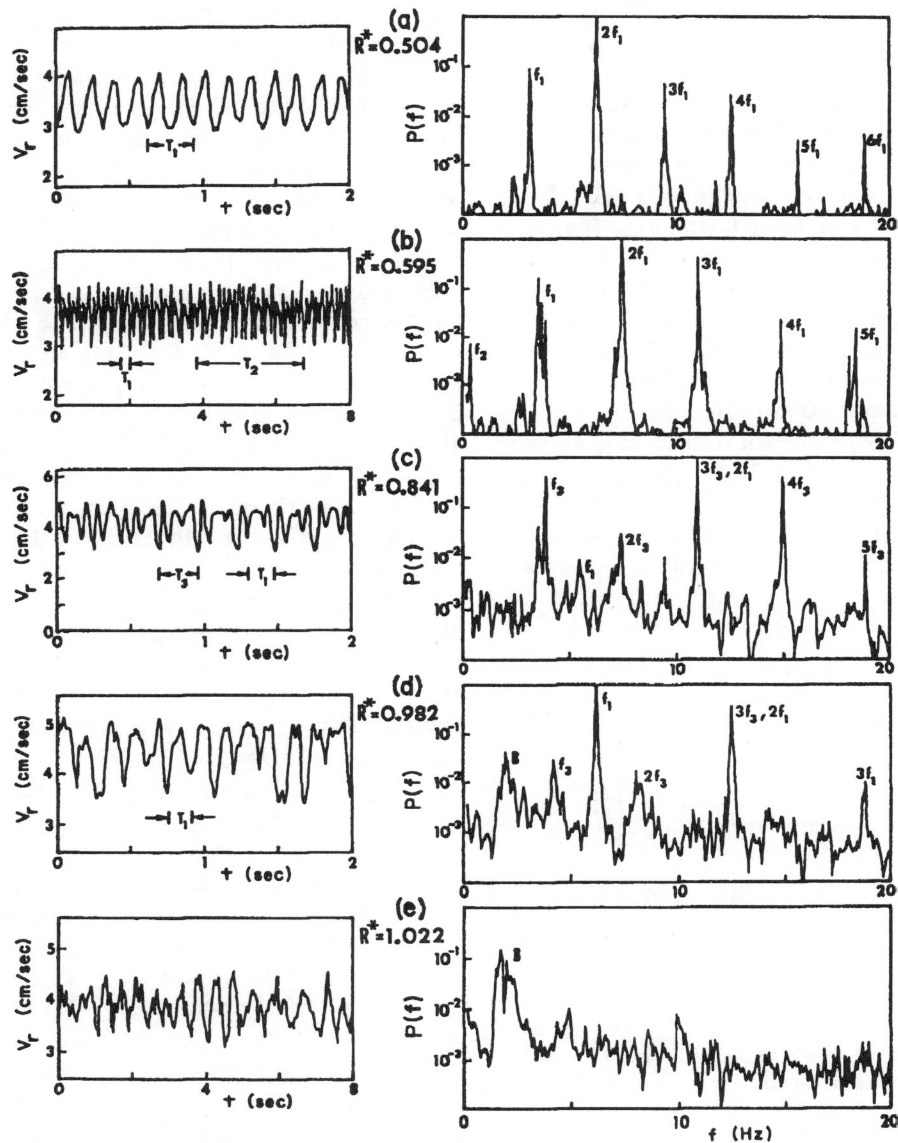

FIG. 1. Time dependence of the radial velocity and corresponding power spectra $P(f)$ [with units cm^2 sec^{-2} Hz^{-1}, normalized so that $\int_0^{25\,\mathrm{Hz}} P(f)\, df = \langle(\Delta V_r)^2\rangle$] for different reduced Reynolds numbers $R^* = R/R_T$.

measured for 1024 adjacent sampling intervals of 5×10^{-4} to 5×10^{-1} sec.

In the discussion to follow the rotation rate is expressed in terms of a reduced Reynolds number $R^*=R/R_T$, where $R=2\pi r_1 d/\nu T$ (ν is the kinematic viscosity) and $R_T=2501$ is the value of R at the onset of aperiodic motion.

We now describe the sequence of transitions which are observed reversibly as R is varied. The first instability (the Taylor instability) occurs at $R^*=0.051$, and results in a time-independent toroidal roll pattern that has been extensively studied.[4-8] The radial velocity is periodic in the axial coordinate z, with wavelength 0.79 cm. Our scattering volume was always positioned at or near one of the maxima in $V_r(z)$, and these locations persisted well into the aperiodic regime.

The first transition to a periodic state occurs at $R^*=0.064$, where transverse waves (with four wavelengths around the annulus) are superimposed on the toroidal vortices.[9] These waves, which have been previously observed visually,[4] manifest themselves as an oscillation at a frequency f_1 in our measurements, as shown in Fig. 1(a) for $R^*=0.504$. The frequency f_1 scales with $R^*\propto T^{-1}$, as Figs. 1(a)–1(d) illustrate; hence the dimensionless frequency $f_1^*\equiv f_1T$ is constant. The range in R^* of this and the subsequent time-dependent states is summarized in Fig. 2. The power spectrum $P(f)$ of the radial velocity, shown on a logarithmic scale in Fig. 1(a), contains strong peaks at the frequency f_1 and its harmonics nf_1, and the 0.05-Hz linewidth of these sharp peaks is determined only by the length of the data segment. The background noise in $P(f)$, which comprises only 2.3% of the total spectral power, is frequency independent to at least 100 Hz. This is mostly instrumental noise, with a magnitude of 10^{-4} cm^2 sec^{-2} Hz^{-1}, which corresponds to $\langle(\Delta V_r)^2\rangle^{1/2}=0.05$ cm/sec.

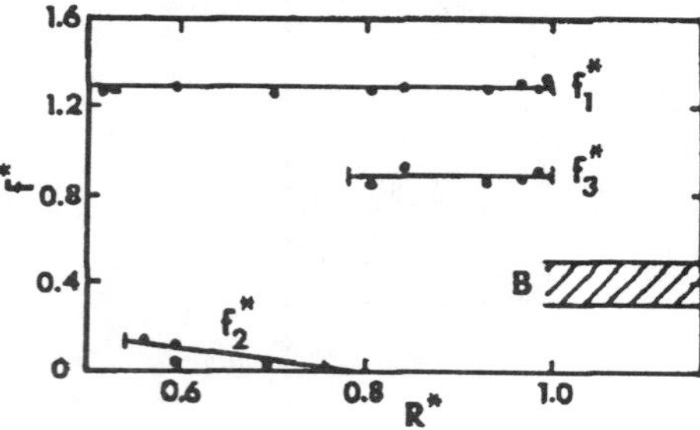

FIG. 2. The dimensionless frequencies $f_i^*=f_iT$ as a function of R^*. The solid lines are to guide the eye, and the vertical bars demarcate the regions in which the f_i are present (except that the lower bound for f_1 is $R^*=0.064$). The fact that $f_1^*=1.30$ and $f_3^*=0.87$ are constant indicates that f_1 and f_3 scale with rotation rate, whereas f_2 does not.

When R^* is increased to 0.54 ± 0.01, a second time-dependent instability occurs, and a new frequency f_2 is visible as a low-frequency modulation of the radial velocity [Fig. 1(b)]. The corresponding power spectrum shows that f_2, though weaker than f_1, is still nearly 2 orders of magnitude above the noise level. The frequency f_2 is a transverse (i.e., axial) disturbance, as is f_1, but its precise nature is unknown. It does interact with the mode at f_1 to produce a splitting of some of the nf_1 lines. Further increase in rotation rate causes f_2 to decrease in frequency until it is no longer visible at $R^*=0.78\pm0.04$ (see Fig. 2). Associated with this decrease is a gradual increase in the background noise level to about 10^{-3} cm^2 sec^{-2} Hz^{-1} in the range $0<f<60$ Hz, as in Fig. 1(c). This background now represents real noise in the fluid, but the peaks, which still contain 90% of the power, remain sharp. A new frequency f_3 (and its harmonics) appears at $R^*=0.78\pm0.03$, and this is also visible in Fig. 1(c). Note that f_3 appears only after f_2 has disappeared. Since f_3 is two-thirds of f_1, the behavior of Fig. 1(c) is periodic except for the additive noise.

Figure 1(d) shows the behavior just below the transition at $R^*=1$. The qualitative features are unchanged, and the reduction in the amplitude of the f_3 peak is caused by a small change in the vertical position of the scattering volume. The peaks in $P(f)$ (which still contain 90% of the spectral power) remain sharp, and the corresponding correlation function $C(\tau)=\langle V_r(t)V_r(t+\tau)\rangle$ oscillates periodically without any detectable decay for time lags τ up to 10 sec.

At $R^*=1$ ($R=R_T=2501$) a dramatic change occurs, as shown in Fig. 1(e). The sharp peaks at nf_1 and nf_3 disappear completely, leaving a broad doublet B that contains 60% of the power. While B was incipient at $R^*=0.982$, it contained only 5% of the power there. Higher-resolution spectra confirm that B is in fact a broad doublet with a total linewidth of about 1 Hz. The disappearance of the sharp peaks is equally apparent in the correlation function, which now decays to zero in a few seconds. The velocity data of Fig. 1(e) show erratic and noisy behavior, but since there is also some noise in 1(d), one must examine $P(f)$ or $C(\tau)$ to see the qualitative distinction. A further increase in rotation rate to $R^* = 1.16$ produced no further qualitative changes, although $P(f)$ broadens substantially.

The transition at $R^* = 1$ is sharp, reversible, and nonhysteretic to within a resolution $\delta R^* = 0.01$. Although the behavior for $R^* > 1$ is independent of the total sample height L, there is some variation with L for $R^* < 1$. However, we always detect three basic frequencies (f_1^*, f_2^*, and f_3^*) followed by a sharp onset of aperiodicity.

Our observation of a sharply defined Reynolds number at which the correlation function $C(\tau)$ decays to zero and the discrete peaks in the power spectrum $P(f)$ disappear represents the first clear demonstration that the Landau picture of the onset of turbulence is wrong. The observed behavior seems to be of the general type described by Ruelle and Takens, in which a few nonlinearly coupled modes are sufficient to produce an aperiodic motion. However, there exists no specific theoretical model applicable to this experiment.

Many questions remain unanswered. The arguments of Ruelle and Takens are quite general, and seem to apply to all systems which exhibit normal bifurcations. For these systems how universal is the behavior we have observed? What physical assumptions are inherent in the arguments of Ruelle and Takens? Finally, what is the sequence of events describing the loss of *spatial* correlation of the velocity fluctuations?

It is a pleasure to acknowledge helpful discussions with H. Z. Cummins, W. Davidon, J. Gersten, J. B. McLaughlin, and W. A. Smith.

---

*Work supported by the National Science Foundation.

†Work performed while on leave from the Physics Department, Haverford College, Haverford, Pa. 19041.

‡J.P.G. gratefully acknowledges the support of the National Oceanic and Atmospheric Administration.

[1]L. Landau, C. R. (Dokl.) Acad. Sci. URSS 44, 311 (1944); L. D. Landau and E. M. Lifshitz, *Fluid Mechanics* (Pergamon, London, England, 1959).

[2]D. Ruelle and F. Takens, Commun. Math. Phys. 20, 167 (1971). Also see R. Bowen and D. Ruelle, to be published; D. Ruelle, to be published.

[3]J. B. McLaughlin and P. C. Martin, Phys. Rev. Lett. 33, 1189 (1974), and Phys. Rev. A 12, 186 (1975).

[4]See, e.g., D. Coles, J. Fluid Mech. 21, 385 (1965); H. A. Snyder, Int. J. Non-Linear Mech. 5, 659 (1970); R. J. Donnelly and R. W. Schwarz, Proc. Roy. Soc. London, Ser. A 283, 531 (1965).

[5]See, e.g., R. Krishnamurti, J. Fluid Mech. 60, 285 (1973); G. E. Willis and J. W. Deardorf, J. Fluid Mech. 44, 661 (1970).

[6]Recent volume-averaged neutron-scattering-intensity measurements on a convecting liquid crystal are provocative but difficult to interpret. See H. B. Møller and T. Riste, Phys. Rev. Lett. 34, 996 (1975).

[7]G. Ahlers, Phys. Rev. Lett. 33, 1185 (1974).

[8]J. P. Gollub and M. H. Freilich, Phys. Rev. Lett. 33, 1465 (1974).

[9]The variety of axial and azimuthal wavelengths observed by Coles were avoided by using a shorter cell and always exceeding $R^* = 1$ before taking data.

Theory of Two-Dimensional Melting

B. I. Halperin and David R. Nelson
Department of Physics, Harvard University, Cambridge, Massachusetts 02138
(Received 17 May 1978)

The consequences of a theory of dislocation-mediated two-dimensional melting are worked out for triangular lattices. Dissociation of dislocation pairs first drives a transition into a "liquid crystal" phase with exponential decay of translational order, but power-law decay of sixfold orientational order. A subsequent dissociation of *disclination* pairs at a higher temperature then produces an isotropic fluid. The critical behavior, as well as the effect of a periodic substrate, is discussed.

Kosterlitz and Thouless[1] have proposed a model of two-dimensional melting, in which the "topological order" of a solid phase is destroyed by the dissociation of dislocation pairs. Similar ideas,[1] with vortices taking the place of dislocations, have led to a rather detailed theory[2] of the superfluid transition in two dimensions. In this Letter, we summarize the consequences of dislocation-mediated melting of triangular lattices, on both smooth and periodic substrates. A more detailed derivation will be given elsewhere.[3]

Consider the properties of a two-dimensional triangular solid on a smooth substrate. By definition, the solid has nonzero long-wavelength elastic constants. The structure factor exhibits[4] power-law singularities, $S(\vec{q}) \sim |\vec{q}-\vec{G}|^{2+\eta_{\vec{G}}}$, near a set of reciprocal lattice vectors $\{\vec{G}\}$, with exponents $\eta_{\vec{G}}$ related to the Lamé elastic constants $\mu_R(T)$ and $\lambda_R(T)$ by $\eta_{\vec{G}} = k_B T|\vec{G}|^2(3\mu_R+\lambda_R)/4\pi\mu_R(2\mu_R+\lambda_R)$. These singularities, which replace the δ-function Bragg peaks found in three-dimensional solids, reflect power-law decay at large distances of the correlation function $\langle\exp\{i\vec{G}\cdot[\vec{u}(\vec{r})-\vec{u}(\vec{0})]\}\rangle$, where $\vec{u}(\vec{r})$ is the lattice displacement at point $\vec{r}$. One can also define an order parameter (analogous to $e^{i\vec{G}\cdot\vec{u}}$) for bond orientations, namely $\psi \equiv e^{6i\theta}$, where $\theta(\vec{r})$ is the orientation, relative to the x axis, of a bond between two nearest-neighbor atoms at $\vec{r}$. In a solid, θ is given in terms of the displacement field, $\theta = \frac{1}{2}(\partial_y u_x - \partial_x u_y)$. The solid phase exhibits long-range orientational order, since $\langle\psi^*(\vec{r})\psi(\vec{0})\rangle$ approaches a nonzero constant at large $\vec{r}$.[5]

If melting is indeed characterized by an unbinding of dislocation pairs at a temperature T_m, one expects that a density $n_f(T)$ of free dislocations above T_m will lead to exponential decay of the translational order parameter $e^{i\vec{G}\cdot\vec{u}}$, with a correlation length $\xi_+(T) \approx n_f^{-2}$. This length diverges as $T \to T_m^+$ [see (6) below]. The structure factor $S(\vec{q})$ is now finite at all Bragg points, and the Lamé coefficients vanish at long wavelengths. We shall see, however, that orientational order persists, in the sense that bond-angle correlations now decay algebraically, $\langle\psi^*(\vec{r})\psi(\vec{0})\rangle \sim 1/r^{\eta_6(T)}$. This phase can be described as a liquid crystal, similar to a two-dimensional nematic, but with a sixfold rather than twofold anisotropy. The exponent $\eta_6(T)$ is related to the Franck constant $K_A(T)$, which is the coefficient of $\frac{1}{2}|\nabla\theta|^2$ in the free-energy density: $\eta_6(T) = 18k_BT/\pi K_A(T)$. We find that K_A is infinite just above T_m, but decreases with increasing temperatures, until a temperature T_i, where dissociation of *disclination* pairs drives a transition into an isotropic phase in which both the translational and orientational order decay exponentially.

The liquid-crystal phase is isomorphic to a two-dimensional superfluid, except that $\pm 60°$ disclinations play the role of vortices. The transition at T_i should belong to the same universality class as the superfluid transition, and we expect, in particular, that[2] $\eta_6(T_i) = \frac{1}{4}$. Although disclination pairs are very tightly bound in the solid phase, screening by a gas of free dislocations produces a weaker logarithmic binding for $T_m < T < T_i$. It is interesting to note that an isolated dislocation can itself be regarded as a tightly bound disclination pair,[6] separated by one lattice constant.

To see the origin of these results, let us decompose the displacement field of a solid into a smoothly varying phonon field $\vec{\varphi}(\vec{r})$, and a part due to dislocations.[1] The Hamiltonian $\mathcal{H}_E$ for the solid, within continuum elasticity theory,[6] then breaks into two parts, $\mathcal{H}_E = \mathcal{H}_0 + \mathcal{H}_D$, with

$$\frac{\mathcal{H}_0}{k_BT} = \frac{1}{2}\int\frac{d^2r}{a_0^2}[2\bar{\mu}\,\varphi_{ij} + \bar{\lambda}\varphi_{ii}^2], \tag{1}$$

$$\frac{\mathcal{H}_D}{k_BT} = -\frac{1}{8\pi}\sum_{\vec{R}\neq\vec{R}'}\left[K_1\vec{b}(\vec{R})\cdot\vec{b}(\vec{R}')\ln\left(\frac{|\vec{R}-\vec{R}'|}{a}\right) + K_2\frac{\vec{b}(\vec{R})\cdot(\vec{R}-\vec{R}')\vec{b}(\vec{R}')\cdot(\vec{R}-\vec{R}')}{|\vec{R}-\vec{R}'|^2}\right] + \frac{E_c}{k_BT}\sum_{\vec{R}}|\vec{b}(\vec{R})|^2. \tag{2}$$

In (1), φ_{ij} is related to the smooth part of the displacement field, $\varphi_{ij}=\frac{1}{2}(\partial_i\varphi_j+\partial_j\varphi_i)$, and $\bar{\mu}$ and $\bar{\lambda}$ are "reduced" elastic constants, given by the usual Lamé coefficients μ and λ multiplied by the square of the lattice spacing, a_0^2, and divided by k_BT. In (2), $\vec{b}(\vec{R})$ is a dimensionless dislocation Burgers vector of the form $\vec{b}(\vec{R}) = m(\vec{R})\vec{e}_1+n(\vec{R})\vec{e}_2$, where $m(\vec{R})$ and $n(\vec{R})$ are integers, and $\vec{e}_1$ and $\vec{e}_2$ are unit vectors spanning the underlying Bravais lattice. The sums in (2) are over, say, a square mesh with spacing a of sites in physical space, and the $\vec{b}(\vec{R})$ must satisfy a vector charge neutrality condition, $\sum_{\vec{R}}\vec{b}(\vec{R})=0$. The quantities K_1 and K_2 are equal, $K_1=K_2\equiv K = 4\bar{\mu}(\bar{\mu}+\bar{\lambda})/(2\bar{\mu}+\lambda)$, and E_c is the core energy of a dislocation.

If dislocations only exist in bound pairs at low temperatures, one expects that they can be ignored, and that the long-wavelength properties of the solid will simply be given by (1), with suitably renormalized elastic constants. The properties of the solid phase quoted above follow directly from this observation.

One of us[7] has studied the properties of (2) in the absence of the dot product or angular terms ($K_2 = 0$). It was found that dislocations are indeed unimportant at low temperatures (large K_1), and that a dislocation-unbinding transition was controlled by the terminus of a fixed *surface*, parametrized by K_1 and $\vec{e}_1\cdot\vec{e}_2$. Here, we restrict ourselves to the triangular lattice ($\vec{e}_1\cdot\vec{e}_2=\frac{1}{2}$), and extend this treatment to the full dislocation Hamiltonian $\mathcal{H}_D$, taking into account the neglected angular terms. Recursion relations for K and $y \equiv \exp(-E_c/k_BT)$ can in fact be obtained rather straightforwardly, by considering the renormalization of elastic constants due to dislocation pairs, in analogy to calculations of the effect of vortices on the superfluid density in a ^{4}He film.[8] Integrating over mesh sizes between a and ae^l, we obtain partially dressed parameters $\bar{\mu}(l)$, $\bar{\lambda}(l)$, $y(l)$, and $K(l)$, which satisfy, to $O(y^2(l))$,

$$\frac{d\bar{\mu}^{-1}}{dl}=3\pi y^2e^{-K/8\pi}I_0\left(\frac{K}{8\pi}\right), \tag{3}$$

$$\frac{d[\bar{\mu}+\bar{\lambda}]^{-1}}{dl}=3\pi y^2e^{-K/8\pi}\left[I_0\left(\frac{K}{8\pi}\right)+I_1\left(\frac{K}{8\pi}\right)\right], \tag{4}$$

$$\frac{dy}{dl}=\left(2-\frac{K}{8\pi}\right)y+2\pi y^2e^{-K/16\pi}I_0\left(\frac{K}{16\pi}\right), \tag{5}$$

where $I_0(x)$ and $I_1(x)$ are modified Bessel functions of the first and second kind. We find that

$$K^{-1}(l)=\tfrac{1}{4}\{\bar{\mu}^{-1}(l)+[\bar{\mu}(l)+\bar{\lambda}(l)]^{-1}\}$$

for all l, so that its recursion relation can be obtained trivially from (3) and (4).

As in Ref. 7, $y(l)$ is driven to zero at large l, for all temperatures below a critical value T_m. Above T_m, $y(l)$ is ultimately driven toward large values and $K(l)$ is driven towards zero, an instability we associate with dislocation-pair unbinding.

Following Kosterlitz[2] and Ref. 7, we determine the behavior near T_m by studying Eqs. (3)–(5) near the critical value $K_c=16\pi$. We identify the correlation length $\xi_+(T)$ with ae^{l^*}, with l^* chosen such that $K(l^*)\approx\frac{1}{2}K_c$. In this way, we find that

$$\xi_+(T)\approx a\exp[b/(T/T_m-1)^{0.44817\cdots}], \tag{6}$$

as $T\to T_m^+$, where b is a constant, and 0.44817... can be expressed in terms of the roots of a quadratic equation with Bessel-function coefficients. The specific heat exhibits only an essential singularity, $C_p\sim\xi_+^{-2}$, while the structure factor at the Bragg points is given by $S(\vec{G})\sim\xi_+^{2-\eta_{\vec{G}}}$. Taking over the discussion for the superfluid density in Ref. 8, we find that the reduced shear modulus in the solid phase is

$$\bar{\mu}_R(T)=\lim_{l\to\infty}\bar{\mu}(l).$$

It follows from Eqs. (3)–(5) that $\mu_R(T)$ approaches a finite limiting value as $T\to T_m^-$. Just below T_m we find $\mu_R(T)=\mu_R(T_m)[1+\text{const}(T_m-T)^{0.44817\cdots}]$, with a similar result for $\lambda_R(T)$. There is a universal relationship involving the elastic constants at T_m,

$$\lim_{T\to T_m^-}\left\{\frac{1}{\mu_R(T)}+\frac{1}{\mu_R(T)+\lambda_R(T)}\right\}=\frac{a_0^2}{4\pi k_BT_m}. \tag{7}$$

This corresponds to the critical value $K_c=16\pi$, and is also suggested by the "entropy argument" of Kosterlitz and Thouless.[1]

The results for orientational correlations above T_m follow from a calculation of the Franck constant K_A:

$$\frac{k_BT}{K_A}=\lim_{q\to 0}q^2\langle\hat{\theta}(q)\hat{\theta}(-q)\rangle$$

$$=\lim_{q\to 0}\frac{q_iq_j}{q^2}\langle\hat{b}_i(q)\hat{b}_j(-g)\rangle, \tag{8}$$

where $\hat{\theta}(q)$ and $\hat{b}_i(q)$ are the Fourier-transformed orientational and Burgers-vector fields, respectively. The second line of (8) follows because the contribution of $\vec{\varphi}(\vec{r})$ to K_A^{-1} is zero and because the dislocation part of $\hat{\theta}(q)$ is just[3,6] $\hat{\theta}(q)=-iq_j\hat{b}_j(q)/q^2$. To estimate K_A just above T_m, we use its transformation properties under the renormalization group, $K_A(K(0), y(0)) = e^{2l}K_A(K(l), y(l))$. Choosing $l = l^* = \ln(\xi_+/a)$, we can evaluate K_A using Debye-Hückel theory, which amounts to treating $\vec{b}(\vec{R})$ as a continuous vector field, rather than restricting it to discrete points on a Bravais lattice. Upon Fourier transformation, $\mathcal{H}_D$ becomes

$$\frac{\mathcal{H}_D}{k_BT} = \frac{1}{2V}\sum_{\vec{q}}\left[\frac{K}{2q^2}\left(\delta_{ij} - \frac{q_iq_j}{q^2}\right) + \frac{2E_ca^2}{k_BT}\delta_{ij}\right]\hat{b}_i(\vec{q})\hat{b}_j(-\vec{q}), \tag{9}$$

where V is the volume. Since the term proportional to the transverse projection operator in (9) does not contribute to (8), one obtains $K_A(K(l^*), y(l^*)) \approx 2E_c(l^*)a^2 = O(k_BT_m)$. It follows that the physical Franck constant is $K_A \sim \xi_+^2(T)$. The algebraic decay of orientational order above T_m and the relationship between $\eta_6(T)$ and $K_A(T)$ are straightforward consequences of this result.

Many experimental investigations of two-dimensional melting are carried out on films adsorbed onto a regular substrate,[9] and so it is important to determine the effect of a periodic potential on our results. One must now distinguish between a "floating solid," characterized by power-law Bragg singularities at reciprocal lattice vectors $\{\vec{G}\}$ which vary continuously with coverage and temperature, and an "epitaxial solid," having δ-function Bragg peaks at a lattice of vectors including the substrate reciprocal lattice $\{\vec{K}\}$ as a proper subset. The floating solid should be rather similar to the solid on a smooth substrate discussed in this paper. Figure 1 shows a schematic phase diagram with fluid, floating solid, and epitaxial phases. A region of two-phase coexistence is also shown, separating epitaxial phase I from a dilute fluid or "vapor."[8,10] We expect an increasing multiplicity and complexity of epitaxial phases with decreasing temperatures.

To understand Fig. 1, consider first the effect of a weak substrate potential commensurate with the lattice of the adsorbed film. Let $\{\vec{M}\}$ be the set of vectors common to $\{\vec{G}\}$ and $\{\vec{K}\}$, and let M_0 be the minimum length of nonzero vectors in $\{\vec{M}\}$. Let us expand the potential on the reciprocal lattice $\{\vec{K}\}$, and focus our attention on

$$\mathcal{H} = \mathcal{H}_E + \sum_{|\vec{M}| = M_0} h_{\vec{M}}\sum_{\vec{r}} e^{i\vec{M}\cdot\vec{u}(\vec{r})}, \tag{10}$$

where $h_{\vec{M}}$ is the potential strength at $\vec{M}$; the terms displayed in (10) are the most important ones for weak potentials. The renormalization-group eigenvalue for $h_{\vec{M}}$ is easily shown to be $\lambda_{\vec{M}} = 2 - \frac{1}{2}\eta_{\vec{G}}|_{\vec{G}=\vec{M}}$, so that any commensurate perturbation becomes relevant at sufficiently low temperatures. If M_0 is sufficiently small ($M_0 \lesssim 8\pi/a_0$), λ_{M_0} remains relevant out to quite high temperatures and a floating solid can never exist; there is then a transition directly from a low-temperature expitaxial phase into a fluid, as shown for epitaxial phases I and III. For large enough M_0 ($M_0 \gtrsim 8\pi/a_0$), however, there is a temperature window where $\lambda_{M_0} < 0$ and $K_R > 16\pi$, indicating that a floating solid is stable to both substrate perturbations and dislocation unbinding. The dotted line in Fig. 1 shows a locus of such points, where the floating solid has the same periodicity as epitaxial phase II, which exists at lower temperatures.

It can be shown[3] that the transitions from floating solid to fluid and from floating solid to epitaxial phase II (marked A and B in Fig. 1) are both describable at long wavelengths by a Hamiltonian of the form (2) with $K_1 \neq K_2$. Indeed, these transitions are essentially dual to each other.[3] The situation is very similar to the "$\cos p\theta$" perturbations discussed by José *et al.*[11] The transition from epitaxial phase II to a floating solid at points other than B may connect two phases with different periodicities; its nature is not yet known. We expect that the transition from floating solid to fluid will be everywhere qualitatively similar to our discussion of dislocation unbinding on a smooth substrate. The orientational bias imposed by the substrate, however, should alter or eliminate the liquid-crystal isotropic transition discussed above.

We wish to emphasize in closing that we have only explored consequences of the dislocation model of melting perturbatively in $y = \exp(-E_c/$

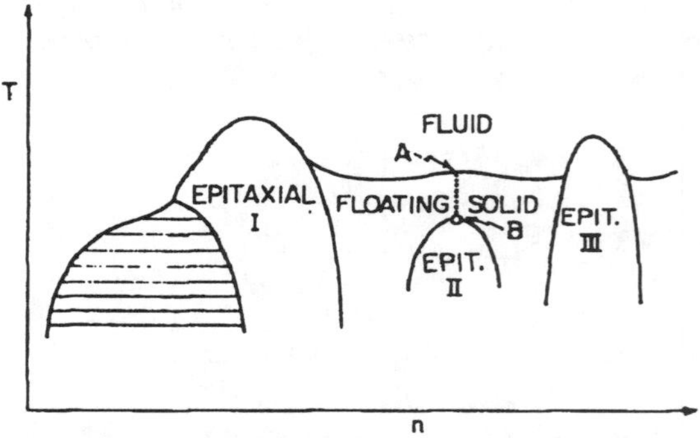

FIG. 1. Hypothetical phase diagram for a submonolayer adsorbed film on a periodic substrate as a function of density n and T.

$k_B T$). Although the theory is stable and self-consistent, we cannot rule out other mechanisms for melting, perhaps leading to a first-order transition. A "premature" unbinding of disclinations (before dislocations dissociate) might constitute such a mechanism.

We have benefitted from discussions with S. Aubry, A. N. Berker, M. Kléman, and R. Peierls. This work was supported by the National Science Foundation, under Grant No. DMR 77-10210, and by the Harvard Materials Research Laboratory program. One of us (D.R.N.) received a Junior Fellowship from the Harvard Society of Fellows.

[1]J. M. Kosterlitz and D. J. Thouless, J. Phys. C 6, 118 (1973), and to be published.

[2]J. M. Kosterlitz, J. Phys. C 7, 1046 (1974); see also J. José, L. P. Kadanoff, S. Kirkpatrick, and D. R. Nelson, Phys. Rev. B 16, 1217 (1977).

[3]D. R. Nelson and B. I. Halperin, to be published.

[4]See, e.g., Y. Imry and L. Gunther, Phys. Rev. B 3, 3939 (1971).

[5]N. D. Mermin, Phys. Rev. 176, 250 (1968).

[6]F. R. N. Nabarro, *Theory of Dislocations* (Clarendon, New York, 1967).

[7]D. R. Nelson, Phys. Rev. B (to be published).

[8]D. R. Nelson and J. M. Kosterlitz, Phys. Rev. Lett. 39, 1201 (1977).

[9]J. G. Dash, *Films on Solid Surface* (Academic, New York, 1975).

[10]See, e.g., A. N. Berker, S. Ostlund, and F. A. Putnam, Phys. Rev. B 17, 3650 (1978).

[11]José *et al.*, Ref. 2.

Chapter 7

GRAVITY PHYSICS AND COSMOLOGY

P. JAMES E. PEEBLES

Introduction

Papers Reprinted in Book

Fritz Zwicky (1898-1974) at California Institute of Technology. (Courtesy of AIP Emilio Segrè Visual Archives.)

P. James Peebles is Professor of Physics at Princeton University, where he has been, with sabbatical time off for good behavior, since he arrived from the University of Manitoba as a graduate student in 1958. He is the author of three books on cosmology.

Robert V. Pound (1919-). (Courtesy of Harvard University News Office, AIP Emilio Segrè Visual Archives.)

Our standard gravity theory, general relativity, was discovered in the second decade of this century. The basis for the now-standard world picture, the relativistic evolving universe, was in place by the end of the third decade. What is new since then is the development of the evidence and physical understanding that established these theories as believable approximations to reality.

When Einstein found general relativity theory, his checks were its consistency with Newtonian gravity as a limiting case and with the Newtonian anomaly in the advance of the perihelion of Mercury. The deflection of light by the Sun was observed soon after, but this is an exceedingly difficult measurement that gives at best a crude check of the predicted factor of 2 difference from the naïve Newtonian calculation. Many steps in the development of the tests of general relativity appear in *The Physical Review*. For example, radio interferometers gave the first 10% measurements showing consistency of the gravitational deflection with the general relativity prediction [Seielstad *et al.*, Phys. Rev. Lett. **24**, 1373 (1970); Muhleman, *et al.*, *ibid.* **24**, 1377 (1970)]. Pound and Rebka [Phys. Rev. Lett. **4**, 337 (1960)] obtained the first laboratory detection of the gravitational redshift, using the Mössbauer effect, and Vessot *et al.* [Phys. Rev. Lett. **45**, 2081 (1980)] used atomic clocks in a suborbital spacecraft to show that the gravitational redshift agrees with the weak equivalence principle to better than one part in 10^4. Shapiro *et al.* [Phys. Rev. Lett. **28**, 1594 (1972)] reported the development of precision planetary radar measurement of the advance in the perihelion of Mercury.

One assesses the significance of the gravity physics tests by comparing the predictions of alternative theories. Thus the precision measurement of spatial isotropy [Hughes *et al.*, Phys. Rev. Lett. **4**, 342 (1960)] tightly constrains two-tensor gravity theories

George Gamow (1904-1968) addressing Junior Academy of Science at George Washington University, Washington, D.C., 1954. (Courtesy of AIP Emilio Segrè Visual Archives.)

Paul Ehrenfest (1880-1933), Eliza Cornelis Wiersma (1901-1944), Mrs. Wiersma, and **Albert Einstein (1879-1955)** in Leiden, The Netherlands. (Courtesy of Uhlenbeck Collection, AIP Emilio Segrè Visual Archives.)

[Peebles and Dicke, Phys. Rev. **127**, 629 (1962)]. The measurements of the gravitational deflection and the perihelion advance push the parameter w in the Brans-Dicke [Phys. Rev. **124**, 925 (1961)] scalar-tensor theory into what is generally considered an uninteresting corner. (The scalar-tensor theory lives on, however, reappearing in inflation scenarios for the early universe.) It is now standard to analyze tests of gravity theories in terms of the post-Newtonian parameters (PPN) originally discussed by Eddington, and introduced to modern physics by Nordvedt [Phys. Rev. **169**, 1017 (1968)]. Will [*Theory and Experiment in Gravity Physics*, 2nd ed. (Cambridge University, Cambridge, 1993)] concludes that the gravity tests within the solar system indicate the corrections to Newtonian gravity theory of PPN order GM/Rc^2 agree with general relativity theory to about 0.2% accuracy. A preview

Fred Hoyle (1915-), Ivor Robinson (1923-), Engelbert L. Schucking (1926-), Alfred E. Schild (1921-1977), Edwin E. Salpeter (1924-), Texas Conference in Dallas, 1973. (Courtesy of E. E. Salpeter Collection, AIP Emilio Segrè Visual Archives.)

Robert H. Dicke (1916-) and **Mark Goldenberg (1935-1979).** (Photograph by Robert P. Matthews. Courtesy of Princeton University.)

Leonard I. Schiff (1915-1971). (Courtesy of Stanford University, AIP Emilio Segrè Visual Archives.)

of the remarkable precision offered by binary pulsars is given by Damour and J. H. Taylor [Phys. Rev. D **45**, 1840 (1992)].

Central to the establishment of the Friedmann-Lemaître relativistic cosmology as the standard world picture was the discovery of the 3 K thermal cosmic background radiation. Gamow [Phys. Rev. **74**, 505 (1948)] collected key pieces of the theory of light-element production in a hot, expanding universe. He pointed out that the predicted rapid expansion of the early universe would make element buildup a dynamical process, rather than the quasistatic situation previously considered. He noted that the key to element production in a hot, rapidly expanding universe is the formation of deuterium, which quickly burns to helium. From the condition that an interesting but not excessive amount of helium be produced, he obtained the entropy per baryon, at a value consistent with the observed background radiation temperature. As Alpher [Phys. Rev. **74**, 1577 (1948)] foresaw, nuclear burning in the hot early universe produces very low abundances of elements heavier than helium. The now-standard model for the origin of the heavy elements employs the Gamow-Alpher neutron capture process in exploding stars rather than in the expanding early universe. For helium the standard picture is production in the early universe, and, so far, this is not challenged by the improving measurements of element abundances.

The thermal cosmic background radiation was discovered and mapped using the Dicke radiometer described in Dicke *et al.*, Phys. Rev. **70**, 340 (1946). The paper established an upper limit of 20 K on the "radiation from cosmic matter" at microwave wavelengths. This is just a factor of 4 above the Gamow entropy, as translated to a temperature by Alpher and Herman, in papers that appeared two years after the Dicke measurement. The connection between the prediction and the method of measurement was only made two decades later, however, and the precision check that the spectrum is a Planck function has been completed only recently, with the magnificent National Air and Space Administration (NASA) Cosmic Background Explorer (COBE) satellite measurements over the peak of the blackbody function.

GRAVITY PHYSICS AND COSMOLOGY

Papers Reproduced on CD-ROM

W. Baade and F. Zwicky. Remarks on supernovae and cosmic rays, *Phys. Rev.* **46**, 76–77(L) (1934)

F. Zwicky. Nebulae as gravitational lenses, *Phys. Rev.* **51**, 290(L) (1937)

J. R. Oppenheimer and G. M. Volkoff. On massive neutron cores, *Phys. Rev.* **55**, 374–381 (1939)

J. R. Oppenheimer and H. Snyder. On continued gravitational contraction, *Phys. Rev.* **56**, 455–459 (1939)

R. H. Dicke, R. Beringer, R. L. Kyhl, and A. B. Vane. Atmospheric absorption measurements with a microwave radiometer, *Phys. Rev.* **70**, 340–348 (1946)

G. Gamow. Expanding universe and the origin of elements, *Phys. Rev.* **70**, 572–573(L) (1946)

W. M. Elsasser. Induction effects in terrestrial magnetism. Part III. Electric modes, *Phys. Rev.* **72**, 821–833 (1947)

M. V. Migeotte. Spectroscopic evidence of methane in the Earth's atmosphere, *Phys. Rev.* **73**, 519–520(L) (1948)

R. A. Alpher, H. Bethe, and G. Gamow. The origin of chemical elements, *Phys. Rev.* **73**, 803–804(L) (1948)

G. Gamow. The origin of elements and the separation of galaxies, *Phys. Rev.* **74**, 505–506(L) (1948)

R. A. Alpher. A neutron-capture theory of the formation and relative abundance of the elements, *Phys. Rev.* **74**, 1577–1589 (1948)

R. A. Alpher and R. C. Herman. Remarks on the evolution of the expanding universe, *Phys. Rev.* **75**, 1089–1095 (1949)

R. A. Alpher, J. W. Follin, Jr., and R. C. Herman. Physical conditions in the initial stages of the expanding universe, *Phys. Rev.* **92**, 1347–1361 (1953)

J. A. Wheeler. Geons, *Phys. Rev.* **97**, 511–536 (1955)

T. Regge and J. A. Wheeler. Stability of a Schwarzschild singularity, *Phys. Rev.* **108**, 1063–1069 (1957)

P. A. M. Dirac. Fixation of coordinates in the Hamiltonian theory of gravitation, *Phys. Rev.* **114**, 924–930 (1959)

J. Weber. Detection and generation of gravitational waves, *Phys. Rev.* **117**, 306–313 (1960)

R. Arnowitt, S. Deser, and C. W. Misner. Canonical variables for general relativity, *Phys. Rev.* **117**, 1595–1602 (1960)

M. D. Kruskal. Maximal extension of Schwarzschild metric, *Phys. Rev.* **119**, 1743–1745 (1960)

G. J. Wasserburg, W. A. Fowler, and F. Hoyle. Duration of nucleosynthesis, *Phys. Rev. Lett.* **4**, 112–114 (1960)

L. I. Schiff. Possible new experimental test of general relativity theory, *Phys. Rev. Lett.* **4**, 215–217 (1960)

R. V. Pound and G. A. Rebka, Jr. Apparent weight of photons, *Phys. Rev. Lett.* **4**, 337–341 (1960)

V. W. Hughes, H. G. Robinson, and V. Beltran-Lopez. Upper limit for the anisotropy of inertial mass from nuclear resonance experiments, *Phys. Rev. Lett.* **4**, 342–344 (1960)

H.-Y. Chiu and P. Morrison. Neutrino emission from black-body radiation at high stellar temperatures, *Phys. Rev. Lett.* **5**, 573–575 (1960)

C. Brans and R. H. Dicke. Mach's principle and a relativistic theory of gravitation, *Phys. Rev.* **124**, 925–935 (1961)

P. J. Peebles and R. H. Dicke. Significance of spatial isotropy, *Phys. Rev.* **127**, 629–631 (1962)

R. P. Kerr. Gravitational field of a spinning mass as an example of algebraically special metrics, *Phys. Rev. Lett.* **11**, 237–238 (1963)

T. S. Jaseja, A. Javan, J. Murray, and C. H. Townes. Test of special relativity or of the isotropy of space by use of infrared masers, *Phys. Rev.* **133**, A1221–A1226 (1964)

K. C. Turner and H. A. Hill. New experimental limit on velocity-dependent interactions of clocks and distant matter, *Phys. Rev.* **134**, B252–B256 (1964)

R. Penrose. Gravitational collapse and space-time singularities, *Phys. Rev. Lett.* **14**, 57–59 (1965)

B. S. DeWitt. Quantum theory of gravity. I. The canonical theory, *Phys. Rev.* **160**, 1113–1148 (1967)

L. B. Kreuzer. Experimental measurement of the equivalence of active and passive gravitational mass, *Phys. Rev.* **169**, 1007–1012 (1968)

K. Nordtvedt, Jr. Equivalence principle for massive bodies. II. Theory, *Phys. Rev.* **169**, 1017–1025 (1968)

I. I. Shapiro, G. H. Pettengill, M. E. Ash, M. L. Stone, W. B. Smith, R. P. Ingalls, and R. A. Brockelman. Fourth test of general relativity: Preliminary results, *Phys. Rev. Lett.* **20**, 1265–1269 (1968)

A. C. Cheung, D. M. Rank, C. H. Townes, D. D. Thornton, and W. J. Welch. Detection of NH_3 molecules in the interstellar medium by their microwave emission, *Phys. Rev. Lett.* **21**, 1701–1705 (1968)

C. W. Misner. Mixmaster universe, *Phys. Rev. Lett.* **22**, 1071–1074 (1969)

J. Weber. Evidence for discovery of gravitational radiation, *Phys. Rev. Lett.* **22**, 1320–1324 (1969)

G. A. Seielstad, R. A. Sramek, and K. W. Weiler. Measurement of the deflection of 9.602-GHz radiation from 3C279 in the solar gravitational field, *Phys. Rev. Lett.* **24**, 1373–1376 (1970)

D. O. Muhleman, R. D. Ekers, and E. B. Fomalont. Radio interferometric test of the general relativistic light bending near the Sun, *Phys. Rev. Lett.* **24**, 1377–1380 (1970)

J. L. Snider. New measurement of the solar gravitational red shift, *Phys. Rev. Lett.* **28**, 853–856 (1972)

I. I. Shapiro, G. H. Pettengill, M. E. Ash, R. P. Ingalls, D. B. Campbell, and R. B. Dyce. Mercury's perihelion advance: Determination by radar, *Phys. Rev. Lett.* **28**, 1594–1597 (1972)

R. Cowsik and J. McClelland. An upper limit on the neutrino rest mass, *Phys. Rev. Lett.* **29**, 669–670 (1972)

GRAVITY PHYSICS AND COSMOLOGY

J. D. Bekenstein. Black holes and entropy, *Phys. Rev. D* **7**, 2333–2346 (1973)

D. P. Woody, J. C. Mather, N. S. Nishioka, and P. L. Richards. Measurement of the spectrum of the submillimeter cosmic background, *Phys. Rev. Lett.* **34**, 1036–1039 (1975)

L. Koester. Verification of the equivalence of gravitational and inertial mass for the neutron, *Phys. Rev. D* **14**, 907–909 (1976)

J. G. Williams, R. H. Dicke, P. L. Bender, C. O. Alley, W. E. Carter, D. G. Currie, D. H. Eckhardt, J. E. Faller, W. M. Kaula, J. D. Mulholland, H. H. Plotkin, S. K. Poultney, P. J. Shelus, E. C. Silverberg, W. S. Sinclair, M. A. Slade, and D. T. Wilkinson. New test of the equivalence principle from lunar laser ranging, *Phys. Rev. Lett.* **36**, 551–554 (1976)

I. I. Shapiro, C. C. Counselman III, and R. W. King. Verification of the principle of equivalence for massive bodies, *Phys. Rev. Lett.* **36**, 555–558 (1976)

B. W. Lee and S. Weinberg. Cosmological lower bound on heavy-neutrino masses, *Phys. Rev. Lett.* **39**, 165–168 (1977)

G. F. Smoot, M. V. Gorenstein, and R. A. Muller. Detection of anisotropy in the cosmic blackbody radiation, *Phys. Rev. Lett.* **39**, 898–901 (1977)

S. Dimopoulos and L. Susskind. Baryon number of the universe, *Phys. Rev. D* **18**, 4500–4509 (1978)

M. Yoshimura. Unified gauge theories and the baryon number of the universe, *Phys. Rev. Lett.* **41**, 281–284 (1978)

S. Tremaine and J. E. Gunn. Dynamical role of light neutral leptons in cosmology, *Phys. Rev. Lett.* **42**, 407–410 (1979)

A. Brillet and J. L. Hall. Improved laser test of the isotropy of space, *Phys. Rev. Lett.* **42**, 549–552 (1979)

S. Weinberg. Cosmological production of baryons, *Phys. Rev. Lett.* **42**, 850–853 (1979)

J. P. Preskill. Cosmological production of superheavy magnetic monopoles, *Phys. Rev. Lett.* **43**, 1365–1368 (1979)

R. F. C. Vessot, M. W. Levine, E. M. Mattison, E. L. Blomberg, T. E. Hoffman, G. U. Nystrom, B. F. Farrel, R. Decher, P. B. Eby, C. R. Baugher, J. W. Watts, D. L. Teuber, and F. D. Wills. Test of relativistic gravitation with a space-borne hydrogen maser, *Phys. Rev. Lett.* **45**, 2081–2084 (1980)

A. H. Guth. Inflationary universe: A possible solution to the horizon and flatness problems, *Phys. Rev. D* **23**, 347–356 (1981)

A. Vilenkin. Cosmological density fluctuations produced by vacuum strings, *Phys. Rev. Lett.* **46**, 1169–1172 (1981)

G. G. Luther and W. R. Towler. Redetermination of the Newtonian gravitational constant G, *Phys. Rev. Lett.* **48**, 121–123 (1982)

A. Albrecht and P. J. Steinhardt. Cosmology for grand unified theories with radiatively induced symmetry breaking, *Phys. Rev. Lett.* **48**, 1220–1223 (1982)

A. H. Guth and S-Y. Pi. Fluctuations in the new inflationary universe, *Phys. Rev. Lett.* **49**, 1110–1113 (1982)

J. P. Turneaure, C. M. Will, B. F. Farrell, E. M. Mattison, and R. F. C. Vessot. Test of the principle of equivalence by a null gravitational red-shift experiment, *Phys. Rev. D* **27**, 1705–1714 (1983)

J. B. Hartle and S. W. Hawking. Wave function of the Universe, *Phys. Rev. D* **28**, 2960–2975 (1983)

P. Sikivie. Experimental tests of the "invisible" axion, *Phys. Rev. Lett.* **51**, 1415–1417 (1983)

R. W. Hellings, P. J. Adams, J. D. Anderson, M. S. Keesey, E. L. Lau, E. M. Standish, V. M. Canuto, and I. Goldman. Experimental test of the variability of G using Viking lander ranging data, *Phys. Rev. Lett.* **51**, 1609–1612 (1983)

J. M. Weisberg and J. H. Taylor. Observations of post-Newtonian timing effects in the binary pulsar PSR 1913+16, *Phys. Rev. Lett.* **52**, 1348–1350 (1984)

PHYSICAL REVIEW VOLUME 70, NUMBERS 5 AND 6 SEPTEMBER 1 AND 15, 1946

Atmospheric Absorption Measurements with a Microwave Radiometer[1]

ROBERT H. DICKE,[2] ROBERT BERINGER,[3] ROBERT L. KYHL, AND A. B. VANE[4]
Massachusetts Institute of Technology, Cambridge Massachusetts
(Received May 18, 1946)

The absorption of microwave radiation in traversing the earth's atmosphere has been measured at three wave-lengths (1.00 cm, 1.25 cm, and 1.50 cm) in the region of a water-vapor absorption line. The measurement employs a sensitive radiometer to detect thermal radiation from the absorbing atmosphere. The theory of such measurements and the connection between absorption and thermal radiation are presented. The measured absorption together with water-vapor soundings of the atmosphere permits the calculation of the absorption coefficients at standard conditions (293°K, 1015 millibar). These are 0.011, 0.026, and 0.014 db/km/g H_2O/m^3 for the wave-lengths 1.00 cm, 1.25 cm, and 1.50 cm, respectively. These values are (50 percent) greater than those given by the theory of Van Vleck. The collision width of the line and its location are in better agreement with the theory and infra-red absorption measurement. It is also found that there is very little (<20°K) radiation from cosmic matter at the radiometer wave-lengths.

I. INTRODUCTION

THE absorption of centimeter, electromagnetic waves in atmospheric gases has received considerable attention recently. There are two known contributions to this absorption: oxygen, which has a band of resonance absorption lines in the region of ½ cm superimposed on a weak continuum extending up to long wave-length and water vapor, which has a weak absorption line at approximately 1.3 cm, and a number of stronger lines below 0.2 cm, the far tails of which contribute to the absorption in the centimeter wave-length region. Both of these were predicted theoretically by Van Vleck[5] in 1942 and their absorption coefficients were calculated subject to experimental determinations of the line widths owing to collision broadening and some uncertainty in the location of the water vapor resonance. More recently, these absorptions have been measured[6] by several methods. The present paper describes atmospheric absorption measurements initiated by one of the authors in 1944 using a microwave radiometer[7] and is principally concerned with the water vapor absorption.

Except for the O_2 band, the atmospheric absorption is too small to be easily measured with path lengths which are available in the laboratory. This has led several observers[6] to employ resonant cavities. The present experiments make use of the entire atmosphere as an absorption path, measuring the thermal radiation which is emitted by the absorbing atmosphere in accordance with the Kirchhoff-Einstein law. The absorption is then calculated from the measured thermal radiation assuming local thermodynamic equilibrium in the atmosphere.

The microwave radiometers consisted of a directional antenna which was pointed at the sky and which was connected to a special microwave receiver which gave an indication of the thermal radiation intercepted by the antenna.[8] It is convenient to consider this radiation as originating in the effective terminating impedance of the antenna and to assign to this impedance a

---

[1] This paper is based on work done for the Office of Scientific Research and Development under contract OEMsr-262 with the Massachusetts Institute of Technology.

[2] Now at Princeton University.

[3] Now at Yale University.

[4] Now at Naval Ordnance Test Station, Inyokern, California.

[5] J. H. Van Vleck, Radiation Laboratory Report 43-2 and Radiation Laboratory Report 664.

[6] J. A. Saxton, British Report RRBS 17, measured the absorption of superheated steam in a resonant cavity. J. W. Miller and R. S. Bender, Radiation Laboratory Report 729 measured the absorption in the atmosphere at 1.25 cm with radar techniques. Kellogg, Phys. Rev. 69, 694A (1946), has measured the absorption in moist air in the range 0.9 cm to 1.7 cm with a cavity resonator. R. Beringer, Radiation Laboratory Report 684, has measured the O_2 absorption in the range 0.5 to 0.6 cm with a wave guide absorption path.

[7] R. H. Dicke, Radiation Laboratory Report 787 and forthcoming article in *The Review of Scientific Instruments*.

[8] Such measurements have been made in the infra-red using somewhat different techniques. See, e.g., J. Strong, J. Opt. Soc. Am. 29, 520 (1938), J. Frank. Inst. 232, 2 (1941).

thermodynamic temperature t (°K) such that the corresponding "Johnson noise" is equal to the observed radiation.[9] If the radiation intercepted by the antenna is P ergs/sec. in the frequency range $\Delta\nu$ cycles/sec., this temperature is $t=P(k\Delta\nu)^{-1}$ °K. This will be called the noise temperature of the antenna or simply the *antenna temperature*. The connection between the antenna temperature and the thermodynamic temperatures of various absorbing systems will be discussed in the following paragraphs.

If the atmosphere is transparent at the signal frequency of the microwave receiver, the effective antenna termination will be the stars and other cosmic matter and the antenna temperature will be characteristic of this matter. At ordinary radio frequencies, the radiation from this matter, the so-called *cosmic noise*, is quite large.[10] However, at the frequencies in question the cosmic noise was found to be very small.

If there is absorption in the atmosphere, this contributes to the antenna temperature and, if the thermodynamic temperature and distribution of absorption in the atmosphere are known, the total fractional absorption in the atmosphere can be computed from a measurement of this contribution to the antenna temperature. In general, this total absorption is made up of contributions from various atmospheric layers, each at a different temperature and pressure. It is possible, however, to reduce these data to some standard condition (say 20°C and 1015 millibar) and to find the corresponding absorption coefficient if the distribution of the absorption in the atmosphere is known. If the variation of this absorption coefficient with wave-length is measured in a region of resonance, the location, strength, and collision-width of the resonance may be deduced.

II. THEORY OF THE MEASUREMENT

Consider a matched[11] antenna connected to a lossless transmission line which is terminated in a matched load (a load impedance equal to the characteristic impedance of the transmission line, Z_0). Let this be imbedded in an absorbing medium in thermodynamic equilibrium at temperature T and of infinite extent, and let Z_0 also be at temperature T. In the frequency range ν to $\nu+\Delta\nu$ the resistor radiates an average power $kT\Delta\nu$ ergs/sec. which in turn is radiated by the antenna. In order that the second law of thermodynamics be obeyed, it is necessary that the antenna intercept an equal amount of power. The absorbing medium, being a blackbody, radiates an amount of power

$$\frac{2h\nu^3}{c^2(e^{h\nu/kt}-1)}\Delta\nu d\Omega \tag{1}$$

into the solid angle $d\nu$ in the frequency range ν to $\nu+\Delta\nu$, as given by Planck's formula. This reduces to the Rayleigh-Jeans expression

$$\frac{2kT\nu^2}{c^2}\Delta\nu d\Omega \tag{2}$$

for $h\nu/k \ll T$ which is the case of interest: (At 2.4×10^{10} cycles/sec., $h\nu/k=1.14$°K as compared with atmospheric temperatures $T\sim300$°K.) The total radiation intercepted by the antenna and in turn absorbed by Z_0 is

$$P=\tfrac{1}{2}\int\frac{2kT\nu^2}{c^2}\Delta\nu\sigma(\theta,\phi)d\Omega \tag{3}$$

in the range $\Delta\nu$, where the factor $\frac{1}{2}$ is introduced because an antenna accepts a single polarization. $\sigma(\phi,\theta)$ is the absorption cross section of the antenna in spherical coordinates. It can be shown from the reciprocity theorem for antennas that

$$\int\sigma(\theta,\phi)d\Omega=c^2/\nu^2. \tag{4}$$

Thus

$$P=kT\Delta\nu, \tag{5}$$

and the thermal radiation emitted by Z_0 is just equal to that absorbed. The temperature T is the antenna temperature in this example. Hence the antenna temperature of any blackbody is just the thermodynamic temperature of that body. Equation (5) will be recognized as a generalization of the Johnson noise formula to a system including antennas.

[9] The available "Johnson noise" power from any impedance is $kT\Delta\nu$ ergs/sec. in the frequency range ν to $\nu+\Delta\nu$ cycles/sec., where k is Boltzmann's constant.

[10] K. G. Jansky, Proc. I.R.E. 20, 1920 (1932); Proc. I.R.E. 21, 1387 (1933); Proc. I.R.E. 23, 1158 (1935); G. Reber; Proc. I.R.E. 28, 68 (1940); Proc. I.R.E. 30, 367 (1942); Astrophys. J. 91, 621 (1940). K. Franz, Hoch. tech. u. Elek: akus. 59, 1943 (1942).

[11] We use *matched* in the usual transmission line sense, e.g., a wave originating in the load and running toward the antenna is radiated into space without reflection.

When an antenna is pointed at an incompletely absorbing medium whose thermodynamic temperature is T, the antenna temperature is, of course, less than T. This is easily shown by the example of Fig. 1. Consider a matched antenna-transmission line-matched load system to be imbedded in a homogeneous, isotropic absorbing medium 1 at temperature T which is bounded at $x=l$ by a similar semi-infinite medium 2 at temperature T'. The antenna pattern, $\sigma(\theta, \phi)$, is assumed to be unidirectional along the x axis. The noise power originating in medium 2 and intercepted by the antenna is

$$kT'\Delta\nu e^{-\alpha l}, \tag{6}$$

where α is the absorption coefficient of medium 1 (defined by $P(x)=e^{-\alpha x}P(x=0)$ for a wave running in the $+x$ direction). If $T'=T$, it is clear from (5) that the total intercepted power is $kT\Delta\nu$. Therefore the noise power originating in medium 1 and intercepted by the antenna is

$$kT\Delta\nu(1-e^{-\alpha l}). \tag{7}$$

In general $T'\neq 0$ and the total noise power intercepted by the antenna is

$$P=k\Delta\nu[T-(T-T')e^{-\alpha l}]. \tag{8}$$

The quantity $[T-(T-T')e^{-\alpha l}]$ is the *antenna temperature*, t, of the system. The quantity $(1-e^{-\alpha l})$ is the *fractional absorption*, a, for the path length l in medium 1 since

$$1-e^{-\alpha l}=\frac{P(x=0)-P(x=1)}{P(x=0)} \tag{9}$$

for a wave running in the $+x$ direction.

The most important case of interest is for $T'=0$. Then the total noise power intercepted by the antenna is

$$P=kT\Delta\nu(1-e^{-\alpha l}) \tag{10}$$

$$=kT\Delta\nu a, \tag{11}$$

and the antenna temperature is

$$t=Ta. \tag{12}$$

Thus if the temperature T of such a medium is known and if the antenna temperature t is measured one immediately has the total fractional absorption a for the path length l by using (12).

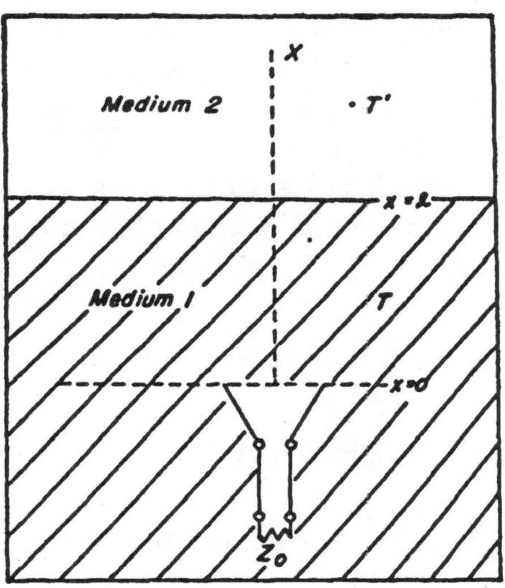

FIG. 1. The measurement of antenna temperatures in a composite absorbing system. Medium 1 is at temperature T(°K) and is bounded at $x=l$ by medium 2 which is at temperature T' and which extends to $x=\infty$.

One can also compute α if l is known. If the absorption is small $a\cong\alpha l$.

In general the absorption coefficient, $\alpha(x)$, and the temperature, $T(x)$, of medium 1 may be functions of x. In the case of $T'\neq 0$ the total noise power intercepted by the antenna is

$$P=k\Delta\nu T'\exp\left[-\int_0^l \alpha(x)dx\right]$$
$$+k\Delta\nu\int_0^l \alpha(x)T(x)dx\exp\left[-\int_0^x \alpha(x)dx\right]. \tag{13}$$

In the measurements reported here the antenna temperature t is measured for the zenith direction and at several angles (48.2°, 60°, 66.5°) from the vertical. This is done for several reasons. First; by reason of its construction the apparatus measures the difference between t and the ambient temperature of the apparatus (say, 300°K). As the absorption is weak, t at the zenith is quite small (say, 30°K), and a small percentage error in the measurement gives a large error in t. The change in t with zenith angle can be measured with considerably greater precision and, under suitable conditions, can be used to compute the absorption. Also, this change in t with angle enables one to evaluate the contribution due to the cosmic noise.

An exact analysis of the antenna temperature as a function of tipping angle is easily made if we assume the atmosphere to be of uniform temperature throughout. Nothing need be assumed about the distribution of the absorption so long as it is horizontally stratified. We shall, for the moment, assume the cosmic noise to be zero. If the absorbing layer is of height h, the antenna temperature at the zenith ($\theta=0$) is, from (13),

$$t_0=T\int_0^h \alpha(x)e^{-\int_0^x \alpha(x)dx}dx, \qquad (14)$$

and at a tipping angle θ, it is

$$t_\theta=T\sec\theta\int_0^h \alpha(x)e^{-\sec\theta\int_0^x \alpha(x)dx}dx. \qquad (15)$$

From (14) and (15)

$$\frac{t_\theta-t_0}{T}=\left(1-\frac{t_0}{T}\right)-\left(1-\frac{t_0}{T}\right)^{\sec\theta} \qquad (16)$$

At $\theta=60°$, (16) can be solved for

$$\frac{t_0}{T}=\frac{1}{2}-\frac{1}{2}\left[1-4\left(\frac{t_{60}-t_0}{T}\right)\right]^{\frac{1}{2}}. \qquad (17)$$

This quantity t_0/T is the fractional absorption a_0 along a path normal to the earth's surface. For $t_0/T\ll 1$, the function (16) is linear in sec θ.

Of course, in practice T is not constant throughout the absorbing region of the atmosphere. Fortunately, this variation is not large nor is it a rapid function of height and a good approximation to $(t_\theta-t_0)$ is obtained by replacing T, in the above formulae, by

$$T_m=\int_0^h T(x)\alpha(x)dx\bigg/\int_0^h \alpha(x)dx. \qquad (18)$$

T_m is called the mean temperature of the atmosphere averaged over the absorption along a vertical path and can be evaluated if the form of $\alpha(x)$ is known.

The cosmic noise is evaluated by comparing t_0 as calculated from (12) with that observed at the zenith. In the experiments such comparisons showed no systematic differences so that the cosmic noise is presumed to be negligible to these frequencies. However, the absolute accuracy of this result was not high (±20°K) for a number of experimental reasons. In any case, a small amount of cosmic noise if distributed uniformly in direction does not introduce much error in (11) for the values of t_0/T relevant here.

III. EXPERIMENTAL METHOD AND DATA

In the experiments three radiometers were operated simultaneously. Each was sensitive at a different wave-length in the region of the water vapor resonance; namely, at 1.00 cm, 1.25 cm, and 1.50 cm. All three had sensitive band widths, (total acceptance band widths to ½-power points), $\Delta\nu$, of 1.6×10^7 cycles/sec. During each run the antenna temperature at the zenith and the antenna temperature change with tipping was observed at the three wave-lengths.

The water vapor contents and temperatures of the atmosphere at various altitudes were taken from balloon and airplane soundings which were made at the time of day and in the geographical regions in which the radiometers were operated. The balloon soundings were kindly furnished to us by the AAF 26th Weather Region at Orlando, Florida. They employed commercially available "Radiosonde" equipment. In addition, the AAF personnel carried out several special flights over our Leesburg, Florida, location using airplanes equipped with the U. S. Army ML-313/AM psychrometer (wet and dry bulb thermometers in a housing which projects into the air-stream). Since these data were taken and analyzed by the AAF personnel we shall not discuss them here.

The microwave radiometer has been described elsewhere.[7] Suffice it to say that it consists of a superheterodyne receiver having a balanced, crystal mixer, a reflex-klystron local-oscillator, an intermediate-frequency amplifier of band width 8×10^6 cycles/sec. centered at 30×10^6 cycles/sec., a vacuum tube detector, and a narrow-band audio amplifier provided with a "lock-in" mixer. The antenna noise is intercepted by a tapered, rectangular horn connecting to a wave guide which carries the received signals to the receiver input. A slot in this wave guide admits a rotating, eccentric absorbing disk which periodically varies the attenuation in the wave guide from essentially zero to a large value. If then, the antenna temperature is different from the disk temperature (room temperature) the noise input to the receiver is modulated at the rotational frequency of the disk (30 cycles/sec.). Thus, the detected noise output from the intermediate-frequency amplifier contains a 30-cycle/sec. component which is amplified and beat against a 30-cycle/sec. signal (obtained from a generator attached to the shaft which drives the disk) in the lock-in mixer, producing a direct-current which is proportional to the difference between the antenna temperature and the disk temperature.

This use of a narrow channel at 30 cycles/sec. for the amplification of the noise signals has two great advantages over a simple scheme in which no modulation is employed and in which the direct-current component of the detected intermediate-frequency noise is observed. In the first place, vacuum tube amplifiers have inherent noise fluctuations which become very large as the frequency approaches zero. The modulation scheme enables one to avoid these large fluctuations and so to measure smaller changes in the antenna temperature. In the second place, the use of a 30-cycle/sec. channel enables one to stabilize the gain of the amplifier system accurately without degeneration of the 30-cycle/sec. signal. This is done by using the direct-current component of the detected noise from the intermediate-frequency amplifier to actuate a feedback system which removes all frequencies of less than 30 cycles/sec. and so stabilizes the amplifier against low frequency gain variations.

The intermediate-frequency band width is made as large as is consistent with small noise contributions by that part of the system, and the time constant of the output meter is made long (4 sec.), since under these conditions the output noise fluctuations are smallest and the useful sensitivity of the system for measuring changes in the antenna temperature is the greatest. The radiometers used had a useful sensitivity (antenna temperature change required to equal the r.m.s. noise fluctuations in the output meter) of 0.4°C. The direct-current output from the lock-in mixer is measured with a recording milliammeter which produces an inked trace which can be analyzed for the best fit with respect to the noise fluctuations over relatively long time intervals (say, 1 minute).

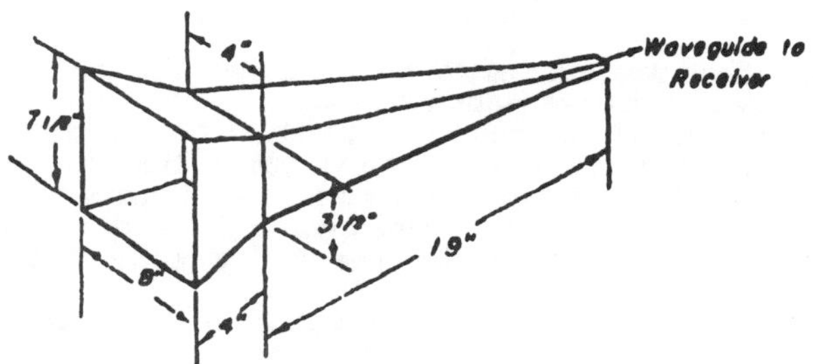

FIG. 2. Tapered, rectangular, horn antenna.

The antenna is shown in Fig. 2. The 4-inch by 3⅛-inch aperture determines the antenna's directional properties. The flared section attached to this is to reduce the interception of radiation from the back hemisphere. This is necessary because such radiation originates from warm terrestrial objects.

The radiometer is calibrated by substituting for the antenna a matched wave guide termination which can be heated to known temperatures by means of a heating coil. This is called a hot load. Such calibrations were made before and after each series of observations on the atmospheric absorption. The sensitivity did not change more than 5 percent between such measurements.

As we have mentioned, the two essential observations are the measurement of the antenna temperature at the zenith and the measurement of the change in antenna temperature with tipping. A systematic procedure was used. First, an observation at the zenith, giving a deflection of say 250°C below the disk temperature. This deflection was then balanced out with a 30-cycle/sec. signal derived from the generator, and the gain of the audio amplifier was increased by a known amount. The apparatus was then tipped successively to 48.2°, 60.0°, and 66.5° remaining at a given position for say 1 minute, and this procedure was repeated in reverse order. The balancing signal was then removed, the zenith temperature measured, and the whole procedure repeated several times.[12] A typical tipping trace is shown in Fig. 3. The tipping deflections were converted to °C with the calibration constant

[12] It should be mentioned that it was found that some cumulus clouds were quite absorbing at the radiometer frequencies. For example, on April 11, 1945, a cloud was observed having absorption of 0.56 db, 0.36 db, and 0.25 db at the three wave-lengths 1.00 cm, 1.25 cm, and 1.50 cm, respectively (assuming a cloud temperature of 10°C, calculated from the ground-level relative humidity and temperature). Consequently, the radiometer antennas were never pointed at clouds during the atmospheric absorption measurements. Also, the sun was avoided for obvious reasons.

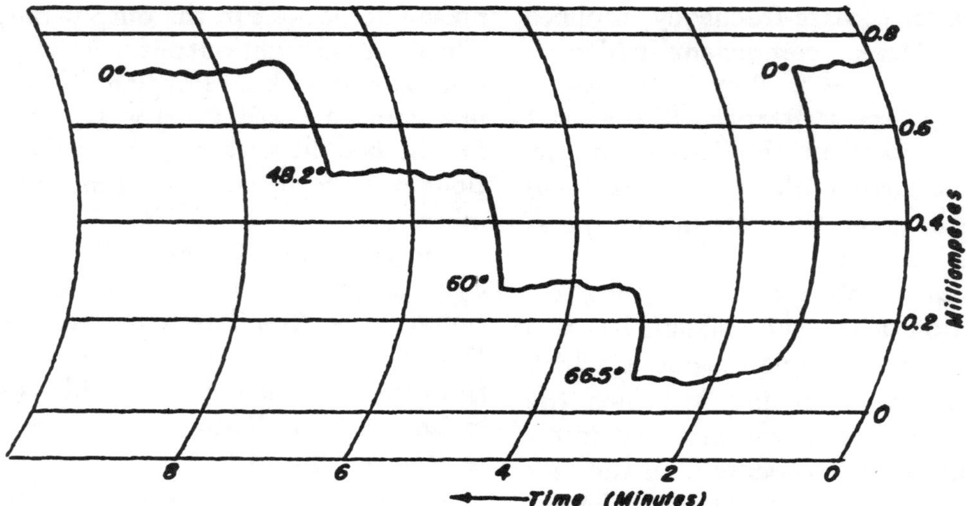

FIG. 3. Tipping trace with the 1.25-cm radiometer. (April 18, 1945.) Trace is labeled with tipping angles (measured from zenith).

deduced from the hot load calibrations[13] and the resulting deflections in °C, namely $(t_\theta-t_0)$, were plotted *vs.* sec θ as in Fig. 4. The best-fit straight line was then drawn through the three points $(t_{48.2}-t_0)$, $(t_{60}-t_0)$, and $(t_{66.5}-t_0)$ and the ordinate of this line at 60° was taken as $(t_{60}-t_0)$ in Eq. (17) from which the fractional absorption at the zenith, t_0/T, was calculated.[14,15] The values of T_m used in Eq. (17) were deduced from Eq. (18) by a method which will be discussed in the next section.[16]

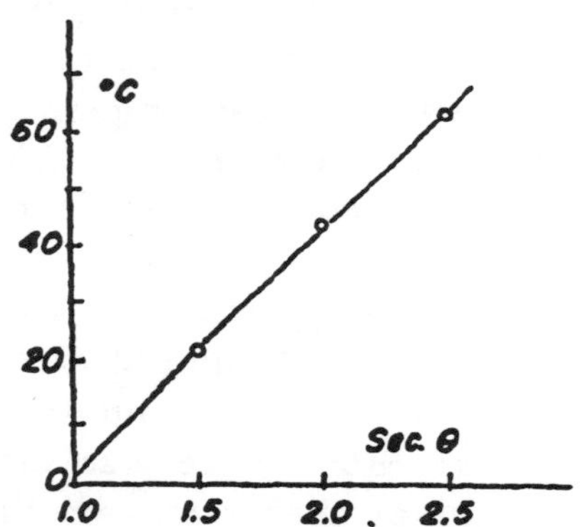

FIG. 4. Tipping deflection for the trace of Fig. 3. $t_\theta-t_0$ is expressed in °C.

The collected absorption data are shown in Table I together with other data which will be discussed in the next section. The fractional absorption is expressed in decibels (db) in Table I. It is, therefore, the total absorption in db along a vertical path through the atmosphere, i.e.,

$$L_{\text{total}}=10\log_{10}(1-t_0/T)^{-1}\text{ db.} \qquad (19)$$

IV. REDUCTION OF DATA

In order to find the absorption coefficients for water vapor (db/km path/g H_2O/m^3/at standard temperature and pressure) from the measured fractional absorptions and the water vapor soundings a rather involved procedure is required. This is because of the following reasons: (a) the fractional absorption contains a contribution due to oxygen, (b) the water vapor is distributed with altitude and hence with pressure, (c) the thermodynamic temperature of the atmosphere is a function of the altitude.

[13] A small correction is applied to the calibration constant because of the difference in the absolute antenna temperature during the calibration and during the tipping. The effect of the feedback gain stabilizing system is to increase the radiometer gain by a small amount for the tipping as compared with that for the calibration, an amount which can be calculated accurately and which did not exceed 2 percent.

[14] From Eq. (11) it is seen that $(t_\theta-t_0)$ is not linear in sec θ for finite values of t_0/T. However, the best fit straight line still passes very near to $(t_{60}-t_0)$. It does, however, have a finite intercept at $\theta=0$. Occasionally this intercept had to be adjusted to have the correct value by displacing the points of Fig. 4 up or down by equal amounts. This troublesome condition was caused by a change in radiometer zero setting in the tipping interval 0° to 48.2°. The effect was eliminated by increasing the mechanical rigidity of the wave guide circuits.

[15] Because of the finite directivity of the experimental antennas, the observed $(t_{60}-t_0)$ values differed from those given by Eq. (12), which was derived for a unidirectional antenna. This was a very small correction and was accurately calculated from the measured directivity of the antenna.

[16] A sufficiently accurate value of T is easily found by inspection of the water vapor sounding data since the range of variation of T is small compared with its mean value.

TABLE I. Total absorption, mean temperature, total water, equivalent water, and absorption coefficients computed from radiometer observations at Leesburg, Florida, in April, 1945.

	Total atmospheric absorption L_t (db) (including oxygen)			Mean temperature of atmosphere T_m (°K)			Total water G	Water vapor absorption L/G (db/kg/m²)			Equivalent water G_{eq} (kg/m²)			Absorption coefficient at S.T.P. $A(P_0)$ (db/km/g/m³)		
Date	λ=1.00 cm	λ=1.25 cm	λ=1.50 cm	λ=1.00 cm	λ=1.25 cm	λ=1.50 cm	kg/m²	λ=1.00 cm	λ=1.25 cm	λ=1.50 cm	λ=1.00 cm	λ=1.25 cm	λ=1.50 cm	λ=1.00 cm	λ=1.25 cm	λ=1.50 cm
Radiosonde runs																
April 8, 1945	0.29	0.77	0.41	285	284	284	26	0.0085	0.028	0.015	23	28	26	0.0095	0.026	0.014
9		.61	.30		284	285	20		.028	.013		21	20		.027	.013
10	.31	.65	.35	286	285	285	24	.010	.026	.013	22	26	25	.011	.024	.013
11	.29	.82	.41	287	287	287	21	.010	.037	.018	19	23	21	.011	.034	.018
12	.35	.85	.41	287	286	286	21	.013	.038	.018	19	24	22	.015	.034	.017
13	.29	.68	.37	289	287	288	26	.0085	.025	.013	23	28	26	.0097	.023	.013
14	.33	.77	.40	289	287	288	32	.0081	.023	.012	28	36	33	.0093	.020	.011
16	.33	.83	.42	292	291	292	27	.0096	.029	.014	24	29	27	.011	.027	.015
18	.44	1.07	.50	289	287	288	38	.0098	.027	.012	32	42	39	.011	.024	.012
20	.25	.52	.30	293	290	291	24	.0075	.020	.011	20	26	24	.0089	.019	.011
21	.29	.51	.28	290	286	288	25	.0088	.019	.010	22	28	25	.010	.017	.0098
23	.45	1.21	.62	289	287	288	36	.011	.032	.016	21	40	37	.012	.029	.016
24	.37	.89	.50	292	290	291	34	.0088	.025	.014	30	38	35	.010	.022	.013
25	.39	.85	.48	294	292	293	28	.011	.029	.016	24	31	28	.013	.026	.016
26	.41	1.10	.56	292	289	291	36	.0094	.029	.015	31	40	36	.011	.026	.015
27	.36	.82	.43	291	289	290	27	.011	.029	.014	24	30	28	.012	.026	.014
28	.23	.64	.33	293	291	292	26	.0062	.023	.012	23	29	26	.0070	.021	.011
													Averages	0.011	0.025	0.014
Airplane soundings																
16	0.33	0.85	0.43	288	287	287					24	29	27	0.011	0.028	0.015
27	.37	.85	.44	287	285	286					26	33	30	.012	.024	.014
28	.23	.66	.33	289	287	288					18	24	22	.0087	.026	.014
													Averages	0.011	0.026	0.014

The Oxygen Contribution

The first step is to subtract the oxygen contribution from the total fractional absorption. This is most easily done by simply plotting the observed fractional absorptions *vs.* the water vapor contents of the atmosphere (G in kg/m²) and to extrapolate this to zero water vapor; the absorption at that point being due to oxygen. This method is good when the experimental variation in G is large. For the Florida data this was not the case and another method was used. The measured L_{total} values at say 1.0 cm are plotted *vs.* those for say 1.25 cm, each point corresponding to a different day of observation. A straight line is drawn through these points. If then, it is assumed that the oxygen absorption varies in the theoretical manner,[5] namely, as $1/\lambda^2$, the intersection of this line with a line of slope $(1.00/1.25)^2$ gives the oxygen absorptions at the wave-length 1.00 cm and 1.25 cm. In principle, this procedure is more accurate than plotting L_{total} *vs.* G at each wave-length since the absolute amount of water vapor is not relevant and the only scatter is introduced by the different pressure variations of the water vapor absorptions at the three wave-lengths. The oxygen absorptions deduced in this manner are shown in Table II. The fractional absorption L for water vapor is obtained by subtracting the values of Table II from L_{total}.

The Absorption Coefficient

The water vapor contribution to the fractional absorption is made up of absorptions in various horizontally stratified layers each at a different pressure and temperature and each containing a different amount of water vapor. Unfortunately, the total absorption of these layers cannot be evaluated independently of the theory.

Before going into this matter, it is well to consider the following approximation. Consider the radio L/G in units db/kg/m². The total water vapor G is the mass of water vapor in a vertical column of one square meter cross section extending from ground level to infinity, and is calculated by simply summing the water vapor densities over the altitude. The ratio L/G, is, therefore, the observed absorption in db along a vertical path per kg/m² in the column. It is also the absorption in db along a 1 km path containing a unit density (1 g/m³) of water vapor. It is, therefore, neglecting the effects of temperature and pressure on the absorption, just the absorption coefficient (db/km/g/m³). Or, one may say that L/G is the absorption coefficient for some average atmospheric conditions. These values are shown in Table I.

The theory must now be invoked in order to reduce the data to an absorption coefficient at standard conditions (20°C, 1015 millibar). This involves assuming that the absorption coefficient depends on pressure in the theoretical manner. It is not necessary to assume the *absolute value* of the theoretical absorption coefficient. Let $A(\lambda, p)$ be the theoretical water vapor absorption coefficient[17] (db/km/g/m³) at a pressure p, a wave-length λ, and a temperature 293°K.[18]

$$A(\lambda, p) = (\text{const})\frac{v}{\lambda^2}\left\{\frac{2.65}{(1/\lambda - 1/\lambda_0)^2 + v^2} + \frac{2.65}{1/\lambda + 1/\lambda_0 + v^2} + 9.65\right\}; \quad (20)$$

$v = pB/p_0$, where p_0 is standard pressure (1015 millibar) and B is one-half of the total line width (cm^{-1}) at half-maximum due to collision broadening at a pressure p_0 and temperature 293°K. The fractional absorption (in db) at wave-length λ(cm) along a vertical path is

$$L(\lambda) = \int_0^{\infty} \rho_w A(\lambda, p)dh \quad (21)$$

$$= \frac{10}{g} \cdot \int_0^{P_0} M(p)A(\lambda, p)dp, \quad (22)$$

where g is the acceleration of gravity (cm/sec.²), ρ_w is the water vapor density (g/meter³) at a height $h(km)$, and $M(p)$ is the corresponding mixing ratio (g H_2O/kg air). Let us define a relative absorption coefficient

$$\sigma(\lambda, p) = A(\lambda, p)/A(\lambda, p_0) \quad (23)$$

at each wave-length λ. Then

$$A(\lambda, p_0) = \frac{L(\lambda)}{\frac{10}{g}\int_0^{P_0} M(p)\sigma(\lambda, p)dp} \quad (24)$$

$$= L(\lambda)/G_{eq}(\lambda), \quad (25)$$

where

$$G_{eq}(\lambda) = \frac{10}{g}\int_0^{P_0} M(p)\sigma(\lambda, p)dp \quad (26)$$

is called the *equivalent water*, being the mass (kg) of water vapor all at pressure p_0 in a column of 1 m² cross section which is required to give the observed absorption. $A(\lambda, p_0)$ is the desired absorption coefficient at standard conditions.

The integration (26) is carried out numerically using the observed distribution of water-vapor in the atmosphere and the function $\sigma(\lambda, p)$ which is calculated from (20) and (23) for assumed values of λ_0 and B. The values assumed were $B = 0.12$ cm^{-1} and $\lambda_0 = 1.33$ cm. We shall see how these may be checked in a following paragraph.

The equivalent water and the absorption coefficient at standard conditions are given in Table I for the Leesburg, Florida, observations. The absorption coefficient can also be expressed as db/kg/m² being the absorption in db along a vertical path which contains 1 kg/m² of water vapor and throughout which the pressure is 1015 millibars.

The Mean Temperature

The mean temperature defined in (18) is seen to reduce to

$$T_m = \frac{\int_{P_0}^{0} T(p)\sigma(\lambda, p)M(p)dp}{\int_{P_0}^{0} \sigma(\lambda, p)M(p)dp}, \quad (27)$$

which can be integrated numerically by use of the meteorological data. These T_m values were used

TABLE II. Deduced O_2 absorption values in db for traversal of the entire atmosphere along the radio vector (Florida data).

λ	L (oxygen)
1.00 cm	0.07 db
1.25	0.04
1.50	0.03

[17] This formula is given in reference 6. The first and second terms give the resonance absorption (line at 1.3 cm) and the last term is the effect of the tails of the infra-red lines.

[18] The variation of $A(\lambda, p)$ with atmospheric temperature is neglected for two reaons. The analytic form of this variation is very complex and the range of temperatures observed at altitudes containing an appreciable amount of water vapor was small and fairly well distributed around 293°K.

TABLE III. Theoretical absorption coefficients at the three-radiometer wave-lengths for a line half-width of 0.12 cm^{-1} located at 1.33 cm.

λ	$A(p_0)$
1.00 cm	0.0071 db/km/g/m^3
1.25	0.017
1.50	0.0092

in (17) in calculating t_0/T and are given in Table I.

The Line Width and Line Position

The assumed values of B and λ_0 can be checked by substituting the calculated values of $A(\lambda, p_0)$ into (20) and solving for B and λ_0. This was done and found to be self-consistent.

B and λ_0 were also solved for by another method. A standard atmosphere ($M(p)$ *vs.* p) was assumed which fitted much of the meteorological data quite well and for which the integrations (22) could be carried out explicitly for arbitrary λ_0 and B. Ratios of $L(\lambda)$ values at the three observation wave-lengths could then be used to find λ_0 and B. The values so found were $B=0.11$ cm^{-1}, $\lambda_0=1.34$ cm.

V. DISCUSSION

It is believed that the principal errors in the absorption coefficients, $A(p_0)$, arise from inaccuracies in the water vapor data, in particular, the "radiosonde" observations. This contention is borne out by the fact that the absorption at the three wave-lengths are relatively in much better agreement than the day to day scatter in the absorption coefficients. Also, the ascents were made at Orlando, Florida, some forty miles from the radiometer location at Leesburg, Florida in a region dotted with ponds and lakes. The airplane soundings were made above the Leesburg location using a psychrometer which is generally regarded as having high precision. It is to be noted, however, that the averages of the radiosonde data agree well with the airplane data.

The absolute agreement of the measured absorption coefficients with the theory[1] is very poor. For the observed line width and position ($\lambda_0=1.33$ cm, $B=0.12$ cm^{-1}) the theory gives the absorption coefficients shown in Table III for $p=1015$ millibar and $T=293°K$. These values include the infra-red contribution. The deduced line-width and position are, however, in good agreement with the values found by Adel and others (unpublished infra-red absorption measurements quoted by Van Vleck[5]). This agreement may have little significance, since we have used the theory generously in reducing our data and have only three points on the absorption line profile.

It is to be noted that the disagreement between the theory and the experiments can be lessened by arbitrarily adjusting the relative contributions of the line absorption and the infra-red absorption. In our experiments this procedure is not justifiable since we have data only at three frequencies all of which is required in finding the three parameters λ_0, B, and the absolute magnitude of the absorption. However, the Columbia[6] experiments on water vapor absorption include data at a number of wave-lengths, permitting such an analysis to be made. This indicates that the infra-red contribution is some four to six times the theoretical value. When this is taken into account the agreement with the theory is very good for the resonant absorption at 1.33 cm.

It is a pleasure to acknowledge the assistance of many members of this laboratory; in particular Dr. E. M. Purcell, and the close cooperation of the AAF in providing facilities at our Leesburg, Florida, location and in supplying the water-vapor data. Mr. A. E. Bent of this laboratory was most helpful in establishing liaison with the AAF.

The Origin of Elements and the Separation of Galaxies

G. GAMOW
George Washington University, Washington, D. C.
June 21, 1948

THE successful explanation of the main features of the abundance curve of chemical elements by the hypothesis of the "unfinished building-up process,"[1,2] permits us to get certain information concerning the densities and temperatures which must have existed in the universe during the early stages of its expansion. We want to discuss here some interesting cosmogonical conclusions which can be based on these informations.

Since the building-up process must have started with the formation of deuterons from the primordial neutrons and the protons into which some of these neutrons have decayed, we conclude that the temperature at that time must have been of the order $T_0 \cong 10^9$ °K (which corresponds to the dissociation energy of deuterium nuclei), so that the density of radiation $\sigma T^4/c^2$ was of the order of magnitude of water density. If, as we shall show later, this radiation density exceeded the density of matter, the relativistic expression for the expansion of the universe must be written in the form:

$$\frac{d}{dt}\lg l=\left(\frac{8\pi G}{3}\frac{\sigma T^4}{c^2}\right)^{\frac{1}{2}} \tag{1}$$

where l is an arbitrary distance in the expanding space, and the term containing the curvature is neglected because of the high density value. Since for the adiabatic expansion T is inversely proportional to l, we can rewrite (1) in the form:

$$-\frac{d}{dt}\lg T=\frac{T^2}{c}\left(\frac{8\pi G\sigma}{3}\right)^{\frac{1}{2}} \tag{2}$$

or, integrating:

$$T^2=\left(\frac{3}{32\pi G\sigma}\right)^{\frac{1}{2}}\cdot\frac{c}{t}. \tag{3}$$

For the radiation density we have:

$$\rho_{\text{rad.}}=\frac{3}{32\pi G}\cdot\frac{1}{t^2}. \tag{4}$$

These formulas show that the time t_0, when the temperature dropped low enough to permit the formation of deuterium, was several minutes. Let us assume that at that time the density of matter (protons plus neutrons) was $\rho_{\text{mat.}}{}^{\circ}$. Since, in contrast to radiation, the matter is conserved in the process of expansion, $\rho_{\text{mat.}}$ was decreasing as $l^{-3}\sim T^3\sim t^{-\frac{3}{2}}$. The value of $\rho_{\text{mat.}}{}^{\circ}$ can be estimated from the fact that during the time period Δt of about 10^2 sec. (which is set by the rate of expansion), about one-half of original particles were combined into deuterons and heavier nuclei. Thus we write:

$$v\Delta t n\sigma \cong 1 \tag{5}$$

where $v = 5\cdot 10^8$ cm/sec. is the thermal velocity of neutrons at 10^9 °K, n is the particle density, and $\sigma \cong 10^{-29}$ cm^2 the capture cross section of fast neutrons in hydrogen. This gives us $n \cong 10^{18}$ cm^{-3} and $\rho_{mat.}{}^{\circ} \cong 10^{-6}$ g/cm^3 substantiating our previous assumption that matter density was negligibly small compared with the radiation density. (Thus we have $\rho_{mat.}{}^{\circ}\cdot\Delta t \cong 10^{-4}$ g$\cdot$cm$^{-3}\cdot$sec. and not 10^{+4} g$\cdot$cm$^{-3}\cdot$sec. as was given incorrectly in the previous paper[2] because of a numerical error in the calculations.)

Since $\rho_{rad.} \sim t^{-2}$ whereas $\rho_{mat.} \sim t^{-3/2}$ the difference by a factor of 10^6 which existed at the time 10^2 sec. must have vanished when the age of the universe was $10^2\cdot(10^6)^2 = 10^{14}$ sec. $\cong 10^7$ years. At that time the density of matter and the density of radiation were both equal to $[(10^6)^2]^{-2} = 10^{-24}$ g/cm^3. The temperature at that epoch must have been of the order $10^9/10^6 \cong 10^3$ °K.

The epoch when the radiation density fell below the density of matter has an important cosmogonical significance since it is only at that time that the Jeans principle of "gravitational instability"[3] could begin to work. In fact, we would expect that as soon as the matter took over the principal role, the previously homogeneous gaseous substance began to show the tendency of breaking up into separate clouds which were later pulled apart by the progressive expansion of the space. The density of these individual gas clouds must have been approximately the same as the density of the universe at the moment of separation, i.e., 10^{-24} g/cm^3. The size of the clouds was determined by the condition that the gravitational potential on their surface was equal to the kinetic energy of the gas particles. Thus we have:

$$\frac{3}{2}kT = \frac{4}{3}\pi R^3 \rho \frac{Gm_H}{R} = \frac{4\pi G m_H \rho}{3} R^2. \tag{6}$$

With $T \cong 10^3$ and $\rho \cong 10^{-24}$ this gives $R \cong 10^{21}$ cm $\cong 10^3$ light years.

The fact that the above-calculated density and radii correspond closely to the observed values for the stellar galaxies strongly suggests that we have here a correct picture of galactic formation. According to this picture the galaxies were formed when the universe was 10^7 years old, and were originally entirely gaseous. This may explain their regular shapes, resembling those of the rotating gaseous bodies, which must have been retained even after all their diffused material was used up in the process of star formation (as, for example, in the elliptic galaxies which consist entirely of stars belonging to the population II).[4]

It may also be remarked that the calculated temperature corresponding to the formation of individual galaxies from the previously uniform mixture of matter and radiation, is close to the condensation points of many chemical elements. Thus we must conclude that some time before or soon after the formation of gaseous galaxies their material separated into the gaseous and the condensed (dusty) phase. The dust particles, being originally uniformly distributed through the entire cloud, were later collected into smaller condensations by the radiation pressure in the sense of the Spitzer-Whipple theory of star formation.[5] In fact, although there were no stars yet, there was still plenty of high intensity radiation which remained from the original stage of expanding universe when the radiation, and not the matter, ruled the things.

In conclusion I must express my gratitude to my astronomical friends, Dr. W. Baade, Dr. E. Hubble, Dr. R. Minkowski, and Dr. M. Schwartzschield for the stimulating discussion of the above topics.

[1] G. Gamow, Phys. Rev. 70, 572 (1946).
[2] R. Alpher, H. Bethe, and G. Gamow, Phys. Rev. 73, 803 (1948).
[3] J. H. Jeans, *Astronomy and Cosmogony* (Cambridge University Press, Teddington 1928).
[4] W. Baade, Astrophys. J. 100, 137 (1944).
[5] L. Spitzer, Jr., Astrophys. J. 95, 329 (1942); F. L. Whipple, Astrophys. J. 104, 1 (1946).

PHYSICAL REVIEW VOLUME 114, NUMBER 3 MAY 1, 1959

Fixation of Coordinates in the Hamiltonian Theory of Gravitation

P. A. M. DIRAC*
Institute for Advanced Study, Princeton, New Jersey
(Received December 10, 1958)

The theory of gravitation is usually expressed in terms of an arbitrary system of coordinates. This results in the appearance of weak equations connecting the Hamiltonian dynamical variables that describe a state at a certain time, leading to supplementary conditions on the wave function after quantization. It is then difficult to specify the initial state in any practical problem.

To remove the difficulty one must eliminate the weak equations by fixing the coordinate system. The general procedure for this elimination is here described. A particular way of fixing the coordinate system is then proposed and its effect on the Poisson bracket relations is worked out.

INTRODUCTION AND NOTATION

THE problem of putting Einstein's equations for the gravitational field into the Hamiltonian form, as a preliminary to quantization, has recently received a good deal of attention, because of the development of mathematical methods sufficiently powerful to make it tractable.

The Hamiltonian form involves the concept of a physical state "at a certain time," which means in a relativistic theory a state on a certain three-dimensional space-like surface in space-time. At first people[1,2] chose the space-like surface independent of the coordinates x^μ, which enabled them to preserve the four-dimensional symmetry of the equations. Later it was realized[3,4] that one could effect a substantial simplification, at the expense of giving up four-dimensional symmetry, by choosing a system of coordinates such that the three-dimensional surfaces $x^0=\text{constant}$ are all space-like and dealing with the physical states on these surfaces.

The main features of the Hamiltonian formalism will be recapitulated here. The notation will be that used by the author,[4] with the exception that the sign of the $g_{\mu\nu}$ will be changed throughout, to make g_{00} negative. Greek suffixes take on the values 0, 1, 2, 3, lower-case Roman suffixes take on the values 1, 2, 3, the determinant of the $g_{\mu\nu}$ is $-J^2$, the determinant of the g_{rs} is K^2, and the reciprocal matrix to g_{rs} is e^{rs}. A lower suffix added to a field quantity denotes an ordinary derivative, while $_{|\mu}$ added to it denotes the covariant derivative.

We shall deal with the gravitational field in interaction with other fields, or possibly particles. Spinor fields are excluded, as they require a special treatment.

* The author's stay at the Institute for Advanced Study was supported by the National Science Foundation.

1 F. A. E. Pirani and A. Schild, Phys. Rev. 79, 986 (1950).

2 Bergmann, Penfield, Schiller, and Zatzkis, Phys. Rev. 80, 81 (1950).

3 Pirani, Schild, and Skinner, Phys. Rev. 87, 452 (1952).

4 P. A. M. Dirac, Proc. Roy. Soc. (London) A246, 333 (1958).

We have an action density of the form

$$\mathcal{L}=\mathcal{L}_G+\mathcal{L}_M,$$

where $\mathcal{L}_G$ is the action density of the gravitational field alone, involving the $g_{\mu\nu}$ and their first derivatives, and $\mathcal{L}_M$ is the action density of the other fields, involving the other field quantities, q_M say, and their first derivatives and involving also the $g_{\mu\nu}$, but not derivatives of the $g_{\mu\nu}$.

The gravitational action density is

$$\mathcal{L}_G=(16\pi\gamma)^{-1}Jg^{\mu\nu}(\Gamma_{\mu\rho}{}^{\sigma}\Gamma_{\nu\sigma}{}^{\rho}-\Gamma_{\mu\nu}{}^{\rho}\Gamma_{\rho\sigma}{}^{\sigma}), \tag{1}$$

where γ is the gravitational constant, occurring in the numerator of Newton's law of force. To save writing, we shall take

$$16\pi\gamma=1. \tag{2}$$

HAMILTONIAN FORM OF GRAVITATIONAL THEORY

We shall deal with the physical state on the surface $x^0=t$ and shall set up Hamiltonian equations of motion to determine how the state varies as t varies. The Hamiltonian is, by the usual definition

$$H=\int\left(g_{\mu\nu 0}\frac{\partial\mathcal{L}_G}{\partial g_{\mu\nu 0}}+\sum q_{M0}\frac{\partial\mathcal{L}_M}{\partial q_{M0}}-\mathcal{L}\right)d^3x, \tag{3}$$

where the sum is over all the nongravitational dynamical coordinates q_M.

It is evident that there must be a good deal of arbitrariness in the equations of motion on account of the arbitrariness in the system of coordinates x^μ. In the first place we see that the $g_{\mu 0}$ can vary with t in an arbitrary way. To describe the geometry of the surface $x^0=t$ and also the system of coordinates x^r in it, we need the g_{rs} at all points on the surface, but we do not need the $g_{\mu 0}$, which refer only to intervals going outside the surface. Different values for the $g_{\mu 0}$ correspond to different choices of a neighboring surface $x^0=t+\epsilon$ and to different systems of coordinates x^r in the neighboring surface, and these are completely arbitrary with a given initial surface $x^0=t$.

We get the simplest form for the equations of motion if we describe the physical state on the surface $x^0=t$ entirely in terms of dynamical variables that are independent of the $g_{\mu 0}$. Let us consider the kind of quantities that can enter into such a description.

Suppose there is a vector field A_μ. The three covariant components A_r on the surface remain invariant under a change of coordinates which leaves the coordinates of each point on the surface invariant. So these A_r will enter into the description. We cannot have A_0, but we have instead the normal component of A, namely $A_\mu l^\mu$, where l^μ is the unit normal. Similarly for a tensor $B_{\mu\nu}$, which may be the covariant derivative $A_{\mu|\nu}$ of A_μ, we have the quantities B_{rs}, $B_{r\nu}l^\nu$, $B_{\mu s}l^\mu$, $B_{\mu\nu}l^\mu l^\nu$. Each of these quantities is unaffected by a change of coordinates which leaves the points on the surface invariant and is thus independent of the $g_{\mu 0}$.

It should be noted that, for a vector A_μ, the ordinary and covariant derivatives A_{rs} and $A_{r|s}$ are both independent of the $g_{\mu 0}$. Their difference, namely $\Gamma_{rs}{}^{\mu}A_\mu$, is thus independent of the $g_{\mu 0}$. We may take A_μ here to be the unit normal, namely

$$l_\mu=g_\mu{}^0/(-g^{00})^{\frac{1}{2}}, \tag{4}$$

and we find that the quantity

$$\Gamma_{rs}{}^0/(-g^{00})^{\frac{1}{2}} \tag{5}$$

is independent of the $g_{\mu 0}$. This quantity may be called the "invariant velocity" of g_{rs}, as it consists of the ordinary velocity g_{rs0} multiplied by a certain factor and with certain terms added on, so as to produce a quantity independent of the choice of coordinate system outside the surface $x^0=t$.

With the physical state described in this way, one easily finds[4] that for a dynamical variable η not involving the $g_{\mu 0}$, $d\eta/dx^0$ is of the form

$$d\eta/dx^0=\int\{(-g^{00})^{-\frac{1}{2}}\xi_L+g_{r0}e^{rs}\xi_s\}d^3x, \tag{6}$$

with ξ_L and ξ_s independent of the $g_{\mu 0}$. We need equations of motion to determine ξ_L, ξ_s for any η. The coefficients $(-g^{00})^{-\frac{1}{2}}$, g_{r0} in (6) are arbitrary and not restricted by the equations of motion.

One gets equations of motion of the form (6) from a Hamiltonian of the form

$$H=\int\{(-g^{00})^{-\frac{1}{2}}\mathcal{H}_L+g_{r0}e^{rs}\mathcal{H}_s\}d^3x, \tag{7}$$

with $\mathcal{H}_L$ and $\mathcal{H}_s$ independent of the $g_{\mu 0}$ and vanishing weakly. It has been shown[4] that the Hamiltonian (3) takes the form (7) provided the dynamical coordinates describing the nongravitational fields are chosen to be independent of the $g_{\mu 0}$, in the way discussed above, and provided one takes for $\mathcal{L}_G$, instead of (1), an expression which differs from (1) by a perfect differential and which does not contain the velocities $g_{\mu 00}$, namely

$$\mathcal{L}_G=Jg^{\mu\nu}(\Gamma_{\mu\rho}{}^{\sigma}\Gamma_{\nu\sigma}{}^{\rho}-\Gamma_{\mu\nu}{}^{\rho}\Gamma_{\rho\sigma}{}^{\sigma}) +(Jg^{00})_0(g^{r0}/g^{00})_r-(Jg^{00})_r(g^{r0}/g^{00})_0. \tag{8}$$

With this $\mathcal{L}_G$, the momenta $p^{\mu 0}$ conjugate to $g_{\mu 0}$ vanish weakly, which results in the degrees of freedom described by $g_{\mu 0}$, $p^{\mu 0}$ dropping out from the Hamiltonian formalism. The weak equations $p^{\mu 0}\approx 0$ give, when one passes to the quantum theory, the conditions $p^{\mu 0}\psi=0$, which show that the wave function ψ does not involve the $g_{\mu 0}$.

The surviving gravitational momenta are

$$p^{rs}=K(e^{ra}e^{sb}-e^{rs}e^{ab})\Gamma_{ab}{}^0/(-g^{00})^{\frac{1}{2}}. \tag{9}$$

They are built up from the invariant velocities (5). The fundamental Poisson bracket (P.b.) relations for them are

$$[g_{ab},p'^{rs}]=\tfrac{1}{2}(g_a{}^r g_b{}^s+g_b{}^r g_a{}^s)\delta(x-x'). \tag{10}$$

The expressions for $\mathcal{H}_L$ and $\mathcal{H}_s$ in (7) are found to be

$$\mathcal{H}_L = K^{-1}(p^{rs}p_{rs} - \tfrac{1}{2}p_r{}^r p_s{}^s) + B + \{K^{-1}(K^2 e^{rs})_r\}_s + \mathcal{H}_{ML}, \quad (11)$$

$$\mathcal{H}_s = p^{ab}g_{abs} - 2(p^{ab}g_{as})_b + \mathcal{H}_{Ms}, \quad (12)$$

where

$$B = \tfrac{1}{4}K g_{rsu} g_{abv}\{(e^{ra}e^{sb} - e^{rs}e^{ab})e^{uv} + 2(e^{ru}e^{ab} - e^{ra}e^{bu})e^{sv}\}, \quad (13)$$

and $\mathcal{H}_{ML}$, $\mathcal{H}_{Ms}$ are the contributions arising from the nongravitational fields. It should be noted that the terms $B + \{K^{-1}(K^2e^{rs})_r\}_s$ are equal to the density of the three-dimensional scalar curvature of the surface $x^0 = t$.

We have the weak equations

$$\mathcal{H}_L \approx 0, \quad \mathcal{H}_s \approx 0. \quad (14)$$

They are χ equations or secondary constraints. To see where they come from, we note that Einstein's field equations are

$$R_\mu{}^\nu - \tfrac{1}{2}g_\mu{}^\nu R_\sigma{}^\sigma = \tfrac{1}{2}T_\mu{}^\nu, \quad (15)$$

where $T_\mu{}^\nu$ is the stress tensor produced by the nongravitational fields. The left-hand side of (15) contains second derivatives of the $g_{\alpha\beta}$ and thus in general contains accelerations $g_{\alpha\beta 00}$. The right-hand side of (15) contains no derivatives of the $g_{\alpha\beta}$. Now the well-known identities

$$(R_\mu{}^\nu - \tfrac{1}{2}g_\mu{}^\nu R_\sigma{}^\sigma)_{|\nu} \equiv 0$$

may be written

$$(R_\mu{}^0 - \tfrac{1}{2}g_\mu{}^0 R_\sigma{}^\sigma)_0 \equiv -(R_\mu{}^r - \tfrac{1}{2}g_\mu{}^r R_\sigma{}^\sigma)_r +, \quad (16)$$

where the + at the end indicates that some further terms, not involving third derivatives of the $g_{\mu\nu}$, must be added on. The right-hand side of (16) evidently does not contain any third time derivatives $g_{\alpha\beta 000}$. Thus the left-hand side cannot involve third time derivatives, so $R_\mu{}^0 - \tfrac{1}{2}g_\mu{}^0 R_\sigma{}^\sigma$ cannot involve accelerations $g_{\alpha\beta 00}$. Thus if we take $\nu = 0$ in (15), we get equations involving only dynamical coordinates and velocities. By substituting for the velocities here in terms of the momenta, we get four equations between dynamical coordinates and momenta only, which yield (14).

The main part of the Hamiltonian is obtained by putting $g_{\mu 0} = -\delta_{\mu 0}$ in (7) and is thus

$$H_{\mathrm{main}} = \int \mathcal{H}_L d^3x = \int \{K^{-1}(p^{rs}p_{rs} - \tfrac{1}{2}p_r{}^r p_s{}^s) + B + \mathcal{H}_{ML}\} d^3x, \quad (17)$$

after removal of a surface integral at infinity. The removal of this surface integral does not disturb the validity of H_{main} for giving equations of motion, but it results in H_{main} not vanishing weakly.

We can write the total Hamiltonian (7) in the form

$$H = H_{\mathrm{main}} + \int \{(-g^{00})^{-\frac{1}{2}} - 1\}\mathcal{H}_L d^3x + \int g_{r0} e^{rs} \mathcal{H}_s d^3x, \quad (18)$$

when it appears as H_{main} with arbitrary linear combinations of $\mathcal{H}_L$ and of $\mathcal{H}_s$, for various values of x^r, added on. These additional terms in the Hamiltonian produce terms in the equations of motion in addition to those produced by H_{main}, corresponding to the surface $x^0 = t$ undergoing arbitrary deformations and having arbitrary changes of its coordinate system x^r as t varies.

NEED FOR FIXATION OF THE COORDINATES

To specify a physical state at a particular time in the classical theory, we must choose numerical values for all the dynamical coordinates and momenta so as to satisfy the constraints (14). This involves solving some differential equations, so it is not such a straightforward matter as specifying a state in particle dynamics.

In the quantum theory the situation is more complicated. The constraints (14) go over into the conditions on the wave function

$$\mathcal{H}_L \psi = 0, \quad (19)$$

$$\mathcal{H}_s \psi = 0. \quad (20)$$

To specify a state at a particular time involves obtaining a solution of Eqs. (19), (20), which are functional equations.

Equation (20) expresses merely that ψ must be invariant under changes of the coordinate system x^r in the surface $x^0 = t$. To get ψ to satisfy this equation is thus not difficult. Equation (19) expresses the requirement that the state shall be specified in a way that is independent of deformations of the surface $x^0 = t$. The treatment of such deformations is essentially as complicated as the treatment of the passage from the surface $x^0 = t$ to a neighboring surface $x^0 = t + \epsilon$, so to get ψ to satisfy (19) is essentially as complicated as solving the equations of motion. Thus we have the situation that we cannot specify the initial state for a problem without solving the equations of motion. The formalism is thus not suitable for dealing with practical problems.

The difficulty does not arise in the weak-field approximation, because then many of the terms in (19) get neglected and the remaining ones, if expressed in terms of Fourier components, are easy to handle.

To obtain a practical formalism of greater accuracy than the weak-field approximation, it is necessary to introduce into the theory some new constraint that fixes the surface $x^0 = t$, so that we no longer have the possibility of making arbitrary deformations in it. Then the supplementary condition (19) gets eliminated. We may also introduce some further constraints that fix the coordinate system x^r in the surface. While not essential for getting a practical formalism, such further constraints serve to simplify the formalism by eliminating the conditions (20), and so making the task of specifying the initial state a trivial one.

The fixation of coordinates is advantageous also in the weak-field approximation, because it leads to some degrees of freedom dropping out from the formalism, the procedure being similar to the elimination of the longitudinal waves in electrodynamics.

When dealing with gravitational waves, people usually restrict the coordinate system by introducing the harmonic conditions

$$(Jg^{\mu\nu})_{,\nu}=0.$$

These conditions would be quite unsuitable in the present formalism because they involve the $g_{\mu 0}$, which the present formalism allows to be completely arbitrary. Any restriction imposed on the $g_{\mu 0}$ would not help one in dealing with Eqs. (14) or (19) and (20). We need some restrictions which affect only the variables involved in (14), namely g_{rs} and p^{rs}, and possibly also the nongravitational variables.

GENERAL METHOD

Let us examine the general principles which come into play when we introduce some new restrictions or constraints on the dynamical variables in a Hamiltonian theory. Suppose we have a number of weak equations $\chi_n\approx 0$ $(n=1, 2,\cdots N)$, which may be either primary or secondary constraints. We are taking N to be finite for definiteness, but the same principles apply with N infinite. Suppose further that these weak equations are all first-class, so that

$$[\chi_n,\chi_{n'}]=0.$$

Now introduce some new restrictions, say the M independent equations

$$Y_m\approx 0,\quad m=1, 2,\cdots, M$$

with $M\leqslant N$. They are, of course, weak equations. Suppose that none of them (and no linear combination of them) has zero P.b. with all the χ's, so that they are all second-class constraints. They will cause M of the χ's to become second-class, while $N-M$ of the χ's (or linear combination of them) remain first-class.

Suppose χ_1, $\chi_2,\cdots\chi_M$ become second-class, while $\chi_{M+1},\cdots,\chi_N$ remain first-class. We now have the $2M$ second-class constraints $\chi_m\approx 0$, $Y_m\approx 0$ $(m=1, 2,\cdots M)$. Let us write $\chi_m=Y_{M+m}$, so that the $2M$ second-class constraints become $Y_s\approx 0$ $(s=1, 2,\cdots, 2M)$.

There is no place for second-class weak equations in the quantum theory, so we have to transform them in some way. We shall see that we can change them into strong equations (holding as equations between operators in the quantum theory) provided we adopt a new definition of P.b.'s, which corresponds to the number of effective degrees of freedom being reduced by M.

In simple cases we can pick out directly the degrees of freedom that have to be dropped and those that survive. Let us take the special case when M of the equation $Y_s\approx 0$ are

$$p_m\approx 0,\quad m=1, 2,\cdots, M. \tag{21}$$

The remaining M of them must then contain all the variables q_m independently, (otherwise the p_m would not all be second-class) and so it must be possible to solve them for the q_m and write them as

$$q_m\approx f_m(q_{M+1}, q_{M+2}\cdots p_{M+1}, p_{M+2}\cdots). \tag{22}$$

We now see that the degrees of freedom associated with q_m, p_m $(m=1, 2,\cdots M)$ cease to play an effective role in the dynamics. We can use Eqs. (21) and (22) to eliminate the variables p_m and q_m from the theory, which implies using these equations as definitions or as strong equations. We then work with P.b.'s that refer only to the other degrees of freedom.

In the general case one retains all the dynamical variables and merely changes their P.b.'s to correspond to the reduction in the number of degrees of freedom. To do this one first sets up the matrix of all the P.b.'s $[Y_s,Y_{s'}]$. It can be shown[5] that this matrix has a nonvanishing determinant, provided there is no linear combination of the Y_s that is first-class. One must then obtain the reciprocal matrix $C_{ss'}$, satisfying

$$C_{ss'}[Y_{s'},Y_{s''}]=\delta_{ss''}. \tag{23}$$

Note that $C_{ss'}$ is a skew matrix, like $[Y_s,Y_{s'}]$. One then defines new P.b.'s by the formula

$$[\xi,\eta]^*=[\xi,\eta]-[\xi,Y_s]C_{ss'}[Y_{s'},\eta]. \tag{24}$$

It can be checked[5] that the new P.b.'s satisfy all the fundamental relations that P.b.'s ought to satisfy.

From (23) and (24) we see at once that $[\xi,Y_s]^*=0$ for any ξ. Thus the Y_s now have zero P.b. with everything, so that we can consider the equations $Y_s=0$ as strong equations and use them before working out P.b.'s.

In applying this method to the gravitational case we desire, of course, that the change in the P.b.'s shall not be too complicated. In particular, we would like to have no change at all in the P.b. of two quantities, neither of which involves the gravitational variables g_{rs}, p^{rs}. This result is ensured provided the two conditions hold: (i) The Y_m $(m=1, 2,\cdots M)$ involve only the gravitational variables; (ii) The P.b.'s $[Y_m,Y_{m'}]$ all vanish. The proof is as follows.

We have already $(\chi_m,\chi_{m'}]\approx 0$ from the assumption that the χ's were originally first-class. With the further condition $[Y_m,Y_{m'}]\approx 0$ we have $[Y_s,Y_{s'}]\approx 0$ except when $1\leqslant s\leqslant M$ and $M+1\leqslant s'\leqslant 2M$ or vice versa. This

[5] P. A. M. Dirac, Can. J. Math. 2, 129 (1950).

leads to $C_{ss'}\approx 0$ except when $1\leqslant s\leqslant M$ and $M+1\leqslant s'\leqslant 2M$ or vice versa. The surviving elements of C are thus $C_{m,\,M+m'}=-C_{M+m',\,m}$. The elements $C_{m,\,M+m'}$ form a matrix of M rows and columns, which is the reciprocal of the matrix $[\chi_{m'},Y_m]$.

The formula (24) now reduces to

$$[\xi,\eta]^*-[\xi,\eta]=-C_{m,\,M+m'}\{[\xi,Y_m][\chi_{m'},\eta]-[\xi,\chi_{m'}][Y_m,\eta]\}. \quad (25)$$

If ξ and η do not involve the gravitational variables, the condition (i) above leads to $[\xi,Y_m]=0$ and $[Y_m,\eta]=0$, so the right-hand side of (25) vanishes.

The introduction of the new constraints into the theory, when combined with the appropriate change in the P.b.'s, leaves the Hamiltonian first-class. It follows that the Hamiltonian equations of motion preserve all the constraints.

FIXATION OF THE SURFACE

To fix the surface $x^0=t$, the natural conditions to take are

$$p_r{}^r=g_{rs}p^{rs}\approx 0. \quad (26)$$

This involves bringing into the theory one Y equation for each point of the surface.

One easily checks that

$$[\mathcal{H}_s,g'_{uv}p'^{uv}]=g_{uv}p^{uv}\delta_{,s}(x-x')\approx 0,$$

so the conditions (26) do not disturb the first-class character of the equations $\mathcal{H}_s\approx 0$. This means that the conditions (26) do not restrict the coordinate system x^r in the surface, a result which is evident from the tensor character of (26). The conditions (26) mean geometrically that the surface shall have a maximum three-dimensional "area." The equations (26) and $\mathcal{H}_L\approx 0$ are now second-class and we can use them to eliminate one degree of freedom at each point of space.

We have

$$[g_{rs},p'_u{}^u]=g_{rs}\delta(x-x').$$

It follows that the ratios of the g_{rs} at any point have zero P.b.'s with $p_u{}^u$ at all points of the surface. Let us put

$$K=\kappa^3,\quad \tilde{g}_{rs}=g_{rs}\kappa^{-2},\quad \tilde{e}^{rs}=e^{rs}\kappa^2. \quad (27)$$

Then $\tilde{g}_{rs}$ involves only such ratios and has zero P.b. with $p_u{}^u$ at all points. There are five independent $\tilde{g}_{rs}$, as their determinant is unity. The $\tilde{e}^{rs}$ form the reciprocal matrix to the matrix $\tilde{g}_{rs}$, and also have the determinant unity.

We have

$$[K^2,p'_u{}^u]=3K^2\delta(x-x'),$$

and so

$$[\ln\kappa,p'_u{}^u]=\tfrac{1}{3}\delta(x-x'). \quad (28)$$

Put

$$\tilde{p}^{rs}=(p^{rs}-\tfrac{1}{3}e^{rs}g_{ab}p^{ab})\kappa^2, \quad (29)$$
$$\tilde{p}_{rs}=\tilde{g}_{ra}\tilde{g}_{sb}\tilde{p}^{ab}.$$

We find that $\tilde{p}^{rs}$ and $\tilde{p}_{rs}$ have zero P.b. with $p_u{}^u$ and κ at all points.

Let us change our basic dynamical coordinates from the six g_{rs} to the five independent $\tilde{g}_{rs}$ and $\ln\kappa$. The momentum conjugate to $\ln\kappa$ is now, from (28), just $2p_u{}^u$, and the momenta conjugate to the $\tilde{g}_{rs}$ are certain functions of the $\tilde{p}^{rs}$ and $\tilde{g}_{rs}$.

The conditions (26) now take the form (21) and we have the equations $\mathcal{H}_L\approx 0$ playing the role of (22). To put them into the form of (22) we must solve them, with the help of (26), to get κ expressed in terms of quantities having zero P.b. with $p_u{}^u$ and κ. Such quantities are the $\tilde{g}_{rs}$, $\tilde{e}^{rs}$, $\tilde{p}^{rs}$, $\tilde{p}_{rs}$, and the nongravitational variables.

From (11), the equation $\mathcal{H}_L\approx 0$ gives,

$$-\{\kappa^{-3}(\kappa^4\tilde{e}^{rs})_{,r}\}_{,s}\approx\kappa^{-3}\tilde{p}^{rs}\tilde{p}_{rs}+B+\mathcal{H}_{ML}, \quad (30)$$

in which we look upon the g_{rs} in B and $\mathcal{H}_{ML}$ as expressed in terms of the $\tilde{g}_{rs}$ and κ. This is a difficult equation to solve generally for κ. However, for gravitational fields that are not too strong, the important terms are those that involve second derivatives of κ, i.e., those on the left-hand side. We can therefore obtain the solution by a method of successive approximation, first putting $\kappa=1$ on the right and solving the resulting simplified equation, then substituting the first approximation for κ on the right and solving to get the second approximation, and so on. We shall consider this equation further in the next section, with reference to a particular system of coordinates, and for the present we shall assume that the solution has been obtained.

Following the method of the preceding section for dealing with the second-class equations (21) and (22), we express H_{main} and $\mathcal{H}_s$ in terms of the variables $\tilde{g}_{rs}$, $\tilde{e}^{rs}$, $\tilde{p}^{rs}$, $\tilde{p}_{rs}$, $p_u{}^u$, and κ, and then eliminate $p_u{}^u$ and κ from them by means of (26) and the solution of (30), which we may now use as strong equations. The elimination from $\mathcal{H}_s$ is trivial, as we get from (12), using (26),

$$\mathcal{H}_s=\tilde{p}^{ab}\tilde{g}_{ab,s}-2(\tilde{p}^{ab}\tilde{g}_{as})_{,b}+\mathcal{H}_{Ms}. \quad (31)$$

If the nongravitational field variables are suitably chosen, $\mathcal{H}_{Ms}$ will not contain κ. The elimination from H_{main} leads to an expression

$$H^*_{\text{main}}=\int(\kappa^{-3}\tilde{p}^{rs}\tilde{p}_{rs}+B+\mathcal{H}_{ML})d^3x, \quad (32)$$

in which κ is understood to have the appropriate value. The integrand here may be considered as the energy density or mass density. The complete Hamiltonian is now

$$H^*_{\text{main}}+\int g_{r0}e^{rs}\mathcal{H}_s d^3x. \quad (33)$$

The term corresponding to the freedom of deformation of the surface, i.e., the middle term of (18), has disappeared.

We now have a Hamiltonian formalism in which the degree of freedom described by $p_u{}^u$ and $\ln\kappa$ has dropped out. The Hamiltonians (32) and (33) are first-class even with the condition (26), so they lead to equations of motion that preserve (26). The procedure of substituting for κ in the derivation of $H^*{}_{\mathrm{main}}$ caused the introduction of the right amount of $\mathcal{H}_L$ into the Hamiltonian to ensure the preservation of (26).

FIXATION OF COORDINATES IN THE SURFACE

To get the theory into a more convenient form, one must also fix the coordinate system x^r in the surface. The most natural conditions to take for this purpose, from the geometrical point of view, are the harmonic conditions in three dimensions:

$$(Ke^{rs})_{,s}\approx 0. \tag{34}$$

However, (34) does not have zero P.b. with (26), so if we adopt (34) together with (26) we must change the P.b. relationships between the nongravitational variables. To avoid this inconvenience, it is better to replace (34) by

$$\tilde{e}^{rs}{}_{,s}\approx 0, \tag{35}$$

which does have zero P.b. with (26).

With the coordinates fixed by (35), Eq. (30) reduces to

$$-4\nabla^2\kappa=\kappa^{-3}\tilde{p}^{rs}\tilde{p}_{rs}+B+\mathcal{H}_{ML}. \tag{36}$$

where ∇^2 denotes the Laplacian operator with respect to the metric $\tilde{g}_{rs}$, namely

$$\nabla^2=\tilde{e}^{rs}\partial^2/\partial x^r\partial x^s. \tag{37}$$

The right-hand side in (36) equals the integrand in (32) and is the mass density. To interpret (36), let us restore the gravitational constant into the theory in accordance with (2). It then becomes

$$-(4\pi\gamma)^{-1}\nabla^2\kappa=16\pi\gamma\kappa^{-3}\tilde{p}^{rs}\tilde{p}_{rs}+(16\pi\gamma)^{-1}B+\mathcal{H}_{ML}. \tag{38}$$

We now see that $\kappa-1$ is the Newtonian potential generated by the mass density in a space with the metric $\tilde{g}_{rs}$. The fact that κ occurs in the right-hand side of (38) can be understood as due to the Newtonian potential itself having some influence on the mass density which generates it.

Let us examine the term with B in (38). The expression (13) for B, written in terms of the new variables, is

$$B=\tfrac{1}{4}\kappa^{-1}(\kappa\tilde{g}_{rsu}+2\kappa_u\tilde{g}_{rs})(\kappa\tilde{g}_{abv}+2\kappa_v\tilde{g}_{ab}) \times\{(\tilde{e}^{ra}\tilde{e}^{sb}-\tilde{e}^{rs}\tilde{e}^{ab})\tilde{e}^{uv}+2(\tilde{e}^{ru}\tilde{e}^{ab}-\tilde{e}^{ra}\tilde{e}^{bu})\tilde{e}^{sv}\}.$$

With the help of the equation

$$\tilde{g}_{rsu}\tilde{e}^{rs}=0,$$

which follows from the determinant of the $\tilde{g}_{rs}$ being unity, and of the equation

$$\tilde{g}_{rsu}\tilde{e}^{ru}=0,$$

which follows from (35), this reduces to

$$B=\tfrac{1}{4}\kappa\tilde{g}_{rsu}\tilde{g}_{abv}(\tilde{e}^{ra}\tilde{e}^{sb}\tilde{e}^{uv}-2\tilde{e}^{ra}\tilde{e}^{bu}\tilde{e}^{sv})-2\kappa^{-1}\kappa_u\kappa_v\tilde{e}^{uv}. \tag{39}$$

The last term here, divided by $16\pi\gamma$, can be interpreted as the mass density (or energy density) of the Newtonian field with the potential $\kappa-1$. It is negative definite, corresponding to the Newtonian force being attractive. The remaining terms of B, together with the first term on the right-hand side of (38), give the energy density of the gravitational waves.

THE NEW POISSON BRACKETS

With the coordinates fixed by (35), the P.b.'s of the gravitational variables with one another and with the nongravitational variables will be altered. The new P.b.'s are given by formula (25) with Y_m replaced by $\tilde{e}^{ru}{}_u$ and $\chi_{m'}$ replaced by $\mathcal{H}'_s$. It thus reads

$$[\xi,\eta]^*-[\xi,\eta]=-\int\int C_r{}^s(x,x')\{[\xi,\tilde{e}^{ru}{}_u][\mathcal{H}'_s,\eta] -[\xi,\mathcal{H}'_s][\tilde{e}^{ru}{}_u,\eta]\}d^3xd^3x'. \tag{40}$$

The coefficient $C_r{}^s(x,x')$ is the reciprocal of the matrix $[\mathcal{H}'_s,\tilde{e}^{ru}{}_u]$ and thus satisfies

$$\int C_v{}^s(x'',x')[\mathcal{H}'_s,\tilde{e}^{ru}{}_u]d^3x'=g_v{}^r\delta(x-x''). \tag{41}$$

Evaluating the P.b. here, we get

$$\int C_v{}^s(x'',x')\{g_s{}^r\tilde{e}^{ab}\delta_{ab}(x-x') +\tfrac{1}{3}\tilde{e}^{ra}\delta_{sa}(x-x')\}d^3x'=g_v{}^r\delta(x-x''),$$

which reduces to

$$\nabla^2C_v{}^r(x',x)+\tfrac{1}{3}\tilde{e}^{ra}C_v{}^s(x',x)_{sa}=g^r{}_v\delta(x-x'), \tag{42}$$

with ∇^2 defined by (37).

This equation may be considered for fixed x', when it is a differential equation for the unknown functions $C_v{}^r(x',x)$ in the variables x. The important domain for x is now the neighborhood of x', since when x is far from x' the functions $C_v{}^r(x',x)$ are small. We can therefore get an approximate solution by considering the space as flat in this domain, so that the $\tilde{e}^{ab}$ are constants. With this approximation we get, on differentiating (42) with respect to x^r,

$$\nabla^2C_v{}^s{}_s=\tfrac{3}{4}\delta_v(x-x'). \tag{43}$$

The solution of this equation is

$$C_v{}^s{}_s=-\frac{1}{4\pi}\times\frac{3}{4}\left(\frac{1}{|x-x'|}\right)_v,$$

where $|x-x'|$ denotes the distance from x to x' with respect to the metric $\tilde{g}_{rs}$,

$$|x-x'|=\{\tilde{g}_{rs}(x^r-x'^r)(x^s-x'^s)\}^{\frac{1}{2}}. \tag{44}$$

Equation (42) now becomes

$$\nabla^2 C_v{}^r = g^r{}_v \delta(x-x') + \frac{1}{16\pi} \bar{e}^{rs} \left(\frac{1}{|x-x'|} \right)_{vs},$$

whose solution is

$$C_v{}^r(x',x) = -\frac{1}{4\pi} g^r{}_v \frac{1}{|x-x'|} + \frac{1}{32\pi} \bar{e}^{rs} |x-x'|_{vs}. \quad (45)$$

One could get the solution of (42) to a higher accuracy by substituting for the $\bar{e}^{ab}$ in the left-hand side of (42), (remembering that $\bar{e}^{ab}$ occurs also in the operator ∇^2,) their Taylor expansions in powers of $x-x'$ and using the first approximation for $C_v{}^r$ in those terms in which it occurs with a factor $x^r - x'^r$. By a process of successive approximation one could get the solution to any desired accuracy.

With the coefficients $C_r{}^s(x,x')$ in (40) determined, the new P.b.'s are determined. It should be noted that the new P.b. of any nongravitational variable with g_{rs} or $\bar{e}^{rs}$ vanishes. However, its new P.b. with $\bar{p}^{rs}$ does not vanish.

QUANTIZATION

To pass over to the quantum theory, we must make all our dynamical variables into operators satisfying commutation relations corresponding to the new P.b.'s. We must then pick out a complete set of commuting observables. We may take these to consist of the $\bar{e}^{rs}$ at all points x^r, together with a complete set of commuting nongravitational observables, ζ say. We can then set up the wave function as a function of these variables,

$$\psi(\bar{e}^{rs}, \zeta).$$

The effective domain of ψ is that for which the $\bar{e}^{rs}$ are restricted to have the determinant unity and to satisfy $\bar{e}^{rs}{}_s = 0$. ψ may be considered as undefined outside this domain. When we operate on ψ with $\bar{p}^{ab}$ or with any dynamical variable in the theory, we get another wave function defined in the same domain, on account of $\bar{p}^{ab}$ commuting with the determinant of the $\bar{e}^{rs}$ and with $\bar{e}^{rs}{}_s$.

There are no supplementary conditions to be imposed on ψ. We can choose it arbitrarily to correspond to the initial state in any problem. There is just one equation for ψ, the Schrödinger equation

$$i\hbar(d\psi/dx^0) = H^*{}_{\text{main}}\psi,$$

which fixes the state at later times.

For the theory to be self-consistent it is necessary that the space-like surface on which the state is defined shall always remain space-like. The condition for this is that K^2, the determinant of the g_{rs}, shall remain always positive. In the present formalism this means $\kappa^6 > 0$, with κ determined by (36). If the mass density is always positive, (36) shows that $\kappa > 1$ and there is no trouble. Difficulties arise only where there is a large negative density. This occurs very close to a point particle, on account of the last term in (39). The gravitational treatment of point particles thus brings in one further difficulty, in addition to the usual ones in the quantum theory.

PHYSICAL REVIEW LETTERS

VOLUME 4 MARCH 1, 1960 NUMBER 5

POSSIBLE NEW EXPERIMENTAL TEST OF GENERAL RELATIVITY THEORY*

L. I. Schiff
Institute of Theoretical Physics, Department of Physics, Stanford University, Stanford, California
(Received February 11, 1960)

In a paper now in process of publication,[1] it is argued that only the planetary orbit precession provides real support for the full structure of the general theory of relativity. The other two of the three "crucial tests," the gravitational red shift and deflection of light, can be inferred correctly from the equivalence principle and the special theory of relativity, both of which are well established by other experimental evidence. It is also pointed out that a terrestrial or satellite experiment that would really test general relativity theory would have either to use particles of finite rest mass in such a way that the equation of motion can be confirmed beyond the Newtonian approximation, or to verify the second-order deviations of the metric tensor from its Minkowski form.

In an attempt to devise a feasible experiment that might accomplish one of these objectives, we have calculated the properties of a spinning test particle (torque-free gyroscope). We start from the covariant equations of Papapetrou[2] for the motion of the center of mass and the spin angular momentum, generalized by inclusion of a nongravitational constraining force $\vec{F}$, and work to lowest order. The motion of the center of mass in the gravitational field of the rotating earth is then described by the Newtonian equation

$$m(d\vec{v}/dt) = -(GmM/r^3)\vec{r} + \vec{F}, \tag{1}$$

where m is the rest mass of the particle, $\vec{r}$ is its coordinate, $\vec{v} = d\vec{r}/dt$ is its velocity, G is the Newtonian gravitational constant, and M is the mass of the earth. The spin angular momentum vector measured by a co-moving observer, $\vec{S}^0$, obeys the equation

$$d\vec{S}^0/dt = \vec{\Omega} \times \vec{S}^0, \tag{2}$$

where

$$\vec{\Omega} = (\vec{F} \times \vec{v})/2mc^2 + (3GM/2c^2r^3)(\vec{r} \times \vec{v}) + (GI/c^2r^3)[3(\vec{\omega}\cdot\vec{r})\vec{r}/r^2 - \vec{\omega}]; \tag{3}$$

$I = 2MR^2/5$ is the moment of inertia of the earth of radius R, assumed to be homogeneous, and $\vec{\omega}$ is its angular velocity vector. The first term on the right side of (3) is the Thomas precession,[3] which is a special relativity effect. The other two are the lowest order effects of general relativity; the second term arises whether or not the earth is rotating, and the third term is the earth rotation effect of Lense and Thirring.[4] While the second term involves the first-order deviations of the metric tensor from its Minkowski form, which can be calculated without the use of general relativity,[1] it also depends on the equation of motion of matter of finite rest mass beyond the Newtonian approximation. It is therefore a genuine consequence of general relativity. The same is true of the third term, since in addition it depends on off-diagonal space-time components of the metric tensor.

Equations (2) and (3) may be obtained either from the standard or the isotropic form of the Schwarzschild line element, and using for the supplementary condition on the angular momentum tensor either that of Corinaldesi and Papapetrou[5] or of Pirani.[6] The equation of motion of the spin in the nonrotating, earth-centered coordinate system looks quite different in these four cases, but they all agree when expressed in terms of the spin measured by a co-moving observer. It should also be remarked that the corrections to Eq. (1) that arise from the spin are unobservably small in any realizable situation.

It follows at once from the form of Eq. (2) that the magnitude of the spin angular momentum measured by a co-moving observer is constant in time. Thus if the moment of inertia of the spinning particle does not change, the angular velocity of rotation is constant, and the spinning particle behaves like a clock which can be set to any desired frequency. This frequency exhibits Doppler and gravitational shifts when observed from outside, just like that of a more conventional clock. It is possible that its frequency stability could be made to compare favorably with those of other types of precision clocks. It also follows from (2) that a number of spinning particles with various magnitudes and directions for their angular momentum vectors maintain fixed angles of these vectors with respect to each other. The vector $\vec{\Omega}$, which in general is not constant, is their common angular velocity of precession with respect to the external "fixed stars"; in comparing their directions with the outside world, a correction must of course be made for aberration whenever $\vec{v} \neq 0$.

If a spinning particle is in free fall, as in a satellite, then $\vec{F}=0$. For an orbit in the earth's equatorial plane, for example,

$$\vec{\Omega}=(3GM/2c^2r)\vec{\omega}_0-(2MGR^2/5c^2r^3)\vec{\omega}, \tag{4}$$

where $\vec{\omega}_0=(\vec{r}\times\vec{v})/r^2$ is the instantaneous orbital angular velocity vector of the particle. The minus sign in Eq. (4) deserves some comment. The third term of Eq. (3) tends to cause a spinning particle to precess in the same direction as the rotating earth at the poles ($\vec{r}$ parallel or antiparallel to $\vec{\omega}$), but in the opposite direction at the equator ($\vec{r}$ perpendicular to $\vec{\omega}$). This is physically reasonable if we think of the moving earth as "dragging" the metric with it to some extent. At the poles, this tends to drag the spin around in the same direction as the rotation of the earth. But at the equator, since the gravitational field falls off with increasing r, the side of the spinning particle nearest the earth is dragged more than the side away from the earth, so that the spin precesses in the opposite direction.

If the center of mass of the spinning particle is constrained to remain at rest with respect to the rotating earth, as in an earth-bound laboratory,[7] then

$$\vec{v}=\vec{\omega}\times\vec{r}, \quad d\vec{v}/dt=\vec{\omega}\times\vec{v}. \tag{5}$$

The required constraining force $\vec{F}$ can then be found from (1) and (5), and substituted into (3). When the particle is at the surface of the earth at latitude λ, the precession angular velocity may be written in the form

$$\vec{\Omega}=[(4gR/5c^2)(1+\cos^2\lambda)-(\omega^2R^2/2c^2)\cos^2\lambda]\vec{\omega} + (4g\sin\lambda/5\omega c^2)(\vec{\omega}\times\vec{v}), \tag{6}$$

where $g=GM/R^2$ is the acceleration of gravity at the surface of the earth. Only the square bracket term in Eq. (6) gives rise to a secular precession of the spin axis, and the second part of it is very small compared to the first. Thus to good approximation, a particle with spin axis perpendicular to the earth's axis precesses at the rate $2\pi(4gR/5c^2)(1+\cos^2\lambda)=3.5\times10^{-9}(1+\cos^2\lambda)$ radians per day. It also follows from Eq. (3) that the corresponding effects caused by the sun and moon are negligibly small in comparison.

A secular precession of 6×10^{-9} radian per day would be very difficult, but perhaps not impossible, to observe. Professor W. M. Fairbank and Professor W. A. Little of this department are exploring the possibility of using for this purpose a gyroscope that consists of a superconducting sphere supported by a static magnetic field.[8] Such a gyroscope would also be of interest as a device for performing experiments in low-temperature physics. If it could be made to operate exceedingly well, it might in addition be used for an experimental test of Mach's principle, by comparing the orientation of its axis with a field of "fixed stars" over a period of a year or so. Most of the experimental difficulties that seem to arise with a high-precision gyroscope are greatly reduced if the gyroscope does not have to be supported against gravity. This, together with the fact that ω_0 is generally much larger than ω, suggests that experiments of this type might be more easily performed in a satellite than in an earth-bound laboratory.

A full account of this work will be submitted shortly for publication in the Proceedings of the National Academy of Sciences.

---

*Supported in part by the U. S. Air Force through the Air Force Office of Scientific Research.

[1]L. I. Schiff, Am. J. Phys. (to be published).
[2]A. Papapetrou, Proc. Roy. Soc. (London) A209, 248 (1951).
[3]L. H. Thomas, Phil. Mag. 3, 1 (1927).
[4]J. Lense and H. Thirring, Phys. Z. 29, 156 (1918).
[5]E. Corinaldesi and A. Papapetrou, Proc. Roy. Soc. (London) A209, 259 (1951).
[6]F. A. E. Pirani, Acta Phys. Polon. 15, 389 (1956). Our Eqs. (2) and (3) are generalizations of some of the results derived by Pirani.
[7]This type of experiment was suggested to the author by W. M. Fairbank.
[8]I. Simon, J. Appl. Phys. 24, 19 (1953); J. Instr. Soc. Am. 1, No. 9, 79 (1954).

PHYSICAL REVIEW LETTERS

VOLUME 4 APRIL 1, 1960 NUMBER 7

APPARENT WEIGHT OF PHOTONS*

R. V. Pound and G. A. Rebka, Jr.
Lyman Laboratory of Physics, Harvard University, Cambridge, Massachusetts
(Received March 9, 1960)

As we proposed a few months ago,[1] we have now measured the effect, originally hypothesized by Einstein,[2] of gravitational potential on the apparent frequency of electromagnetic radiation by using the sharply defined energy of recoil-free γ rays emitted and absorbed in solids, as discovered by Mössbauer.[3] We have already reported[4] a detailed study of the shape and width of the line obtained at room temperature for the 14.4-kev, 0.1-microsecond level in Fe^{57}. Particular attention was paid to finding the conditions required to obtain a narrow line. We found that the line had a Lorentzian shape with a fractional full-width at half-height of 1.13×10^{-12} when the source was carefully prepared according to a prescription developed from experience. We have also investigated the 93-kev, 9.4-microsecond level of Zn^{67} at liquid helium and liquid nitrogen temperatures using several combinations of source and absorber environment, but have not observed a usable resonant absorption. That work will be reported later. The fractional width and intensity of the absorption in Fe^{57} seemed sufficient to measure the gravitational effect in the laboratory.

As a preliminary, we sought possible sources of systematic error that would interfere with measurements of small changes in frequency using this medium. Early in our development of the instrumentation necessary for this experiment, we concluded that there were asymmetries in, or frequency differences between, the lines of given combinations of source and absorber which vary from one combination to another. Thus it is absolutely necessary to measure a change in the relative frequency that is produced by the perturbation being studied. Observation of a frequency difference between a given source and absorber cannot be uniquely attributed to this perturbation. More recently, we have discovered and explained a variation of frequency with temperature of either the source or absorber.[5] We conclude that the temperature difference between the source and absorber must be accurately known and its effect considered before any meaning can be extracted from even a change observed when the perturbation is altered.

The basic elements of the apparatus finally developed to measure the gravitational shift in frequency were a carefully prepared source containing 0.4 curie of 270-day Co^{57}, and a carefully prepared, rigidly supported, iron film absorber. Using the results of our initial experiment, we requested the Nuclear Science and Engineering Corporation to repurify their nickel cyclotron target by ion exchange to reduce cobalt carrier. Following the bombardment, in a special run in the high-energy proton beam of the high-current cyclotron at the Oak Ridge National Laboratory, they electroplated the separated Co^{57} onto one side of a 2-in. diameter, 0.005-in. thick disk of Armco iron according to our prescription. After this disk was received, it was heated to 900°-1000°C for one hour in a hydrogen atmosphere[6] to diffuse the cobalt into the iron foil about 3×10^{-5} cm.

The absorber made by Nuclear Metals Inc., was composed of seven separate units. Each unit consisted of about 80 mg of iron, enriched in Fe^{57} to 31.9%, electroplated onto a polished side of a 3-in. diameter, 0.040-in. thick disk of beryllium. The electroplating technique required considerable development to produce films with absorption lines of width and strength that satisfied our tests. The films finally accepted, resonantly absorbed about 1/3 the recoil-free γ rays from our source. Each unit of the absorber was mounted over the 0.001-in. Al window of a 3 in.×1/4 in. NaI(Tl) scintillation crystal integrally mounted on a Dumont 6363 multiplier phototube. The multiplier supply voltages were separately adjusted to equalize their conversion gains, and their outputs were mixed.

The required stable vertical baseline was conveniently obtained in the enclosed, isolated tower of the Jefferson Physical Laboratory.[7] A statistical argument suggests that the precision of a measurement of the gravitational frequency shift should be independent of the height. Instrumental instability but more significantly the sources of systematic error mentioned above are less critical compared to the larger fractional shifts obtained with an increased height. Our net operating baseline of 74 feet required only conveniently realizable control over these sources of error.

The absorption of the 14.4-kev γ ray by air in the path was reduced by running a 16-in. diameter, cylindrical, Mylar bag with thin end windows and filled with helium through most of the distance between source and absorber. To sweep out small amounts of air diffusing into the bag, the helium was kept flowing through it at a rate of about 30 liters/hr.

The over-all experiment is described by the block diagram of Fig. 1. The source was moved sinusoidally by either a ferroelectric or a moving-coil magnetic transducer. During the quarter of the modulation cycle centered about the time of maximum velocity the pulses from the scintillation spectrometer, adjusted to select the 14.4-kev γ-ray line, were fed into one scaler while, during the opposite quarter cycle, they were fed into another. The difference in counts recorded was a measure of the asymmetry in, or frequency-shift between, the emission and absorption lines. As a precaution the relative phase of the gating pulses and the sinusoidal modulation were displayed continuously. The data were found to be insensitive to phase changes much larger than the drifts of phase observed.

A completely duplicate system of electronics, controlled by the same gating pulses, recorded

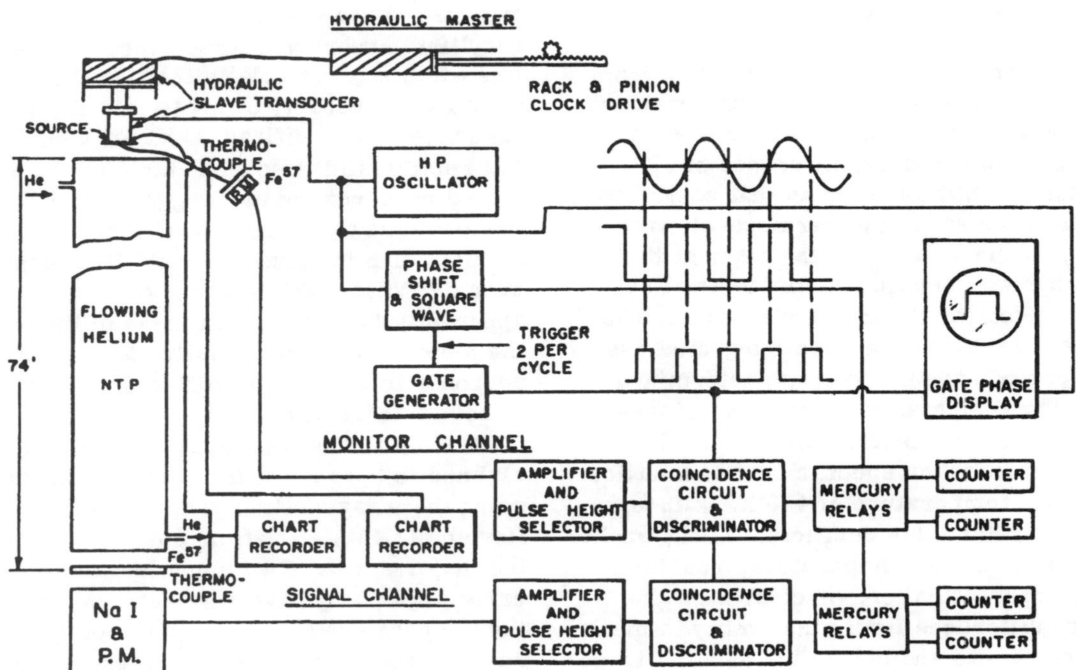

FIG. 1. A block diagram of the over-all experimental arrangement. The source and absorber-detector units were frequently interchanged. Sometimes a ferroelectric and sometimes a moving-coil magnetic transducer was used with frequencies ranging from 10 to 50 cps.

data from a counter having a 1-in. diameter, 0.015-in. thick NaI(Tl) scintillation crystal covered by an absorber similar to the main absorber. This absorber and crystal unit was mounted to see the source from only three feet away. This monitor channel measured the stability of the over-all modulation system, and, because of its higher counting rate, had a smaller statistical uncertainty.

The relation between the counting rate difference and relative frequency shifts between the emission and absorption lines was measured directly by adding a Doppler shift several times the size of the gravitational shift to the emission line. The necessary constant velocity was introduced by coupling a hydraulic cylinder of large bore carrying the transducer and source to a master cylinder of small bore connected to a rack-and-pinion driven by a clock.

Combining data from two periods having Doppler shifts of equal magnitude, but opposite sign, allowed measurement of both sensitivity and relative frequency shift. Because no sacrifice of valuable data resulted, the sensitivity was calibrated about 1/3 of the operating time which was as often as convenient without recording the data automatically. In this way we were able to eliminate errors due to drifts in sensitivity such as would be anticipated from gain or discriminator drift, changes in background, or changes in modulation swing.

The second order Doppler shift resulting from lattice vibrations required that the temperature difference between the source and absorber be controlled or monitored. A difference of 1°C would produce a shift as large as that sought, so the potential difference of a thermocouple with one junction at the source and the other at the main absorber was recorded. An identical system was provided for the monitor channel. The recorded temperature data were integrated over a counting period, and the average determined to 0.03°C. The temperature coefficient of frequency which we have used to correct the data, was calculated from the specific heat of a lattice having a Debye temperature of 420°K. Although at room temperature this coefficient is but weakly dependent on the Debye temperature, residual error in the correction for, or control of, the temperature difference limits the ability to measure frequency shifts and favors the use of a large height difference for the gravitational experiment.

Data typical of those collected are shown in Table I. The right-hand column is the data after correction for temperature difference. All data are expressed as fractional frequency shift $\times 10^{15}$. The difference of the shift seen with γ rays rising and that with γ rays falling should be the result of gravity. The average for the two directions of travel should measure an effective shift of other origin, and this is about four times the difference between the shifts. We confirmed that this shift was an inherent property of the particular combination of source and absorber by measuring the shift for each absorber unit in turn, with temperature correction, when it was six inches from the source. Although this test was not exact because only about half the area of each absorber was involved, the weighted mean shift from this test for the combination of all absorber units agreed well with that observed in the main experiment. The individual fractional frequency shifts found for these, for the monitor absorber, as well as for a 11.7-mg/cm^2 Armco iron foil, are displayed in Table II. The considerable variation among them is as striking as the size of the weighted mean shift. Such shifts could result from differences in a range of about 11% in effective Debye temperature through producing differences in net second order Doppler effect. Other explanations based on hyperfine structure including electric quadrupole interactions are also plausible. Although heat treatment might be expected to change these shifts for the iron-plated beryllium absorbers, experience showed that the line width was materially increased by such treatment, probably owing to interdiffusion. The presence of a significant shift for even the Armco foil relative to the source, both of which had received heat treatments, suggests that it is unlikely one would have, without test, a shift of this sort smaller than the gravitational effect expected in even our "two-way" baseline of 148 feet. The apparently fortuitous smallness of the shift of the monitor absorber relative to our source corresponds to the shift expected for about 30 feet of height difference.

Recently Cranshaw, Schiffer, and Whitehead[8] claimed to have measured the gravitational shift using the γ ray of Fe^{57}. They state that they believe their 43% statistical uncertainty represents the major error. Two much larger sources of error apparently have not been considered: (1) the temperature difference between the source and absorber, and (2) the frequency difference inherent in a given combination of source and absorber. From the above discussion, only 0.6°C of temperature difference would produce a shift

Table I. Data from the first four days of counting. The data are expressed as fractional frequency differences between source and absorber multiplied by 10^{15}, as derived from the appropriate sensitivity calibration. The negative signs mean that the γ ray has a frequency lower than the frequency of maximum absorption at the absorber.

Period	Shift observed	Temperature correction	Net shift
	Source at bottom		
Feb. 22, 5 p.m.	-11.5 ± 3.0	-9.2	-20.7 ± 3.0
	-16.4 ± 2.2[a]	-5.9	-22.3 ± 2.2
	-13.8 ± 1.3	-5.3	-19.1 ± 1.3
	-11.9 ± 2.1[a]	-8.0	-19.9 ± 2.1
	-8.7 ± 2.0[a]	-10.5	-19.2 ± 2.0
Feb. 23, 10 p.m.	-10.5 ± 2.0	-10.6	-21.0 ± 2.0
		Weighted average =	-19.7 ± 0.8
	Source at top		
Feb. 24, 0 a.m.	-12.0 ± 4.1	-8.6	-20.6 ± 4.1
	-5.7 ± 1.4	-9.6	-15.3 ± 1.4
	-7.4 ± 2.1[a]	-7.4	-14.8 ± 2.1
	-6.5 ± 2.1[a]	-5.8	-12.3 ± 2.1
	-13.9 ± 3.1[a]	-7.5	-21.4 ± 3.1
	-6.6 ± 3.0	-5.7	-12.3 ± 3.0
Feb. 25, 6 p.m.	-6.5 ± 2.0[a]	-8.9	-15.4 ± 2.0
	-10.0 ± 2.6	-7.9	-17.9 ± 2.6
		Weighted average =	-15.5 ± 0.8
		Mean shift =	-17.6 ± 0.6
		Difference of averages =	-4.2 ± 1.1

[a]These data were taken simultaneously with a sensitivity calibration.

Table II. Data on asymmetries of various absorbers in apparent fractional frequency shift multiplied by 10^{15}. In the third column we tabulate the Debye temperature increase of the absorber above that of the source which could account for the shift.

Absorber	$(\Delta\nu/\nu)\times 10^{15}$	$\Delta\theta_D$ in °K
No. 1	-8.4 ± 2.5	+15 ± 4
No. 2	-24 ± 3.5	+41 ± 6
No. 3	-28 ± 3.5	+48 ± 6
No. 4	-19 ± 3.5	+33 ± 6
No. 5	-24 ± 3.5	+41 ± 6
No. 6	-17 ± 2.5	+29 ± 4
No. 7	-19 ± 3.5	+33 ± 6
Weighted mean of No. 1-No. 7	-19 ± 3.0	+33 ± 5
Monitor absorber	+0.55 ± 0.15	-0.95 ± 0.26
Armco foil	+10 ± 3.5	-17 ± 6

as large as the whole effect observed. Their additional experiment at the shortened height difference of three meters does not, without concomitant temperature data, resolve the question of inherent frequency difference. Their stated disappointment with the over-all line width observed would seem to add to the probability of existence of such a shift. They mention this broadening in connection with its possible influence on the sensitivity, derived rather than measured, owing to a departure from Lorentzian shape. Clearly such a departure is even more important in allowing asymmetry.

Our experience shows that no conclusion can be drawn from the experiment of Cranshaw et al.

If the frequency-shift inherent in our source-absorber combination is not affected by inversion of the relative positions, the difference between shifts observed with rising and falling γ rays measures the effect of gravity. All data collected since recognizing the need for temperature correction, yield a net fractional shift, $-(5.13\pm 0.51)\times 10^{-15}$. The error assigned is the rms statistical deviation including that of independent sensitivity calibrations taken as representative of their respective periods of operation. The shift observed agrees with -4.92×10^{-15}, the predicted gravitational shift for this "two-way" height difference.

Expressed in this unit, the result is

$$(\Delta\nu)_{\rm exp}/(\Delta\nu)_{\rm theor} = +1.05 \pm 0.10,$$

where the plus sign indicates that the frequency increases in falling, as expected.

These data were collected in about 10 days of operation. We expect to continue counting with some improvements in sensitivity, and to reduce the statistical uncertainty about fourfold. With our present experimental arrangement this should result in a comparable reduction in error in the measurement since we believe we can take adequate steps to avoid systematic errors on the resulting scale. A higher baseline or possibly a narrower γ ray would seem to be required to extend the precision by a factor much larger than this.

We wish to express deep appreciation for the generosity, encouragement, and assistance with details of the experiment accorded us by our colleagues and the entire technical staff of these laboratories during the three months we have been preoccupied with it.

---

[*]Supported in part by the joint program of the Office of Naval Research and the U. S. Atomic Energy Commission and by a grant from the Higgins Scientific Trust.

[1]R. V. Pound and G. A. Rebka, Jr., Phys. Rev. Letters 3, 439 (1959).

[2]A. Einstein, Ann. Physik 35, 898 (1911).

[3]R. L. Mössbauer, Z. Physik 151, 124 (1958); Naturwissenschaften 45, 538 (1958); Z. Naturforsch. 14a, 211 (1959).

[4]R. V. Pound and G. A. Rebka, Jr., Phys. Rev. Letters 3, 554 (1959).

[5]R. V. Pound and G. A. Rebka, Jr., Phys. Rev. Letters 4, 274 (1960).

[6]We wish to thank Mr. F. Rosebury of the Research Laboratory of Electronics, Massachusetts Institute of Technology, for providing his facilities for this treatment.

[7]See E. H. Hall, Phys. Rev. 17, 245 (1903), first paragraph.

[8]T. E. Cranshaw, J. P. Schiffer, and A. B. Whitehead, Phys. Rev. Letters 4, 163 (1960).

Mach's Principle and a Relativistic Theory of Gravitation*

C. Brans† and R. H. Dicke
Palmer Physical Laboratory, Princeton University, Princeton, New Jersey

(Received June 23, 1961)

The role of Mach's principle in physics is discussed in relation to the equivalence principle. The difficulties encountered in attempting to incorporate Mach's principle into general relativity are discussed. A modified relativistic theory of gravitation, apparently compatible with Mach's principle, is developed.

INTRODUCTION

It is interesting that only two ideas concerning the nature of space have dominated our thinking since the time of Descartes. According to one of these pictures, space is an absolute physical structure with properties of its own. This picture can be traced from Descartes vortices[1] through the absolute space of Newton,[2] to the ether theories of the 19th century. The contrary view that the geometrical and inertial properties of space are meaningless for an empty space, that the physical properties of space have their origin in the matter contained therein, and that the only meaningful motion of a particle is motion relative to other matter in the universe has never found its complete expression in a physical theory. This picture is also old and can be traced from the writings of Bishop Berkeley[3] to those of Ernst Mach.[4] These ideas have found a limited expression in general relativity, but it must be admitted that, although in general relativity spatial geometries are affected by mass distributions, the geometry is not uniquely specified by the distribution. It has not yet been possible to specify boundary conditions on the field equations of general relativity which would bring the theory into accord with Mach's principle. Such boundary conditions would, among other things, eliminate all solutions without mass present.

It is necessary to remark that, according to the ideas of Mach, the inertial forces observed locally in an accelerated laboratory may be interpreted as gravitational effects having their origin in distant matter accelerated relative to the laboratory. The imperfect expression of this idea in general relativity can be seen by considering the case of a space empty except for a lone experimenter in his laboratory. Using the traditional, asymptotically Minkowskian coordinate system fixed relative to the laboratory, and assuming a normal laboratory of small mass, its effect on the metric is minor and can be considered in the weak-field approximation. The observer would, according to general relativity, observe normal behavior of his apparatus in accordance with the usual laws of physics. However, also according to general relativity, the experimenter could set his laboratory rotating by leaning out a window and firing his 22-caliber rifle tangentially. Thereafter the delicate gyroscope in the laboratory would continue to point in a direction nearly fixed relative to the direction of motion of the rapidly receding bullet. The gyroscope would rotate relative to the walls of the laboratory. Thus, from the point of view of Mach, the tiny, almost massless, very distant bullet seems to be more important that the massive, nearby walls of the laboratory in determining inertial coordinate frames and the orientation of the gyroscope.[5] It is clear that what is being described here is more nearly an absolute space in the sense of Newton rather than a physical space in the sense of Berkeley and Mach.

The above example poses a problem for us. Apparently, we may assume one of at least three things:

1. that physical space has intrinsic geometrical and inertial properties beyond those derived from the matter contained therein;

2. that the above example may be excluded as nonphysical by some presently unknown boundary condition on the equations of general relativity.

3. that the above physical situation is not correctly described by the equations of general relativity.

These various alternatives have been discussed previously. Objections to the first possibility are mainly philosophical and, as stated previously, go back to the time of Bishop Berkeley. A common inheritance of all present-day physicists from Einstein is an appreciation for the concept of relativity of motion.

As the universe is observed to be nonuniform, it would appear to be difficult to specify boundary conditions which would have the effect of prohibiting unsuitable mass distributions relative to the laboratory *arbitrarily placed*; for could not a laboratory be built near a massive star? Should not the presence of this massive star contribute to the inertial reaction?

The difficulty is brought into sharper focus by con-

* Supported in part by research contracts with the U. S. Atomic Energy Commission and the Office of Naval Research.

† National Science Foundation Fellow; now at Loyola University, New Orleans, Louisiana.

[1] E. T. Whittaker, *History of the Theories of Aether and Electricity* (Thomas Nelson and Sons, New York, 1951).

[2] I. Newton, *Principia Mathematica Philosophiae Naturalis* (1686) (reprinted by University of California Press, Berkeley, California, 1934).

[3] G. Berkeley, *The Principles of Human Knowledge*, paragraphs 111–117, 1710-*De Motu* (1726).

[4] E. Mach, *Conservation of Energy*, note No. 1, 1872 (reprinted by Open Court Publishing Company, LaSalle, Illinois, 1911), and *The Science of Mechanics*, 1883 (reprinted by Open Court Publishing Company, LaSalle, Illinois, 1902), Chap. II, Sec. VI.

[5] Because of the Thirring-Lense effect, [H. Thirring and J. Lense, Phys. Zeits. 19, 156 (1918)], the rotating laboratory would have a weak effect on the axis of the gyroscope.

sidering the laws of physics, including their quantitative aspects, inside a static massive spherical shell. It is well known that the interior Schwarzschild solution is flat and can be expressed in a coordinate system Minkowskian in the interior. Also, according to general relativity all Minkowskian coordinate systems are equivalent and the mass and radius of the spherical shell have no discernible effects upon the laws of physics as they are observed in the interior. Apparently the spherical shell does not contribute in any discernible way to inertial effects in the interior. What would happen if the mass of the shell were decreased, or its radius increased without limit? It might be remarked also that Komar[6] has attempted, without success, to find suitable boundary- and initial-value conditions for general relativity which would bring into evidence Mach's principle.

The third alternative is the subject of this paper. Actually the objectives of this paper are more limited than the formulation of a theory in complete accord with Mach's principle. Such a program would consist of two parts, the formulation of a suitable field theory and the formulation of suitable boundary- and initial-value conditions for the theory which would make the space geometry depend uniquely upon the matter distribution. This latter part of the problem is treated only partially.

At the end of the last section we shall briefly return again to the problem of the rotating laboratory.

A principle as sweeping as that of Mach, having its origins in matters of philosophy, can be described in the absence of a theory in a qualitative way only. A model of a theory incorporating elements of Mach's principle has been given by Sciama.[7] From simple dimensional arguments[8,9] as well as the discussion of Sciama, it has appeared that, with the assumption of validity of Mach's principle, the gravitational constant G is related to the mass distribution in a uniform expanding universe in the following way:

$$GM/Rc^2 \sim 1. \tag{1}$$

Here M stands for the finite mass of the visible (i.e., causally related) universe, and R stands for the radius of the boundary of the visible universe.

The physical ideas behind Eq. (1) have been given in references 7–9 and can be summarized easily. As stated before, according to Mach's principle the only meaningful motion is that relative to the rest of the matter in the universe, and the inertial reaction experienced in a laboratory accelerated relative to the distant matter of the universe may be interpreted equivalently as a gravitational force acting on a fixed laboratory due to the presence of distant accelerated matter.[7] This interpretation of the inertial reaction carries with it an interesting implication. Consider a test body falling toward the sun. In a coordinate system so chosen that the object is not accelerating, the gravitational pull of the sun may be considered as balanced by another gravitational pull, the inertial reaction.[8] Note that the balance is not disturbed by a doubling of all gravitational forces. Thus the acceleration is determined by the mass distribution in the universe, but is independent of the strength of gravitational interactions. Designating the mass of the sun by m_s and its distance by r enables the acceleration to be expressed according to Newton as $a=Gm_s/r^2$ or, from dimensional arguments, in terms of the mass distribution as $a\sim mRc^2/Mr^2$. Combining the two expressions gives Eq. (1).

This relation has significance in a rough order-of-magnitude manner only, but it suggests that either the ratio of M to R should be fixed by the theory, or alternatively that the gravitational constant observed locally should be variable and determined by the mass distribution about the point in question. The first of these two alternatives is of course, in part, simply the limitation of mass distribution which it might be hoped would result from some boundary condition on the field equations of general relativity. The second alternative is not compatible with the "strong principle of equivalence"[10] and general relativity. The reasons for this will be discussed below.

If the inertial reaction may be interpreted as a gravitational force due to distant accelerated matter, it might be expected that the locally observed values of the inertial masses of particles would depend upon the distribution of matter about the point in question. It should be noted, however, that there is a fundamental ambiguity in a statement of this type, for there is no direct way in which the mass of a particle such as an electron can be compared with that of another at a different space-time point. Mass ratios can be compared at different points, but not masses. On the other hand, gravitation provides another characteristic mass

$$(\hbar c/G)^{\frac{1}{2}}=2.16\times10^{-5}\text{ g}, \tag{2}$$

and the mass ratio, the dimensionless number

$$m(G/\hbar c)^{\frac{1}{2}}\cong5\times10^{-23}, \tag{3}$$

provides an unambiguous measure of the mass of an electron which can be compared at different space-time points.

It should also be remarked that statements such as "$\hbar$ and c are the same at all space-time points" are in the same way meaningless within the same context until a method of measurement is prescribed. In fact, it should be noted that $\hbar$ and c may be defined to be constant. A set of physical "constants" may be defined as constant if they cannot be combined to form one or

[6] A. Komar, Ph.D. thesis, Princeton University, 1956 (unpublished).

[7] D. W. Sciama, Monthly Notices Roy. Astron. Soc. 113, 34 (1953); *The Unity of the Universe* (Doubleday & Company, Inc., New York, 1959), Chaps. 7–9.

[8] R. H. Dicke, Am. Scientist 47, 25 (1959).

[9] R. H. Dicke, Science 129, 621 (1959).

[10] R. H. Dicke, Am. J. Phys. 29, 344 (1960).

more dimensionless numbers. The necessity for this limitation is obvious, for a dimensionless number is invariant under a transformation of units and the question of the constancy of such dimensionless numbers is to be settled, not by definition, but by measurements. A set of such independent physical constants which are constant by definition is "complete" if it is impossible to include another without generating dimensionless numbers.

It should be noted that if the number, Eq. (3), should vary with position and $\hbar$ and c are defined as constant, then either m or G, or both, could vary with position. There is no fundamental difference between the alternatives of constant mass or constant G. However, one or the other may be more convenient, for the formal structure of the theory would, in a superficial way, be quite different for the two cases.

To return to Eq. (3), the odd size of this dimensionless number has often been noticed as well as its apparent relation to the large dimensionless numbers of astrophysics. The apparent relation of the square of the reciprocal of this number [Eq. (3)] to the age of the universe expressed as a dimensionless number in atomic time units and the square root of the mass of the visible portion of the universe expressed in proton mass units suggested to Dirac[11] a causal connection that would lead to the value of Eq. (3) changing with time. The significance of Dirac's hypothesis from the standpoint of Mach's principle has been discussed.[8]

Dirac postulated a detailed cosmological model based on these numerical coincidences. This has been criticized on the grounds that it goes well beyond the empirical data upon which it is based.[8] Also in another publication by one of us (R. H. D.), it will be shown that it gives results not in accord with astrophysical observations examined in the light of modern stellar evolutionary theory.

On the other hand, it should be noted that a large dimensionless physical constant such as the reciprocal of Eq. (3) must be regarded as either determined by nature in a completely capricious fashion or else as related to some other large number derived from nature. In any case, it seems unreasonable to attempt to derive a number like 10^{22} from theory as a purely mathematical number involving factors such as $4\pi/3$.

It is concluded therefore, that although the detailed structure of Dirac's cosmology cannot be justified by the weak empirical evidence on which it is based, the more general conclusion that the number [Eq. (3)] varies with time has a more solid basis.

If, in line with the interpretation of Mach's principle being developed, the dimensionless mass ratio given by Eq. (3) should depend upon the matter distribution in the universe, with $\hbar$ and c constant by definition, either the mass m or the gravitational constant, or both, must vary. Although these are alternative descriptions of the same physical situation, the formal structure of the theory would be very different for the two cases. Thus, for example, it can be easily shown that uncharged spinless particles whose masses are position dependent no longer move on geodesics of the metric. (See Appendix I.) Thus, the definition of the metric tensor is different for the two cases. The two metric tensors are connected by a conformal transformation.

The arbitrariness in the metric tensor which results from the indefiniteness in the choice of units of measure raises questions about the physical significance of Riemannian geometry in relativity.[12] In particular the 14 invariants which characterize the space are generally not invariant under a conformal transformation interpreted as a redefinition of the metric tensor in the *same* space.[13] Matters are even worse, for a more general redefinition of the units of measure can be used to reduce all 14 invariants to zero. It should be said that these remarks should not be interpreted as casting doubt on the correctness or usefulness of Riemannian geometry in relativity, but rather that each such geometry is but a particular representation of the theory. It would be expected that the physical content of the theory should be contained in the invariants of the group of position-dependent transformations of units and coordinate transformations. The usual invariants of Riemannian geometry are not invariants under this wider group.

In general relativity the representation is one in which units are chosen so that atoms are described as having physical properties independent of location. It is assumed that this choice is possible!

In accordance with the above, a particular choice of units is made with the realization that the choice is arbitrary and without an invariant significance. The theoretical structure appears to be simpler if one defines the inertial masses of elementary particles to be constant and permits the gravitational constant to vary. It should be noted that this is possible only if the mass ratios of elementary particles are constant. There may be reasonable doubt about this.[9,10] On the other hand, it would be expected that such quantities as particle mass ratios or the fine-structure constant, if they depend upon mass distributions in the universe, would be much less sensitive in their dependence[9] rather than the number given by Eq. (3) and their variation could be neglected in a first crude theory. Also it should be remarked that the requirements of the approximate constancy of the ratio of inertial to passive gravitational mass,[14] and the extremely stringent requirement of spatial isotropy,[15] impose conditions so severe that it has been found to be difficult, if not impossible, to

[11] P. A. M. Dirac, Proc. Roy. Soc. (London) A165, 199 (1938).

[12] E. P. Wigner has questioned the physical significance of Riemannian geometry on other grounds [Relativity Seminar, Stevens Institute, May 9, 1961 (unpublished)].

[13] B. Hoffman, Phys. Rev. 89, 49 (1953).

[14] R. Eötvös, Ann. Physik 68, 11 (1922).

[15] V. W. Hughes, H. G. Robinson, and V. Beltran-Lopez, Phys. Rev. Letters 4, 342 (1960).

construct a satisfactory theory with a variable fine-structure constant.

It should be emphasized that the above argument involving the large dimensionless numbers, Eq. (3), does not concern Mach's principle directly, but that Mach's principle and the assumption of a gravitational "constant" dependent upon mass distributions gives a reasonable explanation for varying "constants."

It would be expected that both nearby and distant matter should contribute to the inertial reaction experienced locally. If the theory were linear, which one does not expect, Eq. (1) would suggest that it is the reciprocal of the gravitational constant which is determined locally as a linear superposition of contributions from the matter in the universe which is causally connected to the point in question. This can be expressed in a somewhat symbolic equation:

$$G^{-1} \sim \sum_i (m_i/r_i c^2), \tag{4}$$

where the sum is over all the matter which can contribute to the inertial reaction. This equation can be given an exact meaning only after a theory has been constructed. Equation (4) is also a relation from Sciama's theory.

It is necessary to say a few words about the equivalence principle as it is used in general relativity and as it relates to Mach's principle. As it enters general relativity, the equivalence principle is more than the assumption of the local equivalence of a gravitational force and an acceleration. Actually, in general relativity it is assumed that the laws of physics, including numerical content (i.e., dimensionless physical constants), as observed locally in a freely falling laboratory, are independent of the location in time or space of the laboratory. This is a statement of the "strong equivalence principle."[9,10] The interpretation of Mach's principle being developed here is obviously incompatible with strong equivalence. The local equality of all gravitational accelerations (to the accuracy of present experiments) is the "weak equivalence principle." It should be noted that it is the "weak equivalence principle" that receives strong experimental support from the Eötvös experiment.

Before attempting to formulate a theory of gravitation which is more satisfactory from the standpoint of Mach's principle than general relativity, the physical ideas outlined above, and the assumptions being made, will be summarized:

1. An approach to Mach's principle which attempts, with boundary conditions, to allow only those mass distributions which produce the "correct" inertial reaction seems foredoomed, for there do exist large localized masses in the universe (e.g., white dwarf stars) and a laboratory could, in principle, be constructed near such a mass. Also it appears to be possible to modify the mass distribution. For example, a massive concrete spherical shell could be constructed with the laboratory in its interior.

2. The contrary view is that locally observed inertial reactions depend upon the mass distribution of the universe about the point of observation and consequently the quantitative aspects of locally observed physical laws (as expressed in the physical "constants") are position dependent.

3. It is possible to reduce the variation of physical "constants" required by this interpretation of Mach's principle to that of a single parameter, the gravitational "constant."

4. The separate but related problem posed by the existence of very large dimensionless numbers representing quantitative aspects of physical laws is clarified by noting that these large numbers involve G and that they are of the same order of magnitude as the large numbers characterizing the size and mass distribution of the universe.

5. The "strong principle of equivalence" upon which general relativity rests is incompatible with these ideas. However, it is only the "weak principle" which is directly supported by the very precise experiments of Eötvös.

A THEORY OF GRAVITATION BASED ON A SCALAR FIELD IN A RIEMANNIAN GEOMETRY

The theory to be developed represents a generalization of general relativity. It is not a completely geometrical theory of gravitation, as gravitational effects are described by a scalar field in a Riemannian manifold. Thus, the gravitational effects are in part geometrical and in part due to a scalar interaction. There is a formal connection between this theory and that of Jordan,[16] but there are differences and the physical interpretation is quite different. For example, the aspect of mass creation[17] in Jordan's theory is absent from this theory.

In developing this theory we start with the "weak principle of equivalence." The great accuracy of the Eötvös experiment suggests that the motion of uncharged test particles in this theory should be, as in general relativity, a geodesic in the four-dimensional manifold.

With the assumption that only the gravitational "constant" (or active gravitational masses) vary with position, the laws of physics (exclusive of gravitation) observed in a freely falling laboratory should be unaffected by the rest of the universe as long as self-gravitational fields are negligible. The theory should be constructed in such a way as to exhibit this effect.

If the gravitational "constant" is to vary, it should be

[16] P. Jordon, *Schwerkraft and Weltall* (Friedrich Vieweg and Sohn, Braunschweig, 1955); Z. Physik 157, 112 (1959). In this second reference, Jordan has taken cognizance of the objections of Fierz (see reference 19) and has written his variational principle in a form which differs in only two respects from that expressed in Eq. (16). See also reference 20.

[17] For a discussion of this, see H. Bondi, *Cosmology*, 2nd edition, 1960.

a function of some scalar field variable. The contracted metric tensor is a constant and devoid of interest. The scalar curvature and the other scalars formed from the curvature tensor are also devoid of interest as they contain gradients of the metric tensor components, and fall off more rapidly than r^{-1} from a mass source. Thus such scalars are determined primarily by nearby mass distributions rather than by distant matter.

As the scalars of general relativity are not suitable, a new scalar field is introduced. The primary function of this field is the determination of the local value of the gravitational constant.

In order to generalize general relativity, we start with the usual variational principle of general relativity from which the equations of motion of matter and nongravitational fields are obtained as well as the Einstein field equation, namely,[18]

$$0=\delta\int[R+(16\pi G/c^4)L](-g)^{\frac{1}{2}}d^4x. \quad (5)$$

Here, R is the scalar curvature and L is the Lagrangian density of matter including all nongravitational fields.

In order to generalize Eq. (5) it is first divided by G, and a Lagrangian density of a scalar field ϕ is added inside the bracket. G is assumed to be a function of ϕ. Remembering the discussion in connection with Eq. (4), it would be reasonable to assume that G^{-1} varies as ϕ, for then a simple wave equation for ϕ with a scalar matter density as source would give an equation roughly the same as (4).

The required generalization of Eq. (6) is clearly

$$0=\delta\int[\phi R+(16\pi/c^4)L-\omega(\phi_{,i}\phi^{,i}/\phi)](-g)^{\frac{1}{2}}d^4x. \quad (6)$$

Here ϕ plays a role analogous to G^{-1} and will have the dimensions $ML^{-3}T^2$. The third term is the usual Lagrangian density of a scalar field, and the scalar in the denominator has been introduced to permit the constant ω to be dimensionless. In any sensible theory ω must be of the general order of magnitude of unity.

It should be noted that the term involving the Lagrangian density of matter in Eq. (6) is identical with that in Eq. (5). Thus the equations of motion of matter in a given externally determined metric field are the same as in general relativity. The difference between the two theories lies in the gravitational field equations which determine g_{ij}, rather than in the equations of motion in a given metric field.

It is evident, therefore, that, as in general relativity, the energy-momentum tensor of matter must have a vanishing covariant divergence,

$$T^{ij}{}_{;j}=0, \quad (7)$$

where

$$T^{ij}=[2/(-g)^{\frac{1}{2}}](\partial/\partial g_{ij})[(-g)^{\frac{1}{2}}L]. \quad (8)$$

It is assumed that L does not depend explicitly upon derivatives of g_{ij}.

Jordan's theory has been criticized by Fierz[19] on the grounds that the introduction of matter into the theory required further assumptions concerning the standards of length and time. Further, the mass creation aspects of this theory and the nonconservation of the energy-momentum tensor raise serious questions about the significance of the energy-momentum tensor. To make it clear that this objection cannot be raised against this version of the theory, we hasten to point out that L is assumed to be the normal Lagrangian density of matter, a function of matter variables and of g_{ij} only, *not* a function of ϕ. It is a well-known result that for *any* reasonable metric field distribution g_{ij} (a distribution which need not be a solution of the field equations of g_{ij}), the matter equations of motion, obtained by varying matter variables in Eq. (6), are such that Eq. (7) is satisfied with T^{ij} defined by Eq. (8). Thus Eq. (7) is satisfied and this theory does not contain a mass creation principle.

The wave equation for ϕ is obtained in the usual way by varying ϕ and $\phi_{,i}$ in Eq. (6). This gives

$$2\omega\phi^{-1}\Box\phi-(\omega/\phi^2)\phi^{,i}\phi_{,i}+R=0. \quad (9)$$

Here the generally covariant d'Alembertian $\Box$ is defined to be the covariant divergence of $\phi^{,i}$:

$$\Box\phi=\phi^{,i}{}_{;i}=(-g)^{-\frac{1}{2}}[(-g)^{\frac{1}{2}}\phi^{,i}]_{,i}. \quad (10)$$

From the form of Eq. (9), it is evident that ϕR and the Lagrangian density of ϕ serves as the source term for the generation of ϕ waves. Remarkably enough, as will be shown below, this equation can be transformed so as to make the source term appear as the contracted energy-momentum tensor of matter alone. Thus, in accordance with the requirements of Mach's principle, ϕ has as its sources the matter distribution in space.

By varying the components of the metric tensor and their first derivatives in Eq. (6), the field equations for the metric field are obtained. This is the analog of the Einstein field equation and is

$$R_{ij}-\tfrac{1}{2}g_{ij}R=(8\pi\phi^{-1}/c^4)T_{ij}+(\omega/\phi^2)(\phi_{,i}\phi_{,j}-\tfrac{1}{2}g_{ij}\phi_{,k}\phi^{,k})+\phi^{-1}(\phi_{,i;j}-g_{ij}\Box\phi). \quad (11)$$

The left side of Eq. (11) is completely familiar and needs no comment. Note that the first term on the right is the usual source term of general relativity, but with the variable gravitational coupling parameter ϕ^{-1}. Note also that the second term is the energy-momentum tensor of the scalar field, also coupled with the gravitational coupling ϕ^{-1}. The third term is foreign and results from the presence of second derivatives of the metric

[18] L. Landau and E. Liftschitz, *Classical Theory of Fields* (Addison-Wesley Publishing Company, Reading, Massachusetts, 1951).

[19] M. Fierz, Helv. Phys. Acta. 29, 128 (1956).

tensor in R in Eq. (6). These second derivatives are eliminated by integration by parts to give a divergence and the extra terms. It should be noted that when the first term dominates the right side of Eq. (11), the equation differs from Einstein's field equation by the presence of a variable gravitational constant only.

While the "extra" terms in Eq. (12) may at first seem strange, their role is essential. They are needed if Eq. (7) is to be consistent with Eqs. (9) and (11). This can be seen by multiplying Eq. (11) by ϕ and then taking the covariant divergence of the resulting equation. The divergence of these two terms cancels the term $\phi_{,i}R_j{}^i=\phi^{,i}R_{ji}$. To show this, use is made of the well-known property of the full curvature tensor that it serves as a commutator for two successive gradient operations applied to an arbitrary vector.

If Eq. (11) is contracted there results

$$-R=(8\pi\phi^{-1}/c^4)T-(\omega/\phi^2)\phi_{,k}\phi^{,k}-3\phi^{-1}\Box\phi. \quad (12)$$

Equation (12) can be combined with Eq. (9) to give a new wave equation for ϕ[20]:

$$\Box\phi=[8\pi/(3+2\omega)c^4]T. \quad (13)$$

With the sign convention

$$ds^2=g_{ij}dx^idx^j \quad \text{and} \quad g_{00}<0,$$

for a fluid

$$T_{ij}=-(p+\epsilon)u_iu_j+pg_{ij}, \quad (14)$$

so that

$$T=-\epsilon+3p, \quad (15)$$

where ϵ is the energy density of the matter in comoving coordinates and p is the pressure in the fluid. With this sign convention and ω positive, the contribution to ϕ from a local mass is positive. Note, however, that there is no direct electromagnetic contribution to T, as the contracted energy-momentum tensor of an electromagnetic field is identically zero. However, bound electromagnetic energy does contribute indirectly through the stress terms in other fields, the stresses being necessary to confine the electromagnetic field.[21] In conclusion, ω must be positive if the contribution to the inertial reaction from nearby matter is to be positive.

THE WEAK FIELD APPROXIMATION

An approximate solution to Eqs. (11) and (13) which is of first order in matter mass densities is now obtained. This weak-field solution plays the same important role that the corresponding solution fills in general relativity.

As in general relativity the metric tensor is written as

$$g_{ij}=\eta_{ij}+h_{ij}, \quad (16)$$

where η_{ij} is the Minkowskian metric tensor

$$\eta_{00}=-1, \quad \eta_{\alpha\alpha}=1, \quad \alpha=1,2,3. \quad (17)$$

h_{ij} is computed to the linear first approximation only. In similar fashion let $\phi=\phi_0+\xi$, where ϕ_0 is a constant and is to be computed to first order in mass densities.

The weak-field solution to Eq. (13) is computed first. In this equation g_{ij} may be replaced by η_{ij}:

$$\Box\phi=\Box\xi=\frac{1}{(-g)^{\frac{1}{2}}}[(-g)^{\frac{1}{2}}g^{ij}\xi_{,i}]_{,j}$$
$$=\nabla^2\xi-\frac{\partial^2\xi}{\partial t^2}=\frac{8\pi T}{(3+2\omega)c^4}. \quad (18)$$

It is evident that a retarded-time solution to Eq. (18) can be written as

$$\xi=-[2/(3+2\omega)]\int Td^3x/rc^4, \quad (19)$$

where T is to be evaluated at the retarded time.

The weak-field solution to Eq. (11) is obtained in a manner similar to that of general relativity by introducing a coordinate condition that simplifies the equation. As a preliminary step let

$$\gamma_{ij}=h_{ij}-\tfrac{1}{2}\eta_{ij}h,$$
$$\sigma_i=\gamma_{ij,k}\eta^{jk}. \quad (20)$$

Equation (11) can be written to first order in h_{ij} and ξ as

$$-\tfrac{1}{2}\{\Box\gamma_{ij}-\sigma_{i,j}-\sigma_{j,i}+\eta_{ij}\sigma_{k,l}\eta^{kl}\}$$
$$=[\xi_{,i,j}-\eta_{ij}\Box\xi]\phi_0^{-1}+\frac{8\pi}{c^4}\phi_0^{-1}T_{ij}. \quad (21)$$

Equation (21) can now be simplified by introducing the four coordinate conditions

$$\sigma_i=\xi_{,i}\phi_0^{-1}, \quad (22)$$

and the notation

$$\alpha_{ij}=\gamma_{ij}-\eta_{ij}\xi\phi_0^{-1}. \quad (23)$$

Equation (21) then becomes

$$\Box\alpha_{ij}=-(16\pi/c^4)\phi_0^{-1}T_{ij}, \quad (24)$$

with the retarded-time solution

$$\alpha_{ij}=(4\phi_0^{-1}/c^4)\int(T_{ij}/r)d^3x. \quad (25)$$

From Eqs. (20) and (23),

$$h_{ij}=\alpha_{ij}-\tfrac{1}{2}\eta_{ij}\alpha-\eta_{ij}\xi\phi_0^{-1}. \quad (26)$$

Thus

$$h_{ij}=\frac{4\phi_0^{-1}}{c^4}\int\frac{T_{ij}}{r}d^3x-\frac{4\phi_0^{-1}}{c^4}\left(\frac{1+\omega}{3+2\omega}\right)\eta_{ij}\int\frac{T}{r}d^3x. \quad (27)$$

[20] There are but two formal differences between the field equations of this theory and those of the particular form of Jordan's theory given in Z. Physik 157, 112 (1959). First, Jordan has defined his scalar field variable reciprocal to ϕ. Thus, the simple wave character of the scalar field equation [Eq. (13)] is not so clear and the physical arguments based on Mach's principle and leading to Eq. (4) have not been satisfied. Second, as a result of its outgrowth from his five-dimensional theory, Jordan has limited his matter variables to those of the electromagnetic field.

[21] C. Misner and P. Putnam, Phys. Rev. 116, 1045 (1959).

For a stationary mass point of mass M these equations become

$$\phi=\phi_0+\xi=\phi_0+2M/(3+2\omega)c^2r, \tag{28}$$

$$g_{00}=\eta_{00}+h_{00}=-1+(2M\phi_0^{-1}/rc^2)[1+1/(3+2\omega)],$$
$$g_{\alpha\alpha}=1+(2M\phi_0^{-1}/rc^2)[1-1/(3+2\omega)], \quad \alpha=1,2,3, \tag{29}$$
$$g_{ij}=0, \quad i\neq j.$$

The above weak-field solution is sufficiently accurate to discuss the gravitational red shift and the deflection of light. However, to discuss the rotation of the perihelion of Mercury's orbit requires a solution good to the second approximation for g_{00}.

The gravitational red shift is determined by g_{00} which also determines the gravitational weight of a body. Thus, there is no anomaly in the red shift. The strange factor $(4+2\omega)/(3+2\omega)$ in g_{00} is simply absorbed into the definition of the gravitational constant

$$G_0=\phi_0^{-1}(4+2\omega)(3+2\omega)^{-1}. \tag{29a}$$

On the other hand, there is an anomaly in the deflection of light. This is determined, not by g_{00} alone, but by the ratio $g_{\alpha\alpha}/g_{00}$. It is easily shown that the light deflection computed from general relativity differs from the value in this theory by the above factor. Thus, the light deflection computed from this theory is

$$\delta\theta=(4G_0M/Rc^2)[(3+2\omega)/(4+2\omega)], \tag{30}$$

where R is the closest approach distance of the light ray to the sun of mass M. It differs from the general relativity value by the factor in brackets. The accuracy of the light deflection observations is too poor to set any useful limit to the size of ω.

On the contrary, there is fair accuracy in the observation of the perihelion rotation of the orbit of Mercury and this does serve to set a limit to the size of ω. In order to discuss the perihelion rotation, an exact solution for a static mass point will be written.

STATIC SPHERICALLY SYMMETRIC FIELD ABOUT A POINT MASS[22]

Expressing the line element in isotropic form gives

$$ds^2=-e^{2\alpha}dt^2+e^{2\beta}[dr^2+r^2(d\theta^2+\sin^2\theta d\phi^2)], \tag{31}$$

where α and β are functions of r only. For $\omega>\frac{3}{2}$ the general vacuum solution can be written in the form

$$e^{2\alpha}=e^{2\alpha_0}[(1-B/r)/(1+B/r)]^{2/\lambda},$$
$$e^{2\beta}=e^{2\beta_0}(1+B/r)^4[(1-B/r)/(1+B/r)]^{2[(\lambda-C-1)/\lambda]}, \tag{32}$$
$$\phi=\phi_0[(1-B/r)/(1+B/r)]^{-C/\lambda},$$

where

$$\lambda=[(C+1)^2-C(1-\tfrac{1}{2}\omega C)]^{\frac{1}{2}}, \tag{33}$$

and α_0, β_0, ϕ_0, B, and C are arbitrary constants. It may be seen by substitution of Eqs. (31) and (32) into Eqs. (13) and (11) that this is the static solution for spherical symmetry when $T_{ij}=0$.

To discuss the perihelion rotation of a planet about the sun requires a specification of the arbitrary constants in Eq. (32) in such a way that this solution agrees in the weak-field limit [first order in $M/(c^2r\phi_0)$] with the previously obtained solution, Eqs. (28) and (29). It may be easily verified that the appropriate choice of constants is

$$\phi_0 \text{ given by Eq. (29a)};$$
$$\alpha_0=\beta_0=0,$$
$$C\cong-1/(2+\omega),$$
$$B\cong(M/2c^2\phi_0)[(2\omega+4)/(2\omega+3)]^{\frac{1}{2}}, \tag{34}$$

with λ given by Eq. (33).

Remembering the previous discussion of Mach's principle, it is clear that the asymptotic Minkowskian character of this solution makes sense only if there is matter at great distance. Second, the matching of the solution to the weak-field solution is permissible only if the sun is a suitable mass distribution for the weak-field approximation. Namely, the field generated by the sun must be everywhere small, including the interior of the sun. With this assumption, the solution, Eqs. (31), (32), (33), and (34), is valid for the sun. It does not, however, justify its use for a point mass.

The question might be raised as to whether a matching of solutions, accurate to first order only in $M/(\phi_0c^2r)$, has a validity to the second order. It should be noted, however, that this matching condition is sufficient to assign sufficiently accurate values to all the adjustable parameters in Eqs. (32) except λB, and that we do not demand that λB be determined in terms of an integration over the matter distribution of the sun; it is determined from the observed periods of the planetary motion.

With the above solution, it is a simple matter to calculate the perihelion rotation. The labor is reduced if $e^{2\alpha}$ is carreid only to second order in $M/(c^2r\phi_0)$, and $e^{2\beta}$ to first order. The result of this calculation is that the relativistic perihelion rotation rate of a planetary orbit is

$$[(4+3\omega)/(6+3\omega)]\times(\text{value of general relativity}). \tag{35}$$

This is a useful result as it sets a limit on permissible values of the constant ω. If it be assumed that the observed relativistic perihelion rotation agrees with an accuracy of 8% or less with the computed result of general relativity, it is necessary for ω in Eq. (35) to satisfy the inequality

$$\omega\geq 6. \tag{36}$$

The observed relativistic perihelion rotation of Mercury (after subtracting off planetary perturbations and other effects presumed known) is $42.6''\pm0.9''$/century.[23] For

[22] This form of solution was suggested to one of us (C. B.) by C. Misner.

[23] G. M. Clemence, Revs. Modern Phys. 19, 361 (1947).

$\omega=6$, the computed relativistic perihelion rotation rate is 39.4″. The difference of 3.2″ of arc per century is 3.3 times the formal probable error. It should also be remarked that Clemence[24] has shown that if some recent data on the general precession constant and the masses of Venus and the Earth-Moon system are adopted, the result is an increase in the discrepancy to 3.7″ while decreasing the formal probable error by a factor of 2.

The formal probable error is thus substantially less than 3.2″ arc, but it may be reasonable to allow this much to take account of systematic errors in observations and future modification of observations, adopted masses, and orbit parameters. Apparently there are many examples in celestial mechanics of quantities changing by substantially more than the formal probable errors. Thus, for example, the following is a list of values which have been assigned to the reciprocal of Saturn's mass (in units of the sun's reciprocal mass) by authors at various times:

$M^{-1}=3501.6\ \pm0.8$, Bessel (1833) from the motion of Saturn's moon Titan;

$=3494.8\ \pm0.3$, Jeffrey (1954) and G. Struve (1924–37) (Titan);

$=3502.2\ \pm0.53$, Hill (1895) Saturn's perturbations of Jupiter;

$=3497.64\pm0.27$, Hertz (1953) Saturn's perturbations of Jupiter;

$=3499.7\ \pm0.4$, Clemence (1960) Saturn's perturbations of Jupiter.

While this example may be atypical, it does suggest that considerable caution be used in judging errors in celestial mechanics.

MACH'S PRINCIPLE

A complete analysis of Mach's principle in relation to the present scalar theory will not be attempted here. However, because of the motivation of this theory by Mach's principle, it is desirable to give a brief discussion. Having formulated the desired field equations, it remains to establish initial-value and boundary conditions to bring the theory in accord with Mach's principle. This will not be attempted in a general way, but in connection with special problems only.

The qualitative discussion in the Introduction suggested that for a static mass shell of radius R and mass M, the gravitational constant in its interior should satisfy the relation

$$GM/Rc^2\sim 1. \tag{37}$$

Equivalently

$$\phi\sim M/Rc^2. \tag{38}$$

It may be noted that in a flat space, with the boundary condition that $\phi=0$ at infinity, Eq. (13) has as a solution for the interior $r<R$

$$\phi=2M/(3+2\omega)Rc^2. \tag{39}$$

This is a hopeful sign and bodes well for Mach's principle within the framework of this theory. One should not be misled by this simple result, however. There are several factors which invalidate Eq. (39) as a quantitative result. First, space is not flat, but is warped by the presence of the mass shell. Second, the asymptotic zero boundary condition may be impossible for the exact static solution to the field equation. Third, it may be impossible to construct such a static massive shell in a universe empty except for the shell, without giving matter nonphysical properties. This third point is not meant to imply a practical limitation of real materials, but rather a fundamental limitation on the stress-energy tensor of matter. In this connection it should be noted that if Eq. (37) is to be satisfied, independent of the size and mass of the shell, the gravitationally induced stresses in the shell are enormous, of the order of magnitude of the energy density of the spherical shell. It is not possible to reduce the stress by decreasing M or increasing R, as the resulting change in the gravitational constant compensates for the change. We ignore here the above third point and assume for the moment that such a shell can be constructed in principle.

To turn now to the massive static shell, consider first the solution to the field equations in the exterior region, $r>R$. This solution is encompassed in the general solution Eqs. (32) and (33). Note that the boundary condition

$$\phi\to 0 \quad \text{as} \quad r\to\infty \tag{40}$$

is not possible.

On the other hand, it is possible to change the sign in the brackets in Eq. (32) and absorb the complex factor into the constant before the bracket. These equations may then be assumed to hold for $r<B$ rather than for $r>B$ as in Eq. (32). The equations now have the form

$$\begin{aligned} e^{2\alpha}&=e^{2\alpha_0}[(B/r-1)/(B/r+1)]^{2/\lambda},\\ e^{2\beta}&=e^{2\beta_0}[1+B/r]^4[(B/r-1)/(B/r+1)]^{2[(\lambda-C-1)/\lambda]},\\ \phi&=\phi_0[(B/r-1)/(B/r+1)]^{C/\lambda}. \end{aligned} \tag{41}$$

It may be noted that this solution, interesting for $r>R$ and $\lambda>0$ only, results in space closure at the radius $r=B$ provided

$$(\lambda-C-1)/\lambda>0. \tag{42}$$

In similar fashion at the closure radius, $\phi\to 0$, provided $C>0$.

Equations (36) and (33) require that

$$C>2/\omega. \tag{43}$$

That this boundary condition is appropriate to Mach's principle can be seen by an application of

[24] G. M. Clemence (private communication). One of us (R.H.D.) is grateful for helpful correspondence and conversations with Dr. Clemence on this and other aspects of celestial mechanics.

Green's theorem. Introduce a Green's function η satisfying

$$\Box\eta=(-g)^{-\frac{1}{2}}[(-g)^{\frac{1}{2}}g^{ij}\eta_{,j}]_{,i}=(-g)^{-\frac{1}{2}}\delta^4(x-x_0), \quad (44)$$

also

$$\Box\phi=[8\pi/(3+2\omega)c^4]T. \quad (45)$$

Combining Eqs. (44) and (45) after the appropriate multiplications gives

$$[(-g)^{\frac{1}{2}}g^{ij}(\eta\phi_{,i}-\phi\eta_{,i})]_{,j} = (-g)^{\frac{1}{2}}[8\pi/(3+2\omega)c^4]T\eta-\phi\delta^4(x-x_0). \quad (46)$$

It is assumed that η is an "advanced-wave" solution to Eq. (44), i.e., $\eta=0$ for all time future to t_0. The condition given by Eq. (42) implies a finite coordinate time for light to propagate from the radius B to R, the radius of the shell, hence to any interior point x_0.

Integrate Eq. (46) over the interior of the closed space ($r<B$) between the time $t_2>t_0$ and the space like surface S_1 so chosen that the η wave starts out at the radius $r=B$ at times lying on this surface and that the normal to the surface at $r=B$ has no component in the r direction. The integral of the left side of Eq. (46), after conversion to a surface integral, vanishes, for η and η_0 both vanish on t_2, and both ϕ and $\phi_{,i}$ vanish on S_1 at $r=B$, with $i\neq 1$.

The integral over the right side of Eq. (46) yields

$$\phi(x_0)=[8\pi/(3+2\omega)c^4]\int\eta T(-g)^{\frac{1}{2}}d^4x, \quad (47)$$

or

$$\phi(x_0)\sim M/Rc^2. \quad (47a)$$

Note that this equation states that ϕ at the point x_0 is determined by an integral over the mass distribution, with each mass element contributing a wavelet which propagates to the point x_0. This is just the interpretation of Mach's principle desired.

COSMOLOGY

A physically more interesting problem to discuss from the standpoint of Mach's principle is the cosmological model derived from this theory. It will be recalled that the assumption of a uniform and isotropic space is supported to some extent by the observations of galaxy distribution. The kinematics of the comoving coordinate system is completely free of dynamical consideration. In spherical coordinates, a form of the line element is

$$ds^2=-dt^2+a^2(t)[dr^2/(1-\lambda r^2)+r^2(d\theta^2+\sin^2\theta d\phi^2)], \quad (48)$$

with $\lambda=+1$ for a closed space, $\lambda=-1$ for open, and $\lambda=0$ for a flat space, and where $r<1$ for the closed space. The Hubble age associated with the rate of expansion of the univese and the galactic red shift is $a/\dot{a}=a/(da/dt)$.

The substitution

$$r=\sin\chi \text{ (closed space, } \lambda=+1)$$

or

$$r=\sinh\chi \text{ (open space, } \lambda=-1) \quad (49)$$

simplifies the expression for line element somewhat:

$$ds^2=-dt^2+a^2[d\chi^2+\sin^2\chi(d\theta+\sin^2\theta d\phi^2)] \quad \text{(closed space).} \quad (50)$$

The most interesting case physically seems to be the closed universe.

Using Eq. (50) for interval and writing the (0,0) component of Eq. (11),

$$R_0{}^0-\tfrac{1}{2}R=-(3/a^2)(\dot{a}^2\pm 1)$$

$$(+\text{, space closed; } -\text{, space open})$$

$$=\frac{8\pi\phi^{-1}}{c^4}T_0{}^0-\frac{\omega}{2\phi^2}\dot{\phi}^2+3\frac{\dot{a}}{a}\frac{\dot{\phi}}{\phi}. \quad (51)$$

Assuming negligible pressure in the universe we have $-T=-T_0{}^0=+\rho c^2$, where the mass density is ρ (observationally ρ seems to satisfy, $\rho>10^{-31}$ g/cm^3).

Again assuming negligible pressure, the energy density times a measure of the volume of the universe is constant. Hence

$$\rho a^3=\rho_0 a_0{}^3=\text{const.} \quad (52)$$

Substituting these results in Eq. (51) yields

$$\left(\frac{\dot{a}}{a}+\frac{1}{2}\frac{\dot{\phi}}{\phi}\right)^2+\frac{\lambda}{a^2}=\tfrac{1}{4}(1+\tfrac{2}{3}\omega)\left(\frac{\dot{\phi}}{\phi}\right)^2+\frac{8\pi}{3\phi}\rho_0\left(\frac{a_0}{a}\right)^3. \quad (53)$$

Here ρ_0 and a_0 refer to values at some arbitrary fixed time t_0. In similar fashion Eq. (13) becomes

$$(d/dt)(\dot{\phi}a^3)=[8\pi/(3+2\omega)]\rho_0 a_0{}^3. \quad (54)$$

After integration, Eq. (54) becomes

$$\dot{\phi}a^3=[8\pi/(3+2\omega)]\rho_0 a_0{}^3(t-t_c). \quad (55)$$

The constant of integration, t_c in Eq. (55), can be evaluated by considerations of Mach's principle.

As before, we introduce Mach's principle into this problem by expressing $\phi(t)$ as an advanced-wave integral over all matter. Equations (46) and (47) require some assumption about the history of matter in the universe. We assume that the universe expands from a highly condensed state. It is possible that in the intense gravitational field of this condensed state, matter is created. For a closed universe, matter from a previous cycle may be regenerated in this high-temperature state. In view of our present state of ignorance, there seems to be little point in speculating about the processes involved. In any case the creation process lies outside the present theory.

We assume, therefore, an initial state at the beginning of the expansion ($t=0$) with $a\cong 0$ and matter already present. Although pressure would certainly be important in such a highly condensed state, with expansion the pressure would rapidly fall and no great harm is done to the model if pressure effects are neglected. In fact, an integration of the initial high-pressure phase for a

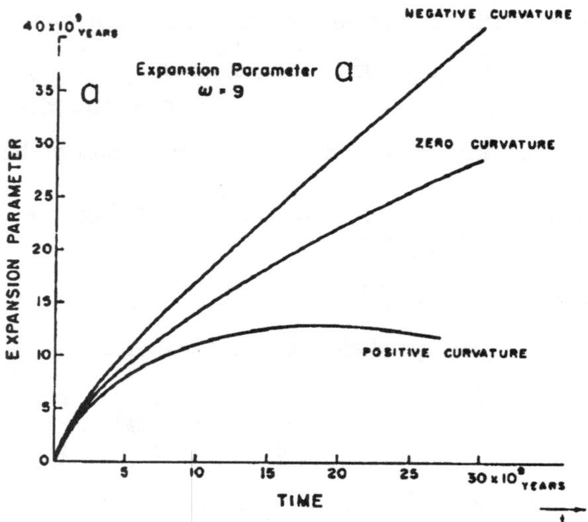

FIG. 1. The expansion parameter a as a function of t for the three cases, closed, open, and flat space with $\omega=9$.

particular cosmological model shows explicitly that it may be neglected to good approximation.

It is assumed that the inertial reaction, and hence ϕ, at time t_0 is determined uniquely by the matter distribution from $t=0$ to $t=t_0$. Hence, if Eq. (46) is integrated over all 3-space from $t=0$ to $t_1>t_0$, the surface integral obtained from the left-hand side should vanish. Initial conditions for Eqs. (53) and (55) in the form of values of a and ϕ at $t=0$ and a value of the constant t_c must be so chosen that the surface integral from Eq. (46) vanishes. In order for this surface integral to be meaningful at $t=0$, the a must be at least infinitesimally positive on the surface, otherwise the metric tensor is singular. If $t_c=0$ and $\phi=0$ on this surface, the surface integral vanishes. This follows because the vanishing factors ϕ and $a^3\phi_{,0}$ [see Eq. (55)] occur in the integral. It is concluded, therefore, that the appropriate initial conditions are $a=\phi=0$ with $t_c=0$. It should be noted that the other surface integral, over the surface $t=t_1$, vanishes since η and all its gradients are zero on this surface (advanced-wave solution).

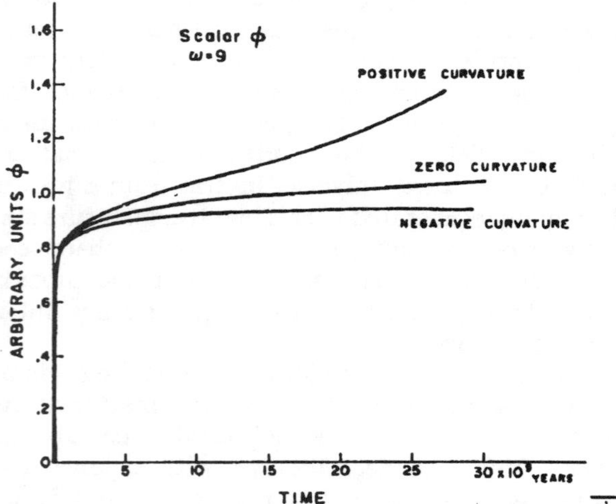

FIG. 2. The scalar ϕ, in arbitrary units, as a function of t for the three cases, closed, open, and flat space, with $\omega=9$.

Letting $t_c=0$ in Eq. (55) and combining with Eq. (53) gives

$$[(\dot{a}/a)+\tfrac{1}{2}(\dot{\phi}/\phi)]^2+\lambda a^{-2} =\tfrac{1}{4}(1+\tfrac{2}{3}\omega)(\dot{\phi}/\phi)^2+(1+\tfrac{2}{3}\omega)(\dot{\phi}/\phi)(1/t), \quad (57)$$

$$\phi a^3=[8\pi/(3+2\omega)]\rho_0 a_0^3 t. \quad (58)$$

It can be seen that for sufficiently small time the term $1/a^2$ in Eq. (57) is negligible and the solution differs only infinitesimally from the flat-space case. The resulting equations can be integrated exactly with the initial conditions

$$\phi=a=0; \quad t=0. \quad (59)$$

As both Eqs. (57) and (58) are now (in this approximation) homogeneous in (a,a_0), the solution is determined within a scale factor in a only.

This solution, good for the early expansion phases (i.e., $a\gg t$), is

$$\phi=\phi_0(t/t_0)^r, \quad a=a_0(t/t_0)^q, \quad (60)$$

with

$$r=2/(4+3\omega), \quad (61)$$

$$q=(2+2\omega)/(4+3\omega), \quad (62)$$

and

$$\phi_0=8\pi[(4+3\omega)/2(3+2\omega)]\rho_0 t_0^2. \quad (63)$$

For the flat-space case, the solution is exact for all $t>0$.

It should be noted that Eq. (63) is compatible with Eq. (1), for in Eq. (1) M is of the order of magnitude of $\rho_0 c^3 t_0^3$ and R is approximately ct_0. Thus, the initial conditions are compatible with Mach's principle as it has been formulated here.

For a nonflat space, the only feasible method of integrating Eqs. (57) and (58) beyond the range of validity of the above solution is numerical integration. An example of an integration is plotted in Figs. 1 and 2, where a and ϕ are plotted as a function of time for the three cases of positive, zero, and negative curvature with $\omega=9$.

It should be noted that for $\omega\geq 6$, and the flat-space solution, the time dependence of a differs only slightly from the corresponding case in general relativity (Einstein-deSitter) where a $\sim t^{2/3}$. Consequently, it would be difficult to distinguish between the two theories on the basis of space geometry only. In similar fashion the mass density required for a particular Hubble age $a/\dot{a}$ (flat space) is the same as for general relativity if $\omega\gg 1$. For $\omega=6$ there is only a 2% difference between the two theories.

On the other hand, stellar evolutionary rates are a sensitive function of the gravitational constant, and this makes an observational test of the theory possible.

This matter is discussed in a companion article by one of us (R. H. D.).[25]

At the beginning of this article a problem was posed, to understand within the framework of Mach's principle the laws of physics seen within a laboratory set rotating within a universe otherwise almost empty. We are now in a position to begin to understand this problem. Consider a laboratory, idealized to a spherical mass shell with a mass m and radius r, and stationary in the comoving coordinate system given by Eqs. (50) with Eqs. (60), (61), (62), and (63) satisfied. Imagine now that the laboratory is set rotating about an axis with an angular velocity α_0. This rotation affects the metric tensor inside the spherical shell in such a way as to cause the gyroscope to precess with an angular velocity[5] [also see Eq. (27)]

$$\alpha=(8m/3rc^2\phi_0)\alpha_0, \tag{64}$$

where ϕ_0 is given by Eq. (63). Equation (64) is valid in the weak-field approximation only for which $m/(rc^2\phi_0)\ll 1$. Substituting Eq. (63) in Eq. (64) gives

$$\alpha=[2(3+2\omega)/3\pi(4+3\omega)]\cdot(m/rc^2\rho_0t_0^2)\alpha_0. \tag{65}$$

It may be noted that if the matter density ρ_0 of the universe is decreased, with t_0 const, α increases. Thus, as the universe is emptied, the Thirring-Lense precession of the gyroscope approaches more closely the rotation velocity α_0 of the laboratory. Unfortunately, the weak-field approximation does not permit a study of the limiting process $\rho_0\rightarrow 0$.

In another publication by one of us (C. B.) other aspects of the theory, including conservation laws, will be discussed.

ACKNOWLEDGMENTS

The authors wish to acknowledge helpful conversations with C. Misner on various aspects of this problem, and one of us (C. B.) is indebted for advice on this and other matters in his thesis. The authors wish also to thank P. Roll and D. Curott for the machine integration of the cosmological solutions [Eqs. (49) and (50)], a small part of which is plotted in Figs. 1 and 2.

APPENDIX

In general relativity the equation of motion of a point particle, without spin, moving in a gravitational field only, may be obtained from the variational principle

$$0=\delta\int m(g_{ij}u^iu^j)^{\frac{1}{2}}ds, \tag{66}$$

or

$$(d/ds)(mu_i)-\tfrac{1}{2}mg_{jk,i}u^ju^k=0. \tag{67}$$

If the mass in Eq. (66) is assumed to be a function of position,

$$m=m_0f(x), \tag{68}$$

an added force term appears and

$$(d/ds)(mu_i)-\tfrac{1}{2}mg_{jk,i}u^ju^k-m_{,i}=0. \tag{69}$$

Both equations are consistent with the constraint condition $u^iu_i=1$. It should be noted that because of the added force term in Eq. (69), the particle does not move on a geodesic of the geometry.

If now the geometry is redefined in such a way that the new metric tensor is (conformal transformation)

$$\bar{g}_{ij}=f^2g_{ij}, \tag{70}$$

and

$$d\bar{s}^2=f^2ds^2,\quad \bar{u}^i=f^{-1}u^i.$$

Equation (69) may be written as

$$(d/d\bar{s})(m_0\bar{u}_i)-\tfrac{1}{2}m_0\bar{g}_{jk,i}\bar{u}^j\bar{u}^k=0. \tag{71}$$

The particle moves on a geodesic of the new geometry. With the new units of length, time, and mass appropriate for this new geometry, the mass of the particle is m_0, a constant.

[25] R. H. Dicke, Revs. Modern Phys. (to be published).

GRAVITATIONAL COLLAPSE AND SPACE-TIME SINGULARITIES

Roger Penrose
Department of Mathematics, Birkbeck College, London, England
(Received 18 December 1964)

The discovery of the quasistellar radio sources has stimulated renewed interest in the question of gravitational collapse. It has been suggested by some authors[1] that the enormous amounts of energy that these objects apparently emit may result from the collapse of a mass of the order of $(10^6\text{-}10^8)M_\odot$ to the neighborhood of its Schwarzschild radius, accompanied by a violent release of energy, possibly in the form of gravitational radiation. The detailed mathematical discussion of such situations is difficult since the full complexity of general relativity is required. Consequently, most exact calculations concerned with the implications of gravitational collapse have employed the simplifying assumption of spherical symmetry. Unfortunately, this precludes any detailed discussion of gravitational radiation—which requires at least a quadripole structure.

The general situation with regard to a spherically symmetrical body is well known.[2] For a sufficiently great mass, there is no final equilibrium state. When sufficient thermal energy has been radiated away, the body contracts and continues to contract until a physical singularity is encountered at $r=0$. As measured by local comoving observers, the body passes within its Schwarzschild radius $r=2m$. (The densities at which this happens need not be enormously high if the total mass is large enough.) To an outside observer the contraction to $r=2m$ appears to take an infinite time. Nevertheless, the existence of a singularity presents a serious problem for any complete discussion of the physics of the interior region.

The question has been raised as to whether this singularity is, in fact, simply a property of the high symmetry assumed. The matter collapses radially inwards to the single point at the center, so that a resulting space-time catastrophe there is perhaps not surprising. Could not the presence of perturbations which destroy the spherical symmetry alter the situation drastically? The recent rotating solution of Kerr[3] also possesses a physical singularity, but since a high degree of symmetry is still present (and the solution is algebraically special), it might again be argued that this is not representative of the general situation.[4] Collapse without assumptions of symmetry[5] will be discussed here.

Consider the time development of a Cauchy hypersurface C^3 representing an initial matter distribution. We may assume Einstein's field equations and suitable equations of state governing the matter. In fact, the only assumption made here about these equations of state will be the non-negative definiteness of Einstein's energy expression (with or without cosmological term). Suppose this matter distribution undergoes gravitational collapse in a way which, at first, qualitatively resembles the spherically symmetrical case. It will be shown that, after a certain critical condition has been fulfilled, deviations from spherical symmetry cannot prevent space-time singularities from arising. If, as seems justifiable, actual physical singularities in space-time are not to be permitted to occur, the conclusion would appear inescapable that inside such a collapsing object at least one of the following holds: (a) Negative local energy occurs.[6] (b) Einstein's equations are violated. (c) The space-time manifold is incomplete.[7] (d) The concept of space-time loses its meaning at very high curvatures—possibly because of quantum phenomena.[2] In fact (a), (b), (c), (d) are somewhat interrelated, the distinction being partly one of attitude of mind.

Before examining the asymmetrical case, consider a spherically symmetrical matter distribution of finite radius in C^3 which collapses symmetrically. The empty region surrounding the matter will, in this case, be a Schwarzschild field, and we can conveniently use the metric $ds^2 = -2drdv + dv^2(1-2m/r) -r^2(d\theta^2+\sin^2\theta d\varphi^2)$, with an advanced time parameter v to describe it.[8] The situation is depicted in Fig. 1. Note that an exterior observer will always see matter outside $r=2m$, the collapse through $r=2m$ to the singularity at $r=0$ being invisible to him.

After the matter has contracted within $r=2m$, a spacelike sphere S^2 (t = const, $2m > r$ = const) can be found in the empty region surrounding the matter. This sphere is an example of what will be called here a trapped surface—defined generally as a closed, spacelike, two-surface T^2 with the property that the two systems of null geodesics which meet T^2 orthogonally converge locally in future directions at T^2. Clearly trapped surfaces will still exist if the matter region has no sharp boundary or if spherical symmetry is dropped, provided that the deviations from the above situation are not too great. Indeed, the Kerr solutions with $m > a$ (angular momentum ma) all possess trapped surfaces, whereas those for which $m \leqslant a$ do not.[9] The argument will be to show that the existence of a trapped surface implies—irrespective of symmetry—that singularities necessarily develop.

The existence of a singularity can never be inferred, however, without an assumption such as completeness for the manifold under consideration. It will be necessary, here, to suppose that the manifold M_+^4, which is the future time development of an initial Cauchy hypersurface C^3 (past boundary of the M_+^4 region), is in fact null complete into the future. The various assumptions are, more precisely, as follows: (i) M_+^4 is a nonsingular (+−−−) Riemannian manifold for which the null half-cones form two separate systems ("past" and "future"). (ii) Every null geodesic in M_+^4 can be extended into the future to arbitrarily large affine parameter values (null completeness). (iii) Every timelike or null geodesic in M_+^4 can be extended

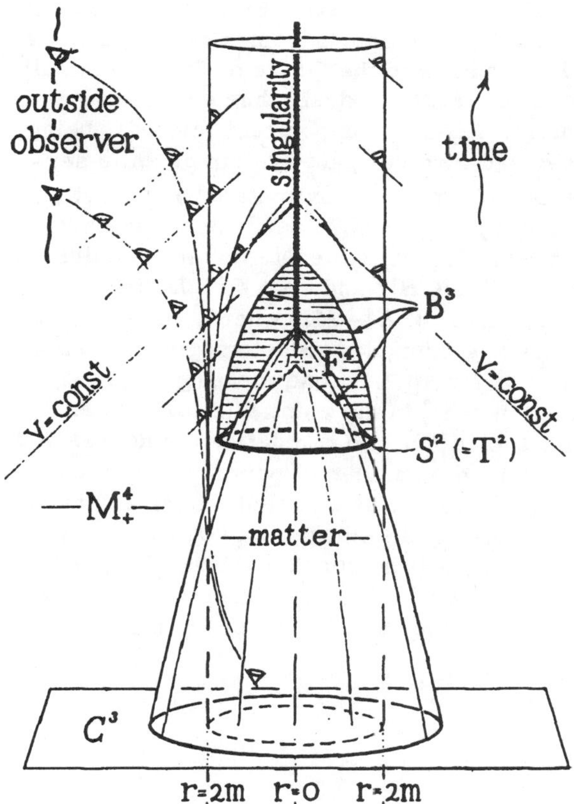

FIG. 1. Spherically symmetrical collapse (one space dimension surpressed). The diagram essentially also serves for the discussion of the asymmetrical case.

into the past until it meets C_3 (Cauchy hypersurface condition). (iv) At every point of $M_+{}^4$, all timelike vectors t^μ satisfy $(-R_{\mu\nu} + \frac{1}{2}Rg_{\mu\nu} - \lambda g_{\mu\nu})t^\mu t^\nu \geq 0$ (non-negativeness of local energy). (v) There exists a trapped surface T^2 in $M_+{}^4$. It will be shown here, in outline, that (i), $\cdots$, (v) are together inconsistent.

Let F^4 be the set of points in $M_+{}^4$ which can be connected to T^2 by a smooth timelike curve leading into the future from T^2. Let B^3 be the boundary of F^4. Local considerations show that B^3 is <u>null</u> where it is nonsingular, being generated by the null geodesic segments which meet T^2 orthogonally at a past endpoint and have a future endpoint if this is a singularity (on a caustic or crossing region) of B^3. Let l^μ (subject to $l^\mu{}_{;\nu}l^\nu = 0$), ρ $(= -\frac{1}{2}l^\mu{}_{;\mu})$, and $|\sigma|$ $\{= [\frac{1}{2}l_{(\mu;\nu)}l^{\mu;\nu} - \frac{1}{4}(l^\mu{}_{;\mu})^2]^{1/2}\}$ be, respectively, a future-pointing tangent vector, the convergence, and the shear for these null geodesics,[10] and let A be a corresponding infinitesimal area of cross section of B^3. Then $[(A^{1/2})_{;\mu}l^\mu]_{;\nu}l^\nu = -(A^{1/2}\rho)_{;\mu}l^\mu = -A^{1/2}(|\sigma|^2 + \Phi) \leq 0$ where $\Phi = -\frac{1}{2}R_{\mu\nu}l^\mu l^\nu$ [≥ 0 by (iv)]. Since T^2 is trapped, $\rho > 0$ at T^2, whence A becomes zero at a finite affine distance to the future of T^2 on each null geodesic. Each geodesic thus encounters a caustic. Hence B^3 is compact (closed), being generated by a compact system of finite segments. We may approximate B^3 arbitrarily closely by a smooth, closed, spacelike hypersurface B^{3*}. Let K^4 denote the set of pairs (P, s) with $P \in B^{3*}$ and $0 \leq s \leq 1$. Define a continuous map μ: $K^4 \to M_+{}^4$ where, for fixed P, $\mu\{(P, s)\}$ is the past geodesic segment normal to B^{3*} at $P = \mu\{(P, 1)\}$ and meeting C^3 [as it must, by (iii)] in the point $\mu\{(P, 0)\}$. At each point Q of $\mu\{K^4\}$, we can define the <u>degree</u> $d(Q)$ of μ to be the number of points of K^4 which map to Q (correctly counted). Over any region not containing the image of a boundary point of K^4, $d(Q)$ will be constant. Near B^{3*}, μ is 1-1, so $d(Q) = 1$. It follows that $d(Q) = 1$ near C^3 also, whence the degree of the map $B^{3*} - C^3$ induced by μ when $s = 0$ must also be unity. The impossibility of this follows from the noncompactness of C^3.

Full details of this and other related results will be given elsewhere.

---

[1]F. Hoyle and W. A. Fowler, Monthly Notices Roy. Astron. Soc. <u>125</u>, 169 (1963); F. Hoyle, W. A. Fowler, G. R. Burbidge, and E. M. Burbidge, Astrophys. J. <u>139</u>, 909 (1964); W. A. Fowler, Rev. Mod. Phys. <u>36</u>, 545 (1964); Ya. B. Zel'dovich and I. D. Novikov, Dokl. Akad. Nauk SSSR <u>155</u>, 1033 (1964) [translation: Soviet Phys.—Doklady <u>9</u>, 246 (1964)]; I. S. Shklovskiĭ and N. S. Kardashev, Dokl. Akad. Nauk SSSR <u>155</u>, 1039 (1964) [translation: Soviet Phys.—Doklady <u>9</u>, 252 (1964)]; Ya. B. Zel'dovich and M. A. Podurets, Dokl. Akad. Nauk SSSR <u>156</u>, 57 (1964) [translation: Soviet Phys.—Doklady <u>9</u>, 373 (1964)]. Also various articles in the <u>Proceedings of the 1963 Dallas Conference on Gravitational Collapse</u> (University of Chicago Press, Chicago, Illinois, 1964).

[2]J. R. Oppenheimer and H. Snyder, Phys. Rev. <u>56</u>, 455 (1939). See also J. A. Wheeler, in <u>Relativity, Groups and Topology</u>, edited by C. deWitt and B. deWitt (Gordon and Breach Publishers, Inc., New York, 1964); and reference 1.

[3]R. P. Kerr, Phys. Rev. Letters <u>11</u>, 237 (1963).

[4]See also E. M. Lifshitz and I. M. Khalatnikov, Advan. Phys. <u>12</u>, 185 (1963).

[5]See also P. G. Bergmann, Phys. Rev. Letters <u>12</u>, 139 (1964).

[6]The negative energy of a "C field" may be invoked to avoid singularities: F. Hoyle and J. V. Narlikar, Proc. Roy. Soc. (London) <u>A278</u>, 465 (1964). However, it is difficult to see how even the presence of negative energy could lead to an effective "bounce" if local causality is to be maintained.

[7]The "I'm all right, Jack" philosophy with regard to the singularities would be included under this heading!

[8]D. Finkelstein, Phys. Rev. <u>110</u>, 965 (1959).

[9]The case $m < a$ is interesting in that here a singularity is "visible" to an outside observer. Whether or not "visible" singularities inevitably arise under appropriate circumstances is an intriguing question not covered by the present discussion.

[10]For the notation, etc., see E. Newman and R. Penrose, J. Math. Phys. <u>3</u>, 566 (1962).

THE

PHYSICAL REVIEW

A journal of experimental and theoretical physics established by E. L. Nichols in 1893

SECOND SERIES, VOL. 169, NO. 5 25 MAY 1968

Experimental Measurement of the Equivalence of Active and Passive Gravitational Mass*

L. B. KREUZER†

Palmer Physical Laboratory, Princeton University, Princeton, New Jersey

(Received 15 January 1968)

A new type of gravitational experiment is reported, and a simple theoretical treatment in terms of Bondi's concepts of active, passive, and inertial mass is used to compare the results with other experiments. A homogeneous Teflon cylinder, which is completely immersed in a mixture of dibromomethane and trichloroethylene prepared to have about the same density as Teflon, was slowly moved back and forth in this liquid. The resultant time-varying gravitational field due to the density difference between the solid Teflon and the displaced liquid was detected by a Cavendish-type torsion balance placed outside of the liquid container. Buoyant forces on the Teflon were measured. The time-varying gravitational field detected by the balance was extrapolated to the condition of neutral buoyancy. The fractional density difference between the Teflon and the liquid required to produce this field was found to be $\Delta\rho/\rho=(1.2\pm4.4)\times10^{-5}$. This upper bound of approximately 5×10^{-5} is compared to 10^{-3}, the best value that can be deduced from other experiments.

INTRODUCTION

THIS paper is divided into three parts. The first describes a gravitational experiment, and the second presents a simple theoretical discussion, which is used to compare this experiment with previously published experiments. The third part presents a summary and conclusion. The work is based on the author's Ph.D. thesis.[1]

I. EXPERIMENTAL

A. Apparatus

A Cavendish-type torsion balance with a beam, suspended by a 2-mil-diam and 20-cm-long tungsten wire, consisting of two pure aluminum cylinders 1.18 in. long and 1.57 in. in diameter, separated by about 10 in., was constructed. The fiber torsion constant was measured to be approximately 1 dyn cm/rad. Pairs of electrodes, with a bias voltage across them, were mounted close to each side of the torsion beam near its ends. They produced an electrostatic torque which was linear in the voltage applied between the electrodes and the torsion beam. A servo used this electrostatic torque to balance other torques on the balance and to prevent beam rotation. This was accomplished by connecting the output of the photoelectric optical lever, which detected beam rotation through appropriate amplifiers and filters, to the electrodes. When the servo was operating, the dynamical properties of the balance were determined, for the most part, by the electrical properties of the servo system rather than by the mechanical properties of the balance suspension. External torques were measured by the size of the electrostatic torque needed to balance their effect on the beam.

The torsion balance was enclosed in a sealed aluminum container which contained air at atmospheric pressure. Operating the torsion balance in a high vacuum could have eliminated unwanted torques produced by thermally generated convection currents in the air surrounding the beam, but it seemed that this advantage would be completely offset by the deleterious effect of removing the air damping of the vibrational modes of the balance. These modes are driven by ground vibrations and, although the torsional motion of the beam is rather insensitive to vibrations, it is driven by nonlinear coupling to other vibrating modes. The air in the balance container reduces the

* Research supported in part by the National Science Foundation and by the U. S. Office of Naval Research.

† Present address: Bell Telephone Laboratories, Murray Hill, N. J.

[1] L. Kreuzer, Ph.D. thesis, Department of Physics, Princeton University, 1966 (unpublished).

coupling of ground vibrations to the torsional motion of the beam by damping these vibrating modes. Temperature gradients which cause convection currents were insulated against by surrounding the beam with three cylindrical, air-spaced aluminum containers.

A tank completely full of a mixture of trichloroethylene and dibromomethane, in which a Teflon[3] cylinder was immersed, was placed next to the balance. The cylinder of Teflon, which had a diameter of 9 in., a length of 9 in., and weighed about 20 kg was attached by fine nylon strings to a motor drive. This was arranged so that the motor could move the cylinder between the ends of the tank. Vertical motion of the cylinder due to buoyant forces was normally restricted by tension in these strings. The size and direction of these bouyant forces was measured by reducing this tension by a fixed amount and noting the direction and extent of the cylinder's vertical motion. The tank which was insulated by a 1-in. thickness of fiber glass and aluminum foil building insulation had a number of thermistors, which were used to measure the temperature of the tank and its contents, fastened in thermal contact to its walls. A uniform winding of electrical resistance wire around the tank could provide heat for temperature control of the tank and its contents.

The proximity of the tank to the balance, as shown in Fig. 1, caused the center of mass of the cylinder, which was periodically moved from one end of the tank to the other, first to be near one end of the torsion beam and then to be near the other. Gravitational attraction between the balance beam and the cylinder then produces a periodic torque on the beam. Since the liquid which is displaced by the immersed cylinder also causes a torque on the beam, the net torque on the beam is proportional to the difference in density between the Teflon and the liquid.

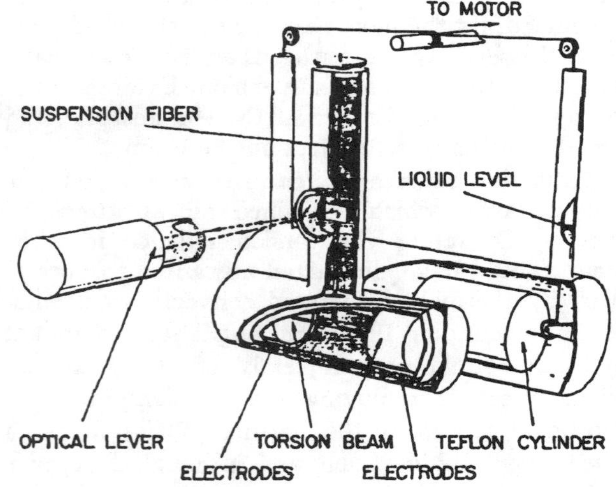

Fig. 1. Cut-away view of tank, Teflon cylinder, and torsion balance.

[3] Trade name for plastic manufactured by E. I. DuPont de Nemours and Company, Inc.

This torque was measured by a synchronous detector which correlated the torque signal from the balance with the position of the Teflon cylinder in the tank. Averaging the output of the synchronous detector over a long period of time allowed the detection of very weak signals.

B. Method

The liquid—a mixture of dibromomethane, which has a density greater than Teflon, and trichloroethylene, which has a density less than Teflon—in which the Teflon cylinder was immersed was prepared to have a density very close to that of Teflon. During the experiment the tank which contained this liquid and the torsion balance were placed in a thermally insulated basement room. The tank temperature changed from time to time, owing to changes in room temperature. The different rates of thermal expansion of Teflon

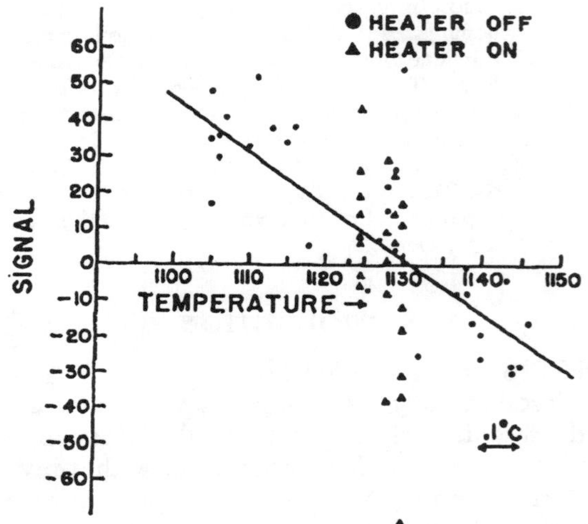

Fig. 2. Torsion balance signal averaged over 3-h periods as a function of liquid temperature.

and the liquid caused the density of the Teflon to be, at different times, both higher and lower than the density of the liquid.

The experimental procedure consisted of measuring synchronous detector output and density difference between the Teflon cylinder and liquid as a function of liquid temperature. The Teflon cylinder was moved back and forth between the ends of the tank with a period of 400 sec. Liquid temperature was recorded continuously on a chart recorder. The density difference between the liquid and the solid was measured twice a day.

Temperature changes of the liquid in which the Teflon cylinder was immersed were measured by recording resistance changes of a thermistor in contact with the liquid. These resistance changes were converted into temperature changes by using a conversion coefficient supplied by the manufacturer. In order to calibrate the experimental results, it was necessary to know the coefficient of thermal expansion of the liquid and that of the Teflon cylinder. The coefficient of expansion of the liquid was measured and that of Teflon was found in the literature.[3]

C. Results

The signal is recorded as a function of liquid temperature in Fig. 2. Each point represents an average of the output of the synchronous detector over a 3-h period. Since it was found that the temperature never changed by more than 0.02°C in any 3-h period, a single temperature has been assigned to each 3-h average. Some of the data were collected with the heater off and the tank in thermal equilibrium with the room, and some were collected with the tank heater on and the tank temperature a few tenths of a degree above room temperature. Although there is some indication of systematic differences between these two groups of data, they have both been retained. The straight line drawn in Fig. 2 is a least-squares fit to the data.

The density difference between the Teflon cylinder and the liquid is plotted as a function of liquid temperature in Fig. 3. The method of measuring density differences, which has been described above, was linear only for small density differences, and for this reason Fig. 3 only contains data taken near neutral bouyancy.

A least-squares fit, of a straight line to the data represented in Fig. 3, gives the temperature of neutral buoyancy to be 1128.7±0.6 Ω. The temperature is given in units of thermistor resistance. The error of ±0.6 Ω is the deviation of the mean of the experimental points from the least-squares fit. The least-squares fit to the signal indicates a signal at this temperature of 3.2±3.4. The units are arbitrary and the error is due to the deviation of the mean about the least-squares fit. This signal may be expressed in terms of an equivalent mass Δm which would have produced this signal if it had been placed in air at the position of the center of the Teflon cylinder and moved back and forth at the same frequency as the Teflon cylinder was moved. Knowledge of the thermal coefficients of volume expansion of the liquid and the Teflon, combined with the slope of the least-squares fit to the data of Fig. 2, provides sufficient information to express the signal at neutral bouyancy in terms of an equivalent mass. If m is the mass of the cylinder, then

$$\Delta m/m=(4.2\pm4.4)\times10^{-5}.$$

The above expression for $\Delta m/m$ contains systematic errors produced by the finite mass of the string used to move the Teflon cylinder. A simple calculation[4] showed that the gravitational attraction between the string

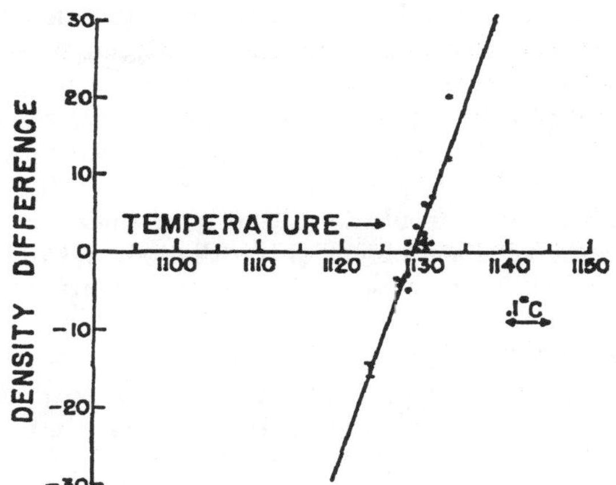

FIG. 3. Density difference between the Teflon cylinder and the liquid as a function of liquid temperature.

and the torsion beam produces an error of

$$\Delta m/m=1.3\times10^{-5}.$$

The difference in density between the nylon string and the Teflon cylinder causes an error in density measurements, which is reflected by an error in the measured temperature of neutral buoyancy. The resultant error introduced into the measured value for $\Delta m/m$ is

$$\Delta m/m=1.7\times10^{-5}.$$

The corrected value of $\Delta m/m$ is then

$$\Delta m/m=(1.2\pm4.4)\times10^{-5}.$$

These experimental results show no detected signal at neutral buoyancy and also set an upper bound to the size of any undetected signal.

A detailed study of the noise sources and possible systematic errors that limited the accuracy of this experiment was not made. It is, however, still possible to make some statements which may be useful in interpreting the results of this experiment and in designing similar types of measurements. The torsion balance was placed on a concrete pier, which went through a hole in the basement floor and penetrated some distance into the ground. Although this did provide some isolation, building vibrations due to machinery and other activities were still detectable on the pier. It seems reasonable to believe that by operating at a location isolated from man-made ground vibrations, a reduction in ground noise by as much as one order of magnitude might be achieved.

Noise was introduced into the measurements by the electrical noise in the photoelectric optical lever and by servo amplifiers. The state of the art of optical lever and electronic amplifier design is such that it would have been possible to use more complex components to achieve lower noise levels.

Mechanical coupling between the tank with the moving Teflon cylinder and the torsion balance could

[3] *Handbook of Chemistry and Physics*, edited by Charles D. Hodgman (Chemical Rubber Publishing Co., Cleveland, Ohio, 1963), 44th ed., p. 1557.

[4] Reference 1, Appendix 3.

have been a source of systematic error. In this experiment this coupling was reduced to a satisfactory level by using a smooth vibration-free drive mechanism and by mechanically isolating the balance and tank by placing the balance on a concrete pier, which penetrated through a hole in the floor on which the tank was supported. It would seem that mechanical isolation will not become a problem until other sources of noise are substantially reduced below the levels of this experiment.

Temperature measurements were made by measuring the resistance of glass bead thermistors with a dc-operated Wheatstone bridge. It is known that these thermistors are very stable and free from drifts, and that more sophisticated means of measuring resistance changes could have increased the precision of the temperature measurements by at least one order of magnitude.

Inhomogeneities in the density of the Teflon cylinder could cause gravitational gradients which would produce systematic errors. This was not a problem in this experiment, and if it became a problem because of the reduction of other sources of noise, the effect could be reduced by rotating the cylinder during the experiment to average out the gradients and by careful preparation of the Teflon to reduce inhomogeneities.

Systematic errors were introduced by the difference in density between the nylon strings used to move the Teflon cylinder and the liquid. This effect could be reduced by using Teflon instead of nylon strings, although the mechanical properties of Teflon are not as desirable.

The torsion balance is sensitive to gravitational gradients produced by objects in its vicinity. In order to reduce this source of noise, it is necessary to conduct the experiment at a reasonable distance from other laboratory activities. The remainder of this paper will be devoted to a theoretical discussion of the above result and to a comparison with other experiments.

II. THEORY

Bondi[5] distinguishes three kinds of mass." Inertial mass is the quantity that enters (and is defined by) Newton's second law; passive gravitational mass is the mass on which the gravitational field acts, that is, it is defined by $F=-m \text{ grad } V$; active mass is the mass that is the source of the gravitational field and is hence the mass that enters Poisson's equation and Gauss's law." Every object has three kinds of mass; or more exactly, there exist three numerically valued measures defined on the set of material objects. Each measure assigns a number to each body. In discussing this concept, it will be assumed that the accelerations are measured with all velocities, relative to the observer, small compared to the velocity of light; thus, relativistic effects are negligible.

Physicists assume today that these three properties of a body are measured by a single quantity. Identities between different types of mass are implied by the postulates of the mechanics of Newton and Einstein. Newton's third law, which states that the sum of the forces in a closed system is zero, implies that the ratio of active to passive mass for a body is a universal constant independent of the size or composition of the body. This may be interpreted to mean that active and passive gravitational mass are measures of the same property of matter and can be made to have identical numerical values by a proper choice of units. The principle of equivalence[6] in general relativity states that, neglecting the effects of gravitational gradients, a body experiences an acceleration in a gravitational field that is independent of its structure. This principle implies that for any body the ratio of passive to inertial mass is a constant independent of size and composition and that these masses are measures of the same property and may have identical numerical values by a proper choice of units.

In order to establish full equivalence between the three types of mass, it is necessary to establish equivalence between two sets of pairs. Roll, Kratkov, and Dicke[7] performed an experiment where the difference between the ratios of inertial to passive mass for gold and aluminum was measured. They concluded that this difference was less than 3×10^{-11}. This indicates that to a high degree of accuracy inertial and passive mass are measures of the same material property, and it will be assumed for the rest of this paper that they are identical. This remarkable degree of precision is in marked contrast to the rather poor experimental confirmation of the equivalence between passive (or inertial) mass and active mass.

The only precise experimental results, other than those reported in this paper, which are relevant to the problem of active gravitational mass come from the Cavendish-type determinations of the gravitational constant. In these experiments a torque is exerted on the torsion beam due to the gravitational attraction between the "small masses" on the torsion beam and the "large masses" which are placed near the balance. By carefully measuring the distance between the masses, the suspension-fiber torque constant, and the weight of the masses, it is possible to calculate the gravitational constant from the beam deflection. If the ratio of active to passive mass for the large masses is a function of their composition, then the value of the gravitational constant will depend on the composition of the large masses. Table I summarizes the results of a

[5] H. Bondi, Rev. Mod. Phys. 29, 423 (1957).

[6] R. H. Dicke, *The Theoretical Significance of Experimental Relativity* (Gordon and Breach Science Publishers, Inc., New York, 1964).

[7] P. G. Roll, R. Krotkov, and R. H. Dicke, Ann. Phys. (N. Y.) 26, 442 (1964).

TABLE I. Summary of experimental determinations of the gravitational constant G.

Experimenter	Date	Small mass	Large mass	G^{a} (10^{-8} dyn cm^2/g^2)	Ref.
Boys	1889–94	Gold	Lead	6.6576±0.002	a
Braun	1887–96	Gilded brass	Brass	6.655	a
Braun	1887–96	Gilded brass	Iron filled with mercury	6.665	a
Poynting	1878–90			6.6984	a
Koning *et al.*	1884–97			6.685 ±0.011	a
Heyle	1930	Gold	Steel	6.678 ±0.003	b
Heyle	1930	Platinum	Steel	6.664 ±0.002	b
Heyle	1942	Glass	Steel	6.674 ±0.002	b
Heyle	1942	Platinum	Steel	6.6755±0.0008	c
Heyle	1942	Platinum	Steel	6.6685±0.0016	c

[a] A. Stanley Mackenzie, *The Laws of Gravitation* (American Book Co., New York, 1900).
[b] P. Heyle, J. Res. Natl. Bur. Std. 5, 1243 (1930).
[c] P. Heyle, J. Res. Natl. Bur. Std. 29, 1 (1942).

number of determinations of the gravitational constant. The errors that are listed are those quoted by the authors and include in most cases only a measure of the statistical spread of the data. This table shows that differences in the gravitational constant as large as 3 parts in 10^3, which depend on the composition of the large mass, could exist without contradiction to the measured values.

Both systematic and random errors contribute to the uncertainty in the results of the Cavendish-type experiments. Measurements of the balance deflection contain noise due to ground vibrations which shake the balance, temperature changes and gradients which cause balance deflections, and random errors produced in measuring the deflection. The size of these errors may be estimated from the statistical properties of the data. Systematic errors are also present and it is much more difficult to estimate their size from published data. They are predominantly due to errors in measuring the distance between and the size and shape of the large and small masses. An estimate of their size may be obtained by comparing the values of G obtained in different experiments. The 1942 experiments of Heyle[8] give two values for G which differ by 0.007, while the errors are claimed to be 0.0008 and 0.0016. This would seem to indicate, since the experiments were almost identical, that these error values are rather optimistic and that 0.007 is some measure of the systematic errors. The experiment reported in this paper was designed to overcome these systematic errors.

The quantity of interest, the difference between the ratios of active to passive mass for two dissimilar substances, is a small or zero difference between two large quantities. By measuring the difference directly, rather than the large quantities themselves and then subtracting them, the need for high precision was eliminated. It is clear from the description of this experiment that the signal, detected by the balance, is proportional to the difference in active mass between the solid Teflon cylinder and an equal volume of the liquid which fills the tank. The measured density difference between the liquid and the Teflon is proportional to the difference in passive mass, since the buoyant forces which are due to the gravitational pull of the earth on the liquid and Teflon are proportional to passive mass. At neutral buoyancy the passive mass densities of the solid and liquid are equal, and any signal detected by the balance must be due to a difference in active mass densities. As described above, the balance is calibrated by use of the coefficients of thermal expansion for Teflon and for the liquid, which do not have to be known accurately.

The data were expressed above in terms of a signal produced by a fraction $\Delta m/m$ of the mass of the cylinder. The gravitational attraction between the cylinder and balance is also proportional to G, the gravitational constant, and if the signal is expressed as a difference ΔG between the liquid and Teflon, then $\Delta G/G=\Delta m/m$. The data of Table I do not exclude the possibility that $\Delta m/m$ between lead and steel is as large as 10^{-3}. If such a difference did exist, it might be a function of nuclear structure, and for this reason the two materials compared in this experiment were selected to have nuclear structures as different as possible. This choice was restricted by the requirements that the solid and liquid have equal densities and that they do not react chemically. Teflon was selected because the solid is homogeneous and chemically inert. It is 76% fluorine by weight. The liquid, a mixture of trichloroethylene and dibromomethane, was 74% bromine by weight. Table II lists the ratio Z/A of protons to nucleons and E/A, the binding energy per nucleon for bromine, fluorine, and the materials used in the experiments of

TABLE II. Binding energy per nuclear (E/A) in MeV and the ratio of protons to nucleons (Z/A).[a]

Element	E/A	Z/A
Lead	7.9	0.39
Iron	8.8	0.46
Brass (copper)	8.7	0.45
Mercury	7.9	0.40
Fluorine	7.8	0.47
Bromine	8.7	0.44

[a] R. Leighton, *Principles of Modern Physics* (McGraw-Hill Book Co., New York, 1959), pp. 736–783.

[8] P. Heyle, J. Res. Natl. Bur. Std. 29, 1 (1942).

Table I. The differences in Z/A and E/A between fluorine and bromine indicate that they are interesting materials to compare. The results of this experiment, given above, indicate that the difference between the ratios of active to passive mass for bromine and fluorine is less than 5×10^{-5}.

III. SUMMARY AND CONCLUSIONS

This experimental technique of comparing the field produced by a homogeneous solid with the field of the fluid which it displaces has made it possible to measure an upper bound for $\Delta m/m$ which is smaller than any value which may be deduced from previous experiments. Although it is difficult to evaluate the possible sources of error in previous experiments and to deduce an upper bound for $\Delta m/m$, the scatter in values for G between various experiments makes it unreasonable to set this upper bound smaller than 10^{-3}. The present experimental result of 5×10^{-5} for an upper bound between fluorine and bromine is both a significant numerical improvement and also a more reliable estimate because it results from a direct measurement of the effect. Improvement by one and possibly two orders of magnitude should be possible by careful application of currently known experimental techniques. To improve the accuracy beyond that point would be very difficult and might require a completely different type of experiment. The present experimental technique would be severely limited by problems of measuring and controlling the temperature of the liquid and solid, by gravitational gradients caused by inhomogeneities in the solid, by noise generated in the balance by thermal effects and ground noise, and by the difficulty of measuring such small density differences.

ACKNOWLEDGMENTS

I am grateful to Professor R. H. Dicke for proposing the experiment and contributing innumerable ideas. Many helpful suggestions were received in conversations with Barry Block, David Curott, Jim Peebles, Peter Roll, Ray Weiss, and David Wilkinson.

Equivalence Principle for Massive Bodies. II. Theory

KENNETH NORDTVEDT, JR.
Department of Physics, Montana State University, Bozeman, Montana
(Received 16 October 1967; revised manuscript received 18 January 1968)

The acceleration of a massive body in an external field for general space-time geometrical gravitational theories is obtained. The condition on the metric is such that $m_g/m_i=1$ is obtained, and we reobtain the result that $m_g/m_i=1$ in Einstein's theory for massive objects with time-independent internal structure. But it is shown that a measurement of m_g/m_i for astronomical bodies would measure space-time metric components which have not been measured in other gravitational experiments. In the scalar-tensor gravitational theory due to Brans and Dicke, it is shown that m_g/m_i differs from 1 by a term of the order of the massive body's gravitational self-energy divided by its total energy.

I. INTRODUCTION

IN another paper,[1] it was shown that the experiments of Eotvos[2] and Dicke[3] which measure the equality of gravitational and inertial masses of bodies to be within a part of 10^{11} indicate nothing about whether the gravitational self-energy of bodies contributes equally to both the gravitational and inertial masses. If the ratio of gravitational to inertial mass for a body is assumed to be

$$\frac{m_g}{m_i}=1+\eta\frac{G}{c^2}\int\rho(x)\rho(x')\frac{d^3xd^3x'}{|\mathbf{x}-\mathbf{x}'|}\bigg/\int\rho(x)d^3x, \quad (1)$$

where η is a dimensionless constant of order 1, G is the gravitational constant, c is the velocity of light, and $\rho(x)$ is the mass density of the body, then the correction term in (1) is of order 10^{-25} for the bodies used by Eotvos[2] and Dicke[3] in their experiments.

For astronomical bodies, the correction term in (1) becomes much larger (10^{-8} for the planet Jupiter, 10^{-6} for the Sun). In I, several experiments were proposed to measure m_g/m_i for astronomical bodies and thereby measure η.

In this paper, gravitational theories will be examined with the purpose of determining what a measurement of η in (1) would reveal about the gravitational theories. Our consideration will be restricted to gravitational theories which can be expressed as geometrical theories, that is, as curved Riemannian space-time geometries in which "test particles" move along geodesics of the geometry. The equivalence principle (EP) is therefore immediately valid for "test particles" which follow geodesics of the geometry, but this paper is concerned with the movement of massive bodies and whether $m_g/m_i=1$ for them also.

Massive bodies will be placed at rest in a space-time also containing a distant external mass source M_e. The acceleration of the massive bodies, which is proportional to

$$\mathbf{g}=-(GM_e/R^3)\mathbf{R}, \quad (2)$$

where $\mathbf{R}$ is the vector from the external mass to the massive body, will be calculated. The ratio of m_g/m_i for the massive body will be obtained by making the identification

$$d^2\mathbf{x}/dt^2=(m_g/m_i)\mathbf{g}. \quad (3)$$

This approach to the problem fits the original domain of application of the EP—weak gravitational fields and slowly moving bodies. Also, by focusing on the Newtonian $1/R^2$ acceleration, several potential complications are bypassed.

(a) A massive body is by necessity an extended body which can sample higher multipoles of the external gravitational field. However, multipole acceleration terms go as R^{-n}, $n>2$, and will not contribute to (3).

(b) Relativistic gravitational theories are known to yield accelerations toward external bodies which deviate from the Newtonian R^{-2} acceleration, but these deviations go as R^{-3}, etc., and also will not contribute to (3).

In this paper we reobtain the results of Fock[4] and Papapetrou,[5] that in Einstein's gravitational theory $m_g/m_i=1$ for a *stationary*, stable massive body. However, this null result is shown to be due to the exact cancellation of several nonzero correction terms in (1), and here we explicitly express our result in terms of the several nonzero terms. Therefore, an experimental measure of η in (1), as proposed in I, would offer an experimental test of metric terms in Einstein's gravitational theory which have not been measured to date.

Also, it is shown that the Brans-Dicke gravitational theory[6] does not fulfill the EP. We obtain the result that $m_g/m_i\neq 1$ for that theory.

II. GENERAL METRIC EXPANSION (SINGLE STATIC SOURCE)

To illustrate the approach to the problem which will be employed in this paper, we review the metric analy-

[1] K. Nordtvedt, Jr., preceding paper, Phys. Rev. 169, 1014 (1968); hereafter referred to as I.
[2] R. V. Eotvos, Ann. Physik 68, 11 (1922).
[3] P. G. Roll, R. Krotkov, and R. H. Dicke, Ann. Phys. (N. Y.) 26, 442 (1964).
[4] V. Fock, *The Theory of Space, Time, and Gravitation* (The Macmillan Co., New York, 1964), 2nd ed., Chap. VI.
[5] A. Papapetrou, Proc. Phys. Soc. (London) 64A, 57 (1951).
[6] C. Brans and R. H. Dicke, Phys. Rev. 124, 925 (1961).

sis of Eddington,[7] Robertson,[8] and Schiff[9] which we will follow and extend. Given a spherically symmetric static source of gravitation of strength

$$m \equiv GM/c^2 \tag{4}$$

(through most of the rest of the work, we will use units in which $G=c=1$), the most general Reimannian space-time exterior geometry can be written as

$$ds^2 = g_{ij}dx^i dx^j, \tag{5}$$

with the metric components given by a general power-series expansion in the sole dimensionless constant of the problem m/r;

$$g_{00} = 1 - 2\alpha(m/r) + 2\beta(m/r)^2 + \cdots, \tag{6a}$$

$$g_{0k} = 0, \tag{6b}$$

$$g_{kk'} = -[1 + 2\gamma(m/r)]\delta_{kk'} + \cdots. \tag{6c}$$

Equations (6a)–(6c) are required to approach the Lorentz metric as $r \to \infty$, 0 indicates the time coordinate, $k=1, 2, 3$ are the three spatial coordinates, r is a radial variable, $r=(x^2+y^2+z^2)^{1/2}$, $\alpha, \beta, \gamma, \cdots$ are dimensionless constants of order 1 which are determined by the assumption of a particular gravitational theory and by field equations for the g_{ij}. The power series Eqs. (6a)–(6c) are assumed to be convergent for sufficiently small m/r. With (4) giving the connection between the source mass M and the parameter m, $\alpha \equiv 1$ in order to obtain Newtonian gravitation as a weak-field limit of the gravitational theory.

An analysis of past and future experimental tests of relativity in terms of the general metric above yields their dependence on the parameters $\alpha, \beta, \gamma, \cdots$.

(a) The frequency shift of spectral lines in a gravitational potential ϕ is[9]

$$\delta\nu/\nu = -\alpha(\phi/c^2). \tag{7}$$

(b) The deflection of light passing at distance d from a source m is[9]

$$\delta\theta = 2(m/d)(\alpha+\gamma). \tag{8}$$

(c) The angular advance of the perihelion position of a planetary orbit of semimajor axis a, period T, and eccentricity e is (per revolution)[9]

$$\theta = [2\alpha(\alpha+\gamma) - \beta]\frac{8\pi^3 a^2}{c^2T^2}\frac{1}{1-e^2}. \tag{9}$$

(d) The change in round-trip radar time between two planets (in circular orbits of radius r_1 and r_2) when the radar path passes close by the Sun at distance d is[10]

$$\delta t = [2(\alpha+\gamma)\ln(4r_1r_2/d^2) - \tfrac{4}{3}(\gamma+2\beta)](m/c). \tag{10}$$

[To obtain (10), it is important to state that the zero-order time must be defined in terms of measurables, i.e., orbital periods of the two planets, not their radial distances r_1 and r_2 which are coordinate-system-dependent.][11]

(e) The geodetic precession of a gyroscope spin axis when the gyroscope is in a circular orbit of angular frequency ω about a central body m, with the gyroscope spin axis in the orbital plane, is[11]

$$\Omega = -[(\alpha+2\gamma)/2](m/r)\omega. \tag{11}$$

For Einstein's theory of gravitation $\alpha=\beta=\gamma=1$, but the value of the general analysis above is that it allows a simple determination of the expected experimental results for any space-time metric. Also the use of the coefficients yields the sensitivity of any experiment on each of the metric components.

III. GENERAL METRIC EXPANSION (SEVERAL MOVING SOURCES)

For the purposes of this paper, Eqs. (6a)–(6c) must be generalized to give the metric for several moving sources. Only g_{00} will be needed beyond the linear order in the source strengths, and g_{00} will only be needed to second order.

Immediately, (6c) generalizes to

$$g_{kk'} = -\left[1 + 2\gamma\left(\frac{m_1}{|\mathbf{r}-\mathbf{r}_1|} + \frac{m_2}{|\mathbf{r}-\mathbf{r}_2|}\right)\right]\delta_{kk'} + \cdots \tag{12a}$$

for two sources, where correction terms due to motion of the sources are not required to be kept in $g_{kk'}$. Equation (12a) is uniquely determined to this approximation by imposing the condition that a two-source metric must become the one-source metric in the limit as either of the source strengths vanishes.

For moving sources, the mixed space-time components of the metric are nonzero and to lowest order in the velocity of the sources are

$$g_{0k} = 4\Delta\left(\frac{m_1}{|\mathbf{r}-\mathbf{r}_1|}\frac{dx_1^k}{dt} + \frac{m_2}{|\mathbf{r}-\mathbf{r}_2|}\frac{dx_2^k}{dt}\right) + 4\Delta'\left(\frac{m_1}{|\mathbf{r}-\mathbf{r}_1|^3}(\mathbf{r}-\mathbf{r}_1)\cdot\frac{d\mathbf{x}_1}{dt}(r-r_1)^k + \frac{m_2}{|\mathbf{r}-\mathbf{r}_2|^3}(\mathbf{r}-\mathbf{r}_2)\cdot\frac{d\mathbf{x}_2}{dt}(r-r_2)^k\right) + \cdots, \tag{12b}$$

[7] A. S. Eddington, *The Mathematical Theory of Relativity* (Cambridge University Press, New York, 1957), p. 105.

[8] H. P. Robertson, in *Space Age Astronomy*, edited by A. J. Deutsch and W. E. Klemperer (Academic Press Inc., New York, 1962), p. 228.

[9] L. I. Schiff, in Proceedings of the 1965 Summer Seminar on Relativity and Astrophysics (unpublished).

[10] D. H. Ross and L. I. Schiff, Phys. Rev. **141**, 1215 (1966).

[11] L. I. Schiff, in Proceedings of International Conferences on Relativity and Gravitation, Warsaw, 1962 (unpublished).

where the new dimensionless constants Δ and Δ' are introduced. Equation (12b) is the most general expression which transforms like a spatial vector under spatial rotations, and which is linear in source strength and source velocity. (Einstein's theory gives $\Delta=1$, $\Delta'=0$, but the Δ' term cannot be ruled out in considering all possible geometries fulfilling our general conditions.)

The g_{00} metric component to lowest order for two sources is

$$g_{00}^{(1)}=1-2\alpha\left(\frac{m_1}{|\mathbf{r}-\mathbf{r}_1|}+\frac{m_2}{|\mathbf{r}-\mathbf{r}_2|}\right). \tag{12c}$$

The next-order general terms which can contribute to g_{00} are given by

$$g_{00}^{(2)}=2\beta\left(\frac{m_1}{|\mathbf{r}-\mathbf{r}_1|}+\frac{m_2}{|\mathbf{r}-\mathbf{r}_2|}\right)^2+2\alpha'\frac{m_1m_2}{|\mathbf{r}_1-\mathbf{r}_2|}\left(\frac{1}{|\mathbf{r}-\mathbf{r}_1|}+\frac{1}{|\mathbf{r}-\mathbf{r}_2|}\right)-\alpha''\left[\frac{m_1}{|\mathbf{r}-\mathbf{r}_1|}\left(\frac{d\mathbf{x}_1}{dt}\right)^2+\frac{m_2}{|\mathbf{r}-\mathbf{r}_2|}\left(\frac{d\mathbf{x}_2}{dt}\right)^2\right]$$
$$+\chi\left[\frac{m_1}{|\mathbf{r}-\mathbf{r}_1|}(\mathbf{r}-\mathbf{r}_1)\cdot\mathbf{a}_1+\frac{m_2}{|\mathbf{r}-\mathbf{r}_2|}(\mathbf{r}-\mathbf{r}_2)\cdot\mathbf{a}_2\right]+\alpha'''\left[\frac{m_1}{|\mathbf{r}-\mathbf{r}_1|^3}\left((\mathbf{r}-\mathbf{r}_1)\cdot\frac{d\mathbf{x}_1}{dt}\right)^2+\frac{m_2}{|\mathbf{r}-\mathbf{r}_1|}\left((\mathbf{r}-\mathbf{r}_2)\cdot\frac{d\mathbf{x}_2}{dt}\right)^2\right]+\cdots, \tag{12d}$$

where $\mathbf{a}=d^2\mathbf{x}/dt^2$. Equation (12c) follows by the superposition principle for linear terms. Equation (12d) is unique (up to the magnitude of the dimensionless coefficients) with the imposition of the following conditions:

(a) g_{00} becomes (6a) as either mass is set equal to zero or the position of either mass goes to infinity and the other mass is at rest.

(b) $g_{00}\rightarrow 1$ as one moves far from all sources. Before the limit 1 is reached, $g_{00}\rightarrow 1-2(M/R)+2\beta(M/R)^2$ [see (6a)], where $M=m_1+m_2$ to lowest order.

(c) The correction terms in g_{00} due to motion of the sources are second order in time derivatives of position. (This is part of a more general condition that g_{0k} be odd in time derivatives of source position, while the other metric components are even in the time derivatives.[12])

(d) g_{00} is symmetric under the interchange of source labels.

(e) g_{00} is a scalar under spatial rotations of the coordinate system.

IV. EQUATION OF MOTION

The massive bodies which will be studied are to be considered an assembly of mass elements, each of which is assumed to follow geodesics of the geometry produced by all the other matter in the space, i.e., each mass element moves in the geometry produced by any external masses plus the other mass elements in the massive body.

The result we possibly expect is that a massive body m of radius a could have an anomalous acceleration in a gravitational field $\mathbf{g}$ of order

$$\mathbf{a}\sim(m/a)\mathbf{g}. \tag{13}$$

If this massive body is divided up into N elements of mass

$$\delta m\sim m/N \tag{14a}$$

and size

$$\delta a\sim a/N^{1/3}, \tag{14b}$$

then the internal effects in each mass element would be expected to cause anomalous accelerations for each element of order

$$\delta\mathbf{a}\sim(\delta m/\delta a)\mathbf{g}\sim(1/N^{2/3})(m/a)\mathbf{g}, \tag{15}$$

which goes to zero as $N\rightarrow\infty$. We can then be justified in considering mass elements as following geodesic paths in the geometry produced by *all other matter*.

The equation of motion for a body with its position x^k given as a function of coordinate time t is desired, so we write the proper time integral

$$s=\int dt(g_{00}+2g_{0k}v^k+g_{kk'}v^kv^{k'})^{1/2}, \tag{16}$$

with $v^k=dx^k/dt$. The proper time s is required to be an extremum for the actual trajectory $x^k(t)$. We write

$$g_{00}=1+h_{00}^{(1)}+h_{00}^{(2)}, \tag{17a}$$

$$g_{0k}=h_{0k}, \tag{17b}$$

$$g_{kk'}=-(1-h_{ss})\delta_{kk'}, \tag{17c}$$

and expand (16) to sufficient approximation to obtain

$$s=\int dt[1-\tfrac{1}{2}v^2-\tfrac{1}{8}v^4+\tfrac{1}{2}h_{00}^{(1)}-\tfrac{1}{8}h_{00}^{(1)2}$$
$$+\tfrac{1}{2}h_{00}^{(2)}+h_{0k}v^k+v^2(\tfrac{1}{2}h_{ss}+\tfrac{1}{4}h_{00}^{(1)})]. \tag{18}$$

Under a variation of the trajectory, $x^k\rightarrow x^k+\delta x^k$ which vanishes on the end points, and integrating by parts, the equation of motion for the mass elements is obtained:

$$\frac{d\mathbf{v}}{dt}+\frac{1}{2}\frac{d}{dt}(v^2\mathbf{v})-\frac{d}{dt}(C\mathbf{v})-\frac{d\mathbf{B}}{dt}$$
$$=-\nabla A-\nabla(\mathbf{B}\cdot\mathbf{v})-\tfrac{1}{2}v^2\nabla C, \tag{19}$$

where

$$A=\tfrac{1}{2}(h_{00}^{(1)}+h_{00}^{(2)})-\tfrac{1}{8}(h_{00}^{(1)})^2, \tag{20a}$$

$$B_{x,y,z}=h_{0x,0y,0z}, \tag{20b}$$

$$C=h_{ss}+\tfrac{1}{2}h_{00}^{(1)}. \tag{20c}$$

[12] This time-symmetry property imposed on the metric implies neglect of gravitational radiation terms. See A. Einstein, L. Infeld, and B. Hoffmann, Ann. Math. 39, 65 (1938).

V. TWO-BODY BOUND SYSTEM (CIRCULAR ORBIT)

Consider two masses m_i and m_j in a circular orbit about each other, and a third distant external mass m_e. This is the simplest massive system that we can construct, using only gravitational forces. We seek that part of the acceleration of the two-mass system toward m_e which is proportional to the inverse square of the distance to m_e.

Operationally, this acceleration can be measured by the following procedure. Any local clock being used by an experimenter is to be calibrated in terms of universal coordinate time t, i.e., the proper time of a clock which is at rest at a large distance from all the masses in the experiment. We place the massive system in a circular orbit about m_e. The radius of the orbit can be calibrated unambiguously by the round-trip time (t_0) for light to travel to the central mass m_e and return. The orbital frequency is measured in terms of t. Then the central acceleration toward m_e can be determined by using Kepler's third law. In particular, we seek the m_g/m_i ratio of massive objects, so

$$m_i\omega^2 r = m_g m_e/r^2$$

or

$$m_g/m_i = \omega^2 r^3/m_e,$$

with ω and r measured as described above and m_e measured by the $\omega^2 r^3$ value of a very small test particle. One can in principle do the experiment with very large ($r\to\infty$) orbits eliminating any order (m_e/r) corrections to the above equations.

In (19), several of the terms contain v^2 of the test particle. We will need v^2 in those terms only to the lowest Newtonian order, so we can use for circular orbits

$$v_i^2 = m_j^2/Mr_{ij}, \tag{21a}$$

$$v_j^2 = m_i^2/Mr_{ij}, \tag{21b}$$

with $M=m_i+m_j$, $r_{ij}=|\mathbf{r}_i-\mathbf{r}_j|$. It is assumed that the two-body system has acquired no significant velocity toward m_e, and it is assumed that m_e is at rest. We now analyze each term in the equation of motion (19):

(a)
$$-\frac{1}{2}\frac{d}{dt}(v^2\mathbf{v}) = -\tfrac{1}{2}v^2\mathbf{a} - \mathbf{v}\cdot\mathbf{a}\mathbf{v}.$$

But we divide the particle's acceleration into an internal ($\mathbf{a}_{\rm int}$) and external ($\mathbf{a}_e$) part with $\mathbf{a}_e$ being proportional to m_e and directed toward m_e. Then this term divides into

$$(-\tfrac{1}{2}v^2\mathbf{a}_e - \mathbf{v}\cdot\mathbf{a}_e\mathbf{v}) + (-\tfrac{1}{2}v^2\mathbf{a}_{\rm int} - \mathbf{v}\cdot\mathbf{a}_{\rm int}\mathbf{v}).$$

We are interested only in the external parts of our acceleration terms, so finally we keep

$$-[\tfrac{1}{2}v^2 + v_{\parallel}^2]\mathbf{a}_e - v_{\parallel}a_e\mathbf{v}_{\perp}, \tag{22}$$

where the acceleration term $\sim\mathbf{v}$ has been divided into a part parallel to $\mathbf{a}_e$ and a part perpendicular to $\mathbf{a}_e$.

(b)
$$\frac{d}{dt}(C\mathbf{v}) = C\mathbf{a} + \frac{\partial C}{\partial t}\mathbf{v} + \mathbf{v}\cdot\nabla C\mathbf{v}.$$

But from (20c), (12a), and (12c)

$$C = -(2\gamma+1)\left(\frac{m_e}{|\mathbf{r}-\mathbf{r}_e|} + \frac{m_i}{|\mathbf{r}-\mathbf{r}_i|}\right). \tag{23}$$

Using our assumption that m_e is at rest, this term gives

$$-(2\gamma+1)\left[\frac{m_e}{|\mathbf{r}-\mathbf{r}_e|}\mathbf{a}_{\rm int} + \frac{m_i}{|\mathbf{r}-\mathbf{r}_i|}\mathbf{a}_e + (\mathbf{v}-\mathbf{v}_i)\cdot\nabla\frac{m_i}{|\mathbf{r}-\mathbf{r}_i|}\mathbf{v} + \mathbf{v}_j\cdot\nabla\frac{m_e}{|\mathbf{r}-\mathbf{r}_e|}\mathbf{v}\right],$$

plus totally internal terms, plus terms proportional to m_e^2. Keeping only accelerations linear in m_e, we have

$$-(2\gamma+1)\left[\frac{m_e}{|\mathbf{r}-\mathbf{r}_e|}\mathbf{g}_{\rm int}(r) + \frac{m_i}{|\mathbf{r}-\mathbf{r}_i|}\mathbf{g}_e + \mathbf{v}\mathbf{v}\cdot\mathbf{g}_e\right], \tag{24}$$

where

$$\mathbf{g}_e = \nabla(m_e/|\mathbf{r}-\mathbf{r}_e|) \tag{25a}$$

and

$$\mathbf{g}_{\rm int}(r) = \nabla(m_i/|\mathbf{r}-\mathbf{r}_i|). \tag{25b}$$

(c)
$$\dot{B}_k = \frac{d}{dt}\left[4\Delta\frac{m_i}{|\mathbf{r}-\mathbf{r}_i|}v_i^k + 4\Delta'\frac{m_i}{|\mathbf{r}-\mathbf{r}_i|^3}(\mathbf{r}-\mathbf{r}_i)\cdot\mathbf{v}_i(r-r_i)^k\right],$$

where we have again used the assumption that m_e is at rest. Keeping only terms proportional to m_e, we have

$$4\Delta\frac{m_i}{|\mathbf{r}-\mathbf{r}_i|}\mathbf{g}_e + 4\Delta'\frac{m_i}{|\mathbf{r}-\mathbf{r}_i|^3}(\mathbf{r}-\mathbf{r}_i)\cdot\mathbf{g}_e(\mathbf{r}-\mathbf{r}_i). \tag{26}$$

(d) This is the usual potential term:

$$-\nabla A = -\nabla\left[(\beta-\tfrac{1}{2})\left(\frac{m_i}{|\mathbf{r}-\mathbf{r}_i|} + \frac{m_e}{|\mathbf{r}-\mathbf{r}_e|}\right)^2 - \frac{m_i}{|\mathbf{r}-\mathbf{r}_i|} - \frac{m_e}{|\mathbf{r}-\mathbf{r}_e|} + \alpha'\frac{m_i m_e}{|\mathbf{r}_i-\mathbf{r}_e|}\left(\frac{1}{|\mathbf{r}-\mathbf{r}_i|} + \frac{1}{|\mathbf{r}-\mathbf{r}_e|}\right) + \tfrac{1}{2}\chi\frac{m_i}{|\mathbf{r}-\mathbf{r}_i|}\mathbf{r}-\mathbf{r}_i\cdot\mathbf{a}_i + \text{(solely internal terms)}\right].$$

Again dividing $\mathbf{a}_i$ into $\mathbf{a}_e+\mathbf{a}_{\rm int}$ and keeping all terms above linear in m_e, we have

$$-\nabla A=\nabla\frac{m_e}{|\mathbf{r}-\mathbf{r}_e|}-(2\beta-1)\left(\frac{m_i}{|\mathbf{r}-\mathbf{r}_i|}\nabla\frac{m_e}{|\mathbf{r}-\mathbf{r}_e|}+\frac{m_e}{|\mathbf{r}-\mathbf{r}_e|}\nabla\frac{m_i}{|\mathbf{r}-\mathbf{r}_i|}\right)-\alpha'\frac{m_e}{|\mathbf{r}_i-\mathbf{r}_e|}\nabla\frac{m_i}{|\mathbf{r}-\mathbf{r}_i|}$$
$$-\tfrac{1}{2}\chi\frac{m_i}{|\mathbf{r}-\mathbf{r}_i|}\mathbf{a}_e+\tfrac{1}{2}\chi\frac{m_i}{|\mathbf{r}-\mathbf{r}_i|^3}(\mathbf{r}-\mathbf{r}_i)\cdot\mathbf{a}_e(\mathbf{r}-\mathbf{r}_i).$$

Regrouping the terms of interest above, then

$$-\nabla A=\mathbf{g}_e\left[1-(2\beta-1+\tfrac{1}{2}\chi)\frac{m_i}{|\mathbf{r}-\mathbf{r}_i|}\right]-\mathbf{g}_{\rm int}(r)\left[\frac{m_e}{|\mathbf{r}-\mathbf{r}_e|}\alpha'+(2\beta-1)\frac{m_e}{|\mathbf{r}-\mathbf{r}_e|}\right]+\tfrac{1}{2}\chi\frac{m_i}{|\mathbf{r}-\mathbf{r}_i|^3}\mathbf{r}-\mathbf{r}_i\cdot\mathbf{g}_e\mathbf{r}-\mathbf{r}_i. \quad (27)$$

(e)
$$\mathbf{B}\cdot\mathbf{v}=4\Delta\frac{m_i}{|\mathbf{r}-\mathbf{r}_i|}\mathbf{v}_i\cdot\mathbf{v}+4\Delta'\frac{m_i}{|\mathbf{r}-\mathbf{r}_i|^3}\mathbf{r}-\mathbf{r}_i\cdot\mathbf{v}\mathbf{r}-\mathbf{r}_i\cdot\mathbf{v}_i.$$

The divergence of $\mathbf{B}\cdot\mathbf{v}$ gives totally an internal acceleration.

(f)
$$-\tfrac{1}{2}v^2\nabla C$$

yields an external term

$$(\gamma+\tfrac{1}{2})v^2\mathbf{g}_e. \quad (28)$$

Combining all of the above results—(22), (24), and (26)–(28)—we arrive at the equation of motion for the particle m_j:

$$\mathbf{a}_j=\mathbf{g}_e\left[1+(4\Delta-2\beta-2\gamma-\tfrac{1}{2}\chi)\frac{m_i}{|\mathbf{r}_j-\mathbf{r}_i|}+\gamma v_j^2-(2\gamma+2)v_{\parallel j}^2\right]+(4\Delta'+\tfrac{1}{2}\chi)\frac{m_i}{|\mathbf{r}_j-\mathbf{r}_i|^3}(\mathbf{r}_j-\mathbf{r}_i)\cdot\mathbf{g}_e(\mathbf{r}_j-\mathbf{r}_i)$$
$$-\mathbf{g}_{\rm int}(r_j)\left[(2\gamma+2\beta)\frac{m_e}{|\mathbf{r}_j-\mathbf{r}_e|}+\alpha'\frac{m_e}{|\mathbf{r}_i-\mathbf{r}_e|}\right]-(2\gamma+2)v_{\parallel j}g_e\mathbf{v}_{\perp j}+\text{(solely internal accelerations)}. \quad (29)$$

Equation (29) can be further simplified by setting

$$\mathbf{r}_j-\mathbf{r}_i=(\mathbf{r}_j-\mathbf{r}_i)_\parallel+(\mathbf{r}_j-\mathbf{r}_i)_\perp$$

(parallel and perpendicular always refer to the direction of $\mathbf{g}_e$). Also a center of mass is defined:

$$\mathbf{R}=(m_i\mathbf{r}_i+m_j\mathbf{r}_j)/M.$$

Then

$$\frac{1}{|\mathbf{r}_j-\mathbf{r}_e|}\cong\frac{1}{|\mathbf{R}-\mathbf{r}_e|}-\frac{(\mathbf{r}_j-\mathbf{R})\cdot(\mathbf{R}-\mathbf{r}_e)}{|\mathbf{R}-\mathbf{r}_e|^3} \quad (30a)$$

and

$$\frac{1}{|\mathbf{r}_i-\mathbf{r}_e|}\cong\frac{1}{|\mathbf{R}-\mathbf{r}_e|}-\frac{(\mathbf{r}_i-\mathbf{R})\cdot(\mathbf{R}-\mathbf{r}_e)}{|\mathbf{R}-\mathbf{r}_e|^3}. \quad (30b)$$

Equation (29) can then be written as

$$\mathbf{a}_j=\mathbf{g}_e\left[1+(4\Delta-2\beta-2\gamma-\tfrac{1}{2}\chi)\frac{m_i}{r_{ij}}+\gamma v_j^2-(2\gamma+2)v_{\parallel j}^2+\left(4\Delta'+\tfrac{1}{2}\chi+\frac{2(\gamma+\beta)m_i-\alpha' m_j}{M}\right)\frac{m_i}{r_{ij}^3}r_{ij\parallel}^2\right]$$
$$-(2\gamma+2)v_{\parallel j}g_e\mathbf{v}_{\perp j}+\left(4\Delta'+\tfrac{1}{2}\chi+\frac{2(\gamma+\beta)m_i-\alpha' m_j}{M}\right)\frac{m_i}{r_{ij}^3}r_{ji\parallel}g_e\mathbf{r}_{ji\perp}+\text{(internal accelerations)}. \quad (31)$$

To obtain the acceleration of the m_i, m_j two-body system we take the combination

$$\mathbf{a}=(m_i\mathbf{a}_i+m_j\mathbf{a}_j)/M. \quad (32)$$

The acceleration $\mathbf{a}_i$ is obtained from (31) by the appropriate inversion of labels $i \leftrightarrow j$. The internal accelerations which we have not been interested in have the expected property

$$m_i \mathbf{g}(\mathbf{r}_i)_{\text{int}} + m_j \mathbf{g}(\mathbf{r}_j)_{\text{int}} = 0. \tag{33}$$

So finally

$$\mathbf{a} = \mathbf{g}_e \left[1 + (8\Delta - 4\beta - 4\gamma - \chi) \frac{m_i m_j}{M r_{ij}} + \gamma \frac{m_i v_i^2 + m_j v_j^2}{M} - (2\gamma + 2) \frac{m_i v_{\parallel i}^2 + m_j v_{\parallel j}^2}{M} + (8\Delta' + \chi + 2\gamma + 2\beta - \alpha') \frac{m_i m_j}{M r_{ij}^3} r_{ij\parallel}^2 \right]$$
$$- (2\gamma + 2) g_e \left[\frac{(m v_\parallel \mathbf{v}_\perp)_i + (m v_\parallel \mathbf{v}_\perp)_j}{M} \right] + (8\Delta' + \chi + 2\gamma + 2\beta - \alpha') \frac{m_i m_j}{M r_{ij}^3} g_e r_{ji\parallel} \mathbf{r}_{ji\perp}. \tag{34}$$

We now consider a circular orbit for m_i and m_j with the orbit plane normal vector making an angle θ with the direction toward m_e. This leads to the following time dependence of the velocity of m_i or m_j (z is the direction toward m_e):

$$v_x = v \sin\omega t \cos\theta\,, \tag{35a}$$

$$v_y = v \cos\omega t\,, \tag{35b}$$

$$v_z = v \sin\omega t \sin\theta\,, \tag{35c}$$

with the magnitude of v given by (21a) and (21b). The interparticle position vector is then

$$(r_j - r_i)_x = r_{ij} \cos\omega t \cos\theta\,, \tag{36a}$$

$$(r_j - r_i)_y = -r_{ij} \sin\omega t\,, \tag{36b}$$

$$(r_j - r_i)_z = r_{ij} \cos\omega t \sin\theta\,. \tag{36c}$$

All the terms in (34) can now be evaluated. The acceleration along $\mathbf{g}_e$ is given by

$$a_z = g_e \{ 1 + (m_i m_j / M r_{ij}) [(8\Delta - 4\beta - 3\gamma - \chi) + \tfrac{1}{2} \sin^2\theta (2\beta + \chi + 8\Delta' - \alpha' - 2)] \} \tag{37}$$

when averaged over the rotation period of m_i and m_j. There is no average acceleration in the y direction. In the x direction, however, there is a time-averaged acceleration

$$a_x = g_e (2\beta + \chi + 8\Delta' - \alpha' - 2)(m_i m_j / M r_{ij}) \times \tfrac{1}{2} \sin\theta \cos\theta\,. \tag{38}$$

Additional acceleration terms which oscillate as $\cos 2\omega t$ and $\sin 2\omega t$ but average to zero over a period of the orbital motion of m_i about m_j will be discussed in Sec. VII.

Equations (37) and (38) can be compared with (1) to give an expression for η, the EP violation coefficient, in terms of the general metric coefficients.

Demanding that $\eta = 0$ for arbitrary orientation of the two-body orbit gives two constraints on the metric expansion coefficients;

$$8\Delta - 4\beta - 3\gamma - \chi = 0 \tag{39a}$$

and

$$2\beta + \chi + 8\Delta' - \alpha' - 2 = 0. \tag{39b}$$

Equation (39b) also guarantees the vanishing of the anomalous a_x acceleration given by (38).

In Einstein's grativational theory $\gamma = \beta = \Delta = \alpha' = \chi = 1$, while $\Delta' = 0$,[12] so both (39a) and (39b) are fulfilled. Note that both (39a) and (39b) contain β, the coefficient of the nonlinear term in g_{00}. This result confirms the suggestion of the previous paper (I), in that the motion of a massive body depends on the gravitational acceleration of gravitational self-energy.

In an appendix to this paper, all of the coefficients above except α' are calculated in the scalar-tensor gravitational theory of Brans and Dicke. We obtain $\beta = \chi = 1$ ($\chi = 1$ is necessary for the metric to be properly retarded), $\gamma = (1+w)/(2+w)$, $\Delta = (3+2w)/(4+2w)$, and $\Delta' = 0$. w is a dimensionless parameter of the Brans-Dicke (BD) theory (as $w \to \infty$ Einstein's theory is reobtained). Equation (39a) is not fulfilled for the BD theory; using the above results,

$$(8\Delta - 4\beta - 3\gamma - \chi)_{\text{BD}} = -1/(2+\omega)\,. \tag{40}$$

VI. MASSIVE GASEOUS SPHERE

The previous computation of the acceleration of a two-body system can be altered to give the acceleration of a massive gas sphere maintained in equilibrium by kinetic gas pressure. We can then apply the results of this work to the examination of the m_g/m_i ratio for normal stars (like the Sun) in which the equilibrium of the star is overwhelmingly produced by the balance of gravitational attraction and particle kinetic pressure.

The two-source metric expansion used previously is generalized to many sources by replacing in all the metric terms the single m_i contributions by a summation ($\sum_i$) over many m_i. Then (31) reads

$$\mathbf{a}_j = \mathbf{g}_e \left[1 + (4\Delta - 2\beta - 2\gamma - \tfrac{1}{2}\chi) \sum_i \frac{m_i}{r_{ij}} + \gamma v_j^2 - (2\gamma + 2) v_{\parallel j}^2 + \sum_i \left(4\Delta + \tfrac{1}{2}\chi + \frac{2(\gamma+\beta) m_i - \alpha' m_j}{M} \right) \frac{m_i}{r_{ij}^3} r_{ij\parallel}^2 \right]$$
$$- (2\gamma + 2) v_{\parallel j} g_e \mathbf{v}_{\perp j} + \sum_i \left[4\Delta + \tfrac{1}{2}\chi + \frac{2(\gamma+\beta) m_i - \alpha' m_j}{M} \right] r_{ji\parallel} g_e \mathbf{r}_{ji\perp}. \tag{41}$$

Taking a directional average by averaging over the kinetic motion and position of many particles in the gas sphere, we have the average quantities

$$\langle v_{ji\parallel}\mathbf{r}_{ji\perp}\rangle=0, \tag{42a}$$

$$\langle v_{\parallel j}\mathbf{v}_{\perp j}\rangle=0, \tag{42b}$$

$$\langle r_{ji\parallel}^2/r_{ij}^3\rangle=\tfrac{1}{3}(r_{ij})^{-1}, \tag{42c}$$

and

$$\langle v_{\parallel}^2\rangle=\tfrac{1}{3}v^2. \tag{42d}$$

Taking the sum

$$\mathbf{a}=\textstyle\sum_j m_j\mathbf{a}_j/M, \tag{43}$$

$$M=\textstyle\sum_j m_j,$$

we obtain the acceleration of the sphere

$$\mathbf{a}=\mathbf{g}_e\left\{1+\left[(4\Delta-2\beta-2\gamma-\tfrac{1}{2}\chi)+\tfrac{1}{6}(\beta+\gamma+4\Delta'+\tfrac{1}{2}\chi-\tfrac{1}{2}\alpha')\right]\sum_{i,j}\frac{m_im_j}{Mr_{ij}}+\tfrac{1}{3}(\gamma-2)\sum_j\frac{m_jv_j^2}{M}\right\}. \tag{44}$$

The v_j^2 in (44) are needed only to the classical Newtonian order, so we can use the usual virial theorem for a system in equilibrium:

$$\sum_j m_jv_j^2=\tfrac{1}{2}\sum_{i,j}\frac{m_im_j}{r_{ij}}. \tag{45}$$

Equation (44) can finally be expressed as

$$\mathbf{a}=\mathbf{g}_e\left\{1+\left[\tfrac{1}{3}(8\Delta-4\beta-3\gamma-\chi)+\tfrac{1}{6}(2\beta+\chi+8\Delta'-\alpha'-2)\right]\sum_{i,j}\frac{m_im_j}{Mr_{ij}}\right\} \tag{46}$$

for the acceleration of a massive gas sphere in an external field $\mathbf{g}_e$.

The two contributions to η in (46) are the terms (39a) and (39b) which have already been shown to vanish in Einstein's theory. For a massive gaseous sphere we can now express the parameter which measures the violation of the EP [η in (1)] in terms of the expansion coefficients of the space-time metric:

$$\eta=4\Delta-(5/3)\beta-\tfrac{3}{2}\gamma+\tfrac{4}{3}\Delta'-\tfrac{1}{6}\alpha'-\tfrac{1}{3}. \tag{47}$$

An experimental measurement of η as proposed in I is therefore seen to be a measurement of a combination of several terms in the space-time metric which have not been measured to date. Only γ and β have been measured, while Δ and Δ' will be measured in the orbiting-gyroscope experiment under development at Stanford University.[11]

VII. ANOMALOUS ACCELERATION OF NONSTATIONARY SYSTEMS

To obtain the results (37) and (38) for the acceleration of a two-body orbiting system in an external field, a time average was performed over a period of the orbiting motion of the two bodies around each other.

There are additional oscillatory terms in the exact expression for the acceleration of the two-body system; they are

$$a_x(\text{osc})=\xi\sin\theta\cos\theta\cos2\omega t, \tag{48a}$$

$$a_y(\text{osc})=-\xi\sin\theta\sin2\omega t, \tag{48b}$$

and

$$a_z(\text{osc})=\xi\sin^2\theta\cos2\omega t, \tag{48c}$$

with

$$\xi=g_e(m_im_j/2Mr_{ij})(8\Delta'+\chi+4\gamma+2\beta+2-\alpha').$$

The magnitude of $\mathbf{a}(\text{osc})$ is

$$|\mathbf{a}(\text{osc})|=\xi\sin\theta. \tag{49}$$

In this section we explore other configurations for massive bodies in order to see if the oscillatory anomalous acceleration found above exists in general.

Next consider a two-body orbit with $\sin\theta=0$, so that the above effects (48a)–(48c) vanish. However, let the orbit be elliptical. Then the conditions (21a) and (21b) will not be valid at all times during the orbital motion, only on a time average over the orbital period.

Specializing (34) to $\sin\theta=0$ gives the simpler expression

$$\mathbf{a}=\mathbf{g}_e\left\{1+(8\Delta-4\beta-2\gamma-\chi)\frac{m_im_j}{Mr_{ij}}+\gamma\left[\frac{m_iv_i^2+m_jv_j^2}{M}\right]\right\}. \tag{50}$$

Letting ϵ be the conserved Newtonian energy of the two-body orbit,

$$\epsilon=\tfrac{1}{2}(m_iv_i^2+m_jv_j^2)-\frac{m_im_j}{r_{ij}}, \tag{51}$$

Eq. (50) yields

$$\mathbf{a}=\mathbf{g}_e\left[1+(8\Delta-4\beta-2\gamma-\chi)\frac{m_im_j}{Mr_{ij}}+\frac{2\epsilon}{M}\gamma\right]. \tag{52}$$

Using Einstein's theory's value for the coefficients gives

$$\mathbf{a}=\mathbf{g}_e\left\{1+\left[\frac{m_im_j}{r_{ij}}+2\epsilon\right]\Big/M\right\}. \tag{53}$$

The anomalous acceleration term in (53) only vanishes when averaged over the orbital period. At orbital perigee, (53) gives

$$\mathbf{a}=\mathbf{g}_e\left(1+e\frac{m_i m_j}{Mr_p}\right), \tag{54a}$$

while at orbital apogee it gives

$$\mathbf{a}=\mathbf{g}_e\left(1-e\frac{m_i m_j}{Mr_a}\right), \tag{54b}$$

with e being the eccentricity of the orbit.

Finally we examine a pulsating gaseous sphere. Our previous results concerning the acceleration of the sphere depended on using an equilibrium condition for the sphere—the virial theorem relating mean kinetic energy to potential energy. Here we assume that the sphere radially pulsates about equilibrium.

When the pulsation is at the extreme condensed state, the kinetic energy of the gas will exceed the requirement of the virial theorem, i.e.,

$$\sum_j m_j v_j^2=\tfrac{1}{2}\sum_{i,j}\frac{m_i m_j}{r_{ij}}+2\delta\epsilon_1. \tag{55}$$

But when the pulsation is at the extreme expanded state, the kinetic energy of the gas is less than required by the virial theorem;

$$\sum_j m_j v_j^2=\tfrac{1}{2}\sum_{i,j}\frac{m_i m_j}{r_{ij}}-2\delta\epsilon_2. \tag{56}$$

Using Einstein's theory's value for the metric coefficients and evaluating expression (44) for the cases (55) and (56), we get, respectively, accelerations of the sphere

$$\mathbf{a}=\mathbf{g}_e\left(1-\frac{2}{3}\frac{\delta\epsilon_1}{M}\right) \tag{57}$$

and

$$\mathbf{a}=\mathbf{g}_e\left(1+\frac{2}{3}\frac{\delta\epsilon_2}{M}\right). \tag{58}$$

These anomalous accelerations will also vanish when time is averaged over the pulsation period.

The common feature of all three systems examined above which showed oscillating anomalous accelerations was that all systems presented a nonstationary, oscillating configuration to the external mass m_e.

In a future paper we will study these oscillating accelerations to see if they produce in principle measurable effects or whether they are simply coordinate-system-dependent anomalies.

APPENDIX

The coefficients which appear in (39a) and (39b) can be obtained for Einstein's theory from the EIH paper.[12] In this paper we calculate these coefficients for the Brans-Dicke (BD) gravitational theory.[6] All of the needed coefficients except α' are calculated, and we hope to obtain α' in a later paper.

The BD field equations for the space-time metric tensor and their scalar field are[6]

$$\square^2\phi=\frac{8\pi}{c^4}\frac{T}{3+2w} \tag{A1}$$

and

$$R_{ij}=-\frac{8\pi}{\phi c^4}\left(T_{ij}-\frac{1+w}{3+2w}Tg_{ij}\right)-\phi_{;i;j}/\phi-w\phi_{,i}\phi_{,j}/\phi^2, \tag{A2}$$

which reduce to Einstein's field equations in the limit $w\rightarrow\infty$. To lowest order, (A1) yields

$$\phi=\phi_0+\frac{2}{c^2}\frac{1}{3+2w}\frac{m_1}{|\mathbf{r}-\mathbf{r}_1|} \tag{A3}$$

for a point source.

In the BD paper, the coefficients γ and β are calculated in order to obtain the advance of planetary perihelion. They obtain[6]

$$\gamma=(1+w)/(2+w) \tag{A4}$$

and

$$\beta=1. \tag{A5}$$

To obtain χ, we need the R_{00} equation to linear order in the source strength:

$$R_{00}=-\frac{8\pi}{\phi_0 c^4}\left(T_{00}-g_{00}T\frac{1+w}{3+2w}\right)-\frac{1}{\phi_0}\frac{d^2}{dt^2}\phi. \tag{A6}$$

Only the part of R_{00} linear in Christoffel symbols is required:

$$R_{00}\cong\Gamma_{k0|0}{}^{k}-\Gamma_{00|k}{}^{k} \tag{A7}$$

or

$$R_{00}\cong-\tfrac{1}{2}\nabla^2 g_{00}+g_{0k|k0}-\tfrac{1}{2}g_{kk|00}. \tag{A8}$$

Equation (A6) then yields the equation

$$-\tfrac{1}{2}\nabla^2 g_{00}=-\frac{4\pi}{\phi_0}\left(\frac{4+2w}{3+2w}\right)\rho-\frac{1}{\phi_0}\frac{d^2}{dt^2}\phi+\tfrac{1}{2}g_{kk|00}-g_{0k|k0}. \tag{A9}$$

Using (A3), and[6]

$$g_{kk'}=-\delta_{kk'}\left(1+\frac{2+2w}{2+w}\frac{m_1}{|\mathbf{r}-\mathbf{r}_1|}\right), \tag{A10}$$

$$g_{0k}=4\Delta\frac{m_1}{|\mathbf{r}-\mathbf{r}_1|}v_1{}^{k}+4\Delta'\frac{m_1}{|\mathbf{r}-\mathbf{r}_1|^3}\mathbf{r}-\mathbf{r}_1\cdot\mathbf{v}_1(\mathbf{r}-\mathbf{r}_1)^{k}, \tag{A11}$$

and[5]

$$[(4+2w)/(3+2w)](1/\phi_0)=1 \quad \text{(in units } G=c=1\text{)}, \tag{A12}$$

Eq. (A9) becomes

$$\nabla^2 g_{00}=8\pi\rho+\left(\frac{8+6w}{2+w}\right)\frac{d^2}{dt^2}\frac{m_1}{|\mathbf{r}-\mathbf{r}_1|} -(8\Delta-8\Delta')\frac{m_1}{|\mathbf{r}-\mathbf{r}_1|}\mathbf{r}-\mathbf{r}_1\cdot\mathbf{a}_1, \tag{A13}$$

which has the solution

$$g_{00}=1-2\frac{m_1}{|\mathbf{r}-\mathbf{r}_1|}+\left(4\Delta-4\Delta'-\frac{4+3w}{2+w}\right) \times\frac{m_1}{|\mathbf{r}-\mathbf{r}_1|}\mathbf{r}-\mathbf{r}_1\cdot\mathbf{a}_1. \tag{A14}$$

Comparing (A14) with (12d) gives

$$\chi=4\Delta-4\Delta'-[(4+3w)/(2+w)]. \tag{A15}$$

In order to obtain Δ and Δ', we use the R_{0k} equation which to linear order is

$$R_{0k}=-\frac{8\pi}{\phi_0 c^4}T_{0k}-\frac{1}{\phi_0}\frac{\partial}{\partial t}\frac{\partial}{\partial x^k}\phi, \tag{A16}$$

with the linearized R_{0k} given by

$$R_{0k}=-\tfrac{1}{2}\nabla^2 g_{0k}+\tfrac{1}{2}g_{0s|sk}-\tfrac{1}{2}g_{ss|0k}+\tfrac{1}{2}g_{ks|0s}. \tag{A17}$$

Therefore we have, using (A12),

$$\nabla^2 g_{0k}-g_{0s|ks}=16\pi\left(\frac{3+2w}{4+2w}\right)T_{0k} +\frac{3+2w}{4+2w}\frac{\partial^2}{\partial t\partial x^k}\phi-g_{ss|0k}+g_{ks|s0} \tag{A18}$$

or

$$\nabla^2 g_{0k}-g_{0k|ks}=16\pi\frac{3+2w}{4+2w}T_{0k}+\frac{6+4w}{2+w} \times\left[\frac{m_1}{|\mathbf{r}-\mathbf{r}_1|^3}v_1^k-3\frac{m_1}{|\mathbf{r}-\mathbf{r}_1|^5}\mathbf{r}-\mathbf{r}_1\cdot\mathbf{v}_1(r-r_1)^k\right], \tag{A19}$$

which yields

$$g_{0k}=4\left(\frac{3+2\omega}{4+2\omega}\right)\frac{m_1}{|\mathbf{r}-\mathbf{r}_1|}v_1^k. \tag{A20}$$

Comparing (A20) with (12b) gives

$$\Delta=(3+2\omega)/(4+2\omega), \tag{A21a}$$

$$\Delta'=0, \tag{A21b}$$

which inserted into (A15) gives

$$\chi=1. \tag{A22}$$

The evaluation of (39a) with the BD coefficients does not yield zero:

$$(8\Delta-4\beta-3\gamma-\chi)_{\rm BD}=-1/(2+w), \tag{A23}$$

indicating that a massive body in the BD theory will possess an anomalous $1/R^2$ acceleration toward an external mass.

MEASUREMENT OF THE DEFLECTION OF 9.602-GHz RADIATION FROM 3C279 IN THE SOLAR GRAVITATIONAL FIELD

G. A. Seielstad,* R. A. Sramek, and K. W. Weiler
Owens Valley Radio Observatory,† California Institute of Technology, Pasadena, California 91109
(Received 30 March 1970)

During its occultation by the sun in October 1969, the position of the radio source 3C279 was interferometrically monitored to determine the deviation of its 9.602-GHz radiation in the solar gravitational field. Rapid instrumental calibration and negligible coronal diffraction enabled the measurement of a general relativity deflection of 1.77″±0.20″ at the limb of the sun. This is in close agreement with Einstein's prediction.

We have used an interferometer at the Owens Valley Radio Observatory consisting of one 27-m antenna and the new 40-m antenna to measure the deflection of electromagnetic radiation in the solar gravitational field predicted by the general theory of relativity.[1] The test was performed by measuring the phase of the interference fringes from 3C279 [assumed $\alpha(1950.0)=12^h53^m35.92^s$, $\delta(1950.0)=-05°31'08.9''$] relative to that of nearby 3C273 [assumed $\alpha(1950.0)=12^h26^m33.20^s$, $\delta(1950.0)=+02°19'41.2''$] many times each day from 30 September through 15 October 1969. The daily vector separations between the sun's center and the two sources are presented in Table I for local sidereal times (LST) between 9.0^h and 16.5^h. The distances ρ are in units of solar radii (using 1 radius ≡ 16′02.0″), and the position angles Θ are in degrees measured from north through east to the vector from sun to source.

The interferometer was conventional, differing only slightly from the system described by Read.[2] Both the common local-oscillator signal delivered to the two elements and the returning i.f. signals were carried on buried low-loss coaxial cables. The so-called "high-frequency reference" within the phase-lock system was the highly amplified output of a crystal-controlled frequency synthesizer at 30.188 679 2 MHz. At each antenna, harmonics of this signal were generated, the 159th then beating in the "special mixers" with klystron oscillators operating at 4.801 GHz. The resultant 1-MHz difference signals were then treated as in Ref. 2.

The only other significant change is that the phase-locked 4.801-GHz klystron outputs passed through frequency doublers just before entering

Table I. Sun-3C273 and sun-3C279 separations.

	3C273		3C279	
Date (1969)	Distance, ρ_3 (Solar Radii)	Position Angle, θ_3 (Degrees E of N)	Distance, ρ_9 (Solar Radii)	Position Angle, θ_9 (Degrees E of N)
Sept. 30	19.2–19.6	1.8–0.0; 0.0–358.7	27.8–26.6	111.7–111.6
Oct. 1	20.8–21.4	352.4–349.7	24.1–23.0	111.4–111.3
2	22.9–23.7	344.4–342.3	20.4–19.3	111.1–110.9
3	25.4–26.2	337.9–336.1	16.8–15.6	110.5–110.3
4	28.1–29.0	332.6–331.1	13.1–12.0	109.7–109.4
5	31.0–32.0	328.2–327.0	9.4– 8.3	108.4–107.7
6	34.1–35.1	324.6–323.6	5.8– 4.6	105.4–103.5
7	37.3–38.3	321.6–320.7	2.2– 1.2	92.4– 72.2
8	40.6–41.6	319.0–318.3	1.8– 2.9	317.3–307.8
9	43.9–45.0	316.8–316.2	5.4– 6.5	300.7–299.2
10	47.3–48.4	314.9–314.3	9.1–10.2	297.3–296.7
11	50.7–51.8	313.2–312.7	12.7–13.9	295.8–295.5
12	54.2–55.3	311.7–311.3	16.4–17.6	294.9–294.7
13	57.7–58.8	310.4–310.0	20.1–21.3	294.3–294.2
14	61.2–62.3	309.2–308.9	23.8–25.0	293.9–293.7
15	64.8–65.9	308.1–307.8	27.5–28.7	293.5–293.4

the mixers preceding the i.f. preamplifiers. Lobe rotation was thus performed at twice the usual rate. Otherwise, the 3-cm wavelength interferometer functioned as described in other Owens Valley Radio Observatory publications.[3] In particular, both sidebands were accepted, so that the phase was insensitive to slight misadjustments of the delay line or to changes in the amplifiers and transmission lines.[2]

The observing frequency of 9.602 GHz was chosen both to achieve high angular resolution and to lessen refraction effects in the solar corona. Assuming the electron distribution of the Allen-Baumbach model,[4] $N=1.55\times10^{14}\rho^{-6}$ m^{-3}, the angular deviation ω of a ray of frequency 9.6 GHz is given by[5] $\omega=82\rho^{-6}$ sec, where ρ is the distance in solar radii of the ray's point of closest approach to the sun's center. If, instead, we use Erickson's coronal model[6] for $\rho=4$-20, $N=5\times10^{11}\rho^{-2}\ \mathrm{m}^{-3}$, the angular deviation is[5] $\omega=0.14\rho^{-2}$ sec. For either model refraction effects are only significant on 7 and 8 October, when $\rho<3$. These days are then useful for studying the corona itself. Two independent analyses of a closely related radar experiment likewise both conclude that the solar corona can be virtually ignored at frequencies of 8-10 GHz.[7]

The baseline of the interferometer[8] was calibrated initially at frequencies of 958.0 and 2881.6 MHz to eliminate any possible lobe ambiguities. Then, the 9.602-GHz baseline was determined repeatedly and in several ways, always during the hours about local midnight to minimize thermal effects. The antenna separation was found to be 3498.0512 ± 0.001 ft (1.066 206 0 km or 34 149.27 wavelengths), and the extended baseline intersected the celestial sphere at hour angle = $06^{\mathrm{h}}01^{\mathrm{m}}38.84^{\mathrm{s}}\pm0.01^{\mathrm{s}}$, declination = $-00°18'20.2''\pm0.1''$. No temporal changes were detected throughout the observing period.

The entire system was tested in several ways. By observing with intentional pointing offsets in October 1968, we demonstrated that the measured phase of small-diameter sources is not noticeably affected by pointing errors. This conclusion is also verified experimentally with the National Radio Astronomy Observatory interferometer.[9] To maintain parallel feeds throughout each day, we rotated both feed and receiver together on the 40-m antenna, which has an altitude-azimuth mounting. Tests in June 1969 demonstrated that the phase was insensitive to this rotation. The phase was likewise insensitive to voltage changes in the phase-lock correction signals. Refocusing either feed, however, produced arbitrarily large phase changes. Therefore, the entire relativity test was peformed with fixed foci. Also in June, we determined the LST dependence of the phase difference between 3C279 and 3C273 at intervals of about 7^{m} when the sun was absent. The standard deviation of a single difference from the curve best fitting all differences was 0.042 lobes. The observed phase scatter could all be attributed to instabilities of the local-oscillator system and fluctuations in the atmospheric water-vapor content, with no attempt made to separate the two. Ionospheric effects at this frequency are negligible.

Most of these tests were repeated in October 1969 with essentially the same results. On 7 October the sun's influence was tested by observing, with nearly identical effective baselines, blank sky at several of the same positions relative to the sun's center as 3C279 had occupied on other days. In no case was a signal significantly exceeding noise detected. At a point less than 3 radii from the center, the measured amplitude was only 0.2% of the flux density of 3C279.

Unpredictable phase changes can arise from instrumental and environmental effects. Examples are differential changes in the structures of the antennas, differential phase shifts in the exposed local oscillator cables to the foci, atmospheric water-vapor fluctuations, instabilities in the local oscillator system, and phase scatter due to receiver noise. By cycling rapidly between 3C273 and 3C279 and considering only phase differences, these and similar effects were minimized. Our actual cycling time was such that over 900 reliable phase differences were obtained in about 100 h of observing.

The phase differences between the two sources did not remain constant throughout the day. Changes having the same LST dependence every day were caused by incorrect source positions, source structure, and incorrect baseline parameters. Dependences which changed daily because of the sun's apparent motion were produced by solar refraction and gravitational deflection. The two dependences were separated by day-to-day comparisons of the data obtained at similar sidereal times.

In determining the phase differences, $\varphi_9-\varphi_3\equiv\Delta$, we linearly interpolated between adjacent measurements of one source to the time of measurement of the other. The subsequent data analysis was performed in three different ways.

All assumed that the gravitational bending was radially directed and varied as $1/\rho$, and all neglected the data from 7 and 8 October.

In two of the methods (carried out independently by G.A.S.), all data from 12-15 October were averaged to obtain an initial calibration curve of phase difference Δ_0 vs LST when both sources were far from the sun. Reasonable agreement with the June results was achieved. This calibration curve was then subtracted from the phase differences for all days. Daily averages of these differences of differences, $\Delta-\Delta_0 \equiv \Box^2$, were formed over four LST intervals.

In one method these means, $\langle\Box^2\rangle$, were compared with values computed for an assumed gravitational deflection of 1.75″ at the solar limb. Differences between the measured and predicted values were used to improve the calibration curve, Δ_0 vs LST. Ratios of the modified measured means to those predicted were then averaged, finally giving measured/predicted $= 1.02 \pm 0.12$, or a deflection at the limb of 1.79″ ±0.21″.

In a second method both the projected baselines (length s in $\sec^{-1}$ and position angle of fringe normal p) and the relative coordinates ρ and Θ at the times appropriate for each mean were calculated for both sources. The quantities $s_9 \cos(p_9-\Theta_9)/\rho_9 \equiv 1/R_9$ and $s_3 \cos(p_3-\Theta_3)/\rho_3 \equiv 1/R_3$ represent the inverse angular distances of 3C279 and 3C273, respectively, as measured in the coordinate system of the interferometer. We expect $\langle\Box^2\rangle = K(1/R_9-1/R_3)$, where K is the gravitational deflection at the solar limb in seconds of arc.

Separately for each of the four LST intervals, a least-squares straight line was fitted to the data. The values of $\langle\Box^2\rangle$ at $1/R_9 = 1/R_3$ were used to improve the initial calibration curve. The corrected data were then combined and a final least-squares fit, illustrated in Fig. 1, gave $K = 1.77'' \pm 0.19''$.

The third and most general method of data analysis (performed independently by K.W.W. and R.A.S.) fits, by the method of least squares, a curve of the form

$$-\Delta = A + B\sin(\mathrm{LST}) + C\cos(\mathrm{LST}) + K(1/R_3 - 1/R_9)$$
$$-D(1/\rho_3 R_3 - 1/\rho_9 R_9) - E(1/\rho_3^{\,5} R_3 - 1/\rho_9^{\,5} R_9)$$

to all 914 individual values of the measured phase differences. The first three terms represent empirically those variations with LST which are the same every day. We find $A = -0.030 \pm 0.018$,

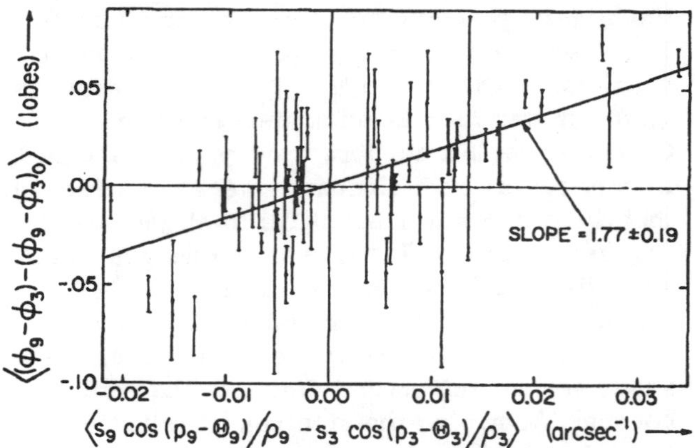

FIG. 1. Mean differences of phase differences, $\langle\Box^2\rangle$, as a function of the difference between inverse angular distances $(1/R_9-1/R_3)$ in the interferometer's coordinate system. There are at most four points per day. The line represents a least-squares fit whose slope, 1.77″±0.19″, is the gravitational deflection at the solar limb.

$B = -0.108 \pm 0.007$, and $C = -0.350 \pm 0.021$ lobes. The fourth term represents the general relativistic effect. We find $K = 1.74'' \pm 0.21''$ for the deflection at the solar limb. The last two terms represent coronal refraction. As asserted previously, this has an insignificant effect in the data considered, for all of which $\rho \gtrsim 5$. For example, holding D fixed at 0.14″ while varying E between 0″ and 100″ changes K by only 0.015″. Likewise, fixing E at 82″ and varying D between 0.0″ and 0.5″ changes K only between 1.72″ and 1.79″. In neither case does A, B, or C change. Figure 2 illustrates the curve fit using $D = 0.14''$ and $E = 82''$ with the A, B, C, and K given above. The general relativity deflection can be clearly seen in the displacement of the curve down for 6 October and up for 9 October as compared with the relatively undisplaced level for the days farther from the sun on 1 October and 14 October.

The experiment does not yield a good determination of D. However, putting $D = 0.14''$ and including the data from 7 and 8 October, days on which the $1/\rho^6$ term dominates, we find $E = 22''$, corresponding to an electron density of $N = 0.42 \times 10^{14}\rho^{-6}\ \mathrm{m}^{-3}$.

We conclude that the gravitational deflection at the limb of the sun is 1.77″ ± 0.20″ (standard deviation) in near exact agreement with Einstein's prediction.[1] This result agrees precisely with that determined in the totally independent experiment at another frequency with a different interferometer by Muhleman, Ekers, and Foma-

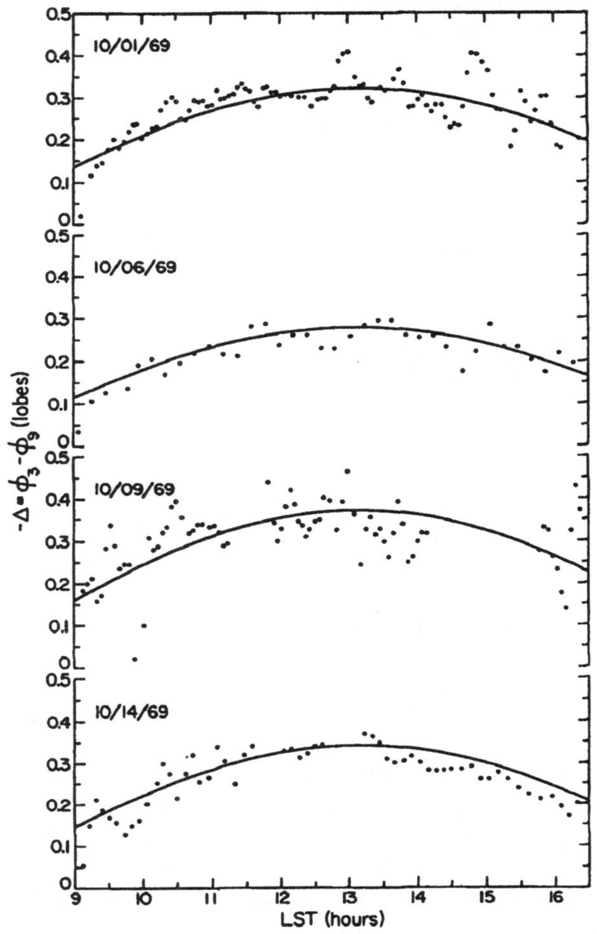

FIG. 2. Examples of four days for the least-squares best-fit curve to the data.

lont, presented in the following Letter.[10]

The scalar-tensor theory of gravitation now predicts a deflection ~0.93 times as large as does Einstein's,[11] and therefore cannot be definitely ruled out by these measurements. However, they seem a significant improvement over most, perhaps all, previous optical determinations (summarized by von Klüber[12]). Not only do these latter have poor internal agreement and even disagreement between different investigators' analyses of the same data, but have indicated that the gravitational deflection appears somewhat larger than the value predicted by the general theory, a result not confirmed by our experiment.

The receivers were designed and tested by G. J. Stanley, who also conceived the local oscillator system. We are especially indebted to C. L. Spencer and R. E. Allen.

---

*Currently on leave of absence as a John Simon Guggenheim Fellow at the Onsala Space Observatory, Chalmers University of Technology, Gothenburg, Sweden.

†Operated with support from the U. S. Office of Naval Research under Contract No. N00014-67-A-0094-0008 and from the National Science Foundation under Contract No. GP15271.

[1]A. Einstein, Ann. Phys. (Leipzig) 49, 769 (1916), and *The Principle of Relativity*, translated by W. Perrett and G. B. Jeffery (Dover, New York, 1923), pp. 109-164.

[2]R. B. Read, Astrophys. J. 138, 1 (1963).

[3]E.g., G. L. Berge, Astrophys. Letters 2, 127 (1968).

[4]C. W. Allen, Monthly Notices Roy. Astron. Soc. 107, 426 (1947).

[5]R. N. Bracewell, V. R. Eshelman, and J. V. Hollweg, Astrophys. J. 155, 367 (1969).

[6]W. C. Erickson, Astrophys. J. 139, 1290 (1964). (At solar maximum, N and ω may be greater by a factor of ~5. Our conclusions will be unaffected.)

[7]I. I. Shapiro, Phys. Rev. Letters 13, 789 (1964); D. O. Muhleman and I. D. Johnston, *ibid.* 17, 455 (1966).

[8]See Ref. 2 for a definition of the baseline and a fuller explanation of its determination. Note also that the azimuth and elevation axes of the 40-m telescope intersect at a point.

[9]*The VLA, A Proposal for a Very Large Array Radio Telescope* (National Radio Astronomy Observatory, Green Bank, W. Va., 1967), Vol. I, p. 5-4.

[10]D. O. Muhleman, R. D. Ekers, and E. B. Fomalont, Phys. Rev. Letters 24, 1377 (1970).

[11]R. H. Dicke, in *Contemporary Physics: Trieste Symposium 1968* (International Atomic Energy Agency, Vienna, Austria, 1969), Vol. 1, pp. 515-531.

[12]H. von Klüber, *Vistas in Astronomy*, edited by A. Beer (Pergamon, New York, 1960), Vol. 3, p. 47.

Mercury's Perihelion Advance: Determination by Radar

Irwin I. Shapiro*† and Gordon H. Pettengill*
Massachusetts Institute of Technology, Cambridge, Massachusetts 02139

and

Michael E. Ash
Massachusetts Institute of Technology Lincoln Laboratory,‡ Lexington, Massachusetts 02173

and

Richard P. Ingalls
Haystack Observatory, Northeast Radio Observatory Corporation, Westford, Massachusetts 01886

and

D. B. Campbell and R. B. Dyce
National Astronomy and Ionosphere Center, Arecibo, Puerto Rico 00612
(Received 10 April 1972)

Measurements of echo delays of radar signals transmitted from Earth to Mercury have yielded an accurate value for the advance of the latter's perihelion position. Given that the Sun's gravitational quadrupole moment is negligible, the result in terms of the Eddington-Robertson parameters is $(2+2\gamma-\beta)/3 \simeq 1.005 \pm 0.007$, where $\gamma=\beta=1$ in general relativity, and where 0.007 represents the statistical standard error. Inclusion of the probable contribution of systematic errors raises the uncertainty to about 0.02.

Interplanetary radar observations, although essentially uncoupled from the "fixed" stars, are nonetheless very sensitive to changes in orbital perihelion positions. For all presently practical purposes, the Sun and planets form a closed dynamical system; the perihelion position of one planet can be determined relative to that of another from the radar measurements of echo delay. Because of the large eccentricity and nearness to the Sun of Mercury's orbit, its non-Newtonian perihelion advance is the easiest to estimate accurately. That radar could provide data for a significant test of the related prediction of general relativity has long been recognized.[1] Here we report the results of an analysis of five years of radar observations of Mercury and Venus as they relate to this well-known test of Einstein's theory.[2] These data were obtained primarily between 1966 and 1971 at the Haystack (Massachusetts) and Arecibo (Puerto Rico) Observatories. All told, there are about 150 Arecibo-Mercury and 200 Haystack-Mercury time-delay measurements; the former were obtained at a radar frequency of 430 MHz and the latter at 7840 MHz. The individual measurement errors are mostly between 5 and 20 μsec. The number of Earth-Venus observations[3] exceeds 500 with many of the recent Haystack time-delay measurements having uncertainties of only 1 μsec. Doppler-shift measurements from Haystack's Mercury and Venus observations were included but have little effect on the results. No optical data at all were included.

In our analysis we considered all potentially important gravitational interactions within the framework of general relativity. As expected, it proved more than sufficient to use for the planets the equations of motion that follow from the Schwarzschild metric for the Sun, supplemented by the Newtonian interplanetary perturbations. To estimate the non-Newtonian perihelion advance explicitly, we parametrized the equations of motion, expressed in harmonic coordinates, such that all non-Newtonian terms were multiplied by the *ad hoc* parameter λ_p. This parametrization does not test the contributions of the individual relativistic terms, but just their cumulative effect. This limitation is of no practical consequence since the existing radar data are very insensitive to all of the predicted relativistic effects on planetary motion save for the non-Newtonian perihelion advance. An estimate of λ_p under these circumstances is therefore equivalent to an estimate of $(2+2\gamma-\beta)/3$, where γ and β are the Eddington-Robertson parameters.[4]

Although the non-Newtonian advance of Mercury's perihelion position is by far the largest relativistic effect, radar observations of Venus play an important role in our estimate of λ_p. The reason is simple. All observations are made from Earth and hence its orbital elements must be determined along with Mercury's when estimating λ_p. Primarily because of the closer approaches of Venus to Earth, much of the corresponding radar echo-delay data are more accurate than the Earth-Mercury data (see above) and serve to determine a more precise orbit for Earth than would be possible with the latter data alone.[5]

In addition to λ_p, therefore, we had to estimate from these data the four "in-plane" orbital parameters[6] for each of the three innermost planets; the light-second equivalent of the astronomical unit (a consequence of our choice of units[6]); the mass of Mercury; the mean equatorial radii of Mercury and Venus; a plasma parameter to account for the interplanetary medium; and two "bias" parameters to disclose possible systematic differences between the Haystack and Arecibo observations of Mercury and Venus.[6] The other orbital elements and masses of the inner planets, as well as the orbits and masses of the Moon and outer planets, are known sufficiently well from other observations so that the uncertainties are too small to affect significantly our estimate for λ_p. A similar comment applies *a fortiori* to the rotation of Earth about its center of mass. The solar corona is of no concern since the data that determine λ_p were obtained mostly near inferior conjunctions where the radio signals do not penetrate inside Mercury's orbit. We also made the assumption that the standardly defined dimensionless parameter J_2, describing the Sun's gravitational quadrupole moment, is zero. The data do not allow a useful result to be obtained from a simultaneous estimate of both λ_p and J_2.

The only other possibly significant source of systematic error is the variation of topography over the equatorial regions of the target planets. Venus's topography presents perhaps the greatest difficulty because of the apparently strong coupling between its spin and the relative orbital motions of Earth and Venus. Nonetheless, significant progress has been made and agreement is reasonable between the results for surface-height variations obtained from echo-delay data and those obtained from absorption effects of the Venus atmosphere on Haystack's X-band radar signal.[7] For the main target, Mercury, the surface-height variations are relatively inconspicuous. None have yet been detected reliably from the echo-delay measurements to the subradar point. But about one third of the equatorial circumference has been mapped by another technique which discloses occasional small (<2 km) peaks and valleys.[8] Aside from these apparently minor variations, there is another reason why Mercury's topography should not affect seriously our estimate of λ_p: The orbital periods of Earth and Mercury are incommensurable and measurements at many different parts of Mercury's orbit have each been made from different parts of Earth's orbit. Thus, although Mercury's spin is coupled to its orbital motion, the effects of topography on the estimate of the orbit will tend to average out over the more than five-year span of the radar data.

Our weighted-least-squares solution for the twenty parameters described above yields

$$(2+2\gamma-\beta)/3 \simeq \lambda_p = 1.006 \pm 0.006,$$

where 0.006 represents the formal standard error and is based on the error assigned to each of the echo delays from a consideration of the signal-to-noise ratio and the relevant systematic errors that might be introduced by the measurement apparatus. Because the lack of a representation of planetary topography is the most serious deficiency in the twenty-parameter theoretical model, we added two-dimensional Fourier series to represent separately the surface-height variations in the equatorial regions of each planet. We considered different numbers of terms in the Fourier series to test more thoroughly the sensitivity of our estimate for λ_p to the surface-height variations. The number of extra parameters involved ranged from 40 to 122 for each planet; the total number in a given solution reached a high of nearly 250. The result for λ_p in no case differed from unity by more than about 0.01. The solution with a typical representation of topography (eighty extra parameters) yielded

$$(2+2\gamma-\beta)/3 \simeq \lambda_p = 1.004 \pm 0.007.$$

The post-fit residuals for the Mercury echo-delay observations for this solution, displayed in Fig. 1 as a function of the longitude of the subradar point,[9] have a weighted-rms value near unity as expected.

The estimates obtained for the other parameters from each of the solutions were all of a nonsurprising nature, with one exception. The bias between the Arecibo and Haystack Mercury observations was consistently in the range from 7 to 10 μsec, with the delays measured at Arecibo being the larger. (The corresponding bias for

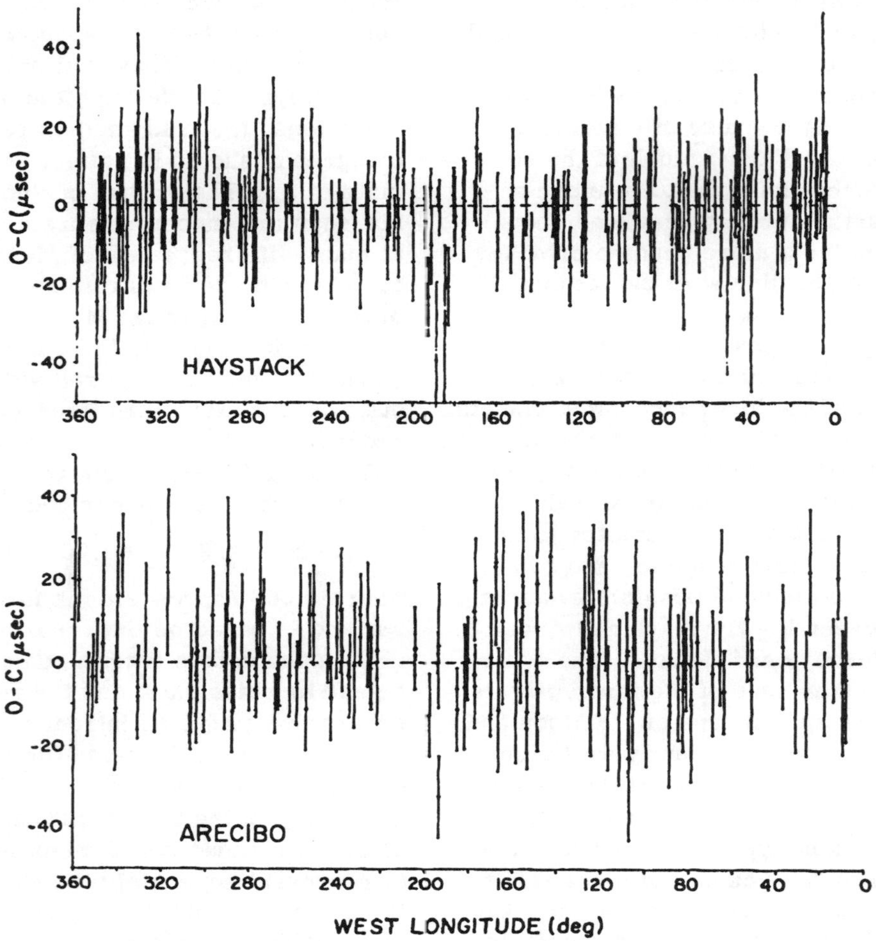

FIG. 1. Post-fit residuals for Earth-Mercury time-delay measurements made at the Haystack and Arecibo Observatories. The residuals are displayed as a function of the longitude of the subradar point on Mercury; low-frequency longitudinal variations in surface heights were removed by means of a Fourier-series parametrization (see text).

the Venus observations was invariably 1 μsec or less.) The possible cause of this discrepancy is still under investigation; it may be related to the radar scattering law since Mercury's surface seems in general to be far coarser on the scale of a wavelength then does Venus's surface. To investigate the possible effect of such a bias on our result for λ_p, we made additional least-squares solutions by alternately suppressing the Haystack and Arecibo data. In some cases the surface-height parameters were also omitted. The results for λ_p were each consistent with a 1.00 value within the somewhat larger bounds of 0.02 and 0.04, for Haystack and Arecibo data, respectively.

In consideration of all of the above, our best judgement is that the radar data alone yield a value for λ_p that does not differ significantly from unity; our estimate of the standard error is about 0.02, mainly attributable to an allowance for the probable contributions of systematic errors.[10] Combining the above with our result for γ[6] yields

$$\gamma \simeq 1.0 \pm 0.1, \quad \beta \simeq 1.1 \mp 0.2$$

for the Eddington-Robertson parameters. Our analysis, as mentioned above, is predicated on the assumption that the solar gravitational quadrupole moment is zero. The result for λ_p can therefore be considered a confirmation of general relativity only insofar as the contribution of J_2 to the orbit of Mercury is negligible.[11] Approximately five years of additional radar observations of the inner planets are required in order to separate usefully the contributions to their orbits of λ_p and J_2. A covariance analysis based on expected improvement in measurement accuracy and in the modeling of planetary topography indicates that the uncertainty of J_2 would be reduced to about 3×10^{-6} and that of λ_p to about 0.3%.

We thank R. Cappallo, R. F. Jurgens, M. A. Slade, and W. B. Smith for important contributions to earlier phases of this study and the staffs of the Arecibo and Haystack Observatories for their aid with the radar observations and data preparation. Research at the Northeast Radio Observatory Corporation Haystack Observatory is supported by the National Science Foundation (Grant No. GP-25865) and the National Aeronautics and Space Administration (Grant No. NGR22-174-003, and Contract No. NAS9-7830). The National Astronomy and Ionosphere Center at Arecibo is operated by Cornell University under a contract with the National Science Foundation.

---

*Department of Earth and Planetary Sciences.

†Department of Physics.

‡Operated with support from the Department of the U. S. Air Force.

[1]I. I. Shapiro, Phys. Rev. Lett. 13, 789 (1964), and Massachusetts Institute of Technology Lincoln Laboratory Technical Report No. 368, DDC No. 614232, 1964 (unpublished).

[2]The radar observations of Mars, at the present state of analysis, cannot contribute usefully to this test.

[3]These include observations from the Millstone Hill radar of the Massachusetts Institute of Technology Lincoln Laboratory and from the Goldstone radars of the Jet Propulsion Laboratory. The latter were kindly sent to us by R. M. Goldstein, J. H. Lieske, and W. G. Melbourne.

[4]See, for example, H. P. Robertson, in *Space Age Astronomy*, edited by A. J. Deutsch and W. E. Klemperer (Academic, New York, 1962), p. 228.

[5]The existing radar observations of Venus do not allow the advance of its orbital perihelion, or Earth's, to be determined with an accuracy useful for testing general relativity.

[6]I. I. Shapiro, M. E. Ash, R. P. Ingalls, W. B. Smith, D. B. Campbell, R. B. Dyce, R. F. Jurgens, and G. H. Pettengill, Phys. Rev. Lett. 26, 1132 (1971). [Note that the result of the first radar time-delay test of general relativity $(1+\gamma)/2\simeq0.9\pm0.2$ was misprinted as 0.09 ± 0.2 in this reference.]

[7]D. B. Campbell, R. B. Dyce, R. P. Ingalls, G. H. Pettengill, and I. I. Shapiro, Science 175, 514 (1972); A. E. E. Rogers, R. P. Ingalls, and L. P. Rainville, Astron. J. 77, 100 (1972).

[8]R. P. Ingalls and L. P. Rainville, Astron. J. 77, 185 (1972).

[9]We use the definition of longitude adopted in *Proceedings of the Fourteenth General Assembly of the International Astronomical Union*, edited by C. De Jager and A. Jappel (D. Reidel, Dordrecht, Holland, 1971), p. 28. In this system, the Sun was above the zero meridian at the time of Mercury's first perihelion passage in 1950.

[10]Prior determinations of the perihelion advance of Mercury's orbit, based solely on optical observations, yielded the equivalent of $\lambda_p\simeq1.00\pm0.01$ [G. M. Clemence, Astron. Papers Amer. Ephemeris Nautical Almanac 11, Part 1 (1943); see also R. L. Duncombe, *ibid.* 16, Part 1 (1959)].

[11]By the same token, given that $\lambda_p\equiv1$, our data show that $J_2<5\times10^{-6}$.

Inflationary universe: A possible solution to the horizon and flatness problems

Alan H. Guth*
Stanford Linear Accelerator Center, Stanford University, Stanford, California 94305

(Received 11 August 1980)

The standard model of hot big-bang cosmology requires initial conditions which are problematic in two ways: (1) The early universe is assumed to be highly homogeneous, in spite of the fact that separated regions were causally disconnected (horizon problem); and (2) the initial value of the Hubble constant must be fine tuned to extraordinary accuracy to produce a universe as flat (i.e., near critical mass density) as the one we see today (flatness problem). These problems would disappear if, in its early history, the universe supercooled to temperatures 28 or more orders of magnitude below the critical temperature for some phase transition. A huge expansion factor would then result from a period of exponential growth, and the entropy of the universe would be multiplied by a huge factor when the latent heat is released. Such a scenario is completely natural in the context of grand unified models of elementary-particle interactions. In such models, the supercooling is also relevant to the problem of monopole suppression. Unfortunately, the scenario seems to lead to some unacceptable consequences, so modifications must be sought.

I. INTRODUCTION: THE HORIZON AND FLATNESS PROBLEMS

The standard model of hot big-bang cosmology relies on the assumption of initial conditions which are very puzzling in two ways which I will explain below. The purpose of this paper is to suggest a modified scenario which avoids both of these puzzles.

By "standard model," I refer to an adiabatically expanding radiation-dominated universe described by a Robertson-Walker metric. Details will be given in Sec. II.

Before explaining the puzzles, I would first like to clarify my notion of "initial conditions." The standard model has a singularity which is conventionally taken to be at time $t=0$. As $t\to 0$, the temperature $T\to\infty$. Thus, no initial-value problem can be defined at $t=0$. However, when T is of the order of the Planck mass ($M_P \equiv 1/\sqrt{G} = 1.22 \times 10^{19}$ GeV)[1] or greater, the equations of the standard model are undoubtedly meaningless, since quantum gravitational effects are expected to become essential. Thus, within the scope of our knowledge, it is sensible to begin the hot big-bang scenario at some temperature T_0 which is comfortably below M_P; let us say $T_0 = 10^{17}$ GeV. At this time one can take the description of the universe as a set of initial conditions, and the equations of motion then describe the subsequent evolution. Of course, the equation of state for matter at these temperatures is not really known, but one can make various hypotheses and pursue the consequences.

In the standard model, the initial universe is taken to be homogeneous and isotropic, and filled with a gas of effectively massless particles in thermal equilibrium at temperature T_0. The initial value of the Hubble expansion "constant" H is taken to be H_0, and the model universe is then completely described.

Now I can explain the puzzles. The first is the well-known horizon problem.[2-4] The initial universe is assumed to be homogeneous, yet it consists of at least $\sim 10^{83}$ separate regions which are causally disconnected (i.e., these regions have not yet had time to communicate with each other via light signals).[5] (The precise assumptions which lead to these numbers will be spelled out in Sec. II.) Thus, one must assume that the forces which created these initial conditions were capable of violating causality.

The second puzzle is the flatness problem. This puzzle seems to be much less celebrated than the first, but it has been stressed by Dicke and Peebles.[6] I feel that it is of comparable importance to the first. It is known that the energy density ρ of the universe today is near the critical value ρ_{cr} (corresponding to the borderline between an open and closed universe). One can safely assume that[7]

$$0.01 < \Omega_p < 10\,, \tag{1.1}$$

where

$$\Omega \equiv \rho/\rho_{cr} = (8\pi/3)G\rho/H^2\,, \tag{1.2}$$

and the subscript p denotes the value at the present time. Although these bounds do not appear at first sight to be remarkably stringent, they, in fact, have powerful implications. The key point is that the condition $\Omega\approx 1$ is unstable. Furthermore, the only time scale which appears in the equations for a radiation-dominated universe is the Planck time, $1/M_P = 5.4\times 10^{-44}$ sec. A typical closed universe will reach its maximum size on the order of this time scale, while a typical open universe will dwindle to a value of ρ much less than ρ_{cr}. A universe can survive $\sim 10^{10}$ years only by extreme fine tuning of the initial values of ρ and H, so that ρ is very near ρ_{cr}. For the initial conditions taken at

$T_0=10^{17}$ GeV, the value of H_0 must be fine tuned to an accuracy of one part in 10^{55}. In the standard model this incredibly precise initial relationship must be assumed without explanation. (For any reader who is not convinced that there is a real problem here, variations of this argument are given in the Appendix.)

The reader should not assume that these incredible numbers are due merely to the rather large value I have taken for T_0. If I had chosen a modest value such as $T_0=1$ MeV, I would still have concluded that the "initial" universe consisted of at least $\sim 10^{22}$ causally disconnected regions, and that the initial value of H_0 was fine tuned to one part in 10^{15}. These numbers are much smaller than the previous set, but they are still very impressive.

Of course, any problem involving the initial conditions can always be put off until we understand the physics of $T \gtrsim M_P$. However, it is the purpose of this paper to show that these puzzles might be obviated by a scenario for the behavior of the universe at temperatures well below M_P.

The paper is organized as follows. The assumptions and basic equations of the standard model are summarized in Sec. II. In Sec. III, I describe the inflationary universe scenario, showing how it can eliminate the horizon and flatness problems. The scenario is discussed in the context of grand models in Sec. IV, and comments are made concerning magnetic monopole suppression. In Sec. V I discuss briefly the key undesirable feature of the scenario: the inhomogeneities produced by the random nucleation of bubbles. Some vague ideas which might alleviate these difficulties are mentioned in Sec. VI.

II. THE STANDARD MODEL OF THE VERY EARLY UNIVERSE

In this section I will summarize the basic equations of the standard model, and I will spell out the assumptions which lead to the statements made in the Introduction.

The universe is assumed to be homogeneous and isotropic, and is therefore described by the Robertson-Walker metric[8]:

$$d\tau^2=dt^2-R^2(t)\left[\frac{dr^2}{1-kr^2}+r^2(d\theta^2+\sin^2\theta\, d\phi^2)\right], \tag{2.1}$$

where $k=+1$, -1, or 0 for a closed, open, or flat universe, respectively. It should be emphasized that any value of k is possible, but by convention r and $R(t)$ are rescaled so that k takes on one of the three discrete values. The evolution of $R(t)$ is governed by the Einstein equations

$$\ddot{R}=-\frac{4\pi}{3}G(\rho+3p)R, \tag{2.2a}$$

$$H^2+\frac{k}{R^2}=\frac{8\pi}{3}G\rho, \tag{2.2b}$$

where $H\equiv\dot{R}/R$ is the Hubble "constant" (the dot denotes the derivative with respect to t). Conservation of energy is expressed by

$$\frac{d}{dt}(\rho R^3)=-p\frac{d}{dt}(R^3), \tag{2.3}$$

where p denotes the pressure. In the standard model one also assumes that the expansion is adiabatic, in which case

$$\frac{d}{dt}(sR^3)=0, \tag{2.4}$$

where s is the entropy density.

To determine the evolution of the universe, the above equations must be supplemented by an equation of state for matter. It is now standard to describe matter by means of a field theory, and at high temperatures this means that the equation of state is to a good approximation that of an ideal quantum gas of massless particles. Let $N_b(T)$ denote the number of bosonic spin degrees of freedom which are effectively massless at temperature T (e.g., the photon contributes two units to N_b); and let $N_f(T)$ denote the corresponding number for fermions (e.g., electrons and positrons together contribute four units). Provided that T is not near any mass thresholds, the thermodynamic functions are given by

$$\rho=3p=\frac{\pi^2}{30}\mathfrak{N}(T)T^4, \tag{2.5}$$

$$s=\frac{2\pi^2}{45}\mathfrak{N}(T)T^3, \tag{2.6}$$

$$n=\frac{\zeta(3)}{\pi^2}\mathfrak{N}'(T)T^3, \tag{2.7}$$

where

$$\mathfrak{N}(T)=N_b(T)+\tfrac{7}{8}N_f(T), \tag{2.8}$$

$$\mathfrak{N}'(T)=N_b(T)+\tfrac{3}{4}N_f(T). \tag{2.9}$$

Here n denotes the particle number density, and $\zeta(3)=1.202\,06\ldots$ is the Riemann zeta function.

The evolution of the universe is then found by rewriting (2.2b) solely in terms of the temperature. Againing assuming that T is not near any mass thresholds, one finds

$$\left(\frac{\dot{T}}{T}\right)^2+\epsilon(T)T^2=\frac{4\pi^3}{45}G\mathfrak{N}(T)T^4, \tag{2.10}$$

where

$$\epsilon(T)=\frac{k}{R^2T^2}=k\left[\frac{2\pi^2}{45}\frac{\mathfrak{N}(T)}{S}\right]^{2/3}, \tag{2.11}$$

where $S\equiv R^3s$ denotes the total entropy in a volume

specified by the radius of curvature R.

Since S is conserved, its value in the early universe can be determined (or at least bounded) by current observations. Taking $\rho < 10\rho_{cr}$ today, it follows that today

$$\left|\frac{k}{R^2}\right| < 9H^2 . \tag{2.12}$$

From now on I will take $k=\pm 1$; the special case $k=0$ is still included as the limit $R\to\infty$. Then today $R>\frac{1}{3}H^{-1}\sim 3\times 10^9$ years. Taking the present photon temperature T_γ as 2.7 °K, one then finds that the photon contribution to S is bounded by

$$S_\gamma > 3\times 10^{85} . \tag{2.13}$$

Assuming that there are three species of massless neutrinos (e, μ, and τ), all of which decouple at a time when the other effectively massless particles are the electrons and photons, then $S_\nu = 21/22 S_\gamma$. Thus,

$$S > 10^{86} \tag{2.14}$$

and

$$|\epsilon| < 10^{-58}\mathfrak{N}^{2/3} . \tag{2.15}$$

But then

$$\left|\frac{\rho-\rho_{cr}}{\rho}\right| = \frac{45}{4\pi^3}\frac{M_P^2}{\mathfrak{N}T^2}|\epsilon| < 3\times 10^{-59}\mathfrak{N}^{-1/3}(M_P/T)^2 . \tag{2.16}$$

Taking $T=10^{17}$ GeV and $\mathfrak{N}\sim 10^2$ (typical of grand unified models), one finds $|\rho-\rho_{cr}|/\rho < 10^{-55}$. This is the flatness problem.

The ϵT^2 term can now be deleted from (2.10), which is then solved (for temperatures higher than all particle masses) to give

$$T^2 = \frac{M_P}{2\gamma t} , \tag{2.17}$$

where $\gamma^2 = (4\pi^3/45)\mathfrak{N}$. (For the minimal SU_5 grand unified model, $N_b = 82$, $N_f = 90$, and $\gamma = 21.05$.) Conservation of entropy implies $RT=$ constant, so $R\propto t^{1/2}$. A light pulse beginning at $t=0$ will have traveled by time t a physical distance

$$l(t) = R(t)\int_0^t dt' R^{-1}(t') = 2t , \tag{2.18}$$

and this gives the physical horizon distance. This horizon distance is to be compared with the radius $L(t)$ of the region at time t which will evolve into our currently observed region of the universe. Again using conservation of entropy,

$$L(t) = [s_p/s(t)]^{1/3} L_p , \tag{2.19}$$

where s_p is the present entropy density and $L_p \sim 10^{10}$ years is the radius of the currently observed region of the universe. One is interested in the ratio of volumes, so

$$\frac{l^3}{L^3} = \frac{11}{43}\left(\frac{45}{4\pi^3}\right)^{3/2}\mathfrak{N}^{-1/2}\left(\frac{M_P}{L_p T_\gamma T}\right)^3$$
$$= 4\times 10^{-89}\mathfrak{N}^{-1/2}(M_P/T)^3 . \tag{2.20}$$

Taking $\mathfrak{N}\sim 10^2$ and $T_0 = 10^{17}$ GeV, one finds $l_0^3/L_0^3 = 10^{-83}$. This is the horizon problem.

III. THE INFLATIONARY UNIVERSE

In this section I will describe a scenario which is capable of avoiding the horizon and flatness problems.

From Sec. II one can see that both problems could disappear if the assumption of adiabaticity were grossly incorrect. Suppose instead that

$$S_p = Z^3 S_0 , \tag{3.1}$$

where S_p and S_0 denote the present and initial values of $R^3 s$, and Z is some large factor.

Let us look first at the flatness problem. Given (3.1), the right-hand side (RHS) of (2.16) is multiplied by a factor of Z^2. The "initial" value (at $T_0 = 10^{17}$ GeV) of $|\rho-\rho_{cr}|/\rho$ could be of order unity, and the flatness problem would be obviated, if

$$Z > 3\times 10^{27} . \tag{3.2}$$

Now consider the horizon problem. The RHS of (2.19) is multiplied by Z^{-1}, which means that the length scale of the early universe, at any given temperature, was smaller by a factor of Z than had been previously thought. If Z is sufficiently large, then the initial region which evolved into our observed region of the universe would have been smaller than the horizon distance at that time. To see how large Z must be, note that the RHS of (2.20) is multiplied by Z^3. Thus, if

$$Z > 5\times 10^{27} , \tag{3.3}$$

then the horizon problem disappears. (It should be noted that the horizon will still exist; it will simply be moved out to distances which have not been observed.)

It is not surprising that the RHS's of (3.2) and (3.3) are approximately equal, since they both correspond roughly to S_0 of order unity.

I will now describe a scenario, which I call the inflationary universe, which is capable of such a large entropy production.

Suppose the equation of state for matter (with all chemical potentials set equal to zero) exhibits a first-order phase transition at some critical temperature T_c. Then as the universe cools through the temperature T_c, one would expect bubbles of the low-temperature phase to nucleate and grow. However, suppose the nucleation rate for this phase transition is rather low. The universe will continue to cool as it expands, and it will then supercool in the high-temperature phase. Suppose that this supercooling continues down to some temperature T_s, many orders of magnitude below T_c. When the phase transition finally takes place at temperature T_s, the latent heat is released. However, this latent heat is characteristic of the energy scale T_c, which is huge relative to T_s. The universe is then reheated to some temperature T_r which is comparable to T_c. The entropy density is then increased by a factor of roughly $(T_r/T_s)^3$ (assuming that the number $\mathfrak{N}$ of degrees of freedom for the two phases are comparable), while the value of R remains unchanged. Thus,

$$Z \approx T_r/T_s . \tag{3.4}$$

If the universe supercools by 28 or more orders of magnitude below some critical temperature, the horizon and flatness problems disappear.

In order for this scenario to work, it is necessary for the universe to be essentially devoid of any strictly conserved quantities. Let n denote the density of some strictly conserved quantity, and let $r \equiv n/s$ denote the ratio of this conserved quantity to entropy. Then $r_p = Z^{-3} r_0 < 10^{-84} r_0$. Thus, only an absurdly large value for the initial ratio would lead to a measurable value for the present ratio. Thus, if baryon number were exactly conserved, the inflationary model would be untenable. However, in the context of grand unified models, baryon number is not exactly conserved. The net baryon number of the universe is believed to be created by CP-violating interactions at a temperature of 10^{13}–10^{14} GeV.[9] Thus, provided that T_c lies in this range or higher, there is no problem. The baryon production would take place after the reheating. (However, strong constraints are imposed on the entropy which can be generated in any phase transition with $T_c \ll 10^{14}$ GeV, in particular, the Weinberg-Salam phase transition.[36])

Let us examine the properties of the supercooling universe in more detail. Note that the energy density $\rho(T)$, given in the standard model by (2.5), must now be modified. As $T \to 0$, the system is cooling not toward the true vacuum, but rather toward some metastable false vacuum with an energy density ρ_0 which is necessarily higher than that of the true vacuum. Thus, to a good approximation (ignoring mass thresholds)

$$\rho(T) = \frac{\pi^2}{30}\mathfrak{N}(T)T^4 + \rho_0 . \tag{3.5}$$

Perhaps a few words should be said concerning the zero point of energy. Classical general relativity couples to an energy-momentum tensor of matter, $T_{\mu\nu}$, which is necessarily (covariantly) conserved. When matter is described by a field theory, the form of $T_{\mu\nu}$ is determined by the conservation requirement up to the possible modification

$$T_{\mu\nu} \to T_{\mu\nu} + \lambda g_{\mu\nu} , \tag{3.6}$$

for any constant λ. (λ *cannot* depend on the values of the fields, nor can it depend on the temperature or the phase.) The freedom to introduce the modification (3.6) is identical to the freedom to introduce a cosmological constant into Einstein's equations. One can always choose to write Einstein's equations without an explicit cosmological term; the cosmological constant Λ is then defined by

$$\langle 0 | T_{\mu\nu} | 0 \rangle = \Lambda g_{\mu\nu} , \tag{3.7}$$

where $|0\rangle$ denotes the true vacuum. Λ is identified as the energy density of the vacuum, and, in principle, there is no reason for it to vanish. Empirically Λ is known to be very small ($|\Lambda| < 10^{-46}$ GeV^4)[10] so I will take its value to be zero.[11] The value of ρ_0 is then necessarily positive and is determined by the particle theory.[12] It is typically of $O(T_c^4)$.

Using (3.5), Eq. (2.10) becomes

$$\left(\frac{\dot{T}}{T}\right)^2 = \frac{4\pi^3}{45}G\mathfrak{N}(T)T^4 - \epsilon(T)T^2 + \frac{8\pi}{3}G\rho_0 . \tag{3.8}$$

This equation has two types of solutions, depending on the parameters. If $\epsilon > \epsilon_0$, where

$$\epsilon_0 = \frac{8\pi^2\sqrt{30}}{45}G\sqrt{\mathfrak{N}\rho_0} , \tag{3.9}$$

then the expansion of the universe is halted at a temperature $T_{\min}$ given by

$$T_{\min}^4 = \frac{30\rho_0}{\pi^2}\left\{\frac{\epsilon}{\epsilon_0} + \left[\left(\frac{\epsilon}{\epsilon_0}\right)^2 - 1\right]^{1/2}\right\}^2 , \tag{3.10}$$

and then the universe contracts again. Note that $T_{\min}$ is of $O(T_c)$, so this is not the desired scenario. The case of interest is $\epsilon < \epsilon_0$, in which case the expansion of the universe is unchecked. [Note that $\epsilon_0 \sim \sqrt{\mathfrak{N}}T_c^2/M_P^2$ is presumably a very small number. Thus $0 < \epsilon < \epsilon_0$ (a closed universe) seems unlikely, but $\epsilon < 0$ (an open universe) is quite plausible.] Once the temperature is low enough for the ρ_0 term to dominate over the other two terms on the RHS of (3.8), one has

$$T(t) \approx \text{const} \times e^{-\chi t} , \tag{3.11}$$

where

$$\chi^2 = \frac{8\pi}{3}G\rho_0 . \tag{3.12}$$

Since $RT = $ const, one has[13]

$$R(t) = \text{const} \times e^{\chi t} . \tag{3.13}$$

The universe is expanding exponentially, in a false

vacuum state of energy density ρ_0. The Hubble constant is given by $H=\dot{R}/R=\chi$. (More precisely, H approaches χ monotonically from above. This behavior differs markedly from the standard model, in which H falls as t^{-1}.)

The false vacuum state is Lorentz invariant, so $T_{\mu\nu}=\rho_0 g_{\mu\nu}$. It follows that $p=-\rho_0$, the pressure is negative. This negative pressure allows for the conservation of energy, Eq. (2.3). From the second-order Einstein equation (2.2a), it can be seen that the negative pressure is also the driving force behind the exponential expansion.

The Lorentz invariance of the false vacuum has one other consequence: The metric described by (3.13) (with $k=0$) does not single out a comoving frame. The metric is invariant under an O(4,1) group of transformations, in contrast to the usual Robertson-Walker invariance of O(4).[14] It is known as the de Sitter metric, and it is discussed in the standard literature.[15]

Now consider the process of bubble formation in a Robertson-Walker universe. The bubbles form randomly, so there is a certain nucleation rate $\lambda(t)$, which is the probability per (physical) volume per time that a bubble will form in any region which is still in the high-temperature phase. I will idealize the situation slightly and assume that the bubbles start at a point and expand at the speed of light. Furthermore, I neglect k in the metric, so $d\tau^2=dt^2-R^2(t)d\vec{x}^2$.

I want to calculate $p(t)$, the probability that any given point remains in the high-temperature phase at time t. Note that the distribution of bubbles is totally uncorrelated except for the exclusion principle that bubbles do not form inside of bubbles. This exclusion principle causes no problem because one can imagine fictitious bubbles which form inside the real bubbles with the same nucleation rate $\lambda(t)$. With all bubbles expanding at the speed of light, the fictitious bubbles will be forever inside the real bubbles and will have no effect on $p(t)$. The distribution of all bubbles, real and fictitious, is then totally uncorrelated.

$p(t)$ is the probability that there are no bubbles which engulf a given point in space. But the number of bubbles which engulf a given point is a Poisson-distributed variable, so $p(t)=\exp[-\overline{N}(t)]$, where $\overline{N}(t)$ is the expectation value of the number of bubbles engulfing the point. Thus[16]

$$p(t)=\exp\left[-\int_0^t dt_1\lambda(t_1)R^3(t_1)V(t,t_1)\right], \qquad (3.14)$$

where

$$V(t,t_1)=\frac{4\pi}{3}\left[\int_{t_1}^t \frac{dt_2}{R(t_2)}\right]^3 \qquad (3.15)$$

is the coordinate volume at time t of a bubble which formed at time t_1.

I will now assume that the nucleation rate is sufficiently slow so that no significant nucleation takes place until $T\ll T_c$, when exponential growth has set in. I will further assume that by this time $\lambda(t)$ is given approximately by the zero-temperature nucleation rate λ_0. One then has

$$p(t)=\exp\left[-\frac{t}{\tau}+O(1)\right], \qquad (3.16)$$

where

$$\tau=\frac{3\chi^3}{4\pi\lambda_0}, \qquad (3.17)$$

and $O(1)$ refers to terms which approach a constant as $\chi t\to\infty$. During one of these time constants, the universe will expand by a factor

$$Z_\tau=\exp(\chi\tau)=\exp\left(\frac{3\chi^4}{4\pi\lambda_0}\right). \qquad (3.18)$$

If the phase transition is associated with the expectation value of a Higgs field, then λ_0 can be calculated using the method of Coleman and Callan.[17] The key point is that nucleation is a tunneling process, so that λ_0 is typically very small. The Coleman-Callan method gives an answer of the form

$$\lambda_0=A\rho_0\exp(-B), \qquad (3.19)$$

where B is a barrier penetration term and A is a dimensionless coefficient of order unity. Since Z_τ is then an exponential of an exponential, one can very easily[18,19,36] obtain values as large as $\log_{10}Z\approx 28$, or even $\log_{10}Z\approx 10^{10}$.

Thus, if the universe reaches a state of exponential growth, it is quite plausible for it to expand and supercool by a huge number of orders of magnitude before a significant fraction of the universe undergoes the phase transition.

So far I have assumed that the early universe can be described from the beginning by a Robertson-Walker metric. If this assumption were really necessary, then it would be senseless to talk about "solving" the horizon problem; perfect homogeneity was assumed at the outset. Thus, I must now argue that the assumption can probably be dropped.

Suppose instead that the initial metric, and the distribution of particles, was rather chaotic. One would then expect that statistical effects would tend to thermalize the particle distribution on a local scale.[20] It has also been shown (in idealized circumstances) that anisotropies in the metric are damped out on the time scale of ~10^3 Planck times.[21] The damping of inhomogeneities in the metric has also been studied,[22] and it is reasonable to expect such damping to occur. Thus, assuming that at least some region of the universe started at temperatures high compared to T_c, one would expect that, by the time the temperature in one of these regions falls to T_c, it will be *locally* homogeneous, isotropic, and in thermal equilibrium. By locally, I am talking about a length scale ξ which is of course less than the horizon distance. It will then be possible to describe this local region of the universe by a Robertson-Walker metric, which will be accurate at distance scales small compared to ξ. When the temperature of such a region falls below T_c, the inflationary scenario will take place. The end result will be a huge region of space which is homogeneous, isotropic, and of nearly critical mass density. If Z is sufficiently large, this region can be bigger than (or much bigger than) our observed region of the universe.

IV. GRAND UNIFIED MODELS AND MAGNETIC MONOPOLE PRODUCTION

In this section I will discuss the inflationary model in the context of grand unified models of elementary-particle interactions.[23,24]

A grand unified model begins with a simple gauge group G which is a valid symmetry at the highest energies. As the energy is lowered, the theory undergoes a hierarchy of spontaneous symmetry breaking into successive subgroups: $G \to H_n \to \cdots \to H_0$, where $H_1 = SU_3 \times SU_2 \times U_1$ [QCD (quantum chromodynamics) × Weinberg-Salam] and $H_0 = SU_3 \times U_1^{EM}$. In the Georgi-Glashow model,[23] which is the simplest model of this type, $G = SU_5$ and $n=1$. The symmetry breaking of $SU_5 \to SU_3 \times SU_2 \times U_1$ occurs at an energy scale $M_X \sim 10^{14}$ GeV.

At high temperatures, it was suggested by Kirzhnits and Linde[25] that the Higgs fields of any spontaneously broken gauge theory would lose their expectation values, resulting in a high-temperature phase in which the full gauge symmetry is restored. A formalism for treating such problems was developed[26] by Weinberg and by Dolan and Jackiw. In the range of parameters for which the tree potential is valid, the phase structure of the SU_5 model was analyzed by Tye and me.[16,27] We found that the SU_5 symmetry is restored at $T > {\sim}10^{14}$ GeV and that for most values of the parameters there is an intermediate-temperature phase with gauge symmetry $SU_4 \times U_1$, which disappears at $T \sim 10^{13}$ GeV. Thus, grand unified models tend to provide phase transitions which could lead to an inflationary scenario of the universe.

Grand unified models have another feature with important cosmological consequences: They contain very heavy magnetic monopoles in their particle spectrum. These monopoles are of the type discovered by 't Hooft and Polyakov,[28] and will be present in any model satisfying the above description.[29] These monopoles typically have masses of order $M_X/\alpha \sim 10^{16}$ GeV, where $\alpha = g^2/4\pi$ is the grand unified fine structure constant. Since the monopoles are really topologically stable knots in the Higgs field expectation value, they do not exist in the high-temperature phase of the theory. They therefore come into existence during the course of a phase transition, and the dynamics of the phase transition is then intimately related to the monopole production rate.

The problem of monopole production and the subsequent annihilation of monopoles, in the context of a second-order or weakly first-order phase transition, was analyzed by Zeldovich and Khlopov[30] and by Preskill.[31] In Preskill's analysis, which was more specifically geared toward grand unified models, it was found that relic monopoles would exceed present bounds by roughly 14 orders of magnitude. Since it seems difficult to modify the estimated annihilation rate, one must find a scenario which suppresses the production of these monopoles.

Kibble[32] has pointed out that monopoles are produced in the course of the phase transition by the process of bubble coalescence. The orientation of the Higgs field inside one bubble will have no correlation with that of another bubble not in contact. When the bubbles coalesce to fill the space, it will be impossible for the uncorrelated Higgs fields to align uniformly. One expects to find topological knots, and these knots are the monopoles. The number of monopoles so produced is then comparable to the number of bubbles, to within a few orders of magnitude.

Kibble's production mechanism can be used to set a "horizon bound" on monopole production which is valid if the phase transition does not significantly disturb the evolution of the universe.[33] At the time of bubble coalescence t_{coal} the size l of the bubbles cannot exceed the horizon distance at that time. So

$$l < 2t_{coal} = \frac{M_P}{\gamma T_{coal}^2} . \tag{4.1}$$

By Kibble's argument, the density n_M of monopoles then obeys

$$n_M \gtrsim l^{-3} > \frac{\gamma^3 T_{coal}^6}{M_P^3} . \tag{4.2}$$

By considering the contribution to the mass density of the present universe which could come from 10^{16} GeV monopoles, Preskill[31] concludes that

$$n_M/n_\gamma < 10^{-24} , \tag{4.3}$$

where n_γ is the density of photons. This ratio changes very little from the time of the phase transition, so with (2.7) one concludes

$$T_{\rm coal} < \left[\frac{10^{-24}\pi^2}{2\zeta(3)}\right]^{1/3} \gamma^{-1} M_P \approx 10^{10}\ {\rm GeV}. \qquad (4.4)$$

If $T_c \sim 10^{14}$ GeV, this bound implies that the universe must supercool by at least about four orders of magnitude before the phase transition is completed.

The problem of monopole production in a strongly first-order phase transition with supercooling was treated in more detail by Tye and me.[16,34] We showed how to explicitly calculate the bubble density in terms of the nucleation rate, and we considered the effects of the latent heat released in the phase transition. Our conclusion was that (4.4) should be replaced by

$$T_{\rm coal} < 2\times 10^{11}\ {\rm GeV}, \qquad (4.5)$$

where $T_{\rm coal}$ refers to the temperature just before the release of the latent heat.

Tye and I omitted the crucial effects of the mass density ρ_0 of the false vacuum. However, our work has one clear implication: If the nucleation rate is sufficiently large to avoid exponential growth, then far too many monopoles would be produced. Thus, the monopole problem seems to also force one into the inflationary scenario.[35]

In the simplest SU_5 model, the nucleation rates have been calculated (approximately) by Weinberg and me.[19] The model contains unknown parameters, so no definitive answer is possible. We do find, however, that there is a sizable range of parameters which lead to the inflationary scenario.[36]

V. PROBLEMS OF THE INFLATIONARY SCENARIO[37]

As I mentioned earlier, the inflationary scenario seems to lead to some unacceptable consequences. It is hoped that some variation can be found which avoids these undesirable features but maintains the desirable ones. The problems of the model will be discussed in more detail elsewhere,[37] but for completeness I will give a brief description here.

The central problem is the difficulty in finding a smooth ending to the period of exponential expansion. Let us assume that $\lambda(t)$ approaches a constant as $t\to\infty$ and $T\to 0$. To achieve the desired expansion factor $Z > 10^{28}$, one needs $\lambda_0/\chi^4 < 10^{-2}$ [see (3.18)], which means that the nucleation rate is slow compared to the expansion rate of the universe. (Explicit calculations show that λ_0/χ^4 is typically much smaller than this value.[18,19,36]) The randomness of the bubble formation process then leads to gross inhomogeneities.

To understand the effects of this randomness, the reader should bear in mind the following facts.

(i) All of the latent heat released as a bubble expands is transferred initially to the walls of the bubble.[17] This energy can be thermalized only when the bubble walls undergo many collisions.

(ii) The de Sitter metric does not single out a comoving frame. The O(4,1) invariance of the de Sitter metric is maintained even after the formation of one bubble. The memory of the original Robertson-Walker comoving frame is maintained by the probability distribution of bubbles, but the local comoving frame can be reestablished only after enough bubbles have collided.

(iii) The size of the largest bubbles will exceed that of the smallest bubbles by roughly a factor of Z; the range of bubble sizes is immense. The surface energy density grows with the size of the bubble, so the energy in the walls of the largest bubbles can be thermalized only by colliding with other large bubbles.

(iv) As time goes on, an arbitrarily large fraction of the space will be in the new phase [see (3.16)]. However, one can ask a more subtle question about the region of space which is in the new phase: Is the region composed of finite separated clusters, or do these clusters join together to form an infinite region? The latter possibility is called "percolation." It can be shown[38] that the system percolates for large values of λ_0/χ^4, but that for sufficiently small values it does *not*. The critical value of λ_0/χ^4 has not been determined, but presumably an inflationary universe would have a value of λ_0/χ^4 below critical. Thus, no matter how long one waits, the region of space in the new phase will consist of finite clusters, each totally surrounded by a region in the old phase.

(v) Each cluster will contain only a few of the largest bubbles. Thus, the collisions discussed in (iii) cannot occur.

The above statements do not quite prove that the scenario is impossible, but these consequences are at best very unattractive. Thus, it seems that the scenario will become viable only if some modification can be found which avoids these inhomogeneities. Some vague possibilities will be mentioned in the next section.

Note that the above arguments seem to rule out the possibility that the universe was ever trapped in a false vacuum state, unless $\lambda_0/\chi^4 \gtrsim 1$. Such a large value of λ_0/χ^4 does not seem likely, but it is possible.[19]

VI. CONCLUSION

I have tried to convince the reader that the standard model of the very early universe requires the assumption of initial conditions which are very implausible for two reasons:

(i) *The horizon problem.* Causally disconnected regions are assumed to be nearly identical; in particular, they are simultaneously at the same temperature.

(ii) *The flatness problem*. For a fixed initial temperature, the initial value of the Hubble "constant" must be fine tuned to extraordinary accuracy to produce a universe which is as flat as the one we observe.

Both of these problems would disappear if the universe supercooled by 28 or more orders of magnitude below the critical temperature for some phase transition. (Under such circumstances, the universe would be growing exponentially in time.) However, the random formation of bubbles of the new phase seems to lead to a much too inhomogeneous universe.

The inhomogeneity problem would be solved if one could avoid the assumption that the nucleation rate $\lambda(t)$ approaches a small constant λ_0 as the temperature $T \to 0$. If, instead, the nucleation rate rose sharply at some T_1, then bubbles of an approximately uniform size would suddenly fill space as T fell to T_1. Of course, the full advantage of the inflationary scenario is achieved only if $T_1 \lesssim 10^{-28} T_c$.

Recently Witten[39] has suggested that the above chain of events may in fact occur if the parameters of the SU_5 Higgs field potential are chosen to obey the Coleman-Weinberg condition[40] (i.e., that $\partial^2 V/\partial\phi^2 = 0$ at $\phi = 0$). Witten[41] has studied this possibility in detail for the case of the Weinberg-Salam phase transition. Here he finds that thermal tunneling is totally ineffective, but instead the phase transition is driven when the temperature of the QCD chiral-symmetry-breaking phase transition is reached. For the SU_5 case, one can hope that a much larger amount of supercooling will be found; however, it is difficult to see how 28 orders of magnitude could arise.

Another physical effect which has so far been left out of the analysis is the production of particles due to the changing gravitational metric.[42] This effect may become important in an exponentially expanding universe at low temperatures.

In conclusion, the inflationary scenario seems like a natural and simple way to eliminate both the horizon and the flatness problems. I am publishing this paper in the hope that it will highlight the existence of these problems and encourage others to find some way to avoid the undesirable features of the inflationary scenario.

ACKNOWLEDGMENTS

I would like to express my thanks for the advice and encouragement I received from Sidney Coleman and Leonard Susskind, and for the invaluable help I received from my collaborators Henry Tye and Erick Weinberg. I would also like to acknowledge very useful conversations with Michael Aizenman, Beilok Hu, Harry Kesten, Paul Langacker, Gordon Lasher, So-Young Pi, John Preskill, and Edward Witten. This work was supported by the Department of Energy under Contract No. DE-AC03-76SF00515.

APPENDIX: REMARKS ON THE FLATNESS PROBLEM

This appendix is added in the hope that some skeptics can be convinced that the flatness problem is real. Some physicists would rebut the argument given in Sec. I by insisting that the equations might make sense all the way back to $t = 0$. Then if one fixes the value of H corresponding to some arbitrary temperature T_*, one always finds that when the equations are extrapolated *backward* in time, $\Omega \to 1$ as $t \to 0$. Thus, they would argue, it is natural for Ω to be very nearly equal to 1 at early times. For physicists who take this point of view, the flatness problem must be restated in other terms. Since H_0 and T_0 have no significance, the model universe must be specified by its conserved quantities. In fact, the model universe is completely specified by the dimensionless constant $\epsilon \equiv k/R^2T^2$, where k and R are parameters of the Robertson-Walker metric, Eq. (2.1). For our universe, one must take $|\epsilon| < 3 \times 10^{-57}$. The problem then is the to explain why $|\epsilon|$ should have such a startlingly small value.

Some physicists also take the point of view that $\epsilon \equiv 0$ is plausible enough, so to them there is no problem. To these physicists I point out that the universe is certainly not described *exactly* by a Robertson-Walker metric. Thus it is difficult to imagine any physical principle which would require a parameter of that metric to be exactly equal to zero.

In the end, I must admit that questions of plausibility are not logically determinable and depend somewhat on intuition. Thus I am sure that some physicists will remain unconvinced that there really is a flatness problem. However, I am also sure that many physicists agree with me that the flatness of the universe is a peculiar situation which at some point will admit a physical explanation.

*Present address: Center for Theoretical Physics, Massachusetts Institute of Technology, Cambridge, Massachusetts 02139.

[1]I use units for which $\hbar = c = k$ (Boltzmann constant) $= 1$. Then 1 m $= 5.068 \times 10^{15}$ GeV^{-1}, 1 kg $= 5.610 \times 10^{26}$ GeV, 1 sec $= 1.519 \times 10^{24}$ GeV^{-1}, and 1 °K $= 8.617 \times 10^{-14}$ GeV.

[2]W. Rindler, Mon. Not. R. Astron. Soc. 116, 663 (1956). See also Ref. 3, pp. 489–490, 525–526; and Ref. 4, pp. 740 and 815.

[3]S. Weinberg, *Gravitation and Cosmology* (Wiley, New York, 1972).

[4]C. W. Misner, K. S. Thorne, and J. A. Wheeler, *Gravitation* (Freeman, San Francisco, 1973).

[5]In order to calculate the horizon distance, one must of course follow the light trajectories back to $t = 0$. This violates my contention that the equations are to be trusted only for $T \lesssim T_0$. Thus, the horizon problem could be obviated if the full quantum gravitational theory had a radically different behavior from the naive extrapolation. Indeed, solutions of this sort have been proposed by A. Zee, Phys. Rev. Lett. 44, 703 (1980) and by F. W. Stecker, Astrophys. J. 235, L1 (1980). However, it is the point of this paper to show that the horizon problem can also be obviated by mechanisms which are more within our grasp, occurring at temperatures below T_0.

[6]R. H. Dicke and P. J. E. Peebles, *General Relativity: An Einstein Centenary Survey*, edited by S. W. Hawking and W. Israel (Cambridge University Press, London, 1979).

[7]See Ref. 3, pp. 475–481; and Ref. 4, pp. 796–797.

[8]For example, see Ref. 3, Chap. 14.

[9]M. Yoshimura, Phys. Rev. Lett. 41, 281 (1978); 42, 746(E) (1979); Phys. Lett. 88B, 294 (1979); S. Dimopoulos and L. Susskind, Phys. Rev. D 18, 4500 (1978); Phys. Lett. 81B, 416 (1979); A. Yu Ignatiev, N. V. Krashikov, V. A. Kuzmin, and A. N. Tavkhelidze, *ibid.* 76B, 436 (1978); D. Toussaint, S. B. Treiman, F. Wilczek, and A. Zee, Phys. Rev. D 19, 1036 (1979); S. Weinberg, Phys. Rev. Lett. 42, 850 (1979); D. V. Nanopoulos and S. Weinberg, Phys. Rev. D 20, 2484 (1979); J. Ellis, M. K. Gaillard, and D. V. Nanopoulos, Phys. Lett. 80B, 360 (1979); 82B, 464 (1979); M. Honda and M. Yoshimura, Prog. Theor. Phys. 62, 1704 (1979); D. Toussaint and F. Wilczek, Phys. Lett. 81B, 238 (1979); S. Barr, G. Segre, and H. A. Weldon, Phys. Rev. D 20, 2494 (1979); A. D. Sakharov, Zh. Eksp. Teor. Fiz. 76, 1172 (1979) [Sov. Phys.—JETP 49, 594 (1979)]; A. Yu Ignatiev, N. V. Krashikov, V. A. Kuzmin, and M. E. Shaposhnikov, Phys. Lett. 87B, 114 (1979); E. W. Kolb and S. Wolfram, *ibid.* 91B, 217 (1980); Nucl. Phys. B172, 224 (1980); J. N. Fry, K. A. Olive, and M. S. Turner, Phys. Rev. D 22, 2953 (1980); 22, 2977 (1980); S. B. Treiman and F. Wilczek, Phys. Lett. 95B, 222 (1980); G. Senjanović and F. W. Stecker, Phys. Lett. B (to be published).

[11]The reason Λ is so small is of course one of the deep mysteries of physics. The value of Λ is not determined by the particle theory alone, but must be fixed by whatever theory couples particles to quantum gravity. This appears to be a separate problem from the ones discussed in this paper, and I merely use the empirical fact that $\Lambda \approx 0$.

[12]S. A. Bludman and M. A. Ruderman, Phys. Rev. Lett. 38, 255 (1977).

[13]The effects of a false vacuum energy density on the evolution of the early universe have also been considered by E. W. Kolb and S. Wolfram, CAL TECH Report No. 79-0984 (unpublished), and by S. A. Bludman, University of Pennsylvania Report No. UPR-0143T, 1979 (unpublished).

[14]More precisely, the usual invariance is O(4) if $k = 1$, O(3,1) if $k = -1$, and the group of rotations and translations in three dimensions if $k = 0$.

[15]See for example, Ref. 3, pp. 385–392.

[16]A. H. Guth and S.-H. Tye, Phys. Rev. Lett. 44, 631 (1980); 44, 963 (1980).

[17]S. Coleman, Phys. Rev. D 15, 2929 (1977); C. G. Callan and S. Coleman, *ibid.* 16, 1762 (1977); see also S. Coleman, in *The Whys of Subnuclear Physics*, proceedings of the International School of Subnuclear Physics, Ettore Majorana, Erice, 1977, edited by A. Zichichi (Plenum, New York, 1979).

[18]A. H. Guth and E. J. Weinberg, Phys. Rev. Lett. 45, 1131 (1980).

[19]E. J. Weinberg and I are preparing a manuscript on the possible cosmological implications of the phase transitions in the SU_5 grand unified model.

[20]J. Ellis and G. Steigman, Phys. Lett. 89B, 186 (1980); J. Ellis, M. K. Gaillard, and D. V. Nanopoulos, *ibid.* 90B, 253 (1980).

[21]B. L. Hu and L. Parker, Phys. Rev. D 17, 933 (1978).

[22]See Ref. 4, Chap. 30.

[23]The simplest grand unified model is the SU(5) model of H. Georgi and S. L. Glashow, Phys. Rev. Lett. 32, 438 (1974). See also H. Georgi, H. R. Quinn, and S. Weinberg, *ibid.* 33, 451 (1974); and A. J. Buras, J. Ellis, M. K. Gaillard, and D. V. Nanopoulos, Nucl. Phys. B135, 66 (1978).

[24]Other grand unified models include the SO(10) model: H. Georgi, in *Particles and Fields—1975*, proceedings of the 1975 meeting of the Division of Particles and Fields of the American Physical Society, edited by Carl Carlson (AIP, New York, 1975); H. Fritzsch and P. Minkowski, Ann. Phys. (N.Y.) 93, 193 (1975); H. Georgi and D. V. Nanopoulos, Phys. Lett. 82B, 392 (1979) and Nucl. Phys. B155, 52 (1979). The E(6) model: F. Gürsey, P. Ramond, and P. Sikivie, Phys. Lett. 60B, 177 (1975); F. Gürsey and M. Serdaroglu, Lett. Nuovo Cimento 21, 28 (1978). The E(7) model: F. Gürsey and P. Sikivie, Phys. Rev. Lett. 36, 775 (1976), and Phys. Rev. D 16, 816 (1977); P. Ramond, Nucl. Phys. B110, 214 (1976). For some general properties of grand unified models, see M. Gell-Mann, P. Ramond, and R. Slansky, Rev. Mod. Phys. 50, 721 (1978). For a review, see P. Langacker, Report No. SLAC-PUB-2544, 1980 (unpublished).

[25]D. A. Kirzhnits and A. D. Linde, Phys. Lett. 42B, 471 (1972).

[26]S. Weinberg, Phys. Rev. D 9, 3357 (1974); L. Dolan and R. Jackiw, *ibid.* 9, 3320 (1974); see also D. A. Kirzhnits and A. D. Linde, Ann. Phys. (N.Y.) 101, 195 (1976); A. D. Linde, Rep. Prog. Phys. 42, 389 (1979). ϵ-expansion techniques are employed by P. Ginsparg, Nucl. Phys. B (to be published).

[27]In the case that the Higgs quartic couplings are comparable to g^4 or smaller (g = gauge coupling), the phase structure has been studied by M. Daniel and C. E. Vayonakis, CERN Report No. TH.2860 1980

(unpublished); and by P. Suranyi, University of Cincinnati Report No. 80-0506 (unpublished).

[28]G. 't Hooft, Nucl. Phys. B79, 276 (1974); A. M. Polyakov, Pis'ma Zh. Eksp. Teor. Fiz. 20, 430 (1974) [JETP Lett. 20, 194 (1974)]. For a review, see P. Goddard and D. I. Olive, Rep. Prog. Phys. 41, 1357 (1978).

[29]If $\Pi_1(G)$ and $\Pi_2(G)$ are both trivial, then $\Pi_2(G/H_0) \approx \Pi_1(H_0)$. In our case $\Pi_1(H_0)$ is the group of integers. For a general review of topology written for physicists, see N. D. Mermin, Rev. Mod. Phys. 51, 591 (1979).

[30]Y. B. Zeldovich and M. Y. Khlopov, Phys. Lett. 79B, 239 (1978).

[31]J. P. Preskill, Phys. Rev. Lett. 43, 1365 (1979).

[32]T. W. B. Kibble, J. Phys. A 9, 1387 (1976).

[33]This argument was first shown to me by John Preskill. It is also described by Einhorn *et al.*, Ref. 34, except that they make no distinction between T_{cool} and T_c.

[34]The problem of monopole production was also examined by M. B. Einhorn, D. L. Stein, and D. Toussaint, Phys. Rev. D 21, 3295 (1980), who focused on second-order transitions. The structure of SU(5) monopoles has been studied by C. P. Dokos and T. N. Tomaras, Phys. Rev. D 21, 2940 (1980); and by M. Daniel, G. Lazarides, and Q. Shafi, Nucl. Phys. B170, 156 (1980). The problem of suppression of the cosmological production of monopoles is discussed by G. Lazarides and Q. Shafi, Phys. Lett. 94B, 149 (1980), and G. Lazarides, M. Magg, and Q. Shafi, CERN Report No. TH.2856, 1980 (unpublished); the suppression discussed here relies on a novel confinement mechanism, and also on the same kind of supercooling as in Ref. 16. See also J. N. Fry and D. N. Schramm, Phys. Rev. Lett. 44, 1361 (1980).

[35]An alternative solution to the monopole problem has been proposed by P. Langacker and S.-Y. Pi, Phys. Rev. Lett. 45, 1 (1980). By modifying the Higgs structure, they have constructed a model in which the high-temperature SU_5 phase undergoes a phase transition to an SU_3 phase at $T \sim 10^{14}$GeV. Another phase transition occurs at $T \sim 10^{1}$ GeV, and below this temperature the symmetry is $SU_3 \times U_1^{EM}$. Monopoles cannot exist until $T < 10^3$ GeV, but their production is negligible at these low temperatures. The suppression of monopoles due to the breaking of U_1^{EM} symmetry at high temperatures was also suggested by S.-H. Tye, talk given at the 1980 Guangzhou Conference on Theoretical Particle Physics, Canton, 1980 (unpublished).

[36]The Weinberg-Salam phase transition has also been investigated by a number of authors: E. Witten, Ref. 41; M. A. Sher, Phys. Rev. D 22, 2989 (1980); P. J. Steinhardt, Harvard report, 1980 (unpublished); and A. H. Guth and E. J. Weinberg, Ref. 18.

[37]This section represents the work of E. J. Weinberg, H. Kesten, and myself. Weinberg and I are preparing a manuscript on this subject.

[38]The proof of this statement was outlined by H. Kesten (Dept. of Mathematics, Cornell University), with details completed by me.

[39]E. Witten, private communication.

[40]S. Coleman and E. J. Weinberg, Phys. Rev. D 7, 1888 (1973); see also, J. Ellis, M. K. Gaillard, D. Nanopoulos, and C. Sachrajda, Phys. Lett. 83B, 339 (1979), and J. Ellis, M. K. Gaillard, A. Peterman, and C. Sachrajda, Nucl. Phys. B164, 253 (1980).

[41]E. Witten, Nucl. Phys. B (to be published).

[42]L. Parker, in *Asymptotic Structure of Spacetime*, edited by F. Esposito and L. Witten (Plenum, New York, 1977); V. N. Lukash, I. D. Novikov, A. A. Starobinsky, and Ya. B. Zeldovich, Nuovo Cimento 35B, 293 (1976).

Observations of Post-Newtonian Timing Effects in the Binary Pulsar PSR 1913+16

J. M. Weisberg and J. H. Taylor
Joseph Henry Laboratories, Physics Department, Princeton University, Princeton, New Jersey 08544
(Received 24 January 1984)

We report the latest results of timing measurements of the binary pulsar PSR 1913+16. Recent high-quality data have enabled us to measure the excess propagation delay of the pulsar signal caused by the gravitational field of the companion star; this result provides strong additional support for the simplest and most straightforward model of the system. The observed rate of orbital period decay is equal to 1.00 ± 0.04 times that expected from gravitational radiation damping.

PACS numbers: 97.60.Gb, 04.80.+z, 95.30.Sf

General relativity predicts that a pair of masses in mutual orbit should gradually spiral closer together as the system loses energy in the form of gravitational radiation. The binary pulsar PSR 1913+16 provides a uniquely suitable system[1] for testing this prediction quantitatively,[2–4] and results already published have shown that the orbital period is decreasing at the rate expected from damping by gravitational radiation.[5–8] In this Letter we report improved measurements of the orbital period decay, based on measurements through August 1983. In addition, we have detected an additional relativistic effect in the PSR 1913+16 system—an excess delay of the pulsar signal, caused by propagation through the gravitational field of the companion star. Measurement of this effect yields another redundant constraint on the already overdetermined parameters of the orbiting pair. The value of the new parameter is in excellent agreement with our earlier prediction.[7] The overall self-consistency of observable parameters now establishes, with a high level of confidence, that the measured decrease in orbital period is the result of gravitational radiation.

Our experimental data, obtained with the 305-m telescope at the Arecibo Observatory in Puerto Rico, consist of 5-min synchronous integrations of the pulsar wave form. Each integration is tagged with the Coordinated Universal Time (UTC) to an accuracy of approximately 1 μs. We determine a "pulse arrival time" for each integrated profile by measuring its phase offset relative to a standard profile (obtained by averaging many hundreds of integrations), and adjusting the nominal UTC accordingly.[7,9] Approximately 3300 such measurements have been accumulated since 1974; the standard errors vary from about 300 μs in the early data to approximately 20 μs in the most recent observations.

The observed pulse arrival times—expressed in units of atomic time as measured by an Earth-based clock—depend on the rotation rate of the neutron star, on the motions of the observatory in the solar system and the pulsar in its orbit, and on relativistic effects involving the accelerations and changes in gravitational potential at both ends of the Earth-pulsar path. The data are used as inputs in a multiparameter least-squares solution for four classes of parameters: (1) the pulsar rotation phase and its time derivatives; (2) the celestial coordinates of the pulsar, measurable because of the large (~ 1000 s) annual variations caused by the Earth's orbital motion; (3) five "classical" orbital elements for the pulsar, essentially determined from the large (~ 4 s) first-order effects of the pulsar's orbital motion; (4) four "relativistic" orbital parameters, measurable in this system because of the extreme conditions $[v/c \sim (GM/c^2R)^{1/2} \sim 10^{-3}]$ that are present.

Classification of the parameters of the orbiting system as "classical" or "relativistic" is done only as a matter of convenience in discussing them. It is well known that a purely Newtonian treatment of data like ours can be used to determine just five orbital parameters; a relativistic analysis, carried out to a precision consistent with present experimental uncertainties, can determine a total of nine. Following essentially the procedure outlined by Blandford and Teukolsky[10] and Epstein,[11] we have adopted a set of nine orbital parameters, the first five of which have the same names, and essentially the same meanings, as their Newtonian counterparts. Measured values of these five parameters are given in the first part of Table I.

The four relativistic parameters listed in Table I are $\langle\dot{\omega}\rangle$, the average rate of rotation of the orbital ellipse within its plane; γ, the amplitude of delays caused by variations in gravitational red shift and time dilation as the pulsar traverses its elliptical or-

TABLE I. Orbital parameters of PSR 1913+16.

(a) "Classical" parameters	
Projected semimajor axis	$a_p \sin i = 2.341\,85 \pm 0.000\,12$ light sec
Eccentricity	$e = 0.617\,127 \pm 0.000\,003$
Orbital period	$P_b = 27\,906.981\,63 \pm 0.000\,02$ s
Longitude of periastron	$\omega_0 = 178.8643 \pm 0.0009$ deg
Julian ephemeris date of periastron and reference time for P_b and ω_0	$T_0 = 2442\,321.433\,208\,4 \pm 0.000\,001\,2$
(b) "Relativistic" parameters	
Mean rate of periastron advance	$\langle\dot{\omega}\rangle = 4.2263 \pm 0.0003$ deg yr^{-1}
Gravitational red shift and time dilation	$\gamma = 0.004\,38 \pm 0.000\,12$ s
Orbital period derivative	$\dot{P}_b = (-2.40 \pm 0.09) \times 10^{-12}$ s s^{-1}
Orbital inclination	$\sin i = 0.76 \pm 0.14$

bit; $\dot{P}_b$, the time derivative of orbital period, which we attribute to the emission of gravitational waves; and $\sin i$, where i is the angle of inclination between the plane of the orbit and the plane of the sky. The last parameter, although cast in terms of geometry, actually specifies the excess delay caused by propagation of the pulsar signal through the gravitational field of the companion. This effect is the phenomenon first postulated by Shapiro[12] and observed in solar system distance measurements.[13] In the PSR 1913+16 system the orbital variation in the gravitational propagation delay as seen from Earth is approximately 25 μs.

Unambiguous detection of the propagation delay term, and consequently measurement of $\sin i$, is made difficult by the large covariances between it and several of the other orbital parameters. Our success was made possible by the high quality of the most recent data and by the recognition of Haugan[14] that the rate of precession of the orbital ellipse in its plane, $\dot{\omega}$, is not constant but varies with the separation of the two stars. The effect on pulse arrival times is of the same order as the gravitational propagation delay and the small post-Newtonian [order $(v/c)^3$] corrections already explicitly included in the analysis.[11] We now include all of these effects in the model, and we believe that our analysis is complete and self-consistent at the microsecond level. Our measurement of $\sin i$ marks the first successful observation of gravitational propagation delay outside the solar system. More importantly, it furnishes a new and independent test of the clean and uncomplicated nature of this binary pulsar system.

The uncertainties listed for the parameter estimates in Table I are between two and four times the formed standard deviations. They are based on a

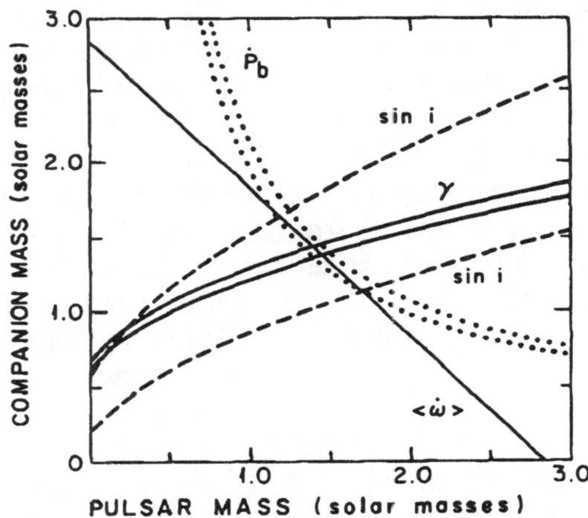

FIG. 1. Families of curves showing how the stellar masses in the PSR 1913+16 system are constrained by the measured values (and estimated errors) of parameters $\langle\dot{\omega}\rangle$, γ, and $\sin i$. (The uncertainty in $\langle\dot{\omega}\rangle$ is less than the width of the sloping straight line.) The fourth pair of curves bracket those mass values that would, according to general relativity, cause the system to emit enough gravitational radiation to explain the observed orbital period derivative $\dot{P}_b$. The simplest model of the system is in good accord with the data if both the pulsar and the compansion star have approximately the Chandrasekhar limiting mass, 1.4 solar masses.

cautious assessment of the range of variation observed for each parameter in a number of different test solutions, and on our semiquantitative judgments about the possible presence of low-level systematic errors in the data. Further details of the error estimation process are contained in Ref. 9.

The simplest model of the PSR 1913+16 system consistent with the observations treats it as a pair of compact masses with negligible quadrupole moments and no significant nonrelativistic dissipative mechanisms. The pulsar itself clearly satisfies these conditions, and the most plausible evolutionary scenarios[4,7] imply that the companion should also be a compact object. If this model is valid, then seven orbital parameters are enough to describe it. Our nine-parameter solution then overdetermines the astrophysical quantities and provides a true test of gravitation theory.

A reasonable choice for a minimal set of parameters would be to replace the last four in Table I with the masses, m_p and m_c, of the pulsar and companion star. The four observable quantities depend on the masses according to the equations (see Ref. 7 and references therein)

$$\dot{\omega} = 3G^{2/3}c^{-2}(P_b/2\pi)^{-5/3}(1-e^2)^{-1}(m_p+m_c)^{2/3} , \tag{1}$$

$$\gamma = G^{2/3}c^{-2}e(P_b/2\pi)^{1/3}m_c(m_p+2m_c)(m_p+m_c)^{-4/3} , \tag{2}$$

$$\sin i = G^{-1/3}c(a_p \sin i/m_c)(P_b/2\pi)^{-2/3}(m_p+m_c)^{2/3} , \tag{3}$$

$$\dot{P}_b = -\frac{192\pi G^{5/3}}{5c^5}\left(\frac{P_b}{2\pi}\right)^{-5/3}(1-e^2)^{-7/2}(1+\tfrac{73}{24}e^2+\tfrac{37}{96}e^4)m_p m_c(m_p+m_c)^{-1/3} \tag{4}$$

Figure 1 shows graphically the values of m_p and m_c that are consistent with the observations and with Eqs. (1)–(4) taken one at a time. It is evident that all four measured parameters are satisfied simultaneously if both masses are close to 1.4 solar masses. At present levels of accuracy the $\sin i$ parameter does not tighten constraints on the masses very much, but it is significant that its measured value is fully consistent with expectations based on the other parameters, using the simplest model of the system.

The most profound result of our observations remains the measurement of orbital period decay at just the rate expected from gravitational radiation damping. In order to state this result succinctly and quantitatively, Eqs. (1)–(3) can be used to solve (by least squares) for the best-fitting values of the masses m_p and m_c. Using the parameter values quoted in Table I, and correctly propagating errors and taking note of the parameter covariances, we obtain the results $m_p = 1.42 \pm 0.03$ and $m_c = 1.40 \pm 0.03$ solar masses. When these values are inserted in Eq. (3), we obtain the predicted rate of orbital period change $\dot{P}_b = (-2.403 \pm 0.002) \times 10^{-12}$ s s^{-1}, in excellent agreement with the observed value $(-2.40 \pm 0.09) \times 10^{-12}$. As we have pointed out before[6,7] most relativistic theories of gravity other than general relativity conflict strongly with our data, and would appear to be in serious trouble in this regard. It now seems an inescapable conclusion that gravitational radiation exists as predicted by the general relativistic quadrupole formula.

Arecibo Observatory is operated by Cornell University under contract with the National Science Foundation.

[1]R. A. Hulse and J. H. Taylor, Astrophys. J. Lett. **195**, L51 (1975).

[2]L. W. Esposito and E. R. Harrison, Astrophys. J. Lett. **196**, L1 (1975).

[3]R. V. Wagoner, Astrophys. J. Lett. **196**, L63 (1975).

[4]L. L. Smarr and R. Blandford, Astrophys. J. **207**, 574 (1976).

[5]J. H. Taylor, L. A. Fowler, and P. M. McCulloch, Nature (London) **277**, 437 (1979).

[6]J. M. Weisberg and J. H. Taylor, Gen. Relativ. Gravit. **13**, 1 (1981).

[7]J. H. Taylor and J. M. Weisberg, Astrophys. J. **253**, 908 (1982).

[8]V. Boriakoff, D. C. Ferguson, M. P. Haugan, Y. Terzian, and S. A. Teukolsky, Astrophys. J. Lett. **261**, L97 (1982).

[9]J. H. Taylor and J. M. Weisberg, to be published.

[10]R. Blandford and S. A. Teukolsky, Astrophys. J. **205**, 580 (1976).

[11]R. Epstein, Astrophys. J. **216**, 92 (1977), and **231**, 644 (1979).

[12]I. I. Shapiro, Phys. Rev. Lett. **13**, 789 (1964).

[13]R. D. Reasenberg, I. I. Shapiro, P. E. MacNeil, R. B. Goldstein, J. C. Breidenthal, J. P. Brenkle, D. L. Cain, T. M. Kaufman, T. A. Komarck, and A. I. Zygielbaum, Astrophys. J. Lett. **234**, L219 (1979).

[14]M. P. Haugan, to be published.

Chapter 8

COSMIC RADIATION

JOHN A. SIMPSON

Introduction

Papers Reprinted in Book

Victor Hess (1883-1964). (Courtesy of Washington University Archives.)

The study of cosmic rays is unique in modern physics for the minuteness of the phenomena, the delicacy of the observations, the adventurous excursions of the observers, the subtlety of the analysis, and the grandeur of the inferences.

K. K. Darrow
Secretary of The American Physical Society, 1933

INTRODUCTION

This overview focuses on the role of *The Physical Review* and *Physical Review Letters* in reporting the development of cosmic-ray physics through 1983 — a field born about ten years after the establishment of *The Physical Review*. Beginning as an inquiry into why gold-leaf electroscopes discharged anomalously rapidly in the laboratory, the discovery of extraterrestrial radiation, now called cosmic rays, by Hess in 1911(a) led to the founding of some

John Alexander Simpson is the Arthur H. Compton Distinguished Service Professor, Emeritus, in the Department of Physics and the Enrico Fermi Institute of The University of Chicago and was a former Director of the Institute. His experiments on more than thirty-two space missions in the interplanetary medium and nine planetary flybys have contributed to an understanding of charged particle acceleration and modulation mechanisms in the heliosphere, at the Sun and in planetary magnetospheres. These studies also included the elemental and isotopic composition of cosmic-ray nuclei. His development of the neutron monitor provides a worldwide record of cosmic-ray intensity. His awards and honors include the Bruno Rossi Prize, the Arctowski Medal and Prize of the National Academy of Sciences, and the United Nations COSPAR Award.

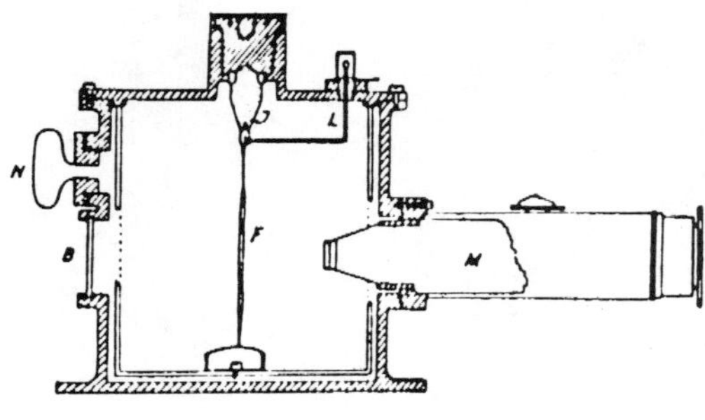

Cross-section view of ionization chamber used by Hess to discover cosmic rays (1911).

of the most diverse new disciplines in 20th century physics. Hess is shown here in a later year reading an electroscope. The subsequent search for the origin of the cosmic radiation initiated many new fields of physics — perhaps the most notable being elementary and strange particle physics [e.g., *The Birth of Particle Physics*, edited by Lillian Hoddeson and Laurie Brown (Cambridge University Press, Cambridge 1983); Panofsky and Trilling, *Elementary Particle Physics Experiments* (this collection)].

In my attempt to portray the explosive development of cosmic-ray research as it evolved in *The Physical Review*, I have focused on some but, due to space limitations, not all of the seminal experimental discoveries that set the pace for inquiry into the origins of the radiations that are also enriching related astrophysical research. Today the search continues into questions of charged particle acceleration processes on all astrophysical scales and sites extending, e.g., from our solar system, to the heliosphere, to supernovae, and to active galactic nuclei.

Cosmic-ray physics is mainly an experimental science driven by observational discoveries. To illustrate the sequence of discoveries and their relationship to new fields of research, I have prepared "The Cosmic Radiation Discovery Time Line" shown in the accompanying chart. Each *Physical Review* or *Physical Review Letters* publication in this time line is indicated by the square symbols for Refs. 1, 2, 3, etc., and in the text by brackets [1], [2], [3], arranged in chronological order. Each of these cited articles is listed in this text, and those through 1983 are included with full text and illustrations in the CD-ROM.

These publications by themselves do not tell the whole story. Therefore, I have added to the discovery time line several publications from other journals in order to better reveal the continuity of intellectual thought and experimental discovery. These contributions from other journals are identified by the circle symbols for Refs. a, b, c, etc., and in the text by parentheses (a), (b), (c), etc. and are also listed in this text.

I decided to allot the available space assigned to me in the book to five of the earliest papers from *The Physical Review* that, together, were important for establishing the foundations of modern cosmic-ray physics and elementary-particle physics. These papers illustrate the diversity of methods and the exploratory characteristics of both the researches and their investigators. Along with all the other papers listed in Table 1, they are also reproduced in the CD-ROM.

A WALK THROUGH THE HISTORY OF COSMIC RADIATION DISCOVERIES

Following Hess's proof (a) of the extraterrestrial origin of the radiation, an extensive period elapsed after World War I in which a variety of investigations were carried out in Europe showing that the radiation at sea level was both highly penetrating and constant in intensity. The names that come to mind include

COSMIC RADIATION

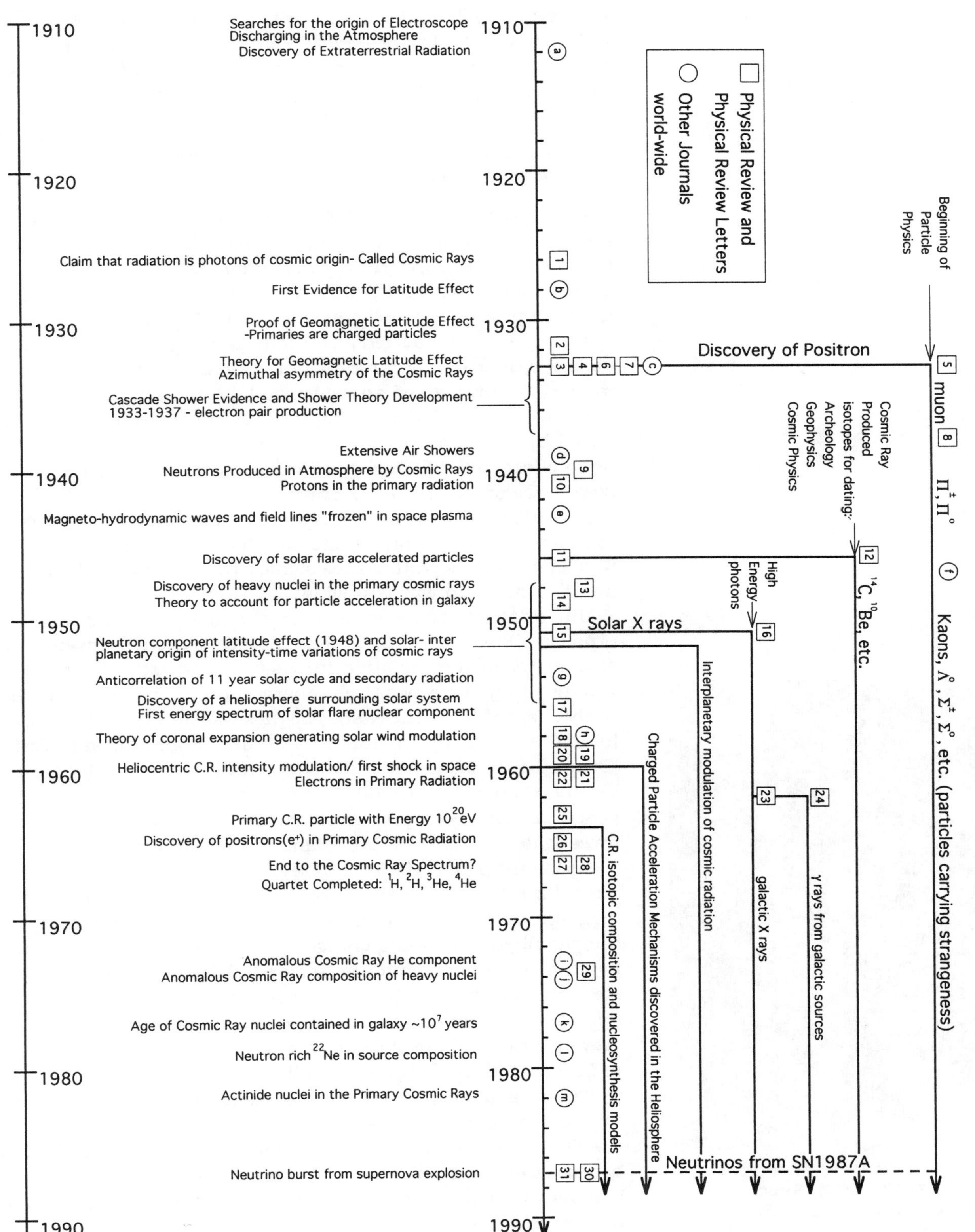

Kolhörster, Bothe, Hess, and Regener, among others. In 1926 Millikan was the first American to publish research [1] confirming the earlier work in Europe and to coin the phrase "cosmic rays," concluding that they

> *...constitute new and quite unambiguous evidence for the existence of very hard ethereal rays of cosmic origin entering the Earth uniformly from all directions.*

He expressed the strong conviction that this was a high-energy neutral radiation from beyond the galaxy — the death cries of atoms. His lectures drew considerable attention from, among others, Eddington and Jeans, who struggled unsuccessfully to describe processes to account for Millikan's claims.

However, Clay reported (b) a geomagnetic latitude effect for the observed intensity by carrying ionization detectors onboard ships in 1927 that traversed an extensive latitude range. This latitude effect was confirmed by Compton, [2] but continued to be disputed by Millikan. There followed a full theory for the geomagnetic effect for high-energy charged particles. [3] To settle the dispute Compton organized a worldwide survey in 1931. The Earth was divided into nine zones and teams, with all investigators using identical ionization chambers. He then reported that there was a latitude effect, that the cosmic rays were charged particles, and that Millikan was wrong! [4]

But what was the sign of these charged particles? Rossi had proposed [Phys. Rev. **36**, 606(L) (1930)] a search for an east-west asymmetry in the radiation intensity at low latitudes to decide whether they were positively charged (from the west) or negatively charged (from the east). The race was on with T. H. Johnson [6] and Alvarez and Compton [7] in Mexico and Rossi in Eritrea [Bruno Rossi, *Moments in the Life of a Scientist* (Cambridge University, Cambridge, 1990)]. The radiation was found to be composed principally of positively charged particles. Three leading investigators of this era are Millikan, Compton, and Rossi.

Bruno Rossi (1905-1994), **Robert Millikan (1868-1953)**, and **Arthur Compton (1892-1962).** (Courtesy of Cambridge University Press.)

Also in the period 1929–1933 several investigators, among them Skobeltzyn and C. D. Anderson, were using cloud chambers in magnetic fields to study the secondary interactions occurring in the chambers, including charged particle showers. These investigations led Anderson to the accidental discovery of the positive electron, the positron. [5] Here at last was the vindication of Dirac's quantum dynamical hole theory where the missing negative-energy electron had now been discovered [e.g., Treiman, *A Century of Particle Theory* (this collection)].

Thus, cosmic-ray investigations opened the major new field of high-energy particle physics (illustrated by the 1933 branch on the chart), which continued into the 1950s, when accelerators became the dominant source. Cosmic rays, through their interactions with matter, were the source for myriad new elementary and strange particles, leading to the rise of modern theories of the nature of matter. Particles of intermediate mass between electrons and protons were discovered [Neddermeyer and C. D. Ander-

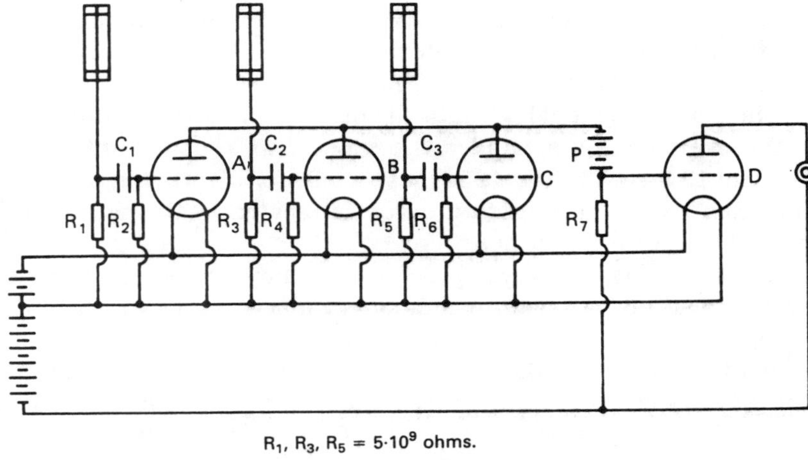

Drawing of the first electronic coincidence circuit. (From Rossi, *Nature*, **125**, 636, 1930.)

son, Phys. Rev. **51**, 884–886(L) (1937); Phys. Rev. **54**, 88 (1938); Street and Stevenson, Phys. Rev. **52**, 1003–1004(L) (1937)]. By 1938 the decay of the meson was established (Euler and Heisenberg) and demonstrated through its atmospheric temperature effect by Blackett. [8]

In the period 1933–1937 shower cascades, shower theory (e.g., Carlson and Oppenheimer and Bhabha and Heitler), and electron-pair production were intensely studied. (c) By 1939 penetrating extensive air showers had been discovered. (d)

With the discovery of the neutron by Chadwick in 1932 it became evident from cosmic-ray interactions in the atmosphere that neutrons would be produced and yield slow neutrons near Earth. [9]

By 1941 high-altitude balloon experiments showed that the positively charged primary particles were protons. [10]

During World War II Alfvén outlined a basic principle by which astrophysical plasmas could transport or "freeze-in" magnetic fields. (e) This concept was later exploited by Fermi to model the acceleration of charged particles in the galaxy to cosmic-ray energies. [14]

Following a lapse in publications due to the war, Libby pointed out the possibility of using the radioactive ^{14}C produced in the atmosphere from neutron interactions with nitrogen as a tool for dating artifacts. [12] This opened the branch field of dating by cosmic-ray isotopes [e.g., ^{10}Be; also W. F. Libby, *Radiocarbon Dating* (University of Chicago Press, Chicago, 1955)], shown in the chart.

The Compton model-C ionization chambers were in continuous operation from the 1930s under the aegis of the Department of Terrestrial Magnetism. Although sudden increases in ionization intensity had been observed earlier by European investigators, Forbush recognized their significance — namely, that these were bursts of particles accelerated to cosmic-ray energies by solar flares. [11]

Unstable elementary particles continued to be the main focus of attention in the 1940s, mainly through the use of nuclear emulsions and cloud chambers, e.g., Ref. (f).

The development of high-altitude plastic balloons, aircraft, and V-2 rockets greatly expanded the experimental opportunities in the late 1940s and early 1950s, especially for identifying primary particles and the elementary particles created in the secondary radiation within the atmosphere.

Although there now was proof of a primary proton component and indications of a helium component in the primary radiation, it came as a surprise in 1948 to discover that the stripped nuclei of many elements in the periodic table were also to be found in the primary cosmic radiation. [13] Furthermore, it was clear that some of the elements — Li, Be, and B — were overabundant by many orders of magnitude (compared with solar system abundances) and therefore must be secondary nuclei from nuclear spallation interactions during the interstellar propagation of primary nuclei in the galaxy. [e.g. Bradt and Peters, Phys. Rev. **74**, 1828 (1948)].

To account for these overabundant secondaries, estimates were obtained for the amount of interstellar matter traversed in the galaxy under the assumption that Li, Be, and B were not present in the cosmic-ray sources. Values ranging from ~3 to ~5 gm-cm^{-2} were obtained, suggesting confinement times in interstellar magnetic fields of order ~10^6 years. {However, Juliusson, Meyer, and Müller [Phys. Rev. Lett. **29**, 445 (1972)] discovered that the ratio of the galactic secondary nuclei to primary nuclei from sources decreased with increasing energy above ~30 GeV nucleon^{-1}.}

Although a solar system origin for the cosmic rays had been promoted by Alfvén and E. Teller, the existence of the primary heavy nuclei and their secondaries presented major difficulties. However, Fermi removed the principal objections to an interstel-

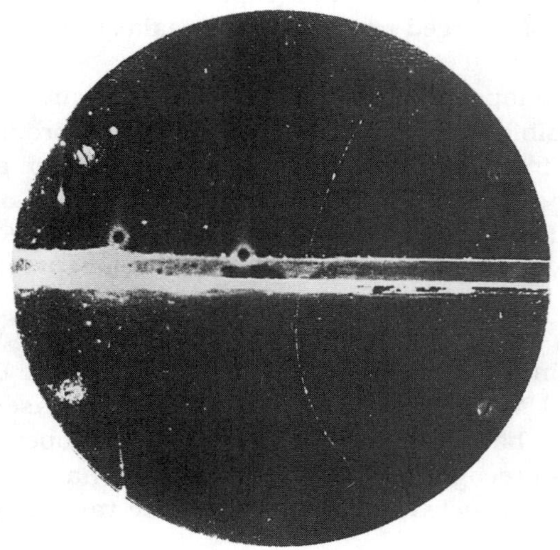

Positron track observed by **C. D. Anderson (1905-1991)**. (Photograph by Carl David Anderson. Courtesy of AIP Emilio Segrè Visual Archives).

lar origin for cosmic-ray particle acceleration [14] by developing a quantitative model for the acceleration of charged particles in the interstellar medium that invoked cosmic rays colliding with moving magnetic fields carried in moving interstellar gas clouds along the lines embodied by Alfvén's ideas, (e), and Alfvén and Herlofson [Phys. Rev. **78**, 616L (1950)].

In work from the late 1930s to the early 1950s Forbush had obtained clear correlations between worldwide geomagnetic storms and sharp decreases in the secondary radiation intensity measured by ion chambers — the so-called Forbush decreases. They were believed to be the result of geomagnetic storm-induced changes in geomagnetic rigidity cutoff. However, following the discovery of the large-scale, low-energy nucleonic component latitude effect in 1948, J. A. Simpson [15] showed from neutron monitor measurements that these intensity variations had their origin in interplanetary dynamical processes controlled by the Sun. The 11-year solar cycle anticorrelation with cosmic-ray intensity discovered by Forbush (g) was therefore due to the solar interplanetary modulation of the galactic cosmic radiation.

The neutron intensity monitors developed in 1948 that measured the time and intensity response of nucleons from the giant solar flare of 23 February 1956 led to the discovery of the heliosphere, an interplanetary magnetic field structure totally enclosing the solar system to great distances beyond the orbit of Earth. [17]

E. N. Parker then quantitatively developed a theory for the coronal expansion of plasma that becomes the solar wind, transporting solar magnetic fields into interplanetary space, to explain the earlier discovery of galactic cosmic-ray modulation by the Sun. [18]

The early experimental and theoretical studies on solar modulation generated widespread research activity on the physics of the solar-terrestrial connections and efforts to deduce, by extrapo-

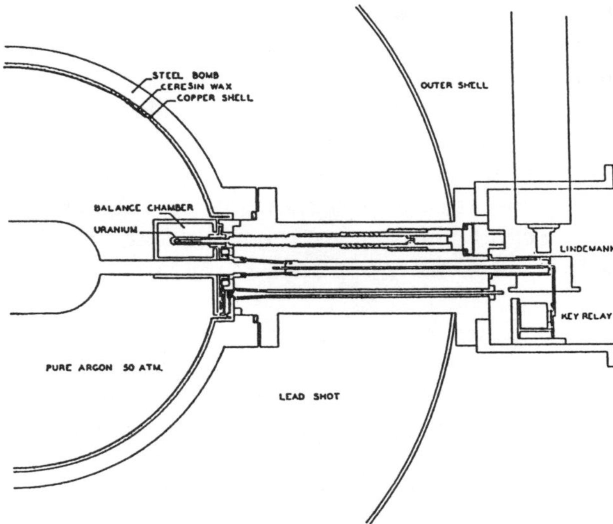

Cross-section drawing of Compton Model-C ionization chamber used in latitude surveys and by Forbush to investigate the secondary ionization component time-intensity variations. (From Compton, Wollan, and Bennett, Rev. Sci. Inst., **5**, 415, 1934.)

Manuel S. Vallarta (1899-1977). (Courtesy of Ciudad Universitaria, Mexico.)

lation, the spectra of cosmic-ray particles in the nearby interstellar medium.

With the launch of Sputnik in 1957 there opened new opportunities for cosmic-ray experiments to measure directly and simultaneously plasmas, magnetic fields, and particles in space.

The discovery of trapped radiation around Earth by Van Allen (h) initiated studies of charged-particle acceleration in space. The first interplanetary shock wave was detected by spacecraft experiments and shown to be the interplanetary origin of the Forbush intensity decreases. [19] The same space probe experiment proved that solar modulation was approximately heliocentric, rather than geocentric. [19]

Colgate and M. H. Johnson invoked hydrodynamic shocks in plasmas [20] to develop a model for cosmic-ray acceleration by supernovae. Later modeling included acceleration by pulsars [e.g. Gunn and Ostriker, Phys. Rev. Lett. **22**, 728, (1969)].

Spacecraft investigations within the heliosphere have led to the opening of a substantial branch of research (noted in the chart) devoted to the discovery of a wide range of acceleration processes, including shocks, in which charged particles, plasmas, and magnetic fields are studied simultaneously. There now exists within the heliosphere a hierarchy of acceleration processes covering a wide scale of energies and a variety of astrophysical sites. At the time of writing this review, PIONEER-10 is approaching a distance of ~60 AU from the Sun and is still inside the heliosphere!

For many years there had been attempts to find electrons in the primary cosmic rays: Success was achieved in 1961. [21], [22] The electrons constituted only a few percent of the charged-particle cosmic-ray flux, but their existence posed new questions regarding their origin: Were they from supernovae or from the decay of $\pi^{\pm}$. This problem was brought into further focus by the discovery that a small portion of the electronic component was primary positrons. [26]

In the postwar years some cosmic-ray physicists turned their attention to the search of x rays using rockets, which provided several minutes of observation beyond the absorbing atmosphere. Friedman and collaborators discovered soft x radiation from the Sun. [16] Later, x-ray sources beyond the solar system were discovered. [23] At about the same time gamma-ray radiation was detected by instruments on spacecraft. [24] This major new field of gamma-ray investigations is shown in the high-energy photon branch in the chart. Subsequent reporting of the exciting discoveries in x and gamma rays appeared mainly in journals on astrophysics.

Although extensive air showers were discovered in 1939, (d) it was not until the 1960s that large-area, ground-based sensors were operated to determine the energies of the ultrahigh-energy primaries. By 1963 primary particle energies as high as ~10^{20} eV were reported. [25]

With the discovery of the cosmic blackbody radiation, the question was raised whether nuclei in excess of 10^{20} eV could escape

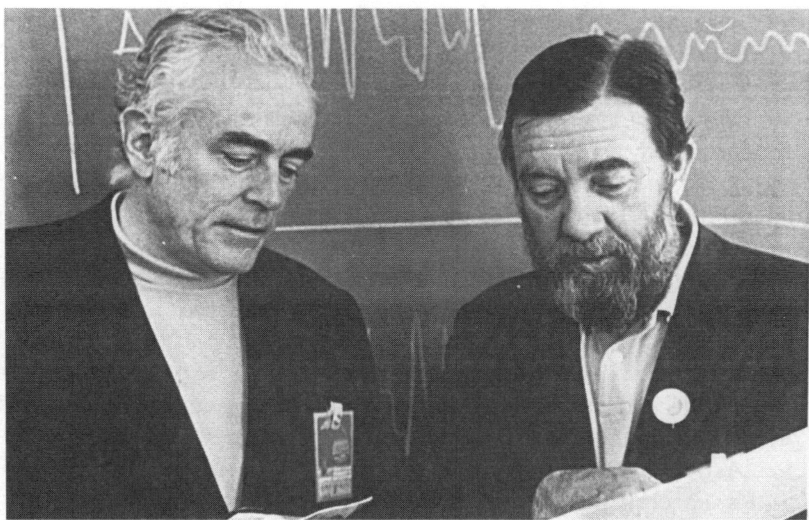

John A. Simpson (1916–) and **James A. Van Allen (1914–).** (Courtesy of NASA, AIP Emilio Segrè Visual Archives.)

annihilation in the galaxy due to collisions with this radiation. [27] However, at the time of writing this report, there appears to be evidence for an event of $\sim 3 \times 10^{20}$ eV.

By 1964 satellites carrying charged-particle telescopes began to determine with precision not only particle energy spectra, but also their elemental and isotopic composition over all modulation levels of the ~11-year solar cycle. The discovery of deuterium [28] in the primary radiation completed the quartet of light isotopes, ^{1}H, ^{2}H, ^{3}He, and ^{4}He, and provided a new tool for exploring cosmic-ray interstellar propagation. An abundance analysis of the ^{7}Be, ^{9}Be, and radioactive ^{10}Be ($\sim 1.6 \times 10^{6}$ years half-life) showed that the containment lifetime in galactic magnetic fields is $\sim 10^{7}$ years (k) and that the mean interstellar density for propagation is ~0.2 atom cm^{-3}.

A new and unanticipated component in the energetic charged-particle population, called anomalous cosmic rays, was discovered to be mixed in with the galactic cosmic rays in the heliosphere. First, a pure ^{4}He component was discovered (i) extending in energy up to ~100 MeV $nucleon^{-1}$. This was followed by a report of a greatly enhanced oxygen component [29] and the finding that both N and O abundances were anomalously high relative to carbon. (j) Current evidence supports a theory proposed by Fisk, Kozlovsky, and Ramaty [Astrophys. J. Lett. **190**, L35 (1974)] in which neutral atoms with high ionization potentials from the interstellar medium are singly ionized by solar UV in the heliosphere, then picked up in the solar wind and accelerated by heliospheric processes, probably at a solar wind termination shock. Current investigations indicate that the anomalous component energy density will equal the galactic cosmic-ray energy density beyond approximately 70–80 AU.

From the early 1960s onward, experiments in spacecraft provided increasingly precise and extended measurements of the energy spectra and elemental and isotopic composition of the

Scott E. Forbush (1904-1984).

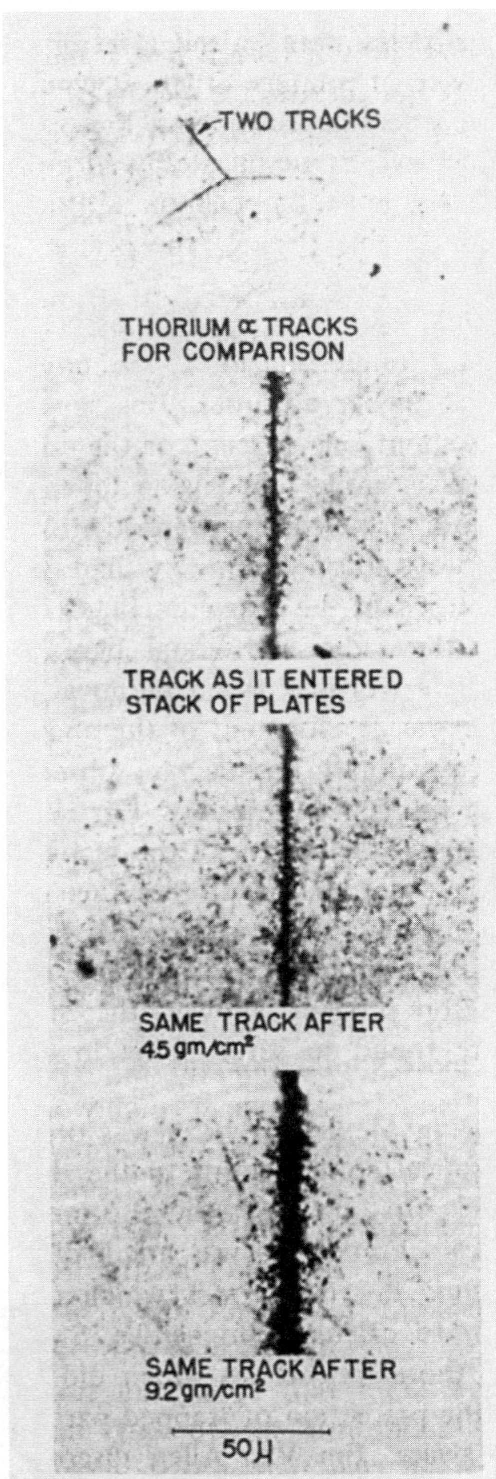

First evidence for very heavy stripped nuclei in the cosmic radiation. (Recorded in Ilford C-2 nucleon emulsion, P. Frier thesis, 1948, University of Minnesota; see also [13].)

cosmic-ray nuclei — the only matter from the galaxy that could be analyzed directly. Thus, the cosmic rays are a "Rosetta stone" for unraveling the nucleosynthesis processes leading to the composition of the source of cosmic rays. The most outstanding isotopic anomaly to challenge nucleosynthesis models is the greatly enhanced, neutron-rich isotope ^{22}Ne. (l) On the other hand, the abundances of the cosmic-ray actinides are not, so far, found to be inconsistent with solar system abundances. (m)

Although a Cosmic-Ray Division of The American Physical Society was established in 1970, the reader will note from the cosmic radiation time line that from about the mid-1960s, the publications of cosmic-ray investigations shifted rapidly from *The Physical Review* to other journals emphasizing astrophysics and physics in space: The most notable were the *Astrophysical Journal, Astrophysical Journal Letters*, and the *Journal of Geophysical Research* (Space Physics) series.

Although this brief walkthrough of experimental cosmic-ray physics has been stipulated to end in 1983, I call attention to a more recent investigation of lasting value. Early in 1987 there occurred an outstanding supernova explosion (SN1987A) in the Large Magellanic Cloud. In coincidence with the visual onset of this event, a burst of neutrinos was observed simultaneously in the U.S. [30] and Japan. [31] By confirming current theories for the evolution of supernovae, these neutrino observations have given an enhanced impetus to future studies of the nuclear processes and dynamics involved in the "death" of stars. The neutrino measurements also confirm that cosmic radiation searches contributing to particle physics are not over — for example, searches for weakly interacting massive particles (WIMPS) and magnetic monopoles continue. The tantalizing but unconfirmed reports of ultrahigh-energy neutral radiation ($\sim 10^{16}$–10^{18} eV) from specific astrophysical sites (such as Cygnus-like objects) have encouraged the building of ground-based detector arrays of ever-increasing size and complexity. Recent discoveries from spacecraft observations of gamma-ray bursts and ultrahigh-energy gamma-ray sources from spacecraft observatories are contributing new insights into the origins of the cosmic radiations. These investigations will continue to be a prime beneficiary of man's entry into the space age well into the 21st century.

Central to the discoveries reported here were the invention of instrumentation including the ionization chamber, counter-controlled cloud chamber, Geiger-Müller (G-M) counter, coincidence circuit, proportional counter, Čerenkov counter, nuclear emulsion, neutron monitor, semiconductor detector, and transition radiation detector, among other important inventions. It is unfortunate that there is no place in this volume to call attention to their development and their role in the discoveries reported here.

I regret that in the space allotted to me I could not cite the many additional publications that deserve mention, especially the wide range of theoretical publications essential to the evolution of cosmic radiation physics. Others may prefer some different selections, but the message would be the same: The discovery of the cosmic radiations and the searches for their origins, as represented in The *Physical Review* and *Physical Review Letters*, have resulted in an explosive and continuing development of our knowledge of particle physics and high-energy astrophysics.

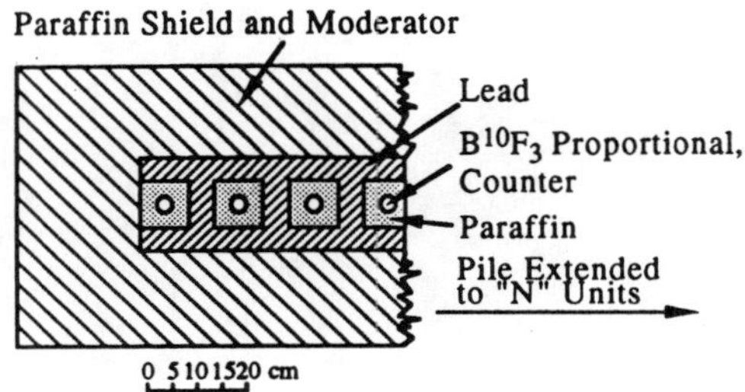

Cross-section drawing of neutron monitor instrument to investigate the interplanetary modulation of galactic cosmic rays. [From Simpson, Fonger, and Treiman, Phys. Rev., **90**, 934 (1953).]

The Physical Review *and* Physical Review Letters *works referenced in* Cosmic Radiation Discovery Time Line. *Papers through 1983 appear in the volume for The Physical Review Centennial (with full texts in the CD-ROM) (□ symbols).*

1. R.A. Millikan and G.H. Cameron, *High Frequency Rays of Cosmic Origin III. Measurements in Snow-Fed Lakes at High Altitudes*, Phys. Rev. **28**, 851–868 (1926).
2. A.H. Compton, *Variation of the Cosmic Rays With Latitude*, Phys. Rev. **41**, 111–113(L) (1932).
3. G. Lemaitre and M.S. Vallarta, *On Compton's Latitude Effect of Cosmic Radiation*, Phys. Rev. **43**, 87–91 (1933).
4. A.H. Compton, *A Geographic Study of Cosmic Rays*, Phys. Rev. **43**, 387–403 (1933).
5. C.D. Anderson, *The Positive Electron*, Phys. Rev. **43**, 491–494 (1933).
6. T.H. Johnson, *The Azimuthal Asymmetry of the Cosmic Radiation*, Phys. Rev. **43**, 834–835(L) (1933).
7. L. Alvarez and A.H. Compton, *A Positively Charged Component of Cosmic Rays*, Phys. Rev. **43**, 835–836(L) (1933).
8. P.M.S. Blackett, *On the Instability of the Barytron and the Temperature Effect of Cosmic Rays*, Phys. Rev. **54**, 973–974(L) (1938).
9. H.A. Bethe, S.A. Korff, and G. Placzek, *On the Interpretation of Neutron Measurements in Cosmic Radiation*, Phys. Rev. **57**, 573–587 (1940).
10. M. Schein, W.P. Jesse, and E.O. Wollan, *The Nature of the Primary Cosmic Radiation and the Origin of the Mesotron*, Phys. Rev. **59**, 615(L) (1941).
11. S.E. Forbush, *Three Unusual Cosmic Ray Increases Probably Due to Charged Particles from the Sun*, Phys. Rev. **70**, 771–772(L) (1946).
12. W.F. Libby, *Atmospheric Helium Three and Radiocarbon from Cosmic Radiation*, Phys. Rev. **69**, 671–672(L) (1946).
13. P. Freier, E.J. Lofgren, E.P. Ney, F. Oppenheimer, H.L. Bradt, and B. Peters, *Evidence for Heavy Nuclei in the Primary Cosmic Radiation*, Phys. Rev. **74**, 213–217 (1948).
14. E. Fermi, *On the Origin of the Cosmic Radiation*, Phys. Rev. **75**, 1169–1174 (1949).
15. J.A. Simpson, *Neutrons Produced in the Atmosphere by Cosmic Radiations*, Phys. Rev. **83**, 1175–1188 (1951).
16. H. Friedman, S.W. Lichtman, and E.T. Byram, *Photon Counter Measurements of Solar X-Rays and Extreme Ultraviolet Light*, Phys. Rev. **83**, 1025–1030 (1951).
17. P. Meyer, E.N. Parker, and J.A. Simpson, *Solar Cosmic Rays of February, 1956 and Their Propagation Through Interplanetary Space*, Phys. Rev. **104**, 768–783 (1956).
18. E.N. Parker, *Cosmic Ray Modulation by Solar Wind*, Phys. Rev. **110**, 1445–1449 (1958).
19. C.Y. Fan, P. Meyer, and J.A. Simpson, *Rapid Reduction of Cosmic Radiation Intensity Measured in Interplanetary Space*, Phys. Rev. Lett. **5**, 269–271 (1960).
20. S.A. Colgate and M.H. Johnson, *Hydrodynamic Origin of Cosmic Rays*, Phys. Rev. Lett. **5**, 235–238 (1960).

21. J. Earl, *Cloud-Chamber Observations of Primary Cosmic-Ray Electrons*, Phys. Rev. Lett. **6**, 125–128 (1961).

22. P. Meyer and R. Vogt, *Electrons in the Primary Cosmic Radiation*, Phys. Rev. Lett. **6**, 193–196 (1961).

23. R. Giacconi, H. Gursky, F.R. Paolini, and B.B. Rossi, *Evidence for X Rays from Sources Outside the Solar System*, Phys. Rev. Lett. **9**, 439–443 (1962).

24. W.L. Kraushaar and G.W. Clark, *Search for Primary Cosmic Gamma Rays with the Satellite Explorer XI*, Phys. Rev. Lett. **8**, 106–109 (1962).

25. J. Linsley, *Evidence for a Cosmic Ray Particle with Energy 10^{20} eV*, Phys. Rev. Lett. **10**, 146–148 (1963).

26. J.A. DeShong, Jr., R.H. Hildebrand, and P. Meyer, *Ratio of Electrons to Positrons in the Primary Cosmic Radiation*, Phys. Rev. Lett. **12**, 3–6 (1964).

27. K. Greisen, *End to the Cosmic Ray Spectrum?*, Phys. Rev. Lett. **16**, 748–750 (1966).

28. C.Y. Fan, G. Gloeckler, and J.A. Simpson, *Galactic Deuterium and its Energy Spectrum Above 20 MeV Per Nucleon*, Phys. Rev. Lett. **17**, 329–333 (1966).

29. D. Hovestadt, O. Vollmer, G. Gloeckler, and C.Y. Fan, *Differential Energy Spectra of Low Energy (<8.5 MeV per Nucleon) Heavy Cosmic Rays During Solar Quiet Times*, Phys. Rev. Lett. **31**, 650–653 (1973).

30. R.M. Bionta, G. Blewitt, C.B. Bratton, D. Casper, A. Ciocio, R. Claus, B. Cortez, M. Crouch, S.T. Dye, S. Errede, G.W. Foster, W. Gajewski, K.S. Ganezar, M. Goldhaber, T.J. Haines, T.W. Jones, D. Kielczewska, W.R. Kropp, J.G. Learned, J.M. LoSecco, J. Matthews, R. Miller, M.S. Mudan, H.S. Park, L.R. Price, F. Reines, J. Schultz, S. Seidel, E. Shumard, D. Sinclair, H.W. Sobel, J.L. Stone, L.R. Sulak, R. Svoboda, G. Thornton, J.C. van der Velde, and C. Wuest, *Observations of a Neutrino Burst in Coincidence with Supernova 1987 in the Large Magellanic Cloud*, Phys. Rev. Lett. **58**, 1494–1496 (1987).

31. K. Hirata, T. Kajita, M. Koshiba, M. Nakahata, Y. Oyama, N. Sato, A. Suzuki, M. Takita, Y. Totsuka, T. Kifune, T. Suda, K. Takahashi, T. Tanimori, K. Miyano, M. Yamada, E.W. Beier, L.R. Feldscher, S.B. Kim, A.K. Mann, F.M. Newcomer, R. Van Berg, W. Zhang, and B.G. Cortez, *Observation of a Neutrino Burst from the Supernova SN1987A*, Phys. Rev. Lett. **58**, 1490–1493 (1987).

Journals other than Physical Review *or* Physical Review Letters *cited for the* Cosmic Radiation Discovery Time Line *[○ symbols].*

a. V. F. Hess, *Über Beobachtungen der durchdringenden Strahlung bei sieben Freiballonfahrten*, Phys. Z. **13**, 1084 (1912).

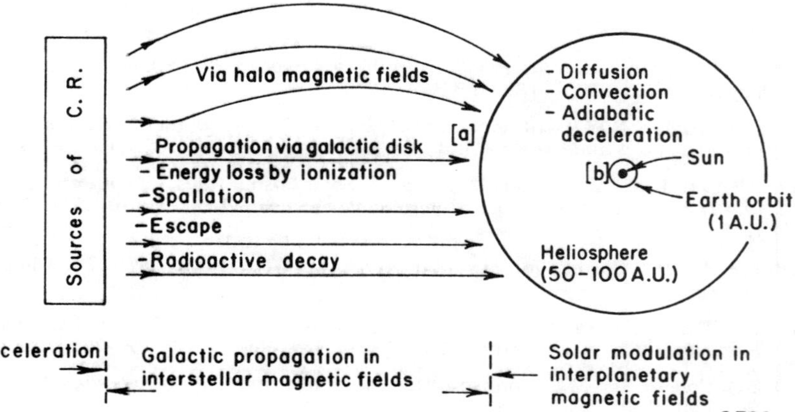

Sketch of history of a cosmic ray nucleon from source to detection at Earth.

b. J. Clay, *Penetrating Radiation*, Proc. R. Acad. Amsterdam **31**, 1091 (1928).

c. P.M.S. Blackett and G.P.S. Occhialini, *Some Photographs of the Tracks of Penetrating Radiation*, Proc. R. Soc. London Ser. A **139**, 699 (1933).

d. P. Auger and R. Maze, *Extension et Pouvoir Pénétrant des Grandes Gerbes de Rayons Cosmiques*, C. R. Acad. Sci. (Institut de France) **208**, 164 (1939).

e. H. Alfvén, *On the Existence of Electromagnetic Hydrodynamic Waves*, Ark. Astron. Fys. **29**B, 1 (1943).

f. G.D. Rochester and C.C. Butler, *Evidence for the Existence of New Unstable Elementary Particles*, Nature **160**, 855 (1947).

g. S.E. Forbush, *World-Wide Cosmic-Ray Variations, 1937-1952*, J. Geophys. Res. **59**, 525 (1954).

h. J.A. Van Allen and L.A. Frank, *Radiation Around the Earth to a Radial Distance of 107,400 km*, Nature **183**, 430, (1959).

i. M. Garcia-Munoz, G.M. Mason, and J.A. Simpson, *New Test for Solar Modulation Theory: The 1972 May-July Low-Energy Galactic Cosmic-Ray Proton and Helium Spectra*, Astrophys. J. **182**, L81 (1973).

j. F.B. McDonald, B.J. Teegarden, J.H. Trainor, and W.R. Webber, *The Anomalous Abundance of Cosmic-Ray Nitrogen and Oxygen Nuclei at Low Energies*, Astrophys. J. **187**, L105 (1974).

k. M. Garcia-Munoz, G.M. Mason, and J.A. Simpson, *The Age of the Galactic Cosmic Rays Derived from the Abundance of ^{10}Be*, Astrophys. J. **217**, 859 (1977).

l. M. Garcia-Munoz, J.A. Simpson, and J.P. Wefel, *The Isotopes of Neon in the Galactic Cosmic Rays*, Astrophys. J. **232**, L95 (1979).

m. W.R. Binns, R.K. Fickle, T.L. Garrard, M.H. Israel, J. Klarmann, E.C. Stone, and C.J. Waddington, *The Abundance of the Actinides in the Cosmic Radiation as Measured on HEAO 3*, Astrophys. J. **261**, L117 (1982).

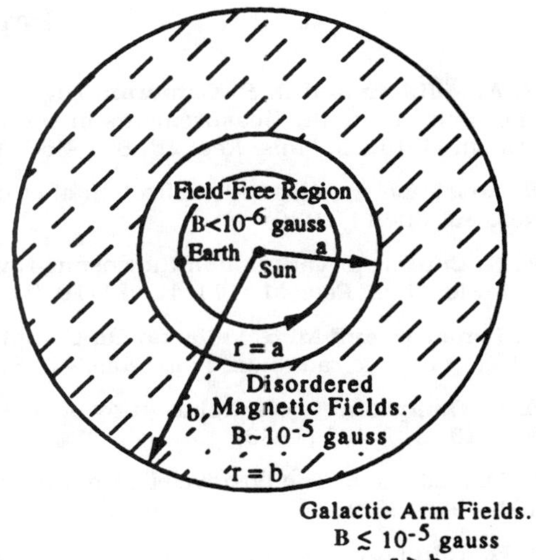

Sketch of first direct experimental evidence for the existence of a dynamical heliosphere [17] (now shown by PIONEER-10 to extend beyond 60 a.u.).

Papers Reproduced on CD-ROM

R. A. Millikan and G. H. Cameron. High frequency rays of cosmic origin. III. Measurements in snow-fed lakes at high altitudes, *Phys. Rev.* **28**, 851–868 (1926)

B. Rossi. On the magnetic deflection of cosmic rays, *Phys. Rev.* **36**, 606(L) (1930)

A. H. Compton. Variation of the cosmic rays with latitude, *Phys. Rev.* **41**, 111–113(L) (1932)

G. Lemaitre and M. S. Vallarta. On Compton's latitude effect of cosmic radiation, *Phys. Rev.* **43**, 87–91 (1933)

A. H. Compton. A geographic study of cosmic rays, *Phys. Rev.* **43**, 387–403 (1933)

C. D. Anderson. The positive electron, *Phys. Rev.* **43**, 491–494 (1933)

T. H. Johnson. The azimuthal asymmetry of the cosmic radiation, *Phys. Rev.* **43**, 834–835(L) (1933)

L. Alvarez and A. H. Compton. A positively charged component of cosmic rays, *Phys. Rev.* **43**, 835–836(L) (1933)

S. H. Neddermeyer and C. D. Anderson. Note on the nature of cosmic-ray particles, *Phys. Rev.* **51**, 884–886 (1937)

J. C. Street and E. C. Stevenson. New evidence for the existence of a particle of mass intermediate between the proton and electron, *Phys. Rev.* **52**, 1003–1004(L) (1937)

S. H. Neddermeyer and C. D. Anderson. Cosmic-ray particles of intermediate mass, *Phys. Rev.* **54**, 88–89(L) (1938)

P. M. S. Blackett. On the instability of the barytron and the temperature effect of cosmic rays, *Phys. Rev.* **54**, 973–974(L) (1938)

B. Rossi, N. Hilberry, and J. B. Hoag. The variation of the hard component of cosmic rays with height and the disintegration of mesotrons, *Phys. Rev.* **57**, 461–469 (1940)

H. A. Bethe, S. A. Korff, and G. Placzek. On the interpretation of neutron measurements in cosmic radiation, *Phys. Rev.* **57**, 573–587 (1940)

M. Schein, W. P. Jesse, and E. O. Wollan. The nature of the primary cosmic radiation and the origin of the mesotron, *Phys. Rev.* **59**, 615(L) (1941)

W. F. Libby. Atmospheric helium three and radiocarbon from cosmic radiation, *Phys. Rev.* **69**, 671–672(L) (1946)

S. E. Forbush. Three unusual cosmic-ray increases possibly due to charged particles from the Sun, *Phys. Rev.* **70**, 771–772(L) (1946)

P. Freier, E. J. Lofgren, E. P. Ney, F. Oppenheimer, H. L. Bradt, and B. Peters. Evidence for heavy nuclei in the primary cosmic radiation, *Phys. Rev.* **74**, 213–217 (1948)

H. L. Bradt and B. Peters. Investigation of the primary cosmic radiation with nuclear photographic emulsions, *Phys. Rev.* **74**, 1828–1837 (1948)

E. Fermi. On the origin of the cosmic radiation, *Phys. Rev.* **75**, 1169–1174 (1949)

H. Alfvén and N. Herlofson. Cosmic radiation and radio stars, *Phys. Rev.* **78**, 616(L) (1950)

K. O. Kiepenheuer. Cosmic rays as the source of general galactic radio emission, *Phys. Rev.* **79**, 738(L) (1950)

H. Friedman, S. W. Lichtman, and E. T. Byram. Photon counter measurements of solar x-rays and extreme ultraviolet light, *Phys. Rev.* **83**, 1025–1030 (1951)

J. A. Simpson. Neutrons produced in the atmosphere by the cosmic radiations, *Phys. Rev.* **83**, 1175–1188 (1951)

P. Meyer, E. N. Parker, and J. A. Simpson. Solar cosmic rays of February 1956 and their propagation through interplanetary space, *Phys. Rev.* **104**, 768–783 (1956)

G. W. Clark and J. Hersil. Polarization of cosmic-ray μ mesons: Experiment, *Phys. Rev.* **108**, 1538–1544 (1957)

E. N. Parker. Cosmic-ray modulation by solar wind, *Phys. Rev.* **110**, 1445–1449 (1958)

S. A. Colgate and M. H. Johnson. Hydrodynamic origin of cosmic rays, *Phys. Rev. Lett.* **5**, 235–238 (1960)

C. Y. Fan, P. Meyer, and J. A. Simpson. Rapid reduction of cosmic-radiation intensity measured in interplanetary space, *Phys. Rev. Lett.* **5**, 269–271 (1960)

J. A. Earl. Cloud-chamber observations of primary cosmic-ray electrons, *Phys. Rev. Lett.* **6**, 125–128 (1961)

P. Meyer and R. Vogt. Electrons in the primary cosmic radiation, *Phys. Rev. Lett.* **6**, 193–196 (1961)

W. L. Kraushaar and G. W. Clark. Search for primary cosmic gamma rays with the satellite Explorer XI, *Phys. Rev. Lett.* **8**, 106–109 (1962)

R. Giacconi, H. Gursky, F. R. Paolini, and B. B. Rossi. Evidence for x rays from sources outside the solar system, *Phys. Rev. Lett.* **9**, 439–443 (1962)

J. Linsley. Evidence for a primary cosmic-ray particle with energy 10^{20} eV, *Phys. Rev. Lett.* **10**, 146–148 (1963)

J. A. De Shong, Jr., R. H. Hildebrand, and P. Meyer. Ratio of electrons to positrons in the primary cosmic radiation, *Phys. Rev. Lett.* **12**, 3–6 (1964)

K. Greisen. End to the cosmic-ray spectrum?, *Phys. Rev. Lett.* **16**, 748–750 (1966)

C. Y. Fan, G. Gloeckler, and J. A. Simpson. Galactic deuterium and its energy spectrum above 20 MeV per nucleon, *Phys. Rev. Lett.* **17**, 329–333 (1966)

J. E. Gunn and J. P. Ostriker. Acceleration of high-energy cosmic rays by pulsars, *Phys. Rev. Lett.* **22**, 728–731 (1969)

E. Juliusson, P. Meyer, and D. Müller. Composition of Cosmic Ray Nuclei at High Energies, *Phys. Rev. Lett.* **29**, 445–448 (1972)

D. Hovestadt, O. Vollmer, G. Gloeckler, and C. Y. Fan. Differential energy spectra of low-energy (<8.5 MeV per nucleon) heavy cosmic rays during solar quiet times, *Phys. Rev. Lett.* **31**, 650–653 (1973)

K. Hirata, T. Kajita, M. Koshiba, M. Nakahata, Y. Oyama, N. Sato, A. Suzuki, M. Takita, Y. Totsuka, T. Kifune, T. Suda, K. Takahashi, T. Tanimori, K. Miyano, M. Yamada, E. W. Beier, L. R. Feldscher, S. B. Kim, A. K. Mann, F. M. Newcomer, R. Van Berg, W. Zhang, and B. G. Cortez. Observation of a neutrino burst from the Supernova SN1987A, *Phys. Rev. Lett.* **58**, 1490–1493 (1987)

R. M. Bionta, G. Blewitt, C. B. Bratton, D. Casper, A. Ciocio, R. Claus, B. Cortez, M. Crouch, S. T. Dye, S. Errede, G. W. Foster, W. Gajewski, K. S. Ganezer, M. Goldhaber, T. J. Haines, T. W. Jones, D. Kielczewska, W. R. Kropp, J. G. Learned, J. M. LoSecco, J. Matthews, R. Miller, M. S. Mudan, H. S. Park, L. R. Price, F. Reines, J. Schultz, S. Seidel, E. Shumard, D. Sinclair, H. W. Sobel, J. L. Stone, L. R. Sulak, R. Svoboda, G. Thornton, J. C. van der Velde, and C. Wuest. Observation of a neutrino burst in coincidence with Supernova 1987A in the Large Megellanic Cloud, *Phys. Rev. Lett.* **58**, 1494–1496 (1987)

1. H. Bethe, **2.** D. Froman, **3.** R. Brode, **4.** A. H. Compton, **5.** E. Teller, **6.** A. Baños, Jr., **7.** G. Groetzinger, **8.** S. Goudsmit, **9.** M. S. Vallarta, **10.** L. Norhdeim, **11.** J. R. Oppenheimer, **12.** C. D. Anderson, **13.** S. Forbush, **14.** Nielsen (of Duke U.), **15.** V. Hess, **16.** V. C. Wilson, **17.** B. Rossi, **18.** W. Bothe, **19.** W. Heisenberg, **20.** P. Auger, **21.** R. Serber, **22.** T. Johnson, **23.** J. Clay (Holland), **24.** W. F. G. Swann, **25.** J. C. Street (Harvard), **26.** J. Wheeler, **27.** S. Neddermeyer, **28.** E. Herzog (?), **29.** M. Pomerantz, **30.** W. Harkins (U. of C.), **31.** H. Beutler, **32.** M. M. Shapiro†, **33.** M. Schein*, **34.** C. Montgomery (Yale), **35.** W. Bostick†, **36.** C. Eckart, **37.** A. Code†, **38.** J. Stearns (Denver?), **39.** J. Hopfield, **40.** E. O. Wollan*, **41.** D. Hughes†, **42.** W. Jesse*, **43.** B. Hoag, **44.** N. Hillberry†, **45.** F. Shonka†, **46.** P. S. Gill†, **47.** A. H. Snell, **48.** J. Schremp, **49.** A. Haas? (Vienna), **50.** E. Dershem, and **51.** H. Jones† at the Cosmic Ray Conference (Symposium on Cosmic Rays, 1939) convened at the University of Chicago in the summer of 1939. (Courtesy of Maurice M. Shapiro.)

**Then research associate of Compton.* †*Then graduate student of Compton.*

Second Series *November, 1926* *Vol. 28, No. 5*

THE

PHYSICAL REVIEW

HIGH FREQUENCY RAYS OF COSMIC ORIGIN III. MEASUREMENTS IN SNOW-FED LAKES AT HIGH ALTITUDES

BY R. A. MILLIKAN AND G. HARVEY CAMERON

ABSTRACT

1. **Absorption experiments in Muir Lake (alt. 11,800 feet).**—The sinking of sealed electroscope No. 3 in Muir Lake showed an ionization decreasing steadily with depth from 13.3 ions per cc per sec. at the surface to 3.6 ions at 50 feet below the surface, below which there was no further decrease. The absorption curve of electroscope No. 3 was in excellent agreement with that of No. 1.

2. **Absorption experiments in Arrowhead Lake (alt. 5,100 feet).**—The electroscope readings in Arrowhead Lake correspond uniformly to readings six feet deeper in Muir Lake. This difference is the exact water equivalent of the absorption of the atmosphere between the two elevations. All readings of both electroscopes fit satisfactorily upon a single curve relating ionization to depth beneath the surface of the atmosphere in equivalent meters of water.

3. **Rays of cosmic origin.**—1 and 2 combined with the failure to detect any systematic diurnal variation, in tests of a number of days duration at high altitudes, constitute new and quite unambiguous evidence for the existence of very hard etherial rays of cosmic origin entering the earth uniformly from all directions.

4. **Spectral distribution of cosmic rays.**—No single absorption coefficient is found to fit the absorption curve, the lower end of which requires a coefficient of .18 per meter of water; the upper end a coefficient, .30 per meter of water. These coefficients correspond, by Compton's equations, to wave-lengths $\lambda = .00038$A and $\lambda = .00063$A. These are fifty times the frequencies of ordinary gamma rays, $\lambda = .025$A, and the former corresponds to an energy of 32,000,000 volts.

5. **Number of pairs of ions due to cosmic rays.**—The observed number of pairs of ions in electroscope No. 1 due to cosmic rays is about 1.4 at sea level, 2.6 at 1600 meters, 4.8 at 3600 meters, 5.9 at 4300 meters.

6. **Stimulated secondary rays.**—Theoretically, cosmic rays of the foregoing energy should not stimulate ether waves of gamma ray hardness, but should produce beta rays capable of penetrating brass walls 5 mm thick. The observations present evidence of rays of about this hardness increasing systematically with altitude in rough proportionality to the intensity of the cosmic rays. This evidence is not completely convincing because of inability thus far to eliminate the effects of the gamma rays from the underlying rocks.

7. **Origin of cosmic rays.**—Evidence is presented that these rays do not result from the union of protons with negative electrons, but they are rather

due to nuclear changes of about one-thirtieth the energy corresponding to such union, taking place throughout the depths of the universe.

I. INTRODUCTION

THE 1922, high altitude, sounding balloon flights reported in Part I[1] had shown that some sort of a penetrating radiation exists in the upper reaches of the atmosphere, though of not more than one fourth the intensity theretofore reported. Again, the 1923 absorption experiments on mountain peaks reported in Part II[2] had shown that there exists at such heights a new radiation of local origin and of something like gamma ray hardness, but they had seemed to prove conclusively that if rays of *cosmic origin* exist at all they must be of somewhat different characteristics from any as yet suggested.

Up to the time of the Pikes Peak observations (September, 1923) the only work which had appeared demanding an absorption coefficient smaller than 5×10^{-3}cm^{-1} for cosmic rays, if they existed, was the aforementioned sounding balloon experiments of Millikan and Bowen performed in April, 1922.[1] However, at the sitting of December 20, 1923, of the Preussischen Akademie der Wissenschaften Dr. Werner Kolhörster presented the results of new experiments, the first of which consisted in sinking electroscopes in different bodies of water at about sea level and observing a slight decrease in the number of ions as compared with the surface value. He attributed the noticeable lack of concordance between the results in the different bodies of water experimented upon to the different influences of the banks, but even with a CO_2 filling of the electroscopes the maximum change produced by sinking in water amounted to 2.1 ions, which would presumably be about 10 percent of the normal surface reading (not recorded in the report).

Dr. Kolhörster's comments upon these observations are as follows. "From the lake-experiments there results the absorption coefficient $\mu = 2\times10^{-3}$cm^{-1}, while my former balloon experiments gave 5×10^{-3}cm^{-1}, a satisfactory agreement in view of the small intensity, about 2 ions, with which the penetrating rays reach the earth," thus indicating that these measurements were not sufficiently certain, in his judgment, to differentiate between $\mu = 5\times10^{-3}$ and $\mu = 2\times10^{-3}$.

He next made observations in crevasses in glaciers at altitudes of 2300 m and 3500 m on the Jungfrau, and obtained in three experiments for μ 1.6×10^{-3}cm^{-1}, 2.6×10^{-3}cm^{-1}, and 2.7×10^{-3}cm^{-1}. Combining

[1] Millikan and Bowen, Carnegie Institution Year Book, No. 21, 386 (1922); also Phys. Rev. 22, 198 (1923) and 27, 353 (1926).

[2] Millikan and Otis, Phys. Rev. 27, 645 (1926).

these three observations on equal footing with those made in water and reported above, he recorded as the rough mean of the four observations $\mu = 2.5 \times 10^{-3}\text{cm}^{-1}$, a figure, however, now so low as to be no longer incompatible with the Kelly Field sounding balloon experiments.[1]

Finally, after having quoted the value of μ for the hardest gamma rays from RaC as $3.9 \times 10^{-2}\text{cm}^{-1}$, and from ThD as $3.3 \times 10^{-2}\text{cm}^{-1}$, he summarizes his paper thus: "To resume, it is to be emphasized that the existence of a hard gamma ray with an absorption coefficient about 1/10 that of the hardest known gamma rays has been demonstrated." The final value chosen was thus about the linear mean of all the observations taken, from 1.6 up to 5.7, namely, about $3.3 \times 10^{-3}\text{cm}^{-1}$. Also, in a very recent paper[3] Dr. Kolhörster again holds that all his observations upon mountains have confirmed the results of his balloon observations, while Hess[4] also holds that the result obtained in Kolhörster's balloon-flights is more trustworthy than that given by Millikan and Bowen's 1922, sounding balloon observations.

The Pike's Peak work of Millikan and Otis[2] had shown, however, (1) that the *mean* absorption coefficient of the rays found on top of the peak was only about that of ThD, and (2) that cosmic rays producing 2 ions per cc at the earth's surface and having an absorption coefficient even as low as $\mu = 2.5 \times 10^{-3}\text{cm}^{-1}$, although no longer in conflict with the sounding balloon experiments would of necessity have produced a 50% larger change inside lead screens in going from sea level to Pikes Peak than that they observed. They concluded, therefore, that cosmic rays *of the assumed characteristics did not exist.* If any of the penetrating rays were of cosmic origin they had to be still harder. The whole of the Pikes Peak data could in fact be explained without them. Accordingly we planned for the summer of 1925 new experiments designed:

(1) To settle definitely the question of the existence or non-existence of a small, very penetrating radiation of cosmic origin—a radiation so hard as to be uninfluenced by, and hence unobservable with the aid of, such screens as we had taken to Pikes Peak; and

(2) To throw light on the cause of the variation with altitude of the radiation of about gamma-ray hardness which our absorption experiments on Pikes Peak showed to be more than twice as copious there as at Pasadena.

The only possible absorbing material obtainable in the immense quantities needed, and of homogeneous and non-radioactive constitution, were the waters of very deep snow-fed lakes—snow-fed because the

[3] Kolhörster, Die Naturwissenschaften 15, 31, 426.
[4] Hess, Phys. Zeits. 27, 405 (1926).

results of under-water experiments which we had previously carried on near Pasadena had been vitiated by our discovery that the waters were appreciably radioactive. We felt that there was much uncertainty as to how much this cause might have affected the European observations in and about glaciers. Further, since the Pikes Peak experiments had demonstrated that if any of the penetrating rays were of *cosmic origin*, the ionization due to them in our electroscope at sea level had to be less than the 2 ions, assumed above, out of the 11.6 observed, the experimental error being, say, half an ion, no crucial tests could possibly be made unless we could find very deep, non-radioactive lakes at very high altitudes where cosmic rays, if they existed, had two or three times the ionizing effect to be expected from them at sea level. We needed at the least three or four ions due to cosmic rays, to vary with absorbing materials, if we were to obtain *unambiguous* evidence.

II. The Electroscopes

The two electroscopes used in these experiments are shown in Fig. 1. Electroscope No. 1 is the same as that used in the experiments described

Fig. 1. Photograph of electroscopes 1 and 3.

in Part II,[2] but with new fibres inserted, while electroscope No. 3 is a new one very much like the first save that it had a greater sensitivity because of a larger volume and a smaller electrical capacity. It was 29.5 cm high and 15 cm in diameter. It was built entirely of brass, side walls 3 mm thick, and had a volume of 3211 cc, 1.69 times that of No. 1. The electrical capacity of No. 3 was 1.10 e.s. units, as measured by the method of mixed capacities, a small condenser the capacity of which (15.85 e.s. units) could be computed accurately from its dimensions being used as a standard. The capacity of No. 1, as remeasured for the purposes of these experiments, was 1.41 electrostatic units. The method of measurement was precisely that described in Part II except that much longer periods of observation, from 5 to 14 hours, were generally used. As described in Part II a calibration curve was drawn for each reading to avoid errors due to changing characteristics of the instrument. That saturation was obtained was shown by the fact that for long observations in which the potential fell from say 200 volts to 50 volts the ionization was not appreciably less than when, with the same external radiation, the fall of potential was from 200 to 150 volts.

III. Experiments in Muir Lake and Arrowhead Lake

The foregoing electroscopes were taken first to Muir Lake, 11,800 feet above sea level, just under the brow of Mount Whitney, the highest peak in the United States, a beautiful snow-fed lake hundreds of feet deep and some 2000 feet in diameter. Here we worked for the last ten days in August, 1925, sinking our electroscopes to various depths down to 67 feet. *Our experiments brought to light altogether unambiguously a radiation of such extraordinary penetrating power that the electroscope-readings kept decreasing down to a depth of 50 feet below the surface.* The atmosphere above the lake was equivalent in absorbing power to 23 feet of water, so that here were rays so penetrating that, if they came from outside the atmosphere, they had the power of passing through $50+23=73$ feet of water, or the equivalent of 6 feet of lead, before being completely absorbed. The most penetrating x-rays that we produce in our hospitals cannot go through half an inch of lead. Here were rays at least a hundred times more penetrating than these. This was in agreement qualitatively with Kolhörster's 1923 contention. The absorption coefficient however came out but one twentieth, instead of "about one-tenth of that of the hardest known gamma rays,"[5] and the number of ions at sea level was but 1.37 (see below).

How unambiguous was now the experimental evidence may be seen from the fact that with the aid of the new electroscope of high sensitivity (because of small capacity and large volume) the change in ions per cc per sec. in going from the surface of Muir Lake to the depth of 15 meters (50 feet) was from 13.3 ions to 3.6 ions, or a decrease to about a fourth value. Since the largest decrease below a surface reading reported by Kolhörster due to sinking electroscopes in water[5] was 2.1 ions, or a decrease of probably about 10%, we seem here to have obtained an altogether new precision of measurement and unambiguity of evidence.

[5] Kolhörster, Sitz.-Ber. Preuss. Akad. Wiss. 34, 366 (1923).

To obtain definite evidence as to whether these very hard rays were however *of cosmic origin*, coming in wholly from above and using the atmosphere merely as an absorbing blanket, we next went to another very deep snow-fed lake, Lake Arrowhead in the San Bernardino mountains, 300 miles farther south and 6700 feet lower in altitude, where the Arrowhead Development Company kindly put all their facilities at our disposal. The atmosphere between the two altitudes has an absorbing power equivalent to about 6 feet of water. *Within the limits of observational error, every reading in Arrowhead Lake corresponded to a reading 6 feet farther down in Muir Lake, thus showing that the rays do come in definitely from above, and that their origin is entirely outside the layer of atmosphere between the levels of the two lakes.* This, taken together with the sounding-balloon data, appears to eliminate completely the idea that the penetrating rays may have their origins in thunder-storms, a possibility recently suggested by C. T. R. Wilson and repeated by Eddington.[6]

The procedure in taking these readings was to take the electroscopes out in a canvass army boat, carried part way up to the lake by pack animals and partly by ourselves, to sink both electroscopes side by side at the chosen depth, and leave them so immersed for a period of from 6 to 14 hours. We could usually obtain but two readings in 24 hours. The process of taking and of treating these readings was precisely that described in II. The elaborate precautions for eliminating leak over the supporting quartz rod were not used because they were found to make no change in the rate of discharge.

Table I

Readings in Lakes Muir and Arrowhead
Electroscope No. 3

Muir Lake									
Depth below surface (m)	0	.45	1.0	2.8	3.0	5.0	10.0	15.0	20.0
Ionization (ions/cc/sec)	13.3	9.7	7.7	6.0	5.45	4.9	4.0	3.6	3.6
	13.2	....	7.8	5.8	....	4.6	4.0	....	3.7
Means	13.25	9.7	7.75	5.9	5.45	4.75	4.0	3.6	3.65
				↙	↙	↙			
Arrowhead									
Depth below surface (m)	0	.7	1.0	1.1	3.0	5.0	....	15.0	....
Ionization (ions/cc/sec)	7.0	5.8	5.5	5.15	4.85	4.4	....	3.7	....
	7.2	....	....	....	4.9	....	....	....	....
	7.5	....	....	....	....	....	....	....	....
	6.9	....	....	....	....	....	....	....	....
	7.2	....	....	....	....	....	....	....	....
		↑	↑		↑				
Means	7.0	5.8	5.5	5.15	4.9	4.4	....	3.7	....

[6] Eddington, Nature Supplement, May 1, 1926, p. 32.

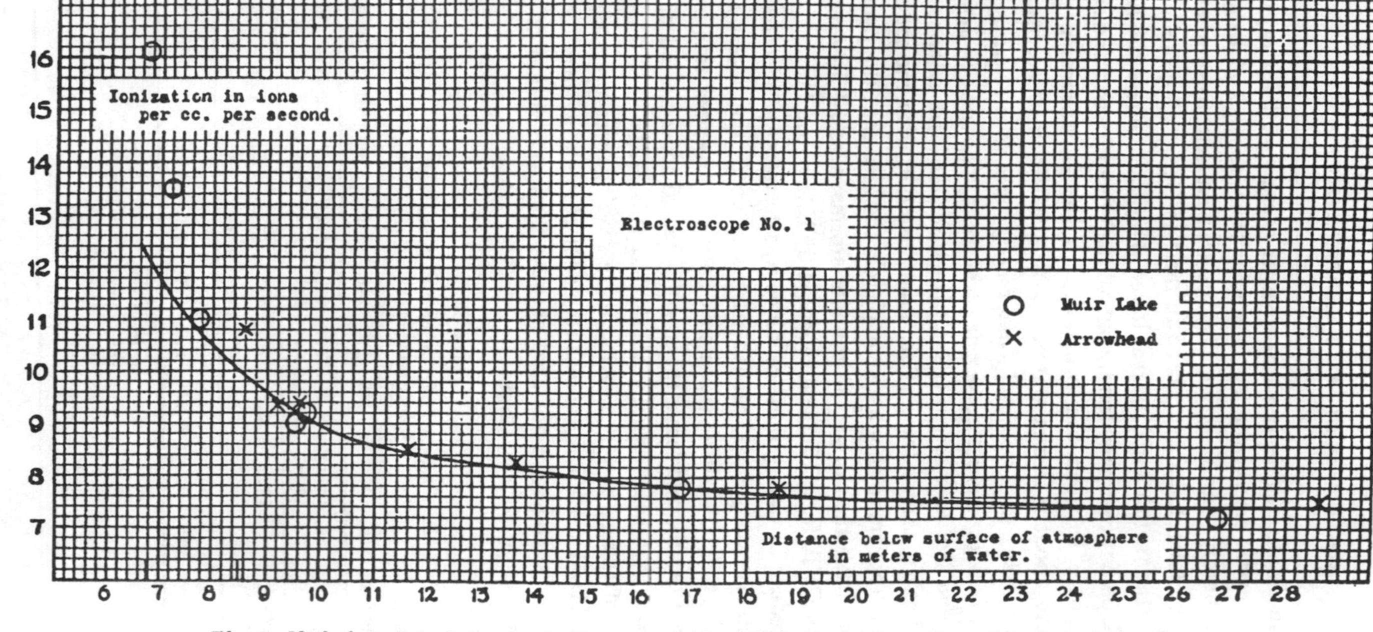

Fig. 3. Variation of the ionization in Electroscope No. 1 with depth below the surface of the atmosphere.

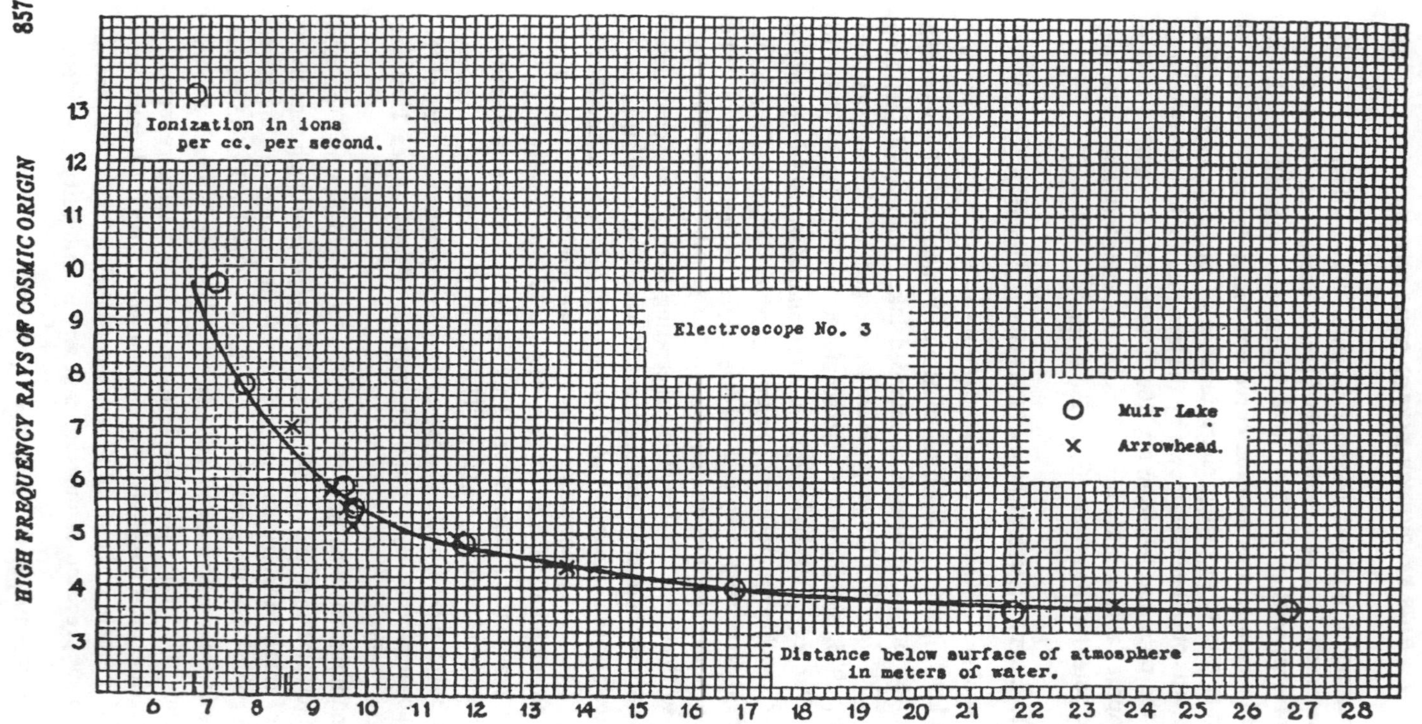

Fig. 2. Variation of the ionization in Electroscope No. 3 with depth below the surface of the atmosphere.

Electroscope No. 1

Muir Lake									
Depth below surface (m)	0	.45	1.0	2.8	3.0	....	10.0	....	20.0
Ionization (ions/cc/sec)	16.5	13.5	11.0	9.0	9.2	....	7.8	....	7.2
	15.8	....	....	....	..	....	....	....	....
	15.9	....	....	....	....	....	....	....	....
Means	16.1	13.5	11.0	9.0	9.2	....	7.8	....	7.2

Arrowhead									
Depth below surface (m)	0	.6	1.0	....	3.0	5.0	10.0		20.0
Ionization (ions/cc/sec)	10.5	9.35	9.6	....	8.6	8.3	7.8	....	7.5
	11.0	....	9.2	....	8.45	....	....	....	....
Means	10.75	9.35	9.4	....	8.5	8.3	7.8	....	7.5

Table I shows all of the readings taken in Lake Muir and Lake Arrowhead. The arrows connect or point toward readings taken at the same depth beneath the top of the atmosphere, and it will be seen that they are all the same within the limits of error whether taken in Lake Muir or Lake Arrowhead.

Figs. 2 and 3 show the curves obtained by plotting all the readings taken in the two lakes as ordinates, and as abscissas the depths in meters beneath the top surface of the atmosphere, reduced to the equivalent depth beneath water. These depths were computed with the aid of the mean temperatures, as a function of altitude, given in the Smithsonian Tables. On these graphs the "depth" beneath the top of the atmosphere of the surface of Muir Lake is 6.75 m, that of Arrowhead Lake 8.6 m, that of Lone Pine 8.5 m, and that of Pasadena 9.98 m. It will be seen that all of the readings corresponding to points more than half a meter beneath the surface of the water fall upon a smooth curve. The fact that readings taken above the surface are all above the curve is due to the presence above the surface in addition to the penetrating radiation of a local radiation of ordinary penetration. Since this latter radiation is all absorbed in a meter or less of water, the points corresponding to depths equal to half a meter are all on the smooth cosmic radiation curve, while those corresponding to readings at the surface are above this curve.

Analysis of these absorption curves shows that the rays are not homogeneous but are hardened as they go through the atmosphere, just as x-rays are hardened by being filtered through a lead screen. Our hardest observed rays have an absorption coefficient of 0.18 per meter of water, and the softest which get down to Muir Lake a coefficient of 0.3 per meter. The sounding balloon experiments of Millikan and Bowen make it improbable that they become very much softer than this at the top of the atmosphere, since otherwise these observers should have obtained larger readings in their very high flight.

Observations carried on by Millikan and Otis day and night for several days on Pikes Peak at an altitude of 14,100 feet, and for two consecutive days on Mount Whitney at an altitude of 13,500 feet had revealed no preferential direction in the heavens from which the rays come. These results were again checked in this work both on Mount Whitney and at Arrowhead Lake. One reading taken in a valley when the Milky Way was practically entirely behind the hills was not at all lower than when the Milky Way was overhead. *Within the limits of our uncertainty of measurement, then, these rays shoot through space equally in all directions.*

IV. Method of Obtaining Absorption Coefficients

In making the foregoing analysis for absorption coefficients the rays were assumed, for the reasons just given, to enter the atmosphere equally from all directions. Then the differential equation for the intensity I at a distance H beneath the surface, in terms of the intensity I_0 coming into the atmosphere from all directions outside its upper surface is

$$dI = 2\pi I_0 \sin\theta d\theta e^{-\mu H \sec\theta} \quad (1)$$

Therefore

$$I/I_0 = 2\pi \int_0^{\pi/2} \sin\theta e^{-\mu H \sec\theta} d\theta. \quad (2)$$

Putting $x = \sec\theta$ this takes the form

$$I/I_0 = 2\pi \int_1^{\infty} \frac{1}{x^2} e^{-\mu H x} dx. \quad (3)$$

Now Gold[7] has published a table of values of an integral of the type in Eq. (3) so that from this table it was possible to obtain the absorption coefficients of rays coming in from all directions. The method of doing this was to select the most reliable observed value near the top of each curve and to see what value of μ in the Gold table best reproduced the portion of the curve near it.

As stated above, however, *no one coefficient would fit the whole curve.* This result is completely new, we think, even as a suggestion, for heretofore it has been the uncertainty of measurement which has made the reported values of μ fluctuate from say .16 up to .57. Here, however, the variation from .18 up to .30 *represents the discrimination of measurement, rather than the uncertainty of reading.* Indeed, no previous observers had worked with the foregoing Gold law for the evaluation of μ since the uncertainty of measurement had theretofore made it useless to attempt to discriminate between rays following a linear absorption law and rays

[7] Gold, Proc. Roy. Soc. A82, 152 (1909).

coming in from all directions. The radiations have clearly become more penetrating with depth. In other words, the radiation is not homogeneous but consists of a spectrum of wave-lengths. For electroscope No. 3 the upper portion of the curve gave, as stated, $\mu = .30$ per meter of water, and the lower end $\mu = .18$. Electroscope No. 1 gave the same result for the lower end of the graph and a value slightly less for the upper end. Electroscope No. 3 was the more dependable since it had about double the sensitivity of No. 1. These coefficients of course characterize the radiation only throughout the region studied. Somewhat softer components are to be expected at greater altitudes.

V. Check Observations with Lead Screens

The same lead screens, 4.8 cm thick, used for making absorption experiments upon the radiations found about the granite rocks on top of Pikes Peak (see Part II) were taken to Muir Lake and observations similar to those there made repeated also upon granite rocks, both at Muir Lake (11,800 feet) and at Lone Pine (5500 feet). Then the instruments were brought back to Pasadena (759 feet) and similar observations made there on a soil consisting of decomposed granite. The lead was adapted for use with electroscope No. 1 alone and was the equivalent in absorbing power of 55 cm of water. The results of all these absorption experiments are collected in Table II.

Table II

Cosmic rays inside 4.8 cm lead

	Pasadena altitude 305 m	Lone Pine altitude 1676 m	Muir Lake altitude 3590 m	Pikes Peak altitude 4298 m
Ions per cc per sec. unshielded	13.0	16.7	20.0	23.7
Shielded with 4.8 cm Pb	9.0	10.1	11.8	12.6
External radiations after screening.	1.6	2.7	4.4	5.2
Cosmic rays (theoretical)	1.3	2.4	4.1	4.9
Cosmic rays (observed)	1.32	2.21	4.08	5.0

In the first and second rows are given the readings without and with the lead shield, respectively. The figures in the third row are obtained by subtracting from those in the second row 7.4, which is seen from Fig. 3 to be the residual ionization in electroscope No. 1 when it is screened from all external radiation by being sunk to a depth of more than 50 feet in water. The third row, therefore, gives the total radiations of all kinds which penetrate at the various altitudes inside the lead screen.

A part of this radiation which gets through the lead is certainly due to the radioactive constituents of the surrounding rocks. Since these rocks were as nearly the same as possible in all the localities, a large variation

in the effect due to them is not to be expected. Kovarik and McKeehan[8] give the ionization on the earth due to γ rays from igneous rocks as about 3 ions. We have therefore taken 3 ions as a probable mean value of the ionization within the unshielded electroscope due to the gamma ray activity of the rocks. About 10 percent of this is able to get through a screen of water 55 cm thick. The figures in the fourth row are therefore obtained by subtracting .3 ions from those in the third row, and what is left, if the assumption as to the constancy of the radioactivity of the rocks is correct, should be, theoretically, the amount of the cosmic rays (see fourth row labelled "cosmic rays (theoretical)" which get through the lead screen in the various localities.

But now we have the possibility of getting these values in another and quite independent way, namely, by taking the readings on the curve of Fig. 3 at each location, for this gives how much of the cosmic radiation actually is present at each elevation, all soft radiations having been screened out by the water. It is then easy to calculate how much of this cosmic radiation will penetrate 55 cm of water, using the coefficient .3 per meter of water (as we compute from our curve the absorption coefficient of the cosmic rays at this depth to be) and the formula $I = I_0 e^{-\mu d}$. This should be the correct formula for this case, since here the shield completely surrounds the electroscope, and most of the radiation goes through it practically perpendicularly. The results are shown in the last row and are labeled "cosmic rays (observed)." The agreement between the observed and computed values in the fourth and fifth rows, respectively, is excellent and shows that after working out the characteristics of the cosmic rays from the observations in water we can actually compute accurately the amount of these cosmic rays which will be found inside a 4.8 cm lead screen with the aid of the assumption that the only other rays which get through the lead screen are the rays from the radioactive constituents found in granite rocks.

VI. Soft Secondary (?) Rays

The agreement in Table II between the cosmic rays actually observed inside the lead (row 5) and those computed (row 4) on the assumption that the only other rays which can get inside the lead are the gamma rays due to the radioactivity of the granite rocks, assumed to be everywhere the same and equal to 3 ions, is apparently good evidence that the considerable amount of other soft radiations of local origin observed by Millikan and Otis on Pikes Peak is unable to penetrate appreciably 4.8

[8] Kovarik and McKeehan, Report of Committee on X-rays and Radioactivity, National Research Council, Washington, D. C., p. 141 (1925).

cm of lead. Before drawing inferences as to the nature and origin of these new soft rays, Table III is presented to show again their existence and distribution with altitude. The first row of this table repeats the total observed ionization inside the unshielded electroscope in the indicated localities, the second row the readings from the cosmic ray curve, Fig. 3. The differences between these two are found in the third row and represent all the rays which have passed through the walls of the unshielded electroscope No. 1 except the cosmic rays, i.e., all the rays of local origin which enter the electroscope from without.

Table III

Soft secondary radiation

	Pasadena 305 m	Lone Pine 1676 m	Muir Lake 3590 m	Pikes Peak 4298 m
Direct observation	13.0	16.7	20.0	23.7
Cosmic rays (from curve)	8.95	10.0	12.2	13.3
Soft rays (observed)	4.05	6.7	7.8	10.4
Stimulated soft rays (assumed)	1.05	3.7	4.8	7.4
Stimulating cosmic rays (observed)	1.55	2.6	4.8	5.9

Since we have just assumed the radioactive rays from the granite rocks to be responsible for 3 of these observed ions per cc per sec., the difference given in the fourth row represents other soft local rays. However, any error in the assumption of the uniform value 3 for the gamma rays of local origin would vitiate badly the fourth row of Table III, whereas it would have had but a small influence on the fourth row of Table II. The reason for this is that the total effect of these gamma rays inside the lead (fourth row, Table II) is but a few tenths of an ion at most, while in the fourth row of Table III it is ten times as much. The fifth row of Table III gives the actual values of the cosmic rays found within the unshielded electroscope. These are obtained from curve 4 for the various altitudes by subtracting the residual ionization, 7.4 ions, from the curve reading. It will be seen that there is a rough proportionality between the stimulating cosmic rays found in row 5 and the new soft rays shown in row 4. This is perhaps sufficiently good, in view of the aforementioned cause of uncertainty as to the values in the fourth row, to furnish evidence that the new soft rays of the fourth row are produced by the cosmic rays of row 5. Row 4 is therefore labeled "Stimulated soft rays." The argument here, however, is not one of certainty. The observed increase with altitude of the soft rays might be explained by making the unlikely assumption that quite accidentally we were observing on rocks of increasing gamma radiation as we progressed upward. Further experiments are needed to settle this point unambiguously.

VII. The Spectral Distribution of the Observed Cosmic Rays

In order to obtain the spectral distribution of rays such as those here found for which the absorption coefficients are as low as 1/20 of those of RaC or ThD and vary from .3 per meter of water to .18 per meter of water there is as yet no altogether infallible guide. But the very recent experimental work of Ahmad[9] has shown that the gamma rays of radium in their absorption by matter obey the same general law as x-rays, and for these the relation between absorption coefficient and frequency is well known. Compton's theory of scattering predicts Ahmad's observational data very satisfactorily. According to the Compton-Ahmad formula the mass absorption coefficients may be calculated from the formula

$$\frac{\mu}{\rho}=\left\{\frac{\sigma_0}{1+2\alpha}+B\lambda^3Z^3\right\}\frac{ZN}{A} \qquad (4)$$

the first term of which represents "Compton scattering" while the last is "true absorption" (ejection of photo-electrons). For absorption in water this last term is nearly negligible even for gamma-ray wave-lengths, so that it must certainly be negligible for the much harder rays here under consideration, so that for these rays

$$\frac{\mu}{\rho}=\left\{\frac{\sigma_0}{1+2\alpha}\right\}\frac{ZN}{A} \qquad (5)$$

where

$$\alpha=\frac{.0242}{\lambda} \text{ and } \sigma_0=6.64\times 10.^{-25} \qquad (6)$$

Making the substitution of $Z/A=10/18$, its value for water, $N=6.06\times 10^{23}$ and the observed range of absorption coefficients, namely, .0030 and .0018 we obtain $\lambda=.000634$A and $\lambda=.00038$A, respectively, or a spectral range of a little less than an octave in a region of frequencies about 50 times higher than that of the shortest measured gamma rays ($\lambda=.02$A). The foregoing reduction of absorption coefficients to wave-length has been given very considerable credentials by Ahmad's experimental proof of the ability of the Compton theory to predict fairly closely his observed results.[10] Also very nearly the same wave-lengths are obtained from Dirac's relativity-quantum-mechanics formula.[11] This yields about 30 percent lower wave-length values.

[9] N. Ahmad, Proc. Roy. Soc. 109, 206 (1925).

[10] A paper has just appeared by Hoffman, Phys. Zeits. 36, 25 (1926) which lends further support to the reliability of Compton's equations for the purpose in question.

[11] Dirac, Proc. Roy. Soc. 111, 405 (1926).

VIII. Nature of Soft Stimulated Radiations

We observe, first, that according to Compton's equations[12] the mean ratio of true scattering to true absorption—this is the ratio between the mean energy in the scattered quant and the energy in the recoil electron—is given by

$$\frac{\sigma_s}{\sigma_a}=\frac{1+\alpha}{\alpha} \tag{7}$$

Since, for the cosmic rays, λ is of the order .0005 the quantity α is large compared to unity and therefore for this case

$$\frac{\sigma_s}{\sigma_a}=1 \tag{8}$$

This means that on the average each particular act of scattering divides the energy of the original quant equally between the new quant and the recoil electron, and the scattered quant has therefore on the average twice the wave-length, or half the frequency of the original one.

Also, according to Compton's equations the average angle of scattering is given by

$$\lambda-\lambda_0=.0242(1-\cos\theta)$$

and since for the cosmic ray, as just shown, $\lambda=2\lambda_0$ we have

$$(1-\cos\theta)=.0005/.024=.02$$

$$\cos\theta=.98 \text{ or } \theta=11°$$

Further, since the original momentum in the direction of the ray is $h\nu_0/c$ and the momentum remaining in the scattered quant, namely $h\nu_0/c$ is very nearly $h\nu_0/2c$, since θ is small, it follows that the momentum imparted to the recoil electron in the direction of the original ray must also be exceedingly close to $h\nu_0/2c$. In other words, *the act of scattering of these very high frequency rays consists merely in taking half the energy of the light quant and transferring it to a recoil electron, both this new light-quant and the electron moving practically straight forward in the direction of the original beam, each with half the original energy.*

Altogether without reference to Compton's theory, the fact that the electrons do actually move more and more nearly straight forward as the frequency of the ray increases is shown directly by the C. T. R. Wilson photographs, so that the qualitative correctness of the foregoing conclusion can scarcely be doubted.

The foregoing equations show that contrary to Eddington's assumption[13] very high frequency cosmic rays do not degenerate in one scattering

[12] A. H. Compton, Phys. Rev. 21, 494 (1923).

[13] Eddington, Nature 117, 31 (1926).

act into rays of gamma-ray frequency as they would if, on the average θ were 90°, but instead, since θ is always very small, they soften to just one-half frequency at each act. Since further, in the process of the degeneration, the energy left in the ether-wave is always proportional to its frequency, when a cosmic ray of wave-length $\lambda=.0005$A has degenerated to an ordinary gamma-ray of wave-length $\lambda=.025$ it has left in it but 1/50th of the original energy. This shows, for example, *that if at a given altitude the cosmic rays produce say 5 ions per cc per sec. there is a wholly negligible ethereal radiation of gamma ray hardness mixed with them.* Indeed, until the radiation has become pretty well absorbed (reduced to less than half its original energy) the bulk of the ionization is due to the primary rays, a smaller part to the secondaries (of half the original frequency) a smaller part still to the tertiaries (of one-fourth the original frequency), and a very small part to all the other members of the series. In other words, while cosmic rays diminish in energy as they go through matter because some of the quanta are removed from the beam by scattering, they soften, or diminish in frequency, very little before the intensity of the beam has been reduced to a small fraction of the original value. Even then it is the secondaries and tertiaries which carry the bulk of the energy, so that the beam has not, on the average, degenerated to anything like gamma-ray hardness. Indeed, a more complete analysis on the basis of the Compton equation shows that *no matter how much the intensity of an originally monochromatic beam has been reduced by passage through matter, of whatever thickness, more than three-fourths of the resultant ether-wave energy is carried by the primaries, secondaries, and tertiaries whose frequencies are respectively 1, 1/2 and 1/4 times the original frequency.* We cannot, therefore, seek the source of the rays "of gamma-ray hardness" found on Pikes Peak in cosmic rays degenerated into actual gamma-rays by Compton scattering.

On the other hand, the following analysis shows that the observed soft rays are in part, at least, the β rays produced by the cosmic rays. For one-half of the incident energy in each cosmic ray goes over in each scattering act into the recoil electron. The highest frequency cosmic ray observed ($\lambda=.00038$) has an energy-value corresponding to the fall of an electron through about 30,000,000 volts. Hence the beta-rays produced by the impact of these with electrons in Compton scattering have an energy of about 15,000,000 volts. The velocity of a 7,500,000 volt beta-ray, in terms of the velocity of light, is .998,[14] and since volts vary as $1/1-\theta^2$, which is proportional, for $\beta\doteq 1$, to $1/\sqrt{2(1-\beta)}$ for 15,000,000 volts rays $\beta=.9995$.

[14] National Research Council, Bulletin on Radioactivity, p. 92 (1925).

Now Bohr[15] has worked out the average range of β-rays as given in Table IV, finding the range proportional to $1/\sqrt{1-\beta^2}$ and Varder[16] has given Bohr's formula experimental verification. It will be seen from the formula and the table that for large velocities the range is proportional to the energy.

TABLE IV

Showing variation of average range with velocity

Velocity of corpuscle / Velocity of light	Average range meters
0.80	0.7
0.85	1.1
0.90	1.9
0.95	3.5
0.99	10.5
0.996	18.0
0.998	26.0
0.9995	52.0

We see, therefore, that beta rays having a range of 52 m in air, equivalent to 5.1 mm of brass, are produced by each act of scattering of the initial cosmic rays. *These are undoubtedly a part, at least, of the soft rays found on Pikes Peak.* In a preceding communication[17] these have been referred to merely as "rays of about gamma-ray hardness," not as actual gamma-rays, though various authors have so understood them.

IX. ORIGIN OF THE COSMIC RAY

It is altogether obvious that any rays of the hardness and distribution indicated, and of cosmic origin, must arise from nuclear changes of some sort going on all about the earth. The energy of the change involved is, however, four times that of any radioactive change thus far on record, being equivalent, for rays of wave-length $\lambda = .00038$ to the fall of an electron through a potential difference of 32,400,000 volts, and for rays of wave-length $\lambda = .000634$ to 19,500,000 volts. The fastest β-ray on record has an energy of 7,500,000 volts.

Both Eddington[18] and Jeans[19] wish to regard these observed cosmic rays as arising from the transmutation of the mass of the proton into radiation by the union of a proton with a negative electron. They regard this process as going on both in the nebulae and in the interior of stars. Such a process, however, would produce a ray of wave-length .000013A, which would be thirty times more energetic and more penetrating than the shortest wave-length which we have observed. This hypothesis does not seem to be tenable if the Compton equations are to be taken as guides, for, as already indicated, rays having that energy could not be softened to an average one-thirtieth of their original frequency by passage through any amount of matter. Their energy could all be dissipated in heat through the beta-rays, but such radiation as got out undissipated would have a considerable fraction of its energy in the original frequency.

Again, if rays thirty times the hardness of the rays observed were present we should not have found in two lakes our electroscopes reaching a constancy of reading at all depths below fifty feet. The reasons adduced by Eddington for assuming that this process is going on seem good, but its seat, if it exists, is presumably in the interiors of stars alone where the energy of the change is all frittered away into heat, through the medium of the beta-rays, before any appreciable part of it has found its way out into space. *The cosmic rays are probably, therefore, not degenerated waves of higher frequency, but are rather generated by nuclear changes having energy values not far from those recorded above.* These changes may be (1) the capture of an electron by the nucleus of a light atom, (2) the formation of helium out of hydrogen, or (3) some new type of nuclear change, such as the condensation of radiation into atoms. The changes are presumably going on not in the stars but in the nebulous matter in space, i.e., throughout the depths of the universe.

SUMMARY

The advances made in these researches seem to us to be

(1) The increased precision, definiteness, and unambiguity with which the properties of the penetrating rays have been brought to light.

(2) The definite proof that some of these rays come from above, the 6700 feet of atmosphere between 11,800 and 5,100 acting merely as a blanket equivalent to six feet of water. This is by far the best evidence found so far for the view that the penetrating rays are partially of cosmic origin.

(3) The bringing forth of evidence for the spectral distribution of cosmic rays and the rough determination of the frequency limits of the spectrum. This is altogether new.

(4) The bringing forth of evidence for the existence of a secondary very penetrating beta radiation stimulated by the primary cosmic rays.

(5) The fixing of the ionization at the earth's surface due to cosmic rays, as measured inside electroscope No. 1 at about 1.4 ions.

The whole of this cosmic ray work has been done with the aid of funds provided by the Carnegie Corporation of New York and administered by the Carnegie Institution of Washington.

NORMAN BRIDGE LABORATORY OF PHYSICS,
CALIFORNIA INSTITUTE OF TECHNOLOGY, PASADENA.

August 7, 1926.

[15] N. Bohr, Phil. Mag. 30, 518 (1915).
[16] Varder, Phil. Mag. 29, 731 (1915).
[17] Millikan, Proc. Natl. Acad. Sci., January (1926).
[18] Eddington, Nature 117, 26 (1926).
[19] Jeans, Nature 116, 861 (1925).

THE PHYSICAL REVIEW

A Journal of Experimental and Theoretical Physics

VOL. 43, No. 2 JANUARY 15, 1933 SECOND SERIES

On Compton's Latitude Effect of Cosmic Radiation

G. LEMAITRE AND M. S. VALLARTA, *University of Louvain and Massachusetts Institute of Technology*

(Received November 18, 1932)

By considering the influence of the earth's magnetic field on the motion of charged particles (electrons, protons, etc.) coming to the earth from all directions in space, it is shown that the experimental variation of cosmic-ray intensity with magnetic latitude, as found by Compton and his collaborators, is fully accounted for. The cosmic radiation must contain charged particles of energy between limits given in the paper. The experimental curve may be represented by a suitable mixture of rays of these energies, but it is not at all excluded that a part of the radiation may consist of photons or neutrons. For predominantly negative particles there must be in the region of rapidly varying intensity a predominant amount of rays coming from the east, and conversely for positive rays. Because of the fact that in regions near the magnetic equator there is a predominance of rays coming nearly horizontally, the absorption by the atmosphere may be increased. Finally the fact that Compton's result definitely shows that the cosmic rays contain charged particles gives some support to the theory of super-radioactive origin of these rays advanced by one of the present authors.

I.

IN the course of a survey of the intensity of cosmic radiation at a large number of stations scattered all over the world, A. H. Compton and his collaborators[1] discovered the remarkable fact that while the intensity is nearly constant for latitudes north of 34° in the American continent and south of 34° in Australasia, it drops sharply between these latitudes to a value about 87 percent as great as that for high latitudes, reaching a minimum at or near the magnetic equator. A close correlation with magnetic latitude was also found. These results are in agreement with those of J. Clay and H. P. Berlage.[2] This discovery rules out the hypothesis that the cosmic radiation consists of photons alone and suggests that it is made up at least partly of electrons, protons or other charged particles. The question as to the origin of these particles remains as yet unanswered; it is very likely bound up with general cosmogonical problems, an hypothesis as to which has already been advanced by one of the present authors.[3]

It is clear that the latitude effect discovered by Compton is attributable to the charged components of the cosmic radiation alone, so that the problem arises as to whether the experimental results can be accounted for by considering the influence of the earth's magnetic field on the motion of such particles.[4] This influence has already been extensively treated by Carl Störmer[5] in connection with his investigations on the origin of the aurora borealis. It will be shown in the present paper that the experimental variation of intensity with latitude is fully accounted for if the cosmic radiation consists at least in part of electrons (or protons) of energy of the order of 10^{10} electron-volts, coming to the earth from all directions in space.

[1] A. H. Compton, Phys. Rev. 41, 111 (1932); 41, 681 (1932); also a paper presented at the Chicago meeting of the American Physical Society, November 25, 1932 (Abstract in Bull. Am. Phys. Soc. 7, 13 (1932)). See also R. D. Bennett, J. L. Dunham, E. H. Bramhall and P. K. Allen, Phys. Rev. 42, 446 (1932).

[2] J. Clay and H. P. Berlage, Naturwiss. 20, 687 (1932). Reference is also made in this letter to a paper by A. Corlin which unfortunately is unavailable to the authors.

[3] G. Lemaitre, Nature 128, 704 (1931).

[4] Qualitatively, the latitude effect was predicted by W. Heisenberg at the end of his paper *Theoretische Überlegungen zur Höhenstrahlung*, Ann. d. Physik 13, 430 (1932).

II.

Since the force is perpendicular to the path of a charged particle moving in a magnetic field, the kinetic energy is constant and the speed is constant. The relativistic mass is therefore also constant and the motion may be treated by the methods of classical dynamics. Since a Hamiltonian exists (the relativistic Hamiltonian for motion in a magnetic field), Liouville's theorem is applicable.[6] If now we assume that the intensity distribution of the cosmic radiation at infinity is homogeneous and isotropic, the intensity in all allowed directions at any point in the earth's magnetic field is, by Liouville's theorem, the same. Thus the question of calculating the intensity at any point on the earth's surface reduces to that of finding out in which directions particles coming from infinity can reach that point. There are, as we shall see, three possibilities: either all directions are forbidden, or all directions are allowed, or only certain directions are allowed and the rest forbidden. At all points belonging to the last category there is a cone which encloses all directions in which trajectories issuing at infinity can reach the point in question. For particles of any given kinetic energy our main problem is thus the determination of this cone, which in the cases of points on the earth's surface belonging to the first two categories may be completely closed or completely open. After the cone is found the intensity at the corresponding point can be immediately calculated by computing the solid angle of the cone.

[5] Carl Störmer, Zeits. f. Astrophys. 1, 237 (1930). References to his previous work are given at the end of this paper. See also Zeits. f. Astrophys. 3, 31 (1931) and a discussion by E. Brucke, Phys. Zeits. 32, 31 (1931).

[6] See for example E. T. Whittaker's *Treatise on Analytical Dynamics*, 2nd edition, p. 284.

III.

We use spherical coordinates r, φ, λ, where r is the distance from the center of the earth, φ the longitude counted positively towards the east, and λ the magnetic latitude. We assume as a first approximation that the earth's magnetic field may be represented by the field of a dipole of moment M at the center of the earth, with its axis towards the magnetic poles.[7] The components of the earth's magnetic field in the direction of r increasing and λ increasing are, respectively,

$$H_r = (2M/r^3)\sin\lambda, \quad H_\lambda = -(M/r^3)\cos\lambda.$$

The equations of motion are then,

$$(m/eM)(r\ddot{\lambda}+2\dot{r}\dot{\lambda}+r\dot{\varphi}^2\sin\lambda\cos\lambda) = -(2/r^2)\dot{\varphi}\sin\lambda\cos\lambda, \quad (1)$$

$$(m/eM)(\ddot{r}-r\dot{\lambda}^2-r\dot{\varphi}^2\cos^2\lambda) = -(\cos^2\lambda/r^2)\dot{\varphi}, \quad (2)$$

$$\frac{m}{eM}\frac{1}{r\cos\lambda}\frac{d}{dt}(r^2\dot{\varphi}\cos^2\lambda) = \frac{2\dot{\lambda}}{r^2}\sin\lambda + \frac{\dot{r}}{r^2}\cos\lambda, \quad (3)$$

where m is the relativistic mass at the constant speed corresponding to the kinetic energy of the particle, e is the charge on the particle and the dots have their usual meaning. The last equation is immediately integrated, yielding as its first integral,

$$-(m/eM)r^2\dot{\varphi}\cos^2\lambda = 2\gamma + (1/r)\cos^2\lambda, \quad (4)$$

where γ is an integration constant which in our physical problem is proportional to the φ-component of the moment of momentum of the particle at infinity. Since the particle may be moving there in any direction, γ may have all values from $-\infty$ to $+\infty$. It follows further that its motion in the magnetic field of the earth can be split up into two motions as already noted by Störmer[5]: a motion in a meridian plane, and a motion of rotation of the meridian plane about the magnetic axis.

The inclination θ of the particle's path with respect to the meridian plane is given by[4]

$$\sin\theta = (r\cos\lambda/v)(d\varphi/dt), \quad (5)$$

[7] The introduction of the real magnetic field of the earth as determined empirically from measurements on the earth's surface, by the method of Gauss, can be made by using the results of A. Schmidt, Zeits. f. Geophys. 2, 38 (1926).

where v is the velocity, and from the conservation of kinetic energy we have,

$$\dot{r}^2+r^2\dot{\lambda}^2+r^2\dot{\varphi}^2\cos^2\lambda=v^2. \quad (6)$$

From (2) and (4) we find, eliminating $\dot{\varphi}$,

$$\tfrac{1}{2}d(r\dot{r})/dt-v^2=(e^2M^2/m^2)(1/r^4)(2\gamma r+\cos^2\lambda) \quad (7)$$

and therefore, by integration,

$$r^2\dot{r}^2-v^2r^2=\frac{e^2M^2}{m^2}\left[-\frac{4\gamma}{r}+2\int\frac{\cos^2\lambda dr}{r^3}-C\right]. \quad (8)$$

From (2), (4), (6) and (8) we have, eliminating $\dot{\varphi}$ and t, and dividing by (8),

$$\frac{r^2d\lambda^2}{dr^2}=\frac{-4\gamma^2/\cos^2\lambda-\cos^2\lambda/r^2+C+2\int\cos^2\lambda dr/r^3}{v^2m^2r^2/e^2M^2-4\gamma/r-C+2\int\cos^2\lambda dr/r^3}, \quad (9)$$

which when integrated by parts gives

$$\frac{r^2d\lambda^2}{dr^2}=\frac{-4\gamma^2/\cos^2\lambda+C+\int\sin 2\lambda d\lambda/r^2}{v^2m^2r^2/e^2M^2-4\gamma/r-C-\int\sin 2\lambda d\lambda/r^2-\cos^2\lambda/r^2}. \quad (10)$$

These equations may be more readily discussed by using a normalized coordinate

$$x=(mv/\pm eM)^{1/2}r, \quad (11)$$

where the sign in the denominator is to be taken either plus or minus according to whether we are dealing with positive or negative particles. In terms of the kinetic energy of the particle measured in electron-volts (11) may be written

$$x=r(V/300McZ)^{1/2}(1+600m_0c^2/\epsilon V)^{1/4}, \quad (12)$$

where V is the potential measured in volts, c the speed of light in vacuum, $Z\epsilon$ the absolute value of the particle charge, ϵ the electronic charge, and m_0 the rest mass of the particle. Placing for r the radius of the earth (6370 km) we obtain a value x_0 which fixes the scale of our normalized coordinate with respect to the earth and is a measure of the energy of the rays.

Likewise it is convenient to use instead of our γ a new γ_1 defined by

$$\gamma_1=-(\pm eM/mv)^{1/2}\gamma, \quad (13)$$

so that (5) now becomes,

$$\mp\sin\theta=-2\gamma_1/x\cos\lambda+\cos\lambda/x^2, \quad (14)$$

where the minus sign refers to positive particles and the plus sign to negative.

IV.

We now examine the three possibilities mentioned above, and investigate first, for any given kinetic energy, those points on the earth where no particles can arrive. For the sake of concreteness we consider the case of electrons, the discussion being similar for protons or other particles. For any given $x_0<1$ and $\gamma_1>1$ no rays coming from infinity can reach the earth because the domain of admissible values of $\sin\theta$ forms a closed region without any connection with infinity. The limiting value λ_1 of λ is therefore given by Eq. (14) with $\sin\theta=1$ and $\gamma_1=1$, and the region where the rays do not come extends from $\lambda=0$ to $\lambda=\lambda_1$. For $x_0>2^{1/2}-1$ there is no region where the rays are completely excluded.

For latitudes greater than λ_1 and values of $x_0<1$ there are trajectories coming from infinity but they have a limit, and this limiting trajectory must be asymptotic to a periodic orbit. Störmer[8] gave the estimate that no periodic orbit exists for $\gamma_1<0.5$. This estimate can be improved by using Eq. (10). A good approximation of the mean value of x for a periodic orbit can be found directly from (9) in agreement with the numerical computations of Störmer. By neglecting the integral and using a mean value for λ, the constant C in the denominator of Eq. (9) must be so chosen that this denominator has a double root. We adopt as the mean value of $\sin^2\lambda$

$$\sin^2\lambda=\sin^2\lambda_m/2, \quad (15)$$

where λ_m is the maximum value of λ, and estimate λ_m as the inclination of the tangent drawn from the origin to the curve $\sin\theta=1$, in Eq. (14). This condition gives,

$$\cos^2\lambda_m=\gamma_1^2. \quad (16)$$

The fourth degree equation fixing the value of x for the periodic orbit is found to be

$$x^4-2\gamma_1x+1-\sin^2\lambda_m/2=0 \quad (17)$$

and C is then given by,

$$C-1=[2(2\gamma_1x_P-1)(1-\gamma_1x_P)+\sin^2\lambda_m/2]/4\gamma_1x_P^3, \quad (18)$$

[8] Reference 5, p. 248.

TABLE I. *Calculated values.*

λ_m	γ_1	x_P	C
0°	1	1	0
10°	0.978	0.984	1.023
20°	0.911	0.934	1.092
30°	0.806	0.810	1.198
31° 40′	0.783	0.736	1.200

where x_P is the root of (17). Table I gives the collected values of λ_m, γ_1, x_P and C.

As the approximation (16) is in good agreement with Störmer's values for $\gamma_1=0.97$ and $\gamma_1=0.8$ we may use it to determine the value of γ_1 at which it becomes impossible to have a double root. Thus we find the value $\gamma_1=0.783$, which must be very close to the limiting value of γ_1 for which periodic orbits disappear. Therefore for values of $\gamma_1<0.783$ there is no limiting trajectory and the earth is reached by rays from all directions. Just as we have proceeded for the first domain ($\gamma_1=1$) we can now determine, using Eq. (14) with $\sin\theta=-1$ and $\gamma_1=0.783$ a limiting value λ_2 such that for values of λ greater than λ_2 the rays of energy corresponding to x_0 will reach the earth from all directions. This applies only to values of x_0 inside of the periodic orbit. We have carried out the numerical integration of Eq. (10) for $\gamma_1=0.911$, corresponding to $\lambda_m=20°$, beginning at $\log x=-0.04$ and $\lambda=14.14°$ and decreasing values of λ. From the result of this integration we have made estimates of the inclination η with the radius vector of the asymptotic family of trajectories passing through points of coordinates $x=0.5$, 0.6, 0.7, 0.8 and 0.9, both for $\lambda=0°$ (equator) and 10°.

TABLE II. $\lambda=0°$.

x_0	θ_1	θ_2	θ_3	η	$I(\%)$
0.5	0°	21°	61°	62°	65
0.6	−34	−15	10	57	32
0.7	−55	−34	−11	71	17
0.8	−70	−46	−26	83	3
0.9	−82	−52	−46	90	1

TABLE III. $\lambda=10°$.

x_0	θ_1	θ_2	θ_3	η'	η''	$I(\%)$
0.5	− 7°	13°	50°	44°	69°	60
0.6	−41	−20	5	42	59	29
0.7	−65	−39	−15	58	68	16
0.8	−90	−51	−26	79	84	—
0.9	−90	−57	−46	90	90	—

For each pair of values of x_0 and λ we know three points of the cone, i.e., the value of θ_1 for $\gamma_1=1$ for which the angle $\eta=0$, θ_2 for $\gamma_1=0.911$ for which we have the estimated values of η as described above, and θ_3 for $\gamma_1=0.783$ for which $\eta=90°$. From these data we have made a graphical integration of the total solid angle of the cone. For $x_0=0.9$ and 0.8, which are greater than 0.736, i.e., the value of x_0 for the limiting periodic orbit we have to replace θ_3 by the values corresponding to $\gamma_1=0.88$ and $\gamma_1=0.80$, respectively, for which the periodic orbit has the value $x_P=0.9$ and $x_P=0.8$, obtained by interpolation from Table I. For $x_0>1$, x_0 is greater than the corresponding value for any periodic orbit and therefore the rays come at every point from every direction.

Our collected results are given in Tables II to V. The last column labelled I gives the percentage intensity and the columns marked η', η'' refer to the north and south, respectively. It is seen

TABLE IV. $\lambda=20°$ *and* $30°$.

	x_0	θ_1	θ_3	$I(\%)$
$\lambda=20°$	0.5	−30°	25°	44
	0.6	−69	−10	15
$\lambda=30°$	0.5	−90	− 9	14

TABLE V. *The latitudes at which the cosmic-ray intensity would become zero* (λ_1) *and would reach maximum value* (λ_2) *for various equivalent energies.*

x_0	λ_1	λ_2
0.1	64.4°	66.7°
0.2	49.0	57.3
0.3	34.8	50.1
0.4	12.0	44.3
0.5	—	39.9
0.6	—	36.3
0.7	—	33.9

TABLE VI. *Equivalent electron voltages corresponding to various values of* x_0.

x_0	Electrons (10^{10} volts)	Protons (10^{10} volts)	α-particles (10^{10} volts)
0.1	0.0596	0.01722	0.01842
0.2	0.238	0.1618	0.2308
0.3	0.536	0.449	0.760
0.4	0.954	0.861	1.564
0.5	1.490	1.397	2.625
0.6	2.145	2.050	3.928
0.7	2.920	2.823	5.46
0.8	3.821	3.719	7.25
0.9	4.830	4.729	9.27
1.0	5.96	5.85	11.52

that there is a slight predominance of the rays coming from the south. Table VI gives the energy measured in electron-volts equivalent to $x_0=0.1$ to $x_0=1.0$ for electrons, protons and alpha-particles.[9] The curves of variation of intensity with magnetic latitude are plotted from these tables in Fig. 1.

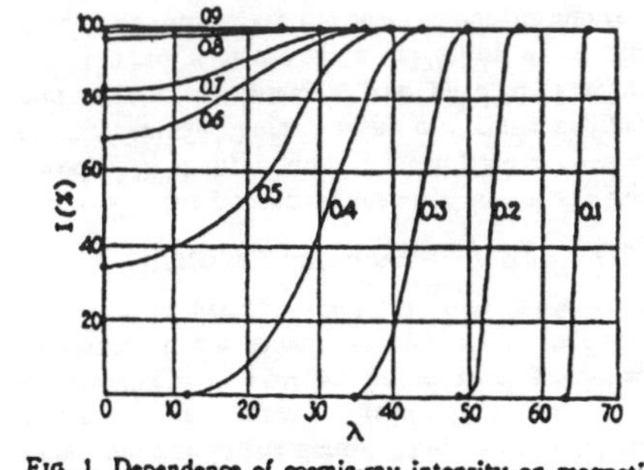

Fig. 1. Dependence of cosmic-ray intensity on magnetic latitude.

V.

From an examination of the curves in Fig. 1 we conclude, first, that the cosmic radiation must contain charged particles of energy between that corresponding to about $x_0=0.3$ and $x_0=0.7$ (see Table VI for equivalent voltages). It seems possible to represent the experimental curve by a suitable mixture of rays of these energies, but it is not at all excluded that a part of the radiation may consist of photons or neutrons.

The sign of the charged particles of the cosmic radiation, or the sign of the predominant part of the rays if they are a mixture of positive and negative particles, is within the reach of experimental detection. For negative particles there must be in the region of rapidly varying intensity a predominant amount of rays coming from the east, and conversely for positive rays. If a large part of the radiation is uncharged this effect may be missed if observations are made too near the magnetic equator.

Because of the fact that in regions near the magnetic equator there is a predominance of rays coming nearly horizontally, the absorption by the atmosphere may be increased. The small southern effect mentioned elsewhere seems to be of the second order and could only be computed by a more refined calculation.

Finally, the fact that Compton's result definitely shows that the cosmic rays contain charged particles gives some experimental support to the theory of super-radioactive origin of the cosmic radiation. In presenting this theory one of the present authors[3] wrote: "I think that a possible test of the theory is that, if I am right, cosmic rays cannot be formed uniquely of photons, but must contain, like the radioactive rays, fast beta-rays and alpha-particles, and even new rays of greater masses and charges. I have shown that the momenta of such rays must be reduced by the expansion (of the universe) in about the same ratio as that of photons."

[9] For the magnetic moment of the earth we take the value 8.04×10^{25} e.m.u. See Handb. d. Physik 15, 288, chapter by G. Angenheister.

THE PHYSICAL REVIEW

A Journal of Experimental and Theoretical Physics

VOL. 43, No. 6 MARCH 15, 1933 SECOND SERIES

A Geographic Study of Cosmic Rays

ARTHUR H. COMPTON, *University of Chicago*

(Received January 30, 1933)

Data are given from measurements of the intensity of cosmic rays by 8 different expeditions at 69 stations distributed at representative points over the earth's surface. Each set of apparatus consisted of a 10 cm spherical steel ionization chamber filled with argon at 30 atmospheres, connected to a Lindemann electrometer, and shielded with 2.5 cm of bronze plus 5.0 cm of lead. Measurements were made by comparing the ionization current due to the cosmic rays with that due to a capsule of radium at a measured distance, the radium standards used with the several sets of apparatus having been intercompared. The method of detecting and correcting for the following disturbing effect is discussed: insulation leak and absorption, local gamma-radiation, radioactive contamination of the ionization chamber, and shielding from cosmic rays by roof and horizon. Intensity *vs.* barometer (altitude) curves are given for various latitudes. These show not only the rapid increase with altitude noted by previous observers, but also the fact that at each altitude the intensity is greater for high latitudes than near the equator. At sea level the intensity at high latitudes is 14±0.6 percent greater than at the equator; at 2000 m elevation, 22 percent greater; and at 4360 m, 33 percent greater. This variation follows the geomagnetic latitude more closely than the geographic or the local magnetic latitude, and is most rapid between geomagnetic latitudes 25 and 40 degrees. Consideration of the conditions necessary for deflection of high-speed electrified particles by the earth's magnetic field indicates that if the cosmic rays are electrons, they must originate not less than several hundred kilometers above the earth. The data can be quantitatively explained on the basis of Lemaitre and Vallarta's theory of electrons approaching the earth from remote space. Acknowledgment is made of the cooperation of more than 60 physicists in this program, 25 of whom are named.

IN the summer of 1930, Professor R. D. Bennett, then of the University of Chicago, Professor J. C. Stearns of the University of Denver, and the writer initiated a coordinated study of the geographical distribution of cosmic rays. While it will be some years before this study is completed, the results already obtained give information whose publication should not be delayed. The present paper is thus a progress report[1] on the findings of our associated expeditions, and gives data which we hope to amplify in subsequent communications.

Previous studies of the relative intensity of the cosmic rays in different parts of the world have been made by J. Clay,[2] who made several trips between Java and Holland, and who found consistently a lower intensity near the equator; Millikan and Cameron,[3] who found but small

[1] Earlier reports of the work of these expeditions have appeared as follows: A. H. Compton, R. D. Bennett and J. C. Stearns, Phys. Rev. 38, 1565, 1566 (1931); ibid. 39, 873; 41, 119 (1932). R. D. Bennett, Technology Review, July, 1932; A. H. Compton, Phys. Rev. 39, 190; 41, 111 and 681; 42, 904 (1932); Scientific Mon., Jan., 1933. A. H. Compton and J. J. Hopfield, Phys. Rev. 41, 539 (1932). J. C. Stearns, W. P. Overbeck and R. D. Bennett, Phys. Rev. 42, 317 (1932). R. D. Bennett, J. L. Dunham, E. H. Bramhall and P. K. Allen, Phys. Rev. 42, 446 (1932).

[2] J. Clay, Proc. Amsterdam Acad. 30, 1115 (1927); 31, 1091 (1928).

[3] R. A. Millikan and G. H. Cameron, Phys. Rev. 31, 163 (1928).

differences between Bolivia and California in their measurements in mountain lakes, and no difference[4] between Pasadena and Churchill, close to the north magnetic pole; Bothe and Kolhörster,[5] who carried a counting tube from Hamburg (53°N) to Spitzbergen (81°N) and back, and who detected no variations in the cosmic rays larger than their rather large experimental error; Kennedy, who under Grant's direction[6] carried similar apparatus from Adelaide, Australia to Antarctica, and likewise found no measurable change; and Corlin,[7] who on going from 50°N to 70°N in Scandinavia found some evidence of a maximum at about 55°N. The prevailing opinion regarding the significance of these measurements has thus been expressed by Hoffmann[8] in a recent summary; "The results so far have on the whole been negative. Most of the observers conclude that within the errors of experiment the intensity is constant and equal, and those authors who do find differences give their results with certain reservations."

In view of the strong indication from the Bothe-Kolhörster double counter experiment[9] that the cosmic rays are high-speed electrical particles, this failure to find a variation of cosmic-ray intensity with latitude was of unusual interest. If the negative results of the experiments could be established with higher precision, it would mean that the cosmic rays could not be electrical particles coming from outside the earth's atmosphere, unless these particles had an unsuspectedly high energy. If, on the other hand, further experiments should confirm the tentative findings of Clay and Corlin, it might be found that a consistent theory of cosmic rays could be built on the assumption of high-speed electrical particles entering the earth's atmosphere.

Organization and Location of the Observing Stations

In order to get as extensive data as possible in the minimum possible time, several expeditions were organized to go into different parts of the world. Seven similar sets of apparatus were constructed, and measurements by the different observers were made with essentially the same procedure. The work has been done not only at sea level, but also at as great a variety of altitudes as possible, in order to learn whether the intensity-altitude relation was independent of the location.

Eight of these associated expeditions have so far reported data. They have been under the direction of:

1. J. C. Stearns and A. H. Compton, working in Colorado and Switzerland.
2. A. H. Compton, in Hawaii, New Zealand, Australia, Panama, Peru, Mexico, northern Canada, Michigan and Illinois.
3. R. D. Bennett, Massachusetts Institute of Technology, in Alaska, California, Colorado and Cambridge.
4. E. O. Wollan, University of Chicago, in Chicago, Spitzbergen and Switzerland.
5. Allen Carpe, of New York City, in Alaska.
6. S. M. Naude, University of Capetown, in South Africa.
7. J. M. Benade, Forman Christian College, Lahore, in India, Ceylon, Malaya, Java, Ladakh.
8. P. G. Ledig, Division of Terrestrial Magnetism, Carnegie Institution of Washington, in South America.

The locations of the major stations where measurements have been made and reported are shown in Fig. 1.

Apparatus

In designing the measuring apparatus, the two main characteristics kept in mind were freedom from sources of systematic error and portability. An ionization chamber was used rather than a counting tube, because of the better reproducibility and the lower statistical error of its readings. The chamber was made small, 10 cm diameter, in order that the weight of the pro-

[4] R. A. Millikan, Phys. Rev. 36, 1595 (1930).

[5] W. Bothe and W. Kolhörster, Berl. Ber. p. 450 (1930).

[6] Kerr Grant, Nature 127, 924 (1931).

[7] A. Corlin, Lund Medd. No. 121 (1930).

[8] G. Hoffmann, Phys. Zeits. 32, 633 (1932). Hoffmann mentions also (without reference) that in their airship flight over the north pole, Malmgrön and Behounek observed the normal cosmic-ray intensity throughout the flight.

[9] W. Bothe and W. Kolhörster, Zeits. f. Physik 56, 751 (1929).

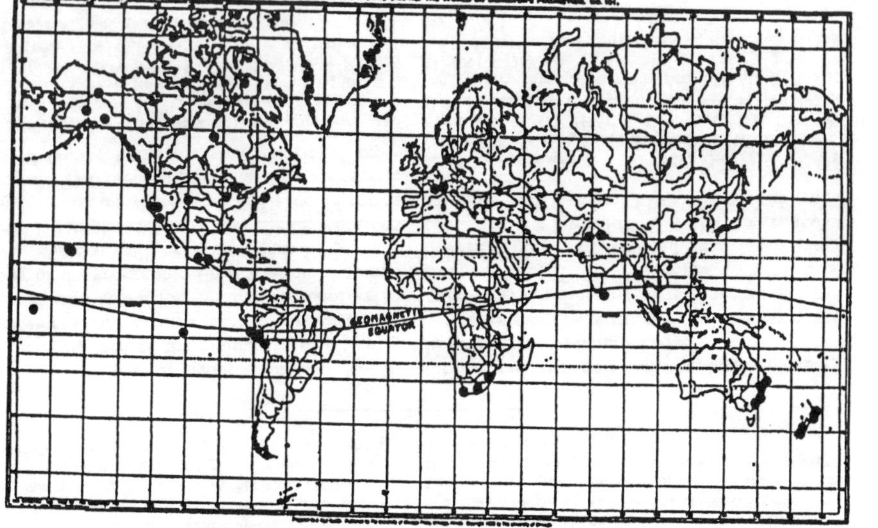

Fig. 1. Map showing location of our major stations for observing cosmic rays.

tecting shields should not be too great for transport by pack trains or porters. By filling the chamber with argon[19] at 30 atmospheres the ionization current was made large enough to be conveniently measurable with a Lindemann electrometer. At each station the ionization due to the cosmic rays was compared with that due to a standard radium capsule. Our measurements were thus independent of the pressure of the gas in the chamber, or of the sensitiveness of our electrical instruments, the reliability of our comparison between different stations depending rather upon the constancy of the radium capsule.

A partly diagrammatic plan of the apparatus is shown in Fig. 2. The ions produced in the argon-filled chamber are collected by a steel rod electrode, which conducts the ionization current to the needle of the electrometer. This electrode is insulated by an amber cone from the brass tube through which it passes. The brass tube serves as a support for the ionization chamber, and remains at ground potential. The chamber containing the key and the connection to the electrometer, though at atmospheric pressure, is kept nearly air-tight with a sponge rubber gasket G, and is dried with a phosphorus pentoxide tube. The electrometer is similarly dried by using a modified form of drying tube. The pressure chambers have shown little if any leak while in service in the field. The various battery potentials indicated in the diagram are supplied by commercial dry batteries. Spherical shells of bronze and of antimony lead encased in thin steel are fitted around the ionization chamber. These shells, each 2.5 cm thick, are sufficient to reduce the intensity of the local gamma-rays to about 5 percent of the cosmic-ray intensity.

For standardizing the instrument, a capsule containing about 1.3 mg of radium enclosed in about 1 cm of lead is placed with its center 1 meter from the center of the ionization chamber. The absolute magnitude of the ionization current due to the radium in this position was measured by the help of a standard cylindrical condenser, which is screwed in place of the grounding key K. This condenser has two pairs of concentric cylinders which may be used alternately. The difference in capacity between the larger and the

[19] Cf. A. H. Compton and J. J. Hopfield, Phys. Rev. 41, 539 (1932).

Fig. 2. Cosmic-ray ionization chamber, electrometer, and electrical connections.

smaller pair can be calculated from the dimensions as 14.95 cm. By comparing the rate of drift of the electrometer needle with and without the condensers, the capacity of apparatus No. 7, for a needle sensitivity of 0.01 volt per division, was found to be 12.72 cm. The ionization current due to the laboratory radium standard at 1 meter produced a potential change of 4.14×10^{-4} volts per second, when the chamber was filled with dry air at 741 mm pressure and 24.5°C, corresponding to 4.626×10^{-4} volts per second filled with air under standard conditions. The volume of the chamber was 430.8 cc. Thus the ionization in standard air due to the laboratory radium standard at 1 meter, through the lead and bronze shields, is,

$$I_s=\frac{1}{300}\times\frac{12.72\times4.626\times10^{-4}}{4.77\times10^{-10}\times430.8}=9.53 \text{ ions cm}^{-3}\text{ sec.}^{-1}. \quad (1)$$

Each of the 7 sets of apparatus has with it its own secondary radium standard, which was compared with the laboratory standard before the equipment was sent out for use. These secondary standards were found to produce respectively in standard air at 1 meter, per cm³ per sec., the values given in Table I.

The Measurements

Each complete set of measurements at a given station consisted of two parts: (1) the determination of the ratio i_2/i_r of the ionization with the radium at a great distance to the ionization with the radium at 1 meter from the center of the ionization chamber, by using both lead shields; and (2) measurement of the ratio i_1/i_2 of the ionization with no radium present when only 1 lead shield surrounds the chamber to that with two lead shields in place. From these two data and the value of I_γ given in Table I can be

Table I. *Ionization by radium capsules.*

Capsule No.	Expedition No.	Ionization (I_γ)
1	3	11.05 ions
2	7	11.8
3	6	12.2
4	5	11.95
5	2 and 8	11.6
6	9	11.55
7	4 and 3	11.85

calculated the ionization due to the cosmic rays (through both lead shields), and the intensity of the local gamma-rays which enter the chamber.

Let i_r be the observed ionization current in arbitrary units, through 2 lead shields, radium at 1 meter; i_1 the ionization through 1 lead shield, radium absent; i_2 the ionization through 2 lead shields, radium absent. The ratio of the ionization due to cosmic and local rays to that due to the gamma-rays alone is then,

$$R_2=i_2/(i_r-i_2)=(i_2/i_r)/(1-i_2/i_r). \quad (2)$$

Also, the ratio of the cosmic rays with 1 shield to the gamma-rays with 2 shields is,

$$R_1=(i_1/i_2)R_2. \quad (3)$$

To correct for the effect of the local radiation we

use the formulas,

$$C=[a/(a-b)](R_2-bR_1)I_\gamma \quad (4)$$

$$L=R_2I_\gamma-C. \quad (5)$$

Here C is the ionization through both lead shields, due to the cosmic rays alone, and I_γ is the ionization by the gamma-rays, so that if I_γ is expressed in ions per cc per second in standard air, C is expressed in the same units.[11] a is the ratio of the ionization due to the cosmic rays when 2 lead shields are used to that when 1 shield is used, and has a value of about 0.9. b is the same ratio for the local gamma-rays, and has a value of 0.286. L represents the ionization due solely to the local rays.

The derivation[12] of Eq. (4) involves the assumption that the ionization due to radioactivity in the chamber itself is negligible. This has been tested with three of our sets of apparatus by placing them in deep tunnels (one in Colorado, one in Thibet and one in Peru). If the assumption is justified, the calculated intensity of the cosmic rays inside the tunnel should be zero. In the Colorado and Thibet measurements it was found to be respectively about 0.2 and 0.1 percent of the intensity outside the tunnel, whereas in the Peruvian measurements there appeared an ionization of about 2 percent, which may have been due to temporary radioactive contamination.

The value of b was determined by using the radium capsule as the source of gamma-rays. In regions where thorium is abundant the resulting value may not be wholly reliable, but approximate tests on the local radiation inside a tunnel in Peru have confirmed the value obtained with radium within experimental error. It was found that the value of a varies so rapidly with altitude that it was necessary to adjust this constant according to the barometric pressure. The values of a used in calculating our results are given in Fig. 3. These values are somewhat arbitrary.

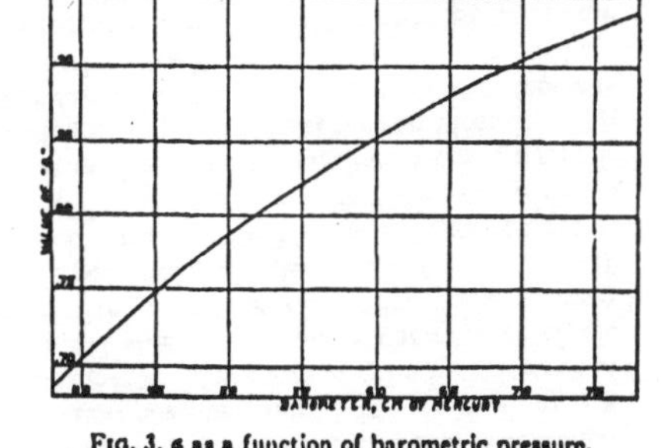

FIG. 3. a as a function of barometric pressure.

Approximate values of a have been determined by measurements of the relative ionization with 1, 2, and 3 shields. Such determinations are however unreliable because the cosmic rays are not exponentially absorbed by the shields. The values given in Fig. 2 are so chosen that the correction for local radiation is on the average about the same for different altitudes. Errors in the value of a will introduce errors in the absolute ionization by the cosmic rays, and in the relative ionization at different altitudes. They will not affect however the relative ionization as calculated at different latitudes.

In any case, since the corrections introduced by Eqs. (3) and (4) for the local radiation are of the order of only 5 percent of the observed intensity, small errors in this correction are not serious.

A typical series of readings consists of: (A) Measurements with the bronze and both lead shields, with and without radium, and with the ionization chamber alternately at +144 and −144 volts. This serves to determine i_2/i_r. (B) Measurements with the bronze and one lead shield, without radium, and with the ionization chamber alternately at +144 and −144 volts. Comparison with A gives i_1/i_2.

Measurements of type A and B are made alternately over a period of from 8 hours to 240 hours, and then averaged. Most of the data reported for a given station represent about 30 hours of readings.

CORRECTIONS

In addition to the correction for local radiation, small corrections were necessary also for the absorption by the roof—if any—protecting the apparatus, and for the shielding due to neighboring mountains or buildings. The roof correction was made by adding to the observed barometric pressure an equivalent to the average weight per cm² of the roof. The maximum correction thus applied amounted to 5.3 mm of mercury. Wherever possible an unprotected site was of course chosen. In no case was the shielding by neighboring buildings significant. In many cases, however, neighboring mountains shielded the apparatus appreciably. A panorama of the altitude of the horizon was then made, and the loss from the shielded zones estimated from Table II. The maximum correction as thus estimated for any station was 10 percent. Since the angular altitude distribution of cosmic rays is known to depend on the elevation, this correction is of course only approximate.

TABLE II.

Degrees from zenith	Normal contribution
0–45°	0.6578
45–60°	0.2105
60–75°	0.1100
75–85°	0.0210
85–90°	0.0007

Several kinds of electrical troubles were encountered. Among these should be mentioned, leakage across the insulation supporting the electrode and the electrometer needle, dielectric absorption by this insulation, ionization in the air filling the electrometer and the tube connecting the electrometer to the ionization chamber, and insulation failures in the auxiliary battery box. Trouble with the batteries could introduce no systematic errors into our measurements, but occasionally reduced their precision. By drying the battery box and connections either with heat or with calcium chloride, the insulation could be kept in satisfactory condition in the dampest weather. Small leaks along the electrometer insulation and ionization in the connecting tube produced effects which average out when a complete series of readings as outlined above is taken. It can be shown that if the resistance of this insulation leak obeys Ohm's law, the average rate from −10 to +10 differs from the rate at 0 by less than 1 percent if the time from 0 to +10 does not exceed that from −10 to 0 by more than 20 percent. Under normal operating conditions the difference was much less than this, and any insulation leak was thus of negligible importance. Tests for such leaks were however a part of our regular routine.

Dielectric absorption showed itself by a faster rate of deflection just after reversing the potential of the sphere than after some minutes had elapsed. The effect could be reduced by careful drying of the air near the electrometer insulation. By using a "null method" of reading, for which each apparatus was wired, the effect of dielectric absorption could be completely eliminated. This method consisted in changing the voltage of the battery C continuously through 6 volts by means of a potentiometer, and thus inducing on the electrode a current which would just balance the ionization current, thus keeping the electrometer needle at zero. In this case the time was noted for which a change of 6 volts was required. This method was however used very little, since the direct deflection method was simpler, and no difference could be detected in the results of the two methods when suitable care was taken.

[11] This statement assumes that the value of R_2 will be independent of the nature and pressure of the gas in the ionization chamber. Its independence of pressure when air or CO_2 is used has been tested and confirmed by Stearns (Phys. Rev. 39, 881 (1932)), Millikan (Phys. Rev. 39, 397 (1932)) and Steinke and Schindler (Naturwiss. 20, 15 (1932)), though Broxon working with nitrogen (Phys. Rev. 40, 327 (1932)) and Hopfield using argon (Phys. Rev. 42, 904A (1932)) find slight differences. If such differences exist, they will require a revision of our *absolute* cosmic-ray intensities, though not of our relative intensities at different altitudes and latitudes.

[12] Let C_1 and L_1 be the intensities of the cosmic rays and the local rays respectively through 1 shell of lead, C_2 and L_2 be the intensities through 2 shells of lead. Then

$$I_1=C_1+L_1 \quad (6)$$

and

$$I_2=C_2+L_2 \quad (7)$$

are the total observed intensities in the two cases, when no radium is present. Also, $a=C_2/C_1$ and $b=L_2/L_1$ are the fractions of the cosmic and the local rays respectively transmitted by the second shell of lead. Thus $C_2=aC_1$ and $L_2=bL_1$. Writing Eq. (7) as

$$I_2=aC_1+bL_1, \quad (8)$$

and combining with (6) we find,

$$C_1=[1/(a-b)](I_2-bI_1).$$

or, since $C_2=aC_1$

$$C_2=[a/(a-b)](I_2-bI_1). \quad (9)$$

However, I_2 is the same as the R_2I_γ, and $I_1=R_1I_\gamma$. Also, C_2 is identical with the C of Eq. (4). Hence,

$$C=[a/(a-b)](R_2-bR_1)I_\gamma, \quad (10)$$

which is Eq. (4) of the text.

We have occasionally noted also that the readings taken immediately after changing the lead and copper shields have shown a slightly higher ionization than occurs after standing for 30 minutes or more. This may be due to a temporary active deposit falling on the outside of the electrically charged chamber when the shields are removed. This may have influenced some of the data taken at Summit Lake in Colorado. The effect is at most a few percent, and seems to be of short duration.

It will be especially noted that the value of the cosmic-ray intensity observed in the field does not depend at all upon the sensitivity of the electrometer nor the precision of our voltmeter, nor, within wide limits, upon the perfection of the insulation. We have relied rather upon the constancy of the radium sample as our only standard.

For measuring the barometric pressure, Paulin barometers and aneroids were chiefly used. Wherever possible, these instruments were compared with mercury barometers, and some boiling point tests were made. On the high mountain measurements, however, it is probable that our barometric errors are more serious than our errors in the cosmic-ray data.

Data

In Table III are recorded the data which have been reported before the end of 1932. All the measurements that have been made are included, except one in Australia, and several near Ceylon and Singapore, for which insulation troubles due to high humidity made the results highly erratic. The tabulated values of i_1/i_r and of i_1/i_2 are merely the averages of the observations, taken as described above, without any corrections. The horizon correction is calculated from Table II, and the values of I_γ are taken from Table I. For expedition 1, the laboratory standard capsule was used at a much shorter distance, thus accounting for the high value of I_γ. The values of C and L are then calculated from these data by using Eqs. (3) and (4). No corrections for radiation from the chamber walls have been applied in the tabulated values of C.

In Fig. 4 are plotted the values of the cosmic-ray intensity C as a function of the barometric pressure, including all the data given in Table III. The circles represent data obtained in the northern hemisphere, and the squares those from the southern hemisphere. The solid dots are values from geomagnetic latitudes[13] higher than 40 degrees, the open dots are between 25°S and 25°N, and the half shaded dots for intermediate latitudes.

In plotting these data, the values of C found by expeditions 1 and 4 have been adjusted so that they can be compared with those from the other expeditions. Expedition 1 used a somewhat larger ionization chamber, filled with air at 30 atmospheres instead of argon, and cylindrical instead of spherical shields. A reliable comparison of the data with the two sets of apparatus is however made possible by comparing the value of $C=6.10$ ions obtained from 240 hours readings at Summit Lake with apparatus number 1 in 1931 with the value of $C=5.70$ ions obtained from a series of readings of equal length at the same location with apparatus number 3. Thus all of the data obtained by expedition 1 have been multiplied by the factor $5.70/6.10=0.935$ to reduce them to the same basis as those obtained by the other expeditions. Expedition 4 used an ionization chamber with which a variety of auxiliary tests had been made over a period of four months. Just before its final filling with argon, it was noted that the residual ionization (radium re-

[13] By geomagnetic latitude we mean the latitude relative to the pole of the earth's uniform magnetization. The north geomagnetic pole, according to Bauer (Terrestrial Magnetism 28, 1 (1923)), is at 78°32′N, 69°08′W. This is not identical with the north "magnetic pole" (90° dip), which Amundsen places at 71°N, 96°W (1903–6). The geomagnetic latitudes here used are calculated by the formula,

$$\sin\lambda=\sin\psi\cos\theta\cos\varphi+\cos\psi\sin\theta, \tag{11}$$

where λ is the geomagnetic latitude, ψ is the colatitude of the north pole of uniform magnetization, θ is the geographic latitude of the point, and $\varphi+69°08'$ is the west longitude of the point.

The "magnetic latitude," as commonly used in papers on terrestrial magnetism, is commonly defined by the formula,

$$\tan\mu=\tfrac{1}{2}\tan\delta, \tag{12}$$

where μ is the magnetic latitude and δ is the inclination or dip of the magnetic needle.

Thus the magnetic latitude is a function of the local magnetic field at the point, whereas the "geomagnetic latitude" depends upon the earth's resultant magnetic moment.

TABLE III. *Cosmic-ray intensities at various locations.*

Expt.	Expd.	Place	Lat.	Long.	Bar. (cm)	i_1/i_r	i_1/i_2	Hor. corr.	I_γ	C	L	C_{76}	C_{60}	C_{45}	Geomag. Lat.	Date
1	1	Mt. Evans	40N	106W	44.7	—	—	1.00	—	7.32	0.13†			6.69	49N	9/31
2	1	Summit Lake	40N	106W	47.4	0.0430	—	.995	137	6.10	.13†			6.71	49N	9/31
3	1	Denver	40N	105W	62.3	—	—	—	—	2.93	—		3.02		49N	9/31
4	1	Jungfraujoch	47N	6E	50.0	.0371	1.344	.944	137	5.45	.16			6.99	49N	10/31
5	2	Haleakala	21N	156W	54.7	.2264	1.504	.99	11.6	3.00	.42		2.42		21N	4/32
6	2	Idlewild	21N	156W	66.0	.1725	1.449	.99	11.6	2.12	.32		2.61		21N	4/32
7	2	Honolulu	21N	158W	76.4	.1304	1.230	1.00	11.6	1.63	.11	1.65			20N	4/32
8	2	SS Aorangi	4S	173W	75.3	.1391	1.500	1.00	11.6	1.65	.20†	1.64			7S	4/32
9	2	Ball Pass	44S	170E	60.2	.2120	1.249	.99	11.6	3.05	.10		3.08		50S	4/32
10	2	Ball Hut	44S	170E	66.3	.1757	1.281	.970	11.6	2.38	.16		3.01		50S	4/32
11	2	Dunedin	46S	170E	76.4	.1460	1.211	1.00	11.6	1.87	.11	1.88			51S	4/32
12	2	Wellington	41S	175E	75.8	.1464	1.235	.99	11.6	1.86	.14	1.86			44S	5/32
13	2	Sydney	34S	151E	76.6	.1614	1.480	1.00	11.6	1.86	.37	1.88			43S	5/32
14	2	Kosciusko	36S	148E	60.6	.2312	1.511	1.00	11.6	2.98	.50		3.06		45S	5/32
15	2	Brisbane*	28S	153E	76.8	.1444	1.510	1.00	11.6	1.61	.35	1.63			37S	5/32
16	2	Guyra*	30S	152E	64.8	.1893	1.232	1.00	11.6	2.60	.11	1.87			39S	5/32
17	2	Auckland	37S	175E	77.5	.1372	1.186	1.00	11.6	1.76	.08	1.81			42S	5/32
18	2	SS Mataroa	13S	106W	76.2	.1300	1.259	1.00	11.6	1.61	.13	1.61			5S	6/32
19	2	Panama	9N	80W	75.5	.1362	1.324	.995	11.6	1.66	.18	1.64			20N	6/32
20	2	Lima	12S	77W	75.0	.1356	1.313	1.00	11.6	1.65	.17	1.62			1S	6/32
21	2	Chosica	12S	77W	68.8	.1541	1.287	.978	11.6	1.99	.16	*1.66			1S	7/32
22	2	Matucana	12S	77W	57.3	.1927	1.243	.924	11.6	2.93	-.06		2.64		1S	7/32
23	2	Chicla	12S	77W	48.6	.2596	1.312	.932	11.6	4.32	.03			5.22	1S	7/32
24	2	Galera	12S	77W	42.8	.337	1.368	.976	11.6	6.03	0			5.24	1S	7/32
25	2	Huantapallacu	13S	74W	43.8	.3221	1.402	.995	11.6	5.41	.12			5.07	2S	7/32
26	2	El Misti	17S	71W	37.7	.4245	1.490	1.00	11.6	8.41	.125			5.10	6S	7/32
27	2	Monte Blanco	17S	71W	42.8	.3500	1.441	.980	11.6	6.15	.21			5.39	6S	7/32
28	2	Arequipa	17S	71W	57.6	.2061	1.300	.995	11.6	2.89	.14		2.63		6S	7/32
29	2	Mollendo	17S	72W	76.8	.1266	1.179	1.00	11.6	1.61	.07	1.63			6S	7/32
30	2	Vera Cruz	19N	96W	76.0	.1306	.998	1.00	11.6	1.68	.06†	1.68			29N	8/32
31	2	Orizaba	19N	97W	66.1	.1619	1.066	.995	11.6	2.19	.06†		2.77		29N	8/32
32	2	Mexico City	19N	99W	58.5	.2039	1.264	1.00	11.6	2.87	.10		2.69		29N	8/32
33	2	Nevado Toluco	19N	100W	46.5	.3204	1.386	.981	11.6	5.44	.13			5.95	29N	8/32
34	2	SS Ocean Eagle	67N	80W	75.3	.1400	1.150	.982	11.6	1.86	.06	1.84			78N	8/32
35	2	Churchill*	59N	94W	76.0	.1386	1.186	.99	11.6	1.80	.09	1.80			69N	9/32
36	2	Otsego Lake	45N	85W	73.2	.1469	1.184	1.00	11.6	1.91	.08	1.80			56N	9/32
37	2	Chicago	42N	88W	74.5	.1514	1.205	1.00	11.6	1.96	.10	1.88			53N	10/32
38	3	Fort Yukon	67N	145W	75.1	.1500	1.190	(1.00)	11.05	1.88	.09	1.84			71N	6/32
39	3	Kennecott	62N	143W	61.0	.2092	1.214	(1.00)	11.05	2.88	.06		3.01		66N	7/32
40	3	Berkeley	38N	122W	75.1	.1465	1.220	(1.00)	11.05	1.81	.11	1.78			46N	7/32
41	3	Tioga Pass	38N	119W	53.0	.2788	1.307	(1.00)	11.05	4.18	.13		3.04		46N	8/32
42	3	Pasadena	34N	118W	73.6	.1537	1.250	(1.00)	11.05	1.88	.14	1.79			42N	8/32
43	3	Denver	40N	105W	62.3	.2026	1.271	(1.00)	11.05	2.69	.14		2.96		49N	8/32
44	3	Summit Lake	40N	106W	48.1	.3466	1.372	(1.00)	11.05	5.70	.20			6.96	49N	9/32
45	4	Chicago	42N	88W	74.5	.1838	1.168	1.00	11.85	2.57	.09	1.80			53N	7/32
46	4	Advent Bay	78N	16E	75.5	.1844	1.152	.98	11.85	2.59	.09	1.86			75N	8/32
47	4	Zurich	47N	9E	71.7	.1930	1.184	.995	11.85	2.74	.10	1.82			49N	10/32
48	4	Eiger Gletcher*	47N	8E	57.4	.2460	1.236	.90	11.85	4.21	.07		3.14		49N	10/32
49	4	Wengen	47N	8E	65.5	.2208	1.195	.99	11.85	3.36	.07		3.20		49N	10/32
50	4	Jungfraujoch	47N	8E	50.1	.3246	1.260	.99	11.85	5.68	.06†			6.84	49N	10/32
51	5	Mt. McKinley*	63N	151W	50.0	.333	1.312	.98	11.95	6.00	.10			8.50	67N	5/32
52	6	Capetown	34S	18E	76.0	.1294	1.234	(1.00)	12.2	1.70	.12	1.70			32S	6/32
53	6	Lootsberg	32S	25E	62.2	.1875	1.292	(1.00)	12.2	2.65	.16		2.91		32S	7/32
54	6	Johannesburg	26S	28E	62.6	.1799	1.224	(1.00)	12.2	2.59	.09		2.88		27S	7/32
55	6	Vinkfontein	27S	30E	63.9	.1749	1.328	(1.00)	12.2	2.38	.20		2.75		28S	7/32
56	6	Pretoria	27S	28E	65.6	.1644	1.252	(1.00)	12.2	2.31	.12		2.81		28S	7/32
57	6	Lodewykslust	27S	31E	67.7	.1560	1.290	(1.00)	12.2	2.09	.16	1.69			28S	7/32
58	6	Nat. Park, Natal	29S	29E	64.9	.1674	1.270	(1.00)	12.2	2.31	.14		2.77		30S	7/32
59	6	Mt-aux-Sources	29S	29E	54.5	.2264	1.300	(1.00)	12.2	3.51	.05		2.80		30S	7/32
60	7	Lahore	31N	74E	74.3	.127	1.20	1.00	11.8	1.63	.08	1.59			21N	8/32
61	7	Lal Tibba	30N	78E	58.2	.184	1.27	.98	11.8	2.62	.09		2.44		20N	8/32
62	7	Rohtang La	32N	77E	47.7	.281	1.375	.98	11.8	4.51	.15			5.24	22N	8/32
63	7	Sarchu	33N	78E	46.1	.297	1.37	(1.00)	11.8	4.87	.12			5.21	23N	8/32
64	7	Lachalung La	33N	78E	42.8	.315	1.41	.96	11.8	5.32	.10			4.65	23N	8/32
65	7	Hara Lacha La	33N	77E	42.5	.343	1.42	.99	11.8	6.09	.13			5.21	23N	8/32
66	7	Telekonka*	33N	78E	41.2	.384	1.43	(1.00)	11.8	7.24	.10			5.60	23N	8/32
67	7	Lanyar La	30N	78E	37.7	.421	1.48	.99	11.8	8.58	.07			5.25	20N	8/32
68	8	Lima	12S	77W	74.6	.1306	1.255	1.00	11.6	1.61	.13	1.58			1S	12/32
69	3	Cambridge	42N	71W	76.8	.184	1.140	1.00	11.85	2.55	.12	1.87			53N	12/32

* Values at these stations have low statistical weight, because of either low reproducibility of the data or a small number of readings.

† These values of L are estimated, because of insufficient data for their reliable calculation.

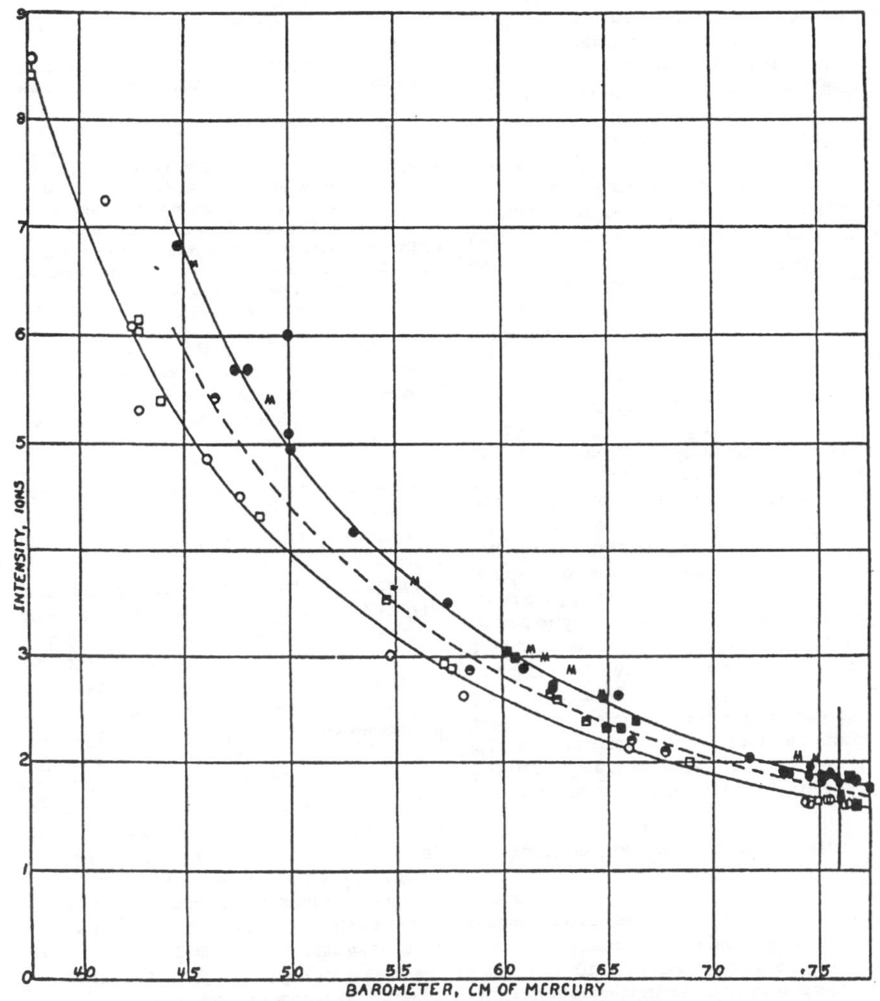

Fig. 4. Cosmic-ray intensity at different elevations. Squares, southern hemisphere. Circles, northern hemisphere. Solid dots, 90° to 40°. Shaded dots, 40° to 25°. Open dots, 0° to 25°.

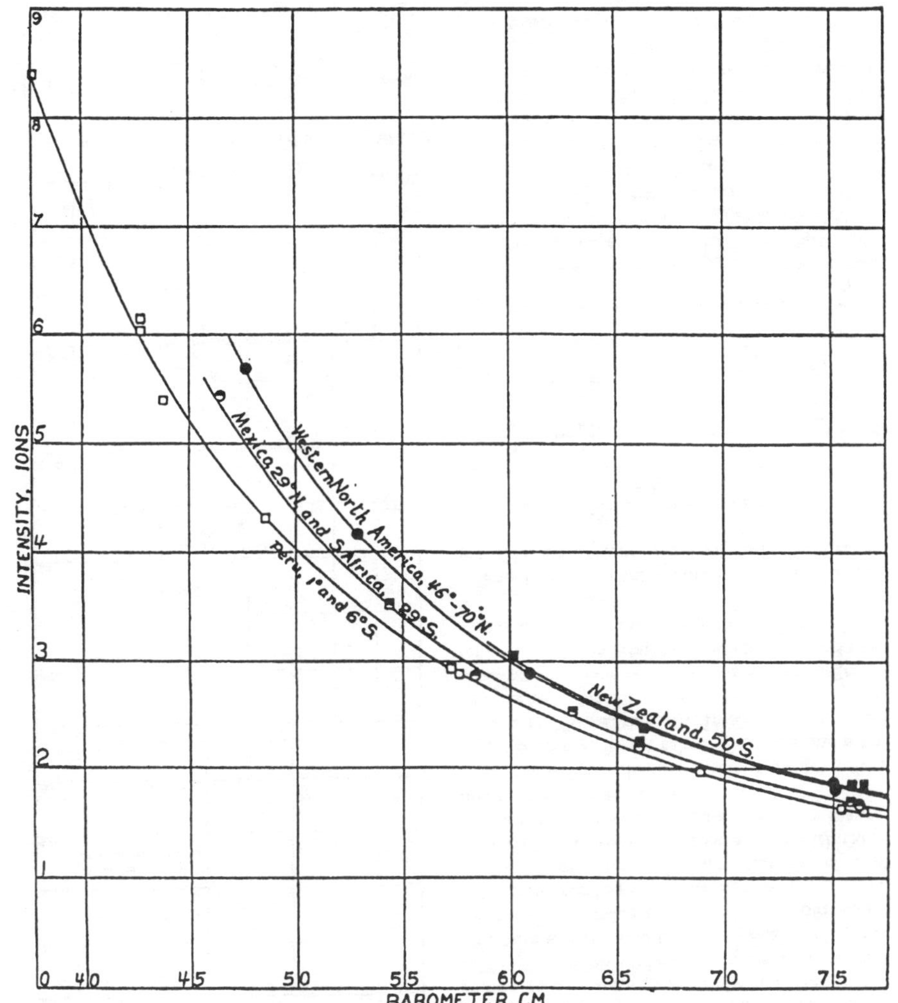

Fig. 5. Typical intensity vs. altitude curves for various latitudes.

moved and shields in place) was greater with air at 1 atmosphere than with air at 100 atmospheres. This could only mean an alpha-ray contamination, whose ionization is reduced by pressure. The magnitude of this ionization from the walls has been estimated by comparing the mean value of C_{76} = 2.55 ions from experiments 45, 46 and 69 (Chicago, Spitzbergen and Cambridge) with the mean value of C_{76} = 1.84 ions obtained for similar latitudes by expeditions 2 and 3. Thus, by subtracting 0.71 ion from the data obtained with apparatus 4, its results become comparable with those from the other sets of equipment.[14]

On the same figure are plotted the data of Millikan and Cameron,[15] indicated by the symbol *M*. The agreement between their values and ours is satisfactory, in view of the differences in the type of apparatus employed.

In Fig. 4 all of the data are included, without choice on the basis of statistical weight. The fact that all of the solid dots lie higher than the open ones over the whole range of barometric pressures is definite evidence that the cosmic rays are more intense at high latitudes than near the equator. This difference is shown more clearly in Fig. 5, which shows the results of six series of intensity *vs.* altitude measurements, made under favorable conditions, at six different latitudes. Those in New Zealand, Mexico and Peru were taken with apparatus number 2, Western North America with number 3, and South Africa with number 6. The excellent agreement between the New Zealand and the North American data, and between the South African and Mexican data indicates the reliability of our method of measurement. This figure shows also how the difference in intensity between the tropic and polar zones becomes much more prominent at the higher elevations.

[14] It is probable that there may be wall corrections of a similar type that should be applied to the other sets of apparatus. If so, however, the corrections are certainly much smaller than those for chamber number 4. Lacking exact information regarding the magnitude of these corrections, we have preferred to make none at all for the other sets of equipment. This procedure is partially justified by the small magnitude of the correction as found for the three sets of equipment that have been tested, and by the consistency of the data as here shown. It is probable that an even better agreement will, however, appear when it becomes possible to determine this zero correction by direct experiment. It will be about a year before such tests can be completed. Of course any possible zero correction cannot affect the differences observed at different latitudes and altitudes when using the same set of apparatus.

[15] R. A. Millikan and G. H. Cameron, Phys. Rev. 37, 235 (1931). The ionization values given in their Table III are divided by the factor 13.80 given by Millikan (Phys. Rev. 39, 397 (1932)) to reduce their measurements at 30 atmospheres air pressure to 1 atmosphere.

Variation of Cosmic-Ray Intensity with Latitude

In order to make a precise comparison of the data for different latitudes, the data plotted in Fig. 4 have been reduced to three standard barometric pressures, by using the smooth curves drawn in this figure as guides in effecting the reduction. All data for pressures over 67.5 cm have been reduced to 76 cm; those between 52.5 cm and 67.5 cm to 60 cm; and those between 37.5 cm and 52.5 cm to 45 cm. The values as thus standardized are given in Table III in columns C_{76}, C_{60} and C_{45}, and are plotted in Fig. 6 as functions of the geomagnetic latitude. In making this graph, the values in brackets in columns C_{76}, C_{60} and C_{45} have been averaged and plotted as a single point, in order to avoid unnecessary complexity. The numbers within the datum points indicate the apparatus with which the different values were obtained.

All the data agree in showing a marked difference between the intensities found within 20 degrees of the magnetic equator as compared with those at a higher latitude than 50° north or south. The increase in going to the higher latitudes averages 14 percent at sea level, 22 percent at 2000 meters (barometer 60 cm), and 33 percent at 4360 meters (barometer 45 cm).

In order to show the sea level values more clearly, they have been plotted in Fig. 7 on a larger scale. There have been added to this figure the data of Millikan,[16] taken at Pasadena and Churchill; and the recent ones of Clay and Berlage,[17] obtained with a set of Steinke recording equipment, compared with a standard at Amsterdam and carried on a ship from Genoa to Singapore. The values in Table IV, read from Clay and Berlage's curve, have been used in

[16] R. A. Millikan, Phys. Rev. 36, 1595 (1930).

[17] J. Clay and H. P. Berlage, Naturwiss. 37, 687 (1932).

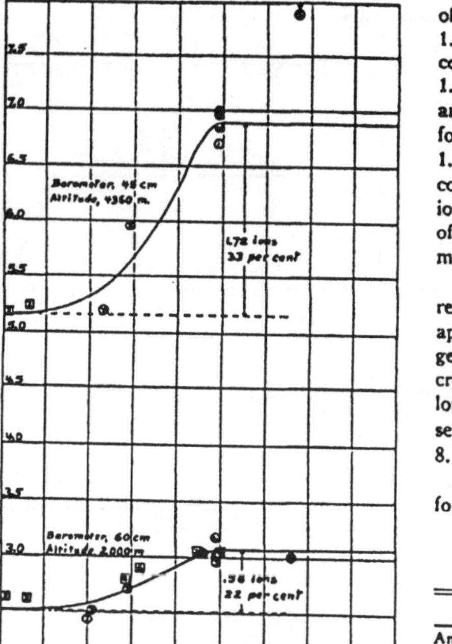

Fig. 6. Intensity *vs.* geomagnetic latitude for different elevations.

Fig. 5. These values have been plotted as crosses without any adjustment of the constants. Millikan's values are given for an ionization chamber pressure of 30 atmospheres; which, as he notes in a later paper,[18] multiplies the ionization by 13.80. With this reduction factor, his value at Pasadena is 2.050 ions for barometer 74.0 cm, and at Churchill is 2.020 ions for barometer 74.8 cm. Reduced to sea level according to the curves of Fig. 4, these intensities become 1.970 and 1.974 respectively. These values have been compared with ours by multiplying by the factor 1.79/1.97, which makes his value at Pasadena and ours at the same place identical. This gives for Millikan's value at Churchill (on our scale) 1.793 ions, plotted as a plus sign, which is to be compared with our value at Churchill of 1.80 ions. Our data are thus in good accord with those of the other observers who have made measurements in the same regions.

[18] R. A. Millikan, Phys. Rev. 39, 397 (1932).

It is significant that most of the datum points recorded in Fig. 7 were taken with the same set of apparatus (expedition 2). It was carried from geomagnetic latitude $\lambda = 53$ N to 51 S. After crossing the equator four times at different longitudes it was brought back to 78 N, and then sent once more across the equator with expedition 8.

The 15 experimental points plotted in Fig. 7 for this instrument were all taken under essentially the same conditions, so that we cannot find any source of systematic error that might affect the results. The readings with instruments 3 and 4 confirm the absence of appreciable variation north of $\lambda = 42°$, and comparison with the data from instruments 6 and 7 confirms both the magnitude of the variation observed by instrument 2 and the latitude at which it occurs.

Table IV. *Cosmic-ray intensities from Clay and Berlage's curve.*

Place	Lat.	Long.	C	Geomag. Lat.
Amsterdam	52 N	5 E	1.87	54 N
Genoa	44 N	9 E	1.79	44 N
At sea	37 N	19 E	1.72	35 N
Suez	30 N	32 E	1.66	27 N
Guardafui	12 N	51 E	1.61	6 N
Colombo	7 N	80 E	1.59	4 S
Singapore	1 N	104 E	1.58	11 S

The average of our 8 datum points taken for geomagnetic latitudes less than 22 degrees is 1.620±0.006 ions. For our 9 datum points at latitudes higher than 48°, the average is 1.839 ±0.006 ions. The difference between the two latitude ranges is thus 0.219±0.0085 ion; that is, more than 25 times the probable error. This probable error is also approximately that estimated also on the basis of statistical fluctuations in the readings taken under uniform conditions.

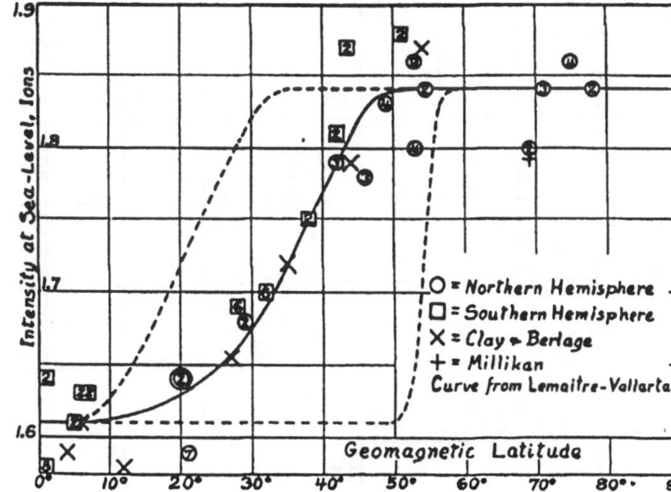

FIG. 7. Intensity *vs.* geomagnetic latitude at sea level, including data of Clay and Millikan.

There is thus no indication of any serious systematic error. From a statistical standpoint, therefore, the probability of the existence of this latitude effect amounts practically to a certainty.

Comparison with Lemaitre-Vallarta Theory

The sharpness of the increase in intensity between geomagnetic latitudes 25° and 45° is a major feature of these data. This shape of curve is just what may be expected from the new theory of Lemaitre and Vallarta,[19] which considers the effect of the earth's magnetic field on the motion of electrons approaching the earth from remote space. To show that this is true, the smooth curves drawn in Figs. 6 and 7 have been calculated from the Lemaitre-Vallarta theory (by graphical interpolation in their family of $F(x)$ *vs.* latitude curves). The arbitrary constants used for Fig. 7 are: 1.605 ions due to rays unaffected by the earth's magnetic field (neutral rays, or electrons of energy over 4×10^{10} electron-volts), and a band of electrons approaching the earth with energies between 0.5×10^{10} and 1.3×10^{10} electron-volts, which reach the earth at latitudes higher than 50°, producing 0.235 ion, but which fail to reach the earth at the geomagnetic equator.[20] The excellent agreement between this curve and the experimental data means that the variation of cosmic-ray intensity with latitude is that to be expected if a considerable portion of the ionization at high latitudes is due to electrons coming from remote space with energies of about 7×10^{9} electron-volts.

At the higher altitudes, as shown in Fig. 6, the latitude effect is so large as to be unquestionable, even without statistical analysis. Within experimental error, the zone over which the cosmic-ray intensity varies with latitude is the same as that found at sea level. There is one datum, of 8.5 ions at geomagnetic latitude 67 degrees, which perhaps indicates a continued increase in intensity at the higher latitudes at high elevations. This datum, however, is starred in Table III, because it represents a single unchecked result obtained by the ill-fated Carpe expedition[21] on Mt. McKinley. Bennett and his

[19] G. Lemaitre and M. S. Vallarta, Phys. Rev. 42, 914 (1932).

[20] In calculating the ionization due to this band of rays, 4 equally ionizing groups were assumed, having respectively the following values of Lemaitre and Vallarta's x: 0.30, 0.35, 0.40 and 0.45.

[21] Both observers, Allen Carpe and Theodore Koven, were killed by falling into a crevasse in the Muldrow Glacier, on which the observation tent was pitched. Some parts of the apparatus, and their notebook, were later recovered, but their Paulin barometer remains on the mountain. Though the cosmic-ray data appear reliable, the unchecked barometer might be in error by enough to bring this point in line with the others.

coworkers (data here recorded under expedition 3) have concluded from their Alaskan measurements at 1840 meters that the shape of the intensity-altitude curve is the same for geomagnetic latitude 66°N as at 48°N.[22] This is further supported, to much higher altitudes, by Millikan's recently reported[23] airplane measurements at 64°N, which showed the same increase with altitude as did experiments at 55°N. We thus seem justified in neglecting Carpe's datum, and in assuming that, as at lower altitudes, the intensity-latitude curve is flat north of 50°. This means, according to the Lemaitre-Vallarta theory, that there is no marked difference in the energy distribution of the particles responsible for the latitude effect as observed at different altitudes up to 4360 meters.

On the other hand, it will be seen from Fig. 6, that the extra component of cosmic rays which appears at the higher latitudes is more rapidly absorbed than is the main body of the rays. This would be anticipated if the portion of the rays that is unaffected by the earth's magnetic field consists of electrons of greater energy. Other interpretations are however possible. Thus, the uniform background may be due to some electrically neutral ray, such as photons, neutrons, or high speed neutral atoms.

Electrified Particles Bombarding the Earth from Remote Space

We have shown that the variation in intensity with latitude, which our experiments have brought to light, can be accounted for satisfactorily on Lemaitre and Vallarta's theory, which assumes that the rays consist of electrons of high energy coming from remote space, but are deflected by the earth's magnetic field. May there not, however, be alternative explanations?

A guide to possible interpretations is given by plotting the cosmic-ray intensity against the different variables of which it may be supposed to be a function. In Fig. 8 the same data as those used in Fig. 5 are plotted against the geographic latitude. Though there is clearly some correlation, in the intermediate latitudes, such as 30°, the scattering is definitely greater than is to be expected from the probable error of the experiments. Thus, the intensity is probably not a direct function of the geographic latitude.

In Fig. 9 the data are plotted against the local "magnetic latitude."[13] Though the correlation in this case is somewhat better than with the geographic latitude, the scattering in the neighborhood of 40°, where the change is most rapid, is again larger than would be expected from the probable error. This result indicates that the latitude variation is not due to the local magnetic field, which presumably would be effective for several hundred kilometers above the ground. It is thus difficult to attribute the latitude variation to any magnetic effect on the rays which occurs primarily within the earth's atmosphere.

The absence of any systematic variation of the atmospheric electric gradient with the latitude makes it hopeless to attempt to correlate the present effect with the electric field of the earth's atmosphere. Also, there is no systematic difference detectable between the data obtained at sea, such as experiments 5, 6, 7, 8, 18, 34, 46; and those taken at corresponding geomagnetic latitudes in mid-continental stations, such as 25, 28, 32 and 60 at the lower latitudes and 14, 37, 39 and 47 at the higher latitudes. This uniformity of the rays with respect to oceanic and continental areas makes it difficult to ascribe the latitude variation to any kind of atmospheric phenomenon.

It can be shown further that this latitude effect cannot be due to any bending of the cosmic-ray particles within the earth's atmosphere. For an electron moving with a speed nearly equal to that of light, the radius of curvature of the path when crossing a magnetic field of strength H is approximately,

$$R = mc^2/eH. \tag{13}$$

With 0.3 gauss as the value of H at the equator, where the magnetic effect is a maximum, this

[22] R. D. Bennett, J. L. Dunham, E. H. Bramhall, P. K. Allen, Phys. Rev. 42, 447 (1932).

[23] R. A. Millikan as quoted in New York Times, Dec. 31, 1932.

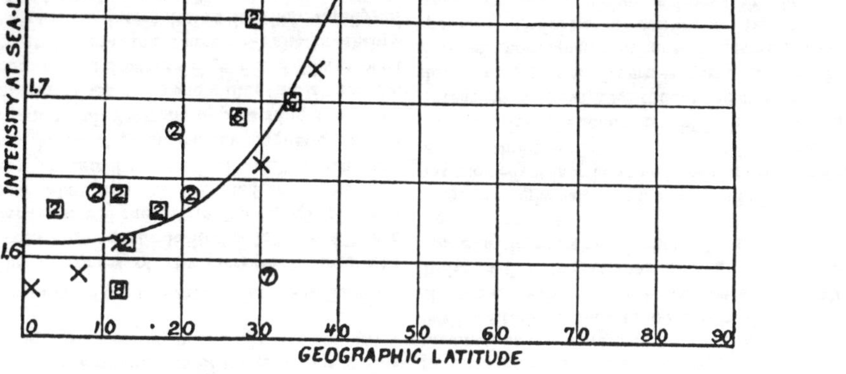

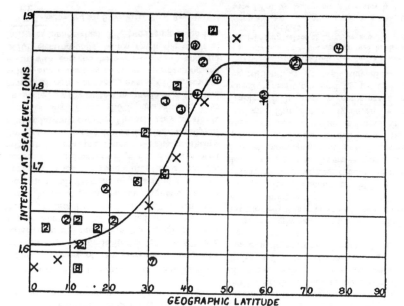

FIG. 8. Intensity *vs.* geographic latitude.

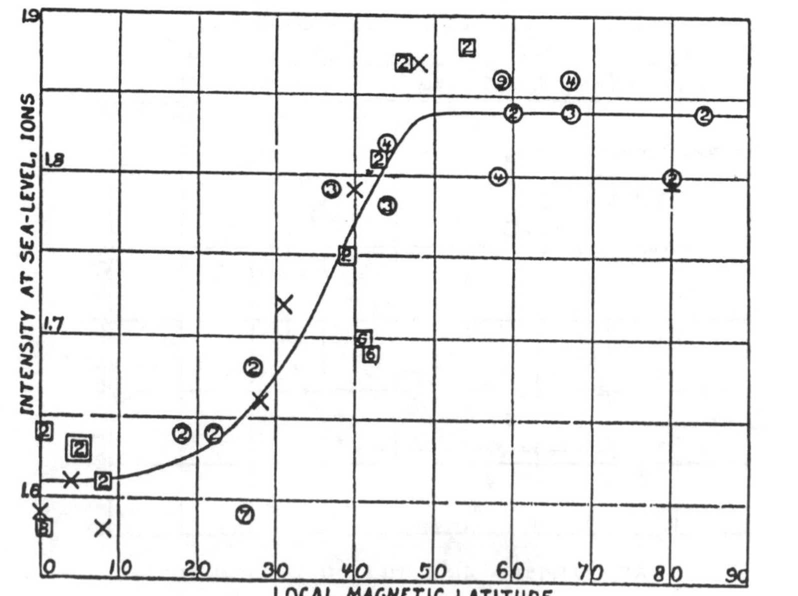

FIG. 9. Intensity *vs.* local magnetic latitude.

becomes,

$$R=6\times10^{30}m \text{ cm}, \qquad (14)$$

where m is the mass of the moving particle expressed in grams. The range of beta-particles has been discussed at length by Rutherford, Chadwick and Ellis,[24] chiefly on the basis of Bohr's theory of beta-ray absorption. From their formulas (10), p. 438; and (7), p. 442, the range of a beta-particle in air at atmospheric pressure can be written approximately as:

$$r=0.2\times10^{30}m \text{ cm}, \qquad (15)$$

if the mass m of the electron in motion is much larger than the rest mass. That is, the radius of curvature in the earth's magnetic field is at atmospheric pressure, $6/0.2=30$ times the particle's range. Thus, at atmospheric pressure the curvature produced in the path of an electron by the earth's magnetic field is negligible. At an altitude of 25 kilometers, where the density of the earth's atmosphere is about 1/30 of that at sea-level, the range of an electron should be about equal to its radius of curvature, and above this altitude the earth's magnetic field should have an appreciable effect on the particle's range. This means that, if the cosmic rays which are affected by the earth's magnetic field are electrons, they must originate at an altitude of more than 25 km. For other electrified particles, protons, alpha-particles, etc., the limiting altitude as thus calculated is still higher.

Supposing that it is such electrons, originating high in the earth's atmosphere, which are detected by our ionization chambers at the earth's surface; how much energy is necessary to penetrate the atmosphere? If in Eq. (15) we place the range $r=8.0\times10^{5}$ cm, which is the equivalent of 1 atmosphere if the air were of uniform density equal to that under standard conditions, we obtain $m=4\times10^{-24}$g. Multiplying by the conversion factor, $300c^2/e$, this means an energy of 2.3×10^{9} electron-volts.[25] This would be the minimum possible energy for an electron passing vertically through the atmosphere without deflection. For such an electron, however, according to Eq. (14), the radius of curvature is 240 kilometers. This means that, if a beta-particle capable of penetrating the earth's atmosphere is appreciably affected by the earth's magnetic field, it must originate not less than some hundreds of kilometers above the earth's surface. This conclusion supports the comparison of Figs. 7 and 9, which indicated that it is the average rather than the local magnetic field of the earth which is responsible for the latitude variation.

It would seem that an effect due to a magnetic field necessarily implies that the rays thus affected are moving charged particles. If so, our data mean that a portion at least of the cosmic rays consists of high speed particles. But we have seen also that these particles must originate high above the earth. It is accordingly not permissible to assume that these cosmic electrons are secondary beta-rays produced within the earth's atmosphere by some form of electrically neutral rays such as photons. Our experiments seem to require rather that the portion of the cosmic rays which varies with the geomagnetic latitude shall consist of electrified particles approaching the earth from distances of not less than some hundreds of kilometers.

The quantitative agreement of the curve taken from Lemaitre and Vallarta's theory, with the data plotted in Fig. 7, strongly supports this conclusion. Though in fitting this curve to the data several arbitrary constants were available, it is by no means true that any arbitrary set of data could thus be fitted. If the cosmic-ray intensities at the equator and the poles are to be respectively 1.61 and 1.84 ions, the Lemaitre-Vallarta theory requires that the intensities at intermediate latitudes shall lie between the two broken curves of this figure. Of these, that for the higher latitude represents the minimum energy (2.3×10^{9} electron-volts) that an electron can have which will penetrate the earth's atmosphere,

[24] E. Rutherford, J. Chadwick, C. D. Ellis, *Radiations from Radioactive Substances* pp. 434–444, 1930.

[25] Neglecting a small correction due to the different ratio of (atomic weight)/(atomic number), this corresponds to an energy loss of 2.5×10^{7} electron-volts per cm of lead traversed by a β-particle. This is in substantial accord with the loss of 3.5×10^{7} electron-volts per cm of lead as reported recently by C. D. Anderson (Bulletin Am. Phys. Soc. 7, No. 7, p. 15, Dec. 7, 1932). With Anderson's value, the energy lost by an electron traversing the atmosphere would be slightly greater than that here estimated.

superposed on a background of radiation that is unaffected by the earth's field. The curve for the lower latitude represents the maximum energy the electrons can have (3.2×10^{10} electron-volts) if the difference in intensity between the equator and the poles is to be of the specified amount. It would be a surprising coincidence that the experimental curve should fall within the rather narrow limits defined by this theory if the latitude variation has some other origin.

The experimental data thus give very strong support to Lemaitre and Vallarta's theory of the variation of cosmic-ray intensity with latitude. This means that this variation seems to be due to the presence in the cosmic rays of charged particles coming into the earth's atmosphere from remote space with an energy, if they are electrons, of about 7×10^{9} electron-volts.

Acknowledgments

In addition to the leaders of the various expeditions mentioned above, the help of the following persons should be specially mentioned: Dr. Marcel Schein, of the University of Zurich, who helped in organizing the two Alpine expeditions; Dr. J. J. Hopfield, Dr. E. O. Wollan and Dr. V. J. Andrew, who have helped with the design, construction and testing of the apparatus at Chicago; Professor Harry Kirkpatrick of the University of Hawaii, who organized the Hawaiian expedition; Professor P. W. Burbidge of Anckland University College, who made the plans for New Zealand and gave valuable help with the measurements; Professor O. U. Vonwiller of the University of Sydney, who planned the Australian program; Dr. Robert Enders, who arranged for the experiments in Panama; Mr. P. G. Ledig and Dr. J. E. I. Cairns of the Huancayo Magnetic Observatory of the Carnegie Institution, without whose help the Peruvian measurements could hardly have been completed; Professor M. S. Vallarta of Massachusetts Institute of Technology, who looked after all arrangements in Mexico; Mr. George Kydd, engineer in charge of construction at Churchill, who made possible the trip to Foxe Channel; Dr. J. L. Dunham, Dr. E. H. Bramhall and Dr. P. K. Allen, who worked with expedition 3 for several months; Mr. Theodore Koven, who with Mr. Allen Carpe gave his life in the effort to secure data on Mt. McKinley; Professor B. F. J. Schonland, of the University of Capetown, who has helped with the experiments in South Africa; Professors Ross Wilson of Lahore and Sharma of Allahabad, who have helped with the Himalayan measurements, and Dr. J. A. Fleming of the Carnegie Institution of Washington, who has supplied valuable information. There have been in addition more than thirty other physicists who have given generously of their time in performing these experiments.

A grant from the Carnegie Foundation made possible a survey on a world-wide scale. Among the other institutions which have contributed support in money or equipment are: the University of Chicago, Massachusetts Institute of Technology, the Rumford Fund, the University of Denver, the University of Hawaii, Auckland University College, the government of New South Wales and the Radium Service Corporation of America. Most of the members of the various expeditions have also borne a large share, or all, of their own expenses while engaged in these experiments. Without such remarkable cooperation, the progress of this survey would have been greatly delayed.

The Positive Electron

Carl D. Anderson, *California Institute of Technology, Pasadena, California*

(Received February 28, 1933)

Out of a group of 1300 photographs of cosmic-ray tracks in a vertical Wilson chamber 15 tracks were of positive particles which could not have a mass as great as that of the proton. From an examination of the energy-loss and ionization produced it is concluded that the charge is less than twice, and is probably exactly equal to, that of the proton. If these particles carry unit positive charge the curvatures and ionizations produced require the mass to be less than twenty times the electron mass. These particles will be called positrons. Because they occur in groups associated with other tracks it is concluded that they must be secondary particles ejected from atomic nuclei.

Editor

On August 2, 1932, during the course of photographing cosmic-ray tracks produced in a vertical Wilson chamber (magnetic field of 15,000 gauss) designed in the summer of 1930 by Professor R. A. Millikan and the writer, the tracks shown in Fig. 1 were obtained, which seemed to be interpretable only on the basis of the existence in this case of a particle carrying a positive charge but having a mass of the same order of magnitude as that normally possessed by a free negative electron. Later study of the photograph by a whole group of men of the Norman Bridge Laboratory only tended to strengthen this view. The reason that this interpretation seemed so inevitable is that the track appearing on the upper half of the figure cannot possibly have a mass as large as that of a proton for as soon as the mass is fixed the energy is at once fixed by the curvature. The energy of a proton of that curvature comes out 300,000 volts, but a proton of that energy according to well established and universally accepted determinations[1] has a total range of about 5 mm in air while that portion of the range actually visible in this case exceeds 5 cm without a noticeable change in curvature. The only escape from this conclusion would be to assume that at exactly the same instant (and the sharpness of the tracks determines that instant to within about a fiftieth of a second) two independent electrons happened to produce two tracks so placed as to give the impression of a single particle shooting through the lead plate. This assumption was dismissed on a probability basis, since a sharp track of this order of curvature under the experimental conditions prevailing occurred in the chamber only once in some 500 exposures, and since there was practically no chance at all that two such tracks should line up in this way. We also discarded as completely untenable the assumption of an electron of 20 million volts entering the lead on one side and coming out with an energy of 60 million volts on the other side. A fourth possibility is that a photon, entering the lead from above, knocked out of the nucleus of a lead atom two particles, one of which shot upward and the other downward. But in this case the upward moving one would be a positive of small mass so that either of the two possibilities leads to the existence of the positive electron.

In the course of the next few weeks other photographs were obtained which could be interpreted logically only on the positive-electron basis, and a brief report was then published[2] with due reserve in interpretation in view of the importance and striking nature of the announcement.

Magnitude of Charge and Mass

It is possible with the present experimental data only to assign rather wide limits to the

[1] Rutherford, Chadwick and Ellis, *Radiations from Radioactive Substances*, p. 294. Assuming $R \propto v^3$ and using data there given the range of a 300,000 volt proton in air S.T.P. is about 5 mm.

[2] C. D. Anderson, Science 76, 238 (1932).

Fig. 1. A 63 million volt positron ($H\rho = 2.1\times10^5$ gauss-cm) passing through a 6 mm lead plate and emerging as a 23 million volt positron ($H\rho = 7.5\times10^4$ gauss-cm). The length of this latter path is at least ten times greater than the possible length of a proton path of this curvature.

magnitude of the charge and mass of the particle. The specific ionization was not in these cases measured, but it appears very probable, from a knowledge of the experimental conditions and by comparison with many other photographs of high- and low-speed electrons taken under the same conditions, that the charge cannot differ in magnitude from that of an electron by an amount as great as a factor of two. Furthermore, if the photograph is taken to represent a positive particle penetrating the 6 mm lead plate, then the energy lost, calculated for unit charge, is approximately 38 million electron-volts, this value being practically independent of the proper mass of the particle as long as it is not too many times larger than that of a free negative electron. This value of 63 million volts per cm energy-loss for the positive particle it was considered legitimate to compare with the measured mean of approximately 35 million volts[3] for negative electrons of 200–300 million volts energy since the rate of energy-loss for particles of small mass is expected to change only very slowly over an energy range extending from several million to several hundred million volts. Allowance being made for experimental uncertainties, an upper limit to the rate of loss of energy for the positive particle can then be set at less than four times that for an electron, thus fixing, by the usual relation between rate of ionization and

[3] C. D. Anderson, Phys. Rev. 43, 381A (1933).

charge, an upper limit to the charge less than twice that of the negative electron. It is concluded, therefore, that the magnitude of the charge of the positive electron which we shall henceforth contract to positron is very probably equal to that of a free negative electron which from symmetry considerations would naturally then be called a negatron.

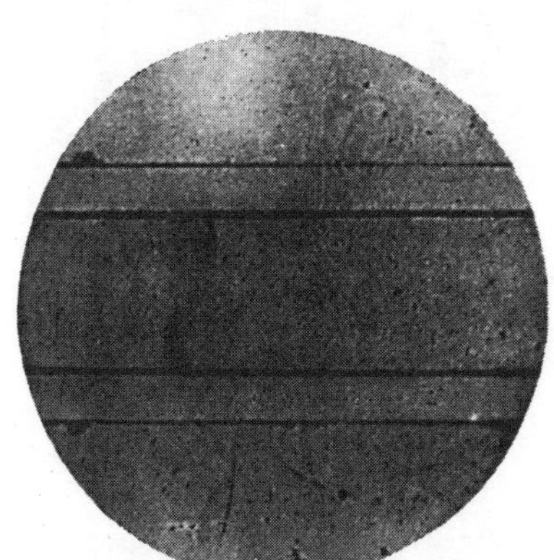

FIG. 2. A positron of 20 million volts energy ($H\rho = 7.1 \times 10^4$ gauss-cm) and a negatron of 30 million volts energy ($H\rho = 10.2 \times 10^4$ gauss-cm) projected from a plate of lead. The range of the positive particle precludes the possibility of ascribing it to a proton of the observed curvature.

It is pointed out that the effective depth of the chamber in the line of sight which is the same as the direction of the magnetic lines of force was 1 cm and its effective diameter at right angles to that line 14 cm, thus insuring that the particle crossed the chamber practically normal to the lines of force. The change in direction due to scattering in the lead,[3] in this case about 8° measured in the plane of the chamber, is a probable value for a particle of this energy though less than the most probable value.

The magnitude of the proper mass cannot as yet be given further than to fix an upper limit to it about twenty times that of the electron mass. If Fig. 1 represents a particle of unit charge passing through the lead plate then the curvatures, on the basis of the information at hand on ionization, give too low a value for the energy-loss unless the mass is taken less than twenty times that of the negative electron mass. Further determinations of $H\rho$ for relatively low energy particles before and after they cross a known amount of matter, together with a study of ballistic effects such as close encounters with electrons, involving large energy transfers, will enable closer limits to be assigned to the mass.

To date, out of a group of 1300 photographs of cosmic-ray tracks 15 of these show positive particles penetrating the lead, none of which can be ascribed to particles with a mass as large as that of a proton, thus establishing the existence of positive particles of unit charge and of mass small compared to that of a proton. In many other cases due either to the short section of track available for measurement or to the high energy of the particle it is not possible to differentiate with certainty between protons and positrons. A comparison of the six or seven hundred positive-ray tracks which we have taken is, however, still consistent with the view that the positive particle which is knocked out of the nucleus by the incoming primary cosmic ray is in many cases a proton.

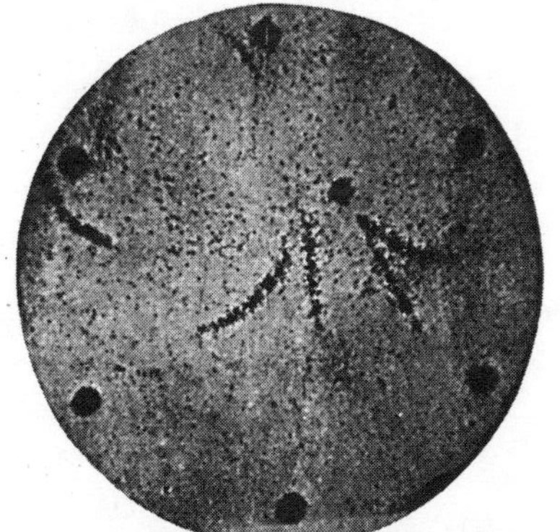

FIG. 3. A group of six particles projected from a region in the wall of the chamber. The track at the left of the central group of four tracks is a negatron of about 18 million volts energy ($H\rho = 6.2 \times 10^4$ gauss-cm) and that at the right a positron of about 20 million volts energy ($H\rho = 7.0 \times 10^4$ gauss-cm). Identification of the two tracks in the center is not possible. A negatron of about 15 million volts is shown at the left. This group represents early tracks which were broadened by the diffusion of the ions. The uniformity of this broadening for all the tracks shows that the particles entered the chamber at the same time.

From the fact that positrons occur in groups associated with other tracks it is concluded that they must be secondary particles ejected from an atomic nucleus. If we retain the view that a nucleus consists of protons and neutrons (and α-

FIG. 4. A positron of about 200 million volts energy ($H\rho = 6.6 \times 10^5$ gauss-cm) penetrates the 11 mm lead plate and emerges with about 125 million volts energy ($H\rho = 4.2 \times 10^5$ gauss-cm). The assumption that the tracks represent a proton traversing the lead plate is inconsistent with the observed curvatures. The energies would then be, respectively, about 20 million and 8 million volts above and below the lead, energies too low to permit the proton to have a range sufficient to penetrate a plate of lead of 11 mm thickness.

particles) and that a neutron represents a close combination of a proton and electron, then from the electromagnetic theory as to the origin of mass the simplest assumption would seem to be that an encounter between the incoming primary ray and a proton may take place in such a way as to expand the diameter of the proton to the same value as that possessed by the negatron. This process would release an energy of a billion electron-volts appearing as a secondary photon. As a second possibility the primary ray may disintegrate a neutron (or more than one) in the nucleus by the ejection either of a negatron or a positron with the result that a positive or a negative proton, as the case may be, remains in the nucleus in place of the neutron, the event occurring in this instance without the emission of a photon. This alternative, however, postulates the existence in the nucleus of a proton of negative charge, no evidence for which exists. The greater symmetry, however, between the positive and negative charges revealed by the discovery of the positron should prove a stimulus to search for evidence of the existence of negative protons. If the neutron should prove to be a fundamental particle of a new kind rather than a proton and negatron in close combination, the above hypotheses will have to be abandoned for the proton will then in all probability be represented as a complex particle consisting of a neutron and positron.

While this paper was in preparation press reports have announced that P. M. S. Blackett and G. Occhialini in an extensive study of cosmic-ray tracks have also obtained evidence for the existence of light positive particles confirming our earlier report.

I wish to express my great indebtedness to Professor R. A. Millikan for suggesting this research and for many helpful discussions during its progress. The able assistance of Mr. Seth H. Neddermeyer is also appreciated.

A Positively Charged Component of Cosmic Rays

The relatively low intensity of cosmic rays at low geomagnetic latitudes, as recently found by our associated expeditions[1] and others,[2] indicates that a part of the cosmic rays consists of electrified particles. When interpreted in terms of Lemaitre and Vallarta's theory[3] of the deflection of electrified particles by the earth's magnetic field, these results indicate that at geomagnetic latitudes higher than about 45° the earth's magnetic field should not alter the direction of the incoming rays as observed at sea level. This is in accord with the sea-level observations of Johnson and Street,[4] which show a symmetrical East-West distribution. At the geomagnetic equator an analysis of our intensity-latitude curves suggests that most of the cosmic rays which are affected by the earth's magnetic field are too strongly deflected to reach the earth's surface. If this is correct, there should be but little asymmetry in the direction of approach of the cosmic rays near the equator. In an intermediate zone, however, where the intensity *vs.* latitude curve is steep, the rays that are being affected by

[1] A. H. Compton, Phys. Rev. **41**, 111 (1932); **43**, 387 (1933).

[2] J. Clay and H. P. Berlage, Naturwiss. **37**, 687 (1932).

[3] G. Lemaitre and M. S. Vallarta, Phys. Rev. **43**, 87 (1933).

[4] T. H. Johnson, J. Frank. Inst. **214**, 689 (1932).

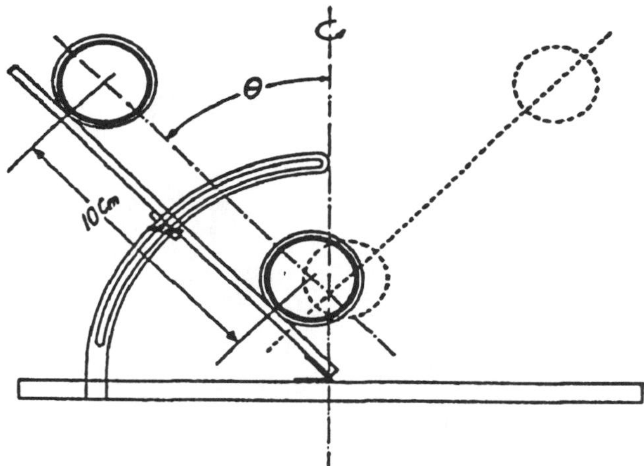

FIG. 1. Arrangement of coincidence counting tubes for studying East-West asymmetry of cosmic rays.

the earth's magnetic field should strike the earth from certain directions but not from others. If the rays are positively charged, they should come mostly from the west, if negatively, predominantly from the east, due to deflection by the earth's magnetic field. From such considerations Vallarta has suggested that Mexico City should be a good place to search for the predicted asymmetry in the direction of the incoming cosmic rays. Besides being in the favorable zone of geomagnetic latitude (29°N), its elevation (2310 meters) is sufficient to avoid some of the disturbing effects of the atmosphere.

In order to observe the direction of the incoming particles we have used a double coincidence counter, as shown diagrammatically in Figs. 1 and 2. Tests made by separating the tubes indicate that chance coincidences occur at the rate of only about 1.5 per hour, so that with a normal counting rate of about 5 per minute these were of negligible importance. The zenith angle θ of the line joining the axes of the tubes with the vertical was measured with the help of a protractor and spirit level. In order to avoid any possible change of conditions, the whole apparatus was mounted on a platform, which was rotated through 180 degrees when the changes between east and west were made. Readings of about a half hour's duration were taken alternately between east and west at the same zenith angle θ. For each angle the final series of readings totaled about fourteen hours on either side. By changing thus back and forth, enough readings were obtained to make a good estimate of the probable errors of the observed counting rates. The errors thus estimated from the consistency of successive readings under similar conditions were but little greater than those calculated as the statistical error from the total number of coincidences in the series. This means that no serious disturbing factor was affecting the readings. Table I summarizes our results.

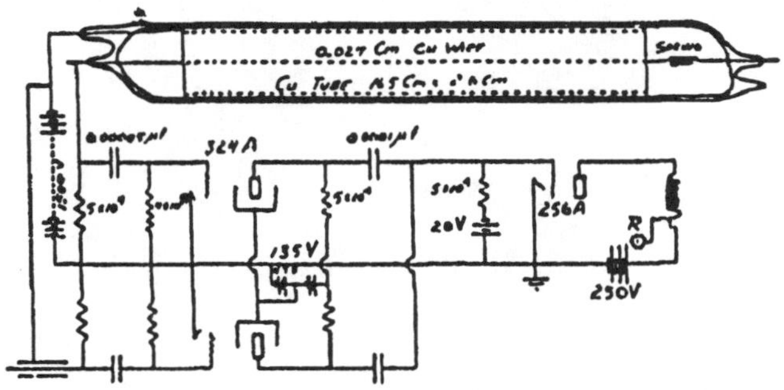

FIG. 2. Circuit used with double coincidence counter.

TABLE I. *East-west measurements at Mexico City, April, 1933.*

Geomagnetic latitude 29°N, elevation 2310 m, barometer, 56.5 cm.

Zenith angle		West	East	West/East
15°	Counts	5370	4856	
	Rate	6.83±0.07	6.64±0.07	1.03±0.02
30°	Counts	4897	4869	
	Rate	5.79±0.06	5.49±0.06	1.055±0.015
45°	Counts	2691	2693	
	Rate	3.70±0.05	3.30±0.05	1.12±0.02

It will be noted that at the larger zenith angles the rate at which the rays come from the west is greater than that from the east by several times the probable error of the measurements. It would appear that the asymmetry thus observed at Mexico City is considerably larger and more definite than that found by Johnson and Street[5] on Mt. Washington, geomagnetic latitude 55°N, elevation 1920 meters. This preponderance of rays from the west seems necessarily to imply the existence of a positively charged component of the cosmic rays.

Since our earlier measurements have shown that the cosmic rays at geomagnetic latitude 29° differ by only about 14 percent from those in high latitudes for this elevation, the difference in counts in the east and west directions is of the order of magnitude to be expected due to the deflection of the particles by the earth's magnetic field. The smallness of the effect confirms our earlier conclusion that most of the rays capable of penetrating the earth's atmosphere are not sufficiently bent by the earth's magnetic field to prevent them from reaching the earth. We may add that these data are consistent with the view that the positively charged component here found consists of Anderson's newly discovered[6] positrons.

We wish to thank Professor R. D. Bennett for suggesting the circuit used, Dr. M. S. Vallarta and Dr. T. H. Johnson for valuable suggestions in carrying on the experiment, and the Carnegie Institution of Washington for financial aid.

LUIS ALVAREZ
ARTHUR H. COMPTON

University of Chicago,
April 22, 1933.

[5] T. H. Johnson and J. C. Street, Phys. Rev. 43, 381 (1933).

[6] C. D. Anderson, Science 76, 238 (1932); Phys. Rev. 43, 491 (1933).

Chapter 9

CONDENSED MATTER

PAUL C. MARTIN

Introduction

Papers Reprinted in Book

John H. Van Vleck (1899-1980). (Courtesy of AIP Meggers Gallery of Nobel Laureates, AIP Emilio Segrè Visual Archives.)

Paul C. Martin is the John H. Van Vleck Professor of Pure and Applied Physics at Harvard University, where he has spent almost all of his professional life. Martin is also Dean of the Division of Applied Sciences and an Associate Dean of the Faculty of Arts and Sciences in which positions he is called upon to predict where physics research is headed. His own research includes work on field theory and quantum electrodynamics, nuclear physics, and fluid dynamics as well as on condensed matter and statistical physics. He has held Sloan and Guggenheim Fellowships and is a member of the National Academy of Sciences and the American Academy of Arts and Sciences.

Leon Cooper (1930-). (Courtesy of AIP Emilio Segrè Visual Archives.)

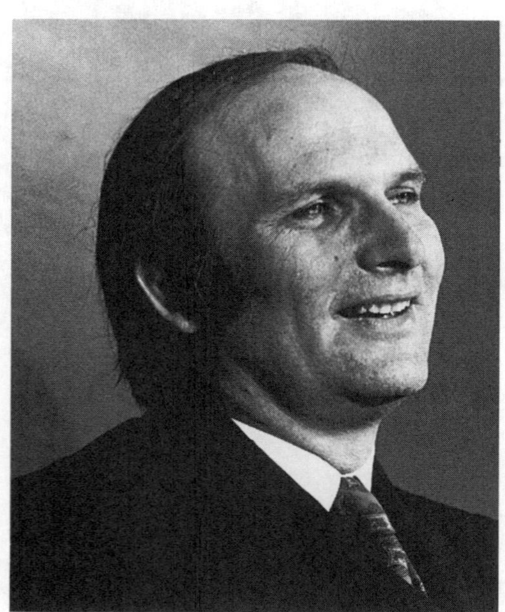

John R. Schrieffer (1931–). (Courtesy of Frank Ross Photography, AIP Emilio Segrè Visual Archives.)

The first 100 years of *The Physical Review* coincide with the golden age of physics. Our civilization has never experienced, nor will it in the future, a century with so many significant and revolutionary advances; the basic laws that underlie all earthly physical phenomena have been established. New insights will deepen our understanding of these laws and give rise to new applications; new methods will yield more accurate values of the parameters they contain; and new discoveries at subquark or supergalactic dimensions may reveal relationships among these parameters. But these insights, methods, and discoveries will not alter the basic principles of relativity and quantum mechanics that describe physical phenomena throughout the universe. Nor will they significantly modify the properties of the particles and fields we encounter, their strong, weak, electromagnetic, and gravitational interactions, or the deterministic and statistical relationships between these interacting particles and fields and the macroscopic workings of the world.

During the last half of this golden age, American physicists and *The Physical Review* have played a dominant role. On the journal's 100th birthday, those who have served the physics community well by editing, publishing, and distributing this mammoth volume of research deserve a hearty cheer. So do we all, for we, the members of the physics community who write, review, read, and support these journals, are primarily responsible for their quality. Moreover, our elected American Physical Society (APS) officials are responsible for assuring that *The Physical Review* does not go the way of most of its for-profit shelfmates.

In view of the "cosmic" significance of the knowledge generated during this period, the fair and objective selection of the "best" contributions from the massive corpus of *The Physical Review* articles would seem an awesome assignment — especially in such a field as condensed matter physics, where most advances have occurred in the past 60 years. At first, some of us making the selections were intimidated, but we have lifted this weight from our shoulders by observing that the project is inherently artificial.

Suppose the officials of Vienna, a city with unparalleled musical traditions, had decided to celebrate 250 years of music by issuing compact disks of operatic, symphonic, and chamber music. Suppose each disk could include only music composed in Vienna and that its playing time could not exceed eight hours. Although anniversary albums for Vienna fashioned under such constraints would contain more outstanding music than similarly constrained anthologies from other cities, they would only be souvenirs. To include a variety of composers, movements of Beethoven symphonies would need to be excised. Part of Mozart's *Marriage of Figaro* might be included, but not *Don Giovanni*, written in Prague. "Top tunes" would — by criteria appropriate for souvenirs — be dictated by fame, length, and popularity. Scholars and lovers of music would surely go elsewhere for serious music and music criticism. Any living composers would recognize the inevitably capricious character of the selection process and would be only mildly chagrined if their work, or that of those who most inspired them, did not appear.

Nominators and authors of contributions not included in this volume should be equally understanding. (Most of the items in this section of the book and the CD-ROM have been selected from lists submitted by individuals, groups, and organization in re-

Mildred S. Dresselhaus (1930–).
(Courtesy of Mildred S. Dresselhaus, MIT)

Ivar Giaever (1929-). (Courtesy of Weber Collection, AIP Emilio Segrè Visual Archives.)

sponse to APS requests. Colleagues, including Nicolaas Bloembergen, Bertrand Halperin, and Michael Tinkham, also have given me advice.)

Brevity was a major criterion in selecting articles. Perhaps the most significant theoretical advance in condensed matter physics is the explanation of superconductivity in terms of a coherent state involving correlated electron pairs [Bardeen, Cooper, and Schrieffer (BCS), Phys. Rev. **108**, 1175 (1958)]. Only the abstract of this fundamental paper appears in the book. The paper itself has been relegated to the CD-ROM because it is 30 pages long. So too have the fundamental papers on nuclear magnetic resonance by Bloch [Phys. Rev. **70**, 460 (1948)] and Bloch, Hansen, and Packard [Phys. Rev. **70**, 474 (1948)] and by Bloembergen, Pound, and Purcell [Phys. Rev. **73**, 679 (1948)]. These four papers alone would have exhausted the original space allotment for solid state physics. A few short articles have been selected as surrogates for these articles, and some others because they played a significant role in otherwise unrepresented fields.

One criterion (highly inappropriate for a constraint-free scholarly anthology but acceptable for a souvenir volume) simplified the choice of articles and letters in this section of the book. With few exceptions, all are seminal contributions by Nobel Prize laureates or, for superconductivity and nuclear magnetic resonance, either seminal precursors or contributions that elucidate and extend research awarded "The Prize."

In the case of superconductivity, neither Cooper's letter [Phys. Rev. **104**, 1189 (1956)], which notes the instability of the Fermi sea to attractive long-range interactions, nor the preliminary letter by Bardeen, Cooper, and Schrieffer [Phys. Rev. **106**, 162 (1957)], which estimates the amount by which the paired state is lowered, shows that the new state with condensed pairs has superconducting magnetic properties (i.e., a Meissner effect); neither describes the finite-temperature properties (i.e., the second-order transition and the exponentially small specific heat observed in weak-coupling superconductors). But both letters foreshadowing the theory are short, as are those demonstrating experimentally by tunneling that a gap exists [Giaever, Phys. Rev. Lett. **5**, 147, 467 (1960)].

The articles on superconductivity highlight why assembling articles from a single journal is not of much value to students and scholars. As the organizer and chair of a special session at the Low Temperature Conference in 1962 at St. Mary's College in London, I shall always remember the spirited debate between Bardeen and Josephson when, five years after BCS, the fundamental significance of the time- and space-varying complex pair wave function was first being recognized. The letters by P. W. Anderson and Rowell [Phys. Rev. Lett. **10**, 230 (1963)] and Jaklevic, Lambe, Silver, and Mercereau [Phys. Rev. Lett. **12**, 159 (1964)] appear in this anthology partly because of their intrinsic importance and partly because they verify Josephson's ideas on ac and dc tunneling and earlier Russian theoretical work on pair wave functions, vortex lattices, and quantized flux; and the letter by W. H. Parker, B. N. Taylor, and Langenberg [Phys. Rev. Lett. **18**, 287 (1967)] because it uses the Josephson effect to measure h/e to such high precision.

The breadth and richness of solid state physics is enormous. To particle physicists and to atomic physicists striving to isolate single particles, the fact that impure materials can be used not only to build practical devices but also to test fundamental quantum phenomena quantitatively is understandably astonishing.

Only 80 years ago W. H. Bragg and W. L. Bragg demonstrated by x-ray diffraction the periodic arrangements of ions in crystalline solids. A few years later, Einstein and then Debye used quantum mechanics to estimate the thermodynamic effects of vibrations of these ions. As x-ray scattering was instrumental in displaying the positions of ions, elastic neutron scattering has been instrumental in displaying their magnetic ordering and inelastic neutron scattering in determining the spectra of the atomic vibrations. Over the years, many experimental studies of phonons in solids have appeared in the literature, but there is little basis for singling out one for inclusion here.

Given the difficulties in accurately computing the properties of systems with two electrons (the helium atom and the hydrogen molecule), in the early days of quantum mechanics many physicists thought quantitatively calculating the electronic properties of real solids and electronic contributions to their binding energies hopelessly ambitious. Curiously, in some quarters, Pauli's characterization of solid state physics as "Schmutzphysik" still persists. Nevertheless, in 1928, very soon after the advent of quantum mechanics, the qualitative effects of periodic potentials were discussed and Bloch states and Brillouin zones introduced. The possibility of and the need for determining the average effect of the interelectronic potential self-consistently, with due regard for Fermi statistics, were recognized by Hartree, Fock, and Slater. In

Léon N. Brillouin (1889-1969). (Courtesy of AIP Emilio Segrè Visual Archives.)

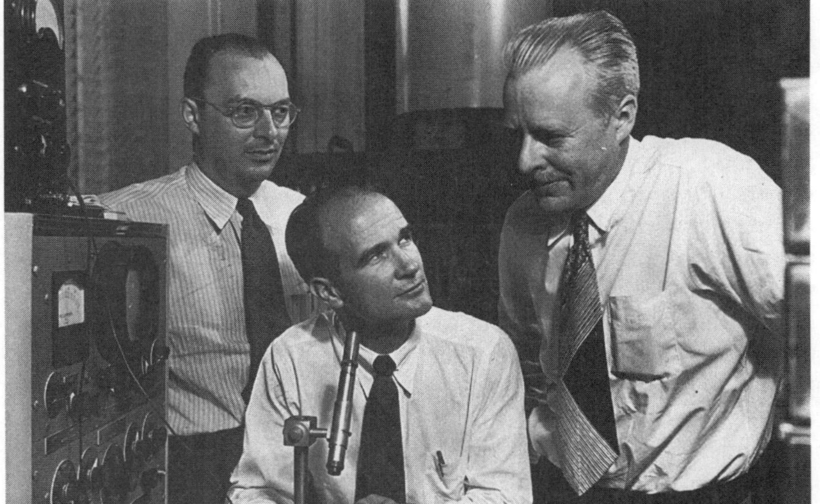

John Bardeen (1908-1991), William Shockley (1910-1989), and **Walter Brattain (1902-1987).** (Courtesy of Bell Telephone Laboratories, AIP Emilio Segrè Visual Archives.)

Philip W. Anderson (1923–) and **P. L. Richards.** (Courtesy of Bell Telephone Laboratories, AIP Emilio Segrè Visual Archives.)

Leo Esaki (1925-). (Courtesy of AIP Meggers Gallery of Nobel Laureates, AIP Emilio Segrè Visual Archives.)

their 1933 paper [Phys. Rev. **43**, 804 (1933)], Wigner and Seitz introduced their cell and showed that the energy of metallic electrons can be estimated by filling it with uncorrelated fermions. Wigner's estimate of the increased binding energy due to correlations [Phys. Rev. **46**, 1002 (1934)] and the general group theory treatment of electronic band structure [Bouckaert, Smoluchowski, and Wigner, Phys. Rev. **50**, 58 (1936)] pointed the way to the quantitative understanding now possible with powerful computers and sophisticated techniques.

The possibility that the microscopic behavior of electrons and ions might some day be quantitatively related to investigations of phase changes under pressure had little influence on the work of Bridgman [Phys. Rev. 2nd Ser., **III**, 126 (1914)] throughout his career. But much of the import of his work and the impetus for its continuation rest upon the effects of pressure on band structure and, consequently, on the phase transformations and properties of crystalline materials.

Few developments in solid state physics over the past 50 years have revolutionized our outlook as much as superconductivity. Some day superconducting magnets and superconducting devices may affect our daily lives; high-T_c superconductors surely increase the prospects. To date, however, the materials that have most profoundly affected our daily lives are semiconductors. Of considerable interest during the 1930s and 1940s, the recognition of their utility in microelectronic devices has led to sensor, computer, and communications technologies. Many such uses have their origin in transistors, the semiconductor triodes invented by Bardeen, Brattain, and Shockley and first discussed in *The Physical Review* [**74**, 230 (1948); **75**, 1208 (1949)]. The development of the integrated-circuit microelectronic devices that have led to the silicon revolution are represented by the work of Esaki [Phys. Rev. **109**, 603 (1958)] on the tunneling diode and the work of Esaki and Chang [Phys. Rev. Lett. **33**, 495 (1974)] on heterostructures.

Many great advances of the past 40 years have been directed toward characterizing or else have serendipitously enabled scientists to characterize materials—not just regular solids, but complex heterogeneous mixtures of solids and fluids. The first of these tools, and perhaps the one that has seen the widest use, is nuclear magnetic resonance. Instruments that determine the local chemical environment of atoms by measuring the effects of that environment on the precessional frequency and relaxation rates of aligned magnetic moments of atomic nuclei permeate not only the laboratories of physicists, chemists, and biologists studying materials and reactions, but also the imaging facilities of medical researchers and practitioners. As precursive "markers" for the papers of Bloembergen, Pound, and Purcell and of Bloch, Hansen, and Packard (cited earlier), the short contributions by Bloch *et al.* [Phys. Rev. **69**, 127 (1946)] and Purcell, Torrey, and Pound [Phys Rev. **69**, 37 (1946)] are included.

Roman Smoluchowski (1910–). (Courtesy of AIP Emilio Segrè Visual Archives.)

A second important technique for studying the properties of electrons and ions is neutron scattering, represented here by the early work of Shull and Smart [Phys. Rev. **76**, 1256 (1949)], whose studies of elastic, magnetically scattered neutrons first exhibited the microscopic magnetic ordering of antiferromagnets.

Many important detection techniques — ranging from electron microscopy to electron resonance techniques, x-ray scattering, positron annihilation, and photoemission — are represented in the CD-ROM articles. Other facets of laser spectroscopy that might have appeared in this section appear elsewhere in this volume. Perhaps the ultimate diagnostic technique is the scanning tunneling microscope, which displays the position of individual atoms and may eventually serve not only as the ultimate microscope but as the ultimate tweezers. This tool and an important application are described by the microscope's inventors, Binnig, Rohrer, Gerber, and Weibel, in their letters [Phys. Rev. Lett. **49**, 57 (1982); **50**, 120 (1983)].

One of the more subtle electronic features of real materials is the distinction between continuum electronic states and conducting electronic states. The fact that continuum electronic states in band tails might be localized was recognized by P. W. Anderson [Phys. Rev. **109**, 1492 (1958)] in a paper that appeared just about the time superconductivity was being explained. The concept of localization and the fundamental differences between localization in two and three dimensions are discussed in the 1979 paper by Abrahams, Anderson, Licciardello, and Ramakrishnan [Phys. Rev. **42**, 673 (1979)].

An unexpected phenomenon closely linked with localization is the quantized Hall resistance, discovered in 1980 by von Klitzing, Dorda, and Pepper [Phys. Rev. Lett. **45**, 494 (1980)]. The theory of this effect, elucidated a year later by Laughlin [Phys. Rev. **23**, 5632 (1981)], displays yet again the intimate connection between

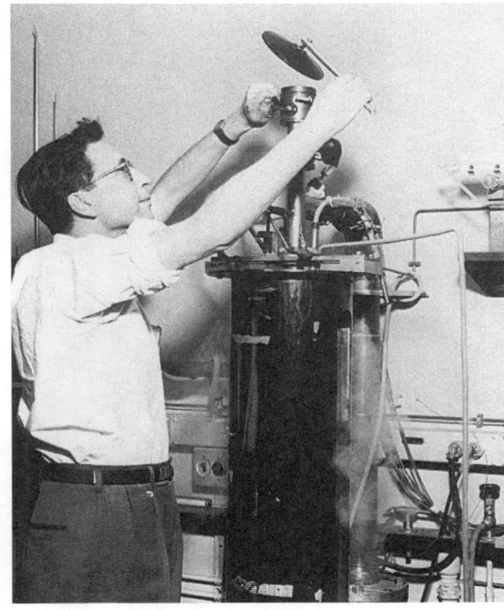

Bernd T. Matthias (1918-1980). (Courtesy of *Physics Today* Collection, AIP Emilio Segrè Visual Archives.)

Clifford G. Shull (1915-). (Courtesy of AIP Emilio Segrè Visual Archives.)

quantum-mechanical electronic motions and macroscopic phenomena. Even more surprising were the discovery by Tsui, Gossard, and Stormer [Phys. Rev. Lett. **48**, 1559 (1982)] of the fractional quantum Hall effect and its explanation by Laughlin [Phys. Rev. Lett. **50**, 1395 (1983)] in terms of unanticipated quantum-mechanical states of the two-dimensional, interacting electron system.

With increasing knowledge and experimental and theoretical sophistication, solid state physicists have been directing their attention toward understanding the subtleties of real materials — not only the simple materials with real impurities that exhibit localization, but more complex materials with more subtle forms of order; smaller chunks of material (clusters and mesoscopic systems, with their universal conductivity fluctuations); and tailor-made materials (ranging from samples sculpted layer by layer or atom by atom for quantum-well devices to composite macroscopic materials with unusual strength and durability). Much of this work has occurred in the last ten years, much is occurring now, and much will occur in the future. Still, as the most recent paper in this section on quasicrystals illustrates [Shechtman, Blech, Gratias, and Cahn, Phys. Rev. Lett. **53**, 1951 (1984)] the possibilities for unanticipated types of complex natural order may not be exhausted even if the vast array of biological materials future generations will study is set aside.

Based on such recent history, the discovery and design of liquids and solids will continue to reveal phenomena that display or exploit the subtle effects of known physical laws in surprising ways — with consequences for mathematics and theoretical speculation on the one hand and for new technology and practical engineering on the other. Let us hope that many of these discoveries will be reported in future volumes of *The Physical Review*.

Martin E. Packard (1921–) and **Russel H. Varian (1898–1959)** holding the first proton free precession magnetometer, 1953. (Courtesy of Varian Associates, AIP Emilio Segrè Visual Archives.)

Papers Reproduced on CD-ROM

P. W. Bridgman. Change of phase under pressure. I. The phase diagram of eleven substances with especial reference to the melting curve, *Phys. Rev.* **3**, 126–141, 189–203 (1914)

P. W. Bridgman. Change of phase under pressure. II. New melting curves with a general thermodynamic discussion of melting, *Phys. Rev.* **6**, 1–3, 94–100, 106–112 (1915)

J. C. Slater. Note on Hartree's method, *Phys. Rev.* **35**, 210–211(L) (1930)

J. Frenkel. On the transformation of light into heat in solids. I., *Phys. Rev.* **37**, 17–44 (1931)

E. Wigner and F. Seitz. On the constitution of metallic sodium, *Phys. Rev.* **43**, 804–810 (1933)

E. Wigner. On the interaction of electrons in metals, *Phys. Rev.* **46**, 1002–1011 (1934)

L. P. Bouckaert, R. Smoluchowski, and E. Wigner. Theory of Brillouin zones and symmetry properties of wave functions in crystals, *Phys. Rev.* **50**, 58–67 (1936)

K. B. Blodgett and I. Langmuir. Built-up films of barium stearate and their optical properties, *Phys. Rev.* **51**, 964–982 (1937)

G. H. Wannier. The structure of electronic excitation levels in insulating crystals, *Phys. Rev.* **52**, 191–197 (1937)

T. Holstein and H. Primakoff. Field dependence of the intrinsic domain magnetization of a ferromagnet, *Phys. Rev.* **58**, 1098–1113 (1940)

R. Smoluchowski. Anisotropy of the electronic work function of metals, *Phys. Rev.* **60**, 661–674 (1941)

J. Bardeen. Surface states and rectification at a metal semi-conductor contact, *Phys. Rev.* **71**, 717–727 (1947)

C. J. Gorter and J. H. van Vleck. The role of exchange interaction in paramagnetic absorption, *Phys. Rev.* **72**, 1128–1129(L) (1947)

J. Bardeen and W. H. Brattain. The transistor, a semi-conductor triode, *Phys. Rev.* **74**, 230–231(L) (1948)

J. H. van Vleck. The dipolar broadening of magnetic resonance lines in crystals, *Phys. Rev.* **74**, 1168–1183 (1948)

G. L. Pearson and J. Bardeen. Electrical properties of pure silicon and silicon alloys containing boron and phosphorus, *Phys. Rev.* **75**, 865–883 (1949)

J. Bardeen and W. H. Brattain. Physical principles involved in transistor action, *Phys. Rev.* **75**, 1208–1225 (1949)

C. G. Shull and J. S. Smart. Detection of antiferromagnetism by neutron diffraction, *Phys. Rev.* **76**, 1256–1257(L) (1949)

W. G. Proctor and F. C. Yu. The dependence of a nuclear magnetic resonance frequency upon chemical compound, *Phys. Rev.* **77**, 717(L) (1950)

C. H. Townes, C. Herring, and W. D. Knight. The effect of electronic paramagnetism on nuclear magnetic resonance frequencies in metals, *Phys. Rev.* **77**, 852–853(L) (1950)

C. A. Reynolds, B. Serin, W. H. Wright, and L. B. Nesbitt. Superconductivity of isotopes of mercury, *Phys. Rev.* **78**, 487(L) (1950)

P. W. Anderson. Antiferromagnetism. Theory of superexchange interaction, *Phys. Rev.* **79**, 350–356 (1950)

R. V. Pound. Nuclear electric quadrupole interactions in crystals, *Phys. Rev.* **79**, 685–702 (1950)

J. R. Haynes and W. Shockley. The mobility and life of injected holes and electrons in germanium, *Phys. Rev.* **81**, 835–843 (1951)

C. Herring and C. Kittel. On the theory of spin waves in ferromagnetic media, *Phys. Rev.* **81**, 869–880 (1951)

W. Shockley, M. Sparks, and G. K. Teal. *p-n* junction transistors, *Phys. Rev.* **83**, 151–162 (1951)

D. Bohm and T. Staver. Application of collective treatment of electron and ion vibrations to theories of conductivity and superconductivity, *Phys. Rev.* **84**, 836–837(L) (1951)

A. W. Overhauser. Polarization of nuclei in metals, *Phys. Rev.* **92**, 411–415 (1953)

H. Y. Carr and E. M. Purcell. Effects of diffusion on free precession in nuclear magnetic resonance experiments, *Phys. Rev.* **94**, 630–638 (1954)

W. Kohn and N. Rostoker. Solution of the Schrödinger equation in periodic lattices with an application to metallic lithium, *Phys. Rev.* **94**, 1111–1120 (1954)

B. T. Matthias, T. H. Geballe, S. Geller, and E. Corenzwit. Superconductivity of Nb_3Sn, *Phys. Rev.* **95**, 1435 (1954)

M. A. Ruderman and C. Kittel. Indirect exchange coupling of nuclear magnetic moments by conduction electrons, *Phys. Rev.* **96**, 99–102 (1954)

J. M. Luttinger and W. Kohn. Motion of electrons and holes in perturbed periodic fields, *Phys. Rev.* **97**, 869–883 (1955)

G. Dresselhaus, A. F. Kip, and C. Kittel. Cyclotron resonance of electrons and holes in silicon and germanium crystals, *Phys. Rev.* **98**, 368–384 (1955)

A. G. Redfield. Nuclear magnetic resonance saturation and rotary saturation in solids, *Phys. Rev.* **98**, 1787–1809 (1955)

J. Bardeen and D. Pines. Electron–phonon interaction in metals, *Phys. Rev.* **99**, 1140–1150 (1955)

W. C. Dash and R. Newman. Intrinsic optical absorption in single-crystal germanium and silicon at 77 °K and 300 °K, *Phys. Rev.* **99**, 1151–1155 (1955)

R. T. Schumacher and C. P. Slichter. Electron spin paramagnetism of lithium and sodium, *Phys. Rev.* **101**, 58–65 (1956)

C. Herring and E. Vogt. Transport and deformation-potential theory for many-valley semiconductors with anisotropic scattering, *Phys. Rev.* **101**, 944–961 (1956)

T. R. Carver and C. P. Slichter. Experimental verification of the Overhauser nuclear polarization effect, *Phys. Rev.* **102**, 975–980 (1956)

G. Feher. Observation of nuclear magnetic resonances via the electron spin resonance line, *Phys. Rev.* **103**, 834–835(L) (1956)

R. E. Glover III and M. Tinkham. Transmission of superconducting films at millimeter-microwave and far infrared frequencies, *Phys. Rev.* **104**, 844–845(L) (1956)

M. Tinkham. Energy gap interpretation of experiments on infrared transmission through superconducting films, *Phys. Rev.* **104**, 845–846(L) (1956)

L. N. Cooper. Bound electron pairs in a degenerate Fermi gas, *Phys. Rev.* **104**, 1189–1190(L) (1956)

J. Bardeen, L. N. Cooper, and J. R. Schrieffer. Microscopic theory of superconductivity, *Phys. Rev.* **106**, 162–164(L) (1957)

K. Yosida. Magnetic properties of Cu–Mn alloys, *Phys. Rev.* **106**, 893–898 (1957)

J. Bardeen, L. N. Cooper, and J. R. Schrieffer. Theory of superconductivity, *Phys. Rev.* **108**, 1175–1204 (1957)

L. Esaki. New phenomenon in narrow germanium *p-n* junctions, *Phys. Rev.* **109**, 603–604(L) (1958)

P. W. Anderson. Absence of diffusion in certain random lattices, *Phys. Rev.* **109**, 1492–1505 (1958)

P. W. Anderson. Coherent excited states in the theory of superconductivity: Gauge invariance and the Meissner effect, *Phys. Rev.* **110**, 827–835 (1958)

W. L. Roth. Magnetic structures of MnO, FeO, CoO, and NiO, *Phys. Rev.* **110**, 1333–1341 (1958)

I. S. Jacobs and R. W. Schmitt. Low-temperature electrical and magnetic behavior of dilute alloys: Mn in Cu and Co in Cu, *Phys. Rev.* **113**, 459–463 (1959)

L. C. Hebel and C. P. Slichter. Nuclear spin relaxation in normal and superconducting aluminum, *Phys. Rev.* **113**, 1504–1519 (1959)

A. G. Anderson and A. G. Redfield. Nuclear spin-lattice relaxation in metals, *Phys. Rev.* **116**, 583–591 (1959)

L. M. Corliss, J. M. Hastings, and R. J. Weiss. Antiphase antiferromagnetic structure of chromium, *Phys. Rev. Lett.* **3**, 211–212 (1959)

W. Cochran. Crystal stability and the theory of ferroelectricity, *Phys. Rev. Lett.* **3**, 412–414 (1959)

I. Giaever. Energy gap in superconductors measured by electron tunneling, *Phys. Rev. Lett.* **5**, 147–148 (1960)

I. Giaever. Electron tunneling between two superconductors, *Phys. Rev. Lett.* **5**, 464–466 (1960)

D. G. Henshaw and A. D. B. Woods. Modes of atomic motions in liquid helium by inelastic scattering of neutrons, *Phys. Rev.* **121**, 1266–1274 (1961)

I. Giaever and K. Megerle. Study of superconductors by electron tunneling, *Phys. Rev.* **122**, 1101–1111 (1961)

P. W. Anderson. Localized magnetic states in metals, *Phys. Rev.* **124**, 41–53 (1961)

L. R. Walker, G. K. Wertheim, and V. Jaccarino. Interpretation of the Fe^{57} isomer shift, *Phys. Rev. Lett.* **6**, 98–101 (1961)

B. S. Deaver, Jr. and W. M. Fairbank. Experimental evidence for quantized flux in superconducting cylinders, *Phys. Rev. Lett.* **7**, 43–46 (1961)

R. Doll and M. Näbauer. Experimental proof of magnetic flux quantization in a superconducting ring, *Phys. Rev. Lett.* **7**, 51–52 (1961)

S. R. Hartmann and E. L. Hahn. Nuclear double resonance in the rotating frame, *Phys. Rev.* **128**, 2042–2053 (1962)

C. P. Bean, M. V. Doyle, and A. G. Pincus. Synthetic high-field, high-current superconductor, *Phys. Rev. Lett.* **9**, 93–94 (1962)

P. W. Anderson. Theory of flux creep in hard superconductors, *Phys. Rev. Lett.* **9**, 309–311 (1962)

P. W. Anderson and J. M. Rowell. Probable observation of the Josephson superconducting tunneling effect, *Phys. Rev. Lett.* **10**, 230–232 (1963)

J. M. Rowell, P. W. Anderson, and D. E. Thomas. Image of the phonon spectrum in the tunneling characteristic between superconductors, *Phys. Rev. Lett.* **10**, 334–336 (1963)

J. R. Schrieffer, D. J. Scalapino, and J. W. Wilkins. Effective tunneling density of states in superconductors, *Phys. Rev. Lett.* **10**, 336–339 (1963)

S. Groves and W. Paul. Band structure of gray tin, *Phys. Rev. Lett.* **11**, 194–196 (1963)

C. N. Berglund and W. E. Spicer. Photoemission studies of copper and silver: Theory, *Phys. Rev.* **136**, A1030–A1044 (1964)

P. Hohenberg and W. Kohn. Inhomogeneous electron gas, *Phys. Rev.* **136**, B864–B871 (1964)

R. C. Jaklevic, J. Lambe, A. H. Silver, and J. E. Mercereau. Quantum interference effects in Josephson tunneling, *Phys. Rev. Lett.* **12**, 159–160 (1964)

W. Kohn and L. J. Sham. Self-consistent equations including exchange and correlation effects, *Phys. Rev.* **140**, A1133–A1138 (1965)

W. L. McMillan and J. M. Rowell. Lead phonon spectrum calculated from superconducting density of states, *Phys. Rev. Lett.* **14**, 108–112 (1965)

A. R. Hutson, A. Jayaraman, A. G. Chynoweth, A. S. Coriell, and W. L. Feldman. Mechanism of the Gunn effect from a pressure experiment, *Phys. Rev. Lett.* **14**, 639–641 (1965)

I. Giaever. Detection of the ac Josephson effect, *Phys. Rev. Lett.* **14**, 904–906 (1965)

I. Giaever. Magnetic coupling between two adjacent type-II superconductors, *Phys. Rev. Lett.* **15**, 825–827 (1965)

N. W. Ashcroft and J. Lekner. Structure and resistivity of liquid metals, *Phys. Rev.* **145**, 83–90 (1966)

CONDENSED MATTER

C. C. Ackerman, B. Bertman, H. A. Fairbank, and R. A. Guyer. Second sound in solid helium, *Phys. Rev. Lett.* **16**, 789–791 (1966)

A. B. Fowler, F. F. Fang, W. E. Howard, and P. J. Stiles. Magneto-oscillatory conductance in silicon surfaces, *Phys. Rev. Lett.* **16**, 901–903 (1966)

P. Soven. Coherent-potential model of substitutional disordered alloys, *Phys. Rev.* **156**, 809–813 (1967)

B. R. Appleton, C. Erginsoy, and W. M. Gibson. Channeling effects in the energy loss of 3–11-MeV protons in silicon and germanium single crystals, *Phys. Rev.* **161**, 330–349 (1967)

J. S. Langer and V. Ambegaokar. Intrinsic resistive transition in narrow superconducting channels, *Phys. Rev.* **164**, 498–510 (1967)

R. E. Slusher, C. K. N. Patel, and P. A. Fleury. Inelastic light scattering from Landau-level electrons in semiconductors, *Phys. Rev. Lett.* **18**, 77–80 (1967)

W. H. Parker, B. N. Taylor, and D. N. Langenberg. Measurement of $2e/h$ using the ac Josephson effect and its implications for quantum electrodynamics, *Phys. Rev. Lett.* **18**, 287–291 (1967)

W. L. McMillan. Transition temperature of strong-coupled superconductors, *Phys. Rev.* **167**, 331–344 (1968)

W. H. Parker, D. N. Langenberg, A. Denenstein, and B. N. Taylor. Determination of e/h, using macroscopic quantum phase coherence in superconductors. I. Experiment, *Phys. Rev.* **177**, 639–664 (1969)

P. Nozières and C. T. De Dominicis. Singularities in the x-ray absorption and emission of metals. III. One-body theory exact solution, *Phys. Rev.* **178**, 1097–1107 (1969)

R. B. Meyer. Piezoelectric effects in liquid crystals, *Phys. Rev. Lett.* **22**, 918–921 (1969)

A. Jayaraman, V. Narayanamurti, E. Bucher, and R. G. Maines. Pressure-induced metal-semiconductor transition and 4*f* electron delocalization in SmTe, *Phys. Rev. Lett.* **25**, 368–370 (1970)

R. C. Zeller and R. O. Pohl. Thermal conductivity and specific heat of noncrystalline solids, *Phys. Rev. B* **4**, 2029–2041 (1971)

L. C. Feldman and D. E. Murnick. Channeling in iron and lattice location of implanted xenon, *Phys. Rev. B* **5**, 1–6 (1972)

W. F. Brinkman, T. M. Rice, P. W. Anderson, and S. T. Chui. Metallic state of the electron-hole liquid, particularly in germanium, *Phys. Rev. Lett.* **28**, 961–964 (1972)

D. B. McWhan, A. Menth, J. P. Remeika, W. F. Brinkman, and T. M. Rice. Metal-insulator transitions in pure and doped V_2O_3, *Phys. Rev. B* **7**, 1920–1931 (1973)

L. Esaki and L. L. Chang. New transport phenomenon in a semiconductor "superlattice," *Phys. Rev. Lett.* **33**, 495–498 (1974)

D. B. McWhan, A. Jayaraman, J. P. Remeika, and T. M. Rice. Metal-insulator transition in $(V_{1-x}Cr_x)_2O_3$, *Phys. Rev. Lett.* **34**, 547–550 (1975)

M. Kastner, D. Adler, and H. Fritzsche. Valence-alternation model for localized gap states in lone-pair semiconductors, *Phys. Rev. Lett.* **37**, 1504–1507 (1976)

R. J. Birgeneau, J. Als-Nielsen, and G. Shirane. Critical behavior of pure and site-random two-dimensional antiferromagnets, *Phys. Rev. B* **16**, 280–292 (1977)

N. P. Ong and P. Monceau. Anomalous transport properties of a linear-chain metal: $NbSe_3$, *Phys. Rev. B* **16**, 3443–3455 (1977)

C. K. Chiang, C. R. Fincher, Jr., Y. W. Park, A. J. Heeger, H. Shirakawa, E. J. Louis, S. C. Gau, and A. G. MacDiarmid. Electrical conductivity in doped polyacetylene, *Phys. Rev. Lett.* **39**, 1098–1101 (1977)

E. Abrahams, P. W. Anderson, D. C. Licciardello, and T. V. Ramakrishnan. Scaling theory of localization: Absence of quantum diffusion in two dimensions, *Phys. Rev. Lett.* **42**, 673–676 (1979)

C. C. Grimes and G. Adams. Evidence for a liquid-to-crystal phase transition in a classical, two-dimensional sheet of electrons, *Phys. Rev. Lett.* **42**, 795–798 (1979)

W. P. Su, J. R. Schrieffer, and A. J. Heeger. Solitons in polyacetylene, *Phys. Rev. Lett.* **42**, 1698–1701 (1979)

G. J. Dolan and D. D. Osheroff. Nonmetallic conduction in thin metal films at low temperatures, *Phys. Rev. Lett.* **43**, 721–724 (1979)

F. Steglich, J. Aarts, C. D. Bredl, W. Lieke, D. Meschede, and W. Franz. Superconductivity in the presence of strong Pauli paramagnetism: $CeCu_2Si_2$, *Phys. Rev. Lett.* **43**, 1892–1896 (1979)

K. v. Klitzing, G. Dorda, and M. Pepper. New method for high-accuracy determination of the fine-structure constant based on quantized Hall resistance, *Phys. Rev. Lett.* **45**, 494–497 (1980)

R. B. Laughlin. Quantized Hall conductivity in two dimensions, *Phys. Rev. B* **23**, 5632–5633 (1981)

J. Orenstein and M. Kastner. Photocurrent transient spectroscopy: Measurement of the density of localized states in α-As_2Se_3, *Phys. Rev. Lett.* **46**, 1421–1424 (1981)

D. E. Moncton, P. W. Stephens, R. J. Birgeneau, P. M. Horn, and G. S. Brown. Synchrotron x-ray study of the commensurate–incommensurate transition of monolayer krypton on graphite, *Phys. Rev. Lett.* **46**, 1533–1536 (1981)

D. C. Tsui, H. L. Stormer, and A. C. Gossard. Two-dimensional magnetotransport in the extreme quantum limit, *Phys. Rev. Lett.* **48**, 1559–1562 (1982)

G. Binnig, H. Rohrer, Ch. Gerber, and E. Weibel. Surface studies by scanning tunneling microscopy, *Phys. Rev. Lett.* **49**, 57–61 (1982)

G. Binnig, H. Rohrer, Ch. Gerber, and E. Weibel. 7×7 reconstruction on Si(111) resolved in real space, *Phys. Rev. Lett.* **50**, 120–123 (1983)

D. R. Nelson. Liquids and glasses in spaces of incommensurate curvature, *Phys. Rev. Lett.* **50**, 982–985 (1983)

R. B. Laughlin. Anomalous quantum Hall effect: An incompressible quantum fluid with fractionally charged excitations, *Phys. Rev. Lett.* **50**, 1395–1398 (1983)

D. Shechtman, I. Blech, D. Gratias, and J. W. Cahn. Metallic phase with long-range orientational order and no translational symmetry, *Phys. Rev. Lett.* **53**, 1951–1953 (1984)

D. Levine and P. J. Steinhardt. Quasicrystals: A new class of ordered structures, *Phys. Rev. Lett.* **53**, 2477–2480 (1984)

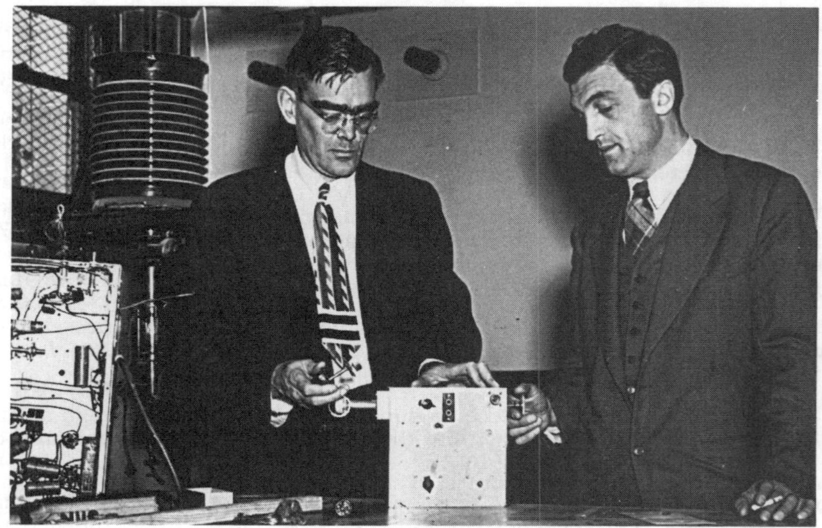

William W. Hansen (1909–1949) and **Felix Bloch (1905–1983).** (Courtesy of Stanford University, AIP Emilio Segrè Visual Archives.)

Edwin H. Land (1909–1991) and **Edward M. Purcell (1912–).** [Courtesy of Paul H. Donaldson (Cruft Laboratory, Harvard Univerisity), AIP Emilio Segrè Visual Archives.]

Note on Hartree's Method

Hartee's method of self-consistent fields, for determining atomic models, has seemed to many persons to stand rather apart from the main current of quantum theory; in spite of the papers of Gaunt and the writer, showing its connection with Schrödinger's equation, it has seemed to contain arbitrary and empirical elements. It appears, however, that it has a very close relation to the variation method. That principle states that, if one has an approximate wave function containing arbitrary parameters or arbitrary functions, one will have the best approximation to a solution of Schrödinger's equation if one chooses the parameters or functions so that the energy is stationary with respect to slight variations of them. Suppose one sets up an approximate wave function for a general problem of the motion of electrons among stationary nuclei, by assuming a product of functions of the various electrons: $u = u_1(x_1) \cdots u_n(x_n)$; suppose further that one apply the variation principle by varying separately each of the functions u_i, leaving the others constant. The n variation equations so obtained prove to be those for the motion of the n electrons, each in a separate electrostatic field; and the field for each electron is obtained by adding the densities u_i^2 for all the other electrons, and finding by electrostatics the field of this charge and of the nuclei. Thus this field is self-consistent in the sense of Hartree; the result is a generalization of his method to more complicated problems than atomic ones.

For atoms, Hartree's procedure differs in the one detail that instead of taking the field so obtained (which would not be quite spherically symmetrical), he averages over all directions, to get a real central field for each electron to move in. This process also can be partly, altnough not entirely, justified by variation methods. If one demands that the functions u_i be solutions of a central field problem—that is, that they be products of spherical harmonics of the angles, by functions of r—and vary only the function of r, then one finds that the potential should be averaged over all orientations; but with a certain weighting function, not used by Hartree, depending on the spherical harmonic occurring in the wave function of the electron for which we are finding the potential. That is, Hartree's method could be slightly changed in this matter, with slight improvement of the results; but, except for this, it is definitely the best method, in the sense of variation principle, using central fields.

Closer examination shows that the potential we have found differs from Hartree's merely in being slightly dependent on the m, as well as the n and l, of the electron one considers. As a result, the energy levels of the different electrons of the same n and l, but different m, will differ slightly, an effect without physical meaning. In an earlier paper of the writer (Phys. Rev. **32**, 343, 1928), it is shown that essentially this same dependence on m results in only a small error, and further, that this effect has a very close connection with the resonance terms, and that by considering these also, the energy again becomes independent of m. It would therefore not be sensible to make this slight improvement in Hartree's method without at the same time making the more important one of considering resonance. If one wished to do this, one could again use variation methods, to derive a new method similar to self-consistent fields: one would demand that the wave function be written not as $u_1(x_1)\cdots u_n(x_n)$, but as a linear combination of such functions with permuted indices, so arranged as to have proper symmetry relations; and one would then vary the u_i's to make the energy a minimum. This would give a method similar to Hartree's, but really an improvement on it. Some preliminary calculations by Drs. Zener and Guillemin indicate that it may really be feasible to carry through this improvement. Their method differs from that sketched here only in that they represent the u_i's by simple analytic forms containing parameters, which they vary, instead of determining them by numerical solutions of a differential equation.

J. C. SLATER

Leipzig, Germany
December 19, 1929.

On the Constitution of Metallic Sodium

E. WIGNER AND F. SEITZ, *Department of Physics, Princeton University*

(Received March 18, 1933)

Previous developments in the theory of metals may be divided clearly into two parts: that based principally upon the hypothesis of free electrons and dealing with conductivity properties, and that based upon calculations of valence forces and dealing with the chemical properties. In the present article an intermediate point of view is adopted and the free-electron picture is employed in an investigation of chemical properties of metallic sodium. The assumption is made that in the metal the K and L shells of an atom are not altered from their form in the free atom. The properties of the wave functions of the electrons are discussed qualitatively, first of all, and it is concluded that the binding energy will be positive even when the Pauli principle is taken account of. This is followed by a quantitative investigation of the energy to be associated with the lowest state. First of all it is shown to what extent the present picture takes account of the interactions of electrons with both parallel and antiparallel spins, and to what extent remaining effects may be neglected. Next a Schroedinger equation is solved in order to determine the lowest energy level for various values of the lattice constant. To this a correction is made to account for the Pauli principle and from the result the lattice constant, binding energy and compressibility are calculated with favorable results.

I.

THE investigations which have been carried out so far on the constitution of metals by quantum mechanics may be divided into two classes, the work on conductivity and related phenomena, carried out chiefly by Bloch, Peierls, Nordheim and Brillouin[1] are mainly based on the hypothesis of free electrons[2] and are concerned with the interaction between the electronic motion and the vibrations of the lattice, which is responsible for the electric resistance. The works of the other class[3] are mainly concerned with the chemical properties and crystal structure of the metals and are based on calculations of valence forces. They encounter great mathematical difficulties because the application of the usual methods to calculate valence forces becomes more and more difficult as the number of atoms increases.

The present work intends to take an intermediate point of view by applying the free electron picture but aiming at a calculation of chemical properties of metallic sodium such as lattice constant, heat of vaporization, compressibility, etc. The method of calculation is the same one as that proposed by Hund for molecules[4] and more recently applied by Lenz and Jensen[5] to ionic lattices, and by Lennard-Jones and H. J. Woods[6] to two dimensional metallic lattices. The electrons are assumed to move freely in a potential field and their interaction is supposed to be contained to a large extent in this field, much as in Hartree's method of the self-consistent field which is actually the field adopted in the calculations of Lenz (not in ours). The initial assumptions which one makes about the statistical connections of positions of different electrons are necessarily rather rough in this picture and should be improved afterwards.

II.

We assume first that the electrons in the K and L shells are not affected by the metallic bond and their wave functions the same as in the

[1] Cf. the comprehensive treatment by L. Brillouin, Die Quantenstatistik. Berlin, 1932.

[2] Cf. W. Pauli, Zeits. f. Physik 41, 81 (1927); A. Sommerfeld, Zeits. f. Physik 47, 1 (1928).

[3] J. C. Slater, Phys. Rev. 35, 509 (1930); E. A. Hylleraas, Zeits. f. Physik 63, 771 (1930); and especially H. S. Taylor, H. Eyring, A. Sherman, J. Chem. Phys. 1, 68 (1933).

[4] F. Hund, Zeits. f. Physik 40, 742 (1927) and applications of this point of view to crystals, Zeits. f. Physik 74, 1 (1932).

[5] W. Lenz, Zeits. f. Physik 77, 713 (1932); H. Jensen, Zeits. f. Physik 77, 722 (1932).

[6] J. E. Lennard-Jones and H. J. Woods, Proc. Roy. Soc. A120, 727 (1928).

free state. This is justified since the corresponding wave functions practically vanish in half the interatomic distance. For the valence electron, however, such an assumption is quite out of the way, since the maximum of the corresponding wave function is (quite necessarily, as we shall see) just about half way between two atoms. Contrary to the conditions which exist in the free state, however, the wave function must not drop to zero after the maximum but can continue periodically through the whole crystal. It will therefore be much smoother than the wave function of the free atom, and the kinetic energy of the corresponding state will consequently be much smaller than that of the electron in the free atom. The potential energy, on the other hand, will be negatively larger in the lattice than in the free state because outside of the above-mentioned maximum the wave will not be under the influence of the nucleus considered originally, but under that of the next nucleus of the lattice, which is nearer. The electron with the wave function just described will have a larger negative energy than that in the free atom and we consider this to be the essence of the metallic state.

Of course, the wave function which one obtains by continuing the atomic wave function periodically in the lattice is not the real wave function of the free electron in the lattice, but the energy of the latter will be even smaller than that of the former. We shall try to find an approximation to the real wave function by actually solving a differential equation.

It must be added that not all the free electrons can be in the state given above, because of the Pauli principle. This reduces the magnitude of the metallic bond, because the electrons must have an additional kinetic energy, which is known as the zero-point energy of a Fermi gas. One sees easily, however, that this additional energy is smaller than the reduction of the kinetic energy which was obtained by continuing the wave function periodically through the whole lattice, so that there certainly remains a positive amount for the metallic bond.

III.

First, we shall calculate the energy of the free electron in the lowest state. We shall do this by numerically solving a Schroedinger equation. It will not be necessary to solve it for the entire lattice, because it will have the same symmetry as the crystal and hence will merely repeat itself a great number of times. Because of this symmetry, the derivative of the wave function at every crystallographic symmetry plane will be zero perpendicular to this plane. This will be used as a boundary condition. The crystallographic symmetry planes which we shall use in this way bisect perpendicularly the lines connecting the second nearest atoms. If we draw lines connecting the nearest atoms and consider the planes bisecting these perpendicularly, we have every atom surrounded by a truncated octahedron (Fig. 1).

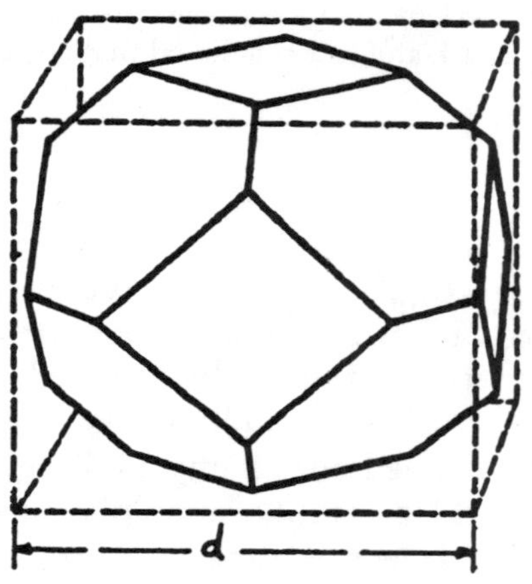

FIG. 1.

The middle points of the planes of the latter possess such symmetry (S_6) that the derivative of the wave function must vanish at these points in every direction. It will be quite a good approximation to replace the polyhedron of Fig. 1 by a sphere of equal volume, and to take as boundary conditions that the derivative of the wave function vanishes at the boundary of this sphere.

The determination of the potential function to be used inside this sphere is more difficult. It would be quite out of the way to use Hartree's method as the density of electrons in the greatest part of the domain is very small. This fundamental difference between metallic and ionic lattices was already pointed out by Lenz.[5] If we assume that two electrons are never around the same ion, every ion may be supposed to be surrounded by a spherical electron cloud which will exactly[7] cancel its potential outside of the sphere. Hence it seems to be the best simple assumption to take the potential as that of the ion inside the sphere mentioned above. The knowledge of this potential will allow us to set down the differential equation for the free electron which will be solved with the boundary condition that the derivative vanish at the boundaries of a sphere. The solution will be obviously spherically symmetric about each atom, which is of course not true for the actual wave function of the free electron but is a consequence of our approximations. It is not very far from the truth, however, since the wave function will actually have the highest crystallographic symmetry (O^h) which is not very far from spherical symmetry, for to every direction there are not less than 47 other equivalent directions.

The justification of the assumption that two electrons are never around the same ion arises from two sources Consider first the statistical connections between the positions of two electrons in an ideal Fermi gas. The complete wave function will be a determinant

$$\Psi(1,2,\cdots n)=\frac{1}{(n!)^{\frac{1}{2}}}\begin{vmatrix}\psi_1(1) & \psi_2(1) & \cdots & \psi_n(1)\\ \psi_1(2) & \psi_2(2) & \cdots & \psi_n(2)\\ \vdots & \vdots & & \vdots\\ \psi_1(n) & \psi_2(n) & \cdots & \psi_n(n)\end{vmatrix} \qquad (1)$$

where the numbers in parentheses represent the Cartesian and spin coordinates of the corresponding electrons and the ψ are the wave functions of the different states.[8] In order to obtain the statistical relations between the positions of two electrons we have to square (1) and integrate it over the coordinates of all electrons except those considered, which will be taken as 1 and 2. Because of the orthogonality relations of the ψ_i the result will be, apart from a constant,

$$\sum_{\kappa=1}^{n}\sum_{\lambda=1}^{n}\left[|\psi_\kappa(1)|^2|\psi_\lambda(2)|^2-\psi_\kappa(1)\psi_\lambda(1)^*\psi_\kappa(2)^*\psi_\lambda(2)\right]. \qquad (2)$$

This still contains the spin coordinates s_1 and s_2 of 1 and 2 and reads more explicitly, if the edge of the cubic-shaped crystal is L,

$$\sum_{\nu_1\nu_2\nu_3}\sum_{\mu_1\mu_2\mu_3}\left[|e^{2\pi i(\nu_1x_1+\nu_2y_1+\nu_3z_1)/L}|^2|e^{2\pi i(\mu_1x_2+\mu_2y_2+\mu_3z_2)/L}|^2\delta_{s_1\sigma_1}\delta_{s_2\sigma_2} - e^{2\pi i[(\nu_1-\mu_1)(x_1-x_2)+(\nu_2-\mu_2)(y_1-y_2)+(\nu_3-\mu_3)(z_1-z_2)]/L}\delta_{s_1\sigma_1}\delta_{s_1\sigma_2}\delta_{s_2\sigma_2}\delta_{s_2\sigma_1}\right]. \qquad (2a)$$

There are really two questions to discuss: the statistical connection between electrons with antiparallel spin ($\sigma_1=-\sigma_2$) and between those with parallel spin ($\sigma_1=\sigma_2$). For a pair of the first kind the second term of (2a) vanishes so that they are statistically independent. For two electrons with parallel spin, on the other hand, we have to evaluate the sum of (2a) after having omitted the spin factors. We shall denote the distance of the two electrons by r, and may assume that the line joining them lies in the X direction, since the probability will not depend on the direction. With these conditions, (2a) becomes

$$\sum_{\nu_1\nu_2\nu_3}\sum_{\mu_1\mu_2\mu_3}(1-e^{2\pi i(\nu_1-\mu_1)r/L}). \qquad (3)$$

Here the summation over ν_2, ν_3, μ_2, μ_3 can be carried out at once. The limitation on the ν_1, ν_2, ν_3 and μ_1, μ_2, μ_3 being

$$\nu_1^2+\nu_2^2+\nu_3^2\leq\nu^2;\quad \mu_1^2+\mu_2^2+\mu_3^2\leq\nu^2$$
$$\nu=(3n/8\pi)^{\frac{1}{3}} \qquad (4)$$

it gives

$$P(r)=\sum_{\nu_1=-\nu}^{\nu}\sum_{\mu_1=-\nu}^{\nu}\pi^2(\nu^2-\nu_1^2)(\nu^2-\mu_1^2)(1-e^{2\pi i(\nu_1-\mu_1)r/L}). \qquad (5)$$

Now the summation over ν_1 and μ_1, after dividing by $\pi^2\nu^2(4\nu^2-1)^2/9$ gives for the probability of the electrons with parallel spin being a distance r apart

[7] There is nothing like exchange forces in our picture.

[8] This consideration is contained implicitly in the work of Uhlenbeck and Gropper for the case of only slightly degenerated gases, Phys. Rev. 40, 1029 (1932). We are interested in the case of complete degeneracy, however.

$$4\pi r^2P(r)=4\pi r^2\left\{1-\left(\frac{3}{2}\frac{\cos(\pi r/L)\sin(2\pi\nu r/L)-2\nu\sin(\pi r/L)\cos(2\pi\nu r/L)}{\nu(4\nu^2-1)\sin^3(\pi r/L)}\right)^2\right\}$$
$$=4\pi^2r^2\left\{1-\left(3\frac{\sin(r/d')-(r/d')\cos(r/d')}{(r/d')^3}\right)^2\right\}. \quad (6)$$

Here $d'=v_0^{\frac{1}{3}}/3^{\frac{1}{3}}\pi^{\frac{2}{3}}$ and $v_0=L^3/n$ is the atomic volume. The function $P(r)$ is sketched in Fig. 2, it vanishes for $r=0$ and approaches 1 as r becomes large compared with the lattice constant. It attains its half-value for $r=1.79\ d'$ or $0.460\ d$ for a body-centered lattice with a cube edge d. The radius of the sphere described above is about the same, namely $(\frac{3}{8}\pi)^{\frac{1}{3}}d=0.492\ d$. So we see that two electrons with parallel spin will be very rarely at the same ion, simply in consequence of the exclusion principle. This will be also true for a Fermi gas subject to a periodic potential, as the potential does not materially alter this argument.

It remains to investigate the case of antiparallel spins somewhat more closely. As it is not possible that three electrons have antiparallel spins, the probability of three electrons being at the same ion will be very small anyway. For two electrons with antiparallel spins and without interaction, however, there is no statistical connection of the positions. If one should take the interaction into account, it would turn out, however, that there is a connection of such a kind that they are but rarely in the neighborhood of each other. This is already indicated in the well-known solution of Hylleraas[9] for He and in the similar solution of Bethe[10] for the negative hydrogen ion. These solutions also show that the connection is of such an order of magnitude that the choice of potential function, which corresponds to our rather rough picture of the metallic bond, is justified to some extent. It must be admitted, however, that the lower limit of the energies of the free electrons, which we calculate in this way, will certainly give too large a binding energy for the lattice, as all of the electrons will not be at different ions with certainty, and also because the terms of Hylleraas and Bethe just discussed will increase the kinetic energy above the value which we obtain by our boundary conditions. We shall not take up this question in more detail this time as it is deeply connected with the interaction problem of the electron and the justification of the notions of the free and bound electrons and we hope to return to it at another time.

IV.

The calculation of the wave function inside the proper spheres of the ions is very simple in principle. The potential function of Prokofjew[11] was used for the purpose. This was obtained by Prokofjew following a method of Kramers[12] in which one employs experimental values of the terms of Na. The differential equation for the radial function $R=r\psi(r)$ is

$$-(h^2/8\pi^2m)(\partial^2R/\partial r^2)+V(r)R=ER(r) \quad (7)$$

and in units of the Bohr radius of H, the quantity $Q(\rho)=-a_0\rho^2V/e^2$, is approximated for various intervals by parabolas, as follows:

ρ = from	to	Q
0.00	0.01	$Q=11\rho$
0.01	0.15	$=-26.4\ \rho^2+11.53\ \rho-0.00264$
0.15	1.00	$=-\ 2.84\ \rho^2+\ 4.46\ \rho+0.5275$
1.00	1.55	$=+\ 1.508\ \rho^2-\ 4.236\ \rho+4.876$
1.55	3.30	$=\ 0.1196\rho^2+\ 0.2072\rho+1.319$
3.30	6.74	$=\ 0.0005\rho^2+\ 0.9933\rho+0.0222$
6.74	∞	$=\rho$

The boundary condition $\partial\psi/\partial r=0$ requires that at the boundary R should satisfy

$$\partial R/\partial r=R/r. \quad (8)$$

[9] E. A. Hylleraas, Zeits. f. Physik 48, 469 (1928).

[10] H. Bethe, Zeits. f. Physik 57, 815 (1929).

[11] W. Prokofjew, Zeits. f. Physik 58, 255 (1929). (Note: Prokofjew's table of $Q(\rho)$ (p. 258) contains obvious errors in two places, one in decimal point and one in sign. These were easily detected by the continuity conditions. The form given here is corrected.)

[12] H. A. Kramers, Zeits. f. Physik 39, 828 (1926).

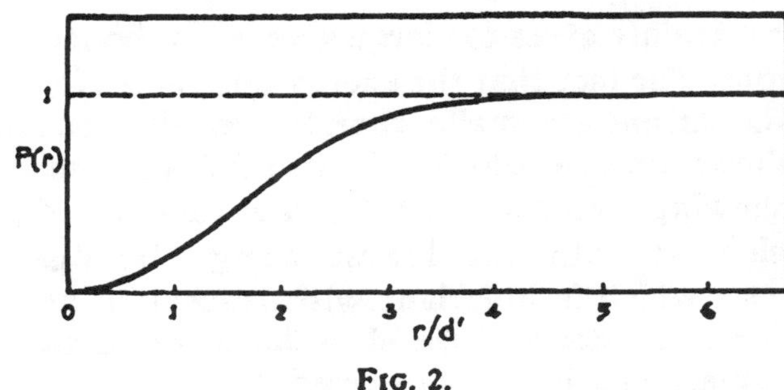

FIG. 2.

Instead of finding the energy value E for different radii of the sphere, the radii corresponding to different energy values were determined. Thus an arbitrary energy value E was taken and the corresponding wave function was obtained from (7) using the method of finite differences employed by Prokofjew. The calculation was started at $r=0$, however, so that every calculated wave function could be used. For $r=0$ the radial function R vanishes and the solution up to $r=0.025$ was calculated by means of a power series in r. After this the method of finite differences was employed, first with differences of 0.005 and then with larger ones when allowable. The largest difference employed was 0.32 ($r>4.6$). The wave function had practically no dependence on E in the neighborhood of the origin so that it was not always necessary to repeat this part of the calculation. As a check, the energy of the electron in the free atom was also determined, the calculated value lies between 0.3820 and 0.3800 Rydberg units, while the experimental value is 0.3778. In Fig. 3, the wave function of the free atom and the wave function for $E=0.500$ are plotted. The numerical tables will be published at another time.

After having the wave function, it was easy to determine the radius of the sphere, for which the boundary condition (8) is satisfied by drawing the tangents to R from the origin. The figure shows, that the boundary conditions are satisfied for two different radii, so that every numerical integration yields two points of the $E(r)$ curve, which gives the energy of the most strongly bound free electron as a function of the lattice constant $d=(8\pi/3)^{\frac{1}{3}}r$. In Fig. 4 the $E(r)$ curve is given (lower line), the unit of energy being the ionization energy of H. For very large r it approaches the ionization energy of atomic Na, possesses a minimum around $r=3$, and rises again for smaller values of r. This latter behavior is due to the fact that a further compression of the lattice would push the valence electron inside the closed L shell, which of course requires energy. The lattice, unlike a similar H lattice,[13] would be stable, therefore, even without taking into account Fermi statistics.

The calculation of a wave function took about two afternoons, and five wave functions were calculated on the whole, giving the ten points of

[13] Cf. E. A. Hylleraas, reference 3.

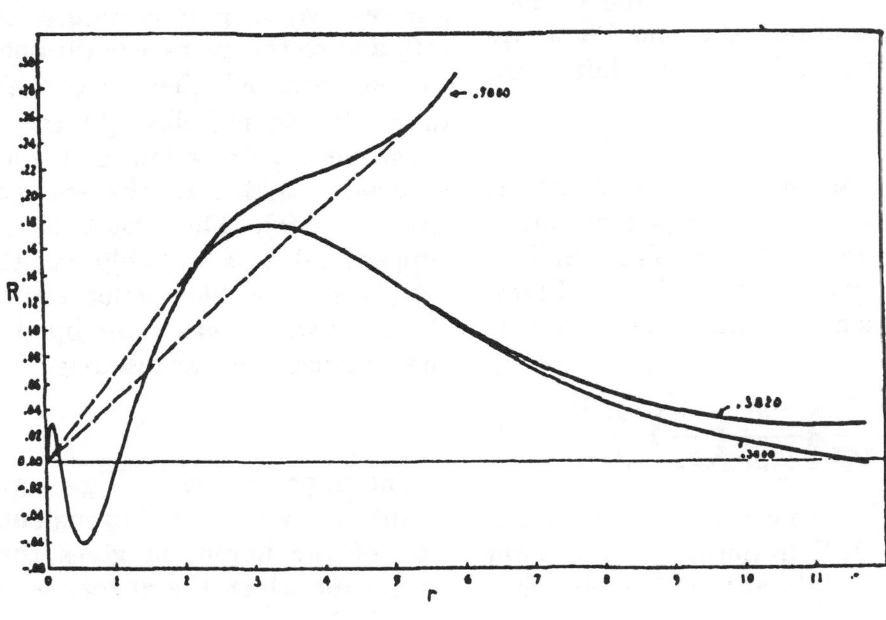

FIG. 3.

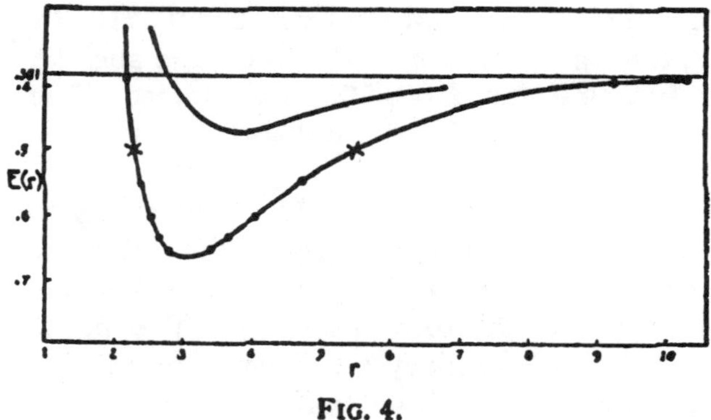

FIG. 4.

the figure. The points of the wave function of Fig. 4 are marked by a cross.

Another point which should be mentioned is that concerning the change in energy of the inner shells. The change is not due chiefly to the change of the boundary conditions as discussed above for the valence electron. This, of course, does increase the binding energy of these electrons, but only by a very small amount. A greater change arises from the increase in the probability of the valence electron being inside the L-electrons, because of the material change of the normalization factor. This decreases the binding energy of the inner electrons and hence lowers the heat of vaporization. A calculation of this effect has been made and shows that the decrease in binding energy is 0.008 Rydberg units per atom, or 2500 small calories per mole. This was obtained by evaluating the change of the potential of the inner electrons in the field due to the valence electron for the free and bound atom. The inner charge distribution was taken to be that given by Hartree.*

V.

The last question we have to investigate is concerned with the additional energy of the other free electrons due to the Fermi distribution. This energy was calculated by the simple Fermi formula and it gives a mean additional energy for every electron

$$\frac{3h^2}{10m}\left(\frac{3}{8\pi}\right)^{2/3}\frac{1}{V_0^{2/3}}=\frac{9h^2}{80\pi m}\left(\frac{3}{2\pi}\right)^{2/3}\frac{1}{r^2} \qquad (9)$$

or $(9\pi/10)(3/2\pi)^{2/3}r^{-2}$ if the energy is measured in Rydberg units and r in Bohr units. As a matter of fact, this formula is valid only for free electrons and it certainly gives too large a value for bound electrons. The fact that the energy differences for bound electrons are smaller than for free electrons was shown first by Bloch.[14] It also follows from the following argument. Let the wave function of the electron with the lowest energy be ψ_{000} (x, y, z), which is invariant with respect to an addition of an identity period to the coordinates. This invariance is not possessed by the wave functions of the other free electrons, and that with the quantum numbers ν_1, ν_2, ν_3, will be multiplied by $e^{2\pi i\nu_1 d/L}$, $e^{2\pi i\nu_2 d/L}$ and by $e^{2\pi i\nu_3 d/L}$ if x, y, or z are increased by d, respectively. Now ψ_{000} gives the lowest possible energy of all wave functions, which are orthogonal to the wave functions of the L and K shells. For the wave function with the quantum numbers ν_1, ν_2, ν_3 this is true if we compare it only with functions which have the same symmetry character, i.e., are multiplied by the same factors if one replaces x, y or z by $x+d$, $y+d$ or $z+d$, respectively. There is, however, the function

$$\psi_{\nu_1\nu_2\nu_3}=e^{2\pi i(\nu_1 x+\nu_2 y+\nu_3 z)/L}\psi_{000}(x, y, z), \qquad (10)$$

which has all the required properties and the energy of which differs from that of ψ_{000} only by $(h^2/2mL^2)(\nu_1^2+\nu_2^2+\nu_3^2)$, the Fermi energy for free electrons. This is easily seen upon calculating $(\psi_{\nu_1\nu_2\nu_3}, H\psi_{\nu_1\nu_2\nu_3})$ for (10) and remembering that ψ_{000} may be assumed to be real. The energy of the real wave function with the quantum numbers ν_1, ν_2, ν_3 is certainly less than that of (10) and so the average additional energy for the free electrons in higher states is also certainly less than (9). Nevertheless (9) was adopted in the subsequent calculation and the corresponding expression added to the energy of the lowest electron $E(r)$. The result is given in Fig. 4 (upper line). It is probably true that the fact that (9) gives a too high value largely compensates the error which was made by the assumption of the free electrons, as discussed in Section III.

VI.

The upper curve in Fig. 4 gives at once all quantities we desire to calculate. The position of the minimum gives the radius of the sphere for which the energy is the smallest and

* Hartree, Proc. Camb. Phil. Soc. 24, 111 (1927).

[14] F. Bloch, Zeits. f. Physik.

when multiplied by $(8\pi/3)^{\frac{1}{3}}$ it should give the lattice constant for the absolute zero point. Similarly the depth of the minimum below the line of the energy for the free atom should give, after subtracting the correction for the energy gained by the inner shells (2.5 kilo cal.) the energy difference between the gas and the solid state (i.e., the heat of vaporization per atom) in Rydberg units at the absolute zero point. Finally the curvature at the minimum r_m is in a simple connection with the compressibility: the energy change for a linear compression in the ratio α per volume v_0 is $v_0\kappa(3\alpha)^2/2$, where κ is the compressibility, and, on the other hand, it is $\frac{1}{2}r_m^2\alpha^2d^2E(r_m)/dr^2$ when calculated from the figure. This gives

$$\kappa=(1/9)(r_m^2/V_0)(d^2E(r_m)/dr^2). \quad (11)$$

The quantity d^2E/dr^2 was calculated as if the lower curve were linear at r_m and all the curvature arose from (9), which is approximately true according to the figure. The final values obtained for the three quantities d, λ and κ are $d=4.2$A, $\lambda=25.6$ kilo cal./mol, $\kappa=1.6\times10^{-11}$ c.g.s. The depth of the minimum of the lower curve below the energy of the free atom is 88.5 kilo cal.; and the depth at the point where the upper curve has its minimum is 70.4. The Fermi correction at this point is 42.3 kilo cal. In order to have a fair comparison with experiment, the experimental values for these quantities must be extrapolated to the absolute zero point. It must be remembered, however, that we did not treat the motion of the nuclei by quantum mechanics and in consequence, the extrapolation should be done in such a way as to neglect the quantum effects. This was done by taking the values for room temperature and correcting them linearly. The three values for room temperatures are[15]: $d=4.30$A; $\lambda=26.00$ kilo cal./mol; $\kappa=1.67\times10^{-11}$ c.g.s. The coefficient of thermal expansion is 62×10^{-6};[16] the corrected value of λ is determined by adding the difference in heat content of solid and gas ((6−3) cal./deg.×300 deg. = 900 cal.) to the room temperature value of λ; and the value of κ at 0°K was obtained by extrapolating values given along with the above. The final values are: $d=4.23$A, $\lambda=26.9$ kilo cal./mol; $\kappa\sim1.0\times10^{-11}$ c.g.s. The theoretical values compare favorably with these, partly, without doubt, as a consequence of compensating errors.

The work on sodium is being extended with particular reference to a more exact determination of the distribution of energy levels in the neighborhood of the lowest one. Moreover, the corresponding calculations on Li, K, Rb, by using Hartree's and Hargreaves' fields,[17] are being undertaken by one of us.

---

[15] d: P. P. Ewald, Hand. d. Phys. XXIV, 331.
λ: J. Sherman, Chem. Rev. 11, 93 (1932).
κ: Landolt Bornstein, Erster Ergaenzungsband 5 auf., 25. Int. Crit. Tables III, 47.

[16] Int. Crit. Tables II, 461.

[17] We wish to offer our thanks, at this time, to Professor J. C. Slater and hence to Dr. Hartree, for the use of the unpublished tables of K^+.

PHYSICAL REVIEW VOLUME 70, NUMBERS 7 AND 8 OCTOBER 1 AND 15, 1946

Nuclear Induction

F. Bloch
Stanford University, California
(Received July 19, 1946)

The magnetic moments of nuclei in normal matter will result in a nuclear paramagnetic polarization upon establishment of equilibrium in a constant magnetic field. It is shown that a radiofrequency field at right angles to the constant field causes a forced precession of the total polarization around the constant field with decreasing latitude as the Larmor frequency approaches adiabatically the frequency of the r-f field. Thus there results a component of the nuclear polarization at right angles to both the constant and the r-f field and it is shown that under normal laboratory conditions this component can induce observable voltages. In Section 3 we discuss this nuclear induction, considering the effect of external fields only, while in Section 4 those modifications are described which originate from internal fields and finite relaxation times.

PHYSICAL REVIEW VOLUME 70, NUMBERS 7 AND 8 OCTOBER 1 AND 15, 1946

The Nuclear Induction Experiment

F. BLOCH, W. W. HANSEN, AND M. PACKARD
Stanford University, California

(Received July 19, 1946)

The phenomenon of nuclear induction has been studied experimentally. The apparatus used is described, both as to principle and detail. Experiments have been carried out in which the signals from protons contained in a variety of substances were observed. The results show the role played by the relaxation time, which was found to vary between about 10^{-6} second and many seconds.

PHYSICAL REVIEW VOLUME 73, NUMBER 7 APRIL 1, 1948

Relaxation Effects in Nuclear Magnetic Resonance Absorption*

N. BLOEMBERGEN,** E. M. PURCELL, AND R. V. POUND,***
Lyman Laboratory of Physics, Harvard University, Cambridge, Massachusetts
(Received December 29, 1947)

The exchange of energy between a system of nuclear spins immersed in a strong magnetic field, and the heat reservoir consisting of the other degrees of freedom (the "lattice") of the substance containing the magnetic nuclei, serves to bring the spin system into equilibrium at a finite temperature. In this condition the system can absorb energy from an applied radiofrequency field. With the absorption of energy, however, the spin temperature tends to rise and the rate of absorption to decrease. Through this "saturation" effect, and in some cases by a more direct method, the *spin-lattice relaxation time* T_1 can be measured. The interaction *among* the magnetic nuclei, with which a characteristic time T_2' is associated, contributes to the width of the absorption line. Both interactions have been studied in a variety of substances, but with the emphasis on liquids containing hydrogen.

Magnetic resonance absorption is observed by means of a radiofrequency bridge; the magnetic field at the sample is modulated at a low frequency. A detailed analysis of the method by which T_1 is derived from saturation experiments is given. Relaxation times observed range from 10^{-4} to 10^{2} seconds. In liquids T_1 ordinarily decreases with increasing viscosity, in some cases reaching a minimum value after which it increases with further increase in viscosity. The line width meanwhile increases monotonically from an extremely small value toward a value determined by the spin-spin interaction in the rigid lattice. The effect of paramagnetic ions in solution upon the proton relaxation time and line width has been investigated. The relaxation time and line width in ice have been measured at various temperatures.

The results can be explained by a theory which takes into account the effect of the thermal motion of the magnetic nuclei upon the spin-spin interaction. The local magnetic field produced at one nucleus by neighboring magnetic nuclei, or even by electronic magnetic moments of paramagnetic ions, is spread out into a spectrum extending to frequencies of the order of $1/\tau_c$, where τ_c is a correlation time associated with the local Brownian motion and closely related to the characteristic time which occurs in Debye's theory of polar liquids. If the nuclear Larmor frequency ω is much less than $1/\tau_c$, the perturbations caused by the local field nearly average out, T_1 is inversely proportional to τ_c, and the width of the resonance line, in frequency, is about $1/T_1$. A similar situation is found in hydrogen gas where τ_c is the time between collisions. In very viscous liquids and in some solids where $\omega\tau_c > 1$, a quite different behavior is predicted, and observed. Values of τ_c for ice, inferred from nuclear relaxation measurements, correlate well with dielectric dispersion data.

Formulas useful in estimating the detectability of magnetic resonance absorption in various cases are derived in the appendix.

The Transistor, A Semi-Conductor Triode

J. BARDEEN AND W. H. BRATTAIN
Bell Telephone Laboratories, Murray Hill, New Jersey
June 25, 1948

A THREE-ELEMENT electronic device which utilizes a newly discovered principle involving a semi-conductor as the basic element is described. It may be employed as an amplifier, oscillator, and for other purposes for which vacuum tubes are ordinarily used. The device consists of three electrodes placed on a block of germanium[1] as shown schematically in Fig. 1. Two, called the emitter and collector, are of the point-contact rectifier type and are placed in close proximity (separation ~.005 to .025 cm) on the upper surface. The third is a large area low resistance contact on the base.

The germanium is prepared in the same way as that used for high back-voltage rectifiers.[2] In this form it is an *N*-type or excess semi-conductor with a resistivity of the order of 10 ohm cm. In the original studies, the upper surface was subjected to an additional anodic oxidation in a glycol borate solution[3] after it had been ground and etched in the usual way. The oxide is washed off and plays no direct role. It has since been found that other surface treatments are equally effective. Both tungsten and phosphor bronze points have been used. The collector point may be electrically formed by passing large currents in the reverse direction.

Each point, when connected separately with the base electrode, has characteristics similar to those of the high back-voltage rectifier. Of critical importance for the operation of the device is the nature of the current in the forward direction. We believe, for reasons discussed in detail in the accompanying letter,[4] that there is a thin layer next to the surface of *P*-type (defect) conductivity. As a result, the current in the forward direction with respect to the block is composed in large part of holes, i.e., of carriers of sign opposite to those normally in excess in the body of the block.

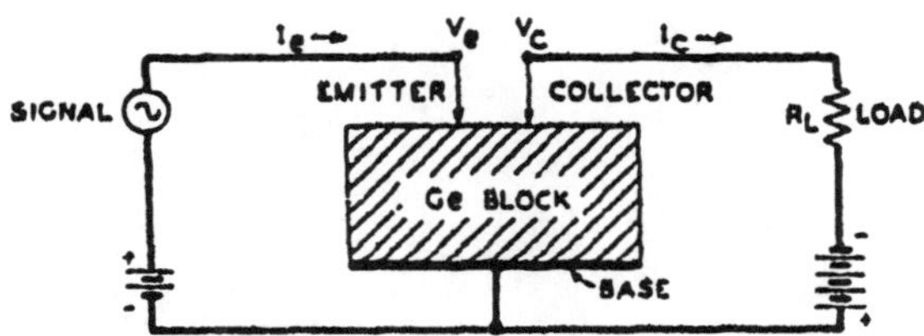

FIG. 1. Schematic of semi-conductor triode.

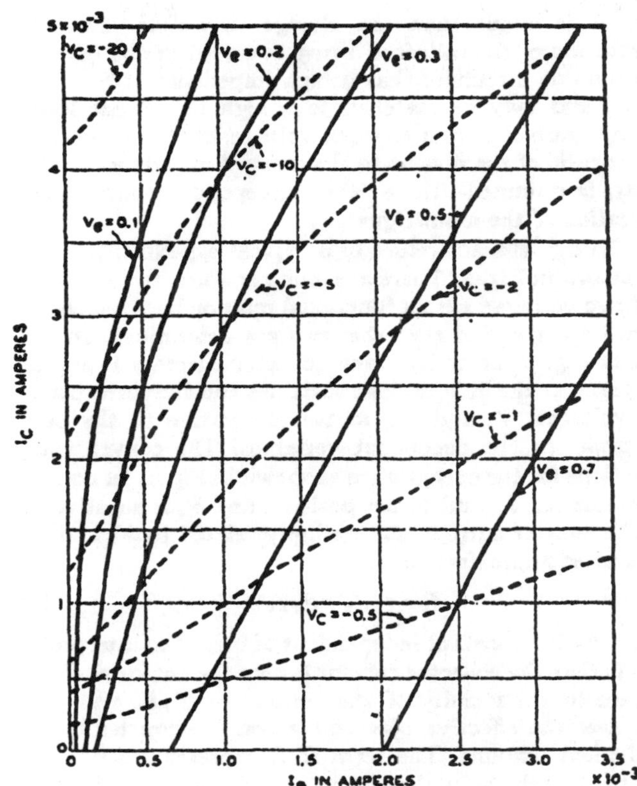

FIG. 2. d.c. characteristics of an experimental semi-conductor triode. The currents and voltages are as indicated in Fig. 1.

When the two point contacts are placed close together on the surface and d.c. bias potentials are applied, there is a mutual influence which makes it possible to use the device to amplify a.c. signals. A circuit by which this may be accomplished in shown in Fig. 1. There is a small forward (positive) bias on the emitter, which causes a current of a few milliamperes to flow into the surface. A reverse (negative) bias is applied to the collector, large enough to make the collector current of the same order or greater than the emitter current. The sign of the collector bias is such as to attract the holes which flow from the emitter so that a large part of the emitter current flows to and enters the collector. While the collector has a high impedance for flow of electrons into the semi-conductor, there is little impediment to the flow of holes into the point. If now the emitter current is varied by a signal voltage, there will be a corresponding variation in collector current. It has been found that the flow of holes from the emitter into the collector may alter the normal current flow from the base to the collector in such a way that the change in collector current is larger than the change in emitter current. Furthermore, the collector, being operated in the reverse direction as a rectifier, has a high impedance (10^4 to 10^6 ohms) and may be matched to a high impedance load. A large ratio of output to input voltage, of the same order as the ratio of the reverse to the forward impedance of the point, is obtained. There is a corresponding power amplification of the input signal.

The d.c. characteristics of a typical experimental unit are shown in Fig. 2. There are four variables, two currents and two voltages, with a functional relation between them. If two are specified the other two are determined. In the plot of Fig. 2 the emitter and collector currents I_e and I_c are taken as the independent variables and the corresponding voltages, V_e and V_c, measured relative to the base electrode, as the dependent variables. The conventional directions for the currents are as shown in Fig. 1. In normal operation, I_e, I_c, and V_e are positive, and V_c is negative.

The emitter current, I_e, is simply related to V_e and I_c. To a close approximation:

$$I_e = f(V_e + R_F I_c), \tag{1}$$

where R_F is a constant independent of bias. The interpretation is that the collector current lowers the potential of the surface in the vicinity of the emitter by $R_F I_c$, and thus increases the effective bias voltage on the emitter by an equivalent amount. The term $R_F I_c$ represents a positive feedback, which under some operating conditions is sufficient to cause instability.

The current amplification factor α is defined as

$$\alpha = (\partial I_c / \partial I_e)_{V_c = \text{const}}.$$

This factor depends on the operating biases. For the unit shown in Fig. 2, α lies between one and two if $V_c < -2$.

Using the circuit of Fig. 1, power gains of over 20 db have been obtained. Units have been operated as amplifiers at frequencies up to 10 megacycles.

We wish to acknowledge our debt to W. Shockley for initiating and directing the research program that led to the discovery on which this development is based. We are also indebted to many other of our colleagues at these Laboratories for material assistance and valuable suggestions.

[1] While the effect has been found with both silicon and germanium, we describe only the use of the latter.

[2] The germanium was furnished by J. H. Scaff and H. C. Theuerer. For methods of preparation and information on the rectifier, see H. C. Torrey and C. A. Whitmer, *Crystal Rectifiers* (McGraw-Hill Book Company, Inc., New York, New York, 1948), Chap. 12.

[3] This surface treatment is due to R. B. Gibney, formerly of Bell Telephone Laboratories, now at Los Alamos Scientific Laboratory.

[4] W. H. Brattain and J. Bardeen, Phys. Rev., this issue.

Detection of Antiferromagnetism by Neutron Diffraction*

C. G. Shull
Oak Ridge National Laboratory, Oak Ridge, Tennessee
AND
J. Samuel Smart
Naval Ordnance Laboratory, White Oak, Silver Spring, Maryland
August 29, 1949

TWO necessary conditions for the existence of ferromagnetism are: (1) the atoms must have a net magnetic moment due to an unfilled electron shell, and (2) the exchange integral J relating to the exchange of electrons between neighboring atoms must be positive. This last condition is required in order that spin states of high multiplicity, which favor ferromagnetism, have the lowest energy. It seems certain that for many of the non-ferromagnetic substances containing a high concentration of magnetic atoms the exchange integrals are negative. In such cases the lowest energy state is the one in which the maximum number of antiparallel pairs occur. An approximate theory of such substances has been developed by Néel,[1] Bitter,[2] and Van Vleck[3] for one specific case and the results are briefly described below.

Consider a crystalline structure which can be divided into two interpenetrating lattices such that atoms on one lattice have nearest neighbors only on the other lattice. Examples are simple cubic and body-centered cubic structures. Let the exchange integral for nearest neighbors be negative and consider only nearest neighbor interactions. Theory then predicts that the structure will exhibit a Curie temperature. Below the Curie temperature the spontaneous magnetization *vs.* temperature curve for one of the sub-lattices is that for an ordinary ferromagnetic material. However, the magnetization directions for the two lattices are antiparallel so that no net spontaneous magnetization exists. At absolute zero all of the atoms on one lattice have their electronic magnetic moments aligned in the same direction and all of the atoms on the other lattice have their moments antiparallel to the first. Above the Curie temperature the thermal energy is sufficient to overcome the tendency of the atoms to lock antiparallel and the behavior is that of a normal paramagnetic substance.

Materials exhibiting the characteristics described above have been designated "antiferromagnetic." Up to the present time the only methods of detecting antiferromagnetism experimentally have been indirect, e.g., determination of Curie points by susceptibility and specific heat anomalies. It has occurred to one of us (J.S.S.) that neutron diffraction experiments might provide a direct means of detecting antiferromagnetism. In an antiferromagnetic material below the Curie temperature a rigid lattice of magnetic ions is formed and the interaction of the neutron magnetic moment with this lattice should result in measurable coherent scattering. Halpern and Johnson[4] have shown that the magnetic and nuclear scattering amplitudes of a paramagnetic atom should be of the same order of magnitude and this result has been qualitatively verified by experimental investigators.[5] At the time of the above suggestion, an experimental program on the determination of the magnetic scattering patterns for various paramagnetic substances (MnO, MnF_2, $MnSO_4$ and Fe_2O_3) was underway at Oak Ridge National Laboratory and room temperature examination had shown (1) a form factor type of diffusion magnetic scattering (no coupling of the atomic moments) to exist for MnF_2 and $MnSO_4$, (2) a liquid type of magnetic scattering (short-range order coupling of oppositely directed magnetic moments) to exist for MnO and (3) the presence of strong coherent magnetic diffraction peaks at forbidden reflection positions for the α-Fe_2O_3 lattice. The latter two observations are in complete accord with the antiferromagnetic notion since the Curie points for MnO and α-Fe_2O_3 are respectively[6] 122°K and 950°K.

Figure 1 shows the neutron diffraction patterns obtained for powdered MnO at room temperature and at 80°K. The room temperature pattern shows coherent nuclear diffraction peaks at the regular face-centered cubic reflection positions and the liquid type of diffuse magnetic scattering in the background. It should be pointed out that the coherent nuclear scattering amplitudes for Mn and O are of opposite sign so that the diffraction pattern is a reversed NaCl type of pattern. The low temperature pattern also shows the same nuclear diffraction peaks, since there is no crystallographic transition in this temperature region,[7] and in addition shows the presence of strong magnetic reflections at positions not allowed on the basis of the chemical unit cell. The magnetic reflections can be indexed, however, making use of a magnetic unit cell twice as large as the chemical unit cell. A complete description of the magnetic structure will be given at a later date.

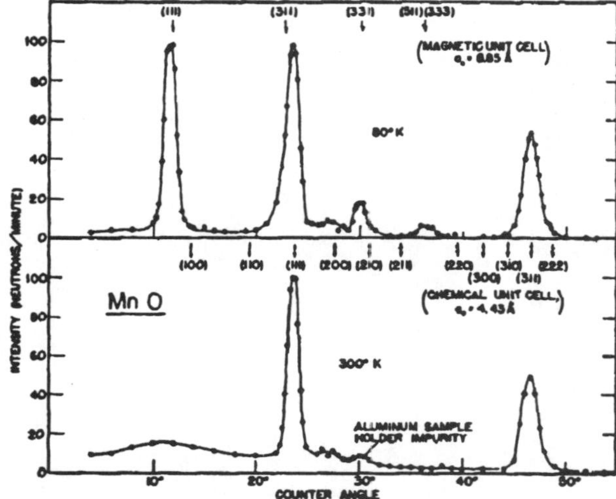

FIG. 1. Neutron diffraction patterns for MnO at room temperature and at 80°K.

In conclusion it appears that neutron diffraction studies of antiferromagnetic materials should provide a new and important method of investigating the exchange coupling of magnetic ions.

* This work was supported in part by the ONR.

[1] L. Néel, Ann. de physique 17, 5 (1932).
[2] F. Bitter, Phys. Rev. 54, 79 (1938).
[3] J. H. Van Vleck, J. Chem. Phys. 9, 85 (1941).
[4] O. Halpern and M. H. Johnson, Phys. Rev. 55, 898 (1939).
[5] Whittaker, Beyer, and Dunning, Phys. Rev. 54, 771 (1938); Ruderman, Havens, Taylor, and Rainwater, Phys. Rev. 75, 895 (1949); and also unpublished work at Oak Ridge National Laboratory.
[6] Bizette, Squire, and Tsai, Comptes Rendus 207, 449 (1938).
[7] B. Ruhemann, Physik. Zeits. Sowjetunion 7, 590 (1935).

Letters to the Editor

PUBLICATION of brief reports of important discoveries in physics may be secured by addressing them to this department. The closing date for this department is five weeks prior to the date of issue. No proof will be sent to the authors. The Board of Editors does not hold itself responsible for the opinions expressed by the correspondents. Communications should not exceed 600 words in length and should be submitted in duplicate.

Bound Electron Pairs in a Degenerate Fermi Gas*

LEON N. COOPER
Physics Department, University of Illinois, Urbana, Illinois
(Received September 21, 1956)

It has been proposed that a metal would display superconducting properties at low temperatures if the one-electron energy spectrum had a volume-independent energy gap of order $\Delta \simeq kT_c$, between the ground state and the first excited state.[1,2] We should like to point out how, primarily as a result of the exclusion principle, such a situation could arise.

Consider a pair of electrons which interact above a quiescent Fermi sphere with an interaction of the kind that might be expected due to the phonon and the screened Coulomb fields. If there is a net attraction between the electrons, it turns out that they can form a bound state, though their total energy is larger than zero. The properties of a noninteracting system of such bound pairs are very suggestive of those which could produce a superconducting state. To what extent the actual many-body system can be represented by such noninteracting pairs will be discussed in a forthcoming paper.

Because of the similarity of the superconducting transition in a wide variety of complicated and differing metals, it is plausible to assume that the details of metal structure do not affect the qualitative features of the superconducting state. Thus, we neglect band and crystal structure and replace the periodic ion potential by a box of volume V. The electrons in this box are free except for further interactions between them which may arise due to Coulomb repulsions or to the lattice vibrations.

In the presence of interaction between the electrons, we can imagine that under suitable circumstances there will exist a wave number q_0 below which the free states are unaffected by the interaction due to the large energy denominators required for excitation. They provide a floor (so to speak) for the possible transitions of electrons with wave number $k_i > q_0$. One can then consider the eigenstates of a pair of electrons with k_1, $k_2 > q_0$.

For a complete set of states of the two-electron system we take plane-wave product functions, $\varphi(\mathbf{k}_1,\mathbf{k}_2;\, \mathbf{r}_1,\mathbf{r}_2) = (1/V)\exp[i(\mathbf{k}_1\cdot\mathbf{r}_1+\mathbf{k}_2\cdot\mathbf{r}_2)]$ which satisfy periodic boundary conditions in a box of volume V, and where $\mathbf{r}_1$ and $\mathbf{r}_2$ are the coordinates of electron one and electron two. (One can use antisymmetric functions and obtain essentially the same results, but alternatively we can choose the electrons of opposite spin.) Defining relative and center-of-mass coordinates, $\mathbf{R}=\frac{1}{2}(\mathbf{r}_1+\mathbf{r}_2)$, $\mathbf{r}=(\mathbf{r}_2-\mathbf{r}_1)$, $\mathbf{K}=(\mathbf{k}_1+\mathbf{k}_2)$ and $\mathbf{k}=\frac{1}{2}(\mathbf{k}_2-\mathbf{k}_1)$, and letting $\mathcal{E}_K+\epsilon_k=(\hbar^2/m)(\frac{1}{4}K^2+k^2)$, the Schrödinger equation can be written

$$(\mathcal{E}_K+\epsilon_k-E)a_k+\sum_{k'} a_{k'}(\mathbf{k}|H_1|\mathbf{k}')\times\delta(\mathbf{K}-\mathbf{K}')/\delta(0)=0 \quad (1)$$

where

$$\Psi(\mathbf{R},\mathbf{r})=(1/\sqrt{V})e^{i\mathbf{K}\cdot\mathbf{R}}\chi(\mathbf{r},K),\qquad \chi(\mathbf{r},K)=\sum_k (a_k/\sqrt{V})e^{i\mathbf{k}\cdot\mathbf{r}}, \quad (2)$$

and

$$(\mathbf{k}|H_1|\mathbf{k}')=\left(\frac{1}{V}\int d\mathbf{r}e^{-i\mathbf{k}\cdot\mathbf{r}}H_1e^{i\mathbf{k}'\cdot\mathbf{r}}\right)_{0\text{ phonons}}$$

We have assumed translational invariance in the metal. The summation over $\mathbf{k}'$ is limited by the exclusion principle to values of k_1 and k_2 larger than q_0, and by the delta function, which guarantees the conservation of the total momentum of the pair in a single scattering. The K dependence enters through the latter restriction.

Bardeen and Pines[3] and Fröhlich[4] have derived approximate formulas for the matrix element $(\mathbf{k}|H_1|\mathbf{k}')$; it is thought that the matrix elements for which the two electrons are confined to a thin energy shell near the Fermi surface, $\epsilon_1\simeq\epsilon_2\simeq\epsilon_F$, are the principal ones involved in producing the superconducting state.[2-4] With this in mind we shall approximate the expressions for $(\mathbf{k}|H_1|\mathbf{k}')$ derived by the above authors by

$$\begin{aligned}(\mathbf{k}|H_1|\mathbf{k}')&=-|F| \quad \text{if} \quad k_0\leqslant k,\, k'\leqslant k_m\\ &=0 \quad \text{otherwise,}\end{aligned} \quad (3)$$

where F is a constant and $(\hbar^2/m)(k_m^2-k_0^2)\simeq 2\hbar\omega\simeq 0.2$ ev. Although it is not necessary to limit oneself so strongly, the degree of uncertainty about the precise form of $(\mathbf{k}|H_1|\mathbf{k}')$ makes it worthwhile to explore the consequences of reasonable but simple expressions.

With these matrix elements, the eigenvalue equation becomes

$$1=-|F|\int_{\epsilon_0}^{\epsilon_m}\frac{N(K,\epsilon)d\epsilon}{E-\epsilon-\mathcal{E}_K}, \quad (4)$$

where $N(K,\epsilon)$ is the density of two-electron states of total momentum K, and of energy $\epsilon=(\hbar^2/m)k^2$. To a very good approximation $N(K,\epsilon)\simeq N(K,\epsilon_0)$. The resulting spectrum has one eigenvalue smaller than $\epsilon_0+\mathcal{E}_K$. while the rest lie in the continuum. The lowest eigenvalue is $E_0=\epsilon_0+\mathcal{E}_K-\Delta$, where Δ is the binding energy of the pair

$$\Delta=(\epsilon_m-\epsilon_0)/(e^{1/\rho}-1), \quad (5)$$

where $\beta=N(K,\epsilon)|F|$. The binding energy, Δ, is independent of the volume of the box, but is strongly dependent on the parameter β.

Following a method of Bardeen,[5] by which the coupling constant for the electron-electron interaction, which is due to phonon exchange, is related to the high-temperature resistivity which is due to phonon absorption, one gets $\beta\simeq\rho n\times10^{-6}$, where ρ is the high-temperature resistivity in esu and n is the number of valence electrons per unit volume. The binding energy displays a sharp change of behavior in the region $\beta\simeq1$ and it is just this region which separates, in almost every case, the superconducting from the nonsuperconducting metals.[5] (Also it is just in this region where the attractive interaction between electrons, due to the phonon field, becomes about equal to the screened Coulomb repulsive interaction.)

The ground-state wave function,

$$\chi_0(r,K)=(\text{const})\int\frac{e^{i\mathbf{k}\cdot\mathbf{r}}N(K,\epsilon(k))}{\mathcal{E}_K+\epsilon(k)-E}\left(\frac{d\epsilon}{dk}\right)d\mathbf{k},\quad(6)$$

represents a true bound state which for large values of r decreases at least as rapidly as const$/r^2$. The average extension of the pair, $[\langle r^2\rangle_{\mathrm{Av}}]^{\frac{1}{2}}$, is of the order of 10^{-4} cm for $\Delta\simeq kT_c$. The existence of such a bound state with nonexponential dependence for large r is due to the exclusion of the states $k<k_0$ from the unperturbed spectrum, and the concomitant degeneracy of the lowest energy states of the unperturbed system. One would get no such state if the potential between the electrons were always repulsive. All of the excited states $\chi_{n>0}(\mathbf{r},K)$ are very nearly plane waves.

The pair described by $\chi_0(r)$ may be thought to have some Bose properties (to the extent that the binding energy of the pair is larger than the energy of interaction between pairs).[6] However, since $N(K,\epsilon)$ is strongly dependent on the total momentum of the pair, K, the binding energy Δ is a very sensitive function of K, being a maximum where $K=0$ and going very rapidly to zero where $K\simeq k_m-k_0$. Thus the elementary excitations of the pair might correspond to the splitting of the pair rather than to increasing the kinetic energy of the pair.

In either case the density of excited states (dN/dE) would be greatly reduced from the free-particle density and the elementary excitations would be removed from the ground state by what amounted to a small energy gap.

If the many-body system could be considered (at least to a lowest approximation) a collection of pairs of this kind above a Fermi sea, we would have (whether or not the pairs had significant Bose properties) a model similar to that proposed by Bardeen which would display many of the equilibrium properties of the superconducting state.

The author wishes to express his appreciation to Professor John Bardeen for his helpful instruction in many illuminating discussions.

* This work was supported in part by the Office of Ordnance Research, U. S. Army.

[1] J. Bardeen, Phys. Rev. **97**, 1724 (1955).

[2] See also, for further references and a general review, J. Bardeen, *Theory of Superconductivity Handbuch der Physik* (Springer-Verlag, Berlin, to be published), Vol. 15, p. 274.

[3] J. Bardeen and D. Pines, Phys. Rev. **99**, 1140 (1955).

[4] H. Fröhlich, Proc. Roy. Soc. (London) **A215**, 291 (1952).

[5] John Bardeen, Phys. Rev. **80**, 567 (1950).

[6] It has also been suggested that superconducting properties would result if electrons could combine in even groupings so that the resulting aggregates would obey Bose statistics. V. L. Ginzburg, Uspekhi Fiz. Nauk **48**, 25 (1952); M. R. Schafroth, Phys. Rev. **100**, 463 (1955).

Microscopic Theory of Superconductivity*

J. BARDEEN, L. N. COOPER, AND J. R. SCHRIEFFER
Department of Physics, University of Illinois, Urbana, Illinois
(Received February 18, 1957)

SINCE the discovery of the isotope effect, it has been known that superconductivity arises from the interaction between electrons and lattice vibrations, but it has proved difficult to construct an adequate theory based on this concept. As has been shown by Fröhlich,[1] and in a more complete analysis by Bardeen and Pines[2] in which Coulomb effects were included, interactions between electrons and the phonon field lead to an interaction between electrons which may be expressed in the form

$$H_I = \sum_{\mathbf{k},\mathbf{k}',s,s'} \frac{\hbar\omega|M_\kappa|^2}{(E_\mathbf{k}-E_{\mathbf{k}'})^2-(\hbar\omega)^2} \times c^*_{\mathbf{k}'-\kappa,s'}c_{\mathbf{k}',s'}c^*_{\mathbf{k}+\kappa,s}c_{\mathbf{k},s}+H_{\mathrm{Coul}}, \quad (1)$$

where $|M_\kappa|^2$ is the matrix element for the electron-phonon interaction for the phonon wave vector κ, calculated for the zero-point amplitude of the vibrations, the c's are creation and destruction operators for the electrons in the Bloch states specified by the wave vector $\mathbf{k}$ and spin s, and H_{Coul} represents the screened Coulomb interaction.

Early attempts[3] to construct a theory were based essentially on the self-energy of the electrons, although it was recognized that a true interaction between electrons probably played an essential role. These theories gave the isotope effect, but contained various difficulties, one of which was that the calculated energy difference between what was thought to represent normal and superconducting states was far too large. It is now believed that the self-energy occurs in the normal state, and results in a slight shift of the energies of the Bloch states and a renormalization of the matrix elements.

The present theory is based on the fact that the phonon interaction is negative for $|E_\mathbf{k}-E_{\mathbf{k}'}|<\hbar\omega$. We believe that the criterion for superconductivity is essentially that this negative interaction dominate over the matrix element of the Coulomb interaction, which for free electrons in a volume Ω is $2\pi e^2/\Omega\kappa^2$. In the Bohm-Pines[4] theory, the minimum value of κ is κ_c, somewhat less than the radius of the Fermi surface. This criterion may be expressed in the form

$$-V = \langle -(|M_\kappa|^2/\hbar\omega)+(4\pi e^2/\Omega\kappa^2)\rangle_{Av}<0. \quad (2)$$

Although based on a different principle, this criterion is almost identical with the one given by Fröhlich.[1,3]

If one has a Hamiltonian matrix with predominantly negative off-diagonal matrix elements, the ground state, $\Psi=\sum\alpha_j\varphi_j$, is a linear combination of the original basic states with coefficients predominantly of one sign. A particularly simple example is one for which the original states are degenerate and each state is connected to n other states by the same matrix element $-V$. The ground state, a sum of the original set with equal coefficients, is lowered in energy by $-nV$. One of the authors made use of this principle to construct a wave function for a single pair of electrons excited above the Fermi surface and found that for a negative interaction a bound state is formed no matter how weak the interaction.[5]

Because of the Fermi-Dirac statistics, difficulties are encountered if one tries to apply this principle directly to (1). Matrix elements of H_I between states specified by occupation numbers (Slater determinants) in general may be of either sign. We want to pick out a subset of these between which matrix elements are always of the same sign. This may be done by occupying the individual particle states in pairs, such that if one of the pair is occupied, the other is also. The pairs should be chosen so that transitions between them are possible, i.e., they all have the same total momentum. To form the ground state, the best choice is $\mathbf{k}\uparrow$, $-\mathbf{k}\downarrow$, since exchange terms reduce the matrix elements between states of parallel spin. To form a state with a net current flow, one might take a pairing $\mathbf{k}\uparrow$, $-\mathbf{k}+\mathbf{q}\downarrow$, where $\mathbf{q}$ is a small wave vector, the same for all $\mathbf{k}$ and such that both states are within the range of energy $\hbar\omega$. The occupation of the pairs may be specified by a single spin-independent occupation number, $m_{\mathbf{k}}=0$ or 1. Nonvanishing matrix elements connect configurations which differ in only one of the occupied pairs.[6] It is often convenient to specify occupation in terms of electron pairs above the Fermi surface and hole pairs below.

The best wave function of this form will be a linear combination

$$\Psi=\sum_{\mathbf{k}_1\cdots\mathbf{k}_n} b(\mathbf{k}_1\cdots\mathbf{k}_n)f(\cdots m_{\mathbf{k}_1}\cdots m_{\mathbf{k}_n}\cdots), \tag{3}$$

where the sum is over all possible configurations. In our calculations, we have made a Hartree-like approximation and replaced b by $b(k_1)b(k_2)\cdots b(k_n)$. We have also assumed an isotropic Fermi surface [so that $b(k)$ depends only on the energy ϵ of the Bloch state involved], and that V is the same for all transitions within a constant energy $\hbar\omega$ of the Fermi surface, $\epsilon=0$. A direct calculation gives for the interaction energy

$$W_I=-4[N(0)]^2V\int_0^{\hbar\omega}\int_0^{\hbar\omega}\Gamma(\epsilon)\Gamma(\epsilon')d\epsilon d\epsilon', \tag{4}$$

where $N(0)$ is the density of states at the Fermi surface. The kinetic energy measured from the Fermi sea is

$$W_K=4N(0)\int_0^{\hbar\omega}g(\epsilon)\epsilon d\epsilon, \tag{5}$$

where $g(\epsilon)$ is the probability that a given state of energy ϵ is occupied by a pair, and

$$\Gamma(\epsilon)=\{g(\epsilon)[1-g(\epsilon)]\}^{\frac{1}{2}}. \tag{6}$$

One may interpret the factor $\Gamma(\epsilon)\Gamma(\epsilon')$ as representing the effect of the exclusion principle on restricting the number of configurations which are connected to a given typical configuration. Matrix elements corresponding to $\mathbf{k}\rightarrow\mathbf{k}'$ are possible only if the state $\mathbf{k}$ is occupied and $\mathbf{k}'$ unoccupied in the initial configuration and $\mathbf{k}'$ occupied and $\mathbf{k}$ unoccupied on the final configuration. The probability that this occurs is

$$g(\epsilon)[1-g(\epsilon')]g(\epsilon')[1-g(\epsilon)]=[\Gamma(\epsilon)]^2[\Gamma(\epsilon')]^2. \tag{7}$$

Since matrix elements have probability amplitudes rather than probabilities, the square root of (7) occurs in (4).

A variational calculation to determine the best $g(\epsilon)$ gives

$$W=W_I+W_K=-\frac{2N(0)(\hbar\omega)^2}{\exp[2/N(0)V]-1}. \tag{8}$$

Thus if there is a net negative interaction, no matter how weak, there is a condensed state in which pairs are virtually excited above the Fermi surface. The product $N(0)V$ is independent of isotopic mass and of volume. The energy W varies as $(\hbar\omega)^2$, in agreement with the isotope effect. It should be noted that (8) cannot be obtained in any finite order of perturbation theory. The energy gain comes from a coherence of the electron wave functions with lattice vibrations of short wavelength, and does not represent a condensation in real space.

Empirically, energies are of the order of magnitude of $N(0)(kT_c)^2$, and of course kT_c is much less than an average phonon energy $\hbar\omega$. According to our theory, this will occur if $N(0)V<1$, a not unreasonable assumption. In this weak-coupling limit, the energy may be expressed simply in terms of the number of electrons, n_c, virtually excited in coherent pairs above the Fermi surface at $T=0°$K

$$W=-\tfrac{1}{2}n_c^2/N(0), \tag{9}$$

where

$$n_c=2N(0)\hbar\omega\exp[-1/N(0)V]. \tag{10}$$

It is a great advantage energy-wise to include in the ground state wave function only pairs with the same total momentum. Suppose that instead one had chosen a random pairing, $\mathbf{k}_1\uparrow$, $\mathbf{k}_2\downarrow$, with $\mathbf{k}_1+\mathbf{k}_2=\mathbf{q}$ and consider a typical matrix element $(\mathbf{k}_1,\mathbf{k}_2|H_I|\mathbf{k}_1'\mathbf{k}_2')$ which vanishes unless $\mathbf{k}_1'+\mathbf{k}_2'=\mathbf{q}'=\mathbf{q}$. We shall assume that the $\mathbf{q}$'s of all pairs are small so that if $\mathbf{k}_1$ and $\mathbf{k}_2$ are both within $\hbar\omega$ of the Fermi surface, so are $\mathbf{k}_1'$ and $\mathbf{k}_2'$. If we construct a wave function made up of a linear combination of states with such virtual excited pairs and determine the interaction energy, we would find an expression similar to (4) but with (7) replaced by the much smaller quantity:

$$g(\epsilon_1)g(\epsilon_2)[1-g(\epsilon_1')][1-g(\epsilon_2')]g(\epsilon_1')\times g(\epsilon_2')[1-g(\epsilon_1)][1-g(\epsilon_2)]. \tag{11}$$

The pairing $\mathbf{k}_2=-\mathbf{k}_1$ corresponds to $\mathbf{q}=0$ for all pairs and insures that if $\mathbf{k}_1'$ is unoccupied, so is $\mathbf{k}_2'$. This is also true if all pairs have the same $\mathbf{q}$.

Wave functions corresponding to individual particle excitations may be made of linear combinations of states in which certain occupation numbers, corresponding to real excited electrons or holes, are specified and the rest are used to make all possible combinations of virtual excitations of $\mathbf{k}\uparrow$, $-\mathbf{k}\downarrow$ pairs. Because of the reduction in phase space available to the pairs, the interaction energy is reduced in magnitude. For small excitations the consequent increase in total energy is proportional to the number of excited electrons. This means that a finite energy is required to excite an electron from the ground state. The same applies to real excited $\mathbf{k}, -\mathbf{k}$ pairs. If $f(\epsilon)$ is the probability that a Bloch state of energy ϵ is occupied by an excited electron above the Fermi sea, and $1-f(-\epsilon)$ the probability that there is a hole below, one finds for the interaction energy an expression similar to (4) but with $[\Gamma(\epsilon)]^2$ replaced by $g(\epsilon)\{1-[f(\epsilon)]^2-g(\epsilon)\}$. For small excitations above $T=0°\text{K}$, the total pair energy may be expressed in the weak-coupling limit as

$$W=-\frac{n_c^2}{2N(0)}\left(1-\frac{4n_e}{n_c}\right),\quad n_e \ll n_c, \tag{12}$$

where n_c is the number of electrons in the virtually excited states at $T=0$ and n_e is the number of actually excited electrons. This leads to an energy gap[7] (i.e., the energy required to create an electron-hole pair):

$$E_G=\partial W/\partial n_e=2n_c/N(0) \quad \text{at} \quad T=0°\text{K}. \tag{13}$$

Taking the empirical $W=-H_c^2/8\pi$ and estimating $N(0)$ from the electronic specific heat, we find $E_G=k \times 13.8°\text{K}$ for tin. This is to be compared with the experimental value of about $k\times 11.2°\text{K}$. Calculations are under way to determine the thermal properties at higher temperatures.

Advantages of the theory are (1) It leads to an energy-gap model of the sort that may be expected to account for the electromagnetic properties.[8] (2) It gives the isotope effect. (3) An order parameter, which might be taken as the fraction of electrons above the Fermi surface in virtual pair states, comes in a natural way. (4) An exponential factor in the energy may account for the fact that kT_c is very much smaller than $\hbar\omega$. (5) The theory is simple enough so that it should be possible to make calculations of thermal, transport, and electromagnetic properties of the superconducting state.

* This work was supported in part by the Office of Ordnance Research, U. S. Army. One of us (J.R.S.) wishes to thank the Corning Glass Works Foundation for a grant which aided in the support of this work.

[1] H. Fröhlich, Phys. Rev. **79**, 845 (1950); Proc. Roy. Soc. (London) **A215**, 291 (1952).

[2] J. Bardeen and D. Pines, Phys. Rev. **99**, 1140 (1955).

[3] For reviews of this work, mainly by Fröhlich and by Bardeen, see J. Bardeen, Revs. Modern Phys. **23**, 261 (1951); *Handbuch der Physik* (Springer-Verlag, Berlin, 1956), Vol. 15, p. 274.

[4] See D. Pines, *Solid State Physics* (Academic Press, Inc., New York, 1955), Vol. 1, p. 367.

[5] L. N. Cooper, Phys. Rev. **104**, 1189 (1956).

[6] This looks deceptively like a one-particle problem, but is not, because of the peculiar statistics involved. The pairs cannot be treated as bosons.

[7] Some references to experimental evidence for an energy gap are R. E. Glover and M. Tinkham, Phys. Rev. **104**, 844 (1956); M. Tinkham, Phys. Rev. **104**, 845 (1956); Blevins, Gordy, and Fairbank, Phys. Rev. **100**, 1215 (1955); Corak, Goodman, Satterthwaite, and Wexler, Phys. Rev. **102**, 656 (1956); W. S. Corak and C. B. Satterthwaite, Phys. Rev. **102**, 662 (1956).

[8] J. Bardeen, Phys. Rev. **97**, 1724 (1955).

PHYSICAL REVIEW VOLUME 108, NUMBER 5 DECEMBER 1, 1957

Theory of Superconductivity*

J. BARDEEN, L. N. COOPER,† AND J. R. SCHRIEFFER‡
Department of Physics, University of Illinois, Urbana, Illinois
(Received July 8, 1957)

A theory of superconductivity is presented, based on the fact that the interaction between electrons resulting from virtual exchange of phonons is attractive when the energy difference between the electrons states involved is less than the phonon energy, $\hbar\omega$. It is favorable to form a superconducting phase when this attractive interaction dominates the repulsive screened Coulomb interaction. The normal phase is described by the Bloch individual-particle model. The ground state of a superconductor, formed from a linear combination of normal state configurations in which electrons are virtually excited in pairs of opposite spin and momentum, is lower in energy than the normal state by amount proportional to an average $(\hbar\omega)^2$, consistent with the isotope effect. A mutually orthogonal set of excited states in one-to-one correspondence with those of the normal phase is obtained by specifying occupation of certain Bloch states and by using the rest to form a linear combination of virtual pair configurations. The theory yields a second-order phase transition and a Meissner effect in the form suggested by Pippard. Calculated values of specific heats and penetration depths and their temperature variation are in good agreement with experiment. There is an energy gap for individual-particle excitations which decreases from about $3.5kT_c$ at $T=0°K$ to zero at T_c. Tables of matrix elements of single-particle operators between the excited-state superconducting wave functions, useful for perturbation expansions and calculations of transition probabilities, are given.

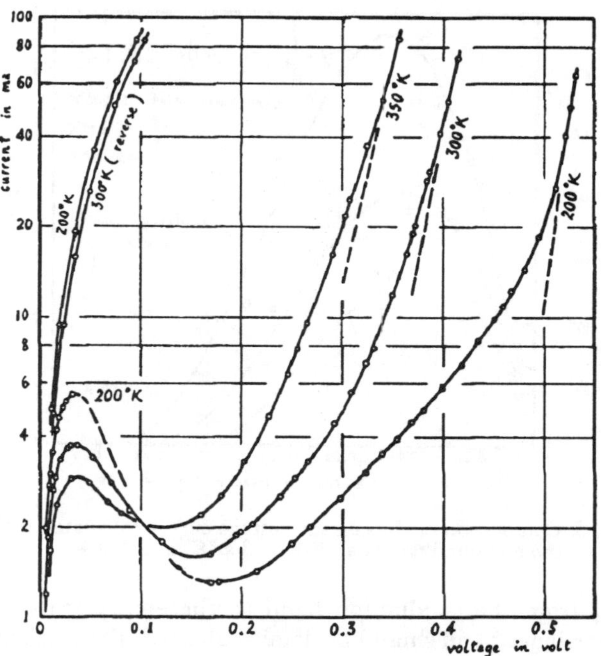

FIG. 1. Semilog plots of the measured current-voltage characteristic at 200°K, 300°K, and 350°K.

New Phenomenon in Narrow Germanium *p-n* Junctions

LEO ESAKI
Tokyo Tsushin Kogyo, Limited, Shinagawa, Tokyo, Japan
(Received October 11, 1957)

IN the course of studying the internal field emission in very narrow germanium *p-n* junctions, we have found an anomalous current-voltage characteristic in the forward direction, as illustrated in Fig. 1. In this *p-n* junction, which was fabricated by alloying techniques, the acceptor concentration in the *p*-type side and the donor concentration in the *n*-type side are, respectively, 1.6×10^{19} cm^{-3} and approximately 10^{19} cm^{-3}. The maximum of the curve was observed at 0.035 ± 0.005 volt in every specimen. It was ascertained that the specimens were reproducibly produced and showed a general behavior relatively independent of temperature. In the range over 0.3 volt in the forward direction, the current-voltage curve could be fitted almost quantitatively by the well-known relation: $I=I_s[\exp(qV/kT)-1]$. This junction diode is more conductive in the reverse direction than in the forward direction. In this respect it agrees with the rectification direction predicted by Wilson, Frenkel, and Joffe, and Nordheim 25 years ago.[1]

The energy diagram of Fig. 2 is proposed for the case in which no voltage is applied to the junction, though the band scheme may be, at best, a poor approximation for such a narrow junction. (The remarkably large values observed in the capacity measurement indicated that the junction width is approximately 150 angstroms, which results in a built-in field as large as 5×10^5 volts/cm.)[2] In the reverse direction and even in the forward direction for low voltage, the current might be carried only by internal field emission and the possibility of an avalanche might be completely excluded because the breakdown occurs at much less than the threshold voltage for electron-hole pair production.[3] Owing to the large density of electrons and holes, their distribution should become degenerate; the Fermi level in the *p*-type side will be 0.06 ev below the top of the valence band, E_v, and that in the *n*-type side will lie above the bottom of the conduction band, E_c. At zero bias, the field emission current $I_{v\to c}$ from the valence band to the empty state of the conduction band and the current

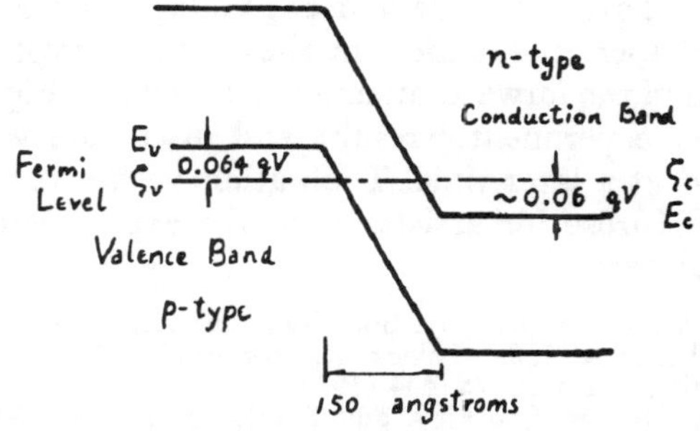

FIG. 2. Energy diagram of the *p-n* junction at 300°K and no bias voltage.

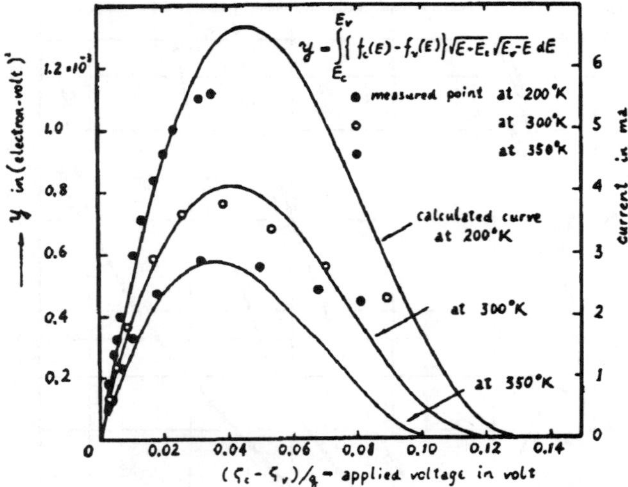

FIG. 3. Comparison of the current-voltage curves calculated with the measured points at 200°K, 300°K, and 350°K.

$I_{c\to v}$ from the conduction band to the empty state of the valence band should be detail-balanced. Expressions for $I_{c\to v}$ and $I_{v\to c}$ might be formulated as follows:

$$I_{c\to v}=A\int_{E_c}^{E_v} f_c(E)\rho_c(E)Z_{c\to v}\{1-f_v(E)\}\rho_v(E)dE,$$

$$I_{v\to c}=A\int_{E_c}^{E_v} f_v(E)\rho_v(E)Z_{v\to c}\{1-f_c(E)\}\rho_c(E)dE,$$

where $Z_{c\to v}$ and $Z_{v\to c}$ are the probabilities of penetrating the gap (these could be assumed to be approximately equal); $f_c(E)$ and $f_v(E)$ are the Fermi-Dirac distribution functions, namely, the probabilities that a quantum state is occupied in the conduction and valence bands, respectively; $\rho_c(E)$ and $\rho_v(E)$ are the energy level densities in the conduction and valence bands, respectively.

When the junction is slightly biased positively and negatively, the observed current I will be given by

$$I=I_{c\to v}-I_{v\to c}=A\int_{E_c}^{E_v}\{f_c(E)-f_v(E)\}Z\rho_c(E)\rho_v(E)dE.$$

From this equation, if Z may be considered to be almost constant in the small voltage range involved, we could calculate fairly well the current-voltage curve at a certain temperature, indicating the dynatron-type characteristic in the forward direction, as shown in Fig. 3.

Further experimental results and discussion will be published at a later time. The author wishes to thank Miss Y. Kurose for assistance in the experiment and the calculations.

[1] A. H. Wilson, Proc. Roy. Soc. (London) **A136**, 487 (1932); J. Frenkel and A. Joffe, Physik. Z. Sowjetunion **1**, 60 (1932); L. Nordheim, Z. Physik **75**, 434 (1932).

[2] McAfee, Ryder, Shockley, and Sparks, Phys. Rev. **83**, 650 (1951); C. Zener, Proc. Roy. Soc. (London) **145**, 523 (1934).

[3] S. L. Miller, Phys. Rev. **99**, 1234 (1955); A. G. Chynoweth and K. G. McKay, Phys. Rev. **106**, 418 (1957).

PHYSICAL REVIEW VOLUME 109, NUMBER 5 MARCH 1, 1958

Absence of Diffusion in Certain Random Lattices

P. W. ANDERSON
Bell Telephone Laboratories, Murray Hill, New Jersey

(Received October 10, 1957)

This paper presents a simple model for such processes as spin diffusion or conduction in the "impurity band." These processes involve transport in a lattice which is in some sense random, and in them diffusion is expected to take place via quantum jumps between localized sites. In this simple model the essential randomness is introduced by requiring the energy to vary randomly from site to site. It is shown that at low enough densities no diffusion at all can take place, and the criteria for transport to occur are given.

I. INTRODUCTION

A NUMBER of physical phenomena seem to involve quantum-mechanical motion, without any particular thermal activation, among sites at which the mobile entities (spins or electrons, for example) may be localized. The clearest case is that of spin diffusion[1,2]; another might be the so-called impurity band conduction at low concentrations of impurities. In such situations we suspect that transport occurs not by motion of free carriers (or spin waves), scattered as they move through a medium, but in some sense by quantum-mechanical jumps of the mobile entities from site to site. A second common feature of these phenomena is randomness: random spacings of impurities, random interactions with the "atmosphere" of other impurities, random arrangements of electronic or nuclear spins, etc.

Our eventual purpose in this work will be to lay the foundation for a quantum-mechanical theory of transport problems of this type. Therefore, we must start with simple theoretical models rather than with the complicated experimental situations on spin diffusion or impurity conduction. In this paper, in fact, we attempt only to construct, for such a system, the simplest model we can think of which still has some expectation of representing a real physical situation reasonably well, and to prove a theorem about the model. The theorem is that at sufficiently low densities, transport does not take place; the exact wave functions are localized in a small region of space. We also obtain a fairly good estimate of the critical density at which the theorem fails. An additional criterion is that the forces be of sufficiently short range—actually, falling off as $r\rightarrow\infty$ faster than $1/r^3$—and we derive a rough estimate of the rate of transport in the $V\propto 1/r^3$ case.

Such a theorem is of interest for a number of reasons: first, because it may apply directly to spin diffusion among donor electrons in Si, a situation in which Feher[3] has shown experimentally that spin diffusion is negligible; second, and probably more important, as an example of a real physical system with an infinite number of degrees of freedom, having no obvious oversimplification, in which the approach to equilibrium is simply impossible; and third, as the irreducible minimum from which a theory of this kind of transport, if it exists, must start. In particular, it re-emphasizes the caution with which we must treat ideas such as "the thermodynamic system of spin interactions" when there is no obvious contact with a real external heat bath.

The simplified theoretical model we use is meant to represent reasonably well one kind of experimental situation: namely, spin diffusion under conditions of

[1] N. Bloembergen, Physica 15, 386 (1949).
[2] A. M. Portis, Phys. Rev. 104, 584 (1956).
[3] G. Feher (private communication).

"inhomogeneous broadening."[4] We assume that we have sites j distributed in some way, regularly or randomly, in three-dimensional space; the array of sites we call the "lattice." We then assume we have entities occupying these sites. They may be spins or electrons or perhaps other particles, but let us call them spins here for brevity. If a spin occupies site j it has energy E_j which (and this is vital) is a stochastic variable distributed over a band of energies completely randomly, with a probability distribution $P(E)dE$ which can be characterized by a width W. Finally, we assume that between the sites we have an interaction matrix element $V_{jk}(r_{jk})$, which transfers the spins from one site to the next. V_{jk} may or may not itself be a stochastic variable with a probability distribution. If one thinks of the mobile entities as up or down electron spins which can occupy various impurity sites, such as color or donor centers, in a crystal, then the random energies E_j are the hyperfine interactions with the surrounding nuclei—Si^{29} for the donors, alkali and halide nuclei for color centers—and $P(E)$ is the line-shape function. In this example, V_{jk} is that part of the interaction which allows an up spin on atom j to flip down while a down spin on k flips up, and the simple process we study is the motion of a single "up" spin among "down" spins.

The "impurity-band" example would again make the sites donors or acceptors, but the E_j's would be energy fluctuations of the donor ground state caused perhaps by Coulomb interactions with randomly placed charged centers; the moving entity would be a single ionized donor. We would have to assume the states of the different donors to be orthogonal, which is no restriction if V is arbitrary. More generally, the situation described by our theorem probably holds in the low-concentration limit and the low-energy tail of almost any model for an impurity band.

One important feature which is missing from our simple model is contact with any external thermal reservoir. When the present theorem holds, some such contact will actually control the transport processes. Our purpose is only to show that the model in itself provides no such reservoir and permits no transport, in spite of its large size and random character; study of the real relaxation and transport processes must come later.

Our basic technique is to place a single "spin" on site n at an initial time $t=0$, and to study the behavior of the wave function thereafter as a function of time. Our fundamental theorem may be restated as: if $V(r_{jk})$ falls off at large distances faster than $1/r^3$, and if the average value of V is less than a certain critical V_c of the order of magnitude of W; then there is actually no transport at all, in the sense that even as $t\rightarrow\infty$ the amplitude of the wave function around site n falls off rapidly with distance, the amplitude on site n itself remaining finite.

One can understand this as being caused by the failure of the energies of neighboring sites to match sufficiently well for V_{jk} to cause real transport. Instead, it causes virtual transitions which spread the state, initially localized at site n, over a larger region of the lattice, without destroying its localized character. More distant sites are not important because the probability of finding one with the right energy increases much more slowly with distance than the interaction decreases.

This theorem leaves two regions of failure (and therefore transport) to be investigated, namely, $V\propto 1/r^3$, as in spin diffusion by dipolar interactions, and $V\sim W$, or the high-concentration limit. In both cases, the methods used to prove the fundamental "nontransport" theorem will probably allow us to outline an approach to the transport problem. In the $1/r^3$ case, we show that transport may be much slower than the estimates of reference 2 would predict. In the case $V\sim W$, we show that transport finally occurs not by single real jumps from one site to another but by the multiplicity of very long paths involving multiple virtual jumps from site to site.

II. SUMMARY OF THE REASONING

Since the mathematical development is fairly complicated and involves lengthy consideration of each of a number of points, we should like to summarize the reasoning rather fully in this section, leaving the proofs and details to later sections. First, then, let us set up the simple model which we study. The equation for the time-dependence of the probability amplitude a_j that a particle is on the site j is:

$$i\dot{a}_j=E_ja_j+\sum_{k\neq j}V_{jk}a_k. \tag{1}$$

Here we measure energies in frequency units, so we can set $\hbar=1$. Equation (1) simply restates the assumptions about the model made in the Introduction.

We study the Laplace transform of the equation (1): let

$$f_j(s)=\int_0^\infty e^{-st}a_j(t)dt, \tag{2}$$

and then

$$i[sf_j(s)-a_j(0)]=E_jf_j+\sum_{k\neq j}V_{jk}f_k. \tag{3}$$

The variable s must be studied as an arbitrary complex variable with positive or zero real part.

The transport problem which interests us is: suppose we know the probability distribution $|a_j(0)|^2$ at time $t=0$, and that it is appreciable only in a certain range of frequency E_j or space $\mathbf{r}_j$. Then how fast, if at all, does this probability distribution diffuse away from this region?

[4] A. M. Portis, Phys. Rev. 91, 1071 (1953).

The simplest question we can ask is to assume $a_0(0)=1$ for a particular atom $j=0$ and inquire how a_j varies with time, or $f_j(s)$ with s. In particular, for very small real part of s we are studying the behavior as $t\to\infty$; for instance, $\lim_{s\to 0+} sf_j(s)$ is in fact $\langle a_j(\infty)\rangle_{\text{Av}}$.

Equation (3) can be written

$$f_j(s)=\frac{i\delta_{0j}}{is-E_j}+\sum_{k\neq j}\frac{1}{is-E_j}V_{jk}f_k(s). \tag{4}$$

In one approach to ordinary transport theory[5] this equation is solved by iteration, in which the equations (4) not involving $f_0(s)$ are solved for $f_j(s)$ in terms of $f_0(s)$:

$$f_j(s)=\frac{1}{is-E_j}V_{j0}f_0(s)+\sum_k\frac{1}{is-E_j}V_{jk}\frac{1}{is-E_k}V_{k0}f_0(s)+\cdots. \tag{5}$$

Then the zeroth equation becomes

$$f_0(s)=\frac{i}{is-E_0}+\sum_k\frac{1}{is-E_0}V_{0k}\left(\frac{V_{k0}}{is-E_k}+\sum_l\frac{1}{is-E_k}V_{kl}\frac{1}{is-E_l}V_{l0}+\cdots\right)f_0(s). \tag{6}$$

Let us call the quantity

$$\sum_k\frac{(V_{0k})^2}{is-E_k}+\sum_{k,l}\frac{V_{0k}V_{kl}V_{l0}}{(is-E_k)(is-E_l)}+\cdots=V_c(0). \tag{7}$$

In many cases the first term suffices. Studying this first term, we see that it can be written

$$V_c(0)=\sum_k(V_{0k})^2\left(\frac{-E_k}{s^2+E_k^2}-\frac{is}{s^2+E_k^2}\right). \tag{8}$$

In the limit as $s\to 0$, the first part of this is obviously just the second-order perturbation $-\Delta E^{(2)}$ of the energy. The second part may be written

$$\lim_{s\to 0}(V_c)=-i\sum_k(V_{0k})^2\delta(E_k)-is\sum_{k,E_k\neq 0}\frac{(V_{0k})^2}{E_k^2}\equiv-\frac{i}{\tau}-isK. \tag{9}$$

Of course, if the first of these terms, the usual transition probability, is finite the second is indeterminate. We include both in order to see what happens if the first term does vanish.

[5] See, for example, W. Kohn and J. M. Luttinger, Phys. Rev. 108, 590 (1957).

Now the equation for $f_0(s)$ is

$$f_0(s)=\frac{i}{is-E_0}+\frac{1}{is-E_0}\left(-\Delta E^{(2)}-\frac{i}{\tau}-isK\right)f_0(s).$$

The solution for f_0 is

$$f_0=\frac{i}{is(1+K)+(i/\tau)-(E_0-\Delta E^{(2)})}. \tag{10}$$

If τ is finite, one gets the usual result of perturbation theory:

$$f_0(s)=\frac{1}{s+(1/\tau)+i(E_0-\Delta E^{(2)})}, \tag{10A}$$

which represents a state of perturbed energy $E_0-\Delta E^{(2)}$ decaying at a rate $e^{-t/\tau}$. If, on the other hand, τ is infinite [$\text{Im}(V_c)\to 0$ as $s\to 0$], then the constant K enters and the amplitude is

$$f_0(s)=\frac{1}{s(1+K)+i(E_0-\Delta E^{(2)})}. \tag{10B}$$

This is a state of the same perturbed energy which does not decay, but has a finite amplitude a_0 $(t\to\infty)$ reduced from unity by the ratio $1/(1+K)$. K is then simply a measure of how much this state has spread to the neighbors by virtual transitions, and has nothing to do with real transport.

Our technique, then, will be to study the behavior of the quantity $V_c(s)$ defined in (7) by an infinite series, just as in the usual transport theory. However, the quantities entering into (7) are completely different in character because of the fact that we have chosen to start with localized states of random energies E_j. In the usual case, there are an infinite number of states j connected to any state 0 by infinitesimal matrix elements V_{0j}, so that within any small range of energies there will be many possible energy-conserving transitions, no one of which takes place with particularly large probability. In such a case the first term of (9) is a meaningful limit of a certain integral. Here, only a few V_{0j}'s are large, and the energies they lead to are stochastically distributed, so that whether or not energy can be conserved is a probability question.

We find that the quantity $V_c(s)$ must be studied as a probability variable: that is, we pick a starting atom 0 and an arbitrary energy E [imaginary part of s; $\text{Re}(s)\to 0$] and study the probability distribution of V_c. Our study then resolves itself into three parts: First, we study the first term (8); second, we discuss the convergence of the series of higher order perturbations. Both of these questions we can resolve in the sense that there is a region in which, with probability unity, $\text{Im}(V_c)\to 0$ as $\text{Re}(s)\to 0$ and the series is convergent. These two parts we shall discuss here briefly, and expand upon in Secs. III and IV. Finally, we must decide whether this kind of convergence in a probability sense is meaningful, and in particular whether the choice of an arbitrary energy is correct. Since this seems reasonable from the start, we shall not go into it further here, but reserve the discussion for Sec. V. We find there that this convergence means that the states are localized, but that it is not easy to assign a correspondence between the perturbed and unperturbed states.

Let us, then, go ahead with the first two questions. The important quantity for the first is

$$\operatorname{Im}(V_c) = -s \sum_k \frac{|V_{0k}|^2}{s^2+E_k^2}.$$

Let

$$\sum_k \frac{|V_{0k}|^2}{s^2+E_k^2} \equiv X(s). \tag{11}$$

We note that X also represents the quantity

$$X(s) = \sum_{j\neq 0} \frac{|f_j(s)|^2}{|f_0(s)|^2} \tag{11'}$$

in first order, as is clear from (5). This points up the interpretation that as $s \to 0$ a finite X means no real transport.

Using the Holtsmark-Markoff[6] method, in the next section we calculate the probability distribution of this sum X, and find that if $V(r) \sim 1/r^{3+\epsilon}$, where $\epsilon > 0$, then $X(0)$ has a distribution law with a perfectly finite most probable value, while the distribution falls off as $1/X^{\frac{3}{2}}$ for large values of X. This latter fact shows, as can also be verified directly, that the mean value of X is infinite. Clearly this is merely the result of an infinitesimal fraction of atoms having a very large value. It can be shown that even these few large values are illusory, and that by suitably redefining the localized states in accordance with multiple-scattering theory[7] the probability of such divergences is greatly reduced.[8]

In reference 2 the transition probability is calculated by taking the mean of (11) over all possible starting atoms. The resulting finite transition probability is therefore meaningless, as discussed above.

The case of $V(r) = A/r^3$, or normal dipolar spin diffusion, is a special one. In this case $X(0) \to \infty$ for all atoms, but the divergence is so extremely weak that $V_c(0)$ is finite. In fact, the distribution of $X(s)$ has the same form, but the most probable value is roughly

$$[X(s)]_{\mathrm{M.P.}} \approx \frac{n^2A^2}{W^2}\left[\sinh^{-1}\left(\frac{W}{2s}\right)\right]^2, \tag{12}$$

where n is the density of sites and W is the width of the distribution of the E_j. This most probable value diverges as the square of $\ln s$ when $s \to 0$.

This case leads to a transport theory with decays slower than exponential. Since the divergence of (12) is caused by large values of r, single, rather long, jumps have an important effect on the transport process. The existence of this transport process for $V \sim 1/r^3$ provides a counter-example for those who may think the present theorem is self-evident for small enough V.

One may roughly estimate transport times by noting that, by the definining relationship (2), that value of s for which $sf(s)$ is of order unity is something like the inverse of a time of decay due to spin diffusion. Equation (12) shows that this time increases exponentially with W/nA for large values—for example, for Si donors at 10^{16} concentration we compute a rate of $\exp(-10^4)$ sec^{-1}.

On the other hand, for $V \sim 1/r^{3+\epsilon}$ (or exponential, as it often is), the first term of perturbation theory leads to a vanishing rate of transport independent of V or W. Thus, if transport is to appear at all it must come in higher terms, and in fact it is easy to convince oneself that it can come only by a divergence of the whole series for V_c. Therefore it is of great importance to our theory to learn how to handle the sums of products

$$V_c(0) = \sum_{j,k,l\cdots m\neq 0} V_{0j}\frac{1}{is-E_j}V_{jk}\frac{1}{is-E_k} \times V_{kl}\frac{1}{is-E_l}\cdots V_{m0}, \tag{13}$$

which represent the possibility of successive virtual transitions until, at possibly some very great distance from site 0, a real, phase-destroying process can occur.

Our method for this problem, set forth in Sec. IV, involves both the idea of calculating a probability distribution rather than a mean for these terms, and also a modification of the multiple-scattering methods of Watson[7] in order to eliminate certain troublesome repeated terms. We must do this elimination first.

Certain terms in (13) are apparently very large because of the fact that there is no prohibition on repeated indices. Suppose, for instance, that V_{jk} is particularly large and that both $is-E_j$ and $is-E_k$ are particularly small so that

$$\left|\frac{1}{is-E_j}V_{jk}\frac{1}{is-E_k}V_{kj}\right| > 1.$$

Then there is the possibility of a term like

$$V_{0l}\frac{1}{is-E_l}\cdots V_{mj}\frac{1}{is-E_j}V_{jk}\frac{1}{is-E_k}V_{kj}\frac{1}{is-E_j}\cdots \times V_{jk}\frac{1}{is-E_j}\cdots\frac{1}{is-E_k}V_{kn}\cdots\frac{1}{is-E_p}V_{p0}.$$

[6] J. Holtsmark, Ann. Physik 58, 577 (1919).

[7] See, for instance, K. M. Watson, Phys. Rev. 105, 1388 (1957), also the references therein.

[8] In this case this is particularly simple: near such an infinity $\operatorname{Re}(V_c)$ will also be large so that the energy will simply be changed to a value at which $\operatorname{Im}(V_c)$ is finite.

Such a term will get larger the more times we repeat V_{jk}. We can represent the terms of our series by closed diagrams through the various sites of the lattice, starting and ending at 0 (see Fig. 1), and this type of diagram involves a "ladder" which repeatedly runs back and forth from j to k. Physically, we can think of it as resulting from a pair of closely coupled atoms.

The technique of Watson[7,9] shows us that we may eliminate all repeated indices in a self-consistent way by including in the energy denominator for atom k the perturbed energy $V_e(k)$ calculated from just such a series of terms as (13). A complicating factor is that $V_e(k)$ must be calculated from a series of diagrams which do not include *any* indices which have previously appeared before in the particular term of $V_e(0)$ we are calculating. That is, if we want the term

$$V_{03}\frac{1}{e_3}V_{32}\frac{1}{e_2}V_{21}\frac{1}{e_1}V_{10},$$

where for brevity we introduce the usual "propagator" notation

$$is-E_j-V_c(j)=e_j, \tag{14}$$

then the propagator e_2, for instance, is given by

$$\begin{aligned} e_2&=is-E_2-V_c^{0,1}(2)\\ &=is-E_2-\sum_{j,k,l\neq 0,1} V_{2j}\frac{1}{e_j}V_{jl}\frac{1}{e_l}\cdots\frac{1}{e_k}V_{k2} \end{aligned} \tag{15}$$

and again, each of the propagators in this series must be appropriately modified not to include either 0, 1, or any of the previous indices in the $V_c^{0,1}(2)$ series.

Thus we may now write

$$V_e(0)=\sum_{\substack{i\neq 0\\ j\neq i,0\\ k\neq i,j,0\\ \cdots\\ l\neq\cdots i,j,k,0}} V_{0l}\cdots\frac{1}{is-E_k-V_c^{0,i,j}(k)}$$

$$\times V_{kj}\frac{1}{is-E_j-V_c^{0,i}(j)}V_{ji}\frac{1}{is-E_i-V_c^{0}(i)}V_{i0}. \tag{16}$$

All of this, of course, involves a self-consistent type of reasoning, since it is only if these series converge that we can find $V_c(j)$ in this way, and therefore that we can define the modified series. We say in defense that clearly we can always make the sum converge for large enough s, and also that the V_c's in the higher terms, since they may have many forbidden indices, are more convergent than those we derive from them.

The prohibition of repeated indices has two useful consequences. The most obvious is to prevent extensive correlations between successive factors V/e of a given product. However, they also introduce a useful and important correlation. Namely, suppose that one factor of our term, $V_{kl}/e_k^{(l\cdots)}$, is particularly large. The V_e of the previous factor, say $V_{lm}/e_l^{(\cdots)}$, will contain the term:

$$|V_{kl}|^2/e_k^{(l\cdots)}.$$

Therefore this previous factor would contain the large factor in its denominator, leading to a tendency to cancel. On a quantitative basis, first think of all V's as having the same order of magnitude. Then, since the other terms of the denominator e_l will all be of order W or less, it is easy to see that we simply decrease the total unless

$$|V_{kl}|^2/e_k\leq W,$$

or

$$|V_{kl}/e_k|\leq W/V. \tag{17}$$

W is the breadth of the distribution $P(E)$.

We shall use the limitation (17) in our later computations. We note that it is meaningless if V is small; but the work of Sec. IV will show that small values of V are never important. In any case the results do not depend sensitively on the existence of this limitation.

Actually, (17) is only the most important of an extensive system of correlations, since similar considerations hold for any group of factors starting from atom k and ending at atom l, if there is a distinct return path to atom k from l which has a finite factor. The fact that isolated large factors are not important,

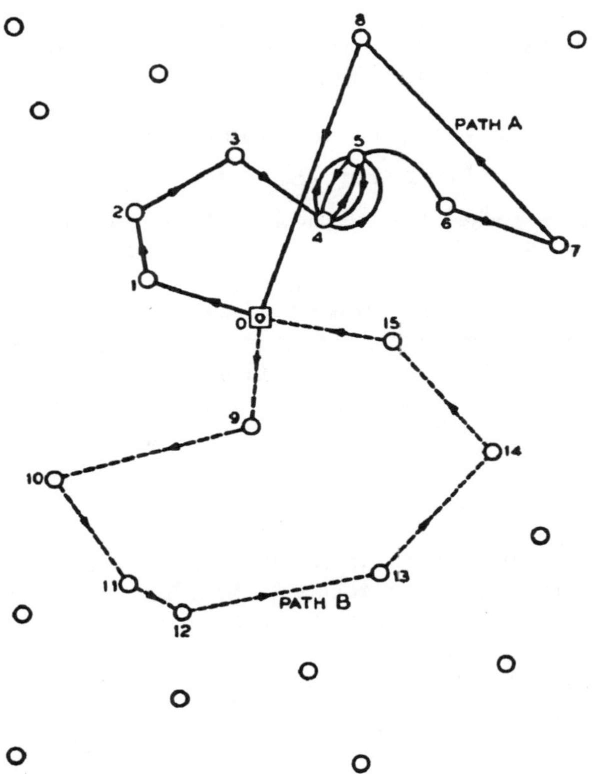

FIG. 1. Diagrams corresponding to terms in the perturbation expansion of V_c. (A) may be large and must be summed over; (B) is a legitimate term.

[9] For this purpose, one could equally well use the method of E. Feenberg, Phys. Rev. 74, 206 (1948).

however, should apply even more strongly to groups of factors. We merely note that we will probably underestimate the limiting density, even using (17).

Our problem now is to study series of the form (16) in which the E's, to a lesser extent the V_c's, and possibly the V_{jk}'s are stochastic variables. We want primarily to find whether such series converge in some sense as $s \to 0$; if so, we have found a self-consistent set of localized wave functions. If the convergence limit on s is finite it may still be possible to calculate some transport properties, but that can be reserved for later work.

In the fourth section we discuss such series. The principle of this discussion is the following: since the terms T_L of the series having a given length (number of denominators, say) L, are themelves random variables, we try to find a distribution function for their values. This distribution can be expressed as a number distribution.

(Average number of terms of length L between

$$T_L \text{ and } T_L+dT_L)=n(T_L)dT_L. \quad (18)$$

(It is necessary to use a number rather than a probability distribution because, when V extends to large r, the total number of terms of any length is infinite.)

The techniques involved in getting such a distribution are closely related to the Markoff method of random walk theory, although we use, for convenience, the Laplace transform and the convolution theory for it. We are able to get $n(T_L)$ explicitly in two cases. In both cases we make the unimportant simplification that the distribution function $P(E)$ of the energies E_j is flat:

$$\begin{aligned} P(E)&=1/W \quad \text{for} \quad -\tfrac{1}{2}W\leq E\leq \tfrac{1}{2}W, \\ P(E)&=0 \quad \text{for} \quad |E|>\tfrac{1}{2}W, \end{aligned} \quad (19)$$

and we neglect the influence of V_c on the frequency denominators except for the limitation (17).

In the first case we assume that V_{jk} is finite only between "nearest neighbors," of which there are some finite number Z; between these neighbors, it has a constant value V. Then the problem simplifies to finding the probability distribution of the product of denominators $P(\Pi_D)d\Pi_D$

$$\Pi_D=\frac{1}{e_1}\frac{1}{e_2}\frac{1}{e_3}\cdots\frac{1}{e_L}. \quad (20)$$

Given $P(\Pi_D)$, we can use the idea of the "connectivity" from the percolation theory of Broadbent and Hammersley.[10] The connectivity K for any given lattice with near-neighbor connections only is defined by exactly the relation we want:

(Number of nonrepeating paths of length L

leading from any given atom)$\sim K^L$. (21)

K is generally of order $Z-1$ to $Z-2$. Then obviously

$$n(T)dT=(K^L/V^L)P(T/V^L)dT. \quad (22)$$

A second case we are able to solve is that of the purely random lattice in which V falls off as some power of r. Unfortunately, we have to ignore the restriction of nonrepeating paths in this case, so that we rather badly overestimate the sum. On the other hand, this at least tells us whether or not large r's are important since this restriction is not important for large jumps. Thus by this case we can show rigorously that $V\sim 1/r^{3+\epsilon}$ is the correct restriction on the range of V.

In each case we come out with an $n(T)$ of the following form:

$$n(T)dT=[F(K,W/V)]^L\frac{dT}{T^2}L(T), \quad (23)$$

where $L(T)$ is a slowly varying function relative to T. This form allows us to make use of the following result, which is implicit in (for instance) the theory of the Holtsmark distribution. The probability distribution of the sum of a collection of random terms of random sign such as (23) is the same as the distribution of the single largest term (a) for values greater than or of the order of the most probable or median value of the sum; and (b) if $L(T)$ is increasing, or decreasing no more rapidly than $T^{-\frac{1}{2}}$. This is essentially the same as the theorem that the force on a dipole in an unpolarized gas, or on an electron in a discharge, comes primarily from the nearest neighbor.

Since $L(T)$ obeys this condition very well in both cases, at least for reasonable values of V/W, etc., we may immediately get the critical values of the parameters from (23). We know now that, if

$$\Sigma=\sum_{n=1}^{K^L}(\pm T_n) \quad (24)$$

(where we use for clarity the case of a finite number of neighbors), then

$$P(\Sigma)d\Sigma\sim F^L(K,W/V)\frac{d\Sigma}{\Sigma}L(\Sigma). \quad (25)$$

First we find $(W/V)_0$ to satisfy

$$F^L(K,(W/V)_0)L(1)=1. \quad (26)$$

If (W/V) has a value even very slightly greater than this, the most probable value of Σ will be small of order e^{-L} and the probability of a value$=1$ will also be of order e^{-L}. Now we consider $L\to\infty$: The number of Σ's only increases as L, while with probability $\sim 1-e^{-L}$ their value is less than e^{-L}. Therefore the series converges almost always if

$$W/V>(W/V)_0, \quad (27)$$

which is the desired criterion. In Sec. IV the critical values will be discussed numerically. A typical estimate

[10] S. R. Broadbent and J. M. Hammersley, Proc. Cambridge Phil. Soc. 53, 629 (1957). This work suggested some features of our approach to the present problem.

would be $(W/V)_0=26$ for $K=4.5$ (about correct for the simple cubic lattice).

With this result the theorem is established.

III. PROBABILITY DISTRIBUTION OF THE FIRST TERM OF V_c

Before Eq. (11) we related the transition rate to a certain quantity $X(s)$, and showed that if $X(s)$ remains finite as $s\rightarrow 0$ no real transport takes place, in the first order of perturbation theory. Now $X(s)$ is a sum over all possible single jumps:

$$X(s)=\sum_k \frac{|V_{0k}|^2}{s^2+E_k^2}. \quad (28)$$

The probability distribution of such a sum is best calculated by the Markoff method,[11] as modified by Holtsmark.[6] In this method we find the Fourier transform of the probability distribution $P(X)$:

$$P(X)=\int_{-\infty}^{\infty} e^{ixX}\varphi(x)dx, \quad (29)$$

where

$$\varphi(x)=\exp\left\{-n\left\langle\int\left[1-\exp\left(ix\frac{V^2(r)}{E^2+s^2}\right)\right]d\tau\right\rangle\right\}. \quad (30)$$

The average is to be taken over the probability distribution of E, $P(E)$, and n is the density of sites.

Let us write out the important integral in the exponent of (30):

$$I=\int_{-\infty}^{\infty}P(E)dE4\pi\int_0^{\infty}r^2dr\left[1-\exp\left(\frac{ixV^2(r)}{E^2+s^2}\right)\right]. \quad (31)$$

The behavior of $P(X)$ for large X depends on the behavior of I for sufficiently small x. Let us first consider the case $s=0$. Now for small enough x, and a finite E (say of order W), the exponential $\exp ixV^2/E^2$ can be expanded in a power series in x, and the integration over r done (so long as V is finite and falls off faster than $r^{-\frac{3}{2}}$) to obtain terms which go as x^1 or higher powers for small x. Thus only the behavior for small E is important, and we can neglect the variation of $P(E)$, replacing it by a constant, $1/W$. Then

$$I\cong\frac{4\pi}{W}\int_0^{\infty}r^2dr\int_{-\infty}^{\infty}dE\left\{1-\exp\left[\frac{ixV^2(r)}{E^2}\right]\right\}$$
$$=\frac{4\pi}{W}(x)^{\frac{1}{2}}\int_0^{\infty}r^2drV(r)\int_{-\infty}^{\infty}du[1-\exp(i/u^2)] \quad (32)$$
$$=2\left(\frac{x}{i}\right)^{\frac{1}{2}}\Gamma(\tfrac{1}{2})\frac{\langle V(r)\rangle}{W}.$$

The integration here depends on the more stringent condition $V\sim 1/r^{3+\epsilon}$ for large r. The probability distribution, which will be valid for large X at least, is familiar from line-broadening theory[12]:

$$P(X)=\frac{n\langle V\rangle}{WX^{\frac{3}{2}}}\exp\left[-\left(2n\Gamma(\tfrac{1}{2})\frac{\langle V\rangle}{W}\right)^2\cdot\frac{1}{X}\right]. \quad (33)$$

For large X, this falls off as $X^{-\frac{3}{2}}$, as stated in Sec. II: while the mean of X is divergent, the probability that X is larger than some value X_0 decreases as $X_0^{-\frac{1}{2}}$. Thus, for any given starting atom n, the renormalization constant K may be large, but the probability that τ is finite is exactly zero.

We see that none of these conclusions are valid for the case $V\propto r^{-3}$ for large r. In this case a finite s must be retained. Again looking at the integral (31), let us substitute $V(r)=A/r^3$ for all r.[13] Then we do the integration over r first:

$$I=\frac{4\pi}{3}\int_{-\infty}^{\infty}P(E)dE\int_0^{\infty}d(r^3)\left\{1-\exp\left[\frac{ixA^2}{r^6(E^2+s^2)}\right]\right\}$$
$$=\frac{4\pi A}{3}\left(\frac{x}{i}\right)^{\frac{1}{2}}\int_{-\infty}^{\infty}\frac{P(E)dE}{(E^2+s^2)^{\frac{1}{2}}}. \quad (34)$$

We see that I indeed has a logarithmic singularity as $s\rightarrow 0$. A simple case for which we can evaluate the integral (34) is the flat distribution (19) of width W for which

$$\frac{W}{2}I=\frac{4\pi A}{3}\left[\sinh^{-1}\left(\frac{W}{2s}\right)\right]\left(\frac{x}{i}\right)^{\frac{1}{2}}. \quad (35)$$

This leads to the probability distribution of the sum X:

$$P(X)=\left(\frac{4\sqrt{2}nA}{3\pi^{\frac{1}{2}}W^{\frac{1}{2}}}\right)\frac{1}{X^{\frac{3}{2}}}\sinh^{-1}\left(\frac{W}{2s}\right)$$
$$\times\exp\left\{-\left[\frac{8\pi nA}{3W}\sinh^{-1}\left(\frac{W}{2s}\right)\right]^2\left(\frac{1}{X}\right)\right\}, \quad (36)$$

which was discussed in Sec. II.

IV. DISTRIBUTIONS OF HIGH-ORDER TERMS IN THE PERTURBATION THEORY

In order to simplify the later manipulations, and to make closer contact with the Watson theory, it will be useful to expand some of the formalism of the problem.

Equation (4) presents our basic equations of motion:

$$f_j(s)=\frac{i\delta_{0j}}{is-E_j}+\sum_k\frac{1}{is-E_j}V_{jk}f_k(s).$$

[11] S. Chandrasekhar, Revs. Modern Phys. 15, 1 (1943). I am indebted to L. R. Walker for suggesting the use of the Markoff method here.

[12] H. Margenau, Phys. Rev. 48, 755 (1935).

[13] We make this substitution since there exists some r_0 beyond which $V-(A/r^3)\simeq 0$. We point out that $\int_0^{r_0}$ leads to terms in I of identical form with those we have already found, and thus independent of s.

Let us consider simultaneously all possible initial starting atoms 0. The $f(s)$'s which result will form a matrix f_{lm}, of which our f's are the particular row f_{j0}. It is simple to introduce as the "Green's function" the matrix

$$(j|W|k)=-i\,f_{jk}(s), \tag{37}$$

which satisfies the equation

$$(j|W|k)=\frac{\delta_{jk}}{is-E_j}+\sum_l \frac{1}{is-E_j}V_{jl}(l|W|k); \tag{38}$$

or, if one introduces the matrix

$$(j|a|k)=\delta_{jk}(is-E_j),$$

then W satisfies

$$W=\frac{1}{a}+\frac{1}{a}VW. \tag{39}$$

The Møller wave matrix Ω is defined also, as

$$\Omega=Wa, \tag{40}$$

and satisfies the equation

$$\Omega=1+\frac{1}{a}V\Omega. \tag{41}$$

The general philosophy of the multiple-scattering theory is to try to separate and identify, in these matrices Ω or W, effects which are "coherent" in the sense that they involve perturbations in which, finally, the system returns coherently to the initial state, from effects which involve real (as opposed to virtual) transitions to other states. Another way to put it is that this is an attempt to define a new set of states, once the perturbation V is applied, which correspond in some sense to the unperturbed states, and thus to have left over only the effects of whatever real transitions may occur. In principle this is exactly what we are attempting in this paper.

The coherent effects are mainly included in a diagonal coherent wave matrix Ω_c, and the rest of the problem is contained in the "model operator" F:

$$\Omega=F\Omega_c, \tag{42}$$

where we want F to satisfy an equation like

$$F=1+\frac{1}{a-V_c}PVF, \tag{43}$$

where P is an operator which prevents in some way the repetition of indices in the perturbation series for F, while V_c is a correction which must therefore be made to the energy a.

It is perhaps simplest just to go ahead and show what must be done. The solution of Eq. (41) may be written out as a perturbation series by direct iteration:

$$(j|\Omega|k)=\delta_{jk}+\frac{1}{a_j}V_{jk}+\sum_l \frac{1}{a_j}V_{jl}\frac{1}{a_l}V_{lk}$$
$$+\sum_{l,m}\frac{1}{a_j}V_{jl}\frac{1}{a_l}V_{lm}\frac{1}{a_m}V_{mk}+\cdots. \tag{44}$$

A very direct way to eliminate repeated indices is as follows: take any given term of (44) and, starting from the right, look through until we find the first repetition of the index k. Between these two repetitions comes a factor:

$$\left(\frac{1}{a_k}V_{kl}\frac{1}{a_l}\cdots V_{mk}\right).$$

Next we continue to look from right to left and find another similar factor, etc., until we find all such factors and no more k indices occur. Now everything which remains to the left will also appear multiplied by the factors we have found in all other possible combinations and repetitions, and in fact by all possible such factors in all such combinations. Summing all such factors, we get for that series which comes to the right of the last repetition of index k:

$$\left[1+\frac{1}{a_k}V_{kk}+\left(\frac{1}{a_k}V_{kk}\right)^2+\cdots+\sum_l \frac{1}{a_{kl}}V_{kl}\frac{1}{a_l}V_{lk}\right.$$
$$\left.+\left(\sum_l \frac{1}{a_k}V_{kl}\frac{1}{a_l}V_{ak}\right)^2+\cdots\right]$$
$$=\left[1-\left(\frac{1}{a_k}V_{kk}+\sum_l \frac{1}{a_k}V_{kl}\frac{1}{a_l}V_{lk}+\cdots\right)\right]^{-1}$$
$$=\frac{a_k}{a_k-V_c(k)}=\frac{a_k}{e_k}=(\Omega_c)_k, \tag{45}$$

where $V_c(k)$ is defined as in (13). The corresponding term of Ω is then in the form

$$\frac{1}{a_j}V_{jl}\frac{1}{a_l}V_{lm}\cdots\frac{1}{a_n}V_{nk}(\Omega_c)_k.$$

We now begin the same process as in (45), looking for repetitions of the index n which appear next. We collect together all such terms, and finally find that we can replace a_n by $a_n-V_c^k(n)$, where

$$V_c^k(n)=V_{nn}+\sum_{l\neq k,n} V_{nl}\frac{1}{a_l}V_{ln}$$
$$+\sum_{l,m\neq k,n} V_{nl}\frac{1}{a_l}V_{lm}\frac{1}{a_m}V_{mn}\cdots. \tag{46}$$

This process may be repeated until we come to the end

of the given term. Thus we have successfully expanded Ω as

$$\Omega = F\Omega_e,$$

$$\Omega_e = a/e,$$

$$F_{j\neq k} = \frac{1}{e_j}V_{jk} + \sum_{l\neq k,j} \frac{1}{e_j}V_{jl}\frac{1}{e_l}V_{lk} + \frac{1}{e_j}\sum_{l\neq m,k,j} V_{jl}\frac{1}{e_l}\sum_{m\neq j,k} V_{jm}\frac{1}{e_m}V_{mk}+\cdots. \quad (47)$$

The final step is obvious: in each of the V_e's in (45) itself we eliminate repeated indices in an exactly similar manner, obtaining expressions like (16) for V_e.[14]

Now the usefulness of such expressions in the usual multiple-scattering theory comes only from the fact that the limitations on the sums are unimportant, since there are an infinite number of V_{jk}'s starting from any j, each being small. In our case we have a different kind of fortunate circumstance which allows us to get around this problem; namely, all of the quantities of the theory are stochastic variables, so that all we really wish to know is the distribution function of the e_j's, which except for the restriction (17) is practically that of the E_j's unless V_e is quite large. Even if V_e were large, one might study it as a stochastic variable.

We now have the problem of calculating probability distributions for products such as the terms of (47). Because the only question is that of convergence we are interested only in the terms of very high order, that is, those of order L with $L\gg 1$. We call such a term T:

$$T_{jk}^L = \frac{1}{e_j}V_{jl}\frac{1}{e_l}V_{lm}\frac{1}{e_m}\cdots V_{nk}$$

$$=\exp\{[\ln V_{jl}+\ln V_{lm}+\cdots(L \text{ terms})] - [\ln e_j+\ln e_l+\cdots(L\text{ terms})]\}. \quad (48)$$

In Sec. II we presented the two cases in which it has been possible to calculate explicitly the number distribution of T. We take up the simplest first: the case in which V is a constant, so that the only stochastic element in (48) is the denominators. In this case

$$T = V^L \prod_{l=1}^{L} \frac{1}{e_l}. \quad (49)$$

To find the convergence limits, we go to $s=0$ immediately. Then clearly T has random sign, which must be taken into account later in summing the T's.

As we discussed before, V_e is important only in that it causes a certain restriction (17) on the magnitude of the separate factors of the product; otherwise we shall make the approximation of neglecting it altogether as an unimportant correction to the stochastic variable E_j. Since we are for simplicity confining ourselves to the flat distribution $P(E)=1/W$ of Eq. (19), we can express (17) semiquantitatively. The smallest denominator e_l which will not seriously affect the probability distribution of subsequent factors we define to be $\Delta/2$. The quantity $\Delta/2$ satisfies the criterion that the contribution to V_e from it must be less than the maximum possible E_j:

$$V^2/(\tfrac{1}{2}\Delta) < \tfrac{1}{2}W,$$

or

$$\Delta > 4V^2/W. \quad (50)$$

Our approximation is simply to take this as a lower limit on e and modify (19) to

$$P(e_j) = \frac{1}{W-\Delta}, \quad \tfrac{1}{2}\Delta < |e_j| < \tfrac{1}{2}W;$$
$$P(e_j)=0, \quad |e_j| < \tfrac{1}{2}\Delta \quad \text{or} \quad |e_j| > \tfrac{1}{2}W. \quad (51)$$

We shall find that the use of (51) changes our convergence limits by less than a factor of 2, in spite of its apparent importance in eliminating singular factors; this is our justification for the crude approximation.

We now take up the question of the probability distribution of T. Let us define the variable S as

$$\left|\prod_{j=1}^{L}\frac{1}{e_j}\right| = e^S/(\tfrac{1}{2}W)^L. \quad (52)$$

The range of S starts from zero, so that we can apply a Laplace transformation to its probability distribution:

$$F_L(p) = \int_0^\infty e^{-pS}P(S)dS$$

$$= \left\langle \prod_{j=1}^{L} \exp\{p[\ln e_j - \ln(\tfrac{1}{2}W)]\}\right\rangle. \quad (53)$$

Since all the e_j's are independent variables, the average in (53) is just the product of the separate averages. Thus

$$F_L(p) = \left[\frac{1-(\Delta/W)^{p+1}}{p+1}\frac{W}{W-\Delta}\right]^L \quad (p>-1). \quad (54)$$

For regions of S in which we may neglect Δ, this is very easily inverted:

$$\Delta\cong 0: \quad F_L(p)\simeq 1/(p+1)^L, \quad P(S)=e^{-S}S^{L-1}/\Gamma(L). \quad (55)$$

Since

$$T = V^L/(\tfrac{1}{2}W)^L e^S,$$

to our order of approximation we have

$$P(T)dT \cong \left(\frac{2eV}{W}\right)^L \left[\frac{\ln T}{L} - \ln\left(\frac{2V}{W}\right)\right]^L \frac{dT}{T^2}, \quad (56)$$

which shows that it is of the form (23).

[14] I am indebted to P. A. Wolff for many helpful discussions on the above, and in particular for pointing out that (43) is true only in the sense of the perturbation series.

It is interesting to note that while (55) is a reasonably narrow distribution, satisfactorily obeying the central limit theorem, the fact that the quantity of interest is exponential in S transfers our attention to what, in (55), appears to be the extreme tail of the distribution. This is a characteristic of this problem. On the other hand, our task is simplified, in that nothing smaller than factors exponential in L affects our results.

Without neglecting Δ, we can get the full distribution with sufficient accuracy from (54) by applying the inversion formula for the Laplace transform,

$$P(S)=\frac{1}{2\pi i}\int_{-i\infty}^{i\infty} F_L(p)e^{pS}dp, \tag{57}$$

and using the method of steepest descents. Writing out (57) in appropriate form for this method, we obtain

$$P(S)=\frac{1}{2\pi}\int_{-i\infty}^{i\infty} \exp\left\{pS-L\left[\ln(p+1)+\ln\left(1-\frac{\Delta}{W}\right)-\ln\left[1-\left(\frac{\Delta}{W}\right)^{1+p}\right]\right]\right\}. \tag{58}$$

First notice that as p becomes large and positive, the exponent will approach $+\infty$ unless $S<0$. If $S<0$, we can find a path via $p\rightarrow+\infty$ and $P(S)=0$, as it should be. Similarly, as $p\rightarrow-\infty$, the S term dominates if $S>L\ln(W/\Delta)$, so that, correctly, $P(S>L\ln(W/\Delta))=0$. Within these limits, there is a saddle point for finite p which may be found by differentiating the exponent.

At the point $1+p=0$, the exponent changes character, from not depending on Δ/W above this point to depending primarily on it below. It is instructive to expand the exponent about this point:

$$-L\left\{\ln(1+p)+\ln\left(1-\frac{\Delta}{W}\right)-\ln\left[1-\left(\frac{\Delta}{W}\right)^{1+p}\right]\right\}\cong-L\left\{\ln\left(1-\frac{\Delta}{W}\right)-\ln\left[\ln\frac{W}{\Delta}-\frac{1+p}{2}\left(\ln\frac{W}{\Delta}\right)^2\cdots\right]\right\}. \tag{59}$$

As we see, this point is actually not a singularity. Taking derivatives, we find that the condition that the saddle be here is:

$$S|_{p_0=-1}=\tfrac{1}{2}L\ln(W/\Delta),$$

and that at this point

$$P\left(\tfrac{1}{2}L\ln\frac{W}{\Delta}\right)\cong e^{-S}\left(\frac{\ln(W/\Delta)}{1-\Delta/W}\right)^L. \tag{60}$$

As S gets appreciably larger than this value, $F_L(p)$ and the exponent again become simple when we can neglect 1 relative to $(\Delta/W)^{p+1}$. Here it becomes necessary to continue the logarithms in (58) to negative values of their arguments, but this can be done by referring to (54) and noticing that we must change the sign of $1+p$ and $-(\Delta/W)^{p+1}$ simultaneously. Then the saddle-point condition is

$$S-\frac{L}{1+p_0}+L\ln\frac{\Delta}{W}=0, \tag{61}$$

and the probability of such values of S is

$$P(S)\cong e^{L-S}\left(\ln\frac{W}{\Delta}-\frac{S}{L}\right)^L\Big/\left(1-\frac{\Delta}{W}\right)^L. \tag{62}$$

The results (56), (60), and (62) may be summarized in the following way:

$$P(S)=\left(\frac{e}{1-\Delta/W}\right)^L e^{-S}[\psi(S/L)]^L, \tag{63}$$

where ψ is a slowly varying function which may be estimated in each of the three regions:

$$\begin{aligned}&\text{I: } \tfrac{1}{2}L\ln(W/\Delta)\gg S>0, && \psi(S/L)\simeq S/L;\\ &\text{II: } S\simeq\tfrac{1}{2}L\ln(W/\Delta), && \psi(S/L)\simeq\ln(W/\Delta)/e;\\ &\text{III: } L\ln(W/\Delta)>S\gg\tfrac{1}{2}L\ln(W/\Delta), && \\ & && \psi(S/L)\cong\ln(W/\Delta)-(S/L).\end{aligned} \tag{64}$$

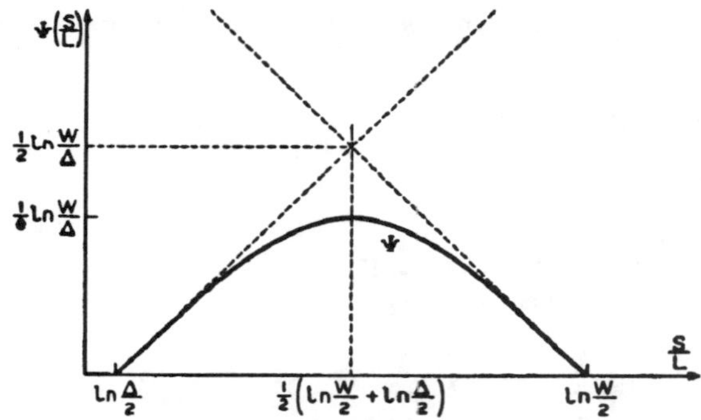

FIG. 2. The function ψ of Eq. (64), giving the slowly varying part of $P(S)$.

The function ψ is easily plotted approximately from these results and is shown in Fig. 2. For the probability distribution of T, we obtain

$$P(T)dT=\frac{dT}{T^2}\left(\frac{2eV}{W}\right)^L\times\psi^L\left[\frac{\ln T}{L}-\ln\frac{2V}{W}\right]\Big/\left(1-\frac{\Delta}{W}\right)^L. \tag{65}$$

Except for the most remote parts of region III, (65) is of the form (23) and satisfies our condition.

We have not explored very carefully the corrections to the saddle-point method, both because we need only terms of the order e^L, and because the results are clearly in accordance with expectations.

Before studying (65) numerically, let us go on to the second case which can be solved; namely,

$$V = V_0/r^{3N}. \quad (66)$$

We find a solution only in the limited sense that various crude approximations are made for small r; what we try to do is to assure ourselves that the region of large r and small V is not important, in spite of the extra mathematical difficulties to which it leads.

These "crude approximations" are threefold: (a) We ignore the fact that a path leading to an atom from its nearest neighbor must leave it via a further neighbor—i.e., in each factor we allow V to be randomly distributed, ignoring the favorable correlations caused by the restriction to nonrepeating paths. (b) We ignore (17) and thus can use (55) and (56) as the distribution function of the denominators, again overestimating the effects of large V's without disturbing the small V region. (c) We limit $|V|$ simply by introducing a minimum radius a; the maximum V is then

$$V < V_m = V_0/a^{3N}. \quad (67)$$

Now we study the distribution of terms of the form

$$T = \prod_{j=1}^{L} V_j \Big/ \prod_{j=1}^{L} e_j. \quad (68)$$

The distribution of the V's may be approximated by using a perfectly random distribution of neighbor distances:

$$n(r)dr = 4\pi\rho r^2 dr,$$

where ρ is the density of sites. From this and (67) the distribution of V is immediately

$$n(V)dV = \frac{4\pi\rho}{3N} V_0^{1/N} \frac{dV}{V^{1+1/N}}, \quad V < V_m$$
$$= 0, \quad V > V_m. \quad (69)$$

The numerator in (68) may take on values from V_m^L to zero; the inverse of the denominator, from ∞ to a certain minimum. Again the mathematics is more familiar if we study the logarithm, which is a variable extending from $-\infty$ to ∞. Thus it is necessary to use now the bilateral Laplace transform. This causes no difficulty in (53), (55), and (56) since $S>0$ anyhow; we simply reinterpret these as the corresponding bilateral formulas.

The logarithmic variable Σ, corresponding to S, we define by

$$\prod_{j=1}^{L} V_j = V_m^L e^{\Sigma}, \quad (70)$$

and the quantities (68) whose distribution we want are

$$T = \frac{V_m^L}{(W/2)^L} e^{S+\Sigma}. \quad (71)$$

Thus, once we have the distribution of Σ, we can find the distribution of $S+\Sigma$ by convolution of the bilateral Laplace transforms, and thence find that of T.

The required Laplace transform is

$$\varphi_L(p) = \int_{-\infty}^{\infty} P(\Sigma) e^{-p\Sigma} d\Sigma$$
$$= \left\{ \frac{4\pi\rho a^3}{3N} \int_0^1 \frac{(V/V_m)^{-p} d(V/V_m)}{(V/V_m)^{1+1/N}} \right\}^L, \quad (72)$$

$$\varphi_L(p) = \left(\frac{a^3}{N r_s^3} \right)^L \left\{ -\frac{1}{p+1/N} \right\}^L, \quad p < -1/N. \quad (73)$$

Here we have defined

$$\tfrac{4}{3}\pi r_s^3 = 1/\rho. \quad (74)$$

The transform of the denominators, $F_L(p)$, was convergent for $p > -1$ [Eq. (55)]. Thus for $N>1$, or forces falling off more rapidly than $1/r^3$, the two transforms have a common strip of convergence. This is the criterion on range we already expected from Sec. III. Now let

$$X = S + \Sigma,$$
$$\int_{-\infty}^{\infty} n(X) dX e^{-pX} = \psi_L(p). \quad (75)$$

Also, by the convolution theorem,

$$\psi_L(p) = \varphi_L(p) F_L(p), \quad -1 < p < -1/N$$
$$= \left(\frac{a^3}{N r_s^3} \right)^L \left\{ -\frac{1}{(1+p)(p+1/N)} \right\}^L. \quad (76)$$

The inversion of this is simple if we apply the shifting operator to p, bringing the origin to the center of the convergence strip:

$$p' = p + \tfrac{1}{2}(1+1/N),$$

$$\psi_L(p') = \int_{-\infty}^{\infty} \exp(-p'X) n'(X) dX, \quad (77)$$

$$n'(X) = n(X) e^{\frac{1}{2}X(1+1/N)}.$$

Then

$$\psi_L(p') = \left[\frac{a^3}{N r_s^3} \frac{1}{\frac{1}{4}(1-1/N)^2 - p'^2} \right]^L.$$

The standard inversion formula now gives us

$$n'(X) = \frac{1}{2\pi i} \int_{-i\infty}^{i\infty} \frac{dp' \exp(p'X)}{[\frac{1}{4}(1-1/N)^2 - p'^2]^L} \left[\frac{a^3}{N r_s^3} \right]^L. \quad (78)$$

This integral can be expressed in terms of the Bessel function $K_{L-\frac{1}{2}}$ (which is actually a polynomial in elementary functions) by Basset's formula,[15] which in

[15] G. N. Watson, *Bessel Functions* (Cambridge University Press, Cambridge, 1944), p. 172.

turn can be expanded by the well-known asymptotic expansion for large order.[16] The result, for our purposes, could be foreseen by realizing that in (78) practically all the contribution will come from $p'=0$. Using (77), and neglecting unimportant constant factors, it is

$$n(X)=\left[\frac{4a^3}{Nr_s^3(1-1/N)^2}\right]^L \exp[-\tfrac{1}{2}(1+1/N)X]. \quad (79)$$

Now we get the number distribution of the terms T, using (71):

$$n(T)dT=\frac{dT}{T^{1+1/2N}}\left[\frac{4a^3}{Nr_s^3(1-1/N)^2}\right]^L\left[\frac{2V_m}{W}\right]^{\frac{1}{2}L(1+1/N)}. \quad (80)$$

Again we find a distribution of the form (23), satisfying very well the condition that the distribution of large values be that of the largest term.

Now the only remaining task to complete the discussion is to study the criterion for localization numerically, using (23) and (26). First we shall deal very briefly with the unrealistic second case. Upon using (80), Eq. (26) becomes

$$1=\left[\frac{4a^3}{Nr_s^3(1-1/N)^2}\right]^L\left[\frac{2V_m}{W}\right]^{\frac{1}{2}L(1+1/N)}.$$

It is interesting to put this in the following form:

$$\frac{V_0}{\left\{r_s^2a^{N-1}\left[\dfrac{2}{N^{\frac{1}{2}}-1/N^{\frac{1}{2}}}\right]^{4/3}\right\}^{3N/(N+1)}}=\frac{W}{2}. \quad (81)$$

The $3N$th root of the denominator on the left can be thought of as an effective radius of interaction; namely, contributions from much greater distances are certainly of no importance. Except when N is very close to unity this radius is smaller than the mean nearest-neighbor distance ($\simeq 0.8r_s$) and strongly dependent upon a. This tells us two things: first, that the infinite range of the potential is, unless $N\simeq 1$, of no importance, and the important interactions are with close neighbors; and second, that this particular calculation, which did not take into account the important correlations of near neighbors introduced by the restriction to nonrepeating paths, is of no direct value.

An order-of-magnitude guess at the correct answer to this problem might be gotten by inserting something like r_s for a in (81); this eliminates the false effect of single, very close neighbors, which by the restriction to nonrepeating paths should make no contribution. We see then that $r_{\rm eff}\simeq r_s$, or approximately the mean nearest-neighbor distance. The resulting V/W is rather surprisingly large and tends to explain Feher's result that even when a considerable fraction of the atoms have a close neighbor for which $V>W$, there is still little or no spin diffusion.

These difficulties with correlations confine our quantitative work to the one case in which this problem is solved for us, at least in principle, by Hammersley's work: The regular lattice with a finite number Z of equal interactions.

Here we apply very directly the discussion of Eqs. (22)–(26): that is, we find the probability distribution of the sum of K^L terms from that of the largest term, and thence find a critical $(W/V)_0$ below which all the higher terms are exponentially small with exponentially large probability. The criterion for applicability of the basic theorem in the case of the present distribution (65) turns out to be

$$\ln\frac{2V}{\Delta}-\frac{\ln T}{L}<1, \quad (82)$$

which we can show to be true at the critical V/W in all cases.

Equation (26) for the critical W/V is, when one uses (65) and takes the Lth root,

$$1=\frac{2eK}{1-(\Delta/W)_c}(V/W)_c\psi\{-\ln[2(V/W)_c]\}. \quad (83)$$

Since we are really quite uncertain as to exactly how to take the correlations into account, we shall solve (83) first in the case of no correlation, which gives us an upper limit on $(W/V)_0$, as well as in the case of a finite Δ given by (50).

I. *Upper limit.*—Here we assume that $\Delta=0$ and that ψ for region I is always correct, so that

$$eK\ln(W/2V)_{\rm u.l.}=(W/2V)_{\rm u.l.}. \quad (84)$$

A rough solution by iteration is clearly $(W/2V)_{\rm u.l.}=eK\ln(eK)$. More accurate solutions can be obtained by plotting (84) and are given in Fig. 3. The values are rather large; for instance, at $K=4.5$, approximately correct for the simple cubic lattice, $(W/2V)_{\rm u.l.}=45$. The rather large effect of increasing connectivity is interesting.

II. *Lower limit.*—Next let us do the calculation taking into account the approximate lower limit (50). It turns out that in this case the T of interest is in region II. In fact, with the definition (50),

$$-\ln(2V/W)_c=\tfrac{1}{2}\ln(W/\Delta),$$

so that S is exactly correct for region II. This leads to

$$\frac{2K\ln(W/2V)_c}{1-(4V^2/W^2)_c}=(W/2V)_c. \quad (85)$$

Except for the correction in the denominator, which is negligible for all but the smallest values of K, this is the same equation as before with 2 replacing e. Its

[16] J. W. Nicholson, Phil. Mag. 20, 938 (1910).

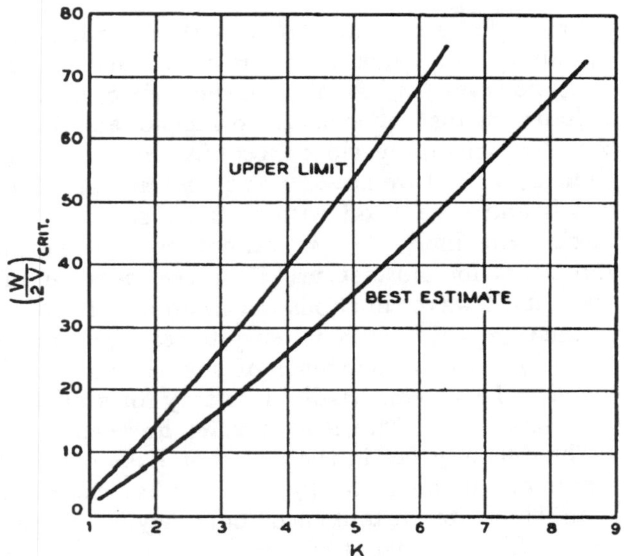

FIG. 3. Numerical estimates for the critical $W/2V$, the ratio of line width to interaction, for transport, plotted against connectivity K. The upper curve is a quasi-exact upper limit; the lower one is our best estimate.

solution is the lower curve of Fig. 3. It is very unlikely that (85) is accurate for $K\sim 1$, so no plot has been made in this region. This concludes our estimates of the critical ratio for transport.

V. MEANING OF CONVERGENCE OF THE PERTURBATION THEORY

The results of Secs. III and IV may be stated simply as follows:

(1) The first few terms of perturbation theory are convergent in the sense that $\text{Im}(V_c)$ is finite with probability unity, at any particular randomly chosen point on the energy axis.

(2) The terms of order L or higher are, at any particular random point on the energy axis, smaller than $e^{-\epsilon L}$ with probability $1-e^{-\epsilon L}$, under the appropriate conditions.

We discuss here the question of what this tells us about the actual perturbed energy states. Our conclusion is that we can show that a typical perturbed state is localized with unity probability; but that we cannot prove that it is possible to assign localized perturbed states a one-to-one correspondence with localized unperturbed states in any obvious way, so that perhaps with very small probability a few states may not be localized in any clear sense.[17]

Equation (37) introduces the "Green's function" $(j|W(s)|k)=-if_{jk}(s)$. Let us define

$$s=\sigma+iE. \tag{86}$$

According to the meaning of $f_{jk}(s)$, we can find the final amplitudes on the various atoms by the prescription:

$$\lim_{\sigma\to 0} i\sigma(j|W|k)=\bar{a}_{jk}(E), \tag{87}$$

where $\bar{a}_{jk}(E)$ is the amplitude at energy E and on atom j of the wave function which initially was unity on atom k. The total amplitude on atom j is the sum of squares of (87) over all energies E which are exact eigenenergies of the problem. Finally, W obeys the usual definition of the Green's function:

$$W=\frac{1}{is-H_0-V}=\frac{1}{i\sigma+E-H}. \tag{88}$$

[This is just (39).]

The scheme of the multiple-scattering method is to replace V in (88) by a number V_c, which can, of course, only be done in the diagonal elements:

$$(j|W|j)=\frac{1}{i\sigma+E-E_j-V_c{}^j(i\sigma+E)}. \tag{89}$$

Then the general elements of W are obtained by using the model operator F of (42) and (47):

$$(j|W|k)=(j|F|k)(k|W|k). \tag{90}$$

It is important that in the expansion (47) of F the particular denominator occurring in (89) never appears. Let us start by thinking of a large but finite system, so that we know the energy levels are not a continuum. Then the exact energies of the problem are clearly the poles of W, which occur at

$$E-E_j-V_c{}^{(j)}(E)=0. \tag{91}$$

This is just the expression one gets in Brillouin-Wigner perturbation theory. At such an energy, the amplitude at $t\to\infty$ is given by

$$\bar{a}_{jj}(E)=\lim_{\sigma\to 0}\frac{i\sigma}{i\sigma-i\,\text{Im}[V_c{}^{(j)}(i\sigma+E)]}, \tag{92}$$

so that if Im V_c converges and this convergence remains as $N\to\infty$, this is finite, not zero as in usual transport theory.

Thus our proof that, at an arbitrarily chosen point E, ImV_c, or any of the quantities of the theory, converge, and (as we could also show) $(j|F|k)$ falls off rapidly with distance, would be complete if (91) furnished us with only one solution per E_j, and if this solution were displaced by a finite and indeterminate amount from the poles of V_c or F. Unfortunately, this is only true of some of the solutions of (91), those which correspond to nearby localized functions; but because (91) is a Brillouin-Wigner result it also contains all the other eigenstates as well. We have to show, of those distant from atom j, that their energies are *not* random relative to the poles of $(V_c{}^{(j)})$, but instead are specially related to V_c so that $\bar{a}_{jj}$ and $\bar{a}_{jk}$ are decreasing functions of distance.

[17] Our criterion for nontransport is the same as that of L. Van Hove [Physica 23, 441 (1957); see especially Sec. 4] translated into probability language. Van Hove's resolvent R_l is our Green's function and his G_l our V_c.

Why this is so is just as apparent from the simplest second-order theory as from the whole sum. Let us write (91) to second order:

$$E-E_j-\sum_k \frac{|V_{jk}|^2}{E-E_k}=0.$$

Clearly this has a solution $E_0(j)$ near E_j at which the sum takes on the value

$$\sum_k \frac{|V_{jk}|^2}{E_j-E_k},$$

approximately. There is no close relation between $E_0(j)$ and any of the poles of V_c to this order. But even for very small V_{jk}'s it also has a solution $E_0(k)$ near E_k, given by

$$E_0(k)-E_k\cong\frac{(V_{jk})^2}{E_k-E_j}=(\delta E)_k. \quad (93)$$

This solution is closer to E_k, the weaker the coupling with j.

Let us now compute $\bar{a}_{jj}(E)$ at this energy, from (92). It is

$$\bar{a}_{jj}(E_0(k))\cong\lim_{\sigma\to 0}\left\{1-\frac{1}{\sigma}\operatorname{Im}\left[\frac{(V_{jk})^2}{(\delta E)_k+i\sigma}\right]\right\}^{-1}$$

$$=\frac{|V_{jk}|^2}{(V_{jk})^2+(E_j-E_k)^2}\ll 1. \quad (94)$$

In a similar way, one can show that

$$(j|F|k)\simeq(E_j-E_k)/|V_{jk}|,$$

so that

$$\lim_{V_{jk}\to 0}\bar{a}_{jk}(E_0(k))=0 \quad (95)$$

also.

In the finite random lattice with all orders of perturbations the same considerations obviously hold; now, however, E_k is a perturbed energy, and V_{jk} includes virtual effects. In principle all the same considerations apply, since except for the exact localized state E_k of (93) we have proved that all contributions to V's and F's fall off with distance sufficiently fast. Taking the limit $N\to\infty$ cannot change the above arguments for most states. However, because one of the distant wave functions may pop up at any point on the energy axis, it is not easy to see a way to assign a one-to-one correspondence of sites j and perturbed energies. This seems hardly necessary for a qualitative understanding of what is happening, however.

The difficulty lies in the fact that V_c is not really a continuous function in any sense. What can be done is to eliminate distant neighbors beyond a certain radius R. Then we find an appropriate perturbed energy $E_j'(R)$. The contribution to V_c, $\delta V_c(R)$, from beyond R is a probability variable which can be made to have as narrow a distribution as desired, but at all R may be large at a few E. Thus it is probable, but not certain, that a state localized around j has an energy within a predetermined $\delta V_c(R)$ of $E_j'(R)$. However, as we increase the size of the system we can never find an R beyond which $E_j'(R)$ is always as close as we like to the correct value. The fault lies in the Brillouin-Wigner technique, and is similar to problems as yet unsolved in other theories using this technique.

VI. ACKNOWLEDGMENTS

Aside from the help of Dr. L. R. Walker and Dr. P. A. Wolff already mentioned, I should like to thank a number of my colleagues for discussing this work with me; in particular, Dr. E. Abrahams, Dr. G. Feher, Dr. C. Herring, Dr. D. Pines, and Dr. G. H. Wannier.

ENERGY GAP IN SUPERCONDUCTORS MEASURED BY ELECTRON TUNNELING

Ivar Giaever
General Electric Research Laboratory, Schenectady, New York
(Received July 5, 1960)

If a potential difference is applied to two metals separated by a thin insulating film, a current will flow because of the ability of electrons to penetrate a potential barrier. The fact that for low fields the tunneling current is proportional to the applied voltage[1] suggested that low-voltage tunneling experiments could reveal something of the electronic structure of superconductors.

Aluminum/aluminum oxide/lead sandwiches were prepared by vapor-depositing aluminum on glass slides in vacuum, oxidizing the aluminum in air for a few minutes at room temperature, and then vapor-depositing lead over the aluminum oxide. The oxide layer separating aluminum and lead is thought to be about 15-20A thick.

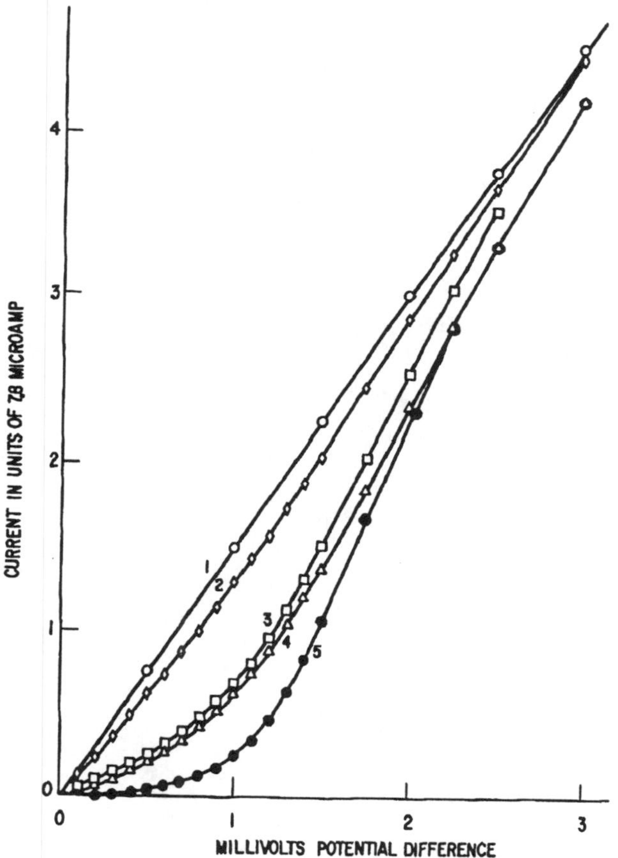

FIG. 1. Tunnel current between Al and Pb through Al_2O_3 film as a function of voltage. (1) $T=4.2°K$ and $1.6°K$, $H=2.7$ koe (Pb normal). (2) $T=4.2°K$, $H=0.8$ koe. (3) $T=1.6°K$, $H=0.8$ koe. (4) $T=4.2°K$, $H=0$ (Pb superconducting). (5) $T=1.6°K$, $H=0$ (Pb superconducting).

At liquid helium temperature, in the presence of a magnetic field applied parallel to the film and sufficiently strong to keep the lead in the normal state, the tunnel current is linear in the voltage. However, when the magnetic field is removed, and lead becomes superconducting, the tunnel current is very much reduced at low voltages as shown in Fig. 1. There is no influence of polarity, identical results being obtained with both directions of current flow.

The slope dI/dV of the curve in Fig. 1 where $H=0$, $T=1.6°K$, divided by dI/dV for normal lead, is plotted in Fig. 2. On the naive picture that tunneling is proportional to density of states,[2] this curve expresses the density of states in superconducting lead relative to the density of states when lead is in its normal state, as a function of energy measured from the Fermi energy. It seems clear that the density of states at the Fermi level is drastically changed when a metal becomes a superconductor, the change being symmetric with respect to the Fermi level. The curve resembles the Bardeen-Cooper-Schrieffer[3] density of states for quasi-particle excitations. There is a broadening of the peak that decreases with decreasing temperature. An approximate measure of half the energy gap

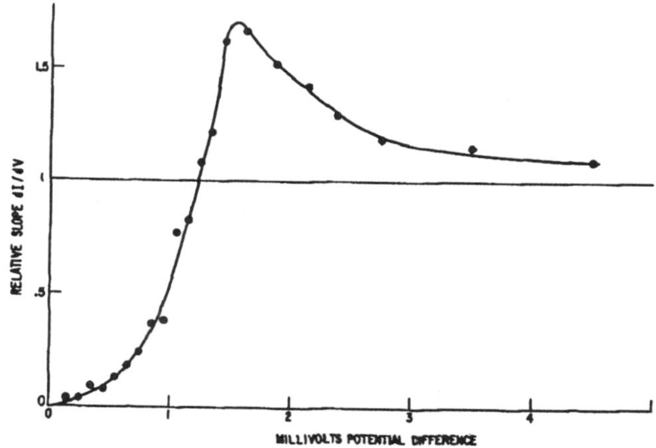

FIG. 2. From Fig. 1, slope dI/dV of curve 5 relative to slope of curve 1.

is given by the point at which the relative slope $dI/dV = 1$. On this basis the gap width for lead is $(4.2 \pm 0.1)kT_c$.

The experiment has been repeated with tin and indium giving entirely similar results; the gap in each case is approximately $4kT_c$. These results are of a preliminary nature, and experiments at lower temperatures will make them more precise.

I wish to thank C. P. Bean and J. C. Fisher for their interest and encouragement, and P. E. Lawrence for his help in performing the experiments.

---

[1]J. C. Fisher and I. Giaever (to be published).

[2]W. A. Harrison (private communication) has pointed out that the tunnel current is not proportional to the density of states except in the limiting case of a low density of states.

[3]J. Bardeen, L. N. Cooper, and J. R. Schrieffer, Phys. Rev. 108, 1175 (1957).

ELECTRON TUNNELING BETWEEN TWO SUPERCONDUCTORS

Ivar Giaever
General Electric Research Laboratory, Schenectady, New York
(Received October 31, 1960)

When two metals are separated by a thin insulating film, electrons can flow between the two conductors due to the quantum mechanical tunnel effect. If a small potential difference is applied between the two metals, the current through the film will vary linearly with the applied voltage, as long as the density of states in the two metals is constant over the applied voltage range,[1] as it is for most metals. In a superconductor, however, the density of states changes rapidly in a narrow energy range centered at the Fermi level, so that the voltage-current characteristic becomes nonlinear.[2] It is relatively easy to correlate the change from linearity with the variation in the density of states. Under the assumption that the tunnel current is proportional to the density of states, the current between normal and superconducting metals is in good agreement with the density of states calculated for a superconductor by the Bardeen-Cooper-Schrieffer[3] theory.[4]

A more direct measure of the energy gap is possible when electrons tunnel between two superconductors, as may be understood from a one-particle model of a superconductor as shown in Fig. 1. All the observed phenomena of tunneling into superconductors can be understood both qualitatively and quantitatively if we are willing to accept this model, which actually guided the experiments.

The samples were prepared by vapor-depositing aluminum on ordinary glass slides and allowing the surface of the aluminum film to oxidize. After a suitable oxide layer had formed, lead, indium, or aluminum was vapor-deposited over it to form a metal-oxide-metal sandwich. The oxide layer is thought to be 15-20 A thick.

In Fig. 2 are shown some typical voltage-current characteristics for the three different metal-oxide-metal sandwiches tested. The voltage scale is in millivolts while the current scale is in arbitrary units. An X-Y recorder was used in taking the data. The sandwich involving the lead behaves exactly as predicted from the model in Fig. 1. Actually to obtain the curve it was necessary to shunt the sample with an RC network to damp out self-induced oscillations. The sandwich involving the indium shows basically the same characteristics, although for indium the unstable region was not traced out. Also, as is apparent from the low-current behavior of this sample, the oxide film is pierced by a superconductive bridge. When the current is increased the bridge goes normal, and its conductivity is too low to affect the general characteristics of the tunneling. When the current is decreased, the bridge remains normal at a lower current due to Joule heating. Finally the sandwich involving the aluminum is a little different, as here the energy gaps on either side of the oxide are equal, and at this temperature the Fermi tail is of the same order of magnitude as half the gap width.

The energy gaps obtained from these experi-

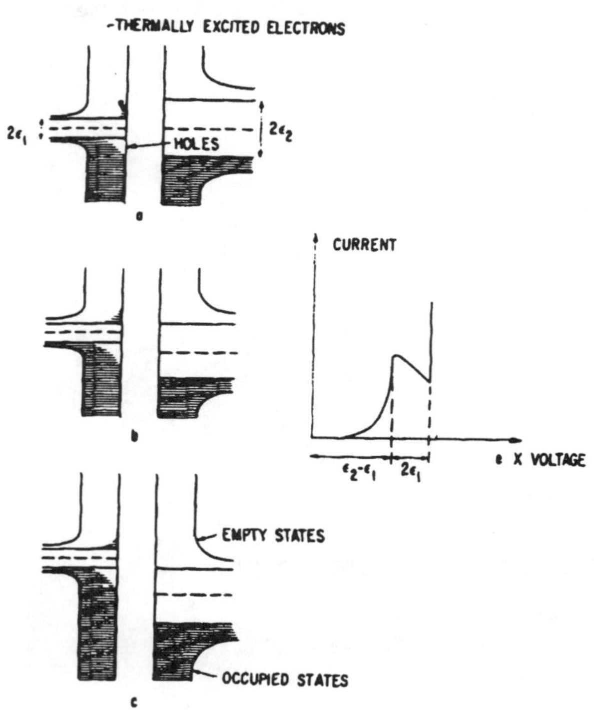

FIG. 1. Analysis of the current-voltage characteristic of two superconductors separated by a thin film. (a) The two superconductors with no voltage applied. Thermally excited electrons and holes are shown for the smaller gap, while for the larger gap there will be relatively few thermally excited electrons. (b) When a voltage is applied, a current will flow and will increase with voltage, because more and more of the thermally excited electrons in the left-hand superconductor are raised above the forbidden gap in the right-hand superconductor, and can tunnel. When the applied voltage corresponds to half the difference of the two energy gaps, $\epsilon_2-\epsilon_1$, it has become energetically possible for all the thermally excited electrons to tunnel through the film. (c) When the voltage is increased further, only the same number of electrons can tunnel, and since they now face a less favorable (lower) density of states, the current will decrease. Finally, when a voltage greater than half the sum of the two energy gaps, $\epsilon_2+\epsilon_1$, is applied, the current will increase rapidly because electrons below the gap can begin to flow.

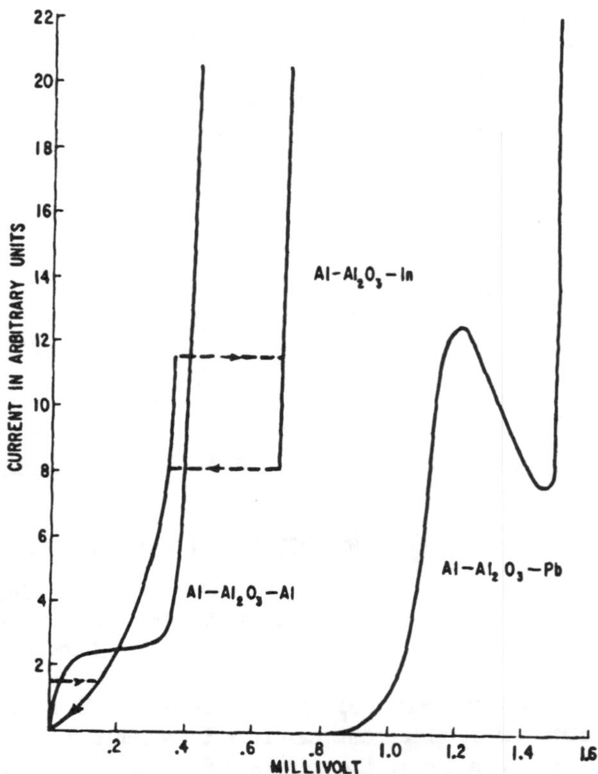

FIG. 2. Characteristic curves for tunneling between two superconductors, showing agreement with the analysis of Fig. 1. The curves $Al-Al_2O_3-In$ and $Al-Al_2O_3-Al$ are taken at $T\sim 1.1°K$, while the curve $Al-Al_2O_3-Pb$ is at $T\sim 1.0°K$.

ments are

$$2\epsilon_{Pb}=(2.68\pm0.06)\times10^{-3}\text{ electron volt},$$

$$2\epsilon_{In}=(1.05\pm0.03)\times10^{-3}\text{ electron volt},$$

$$2\epsilon_{Al}=(0.32\pm0.03)\times10^{-3}\text{ electron volt},$$

For indium and lead, these gaps should not be significantly different at absolute zero, and we obtain

$$2\epsilon_{Pb}=(4.33\pm0.10)kT_c,$$

$$2\epsilon_{In}=(3.63\pm0.10)kT_c,$$

where T_c is the bulk transition temperature. This direct measurement of the energy gap for lead is a little smaller than what was obtained by fitting the experimental results with the BCS theory, where the best fit was obtained with an energy gap of $4.5kT_c$.[4] It should be noted that quite a large spread in the transition temperature of the aluminum films has been found; a transition temperature as high as 1.8°K has been observed. Whether this is true for the lead and indium films as well is not known.

I wish to thank J. C. Fisher and C. P. Bean for their interest and encouragement, L. B. Nesbitt for his advice concerning low-temperature physics, and K. R. Megerle for his help in performing the experiments.

[1]J. C. Fisher and I. Giaever, J. Appl. Phys. (to be published).

[2]I. Giaever, Phys. Rev. Letters 5, 147 (1960).

[3]J. Bardeen, L. N. Cooper, and J. R. Schrieffer, Phys. Rev. 108, 1175 (1957).

[4]I. Giaever, Proceedings of the Seventh International Conference on Low-Temperature Physics (to be published).

EXPERIMENTAL EVIDENCE FOR QUANTIZED FLUX IN SUPERCONDUCTING CYLINDERS*

Bascom S. Deaver, Jr., and William M. Fairbank
Department of Physics, Stanford University, Stanford, California
(Received June 16, 1961)

We have observed experimentally quantized values of magnetic flux trapped in hollow superconducting cylinders. That such an effect might occur was originally suggested by London[1] and Onsager,[2] the predicted unit being hc/e. The quantized unit we find experimentally is not hc/e, but $hc/2e$ within experimental error.[3]

Although the unit of quantized flux is small ($hc/2e = 2.07\times10^{-7}$ gauss cm^2), it can be produced by a magnetic field easily measured and controlled in the laboratory if the area to which it is confined is sufficiently small. For our samples, one flux unit corresponds to a magnetic field of the order of 0.1 gauss. Measurements were made on two hollow tin cylinders. Cylinder No. 1 was 0.8 cm long, 2.33×10^{-3} cm outside diameter and 1.33×10^{-3} cm inside diameter. Cylinder No. 2 was 0.9 cm long, 1.64×10^{-3} cm outside diameter and 1.35×10^{-3} cm inside diameter. These were fabricated by electroplating tin on a one-centimeter portion of a No. 56 copper wire. The sample, plus protruding wire, was jacketed for protection and strength with electroplated copper to an approximate outside diameter of 8×10^{-3} cm.

A field-free region ($H = 0 \pm 0.001$ gauss) is prepared using three orthogonal Helmholtz coils. The tin cylinder is placed in this region and cooled through the superconducting transition in the presence of a known applied axial magnetic field. The net flux in the cylinder is measured both with the field on and after the field is turned off. The measurement is made by moving the tin cylinder up and down one hundred times per second with an amplitude of one millimeter and observing the electrical pickup in two small coils, each of ten thousand turns, surrounding the ends of the cylinder. The instrument is similar in concept to that described by Foner.[4] The induced emf measures the difference in the flux contained within the area of the cylinder and that which would have been in the same area if the cylinder were absent (or in the normal state). The system is calibrated by cooling the sample below the superconducting transition in zero field and observing the signal from the completely diamagnetic cylinder when a known magnetic field is applied. From the value of the field and the measured cross-section area of the cylinder, the absolute value of the flux for a given signal is calculated.

The diameter of each cylinder was measured with a microscope equipped with a micrometer eyepiece. X-ray photographs verified the dimensions of the tin cylinder after the application of the copper jacket. For the purpose of calculating flux, the measured radii of the cylinders are reduced by 0.6 micron due to an expected loss of superconducting properties on the surface of the tin in contact with the copper.[5] That this correction was approximately valid is indicated by a 0.2°K decrease in the value of the transition temperature for the sample No. 2 whose cylindrical walls were 1.5 microns thick, leaving, we believe, only 0.3 micron of superconducting material in the center after allowance for the effect of the center copper wire and the outside copper jacket. This is in agreement with the experimental results on electroplated tin.[5]

With this adjustment, the area used for the diamagnetic calibration of cylinder No. 1 is 3.84×10^{-6} cm^2, and the area of the hole is 1.65×10^{-6} cm^2. For cylinder No. 2 the corresponding quantities are 1.81×10^{-6} cm^2 and 1.70×10^{-6} cm^2.

Data on sample No. 1 are shown in Fig. 1, and on sample No. 2 in Fig. 2. The diagonal line through the origin represents the calibration. It is the signal corresponding to zero flux in the cylinder and hole in the presence of the applied field as described above. The experimental points shown on the graph represent two types of data for each value of the applied field. The points on the bottom half of each graph represent the signal in the presence of the applied field after cooling through the transition in that field. The points in the upper half represent the trapped flux after the field is subsequently reduced to zero. For each point in the lower curve there is a corresponding point in the upper curve. The solid lines represent calculated integral values of $hc/2e$.

It can be seen that certain features of the data are common to both samples. (1) Below a certain value of applied field, the total cross section of the cylinder acts as a perfect diamagnet, excluding all the flux, and no flux is trapped when the applied field is turned off. (We believe this

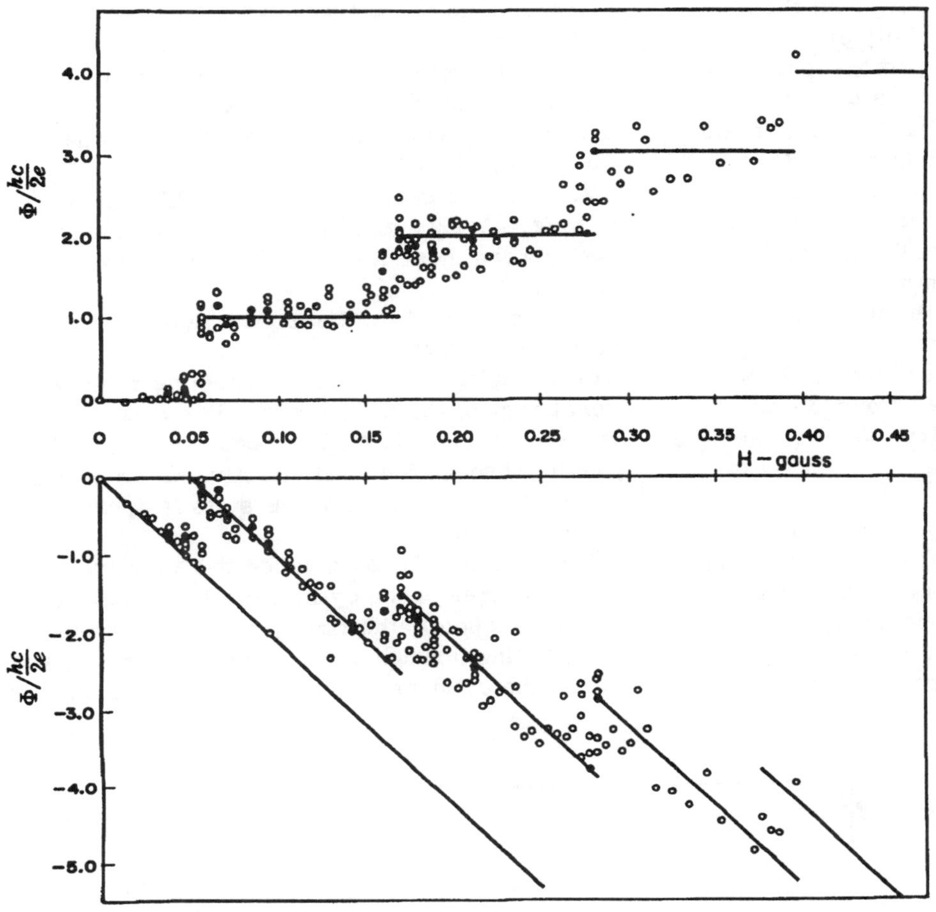

FIG. 1. (Upper) Trapped flux in cylinder No. 1 as a function of magnetic field in which the cylinder was cooled below the superconducting transition temperature. The open circles are individual data points. The solid circles represent the average value of all data points at a particular value of applied field including all the points plotted and additional data which could not be plotted due to severe overlapping of points. Approximately two hundred data points are represented. The lines are drawn at multiples of $hc/2e$. (Lower) Net flux in cylinder No. 1 before turning off the applied field in which it was cooled as a function of the applied field. Open and solid circles have the same significance as above. The lower line is the diamagnetic calibration to which all runs have been normalized. The other lines are translated vertically by successive steps of $hc/2e$.

provides a way of obtaining a truly zero-magnetic-field region.) (2) When the applied field exceeds a certain value, flux is trapped both with the field on, and after the applied field is turned off. The amount of this trapped flux within the experimental accuracy of the data is $hc/2e$.

The amount of trapped flux remains constant as the applied field is increased until a value approximately three times that for the initial trapping is reached, at which point the trapped flux increases to about twice the original amount. There appears to be evidence for additional changes at five and seven times the field for the first trapping.

Fluctuations in the data are caused by variations in the zero of the magnetic field, changes in the gain, vibration amplitude, drift, and random noise in the detection system. The approximately two hundred data points for sample No. 1 were taken over a three-week period during which the drift and noise were gradually improved. The fluctuations of the data around the values 0 and $hc/2e$ represent, we believe, expected scatter from drift and noise. This scatter has been greatly improved for sample No. 2. For both samples the data are consistent with values 0 and $hc/2e$ for the trapped flux as described above.

Near the transition to the second and third steps the fluctuations in the data are greater, and in addition points lie between the steps. Some increased scatter is expected since the absolute fluctuations due to changes in gain and vibration amplitude are proportional to the size of the signal. The points between the steps do not necessarily indicate trapping of nonintegral values of flux. Since the observed signal is the sum of the emf's from coils at the two ends of the sample, a flux line passing out of the cylinder at some point other than the end may produce different signals in the two coils. The two ends of the cylinder are not quite identical; so near the transition region it is probable that the two ends might trap a different number of units of flux, the extra unit being shoved out the side of the cylinder. This is especially probable for sample No. 1 since the x-ray photograph showed a break in the tin coating near the middle of the cylinder. Also, it is known that flux can create a normal region in a superconductor by shrinking in size until the critical field is exceeded. In this experiment we were unable to measure independently the signals from the two coils. However, in future experiments this will be done to remove this ambiguity. It is interesting to note that no intermediate points are found outside the expected scatter of the data near the first step. One point for which no flux was trapped was found near the center of the first step with sample No. 1.

In conclusion, we find:

1. The flux trapped in a superconducting cylinder both in the presence and absence of an applied magnetic field is not continuous but exhibits a step behavior, the first step occurring for $\Phi = hc/2e$, within experimental error in the data. Considering all sources of error, we estimate that the value of the trapped flux at the first step is $hc/2e \pm 20\,\%$. If the correction to the size of the cylinder due to the presence of the copper should prove invalid, an additional 11 % error could arise for the large cylinder and 17 % for the small cylinder.

2. The data seem to indicate additional steps at hc/e, $3hc/2e$, and $2hc/e$. The points appearing between these levels will be investigated further.

3. The ratio of the fields at which the steps occur are approximately 1, 3, 5, and 7. In the first cylinder (for which the effective cross-sectional area of the cylinder is 2.33 times the area of the hole), the first jump occurs when the flux passing through the total effective cross section of the cylinder in the normal state is approximately $hc/2e$.

For cylinder No. 2 (in which the effective cross-sectional area of the cylinder is 1.1 times the area of the hole), the first jump occurs when the flux passing through the total effective cross section of the cylinder in the normal state is approxi-

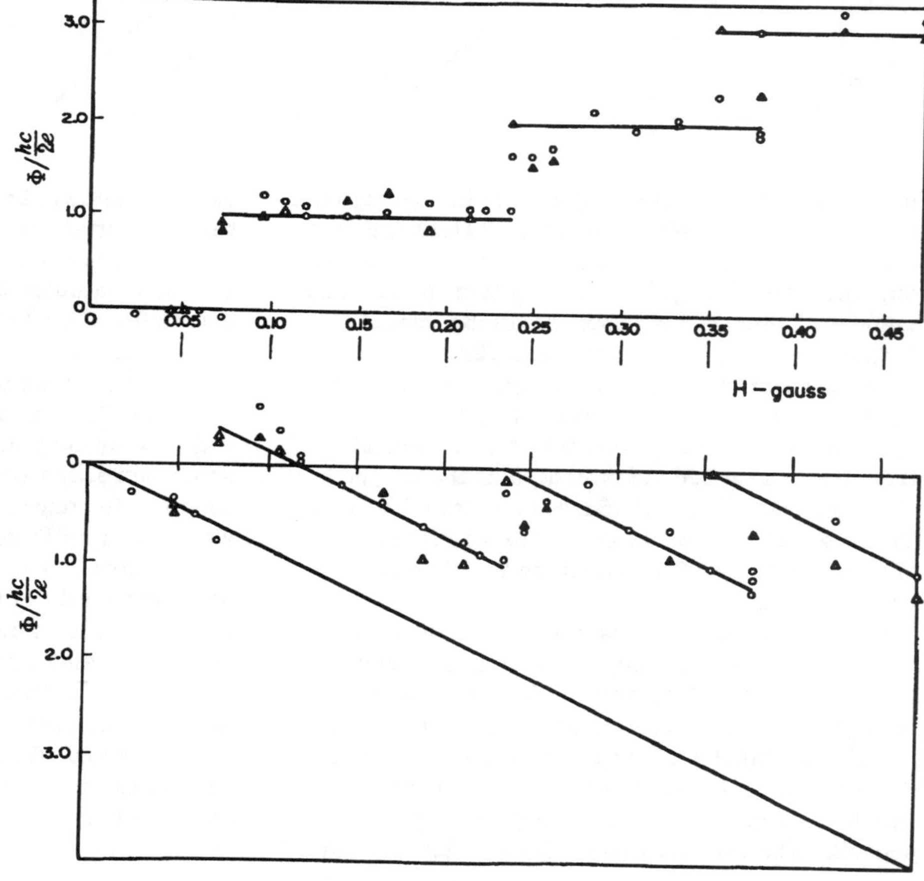

FIG. 2. (Upper) Trapped flux in cylinder No. 2 as a function of magnetic field in which the cylinder was cooled below the superconducting transition temperature. The circles and triangles indicate points for oppositely directed applied fields. Lines are drawn at multiples of $hc/2e$. (Lower) Net flux in cylinder No. 2 before turning off the applied field as a function of the applied field. The circles and triangles are points for oppositely directed applied fields. The lower line is the diamagnetic calibration to which all runs have been normalized. The other lines are translated vertically by successive steps of $hc/2e$.

mately $0.6hc/2e$. In a following Letter, Byers and Yang[6] conclude that in a thin ring the first jump should occur at $0.5hc/2e$.

4. Since the time constant of our measuring circuit is 25 seconds, this experiment gives only a large upper limit for the time involved in reaching these quantized flux values. Mercereau and Vant-Hull[7] have reported a negative experiment designed to observe quantized flux in a 1-mm ring cooled 6000 times per second through the superconducting transition in a small magnetic field. It is possible that the difference in their results and the results of our experiment are due to a minimum time necessary to establish equilibrium. We are planning to investigate this relaxation time.

We have had the pleasure of discussing the results of this experiment with N. Byers, C. N. Yang, and L. Onsager, whose interpretation of these results appear in the following Letters.[6,8] One of us (WMF) also wishes to acknowledge his indebtedness to F. London and M. J. Buckingham who greatly influenced his concept of the superfluid state. We also wish to thank F. Bloch, L. I. Schiff, and J. D. Bjorken for many stimulating discussions of the experiment. We wish to acknowledge the invaluable assistance of M. B. Goodwin.

---

*Work supported in part by grants from the National Science Foundation, the Office of Ordnance Research (U. S. Army), and the Linde Company.

[1]F. London, Superfluids (John Wiley & Sons, New York, 1950), p. 152.

[2]L. Onsager, Proceedings of the International Conference on Theoretical Physics, Kyoto and Tokyo, September, 1953 (Science Council of Japan, Tokyo, 1954), pp. 935-6.

[3]Such a possibility was mentioned by Lars Onsager to one of us (WMF) at the conference on superconductivity in Cambridge, England, 1959 (unpublished).

[4]S. Foner, Rev. Sci. Instr. 30, 548 (1959).

[5]E. Burton, H. Grayson-Smith, and J. Wilhelm, Phenomena at the Temperature of Liquid Helium (Reinhold Publishing Corporation, New York, 1940), p. 120.

[6]N. Byers and C. N. Yang, following Letter [Phys. Rev. Letters 7, 46 (1961)].

[7]J. E. Mercereau and L. L. Vant-Hull, Bull. Am. Phys. Soc. 6, 121 (1961).

[8]L. Onsager, this issue [Phys. Rev. Letters 7, 50 (1961)].

EXPERIMENTAL PROOF OF MAGNETIC FLUX QUANTIZATION IN A SUPERCONDUCTING RING*

R. Doll and M. Näbauer
Kommission für Tieftemperaturforschung der Bayerischen Akademie der Wissenschaften, Herrsching/Ammersee, Germany
(Received June 19, 1961)

From theoretical considerations, based on wave mechanics, London[1] concluded that the magnetic flux frozen in in a twofold-connected superconducting body (ring or tube) should not have any arbitrary value, but only such values which are integer multiples of a basic unit ϕ_0,

$$\phi_0 = hc/e = 4.12\times10^{-7} \text{ gauss cm}^2. \quad (1)$$

That means the magnetic flux should be quantized. Bardeen and Schrieffer[2] also agreed with this conclusion.

In order to verify a possible flux quantization, the mechanical torque exerted by a magnetic field H_x on a small superconducting lead tube with frozen-in magnetic flux has been measured.

From the following equation, one obtains the magnetic field H_{y0} necessary to freeze in just one flux unit ϕ_0:

$$H_{y0} = \phi_0 \times (4/\pi d^2), \quad (2)$$

where d is the diameter of the tube. The diameter has to be chosen sufficiently small that the fluctuations of the earth's magnetic field can be neglected relative to the magnetic field adequate to freeze in one flux unit. This has been achieved with a tube of 10.3-micron diameter and 0.6-mm length (see Fig. 1), which gives a magnetic field of $H_{y0} = 0.5$ oe, according to Eq. (2).

The sample consists of a small lead cylinder, prepared by evaporating lead on a quartz fiber (of about 10-micron diameter and with a length of about 1 mm).

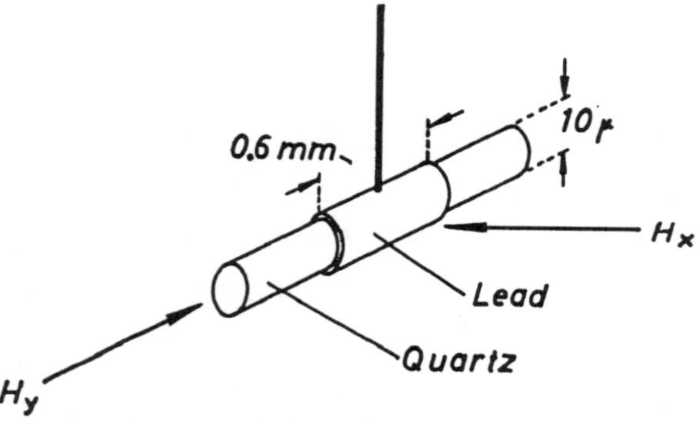

FIG. 1. Schematic diagram of the sample with the directions of the applied field H_y to be frozen in, and the measuring field H_x.

The very small torque, proportional to the frozen-in flux and to the measuring field H_x (normal to H_y), can be observed by an already known autoresonance method.[3,4] The sample is suspended on a thin torsion fiber (normal to H_y and H_x) inside a coil. In connection with a mirror for recording the oscillation amplitude it represents a system of damped torsion oscillation. The oscillation of the system can be kept at a constant amplitude by the alternating torque, caused by periodically reversing the magnetic field H_x of the coil. The oscillating system itself controls the time of switching the field by means of a photoelectric device. The damping of the system being known, the constant resonance amplitude is a measure for the torque which acts on the sample.

Each experimental value was obtained in the following manner: 1. The sample was heated above the transition temperature; then a defined field to be frozen in, H_y, was applied. 2. After recooling below the transition temperature, H_y was switched off. 3. The resonance amplitude was measured as described above.

The resonance amplitude is proportional to the product (measuring field H_x) times (magnetic moment of the sample). The latter in turn is proportional to the frozen-in flux. Figure 2 shows the measured resonance amplitude divided by the driving field H_x as a function of the field H_y. (H_x has always been about 10 oe.)

As Fig. 2 shows, it is impossible to freeze in any flux between $H_y = -0.1$ and $+0.1$ oe. Near ±0.1 oe there occur marked steps. Upon increasing the magnetic field H_y, the frozen-in fluxes remain nearly constant between 0.1 and 0.3 oe. At 0.3 oe another step occurs, again followed by a series of constant values.

This is exactly what is expected of a quantized magnetic flux in a twofold-connected superconducting body. If an arbitrary flux could be frozen in, the relation between magnetic flux and field H_y would be as shown in Fig. 2 by the dashed line. This has been obtained by measurements at comparatively high fields ($H_y = 10$ oe), in which case the value of one flux unit is already very small compared with the entire frozen-in flux.

With the microscopically measured diameter of

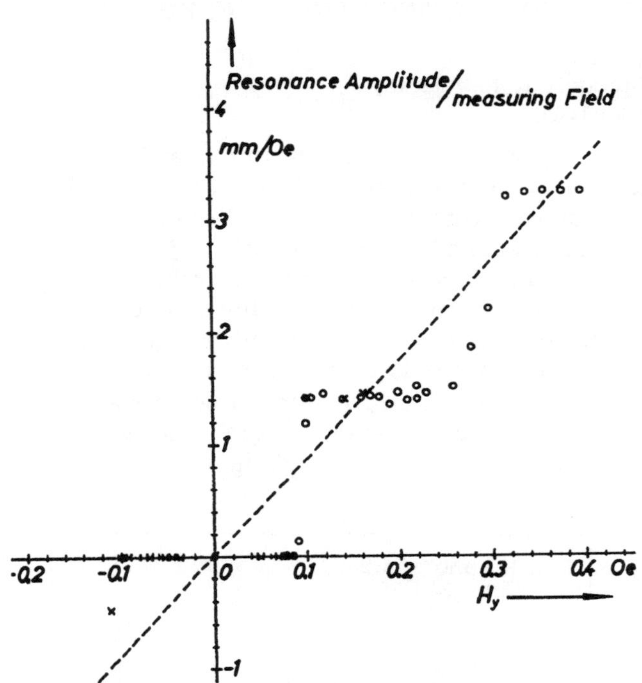

FIG. 2. Resonance amplitude divided by measuring field H_x as a function of the applied field H_y. The ordinate is proportional to the frozen-in flux. ×— First run; o— second run.

the lead tube, Eq. (2) predicts for the interval of the magnetic field strength corresponding to one flux unit a value of $H_y = 0.5$ oe. The experimentally observed interval, however, reaches only 0.2 oe, that is about 40% of the calculated value. So far the reason for this discrepancy is not clear. For example, an error of 60% in the determination of the lead tube's diameter would explain the difference, but such an error is improbable.

The experiments are being continued with higher fields H_y and other superconductors of various diameters.

Mercereau and Vant-Hull[5] also tried to verify London's postulate of the quantization of magnetic flux in a superconducting ring. The result of their experiments was negative.

The authors are indebted to Professor W. Meissner who made possible and promoted this work. The authors would further like to thank Professor F. X. Eder for encouragement and helpful discussions.

---

*Presented at the Conference on Fundamental Research in Superconductivity, IBM Research Center, Yorktown Heights, New York, June 15-17 (1961).

[1]F. London, Superfluids (John Wiley & Sons, New York, 1950), Vol. I, p. 152.

[2]J. Bardeen and I. R. Schrieffer, Progress in Low-Temperature Physics (North-Holland Publishing Company, Amsterdam, 1961), Vol. III, p. 182.

[3]A. Einstein and W. J. de Haas, Verhandl. deut. physik. Ges. 17, 152 (1915).

[4]R. Doll, Z. Physik 153, 207 (1958).

[5]J. E. Mercereau and L. L. Vant-Hull, Bull. Am. Phys. Soc. 6, 121 (1961).

PROBABLE OBSERVATION OF THE JOSEPHSON SUPERCONDUCTING TUNNELING EFFECT

P. W. Anderson and J. M. Rowell
Bell Telephone Laboratories, Murray Hill, New Jersey
(Received 11 January 1963)

Josephson[1] has recently predicted theoretically a new kind of tunneling current through an insulating barrier between two superconducting metals. This anomalous current behaves like a direct tunneling of condensed electron pairs between the Fermi surfaces of the two metals. When the voltage difference across the barrier is zero, the current should be dc but may range between limits of the order of magnitude of the usual tunnel current above the gap, depending on the relative value of the phase of the energy gap function or "second Green's function F" on the two sides, while at a nonzero voltage difference ΔV it is alternating at a frequency $2e\Delta V/h$.

We have observed an anomalous dc tunneling current at or near zero voltage in very thin tin oxide barriers between superconducting Sn and Pb, which we cannot ascribe to superconducting leakage paths across the barrier, and which behaves in several respects as the Josephson current might be expected to.

Figure 1 shows an X - Y recorder plot of the tunneling current vs voltage for one of these structures at ~1.5°K. The lead and tin films are both approximately 2000Å thick, and the junction has dimensions 0.025×0.065 cm^2 and a resistance (both metals normal) of 0.4 Ω. Voltage is applied to two arms of the junction from a 1-kΩ potentiometer and the resulting current flow is measured as voltage across a series resistor of 10 Ω. The voltage appearing across the barrier is taken directly from the other two arms of the junction. Figure 2 shows the plot with current scale expanded to show the anomalous region near the origin. The current

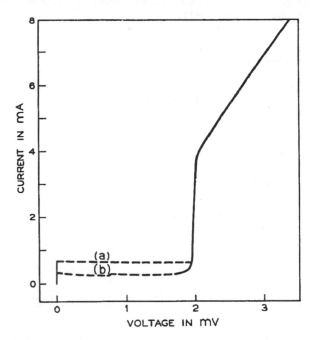

FIG. 1. Current-voltage characteristic for a tin-tin oxide-lead tunnel structure at ~1.5°K, (a) for a field of 6×10^{-3} gauss and (b) for a field 0.4 gauss.

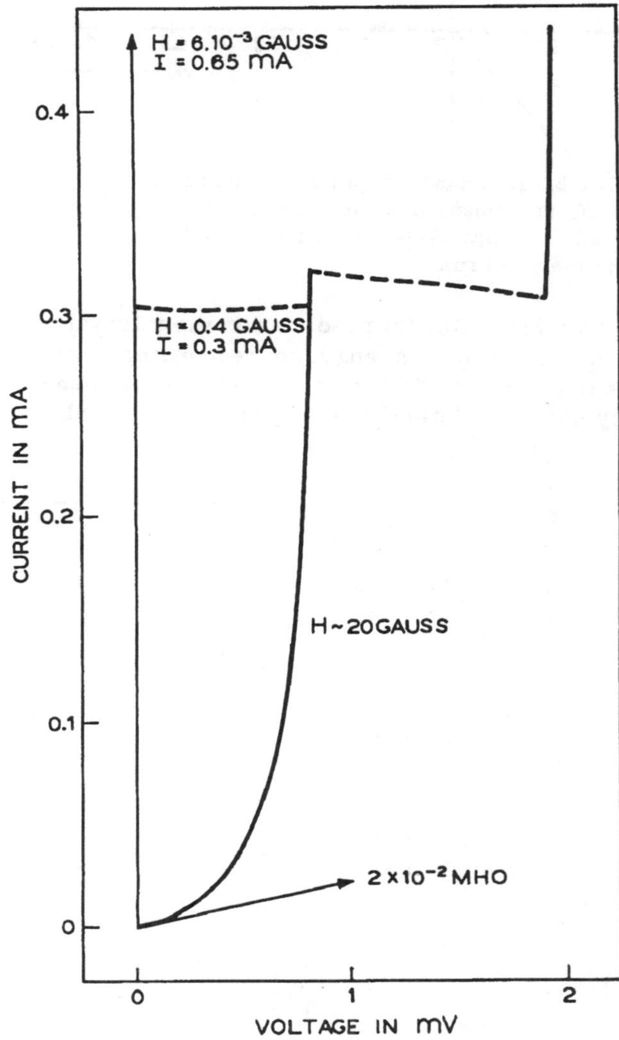

FIG. 2. Current-voltage characteristic for structure of Fig. 1 with expanded current scale. Note the conductance at low voltages in a field ~20 gauss.

at first increases up to a value of 0.3 mA with no voltage appearing across the barrier. At this point the junction becomes unstable and may fluctuate back and forth between the vertical characteristic and the expected "two-superconductor" characteristic. With a small increase in current, the junction settles stably on the latter. Expansion of the voltage scale showed some fluctuations even at lower currents in many cases.

One possible explanation that will be suggested is, of course, that in such thin junctions we have not avoided small superconducting shorts across the barrier. There are, however, four experimental points suggesting that this is indeed the Josephson effect.

(1) As pointed out in Josephson's Letter,[1] the effect should be quite sensitive to magnetic fields. Since the proposed proportionality to $\mathrm{Im}(\Delta_{\mathrm{Sn}}{}^*\Delta_{\mathrm{Pb}})$ is not gauge invariant, we expect an additional dependence on $\sin\int_1^2(2e/\hbar c)\vec{A}\cdot d\vec{l}$, where $\vec{A}$ is the vector potential, which will lead to cancellation of currents in various parts of the barrier if the magnetic flux flowing between the superconductors reaches one or two quanta. With an area of $\sim 10^{-7}\ \mathrm{cm}^2$, this corresponds to about 1 gauss. We have found (see Fig. 2) that when the junction was carefully shielded by a mu-metal can with a measured interior field of 6×10^{-3} gauss, the vertical characteristic reached 0.65 mA, with no shielding (0.4 gauss), 0.30 mA, and when less than 20 gauss was applied the anomalous behavior was not observed. Fine superconducting filaments should show anomalously high, not low, critical fields.

(2) The effect can only occur if both metals are superconducting and should be proportional to $\sim|\Delta_{\mathrm{Sn}}\Delta_{\mathrm{Pb}}|$. On cooling the junction we find that the vertical characteristic appears at the tin transition (as measured by the negative resistance region of width Δ_{Sn} first appearing at Δ_{Pb}) within experimental error. It seems unlikely that the "superleaks" would have precisely the same T_c as the tin film.

(3) Critical currents of finely divided specimens never exceed $10^7\ \mathrm{A/cm}^2$. Our observed maximum current of 0.65 mA corresponds, then, to an area A of superleaks of not less than 6×10^{-11} cm^2. Assuming a rather poor conductivity of 10^4 $\mathrm{ohm}^{-1}\ \mathrm{cm}^{-1}$ for the filaments, this leads to a conductance, which should be observed even when the filaments are normal, of

$$Y \geqslant \frac{10^4\times6\times10^{-11}}{10^{-7}} = 6\ \mathrm{mho},$$

whereas (Fig. 2) we observe with an applied field at most 2×10^{-2} mho, most of which is probably thermally excited quasi-particle tunneling.

(4) We attempt to "burn out" the leaks by passing increasing currents through the junction, checking the anomalous current between each step. We observed no change in the vertical characteristic for a number of junctions ($\frac{1}{10}$ to 1 Ω) until (at a voltage usually between 0.3 and 0.6 volt) destruction of the junction resulted. Metallic leaks should burn out before the junction as a whole does.

These last arguments seem to us nearly to exclude the conducting leak hypothesis.

The maximum current we observe is still, even in the best units, about an order of magnitude less than predicted. Other questions concern the fluctuation effects and the fact that thin junctions are necessary to see the effect.

We believe that these questions are probably all best elucidated by looking at the effect as related to a coupling energy between the phases of the gap functions on the two sides. Calculations which will be reported elsewhere show that this energy is proportional to the negative cosine of the phase difference, and in magnitude is

$$\Delta E = (\hbar/e)J_1,$$

where J_1 is the maximum amplitude of Josephson's predicted current. This energy coupling is reduced both by the presence of magnetic fields and by driving current through the unit. In order to observe the dc Josephson effect, this energy must be large enough to keep the phases on the two sides coupled against thermal or other fluctuations.

The total magnitude of ΔE for the entire unit is rather small—of order 1 or a few eV in the thin barriers, 10^{-2} eV in the more normal units. Obviously the thicker units are completely unstable against thermal fluctuations (remember that most of the circuit is at room temperature or higher). The thin barriers are more stable, but as we drive larger currents through them, or apply stray magnetic fields, we lower the coupling energy, and random fluctuations will eventually destroy the dc current and replace it with noise.

A perfectly rigorous way to think of this process is that the energy ΔE serves as a barrier against the passage of quantized flux lines through the unit. It will act as a superconductor only if fluctuations over the applied magnetic stresses cannot drive lines through the barrier region at an appreciable rate.

We wish to acknowledge the assistance of L. Kopf in the preparation of the tunnel units, and discussions with V. Ambegaokar. B. D. Josephson informs us that he has independently reached some of the theoretical conclusions of the last few paragraphs.

---

[1]B. D. Josephson, Phys. Letters 1, 251 (1962).

PHYSICAL REVIEW LETTERS

VOLUME 12 | 17 FEBRUARY 1964 | NUMBER 7

QUANTUM INTERFERENCE EFFECTS IN JOSEPHSON TUNNELING

R. C. Jaklevic, John Lambe, A. H. Silver, and J. E. Mercereau
Ford Scientific Laboratories, Dearborn, Michigan
(Received 16 January 1964)

Current flow through a Josephson junction has been shown[1] to depend periodically upon the magnetic flux contained within the junction—the period in contained flux being $(h/2e)$. This concept has been extended to include multiple junctions in parallel connected by superconducting links. Such a parallel configuration of two junctions leads to the expectation of two periodicities of the current with flux. One period is again associated with the flux contained in a single junction, but now another period arises associated with the flux enclosed in the area between junctions. This second period involves a quantum mechanical interference between the currents flowing through separate junctions in direct analogy with double-slit electron beam interference effects. The purpose of this Letter is to report experimental observation of such interference effects.

Current flow through a Josephson junction is given by[2]

$$j_J = j_0 \sin\left(\gamma_a - \gamma_b - \frac{2e}{\hbar}\int_a^b A dx\right), \qquad (1)$$

where γ_a and γ_b are the phases of the wave function at superconductors a and b separated by the Josephson junction. Integrating (1) over the current-carrying area of the junction gives a total current having a functional form typical of "diffraction" effects:

$$I = I_0 \frac{|\sin(\Phi_j e/\hbar)|}{(\Phi_j e/\hbar)} |\sin(\Delta\gamma)|,$$

where Φ_j is the flux contained in the current-carrying area of the junction and $\Delta\gamma = \gamma_1 - \gamma_2$. If (1) is applied to two identical junctions connected in parallel by superconducting links, a double periodicity results. The integration of (1) over both junctions, keeping account of the relative phase between the separate junctions, leads to additional interference effects:

$$I \simeq I_0 \frac{|\sin(\Phi_j e/\hbar)|}{(\Phi_j e/\hbar)} |\sin(\Delta\gamma - \Phi_T e/\hbar)|,$$

where Φ_T is the total flux enclosed between junctions.

Double junctions were fabricated as shown in Fig. 1. The normal resistance of the junctions used was about $\frac{1}{2}$ ohm. Junctions were spaced ap-

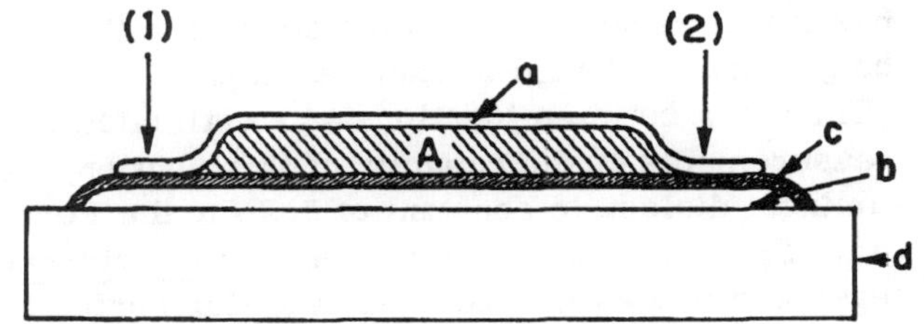

FIG. 1. Cross section of a Josephson junction pair vacuum-deposited on a quartz substrate (d). A thin oxide layer (c) separates thin (~1000Å) tin films (a) and (b). The junctions (1) and (2) are connected in parallel by superconducting thin film links forming an enclosed area (A) between junctions. Current flow is measured between films (a) and (b).

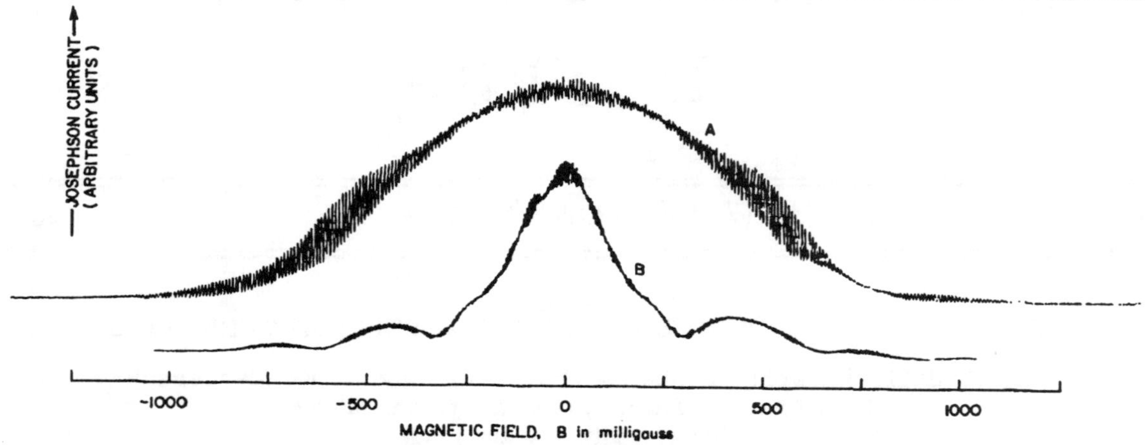

FIG. 2. Josephson current vs magnetic field for two junctions in parallel showing interference effects. Magnetic field applied normal to the area between junctions. Curve (*A*) shows interference maxima spaced at $\Delta B = 8.7 \times 10^{-3}$ G, curve (*B*) spacing $\Delta B = 4.8 \times 10^{-3}$ G. Maximum Josephson current indicated here is approximately 10^{-3} A.

proximately 3.5 mm apart forming an area between junctions ranging from 10^{-4} to 10^{-5} cm^2 for the data given herein. The junctions were tin-(tin oxide)-tin separated by a Formvar spacer.

Typical experimental results for two different junction pairs are given in Fig. 2. In the upper trace (*A*) the individual junctions are narrow showing only the central maximum of the single-slit diffraction in this field span but also clearly showing the interference effects between junctions. The lower trace (*B*) with somewhat wider junctions shows both the single-slit diffraction with side peaks[1] and the expected interference effects. As the magnetic field is rotated in the plane of the junction away from normal to the area between junctions, the field spacing between interference maxima increases as expected corresponding to the geometric change in enclosed flux.

This area between junctions was estimated from measurements of the capacity of this section, assuming a dielectric constant of 3.2 for the Formvar. From the field spacing between interference peaks and this estimated area, the flux period for the junction pair (*A*) is 2.7×10^{-7} G cm^2, while for junction pair (*B*) the period is 2.4×10^{-7} G cm^2. The flux period associated with the diffraction minima in junction pair (*B*) was found to be 2.5×10^{-7} G cm^2, using as area the width (0.8 mm) multiplied by twice the penetration depth (510 Å). This width (0.8 mm) is somewhat larger than the expected effective width for this junction assuming a "Josephson-length" effect. The effective width[1] for this junction is about 0.7 mm, yielding a flux period of 2.2×10^{-7} G cm^2.

This flux period from the diffraction minima is in reasonable agreement with previous work[1] and the theoretical value ($h/2e$). The somewhat larger period determined from the interference effects in these two junction pairs probably reflects the inadequacy of the area determination technique.

Similar interference effects have been observed in all Josephson pairs we have examined. We believe these data demonstrate interference effects (and thus phase coherence) in the quantum wave function in solids at distances, in these junctions, of up to 3.5 mm. The obvious experiment to measure the effects of vector potential alone is in progress.

---

[1]J. M. Rowell, Phys. Rev. Letters 11, 200 (1963).

[2]P. W. Anderson and J. M. Rowell, Phys. Rev. Letters 10, 230 (1963).

MEASUREMENT OF $2e/h$ USING THE ac JOSEPHSON EFFECT AND ITS IMPLICATIONS FOR QUANTUM ELECTRODYNAMICS*

W. H. Parker
Department of Physics and Laboratory for Research on the Structure of Matter, University of Pennsylvania, Philadelphia, Pennsylvania

and

B. N. Taylor†
RCA Laboratories, Princeton, New Jersey

and

D. N. Langenberg
Department of Physics and Laboratory for Research on the Structure of Matter, University of Pennsylvania, Philadelphia, Pennsylvania

(Received 23 January 1967)

Using the ac Josephson effect, we have determined that $2e/h = 483.5912 \pm 0.0030$ MHz/μV. The implications of this measurement for quantum electrodynamics are discussed as well as its effect on our knowledge of the fundamental physical constants.

In this Letter, we report a high-accuracy measurement of $2e/h$ using the ac Josephson effect (here, e is the electron charge and h is Planck's constant). When combined with the measured values of other fundamental constants, this measurement yields a new value for the fine-structure constant α which differs by 21 ppm from the presently accepted value. This change in α removes the present discrepancy between the theoretical and experimental values of the hyperfine splitting in the ground state of atomic hydrogen, one of the major unsolved problems of quantum electrodynamics today. We also discuss the effect of this change on our present knowledge of the fundamental physical constants.

The phenomenon used in these experiments was first predicted by Josephson in 1962.[1] He showed theoretically that when two weakly coupled superconductors are maintained at a potential difference V, an ac supercurrent of frequency

$$\nu = (2e/h)V \tag{1}$$

flows between them. This equation, known as the Josephson frequency-voltage relation, can be shown to follow from quite general assumptions concerning superconductivity,[1,2] and is believed to be exact. In a recent publication[3] we reported an experimental test of this relation which verified that the frequency-voltage ratio was equal to the then best value of $2e/h$ to within the 60-ppm uncertainty of the measurements, and was completely independent of all of the experimental variables tested. This work also demonstrated that the main experimental difficulty with using the ac Josephson effect to make a high-accuracy determination of $2e/h$ was the calibration of the voltage-measuring system. Since then, we have acquired a completely self-calibrating system, and have carried out a new series of measurements with an order-of-magnitude improvement in accuracy.

In these new experiments, measurements were performed on two types of "junctions" which show ac Josephson-effect phenomena—evaporated thin-film tunnel junctions, and point-contact weak links.[3] The particular phenomenon used was that of microwave-induced, constant-voltage current steps[1,2] of the type first observed by Shapiro.[4] Such steps appear as excess dc currents in the current-voltage characteristics of Josephson junctions when they are irradiated with microwave radiation of frequency ν. Physically, this effect arises from the presence of a microwave-induced ac voltage across the junction which frequency-modulates the ac Josephson current. The constant-voltage current steps simply correspond to zero-frequency (dc) sidebands. The relationship between the voltages V_n at which these steps occur and the frequency of the applied radiation is $2eV_n = nh\nu$, where n is the number of the step. A determination of $2e/h$ can thus be made simply by measuring the frequency of the applied microwave radiation and the absolute voltage at which the current steps occur. No other measurements are required except those necessary in calibrating the equipment, and in this sense it is a remarkably straightforward fundamental constant experiment.

The applied microwave radiation was generated by an X-band (8- to 12.4-GHz) oscillator with a stability of one part in 10^8 per hour when phase-locked to a quartz crystal reference. The frequency of the radiation was measured to an accuracy of one part in 10^8 by using an electronic counter and a microwave frequency converter. The reference time base of the counter was maintained to an accuracy of better than one part in 10^8 by regular phase comparisons with the U. S. frequency standard as broadcast by radio station WWVB, Fort Collins, Colorado. Thus, the frequency measurement contributed negligible error (~0.01 ppm) to the measured value of $2e/h$. The over-all accuracy was limited by the voltage-measuring system, i.e., potentiometer and standard reference voltage.

The reference voltage used was the mean voltage of a set of six standard cells, calibrated by the U. S. National Bureau of Standards (NBS), in a constant-temperature air bath. This voltage was known to 1 ppm in terms of the NBS legal volt, the uncertainty being an estimate of the possible changes in the emf of the standard cells due to transporting them from NBS to our laboratory. In order to obtain a value of $2e/h$ in absolute units, it is necessary to convert from the legal volt to absolute volts. The present best value of this conversion factor is 1 NBS legal volt $= 1.000\,012 \pm 0.000\,004$ absolute volts, where the 4-ppm uncertainty is intended to represent a 50% confidence level.[5]

The potentiometer used was the Julie Research Laboratories PVP 1001.[6] This nanovolt instrument is self-calibrating in that it has provisions which enable the operator to measure all factors which contribute to the accuracy of a voltage measurement and to make any necessary corrections. Using techniques developed by Julie Research Laboratories[7] and NBS,[8] the 1-mV full-scale range can be calibrated with an rms uncertainty of between 3 and 4 ppm (this was the range normally used since the voltages of the induced current steps rarely exceeded 1 mV). The null detector used with the potentiometer consisted of a photocell amplifier and galvanometer and had a resolution of 1 nV.

In making accurate measurements of such small voltages, great effort is necessary to eliminate or correct for spurious voltages in the measuring circuit. In the measurements described here, the effect of voltages which do not reverse when the current is reversed (thermoelectric voltages, for example) was eliminated by measuring constant-voltage current steps of both polarities. Voltages which reverse with current (those from Ohmic sources, for example) were shown to be negligible by observing that the voltage in the measuring circuit was constant over the full range of the zero-voltage current arising from the dc Josephson effect.[1] Spurious voltages due to any rectification of the microwaves were also shown to be negligible by observing that the measured value of $2e/h$ was independent of microwave power over a range of 10 dB for a given sample and of 20 dB from sample to sample.

The results of measurments on several thin-film tunnel junctions and point-contact weak links are given in Table I. The standard deviation of a set of measurements obtained during one run[9] on any particular junction, typically 2 ppm, is due to the 1-nV resolution of the null detector, the stability and linearity of the potentiometer, and the stability of the thermoelectric voltages in the measuring circuit (usually of order 100 nV). Within this 2-ppm standard deviation, the measured value of $2e/h$ was found to be independent of a wide variety of experimental conditions, including step number up to $n = 40$, magnetic field from 0 to 10 G, microwave frequency from 8 to 12 GHz, and microwave power. All of the measurements were carried out between 1.2 and 1.6°K.

For most of the current steps from which the data were obtained, the voltage was constant to within 1 nV (the resolution of the null detector) over the full range of the step. However, for three of the point contacts (marked by an asterisk in Table I), the voltage was found to increase by 5-10 nV as the current was increased over the range of the step. It was observed that in higher resistance point contacts (several tenths of an ohm rather than several hundredths of an ohm), where the voltage variation was as much as 200 nV, the midcurrent point of the step gave a value of $2e/h$ equal to the average of all the data obtained on the constant-voltage steps. As the resistance was decreased, the voltage variation decreased and the midpoint continued to give a value of $2e/h$ in agreement with the constant-voltage step data. Extrapolating this behavior to contacts with 5- to 10-nV variation, we assume that the midcurrent point corresponds to the voltage at which the step would occur if it were constant. Thus, all measurements on such steps were made at this midpoint.

The average of all of the data in Table I, weighted as the inverse square of the rms uncertainties, gives

$$2e/h = 483.5912 \pm 0.0030 \text{ MHz}/\mu\text{V},$$

or in more conventional terms,

$$h/e = 4.135\,725 \pm 0.000\,026 \times 10^{-15} \text{ J sec/C}$$
$$= 1.379\,529 \pm 0.000\,008 \times 10^{-17} \text{ erg sec/esu}.$$

The quoted uncertainty, about 6 ppm (70% confidence level), is an rms sum of all known sources of error, either systematic or random, and includes the uncertainty in the calibration of the potentiometer, the standard deviation

Table I. Summary of experimental data. Junctions of the form Sn-SnO-X are evaporated thin-film tunnel junctions while the others are point-contact weak links. The table entries are in chronological order and the decreasing uncertainty in the potentiometer calibration results from improved techniques.

JUNCTION	UNCORRECTED $n\nu/V$ WITH STD. DEV. MHz/μV	POTENTIOMETER CORRECTION WITH rms UNCERTAINTY ppm	STD. CELL TEMPERATURE CORRECTION ppm	CORRECTED $n\nu/V$ WITH rms UNCERTAINTY MHz/μV
Sn-SnO-Sn	483.610 ± .002	−30 ± 10	—	483.596 ± .005
Sn-Sn	483.6126 ± .0011	−29 ± 7	—	483.5986 ± .0032
Ta-Ta	483.6164 ± .0009	−39 ± 4	−0.5	483.5973 ± .0021
Sn-SnO-Sn	483.6156 ± .0008	−39 ± 4	−0.5	483.5965 ± .0021
Sn-SnO-Sn	483.6158 ± .0007	−42 ± 4	−0.3	483.5955 ± .0021
Sn-SnO-Pb	483.6174 ± .0013	−43 ± 4	−1.0	483.5962 ± .0023
Nb-Ta*	483.6195 ± .0007	−42 ± 4	−1.0	483.5987 ± .0021
Ta-Nb_3Sn*	483.6185 ± .0011	−44 ± 4	−0.5	483.5975 ± .0022
Ta-Ta*	483.6185 ± .0016	−46 ± 4	−0.4	483.5961 ± .0025
Sn-SnO-Sn	483.6194 ± .0005	−45 ± 4	−0.0	483.5976 ± .0020
WEIGHTED AVERAGE OF DATA IN TERMS OF NBS VOLT MHz/μV_{NBS}				483.5971 ± .0022
NBS VOLT TO ABSOLUTE VOLT CONVERSION				−0.0058 ± .0019
FINAL VALUE FOR $2e/h$ IN ABSOLUTE UNITS MHz/μV				483.5912 ± .0030

of a set of measurements, the uncertainty in the absolute value of the NBS legal volt, and the uncertainty in transferring the NBS volt to our laboratory. It should be noted that the standard deviation of the eight most accurate measurements, made over a period of several months, is only 2 ppm, an indication of the high precision of the measurements. To within this 2-ppm precision, the measured value of $2e/h$ is independent of the material and type of junction used.

A value of the fine-structure constant can be derived from our value of $2e/h$ and other directly measured quantities by use of the equation

$$\alpha^{-1}=\left[\frac{c}{4R_\infty\gamma_p}\frac{\mu_p}{\mu_0}\frac{2e}{h}\right]^{1/2}, \qquad (2)$$

where c is the velocity of light, R_∞ is the Rydberg constant for infinite mass, γ_p is the gyromagnetic ratio of the proton, and μ_p/μ_0 is the magnetic moment of the proton in units of the Bohr magneton. Taking the best values for these quantities,[10] $c=2.997925\times10^8$ m sec^{-1} ±0.3 ppm, $R_\infty=1.0973731\times10^7$ m$^{-1}\pm0.1$ ppm, $\mu_p/\mu_0=1.5210325\times10^{-3}\pm0.5$ ppm, $\gamma_p=2.675192\times10^8$ rad sec^{-1} T$^{-1}\pm3$ ppm, and $2e/h=4.835912\times10^{14}$ Hz V$^{-1}\pm5$ ppm, Eq. (2) gives[11]

$$(\alpha^{-1})_{2e/h}=137.0359\pm0.0004.$$

This value is 21 ± 5 ppm less than the presently accepted value derived from the fine-structure splitting (fs) in deuterium as measured by Treibwasser, Dayhoff, and Lamb,[12,13]

$$(\alpha^{-1})_{\rm fs}=137.0388\pm0.0006.$$

The new value of α derived here entirely removes the apparent discrepancy between the theoretical and experimental values for the hyperfine splitting (hfs) in the ground state of atomic hydrogen. This splitting ($\nu_{\rm hfs}$) has been measured to the extraordinary accuracy of 2 parts in 10^{11} by Crampton, Kleppner and Ramsey.[14] The quantum-electrodynamic expression for the splitting, which includes all theoretical effects other than the dynamic polarizability of the proton, is believed to be accurate to a few ppm.[15] When this expression is evaluated using $(\alpha^{-1})_{\rm fs}$, it predicts

$$\frac{\nu_{\rm hfs}({\rm expt})-\nu_{\rm hfs}({\rm theory})}{\nu_{\rm hfs}({\rm expt})}=43\pm12\ {\rm ppm}.$$

[The quoted errors include an uncertainty of 2 ppm in the estimate of form factors and 5 ppm in $(\alpha^{-1})_{\rm fs}$.] If the theoretical expression is evaluated using $(\alpha^{-1})_{2e/h}$, it predicts

$$\frac{\nu_{\rm hfs}({\rm expt})-\nu_{\rm hfs}({\rm theory})}{\nu_{\rm hfs}({\rm expt})}=0\pm8\ {\rm ppm}.$$

Although at present it is impossible to calculate the proton polarizability exactly, the best estimates indicate that it would increase $\nu_{\rm hfs}$(theory) by less than 10 ppm.[15] Thus, unless the proton polarizability is much larger than is presently believed, $(\alpha^{-1})_{\rm fs}$ suggests a breakdown of quantum electrodynamics, while $(\alpha^{-1})_{2e/h}$ is consistent with both quantum electrodynamics and a small proton polarizability.

The change in α implied here is also important because in the 1963 adjustment of the fundamental physical constants by Cohen and DuMond, the value $(\alpha^{-1})_{\rm fs}$ was used as an input datum. Because of the pivotal role played by α in this adjustment,[16] any change in α will cause large changes in the values of the other fundamental constants. In Table II, we give the values of some of the more important constants which would have resulted if $(\alpha^{-1})_{2e/h}$ had been used as an input datum in the 1963 adjustment.

We might also point out that with this new experimental value for $2e/h$, a new and more reliable value for the x-ray wavelength conversion factor Λ can be obtained. Using the recent experimental data of Spijkerman and Bearden[17] for the voltage-to-wavelength conversion factor, $V\lambda_s$, we find that $\Lambda=1.002067\pm0.000023$ Å/kxu based on Bearden's new definition of the x unit.[18]

Although we believe that the results reported here are highly reliable, work is continuing to ensure that there is no unknown systematic error in the measurements. To this end, experiments of the type described here are being carried out at higher frequencies (~70 GHz) as well as experiments involving the measurement of the frequency of the radiation emitted by a Josephson tunnel junction when biased to a known voltage. Preliminary results from experiments of the latter type are in complete agreement with the results presented here.

We should like to thank Mr. Loebe Julie for helpful discussions concerning the potentiometer, Mr. A. G. McNish for arranging the cal-

Table II. Changes in some of the fundamental physical constants resulting from substituting $(\alpha^{-1})_{2e/h}$ for $(\alpha^{-1})_{fs}$ in the input data of the 1963 adjustment. (N is Avogadro's number and m is the rest mass of the electron.) The numbers in parentheses are the one-standard-deviation errors in ppm.

Quantity	Units	Value given by 1963 adjustment	Value implied by this measurement	Change (ppm)
α^{-1}	-	137.0388(4)	137.0359(3)	−21
e	10^{-19} C	1.602 10(13)	1.602 20(13)	+63
	10^{-10} esu	4.802 98(13)	4.803 28(13)	
h	10^{-34} J sec	6.625 59(24)	6.626 28(24)	+105
m_e	10^{-31} kg	9.109 08(14)	9.109 65(14)	+63
N	10^{26} kmole^{-1}	6.022 52(15)	6.022 14(15)	−63

ibration of the standard cells, Dr. E. Richard Cohen and Dr. Jesse W. M. DuMond for their interest and encouragement, Professor S. A. Bludman and Professor D. J. Scalapino for helpful discussions, and Mr. A. Denestein for his excellent technical assistance.

---

*A contribution from the Laboratory for Research on the Structure of Matter, University of Pennsylvania, covering research sponsored by the National Science Foundation and Advanced Research Projects Agency.

†Part of this work was performed while the author was at the University of Pennsylvania and part while at RCA Laboratories.

[1]B. D. Josephson, Phys. Letters 1, 251 (1962); Advan. Phys. 14, 419 (1965).

[2]P. W. Anderson, Rev. Mod. Phys. 38, 298 (1966); Progress in Low Temperature Physics, edited by C. J. Gorter (North-Holland Publishing Company, Amsterdam, 1964), Vol. IV.

[3]D. N. Langenberg, W. H. Parker, and B. N. Taylor, Phys. Rev. 150, 186 (1966).

[4]S. Shapiro, Phys. Rev. Letters 11, 80 (1963); S. Shapiro, A. R. Janus, and S. Holly, Rev. Mod. Phys. 36, 223 (1964).

[5]F. K. Harris, "Electric Units" (to be published); and private communication.

[6]L. Julie, "An Unusually Accurate Universal Potentiometer for the Range from 1 Nanovolt to 10 Volts" (unpublished).

[7]L. Julie, "A Ratiometric Method for Precise Calibration of Volt Boxes" (to be published).

[8]R. F. Dziuba and T. M. Souders, "A Method for Calibrating Volt Boxes with Analysis of Volt-Box Self-Heating Characteristics", IEEE International Convention Record, March 1966, Pt. 10.

[9]A typical run might consist of 40 separate measurements on about 8 different steps.

[10]E. R. Cohen and J. W. M. DuMond, Rev. Mod. Phys. 37, 537 (1965).

[11]Throughout this Letter, all uncertainties correspond to one standard deviation unless otherwise stated. The uncertainties for γ_p and $2e/h$ do not include the uncertainty in the conversion factor from NBS "as maintained" electrical units to absolute units because the conversion factor for each of these quantities is essentially the same and enters Eq. (2) in such a way as to cancel out. Thus, we are able to completely bypass the problem of the relationship between NBS legal units and absolute units.

[12]S. Triebwasser, E. S. Dayhoff, and W. E. Lamb, Jr., Phys. Rev. 89, 98 (1953).

[13]The recent measurements of the hyperfine splitting in muonium by W. E. Cleland et al., Phys. Rev. Letters 13, 202 (1964), after correction by M. A. Ruderman, Phys. Rev. Letters 17, 794 (1966), as well as the measurements of R. T. Robiscoe and B. L. Cosens of the Lamb shift in H, Phys. Rev. Letters 17, 69 (1966), indicate that $(\alpha^{-1})_{fs}$ may be in error. Also see Ref. 16.

[14]S. B. Crampton, D. Kleppner, and N. Ramsey, Phys. Rev. Letters 11, 338 (1963).

[15]For an up-to-date review of the theoretical situation and the original references, see S. D. Drell and J. D. Sullivan, Phys. Rev. (to be published).

[16]J. W. M. DuMond, Z. Naturforsch. 21a, 70 (1966).

[17]J. J. Spijkerman and J. A. Bearden, Phys. Rev. 134, A871 (1964).

[18]J. A. Bearden et al., X-Ray Wavelengths, Report No. NYO-10586 (Clearing House for Federal Scientific and Technical Information, Springfield, Virginia, 1964)

New Transport Phenomenon in a Semiconductor "Superlattice"*

L. Esaki and L. L. Chang
IBM Thomas J. Watson Research Center, Yorktown Heights, New York 10598
(Received 1 July 1974)

We report electronic transport properties in a GaAs-AlAs periodic structure known as a "superlattice" prepared by a molecular-beam epitaxy. Its differential conductance in the superlattice direction first gradually decreases, followed by a rapid drop to negative values, then, at high fields, exhibits an oscillatory behavior with respect to applied voltages. This observation is interpreted in terms of the formation and expansion of a high-field domain. The voltage period of the oscillation provides the energy of the first-excited band which is in good agreement with that predicted by the theory.

It has been proposed[1,2] that quantum states with desirable energies or bandwidths can be created in monocrystalline semiconductors, once a well-defined structure with extremely narrow potential barriers and wells is achieved in a controlled manner: The transport of electrons in the structure is then expected to be largely governed by such quantum states.

In this Letter, we report transport properties in a periodic structure known as a "superlattice."

The current and conductance as a function of applied voltage show nonlinear characteristics and exhibit an oscillatory behavior beyond a certain threshold voltage. The observed period of the oscillation in terms of applied voltage appears to coincide with the energy difference between the quantized states or bands. This result, we believe, not only suggests the formation of a high-field domain but also confirms the creation of the ground and first-excited bands in the superlattice.

The superlattice used in the present experiments typically comprises fifty periods. Each period has a thickness of 85 Å, consisting of 45 Å GaAs and 40 Å AlAs. The latter provides a potential barrier of a height of 0.4–0.5 eV. The structure has been grown on GaAs substrates by the advanced facilities of computer-controlled molecular-beam epitaxy.[3] The conduction-electron density is $\sim 10^{17}$ cm^{-3} in the superlattice region and $\sim 10^{18}$ cm^{-3} in contacting GaAs regions, the substrate on one side and an overlaid layer on the other. The formation of such a superfine structure has been verified by a combined technique of Auger electron spectroscopy and argon-ion sputter etching,[4] and more recently by small- and large-angle x-ray scattering techniques.[5]

The allowed energy bands in a superlattice can be calculated from the following expression, assuming a one-dimensional, periodic square-well potential.[6]

$$-1 \leq \cos\frac{a(2mE)^{1/2}}{\hbar}\cosh\frac{b[2m(V-E)]^{1/2}}{\hbar}+\left(\frac{V}{2E}-1\right)\left(\frac{V}{E}-1\right)^{-1/2}\sin\frac{a(2mE)^{1/2}}{\hbar}\sinh\frac{b[2m(V-E)]^{1/2}}{\hbar}-1,$$

where E is the electron energy in the superlattice direction, V the barrier height (0.4 eV), a the well width, b the barrier width, and m the effective mass ($0.1m_0$). The hatched region in Fig. 1 indicates the two energy bands, E_1 and E_2, as a function of well width, with the barrier width constant at 40 Å. In the present case with a well width of 45 Å, the widths of the ground and first-excited bands, E_1 and E_2, are as narrow as 5 and 40 meV, respectively. Such narrow bandwidths are a result of the relatively tight-binding potential in the present structure,[7] as compared with that previously investigated.[8] Thus the locations of E_1 and E_2 in this case are essentially the same as those of discrete energy levels in a single square well. Since the scattering time τ is estimated to be of the order of 10^{-13} sec in our epitaxial structure,[8] an inherent broadening even for these discrete levels is expected to be 6–7 meV. Therefore, in order to test these theoretical curves, we have employed the method of the previously reported resonant tunneling in double barriers.[9] The energies, E_1 and E_2, for three different well widths, 40, 50, and 65 Å, were obtained from a large number of double barriers. Although there is some understandable spread, as shown in Fig. 1, because of thickness fluctuation and other imperfections, the measured values generally fall on the calculated curves, particularly well for the ground state.

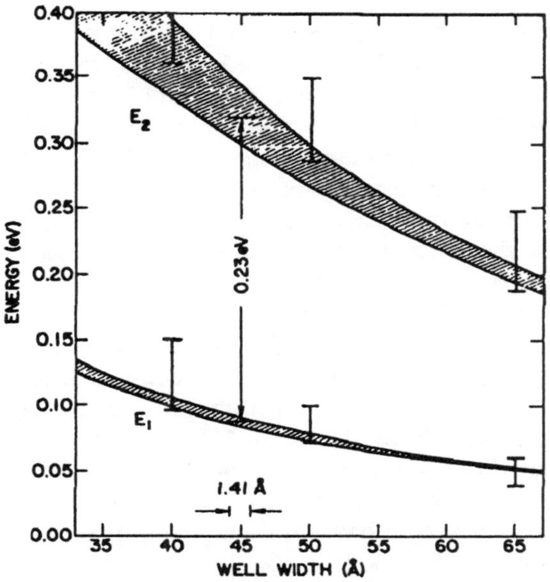

FIG. 1. Two superlattice energy bands, E_1 and E_2, calculated as a function of well width with a barrier width of 40 Å. Each range indicates some spread of experimental values taken from double barriers. Distance of 1.41 Å represents the spacing between (100) atomic planes in GaAs or AlAs.

Transport measurements were made on the present superlattice with two-terminal specimens of an area of about 10^{-7} cm^2. The current is found to increase smoothly with applied voltages up to a few tenths of a volt. Subsequently it starts to saturate or decrease and at high voltages a fine oscillatory behavior is developed. Results of 50-nsec pulsed measurements remain essentially the same. Figure 2 illustrates the differential conductance as a function of applied voltage at four different temperatures: 65, 125, 210, and 300°K. At room temperature, the conductance neither becomes negative nor shows any

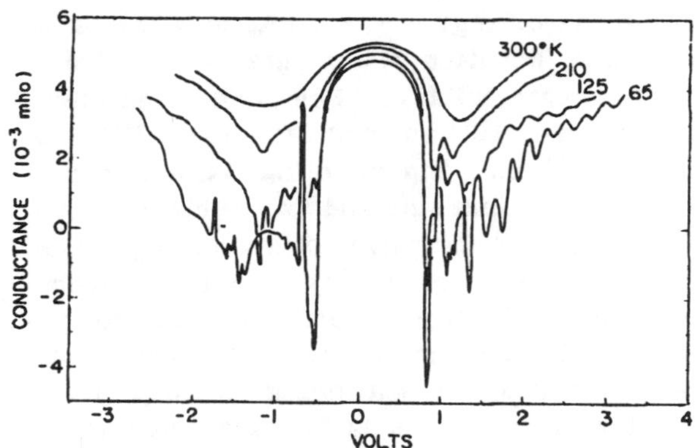

FIG. 2. Differential conductance versus applied voltage in a superlattice at four specified temperatures.

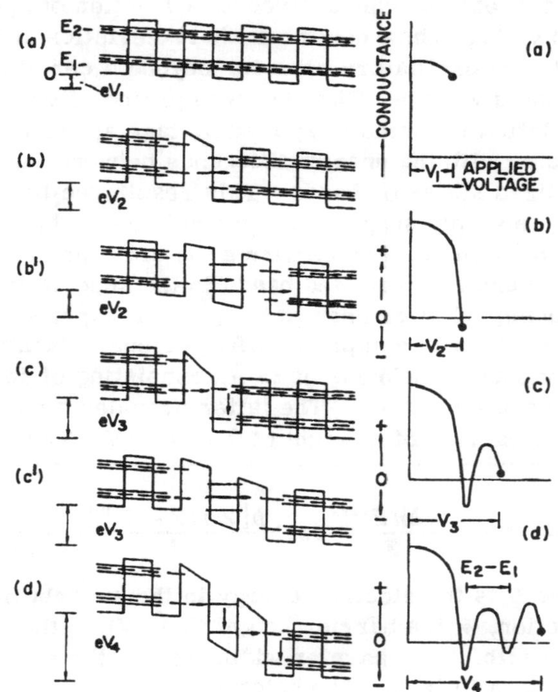

FIG. 3. Schematic energy diagrams (left-hand side) and corresponding conductances (right-hand side): (a) band conduction; (b) and (b′) spontaneous generation of a domain; (c), (c′), and (d) development of domain expansion.

fine structure. The oscillatory behavior is observable at 210°K and becomes increasingly pronounced at lower temperatures. On the other hand, both the conductance near zero bias and a threshold voltage where the conductance begins to drop sharply are relatively insensitive to the temperature variation. Instabilities sometimes exist in the negative-conductance range which give rise to spurious conductance curves. The asymmetric characteristic with respect to polarity, as seen in Fig. 2, is fairly common in many specimens, and is likely due to an asymmetry in the potential profile, although its origin is not clear at this moment. In measuring over 20 specimens, the general features of the conductance, however, are quite reproducible: The onset voltage for the negative differential conductance ranges from 0.4 to 0.8 V and the period of the oscillation falls between 0.21 and 0.24 V.

The observed characteristics are interpreted as schematically illustrated in Fig. 3. At low applied fields, a marginal band-type conduction probably governs the electron transport. The mobility deduced from the conductance near zero voltage is of the order of 50 cm^2/V sec in the direction of the superlattice. After showing a slow decrease as in Fig. 3(a), the conductance starts to drop rapidly at a field of $\sim 10^4$ V/cm: The voltage drop per unit cell (per one period) approaches about 8 meV, if a uniform field distribution is assumed. Beyond this point it seems reasonable to assume that the band conduction fails to be sustained throughout the entire superlattice region and a narrow high-field domain is spontaneously generated. Although the domain region may initially extend only to one barrier [Fig. 3(b)] or two [Fig. 3(b′)], a substantial fraction of the total voltage will be applied across it. This will leave intact the band conduction in the rest of the superlattice region. It should be pointed out that the high-field domain formation, initiated possibly by a random noise fluctuation or more likely an unavoidable nonuniformity in the superlattice structure, is indeed an inherent feature of the voltage-controlled negative-conductance medium.[10]

We proceed to analyze tunneling characteristics across the high-field domain, because it will dominate the total conductance characteristic of the superlattice system. As shown in Figs. 3(b) and 3(c) or in Figs. 3(b′) and 3(c′), one may realize that this situation is somewhat analogous to the double-barrier tunneling.[9] In the most simple case, a single barrier is sufficient to provide a current peak or a negative conductance arising from matching or mismatching of energy levels on both sides of the domain, as shown in Fig. 3(c), because all involved electrons here are two dimensional in nature. In this domain tunneling, the current peak or dip is expected to be just as sharp as that in the case of the double-barrier where electrode electrons are three dimensional, if the transverse-momentum conservation of electrons is invoked.[11] The tunneling current across the domain will increase at the matching condition when the applied voltage across it reaches a value corresponding to the difference between E_1 and E_2, as shown in Figs. 3(c) and 3(c′). With further increase in applied voltage, the domain region will be expanded to the adjacent barriers, one by one, and, correspondingly, off- and on-matching conditions will be alternately repeated. Figure 3(d) shows that both barriers in the domain are simultaneously at the matching condition. This discrete nature in the domain expansion will give the oscillatory behavior in the conductance curve, as illustrated sequentially from top to bottom on the right-hand side of Fig. 3. Therefore, the observed voltage period (0.21–0.24 V) will provide the experimental value for $E_2 - E_1$, which is in good agreement with the estimated value (0.23 eV) shown in Fig. 1.

The observed negative conductance is not due to the Gunn effect, because the onset voltage (0.4–0.8 V) for the negative conductance, the electron mobility (50 cm^2/V sec), and the electron mean free path (a few hundred angstroms) in the structure are much too low to activate the electron transfer mechanism.[12] Furthermore, the observation of the oscillatory effect will exclude the possibility of the involvement of hot electrons. The involvement of the optical phonons, $\hbar\omega_0$, is also clearly denied because, in that case, the oscillation peaks are predicted to occur at fields F given by $eFd = \hbar\omega_0/n$, where n is an integer and d is the superlattice period.[13]

In the proposed model, we have used simplified assumptions[14] such as a square-well potential profile, a constant effective mass, ignoring a number of possible effects such as the upper valley in GaAs, band bending, junction problems between the superlattice region and the heavily doped contact regions, etc. These, however, are believed to play rather minor roles and will be required only for further refinement of the theory. The suprisingly good agreement between the calculated and experimental values of the location of the first-excited band seems to support our model and elucidates the most fundamental aspects of the transport characteristics in the superlattice.

We are grateful to R. Tsu, A. Koma, and L. F. Alexander for their cooperation in the project. Technical contributions of L. E. Osterling, C. C. Periu, and M. S. Christie are also acknowledged.

---

*Research sponsored in part by the U. S. Army Research Office (Durham).

[1] L. Esaki and R. Tsu, IBM J. Res. Develop. 14, 61 (1970).

[2] R. Tsu and L. Esaki, Appl. Phys. Lett. 22, 562 (1973).

[3] L. L. Chang, L. Esaki, W. E. Howard, R. Ludeke, and G. Schul, J. Vac, Sci. Technol. 10, 655 (1973); L. Esaki, J. Jpn. Soc. Appl. Phys., Suppl. 43, 452 (1974).

[4] R. Ludeke, L. Esaki, and L. L. Chang, Appl. Phys. Lett. 24, 417 (1974).

[5] A. Segmüller, private communication.

[6] I. I. Gol'dman and V. Krivchenkov, *Problems in Quantum Mechanics* (Addison-Wesley, Reading, Mass., 1961), p. 60.

[7] R. F. Kazarinov and R. A. Suris, Fiz. Tekh. Poluprov. 5, 797 (1971) [Sov. Phys. Semicond. 5, 707 (1971)].

[8] L. Esaki, L. L. Chang, W. E. Howard, and V. L. Rideout, in *Proceedings of the Eleventh International Conference on the Physics of Semiconductors, Warsaw, Poland, 1972*, edited by The Polish Academy of Sciences (PWN-Polish Scientific Publishers, Warsaw, Poland, 1972), p. 431.

[9] L. L. Chang, L. Esaki, and R. Tsu, Appl. Phys. Lett. 24, 593 (1974).

[10] For instance, see S. M. Sze, *Physics of Semiconductor Devices* (Wiley, New York, 1969), p. 732.

[11] For instance, see L. Esaki, Science 183, 1149 (1974), Fig. 12 and its explanation.

[12] For instance, see Ref. 10, p. 743.

[13] M. Saitoh, J. Phys. C: Proc. Phys. Soc., London 5, 914 (1972).

[14] The barrier height is taken to be 0.4 eV. The choice of the value in that neighborhood is not critical for calculated results on the location of energy levels. As seen in Ref. 8, about 80% of the band-gap mismatch at the heterostructure is considered to be accommodated in the conduction band, although the actual situation of GaAs-AlAs is somewhat complicated because of the Γ-X band crossing.

Scaling Theory of Localization: Absence of Quantum Diffusion in Two Dimensions

E. Abrahams
Serin Physics Laboratory, Rutgers University, Piscataway, New Jersey 08854

and

P. W. Anderson,[a] D. C. Licciardello, and T. V. Ramakrishnan[b]
Joseph Henry Laboratories of Physics, Princeton University, Princeton, New Jersey 08540
(Received 7 December 1978)

Arguments are presented that the $T=0$ conductance G of a disordered electronic system depends on its length scale L in a universal manner. Asymptotic forms are obtained for the scaling function $\beta(G)=d\ln G/d\ln L$, valid for both $G \ll G_c \simeq e^2/\hbar$ and $G \gg G_c$. In three dimensions, G_c is an unstable fixed point. In two dimensions, there is no true metallic behavior; the conductance crosses over smoothly from logarithmic or slower to exponential decrease with L.

Scaling theories of localization have been discussed by Thouless and co-authors[1-3] and by Wegner.[4] Recently Schuster,[5] using methods related to those of Aharony and Imry,[6] has proposed a close relationship of the localization problem to a dirty XY model of the same dimensionality. Wegner has proposed a scaling for dimensionality $d \neq 2$ of the conductivity

$$\sigma \sim (E-E_c)^{(d-2)\nu}, \tag{1}$$

while Schuster identifies ν as the correlation-length exponent of the XY model $\to \frac{1}{2}$ at $d>4$. The latter proposes a universal jump of conductivity for $d=2$ given by $e^2/\hbar\pi^2$. This is not inconsistent with the results of Wegner[4] and is in rough agreement with the early calculations of Ref. 2. It has not been clear how (1) could be reconciled with the physical ideas of Mott[7] as related to the beginning of a scaling theory by Thouless.[3] We here develop a renormalization-group scheme based on the Mott-Thouless arguments, which in many essential ways agrees with Wegner's results, and in other ways severely disagrees. In particular, we recover (1) for $d>2$, where ν is the localization-length exponent below E_c. This is in clear contradiction to Mott,[7] who argues that in all cases the conductivity jumps to zero at E_c. At $d=2$, we find no jump in σ but a steep crossover from exponential to very slow dependence on size. There is *no* true metallic conductivity. These results were presaged by Thouless and coworkers[8,9] to some extent, with Ref. 8 indicating a transition region for three dimensions, and Ref. 9 a size-dependent minimum metallic conductivity.

Our ideas are based on the relationship[1] between conductance as determined by the Kubo-Greenwood formula and the response to perturbation of boundary conditions in a finite sample described by Thouless and co-workers[3]

$$\frac{\text{“}V\text{”}}{W}=\frac{\Delta E}{dE/dN}=\frac{2\hbar}{e^2}C=\frac{2\hbar}{e^2}\sigma L^{d-2}. \tag{2}$$

Here G is the conductance (*not* conductivity σ) of a hypercube of size L^d [here $L \ggg L_0$ (L_0 = microscopic size)], dE/dN is the mean spacing of its energy levels, and ΔE is the geometric mean of the fluctuation in energy levels caused by replacing periodic by antiperiodic boundary conditions. Actually, when "V/W" is relatively large, it is hard to match the energy levels and, in fact, ΔE is defined using the curvature for small χ when we replace periodic $\psi(n+1)=\psi(1)$ by $\psi(n+1)=e^{i\chi}\times\psi(1)$ boundary conditions. This procedure is valid throughout the range of interest.[3] We will comment on the validity of (2) in a fuller paper, but here we add the following remarks. The equivalence of the Kubo-Greenwood formula and the breadth ν of the distribution of ΔE as described in Ref. 3a is not quite precisely provable but does not depend as stated in that reference on independence of momentum matrix elements $p_{\alpha\beta}$ and energy difference $E_\alpha - E_\beta$ between two states, only on a uniform distribution of those $E_\alpha - E_\beta$ which have large $p_{\alpha\beta}$.

Our scaling theory depends on the following ideas.

(I) We define a generalized dimensionless conductance which we call the "Thouless number" as a function of scale L:

$$g(L)=\frac{\Delta E(L)}{dE(L)/dN}\left(=\frac{G(L)}{e^2/2\hbar}\right), \tag{3}$$

where we now contemplate a small *finite* hypercube of size L. In the case $L \gg l$, the mean free path, we may use (2) to define a conductance $G(L)$ which is not related directly to the macroscopic conductivity but is a function of L, and is defined by (3) from the average of the Thouless energy-level differences at scale L. When $L<l$, there is phase coherence on a scale L and g is no longer given by (3) but it can be shown that $(e^2/\hbar)g=G$ can be defined as the conductance of a hypercube imbedded in a perfect crystal.

(II) We remark that $g(L)$ is the relevant dimensionless ratio which determines the change of energy levels when two hypercubes are fitted together. This is the hypothesis of Thouless and can be justified in several ways on physical grounds. For instance, once l > mean free path, the phase relationships for an arbitrary integration of the wave equation across the cube are as random from one side to another as those between wave functions on different cubes. This could be shown to be related to Wegner's "neglect of eigenvalues far from E_F" by a scaling argument. In this limit $g(L)$ represents [as indicated in (2)] the "V/W" of an equivalent many-level Anderson model where each block has $(L/L_0)^d$ energy levels and a width of spectrum

$$W=(dE/dN)(L/L_0)^d.$$

We cannot see how any statistical feature of the energy levels other than this coupling/granularity ratio can be relevant.

(III) We then contemplate combining b^d cubes into blocks of side bL and computing the new $\Delta E'/(dE/dN)'$ at the resulting scale bL. The result will be

$$g(bL)=f(b,g(L)), \tag{4}$$

or in continuous terms

$$d\ln g(L)/d\ln L=\beta(g(L)). \tag{5}$$

The scaling trajectory has only one parameter, g.

(IV) At large and small g we can get the asymptotics of β from general physical arguments. For large g, macroscopic transport theory is correct and, as in (2),

$$G(L)=\sigma L^{d-2},$$

so that

$$\lim_{g\to\infty}\beta_d(g)=d-2. \tag{6}$$

For small g ($V/W\ll 1$), exponential localization is surely valid and therefore g falls off exponentially:

$$g=g_a e^{-\alpha L}.$$

Thence

$$\lim_{g\to\infty}\beta_d(g)=\ln[g/g_a(d)]. \tag{7}$$

Here g_a is a dimensionless ratio of order unity.

From the asymptotics (6) and (7), we may sketch the universal curve $\beta_d(g)$ in $d=1,2,3$ dimensions (Fig. 1). The central assumption of Fig. 1 is continuity: Since β represents the blocking of finite groups of sites, it can have no built-in singularity, and hence it would be unreasonable for it to have the cusp indicated by the dashed line: This is the curve which would be required to give the Mott-Schuster jump in conductivity for $d=2$. The only singularities then, must be fixed points $\beta=0$. Physically, it is also certain that β is monotonic in g, since smaller V/W surely always means more localization.

In constructing Fig. 1, we have used perturbation theory in V/W which shows that the first deviation of β from $\ln(g/g_a)$ is *positive*, with

$$\beta=\ln(g/g_a)[1+\alpha g+\sim g^2+\dots], \tag{8}$$

since this is essentially just the "locator" perturbation series first discussed by Anderson.[10] The steepening of the slope of β given by (8) makes $\nu\lesssim 1$, as we shall see, for $d=3$ or greater.

For large g, we suppose that β may be calculated as a perturbation series in $W/V=g^{-1}$:

$$dg/d\ln L=g(d-2-a/g+\dots). \tag{9}$$

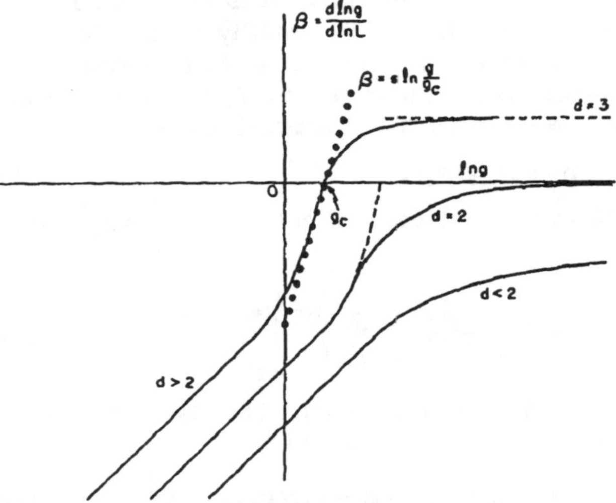

FIG. 1. Plot of $\beta(g)$ vs $\ln g$ for $d>2$, $d=2$, $d<2$. $g(L)$ is the normalized "local conductance." The approximation $\beta=s\ln(g/g_c)$ is shown for $g>2$ as the solid-circled line; this unphysical behavior necessary for a conductance jump in $d=2$ is shown dashed.

The first correction term in this series may be estimated in perturbation theory by considering backscattering processes of the sort first discussed by Langer and Neal[11] in their analysis of the dependence of resistivity on impurity concentration.

The use of Langer and Neal involves a rather subtle question. Converting the calculation of Langer and Neal to dimensionality 2 in particular, one obtains

$$g(L) = g_0 - g_1 \ln\Lambda,$$

where Λ is a length cutoff for a certain divergent integral of second order in the density of scatterers. Langer and Neal assume this cutoff is l, the mean free path; for scales $L<l$, it should, of course, be L and we obtain just the result expected from (9), $d\ln g/d\ln L \sim -g^{-1}$. On the other hand, our universality argument seems to require $L>l$. We have restudied the cutoff question and will show in a fuller paper that their cutoff is not correct.

On the other hand, we have been unable to show definitively that the mean free path does *not* represent a relevant scale for the problem, since once $L>l$, we find, for example, that the coefficient of $\ln L$ depends on l. We must rely rather on our general arguments from continuity and regularity, and an intuition that only g is relevant, to suppose that a series development of β in g^{-1} should exist, once $L \gg l$.

The consequences of Fig. 1 and Eqs. (8) and (9) are as follows: For $d>2$, the β function has a zero at g_c of order unity. It is an unstable fixed point which signals the mobility edge. The critical behavior can be estimated by integrating β starting from a microscopic L_0 and with g_0 near g_c. We use the linear approximation

$$\beta = s \ln(g/g_c), \qquad (10)$$

where $s>1$, since $\alpha>0$ in (8). For $g_0>g_c$, we obtain

$$\sigma = A\frac{e^2}{\hbar}\frac{g_c}{L_0^{d-2}}\left(\ln\frac{g_0}{g_c}\right)^{(d-2)/s}, \qquad (11)$$

where A is of order unity. The distance to the mobility edge is measured by

$$\epsilon = \ln(g_0/g_c) \approx (g_0-g_c)/g_c, \qquad (12)$$

and the factor $Ae^2/\hbar L_0^{d-2}$ in (11) is the Mott conductivity which here appears in the scaling form proposed by Wegner.

In the localized regime ($g_0<g_c$), we get

$$g \approx g_c \exp(-A|\epsilon|^{1/s}L/L_0), \qquad (13)$$

so that the exponent of the localization length is the inverse slope of β at g_c,

$$\nu = 1/s. \qquad (14)$$

These results again agree with those of Wegner.

In two dimensions, we have a strikingly different picture (see Fig. 1). Instead of a sharp mobility edge there is *no* critical g_c where $\beta(g_c)=0$, but β is *always negative* so that in all cases $g(L\to\infty)=0$. Instead of a sharp universal minimum metallic conductivity, there is a universal crossover from logarithmic to exponential behavior which for many experimental purposes may resemble a sharp mobility edge fairly closely. If we extrapolate the form we would deduce from Langer's perturbation-theory calculation, on the "extended" side of the crossover

$$g \simeq g_0 - Ag_c\ln(L/L_0) \qquad (15)$$

the conductivity decreases logarithmically with scale until $g=g_c$ at the scale L_l, where

$$\frac{L_l}{L_0} = \exp\left[\frac{1}{A}\left(\frac{g_0}{g_c}-1\right)\right]. \qquad (16)$$

From this point on g decreases exponentially with L, the localization length being of order L_l as given by (16). This type of behavior was already anticipated from computer studies,[9] but the nature of the actual solution is surprising to say the least, as well as the fact that it appears to have been anticipated in terms of a divergence of perturbation theory in the weak-coupling limit by Langer and Neal.[11]

This work is supported in part by the National Science Foundation Grants No. DMR 78-03015 and No. DMR 76-23330-A-1, and by the U. S. of Naval Research.

[(a)]Also at Bell Laboratories, Murray Hill, N. J. 07974.

[(b)]On leave from Indian Institute of Technology, Kanpur, India.

[1]J. T. Edwards and D. J. Thouless, J. Phys. C 5, 807 (1972).

[2]D. C. Licciardello and D. J. Thouless, J. Phys. C 8, 4157 (1975).

[3a]D. J. Thouless, Phys. Rep. 13C, 93 (1974).

[3b]D. J. Thouless, to be published.

[4]F. J. Wegner, Z. Phys. 25, 327 (1976).
[5]H. G. Schuster, Z. Phys. 31, 99 (1978).
[6]A. Aharony and Y. Imry, J. Phys. C 10, L487 (1977).
[7]N. F. Mott, *Metal Insulator Transitions* (Taylor and Francis, London, 1974).
[8]B. J. Last and D. J. Thouless, J. Phys. C 7, 699 (1974).
[9]D. C. Licciardello and D. J. Thouless, J. Phys. C 11, 925 (1978).
[10]P. W. Anderson, Phys. Rev. 109, 1492 (1958).
[11]J. S. Langer and T. Neal, Phys. Rev. Lett. 16, 984 (1966).

New Method for High-Accuracy Determination of the Fine-Structure Constant Based on Quantized Hall Resistance

K. v. Klitzing
Physikalisches Institut der Universität Würzburg, D-8700 Würzburg, Federal Republic of Germany, and Hochfeld-Magnetlabor des Max-Planck-Instituts für Festkörperforschung, F-38042 Grenoble, France

and

G. Dorda
Forschungslaboratorien der Siemens AG, D-8000 München, Federal Republic of Germany

and

M. Pepper
Cavendish Laboratory, Cambridge CB3 0HE, United Kingdom
(Received 30 May 1980)

Measurements of the Hall voltage of a two-dimensional electron gas, realized with a silicon metal-oxide-semiconductor field-effect transistor, show that the Hall resistance at particular, experimentally well-defined surface carrier concentrations has fixed values which depend only on the fine-structure constant and speed of light, and is insensitive to the geometry of the device. Preliminary data are reported.

PACS numbers: 73.25.+i, 06.20.Jr, 72.20.My, 73.40.Qv

In this paper we report a new, potentially high-accuracy method for determining the fine-structure constant, α. The new approach is based on the fact that the degenerate electron gas in the inversion layer of a MOSFET (metal-oxide-semiconductor field-effect transistor) is fully quantized when the transistor is operated at helium temperatures and in a strong magnetic field of order 15 T.[1] The inset in Fig. 1 shows a schematic diagram of a typical MOSFET device used in this work. The electric field perpendicular to the surface (gate field) produces subbands for the motion normal to the semiconductor-oxide interface, and the magnetic field produces Landau quantization of motion parallel to the interface. The density of states $D(E)$ consists of broadened δ functions[2]; minimal overlap is achieved if the magnetic field is sufficiently high. The number of states, N_L, within each Landau level is given by

$$N_L = eB/h, \tag{1}$$

where we exclude the spin and valley degeneracies. If the density of states at the Fermi energy, $N(E_F)$, is zero, an inversion layer carrier cannot be scattered, and the center of the cyclotron orbit drifts in the direction perpendicular to the electric and magnetic field. If $N(E_F)$ is finite but small, an arbitrarily small rate of scattering cannot occur and localization produced by the long lifetime is the same as a zero scattering rate, i.e., the same absence of current-carrying states occurs.[3] Thus, when the Fermi level is between

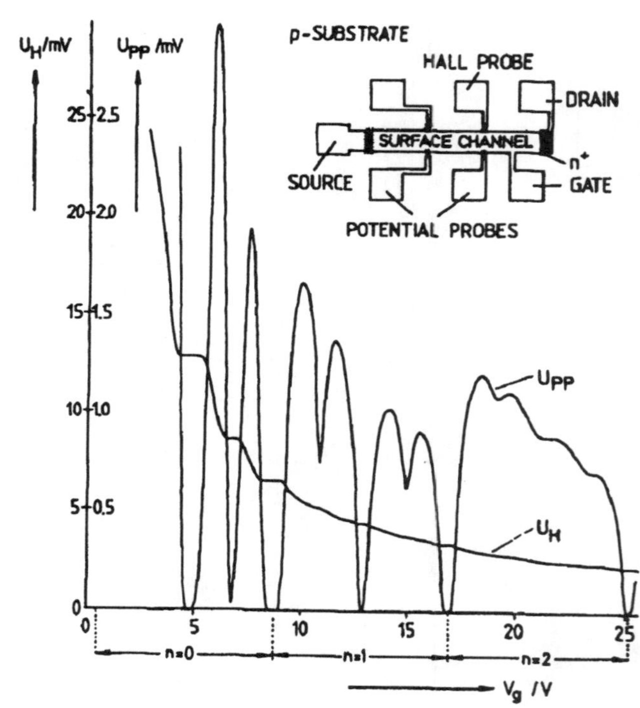

FIG. 1. Recordings of the Hall voltage U_H, and the voltage drop between the potential probes, U_{pp}, as a function of the gate voltage V_g at $T = 1.5$ K. The constant magnetic field (B) is 18 T and the source drain current, I, is 1 μA. The inset shows a top view of the device with a length of $L = 400$ μm, a width of $W = 50$ μm, and a distance between the potential probes of $L_{pp} = 130$ μm.

Landau levels the device current is thermally activated and the minima in σ_{xx}, $\sigma_{xx}^{\min}$, can be less than $10^{-7}\sigma_{xx}^{\max}$.[4] Increasing the magnetic field and decreasing the temperature, further decreases $\sigma_{xx}^{\min}$. The Hall conductivity σ_{xy}, which is usually a complicated function of the scattering process, becomes very simple in the absence of scattering and is given by[2]

$$\sigma_{xy} = -Ne/B, \tag{2}$$

where N is the carrier concentration.

The correction term to the above relation, $\Delta\sigma_{xy}$, is of the order of $\sigma_{xx}/\omega\tau$, where ω is the cyclotron frequency and τ is the relaxation time of the conduction electrons; $\omega\tau \gg 1$ in strong magnetic fields. When the Fermi energy is between Landau levels, and $\sigma_{xx}^{\min} \sim 10^{-7}\sigma_{xx}^{\max}$, the correction $\Delta\sigma_{xy}/\sigma_{xy} < 10^{-8}$. Subject to any error imposed by $\Delta\sigma_{xy}$, when a Landau level is fully occupied and $N = N_L i$ ($i = 1, 2, 3, \ldots$), σ_{xy} is immediately given from Eqs. (1) and (2):

$$-\sigma_{xy} = e^2 i/h. \tag{3}$$

The Hall resistivity $\rho_{xy} = -\sigma_{xy}/(\sigma_{xx}^2 + \sigma_{xy}^2) \approx -\sigma_{xy}^{-1}$ is defined by E_H/j (E_H = Hall field, j = current density) and can be rewritten R_H/I, where R_H is the Hall resistance, U_H the Hall voltage and I the current. Thus, $R_H = h/e^2 i$, which may finally be written as[5]

$$R_H = \alpha^{-1}\mu_0 c/2i, \tag{4}$$

where μ_0 is the permeability of vacuum and exactly equal to $4\pi \times 10^{-7}$ H m^{-1}, c is the speed of light in vacuum and equal to 299 792 458 m s^{-1} with a current uncertainty[5] of 0.004 ppm and $\alpha \approx \frac{1}{137}$ is the fine-structure constant. It is clear from Eq. (4) that a high-accuracy measurement of the Hall resistance in SI units will give a value of α with essentially the same accuracy. Since resistances can be determined in SI units to a few parts in 10^8 by means of the so-called calculable cross capacitor by Thompson and Lampard,[6] the question of absolute units versus as-maintained units is much less of a problem than in the determination of e/h from the ac Josephson effect. Furthermore, the magnitude of R_H falls within a relatively convenient range: $R_H \approx (25\,813\ \Omega)/i$, with i typically between 2 and 8. Finally, we note that if α is assumed to be known from some other experiment (for example, from $2e/h$ and the proton gyromagnetic ratio γ_p), Eq. (4) may be used to derive a known standard resistance.

Two well-known corrections in the low-field Hall effect become unimportant. The first is the correction due to the shorting of the Hall voltage by the source and drain contacts.[7] This is important at low fields for samples with length-to-width ratio, L/W, less than 4, but becomes negligible when the Hall angle is 90°, i.e., $\sigma_{xx} = 0$.[8] The second correction which becomes unimportant is that due to an inexact alignment of the Hall probes, i.e., they are not exactly opposite: This is irrelevant, as the voltage drop along the sample vanishes when $\sigma_{xx} = 0$.[9]

The experiments were carried out on MOS devices with a range of oxide thicknesses ($d_{ox} = 100$ nm–400 nm), and length-to-width ratios ranging from $L/W = 25$ to $L/W = 0.65$. All the transistors were fabricated on the (100) surface orientation and, typically, the p-type substrate had room temperature resistivity of 10 Ω cm. The resistivity at helium temperature was higher than 10^{13} Ω cm, and no current flow between source and drain around the channel could be measured. The long devices ($L/W > 8$) had potential probes in addition to the Hall probes.

A typical recording of the measured Hall voltage U_H, and the voltage between the potential probes U_{pp}, as a function of the gate voltage is shown in Fig. 1. These results were obtained at a constant magnetic field of $B = 18$ T, a temperature of 1.5 K and a constant source drain current of $I = 1$ μA. Relevant device parameters were $L = 400$ μm, $W = 50$ μm and the distance between potential probes was about 130 μm.

The measured voltage U_{pp} is proportional to the resistivity component $\rho_{xx} = \sigma_{xx}[\rho_{xx}^2 + \rho_{xy}^2]$. At gate voltages where the E_F is in the energy gap between Landau levels, minima in both σ_{xx} and ρ_{xx} are observed.[9] Such minima are clearly visible, and are identified, in Fig. 1; the minima due to the lifting of the spin and the (twofold) valley degeneracy are also apparent. The Hall voltage clearly levels off at those values of carrier concentration where σ_{xx} and ρ_{xx} are zero. The values of U_H obtained in the regions are in good agreement with the predicted values, Eq. (4), if the error due to the 1-MΩ input impedance of the X-Y recorder is taken into account. It was found that the value of U_H in the "steps" was, for instance current, independent of sample geometry and direction of magnetic field, provided that σ_{xx} was zero.

An area of possible criticism of the theoretical basis of this experiment, is the role of carriers which are localized outside the main Landau level. Here we do not specify the localization mechanism, but the presence of localized carriers will invalidate both the relation $N = N_L$ and Eq. (4). However, the experimental results strongly suggest that such carriers do not invalidate Eq. (4). At present there is both theoretical and experimental investigation of this type of localization.[3,4,9-12] Ando[2] has suggested that the electrons in impurity bands, arising from short range scatterers, do not contribute to the Hall current; whereas the electrons in the Landau level give rise to the same Hall current as that obtained when all the electrons are in the level and can move freely. Clearly this process must be occuring but its range of validity must be carefully examined as an accompaniment to highly accurate measurements of Hall resistance.

For high-precision measurements we used a normal resistance R_0 in series with the device. The voltage drop, U_0, across R_0, and the voltages U_H and U_{pp} across and along the device was measured with a high impedance voltmeter ($R > 2\times10^{10}$ Ω). The resistance R_0 was calibrated by the Physikalisch Technische Bundesanstalt, Braunschweig, and had a value of $R_0 = 9999.69$ Ω at a temperature of 20 °C. A typical result of the measured Hall resistance $R_H = U_H/I = U_H R_0/U_0$, and the resistance, $R_{pp} = U_{pp} R_0/U_0$, between the potential probes of the device is shown in Fig. 2 ($B = 13$ T, $T = 1.8$ K). The minimum in σ_{xx} at $V_g = 23.6$ V corresponds to the minimum at $V_g = 8.7$ V in Fig. 1, because the thicknesses of the gate oxides of these two samples differ by a factor of 3.6. Our experimental arrangement was not sensitive enough to measure a value of R_{pp} of less than 0.1 Ω which was found in the gate-voltage region 23.40 V $< V_g <$ 23.80 V. The Hall resistance in this gate voltage region had a value of 6453.3 ± 0.1 Ω. This inaccuracy of ± 0.1 Ω was due to the limited sensitivity of the voltmeter. We would like to mention that most of the samples, especially devices with a small length-to-width ratio, showed a minimum in the Hall voltage as a function of V_g at gate voltage close to the left side of the plateau. In Fig. 2, this minimum is relatively shallow and has a value of 6 452.87 Ω at $V_g = 23.30$ V.

In order to demonstrate the insensitivity of the Hall resistance on the geometry of the device, measurements on two samples with a length-to-width ratio of $L/W = 0.65$ and $L/W = 25$, respectively, are plotted in Fig. 3. The gate-voltage scale

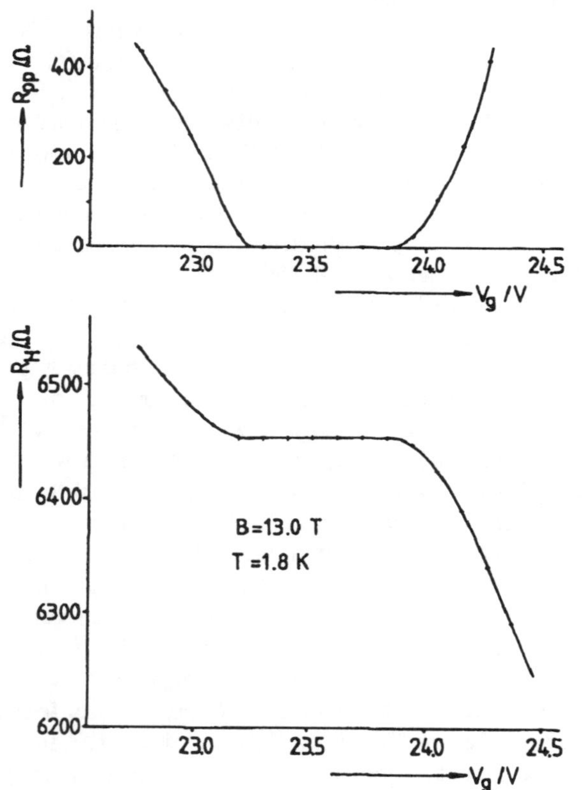

FIG. 2. Hall resistance R_H, and device resistance, R_{pp}, between the potential probes as a function of the gate voltage V_g in a region of gate voltage corresponding to a fully occupied, lowest ($n = 0$) Landau level. The plateau in R_H has a value of 6453.3 ± 0.1 Ω. The geometry of the device was $L = 400$ μm, $W = 50$ μm, and $L_{pp} = 130$ μm; $B = 13$ T.

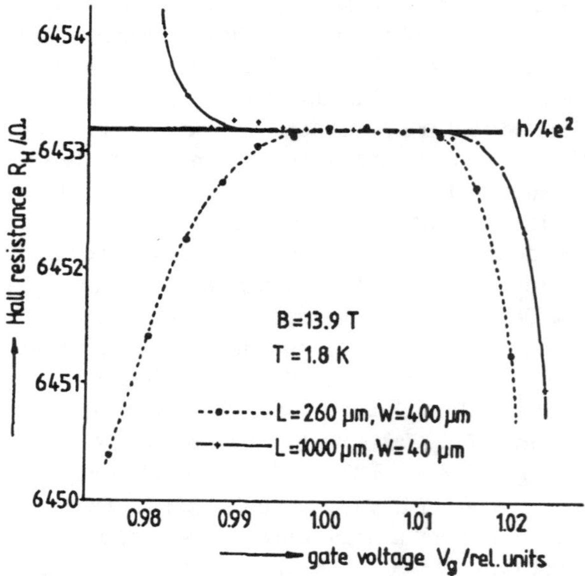

FIG. 3. Hall resistance R_H for two samples with different geometry in a gate-voltage region V_g where the $n = 0$ Landau level is fully occupied. The recommended value $h/4e^2$ is given as 6453.204 Ω.

is given in arbitrary units, and is different for the two samples because the thicknesses of the gate oxides are different. A gate voltage $V_g = 1.00$ corresponds, approximately, to a surface carrier concentration where the first fourfold-degenerate Landau level, $n = 0$, is completely filled. Within the experimental accuracy of 0.1 Ω, the same value for the plateau in the Hall resistance is measured. The value for $h/4e^2 = 6\,453.204 \pm 0.005$ Ω based on the recommended value for the fine-structure constant[5] is plotted in this figure, too. The decrease of the Hall resistance with decreasing gate voltage for the sample with $L/W = 0.65$ originates mainly from the shorting of the Hall voltage at the contacts. This effect is most pronounced when the Hall angle becomes smaller than 90°. In the limit of small Hall angles, the Hall voltage is reduced by a factor of 2 for the sample with $L/W = 0.65$.[7]

The mean value of the Hall resistance for all samples investigated was 6453.22 ± 0.10 Ω for measurements in the energy gap between the Landau levels $n = 0$ and $n = 1$ (corresponding to $i = 4$ in Eq. 4), 3226.62 ± 0.10 Ω for measurements in the energy gap between Landau levels $n = 1$ and $n = 2$ ($i = 8$), and 12 906.5 ± 1.0 Ω for measurements in the energy gap between the spin split levels with $n = 0$ ($i = 2$). These resistances agree very well with the calculated values $h/e^2 i$ based on the recently reported[13] highly accurate value of $\alpha^{-1} = 137.035\,963(15)$ (0.11 ppm).

Measurements with a voltmeter with higher resolution and a calibrated standard resistor with a vanishing small temperature coefficient at $T = 25\,°C$ yield a value of $h/4e^2 = 6453.17 \pm 0.02$ Ω corresponding to a fine-structure constant of $\alpha^{-1} = 137.0353 \pm 0.0004$.

We would like to thank the Physikalisch Technische Bundesanstalt, Braunschweig, for experimental support and E. R. Cohen, Th. Englert, V. Kose, G. Landwehr, and B. N. Taylor for valuable discussions. One of us (M.P.) would like to thank the European Research Office of the U. S. Army for partial support.

---

[1]For a review see for example: F. Stern, Crit. Rev. Solid State Sci. 5, 499 (1974); G. Landwehr, in *Advances in Solid State Physics: Festkörperprobleme*, edited by H. J. Queisser (Pergamon, New York-Vieweg, Braunschweig, 1975), Vol. 15, p. 48.

[2]T. Ando, J. Phys. Soc. Jpn. 37, 622 (1974).

[3]H. Aoki and H. Kamimura, Solid State Commun. 21, 45 (1977).

[4]R. J. Nicholas, R. A. Stradling, and R. J. Tidey, Solid State Commun. 23, 341 (1977).

[5]E. R. Cohen and B. N. Taylor, J. Phys. Chem. Ref. Data 2, 633 (1973).

[6]A. M. Thompson and D. G. Lampard, Nature (London) 177, 888 (1956).

[7]I. Isenberg, B. R. Russel, and F. R. Greene, Rev. Sci. Instrum. 19, 685 (1948).

[8]R. F. Wick, J. Appl. Phys. 25, 741 (1954).

[9]Th. Englert and K. V. Klitzing, Surf. Sci. 73, 71 (1978).

[10]S. Kawaji and J. Wakabayashi, Surf. Sci. 58, 238 (1976).

[11]M. Pepper, Philos. Mag. 37B, 83 (1978).

[12]S. Kawaji, Surf. Sci. 73, 46 (1978).

[13]E. R. Williams and P. T. Olsen, Phys. Rev. Lett. 42, 1575 (1979).

Quantized Hall conductivity in two dimensions

R. B. Laughlin
Bell Laboratories, Murray Hill, New Jersey 07974
(Received 20 January 1981)

It is shown that the quantization of the Hall conductivity of two-dimensional metals which has been observed recently by Klitzing, Dorda, and Pepper and by Tsui and Gossard is a consequence of gauge invariance and the existence of a mobility gap. Edge effects are shown to have no influence on the accuracy of quantization. An estimate of the error based on thermal activation of carriers to the mobility edge is suggested.

There has been considerable interest in the remarkable observation made recently by von Klitzing, Dorda, and Pepper[1,2] and by Tsui and Gossard[2] that, under suitable conditions, the Hall conductivity of an inversion layer is quantized to better than one part in 10^5 to integral multiples of e^2/h. The singularity of the result lies in the apparent total absence of the usual dependence of this quantity on the density of mobile electrons, a sample-dependent parameter. As it has been proposed[1] to use this effect to define a new resistance standard or to refine the known value of the fine-structure constant, an important issue at present is to what accuracy the quantization is exact, particularly in the regime of high impurity density. Some light has been shed on this question by the renormalized weak-scattering calculations of Ando,[3] who has shown that the presence of an isolated impurity does not affect the Hall current. A similar result has been obtained recently by Prange,[4] who has shown that an isolated δ-function impurity does not affect the Hall conductivity to lowest order in the drift velocity $v = cE/H$, even though it binds a localized state, because the remaining delocalized states carry exactly enough extra current to compensate for its loss. The exactness of these results and their apparent insensitivity to the type or location of the impurity suggest that the effect is due, ultimately, to a fundamental principle. In this communication, we point out that it is, in fact, due to the long-range phase rigidity characteristic of a supercurrent, and that quantization can be derived from gauge invariance and the existence of a mobility gap.

We consider the situation illustrated in Fig. 1, of a ribbon of two-dimensional metal bent into a loop of circumference L, and pierced everywhere by a magnetic field H_0 normal to its surface. The density of states of this system, also illustrated in Fig. 1, consists, in the absence of disorder, of a sequence of δ functions, one for each Landau level. These broaden, in the presence of disorder, into bands of extended states separated by tails of localized ones. We consider the disordered case with the Fermi level in a mobility gap, as shown.

We wish to relate the total current I carried around the loop to the potential drop V from one edge to another. This current is equal to the adiabatic derivative of the total electronic energy U of the system with respect to the magnetic flux ϕ through the loop. This may be obtained by differentiating with respect to a uniform vector potential A pointing around the loop, in the manner

$$I = c\frac{\partial U}{\partial \phi} = \frac{c}{L}\frac{\partial U}{\partial A} \quad . \tag{1}$$

This derivative is nonzero only by virtue of the phase coherence of the wave functions around the loop. If, for example, all the states are localized then the only effect of A is to multiply each wave function by $\exp(ieAx/\hbar c)$, where x is the coordinate around the loop, and the energy change and current are zero. If a state is extended, on the other hand, such a gauge transformation is illegal unless

$$A = n\frac{hc}{eL} \quad . \tag{2}$$

In the case on noninteracting electrons, phase coherence enables a vector potential increment to

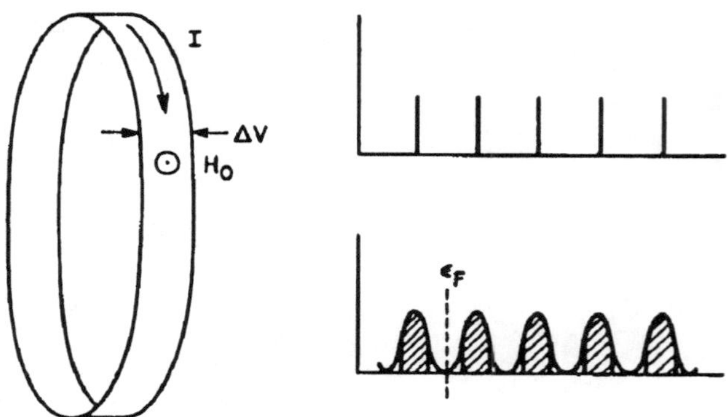

FIG. 1. Left: Diagram of metallic loop. Right: Density of states without (top) and with (bottom) disorder. Regions of delocalized states are shaded. The dashed line indicates the Fermi level.

change the total energy by forcing the filled states toward one edge of the ribbon. Specifically, if one adopts the usual isotropic effective-mass Hamiltonian,

$$H = \frac{1}{2m^*}\left(\vec{p} - \frac{e}{c}\vec{A}\right)^2 + eE_0 y \quad , \tag{3}$$

where E_0 is the electric field across the ribbon, and adopts Landau gauge

$$\vec{A} = H_0 y \hat{x} \quad , \tag{4}$$

then the wave functions, given by

$$\psi_{k,n} = e^{ikx}\phi_n(y - y_0) \quad , \tag{5}$$

where ϕ_n is the solution to the harmonic-oscillator equation

$$\left[\frac{1}{2m^*}p_y^2 + \frac{1}{2m^*}\left(\frac{eH_0}{c}\right)^2 y^2\right]\phi_n = (n + \tfrac{1}{2})\hbar\omega_c\phi_n \quad , \tag{6}$$

and y_0 is related to k by

$$y_0 = \frac{1}{\omega_c}\left(\frac{\hbar k}{m^*} - \frac{cE_0}{H_0}\right) \quad , \tag{7}$$

are affected by a vector potential increment $\Delta A\hat{x}$ only through the location of their centers, in the manner

$$y_0 \to y_0 - \frac{\Delta A}{H_0} \quad . \tag{8}$$

The energy of the state, still given by

$$\epsilon_{n,k} = (n + \tfrac{1}{2})\hbar\omega_c + eE_0 y_0 + \tfrac{1}{2}m^*(cE_0/H_0)^2 \tag{9}$$

thus changes linearly with ΔA. This gives rise to the derivative in Eq. (1), which may be conveniently evaluated via the substitution

$$\frac{\partial U}{\partial \phi} \to \frac{\Delta U}{\Delta \phi} \tag{10}$$

with $\Delta\phi = hc/e$ a flux quantum. Since, by gauge invariance (2), adding $\Delta\phi$ maps the system back into itself, the energy increase due to it results from the net transfer of n electrons (no spin degeneracy) from one edge to the other. The current is thus

$$I = c\frac{neV}{\Delta\phi} = \frac{ne^2 V}{h} . \tag{11}$$

We now consider the dirty interacting system. As in the ideal case, gauge invariance is an exact symmetry forcing the addition of a flux quantum to result only in excitation or deexcitation of the original system. Also as in the ideal case, there is a gap, although the gap now exists between the electrons and holes affected by the perturbation, those contiguous about the loop, rather than in the density of states. Since adiabatic change of the many-body Hamiltonian cannot excite quasiparticles across this gap, it can only produce an excitation of the charge-transfer variety discussed in the ideal case, although the number of electrons transferred need not be the ideal number, and can be zero, as is the case for most systems with gaps. Therefore, Eq. (11) is always true, as a bulk property, for some integer n whenever the local Fermi level lies everywhere in a gap in the extended-state spectrum.

At the edges of the ribbon, the effective gap collapses and communication between the extended states and the local Fermi level is reestablished. Particles injected into this region rapidly thermalize to the Fermi level, in the process losing all memory of having been mapped adiabatically. This would be a significant source of error in Eq. (11) were it not for the fact that *isothermal* differentiation with respect to ϕ, the thermodynamically correct procedure for obtaining I, is equivalent to adiabatic differentiation in the sample interior and is reversible. Thus, slow addition of $\Delta\phi$ physically removes a particle from the local Fermi level at one edge of the ribbon and injects it at the local Fermi level of the other, acting as a pump. Since the Fermi energy is defined as the change in V resulting from the injection of a particle, and since eV is defined to be the Fermi-level difference, edge effects are not a source of error in Eq. (11).

Several other sources remain to be investigated, including possible ϕ dependence, the effect of substituting the ring geometry of Fig. 1 for the usual strip geometry, and effects of tunneling. However, we find it intuitively appealing that the quantum effect should go hand in hand with the persistence of currents, and thus that the physically significant source of error should be thermal activation of carriers to the mobility edge. These carriers produce a large, but finite, normal resistance per square R, which in the steady-state strip geometry, results in a Hall resistance too small in the amount

$$\left|\frac{\Delta R_H}{R_H}\right| = \left(\frac{R_H}{R}\right)^2 . \tag{12}$$

In summary, we have shown that the quantum Hall effect is intimately related to the extended nature of the states near the center of the disorder-broadened Landau level, and that edge effects do not influence the accuracy of the quantization. We speculate that the only significant source of error is thermal activation of carriers to the mobility edge.

ACKNOWLEDGMENTS

I am grateful to P. A. Lee, D. C. Tsui, R. E. Prange, and H. Störmer for helpful discussions.

[1]K. V. Klitzing, G. Dorda and M. Pepper, Phys. Rev. Lett. 45, 494 (1980).

[2]Identical behavior has been seen in GaAs-Al$_x$Ga$_{1-x}$As heterostructures. D. C. Tsui and A. C. Gossard (unpublished).

[3]T. Ando, J. Phys. Soc. Jpn. 37, 622 (1974).

[4]R. E. Prange (unpublished).

Two-Dimensional Magnetotransport in the Extreme Quantum Limit

D. C. Tsui,[a),(b)] H. L. Stormer,[(a)] and A. C. Gossard
Bell Laboratories, Murray Hill, New Jersey 07974
(Received 5 March 1982)

A quantized Hall plateau of $\rho_{xy} = 3h/e^2$, accompanied by a minimum in ρ_{xx}, was observed at $T < 5$ K in magnetotransport of high-mobility, two-dimensional electrons, when the lowest-energy, spin-polarized Landau level is $\frac{1}{3}$ filled. The formation of a Wigner solid or charge-density-wave state with triangular symmetry is suggested as a possible explanation.

PACS numbers: 72.20.My, 71.45.-d, 73.40.Lq, 73.60.Fw

In the presence of an intense perpendicular magnetic field B, a system of two-dimensional (2D) electrons is expected to form a Wigner solid[1,2] at low temperatures (T). In the infinite-B limit, an analogy can be drawn to the classical electron gas on the surface of liquid helium, which crystallizes into a solid[3] when the ratio of the electron's average potential energy to thermal energy $\Gamma \equiv e^2\pi^{1/2}n/\epsilon kT = 137$ (n is the electron areal density). At finite B, quantum effects become important and it has been suggested that a charge-density-wave (CDW) state[4] may be possible at considerably higher T as a precursor to Wigner crystallization. Early experiments were carried out on the Si inversion layer at the Si-SiO_2 interface. Kawaji and Wakabayashi[5] and Tsui[6] made high-B magnetoconductivity measurements and observed structures and electric field dependences which cannot be explained by the independent-electron theory of Ando and Uemura.[7] Subsequently, Kennedy *et al.*[8] observed a shift in cyclotron resonance, concomitant with a drastic line narrowing, in the high-B limit, when the average electron separation exceeds the cyclotron diameter. Wilson, Allen, and Tsui[9] studied the dependence of this phenomenon on the Landau-level filling factor ($\nu = nh/eB$), and found that a pinned-CDW model[10] gave the most satisfactory account of the cyclotron resonance data. However, in the range of n at which these experiments were performed, localization due to disorder at the Si-SiO_2 interface is known to be important even in the absence of B, and consequently, it has not been possible to discern true Coulomb effects from those due to disorder.

In this Letter, we report some striking, new results on the transport of high-mobility, 2D electrons, in GaAs-AlGaAs heterojunctions, in the extreme quantum limit ($\nu < 1$), when the lowest-energy, spin-polarized Landau level is partially filled. We found that at temperatures $T < 5$ K, the diagonal part ρ_{xx} of the resistivity tensor shows a dip at $\nu = \frac{1}{3}$, which becomes stronger at lower T. For $\nu < \frac{1}{3}$, ρ_{xx} follows an approximately exponential increase with inverse T. The Hall resistivity ρ_{xy}, on the other hand, approaches a step of $3h/e^2$ at $\nu = \frac{1}{3}$ as T decreases, but remains essentially independent of T away from this Hall plateau. These features of the data resemble those of the quantized Hall resistance and the zero-resistance state expected exclusively for integral values of ν. We suggest that these striking results are evidence for a new electronic state at $\nu = \frac{1}{3}$. They are consistent with the notion that a Wigner solid, or a CDW state with triangular crystal symmetry, is favored at $\nu = \frac{1}{3}$ when the unit cell area of the lattice is a multiple of the area of a magnetic flux quantum.

The samples, consisting of 1-μm undoped GaAs, 500-Å undoped $Al_{0.3}Ga_{0.7}As$, 600-Å Si-doped $Al_{0.3}Ga_{0.7}As$, and 200-Å Si-doped GaAs single crystals, were sequentially grown on insulating GaAs substrates using molecular-beam-epitaxy techniques.[11] The 2D electron gas, resulting from ionized donors placed 500 Å inside AlGaAs,[12] is established at the undoped GaAs side of the GaAs-AlGaAs heterojunction. Samples were cut into standard Hall bridges and Ohmic contacts to the electron layer were made with In at 400 °C. Low-field transport measurements were used to determine n and μ. Our samples have n from 1.1×10^{11} to 1.4×10^{11} cm^{-2} and μ from 80 000 to 100 000 cm^2/V sec. The high-B measurements were performed at the Francis Bitter National Magnet Laboratory, Cambridge, Mass.

Figure 1 shows ρ_{xy} and ρ_{xx} of one specimen as a function of B at four different temperatures. The scale at the top of the figure shows the Landau level filling factor ν, which gives the number of occupied levels. At integral values of ν, the data show the characteristic features of the quantized Hall plateaus and the vanishing of ρ_{xx},[13-15] when the Fermi energy E_F is pinned in the gap between two adjacent levels. Removal of spin de-

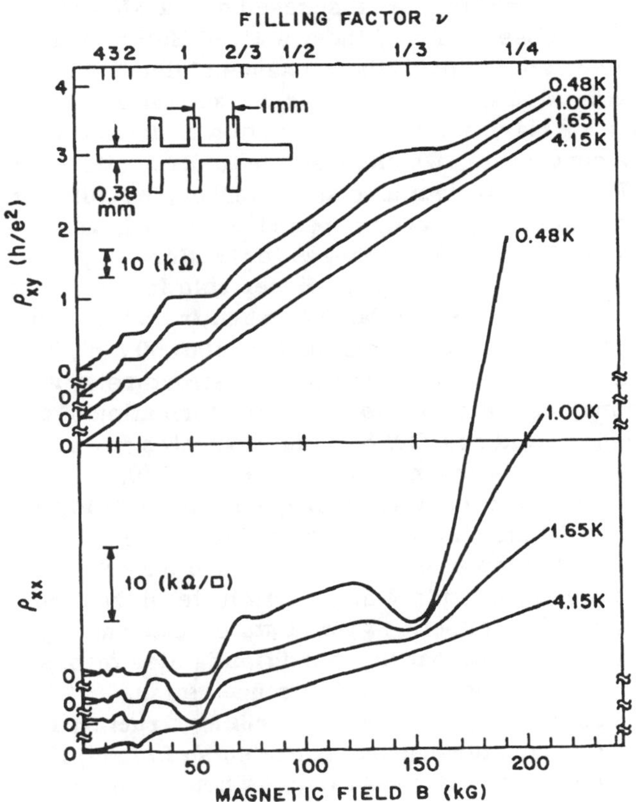

FIG. 1. ρ_{xy} and ρ_{xx} vs B, taken from a GaAs-$Al_{0.3}$-$Ga_{0.7}$As sample with $n = 1.23 \times 10^{11}/cm^2$, $\mu = 90\,000\ cm^2/V\ sec$, using $I = 1\ \mu A$. The Landau level filling factor is defined by $\nu = nh/eB$.

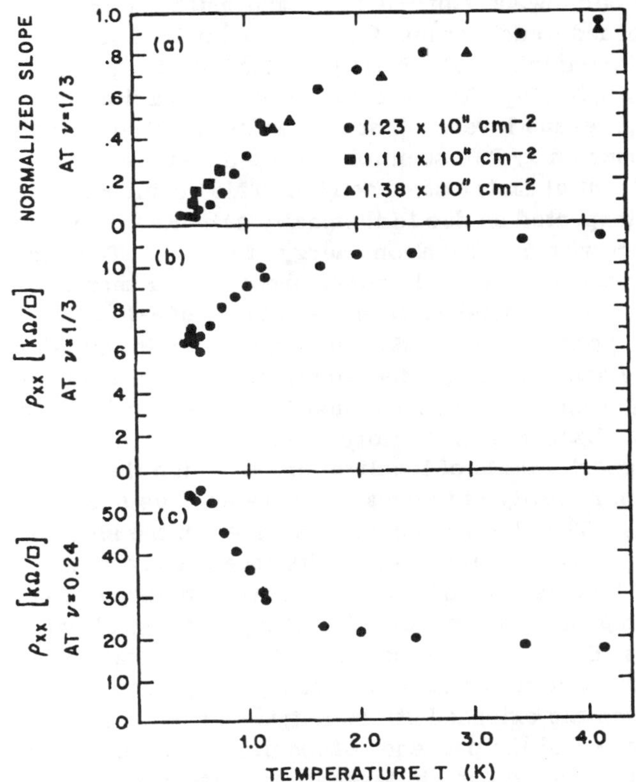

FIG. 2. T dependence of (a) the slope of ρ_{xy} at $\nu = \frac{1}{3}$, normalized to the slope at ~ 30 K, (b) ρ_{xx} at $\nu = \frac{1}{3}$, and (c) ρ_{xx} at $\nu = 0.24$.

generacy[16] is seen in the appearance of these features at odd-integer values of ν. As observed earlier,[17] the plateaus in ρ_{xy} as well as the vanishing of ρ_{xx} become increasingly pronounced as T is decreased.

In the extreme quantum limit, $\nu < 1$, only the lower spin state of the lowest Landau level, i.e., the $(0, \uparrow)$ level, remains partially occupied. In this regime (i.e., $B > 50$ kG in Fig. 1), the system is completely spin polarized. For $T > 4.2$ K, $\rho_{xy} = B/ne$, and ρ_{xx} shows also nearly linear dependence on B, as expected from the free-electron theory of Ando and Uemura.[7,16] At lower T, ρ_{xy} deviates from $\rho_{xy} = B/ne$ at $\nu \sim \frac{1}{3}$. This deviation becomes more pronounced as T decreases and approaches a plateau of $\rho_{xy} = 3h/e^2$, within an accuracy better than 1% at 0.42 K. The appearance of this plateau is accompanied by a minimum in ρ_{xx}, as apparent in the lower panel of Fig. 1. The development of these features is similar to that of the quantized Hall resistance and the concomitant vanishing of ρ_{xx}, observed at integral values of ν at higher T. Moreover, for $\nu < \frac{1}{3}$ and away from the plateau region, ρ_{xx} shows strong increase with decreasing T, while ρ_{xy} shows very weak decrease or essentially independence of T. This behavior has been seen to $\nu = 0.21$, the smallest ν attained in this experiment.

Figure 2 illustrates the development of ρ_{xx} and ρ_{xy} at fixed B as a function of T. Figure 2(a) shows the slope of ρ_{xy} at $\nu = \frac{1}{3}$, normalized to the slope at high T (~ 30 K), for three samples with slightly different n. Figure 2(b) shows the accompanying ρ_{xx} minimum (at $\nu = \frac{1}{3}$), and Fig. 2(c) shows ρ_{xx} at $\nu = 0.24$ to illustrate its T dependence for $\nu < \frac{1}{3}$, away from the Hall plateau. Several points should be noted. First, the slope of ρ_{xy} at $\nu = \frac{1}{3}$ approaches zero at $T \sim 0.4$ K, indicative of a true quantized Hall plateau. Second, replotting the data in Fig. 2(a) on logarithmic slope versus inverse T scale shows a linear portion for data taken at $T \geq 1.1$ K. This fact allows us to extrapolate the normalized slope to 1 at $T_0 = 5$ K, which we identify as the temperature for the onset of this phenomenon. Third, ρ_{xx} at $\nu = \frac{1}{3}$ is ~ 6 kΩ/□

at our lowest T of ~0.42 K, and much lower T is needed to determine if ρ_{xx} at $\nu = \frac{1}{3}$ indeed vanishes with vanishing T. Finally, if the data for ρ_{xx} at $\nu = 0.24$ [Fig. 2(c)] are replotted on logarithmic ρ_{xx} versus inverse T scale, an exponential dependence on $1/T$ is seen for $T \geqslant 0.6$ K, with a preexponential factor of ~13kΩ/□. This result may be interpreted as due to thermally activated transport with an activation energy of 0.94 K. The preexponential factor is considerably lower than the maximum metallic sheet resistance of ~40 kΩ/□, predicted for Anderson localization in the tails of Landau levels,[18] but comparable to that signifying a metal-insulator transition for 2D systems in the absence of B.[19] Moreover, the data for $\nu < \frac{1}{3}$ suggest a state of localization, in which the electron mobility is thermally activated,[17] as seen in ρ_{xx}. The electron density, as seen in the slope of ρ_{xy} vs B, remains essentially independent of T.

The existence of quantized Hall resistance accompanied by the vanishing of ρ_{xx} at integral values of ν is now well known. Their observation is attributed to the existence of an energy gap between the extended states in two adjacent Landau levels and the presence of localized states, which pin E_F in the gap, to keep all the extended states in finite numbers of Landau levels completely occupied for finite regions of B or n. Laughlin's argument[20] based on gauge invariance demonstrated that the quantized Hall resistance, given by $\rho_{xy} = h/ie^2$ ($i = 1, 2, \ldots$), results from complete occupation of all the extended states in the Landau level, regardless of the presence of localization. Our observation of a quantized Hall resistance of $3h/e^2$ at $\nu = \frac{1}{3}$ is a case where Laughlin's argument breaks down. If we attribute it to the presence of a gap at E_F when $\frac{1}{3}$ of the lowest Landau level is occupied, his argument will lead to quasiparticles with fractional electronic charge of $\frac{1}{3}$, as has been suggested for $\frac{1}{3}$-filled quasi one-dimensional systems.[21]

At the present, there is no satisfactory explanation for all of our observations. The fact that this phenomenon always occurs at $\nu = \frac{1}{3}$ and that it is most striking in samples with the highest electron mobility suggest the formation of a new spin-polarized electronic state, such as Wigner solid or CDW, with a triangular symmetry,[22] which is favored at $\nu = \frac{1}{3}$. In this picture, the observed features of ρ_{xx} and ρ_{xy} may be attributed to transport of the collective ground state. At $T = 0$, the transport is free of dissipation and ρ_{xx} is expected to vanish. Since the number of electrons in this ground state is $n = eB/3h$, the Hall resistivity is $\rho_{xy} = B/ne = 3h/e^2$. As discussed by Baraff and Tsui,[23] observation of the quantized Hall plateau may be attributed to the presence of donor states inside $Al_xGa_{1-x}As$. The thermal activation of ρ_{xx} at $\nu = \frac{1}{3}$ may result from activation of defects in the condensate, which give rise to dissipation.

Finally, our data also show weaker, but similar, structures in ρ_{xx} near $\nu = \frac{2}{3}$ and near $\nu = \frac{3}{2}$, accompanied by slight changes in the slope of ρ_{xy}. These structures, though discernible in Fig. 1, are well resolved in the data taken from the sample with $n = 1.4 \times 10^{11}/\mathrm{cm}^2$ and $\mu = 100\,000$ cm²/V sec at 1.2 K. In our picture, the structure at $\nu = \frac{2}{3}$ may be identified as due to the formation of a Wigner solid or CDW of holes, expected from electron-hole symmetry. At $\nu = \frac{3}{2}$, the (0, ↑) Landau level is completely filled, but the (0, ↓) level is half filled. Consequently, $\frac{2}{3}$ of all the electrons occupy the low-lying (0, ↑) level, and only the remaining $\frac{1}{3}$ with spin ↓ may participate in the formation of a collective ground state. Our data appear to suggest that this condition is also favored for a Wigner solid or CDW ground state.

In summary, we observed striking structures in the magnetotransport coefficients of high-μ, 2D electrons in GaAs-$Al_xGa_{1-x}As$ heterojunctions at $\nu = \frac{1}{3}$, and similar, but much weaker, structures at $\nu = \frac{2}{3}$ and $\frac{3}{2}$. Their development as a function of T is reminiscent of the quantized ρ_{xy} and the concomitant vanishing of ρ_{xx}, expected only for integral values of ν. We suggest as a possible explanation the formation of a new electronic state, such as a Wigner solid or CDW state with a triangular symmetry.

We thank P. M. Tedrow for the He^3 refrigerator; R. B. Laughlin, P. A. Lee, V. Narayanamurti, and P. M. Platzman for discussions; and K. Baldwin, G. Kaminsky, and W. Wiegmann for technical assistance. This work was supported in part by the National Science Foundation.

---

[a]Visiting scientist at the Francis Bitter National Magnet Laboratory, Cambridge, Mass. 02139.

[b]Present address: Department of Electrical Engineering and Computer Science, Princeton University, Princeton, N.J. 08544.

[1]E. P. Wigner, Phys. Rev. 46, 1002 (1934).

[2]Y. E. Lozovik and V. I. Yudson, Pis'ma Zh. Eksp. Teor. Fiz. 22, 26 (1975) [JETP Lett. 22, 11 (1975)].

[3]C. C. Grimes and G. Adams, Phys. Rev. Lett. 42, 795 (1979).

[4]H. Fukuyama, P. M. Platzman, and P. W. Anderson, Phys. Rev. B 19, 5211 (1979).

[5]S. Kawaji and J. Wakabayashi, Solid State Commun. 22, 87 (1977).
[6]D. C. Tsui, Solid State Commun. 21, 675 (1977).
[7]T. Ando and Y. Uemura, J. Phys. Soc. Jpn. 36, 959 (1974).
[8]T. A. Kennedy, R. J. Wagner, B. D. McCombe, and D. C. Tsui, Solid State Commun. 22, 459 (1977).
[9]B. A. Wilson, S. J. Allen, and D. C. Tsui, Phys. Rev. Lett. 44, 479 (1980).
[10]H. Fukuyama and P. A. Lee, Phys. Rev. B 18, 6245 (1978).
[11]A. Y. Cho and J. R. Arthur, Prog. Solid State Chem. 10, 157 (1975).
[12]H. L. Stormer, A. Pinczuk, A. C. Gossard, and W. Wiegmann, Appl. Phys. Lett. 38, 691 (1981); T. J. Drummond, H. Morkoc, and A. Y. Cho, J. Appl. Phys. 52, 1380 (1981).
[13]K. von Klitzing, G. Dorda, and M. Pepper, Phys. Rev. Lett. 45, 494 (1980).
[14]D. C. Tsui and A. C. Gossard, Appl. Phys. Lett. 37, 550 (1981).
[15]D. C. Tsui, H. L. Stormer, and A. C. Gossard, Phys Rev. B 25, 1405 (1982).
[16]D. C. Tsui, H. L. Stormer, A. C. Gossard, and W. Wiegmann, Phys. Rev. 21, 1589 (1980); Th. Englert, D. C. Tsui, and A. C. Gossard, Surface Sci. 113, 295 (1982).
[17]M. A. Paalanen, D. C. Tsui, and A. C. Gossard, Phys. Rev. B 25, 5566 (1982).
[18]H. Aoki and H. Kamimura, Solid State Commun. 21, 45 (1977).
[19]D. J. Bishop, D. C. Tsui, and R. C. Dynes, Phys. Rev. Lett. 44, 1153 (1980).
[20]R. B. Laughlin, Phys. Rev. B 23, 5632 (1981).
[21]W. P. Su and J. R. Schrieffer, Phys. Rev. Lett. 46, 738 (1981).
[22]D. Yoshioka and H. Fukuyama, J. Phys. Soc. Jpn. 47, 394 (1979), and 50, 1560 (1981).
[23]G. A. Baraff and D. C. Tsui, Phys. Rev. B 24, 2274 (1981).

Surface Studies by Scanning Tunneling Microscopy

G. Binnig, H. Rohrer, Ch. Gerber, and E. Weibel
IBM Zurich Research Laboratory, 8803 Rüschlikon-ZH, Switzerland
(Received 30 April 1982)

Surface microscopy using vacuum tunneling is demonstrated for the first time. Topographic pictures of surfaces on an *atomic scale* have been obtained. Examples of resolved monoatomic steps and surface reconstructions are shown for (110) surfaces of $CaIrSn_4$ and Au.

PACS numbers: 68.20.+t, 73.40.Gk

In two previous reports,[1,2] we demonstrated the experimental feasibility of controlled vacuum tunneling. The tunnel current flowed from a W tip to a Pt surface at some 10 Å distance from each other. The tunnel distance could be stabilized within 0.2 Å. These experiments were the first step towards the development of scanning tunneling microscopy. Previous developments were unsuccessful for various reasons.[3]

The present Letter contains the first experimental results on surface topography obtained with this novel technique. They demonstrate an unprecedented resolution of the *scanning tunneling microscope* (STM) and should give a taste of its fascinating possibilities for surface characterization.

The principle of the STM is straightforward. It consists essentially in scanning a metal tip over the surface at *constant* tunnel current as shown in Fig. 1. The displacements of the metal tip given by the voltages applied to the piezodrives then yield a topographic picture of the surface. The very high resolution of the STM rests on the strong dependence of the tunnel current on the distance between the two tunnel electrodes, i.e., the metal tip and the scanned surface. The tunnel current through a planar tunnel barrier of average height φ and width s is given by[4]

$$J_T \propto \exp(-A\varphi^{1/2}s)\,, \tag{1}$$

where $A=(4\pi/h)2m)^{1/2}=1.025$ Å^{-1} eV$^{-1/2}$, with m the free-electron mass, appropriate for a vacuum tunnel barrier. With barrier heights (work functions) of a few electronvolts, a change of the tunnel barrier width by a single atomic step (~2–5 Å) changes the tunnel current up to three orders of magnitude. Using only the distance dependence as given by Eq. (1), and a spherical tip of radius R, one estimates a lateral spread δ of a surface step as $\delta \approx 3r_0 = 3(2R/A\varphi^{1/2})^{1/2}$, i.e., $\delta(\text{Å}) \approx 3[R(\text{Å})]^{1/2}$. Thus, a lateral resolution considerably below 100 Å requires tip radii of the order of 100 Å. Such tips are standard in field-emission microscopy. However, since suppression of vibrations is evidently more vital for the STM, long and narrow field-emission tips might not be satisfactory. Instead, we used solid metal rods of 1 mm diameter, and ground 90° tips with a conventional grinding machine. This yielded overall tip radii of only some thousand angstroms to 1 μm, but with some rather sharp minitips. The extreme sensitivity of the tunnel current on gap width then selects the longest of the minitips for operation of the STM. The lateral resolution could be increased further by gently touching the surface with the tip and subsequently retracting it. This "mini-spot-welding" procedure created very fine tips, such that monoatomic steps could be resolved within 10 Å laterally.

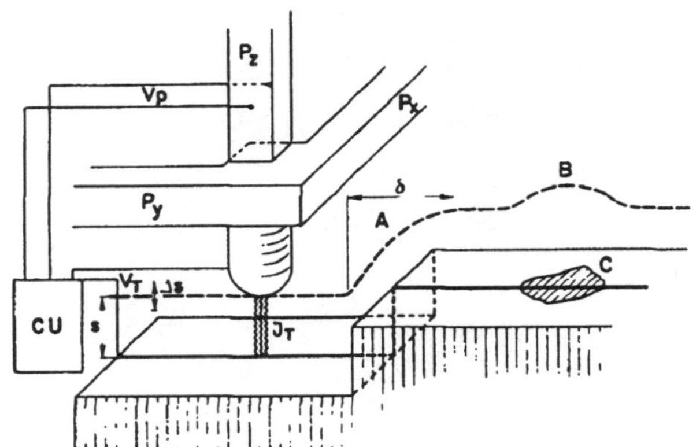

FIG. 1. Principle of operation of the scanning tunneling microscope. (Schematic: distances and sizes are not to scale.) The piezodrives P_x and P_y scan the metal tip M over the surface. The control unit (CU) applies the appropriate voltage V_p to the piezodrive P_z for constant tunnel current J_T at constant tunnel voltage V_T. For constant work function, the voltages applied to the piezodrives P_x, P_y, and P_z yield the topography of the surface directly, whereas modulation of the tunnel distance s by Δs gives a measure of the work function as explained in the text. The broken line indicates the z displacement in a y scan at (A) a surface step and (B) a contamination spot, C, with lower work function.

Scanning the tunnel tip at constant tunnel current implies $\varphi^{1/2}s=\text{const}$. Thus, the z displacement of the tunnel tip gives the surface topography only for constant work function φ, and therefore constant gap width s, as shown in Fig. 1 at A. On the other hand at B, the z displacement is caused by a change of work function on a structureless part of the surface. However, true surface structures and work-function–mimicked structures can be separated by modulating the gap width s while scanning, at a frequency higher than the cutoff frequency of the control unit. In a simple situation, as depicted in Fig. 1, the modulation signal gives the square root of the work function $\varphi^{1/2}=\Delta(\ln J_T)/\Delta s$, directly, A in Eq. (1) being nearly 1. For general surface topographies, and work-function profiles, the separation process becomes rather involved. Then, the modulation Δs of the gap with s is no longer equal to the length modulation Δz of the piezodrive P_z. Essentially, $\Delta s=\Delta z\cos\psi$, where ψ is the angle between the tunnel-surface element and the z direction. In turn, the modulation signal is no longer constant at true surface structures even for constant work function φ. However, since V_p and the modulation signal contain φ and s in a different way, their separation is, in principle, still possible even for involved structures and work-function profiles. In the following, we present topographic pictures of (110) surfaces of $CaIrSn_4$ and Au. Work-function profiles have not yet been studied in detail. They were used rather to get an overall picture of the surface condition.

$CaIrSn_4$.—The flux-grown single crystals exhibited shiny, natural faces after solving the remaining flux in HCl. Solvent etching probably stops at Ir layers, which appear to be rather

inert.[5] Therefore, they were good candidates for testing the operation of the STM at moderate vacuum ($\approx 10^{-6}$ Torr). Figure 2(a) shows a STM picture of a (110) surface obtained at room temperature without further surface treatment. We take the large flat parts (flat on an atomic scale) as an indication for a weak and homogeneous surface contamination. (No provision for simultaneous recording of work function and topography existed at the time of these experiments.) The pronounced structure on the left is the beginning of a growth spiral. Such spirals could be observed with both light and scanning electron microscopes. In the flat region, some monoatomic steps are clearly seen. Two scans with monoatomic, double-atomic, and triple-atomic steps are shown in Fig. 2(b). From all the steps observed, we obtained 6.7 Å as the average spacing of the Ir(110) planes. The piezodrives were calibrated by relacing the tip and sample with capacitor plates, giving a sensitivity of 2.0(±0.2) Å/V in each direction. This step height agrees well with the 6.87 Å inferred from crystallographic data.[6] Moreover, the form of the large steps is in qualitative agreement with that expected from simple calculations: a relatively sharp edge at the beginning of the step and considerable smearing out at the end (as sketched in Fig. 1 at *A*).

Au.—The Au pictures were taken with a new, improved tunnel unit with considerably increased stability. The piezodrive material was calibrated in a conventional capacitance dilatometer within 2%, giving an accuracy of the sensitivity of the whole piezodrive of about 5%. The untreated (110) surface appeared structureless and mostly atomically flat. After Ar sputtering and subsequent annealing at 600 °C in (2 to 7)$\times 10^{-10}$ Torr [a standard procedure for inducing reconstructions of Au (110) surfaces[7-9]], the surface appeared gently corrugated in the [001] direction as shown in Fig. 3(a). The work function was practically constant. The modulated signal showed variations of the order of a percent which reflect the surface corrugation rather than a true variation of the work function, as explained above. Repetition of the cleaning procedure led to qualitatively the same result. The corrugation is not strictly periodic; it varies from 20 to 100 Å in length and from some tenths to 2 Å in height, but with only small local variation in

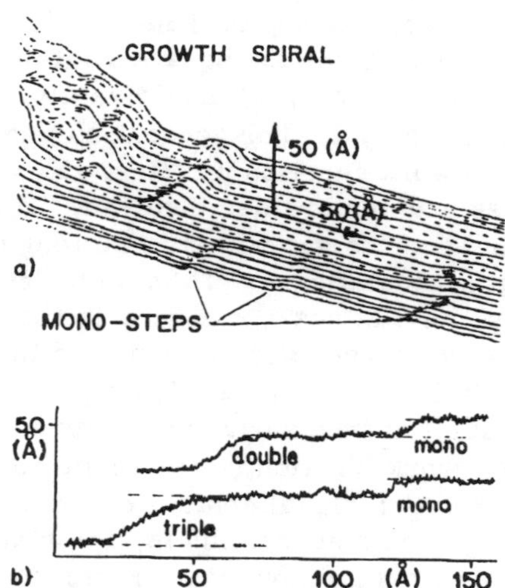

FIG. 2. Topography of a $CaIrSn_4$ (110) surface. (a) Overall view of a flat part with single atomic steps (right) and start of a growth spiral (left). For better visualization of the topography of the surface, some additional lines have been interpolated (broken) between the smoothed scans. The bottom line is as measured. The distance between the scans is uncalibrated. (b) Two individual scans, exhibiting triple, double, and monoatomic steps. The broken lines indicate (110) faces, with the proper distance.

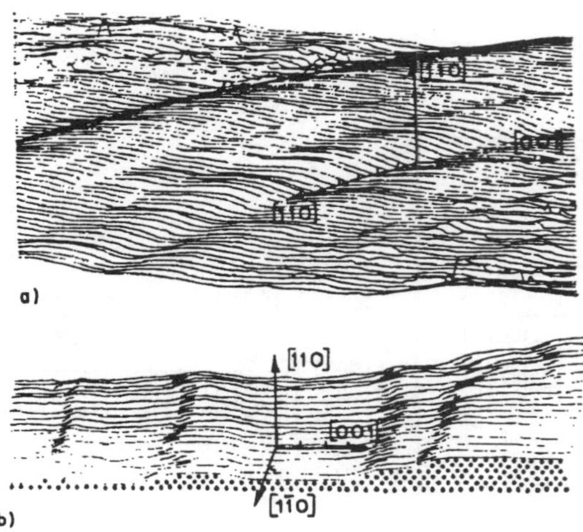

FIG. 3. Two examples of scanning tunneling micrographs of a Au (110) surface, taken at (a) room temperature, and (b) 300 °C after annealing for 20 h at the same temperature (and essentially constant work function). The sensitivity is 10 Å/div everywhere. Because of a small thermal drift, there is some uncertainty in the crystal directions in the surface. In (a), the surface is gently corrugated in the [001] direction, except for a step of four atomic layers (≅ 2 atomic radii) along the [1$\bar{1}$0] direction, as indicated by the discontinuity of the shaded ribbon. The steps in (b), which were always found along the [1$\bar{1}$0] direction, are visualized by the possible positions of the Au atoms (dots).

periodicity and height. A small corrugation of about 100 Å length in the [1$\bar{1}$0] direction could be induced by rapidly cooling the sample to room temperature after annealing at 600 °C. Atomic steps could not usually be observed, and the step of 6 Å [equal to four (110) spacings or two atomic radii] shown in Fig. 3(a) is an exception. However, double or monoatomic steps were easily found at 300 °C [Fig. 3(b)]. An independent indication of an increasing step density with increasing temperature was recently obtained from an analysis of He-diffraction line shapes for Ni(100).[10]

Disorder along the [001] direction of Au(110) surfaces has been inferred from low-energy electron diffraction (LEED) experiments.[8] The varying wavelength of the corrugation found in the present experiments induces such an anisotropic disorder. In view of the resolution demonstrated [see the steps in Fig. 3(b)], the corrugation is too smooth and flat to be explained in terms of some sequence of unrelaxed steps or a disordered 2×1 reconstruction of the missing-row type.[8] It rather indicates a more continuous vertical displacement of the Au atoms. Surface buckling has been conjectured[11] for the Au(100) surface as a consequence of a mismatch of a topmost hexagonal layer with the underlying fcc structure. Reconstructions of the (110) surface are subject to quite some controversy. In particular, spin-polarized LEED experiments seem to rule out any of the proposed models containing a mirror plane perpendicular to the [1$\bar{1}$0] chains or twofold rotations.[12] The distorted hexagonal topmost layer model[13] is compatible with the symmetry requirements of the spin-polarized LEED results. Although the present experiment did not reveal the double periodicity in the [100] direction, some distorted hexagonal topmost-layer structure appears to be an attractive explanation for the long-wave buckling. Even more, nonobservation of the 2×1 structure in the present experiment could be considered as support of this model. However, it is not certain whether a 2×1 reconstruction was indeed present, although it had been previously observed in the same crystal by TEAMS experiments.[7] Combined LEED and tunnel experiments are planned to clear this point. Finally, it is interesting to note that the step in Fig. 3(a) separates a smooth portion of the surface (on the right) from an atomically rough one.

In summary, we have shown that scanning tunneling microscopy yields a true three-dimensional topography of surfaces on an atomic scale, i.e., a resolution orders of magnitude better than scanning electron microscopy, with the possibility of extending it to work-function profiles (fourth dimension). The technique is nondestructive (energy of the tunnel "beam" 1 meV up to 4 eV), and uses fields down to three orders of magnitude less than field-ionization microscopy. The high current densities of 10^3 to 10^4 A/cm^2 appear to be no problem, and the technique has already been successfully extended to low-doped semiconductors.[14] The significance of vacuum tunneling to surface studies and many other fields like space-resolved tunneling spectroscopy, microscopy of adsorbed molecules, and crystal growth, as well as for fundamental aspects of tunneling, especially in small geometries, is evident.

We thank R. Gambino and K. H. Rieder for providing the $CaIrSn_4$ and Au samples, respectively, H. R. Ott for calibrating the piezodrive material, B. Reihl and K. H. Rieder for discussions on surface aspects, and E. Courtens, K. A. Müller, and H. J. Scheel for their active interest in the STM.

[1]G. Binnig, H. Rohrer, Ch. Gerber, and E. Weibel, Physica (Utrecht) 107B+C, 1335 (1981), Proceedings of the Sixteenth International Conference on Low-Temperature Physics, Los Angeles, 19–25 August 1981.

[2]G. Binnig, H. Rohrer, Ch. Gerber, and E. Weibel, Appl. Phys. Lett. 40, 178 (1981).

[3]R. D. Young, J. Ward, and F. Scire, Rev. Sci. Instrum. 43, 999 (1972).

[4]R. H. Fowler and L. Nordheim, Proc. Roy. Soc. London, Ser. A 119, 173 (1928); J. Frenkel, Phys. Rev. 36, 1604 (1930).

[5]R. Gambino, private communication.

[6]A. S. Cooper, Mater. Res. Bull. 15, 799 (1980).

[7]K. H. Rieder, T. Engel, and N. Garcia, in *Proceedings of the Fourth International Conference on Solid Surfaces, and Third European Conference on Surface Science, Cannes 1980*, Supplement to Revue Le Vide, Les Couche Minces, No. 201 (Societé Francaise du Vide, Paris, 1980), p. 861.

[8]D. Wolf, H. Jagodzinski, and M. Moritz, Surf. Sci. 77, 265, 283 (1978).

[9]J. R. Noonan and H. J. Davis, J. Vac. Sci. Technol. 16, 587 (1979).

[10]K. H. Rieder and H. Wilsch, private communication.

[11]M. A. van Hove, R. J. Koestner, P. C. Stair, J. P. Bibérian, L. L. Kesmodel, I. Bartos, and G. A. Somorjai, Surf. Sci. 103, 181, 218 (1981).

[12]B. Reihl and B. I. Dunlap, Appl. Phys. Lett. 37, 941 (1980).

[13]E. Lang, K. Heinz, and K. Müller, Verh. Dtsch.

Phys. Ges. 1, 278 (1978); E. Lang, private communication.
[14]G. Binnig and H. Rohrer, Europhys. Conf. Abstr. 6A, 210 (1982), and Verh. Dtsch. Phys. Ges. 6, 999 (1982), and in Proceedings of the Société Suisse de Physique Réunion de Printemps, 1982 (unpublished).

7 × 7 Reconstruction on Si(111) Resolved in Real Space

G. Binnig, H. Rohrer, Ch. Gerber, and E. Weibel
IBM Zurich Research Laboratory, 8803 Rüschlikon-ZH, Switzerland
(Received 17 November 1982)

The 7×7 reconstruction on Si(111) was observed in real space by scanning tunneling microscopy. The experiment strongly favors a modified adatom model with 12 adatoms per unit cell and an inhomogeneously relaxed underlying top layer.

PACS numbers: 68.20.+t, 73.40.Gk

The 7×7 reconstruction of the Si(111) surface is one of the most intriguing problems in surface science. In recent years, most impressive experimental and theoretical efforts have dealt with the structure of this reconstruction.[1,2] However, the complexity of the large basic unit cell with 49 atoms is a serious handicap to deriving a structural model unambiguously even from an abundant set of experiments. Models in accordance with one class of experiments are in conflict with others.[1] The ever-increasing number of models and their variations appear to confuse rather than clarify the issue. In order to make significant progress, some basically new approach is required. Such a new approach is the *scanning tunneling microscopy* recently introduced by the authors.[3]

In the following, we report on the first *real-space* determination of the Si(111) 7×7 reconstruction. The scanning tunneling micrographs yield the principal structural features of the 7×7 unit cell.

The principle of the scanning tunneling microscope (STM) is explained in Ref. 3. It consists essentially in scanning a metal tip at constant tunnel current over the surface to be investigated. The corrugation monitored by the vertical motion of the tunnel tip reflects qualitatively the topography of the surface.

The 7×7 reconstruction was generated as follows. After etching the oxide with an HF solution the Si wafer was immediately transferred to the STM in the UHV chamber. Repeated heating to 900 °C in a vacuum not exceeding 3×10^{-8} Pa effects sublimation of the SiO layer grown during the transfer. This procedure is known to give fairly clean surfaces,[4] e.g., less than $\frac{1}{25}$ of a monolayer of carbon. However, the condition of the surface could not yet be tested by another contamination-sensitive method.

The micrographs were taken at 2.9 V (tip positive) since tunnel voltages below 2.5 V lead to direct contact between tip and sample. The voltage drop across the vacuum gap is smaller by an unknown amount because of band bending in the semiconductor. Only unidirectional scans (3 deg off the [$\bar{2}$11] direction) were recorded to avoid small but noticeable hysteresis effects of the scanning piezodrives.

The original recordings of two complete 7×7 unit cells are presented in Fig. 1 in relief form. The rhombohedral 7×7 unit cell is clearly bounded by the lines of minima with deep corners. Inside each cell, twelve maxima appear. The diagonals are 46 ± 1 and 29 ± 4 Å, in agreement with the crystallographic values of 46.56 and 26.88 Å, respectively. The short diagonal is less accurately determined since thermal drifts are more noticeable (scanning along the long diagonal with 2 sec/scan). For comparison with models, the cell size is adjusted to its proper value. Figure 2, a top view of Fig. 1, shows the sixfold rotational symmetry of the positions of the maxima around each corner. The minima pattern, however, ap-

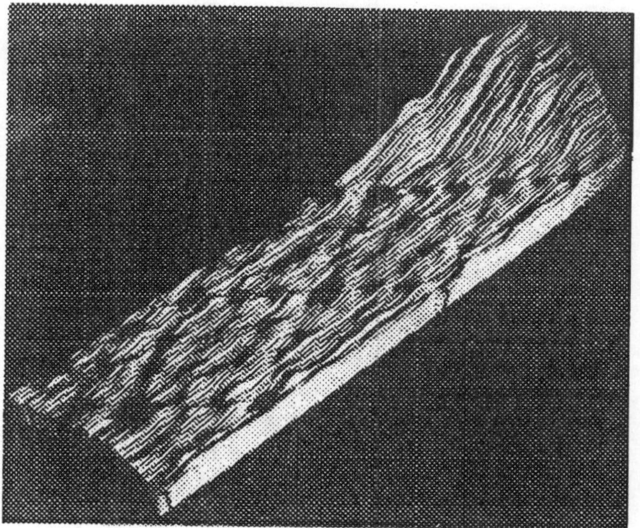

FIG. 1. Relief of two complete 7×7 unit cells, with nine minima and twelve maxima each, taken at 300 °C. Heights are enhanced by 55%; the hill at the right grows to a maximal height of 15 Å. The [$\bar{2}$11] direction points from right to left, along the long diagonal.

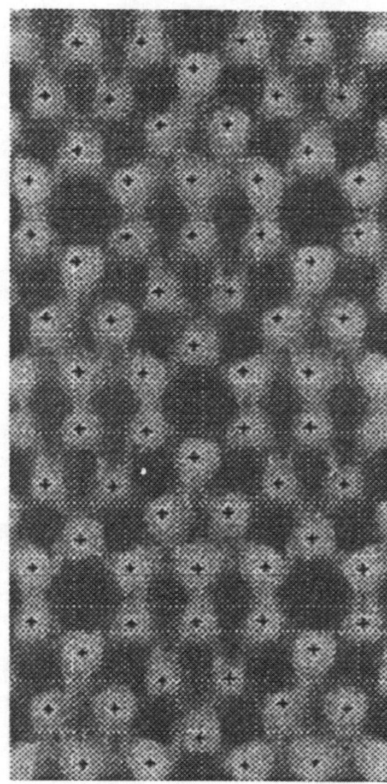

FIG. 2. Top view of the relief shown in Fig. 1 (the hill at the right is not included) clearly exhibiting the sixfold rotational symmetry of the maxima around the rhombohedron corners. Brightness is a measure of the altitude, but is not to scale. The crosses indicate adatom positions of the modified adatom model (see Fig. 3) or "milk-stool" positions (Ref. 5).

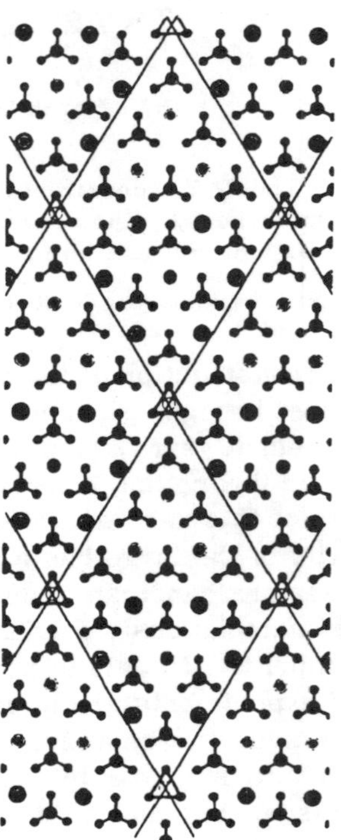

FIG. 3. Modified adatom model. The underlying top-layer atom positions are shown by dots, and the rest atoms with unsatisfied dangling bonds carry circles, whose thickness indicates the depth measured as discussed in the text. The adatoms are represented by large dots with corresponding bonding arms. The empty potential adatom position is indicated by an empty circle in the triangle of adjacent rest atoms. The grid indicates the 7×7 unit cells.

pears to have no rotational symmetry since it is deeper (on the average -0.26 Å) along the edges than along the short diagonal. A somewhat smaller height difference in the maxima is visually not as clearly evident in Fig. 2. However, we believe that these differences are experimental artifacts due to a slight overshoot when recording a minimum after a maximum (or vice versa). If we make the appropriate small corrections for this overshoot, the height of the maxima is uniformly 0.7 ± 0.1 Å with respect to an average level of zero for the unit cell. Likewise, the depth of the corner minimum becomes 2.1 ± 0.2 Å, and those along the edges and the short diagonal, 0.9 ± 0.1 Å. The depth of the single minimum in the left half of the cell is estimated at 0.5 ± 0.2 Å. Finally, a very shallow minimum with three arms of depth nearly zero lies in the right half of the cell. Then, the corrected minima pattern shows threefold rotational symmetry.

The maxima pattern is congruent with the positions of the "milk stools" in the model of Snyder, Wassermann, and Moskowitz[5] or with those of adatoms of a slight modification (missing adatom at the cell corner) of Harrison's model[6] (see Fig. 2). In both models, 36 dangling bonds of the truncated bulk surface are saturated by the adatoms or milk-stool atoms, leaving 13 (rest atoms) with unsatisfied bonds. The virtue of the milk-stool model is dangling bonds on neighboring sites (an important element also in the π-bonding model of Pandey[7]), and that of the adatom model, the reduction of dangling bonds. Our modified adatom model is shown in Fig. 3. In the milk-stool model, the adatom is replaced by a three-membered ring of atoms. Since tunneling is expected to occur predominantly from the dangling bonds, the maxima observed should reflect the dangling-bond positions of the topmost atoms. Then, the milk-stool model can hardly be reconciled with the experiment, since either 36 single maxima or a substantially different structure (e.g., a "doughnut" around the corner instead of six distinct maxima) should be observed, depending on resolution.

In Fig. 4, we compare the lateral positions of the minima with those of the rest atoms of Fig. 3. The two patterns are again congruent within the experimental uncertainty of ±0.5 Å when the three adjacent rest atoms are assigned to the deep corner minimum, and the three central rest atoms in the right half of the cell to the shallow three-armed minimum. In the model of Fig. 3, the adatoms sit on top of an "empty" triangular surface site (no atom below the adatom in the second layer). Shifting the adatoms to "filled" triangular surface sites (with an atom below the adatom in the second layer) corresponds to a reflection about the short diagonal. In that case, rest-atom sites and minima clearly do not match. Thus, the experiment unambiguously confirms the "empty" positions expected for the adatoms. Note also that the proposed milk-stool positions[5] correspond to "filled" sites.

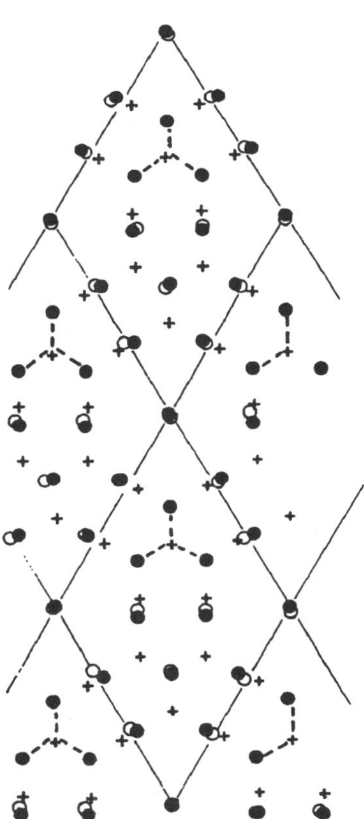

FIG. 4. Comparison of the lateral position of the minima (open circles, and three-armed star for the shallow minimum) with the rest-atom positions (filled circles for model of Fig. 3, crosses for reflection about the short diagonal). The diameter of the circles corresponds to 1.4 Å. The grid indicates the 7× 7 unit cells.

The excellent agreement of the lateral positions of maxima and minima with those of the adatoms at "empty" sites, and of rest atoms, respectively, strongly supports, of the models already proposed, a slight modification of the adatom model[6] for the structure of the 7×7 unit cell. Of course, many interesting questions still remain open, two of which we briefly address below.

The first remark concerns the different depths of the minima. The deep and inhomogeneous corrugation observed cannot be explained by an unrelaxed adatom model. Nonuniform relaxation can, however, enhance the corrugation in a twofold manner: by the shift of the core positions and by relative changes of the energy of the dangling-bond states and/or their occupation. An appreciable tunneling corrugation by a change in bond occupation would require substantial charge transfer costing too much energy.[7] On the other hand, lowering the bond energy would substantially decrease the tunneling current through the shorter decay length into the vacuum gap. When shifted below the Fermi level of the counterelectrode, the state no longer even contributes to the tunnel current. A quantitative interpretation of the tunneling corrugation amplitudes found requires, however, a microscopic tunneling theory, which is still lacking.

A second point concerns the origin of the 7×7 structure. The adatom model with the lowest dangling-bond density is a $\sqrt{3}\times\sqrt{3}$ structure with no rest atoms. It appears that rest atoms are needed to provide (by inward relaxation) the charge transfer to the adatoms. However, if it were merely a matter of charge compensation, then a 2×2 reconstruction with equal number of adatoms and rest atoms would be most appealing. Therefore, we believe that the secret lies in the special nature of the corner, although other unit cells can be imagined with this key site. The smallest two, a $2\sqrt{3}\times 2\sqrt{3}$ and a 3×3, consist only of such key sites and have never been observed. The 5×5, also containing other sites, however, has been reported.[8] The unique character of the corner is not merely a matter of the three adjacent rest atoms. It is also the vertex of different kinds of regions. The 7×7 unit cell can be viewed as being composed of two equilateral triangles. The right one (see Fig. 3) contains practically unrelaxed rest atoms; in the other one the rest atoms are relaxed. These two kinds of triangles meet alternately at each corner. Preference for the 7×7 structure is likely the result of optimal balance of charge transfer and stress.

In spite of the various interesting open questions, we trust that this Letter has clarified the basic aspects of a very intriguing problem, and will contribute to a better understanding of Si surfaces.

We thank Hartwig Thomas for image processing, F. Himpsel for hints concerning sample preparation, and K. C. Pandey and K. H. Rieder for illuminating discussions on surface problems.

---

[1]Reviews of activities prior to 1980 are found in D. E. Eastman, J. Vac. Sci. Technol. 17, 492 (1980); D. J. Chadi, J. Phys. Soc. Jpn. 49, Suppl. A, 1035 (1980); D. J. Miller and D. Haneman, Surf. Sci. 104, L237 (1980).

[2]J. C. Phillips, Phys. Rev. Lett. 45, 905 (1980); R. J. Culbertson, L. C. Feldman, and P. J. Silverman, Phys. Rev. Lett. 45, 2043 (1980); G. Le Lay, Surf. Sci. 108, L429 (1981); C. Y. Su, P. R. Skeath, I. Lindau, and W. E. Spicer, Surf. Sci. 107, L355 (1981); M. J. Cardillo, Chem. Phys. 21, 54 (1982); N. Garcia, Solid State Commun. 40, 719 (1981); F. J. Himpsel, D. E. Eastman, P. Heimann, B. Reihl, C. W. White, and D. M. Zehner, Phys. Rev. B 24, 1120 (1980); J. Pollmann, Phys. Rev. Lett. 49, 1649 (1982); a trimer model has been proposed by F. J. Himpsel, to be published.

[3]G. Binnig, H. Rohrer, Ch. Gerber, and E. Weibel, Appl. Phys. Lett. 40, 178 (1982), and Phys. Rev. Lett. 49, 57 (1982); G. Binnig and H. Rohrer, to be published.

[4]F. J. Himpsel, private communication; see also P. A. Bennet and M. W. Webb, Surf. Sci. 104, 74 (1981).

[5]L. C. Snyder, Z. Wassermann, and J. W. Moskowitz, J. Vac. Sci. Technol. 16, 1266 (1979).

[6]W. A. Harrison, Surf. Sci. 55, 1 (1976).

[7]K. C. Pandey, Phys. Rev. Lett. 49, 223 (1982).

[8]J. Lander, in *Progress in Solid State Chemistry*, edited by H. Reiss (Pergamon, Oxford, 1965), Vol. 2, p. 26. This work is disputed; see F. Jona, IBM J. Res. Dev. 9, 357 (1965). On the other hand, Sn on Ge leads to both a 5×5 and a 7×7 structure: T. Ichikawa and S. Ino, Solid State Commun. 27, 483 (1978).

Anomalous Quantum Hall Effect: An Incompressible Quantum Fluid with Fractionally Charged Excitations

R. B. Laughlin

Lawrence Livermore National Laboratory, University of California, Livermore, California 94550

(Received 22 February 1983)

This Letter presents variational ground-state and excited-state wave functions which describe the condensation of a two-dimensional electron gas into a new state of matter.

PACS numbers: 71.45.Nt, 72.20.My, 73.40.Lq

The "$\frac{1}{3}$" effect, recently discovered by Tsui, Störmer, and Gossard,[1] results from the condensation of the two-dimensional electron gas in a GaAs-Ga_xAl_{1-x}As heterostructure into a new type of collective ground state. Important experimental facts are the following: (1) The electrons condense at a particular density, $\frac{1}{3}$ of a full Landau level. (2) They are capable of carrying electric current with little or no resistive loss and have a Hall conductance of $\frac{1}{3}e^2/h$. (3) Small deviations of the electron density do not affect either conductivity, but large ones do. (4) Condensation occurs at a temperature of ~ 1.0 K in a magnetic field of 150 kG. (5) The effect occurs in some samples but not in others. The purpose of this Letter is to report variational ground-state and excited-state wave functions that I feel are consistent with all the experimental facts and explain the effect. The ground state is a new state of matter, a quantum fluid the elementary excitations of which, the quasielectrons and quasiholes, are fractionally charged. I have verified the correctness of these wave functions for the case of small numbers of electrons, where direct numerical diagonalization of the many-body Hamiltonian is possible. I predict the existence of a sequence of these ground states, decreasing in density and terminating in a Wigner crystal.

Let us consider a two-dimensional electron gas in the x-y plane subjected to a magnetic field H_0 in the z direction. I adopt a symmetric gauge vector potential $\vec{A} = \frac{1}{2}H_0[x\hat{y} - y\hat{x}]$ and write the eigenstates of the ideal single-body Hamiltonian $H_{sp} = |(\hbar/i)\nabla - (e/c)\vec{A}|^2$ in the manner

$$|m,n\rangle = (2^{m+n+1}\pi m!\,n!)^{-1/2}\exp[\tfrac{1}{4}(x^2+y^2)]\left(\frac{\partial}{\partial x}+i\frac{\partial}{\partial y}\right)^m\left(\frac{\partial}{\partial x}-i\frac{\partial}{\partial y}\right)^n\exp[-\tfrac{1}{2}(x^2+y^2)], \quad (1)$$

with the cyclotron energy $\hbar\omega_c = \hbar(eH_0/mc)$ and the magnetic length $a_0 = (\hbar/m\omega_c)^{1/2} = (\hbar c/eH_0)^{1/2}$ set to 1. We have

$$H_{sp}|m,n\rangle = (n+\tfrac{1}{2})|m,n\rangle. \quad (2)$$

The manifold of states with energy $n+\frac{1}{2}$ constitutes the nth Landau level. I abbreviate the states of the lowest Landau level as

$$|m\rangle = (2^{m+1}\pi m!)^{-1/2} z^m \exp(-\tfrac{1}{4}|z|^2), \quad (3)$$

where $z = x+iy$. $|m\rangle$ is an eigenstate of angular momentum with eigenvalue m. The many-body Hamiltonian is

$$H = \sum_j\{|(\hbar/i)\nabla_j - (e/c)\vec{A}_j|^2 + V(z_j)\} + \sum_{j>k} e^2/|z_j - z_k|, \quad (4)$$

where j and k run over the N particles and V is a potential generated by a uniform neutralizing background.

I showed in a previous paper[2] that the $\frac{1}{3}$ effect could be understood in terms of the states in the lowest Landau level solely. With $e^2/a_0 \lesssim \hbar\omega_c$, the situation in the experiment, quantization of interelectronic spacing follows from quantization of angular momentum: The only wave functions composed of states in the lowest Landau level which describe orbiting with angular momentum m about the center of mass are of the form

$$\psi = (z_1 - z_2)^m (z_1 + z_2)^n \exp[-\tfrac{1}{4}(|z_1|^2 + |z_2|^2)]. \quad (5)$$

My present theory generalizes this observation to N particles.

I write the ground state as a product of Jastrow functions in the manner

$$\psi = \{\prod_{j<k} f(z_j - z_k)\}\exp(-\tfrac{1}{4}\textstyle\sum_l |z_l|^2), \quad (6)$$

and minimize the energy with respect to f. We

observe that the condition that the electrons lie in the lowest Landau level is that $f(z)$ be polynomial in z. The antisymmetry of ψ requires that f be odd. Conservation of angular momentum requires that $\prod_{j<k} f(z_j - z_k)$ be a homogeneous polynomial of degree M, where M is the total angular momentum. We have, therefore, $f(z) = z^m$, with m odd. To determine which m minimizes the energy, I write

$$|\psi_m|^2 = |\{\prod_{j<k}(z_j - z_k)^m\}\exp(-\tfrac{1}{4}\sum_l |z_l|^2)|^2 = e^{-\beta\Phi}, \tag{7}$$

where $\beta = 1/m$ and Φ is a classical potential energy given by

$$\Phi = -\sum_{j<k} 2m^2 \ln|z_j - z_k| + \tfrac{1}{2} m \sum_l |z_l|^2. \tag{8}$$

Φ describes a system of N identical particles of charge $Q = m$, interacting via logarithmic potentials and embedded in a uniform neutralizing background of charge density $\sigma = (2\pi a_0^2)^{-1}$. This is the classical one-component plasma (OCP), a system which has been studied in great detail. Monte Carlo calculations[3] have indicated that the OCP is a hexagonal crystal when the dimensionless plasma parameter $\Gamma = 2\beta Q^2 = 2m$ is greater than 140 and a fluid otherwise. $|\psi_m|^2$ describes a system uniformly expanded to a density of $\sigma_m = m^{-1}(2\pi a_0^2)^{-1}$. It minimizes the energy when σ_m equals the charge density generating V.

In Table I, I list the projection of ψ_m for three particles onto the lowest-energy eigenstate of angular momentum $3m$ calculated numerically. These are all nearly 1. This supports my assertion that a wave function of the form of Eq. (6) has adequate variational freedom. I have done a similar calculation for four particles with Coulombic repulsions and find projections of 0.979 and 0.947 for the $m = 3$ and $m = 5$ states.

ψ_m has a total energy per particle which for small m is more negative than that of a charge-density wave (CDW).[4] It is given in terms of the radial distribution function $g(r)$ of the OCP by

$$U_{\text{tot}} = \pi \int_0^\infty \frac{e^2}{r}[g(r) - 1]\, r\, dr. \tag{9}$$

In the limit of large Γ, U_{tot} is approximated

TABLE I. Projection of variational three-body wave functions ψ_m in the manner $\langle\psi_m|\Phi_m\rangle/(\langle\psi_m|\psi_m\rangle\langle\Phi_m|\times\Phi_m\rangle)^{1/2}$. Φ_m is the lowest-energy eigenstate of angular momentum $3m$ calculated with $V = 0$ and an interelectronic potential of either $1/r$, $-\ln(r)$, or $\exp(-r^2/2)$.

m	$1/r$	$-\ln(r)$	$\exp(-r^2/2)$
1	1	1	1
3	0.999 46	0.996 73	0.999 66
5	0.994 68	0.991 95	0.999 39
7	0.994 76	0.992 95	0.999 81
9	0.995 73	0.994 37	0.999 99
11	0.996 52	0.995 42	0.999 96
13	0.997 08	0.996 15	0.999 85

within a few percent by the ion disk energy:

$$U_{\text{tot}} \simeq -\sigma_m \int \frac{e^2}{|r|} d^2 r + \frac{\sigma_m^2}{2} \iint \frac{e^2}{|r_{12}|} dr_1^2\, dr_2^2 = (4/3\pi - 1) 2e^2/R, \tag{10}$$

where the integration domain is a disk of radius $R = (\pi\sigma_m)^{-1/2}$. At $\Gamma = 2$ we have the exact result[5] that $g(r) = 1 - \exp[-(r/R)^2]$, giving $U_{\text{tot}} = -\frac{1}{2}\pi^{1/2} e^2/R$. At $m = 3$ and $m = 5$ I have reproduced the Monte Carlo $g(r)$ of Caillol *et al.*[3] using the modified hypernetted chain technique described by them. I obtain $U_{\text{tot}} = (-0.4156 \pm 0.0012)e^2/a_0$ and $U_{\text{tot}}(5) = (-0.3340 \pm 0.0028)e^2/a_0$. The corresponding values for the charge-density wave[4] are $-0.389e^2/a_0$ and $-0.322e^2/a_0$. U_{tot} is a smooth function of Γ. I interpolate it crudely in the manner

$$U_{\text{tot}}(m) \simeq \frac{0.814}{\sqrt{m}}\left(\frac{0.230}{m^{0.64}} - 1\right)\frac{e^2}{a_0}. \tag{11}$$

This interpolation converges to the CDW energy near $m = 10$. The actual crystallization point cannot be determined from that of the OCP since the CDW has a lower energy than the crystal described by ψ_m for $m > 71$.

I generate the elementary excitations of ψ_m by piercing the fluid at z_0 with an infinitely thin solenoid and passing through it a flux quantum $\Delta\varphi = hc/e$ adiabatically. The effect of this operation on the single-body wave functions is

$$(z - z_0)^m \exp(-\tfrac{1}{4}|z|^2) \to (z - z_0)^{m+1}\exp(-\tfrac{1}{4}|z|^2). \tag{12}$$

Let us take as approximate representations of these excited states

$$\psi_m^{+z_0} = A_{z_0}\psi_m = \exp(-\tfrac{1}{4}\sum_l |z_l|^2)\{\prod_i (z_i - z_0)\}\{\prod_{j<k}(z_j - z_k)^m\}, \tag{13}$$

and

$$\psi_m^{-z_0} = A_{z_0}^{\dagger}\psi_m = \exp(-\tfrac{1}{4}\sum_l |z_l|^2)\left\{\prod_i\left(\frac{\partial}{\partial z_i} - \frac{z_0}{a_0^2}\right)\right\}\{\prod_{j<k}(z_j - z_k)^m\}, \tag{14}$$

for the quasihole and quasielectron, respectively. For four particles, I have projected these wave functions onto the analogous ones computed numerically. I obtain 0.998 for ψ_3^{-0} and 0.994 for ψ_5^{-0}. I obtain 0.982 for $\bar{\psi}_3^{+\bar{0}} = \{\prod_i (z_i - \bar{z})\}\psi_3$, which is ψ_3^{+0} with the center-of-mass motion removed.

These excitations are particles of charge $1/m$. To see this let us write $|\psi^{+z_0}|^2$ as $e^{-\beta\Phi'}$, with $\beta = 1/m$ and

$$\Phi' = \Phi - 2\sum_i \ln|z_i - z_0|. \tag{15}$$

Φ' describes an OCP interacting with a phantom point charge at z_0. The plasma will completely screen this phantom by accumulating an equal and opposite charge near z_0. However, since the plasma in reality consists of particles of charge 1 rather than charge m, the real accumulated charge is $1/m$. Similar reasoning applies to ψ^{-z_0} if we approximate it as $\prod_j (z_j - z_0)^{-1} P_{z_0}\psi_3$, where P_{z_0} is a projection operator removing all configurations in which any electron is in the single-body state $(z - z_0)^0 \exp(-\frac{1}{4}|z|^2)$. The projection of this approximate wave function onto $\psi_3^{-z_0}$ for four particles is 0.922. More generally, one observes that far away from the solenoid, adiabatic addition of $\Delta\varphi$ moves the fluid rigidly by exactly one state, per Eq. (12). The charge of the particles is thus $1/m$ by the Schrieffer counting argument.[6]

The size of these particles is the distance over which the OCP screens. Were the plasma weakly coupled ($\Gamma \leq 2$) this would be the Debye length $\lambda_D = a_0/\sqrt{2}$. For the strongly coupled plasma, a better estimate is the ion-disk radius associated with a charge of $1/m$: $R = \sqrt{2}\, a_0$. From the size we can estimate the energy required to make a particle. The charge accumulated around the phantom in the Debye-Hückel approximation is

$$\delta\rho = \frac{e/m}{2\pi\lambda_D^2} K_0(r/\lambda_D),$$

where K_0 is a modified Bessel function of the second kind. The energy required to accumulate it is

$$\Delta_{\text{Debye}} = \tfrac{1}{2}\iint \frac{\delta\rho\,\delta\rho}{|r_{12}|} = \frac{\pi}{4\sqrt{2}} \frac{1}{m^2} \frac{e^2}{a_0}. \tag{16}$$

This estimate is an upper bound, since the plasma is strongly coupled. To make a better estimate let $\delta\rho = \sigma_m$ inside the ion disk and zero outside, to obtain

$$\Delta_{\text{disk}} = \frac{3}{2\sqrt{2}\pi} \frac{1}{m^2} \frac{e^2}{a_0}. \tag{17}$$

For $m = 3$, these estimates are $0.062e^2/a_0$ and $0.038e^2/a_0$. This compares well with the value $0.033e^2/a_0$ estimated from the numerical four-particle solution in the manner

$$\Delta \simeq \tfrac{1}{2}\{E(\psi_3^{-0}) + E(\bar{\psi}_3^{+\bar{0}}) - 2E(\psi_3)\}, \tag{18}$$

where $E(\psi_3)$ denotes the eigenvalue of the numerical analog of ψ_3. This expression averages the electron and hole creation energies while subtracting off the error due to the absence of V. I have performed two-component hypernetted chain calculations for the energies of $\psi_3^{+z_0}$ and $\psi_3^{-z_0}$. I obtain $(0.022 \pm 0.002)e^2/a_0$ and $(0.025 \pm 0.005)e^2/a_0$. If we assume a value $\epsilon = 13$ for the dielectric constant of GaAs, we obtain $0.02e^2/\epsilon a_0 \simeq 4$ K when $H_0 = 150$ kG.

The energy to make a particle does not depend on z_0, so long as its distance from the boundary is greater than its size. Thus, as in the single-particle problem, the states are degenerate and there is no kinetic energy. We can expand the creation operator as a power series in z_0:

$$A_{z_0} = \sum_{j=0}^{N} A_j(z_1 \cdots, z_N) z_0^{N-j}. \tag{19}$$

These A_j are the elementary symmetric polynomials,[7] the algebra of which is known to span the set of symmetric functions. Since every antisymmetric function can be written as a symmetric function times ψ_1, these operators and their adjoints generate the entire state space. It is thus appropriate to consider them N linearly independent particle creation operators.

The state described by ψ_m is incompressible because compressing or expanding it is tantamount to injecting particles. If the area of the system is reduced or increased by δA the energy rises by $\delta U = \sigma_m \Delta|\delta A|$. Were this an elastic solid characterized by a bulk modulus B, we would have $\delta U = \frac{1}{2}B(\delta A)^2/A$. Incompressibility causes the longitudinal collective excitation roughly equivalent to a compressional sound wave to be absent, or more precisely, to have an energy $\sim\Delta$ in the long-wavelength limit. This facilitates current conduction with no resistive loss at zero temperature. Our prototype for this behavior is full Landau level ($m = 1$) for which this collective excitation occurs at $\hbar\omega_c$. The response of this system to compressive stresses is analogous to the response of a type-II superconductor to the application of a magnetic field. The system first generates Hall currents without compressing, and then at a critical stress collapses by an area quantum $m2\pi a_0^2$

and nucleates a particle. This, like a flux line, is surrounded by a vortex of Hall current rotating in a sense opposite to that induced by the stress.

The role of sample impurities and inhomogeneities in this theory is the same as that in my theory of the ordinary quantum Hall effect.[8] The electron and hole bands, separated in the impurity-free case by a gap 2Δ, are broadened into a continuum consisting of two bands of extended states separated by a band of localized ones. Small variations of the electron density move the Fermi level within this localized state band as the extra quasiparticles become trapped at impurity sites. The Hall conductance is $(1/m)\times(e^2/h)$ because it is related by gauge invariance to the charge of the quasiparticles e^* by $\sigma_{\rm Hall}=e^*e/h$, whenever the Fermi level lies in a localized state band. As in the ordinary quantum Hall effect, disorder sufficient to localize all the states destroys the effect. This occurs when the collision time τ in the sample in the absence of a magnetic field becomes smaller than $\tau<\hbar/\Delta$.

I wish to thank H. DeWitt for calling my attention to the Monte Carlo work and D. Boercker for helpful discussions. I also wish to thank P. A. Lee, D. Yoshioka, and B. I. Halperin for helpful criticism. This work was performed under the auspices of the U. S. Department of Energy by Lawrence Livermore National Laboratory under Contract No. W-7405-Eng-48.

---

[1]D. C. Tsui, H. L. Störmer, and A. C. Gossard, Phys. Rev. Lett. 48, 1559 (1982).

[2]R. B. Laughlin, Phys. Rev. B 27, 3383 (1983).

[3]J. M. Caillol, D. Levesque, J. J. Weis, and J. P. Hansen, J. Stat. Phys. 28, 325 (1982).

[4]D. Yoshioka and H. Fukuyama, J. Phys. Soc. Jpn. 47, 394 (1979); D. Yoshioka and P. A. Lee, Phys. Rev. B 27, 4986 (1983), and private communication.

[5]B. Jancovici, Phys. Rev. Lett. 46, 386 (1981). $\Gamma=2$ corresponds to a full Landau level, for which the total energy equals the Hartree-Fock energy $-\sqrt{\pi/8}\,e^2/a_0$. This correspondence may be viewed as the underlying reason an exact solution at $\Gamma=2$ exists.

[6]W. P. Su and J. R. Schrieffer, Phys. Rev. Lett. 46, 738 (1981).

[7]S. Lang, *Algebra* (Addison-Wesley, Reading, Mass., 1965), p. 132.

[8]R. B. Laughlin, Phys. Rev. B 23, 5632 (1981).

Metallic Phase with Long-Range Orientational Order and No Translational Symmetry

D. Shechtman and I. Blech
Department of Materials Engineering, Israel Institute of Technology–Technion, 3200 Haifa, Israel

and

D. Gratias
Centre d'Etudes de Chimie Métallurgique, Centre National de la Recherche Scientifique, F-94400 Vitry, France

and

J. W. Cahn
Center for Materials Science, National Bureau of Standards, Gaithersburg, Maryland 20760
(Received 9 October 1984)

We have observed a metallic solid (Al–14-at.%-Mn) with long-range orientational order, but with icosahedral point group symmetry, which is inconsistent with lattice translations. Its diffraction spots are as sharp as those of crystals but cannot be indexed to any Bravais lattice. The solid is metastable and forms from the melt by a first-order transition.

PACS numbers: 61.50.Em, 61.55.Hg, 64.70.Ew

We report herein the existence of a metallic solid which diffracts electrons like a single crystal but has point group symmetery $m\overline{3}\overline{5}$ (icosahedral) which is inconsistent with lattice translations. If the specimen is rotated through the angles of this point group (Fig. 1), selected-area electron diffraction patterns clearly display the six fivefold, ten threefold, and fifteen twofold axes characteristic[1] of icosahedral symmetry (Fig. 2). Grains up to 2 μm in size with this structure form in rapidly cooled alloys of Al with 10–14 at.% Mn, Fe, or Cr. We will refer to the phase as the icosahedral phase. Microdiffraction from many different volume elements of a grain and dark-field imaging from various diffraction spots confirm that entire grains have long-range orientational order. If the orientational order decays with distance, its correlation length is far greater than the grain size. We have thus a solid metallic phase with no translational order and with with long-range orientational order.

The remarkable sharpness of the diffraction spots (Fig. 2) indicates a high coherency in the spatial interference, comparable to the one usually encountered in crystals. The diffraction data are qualitatively well fitted by a model consisting of a random packing of nonoverlapping parallel icosahedra attached by edges.[2] The invariance of the local orientational symmetry from site to site and the finite number of possible translations between two adjacent icosahedra seem to be sufficient for insuring highly coherent interferences. Icosahedra are a common packing unit in intermetallic crystals with the smaller transition element at the center surrounded by twelve larger atoms arranged like the corners of an icosahedron.[3] The symmetries of the crystals dictate that the several icosahedra in a unit cell have different orientations and allow them to be distorted, leaving the overall crystal consistent with the well-known crystallographic point and space groups. Even though icosahedral symmetry is of great importance as an approximate site symmetry in crystals, it cannot survive the imposition of lattice translations: Crystals cannot and do not exhibit the icosahedral point group symmetry.

Elementary crystallography indicates that fivefold

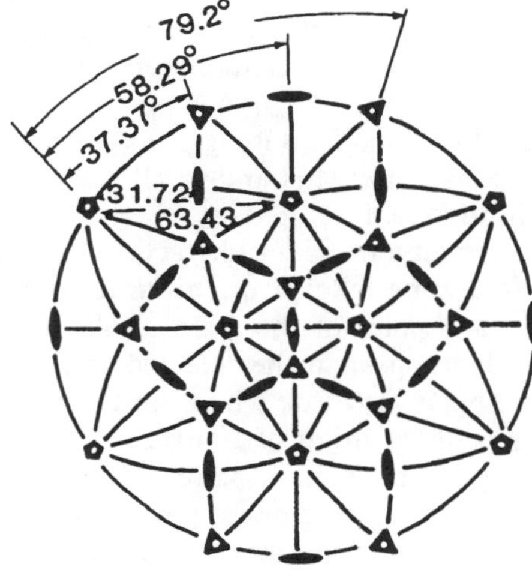

FIG. 1. Stereographic projection of the symmetry elements of the icosahedral group $m\overline{3}\overline{5}$.

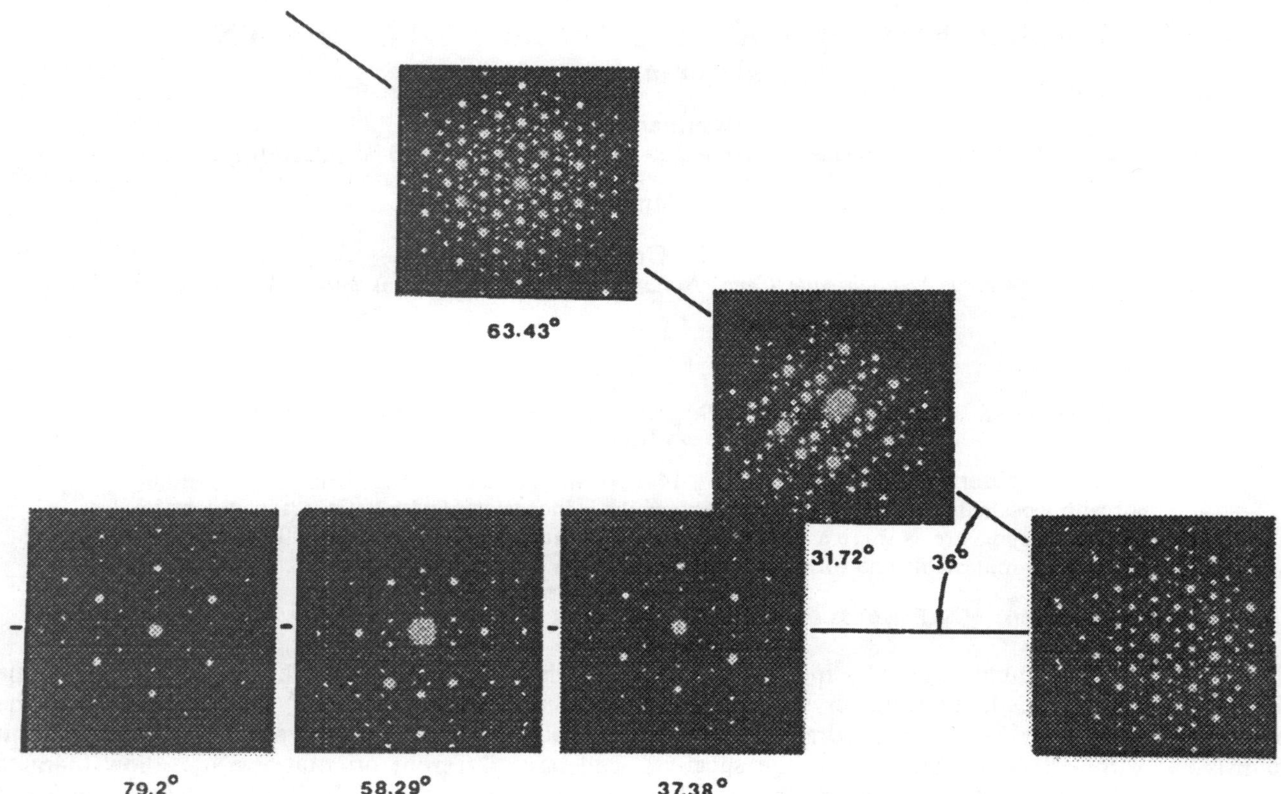

FIG. 2. Selected-area electron diffraction patterns taken from a single grain of the icosahedral phase. Rotations match those in Fig. 1.

axes are inconsistent with translational order.[4] Crystals with an apparent fivefold axis do occur as multiple twins.[5] Icosahedral symmetry could conceivably occur by multiple twinning, but at least five twins are required in order to obtain an icosahedral diffraction set from a conventional crystal. Experiments in the electron microscope and with x-ray diffraction contradict the twin hypothesis:

(1) A set of dark-field images taken from any reflection reveals no twins and in all cases the whole grain is illuminated.

(2) A convergent-beam diffraction pattern along the fivefold zone taken from an area 20 nm in diameter at any point of the grain displays all the reflections that appear in the selected-area diffraction pattern. This is also true when the thickness of the specimen is of the order of 10 nm.

(3) An x-ray diffraction pattern (Cu $K\alpha$ source) was taken from a single-phase sample of the material containing many grains of various orientations. Had the phase consisted of a multiply twinned crystalline structure, it should have been possible to index the powder pattern regardless of the twins. The pattern obtained from the icosahedral phase could not be indexed to any Bravais lattice.

On the basis of these experiments we conclude that the icosahedral phase does not consist of multiply twinned regular crystal structures.

The icosahedral phase forms during rapid cooling of the melt by a nucleation and growth mechanism. This mechanism is characteristic of a first-order transition because the two phases coexist along a moving interface. Each particle nucleates at a center and grows out from there. The atomic rearrangements that result in the orientational order of the icosahedral phase occur at this interface, and the two adjoining phases differ in entropy and, for some alloys, in composition. If the transition were higher order, ordering would occur everywhere in the liquid instead of being confined to interfaces. Our evidence for the first-order character is morphological. We examined alloys with 10–14 at.% Mn. Samples with 10% to 12% Mn showed many nodular grains separated from each other by crystalline films of fcc Al. The grains were approximately spherical in shape but were deeply indented with radial streaks composed of fcc Al crystals. The morphology is similar to a commonly observed one in rapid solidification in which crystals nucleate at many centers and grow until the remaining liquid solidifies, except that instead of isolated crystals we have isolated icosahedral grains embedded in solidified Al.

Another aspect of the first-order character of the transition is that segregation accompanies the growth of the icosahedral phase. In the 10%- and 12%-Mn samples the growing phase rejects Al into the liquid. The indentations in the nodules are characteristic of the cellular morphology caused by an instability in the diffusion layer surrounding the solid resulting from segregation during growth.[6] In the 14%-Mn alloys the icosahedral grains occupy almost all of the volume of the specimen. Only small amounts of fcc Al occur along some grain boundaries where the grains grew to impingement. Some radial streaking still points to the nucleation centers from which the grains grew. We conclude that little segregation occurs in this alloy and that the icosahedral phase contains 14% Mn.

The obvious cellular morphology indicates that the growth of the icosahedral phase is slow enough to permit diffusional segregation on the scale of 1 μm. With the diffusion coefficient in the liqud of order 10^{-8} m^2/s this indicates interface velocities of order 10^{-2} m/s and formation times of 10^{-4} s. This is 2 to 3 orders of magnitude slower than the fastest known crystallization velocities of a metallic liquid during rapid cooling, but it is a typical upper limit of velocity of crystallization with composition changes.[7] There is thus plenty of time for atomic rearrangements to occur at the interface between the melt and the icosahedral phase.

The icosahedral phase in rapidly solidified Al-Mn alloys is remarkably resistant to crystallization. Prolonged heating of 6 h at 300 °C and 1 h at 350 °C produced no detectable crystallization, but 1 h at 400 °C caused crystallization to the stable Al_6Mn phase. We conclude that the icosahedral phase is a truly metastable phase which nucleates and grows for a range of cooling rates which are slow enough to permit its formation but rapid enough to prevent crystallization, either from the melt or from the icosahedral phase after its formation.

The icosahedral phase has symmetries intermediate between those of a crystal and a liquid. It differs from other intermediate phases in that it is both solid, like a metallic glass, and that it has long-range orientational order. Many intermediate phases do have orientational order, but usually it is only local, and the transition to such phases is continuous.[8,9] The possibility of an icosahedral phase with long range order was inferred from a computer simulation,[10,11] and a first-order liquid-to-icosahedral phase transition has been predicted from a mean-field theory.[11,12]

We thank F. S. Biancaniello for spinning the alloys, C. R. Hubbard for x-ray experiments, and B. Burton for a critical review of the manuscript. This work was sponsored by the Defense Advanced Research Projects Agency and the National Science Foundation. This work was performed while one of us (D.S.) was a guest worker at the National Bureau of Standards and while another of us (D.G.) was at the Institute for Theoretical Physics, University of California at Santa Barbara.

[1] *International Tables for Crystallography* (Reidel, Higham, Mass., 1983), Vol. A, p. 179.

[2] D. Shechtman and I. Blech, to be published.

[3] F. C. Frank and J. S. Kasper, Acta Crystallogr. **11**, 184 (1958), and **12**, 483 (1959).

[4] M. J. Buerger, *Elementary Crystallography* (MIT Press, Cambridge, 1978).

[5] G. Friedel, *Lecon de Crystallographie* (Masson, Paris, 1921).

[6] W. W. Mullins and R. F. Sekerka, J. Appl. Phys. **35**, 444 (1963).

[7] W. J. Boettinger, F. S. Biancaniello, G. Kalonji, and J. W. Cahn, in *Rapid Solidification Processing II*, edited by R. Mehrabian *et al.* (Claitors, Baton Rouge, 1980), p. 50.

[8] D. R. Nelson, Phys. Rev. B **28**, 5515 (1983).

[9] J. Sadoc, J. Phys. (Paris), Colloq. **41**, C8-326 (1980).

[10] P. J. Steinhardt, D. R. Nelson and M. Rouchetti, Phys. Rev. Lett. **47**, 1297 (1981).

[11] P. J. Steinhardt, D. R. Nelson and M. Rouchetti, Phys. Rev. B **28**, 784 (1983).

[12] A. D. J. Haymet, Phys. Rev. B **27**, 1725 (1983).

Chapter 10
PLASMA PHYSICS
MARSHALL N. ROSENBLUTH

Introduction

Papers Reprinted in Book

Marshall N. Rosenbluth (1927-). (Courtesy of *Physics Today* Collection, AIP Emilio Segrè Visual Archives.)

Marshall N. Rosenbluth, a theoretical plasma physicist, is now a Professor Emeritus at the University of California, San Diego, working as a Joint Central Team scientist at the International Thermonuclear Experimental Reactor (ITER). He has also worked at Stanford, Los Alamos, General Atomics, the Institute for Advanced Study, and the University of Texas. Honors include the Lawrence Prize, the Einstein Prize, and the Enrico Fermi Award.

Harold P. Furth (1930–). (Courtesy of Carol A. Phillips, Plasma Physics Laboratory, Princeton University.)

Bruno Coppi (1935-), George Bekefi (1925-) and **Abraham Bers (1930-)** taken in front of MIT's Building 20, where radar was developed in World War II. (Photograph by John F. Cook. Courtesy of MIT/RLE.)

Like many of the disciplines represented in this volume, plasma physics did not exist until about 65 years ago, although it is of course firmly rooted in the classical disciplines of the 19th century — electromagnetic theory, classical mechanics, and kinetic theory. Indeed, it not only represents a synthesis of these disciplines, but has ramifications for so many areas of physics and technology as to present serious problems in selecting highlights from *The Physical Review* and *Physical Review Letters*.

Since 1960 most U.S. fusion and fluid plasma physics articles have been published in *Physics of Fluids* or *Nuclear Fusion* as the journals of choice, space and geophysics in *Journal of Geophysical Research*, astrophysics in *Astrophysical Journal*, and many applied topics in *Journal of Applied Physics*. Hence, as can be seen from the printed selections, what remains is primarily highlights and first accomplishments suitable for publication in *Physical Review Letters*. Let us thank the editors and reviewers for enforcing the admonishments to keep the *Letters* to important accomplishments of general interest presented in a comprehensible and jargon-free manner. In view of the very limited space available in the book, I have decided to relegate the excellent longer, and usually older, articles to the CD in order that the book convey a sense of the rich diversity of research in plasma physics.

The field of plasma physics really originated with the discovery of plasma oscillations,[1] at which time its name was coined to describe the nature of a complex, jiggling medium, prone to waves and instabilities. Shortly thereafter it was natural to consider tam-

Henry N. Russell (1887-1957) and **Lyman Spitzer (1919-).** (Courtesy of Edmondson Collection, AIP Emilio Segrè Visual Archives.)

ing the medium with magnetic fields.[2] This has obviously been a continuing theme of much of fusion research as well as astrophysical and geophysical plasma research.

In order to have a framework for calculation of the simplest behavior, it was required to develop a classical transport theory,[3,4] which could be done fairly completely in the weak-plasma limit–scattering mean free path long compared to the Debye-shielding distance, or equivalently, interparticle potential energy much less than kT. Extensions to the strong coupling limit[5] are not so easy, and even the statistical mechanics, when quantum effects are considered,[6] is an incomplete topic requiring extensive computation.

One of the themes running through much of plasma research is the dominant constraint that in the ideal limit—small gyroradius and no dissipation—plasma remains perfectly frozen to magnetic-field lines. If this constraint were exact, plasma physics would be quite dull, controlled fusion would be easy, and there would be no aurora or even perhaps cosmic rays. In the early 1960s the discovery of "tearing" modes provided a basis for understanding one way this constraint could be violated in singular boundary layers. Many applications to geophysics[7] emerged, and the original fluid-type theory was extended to take into account complex kinetic[8] structures and the chaos-like effects that could occur when the reconnection led to magnetic-field lines becoming tangled.[9] Practical applications[10] to the understanding of the behavior of magnetically confined plasmas are critical.

While the plasma medium is, unfortunately, usually turbulent and complex, clever experiments under quiescent conditions were nonetheless able to verify much underlying plasma theory.[11] On the other hand, the necessity of understanding nonlinear behavior[12] has made plasma physics a key contributor to modern nonlinear physics. Only a beginning has been made in understand-

Melvin B. Gottlieb (1917–). (Courtesy of Carol A. Phillips, Plasma Physics Laboratory, Princeton University.)

Lewi Tonks (1897–1971). (Courtesy of Tonks Collection, AIP Emilio Segrè Visual Archives.)

ing the complex and transient structures[13] that can evolve from the equations of plasma physics.

Of course the bulk of plasma physics research in the last 40 years, at least to judge from the numbers of scientists involved and papers published, has been devoted to the control of fusion energy to serve as an inexhaustible energy source. Much of the magnetic fusion effort was classified in its early years (1945–1958), and much of inertial (laser) fusion remains classified up to now. However, some fascinating experiments and theory[14] concerning the interactions of plasmas with intense lasers are in the public domain.

Most magnetic fusion research has been devoted to improving understanding and performance of what is arguably the simplest, and certainly by far the most completely explored, confinement scheme, the Tokamak. However, a number of interesting propos-

Livermore fusion program Nova laser. (Courtesy of University of California, Lawrence Livermore National Laboratory.)

als[15,16] based on ingenious, but necessarily oversimplified, physics continue to be made for alternative confinement concepts, which deserve to be explored further.

In Tokamak research, improvements in technology, increases in size, and a wealth of semiempirical understanding have led to accomplishments[17] over the years such that at present there is little doubt that a 1-GW-sized reactor can be built, although many optimization, cost, reliability, and environmental issues remain open. While the large-scale physics of Tokamaks is reasonably well understood, discovering the detailed nature of the microturbulence leading to anomalous leakage of energy and fuel remains a principal focus of scientific research. It is clear that such research,[18] characterized by physical intuition, lies between experiment and nonlinear theory, while increasing use of computers is required for understanding and, eventually, better control of thermonuclear plasmas.

One can safely predict that plasma physics will remain a vital discipline for many decades, perhaps for the next 100 years. Fundamental issues such as the Earth's dynamo and the effects of magnetic fields on cosmological development remain open. Plasma-material interfaces are of increasing technological importance. The detailed predictive understanding of the turbulence that underlies these applications as well as the behavior of fusion plasmas are intellectual challenges of great importance in physics.

Martin D. Kruskal (1925–). (Courtesy of Princeton University, AIP Emilio Segrè Visual Archives.)

REFERENCES

1. L. Tonks and I. Langmuir, *Oscillations in Ionized Gases*, Phys. Rev. **33**, 195–210 (1929).
2. W. H. Bennett, *Magnetically Self-Focussing Streams*, Phys. Rev. **45**, 890–897 (1934).
3. L. Spitzer and R. Härm, *Transport Phenomena in a Completely Ionized Gas*, Phys. Rev. **89**, 977–981 (1953).
4. M.N. Rosenbluth, W.M. MacDonald, and D.L. Judd, *Fokker-Planck Equation for an Inverse-Square Force*, Phys. Rev. **107**, 1–6 (1957).
5. S. Ichimaru, S. Tanaka, and H. Iyetomi, *Screening Potential and Enhancement of the Thermonuclear Reaction Rate in Dense Plasmas*, Phys. Rev. A **29**, 2033–2035 (1984).
6. D.M. Ceperley and B.J. Alder, *Ground State of the Electron Gas by a Stochastic Method*, Phys. Rev. Lett. **45**, 566–569 (1980).
7. B. Coppi, G. Laval, and R. Pellat, *Dynamics of the Geomagnetic Tail*, Phys. Rev. Lett. **16**, 1207–1210 (1966).
8. S.M. Mahajan, R.D. Hazeltine, H.R. Strauss, and D.W. Ross, *Collisionless "Current-Channel" Tearing Modes*, Phys. Rev. Lett. **41**, 1375–1378 (1979).
9. A.B. Rechester and T.H. Stix, *Magnetic Braiding Due to Weak Asymmetry*, Phys. Rev. Lett. **36**, 587–591 (1976).
10. A.H. Glasser, H.P. Furth, and P.H. Rutherford, *Stabilization of Resistive Kink Modes in the Tokamak*, Phys. Rev. Lett. **38**, 234–237 (1977).
11. R.W. Gould, T.M. O'Neil, and J.H. Malmberg, *Plasma Wave Echo*, Phys. Rev. Lett. **19**, 219–222 (1967).
12. C.S. Gardner, J.M. Greene, M.D. Kruskal, and R.M. Miura, *Method for Solving the Korteweg—de Vries Equation*, Phys. Rev. Lett. **19**, 1095–1097 (1967).

Tihiro Ohkawa (1928–). (Courtesy of Douglas M. Fouquet, Fusion Research, General Atomics.)

John M. Dawson (1930–). (Photograph by UCLA Campus Studio. Courtesy of John M. Dawson, Computational Plasma Physics.)

13. R.H. Berman, D.J. Tetreault, T.H. Dupree, and T. Boutros-Ghali, *Computer Simulation of Nonlinear Ion-Electron Instability*, Phys. Rev. Lett. **48**, 1249–1252 (1983).

14. K. Estabrook and W.L. Kruer, *Threshold of Convective Raman Backscattering: Effects of* ∇n, *Thomson Scattering, Noise, Collisions and Landau Damping*, Phys. Rev. Lett. **53**, 465–468 (1984).

15. D.E. Baldwin and B.G. Logan, *Improved Tandem Mirror Fusion Reactor*, Phys. Rev. Lett. **43**, 1318–1321 (1979).

16. R. Sudan and P. Kaw, *Stabilizing Effect of Finite-Gyroradius Beam Particles on the Tilting Mode of Spheromaks*, Phys. Rev. Lett. **47**, 575–578 (1981).

17. H. Eubank, R. Goldston, *et al.*, *Neutral-Beam-Heating Results from the Princeton Large Torus*, Phys. Rev. Lett. **43**, 270–274 (1978).

18. B.A. Carreras, P.H. Diamond, M. Murakami, J.L. Dunlap, J.D. Bell, H.R. Hicks, J.A. Holmes, E.A. La Zarus, V.K. Paré, P. Similon, C.E. Thomas, and R. M. Wieland, *Transport Effects Induced by Resistive Ballooning Modes and Comparison with High-*β_p *ISX-B Tokamak Confinement*, Phys. Rev. Lett. **50**, 503–506 (1983).

Donald W. Kerst (1911–1993). (Fusion Research, General Atomics and Plasma Physics, University of Wisconsin. Courtesy of Ken Maas.)

Sydney Chapman (1880–1970) and **Harold Grad (1923–1987).** (Courtesy of Eli Aaron, AIP Emilio Segrè Visual Archives.)

PLASMA PHYSICS

Papers Reproduced on CD-ROM

L. Tonks and I. Langmuir. Oscillations in ionized gases, *Phys. Rev.* **33**, 195–210 (1929)

W. H. Bennett. Magnetically self-focussing streams, *Phys. Rev.* **45**, 890–897 (1934)

T. Holstein. Imprisonment of resonance radiation in gases, *Phys. Rev.* **72**, 1212–1233 (1947)

D. Bohm and E. P. Gross. Theory of plasma oscillations. A. Origin of medium-like behavior, *Phys. Rev.* **75**, 1851–1876 (1949)

P. W. Anderson. Pressure broadening in the microwave and infra-red regions, *Phys. Rev.* **76**, 647–661 (1949)

R. S. Cohen, L. Spitzer, Jr. and P. McR. Routly. The electrical conductivity of an ionized gas, *Phys. Rev.* **80**, 230–238 (1950)

D. Pines and D. Bohm. A collective description of electron interactions: II. Collective *vs* individual particle aspects of the interactions, *Phys. Rev.* **85**, 338–353 (1952)

L. Spitzer, Jr. and R. Härm. Transport phenomena in a completely ionized gas, *Phys. Rev.* **89**, 977–981 (1953)

T. H. Stix. Oscillations of a cylindrical plasma, *Phys. Rev.* **106**, 1146–1150 (1957)

M. N. Rosenbluth, W. M. MacDonald, and D. L. Judd. Fokker–Planck equation for an inverse-square force, *Phys. Rev.* **107**, 1–6 (1957)

I. B. Bernstein, J. M. Greene, and M. D. Kruskal. Exact nonlinear plasma oscillations, *Phys. Rev.* **108**, 546–550 (1957)

I. B. Bernstein. Waves in a plasma in a magnetic field, *Phys. Rev.* **109**, 10–21 (1958)

O. Buneman. Instability, turbulence, and conductivity in current-carrying plasma, *Phys. Rev. Lett.* **1**, 8–11, 119(E) (1958)

T. G. Northrop and E. Teller. Stability of the adiabatic motion of charged particles in the Earth's field, *Phys. Rev.* **117**, 215–225 (1960)

H. Dreicer. Electron velocity distributions in a partially ionized gas, *Phys. Rev.* **117**, 343–354 (1960)

E. E. Salpeter. Electron density fluctuations in a plasma, *Phys. Rev.* **120**, 1528–1535 (1960)

R. F. Post, R. E. Ellis, F. C. Ford, and M. N. Rosenbluth. Stable confinement of a high-temperature plasma, *Phys. Rev. Lett.* **4**, 166–170 (1960)

R. Bowers, C. Legendy, and F. Rose. Oscillatory galvanomagnetic effect in metallic sodium, *Phys. Rev. Lett.* **7**, 339–341 (1961)

H. R. Griem. Theory of wing broadening of the hydrogen Lyman-α line by electrons and ions in a plasma, *Phys. Rev.* **140**, A1140–A1154 (1965)

T. H. Stix. Radiation and absorption via mode conversion in an inhomogeneous collision-free plasma, *Phys. Rev. Lett.* **15**, 878–882 (1965)

B. Coppi, G. Laval, and R. Pellat. Dynamics of the geomagnetic tail, *Phys. Rev. Lett.* **16**, 1207–1210 (1966)

R. W. Gould, T. M. O'Neil, and J. H. Malmberg. Plasma wave echo, *Phys. Rev. Lett.* **19**, 219–222 (1967)

C. S. Gardner, J. M. Greene, M. D. Kruskal, and R. M. Miura. Method for solving the Korteweg–deVries equation, *Phys. Rev. Lett.* **19**, 1095–1097 (1967)

D. L. Albritton, T. M. Miller, D. W. Martin, and E. W. McDaniel. Mobilities of mass-identified H_3^+ and H^+ ions in hydrogen, *Phys. Rev.* **171**, 94–102 (1968)

F. H. Coensgen, W. F. Cummins, B. G. Logan, A. W. Molvik, W. E. Nexsen, T. C. Simonen, B. W. Stallard, and W. C. Turner. Stabilization of a neutral-beam–sustained, mirror-confined plasma, *Phys. Rev. Lett.* **35**, 1501–1503 (1975)

A. B. Rechester and T. H. Stix. Magnetic braiding due to weak asymmetry, *Phys. Rev. Lett.* **36**, 587–591 (1976)

A. H. Glasser, H. P. Furth, and P. H. Rutherford. Stabilization of resistive kink modes in the tokamak, *Phys. Rev. Lett.* **38**, 234–237 (1977)

B. Coppi. Topology of ballooning modes, *Phys. Rev. Lett.* **39**, 939–942 (1977)

N. J. Fisch. Confining a tokamak plasma with rf-driven currents, *Phys. Rev. Lett.* **41**, 873–876 (1978)

S. M. Mahajan, R. D. Hazeltine, H. R. Strauss, and D. W. Ross. Collisionless "current-channel" tearing modes, *Phys. Rev. Lett.* **41**, 1375–1378 (1979)

A. B. Rechester, M. N. Rosenbluth, and R. B. White. Calculation of the Kolmogorov entropy for motion along a stochastic magnetic field, *Phys. Rev. Lett.* **42**, 1247–1250 (1979)

H. Eubank, R. Goldston, V. Arunasalam, M. Bitter, K. Bol, D. Boyd, N. Bretz, J.-P. Bussac, S. Cohen, P. Colestock, S. Davis, D. Dimock, H. Dylla, P. Efthimion, L. Grisham, R. Hawryluk, K. Hill, E. Hinnov, J. Hosea, H. Hsuan, D. Johnson, G. Martin, S. Medley, E. Meservey, N. Sauthoff, G. Schilling, J. Schivell, G. Schmidt, F. Stauffer, L. Stewart, W. Stodiek, R. Stooksberry, J. Strachan, S. Suckewer, H. Takahashi, G. Tait, M. Ulrickson, S. von Goeler, and M. Yamada. Neutral-beam—Heating results from the Princeton Large Torus, *Phys. Rev. Lett.* **43**, 270–274 (1979)

D. E. Baldwin and B. G. Logan. Improved tandem mirror fusion reactor, *Phys. Rev. Lett.* **43**, 1318–1321 (1979)

D. M. Ceperley and B. J. Alder. Ground state of the electron gas by a stochastic method, *Phys. Rev. Lett.* **45**, 566–569 (1980)

R. L. McCrory, L. Montierth, R. L. Morse, and C. P. Verdon. Nonlinear evolution of ablation-driven Rayleigh–Taylor instability, *Phys. Rev. Lett.* **46**, 336–339 (1981)

R. N. Sudan and P. K. Kaw. Stabilizing effect of finite-gyroradius beam particles on the tilting mode of spheromaks, *Phys. Rev. Lett.* **47**, 575–578 (1981)

PLASMA PHYSICS

A. H. Boozer and R. B. White. Particle diffusion in tokamaks with partially destroyed magnetic surfaces, *Phys. Rev. Lett.* **49**, 786–789 (1982)

R. H. Berman, D. J. Tetreault, T. H. Dupree, and T. Boutros-Ghali. Computer simulation of nonlinear ion-electron instability, *Phys. Rev. Lett.* **48**, 1249–1252 (1983)

B. A. Carreras, P. H. Diamond, M. Murakami, J. L. Dunlap, J. D. Bell, H. R. Hicks, J. A. Holmes, E. A. Lazarus, V. K. Paré, P. Similon, C. E. Thomas, and R. M. Wieland. Transport effects induced by resistive ballooning modes and comparison with high-β_p ISX-*B* tokamak confinement, *Phys. Rev. Lett.* **50**, 503–506 (1983)

S. Ichimaru, S. Tanaka, and H. Iyetomi. Screening potential and enhancement of the thermonuclear reaction rate in dense plasmas, *Phys. Rev. A* **29**, 2033–2035 (1984)

M. C. Zarnstorff and S. C. Prager. Experimental observation of neoclassical currents in a plasma, *Phys. Rev. Lett.* **53**, 454–457 (1984)

K. Estabrook and W. L. Kruer. Thresholds of convective Raman backscattering: Effects of ∇n, Thomson scattering, noise, collisions, and Landau damping, *Phys. Rev. Lett.* **53**, 465–468 (1984)

E. Ott, T. M. Antonsen Jr., and J. D. Hanson. Effect of noise on time-dependent quantum chaos, *Phys. Rev. Lett.* **53**, 2187–2190 (1984)

D. W. Forslund, J. M. Kindel, W. B. Mori, C. Joshi, and J. M. Dawson. Two-dimensional simulations of single-frequency and beat-wave laser-plasma heating, *Phys. Rev. Lett.* **54**, 558–561 (1985)

R. E. Waltz. Subcritical magnetohydrodynamic turbulence, *Phys. Rev. Lett.* **55**, 1098–1101 (1985)

Tokamak Fusion Test Reactor (TFTR) at the Princeton University Plasma Physics Laboratory. (Courtesy of Carol A. Phillips, Plasma Physics Laboratory, Princeton University.)

View inside **D-III D fusion device** at General Atomics, San Diego, California. (Courtesy of Douglas M. Fouquet, General Atomics.)

JUNE 15, 1934 PHYSICAL REVIEW VOLUME 45

Magnetically Self-Focussing Streams

WILLARD H. BENNETT, *Ohio State University*

(Received January 13, 1933)

Streams of fast electrons which can accumulate positive ions in sufficient quantity to have a linear density of positives about equal to the linear density of electrons, along the stream, become magnetically self-focussing when the current exceeds a value which can be calculated from the initial stream conditions. Focussing conditions obtain when breakdown occurs in cold emission. The characteristic features of breakdown are explained by the theory. Failure of high voltage tubes is also discussed.

INTRODUCTION

ALTHOUGH the focussing effect which residual gas can have on low voltage electron streams due to the fact that the ions have much smaller velocities than the electrons freed by ionization of the residual gas, has been discussed,[1] not much attention seems to have been given to the focussing effect on electron streams due to the effect of magnetic attractions between the parts of the stream. Calculation shows that such focussing can be very important at high voltages and the result explains some phenomena which occur in high voltage tubes, and which have never been satisfactorily explained. The result has a bearing on the phenomenon of breakdown in cold emission.

How such focussing can take place can be seen qualitatively by considering a stream of high velocity electrons with velocity, u, all moving parallel to a direction which we may choose as the Z-axis, and positive ions moving with velocity, v, in the opposite direction. The density of electrons, positives, and residual gas are assumed to be small and collisions infrequent. The force acting between any two electrons has components[2]

$$F_x=(1-u^2/c^2)E_x;\quad F_y=(1-u^2/c^2)E_y;\quad F_z=E_z;\quad (1)$$

where E_x, E_y and E_z are the components of the force which would exist if the electrons were not moving, i.e., the familiar Coulomb force. The force acting between any two positives has components

$$F_x'=(1-v^2/c^2)E_x';\ F_y'=(1-v^2/c^2)E_y';\quad F_z'=E_z';\quad (2)$$

and the force between any electron and positive

$$F_x''=(1+uv/c^2)E_x'';\ F_y''=(1+uv/c^2)E_y'';\quad F_z''=E_z''.\quad (3)$$

If the density of positives everywhere equals the density of electrons, the static attractions and repulsions cancel, but the additional magnetic force in every case is an attraction so that the charge in every element of volume attracts the charge in every other element of volume. In a cylindrically symmetric stream which is of uniform composition along its length, and which is long compared with its diameter, the attraction between each small element of volume and any long thin element of the stream parallel to the axis, varies inversely as the radial distance from the first small element to the axis of the long thin element, and so the potential energy of the first small element is proportional to the integral of the logarithm of the radial distance to the various long thin elements of volume.

The stream could not spread out indefinitely because in so doing, the potential energy of the stream would approach a logarithmic infinity. No single electron or positive could leave the stream indefinitely far because its individual potential energy would approach a logarithmic infinity. Hence positives in such a stream could

[1] Johnson, J.O.S.A. and R.S.I. 6, 701 (1922) and Buchta, J.O.S.A. and R.S.I. 10, 581 (1925).

[2] See Mason and Weaver, *Electromagnetic Field*, p. 299.

not leave it except by moving out of one of the ends, unless there were more positives than electrons, in which case, only enough would leave the stream to equalize the numbers of each kind of particle per unit length of stream.

The hypothesis of positives at the same density as electrons is not so improbable of realization as might have appeared at first sight, both because the velocity of a positive is much smaller than that of an electron formed by the same collision, and also because the positives once formed in the stream have to pass down the stream clear to the end. An electron freed by the same collision that formed a positive, is acted upon by an attractive force which is much smaller than that acting on the high velocity electrons in the stream, as may be seen from the form of expressions (1). The force between the slow electron the Z-component of whose velocity is u', and each fast electron with velocity u, has components

$$F_x'''=(1-uu'/c^2)E_x'''; \quad F_y'''=(1-uu'/c^2)E_y'''; \quad F_z'''=E_z'''.$$

Since u' is much smaller than u, the attractive force which is proportional to $u\cdot u'/c^2$ is much smaller than the attractive force on a fast electron at the same position. A strong selection thus acts to allow the slow electrons so freed to leave the stream radially with whatever radial velocity they had after the collisions, rather than to allow them to expel the high velocity electrons. Especially in the case of high voltage streams in regions of low field in the direction of the stream will the stream tend to collect positives at high density. This hypothesis will be treated more in detail, later, after a much more definite and quantitative treatment has been given the problem.

Equilibrium Distribution

It will be convenient for the later discussion of actual streams to describe first a special distribution of the particles which is in dynamic equilibrium and which can be designated as the equilibrium distribution. This is to be a cylindrically symmetric stream having the Z-axis as the axis of symmetry. The stream consists of electrons with velocity, u, in the positive Z-direction, and positive ions having velocity, v, in the opposite direction. Superimposed on the relatively large velocity, u, the electrons are supposed to have a Maxwellian distribution of X- and Y-components of velocity so that to an observer moving parallel to the stream with a velocity, u, the number of electrons per unit length of stream with X-components of velocity between U and $U+dU$, and with Y-components between V and $V+dV$, is

$$\lambda_1\left(\frac{m}{2\pi kT_1}\right)\exp\left[-\frac{m}{2kT_1}(U^2+V^2)\right]dU\cdot dV$$

where λ_1 is the number of all electrons per unit length of stream, m is the mass of an electron, and T_1 is a constant which would be called the temperature if the Z-components of velocity were also being considered. It can be called the temperature of the electrons if it is kept in mind that temperature in this discussion means the two-dimensional temperature so defined. The distribution of Z-components of velocity does not make any difference in this problem because small variations in the kinetic energies due to Z-components of velocity cannot affect appreciably the radial distribution of the particles in the stream. Similarly, the positives are supposed to have a Maxwellian distribution of X- and Y-components of velocity superimposed on the velocity, v, in the negative Z-direction, so that to an observer moving with a velocity, v, the number of positives with X-components between U and $U+dU$, and Y-components between V and $V+dV$, is

$$\lambda_2\left(\frac{M}{2\pi kT_2}\right)\exp\left[-\frac{\mu}{2kT_2}(U^2+V^2)\right]dU\cdot dV$$

where λ_2 is the number of all positives per unit length of stream, M is the mass of a positive, and T_2 is the (two-dimensional) temperature of the positives.

Although collisions are assumed to be infrequent, and hence no thermodynamical equilibrium is being approached, it can be shown[3] that a system of particles moving in a field where the potential energy of a particle is $X(x, y)$, and having a distribution such that the number of particles between x and $x+dx$, y and $y+dy$,

[3] See Jeans, *Dynamical Theory of Gases*, p. 89.

z and $z+dz$ and with velocities between U and $U+dU$, V and $V+dV$, is

$$\rho_0\left(\frac{m}{2\pi kT}\right)\exp\left[-\frac{X(x,y)}{kT}\right]$$

$$\exp\left[-\frac{m}{2kT}(U^2+V^2)\right]dx\cdot dy\cdot dz\cdot dU\cdot dV$$

will retain this distribution. The familiar derivation of this expression[4] shows that this kind of distribution is retained in spite of collisions and so it is obvious that such a distribution is retained when collisions are negligible.

A solution will now be obtained for a distribution of electrons and positives each of which has this kind of distribution, and this will be called the equilibrium distribution. This name need not include any ideas concerning how this distribution was arrived at, provided this distribution will be retained by the stream.

To an observer moving along with the electrons, the density of electrons is

$$\rho_{11}=\rho_{110}\cdot e^{-X_{11}(r)/kT_{11}}, \tag{4}$$

where ρ_{110} is the density of electrons at a radial distance where the potential energy of an electron is zero. Similarly, to an observer moving along with the positives, the density of positives is

$$\rho_{22}=\rho_{220}\cdot e^{-X_{22}(r)/kT_{22}}. \tag{5}$$

By applying Poisson's equation in the first instance, since

$$X_{11}=-e\cdot V_1$$

where V_1 is the electric potential in the first system of coordinates,

$$\nabla^2 V_1=-4\pi[-e\cdot\rho_{11}+e\cdot\rho_{21}],$$

where ρ_{11} is the density of electrons in the first system of coordinates, and ρ_{21} is the density of positives in the first system of coordinates. Substituting from Eq. (4),

$$\nabla^2\log\rho_{11}=4\pi e^2[\rho_{11}-\rho_{21}]/kT_{11}. \tag{6}$$

Analogously, on a second system of coordinates moving with a velocity, v, with the positives,

$$\nabla^2 V_{22}=-4\pi[-e\cdot\rho_{12}+e\cdot\rho_{22}]$$

and substituting from Eq. (5)

$$\nabla^2\log\rho_{22}=4\pi e^2[\rho_{22}-\rho_{12}]/kT_{22}, \tag{7}$$

where ρ_{22} is the density of positives in the second system of coordinates and ρ_{12} is the density of electrons in this system. Writing

$$4\pi e^2/kT_{11}=\alpha_1 \quad\text{and}\quad 4\pi e^2/kT_{22}=\alpha_2$$

the equations can be transformed to equations in the rest system of coordinates[5] by writing

$$\rho_{11}=\rho_1/\beta_1;\quad \rho_{12}=\beta_2[1+uv/c^2]\cdot\rho_1;\quad \rho_{21}=\beta_1[1+uv/c^2]\cdot\rho_2;\quad \rho_{22}=\rho_2/\beta_2,$$

where

$$\beta_1=(1-u^2/c^2)^{-\frac{1}{2}};\quad \beta_2=(1-v^2/c^2)^{-\frac{1}{2}}$$

and Eqs. (6) and (7) become

$$\nabla^2\log\rho_1=\alpha_1\beta_1[1-u^2/c^2]\cdot\rho_1-\alpha_1\beta_1[1+uv/c^2]\cdot\rho_2, \tag{8}$$

$$\nabla^2\log\rho_2=\alpha_2\beta_2[1-v^2/c^2]\cdot\rho_2-\alpha_2\beta_2[1+uv/c^2]\cdot\rho_1. \tag{9}$$

These are exactly the same equations as are arrived at, if, instead of using Poisson's equation in each of two moving systems of coordinates, we calculate the potential energy of each of the two kinds of particle in the rest system of coordinates, using expressions (1), (2) and (3) for the forces between particles given in the Introduction, *viz.*,

$$X_1=\int_R^r 2e\left[1+\frac{uv}{c^2}\right]\frac{d\lambda}{\lambda}\int_0^\lambda e\cdot\rho_2(\xi)\cdot 2\pi\xi\cdot d\xi-\int_R^r 2e\left[1-\frac{u^2}{c^2}\right]\frac{d\lambda}{\lambda}\int_0^\lambda e\cdot\rho_1(\xi)\cdot 2\pi\xi\cdot d\xi,$$

$$X_2=\int_R^r 2e\left[1+\frac{uv}{c^2}\right]\frac{d\lambda}{\lambda}\int_0^\lambda e\cdot\rho_1(\xi)\cdot 2\pi\xi\cdot d\xi-\int_R^r 2e\left[1-\frac{v^2}{c^2}\right]\frac{d\lambda}{\lambda}\int_0^\lambda e\cdot\rho_2(\xi)\cdot 2\pi\xi\cdot d\xi,$$

[4] The derivation usually includes the distribution in the Z-component of velocity, too, but since the potential function is not a function of z, the distribution in the Z-component of velocity cannot affect the density.

[5] A discussion of the relativistic transformation of density can be found in Eddington's *Mathematical Theory of Relativity*, p. 33, *et seq.*

where R is the radial distance from the axis at which the potential energy is zero. These can be substituted in the Boltzmann expressions (4) and (5) (taking logarithms and derivatives to put the equations into the form of (8) and (9)) if we keep in mind that the temperatures in the rest system must be relativistically transformed from those in the moving systems by

$$T_1=\beta_1\cdot T_{11};\quad T_2=\beta_2\cdot T_{22}$$

because the two-dimensional temperatures used here are proportional to the transverse kinetic energies in which the mass and transverse velocities must be transformed.

Since exactly the same equations are obtained regardless of which method of approach to the problem is used, even to the extent of being relativistically invariant, we are justified in using expressions (1), (2) and (3) for the forces on the particles in the later treatment of non-equilibrium cases.

A particular solution of (8) and (9) is

$$\rho_1=\rho_0/[1+b\rho_0r^2]^2$$
$$\rho_2=\frac{\alpha_1\beta_1[1-u^2/c^2]+\alpha_2\beta_2[1+uv/c^2]}{\alpha_2\beta_2[1-v^2/c^2]+\alpha_1\beta_1[1+uv/c^2]}\cdot\rho_1, \tag{10}$$

where

$$b=\frac{\alpha_1\alpha_2\beta_1\beta_2}{8c^2}\cdot\frac{(u+v)^2}{\alpha_1\beta_1[1+uv/c^2]+\alpha_2\beta_2[1-v^2/c^2]}.$$

This solution is of course relativistically invariant.

If the ratio of the density of electrons to the density of positives in each of any two elements of volume in the stream, swept out by passing two small elements of cross-sectional area along the entire stream parallel to the axis, is greater than

$$\frac{1+v/c}{1-u/c}\quad\text{or less than}\quad\frac{1-v/c}{1+u/c}$$

it is seen from the application of expressions (1), (2) and (3) given for the forces between particles, in the Introduction, that the charge in any two such elements would repel and an equilibrium distribution with such ratios present is impossible. Thus, although expressions (10), are only a particular solution, they are a very good approximation to the complete solution for the distribution when the velocities u and v are small compared with the velocity of light, c, and in fact no serious error will be introduced by setting the coefficient in Eqs. (10) equal to unity. This makes the approximate solution

$$\rho_1=\rho_2=\rho_0/[1+b\rho_0r^2]^2,$$

where

$$b=\frac{\pi e^2}{2kc^2}\cdot\frac{(u+v)^2}{T_1+T_2}. \tag{11}$$

The number of either kind of particle per unit length of stream is

$$\lambda_0=[c^2\cdot 2K(T_1+T_2)]/e^2(u+v)^2 \tag{12}$$

and the current is

$$i_0=[c^2\cdot 2K(T_1+T_2)]/e(u+v),$$

which is rigidly fixed by the values of u, v, T_1 and T_2. Thus an equilibrium distribution of the kind just described can exist only for one special value of the current. This will be designated as the critical current.

Sources and Distributions of Positives

At electron velocities corresponding to potential drops of the order of 5000 volts and larger, the probability that a collision with a neutral molecule will result in ionization is to a good approximation (except when the velocities approach that of light) inversely proportional to the voltage of the primary electron.[6] The number of ions formed by each electron per centimeter of path from gases more apt to be present in tubes at moderately high vacuum, can be esti-

[6] See J. J. Thomson, *Conduction of Electricity through Gases*, Vol. 2, 3rd Ed., p. 96, *et seq.*

mated at least for order of magnitude as $200p/V$, where p is the pressure in mm of mercury, and V is the potential of the electrons in e.s.u. If the current in e.s.u. in a stream of fast electrons is i, the number of ions formed per centimeter length of stream, per second, is $200pi/Ve$. If it is supposed that these ions are formed in a region having a uniform field intensity, E, the number of ions per centimeter length of stream coming from along the stream back a distance, d, is

$$200\frac{p}{V}\cdot\frac{i}{e}\cdot\left(\frac{2Md}{Ee}\right)^{\frac{1}{2}},$$

where M is the mass of an ion. For this to be equal to or greater than the number of fast electrons per centimeter length of stream

$$\frac{i}{e}\left(\frac{m}{2Ve}\right)^{\frac{1}{2}}\leq 200\frac{p}{V}\cdot\frac{i}{e}\left(\frac{2Md}{Ee}\right)^{\frac{1}{2}}$$

and so the pressure of the gas must be, approximately

$$p\geq 2\cdot 10^{-7}(VE/Wd)^{\frac{1}{2}}, \qquad (13)$$

where V and E are in volts, and volts per centimeter, respectively, and W is the molecular weight of the gas from which the ions are formed. If the actual gas pressure exceeds the value calculated from expression (13) for the conditions at any part of the stream, the number of positives per unit length of stream will exceed the number of fast electrons at that part of the stream.

In addition to the residual gas as a source of ions, we must keep in mind that it has been found in numerous cases[7] that large amounts of the anode material can be removed from the anode by high voltage streams. This necessarily increases the density of matter in the vicinity of the stream while large electron currents are passing.

In order to see what is to be expected to be the kinetic energy of the positives after they are formed, some experiments by Ishino[8] can be considered in which it was found that when fast electrons are passed through a rarefied gas, the electrons freed in ionizations have transverse kinetic energies distributed so that about ninety percent of the electrons have energies less than forty volts and about ten percent have energies greater than this value. Since the rapidly moving ionizing electron acts on both the atomic electron and the rest of the atom during the same interval of time, the momentum given the two must be equal and opposite so at least ninety percent of the ions must have transverse energies less than $0.022/W$ electron-volts, where W is the molecular weight. If the ions are formed from residual gas at room temperature, their velocities are not greatly affected by ionization and the average kinetic energy is approximately the energy corresponding to that temperature.

If positive ions are present at densities greater than those of the high velocity electrons, only enough of the low velocity electrons freed in the formation of positives will move directly out of the stream, to leave an average charge density only slightly positive in the stream. The static focussing field which can be obtained in this way is a small field (order of magnitude of a few volts per centimeter) similar in nature to the focussing field characteristic of the low voltage streams in the presence of gas at much higher pressure, mentioned earlier.[1]

If the positive ions have velocities much smaller than the fast electrons, as they do have except in extreme proximity to the cathode, it is apparent from the form of expressions (1), (2) and (3) that the excess of positive ions over the fast electrons, together with the neutralizing space charge of slow electrons, do not appreciably affect the motions of the fast electrons magnetically and consequently have a negligible effect on the stream. Thus in any section where there are at least as many positives as fast electrons, we are concerned with the dynamics of a stream consisting of equal numbers of each kind of particle per unit length of stream.

Stability

In order to extend this treatment to streams under conditions apt to occur experimentally, a function whose value will increase as the particles spread out from the axis can be used as a criterion of how much the stream has spread:

$$F=\int_0^\infty r^2\cdot\rho\cdot 2\pi r\cdot dr.$$

[7] Hull and Burger, Phys. Rev. 31, 1121 (1928); Snoddy, Phys. Rev. 37, 1878 (1931); Newman, Phil. Mag. 14, 712 (1932); Bennett, Phys. Rev. 37, 590 (1931).

[8] Ishino, Phil. Mag. 32, 202 (1916).

A maximum possible value for this function can be calculated subject to the restriction that the potential energy of the stream per unit length of stream shall not exceed a value, H, which is the initial potential energy plus the sum of the kinetic energies due to components of velocity transverse to the axis, of all the particles.

Although the density of positives will not in general equal the density of electrons at all radial distances from the axis, the maximum spread occurs when the two densities are everywhere equal, so that in order to find the maximum value of F, we can write the restriction as

$$\tfrac{1}{2}\cdot(2e^2(u+v)^2/c^2)\int_0^\infty \log(r/R)\cdot 2\pi r\cdot\rho(r)\cdot dr\int_0^r \rho(\xi)\cdot 2\pi\xi\cdot d\xi = H.$$

This calculation can be simplified by using as the independent variable, n, the total number of either kind of particle from the axis out to a radial distance, r, and letting the distance, r, be dependent on n. Then

$$F=\int_0^N r^2\cdot dn \quad\text{and}\quad H=g\int_0^N n\cdot\log(r/R)dn,$$

where $g=e^2(u+v)^2/c^2$ and N is the total number of either kind of particle per unit length of stream. In order to find the maximum spread, set $\delta\{F+\lambda H\}=0$, where λ is an arbitrary multiplier. This gives a distribution with uniform density ρ' from the axis out to a radial distance, r_0, and zero beyond, where

$$\pi r_0^2\rho'=N;\quad \pi r^2\rho'=n;\quad r=r_0(n/N)^{\frac{1}{2}}.$$

If the stream is initially distributed uniformly inside a radial distance from the axis, s_0, and has a total kinetic energy due to transverse components of velocity, K, which can be related to another constant, T, by $K=2N\cdot kT$, the maximum spread is a uniform distribution out to a radial distance, r_0, given by

$$g\int_0^N n\left(\log\frac{r_0}{R}\cdot\left(\frac{n}{N}\right)^{\frac{1}{2}}\right)\cdot dn=g\int_0^N n\left(\log\frac{s_0}{R}\left(\frac{n}{N}\right)^{\frac{1}{2}}\right)\cdot dn+2N\cdot kT,$$

i.e.,

$$r_0=s_0\cdot\exp\left[\frac{c^2\cdot 2\cdot 2kT}{e^2(u+v)^2}\cdot\frac{1}{N}\right]. \tag{14}$$

More than half of the particles must always lie nearer the axis than $1/2^{\frac{1}{2}}$ times this value of r_0.

If the kinetic energy of the particles due to transverse components of velocity could be redistributed among the particles so that the transverse components of velocity would have Maxwellian distributions, the (two-dimensional) temperature would be T given by $K=2N\cdot kT$ and the linear density λ_0 of the stream which could have an equilibrium distribution such as given by Eqs. (12), can be substituted in expression (14). Thus more than half the particles in the stream must always lie nearer the axis than a radial distance given by

$$(s_0/2^{\frac{1}{2}})\cdot e^{\lambda_0/N}=(s_0/2^{\frac{1}{2}})\cdot e^{i_0/I}, \tag{15}$$

where i_0 is the critical current (expression (12)) and I is the actual current.

If half of the particles get outside this radial distance, all of the transverse velocities would be simultaneously reduced to zero which is obviously impossible. Thus this expression is an exaggerated outside limit of the radial distance inside which at least half of the particles must always remain. In fact, calculation of the directions in which most of the particles would begin to drift for several special distributions show that the ratio i_0/I distinguishes between streams which begin to contract towards the axis, and those which begin to expand, the contraction occurring when $I>i_0$.

From the form of expression (15), it is seen that if the actual current is one-tenth the critical current, the radial distance inside which at least half the particles must always remain, is liable to increase to more than 10,000 times the initial value for the outer radius of the stream, which for most experimental cases means no observable focussing action. In this sense, it may be said that the critical current given by expression (12) rather sharply distinguishes between streams which focus and those which do not. This critical current may be written as

$$i_0 = 2.5 \cdot 10^{-3} T / V^{\frac{1}{2}}, \tag{16}$$

where i_0 is in amperes, V is the potential through which the fast electrons have fallen, in volts (except in extreme proximity to the cathode), and T is in equivalent degrees absolute, and is a constant proportional to the total transverse energy of the stream per unit length.

Applications

In an earlier communication,[9] evidence was presented supporting the contention that breakdown in cold emission is caused by positive ions liberated at the anode by small currents, and travelling back along the electron stream to the cathode. A difficulty with this idea is that even if positives do bombard the cathode in this way, still at such low densities they might be expected to eliminate any emitting areas where they strike, instead of producing them. Now this is just what has been observed to occur except when the current rises to above the order of 10^{-2} amperes. It was not clear then why this order of magnitude of the current should be so critical in changing from elimination of emitting areas to severe rupture of the cathode surface.

Focussing the positives on a small area of the cathode cannot be produced in the analogous way to the focussing of low voltage electron streams in gases at relatively higher pressures[1] by interchanging the rôles of positives and electrons, because the electrons are not produced by the positives from the residual gas as would be required by that kind of explanation.

On the other hand, if the current is momentarily large enough to give magnetic self-focussing over any considerable part of the stream, the electrons towards the anode are brought nearer the axis by this focussing action, and then since the initial total energy of positives in any part of the stream depends on the position with respect to the axis at which they were formed, and since bringing electrons nearer the axis in parts of the stream towards the anode reacts to decrease the total transverse energy of the stream per unit length in the central part where self-focussing conditions obtain, the effect of the magnetic self-focussing is to bring all particles nearer the axis over an increasingly long segment of the stream. In an analogous way, bringing the positives nearer the axis of the stream very near the cathode increases the emission from the part of the cathode delivering electrons along the axis, and this serves to decrease the total transverse energy per unit length in the central part of the stream.

This kind of process can go on indefinitely provided only that conditions for magnetic self-focussing obtain over any segment of the stream long compared with its diameter. This kind of focussing results in concentrating the positive ion current at the cathode until the current density of the positive ions is a far higher order of magnitude than the value which is known to eliminate emitting areas instead of producing them. At these higher positive ion current densities, the positives would certainly be expected to dig out the small craters which it is well known are produced during breakdown. In this way, we have an explanation of why the order of magnitude of current given by expression (16) distinguishes between breakdown and no breakdown.

It seems probable that the entire process just described takes place in a very small interval of time (order of 10^{-4} or 10^{-5} seconds) after which the streams are no longer self-focussing, because the transverse energies of the electrons arising at the ragged edges of the crater are too great, or that the density of matter near the cathode rises to so high a value that collisions can no longer be neglected with respect to the dynamics of the stream. The point of the discussion has been to show that structural defects in the cathode surface are not necessary for breakdown, but that rather, we must look to stream conditions to explain the many characteristic features of breakdown.

Any explanation of breakdown in cold emission to agree with experimental fact, should explain

[9] Bennett, Phys. Rev. 40, 416 (1932).

the following general observations: (1) breakdown is much more abrupt when the electrodes are unoutgassed than when they are outgassed,[10] i.e., the rise of current to the maximum value permitted by the resistance in series with the source of potential is from an initially much lower value in the former than in the latter case; (2) the fineness of the polish on the surface of the cathode does not affect the value of the field intensity at which breakdown occurs[11] but only affects the fields at which the first small currents pass; (3) with unoutgassed electrodes, after breakdown has occurred, if a field intensity a little less than the field at which breakdown occurred is applied to the cathode, large sudden surges of current pass through the tube, similar in every observable respect to the surge of current which accompanied the original breakdown, but the frequency of occurrence of these surges decreases with time; (4) breakdown can be prevented, or at least its severity greatly diminished either by outgassing the electrodes (both are necessary) or by increasing the resistance in series with the source of potential[12] to a value which will limit the maximum value to which the current can rise, to the order of 10^{-3} or 10^{-4} or less; (5) before breakdown, the current is very erratic, rising abruptly, and dropping sometimes abruptly, and sometimes gradually; these erratic fluctuations while the current is small occur at values of field much higher than the values of the field for the same order of magnitude of current after breakdown; but once the value of the current has increased to more than the order of 10^{-2} amperes, the resistance of the tube decreases and the surface of the cathode becomes permanently altered to give much higher order currents at a given field than before; (6) when breakdown occurs, small flashes of light are observed both at the anode and at the cathode.

[10] Millikan and Shackelford, Phys. Rev. 15, 239 (1920); Eyring, Mackeown and Millikan, Phys. Rev. 31, 900 (1928); Bennett, Phys. Rev. 37, 582 (1931).

[11] Only superficial significance was given this fact on p. 584 of the paper in Phys. Rev. 37, 582 (1931).

[12] Unpublished work. See R. W. Mebs, Phys. Rev. 43, 1058 (1933).

The explanation which writers have usually used in the past, *viz.*, that a particle of impurity or a piece of the metal of the cathode is torn bodily out of the cathode by the field, fails to explain the above observations, but their explanation in terms of the idea of magnetically self-focussing streams follows easily from the discussion of streams already given.

The theory of streams also explains the fact that even after electrodes have been fatigued or "conditioned" and are giving steady field-current characteristics, if the residual gas pressure is allowed to rise to the order of magnitude corresponding to expression (13) the current becomes unsteady, and the surges of current which pass can be prevented from entirely disfiguring the cathode only by increasing the resistance in series with the source of potential.

The fact that the field intensity for breakdown in mercury obtained recently by Beams[13] agrees so well with the results of the writer for other metals with a similar arrangement of electrodes shows that surface impurities can at most only serve to give the first small electron currents, but that an entirely different explanation is necessary in order to explain breakdown.

The theory of streams explains the fact that the experience with tubes to which extremely high potentials are applied[14] has been that destructive cold emission occurs at fields of the order of 100,000 volts per centimeter, but that the application of fields of the order of 1,000,000 volts per centimeter to surfaces prepared in the same way, but held at much shorter distances from the anode were found by the writer to be necessary to produce measurable emission. In the case of high voltage tubes, the longer distances and the higher voltages both aid in the formation of self-focussing streams.

In conclusion, the writer wishes to thank Professor L. H. Thomas of this laboratory, for the very generous assistance which he has given in discussions of mathematical methods.

[13] J. W. Beams, Phys. Rev. 44, 803 (1933); Bennett, Phys. Rev. 37, 582 (1931).

[14] Lauritsen and R. D. Bennett, Phys. Rev. 32, 850 (1928); Breit, Tuve and Dahl, Phys. Rev. 35, 51 (1930); Lauritsen and Cassen, Phys. Rev. 36, 988 (1930); Lauritsen, Phys. Rev. 36, 1680 (1930).

PHYSICAL REVIEW VOLUME 89, NUMBER 5 MARCH 1, 1953

Transport Phenomena in a Completely Ionized Gas*

LYMAN SPITZER, JR., AND RICHARD HÄRM
Princeton University Observatory, Princeton, New Jersey

(Received November 10, 1952)

The coefficients of electrical and thermal conductivity have been computed for completely ionized gases with a wide variety of mean ionic charges. The effect of mutual electron encounters is considered as a problem of diffusion in velocity space, taking into account a term which previously had been neglected. The appropriate integro-differential equations are then solved numerically. The resultant conductivities are very close to the less extensive results obtained with the higher approximations on the Chapman-Cowling method, provided the Debye shielding distance is used as the cutoff in summing the effects of two-body encounters.

I. INTRODUCTION

A PREVIOUS paper by Cohen, Spitzer, and Routly,[1] referred to hereafter as CSR, presented a new approach to the problem of transport phenomena in a completely ionized gas. In effect, the influence of mutual electron encounters on the velocity distribution function for electrons was considered as a problem of diffusion in velocity space. In particular, the electrical conductivity of an electron-proton gas was computed in this way. However, the results were not exact, since one term in the diffusion equation was neglected. In the present paper, a solution of the complete diffusion equation is obtained, and the results are extended to completely ionized gases with different mean nuclear charges. Computations are carried out for the thermal as well as the electrical conductivity.

In the first section below the general principles are explained and justified. Subsequent sections outline the derivation of the equations, the method of solution, and the results obtained.

* This work has been supported in part by the U. S. Atomic Energy Commision.

[1] Cohen, Spitzer, and Routly, Phys. Rev. 80, 230 (1950).

II. GENERAL PRINCIPLES

The velocity distribution function $f_r(v)$ for particles of type r is determined by the familiar Boltzmann equation, basic in all studies of this sort,

$$\frac{\partial f_r}{\partial t}+\sum_i v_{ri}\frac{\partial f_r}{\partial x_i}+\sum_i F_{ri}\frac{\partial f_r}{\partial v_{ri}}=\sum_s\left(\frac{\partial_s f_r}{\partial t}\right)_s, \qquad (1)$$

where the notation in CSR has been followed. The complexity of the problem arises entirely from the term $(\partial_s f_r/\partial t)_s$, which gives the change in f_r produced by encounters of r particles with particles of type s.

To visualize the physical situation more accurately, let us follow a single electron as it moves through the gas. The random electrical fields encountered by the electron will produce deflections and changes in velocity. To some extent these fields can be described in terms of separate two-body encounters; let b be the impact parameter for such an encounter—the distance of closest approach between the two particles in the absence of any mutual force. The situation is characterized by the values of the following four distances: d, the mean distance from an electron to its nearest neighbor; b_0, the value of the collision parameter for

TABLE I. Values of the velocity distribution function $D(x)$ when an electrical field is present.

x	$ZD(x)/A$ $Z=1$	$Z=2$	$Z=4$	$Z=16$	$Z=\infty$
0.10	0.0008093	0.0001340	···	···	0.0001000
0.11	0.001300	0.0002262	···	···	0.0001464
0.12	0.001970	0.0003630	···	···	0.0002074
0.13	0.002847	0.0005582	···	···	0.0002856
0.14	0.003955	0.0008262	···	···	0.0003842
0.15	0.005317	0.001183	···	···	0.0005062
0.16	0.006955	0.001645	···	···	0.0006554
0.17	0.008886	0.002228	···	···	0.0008352
0.18	0.01113	0.002952	···	···	0.001050
0.19	0.01370	0.003832	···	···	0.001303
0.20	0.01660	0.004884	0.002163	0.001645	0.001600
0.22	0.02347	0.007576	0.003373	0.002432	0.002343
0.24	0.03180	0.01116	0.005044	0.003477	0.003318
0.26	0.04165	0.01575	0.007280	0.004833	0.004570
0.28	0.05304	0.02146	0.01018	0.006560	0.006147
0.30	0.06601	0.02840	0.01386	0.008721	0.008100
0.32	0.08057	0.03666	0.01842	0.01139	0.01049
0.34	0.09672	0.04632	0.02398	0.01463	0.01336
0.36	0.1145	0.05746	0.03063	0.01853	0.01680
0.38	0.1338	0.07012	0.03849	0.02317	0.02085
0.40	0.1548	0.08440	0.04764	0.02866	0.02560
0.44	0.2015	0.1180	0.07028	0.04249	0.03748
0.48	0.2545	0.1586	0.09924	0.06082	0.05308
0.52	0.3137	0.2066	0.1352	0.08443	0.07312
0.56	0.3792	0.2620	0.1789	0.1142	0.09834
0.60	0.4508	0.3254	0.2309	0.1511	0.1296
0.64	0.5285	0.3968	0.2917	0.1958	0.1678
0.68	0.6123	0.4764	0.3618	0.2494	0.2138
0.72	0.7023	0.5646	0.4416	0.3127	0.2687
0.76	0.7983	0.6612	0.5324	0.3868	0.3336
0.80	0.9005	0.7668	0.6336	0.4722	0.4096
0.88	1.123	1.005	0.8704	0.6819	0.5997
0.96	1.371	1.281	1.156	0.9499	0.8493
1.04	1.645	1.596	1.494	1.287	1.170
1.12	1.945	1.952	1.889	1.702	1.574
1.20	2.273	2.352	2.344	2.204	2.074
1.28	2.630	2.796	2.864	2.804	2.684
1.36	3.017	3.290	3.455	3.510	3.421
1.44	3.435	3.836	4.120	4.331	4.300
1.52	3.887	4.440	4.868	5.281	5.338
1.60	4.375	5.096	5.700	6.366	6.554
1.76	5.465	6.604	7.660	8.996	9.595
1.92	6.728	8.406	10.06	12.34	13.59
2.08	8.190	10.54	12.96	16.54	18.72
2.24	9.880	13.05	16.46	21.74	25.18
2.40	11.83	16.00	20.64	28.10	33.18
2.56	14.06	19.40	25.60	35.80	42.95
2.72	16.62	23.3	31.4	45.0	54.74
2.88	19.53	27.7	38.2	56.0	68.80
3.04	22.74	32.7	46.0	68.8	85.41
3.20	26.00	38.5	54.6	83.9	104.9

which an electron is deflected 90° in an encounter with a stationary positive ion; h, the Debye shielding distance; and λ, the mean free path for a net deflection of 90°. It is readily verified that for virtually all situations of interest,

$$b_0 \ll d \ll h \ll \lambda. \tag{2}$$

It is clear that encounters for which $b \ll d$ can be described adequately in terms of successive two-body encounters, since usually an encounter with one particle will be effectively over before another particle approaches to a distance less than d. These successive encounters may be divided into two classes. Those with $b \leqslant b_0$ produce large deflections, and will be termed "close" encounters. Those with $b_0 < b < d$ produce relatively small deflections, and will be called "distant" encounters. As shown in CSR and elsewhere, the cumulative effect of many distant encounters outweighs the effect of the less frequent close encounters, in the special case of inverse-square forces between the particles.

Encounters for which $d \leqslant b < h$ cannot be regarded as independent, since several such encounters will be taking place at the same time. More correctly, the deflection of a particular electron caused by such "encounters" must be attributed to statistical fluctuations of the electron density in a sphere of radius b. As shown in CSR, however, the mean square change of electron velocity produced by such fluctuations is correctly given if the formulas derived for successive two-body encounters are applied for $b > d$.

Particles passing at a distance large compared to h produce a negligible effect. From the standpoint of the Debye shielding theory, the effective field of a charge in a plasma varies as e^{-hr}/r, where h is given by

$$h^2 = \frac{kT}{4\pi n_e e^2(1+Z)}. \tag{3}$$

If one considers rather the statistical fluctuations in electron density, Pines and Bohm[2] have shown that collective phenomena in a plasma reduce markedly the statistical fluctuations in electron density with wavelengths large compared to h, thus justifying the neglect of encounters such that $b > h$. There is some interaction between a single electron and the organized oscillations of the plasma—see Eq. (59a) of Pines and Bohm. However, comparison of their Eqs. (59a) and (59b) shows that for thermal electrons, with mean kinetic energies of the order kT, the rate of energy loss due to this process is less by a factor $1/\ln(h/b_0)$ than the energy loss due to random encounters such that $b < h$. The generation of plasma oscillations by a single thermal electron may therefore be neglected, together with a number of other terms of the same order.[3] Hence, we may neglect all interactions between electrons for which the distance of closest approach exceeds h.

Since λ is much greater than h, it is evident that many small deflections will be experienced by a particle traversing its mean free path. It is also clear that these deflections are essentially independent of each other. Inasmuch as collective phenomena (oscillations) have been neglected, the random electrical fields encountered by an electron in one region will be completely independent from the fields in a similar region separated by a distance appreciably greater than h. Hence, the successive changes in velocity represent a Markoff process, and the change of the velocity distribution function may be found from the Fokker-Planck equation.[4] This equation neglects the close encounters; the relative error introduced is again of the order $1/\ln(h/b_0)$.

III. DERIVATION OF EQUATIONS

Equations (10), (23), and (24) of CSR give the basic equations of the problem on the following assumptions: (a) the Fokker-Planck equation may be used to give $\partial f_e/\partial t$; (b) a steady state is established; (c) the velocity

[2] D. Pines and D. Bohm, Phys. Rev. 85, 338 (1952).

[3] The various terms of relative magnitude $1/\ln(d/b_0)$ have been called "nondominant" by Chandrasekhar, and are usually neglected in the computation of diffusion coefficients—see S. Chandrasekhar, *Principles of Stellar Dynamics* (University of Chicago Press, Chicago, 1939).

[4] See S. Chandrasekhar, Revs. Modern Phys. 15, 1 (1943).

distribution function may be expressed as the sum of a Maxwellian function $f^{(0)}$ plus a small term $f^{(1)}$ whose square may be neglected. The values of the diffusion coefficients have now been recomputed, using a straightforward and conceptually very simple method. A new integro-differential equation has then been obtained for $D(lv)$, the function which gives the dependence of the ratio $f^{(1)}/f^{(0)}$ on the velocity v; $1.5/l^2$ is the mean square electron velocity.

The electron-ion interaction is relatively simple to consider. Equation (28) of CSR must be modified to include a factor Z, the mean ionic charge, defined by

$$Z \equiv \sum_i n_i Z_i^2 / n_e, \tag{4}$$

summed over all positive ions, each of charge $Z_i e$ and of particle density n_i. We find

$$K(ff_i) = [3ZLf^{(0)}D(lv)\cos\theta]/2v^3, \tag{5}$$

where all the symbols have the same meaning as in CSR, except that to avoid confusion with the current density, j has been replaced by l.

The quantity $K(ff)$, giving the contribution of electron-electron interactions to $\partial_e f/\partial t$, is much more complicated. To evaluate the diffusion coefficients needed, the values of Δ_ξ, Δ_η, and Δ_ζ, the changes of velocity in a single collision, along coordinates parallel and perpendicular, respectively, to the original velocity, were first determined. In this computation, the velocity changes were first taken in a frame of reference moving with the center of gravity of the two particles. If the orbit lies in the xy plane, and the x axis is taken to be in the direction in which the original particle is moving, the velocity changes become very simple. The components of the vector change in velocity along the ξ, η, and ζ axes may then be found by successive rotations of the coordinate axes.

Next the values of the velocity changes, together with their products and squares, are averaged over all collisions, in accordance with Eq. (8) of CSR. The final values for $\langle\Delta_\xi\rangle$, $\langle\Delta_\xi^2\rangle$, and $\langle\Delta_\eta^2\rangle$ are the same as given in Eqs. (31) through (35) of CSR. For the two remaining coefficients we find

$$\langle\Delta_\eta\rangle = -\frac{4Ll^2}{\pi^{\frac{1}{2}}}\sin\theta\{I_0(\infty) - I_0(x) + I_3(x)/x^2\}, \tag{6}$$

$$\langle\Delta_\xi\Delta_\eta\rangle = -\frac{2Ll}{\pi^{\frac{1}{2}}}\sin\theta\{0.4x(I_0(\infty) - I_0(x)) + I_3(x)/x^2 - 0.6I_5(x)/x^4\}, \tag{7}$$

where L and $I_n(x)$ are defined in Eqs. (29) and (37) of CSR.

If these values of the diffusion coefficients are substituted into $K(ff)$, Eq. (1) yields, after considerable algebra, the equation

$$D''(x) + P(x)D'(x) + Q(x)D(x) = R(x) + S(x), \tag{8}$$

exactly as in Eq. (41) of CSR; x is defined as lv. While $P(x)$ is unchanged from the form given in Eq. (42) of CSR, $Q(x)$ and $S(x)$ become

$$Q(x) = \frac{1}{x^2} - 2\frac{Z + \Phi(x) - 2x^3\Phi'(x)}{A}, \tag{9}$$

$$S(x) = \frac{16}{3\pi^{\frac{1}{2}}A}\{xI_3(x) - 1.2xI_5(x) - x^4I_0(x)(1 - 1.2x^2)\}, \tag{10}$$

where $\Phi(x)$ is the error function, while $A(x)$ is defined in Eq. (46) of CSR. If an electrical field is present,

$$R(x) = -(2x^4A/ZA)\{Z - 1 + 1.2x^2\}, \tag{11}$$

where A is again given by Eq. (40) of CSR. If a temperature gradient is present, we have

$$R(x) = -(x^4B/A)\{2.5 - x^2\}, \tag{12}$$

where

$$B = \frac{2k^2T|\nabla T|}{\pi n_e e^4 \ln(qC^2)}; \tag{13}$$

qC^2 is given in Eq. (65), CSR.

TABLE II. Values of the velocity distribution function $D(x)$ when a temperature gradient is present.

x	$ZD(x)/B$ $Z=1$	$Z=2$	$Z=4$	$Z=16$	$Z=\infty$
0.10	0.0005906	0.0002306	...	...	0.0001245
0.11	0.0009023	0.0003630	...	...	0.0001521
0.12	0.001309	0.0005460	...	...	0.0002577
0.13	0.001821	0.0007914	...	...	0.0003546
0.14	0.002448	0.001110	...	...	0.0004764
0.15	0.003197	0.001515	...	...	0.0006271
0.16	0.004074	0.002016	...	...	0.0008108
0.17	0.005082	0.002624	...	...	0.001032
0.18	0.006225	0.003348	...	...	0.001295
0.19	0.007504	0.004200	...	...	0.001605
0.20	0.008922	0.005188	0.003095	0.002146	0.001968
0.22	0.01217	0.007601	0.004668	0.003156	0.002872
0.24	0.01596	0.01064	0.006744	0.004487	0.004052
0.26	0.02027	0.01434	0.009400	0.006199	0.005558
0.28	0.02508	0.01874	0.01270	0.008356	0.007442
0.30	0.03035	0.02384	0.01673	0.01103	0.009760
0.32	0.03607	0.02966	0.02152	0.01428	0.01257
0.34	0.04218	0.03616	0.02713	0.01818	0.01593
0.36	0.04865	0.04338	0.03360	0.02281	0.01991
0.38	0.05545	0.05126	0.04092	0.02824	0.02456
0.40	0.06254	0.05978	0.04916	0.03453	0.02995
0.44	0.07746	0.07866	0.06840	0.04995	0.04322
0.48	0.09302	0.09966	0.09116	0.06955	0.06024
0.52	0.1090	0.1224	0.1174	0.09370	0.08161
0.56	0.1250	0.1466	0.1468	0.1226	0.1075
0.60	0.1407	0.1717	0.1790	0.1565	0.1387
0.64	0.1558	0.1973	0.2134	0.1952	0.1754
0.68	0.1700	0.2228	0.2497	0.2386	0.2178
0.72	0.1829	0.2476	0.2871	0.2864	0.2663
0.76	0.1944	0.2714	0.3249	0.3384	0.3207
0.80	0.2036	0.2934	0.3624	0.3940	0.3809
0.88	0.2151	0.3296	0.4324	0.5135	0.5174
0.96	0.2139	0.3506	0.4896	0.6384	0.6703
1.04	0.1965	0.3502	0.5244	0.7584	0.8297
1.12	0.1587	0.3210	0.5252	0.8565	0.9800
1.20	0.0957	0.2546	0.4796	0.9096	1.099
1.28	+0.0021	+0.1414	0.3717	0.8858	1.156
1.36	−0.1289	−0.0302	+0.1834	0.7474	1.113
1.44	−0.3041	−0.2734	−0.1067	+0.4525	0.9167
1.52	−0.5339	−0.6040	−0.5240	−0.1736	+0.5060
1.60	−0.8268	−1.040	−1.098	−0.7999	−0.1966
1.76	−1.657	−2.320	−2.861	−3.313	−2.867
1.92	−2.921	−4.330	−5.752	−7.691	−8.061
2.08	−4.774	−7.366	−10.25	−14.79	−17.09
2.24	−7.448	−11.81	−16.98	−25.78	−31.69
2.40	−11.23	−18.18	−26.78	−42.20	−54.08
2.56	−16.5	−27.1	−40.6	−66.0	−87.05
2.72	−23.5	−39.3	−59.9	−99.4	−134.1
2.88	−33.2	−55.9	−86.0	−145.	−199.3
3.04	−46.0	−77.8	−121.	−207.	−287.9
3.20	−62.0	−106.	−166.	−290.	−405.8

From the principle of conservation of momentum the integral $I_0(\infty)$ can be evaluated in simple form. In the case of an electrical field, Eq. (39) of CSR becomes modified if Z differs from unity, and we have

$$I_0(\infty)=3\pi^{\frac{1}{2}}A/8Z. \quad (14)$$

If a temperature gradient is considered, then we have instead

$$I_0(\infty)=0. \quad (15)$$

IV. SOLUTION OF THE EQUATIONS

The method of numerical solution followed is, in principle, identical with that employed in CSR. For values of Z different from unity the functions $Q_{01}(x)$ and $Q_{02}(x)$ differed from those used previously, with somewhat different forms for the solutions $U(x)$ and $V(x)$. No changes in $P_{01}(x)$ or $P_{02}(x)$ were required. In each specific range the substitutions given below were the same for both electrical and thermal conductivity.

In the range $x\leqslant 0.80$, we have

$$Q_{01}(x)=(-3Z\pi^{\frac{1}{2}}/2x^3)-2/x^2, \quad (16)$$

$$U_1=(Z/x)^{\frac{1}{2}}I_3(Z^{\frac{1}{2}}\alpha/x^{\frac{1}{2}}), \quad (17)$$

$$V_1=(Z/x)^{\frac{1}{2}}K_3(Z^{\frac{1}{2}}\alpha/x^{\frac{1}{2}}), \quad (18)$$

where α^2 is again equal to $6\pi^{\frac{1}{2}}$. For the range $0.80\leqslant x<3.20$, we have

$$Q_{02}=-2(Z+1), \quad (19)$$

$$U_2=D_Z(i2^{\frac{1}{2}}x)\exp(x^2/2), \quad (20)$$

$$V_2=D_{-1-Z}(2^{\frac{1}{2}}x)\exp(x^2/2). \quad (21)$$

In Eqs. (20) and (21) $D_n(y)$ is the parabolic cylinder function, which, for any integral value of n, may be expressed in terms of the error function $\Phi(x)$ by means of the recursion formula.[5]

As before, two independent numerical solutions were computed, each satisfying the boundary condition for small x. A linear combination of these was taken to satisfy the boundary condition at $x=3.2$. The resulting values of the velocity distribution function are given in Table I for electrical conductivity and in Table II for thermal conductivity. For $Z=4$ the integration was started at $x=0.20$, and no values were computed for smaller x.

As Z becomes very large, the mutual electron interactions become unimportant, and the following simple formulas for a so-called Lorentz gas are applicable.

$$D(x)=x^4A/Z \text{ for electrical conductivity,} \quad (22)$$

$$D(x)=x^4B(2.5-x^2)/2Z \text{ for thermal conductivity.} \quad (23)$$

Hence, for ready comparison with the Lorentz gas Table I gives values of $ZD(x)/A$, and Table II gives $ZD(x)/B$.

Also listed in Tables I and II are values of the velocity distribution function for Z equal to 16 and ∞. The latter values were found directly from Eqs. (22) and (23). Those for $Z=16$ were determined from an asymptotic series for $D(x)$ in increasing powers of $1/Z$, which we may write

$$D(x)=\sum_{n=1}^{\infty}\frac{d_n(x)}{Z^n}. \quad (24)$$

If Eq. (24) is substituted into Eq. (8) and coefficients of each power of Z are separately set equal to zero, each d_n is given as a function of d_{n-1}, the first and second derivatives of d_{n-1}, and certain integrals over d_{n-1}. The quantity d_1 is simply the value of $D(x)$ for a Lorentz gas, given by Eqs. (22) or (23); successive functions d_n were found by straightforward computation up to $n=4$.

V. RESULTS

In the presence of a weak electrical field $\mathbf{E}$ and a small temperature gradient ∇T, the current density $\mathbf{j}$ and the rate of flow of heat $\mathbf{Q}$ per unit area are given by

$$\mathbf{j}=\sigma\mathbf{E}+\alpha\nabla T, \quad (25)$$

$$\mathbf{Q}=-\beta\mathbf{E}-K\nabla T. \quad (26)$$

In terms of the velocity distribution function $D(x)$, j, summed over all electrons—see Eq. (61) of CSR—is given by

$$j=-2\pi^{-\frac{1}{2}}en_eC(2/3)^{\frac{1}{2}}I_3(\infty), \quad (27)$$

while for the heat flow Q we have

$$Q=\pi^{-\frac{1}{2}}mn_eC^3(2/3)^{5/2}I_5(\infty), \quad (28)$$

where e and m are the electron charge and mass, and C is the root mean square electron velocity. From the numerical values found for the integrals $I_3(\infty)$ and $I_5(\infty)$, values of the coefficients σ, α, β, and K may be determined.

It is convenient to express these transport coefficients in terms of their values in a Lorentz gas. In the case of an electrical field, we define

$$\gamma_E=ZI_3(\infty)/3A, \quad (29)$$

$$\delta_E=ZI_5(\infty)/12A. \quad (30)$$

When $Z=1$, γ_E is identical with the quantity γ introduced in CSR. In the corresponding case where a

TABLE III. Values of transport coefficients.

	$Z=1$	$Z=2$	$Z=4$	$Z=16$	$Z=\infty$
γ_E	0.5816	0.6833	0.7849	0.9225	1.000
γ_T	0.2727	0.4137	0.5714	0.8279	1.000
δ_E	0.4652	0.5787	0.7043	0.8870	1.000
δ_T	0.2252	0.3563	0.5133	0.7907	1.000
ϵ	0.4189	0.4100	0.4007	0.3959	0.4000

[5] See section 16.7, E. T. Whittaker and G. N. Watson, *Modern Analysis* (Cambridge University Press, Cambridge, 1950).

temperature gradient is present, we write

$$\gamma_T = -4ZI_3(\infty)/9B, \tag{31}$$

$$\delta_T = -ZI_5(\infty)/15B. \tag{32}$$

From Eqs. (22) and (23) above the definition of $I_n(\infty)$—see Eq. (37) of CSR—it is readily verified that for a Lorentz gas all four of these coefficients are unity. On elimination of A and B by means of Eq. (40) of CSR and Eq. (13) above, we obtain, after some substitutions,

$$\sigma = \frac{2mC^3}{e^2 Z \ln(qC^2)}\left(\frac{2}{3\pi}\right)^{\frac{1}{2}}\gamma_E, \tag{33}$$

$$\alpha = \frac{3mkC^3}{e^3 Z \ln(qC^2)}\left(\frac{2}{3\pi}\right)^{\frac{1}{2}}\gamma_T, \tag{34}$$

$$\beta = \frac{8m^2C^5}{3e^3 Z \ln(qC^2)}\left(\frac{2}{3\pi}\right)^{\frac{1}{2}}\delta_E, \tag{35}$$

$$K = \frac{20m^2kC^5}{3e^4 Z \ln(qC^2)}\left(\frac{2}{3\pi}\right)^{\frac{1}{2}}\delta_T. \tag{36}$$

Values of the four transport coefficients are given in Table III for various values of Z.

The quantity qC^2, which is essentially h/b_0, is given in Eq. (65) of CSR, while values of $\ln(qC^2)$ are tabulated in Table IV of the same paper. The electron charge e has here been taken in electrostatic units throughout. To obtain the conductivity in practical units (mho) the value found from Eq. (33) must be divided by 9×10^{11}.

It should be noted that if a temperature gradient is present in a steady state, but no steady current is flowing, the electrostatic field E will build up to such a value that j vanishes. This field then reduces the flow of heat, and K', the effective coefficient of heat conductivity is readily shown to be

$$K' = \epsilon K, \tag{37}$$

where

$$\epsilon = 1 - 3\delta_E\gamma_T/(5\delta_T\gamma_E). \tag{38}$$

Values of ϵ are also given in Table III.

It remains to compare these results with those of previous workers. When this work was undertaken, the best available results for the electrical conductivity of a completely ionized gas were those of Chapman and Cowling[6] and of Cowling,[7] who had obtained first and second approximations, respectively, for the conductivity. In terms of the present notation, the Chapman-Cowling method utilizes the expansion

$$D(x) = \frac{3\pi^{\frac{1}{2}}x}{4}\sum_{j=0}^{n-1}(-1)^j\frac{\Delta_{0j}^{(n)}}{\Delta^{(n)}}L_j^{(3/2)}(x^2), \tag{39}$$

where $L_j^{(3/2)}$ is a Laguerre polynomial, and the ratios $\Delta_{0j}^{(n)}/\Delta^{(n)}$ are determined from encounter theory and the Boltzmann equation; n is the order of the approximation used. Since the value of σ found by Cowling[7] with $n=2$ was about twice the value obtained by Chapman and Cowling[6] with $n=1$, it appeared that the present treatment, equivalent to letting $n=\infty$ in Eq. (39), might give a markedly different value.

TABLE IV. Comparison with results obtained by Landshoff for $Z=1$.

n	1	2	3	4	∞
γ_E	0.2945	0.5693	0.5743	0.5777	0.5816

More recently, this same problem has been considered by Landshoff,[8] using the Chapman-Cowling method, but with values of n up to 4. From the values of $\Delta_{0j}^{(n)}/\Delta^{(n)}$ which he gives for $Z=1$, the constant γ_E has been computed, and is given in Table IV, together with the value found in the present work ($n=\infty$).

In view of the large difference between the first and second approximation, it is rather remarkable how close to the truth is the second approximation for γ_E. For thermal conductivity the convergence is somewhat less rapid, with the fourth approximation in close agreement with the value for $n=\infty$. Evidently the present results agree with Cowling's[7] second approximation for the electrical conductivity, provided that for the cutoff in the integration over the impact parameter b, we take the Debye shielding distance h rather than the electronic separation d taken by Cowling. The value 0.490 obtained for γ_E in CSR, in disagreement with Cowling's value 0.578, was the consequence of the neglect of the $\langle\Delta_\xi\Delta_\eta\rangle$ term in $K(ff)$; inclusion of this term removes virtually all the disagreement in γ_E.

It should be emphasized that the present theory considers only those terms in $\partial f_e/\partial t$ which are of order $\ln(h/b_0)$, and a variety of terms of order unity have been neglected, including, for example, the interaction between a high speed electron and its wake of plasma oscillation, an effect explored by Pines and Bohm.[2] Thus, the relative accuracy of the present theory does not exceed $1/\ln(h/b_0)$, or some 5 to 10 percent for most conditions of astrophysical interest. In view of the lack of observational data in this field, development of a more refined theory does not seem worth the very considerable effort required.

[6] S. Chapman and T. G. Cowling, *The Mathematical Theory of Non-Uniform Gases* (Cambridge University Press, Cambridge, 1939).

[7] T. G. Cowling, Proc. Roy. Soc. (London) A183, 453 (1945).

[8] R. Landshoff, Phys. Rev. 82, 442 (1951).

THE

PHYSICAL REVIEW

A journal of experimental and theoretical physics established by E. L. Nichols in 1893

SECOND SERIES, VOL. 107, NO. 1 JULY 1, 1957

Fokker-Planck Equation for an Inverse-Square Force*

MARSHALL N. ROSENBLUTH,† WILLIAM M. MACDONALD,†† AND DAVID L. JUDD
Radiation Laboratory, University of California, Berkeley, California

(Received August 31, 1956)

The contribution to the Fokker-Planck equation for the distribution function for gases, due to particle-particle interactions in which the fundamental two-body force obeys an inverse square law, is investigated. The coefficients in the equation, $\langle\Delta\mathbf{v}\rangle$ (the average change in velocity in a short time) and $\langle\Delta\mathbf{v}\Delta\mathbf{v}\rangle$, are obtained in terms of two fundamental integrals which are dependent on the distribution function itself. The transformation of the equation to polar coordinates in a case of axial symmetry is carried out. By expanding the distribution function in Legendre functions of the angle, the equation is cast into the form of an infinite set of one-dimensional coupled nonlinear integro-differential equations. If the distribution function is approximated by a finite series, the resultant Fokker-Planck equations may be treated numerically using a computing machine. Keeping only one or two terms in the series corresponds to the approximations of Chandrasekhar, and Cohen, Spitzer and McRoutly, respectively.

I. INTRODUCTION

IN dealing with the nonequilibrium properties of systems whose particles obey an inverse-square law of interaction, it is convenient to make use of the fact that under most circumstances small-angle collisions are much more important than collisions resulting in large momentum changes.[1] This leads to the method often used for treating such systems, in which one expands the collision integrand of the Boltzmann equation in powers of the momentum change per collision.

A more generally valid approach to the problem of treating changes in a distribution function resulting from frequently occurring "events," each of which produces a small change in the momentum of a particle, is to use the Fokker-Planck equation, which has been discussed by Chandrasekhar.[2] He has used the formalism of this equation to evaluate the collision terms of the Boltzmann equation under the assumptions that (a) the events producing changes in particle momenta are two-body interactions described by the associated differential scattering cross sections, and (b) that the distribution function is isotropic in space and velocity space. Spitzer and collaborators[3,4] have extended this calculation to the case in which the distribution function is of the form $f^{(0)}+\mu f^{(1)}$, where $f^{(0)}$ and $f^{(1)}$ are isotropic and μ is the direction cosine between the particle trajectory and some preferred direction in space, and $f^{(1)}$ is assumed to be small.

It is the purpose of this paper to present the mechanics of two-body collisions in a somewhat simplified form, and to derive the Fokker-Planck equation for an arbitrary distribution function. From this general equation such special cases as those of Chandrasekhar and Spitzer may easily be obtained. For more complex situations the equation is suitable for integration by an electronic computer.

II. FORMULATION OF THE PROBLEM

The Boltzmann equation for the change of the molecular distribution function is given by

$$\frac{\partial f_a}{\partial t}+v^{\mu}\frac{\partial f_a}{\partial x^{\mu}}+\frac{F^{\mu}}{m}\frac{\partial f_a}{\partial v^{\mu}}=\left(\frac{\partial f_a}{\partial t}\right)_c, \tag{1}$$

* This work was done under the auspices of the U. S. Atomic Energy Commission.

† Present address: General Atomic, San Diego, California.

†† Present address: Physics Department, University of Maryland, College Park, Maryland.

[1] S. Chapman and T. G. Cowling, *Mathematical Theory of Non-Uniform Gases* (Cambridge University Press, London, 1952), second edition, pp. 178–179.

[2] S. Chandrasekhar, Revs. Modern Phys. 15, 1 (1943).

[3] Cohen, Spitzer, and McRoutly, Phys. Rev. 80, 230 (1950). A more complete list of references is given in this paper.

[4] L. Spitzer and R. Harm, Phys. Rev. 89, 977 (1953).

where f_a is the number of molecules of type a per unit volume in the phase space of configuration and velocity, and $\mathbf{F}$ is an external force field. The summation on repeated Greek indices is understood in this paper, while sums on Roman letters denoting molecular species are explicitly indicated. The quantity $(\partial f_a/\partial t)_c$ represents the change in the distribution function produced by collisions, and it is this term with which we are concerned. Since the interactions take place between molecules within the same small region in space, we need only consider the velocity dependence of the distribution function in evaluating this term.

The Fokker-Planck equation, which is simply a conservation equation, gives the time rate of change of f due to collisions as

$$\left(\frac{\partial f_a}{\partial t}\right)_c = -\frac{\partial}{\partial v^\mu}(f_a\langle\Delta v^\mu\rangle_a)+\frac{1}{2}\frac{\partial^2}{\partial v^\mu\partial v^\nu}(f_a\langle\Delta v^\mu\Delta v^\nu\rangle_a), \quad (2)$$

where v^μ is the component of particle velocity in Cartesian coordinates and $\langle\Delta v^\mu\rangle_a$ is the average increment per unit time of the μth component of velocity of a molecule of type a. The derivation of this equation rests on the approximation that small changes in v^μ are the most probable and that terms involving higher powers of Δv^μ contribute negligibly to $(\partial f_a/\partial t)_c$.[2] In the next section we give a more precise statement of the approximation made here.

In calculating the average values $\langle\Delta v^\mu\rangle$ and $\langle\Delta v^\mu\Delta v^\nu\rangle$, we make the usual assumption that changes in velocity v^μ result from two-particle interactions, or collisions during which spatial correlation effects (polarization effects or multiple collisions) are unimportant. For many situations this assumption is believed to be justified, as is indicated by the work of Chapman, Ferraro, and Persico,[1] and more recently, Gasiorowicz, Neuman, and Riddell.[5] The expression for $\langle\Delta v^\mu\rangle_a$ becomes

$$\langle\Delta v^\mu\rangle_a=\sum_b \int d\mathbf{v}' f_b(v'^\mu)\int d\Omega\sigma(u,\Omega)u\Delta v^\mu, \quad (3)$$

where u is the magnitude of the relative velocity $|v'^\mu - v^\mu|$, $\sigma(u)$ is the differential scattering cross section, Ω is the scattering solid angle, and Δv^μ is the change in v'^μ resulting from the collision. The increment Δv^μ has been integrated over all scattering angles, all velocities v^μ of the scattering particle, and has been summed over all the species of particles. Similarly the average value $\langle\Delta v^\mu\Delta v^\nu\rangle_a$ is given by

$$\langle\Delta v^\mu\Delta v^\nu\rangle_a=\sum_b \int d\mathbf{v}' f_b(v'^\mu)\int d\Omega\sigma(u)u\Delta v^\mu\Delta v^\nu. \quad (4)$$

The differential scattering cross section that we use in Eqs. (3) and (4) is that for an inverse-square law of force,

$$\sigma(\Omega)=(e^4/4m_{ab}^2u^4)[\sin(\theta/2)]^{-4}, \quad (5)$$

where $m_{ab}=m_am_b/(m_a+m_b)$ is the reduced mass of the colliding particles and θ is the scattering angle in the center-of-mass system.

III. DERIVATION OF THE EQUATION

We first discuss the kinematics of the collision between a molecule of type a and belocity $\mathbf{v}$ and a molecule of type b and velocity $\mathbf{v}'$. The relation between $\mathbf{v}_a$, the velocity $\mathbf{V}$ of the center of mass, and the relative velocity $\mathbf{u}=\mathbf{v}-\mathbf{v}'$ is

$$\mathbf{v}_a=\mathbf{V}+\frac{m_b}{m_a+m_b}\mathbf{u}.$$

The change in the μth component of $\mathbf{v}_a$ is given by

$$\Delta v^\mu=\frac{m_b}{m_a+m_b}\Delta u^\mu. \quad (6)$$

We find it convenient to introduce a local Cartesian coordinate system with unit vectors $\mathbf{e}_1'$, $\mathbf{e}_2'$, $\mathbf{e}_3'$ whose relation to the fixed system $\mathbf{e}_1$, $\mathbf{e}_2$, $\mathbf{e}_3$ is given by

$$\mathbf{e}_1'=\frac{\mathbf{u}}{u},\quad \mathbf{e}_2'=\frac{\mathbf{e}_3\times\mathbf{u}}{[(u^1)^2+(u^2)^2]^{\frac{1}{2}}},\quad \mathbf{e}_3'=\mathbf{e}_1'\times\mathbf{e}_2', \quad (7)$$

and in which the relative velocity has components u_L^μ. The changes in the components of u_L^μ produced by a collision are easily calculated in the local Cartesian coordinates, since the relative velocity vector simply undergoes a rotation through an angle θ,

$$\begin{aligned}\Delta u_L^1&=-2u\sin^2(\theta/2),\\ \Delta u_L^2&=2\mu\sin(\theta/2)\cos(\theta/2)\cos\phi,\\ \Delta u_L^3&=2u\sin(\theta/2)\cos(\theta/2)\sin\phi.\end{aligned} \quad (8)$$

A diagram of the scattering is shown in Fig. 1. The changes in the components of the relative velocity $\mathbf{u}$ in the fixed coordinate system are related to these changes in the local system by

$$\begin{aligned}\Delta u^\mu&=(\mathbf{e}_\mu\cdot\mathbf{e}_\nu')\Delta u_L^\nu,\\ \Delta u^\mu\Delta u^\nu&=(\mathbf{e}_\mu\cdot\mathbf{e}_\sigma')(\mathbf{e}_\nu\cdot\mathbf{e}_\omega')\Delta u_L^\sigma\Delta u_L^\omega.\end{aligned} \quad (9)$$

We can next calculate the change of relative velocity in the local system for all collisions by integrating over the scattering angles θ, ϕ, which will be denoted as follows:

$$\{\Delta u_L^\mu\}\equiv\int d\Omega\sigma u(\Delta u_L^\mu). \quad (10)$$

Using Eqs. (5) and (8), we have

$$\{\Delta u_L^1\}=-\left(\frac{\pi e^4}{m_{ab}^2u^2}\right)\int_0^\pi \frac{\sin^2(\theta/2)\sin\theta}{\sin^4(\theta/2)}d\theta, \quad (11)$$

[5] Gasiorowicz, Neuman, and Riddell, Phys. Rev. 101, 922 (1956).

where we have performed the integration over ϕ. The integral diverges logarithmically at small angles, and we therefore introduce a cutoff at $\theta_{\min}$ to obtain

$$\{\Delta u_L^1\} \simeq -(4\pi e^4/m_{ab}^2u^2)\ln(2/\theta_{\min}). \qquad (12)$$

The small-angle deflections correspond to scatterings with very large impact parameters, and the divergence arises from the long-range nature of the Coulomb forces. The divergence is eliminated, however, when we take into account the shielding that arises from the polarization of charge surrounding the scatterer.[6] The polarization screens the scattering particle and provides a natural cutoff on the maximum impact parameter of the order of a Debye length $\lambda_D=(kT/4\pi ne^2)^{\frac{1}{2}}$, and a value for the logarithm in Eq. (12) of

$$\ln(2/\theta_m)=\ln D=\ln(\tfrac{1}{2}m_{ab}u^2)(e^2/\lambda_D)^{-1},$$

$$\lambda_D=(kT/4\pi ne^2Z_{\rm eff})^{\frac{1}{2}}.$$

In this equation kT is proportional to the average kinetic energy, n is the number of particles per unit volume, and e is the electronic charge. The quantity D, which is the ratio of the Debye length λ_D to the classical distance of closest approach $(\frac{1}{2}mu^2/e^2)$ for two particles of relative velocity u, in most cases of interest will be a very large number so that $\ln D\gg 1$. From Eqs. (3), (5), and (8) one can easily see that terms of higher order in Δv^μ, like $\langle\Delta v^\mu\Delta v^\nu\Delta v^\omega\rangle$, will not contain $\ln D$, and that the neglect of these terms in the Fokker-Planck equation is therefore justified. The insensitivity of $\ln D$ to the precise value of u means that we can simplify our further development by neglecting the weak dependence on u and using the value for a Maxwell-Boltzmann distribution of $\frac{1}{2}mu^2\cong\frac{3}{2}kT$. It would probably not be justified in any event to consider the argument of the logarithm as better determined than this.

The remaining integrations yield

$$\{\Delta u_L^2\}_a=\{\Delta u_L^3\}_a=0, \qquad (13)$$

and

$$\{(\Delta u_L^1)^2\}=0,$$

$$\{(\Delta u_L^2)^2\}=\{(\Delta u_L^3)^2\}=(4\pi e^4/m_{ab}^2u^2)\ln D, \qquad (14)$$

with all other second-order terms zero (compared to $\ln D$).

Using these results with Eqs. (6), (7), and (9), we can immediately write down the integrals in the fixed coordinate system,

$$\{\Delta v^\mu\}_a=-[4\pi e^4(\ln D)/m_{ab}m_au^3]u^\mu,$$

$$\{\Delta v^\mu\Delta v^\nu\}_a=[4\pi e^4(\ln D)/m_a^2]\{\delta^{\mu\nu}-u^\mu u^\nu/(u)^2\}. \qquad (15)$$

These equations can be simplified by noting that

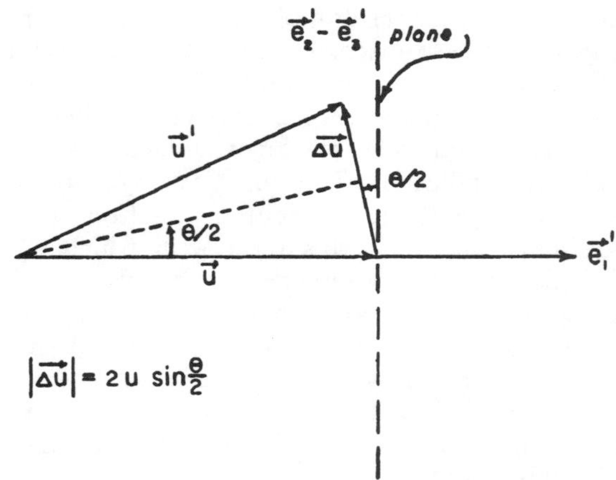

Fig. 1. Diagram showing kinematics of an elastic scatter in the local Cartesian system.

$u=[(v^\mu-v'^\mu)(v^\mu-v'^\mu)]^{\frac{1}{2}}$, so that we have

$$\{\Delta v^\mu\}=\Gamma_a\frac{m_a}{m_{ab}}\frac{\partial}{\partial v^\mu}\frac{1}{u},\quad \{\Delta v^\mu\Delta v^\nu\}=\Gamma_a\frac{\partial^2(u)}{\partial v^\mu\partial v^\nu}, \qquad (16)$$

$$\Gamma_a\equiv(4\pi e^4\ln D/m_0^2).$$

Finally, from Eqs. (3), (4), and (16), we obtain

$$\langle\Delta v^\mu\rangle_a=\sum_b\int d\mathbf{v}'f(\mathbf{v}')\{\Delta v^\mu\}_a=\Gamma_a(\partial h_a/\partial v^\mu), \qquad (17)$$

$$\langle\Delta v^\mu\Delta v^\nu\rangle_a=\Gamma_a(\partial^2g/\partial v^\mu\partial v^\nu), \qquad (18)$$

where

$$h_a(\mathbf{v})=\sum_b\frac{m_a+m_b}{m_b}\int d\mathbf{v}'f_b(\mathbf{v}')|\mathbf{v}-\mathbf{v}'|^{-1},$$

$$g(\mathbf{v})=\sum_b\int d\mathbf{v}'f_b(\mathbf{v}')|\mathbf{v}-\mathbf{v}'|. \qquad (19)$$

It is interesting to note a formal similarity with potential theory,

$$\nabla_v^2h_a\equiv(\partial^2/\partial v^\mu\partial v^\mu)h_a=-4\pi\sum_b\frac{m_a+m_b}{m_b}f_b(v),$$

$$\nabla_v^4g=(\partial^4/\partial v^\mu\partial v^\nu\partial v^\mu\partial v^\nu)g=-8\pi\textstyle\sum_b f_b. \qquad (20)$$

Substituting Eq. (18) into Eq. (2), we obtain the Fokker-Planck equation for an arbitrary distribution function:

$$\frac{1}{\Gamma_a}\left(\frac{\partial f_a}{\partial t}\right)_c=-\frac{\partial}{\partial v^i}\left(f_a\frac{\partial h_a}{\partial v^i}\right)+\frac{1}{2}\frac{\partial^2}{\partial v^i\partial v^j}\left(f_a\frac{\partial^2g}{\partial v^i\partial v^j}\right). \qquad (21)$$

In the general case this fourth-order, three-dimensional, time-dependent, and nonlinear partial differential equation seems quite difficult to handle. In many cases, however, there are simplifications which result when a coordinate system is adopted that embodies the natural symmetries of a problem. For example, in many

[6] The choice of a cutoff is discussed at length in reference 3.

problems there will be a preferred direction in space, such as the direction of an external applied field, with azimuthal symmetry about this direction. Polar coordinates seem especially suitable for such a problem.

IV. TRANSFORMATION OF THE EQUATION

Although it is possible to transform Eq. (21) by a straightforward change of variables, the procedure is tedious and unnecessarily involved. A much simpler and more direct procedure is to write the equation in a covariant form valid in any set of curvilinear coordinates q^1, q^2, and q^3. Let the expression for distance between two points whose coordinates differ by dq^1, dq^2, and dq^3 be

$$(ds)^2 = a_{\mu\nu} dq^\mu dq^\nu,$$

where $a_{\mu\nu}$ is a metric tensor, and let $a^{\mu\nu}\det a$ be the cofactor of $a_{\mu\nu}$ in the matrix $a=(a_{\mu\nu})$, i.e., $a^{\mu\omega}a_{\omega\nu}=\delta_\nu{}^\mu$. We observe that the quantities $T_a{}^\mu = \Gamma_a{}^{-1}\langle\Delta v_a{}^\mu\rangle$ and $S_a{}^{\mu\nu}=\Gamma_a{}^{-1}\langle\Delta v^\mu \Delta v^\nu\rangle_a$ transform like a contravariant vector and tensor, respectively, between different Cartesian coordinate systems. The appropriate tensor extension of Eq. (2) is therefore

$$\Gamma_a{}^{-1}(\partial f_a/\partial t)_c = -(fT_a{}^\mu)_{,\mu} + \tfrac{1}{2}(fS_a{}^{\mu\nu})_{,\mu\nu}, \quad (22)$$

where the commas indicate covariant derivatives with respect to the q^μ. In any Cartesian coordinate system Eq. (22) has precisely the form of Eq. (2). We can now write Eq. (18) in a covariant form,

$$T_a{}^\mu = a^{\mu\nu}(h_a)_{,\nu}, \quad S^{\mu\nu} = a^{\mu\omega}a^{\nu\tau}(g_{,\omega\tau}). \quad (23)$$

The two covariant derivatives that appear in Eq. (23) can be found from

$$(h_a)_{,\nu} = \partial h_a/\partial q^\nu, \quad g_{,\omega\tau} = \partial^2 g/\partial q^\omega \partial q^\tau - \left\{ \begin{matrix} \sigma \\ \omega\tau \end{matrix} \right\} (\partial g/\partial q^\sigma), \quad (24)$$

where $\left\{ \begin{matrix} \sigma \\ \omega\tau \end{matrix} \right\}$ is a Christoffel symbol of the second kind defined by

$$\left\{ \begin{matrix} \nu \\ \omega\mu \end{matrix} \right\} = a^{\nu\tau}[\omega\mu,\tau] = \tfrac{1}{2}a^{\nu\tau}(\partial a_{\omega\tau}/\partial q^\mu + \partial a_{\mu\tau}/\partial q^\omega - \partial a_{\omega\mu}/\partial q^\tau). \quad (25)$$

The covariant derivative $(fT_a{}^\mu)_{,\mu}$ can be simply expressed

$$(fT_a{}^\mu)_{,\mu} = a^{-\frac{1}{2}}(\partial/\partial q^\mu)(a^{\frac{1}{2}} f T_a{}^\mu), \quad a=\det(a_{\mu\nu}), \quad (26)$$

and for $(fS^{\mu\nu})_{,\mu\nu}$,

$$(fS^{\mu\nu})_{,\mu\nu} = a^{-\frac{1}{2}}(\partial^2/\partial q^\mu \partial q^\nu)(a^{\frac{1}{2}} f S^{\mu\nu}) + a^{-\frac{1}{2}}(\partial/\partial q^\nu)\left[a^{\frac{1}{2}}\left\{ \begin{matrix} \nu \\ \omega\mu \end{matrix} \right\} f S^{\mu\omega}\right]. \quad (27)$$

The writing of Eq. (22) in arbitrary curvilinear coordinates now becomes a straightforward application of Eqs. (23), (24), (26), and (27), in that order.

As an example we can easily write down the equation in spherical polar coordinates in *velocity space*, assuming azimuthal symmetry about the $\theta=0$ symmetry axis, so that we have $f(v,\mu)$, where $\mu=\cos\theta$. In these coordinates we have

$$\begin{aligned}
&q^1=v, \quad q^2=\mu, \quad q^3=\phi,\\
&ds^2 = dv^2 + v^2(1-\mu^2)^{-1}(d\mu)^2 + v^2(1-\mu^2)(d\phi)^2,\\
&a_{11}=1, \quad a_{22}=v^2(1-\mu^2)^{-1}, \quad a_{33}=v^2(1-\mu^2),\\
&\qquad\qquad a_{ij}=0 \quad \text{if } i\neq j, \qquad (28)\\
&a^{11}=1, \quad a^{22}=v^{-2}(1-\mu^2), \quad a^{33}=v^{-2}(1-\mu^2)^{-1},\\
&\qquad\qquad a^{ij}=0 \quad \text{if } i\neq j,\\
&a=\det(a_{\mu\nu})=v^4.
\end{aligned}$$

From Eqs. (23), (24), and (26) we obtain

$$\begin{aligned}
&T_a{}^1 = (\partial h_a/\partial v), \quad T_a{}^2 = v^{-2}(1-\mu^2)(\partial h/\partial \mu), \quad T_a{}^3=0,\\
&(fT_a{}^\mu)_{,\mu} = v^{-2}(\partial/\partial v)(fv^2\partial h_a/\partial v) \qquad (29)\\
&\qquad + v^{-2}(\partial/\partial\mu)[f(1-\mu^2)\partial h_a/\partial\mu].
\end{aligned}$$

The second-rank tensor $S^{\mu\nu}$ follows in the same way:

$$\begin{aligned}
&S^{11} = \partial^2 g/\partial v^2,\\
&S^{22} = v^{-4}(1-\mu^2)^2[\partial^2 g/\partial\mu^2 + v(1-\mu^2)^{-1}(\partial g/\partial v)\\
&\qquad - \mu(1-\mu^2)^{-1}\partial g/\partial\mu],\\
&S^{13}=S^{23}=0, \qquad (30)\\
&S^{12} = v^{-2}(1-\mu^2)[\partial^2 g/\partial v\partial\mu - v^{-1}\partial g/\partial\mu],\\
&S^{33} = v^{-4}(1-\mu^2)^{-1}[v\partial g/\partial v - \mu\partial g/\partial\mu].
\end{aligned}$$

Using Eq. (26) we calculate the second covariant derivative of $(fS^{\mu\nu})_{,\mu\nu}$ and can then write down Eq. (22) as

$$\begin{aligned}
&\Gamma_a{}^{-1}(\partial f_a/\partial t)_c\\
&= -v^{-2}(\partial/\partial v)(f_a v^2 \partial h_a/\partial v) - v^{-2}(\partial/\partial\mu)\\
&\quad \times[f_a(1-\mu^2)\partial h_a/\partial\mu] + (2v^2)^{-1}(\partial^2/\partial v^2)\\
&\quad \times(f_a v^2 \partial^2 g/\partial v^2) + (2v^2)^{-1}(\partial^2/\partial\mu^2)\{f_a[v^{-2}(1-\mu^2)^2\\
&\quad \times(\partial^2 g/\partial\mu^2) + v^{-1}(1-\mu^2)(\partial g/\partial v) - v^{-2}\mu(1-\mu^2)\\
&\quad \times(\partial g/\partial\mu)]\} + v^{-2}(\partial^2/\partial\mu\partial v)\{f_a(1-\mu^2)\\
&\quad \times[(\partial^2 g/\partial\mu\partial v) - v^{-1}(\partial g/\partial\mu)]\} + (2v^2)^{-1}(\partial/\partial v)\\
&\quad \times\{f_a[-v^{-1}(1-\mu^2)(\partial^2 g/\partial\mu^2) - 2(\partial g/\partial v)\\
&\quad + 2\mu v^{-1}(\partial g/\partial\mu)]\} + (2v^2)^{-1}(\partial/\partial\mu)\\
&\quad \times\{f_a[v^{-2}\mu(1-\mu^2)(\partial^2 g/\partial\mu^2) + 2\mu v^{-1}(\partial g/\partial v)\\
&\qquad + 2v^{-1}(1-\mu^2)(\partial^2 g/\partial\mu\partial v) - 2v^{-2}(\partial g/\partial\mu]\}. \qquad (31)
\end{aligned}$$

The equation that describes a system of particles interacting according to an inverse-square law of force when there exists an axis of symmetry is now obtained by combining Eqs. (1), (19), and (31). The quantities h_a and g can be given in terms of two-dimensional integrals,

$$h_a(v,\mu)=\sum_b \frac{m_a+m_b}{m^b}\int_0^\infty dv'v'^2\int_{-1}^1 d\mu' \times f_b(v',\mu')\Lambda(v',\mu';v,\mu), \quad (32)$$

$$g(v,\mu)=\sum_b \int_0^\infty dv'v'^2\int_{-1}^1 d\mu' f_b(v',\mu')\Omega(v',\mu';v,u),$$

with Λ and Ω defined in terms of the complete elliptic integrals K and E as follows:

$$\Lambda=4[v^2+v'^2-2vv'(\mu\mu'-[(1-\mu^2)(1-\mu'^2)]^{\frac{1}{2}})]^{-\frac{1}{2}} \times K\left(\left\{\frac{4vv'[(1-\mu^2)(1-\mu'^2)]^{\frac{1}{2}}}{v^2+v'^2-2vv'[(1-\mu^2)(1-\mu'^2)]^{\frac{1}{2}}}\right\}^{\frac{1}{2}}\right),$$

$$\Omega=4[v^2+v'^2-2vv'(\mu\mu'-[(1-\mu^2)(1-\mu'^2)]^{\frac{1}{2}})]^{\frac{1}{2}} \times E\left(\left\{\frac{4vv'[(1-\mu^2)(1-\mu'^2)]^{\frac{1}{2}}}{v^2+v'^2-2vv'(\mu\mu'-[(1-\mu^2)(1-\mu'^2)]^{\frac{1}{2}}}\right\}^{\frac{1}{2}}\right). \quad (33)$$

The spatially homogeneous two-dimensional time-dependent Eq. (31) is not too complex for electronic digital computers. Moreover, Eq. (31) forms a useful starting point for developing an approximate distribution function when axial symmetry exists. A method for reducing the Eq. (31) to a coupled set of one-dimensional nonlinear integrodifferential equations which can be treated quite simply numerically will be given.

V. REDUCTION OF THE FOKKER-PLANCK EQUATION FOR AXIAL SYMMETRY

The solution to Eq. (31) can be expanded in a series of Legendre polynomials:

$$f_a(v,\mu)=\sum_{n=0}^{\infty} a_n^{(a)}(v)P_n(\mu). \quad (34)$$

This expansion provides an expansion of the two functions $h_a(v,\mu)$ and $g(v,\mu)$, which can be obtained from Eq. (19). We first evaluate the integral appearing in the definition of $h_a(v,\mu)$. Let us define

$$P_n(\mu)A_n^{(a)}(v,\mu)\equiv\int d\mathbf{v}'a_n^{(a)}(v')P_n(\mu')|\mathbf{v}-\mathbf{v}'|^{-1}. \quad (35)$$

Then, inserting

$$|\mathbf{v}-\mathbf{v}'|^{-1}=(2\pi^2)^{-1}\int d\mathbf{k}e^{i\mathbf{k}\cdot(\mathbf{v}-\mathbf{v}')}k^{-2}$$

into Eq. (35), we have

$$P_n(\mu)A_n^{(a)}(v,\mu) =(2\pi^2)^{-1}\int d\mathbf{v}'a_n^{(a)}(v')\int d\mathbf{k}e^{i\mathbf{k}\cdot(\mathbf{v}-\mathbf{v}')}k^{-2}P_n(\mu'). \quad (36)$$

Writing

$$\mathbf{k}\cdot(\mathbf{v}-\mathbf{v}')=\mathbf{k}\cdot\mathbf{v}-kv'[\mu''\mu'+(1-\mu'^2)^{\frac{1}{2}}(1-\mu''^2)^{\frac{1}{2}}\cos\phi],$$

where $\mu''=\cos(\mathbf{k},\mathbf{e}_z)$, $\mu'=\cos(\mathbf{v}',\mathbf{e}_z)$, and ϕ is the angle between the plane of $\mathbf{v}'$ and $\mathbf{e}_z$ and the plane of $\mathbf{k}$ and $\mathbf{e}_z$, we have

$$\int e^{-i\mathbf{k}\cdot\mathbf{v}'}P_n(\mu')d\mu'd\phi'=2\pi\int e^{-ikv'\mu''\mu'} \times J_0(kv'[(1-\mu''^2)(1-\mu'^2)]^{\frac{1}{2}})^{\frac{1}{2}}P_n(\mu')d\mu'. \quad (37)$$

Using a formula given by Watson[7] (12.14) we can finally integrate this to

$$\int e^{-i\mathbf{k}\cdot\mathbf{v}'}P_n(\mu')d\mu'd\phi'=2\pi(2\pi/kv')^{\frac{1}{2}}(-i)^n \times P_n(\mu'')J_{n+\frac{1}{2}}(kv'). \quad (38)$$

If we write $\mathbf{k}\cdot\mathbf{v}=kv\{\mu\mu''+[(1-\mu^2)(1-\mu''^2)]^{\frac{1}{2}}\cos\gamma\}$, where $\mu=\cos(\mathbf{v},\mathbf{e}_z)$ and γ is the angle between the plane of $(\mathbf{k},\mathbf{e}_z)$ and $(\mathbf{v},\mathbf{e}_z)$, we can employ the same formula to integrate Eq. (36) with respect to k, obtaining

$$A_n^{(a)}(v,\mu)=4\pi\int_0^\infty dv'v'^2a_n^{(a)}(v')\int_0^\infty dkJ_{n+\frac{1}{2}}(kv) \times J_{n+\frac{1}{2}}(kv')k^{-1}(vv')^{-\frac{1}{2}}. \quad (39)$$

The integral over k is found in Watson[7] (13.42) also:

$$\int_0^\infty dkJ_{n+\frac{1}{2}}(kv)J_{n+\frac{1}{2}}(kv')k^{-1}=(2n+1)^{-1}(v_</v_>)^{n+\frac{1}{2}},$$

where $v_<$ is the smaller of v, v', and $v_>$ is the greater. Thus the final result is

$$a_n^{(a)}(v,\mu)=4\pi(2n+1)^{-1}\left[\int_0^v dv'\frac{(v')^{n+2}}{v^{n+1}}a_n^{(a)}(v') +\int_v^\infty dv'\frac{v^n}{(v')^{n-1}}a_n^{(a)}(v')\right]. \quad (40)$$

The expansion for $h_a(v,\mu)$ follows from Eqs. (19) and (35):

$$h_a(v,\mu)=\sum_{n=0}\sum_b(m_a+m_b)m_b^{-1}a_n^{(b)}(v,\mu)P_n(\mu). \quad (41)$$

The expansion for $g(v,\mu)$ can be found in the same way by first using

$$|\mathbf{v}-\mathbf{v}'|=-\pi^{-2}\int d\mathbf{k}e^{i\mathbf{k}\cdot(\mathbf{v}-\mathbf{v}')}k^{-4}.$$

If we define

$$P_n(\mu)B_n^{(a)}(v)\equiv\int d\mathbf{v}'a_n^{(a)}(\mathbf{v}')\int d\mathbf{k}e^{i\mathbf{k}\cdot(\mathbf{v}-\mathbf{v}')}P_n(\mu')k^{-4}, \quad (42)$$

[7] G. N. Watson, *Theory of Bessel Functions* (Cambridge University Press, London, 1944), second edition.

the same steps followed above give

$$B_n^{(a)}(v)=2\pi\int_0^\infty dv'(v')^2a_n^{(a)}(v')\int_0^\infty dkJ_{n+\frac{1}{2}}(kv)\times J_{n+\frac{1}{2}}(kv')k^{-2}(vv')^{-\frac{1}{2}}. \quad (43)$$

The k integral can be evaluated in terms of the hypergeometric function ${}_2F_1(a,b;c;z)$ (reference 7, Sec. 13.4):

$$\int_0^\infty dkJ_{n+\frac{1}{2}}(kv)J_{n+\frac{1}{2}}(kv')k^{-2}=\tfrac{1}{2}(v_<)^{n+\frac{1}{2}}(v_>)^{-n+\frac{1}{2}}(n^2-\tfrac{1}{4})^{-1}\times{}_2F_1(n-\tfrac{1}{2},-1;n+\tfrac{3}{2};v_<^2/v_>^2). \quad (44)$$

The hypergeometric function appearing here is actually a polynomial, and the result for $B_n^{(a)}(v)$ is

$$B_n^{(a)}(v)=-4\pi(4n^2-1)^{-1}\times\left[\int_0^v dv'a_n^{(a)}(v')\frac{(v')^{n+2}}{v^{n-1}}\left(1-\frac{n-\frac{1}{2}}{n+\frac{3}{2}}\frac{(v'^2)}{v^2}\right)+\int_v^\infty dv'a_n^{(a)}(v')\frac{v^n}{(v')^{n-3}}\left(1-\frac{n-\frac{1}{2}}{n+\frac{3}{2}}\frac{v^2}{(v')^2}\right)\right]. \quad (45)$$

The expansion for $g(v,\mu)$ follows from Eqs. (19) and (42):

$$g(v,\mu)=\sum_{n=0}^\infty\sum_b B_n^{(b)}(v)P_n(\mu). \quad (46)$$

The procedure for obtaining an approximate solution to Eq. (31) is to retain terms in $f_a(v,\mu)$ to some order N.

$$f_a(v,\mu)\simeq\sum_{n=0}^N a_n^{(a)}(v)P_n(\mu), \quad (47)$$

and obtain the corresponding expansions of $h_a(v,\mu)$ and $g_a(v,\mu)$, which also are to order N. These expressions are now inserted in Eq. (31) and the result expressed as a series in Legendre polynomials. Of use for this purpose is

$$P_{l'}(\mu)P_l(\mu)=\sum_{k=0}^\infty C_{ll'k}P_k(\mu), \quad (48)$$

where the $C_{ll'k}$ are given in Condon and Shortley.[8] Assuming spatial homogeneity, we find that the velocity-dependent term $v_\mu(\partial f_a/\partial v_\mu)$ of the Boltzmann differential operator Eq. (1) can also be expanded in Legendre polynomials. Equating coefficients of Legendre polynomials of the same order in the expansions of Eqs. (1) and (31), one now obtains a system of coupled one-dimensional nonlinear integro-differential equations.

The two simplest approximations are the following (*a*) $f_a(v,\mu)$ in $h_a(v,\mu)$ and $g(v,\mu)$ is isotropic and Eq. (31) is the equation given by Chandrasekhar; (*b*) $f_a(v,\mu)=a_0(v)+a_1(v)P_1(\mu)$, and Eq. (31) is the equation used by Spitzer and collaborators.[3,4]

This work was begun while the first two authors were at the University of California Radiation Laboratory.

[8] E. U. Condon and G. H. Shortley, *Theory of Atomic Spectra* (Cambridge University Press, London, 1935), Sec. 967.

DYNAMICS OF THE GEOMAGNETIC TAIL

B. Coppi, G. Laval, and R. Pellat
International Centre for Theoretical Physics, Trieste, Italy
(Received 13 January 1966)

In a paper recently published,[1] Ness has reported the experimental evidence of a magnetically neutral sheet behind the earth at a distance of (20-30)R_e (earth radii). This sheet separates regions of oppositely directed magnetic fields in the magnetic tail (Fig. 1). It has been proposed by Ness[1] and subsequent authors that the dynamics of this sheet may have an essential role in geomagnetic phenomena. This suggests as a model the analysis of the stability of a pinch containing a neutral sheet.[2] In fact, theory and laboratory experiments[3-5] clearly show that this configuration is violently unstable as it breaks up into separate pinches (Fig. 2), lying on the neutral sheet, which tend to repel each other. Considering the order of magnitude of the sheet thickness (600 km), and the energy of the electrons there contained, we see immediately that collisional effects (such as resistivity) cannot play a role in the dynamics of the neutral sheet. For this, we can take up the stability analysis of a collisionless pinch,[2,6] with the intention

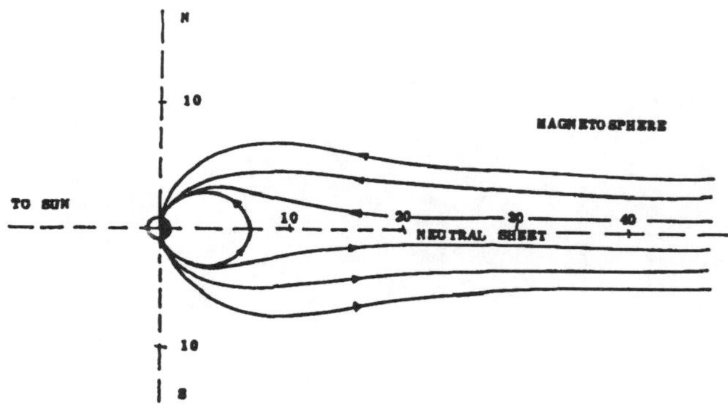

FIG. 1. Projection of magnetic field topology on noon-midnight meridian plane in the vicinity of the neutral sheet (unperturbed configuration). Distance in earth radii.

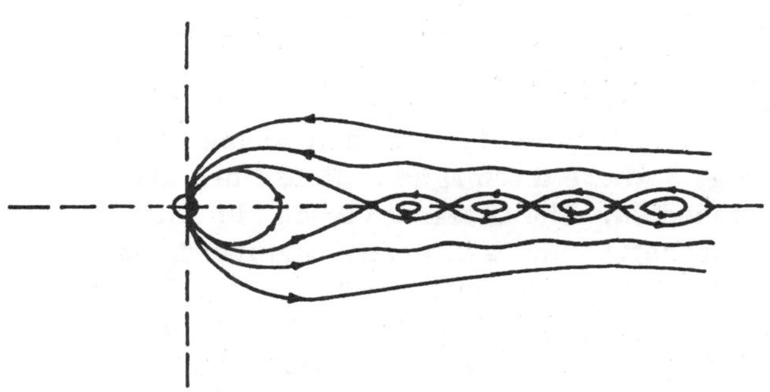

FIG. 2. Effects of the instability on magnetic field topology around the neutral sheet.

to show that the relevant instability has macroscopic effects,[7] as it transforms magnetic energy into kinetic energy, and that it can be suitable to explain the characteristic times of evolution of phenomena observed during auroral events, in the auroral regions, and in the magnetic tail.

We consider for this a one-dimensional model as in Fig. 1 where we set the X axis along the S-N direction, the Z axis on the ecliptic plane pointing from the earth to the sun, and the Y axis perpendicularly to them. The relevant perturbations from this (equilibrium) configuration are constant along the equilibrium current lines, in the Y direction, and have wavelengths parallel to the magnetic field lines, in the Z direction, of the order of the sheet thickness or larger. The result of the instability is a reconnection of initially antiparallel magnetic field lines. This occurs through the transfer of macroscopic energy of the plasma to a relatively small number of electrons through microscopic particle-wave resonant processes. The heated electrons lie on a thin sheet around the plane of zero magnetic field.

All these detailed features have not been observed as yet, but we must remember that the relatively small transverse scale (in the X direction) and the short duration of these phenomena make their direct detection by satellite difficult. Now, we show, in order to prove our point, that by computing the rate at which magnetic energy is transferred to the particles we obtain a growth time in agreement with the one which can be formally derived by the full theory.[6]

Take as equilibrium distribution functions for each species of particle

$$f_j^{\,0} \approx n_0 \exp\{-\Theta_j^{-1}[\tfrac{1}{2}m_j v^2 - V_j(m_j v_y + q_j A_y^{\,0})]\},$$

where v is the total particle velocity, m_j and q_j the mass and the charge, $\Theta_j = kT_j$ represents the temperature, V_j the flow velocity, n_0 the maximum density, and $A_y^{\,0}$ is the only component of the magnetic vector potential. The electric potential is zero as we choose the coordinate frame where $V_i/\Theta_i = -V_e/\Theta_e$. Then using Maxwell's equations, we find for the equilibrium magnetic field $B_y^{\,0} = B_0 \tanh(x/\lambda)$, and for the density $n = n_0/\cosh^2(x/\lambda)$, where $B_0^{\,2}/8\pi = n_0(\Theta_i + \Theta_e)$, $\lambda^{-1} = qB_0(V_i - V_e)/2c(\Theta_i + \Theta_e)$, c being the velocity of light and λ the sheet thickness. For the mode we consider, the perturbed magnetic potential A^1 has the form $A^1 \approx A_y^{\,1}(x)\exp(i\omega t + ikz)$. Then, if W is the kinetic energy, we have

$$\frac{dW}{dt} = \sum_j q_j \int dx\, E_y^{\,1} v_y f_j^{\,1} d^3v,$$

where $E_y^{\,1}$ is the perturbed electric field $(-1/c) \times (\partial A_y^{\,1}/\partial t)$. In the neutral sheet the particle motion can be considered as free and determined by two reflecting walls with distance $d \simeq (r_{\rm L}\lambda)^{1/2}$, $r_{\rm L}$ being the mean-particle Larmor radius. Using the linearized collisionless Boltzmann equation, we can write the approximate form of the perturbed distribution function in this region as

$$f_j^{\,1} = (q_j f_j^{\,0}/\Theta_j)\{V_j A_y^{\,1} - iE_y^{\,1} v_y(\omega + kv_z)^{-1}\},$$

where the first term on the right-hand side represents the influence of the perturbed magnetic field and the second one represents the interaction between wave and particles. The latter contribution is negligible outside the neutral sheet $(x > d)$ as the particle motion becomes adiabatic and there is no electric field along the magnetic lines of force. Then, since it is physically interesting to consider a low-frequency mode, we assume $\omega \ll kv_{{\rm th}i}$ $(v_{{\rm th}i}^{\,2} \equiv \Theta_i 2/m)$ and take $(\omega/k + v_z)^{-1} \approx i\pi\delta(v_z) + {\rm P}(1/v_z)$. Noticing that the integration of the principal part gives no contribution we find

$$\frac{dW}{dt} = \frac{\pi}{k}\sum_j \frac{q_j}{\Theta_j}\int_{d_j}^{d_j} dx \int d^3v\, f_j^{\,0}\delta(v_z)(E_y^{\,1}v_y)^2 - \frac{1}{4\pi}\frac{d}{dt}\int dx \frac{n(x)}{n(0)}\left(\frac{A_y^{\,1}}{\lambda}\right)^2 \equiv \sum_j \frac{d}{dt}W_j^{\,R} - \frac{d}{dt}W^M. \quad (1)$$

Here W_j^R represents the increment of kinetic energy for the resonant particles of species j, contained in the region $|x|<d$, and W^M is the decrement of the macroscopic kinetic energy of the remaining particles.

One can see that the kinetic-ion resonant term is smaller than the electron one by the factor $(r_{Le}/r_{Li})^{1/2}$. On the other hand, by conservation of the total energy,

$$\frac{dW}{dt}=-\frac{1}{8\pi}\frac{d}{dt}\int dx\,[B^1]^2. \tag{2}$$

Then comparing the two expressions (1) and (2) we have

$$W_e^R=\frac{1}{\tau}\int_{-d_e}^{d_e}dx\int d^3v f_e^{\,0}\delta(v_z)\,|(A_y^{\,1})^2 v_y|^2$$

$$=-\frac{k\Theta_e}{8\pi e^2}\int dx\left\{\left|\frac{\partial A_y^{\,1}}{\partial x}\right|^2+|A_y^{\,1}|^2\left[k^2-\frac{2}{\lambda^2\cosh^2(x/\lambda)}\right]\right\}$$

$$=W^M-\frac{1}{8\pi}\int dx(B^1)^2, \tag{3}$$

where $\tau=-i/\omega$ represents the growth time.

We see from Eq. (3) that the instability is purely growing and that it exists for long wavelengths, such that $k^2\lambda^2<1$. Also, these features are typical of an instability having macroscopic effects. In particular, the growth time is of the order of $(\lambda/r_{Le})^2 d_e/v_{the}$ in agreement with the formal result of Ref. 6. Therefore, Eq. (3) proves clearly that free macroscopic energy is transferred to resonant electrons contained in a slab of thickness $d_e\approx(r_{Le}\lambda)^{1/2}$ on the ecliptic plane. Using the exact form $\tau=\pi^{-1/2}(2\lambda/r_{Le})^{3/2}(\lambda/v_{the})\Theta_e/(\Theta_e+\Theta_i)$ and the results of measurements by the IMP-1 satellite,[1] we take $2\lambda=600$ km, $B_0\simeq1.6\times10^{-4}$ G, $\Theta_i\simeq1$ keV, and obtain

$$\tau\approx15\text{ sec if }\Theta_e\simeq10\text{ eV}:$$

$$\tau\approx5\text{ sec if }\Theta_i\simeq1\text{ keV},$$

Θ_e being the electron temperature previous to the perturbation. In particular we see that the growth time is not very sensitive to the electron temperature. The energy to which particles are accelerated by the instability can be estimated by evaluating the electric field as resulting from the variation of magnetic flux across a contour of height λ. In a time of a few seconds, as consistent with the growth time, the electrons can attain an energy of the order of 10 keV, as observed during auroral events.[8]

This acceleration is maximum at the nodes of Fig. 2 where the magnetic field lines are reconnected. The relevant region, in which the magnetic field is violently perturbed, has a width of the order of 20 earth radii (1.2×10^5 km along the Y axis) and a height of the order of $d_e\approx50$ km corresponding to $\Theta_e\simeq1$ keV. The corresponding area, at the earth's surface, crossed by the same tubes of force is $\approx1.8\times10^3$ km^2, for a magnetic field strength ratio of 3×10^{-4}. If the longitudinal extent of this region is $\approx6\times10^3$ km then the latitude width becomes 0.3 km. Thus the above-mentioned region maps into one, near the earth, which is extended in longitude and very narrow in latitude, reminiscent of auroral arcs,[9,10] that have the same geometrical features.

To support this interpretation we can quote the fact that[10] the oval belt of the aurora coincides roughly with the intersection curve between the ionosphere and the outer magnetosphere. The lines of force of the magnetic tail terminate inside the oval belt. After a violent solar flare, the inner magnetosphere shrinks and correspondingly one observes a displacement of the oval belt toward the equator. During the subsequent auroral substorm, the arcs increase in intensity and move polewards, indicating a reconnection of the magnetic field lines which are the closest to the ecliptic plane, as predicted by the theory.

Our simple analysis cannot explain the presence of energetic electron spikes (electron islands) detected far from the neutral sheet in the magnetic tail. If one attributes their presence to large-amplitude magnetic perturbations[11] (solitons) propagating at an angle with the magnetic field, one can ask the question whether these nonlinear perturbations come from the boundary of the outer magnetosphere or from the instability of the neutral sheet which gives rise to macroscopic fluctuations of the magnetic field.

Finally, if the auroral substorm phenomenon[9,10] can be associated with the instability of the neutral sheet, the recovery time of the substorm may be related to the time to re-form the sheet. For this we can suppose that the particles on the night side have sufficient pressure to blow the outer reconnected magnetic field lines farther up to the stage where a sheet with opposite lines of force is formed.

It is a pleasure to thank Dr. C. F. Kennel and Dr. N. F. Ness for their profitable suggestions, and Professor Abdus Salam and the International Atomic Energy Agency for hospitality at the International Centre for Theoretical Physics, Trieste, Italy.

[1]N. F. Ness, J. Geophys. Res. 70, 2989 (1965).

[2]H. P. Furth, in Advanced Plasma Theory, edited by M. N. Rosenbluth (Academic Press, New York, 1964).

[3]R. H. Lovberg, J. Beuford, H. Davis, D. Nyman, and R. Siemon, Seventh Annual Meeting of the Division of Plasma Physics of The American Physical Society, San Francisco, California (to be published).

[4]H. A. B. Bodin, Nucl. Fusion 3, 215 (1963).

[5]A. Eberhagen and H. Glaser, Nucl. Fusion 4, 296 (1964).

[6]G. Laval, R. Pellat, and M. Vuillemin, in Proceedings of the Second International Conference on Plasma Physics and Thermonuclear Fusion (International Atomic Energy Agency, Vienna, 1965), Paper CN-21/71.

[7]Considerations based on instabilities of microscopic type had been used previously to explain auroral precipitation [J. W. Chamberlain, J. Geophys. Res. 68, 5667 (1963)] but they were subject to criticism based on the amount of energy and the number of particles involved in an auroral event [B. Coppi, Nature 201, 998 (1965)].

[8]B. J. O'Brien, Science 148, 449 (1965).

[9]S. I. Akasofu, Planetary Space Sci. 12, 273 (1964).

[10]S. I. Akosofu, Sci. Am. 213, No. 12, 54 (1965).

[11]M. A. Ginzburg, Phys. Rev. Letters 16, 326 (1966).

PLASMA WAVE ECHO*

R. W. Gould
California Institute of Technology, Pasadena, California

and

T. M. O'Neil and J. H. Malmberg
General Atomic Division of General Dynamics Corporation,
John Jay Hopkins Laboratory for Pure and Applied Science, San Diego, California
(Received 29 May 1967)

It is shown that if a longitudinal wave is excited in a collision-free plasma and Landau-damps away, and a second wave is excited and also damps away, then a third wave (i.e., the echo) will spontaneously appear in the plasma.

It has long been recognized that electron plasma waves can be damped, even in the absence of collisions.[1] Collisionless damping (Landau damping) has been the subject of extensive theoretical treatments in recent years and is now believed to play an important role in many related, but more complicated, oscillation and instability phenomena. Only recently has Landau damping been demonstrated experimentally.[2] Landau's treatment shows that macroscopic quantities such as the electric field and the charge density are damped exponentially, but the perturbations in the electron phase-space distribution $f(x,v,t)$ oscillate indefinitely. Since the electron density is given by $n_e = \int f(x,v,t)dv$, one may think of the damping as arising out of the phase mixing of various parts of the distribution function. In this paper, we will show how the direction of the phase evolution of the perturbed distribution function can be reversed by the application of a second electric field. This results in the subsequent reappearance of a macroscopic field (i.e., the echo), many Landau-damping periods after the application of the second pulse. The plasma echo is related to other known echo phenomena[3] in that the decay of a macroscopic physical quantity of the system, caused by phase mixing of rapidly oscillating microscopic elements in the system, is reversed by reversing the direction of phase evolution of the microscopic elements.

The basic mechanism behind the plasma echo can easily be understood. When an electric field of spatial dependence e^{-ik_1x} is excited in a plasma and then Landau-damps away, it modulates the distribution function leaving a perturbation of the form[1] $f_1(v)\exp[-ik_1x+ik_1vt]$. For large t, there is no electric field associated with this perturbation, since a velocity integral over it will phase mix to 0. If after a time τ a wave of spatial dependence e^{ik_2x} is excited and then damps away, it will modulate the unperturbed part of the distribution leaving a first-order term of the form $f_2(v)\exp[ik_2x-ik_2v(t-\tau)]$, but it will also modulate the perturbation due to the first wave leaving a second-order term of the form

$$f_1(v)f_2(v)\exp[i(k_2-k_1)x+ik_2v\tau-i(k_2-k_1)vt].$$

The coefficient of v in this exponential will vanish when $t=\tau[k_2/(k_2-k_1)]$; so at this time a velocity integral over this term will not phase mix to 0, and an electric field will reappear in the plasma. If τ is long compared with a collisionless damping period and $[k_2/(k_2-k_1)]$ is of order unity, then this third electric field will appear long after the first two waves have damped away (i.e., it will be an echo).

This echo phenomenon can be rigorously derived from the collisionless Boltzmann equation and Poisson's equation. For the sake of simplicity, we limit the presentation to one dimension and treat the ions as a uniform positive background charge. If we assume that the electron distribution is initially spatially homogeneous, $f(x,v,t=0)=f_0(v)$, and that the two externally applied pulses are given by[4]

$$\varphi_{\text{ext}} = \Phi_{k_1}\cos(k_1x)\delta[\omega_pt] + \Phi_{k_2}\cos(k_2x)\delta[\omega_p(t-\tau)],$$

then the Fourier transform of the spatial dependence and Laplace transform of the time dependence of the Boltzmann equation and Poisson's equation can be written as

$$(p+ikv)\tilde{f}_k(v,p)=\frac{e}{m}ik\tilde{\varphi}_k(p)\frac{\partial f_0}{\partial v}+\frac{e}{m}\sum_q{}'\int\frac{dp'}{2\pi}i(k-q)\tilde{\varphi}_{k-q}(p-p')\frac{\partial f_q}{\partial v}(p'), \tag{1}$$

$$k^2\tilde{\varphi}_k(p)=4\pi ne\int dv\tilde{f}_k(v,p)+\frac{k_1{}^2\Phi_{k_1}}{2\omega_p}[\delta_{k,k_1}+\delta_{k,-k_1}]+\frac{k_2{}^2\Phi_{k_2}}{2\omega_p}[\delta_{k,k_2}+\delta_{k,-k_2}]e^{-p\tau}, \tag{2}$$

where $\tilde{\varphi}_k(p)$ and $\tilde{f}_k(v,p)$ are the transformed electric potential and distribution function, and the prime on the sigma in Eq. (1) indicates that the $q=0$ term is being treated separately in the manner usual for mode coupling calculations.[5]

To solve Eqs. (1) and (2), we expand in terms of the applied potentials Φ_{k_1} and Φ_{k_2}. The first-order (or linear) solution just describes a Landau-damped plasma wave following each pulse.[1] The second-order solution associated with wave number[6] $k_3\equiv k_2-k_1$ can be written as

$$\varphi_{k_3}{}^{(2)}(t)=\frac{e}{m}\frac{\Phi_{k_1}\Phi_{k_2}k_1k_2}{4k_3{}^2}\int_{\infty}^{+\infty}dv\int_{-i\infty+\sigma}^{i\infty+\sigma}\frac{dp}{2\pi i}\int_{-i\infty+\sigma'}^{i\infty+\sigma'}\frac{dp'}{2\pi i}\frac{ik_3}{\epsilon(k_3p)(p+ik_3v)^2}\frac{\partial f_0}{\partial v}$$

$$\times\left[\frac{e^{pt}e^{-p\tau'}}{\epsilon(k_2,p')\epsilon(-k_1,p-p')(p'+ik_2v)}+\frac{e^{p(t-\tau)}e^{p'\tau}}{\epsilon(k_2,p-p')\epsilon(-k_1,p')(p'-ik_1v)}\right], \tag{3}$$

where $\epsilon(k,p)\equiv 1-\omega_p{}^2/k^2\int dv(\partial f_0/\partial v)[v+p/ik]^{-1}$ is the Landau dielectric function[1] and the p and p' contours are defined by requiring that $0<\sigma'<\sigma$. To carry out the p and p' integrations, we use the Cauchy residue method, closing the contours on the side which produces vanishingly small exponentials in the numerator. If we assume that τ is long compared with a collisionless damping period and that the time between the second pulse and the echo is same order as τ {i.e., that $|\gamma(k_1)\tau|$, $|\gamma(k_2)\tau|$, $|\gamma(k_3)\tau|\gg 1$ and that $[k_1/k_3]\simeq 1$}, then the residues at the poles arising from the roots of the dielectric constants will all be exponentially small and we may neglect them. Picking up the poles at $p'=ik_1v$ and $p=-ik_3v$ yields the result

$$\varphi_{k_3}{}^{(2)}(t)=\frac{e}{m}\frac{\Phi_{k_1}\Phi_{k_2}k_1k_2ik_1\tau}{4k_3{}^2}\int_{-\infty}^{+\infty}dv\frac{\partial f_0}{\partial v}\left\{\frac{e^{iv[k_1\tau-k_3(t-\tau)]}}{\epsilon(-k_1,ik_1v)\epsilon(k_2,-ik_2v)\epsilon(k_3,-ik_3v)}\right\}. \tag{4}$$

This integral does not phase mix to 0 when $k_3(t-\tau)\simeq k_1\tau$ [i.e., when $t\simeq\tau'\equiv\tau(k_2/k_3)$], and this results in the echo. One recognizes the various dielectric functions in the denominator as resulting from the effect of the self-consistent fields associated with first and second pulses and the echo. By setting these dielectric functions equal to unity, one recovers the result for weakly interacting or free streaming particles [an easy limit in which to check Eq. (4)].

When $t<\tau'$, we can evaluate the integral in Eq. (4) by closing the contour in the upper half v plane.[7] In this region of the v plane, we pick up poles from the Landau roots of $\epsilon(-k_1,ik_1v)$. On the other hand, when $t>\tau'$, we must close the contour in the lower half v plane and we pick up poles from the Landau roots of $\epsilon(k_3,-ik_3v)$ and $\epsilon(k_2,-ik_2v)$. However, we may neglect the latter compared with the former, since our assumption that $k_1/k_3\simeq 1$ implies that $k_2\simeq 2k_3$ and that $|\gamma(k_2)|\gg|\gamma(k_3)|$. Carrying out the integrations for these two cases and using a Maxwellian of mean thermal energy T to evaluate the dielectric functions yields the following time-asymptotic solution (i.e., $|t-\tau'|\omega_p>1$):

$$\varphi_{k_3}{}^{(2)}(t)=\Phi_{k_1}(\omega_p\tau)\frac{e\Phi_{k_2}}{T}\frac{k_1{}^4k_2L_D{}^2}{k_3(k_1+k_3)^2}$$

$$\times\frac{-(k_3/k_1)\gamma(k_1)e^{\gamma(k_1)k_3/k_1(\tau'-t)}\cos[\omega(k_1)(k_3/k_1)(\tau'-t)+\delta]}{\{\omega_p[(k_3-k_1)/(k_3+k_1)]^2+\gamma(k_1)^2\}^{1/2}},\quad\text{for } t<\tau'$$

$$\times\frac{\gamma(k_3)e^{\gamma(k_3)(t-\tau')}\cos[\omega(k_3)(t-\tau')+\delta']}{\{\omega_p{}^2[(k_3-k_1)/(k_3+k_1)]^2+\gamma(k_3)^2\}^{1/2}},\quad\text{for } t>\tau', \tag{5}$$

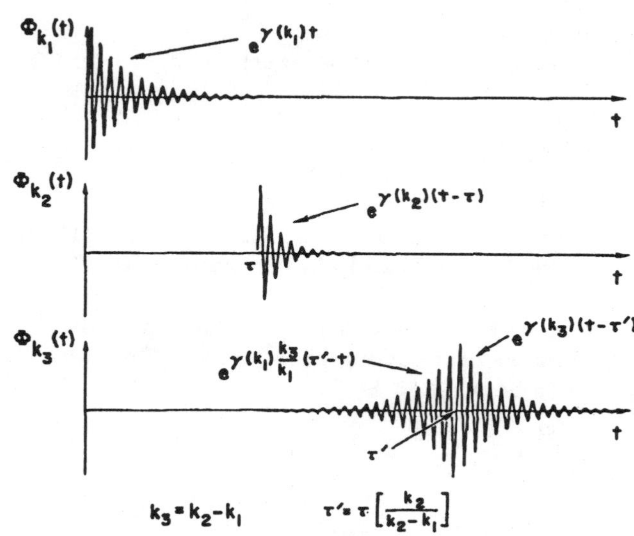

FIG. 1. Approximate variation of the principal Fourier coefficients of the self-consistent field for the case $k_3 \cong k_1 \cong \frac{1}{2}k_2$. Upper line: response to the first pulse; middle line: response to the second pulse; lower line: echo.

where

$$\tan\delta = \gamma(k_1)(k_3 - k_1)/\omega_p(k_3 + k_1)$$

and

$$\tan\delta' = \gamma(k_3)(k_1 - k_3)/\omega_p(k_1 + k_3).$$

It is interesting to note that the echo is not symmetric in that it grows up at the rate $\exp[\gamma(k_1)k_3/k_1(\tau'-t)]$ and damps away at the rate $\exp[\gamma(k_3)\times(t-\tau')]$.

The results of both the first- and second-order calculations are summarized in Fig. 1. The exponentials written in this figure indicate the general dependence of the envelopes of the oscillating curves, which have actually been drawn for the case where $k_1 \simeq k_3$.

The above calculation was based on the collisonless Boltzmann equation and is invalidated if collisions are strong enough to destroy the phase information before the echo can appear. Small angle Coulomb collisions are particularly effective in this regard, since the Fokker-Planck operator representing these collisions enhances the collision rate by a factor $(kv\tau)^2 \simeq (\omega_p\tau)^2$ when operating on a perturbation of the form $e^{ikv\tau}$. By working in a marginal range, one might be able to use this effect as a tool to measure the Coulomb collision rate, even though the neutral collision rate is somewhat higher.

We have considered several variations on the above calculation. Although in this paper we have discussed explicitly only second-order echoes, higher order echoes are also possible. For example, a third-order echo is produced when the velocity space perturbation from the first pulse is modulated by a spatial harmonic of the electric field from the second pulse. The echo then occurs at $t = \tau 2k_2/(2k_2 - k_1)$ or $t = 2\tau$ when $k_2 = k_1$. This result is more closely related to echoes of other types[3] which are also third order for small amplitudes.

It is possible also to have spatial echoes, and these will probably be easier to observe experimentally than the temporal echoes described above. If an electric field of frequency ω_1 is continuously excited at one point in a plasma and an electric field of frequency $\omega_2 > \omega_1$ is continuously excited at a distance l from this point, then a spatial echo of frequency $\omega_2 - \omega_1$ will appear at a distance $l\omega_1/(\omega_2 - \omega_1)$ from the point where the second field is excited.

Finally, although our discussion has been entirely in terms of electron wave echoes, it is clear that the above treatment can be extended in a straightforward manner to include ion dynamics, and this leads to temporal as well as spatial ion wave echoes.

An observation of plasma echoes would be of fundamental interest, since it would experimentally verify the reversible nature of collisionless damping. The analogy with spin echoes[3] strongly suggests the possible use of the ehco technique as a means for studying collisional relaxation phenoma in plasmas.

---

*This research was sponsored in part by the Office of Naval Research under Contract No. Nonr-220(50), and in part by the Defense Atomic Support Agency under Contract No. DA-49-146-XZ-486.

[1]L. Landau, J. Phys. USSR 10, 45 (1946).

[2]A. Y. Wong, N. D'Angelo, and R. W. Motley, Phys. Rev. 133, A436 (1964); J. H. Malmberg and C. B. Wharton, Phys. Rev. Letters 6, 184 (1964); J. H. Malmberg, C. B. Wharton, and W. E. Drummond, in Proceedings of the 1965 Culham Conference (International Atomic Energy Agency, Vienna, 1966), Vol. I, 485.

[3]E. L. Hahn, Phys. Rev. 80, 580 (1950); R. M. Hill and D. E. Kaplan, Phys. Rev. Letters 14, 1062 (1965); R. W. Gould, Phys. Letters 19, 477 (1965); I. D. Abella, N. A. Kurnit, and S. R. Hartman, Phys. Rev. 141, 391 (1966).

[4]Φ_{k_1} and Φ_{k_2} have the dimensions of electric potential owing to our inclusions of ω_p in the arguments of the delta functions.

[5]W. E. Drummond and D. Pines, Nucl. Fusion Suppl. 3, 1049 (1962).

[6]Of course, $\Phi_{k_1+k_2}$, $\Phi_{k_1+k_1}$, and $\Phi_{k_2+k_2}$ all have second-order terms, but there is no echo associated with these terms.

[7]The $\partial f_0/\partial v$ term in Eq. (4) looks like it makes the integrand diverge for large imaginary v, but this term is actually canceled by a similar term hidden in $\epsilon(-k_1, ik_1 v)$.

METHOD FOR SOLVING THE KORTEWEG-deVRIES EQUATION*

Clifford S. Gardner, John M. Greene, Martin D. Kruskal, and Robert M. Miura
Plasma Physics Laboratory, Princeton University, Princeton, New Jersey
(Received 15 September 1967)

A method for solving the initial-value problem of the Korteweg-deVries equation is presented which is applicable to initial data that approach a constant sufficiently rapidly as $|x| \to \infty$. The method can be used to predict exactly the "solitons," or solitary waves, which emerge from arbitrary initial conditions. Solutions that describe any finite number of solitons in interaction can be expressed in closed form.

For a large class of physical systems, nonlinear and dispersive processes compete while dissipation is negligible. In particular, the Korteweg-deVries (KdV) equation,

$$u_t - 6uu_x + u_{xxx} = 0 \tag{1}$$

(subscripts x and t denoting partial differentiations), has been shown to describe the asymptotic development of small- but finite-amplitude shallow-water waves,[1] hydromagnetic waves in a cold plasma,[2] ion-acoustic waves,[3] and acoustic waves in an anharmonic crystal.[4]

The quantities u, x, and t can be rescaled to produce any desired coefficients for the terms of Eq. (1). The present choice is convenient for this paper. Note that u is reversed in sign from previous work since the coefficient of the second term is negative. Further, the KdV equation is Galilean-invariant so that $u(x-6Vt, t)-V$ forms a one-parameter family of solutions.

Previous numerical computations,[5] as well as more recent ones,[6] indicate that for large t the solution of an initial-value problem consists of a finite train of "solitons," or solitary waves, traveling to the right, and an oscillatory train or "tail" spreading to the left. The solitons exhibit a remarkable stability in that their identity is preserved through nonlinear interactions. This property of solitons, which was discovered numerically[5] and justifies the name suggestive of particles, has been proved by Lax[7] for two of them, and can be demonstrated for any number using the solution described below.

We now sketch a general method of solution that can be used to establish these results rigorously. It is applicable to initial data that approach a constant sufficiently rapidly as $|x| \to \infty$. The Galilean invariance described above permits us to set this constant equal to zero.

First consider the differential equation[8]

$$\psi_{xx}-(u-\lambda)\psi=0, \tag{2}$$

where $u(x,t)$ is a solution of Eq. (1), so that $\psi(x,t)$ and $\lambda(t)$ depend parametrically on t. Solving Eq. (2) for u and inserting the result in Eq. (1) yields

$$\lambda_t\psi^2+[\psi Q_x-\psi_x Q]_x=0, \tag{3}$$

with

$$Q\equiv\psi_t+\psi_{xxx}-3(u+\lambda)\psi_x, \tag{4}$$

for the time development of the solutions of Eq. (2). If ψ vanishes as $|x|\to\infty$, the second term of Eq. (3) vanishes on integration over the interval $(-\infty,\infty)$. Therefore $\lambda_t=0$, i.e., the discrete eigenvalues of Eq. (2) are constant when u evolves according to the KdV equation.

Dropping the first term, we can integrate Eq. (3) twice to yield

$$\psi_t+\psi_{xxx}-3(u+\lambda)\psi_x=C\psi+D\varphi. \tag{5}$$

Here $C(t)$ and $D(t)$ are the constants of integration, and φ is a solution of Eq. (2) that is linearly independent of ψ. Thus $\varphi\equiv\psi\int^x dx/\psi^2$.

It is now straightforward to compute the evolution of ψ in regions where u vanishes, and, in particular, asymptotically for $|x|\to\infty$. For a (time-independent) discrete eigenvalue $\lambda_n<0$, $D=0$ because the corresponding ψ_n satisfies Eq. (5) and vanishes exponentially as $|x|\to\infty$, and $C=0$ because we are assuming the normalization $\int\psi_n^2dx=1$. Then inserting

$$\psi_n\approx c_n(t)\exp(-\kappa_n x) \text{ for } x\to\infty, \tag{6}$$

with $\kappa_n=(-\lambda_n)^{1/2}>0$ into Eq. (5), we find

$$c_n(t)=c_n(0)\exp(4\kappa_n^3t). \tag{7}$$

The analogous coefficients for large negative x decay exponentially in time.

For $\lambda=k^2>0$, a solution of Eq. (2) for large $|x|$ is a linear combination of $\exp(\pm ikx)$. We impose on ψ the boundary conditions

$$\psi\approx\exp(-ikx)+b\exp(ikx),\quad x\to\infty, \tag{8}$$

$$\psi\approx a\exp(-ikx),\quad x\to-\infty. \tag{9}$$

In the frequent interpretation of Eq. (2) as describing the normal modes of a wave equation, the coefficients of unity in Eq. (8) and (implied) zero in Eq. (9) indicate prescribed steady radiation arriving from $+\infty$ only. The coefficients of transmission $a(k,t)$ and reflection $b(k,t)$ can be shown to satisfy $|a|^2+|b|^2=1$.

The spectrum for $\lambda>0$ is continuous and we may choose λ constant, so that Eq. (5) is again valid. Inserting Eqs. (8) and (9) into Eq. (5) and equating the coefficients of the two independent solutions at $+\infty$ and at $-\infty$, we find $D=0$, $C=4ik^3$, and two equations which integrate trivially to yield

$$a(k,t)=a(k,0), \tag{10}$$

$$b(k,t)=b(k,0)\exp(8ik^3t). \tag{11}$$

This information on the development of ψ is sufficient to reconstruct u for any value of time! Specifically, given the reflection coefficient $b(k)$ and the κ_n and c_n, let $K(x,y)$ for $y\geq x$ be the solution of the Gel'fand-Levitan equation,[9,10]

$$K(x,y)+B(x+y)+\int_x^\infty K(x,z)B(y+z)dz=0, \tag{12}$$

with

$$B(\xi)\equiv\frac{1}{2\pi}\int_{-\infty}^{\infty}b(k)\exp(ik\xi)dk+\sum_n c_n^2\exp(\kappa_n\xi). \tag{13}$$

Then

$$u(x,t)=2(d/dx)K(x,x). \tag{14}$$

The evolution of $u(x,t)$ is obtained from the explicit dependence on time of $b(k)$ and the c_n given by Eqs. (11) and (7). [In all these formulas the signs of x, y, and z have been reversed from Kay's[10] usage, thus the reference end in Eqs. (6), (8), and (9) is $+\infty$.] Note that $K(x, x)$ as determined by Eq. (12) is independent of values of $B(\xi)$ for $\xi < 2x$.

A number of results can be established by further elaboration of this method, which we mention without going into details.

When u represents a single soliton, there is perfect transmission $[b(k) \equiv 0]$ and exactly one discrete eigenvalue $\lambda_1 = \frac{1}{2}u_{\min}$. More generally, Kay and Moses[11] have given the general solution of Eq. (12) with $b(k) \equiv 0$ in closed form in terms of exponentials. This includes all cases where u decomposes exactly into solitons.

It is more difficult to find exact solutions when $b(k)$ does not vanish. The time dependence of $b(k)$ indicates a strong phase mixing in the integral of Eq. (13) as $t \to \infty$ for positive ξ. The behavior for negative ξ is more complicated since the integrand then has points of stationary phase. This is reflected (in computer studies) by the "tail" moving toward the left.

Since the c_n grow exponentially, as long as there is at least one of them $B(\xi)$ can be approximated by the summation when Eq. (12) is to be solved for $x > 0$ and $t \to \infty$. The solution then reduces to that found by Kay and Moses[11] described above. Thus the magnitude, velocity, and position of each soliton can be found in the limit of large time. Furthermore, the solitons for large negative time can be found from the usual version[10] of Eq. (12) where the reference end is $-\infty$.

A fuller treatment together with other applications and generalizations will be published subsequently.

---

*This work was performed under the auspices of the U. S. Air Force Office of Scientific Research, Contract No. AF49(638)-1555.

[1]D. J. Korteweg and G. deVries, Phil. Mag. 39, 422 (1895).

[2]C. S. Gardner and G. K. Morikawa, Courant Institute of Mathematical Sciences, New York University, Report No. NYO 9082, 1960 (unpublished).

[3]H. Washimi and T. Taniuti, Phys. Rev. Letters 17, 966 (1966).

[4]N. J. Zabusky, in Mathematical Models in Physical Sciences, edited by S. Drobot (Prentice-Hall, Inc., Englewood Cliffs, New Jersey, 1963).

[5]N. J. Zabusky and M. D. Kruskal, Phys. Rev. Letters 15, 240 (1965).

[6]N. J. Zabusky, to be published.

[7]P. D. Lax, private communication.

[8]R. M. Miura, C. S. Gardner, and M. D. Kruskal, to be published.

[9]I. M. Gel'fand and B. M. Levitan, Izv. Akad. Nauk SSSR, Ser. Mat. 15, 309 (1951) [translated in American Mathematical Society Translations (American Mathematical Society, Providence, Rhode Island, 1955), Ser. 2, Vol. 1, p. 253].

[10]I. Kay, Courant Institute of Mathematical Sciences, New York University, Report No. EM-74, 1955 (unpublished); I. Kay and H. E. Moses, Nuovo Cimento 3, 276 (1956).

[11]I. Kay and H. E. Moses, J. Appl. Phys. 27, 1503 (1956).

Stabilization of Resistive Kink Modes in the Tokamak*

A. H. Glasser, H. P. Furth, and P. H. Rutherford
Plasma Physics Laboratory, Princeton University, Princeton, New Jersey 08540
(Received 24 September 1976)

Optimized current profiles are shown to be capable of providing simultaneous stability against all resistive kink modes in the tokamak.

The efficient utilization of the magnetic field in a tokamak increases with the ratio B_θ/B_φ of poloidal to toroidal field strength. In order to avoid unstable helical magnetohydrodynamics (MHD) perturbations (kink modes) of the form $\exp i(m\theta - n\varphi)$, the "safety factor" $q(r) \equiv 2\pi/\iota(r) = rB_\varphi/RB_\theta$ must, however, be restricted.[1] If there is a radial range wherein $q(r)<1$, then the fundamental mode with $m=1$ and $n=1$ is unstable,[1] whether or not the plasma is perfectly conducting at the point where $q(r)=1$. Higher modes (with $m>1$) can be unstable only as resistive kink, or tearing, modes[2] for which the singular point, where $q(s) = m/n$, falls into a resistive region.

The object of this Letter is to demonstrate, within a constant-resistivity, cylindrical theory, the existence of q profiles that provide simultaneous stability against all the low-m kink modes, while minimizing the limiter value $q_a \equiv q(a)$. The principle is contained in a comparison theorem[2] that states the following: For two profiles of the rotational transform ι having the same shear $(d\iota/dr)_s$ and the same transform ι_s at the singular point of a given mode, if the two profiles everywhere satisfy $|\iota_1(r)-\iota_s| > |\iota_2(r)-\iota_s|$, then $\iota_1(r)$ is more stable against the given mode than $\iota_2(r)$. Resistive kink instabilities can also be eliminated[3] (in sufficiently hot plasmas) by a local pressure gradient at s due to favorable average toroidal curvature,[4] or by the proximity of a perfectly conducting exterior shell.

To illustrate optimum profiles, we will consider two cases: case A, a profile with $q_a > 2$, giving stability against all finite-m modes, without need of a conducting shell or of toroidal-curvature effects; and case B, a similar profile, but adding a conducting shell to achieve stability at $q_a < 2$.

We consider first a straight cylindrical configuration, and neglect pressure-gradient effects. The magnetic perturbations outside the resistive layer satisfy the equation for a marginal MHD mode, namely,[1,2]

$$\frac{1}{r^2}\frac{\partial}{\partial r}\left(r^2\frac{\partial\psi}{\partial r}\right) - \frac{m^2}{r^2}\psi = \frac{m}{rF}\frac{dj}{dr}\psi, \tag{1}$$

where $\psi = irB_r/m$ is the perturbed poloidal flux function, $j = d(rB_\theta)/rdr$ is the equilibrium longitudinal current density, and

$$F = \vec{k}\cdot\vec{B} = (m - nq)B_\theta/r. \tag{2}$$

For a *marginal* resistive mode, Eq. (1) is satisfied everywhere, i.e., there is no discontinuity at the singular surface.

Analytic solutions of Eq. (1) can readily be found for the model of $j(r)$ shown in Fig. 1. In this model, there is a central current channel of radius $r=c$ with uniform current density $j(r)=j_1$, surrounding by a "pedestal" of radius $r=a$ with uniform, but lower, current density $j(r)=j_2$. For $r>a$, the current density vanishes, so that the limiter could be placed just outside $r=a$, with a conducting shell at $r=b$.

Within $0<r<c$, the solution of Eq. (1) is given by

$$\psi/\psi_c = (r/c)^m. \tag{3}$$

Across $r=c$, the matching conditions are that $\psi = \psi_c$ be continuous, and that

$$\frac{[\psi']_c}{\psi_c} = \frac{m}{cF_c}[j]_c = -\frac{2m(1-p)}{c(m-nq_c)}, \tag{4}$$

where the square brackets $[\,]_c$ denote the discontinuity of the variable across $r=c$, and $p = j_2/j_1$ is a factor describing the height of the pedestal. (Given q_c and q_a, values of p are possible within

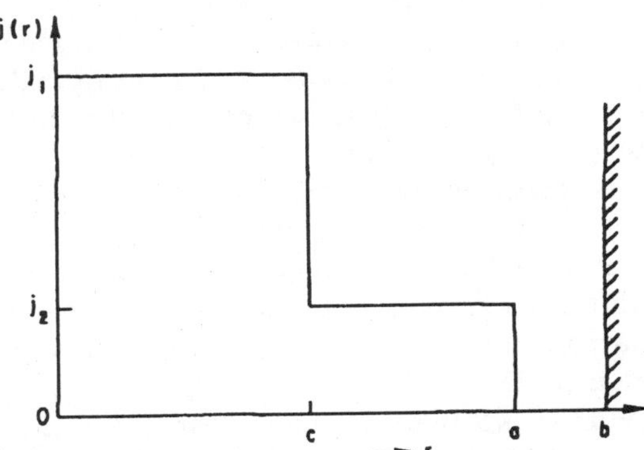

FIG. 1. Simple model of $j(r)$ used in the analytic calculation.

the range $0<p<q_c/q_a$; the relation

$$\frac{c^2}{a^2}=\frac{q_c/q_a-p}{1-p} \tag{5}$$

then determines the ratio c/a.) Applying these matching conditions at $r=c$, the solution of Eq. (1) within $c<r<a$ is given by

$$\frac{\psi}{\psi_c}=\left(\frac{r}{c}\right)^m-\frac{1-p}{m-nq_c}\left[\left(\frac{r}{c}\right)^m-\left(\frac{c}{r}\right)^m\right]. \tag{6}$$

Across $r=a$, the matching conditions are again that $\psi=\psi_a$ be continuous and that

$$\frac{[\psi']_a}{\psi_a}=\frac{m}{aF_a}[j]_c=-\frac{2mq_a p}{aq_c(m-nq_a)}. \tag{7}$$

Applying these matching conditions at $r=a$, and using Eq. (6) at $r=a$ for ψ_a, the solution of Eq. (1) for $r>a$ is given by

$$\frac{\psi}{\psi_c}=\left(\frac{r}{c}\right)^m-\frac{1-p}{m-nq_c}\left[\left(\frac{r}{c}\right)^m-\left(\frac{c}{r}\right)^m\right]-\frac{q_a p}{q_c(m-nq_a)}\left[\left(\frac{r}{a}\right)^m-\left(\frac{a}{r}\right)^m\right]\left\{\left(\frac{a}{c}\right)^m-\frac{1-p}{m-nq_c}\left[\left(\frac{a}{c}\right)^m-\left(\frac{c}{a}\right)^m\right]\right\}. \tag{8}$$

If the conducting shell is absent ($b\to\infty$), the stability condition (i.e., the condition that no marginal mode exists) is that the coefficient of the r^m term be positive, i.e.,

$$1-\frac{1-p}{m-nq_c}>\frac{q_a p}{q_c(m-nq_a)}\left\{1-\frac{1-p}{m-nq_c}\left[1-\left(\frac{c}{a}\right)^{2m}\right]\right\}. \tag{9}$$

Suppose, first, that the pedestal is entirely absent, i.e., $p=0$. In this case, instability occurs if $0<m-nq_c<1$. For the $(m,n)=(2,1)$ mode to be stable, it is clearly necessary to have $q_c<1$, in which case the $(m,n)=(1,1)$ mode is unstable. Moreover, if q_c is just above 1, the entire sequence of modes, (2,1), (3,2), (4,3), etc., is unstable. Even if the $m=2$ mode were stabilized by means of a fairly close conducting shell, the higher-m modes of this sequence would typically remain unstable, since the effect of the shell falls off rapidly with rising m. With the $m=2$ mode stabilized by a conducting shell, one might consider setting q_c just above 1.5, so that the modes, (3,2), (4,3), etc., become stable. However, in this case, the sequence of modes, (5,3), (8,5), etc., would be unstable. It is, thus, of considerable interest to determine whether, in either case, a current profile with a nonzero pedestal can provide simultaneous stability against all modes.

Let us consider case A, in which q_c is just above 1, and q_a is just above 2, with a finite value of p in the range $0<p<0.5$. If q_c is *infinitesimally* above 1, Eq. (9) shows clearly that the "inner" sequence of modes (2,1), (3,2), (4,3), etc., whose singular surfaces fall into the pedestal region, is positively stable, since in each case the left-hand side of Eq. (9) is positive and the right-hand side is negative. We must also, however, demonstrate the stability of the "outer" sequence of modes (3,1), (5,2), (7,3), etc., whose singular surfaces fall outside the pedestal region. A condition stronger than (9) would result from replacing $(c/a)^{2m}$ by $(c/a)^2$; we do this, and substitute Eq. (5) for c/a, to obtain the sufficient stability condition $(1-2p)(m-nq_c-1)>0$, after using $m-nq_a=1$ and $q_a/q_c=2$. Since $p<q_c/q_a=0.5$, this condition is always satisfied by the modes of the "outer" sequence, which have $m-nq_c\geq 2$.

If q_c and q_a exceed 1 and 2, respectively, by small but *finite* increments, a reasonable number of the modes in both the "inner" and "outer" seqences can be made positively stable. This is illustrated in Fig. 2, for the case where $q_c=1.05$ and $q_a=2.1$, and for various values of the pedestal p. We see that, in this case, the optimum value for p, in the sense of stabilizing the greatest range of low m values ($m<8$), is about 0.3.

Let us now consider case B, which requires a conducting shell to stabilize the $m=2$ mode, but offers the advantage that the limiter q value can be dropped below 2. Here q_c is again just above 1, but q_a is chosen to be just above 1.5, with a value of p in the range $0<p<\frac{2}{3}$. As before, if q_c is *infinitesimally* above 1, Eq. (9) shows that the "inner" sequence of modes, with $(m,n)=(3,2)$, (4,3), etc., is positively stable, for any finite value of p in the above range. We must also, however, demonstrate the stability of the "outer" sequence of modes (2,1), (5,3), (8,5), etc. For q_a infinitesimally above 1.5, we find that the stability condition is never satisfied for the (2,1) mode, but it can be satisfied for the (5,3) mode, and all higher modes of this "outer" sequence, provided $p<0.32$. The (2,1) mode can, however, be stabilized by means of a conducting shell. The requirement on the radius b of the shell can be

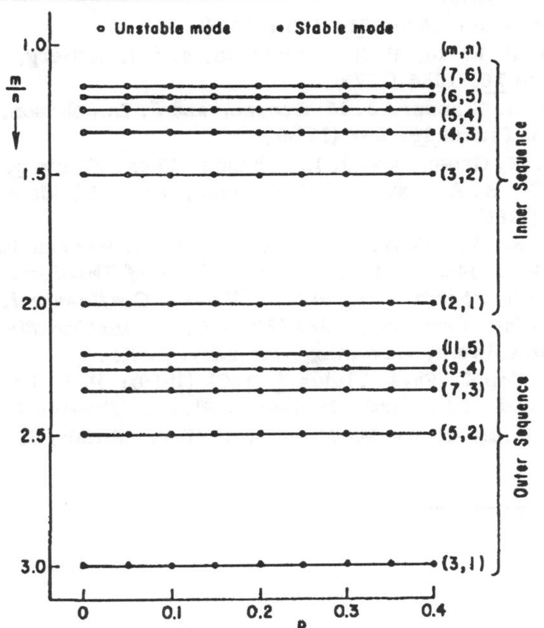

FIG. 2. Stability diagram for modes (m,n) in case A with use of the current profile of Fig. 1, with $q_c=1.05$, $q_a=2.1$, $b=\infty$, and various values of the pedestal p.

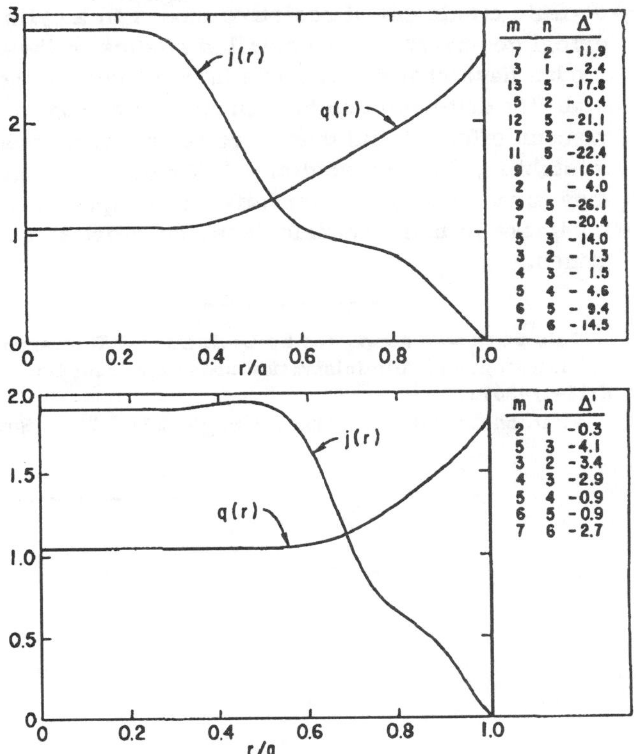

FIG. 3. (a) Example of a stable profile similar to case A, with no conducting shell; the computed values of Δ' show that all the indicated modes are stable. (b) Example of a stable profile similar to case B, with a conducting shell at $r/a=1.2$.

determined from Eq. (8) by setting $\psi/\psi_c>0$ at $r=b$. Employing $q_c=1$, $q_a=1.5$, Eq. (5) for c/a, and $(m,n)=(2,1)$, we obtain

$$b/a<(4/3p)^{1/4}. \qquad (10)$$

For $p\simeq 0.3$, this gives $b/a<1.45$, a requirement that could be met rather easily.

The simple analytic treatment given above has the advantage of clarifying the role of a current pedestal in stabilizing low-m modes. However, as we have seen, the use of a discontinuous function for $j(r)$ has the disadvantage of exciting high-m modes. In order to investigate the possibility of stabilizing *all* kink modes simultaneously, we have employed a computer program that determines the stability of arbitrary *smooth* current profiles by calculating the quantities Δ' that measure the potential-energy perturbations for the various modes.

Figures 3(a) and 3(b) show two examples of "realistic" current profiles resembling the analytic cases A and B. In both cases, we see that the entire spectrum of modes is stabilized ($\Delta'<0$), the higher-m modes apparently being suppressed by the smoothing of $j(r)$. The corresponding limiter q values are 2.6 and 1.8, respectively. In Fig. 3(b) a fairly close conducting shell was needed ($b/a=1.2$); alternatively, one could invoke toroidal-curvature stabilization[3] of the weakly unstable higher-m modes of the "inner" sequence, thus permitting a lower pedestal and a larger value of b/a.

Our results are in accord with the experimentally observed destabilizing effects of limiter q values that approach 2, or high levels of impurity influx. In either case, the outer plasma region would be cooled, so that the pedestal on the current profile would tend to be truncated short of the $q(r)=2$ point. In larger tokamaks, it may be possible to achieve better control over the current distribution, so that profiles resembling Fig. 3(a) could be approximated.

Experiments on conducting-shell stabilization[5] proved successful in suppressing the $(m,n)=(2,1)$ mode, thus obtaining gross stability at $q_a<2$. There was, however, evidence of a deterioration in confinement, particularly for $q_a\simeq 1.5$. Our results for case B show that this could be explained in terms of the truncation of the pedestal on the current profile short of the point where $q(r)=1.5$.

We conclude with a word about our approximations. Our calculation is based upon a collisional-fluid treatment of resistive modes in a cylindrical geometry. More detailed studies of these modes have shown that there is no change in the stability criterion due to diamagnetic and gyroviscous effects,[6] and that there is an improvment in stability due to toroidicity.[3] Thus, our results may turn out to be conservative when applied to tokamaks in more collisionless, reactorlike regimes.

*This work was supported by U. S. Energy Research and Development Administration under contract No. E(11-1)-3073.

[1]V. D. Shafranov, Zh. Tekh. Fiz. 40, 241 (1970) [Sov. Phys. Tech. Phys. 15, 175 (1970)].

[2]H. P. Furth, P. H. Rutherford, and H. Selberg, Phys Fluids 16, 1054 (1973).

[3]A. H. Glasser, J. M. Greene, and J. L. Johnson, Phys. Fluids 19, 567 (1976).

[4]J. M. Greene and J. L. Johnson, Phys. Fluids 5, 510 (1962); A. A. Ware and F. A. Haas, Phys. Fluids 9, 956 (1966).

[5]V. S. Vlasenkov, V. M. Leonov, V. G. Merezhkin, and V. S. Mukhovatov, in *Proceedings of Third International Symposium on Toroidal Plasma Confinement, Garching, Germany, 1973* (Max-Planck-Institut für Plasmaphysik, Garching, Germany, 1973).

[6]B. Coppi, Phys. Fluids 7, 1501 (1969); P. H. Rutherford and H. P. Furth, Princeton Plasma Physics Laboratory Report No. MATT-872, 1971 (unpublished).

Collisionless "Current-Channel" Tearing Modes

S. M. Mahajan, R. D. Hazeltine, H. R. Strauss, and D. W. Ross
Fusion Research Center, The University of Texas at Austin, Austin, Texas 78712
(Received 7 August 1978)

Analytical and numerical studies of collisionless tearing modes that are wider than the "current channel" are presented. The $m=1$ type of inertial mode is shown to be strongly unstable for typical tokamak shear and β_e. Large spatial extension and large growth rate make it a possible candidate for explaining plasma disruption.

We report, in this Letter, analytical and numerical studies of tearing modes in a collisionless plasma. The modes we discuss are characterized by a mode width λ_w greater than the width of the electron layer or the "current channel" $x_e = |\omega/k_\parallel' v_e|$. Here $k_\parallel' = k_y/L_s$, k_y is the azimuthal mode number, L_s is the shear length, v_e is the electron thermal speed, and λ_w measures the region in which the parallel electric field is nonzero. Laval, Pellat, and Vuillemin[1] were the first to present an $m \geq 2$ type of current-channel tearing mode; their result is instructively discussed by Drake and Lee.[2] Later, Chen, Rutherford, and Tang[3] found trapped-particle modification to the mode. Although, these results are easily recovered in the appropriate limit of our dispersion relation, we emphasize here the $m=1$ type of inertial tearing mode first pointed out by Hazeltine and Strauss.[4] Our analytical dispersion relation, which has been verified in detail numerically, modifies the previous results for $\beta_e > m_e/m_i$. We also clarify its relationship to other instabilities. Most importantly, we confirm the potentially rapid growth and wide parameter range for instability of inertial tearing modes with $m=1$ character (large Δ'). Whenever it is consistent to treat the plasma in a collisionless approximation, this mode would be a serious candidate to explain plasma disruption because of its large spatial extension and large growth rate.

The slab geometry formulation of the electromagnetic eigenvalue problem is now standard.[5,6] If the radial wavelength is shorter than the azimuthal wavelength, we have

$$\varphi'' = (\sigma/x_A^2)[(\psi/x) - \varphi], \tag{1}$$

$$\psi'' = (\sigma/x)[(\psi/x) - \varphi], \tag{2}$$

where φ is the electrostatic potential, $\psi = \omega A_\parallel / k_\parallel' c$ is proportional to the vector potential $A_\parallel$, a prime denotes differentiation with respect to the radial coordinate x, $x_A^2 = \omega(\omega+\omega_{i*})/k_\parallel' v_A)^2$, $v_A = (B_0^2/4\pi n_0 m_i)^{1/2}$ is the Alfvén speed, ω_{i*} is the ion diamagnetic drift frequency, and all lengths are normalized to $\rho_s = (T_e/m_i)^{1/2}\Omega_i^{-1}$. In Eqs. (1) and (2), σ is the dimensionless measure of the generalized "conductivity." Although our analytical method can handle fairly complicated models of σ (i.e., with effects of temperature gradients and trapped particles included[3]), we keep here, for illustrative purposes, to the simplest representative collisionless model,

$$\sigma = -x_A^2 \frac{\omega - \omega_{e*}}{\omega + \omega_{i*}}\left[1 + \frac{x_e}{|x|} Z\left(\frac{x_e}{|x|}\right)\right], \tag{3}$$

where $\omega_{e*} = k_y c_s / L_n$ is the electron diagmagnetic drift frequency, $c_s = (T_e/m_i)^{1/2}$ is the sound speed, $L_n = |n_0^{-1}\partial n_0/\partial x|$ is the density scale length, and $x_e = \omega/k_\parallel' v_e$ measures the width of the current channel. It is customary, at this stage, to combine Eqs. (1) and (2) to obtain a single second-order differential equation in $E_x = -\partial\varphi/\partial x$.[6] We depart here from the conventional approach, and instead obtain an equivalent equation for $Q = (\psi/x) - \varphi = -(x_A^2 E_x'/\sigma)\alpha E_\parallel / x$,

$$\frac{d}{dx}\left(\frac{x_A^2 x^2\, dQ/dx}{x^2 - x_A^2}\right) + \sigma Q = -\frac{2xE_0 x_A^2}{(x^2 - x_A^2)^2}, \tag{4}$$

where E_0 is related to Δ', the stability parameter of the kink-tearing mode theory,[5,7] by

$$E_0 = -(\Delta')^{-1}\int_{-\infty}^{+\infty}[\sigma Q/x]\,dx. \tag{5}$$

We have chosen to write our equation in terms of Q $(E_\parallel/x)$ instead of E_x, because it is the behavior of $E_\parallel$ which determines whether the mode has a current channel or not. We notice from Eq. (1) that while the $E_\parallel$ profile can be much broader than the σ profile, i.e., when $\lambda_\omega > |x_e|$, E_x is constrained to follow the σ profile. This makes E_x an unsuitable variable for the study of current-channel modes; and presumably explains the less accurate results of Ref. 4.

We now solve Eqs. (4) and (5) by setting up a variational principle for Q. Following standard methods,[5] we can show that the functional $(\langle\cdots\rangle$ denotes $\int_{-\infty}^{+\infty}\cdots dx)$

$$S = \left[\Delta' + \frac{\epsilon\pi i}{x_A}\right]\left\{\left\langle Q\frac{d}{dx}\left[\frac{x^2 x_A^2\, dQ/dx}{x^2 - x_A^2}\right]\right\rangle + \langle\sigma Q^2\rangle\right\} + 2x_A^2\left\langle\frac{Qx}{x^2 - x_A^2}\right\rangle^2 \tag{6}$$

is variational, in that $\delta S = 0$, generates Eq. (4) with the constraint Eq. (5). To do that we need to use the relation

$$(\Delta' + \epsilon\pi i/x_A)E_0 = 2x_A^2\int_{-\infty}^{+\infty}[Qx/(x^2 - x_A^2)^2]\,dx,$$

which is obtained by multiplying Eq. (4) with $(1/x)$ and integrating over all x. In Eq. (6), $\epsilon = 1$ if $\mathrm{Im}\,x_A > 0$, and $\epsilon = -1$ if $\mathrm{Im}\,x_A < 0$. The extremal value of $S \equiv S^* = 0$ yields the dispersion relation. Since we are looking for localized solutions of $E_\parallel$ which tear the magnetic surfaces, the appropriate trial function should be even in $E_\parallel$. Recalling that $Q = E_\parallel/x$, we choose the trial function

$$Q = x^{-1}\exp(-\alpha x^2/2), \quad \mathrm{Re}\,\alpha > 0, \tag{7}$$

where α is a variational parameter, to evaluate S. All the integrals involved can be evaluated exactly.[8] A general analysis of S will be presented elsewhere. In this Letter, however, we are going to concentrate on recovering the modes which are much wider than the current channel, i.e., for which $|x_e\alpha^{1/2}| < 1$. For further simplification, we also assume that $|x_A\alpha^{1/2}| < 1$. With these approximations, and for $\mathrm{Im}\,x_A > 0$, we can write

$$S \propto S_0 + S_1\alpha^{1/2} + S_2\alpha, \tag{8}$$

where

$$S_0 = \Delta'\left[-\frac{i\pi}{x_A} - \frac{x_A^2}{x_e}\frac{\omega - \omega_{e*}}{\omega + \omega_{i*}} i\sqrt{\pi}\right] + \pi^{3/2}\frac{\omega - \omega_{e*}}{\omega + \omega_{i*}}\frac{x_A}{x_e},$$

$$S_1 = \left(\Delta' + \frac{i\pi}{x_A}\right)2\sqrt{\pi}\, x_A^2\frac{\omega - \omega_{e*}}{\omega + \omega_{i*}}, \tag{9}$$

$$S_2 = -2i\pi x_A\Delta'.$$

A simultaneous solution of $\partial s/\partial\alpha = 0$, $S^* = 0$, gives $\alpha^{1/2} = -S_1/2S_2$ and leads to the dispersion relation

$$S_0 = S_2\alpha = S_1^2/4S_2 \tag{10}$$

which must be solved for the eigenvalue ω. Of course, the acceptable solution must satisfy the consistency criteria $\mathrm{Re}\,\alpha > 0$, $|x_e\alpha^{1/2}| < 1$, $|x_A\alpha^{1/2}| < 1$, and $\mathrm{Im}\,x_A > 0$. The first of these assures a localized solution, the second, a solution with a current channel, and the third is simply for convenience, and is not essential to the analysis. To make further progress, we consider the infinite-mode-width limit, $\alpha \to 0$, which implies that $E_\parallel$ is essentially constant. Since $E_\parallel = \psi - x\varphi$, $\alpha \to 0$ is equivalent to the constant ψ approximation if $x\varphi$ is neglected. Indeed, we recover, in the zeroth-order dispersion relation $S = 0$, the mode of Laval, Pellat and Vuillemin[1] which was derived making use of the constant-ψ approximation. However, we also find an unstable mode when $(\Delta')^{-1}$ is zero. We use $(\Delta')^{-1} = 0$ as a definition of $m = 1$ modes. For simplicity, let us put $\omega_{i*} = 0$, then $S = 0$ leads to

$$(\omega - \omega_{e*})[1 - (i\Delta'\omega/\pi k_\parallel' v_A)] = i\gamma_T, \tag{11}$$

where $\gamma_T = (\Delta' k_\parallel' v_A^2/\sqrt{\pi} v_e)$. Notice that for $\omega_{i*} = 0$, the constraint $\mathrm{Im}\, x_A > 0$ is satisfied for any growing mode. For small Δ', the second term in the square brackets in Eq. (11) is small, and the resulting dispersion relation $\omega = \omega_{e*} + i\gamma_T$ describes the collisionless tearing mode of Ref. 1. For $(\Delta')^{-1} = 0$, Eq. (11) is the dispersion relation for $m = 1$ type of modes, and the solution is

$$\omega = \frac{\omega_{e*}}{2} \pm i\left[k_\parallel'^2 v_A^2 \frac{v_A}{v_e}\sqrt{\pi} - \frac{\omega_{e*}^2}{4}\right]^{1/2} \tag{12}$$

which has a growing root if $\beta_e^{3/2} > 4(L_n/L_s)^2(\pi m_e/m_i)^{1/2}$, where expressions for v_A and ω_{e*} have been used. This restriction is a consequence of evaluating the zeroth-order dispersion relation, and also of expanding the integrals for small $|x_A\alpha^{1/2}|$. A numerical solution of the variational dispersion relation allows us to handle large values of β_e. For $|x_A\alpha^{1/2}| < 1$, the current-channel condition $|x_e\alpha^{1/2}| < 1$ simply requires that $\beta_e > m_e/m_i$. It can be easily seen that for the growing root of Eq. (12), all the consistency conditions are readily satisfied. Since $\beta_e > m_e/m_i$ is required, it is clear that the "current-channel" inertial tearing mode has no electrostatic limit.

We have also solved the Eqs. (1)–(3) by direct numerical integration using a code developed by Miner.[9] This code carries out a finite-element Galerkin procedure employing basic cubic splines. The boundary conditions are $\psi'(0) = \varphi(0) = 0$, and $\psi'(x_b) = \psi(x_b)(x_b + 2/\Delta')^{-1}$, $\varphi(x_b) = \psi(x_b)/x_b$. Since the latter condition is $E_\parallel = 0$, the boundary point is chosen outside the $E_\parallel$ layer, $|\alpha^{1/2}x_b| \gg 1$.

An analysis of Eq. (11) reveals that a positive nonzero $(\Delta')^{-1}$ decreases the growth rate of the $m = 1$ mode. As $(\Delta')^{-1}$ is increased further, the mode smoothly goes over to the Laval mode, which has a smaller growth rate than the $m = 1$ mode. A negative $(\Delta')^{-1}$, on the other hand, enhances the growth rate of the mode, and as Δ' becomes small, γ approaches the large magnetohydrodynamic growth rate $\pi k_\parallel' v_A/|\Delta'|$. The same circumstance pertains in collision-dominated regimes.[6] All of the above-mentioned features of the dispersion relation Eq. (11) have been verified in detail by carrying out extensive numerical experiments in various parameter regimes. Thus we have verified the mode of Laval, Pellat, and Vuillemin which is relevant for small Δ'. We present here the comparison

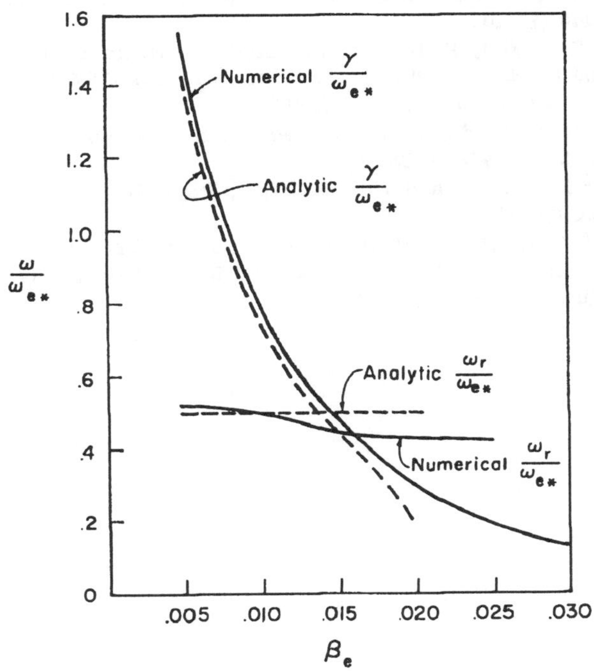

FIG. 1. Normalized eigenfrequency ω/ω_{e*} vs β_e for $\omega_{i*} = 0$, $\Delta' = \infty$, and $L_n/L_s = 0.1$.

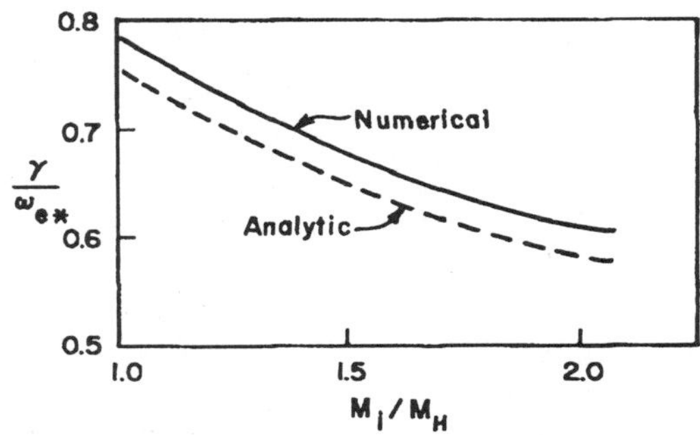

FIG. 2. Normalized growth rate γ/ω_{e*} vs the ion mass m_i/m_H, where $m_H = 1836 m_e$, for $\beta_e = 0.01$ and $L_n/L_s = 0.1$.

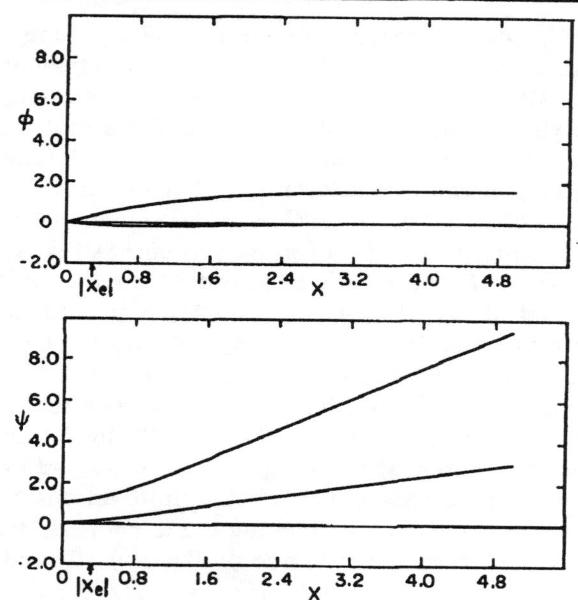

FIG. 3. φ and ψ vs x/ρ_s for $\beta_e = 0.005$, $L_n/L_s = 0.1$, $\omega_{i*} = 0$, and $\omega = 0.52 + 1.54i$. $|x_e|$ denotes the current-channel width.

between our analytical and numerical results for the $m=1$ ($\Delta' = \infty$) inertial tearing mode. In Fig. 1, we plot the real part of the frequency ω_r and the growth rate γ as a function of β_e. The analytical [Eq. (12)] and numerical results are clearly in excellent agreement in the limit of validity of Eq. (12). The numerical solution has been extended to higher values of β_e, and the instability persists. To check the scaling of the growth rate further, we plot γ as a function of the ion mass in Fig. 2, and the agreement is again excellent. In Fig. 3, we show a typical plot of φ and ψ as a function of x/ρ_s. The mode width, i.e., the region in which $E_\parallel$ remains finite is $\simeq 2.5\rho_s$, while the current-channel width is $\simeq 0.25\rho_s$. Thus the mode is indeed much wider than the current channel.

Therefore, we have demonstrated the existence of an $m=1$ ($\Delta' = \infty$) "current-channel" tearing mode in a collisionless plasma by analytical and numerical methods. The mode has a growth rate $\sim \omega_{e*}$, which makes the growth time comparable to typical disruption time. The mode remains unstable for realistic shear $L_n/L_s \sim 0.1$, and for a wide range of β_e: greater than m_e/m_i and up to a few percent. Its large growth rate coupled with its large spatial extension make this mode very important for high-temperature, moderate-density plasmas which can be treated in a collisionless approximation.

This work was supported by the U. S. Department of Energy, Contract No. EY-77-C-05-4478.

[1]G. Laval, R. Pellat, and M. Vuillemin, in *Proceedings of the Second Conference on Plasma Physics and Controlled Nuclear Fusion Research, Culham, England, 1965* (International Atomic Energy Agency, Vienna, Austria, 1965), Vol. II, p. 259.

[2]J. F. Drake and Y. C. Lee, Phys. Fluids 20, 1341 (1977).

[3]Liu Chen, P. H. Rutherford, and W. M. Tang, Phys. Rev. Lett. 39, 460 (1977).

[4]R. D. Hazeltine and H. R. Strauss, Phys. Fluids 21, 1007 (1978).

[5]R. D. Hazeltine and D. W. Ross, Phys. Fluids 21, 1140 (1978).

[6]B. Coppi, R. Galvao, R. Pellat, M. N. Rosenbluth, and D. H. Rutherford, Fiz. Plasmy 2, 961 (1976) [Sov. J. Plasma Phys. 2, 533 (1976)].

[7]H. P. Furth, J. Killeen, and M. N. Rosenbluth, Phys. Fluids 6, 459 (1963).

[8]D. W. Ross and S. M. Mahajan, Phys. Rev. Lett. 40, 324 (1978).

[9]W. H. Miner, Ph.D. dissertation, University of Texas (unpublished); D. W. Ross and W. H. Miner, Phys. Fluids 20, 1957 (1977).

Neutral-Beam-Heating Results from the Princeton Large Torus

H. Eubank, R. Goldston, V. Arunasalam, M. Bitter, K. Bol, D. Boyd,[a] N. Bretz, J.-P. Bussac,[b] S. Cohen, P. Colestock, S. Davis, D. Dimock, H. Dylla, P. Efthimion, L. Grisham, R. Hawryluk, K. Hill, E. Hinnov, J. Hosea, H. Hsuan, D. Johnson, G. Martin, S. Medley, E. Meservey, N. Sauthoff, G. Schilling, J. Schivell, G. Schmidt, F. Stauffer,[a] L. Stewart,[c] W. Stodiek, R. Stooksberry,[d] J. Strachan, S. Suckewer, H. Takahashi, G. Tait,[a] M. Ulrickson, S. von Goeler, and M. Yamada
Plasma Physics Laboratory, Princeton University, Princeton, New Jersey 08544

and

C. Tsai, W. Stirling, W. Dagenhart, W. Gardner, M. Menon, and H. Haselton
Oak Ridge National Laboratory, Oak Ridge, Tennessee 37830

(Received 1 March 1979)

Experimental results from high-power neutral-beam-injection experiments on the Princeton Large Torus tokamak are reported. At the highest beam powers (2.4 MW) and lowest plasma densities $[n_e(0)=5\times10^{13}\text{ cm}^{-3}]$, ion temperatures of 6.5 keV are achieved. The ion collisionality ν_i^* drops below 0.1 over much of the radial profile. Electron heating of $\Delta T_e/T_e\approx50\%$ has also been observed, consistent with the gross energy-confinement time of the Ohmically heated plasma, but indicative of enhanced electron-energy confinement in the core of the plasma.

The purpose of the Princeton Large Torus (PLT) tokamak neutral-beam-injection experiments is to produce collisionless high-temperature tokamak plasmas in which to study ion and electron thermal transport. In this paper we present data from recent neutral-beam-heating experiments on the PLT tokamak, extending the results of previous injection-heating experiments,[1-6] to the better confinement conditions associated with large tokamaks, and also extending our previous results[7] to injection power levels of up to 2.4 MW and ion temperatures of 6.5 keV. The PLT beam-injection system consists of four tangentially aimed beam lines and 40-keV ion sources.[8] Two sources inject parallel to the plasma current and two inject antiparallel.

Ion heating.—The techniques for measuring ion temperature on PLT fall into three categories: mass and energy analysis of the fast neutrals generated by charge exchange, measurements of the Doppler broadening of impurity line radiation in the x-ray and ultraviolet, and thermonuclear-neutron–emission measurements for H^0 injection into D^+ plasmas. The nature of the respective uncertainties and the extent to which the proper correction can be applied to each specific measurement technique has been discussed previously.[7] Generally the ion-temperature assessments agree to within ~10%.

With 2.4 MW of D^0 injection into an H^+ plasma, we have achieved ion temperatures up to 6.6 keV recorded by an analysis of charge-exchange neutrals as shown in Fig. 1. A supportive diagnostic for this ion-temperature measurement is Doppler broadening of Fe XXIV. At this power level, ion temperature and central density $[n_e(0) = 5\times 10^{13}$ $\mathrm{cm}^{-3}]$, Fe XXIV is strongly heated by the beam ions and can be raised to temperatures well above the thermal H^+ plasma. We calculate, for this case, that the temperature of Fe XXIV should exceed that of the thermal protons by 1700 eV. The

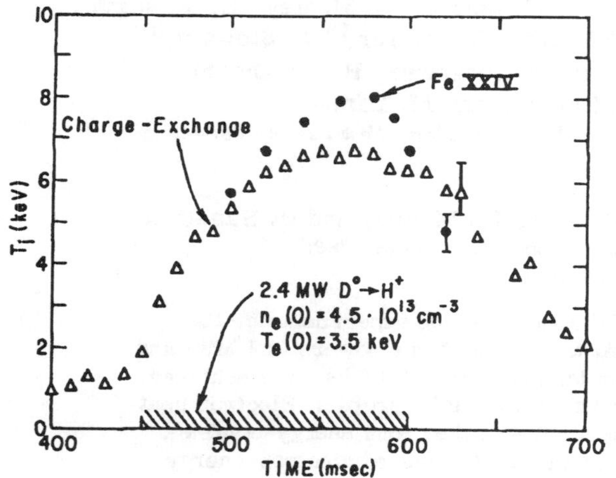

FIG. 1. Ion temperature (H^+) vs time as measured by analysis of charge-exchange neutrals and Doppler broadening of Fe XXIV for 2.4-MW D^0 injection into a H^+ plasma. Calculations indicate that the maximum temperature of Fe XXIV should exceed that of H^+ by 1700 eV.

Fe XXIV temperature shown in Fig. 1, which reaches 8 keV, is thus in good agreement with the charge-exchange data.

While the achievement of an ion temperature in excess of that required for ignition in an ideal D-T fusion reactor (~4 keV) is noteworthy, the true significance of the data lies in the linear relationship of ion temperature to beam power as the temperature moves into the collisionless regime, where trapped-particle modes were predicted to produce enhanced energy transport. (See Fig. 2.) At $T_i = 6.5$ keV, $Z_{eff} \approx 3.5$, and $n_e(0) = 5\times 10^{13}$ cm^{-3}, the ion collisionality parameter ν^* reaches a minimum of 2×10^{-2} and is below unity out to $r = 30$ cm (limiter radius = 40 cm). This represents an ion thermal component as deep within the banana regime as required for many tokamak reactor designs. As we proceed into the collisionless regime by increasing the ion temperature, there is a strong enhancement in the level of density fluctuations as measured by microwave scattering. So far, however, no observable effect on the ion energy balance nor on the circulating fast beam particles has been seen. The scattering volume is wave-number dependent but encompasses about one-half of the minor radius and is centered at $r/a \cong \frac{1}{2}$. The spatial profile of the fluctuations is not known, the rather rapid buildup of the fluctuation intensity occurs at $T_i \gtrsim 4$ keV for these low-density discharges. Although the observed frequency spectrum is characteristic of drift waves, neither the nature of the fluctuation nor the driving source is known at this time.

Injection of 2.2 MW of ~40-keV D^0 into D^+ plasmas, has produced a flux of $1.6\times 10^{14} n/\mathrm{sec}$ or 2

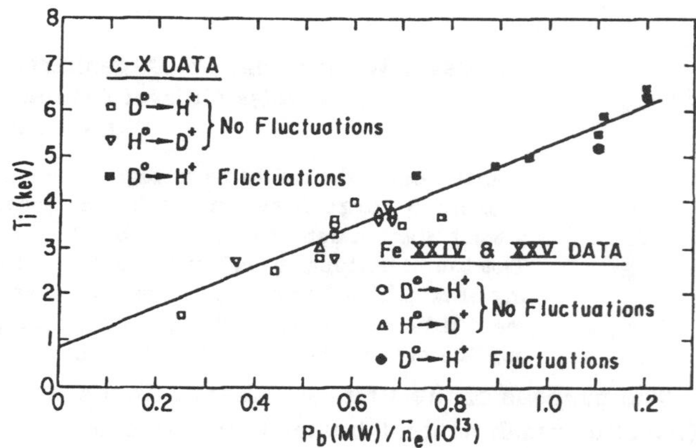

FIG. 2. Ion-temperature increase vs beam power per unit line-average plasma density, illustrating linearity despite the onset of strong density fluctuations.

$\times 10^{13} n$/pulse in good agreement with calculations (Fig. 3). We calculate that these neutrons arise about equally from beam-plasma and beam-beam interactions with less than 10% contribution from the thermal particles by themselves.

Electron heating.—The electron heating obtained in PLT with neutral injection is very sensitive to the choice of limiter material and to the conditions of the vacuum-vessel wall. In early experiments with tungsten limiters and even in more recent experiments with steel limiters, counterinjection into low-density plasmas has consistently resulted in metallic line radiation from the plasma core of ~1 W cm^{-3} (greater than the input beam power), quenching any significant electron-temperature rise. At low densities only with carbon limiters have we been able to routinely maintain a discharge relatively free of metallic impurities, with central radiation ≤ 200 mW cm^{-3} and obtain strong electron heating with beam injection. With titanium gettering onto the vacuum-vessel wall, we have also been able to limit the density increase to $\Delta n_e(0) = 2.5 \times 10^{13}$ cm^{-3} with four-beam injection into a plasma which starts at $n_e(0) = 2.5 \times 10^{13}$ cm^{-3}. At higher densities $[n_e(0) > 5 \times 10^{13}$ $cm^{-3}]$ metallic radiation drops to negligible levels and we are also able to hold the density nearly constant during injection through feedback control of the pulse-valve gas feed. In addition, gettering results in a decreased Z_{eff} and a wider operating range in density.[9] Ion heating has been obtained for all of the limiter materials employed in PLT, namely C, Fe, and W. The efficiency is, of course, affected by the ion-electron coupling and thus the electron temperature. The best ion-heating results have therefore been obtained with carbon limiters.

Figure 4 shows the electron-temperature increase which we obtain with 2.4 MW of D^0 beams into a low-density H^+ plasma as determined by electron-cyclotron emission measurements at the first harmonic. These results are consistent with Thomson scattering measurements of T_e. While the overall radiation level increases with injection, Z_{eff} remains nearly constant. The overshoot in T_e at the end of the injection pulse which is due to ending the cold-electron influx associated with the neutral beam, is reproduced by transport-code calculations.[10]

Ion and electron power balance.—Numerical calculations of the ion and electron radial power flows have been made for a number of discharges both with and without neutral injection and will be discussed in detail elsewhere. Here we summarize only the important results.

For the ions, we calculate $T_i(r)$ on the basis of a Monte Carlo simulation of the beam-heating process, and numerical calculation of energy

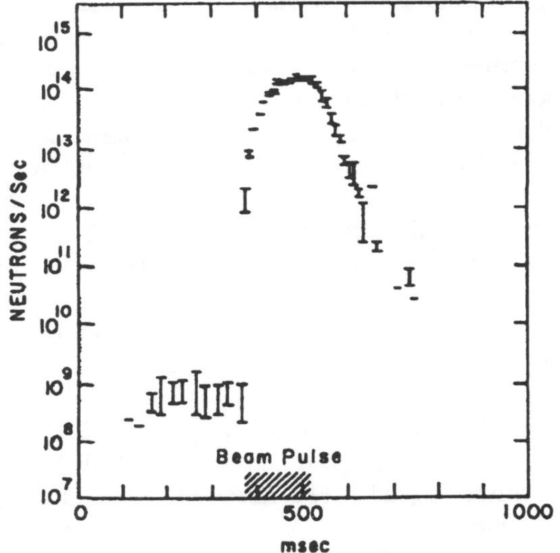

FIG. 3. Neutron emission vs time for 2.2-MW D^0 injection into a D^+ plasma. Peak fusion output is 170 W. D-T equivalent power is 50 kW or $Q = 2\%$. Before injection, $T_i(0) = 1$ keV and $\langle W_i \rangle = 1.5$ keV. During injection, beam ions have $\langle W_i \rangle = 20$ keV and comprise 30% of $n_i(0)$. Bulk ions have $T_i(0) = 6.6$ keV and $\langle W_i \rangle = 10$ keV and are 70% of $n_i(0)$. Thus $\langle W_i \rangle = 13$ keV.

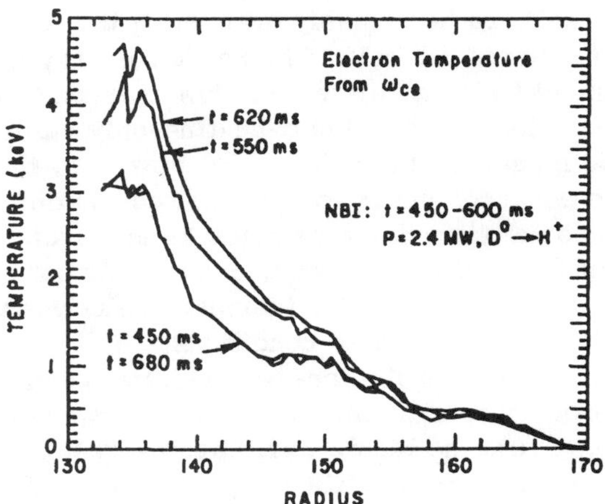

FIG. 4. Electron-temperature profiles during beam heating. $T_e(r)$ has nearly saturated at 550 ms. The short increase following beam turnoff, exemplified by $T_e(r)$ at 620 ms, is due to ending the cold-electron influx associated with the neutral beam, while beam ions circulating inside the plasma continue to heat the electrons.

loss by neoclassical thermal conduction, ion-electron coupling, empirical thermal convection, and charge-exchange, as in Stott.[11] Because of the uncertainties in $Z_{eff}(r)$, $q(r)$, and $n_0(r)$, we find, for a case of 2.1-MW D^0 injection into a low-density $[n_e(0)=4.5\times10^{13}\ cm^{-3}]$ H^+ plasma, a range of predictions for $T_i(0)$ extending from 4 to 7.5 keV, to be compared to the measured value of 5.5 keV. The volume-integrated net ion-energy confinement time τ_{Ei} is 25 msec for this case or about half the value of Ohmic heating alone for the same discharge. In the numerical simulation this difference in τ_{Ei} is primarily due to the enhanced role of charge-exchange and empirical convective losses at the high ion temperatures and steep temperature gradients which occur with beam heating. As neoclassical thermal conduction is only a small fraction of the total ion-energy flow, we cannot rule out, on the basis of the measured temperature, as much as a fivefold enhancement of the thermal conduction over the neoclassical value, which is itself uncertain by almost as large a factor.

In cases of neutral injection into higher-density plasmas, charge-exchange and convective losses are reduced and ion-electron coupling becomes the dominant term in the ion-energy flow. Neoclassical thermal conduction remains a small term in the total ion-power-balance picture, in part because of the reduction of Z_{eff} at higher densities. The neoclassical prediction for ion temperature with injection of 2.0 MW of D^0 into a H^+ plasma of $n_e(0)=7.5\times10^{13}\ cm^{-3}$, is $T_i(0)=3.1\pm0.5$ keV, rather closely bracketing the measured value of 3.2 keV. The net ion-energy confinement time calculated for this case was 95 msec, close to the Ohmic-heating-only value of ≤ 120 msec. Even in this case, however, the possible enhancement of ion thermal conduction over the neoclassical value remains as much as a factor of 3–4. We, therefore, cannot rule out appreciable anomalous ion thermal conduction under these beam-heated conditions.

On the basis of the ion-power-flow calculations, we proceed to examine the electron power balance. From the Monte Carlo beam-orbit code, we get the radial profile of P_{be}, the power flowing directly from the beam ions to the electrons; and from the experimental measurements of $T_e(0)$, coupled with the neoclassical ion-temperature profile calculations, we find $P_{ie}(r)$, the power flowing from the bulk ions to the electrons. In addition, the beam-orbit code evaluates $P_{he}(r)$, the power required to thermalize the cold electrons entering the plasma from the neutral beam. The Ohmic input power is calculated from the experimental $T_e(r)$ profile, assuming $Z_{eff}(r)=$ const and $\sigma\propto T_e^{3/2}$. Finally, we subtract the bolometrically measured radiated power, and find the radial power flow due to electron thermal transport which is required to complete the power balance. Dividing the result by $n_e dT_e/dr$, we arrive at an experimentally determined anomalous electron thermal transport coefficient $\chi_e(r)$.

The intriguing result of applying this analysis to our beam-heating measurements is that, although the volume-integrated net electron-energy confinement time is approximately unchanged during injection, χ_e in the core of the plasma appears to fall roughly as $(n_e T_e)^{-1}$ when the temperature and density are increased with neutral injection. This effect is reflected in a 50% rise in T_e on axis, in moderate density cases, equal to the fractional increase at $r/a=\frac{1}{2}$, even though P_{be} and P_{ie} are relatively much smaller in the hot central core. It seems unlikely, however, that the evident reduction in thermal transport in the central plasma region is due to a simple scaling of χ_e with T_e, since χ_e was not reduced in the outer regions of the plasma, which are also heated.

The results of high-power neutral-beam–heating experiments on the PLT tokamak are very encouraging. We have observed that the thermal-ion component in a tokamak plasma can be driven deep into the collisionless regime without a large enhancement of thermal transport. In addition, we have observed the surprisingly optimistic effect that the anomalous electron transport always observed in tokamak plasmas appears to be reduced in the hot core region of the plasma when T_e is increased by neutral injection. These two results together generate very encouraging predictions for the Tokamak Fusion Test Reactor under construction at Princeton Plasma Physics Laboratory, and for future tokamak fusion devices.

The continuing support of Dr. M. B. Gottlieb, Dr. E. A. Frieman, and Dr. H. P. Furth is gratefully acknowledged. The authors express their appreciation to Dr. D. Post, Dr. J. Ogden, Dr. D. Jassby, and Dr. H. Towner for contributing modeling calculations and to Dr. T. Stix for his significant contributions to the early design work and expected performance evaluations. This work was supported by the U. S. Department of Energy Contracts No. EY-76-C-02-3073 and No. W 7405-ENG-26.

[a]Permanent address: Department of Physics and Astronomy, University of Maryland, College Park, Md. 20742.

[b]On leave from Centre d'Etudes Nucléaires de Fontenay-aux-Roses, B. P. No. 6, Fontenay-aux-Roses, France.

[c]Consultant. Permanent address: Exxon Nuclear Company, Bellevue, Wash.

[d]Consultant. Permanent address: Fusion Power Systems Division, Westinghouse Electric Company, Pittsburgh, Pa.

[1]J. G. Cordey *et al.*, Nucl. Fusion 14, 441 (1973).

[2]K. Bol *et al.*, in *Proceedings of the Fifth International Conference on Plasma Physics and Controlled Nuclear Fusion Research, Tokyo, Japan, 1974* (International Atomic Energy Agency, Vienna, Austria, 1975), Vol. 1, p. 77.

[3]L. A. Berry *et al.*, in *Proceedings of the Sixth International Conference on Plasma Physics and Controlled Nuclear Fusion Research, Berchtesgaden, West Germany, 1976* (International Atomic Energy Agency, Vienna, 1977), Vol. 1, p. 49.

[4]Équipe TFR, see Ref. 3, Vol. I, p. 69.

[5]J. W. M. Paul *et al.*, in Proceedings of the Eighth European Conference on Controlled Fusion and Plasma Physics, Prague, Czechoslovakia, 1977, Vol. II (to be published).

[6]V. S. Vlasenkov *et al.*, see Ref. 3, Vol. 1, p. 85.

[7]H. Eubank *et al.*, in *Proceedings of the Seventh International Conference on Plasma Physics and Controlled Nuclear Fusion Research, Innsbruck, Austria, 1978* (International Atomic Energy Agency, Vienna, Austria, 1979).

[8]J. Kim *et al.*, in *Proceedings of the Second Topical Meeting on Technology of Controlled Fusion, Richland, Washington, 1976* (National Technical Information Service, Springfield, Va., 1976), Vol. 4, p. 1213.

[9]K. Bol *et al.*, see Ref. 7.

[10]D. Post *et al.*, see Ref. 7.

[11]P. E. Stott, Plasma Phys. 18, 251 (1976).

Improved Tandem Mirror Fusion Reactor

D. E. Baldwin and B. G. Logan
Lawrence Livermore Laboratory, University of California, Livermore, California 94550
(Received 22 May 1979)

It is shown that the introduction of barrier potentials between the plugs and solenoid of a tandem mirror substantially reduce ion energy and density required in end plugs. Several means for creating barriers and some of the important physics issues are discussed.

In tandem mirror (TM) confinement of fusion plasma, ions in a magnetized solenoid are confined axially (plugged) by electrostatic potentials of denser mirror-confined plasma.[1,2] For a uniform electron temperature T_e, an ion-confining potential φ_c requires a plug-to-central-cell density ratio[1]

$$n_p/n_c = \exp(\varphi_c/T_e). \quad (1)$$

φ_c varies only logarithmically in the density ratio, whereas the ratio of central-cell fusion power density to the injection power density required to maintain the plugs varies as n_c^2/n_p^2. Increasing T_e by auxiliary heating permits a decrease of n_p/n_c for fixed φ_c, improving the reactor picture. Even with this, however, a conceptual TM reactor has severe technological requirements.[3] To plug a central cell of density $\approx 10^{14}$ cm^{-3}, temperature ≈ 40 keV, and magnetic field ≈ 2 T requires plugs of density $\approx 10^{15}$ cm^{-3} having peak fields ≈ 17 T and neutral-beam injection energies of order 600 keV with, or of order 1–2 MeV without, auxiliary electron heating. In the following, we describe a means by which, for the same central-cell conditions, the density of the plugs might be reduced to a few 10^{13} cm^{-3} requiring peak fields $\lesssim 10$ T and beam injection energies as low as 200 keV.

The essential idea is to raise the plug-electron temperature T_{ep} above the central-cell electron temperature T_{ec} by auxiliary electron heating in the plugs alone. Consider the magnetic field, potential, and density profiles shown in Fig. 1. Electrons from the central cell pass through the plug and mix by weak collisions with those trapped in the higher potential. In TM reactors as previously conceived, the mixing is sufficient to allow only relatively small electron temperature differences between plug and central cell, even though considerable neutral-beam and auxiliary neutral-beam and auxiliary heating are applied to the plug. Introduction of a potential dip $\varphi_b \gtrsim T_{ec}$ markedly increases achievable temperature differences by having φ_b act as a thermal barrier between the plug and solenoid electrons. Because plug electrons are then confined by a potential $\varphi_b + \varphi_c$, the power per volume transferred between the plug and transiting central-cell electrons can be estimated[4] to be

$$P_{cp} \approx G_e(n_p^2/n\tau_{ee})(T_{ep} - T_{ec}) \times \exp[-(\varphi_b + \varphi_c)/T_{ep}], \quad (2)$$

where

$$n\tau_{ee} = (m^{1/2})4\pi e^4 \ln\Lambda)(2T_{ep})^{3/2} = (8.2\times10^9/\ln\Lambda)T_{ep}^{3/2}\ \text{cm}^{-3}\ \text{s},$$

where T_{ep} is expressed in keV and G_e of order unity is a weak function of potential and mirror ratio. (Accurate determination of this transfer rate is important for detailed reactor calculations and is being pursued by analytical and numerical means.) The power applied to the plug electrons transfers to the central-cell electrons and contributes to their total power balance.

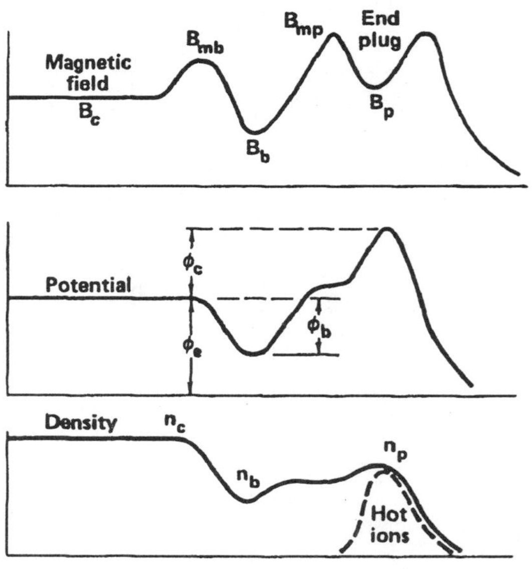

FIG. 1. Sample magnetic field, potential, and density profiles for a thermal barrier.

The barrier density n_b is related to n_c and n_p by

$$n_b = n_c \exp(-\varphi_b/T_{ec}) \tag{3a}$$

$$\simeq n_p \exp[-(\varphi_b+\varphi_c)/T_{ep}] \tag{3b}$$

or, replacing Eq. (1), there results

$$n_p/n_c \simeq \exp[-\varphi_b/T_{ec}+(\varphi_b+\varphi_c)/T_{ep}]. \tag{4}$$

[The approximate equality in Eqs. (3b) and (4) is used because the potential depth $\varphi_b+\varphi_c$ is to be determined by equality of the trapping and entrapping of electrons in the plugs, rather than by a strict Boltzmann relation.]

The depression in the barrier density required in Eq. (3a) is limited by the minimum density n_b(passing) of central-cell ions streaming through the locally negative potential. Assuming that this drop in potential is accompanied by a drop in the magnetic field by a factor $R_b + B_{mb}/B_b$ and that $\varphi_b > T_{ic}$, we obtain

$$n_b(\text{passing}) \approx (n_c/R_b)(T_{ic}/\pi\varphi_b)^{1/2}. \tag{5}$$

Added to this will be the density of ions trapped in φ_b, and it will be most important that their accumulation be prevented. If the total barrier-ion density n_b(passing plus trapped) is normalized to the density of passing ions, i.e., $n_b = g_b n_b$(passing), then from Eqs. (3a) and (5) we find

$$(T_{ic}/\varphi_b)^{1/2}\exp(\varphi_b/T_{ec}) = R_b \pi^{1/2}/g_b. \tag{6}$$

This result points up the importance of a large mirror ratio R_L, and a nearly complete pumpout of the trapped thermal ions from this region (for which $g_b - 1$). In assessing reasonable estimates for g_b, two considerations are important: (i) the degree to which specific pumpout mechanisms can compete with collisional trapping, and (ii) the questions of microstability of the resulting distributions.

Several schemes for maintaining a thermal barrier have been considered. The first three described require strengthening the field at the end of the solenoid to generate a local mirror-field peak. The barrier region would then lie between this peak and the end plug [see Fig. 1(a)]. The field peak throttles flow toward the plug; without collisions the density and potential in the barrier would decrease with B giving Eq. (5).

As a first pumpout means, we propose transit-time pumping of locally trapped ions by means of a parallel force applied at the ion bounce frequency. Examples are a parallel electric field (difficult at high density) or small oscillation of the position, depth, or axial extent of the local magnetic well. The former was used to eject electrons from the Phoenix mirror machine,[5] creating a rise in the ambipolar potential. The absorbed power density for this heating, which ends up as heating to the whole central cell, is that which competes with the trapping rate for passing central-cell ions into the potential well:

$$P_{\text{pump}} \approx G_i \frac{n_b^2(\text{passing})(B_{mb}/B_b)}{(n\tau)_{ii}} T_{ic}, \tag{7}$$

where $(n\tau)_{ii} = 5\times10^{11}\, T_{ic}^{3/2}/\ln\Lambda$ cm^{-3} sec, and $G_i \approx 3$. Determination of the degree to which thermal ions can be pumped out by this means against collisional filling and the amplitude of the required fields depends upon details of the electrostatic and magnetic well shapes, resonance frequencies, their widths and overlap, the applied frequency spectrum, island formation in phase space, etc. Modeling of this process in both the diffusion, or Fokker-Planck, limit and by single particles with a Monte Carlo collision process is underway at Livermore.

In a second pumpout scheme, the minimum of the barrier region in Fig. 1(a) would periodically be raised to the peak mirror value, so that all trapped thermal ions would escape. When returned to its minimum value, the barrier regions would remain empty of trapped ions for a fraction of a collision time, at which time the cycle would be repeated. The potential barrier offered by such a barrier region would not be constant in time, so that this method might require a pair of such barriers, operating out of phase.

A third pumpout method would be the injection of a neutral beam of energy $< \varphi_b$ in to the barrier, at an angle $\theta < \arcsin(1-\sqrt{R_b})$ with respect to the magnetic axis, so that ionization of beam atoms by charge-exchange, ion-impact, and electron-impact collisions would form passing ions. Charge-exchange collisions between beam atoms and trapped barrier ions would convert those trapped ions to neutrals, permitting their escape from the barrier. The beam atoms thereby converted to passing ions would then contribute to fueling the center-cell ion losses.

Microstability due to beam-type modes does not appear to pose a serious limit of the density in the barrier region. We have examined ion-ion two-stream, ion-acoustic, and ion-beam–cyclotron modes[6] and find that $g_b \gtrsim 2$ is sufficient for stability. Accordingly, we have taken $g_b = 2$ in reactor evaluations.

Potential magnetohydrodynamics (MHD) modes in the barrier region are the firehose, requiring for stability $\beta_\parallel - \beta_\perp \lesssim 2$, and the flute interchange, requiring for low-β stability that

$$\int \frac{ds(p_\perp + p_\parallel)K_\psi}{B^2} > 0, \tag{8}$$

where K_ψ is the component of the line curvature normal to the constant-pressure flux surface and the integration runs the entire TM length. We rely on higher-pressure, positive-curvature plugs to stabilize the negative-curvature regions joining the plugs with the uniform central cell, as in the conventional TM.[7] Of possible concern is the added destabilization of the barrier region. Using model fields and pressure profiles, we find that minimum-B plugs of mirror ratio 2 and ellipticity 30 can line-average stabilize a mirror-ratio-10 barrier region of $\beta_{\parallel b} \approx 2\beta_{\perp b}$. This curvature constraint is more stringent than the firehose, but is easily satisfied. Local ballooning ultimately sets a β limit,[7] and is currently being evaluated.

In the stream-stabilized mode of 2XIIB operation,[8] it frequently occurred that the density outside the mirror on the upstream side exceeded the hot-ion density between the mirrors by a factor of 3 or more and the density in the mirror throat by a factor of 10 to 20, with T_e inside twice that outside. (The potential profile was not measured.) We believe residual ion-cyclotron fluctuations precluded accumulation of ions at the magnetic maximum by $\vec{\mu}\cdot\nabla B$ forces induced by perpendicular ion heating. If this interpretation proves correct, it might be possible to extend this technique to reactor conditions, either by internally or externally generated rf, at either the plug or auxiliary mirror peak.

Although a complete TM reactor design employing thermal barriers awaits a more careful evaluation of the efficiency and stability of specific ion-pumping mechanisms, we can get a rough idea of the impact of thermal barriers by estimating the densities, fields, and powers for the plugs and the barriers for $g_b = 2$. Let us take a central-cell density $n_c = 10^{14}$ cm^{-3}, temperature $T_{ic} = T_{ec} = 40$ keV, field $B_c = 2.1$ T ($\beta_c = 0.7$), and radius $r_c = 100$ cm (giving $\simeq 20$ MW fusion power/m). At $\varphi_c \simeq 3T_{ic}$ and $\varphi_e \simeq 7T_{ec}$, fusion α particles will sustain central-cell energy losses at $n_c\tau_{\rm loss} \simeq 10^{15}$ cm^{-3} s obtainable by electrostatic confinement,[9] neglecting radial loss. For a maximum MHD-stable barrier mirror ratio $R_b = 10$, Eqs. (3) and (5) give $n_b = n_c/14 \simeq 7\times10^{12}$ cm^{-3} at $\varphi_b = 2.6T_{ec}$. Choosing the maximum $T_{ep} \simeq \varphi_b + \varphi_c \simeq 230$ keV consistent with our assumption of plug electrons being Maxwellian up to their confining potential, we have $n_p/n_b = 2.7$, or $n_p = 2\times10^{13}$ cm^{-3}, 5 times smaller than n_c. For plug ions to be mirror confined they must be injected above $(\varphi_e + \varphi_c)/(R_p - 1)$, which equals 150 keV for plug mirror ratios $R_p = B_{mp}/B_p(1-\beta_p)^{1/2} = 3.7$; so take $E_{\rm inj} = 200$ keV for adequate confinement. Ion scattering with small electron drag (high T_{ep}) leads to mean plug-ion energies $E_p \simeq 2E_{\rm inj} \simeq 400$ keV, requiring plug fields $B_p \simeq 2.7$ T at $\beta_p \simeq 0.7$, mirror fields $B_{mp} \simeq 5.4$ T for $R_p = 3.7$, and maximum conductor fields $B_{\rm max} \simeq 8$ T. Flux conservation gives $r_p = r_c(2.1/2.7)^{1/2} \simeq 90$ cm; so with spherical plugs $V_p \simeq 3\times10^6$ cm^3, and using Eq. (2), we find an auxiliary electron heating power $P_{cp}V_p \simeq 9.4$ MW per plug. The required injected-neutral-beam power per unit volume is

$$P_{\rm NB} \simeq n_p{}^2 E_{\rm inj}/(n\tau)_p \tag{9}$$

with[10]

$$(n\tau)_p \simeq 6\times10^{10} E_{\rm inj}{}^{3/2} \log_{10}\left[\frac{R_p}{1+(\varphi_e+\varphi_c)/E_{\rm inj}}\right] \simeq 1.5\times10^{13}\ \text{cm}^{-3}\ \text{s} \tag{10}$$

being the particle confinement for hydrogen plug ions with the above parameters, giving a neutral-beam power $P_{\rm NB}V_p = 2.5$ MW per plug. For the barrier parameters we have a barrier field $B_b = 0.5$ T for an MHD-stability-limited $\beta_{\parallel b} \simeq 1.4$ with $\beta_{\perp b} \simeq 0.5$ ($T_{i\perp} \simeq T_{ic}/R_b$ in the barrier); so at $R_b = 10$, $B_b(1-\beta_b)^{1/2} \simeq 0.35$ T, and $B_{mb} = 3.5$ T. Flux conservation gives $r_b = 1.8\,r_c \simeq 180$ cm; approximating the barrier volume as a sphere of this radius and using Eq. (7), we have $P_{\rm pump}V_b = 0.6$ MW per barrier. Thus the total required plasma input power $2(P_{cp}V_p + P_{\rm NB}V_p + P_{\rm pump}V_b)$ would be in the range of 25 MW. Since the center cell produces 20 MW fusion power/m, the TM reactor Q (central-cell fusion power)/(plug + barrier input power) would be approximately $Q = 40$ for a central-cell length of 50 m, and would increase proportionately at higher powers and lengths.

An addition that might be necessary in some situations is the confinement of anisotropic, hot electrons in the barrier region. When added to the right side of Eq. (3a), the presence of such electrons would give a larger φ_b for a given n_b, further reducing the power transfer between the plug and central-cell electrons. However, the power necessary to sustain these hot electrons can be comparable to the bulk plug-electron heating estimated above.

Finally, we see a number of issues concerning conventional TM confinement to be little affected by the addition of a barrier cell. This would certainly be true for the drift-cyclotron-loss-cone mode[11] in the plugs. Various drift modes in the solenoid would be forced to fit parallel wavelengths between the barrier regions. Neoclassical and related transport[12] would be reduced, roughly by a factor $B_c(1-\beta_c)^{1/2}/B_{mb}$, due to that fraction of the solenoidal ions being confined by an axisymmetric field.

In summary, thermal barrier potentials can substantially reduce plug power in TM reactors. Of crucial importance are the questions of how the barrier is formed and how completely the accumulation of thermal ions can be prevented. Methods of forming thermal barriers may represent rather small changes in the conventional TM geometry. We seek to apply electron heating power selectivity in such a way as to improve confinement. The degree to which this can be done will take time to evaluate, and probably we have not yet found the best scheme for efficiently pumping thermal ions. The potential improvements to be gained from this approach are so great as to warrant a thorough theoretical and experimental study.

This work was performed under the auspices of the U.S. Department of Energy by the Lawrence Livermore Laboratory under Contract No. W-7405-ENG-48.

[1]G. I. Dimov, V. V. Zakaidakov, and M. E. Kishinevsky, Fiz. Plasmy 2, 597 (1976) [Sov. J. Plasma Phys. 2, 326 (1976)], in *Proceedings of the Sixth International Conference on Plasma Physics and Controlled Nuclear Fusion Research, Berchtesgaden, West Germany, 1976* (International Atomic Energy Agency, Vienna, 1977). Paper No. C4.

[2]T. K. Fowler and B. G. Logan, Comments Plasma Phys. Controlled Fusion 2, 167 (1977).

[3]G. A. Carlson *et al.*, Lawrence Livermore Laboratory, Report No. UCID-18158, 1979 (unpublished).

[4]J. J. Dorning and R. H. Cohen, Bull. Am. Phys. Soc. 23, 776 (1978).

[5]E. Thompson *et al.*, in Proceedings of the Tenth Meeting of the Division of Plasma Physics, American Physical Society, Miami Beach, November, 1968 (unpublished), Paper No. 3D-4.

[6]W. E. Drummond and M. N. Rosenbluth, Phys. Fluids 5, 1507 (1962).

[7]D. E. Baldwin *et al.*, in *Proceedings of the Seventh International Conference on Plasma Physics and Controlled Nuclear Fusion Research, Innsbruck, Austria, 1978* (International Atomic Energy Agency, Vienna, Austria, 1979), Paper No. CN-37-J-4.

[8]For a description of the MX project, see F. H. Coensgen *et al.*, Lawrence Livermore Laboratory, Report No. LLL-Prop-142, 1976 (unpublished).

[9]R. H. Cohen, M. E. Rensink, T. A. Cutler, and A. A. Mirin, Nucl. Fusion 18, 1229 (1978).

[10]D. E. Baldwin, Rev. Mod. Phys. 49, 317 (1977).

[11]R. F. Post and M. N. Rosenbluth, Phys. Fluids 9, 730 (1966).

[12]D. D. Ryutov and G. V. Stupakov, Fiz. Plazmy 4, 501 (1978) [Sov. J. Plasma Phys. 4, 278 (1978)].

Ground State of the Electron Gas by a Stochastic Method

D. M. Ceperley
National Resource for Computation in Chemistry, Lawrence Berkeley Laboratory, Berkeley, California 94720

and

B. J. Alder
Lawrence Livermore Laboratory, University of California, Livermore, California 94550

(Received 16 April 1980)

An exact stochastic simulation of the Schroedinger equation for charged bosons and fermions has been used to calculate the correlation energies, to locate the transitions to their respective crystal phases at zero temperature within 10%, and to establish the stability at intermediate densities of a ferromagnetic fluid of electrons.

PACS numbers: 67.90.+i, 71.45.Gm

The properties of the ground state of the electron gas, also referred to as the fermion one-component plasma and jellium, have rigorously only been established in the limit of high densities[1] where the system approaches a perfect gas and at low density[2] where the electrons crystallize. Furthermore, Hartree-Fock calculations[3] and variational calculations[4] suggest that at intermediate densities, the spin-aligned state of the electrons will be more stable than the normal, unpolarized state. Precise calculations of this many-fermion system are required to establish the regions of stability of the various phases because of the small energy differences among them. This note outlines a Monte Carlo method that, if run long enough on a computer, can give as precise a solution for the ground state of a given fermion system as desired.

In practice, the precision of such a calculation is limited to about two orders of magnitude smaller than that of an approximate trial wave function that is introduced as an importance function in the Monte Carlo process. That the introduction of such an importance function is essential was previously demonstrated for the many-boson problem.[5] The extension of this boson calculation to fermions requires dealing with antisymmetric functions whose nodes are unknown. This leads to two related complications; namely, the probability density of a random walk cannot be chosen everywhere positive, and unless prevented the random walk will always converge to the all positive, boson ground state. It is demonstrated here that by representing the wave function by the difference between two probability densities, the effect of this inherent instability becomes serious, and it is possible to extract the properties of the lowest antisymmetric state. A more general procedure which removes the effects of the instability has yet to be perfected.

The solution of the fermion problem was carried out in two steps. In the first step the nodes, the places where the trial function vanishes, act as fixed absorbing barriers to the diffusion process. Inside a connected nodal region the wave function is everywhere positive and vanishes at the boundaries. With these boundary conditions, the fermion problem is equivalent to a boson problem. The energy calculated with this procedure, which we will refer to as the "fixed-node" energy, is an upper bound to the exact fermion ground-state energy and generally very close to it. In principle one could next vary the nodal loca tions to obtain the best upper bound, for example, by varying the functions used as elements in the Slater determinant of the trial wave function. In practice, the highly dimensional nodal surfaces are difficult to parametrize in a systematic fashion.

The second step, called "nodal relaxation," begins with the population of walks from the "fixed-node" approximation. In this second proce dure, if a random walk strays across the node of the trial function it is not terminated, but the sign of its contribution to any average is reversed. At any stage of the random walk there is a population of positive walks (those that remained in the same nodal region or crossed an even number of nodes) and a population of negative walks (those that crossed an odd number of nodes). The importance function used in this process is the absolute value of the trial function. It can be easily shown that the difference population converges to the antisymmetric eigenfunction. However, both the positive and negative populations grow geometrically with a rate equal to the difference between the Fermi and Bose energies. If the relaxation time from the fixed-node distribution times this energy difference is less than unity, the fermion energy can be reliably extracted. We have found that for the electron gas this condition is satisfied if the nodes of the Hartree-Fock wave function are used.

Our simulation method is a simpler version of the Green's function Monte Carlo method of Kalos, Levesque, and Verlet.[5] However, it requires numerical truncation. A trial wave function $\Psi_T(R)$ of the Bijl-Jastrow-Slater type[4] and an ensemble of about 100 systems are selected from a variational Monte Carlo calculation, where R represents the $3N$ spatial coordinates of the system of N electrons. Let the probability density of finding a random walk in dR^{3N} at time t be given by $f(R,t)dR^{3N}$. Then the value of f at $t=0$ is given by $|\Psi_T(R)|^2$ properly normalized. The diffusion equation for $f(R,t)$ is

$$\frac{\partial f}{\partial t}=\frac{\hbar^2}{2m}\left[\sum_{i=1}^{N}\nabla_i^2 f-\nabla_i(f\nabla_i \ln|\Psi_T|^2)\right]-\left[\frac{H\Psi_T}{\Psi_T}-E_{\text{ref}}\right]f, \quad (1)$$

where H is the Hamiltonian

$$H=\left(\frac{\hbar^2}{2m}\right)\sum_{i=1}^{N}\nabla_i^2-\sum_{i<j}e^2/r_{ij}. \quad (2)$$

It is easily verified that for large times, $f(r,t) = \Psi_T\varphi_0\exp[-t(E_{\text{ref}}-E_0)]$, where E_0 and φ_0 are the exact eigenvalue and eigenfunction. The above equation for $f(R,t)$ has a simple interpretation as a stochastic process. Each member of the ensemble of systems undergoes (i) random diffusion caused by the zero-point motion, (ii) biasing or drift by the trial quantum force, $\nabla\ln|\Psi_T|^2$, and (iii) branching with probability given by the time step times the difference between the local trial energy, $E_T=H\Psi_T/\Psi_T$, and the arbitrarily chosen reference energy, E_{ref}. By "branching", it is meant that a particular system is either eliminated from the ensemble (if the local energy is greater than the reference energy) or duplicated in the ensemble (otherwise). A steady-state population of the ensemble requires that the reference energy equal the lowest eigenvalue. This is one way of determining the eigenvalue.

The trial wave function employed in the present calculations is identical with those used in an earlier Monte Carlo variational calculation.[4] This trial function is a product of two-body correlation factors times a Slater determinant of single-particle orbitals. The two-body correlation factors are chosen such that they remove exactly the singularities in the local energy when two electrons approach each other, thus reducing tremendously the variance of the estimate of the ground-state energy. For the fluid phase the single-particle orbitals are plane waves, with the wave vector lying within the Fermi sea. For the polarized state, where there is only one spin for each spatial state, as opposed to two for the normal unpolarized state, the Fermi wave vector has been increased to allow for twice as many spatial orbitals. In the crystal phase, the orbitals are Gaussians centered around body-centered cubic lattice sites with a width chosen variationally.

Figure 1 shows that the relaxation from the unpolarized nodes to the ground state is rapid with a small lowering of the energy. A less accurate trial wave function with different nodes obtained from a linear combination of polarized and unpolarized Slater determinants is nevertheless shown to lead to similar energies with a somewhat larger relaxation time. This suggests that the results are insensitive to the original location of the nodes, although a longer calculation beyond this relaxation time would be possible and desirable. Since at all densities the relaxation from the Hartree-Fock nodes was rapid, the ground-state energy of the electron gas by the method employed could be obtained with very little uncertainty.

The largest uncertainty in the results is, in fact, due to the number dependence. Because of

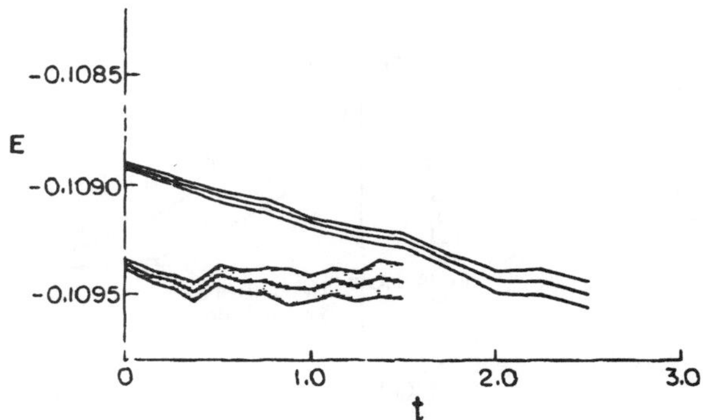

FIG. 1. The energy in rydbergs per particle of a 38-electron system at the density $r_s = 10$ vs diffusion time (in inverse Rydbergs) from removal of the fixed nodes. The lower curve is the relaxation of an ensemble of 1.6×10^4 systems from the nodes of the unpolarized determinant of plane waves. The upper curve is the relaxation of 1.0×10^5 systems from the nodes of a linear combination of polarized and unpolarized determinants.

TABLE I. The ground-state energy of the charged Fermi and Bose systems. The density parameter r_s is the Wigner-sphere radius in units of Bohr radii. The energies are rydbergs and the digits in parentheses represent the error bar in the last decimal place. The four phases are paramagnetic or unpolarized Fermi fluid (PMF); the ferromagnetic or polarized Fermi fluid (FMF); the Bose fluid (BF); and the Bose crystal with a bcc lattice.

r_s	E_{PMF}	E_{FMF}	E_{BF}	E_{bcc}
1.0	1.174(1)	...	...	...
2.0	0.0041(4)	0.2517(6)	-0.4531(1)	...
5.0	-0.1512(1)	-0.1214(2)	-0.216 63(6)	...
10.0	-0.106 75(5)	-0.1013(1)	-0.121 50(3)	...
20.0	-0.063 29(3)	-0.062 51(3)	-0.066 66(2)	...
50.0	-0.028 84(1)	-0.028 78(2)	-0.029 27(1)	-0.028 76(1)
100.0	-0.015 321(5)	-0.015 340(5)	-0.015 427(4)	-0.015 339(3)
130.0	...	...	-0.012 072(4)	-0.012 037(2)
200.0	...	...	-0.008 007(3)	-0.008 035(1)

the high accuracy of the results derived from employing a good trial wave function and the consequent small statistical error, the number dependence, which was empirically established for systems ranging from 38 to 246 particles, is an order of magnitude larger than the statistical error. Extrapolation to infinite-particle results was carried out at each density on the basis of $E(N)=E_0+E_1/N+E_2\Delta_N$, where the coefficients E_0, E_1, and E_2 were empirically determined from the simulations. The E_1 term arises from the potential energy and is due to the correlation between a particle and its images in the periodically extended space that is used in the Ewald summation procedure[4] to eliminate the major surface effects. The E_2 term comes from the discrete nature of the Fermi sea for finite systems, and Δ_N is the size dependence of an ideal Fermi sys-

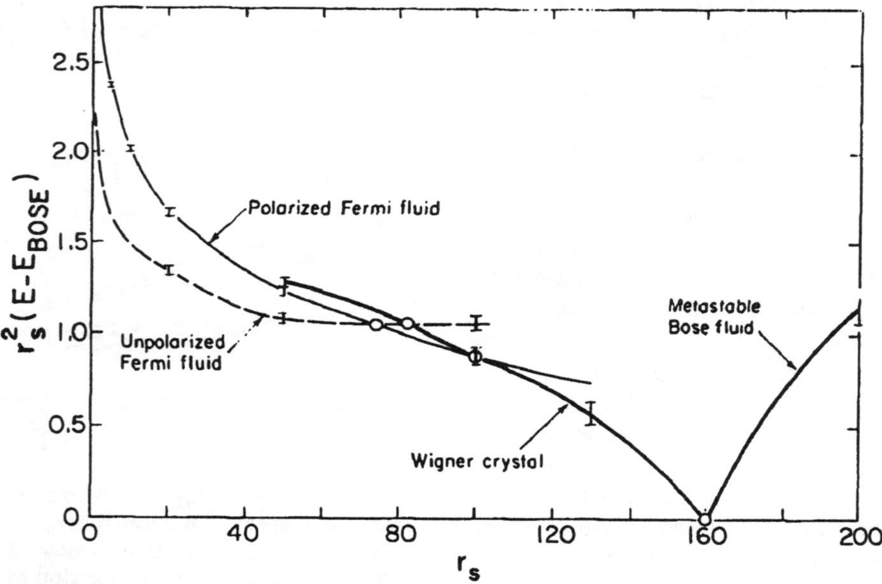

FIG. 2. The energy of the four phases studied relative to that of the lowest boson state times r_s^2 in rydbergs vs r_s in Bohr radii. Below $r_s = 160$ the Bose fluid is the most stable phase, while above, the Wigner crystal is most stable. The energies of the polarized and unpolarized Fermi fluid are seen to intersect at $r_s = 75$. The polarized (ferromagnetic) Fermi fluid is stable between $r_s = 75$ and $r_s = 100$, the Fermi Wigner crystal above $r_s = 100$, and the normal paramagnetic Fermi fluid below $r_s = 75$.

tem at the same density. That term is absent for bosons. In addition, the energies have been extrapolated to zero time step by empirically establishing the validity of linear extrapolation. This correction is quite small, on the order of the statistical error for the time steps used. However, this correction can be completely avoided by using an integral formulation of Eq. (1).[5]

The results for the energy of the plasma in four different phases is given in Table I and plotted in Fig. 2. The boson system undergoes Wigner[6] crystallization at $r_s = 160 \pm 10$. The fermion system has two phase transitions, crystallization at $r_s = 100 \pm 20$ and depolarization at $r_s = 75 \pm 5$. We have found that the difference in energy between a boson crystal and a fermion crystal is less than $1.0 \times 10^{-6} R$ at $r_s = 100$. The energies of the three Fermion states are sufficiently close in the low-density regime that still more accurate calculations on larger systems would be desirable to confirm these results. Although the Bijl-Jastrow-Slater results are quite accurate,[4] the error is different for the different phases, changing their relative stability. This demonstrates how essential it is to perform exact simulations to calculate reliably phase-transition densities.

The authors would like to thank M. H. Kalos for numerous useful discussions and for inspiring the present work. We thank Mary Ann Mansigh for computational assistance.

This work was supported in part by the National Resource for Computation in Chemistry under Contract No. CHE-7721305 from the National Science Foundation, and by the Basic Energy Sciences Division of the U. S. Department of Energy under Contract No. W-7405-ENG-48.

[1]M. Gell-Mann and K. A. Brueckner, Phys. Rev. 106, 364 (1957).

[2]W. J. Carr, Phys. Rev. 122, 1437 (1961).

[3]F. Bloch, Z. Phys. 57, 549 (1929).

[4]D. Ceperley, Phys. Rev. B 18, 3126 (1978).

[5]M. H. Kalos, D. Levesque, and L. Verlet, Phys. Rev. A 9, 2178 (1974); D. M. Ceperley and M. H. Kalos, in *Monte Carlo Methods in Statistical Physics*, edited by K. Binder (Springer-Verlag, New York, 1979), p. 145.

[6]E. P. Wigner, Phys. Rev. 46, 1002 (1934), and Trans. Faraday Soc. 34, 678 (1938).

Stabilizing Effect of Finite-Gyroradius Beam Particles on the Tilting Mode of Spheromaks

R. N. Sudan
Laboratory of Plasma Studies, Cornell University, Ithaca, New York 14853

and

P. K. Kaw
Plasma Physics Laboratory, Princeton University, Princeton, New Jersey 08540
(Received 29 December 1980)

The equilibrium shape of a low-pressure spheromak plasma with a small component of toroidal current carried by finite-gyroradius particles is computed. The stabilizing influence of this current on the tilting mode is determined by employing an energy principle that includes gyroscopic and finite-gyroradius effects.

PACS numbers: 52.55.-s, 52.20.Dq, 52.35.Py

The favorable characteristics of nearly force-free, spherical magnetic configurations[1] dubbed "spheromaks" have led to an enthusiastic vision of a fusion reactor[2] with engineering advantages superior to that of tokamaks. However, Rosenbluth and Bussac[3] have shown that these configurations are subject to the "tilting mode" in which the entire plasma tilts unstably about an axis through the center. The remedy[3] for this mode appears to lie in modifying the plasma shape to one which is oblate and, more importantly, in employing a conducting shell very close to the plasma. Unfortunately, this last requirement detracts considerably from the reactor scenario.

In this paper we show that the inclusion of a component of azimuthal current carried by energetic ions with large gyroradius in the spheromak system leads to (1) an oblateness in the shape of the plasma so that by adjusting this current one could optimize the shape for magnetohydrodynamic stability, and (2) an additional stabilizing effect due to the gyroscopic motion and large angular momentum of these particles. This stabilizing influence can be employed to increase the distance between the plasma and the conducting shroud to improve the reactor prospects.

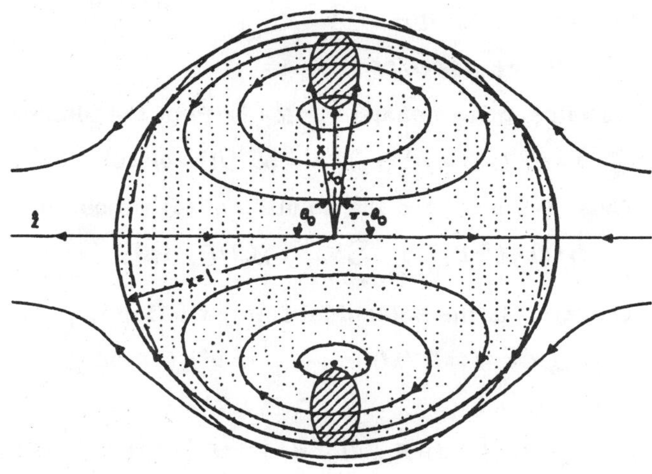

FIG. 1. Magnetic configuration of a spheromak; the cross-hatched region is occupied by the energetic particle current, $\Omega/\Omega_* \sim 1.2$ ($x = r/R$). Dashed line shows perturbed surface $R + \delta r(\theta)$.

The analytic investigation is a perturbation expansion about the spheromak configuration by including a small energetic particle current j_b. Thus

$$\nabla \times \vec{B} = k\vec{B} + 4\pi j_b \hat{\varphi}, \tag{1}$$

with $j_b = q\int d^3v\, v_\varphi f(H + \Omega P_\varphi)$; H, the particle energy, and P_φ, the canonical angular momentum, are constants of motion, Ω is a positive constant, q is the particle charge, and the system is considered to be axisymmetric. The rigid-rotor distribution function is chosen because it maximizes the entropy leading to favorable stability properties. Furthermore, $|I_b/RB_\varphi| \equiv \epsilon \ll 1$ is regarded as an expansion parameter, where I_b is the total particle current. The first-order (in ϵ) perturbed fields are obtained from (spherical coordinates r, θ, φ)

$$\nabla \times \vec{B}^{(1)} = k\vec{B}^{(1)} + 4\pi j_b \hat{\varphi} = k\vec{B}^{(1)} - \hat{\varphi} 4\pi q\Omega m^{-2} 2\pi \int_{-\infty}^{\infty} dP_\varphi \int_V^\infty dH f_0(H + \Omega P_\varphi), \tag{2}$$

where $V = (P_\varphi - q\psi_0)^2/2mr^2\sin^2\theta$. The zero-order fields $\vec{B}^{(0)}$ and poloidal flux ψ_0 are those for the spheromak equilibrium.[3] We choose, for calculational convenience, $f = A$ (constant) for $0 < H + \Omega P_\varphi < \frac{1}{2}mR^2\Omega^2$ and zero everywhere else. This choice ensures that j_b vanishes at $r = R$, and with $B^{(0)} \to B_* \hat{z}$ for $r \gg R$,

$$j_b = -A(4\pi q/3)R^4\Omega^4 x\,|\sin^2\theta|\,[x^2\sin^2\theta(1+2Y) - 1]^{3/2} \tag{3a}$$

where $x = r/R$, $Y = 1.54(\Omega_*/\Omega) j_1(kRx)/x$, $j_m(kRx)$ denotes the spherical Bessel function of order m, and $\Omega_* = qB_*/m$. In evaluating j_b which is a quantity of order ϵ, we have used zero-order poloidal flux ψ_0 in f_0. The current density is nonvanishing only between $\theta_0(x)$ and $\pi - \theta_0(x)$ and between x_0 and 1, where $x^2\sin^2\theta_0 = (1+2Y)^{-1}$ and $x_0^2 = (1+2Y)^{-1}$ (see Fig. 1). The constant A can be expressed in terms of the magnitude of the total particle current

$$I_b = \int d^3r\, |j_b| = A(4\pi q/3)R^6\Omega^4 \int_{x_0}^{1} dx\, x^2 \int_{\theta_0}^{\pi-\theta_0} d\theta \sin\theta\,[x^2\sin^2\theta(1+2Y) - 1]^{3/2},$$

$$\equiv A(4\pi q/3)R^6\Omega^4 M(\Omega/\Omega_*). \tag{3b}$$

The solution of (2) is obtained from

$$\nabla^2\Phi + k^2\Phi = \begin{cases} -\int^\theta d\theta\, 4\pi j_b & \text{for } 0 < x < 1, \\ 0 & \text{for } x > 1, \end{cases} \tag{4}$$

with $\vec{B}^{(1)} = k\vec{r} \times \nabla\Phi + \nabla \times \vec{r} \times \nabla\Phi$. Furthermore, it can be shown for axisymmetric systems that the perturbed poloidal flux

$$\psi^{(1)} = r \sin\theta\, \partial\Phi/\partial\theta. \qquad (5)$$

The perturbed shape $r_s = R + \delta r(\theta)$ of the plasma which coincides with the separatrix is obtained from

$$0 = \psi^{(0)}(r_s) + \psi^{(1)}(r_s) \cdots = \psi^{(0)}(R) + \delta r\, \partial\psi^{(0)}/\partial r|_{r=R} + \psi^{(1)}(R). \qquad (6)$$

Thus, from (5) and the solution of (4) we obtain for the perturbation of the surface ($\mu \equiv \cos\theta$)

$$\delta r(\theta) = (\tfrac{2}{3}B_*) \sum_{m=1,3,5,\ldots} G_m(R) P_m'(\mu), \qquad (7)$$

where B_* is the solenoidal field at infinity,

$$G_m(R) = -(4\pi I_b/M)(m+\tfrac{1}{2})\int_0^1 dx\, x^3 j_m(kRx) C_m(x)/kRj_{m-1}(kR),$$

and

$$C_m(x) \equiv \int_{-1}^{1} d\mu\, P_m(\mu) \int^\theta d\theta\, |\sin\theta| [x^2 \sin^2\theta (1+2Y) - 1]^{3/2},$$

P_m are the Legendre polynomials, and the prime on P_m denotes a derivative. Note that δr is independent of the sign of k. The $m=1$ term is positive and independent of θ showing a uniform expansion; the $m=3$ term is proportional to $-(15\cos^2\theta - 3)$ and leads to an oblateness of the spheromak. The terms alternate in sign but get smaller and smaller.

To establish the stability of this configuration we employ the energy principle of Sudan and Rosenbluth[4] in the form given by Finn and Sudan[5] and Finn,[6]

$$-\omega^2 T + \omega L + \delta W_p + \delta W_1 + \delta W_2 = 0, \qquad (8)$$

where

$$L = \tfrac{1}{2} iq \int d^3r\, n_b \vec{B}\cdot\vec{\xi}^* \times \vec{\xi}, \quad T = \int d^3r\, n_p m_p |\vec{\xi}|^2,$$

$$\delta W_1 = \tfrac{1}{2} q^2 \Omega^2 \int d^3r\, d^3v (\partial f_0/\partial H) \rho^2 |\hat{\varphi}\cdot\vec{\xi}\times\vec{B}|^2,$$

$$\delta W_2 = \tfrac{1}{2} i q^2 (\omega - l\Omega) \int d^3r\, d^3v\, (\partial f_0/\partial H) g\, dg^*/dt,$$

$$\delta W_p = \tfrac{1}{2}\int d^3r (|Q|^2/4\pi - \vec{\xi}^*\cdot\vec{J}_p\times\vec{Q} + \vec{\xi}\cdot\vec{J}_p\times\vec{B}\nabla\cdot\vec{\xi}^* + \gamma p|\nabla\cdot\vec{\xi}|^2),$$

$g = \int_{-\infty}^t dt'\, \vec{\xi}\cdot\vec{v}\times\vec{B}$ is an integral over the orbits of the energetic particles, ω is the eigenfrequency, $\vec{\xi}$ is the plasma displacement $\vec{Q} = \nabla\times\vec{\xi}\times\vec{B}$; in δW_p, $\vec{J}_p$, n_p, and p refer to the plasma current, density, and pressure, respectively, and in a cylindrical coordinate system (ρ, φ, z), $\rho = r\sin\theta$.

Following an argument employed by Rosenbluth and Bussac,[3] we let $\vec{\xi} = \vec{\xi}_0 + \epsilon\vec{\xi}^{(1)}$, where $\vec{\xi}_0 = \theta\vec{r}\times\hat{x}$, a rigid rotation of the configuration about the x axis, is an eigenfunction of the unperturbed spheromak while $\vec{\xi}^{(1)}$ depends upon the distortion caused by j_b; θ is the amplitude of the tilt. If a conducting wall surrounds the plasma at $r = R + \delta r_w(\theta)$, then it has been shown for a force-free spheromak by Rosenbluth and Bussac[3] and under more general conditions by Hammer[7] that $\vec{\xi}^{(1)}$ does not appear in δW_p to first order in ϵ. Omitting the details, we obtain

$$\delta W_p = -\tfrac{3}{4} B_*^2 R^3 \theta^2 [(RB_*)^{-1} \sum_{m=3,5,\ldots} G_m(R) + \delta r_w/R]. \qquad (9)$$

We now proceed to compute the remaining terms in (8) involving gyroscopic and finite-Larmor-radius effects. It is immediately evident that to first order in ϵ these terms can be evaluated with $\vec{\xi}_0$ and $\vec{B}_0$, instead of $\vec{\xi}$ and $\vec{B}$. With $\xi_0 = \theta(-iz, z, i\rho)\exp[i(\varphi - \omega t)]$ in cylindrical coordinates (ρ, φ, z), the orbit integral g is evaluated as follows:

$$g = \int_{-\infty}^t dt'\, \vec{\xi}_0\cdot\vec{v}\times\vec{B}_0 = (m/q)\int_{-\infty}^t dt'\, \vec{\xi}_0\cdot d\vec{v}/dt' = (m/q)[\vec{\xi}_0\cdot\vec{v} + i\omega\int_{-\infty}^t dt'\, \vec{\xi}_0\cdot\vec{v}].$$

The $i\omega$ term is of order $\omega/\Omega < v_A/R\Omega \ll 1$ and can, therefore, be neglected; v_A is the typical Alfvén speed. Thus, after some lengthy algebra we obtain

$$\delta W_1 + \delta W_2 = (9.24\pi I_b B_* R^2\theta^2/M)\int_{x_0}^1 dx\, x^3 j_1(kRx)\int_{\theta_0}^{\pi-\theta_0} d\theta\, |\sin\theta| \cos^2\theta [x^2\sin^2\theta(1+2Y) - 1]^{1/2} \times [x^2\sin^2\theta(\tfrac{5}{2} - Y) - 1]. \qquad (10)$$

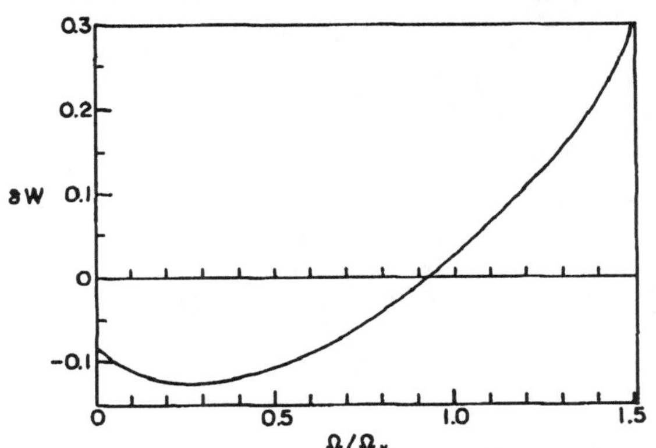

FIG. 2. Plot of $\delta W \equiv (\delta W_1 + \delta W_2)/9.24\pi I_B B_* R^2\theta^2$ as a function of Ω/Ω_*.

The system is stable[8] provided

$$L^2/4T + \delta W_p + \delta W_1 + \delta W_2 > 0. \tag{11}$$

Although $L^2/4T$ provides a stabilizing influence, in our calculations here it is a quantity of $O(\epsilon^2)$ and will, therefore, be neglected. In Fig. 2, $(\delta W_1 + \delta W_2)/9.24\pi I_b B_* R^2\theta^2$ is plotted as a function of Ω/Ω_*. We keep the uniform field B_*, spheromak radius R, and the total particle current I_b constant. We note that $x_0 \to 1$ as $\Omega/\Omega_* \to \frac{3}{2}$, i.e., if the ions have too large an angular velocity they are unconfined. This follows from the fact that ions of energy $\frac{1}{2}mR^2\Omega^2$ are barely confined and their gyrofrequency at $\theta = \pi/2$ is $\frac{3}{2}qB_*/m = \frac{3}{2}\Omega_*$. Thus, for equilibrium to exist $\Omega/\Omega_* \leq \frac{3}{2}$. For $\Omega/\Omega_* < 0.93$, $\delta W_1 + \delta W_2$ becomes negative, i.e., the destabilizing influence of the centrifugal force (δW_1) overcomes the stabilizing effect of the spread in betatron motion (δW_2). Note that $\delta W_1 + \delta W_2$ rises rapidly as $\Omega/\Omega_* \to \frac{3}{2}$. In the distribution f, the spread in $H + \Omega P_\varphi$ is $-\frac{1}{2}mR^2\Omega^2$, while the mean value is $-\frac{1}{4}mR^2\Omega^2$. It is noteworthy that only the radial field B_r appears in the expression for $\delta W_1 + \delta W_2$ while B_θ determines δW_p. Thus, the tilting mode is independent of the toroidal field B_φ. If δr_w is adjusted to make δW_p vanish, then $\delta W_1 + \delta W_2$ provides a margin of stability. Alternatively δr_w can be increased to take into account the stabilizing effect of $\delta W_1 + \delta W_2$ for $0.93 < \Omega/\Omega_* < \frac{3}{2}$. In passing we mention that the calculation for stabilizing Hill's vortex $(B_\varphi = 0)$ proceeds in a very similar fashion and the expression for $\delta W_1 + \delta W_2$ for this case can be obtained from Eq. (10) by replacing $j_1(kRx)/x$ by $\frac{1}{2}(1 - x^2)$.

The practical situation one looks forward to would probably require the particle current to be of the order of magnitude of the plasma current, i.e., a hybrid particle ring–compact torus. Such equilibria have been obtained numerically[9] and their stability would also have to be tested numerically. It must be noted, however, that for $4\pi I_b \sim B_* R$ the rigid tilt may not be the preferred eigenmode. Numerical simulations of non-field-reversed ion rings[10] show that the ring displacement is mostly in the direction of the external field while it varies in azimuth as $\exp i\varphi$.

We are indebted to Dr. A. Turnbull for numerical calculations leading to Fig. 2, and we acknowledge the encouragement of Dr. William C. Condit and our indebtedness to Dr. J. H. Hammer for providing us with an advance copy of his analysis.

[1]G. Morikawa, Phys. Fluids 12, 164 (1967).

[2]M. N. Bussac, H. P. Furth, M. Okabayashi, M. N. Rosenbluth, and A. M. Todd, in *Proceedings of the Seventh International Conference on Plasma Physics and Controlled Nuclear Fusion Research, Innsbruck, Austria, 1978* (International Atomic Energy Agency, Vienna, Austria, 1979), Vol. 3.

[3]M. N Rosenbluth and M. N. Bussac, Nucl. Fusion 19, 489 (1979).

[4]R. N. Sudan and M. N. Rosenbluth, Phys. Rev. Lett. 36, 972 (1976), and Phys. Fluids 22, 282 (1979).

[5]J. M. Finn and R. N. Sudan, Phys. Rev. Lett. 41, 695 (1978), and Phys. Fluids 22, 1148 (1979).

[6]J. M. Finn, Phys. Fluids 22, 1770 (1979).

[7]J. H. Hammer, Bull. Am. Phys. Soc. 25, 862 (1980), and private communication.

[8]Stability here refers only to modes growing on a fast time scale. Because of the neglect of terms higher order in ω/Ω, the system may still have residual instabilities on a long time scale.

[9]A. Mankofsky, R. N. Sudan, and J. Denavit, in Proceedings of the Ninth Conference on Numerical Simulation of Plasmas, Evanston, Illinois, 1980 (unpublished), Paper No. PA-4.

[10]A. Friedman, Ph.D. thesis, Cornell University, 1980 (unpublished).

Computer Simulation of Nonlinear Ion-Electron Instability

R. H. Berman, D. J. Tetreault, T. H. Dupree, and T. Boutros-Ghali
Massachusetts Institute of Technology, Cambridge, Massachusetts 02139
(Received 14 December 1981)

A computer simulation of a one-dimensional electron-ion plasma is described. The results differ substantially from the predictions of the conventional theory of this system. In particular, an instability is observed whose onset occurs for electron drifts well below the threshold of the linear ion acoustic instability and which ultimately dominates the nonlinear evolution of the linear instability.

PACS numbers: 52.25.Gj, 52.35.Py, 52.35.Ra, 52.65.+z

We report numerical experiments for an ion-electron plasma that show an instability well below the threshold predicted by linear theory. We studied a simulation plasma with mass ratio $m_i/m_e=4$ and temperature ratio $T_e/T_i=1$ in which, initially, the average electron distribution $\{\langle f_e\rangle = (2\pi v_e^2)^{-1/2}\exp[-(v-v_d)^2/2v_e^2]\}$ drifts relative to the average ion distribution $\{\langle f_i\rangle=(2\pi v_i^2)^{-1/2} \times\exp(-v^2/2v_i^2)\}$. According to linear theory such a plasma is unstable (the ion acoustic instability for physical mass ratios) for drifts exceeding a certain threshold, $v_d=3.924v_i$. We observed an instability for $v_d \geq 1.5v_i$. The observation of this instability, which is completely at variance with linear theory, is the principal result of this Letter. We describe several diagnostics that elucidate the nature of this instability. Finally, we point out that these results are consistent with recent theoretical predictions of the theory of clumps and holes. This agreement suggests that the results reported here may be the first observations of an important new type of plasma instability.

For our simulations we used a highly optimized one-dimensional, electrostatic code with $N_p=102\,400$ particles per species. We treated a periodic system of length $L=32.42\lambda_d$, where λ_d $(\equiv v_i/\omega_{pi})$ is the Debye length and ω_{pi} is the ion plasma frequency. Various diagnostics were performed which provide information about the fluctuations δf of the distribution function: $\delta f=f-\langle f\rangle$. For our spatially periodic and homogeneous system, the ensemble average $\langle\ldots\rangle$ was approximated by a spatial average. The two basic diagnostics we used were the mean square electric field and the mean square fluctuation $\langle(\delta N)^2\rangle$ of the number of particles, N (electron or ion), in a phase-space cell of size $\Delta x, \Delta v$. Here, $\delta N=N-\langle N\rangle$, where $\langle N\rangle$ is the mean number of particles in a cell. We would have preferred to measure the correlation function $\langle\delta f(1)\delta f(2)\rangle = \langle\delta f(x_1,v_1)\delta f(x_2,v_2)\rangle$ directly, for small $x_-=x_1-x_2$ and small $v_-=v_1-v_2$. Unfortunately, our value of $n_0\lambda_d=3259.5$ $(n_0=N_p/L)$ was not large enough to provide adequate statistical accuracy. Our diagnostics, however, derive from the correlation function since the mean square electric field involves velocity integrals over $\langle\delta f\,\delta f\rangle$, while $\langle(\delta N)^2\rangle$ and $\langle\delta f\,\delta f\rangle$ are related through[1]

$$\langle(\delta N)^2\rangle = \langle[n_0\int_x^{x+\Delta v} dx\int_v^{v+\Delta v} dv\,\delta f(x,v)]^2\rangle = n_0^2\int_{-\Delta x}^{\Delta x} dx_-(\Delta x-|x_-|)\int_{-\Delta v}^{\Delta v} dv_-(\Delta v-|v_-|)\langle\delta f(1)\delta f(2)\rangle. \tag{1}$$

Since $\langle(\delta N)^2\rangle$ is a double integral of $\langle\delta f\delta f\rangle$, it is less sensitive to statistical error. The latter point is clearer when one realizes that Eq. (1) can be solved for the correlation function in terms of a fourth derivative of $\langle(\delta N)^2\rangle$.

We can write, for each species,

$$\langle\delta f(1)\delta f(2)\rangle = n_0^{-1}\delta(x_-)\delta(v_-)\langle f(1)\rangle + g_d(1,2)+g_v(1,2), \tag{2}$$

where the first term of Eq. (2) is the discrete-particle self-correlation function and $g_d(1,2)$ accounts for correlated fluctuations which shield the discrete particles. $g_v(1,2)$ represents the effects of fluctuations over and above this level. Using the first term of Eq. (2) in Eq. (1), we find that the self-correlation contribution to $\langle(\delta N)^2\rangle$ is $\langle N\rangle$—the value for randomly located discrete particles. This value is modified by the contribution from g_d; in particular, as the linear stability threshold is approached, the zeros of the dielectric function will enhance $\langle(\delta N)^2\rangle$ through the emission and absorption of weakly damped waves. We have calculated[2] this contribution to $\langle(\delta N)^2\rangle$ and have found it to be, consistently, much lower than the values of $\langle(\delta N)^2\rangle$ observed in the simulations (cf. Fig. 1). Our observations of $\langle(\delta N)^2\rangle$ are,

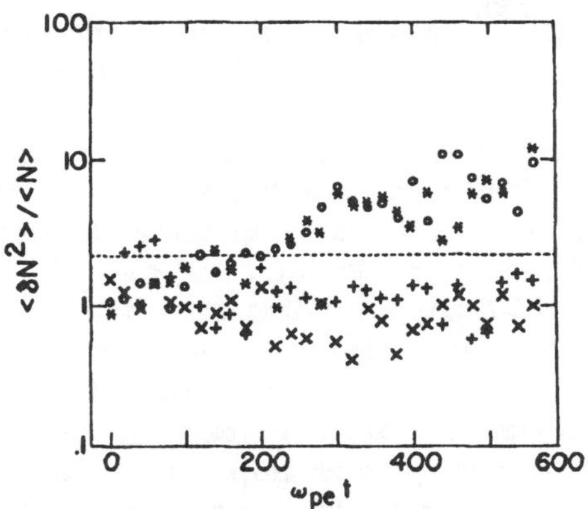

FIG. 1. Ion $\langle(\delta N)^2\rangle/\langle N\rangle$ for a phase-space cell of dimensions $\Delta x = 1.963\lambda_d$, $\Delta v = 0.1v_i$ vs time for $v_d = 2.5v_i$. The points are the simulation values measured at different velocities: crosses, $-1v_i$; plusses, $0v_i$; asterisks, v_i; and circles, $2v_i$. The dashed line is the shielded discrete-test-particle level.

therefore, evidence for collective fluctuations $g_v(1,2)$ well above the dressed test-particle level.

We measured $\langle(\delta N)^2\rangle$ in cells of size $0.05v_i \leq \Delta v \leq 3v_i$ by $0.2\lambda_d \leq \Delta x \leq 3\lambda_d$. The characteristic sizes of the fluctuations in space $(\Delta x)_c$ and in velocity $(\Delta v)_c$ were inferred from the dependence of $\langle(\delta N)^2\rangle/\langle N\rangle$ on $\Delta x, \Delta v$, given by Eq. (1). For Δx, Δv less than the characteristic size, $\langle(\delta N)^2\rangle/\langle N\rangle$ increases with $\Delta x, \Delta v$ since $\langle(\delta N)^2\rangle \simeq n_0^2\langle\delta f^2\rangle(\Delta x \times \Delta v)^2$ and $\langle N\rangle = n_0\langle f\rangle\Delta x\,\Delta v$. For $\Delta x, \Delta v$ greater than the characteristic size,

$$\langle(\delta N)^2\rangle \simeq n_0^2\Delta x\,\Delta v \int_{-\infty}^{\infty} dx_- \int_{-\infty}^{\infty} dv_- \langle\delta f(1)\delta f(2)\rangle$$

so that $\langle(\delta N)^2\rangle/\langle N\rangle$ approaches a constant value. For $\Delta x < (\Delta x)_c$ and $\Delta v > (\Delta v)_c$, and vice versa, $\langle(\delta N)^2\rangle$ becomes proportional to Δx and Δv, respectively. Figure 2 is a typical plot of $\langle(\delta N)^2\rangle/\langle N\rangle$, for the ions, versus cell size. It indicates that the ion fluctuation curves stopped increasing and turned over when $\Delta x \simeq 4\lambda_d$ and $\Delta v \simeq 0.1v_i$. Thus, these values of Δx and Δv represent the characteristic sizes $(\Delta v)_c$ and $(\Delta x)_c$.

Figure 1 shows the time dependence of $\langle(\delta N)^2\rangle/\langle N\rangle$ for the ions for a run with $v_d = 2.5v_i$. After starting at unity, $\langle(\delta N)^2\rangle/\langle N\rangle$ has risen by $\omega_{pe}t = 150$ to the shielded discrete-particle level, which we calculated to be 2.2. Subsequently, growth occurred until $\omega_{pe}t = 300$. Following this unstable growth phase, the instability saturated when the electron distribution function formed a plateau in velocity space. We measured the fluctuation levels at $-v_i$, 0, v_i, and $2v_i$. Fluctuations with phase velocities of v_i and $2v_i$ grew, while those at 0 and $-v_i$ decayed. This would seem to

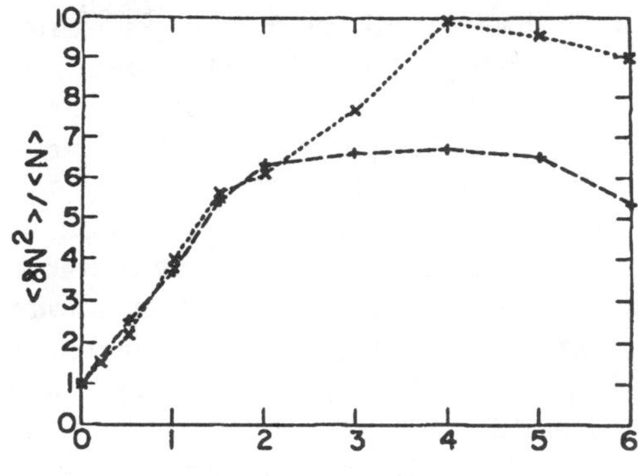

FIG. 2. Ion $\langle(\delta N)^2\rangle/\langle N\rangle$ for a phase-space cell of dimensions $\Delta x, \Delta v$. The points are the simulation values: plusses, $\langle(\delta N)^2\rangle/\langle N\rangle$ vs $20\Delta v/v_i$ at fixed $\Delta x = 0.98\lambda_d$; crosses, $\langle(\delta N)^2\rangle/\langle N\rangle$ vs $\Delta x/\lambda_d$ at fixed $\Delta v = 0.1v_i$.

indicate that the instability grows in regions of opposing velocity gradients of $\langle f_e\rangle$ and $\langle f_i\rangle$. Furthermore, we note that the instability occurred in regions of large negative $\partial\langle f_i\rangle/\partial v \neq 0$—a region where linear theory would predict strong damping of the fluctuations. Figure 3 shows that the fluctuations are unstable over a wide region of velocity space $(0 \leq v \leq 4v_i)$. This is evidenced by the ion tail and distorted electron distribution at $400\omega_{pe}^{-1}$. These distortions in $\langle f_i\rangle$ and $\langle f_e\rangle$ are not the result of discrete-particle collisions, since the collisional relaxation time is substantially larger than the run time of the simulation. Moreover, ion acoustic waves cannot be responsible, since, according to linear theory, they are stable.

Numerous runs were made to study various

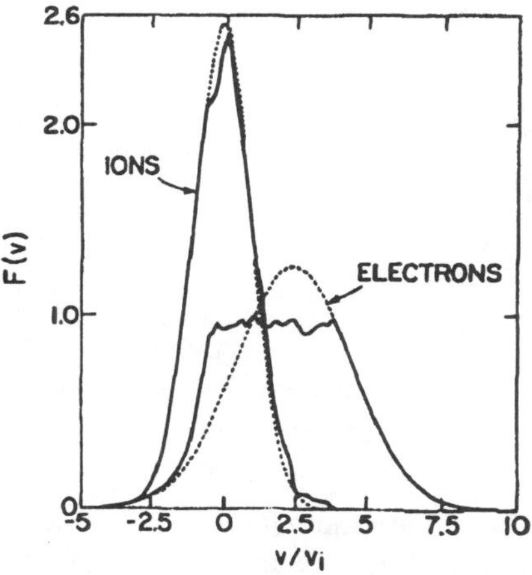

FIG. 3. The spatially averaged distribution functions for the ions and electrons for $v_d = 2.5v_i$ at $\omega_{pe}t = 400$ (solid curve) and at 0 (dashed curve).

features of the instability: First, we examined the effect of using different initial conditions to start the simulation. These included quiet starts with $\langle(\delta N)^2\rangle/\langle N\rangle \ll 1$; "thermal-level" starts with $\langle(\delta N)^2\rangle/\langle N\rangle \approx 1$; and "noisy" starts with $\langle(\delta N)^2\rangle/\langle N\rangle \approx 4$. In all cases an instability was observed for $v_d \geq 1.5v_i$. Thus, to trigger the instability we did not require large-amplitude fluctuations. In fact, the amplitude $\langle(\delta N)^2\rangle/\langle N\rangle \simeq 1$ corresponds to $e\langle\varphi^2\rangle^{1/2}/m_i v_i^2 \simeq 10^{-2}$, where $\langle\varphi^2\rangle$ is the mean square potential.

Second, we investigated the dependence of the instability on v_d. Aside from the growth rate, there were no apparent qualitative differences between the runs for $1.5v_i \leq v_d \leq 4.5v_i$. After an initial growth stage, the instability saturated by forming a quasilinear plateau. The ion distribution function developed a tail and the electron distribution function became significantly flattened as indicated in Fig. 3. We emphasize that these distortions were evident for v_d both below and above the linear stability threshold. The dependence of γ, the observed growth rate, on v_d is illustrated in Fig. 4. The measurements were made, for the ions, at approximately the same amplitude $\langle(\delta N)^2\rangle/\langle N\rangle$ for each run [the electron $\langle(\delta N)^2\rangle/\langle N\rangle$ gave similar growth rates]. The error bars indicate the spread in the measured values of the growth rate when different methods were used to obtain γ. These methods consisted of measuring γ from the mean square electrostatic field, and from the time dependence of

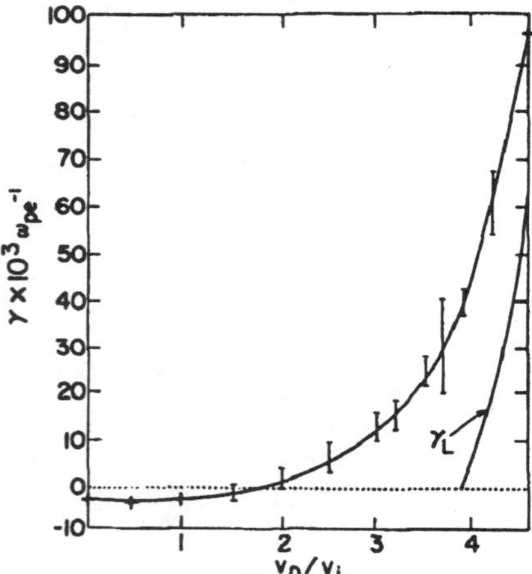

FIG. 4. The simulation values of the growth rate γ vs the electron drift v_d. Plusses denote single measurements while the error bars include several measurements. The growth rate γ_L obtained from linear stability theory is also plotted.

$\langle(\delta N)^2\rangle/\langle N\rangle$ at different phase velocities ($v/v_i = 0.5, 1, 2$), and different cell sizes ($\Delta x/\lambda_d = 1, 2$, $\Delta v/v_i = 0.1, 0.2$). Also shown in Fig. 4 is the linear growth rate γ_L for the most unstable wave number. It is clear that the nonlinear effects dominated in the linearly unstable, as well as the linearly stable, region since $\gamma_L < \gamma$.

The third feature we examined was the effect of varying the parameter $n_0\lambda_d$. When we decreased $n_0\lambda_d$ to 815, no change in γ occurred. From this, we concluded that discrete-particle collisions apparently did not play an important role. Below the nonlinear threshold, we saw the fluctuations decay. For $v_d = 0$, we measured the same decay rate observed in a series of one-species calculations with $n_0\lambda_d = 65\,190$ reported elsewhere.[3]

The measured characteristics of the instability are consistent with the nonlinear theory of ion and electron "clump" regeneration.[4] In a Vlasov plasma, clumps result from the mixing of the incompressible phase-space density by turbulent fluctuations.[5,6] If the clump production rate is equal to their destruction rate (through velocity streaming and the turbulent electric fields) the clump spectrum will regenerate. Overregeneration implies an instability. In Ref. 4 it was theoretically predicted, for the parameters of this simulation, that a clump spectrum would regenerate in regions of opposing velocity gradients for $v_d \geq 2.5v_i$. In addition, the spectrum would exhibit a wide phase-velocity spread (of the order of v_i) and characteristic scales similar to those observed in this simulation. Moreover, approximate calculations[7] show that for a mass ratio of 4, the clump instability has an amplitude-dependent growth rate proportional to the inverse ion trapping time $[\tau_{tr} \equiv (\Delta x)_c/(\Delta v)_c]$. The constant of proportionality (which is of the order of unity) increases rapidly with v_d and $-(\partial\langle f_e\rangle/\partial v)\partial\langle f_i\rangle/\partial v$. The observed values of $(\Delta x)_c$ and $(\Delta v)_c$ lead to a trapping time of the order of $80\omega_{pe}^{-1}$ which is consistent with the measured growth rates (cf. Fig.4).

The existing calculations of clump regeneration omit a number of terms that describe the effects of self-binding of the fluctuations. Clumps are enhancements ($\delta f \geq 0$) or depletions ($\delta f \leq 0$) in the local phase-space density. The depletions, or "holes," have the property of being self-binding.[8] Such an effect would decrease the destruction rate of the fluctuations and therefore reduce the predicted drift-velocity threshold. This would also be consistent with the slow (about $0.1\tau_{tr}^{-1}$) decay rate observed at $v_d = 0$. It is interesting to note that a single, isolated phase-space hole has been shown to be unstable for all $v_d > 0$.[9] This isolated-hole instability is driven by opposing velocity gradients of $\langle f_i \rangle$ and $\langle f_e \rangle$ and is the analog to the turbulent clump instability.

Although the nonlinear instability discussed in this Letter is one dimensional and driven by velocity gradients, we believe that it is representative of an important new class of instabilities since clump and hole phenomena are predicted to occur in three dimensions with a magnetic field. For instance, it recently has been shown that a single phase-space hole in a magnetic field is unstable to a spatial density gradient.[9] This result implies that the clump instability will be driven by a spatial density gradient. Furthermore, our simulation indicates that large amplitudes are not necessary for its onset. Indeed, we have observed the instability growing out of thermal-level fluctuations.

This work is supported by the National Science Foundation and the Department of Energy. Parts of these computations were performed using MACSYMA and LISP machines at Massachusetts Institute of Technology.

---

[1]T. H. Dupree, C. E. Wagner, and W. M. Manheimer, Phys. Fluids 18, 1167 (1975).

[2]The calculation is for the case $\Delta x, \Delta v, > (\Delta x)_c, (\Delta v)_c$ For smaller-size windows $\langle (\delta N)^2 \rangle$ would be smaller.

[3]R. H. Berman, D. J. Tetreault, and T. H. Dupree, Bull. Am. Phys. Soc. 25, 1035 (1980), and in Proceedings of the Ninth Conference on Numerical Simulation of Plasmas, Chicago, 1980 (to be published).

[4]T. Boutros-Ghali and T. H. Dupree, to be published.

[5]T. H. Dupree, Phys. Fluids 15, 334 (1972).

[6]T. Boutros-Ghali and T. H. Dupree, Phys. Fluids 24, 1839 (1981).

[7]D. J. Tetreault, "Growth rate of the Clump Instability" (to be published).

[8]T. H. Dupree, Phys. Fluids 25, 277 (1982).

[9]T. H. Dupree, Bull. Am. Phys. Soc. 26, 1060 (1981).

Transport Effects Induced by Resistive Ballooning Modes and Comparison with High-β_p ISX-B Tokamak Confinement

B. A. Carreras, P. H. Diamond, M. Murakami, J. L. Dunlap, J. D. Bell, H. R. Hicks, J. A. Holmes, E. A. Lazarus, V. K. Paré, P. Similon, C. E. Thomas, and R. M. Wieland

Oak Ridge National Laboratory, Oak Ridge, Tennessee 37830, and Institute for Fusion Studies, University of Texas, Austin, Texas 78712, and Computer Sciences, Nuclear Division, Union Carbide Corporation, Oak Ridge National Laboratory, Oak Ridge, Tennessee 37830

(Received 2 July 1982)

The transport effects induced by resistive ballooning modes are estimated from a theory, and are found to be mainly thermal electron conduction losses. An expression for electron thermal diffusivity χ_e is derived. The theoretical predictions agree well with experimental values of χ_e obtained from power balance for the ISX-B plasmas at high poloidal beta.

PACS numbers: 52.25.Fi, 52.30.+r, 52.55.Gb

A deterioration in confinement is observed in ISX-B tokamak experiments[1,2] with high neutral injection power at high poloidal plasma beta (β_p). From a theoretical point of view, resistive pressure-driven ballooning modes are a possible cause of this deterioration, linked to high-β_p plasmas. There have been several linear studies[3-5] of these instabilities in the past. Recently, numerical and analytical work has been done[6] to understand the linear and nonlinear properties of resistive ballooning modes in the framework of the incompressible resistive magnetohydrodynamic (MHD) equations. Below and near the critical β for ideal instabilities ($\beta_p \gtrsim 1$), the fastest growing mode, with a given torodial mode number n, has a growth rate

$$\gamma_n \cong (n^2/S)^{1/3}[\beta_0 q^2/(\epsilon \rho L_p)]^{2/3}\tau_{hp}^{-1},$$

where S is the ratio of resistive time τ_R to poloidal Alfvén time τ_{hp}, $\beta_0 = 2p(0)\mu_0/B_T^2$, p is the pressure, q is the safety factor, B_T is the toroidal magnetic field, ϵ is the inverse aspect ratio, $L_p = [(-dp/d\rho)/p(0)]^{-1}$, and ρ (with $0 \leq \rho \leq 1$) is a flux surface label. These modes are extended greatly along magnetic field lines, with a characteristic width given by

$$W_n = [q^4\hat{S}^2n^2\gamma_n\tau_{hp}^{-1}/(\rho^2 S)]^{-1/4},$$

where $\hat{S} = [\rho(dq/d\rho)/q]$ and a is the minor radius. Their linear properties are similar to resistive interchanges.[7]

With use of the nonlinear resistive MHD equation in the ballooning representation, a calculation of the renormalized response has been performed.[6] This calculation shows that the dominant nonlinear effect is due to the pressure-convective nonlinearity, which reduces the turbulent pressure response $\tilde{p}$ to $\tilde{\varphi}$, the electrostatic perturbation. This causes a reduction of the interchange destabilizing term, without changing the basic structure of the eigenfunction. A physical interpretation is that the resistive ballooning modes saturate when the pressure fluctuation mixes $dp/d\rho$ over the radial extent Δ of each poloidal subharmonic; thus, $\tilde{p} \simeq \Delta\, dp/d\rho$. Since the pressure is mainly convected, $\tilde{p} \simeq inq\tilde{\varphi}(dp/d\rho)/\rho\gamma_n$. Therefore, the kinetic energy of these modes at saturation is $E_k = \langle |(nq/\rho)\tilde{\varphi}|^2\rangle \approx \Delta^2 \times(\gamma_n\tau_{hp}^{-1})^2$. Details on the stability theory of these modes will be given elsewhere. In this Letter we limit ourselves to the consideration of the induced transport effects and consequences for confinement in ISX-B. We consider two main effects of the saturated modes, the diffusion induced by the convective nonlinearity and the anomalous electron conduction induced by the magnetic field-line stochastization.

The radial diffusion induced by the convective term is given by

$$D \approx \langle |(n'q/\rho)\tilde{\varphi}_{n'}|^2/\gamma_{n-n'}\rangle = \gamma_n\Delta^2.$$

Using the reciprocity of y (ballooning) space and position space, $\Delta \approx (nq\hat{S}W_n/\rho)^{-1}$, we can estimate the convection losses, which are $D \approx \hat{S}^{-2}(\beta_0 q^2/\epsilon L_p)a^2/\tau_R$. This effect is on the order of resistive diffusion and is therefore negligible.

To calculate the induced electron conduction losses, we first calculate the magnetic-field-line diffusion coefficient D_M. It is obtained from the static-field, zero-frequency electron-drift kinetic equation for toroidal geometry, by an iterative procedure of renormalized quasilinear theory. The magnetic nonlinearity and slow radial gradients have been retained. Consistent with conditions of the strongly turbulent regime, a renormalized propagator has been used in calculating D_M. Thus,

$$D_M = \sum_{n'}\sum_m (n'q/\rho)^2 \hat{\psi}_{n'}^*(y+2\pi m) L_{n'} \hat{\psi}_{n'}(y+2\pi m)$$

where $\hat{\psi}_{n'}(y)$ is related to the poloidal flux function ψ, through the ballooning representation. The propagator is

$$L_{n'}^{-1} = i\omega_{De}/v_{\parallel} + d_{n'} + (1/Rq)\partial/\partial y,$$

with

$$d_{n'} = \sum_{n''}\sum_{m'} (n'n''q^2/\rho^2)^2 (2\pi m')^2 \hat{S}^2 \psi_{n''}^* \times (y+2\pi m') L_{n'+n''} \hat{\psi}_{n''}(y+2\pi m').$$

The propagator L_n can be simplified by noting that $\omega_{De}/v_{\parallel} < d_{n'}$, $d_{n'} > (RqW_n)^{-1}$ and that the rapidly oscillating pieces of $(\partial/\partial y)^{-1}\hat{\psi}_{n'}(y+2\pi m)$ make an insignificant contribution to D_M. Therefore, $L_{n'} \approx L_{\parallel n'} \equiv d_{n'}^{-1}$, where $L_{\parallel n'}$ is the parallel correlation length for the field-line propagator $L_{n'}$. Note that in contrast to the familiar slab-model result $L_{\parallel n'} \propto D_M^{-1/3}$, here $L_{\parallel n'} \propto D_M^{-1}$.

In order to estimate D_M, it is necessary to derive an approximate expression for the relationship of $d_{n'}$ to D_M. It follows that $d_n \approx (nq/\rho)^2 \hat{S}^2 W_n^2 D_M$. Thus,

$$D_M \approx \left(\sum_{n'}\sum_m |\hat{\psi}_{n'}(y+2\pi m)|^2/\hat{S}^2 W_n^2\right)^{1/2}.$$

Using Ohm's law to express $\hat{\psi}_{n'}$ in terms of $\hat{\varphi}_{n'}$ and keeping the ordering $\gamma_n < (nq/\rho)^2 \hat{S}^2 W_n^2/S$ gives

$$D_M \approx (SRq)^{-1} \langle (nq\hat{S}/\rho)^4 \rangle^{-1} \gamma_n W_n^{-5},$$

where the saturated value of the kinetic energy has been used to estimate $|\hat{\varphi}_n|^2$.

For ISX-*B*, $L_{\parallel n'}$ is smaller than the electron mean free path. Hence, an approximate form for the anomalous electron thermal conductivity is $\chi_e \approx \frac{3}{2} v_{Te} D_M$. Using the explicit expressions for γ and W_n, one finally finds that

$$\chi_e^{\mathrm{Th}} \approx \tfrac{3}{2}(v_{Te}a) q \frac{1}{S}\left(\frac{\beta_0}{\epsilon}\frac{q^2}{\hat{S}}\frac{1}{L_p}\right)^{3/2}. \qquad (1)$$

Theoretical predictions based on the present model compare favorably with the results of the ISX-*B* beta-scaling experiments.[8] The points of agreement between the experiment and the theory are as follows: (1) Electron heat conduction is the dominant loss channel at the high values of β_p obtained in the experiment with high-power neutral beam injection, and (2) the theoretical predictions of the electron thermal diffusivity (χ_e^{Th}) agree well over a large range of parameter variations with the values and shape of the experimental thermal diffusivity,

$$\chi_e^{\exp}(\rho) = \frac{P_{be} + P_{\mathrm{OH}} - P_{ei} - P_{\mathrm{rad}} - \frac{5}{2} T_e \Gamma_e A_\rho}{(n_e A_\rho/a)|\partial T_e/\partial \rho|}, \qquad (2)$$

based on the power balance considerations as discussed below.

The power balance is carried out in a magnetic geometry consistent with both the MHD equilibrium theory and with the available experimental observations. This is accomplished in the data analysis code ZORNOC[9] with use of the variational moment analysis[10] which incorporates (1) the experimental pressure profile, consisting of those of electrons, thermal ions, and fast ions; (2) the plasma current profile, modeled to be consistent with the radius of the $q = 1$ surface observed with the soft x-ray diagnostics[11]; and (3) the form of the outermost magnetic surface determined from the poloidal magnetic measurements.[12] The total input power consists of a large beam-power input to electrons P_{be} (calculated from classical treatments of beam deposition and slowing down) and much smaller (by factors of 5–10) Ohmic heating power P_{OH}. The ion temperature profile T_i calculated from ion power balance (with a radially constant enhancement factor for neoclassical heat conductivity[13] adjusted to match the measured central ion temperature) is not significantly different from the electron temperature profile $T_e(\rho)$ under typical conditions, making the electron-ion heat transfer P_{ei} a small term. The radiative loss P_{rad} is small except near the edge.[14] The convection loss ($\frac{5}{2} T_e \Gamma_e A_\rho$, where A_ρ is a factor which reduces to the surface area in the low-β and high-aspect-ratio limit) due to particle flux Γ_e is evaluated from particle balance with a standard neutral model,[8] and is a small fraction of the total input. Then, the remaining power is assigned to the electron thermal conduction loss, which is by far the largest fraction of the input power. The electron temperature and density profile needed to calculate $\chi_e^{\exp}(\rho)$ are based on Thomson scattering measurements along a major radius, usually at twelve separate radial positions at 3 cm intervals. Considering the uncertainties and assumptions made in the analysis, the accuracy of determining $\chi_e(\rho)$ is estimated to be within a factor of 2, which is sufficient to assess gross trends. The accuracy of evaluating χ_e^{Th} based on local experimental parameters of

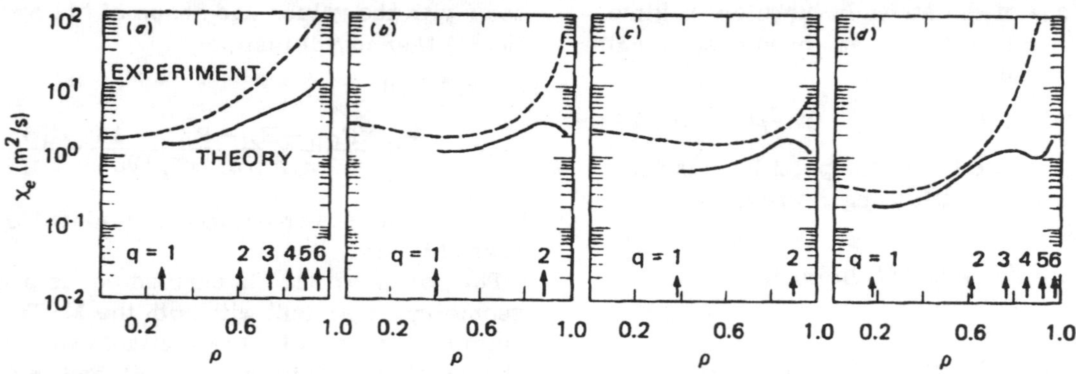

FIG. 1. Comparison of radial dependence of χ_e^{Th} and χ_e^{exp} for (a) $P_b = 2$ MW, $I_p = 83$ kA, $\beta_I = 1.7$; (b) $P_b = 2$ MW, $I_p = 192$ kA, $\beta_I = 1.05$; (c) $P_b = 1$ MW, $I_p = 184$ kA, $\beta_I = 1.05$; (d) $P_b = 0.6$ MW, $I_p = 143$ kA, $\beta_I = 0.85$.

specific ISX-*B* discharges is estimated to be of the same order.

Figure 1 shows the radial dependence of χ_e^{exp} and χ_e^{Th} for several typical high-β_p ISX-*B* discharges. The radial dependence of χ_e^{exp} is conveniently described by a three-plasma-region model: a central core region $[\rho \leq \rho(q=1)]$, dominated by $(m=1, n=1)$ mode and its driven modes[11]; a confinement region outside the core where a large pressure gradient is sustained; and a plasma edge region ($\rho \gtrsim 0.8$) dominated by atomic physics effects and/or recycling of plasma particles. As long as the central core region is not too large (which is the case for ISX-*B* plasmas with $q_\psi > 2.5$) and the plasma is clean, heat conduction in the confinement region determines the confinement. Indeed, we observe that $1/\chi_e^{exp}$ in the confinement region correlates well with τ_{Ee}, the electron energy confinement time, which in turn correlates with global energy confinement time. Since the global confinement time for high-power injection plasmas in ISX-*B* scales differently from that in Ohmic discharges[2], it is not surprising that standard OH models of χ_e, all of which fit Ohmic discharges reasonably well, do not come close to χ_e^{exp} in beam-heated high-β_p discharges in either magnitude or scaling. On the other hand, the present theoretical model χ_e^{Th} fits closely the magnitude and shape of χ_e^{exp} over the large parameter ranges. Figure 2 illustrates this agreement more clearly by directly comparing χ_e^{exp} and χ_e^{Th} for scans of plasma current (at $P_b = 2$ MW) and toroidal field (at $P_b = 0.6$ MW).[2] In this figure, three χ_e values are given for the confinement region of each discharge, those at $\rho = 0.50$, 0.67, and 0.75; χ_e^{Th} agrees well with χ_e^{exp} over nearly two decades. Comparison of χ_e's in a larger ISX-*B* data base indicates that the theoretical values are much smaller than most experimental values at low β_p ($\beta_p < 0.9$). This is expected since the ballooning

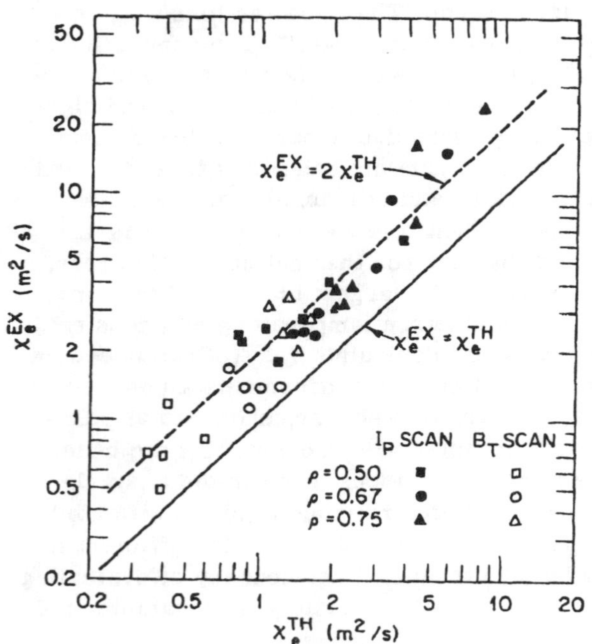

FIG. 2. Comparison of values of χ_e^{Th} and χ_e^{exp}.

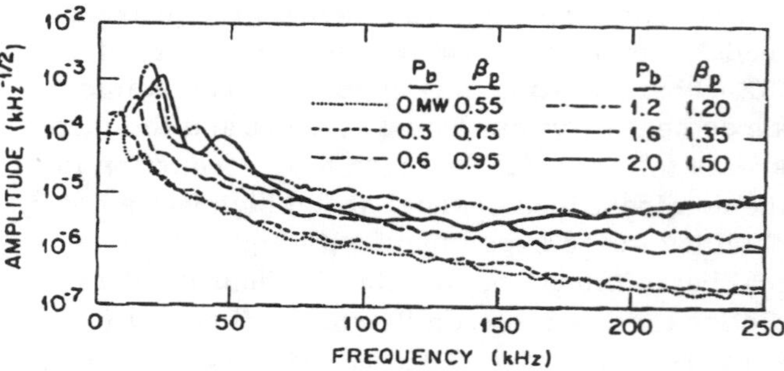

FIG. 3. Amplitude spectra of $\tilde{B}_p/B_p$ of discharges differing in neutral-beam power.

modes are reduced at low β_p, and competing processes could easily dominate confinement. For relevant mode numbers n, the simple collisional model is applicable since γ/ω_{*e} is large. In the I_p scan, at $\rho = 0.67$, the value of ω_{*e} ranges from $5\times10^3 n$ sec^{-1} to $7\times10^3 n$ sec^{-1} while $\gamma = 5\times10^4 n^{2/3}$ sec^{-1} to $\gamma = 2.7\times10^4 n^{2/3}$ sec^{-1}.

Studies of the fluctuations in magnetic field near the edge of the plasma are being conducted, with Mirnov coil diagnostics, in an effort to confirm the presence of these theoretically predicted modes. The $\tilde{B}_p$ spectra (Fig. 3) extend far beyond the range, typically 5 to 25 kHz, of the ($m = 1$, $n = 1$) mode.[11] The high-frequency tail is present independently of the ($m = 1, n = 1$).[6] Its amplitude level increases with neutral-beam heating, and the increases develop after beam turnon in a fashion similar to the development of β_p. Details of the spectra are complex, and as yet there are no specific features that can clearly be correlated with plasma confinement. These studies continue, and similar studies are being undertaken with collimated soft x-ray diagnostics.

We acknowledge with appreciation the support of our many colleagues in the ISX-*B* group and in the engineering staff. We also wish to thank H. C. Howe and M. N. Rosenbluth for many useful discussions. This research was sponsored by the Office of Fusion Energy, U. S. Department of Energy, under Contracts No. W-7405-eng-26 and No. DE-FG05-80ET-53088.

[1]D. W. Swain *et al.*, Nucl. Fusion 21, 1409 (1981).

[2]H. G. Neilson *et al.*, to be published.

[3]H. P. Furth *et al.*, in *Proceedings of the Second International Conference on Plasma Physics and Controlled Nuclear Fusion Research, Culham, England, 1965* (International Atomic Energy Agency, Vienna, Austria, 1966), Vol. 1, p. 103.

[4]M. S. Chance *et al.*, in *Proceedings of the Seventh International Conference on Plasma Physics and Controlled Nuclear Fusion Research, Innsbruck, Austria, 1978* (International Atomic Energy Agency, Vienna, Austria, 1979), Vol. 1, p. 677; G. Bateman and D. B. Nelson, Phys. Rev. Lett. 41, 1804 (1978).

[5]H. R. Strauss, Phys. Fluids 24, 2004 (1981).

[6]B. A. Carreras *et al.*, in Proceedings of the Ninth International Conference on Plasma Physics and Controlled Nuclear Fusion Research, Baltimore, Maryland, September 1982 (to be published), International Atomic Energy Agency Report No. CN-41/P-4.

[7]H. P. Furth, J. Killeen, and M. N. Rosenbluth, Phys. Fluids 6, 459 (1963); B. Coppi, J. M. Green, and J. L. Johnson, Nucl. Fusion 6, 101 (1966).

[8]M. Murakami *et al.*, in Proceedings of the Ninth International Conference on Plasma Physics and Controlled Nuclear Fusion Research, Baltimore, Maryland, September 1982 (to be published), International Atomic Energy Agency Report No. CN-41/A-4.

[9]R. M. Wieland *et al.*, to be published.

[10]L. L. Lao *et al.*, Phys. Fluids 24, 1431 (1981).

[11]J. L. Dunlap *et al.*, Phys. Rev. Lett. 48, 538 (1982).

[12]D. W. Swain and G. H. Neilson, Nucl. Fusion 22, 1015 (1982).

[13]F. L. Hinton and R. D. Hazeltine, Rev. Mod. Phys. 48, 239 (1976).

[14]C. E. Bush *et al.*, to be published.

Screening potential and enhancement of the thermonuclear reaction rate in dense plasmas

Setsuo Ichimaru, Shigenori Tanaka, and Hiroshi Iyetomi
Department of Physics, University of Tokyo, Bunkyo-ku, Tokyo 113, Japan
(Received 16 January 1984)

We present the first quantitative study of the screening potential and the effective cross section for thermonuclear reaction in dense plasmas relevant to the inertially confined fusion experiments and the interior of main-sequence stars. It is shown that the enhancement of the thermonuclear reaction rate due to strong Coulomb correlations has a considerable effect in such a dense plasma.

Strong enhancement of the thermonuclear reaction rate takes place in dense stellar matter owing to the interparticle Coulomb correlations, which act to screen the effective interionic repulsive potential.[1–3] For a plasma with $\Gamma \geq 10$ and $r_s \leq 0.1$, appropriate to the state of matter in degenerate cores,[4] the rate of such an enhancement has been well studied and understood in terms of both the strong ion-ion correlations[1] and the screening arising from relativistic degenerate electrons.[2] Here

$$\Gamma \equiv (Ze)^2/ak_BT$$

refers to the Coulomb coupling constant of the ion system with electric charge Ze, the Wigner-Seitz radius

$$a=(4\pi n/3)^{-1/3} ,$$

and the temperature T;

$$r_s \equiv (4\pi n_e/3)^{-1/3} me^2/\hbar^2$$

is the dimensionless density parameter of the electron system with number density n_e $(=Zn)$.

The screening potential and resulting enhancement of the thermonuclear reaction rate are essential problems also for those plasmas expected in the inertially confined fusion experiments or in the interior of main-sequence stars. Here the ion system forms a classical plasma of intermediate to weak coupling ($\Gamma < 10$); the electron degeneracy parameter $\zeta \equiv k_BT/E_F$, where E_F is the Fermi energy at $T=0$, varies widely. The treatment of ion-ion correlations with the inclusion of the electronic screening effect in this parameter domain is a fairly complex problem, owing to both partial degeneracy and local-field corrections (LFC); no quantitatively dependable calculations of the screening potential have been carried out thus far.

The screening potential $H(r)$ is defined in terms of the radial distribution function $g(r)$ of the ions as[3]

$$H(r)=\frac{(Ze)^2}{rk_BT}+\ln[g(r)] . \tag{1}$$

In this paper, we present the first quantitative study of the screening potential and the resulting enhancement factor $\exp[H(0)]$ for the thermonuclear reaction rate over the entire parameter domain relevant to the inertially confined fusion experiments and the interior of main-sequence stars. The thermal average $\langle \sigma v \rangle$ of the product between reaction cross section and relative velocity is then calculated explicitly for deuterium plasmas; it will be shown that the enhancement of reaction rate by Coulomb correlations is a substantial effect which needs to be taken into serious consideration in high-density plasmas.

The present study relies on the observation that the hypernetted chain (HNC) scheme[5] offers an extremely accurate description of corelations in an intermediate to weakly coupled ($\Gamma < 10$) classical one-component plasma (OCP).[6] For $\Gamma \leq 1$, the HNC results are virtually exact; even for $1 < \Gamma < 10$, then HNC scheme reproduces correlation energies of OCP with relative errors much less than 1%.[7] In the present case, the ions, interacting via screened Coulomb forces, cannot be regarded as an OCP; the screening effect then acts to widen the range of Γ for the validity of the HNC calculation.[8]

In the HNC scheme, the ion-ion correlations are analyzed through the HNC equation,

$$g(r)=\exp\{-[\phi(r)/k_BT]+h(r)-c(r)\} , \tag{2}$$

coupled with the Ornstein-Zernike relation

$$h(r)=c(r)+n\int d\vec{r}\,'c(|\vec{r}-\vec{r}\,'|)h(r') , \tag{3}$$

where $h(r)=g(r)-1$. The interionic potential $\phi(r)$ is given by

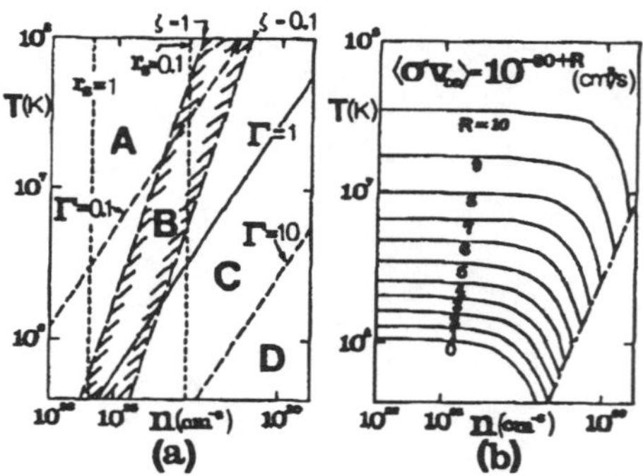

FIG. 1. (a) Density vs temperature plane for plasmas with $Z=1$. Γ, r_s, and ζ are the Coulomb coupling constant for the ions, the dimensionless electron-density parameter, and the electron degeneracy parameter, respectively, defined in the text. (b) n-T contours for deuterium plasmas along which $\langle \sigma v_{\rm DD} \rangle$ takes on constant values. The chain line represents $3\Gamma/\tau=1$.

$$\phi(r)=\frac{(Ze)^2}{2\pi^2}\int d\vec{q}[q^2\epsilon(q)]^{-1}\exp(i\vec{q}\cdot\vec{r}) , \quad (4)$$

where

$$\epsilon(q)=1-\frac{v(q)\chi_0(q)}{1+v(q)G(q)\chi_0(q)} \quad (5)$$

is the (static) screening function of the electrons,[3] $\chi_0(q)$ is the free-electron polarizability, $G(q)$ is LFC, and $v(q)=4\pi e^2/q^2$.

The expression for $\epsilon(q)$ is not known uniformly over the parameter domain illustrated in Fig. 1(a). To circumvent the difficulty we divide it into a number of subdomains: In A, bounded by $\zeta=1$ from the right, the electrons obey the classical statistics; $\chi_0(q)=-n_e/k_BT$, the Debye-Hückel value. The LFC $G(q)$ is calculated from the solution to the HNC equation for the electron system regarded as a classical OCP.[9] B is the interpolation domain. In C, bounded between $\zeta=0.1$ and $\Gamma=10$, the electrons can be regarded as completely degenerate; $\chi_0(q)$ takes the Lindhard expression.[3,10] An adequate, analytic expression for $G(q)$ is available.[3,11] In D, bounded by $\Gamma=10$ from the left, the ions are strongly coupled and the degenerate electrons with $r_s<10^{-2}$ require a relativistic treatment. The screening potential in this subdomain has been well investigated.[1,2]

We thus solve the HNC scheme, (2) and (3), for 23 combinations of n and T in the subdomain A and for 16 combinations in C, to calculate $H(r)$ with $Z=1$. In the calculation, we take account of three different choices for $\epsilon(q)$ in the potential (4): (i) Equation (5) with LFC; (ii) Equation (5) without LFC, i.e., a random-phase approximation (RPA) expression assuming $G(q)=0$; and (iii) $\epsilon(q)=1$, i.e., an OCP treatment for the ions. The calculated values of $H(r)$ decrease as (i)→(ii)→(iii), an indication consistent with earlier observations in analogous problems.[8,12] In either of the subdomains A and C, we find that the difference between (i) and (ii) is small and does not exceed a few percent. The difference between (i) and (iii), however, is considerable and reaches 20–50 % particularly in A.

We then interpolate the results obtained with (i) for the 39 parameter combinations in A and C analytically into the connecting subdomain B. Adequacy of the interpolated results has been examined and assured through comparison with the HNC results derived separately by using an RPA screening function with arbitrary degeneracy.[13]

The parametrized expression for $H(0)$ obtained through the interpolation procedure is

$$H(0)=\Gamma^{3/2}\frac{1.732+0.7174\xi+(1.644-0.1039\xi)\Gamma^{1/2}}{1+(1.096-0.5286\xi)\Gamma^{1/2}+(1.416-0.3238\xi)\Gamma} , \quad (6)$$

where $\xi=\exp(-1/5\zeta)$. This formula reproduces all the computed values of $H(0)$ for the 39 cases in A and C with relative errors less than 4%.[14] In the weak coupling limit ($\Gamma \ll 1$) with $\xi \to 1$, Eq. (6) approaches $\sqrt{6}\Gamma^{3/2}$, the Debye-Hückel value. In the strong coupling limit ($\Gamma \gg 1$) with $\xi \to 0$, Eq. (6) approaches 1.161Γ; this is slightly above the OCP value, either 1.057Γ in the ion-sphere model[3] or 1.053Γ in Jancovici's evaluation,[15] accounting correctly for the additional screening contribution of the electrons.[2] The formula (6) thus connects smoothly with the known results[1,2] in the strong coupling regime.

The thermal average $\langle\sigma v_{DD}\rangle$ for the deuterium plasma is then calculated as

$$\langle\sigma v_{DD}\rangle=3.25\times10^{-14}\left[\frac{T}{10^7\ \mathrm{K}}\right]^{-3/2}[-\tau+H(0)] \quad (7)$$

(in cm^3/s), where

$$\tau=[(27\pi^2/4)M(Ze)^4/\hbar^2k_BT]^{1/3}$$

and M denotes twice the reduced mass of the reacting nuclei. If one ignores the Coulomb-correlation effects altogether and thereby sets $H(0)=0$ in (7), the result turns into the well-known Gamow expression.[16]

In Fig. 1(b) we show on the same parameter domain as in Fig. 1(a) a set of contours along which $\langle\sigma v_{DD}\rangle$ takes on constant values. The contours are terminated at $3\Gamma/\tau=1$, where the classical turning radius at the Gamow peak reaches the Wigner-Seitz radius.[1] Note that the separation between a pair of adjacent curves implies a difference in $\langle\sigma v_{DD}\rangle$ by a factor of 10. If the Gamow expression were adopted, those contours would have been horizontal lines starting with the low-density values. We observe in Fig. 1(b) that the deviation from the Gamow expression is substantial in the high-density regime; Coulomb correlations are found to represent a major effect in the determination of the effective cross section for thermonuclear reaction.

In conclusion we have presented the first quantitative calculation of the effective cross section for thermonuclear reaction in dense plasmas in which the screening effects arising from Coulomb correlations are accurately taken into account. The resulting analytic expressions such as (6) should be of use in an optimization consideration for inertially confined fusion experiments and in an evolution analysis of main-sequence stars. Details of the calculation will be published elsewhere.

This work was supported in part through Grants-in-Aid for Scientific Research provided by the Japanese Ministry of Education, Science, and Culture.

[1]E. E. Salpeter, Aust. J. Phys. 7, 373 (1954); E. E. Salpeter and H. M. Van Horn, Astrophys. J. 155, 183 (1969); H. E. DeWitt, H. C. Graboske, and M. S. Cooper, *ibid.* 181, 439 (1973); H. E. Mitler, *ibid.* 212, 513 (1977); N. Itoh, H. Totsuji, and S. Ichimaru, *ibid.* 218, 477 (1977); A. Alastuey and B. Jancovici, *ibid.* 226, 1034 (1978); N. Itoh, H. Totsuji, S.

Ichimaru, and H. E. DeWitt, *ibid.* 234, 1079 (1979).
[2]S. Ichimaru and K. Utsumi, Astrophys. J. (Lett.) 269, L51 (1983); Astrophys. J. (to be published).
[3]S. Ichimaru, Rev. Mod. Phys. 54, 1017 (1982).
[4]R. G. Couch and W. D. Arnett, Astrophys. J. 196, 791 (1975).
[5]J. M. J. van Leeuwen, J. Groeneveld, and J. De Boer, Physica (Utrecht) 25, 792 (1959); T. Morita, Prog. Theor. Phys. 23, 829 (1960).
[6]J. F. Springer, M. A. Pokrant, and F. A. Stevens, J. Chem. Phys. 58, 4863 (1973); K. C. Ng, *ibid.* 61, 2680 (1974).
[7]H. Iyetomi and S. Ichimaru, Phys. Rev. A 27, 3241 (1983).
[8]S. Galam and J. P. Hansen, Phys. Rev. A 14, 816 (1976); H. Iyetomi, K. Utsumi, and S. Ichimaru, J. Phys. Soc. Jpn. 50, 3769 (1981).
[9]See Eq. (2.4) in Ref. 3.
[10]J. Lindhard, K. Dan. Vidensk. Selsk. Mat.-Fys. Medd. 28 (8), 1 (1954).
[11]S. Ichimaru and K. Utsumi, Phys. Rev. B 24, 7385 (1981); K. Utsumi and S. Ichimaru, Phys. Rev. A 26, 603 (1982).
[12]S. Ichimaru, H. Iyetomi, and K. Utsumi, Phys. Rev. B 25, 1374 (1982).
[13]See, e.g., U. Gupta and A. K. Rajagopal, J. Phys. B 12, L703 (1979); Phys. Rep. 87, 259 (1982).
[14]More precisely, only in six of the total 39 cases of the n-T combinations studied, we find that the parametrized $H(0)$ values of (6) deviate from the HNC calculations by an amount falling in the range 3.0—3.9 %, all in the direction of underestimating the HNC values. On the other hand, it is generally known that the HNC scheme systematically overestimates the true values of $H(0)$ because of its neglect of the bridge-diagram contributions (Ref. 7). For $\Gamma \simeq 10$, where we expect the largest inaccuracy in the subdomains A and C, the estimated HNC error in $H(0)$ is approximately 5%. The overestimation inherent in the HNC scheme is thus compensated by the underestimation contained in (6). The parametrized values (6) are thus closer to the true values than are the original HNC values.
[15]B. Jancovici, J. Stat. Phys. 17, 357 (1977).
[16]G. Gamow and E. Teller, Phys. Rev. 53, 608 (1938); R. F. Post, Rev. Mod. Phys. 28, 338 (1956).

Thresholds of Convective Raman Backscattering: Effects of ∇n, Thomson Scattering, Noise, Collisions, and Landau Damping

Kent Estabrook and W. L. Kruer
Lawrence Livermore National Laboratory, University of California, Livermore, California 94550
(Received 7 November 1983)

Using kinetic simulations we find that Raman backscattering can exist for laser intensities a factor 2π lower than the density gradient (∇n) threshold and that Thomson scattering can account for backscattering signals even lower. In addition, kinetic, collisional simulations show that laser intensities $\gtrsim 3-10$ times the electron-ion collisional threshold are needed for significant Raman absorption. Collisions can also reduce the heated electron temperatures for $\nabla n \neq 0$.

PACS numbers: 52.50.Jm, 52.25.Ps, 52.40.Db, 52.65.+z

Stimulated Raman scattering (SRS)[1-10] in laser-produced plasmas can generate high-energy electrons, which can reduce the energy gain of laser-fusion targets. SRS is the parametric decay of an incident photon into a scattered photon plus a longitudinal electron plasma wave. We describe theories of Raman backscattering thresholds as a function of the density gradient[1] (∇n), plasma noise, electron-ion collisions, and Landau damping, and give the results of electromagnetic, relativistic, one-dimensional, kinetic (particle) simulations[3,6] with ions fixed to isolate Raman backscattering. We find the following: (1) The low-level "SRS" seen in some experiments may be Thomson scattering. (2) True Raman backscattering can exist for $I\lambda_0^2$ lower by a factor of 2π than the traditional ∇n threshold value.[1] Here I and λ_0 are the laser intensity and wavelength. (3) For $\nabla n = 0$, electron-ion collisions (of frequency ν_{ei})[11] can reduce Raman absorption by an order of magnitude for $\nu_{ei}/\gamma_0 = 0.4$ to 1 (depending on gain) even though the Raman growth rate γ_0 is a factor of 2–3 above the collisional threshold. (4) For $\nabla n \neq 0$, electron-ion collisions reduce the temperature (T_{hot}) of the suprathermal electrons produced by Raman scattering.

Convective SRS has been observed[2] much below the traditional density gradient threshold. Figure 1 shows the Raman backscattered spectrum from a numerical simulation in which $I\lambda_0^2 = 3\times10^{13}$ W μm^2/cm^2 is a factor of 30 below the ∇n threshold; the signal amplitude is roughly an order of magnitude above the noise. This spectrum is light scattering from electron-plasma-wave noise (Thomson scattering). The scattered frequency, ω_s, obeys the wave-number– and frequency–matching conditions $k_0 = k_s + k_{epw}$ and $\omega_0 = \omega_s + \omega_{epw}$, as in SRS (the subscripts 0, s, and epw refer to the pump, scattered, and electron plasma wave, respectively). We note that $k_s < 0$: see Liu, Rosenbluth, and White[4] and Ref. 10. The reflected spectrum shows a similar peak at $\omega_0 + \omega_{epw}$, and the transmitted spectrum shows similar (but weaker) peaks at $\omega_0 \pm \omega_{epw}$. A control simulation without the pump shows no such peak. Additional evidence that Fig. 1 is Thomson scattering is that the amplitude of k_{epw} does not rise above noise in the simulation wave-number spectra, even with time averaging. Theoretical estimates of electromagnetic scatter from the measured density fluctuation $\delta n_e(k_{epw})$ over two mismatch lengths[1] agree reasonably well with the Fig. 1 result. Figueroa *et al.*[8] find that Thomson scattering dominates bremsstrahlung as a SRS noise source in an experimental situation. Simon and Short[4] note that the plasma waves in experiments may need to be enhanced over thermal noise for Thomson scattering to account for the

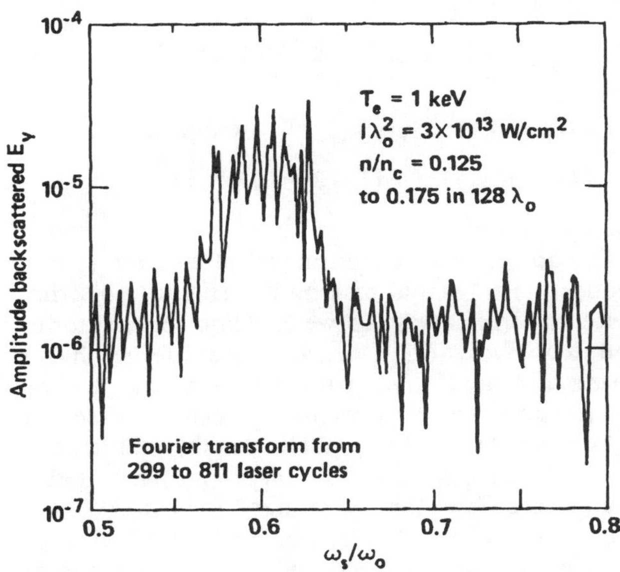

FIG. 1. Fourier transform in time of Thomson backscattered light for which $I\lambda_0^2$ is a factor of 30 below the classical ∇n threshold.

low-level signals observed in University of Rochester experiments.

One reason why experimentalists observe SRS below threshold,[1] $\gamma_0^2/\kappa' v_{gepw} v_{gs} = 1$, is the threshold definition. Equation (4) of Ref. 1 is the amplification:

$$I = I_0 \exp[2\pi(\gamma_0^2/\kappa' v_{gepw} v_{gs})], \tag{1}$$

where I (I_0) are the scattered (noise) light intensities at frequency ω_s, and v_{gepw} and $v_{gs} = c[1-(\omega_p/\omega_s)^2]^{1/2}$ are the group velocities of the electron plasma wave and the scattered wave. The threshold corresponds to 2π exponential growths [$\exp(2\pi) = 535$] as Eq. (1) and Fig. 2 show. This 2π has been pointed out by others in Refs. 1–10.

Following Liu, Rosenbluth, and White,[4] we derive an expression for the threshold in the $k\lambda_D << 1$ and collisionless limit in the convective regime from $\gamma_0^2/\kappa' v_{gepw} v_{gs} > 1$ using the growth rate

$$\gamma_0 = \frac{k_{epw}}{4} v_{osc} \left(\frac{\omega_p}{\omega_s} \frac{\omega_p}{\omega_{epw}}\right)^{1/2} \left(1 - \frac{n}{n_c}\right)^{-1/4}, \tag{2}$$

where $\kappa' = (d/dx)(k_0 - k_s - k_{epw})$. The quantity $\kappa' v_{gepw}$ may be expressed[4] by $\omega_p/2L$, where the density gradient L is $(\Delta x/\Delta n)n$ and ω_p is the plasma frequency at the density n of interest; n changes by Δn in a distance Δx. With use of

$$v_{osc} = eE_0/m_e\omega_0 = 8.55\times 10^{-10}\lambda_0(\mu\text{m})c[I(\text{W/cm}^2)]^{1/2},$$

where E_0 is the laser peak vacuum electric field, $\gamma_0^2/\kappa' v_{gepw} v_{gs} > 1$ becomes

$$I\lambda_0^2 > 1.5\times 10^{18} \frac{k_s c}{\omega_0} \frac{\lambda_0}{\Delta x} \frac{\Delta n}{n} \left(\frac{k_{epw} c}{\omega_0}\right)^{-2} \tag{3}$$

for $n/n_c \lesssim 0.15$. Here $k_{epw}c/\omega_0$ is approximately

$$[1-(n/n_c)]^{1/2} + [1 - 2(n/n_c)^{1/2}]^{1/2}$$

for $k_{epw}\lambda_D << 1$.

Of course, other experimental plasma effects can reduce the threshold making it difficult to discriminate experimentally between Thomson scattering and low-level SRS. Raman sidescattering has a lower threshold[4] than backscattering and is commonly observed. Filamentation can increase the light intensity,[7] and a parabolic density profile has a lower threshold than a linear profile.[8] Other sources are Compton scattering[3,9] for $n > n_{cr}/4$ and flat spots in the density profile.[12] Any electron plasma wave in a density gradient can radiate in an inverse resonant absorption process.[13]

As experiments tend toward higher-frequency lasers, electron-ion collisions can become important

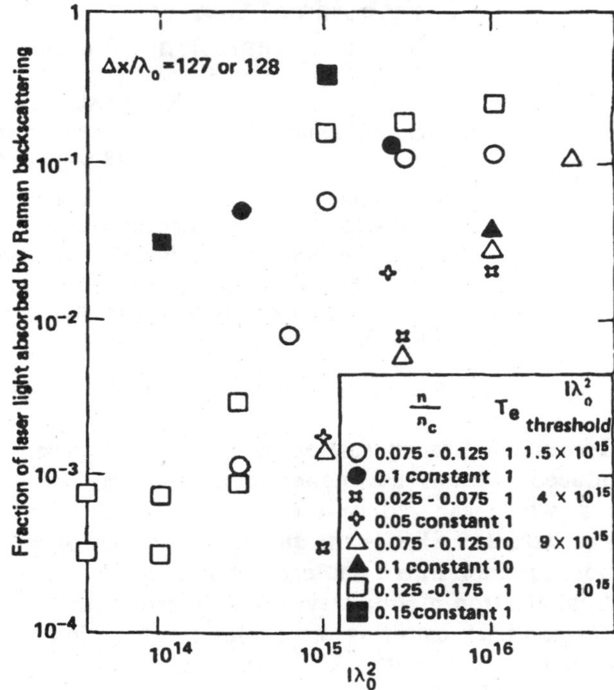

FIG. 2. Fraction of laser light absorbed by Raman backscattering, measured at the end of the run, as a function of laser intensity, wavelength, plasma density, and gradient. The high-level Raman absorption varies in time (Ref. 3). In low-level cases, absorption is inferred from the scattered light spectrum by use of the Manley-Rowe relation (Ref. 6). The time scales of the simulation were from several hundred to 2000 whole laser periods. The last column in the legend is $I[\lambda_0/(1.06\ \mu\text{m})]^2$ from the ∇n threshold [Eq. (4)] and corresponds to 2π e foldings in gain. The effect of doubling the number of electrons from 80 000 to 160 000 can be seen in the lower open squares: The absorption is lower because of lower noise. The three regimes of Thomson scattering, exponential Raman growth, and saturation can be seen in the open square points.

in suppressing SRS. Electron-ion collisions reduce SRS in three ways: (1) by absorbing the incident laser light directly, (2) by directly damping the electron plasma waves (as Landau damping does), and (3) by absorbing the scattered light wave that contributes to the amplification.

Nishikawa and Liu[10] have derived a Raman threshold in the heavily damped limit (Ref. 10, p. 44), $\Gamma_{epw}/v_{gepw} >> 1/(\text{gain length}^{3,6})$ or

$$\frac{\Gamma_{epw}}{\omega_p} >> \left(\frac{3k_{epw}}{4k_s}\right)^{1/2} \frac{v_{osc}}{c} k\lambda_D,$$

where $\Gamma_{epw} = \omega_i + \nu_{ei}/2$. They include the effects of ν_{ei} on the scattered wave (but not on the incident wave), and obtain

$$I\lambda_0^2 > 3\times10^{18}\left(\frac{k_s c}{\omega_0}\right)\left(\frac{\Delta n}{n_c}\right)\left[1+\frac{\nu_{ei}}{\omega_0}\frac{\Delta x}{\lambda_0}\frac{\omega_p}{\omega_s}\frac{\omega_p}{\omega_0}\left(\frac{k_s c}{\omega_0}\right)^{-1}\right]\frac{\omega_{epw}}{\omega_p}\left[\left(\frac{k_{epw}c}{\omega_0}\right)^2\frac{\Delta x}{\lambda_0}\frac{n}{n_c}\arctan(\alpha)\right]^{-1}, \tag{4}$$

where

$$\alpha = \frac{\Delta n}{n_c}\left[4\left(\frac{\omega_i}{\omega_0}+\frac{\nu_{ei}}{2\omega_0}\right)\frac{\omega_p}{\omega_0}\right]^{-1}.$$

Note that Eq. (4) reduces to Eq. (3) as ν_{ei} and ω_i approach 0.

In a plasma for which $\nabla n = 0$, we find that the heavily damped limit is equivalent to the convective limit, $\gamma_0 < (\Gamma_{epw}/2)(v_{gs}/v_{gepw})^{1/2}$ (Ref. 10, Eq. III-48). In this limit the Raman gain for $\nabla n = 0$ (Ref. 6 and Eq. III-58 from Ref. 10) is

$$\frac{k_{epw}^2\Delta x}{8k_s}\left(\frac{v_{osc}}{c}\right)^2\frac{\omega_p}{\Gamma_{epw}}\frac{\omega_p}{\omega_{epw}}\left(1-\frac{n}{n_c}\right)^{-1/2} - 2\pi\frac{\nu_{ei}}{\omega_0}\left(\frac{\omega_p}{\omega_s}\right)^2\frac{\Delta x}{\lambda_0}\left[1-\left(\frac{\omega_p}{\omega_s}\right)^2\right]^{-1/2}. \tag{5}$$

We note that the heavily damped (convective) limit is not always valid; when it is not, the theory will underestimate the scattering.

We model electron-ion collisions by randomly rotating the electron velocity vector.[14] For simplicity, we make ν_{ei} a function only of density and not of temperature. Raman scattering is strongly reduced in Figs. 3 and 4 for γ_0 ($I\lambda_0^2$) a factor of 2–3 (4–9) above the $\nabla n = 0$ threshold[5]:

$$\gamma_0^2 > (\nu_{ei}/2)(\omega_p/\omega_s)^2(\nu_{ei}/2+\omega_i)$$

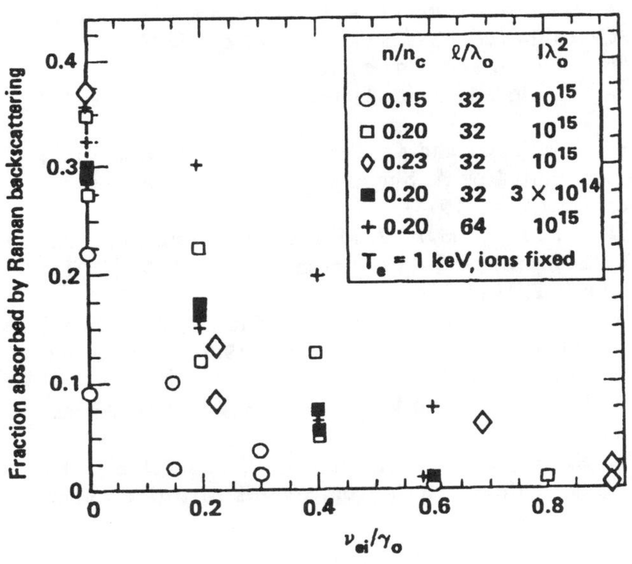

FIG. 3. Fraction absorbed by Raman backscattering in a plasma with $\nabla n = 0$ for a variety of collision frequencies, where ν_{ei} is defined at the n/n_c indicated. The higher symbol is measured at the peak absorption averaged over at least 100 laser periods; the lower symbol is the long-time absorption. The measured electromagnetic noise level of Fig. 3 was $eB_z(k_s)/m_e\omega_0 c \sim 5\times10^{-5}$ independent of ν_{ei}. This is the numerical noise inherent in a particle simulation.

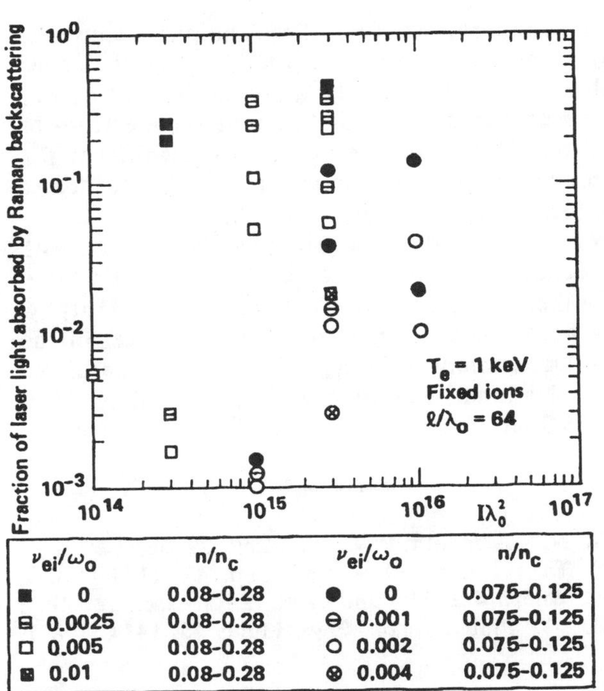

FIG. 4. Fraction of light absorbed by Raman backscattering as a function of $I\lambda_0^2$ for $\nabla n \neq 0$. The higher symbol of each vertical pair is measured at the peak absorption averaged over at least 100 laser periods; the lower symbol is the long-time absorption. Equation (4) predicts the threshold for the circles to be $\sim 3\times10^{15}$. For the squares, ν_{ei} was determined for $n = n_c/4$; for the circles, $n = n_c/10$ was used.

or

$$I > 3\times10^{-12}\left(\frac{k_{\rm epw}c}{\omega_0}\right)^{-2}\frac{\omega_{\rm epw}}{\omega_s}\nu_{\rm ei}(\nu_{\rm ei}/2+\omega_i)\ \frac{\rm W}{\rm cm^2}\propto\lambda_0^{-3}. \tag{6}$$

For simulations with high power, large ν_{ei}, and long run times, the theory is not rigorously applicable because ω_i becomes comparable to ν_{ei} by the inverse bremsstrahlung heating, and by linear and nonlinear Landau damping. However, the $I\lambda_0^2 = 3\times10^{14}$ W/cm^2, $\nu_{ei}\neq 0$ runs in Fig. 3 have low enough intensity that the linear and nonlinear $\omega_i \ll \nu_{ei}$ for the entire run. As a typical experimental example for gold, for $T_e = 3$ keV, $\bar{z}$ (the average ionization state) $=50$, $I = 5\times10^{14}$ W/cm^2, $\lambda_0 = 0.53$ μm, and $n/n_c = 0.15$, we find $\nu_{ei}/\omega_i \sim 1.6$ and $\nu_{ei}/\gamma_0 = 0.6$.

Our simulations also indicate that electron-ion collisions reduce the heated electron temperature ($T_{\rm hot}$) produced by Raman backscattering in a density gradient. For $\langle n\rangle = n_c/4$, we find $T_{\rm hot}\sim 120$, 62, 48 keV for $\nu_{ei}/\omega_0 = 0$, 0.0025, and 0.005, respectively; for $\langle n\rangle = n_c/10$, we find $T_{\rm hot}\sim 16$, 10, 9 keV for $\nu_{ei}/\omega_0 = 0$, 0.001, and 0.002 (data from runs of Fig. 4).

In summary, we find that Thomson scattering may account for Raman backscatter signals considerably below the density gradient threshold, and that laser intensities $\gtrsim 3$–10 above the electron-ion collisional threshold are needed for significant Raman absorption. Collisions can reduce the heated electron temperatures for $\nabla n \neq 0$.

We acknowledge valuable conversations with W. Seka, C. Randall, B. Lasinski, D. Phillion, P. Drake, R. Turner, M. Campbell, and P. Murphy. This work was performed under the auspices of the U. S. Department of Energy by the Lawrence Livermore National Laboratory under contract No. W7405-ENG-48.

[1]M. N. Rosenbluth, Phys. Rev. Lett. **29**, 565 (1972).

[2]K. Tanaka *et al.*, Phys. Rev. Lett. **48**, 1179 (1982); C. C. Shepard *et al.*, Bull. Am. Phys. Soc. **28**, 1058 (1983); D. Phillion *et al.*, Phys. Fluids **25**, 1434 (1982), and Phys. Rev. Lett. **49**, 1405 (1982).

[3]Kent Estabrook and W. L. Kruer, Phys. Fluids **26**, 1892 (1983).

[4]C. S. Liu, M. N. Rosenbluth, and R. B. White, Phys. Fluids **17**, 1211 (1974); see also D. W. Forslund *et al.*, Phys. Fluids **18**, 1002 (1975), and Phys. Rev. Lett. **30**, 739 (1973); B. I. Cohen *et al.*, Phys. Rev. Lett. **29**, 581 (1972); W. Manheimer and H. Klein, Phys. Fluids **17**, 1889 (1974); D. Biskamp and H. Welter, Phys. Rev. Lett. **34**, 312 (1975); J. J. Thomson, Phys. Fluids **21**, 2082 (1978); T. Tajima and J. M. Dawson, Phys. Rev. Lett. **43**, 267 (1979); W. Rozmus *et al.*, Phys. Fluids **26**, 1071 (1983); B. Amini and F. F. Chen, private communication; E. A. Williams and R. Short, in Proceedings of the Thirteenth Anomalous Absorption Conference, Banff, 1983 (unpublished); G. Picard and T. W. Johnston, to be published; D. R. Nicholson, Phys. Fluids **19**, 889 (1976); H. C. Barr and F. F. Chen, to be published; A. A. Offenberger *et al.*, Phys. Rev. Lett. **49**, 371 (1982); A. Simon and R. W. Short, to be published; K. Estabrook and W. L. Kruer, Bull. Am. Phys. Soc. **28**, 1165 (1983).

[5]Derived for Brillouin scattering by S. E. Bodner and J. L. Eddleman, University of California Radiation Laboratory Report No. UCRL 73378, 1971 (unpublished).

[6]K. Estabrook *et al.*, Phys. Rev. Lett. **45**, 1399 (1980).

[7]C. Joshi *et al.*, Phys. Rev. Lett. **48**, 874 (1982).

[8]H. Figueroa *et al.*, to be published.

[9]A. T. Lin and J. M. Dawson, Phys. Fluids **18**, 201 (1975).

[10]K. Nishikawa and C. S. Liu, in *Advances in Plasma Physics*, edited by A. Simon and W. Thomson (Interscience, New York, 1976).

[11]T. W. Johnston and J. M. Dawson, Phys. Fluids **16**, 722 (1973); $\nu_{ei}\approx 9\times10^{-11}n\,({\rm cm}^{-3})\bar{z}[T_e({\rm keV})]^{-3/2}[31-\ln(n^{1/2}/T_e)]$ sec^{-1}.

[12]L. V. Powers *et al.*, Nucl. Fusion **19**, 659 (1979); O. Willi *et al.*, Phys. Fluids **23**, 2061 (1980); C. R. Gwinn *et al.*, Phys. Fluids **26**, 275 (1983).

[13]V. L. Ginzburg and V. V. Zheleznyakov, Sov. Astron. AJ **2**, 653 (1958).

[14]R. Shanny *et al.*, Phys. Fluids **10**, 1281 (1967); the mapping of three velocity components to two was done by J. Denavit.

Chapter 11

ELEMENTARY PARTICLE PHYSICS EXPERIMENTS

WOLFGANG K. H. PANOFSKY

GEORGE H. TRILLING

Introduction

Papers Reprinted in Book

Jack Steinberger (1921-). (Courtesy of AIP Emilio Segrè Visual Archives.)

Wolfgang K. H. Panofsky is Director Emeritus at the Stanford Linear Accelerator Center (SLAC), California, and has been its Director for more than twenty-five years. His work on the absorption of negative pions in hydrogen and deuterium resulted in the term Panofsky ratio and established the intrinsic parity of the pion. His awards and honors include the National Medal of Science, the Franklin Medal, the Ernest O. Lawrence Medal, the Leo Szilard Award, the Enrico Fermi Award, and the Richtmeyer Lecture by the American Association of Physics Teachers.

George H. Trilling is at the University of California in Berkeley and at the Lawrence Berkeley Laboratory. He has served as Chairman of the University's Physics Department, and as Director of the LBL Physics Division. His research over the past 40 years has focused on accelerator-based experimental high energy physics.

ELEMENTARY PARTICLE PHYSICS EXPERIMENTS

Wolfgang K. H. Panofsky (1919–), adjusting the entrance slits on an analyzing magnet used in meson experiments performed with the 1-GeV linear electron accelerator at the High Energy Physics Lab, Stanford, CA, in 1959. (Courtesy of Stanford University, AIP Emilio Segrè Visual Archives.)

Emilio Segrè, (1905-1989). (Courtesy of Lawrence Berkeley Laboratory, AIP Emilio Segrè Visual Archives.)

ELEMENTARY PARTICLE PHYSICS EXPERIMENTS

Elementary particle physics has undergone dramatic changes in content during the 100-year history of *The Physical Review*, but it has maintained a unity of purpose: to understand those particles that, at the relevant time, were considered to be "elementary," that is, without known subconstituents. As the feasible energy of investigation advanced, substructures became evident, and therefore the term elementary became subject to repeated reinterpretation. This transformation creates a potential for overlap between this overview and those on nuclear physics, atomic physics, foundations of physics, cosmic rays, and astrophysics. For purposes of this overview, we assume that the pre-World War II "elementary particle" work in atomic physics, starting with the discovery of the electron and then passing through the important prewar period in which deep insights on elementary processes were guided by x-ray investigations, is covered by other authors. The understanding of the nucleon-nucleon forces that before the war was justly considered elementary particle physics is covered under the nuclear physics heading. Also events based on cosmic-ray experiments, such as the discoveries of the positron and charged pion, the observation of photon fluxes predating the discovery of the neutral pion, and the initial studies of strange particles, are covered in the cosmic-ray section. In addition, any understanding of the development of particle experiment requires close familiarity with simultaneous theoretical advances; yet we shall leave the discussion of the relevant theoretical work to another chapter. Finally, we must make one other important qualification: The development of particle physics has been very much an international endeavor, and we shall occasionally allude to work published in non-U.S. journals. However, because our emphasis in this report must be on the published work in *The Physical Review* and *Physical Review Letters*, our

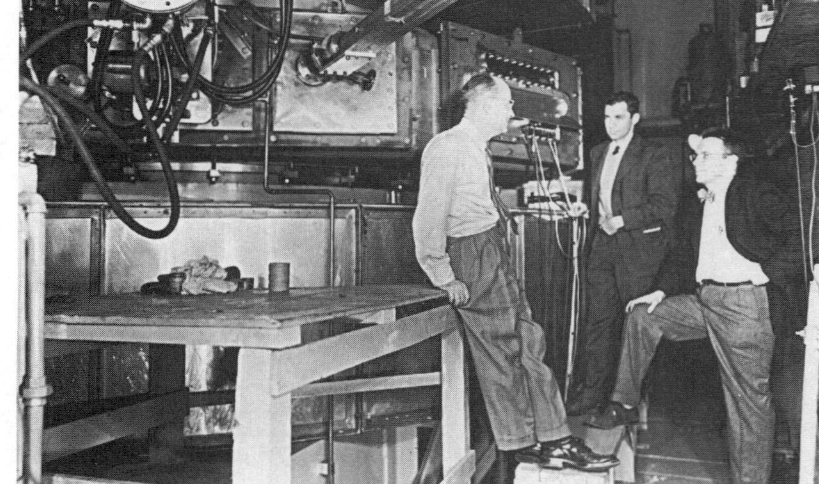

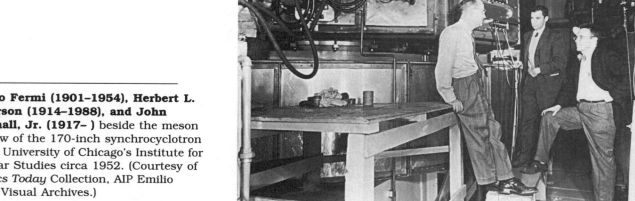

Enrico Fermi (1901–1954), Herbert L. Anderson (1914–1988), and John Marshall, Jr. (1917–) beside the meson window of the 170-inch synchrocyclotron at the University of Chicago's Institute for Nuclear Studies circa 1952. (Courtesy of *Physics Today* Collection, AIP Emilio Segrè Visual Archives.)

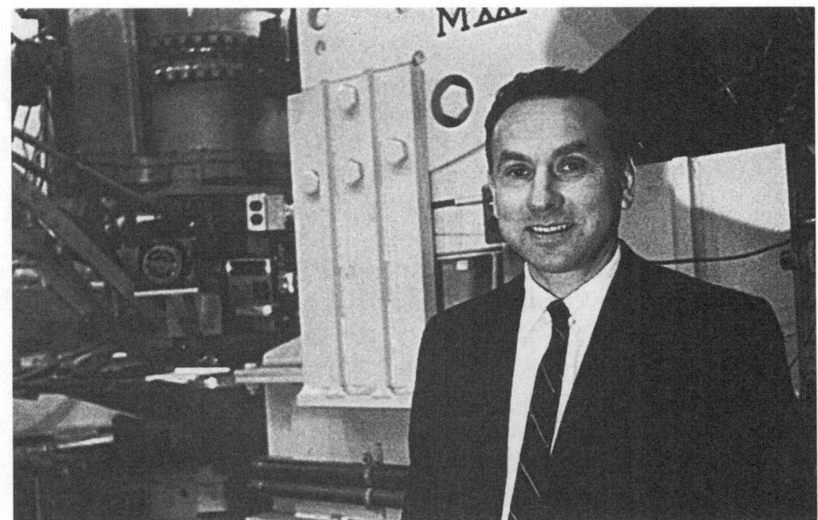

Robert Hofstadter (1915–1990). (Courtesy of Stanford University, AIP Emilio Segrè Visual Archives.)

comments will not represent a completely balanced account of the development of this field.

The papers covered here comprise in essence the milestones passed in creating what is now known as the Standard Model, via those experimental investigations that used the postwar particle accelerators. The development of the Standard Model can be characterized as one of the great success stories of scientific thought. Its theoretical components are quantum electrodynamics (QED), the unification of QED with weak interaction theory, and the formulation of quantum chromodynamics (QCD). These theories provide an almost complete description of what is currently known about elementary particle physics. The building blocks of the Standard Model are families of point-like quarks and leptons, each carrying appropriate charge and flavor, which link to the carriers of the interactions between them. However, the Standard Model requires that the theory incorporate a large number of arbitrary parameters that define the masses of the quarks and leptons, the masses of the carriers of the forces between them, the strengths of the interactions, and the "mixing angles" among the forces that constitute the basic interactions. The very essence of such a large number of parameters makes it clear that the Standard Model cannot be the "last word" or "theory of everything," quite apart from the fact that gravity has not yet found a place within the Standard Model.

PIONS

The experimental path to the Standard Model started with pion physics using the postwar accelerators. Prior to the war, Yukawa had postulated the existence of the meson, or mesotron, as the carrier of the nuclear force. Initially, the Yukawa particle was identified with the muon, the particle constituting the penetrating component of cosmic rays, because, as predicted for Yukawa's

Leon Lederman (1922–). (Courtesy of AIP Meggers Gallery of Nobel Laureates, AIP Emilio Segrè Visual Archives.)

meson, its mass lay between those of the proton and electron. This illusion was shattered by the discovery of Conversi *et al.*[1] demonstrating that the muon could not be the carrier of the strong force. This discovery was confirmed by subsequent investigations of scattering of cosmic-ray muons, which also could not be interpreted in terms of strong interactions. The real Yukawa particle turned out to be the pion.

The charged pions were first detected as tracks in thick photographic emulsions exposed to cosmic rays by Perkins and by Lattes and colleagues in Bristol. It was found that a stopping positive pion decays into a muon, which in turn decays into an electron. A negative pion stopping in an emulsion is absorbed by a nucleus and generates a nuclear breakup or "star," a direct proof of its strong interaction. Subsequently, the charged pions were produced in large quantity with the Berkeley 184-in. proton synchrocyclotron by Gardner and Lattes, and their properties were studied extensively.

The neutral pion was discovered through identification of its decay into two photons. The initial observations involved study of the photon fluxes from targets exposed to high-energy proton beams. More conclusive proof came through observations of two-photon coincidences from targets bombarded by energetic photon beams.[2] Thus the existence of the pion in all three charge states was demonstrated.

The spin of the charged pion was determined to be zero by comparing the cross section for pion-plus-deuteron production in nucleon-nucleon collisions with the cross section for the inverse process, and applying detailed balance arguments.[3] The intrinsic parity of the pion was determined to be odd from the observation of two-neutron final states (forbidden for even parity) produced in the absorption of negative pions in deuterium.[4]

The combination of all the experiments thus determined that the pion in all its then-known intrinsic properties exhibited the characteristics of the Yukawa particle—that is, the mediator of the nuclear force. In addition the masses and decay patterns of the pions were established. This postwar leap in our understanding of the carrier of the nuclear force, and the insights into elementary particle physics still to come, would have been impossible without major developments in accelerator physics and in the technology of particle detectors. However, since these achievements are generally published in journals other than *The Physical Review*, they shall not be considered here.

The decay of the charged pion into a muon plus a neutrino, occurring with a mean life of about 25 ns, is a weak decay, and it represents the first such decay beyond the well-known nuclear beta decay. Similarly, the muon decay into an electron plus two neutrinos (the electron's spectrum is continuous) with a mean life of about 2 μs is also a weak decay. In both cases, as in beta decay, the decay products include at least one neutrino, a particle whose only interactions are weak.

STRANGE PARTICLES

A new phenomenon surfaced with the discovery of particles produced with large cross sections, hence strongly interacting, and yet decaying into pairs or triplets of other strongly interacting particles with the long lifetimes, 10^{-8} to 10^{-10} s, typical of weak decays. These new particles were first observed in cosmic-ray experiments by Rochester and Butler, and they were subsequently produced in the newly built Cosmotron and Bevatron accelerators. These observations led Gell-Mann to postulate a new additive quantum number S (for strangeness — because of their unusual behavior the new particles were called "strange" particles), conserved in strong and electromagnetic interactions and not conserved in weak interactions. One consequence was the associated production of strange particles, experimentally verified at the Cosmotron.[5] Both strange baryons (particles for which $S \neq 0$ and whose decay chains must eventually terminate in a proton or neutron) and strange mesons of various masses and decay channels were discovered. The charged K mesons, or kaons ($S=\pm 1$), decay by a variety of channels; but, by showing that all these channels correspond to the same lifetime, it was established that there was but a single species of charged kaon.

HADRON RESONANCES

Subsequent to the elucidation of their properties, charged pions produced in particle accelerators were used as beam particles to bombard hydrogen targets, leading to the observations of baryon-excited states. These manifest themselves as peaks in the pion-proton interaction cross sections.[6] With the development of both the hydrogen bubble chamber and particle beams consisting almost exclusively of positive or negative kaons, it became possible to observe strange baryon resonances (i.e., with strangeness $S \neq 0$).[7]

30-GeV proton proton elastic scattering at the BNL AGS. Scattered and recoil protons go off at equal angles to the left and right through the magnets shown to counter telescopes outside the shield wall. (Courtesy of Brookhaven National Laboratory, AIP Emilio Segrè Visual Archives.)

Frederick Reines (1918-). (Photograph by Ed Nano. Courtesy of AIP Emilio Segrè Visual Archives.)

By exposing the hydrogen bubble chambers immersed in magnetic fields to various charged-particle beams, one could measure particle momenta and, taking account of momentum-energy conservation, fully and precisely reconstruct individual interaction events. By detecting population peaks at particular values of invariant mass for various combinations of outgoing particles ($\pi\pi$, $\pi\pi\pi$, πK, etc.), one also discovered meson resonances (or excited meson states).[8] Eventually an extensive spectroscopy of excited meson and baryon states, with measured spins and parities, was uncovered and systematized into SU(3) multiplets, with each member of a multiplet characterized by particular values of the isospin and hypercharge quantum numbers and with all members of the same multiplet having the same spin and parity. The culmination of this effort was the prediction and subsequent discovery at Brookhaven National Laboratory (BNL) of the nearly stable Ω^- baryon.[9] Spectroscopy for atoms and nuclei has been understood in terms of the dynamics of the relevant constituents, electrons for atoms and nucleons for nuclei. Gell-Mann and Zweig showed that observed baryon and meson states could be understood in terms of constituents that Gell-Mann called quarks. With three quarks, u, d, and s (standing for up, down, and strange), one could account for observed meson states as bound levels of a quark and an antiquark, and baryon states as bound levels of three quarks. At this point quarks were more a useful mathematical device than a real physical entity, and no free quark was ever observed.

ANTIBARYONS

From charge conjugation invariance, one would expect for each particle an oppositely charged antiparticle of equal mass. This had long been established for electrons and muons; but, until completion of the Berkeley Bevatron, no accelerator had sufficiently high beam energy to create the baryon-antibaryon pairs mandated by baryon number conservation. The discovery of the antiproton by Chamberlain *et al.* came soon after Bevatron turnon,[10] followed by the discovery of the antineutron[11] by Cork *et al.* Antiproton-proton annihilation was observed in photographic emulsions[12] and bubble chambers, providing conclusive proof of the antiproton identification. The antiproton was to have a particularly important application in particle physics some 25 years after its discovery: its use in proton-antiproton colliders to reach ultrahigh collision energies and, ultimately, to discover experimentally the W and Z bosons.

PARITY NONCONSERVATION

We now return to the weak interactions. Experimental study of the charged-kaon decay modes into two pions and into three pions indicated that the extra pion carried a negative intrinsic parity uncompensated by the relevant spatial parities, and hence that two different decay modes from the same parent carried opposite parities. This remarkable observation was interpreted by Lee and Yang as evidence that mirror symmetry (manifested by parity conservation) was not a symmetry of the weak interactions. This was confirmed by the Wu *et al.* observation[13] of the angular asymmetry of beta particles emitted from polarized Co^{60}, and by the Garwin *et al.* measurement[14] of the asymmetry in the electron angular distribution from muon decay (the muons are polarized through parity nonconservation in pion decay). These observations also established the violation in weak interactions of charge conjugation invariance. Symmetry with respect to the combined operations of mirror reflection and charge conjugation (CP) appeared preserved; although, as will be discussed later, small violations of CP symmetry were eventually observed in weak decays. Parity nonconservation was subsequently detected in weak decays of strange particles. It is worth mentioning that the parity-violating asymmetry in muon decay, which permits easy measurement of the spin direction, allowed increasingly precise measurements of the muon magnetic moment. These measurements, in agreement with much other evidence, made it clear at a very precise level that, in all properties except those related to its mass, the muon appears identical to the electron.

Val L. Fitch (1923–). (Courtesy of Clem Fiori, AIP Emilio Segrè Visual Archives.)

NEUTRINO/ANTINEUTRINO BEAMS

A major milestone in weak interaction experiments was the detection of collisions initiated by incident neutrinos or antineutrinos, enormously difficult because of the minute cross sections involved. In the first such experiment, Reines and Cowan[15] detected positrons produced by collisions on hydrogen of antineutrinos originating at a nuclear reactor. The neutrino and antineutrino collision cross sections increase with energy, and accelerator experiments with high-energy neutrinos/antineutrinos produced from kaon and pion decay were initiated in the early 1960s. The first such successful experiment,[16] done by Schwartz, Lederman, Steinberger, and colleagues, demonstrated that neutrinos arising from pion or kaon decay into muons would, after collision, produce only muons, rather than the positrons observed in the earlier reactor experiments. Hence the neutrinos associated with muons are different from the neutrinos associated with beta decay, and lepton number is conserved. This early accelerator experiment was the precursor of an extensive program of collision experiments with neutrino and antineutrino beams that continues to the present time.

NEUTRAL KAON PHENOMENA AND CP VIOLATION

Returning to consideration of kaons, the K^0 and $\bar{K}^0$ differ from each other in having opposite strangeness, but they have the common dominant 2π-decay modes. It was pointed out theoretically by Gell-Mann and Pais that this remarkable situation led one to the expectation that the neutral kaon eigenstates of definite mass and lifetime should be linear combinations of K^0 and $\bar{K}^0$, one of which, K_S, decays with short lifetime into the 2π modes, and the other, K_L, decays into higher multiplicity modes with much longer lifetime. The existence of long-lived neutral kaons with decay modes other than 2π was first shown in a Brookhaven Cosmotron experiment.[17] The regeneration of K_S from a K_L beam by interaction in matter was subsequently demonstrated. In 1964 observation by Cronin, Fitch, and colleagues[18] of a small rate of 2π decays from a K_L beam demonstrated for the first time CP violation in weak interactions. This interpretation has been confirmed through observation of K_S-K_L interference in 2π decay, and charge asymmetries in K_L semileptonic decay. Quantitative measurements of CP violation in K decay continue to the present day, and there has been no laboratory observation of CP violation anywhere other than in the neutral kaon system.

James W. Cronin (1931–). (Photograph by Mike Weinstein. Courtesy of AIP Emilio Segrè Visual Archives.)

HADRONIC STRUCTURE

In the discussion of resonances, the potential relevance of nucleon substructure in the form of quarks was noted. A different approach to the study of nucleon substructure came from the use of high energy electrons as probes in experiments at the Stanford High Energy Physics Laboratory and the Stanford Linear Accelerator Center (SLAC). Hofstadter and colleagues used electron elastic scattering measurements to establish a finite size of the proton.[19] A SLAC-MIT collaboration pursued the study of inelastic electron scattering[20] and discovered that the cross sections at large momentum transfers fell off very slowly with momentum transfer and exhibited a scaling behavior. This could be naturally interpreted in terms of point-like constituents of the nucleon. These experiments were the first in a fruitful series carried out in various accelerators and using as probe particles electrons, muons, and neutrinos. The point-like constituents could be readily interpreted as quarks and gluons. Deviations from the scaling behavior were interpreted in terms of quantum chromodynamics and the emission of gluons from quarks (roughly analogous to the emission of photons from electrons). These experiments gave new dynamical reality to the quarks as constituents of the hadrons.

WEAK NEUTRAL CURRENTS

The late 1960s and early 1970s saw the development of the standard electroweak model by Glashow, Weinberg, and Salam. One of its crucial predictions was the existence of weak neutral currents, in addition to the well-known weak charged currents responsible for all observed weak decays. Search for neutral currents in such decays as $K_L \rightarrow \mu^+\mu^-$ was unsuccessful, but in 1973 the neutral currents were observed in interactions of neutrinos and antineutrinos leading to final states without muons or electrons, and where the neutrino or antineutrino is reemitted. The initial experiment was done at the European Laboratory for Particle Physics (CERN) by a European collaboration. The final triumph of the standard electroweak model came in the discovery of the W and Z bosons, the mediators of charged and neutral currents, respectively, at the theoretically expected mass values. This was done at CERN, using high-energy counterrotating beams of protons and antiprotons to produce collisions of sufficiently high energy to create W and Z bosons.

Sam Ting (1936–) and his team. (Courtesy of Brookhaven National Laboratory, AIP Emilio Segrè Visual Archives.)

THE CHARM QUARK

The very small observed rate of $K_L \rightarrow \mu\mu$ decays mentioned above was interpreted by Glashow and colleagues as evidence for the existence of a fourth quark, which they called the charm or c quark. The effect of the additional c quark was to produce a cancellation in the $K_L \rightarrow \mu\mu$ amplitude. There would then be two complete families of quarks and leptons: (1) u and d quarks, e and ν_e leptons and (2) c and s quarks, μ and ν_μ leptons. The discovery of the c quark came in 1974 through the observation of the J/ψ meson, a bound state of charm quark and charm antiquark, in two completely different experiments, one at Brookhaven[21] led by Ting and the other at SLAC[22] led by Richter. The SLAC experiment took advantage of the new electron-positron collider, Stanford Positron Electron Asymmetric Ring (SPEAR). The interpretation of the J/ψ as a $c\bar{c}$ bound state was strongly supported by the observation of excited levels in accord with expectations. The final compelling evidence for c quarks came from the observation of mesons or baryons that are low-lying bound states of a c quark and u, d, or s quarks or antiquarks. For such particles, carrying nonzero charm quantum number, only weak decay modes would be allowed. Although evidence for such charmed particles came from several experiments, the definitive observations of

Burton Richter (1931–). (Courtesy of AIP Gallery of Member Society Presidents, AIP Emilio Segrè Visual Archives.)

states with just the properties expected from charmed meson decays came from G. Goldhaber *et al.* [23] at SPEAR.

HADRON JETS

We have already mentioned the role of point-like quarks and gluons in understanding deep inelastic scattering of electrons, muons, and neutrinos by nucleons. The discovery of the charm quark gave additional reality to the quarks. The process of high-energy electron-positron annihilation into hadrons is best understood in terms of these same quarks. Specifically, such annihilation leads via an electromagnetic interaction to the production of quark-antiquark pairs. The quarks are not directly detectable, but they "hadronize" and produce, at high energy, "jets" of mesons and sometimes baryons, in which most of the energy flows along the original quark direction. Jet structure has been seen in many types of high-energy interactions, but its most compelling evidence initially came from two-jet structure observed in electron-positron annihilation at 7 GeV total energy at SPEAR.[24] The angular distribution of the jet axes with respect to the incident beams was consistent with expectations from the production by annihilation of a spin 1/2 point-like charged particle and its antiparticle (such as a quark-antiquark pair). These results have been confirmed at higher energies for which the outgoing quarks can radiate fast gluons, leading to an expected three-jet pattern from annihilation. This was indeed observed in experiments in Europe at the Deutshes Elektronen Synchrotron (DESY) Laboratory electron-positron collider, Positronen Elektronen Tandem Ring Anlage (PETRA).

THE THIRD GENERATION

Even before the existence for the charm quark and the two quark-lepton families had been fully established, evidence for a third family had appeared. In particular, Perl and his colleagues[25] in the SLAC-LBL (Lawrence Berkeley Laboratory) collaboration at SPEAR established the existence of annihilation events producing only an electron, a muon, and missing energy in the form of undetected neutral particles. These were the first evidence for what was later definitively established to be a new charged lepton, the τ lepton, differing from its lighter sisters, the electron and muon, only through its higher mass. The other members of this third family would be the associated ν_τ and the two quarks t (for top) and b (for bottom). Compelling evidence for the b quark was obtained through the discovery of the Υ, the lowest bound state of a b quark and antiquark, by Lederman and his colleagues[26] in an experiment searching for high-mass dimuon states at the Fermi National Accelerator Laboratory (Fermilab). Strong confirmation and detailed information has come from observations of Υ states produced by electron-positron annihilation at the Doppel Ring Speicher (DORIS) storage ring in DESY and the Cornell Electron-Positron Storage Ring (CESR).[27] As in the case of charm,

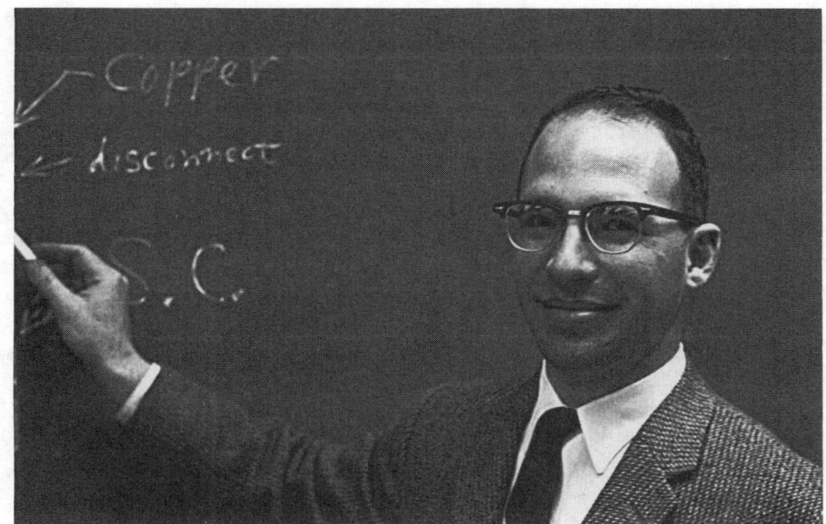

Richard L. Garwin (1928–) (Courtesy of AIP Emilio Segrè Visual Archives.)

excited $b\bar{b}$ states were observed, and the final demonstration of the existence of a new quark came from the observation of weakly decaying mesons consisting of a b quark plus lighter antiquarks (weakly decaying states with nonzero b quantum number). This was achieved in both DORIS and CESR[28] by 1983. As of this writing, the other two members of the third family, the ν_τ and the t quark, have not yet been conclusively detected, although there is little doubt of their existence. A major experimental effort at the Fermilab Tevatron Collider has established evidence by direct observation for the existence of the t quark.

REFERENCES

1. M. Conversi *et al.*, *On the Disintegration of Negative Mesons*, Phys. Rev. **71**, 209–210(L) (1947).
2. J. Steinberger *et al.*, *Evidence for the Production of Neutral Mesons by Photons*, Phys. Rev. **78**, 802–805 (1950).
3. R. Durbin, H. Loar, and J. Steinberger, *The Spin of the Pion via the Reaction* π^++d⇋p+p, Phys. Rev. **83**, 646–648 (1951).
4. W. K. H. Panofsky *et al.*, *The Gamma-Ray Spectrum Resulting from Capture of Negative* π*-Mesons in Hydrogen and Deuterium*, Phys. Rev. **81**, 565–574 (1951).
5. W. B. Fowler *et al.*, *Production of Heavy Unstable Particles by Negative Pions*, Phys. Rev. **93**, 861–867 (1954).
6. H. L. Anderson *et al.*, *Total Cross Section of Positive Pions in Hydrogen*, Phys. Rev. **85**, 936(L) (1952).
7. M. Alston *et al.*, *Resonance in the* $\Lambda\pi$ *System*, Phys. Rev. Lett. **5**, 520–524 (1960).
8. A. R. Erwin *et al.*, *Evidence for a* π-π *Resonance in the I=1, J=1 State*, Phys. Rev. Lett. **6**, 628–630 (1961).
9. V. E. Barnes *et al.*, *Observation of a Hyperon with Strangeness Minus Three*, Phys. Rev. Lett. **12**, 204–206 (1964).
10. O. Chamberlain *et al.*, *Observation of Antiprotons*, Phys. Rev. **100**, 947–950(L) (1955).
11. B. Cork *et al.*, *Antineutrons Produced from Antiprotons in Charge-Exchange Collisions*, Phys. Rev. **104**, 1193–1197(L) (1957).

Stanford Linear Accelerator Center, aerial view. (Courtesy of Stanford University, AIP Emilio Segrè Visual Archives.)

12. O. Chamberlain *et al.*, *Example of an Antiproton-Nucleon Annihilation*, Phys. Rev. **102**, 921–923(L) (1956).
13. C. S. Wu *et al.*, *Experimental Test of Parity Conservation in Beta Decay*, Phys. Rev. **105**, 1413–1415(L) (1957).
14. R. Garwin *et al.*, *Observation of the Failure of Conservation of Parity and Charge Conjugation in Meson Decays: The Magnetic Moment of the Free Muon*, Phys. Rev. **105**, 1415–1417 (1957).
15. F. Reines and C. L. Cowan, *Free Antineutrino Absorption Cross Section. I. Measurement of the Free Antineutrino Absorption Cross Section by Protons*, Phys. Rev. **113**, 273–279 (1959).
16. G. Danby *et al.*, *Observation of High-Energy Neutrino Reactions and the Existence of Two Kinds of Neutrinos*, Phys. Rev. Lett. **9**, 36–44 (1962).
17. K. Lande *et al.*, *Observation of Long-Lived Neutral V Particles*, Phys. Rev. **103**, 1901–1904(L) (1956).
18. J. H. Christenson *et al.*, *Evidence for the* 2π *Decay of the* K^0_2 *Meson*, Phys. Rev. Lett. **13**, 138–140 (1964).
19. R. Hofstadter and R. W. McAllister, *Electron Scattering from the Proton*, Phys. Rev. **98**, 217–218(L) (1955).
20. M. Breidenbach *et al.*, *Observed Behavior of Highly Inelastic Electron-Proton Scattering*, Phys. Rev. Lett. **23**, 935–939 (1969).
21. J. J. Aubert *et al.*, *Experimental Observation of a Heavy Particle J*, Phys. Rev. Lett. **33**, 1404–1406 (1974).
22. J. E. Augustin *et al.*, *Discovery of a Narrow Resonance in* e^+e^- *Annihilation*, Phys. Rev. Lett. **33**, 1406–1408 (1974).
23. G. Goldhaber *et al.*, *Observation in* e^+e^- *Annihilation of a Narrow State at 1865 MeV/*c^2 *Decaying to* $K\pi$ *and* $K\pi\pi\pi$, Phys. Rev. Lett. **37**, 255–259 (1976).
24. G. Hanson *et al.*, *Evidence for Jet Structure in Hadron Production by* e^+e^- *Annihilation*, Phys. Rev. Lett. **35**, 1609–1612 (1975).
25. M. L. Perl *et al.*, *Evidence for Anomalous Lepton Production in* e^+-e^- *Annihilation*, Phys. Rev. Lett. **35**, 1489–1492 (1975).
26. S. W. Herb *et al.*, *Observation of a Dimuon Resonance at 9.5 GeV in 400-GeV Proton-Nucleus Collisions*, Phys. Rev. Lett. **39**, 252–255 (1977).
27. T. Böhringer *et al.*, *Observations of* Υ, Υ', *and* Υ'' *at the Cornell Electron Storage Ring*, Phys. Rev. Lett. **44**, 1111–1114 (1980).
28. S. Behrends *et al.*, *Observation of Exclusive Decay Modes of b-Flavored Mesons*, Phys. Rev. Lett. **50**, 881–884 (1983).

ELEMENTARY PARTICLE PHYSICS EXPERIMENTS

Papers Reproduced on CD-ROM

D. W. Kerst. Acceleration of electrons by magnetic induction, *Phys. Rev.* **58**, 841(L) (1940)

D. Iwanenko and I. Pomeranchuk. On the maximal energy attainable in a betatron, *Phys. Rev.* **65**, 343(L) (1944)

J. P. Blewett. Radiation losses in the induction electron accelerator, *Phys. Rev.* **69**, 87–95 (1946)

M. Conversi, E. Pancini, and O. Piccioni. On the disintegration of negative mesons, *Phys. Rev.* **71**, 209–210(L), 557(E) (1947)

J. Steinberger. On the range of the electrons in meson decay, *Phys. Rev.* **75**, 1136–1143 (1949)

J. Steinberger, W. K. H. Panofsky, and J. Steller. Evidence for the production of neutral mesons by photons, *Phys. Rev.* **78**, 802–805 (1950)

W. K. H. Panofsky, R. L. Aamodt, and J. Hadley. The gamma-ray spectrum resulting from capture of negative π-mesons in hydrogen and deuterium, *Phys. Rev.* **81**, 565–574 (1951)

R. Durbin, H. Loar, and J. Steinberger. The spin of the pion via the reaction $\pi^{+}+d \rightleftharpoons p+p$, *Phys. Rev.* **83**, 646–648 (1951)

H. L. Anderson, E. Fermi, E. A. Long, and D. E. Nagle. Total cross sections of positive pions in hydrogen, *Phys. Rev.* **85**, 936(L) (1952)

D. A. Glaser. Some effects of ionizing radiation on the formation of bubbles in liquids, *Phys. Rev.* **87**, 665(L) (1952)

E. D. Courant, M. S. Livingston, and H. S. Snyder. The strong-focusing synchroton—A new high energy accelerator, *Phys. Rev.* **88**, 1190–1196 (1952)

R. W. Thompson, A. V. Buskirk, L. R. Etter, C. J. Karzmark, and R. H. Rediker. The disintegration of V^0 particles, *Phys. Rev.* **90**, 329–330(L) (1953)

D. A. Glaser. Bubble chamber tracks of penetrating cosmic-ray particles, *Phys. Rev.* **91**, 762–763(L) (1953)

W. B. Fowler, R. P. Shutt, A. M. Thorndike, and W. L. Whittemore. Production of $V_1{}^0$ particles by negative pions in hydrogen, *Phys. Rev.* **91**, 1287(L) (1953)

F. Reines and C. L. Cowan, Jr.. Detection of the free neutrino, *Phys. Rev.* **92**, 830–831(L) (1953)

W. B. Fowler, R. P. Shutt, A. M. Thorndike, and W. L. Whittemore. Production of heavy unstable particles by negative pions, *Phys. Rev.* **93**, 861–867 (1954)

F. Reines, C. L. Cowan, Jr., and M. Goldhaber. Conservation of the number of nucleons, *Phys. Rev.* **96**, 1157–1158(L) (1954)

R. Hofstadter and R. W. McAllister. Electron scattering from the proton, *Phys. Rev.* **98**, 217–218(L) (1955)

O. Chamberlain, E. Segrè, C. Wiegand, and T. Ypsilantis. Observation of antiprotons, *Phys. Rev.* **100**, 947–950(L) (1955)

D. W. Kerst, F. T. Cole, H. R. Crane, L. W. Jones, L. J. Laslett, T. Ohkawa, A. M. Sessler, K. R. Symon, K. M. Terwilliger, and N. V. Nilsen. Attainment of very high energy by means of intersecting beams of particles, *Phys. Rev.* **102**, 590–591(L) (1956)

O. Chamberlain, W. W. Chupp, A. G. Ekspong, G. Goldhaber, S. Goldhaber, E. J. Lofgren, E. Segrè, C. Wiegand, E. Amaldi, G. Baroni, C. Castagnoli, C. Franzinetti, and A. Manfredini. Example of an antiproton–nucleon annihilation, *Phys. Rev.* **102**, 921–923(L) (1956)

G. K. O'Neill. Storage-ring synchrotron: Device for high-energy physics research, *Phys. Rev.* **102**, 1418–1419(L) (1956)

R. Budde, M. Chretien, J. Leitner, N. P. Samios, M. Schwartz, and J. Steinberger. Properties of heavy unstable particles produced by 1.3-Bev π^- mesons, *Phys. Rev.* **103**, 1827–1836 (1956)

K. Lande, E. T. Booth, J. Impeduglia, and L. M. Lederman. Observation of long-lived neutral *V* particles, *Phys. Rev.* **103**, 1901–1904(L) (1956)

B. Cork, G. R. Lambertson, O. Piccioni, and W. A. Wenzel. Antineutrons produced from antiprotons in charge-exchange collisions, *Phys. Rev.* **104**, 1193–1197(L) (1956)

R. L. Garwin, L. M. Lederman, and M. Weinrich. Observations of the failure of conservation of parity and charge conjugation in meson decays: The magnetic moment of the free muon, *Phys. Rev.* **105**, 1415–1417(L) (1957)

J. I. Friedman and V. L. Telegdi. Nuclear emulsion evidence for parity nonconservation in the decay chain π^{+}-μ^{+}-e^{+}, *Phys. Rev.* **105**, 1681–1682(L) (1957)

F. Eisler, R. Plano, A. Prodell, N. Samios, M. Schwartz, J. Steinberger, P. Bassi, V. Borelli, G. Puppi, G. Tanaka, P. Woloschek, V. Zoboli, M. Conversi, P. Franzini, I. Mannelli, R. Santangelo, V. Silvestrini, D. A. Glaser, C. Graves, and M. L. Perl. Demonstration of parity nonconservation in hyperon decay, *Phys. Rev.* **108**, 1353–1355(L) (1957)

M. Goldhaber, L. Grodzins, and A. W. Sunyar. Helicity of neutrinos, *Phys. Rev.* **109**, 1015–1017(L) (1958)

F. Reines and C. L. Cowan, Jr. Free antineutrino absorption cross section. I. Measurement of the free antineutrino absorption cross section by protons, *Phys. Rev.* **113**, 273–279 (1959)

F. Reines, C. L. Cowan, Jr., F. B. Harrison, A. D. McGuire, and H. W. Kruse. Detection of the free antineutrino, *Phys. Rev.* **117**, 159–173 (1960)

M. Schwartz. Feasibility of using high-energy neutrinos to study the weak interactions, *Phys. Rev. Lett.* **4**, 306–307 (1960)

M. Alston, L. W. Alvarez, P. Eberhard, M. L. Good, W. Graziano, H. K. Ticho, and S. G. Wojcicki. Resonance in the $\Lambda\pi$ system, *Phys. Rev. Lett.* **5**, 520–524 (1960)

ELEMENTARY PARTICLE PHYSICS EXPERIMENTS

A. R. Erwin, R. March, W. D. Walker, and E. West. Evidence for a π-π resonance in the $I=1$, $J=1$ state, *Phys. Rev. Lett.* **6**, 628–630 (1961)

G. Danby, J-M. Gaillard, K. Goulianos, L. M. Lederman, N. Mistry, M. Schwartz, and J. Steinberger. Observation of high-energy neutrino reactions and the existence of two kinds of neutrinos, *Phys. Rev. Lett.* **9**, 36–44 (1962)

L. Bertanza, V. Brisson, P. L. Connolly, E. L. Hart, I. S. Mittra, G. C. Moneti, R. R. Rau, N. P. Samios, I. O. Skillicorn, S. S. Yamamoto, M. Goldberg, L. Gray, J. Leitner, S. Lichtman, and J. Westgard. Possible resonances in the $\Xi\pi$ and $K\bar{K}$ systems, *Phys. Rev. Lett.* **9**, 180–183 (1962)

R. L. Fleischer, P. B. Price, and R. M. Walker. Track registration in various solid-state nuclear track detectors, *Phys. Rev.* **133**, A1443–A1449 (1964)

V. E. Barnes, P. L. Connolly, D. J. Crennell, B. B. Culwick, W. C. Delaney, W. B. Fowler, P. E. Hagerty, E. L. Hart, N. Horwitz, P. V. C. Hough, J. E. Jensen, J. K. Kopp, K. W. Lai, J. Leitner, J. L. Lloyd, G. W. London, T. W. Morris, Y. Oren, R. B. Palmer, A. G. Prodell, D. Radojičić, D. C. Rahm, C. R. Richardson, N. P. Samios, J. R. Sanford, R. P. Shutt, J. R. Smith, D. L. Stonehill, R. C. Strand, A. M. Thorndike, M. S. Webster, W. J. Willis, and S. S. Yamamoto. Observation of a hyperon with strangeness minus three, *Phys. Rev. Lett.* **12**, 204–206 (1964)

J. H. Christenson, J. W. Cronin, V. L. Fitch, and R. Turlay. Evidence for the 2π decay of the K_2^0 meson, *Phys. Rev. Lett.* **13**, 138–140 (1964)

T. Janssens, R. Hofstadter, E. B. Hughes, and M. R. Yearian. Proton form factors from elastic electron–proton scattering, *Phys. Rev.* **142**, 922–931 (1966)

M. Breidenbach, J. I. Friedman, H. W. Kendall, E. D. Bloom, D. H. Coward, H. DeStaebler, J. Drees, L. W. Mo, and R. E. Taylor. Observed behavior of highly inelastic electron–proton scattering, *Phys. Rev. Lett.* **23**, 935–939 (1969)

J. J. Aubert, U. Becker, P. J. Biggs, J. Burger, M. Chen, G. Everhart, P. Goldhagen, J. Leong, T. McCorriston, T. G. Rhoades, M. Rohde, S. C. C. Ting, S. L. Wu, and Y. Y. Lee. Experimental observation of a heavy particle J, *Phys. Rev. Lett.* **33**, 1404–1406 (1974)

J. -E. Augustin, A. M. Boyarski, M. Breidenbach, F. Bulos, J. T. Dakin, G. J. Feldman, G. E. Fischer, D. Fryberger, G. Hanson, B. Jean-Marie, R. R. Larsen, V. Lüth, H. L. Lynch, D. Lyon, C. C. Morehouse, J. M. Paterson, M. L. Perl, B. Richter, P. Rapidis, R. F. Schwitters, W. M. Tanenbaum, F. Vannucci, G. S. Abrams, D. Briggs, W. Chinowsky, C. E. Friedberg, G. Goldhaber, R. J. Hollebeek, J. A. Kadyk, B. Lulu, F. Pierre, G. H. Trilling, J. S. Whitaker, J. Wiss, and J. E. Zipse. Discovery of a narrow resonance in e^+e^- annihilation, *Phys. Rev. Lett.* **33**, 1406–1408 (1974)

G. S. Abrams, D. Briggs, W. Chinowsky, C. E. Friedberg, G. Goldhaber, R. J. Hollebeek, J. A. Kadyk, A. Litke, B. Lulu, F. Pierre, B. Sadoulet, G. H. Trilling, J. S. Whitaker, J. Wiss, J. E. Zipse, J. -E. Augustin, A. M. Boyarski, M. Breidenbach, F. Bulos, G. J. Feldman, G. E. Fischer, D. Fryberger, G. Hanson, B. Jean-Marie, R. R. Larsen, V. Luth, H. L. Lynch, D. Lyon, C. C. Morehouse, J. M. Paterson, M. L. Perl, B. Richter, P. Rapidis, R. F. Schwitters, W. Tanenbaum, and F. Vannucci. Discovery of a second narrow resonance in e^+e^- annihilation, *Phys. Rev. Lett.* **33**, 1453–1455 (1974)

E. G. Cazzoli, A. M. Cnops, P. L. Connolly, R. I. Louttit, M. J. Murtagh, R. B. Palmer, N. P. Samios, T. T. Tso, and H. H. Williams. Evidence for $\Delta S=-\Delta Q$ currents or charmed-baryon production by neutrinos, *Phys. Rev. Lett.* **34**, 1125–1128 (1975)

M. L. Perl, G. S. Abrams, A. M. Boyarski, M. Breidenbach, D. D. Briggs, F. Bulos, W. Chinowsky, J. T. Dakin, G. J. Feldman, C. E. Friedberg, D. Fryberger, G. Goldhaber, G. Hanson, F. B. Heile, B. Jean-Marie, J. A. Kadyk, R. R. Larsen, A. M. Litke, D. Lüke, B. A. Lulu, V. Lüth, D. Lyon, C. C. Morehouse, J. M. Paterson, F. M. Pierre, T. P. Pun, P. A. Rapidis, B. Richter, B. Sadoulet, R. F. Schwitters, W. Tanenbaum, G. H. Trilling, F. Vannucci, J. S. Whitaker, F. C. Winkelmann, and J. E. Wiss. Evidence for anomalous lepton production in e^+–e^- annihilation, *Phys. Rev. Lett.* **35**, 1489–1492 (1975)

G. Hanson, G. S. Abrams, A. M. Boyarski, M. Breidenbach, F. Bulos, W. Chinowsky, G. J. Feldman, C. E. Friedberg, D. Fryberger, G. Goldhaber, D. L. Hartill, B. Jean-Marie, J. A. Kadyk, R. R. Larsen, A. M. Litke, D. Lüke, B. A. Lulu, V. Lüth, H. L. Lynch, C. C. Morehouse, J. M. Paterson, M. L. Perl, F. M. Pierre, T. P. Pun, P. A. Rapidis, B. Richter, B. Sadoulet, R. F. Schwitters, W. Tanenbaum, G. H. Trilling, F. Vannucci, J. S. Whitaker, F. C. Winkelmann, and J. E. Wiss. Evidence for jet structure in hadron production by e^+e^- annihilation, *Phys. Rev. Lett.* **35**, 1609–1612 (1975)

G. Goldhaber, F. M. Pierre, G. S. Abrams, M. S. Alam, A. M. Boyarski, M. Breidenbach, W. C. Carithers, W. Chinowsky, S. C. Cooper, R. G. DeVoe, J. M. Dorfan, G. J. Feldman, C. E. Friedberg, D. Fryberger, G. Hanson, J. Jaros, A. D. Johnson, J. A. Kadyk, R. R. Larsen, D. Lüke, V. Lüth, H. L. Lynch, R. J. Madaras, C. C. Morehouse, H. K. Nguyen, J. M. Paterson, M. L. Perl, I. Peruzzi, M. Piccolo, T. P. Pun, P. Rapidis, B. Richter, B. Sadoulet, R. H. Schindler, R. F. Schwitters, J. Siegrist, W. Tanenbaum, G. H. Trilling, F. Vannucci, J. S. Whitaker, and J. E. Wiss. Observation in e^+e^- annihilation of a narrow state at 1865 MeV/c^2 decaying to $K\pi$ and $K\pi\pi\pi$, *Phys. Rev. Lett.* **37**, 255–259 (1976)

M. J. Alguard, W. W. Ash, G. Baum, J. E. Clendenin, P. S. Cooper, D. H. Coward, R. D. Ehrlich, A. Etkin, V. W. Hughes, H. Kobayakawa, K. Kondo, M. S. Lubell, R. H. Miller, D. A. Palmer, W. Raith, N. Sasao, K. P. Schüler, D. J. Sherden, C. K. Sinclair, and P. A. Souder. Deep inelastic scattering of polarized electrons by polarized protons, *Phys. Rev. Lett.* **37**, 1261–1265 (1976)

S. W. Herb, D. C. Hom, L. M. Lederman, J. C. Sens, H. D. Snyder, J. K. Yoh, J. A. Appel, B. C. Brown, C. N. Brown, W. R. Innes, K. Ueno, T. Yamanouchi, A. S. Ito, H. Jöstlein, D. M. Kaplan, and R. D. Kephart. Observation of a dimuon resonance at 9.5 GeV in 400-GeV proton–nucleus collisions, *Phys. Rev. Lett.* **39**, 252–255 (1977)

D. M. Kaplan, R. J. Fisk, A. S. Ito, H. Jöstlein, J. A. Appel, B. C. Brown, C. N. Brown, W. R. Innes, R. D. Kephart, K. Ueno, T. Yamanouchi, S. W. Herb, D. C. Hom, L. M. Lederman, J. C. Sens, H. D. Synder, and J. K. Yoh. Study of the high-mass dimuon continuum in 400-GeV proton–nucleus collisions, *Phys. Rev. Lett.* **40**, 435–438 (1978)

T. Böhringer, F. Constantini, J. Dobbins, P. Franzini, K. Han, S. W. Herb, D. M. Kaplan, L. M. Lederman, G. Mageras, D. Peterson, E. Rice, J. K. Yoh, G. Finocchiaro, J. Lee-Franzini, G. Giannini, R. D. Schamberger, Jr., M. Sivertz, L. J. Spencer, and P. M. Tuts. Observation of Υ, Υ′, and Υ″ at the Cornell Electron Storage Ring, *Phys. Rev. Lett.* **44**, 1111–1114 (1980)

S. Behrends, K. Chadwick, J. Chauveau, P. Ganci, T. Gentile, J. M. Guida, J. A. Guida, R. Kass, A. C. Melissinos, S. L. Olsen, G. Parkhurst, D. Peterson, R. Poling, C. Rosenfeld, G. Rucinski, E. H. Thorndike, J. Green, R. G. Hicks, F. Sannes, P. Skubic, A. Snyder, R. Stone, A. Chen, M. Goldberg, N. Horwitz, A. Jawahery, M. Jibaly, P. Lipari, G. C. Moneti, C. G. Trahern, H. van Hecke, M. S. Alam, S. E. Csorna, L. Garren, M. D. Mestayer, R. S. Panvini, D. Andrews, P. Avery, C. Bebek, K. Berkelman, D. G. Cassel, J. W. DeWire, R. Ehrlich, T. Ferguson, R. Galik, M. G. D. Gilchriese, B. Gittelman, M. Halling, D. L. Hartill, D. Herrup, S. Holzner, M. Ito, J. Kandaswamy, V. Kistiakowsky, D. L. Kreinick, Y. Kubota, N. B. Mistry, F. Morrow, E. Nordberg, M. Ogg, R. Perchonok, R. Plunkett, A. Silverman, P. C. Stein, S. Stone, R. Talman, D. Weber, R. Wilcke, A. J. Sadoff, R. Giles, J. Hassard, M. Hempstead, J. M. Izen, K. Kinoshita, W. W. MacKay, F. M. Pipkin, J. Rohlf, R. Wilson, and H. Kagan. Observation of exclusive decay modes of *b*-flavored mesons, *Phys. Rev. Lett.* **50**, 881–884 (1983)

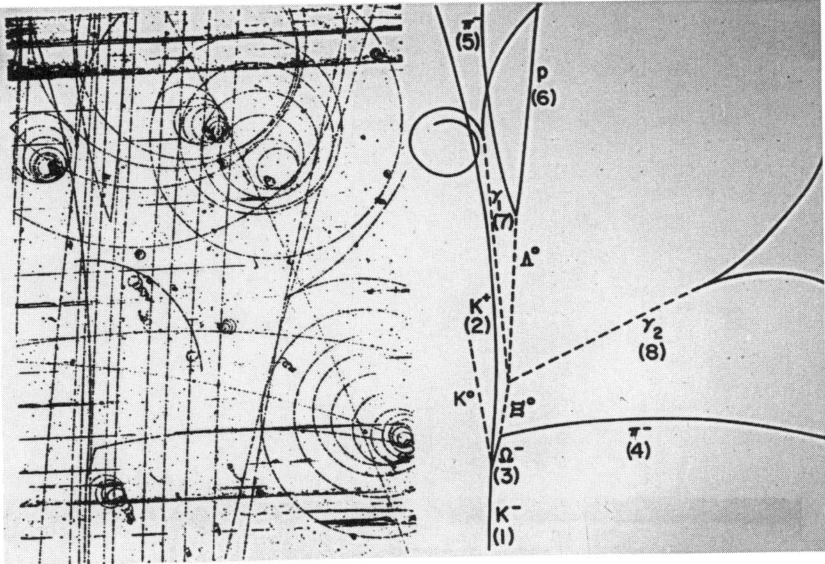

Brookhaven National Laboratory bubble chamber photograph of the production of the Omega Minus (white background) with diagram. (Courtesy of Brookhaven National Laboratory, AIP Emilio Segrè Visual Archives.)

Letters to the Editor

PUBLICATION of brief reports of important discoveries in physics may be secured by addressing them to this department. The closing date for this department is, for the issue of the 1st of the month, the 8th of the preceding month and for the issue of the 15th, the 23rd of the preceding month. No proof will be sent to the authors. The Board of Editors does not hold itself responsible for the opinions expressed by the correspondents. Communications should not exceed 600 words in length.

On the Disintegration of Negative Mesons

M. CONVERSI, E. PANCINI, AND O. PICCIONI*
Centro di Fisica Nucleare del C. N. R. Istituto di Fisica dell'Università di Roma, Italia

December 21, 1946

IN a previous Letter to the Editor,[1] we gave a first account of an investigation of the difference in behavior between positive and negative mesons stopped in dense materials. Tomonaga and Araki[2] showed that, becuase of the Coulomb field of the nucleus, the capture probability for negative mesons at rest would be much greater than their decay probability, while for positive mesons the opposite should be the case. If this is true, then practically all the decay processes which one observes should be owing to positive mesons.

Several workers[3] have measured the ratio η between the number of the disintegration electrons and the number of mesons stopped in dense materials. Using aluminum, brass, and iron, these workers found values of η close to 0.5 which, if one assumes that the primary radiation consists of approximately equal numbers of positive and negative mesons, support the above theoretical prediction. Auger, Maze, and Chaminade,[4] on the contrary, found η to be close to 1.0, using aluminum as absorber.

Last year we succeeded in obtaining evidence of different behavior of positive and negative mesons stopped in 3 cm of iron as an absorber by using magnetized iron plates to concentrate mesons of the same sign while keeping away mesons of the opposite sign (at least for mesons of such energy that would be stopped in 3 cm of iron). We obtained results in agreement with the prediction of Tomonaga and Araki. After some improvements intended to increase the counting rate and improve our discrimination against the "mesons of the opposite sign," we continued the measurements using, successively, iron and carbon as absorbers. The recording equipment was one which two of us had previously used in a measurement of the meson's mean life.[5] It gave threefold (III) and fourfold (IV) delayed coincidences. The difference (III)−(IV) (after applying a slight correction for the lack of efficiency of the fourfold coincidences) was owing to mesons stopped in the absorber and ejecting a disintegration electron which produced a delayed coincidence. The minimum detected delay was about 1 μsec. and the maximum about 4.5 μsec. Our calculations of the focusing properties of the magnetized plates (20 cm high; $\beta = 15{,}000$ gauss) and including roughly the effects of scattering, showed that we should expect almost complete cut-off for the "mesons of the opposite sign." This is confirmed by our results, since otherwise it would be very hard to explain the almost complete dependence on the sign of the meson observed in the case of iron.

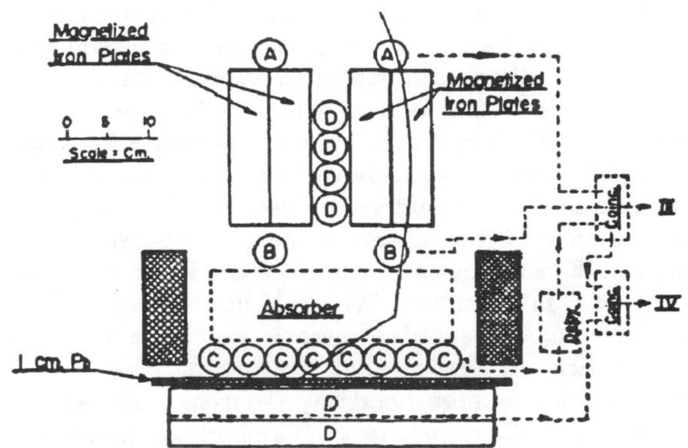

FIG. 1. Disposition of counters, absorber, and magnetized iron plates. All counters "D" are connected in parallel.

TABLE I. Results of measurements on β-decay rates for positive and negative mesons.

Sign	Absorber	III	IV	Hours	M/100 hours
(a) +	5 cm Fe	213	106	155.00′	67±6.5
(b) −	5 cm Fe	172	158	206.00′	3
(c) −	none	71	69	107.45′	−1
(d) +	4 cm C	170	101	179.20′	36±4.5
(e) −	4 cm C+5 cm Fe	218	146	243.00′	27±3.5
(f) −	6.2 cm Fe	128	120	240.00′	0

The results of our last measurements with two different absorbers are given in Table I. In this table "Sign" refers to the sign of the meson concentrated by the magnetic field. $M = (\text{III}) - (\text{IV}) - P(\text{IV})$, the number of decay electrons, is corrected for the lack of efficiency (p) in our fourfold coincidences (∼0.046).

The value M_- (5 cm Fe) is but slightly greater than the correction for the lack of efficiency in our counting, so that we can say that perhaps no negative mesons and, at most, only a few (∼5) percent undergo β-decay with the accepted half-life.

The results with carbon as absorber turn out to be quite inconsistent with Tomonaga and Araki's prediction. We used cylindrical graphite rods having a mean effective thickness of 4 cm because we were unable to procure a graphite plate. In addition, when concentrating negative mesons, we placed above the graphite a 5-cm thick plate of iron to guard against the scattering of very low energy mesons which might destroy the concentrating effect of our magnets. We alternated the following three measurements:

A. Negative mesons with 4 cm C and 5 cm Fe,
B. Negative mesons with 6.2 cm Fe (6.2 cm Fe is approximately equivalent to 4 cm C+5 cm Fe as far as energy loss is concerned.
C. Positive mesons with 4 cm C.

The comparison between A and B gave the difference in behavior between Fe and C, once we had established the fact that practically no disintegration electrons came from negative mesons in the 5-cm iron plate. The comparison between A and C gives the difference in behavior between negative and positive mesons in carbon. This must be considered as a qualitative comparison because of the slightly different action of the magnetic field in concentrating mesons of different ranges (4 cm C+5 cm Fe in one case and 4 cm of C in the other). We could not, of course, add 5 cm of Fe for the positive mesons too, since positive mesons do decay in Fe.

The great yield of negative decay electrons from carbon shows a marked difference between it and iron as absorbers. Tomonaga and Araki's calculation also give for carbon a much higher ratio of capture to decay probability for negative mesons, so we are forced to doubt their estimation. It is possible that a suitable dependence of the capture cross section, σ_c, on the nuclear charge, Z, might explain these results; however, if the ratio of the capture to decay probability also depends on the density as Tomonaga and Araki pointed out, then it would require a very irregular dependence on Z to also explain the cloud-chamber pictures of some authors[6] showing negative mesons stopped in the chamber without any decay electrons coming out.

Concerning the difference between M_+ and M_- in carbon, we should like to point out that it is not necessary to assume that σ_c for carbon has an appreciable value for negative mesons. A positive excess, $(H_+-H_-)/(H_++H_-)$ of 20 percent in the hard component, as it seems to be[7] is sufficient to explain our results since this gives $H_+/H_-=1.5$ which is greater than M_+/M_- for carbon. Impurities in the graphite could also explain some preference for M_+, with a suitable dependence of σ_c on Z.

Further experiments on this subject are now in progress, in an attempt to calculate the capture cross section, and to know how it depends on Z.

* Now Visiting Research Associate at Massachusetts Institute of Technology, Cambridge, Massachusetts.

[1] M. Conversi, E. Pancini, and O. Piccioni, Phys. Rev. 68, 232 (1945).

[2] S. Tomonaga and G. Araki, Phys. Rev. 58, 90 (1940).

[3] F. Rasetti, Phys. Rev. 60, 198 (1941); B. Rossi and N. Nereson, Phys. Rev. 62, 417 (1942); M. Conversi and O. Piccioni, Nuovo Cimento 2, 71 (1944); Phys. Rev. 70, 874 (1946).

[4] P. Auger, R. Maze, and Chaminade, Comptes rendus 213, 381 (1941).

[5] M. Conversi and O. Piccioni, Nuovo Cimento 2, 40 (1944); Phys. Rev. 70, 859 (1946).

[6] Y. Nishina, M. Takeuchi, and T. Ichimiya, Phys. Rev. 55, 585 (1939); H. Maier-Leibnitz, Zeits f. Physik 112, 569 (1939); T. H. Johnson and R. P. Shutt, Phys. Rev. 61, 380 (1942).

[7] H. Jones, Rev. Mod. Phys. 11, 235 (1939); D. J. Hughes, Phys. Rev. 57, 592 (1940); G. Bernardini, M. Conversi, E. Pancini, E. Scrocco, and G. C. Wick, Phys. Rev. 68, 109 (1945).

Erratum: On the Disintegration of Negative Mesons

[Phys. Rev. 71, 209 (1947)]

M. CONVERSI, E. PANCINI, AND O. PICCIONI
Centro di Fisica, Nucleare del C. N. R, Istituto di Fisica dell'Università di Roma, Italia

IN the letter to the Editor with the above title, Fig. 1, the block diagram of the registering set, shows the "delay" in a wrong place, that is between counters "*C*" and "coincidences III." Actually the "delay" must be understood to be between counters "*A*" and "coincidences III" and between counters "*B*" and "coincidences III." No delay is inserted between counters "*C*" and "coincidences III." For more details, see M. Conversi and O. Piccioni, Phys. Rev. 70, 859 (1946).

Evidence for the Production of Neutral Mesons by Photons*

J. STEINBERGER, W. K. H. PANOFSKY, AND J. STELLER
Radiation Laboratory, Department of Physics, University of California, Berkeley, California
(Received April 28, 1950)

In the bombardment of nuclei by 330-Mev x-rays, multiple gamma-rays are emitted. From their angular correlation it is deduced that they are emitted in pairs in the disintegration of neutral particles moving with relativistic velocities and therefore of intermediate mass. The neutral mesons are produced with cross sections similar to those for the charged mesons and with an angular distribution peaked more in the forward direction. The production cross section in hydrogen and the production cross section per nucleon in C and Be are comparable.

I. INTRODUCTION

NEUTRAL mesons which are coupled strongly to nuclei must be expected to be unstable against decay into two or more gamma-rays. The modes of decay, and expected lifetimes, have been discussed extensively.[1] These gamma-rays are then supposed to be responsible for the soft showers which often accompany energetic cosmic-ray nuclear events.[2] The evidence in favor of the existence of the neutral meson has recently been greatly strengthened by the discovery at Berkeley[3] of gamma-rays which behave in all ways as if they were due to the disintegration of a neutral meson. They are produced by proton bombardment of various nuclei and have a production cross section which depends on proton energy much like that of charged mesons. Their energy is approximately 70 Mev on the average, half that of the charged π-meson, and the energy spread is in agreement with the Doppler shift due to the velocity of the parent mesons. The lifetime of the mesons is less than 10^{-13} sec., which is in agreement with the theoretical expectations.

The evidence is therefore already much in favor of the existence of a gamma-unstable neutral meson. However, until now, coincidences between the two gamma-rays have never been observed. We report here the detection of such coincidences, produced by the bombardment of various nuclei in the x-ray beam of the Berkeley synchrotron. This must be regarded as strong additional evidence supporting the existence of the neutral meson.

II. EXPERIMENTAL ARRANGEMENT

The apparatus is sketched in Fig. 1. The synchrotron x-ray beam of 330-Mev maximum energy is collimated in two successive collimators. The second collimator serves only to intercept some of the electrons produced at the edge of the first collimator. The beam then strikes a target, which, for most of the experiment, is a cylinder of beryllium, $1\frac{1}{2}$ inches long and 2 inches in diameter. The particles produced in the target are detected in two telescopes, each consisting of three scintillation counters. A converter, usually $\frac{1}{4}$ inch of lead, is inserted between the two crystals nearest the target in each telescope. An event is recorded if simultaneous (resolving time 10^{-7} sec.) pulses are recorded in the outer four crystals, but none in the two near the target. That is, we require that there be two particles, one in each telescope, neutral at first which are converted into charged particles by the lead, and which penetrate one crystal and enter the next. With a beam intensity of about 10^{11} Mev/min. the counting rate for such coincidences at favorable orientations of the telescopes is about 10 counts/min.

III. NATURE OF THE COINCIDENCES

Let us first describe the experiments which identify the particles as gamma-rays, indicate their energy and show that their origin is the nuclear rather than the Coulomb field. In Table I we list the relative detection

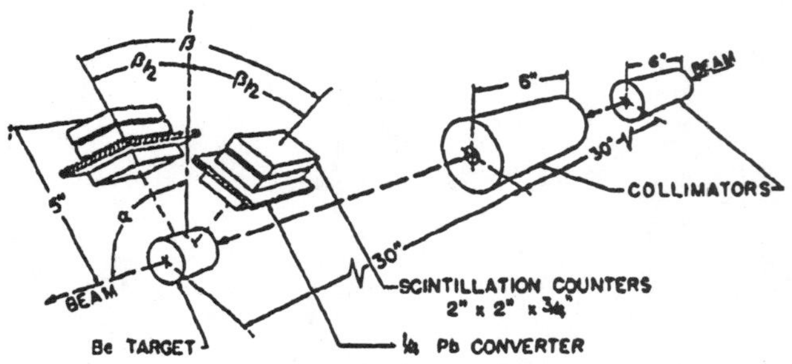

FIG. 1. Experimental arrangement

* This work was performed under the auspices of the AEC.

[1] Y. Tanikawa, Proc. Phys. Math. Soc. Japan 24, 610 (1940). R. J. Finkelstein, Phys. Rev. 72, 414 (1947). H. Fukuda and Y. Miamoto, Prog. Theor. Phys. 4, 347 (1949). Ozaki, Oneda, and Sasaki, Prog. Theor. Phys. 4, 524 (1949). J. Steinberger, Phys. Rev. 76, 1180 (1949). C. N. Yang, Phys. Rev. 77, 243 (1950).

[2] The implications of the gamma-decay of neutral mesons for the soft component in the cosmic radiation were pointed out by J. R. Oppenheimer (Phys. Rev. 71, 462 (T) (1947). It was assumed that in high energy nuclear events neutral mesons are emitted with multiplicities similar to those for charged mesons. The neutral mesons decay into photons and account for the early development of extensive showers, as well as the large total amount of soft radiation. These bursts of soft radiation accompanying energetic nuclear events were then actually observed in the cloud chamber by W. Fretter, Phys. Rev. 73, 41 (1948), 76, 511 (1949); C. Y. Chao, Phys. Rev. 75, 581 (1949); Gregory, Rossi, and Tinlot, Phys. Rev. 77, 299 (1949); and J. Green, Thesis, University of California, 1950. They were found in photographic plates by Kaplan, Peters, and Bradt, Phys. Rev. 76, 1735 (1949). Both the cloud-chamber pictures and the photographic star show that the showers begin with gamma-rays rather than electrons.

[3] Bjorklund, Crandall, Moyer, and York, Phys. Rev. 77, 213 (1950).

efficiency for various converter materials and thicknesses. Without converters the counting rate is almost zero, then increases as the converter thickness in each arm is increased to $\frac{1}{4}$ inch of lead, and only slightly from $\frac{1}{4}$ inch to $\frac{1}{2}$ inch. This is as expected from shower theory for about 100-Mev photons. Copper of $\frac{1}{4}$ inch thickness has approximately the same conversion efficiency as has $\frac{1}{16}$ inch of lead, again in agreement with shower theory, since the number of shower units is the same for these thicknesses.

The coincidences are attenuated by a factor of four when $\frac{1}{4}$ inch of lead is inserted between the target and the anticoincidence crystals. This again is as expected for photons. Furthermore, it can be seen from Table I that both telescopes require converters, so that both particles must be photons.

To measure the energy of the conversion electrons, aluminum absorbers were inserted between the last two crystals of one of the telescopes. Unfortunately, at these energies the radiation losses are important, and therefore the straggling large. We have plotted in Fig. 2 the coincidence counting rate as a function of the average energy required to traverse the telescope. Because the photons originate in moving mesons, the average gamma-ray energy is expected to be approximately 100 Mev, and the average electron energy 50 Mev, quite in agreement with the observed attenuation.

The nuclear origin of the photons is demonstrated by the fact that the cross section for these coincidences is only six times as big for a lead nucleus as for beryllium, which is less than the ratio of the nuclear areas. On the other hand, ordinary shower cross sections increase by a factor of 400.

Finally, we have looked for coincidences with the beam energy reduced to about 175 Mev with angles α and β of the telescope both 90°. The cross section per Q (the number Q for a bremsstrahlung beam is equal to the total energy divided by the maximum energy of the spectrum) is at least 50 times smaller here than at 330 Mev. This steep excitation function is also observed for charged meson production.

We believe, therefore, that it is demonstrated that the observed coincidences are caused by gamma-rays of about 100-Mev average energy, of non-Coulombic origin, and with a threshold similar to that for charged mesons.

TABLE I. Relative detection efficiency as a function of absorber material and thickness.

Converter in telescope 1	Converter in telescope 2	Relative counting rate $\alpha=\beta=90°$
none	none	0.01 ± 0.005
$\frac{1}{32}$-in. Pb	$\frac{1}{32}$-in. Pb	0.17 ± 0.013
$\frac{1}{16}$-in. Pb	$\frac{1}{16}$-in. Pb	$0.3\ \pm0.02$
$\frac{1}{8}$-in. Pb	$\frac{1}{8}$-in. Pb	0.67 ± 0.08
$\frac{1}{4}$-in. Pb	$\frac{1}{4}$-in. Pb	1.00 ± 0.06
$\frac{1}{4}$-in. Cu	$\frac{1}{4}$-in. Cu	0.39 ± 0.03
none	$\frac{1}{4}$-in. Pb	0.15 ± 0.05
$\frac{1}{16}$-in. Pb	$\frac{1}{4}$-in. Pb	0.62 ± 0.07
$\frac{1}{2}$-in. Pb	$\frac{1}{4}$-in. Pb	1.07 ± 0.1
$\frac{1}{4}$-in. Pb	$\frac{1}{4}$-in. Pb	0.28 ± 0.05

$\frac{1}{4}$-in. Pb absorbers placed in front of both telescopes.

IV. ANGULAR CORRELATION AND DISTRIBUTION OF THE GAMMA-RAYS

To study further the properties of these coincidences, we have measured their rate as a function of the angle, α, between the beam direction and the plane of the telescopes and of the correlation angle β (see Fig. 1). Consider first the variation with β at a fixed α, say 90°. 180° coincidences are rare. The counting rate increases with decreasing β to a maximum at 90°, and then drops sharply. This behavior must actually be expected of gamma-rays which are the decay products of neutral mesons, because of the motion of the decaying mesons. A meson at rest decaying into two gamma-rays, emits them in opposite direction. But when this is seen from a system in which the meson has a total energy E, then the included angle β varies between π and $2\sin^{-1}(1/E)$ with a probability which favors the small angles tremendously. The median angle is $2\sin^{-1}[2/(3E^2+1)^{\frac{1}{2}}]$. E is the total meson energy in units of its rest energy.

For 70-Mev mesons the minimum angle of β is 84° and the median angle 92°. Since the distribution is so heavily peaked, not much error is introduced if one assumes, as is done in the following, that to an angle β corresponds a unique energy, that of the median angle. Therefore a measurement of the distribution in β is a measure of the distribution in energy of the neutral mesons, although the angular resolution of our telescopes is insufficient to give more than a glimpse of the energy distribution. We have included in Fig. 3 curves in which the observed[4] energy distributions of the π^+-meson made by the same x-rays on hydrogen are transformed into distributions in β and arbitrarily normalized. All corrections due to scattering and angular resolution are omitted. The general shape of the curves is certainly well reproduced by the experiment. It is

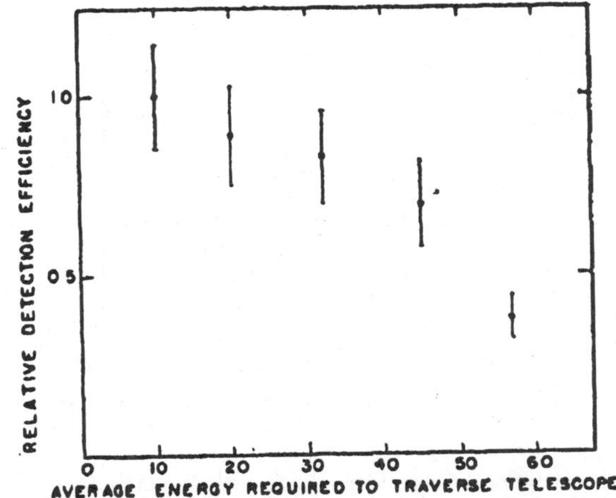

FIG. 2. Absorption of conversion electrons in aluminum. The energy includes the average radiation loss.

[4] Bishop, Cook, and Steinberger (to be published).

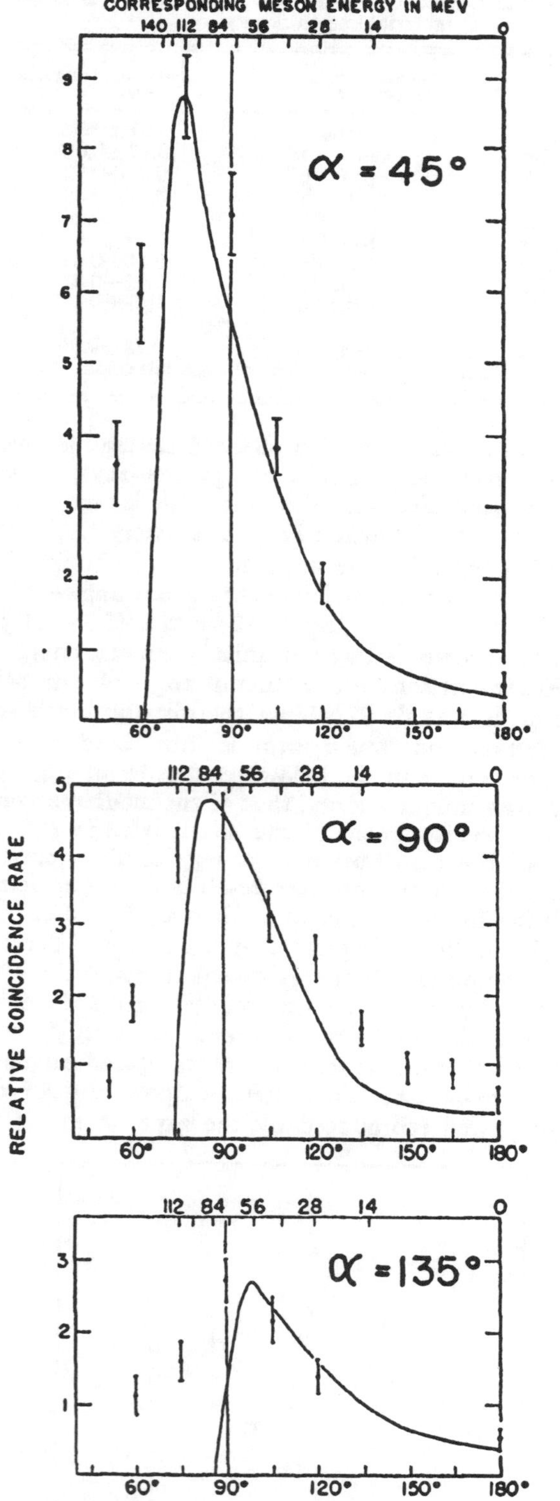

FIG. 3. Variation of coincidence rate with the included angle β between the two arms of the telescope. The curves are those expected on the assumption that the gamma-rays are the decay products of a neutral meson, emitted with the same energy distribution as are π^+-mesons from hydrogen. The curves are arbitrarily normalized for each angle α.

therefore clear that if the gamma-rays are the decay product of intermediate particles, these particles must move with velocities of the order of $v/c \simeq 0.8$. Excited nucleons of this velocity cannot be produced by x-rays of 330 Mev; the particles must therefore have an intermediate mass. Furthermore, it is possible to see that the decay must be into only 2 photons, since the expected angular distributions for a decay into more than two photons would not show a valley for small angles β.

The distribution in the angle α of the beam with the plane of the telescope shown in Fig. 4, is interesting chiefly because of the difference between this distribution and the angular distribution of π^+-photo-mesons[5] from either carbon or hydrogen targets. This is not particularly surprising, since various theories also give quite different results for charged and neutral mesons.

V. HYDROGEN CROSS SECTION AND TOTAL CROSS SECTION

At one setting of the telescopic angles, $\alpha = \beta = 90°$, we have compared the cross sections of hydrogen and carbon. This was done by comparing the count from a polyethylene (CH_2) block and a perforated carbon block of the same size and carbon content as the CH_2. The result is: $\sigma H_{\pi 0}/\sigma C_{\pi 0} = 0.12 \pm 0.03$. This again differs from the results for positive mesons, where $\sigma H_{\pi^+}/\sigma C_{\pi^+} \simeq 0.55$. The difference is probably in part caused by the fact that both neutrons and protons can contribute to neutral meson production, but only protons to π^+-production. In part, it may also be possible to ascribe this to the same phenomenon which, according to Chew,[6] is responsible for the large hydrogen-carbon ratio for the positive mesons. In the case of π^+-production, the reaction is inhibited by the fact that, when the proton is changed into a neutron, there is an oversupply of neutrons in the immediate neighborhood and the number of states available to it is small because of the Pauli principle. This is not significant in the neutral case because the nucleon's charge does not change.

The curves in Fig. 3 can be integrated to yield a total cross section for beryllium. $\sigma_{Be} = 7.5 \times 10^{-28}$ cm² per Q, while for carbon and hydrogen, assuming the same angular distribution, $\sigma_C = 10 \times 10^{-28}$ and $\sigma_H = 1.3 \times 10^{-28}$ cm² per Q. The absolute x-ray intensity is known[7] to about 10 percent, but the efficiency of the detecting system only to within a factor of two, so that there is a corresponding error in the above cross sections. The hydrogen cross section is approximately the same as that for π^+-production;[5] those for carbon and beryllium are somewhat higher.[8]

One might assume that the charge of the meson would play an important role in the production of mesons in the electromagnetic field of the photon. This

[5] J. Steinberger and A. S. Bishop, Phys. Rev. 78, 493 (1950).
[6] G. Chew (private communication).
[7] Blocker, Kenney, and Panofsky (to be published).
[8] McMillan, Peterson, and White, Science 110, 579 (1949).

is contradicted by the observed angular distribution of π^+-mesons produced by photons in H_2. The angular distribution indicates that the principal process responsible for charged meson production is the interaction of the photon with the spin of the nucleon. If the neutral meson has the same transformation properties as the charged, it then appears plausible that the production cross sections in hydrogen should be comparable, as seems to be the case. However, actual calculations on the basis of pseudoscalar theory, both in the classical and in the perturbation theory approximation, which give a reasonable angular distribution for the π^+-production, give smaller values for neutral meson production. Whether or not this is a new difficulty in a theory which already has several, is not clear. From a less restricted point of view it is not a surprising result.

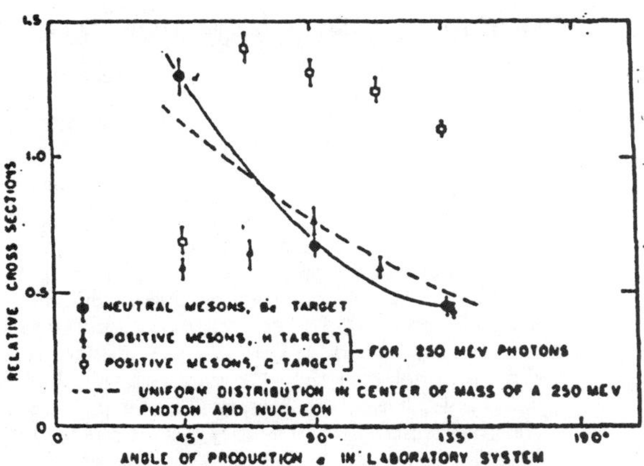

FIG. 4. Variation with the angle α between the plane of the telescope and the beam. Each point represents an integral over the angle β.

VI. SUMMARY

In the bombardment of various nuclei by 330-Mev x-rays, photons with the following properties are emitted:

(1) At least two are emitted in coincidence.
(2) They each have an average energy of about 100 Mev.
(3) The Z dependence of the production indicates that they have their origin in a nuclear interaction, and not in the Coulomb field.
(4) The threshold for their production is at least 150 Mev.
(5) The angular correlation of the photons shows that they are emitted in pairs as the only decay products of particles moving with velocities of the order of $v/c=0.8$, and therefore of intermediate mass.
(6) The total cross section for production from hydrogen is about the same as that for production of π^+-mesons; other light nuclei cross sections are somewhat higher than those for the positive mesons.

It is clear from these properties that the gamma-rays are the decay products of neutral mesons. Since spin $\frac{1}{2}$, and spin 1 mesons are forbidden to decay into two photons,[1] the spin must be zero, excluding the possibility of very high intrinsic angular momenta. It seems reasonable, and it is in good agreement with all observations, to assume that both charged and neutral mesons are of the same type. It then follows from the angular distribution of the x-ray produced π^+-mesons, and the high cross sections for making neutral mesons by x-rays, that the π-meson is a pseudoscalar. This remark applies, of course, only to the character of the meson, and not to any particular field theory for the interaction of mesons with nucleons.

All phases of this experiment have been discussed with Professor Edwin McMillan and his advice has been of great help. The bombardments were carried out by the synchrotron crew under the direction of W. Gibbons.

PHYSICAL REVIEW VOLUME 81, NUMBER 4 FEBRUARY 15, 1951

The Gamma-Ray Spectrum Resulting from Capture of Negative π-Mesons in Hydrogen and Deuterium

WOLFGANG K. H. PANOFSKY, R. LEE AAMODT, AND JAMES HADLEY
*Radiation Laboratory, Department of Physics, University of California, Berkeley, California**

(Received October 6, 1950)

π^- mesons produced in an internal wolfram target bombarded by 330-Mev protons in the 184-inch cyclotron are absorbed in a high pressure hydrogen target. The resulting gamma-ray spectrum is analyzed outside the shielding of the cyclotron by means of a 30-channel electron-positron pair spectrometer. The principal results are as follows. (1) The gamma-rays result from two competing reactions: $\pi^-+p\rightarrow n+\gamma$ and $\pi^-+p\rightarrow n+\pi^0$; $\pi^0\rightarrow 2\gamma$. (2) The ratio between the π^0 yield to the single gamma-ray yield is $=0.94\pm0.20$. (3) The mass difference between the π^- meson and the π^0 meson is given by 10.6 ± 2.0 electron masses. (4) The π^- mass is 275.2 ± 2.5 electron masses. The large mass difference between π^- and π^0 precludes the conclusion that the unexpectedly small π^0 to γ ratio is due to the small amount of momentum space available for π^0 emission. It rather indicates that π^0 emission is slowed down by the nature of the coupling of the π^0 field to the nucleons. The experiment has been repeated by substituting D_2 for H_2 in the vessel. The result is that the reaction $\pi^-+D\rightarrow 2n$ and $\pi^-+D\rightarrow 2n+\gamma$ compete in the ratio 2 : 1. The reaction $\pi^-+D\rightarrow 2n+\pi^0$ is absent.

I. INTRODUCTION

THE classic experiments of Conversi, *et al.*[1] on the absorption of negative μ-mesons in matter gave the first information on the fact that the coupling of μ-mesons with nuclei is weak. On the other hand, experiments on the absorption of π^- mesons[2,3] have confirmed the fact that π-mesons are strongly coupled and that they have integral spin. This is evidenced by the fact that the process

$$\pi^-+p\rightarrow n^* \tag{1}$$

taking place within a nucleus could not give rise to the large nuclear stars observed if an additional light particle of half-integral spin had to be emitted, as in the absorption of μ^- mesons. Other than confirming this qualitative fact, the absorption of slow π^- mesons in matter has led to no quantitative information as to the nature of the coupling of π-mesons to nuclei, since the dominant time in the capture process is the slow-down time by ionization; once a π^- meson has arrived in the K shell of an atom, absorption is presumed to take place in about 10^{-14} sec.

The case of π^- absorption in hydrogen is a singular one since clearly process (1) is forbidden by the conservation laws in the case of a free proton. Absorption of a π^- meson in hydrogen must thus lead to the emission of one or more additional particles. Excluding processes involving several spin ½ particles which are clearly too slow to compete, the possible processes which might result are:

$$\pi^-+p\rightarrow n+\gamma, \tag{2}$$

$$\pi^-+p\rightarrow n+2\gamma, \tag{3}$$

$$\pi^-+p\rightarrow n+\pi^0. \tag{4}$$

Prior to this work the possibility for process (4) rested, of course, on the possibility that the π^- might be sufficiently heavier than the π^0 to make the process energetically possible. Evidence from direct gamma-ray production in a cyclotron target bombarded by 350-Mev protons[4] points to the existence of a π^0 of mass of the order of the π^- mass, but the center of the gamma-ray spectrum cannot be localized with sufficient accuracy to decide the sign of the mass difference. Cosmic-ray evidence[5-8] and particularly the observations of gamma-gamma coincidences observed from targets bombarded in the x-ray beam of the Berkeley synchrotron[9] have shown conclusively that a π^0 exists and that it disintegrates into two gamma-rays and thus cannot have spin one. Recently, Carlson, Cooper, and King[10] have succeeded in analyzing positron-electron pairs observed in nuclear emulsions exposed at 70,000 ft in terms of neutral mesons. They show that the observed energy spectrum of such pairs is compatible with their origin from gamma-rays from a π^0 meson of mass 295 ± 20 electron masses, where only the statistical error is included in the mass estimate. Carlson, Cooper, and King also deduce the mean life, τ, of the π^0 meson to be $\tau<5\times10^{-14}$ sec.

Preliminary reports of the present experiment[11] have indicated qualitatively that both processes (2) and (4) exist. However, no accurate mass determination of the π^0 was possible and thus no very significant branching ratio between the processes could be inferred.

The evidence presented here excludes any appreciable competition from process (3). The reason is firstly a

* This work was done under the auspices of the AEC.
[1] Conversi, Pancini, and Piccioni, Phys. Rev. 68, 232 (1945).
[2] Manon, Muirhead, and Rachat, Phil. Mag. 41, 583 (1950).
[3] F. Adelman and S. Jones, Science 111, 226 (1950); W. Cheston and L. Goldfarb, Phys. Rev. 78, 320A (1950).
[4] Bjorklund, Crandall, Moyer, and York, Phys. Rev. 77, 213 (1950).
[5] W. Fretter, Phys. Rev. 73, 41 (1948); Phys. Rev. 76, 213 (1949).
[6] C. Y. Chao, Phys. Rev. 75, 581 (1949).
[7] Gregory, Rossi, and Tinlot, Phys. Rev. 77, 299 (1949).
[8] Kaplon, Peters, and Bradt, Phys. Rev. 76, 1735 (1949).
[9] Steinberger, Panofsky, and Steller, Phys. Rev. 78, 802 (1950).
[10] Carlson, Cooper, and King, Phil. Mag. 41, 701 (1950).
[11] W. K. H. Panofsky, Post Deadline Paper, New York Meeting A.P.S., 1950; Panofsky, Aamodt, and York, Phys. Rev. 78, 825 (1950).

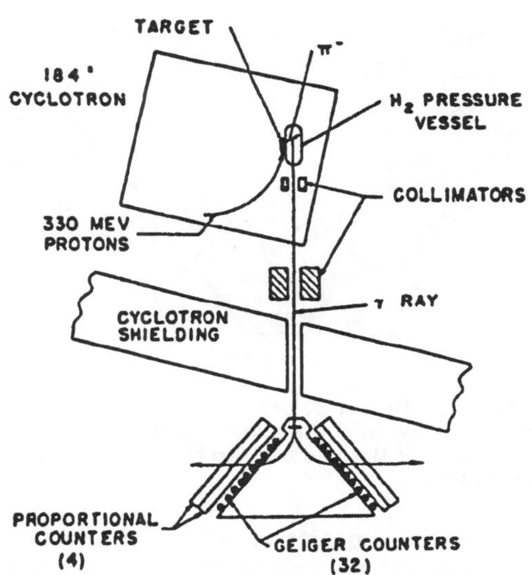

FIG. 1. Geometrical arrangement of π^- capture experiment. π^- mesons produced in a primary wolfram target of the 184-inch cyclotron are captured in the H_2 pressure vessel. The resultant gamma-rays are collimated and leave the cyclotron shielding through a hole tapering from 2 in. to 3 in. in diameter. The gamma-rays are then analyzed by a pair spectrometer.

theoretical one. It appears to be difficult to construct a selection rule which would make double gamma-emission compete effectively with single gamma-emission. Secondly, the double peaked energy distribution (see Fig. 12) of the emitted radiation practically excludes a two-gamma-process.

The details of the slowdown process of π^- in hydrogen have been discussed in considerable detail by Wightman.[12] The significant sequence of the process is as follows. (1) Slowdown of the fast meson by the ordinary stopping power mechanism ($\sim 10^{-10}$ sec). (2) Slowdown by collisions with orbital electrons of velocities comparable with that of the meson ($\sim 10^{-12}$ sec). (3) Capture of the mesons in an outer orbit leading to an excited $\pi^- - H^+$ system.[13] (4) Reduction of energy of the neutral $\pi^- - H^+$ system to the lowest quantum state. This latter process principally is not radiative but is due to collisions of the neutral system with hydrogen molecules which leads by various mechanisms to the emission of an Auger electron ($\sim 10^{-10} - 10^{-9}$ sec). In liquid or high pressure hydrogen the over-all time to enter the K shell is thus sufficiently short to compete effectively with the $\pi-\mu$ decay time.[14] This is true, however, only if the H_2 density is sufficient. This point has been verified experimentally (see Sec. VII).

Capture in flight[15] corresponds to a lifetime of the order of 10^{-4} to 10^{-6} sec, depending on assumptions as to the interaction.

[12] A. S. Wightman, thesis, Princeton University, Princeton, New Jersey (June, 1949), and Phys. Rev. 77, 521 (1950).

[13] E. Fermi and E. Teller, Phys. Rev. 72, 399 (1947).

[14] J. R. Richardson, Phys. Rev. 74, 1720 (1948); E. A. Martinelli and W. K. H. Panofsky, Phys. Rev. 77, 465 (1950); Chamberlain, Mozley, Steinberger, and Wiegand, Phys. Rev. 79, 394 (1950); M. Jakobsen, private communication.

[15] R. Marshak and A. S. Wightman, Phys. Rev. 76, 114 (1949).

We can therefore conclude that at densities approximating that of liquid hydrogen all but of the order of 10^{-3} of the secondary radiations resulting from the capture of π^- in H_2 result either from absorption of the π^- meson from an inner shell or from $\pi-\mu$-decay; the $\pi-\mu$-decay branching is small.

II. GEOMETRICAL ARRANGEMENT; THE HYDROGEN SYSTEM

The geometrical layout of the experiment is shown in Fig. 1. The 330-Mev protons circulating in the internal beam of the 184-inch cyclotron strike a wolfram target $\frac{1}{2}$ in. deep (parallel to the beam) and 0.040 in. thick (transverse to the beam). Wolfram was chosen, since the π^- cross section measurements of Weissbluth[16] showed that a heavy element favors π^- production. Also, auxiliary measurements on beam penetration and "scattering out" showed that a high density target was desirable here from the point of view of total meson yield. Finally, as discussed later, the background in this experiment is principally produced by high energy protons elastically scattered and striking the hydrogen vessel. At this energy scattering at the angles in question is principally diffraction scattering; heavy elements produce a smaller diffraction angle.

The π^- mesons enter the high pressure hydrogen through the walls of the pressure vessel shown in Fig. 2. To produce maximum yield the wall thickness is limited in order that the mesons absorbed in the hydrogen are those produced at a sufficiently low energy to correspond to a rising portion of the meson yield curve as a function of meson energy. Such considerations limit the wall thickness to ~ 4 g/cm². However, an H_2 density near that of the liquid phase is desirable by the capture considerations already given. The use of liquid H_2 was

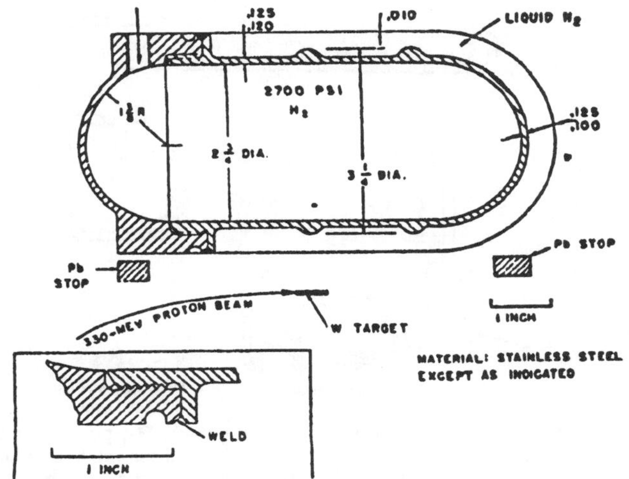

FIG. 2. High pressure vessel used for absorption of H_2 and D_2. The vessel is constructed of stainless steel machined as indicated. The load is carried by threads with the weld serving as a seal only. The outer stainless steel liquid N_2 jacket (0.010 in. thick) is soft soldered to the thick portion of the main vessel.

[16] M. Weissbluth, Phys. Rev. 78, 86A (1950).

not advisable here owing to the difficulty of cooling a long horizontal filling tube required by the geometry of the cyclotron. The vessel shown in Fig. 2 operates at a factor of safety of about 2.5 when maintained at 2700 psi and at liquid N_2 temperature. The specific gravity[17] under these conditions is 0.046. The aforementioned factor of safety makes use of the appreciable increase in strength of stainless steel at low temperature.[18] The outer jacket is fed by an external liquid N_2 Dewar vessel. The pressure vessel is filled by an external oil piston pressure pump[18a] fed by commercial H_2, dried and purified in a liquid N_2 trap. A flow diagram of the arrangement is shown in Fig. 3. A similar system is used for deuterium with certain modifications to permit recovery.

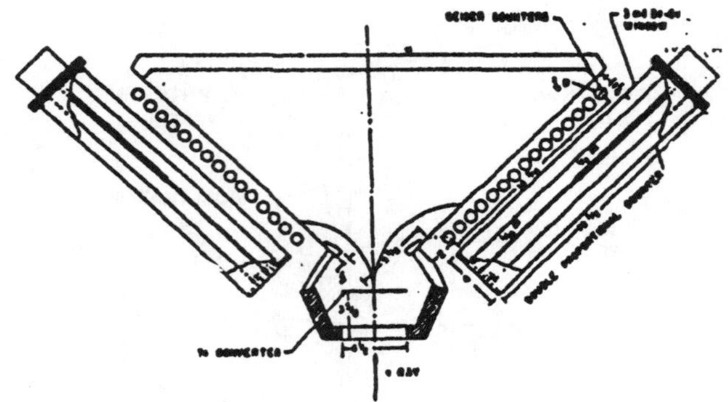

Fig. 4. Outline diagram of pair spectrometer. The converter, Geiger counter array, and the proportional counters are shown. Note the geometry of the pole piece to give a uniform field in the area of the converter.

III. THE PAIR SPECTROMETER

Since the expected pair spectra from processes (2) and (4) exhibit discontinuities, satisfactory analysis and also good signal-to-background ratio requires a spectrometer with a large number of channels. Since the counting rates in this experiment are limited entirely by absolute available intensity and not by errors introduced by accidental coincidences, etc., Geiger counters as used by Lawson[19] seem to offer the best solution to the multiple channel problem. The geometrical layout of the pair spectrometer is shown in Fig. 4. The magnet has a useful gap of 3.5 in. and a maximum field of 14,000 gauss although for this experiment only fields of the order of 5000 to 10,000 gauss were used. The pole piece is in the shape of a 90° triangle, the hypotenuse of the triangle being 30 in. The pole is widened near the converter position to improve the uniformity of the field near the converter; field uniformity is not as essential near the counters as it is near the converter. There is no specific advantage to the choice of 90° for the triangle apex angle; the angle was defined principally by arguments of size and weight. A 90° apex angle (and hence a 90° deflection angle!) provides for first-order horizontal focusing for particles of the same energy originating in different parts of the converter; this is, however, not an essential consideration if one is interested only in the sum of the energy of the two pair fragments.

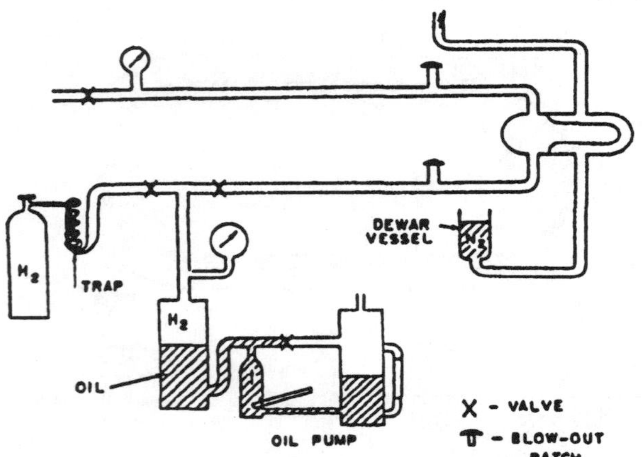

Fig. 3. Flow diagram of H_2 pressure system. H_2, purified in a liquid N_2 trap, is admitted into the pressure chamber above the pump and at tank pressure. The pressure is then raised by displacing the H_2 in the pressure chamber with oil pumped as shown.

The resolving power of a pair spectrometer of this type is essentially defined by three factors: (1) channel width; (2) multiple scattering in the converter; and (3) radiation straggling of the pair members. A feature of the triangular design is the fact that the energy width due to multiple scattering is constant. The converter thickness has been chosen such that the combination of the latter two widths approximately matches the first. This condition can only be approximately achieved, since the ratio of the radiation error to the scattering error varies as the pair energy. The choice of converter material is not critical, since pair conversion efficiency, multiple scattering, and radiation straggling depend only on the number of radiation lengths of converter used. Tantalum converters were used in this experiment. The choice of triangular shape of magnet and the analysis of the resolution of such a magnet are due to Professor Edwin McMillan. The initial design of the pair spectrometer magnet was carried out by Herbert F. York and the mechanical engineering design is due to Robert Meuser; the authors are greatly indebted to them for their contributions.

Because of the high singles rates (approximately 3 counts/sec) of the Geiger counters (Victoreen type 1B85) and the low true pair rate (approximately 30 c/hr), additional selection of events is necessary. This is provided by 4 large proportional counters (Fig. 4) backing up the counter arrangements. A pair event is selected by a quadruple coincidence count in the proportional counters; this quadruple coincidence opens a gate which passes the amplified Geiger pulses into a recorder. This recorder consists of 30 pens mark-

[17] Johnston, Berman, Rubin, Rifkin, Swanson, and Corak, MDDC-850 (to be published).
[18] E. E. Thum, *The Book of Stainless Steels* (American Society for Metals, Cleveland, 1935), second edition, p. 376.
[18a] Manufactured by the American Instrument Company, Superpressure Division, Silver Spring, Maryland.
[19] J. L. Lawson, Phys. Rev. 75, 433 (1949).

ing voltage-sensitive paper on a rotating drum. A typical event thus appears as two dots in the appropriate channels. The arrangement of the electronic components is shown in Fig. 5. The counting rate is sufficiently slow to permit this mechanical method of recording. The proportional counter gate width is 1.5 μsec; accidental counts are entirely negligible; the counting rate loss due to Geiger counter dead time is estimated at less than 2 percent.

The magnet is fed by a motor generator set, electronically regulated; the magnetic field has been calibrated against the magnetic moment of the proton; during runs the fields are monitored by current readings with a shunt and potentiometer or with a proton moment apparatus if the accuracy is needed.

The sensitivity of a pair spectrometer is not constant over the energy range covered by the instrument. This is due to (a) the variability of number of channels available to record the pair fragments of a given total energy, (b) the variation of pair production cross section with energy, and (c) the variability of loss by scattering. It can be shown easily that (c) can be neglected in this geometry. The correction curve of the instrument due to causes (a) and (b) is given in Fig. 6.

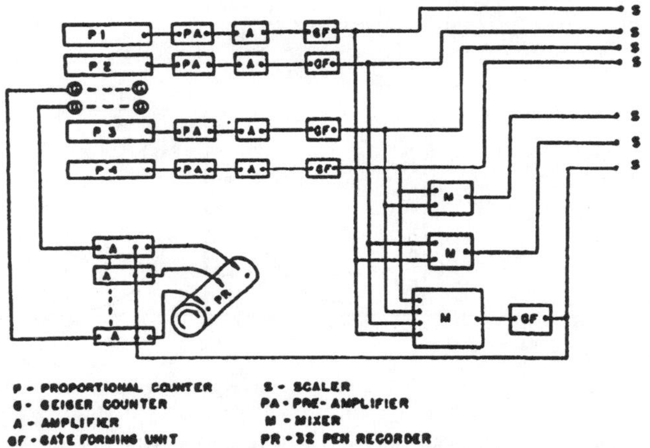

FIG. 5. Block diagram of electronic components.

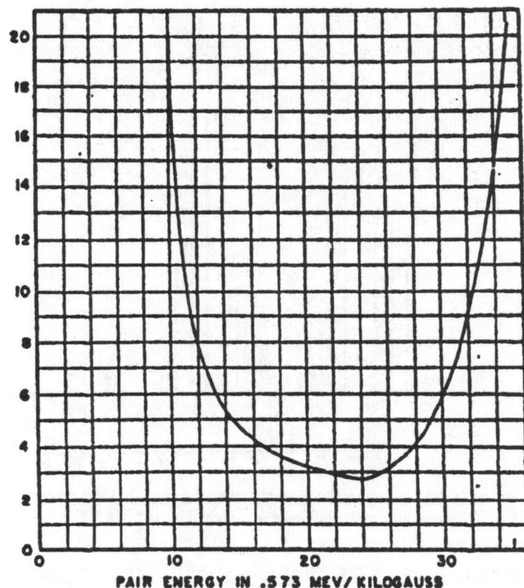

FIG. 6. Multiplication factor to be used to reduce observed counting rates to gamma-ray intensity. This factor arises from (a) variation of pair production cross section with energy and (b) number of available channels into which a given total pair energy can divide.

IV. OPERATION OF RUNS

Before every run the spectrometer is checked by using gamma-rays directly produced in the cyclotron target.[4] The yield of the direct gamma-rays is sufficient to permit the plateaus of all counters to be checked with good statistics. Also, all Geiger channels are checked for singles rates by removing the gate formed by the proportional counters. The target and pressure vessel are then moved so that the primary target is well shielded from the spectrometer and the hydrogen vessel is aligned with the collimators and the spectrometer (Fig. 1).

Readings are made with the pressure vessel either evacuated or pressurized with H_2 or D_2. If the vacuum runs are to represent the true background it is necessary that no other process (e.g., scattering of protons from the target) produce gamma-rays in hydrogen. This assumption receives support from experiments by Crandall, Hildebrand, Moyer, and York,[20] indicating that the production cross section for gamma-ray production by bombardment with 345-Mev protons in hydrogen is less than 2 percent of the cross section in carbon. This means that if a sufficient number of primary protons were scattered into the hydrogen to produce gamma-rays of significant intensity, then the background due to gamma-rays hitting the steel vessel would be much higher. It therefore appears to be certain that the gamma-rays depending on the introduction of the H_2 are not produced by fast particle collisions in H_2. This argument is not as significant in the case of D_2. As a further link in the qualitative interpretation of the experiment it was shown that no statistically significant gamma-ray counts beyond background were produced by the introduction of He into the pressure vessel. This check was done with relatively poor statistics; a positive helium effect in the form of a broad gamma-ray spectrum of 10 percent total intensity of that observed in H_2 is not excluded by the data.

The background has the general character of the gamma-ray spectra observed by Bjorklund, Crandall, Moyer, and York[4] at 180° from the target. The background is almost certainly due to protons scattered by the primary target onto the steel jacket of the pressure vessel and other parts of the cyclotron. The background is negligible in the 130-Mev region but is of the same order of intensity as the H-capture gamma-rays in the 70-Mev region.

[20] Crandall, Hildebrand, Moyer, and York, private communication.

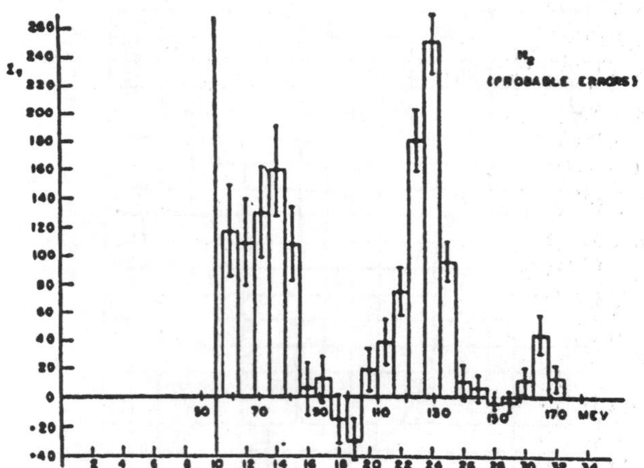

FIG. 7. Pair spectrum of gamma-rays produced by capture of π^- in H_2. Center of spectrometer set near 130 Mev.

V. GAMMA-RAY SPECTRA FROM HYDROGEN

As can be seen from the spectrometer response curve (Fig. 6) it is not possible to cover the entire spectrum of the spectrometer with good efficiency. Accordingly, different runs were made with the spectrometer set with its central point at (a) single gamma-ray peak, (b) center between the two peaks to permit easy relative area measurements, and (c) π^0 peak.

(a) The High Energy Peak; The π^- Mass

Figure 7 shows the spectrum observed with the spectrometer maximum set near the high energy peak. Note that the "π^0 peak" also appears clearly.

Since the position of the single gamma peak gives a precise measurement of the π^- mass, it is a matter of considerable interest to analyze the observed peak accurately in terms of resolving power of the pair spectrometer. Figure 8 shows plotted individually the resolving power curves due to the three principal causes of finite resolving power. The first cause is the finite channel width. This gives rise to a triangular resolution $R_1(E)$ of base equal to twice the channel width, which is 5.36 Mev. The second cause is the multiple Coulomb scattering in the converter. One can show easily that if $\langle\theta\rangle$ is the root mean square plane projected scattering angle[21] of an electron of energy of one-half the gamma-energy after having passed through the full converter thickness, then the fractional rms error in gamma-energy is given by

$$\delta = \delta E_\gamma/E_\gamma = (\sqrt{2}\langle\theta\rangle)/3. \qquad (5)$$

This gives a resolving power $R_2(E)$ given by

$$R_2(E) = \exp[-(E-E_\gamma)^2/(2\delta E_\gamma^2)] \qquad (6)$$

plotted in Fig. 8b.

The third cause is radiation straggling of the outgoing pair represented by a resolving power curve $R_4(E)$. Let, in the notation used by Heitler,[22] $w(y)dy$ be the probability that the energy of a single electron has decreased to e^{-y} times its initial value after traversing a thickness t. Let $W(p)dp$ be the resultant probability that an outgoing pair fragment has retained a fraction between p and $p+dp$ of its energy of formation. $W(p)$ can be generated by averaging $w(y)$ over the converter thickness. Let $P(E_1, E)dE_1$ be the probability that a pair fragment have an energy between E_1 and E_1+dE_1 for a total pair energy E_γ. It can be shown that the probability $\pi(f)df$ that the resultant pair shall have retained a fraction between f and $f+df$ of its initial energy is

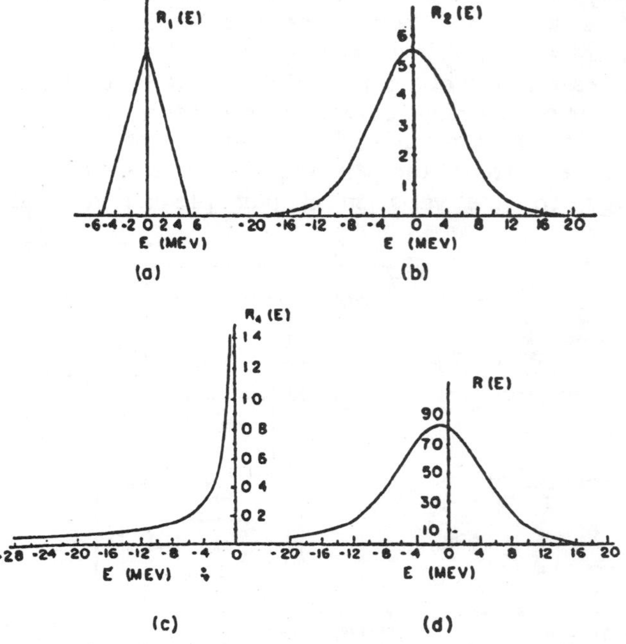

FIG. 8. Curves showing the components of the resolving power curve of the spectrometer. (a) Resolving power due to finite channel width. (b) Resolving power due to multiple scattering of pairs in converter. (c) Resolving power due to radiation straggling of outgoing pair. (d) "Fold" of a, b, and c giving total resolving power.

given by:

$$\pi(f)df = df\int_{E_A}^{E_B} dE_1\left\{\int_{1-(E_\gamma/E_1)(1-f)}^{1}\left(\frac{E_\gamma}{E_\gamma-E_1}\right)\right.$$
$$\left.\times W(p)W\left(\frac{E_\gamma f - pE_1}{E_\gamma - E_1}\right)P(E_1, E_\gamma)dp\right\}. \qquad (7)$$

Here E_A and E_B are the pair fragment energy limits defined by the spectrometer. This integral has been evaluated using the forms $P(E_1, E_\gamma)dE_1 = dE_1/E_\gamma$ and $W(p) = K/(1-p)^{1-\alpha}$, where α was fitted to the computed radiation curves. The result shows that the resolving power has the approximate form:

$$R_4(E) = 1/(E_\gamma - E)^{1-2\alpha}, \quad E<E_\gamma$$
$$= 0, \quad E>E_\gamma. \qquad (8)$$

[21] B. Rossi and K. Greisen, Rev. Mod. Phys. 13, 240 (1941).

[22] W. Heitler, *The Quantum Theory of Radiation* (Oxford University Press, Oxford, 1936), p. 223 ff.

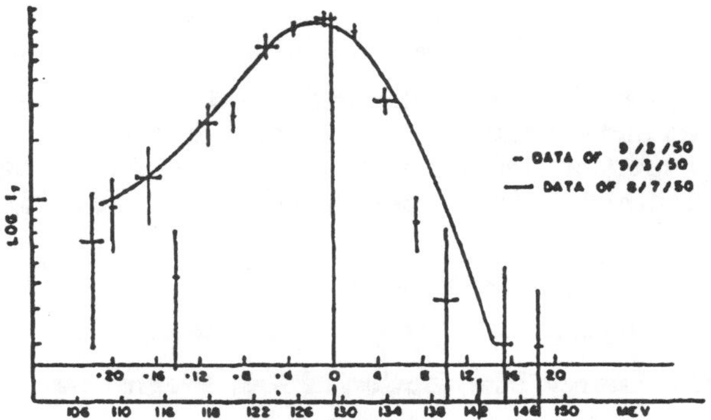

FIG. 9. Pair spectra of gamma-rays from the process $\pi^-+H\rightarrow n+\gamma$ plotted on a logarithmic scale. Plotted (solid line) also is the theoretical resolving power curve adjusted for best fit. The origin of the resolving power curve marks the energy value of the gamma-ray on the abscissa of the pair spectrum.

This is plotted in Fig. 8c for $\alpha=0.081$, $E_\gamma=132$ Mev. This corresponds to a 0.020-in. tantalum radiator. The three resolving powers are then combined numerically by a successive folding[22a] process; the resultant curve is shown in Fig. 8d.

Figure 9 shows both the final resolving power curve and the experimental data superimposed to give optimum fit. A logarithmic scale is chosen for the intensity to permit satisfactory normalization. It is observed that the fit is quite satisfactory.

It is estimated that the probable error in fitting the curves is ±0.8 percent. The remaining errors deal with the establishment of the energy scale.

The magnetic field was monitored continuously during operation by means of a magnetic moment of the proton apparatus. The probable error in magnetic field measurement of a chosen reference point is estimated at ±0.03 percent. The magnetic field was mapped with a flip coil accurate in relative measure to an accuracy of 0.1 percent probable error. Trajectories were laid out to determine the small corrections for field nonuniformity. It is estimated that the error due to uncertainty in trajectory layout is a ±0.3 percent probable error. The geometry was laid out accurately to a ±0.1 percent probable error. As a result the over-all probable error is ±0.9 percent, the principal contribution being the accuracy of curve fitting. We thus obtain

$$M_{\pi^-}=275.2\pm2.5 \text{ electron masses.}$$

A more accurate measurement based on this method is planned.

The excellent agreement of this result with the photographic work[23,24] confirms also the argument that we are, in fact, observing process (2).

(b) The Low Energy Peak

The spectrum in the neighborhood of the low energy peak is shown in Fig. 10. The resolving power at this energy is defined principally by multiple scattering of the pair fragments. The resolving power is again calculated as before and has been plotted in Fig. 10. Note that the resolution is sufficient to assure that the width of the curve of Fig. 10 is real rather than instrumental. Experimentally we take the lower and upper limits at $W_1=53.6\pm2.8$ Mev and $W_2=85\pm2.8$ Mev, respectively. The probable errors are estimated on the basis of the uncertainty in fit of these computed curves to the experimental data.

Analysis of the process

$$\pi^-+p\rightarrow n+\pi^0 \quad \pi^0\rightarrow 2\gamma \tag{9}$$

is based on the following physical picture. Since essentially all the observed radiation results from π^- mesons of initial velocity $\beta\sim1/137$, we can assume that the kinetic energy of the π^0 plus that of the neutron in process (9) are essentially equal to the mass differences involved. Since the Doppler width of the emitted gamma-ray are proportional to the momentum of the π^0, the following equation can be deduced:

$$\delta=M_{\pi^-}-(M_{\pi^-}-\Delta)\left[1-\frac{2M_n\alpha(M_{\pi^-}+M_p)}{(M_{\pi^-}-\Delta)^2}\right]^{\frac{1}{2}}, \tag{10}$$

where

$$\alpha=(1/M_n)\{(M_n^2+\Delta W^2)^{\frac{1}{2}}-M_n\}\approx\Delta W^2/2\cdot M_n^2.$$

Here δ is the mass difference between π^- and π^0, Δ is the neutron-proton mass difference,[25] and ΔW is the width of the peak: $(\delta-\Delta)$ depends thus essentially on the width of the peak of Fig. 10.

It can easily be shown statistically that the expected distribution of γ-energies is uniform on an energy scale between the two Doppler limits. The sum of the lower and upper limits of the γ-peak represents the total relativistic energy of the π^0; it is thus equal to the π^- mass less the neutron-proton mass difference and the neutron recoil energy; thus, if S is the sum of the upper and lower spectral limits,

$$S\doteq M_{\pi^-}-\Delta-[(\Delta W)^2/2M_n]. \tag{11}$$

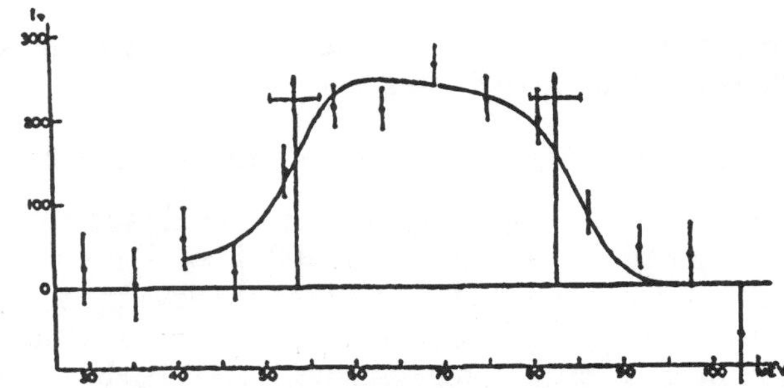

FIG. 10. Pair spectrum of gamma-rays presumably due to the process $\pi^-+H\rightarrow n+\pi^0\rightarrow n+2\gamma$. Plotted also is the theoretical spectral shape assuming that the spectrum lies between the limits of 53.6 Mev and 85 Mev. The estimated probable errors of these limits are indicated.

[22a] The "fold," $f(x)$, of two functions $g(x)$ and $h(x)$ is defined by the formula $f(x)=\int_{-\infty}^{+\infty}g(t-x)h(t)dt$.

[23] Gardner, Barkas, Smith, and Bradner, Science 111, 191 (1950).

[24] F. Smith, private communication.

[25] R. E. Bell and L. G. Elliott, Phys. Rev. 74, 1552 (1948).

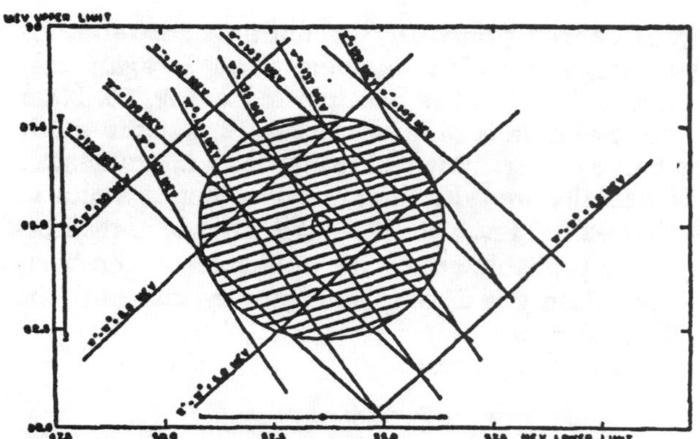

FIG. 11. Graphical representation of the relations defining the meson masses in terms of the lower and upper limit of the spectrum of Fig. 10. The ordinate and abscissa are the upper and lower limits, respectively. Plotted on this graph are: (a) The experimental values and the probable errors of the lower and upper limits. (b) The ellipse (a circle in this case) in the coordinate plane representing the area of 50 percent probability for the quoted masses. (c) The lines of constant π^- mass. (d) The lines of constant $\pi^- - \pi^0$ mass difference. (e) The curves of constant π^0 mass.

Figure 11 shows graphically the relation between the energy limits and the masses involved. The ordinates and abscissae of Fig. 11 are W_2 and W_1, the upper and lower limits, respectively, of the spectrum of Fig. 10, and the measured values are indicated. Lines of constant $\pi^- - \pi^0$ mass difference (δ) are practically of the form $W_2 - W_1 = \text{const.}$, while lines of constant π^- mass are of the form $W_2 + W_1 = \text{const.}$ Both of these functions have been plotted, in addition to the lines of constant π^0 mass. The measured values of W_1 and W_2 and their probable errors generate an ellipse of error in the $W_2 - W_1$ plane which is given here. Accordingly,

$$\delta = M_{\pi^-} - M_{\pi^0} = 5.4^{+1.4}_{-0.5}\ \text{Mev} \cong 5.4 \pm 1.0\ \text{Mev} = 10.6 \pm 2.0\ \text{electron masses.} \quad (12)$$

From the diagram:

$$M_{\pi^0} = 135 \pm 4\ \text{Mev} = 265 \pm 8\ \text{electron masses.}$$

Using the mass values already determined for the π^- mass, Eq. (12) leads to the π^0 mass:

$$M_{\pi^0} = 264.6 \pm 3.2\ \text{e.m.}$$

Further reduction of the probable errors of the π^- mass is, however, anticipated.

(c) Branching Ratio Between π^0 and γ-Emission

A run was made with the peak of the pair spectrometer intensity curve located at 100 Mev. This gives somewhat poorer statistics on either the low energy or the high energy peaks but permits a mesaurement of the branching ratio. The resulting curve is shown in Fig. 12. Figure 12 also shows a rectangular profile of area equal to the total intensity of the low energy group and of width equal to the width of the low energy peak in Fig. 10. Considering the fact that two photons are emitted per π^0 disintegration, we obtain a branching ratio of

$$\Gamma_{\pi^0}/\Gamma_\gamma = 0.94 \pm 0.20. \quad (13)$$

The probable error quoted is due to the statistics of the data and a reasonable allowance for the uncertainty in the spectrometer sensitivity curve. In interpreting this branching ratio it should be noted that, from Fig. 10,

$$p/M_{\pi}c = 0.23 \pm 0.03 \quad (14)$$

gives the momentum of the π^0 in process (4).

(d) Energy Distribution of Pair Fragments

The pair fragment energy of the pairs entering into the gamma-ray spectrum of Fig. 10 has been tabulated as a check on the performance of the pair spectrometer. Although the statistics are insufficient to provide data of interest to pair theory, it is relevant to show that the probability of division of a pair is essentially constant as a function of the division percentage.[23] Figure 13 shows a graphical representation of the number of pairs of fragment energy E_1 and E_2 plotted as a function of the division fraction $E_1/(E_1+E_2)$. The instrument does not permit recording of all possible division ratios, since (Fig. 4) the counter arrangement does not reach to the converter and is limited in radial extent. Accordingly, the scale of intensities of Fig. 13 is weighted inversely to the gamma-ray energy interval which contributes to the particular division ratio. Division ratios are given only in the range $0.2 < E_1/(E_1+E_2) < 0.8$. It is seen that within statistics the distribution is uniform as predicted theoretically.

VI. RUNS WITH MATERIALS OTHER THAN HYDROGEN OR DEUTERIUM

In the beginning stages of these experiments hydrogenous compounds were used in place of the high pressure H_2 vessel.[24] The pair spectrometer used had only a

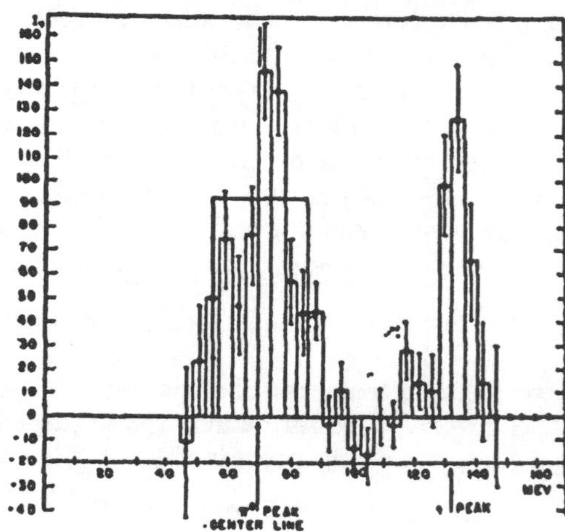

FIG. 12. Pair spectrum resulting from the absorption of π^- mesons in hydrogen. The center of the spectrometer is set near 100 Mev. The spectrum clearly shows the separation between processes (2) and (4). The branching ratio between these reactions can be derived from this spectrum.

[24] W. K. H. Panofsky and H. York, Phys. Rev. 78, 89(A) (1950).

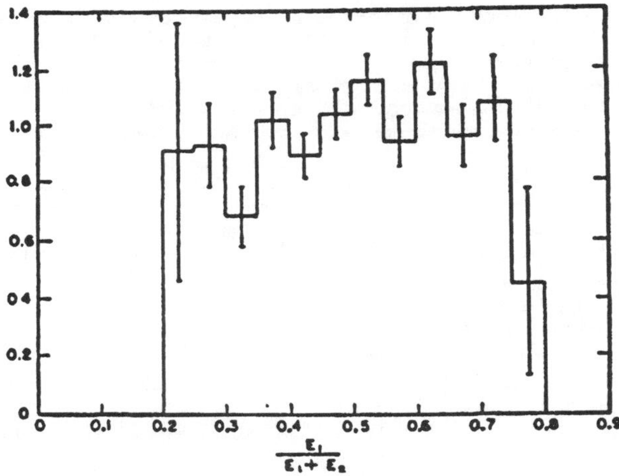

FIG. 13. Energy distribution of pair fragments. The frequency of occurrence of a given energy division is plotted against the energy fraction $E_1/(E_1+E_2)$; the measurements cover the range $0.2<E_1/(E_1+E_2)<0.8$. The ordinate does not represent the actual count but is divided by the gamma-ray energy value contributing to the particular division ratio interval. It is seen that the distribution is uniform within statistics.

single channel 30 percent, wide but the instrumentation was sufficient to show the presence or absence of gamma-ray yield within certain limits.

Let f = fraction of π^- absorbed which is finally absorbed by a hydrogen nucleus. The expected and observed intensities are tabulated in Table I.

The "expected count" was computed from evaluation of the geometry and the conversion yields. The calculations were checked by observing the gamma-rays originating directly from the target.[4] The expected count is estimated to be certainly correct to within a factor of 3. We can conclude, therefore, that $f<2\times10^{-3}$ for CH_2 and $f<6\times10^{-3}$ for LiH. It is thus clear that the absorption probability is not simply proportional[12] to Z, but that a special mechanism favors the absorption on a higher Z nucleus. This mechanism operates as follows. Once a π^- meson is captured by a hydrogen nucleus in an outer Bohr orbit, the atom loses its electron and hence its chemical binding. The lifetime of the resulting excited π^-—H system toward radiation into the K shell is long compared with the collision time with other nuclei, during the collision of the π^-—H system with a heavier nucleus, the probability is large that the π^- be captured by the heavier nucleus with the consequent production of a nuclear star, rather than a gamma-ray. Approximate estimates of the value of f have been made on the basis of this mechanism and are not inconsistent with the observations.

It has been suggested by Barkas[27] that π^0 emission might compete with nuclear stars induced by π^- capture. This is energetically possible if the mass of the capturing nucleus (Z, A) is less than that of the resulting $(Z-1, A)$ by less than the $\pi^- - \pi^0$ mass difference, δ. In particular if the corresponding β transition $(Z-1)\rightarrow Z$ is allowed, the gamma-rays from the disintegration of the resulting π^0 might be observable. This possibility is now being studied. A special case in this class is absorption in deuterium, studied theoretically by Marshak and his co-workers;[28] we shall discuss the process in the next section.

TABLE I. Tabulation of gamma-ray yield from the absorption of π^--mesons in hydrogeneous materials. (f = fraction of π^--mesons absorbed ultimately by a proton.)

Absorbing material	Gamma-ray energy (Mev)	Expected count (c/hr)	Observed count (c/hr)
CH_2	135	$5000 f$	0 ± 2
LiH	135	$1500 f$	0.5 ± 2
LiH	68	$1500 f$	-1 ± 4

VII. ABSORPTION IN DEUTERIUM

(a) Discussion of the Process

The absorption of π^- mesons in deuterium can lead to processes analogous to those previously outlined, in addition to pure heavy particle emission in the form of two neutrons. We consider thus the processes

$$\pi^- + D \rightarrow 2n, \tag{15}$$

$$\pi^- + D \rightarrow 2n + \gamma, \tag{16}$$

$$\pi^- + D \rightarrow 2n + \pi^0. \tag{17}$$

Process (15) will lead to monoenergetic neutrons. At first sight one might expect that process (16) would lead to a very broad gamma-ray spectrum. This is actually not so, since if the gamma-ray has low energy the conservation laws are satisfied only if the neutrons are emitted nearly exactly in opposite directions. Consequently large gamma-ray energies will be favored. One can show easily that, in the absence of angular correlation between the particles, the gamma-ray distribution is given by

$$N_\gamma dE = (E_0 - E - E^2/4M_nc^2)1/2^{\frac{1}{2}}E^2dE, \tag{18}$$

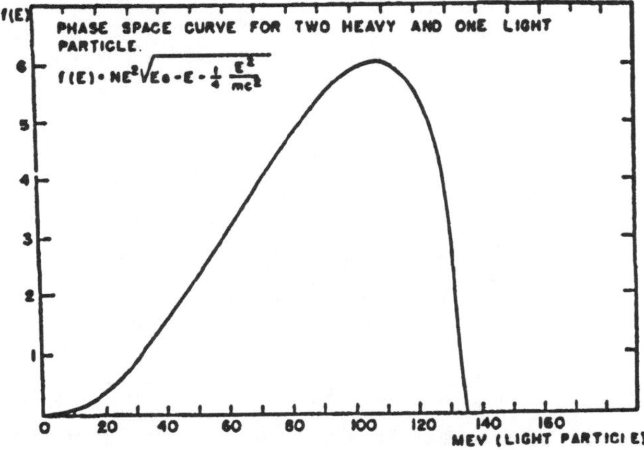

FIG. 14. Plot of the function giving the energy distribution of gamma-rays from the process $\pi^- + D \rightarrow 2n + \gamma$ for a constant matrix element.

[27] W. H. Barkas, private communication.

[28] S. Tamor, Phys. Rev. 79, 221 (1950); R. E. Marshak, private communication.

where $E_0 = M_{\pi^-} + M_D - 2M_n$. (See Fig. 14.) Actually, we shall see that the spectrum is still considerably narrower.

The spectrum of process (17) is expected again to lie between boundaries defined by the Doppler effect of the moving π^0. Because of the extra energy required in dissociating the deuteron, the peak is expected to be narrower. Using the foregoing value for the π^0 mass, the kinetic energy of the π^0 is approximately 1.9 ± 1.0 Mev. Process (17) is thus energetically permitted nominally although the π^0 mass accuracy is not sufficient to establish this fact beyond reasonable doubt.

(b) Experimental Results

Two experimental runs on deuterium were performed in the geometry, pressure, temperature, and spectrometer setting identical to the conditions under which the high energy spectrum of the absorption of π^- in H_2 was taken (Fig. 7). Accordingly, we feel safe in comparing the yields of the two processes by assuming that the same number of π^- mesons reach the K shell in the two cases for a given proton bombardment of the primary target. There might be two qualifications to this statement. Both with H_2 and D_2 we are dealing with the same number of electrons/cm^3 and thus comparable stopping power. The time required to reach an outer Bohr orbit is thus identical in the two cases. However, the time required for transition to lower states depends on collision processes;[12] the situation here is thus not quite identical in the two cases. Furthermore, possible capture from an orbit other than an S orbit depends, of course, on the process in question.

An experiment was carried out to determine whether an appreciable number of π^- mesons were lost by $\pi-\mu$ decay before reaching the K shell. If the time of the energy reduction process by collision were comparable to the $\pi-\mu$ decay time (contrary to the calculation by Wightman[12]), then the yield of gamma-ray from H_2 would fall off more rapidly with reduced density than the density itself. Table II shows the observed gamma-ray intensities at the usual operating pressure (2700 psi) and at reduced pressure.

Clearly, no significant nonlinear decrease is observed. Accordingly, we conclude that in agreement with Wightman,[12] no significant $\pi-\mu$ loss occurs; and hence, the intensity comparison between H_2 and D_2 γ-yields are valid. The evidence for process (15) (see Table III) thus rests on a good quantitative basis. A separate experiment now in progress[29] also tentatively confirms the existence of fast neutrons correlated to the presence of deuterium in the pressure vessel.

The spectrum corresponding to the spectrum of Fig. 7 is shown in Fig. 15. The first conclusion is that its spectrum does not conform to the momentum space function plotted in Fig. 14. The reason is that the two slow neutrons involved cannot be considered free but will interact to favor a small relative velocity between

[29] K. Crowe and H. F. York (private communication).

TABLE II. Tabulation of the relative gamma-ray yields from H_2 at 2700 psi and 1300 psi. Intensities are tabulated in terms of counts/minute. Total quadruple coincidences in the proportional counters, total number of recorded pairs, and the pairs corresponding to the high energy peak only are tabulated.

	Intensity in counts/minute 2700 psi	1300 psi	Ratio
Total number of quadruple coincidences	0.853 ± 0.033	0.523 ± 0.042	1.63 ± 0.14
Total number of gamma-rays recorded on multiple channel unit	0.302 ± 0.021	0.190 ± 0.027	1.60 ± 0.25
Gamma-rays recorded in the high energy peak only	0.213 ± 0.013	0.121 ± 0.017	1.74 ± 0.27
Density	0.046	0.028	1.65

the neutrons. This will result in a spectrum peaked toward high energy. The effect is analogous to the high energy peak observed in the π^+ spectrum formed in $p-p$ collisions,[30] there the $n-p$ interaction favors a high π^+ energy. The low internal energy of the deuteron will also favor low relative velocities between the two neutrons.

The second result is the apparent absence of the π^0 peak. A separate run was made with the spectrometer centered at 70 Mev in order to place more rigid limits on the π^0 intensity. The observed counts are given in Table III. Since the H_2 spectrum shown in Fig. 7 was obtained under comparable conditions, we can directly compare the gamma-yields. This comparison is given in Table III. The intensity of the fast neutron yield (process 15) is only inferred from the intensity balance with hydrogen and does not represent direct observation.

VIII. CONCLUSIONS

In a qualitative sense the results reported here confirm some of the already reasonably well-established facts concerning π-mesons. (1) The existence of the monochromatic high energy peak from hydrogen proves that the π^- meson is a boson. (2) A π^0 meson exists[6,10]

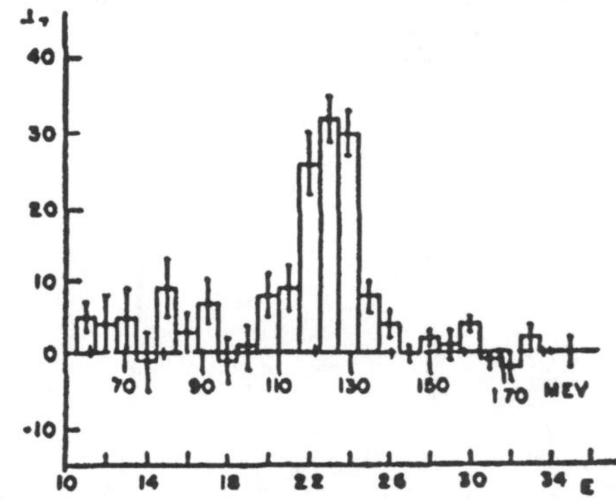

FIG. 15. Pair spectrum resulting from π^- capture in deuterium. The spectrometer center is set near 130 Mev.

[30] Cartwright, Richman, Whitehead, and Wilcox, Phys. Rev. 78, 823 (1950); V. Z. Peterson, Phys. Rev. 79, 407 (1950).

TABLE III. Summary of relative intensities of various processes obtained as a result of π^- capture in various materials. Tabulated values are given in counts per minute. In the case of single gamma-ray processes this represents the counting rate of the spectrometer; in the case of π^0 emission it represents $\frac{1}{2}$ times the spectrometer rate. The "two fast neutrons" count does not represent direct observation but only the intensity inferred by balancing counts between hydrogen and deuterium absorption.

Particle emitted \ Absorber	H	D
π^0	0.45 ± 0.09 c/m	-0.007 ± 0.020 c/m
Single γ	0.470 ± 0.046 c/m	0.275 ± 0.034 c/m
Two fast neutrons		0.65 ± 0.11

and it disintegrates into two photons; it thus must be a spin zero particle. (3) The electrostatic self-energy or other causes make the π^- heavier than the π^0 by about 11 electron masses. (4) As long as emission of the π^0 can be in an S state only, the π^0 and π^- must be particles of identical parity properties. Considering the large kinetic energy of the π^0 emitted in process (4), this conclusion is no longer rigorous.

A quantitative result which might permit interpretation at the present time is the branching ratio between π^0 and γ-emission and the ratio between $2n$ and γ-emission in deuteron capture. By elementary notions mesons are strongly coupled to nuclei while photons are weakly coupled; therefore a branching ratio close to unity seems paradoxical, since the π^0 phase space factor $p/M_{\pi^0}c$ is as large as $\frac{1}{2}$. Processes (2) and (4) are essentially the inverse of photo meson production and of "charge exchange" scattering of π-mesons on nucleons. By a detailed balancing argument the ratio of cross section for such processes would be of the order of unity. Actually, at high energies it appears as though the photo production cross sections for mesons are well below nuclear interaction cross sections of mesons; but consideration should be given to the fact that the energies involved in the capture experiment and the inverse processes mentioned are dissimilar.

A very analogous difficulty appears in the case of the deuterium results. Process (15) is essentially the inverse of meson production in like particle collisions[20] and process (16) is essentially the inverse of the photo-production of π^- mesons;[21] experimentally, the latter process has a very much smaller cross section than the production cross section in like particle collisions. Again a dependence of the matrix elements on π^- energy could remove the contradiction.

A definite result which can be deduced from the inferred presence of the fast neutron yield is the fact that the π^- is not a scalar particle. This is clear, since the process

$$\begin{array}{ccc} \pi^- & +D\rightarrow & 2n \\ S\text{-orbit} & {}^3S & {}^1S \text{ or } {}^3P \end{array} \qquad (19)$$

violates either parity or angular momentum conservation for a scalar π^-. Capture from a p-state might weaken this selection rule; however, calculations by Bruecker and Watson[22] indicate that the lifetime for radiation from a p-orbit is very short compared with the capture time, so that this effect can be neglected.

The absence of the π^0 peak in the case of π^- capture by deuterium is not surprising. If the π^- and π^0 are particles of equal parity, then in the process

$$\begin{array}{cccc} \pi^- & +D\rightarrow 2n+ & \pi^0 & \\ S\text{-orbit} & {}^3S \quad {}^3P & P\text{-wave} \end{array} \qquad (20)$$

both the π^0 and the two neutrons must be emitted in odd states of angular momentum as indicated, in order to obey conservations of parity and angular momentum and the exclusion principle. This effect[23] produces a greatly retarded yield at the small π^0 energy available.

Direct calculations of the branching ratio based on various combinations of meson character and coupling have been made by several authors.[21,22] The formal perturbation calculations show that the number of possibilities of meson character and coupling is greatly reduced by the results of this experiment.

It has been shown quantitatively by Brueckner, Serber, and Watson[21] that the variability of the matrix element, predicted by the comparison of branching ratio measurements reported here with the inverse processes, is subject to experimental check by measurements of the excitation function of meson production in p–p collisions.

As a further remark it might be mentioned that there exists here, of course, no positive proof that the π^0 mesons as observed here are identical with the π^0 observed as produced by nuclear collisions[4–8, 10] and photo-production,[19] but the inference appears to be justified.

The authors have benefited greatly by the active cooperation of many members of this Laboratory. In particular the help of Dr. Herbert F. York during the first phases of the experiment has been indispensable. We are also indebted to Robert Meuser and Hugh Smith for mechanical design, to Mr. Alex Stripeika for assistance in electronics, and to the cyclotron crew under Mr. James Vale for the bombardments and assistance in the adjustments. The authors have also gained valuable information by discussions with Doctors Marshak, Serber, Wick, Brueckner, and Watson. Mr. Robert Phillips has actively participated in the execution of the experiment and has been of invaluable assistance.

[21] Peterson, McMillan, and White, Science 110, 579 (1949).

[22] Brueckner, Serber, and Watson, Phys. Rev. 81, 575 (1951).

The Spin of the Pion via the Reaction $\pi^+ + d \rightleftharpoons p + p$

R. DURBIN, H. LOAR, AND J. STEINBERGER
*Columbia University, New York, New York**

(Received June 21, 1951)

It is possible to determine the spin of the pion by comparing the forward and backward rates of the reaction $\pi^+ + d \rightleftharpoons p + p$. The backward rate has been measured in Berkeley. We have measured the forward rate. Comparison of the two results shows the spin to be zero. In the light of other recent experimental results the meson is then pseudoscalar.

DURING the past few years there has accumulated an increasing amount of evidence that the pion is pseudoscalar. This consists chiefly of the following:

(a) The pion has integral spin. This follows from star formation in the capture of π^- mesons in photographic emulsions, as well as from angular momentum conservation in such reactions as $p+p \rightarrow \pi^+ + d$; $p + h\nu \rightarrow \pi^+ + n$.

(b) The neutral pion does not have spin one, since it decays into two γ-rays and such a transition is forbidden for systems with angular momentum $\hbar$.[1]

(c) Because of their similar masses and their similar nuclear production cross sections, as well as from the evidence on charge independence of nuclear forces, it is likely that neutral and charged mesons have the same transformation properties, so that charged pions also cannot have spin one.

(d) The pion is not scalar. This follows from the experiment on the capture of stopped negative pions in deuterium,[2] as well as the results on the production of pions, both charged and uncharged, by γ-rays.[3] Both experiments give best theoretical agreement in the pseudoscalar meson theory. It has therefore appeared quite probable that the pion is pseudoscalar; but the evidence, especially against a spin of two or greater, is poor.

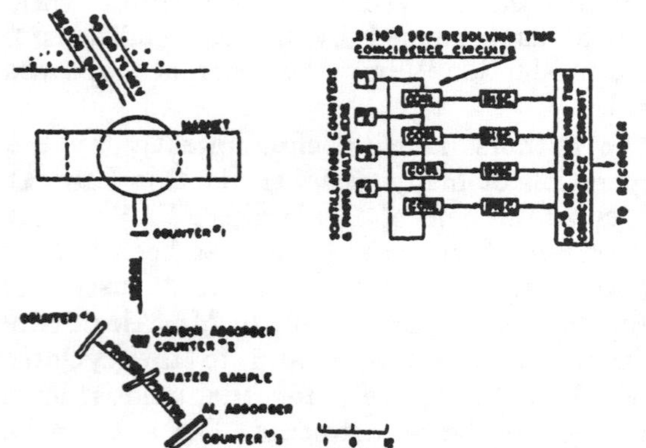

FIG. 1. Arrangement of the beam collimation, water sample, and detectors. Block diagram of circuits.

It has been pointed out by Cheston and Marshak[4] that the reaction $\pi^+ + d \rightleftharpoons p + p$ lends itself to a determination of the spin of the pion. The forward and backward reactions are related by a detailed balancing argument, if one assumes initially unpolarized particles, so that

$$\frac{d\sigma(\rightarrow)}{d\Omega} \bigg/ \frac{d\sigma(\leftarrow)}{d\Omega} = \frac{4}{3} \frac{p^2}{q^2(2s+1)},$$

where s is the spin of the meson, and p, q, are the momenta of the proton and meson in the center-of-mass system. The cross sections, of course, are also in the center-of-mass system. The argument is rigorous, independent of meson theory, and in this rests its chief contribution. The reaction $p+p \rightarrow \pi^+ + d$ has been measured by Cartwright, Richman, Whitehead, and Wilcox[5] for 340-Mev protons, corresponding to a meson energy of 21 Mev in the center-of-mass system. We present here results on the inverse reaction.

The experimental arrangement is shown in Fig. 1. The Nevis cyclotron delivers a beam of approximately 20 positive mesons per square centimeter per second outside the concrete shielding. These are produced when the 380-Mev protons strike an internal Be target. The mesons are magnetically analyzed in the fringing field of the cyclotron, and by a small magnet outside the shielding. The energy resolution is ± 4 Mev at 75 Mev and the composition of the beam is 90 percent π^+, and 10 percent μ^+ mesons.[6] The beam is defined by the

---

* This research has been supported by a joint program of the ONR and AEC.

[1] Steinberger, Panofsky, and Steller, Phys. Rev. 78, 802 (1950); C. N. Yang, Phys. Rev. 77, 242 (1950).

[2] Panofsky, Aamodt, and Hadley, Phys. Rev. 82, 97 (1951); Brueckner, Serber, and Watson, Phys. Rev. 81, 575 (1951).

[3] Bishop, Steinberger, and Cook, Phys. Rev. 80, 291 (1950).

[4] R. E. Marshak, Phys. Rev. 82, 313 (1951); W. B. Cheston, to be published.

[5] Cartwright, Richman, Whitehead, and Wilcox, Phys. Rev. 81, 652 (1951); V. Peterson, Phys. Rev. 79, 407 (1950); C. Richman and M. H. Whitehead, to be published. We are most indebted to Professors Richman and Wilcox, Mr. Cartwright, and Miss Whitehead for the privilege of quoting as yet unpublished results, and in particular, the fine meson spectrum shown in Fig. 5.

[6] The beam is analyzed in the following way: Heavy particles and electrons are detected in a measurement of the velocity distribution of the beam particles by means of a time of flight measurement. The mesons have velocities $\sim 0.7c$, the electrons which penetrate the counters have the velocity of light, and heavy particles have smaller velocities. The μ-mesons are measured at the end of their range by means of the delayed coincidences of their electron decay product. This is only possible in the π^- beam, since π^+ mesons at the end of their range are not captured but produce μ^+ mesons and interfere. We have assumed that, since the μ-mesons in the beam are the result of the decay in flight of the π-meson and since π^- and π^+ mesons have at least approximately the same lifetime [Lederman, Booth, Byfield, and Kessler,

crystal scintillation counters 1 and 2, and the energy reduced in the carbon absorber. The liquid scintillation counters 3 and 4 are set with respect to the water sample so that protons emitted 180° apart in the center-of-mass system of the meson and deuteron will be detected.[7] The water sample is 2.5 g/cm^2 thick. The aluminum absorber in front of counter No. 3 is thin enough to transmit the protons emitted in the meson absorption under study, but stops scattered protons or deuterons. The difference in counting rate using heavy water and light water targets is entirely due to the reaction $\pi^+ + d \rightarrow p + p$. It is in principle quite easy to check that this is so. In actuality, the counting rate is small and only a limited number of checks have been made, to wit: (1) Plateau. In the beginning of each experiment the counters are placed so that the meson beam penetrates all four, and the voltages (amplification) of the phototubes adjusted so that mesons are detected with full efficiency. But the proton pulses in counters 3 and 4 should be larger by a factor two or three than the

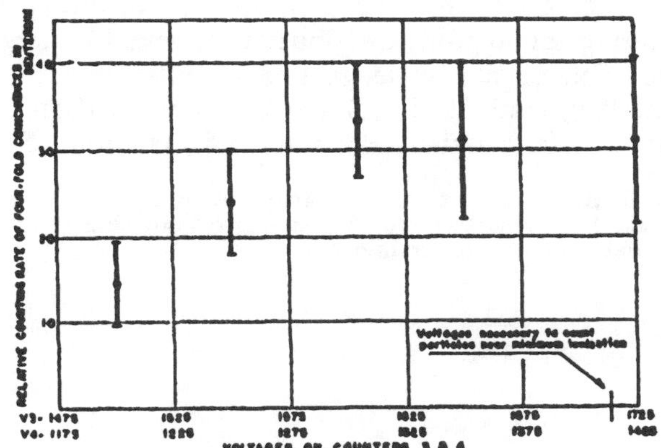

Fig. 2. Counting rate of fourfold events after subtraction, as a function of phototube voltage (amplification) in the proton detection counters 3 and 4.

meson pulses. Figure 2 shows that the pulses responsible for the subtracted fourfold coincidences are large. The $D_2O - H_2O$ difference is counted with full efficiency at voltages 100 volts (i.e., a factor of two in gain) below those necessary to count mesons. (2) When counters 3 and 4 were moved out of line to angles improper for the detection of protons with 180° c.m. angular correlation, no events were observed within statistical accuracy.

Figure 3 shows the energy dependence of the reaction at 45° in the c.m. system. It is sufficiently flat that errors due to energy spread of the meson beam and finite target thickness are small.

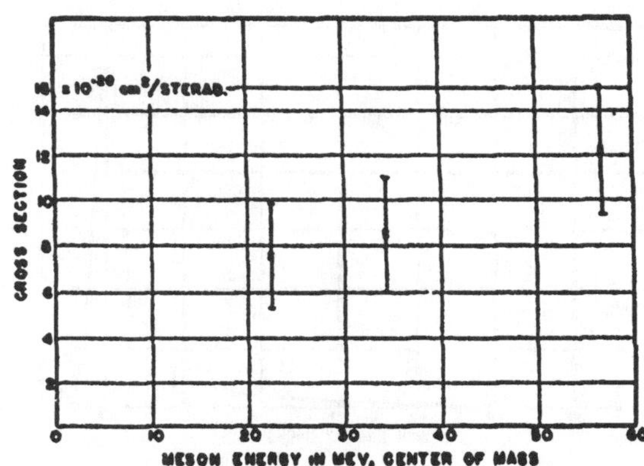

Fig. 3. Differential cross section for the emission of a proton at 45° (and one at 135°) to the meson beam in the c.m. system.

The results on the differential cross section at three angles are shown in Fig. 4. They are the combined results of three determinations, under conditions which varied somewhat. For instance, counters No. 3 and No. 4 were sometimes $4\frac{1}{2}$ in. and at other times 8 in. in diameter. The average meson energy in the target for the three runs was 28 Mev. The data are corrected for the geometrical efficiency of the detecting system which varied from 0.48 to 0.84, for the beam composition (90 percent π^+, 10 percent μ^+), for the nuclear absorption of the mesons and the protons in the target (7 percent), and for an inefficiency of 8 percent in the circuits due to blocking.[7]

The cross sections expected under the assumption of spins zero and one, on the basis of the results of Cartwright, Richman, Whitehead, and Wilcox, and of Peterson[5] are also shown. The Berkeley group has measured the energy distribution of mesons produced in the bombardment of hydrogen by 340 Mev protons.

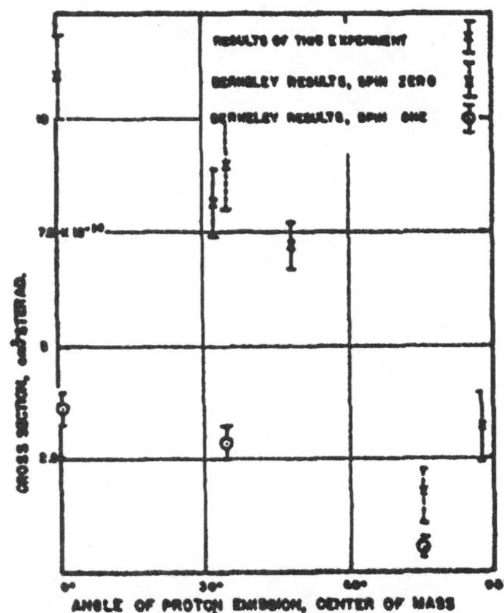

Fig. 4. Differential cross section of the reaction $\pi^+ + d \rightleftharpoons p + p$ at three angles of emission of the proton in the c.m. system. The average meson energy is 28 Mev in the c.m. system. The dotted points show the cross sections expected for spin one and spin zero pions on the basis of the Berkeley results (see reference 5).

Phys. Rev. 83, 686 (1951)], the μ-contamination is independent of charge.

[7] The center-of-mass transformation is small, approximately 0.05c. The correlation angles in the laboratory system do not differ from 180° by more than 15°. Angular distribution measurements are easier in this reaction than in the inverse, where the large center-of-mass transformation results in a large variation of experimental conditions with angle.

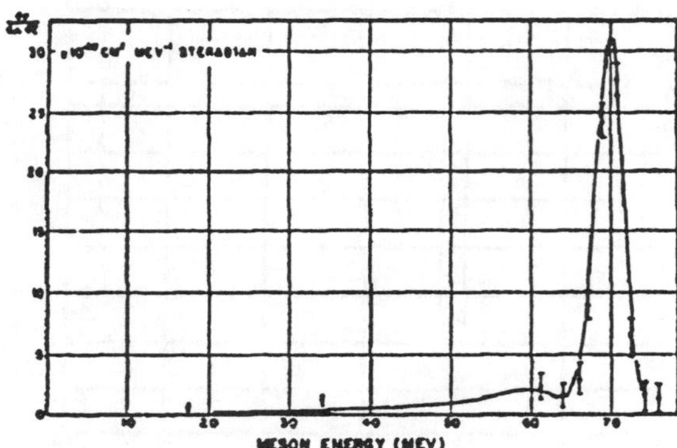

FIG. 5. The spectrum of mesons produced in the collision of 340 Mev protons in the forward direction. This experiment has been performed by Cartwright, Richman, Whitehead, and Wilcox (see reference 5).

(See Fig. 5.) This spectrum consists of a continuum due to the reaction $p+p\rightarrow\pi^{+}+n+p$, and a sharp peak at an energy which exceeds the theoretical limit of the continuum and is due to the reaction $p+p\rightarrow\pi^{+}+d$.[8] The cross section is obtained by integrating the energy spectrum under the peak.

Comparison of the two results shows that the π^{+} meson spin is zero, quite outside the possible limits of error in the two experiments. Combining this result with those of Panofsky[2] and those on the photomeson production,[3] the meson is very likely pseudoscalar.

It is necessary to point out that the same ratio of cross sections could be obtained also for non-zero spin mesons provided that pions were completely polarized both in the $p+$Be and $p+p$ meson production reactions. Such polarization is theoretically possible, but only in the longitudinal mode, that is, in the mode with zero component of angular momentum along the propagation axis. This seems a rather remote possibility.

A similar experiment has been performed by D. L. Clark, A. Roberts, and R. Wilson, who have reached the same conclusions. We are indebted to them for a pre-publication copy of their results.

The experiment is being continued. The reaction is interesting also in other connections. As Bethe[9] has pointed out, the angular distribution is quite perplexing. Furthermore, the energy dependence will shed some light on the momentum dependence of the meson nucleon interaction. We are therefore in the process of measuring the angular distribution at several energies.

We wish to acknowledge our indebtedness to the engineering staff of the Nevis Cyclotron Laboratory, especially Mr. Harrison Edwards and Mr. Julius Spiro.

[8] The actual production of deuterons in coincidence with mesons has been observed by Crawford, Crowe, and Stevenson, Phys. Rev. 82, 97 (1951).

[9] H. A. Bethe, letter to R. E. Marshak with copies to C. Richman and J. Steinberger. We wish to express our thanks to Professor Bethe for this communication.

Total Cross Sections of Positive Pions in Hydrogen*

H. L. Anderson, E. Fermi, E. A. Long,† and D. E. Nagle
Institute for Nuclear Studies, University of Chicago, Chicago, Illinois
(Received January 21, 1952)

In a previous letter,[1] measurements of the total cross sections of negative pions in hydrogen were reported. In the present letter, we report on similar experiments with positive pions.

The experimental method and the equipment used in this measurement was essentially the same as that used in the case of negative pions. The main difference was in the intensity, which for the positives was much less than for the negatives, the more so the higher the energy. This is due to the fact that the positive pions which escape out of the fringing field of the cyclotron magnet are those which are emitted in the backward direction with respect to the proton beam, whereas the negative pions are those emitted in the forward direction. The difficulty of the low intensity was in part compensated by the fact that the cross section for positive pions turned out to be appreciably larger than for negative pions. The results obtained thus far are summarized in Table I.

In Fig. 1 the total cross sections of positive and negative pions are collected. It is quite apparent that the cross section of the positive particles is much larger than that of the negative particles, at least in the energy range from 80 to 150 Mev.

In this letter and in the two preceding ones,[1,2] the three processes: (1) scattering of positive pions, (2) scattering of negative pions with exchange of charge, and (3) scattering of negative pions without exchange of charge have been investigated. It appears that over a rather wide range of energies, from about 80 to 150 Mev, the cross section for process (1) is the largest, for process (2) is intermediate, and for process (3) is the smallest. Furthermore, the cross sections of both positive and negative pions increase rather rapidly with the energy. Whether the cross sections level off at a high value or go through a maximum, as might be expected if there should be a resonance, is impossible to determine from our present experimental evidence.

Brueckner[3] has recently pointed out that the existence of a broad resonance level with spin 3/2 and isotopic spin 3/2 would give an approximate understanding of the ratios of the cross sections for the three processes (1), (2), and (3). We might point out in this connection that the experimental results obtained to date are also compatible with the more general assumption that in the energy interval in question the dominant interaction responsible for the scattering is through one or more intermediate states of isotopic spin 3/2, regardless of the spin. On this assumption, one finds that the ratio of the cross sections for the three

Table I. Total cross sections of positive pions in hydrogen.

Energy (Mev)	Cross section (10^{-27} cm²)
56±8	20±10
82±7	50±13
118±6	91± 6
136±6	152±14

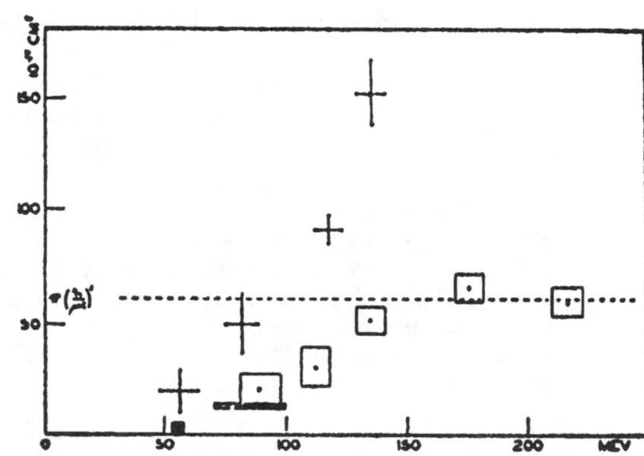

Fig. 1. Total cross sections of negative pions in hydrogen (sides of the rectangle represent the error) and positive pions in hydrogen (arms of the cross represent the error). The cross-hatched rectangle is the Columbia result. The black square is the Brookhaven result and does not include the charge exchange contribution.

processes should be (9:2:1), a set of values which is compatible with the experimental observations. It is more difficult, at present, to say anything specific as to the nature of the intermediate state or states. If there were one state of spin 3/2, the angular distribution for all three processes should be of the type $1+3\cos^2\theta$. If the dominant effect were due to a state of spin 1/2, the angular distribution should be isotropic. If states of higher spin or a mixture of several states were involved, more complicated angular distributions would be expected. We intend to explore further the angular distribution in an attempt to decide among the various possibilities.

Besides the angular distribution, another important factor is the energy dependence. Here the theoretical expectation is that, if there is only one dominant intermediate state of spin 3/2 and isotopic spin 3/2, the total cross section of negative pions should at all points be less than $(8/3)\pi\lambda^2$. Apparently, the experimental cross section above 150 Mev is larger than this limit, which indicates that other states contribute appreciably at these energies. Naturally, if a single state were dominant, one could expect that the cross sections would go through a maximum at an energy not far from the energy of the state involved. Unfortunately, we have not been able to push our measurements to sufficiently high energies to check on this point.

Also very interesting is the behavior of the cross sections at low energies. Here the energy dependence should be approximately proportional to the 4th power of the velocity if only states of spin 1/2 and 3/2 and even parity are involved and if the pion is pseudoscalar. The experimental observations in this and other laboratories seem to be compatible with this assumption, but the cross section at low energy is so small that a precise measurement becomes difficult.

* Research sponsored by the ONR and AEC.
† Institute for the Study of Metals, University of Chicago.
[1] Anderson, Fermi, Long, Martin, and Nagle, Phys. Rev., this issue.
[2] Fermi, Anderson, Lundby, Nagle, and Yodh, preceding Letter, this issue, Phys. Rev.
[3] K. A. Brueckner (private communication).

Production of Heavy Unstable Particles by Negative Pions*

W. B. Fowler, R. P. Shutt, A. M. Thorndike, and W. L. Whittemore
Brookhaven National Laboratory, Upton, New York

(Received November 10, 1953)

In addition to two previously discussed cloud-chamber examples of V-particle production by 1.5-Bev π^- mesons from the Cosmotron, four further examples are discussed here. In two of the new examples a $\Lambda^0(V_1^0)$ and a $\vartheta^0(V_2^0)$ are seen to decay in a geometry indicating that they were produced together in a $\pi^- - p$ collision. A third example is best interpreted as production of a $\Lambda^-(V_1^-)$ together with a $K^+(V_2^+)$ by a π^- colliding with a proton. A fourth example shows a probable Λ^- decaying into a π^- and a neutron with a Q value of about 130 Mev. A cross section of $\sim$1 millibarn for V-particle production is inferred from the number of $\pi^- - p$ collisions observed.

In a previous letter[1] we have reported two examples of Λ^0 particles[2] produced in hydrogen by negative pions (π^-) of 1.5-Bev kinetic energy. For both examples it was shown that if there were only two resulting particles the second was a K^0 particle of a mass of about 650 Mev. Since, however, the K^0 was not seen to decay it was also possible to balance energies and momenta by assuming two lighter neutral particles (including π^0) instead of one K^0 in addition to the Λ^0.

Observations on heavy unstable particles in cosmic radiation indicate lifetimes as long as 10^{-10} to 10^{-9} second with production cross sections at least 10^{-2} times those for π mesons. These facts can be reconciled theoretically[3] if it is assumed that the particles must be produced doubly (two at a time). Different particles may be produced simultaneously, in addition to pairs of identical ones. It appears that no evidence for double production of such particles, exceeding purely statistical coincidence, has been reported in cosmic radiation although a few examples might perhaps be interpreted in this manner.[4] The two examples of production in hydrogen reported by this group indicate multiple production. This interpretation is uncertain, however, since the K^0 particles are not observed to decay and the computed masses are higher than the value of about 500 Mev most commonly found for K particles.

We now wish to describe three additional cases where double production of heavy unstable particles by 1.5-Bev π^- is indicated with more certainty then in the previous two examples. We shall refer to the previous examples as cases A and B. Case C shows a Λ^0 together with a ϑ^0 probably produced in a heavy nucleus. Case D shows the same combination produced in hydrogen. Case E shows what may well be a Λ^- together with a K^+ with mass of about 500 Mev produced in hydrogen. A few cases of Λ^+ have been observed in cosmic radiation[5,6] but evidence for Λ^- is less certain. In a fourth case, called F, we shall describe an additional example best interpreted as a Λ^-.

Details of the experimental method will be described in a later article. A diffusion cloud chamber filled with 20 atmospheres of hydrogen was exposed to a 1.5-Bev π^- beam produced in a carbon target by the 2.2-Bev circulating proton beam in the Cosmotron. The pions were selected and collimated by the field of the Cosmotron magnet and a channel in the concrete shielding around the machine. The beam thus obtained is quite monoenergetic with a spread probably less than ± 0.1 Bev. The beam was finally deflected by a magnet into a concrete house containing the cloud chamber mounted between the pole faces of a magnet providing an average field of 10 500 gauss. This magnetic field has been calibrated to 1 percent accuracy and its vertical and horizontal components have been mapped throughout the chamber.

The examples to be described were obtained out of a total of 26 000 photographs obtained at a rate of one every 7 to 8 seconds. In addition many examples of V events were found for which no associated V event or nuclear interaction in the gas was observed. Finally, we have observed about 170 pion interactions in hydrogen, most of which lead to single and multiple production of pions. These will be described in a later article.

CASE C: Λ^0 WITH ϑ^0

Figure 1 (case C) shows a photograph of two V events, of which (a) is considered to be a Λ^0, and (b) a ϑ^0. Data which lead to this identification are given in Table I. Row 2 gives the momenta as measured on the 35-mm film by means of a Cooke microscope. The measure-

* Work performed under the auspices of the U. S. Atomic Energy Commission.

[1] Fowler, Shutt, Thorndike, and Whittemore, Phys. Rev. 91, 1287 (1953).

[2] We are using here the nomenclature suggested for V events at the International Congress on Cosmic Radiation, Bagnères-de-Bigorre, France. Accordingly $\Lambda^{0,+,-} \rightarrow \text{nucleon} + \text{pion} + Q_\Lambda$; $K^{0,+,-}$ is any particle whose mass falls between those of pion and proton; for example, $\vartheta^0(\rightarrow \pi^+ + \pi^- + Q_\vartheta)$ is a K^0.

[3] A. Pais, Phys. Rev. 86, 663 (1952). References to previous work are cited in this article. A. Pais, Proceedings of the Lorentz-Kamerlingh Onnes Conference, Physica (to be published); M. Goldhaber, Phys. Rev. 92, 1279 (1953); M. Gell-Mann, Phys. Rev. 92, 833 (1953).

[4] Leighton, Wanlass, and Anderson, Phys. Rev. 89, 148 (1953); Fretter, May, and Nakada, Phys. Rev. 89, 168 (1953); G. D. Rochester and C. C. Butler, Repts. Progr. in Phys. 16, 364 (1953), give a comprehensive review of cosmic-ray results.

[5] York, Leighton, and Bjornerud, Phys. Rev. 90, 167 (1953).

[6] International Congress on Cosmic Radiation, Bagnères-de-Bigorre, France (unpublished).

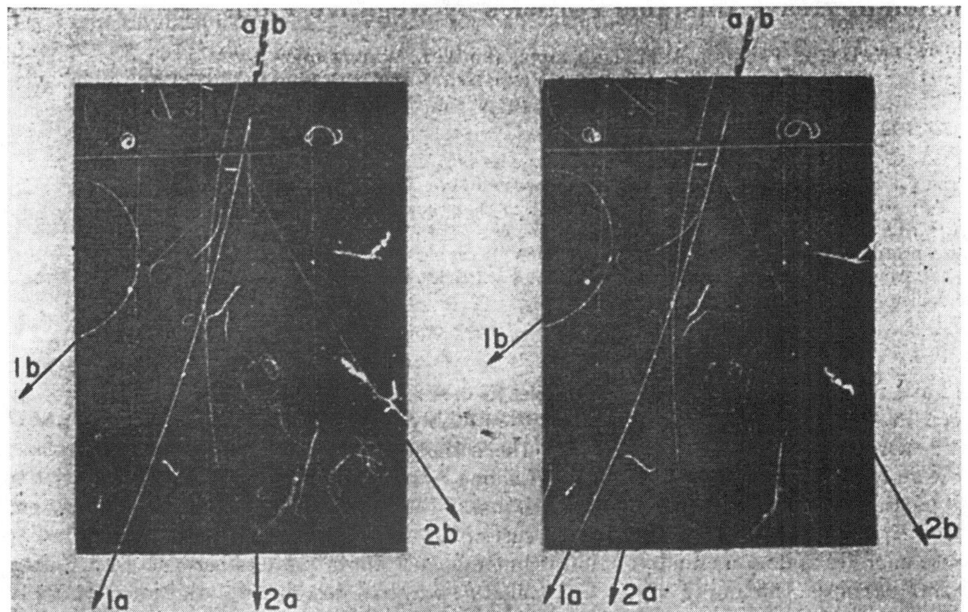

Fig. 1. Case C. Diffusion cloud-chamber photograph of two neutral V particles (a) and (b), whose lines of flight are almost colinear. (a) is believed to be a Λ^0 decaying into a proton (1a) and a negative π meson (2a). Tracks 1a and 2a practically coincide in the right view. (b) is probably a ϑ^0 decaying into π^+ (1b) and π^- (2b).

ments have been corrected for dip angle and space variations of magnetic field and magnification. The mass limits given in row 4 were deduced from measured momenta and estimated ionization densities. Row 5 shows what particle was assumed for further calculation. The angles between tracks in row 6 were obtained from a 3-dimensional reprojecting system as well as by geometrical reconstruction and calculation. The Q value for (a) is, within the given error, consistent with the usually accepted value of 37 Mev for a Λ^0. The Q value for (b) is rather large compared to the best value of 214 Mev given in the literature[7] for a ϑ^0.

From the measured momenta and angles in space for each track one can infer that the lines of flight of the Λ^0 and ϑ^0 practically coincide. From the total number of Λ^0 and ϑ^0 (originating in the walls) found in the 26 000 pictures one concludes that the probability for random association of the two particles in such a geometry is $\sim 10^{-4}$. We therefore should be justified in assuming that the Λ^0 and ϑ^0 were produced in one act. Since no incident particle is visible in the chamber and since the Λ^0 and ϑ^0 travel at an angle of about 24° upwards from the bottom of the chamber, we assume that they were produced in or below the bottom glass plate, though probably not much further down since the pion beam was fairly well collimated. The origin must, of course, lie on the line of intersection of the decay planes of the Λ^0 and ϑ^0. Both particles travel in directions between 0.5° and 1.5° to this line for a range of reasonable choices for the position of the origin.

The following procedure can now be used to determine the Q values more accurately than before. Since we know the directions of the lines of flight of Λ^0 and ϑ^0 with fair certainty we can deduce the momentum of one decay product in each V event from the momentum of the other. As row 2, Table I, shows, the momenta of 1a and 2b are known much better than those of 1b and 2a. We therefore have deduced the momenta of 1b and 2a. For the momenta of 1a and 2b we have chosen their lowest values within the given experimental errors. This choice will result in somewhat lower Q values than calculated above. The new momenta are given in row 8 and the recalculated Q values in row 9. The indicated remaining errors are now mainly due to the mentioned uncertainty in the directions of flight of Λ^0 and ϑ^0. The Q value for the ϑ^0 is still not in good agreement with the usual value of 214 Mev.

From the momenta in row 8 and the given angles one finds the momenta of Λ^0 and ϑ^0 given in row 10. Assuming that the particles were produced in a collision between a π^- and a nucleon and that no additional particle was involved ($\pi^- + p \rightarrow \Lambda^0 + \vartheta^0$) one calculates from the resultant of the momenta of Λ^0 and ϑ^0 that the total energy of incident pion and nucleon must have been 2050±25 Mev before the collision. On the other hand, with the mases of 1119±10 and 546±20 Mev for Λ^0 and ϑ^0, respectively, one finds for the sum of their energies a value of 2068±30 Mev. This value must be equal to the initial energy just calculated, if no other particles are involved in the collision, and one indeed finds agreement. Therefore the Λ^0 and ϑ^0 could have been produced in a $\pi^- - p$ collision, the π^- previously having been scattered upward by 24°, with the required energy. Such repeated interactions could take place in a heavy nucleus. For this interpretation the almost colinear flight of the Λ^0 and ϑ^0 would mean that in the

[7] Thompson, Buskirk, Etter, Karzmark, and Rediker, Phys. Rev. 90, 1122 (1953).

TABLE I. Measurements and results for case *C*.

	Track	Event a 1a		Event a 2a	Event b 1b		Event b 2b
1	Sign of charge	+		−	+		−
2	Measured momenta (Mev/*c*)	272±8		205±40	451±70		391±15
3	Estimated ionization density	>5×min		<1.5×min	<1.5×min		<1.5×min
4	Mass limit (Mev)	>780		<170	<380		<330
5	Assumed particle	proton		pion	pion		pion
6	Angle between tracks (degrees)		31.1±1			70±1	
7	*Q* values calculated from above data (Mev)		54±20			271±30	
8	Momenta for (2a) and (1b) calculated from direction of line of flight, assuming lowest momenta for (1a) and (2b) (Mev/*c*)	264		189±20	441±15		376
9	*Q* values recalculated (Mev)		49±10			266±20	
10	Momentum of unstable particle (Mev/*c*)		437			673	

center-of-mass system (c.m.s.) of π^- and nucleon the Λ^0 went almost straight backward. The same will be found for the next example to be discussed. From the observed momenta and angles a π^- of the beam colliding with a nucleon could *not* have produced the Λ^0 and ϑ^0 traveling in the observed direction, with simultaneous production of some other particle going downward to balance transverse momenta.

CASE *D*: Λ^0 AND θ^0 PRODUCED IN $\pi^- - p$ COLLISION

Figure 2 (case *D*) shows a photograph of a π^- track (marked π^-) disappearing abruptly in the hydrogen, and two nearby *V* events of which (a) is considered to be a Λ^0 and (b) a ϑ^0. Data which lead to this identification are given in Table II. The measured momentum of the incident particle (row 2) agrees well with the momentum of 1630 Mev/*c* for a 1.5-Bev π^-. The other momenta are not well determined, mostly because the tracks are of short length. The longest track (2b), furthermore, seems to show a slight deflection, perhaps due to scattering or to $\pi-\mu$ decay, which makes its momentum also unreliable. The given mass limits (row 4), however, seem to justify the assumed choices for the individual particles (row 5). In particular, since in a collision in hydrogen only one proton was present and since this proton is probably found in 1a, 2b could hardly be another proton in spite of the rather high upper mass limit.

The decay planes of both particles contain the end point of the incident track. The directions of flight of Λ^0 and ϑ^0 and incident π^- appear coplanar, though this is not accurately determined because of the small angle between π^- and ϑ^0 (b). Therefore no additional neutral particle has to be assumed, although calculation shows that kinematically this would be possible. One is therefore justified in assuming the same reaction as in case *C* ($\pi^- + p \rightarrow \Lambda^0 + \vartheta^0$). All pertinent angles are given in rows 6 and 7. From these angles and the momentum of the incident π^- one can calculate the momenta of the decay products directly. The results (row 9) agree with the measured values (row 2) within the errors. With the momenta from row 9, *Q* values have been calculated (row 10). Two experimental errors are given separately with each *Q* value. The first was determined from the probable variation of *Q* with the possible variations of all of the angles (rows 6 and 7) used for the computation. The second error was found from the variation of *Q* with a possible uncertainty of ±100 Mev/*c* for the momentum of the incident π^-. One sees that within the errors both *Q*'s may agree with the usual values of 37 and 214 Mev, respectively, although the *Q* for the ϑ^0 is again rather high and agrees with the value found for case *C*. It should be pointed out that the error for the *Q* for the Λ^0 depends most strongly on the measurement

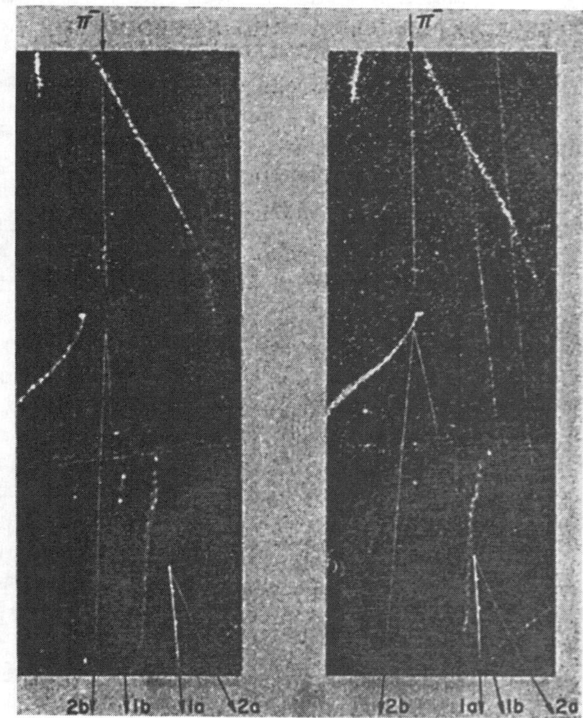

FIG. 2. Case D. Photograph of a 1.5-Bev π^- producing two neutral *V* particles in a collision with a proton. Tracks 1a and 2a, believed to be proton and π^-, respectively, are the decay products of a Λ^0. A ϑ^0 is probably seen to decay into π^+ (1b) and π^- (2b). Because of the rather "foggy" quality of this picture tracks 1b, 2a, and 2b have been retouched for better reproduction.

TABLE II. Data for case D.

		Incident track	Event a 1a	Event a 2a	Event b 1b	Event b 2b
1	Sign of charge	−	+	−	+	−
2	Measured momenta (Mev/c)	1620±160	210^{+210}_{-70}	140^{+300}_{-60}	210^{+210}_{-70}	840±300
3	Estimated ionization density	<1.5×min	>5×min	<1.5×min	<1.5×min	<1.5×min
4	Mass limit (Mev)		>400	<370	<350	<870
5	Assumed particle	pion	proton	pion	pion	pion
6	Angle between incident π^- and direction of flight (degrees)		11.1±1		2.3±0.5	
7	Angle between direction of flight and decay products (degrees)		8.8±1	11.3±1.5	29.3±2	11.0±2
8	Momentum of (a) and (b) calculated from incident momentum and angles (Mev/c)	1630	282		1357	
9	Momenta of decay products from momenta of (a) and (b) and angles (Mev/c)		160	125	400	1027
10	Q values calculated from rows 7 and 9		27±11±3		258±35±21	

of the small angle (2.3°) between incident π^- and ϑ^0 (b). Modifying this angle to make Q_{Λ^0} equal to 37 Mev would reduce Q_{ϑ^0}, which depends on this angle less strongly, to a value of 244 Mev.

From the momenta (row 8) of the Λ^0 and ϑ^0 and their masses of 1097±12 and 538±40 Mev, respectively, one calculates that the kinetic energy of the incident π^- must have been 1.52±0.04 Bev which is quite consistent with the 1.5-Bev beam energy.

CASE E: POSSIBLE Λ^- AND K^+ PRODUCED IN $\pi^- - p$ COLLISION

Figure 3 (case E) shows a photograph of what on first sight appears to be a π^- scattered forward by a proton with a subsequent decay of the π^-. Closer inspection shows that the momentum of the decay product

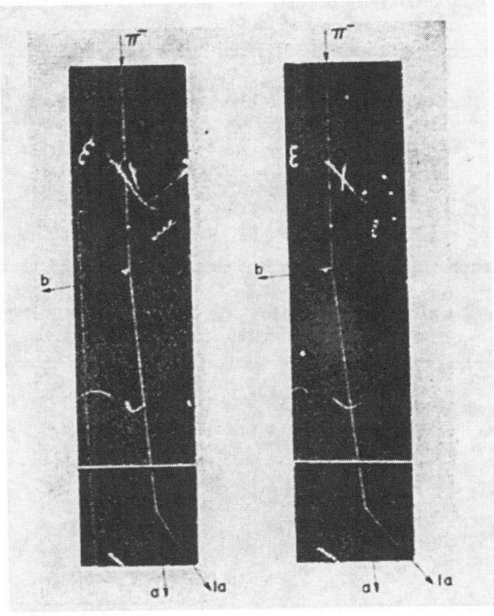

FIG. 3. Case E. Photograph of a $\pi^- - p$ collision event possibly resulting in a Λ^- (a) with a π^- (1a) as a decay product and a K^+ whose decay is not seen.

(1a) and its angle with respect to the track of the decaying particle (a) are much too large for a $\pi-\mu$ or $\mu-\beta$ decay. Therefore (a) must have been a heavy unstable particle.

As far as can be ascertained from the "scattering" event, particle (a) might have been produced at the Cosmotron target or at the cloud chamber wall, and scattered by a proton in the cloud chamber. Certainly the rate of production of heavy mesons would have to be large (~10 percent of that for pions) and their decay lifetime long (~10^{-8} sec) for the beam to contain an appreciable contamination of heavy mesons. No beam particles have shown a decay resembling that of (a). If (a) were produced in the target, it would be quite remarkable that it lives until it reaches the cloud chamber, is scattered, and then decays within the chamber. Such an origin is possible but seems unlikely. If (a) were a particle produced in the wall and scattered in the chamber, it would be remarkable that the incident track has both direction and momentum characteristic of beam tracks. For these reasons we assume that the incident particle is a beam π^- producing a charged unstable particle in a collision with a proton.

Data are given in Table III. Since the tracks are short and the momenta are high, the latter are not well determined. We therefore assume that the incident particle had the beam energy of 1.5 Bev, and compute the momenta of (a) and (b) by assuming that no additional neutral particle was produced. This assumption is very probably justified because (a), (b), and incident track are coplanar. One finds the momentum values given in row 8. If the mass of (a) is given, that of (b) is determined so that their total energies equal that of the incident π^-. In Table IV a number of such consistent values are given. The errors on the mass values of (b) include the uncertainties given for the angle measurements as well as an improbably large uncertainty of ±300 Mev for the energy of the incident π^-. To obtain a mass of 930 Mev (proton) for (b) does not seem to be possible unless (a) is a pion, without making

TABLE III. Data for case *E*.

		Incident track	Track a	Track 1a	Track b
1	Sign of charge	−	−	−	+
2	Measured momenta (Mev/*c*)	1800±600	1400±400	>120[a]	−
3	Estimated ionization density	<1.5×min	<1.5×min	<1.5×min	3 to 6×min
4	Mass limit (Mev)		<1500		
5	Assumed particle	pion		pion	
6	Angle between incident track and collision products (degrees)		8±1		79±3
7	Angle between (a) and its decay product (degrees)			36±1	
8	Momenta from incident momentum and angles (Mev/*c*)	1630	1604		227

[a] Track 1a is slightly distorted. Therefore only a lower limit can be given.

quite unreasonable assumptions for the uncertainties of the measured angles and of the incident momentum.[8] In the cosmic radiation Λ^+ particles have been found[6] with Q values of about 130 Mev, leading to a mass of 1200 Mev. Row 5 of Table IV shows that such a mass for (a) would lead to a mass for (b) which is quite consistent with that usually found for K particles. Case E can therefore be interpreted as the charged counterpart of case C and case D in which Λ^0 and K^0 were produced. For this interpretation the present photograph is an example of the reaction $\pi^-+p\rightarrow\Lambda^-+K^+$.

A calculation has been performed to investigate whether a neutrino (ν) or π^0 could have been produced in addition, which might then change the above conclusions. One finds that only the combinations $[\Lambda^-, K^+, \pi^0]$ or $[\Lambda^-, K^+, \nu]$ are possible (though unlikely because of the observed coplanarity) while the combinations $[\Lambda^+, K^-, \pi^0]$, $[\Lambda^+, K^-, \nu]$, $[p, K^-, \pi^0]$, or $[p, K^-, \nu]$ are kinematically not possible.

From the mass of (b) of 520 Mev and its momentum of 227 Mev/c one finds that (b) should show an ionization density of $\sim$4×min which agrees with the estimated ionization density given in row 3.

Assuming a decay of $\Lambda^-\rightarrow n+\pi^-$, the calculated momentum for 1a of 1604 Mev/c and measured momentum for (b) of $\geqslant$120 Mev/c lead to a Q value $\geqslant$50 Mev. Track 1a may be somewhat distorted so that these figures probably represent lower limits only. If much of the apparent curvature of 1a is due to distortion, the Q value may be as high as 130 Mev, corresponding to a momentum of 440 Mev/c for 1a, as has been assumed in the previous discussion.

[8] For the present discussion it is of particular importance to exclude the possibility that the mass of (b), M_b, could be that of a proton (930 Mev) for the assumption that (a) is a K^- of mass 500 Mev. One finds that $\partial M_b/\partial p_0=8.1$ Mev/(100 Mev/c), where p_0 is the momentum of the incident π^- (so far assumed to be the beam momentum of 1630 Mev/c). Furthermore $\partial M_b/\partial\alpha_a=-11.3$ Mev/degree and $\partial M_b/\partial\alpha_b=-3.7$ Mev/degree, where α_a and α_b are the angles recorded in Table III, row 6, for (a) and (b), respectively. It is very hard to conceive how π^- of momenta >1700 Bev/c could be contained in the beam. With this upper limit and the uncertainties for α_a and α_b given in Table IV one sees that $M_b\leq890$ Mev if all uncertainties are assumed to act together in a direction to increase M_b. Only by decreasing α_a by 3° and α_b by 9°, for example, could one obtain a value for M_b of 930 Mev. The values for M_a and M_b noted in rows 2 and 3 of Table IV are then not possible without destroying a nucleon.

CASE F: EXAMPLE OF A Λ^-

The photograph shown in Fig. 4 (case F) shows another V event which may be interpreted as a Λ^-. The picture shows a negative particle (a) of momentum 1190±170 Mev/c and of estimated ionization density $\leqslant$1.5×min, apparently produced in the wall of the chamber. The angle between (a) and the beam direction is 8°. Particle (a) decays into a negative particle (1a) of momentum 83±3 Mev/c and estimated ionization density 2 to 3×minimum.[9] The mass of the decay product thus lies between 110 and 150 Mev, identifying it as a π^-. The angle between (a) and 1a is 76°. If one additional neutral decay product is assumed one calculates Q values and mass values for (a) as given in Table V. One sees that only the assumption of a neutron (row 1) leads to Q and mass values compatible with those found in cosmic radiation. Particle (a) can also not be identified as a $\tau^-\rightarrow2\pi^0+\pi^-+70$ Mev because 1a alone would have an energy of 230 Mev in the rest system of the τ^-. The assumption in row 4 ($K^-\rightarrow\mu^-+2\nu$) is unlikely because the decay product is most probably

TABLE IV. Consistent masses for particle (a) and particle (b) of case *E*.

	Particle (a)	Mass of (a) (Mev)	Particle (b)	Mass of (b) (Mev)
1	π^-	140	proton	934±6
2	K^-	500	?	860±20
3	?	760	?	760±30
4	Λ^-(Q=37 Mev)	1107	K^+	570±50
5	Λ^-(Q=130 Mev)	1200	K^+	520±50

TABLE V. Q values and masses for case F, assuming different masses for the neutral decay product.

	Charged decay product	Neutral decay product	Mass of neutral decay prod. (Mev)	Q for (a) (Mev)	Mass of (a) (Mev)
1	π^-	n	930	130^{+25}_{-15}	1200
2	π^-	π^0	140	430 ±70	710
3	π^-	K^0	500	230±35	870
4	μ^-	2ν	0	>520±80	>620

[9] For the assumption of a Λ^0 traveling backwards the momentum of (a) is much too high.

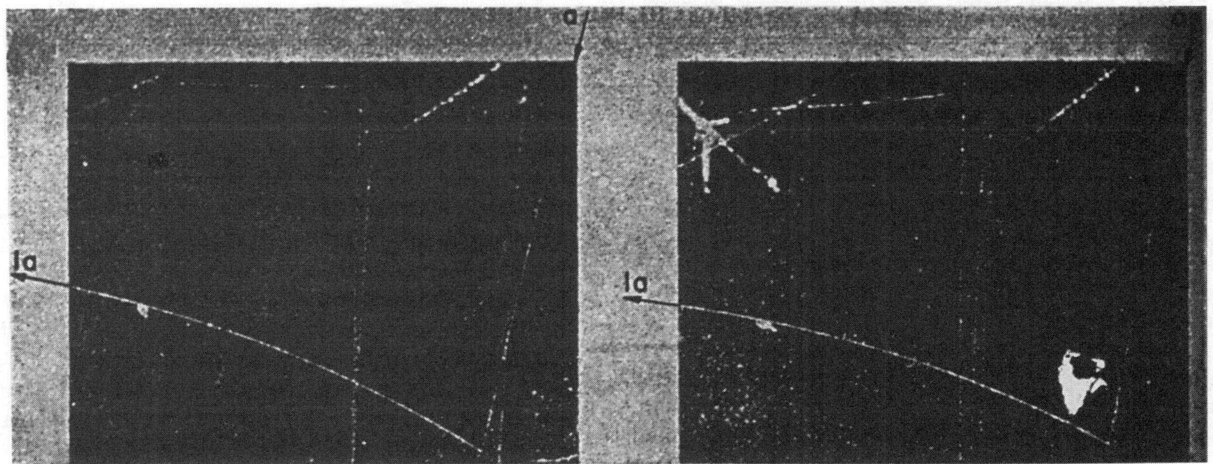

FIG. 4. Case F. Photograph of a negative unstable particle (a) best interpreted as a Λ^-. The decay product (1a) is identified as a π^- from momentum and ionization density.

identified as a π^-. We therefore are left with the conclusion that the present photograph shows the decay of a $\Lambda^- \to n+\pi^-+Q$, where $Q=130_{-15}^{+25}$ Mev. The errors on the latter are due to the uncertainty of the measured momentum of (a).

It has been pointed out[10] that if the hypothesis of charge independence applies to $\Lambda^{+,-,0}$, then the Λ^- must have isotopic spin z component $(-3/2)$ with isotopic spin 3/2, in which case a doubly-charged Λ^{++} should also exist. No Λ^{++} decays have been reported to date, but the number of observed Λ^- and Λ^+ is so small that no significant discrepancy exists. A track which might be interpreted as a Λ^{++} has been reported by Ascoli.[11]

LIFETIMES, CROSS SECTION, AND ANGULAR RELATIONSHIPS

The role played by chance in apparent lifetimes, cross sections, and angular relationships deduced from a small number of cases is very large. Nevertheless the few available observations will be compiled in the following paragraphs.

Table VI shows the lifetimes. All values are consistent with mean lifetimes of 10^{-10} to 3×10^{-10} sec cited in the literature.[4–6]

TABLE VI. Observed lifetimes of all particles in units of 10^{-10} sec.

Particle	A	B	C	D	E	F
Λ^0	0.4	0.3	6	9		
Λ^-					2	3
ϑ^0			2	0.1		
K^0	>4[a]	>3[a]				
K^+					>0.7	

[a] Not taking into account that decay may result in two neutral particles and thus be invisible.

[10] D. C. Peaslee, Phys. Rev. 86, 127 (1952).
[11] G. Ascoli, Phys. Rev. 90, 1079 (1953).

For a mean lifetime of 3×10^{-10} sec for the Λ^0 and 1.5×10^{-10} for the ϑ^0, the given cloud-chamber geometry, a π^--beam energy of 1.5 Bev, and isotropic angular distributions (in the c.m.s.) of Λ and K particles one can estimate that 60 percent of the Λ and 50 percent of of the K should be seen to decay inside the chamber. One can conclude that for 80 percent of all occurring cases one should see at least one of the two particles decay and for 20 percent of all cases one should see both decaying. (For shorter lifetimes or angular distributions peaked forwards or backwards the probabilities are even larger.) Therefore the 4 cases of unstable particle production in hydrogen (A, B, D, E) observed here may correspond to 5 ± 3 cases that actually happened.[12] This number can be compared with the other 170 $\pi^- - p$ interactions observed, including events with two and four outgoing prongs. Making use of Fermi's statistical theory of meson production[13] and of the isotopic spin formalism, Fermi[14] has calculated the probabilities for the different combinations. One finds that combinations resulting in no outgoing prongs are expected to occur in only 12 percent of all interactions. Therefore our 170 observed cases may correspond to 190 actual interactions. The total interaction cross section has been found to be 34 ± 3 millibarns.[15] Therefore the cross section for heavy unstable particle production by 1.5-Bev π^- in hydrogen is ~1 millibarn.

Table VII gives the angles between the decay planes of the particles and their production planes (the plane formed by the incident track and the line of flight of the unstable particle.) The absence of large angles for the Λ particles is surprising and might be taken as an indication for a large spin for these particles. In fact the

[12] Decays of Λ^0 and ϑ^0 into neutral particles cannot be observed in this experiment and are not taken into account here.
[13] E. Fermi, Progr. Theoret. Phys. (Japan) 5, 570 (1950).
[14] E. Fermi (private communication).
[15] Cool, Madansky, and Piccioni, Phys. Rev. 93, 637 (1954).

separation between the decay and production planes seems smaller than would be expected unless the spin is taken to be unreasonably large.

Table VIII shows the angles between incident π^- and direction of emission of Λ^0 in the c.m.s. Again the result is surprising because of the absence of angles near 90° for which the solid angle is largest. The solid angle between 170° and 180° amounts to only 1.5 percent of the hemisphere, yet both C and D fall into this region. This may indicate that large angular momentum states are involved in the production of these particles, which might be consistent with the possibility

TABLE VII. Angles between decay plane and production plane.

Particle	*A*	*B*	*C*	*D*	*E*
Λ^0	5°±5°	30°±20°	18°±7°	27°±10°	
Λ^-					7°±5°
ϑ^0			60°±6°	70°±5°	

of a large spin of the Λ^0. Since $\lambda\!\!\!^-_0 \approx 2\times10^{-14}$ cm in the c.m.s. for a π^- of 1.5-Bev laboratory energy, angular momentum states up to $L=6$ may be possible.

SUMMARY

These examples of the production of heavy unstable particles in $\pi^- - p$ collisions have been shown to be consistent with a double production process,

$$\pi^- + p \rightarrow \Lambda + K,$$

occurring with a cross section of about 1 millibarn for

TABLE VIII. Angle between incident π^- and direction of emission of $\Lambda^{0,-}$ in c.m.s.

Particle	*A*	*B*	*C*	*D*	*E*
Λ^0	141°	125°	177°	174°	
Λ^-					30°

1.5-Bev π^-. Further work is required to determine whether production is *always* double in these and nucleon-nucleon collisions.

In four cases the Λ is a Λ^0, and in two of these the K^0 is observed to decay and can be considered to be a ϑ^0, although the Q values of about 260 Mev are not quite consistent with the cosmic-ray value of 214 Mev. One case is interpreted as a Λ^-, and an additional charged decay originating outside the chamber is thought to be a Λ^- with Q value of 130 Mev.

The data suggest an angular correlation between Λ-decay planes and production planes, which may mean that Λ particles have large spin, and a preferred backward (or forward) emission in the c.m.s., which may mean that Λ particles are produced in states of high angular momentum. Many more data are needed to determine such angular relationships.

We wish to express our thanks to the many members of the Cosmotron Department whose efforts have created the opportunity for this work and whose cooperation is enabling us to pursue it. Our thanks are also due to the other members of the cloud-chamber group without whose support this work could not be continued.

We have found that deviations in excess of Mott scattering are readily apparent at large scattering angles. The early results (reported at the Seattle meeting, July, 1954) at smaller angles showed the expected agreement with the Mott formula within experimental error. Deviations from the Mott formula such as we have found may be anticipated at large angles because of additional scattering from the magnetic moment of the proton.[3] We have observed this additional scattering but in an amount smaller than predicted by theory.

The experimental curve at 188 Mev is given in Fig. 1. It may be observed that the experimental points do not fit either the Mott curve or the theoretical curve of Rosenbluth,[3] computed for a point charge and point (anomalous) magnetic moment of the proton. Furthermore, the experimental curve does not fit a Rosenbluth curve with the Dirac magnetic moment and a point charge. The latter curve would lie close to the Mott curve and slightly above it. Similar behavior is observed at 236 Mev.

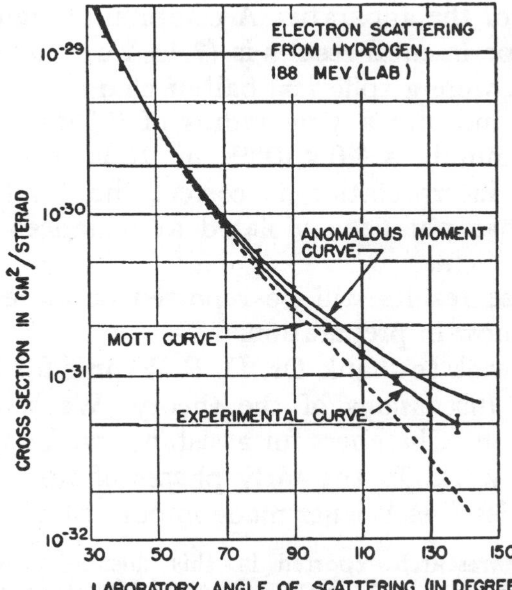

FIG. 1. The figure shows the experimental curve, the Mott curve, and the point-charge, point-magnetic-moment curve. The experimental curve passes through the points with the attached margins of error. The margins of error are not statistical; statistical errors would be much smaller than the errors shown. The limits of error are, rather, the largest deviations observed in the many complete and partial runs taken over a period of several months. Absolute cross sections given in the ordinate scale were not measured experimentally but were taken from theory. The radiative corrections of Schwinger have been ignored since they affect the angular distribution hardly at all. The radiative corrections do influence the absolute cross sections. Experimental points in the figure refer to areas under the elastic peaks taken over an energy interval of ± 1.5 Mev centering about the peak. The data at the various points are unchanged in relation to each other when the energy interval is increased to ± 2.5 Mev about the peak; the latter widths include essentially all the area under the peak.

Electron Scattering from the Proton*†‡

ROBERT HOFSTADTER AND ROBERT W. MCALLISTER
Department of Physics and High-Energy Physics Laboratory, Stanford University, Stanford, California
(Received January 24, 1955)

WITH apparatus previously described,[1,2] we have studied the elastic scattering of electrons of energies 100, 188, and 236 Mev from protons initially at rest. At 100 Mev and 188 Mev, the angular distributions of scattered electrons have been examined in the ranges 60°–138° and 35°–138°, respectively, in the laboratory frame. At 236 Mev, because of an inability of the analyzing magnet to bend electrons of energies larger than 192 Mev, we have studied the angular distribution between 90° and 138° in the laboratory frame. In all cases a gaseous hydrogen target was used.

The correct interpretation of these results will require a more elaborate explanation (probably involving a good meson theory) than can be given at the moment, although Rosenbluth already has made weak-coupling calculations in meson theory which predict an effect of the kind we have observed.[4]

Nevertheless, if we make the naive assumption that the proton charge cloud and its magnetic moment are both spread out in the same proportions we can calculate simple form factors for various values of the proton "size." When these calculations are carried out we find that the experimental curves can be represented very well by the following choices of size. At 188 Mev, the data are fitted accurately by an rms radius of $(7.0\pm2.4)\times10^{-14}$ cm. At 236 Mev, the data are well fitted by an rms radius of $(7.8\pm2.4)\times10^{-14}$ cm. At 100 Mev the data are relatively insensitive to the radius but the experimental results are fitted by both choices given above. The 100-Mev data serve therefore as a valuable check of the apparatus. A compromise value fitting all the experimental results is $(7.4\pm2.4)\times10^{-14}$ cm. If the proton were a spherical ball of charge, this rms radius would indicate a true radius of 9.5×10^{-14} cm, or in round numbers 1.0×10^{-13} cm. It is to be noted that if our interpretation is correct the Coulomb law of force has not been violated at distances as small as 7×10^{-14} cm.

These results will be reported in more detail in a paper now in preparation.

We wish to thank Dr. D. R. Yennie for his generous aid in discussions of the theory. We wish to thank Mr. E. E. Chambers for assistance with several phases of the work. In the early phases of this research, the late Miss Eva Wiener made important contributions.

* The research reported in this document was supported jointly by the U. S. Navy (Office of Naval Research) and the U. S. Atomic Energy Commission, and the U. S. Air Force through the Office of Scientific Research of the Air Research and Development Command.

† Aided by a grant from the Research Corporation.

‡ Early results were reported at the Seattle Meeting of the American Physical Society [Phys. Rev. **96**, 854(A) (1954)]. More recent results were presented at the Berkeley meeting of the American Physical Society [Bull. Am. Phys. Soc. **29**, No. 8, 29 (1954)].

[1] Hofstadter, Fechter, and McIntyre, Phys. Rev. **92**, 978 (1953).

[2] Hofstadter, Hahn, Knudsen, and McIntyre, Phys. Rev. **95**, 512 (1954).

[3] M. N. Rosenbluth, Phys. Rev. **79**, 615 (1950).

[4] See also the classical calculation of L. I. Schiff reported in Rosenbluth's paper.

Observation of Antiprotons*

OWEN CHAMBERLAIN, EMILIO SEGRÈ, CLYDE WIEGAND, AND THOMAS YPSILANTIS

Radiation Laboratory, Department of Physics, University of California, Berkeley, California

(Received October 24, 1955)

ONE of the striking features of Dirac's theory of the electron was the appearance of solutions to his equations which required the existence of an antiparticle, later identified as the positron.

The extension of the Dirac theory to the proton requires the existence of an antiproton, a particle which bears to the proton the same relationship as the positron to the electron. However, until experimental proof of the existence of the antiproton was obtained, it might be questioned whether a proton is a Dirac particle in the same sense as is the electron. For instance, the anomalous magnetic moment of the proton indicates that the simple Dirac equation does not give a complete description of the proton.

The experimental demonstration of the existence of antiprotons was thus one of the objects considered in the planning of the Bevatron. The minimum laboratory kinetic energy for the formation of an antiproton in a nucleon-nucleon collision is 5.6 Bev. If the target nucleon is in a nucleus and has some momentum, the threshold is lowered. Assuming a Fermi energy of 25 Mev, one may calculate that the threshold for formation of a proton-antiproton pair is approximately 4.3 Bev. Another, two-step process that has been considered by Feldman[1] has an even lower threshold.

TABLE I. Characteristics of components of the apparatus.

$S1$, $S2$	Plastic scintillator counters 2.25 in. diameter by 0.62 in. thick.
$C1$	Čerenkov counter of fluorochemical 0–75, ($C_8F_{16}O$); $\mu_D=1.276$; $\rho=1.76$ g cm^{-3}. Diameter 3 in.; thickness 2 in.
$C2$	Čerenkov counter of fused quartz: $\mu_D=1.458$; $\rho=2.2$ g cm^{-3}. Diameter 2.38 in.; length 2.5 in.
$Q1$, $Q2$	Quadrupole focusing magnets: Focal length 119 in.; aperture 4 in.
$M1$, $M2$	Deflecting magnets 60 in. long. Aperture 12 in. by 4 in. $B\cong 13\,700$ gauss.

There have been several experimental events[2–4] recorded in cosmic-ray investigations which might be due to antiprotons, although no sure conclusion can be drawn from them at present.

With this background of information we have performed an experiment directed to the production and detection of the antiproton. It is based upon the determination of the mass of negative particles originating at the Bevatron target. This determination depends on the simultaneous measurement of their momentum and velocity. Since the antiprotons must be selected from a heavy background of pions it has been necessary to measure the velocity by more than one method. To date, sixty antiprotons have been detected.

Figure 1 shows a schematic diagram of the apparatus. The Bevatron proton beam impinges on a copper target and negative particles scattered in the forward direction with momentum 1.19 Bev/c describe an orbit as shown in the figure. These particles are deflected 21° by the field of the Bevatron, and an additional 32° by magnet $M1$. With the aid of the quadrupole focusing magnet $Q1$ (consisting of 3 consecutive quadrupole magnets) these particles are brought to a focus at counter $S1$, the first scintillation counter. After passing through counter $S1$, the particles are again focused (by $Q2$), and deflected (by $M2$) through an additional angle of 34°, so that they are again brought to a focus at counter $S2$.

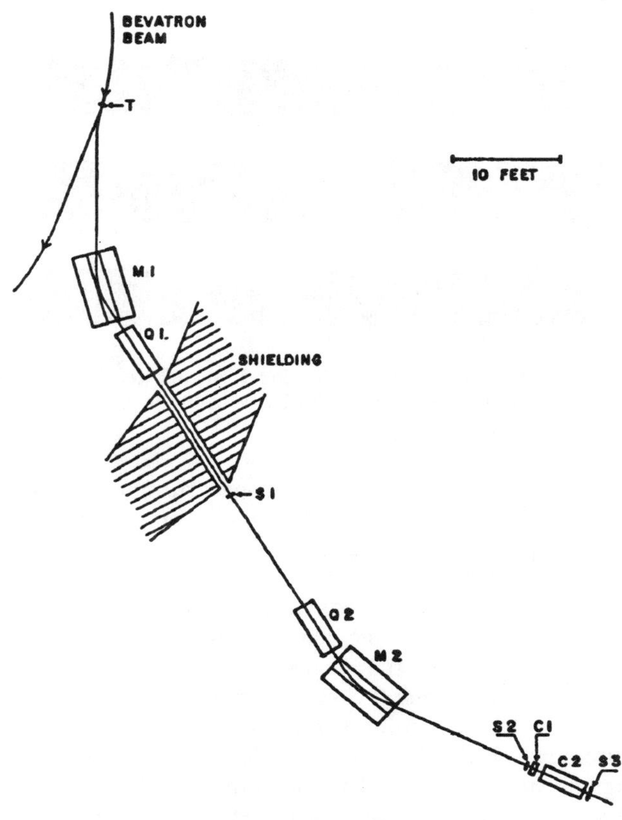

FIG. 1. Diagram of experimental arrangement. For details see Table I.

The particles focused at $S2$ all have the same momentum within 2 percent.

Counters $S1$, $S2$, and $S3$ are ordinary scintillation counters. Counters $C1$ and $C2$ are Čerenkov counters. Proton-mass particles of momentum 1.19 Bev/c incident on counter $S2$ have $v/c=\beta=0.78$. Ionization energy loss in traversing counters $S2$, $C1$, and $C2$ reduces the average velocity of such particles to $\beta=0.765$. Counter $C1$ detects all charged particles for which $\beta>0.79$. $C2$ is a Čerenkov counter of special design that counts only particles in a narrow velocity interval, $0.75<\beta<0.78$. This counter will be described in a separate publication. In principle, it is similar to some of the counters described by Marshall.[5] The requirement that a particle be counted in this counter represents one of the determinations of velocity of the particle.

The velocity of the particles counted has also been determined by another method, namely by observing the time of flight between counters $S1$ and $S2$, separated by 40 ft. On the basis of time-of-flight measurement the separation of π mesons from proton-mass particles is quite feasible. Mesons of momentum 1.19 Bev/c have $\beta=0.99$, while for proton-mass particles of the same momentum $\beta=0.78$. Their respective flight times over the 40-ft distance between $S1$ and $S2$ are 40 and 51 millimicroseconds.

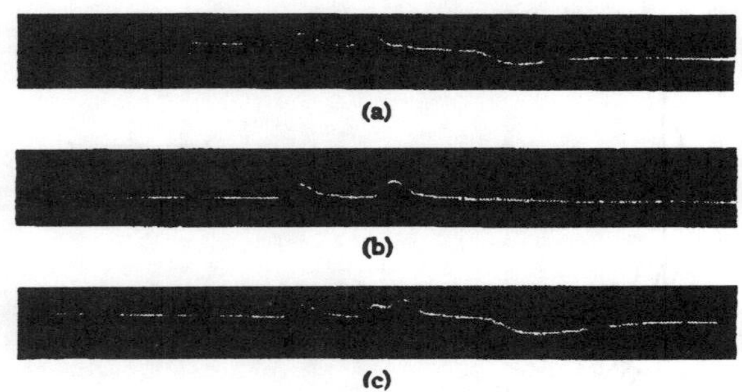

FIG. 2. Oscilloscope traces showing from left to right pulses from $S1$, $S2$, and $C1$. (a) meson, (b) antiproton, (c) accidental event.

The beam that traverses the apparatus consists overwhelmingly of π^- mesons. One of the main difficulties of the experiment has been the selection of a very few antiprotons from the huge pion background. This has been accomplished by requiring counters $S1$, $S2$, $C2$, and $S3$ to count in coincidence. Coincidence counts in $S1$ and $S2$ indicate that a particle of momentum 1.19 Bev/c has traversed the system with a flight time of approximately 51 millimicroseconds. The further requirement of a coincidence in $C2$ establishes that the particle had a velocity in the interval $0.75<\beta<0.78$. The latter requirement of a count in $C2$ represents a measure of the velocity of the particle which is essentially independent of the cruder electronic time-of-flight measurement. Finally, a coincident count in counter $S3$ was required in order to insure that the particle traversed the quartz radiator in $C2$ along the axis and suffered no large-angle scattering.

As outlined thus far, the apparatus has some shortcomings in the determination of velocity. In the first place, accidental coincidences of $S1$ and $S2$ cause some mesons to count, even though a single meson would be completely excluded because its flight time would be too short. Secondly, the Čerenkov counter $C2$ could be actuated by a meson (for which $\beta=0.99$) if the meson suffered a nuclear scattering in the radiator of the counter. About 3 percent of the mesons, which ideally should not be detected in $C2$, are counted in this manner. Both of these deficiencies have been eliminated by the insertion of the guard counter $C1$, which records all particles of $\beta>0.79$. A pulse from $C1$ indicates a particle (meson) moving too fast to be an antiproton of the selected momentum and indicates that this event should be rejected. In Table I, the characteristics of the components of the apparatus are summarized.

The pulses from counters $S1$, $S2$, and $C1$ were displayed on an oscilloscope trace and photographically recorded. From the separation of pulses from $S1$ and $S2$ the flight time of the particle could be measured with an accuracy of 1 millimicrosecond, and the pulse in the guard counter $C1$ could be measured. Figure 2 shows three oscilloscope traces, with the pulses from $S1$, $S2$, and $C1$ appearing in that order. The first trace (a) shows the pulses due to a meson passing through the system. It was recorded while the electronic circuits were adjusted for meson time of flight for calibration purposes. The second trace, Fig. 2(b), shows the pulses resulting from an antiproton. The separation of pulses from $S1$ and $S2$ indicates the correct antiproton time of

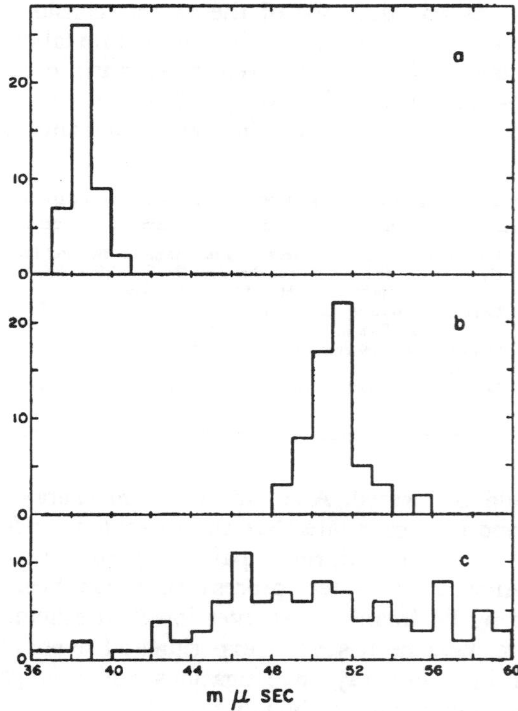

FIG. 3. (a) Histogram of meson flight times used for calibration. (b) Histogram of antiproton flight times. (c) Apparent flight times of a representative group of accidental coincidences. Times of flight are in units of 10^{-9} sec. The ordinates show the number of events in each 10^{-10}-sec intervals.

flight, and the absence of the $C1$ pulse shows that no meson passed through $C1$. The third trace, Fig. 2(c), shows the accidental coincidence of two mesons with a difference of time such as to register in the electronic circuits. Either the presence of a pulse from $C2$ or the presence of multiple pulses from $S1$ or $S2$ would be sufficient to identify the trace as due to one or more mesons.

An over-all test of the apparatus was obtained by changing the position of the target in the Bevatron, inverting the magnetic fields in $M1$, $M2$, $Q1$, and $Q2$, and detecting positive protons.

Each oscilloscope sweep of the type shown in Fig. 2 can be used to make an approximate mass measurement for *each* particle, since the magnetic fields determine the momentum of the particle and the separation of pulses $S1$ and $S2$ determine the time of flight. For protons of our selected momentum the mass is measured to about 10 percent, using this method only.

The observed times of flight for antiprotons are made more meaningful by the fact that the electronic gate time is considerably longer than the spread of observed antiproton flight times. The electronic equipment accepts events that are within ± 6 millimicroseconds of the right flight time for antiprotons, while the actual antiproton traces recorded show a grouping of flight times to ± 1 or 2 millimicroseconds. Figure 3(a) shows a histogram of meson flight times; Fig. 3(b) shows a similar histogram of antiproton flight times. Accidental coincidences account for many of the sweeps (about $\frac{2}{3}$ of the sweeps) during the runs designed to detect antiprotons. A histogram of the apparent flight times of accidental coincidences is shown in Fig. 3(c). It will be noticed that the accidental coincidences do not show the close grouping of flight times characteristic of the antiproton or meson flight times.

Mass measurement.—A further test of the equipment has been made by adjusting the system for particles of different mass, in the region of the proton mass. A test for the reality of the newly detected negative particles is that there should be a peak of intensity at the proton mass, with small background at adjacent mass settings. By changing only the magnetic field values of $M1$, $M2$, $Q1$, and $Q2$, particles of different momentum may be chosen. Providing the velocity selection is left completely unchanged, the apparatus is then set for particles of a different mass. These tests have been made for both positive and negative particles in the vicinity of the proton mass. Figure 4 shows the curve obtained using positive protons, which is the mass resolution curve of the instrument. Also shown in Fig. 4 are the experimental points obtained with antiprotons. The observations show the existence of a peak of intensity at the proton mass, with no evidence of background when the instrument is set for masses appreciably greater or smaller than the proton mass. This test is considered one of the most important for the establishment of the reality of these observations, since background, if present, could be expected to appear at any mass setting of the instrument. The peak at proton mass may further be used to say that the new particle has a mass within 5 percent of that of the proton mass. It is mainly on this basis that the new particles have been identified as antiprotons.

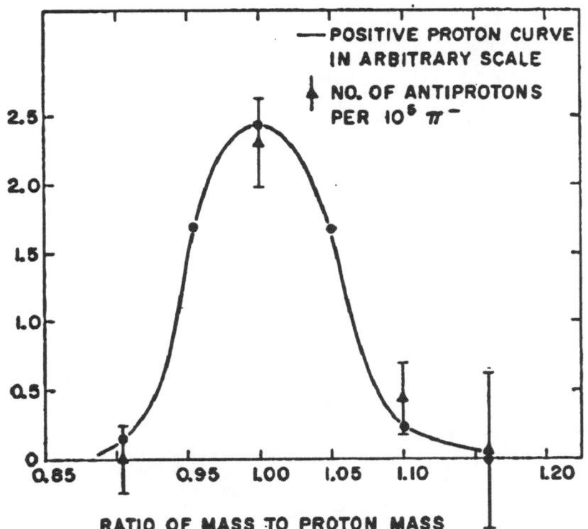

FIG. 4. The solid curve represents the mass resolution of the apparatus as obtained with protons. Also shown are the experimental points obtained with antiprotons.

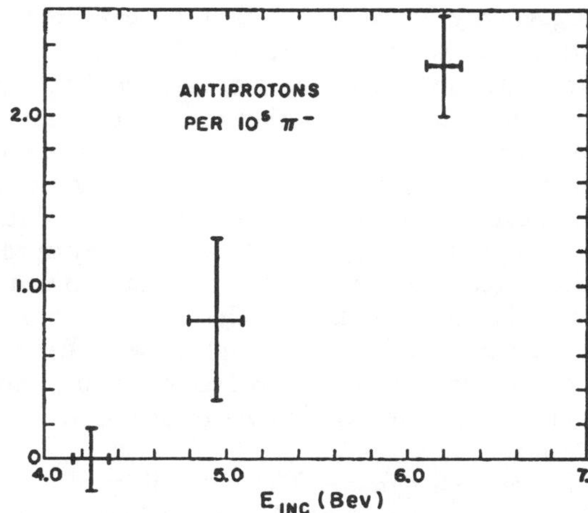

FIG. 5. Excitation curve for the production of antiprotons relative to meson production as a function of Bevatron beam energy.

Excitation function.—A very rough determination has been made of the dependence of antiproton production cross section on the energy of the Bevatron proton beam. A more exact determination will be attempted in the future, but up to the present it has not been possible to monitor reliably the amount of beam actually striking the target. Furthermore, the solid angle of acceptance of the detection apparatus may not be independent of Bevatron energy since the shape of the orbit on which the antiprotons emerge depends somewhat on the magnetic field strength within the Bevatron magnet. It has, however, been possible to measure the ratio of antiprotons to mesons (both at momentum 1.19 Bev/c) emitted in the forward direction from the target as a function of Bevatron energy. The resulting approximate excitation function is shown in the form of three experimental points in Fig. 5.

Even at 6.2 Bev, the antiprotons appear only to the extent of one in 44 000 pions. Because of the decay of pions along the trajectory through the detecting apparatus, this number corresponds to one antiproton in 62 000 mesons generated at the target. It will be seen from Fig. 5 that there is no observed antiproton production at the lowest energy. Although the production of antiprotons does not seem to rise as sharply with increasing energy as might at first be expected, the data indicate a reasonable threshold for production of antiprotons. It must again be emphasized that Fig. 5 shows only the excitation function relative to the meson excitation function, hence the true excitation function is not known at this time. If and when detailed meson production excitation functions become known, data of the type shown in Fig. 5 may allow a true antiproton production excitation function to be determined. It should also be mentioned that the angle of emission from the target actually varies slightly with Bevatron energy. At 6.2 Bev, it is 3°, at 5.1 Bev it is 6°, and at 4.2 Bev it is 8° from the forward direction at the Bevatron target.

Possible spurious effects.—The possibility of a negative hydrogen ion being mistaken for an antiproton is ruled out by the following argument: It is extremely improbable that such an ion should pass through all the counters without the stripping of its electrons. It may be added that except for a few feet near the target the whole trajectory through the apparatus is though gas at atmospheric pressure, either in air or, near the magnetic lenses, in helium gas introduced to reduce multiple scattering.

None of the known heavy mesons or hyperons have the proper mass to explain the present observations. Moreover, no such particles are known that have a mean life sufficiently long to pass through the apparatus without a prohibitive amount of decay since the flight time through the apparatus of a particle of proton mass is 10.2×10^{-8} sec. However, this possibility cannot be strictly ruled out. In the description of the new particles as antiprotons, a reservation must be made for the possible existence of previously unknown negative particles of mass very close to 1840 electron masses.

The observation of pulse heights in counters $S1$ and $S2$ indicates that the new particles must be singly charged. No multiply charged particle could explain the experimental results.

Photographic experiments directed toward the detection of the terminal event of an antiproton are in progress in this laboratory and in Rome, Italy, using emulsions irradiated at the Bevatron, but to this date no positive results can be given. An experiment in conjunction with several other physicists to observe the energy release upon the stopping of an antiproton in a large lead-glass Čerenkov counter is in progress and its results will be reported shortly. It is also planned to try to observe the annihilation process of the antiproton in a cloud chamber, using the present apparatus for counter control.

The whole-hearted cooperation of Dr. E. J. Lofgren, under whose direction the Bevatron has been operated, has been of vital importance to this experiment. Mr. Herbert Steiner and Mr. Donald Keller have been very helpful throughout the work. Dr. O. Piccioni has made very useful suggestions in connection with the design of the experiment. Finally, we are indebted to the operating crew of the Bevatron and to our colleagues, who have cheerfully accepted many weeks' postponement of their own work.

* This work was done under the auspices of the U. S. Atomic Energy Commission.

[1] G. Feldman, Phys. Rev. 95, 1967 (1954).
[2] Evans Hayward, Phys. Rev. 72, 937 (1947).
[3] Amaldi, Castagnoli, Cortini, Franzinetti, and Manfredini, Nuovo cimento 1, 492 (1955).
[4] Bridge, Courant, DeStaebler, and Rossi, Phys. Rev. 95, 1101 (1954).
[5] J. Marshall, Ann. Rev. Nuc. Sci. 4, 141 (1954).

expected, assuming a geometric interaction cross section for antiprotons in copper. It has now been found[3] that the cross section in copper is about twice geometric, which explains this low yield.

In view of this result a new irradiation was planned in which (1) no absorbing material preceded the stack,

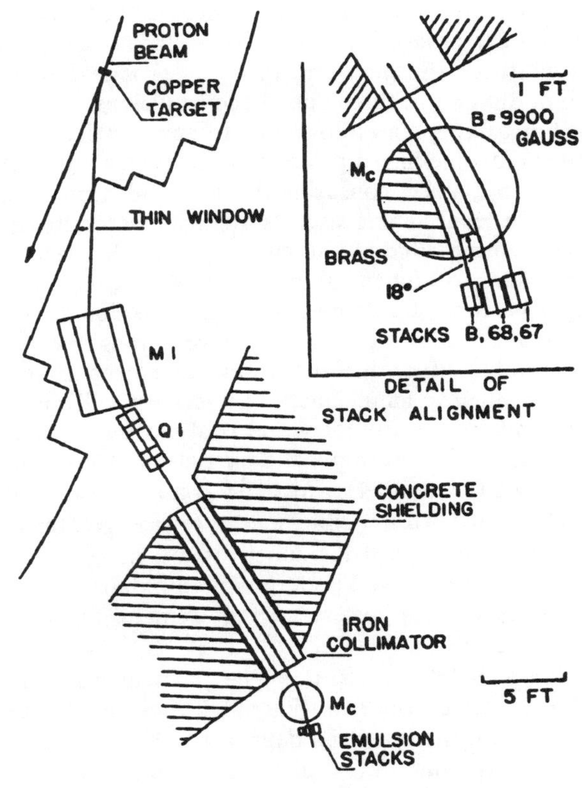

FIG. 1. Plan of the irradiation.

(2) the range of the antiprotons ended in the stack, and (3) antiprotons and mesons were easily distinguishable by grain density at the entrance of the stack. In order to achieve these three results it was necessary to select antiprotons of lower momentum, even if these should be admixed with a larger number of π^- than at higher momenta.

In the present experiment we exposed a stack in the same beam used previously, adjusted for a momentum of 700 Mev/c instead of 1090 Mev/c. Since the previous work[2] had indicated that the most troublesome background was due to ordinary protons, the particles were also passed through a clearing magnetic field just prior to their entrance into the emulsion stacks. The clearing magnet (M_c) had $B=9900$ gauss, circular pole faces of diameter 76 cm, and a gap of 18 cm, so particles scattered from the pole faces of the clearing magnet could be ignored on the basis of their large dip angle in the emulsions. With this arrangement we have achieved conditions in which the negative particles enter the emulsions at a well-defined angle, and extremely few positive particles enter the emulsions within the same

Example of an Antiproton-Nucleon Annihilation

O. CHAMBERLAIN, W. W. CHUPP, A. G. EKSPONG, G. GOLDHABER, S. GOLDHABER, E. J. LOFGREN, E. SEGRÈ, AND C. WIEGAND, *Radiation Laboratory and Department of Physics, University of California, Berkeley, California*

AND

E. AMALDI, G. BARONI, C. CASTAGNOLI, C. FRANZINETTI, AND A. MANFREDINI, *Istituto Fisico dell'Università Roma, Italy and Istituto Nazionale di Fisica Nucleare Sez. di Roma, Italy*

(Received March 8, 1956)

THE existence of antiprotons has recently been demonstrated at the Berkeley Bevatron by a counter experiment.[1] The antiprotons were found among the momentum-analyzed (1190 Mev/c) negative particles emitted by a copper target bombarded by 6.2-Bev protons. Concurrently with the counter experiment, stacks of nuclear emulsions were exposed in the beam adjusted to accept 1090-Mev/c negative particles in an experiment designed to observe the properties of antiprotons when coming to rest. This required a 132-g/cm^2 copper absorber to slow down the antiprotons sufficiently to stop them in the emulsion stack. Only one antiproton was found[2] in stacks in which seven were range of angles. For the first time we have obtained an exposure in which more antiprotons than protons enter the stacks with the proper entrance angles. Under these conditions it is relatively easy to find antiprotons in these stacks even though approximately 5×10^5 negative π mesons at minimum ionization accompany one antiproton. The exposure arrangement is shown in Fig. 1. The beam collimation was such that at any given position at the leading edge of the stacks the angular half-width of the pion entrance angles is less than 1° both in dip and in the plane of the emulsions. This very small angular spread allowed us to apply strict angular criteria for picking up antiproton tracks, and thus helped to reduce confusing background tracks to a negligible level. The antiproton tracks were picked up at the leading edge of the emulsions on the basis of a grain count (~twice minimum) and angular criteria (angle between track and average direction of pions less than 5°) and were then followed along the track.

A number of antiproton stars have been observed in these nuclear emulsions.[4] The one we will describe here was found in Berkeley and is of particular interest since it is the first example of a particle of protonic mass ($m/M_p=1.013\pm0.034$) which on coming to rest gave rise to a star with a visible energy release greater than M_pc^2. This example thus constitutes a proof that the particles here observed undergo an annihilation process with a nucleon, a necessary requirement for Dirac's antiproton.

Description of the Event.—The particle marked P^- in Fig. 2 entered the emulsion stack at an angle of less than 1° from the direction defined by the π^- mesons in the beam. It came to rest in the stack and produced an 8-prong star. Its total range was $R=12.13\pm0.14$ cm. Table I gives the results of three independent mass measurements on the incoming particle. The first two methods listed in Table I use measurements made entirely in the emulsion stack. The third combines the range, as measured in the stack, with the momentum as determined by magnetic field measurements. For the position and entrance angle of this particle into the stack the momentum is $p=696$ Mev/c with an estimated 2% error. All three methods are in good agreement and give a mass of $m=1.013\pm0.034$ in proton mass units.

TABLE I. Mass measurements.

Method	Residual range cm of emulsion	Mass M/M_p
Ionization–range	2, 5.5, and 12	0.97 ±0.10
Scattering–range (constant sagitta)	0–1	0.93 ±0.14
Momentum–range	12.13+0.12 (air and helium equivalent)	1.025±0.037
Weighted mean		1.013±0.034

Of the particles forming the star, five came to rest in the emulsion stack, two left the stack (tracks numbered 4 and 8), and one disappeared in flight (track number 3). The tracks numbered 1, 4, and 6 in Fig. 2 were caused by heavy particles. Particle 4 was near the end of its range ($R_{res}=2$ mm) when it left the stack. Tracks 1 and 4 are probably due to protons and track 6 to a triton. However, owing to the large dip angles the assignments for tracks 1, 4, and 6 are not certain. Track 2 has the characteristics of a π meson and on coming to rest gives a 2-prong σ star. It is thus a negative π meson. Particle 5 came to rest and gave the typical π-μ-e decay, and was thus a positive π meson. From the measured range its energy would have been 18 Mev; however, after

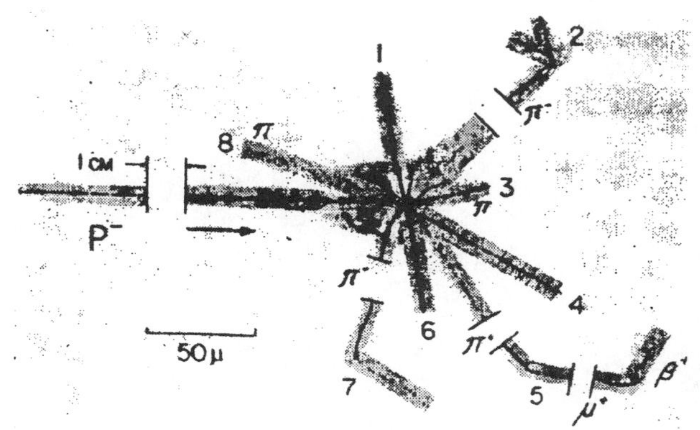

FIG. 2. Reproduction of the P^- star. The description of the prongs is given in Table II. The star was observed by A. G. Ekspong and the photomicrograph was made by D. H. Kouns.

0.22 mm it underwent a 22° scattering that appears to be inelastic. The initial energy as estimated from the grain density change was 30±6 Mev. Track 7 is very steep (dip angle=83.5°). The particle came to rest as a typical light ρ meson after traversing 30 emulsions. At the end of the track there is a blob and possibly an associated slow electron. The most probable assignment is a negative π meson, although a negative μ meson cannot be ruled out.

In addition to the three stopping π mesons there are two other tracks which we know were caused by light particles, presumably π mesons. Track 8 had $p\beta=190$ ±30 Mev/c and $g/g_0=1.10\pm0.04$, which is consistent with a π meson of 125±25 Mev energy, but is not consistent with a much heavier particle. After 16 mm it shows a 17° scattering with no detectable change in energy. Track 3 is very steep (dip angle=73.5°) and its ionization is about minimum. The particle traversed 81 plates and disappeared in flight after an observed range of 50 mm. The $p\beta$ has been determined by a new modification of the multiple scattering technique to be 250±45 Mev/c. The new method, which is applicable to steep tracks, is based on measurements of the coordinates of the exit point of the track in each emulsion with reference to a well-aligned millimeter grid[5] printed on each pellicle in the stack. A detailed description of this method will be given in a subsequent paper.

TABLE II. Measurements and data on the eight prongs of the P^- star shown in Fig. 2.

Track number	Range mm	Number of plates traversed	Dip angle	Projected angle	$p\beta$ Mev/c	Ionization g/g_0	Identity	E_{kin} Mev	Total energy Mev
1	0.59	2	−56.5°	103°			p(?)	10	18
2	27.9	11	+6.5°	61.5°			π^-	43	183
3	>50	81	−73.5°	14.5°	250±45	1.10±0.04	π(?)	174±40	314±40
4	>14.2	16	+53°	318.5°			p(?)	70±5	78±5
5	6.2	3	+4°	305.5°			π^+	30±6	170±6
6	9.5	15	−63.5°	281°			T(?)	82	98
7	18.6	30	−83.5°	255°			π^-	34	174
8	>22.3	16	+33°	163°	190±30	∼1	π(?)	125±25	265±25

Total visible energy[a]: 1300±50 Mev
For momentum balance: ≳100 Mev
Total energy release: ≳1400±50 Mev

[a] To obtain the minimum possible value of the visible energy release, still consistent with our observations, one has to make the very unlikely assumptions about the identity of tracks 3, 6, 7, and 8: that tracks 3 and 8 are due to electrons, track 6 to a proton, and track 7 to a μ^- meson. The total visible energy release in this case becomes 1084±55 Mev. To this must be added at least 50 Mev to balance momentum, bringing the total energy release to ≳1134±55 Mev.

The observations do not allow us to rule out the possibility that tracks 3 and 8 are due to electrons. It is, however, very unlikely that a fast electron could travel 50 mm (1.7 radiation lengths) in the emulsion (as does track 3) without a great loss of energy due to bremsstrahlung. The energy (particle 3) deduced from the measured $p\beta(E=250$ Mev) must be considered a lower limit.

In Table II, the pertinent data on the eight prongs are summarized. The last column gives the total visible energy per particle (E_{kin}+8-Mev binding energy per nucleon, or E_{kin}+140-Mev rest energy per π meson) for the most probable assignments as discussed above. The total visible energy is 1300±50 Mev, and the momentum unbalance is 750 Mev/c. To balance momentum, an energy of at least 100 Mev is required in neutral particles (i.e., about 5 neutrons with parallel and equal momenta), which brings the lower limit for the observed energy release to 1400±50 Mev.

However, as some of the identity assignments to the star prongs are not certain, we have also computed the energy release for the extreme and very unlikely assignments, given at the foot of Table II, which are chosen to give the minimum energy release. In this case the total visible energy is 1084±55 Mev and the resultant momentum is 380 Mev/c, which to be balanced requires at least 50 Mev in neutral particles (three or four neutrons). In this unrealistic case the lower limit for the observed energy release is 1134±55 Mev, which still exceeds the rest energy of the incoming particle by about three standard deviations.

We conclude that the observations made on this reaction constitute a conclusive proof that we are dealing with the antiparticle of the proton.

A second important observation is the high multiplicity of charged π mesons (one π^+, two π^-, and two π mesons with unknown charge). The fact that so many π mesons escaped from the nucleus where the annihilation took place, together with the low number of heavy particles emitted (three), may indicate that the struck nucleus was one of the light nuclei of the emulsion (C, N, O). Two of the outgoing heavy prongs carried rather high energies (70 Mev for the proton, 82 Mev for the triton), and they may have resulted from the reabsorption of another two π mesons.

We are greatly indebted to the Bevatron crew for their assistence in carrying out the exposure. We also wish to thank Mr. J. E. Lannutti for help with measurements and the analysis of the event.

* This work was performed under the auspices of the U. S. Atomic Energy Commission.

[1] Chamberlain, Segrè, Wiegand, and Ypsilantis, Phys. Rev. **100**, 947 (1955).

[2] Chamberlain, Chupp, Goldhaber, Segrè, Wiegand, Amaldi, Baroni, Castagnoli, Franzinetti, and Manfredini, Phys. Rev. **101**, 909 (1956), and Nuovo cimento (to be published).

[3] Chamberlain, Keller, Segrè, Steiner, Wiegand, and Ypsilantis, Phys. Rev. (to be published).

[4] Several stacks exposed in the 700 Mev/c beam are being studied in Berkeley by A. G. Ekspong and G. Goldhaber; W. W. Chupp and S. Goldhaber; R. Birge, D. H. Perkins, D. Stork, and L. van Rossum; W. Barkas, H. Heckman, and F. Smith; and in Rome by E. Amaldi, G. Baroni, C. Castagnoli, C. Franzinetti, and A. Manfredini.

[5] Goldhaber, Goldsack, and Lannutti, University of California Radiation Laboratory Report UCRL-2928 (unpublished).

Observation of Long-Lived Neutral *V* Particles*

K. Lande, E. T. Booth, J. Impeduglia, and L. M. Lederman,
Columbia University, New York, New York
and
W. Chinowsky, *Brookhaven National Laboratory, Upton, New York*

(Received July 30, 1956)

THE application of rigorous charge conjugation invariance to strange particle interactions has led to the prediction of rather startling properties for the θ^0-meson state.[1] Some of these are: (I) the existence of a second neutral particle, $\theta_2{}^0$, for which two-pion decay is prohibited; (II) the consequent existence of a second

TABLE I. Data on V^0 events.

Event number	P_+ Mev/c	I_+[a]	P_- Mev/c	I_-[a]	θ[b]	$Q^*_{\pi\pi}$[c] Mev	Comment
1	360±10	<2	206±37	<2	52°	96±17	Not τ^0[d]
2	...		117±50	<2	151°	...	(−) track short
3	>100	<2	>100	<2	140°	...	Both tracks short
4	224±5	<2	58.5±4	<2	91.5°	66±2	Not τ^0; probable e^-
5	147±23	<2	197±6	<2	113.6°	121±12	Not τ^0
6	83±5	<2	137±3	<2	81.0°	40±2	$I_- > I_+$, probable $\pi^-\mu^+$, π^-e^+, or μ^-e^+
7	142±13	<2	255±5	<2	124°	163±10	Not τ^0
8	197±25	<2	234±9	<2	97°	147±14	Not τ^0
9	241±33	<2	67±20	<2	142°	109±18	Not τ^0, probable e^-
10	194±8	<2	223±4	<2	140°	140±5	Not τ^0
11	111±4	<2	114±5	<2	77.5°	34±1.4	$I_- > I_+$, like No. 6
12	249±5	<2	89±1	2–3	42.0°	38±1.2	Probable π^-
13	290±25	<2	86±25	<2	92°	103±12	μ^- or e^-, ∴ not τ^0
14	183±5	<2	44±5	3–4	102.7	52±2	π^- or μ^-
15	>150	<2	62±7	<2	99°	...	3° deflection in +. $P_{sec}=287\pm30$. Possible $\pi-\mu$ decay. Probable e^-
16	508±18	<2	150±2	<2	65.9°	160±7	Not τ^0, $I_- > I_+$
17	136±8	1.5–2	164±5	<2	118.6°	101±5	$I_+ > I_-$. Probable π^+, not τ^0
18	251±15	<2	201±50	<2	65.9°	93±15	Not τ^0
19	327±15	<2	112±10	<2	38.3°	51±4	$\pi^+\rightarrow\mu^+$, $P_\mu^*=20\pm10$ Mev
20	167±3	<2	114±3	<2	65.5°	40±2	
21	152±5	<2	120±24	<2	112.5°	79±8	
22	283±10	<2	222±10	<2	50.1°	72±6	Coplanar, $P_\perp=59\pm30$ Mev/c
23	89±1	<2	272±9	<3	128.5°	134±10	Not τ^0

[a] This is a visual estimate of the ionization, in units of minimum ionization as determined from nearby light tracks of $P<50$ Mev/c.
[b] Angle errors have not been computed. An average error of 3° has been used.
[c] $Q^*_{\pi\pi}$ for a normal θ^0 is 214 Mev.
[d] τ^0 is defined as $\rightarrow\pi^+ + \pi^- + \pi^0$ and is excluded by Q value or transverse momentum.

lifetime, considerably longer than that for two-pion decay of the $\theta_1^0(\sim 1\times 10^{-10}$ sec); (III) a complicated time dependence for the nuclear interaction properties.[2] The only additional assumption in this "particle mixture" theory is the nonidentity of θ^0 and its antiparticle.

These theoretical considerations have stimulated us to undertake a search for long-lived neutral particles. To this end, the Columbia 36-in. magnet cloud chamber was exposed to the neutral radiation emitted from a copper target at an angle of 68° to the 3-Bev external proton beam of the Brookhaven Cosmotron.[3] Charged particles are eliminated by the combination of a 4-ft long Pb collimator and a 4×10^5 gauss-inch sweeping magnet. The 6-meter flight path from target to chamber represents ~100 mean lives for the well-known Λ^0 and θ^0 particles which are produced at this energy.[4] To date twenty-six V^0 events have been observed. All of these events have anomalous Q values for two-pion decay, all but one are noncoplanar with the line of flight, and all but one demand at least one neutral secondary to balance transverse momentum.

The cloud chamber operates at a pressure of 0.91 atmos of He and 0.10 atmos of argon. The only additional matter in the direct path of the neutral radiation is the 1-cm thick Lucite chamber wall. A 1.5-in. thick lead filter was placed at the entrance to the collimator to reduce the γ-ray flux reaching the chamber. The aperture (5 in.×1.5 in.) defined a solid angle of 0.002 steradian at 68° to the incident protons. The arrangement yielded readable photographs at a beam intensity of $\sim 10^8$ protons per pulse, although the flux through the chamber was estimated to be $\sim 10^4$ neutrons. The latter fact points up the virtue of the technique employed.

The relevant primary data on 23 measured V^0 events, found in a run of 1200 pictures, are listed in Table I. We have considered various background effects which could possibly simulate V^0 events:

(1) Production of meson pairs in the gas by neutrons or photons, the nuclear recoil track being too short to observe. However, the number of neutrons above meson production threshold energy at 68° was expected to be quite small. This was verified experimentally by the fact that no negative prongs (i.e., π^- mesons) were observed to emerge from 1218 neutron-induced stars in the gas.

(2) Decay of π^0 mesons, produced in the gas without recoil, into the alternate mode $e^+e^-\gamma$. This is ruled out kinematically for 16 of the events. The argument in (1) also applies.

(3) Production of large-angle electron pairs in the gas by photons.

(4) Bremsstrahlung or scattering of backward-moving particles with consequent large-angle deflections.

These possibilities lead to the prediction of thousands of smaller angle events and to the necessity for large fluxes of backward-moving particles. Neither of these is observed. These arguments will be detailed in a more complete report. They lead to the conclusion that

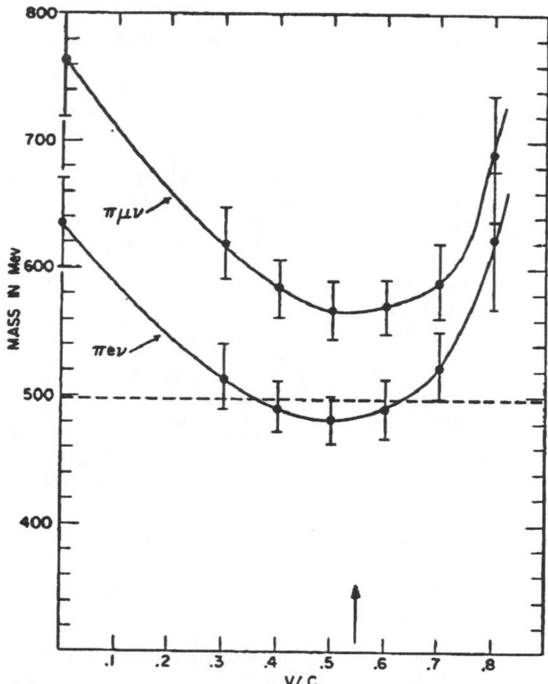

FIG. 1. Average calculated primary mass *vs* velocity of primary particle for assumed decay schemes $\pi^+e^-\nu^0$ and $\pi^+\mu^-\nu^0$. The arrow indicates the peak of the phase space spectra (reference 5). The vertical bars are average deviations of the mean.

the events listed in Table I are indeed examples of the disintegration of a long-lived neutral particle.

A preliminary analysis of the data yields some information on the properties of the new particle.

(1) All but three of the forty-six secondaries are determined to be lighter in mass than the K meson. None can be protons. We have assumed that all are pions, muons, or electrons. The identification of several of the decay products as pions or electrons is indicated in the table.

(2) We have considered various three-body decay schemes, motivated by the observed charged K-meson modes. In Fig. 1, we plot for assumed decay products $\pi^\pm e^\mp\nu^0$ and for $\pi^\pm\mu^\mp\nu^0$, the variation of the average computed mass (15 events were available) of the incoming primary as a function of its assumed velocity. Permutations of the relevant combinations of π's, μ's, e's, and ν's yield similar results. For example $\pi^+\mu^-\nu^0$, $\mu^+\pi^-\nu^0$, $\mu^+e^-\pi^0$, and $\pi^+e^-\pi^0$ are almost coincident. These graphs emphasize the conclusion that the resultant incoming velocity distribution is kinematically sensible only for primary masses near the K mass of 500 Mev.[5] One may also infer that, for a K mass primary, $\pi e\nu$ secondaries are more frequent than $\pi\mu\nu$ or, say, $\mu e\pi$.

(3) All but two of the events are kinematically inconsistent with a Λ^0-mass particle decaying into $\mu^\pm e^\mp N$ or $e^\pm e^\mp N$.[6]

(4) Figure 2 illustrates the detection sensitivity as a function of lifetime for a K-mass particle. Although

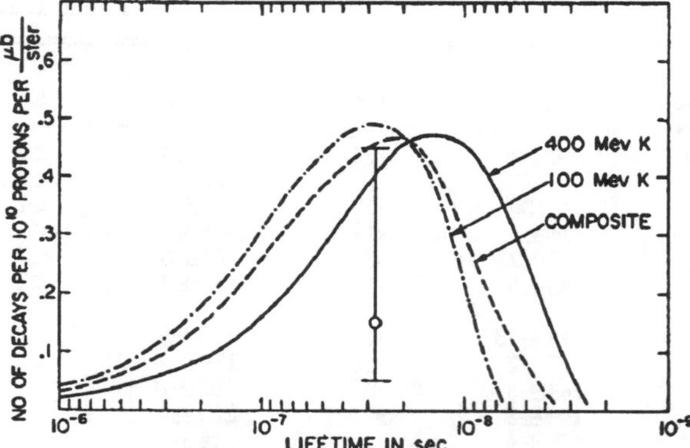

FIG. 2. Detection sensitivity for K mesons as function of lifetime. The composite curve is obtained with the spectra of reference 5. The point indicates the observed yield with a production cross section of $\sim$20 μb/sterad.

the production cross section for K^0 mesons[7] has a large uncertainty, comparison with the observed yield serves to limit the lifetime to the range 10^{-6} sec$>\tau>3\times10^{-9}$ sec. The observed uniform distribution of events in the chamber, together with Fig. 1 also sets a lower limit: $\tau>1\times10^{-9}$ sec. If the lifetime is on the short side of the above interval, then it is likely that many of the anomalous V^0's observed in cosmic rays are examples of this particle, and not alternate decay modes of the $\theta_1{}^0$.[8]

At the present stage of the investigation one may only conclude that Table I, Fig. 2, and Q^* plots are consistent with a K^0-type particle undergoing three-body decay. In this case the mode $\pi e\nu$ is probably prominent,[9] the mode $\pi\mu\nu$ and perhaps other combinations may exist but are more difficult to establish, and $\pi^+\pi^-\pi^0$ is relatively rare. Although the Gell-Mann-Pais predictions (I) and (II) have been confirmed, long lifetime and "anomalous" decay mode are not sufficient to identify the observed particle with $\theta_2{}^0$. In particular, a neutral τ meson, if three-pion decay has a small branching ratio, may have these properties. A much stronger test of particle mixtures must await the observation of nuclear interactions or of the striking interference effects which are also predicted by Pais and Piccioni,[2] Treiman and Sachs,[2] and Serber.[10]

The authors are indebted to Professor A. Pais whose elucidation of the theory directly stimulated this research. The effectiveness of Cosmotron staff collaboration is evidenced by the successful coincident operation of six magnets and the Cosmotron with the cloud chamber.

* Supported by the U. S. Atomic Energy Commission and the U. S. Atomic Energy Commission-Office of Naval Research Joint Program.

[1] M. Gell-Mann and A. Pais, Phys. Rev. 97, 1387 (1955).

[2] Further discussion of particle mixtures have been given by A. Pais and O. Piccioni, Phys. Rev. 100, 932 (1955); G. Snow, Phys. Rev. 103, 1111 (1956); S. Treiman and R. G. Sachs, Phys. Rev. 103, 1545 (1956); K. Case, Phys. Rev. 103, 1449 (1956).

[3] See Piccioni, Clark, Cool, Friedlander, and Kassner, Rev. Sci. Instr. 26, 232 (1955). The ejected beam is focused by a quadrupole magnet pair to a 3-in. diameter circle. Two bending magnets

were used to steer the beam onto the 1.5 in.×4 in.×5 in. long target.

[4] Blumenfeld, Booth, Lederman, and Chinowsky, Phys. Rev. 102, 1184 (1956).

[5] We are grateful to R. Sternheimer for computing the energy spectrum of K mesons emitted at 68° under various assumptions as to the collision mechanism. These calculations yield similar spectra, all of which peak near 100 Mev. See Block, Harth, and Sternheimer, Phys. Rev. 100, 324 (1956).

[6] For example, one member of a Λ^0 parity doublet may have a long lifetime. See T. D. Lee and C. N. Yang, Phys. Rev. 102, 290 (1956).

[7] Collins, Fitch, and Sternheimer (private communication).

[8] Kadyk, Trilling, Leighton, and Anderson, Bull. Am. Phys. Soc. Ser. II, 1, 251 (1956). For a recent summary see Ballam, Grisaru, and Treiman, Phys. Rev. 101, 1438 (1956).

[9] Examples of this decay mode have been reported by Slaughter, Block, and Harth, Bull. Am. Phys. Soc. Ser. II, 1, 186 (1956). A particularly clear event has been observed by the Ecole Polytechnique group. We are indebted to J. Tinlot and B. Gregory for this data and for helpful discussions on anomalous V^0's.

[10] R. Serber (private communication).

Antineutrons Produced from Antiprotons in Charge-Exchange Collisions*

BRUCE CORK, GLEN R. LAMBERTSON, ORESTE PICCIONI,† AND WILLIAM A. WENZEL
Radiation Laboratory, University of California, Berkeley, California
(Received October 3, 1956)

THE principle of invariance under charge conjugation gained strong support when it was found that the Bevatron produces antiprotons.[1–3] Another prediction of the same theory which could be tested experimentally was the existence of the antineutron. Additional interest arises from the fact that charge conjugation has somewhat less obvious consequences when applied to neutral particles than it has when applied to particles with electric charge.

The purpose of this experiment was to detect the annihilation of antineutrons produced by charge exchange from antiprotons. Because the yield of antineutrons was expected to be low, a relatively large flux of antiprotons was required. Protons of 6.2-Bev energy bombarded an internal beryllium target of the Bevatron (Fig. 1). With a system of two deflecting magnets and five magnetic lenses a beam of 1.4-Bev/c negative particles was obtained. Six scintillation counters connected in coincidence distinguished antiprotons from negative mesons by time of flight. In Figs. 1 and 2, F is the last counter of this system, which counted 300 to 600 antiprotons per hour. Antiprotons interacting in the thick converter, X (Fig. 2), sometimes produce antineutrons which pass through the scintillators S_1 and S_2 without detection and finally interact in the lead glass Čerenkov Counter C, producing there a pulse of light so large as to indicate the annihilation of a nucleon and an antinucleon.

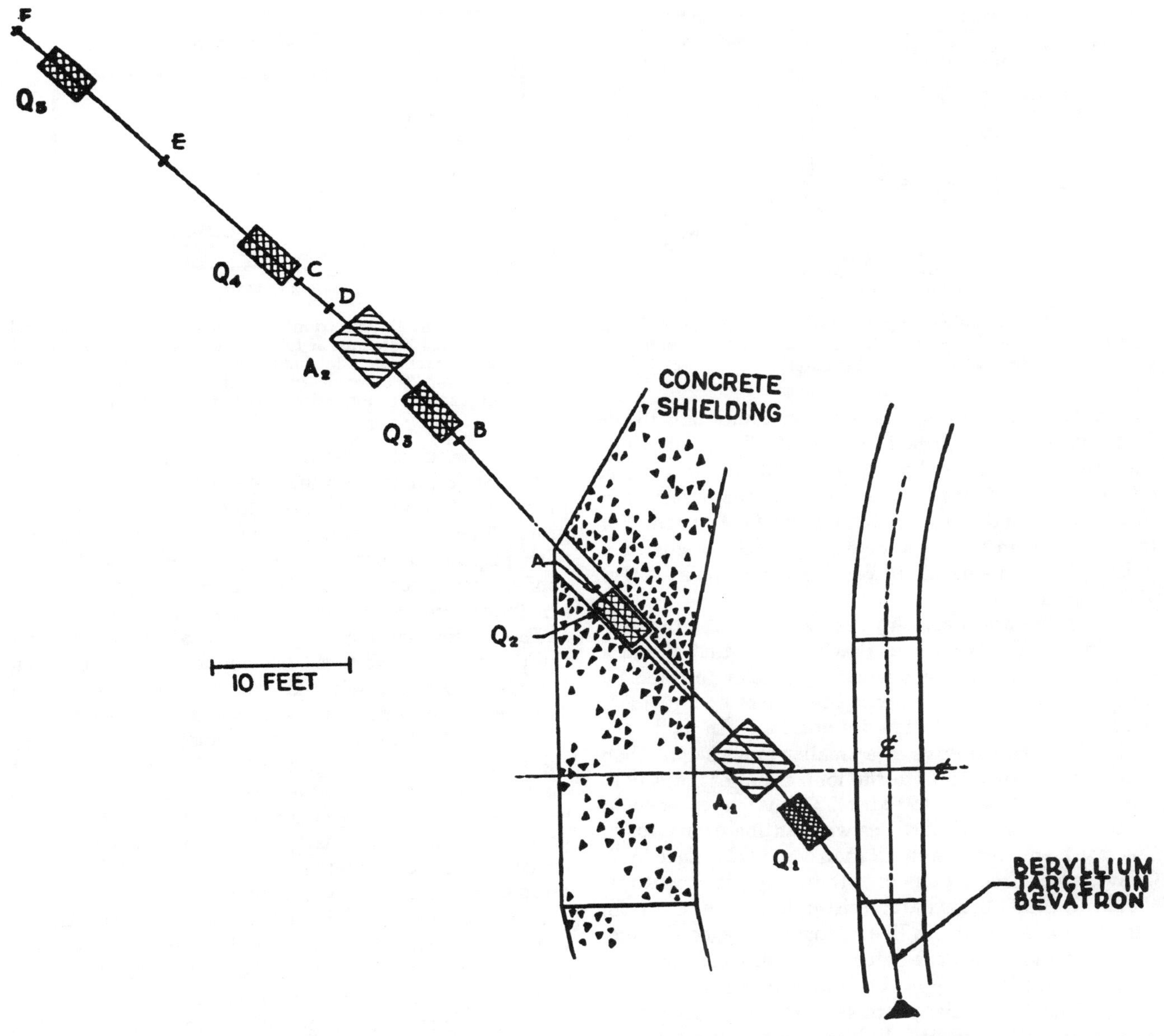

FIG. 1. Antiproton-selecting system. Q_1 through Q_5 are focusing quadrupoles. A_1 and A_2 are analyzing magnets. A through F are 4-by-4-by-¼-inch scintillators.

The Čerenkov Counter C is a piece of lead glass, 13 by 13 by 14 in., density=4.8, index of refraction =1.8, viewed by 16 RCA 6655 photomultipliers. This instrument is similar to the one used in a previous experiment on antiprotons.[2] A 1-in. lead plate is placed between S_1 and S_2 to convert high-energy gamma rays which could otherwise be confused with the antineutrons. Ordinary neutrons and neutral mesons (heavier than pions) can be detected by the Čerenkov counter but their average light pulse is much smaller than that from the annihilation of an antineutron. However, a relatively small background of these neutral secondaries would distort the apparent spectrum of antineutrons.

To discriminate against these neutral secondaries, the charge-exchange converter, X, was made of a scintillating toluene-terphenyl solution, viewed by four photomultipliers connected in parallel. In this way neutral particles producing pulses in the Čerenkov counter ("neutral events") could be separated according to whether they originated in an annihilation, indicated by a large pulse in X, or in the less violent process expected to accompany the charge-exchange production of an antineutron. A quantitative criterion for this separation is derived from a comparison between the pulse-height spectra in X, shown in Fig. 3. The dashed curve, obtained in a separate experiment, is the spectrum produced by antiprotons passing through but not interacting in X. The sharp peak in the spectrum provides the calibration of X; the ionization loss for

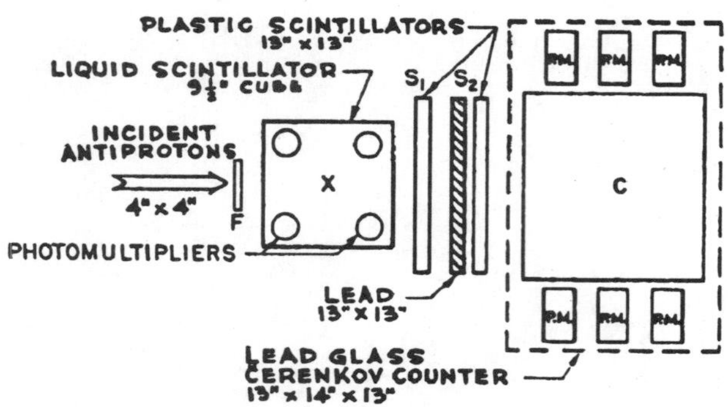

FIG. 2. Antineutron-detecting system. X is the charge-exchange scintillator; S_1 and S_2 are scintillation counters; C is a lead-glass Čerenkov counter (later a large scintillator).

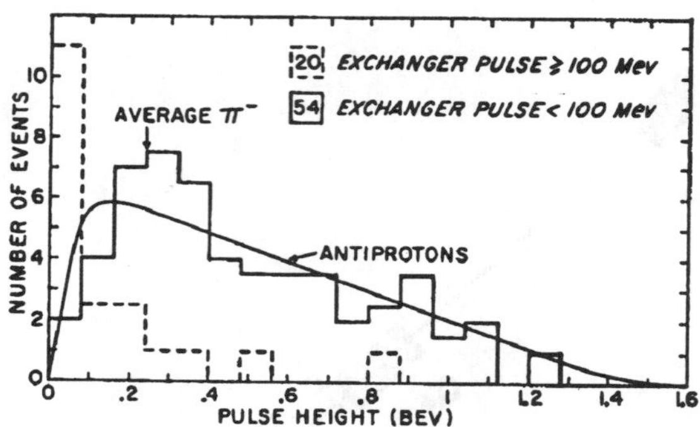

FIG. 4. Pulse-height spectrum in lead glass counter for neutral events. The solid histogram is for 54 antineutron events (energy loss in charge-exchange scintillator less than 100 Mev). Dashed histogram is for 20 other neutral events. Smooth solid curve is for antiprotons and is normalized to the solid histogram.

transmitted antiprotons is readily computed to be 50 Mev. The smooth solid curve of Fig. 3, obtained with the geometry of Fig. 2, represents all antiproton interactions in X from which no pulse was observed in S_1 and S_2, whether or not a pulse in C occurred. For those events in which a neutral particle produced a pulse in C, the histogram of Fig. 3 gives the pulse-height distribution.

The difference between the solid curve and the histogram is remarkable in that it shows that the rare interactions that produce neutral particles detected by the Čerenkov counter release much less energy in X than the other unselected interactions. In fact, the peak of the histogram is at a smaller pulse height than that which corresponds to the ionization loss of a noninteracting antiproton (50 Mev). This is what we should expect if the neutral particles were antineutrons, for in this case no nucleonic annihilation could take place in X. Conversely, production of other energetic neutrals should exhibit the characteristic large pulse of an annihilation event in X. The histogram suggests, therefore, that the apparatus detects a small background of events of this latter type. The pulse height of 100 Mev in Fig. 3 has been selected to separate this background from antineutron events. Figure 4 shows the separate pulse-height distributions in the Čerenkov counter for the events which produced in X a pulse less than 100 Mev (solid histogram), and for the events which produced a pulse larger than 100 Mev (dashed histogram). The great difference between the two histograms with respect to both average pulse height and shape confirms the interpretation by which the neutral events are divided into antineutrons and background.

The energy scale in Fig. 4 is obtained by relating the pulse height produced by π mesons going through the glass to the computed ionization energy loss of 240 Mev. This calibration was repeated every day.

The standard for annihilation pulses is provided by the smooth curve of Fig. 4, which is the pulse-height distribution for antiprotons entering the lead glass when S_1, S_2, and the lead plate are removed. Comparison of the solid histogram with this antiproton curve justifies our interpretation that the solid histogram is produced by antineutrons.

For comparison with the annihilation spectra of Fig. 4, Fig. 5 shows the spectra obtained with 750-Mev positive protons (solid curve) and with 600-Mev negative pions incident on the glass Counter C. These spectra indicate that large pulses are rarely produced by particles of such energies. From this it is apparent that even high-energy neutrons could not produce a spectrum like the solid histogram of Fig. 4.

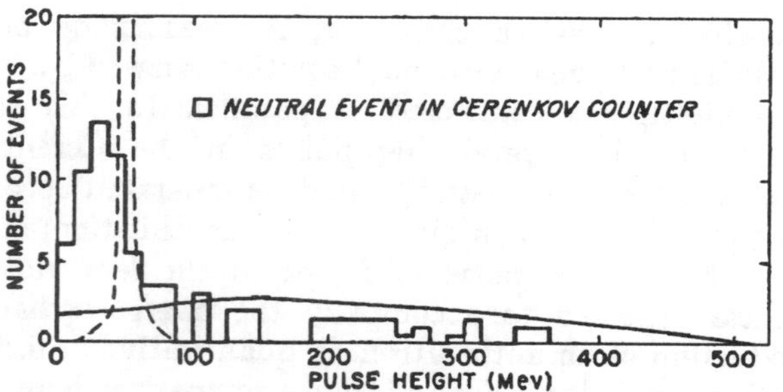

FIG. 3. Pulse-height spectrum in charge-exchange scintillator for 74 neutral events in lead glass. Histogram is for all neutral events. The smooth solid curve is for calibrating antiprotons for which no pulse occurred in S_1 or S_2. Smooth dashed curve is for noninteracting antiprotons. Smooth curves are each normalized to histogram.

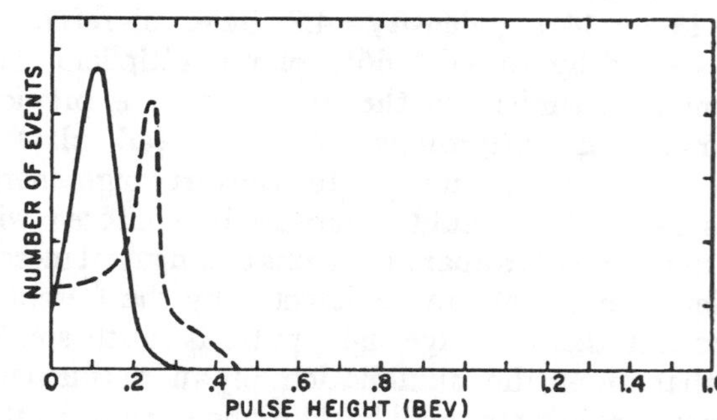

FIG. 5. Pulse-height spectrum in lead glass counter for π mesons (dashed curve) and for positive protons (solid curve). The curves are normalized.

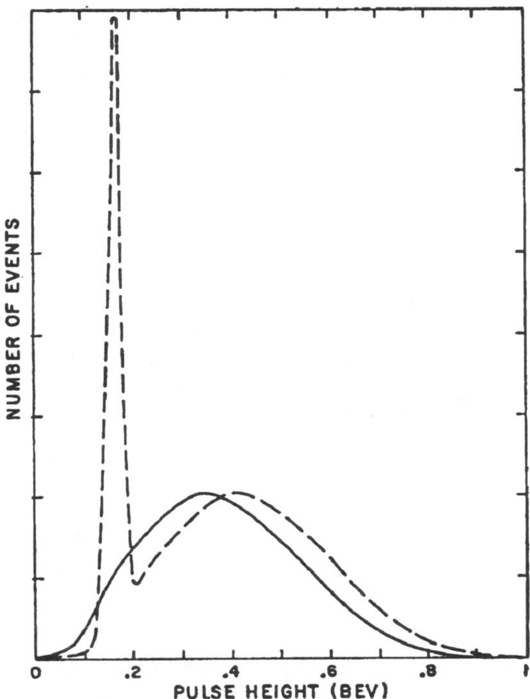

FIG. 6. Pulse-height spectrum of antiprotons in large scintillation counter. The dashed curve is for all incident antiprotons. The solid curve has had noninteracting antiprotons removed and includes a correction to permit comparison with antineutrons.

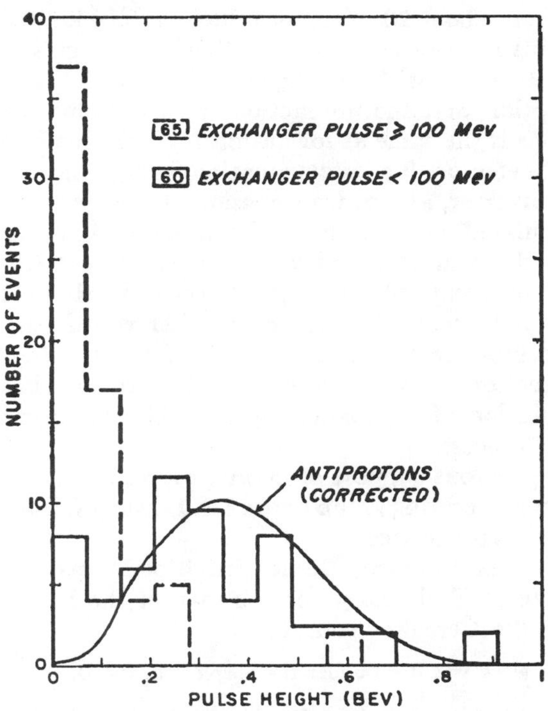

FIG. 7. Pulse-height spectrum in large scintillation counter for neutral events. Solid histogram is for 60 antineutrons (energy loss in charge-exchange scintillator less than 100 Mev). Dashed histogram is for 65 other neutral events. The smooth solid curve is the corrected antiproton curve from Fig. 6.

To determine the number of γ rays incident on S_1, the lead between S_1 and S_2 was removed. The number of neutral events per incident antiproton increased by a factor of 7. From the known probability that a single high-energy γ ray would be transmitted through 1 in. of lead without converting (3% for a γ-ray energy of 300 Mev), this observed increase shows that our neutral events contain at most 20% of γ-ray background before selection on the basis of pulse height in X.

The lead glass Counter C is very sensitive to γ rays and insensitive to ionization losses by slow particles. The desirability of comparing the spectra of antineutrons and antiprotons obtained with an entirely different type of detector led us to repeat the experiment with Counter C replaced by a liquid scintillator. This scintillator, 28 in. thick and 5 ft^3 in volume, was large enough to detect a substantial part of the energy of an annihilation event. For this experiment the thickness of the lead converter between S_1 and S_2 was increased to 1.5 in. As before, the antineutron detector was calibrated with antiprotons. The pulse-height distribution of antiprotons in the large scintillator is given by Fig. 6. The noninteracting antiprotons produce the sharp peak.

The solid smooth curve in Fig. 7 is the solid curve of Fig. 6, obtained by correcting the pulse height by 70 Mev toward lower energy because antiprotons ionize before interacting in the scintillator. Sixty neutral events were obtained (Fig. 7) after selection with the same criterion as before on the pulse height in X. Again the selected neutral spectrum and the antiproton spectrum are in agreement, although not so strikingly as with the lead glass. The sixty selected events apparently include some contamination. This interpretation is confirmed by the shape of the spectrum in X for all neutral events (Fig. 8). There are now many more neutral secondaries from inelastic collisions of antiprotons than there were in the experiment with the lead glass, and the separation between antineutrons and background is therefore not so good. The larger number of neutral secondaries is probably to be attributed to the greater sensitivity of the scintillator to neutrons.

The lead glass and the scintillator are of nearly equal efficiency in detecting the antineutrons. The observed

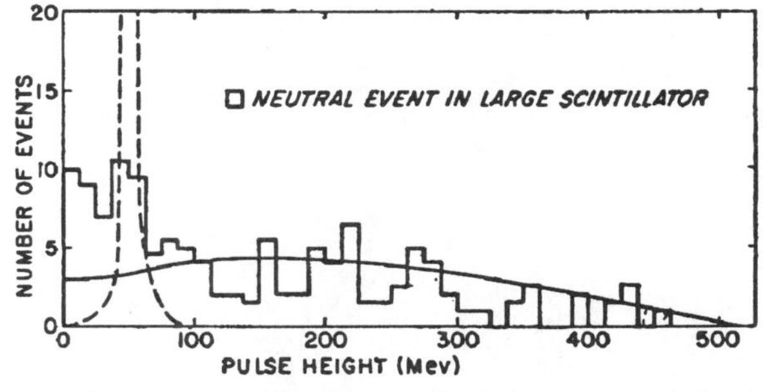

FIG. 8. Pulse-height spectrum in charge-exchange scintillator for 125 neutral events in large scintillator. The smooth curves are the same as in Fig. 3, each normalized to histogram.

yield from about 20 g/cm^2 of toluene is 0.0030 ± 0.0005 antineutrons per antiproton with the lead glass, and 0.0028 ± 0.0005 with the liquid scintillator. With the assumption that the interaction cross section for antineutrons is the same as for antiprotons, the inefficiency of the detector due to attenuation in S_1, S_2, and the lead converter, and to transmission of the detector can be calculated, and is found to be about 50%. From the observed antineutron yield the mean free path for charge exchange of the type detected is about 2300 g/cm^2 of toluene (C_7H_8); or, in other words, the exchange cross section is about 2% of the annihilation cross section for this material. This corresponds to a cross section of approximately 8 millibarns in carbon for this process.

The generous support of many groups, including the Bevatron operating group under Dr. Edward J. Lofgren, is greatly appreciated.

We thank Professor David Frisch of Massachusetts Institute of Technology for the loan of the lead glass used in the Čerenkov counter.

* This work was done under the auspices of the U. S. Atomic Energy Commission.

† On leave of absence from Brookhaven National Laboratory, Upton, New York.

[1] Chamberlain, Segrè, Wiegand, and Ypsilantis, Phys. Rev. **100**, 947 (1955).

[2] Brabant, Cork, Horwitz, Moyer, Murray, Wallace, and Wenzel, Phys. Rev. **101**, 498 (1956).

[3] Chamberlain, Chupp, Ekspong, Goldhaber, Goldhaber, Lofgren, Segrè, Wiegand, Amaldi, Baroni, Castagnoli, Franzinetti, and Manfredini, Phys. Rev. **102**, 921 (1956).

Observations of the Failure of Conservation of Parity and Charge Conjugation in Meson Decays: the Magnetic Moment of the Free Muon*

RICHARD L. GARWIN,† LEON M. LEDERMAN, AND MARCEL WEINRICH

Physics Department, Nevis Cyclotron Laboratories, Columbia University, Irvington-on-Hudson, New York, New York

(Received January 15, 1957)

LEE and Yang[1–3] have proposed that the long held space-time principles of invariance under charge conjugation, time reversal, and space reflection (parity) are violated by the "weak" interactions responsible for decay of nuclei, mesons, and strange particles. Their hypothesis, born out of the $\tau-\theta$ puzzle,[4] was accompanied by the suggestion that confirmation should be sought (among other places) in the study of the successive reactions

$$\pi^+\rightarrow\mu^++\nu, \qquad (1)$$

$$\mu^+\rightarrow e^++2\nu. \qquad (2)$$

They have pointed out that parity nonconservation implies a polarization of the spin of the muon emitted from stopped pions in (1) along the direction of motion and that furthermore, the angular distribution of electrons in (2) should serve as an analyzer for the muon polarization. They also point out that the longitudinal polarization of the muons offers a natural way of determining the magnetic moment.[5] Confirmation of this proposal in the form of preliminary results on β decay of oriented nuclei by Wu *et al.* reached us before this experiment was begun.[6]

By stopping, in carbon, the μ^+ beam formed by forward decay in flight of π^+ mesons inside the cyclotron, we have performed the meson experiment, which establishes the following facts:

I. A large asymmetry is found for the electrons in (2), establishing that our μ^+ beam is strongly polarized.

II. The angular distribution of the electrons is given by $1+a\cos\theta$, where θ is measured from the velocity vector of the incident μ's. We find $a=-\frac{1}{3}$ with an estimated error of 10%.

III. In reactions (1) and (2), parity is not conserved.

IV. By a theorem of Lee, Oehme, and Yang,[2] the observed asymmetry proves that invariance under charge conjugation is violated.

V. The g value (ratio of magnetic moment to spin) for the (free) μ^+ particle is found to be $+2.00\pm0.10$.

VI. The measured g value and the angular distribution in (2) lead to the very strong probability that the spin of the μ^+ is $\frac{1}{2}$.[7]

VII. The energy dependence of the observed asymmetry is not strong.

VIII. Negative muons stopped in carbon show an asymmetry (also leaked backwards) of $a\sim-1/20$, i.e., about 15% of that for μ^+.

IX. The magnetic moment of the μ^-, bound in carbon, is found to be negative and agrees within limited accuracy with that of the μ^+.[8]

X. Large asymmetries are found for the e^+ from polarized μ^+ beams stopped in polyethylene and calcium. Nuclear emulsion (as a target in Fig. 1) yields an asymmetry of about half that observed in carbon.

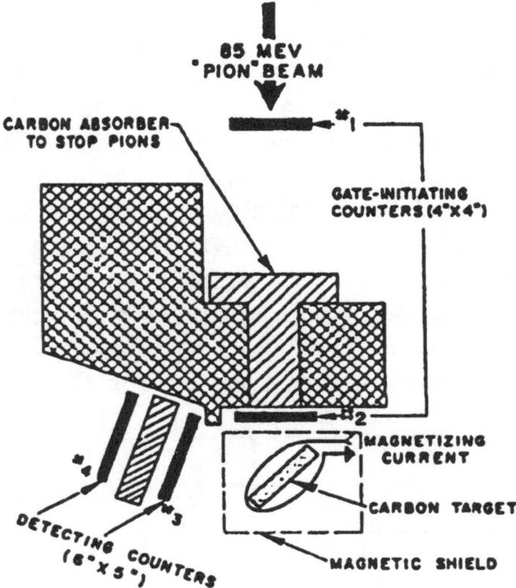

FIG. 1. Experimental arrangement. The magnetizing coil was close wound directly on the carbon to provide a uniform vertical field of 79 gauss per ampere.

The experimental arrangement is shown in Fig. 1. The meson beam is extracted from the Nevis cyclotron in the conventional manner, undergoing about 120° of magnetic deflection in the cyclotron fringing field and about −30° of deflection and mild focusing upon emerging from the 8-ft shielding wall. The positive beam contains about 10% of muons which originate principally in the vicinity of the cyclotron target by pion decay-in-flight. Eight inches of carbon are used in the entrance telescope to separate the muons, the mean range of the "85-"Mev pions being ~5 in. of carbon. This arrangement brings a maximum number of muons to rest in the carbon target. The stopping of a muon is signalled by a fast 1–2 coincidence count. The subsequent beta decay of the muon is detected by the electron telescope 3–4 which normally requires a particle of range >8 g/cm²(~25-Mev electrons) to register. This arrangement has been used to measure the lifetimes of μ^+ and μ^- mesons in a vast number of elements.[9] Counting rates are normally ~20 electrons/

min in the μ^+ beam and ~150 electrons/min in the μ^- beam with background of the order of 1 count/min.

In the present investigation, the 1–2 pulse initiates a gate of duration $T=1.25$ μsec. This gate is delayed by $t_1=0.75$ μsec and placed in coincidence with the electron detector. Thus the system counts electrons of energy >25 Mev which are born between 0.75 and 2.0 μsec after the muon has come to rest in carbon. Consider now the possibility that the muons are created in reaction (1) with large polarization in the direction of motion. If the gyromagnetic ratio is 2.0, these will maintain their polarization throughout the trajectory. Assume now that the processes of slowing down, stopping, and the microsecond of waiting do not depolarize the muons. In this case, the electrons emitted from the target may have an angular asymmetry about the polarization direction, e.g., for spin $\frac{1}{2}$ of the form $1+a\cos\theta$. In the absence of any vertical magnetic field, the counter system will sample this distribution at $\theta=100°$. We now apply a small vertical field in the magnetically shielded enclosure about the target, which causes the muons to precess at a rate of $(\mu/s\hbar)H$ radians per sec. The probability distribution in angle is carried around with the μ-spin. In this manner we can, with a fixed counter system, sample the entire distribution by plotting counts as a function of magnetizing current for a given time delay. A typical run is shown in Fig. 2. As an example of a systematic check, we have

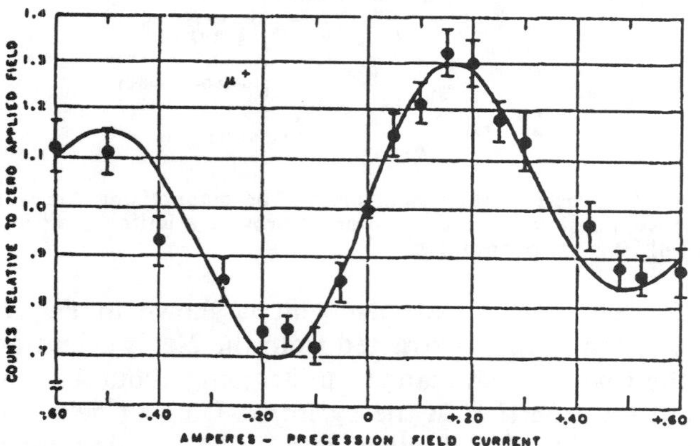

FIG. 2. Variation of gated 3–4 counting rate with magnetizing current. The solid curve is computed from an assumed electron angular distribution $1-\frac{1}{3}\cos\theta$, with counter and gate-width resolution folded in.

reduced the absorber in the telescope to 5 in. so that the end-of-range of the main pion beam occurred at the carbon target. The electron rate rose accordingly by a factor of 10, indicating that now electrons were arising from muons isotropically emitted by pions at rest in the carbon. No variation in counting rate with magnetizing current was then observed, the ratio of the rate for $I=+0.170$ amp to that for $I=-0.150$ amp, for example, being 0.989 ± 0.028. The highest field produced at the target was ~50 gauss which generates a stray field outside of the magnetic shield of $<\frac{1}{10}$ the cyclotron fringing field of 20 gauss. The only conceivable effect of the magnetizing current is the precession of muon spins and we are, therefore, led to conclusions I–IV as necessary consequences of these observations.

The solid curve in Fig. 2 is a theoretical fit to a distribution $1-\frac{1}{3}\cos\theta$, where

(1) the gyromagnetic ratio is taken to be +2.00;[10]

(2) the angular breadth of the electron telescope and the gate-width smearing are folded in, as well as (to first order) the exponential decay rate of muons within the gate;

(3) the small residual cyclotron stray field (*up* for Fig. 2, the positive magnetizing current producing a *down* field) is included. This has the accidental effect of converting the 100° initial angle ($H=0$) to 89° as in Fig. 2. We note that this experiment establishes only a lower limit to the magnitude of a, since the percent polarization at the time of decay is not known. If polarization is complete, $a=-0.33\pm0.03$.

Proof of the 2π symmetry of the distribution and the sign of the moment was obtained by shifting the electron counters to 65° with respect to the incident muon direction. The repetition of a magnetizing run yielded a curve as in Fig. 2 but shifted to the right by 0.075 ampere (5.9 gauss) corresponding to a precession angle of 37°, in agreement with the spatial rotation of the counter system. Thus we are led to conclusions V and VI.

A specific model, the two-component neutrino theory, has been proposed by Lee and Yang[8] in an attempt to introduce parity nonconservation naturally into elementary particle theory. This theory predicts, for our experimental arrangement and on the basis of 1.86 for the integrated spectrum (Fig. 2), a ratio of the order of 2.5 for energies greater than 35 Mev. We have increased the amount of absorber in the electron telescope to exclude electrons of less than ~35 Mev. The resulting peak-to-valley ratio was then observed to be 1.92 ± 0.19.[11]

We have also detected asymmetry in negative muon decay and have verified that the moment is negative and roughly equal to that of the positive muon.[7] The asymmetry in this case is also peaked backwards.

Various other materials were investigated for μ^+ mesons. Nuclear emulsion as a target was found to have a significantly weaker asymmetry (peak-to-valley ratio of 1.40 ± 0.07) and it is interesting to note that this did not increase with reduced delay and gate width. Neither was there any evidence for an altered moment. It seems possible that polarized positive and negative muons will become a powerful tool for exploring magnetic fields in nuclei (even in Pb, 2% of the μ^- decay into electrons[9]), atoms, and interatomic regions.

The authors wish to acknowledge the essential role of Professor Tsung-Dao Lee in clarifying for us the papers of Lee and Yang. We are also indebted to Professor C. S. Wu[6] for reports of her preliminary results in the Co^{60} experiment which played a crucial part in the

Columbia discussions immediately preceding this experiment.

* Research supported in part by the joint program of the Office of Naval Research and the U. S. Atomic Energy Commission.

† Also at International Business Machines, Watson Scientific Laboratories, New York, New York.

[1] T. D. Lee and C. N. Yang, Phys. Rev. 104, 254 (1956).

[2] Lee, Oehme, and Yang, Phys. Rev. (to be published).

[3] T. D. Lee and C. N. Yang, Phys. Rev. (to be published).

[4] R. Dalitz, Phil. Mag. 44, 1068 (1953).

[5] T. D. Lee and C. N. Yang (private communication).

[6] Wu, Ambler, Hudson, Hoppes, and Hayward, Phys. Rev. 105, 1413 (1957), preceding Letter.

[7] The Fierz-Pauli theory for spin $\frac{3}{2}$ particles predicts a g value of $\frac{2}{3}$. See F. J. Belinfante, Phys. Rev. 92, 997 (1953).

[8] V. Fitch and J. Rainwater, Phys. Rev. 92, 789 (1953).

[9] M. Weinrich and L. M. Lederman, *Proceedings of the CERN Symposium, Geneva, 1956* (European Organization of Nuclear Research, Geneva, 1956).

[10] The field interval, ΔH, between peak and valley in Fig. 2 gives the magnetic moment directly by $(\mu\Delta H/s\hbar)(t_1+\frac{1}{2}T)\delta=\pi$, where $\delta=1.06$ is a first-order resolution correction which takes into account the finite gate width and muon lifetime. The 5% uncertainty comes principally from lack of knowledge of the magnetic field in carbon. Independent evidence that $g=2$ (to $\sim$10%) comes from the coincidence of the polarization axis with the velocity vector of the stopped μ's. This implies that the spin precession frequency is identical to the μ cyclotron frequency during the 90° net magnetic deflection of the muon beam in transit from the cyclotron to the 1–2 telescope. We have designed a magnetic resonance experiment to determine the magnetic moment to $\sim$0.03%.

[11] *Note added in proof.*—We have now observed an energy dependence of a in the $1+a\cos\theta$ distribution which is somewhat less steep but in rough qualitative agreement with that predicted by the two-component neutrino theory $(\mu\rightarrow e+\nu+\bar{\nu})$ without derivative coupling. The peak-to-valley ratios for electrons traversing 9.3 g/cm^2, 15.6 g/cm^2, and 19.8 g/cm^2 of graphite are observed to be 1.80 ± 0.07, 1.84 ± 0.11, and 2.20 ± 0.10, respectively.

Free Antineutrino Absorption Cross Section. I. Measurement of the Free Antineutrino Absorption Cross Section by Protons*

FREDERICK REINES AND CLYDE L. COWAN, JR.†
Los Alamos Scientific Laboratory, University of California, Los Alamos, New Mexico

(Received September 8, 1958)

The cross section for the reaction $p(\bar{\nu},\beta^+)n$ was measured using antineutrinos ($\bar{\nu}$) from a powerful fission reactor at the Savannah River Plant of the United States Atomic Energy Commission. Target protons were provided by a 1.4×10^3 liter liquid scintillation detector in which the scintillator solution (triethylbenzene, terphyenyl, and POPOP) was loaded with a cadmium compound (cadmium octoate) to allow the detection of the reaction by means of the delayed coincidence technique. The first pulse of the pair was caused by the slowing down and annihilation of the positron (β^+), the second by the capture of the neutron (n) in cadmium following its moderation by the scintillator protons. A second giant scintillation detector without cadmium loading was used above the first to provide an anticoincidence signal against events induced by cosmic rays. The antineutrino signal was related to the reactor by means of runs taken while the reactor was on and off. Reactor radiations other than antineutrinos were ruled out as the cause of the signal by a differential shielding experiment. The signal rate was 36 ± 4 events/hr and the signal-to-noise ratio was $\frac{1}{3}$, where half the noise was correlated and cosmic-ray associated and about half was due to non-reactor-associated accidental coincidences. The cross section per fission $\bar{\nu}$ (assuming 6.1 $\bar{\nu}$ per fission) for the inverse beta decay of the proton was measured to be $(11\pm2.6)\times10^{-44}$ cm$^2/\bar{\nu}$ or $(6.7\pm1.5)\times10^{-43}$ cm^2/fission. These values are consistent with prediction based on the two-component theory of the neutrino.

I. INTRODUCTION

A DETERMINATION of the cross section for the reaction: antineutrino ($\bar{\nu}$) on a proton (p^+) to yield a positron (β^+) and a neutron (n),

$$\bar{\nu}+p^+\rightarrow\beta^++n, \tag{1}$$

permits a check to be made on the combination of fundamental parameters on which the cross section depends. Implicit in a theoretical prediction of the cross section are (1) the principle of microscopic reversibility, (2) the spin of the $\bar{\nu}$, (3) the particular neutrino theory employed: e.g., two- or four-component, (4) the neutron half-life and its decay electron spectrum, and (5) the spectrum of the incident $\bar{\nu}$'s.

An experiment which was performed to identify antineutrinos from a fission reactor[1] yielded an approximate value for this cross section. Following this work, however (and prior to the parity developments involved in point 3), the equipment was modified in order to obtain a better value of the cross section. The modification consisted in the addition of a cadmium salt of 2-ethylhexanoic acid to the scintillator solution[2] of one of the detectors of reference 1, utilizing the protons of the solution as targets for antineutrinos, and making the necessary changes in circuitry to observe both positrons and neutron captures in the detector resulting from antineutrino-induced beta decay in the detector. In addition, a second detector used in the experiment of reference 1 was now used as an anticoincidence shield against cosmic-ray-induced backgrounds, and static shielding was increased by provision of a water tank about 12-inches thick below the target detector. The delayed-coincidence count rate resulting from the positron pulse followed by the capture of the neutron was observed as a function of reactor power, and an analysis of the reactor-associated signal yielded, in addition to an independent identification of the free antineutrino, a measure of the cross section for the reaction and a spectrum of first-pulse (or $\bar{\nu}$) energies. Since the antineutrino spectrum is simply related to the β^+ spectrum, the measurement yields an antineutrino spectrum above the 1.8-Mev reaction threshold. The spectrum is, however, seriously degraded by edge effects in the detector.

This experiment was identical in principle with that performed at Hanford in 1953.[3] It was, however, definitive from the point of view of antineutrino identification (whereas the Hanford experiment was not) because of a series of technical improvements, coupled with the better shielding against cosmic rays achieved by going underground. The improvements consisted in the use of an isolated power supply to diminish electrical noise from nearby machinery, better shielding from the reactor gamma-ray and neutron background, a more complete anticoincidence shield against charged cosmic rays through the use of a liquid scintillation detector, and use of a large detector containing 6.5 times as many proton targets.[4] In addition, oscilloscopic presentation and photographic recording of the data assisted materially in analyzing the signals and rejecting electrical noise.

* Work performed under the auspices of the U. S. Atomic Energy Commission.

† Now at the Department of Physics, George Washington University, Washington, D. C.

[1] Cowan, Reines, Harrison, Kruse, and McGuire, Science 124, 103 (1956).

[2] Ronzio, Cowan, and Reines, Rev. Sci. Instr. 29, 146 (1958), describe the preparation and handling of liquid scintillators developed for the Los Alamos neutrino program.

[3] F. Reines and C. L. Cowan, Jr., Phys. Rev. 90, 492 (1953).

[4] The gain, times 6.5, due to the increase in target protons was largely balanced by a decrease in the neutron detection efficiency, times $\frac{1}{3}$, made necessary by other experimental considerations.

II. THE EXPERIMENT

Figure 1 represents schematically the sequence of events which occur when an antineutrino is captured by a proton. The cross section $\bar{\sigma}$ for the process for an average fission $\bar{\nu}$ is determined from the relation

$$\bar{\sigma}=\frac{R}{3600 f n \epsilon_{\beta^+} \epsilon_n}\ \text{cm}^2, \qquad (2)$$

where R = the observed signal rate in counts/hr; n = the number of target protons $= 8.3\times10^{28}$; f = the antineutrino flux at the detector in $\bar{\nu}/\text{cm}^2\,\text{sec} = 1.3\times10^{13}$, assuming $N=6.1\ \bar{\nu}$/fission; ϵ_{β^+} = the positron detection efficiency, and ϵ_n = the neutron detection efficiency.

Note that the mean cross section per fission ($N\bar{\sigma}$) is independent of the number of antineutrinos assumed emitted by the fission-fragments per fission (N). Uncertainties in the $\bar{\nu}$ flux f (5 to 10%) arise from imprecise knowledge of reactor power, uncertainty concerning energy released per fission and the number of $\bar{\nu}$ per fission, and incomplete knowledge of the fission-fragment distribution in the reactor. The $\bar{\nu}$ energy spectrum is determined from a measured β^+ spectrum (or the first-pulse spectrum of the antineutrino-produced delayed coincidences after appropriate energy-resolution corrections). The energy $E_{\bar{\nu}}$ of the $\bar{\nu}$ is related to E_{β^+}, the kinetic energy of the product β^+, by the equation

$$E_{\bar{\nu}}=3.53+E_{\beta^+}(mc^2 \text{ units}). \qquad (3)$$

We have neglected the few-kilovolt recoil energy of the product neutron.

With these quantities in mind we will describe the experiment in conjunction with a schematic diagram of the equipment (Fig. 2).[5] Assume that an antineutrino-

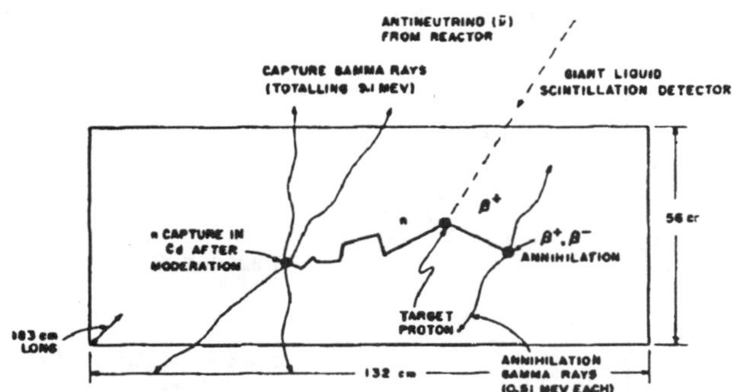

FIG. 1. Schematic of antineutrino detector. This 1.4×10^3-liter detector is filled with a mixture which consists primarily of triethylbenzene (TEB) with small amounts of p-terphenyl (3 g/liter), POPOP wavelength shifter (0.2 g/liter and cadmium (1.8 g/liter) as cadmium octoate. An antineutrino is shown transmuting a proton to produce a neutron and positron. The positron slows down and annihilates, producing annihilation radiation. The neutron is moderated by the hydrogen of the scintillator and is captured by the cadmium, producing capture gamma rays.

induced reaction occurs in the detector. The β^- signal is seen by each of two interleaved banks of 55, 5-inch Dumont photomultiplier tubes in prompt coincidence within the 0.2-μsec resolving time of the equipment. The signals are added by preamplifiers whose gains have been balanced to allow for slight differences in the response of the two photomultiplier banks, amplified further and sent via a 30-μsec delay line to the deflection plates of a recording oscilloscope. At the same time the two signals are sent separately to a prompt coincidence unit (marked β^+) which accepts them if they correspond to pulse-height amplitudes between 1.5 and 8 Mev. On receipt of an acceptable signal, the β^+ scaler is tripped, and a gating pulse is sent to the second coincidence unit (marked n). If during a prescribed time (0.75 to 25.75 μsec) following the β^+ pulse, a neutron pulse corresponding to an energy deposition of 3 to 10 Mev in the antineutrino detector occurs (again in prompt coincidence from the two interleaved photomultiplier banks), the neutron coincidence unit signals a delayed coincidence. This delayed coincidence is registered by a scaler and triggers the scope sweep, allowing the entire sequence which has been stored in in the 30-μsec delay lines to be displayed and photographed. The neutron prompt coincidences are also recorded by a scaler.[6] Therefore the raw data obtained for analysis are the following: the rates in the positron and neutron gates, delayed coincidence rate, scope trigger rate, pulse amplitudes, and time intervals between pulses as seen on the recording oscilloscope. These data are obtained with the reactor on and off and with gross changes in bulk shielding provided by bags of wet sawdust.

In addition to the above arrangement there is provision for the reduction of cosmic-ray-associated background by means of an anticoincidence detector placed above the antineutrino detector as shown in Fig. 2. If, for example, a pulse occurs in the anticoincidence detector of amplitude >0.5 Mev in coincidence with otherwise acceptable β^+-like pulses, the event is not accepted by the β^+ coincidence-anticoincidence unit and hence is not recorded by the oscilloscope. This is a reasonable criterion since the annihilation radiation which might reach the anticoincidence detector for a bonafide $\bar{\nu}$ event is at most 0.5 Mev. In order to reduce the background from events secondary to the passage of high-energy (>8 Mev) charged cosmic rays and delayed in time, the β^+ coincidence unit had incorporated into it a long gate which rendered the system insensitive for ~60 μsec following such a pulse. Pulses triggered by electrical noise are also eliminated by means of the distinctive visual record.

[5] Photographs of the detectors and some associated equipment may be found in F. Reines and C. L. Cowan, Jr., Physics Today 10, No. 8, 12 (1957). Detector details are described in an article by F. Reines in the forthcoming book, *Methods of Experimental Physics,* edited by L. C. Yuan and C. S. Wu [Academic Press, Inc., New York (to be published)], Vol. 5.

[6] To be precise, we should use the phrases "β^+-like" and "n-like" to describe the pulses because pulses in these energy ranges are produced by other particles as well.

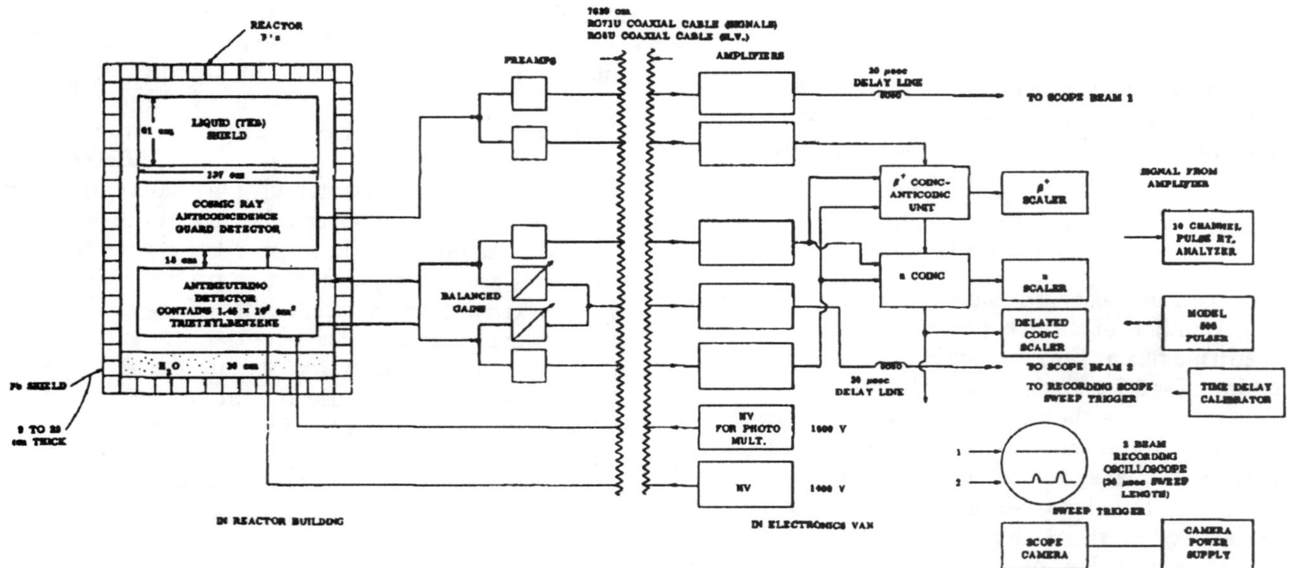

FIG. 2. Schematic of experimental arrangement.

A. Calibration

Calibrations of energy and time-interval response were made periodically. The first was accomplished by employing the μ-meson "through-peak" energy, the second by means of standard time markers put directly on the oscilloscope traces by a crystal oscillator. Gate lengths were checked against a time-delay calibrator designed for the purpose. Figure 3 shows the through-peak, a pulse-amplitude distribution resulting from the vertical passage through the tank of penetrating cosmic-ray μ mesons, taken before and after the present experiment for each of the two interleaved photomultiplier tube banks. Since most of the mesons are minimally ionizing, and the depth of the liquid is 60 cm, the specific energy loss[7] in the liquid of 1.57 Mev/cm gives the location of the peak as 100 Mev. The peak represents a slightly higher energy than that calculated from the energy loss/cm times the tank depth because of the finite lateral extent of the tank and the angular distribution of the cosmic rays. The peak is located to an accuracy of $\pm 5\%$. Since the detector response is proportional to the energy deposited in it, a standard linear pulser was calibrated with the through-peak amplitude and then used to calibrate the system in turn and set the appropriate gates. Based on measurements using artificial radioactive sources and the end point of the electron spectrum from μ-meson decay in our large liquid scintillation detectors, the error in energy calibration is believed to be less than $\pm 10\%$.

B. Determination of the Signal Rate R

The signal rate R was determined from the four series of measurements summarized in Table I. In principle the procedure is straightforward: the accidental background rate A (hr^{-1}) is determined for each run from the relation

$$A = 3600 \alpha \bar{n} \bar{\beta}^{+} \tau \text{ hr}^{-1}, \tag{4}$$

where τ = delayed-coincidence gate length in sec $= 25 \times 10^{-6}$; α = overlap factor for counts in n and β^+ gates; $\bar{n}$ and $\bar{\beta}^+$ = the rates in the neutron and positron gates averaged over each run as measured by the scalers.

We see from a comparison of the delayed-coincidence rate as given by the scalers and film analysis, however, that about $\frac{1}{3}$ of the scaler rate is rejected as unsuitable on inspection of the film traces. This means that the accidental background rate calculated from the n and β^+ scalers is too high by a factor of about $\frac{1}{3}$. In addition, the energies in the n and β^+ gates

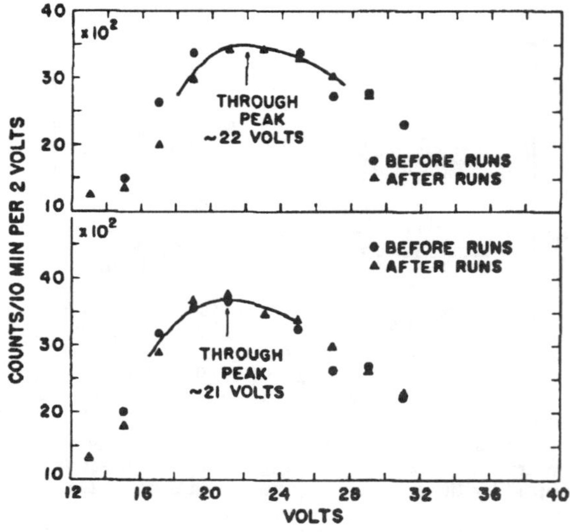

FIG. 3. Ungated meson through-peaks for energy calibration and stability check.

[7] This value is obtained from Fig. 2.91 of the book by B. Rossi, *High-Energy Particles* (Prentice-Hall, Inc., Englewood Cliffs, New Jersey, 1952), using a carbon density of 0.88 g/cm^3.

TABLE I. Summary of runs (a).

Run No.[a]	Comments	Run length (hr)	Scaler readings			Total scope film accept. rate (hr^{-1}) (in 25 μsec)
			n (sec^{-1})	$\bar{\beta}^+$ (sec^{-1})	Del. coinc (hr^{-1}) (25 μsec gate)	
232	Reactor *ON*, wet saw-	2.05	15.9	68.5	307.8	212.6
233	dust shield in place.	14.43	15.8	67.5	304.5	220.3
234	(Category *A*)	8.0	15.6	66.6	302.6	209.4
235		13.30	15.6	66.8	297.8	205.3
236		9.5	15.4	67.0	299.5	214.8
237	Reactor *OFF*, sawdust	12.47	14.3	65.0	251.0	165.6
238	shield in place.	9.37	14.0	63.6	249.0	170.9
240	(Category *B*)	9.54	14.2	68.0	256.6	173.7
241		14.43	14.5	66.3	251.8	170.1
243		6.77	14.4	65.4	251.1	163.5
246	Reactor *ON*, sawdust	12.20	16.3	71.6	313.4	228.3
247	shield in place.	2.00	16.2	71.1	300.0	213.5
248	(Category *C*)	11.12	16.1	71.1	314.5	224.4
249		9.53	16.2	71.2	327.2	236.6
251		10.53	16.5	72.4	320.9	226.2
252		11.67	16.2	71.5	324.7	232.6
253		8.92	16.3	71.6	316.3	222.5
255	Reactor *ON*, sawdust	6.48	17.2	75.2	334.4	250.5
256	removed.	10.38	17.3	76.0	331.1	240.2
	(Category *D*)					

[a] Runs are listed in chronological sequence. Missing runs were omitted either because they were incomplete or were not a relevant part of this series.

overlap, and judging by the rates in these gates, $1.23>\alpha>1.00$. Basing our calculations on the scope films, we find the net rates (total less accidental) for the four categories of runs which we list in Table II. Since $\alpha<1.23$ and the truth is between (a) and (c), we quote $R=36\pm4$ hr^{-1}, where ±4 includes the statistical error listed in column (a) and an allowance for the drift in the energy calibration, which analysis of the data shows likely to have occurred in the period between the series of runs *A* and *C*. The ratio of the n/β^+ rates is lower for runs *C* than for *A*. This is consistent with an increase in the over-all gain of the system, since the background spectrum decreases monotonically with increasing energy, and an increase in gain would bring in relatively more low-energy pulses. Runs *D* were made to demonstrate that the sawdust shield, though effective in reducing neutron signals from an Am-Be source (and hence reactor neutrons) by a factor of 15 and gammas by a factor of 2, had no effect on the antineutrino signal. The antineutrino flux during *D* was up by 10% because of a change in reactor power which happened to coincide with these runs. When corrected for this rise in reactor power, the results from *D* are consistent with the other runs.

C. Signal-to-Background Ratio

From Tables I and II we conclude that the signal-to-total-background ratio is approximately $\frac{1}{5}$, with the background about equally divided between correlated and accidental events. Correlated events arise primarily from fast neutrons produced by μ-meson capture in the vicinity of the detector: the first pulse is produced as a proton recoil, the second by the capture of the neutron in the scintillator cadmium. The correlated reactor-associated background is deduced from the absence of an observable effect due to the 75-cm sawdust shield (density, 0.5 g/cm^3; neutron shielding factor, 15) to be $<1/10$ the signal. An accidental background increase of 15 hr^{-1} was associated with the reactor so that the ratio of signal to accidental reactor-associated background was about 2/1.

TABLE II. Summary of results (b).

Run category	Net rate (hr^{-1}) = Gross rate less calculated accidental background (a) Bkd. reduced by signal ratio and $\alpha=1$	(b) No correction "$\frac{1}{5}$" factor and $\alpha=1$	(c) Correction "$\frac{1}{5}$" factor and $\alpha=1$	Results Reactor associated signal $=[(C-B)66+(A-B)47.3]/113.3$
A (47.3 hr)	146.2±1.7	118.4	130.8	(a) 38±3 (hr^{-1})
B (52.6 hr)	112.2±1.5	84.4	99.1	(b) 37±3
C (66.0 hr)	153.0±1.4	123.3	135.8	(c) 35±3
D (16.9 hr)	157.9±2	126.6	138.1	

D. Efficiency Estimates

In order to evaluate the cross section, we require the efficiencies ϵ_{β^+} and ϵ_n. Since these quantities were inferred rather than measured directly, some discussion of the efficiency evaluation procedure employed is in order.

ε_{β^+}

It is evident that the β^+-detection efficiency is high because of the small probability of β^+ leakage from the detector. The problem is to determine the probability that an event will fall within the energy gates employed, i.e., 1.5 to 8 Mev. To estimate this probability, plots were made of the first-pulse spectrum with the reactor on and off as measured in runs *A*, *B*, and *C*. Figure 4 shows the spectrum of first pulses scaled to run time of 47.3 hr. The lowest energy points are seen to drop sharply, a fact attributed to the effect of energy gates cutting into the spectrum. Since the background spectrum should continue to rise with decreasing energy,

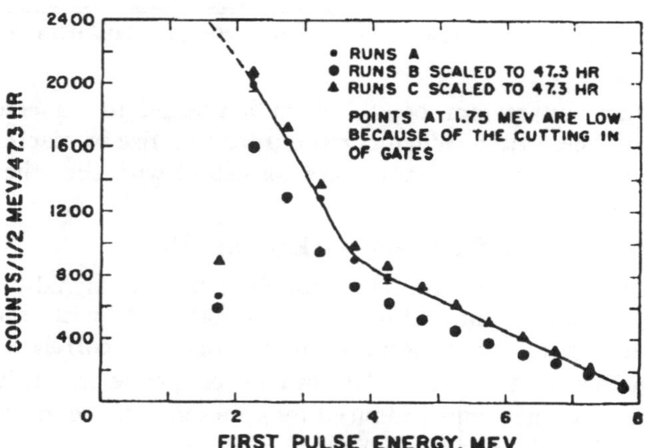

FIG. 4. First-pulse spectrum.

the reactor on-off difference was scaled up by a factor determined from an extrapolation to lower energies and is shown on the first-pulse difference curve on Fig. 5. In deriving the difference curve, no account was taken of the increase in accidental background associated with the reactor, and so the curve rises more sharply at lower energies than does the true β^+ spectrum. The β^+ detection efficiency was deduced from this curve by extrapolating to the origin and measuring the fraction of the area in the experimental *vs* the extrapolated curve. This procedure underestimates the efficiency somewhat because a subsequent measurement of the ungated spectrum seen from a Cu^{64} β^+ source dissolved in the scintillator showed no pulses of energy <0.45 Mev, whereas we have here assumed pulses down to 0 Mev. Accordingly, the β^+ efficiency estimate from Fig. 5 (0.81) is raised slightly and taken to be $\epsilon_{\beta^+}=0.85\pm0.05$, where 0.05 is meant to indicate the limits of error in ϵ_{β^+}.

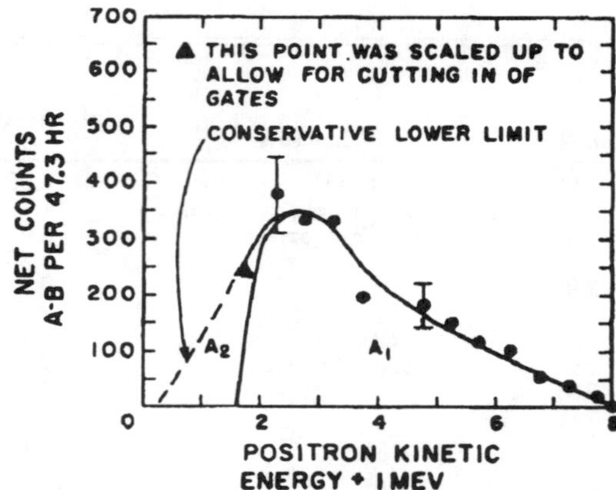

FIG. 5. Positron spectrum, $A_1/(A_1+A_2)=0.81$. No correction is made here for reactor-associated background. This background raises the lower-energy part of the spectrum and hence $\epsilon_{\beta^+}>0.81$. Take $\epsilon_{\beta^+}=0.85\pm0.05$.

ε_n

The neutron-detection efficiency is somewhat more difficult to estimate. This efficiency is given as the product of three factors:

$$\epsilon_n=\epsilon_{n1}\epsilon_{n2}\epsilon_{n3}, \tag{5}$$

where ϵ_{n1}=probability that the neutron will not leak out of the system, ϵ_{n2}=probability that the neutron will be captured in the scintillator cadmium in the 25-μsec time interval (0.75 to 25.75 μsec) after its birth, and ϵ_{n3}=probability that the neutron capture gamma rays will produce a signal which falls within the chosen energy gates, 3 to 10 Mev.

We estimate ϵ_{n1} from a consideration of the detector-volume fraction within an antineutrino-produced neutron mean free path of the detector surface. From the conservation laws applied to reaction (1) the neutron energy is $\lesssim$10 kev and therefore has a mean free path in the scintillator of about 1 cm. The fraction of the detector volume within 1 cm of the edge is about 6% and approximately $\frac{1}{2}$ (or 3%) of the neutrons born in this region will be travelling outward; hence $\epsilon_{n1}=0.97$.

The least certain of the factors involved in the neutron-detection efficiency is ϵ_{n2}. It was estimated in

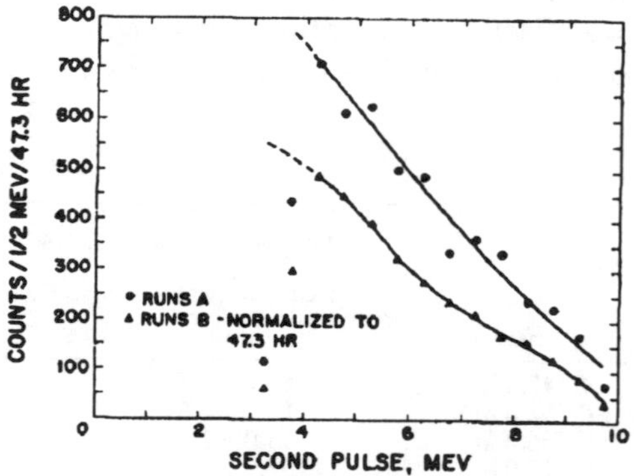

FIG. 6. Second-pulse spectrum.

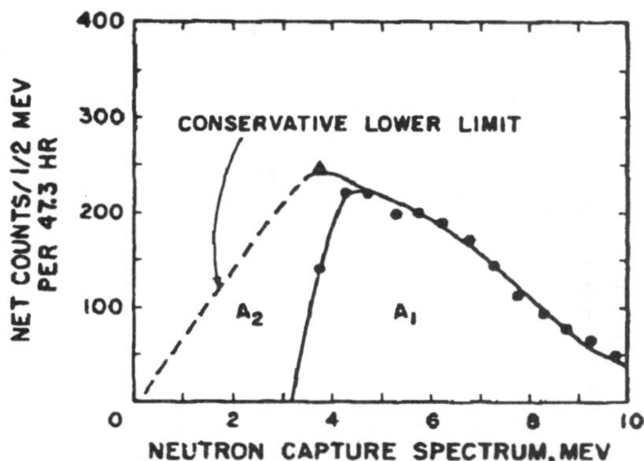

FIG. 7. Neutron-capture gamma spectrum. $A_1/(A_1+A_2)=0.68$. No correction is made here for reactor-associated background. This background raises the lower-energy part of the spectrum and hence $\epsilon_{n3}>0.68$. Take $\epsilon_{n3}=0.75\pm0.05$.

two ways: by an interpolation of the curves calculated via the Monte Carlo method for cases[8] involving higher Cd/H ratios than the one used in this experiment (here Cd/H=0.000145) and by integration of the cadmium-capture probability for thermal neutrons from 0.75 to 25.75 μsec after their introduction into the scintillator. The interpolation gives $\tau_{n2}=0.15$ with a ±0.02 uncertainty. If capture competition by the scintillator hydrogen is neglected, the mean time for capture $\tau_{Cd}=161$ μsec, and the capture probability is calculated to be 0.142. Since the mean capture time in scintillator hydrogen $\tau_H=235$ μsec, the hydrogen captures in this period reduce the number of captures in Cd so that $\epsilon_{n2}=0.135$.

We estimated ϵ_{n3} in much the same way as ϵ_{β^+}. Figure 6 shows the first-pulse spectra in runs A and B normalized to 47.3 hr. Figure 7 shows the difference spectrum and $A_1/(A_1+A_2)=0.68$. Since, as with the first-pulse spectrum, no allowance was made for the accidental background, we take

$$\epsilon_{n3}=0.75\pm0.05,$$

where ±0.05 is meant to indicate the limits of error in ϵ_{n3}.

To summarize, $\epsilon_n=0.97\times0.75\times0.14=0.10$. It seems reasonable to assign error limits of $\pm20\%$ to this efficiency. An experimental attempt to measure the neutron-detection efficiency succeeded only in setting a lower limit of 6%. Figures 8 and 9 show the distribution of time-delay intervals between the pairs of pulses comprising the delayed coincidences. The curves are characteristic of neutron captures in the scintillator.[8]

E. The Cross Section

Inserting the efficiency numbers, etc., into Eq. (2) we find the cross section for fission antineutrino absorption by protons:

$$\bar{\sigma}=\frac{36\pm4}{3600\times1.3\times10^{13}\times8.3\times10^{28}\times(0.85\pm0.05)\times(0.10\pm0.02)}=11\pm2.6\times10^{-44}\ \mathrm{cm}^2/\bar{\nu},$$

or

$$N\bar{\sigma}=6.7\pm1.5\times10^{-43}\ \mathrm{cm}^2/\mathrm{fission},$$

where we have quoted the root-mean-square error.

This value of the cross section is consistent with predictions based on the two-component theory of the neutrino.[9]

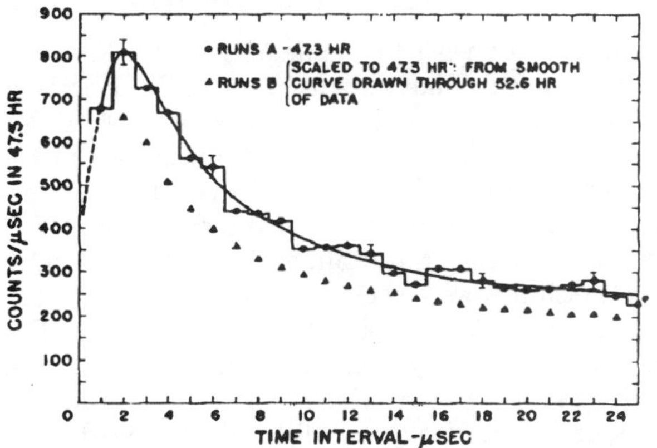

FIG. 8. Distribution of time intervals between pulses. ● Runs A (47.3 hr). ▲ Runs B (scaled to 47.3 hr from smooth curve drawn through 52.6 hr of data).

F. The $\bar{\nu}$ Spectrum from Fission Fragments

It is possible to deduce the fission fragment $\bar{\nu}$ spectrum from a measurement of the β^+ energies in reaction (1) and a knowledge of the cross section for the process. Because of the large experimental error involved in our determination of the β^+ spectra, the resultant $\bar{\nu}$ spectrum is very poorly determined. Nonetheless it seems worthwhile to make such a deduction. Apart from statistical fluctuations, the major uncertainty is in the energy resolution of the system. This uncertainty is due to leakage of one 0.511-Mev annihilation gamma ray from the detector and the poor energy resolution ($\pm25\%$) for 1 Mev deposited in the large detector. The effect of gamma-ray leakage on the spectrum was checked by dissolving a Cu^{64}-octoate source in the scintillator and measuring the energy

[8] Reines, Cowan, Harrison, and Carter, Rev. Sci. Instr. 25, 1061 (1954).

[9] The cross sections for monoenergetic $\bar{\nu}$ and for the measured $\bar{\nu}$ spectrum for fission fragments are given by Carter, Reines, Wagner, and Wyman, Phys. Rev. 113, 280 (1959), following paper.

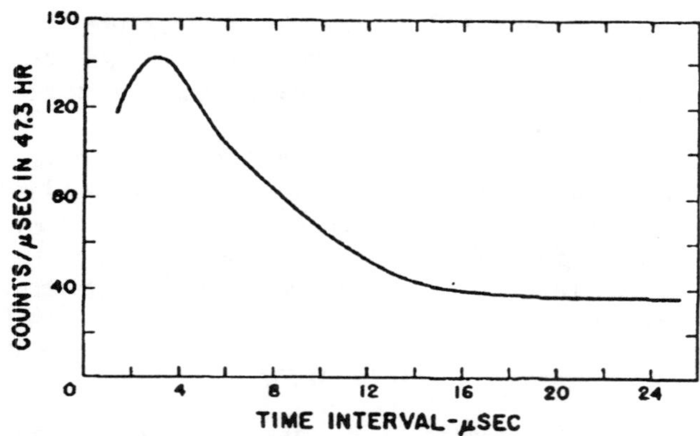

FIG. 9. Reactor-associated distribution of time intervals between pulses. Runs *A-B* (normalized to 47.3 hr).

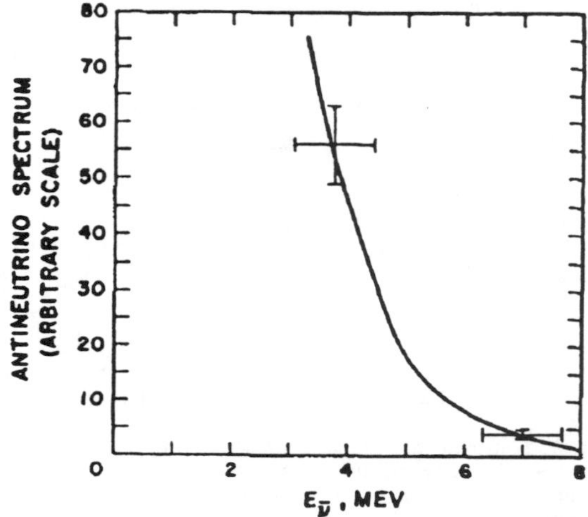

FIG. 10. Antineutrino spectrum from fission fragments deduced from β^+ spectrum in reaction $\bar{\nu}(p^+,n)\beta^+$. Crosses denote estimates of uncertainties.

spectrum. It was somewhat distorted, but the main effect was to drop the energy by about 0.5 Mev.

The β^+ spectrum $n(E_{\beta^+})$, the cross section for monoenergetic $\bar{\nu}$, $\sigma(E_{\bar{\nu}})$ and the $\bar{\nu}$ spectrum $m(E_{\bar{\nu}})$ are related[9] by the equation

$$m(E_{\bar{\nu}})=n(E_{\beta^+})/\sigma(E_{\bar{\nu}}). \quad (6)$$

Figure 10 shows $m(E_{\bar{\nu}})$. Gamma-ray leakage was considered in that the β^+ curve was shifted to the right by 0.5 Mev prior to the calculation. The data are not considered sufficiently accurate to warrant estimation of the distortion due to the energy resolution of the detector.

III. REMARKS CONCERNING AN IMPROVED MEASUREMENT

At least two major improvements could be made in this measurement to increase the counting rate and improve the energy resolution. The neutron-detection efficiency can be raised from its present 10% to about 80% by increasing the cadmium content of the scintillator by a factor of 20. This can be done without unreasonable reduction in the light transmission and scintillator efficiency by using the highly purified cadmium octoate recently developed by Ronzio.[10]

In addition, redesign of the detector using an inner "cadmiated" region enclosed in a noncadmium-bearing scintillator would minimize end effects due to gamma-ray leakage from the detector. Such an increase in detection efficiency would permit a factor of ten reduction in the size of the antineutrino-sensitive (or cadmiated) volume without undue sacrifice in signal rate. At presently available antineutrino fluxes, a signal rate of 30 hr^{-1} or more would result from the smaller improved detector. The signal-to-accidental-background ratio would be raised by a factor of about ten for a 25-μsec delayed-coincidence gate because of the increase in the signal rate and the decrease in detector size. The uncadmiated scintillator blanket should help shield against cosmic-ray-correlated events which are due to a neutron produced by μ-meson capture in the vicinity of the detector. A cylindrical shape with photomultiplier tubes placed around the cylinder wall would make for a uniform light collection and hence improved energy resolution. This detector could be shielded against cosmic rays with the anti-coincidence detector as before, and much the same electronics could be used.

IV. ACKNOWLEDGMENTS

The authors wish to thank Mr. H. W. Kruse for his analysis of the film records. Dr. F. B. Harrison and Dr. A. D. McGuire and R. P. Jones and M. P. Warren helped set up the equipment. Many groups in the Los Alamos Scientific Laboratory have given support in the design and construction of the equipment. We wish especially to thank the Shops Department under Mr. G. H. Schultz and the Electronics Group under Mr. John Lamb. The hospitality of the E. I. duPont de Nemours Company, which operates the Savannah River Plant of the United States Atomic Energy Commission, is also gratefully acknowledged.

[10] A. R. Ronzio (private communication). Another solution, cadmium propionate in toluene with the suggested Cd/H ratio of 0.003, was used in the Hanford work[3,8] but its generally undesirable characteristics, e.g., fire hazard and toxicity, militated against its use in the present experiment.

RESONANCE IN THE $\Lambda\pi$ SYSTEM*

Margaret Alston, Luis W. Alvarez, Philippe Eberhard,† Myron L. Good,‡
William Graziano, Harold K. Ticho,‖ and Stanley G. Wojcicki
Lawrence Radiation Laboratory and Department of Physics, University of California, Berkeley, California
(Received October 31, 1960)

We report a study of the reaction

$$K^- + p \to \Lambda^0 + \pi^+ + \pi^- \qquad (1)$$

produced by 1.15-Bev/c K^- mesons and observed in the Lawrence Radiation Laboratory's 15-in. hydrogen bubble chamber. A preliminary report of these results was presented at the 1960 Rochester Conference.[1] The beam was purified by two velocity spectrometers.[2] A Ξ^0 hyperon observed during the run[3] and the preliminary cross sections[4] for various K^- reactions at 1.15 Bev/c have been reported previously. Reaction (1) was the first one selected for detailed study, because it appeared to take place with relatively large probability and because the event, a 2-prong interaction accompanied by a V, was easily identified. In a volume of the chamber sufficiently restricted so that the scanning efficiency was near 100%, 255 such events were found. These events were measured, and the track data supplied to a computer which tested each event for goodness of fit to various kinematic hypotheses. The possible reactions, the distribution of events, and the corresponding cross sections are given in Table I. An event was placed in a given category of Table I if the χ^2 probability for the other hypotheses was <1%. It appears likely that the majority of the events in group (e) are also reactions of type (1). This belief is based on the following arguments:

1. Since the kinematics of a $\Lambda\pi\pi$ fit (four constraints) are more overdetermined than those of a $\Sigma^0\pi\pi$ fit (two constraints), it is relatively easy for a $\Lambda\pi\pi$ reaction to fit a $\Sigma^0\pi\pi$ reaction, but only

Table I. Distribution of events among different reactions.

Reaction	No. of events	Cross section (mb)
(a) $K^- + p \to \bar{K}^0 + p + \pi^-$	48	2.0 ± 0.3
(b) $K^- + p \to (\Lambda \text{ or } \Sigma^0) + \pi^+ + \pi^- + \pi^0$	39	1.1 ± 0.2
(c) $K^- + p \to \Sigma^0 + \pi^+ + \pi^-$	27	
(d) $K^- + p \to \Lambda + \pi^+ + \pi^-$	49	4.1 ± 0.4
(e) $K^- + p \to (\Lambda \text{ or } \Sigma^0) + \pi^+ + \pi^-$	92	
Total	255	7.2 ± 0.5

very few Σ^0 configurations can fit the $\Lambda\pi\pi$ reactions.

2. The events of group (e) when treated as $\Sigma^0\pi\pi$ reactions give a χ^2 distribution which is much worse than that obtained when they are treated as $\Lambda\pi\pi$ reactions.

In what follows, the 141 events of groups (d) and (e) are treated as examples of reaction (1). We estimate that 10 to 15% are actually Σ^0 events.

The energy distribution of the two pions in the K^--p barycentric system is shown in Fig. 1. If the cross section were dominated by phase space alone, the distribution of the points on the two-dimensional plot of Fig. 1 should be uniform. This is clearly not the case. On the contrary, both the π^+ and the π^- distributions have peaks near 285 Mev, such as would be expected from a quasi-two-body reaction of the type

$$K^- + p \to Y^{*\pm} + \pi^{\mp}, \tag{2}$$

the Y^* having a mass spectrum peaking at ~1380 Mev. If the Y^* of mass 1380 Mev breaks up according to

$$Y^{*\pm} \to \Lambda^0 + \pi^{\pm}, \tag{3}$$

the pions from this breakup are expected to have energies ranging from 58 to 175 Mev in the K^-p rest system. Those pions from (3) are well separated from the pions arising from reaction (2) in the energy histograms.

The isotopic spin of this excited hyperon must be one, since it breaks up into a Λ and a π. Since the Y^* is produced with a pion, also of isotopic spin one, the reaction could proceed either in the $I=0$ or the $I=1$ state. Therefore the ratio of Y^{*+} to Y^{*-} will depend on the relative magnitude and phase of the two isotopic-spin amplitudes and thus could differ from unity. We observed 59 Y^{*+} events and 82 Y^{*-} events, using the criterion for separation that the high-momentum π meson is the pion from reaction (2).

Figure 2 shows the distribution in mass of the Y^* state (both Y^{*+} and Y^{*-}) including all 141 events, again using the higher energy pion in each event to calculate the Y^* mass. The experimental uncertainty in the mass for each event is small compared to the observed width of the curve. The curves of Fig. 2 are discussed later.

Figure 3 shows production angular distributions for Y^{*+} and Y^{*-} in the K^-p rest system. Partial waves with $l > 0$ appear to be present, as would be expected since $\hbar k/m_\pi c$ approximately equals 3. The difference between the Y^{*+} and Y^{*-} angular distributions may reflect the different superposi-

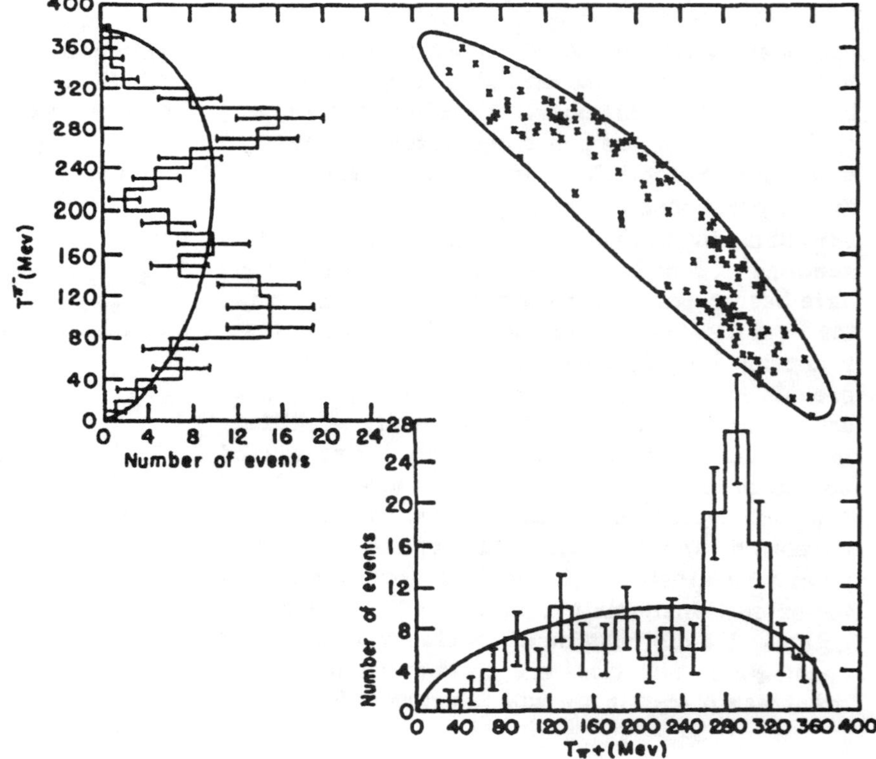

FIG. 1. Energy distribution of the two pions from the reaction $K^- + p \to \Lambda + \pi^+ + \pi^-$. Each event is plotted only once on the Dalitz plot, which should be uniformly populated if phase space dominated the reaction. The two energy histograms are merely one-dimensional projections of the two-dimensional plot, and each event is represented once on each histogram. The solid lines superimposed over the histograms are the phase-space curves.

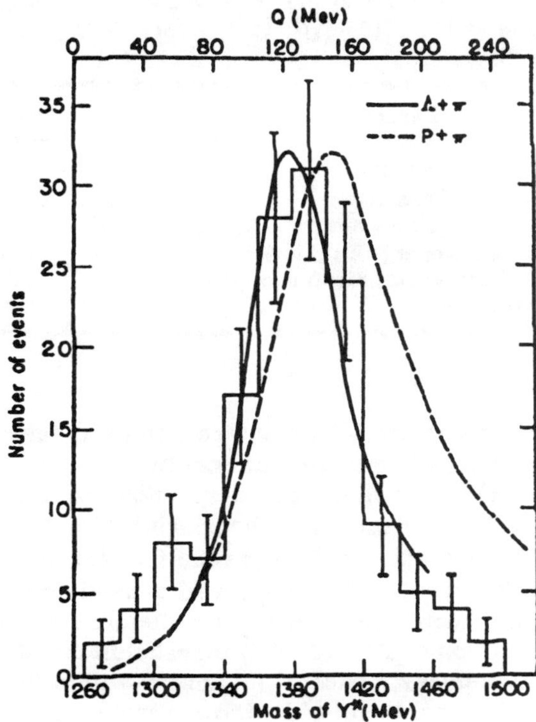

FIG. 2. Mass distribution for Y^* and fitted curves for $\pi\Lambda$ and πp resonances. The lower scale refers only to the $\pi\Lambda$ resonance. Q is the kinetic energy released when either isobar dissociates. The curve for the $\pi\Lambda$ resonances is fitted to the center eight histogram intervals of our data. The πp curve is the fit obtained by Gell-Mann and Watson,[7] to πp scattering data. Both fits are to the formula $\sigma \propto \lambda\!\!\!^{-2}\Gamma^2/[(E-E_0)^2+\frac{1}{4}\Gamma^2]$, where $\Gamma = 2b(a/\lambda\!\!\!^{-})^3/[1+(a/\lambda\!\!\!^{-})^2]$.

tions of the isotopic-spin zero and one amplitudes for the two cases.

The following two methods were used in an effort to determine the spin of Y^*.

(a) The angular momentum of Y^* was investigated by means of an Adair analysis.[8] We first restricted ourselves to production angles with $|\cos\theta| \geq 0.8$. For this angular range the Adair analysis should be valid if only S and P waves are present in the production process. We then computed η for each event, where

$$\eta = \vec{P}_{K^-} \cdot \vec{P}_\Lambda / (|\vec{P}_{K^-}||\vec{P}_\Lambda|).$$

Of the 29 events with $|\cos\theta| \geq 0.8$, the fraction 0.62 ± 0.09 has $|\eta| \geq 0.5$. If the above-mentioned restriction on the angular interval is sufficient to insure the validity of the Adair analysis, this ratio is expected to be 0.50 for $j = 1/2$ and 0.73 for $j = 3/2$. The experimental result is thus ~1.3 standard deviations from both possibilities, and no conclusion may be drawn from the data. Similar results were obtained for several larger values of the cutoff angle. Presence of D waves, however, cannot be excluded by the production angular distributions (Fig. 3). If they are present, indeed, then none of these choices of angle would be sufficiently restrictive to guarantee the success of the Adair analysis.

(b) Since Y^* may be polarized perpendicular to its plane of production, correlations can exist between the decay angle of the Y^* and the polarization of the resulting Λ. Also, a net Λ polarization

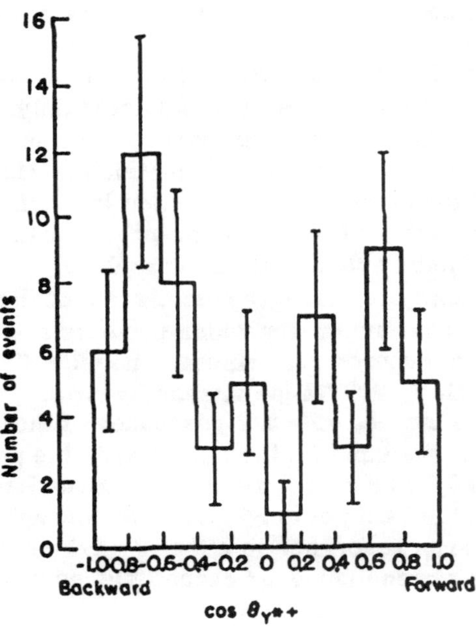

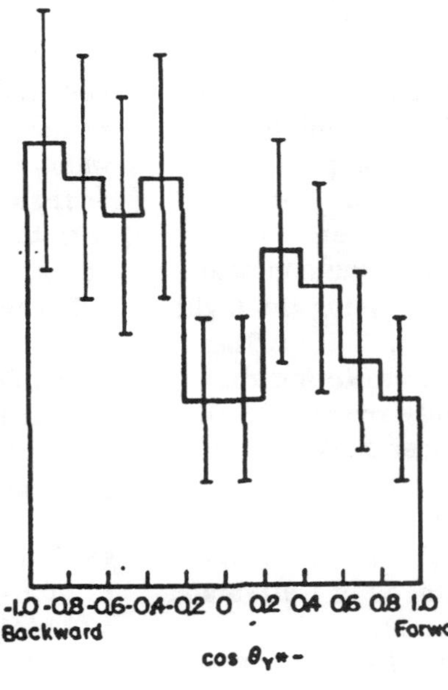

FIG. 3. Angular distribution of $Y^{*\pm}$ in the K^-p barycentric system for the reactions $K^- + p \rightarrow Y^{*\pm} + \pi^{\mp}$.

can result. With our limited data, we see no statistically significant Λ polarization or angular correlations.

However, one can also look for anisotropy, i.e., a polar-to-equatorial ratio, in the decay angle of Y^* with respect to the normal to the plane of production. For spin 3/2, the distribution must be of the form $A+B\xi^2$ by the Sachs-Eisler theorem,[6] independent of the Y^* parity, where we have

$$\xi=(\vec{P}_{K^-}\times\vec{P}_{Y^*})\cdot\vec{P}_\Lambda/(|\vec{P}_{K^-}\times\vec{P}_{Y^*}||\vec{P}_\Lambda|),$$

and $\vec{P}_a$ is the momentum of the particle a in the K^-p barycentric system.

Since the coefficient B is a function of the production angle, we want to restrict ourselves to that range of the solid angle where the polar-to-equatorial anisotropy is probably greatest along the normal to the production plane. For production angles near 0 deg and 180 deg (Adair-analysis region), one expects the polar-to-equatorial ratio to be most different from unity in another direction (namely along the direction of the beam). Thus the equatorial region of production angles is more likely to show a large anisotropy along the direction in question. Therefore the production-angle range $\sin\theta \geq 0.866$ was selected for study. We find the ratio of events with $|\xi|>0.5$ to all events is 0.355. If the distribution is isotropic, as is required for spin 1/2, we expect 0.500 ± 0.063 for our 62 events. The result is thus 2.3 standard deviations from isotropy. The 45-to-1 odds against isotropy overstate the case for higher spin because this is the fourth anisotropy looked for.

Since Y^* may be regarded as a hyperon isobar, which decays into a π and a Λ, it evidently corresponds to a resonance in pion-hyperon scattering. The mass distribution of Fig. 2 then invites a comparison to the cross section for pion-nucleon scattering in the 3/2 - 3/2 state. For this purpose a p-wave resonance formula employed by Gell-Mann and Watson[7] for pion-nucleon scattering was fitted to our $\pi\Lambda$ data by using the eight central histogram intervals of Fig. 2. In fitting the curve, it was found that the interaction radius (a) could be varied over a wide range without changing the goodness of fit appreciably, provided that the reduced width (b) was also changed appropriately. The radius parameter was therefore fixed arbitrarily at $\hbar/m_\pi c$. Table II summarizes our results for Y^*, along with those of Gell-Mann and Watson for the 3-3 resonance.

Even if Y^* does turn out to be a p-wave resonance, there are still many reasons why the π - Λ resonance parameters must not be taken too literally: (a) There is a small contamination of $\Sigma^0\pi\pi$ events in our data. (b) A nonresonant background may be present. (c) The production matrix element for reaction (2) might well depend on the outgoing momentum, and hence distort the mass distribution of Y^*. (d) Two thresholds for other possible decay modes of Y^* appear within the mass interval covered by the resonance curve; i.e., the $\Sigma\pi$ mode threshold around 1330 Mev and the $\bar{K}N$ threshold around 1435 Mev. This must have some effect on the shape of the mass spectrum as observed via the $\Lambda\pi$ decay mode. (e) Final-state pion-pion interaction could disturb the spectrum. (f) Even when the two resonances, Y^{*+} and Y^{*-}, are well resolved in terms of intensity—as in our experiment—there can still be an appreciable interference between the amplitude in which the π^+ arises from reaction (2) and the π^- from reaction (3) and the amplitude in which the roles of the two pions are reversed.

Table II. Parameters for π - Λ and π - p resonance fitted to $\sigma\propto\lambda\!\!\!^-{}^2\Gamma^2/[(E-E_0)^2+\frac{1}{4}\Gamma^2]$, where $\Gamma=2b(a/\lambda\!\!\!^-)^3/[1+(a/\lambda\!\!\!^-)^2]$.

Parameter	π - p	π - Λ
Interaction radius a in units of $\hbar/m_\pi c$	0.88	1
Reduced width b (Mev)	58	33.4
Resonance energy E_0 (Mev)	159	129.3
Full width at half maximum (Mev)	100	64
Lifetime (sec)		$\sim 10^{-23}$

If we bear all these uncertainties in mind, the resemblance to the 3-3 resonance is certainly remarkable (Fig. 2). The resonance energies when expressed in terms of barycentric kinetic energies differ by only 30 Mev, which is much less than the width of either resonance. Furthermore, the widths are at least comparable.

These results are strongly reminiscent of the concept of global symmetry which predicts two spin 3/2 pion-hyperon resonances, one with $T=1$, the other with $T=2$.[8] These are the hyperon counterparts of the $J=T=3/2$ resonance of the pion-nucleon system. On the other hand, the possibility that Y^* is a $J=1/2$ resonance cannot be excluded on the basis of our data. The concept of pion-hyperon resonance in either $J=1/2$ or 3/2 state has been discussed recently by several authors.[9]

A study of $\Sigma^{\mp}\pi^{\pm}\pi^{0}$ events in our experiment is under way at present. The results, however, are too incomplete for us to be able to draw any definite conclusions.

The authors are greatly indebted to the bubble chamber crew under the direction of James D. Gow for their fine job in operating the chamber, especially Robert D. Watt and Glen J. Eckman for their invaluable help with the velocity spectrometers. We also gratefully acknowledge the cooperation of Dr. Edward J. Lofgren and the Bevatron crew, as well as the skilled work and cooperation of our scanning and measuring staff. Special thanks are due the many colleagues in our group who developed the PANG and KICK computer programs—especially Dr. Arthur H. Rosenfeld, and to Dr. Frank Solmitz for many helpful discussions.

One of us (P.E.) is grateful to the Philippe's Foundation Inc. and to the Commisariat à l'Energie Atomique for a fellowship.

---

*This work was done under the auspices of the U. S. Atomic Energy Commission.

†Presently at Laboratoire de Physique Atomique, College de France, Paris, France.

‡Presently at University of Wisconsin, Madison, Wisconsin.

‖Presently at University of California at Los Angeles, Los Angeles, California.

[1]Margaret Alston, L. W. Alvarez, P. Eberhard, M. L. Good, W. Graziano, H. K. Ticho, and S. Wojcicki, paper presented at the Tenth Annual Rochester Conference on High-Energy Nuclear Physics, 1960 (to be published).

[2]P. Eberhard, M. L. Good, and H. K. Ticho, Lawrence Radiation Laboratory Report UCRL-8878 Rev, December, 1959 (unpublished); also Rev. Sci. Instr. (to be published).

[3]L. W. Alvarez, P. Eberhard, M. L. Good, W. Graziano, H. K. Ticho, and S. Wojcicki, Phys. Rev. Letters 2, 215 (1959).

[4]L. W. Alvarez, in Proceedings of the 1959 International Conference on High-Energy Physics at Kiev (unpublished); also Lawrence Radiation Laboratory Report UCRL-9354, August, 1960 (unpublished).

[5]R. K. Adair, Phys. Rev. 100, 1540 (1955).

[6]E. Eisler and R. G. Sachs, Phys. Rev. 72, 680 (1947).

[7]M. Gell-Mann and K. Watson, Annual Review of Nuclear Science (Annual Reviews, Inc., Palo Alto, California, 1954), Vol. 4.

[8]M. Gell-Mann, Phys. Rev. 106, 1297 (1957).

[9]R. H. Capps, Phys. Rev. 119, 1753 (1960); R. H. Capps and M. Nauenberg, Phys. Rev. 118, 593 (1960); R. H. Dalitz and S. F. Tuan, Ann. Phys. 10, 307 (1960); M. Nauenberg, Phys. Rev. Letters 2, 351 (1959); A. Komatsuzawa, R. Sugano, and Y. Nogami, Progr. Theoret. Phys. (Kyoto) 21, 151 (1959); Y. Nogami, Progr. Theoret. Phys. (Kyoto) 22, 25 (1959); D. Amati, A. Stanghellini, and B. Vitale, Nuovo cimento 13, 1143 (1959); L. F. Landovitz and B. Margolis, Phys. Rev. Letters 2, 318 (1959); M. H. Ross and C. L. Shaw, Ann. Phys. (to be published).

EVIDENCE FOR A π-π RESONANCE IN THE $I=1$, $J=1$ STATE*

A. R. Erwin, R. March, W. D. Walker, and E. West
Brookhaven National Laboratory, Upton, New York and University of Wisconsin, Madison, Wisconsin
(Received May 11, 1961)

Since the earliest data became available on pion production by pions, certain features have been quite clear. The main feature which is strongly exhibited above energies of 1 Bev is that collisions are preferred in which there is a small momentum transfer to the nucleon.[1] This is shown by the nucleon angular distributions which are sharply peaked in the backward direction. These results suggest that large-impact-parameter collisions are important in such processes. The simplest process that could give rise to such collisions is a pion-pion collision with the target pion furnished in a virtual state by the nucleon. The quantitative aspects of such collisions have been discussed by a number of authors. Goebel, Chew and Low, and Salzman and Salzman[2] discussed means of extracting from the data the π-π cross section.

Holladay and Frazer and Fulco[3] deduced from electromagnetic data that indeed there must be a strong pion-pion interaction. In particular, Frazer and Fulco deduced that there probably was a resonance in the $I=1$, $J=1$ state. A qualitative set of π-p phase shifts in the 400-600 Mev[4] region were used by Bowcock et al.[5] to deduce an energy of about 660 Mev in the π-π system for the resonance. The work of Pickup et al.[6] showed an indication of a peak in the π-π spectrum at an energy of about 600 Mev.

The present experiment was designed to explore the π-π system up to an energy of about 1 Bev. The π^- beam was produced by the external proton beam No. 1 at the Cosmotron. A suitable set of quadrupole and bending magnets focussed the pion beam on a Hevimet slit about 10 ft from the Adair-

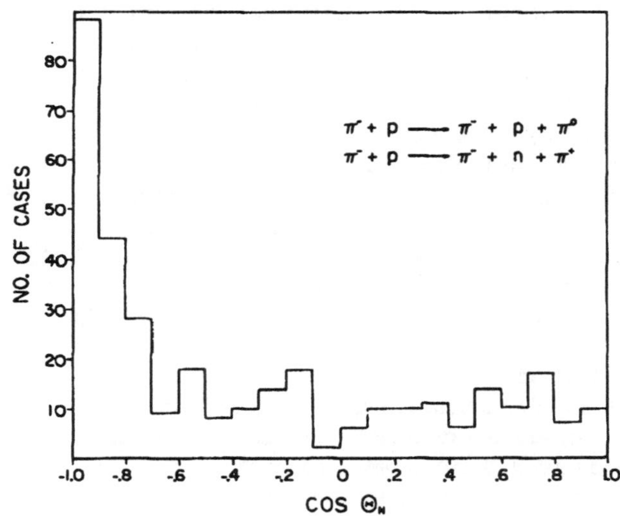

FIG. 1. The angular distribution of the nucleons from the processes $\pi^- + p \to \pi^- + \pi^0 + p$ and $\pi^- + p \to \pi^- + \pi^+ + n$.

Leipuner 14-in. H_2 bubble chamber. The pions were guided into the chamber by another bending magnet. The measured momentum was 1.89 ± 0.07 Bev/c.

Events selected for measurement were taken in a fiducial volume of the chamber. The forward-going track was required to be at least 10 cm long. Measurements were made on a digitized system and the output was analyzed by use of an IBM-704. The events were analyzed by means of a program based on the "Guts" routine written by members of the Alvarez bubble chamber group.

Figure 1 shows the combined angular distribution for the nucleons from the two processes, $\pi^- + p \to \pi^- + \pi^0 + p$ and $\pi^- + p \to \pi^- + \pi^+ + n$, which appear to be identical within statistics. The results indicate a large number of events with small momentum transfer to the nucleons.

We concentrate our interest on those events with small momentum transfer since these events satisfy the qualitative criterion of being examples of π-π collisions. Somewhat arbitrarily, we center our attention on cases in which the momentum transfer to the nucleon is less than 400 Mev/c. Table I gives the ratios of the three possible final states $\pi^-\pi^+n$, $\pi^-\pi^0p$, and $\pi^0\pi^0n$, assuming the π-π scattering to be dominated, respectively, by the $I=0$, 1, 2 scattering states of the π-π system.

The experimental results in the last column indicate a strong domination by $I=1$ state. For the $I=1$ state the basic π-π scattering cross sections $\sigma(\pi^-\pi^0 \to \pi^-\pi^0)$ and $\sigma(\pi^-\pi^+ \to \pi^-\pi^+)$ are equal.

Table I. Ratios of final states.

	$I=0$	$I=1$	$I=2$	Experiment ($\Delta \leq 400$ Mev/c)
$\pi^-\pi^+n$	2	2	2/9	1.7 ± 0.3
$\pi^-\pi^0p$	0	1	1	1
$\pi^0\pi^0n$	1	0	4/9	$< 0.25 \pm 0.25$

The nucleon four-momentum transfer spectrum seems to be in qualitative agreement with the theory for the process in which a π is knocked out of the cloud. Figure 2 shows ideograms for the mass spectrum of the di-pions for cases with $\Delta \leq 400$ Mev/c and $\Delta > 400$ Mev/c, where Δ is the four-momentum transfer to the nucleon. The curve for $\Delta \leq 400$ Mev/c clearly shows a peak at 765 Mev/c. In the ideogram for $\Delta > 400$ Mev/c the peak is still present but seems to be smeared to higher values of the di-pion mass, m^*. One worries that diagrams other than the one involving

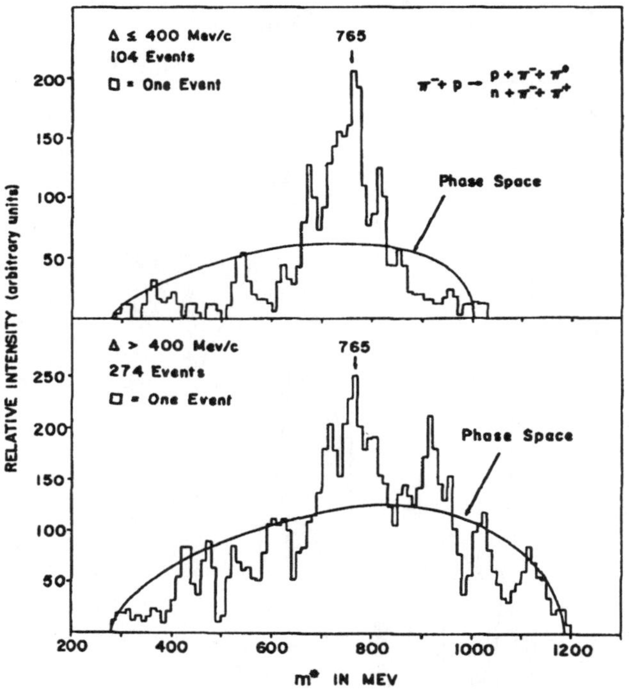

FIG. 2. The combined mass spectrum for the $\pi^-\pi^0$ and $\pi^-\pi^+$ system. The smooth curve is phase space as modified for the included momentum transfer and normalized to the number of events plotted. Events used in the upper distribution are not contained in the lower distribution.

one-pion exchange might be contributing to the observed peak in this m^* spectrum. In particular, an important contribution at lower energies comes from a diagram in which one of the π's rescatters off the nucleon and ends up in the 3-3 state with respect to the nucleon. If one restricts the data to cases with $\Delta \leq 400$ Mev/c this diagram does not seem to be very important, but if one takes cases with $\Delta > 400$ Mev/c many cases consistent with rescattering are found.

In order to deduce values of the π-π cross section, we use the formula[2]

$$\frac{d^2\sigma}{dm^* d\Delta^2} = \frac{3f^2}{\pi} \frac{\Delta^2}{(\Delta^2+1)^2} \left(\frac{m^*}{q_{iL}}\right)^2 K\bar{\sigma}_{\pi-\pi},$$

where $\bar{\sigma}_{\pi-\pi}$ is the mean of $\sigma(\pi^-\pi^0 \to \pi^-\pi^0)$ and $\sigma(\pi^-\pi^+ \to \pi^-\pi^+)$. In the above formula all momenta and energies are measured in units of pion masses. q_{iL} = momentum of the incoming pion measured in units of the pion mass. K = momentum of the pions in the di-pion center-of-mass system. Then

$$\delta\sigma = \frac{3f^2}{\pi} \left(\frac{m^*}{q_{iL}}\right)^2 K\bar{\sigma}_{\pi-\pi} \delta m^* \int_{\Delta_{\min}(m^*)}^{\Delta_{\max}} \frac{\Delta^2 d\Delta^2}{(\Delta^2+1)^2}.$$

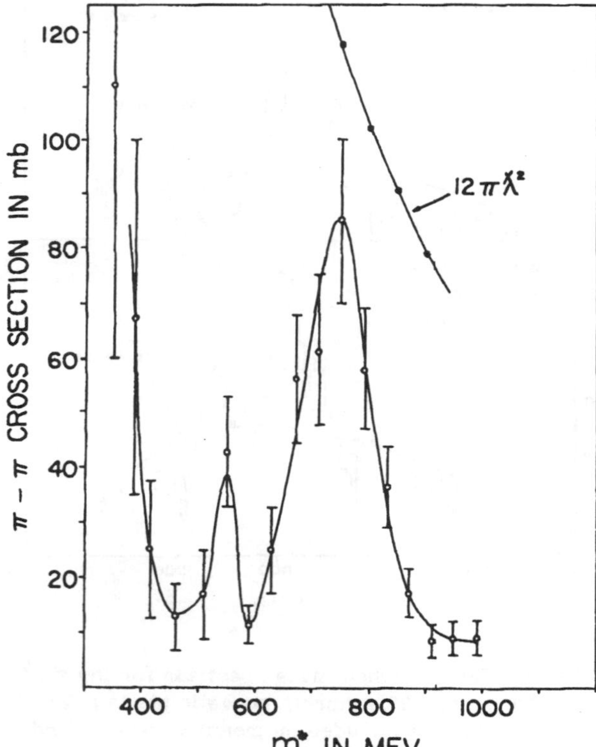

FIG. 3. The π-π cross section as deduced from cases with the four-momentum transfer less than 400 Mev/c.

The results of this calculation using the experimentally determined $\delta\sigma$'s are shown in Fig. 3. The results indicate a peak in the neighborhood of 750 Mev with a width of 150-200 Mev, which is about 3/4 of what it would be ($12\pi\lambda\!\!\!^{-2}$) for a resonance in the $I=1$, $J=1$ state. Since this cross section was determined off the energy shell, it is difficult to estimate the effect of the interference of other diagrams and also the effect of line broadening.[7] Whether or not the other peak and the S-wave scattering indicated in Fig. 3 are real will have to await better statistics for verification.

We wish to acknowledge with gratitude the help and cooperation of R. K. Adair and L. Leipuner in the use of their bubble chamber, and to the latter also for his assistance in adapting the "Guts" routine to our use. We also acknowledge the help of J. Boyd, J. Bishop, P. Satterblom, R. P. Chen, C. Seaver, and K. Eggman in measuring, scanning, and tabulating. We were greatly aided by Dr. J. Ballam and Dr. H. Fechter in setting up the beam. We have had helpful conversations with Dr. R. K. Adair, Dr. C. J. Goebel, Dr. M. L. Good, and in particular Dr. G. Takeda.

---

*Work supported in part by the U. S. Atomic Energy Commission and Wisconsin Alumni Research Foundation.

[1]L. M. Eisberg, W. B. Fowler, R. M. Lea, W. D. Shephard, R. P. Shutt, A. M. Thorndike, and W. L. Whittemore, Phys. Rev. 97, 797 (1955); W. D. Walker and J. Crussard, Phys. Rev. 98, 1416 (1955).

[2]C. Goebel, Phys. Rev. Letters 1, 337 (1958); G. F. Chew and F. E. Low, Phys. Rev. 113, 1640 (1959); F. Salzman and G. Salzman, Phys. Rev. 120, 599 (1960).

[3]W. Holladay, Phys. Rev. 101, 1198 (1956); W. Frazer and J. Fulco, University of California Radiation Laboratory Report UCRL-8880, August, 1959 (unpublished).

[4]W. D. Walker, J. Davis, and W. D. Shephard, Phys. Rev. 118, 1612 (1960).

[5]J. Bowcock, N. Cottingham, and D. Lurie, Nuovo cimento 19, 142 (1961), and Phys. Rev. Letters 5, 386 (1960).

[6]E. Pickup, F. Ayer, and E. O. Salant, Phys. Rev. Letters 5, 161 (1960); Proceedings of the Tenth Annual International Conference on High-Energy Physics at Rochester (Interscience Publishers, Inc., New York, 1960); see also F. Bonsignori and F. Selleri, Nuovo cimento 15, 465 (1960).

[7]Deductions made by extrapolations seem to give different values of the cross section. Also the position of the maximum is not determined very well. J. A. Anderson, Vo. X. Bang, P. G. Burke, D. D. Carmony, and N. Schmitz, Phys. Rev. Letters 6, 365 (1961).

OBSERVATION OF HIGH-ENERGY NEUTRINO REACTIONS AND THE EXISTENCE OF TWO KINDS OF NEUTRINOS*

G. Danby, J-M. Gaillard, K. Goulianos, L. M. Lederman, N. Mistry, M. Schwartz,† and J. Steinberger†

Columbia University, New York, New York and Brookhaven National Laboratory, Upton, New York
(Received June 15, 1962)

In the course of an experiment at the Brookhaven AGS, we have observed the interaction of high-energy neutrinos with matter. These neutrinos were produced primarily as the result of the decay of the pion:

$$\pi^{\pm} \to \mu^{\pm} + (\nu/\bar{\nu}). \qquad (1)$$

It is the purpose of this Letter to report some of the results of this experiment including (1) demonstration that the neutrinos we have used produce μ mesons but do not produce electrons, and hence are very likely different from the neutrinos involved in β decay and (2) approximate cross sections.

Behavior of cross section as a function of energy. The Fermi theory of weak interactions which works well at low energies implies a cross section for weak interactions which increases as phase space. Calculation indicates that weak interacting cross sections should be in the neighborhood of 10^{-38} cm^2 at about 1 BeV. Lee and Yang[1] first calculated the detailed cross sections for

$$\nu+n \to p+e^-,$$

$$\bar{\nu}+p \to n+e^+, \quad (2)$$

$$\nu+n \to p+\mu^-,$$

$$\bar{\nu}+p \to n+\mu^+, \quad (3)$$

using the vector form factor deduced from electron scattering results and assuming the axial vector form factor to be the same as the vector form factor. Subsequent work has been done by Yamaguchi[2] and Cabbibo and Gatto.[3] These calculations have been used as standards for comparison with experiments.

Unitarity and the absence of the decay $\mu \to e+\gamma$. A major difficulty of the Fermi theory at high energies is the necessity that it break down before the cross section reaches $\pi\lambda\!\!\!^-{}^2$, violating unitarity. This breakdown must occur below 300 BeV in the center of mass. This difficulty may be avoided if an intermediate boson mediates the weak interactions. Feinberg[4] pointed out, however, that such a boson implies a branching ratio $(\mu \to e+\gamma)/(\mu \to e+\nu+\bar{\nu})$ of the order of 10^{-4}, unless the neutrinos associated with muons are different from those associated with electrons.[5] Lee and Yang[6] have subsequently noted that any general mechanism which would preserve unitarity should lead to a $\mu \to e+\gamma$ branching ratio not too different from the above. Inasmuch as the branching ratio is measured to be $\lesssim 10^{-8}$,[7] the hypothesis that the two neutrinos may be different has found some favor. It is expected that if there is only one type of neutrino, then neutrino interactions should produce muons and electrons in equal abundance. In the event that there are two neutrinos, there is no reason to expect any electrons at all.

The feasibility of doing neutrino experiments at accelerators was proposed independently by Pontecorvo[8] and Schwartz.[9] It was shown that the fluxes of neutrinos available from accelerators should produce of the order of several events per day per 10 tons of detector.

The essential scheme of the experiment is as follows: A neutrino "beam" is generated by decay in flight of pions according to reaction (1). The pions are produced by 15-BeV protons striking a beryllium target at one end of a 10-ft long straight section. The resulting entire flux of particles moving in the general direction of the detector strikes a 13.5-m thick iron shield wall at a distance of 21 m from the target. Neutrino interactions are observed in a 10-ton aluminum spark chamber located behind this shield.

The line of flight of the beam from target to detector makes an angle of 7.5° with respect to the internal proton direction (see Fig. 1). The operating energy of 15 BeV is chosen to keep the muons penetrating the shield to a tolerable level.

The number and energy spectrum of neutrinos from reaction (1) can be rather well calculated, on the basis of measured pion-production rates[10] and the geometry. The expected neutrino flux from π decay is shown in Fig. 2. Also shown is

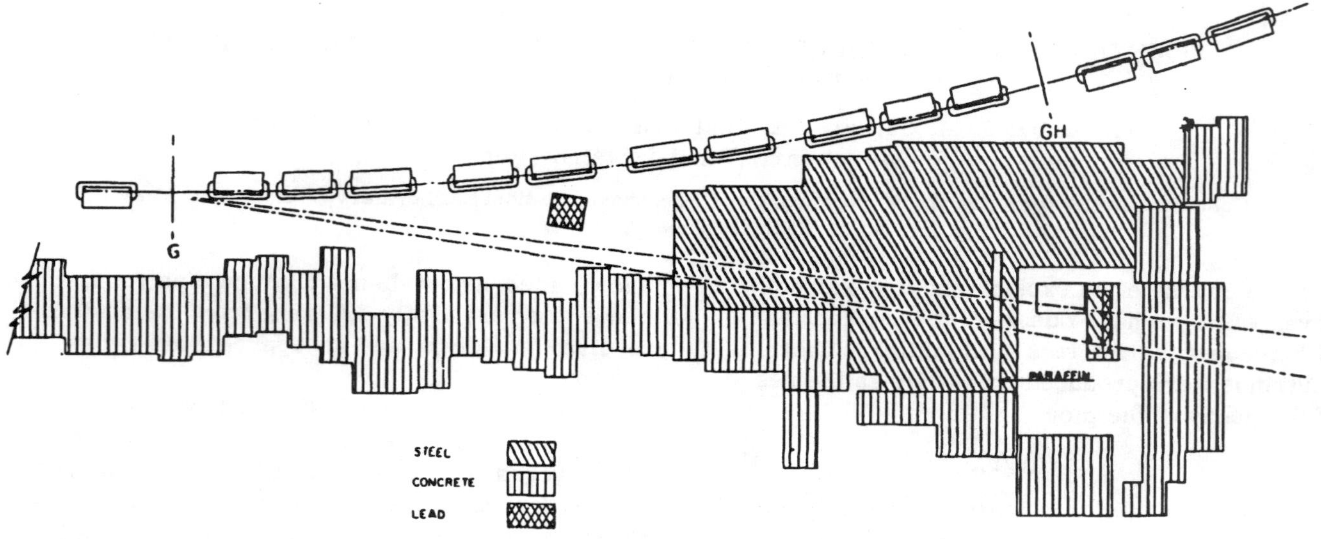

FIG. 1. Plan view of AGS neutrino experiment.

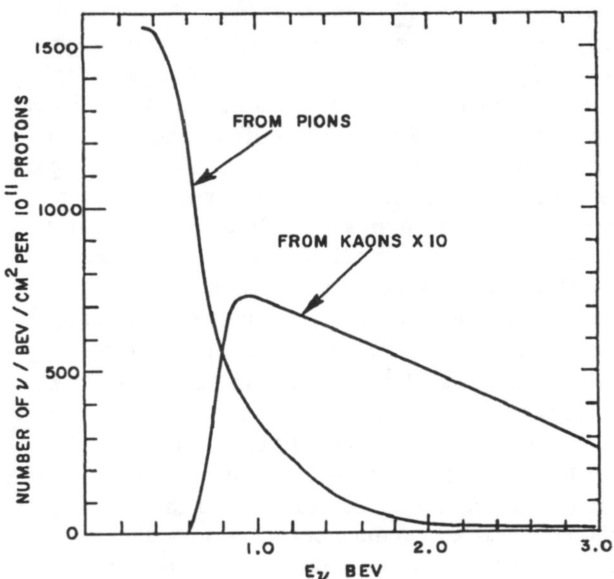

FIG. 2. Energy spectrum of neutrinos expected in the arrangement of Fig. 1 for 15-BeV protons on Be.

an estimate of neutrinos from the decay $K^{\pm} \rightarrow \mu^{\pm} + \nu(\overline{\nu})$. Various checks were performed to compare the targeting efficiency (fraction of circulating beam that interacts in the target) during the neutrino run with the efficiency during the beam survey run. (We believe this efficiency to be close to 70%.) The pion-neutrino flux is considered reliable to approximately 30% down to 300 MeV/c, but the flux below this momentum does not contribute to the results we wish to present.

The main shielding wall thickness, 13.5 m for most of the run, absorbs strongly interacting particles by nuclear interaction and muons up to 17 BeV by ionization loss. The absorption mean free path in iron for pions of 3, 6, and 9 BeV has been measured to be less than 0.24 m.[11] Thus the shield provides an attenuation of the order of 10^{-24} for strongly interacting particles. This attenuation is more than sufficient to reduce these particles to a level compatible with this experiment. The background of strongly interacting particles within the detector shield probably enters through the concrete floor and roof of the 5.5-m thick side wall. Indications of such leaks were, in fact, obtained during the early phases of the experiment and the shielding subsequently improved. The argument that our observations are not induced by strongly interacting particles will also be made on the basis of the detailed structure of the data.

The spark chamber detector consists of an array of 10 one-ton modules. Each unit has 9 aluminum plates 44 in. ×44 in. ×1 in. thick, separated by $\frac{3}{8}$-in. Lucite spacers. Each module is driven by a specially designed high-pressure spark gap and the entire assembly triggered as described below. The chamber will be more fully described elsewhere. Figure 3 illustrates the arrangement of coincidence and anticoincidence counters. Top, back, and front anticoincidence sheets (a total of 50 counters, each 48 in. ×11 in. ×$\frac{1}{2}$ in.) are provided to reduce the effect of cosmic rays and AGS-produced muons which penetrate the shield. The top slab is shielded against neutrino events by 6 in. of steel and the back slab by 3 ft of steel and lead.

Triggering counters were inserted between adjacent chambers and at the end (see Fig. 3). These consist of pairs of counters, 48 in. ×11 in. ×$\frac{1}{2}$ in., separated by $\frac{3}{4}$ in. of aluminum, and in fast coincidence. Four such pairs cover a chamber; 40 are employed in all.

The AGS at 15 BeV operates with a repetition period of 1.2 sec. A rapid beam deflector drives the protons onto the 3-in. thick Be target over a period of 20-30 μsec. The radiation during this interval has rf structure, the individual bursts being 20 nsec wide, the separation 220 nsec. This structure is employed to reduce the total "on" time and thus minimize cosmic-ray background. A Čerenkov counter exposed

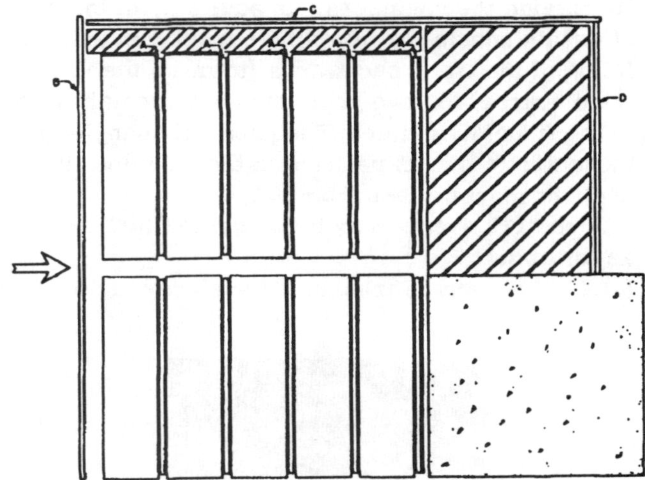

FIG. 3. Spark chamber and counter arrangement. A are the triggering slabs; B, C, and D are anticoincidence slabs. This is the front view seen by the four-camera stereo system.

to the pions in the neutrino "beam" provides a train of 30-nsec gates, which is placed in coincidence with the triggering events. The correct phasing is verified by raising the machine energy to 25 BeV and counting the high-energy muons which now penetrate the shield. The tight timing also serves the useful function of reducing sensitivity to low-energy neutrons which diffuse into the detector room. The trigger consists of a fast twofold coincidence in any of the 40 coincidence pairs in anticoincidence with the anticoincidence shield. Typical operation yields about 10 triggers per hour. Half the photographs are blank, the remainder consist of AGS muons entering unprotected faces of the chamber, cosmic rays, and "events." In order to verify the operation of circuits and the gap efficiency of the chamber, cosmic-ray test runs are conducted every four hours. These consist of triggering on almost horizontal cosmic-ray muons and recording the results both on film and on Land prints for rapid inspection (see Fig. 4).

A convenient monitor for this experiment is the number of circulating protons in the AGS machine. Typically, the AGS operates at a level of $2-4\times10^{11}$ protons per pulse, and 3000 pulses per hour. In an exposure of 3.48×10^{17} protons, we have counted 113 events satisfying the following geometric criteria: The event originates within a fiducial volume whose boundaries lie 4 in. from the front and back walls of the chamber and 2 in. from the top and bottom walls. The first two gaps must not fire, in order to exclude events whose origins lie outside the chambers. In addition, in the case of events consisting of a single track, an extrapolation of the track backwards (towards the neutrino source) for two gaps must also remain within the fiducial volume. The production angle of these single tracks relative to the neutrino line of flight must be less than 60°.

These 113 events may be classified further as follows:

(a) 49 short single tracks. These are single tracks whose visible momentum, if interpreted as muons, is less than 300 MeV/c. These presumably include some energetic muons which leave the chamber. They also include low-energy neutrino events and the bulk of the neutron produced background. Of these, 19 have 4 sparks or less. The second half of the run (1.7×10^{17} protons) with improved shielding yielded only three tracks in this category. We will not consider these as acceptable "events."

(b) 34 "single muons" of more than 300 MeV/c. These include tracks which, if interpreted as muons, have a visible range in the chambers such that their momentum is at least 300 MeV/c. The origin of these events must not be accompanied by more than two extraneous sparks. The latter requirement means that we include among "single tracks" events showing a small recoil. The 34 events are tabulated as a function of momentum in Table I. Figure 5 illustrates 3 "single muon" events.

(c) 22 "vertex" events. A vertex event is one whose origin is characterized by more than one track. All of these events show a substantial energy release. Figure 6 illustrates some of these.

(d) 8 "showers." These are all the remaining events. They are in general single tracks, too irregular in structure to be typical of μ mesons, and more typical of electron or photon showers. From these 8 "showers," for purposes of comparison with (b), we may select a group of 6 which are so located that their potential range within the chamber corresponds to μ mesons in excess of 300 MeV/c.

In the following, only the 56 energetic events of type (b) (long μ's) and type (c) (vertex events) will be referred to as "events."

Arguments on the neutrino origin of the ob-

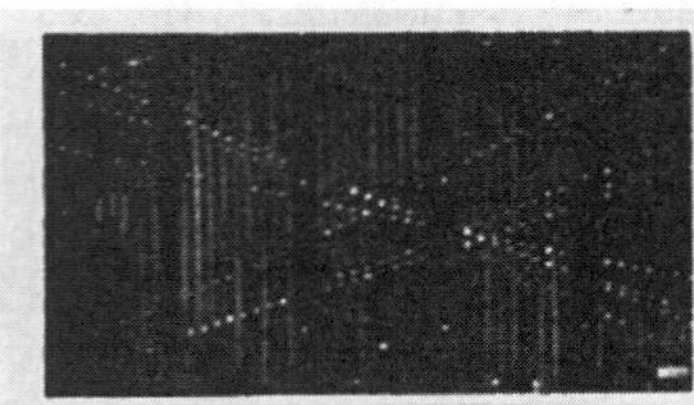

FIG. 4. Land print of Cosmic-ray muons integrated over many incoming tracks.

Table I. Classification of "events."

Single tracks			
$p_\mu < 300$ MeV/c [a]	49	$p_\mu > 500$	8
$p_\mu > 300$	34	$p_\mu > 600$	3
$p_\mu > 400$	19	$p_\mu > 700$	2
Total "events" 34			
Vertex events			
Visible energy released < 1 BeV	15		
Visible energy released > 1 BeV	7		

[a]These are not included in the "event" count (see text).

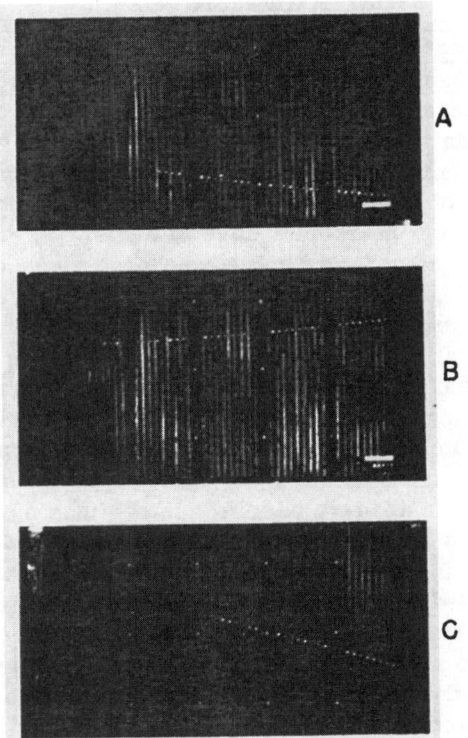

FIG. 5. Single muon events. (A) $p_\mu > 540$ MeV and δ ray indicating direction of motion (neutrino beam incident from left); (B) $p_\mu > 700$ MeV/c; (C) $p_\mu > 440$ with δ ray.

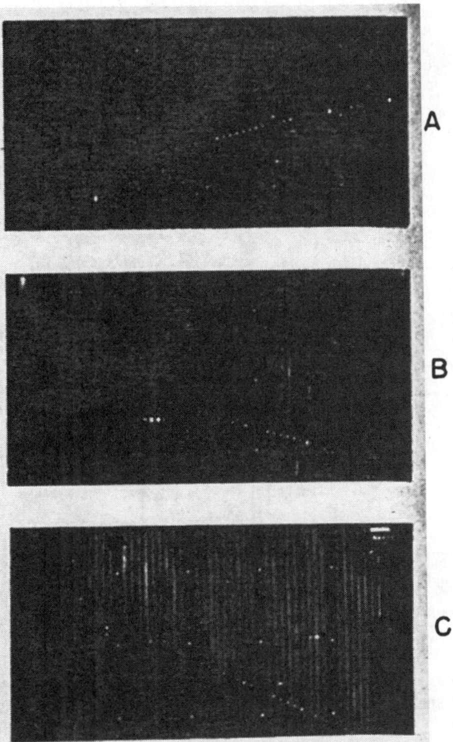

FIG. 6. Vertex events. (A) Single muon of $p_\mu > 500$ MeV and electron-type track; (B) possible example of two muons, both leave chamber; (C) four prong star with one long track of $p_\mu > 600$ MeV/c.

served "events."

1. The "events" are not produced by cosmic rays. Muons from cosmic rays which stop in the chamber can and do simulate neutrino events. This background is measured experimentally by running with the AGS machine off on the same triggering arrangement except for the Čerenkov gating requirement. The actual triggering rate then rises from 10 per hour to 80 per second (a dead-time circuit prevents jamming of the spark chamber). In 1800 cosmic-ray photographs thus obtained, 21 would be accepted as neutrino events. Thus 1 in 90 cosmic-ray events is neutrino-like. Čerenkov gating and the short AGS pulse effect a reduction by a factor of $\sim 10^{-6}$ since the circuits are "on" for only 3.5 μsec per pulse. In fact, for the body of data represented by Table I, a total of 1.6×10^6 pulses were counted. The equipment was therefore sensitive for a total time of 5.5 sec. This should lead to $5.5 \times 80 = 440$ cosmic-ray tracks which is consistent with observation. Among these, there should be 5 ± 1 cosmic-ray induced "events." These are almost evident in the small asymmetry seen in the angular distributions of Fig. 7. The remaining 51 events cannot be the result of cosmic rays.

2. The "events" are not neutron produced. Several observations contribute to this conclusion.

(a) The origins of all the observed events are uniformly distributed over the fiduciary volume, with the obvious bias against the last chamber induced by the $p_\mu > 300$ MeV/c requirement. Thus there is no evidence for attenuation, although the mean free path for nuclear interaction in aluminum is 40 cm and for electromagnetic interaction 9 cm.

(b) The front iron shield is so thick that we can expect less than 10^{-4} neutron induced reactions in the entire run from neutrons which have penetrated this shield. This was checked by removing 4 ft of iron from the front of the thick shield. If our events were due to neutrons in line with the target, the event rate would have increased by a factor of one hundred. No such effect was observed (see Table II). If neutrons penetrate the shield, it must be from other di-

10

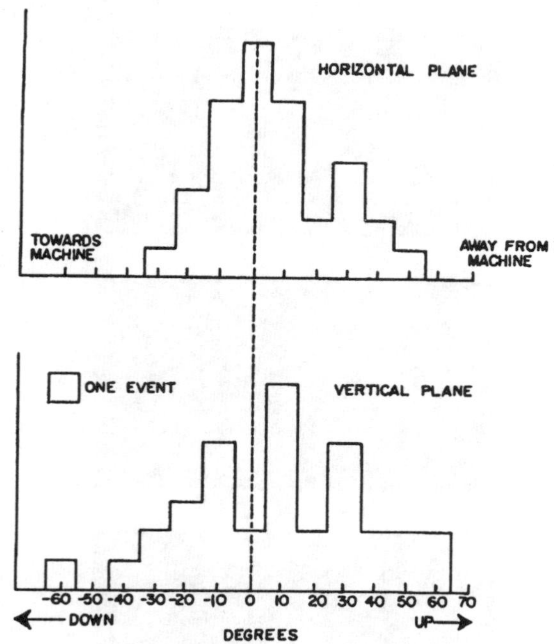

FIG. 7. Projected angular distributions of single track events. Zero degree is defined as the neutrino direction.

rections. The secondaries would reflect this directionality. The observed angular distribution of single track events is shown in Fig. 7. Except for the small cosmic-ray contribution to the vertical plane projection, both projections are peaked about the line of flight to the target.

(c) If our 29 single track events (excluding cosmic-ray background) were pions produced by neutrons, we would have expected, on the basis of known production cross sections, of the order of 15 single π^0's to have been produced. No cases of unaccompanied π^0's have been observed.

Table II. Event rates for normal and background conditions.

	Circulating protons $\times 10^{16}$	No. of Events	Calculated cosmic-ray[c] contribution	Net rate per 10^{16}
Normal run	34.8	56	5	1.46
Background I[a]	3.0	2	0.5	0.5
Background II[b]	8.6	4	1.5	0.3

[a] 4 ft of Fe removed from main shielding wall.
[b] As above, but 4 ft of Pb placed within 6 ft of Be target and subtending a horizontal angular interval from 4° to 11° with respect to the internal proton beam.
[c] These should be subtracted from the "single muon" category.

3. The single particles produced show little or no nuclear interaction and are therefore presumed to be muons. For the purpose of this argument, it is convenient to first discuss the second half of our data, obtained after some shielding improvements were effected. A total traversal of 820 cm of aluminum by single tracks was observed, but no "clear" case of nuclear interaction such as large angle or charge exchange scattering was seen. In a spark chamber calibration experiment at the Cosmotron, it was found that for 400-MeV pions the mean free path for "clear" nuclear interactions in the chamber (as distinguished from stoppings) is no more than 100 cm of aluminum. We should, therefore, have observed of the order of 8 "clear" interactions; instead we observed none. The mean free path for the observed single tracks is then more than 8 times the nuclear mean free path.

Included in the count are 5 tracks which stop in the chamber. Certainly a fraction of the neutrino secondaries must be expected to be produced with such small momentum that they would stop in the chamber. Thus, none of these stoppings may, in fact, be nuclear interactions. But even if all stopping tracks are considered to represent nuclear interactions, the mean free path of the observed single tracks must be 4 nuclear mean free paths.

The situation in the case of the earlier data is more complicated. We suspect that a fair fraction of the short single tracks then observed are, in fact, protons produced in neutron collisions. However, similar arguments can be made also for these data which convince us that the energetic single track events observed then are also noninteracting.[12]

It is concluded that the observed single track events are muons, as expected from neutrino interactions.

4. The observed reactions are due to the decay products of pions and K mesons. In a second background run, 4 ft of iron were removed from the main shield and replaced by a similar quantity of lead placed as close to the target as feasible. Thus, the detector views the target through the same number of mean free paths of shielding material. However, the path available for pions to decay is reduced by a factor of 8. This is the closest we could come to "turning off" the neutrinos. The results of this run are given in terms of the number of events per 10^{16} circulating protons in Table II. The rate of "events" is reduced from 1.46 ± 0.2 to 0.3 ± 0.2 per 10^{16} in-

cident protons. This reduction is consistent with that which is expected for neutrinos which are the decay products of pions and K mesons.

Are there two kinds of neutrinos? The earlier discussion leads us to ask if the reactions (2) and (3) occur with the same rate. This would be expected if ν_μ, the neutrino coupled to the muon and produced in pion decay, is the same as ν_e, the neutrino coupled to the electron and produced in nuclear beta decay. We discuss only the single track events where the distinction between single muon tracks of $p_\mu > 300$ MeV/c and showers produced by high-energy single electrons is clear. See Figs. 8 and 4 which illustrate this difference.

We have observed 34 single muon events of which 5 are considered to be cosmic-ray background. If $\nu_\mu = \nu_e$, there should be of the order of 29 electron showers with a mean energy greater than 400 MeV/c. Instead, the only candidates which we have for such events are six "showers" of qualitatively different appearance from those of Fig. 8. To argue more precisely, we have exposed two of our one-ton spark chamber modules to electron beams at the Cosmotron. Runs were taken at various electron energies. From these we establish that the triggering efficiency for 400-MeV electrons is 67%. As a quantity characteristic of the calibration showers, we have taken the total number of observed sparks. The mean number is roughly linear with electron energy up to 400 MeV/c. Larger showers saturate the two chambers which were available. The spark distribution for 400 MeV/c showers is plotted in Fig. 9, normalized to the $\frac{2}{3} \times 29$ expected showers. The six "shower" events are also plotted. It is evident that these are not consistent with the prediction based on a universal theory with $\nu_\mu = \nu_e$. It can perhaps be argued that the absence of electron events could be understood in terms of the coupling of a single neutrino to the electron which is much weaker than that to the muon at higher momentum transfers, although at lower momentum transfers the results of β decay, μ capture, μ decay, and the ratio of $\pi \rightarrow \mu + \nu$ to $\pi \rightarrow e + \nu$ decay show that these couplings are equal.[13] However, the most plausible explanation for the absence of the electron showers, and the only one which preserves universality, is then that $\nu_\mu \neq \nu_e$; i.e., that there are at least two types of neutrinos. This also resolves the problem raised by the forbiddenness of the $\mu^+ \rightarrow e^+ + \gamma$ decay.

It remains to understand the nature of the 6 "shower" events. All of these events were obtained in the first part of the run during conditions in which there was certainly some neutron background. It is not unlikely that some of the events are small neutron produced stars. One or two could, in fact, be μ mesons. It should also be remarked that of the order of one or two electron events are expected from the neutrinos produced in the decays $K^+ \rightarrow e^+ + \nu_e + \pi^0$ and

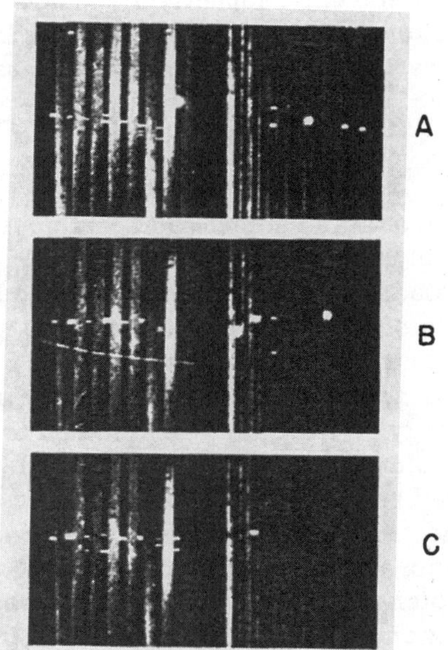

FIG. 8. 400-MeV electrons from the Cosmotron.

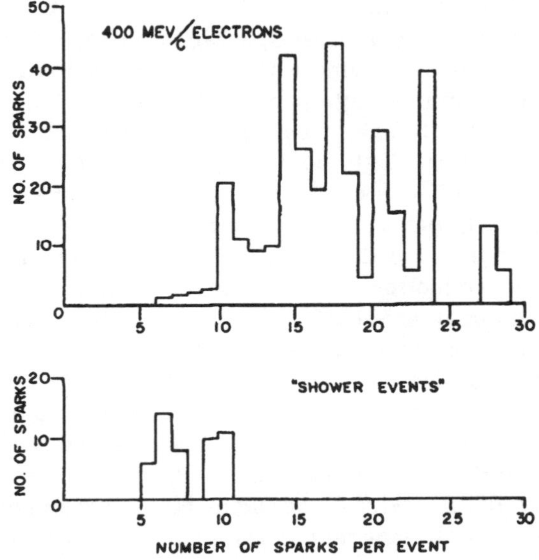

FIG. 9. Spark distribution for 400-MeV/c electrons normalized to expected number of showers. Also shown are the "shower" events.

$K_2^0 \to e^{\pm} + \nu_e + \pi^{\mp}$.

The intermediate boson. It has been pointed out[1] that high-energy neutrinos should serve as a reasonable method of investigating the existence of an intermediate boson in the weak interactions. In recent years many of the objections to such a particle have been removed by the advent of V-A theory[14] and the remeasurement of the ρ value in μ decay.[15] The remaining difficulty pointed out by Feinberg,[4] namely the absence of the decay $\mu \to e + \gamma$, is removed by the results of this experiment. Consequently it is of interest to explore the extent to which our experiment has been sensitive to the production of these bosons.

Our neutrino intensity, in particular that part contributed by the K-meson decays, is sufficient to have produced intermediate bosons if the boson had a mass m_w less than that of the mass of the proton (m_p). In particular, if the boson had a mass equal to $0.6 m_p$, we should have produced ~20 bosons by the process $\nu + p \to w^+ + \mu^- + p$. If $m_w = m_p$, then we should have observed 2 such events.[16]

Indeed, of our vertex events, 5 are consistent with the production of a boson. Two events, with two outgoing prongs, one of which is shown in Fig. 6(B), are consistent with both prongs being muons. This could correspond to the decay mode $w^+ \to \mu^+ + \nu$. One event shows four outgoing tracks, each of which leaves the chamber after traveling through 9 in. of aluminum. This might in principle be an example of $w^+ \to \pi^+ + \pi^- + \pi^+$. Another event, by far our most spectacular one, can be interpreted as having a muon, a charged pion, and two gamma rays presumably from a neutral pion. Over 2 BeV of energy release is seen in the chamber. This could in principle be an example of $w^+ \to \pi^+ + \pi^0$. Finally, we have one event, Fig. 6(A), in which both a muon and an electron appear to leave the same vertex. If this were a boson production, it would correspond to the boson decay mode $w^+ \to e^+ + \nu$. The alternative explanation for this event would require (i) that a neutral pion be produced with the muon; and (ii) that one of its gamma rays convert in the plate of the interaction while the other not convert visibly in the chamber.

The difficulty of demonstrating the existence of a boson is inherent in the poor resolution of the chamber. Future experiments should shed more light on this interesting question.

Neutrino cross sections. We have attempted to compare our observations with the predicted cross sections for reactions (2) using the theory.[1-3] To include the fact that the nucleons in (2) are, in fact, part of an aluminum nucleus, a Monte Carlo calculation was performed using a simple Fermi model for the nucleus in order to evaluate the effect of the Pauli principle and nucleon motion. This was then used to predict the number of "elastic" neutrino events to be expected under our conditions. The results agree with simpler calculations based on Fig. 2 to give, in terms of number of circulating protons,

from $\pi \to \mu + \nu$, 0.60 events/10^{16} protons,

from $K \to \mu + \nu$, 0.15 events/10^{16} protons,

Total 0.75 events/10^{16} ± ~30%.

The observed rates, assuming all single muons are "elastic" and all vertex events "inelastic" (i.e., produced with pions) are

"Elastic": 0.84 ± 0.16 events/10^{16} (29 events),

"Inelastic": 0.63 ± 0.14 events/10^{16} (22 events).

The agreement of our elastic yield with theory indicates that no large modification to the Fermi interaction is required at our mean momentum transfer of 350 MeV/c. The inelastic cross section in this region is of the same order as the elastic cross section.

Neutrino flip hypothesis. Feinberg, Gursey, and Pais[17] have pointed out that if there were two different types of neutrinos, their assignment to muon and electron, respectively, could in principle be interchanged for strangeness-violating weak interactions. Thus it might be possible that

$$\begin{matrix} \pi^+ \to \mu^+ + \nu_1 & & K^+ \to \mu^+ + \nu_2 \\ & \text{while} & \\ \pi^+ \to e^+ + \nu_2 & & K^+ \to e^+ + \nu_1. \end{matrix}$$

This hypothesis is subject to experimental check by observing whether neutrinos from $K_{\mu 2}$ decay produce muons or electrons in our chamber. Our calculation of the neutrino flux from $K\mu 2$ decay indicates that we should have observed 5 events from these neutrinos. They would have an average energy of 1.5 BeV. An electron of this energy would have been clearly recognizable. None have been seen. It seems unlikely therefore that the neutrino flip hypothesis is correct.

The authors are indebted to Professor G. Feinberg, Professor T. D. Lee, and Professor C. N. Yang for many fruitful discussions. In particular, we note here that the emphasis by Lee and Yang on the importance of the high-energy behavior of weak interactions and the likelihood of the existence of two neutrinos played an important part in stimulating this research.

We would like to thank Mr. Warner Hayes for technical assistance throughout the experiment. In the construction of the spark chamber, R. Hodor and R. Lundgren of BNL, and Joseph Shill and Yin Au of Nevis did the engineering. The construction of the electronics was largely the work of the Instrumentation Division of BNL under W. Higinbotham. Other technical assistance was rendered by M. Katz and D. Balzarini. Robert Erlich was responsible for the machine calculations of neutrino rates, M. Tannenbaum assisted in the Cosmotron runs.

The experiment could not have succeeded without the tremendous efforts of the Brookhaven Accelerator Division. We owe much to the cooperation of Dr. K. Green, Dr. E. Courant, Dr. J. Blewett, Dr. M. H. Blewett, and the AGS staff including J. Spiro, W. Walker, D. Sisson, and L. Chimienti. The Cosmotron Department is acknowledged for its help in the initial assembly and later calibration runs.

The work was generously supported by the U. S. Atomic Energy Commission. The work at Nevis was considerably facilitated by Dr. W. F. Goodell, Jr., and the Nevis Cyclotron staff under Office of Naval Research support.

---

*This research was supported by the U. S. Atomic Energy Commission.

†Alfred P. Sloan Research Fellow.

[1]T. D. Lee and C. N. Yang, Phys. Rev. Letters 4, 307 (1960).

[2]Y. Yamaguchi, Progr. Theoret. Phys. (Kyoto) 6, 1117 (1960).

[3]N. Cabbibo and R. Gatto, Nuovo cimento 15, 304 (1960).

[4]G. Feinberg, Phys. Rev. 110, 1482 (1958).

[5]Several authors have discussed this possibility. Some of the earlier viewpoints are given by: E. Konopinski and H. Mahmoud, Phys. Rev. 92, 1045 (1953); J. Schwinger, Ann. Phys. (New York) 2, 407 (1957); I. Kawakami, Progr. Theoret. Phys. (Kyoto) 19, 459 (1957); M. Konuma, Nuclear Phys. 5, 504 (1958); S. A. Bludman, Bull. Am. Phys. Soc. 4, 80 (1959); S. Oneda and J. C. Pati, Phys. Rev. Letters 2, 125 (1959); K. Nishijima, Phys. Rev. 108, 907 (1957).

[6]T. D. Lee and C. N. Yang (private communications). See also Proceedings of the 1960 Annual International Conference on High-Energy Physics at Rochester (Interscience Publishers, Inc., New York, 1960), p. 567.

[7]D. Bartlett, S. Devons, and A. Sachs, Phys. Rev. Letters 8, 120 (1962); S. Frankel, J. Halpern, L. Holloway, W. Wales, M. Yearian, O. Chamberlain, A. Lemonick, and F. M. Pipkin, Phys. Rev. Letters 8, 123 (1962).

[8]B. Pontecorvo, J. Exptl. Theoret. Phys. (U.S.S.R.) 37, 1751 (1959) [translation: Soviet Phys.—JETP 10, 1236 (1960)].

[9]M. Schwartz, Phys. Rev. Letters 4, 306 (1960).

[10]W. F. Baker et al., Phys. Rev. Letters 7, 101 (1961).

[11]R. L. Cool, L. Lederman, L. Marshall, A. C. Melissinos, M. Tannenbaum, J. H. Tinlot, and T. Yamanouchi, Brookhaven National Laboratory Internal Report UP-18 (unpublished).

[12]These will be published in a more complete report.

[13]H. L. Anderson, T. Fujii, R. H. Miller, and L. Tau, Phys. Rev. 119, 2050 (1960); G. Culligan, J. F. Lathrop, V. L. Telegdi, R. Winston, and R. A. Lundy, Phys. Rev. Letters 7, 458 (1961); R. Hildebrand, Phys. Rev. Letters 8, 34 (1962); E. Bleser, L. Lederman, J. Rosen, J. Rothberg, and E. Zavattini, Phys. Rev. Letters 8, 288 (1962).

[14]R. Feynman and M. Gell-Mann, Phys. Rev. 109, 193 (1958); R. Marshak and E. Sudershan, Phys. Rev. 109, 1860 (1958).

[15]R. Plano, Phys. Rev. 119, 1400 (1960).

[16]T. D. Lee, P. Markstein, and C. N. Yang, Phys. Rev. Letters 7, 429 (1961).

[17]G. Feinberg, F. Gursey, and A. Pais, Phys. Rev. Letters 7, 208 (1961).

OBSERVATION OF A HYPERON WITH STRANGENESS MINUS THREE*

V. E. Barnes, P. L. Connolly, D. J. Crennell, B. B. Culwick, W. C. Delaney, W. B. Fowler, P. E. Hagerty,† E. L. Hart, N. Horwitz,† P. V. C. Hough, J. E. Jensen, J. K. Kopp, K. W. Lai, J. Leitner,† J. L. Lloyd, G. W. London,‡ T. W. Morris, Y. Oren, R. B. Palmer, A. G. Prodell, D. Radojičić, D. C. Rahm, C. R. Richardson, N. P. Samios, J. R. Sanford, R. P. Shutt, J. R. Smith, D. L. Stonehill, R. C. Strand, A. M. Thorndike, M. S. Webster, W. J. Willis, and S. S. Yamamoto
Brookhaven National Laboratory, Upton, New York
(Received 11 February 1964)

It has been pointed out[1] that among the multitude of resonances which have been discovered recently, the $N_{3/2}^*(1238)$, $Y_1^*(1385)$, and $\Xi_{1/2}^*(1532)$ can be arranged as a decuplet with one member still missing. Figure 1 illustrates the position of the nine known resonant states and the postulated tenth particle plotted as a function of mass and the third component of isotopic spin. As can be seen from Fig. 1, this particle (which we call Ω^-, following Gell-Mann[1]) is predicted to be a negatively charged isotopic singlet with strangeness minus three.[2] The spin and parity should be the same as those of the $N_{3/2}^*$, namely, $3/2^+$. The 10-dimensional representation of the group SU_3 can be identified with just such a decuplet. Consequently, the existence of the Ω^- has been cited as a crucial test of the theory of unitary symmetry of strong interactions.[3,4] The mass is predicted[5] by the Gell-Mann–Okubo mass formula to be about 1680 MeV/c^2. We wish to report the observation of an event which we believe to be an example of the production and decay of such a particle.

The BNL 80-in. hydrogen bubble chamber was exposed to a mass-separated beam of 5.0-BeV/c K^- mesons at the Brookhaven AGS. About 100 000 pictures were taken containing a total K^- track length of ~10^6 feet. These pictures have been partially analyzed to search for the more characteristic decay modes of the Ω^-.

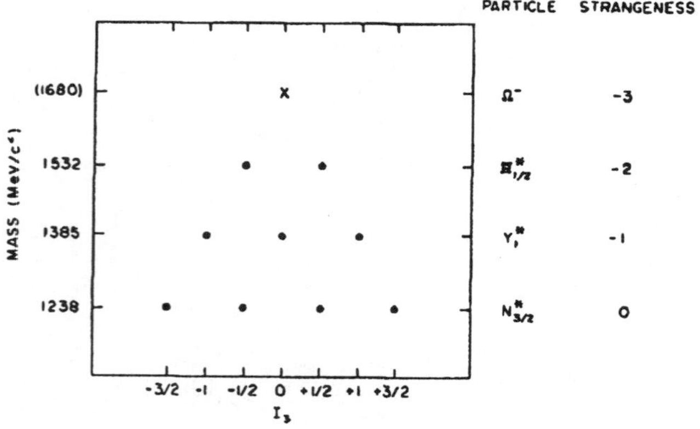

FIG. 1. Decuplet of $\frac{3}{2}^+$ particles plotted as a function of mass versus third component of isotopic spin.

The event in question is shown in Fig. 2, and the pertinent measured quantities are given in Table I. Our interpretation of this event is

$$\begin{array}{l} K^- + p \to \Omega^- + K^+ + K^0 \\ \quad \llcorner \Xi^0 + \pi^- \\ \qquad \llcorner \Lambda^0 + \pi^0 \\ \qquad\quad \llcorner \gamma_1 + \gamma_2 \\ \qquad\qquad \llcorner e^+ + e^- \\ \qquad\qquad \llcorner e^+ + e^- \\ \qquad \llcorner \pi^- + p. \end{array} \qquad (1)$$

From the momentum and gap length measurements, track 2 is identified as a K^+. (A bubble density of 1.9 times minimum was expected for this track while the measured value was 1.7 ± 0.2.) Tracks 5 and 6 are in good agreement with the decay of a Λ^0, but the Λ^0 cannot come from the primary interaction. The Λ^0 mass as calculated from the measured proton and π^- kinematic quantities is 1116 ± 2 MeV/c^2. Since the bubble density from gap length measurement of track 6 is 1.52 ± 0.17, compared to 1.0 expected for a π^+ and 1.4 for a proton, the interpretation of the V as a K^0 is unlikely. In any case, from kinematical considerations such a K^0 could not come from the production vertex. The Λ^0 appears six decay lengths from the wall of the bubble chamber, and there is no other visible origin in the chamber.

The event is unusual in that two gamma rays, apparently associated with it, convert to electron-positron pairs in the liquid hydrogen. From measurements of the electron momenta and angles, we determine that the effective mass of the two gamma rays is 135.1 ± 1.5 MeV/c^2, consistent with a π^0 decay. In a similar manner, we have used the calculated π^0 momentum and angles, and the values from the fitted Λ^0 to deter-

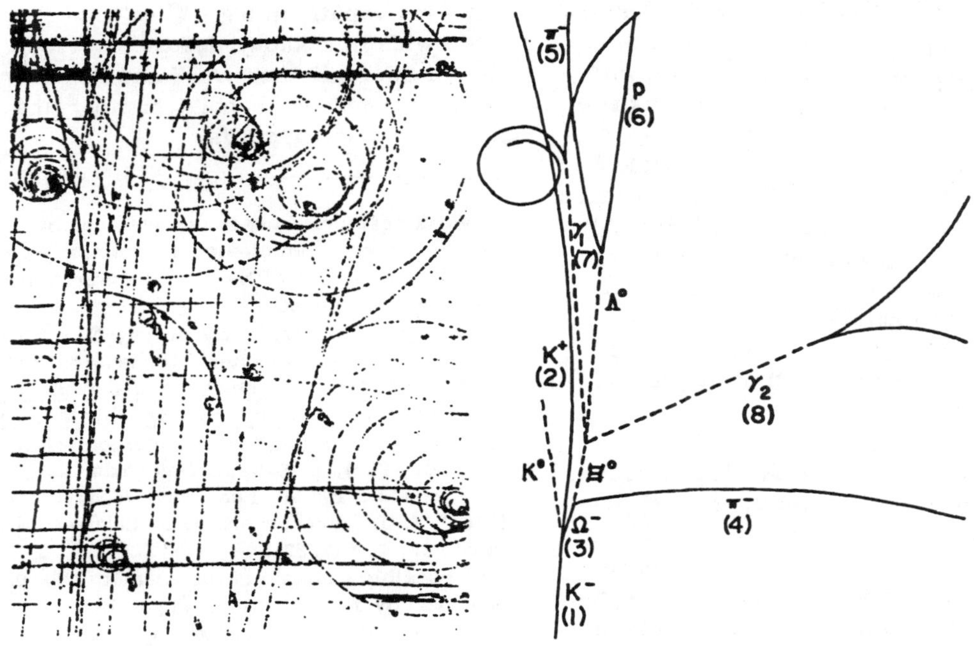

FIG. 2. Photograph and line diagram of event showing decay of Ω^-.

mine the mass of the neutral decaying hyperon to be 1316 ± 4 MeV/c^2 in excellent agreement with that of the Ξ^0. The projections of the lines of flight of the two gammas and the Λ^0 onto the XY plane (parallel to the film) intersect within 1 mm and in the XZ plane within 3 mm. The calculated momentum vector of the Ξ^0 points back to the decay point of track 3 within 1 mm and misses the production vertex by 5 mm in the XY plane. The length of the Ξ^0 flight path is 3 cm with a calculated momentum of 1906 ± 20 MeV/c. The transverse momenta of the Ξ^0 and of track 4 balance within the errors, indicating that no other particle is emitted in the decay of particle 3.

We will now discuss the decay of particle 3. From the momentum and gap length measurements on track 4, we conclude that its mass is less than that of a K. Using the Ξ^0 momentum and assuming particle 4 to be a π^-, the mass of particle 3 is computed to be 1686 ± 12 MeV/c^2 and its momentum to be 2015 ± 20 MeV/c. Note that the measured transverse momentum of track 4, 248 ± 5 MeV/c, is greater than the maximum momentum for the possible decay modes of the known particles (given in Table II), except for $\Xi^- \to e^- + n + \nu$. We reject this hypothesis not only because it involves $\Delta S = 2$, but also because it disregards the previously established associations of the Λ and two gammas with the event.

Table I. Measured quantities.

Track	Azimuth (deg)	Dip (deg)	Momentum (MeV/c)
1	4.2 ± 0.1	1.1 ± 0.1	4890 ± 100
2	6.9 ± 0.1	3.3 ± 0.1	501 ± 5.5
3	14.5 ± 0.5	-1.5 ± 0.6	...
4	79.5 ± 0.1	-2.7 ± 0.1	281 ± 6
5	344.5 ± 0.1	-12.0 ± 0.2	256 ± 3
6	9.6 ± 0.1	-2.5 ± 0.1	1500 ± 15
7	357.0 ± 0.3	3.9 ± 0.4	82 ± 2
8	63.3 ± 0.3	-2.4 ± 0.2	177 ± 2

Table II. Maximum transverse momentum of the negative decay product for various particle decays.

Decay modes	Maximum transverse momentum (MeV/c)
$\pi^- \to \mu^- + \nu$	30
$K^- \to \mu^- + \nu$	236
$K^- \to \pi^- + \pi$	205
$K^- \to e^- + \pi^0 + \nu$	229
$\Sigma^- \to \pi^- + n$	192
$\Sigma^- \to e^- + \Lambda^0 + \nu$	78
$\Sigma^- \to e^- + n + \nu$	229
$\Xi^- \to \pi^- + \Lambda^0$	139
$\Xi^- \to e^- + \Lambda^0 + \nu$	190
$\Xi^- \to e^- + n + \nu$	327

The proper lifetime of particle 3 was calculated to be 0.7×10^{-10} sec; consequently we may assume that it decayed by a weak interaction with $\Delta S=1$ into a system with strangeness minus two. Since a particle with $S=-1$ would decay very rapidly into $Y+\pi$, we may conclude that particle 3 has strangeness minus three. The missing mass at the production vertex is calculated to be 500 $\pm$ 25 MeV/c^2, in good agreement with the K^0 assumed in Reaction (1). Production of the event by an incoming π^- is excluded by the missing mass calculated at the production vertex, and would not alter the interpretation of the decay chain starting with track 3.

In view of the properties of charge ($Q=-1$), strangeness ($S=-3$), and mass ($M=1686\pm12$ MeV/c^2) established for particle 3, we feel justified in identifying it with the sought-for Ω^-. Of course, it is expected that the Ω^- will have other observable decay modes, and we are continuing to search for them. We defer a detailed discussion of the mass of the Ω^- until we have analyzed further examples and have a better understanding of the systematic errors.

The observation of a particle with this mass and strangeness eliminates the possibility which has been put forward[6] that interactions with $\Delta S=4$ proceed with the rates typical of the strong interactions, since in that case the Ω^- would decay very rapidly into $n+K^0+\pi^-$.

We wish to acknowledge the excellent cooperation of the staff of the AGS and the untiring efforts of the 80-in. bubble chamber and scanning and programming staffs.

---

*Work performed under the auspices of the U. S. Atomic Energy Commission and partially supported by the U. S. Office of National Research and the National Science Foundation.

†Syracuse University, Syracuse, New York.

‡University of Rochester, Rochester, New York.

[1]M. Gell-Mann, Proceedings of the International Conference on High-Energy Nuclear Physics, Geneva, 1962 (CERN Scientific Information Service, Geneva, Switzerland, 1962), p. 805; R. Behrends, J. Dreitlein, C. Fronsdal, and W. Lee, Rev. Mod. Phys. 34, 1 (1962); S. L. Glashow and J. J. Sakurai, Nuovo Cimento 25, 337 (1962).

[2]A possible example of the decay of this particle was observed by Y. Eisenberg, Phys. Rev. 96, 541 (1954).

[3]M. Gell-Mann, Phys. Rev. 125, 1067 (1962); Y. Ne'eman, Nucl. Phys. 26, 222 (1961).

[4]See, however, R. J. Oakes and C. N. Yang, Phys. Rev. Letters 11, 174 (1963).

[5]M. Gell-Mann, Synchrotron Laboratory, California Institute of Technology, Internal Report No. CTSL-20, 1961 (unpublished); S. Okubo, Progr. Theoret. Phys. (Kyoto) 27, 949 (1962).

[6]G. Racah, Nucl. Phys. 1, 302 (1956); H. J. Lipkin, Phys. Letters 1, 68 (1962).

EVIDENCE FOR THE 2π DECAY OF THE K_2^0 MESON*†

J. H. Christenson, J. W. Cronin,‡ V. L. Fitch,‡ and R. Turlay§
Princeton University, Princeton, New Jersey
(Received 10 July 1964)

This Letter reports the results of experimental studies designed to search for the 2π decay of the K_2^0 meson. Several previous experiments have served[1,2] to set an upper limit of 1/300 for the fraction of K_2^0's which decay into two charged pions. The present experiment, using spark chamber techniques, proposed to extend this limit.

In this measurement, K_2^0 mesons were produced at the Brookhaven AGS in an internal Be target bombarded by 30-BeV protons. A neutral beam was defined at 30 degrees relative to the circulating protons by a $1\frac{1}{2}$-in.×$1\frac{1}{2}$-in.×48-in. collimator at an average distance of 14.5 ft. from the internal target. This collimator was followed by a sweeping magnet of 512 kG-in. at ~20 ft. and a 6-in.×6-in.×48-in. collimator at 55 ft. A $1\frac{1}{2}$-in. thickness of Pb was placed in front of the first collimator to attenuate the gamma rays in the beam.

The experimental layout is shown in relation to the beam in Fig. 1. The detector for the decay products consisted of two spectrometers each composed of two spark chambers for track delineation separated by a magnetic field of 178 kG-in. The axis of each spectrometer was in the horizontal plane and each subtended an average solid angle of 0.7×10^{-3} steradians. The spark chambers were triggered on a coincidence between water Cherenkov and scintillation counters positioned immediately behind the spectrometers. When coherent K_1^0 regeneration in solid materials was being studied, an anticoincidence counter was placed immediately behind the regenerator. To minimize interactions K_2^0 decays were observed from a volume of He gas at nearly STP.

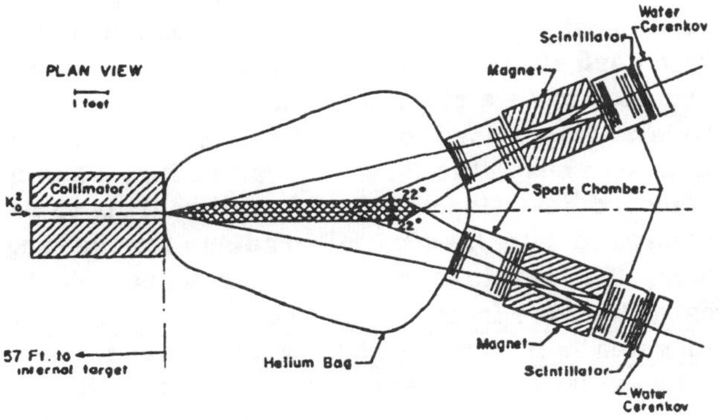

FIG. 1. Plan view of the detector arrangement.

The analysis program computed the vector momentum of each charged particle observed in the decay and the invariant mass, m^*, assuming each charged particle had the mass of the charged pion. In this detector the K_{e3} decay leads to a distribution in m^* ranging from 280 MeV to ~536 MeV; the $K_{\mu3}$, from 280 to ~516; and the $K_{\pi3}$, from 280 to 363 MeV. We emphasize that m^* equal to the K^0 mass is not a preferred result when the three-body decays are analyzed in this way. In addition, the vector sum of the two momenta and the angle, θ, between it and the direction of the K_2^0 beam were determined. This angle should be zero for two-body decay and is, in general, different from zero for three-body decays.

An important calibration of the apparatus and data reduction system was afforded by observing the decays of K_1^0 mesons produced by coherent regeneration in 43 gm/cm^2 of tungsten. Since the K_1^0 mesons produced by coherent regeneration have the same momentum and direction as the K_2^0 beam, the K_1^0 decay simulates the direct decay of the K_2^0 into two pions. The regenerator was successively placed at intervals of 11 in. along the region of the beam sensed by the detector to approximate the spatial distribution of the K_2^0's. The K_1^0 vector momenta peaked about the forward direction with a standard deviation of 3.4 ± 0.3 milliradians. The mass distribution of these events was fitted to a Gaussian with an average mass 498.1 ± 0.4 MeV and standard deviation of 3.6 ± 0.2 MeV. The mean momentum of the K_1^0 decays was found to be 1100 MeV/c. At this momentum the beam region sensed by the detector was 300 K_1^0 decay lengths from the target.

For the K_2^0 decays in He gas, the experimental distribution in m^* is shown in Fig. 2(a). It is compared in the figure with the results of a Monte Carlo calculation which takes into account the nature of the interaction and the form factors involved in the decay, coupled with the detection efficiency of the apparatus. The computed curve shown in Fig. 2(a) is for a vector interaction, form-factor ratio $f^-/f^+=0.5$, and relative abundance 0.47, 0.37, and 0.16 for the K_{e3}, $K_{\mu3}$, and $K_{\pi3}$, respectively.[3] The scalar interaction has been computed as well as the vector interaction

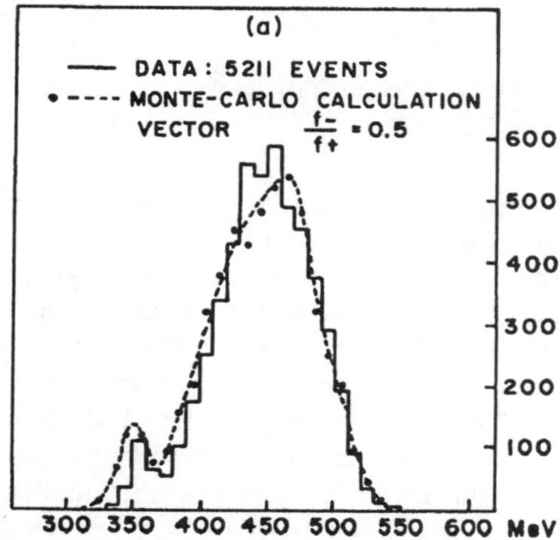

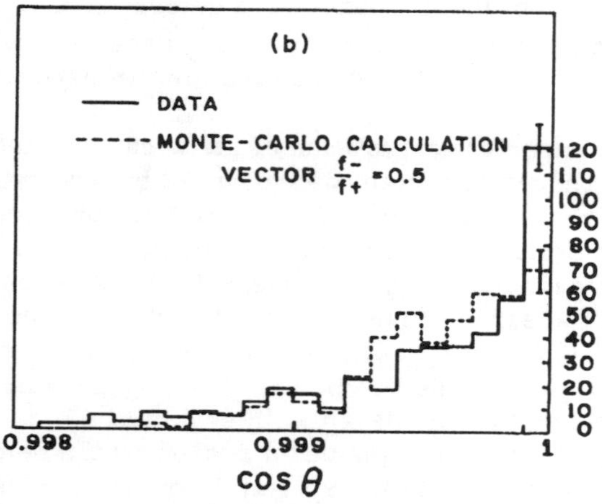

FIG. 2. (a) Experimental distribution in m^* compared with Monte Carlo calculation. The calculated distribution is normalized to the total number of observed events. (b) Angular distribution of those events in the range 490 < m^* < 510 MeV. The calculated curve is normalized to the number of events in the complete sample.

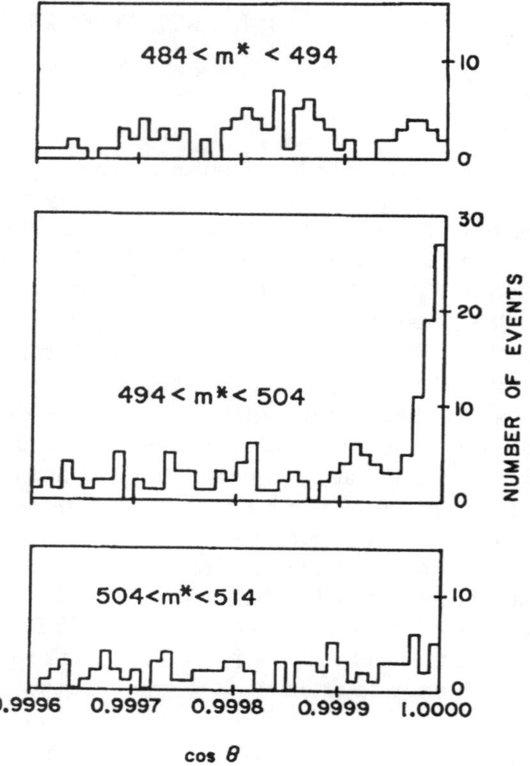

FIG. 3. Angular distribution in three mass ranges for events with cosθ > 0.9995.

with a form-factor ratio $f^-/f^+ = -6.6$. The data are not sensitive to the choice of form factors but do discriminate against the scalar interaction.

Figure 2(b) shows the distribution in cosθ for those events which fall in the mass range from 490 to 510 MeV together with the corresponding result from the Monte Carlo calculation. Those events within a restricted angular range (cosθ > 0.9995) were remeasured on a somewhat more precise measuring machine and recomputed using an independent computer program. The results of these two analyses are the same within the respective resolutions. Figure 3 shows the results from the more accurate measuring machine. The angular distribution from three mass ranges are shown; one above, one below, and one encompassing the mass of the neutral K meson.

The average of the distribution of masses of those events in Fig. 3 with cosθ > 0.999 99 is found to be 499.1 ± 0.8 MeV. A corresponding calculation has been made for the tungsten data resulting in a mean mass of 498.1 ± 0.4. The difference is 1.0 ± 0.9 MeV. Alternately we may take the mass of the K^0 to be known and compute the mass of the secondaries for two-body decay. Again restricting our attention to those events with cosθ > 0.999 99 and assuming one of the secondaries to be a pion, the mass of the other particle is determined to be 137.4 ± 1.8. Fitted to a Gaussian shape the forward peak in Fig. 3 has a standard deviation of 4.0 ± 0.7 milliradians to be compared with 3.4 ± 0.3 milliradians for the tungsten. The events from the He gas appear identical with those from the coherent regeneration in tungsten in both mass and angular spread.

The relative efficiency for detection of the three-body K_2^0 decays compared to that for decay to two pions is 0.23. We obtain 45 ± 9 events in the forward peak after subtraction of background out of a total corrected sample of 22 700 K_2^0 decays.

Data taken with a hydrogen target in the beam also show evidence of a forward peak in the $\cos\theta$ distribution. After subtraction of background, 45 ± 10 events are observed in the forward peak at the K^0 mass. We estimate that ~10 events can be expected from coherent regeneration. The number of events remaining (35) is entirely consistent with the decay data when the relative target volumes and integrated beam intensities are taken into account. This number is substantially smaller (by more than a factor of 15) than one would expect on the basis of the data of Adair et al.[4]

We have examined many possibilities which might lead to a pronounced forward peak in the angular distribution at the K^0 mass. These include the following:

(i) K_1^0 coherent regeneration. In the He gas it is computed to be too small by a factor of $\sim 10^6$ to account for the effect observed, assuming reasonable scattering amplitudes. Anomalously large scattering amplitudes would presumably lead to exaggerated effects in liquid H_2 which are not observed. The walls of the He bag are outside the sensitive volume of the detector. The spatial distribution of the forward events is the same as that for the regular K_2^0 decays which eliminates the possibility of regeneration having occurred in the collimator.

(ii) $K_{\mu 3}$ or K_{e3} decay. A spectrum can be constructed to reproduce the observed data. It requires the preferential emission of the neutrino within a narrow band of energy, ±4 MeV, centered at 17 ± 2 MeV ($K_{\mu 3}$) or 39 ± 2 MeV (K_{e3}). This must be coupled with an appropriate angular correlation to produce the forward peak. There appears to be no reasonable mechanism which can produce such a spectrum.

(iii) Decay into $\pi^+\pi^-\gamma$. To produce the highly singular behavior shown in Fig. 3 it would be necessary for the γ ray to have an average energy of less than 1 MeV with the available energy extending to 209 MeV. We know of no physical process which would accomplish this.

We would conclude therefore that K_2^0 decays to two pions with a branching ratio $R = (K_2 \to \pi^+ + \pi^-)/(K_2^0 \to \text{all charged modes}) = (2.0 \pm 0.4) \times 10^{-3}$ where the error is the standard deviation. As emphasized above, any alternate explanation of the effect requires highly nonphysical behavior of the three-body decays of the K_2^0. The presence of a two-pion decay mode implies that the K_2^0 meson is not a pure eigenstate of CP. Expressed as $K_2^0 = 2^{-1/2}[(K_0 - \bar{K}_0) + \epsilon(K_0 + \bar{K}_0)]$ then $|\epsilon|^2 \cong R_T \tau_1 \tau_2$ where τ_1 and τ_2 are the K_1^0 and K_2^0 mean lives and R_T is the branching ratio including decay to two π^0. Using $R_T = \frac{3}{2}R$ and the branching ratio quoted above, $|\epsilon| \cong 2.3 \times 10^{-3}$.

We are grateful for the full cooperation of the staff of the Brookhaven National Laboratory. We wish to thank Alan Clark for one of the computer analysis programs. R. Turlay wishes to thank the Elementary Particles Laboratory at Princeton University for its hospitality.

---

*Work supported by the U. S. Office of Naval Research.

†This work made use of computer facilities supported in part by National Science Foundation grant.

‡A. P. Sloan Foundation Fellow.

§On leave from Laboratoire de Physique Corpusculaire à Haute Energie, Centre d'Etudes Nucléaires, Saclay, France.

[1]M. Bardon, K. Lande, L. M. Lederman, and W. Chinowsky, Ann. Phys. (N.Y.) 5, 156 (1958).

[2]D. Neagu, E. O. Okonov, N. I. Petrov, A. M. Rosanova, and V. A. Rusakov, Phys. Rev. Letters 6, 552 (1961).

[3]D. Luers, I. S. Mittra, W. J. Willis, and S. S. Yamamoto, Phys. Rev. 133, B1276 (1964).

[4]R. Adair, W. Chinowsky, R. Crittenden, L. Leipuner, B. Musgrave, and F. Shively, Phys. Rev. 132, 2285 (1963).

OBSERVED BEHAVIOR OF HIGHLY INELASTIC ELECTRON-PROTON SCATTERING

M. Breidenbach, J. I. Friedman, and H. W. Kendall
Department of Physics and Laboratory for Nuclear Science,*
Massachusetts Institute of Technology, Cambridge, Massachusetts 02139

and

E. D. Bloom, D. H. Coward, H. DeStaebler, J. Drees, L. W. Mo, and R. E. Taylor
Stanford Linear Accelerator Center,† Stanford, California 94305
(Received 22 August 1969)

Results of electron-proton inelastic scattering at 6° and 10° are discussed, and values of the structure function W_2 are estimated. If the interaction is dominated by transverse virtual photons, νW_2 can be expressed as a function of $\omega = 2M\nu/q^2$ within experimental errors for $q^2 > 1$ $(\mathrm{GeV}/c)^2$ and $\omega > 4$, where ν is the invariant energy transfer and q^2 is the invariant momentum transfer of the electron. Various theoretical models and sum rules are briefly discussed.

In a previous Letter,[1] we have reported experimental results from a Stanford Linear Accelerator Center–Massachusetts Institute of Technology study of high-energy inelastic electron-proton scattering. Measurements of inelastic spectra, in which only the scattered electrons were detected, were made at scattering angles of 6° and 10° and with incident energies between 7 and 17 GeV. In this communication, we discuss some of the salient features of inelastic spectra in the deep continuum region.

One of the interesting features of the measurements is the weak momentum-transfer dependence of the inelastic cross sections for excitations well beyond the resonance region. This weak dependence is illustrated in Fig. 1. Here we have plotted the differential cross section divided by the Mott cross section, $(d^2\sigma/d\Omega dE')/(d\sigma/d\Omega)_{\mathrm{Mott}}$, as a function of the square of the four-momentum transfer, $q^2 = 2EE'(1-\cos\theta)$, for constant values of the invariant mass of the recoiling target system, W, where $W^2 = 2M(E-E') + M^2 - q^2$. E is the energy of the incident electron, E' is the energy of the final electron, and θ is the scattering angle, all defined in the laboratory system; M is the mass of the proton. The cross section is divided by the Mott cross section

$$\left(\frac{d\sigma}{d\Omega}\right)_{\mathrm{Mott}} = \frac{e^4}{4E^2}\frac{\cos^2\frac{1}{2}\theta}{\sin^4\frac{1}{2}\theta}$$

in order to remove the major part of the well-known four-momentum transfer dependence arising from the photon propagator. Results from both 6° and 10° are included in the figure for each value of W. As W increases, the q^2 dependence appears to decrease. The striking difference between the behavior of the inelastic and elastic cross sections is also illustrated in Fig. 1, where the elastic cross section, divided by the Mott cross section for $\theta = 10°$, is included. The q^2 dependence of the deep continuum is also consider-

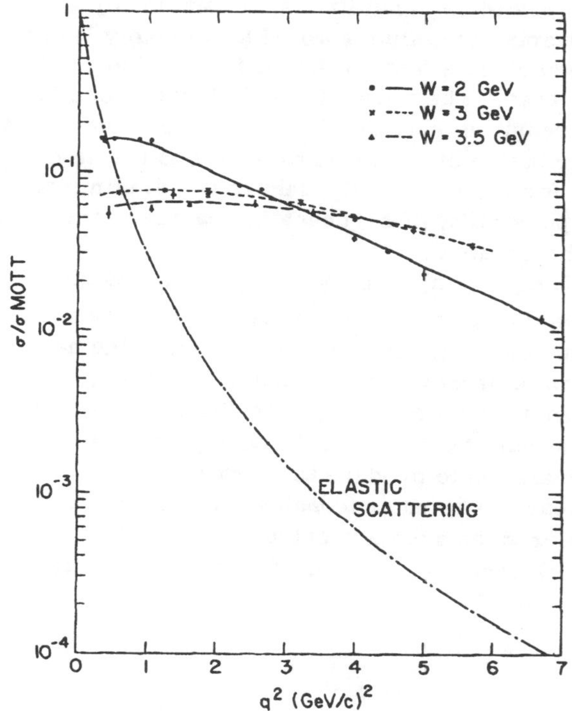

FIG. 1. $(d^2\sigma/d\Omega dE')/\sigma_{\mathrm{Mott}}$, in GeV^{-1}, vs q^2 for $W = 2$, 3, and 3.5 GeV. The lines drawn through the data are meant to guide the eye. Also shown is the cross section for elastic e-p scattering divided by σ_{Mott}, $(d\sigma/d\Omega)/\sigma_{\mathrm{Mott}}$, calculated for $\theta = 10°$, using the dipole form factor. The relatively slow variation with q^2 of the inelastic cross section compared with the elastic cross section is clearly shown.

ably weaker than that of the electroexcitation of the resonances,[2] which have a q^2 dependence similar to that of elastic scattering for $q^2 > 1$ (GeV/c)^2.

On the basis of general considerations, the differential cross section for inelastic electron scattering in which only the electron is detected can be represented by the following expression[3]:

$$\frac{d^2\sigma}{d\Omega dE'} = \left(\frac{d\sigma}{d\Omega}\right)_{\text{Mott}} (W_2 + 2W_1 \tan^2\tfrac{1}{2}\theta).$$

The form factors W_2 and W_1 depend on the properties of the target system, and can be represented as functions of q^2 and $\nu = E - E'$, the electron energy loss. The ratio W_2/W_1 is given by

$$\frac{W_2}{W_1} = \left(\frac{q^2}{\nu^2 + q^2}\right)(1 + R), \quad R \geq 0,$$

where R is the ratio of the photoabsorption cross sections of longitudinal and transverse virtual photons, $R = \sigma_S/\sigma_T$.[4]

The objective of our investigations is to study the behavior of W_1 and W_2 to obtain information about the structure of the proton and its electromagnetic interactions at high energies. Since at present only cross-section measurements at small angles are available, we are unable to make separate determinations of W_2 and W_1. However, we can place limits on W_2 and study the behavior of these limits as a function of the invariants ν and q^2.

Bjorken[5] originally suggested that W_2 could have the form

$$W_2 = (1/\nu)F(\omega),$$

where

$$\omega = 2M\nu/q^2.$$

$F(\omega)$ is a universal function that is conjectured to be valid for large values of ν and q^2. This function is universal in the sense that it manifests scale invariance, that is, it depends only on the ratio ν/q^2. Since

$$\nu W_2 = \frac{\nu d^2\sigma/d\Omega\, dE'}{(d\sigma/d\Omega)_{\text{Mott}}}\left[1 + 2\frac{1}{1+R}\left(1 + \frac{\nu^2}{q^2}\right)\tan^2\tfrac{1}{2}\theta\right]^{-1},$$

the value of νW_2 for any given measurement clearly depends on the presently unknown value of R. It should be noted that the sensitivity to R is small when $2(1+\nu^2/q^2)\tan^2\frac{1}{2}\theta \ll 1$. Experimental limits on νW_2 can be calculated on the basis of the extreme assumptions $R = 0$ and $R = \infty$. In Figs. 2(a) and 2(b) the experimental values of νW_2

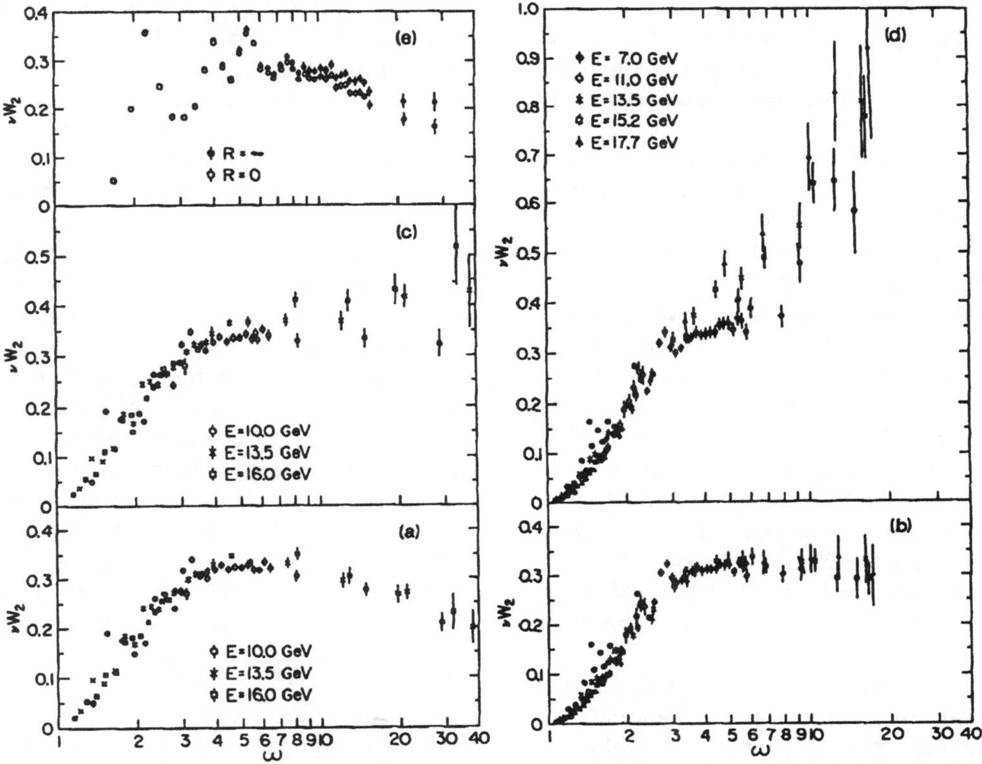

FIG. 2. νW_2 vs $\omega = 2M\nu/q^2$ is shown for various assumptions about $R = \sigma_S/\sigma_T$. (a) 6° data except for 7-GeV spectrum for $R = 0$. (b) 10° data for $R = 0$. (c) 6° data except for 7-GeV spectrum for $R = \infty$. (d) 10° data for $R = \infty$. (e) 6°, 7-GeV spectrum for $R = 0$ and $R = \infty$.

from the 6° and 10° data for $q^2 > 0.5$ (GeV/c)^2 are shown as a function of ω for the assumption that $R=0$. Figures 2(c) and 2(d) show the experimental values of νW_2 calculated from the 6° and 10° data with $q^2>0.5$ (GeV/c)^2 under the assumption $R=\infty$. The 6°, 7-GeV results for νW_2, all of which have values of $q^2 \leq 0.5$ (GeV/c)^2, are shown for both assumptions in Fig. 2(e). The elastic peaks are not displayed in Fig. 2.

The results shown in these figures indicate the following:

(1) If $\sigma_T \gg \sigma_S$, the experimental results are consistent with a universal curve for $\omega \gtrsim 4$ and $q^2 \gtrsim 0.5$ (GeV/c)^2. Above these values, the measurements at 6° and 10° give the same results within the errors of measurements. The 6°, 7-GeV measurements of νW_2, all of which have values of $q^2 \leq 0.5$ (GeV/c)^2, are somewhat smaller than the results from the other spectra in the continuum region.

The values of νW_2 for $\omega \gtrsim 5$ show a gradual decrease as ω increases. In order to test the statistical significance of the observed slope, we have made linear least-squares fits to the values of νW_2 in the region $6 \leq \omega \lesssim 25$. These fits give $\nu W_2 = (0.351 \pm 0.023) - (0.00386 \pm 0.00088)\omega$ for data with $q^2 > 0.5$ (GeV/c)^2 and $\nu W_2 = (0.366 \pm 0.024) - (0.0045 \pm 0.0019)\omega$ for $q^2 > 1$ (GeV/c)^2. The quoted errors consist of the errors from the fit added in quadrature with estimates of systematic errors.

Since $\sigma_T + \sigma_S \simeq 4\pi^2\alpha\nu W_2/q^2$ for $\omega \gg 1$, our results can provide information about the behavior of σ_T if $\sigma_T \gg \sigma_S$. The scale invariance found in the measurements of νW_2 indicates that the q^2 dependence of σ_T is approximately $1/q^2$. The gradual decrease exhibited in νW_2 for large ω suggests that the photoabsorption cross section for virtual photons falls slowly at constant q^2 as the photon energy ν increases.

The measurements indicate that νW_2 has a broad maximum in the neighborhood of $\omega = 5$. The question of whether this maximum has any correspondence to a possible quasielastic peak[6] requires further investigation.

It should be emphasized that all of the above conclusions are based on the assumption that $\sigma_T \gg \sigma_S$.

(2) If $\sigma_S \gg \sigma_T$, the measurements of νW_2 do not follow a universal curve and have the general feature that at constant $2M\nu/q^2$, the value of νW_2 increases with q^2.

(3) For either assumption, νW_2 shows a threshold behavior in the range $1 \leq \omega \lesssim 4$. W_2 is constrained to be zero at inelastic threshold which corresponds to $\omega \simeq 1$ for large q^2. In the threshold region of νW_2, W_2 falls rapidly as q^2 increases at constant ν. This qualitatively different from the weak q^2 behavior for $\omega > 4$. For $q^2 \approx 1$ (GeV/c)^2, the threshold region contains the resonances excited in electroproduction. As q^2 increases, the variations due to these resonances damp out and the values of νW_2 do not appear to vary rapidly with q^2 at constant ω.

It can be seen from a comparison of Figs. 2(a) and 2(c) that the 6° data provide a measurement of νW_2 to within 10% up to a value of $\omega \approx 6$, irrespective of the values of R.

There have been a number of different theoretical approaches in the interpretation of the high-energy inelastic electron-scattering results. One class of models,[6-9] referred to as parton models, describes the electron as scattering incoherently from pointlike constituents within the proton. Such models lead to a universal form for νW_2, and the point charges assumed in specific models give the magnitude of νW_2 for $\omega > 2$ to within a factor of 2.[6] Another approach[10,11] relates the inelastic scattering to off-the-mass-shell Compton scattering which is described in terms of Regge exchange using the Pomeranchuk trajectory. Such models lead to a flat behavior of νW_2 as a function of ν but do not require the weak q^2 dependence observed and do not make any numerical predictions at this time. Perhaps the most detailed predictions made at present come from a vector-dominance model which primarily utilizes the ρ meson.[12] This model reproduces the gross behavior of the data and has the feature that νW_2 asymptotically approaches a function of ω as $q^2 \to \infty$. However, a comparison of this model with the data leads to statistically significant discrepancies. This can be seen by noting that the prediction for $d^2\sigma/d\Omega dE'$ contains a parameter ξ, the ratio of the cross sections for longitudinally and transversely polarized ρ mesons on protons, which is expected to be a function of W but which should be independent of q^2. For values of $W \geq 2$ GeV, the experimental values of ξ increase by about (50 ± 5)% as q^2 increases from 1 to 4 (GeV/c)^2. This model predicts that

$$\sigma_S/\sigma_T = \xi(W)(q^2/m_\rho^2)[1 - q^2/2m\nu],$$

which will provide the most stringent test of this approach when a separation of W_1 and W_2 can be made.

The application of current algebra[13-17] and the use of current commutators leading to sum rules and sum-rule inequalities provide another way of comparing the measurements with theory. There have been some recent theoretical considerations[18-20] which have pointed to possible ambiguity in these calculations; however, it is still of considerable interest to compare them with experiment.

In general, W_2 and W_1 can be related to commutators of electromagnetic current densities.[6,16] The experimental value of the energy-weighted sum $\int_1^\infty (d\omega/\omega^2)(\nu W_2)$, which is related to the equal-time commutator of the current and its time derivative, is 0.16 ± 0.01 for $R=0$ and 0.20 ± 0.03 for $R=\infty$. The integral has been evaluated with an upper limit $\omega = 20$. This integral is also important in parton theories where its value is the mean square charge per parton.

Gottfried[21] has calculated a constant-q^2 sum rule for inelastic electron-proton scattering based on a nonrelativistic quark model involving pointlike quarks. The resulting sum rule is

$$\int_1^\infty \frac{d\omega}{\omega}(\nu W_2) = \int_{q^2/2M}^\infty d\nu W_2$$

$$= 1 - \frac{G_{Ep}{}^2 + (q^2/4M^2)G_{Mp}}{1+q^2/4M^2},$$

where G_{Ep} and G_{Mp} are the electric and magnetic form factors of the proton. The experimental evaluation of this integral from our data is much more dependent on the assumption about R than the previous integral. We will thus use the 6° measurements of W_2 which are relatively insensitive to R. Our data for a value of $q^2 \simeq 1$ (GeV/c)^2, which extend to a value of ν of about 10 GeV, give a sum that is 0.72 ± 0.05 with the assumption that $R=0$. For $R=\infty$, its value is 0.81 ± 0.06. An extrapolation of our measurements of νW_2 for each assumption suggests that the sum is saturated in the region $\nu \simeq 20$-40 GeV. Bjorken[13] has proposed a constant-q^2 sum-rule inequality for high-energy scattering from the proton and neutron derived on the basis of current algebra. His result states that

$$\int_1^\infty \frac{d\omega}{\omega}\nu(W_{2p}+W_{2n}) = \int_{q^2/2M}^\infty d\nu(W_{2p}+W_{2n}) \geq \tfrac{1}{2},$$

where the subscripts p and n refer to the proton and neutron, respectively. Since there are presently no electron-neutron inelastic scattering results available, we estimate W_{2n} in a model-dependent way. For a quark model[22] of the proton, $W_{2n} \simeq 0.8 W_{2p}$ whereas in the model[8] of Drell and co-workers, W_{2n} rapidly approaches W_{2p} as ν increases. Using our results, this inequality is just satisfied at $\omega \simeq 4.5$ for the quark model and at $\omega \simeq 4.0$ for the other model for either assumption about R. For example, this corresponds to a value of $\nu \simeq 4.5$ GeV for $q^2 = 2$ (GeV/c)^2. Bjorken[23] estimates that the experimental value of the sum is too small by about a factor of 2 for either model, but is should be noted that the q^2 dependence found in the data is consistent with the predictions of this calculation.

---

*Work supported in part through funds provided by the U. S. Atomic Energy Commission under Contract No. AT(30-1)2098.

†Work supported by the U. S. Atomic Energy Commission.

[1]E. Bloom et al., preceding Letter [Phys. Rev. Letters 23, 930 (1969)].

[2]Preliminary results from the present experimental program are given in the report by W. K. H. Panofsky, in Proceedings of the Fourteenth International Conference On High Energy Physics, Vienna, Austria, 1968 (CERN Scientific Information Service, Geneva, Switzerland, 1968), p. 23.

[3]R. von Gehlen, Phys. Rev. 118, 1455 (1960); J. D. Bjorken, 1960 (unpublished); M. Gourdin, Nuovo Cimento 21, 1094 (1961).

[4]See L. Hand, in Proceedings of the Third International Symposium on Electron and Photon Interactions at High Energies, Stanford Linear Accelerator Center, Stanford, California, 1967 (Clearing House of Federal Scientific and Technical Information, Washington, D. C., 1968), or F. J. Gilman, Phys. Rev. 167, 1365 (1968).

[5]J. D. Bjorken, Phys. Rev. 179, 1547 (1969).

[6]J. D. Bjorken and E. A. Paschos, Stanford Linear Accelerator Center, Report No. SLAC-PUB-572, 1969 (to be published).

[7]R. P. Feynman, private communication.

[8]S. J. Drell, D. J. Levy, and T. M. Yan, Phys. Rev. Letters 22, 744 (1969).

[9]K. Huang, in Argonne National Laboratory Report No. ANL-HEP 6909, 1968 (unpublished), p. 150.

[10]H. D. Abarbanel and M. L. Goldberger, Phys. Rev. Letters 22, 500 (1969).

[11]H. Harari, Phys. Rev. Letters 22, 1078 (1969).

[12]J. J. Sakurai, Phys. Rev. Letters 22, 981 (1969).

[13]J. D. Bjorken, Phys. Rev. Letters 16, 408 (1966).

[14]J. D. Bjorken, in Selected Topics in Particle Physics, Proceedings of the International School of Physics "Enrico Fermi," Course XLI, edited by J. Steinberger (Academic Press, Inc., New York, 1968).

[15]J. M. Cornwall and R. E. Norton, Phys. Rev. 177, 2584 (1969).

[16]C. G. Callan, Jr., and D. J. Gross, Phys. Rev. Letters 21, 311 (1968).

[17]C. G. Callan, Jr., and D. J. Gross, Phys. Rev. Letters 22, 156 (1969).

[18]R. Jackiw and G. Preparata, Phys. Rev. Letters 22, 975 (1969).

[19]S. L. Adler and W.-K. Tung, Phys. Rev. Letters 22, 978 (1969).

[20]H. Cheng and T. T. Wu, Phys. Rev. Letters 22, 1409 (1969).

[21]K. Gottfried, Phys. Rev. Letters 18, 1174 (1967).

[22]J. D. Bjorken, Stanford Linear Accelerator Center, Report No. SLAC-PUB-571, 1969 (unpublished).

[23]J. D. Bjorken, private communication.

Experimental Observation of a Heavy Particle J†

J. J. Aubert, U. Becker, P. J. Biggs, J. Burger, M. Chen, G. Everhart, P. Goldhagen, J. Leong, T. McCorriston, T. G. Rhoades, M. Rohde, Samuel C. C. Ting, and Sau Lan Wu
Laboratory for Nuclear Science and Department of Physics, Massachusetts Institute of Technology, Cambridge, Massachusetts 02139

and

Y. Y. Lee
Brookhaven National Laboratory, Upton, New York 11973
(Received 12 November 1974)

We report the observation of a heavy particle J, with mass $m = 3.1$ GeV and width approximately zero. The observation was made from the reaction $p + \mathrm{Be} \rightarrow e^+ + e^- + x$ by measuring the e^+e^- mass spectrum with a precise pair spectrometer at the Brookhaven National Laboratory's 30-GeV alternating-gradient synchrotron.

This experiment is part of a large program to study the behavior of timelike photons in $p+p \rightarrow e^+ + e^- + x$ reactions[1] and to search for new particles which decay into e^+e^- and $\mu^+\mu^-$ pairs.

We use a slow extracted beam from the Brookhaven National Laboratory's alternating-gradient synchrotron. The beam intensity varies from 10^{10} to 2×10^{12} p/pulse. The beam is guided onto an extended target, normally nine pieces of 70-mil Be, to enable us to reject the pair accidentals by requiring the two tracks to come from the same origin. The beam intensity is monitored with a secondary emission counter, calibrated daily with a thin Al foil. The beam spot size is 3×6 mm^2, and is monitored with closed-circuit television. Figure 1(a) shows the simplified side view of one arm of the spectrometer. The two arms are placed at 14.6° with respect to the incident beam; bending (by $M1$, $M2$) is done vertically to decouple the angle (θ) and the momentum (p) of the particle.

The Cherenkov counter C_0 is filled with one atmosphere and C_e with 0.8 atmosphere of H_2. The counters C_0 and C_e are decoupled by magnets $M1$ and $M2$. This enables us to reject knock-on electrons from C_0. Extensive and repeated calibra-

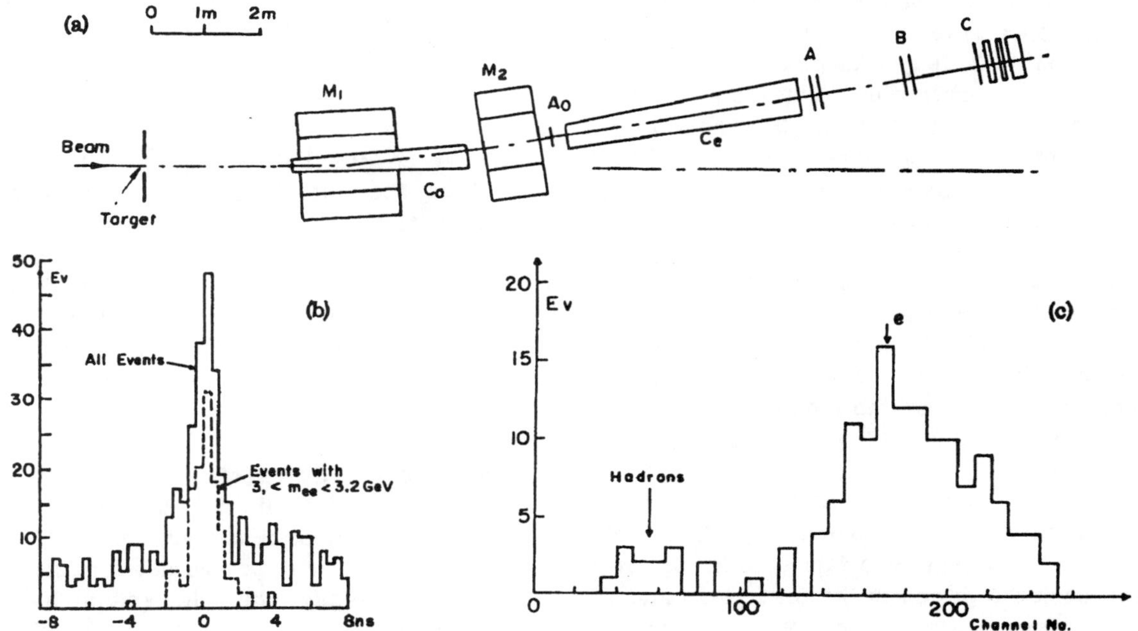

FIG. 1. (a) Simplified side view of one of the spectrometer arms. (b) Time-of-flight spectrum of e^+e^- pairs and of those events with $3.0 < m < 3.2$ GeV. (c) Pulse-height spectrum of e^- (same for e^+) of the e^+e^- pair.

tion of all the counters is done with approximately 6-GeV electrons produced with a lead converter target. There are eleven planes ($2\times A_0$, $3\times A$, $3\times B$, $3\times C$) of proportional chambers rotated approximately 20° with respect to each other to reduce multitrack confusion. To further reduce the problem of operating the chambers at high rate, eight vertical and eight horizontal hodoscope counters are placed behind chambers A and B. Behind the largest chamber C (1 m×1 m) there are two banks of 25 lead glass counters of 3 radiation lengths each, followed by one bank of lead-Lucite counters to further reject hadrons from electrons and to improve track identification. During the experiment all the counters are monitored with a PDP 11-45 computer and all high voltages are checked every 30 min.

The magnets were measured with a three-dimensional Hall probe. A total of 10^5 points were mapped at various current settings. The acceptance of the spectrometer is $\Delta\theta = \pm 1°$, $\Delta\varphi = \pm 2°$, $\Delta m = 2$ GeV. Thus the spectrometer enables us to map the e^+e^- mass region from 1 to 5 GeV in three overlapping settings.

Figure 1(b) shows the time-of-flight spectrum between the e^+ and e^- arms in the mass region $2.5 < m < 3.5$ GeV. A clear peak of 1.5-nsec width is observed. This enables us to reject the accidentals easily. Track reconstruction between the two arms was made and again we have a clear-cut distinction between real pairs and accidentals. Figure 1(c) shows the shower and lead-glass pulse height spectrum for the events in the mass region $3.0 < m < 3.2$ GeV. They are again in agreement with the calibration made by the e beam.

Typical data are shown in Fig. 2. There is a clear sharp enhancement at $m = 3.1$ GeV. Without folding in the 10^5 mapped magnetic points and the radiative corrections, we estimate a mass resolution of 20 MeV. As seen from Fig. 2 the width of the particle is consistent with zero.

To ensure that the observed peak is indeed a real particle ($J \rightarrow e^+e^-$) many experimental checks were made. We list seven examples:

(1) When we decreased the magnet currents by 10%, the peak remained fixed at 3.1 GeV (see Fig. 2).

(2) To check second-order effects on the target, we increased the target thickness by a factor of 2. The yield increased by a factor of 2, not by 4.

(3) To check the pileup in the lead glass and shower counters, different runs with different voltage settings on the counters were made. No effect was observed on the yield of J.

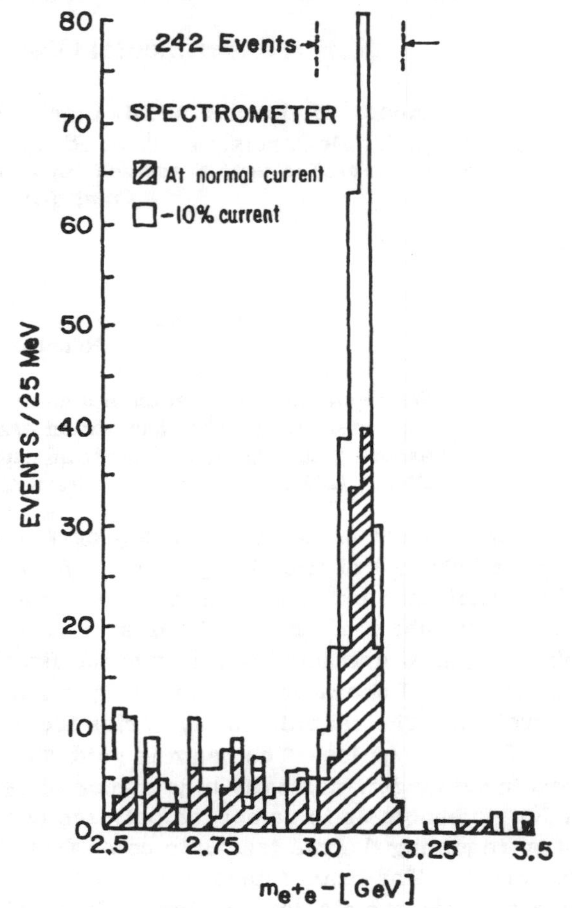

FIG. 2. Mass spectrum showing the existence of J. Results from two spectrometer settings are plotted showing that the peak is independent of spectrometer currents. The run at reduced current was taken two months later than the normal run.

(4) To ensure that the peak is not due to scattering from the sides of magnets, cuts were made in the data to reduce the effective aperture. No significant reduction in the J yield was found.

(5) To check the read-out system of the chambers and the triggering system of the hodoscopes, runs were made with a few planes of chambers deleted and with sections of the hodoscopes omitted from the trigger. No effect was observed on the J yield.

(6) Runs with different beam intensity were made and the yield did not change.

(7) To avoid systematic errors, half of the data were taken at each spectrometer polarity.

These and many other checks convinced us that we have observed a real massive particle $J \rightarrow ee$.

If we assume a production mechanism for J to be $d\sigma/dp_\perp \propto \exp(-6p_\perp)$ we obtain a yield of J of approximately 10^{-34} cm^2.

The most striking feature of J is the possibility that it may be one of the theoretically suggested charmed particles[2] or a's[3] or Z_0's,[4] etc. In order to study the real nature of J,[5] measurements are now underway on the various decay modes, e.g., an $e\pi\nu$ mode would imply that J is weakly interacting in nature.

It is also important to note the absence of an e^+e^- continuum, which contradicts the predictions of parton models.[6]

We wish to thank Dr. R. R. Rau and the alternating-gradient synchrotron staff who have done an outstanding job in setting up and maintaining this experiment. We thank especially Dr. F. Eppling, B. M. Bailey, and the staff of the Laboratory for Nuclear Science for their help and encouragement. We thank also Ms. I. Schulz, Ms. H. Feind, N. Feind, D. Osborne, G. Krey, J. Donahue, and E. D. Weiner for help and assistance. We thank also M. Deutsch, V. F. Weisskopf, T. T. Wu, S. Drell, and S. Glashow for many interesting conversations.

---

†Accepted without review under policy announced in Editorial of 20 July 1964 [Phys. Rev. Lett. 13, 79 (1964)].

[1]The first work on $p + p \rightarrow \mu^+ + \mu^- + x$ was done by L. M. Lederman *et al.*, Phys. Rev. Lett. 25, 1523 (1970).

[2]S. L. Glashow, private communication.

[3]T. D. Lee, Phys. Rev. Lett. 26, 801 (1971).

[4]S. Weinberg, Phys. Rev. Lett. 19, 1264 (1967), and 27, 1688 (1971), and Phys. Rev. D 5, 1412, 1962 (1972).

[5]After completion of this paper, we learned of a similar result from SPEAR. B. Richter and W. Panofsky, private communication; J.-E. Augustin *et al.*, following Letter [Phys. Rev. Lett. 33, 1404 (1974)].

[6]S. D. Drell and T. M. Yan, Phys. Rev. Lett. 25, 316 (1970). An improved version of the theory is not in contradiction with the data.

Discovery of a Narrow Resonance in e^+e^- Annihilation*

J.-E. Augustin,† A. M. Boyarski, M. Breidenbach, F. Bulos, J. T. Dakin, G. J. Feldman, G. E. Fischer, D. Fryberger, G. Hanson, B. Jean-Marie,† R. R. Larsen, V. Lüth, H. L. Lynch, D. Lyon, C. C. Morehouse, J. M. Paterson, M. L. Perl, B. Richter, P. Rapidis, R. F. Schwitters, W. M. Tanenbaum, and F. Vannucci‡

Stanford Linear Accelerator Center, Stanford University, Stanford, California 94305

and

G. S. Abrams, D. Briggs, W. Chinowsky, C. E. Friedberg, G. Goldhaber, R. J. Hollebeek, J. A. Kadyk, B. Lulu, F. Pierre,§ G. H. Trilling, J. S. Whitaker, J. Wiss, and J. E. Zipse

Lawrence Berkeley Laboratory and Department of Physics, University of California, Berkeley, California 94720

(Received 13 November 1974)

We have observed a very sharp peak in the cross section for $e^+e^- \to$ hadrons, e^+e^-, and possibly $\mu^+\mu^-$ at a center-of-mass energy of 3.105 ± 0.003 GeV. The upper limit to the full width at half-maximum is 1.3 MeV.

We have observed a very sharp peak in the cross section for $e^+e^- \to$ hadrons, e^+e^-, and possibly $\mu^+\mu^-$ in the Stanford Linear Accelerator Center (SLAC)–Lawrence Berkeley Laboratory magnetic detector[1] at the SLAC electron-positron storage ring SPEAR. The resonance has the parameters

$$E = 3.105 \pm 0.003 \text{ GeV},$$

$$\Gamma \leqslant 1.3 \text{ MeV}$$

(full width at half-maximum), where the uncertainty in the energy of the resonance reflects the uncertainty in the absolute energy calibration of the storage ring. [We suggest naming this structure $\psi(3105)$.] The cross section for hadron production at the peak of the resonance is $\geqslant 2300$ nb, an enhancement of about 100 times the cross section outside the resonance. The large mass, large cross section, and narrow width of this structure are entirely unexpected.

Our attention was first drawn to the possibility of structure in the $e^+e^- \to$ hadron cross section during a scan of the cross section carried out in 200-MeV steps. A 30% (6 nb) enhancement was observed at a c.m. energy of 3.2 GeV. Subsequently, we repeated the measurement at 3.2 GeV and also made measurements at 3.1 and 3.3 GeV. The 3.2-GeV results reproduced, the 3.3-GeV measurement showed no enhancement, but the 3.1-GeV measurements were internally inconsistent—six out of eight runs giving a low cross section and two runs giving a factor of 3 to 5 higher cross section. This pattern could have been caused by a very narrow resonance at an energy slightly larger than the nominal 3.1-GeV setting of the storage ring, the inconsistent 3.1-GeV cross sections then being caused by setting errors in the ring energy. The 3.2-GeV enhancement would arise from radiative corrections which give a high-energy tail to the structure.

We have now repeated the measurements using much finer energy steps and using a nuclear magnetic resonance magnetometer to monitor the ring energy. The magnetometer, coupled with measurements of the circulating beam position in the storage ring made at sixteen points around the orbit, allowed the relative energy to be determined to 1 part in 10^4. The determination of the absolute energy setting of the ring requires the knowledge of $\int B\,dl$ around the orbit and is accurate to $\pm 0.1\%$.

The data are shown in Fig. 1. All cross sections are normalized to Bhabha scattering at 20 mrad. The cross section for the production of hadrons is shown in Fig. 1(a). Hadronic events are required to have in the final state either ≥ 3 detected charged particles or 2 charged particles noncoplanar by $> 20°$.[2] The observed cross section rises sharply from a level of about 25 nb to a value of 2300 ± 200 nb at the peak[3] and then exhibits the long high-energy tail characteristic of radiative corrections in e^+e^- reactions. The detection efficiency for hadronic events is 45% over the region shown. The error quoted above includes both the statistical error and a 7% contribution from uncertainty in the detection efficiency.

Our mass resolution is determined by the energy spread in the colliding beams which arises from quantum fluctuations in the synchrotron radiation emitted by the beams. The expected Gaussian c.m. energy distribution ($\sigma = 0.56$ MeV), folded with the radiative processes,[4] is shown as the dashed curve in Fig. 1(a). The width of the resonance must be smaller than this spread; thus an upper limit to the full width at half-maximum is 1.3 MeV.

Figure 1(b) shows the cross section for e^+e^- final states. Outside the peak this cross section

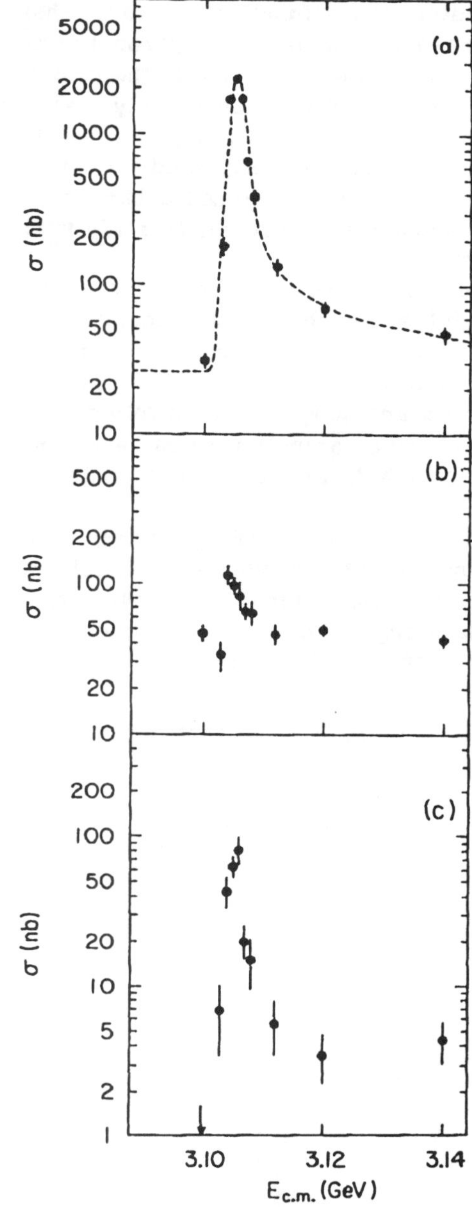

FIG. 1. Cross section versus energy for (a) multi-hadron final states, (b) e^+e^- final states, and (c) $\mu^+\mu^-$, $\pi^+\pi^-$, and K^+K^- final states. The curve in (a) is the expected shape of a δ-function resonance folded with the Gaussian energy spread of the beams and including radiative processes. The cross sections shown in (b) and (c) are integrated over the detector acceptance. The total hadron cross section, (a), has been corrected for detection efficiency.

is equal to the Bhabha cross section integrated over the acceptance of the apparatus.[1]

Figure 1(c) shows the cross section for the production of collinear pairs of particles, excluding electrons. At present, our muon identifications system is not functioning and we therefore cannot separate muons from strongly interacting particles. However, outside the peak the data are consistent with our previously measured μ-pair cross section. Since a large $\pi\pi$ or KK branching ratio would be unexpected for a resonance this massive, the two-body enhancement observed is *probably* but not *conclusively* in the μ-pair channel.

The $e^+e^- \rightarrow$ hadron cross section is presumed to go through the one-photon intermediate state with angular momentum, parity, and charge conjugation quantum numbers $J^{PC} = 1^{--}$. It is difficult to understand how, without involving new quantum numbers or selection rules, a resonance in this state which decays to hadrons could be so narrow.

We wish to thank the SPEAR operations staff for providing the stable conditions of machine performance necessary for this experiment. Special monitoring and control techniques were developed on very short notice and performed excellently.

---

*Work supported by the U. S. Atomic Energy Commission.

†Present address: Laboratoire de l'Accélérateur Linéaire, Centre d'Orsay de l'Université de Paris, 91 Orsay, France.

‡Permanent address: Institut de Physique Nucléaire, Orsay, France.

§Permanent address: Centre d'Etudes Nucléaires de Saclay, Saclay, France.

[1]The apparatus is described by J.-E. Augustin *et al.*, to be published.

[2]The detection-efficiency determination will be described in a future publication.

[3]While preparing this manuscript we were informed that the Massachusetts Institute of Technology group studying the reaction $pp \rightarrow e^+e^- + x$ at Brookhaven National Laboratory has observed an enhancement in the e^+e^- mass distribution at about 3100 MeV. J. J. Aubert *et al.*, preceding Letter [Phys. Rev. Lett. 33, 1402 (1974)].

[4]G. Bonneau and F. Martin, Nucl. Phys. B27, 381 (1971).

Evidence for Anomalous Lepton Production in e^+-e^- Annihilation*

M. L. Perl, G. S. Abrams, A. M. Boyarski, M. Breidenbach, D. D. Briggs, F. Bulos, W. Chinowsky, J. T. Dakin,† G. J. Feldman, C. E. Friedberg, D. Fryberger, G. Goldhaber, G. Hanson, F. B. Heile, B. Jean-Marie, J. A. Kadyk, R. R. Larsen, A. M. Litke, D. Lüke,‡ B. A. Lulu, V. Lüth, D. Lyon, C. C. Morehouse, J. M. Paterson, F. M. Pierre,§ T. P. Pun, P. A. Rapidis, B. Richter, B. Sadoulet, R. F. Schwitters, W. Tanenbaum, G. H. Trilling, F. Vannucci,‖ J. S. Whitaker, F. C. Winkelmann, and J. E. Wiss

Lawrence Berkeley Laboratory and Department of Physics, University of California, Berkeley, California 94720, and Stanford Linear Accelerator Center, Stanford University, Stanford, California 94305

(Received 18 August 1975)

We have found events of the form $e^+ + e^- \rightarrow e^\pm + \mu^\mp$ + missing energy, in which no other charged particles or photons are detected. Most of these events are detected at or above a center-of-mass energy of 4 GeV. The missing-energy and missing-momentum spectra require that at least two additional particles be produced in each event. We have no conventional explanation for these events.

We have found 64 events of the form

$$e^+ + e^- \rightarrow e^\pm + \mu^\mp + \geq 2 \text{ undetected particles} \quad (1)$$

for which we have no conventional explanation. The undetected particles are charged particles or photons which escape the 2.6π sr solid angle of the detector, or particles very difficult to detect such as neutrons, K_L^0 mesons, or neutrinos. Most of these events are observed at center-of-mass energies at, or above, 4 GeV. These events were found using the Stanford Linear Accelerator Center–Lawrence Berkeley Laboratory (SLAC-

LBL) magnetic detector at the SLAC colliding-beams facility SPEAR.

Events corresponding to (1) are the signature for new types of particles or interactions. For example, pair production of heavy charged leptons[1-4] having the decay modes $l^- \to \nu_l + e^- + \bar{\nu}_e$, $l^+ \to \bar{\nu}_l + e^+ + \nu_e$, $l^- \to \nu_l + \mu^- + \bar{\nu}_\mu$, and $l^+ \to \bar{\nu}_l + \mu^+ + \nu_\mu$ would appear as such events. Another possibility is the pair production of charged bosons with decays $B^- \to e^- + \bar{\nu}_e$, $B^+ \to e^+ + \nu_e$, $B^- \to \mu^- + \bar{\nu}_\mu$, and $B^+ \to \mu^+ + \nu_\mu$. Charmed-quark theories[5,6] predict such bosons. Intermediate vector bosons which mediate the weak interactions would have similar decay modes, but the mass of such particles (if they exist at all) is probably too large[7] for the energies of this experiment.

The momentum-analysis and particle-identifier systems of the SLAC-LBL magnetic detector[8] cover the polar angles $50° \leq \theta \leq 130°$ and the full 2π azimuthal angle. Electrons, muons, and hadrons are identified using a cylindrical array of 24 lead-scintillator shower counters, the 20-cm-thick iron flux return of the magnet, and an array of magnetostrictive wire spark chambers situated outside the iron. Electrons are identified solely by requiring that the shower-counter pulse height be greater than that of a 0.5-GeV e. Incidently, the e's in the e-μ events thus selected give no signal in the muon chambers; and their shower-counter pulse-height distribution is that expected of electrons. Also the positions of the e's in the shower counters as determined from the relative pulse heights in the photomultiplier tubes at each end of the counters agree within measurement errors with the positions of the e tracks. Hence the e's in the e-μ events are not misidentified combinations of $\mu+\gamma$ or $\pi+\gamma$ in a single shower counter, except possibly for a few events already contained in the background estimates. Muons are identified by two requirements. The μ must be detected in one of the muon chambers after passing through the iron flux return and other material totaling 1.67 absorption lengths for pions. And the shower-counter pulse height of the μ must be small. All other charged particles are called hadrons. The shower counters also detect photons (γ). For γ energies above 200 MeV, the γ detection efficiency is about 95%.

To illustrate the method of searching for events corresponding to Reaction (1), we consider our data taken at a total energy ($\sqrt{s}$) of 4.8 GeV. This sample contains 9550 three-or-more-prong events and 25 300 two-prong events which include $e^+ + e^- \to e^+ + e^-$ events, $e^+ + e^- \to \mu^+ + \mu^-$ events, two-prong hadronic events, and the e-μ events described here. To study two-prong events we define a coplanarity angle

$$\cos\theta_{\text{copl}} = -(\vec{n}_1 \times \vec{n}_{e^+}) \cdot (\vec{n}_2 \times \vec{n}_{e^+}) / |\vec{n}_1 \times \vec{n}_{e^+}| |\vec{n}_2 \times \vec{n}_{e^+}|, \quad (2)$$

where $\vec{n}_1$, $\vec{n}_2$, and $\vec{n}_{e^+}$ are unit vectors along the directions of particles 1, 2, and the e^+ beam. The contamination of events from the reactions $e^+ + e^- \to e^+ + e^-$ and $e^+ + e^- \to \mu^+ + \mu^-$ is greatly reduced if we require $\theta_{\text{copl}} > 20°$. Making this cut leaves 2493 two-prong events in the 4.8-GeV sample.

To obtain the most reliable e and μ identification[9] we require that each particle have a momentum greater than 0.65 GeV/c. This reduces the 2493 events to the 513 in Table I. The 24 e-μ events with no associated photons, called the signature events, are candidates for Reaction (1). The e-μ events can come conventionally from the two-virtual-photon process[10] $e^+ + e^- \to e^+ + e^- + \mu^+ + \mu^-$. Calculations indicate that this source is negligible, and the absence of e-μ events with charge 2 proves this point since the number of charge-2 e-μ events should equal the number of charge-0 e-μ events from this source.

We determine the background from hadron misidentification or decay by using the 9550 three-or-more-prong events and assuming that every particle called an e or a μ by the detector either was a misidentified hadron or came from the decay of a hadron. We use $P_{h\to l}$ to designate the sum of the probabilities for misidentification or decay causing a hadron h to be called a lepton l. Since the P's are momentum dependent[9] we use all the

TABLE I. Distribution of 513 two-prong events, obtained at $E_{\text{c.m.}} = 4.8$ GeV, which meet the criteria $|\vec{p}_1| > 0.65$ GeV/c, $|\vec{p}_2| > 0.65$ GeV/c, and $\theta_{\text{copl}} > 20°$. Events are classified according to the number N_γ of photons detected, the total charge, and the nature of the particles. All particles not identified as e or μ are called h for hadron.

Particles \ N_γ	0	1	>1	0	1	>1
	Total charge = 0			Total charge = ±2		
e-e	40	111	55	0	1	0
e-μ	24	8	8	0	0	3
μ-μ	16	15	6	0	0	0
e-h	20	21	32	2	3	3
μ-h	17	14	31	4	0	5
h-h	14	10	30	10	4	6

e-h, μ-h, and h-h events in column 1 of Table I to determine a "hadron" momentum spectrum, and weight the P's accordingly. We obtain the momentum-averaged probabilities $P_{h\to e}=0.183\pm0.007$ and $P_{h\to\mu}=0.198\pm0.007$. Collinear e-e and μ-μ events are used to determine $P_{e\to h}=0.056\pm0.02$, $P_{e\to\mu}=0.011\pm0.01$, $P_{\mu\to h}=0.08\pm0.02$, and $P_{\mu\to e}<0.01$.

Using these probabilities and assuming that all e-h and μ-h events in Table I result from particle misidentifications or particle decays, we calculate for column 1 the contamination of the e-μ sample to be 1.0 ± 1.0 event from misidentified e-e,[11] <0.3 event from misidentified μ-μ,[11] and 3.7 ± 0.6 events from h-h in which the hadrons were misidentified or decayed. The total e-μ background is then 4.7 ± 1.2 events.[12,13] The statistical probability of such a number yielding the 24 signature e-μ events is very small. The same analysis applied to columns 2 and 3 of Table I yields 5.6 ± 1.5 e-μ background events for column 2 and 8.6 ± 2.0 e-μ background events for column 3, both consistent with the observed number of e-μ events.

Figure 1(a) shows the momentum of the μ versus the momentum of the e for signature events.[14] Both p_μ and p_e extend up to 1.8 GeV/c, their average values being 1.2 and 1.3 GeV/c, respectively. Figure 1(b) shows the square of the invariant e-μ mass (M_i^2) versus the square of the missing mass (M_m^2) recoiling against the e-μ system. To explain Fig. 1(b) at least two particles must escape detection. Figure 1(c) shows the distribution in collinearity angle between the e and μ ($\cos\theta_{\text{coll}}=-\vec{p}_e\cdot\vec{p}_\mu/|\vec{p}_e||\vec{p}_\mu|$). The dip near $\cos\theta_{\text{coll}}=1$ is a consequence of the coplanarity cut; however, the absence of events with large θ_{coll} has dynamical significance.

Figure 2 shows the *observed* cross section in the range of detector acceptance for signature e-μ events versus center-of-mass energy with the background subtracted at each energy as described above.[9] There are a total of 86 e-μ events summed over all energies, with a calculated background of 22 events.[12] The corrections to obtain the true cross section for the angle and momentum cuts used here depend on the hypothesis as to the origin of these e-μ events, and the corrected cross section can be many times larger than the observed cross section. While Fig. 2 shows an apparent threshold at around 4 GeV, the statistics are small and the correction fac-

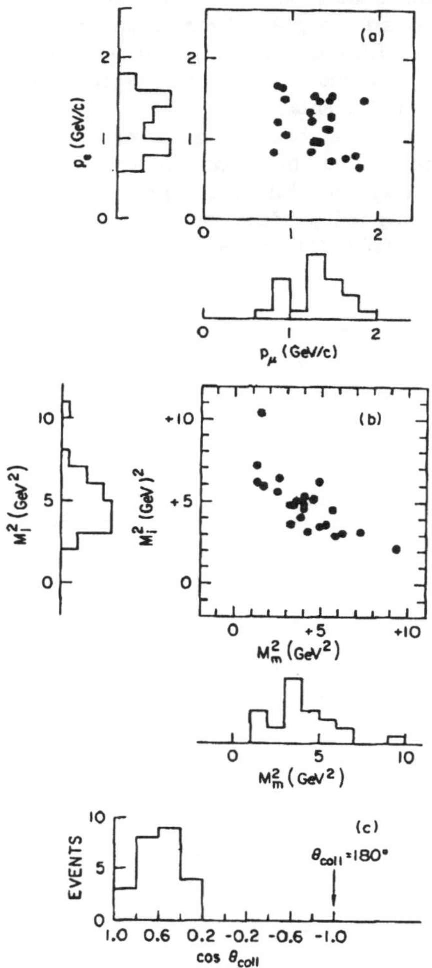

FIG. 1. Distribution for the 4.8-GeV e-μ signature events of (a) momenta of the e (p_e) and μ (p_μ); (b) square of the invariant mass (M_i^2) and square of the missing mass (M_m^2); and (c) $\cos\theta_{\text{coll}}$.

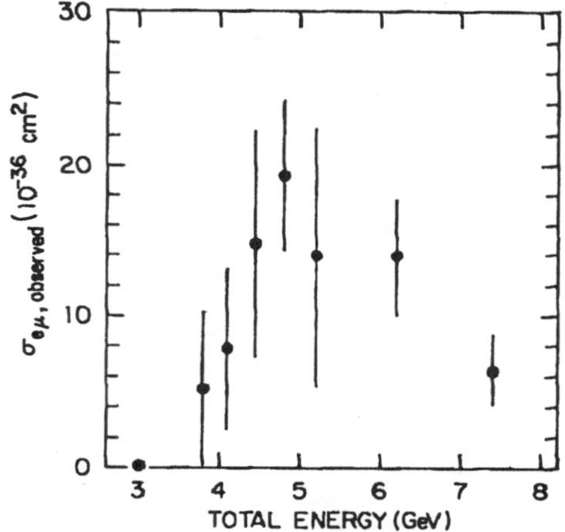

FIG. 2. The *observed* cross section for the signature e-μ events.

tors are largest for low $\sqrt{s}$. Thus, the apparent threshold may not be real.

We conclude that the signature e-μ events cannot be explained either by the production and decay of any presently known particles or as coming from any of the well-understood interactions which can conventionally lead to an e and a μ in the final state. A possible explanation for these events is the production and decay of a pair of new particles, each having a mass in the range of 1.6 to 2.0 GeV/c^2.

---

*Work supported by the U. S. Energy Research and Development Administration.

†Present address: Department of Physics and Astronomy, University of Massachusetts, Amherst, Mass. 01002.

‡Fellow of Deutsche Forschungsgemeinschaft.

§Centre d'Etudes Nucléaires de Saclay, Saclay, France.

‖Institut de Physique Nucléaire, Orsay, France.

[1]M. L. Perl and P. A. Rapidis, SLAC Report No. SLAC-PUB-1496, 1974 (unpublished).

[2]J. D. Bjorken and C. H. Llewellyn Smith, Phys. Rev. D 7, 887 (1973).

[3]Y. S. Tsai, Phys. Rev. D 4, 2821 (1971).

[4]M. A. B. Beg and A. Sirlin, Annu. Rev. Nucl. Sci. 24, 379 (1974).

[5]M. K. Gaillard, B. W. Lee, and J. L. Rosner, Rev. Mod. Phys. 47, 277 (1975).

[6]M. B. Einhorn and C. Quigg, Phys. Rev. D (to be published).

[7]B. C. Barish *et al.*, Phys. Rev. Lett. 31, 180 (1973).

[8]J.-E. Augustin *et al.*, Phys. Rev. Lett. 34, 233 (1975); G. J. Feldman and M. L. Perl, Phys. Rep. 19C, 233 (1975).

[9]See M. L. Perl, in Proceedings of the Canadian Institute of Particle Physics Summer School, Montreal, Quebec, Canada, 16–21 June 1975 (to be published).

[10]V. M. Budnev *et al.*, Phys. Rep. 15C, 182 (1975); H. Terazawa, Rev. Mod. Phys. 45, 615 (1973).

[11]These contamination calculations do not depend upon the source of the e or μ; anomalous sources lead to overestimates of the contamination.

[12]Using *only* events in column 1 of Table I we find at 4.8 GeV $P_{h\to e}=0.27\pm0.10$, $P_{h\to\mu}=0.23\pm0.09$, and a total e-μ background of 7.9 ± 3.2 events. The same method yields a total e-μ background of 30 ± 6 events summed over all energies. This method of background calculation (Ref. 9) allows the hadron background in the two-prong, zero-photon events to be different from that in other types of events.

[13]Our studies of the two-prong and multiprong events show that there is *no* correlation between the misidentification or decay probabilities; hence the background is calculated using independent probabilities.

[14]Of the 24 events, thirteen are $e^+ + \mu^-$ and eleven are $e^- + \mu^+$

Evidence for Jet Structure in Hadron Production by e^+e^- Annihilation*

G. Hanson, G. S. Abrams, A. M. Boyarski, M. Breidenbach, F. Bulos, W. Chinowsky, G. J. Feldman, C. E. Friedberg, D. Fryberger, G. Goldhaber, D. L. Hartill,† B. Jean-Marie, J. A. Kadyk, R. R. Larsen, A. M. Litke, D. Lüke,‡ B. A. Lulu, V. Lüth, H. L. Lynch, C. C. Morehouse, J. M. Paterson, M. L. Perl, F. M. Pierre,§ T. P. Pun, P. A. Rapidis, B. Richter, B. Sadoulet, R. F. Schwitters, W. Tanenbaum, G. H. Trilling, F. Vannucci,‖ J. S. Whitaker, F. C. Winkelmann, and J. E. Wiss

Lawrence Berkeley Laboratory and Department of Physics, University of California, Berkeley, California 94720, and Stanford Linear Accelerator Center, Stanford University, Stanford, California 94305

(Received 8 October 1975)

We have found evidence for jet structure in $e^+e^- \rightarrow$ hadrons at center-of-mass energies of 6.2 and 7.4 GeV. At 7.4 GeV the jet-axis angular distribution integrated over azimuthal angle was determined to be proportional to $1+(0.78\pm0.12)\cos^2\theta$.

In quark-parton constituent models of elementary particles, hadron production in e^+e^- annihilation reactions proceeds through the annihilation of the e^+ and e^- into a virtual photon which subsequently produces a quark-parton pair, each member of which decays into hadrons. At sufficiently high energy the limited transverse-momentum distribution of the hadrons with respect to the original parton production direction, characteristic of all strong interactions, results in oppositely directed jets of hadrons.[1-4] The spins of the constituents can, in principle, be determined from the angular distribution of the jets. In this Letter we report the evidence for the existence of jets and the angular distribution of the jet axis.

The data were taken with the Stanford Linear Accelerator Center–Lawrence Berkeley Laboratory magnetic detector at the SPEAR storage ring of the Stanford Linear Accelerator Center. Hadron production, muon pair production, and Bhabha scattering data were recorded simultaneously. The detector and the selection of events have been described previously.[5,6] The detector subtended $0.65\times4\pi$ sr with full acceptance in azimuthal angle and acceptance in polar angle from

50° to 130°. We have used the large blocks of data at center-of-mass energies ($E_{c.m.}$) of 3.0, 3.8, 4.8, 6.2, and 7.4 GeV. We included only those hadronic events in which three or more particles were detected in order to avoid background contamination in events with only two charged tracks due to beam-gas interactions and photon-photon processes.

To search for jets we find for each event that direction which minimizes the sum of squares of transverse momenta.[7] For each event we calculate the tensor

$$T^{\alpha\beta}=\sum_i (\delta^{\alpha\beta}\vec{p}_i^{\,2}-p_i^{\alpha}p_i^{\beta}), \tag{1}$$

where the summation is over all detected particles and α and β refer to the three spatial components of each particle momentum $\vec{p}_i$. We diagonalize $T^{\alpha\beta}$ to obtain the eigenvalues λ_1, λ_2, and λ_3 which are the sums of squares of transverse momenta with respect to the three eigenvector directions. The smallest eigenvalue (λ_3) is the minimum sum of squares of transverse momenta. The eigenvector associated with λ_3 is defined to be the reconstructed jet axis. In order to determine how jetlike an event is, we calculate a quantity which we call the sphericity (S):

$$S=\frac{3\lambda_3}{\lambda_1+\lambda_2+\lambda_3}=\frac{3(\sum_i p_{\perp i}^{\,2})_{\min}}{2\sum_i \vec{p}_i^{\,2}}. \tag{2}$$

S approaches 0 for events with bounded transverse momenta and approaches 1 for events with large multiplicity and isotropic phase-space particle distributions.

The data at each energy were compared to Monte Carlo simulations which were based on either an isotropic phase-space (PS) model or a jet model. In both models only pions (charged and neutral) were produced. The total multiplicity was given by a Poisson distribution. The jet model modified phase space according to the square of a matrix element of the form

$$M^2=\exp(-\sum_i p_{\perp i}^{\,2}/2b^2), \tag{3}$$

where $p_\perp$ is the momentum perpendicular to the jet axis.

The angular distribution for the jet axis is expected to have the form

$$d\sigma/d\Omega\propto 1+\alpha\cos^2\theta+P^2\alpha\sin^2\theta\cos(2\varphi), \tag{4}$$

where θ is the polar angle of the jet axis with respect to the incident positron direction, φ is the azimuthal angle with respect to the plane of the storage ring, $\alpha=(\sigma_T-\sigma_L)/(\sigma_T+\sigma_L)$ with σ_T and σ_L the transverse and longitudinal production cross sections, and P is the polarization of each beam. (The polarization term will be discussed later.) The angular distribution given by Eq. (4) was used in the jet-model simulation. The simulations included the geometric acceptance, the trigger efficiency, and all other known characteristics of the detector. The total multiplicity and the charged-neutral multiplicity ratio for both models were obtained by fitting to the observed charged-particle mean multiplicity and mean momentum at each energy. In the jet model the parameter b was determined by fitting to the observed mean S at the highest energy (7.4 GeV). For lower energies the value of b was determined by requiring that the mean $p_\perp$ in the jet model be the same (315 MeV/c) as at 7.4 GeV.

Figure 1 shows the observed mean S and the model predictions. Both models are consistent with the data in the 3–4-GeV region. At higher energies the data have significantly lower mean S than the PS model and agree with the jet model. Figure 2 shows the S distributions at several energies. At 3.0 GeV the data agree with either the PS or the jet model [Fig. 2(a)]. At 6.2 and 7.4 GeV the data are peaked toward low S, favoring

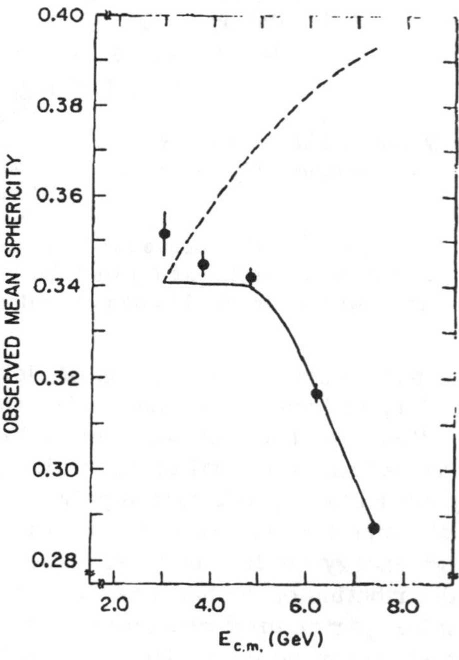

FIG. 1. Observed mean sphericity versus center-of-mass energy $E_{c.m.}$ for data, jet model with $\langle p_\perp\rangle$ = 315 MeV/c (solid curve), and phase-space model (dashed curve).

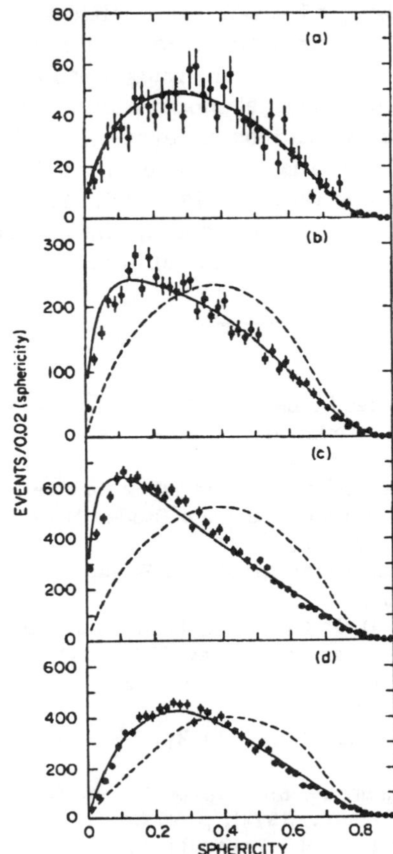

FIG. 2. Observed sphericity distributions for data, jet model with $\langle p_\perp \rangle = 315$ MeV/c (solid curves), and phase-space model (dashed curves) for (a) $E_{c.m.} = 3.0$ GeV; (b) $E_{c.m.} = 6.2$ GeV; (c) $E_{c.m.} = 7.4$ GeV; and (d) $E_{c.m.} = 7.4$ GeV, events with largest $x < 0.4$. The distributions for the Monte Carlo models are normalized to the number of events in the data.

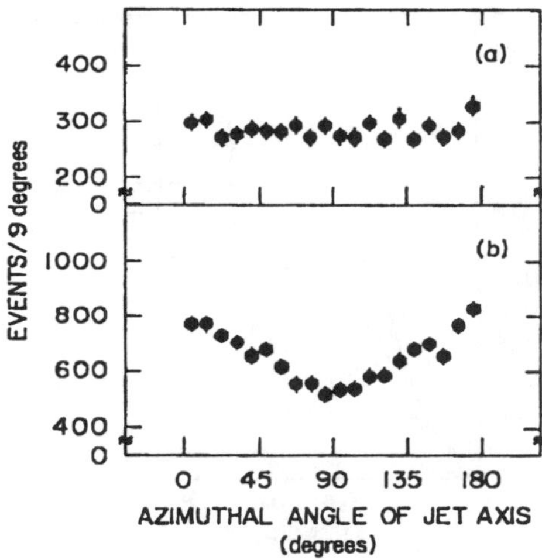

FIG. 3. Observed distributions of jet-axis azimuthal angles from the plane of the storage ring for jet axes with $|\cos\theta| \leq 0.6$ for (a) $E_{c.m.} = 6.2$ GeV and (b) $E_{c.m.} = 7.4$ GeV.

the jet model [Figs. 2(b) and 2(c)]. At the highest two energies, the PS model poorly reproduces the single-particle momentum spectra, having fewer particles with $x > 0.4$ ($x = 2p/E_{c.m.}$ and p is the particle momentum) than the data.[8] The jet-model x distributions are in better agreement. For $x < 0.4$ the x distributions for both models agree with the data. Therefore, we show in Fig. 2(d) the S distributions at 7.4 GeV for those events in which *no* particle has $x > 0.4$. The jet model is still preferred.

At $E_{c.m.} = 7.4$ GeV the electron and positron beams in the SPEAR ring are transversely polarized, and the hadron inclusive distributions show an azimuthal asymmetry.[9] The φ distributions of the jet axis for jet axes with $|\cos\theta| \leq 0.6$ are shown in Fig. 3 for 6.2 and 7.4 GeV.[10] At 6.2 GeV, the beams are unpolarized[9] and the φ distribution is flat, as expected. At 7.4 GeV, the φ distribution of the jet axis shows an asymmetry with maxima and minima at the same values of φ as for $e^+e^- \to \mu^+\mu^-$.

The φ distribution shown in Fig. 3(b) and the value for P^2 (0.47 ± 0.05) measured simultaneously by the reaction[9] $e^+e^- \to \mu^+\mu^-$ were used to determine the parameter α of Eq. (4). The value obtained for the *observed* jet axis is $\alpha = 0.45 \pm 0.07$. This observed value of α will be less than the true value which describes the production of the jets because of the incomplete acceptance of the detector, the loss of neutral particles, and our method of reconstructing the jet axis. We have used the jet-model Monte Carlo simulation to estimate the ratio of observed to produced values of α and find this ratio to be 0.58 at 74 GeV. Thus the value of α describing the *produced* jet-axis angular distribution is $\alpha = 0.78 \pm 0.12$ at $E_{c.m.} = 7.4$ GeV. The error in α is statistical only; we estimate that the systematic errors in the observed α can be neglected. However, we have not studied the model dependence of the correction factor relating observed to produced values of α.

The sphericity and the value of α as determined above are properties of whole events. The simple jet model used for the sphericity analysis can also be used to predict the single-particle inclusive angular distributions for all values of the secondary particle momentum. In Fig.

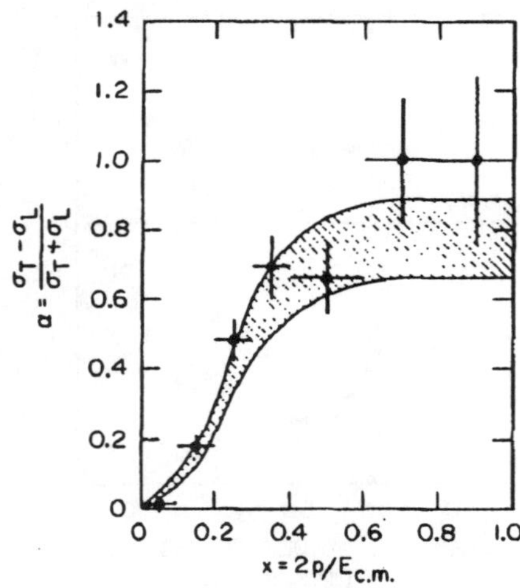

FIG. 4. Observed inclusive α versus x (from Ref. 9) for particles with $|\cos\theta| \leq 0.6$ in hadronic events at $E_{c.m.} = 7.4$ GeV. The prediction of the jet-model Monte Carlo simulation for jet-axis angular distribution with $\alpha = 0.78 \pm 0.12$ is represented by the shaded band.

4 values for the inclusive hadron α as a function of x at 7.4 GeV[9] are compared with the jet-model calculation. The model assumed the value $\alpha = 0.78 \pm 0.12$ for the jet-axis angular distribution. The prediction agrees well with the data for all values of x.

We conclude that the data strongly support the jet hypothesis for hadron production in e^+e^- annihilation. The data show a decreasing mean sphericity with increasing $E_{c.m.}$ and the sphericity distributions peak more strongly at low values as $E_{c.m.}$ increases. Both of these trends agree with a jet model and disagree with an isotropic PS model. The mean transverse momentum relative to the jet axis obtained using the jet-model Monte Carlo simulation was found to be 315 ± 2 MeV/c. At $E_{c.m.} = 7.4$ GeV the coefficient α for the jet-axis angular distribution in Eq. (4) has been found to be nearly +1 giving a value for σ_L/σ_T of 0.13 ± 0.07. The jet model also reproduces well the inclusive hadron α versus x. All of this indicates not only that there are jets but also that the helicity along the jet axis is ±1. In the framework of the quark-parton model, the partons must must have spin $\frac{1}{2}$ rather than spin 0.

---

*Work supported by the U. S. Energy Research and Development Administration.

†Alfred P. Sloan Fellow.

‡Fellow of Deutsche Forschungsgemeinschaft.

§Permanent address: Centre d'Etudes Nucléaires de Saclay, Saclay, France

‖Permanent address: Institut de Physique Nucléaire, Orsay, France.

[1]S. D. Drell, D. J. Levy, and T. M. Yan, Phys. Rev. 187, 2159 (1969), and Phys. Rev. D 1, 1617 (1970).

[2]N. Cabibbo, G. Parisi, and M. Testa, Lett. Nuovo Cimento 4, 35 (1970).

[3]J. D. Bjorken and S. J. Brodsky, Phys. Rev. D 1, 1416 (1970).

[4]R. P. Feymann, *Photon-Hadron Interactions* (Benjamin, Reading, Mass., 1972), p. 166.

[5]J.-E. Augustin *et al.*, Phys. Rev. Lett. 34, 233 (1975).

[6]J.-E. Augustin *et al.*, Phys. Rev. Lett. 34, 764 (1975).

[7]It is impossible to determine the jet axis exactly, even with perfect detection efficiency; the method described here, which was suggested in Ref. 3, is the best approximation known to us.

[8]The momentum distributions will be discussed in a subsequent paper.

[9]R. F. Schwitters *et al.*, Phys. Rev. Lett. 35, 1320 (1975).

[10]Since the jet axis is a symmetry axis, the azimuthal angle $\varphi + 180°$ is equivalent to the azimuthal angle φ.

Observation in e^+e^- Annihilation of a Narrow State at 1865 MeV/c^2 Decaying to $K\pi$ and $K\pi\pi\pi$ †

G. Goldhaber,* F. M. Pierre,‡ G. S. Abrams, M. S. Alam, A. M. Boyarski, M. Breidenbach, W. C. Carithers, W. Chinowsky, S. C. Cooper, R. G. DeVoe, J. M. Dorfan, G. J. Feldman, C. E. Friedberg, D. Fryberger, G. Hanson, J. Jaros, A. D. Johnson, J. A. Kadyk, R. R. Larsen, D. Lüke,§ V. Lüth, H. L. Lynch, R. J. Madaras, C. C. Morehouse,‖ H. K. Nguyen,** J. M. Paterson, M. L. Perl, I. Peruzzi,†† M. Piccolo,†† T. P. Pun, P. Rapidis, B. Richter, B. Sadoulet, R. H. Schindler, R. F. Schwitters, J. Siegrist, W. Tanenbaum, G. H. Trilling, F. Vannucci,‡‡ J. S. Whitaker, and J. E. Wiss

Lawrence Berkeley Laboratory and Department of Physics, University of California, Berkeley, California 94720, and Stanford Linear Accelerator Center, Stanford University, Stanford, California 94305
(Received 14 June 1975)

We present evidence, from a study of multihadronic final states produced in e^+e^- annihilation at center-of-mass energies between 3.90 and 4.60 GeV, for the production of a new neutral state with mass 1865 ± 15 MeV/c^2 and decay width less than 40 MeV/c^2 that decays to $K^\pm\pi^\mp$ and $K^\pm\pi^\mp\pi^\pm\pi^\mp$. The recoil-mass spectrum for this state suggests that it is produced only in association with systems of comparable or larger mass.

We have observed narrow peaks near 1.87 GeV/c^2 in the invariant-mass spectra for neutral combinations of the charged particles $K^\pm\pi^\mp$ ($K\pi$) and $K^\pm\pi^\mp\pi^\pm\pi^\mp$ ($K3\pi$) produced in e^+e^- annihilation. The agreement in mass, width, and recoil-mass spectrum for these peaks strongly suggests they represent different decay modes of the same object. The mass of this state is 1865 ± 15 MeV/c^2 and its decay width (full width at half-maximum) is less than 40 MeV/c^2 (90% confidence level). The state appears to be produced only in association with systems of comparable or higher mass.

Our results are based on studies of multihadronic events recorded by the Stanford Linear Accelerator Center–Lawrence Berkeley Laboratory magnetic detector operating at the colliding-beam facility SPEAR. Descriptions of the detector and event-selection procedures have been published.[1,2]

A new feature in our analysis is the use of time of flight (TOF) information to help identify hadrons. The TOF system includes 48 2.5 cm×20 cm×260 cm Pilot Y scintillation counters arranged in a cylindrical array immediately outside the tracking spark chambers at a radius of 1.5 m from the beam axis. Both ends of each counter are viewed by Amperex 56DVP photomultiplier tubes (PM); anode signals from each PM are sent to separate time-to-digital converters (TDC's), analog-to-digital converters, and latches. Pulse-height information is used to correct times given by the TDC's. The collision time is derived from a pickup electrode that senses the passage of the 0.2-nsec-long beam pulses; the period between successive collisions is 780 nsec. Run-to-run calibrations of the TOF system are performed with Bhabha scattering ($e^+e^- \to e^+e^-$) events. The rms resolution of the TOF system is 0.4 nsec.

Evidence for a new state in the $K\pi$ system was found among 29 000 hadronic events collected at center-of-mass (c.m.) energies between 3.90 and 4.60 GeV. As shown by the top row of Fig. 1, a

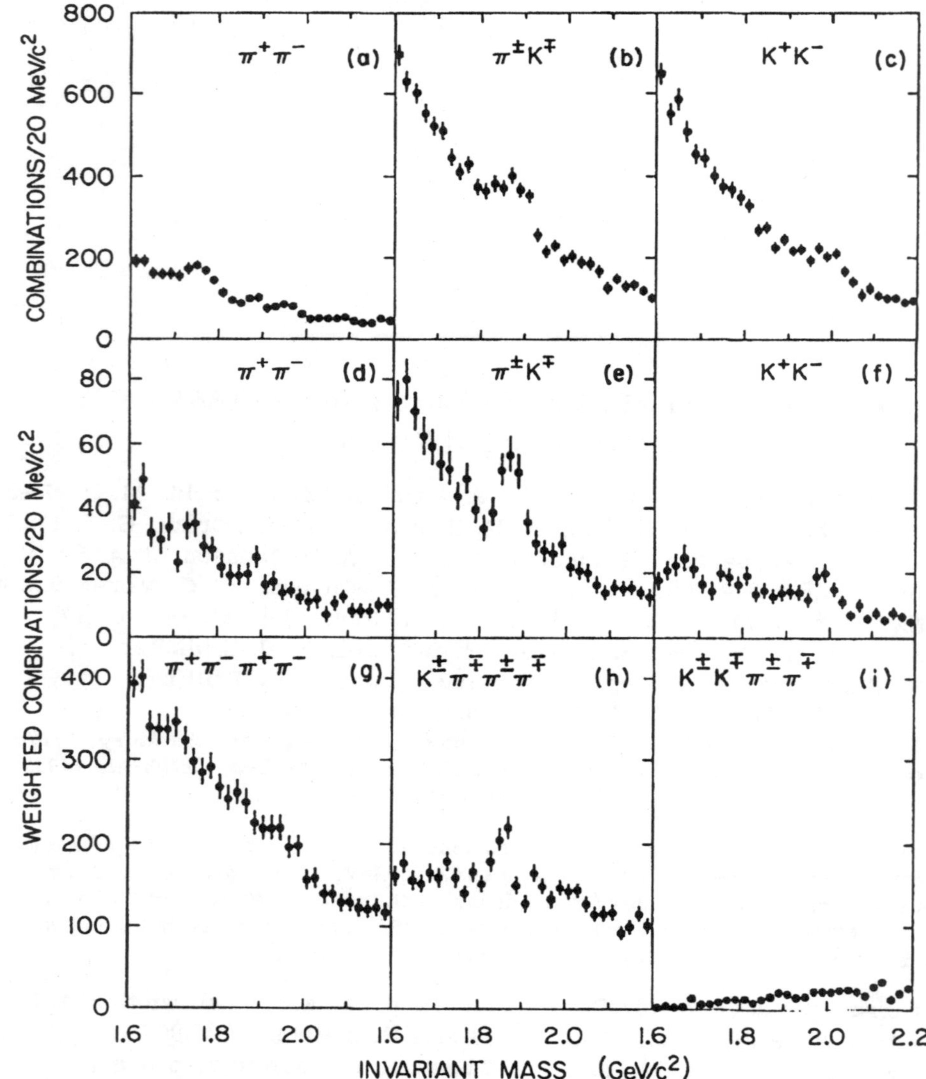

FIG. 1. Invariant-mass spectra for neutral combinations of charged particle. (a) $\pi^+\pi^-$ assigning π mass to all tracks, (b) $K^\mp\pi^\pm$ assigning K and π masses to all tracks, (c) K^+K^- assigning K mass to all tracks, (d) $\pi^+\pi^-$ weighted by $\pi\pi$ TOF probability, (e) $K^\pm\pi^\mp$ weighted by $K\pi$ TOF probability, (f) K^+K^- weighted by KK TOF probability, (g) $\pi^+\pi^-\pi^+\pi^-$ weighted by 4π TOF probability, (h) $K^\pm\pi^\mp\pi^\pm\pi^\mp$ weighted by $K3\pi$ TOF probability, (i) $K^+K^-\pi^+\pi^-$ weighted by $KK\pi\pi$ TOF probability.

significant signal[3] appears when we simply consider invariant-mass spectra for all possible neutral combinations of two charged particles *assuming* both π and K masses for the particles as was done in our previous search for the production of narrow peaks.[4] Through kinematic reflections, the signal appears near 1.74 GeV/c^2 for the $\pi^+\pi^-$ hypothesis [Fig. 1(a)], 1.87 GeV/c^2 in the case of $K^+\pi^-$ or $K^-\pi^+$ [Fig. 1(b)], and 1.98 GeV/c^2 for K^+K^- [Fig. 1(c)].

To establish the correct choice of final-state particles associated with these peaks, we use the TOF information. Because the typical time difference between a π and a K in the $K\pi$ signal is only about 0.5 nsec, we have used the following technique to extract maximal information on particle identity. First, tracks are required to have good timing information from both PM's, consistent with the extrapolated position of the track in the counter. Next, each track is assigned probabilities that it is a π or K; they are determined from the measured momentum and TOF assuming a Gaussian probability distribution with standard deviation 0.4 nsec. Tracks with net (π plus K) probability less than 1% are rejected.[5] Then, the relative π-K probabilities are renormalized so that their sum is unity, and two-particle combinations are weighted by the joint probability that the particles satisfy the particular π or K hypothesis assigned to them. In this way, the total weight assigned to all $\pi\pi$, $K\pi$, and KK combinations equals the number of two-body combinations and no double counting occurs.

Invariant-mass spectra weighted by the above procedure are presented in the second row of Fig. 1. We see that the $K\pi$ hypothesis [Fig. 1(e)] for the peak at the $K\pi$ mass 1.87 GeV/c^2 is clearly preferred over either $\pi^+\pi^-$ [Fig. 1(d)] or K^+K^- [Fig. 1(f)]. The areas under the small peaks remaining in the $\pi^+\pi^-$ and K^+K^- channels are consistent with the entire signal being $K\pi$ and the resulting misidentification of true $K\pi$ events expected for our TOF system. From consideration of possible residual uncertainties in the TOF calibration, we estimate that the confidence level for this signal to arise only from $\pi^+\pi^-$ or K^+K^- is less than 1%. Assuming the entire signal in Figs. 1(d)–1(f) to be in the $K\pi$ channel, we find a total of 110± 25 decays of the new state; the significance of the peak in Fig. 1(e) is greater than 5 standard deviations. No signal occurs in the corresponding doubly charged channels.

Evidence for the decay of this state to neutral combinations of a charged K and three charged π's is presented in the third row of Fig. 1. Again, we employ the TOF weighting technique discussed above; the hadron event sample is the same as that used for the $K\pi$ study. Four-body mass combinations are weighted by their joint π-K probabilities. In order to recover tracks when an extra particle is present in the TOF counter, or when they miss a counter, all tracks failing the timing quality criteria are called π.

As can be seen in Fig. 1(h), a clear signal is obtained in the $K3\pi$ system at a mass near 1.86 GeV/c^2. No corresponding signal is evident at this mass or the appropriate kinematically reflected mass for either the $\pi^+\pi^-\pi^+\pi^-$ or $K^+K^-\pi^+\pi^-$ systems. We estimate the number of $K3\pi$ decays in the 1.86-GeV/c^2 peak to be 124± 21, an effect of more than 5 standard deviations. Again, there is no signal in the corresponding doubly charged channel.

To determine the masses and widths of the peaks in the $K\pi$ and $K3\pi$ mass spectra, we have fitted the data represented by Fig. 1 with a Gaussian for the peak and linear and quadratic background terms under various conditions of bin size, event-selection criteria, and kinematic cuts. Masses for the $K\pi$ signal center at 1870 MeV/c^2; those for the $K3\pi$ signal center at 1860 MeV/c^2. The spread in central-mass values for the various fits is ± 5 MeV/c^2. Within the statistical errors of ± 3 to 4 MeV/c^2, the widths obtained by these fits agree with those expected from experimental resolution alone. From Monte Carlo calculations we expect a rms mass resolution of 25 MeV/c^2 for the $K\pi$ system and 13 MeV/c^2 for the $K3\pi$ system. Systematic errors in momentum measurement are estimated to contribute a ± 10-MeV/c^2 uncertainty in the absolute mass determination, and can account for the 10-MeV/c^2 mass difference observed between the $K\pi$ and $K3\pi$ systems. Thus, both signals are consistent with being decays of the same state and, from our mass resolution, we deduce a 90%-confidence-level upper limit of 40 MeV/c^2 for the decay width of this state.

In Fig. 2, we show the spectra of masses recoiling against neutral $K\pi$ and $K3\pi$ systems in the signal region. The entries are weighted by the TOF likelihood as discussed above. Background estimates are obtained by plotting smooth curves corresponding to the recoil spectra for $K\pi$ and $K3\pi$ invariant-mass combinations in bands on either side of the signal region. The normalizations of these curves are fixed by the areas of the respective control regions.

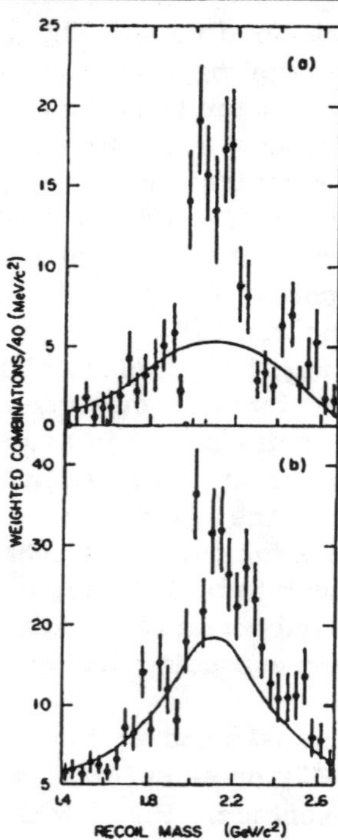

FIG. 2. Recoil-mass spectra for combinations in the $K\pi$ and $K3\pi$ peaks. Smooth curves are estimates of the background obtained from combinations whose invariant masses are on either side of the peak mass region. (a) $K^{\pm}\pi^{\mp}$, peak mass region of 1.84 to 1.90 GeV/c^2 and background mass regions of 1.70 to 1.82 GeV/c^2 and 1.92 to 2.04 GeV/c^2. (b) $K^{\pm}\pi^{\mp}\pi^{\pm}\pi^{\mp}$, peak mass region of 1.84 to 1.88 GeV/c^2 and background mass regions of 1.74 to 1.82 GeV/c^2 and 1.90 to 1.98 GeV/c^2.

From Fig. 2 we find no evidence for the production of recoil systems having masses less than or equal to 1.87 GeV/c^2 in either spectrum. The $K\pi$ data of Fig. 2(a) show a large signal for recoil masses in the range 1.96 to 2.20 GeV/c^2 with contributions up to 2.5 GeV/c^2. The $K3\pi$ recoil-mass spectrum [Fig. 2(b)] has more background, but appears to be consistent with the $K\pi$ spectrum. These spectra suggest that the $K\pi$ and $K3\pi$ systems are produced with thresholds occuring above 3.7-GeV c.m. energy; more detailed interpretations of Fig. 2 are made difficult by the broad range of c.m. energies over which this data sample was collected.

As a further test of this apparent threshold behavior, we have examined 150 000 multihadronic events collected at the ψ mass ($E_{\text{c.m.}} = 3.1$ GeV) and 350 000 events at the ψ' mass ($E_{\text{c.m.}} = 3.7$ GeV) for $K\pi$ and $K3\pi$ signals near 1.87 GeV/c^2. Because of the large cascade decay rate[6] of ψ' to ψ and the large second-order electromagnetic decay rate[7] of the ψ, the resonance events contain 72 000 examples of hadron production by a virtual photon of c.m. energy 3.1 GeV. From fits to invariant-mass spectra (with the signal mass near 1.87 GeV/c^2) we find no $K\pi$ signal larger than 0.3 standard deviations and no $K3\pi$ signal larger than 1.2 standard deviations in this large sample of events. The upper limits (90% confidence level) are 60 events for the $K\pi$ signal and 200 events for the $K3\pi$ signal.

The threshold behavior noted above as well as the narrow widths argue against the interpretation of the structure in Fig. 1 as being a conventional K^*, e.g., the strange counterpart of the $g(1680)$.

Preliminary Monte Carlo calculations to estimate detection efficiencies for two modes have been performed; present systematic uncertainties in these detection efficiencies could be as large as ± 50%. Our estimate of the cross section times branching ratio σB (errors quoted are statistical) averaged over our 3.9–4.6-GeV c.m. energy data is 0.20± 0.05 nb for the $K\pi$ mode and 0.67± 0.11 nb for the $K3\pi$ mode. These are to be compared with the average total hadronic cross section σ_T in this energy region[8] of 27± 3 nb. We have also searched for these signals in the events at higher c.m. energies. In our previous search for the production of narrow peaks[4] at 4.8 GeV, there was a small $K\pi$ signal at 1.87 GeV/c^2 corresponding to a σB of 0.10± 0.07 nb. This signal set the upper limit quoted in the paper ($\sigma B < 0.18$ nb for the $K\pi$ system of mass between 1.85 and 2.40 GeV/c^2) but lacked the statistical significance necessary to be considered a convincing peak. The value of σ_T at 4.8 GeV is 18± 2 nb.[2] In the c.m. energy range 6.3 to 7.8 GeV the $K\pi$ σB is 0.04± 0.03 nb and the average σ_T is 10± 2 nb.

In summary, we have observed significant peaks in the invariant-mass spectra of $K^{\pm}\pi^{\mp}$ and $K^{\pm}\pi^{\mp}\pi^{+}\pi^{-}$ that we associate with the decay of a state of mass 1865± 15 MeV/c^2 and width less than 40 MeV/c^2. The recoil-mass spectra indicate that this state is produced in association with systems of comparable or larger mass.

We find it significant that the threshold energy for pair-producing this state lies in the small interval between the very narrow ψ' and the broader structures present in e^+e^- annihilation near 4 GeV.[8] In addition, the narrow width of this state, its production in association with systems of even greater mass, and the fact that the decays we observe involve kaons form a pattern of observation that would be expected for a state possessing the proposed new quantum number charm.[9,10]

---

†Work supported by the U. S. Energy Research and Development Administration.

*Miller Institute for Basic Research in Science, Berkeley, Calif. (1975–1976).

‡Permanent address: Centre d'Etudes Nucléaires de Saclay, Saclay, France.

§Fellow of Deutsche Forschungsgemeinschaft.

‖Permanent address: Varian Associates, Palo Alto, California.

**Permanent address: Laboratoire de Physique Nucléaire et Haute Energie, Université Paris VI, Paris, France.

††Permanent address: Laboratori Nazionali, Frascatti, Rome, Italy.

‡‡Permanent address: Institut de Physique Nucléaire, Orsay, France.

[1]J.-E. Augustin *et al.*, Phys. Rev. Lett. 34, 233 (1975).

[2]J.-E. Augustin *et al.*, Phys. Rev. Lett. 34, 764 (1975).

[3]The only other feature in the $K\pi$ system that we observe in this data sample is the $K^*(890)$.

[4]A. M. Boyarski *et al.*, Phys. Rev. Lett. 35, 196 (1975).

[5]This cut rejects most nucleons (p and $\bar{p}$) as well as tracks accompanied by extra particles in the TOF counter.

[6]G. S. Abrams *et al.*, Phys. Rev. Lett. 34, 1181 (1975).

[7]A. M. Boyarski *et al.*, Phys. Rev. Lett. 34, 1357 (1975).

[8]J. Siegrist *et al.*, Phys. Rev. Lett. 36, 700 (1976).

[9]S. L. Glashow, J. Iliopoulos, and L. Maiani, Phys. Rev. D 2, 1285 (1970); S. L. Glashow, in *Experimental Meson Spectroscopy—1974*, AIP Conference Proceedings No. 21, edited by D. A. Garelick (American Institute of Physics, New York, 1974), p. 387.

[10]Other indications of possible charmed-particle production have come from experiments involving neutrino interactions. See, for example, A. Benvenuti *et al.*, Phys. Rev. Lett. 34, 419 (1975); E. G. Cazzoli *et al.*, Phys. Rev. Lett. 34, 1125 (1975); J. Bleitschau *et al.*, Phys. Lett. 60B, 207 (1976); J. von Krogh *et al.*, Phys. Rev. Lett. 36, 710 (1976); B. C. Barish *et al.*, Phys. Rev. Lett. 36, 939 (1976).

Observation of a Dimuon Resonance at 9.5 GeV in 400-GeV Proton-Nucleus Collisions

S. W. Herb, D. C. Hom, L. M. Lederman, J. C. Sens,[a] H. D. Snyder, and J. K. Yoh
Columbia University, New York, New York 10027

and

J. A. Appel, B. C. Brown, C. N. Brown, W. R. Innes, K. Ueno, and T. Yamanouchi
Fermi National Accelerator Laboratory, Batavia, Illinois 60510

and

A. S. Ito, H. Jöstlein, D. M. Kaplan, and R. D. Kephart
State University of New York at Stony Brook, Stony Brook, New York 11974

(Received 1 July 1977)

Accepted without review at the request of Edwin L. Goldwasser under policy announced 26 April 1976

Dimuon production is studied in 400-GeV proton-nucleus collisions. A strong enhancement is observed at 9.5 GeV mass in a sample of 9000 dimuon events with a mass $m_{\mu^+\mu^-}$ > 5 GeV.

We have observed a strong enhancement at 9.5 GeV in the mass spectrum of dimuons produced in 400-GeV proton-nucleus collisions. Our conclusions are based upon an analysis of 9000 dimuon events with a reconstructed mass $m_{\mu^+\mu^-}$ greater than 5 GeV corresponding to 1.6×10^{16} protons incident on Cu and Pt targets:

$$p + (\mathrm{Cu}, \mathrm{Pt}) \rightarrow \mu^+ + \mu^- + \text{anything}.$$

The produced muons are analyzed in a double-arm magnetic-spectrometer system with a mass resolution $\Delta m/m$ (rms) $\approx 2\%$.

The experimental configuration (Fig. 1) is a modification of an earlier dilepton experiment in the Fermilab Proton Center Laboratory.[1-3] Narrow targets (~ 0.7 mm) with lengths corresponding to 30% of an interaction length are employed.

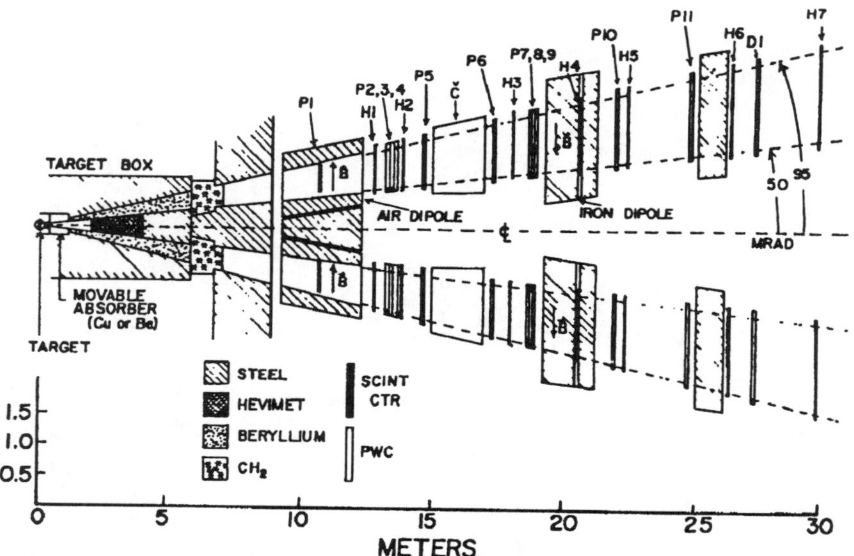

FIG. 1. Plan view of the apparatus. Each spectrometer arm includes eleven PWC's P1–P11, seven scintillation counter hodoscopes H1–H7, a drift chamber D1 and a gas-filled threshold Čerenkov counter Č. Each arm is up/down symmetric and hence accepts both positive and negative muons.

TABLE I. Yield of muon pairs for various running conditions.

Analyzing-magnet current (A)	Target	First 30 cm absorber	No. of incident protons	No. of events $m_{\mu^+\mu^-} \gtrsim 5$ GeV
1500	Cu	Cu	8.1×10^{15}	4093
1500	Pt	Cu	4.1×10^{15}	2076
1250	Cu	Cu	2.5×10^{15}	1891
1250	Pt	Be	1.6×10^{15}	911

Beryllium (18 interaction lengths) is used as a hadron filter, covering the 50–95-mrad (70–110° c.m.) horizontal and ± 10-mrad vertical aperture in each arm. The Be is closely packed against steel and tungsten which minimize particle leakage from outside the aperture, especially from the tungsten beam dump located 2.2 m downstream of the target. Polyethylene (1.5 m) and a 2.2-m steel collimator complete the shielding. The first 30 cm of beryllium (starting 13 cm downstream of the target center) can be remotely exchanged for 30 cm of copper.

The spectrometer dipoles deflect vertically, decoupling the production angle of each muon from its momentum determination. At full excitation (1500 A), the magnets provide a transverse momentum kick $p_t \approx 1.2$ GeV. In order to maximize the usable luminosity, no detectors are placed upstream of the magnet. Conventional proportional wire chambers (PWC's) and scintillation hodoscopes serve to define the muon trajectory downstream of the air dipole. Following the PWC's is a solid iron magnet (1.8 m long, energized to 20 kG) used to refocus partially the muons vertically and to redetermine the muon momentum to ± 15%. A threshold Čerenkov counter on each arm also helps prevent possible low-momentum muon triggers. The apparatus is ar-

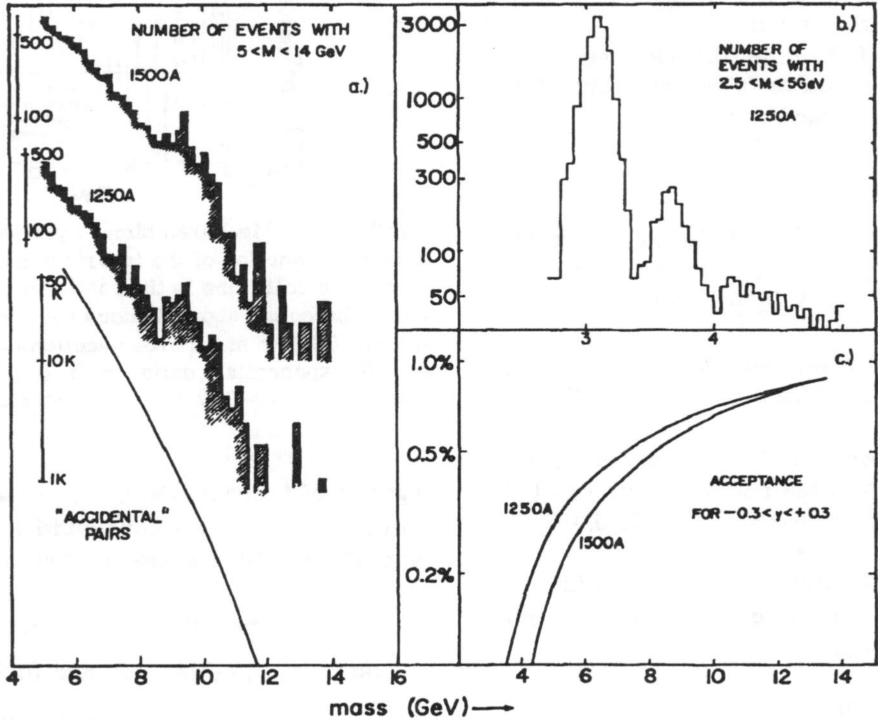

FIG. 2. (a) Dimuon yield at 1500 and 1250 A; the data with Cu and Pt targets have been combined. Also shown is the mass spectrum generated by combining two muons from different events. (b) Excess of opposite-sign over equal-sign muon pairs in the ψ, ψ' region. (c) Dimuon mass acceptance for the two excitations of the air dipole.

ranged symmetrically with respect to the horizontal median plane in order to detect both μ^+ and μ^- in each arm.

The data sets presented here are listed in Table I. Low-current runs produced ~ 15 000 J/ψ and 1000 ψ' particles which provide a test of resolution, normalization, and uniformity of response over various parts of the detector. Figure 2(b) shows the 1250-A J/ψ and ψ' data. The yields are in reasonable agreement with our earlier measurements.[2]

High-mass data (1250 and 1500 A) were collected at a rate of 20 events/h for $m_{\mu^+\mu^-} > 5$ GeV using $(1.5-3)\times 10^{11}$ incident protons per accelerator cycle. The proton intensity is limited by the requirement that the singles rate at any detector plane not exceed 10^7 counts/sec. The copper section of the hadron filter has the effect of lowering the singles rates by a factor of 2, permitting a corresponding increase in protons on target. The penalty is an ~15% worsening of the resolution at 10 GeV mass. Figure 2(a) shows the yield of muon pairs obtained in this work.

At the present stage of the analysis, the following conclusions may be drawn from the data [Fig. 3(a)]:

(1) A statistically significant enhancement is observed at 9.5-GeV $\mu^+\mu^-$ mass.

(2) By exclusion of the 8.8–10.6-GeV region, the continuum of $\mu^+\mu^-$ pairs falls smoothly with mass. A simple functional form,

$$[d\sigma/dmdy]_{y=0} = Ae^{-bm},$$

with $A = (1.89\pm 0.23)\times 10^{-33}$ cm^2/GeV/nucleon and $b = 0.98\pm 0.02$ GeV^{-1}, gives a good fit to the data for 6 GeV $< m_{\mu^+\mu^-} <$ 12 GeV ($\chi^2 = 21$ for 19 degrees of freedom).[4,5]

(3) In the excluded mass region, the continuum fit predicts 350 events. The data contain 770 events.

(4) The observed width of the enhancement is greater than our apparatus resolution of a full width at half-maximum (FWHM) of 0.5±0.1 GeV. Fitting the data minus the continuum fit [Fig. 3(b)] with a simple Gaussian of variable width yields the following parameters (B is the branching ratio to two muons):

$$\text{Mass} = 9.54\pm 0.04 \text{ GeV},$$

$$[Bd\sigma/dy]_{y=0} = (3.4\pm 0.3)\times 10^{-37} \text{ cm}^2/\text{nucleon},$$

with FWHM = 1.16 ± 0.09 GeV and $\chi^2 = 52$ for 27

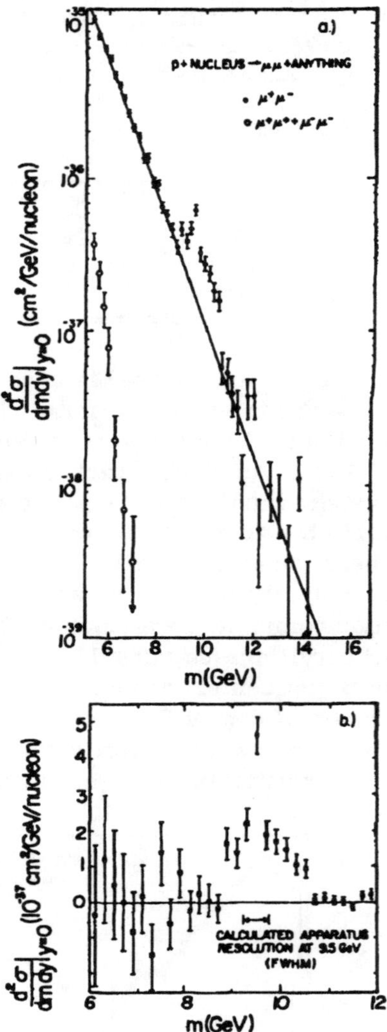

FIG. 3. (a) Measured dimuon production cross sections as a function of the invariant mass of the muon pair. The solid line is the continuum fit outlined in the text. The equal-sign–dimuon cross section is also shown. (b) The same cross sections as in (a) with the smooth exponential continuum fit subtracted in order to reveal the 9–10-GeV region in more detail.

degrees of freedom (Ref. 5). An alternative fit with two Gaussians whose widths are fixed at the resolution of the apparatus yields

$$\text{Mass} = 9.44\pm 0.03 \text{ and } 10.17\pm 0.05 \text{ GeV},$$

$$[Bd\sigma/dy]_{y=0} = (2.3\pm 0.2) \text{ and } (0.9\pm 0.1) \times 10^{-37} \text{ cm}^2/\text{nucleon},$$

with $\chi^2 = 41$ for 26 degrees of freedom (Ref. 5).

The Monte Carlo program used to calculate the acceptance [see Fig. 2(c)] and resolution of the apparatus assumed a mass, p_t, rapidity and decay angular distribution of the $\mu^+\mu^-$ pair consistent with these data and previously published dilepton searches. It also included all multiple-scattering effects in the hadron absorber and detector resolutions. The conclusions stated above are insensitive to these assumptions. In particular we note that the acceptance is relatively flat in the 9–10-GeV region.

The following checks have been made to verify the validity of the conclusions reached above:

(1) The spectrum of $\mu^+\mu^+$ and $\mu^-\mu^-$ events in Fig. 3(a) constitute an upper limit on the combined effects of accidental coincidences and hadronic decays. Misidentified $\psi \to \mu^+ + \mu^-$ decays are prevented from producing background at high mass by the remeasurement of the muon momenta both downstream by the second magnet and also by the PWC at the center of the first magnet. This is confirmed by the clean separation of the ψ and ψ' peaks in Fig. 2(a). Their widths agree with the calculated apparatus resolution.

(2) Various subsets of the data were studied in order to search for possible apparatus bias. In addition to the subsets shown in Table I, data were studied as a function of magnet polarity and magnetic bend direction. All fits showed enhancements consistent with the values quoted above.

(3) To check our analysis software (and as a further check of the apparatus), we mixed muons from different events, yielding the smooth mass spectrum shown in Fig. 2(a). The geometrical distribution of events in the 9–10-GeV region at the various detector planes in the apparatus is consistent with that of events in neighboring mass bins.

(4) The longitudinal distribution of muon-pair vertices at the target (FWHM = 16 cm) is cleanly separated from events generated in the beam dump, 220 cm downstream. A separate target-out run with 6×10^{14} incident protons produced no acceptable $\mu^+\mu^-$ candidates above 6 GeV (an equivalent run with a Cu or Pt target would have yielded about 200 events with 25 of these in the 9–10-GeV region).

In conclusion, the measured spectrum of $\mu^+\mu^-$ pairs produced in proton-nucleus collisions shows significant structure[6] in the 9–10-GeV region on an exponentially falling continuum. The structure is wider than the apparatus resolution. The 9.5-GeV enhancement and the continuum are in agreement with our previous measurements.[7]

We owe much to Ken Gray, Karen Kephart, Frank Pearsall, and S. Jack Upton for technical assistance, and to F. William Sippach for our electronic systems design. We also thank Brad Cox, William Thomas, and the staffs of the Fermilab Proton Department and Accelerator Division for their efforts. The work at Columbia University and at the State University of New York at Stony Brook was supported by the National Science Foundation, and that at the Fermi National Accelerator Laboratory by the U. S. Energy Research and Development Administration.

---

[a]Permanent address: Foundation for Fundamental Research on Matter, The Netherlands.

[1]D. C. Hom *et al.*, Phys. Rev. Lett. 36, 1236 (1976).

[2]H. D. Snyder *et al.*, Phys. Rev. Lett. 36, 1415 (1976).

[3]D. C. Hom *et al.*, Phys. Rev. Lett. 37, 1374 (1976).

[4]Cu and Pt yields were reduced to cross sections per nucleon by assuming an atomic-number dependence of $A^{1.0}$.

[5]The errors quoted on the magnitude of the continuum and resonance cross sections and the resonance masses are statistical only. Systematic normalization effects are probably less than 25% and do not affect the conclusions drawn here. Systematic errors on the mass calibration are probably less than 1%.

[6]Following Ref. 1, a reasonable designation for this enhancement is Υ(9.5).

[7]We note that the 9–10-GeV mass bin in the e^+e^- and $\mu^+\mu^-$ spectra previously published by this group (Ref. 3) shows an excess of events, consistent with the statistically more significant results here. If we add our preliminary unpublished e^+e^- data to our published e^+e^- yield (Ref. 1), the spectrum contains a cluster of 6 e^+e^- events near 9.5 GeV where ~5 events would be expected on the assumption of μ-e universality.

Observation of Υ, Υ′, and Υ″ at the Cornell Electron Storage Ring

T. Böhringer, F. Costantini,[a] J. Dobbins, P. Franzini, K. Han, S. W. Herb, D. M. Kaplan, L. M. Lederman,[b] G. Mageras, D. Peterson, E. Rice, and J. K. Yoh
Columbia University, New York, New York 10027

and

G. Finocchiaro, J. Lee-Franzini, G. Giannini, R. D. Schamberger, Jr., M. Sivertz, L. J. Spencer, and P. M. Tuts
The State University of New York at Stony Brook, Stony Brook, New York 11794
(Received 15 February 1980)

The Υ, Υ′, and Υ″ states have been observed at the Cornell Electron Storage Ring as narrow peaks in $\sigma(e^+e^- \to \text{hadrons})$ versus beam energy. Data were collected during a run with integrated luminosity of 1000 nb^{-1}, using the Columbia University–Stony Brook segmented NaI detector. The measured mass differences are $M(\Upsilon') - M(\Upsilon) = 559 \pm 1(\pm 3)$ MeV and $M(\Upsilon'') - M(\Upsilon) = 889 \pm 1(\pm 5)$ MeV, where the errors in parentheses represent systematic uncertainties. Preliminary values for the leptonic width ratios were also obtained.

PACS numbers: 13.65.+i

The discovery at Fermilab[1] of narrow enhancements in the dimuon spectrum near 10 GeV invariant mass was considered evidence for the existence of a new heavy quark. Two of these states, Υ and Υ′, were later observed with much better resolution at the electron-positron storage ring DORIS[2,3] through the process $e^+e^- \to$ hadrons. The leptonic decay widths inferred from the DORIS measurements were consistent with models[4-6] describing the Υ and Υ′ as the 1^3S_1 and

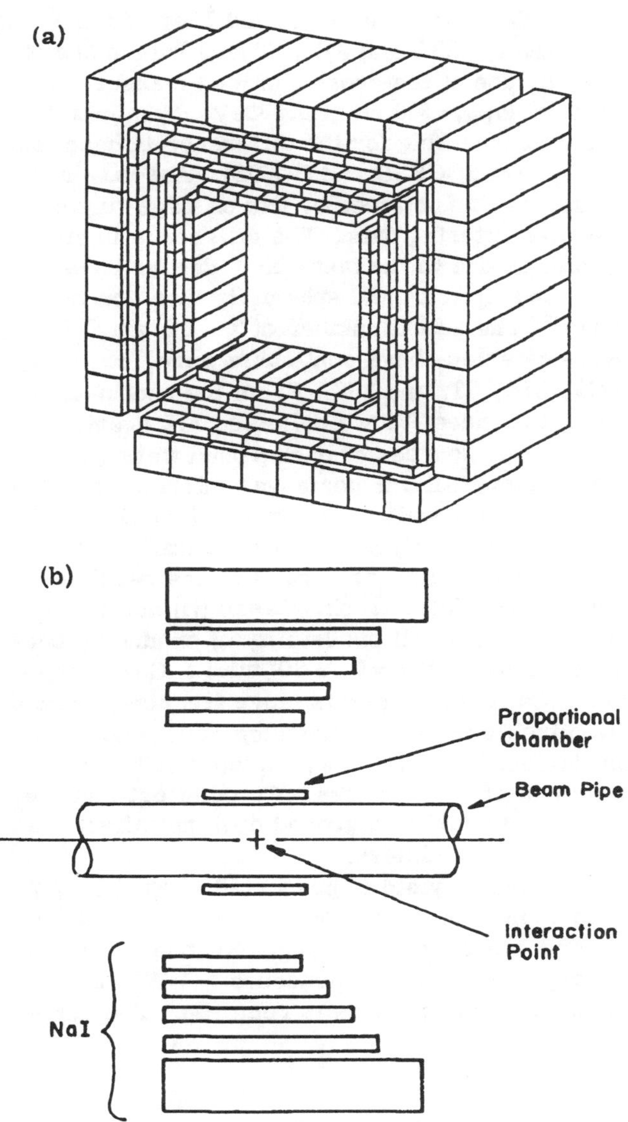

FIG. 1. (a) An isometric view of the NaI array as used in the present run and (b) a side view showing the positioning of the array relative to the interaction point and the positions of proportional chambers.

2^3S_1 bound states of a quark-antiquark pair, $b\bar{b}$, where the b quark has a mass of about 5 GeV and carries $\frac{1}{3}$-integer charge. The results from the Fermilab experiment[7] included evidence for a third state near 10.4 GeV. We report here a new observation of the Υ, Υ', and Υ'' states in e^+e^- collisions with improved energy resolution which confirms the existence of the Υ'' and gives, for the first time, a precise measurement of both level spacings. Preliminary values for the relative leptonic widths are also given.

These results were obtained during the *first* energy scan in the 10-GeV region performed at the Cornell Electron Storage Ring (CESR). This scan lasted ~5 weeks yielding an integrated luminosity of ~1000 nb^{-1} of which ~600 nb^{-1} were concentrated in three intervals around 9.5, 10.0, and 10.3 GeV. The additional 400 nb^{-1} were used for a scan around 10.6 GeV.

Our detector is located at one of the two interaction regions at CESR (the "North Area") and is designed primarily for the study of photons and electrons. The detector as designed consists of a segmented array of NaI, 8 radiation lengths (r.l.) thick, followed by 7 r.l. of lead glass. Drift chambers before the NaI, and cathode readout proportional chambers ("strip chambers") within the NaI array, provide tracking. The NaI is divided into 32 azimuthal sectors and 2 polar sectors. This provides complete azimuthal coverage in the region $45° < \theta < 135°$. In addition, it is subdivided radially into five layers. At normal incidence, the inner four layers are 1 r.l. each and the last layer is 4 r.l. For the scan reported here, only one NaI half array was used, centered optimally over the interaction point. Four strip chambers at the beam pipe gave the actual position of the interaction point. Complete azimuthal coverage was maintained [Fig. 1(a)]. The polar acceptance was $70° < \theta < 110°$ with some asymmetry [Fig. 1(b)].

Charged particles originating at the interaction point deposit energy in each layer, yielding five independent dE/dx measurements. This signature helped us to eliminate events resulting from beam-gas and beam-wall interactions, although such events were already suppressed since our region of acceptance was located at 90° to the beam.

An absolute energy scale for each NaI crystal was set with γ rays from Cs^{137} and Co^{60}. Full-scale settings ranged from 0.5 to 3.0 GeV. Photomultiplier tube (PM) stability was monitored with light from a spark in argon.

All signals from PM's and strips were integrated every beam crossing (every 2.56 μs) while a trigger decision was made. If no trigger was present, all integrators were reset to be ready for the next crossing. Only a total-energy trigger was used for the data presented. This was generated by adding all signals from the three outer layers of NaI and requiring this sum to exceed a threshold equivalent to 420 MeV and to be coincident with the beam. If the trigger was produced, all signals were digitized and recorded on tape. This trigger gave an event rate of 0.3 Hz for a luminosity of 1 $\mu b^{-1} s^{-1}$. A typical fill of CESR lasts 3 to 5 hours yielding an integrated luminosity of up to ~15 nb^{-1}. The integrated luminosity for each run was measured by detecting and counting small-angle (40 to 80 mrad) collinear Bhabha scatters with lead-scintillator sandwich shower detectors. The long-term stability of the luminosity monitor is confirmed by the yield of large-angle Bhabha scattering events in the NaI array.

Because of the limited solid angle of the NaI array as used, a major fraction of the hadronic e^+e^- annihilations gave very few particles in the detector. Rather than trying to identify all hadronic events, which would result in an unacceptable amount of background, our aim in the analysis was to obtain a clean sample through the use of strict event-selection criteria. Fundamental in all criteria used was the identification of minimum-ionizing hadrons. At normal incidence, minimum-ionizing particles deposit 15 MeV in the first four NaI layers and ~68 MeV in the last layer of a single sector. In all scans one unambiguous and isolated minimum-ionizing track plus at least two other tracks or showers were required. All data were scanned by physicists and with computer programs. The acceptance criteria for data presented were determined by maximizing detection efficiency while maintaining the background level well below 10% of the continuum cross section. The overall efficiencies for detecting continuum and Υ events are, respectively, 28% and 37%. These values are obtained by use of the cross sections measured at DORIS[2,3] (σ_{cont} = 3.8 nb at 9.4 GeV, $\sigma_{\Upsilon peak}$ = 18.5 nb after correcting for the difference in beam energy spread at CESR and DORIS). Absolute normalization was obtained by use of large-angle Bhabha-scattering data. The difference in efficiencies is due to the fact that Υ decays have higher multiplicity and sphericity than continuum events.[2] The actual number of Υ, Υ', and Υ'' events detected above continuum were, respectively, 214, 53, and 133. From the continuum around the three Υ's we collected 272 events.

The major sources of background were (i) far single beam-wall and beam-gas interactions, (ii) close beam-wall interactions, (iii) close beam-gas interactions, and (iv) cosmic rays. Case (i) was trivially removed by the requirement of an isolated track. Cases (ii) and (iii) occur with very small probability of producing penetrating hadrons at $\theta = 90° \pm 30°$ with 5-GeV electrons. Case (ii), which is more frequent, is also recognizable by tracks crossing azimuthal sector boundaries. Case (iv) was rejected by the requirement of three tracks. We point out that the minimal residual background does not affect the results presented here.

The hadronic yield is presented in Fig. 2, plotted in arbitrary units proportional to the ratio of detected events to small-angle Bhabha yield. In this way, the energy dependence ($\sim 1/E^2$) of the single-photon processes is removed. The hori-

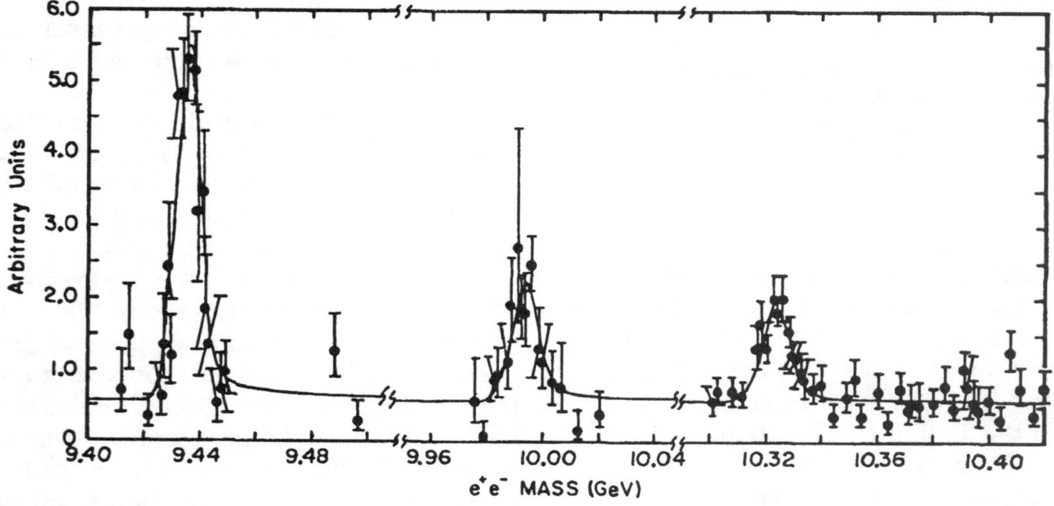

FIG. 2. The number of hadronic events, normalized to the small-angle Bhabha yield. The solid line indicates a fit described in the text.

zontal scale is $M(e^+e^-)$, twice the nominal machine energy. Mass values of the three resonances are determined by fitting the data with a constant continuum plus three radiatively corrected[8] Gaussians, with widths representing the machine energy spread. This fit is shown as a solid line in Fig. 2. A single free width parameter is used for all three resonances, allowing for scaling as E^2 as predicted for the stored-beam energy spread. Our fitted e^+e^- invariant-mass spread at 9.5 GeV is 4.0± 0.3 MeV rms, in agreement with the computed value for CESR. Our mass values for Υ and Υ′ are ~0.3% lower than the DORIS results. This difference is consistent with the accuracy of the CESR energy calibration.

More relevant to model calculations of the $q\bar{q}$ bound states are the mass differences. Our result for $M(\Upsilon') - M(\Upsilon)$ is 559± 1 MeV. To this purely statistical error one should add an estimated systematic uncertainty of ± 3 MeV due to the machine energy calibration. This result is in good agreement with the DORIS results.[2,3] The mass differences $M(\Upsilon'') - M(\Upsilon)$ has been measured accurately for the first time in this experiment, and at the same time by the CLEO collaboration.[9] Our result is 889± 1 MeV, with an additional systematic uncertainty of ± 5 MeV. Table I contains a summary of all measured parameters obtained from the fit described.

Another quantity of interest for the phenomenology of the $q\bar{q}$ bound states is the ratio of the leptonic widths for the three resonances. Here the lack of information on decay details and our limited solid angle introduces severe uncertainties on such quantities. Our best estimates, computed without correcting for possible differences in decay multiplicities and angular distributions at the three resonances, are $\Gamma_{ee}(\Upsilon')/\Gamma_{ee}(\Upsilon) = 0.39 \pm 0.06$ and $\Gamma_{ee}(\Upsilon'')/\Gamma_{ee}(\Upsilon) = 0.32 \pm 0.04$. These errors represent only the statistical uncertainty. An estimate of the *systematic errors* for the partial-width ratios given above was obtained by relaxing the acceptance criteria which increased the number of accepted events for the three resonances by ~50%. In this way, we concluded that the systematic errors are smaller than the quoted statistical errors. The results represented here are in good agreement with many predictions,[4-6] reinforcing the validity of the interpretation of the Υ family as $b\bar{b}$ bound states.

TABLE I. Results from the fit[a].

σ_{machine}	4.0± 0.3 MeV
$\Gamma_{ee}(\Upsilon')/\Gamma_{ee}(\Upsilon)$	0.39± 0.06
$\Gamma_{ee}(\Upsilon'')/\Gamma_{ee}(\Upsilon)$	0.32± 0.04
$M(\Upsilon)$	9.4345± 0.0004 GeV
$M(\Upsilon')$	9.9930± 0.0010 GeV
$M(\Upsilon'')$	10.3232± 0.0007 GeV

[a]Only statistical errors are given.

We wish to acknowledge the tremendous effort of the whole CESR staff in bringing the storage ring successfully into operation. We thank B. McDaniel, M. Tigner, and members of the Cornell Laboratory of Nuclear Studies for help during many months of installation. Part of the electronics design and construction and most of the mechanical design and construction were done by the Nevis Laboratories and Stony Brook shops. We thank Paula Franzini for help in running, and Carl Brown, Tom Regan, and Bill Marterer for help during assembly and testing of the detector at CESR. This research was supported in part by the National Science Foundation.

(a)On leave from University of Pisa, I-56100 Pisa, Italy, and Istituto Nazionale di Fisica Nucleare, I-56100 Pisa, Italy.

(b)Also at Fermi National Accelerator Laboratory, Batavia, Ill. 60510.

[1]S. W. Herb *et al.*, Phys. Rev. Lett. 39, 252 (1977); W. R. Innes *et al.*, Phys. Rev. Lett. 39, 1240, 1640(E) (1977).

[2]Ch. Berger *et al.*, Phys. Lett 76B, 243 (1978); Ch. Berger *et al.*, Phys. Lett. 78B, 176 (1978); C. W. Darden *et al.*, Phys. Lett. 76B, 246 (1978); F. H. Heimlich *et al.*, Phys. Lett. 86B, 399 (1979).

[3]J. K. Bienlein *et al.*, Phys. Lett. 78B, 360 (1978); C. W. Darden *et al.*, Phys. Lett. 78B, 364 (1978).

[4]C. Quigg and J. L. Rosner, Phys. Lett. 71B, 153 (1977).

[5]E. Eichten, K. Gottfried, T. Kinoshita, K. Lane, and T.-M. Yan, Phys. Rev. D 17, 3090 (1978), and 21, 203 (1980).

[6]G. Bhanot and S. Rudaz, Phys. Lett. 78B, 119 (1978).

[7]K. Ueno *et al.*, Phys. Rev. Lett. 42, 486 (1979).

[8]J. D. Jackson and D. L. Scharre, Nucl. Instrum. Methods 128, 13 (1975).

[9]D. Andrews *et al.*, preceding Letter [Phys. Rev. Lett. 44, 1108 (1980)].

Observation of Exclusive Decay Modes of b-Flavored Mesons

S. Behrends, K. Chadwick, J. Chauveau,[(a)] P. Ganci, T. Gentile, Jan M. Guida, Joan A. Guida, R. Kass, A. C. Melissinos, S. L. Olsen, G. Parkhurst, D. Peterson, R. Poling, C. Rosenfeld, G. Rucinski, and E. H. Thorndike
University of Rochester, Rochester, New York 14627

and

J. Green, R. G. Hicks, F. Sannes, P. Skubic,[(b)] A. Snyder, and R. Stone
Rutgers University, New Brunswick, New Jersey 08854

and

A. Chen, M. Goldberg, N. Horwitz, A. Jawahery, M. Jibaly, P. Lipari, G. C. Moneti, C. G. Trahern, and H. van Hecke
Syracuse University, Syracuse, New York 13210

and

M. S. Alam, S. E. Csorna, L. Garren, M. D. Mestayer, and R. S. Panvini
Vanderbilt University, Nashville, Tennessee 37235

and

D. Andrews,[(c)] P. Avery, C. Bebek, K. Berkelman, D. G. Cassel, J. W. DeWire, R. Ehrlich, T. Ferguson, R. Galik, M. G. D. Gilchriese, B. Gittelman, M. Halling, D. L. Hartill, D. Herrup,[(d)] S. Holzner, M. Ito, J. Kandaswamy, V. Kistiakowsky,[(e)] D. L. Kreinick, Y. Kubota, N. B. Mistry, F. Morrow, E. Nordberg, M. Ogg, R. Perchonok, R. Plunkett,[(f)] A. Silverman, P. C. Stein, S. Stone, R. Talman, D. Weber, and R. Wilcke
Cornell University, Ithaca, New York 14853

and

A. J. Sadoff
Ithaca College, Ithaca, New York 14850

and

R. Giles, J. Hassard, M. Hempstead, J. M. Izen,[(g)] K. Kinoshita, W. W. MacKay, F. M. Pipkin, J. Rohlf, and Richard Wilson
Harvard University, Cambridge, Massachusetts 02138

and

H. Kagan
Ohio State University, Columbus, Ohio 43210
(Received 24 January 1983)

B-meson decays to final states consisting of a D^0 or D^{*+} and one or two charged pions have been observed. The charged-B mass is $5270.8 \pm 2.3 \pm 2.0$ MeV and the neutral-B mass is $5274.2 \pm 1.9 \pm 2.0$ MeV.

PACS numbers: 14.40.Jz, 13.25.+m, 13.65.+i

The upsilon states[1] are interpreted as q-q resonances of a new quark, the b quark. The first three resonances are narrow,[2,3] implying bound b flavor and a suppressed strong decay. The large width of the $\Upsilon(4S)$ resonance discovered at the e^+e^- storage ring CESR[4] (Cornell Electron Storage Ring) and the observation that the decay products from the $\Upsilon(4S)$ include high-momentum leptons[5] imply that the $\Upsilon(4S)$ decays strongly into $B\bar{B}$ meson pairs, which then decay weakly. Until now, however, the b-flavored mesons themselves had not been found. Here we report that discovery.

The b quark has been shown to decay predominantly to the c quark.[6] Thus the principal decay mode of the B meson will be to a charmed meson plus pions. Since the high multiplicity in $\Upsilon(4S)$ decay[7] leads to large combinatorial background, we have restricted our search to low-multiplicity decay modes, D^0 or $D^{*\pm}$ plus one or two charged pions.

The data sample used is 40.7 pb^{-1} of $\Upsilon(4S)$ data and 19.6 pb^{-1} of continuum data taken with the CLEO detector at CESR. The $\Upsilon(4S)$ cross section is a 1.0-nb enhancement above a 2.5-nb continuum contribution. The detector has been described in detail elsewhere.[8] In this work we have used the cylindrical drift chamber inside a 1.0-T solenoid magnet to determine momenta of charged particles. In addition we have used the dE/dx-measuring wire proportional chambers and the time-of-flight scintillation counters located outside the solenoid magnet to identify charged kaons over a momentum range from 0.45 to 1.0 GeV/c.

The two-body decay modes, $D^0 \to K^-\pi^+$ and its charge conjugate, were used to find D^0 mesons. Identified kaons were paired with each oppositely charged particle in the event (assumed to be a pion). The combination was kept only if its momentum was below 2.6 GeV/c, since D^0's from B decay cannot exceed this momentum. The resulting $K^\pm\pi^\mp$ mass distribution is shown in Fig. 1(a). Mass combinations within $\pm$ 40 MeV of the D^0 mass were kept as D^0 candidates.

We looked for charged D^* mesons through the cascade $D^{*+} \to D^0\pi^+$, $D^0 \to K^-\pi^+$ and its charge conjugate. We did not require that the charged kaon be identified as such. Rather we first formed mass combinations of all pairs of oppositely charged particles in an event, assuming that each particle in turn is a kaon. We then added an additional particle (assumed to be a pion) of charge opposite to that of the assumed kaon. We kept as D^* candidates the combinations for which the $[K\pi\pi, K\pi]$ mass difference was within $\pm$ 3.0 MeV of the $[D^{*\pm}, D^0]$ mass difference of 145.4 MeV,[9] and $K\pi$ masses within $\pm$ 80 MeV of the D^0 mass. We further required that the D^* candidate have momentum below 2.6 GeV/c, eliminating the high-momentum D^* contribution from the continuum. With these requirements the $D^{*\pm}$ signal is hidden under considerable background. By demanding a more restrictive set of conditions [see Fig. 1(b)], we demonstrate that the $\Upsilon(4S)$ decays contain a $D^{*\pm}$ signal. Bécause these latter restrictions lower

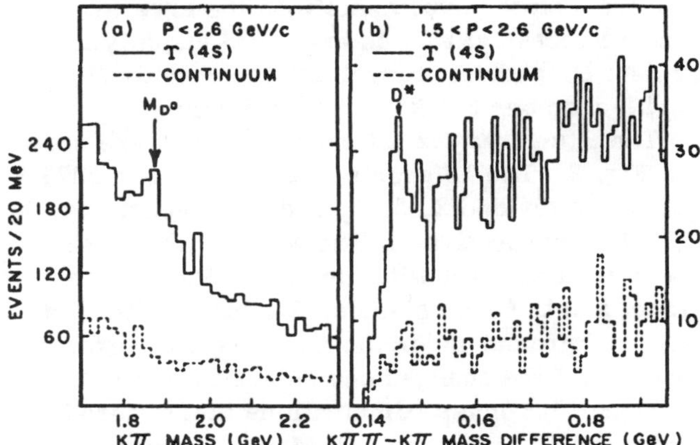

FIG. 1. (a) Mass distribution of $K^\pm\pi^\mp$, for $K\pi$ momenta below 2.6 GeV/c, using identified kaons. The solid line shows data from 40.7 pb^{-1} of $\Upsilon(4S)$ running; the dashed line is from 19.6 pb^{-1} of continuum running at energies just below the $\Upsilon(4S)$. A D^0 signal at 1.86 GeV is evident in the 4S data. (b) $(K^\pm\pi^\mp\pi^\mp)-(K^\pm\pi^\mp)$ mass-difference distribution, for $K\pi\pi$ momenta between 1.5 and 2.6 GeV/c and $K\pi$ masses within 20 MeV of the D^0 mass. Kaons were not directly identified. Curves are as in (a). The signal at the $D^{*\pm}-D^0$ mass difference (145.4 MeV) is evidence of $D^{*\pm}$ production in $\Upsilon(4S)$ decay.

the $D^{*\pm}$ detection efficiency they were not used in the search for B mesons.

Each event containing a D^0 or $D^{*\pm}$ candidate was fitted to the following hypotheses[10] (or their charge conjugates):

$$B^- \to D^0\pi^-, \tag{1}$$

$$\bar{B}^0 \to D^0\pi^+\pi^-, \tag{2}$$

$$\bar{B}^0 \to D^{*+}\pi^-, \tag{3}$$

$$B^- \to D^{*+}\pi^-\pi^-. \tag{4}$$

We considered only these charge combinations, since they preserve the quark decay scheme $b \to c \to s$. In making the fit, we constrained the B-meson energy to the beam energy and constrained the D^0 or $D^{*\pm}$ decay products to the known D^0, $D^{*\pm}$ masses, respectively. This fitting procedure measures the B mass relative to the CESR beam energy, which is scaled to agree with the VEPP4 measurement of the $\Upsilon(1S)$ mass.[11] Since the threshold for $B\bar{B}$ production is known to lie between the $\Upsilon(3S)$ and $\Upsilon(4S)$ resonances, we considered B-meson mass combinations between half the $\Upsilon(3S)$ and $\Upsilon(4S)$ resonance masses (i.e., 5180 and 5290 MeV). Candidate fits were required to have a χ^2 value less than 14. If an event had two acceptable fits in this mass interval we took the hypothesis with the lower χ^2. Each successful fit was examined visually to reject B candidates involving incorrectly fitted drift-chamber tracks (a 15% rejection).

The B masses for all successful fits are shown in Fig. 2. The 18 events in the peak near 5275 MeV are divided 2, 5, 5, and 6 for reactions (1)–(4), respectively. The width of the mass peak is consistent with the resolution expected from Monte Carlo studies. We have estimated the background under the mass peak in several ways. (1) We changed our selection criteria to accept $K^{\mp}\pi^{\pm}$ mass combinations that differed from the D^0 mass by ± 200 MeV. The spectrum of reconstructed "B-meson" masses for this "sideband" search is shown in Fig. 3(a). (2) We considered wrong charge combinations, which corresponded to doubly charged B's, or corresponded to decay sequences other than $b \to c \to s$. The mass spectrum for wrong charge combinations is shown in Fig. 3(b). Both distributions in Fig. 3 have been normalized so that the vertical scales are directly comparable with Fig. 2. (3) We performed Monte Carlo studies of background from $B\bar{B}$ events and from continuum events, to determine how the background in Fig. 2 should be extrapolated from lower masses to the peak region. (4) We searched 19.6 pb^{-1} of data accumulated just below the $\Upsilon(4S)$ for apparent B's, finding two in the region of the mass peak. These studies lead to estimates of the background under the peak at 5275 MeV which lie between 4 and 7 events.

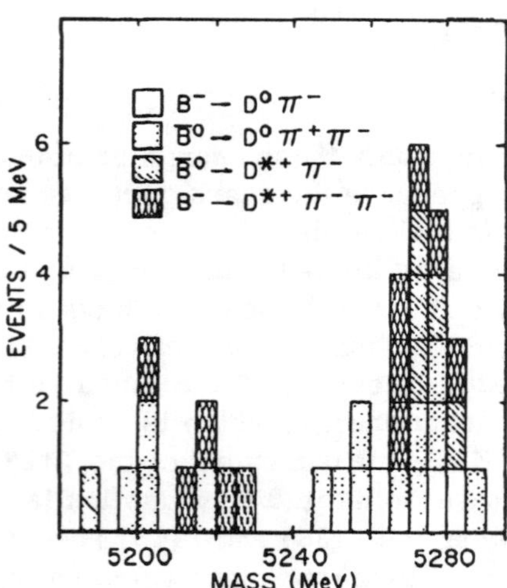

FIG. 2. Mass distribution of B-meson candidates. The $B \to$ final-state decay labels should be interpreted as including the charge-conjugate reaction.

If a B decay contains a low-energy particle that escapes detection, the remaining particles from that B may still be consistent with the beam-energy constraint and give an acceptable fit. We frequently cannot distinguish reactions (1) and (2) from similar reactions with the D^0 replaced by D^{*0}, where $D^{*0} \to D^0\pi^0$ or $D^0\gamma$. Similarly, the decay $\bar{B}^0 \to D^{*+}\pi^-$, $D^{*+} \to D^0\pi^+$ (π^+ not detected), can masquerade as $B^- \to D^0\pi^-$, causing us to assign an incorrect charge to the B. Monte Carlo studies show that the reconstructed mass is shifted down a few megaelectronvolts from the true B mass, and the mass resolution is worsened slightly. The problem of missed low-energy particles is not important for reactions (3) and (4), and therefore we use only these to determine the B mass.

We find a mass of $5274.2 \pm 1.9 \pm 2.0$ MeV for the neutral B, and $5270.8 \pm 2.3 \pm 2.0$ MeV for the charged B, where the first error is statistical and the second error systematic. The $[\bar{B}^0, B^-]$ mass difference is $3.4 \pm 3.0 \pm 2.0$ MeV, consistent

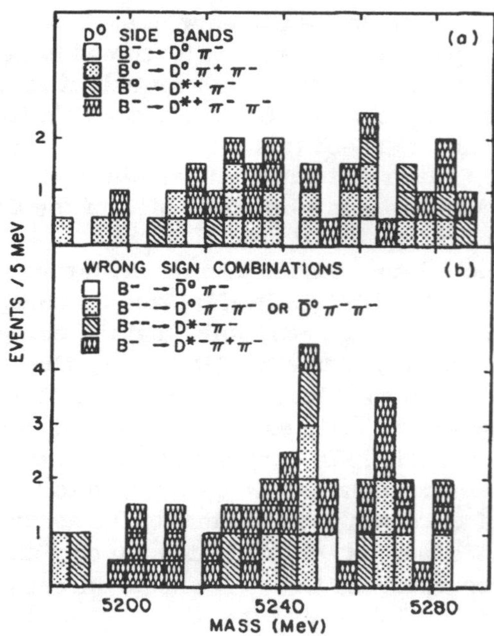

FIG. 3. Mass distribution for two estimates of the background to the B-meson candidates of Fig. 2. (a) D^0's chosen from sidebands. The events shown are plotted with a weight of $\frac{1}{2}$ event, since there are approximately twice as many events in the sidebands as in the D region. (b) Wrong charge combinations. The events shown in this distribution have been scaled to account for the difference in the number of combinations leading to a wrong-sign B compared to those leading to the correct-sign charged B.

with the theoretical prediction[12] of 4.4 MeV. The average of charged and neutral B masses is 5272.3± 1.5± 2.0 MeV. This corresponds to a mass difference ΔM of 32.4± 3.0± 4.0 MeV between the mass of the $\Upsilon(4S)$ and twice the B-meson mass. If the $[B^*, B]$ mass difference is ~ 50 MeV as expected theoretically,[12] the $\Upsilon(4S)$ must decay exclusively to $B\bar{B}$, with no contribution from $B^*\bar{B}$. Previous experimental information on ΔM comes from the fact that Schamberger *et al.* do not observe monochromatic photons from $B^* \to \gamma B$ decay.[13] Their experiment sets an upper limit of 50 MeV on ΔM. Theoretical calculations of ΔM using the width and the height of the $\Upsilon(4S)$ fall either above[12] or below[14] our result. Using our measured value for ΔM and the theoretical value of 4.4 MeV for the $[\bar{B}^0, B^-]$ mass difference, we obtain the branching fractions $B(\Upsilon(4S) \to B^+B^-) = 0.60 \pm 0.02$ and $B(\Upsilon(4S) \to B^0\bar{B}^0) = 0.40 \pm 0.02$. We estimate branching ratios of 4.2± 4.2%, 13± 9%, 2.6± 1.9%, and 4.8± 3.0% for reactions (1)–(4), respectively.[15,16]

In conclusion, we have explicitly demonstrated the existence of the B meson through its decay into exclusive final states and have measured its mass.

We gratefully acknowledge the efforts of the CESR staff who made this work possible. We thank H. Tye for useful discussions. This work was supported in part by the National Science Foundation and the U. S. Department of Energy.

(a)Permanent address: Laboratoire de Physique Corpusculaire, College de France, F-75231 Paris, France.

(b)Present address: University of Oklahoma, Norman, Okla. 73019.

(c)Present address: AIRCO Superconductors, Carteret, N.J. 07008.

(d)Present address: Lawrence Berkeley Laboratory, Berkeley, California 94720.

(e) Permanent address: Massachusetts Institute of Technology, Cambridge, Mass. 02139.

(f)Present address: CERN, EP Division, CH-1211 Geneva 23, Switzerland.

(g)Present address: University of Wisconsin, Madison, Wisc. 53706.

[1]S. W. Herb *et al.*, Phys. Rev. Lett. 39, 252 (1977).

[2]C. Berger *et al.*, Phys. Lett. 76B, 243 (1978); C. W. Darden *et al.*, Phys. Lett. 76B, 246 (1978), and 78B, 364 (1978); J. Bienlein *et al.*, Phys. Lett. 78B, 360 (1978).

[3]D. Andrews *et al.*, Phys. Rev. Lett. 44, 1108 (1980); T. Bohringer *et al.*, Phys. Rev. Lett. 44, 222 (1980).

[4]D. Andrews *et al.*, Phys. Rev. Lett. 45, 219 (1980); G. Finocchiaro *et al.*, Phys. Rev. Lett. 45, 222 (1980).

[5]C. Bebek *et al.*, Phys. Rev. Lett. 46, 84 (1981); K. Chadwick *et al.*, Phys. Rev. Lett. 46, 88 (1981); L. J. Spencer *et al.*, Phys. Rev. Lett. 47, 771 (1981).

[6]A. Brody *et al.*, Phys. Rev. Lett. 48, 1070 (1982); D. Andrews *et al.*, Cornell University Report No. CLNS 82/547, 1982 (to be published), p. 14; Spencer *et al.*, Ref. 5.

[7]M. S. Alam *et al.*, Phys. Rev. Lett. 49, 357 (1982).

[8]D. Andrews *et al.*, in Cornell University Report No. CLNS 82/538 (to be published).

[9]M. Roos *et al.* (Particle Data Group), Phys. Lett. 111B, 1 (1982).

[10]Our notation for neutral B's follows the convention used for neutral kaons, i.e., $\bar{B}^0 = (b\bar{d})$, $B^0 = (\bar{b}d)$.

[11]A. S. Artamonov *et al.*, Institute of Nuclear Physics, Academy of Science of the U.S.S.R., Novosibirsk Report No. 82-94 (to be published).

[12]E. Eichten, Phys. Rev. D 22, 1819 (1980).

[13]R. D. Schamberger *et al.*, Phys. Rev. D 26, 720 (1982).

[14]M. Gronau *et al.*, Phys. Rev. D 25, 3100 (1982); I. I. Bigi and S. Ono, Nucl. Phys. B189, 229 (1981).

[15]D^0 branching ratios are from R. H. Schindler *et al.*, Phys. Rev. D 24, 78 (1981).

[16]As noted earlier, the branching ratio for reaction (1) includes a contamination from $B^- \to D^{*0}\pi^-$ and $\bar{B}^0 \to D^{*+}\pi^-$, while reaction (2) includes $\bar{B}^0 \to D^{*0}\pi^+\pi^-$.

Chapter 12

PARTICLE THEORY

SAM TREIMAN

A. Henri Becquerel (1852–1908). (Courtesy of William G. Myers Collection, AIP Emilio Segrè Visual Archives.)

Introduction

Papers Reprinted in Book

Sam Treiman is Professor of Physics at Princeton University, past chairman of the Physics Department, and, currently, chair of the University Research Board. He has served on numerous high energy physics advisory panels and was instrumental during the early days of Fermilab in helping to establish a theory group there. His work has ranged over various aspects of particle theory. His honors include the Oersted Medal of the American Association of Physics Teachers.

Hans A. Bethe (1906-). (Courtesy of Edith Michaelis, 1954.)

The Physical Review was born just in time for the beginnings of "modern" physics. Before the journal was three years old, Roentgen and Becquerel had respectively discovered x rays and natural radioactivity (the medical implications of x rays drew immediate worldwide attention); and within another three years, Thomson and others pretty well got the electron nailed down. The photon had a more protracted birth. It entered the world scarcely recognizable when Planck introduced the quantum hypothesis for blackbody radiation. It took firmer shape, regarded as a packet of energy, in Einstein's hands five years later, and it became indubitably particle-like in the Compton experiments 20 years after that. *The Physical Review* watched the early developments in modern physics mostly from a distance. But by the time of the golden decades following World War II *The Physical Review* had become the dominant medium for publication in particle physics (and much else).

The electron preceded the quantum. It could be — and was — conceived of in classical terms. It was an unusually interesting classical particle, to be sure: It evidenced a substructure of atoms, it was a carrier of electric charge, and it was, surprisingly, so very much less massive than atoms. Still, it was born as a classical particle, presumably immutable throughout all its wanderings. The photon was harder to swallow in these terms. It was somehow related to the classical electromagnetic *field*; and, as a particle, it could easily be created or destroyed *de novo*. From the beginning, that is, the photon was intrinsically a quantum mechanical object.

Differences between the treatment of electrons (handled nonrelativistically) and photons (always ultrarelativistic, of course) persisted for a while after the principles of quantum theory were laid down in the middle 1920s. For electrons (or other nonrelativistic particles) one started with classical *particle* dynamics and switched to quantum dynamics by converting the position and momentum observables into operators (spin could easily be grafted on). In the quantum version of the dynamics, one was still dealing with systems having a finite number of degrees of freedom, and the electron was still immutable. In the case of the photon, one starts not with a classical particle but with the classical Maxwell *field*; it is the field dynamics, with its infinite number of degrees of freedom, that is quantized. The photons are not put in; they *emerge* as quanta of the quantum field theory. The combined dynamics of (nonrelativistic) quantum electrons and the quantum electromagnetic field provided the necessary ingredients for dealing with processes such as photon emission and absorption in atomic transitions, (nonrelativistic) Compton scattering, etc.

The distinction between particle and field quantum dynamics begins to fade when one attempts to generalize to relativistic particle dynamics, which is what Dirac did for the electron with his celebrated equation.[1] There he ran into the dread negative-energy states of the free electron. To cope, he devised his now-familiar hole theory.[2] According to this, the sea of negative-energy states is normally filled, a hole in that sea, i.e., a missing negative-energy electron behaving like a particle of positive energy and positive charge, i.e., behaving like a positron. Hole theory made it possible to handle processes such as positron-electron pair production in the collision, say, of a photon with an atom. In hole theory, the electron is still in a sense immutable, but when the photon lifts an electron from the negative sea to a positive-energy state, the hole left behind is identified as a positron. However, the need for an infinite sea signals that in dealing with the Dirac electron, one is facing a system with infinitely many degrees of freedom. Indeed, through the technique of "second quantization" that was taken up widely soon after hole theory was devised, one could introduce a field theory description in which electrons and positrons emerge as field quanta.

A new threshold was crossed in the early 1930s following the discovery of the neutron, when it became plausible to interpret nuclear beta decay as the spontaneous transformation of a neutron into proton, electron, and neutrino. Here was particle creation and destruction with a vengeance, beyond any interpretative help from hole theory: No more immutability. Fermi introduced a field theory treatment of beta decay, patterning it on quantum electrodynamics.[3] Quantum field theory had taken its place as the proper dynamical basis for particle physics. It was in

P. A. M. Dirac (1902-1984). (Courtesy of AIP Emilio Segrè Visual Archives.)

Julian S. Schwinger (1918-1994). (Courtesy of AIP Meggers Gallery of Nobel Laureates, AIP Emilio Segrè Visual Archives.)

this framework that Yukawa, in the mid-1930s, suggested that nuclear forces may be mediated by charged "heavy particles," the pi mesons, as we now call them.[4] The 1930s were much taken up with quantum field theories. For beta decay there were various coupling types and coupling constants to be determined; for the nuclear force problem, various meson types (scalar, vector, etc.), coupling types, and coupling constants to be considered as possibilities. But for quantum electrodynamics (QED) — the physics of electrons and photons — everything seemed well specified. An abundance of experimental information was available on a wide variety of processes, and there was a definite theory to be tested. The theory came off well, as long as one computed only to lowest order in the small fine structure constant ($\alpha = 1/137$) that serves as the perturbative expansion parameter for QED. The trouble was that the higher order "corrections" were plagued with infinities, a problem for quantum field theories in general.

This brings us to the early postwar years. Yukawa's pi meson was finally discovered and disentangled from the mu meson. Other new particles began to turn up as well, totally unforeseen. The Berkeley 184-in. cyclotron came into operation, and other high-energy accelerators followed. Particle physics was emerging out of atomic and nuclear physics as a distinct discipline, although those parent disciplines have continued to intermingle with it and provide it with critical testbeds (e.g., QED and the hydrogen Lamb shift, parity violation and the cobalt-60 nucleus).

At the Shelter Island Conference of June 1947, Lamb reported his findings on the level structure of hydrogen: To wit, the $2S_{1/2}$ level sits above the $2P_{1/2}$ level with a separation of about 1000 MHz.[5] A "vacuum polarization" contribution to this separation had already been worked out in the prewar years, but it is far too small (and has the wrong sign as well) to account for the observed effect.[6] As for other radiative corrections, where the infinities are encountered, the ideas of renormalization had similarly been introduced in the prewar years; but around the period of Shelter Island, techniques for sharpening those ideas and for implementing them were being developed independently by Feynman, Schwinger, Tomonaga, and others. Bethe, who was at the conference, didn't wait for all the niceties to be clarified. A few days after the meeting, he sent around a preprint on the Lamb shift. In it he captured the essence of mass renormalization in quantitative terms. Within his nonrelativistic treatment of the atomic electron, he encountered a mild (logarithmic) singularity. Undaunted, he simply introduced a cutoff at what he took to be a reasonable value. It *was* reasonable. His finding for the Lamb shift was in remarkable agreement with experiment.

The selectors for the particle theory section of this centennial volume begin with Bethe's wonderful paper.[7] Within a year or so after its publication, various technical confusions surrounding attempts to provide a more accurate treatment of the (lowest order) Lamb shift were surmounted.[8] Similarly, computation of the lowest order correction to the electron magnetic moment was ac-

Charles G. Darwin (1887-1962), Llewellyn H. Thomas (1903-1992), and **Gregory Breit (1899-1981).** (Courtesy of Goudsmit Collection, AIP Emilio Segrè Visual Archives.)

complished.[9] Dyson could soon establish the equivalence of the Feynman and Schwinger methods, and he could also demonstrate well enough that QED is renormalizable to all orders (refined proofs by others were produced over time).[10] Feynman's techniques — Feynman diagrams — took over rather quickly.[11] QED settled in, and in its higher order usages it stands today as the most quantitatively accurate of all our theories of nature.

Once the pions came along, theorists set out to master the strong interactions with Yukawa field theory, just as they had mastered the electromagnetic interactions with QED. Abysmal failure, on every front! The coupling constant was simply too big for a perturbative approach. Perturbation theory was not going to reveal the Δ resonance, for example, or any of the other pi-nucleon and multipion resonances that came along later on. Therefore, who knows, maybe Yukawa theory was wrong, even when gener-

Igor Tamm (1895-1971), Freeman Dyson (1923-), Rudolf E. Peierls (1907-), and **V. L. Ginsburg (1916-)** in Moscow, 1956. (Courtesy of Rudolf Peierls Collection, AIP Emilio Segrè Visual Archives.)

Gian Carlo Wick (1909-1992). (Courtesy of AIP Emilio Segrè Visual Archives.)

alized to accommodate the newly discovered strange particles and resonances. In time the distinction between "particles" and "resonances" began to fade, although it is hard to put a starting date to this. By the early 1960s, certainly, as the ρ, ω, and η mesons were being discovered in quick succession, they were generally regarded from the start as being no less particle-like than the pi mesons. Indeed, with the growing proliferation of particles/resonances, it became very difficult to believe that they could all be "fundamental" in any meaningful sense of that word. It was also hard to find reasons why any particular subset of hadrons should be regarded as more fundamental than the rest. The quark hypothesis, when introduced in 1964, fell on fertile ground.[12]

The failures of perturbative Yukawa theory, already well evident by the early 1950s, produced only limited despair. There was so much happening on the experimental front and so much phenomenology to be rummaged amidst in search of partial theoretical insights that transcend the details of the underlying quantum dynamics. So much in fact that in a short review of this sort, one can scarcely go beyond the most highly abbreviated sampling of topics. One area that the selectors have bypassed, an area that we can mention only fleetingly here, is analytic S-matrix theory. It was a very prominent topic for a decade or more starting in the mid 1950s. Broadly speaking, it seeks to extract all that can be extracted from certain general principles, notably, unitarity, causality (which has implications for analyticity), and crossing symmetry. One of the solid achievements was the forward-scattering dispersion relation for pion-nucleon scattering. This was first written down, persuasively but somewhat heuristically, by Goldberger,[13] and then derived more rigorously later on.[14] In time, additional but more speculative properties were attributed to the S matrix. Some were at least provable in the context of nonrelativistic potential scattering: notably, the double dispersions relation of Mandelstam[15] and Regge poles.[16] A considerable phenomenology got built up around Regge poles. But we must now leave this subject.

With the arrival of the pions, isotopic spin symmetry came into its own as a successful symmetry principle of the strong interactions. Concerning the newly discovered particles that we now recognize as carrying strangeness, in the early 1950s the first thing to be understood was why they decay so slowly while being produced so copiously in nucleon-nucleon collisions. The correct direction was pointed out by Pais, who introduced the notion of associated production.[17] Subsequently, Gell-Mann introduced a selection rule based on a new quantum number, strangeness, assumed to be conserved in the strong and electromagnetic interactions but violated in the weak ones.[18] It was not until some years later that this quantum number got incorporated along with isospin into a global symmetry of the strong interactions, SU(3).[19] This culminated a long search aimed at organizing the growing array of particles into multiplets of some higher symmetry beyond isospin.

It is possible to have personal favorites among the hordes of hadrons. Many will choose the K mesons, in particular the neutral K meson complex composed of K^0 and its distinct antiparticle $\bar{K}^0$. The one has strangeness quantum number S = 1, the other S = -1. They cannot interconnect under the strong and electromagnetic interactions, but they *can* do so through the strangeness-violating weak interactions. As pointed out by Gell-Mann and Pais,[20] this means that the entities with definite lifetimes and masses are not K^0 and $\bar{K}^0$ but rather the linear combinations $K_1^0 = (K^0 + \bar{K}^0) / \sqrt{2}$ and $K_2^0 = (K^0 - \bar{K}^0) / \sqrt{2}$. All kinds of interesting interference effects in decay and regeneration follow, as does forbiddenness of K_2^0 decay into two pions—assuming, as one did in that era, that C invariance holds, though CP invariance in fact suffices to forbid this decay. But the K_2^0 *does* decay into two pions, with small but nonvanishing branching ratio, as was discovered in the celebrated experiment of Christenson, Cronin, Fitch, and Turlay.[21] Through the CPT theorem, this implies the breakdown of time-reversal invariance. The experiment took place some years after the famous parity revolution of the 1950s. That revolution was initiated by mounting indications that the charged K meson can decay in both a two-pion mode ("θ mode") and a three-pion mode ("τ mode"). Initially, one could imagine that the parent particles corresponding to these two modes are distinct, but the accumulating evidence suggested that they were not distinct—except for their opposite parities. This led to an investigation by Lee and Yang, who showed, to the surprise and consternation of many, that parity invariance had never really been tested experimentally in weak processes and who also proposed the kinds of experimental tests that might be most critical.[22] This was followed shortly by the discovery of parity violation in Co^{60} beta decay[23] and in pion and muon decays,[24] then elsewhere in the weak interactions. On the leptonic side, the effects were essentially maxi-

Murray Gell-Mann (1929-), Lev Landau (1908-1968), and **Robert E. Marshak (1916-1992)** at the 1956 Moscow Conference. (Courtesy of Marshak Collection, AIP Emilio Segrè Visual Archives.)

Marvin L. Goldberger (1922–). (Courtesy of AIP Emilio Segrè Visual Archives.)

mal, pointing to a two-component theory of the neutrino.[25] Confusion about the coupling types in beta decay were also resolved in this period.

What emerged was a picture in which the weak interactions have a current-current structure, with charge-changing currents that contain both hadronic and leptonic parts, each transforming like vector minus axial vector (V – A). It was proposed, moreover, that the (strangeness-conserving) hadronic vector current belongs to the same multiplet as the isovector part of the electromagnetic current.[26] This hypothesis—subsequently confirmed—tied together two hitherto disparate branches of particle physics. All of these things were in time generalized to include strangeness-changing hadronic currents. In 1964 an ingenious set of commutation relations involving the V and A parts of the hadronic current was proposed by Gell-Mann on the basis of a model in which the currents are constructed out of quark fields.[27] Here was a bit of quantum field theory slipping back in, although one was entitled to adopt the commutation relations and throw away the quark model that suggested them. This "algebra of currents" achieved a number of successes, especially when combined with the so-called partially conserved axial current (PCAC) hypothesis—the notion that the divergence of the axial vector current acts (in a certain effective sense) like the field that creates and destroys pions.[28] The PCAC hypothesis provided an alternate route to a theoretical connection, remarkably well supported by experiment, that was stumbled upon years earlier through some dubious dispersion-theoretic reasoning: namely, a formula connecting the strong pi-nucleon coupling constant and the effective axial vector coupling constant of neutron beta decay.[29] As the 1960s moved on, the quarks were being taken more and more seriously as the building blocks of matter. Certain difficulties that had been encountered could be dealt with if one was prepared to suppose that each quark "flavor" (up, down, strange) is in fact a trio of quarks (distinguished terminologically by one's favorite set of three "colors"), corresponding to a triplet under a new "color" SU(3).[30]

In the middle of the 1960s a series of experiments got underway at SLAC on inelastic scattering of very-high-energy electrons off nucleons ("deep inelastic electron scattering"): e + nucleon →e + hadrons. One expected that the cross section would fall off precipitously at large angles, or better put, at large electron momentum transfers. It turned out otherwise, quite momentously.[31] The cross section fell off slowly; indeed, to a good approximation it displayed an interesting "scaling" behavior—namely, the form factors that characterize the cross section turned out to depend essentially only on the ratio of momentum transfer to energy transfer. The unexpected frequency of large-angle scatterings was not unlike the surprise experienced by Rutherford when he detected large-angle events in the scattering of alpha particles off an atom. That finding was indicative of hard scattering off an essentially point-like ingredient (the nucleus) within the atom. Similarly, the SLAC results soon led to the "parton" model, developed by

Feynman[32] and Bjorken,[33] in which the nucleon is pictured as being made up of small point-like ingredients that act independently in their scattering interactions with the electron. It was natural to identify the partons with quarks. The quarks were becoming ever more real.

In our terse review of some of the major advances of the 1950s and 1960s, we have concentrated so far on developments that do not rely much (or at all) on possession of the correct underlying field theory. All the while, however, theoretical investigations were going forward in various directions on the field theory front. With hindsight, one can now look back and identify those lines that turned out to be on the main track toward the great syntheses achieved in the late 1960s and early 1970s: electroweak unification and quantum chromodynamics (QCD)—jointly, the Standard Model. Fundamental to both are the ideas of non-Abelian gauge theory, developed in the 1954 papers of Yang and Mills.[34] In gauge theory, global symmetries are replaced by local symmetries. QED is an example of an Abelian gauge theory. Here the gauge boson, the photon, couples to charged particles, but the photon itself does not carry charge. In non-Abelian gauge theories, the gauge bosons themselves carry the conserved "charges." Another critical advance in the preunification era had to do with the idea of broken symmetry. On this, work done by Higgs was especially well suited for the later developments.[35] Here one deals with situations in which the dynamics is gauge invariant but where there is a continuum of degenerate vacuums. Selection of a particular vacuum breaks the symmetry and introduces the possibility of giving mass to the gauge particles and to other particles of the theory; the unbroken symmetry would have required masslessness. Also crucial for the Standard Model were developments that had taken place in the theory of the renormalization group. What this subject addresses in the field theory context is the connection between renormalizability and scale changes. The Callan-Symanzik formulation was especially well suited for what was to come.[36]

In 1967 Weinberg[37] and Salam[38] independently proposed to unify the weak and electromagnetic interactions in the framework of an $SU(2) \times U(1)$ non-Abelian gauge theory, with the massless photon, a massive charged pair of bosons W^+ and W^-, and a massive neutral boson Z^0 as the vector gauge particles. The Higgs broken-symmetry mechanism was invoked to generate masses. In 1971 't Hooft importantly proved that the theory is renormalizable;[39] at about that same time Weinberg took up the postulate of a fourth quark ("charm") and incorporated it with the other quarks into the model (which hitherto had treated only the leptons).[40] The charm quark idea had had a considerable history of its own, but it received an especially critical boost in the 1970 paper of Glashow, Iliopoulos, and Maiani,[41] who pointed to a number of its virtues. Taken altogether, this constellation of happenings coalesced into a serious model of electroweak unification. It absorbed the phenomenology that had gone before, and

Richard P. Feynman (1918–1988). (Courtesy of *Physics Today* Collection, AIP Emilio Segrè Visual Archives.)

it famously predicted the existence of a new quantum number, charm, as well as the existence of weak neutral current phenomena such as $\nu + p \rightarrow \nu$ + hadrons, $\nu + e \rightarrow \nu + e$, etc.

Neutral current events were discovered in 1973.[42] "Hidden charm" was revealed via the discovery of the J/Ψ in the November Revolution of 1974;[43] "naked charm" was detected in 1976.[44] A new "family" announced itself in 1975 with the discovery of the τ lepton.[45] The bottom-quark family member appeared in 1975.[46] The weak gauge bosons W and Z were discovered in 1983.[47] Some decade!

On the strong-interaction front, the experimental evidence for scaling and success of the quark model led to the search for field theories, if any, with the requisite property of "asymptotic freedom": roughly, the property that the strong forces become weak at short distances (large momentum transfers). In 1973 Gross and Wilczek[48] and, independently, Politzer[49] discovered that non-Abelien gauge theories have the wanted asymptotic behavior. They fit this into a color SU(3) gauge model that had already been under discussion.[30] The gauge bosons of the model — the "gluons" — form a color octet. Asymptotic freedom gave an excellent account of scaling or, better put, of the small but characteristic departures from exact scaling that subsequent experiments revealed with increasing precision. In later years hadron "jets" unmistakably associated with quarks and gluons were detected in high-energy nucleon-nucleon collisions.

The Standard Model has met with one success after another. The selectors for this centennial volume have carried their samplings only through the birth of the model. They have not gone beyond. One can't cover everything, and that's not a bad place to stop [for a splendid account of the whole century, extending to the middle of 1980s, see Pais, *Inward Bound* (Oxford University, New York, 1986)]. But of course, theoretical work has by no means

Hideki Yukawa (1907–1981) and Abraham Pais (1918–). (Courtesy of University of Rochester, AIP Emilio Segrè Visual Archives.)

come to an end. It continues within the Standard Model with the goal of extracting its phenomenological consequences, something that remains difficult wherever the strong interactions are strong (i.e., nonasymptotic). Even more intense are the efforts to go beyond the Standard Model. There is plenty of theoretical motivation for this. Although very likely correct over a very considerable range of particle phenomena, the Standard Model is almost certainly incomplete; there are simply too many free parameters, unification surely has not gone far enough, etc. Indeed, attempts to extend unification to include the strong interactions got underway soon after the birth of electroweak unification, and a vast literature has grown up around this. There are various exciting but still speculative predictions: proton instability; neutrino masses and neutrino oscillation; the existence of new particles—in particular, supersymmetric partners to known particles, reflecting a kind of symmetry between fermions and bosons; etc.

But one can't cover everything in the limited space available. In particular, there is no space here for the "theory of everything," an immodest-sounding title but one that is fair enough if appropriately understood: The reference is to string theory, an attempt to incorporate gravity along with the other forces in a grand quantum-mechanical theory. String theory, whose roots go back to analytic S-matrix theory, has been under very intense study and development for a number of years. When it comes time for the next centennial volume (but very, very much sooner, one hopes), it may be the centerpiece.

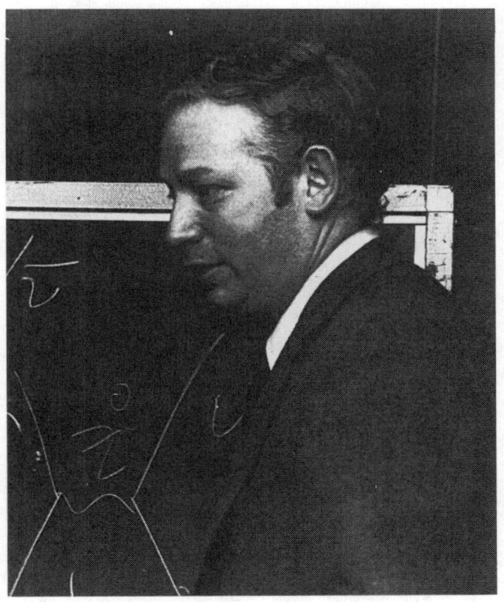

Steven Weinberg (1933–). (Courtesy of AIP Emilio Segrè Visual Archives.)

REFERENCES

1. P. A .M. Dirac, *The Quantum Theory of the Electron*, Proc. R. Soc. London Ser. A **117**, 610–624 (1928).
2. P. A. M. Dirac, *A Theory of Electrons and Protons*, Proc. R. Soc. London Ser. A **126**, 360–365 (1928).
3. E. Fermi, *Theory of β-Rays*, Nuovo Cimento **11**, 1–19 (1934); *Versuch einer Theorie der β-Strahlen. I.*, Z. Phys. **88**, 161–177 (1934).
4. H. Yukawa, *Interaction of Elementary Particles*, Proc. Phys. Math Soc. J. **17**, 48–57 (1935).
5. W. E. Lamb and R. C. Retherford, *Fine Structure of the Hydrogen Atom by a Microwave Method*, Phys. Rev. **72**, 241–243 (1947).
6. E. Uehling, *Polarization Effects in the Positron Theory*, Phys. Rev. **48**, 55–63 (1935).
7. H. A. Bethe, *The Electromagnetic Shift of Energy Levels*, Phys. Rev. **72**, 339–341 (1947).
8. H. Fukuda, Y. Miyamoto, and S. Tomonaga, *A Self-Consistent Subtraction Method in the Quantum Field Theory. II.*, Prog. Theor. Phys. **4**, 47–59 (1949); N. M. Kroll and W. E. Lamb, *On the Self-Energy of a Bound Electron*, Phys. Rev. **75**, 388–398 (1949); J. Schwinger, *On Radiative Corrections to Electron Scattering*, *ibid.* **75**, 898–899(L) (1949); R. P. Feynman, *Space-Time Approach to Quantum Electrodynamics*, *ibid.* **76**, 769–789 (1949).
9. J. Schwinger, *On Quantum Electrodynamics and the Magnetic Moment of the Electron*, Phys. Rev. **73**, 416–417(L) (1948).
10. F. J. Dyson, *The Radiative Theories of Tomonaga, Schwinger and Feynman*, Phys. Rev. **75**, 486–502 (1949).
11. R. P. Feynman, *The Theory of Positrons*, Phys. Rev. **76**, 749–759 (1949).

Abdus Salam (1926–). (Courtesy of Weber Collection, AIP Emilio Segrè Visual Archives.)

12. M. Gell-Mann, *A Schematic Model of Baryons and Mesons*, Phys. Lett. **8**, 214–215 (1964); G. Zweig [CERN preprint 8182/Th 401 (1964) (unpublished)].

13. M. L. Goldberger, *Causality Conditions and Dispersion Relations. I. Boson Fields*, Phys. Rev. **99**, 979–985 (1955); M. L. Goldberger, H. Miyazawa, and R. Oehme, *Application of Dispersion Relations to Pion-Nucleon Scattering, ibid.* **99**, 986–988 (1955).

14. H. Lehmann, *Scattering Matrix and Field Operators*, Nuovo Cimento Suppl. **14**, 153–176 (1959).

15. S. Mandelstam, *Determination of the Pion-Nucleon Scattering Amplitude from Dispersion Relations and Unitarity. General Theory*, Phys. Rev. **112**, 1344–1360 (1958); for potential scattering, see R. Blankenbecler, M. L. Goldberger, N. N. Khuri, and S. B. Treiman, *Mandelstam Representations for Potential Scattering*, Ann. Phys. **10**, 62–93 (1960).

16. T. Regge, *Introduction to Complex Orbital Momenta*, Nuovo Cimento **14**, 951–976 (1959); *Bound States, Shadow States and Mandelstam Representation, ibid.* **18**, 947–956 (1960).

17. A. Pais, *Some Remarks on the V Particles*, Phys. Rev. **86**, 663–672 (1952).

18. M. Gell-Mann, *Isotopic Spin and New Unstable Particles*, Phys. Rev. **92**, 833–834(L) (1953).

19. See M. Gell-Mann and Y. Ne'eman, *The Eightfold Way* (Benjamin, New York, 1964).

20. M. Gell-Mann and A. Pais, *Behavior of Neutral Particles under Charge Conjugation*, Phys. Rev. **97**, 1387–1389 (1953).

21. J. Christenson, J. W. Cronin, V. L. Fitch, and R. Turlay, *Evidence for the 2π Decay of the K Meson*, Phys. Rev. Lett. **13**, 138–140 (1964).

22. T. D. Lee and C. N. Yang, *Question of Parity Conservation in Weak Interactions*, Phys. Rev. **104**, 254–258 (1956).

23. C. S. Wu, E. Ambler, R. W. Hayward, D. D. Hoppes, and R. P. Hudson, *Experimental Test of Parity Conservation in Beta Decay*, Phys. Rev. **105**, 1413–1415(L) (1957).

24. R. L. Garwin, L. M. Lederman, and M. Weinrich, *Observation of the Failure of Conservation of Parity and Charge Conjugation in Meson Decays: The Magnetic Moment of the Free Muon*, Phys. Rev. **105**, 1415–1417 (1957); J. I. Friedman and V. L. Telegdi, *Nuclear Emulsion Evidence for Parity Nonconservation in the Decay Chain π^+-μ^+-e^+*, *ibid.* **105**, 1681–1682(L) (1957).

25. T. D. Lee and C. N. Yang, *Parity Nonconservation and a Two-Component Theory of the Neutrino*, Phys. Rev. **105**, 1671–1675 (1957).

26. M. Gell-Mann and R. P. Feynman, *Theory of the Fermi Interaction*, Phys. Rev. **109**, 193–198 (1958).

27. M. Gell-Mann, *The Symmetry Group of Vector and Axial Vector Currents*, Physics **1**, 63–75 (1964).

28. For a review, see S. B. Treiman, R. Jackiw, and D. J. Gross, *Lectures on Current Algebra and Its Applications* (Princeton University Press, Princeton, 1972).

29. M. L. Goldberger and S. B. Treiman, *Decay of the Pi Meson*, Phys. Rev. **110**, 1178–1184 (1958).

30. M. Y. Han and Y. Nambu, *Three-Triplet Model with Double SU(3) Symmetry*, Phys. Rev. **139**, B1006–B1010 (1965).

31. E. D. Bloom *et al.*, *High-Energy Inelastic e-p Scattering at* 6° and 10°, Phys. Rev. Lett. **23**, 930–934 (1969); M. Breidenbach *et al.*, *Observed Behavior of Highly Inelastic Electron-Proton Scattering, ibid.* **23**, 935–939 (1969).

32. R. P. Feynman, *Very High-Energy Collisions of Hadrons*, Phys. Rev. Lett. **23**, 1415–1417 (1969).

33. J. D. Bjorken, *Asymptotic Sum Rules at Infinite Momentum*, Phys. Rev. **179**, 1547–1553 (1969).

34. C. N. Yang and R. L. Mills, *Isotopic Spin Conservation and a Generalized Gauge Invariance*, Phys. Rev. **95**, 631(A) (1954); *Conservation of Isotopic Spin and Isotopic Gauge Invariance, ibid.* **96**, 191–195 (1954).

35. P. W. Higgs, *Broken Symmetries, Massless Particles and Gauge Fields*, Phys. Lett. **12**, 132–133 (1964); *Spontaneous Symmetry Breakdown without Massless Bosons*, Phys. Rev. **145**, 1156–1163 (1966).

36. C. G. Callan, *Broken Scale Invariance in Scalar Field Theory*, Phys. Rev. D **2**, 1541–1547 (1970); K. Symanzik, *Small Distance Behavior in Field Theory and Power Counting*, Commun. Math, Phys. **18**, 227–246 (1970).

37. S. Weinberg, *A Model of Leptons*, Phys. Rev. Lett. **19**, 1264–1266 (1967).

38. A Salam, in *Elementary Particle Theory*, edited by N. Svarthholm (Almqvist and Wiskell, Stockholm, 1968), p. 1367.

39. G. 't Hooft, *Renormalizable Lagrangians for Massive Yang-Mills Fields*, Nucl. Phys. **35B**, 167–188 (1971).

40. S. Weinberg, *Physical Processes in a Convergent Theory of the Weak and Electromagnetic Interactions*, Phys. Rev. Lett. **27**, 1688–1691 (1971); *Effects of a Neutral Intermediate Boson in Semileptonic Processes, ibid.* **5**, 1412–1417 (1972).

41. S. L. Glashow, J. Iliopoulos, and L. Maiani, *Weak Interactions with Lepton-Hadron Symmetry*, Phys. Rev. D **2**, 1285–1292 (1970).

42. F. J. Hasert *et al.*, *Observation of Neutrino-Like Interactions without Muon or Electron in the Gargamelle Neutrino Experiment*, Phys. Lett. B **46**, 138–140 (1973); A. Benvenuti *et al.*, *Observation of Muonless Neutrino-Induced Inelastic Interactions*, Phys. Rev. Lett. **32**, 800–803 (1974); B. Aubert *et al.*, *Further Observation of Muonless Neutrino-Induced Inelastic Interactions, ibid.* **32**, 1454–1457 (1974).

43. J. J. Aubert *et al.*, *Experimental Observation of a Heavy Particle J*, Phys. Rev. Lett. **33**, 1404–1406 (1974); J. E. Augustin *et al.*, *Discovery of a Narrow Resonance in $e^+ e^-$ Annihilation, ibid.* **33**, 1406–1408 (1974).

44. G. Goldhaber *et al.*, *Observation in e^+e^- Annihilation of a Narrow State at 1865 MeV/c^2 Decaying to $K\pi$ and $K\pi\pi\pi$*, Phys. Rev. Lett. **37**, 255–259 (1976); I. Peruzzi *et al.*, *Observation of a Narrow Charged State at 1876 MeV/c^2 Decaying to an Exotic Combination of $K\pi\pi$, ibid.* **37**, 569–571 (1976).

45. M. L. Perl *et al.*, *Evidence for Anomalous Lepton Production in e^+-e^- Annihilation*, Phys. Rev. Lett. **35**, 1489–1492 (1975).

46. S. W. Herb *et al.*, *Observation of a Dimuon Resonance at 9.5 GeV in 400-GeV Proton Nucleus Collisions*, Phys. Rev. Lett. **39**, 252–255 (1977).

47. G. Arnison *et al.*, *Experimental Observation of Isolated Large Transverse Energy Electrons with Associated Missing Energy at s = 540 GeV*, Phys. Lett. B **122**, 103–116 (1983); M. Banner *et al.*, *Observation of Single Isolated Electrons of High Transverse Momentum in Events with Missing Transverse Energy at the CERN $\bar{p}p$ Collider, ibid.* **122**, 476–485 (1983); G. Arnison *et al.*, *Experimental Observation of Lepton Pairs of Invariant Mass around 95 GeV/c^2 at the CERN SPS Collider, ibid.* **126**, 398–410 (1983); P. Bagnaia *et al.*, *Evidence for $Z^0 \rightarrow e^+e^-$ at the CERN $\bar{p}p$ Collider, ibid.* **129**, 130–140 (1983).

48. D. J. Gross and F. Wilczek, *Ultraviolet Behavior of Non-Abelian Gauge Theories*, Phys. Rev. Lett. **30**, 1343–1346 (1973); *Asymptotically Free Gauge Theories. I.*, Phys. Rev. D **8**, 3633–3652 (1973).

49. H. D. Politzer, *Reliable Perturbative Results for Strong Interactions?*, Phys. Rev. Lett. **30**, 1346–1349 (1973).

Sheldon L. Glashow (1932–). (Courtesy of Harvard University, AIP Emilio Segrè Visual Archives.)

PARTICLE THEORY

Papers Reproduced on CD-ROM

J. R. Oppenheimer. Note on the theory of the interaction of field and matter, *Phys. Rev.* **35**, 461–477 (1930)

E. C. G. Stueckelberg. On the existence of heavy electrons, *Phys. Rev.* **52**, 41–42(L) (1937)

F. Bloch and A. Nordsieck. Note on the radiation field of the electron, *Phys. Rev.* **52**, 54–59 (1937)

V. F. Weisskopf. On the self-energy and the electromagnetic field of the electron, *Phys. Rev.* **56**, 72–85 (1939)

W. Rarita and J. Schwinger. On a theory of particles with half-integral spin, *Phys. Rev.* **60**, 61(L) (1941)

H. A. Bethe. The electromagnetic shift of energy levels, *Phys. Rev.* **72**, 339–341 (1947)

R. E. Marshak and H. A. Bethe. On the two-meson hypothesis, *Phys. Rev.* **72**, 506–509 (1947)

G. F. Chew and M. L. Goldberger. High energy neutron–proton scattering, *Phys. Rev.* **73**, 1409(L) (1948)

S-I. Tomonaga. On infinite field reactions in quantum field theory, *Phys. Rev.* **74**, 224–225(L) (1948)

P. A. M. Dirac. The theory of magnetic poles, *Phys. Rev.* **74**, 817–830 (1948)

J. Schwinger. Quantum electrodynamics. I. A covariant formulation, *Phys. Rev.* **74**, 1439–1461 (1948)

F. J. Dyson. The radiation theories of Tomonaga, Schwinger, and Feynman, *Phys. Rev.* **75**, 486–502 (1949)

J. Schwinger. Quantum electrodynamics. II. Vacuum polarization and self-energy, *Phys. Rev.* **75**, 651–679 (1949)

J. Schwinger. On radiative corrections to electron scattering, *Phys. Rev.* **75**, 898–899(L) (1949)

F. J. Dyson. The S matrix in quantum electrodynamics, *Phys. Rev.* **75**, 1736–1755 (1949)

R. P. Feynman. The theory of positrons, *Phys. Rev.* **76**, 749–759 (1949)

R. P. Feynman. Space-time approach to quantum electrodynamics, *Phys. Rev.* **76**, 769–789 (1949)

J. Schwinger. Quantum electrodynamics. III. The electromagnetic properties of the electron—radiative corrections to scattering, *Phys. Rev.* **76**, 790–817 (1949)

J. C. Ward. The scattering of light by light, *Phys. Rev.* **77**, 293(L) (1950)

J. C. Ward. An identity in quantum electrodynamics, *Phys. Rev.* **78**, 182(L) (1950)

E. M. Purcell and N. F. Ramsey. On the possibility of electric dipole moments for elementary particles and nuclei, *Phys. Rev.* **78**, 807(L) (1950)

G. C. Wick. The evaluation of the collision matrix, *Phys. Rev.* **80**, 268–272 (1950)

R. P. Feynman. Mathematical formulation of the quantum theory of electromagnetic interaction, *Phys. Rev.* **80**, 440–457 (1950)

H. A. Bethe and E. E. Salpeter. A relativistic equation for bound state problems, *Phys. Rev.* **82**, 309–310(A) (1951)

J. Schwinger. On gauge invariance and vacuum polarization, *Phys. Rev.* **82**, 664–679 (1951)

M. Gell-Mann and F. Low. Bound states in quantum field theory, *Phys. Rev.* **84**, 350–354 (1951)

A. Pais. Some remarks on the V-particles, *Phys. Rev.* **86**, 663–672 (1952)

M. Gell-Mann. Isotopic spin and new unstable particles, *Phys. Rev.* **92**, 833–834(L) (1953)

R. H. Dalitz. Decay of τ mesons of known charge, *Phys. Rev.* **94**, 1046–1051 (1954)

M. Gell-Mann and F. E. Low. Quantum electrodynamics at small distances, *Phys. Rev.* **95**, 1300–1312 (1954)

C. N. Yang and R. L. Mills. Conservation of isotopic spin and isotopic gauge invariance, *Phys. Rev.* **96**, 191–195 (1954)

M. Gell-Mann and A. Pais. Behavior of neutral particles under charge conjugation, *Phys. Rev.* **97**, 1387–1389 (1955)

F. E. Low. Boson–Fermion scattering in the Heisenberg representation, *Phys. Rev.* **97**, 1392–1398 (1955)

A. Pais and O. Piccioni. Note on the decay and absorption of the θ^0, *Phys. Rev.* **100**, 1487–1489 (1955)

R. E. Behrends, R. J. Finkelstein, and A. Sirlin. Radiative corrections to decay processes, *Phys. Rev.* **101**, 866–873 (1956)

T. D. Lee and C. N. Yang. Mass degeneracy of the heavy mesons, *Phys. Rev.* **102**, 290–291 (1956)

T. D. Lee and C. N. Yang. Question of parity conservation in weak interactions, *Phys. Rev.* **104**, 254–258 (1956)

T. D. Lee and C. N. Yang. Parity nonconservation and a two-component theory of the neutrino, *Phys. Rev.* **105**, 1671–1675 (1957)

R. P. Feynman and M. Gell-Mann. Theory of Fermi interaction, *Phys. Rev.* **109**, 193–198 (1958)

N. F. Ramsey. Time reversal, charge conjugation, magnetic pole conjugation, and parity, *Phys. Rev.* **109**, 225–226(L) (1958)

E. C. G. Sudarshan and R. E. Marshak. Chirality invariance and the universal Fermi interaction, *Phys. Rev.* **109**, 1860–1862(L) (1958)

M. L. Goldberger and S. B. Treiman. Decay of the pi meson, *Phys. Rev.* **110**, 1178–1184 (1958)

J. Schwinger. Field theory commutators, *Phys. Rev. Lett.* **3**, 296–297 (1959)

T. D. Lee and C. D. Yang. Implications of the intermediate boson basis of the weak interactions: Existence of a quartet of intermediate bosons and their dual isotopic spin transformation properties, *Phys. Rev.* **119**, 1410–1419 (1960)

PARTICLE THEORY

Y. Nambu. Axial vector current conservation in weak interactions, *Phys. Rev. Lett.* **4**, 380–382 (1960)

Y. Nambu and G. Jona-Lasinio. Dynamical model of elementary particles based on an analogy with superconductivity. I., *Phys. Rev.* **122**, 345–358 (1961)

M. Gell-Mann. Symmetries of baryons and mesons, *Phys. Rev.* **125**, 1067–1084 (1962)

J. Goldstone, A. Salam, and S. Weinberg. Broken symmetries, *Phys. Rev.* **127**, 965–970 (1962)

J. Schwinger. Gauge invariance and mass. II., *Phys. Rev.* **128**, 2425–2429 (1962)

L. Rosenberg. Electromagnetic interactions of neutrinos, *Phys. Rev.* **129**, 2786–2788 (1963)

N. Cabibbo. Unitary symmetry and leptonic decays, *Phys. Rev. Lett.* **10**, 531–533 (1963)

T. D. Lee and M. Nauenberg. Degenerate systems and mass singularities, *Phys. Rev.* **133**, B1549–B1562 (1964)

N. Cabibbo. Unitary symmetry and nonleptonic decays, *Phys. Rev. Lett.* **12**, 62–63 (1964)

F. Englert and R. Brout. Broken symmetry and the mass of gauge vector mesons, *Phys. Rev. Lett.* **13**, 321–323 (1964)

P. W. Higgs. Broken symmetries and the masses of gauge bosons, *Phys. Rev. Lett.* **13**, 508–509 (1964)

O. W. Greenberg. Spin and unitary-spin independence in a paraquark model of baryons and mesons, *Phys. Rev. Lett.* **13**, 598–602 (1964)

M. Y. Han and Y. Nambu. Three-triplet model with double *SU*(3) symmetry, *Phys. Rev.* **139**, B1006–B1010 (1965)

W. I. Weisberger. Renormalization of the weak axial-vector coupling constant, *Phys. Rev. Lett.* **14**, 1047–1051 (1965)

S. L. Adler. Calculation of the axial-vector coupling constant renormalization in β decay, *Phys. Rev. Lett.* **14**, 1051–1055 (1965)

S. L. Adler. Sum rules giving tests of local current commutation relations in high-energy neutrino reactions, *Phys. Rev.* **143**, 1144–1155 (1966)

P. W. Higgs. Spontaneous symmetry breakdown without massless bosons, *Phys. Rev.* **145**, 1156–1163 (1966)

J. D. Bjorken. Inequality for electron and muon scattering from nucleons, *Phys. Rev. Lett.* **16**, 408 (1966)

S. Weinberg. Pion scattering lengths, *Phys. Rev. Lett.* **17**, 616–621 (1966)

S. Coleman and J. Mandula. All possible symmetries of the *S* matrix, *Phys. Rev.* **159**, 1251–1256 (1967)

S. Weinberg. Precise relations between the spectra of vector and axial-vector mesons, *Phys. Rev. Lett.* **18**, 507–509 (1967)

S. Weinberg. A model of leptons, *Phys. Rev. Lett.* **19**, 1264–1266 (1967)

J. Glimm and A. Jaffe. A $\lambda\varphi^4$ quantum field theory without cutoffs. I., *Phys. Rev.* **176**, 1945–1951 (1968)

S. L. Adler. Axial–vector vertex in spinor electrodynamics, *Phys. Rev.* **177**, 2426–2438 (1969)

K. G. Wilson. Non-Lagrangian models of current algebra, *Phys. Rev.* **179**, 1499–1512 (1969)

J. D. Bjorken. Asymptotic sum rules at infinite momentum, *Phys. Rev.* **179**, 1547–1553 (1969)

W. A. Bardeen. Anomalous Ward identities in spinor field theories, *Phys. Rev.* **184**, 1848–1857 (1969)

J. D. Bjorken and E. A. Paschos. Inelastic electron–proton and γ-proton scattering and the structure of the nucleon, *Phys. Rev.* **185**, 1975–1982 (1969)

R. P. Feynman. Very high-energy collisions of hadrons, *Phys. Rev. Lett.* **23**, 1415–1417 (1969)

S. L. Glashow, J. Iliopoulos, and L. Maiani. Weak interactions with lepton–hadron symmetry, *Phys. Rev. D* **2**, 1285–1292 (1970)

C. G. Callan, Jr. Broken scale invariance in scalar field theory, *Phys. Rev. D* **2**, 1541–1547 (1970)

S. D. Drell and T-M Yan. Massive lepton-pair production in hadron–hadron collisions at high energies, *Phys. Rev. Lett.* **25**, 316–320 (1970)

S. J. Brodsky, T. Kinoshita, and H. Terazawa. Dominant colliding-beam cross sections at high energies, *Phys. Rev. Lett.* **25**, 972–975 (1970)

M. G. G. Laidlaw and C. Morette DeWitt. Feynman functional integrals for systems of indistinguishable particles, *Phys. Rev. D* **3**, 1375–1378 (1971)

S. Weinberg. Effects of a neutral intermediate boson in semileptonic processes, *Phys. Rev. D* **5**, 1412–1417 (1972)

T. Kinoshita and P. Cvitanovic. Sixth-order radiative corrections to the electron magnetic moment, *Phys. Rev. Lett.* **29**, 1534–1537 (1972)

S. Coleman and E. Weinberg. Radiative corrections as the origin of spontaneous symmetry breaking, *Phys. Rev. D* **7**, 1888–1910 (1973)

D. J. Gross and F. Wilczek. Ultraviolet behavior of non-Abelian gauge theories, *Phys. Rev. Lett.* **30**, 1343–1346 (1973)

H. D. Politzer. Reliable perturbative results for strong interactions?, *Phys. Rev. Lett.* **30**, 1346–1349 (1973)

S. J. Brodsky and G. R. Farrar. Scaling laws at large transverse momentum, *Phys. Rev. Lett.* **31**, 1153–1156 (1973)

D. J. Gross and F. Wilczek. Asymptotically free gauge theories. II., *Phys. Rev. D* **9**, 980–993 (1974)

L. Dolan and R. Jackiw. Symmetry behavior at finite temperature, *Phys. Rev. D* **9**, 3320–3341 (1974)

A. Chodos, R. L. Jaffe, K. Johnson, C. B. Thorn, and V. F. Weisskopf. New extended model of hadrons, *Phys. Rev. D* **9**, 3471–3495 (1974)

J. Kogut and L. Susskind. Vacuum polarization and the absence of free quarks in four dimensions, *Phys. Rev. D* **9**, 3501–3512 (1974)

K. G. Wilson. Confinement of quarks, *Phys. Rev. D* **10**, 2445–2459 (1974)

R. Balian, J. M. Drouffe, and C. Itzykson. Gauge fields on a lattice. I. General outlook, *Phys. Rev. D* **10**, 3376–3395 (1974)

H. Georgi and S. L. Glashow. Unity of all elementary-particle forces, *Phys. Rev. Lett.* **32**, 438–441 (1974)

H. Georgi, H. R. Quinn, and S. Weinberg. Hierarchy of interactions in unified gauge theories, *Phys. Rev. Lett.* **33**, 451–454 (1974)

D. Z. Freedman, P. van Nieuwenhuizen, and S. Ferrara. Progress toward a theory of supergravity, *Phys. Rev. D* **13**, 3214–3218 (1976)

R. Jackiw and C. Rebbi. Solitons with fermion number 1/2, *Phys. Rev. D* **13**, 3398–3409 (1976)

R. Jackiw and C. Rebbi. Conformal properties of a Yang–Mills pseudoparticle, *Phys. Rev. D* **14**, 517–523 (1976)

S. J. Brodsky and B. T. Chertok. Asymptotic form factors of hadrons and nuclei and the continuity of particle and nuclear dynamics, *Phys. Rev. D* **14**, 3003–3020 (1976)

R. D. Field and R. P. Feynman. Quark elastic scattering as a source of high-transverse-momentum mesons, *Phys. Rev. D* **15**, 2590–2616 (1977)

R. D. Peccei and H. R. Quinn. Constraints imposed by *CP* conservation in the presence of pseudoparticles, *Phys. Rev. D* **16**, 1791–1797 (1977)

R. D. Peccei and H. R. Quinn. *CP* conservation in the presence of pseudoparticles, *Phys. Rev. Lett.* **38**, 1440–1443 (1977)

R. P. Feynman, R. D. Field, and G. C. Fox. Quantum-chromodynamic approach for the large-transverse-momentum production of particles and jets, *Phys. Rev. D* **18**, 3320–3343 (1978)

M. Creutz, L. Jacobs, and C. Rebbi. Experiments with a gauge-invariant Ising system, *Phys. Rev. Lett.* **42**, 1390–1391 (1979)

M. Creutz. Confinement and the critical dimensionality of space-time, *Phys. Rev. Lett.* **43**, 553–556 (1979)

A. Sirlin. Radiative corrections in the $SU(2)_L \times U(1)$ theory: A simple renormalization framework, *Phys. Rev. D* **22**, 971–981 (1980)

M. Creutz. Asymptotic-freedom scales, *Phys. Rev. Lett.* **45**, 313–316 (1980)

S. Deser, R. Jackiw, and S. Templeton. Three-dimensional massive gauge theories, *Phys. Rev. Lett.* **48**, 975–978 (1982)

J. D. Bjorken. Highly relativistic nucleus–nucleus collisions: The central rapidity region, *Phys. Rev. D* **27**, 140–151 (1983)

E. J. Eichten, K. D. Lane, and M. E. Peskin. New tests for quark and lepton substructure, *Phys. Rev. Lett.* **50**, 811–814 (1983)

Werner Heisenberg (1901–1976) and **Sin-itiro Tomonaga (1906–1979).** (Courtesy of University of Isukuba, AIP Emilio Segrè Visual Archives.)

PHYSICAL REVIEW VOLUME 72, NUMBER 4 AUGUST 15, 1947

The Electromagnetic Shift of Energy Levels

H. A. Bethe
Cornell University, Ithaca, New York

(Received June 27, 1947)

By very beautiful experiments, Lamb and Retherford[1] have shown that the fine structure of the second quantum state of hydrogen does not agree with the prediction of the Dirac theory. The 2*s* level, which according to Dirac's theory should coincide with the $2p_{\frac{1}{2}}$ level, is actually higher than the latter by an amount of about 0.033 cm^{-1} or 1000 megacycles. This discrepancy had long been suspected from spectroscopic measurements.[2,3] However, so far no satisfactory theoretical explanation has been given. Kemble and Present, and Pasternack[4] have shown that the shift of the 2*s* level cannot be explained by a nuclear interaction of reasonable magnitude, and Uehling[5] has investigated the effect of the "polarization of the vacuum" in the Dirac hole theory, and has found that this effect also is much too small and has, in addition, the wrong sign.

Schwinger and Weisskopf, and Oppenheimer have suggested that a possible explanation might be the shift of energy levels by the interaction of the electron with the radiation field. This shift comes out infinite in all existing theories, and has therefore always been ignored. However, it is possible to identify the most strongly (linearly) divergent term in the level shift with an electromagnetic *mass* effect which must exist for a bound as well as for a free electron. This effect should

[1] Phys. Rev. **72**, 241 (1947).
[2] W. V. Houston, Phys. Rev. **51**, 446 (1937).
[3] R. C. Williams, Phys. Rev. **54**, 558 (1938).
[4] E. C. Kemble and R. D. Present, Phys. Rev. **44**, 1031 (1932); S. Pasternack, Phys. Rev. **54**, 1113 (1938).
[5] E. A. Uehling, Phys. Rev. **48**, 55 (1935).

properly be regarded as already included in the observed mass of the electron, and we must therefore subtract from the theoretical expression, the corresponding expression for a free electron of the same average kinetic energy. The result then diverges only logarithmically (instead of linearly) in non-relativistic theory: Accordingly, it may be expected that in the hole theory, in which the *main* term (self-energy of the electron) diverges only logarithmically, the result will be *convergent* after subtraction of the free electron expression.[6] This would set an effective upper limit of the order of mc^2 to the frequencies of light which effectively contribute to the shift of the level of a bound electron. I have not carried out the relativistic calculations, but I shall assume that such an effective relativistic limit exists.

The ordinary radiation theory gives the following result for the self-energy of an electron in a quantum state m, due to its interaction with transverse electromagnetic waves:

$$W = -(2e^2/3\pi hc^3)$$

$$\times \int_0^K kdk \sum_n |\mathbf{v}_{mn}|^2/(E_n - E_m + k), \quad (1)$$

where $k = h\omega$ is the energy of the quantum and $\mathbf{v}$ is the velocity of the electron which, in non-relativistic theory, is given by

$$\mathbf{v} = \mathbf{p}/m = (h/im)\nabla. \quad (2)$$

Relativistically, $\mathbf{v}$ should be replaced by $c\alpha$ where α is the Dirac operator. Retardation has been neglected and can actually be shown to make no substantial difference. The sum in (1) goes over all atomic states n, the integral over all quantum energies k up to some maximum K to be discussed later.

For a free electron, $\mathbf{v}$ has only diagonal elements and (1) is replaced by

$$W_0 = -(2e^2/3\pi hc^3) \int kdk\mathbf{v}^2/k. \quad (3)$$

This expression represents the change of the kinetic energy of the electron for fixed momentum, due to the fact that electromagne[tic] mass is added to the mass of the electron. T[he] electromagnetic mass is already contained in t[he] experimental electron mass; the contribution to the energy should therefore be disregard[ed]. For a bound electron, $\mathbf{v}^2$ should be replaced by [its] expectation value, $(\mathbf{v}^2)_{mm}$. But the matrix e[le]ments of $\mathbf{v}$ satisfy the sum rule

$$\sum_n |\mathbf{v}_{mn}|^2 = (\mathbf{v}^2)_{mm}.$$

Therefore the relevant part of the self-ene[rgy] becomes

$$W' = W - W_0 = +\frac{2e^2}{3\pi hc^3}$$

$$\times \int_0^K dk \sum_n \frac{|\mathbf{v}_{mn}|^2(E_n - E_m)}{E_n - E_m + k}.$$

This we shall consider as a true shift of the le[vel] due to radiation interaction.

It is convenient to integrate (5) first ove[r k]. Assuming K to be large compared with all ene[rgy] differences $E_n - E_m$ in the atom,

$$W' = \frac{2e^2}{3\pi hc^3} \sum_n |\mathbf{v}_{mn}|^2 (E_n - E_m) \ln \frac{K}{|E_n - E_m|}.$$

(If $E_n - E_m$ is negative, it is easily seen that principal value of the integral must be take[n] was done in (6).) Since we expect that relati[vistic] theory will provide a natural cut-off for frequency k, we shall assume that in (6)

$$K \approx mc^2.$$

(This does not imply the same limit in Eqs and (3).) The argument in the logarithm in (therefore very large; accordingly, it seems missible to consider the logarithm as cons[tant] (independent of n) in first approximation.

We therefore should calculate

$$A = \sum_n A_{nm} = \sum_n |\mathbf{p}_{nm}|^2 (E_n - E_m).$$

This sum is well known; it is

$$A = \sum |\mathbf{p}_{nm}|^2 (E_n - E_m)$$

$$= -h^2 \int \psi_m^* \nabla V \cdot \nabla \psi_m d\tau$$

$$= \tfrac{1}{2} h^2 \int \nabla^2 V \psi_m^2 d\tau = 2\pi h^2 e^2 Z \psi_m^2(0),$$

[6] It was first suggested by Schwinger and Weisskopf that hole theory must be used to obtain convergence in this problem.

for a nuclear charge Z. For any electron with angular momentum $l \neq 0$, the wave function vanishes at the nucleus; therefore, the sum $A=0$. For example, for the $2p$ level the negative contribution $A_{1S,2P}$ balances the positive contributions from all other transitions. For a state with $l=0$, however,

$$\psi_m^2(0)=(Z/na)^3/\pi, \tag{10}$$

where n is the principal quantum number and a is the Bohr radius.

Inserting (10) and (9) into (6) and using relations between atomic constants, we get for an S state

$$W_{ns}'=\frac{8}{3\pi}\left(\frac{e^2}{hc}\right)^3 \text{Ry}\frac{Z^4}{n^3}\ln\frac{K}{\langle E_n-E_m\rangle_{Av}}, \tag{11}$$

where Ry is the ionization energy of the ground state of hydrogen. The shift for the $2p$ state is negligible; the logarithm in (11) is replaced by a value of about -0.04. The average excitation energy $\langle E_n-E_m\rangle_{Av}$ for the $2s$ state of hydrogen has been calculated numerically[7] and found to be 17.8 Ry, an amazingly high value. Using this figure and $K=mc^2$, the logarithm has the value 7.63, and we find

$$W_{ns}'=136\ln[K/(E_n-E_m)]$$
$$=1040 \text{ megacycles}. \tag{12}$$

[7] I am indebted to Dr. Stehn and Miss Steward for the numerical calculations.

This is in excellent agreement with the observed value of 1000 megacycles.

A relativistic calculation to establish the limit K is in progress. Even without exact knowledge of K, however, the agreement is sufficiently good to give confidence in the basic theory. This shows

(1) that the level shift due to interaction with radiation is a real effect and is of finite magnitude,

(2) that the effect of the infinite electromagnetic mass of a point electron can be eliminated by proper identification of terms in the Dirac radiation theory,

(3) that an accurate experimental and theoretical investigation of the level shift may establish relativistic effects (e.g., Dirac hole theory). These effects will be of the order of unity in comparison with the logarithm in Eq. (11).

If the present theory is correct, the level shift should increase roughly as Z^4 but not quite so rapidly, because of the variation of $\langle E_n-E_m\rangle_{Av}$ in the logarithm. For example, for He^+, the shift of the $2s$ level should be about 13 times its value for hydrogen, giving 0.43 cm^{-1}, and that of the $3s$ level about 0.13 cm^{-1}. For the x-ray levels LI and LII, this effect should be superposed upon the effect of screening which it partly compensates. An accurate theoretical calculation of the screening is being undertaken to establish this point.

This paper grew out of extensive discussions at the Theoretical Physics Conference on Shelter Island, June 2 to 4, 1947. The author wishes to express his appreciation to the National Academy of Science which sponsored this stimulating conference.

PHYSICAL REVIEW VOLUME 75, NUMBER 3 FEBRUARY 1, 1949

The Radiation Theories of Tomonaga, Schwinger, and Feynman

F. J. DYSON
Institute for Advanced Study, Princeton, New Jersey

(Received October 6, 1948)

A unified development of the subject of quantum electrodynamics is outlined, embodying the main features both of the Tomonaga-Schwinger and of the Feynman radiation theory. The theory is carried to a point further than that reached by these authors, in the discussion of higher order radiative reactions and vacuum polarization phenomena. However, the theory of these higher order processes is a program rather than a definitive theory, since no general proof of the convergence of these effects is attempted.

The chief results obtained are (a) a demonstration of the equivalence of the Feynman and Schwinger theories, and (b) a considerable simplification of the procedure involved in applying the Schwinger theory to particular problems, the simplification being the greater the more complicated the problem.

I. INTRODUCTION

As a result of the recent and independent discoveries of Tomonaga,[1] Schwinger,[2] and Feynman,[3] the subject of quantum electrodynamics has made two very notable advances. On the one hand, both the foundations and the applications of the theory have been simplified by being presented in a completely relativistic way; on the other, the divergence difficulties have been at least partially overcome. In the reports so far published, emphasis has naturally been placed on the second of these advances; the magnitude of the first has been somewhat obscured by the fact that the new methods have been applied to problems which were beyond the range of the older theories, so that the simplicity of the methods was hidden by the complexity of the problems. Furthermore, the theory of Feynman differs so profoundly in its formulation from that of Tomonaga and Schwinger, and so little of it has been published, that its particular advantages have not hitherto been available to users of the other formulations. The advantages of the Feynman theory are simplicity and ease of application, while those of Tomonaga-Schwinger are generality and theoretical completeness.

The present paper aims to show how the Schwinger theory can be applied to specific problems in such a way as to incorporate the ideas of Feynman. To make the paper reasonably self-contained it is necessary to outline the foundations of the theory, following the method of Tomonaga; but this paper is not intended as a substitute for the complete account of the theory shortly to be published by Schwinger. Here the emphasis will be on the application of the theory, and the major theoretical problems of gauge-invariance and of the divergencies will not be considered in detail. The main results of the paper will be general formulas from which the radiative reactions on the motions of electrons can be calculated, treating the radiation interaction as a small perturbation, to any desired order of approximation. These formulas will be expressed in Schwinger's notation, but are in substance identical with results given previously by Feynman. The contribution of the present paper is thus intended to be twofold: first, to simplify the Schwinger theory for the benefit of those using it for calculations, and second, to demonstrate the equivalence of the various theories within their common domain of applicability.*

[1] Sin-itiro Tomonaga, Prog. Theoret. Phys. 1, 27 (1946); Koba, Tati, and Tomonaga, Prog. Theoret. Phys. 2, 101 198 (1947); S. Kanesawa and S. Tomonaga, Prog. Theoret. Phys. 3, 1, 101 (1948); S. Tomonaga, Phys. Rev. 74, 224 (1948).

[2] Julian Schwinger, Phys. Rev. 73, 416 (1948); Phys. Rev. 74, 1439 (1948). Several papers, giving a complete exposition of the theory, are in course of publication.

[3] R. P. Feynman, Rev. Mod. Phys. 20, 367 (1948); Phys. Rev. 74, 939, 1430 (1948); J. A. Wheeler and R. P. Feynman, Rev. Mod. Phys. 17, 157 (1945). These articles describe early stages in the development of Feynman's theory, little of which is yet published.

* After this paper was written, the author was shown a letter, published in Progress of Theoretical Physics 3, 205 (1948) by Z. Koba and G. Takeda. The letter is dated May 22, 1948, and briefly describes a method of treatment of radiative problems, similar to the method of this paper.

II. OUTLINE OF THEORETICAL FOUNDATIONS

Relativistic quantum mechanics is a special case of non-relativistic quantum mechanics, and it is convenient to use the usual non-relativistic terminology in order to make clear the relation between the mathematical theory and the results of physical measurements. In quantum electrodynamics the dynamical variables are the electromagnetic potentials $A_\mu(\mathbf{r})$ and the spinor electron-positron field $\psi_\alpha(\mathbf{r})$; each component of each field at each point $\mathbf{r}$ of space is a separate variable. Each dynamical variable is, in the Schrödinger representation of quantum mechanics, a time-independent operator operating on the state vector Φ of the system. The nature of Φ (wave function or abstract vector) need not be specified; its essential property is that, given the Φ of a system at a particular time, the results of all measurements made on the system at that time are statistically determined. The variation of Φ with time is given by the Schrödinger equation

$$i\hbar[\partial/\partial t]\Phi=\left\{\int H(\mathbf{r})d\tau\right\}\Phi, \tag{1}$$

where $H(\mathbf{r})$ is the operator representing the total energy-density of the system at the point $\mathbf{r}$. The general solution of (1) is

$$\Phi(t)=\exp\left\{[-it/\hbar]\int H(\mathbf{r})d\tau\right\}\Phi_0, \tag{2}$$

with Φ_0 any constant state vector.

Now in a relativistic system, the most general kind of measurement is not the simultaneous measurement of field quantities at different points of space. It is also possible to measure independently field quantities at different points of space at different times, provided that the points of space-time at which the measurements are made lie outside each other's light cones, so that the measurements do not interfere with each other. Thus the most comprehensive general type of measurement is a measurement of field quantities at each point $\mathbf{r}$ of space at a time $t(\mathbf{r})$, the locus of the points $(\mathbf{r}, t(\mathbf{r}))$ in space-time forming a 3-dimensional surface σ which is space-like (i.e., every pair of points on it is separated by a space-like interval). Such a measurement will be called "an observation of the system on σ." It is easy to see what the result of the measurement will be. At each point $\mathbf{r}'$ the field quantities will be measured for a state of the system with state vector $\Phi(t(\mathbf{r}'))$ given by (2). But all observable quantities at $\mathbf{r}'$ are operators which commute with the energy-density operator $H(\mathbf{r})$ at every point $\mathbf{r}$ different from $\mathbf{r}'$, and it is a general principle of quantum mechanics that if B is a unitary operator commuting with A, then for any state Φ the results of measurements of A are the same in the state Φ as in the state $B\Phi$. Therefore, the results of measurement of the field quantities at $\mathbf{r}'$ in the state $\Phi(t(\mathbf{r}'))$ are the same as if the state of the system were

$$\Phi(\sigma)=\exp\left\{-[i/\hbar]\int t(\mathbf{r})H(\mathbf{r})d\tau\right\}\Phi_0, \tag{3}$$

which differs from $\Phi(t(\mathbf{r}'))$ only by a unitary factor commuting with these field quantities. The important fact is that the state vector $\Phi(\sigma)$ depends only on σ and not on $\mathbf{r}'$. The conclusion reached is that observations of a system on σ give results which are completely determined by attributing to the system the state vector $\Phi(\sigma)$ given by (3).

The Tomonaga-Schwinger form of the Schrödinger equation is a differential form of (3). Suppose the surface σ to be deformed slightly near the point $\mathbf{r}$ into the surface σ', the volume of space-time separating the two surfaces being V. Then the quotient

$$[\Phi(\sigma')-\Phi(\sigma)]/V$$

tends to a limit as $V\rightarrow 0$, which we denote by $\partial\Phi/\partial\sigma(\mathbf{r})$ and call the functional derivative of Φ with respect to σ at the point $\mathbf{r}$. From (3) it follows that

$$i\hbar c[\partial\Phi/\partial\sigma(\mathbf{r})]=H(\mathbf{r})\Phi, \tag{4}$$

and (3) is, in fact, the general solution of (4).

The whole meaning of an equation such as (4) depends on the physical meaning which is attached to the statement "a system has a constant state vector Φ_0." In the present context, this statement means "results of measurements of

Results of the application of the method to a calculation of the second-order radiative correction to the Klein-Nishina formula are stated. All the papers of Professor Tomonaga and his associates which have yet been published were completed before the end of 1946. The isolation of these Japanese workers has undoubtedly constituted a serious loss to theoretical physics.

field quantities at any given point of space are independent of time." This statement is plainly non-relativistic, and so (4) is, in spite of appearances, a non-relativistic equation.

The simplest way to introduce a new state vector Ψ which shall be a relativistic invariant is to require that the statement "a system has a constant state vector Ψ" shall mean "a system consists of photons, electrons, and positrons, traveling freely through space without interaction or external disturbance." For this purpose, let

$$H(\mathbf{r})=H_0(\mathbf{r})+H_1(\mathbf{r}), \tag{5}$$

where H_0 is the energy-density of the free electromagnetic and electron fields, and H_1 is that of their interaction with each other and with any external disturbing forces that may be present. A system with constant Ψ is, then, one whose H_1 is identically zero; by (3) such a system corresponds to a Φ of the form

$$\Phi(\sigma)=T(\sigma)\Phi_0,$$

$$T(\sigma)=\exp\left\{-[i/\hbar]\int t(\mathbf{r})H_0(\mathbf{r})d\tau\right\}. \tag{6}$$

It is therefore consistent to write generally

$$\Phi(\sigma)=T(\sigma)\Psi(\sigma), \tag{7}$$

thus defining the new state vector Ψ of any system in terms of the old Φ. The differential equation satisfied by Ψ is obtained from (4), (5), (6), and (7) in the form

$$i\hbar c[\partial\Psi/\partial\sigma(\mathbf{r})]=(T(\sigma))^{-1}H_1(\mathbf{r})T(\sigma)\Psi. \tag{8}$$

Now if $q(\mathbf{r})$ is any time-independent field operator, the operator

$$q(x_0)=(T(\sigma))^{-1}q(\mathbf{r})T(\sigma)$$

is just the corresponding time-dependent operator as usually defined in quantum electrodynamics.[4] It is a function of the point x_0 of space-time whose coordinates are $(\mathbf{r}, ct(\mathbf{r}))$, but is the same for all surfaces σ passing through this point, by virtue of the commutation of $H_1(\mathbf{r})$ with $H_0(\mathbf{r}')$ for $\mathbf{r}'\neq\mathbf{r}$. Thus (8) may be written

$$i\hbar c[\partial\Psi/\partial\sigma(x_0)]=H_1(x_0)\Psi, \tag{9}$$

where $H_1(x_0)$ is the time-dependent form of the energy-density of interaction of the two fields with each other and with external forces. The left side of (9) represents the degree of departure of the system from a system of freely traveling particles and is a relativistic invariant; $H_1(x_0)$ is also an invariant, and thus is avoided one of the most unsatisfactory features of the old theories, in which the invariant H_1 was added to the non-invariant H_0. Equation (9) is the starting point of the Tomonaga-Schwinger theory.

III. INTRODUCTION OF PERTURBATION THEORY

Equation (9) can be solved explicitly. For this purpose it is convenient to introduce a one-parameter family of space-like surfaces filling the whole of space-time, so that one and only one member $\sigma(x)$ of the family passes through any given point x. Let $\sigma_0, \sigma_1, \sigma_2, \cdots$ be a sequence of surfaces of the family, starting with σ_0 and proceeding in small steps steadily into the past. By

$$\int_{\sigma_1}^{\sigma_0} H_1(x)dx$$

is denoted the integral of $H_1(x)$ over the 4-dimensional volume between the surfaces σ_1 and σ_0; similarly, by

$$\int_{-\infty}^{\sigma_0} H_1(x)dx, \quad \int_{\sigma_0}^{\infty} H_1(x)dx$$

are denoted integrals over the whole volume to the past of σ_0 and to the future of σ_0, respectively. Consider the operator

$$U=U(\sigma_0)=\left(1-[i/\hbar c]\int_{\sigma_1}^{\sigma_0} H_1(x)dx\right)$$

$$\times\left(1-[i/\hbar c]\int_{\sigma_2}^{\sigma_1} H_1(x)dx\right)\cdots, \tag{10}$$

the product continuing to infinity and the surfaces $\sigma_0, \sigma_1, \cdots$ being taken in the limit infinitely close together. U satisfies the differential equation

$$i\hbar c[\partial U/\partial\sigma(x_0)]=H_1(x_0)U, \tag{11}$$

and the general solution of (9) is

$$\Psi(\sigma)=U(\sigma)\Psi_0, \tag{12}$$

with Ψ_0 any constant vector.

[4] See, for example, Gregor Wentzel, *Einführung in die Quantentheorie der Wellenfelder* (Franz Deuticke, Wien, 1943), pp. 18–26.

Expanding the product (10) in ascending powers of H_1 gives a series

$$U=1+(-i/\hbar c)\int_{-\infty}^{\sigma_0} H_1(x_1)dx_1+(-i/\hbar c)^2 \times\int_{-\infty}^{\sigma_0} dx_1\int_{-\infty}^{\sigma(x_1)} H_1(x_1)H_1(x_2)dx_2+\cdots. \quad (13)$$

Further, U is by (10) obviously unitary, and

$$U^{-1}=\bar{U}=1+(i/\hbar c)\int_{-\infty}^{\sigma_0} H_1(x_1)dx_1+(i/\hbar c)^2 \times\int_{-\infty}^{\sigma_0} dx_1\int_{-\infty}^{\sigma(x_1)} H_1(x_2)H_1(x_1)dx_2+\cdots. \quad (14)$$

It is not difficult to verify that U is a function of σ_0 alone and is independent of the family of surfaces of which σ_0 is one member. The use of a finite number of terms of the series (13) and (14), neglecting the higher terms, is the equivalent in the new theory of the use of perturbation theory in the older electrodynamics.

The operator $U(\infty)$, obtained from (10) by taking σ_0 in the infinite future, is a transformation operator transforming a state of the system in the infinite past (representing, say, converging streams of particles) into the same state in the infinite future (after the particles have interacted or been scattered into their final outgoing distribution). This operator has matrix elements corresponding only to real transitions of the system, i.e., transitions which conserve energy and momentum. It is identical with the Heisenberg S matrix.[6]

IV. ELIMINATION OF THE RADIATION INTERACTION

In most of the problem of electrodynamics, the energy-density $H_1(x_0)$ divides into two parts—

$$H_1(x_0)=H^i(x_0)+H^e(x_0), \quad (15)$$

$$H^i(x_0)=-[1/c]j_\mu(x_0)A_\mu(x_0), \quad (16)$$

the first part being the energy of interaction of the two fields with each other, and the second part the energy produced by external forces. It is usually not permissible to treat H^e as a small perturbation as was done in the last section. Instead, H^i alone is treated as a perturbation, the aim being to eliminate H^i but to leave H^e in its original place in the equation of motion of the system.

Operators $S(\sigma)$ and $S(\infty)$ are defined by replacing H_1 by H^i in the definitions of $U(\sigma)$ and $U(\infty)$. Thus $S(\sigma)$ satisfies the equation

$$i\hbar c[\partial S/\partial\sigma(x_0)]=H^i(x_0)S. \quad (17)$$

Suppose now a new type of state vector $\Omega(\sigma)$ to be introduced by the substitution

$$\Psi(\sigma)=S(\sigma)\Omega(\sigma). \quad (18)$$

By (9), (15), (17), and (18) the equation of motion for $\Omega(\sigma)$ is

$$i\hbar c[\partial\Omega/\partial\sigma(x_0)]=(S(\sigma))^{-1}H^e(x_0)S(\sigma)\Omega. \quad (19)$$

The elimination of the radiation interaction is hereby achieved; only the question, "How is the new state vector $\Omega(\sigma)$ to be interpreted?," remains.

It is clear from (19) that a system with a constant Ω is a system of electrons, positrons, and photons, moving under the influence of their mutual interactions, but in the absence of external fields. In a system where two or more particles are actually present, their interactions alone will, in general, cause real transitions and scattering processes to occur. For such a system, it is rather "unphysical" to represent a state of motion including the effects of the interactions by a constant state vector; hence, for such a system the new representation has no simple interpretation. However, the most important systems are those in which only one particle is actually present, and its interaction with the vacuum fields gives rise only to virtual processes. In this case the particle, including the effects of all its interactions with the vacuum, appears to move as a free particle in the absence of external fields, and it is eminently reasonable to represent such a state of motion by a constant state vector. Therefore, it may be said that the operator,

$$H_T(x_0)=(S(\sigma))^{-1}H^e(x_0)S(\sigma), \quad (20)$$

on the right of (19) represents the interaction of a physical particle with an external field, including radiative corrections. Equation (19) describes the extent to which the motion of a

[6] Werner Heisenberg, Zeits. f. Physik 120, 513 (1943), 120, 673 (1943), and Zeits. f. Naturforschung 1, 608 (1946).

single physical particle deviates, in the external field, from the motion represented by a constant state-vector, i.e., from the motion of an observed "free" particle.

If the system whose state vector is constantly Ω undergoes no real transitions with the passage of time, then the state vector Ω is called "steady." More precisely, Ω is steady if, and only if, it satisfies the equation

$$S(\infty)\Omega=\Omega. \tag{21}$$

As a general rule, one-particle states are steady and many-particle states unsteady. There are, however, two important qualifications to this rule.

First, the interaction (20) itself will almost always cause transitions from steady to unsteady states. For example, if the initial state consists of one electron in the field of a proton, H_T will have matrix elements for transitions of the electron to a new state with emission of a photon, and such transitions are important in practice. Therefore, although the interpretation of the theory is simpler for steady states, it is not possible to exclude unsteady states from consideration.

Second, if a one-particle state as hitherto defined is to be steady, the definition of $S(\sigma)$ must be modified. This is because $S(\infty)$ includes the effects of the electromagnetic self-energy of the electron, and this self-energy gives an expectation value to $S(\infty)$ which is different from unity (and indeed infinite) in a one-electron state, so that Eq. (21) cannot be satisfied. The mistake that has been made occurred in trying to represent the observed electron with its electromagnetic self-energy by a wave field with the same characteristic rest-mass as that of the "bare" electron. To correct the mistake, let δm denote the electromagnetic mass of the electron, i.e., the difference in rest-mass between an observed and a "bare" electron. Instead of (5), the division of the energy-density $H(\mathbf{r})$ should have taken the form

$$H(\mathbf{r})=(H_0(\mathbf{r})+\delta mc^2\psi^*(\mathbf{r})\beta\psi(\mathbf{r}))+(H_1(\mathbf{r})-\delta mc^2\psi^*(\mathbf{r})\beta\psi(\mathbf{r})).$$

The first bracket on the right here represents the energy-density of the free electromagnetic and electron fields with the observed electron rest-mass, and should have been used instead of $H_0(\mathbf{r})$ in the definition (6) of $T(\sigma)$. Consequently, the second bracket should have been used instead of $H_1(\mathbf{r})$ in Eq. (8).

The definition of $S(\sigma)$ has therefore to be altered by replacing $H^i(x_0)$ by[6]

$$H^I(x_0)=H^i(x_0)+H^S(x_0)=H^i(x_0)-\delta mc^2\bar{\psi}(x_0)\psi(x_0). \tag{22}$$

The value of δm can be adjusted so as to cancel out the self-energy effects in $S(\infty)$ (this is only a formal adjustment since the value is actually infinite), and then Eq. (21) will be valid for one-electron states. For the photon self-energy no such adjustment is needed since, as proved by Schwinger, the photon self-energy turns out to be identically zero.

The foregoing discussion of the self-energy problem is intentionally only a sketch, but it will be found to be sufficient for practical applications of the theory. A fuller discussion of the theoretical assumptions underlying this treatment of the problem will be given by Schwinger in his forthcoming papers. Moreover, it must be realized that the theory as a whole cannot be put into a finally satisfactory form so long as divergencies occur in it, however skilfully these divergencies are circumvented; therefore, the present treatment should be regarded as justified by its success in applications rather than by its theoretical derivation.

The important results of the present paper up to this point are Eq. (19) and the interpretation of the state vector Ω. The state vector Ψ of a system can be interpreted as a wave function giving the probability amplitude of finding any particular set of occupation numbers for the various possible states of free electrons, positrons, and photons. The state vector Ω of a system with a given Ψ on a given surface σ is, crudely speaking, the Ψ which the system would have had in the infinite past if it had arrived at the given Ψ on σ under the influence of the interaction $H^I(x_0)$ alone.

The definition of Ω being unsymmetrical between past and future, a new type of state vector Ω' can be defined by reversing the direction of time in the definition of Ω. Thus the Ω' of a system with a given Ψ on a given σ is the Ψ

[6] Here Schwinger's notation $\bar{\psi}=\psi^*\beta$ is used.

which the system would reach in the infinite future if it continued to move under the influence of $H^I(x_0)$ alone. More simply, Ω' can be defined by the equation

$$\Omega'(\sigma)=S(\infty)\Omega(\sigma). \tag{23}$$

Since $S(\infty)$ is a unitary operator independent of σ, the state vectors Ω and Ω' are really only the same vector in two different representations or coordinate systems. Moreover, for any steady state the two are identical by (21).

V. FUNDAMENTAL FORMULAS OF THE SCHWINGER AND FEYNMAN THEORIES

The Schwinger theory works directly from Eqs. (19) and (20), the aim being to calculate the matrix elements of the "effective external potential energy" H_T between states specified by their state vectors Ω. The states considered in practice always have Ω of some very simple kind, for example, Ω representing systems in which one or two free-particle states have occupation number one and the remaining free-particle states have occupation number zero. By analogy with (13), $S(\sigma_0)$ is given by

$$S(\sigma_0)=1+(-i/\hbar c)\int_{-\infty}^{\sigma 0}H^I(x_1)dx_1+(-i/\hbar c)^2 \times\int_{-\infty}^{\sigma 0}dx_1\int_{-\infty}^{\sigma(x_1)}H^I(x_1)H^I(x_2)dx_2+\cdots, \tag{24}$$

and $(S(\sigma_0))^{-1}$ by a corresponding expression analogous to (14). Substitution of these series into (20) gives at once

$$H_T(x_0)=\sum_{n=0}^{\infty}(i/\hbar c)^n\int_{-\infty}^{\sigma(x_0)}dx_1\int_{-\infty}^{\sigma(x_1)}dx_2\cdots \times\int_{-\infty}^{\sigma(x_n-1)}dx_n\times[H^I(x_n),[\cdots,[H^I(x_2),[H^I(x_1),H^e(x_0)]]\cdots]]. \tag{25}$$

The repeated commutators in this formula are characteristic of the Schwinger theory, and their evaluation gives rise to long and rather difficult analysis. Using the first three terms of the series, Schwinger was able to calculate the second-order radiative corrections to the equations of motion of an electron in an external field, and obtained satisfactory agreement with experimental results. In this paper the development of the Schwinger theory will be carried no further: in principle the radiative corrections to the equations of motion of electrons could be calculated to any desired order of approximation from formula (25).

In the Feynman theory the basic principle is to preserve symmetry between past and future. Therefore, the matrix elements of the operator H_T are evaluated in a "mixed representation;" the matrix elements are calculated between an initial state specified by its state vector Ω_1 and a final state specified by its state vector Ω_2'. The matrix element of H_T between two such states in the Schwinger representation is

$$\Omega_2^*H_T\Omega_1=\Omega_2'^*S(\infty)H_T\Omega_1, \tag{26}$$

and therefore the operator which replaces H_T in the mixed representation is

$$H_F(x_0)=S(\infty)H_T(x_0) =S(\infty)(S(\sigma))^{-1}H^e(x_0)S(\sigma). \tag{27}$$

Going back to the original product definition of $S(\sigma)$ analogous to (10), it is clear that $S(\infty)\times(S(\sigma))^{-1}$ is simply the operator obtained from $S(\sigma)$ by interchanging past and future. Thus,

$$R(\sigma)=S(\infty)(S(\sigma))^{-1}=1+(-i/\hbar c)\times\int_{\sigma}^{\infty}H^I(x_1)dx_1+(-i/\hbar c)^2\int_{\sigma}^{\infty}dx_1 \times\int_{\sigma(x_1)}^{\infty}H^I(x_2)H^I(x_1)dx_2+\cdots. \tag{28}$$

The physical meaning of a mixed representation of this type is not at all recondite. In fact, a mixed representation is normally used to describe such a process as bremsstrahlung of an electron in the field of a nucleus when the Born approximation is not valid; the process of bremsstrahlung is a radiative transition of the electron from a state described by a Coulomb wave function, with a plane ingoing and a spherical outgoing wave, to a state described by a Coulomb wave function with a spherical ingoing and a plane outgoing wave. The initial and final states here belong to different orthogonal systems of wave functions, and so the transition matrix elements are calculated in a mixed representation. In the Feynman theory the situation is analogous; only the roles of the radiation interaction and the external (or Coulomb) field are interchanged; the radiation interaction is used instead of the Coulomb field to modify the state vectors (wave functions) of the initial and final states, and the external field instead of the radiation interaction causes transitions between these state vectors.

In the Feynman theory there is an additional simplification. For if matrix elements are being calculated between two states, either of which is steady (and this includes all cases so far considered), the mixed representation reduces to an ordinary representation. This occurs, for example, in treating a one-particle problem such as the radiative correction to the equations of motion of an electron in an external field; the operator $H_F(x_0)$, although in general it is not even Hermitian, can in this case be considered as an effective external potential energy acting on the particle, in the ordinary sense of the words.

This section will be concluded with the derivation of the fundamental formula (31) of the Feynman theory, which is the analog of formula (25) of the Schwinger theory. If

$$F_1(x_1), \quad \cdots, \quad F_n(x_n)$$

are any operators defined, respectively, at the points $x_1, \cdots, x_n$ of space-time, then

$$P(F_1(x_1), \quad \cdots, \quad F_n(x_n)) \tag{29}$$

will denote the product of these operators, taken in the order, reading from right to left, in which the surfaces $\sigma(x_1), \cdots, \sigma(x_n)$ occur in time. In most applications of this notation $F_i(x_i)$ will commute with $F_j(x_j)$ so long as x_i and x_j are outside each other's light cones; when this is the case, it is easy to see that (29) is a function of the points $x_1, \cdots, x_n$ only and is independent of the surfaces $\sigma(x_i)$. Consider now the integral

$$I_n = \int_{-\infty}^{\infty} dx_1 \cdots \int_{-\infty}^{\infty} dx_n P(H^e(x_0), \\ H^I(x_1), \quad \cdots, \quad H^I(x_n)).$$

Since the integrand is a symmetrical function of the points $x_1, \cdots, x_n$, the value of the integral is just $n!$ times the integral obtained by restricting the integration to sets of points $x_1, \cdots, x_n$ for which $\sigma(x_i)$ occurs after $\sigma(x_{i+1})$ for each i. The restricted integral can then be further divided into $(n+1)$ parts, the j'th part being the integral over those sets of points with the property that $\sigma(x_0)$ lies between $\sigma(x_{j-1})$ and $\sigma(x_j)$ (with obvious modifications for $j=1$ and $j=n+1$). Therefore,

$$I_n = n! \sum_{j=1}^{n+1} \int_{-\infty}^{\sigma(x_0)} dx_j \cdots \int_{-\infty}^{\sigma(x_{n-1})} dx_n \\ \times \int_{\sigma(x_0)}^{\infty} dx_{j-1} \cdots \int_{\sigma(x_2)}^{\infty} dx_1 \times H^I(x_1) \cdots \\ H^I(x_{j-1}) H^e(x_0) H^I(x_j) \cdots H^I(x_n). \tag{30}$$

Now if the series (24) and (28) are substituted into (27), sums of integrals appear which are precisely of the form (30). Hence finally

$$H_F(x_0) = \sum_{n=0}^{\infty} (-i/\hbar c)^n [1/n!] I_n \\ = \sum_{n=0}^{\infty} (-i/\hbar c)^n [1/n!] \int_{-\infty}^{\infty} dx_1 \cdots \int_{-\infty}^{\infty} dx_n \\ \times P(H^e(x_0), \; H^I(x_1), \; \cdots, \; H^I(x_n)). \tag{31}$$

By this formula the notation $H_F(x_0)$ is justified, for this operator now appears as a function of the point x_0 alone and not of the surface σ. The further development of the Feynman theory is mainly concerned with the calculation of matrix elements of (31) between various initial and final states.

As a special case of (31) obtained by replacing H^e by the unit matrix in (27),

$$S(\infty) = \sum_{n=0}^{\infty} (-i/\hbar c)^n [1/n!] \int_{-\infty}^{\infty} dx_1 \cdots \int_{-\infty}^{\infty} dx_n \\ \times P(H^I(x_1), \; \cdots, \; H^I(x_n)). \tag{32}$$

VI. CALCULATION OF MATRIX ELEMENTS

In this section the application of the foregoing theory to a general class of problems will be explained. The ultimate aim is to obtain a set of rules by which the matrix element of the operator (31) between two given states may be written down in a form suitable for numerical evaluation, immediately and automatically. The fact that such a set of rules exists is the basis of the Feynman radiation theory; the derivation in this section of the same rules from what is fundamentally the Tomonaga-Schwinger theory constitutes the proof of equivalence of the two theories.

To avoid excessive complication, the type of matrix element considered will be restricted in two ways. First, it will be assumed that the external potential energy is

$$H^e(x_0) = -[1/c]j_\mu(x_0)A_\mu{}^e(x_0), \quad (33)$$

that is to say, the interaction energy of the electron-positron field with electromagnetic potentials $A_\mu{}^e(x_0)$ which are given numerical functions of space and time. Second, matrix elements will be considered only for transitions from a state A, in which just one electron and no positron or photon is present, to another state B of the same character. These restrictions are not essential to the theory, and are introduced only for convenience, in order to illustrate clearly the principles involved.

The electron-positron field operator may be written

$$\psi_\alpha(x) = \sum_u \phi_{u\alpha}(x)a_u, \quad (34)$$

where the $\phi_{u\alpha}(x)$ are spinor wave functions of free electrons and positrons, and the a_u are annihilation operators of electrons and creation operators of positrons. Similarly, the adjoint operator

$$\bar\psi_\alpha(x) = \sum_u \bar\phi_{u\alpha}(x)\bar a_u, \quad (35)$$

where $\bar a_u$ are annihilation operators of positrons and creation operators of electrons. The electromagnetic field operator is

$$A_\mu(x) = \sum_v (A_{v\mu}(x)b_v + A_{v\mu}{}^*(x)\bar b_v), \quad (36)$$

where b_v and $\bar b_v$ are photon annihilation and creation operators, respectively. The charge-current 4-vector of the electron field is

$$j_\mu(x) = iec\bar\psi(x)\gamma_\mu\psi(x); \quad (37)$$

strictly speaking, this expression ought to be antisymmetrized to the form[7]

$$j_\mu(x) = \tfrac{1}{2}iec\{\bar\psi_\alpha(x)\psi_\beta(x) - \psi_\beta(x)\bar\psi_\alpha(x)\}(\gamma_\mu)_{\alpha\beta}, \quad (38)$$

but it will be seen later that this is not necessary in the present theory.

Consider the product P occurring in the n'th integral of (31); let it be denoted by P_n. From (16), (22), (33), and (37) it is seen that P_n is a sum of products of $(n+1)$ operators ψ_α, $(n+1)$ operators $\bar\psi_\alpha$, and not more than n operators A_μ, multiplied by various numerical factors. By Q_n may be denoted a typical product of factors ψ_α, $\bar\psi_\alpha$, and A_μ, not summed over the indices such as α and μ, so that P_n is a sum of terms such as Q_n. Then Q_n will be of the form (indices omitted)

$$Q_n = \bar\psi(x_{i_0})\psi(x_{i_0})\bar\psi(x_{i_1})\psi(x_{i_1})\cdots\bar\psi(x_{i_n})\psi(x_{i_n}) \times A(x_{j_1})\cdots A(x_{j_m}), \quad (39)$$

where $i_0, i_1, \cdots, i_n$ is some permutation of the integers $0, 1, \cdots, n$, and $j_1, \cdots, j_m$ are some, but not necessarily all, of the integers $1, \cdots, n$ in some order. Since none of the operators $\bar\psi$ and ψ commute with each other, it is especially important to preserve the order of these factors. Each factor of Q_n is a sum of creation and annihilation operators by virtue of (34), (35), and (36), and so Q_n itself is a sum of products of creation and annihilation operators.

Now consider under what conditions a product of creation and annihilation operators can give a non-zero matrix element for the transition $A \rightarrow B$. Clearly, one of the annihilation operators must annihilate the electron in state A, one of the creation operators must create the electron in state B, and the remaining operators must be divisible into pairs, the members of each pair respectively creating and annihilating the same particle. Creation and annihilation operators referring to different particles always commute or anticommute (the former if at least one is a photon operator, the latter if both are electron-positron operators). Therefore, if the two single operators and the various pairs of operators in the product all refer to different particles, the order of factors in the product can be altered so as to bring together the two single operators and the two members of each pair, without changing the value of the product except for a change of sign if the permutation made in the order of the electron and positron operators is odd. In the case when some of the single operators and pairs of operators refer to the same particle, it is not hard to verify that the same change in order of factors can be made, provided it is remembered that the division of the operators into pairs is no longer unique, and the change of order is to

[7] See Wolfgang Pauli, Rev. Mod. Phys. 13, 203 (1941), Eq. (96), p. 224.

be made for each possible division into pairs and the results added together.

It follows from the above considerations that the matrix element of Q_n for the transition $A\rightarrow B$ is a sum of contributions, each contribution arising from a specific way of dividing the factors of Q_n into two single factors and pairs. A typical contribution of this kind will be denoted by M. The two factors of a pair must involve a creation and an annihilation operator for the same particle, and so must be either one $\bar{\psi}$ and one ψ or two A; the two single factors must be one $\bar{\psi}$ and one ψ. The term M is thus specified by fixing an integer k, and a permutation $r_0, r_1, \cdots, r_n$ of the integers $0, 1, \cdots, n$, and a division (s_1,t_1), (s_2,t_2), $\cdots$, (s_h,t_h) of the integers $j_1, \cdots, j_m$ into pairs; clearly $m=2h$ has to be an even number; the term M is obtained by choosing for single factors $\bar{\psi}(x_k)$ and $\psi(x_{r_k})$, and for associated pairs of factors $(\bar{\psi}(x_i),\psi(x_{r_i}))$ for $i=0, 1, \cdots, k-1, k+1, \cdots, n$ and $(A(x_{s_i}),A(x_{t_i}))$ for $i=1, \cdots, h$. In evaluating the term M, the order of factors in Q_n is first to be permuted so as to bring together the two single factors and the two members of each pair, but without altering the order of factors within each pair; the result of this process is easily seen to be

$$Q_n'=\epsilon P(\bar{\psi}(x_0),\psi(x_{r_0}))\cdots P(\bar{\psi}(x_n),\psi(x_{r_n}))\times P(A(x_{s_1}),A(x_{t_1}))\cdots P(A(x_{s_h}),A(x_{t_h})), \quad (40)$$

a factor ϵ being inserted which takes the value ± 1 according to whether the permutation of $\bar{\psi}$ and ψ factors between (39) and (40) is even or odd. Then in (40) each product of two associated factors (but not the two single factors) is to be independently replaced by the sum of its matrix elements for processes involving the successive creation and annihilation of the same particle.

Given a bilinear operator such as $A_\mu(x)A_\nu(y)$, the sum of its matrix elements for processes involving the successive creation and annihilation of the same particle is just what is usually called the "vacuum expectation value" of the operator, and has been calculated by Schwinger. This quantity is, in fact (note that Heaviside units are being used)

$$\langle A_\mu(x)A_\nu(y)\rangle_0=\tfrac{1}{2}\hbar c\delta_{\mu\nu}\{D^{(1)}+iD\}(x-y),$$

where $D^{(1)}$ and D are Schwinger's invariant D functions. The definitions of these functions will not be given here, because it turns out that the vacuum expectation value of $P(A_\mu(x),A_\nu(y))$ takes an even simpler form. Namely,

$$\langle P(A_\mu(x),A_\nu(y))\rangle_0=\tfrac{1}{2}\hbar c\delta_{\mu\nu}D_F(x-y), \quad (41)$$

where D_F is the type of D function introduced by Feynman. $D_F(x)$ is an even function of x, with the integral expansion

$$D_F(x)=-[i/2\pi^2]\int_0^\infty \exp[i\alpha x^2]d\alpha, \quad (42)$$

where x^2 denotes the square of the invariant length of the 4-vector x. In a similar way it follows from Schwinger's results that

$$\langle P(\bar{\psi}_\alpha(x),\psi_\beta(y))\rangle_0=\tfrac{1}{2}\eta(x,y)S_{F\beta\alpha}(x-y), \quad (43)$$

where

$$S_{F\beta\alpha}(x)=-(\gamma_\mu(\partial/\partial x_\mu)+\kappa_0)_{\beta\alpha}\Delta_F(x), \quad (44)$$

κ_0 is the reciprocal Compton wave-length of the electron, $\eta(x,y)$ is -1 or $+1$ according as $\sigma(x)$ is earlier or later than $\sigma(y)$ in time, and Δ_F is a function with the integral expansion

$$\Delta_F(x)=-[i/2\pi^2]\int_0^\infty \exp[i\alpha x^2-i\kappa_0^2/4\alpha]d\alpha. \quad (45)$$

Substituting from (41) and (44) into (40), the matrix element M takes the form (still omitting the indices of the factors $\bar{\psi}$, ψ, and A of Q_n)

$$M=\epsilon\prod_{i\neq k}(\tfrac{1}{2}\eta(x_i,x_{r_i})S_F(x_i-x_{r_i}))\times\prod_j(\tfrac{1}{2}\hbar cD_F(x_{s_j}-x_{t_j}))P(\bar{\psi}(x_k),\psi(x_{r_k})). \quad (46)$$

The single factors $\bar{\psi}(x_k)$ and $\psi(x_{r_k})$ are conveniently left in the form of operators, since the matrix elements of these operators for effecting the transition $A\rightarrow B$ depend on the wave functions of the electron in the states A and B. Moreover, the order of the factors $\bar{\psi}(x_k)$ and $\psi(x_{r_k})$ is immaterial since they anticommute with each other; hence it is permissible to write

$$P(\bar{\psi}(x_k),\psi(x_{r_k}))=\eta(x_k,x_{r_k})\bar{\psi}(x_k)\psi(x_{r_k}).$$

Therefore (46) may be rewritten

$$M=\epsilon'\prod_{i\neq k}(\tfrac{1}{2}S_F(x_i-x_{r_i}))\prod_j(\tfrac{1}{2}\hbar cD_F(x_{s_j}-x_{t_j}))\times\bar{\psi}(x_k)\psi(x_{r_k}), \quad (47)$$

with

$$\epsilon'=\epsilon\prod_i\eta(x_i,x_{r_i}). \quad (48)$$

Now the product in (48) is $(-1)^p$, where p is the number of occasions in the expression (40) on which the ψ of a P bracket occurs to the left of the $\bar{\psi}$. Referring back to the definition of ϵ after Eq. (40), it follows that ϵ' takes the value $+1$ or -1 according to whether the permutation of $\bar{\psi}$ and ψ factors between (39) and the expression

$$\bar{\psi}(x_0)\psi(x_{r_0})\cdots\bar{\psi}(x_n)\psi(x_{r_n}) \tag{49}$$

is even or odd. But (39) can be derived by an even permutation from the expression

$$\bar{\psi}(x_0)\psi(x_0)\cdots\bar{\psi}(x_n)\psi(x_n), \tag{50}$$

and the permutation of factors between (49) and (50) is even or odd according to whether the permutation $r_0, \cdots, r_n$ of the integers $0, \cdots, n$ is even or odd. Hence, finally, ϵ' in (47) is $+1$ or -1 according to whether the permutation $r_0, \cdots, r_n$ is even or odd. It is important that ϵ' depends only on the type of matrix element M considered, and not on the points $x_0, \cdots, x_n$; therefore, it can be taken outside the integrals in (31).

One result of the foregoing analysis is to justify the use of (37), instead of the more correct (38), for the charge-current operator occurring in H^e and H^i. For it has been shown that in each matrix element such as M the factors $\bar{\psi}$ and ψ in (38) can be freely permuted, so that (38) can be replaced by (37), except in the case when the two factors form an associated pair. In the exceptional case, M contains as a factor the vacuum expectation value of the operator $j_\mu(x_i)$ at some point x_i; this expectation value is zero according to the correct formula (38), though it would be infinite according to (37); thus the matrix elements in the exceptional case are always zero. The conclusion is that only those matrix elements are to be calculated for which the integer r_i differs from i for every $i\neq k$, and in these elements the use of formula (37) is correct.

To write down the matrix elements of (31) for the transition $A\rightarrow B$, it is only necessary to take all the products Q_n, replace each by the sum of the corresponding matrix elements M given by (47), reassemble the terms into the form of the P_n from which they were derived, and finally substitute back into the series (31). The problem of calculating the matrix elements of (31) is thus in principle solved. However, in the following section it will be shown how this solution-in-principle can be reduced to a much simpler and more practical procedure.

VII. GRAPHICAL REPRESENTATION OF MATRIX ELEMENTS

Let an integer n and a product P_n occurring in (31) be temporarily fixed. The points $x_0, x_1, \cdots, x_n$ may be represented by $(n+1)$ points drawn on a piece of paper. A type of matrix element M as described in the last section will then be represented graphically as follows. For each associated pair of factors $(\bar{\psi}(x_i),\psi(x_{r_i}))$ with $i\neq k$, draw a line with a direction marked in it from the point x_i to the point x_{r_i}. For the single factors $\bar{\psi}(x_k)$, $\psi(x_{r_k})$, draw directed lines leading out from x_k to the edge of the diagram, and in from the edge of the diagram to x_{r_k}. For each pair of factors $(A(x_{s_i}),A(x_{t_i}))$, draw an undirected line joining the points x_{s_i} and x_{t_i}. The complete set of points and lines will be called the "graph" of M; clearly there is a one-to-one correspondence between types of matrix element and graphs, and the exclusion of matrix elements with $r_i=i$ for $i\neq k$ corresponds to the exclusion of graphs with lines joining a point to itself. The directed lines in a graph will be called "electron lines," the undirected lines "photon lines."

Through each point of a graph pass two electron lines, and therefore the electron lines together form one open polygon containing the vertices x_k and x_{r_k}, and possibly a number of closed polygons as well. The closed polygons will be called "closed loops," and their number denoted by l. Now the permutation $r_0, \cdots, r_n$ of the integers $0, \cdots, n$ is clearly composed of $(l+1)$ separate cyclic permutations. A cyclic permutation is even or odd according to whether the number of elements in it is odd or even. Hence the parity of the permutation $r_0, \cdots, r_n$ is the parity of the number of even-number cycles contained in it. But the parity of the number of odd-number cycles in it is obviously the same as the parity of the total number $(n+1)$ of elements. The total number of cycles being $(l+1)$, the parity of the number of even-number cycles is $(l-n)$. Since it was seen earlier that the ϵ' of Eq. (47) is determined just by the parity of the permutation $r_0, \cdots, r_n$, the above argument yields the simple formula

$$\epsilon' = (-1)^{l-n}. \tag{51}$$

This formula is one result of the present theory which can be much more easily obtained by intuitive considerations of the sort used by Feynman.

In Feynman's theory the graph corresponding to a particular matrix element is regarded, not merely as an aid to calculation, but as a picture of the physical process which gives rise to that matrix element. For example, an electron line joining x_1 to x_2 represents the possible creation of an electron at x_1 and its annihilation at x_2, together with the possible creation of a positron at x_2 and its annihilation at x_1. This interpretation of a graph is obviously consistent with the methods, and in Feynman's hands has been used as the basis for the derivation of most of the results, of the present paper. For reasons of space, these ideas of Feynman will not be discussed in further detail here.

To the product P_n correspond a finite number of graphs, one of which may be denoted by G; all possible G can be enumerated without difficulty for moderate values of n. To each G corresponds a contribution $C(G)$ to the matrix element of (31) which is being evaluated.

It may happen that the graph G is disconnected, so that it can be divided into subgraphs, each of which is connected, with no line joining a point of one subgraph to a point of another. In such a case it is clear from (47) that $C(G)$ is the product of factors derived from each subgraph separately. The subgraph G_1 containing the point x_0 is called the "essential part" of G, the remainder G_2 the "inessential part." There are now two cases to be considered, according to whether the points x_k and x_{rk} lie in G_2 or in G_1 (they must clearly both lie in the same subgraph). In the first case, the factor $C(G_2)$ of $C(G)$ can be seen by a comparison of (31) and (32) to be a contribution to the matrix element of the operator $S(\infty)$ for the transition $A\rightarrow B$. Now letting G vary over all possible graphs with the same G_1 and different G_2, the sum of the contributions of all such G is a constant $C(G_1)$ multiplied by the total matrix element of $S(\infty)$ for the transition $A\rightarrow B$. But for one-particle states the operator $S(\infty)$ is by (21) equivalent to the identity operator and gives, accordingly, a zero matrix element for the transition $A\rightarrow B$. Consequently, the disconnected G for which x_k and x_{rk} lie in G_2 give zero contribution to the matrix element of (31), and can be omitted from further consideration. When x_k and x_{rk} lie in G_1, again the $C(G)$ may be summed over all G consisting of the given G_1 and all possible G_2; but this time the connected graph G_1 itself is to be included in the sum. The sum of all the $C(G)$ in this case turns out to be just $C(G_1)$ multiplied by the expectation value in the vacuum of the operator $S(\infty)$. But the vacuum state, being a steady state, satisfies (21), and so the expectation value in question is equal to unity. Therefore the sum of the $C(G)$ reduces to the single term $C(G_1)$, and again the disconnected graphs may be omitted from consideration.

The elimination of disconnected graphs is, from a physical point of view, somewhat trivial, since these graphs arise merely from the fact that meaningful physical processes proceed simultaneously with totally irrelevant fluctuations of fields in the vacuum. However, similar arguments will now be used to eliminate a much more important class of graphs, namely, those involving self-energy effects. A "self-energy part" of a graph G is defined as follows; it is a set of one or more vertices not including x_0, together with the lines joining them, which is connected with the remainder of G (or with the edge of the diagram) only by two electron lines or by one or two photon lines. For definiteness it may be supposed that G has a self-energy part F, which is connected with its surroundings only by one electron line entering F at x_1, and another leaving F at x_2; the case of photon lines can be treated in an entirely analogous way. The points x_1 and x_2 may or may not be identical. From G a "reduced graph" G_0 can be obtained by omitting F completely and joining the incoming line at x_1 with the outgoing line at x_2 to form a single electron line in G_0, the newly formed line being denoted by λ. Given G_0 and λ, there is conversely a well determined set Γ of graphs G which are associated with G_0 and λ in this way; G_0 itself is considered also to belong to Γ. It will now be shown that the sum $C(\Gamma)$ of the contributions $C(G)$ to the matrix element of (31) from all the graphs G of Γ reduces to a single term $C'(G_0)$.

Suppose, for example, that the line λ in G_0 leads from a point x_3 to the edge of the diagram. Then $C(G_0)$ is an integral containing in the integrand the matrix element of

$$\bar{\psi}_\alpha(x_3) \tag{52}$$

for creation of an electron into the state B. Let the momentum-energy 4-vector of the created electron be p; the matrix element of (52) is of the form

$$Y_\alpha(x_3)=a_\alpha \exp[-i(p\cdot x_3)/\hbar] \tag{53}$$

with a_α independent of x_3. Now consider the sum $C(\Gamma)$. It follows from an analysis of (31) that $C(\Gamma)$ is obtained from $C(G_0)$ by replacing the operator (52) by

$$\sum_{n=0}^{\infty}(-i/\hbar c)^n[1/n!]\int_{-\infty}^{\infty}dy_1\cdots\int_{-\infty}^{\infty}dy_n \times P(\bar{\psi}_\alpha(x_3),\, H^I(y_1),\, \cdots,\, H^I(y_n)). \tag{54}$$

(This is, of course, a consequence of the special character of the graphs of Γ.) It is required to calculate the matrix element of (54) for a transition from the vacuum state O to the state B, i.e., for the emission of an electron into state B. This matrix element will be denoted by Z_α; $C(\Gamma)$ involves Z_α in the same way that $C(G_0)$ involves (53). Now Z_α can be evaluated as a sum of terms of the same general character as (47); it will be of the form

$$Z_\alpha=\sum_i\int_{-\infty}^{\infty}K_i^{\alpha\beta}(y_i-x_3)\,Y_\beta(y_i)dy_i,$$

where the important fact is that K_i is a function only of the coordinate differences between y_i and x_3. By (53), this implies that

$$Z_\alpha=R_{\alpha\beta}(p)\,Y_\beta(x_3), \tag{55}$$

with R independent of x_3. From considerations of relativistic invariance, R must be of the form

$$\delta_{\beta\alpha}R_1(p^2)+(p_\mu\gamma_\mu)_{\beta\alpha}R_2(p^2),$$

where p^2 is the square of the invariant length of the 4-vector p. But since the matrix element (53) is a solution of the Dirac equation,

$$p^2=-\hbar^2\kappa_0^2,\quad (p_\mu\gamma_\mu)_{\beta\alpha}Y_\beta=i\hbar\kappa_0 Y_\alpha,$$

and so (55) reduces to

$$Z_\alpha=R_1Y_\alpha(x_3),$$

with R_1 an absolute constant. Therefore the sum $C(\Gamma)$ is in this case just $C'(G_0)$, where $C'(G_0)$ is obtained from $C(G_0)$by the replacement

$$\bar{\psi}(x_3)\rightarrow R_1\bar{\psi}(x_3). \tag{56}$$

In the case when the line λ leads into the graph G_0 from the edge of the diagram to the point x_3, it is clear that $C(\Gamma)$ will be similarly obtained from $C(G_0)$ by the replacement

$$\psi(x_3)\rightarrow R_1^*\psi(x_3). \tag{57}$$

There remains the case in which λ leads from one vertex x_3 to another x_4 of G_0. In this case $C(G_0)$ contains in its integrand the function

$$\tfrac{1}{2}\eta(x_3,x_4)S_{F\beta\alpha}(x_3-x_4), \tag{58}$$

which is the vacuum expectation value of the operator

$$P(\bar{\psi}_\alpha(x_3),\psi_\beta(x_4)) \tag{59}$$

according to (43). Now in analogy with (54), $C(\Gamma)$ is obtained from $C(G_0)$ by replacing (59) by

$$\sum_{n=0}^{\infty}(-i/\hbar c)^n[1/n!]\int_{-\infty}^{\infty}dy_1\cdots\int_{-\infty}^{\infty}dy_n \times P(\bar{\psi}_\alpha(x_3),\, \psi_\beta(x_4),\, H^I(y_1),\, \cdots,\, H^I(y_n)), \tag{60}$$

and the vacuum expectation value of this operator will be denoted by

$$\tfrac{1}{2}\eta(x_3,x_4)S'_{F\beta\alpha}(x_3-x_4). \tag{61}$$

By the methods of Section VI, (61) can be expanded as a series of terms of the same character as (47); this expansion will not be discussed in detail here, but it is easy to see that it leads to an expression of the form (61), with $S_F'(x)$ a certain universal function of the 4-vector x. It will not be possible to reduce (61) to a numerical multiple of (58), as Z_α was in the previous case reduced to a multiple of Y_α. Instead, there may be expected to be a series expansion of the form

$$S_{F\beta\alpha}(x)=(R_2+a_1(\Box^2-\kappa_0^2)+a_2(\Box^2-\kappa_0^2)^2+\cdots)S_{F\beta\alpha}(x)+(b_1+b_2(\Box^2-\kappa_0^2)+\cdots)\times(\gamma_\mu[\partial/\partial x_\mu]-\kappa_0)_{\beta\gamma}S_{F\gamma\alpha}(x), \tag{62}$$

where $\Box^2$ is the Dalembertian operator and the a, b are numerical coefficients. In this case $C(\Gamma)$ will be equal to the $C'(G_0)$ obtained from $C(G_0)$ by the replacement

$$S_F(x_3-x_4)\rightarrow S_F'(x_3-x_4). \tag{63}$$

Applying the same methods to a graph G with a self-energy part connected to its surroundings by two photon lines, the sum $C(\Gamma)$ will be obtained as a single contribution $C'(G_0)$ from the reduced graph G_0, $C'(G_0)$ being formed from $C(G_0)$ by the replacement

$$D_F(x_3-x_4)\to D_F'(x_3-x_4). \quad (64)$$

The function D_F' is defined by the condition that

$$\tfrac{1}{2}\hbar c\delta_{\mu\nu}D_F'(x_3-x_4) \quad (65)$$

is the vacuum expectation value of the operator

$$\sum_{n=0}^{\infty}(-i/\hbar c)^n[1/n!]\int_{-\infty}^{\infty}dy_1\cdots\int_{-\infty}^{\infty}dy_n$$

$$\times P(A_\mu(x_3), A_\nu(x_4), H^I(y_1), \cdots, H^I(y_n)), \quad (66)$$

and may be expanded in a series

$$D_F'(x) = (R_3+c_1\Box^2+c_2(\Box^2)^2+\cdots)D_F(x). \quad (67)$$

Finally, it is not difficult to see that for graphs G with self-energy parts connected to their surroundings by a single photon line, the sum $C(\Gamma)$ will be identically zero, and so such graphs may be omitted from consideration entirely.

As a result of the foregoing arguments, the contributions $C(G)$ of graphs with self-energy parts can always be replaced by modified contributions $C'(G_0)$ from a reduced graph G_0. A given G may be reducible in more than one way to give various G_0, but if the process of reduction is repeated a finite number of times a G_0 will be obtained which is "totally reduced," contains no self-energy part, and is uniquely determined by G. The contribution $C'(G_0)$ of a totally reduced graph to the matrix element of (31) is now to be calculated as a sum of integrals of expressions like (47), but with a replacement (56), (57), (63), or (64) made corresponding to every line in G_0. This having been done, the matrix element of (31) is correctly calculated by taking into consideration each totally reduced graph once and once only.

The elimination of graphs with self-energy parts is a most important simplification of the theory. For according to (22), H^I contains the subtracted part H^S, which will give rise to many additional terms in the expansion of (31). But if any such term is taken, say, containing the factor $H^S(x_i)$ in the integrand, every graph corresponding to that term will contain the point x_i joined to the rest of the graph only by two electron lines, and this point by itself constitutes a self-energy part of the graph. Therefore, all terms involving H^S are to be omitted from (31) in the calculation of matrix elements. The intuitive argument for omitting these terms is that they were only introduced in order to cancel out higher order self-energy terms arising from H^i, which are also to be omitted; the analysis of the foregoing paragraphs is a more precise form of this argument. In physical language, the argument can be stated still more simply; since δm is an unobservable quantity, it cannot appear in the final description of observable phenomena.

VIII. VACUUM POLARIZATION AND CHARGE RENORMALIZATION

The question now arises: What is the physical meaning of the new functions D_F' and S_F', and of the constant R_1? In general terms, the answer is clear. The physical processes represented by the self-energy parts of graphs have been pushed out of the calculations, but these processes do not consist entirely of unobservable interactions of single particles with their self-fields, and so cannot entirely be written off as "self-energy processes." In addition, these processes include the phenomenon of vacuum polarization, i.e., the modification of the field surrounding a charged particle by the charges which the particle induces in the vacuum. Therefore, the appearance of D_F', S_F', and R_1 in the calculations may be regarded as an explicit representation of the vacuum polarization phenomena which were implicitly contained in the processes now ignored.

In the present theory there are two kinds of vacuum polarization, one induced by the external field and the other by the quantized electron and photon fields themselves; these will be called "external" and "internal," respectively. It is only the internal polarization which is represented yet in explicit fashion by the substitutions (56), (57), (63), (64); the external will be included later.

To form a concrete picture of the function D_F', it may be observed that the function $D_F(y-z)$ represents in classical electrodynamics the retarded potential of a point charge at y acting upon a point charge at z, together with the retarded potential of the charge at z acting on the charge at y. Therefore, D_F may be spoken of loosely as "the electromagnetic interaction between two point charges." In this semiclassical picture, D_F' is then the electromagnetic interaction between two point charges, including the effects of the charge-distribution which each charge induces in the vacuum.

The complete phenomenon of vacuum polarization, as hitherto understood, is included in the above picture of the function D_F'. There is nothing left for S_F' to represent. Thus, one of the important conclusions of the present theory is that there is a second phenomenon occurring in nature, included in the term vacuum polarization as used in this paper, but additional to vacuum polarization in the usual sense of the word. The nature of the second phenomenon can best be explained by an example.

The scattering of one electron by another may be represented as caused by a potential energy (the Møller interaction) acting between them. If one electron is at y and the other at z, then, as explained above, the effect of vacuum polarization of the usual kind is to replace a factor D_F in this potential energy by D_F'. Now consider an analogous, but unorthodox, representation of the Compton effect, or the scattering of an electron by a photon. If the electron is at y and the photon at z, the scattering may be again represented by a potential energy, containing now the operator $S_F(y-z)$ as a factor; the potential is an exchange potential, because after the interaction the electron must be considered to be at z and the photon at y, but this does not detract from its usefulness. By analogy with the 4-vector charge-current density j_μ which interacts with the potential D_F, a spinor Compton-effect density u_α may be defined by the equation

$$u_\alpha(x) = A_\mu(x)(\gamma_\mu)_{\alpha\beta}\psi_\beta(x),$$

and an adjoint spinor by

$$\bar{u}_\alpha(x) = \bar{\psi}_\beta(x)(\gamma_\mu)_{\beta\alpha}A_\mu(x).$$

These spinors are not directly observable quantities, but the Compton effect can be adequately described as an exchange potential, of magnitude proportional to $S_F(y-z)$, acting between the Compton-effect density at any point y and the adjoint density at z. The second vacuum polarization phenomenon is described by a change in the form of this potential from S_F to S_F'. Therefore, the phenomenon may be pictured in physical terms as the inducing, by a given element of Compton-effect density at a given point, of additional Compton-effect density in the vacuum around it.

In both sorts of internal vacuum polarization, the functions D_F and S_F, in addition to being altered in shape, become multiplied by numerical (and actually divergent) factors R_3 and R_2; also the matrix elements of (31) become multiplied by numerical factors such as $R_1R_1^*$. However, it is believed (this has been verified only for second-order terms) that all n'th-order matrix elements of (31) will involve these factors only in the form of a multiplier

$$(eR_2R_3^{\frac{1}{2}})^n;$$

this statement includes the contributions from the higher terms of the series (62) and (67). Here e is defined as the constant occurring in the fundamental interaction (16) by virtue of (37). Now the only possible experimental determination of e is by means of measurements of the effects described by various matrix elements of (31), and so the directly measured quantity is not e but $eR_2R_3^{\frac{1}{2}}$. Therefore, in practice the letter e is used to denote this measured quantity, and the multipliers R no longer appear explicitly in the matrix elements of (31); the change in the meaning of the letter e is called "charge renormalization," and is essential if e is to be identified with the observed electronic charge. As a result of the renormalization, the divergent coefficients R_1, R_2, and R_3 in (56), (57), (62), and (67) are to be replaced by unity, and the higher coefficients a, b, and c by expressions involving only the renormalized charge e.

The external vacuum polarization induced by the potential A_μ^e is, physically speaking, only a special case of the first sort of internal polarization; it can be treated in a precisely similar manner. Graphs describing external polarization effects are those with an "external polarization part," namely, a part including the point x_0 and connected with the rest of the graph by only a single photon line. Such a graph is to be "reduced" by omitting the polarization part entirely and renaming with the label x_0 the point at the further end of the single photon line. A discussion similar to those of Section VII leads to the conclusion that only reduced graphs need be considered in the calculation of the matrix element of (31), and that the effect of external polarization is explicitly represented if in the contributions from these graphs a replacement

$$A_\mu^e(x) \rightarrow A_\mu^{e\prime}(x) \tag{68}$$

is made. After a renormalization of the unit of potential, similar to the renormalization of charge, the modified potential $A_\mu^{e\prime}$ takes the form

$$A_\mu^{e\prime}(x) = (1 + c_1\Box^2 + c_2(\Box^2)^2 + \cdots)A_\mu^e(x), \tag{69}$$

where the coefficients are the same as in (67).

It is necessary, in order to determine the functions D_F', S_F', and $A_\mu^{e\prime}$, to go back to formulas (60) and (66). The determination of the vacuum expectation values of the operators (60) and (66) is a problem of the same kind as the original problem of the calculation of matrix elements of (31), and the various terms in the operators (60) and (66) must again be split up, represented by graphs, and analyzed in detail. However, since D_F' and S_F' are universal functions, this further analysis has only to be carried out once to be applicable to all problems.

It is one of the major triumphs of the Schwinger theory that it enables an unambiguous interpretation to be given to the phenomenon of vacuum polarization (at least of the first kind), and to the vacuum expectation value of an operator such as (66). In making this interpretation, profound theoretical problems arise, particularly concerned with the gauge invariance of the theory, about which nothing will be said here. For Schwinger's solution of these problems, the reader must refer to his forthcoming papers. Schwinger's argument can be transferred without essential change into the framework of the present paper.

Having overcome the difficulties of principle, Schwinger proceeded to evaluate the function D_F' explicitly as far as terms of order $\alpha = (e^2/4\pi\hbar c)$ (heaviside units). In particular, he found for the coefficient c_1 in (67) and (69) the value $(-\alpha/15\pi\kappa_0^2)$ to this order.[8] It is hoped to publish in a sequel to the present paper a similar evaluation of the function S_F'; the analysis involved is too complicated to be summarized here.

[8] Schwinger's results agree with those of the earlier, theoretically unsatisfactory treatment of vacuum polarization. The best account of the earlier work is V. F. Weisskopf, Kgl. Danske Sels. Math.-Fys. Medd. 14, No. 6 (1936).

IX. SUMMARY OF RESULTS

In this section the results of the preceding pages will be summarized, so far as they relate to the performance of practical calculations. In effect, this summary will consist of a set of rules for the application of the Feynman radiation theory to a certain class of problems.

Suppose an electron to be moving in an external field with interaction energy given by (33). Then the interaction energy to be used in calculating the motion of the electron, including radiative corrections of all orders, is

$$\begin{aligned} H_E(x_0) &= \sum_{n=0}^{\infty} (-i/\hbar c)^n [1/n!] J_n \\ &= \sum_{n=0}^{\infty} (-i/\hbar c)^n [1/n!] \int_{-\infty}^{\infty} dx_1 \cdots \int_{-\infty}^{\infty} dx_n \\ &\quad \times P(H^e(x_0), H^i(x_1), \cdots, H^i(x_n)), \end{aligned} \tag{70}$$

with H^i given by (16), and the P notation as defined in (29).

To find the effective n'th-order radiative correction to the potential acting on the electron, it is necessary to calculate the matrix elements of J_n for transitions from one one-electron state to another. These matrix elements can be written down most conveniently in the form of an operator K_n bilinear in $\bar\psi$ and ψ, whose matrix elements for one-electron transitions are the same as those to be determined. In fact, the operator K_n itself is already the matrix element to be determined if the $\bar\psi$ and ψ contained in it are regarded as one-electron wave functions.

To write down K_n, the integrand P_n in J_n is first expressed in terms of its factors $\bar\psi$, ψ, and A, all suffixes being indicated explicitly, and the expression (37) used for j_μ. All possible graphs G with $(n+1)$ vertices are now drawn as described in Section VII, omitting disconnected graphs, graphs with self-energy parts, and graphs with external vacuum polarization parts as defined in Section VIII. It will be found that in each graph there are at each vertex two electron lines and one photon line, with the exception of x_0 at which there are two electron lines only; further, such graphs can exist only for even n. K_n is the sum of a contribution $K(G)$ from each G.

Given G, $K(G)$ is obtained from J_n by the following transformations. First, for each photon line joining x and y in G, replace two factors $A_\mu(x)A_\nu(y)$ in P_n (regardless of their positions) by

$$\tfrac{1}{2}\hbar c\delta_{\mu\nu}D_F'(x-y), \qquad (71)$$

with D_F' given by (67) with $R_3=1$, the function D_F being defined by (42). Second, for each electron line joining x to y in G, replace two factors $\bar{\psi}_\alpha(x)\psi_\beta(y)$ in P_n (regardless of positions) by

$$\tfrac{1}{2}S'_{F\beta\alpha}(x-y) \qquad (72)$$

with S_F' given by (62) with $R_2=1$, the function S_F being defined by (44) and (45). Third, replace the remaining two factors $P(\bar{\psi}_\gamma(z),\psi_\delta(w))$ in P_n by $\bar{\psi}_\gamma(z)\psi_\delta(w)$ in this order. Fourth, replace $A_\mu{}^e(x_0)$ by $A_\mu{}^{e\prime}(x_0)$ given by

$$A_\mu{}^{e\prime}(x)=A_\mu{}^e(x)-[\alpha/15\pi\kappa_0{}^2]\Box^2A_\mu{}^e(x) \qquad (73)$$

or, more generally, by (69). Fifth, multiply the whole by $(-1)^l$, where l is the number of closed loops in G as defined in Section VII.

The above rules enable K_n to be written down very rapidly for small values of n. It should be observed that if K_n is being calculated, and if it is not desired to include effects of higher order than the n'th, then D_F', S_F', and $A_\mu{}^{e\prime}$ in (71), (72), and (73) reduce to the simple functions D_F, S_F, and $A_\mu{}^e$. Also, the integrand in J_n is a symmetrical function of $x_1, \cdots, x_n$; therefore, graphs which differ only by a relabeling of the vertices $x_1, \cdots, x_n$ give identical contributions to K_n and need not be considered separately.

The extension of these rules to cover the calculation of matrix elements of (70) of a more general character than the one-electron transitions hitherto considered presents no essential difficulty. All that is necessary is to consider graphs with more than two "loose ends," representing processes in which more than one particle is involved. This extension is not treated in the present paper, chiefly because it would lead to unpleasantly cumbersome formulas.

X. EXAMPLE—SECOND-ORDER RADIATIVE CORRECTIONS

As an illustration of the rules of procedure of the previous section, these rules will be used for writing down the terms giving second-order radiative corrections to the motion of an electron in an external field. Let the energy of the external field be

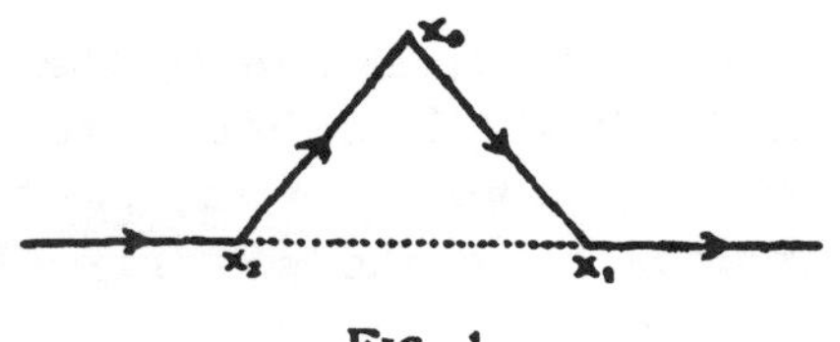

FIG. 1.

$$-[1/c]j_\mu(x_0)A_\mu{}^e(x_0). \qquad (74)$$

Then there will be one second-order correction term

$$U=[\alpha/15\pi\kappa_0{}^2][1/c]j_\mu(x_0)\Box^2A_\mu{}^e(x_0)$$

arising from the substitution (73) in the zero-order term (74). This is the well-known vacuum polarization or Uehling term.[9]

The remaining second-order term arises from the second-order part J_2 of (70). Written in expanded form, J_2 is

$$J_2=ie^3\int_{-\infty}^{\infty}dx_1\int_{-\infty}^{\infty}dx_2P(\bar{\psi}_\alpha(x_0)(\gamma_\lambda)_{\alpha\beta}\psi_\beta(x_0)A_\lambda{}^e(x_0),$$
$$\bar{\psi}_\gamma(x_1)(\gamma_\mu)_{\gamma\delta}\psi_\delta(x_1)A_\mu(x_1),$$
$$\bar{\psi}_\epsilon(x_2)(\gamma_\nu)_{\epsilon\zeta}\psi_\zeta(x_2)A_\nu(x_2)).$$

Next, all admissable graphs with the three vertices x_0, x_1, x_2 are to be drawn. It is easy to see that there are only two such graphs, that G shown in Fig. 1, and the identical graph with x_1 and x_2 interchanged. The full lines are electron lines, the dotted line a photon line. The contribution $K(G)$ is obtained from J_2 by substituting according to the rules of Section IX; in this case $l=0$, and the primes can be omitted from (71), (72), (73) since only second-order terms are required. The integrand in $K(G)$ can be reassembled into the form of a matrix product, suppressing the suffixes $\alpha, \cdots, \zeta$. Then, multiplying by a factor 2 to allow for the second graph, the complete second-order correction to (74) arising from J_2 becomes

$$L=-i[e^3/8\hbar c]\int_{-\infty}^{\infty}dx_1\int_{-\infty}^{\infty}dx_2D_F(x_1-x_2)A_\mu{}^e(x_0)$$
$$\times\bar{\psi}(x_1)\gamma_\nu S_F(x_0-x_1)\gamma_\mu S_F(x_2-x_0)\gamma_\nu\psi(x_2).$$

[9] Robert Serber, Phys. Rev. **48**, 49 (1935); E. A. Uehling, Phys. Rev. **48**, 55 (1935).

This is the term which gives rise to the main part of the Lamb-Retherford line shift,[10] the anomalous magnetic moment of the electron,[11] and the anomalous hyperfine splitting of the ground state of hydrogen.[12]

The above expression L is formally simpler than the corresponding expression obtained by Schwinger, but the two are easily seen to be equivalent. In particular, the above expression does not lead to any great reduction in the labor involved in a numerical calculation of the Lamb shift. Its advantage lies rather in the ease with which it can be written down.

In conclusion, the author would like to express his thanks to the Commonwealth Fund of New York for financial support, and to Professors Schwinger and Feynman for the stimulating lectures in which they presented their respective theories.

Notes added in proof (To Section II). The argument of Section II is an over-simplification of the method of Tomonaga,[1] and is unsound. There is an error in the derivation of (3); derivatives occurring in $H(r)$ give rise to non-commutativity between $H(r)$ and field quantities at r' when r is a point on σ infinitesimally distant from r'. The argument should be amended as follows. Φ is defined only for flat surfaces $t(r)=t$, and for such surfaces (3) and (6) are correct. Ψ is defined for general surfaces by (12) and (10), and is verified to satisfy (9). For a flat surface, Φ and Ψ are then shown to be related by (7). Finally, since H_1 does not involve the derivatives in H, the argument leading to (3) can be correctly applied to prove that for general σ the state-vector $\Psi(\sigma)$ will completely describe results of observations of the system on σ.

(To Section III). A covariant perturbation theory similar to that of Section III has previously been developed by E. C G. Stueckelberg, Ann. d. Phys. **21**, 367 (1934); Nature, **153**, 143 (1944).

(To Section V). Schwinger's "effective potential" is not H_T given by (25), but is $H_T'=QH_TQ^{-1}$. Here Q is a "square-root" of $S(\infty)$ obtained by expanding $(S(\infty))^{\frac{1}{2}}$ by the binomial theorem. The physical meaning of this is that Schwinger specifies states neither by Ω nor by Ω', but by an intermediate state-vector $\Omega''=Q\Omega=Q^{-1}\Omega'$, whose definition is symmetrical between past and future. H_T' is also symmetrical between past and future. For one-particle states, H_T and H_T' are identical.

Equation (32) can most simply be obtained directly from the product expansion of $S(\infty)$.

(To Section VII). Equation (62) is incorrect. The function S_F' is well-behaved, but its fourier transform has a logarithmic dependence on frequency, which makes an expansion precisely of the form (62) impossible.

(To Section X). The term L still contains two divergent parts. One is an "infra-red catastrophe" removable by standard methods. The other is an "ultraviolet" divergence, and has to be interpreted as an additional charge-renormalization, or, better, cancelled by part of the charge-renormalization calculated in Section VIII.

[10] W. E. Lamb and R. C. Retherford, Phys. Rev. **72**, 241 (1947).

[11] P. Kusch and H. M. Foley, Phys. Rev. **74**, 250 (1948).

[12] J. E. Nafe and E. B. Nelson, Phys. Rev. **73**, 718 (1948); Aage Bohr, Phys. Rev. **73**, 1109 (1948).

PHYSICAL REVIEW VOLUME 76, NUMBER 6 SEPTEMBER 15, 1949

Space-Time Approach to Quantum Electrodynamics

R. P. FEYNMAN
Department of Physics, Cornell University, Ithaca, New York

(Received May 9, 1949)

In this paper two things are done. (1) It is shown that a considerable simplification can be attained in writing down matrix elements for complex processes in electrodynamics. Further, a physical point of view is available which permits them to be written down directly for any specific problem. Being simply a restatement of conventional electrodynamics, however, the matrix elements diverge for complex processes. (2) Electrodynamics is modified by altering the interaction of electrons at short distances. All matrix elements are now finite, with the exception of those relating to problems of vacuum polarization. The latter are evaluated in a manner suggested by Pauli and Bethe, which gives finite results for these matrices also. The only effects sensitive to the modification are changes in mass and charge of the electrons. Such changes could not be directly observed. Phenomena directly observable, are insensitive to the details of the modification used (except at extreme energies). For such phenomena, a limit can be taken as the range of the modification goes to zero. The results then agree with those of Schwinger. A complete, unambiguous, and presumably consistent, method is therefore available for the calculation of all processes involving electrons and photons.

The simplification in writing the expressions results from an emphasis on the over-all space-time view resulting from a study of the solution of the equations of electrodynamics. The relation of this to the more conventional Hamiltonian point of view is discussed. It would be very difficult to make the modification which is proposed if one insisted on having the equations in Hamiltonian form.

The methods apply as well to charges obeying the Klein-Gordon equation, and to the various meson theories of nuclear forces. Illustrative examples are given. Although a modification like that used in electrodynamics can make all matrices finite for all of the meson theories, for some of the theories it is no longer true that all directly observable phenomena are insensitive to the details of the modification used.

The actual evaluation of integrals appearing in the matrix elements may be facilitated, in the simpler cases, by methods described in the appendix.

THIS paper should be considered as a direct continuation of a preceding one[1] (I) in which the motion of electrons, neglecting interaction, was analyzed, by dealing directly with the *solution* of the Hamiltonian differential equations. Here the same technique is applied to include interactions and in that way to express in simple terms the solution of problems in quantum electrodynamics.

For most practical calculations in quantum electrodynamics the solution is ordinarily expressed in terms of a matrix element. The matrix is worked out as an expansion in powers of $e^2/\hbar c$, the successive terms corresponding to the inclusion of an increasing number of virtual quanta. It appears that a considerable simplification can be achieved in writing down these matrix elements for complex processes. Furthermore, each term in the expansion can be written down and understood directly from a physical point of view, similar to the space-time view in I. It is the purpose of this paper to describe how this may be done. We shall also discuss methods of handling the divergent integrals which appear in these matrix elements.

The simplification in the formulae results mainly from the fact that previous methods unnecessarily separated into individual terms processes that were closely related physically. For example, in the exchange of a quantum between two electrons there were two terms depending on which electron emitted and which absorbed the quantum. Yet, in the virtual states considered, timing relations are not significant. Olny the order of operators in the matrix must be maintained. We have seen (I), that in addition, processes in which virtual pairs are produced can be combined with others in which only positive energy electrons are involved. Further, the effects of longitudinal and transverse waves can be combined together. The separations previously made were on an unrelativistic basis (reflected in the circumstance that apparently momentum but not energy is conserved in intermediate states). When the terms are combined and simplified, the relativistic invariance of the result is self-evident.

We begin by discussing the solution in space and time of the Schrödinger equation for particles interacting instantaneously. The results are immediately generalizable to delayed interactions of relativistic electrons and we represent in that way the laws of quantum electrodynamics. We can then see how the matrix element for any process can be written down directly. In particular, the self-energy expression is written down.

So far, nothing has been done other than a restatement of conventional electrodynamics in other terms. Therefore, the self-energy diverges. A modification[2] in interaction between charges is next made, and it is shown that the self-energy is made convergent and corresponds to a correction to the electron mass. After the mass correction is made, other real processes are finite and insensitive to the "width" of the cut-off in the interaction.[3]

Unfortunately, the modification proposed is not completely satisfactory theoretically (it leads to some difficulties of conservation of energy). It does, however, seem consistent and satisfactory to define the matrix

[1] R. P. Feynman, Phys. Rev. 76, 749 (1949), hereafter called I.

[2] For a discussion of this modification in classical physics see R. P. Feynman, Phys. Rev. 74 939 (1948), hereafter referred to as A.

[3] A brief summary of the methods and results will be found in R. P. Feynman, Phys. Rev. 74, 1430 (1948), hereafter referred to as B.

element for all real processes as the limit of that computed here as the cut-off width goes to zero. A similar technique suggested by Pauli and by Bethe can be applied to problems of vacuum polarization (resulting in a renormalization of charge) but again a strict physical basis for the rules of convergence is not known.

After mass and charge renormalization, the limit of zero cut-off width can be taken for all real processes. The results are then equivalent to those of Schwinger[4] who does not make explicit use of the convergence factors. The method of Schwinger is to identify the terms corresponding to corrections in mass and charge and, previous to their evaluation, to remove them from the expressions for real processes. This has the advantage of showing that the results can be strictly independent of particular cut-off methods. On the other hand, many of the properties of the integrals are analyzed using formal properties of invariant propagation functions. But one of the properties is that the integrals are infinite and it is not clear to what extent this invalidates the demonstrations. A practical advantage of the present method is that ambiguities can be more easily resolved; simply by direct calculation of the otherwise divergent integrals. Nevertheless, it is not at all clear that the convergence factors do not upset the physical consistency of the theory. Although in the limit the two methods agree, neither method appears to be thoroughly satisfactory theoretically. Nevertheless, it does appear that we now have available a complete and definite method for the calculation of physical processes to any order in quantum electrodynamics.

Since we can write down the solution to any physical problem, we have a complete theory which could stand by itself. It will be theoretically incomplete, however, in two respects. First, although each term of increasing order in $e^2/\hbar c$ can be written down it would be desirable to see some way of expressing things in finite form to all orders in $e^2/\hbar c$ at once. Second, although it will be physically evident that the results obtained are equivalent to those obtained by conventional electrodynamics the mathematical proof of this is not included. Both of these limitations will be removed in a subsequent paper (see also Dyson[4]).

Briefly the genesis of this theory was this. The conventional electrodynamics was expressed in the Lagrangian form of quantum mechanics described in the Reviews of Modern Physics.[5] The motion of the field oscillators could be integrated out (as described in Section 13 of that paper), the result being an expression of the delayed interaction of the particles. Next the modification of the delta-function interaction could be made directly from the analogy to the classical case.[2] This was still not complete because the Lagrangian method had been worked out in detail only for particles obeying the non-relativistic Schrödinger equation. It was then modified in accordance with the requirements of the Dirac equation and the phenomenon of pair creation. This was made easier by the reinterpretation of the theory of holes (I). Finally for practical calculations the expressions were developed in a power series in $e^2/\hbar c$. It was apparent that each term in the series had a simple physical interpretation. Since the result was easier to understand than the derivation, it was thought best to publish the results first in this paper. Considerable time has been spent to make these first two papers as complete and as physically plausible as possible without relying on the Lagrangian method, because it is not generally familiar. It is realized that such a description cannot carry the conviction of truth which would accompany the derivation. On the other hand, in the interest of keeping simple things simple the derivation will appear in a separate paper.

The possible application of these methods to the various meson theories is discussed briefly. The formulas corresponding to a charge particle of zero spin moving in accordance with the Klein Gordon equation are also given. In an Appendix a method is given for calculating the integrals appearing in the matrix elements for the simpler processes.

The point of view which is taken here of the interaction of charges differs from the more usual point of view of field theory. Furthermore, the familiar Hamiltonian form of quantum mechanics must be compared to the over-all space-time view used here. The first section is, therefore, devoted to a discussion of the relations of these viewpoints.

1. COMPARISON WITH THE HAMILTONIAN METHOD

Electrodynamics can be looked upon in two equivalent and complementary ways. One is as the description of the behavior of a field (Maxwell's equations). The other is as a description of a direct interaction at a distance (albeit delayed in time) between charges (the solutions of Lienard and Wiechert). From the latter point of view light is considered as an interaction of the charges in the source with those in the absorber. This is an impractical point of view because many kinds of sources produce the same kind of effects. The field point of view separates these aspects into two simpler problems, production of light, and absorption of light. On the other hand, the field point of view is less practical when dealing with close collisions of particles (or their action on themselves). For here the source and absorber are not readily distinguishable, there is an intimate exchange of quanta. The fields are so closely determined by the motions of the particles that it is just as well not to separate the question into two problems but to consider the process as a direct interaction. Roughly, the field point of view is most practical for problems involv-

[4] J. Schwinger, Phys. Rev. 74, 1439 (1948), Phys. Rev. 75, 651 (1949). A proof of this equivalence is given by F. J. Dyson, Phys. Rev. 75, 486 (1949).

[5] R. P. Feynman, Rev. Mod. Phys. 20, 367 (1948). The application to electrodynamics is described in detail by H. J. Groenewold, Koninklijke Nederlandsche Akademia van Weteschappen. Proceedings Vol. LII, 3 (226) 1949.

ing real quanta, while the interaction view is best for the discussion of the virtual quanta involved. We shall emphasize the interaction viewpoint in this paper, first because it is less familiar and therefore requires more discussion, and second because the important aspect in the problems with which we shall deal is the effect of virtual quanta.

The Hamiltonian method is not well adapted to represent the direct action at a distance between charges because that action is delayed. The Hamiltonian method represents the future as developing out of the present. If the values of a complete set of quantities are known now, their values can be computed at the next instant in time. If particles interact through a delayed interaction, however, one cannot predict the future by simply knowing the present motion of the particles. One would also have to know what the motions of the particles were in the past in view of the interaction this may have on the future motions. This is done in the Hamiltonian electrodynamics, of course, by requiring that one specify besides the present motion of the particles, the values of a host of new variables (the coordinates of the field oscillators) to keep track of that aspect of the past motions of the particles which determines their future behavior. The use of the Hamiltonian forces one to choose the field viewpoint rather than the interaction viewpoint.

In many problems, for example, the close collisions of particles, we are not interested in the precise temporal sequence of events. It is not of interest to be able to say how the situation would look at each instant of time during a collision and how it progresses from instant to instant. Such ideas are only useful for events taking a long time and for which we can readily obtain information during the intervening period. For collisions it is much easier to treat the process as a whole.[6] The Møller interaction matrix for the the collision of two electrons is not essentially more complicated than the non-relativistic Rutherford formula, yet the mathematical machinery used to obtain the former from quantum electrodynamics is vastly more complicated than Schrödinger's equation with the e^2/r_{12} interaction needed to obtain the latter. The difference is only that in the latter the action is instantaneous so that the Hamiltonian method requires no extra variables, while in the former relativistic case it is delayed and the Hamiltonian method is very cumbersome.

We shall be discussing the solutions of equations rather than the time differential equations from which they come. We shall discover that the solutions, because of the over-all space-time view that they permit, are as easy to understand when interactions are delayed as when they are instantaneous.

As a further point, relativistic invariance will be self-evident. The Hamiltonian form of the equations develops the future from the instantaneous present. But for different observers in relative motion the instantaneous present is different, and corresponds to a different 3-dimensional cut of space-time. Thus the temporal analyses of different observers is different and their Hamiltonian equations are developing the process in different ways. These differences are irrelevant, however, for the solution is the same in any space time frame. By forsaking the Hamiltonian method, the wedding of relativity and quantum mechanics can be accomplished most naturally.

We illustrate these points in the next section by studying the solution of Schrödinger's equation for non-relativistic particles interacting by an instantaneous Coulomb potential (Eq. 2). When the solution is modified to include the effects of delay in the interaction and the relativistic properties of the electrons we obtain an expression of the laws of quantum electrodynamics (Eq. 4).

2. THE INTERACTION BETWEEN CHARGES

We study by the same methods as in I, the interaction of two particles using the same notation as I. We start by considering the non-relativistic case described by the Schrödinger equation (I, Eq. 1). The wave function at a given time is a function $\psi(\mathbf{x}_a, \mathbf{x}_b, t)$ of the coordinates $\mathbf{x}_a$ and $\mathbf{x}_b$ of each particle. Thus call $K(\mathbf{x}_a, \mathbf{x}_b, t; \mathbf{x}_a', \mathbf{x}_b', t')$ the amplitude that particle a at $\mathbf{x}_a'$ at time t' will get to $\mathbf{x}_a$ at t while particle b at $\mathbf{x}_b'$ at t' gets to $\mathbf{x}_b$ at t. If the particles are free and do not interact this is

$$K(\mathbf{x}_a, \mathbf{x}_b, t; \mathbf{x}_a', \mathbf{x}_b', t') = K_{0a}(\mathbf{x}_a, t; \mathbf{x}_a', t')K_{0b}(\mathbf{x}_b, t; \mathbf{x}_b', t')$$

where K_{0a} is the K_0 function for particle a considered as free. In *this* case we can obviously define a quantity like K, but for which the time t need not be the same for particles a and b (likewise for t'); e.g.,

$$K_0(3, 4; 1, 2) = K_{0a}(3, 1)K_{0b}(4, 2) \tag{1}$$

can be thought of as the amplitude that particle a goes from $\mathbf{x}_1$ at t_1 to $\mathbf{x}_3$ at t_3 and that particle b goes from $\mathbf{x}_2$ at t_2 to $\mathbf{x}_4$ at t_4.

When the particles do interact, one can only define the quantity $K(3, 4; 1, 2)$ precisely if the interaction vanishes between t_1 and t_2 and also between t_3 and t_4. In a real physical system such is not the case. There is such an enormous advantage, however, to the concept that we shall continue to use it, imagining that we can neglect the effect of interactions between t_1 and t_2 and between t_3 and t_4. For practical problems this means choosing such long time intervals t_3-t_1 and t_4-t_2 that the extra interactions near the end points have small relative effects. As an example, in a scattering problem it may well be that the particles are so well separated initially and finally that the interaction at these times is negligible. Again energy values can be defined by the average rate of change of phase over such long time intervals that errors initially and finally can be neglected. Inasmuch as any physical problem can be defined in terms of scattering processes we do not lose much in

[6] This is the viewpoint of the theory of the S matrix of Heisenberg.

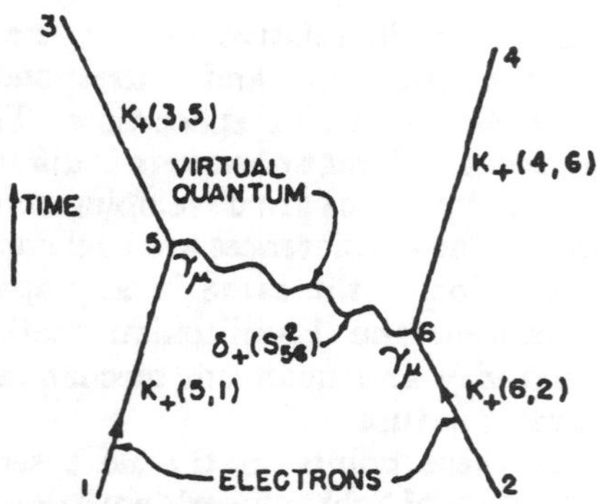

FIG. 1. The fundamental interaction Eq. (4). Exchange of one quantum between two electrons.

a general theoretical sense by this approximation. If it is not made it is not easy to study interacting particles relativistically, for there is nothing significant in choosing $t_1=t_3$ if $\mathbf{x}_1\neq\mathbf{x}_3$, as absolute simultaneity of events at a distance cannot be defined invariantly. It is essentially to avoid this approximation that the complicated structure of the older quantum electrodynamics has been built up. We wish to describe electrodynamics as a delayed interaction between particles. If we can make the approximation of assuming a meaning to $K(3,4;1,2)$ the results of this interaction can be expressed very simply.

To see how this may be done, imagine first that the interaction is simply that given by a Coulomb potential e^2/r where r is the distance between the particles. If this be turned on only for a very short time Δt_0 at time t_0, the first order correction to $K(3,4;1,2)$ can be worked out exactly as was Eq. (9) of I by an obvious generalization to two particles:

$$K^{(1)}(3,4;1,2)=-ie^2\int\int K_{0a}(3,5)K_{0b}(4,6)r_{56}^{-1}\times K_{0a}(5,1)K_{0b}(6,2)d^3\mathbf{x}_5d^3\mathbf{x}_6\Delta t_0,$$

where $t_5=t_6=t_0$. If now the potential were on at all times (so that strictly K is not defined unless $t_4=t_3$ and $t_1=t_2$), the first-order effect is obtained by integrating on t_0, which we can write as an integral over both t_5 and t_6 if we include a delta-function $\delta(t_5-t_6)$ to insure contribution only when $t_5=t_6$. Hence, the first-order effect of interaction is (calling $t_5-t_6=t_{56}$):

$$K^{(1)}(3,4;1,2)=-ie^2\int\int K_{0a}(3,5)K_{0b}(4,6)r_{56}^{-1}\times\delta(t_{56})K_{0a}(5,1)K_{0b}(6,2)d\tau_5d\tau_6, \quad (2)$$

where $d\tau=d^3\mathbf{x}dt$.

We know, however, in classical electrodynamics, that the Coulomb potential does not act instantaneously, but is delayed by a time r_{56}, taking the speed of light as unity. This suggests simply replacing $r_{56}^{-1}\delta(t_{56})$ in (2) by something like $r_{56}^{-1}\delta(t_{56}-r_{56})$ to represent the delay in the effect of b on a.

This turns out to be not quite right,[7] for when this interaction is represented by photons they must be of only positive energy, while the Fourier transform of $\delta(t_{56}-r_{56})$ contains frequencies of both signs. It should instead be replaced by $\delta_+(t_{56}-r_{56})$ where

$$\delta_+(x)=\int_0^\infty e^{-i\omega x}d\omega/\pi=\lim_{\epsilon\to 0}\frac{(\pi i)^{-1}}{x-i\epsilon}=\delta(x)+(\pi ix)^{-1}. \quad (3)$$

This is to be averaged with $r_{56}^{-1}\delta_+(-t_{56}-r_{56})$ which arises when $t_5<t_6$ and corresponds to a emitting the quantum which b receives. Since

$$(2r)^{-1}(\delta_+(t-r)+\delta_+(-t-r))=\delta_+(t^2-r^2),$$

this means $r_{56}^{-1}\delta(t_{56})$ is replaced by $\delta_+(s_{56}^2)$ where $s_{56}^2=t_{56}^2-r_{56}^2$ is the square of the relativistically invariant interval between points 5 and 6. Since in classical electrodynamics there is also an interaction through the vector potential, the complete interaction (see A, Eq. (1)) should be $(1-(\mathbf{v}_5\cdot\mathbf{v}_6)\delta_+(s_{56}^2)$, or in the relativistic case,

$$(1-\alpha_a\cdot\alpha_b)\delta_+(s_{56}^2)=\beta_a\beta_b\gamma_{a\mu}\gamma_{b\mu}\delta_+(s_{56}^2).$$

Hence we have for electrons obeying the Dirac equation,

$$K^{(1)}(3,4;1,2)=-ie^2\int\int K_{+a}(3,5)K_{+b}(4,6)\gamma_{a\mu}\gamma_{b\mu}\times\delta_+(s_{56}^2)K_{+a}(5,1)K_{+b}(6,2)d\tau_5d\tau_6, \quad (4)$$

where $\gamma_{a\mu}$ and $\gamma_{b\mu}$ are the Dirac matrices applying to the spinor corresponding to particles a and b, respectively (the factor $\beta_a\beta_b$ being absorbed in the definition, I Eq. (17), of K_+).

This is our fundamental equation for electrodynamics. It describes the effect of exchange of one quantum (therefore first order in e^2) between two electrons. It will serve as a prototype enabling us to write down the corresponding quantities involving the exchange of two or more quanta between two electrons or the interaction of an electron with itself. It is a consequence of conventional electrodynamics. Relativistic invariance is clear. Since one sums over μ it contains the effects of both longitudinal and transverse waves in a relativistically symmetrical way.

We shall now interpret Eq. (4) in a manner which will permit us to write down the higher order terms. It can be understood (see Fig. 1) as saying that the amplitude for "a" to go from 1 to 3 and "b" to go from 2 to 4 is altered to first order because they can exchange a quantum. Thus, "a" can go to 5 (amplitude $K_+(5,1)$)

[7] It, and a like term for the effect of a on b, leads to a theory which, in the classical limit, exhibits interaction through half-advanced and half-retarded potentials. Classically, this is equivalent to purely retarded effects within a closed box from which no light escapes (e.g., see A, or J. A. Wheeler and R. P. Feynman, Rev. Mod. Phys. 17, 157 (1945)). Analogous theorems exist in quantum mechanics but it would lead us too far astray to discuss them now.

emit a quantum (longitudinal, transverse, or scalar $\gamma_{a\mu}$) and then proceed to 3 ($K_+(3, 5)$). Meantime "b" goes to 6 ($K_+(6, 2)$), absorbs the quantum ($\gamma_{b\mu}$) and proceeds to 4 ($K_+(4, 6)$). The quantum meanwhile proceeds from 5 to 6, which it does with amplitude $\delta_+(s_{56}^2)$. We must sum over all the possible quantum polarizations μ and positions and times of emission 5, and of absorption 6. Actually if $t_5 > t_6$ it would be better to say that "a" absorbs and "b" emits but no attention need be paid to these matters, as all such alternatives are automatically contained in (4).

The correct terms of higher order in e^2 or involving larger numbers of electrons (interacting with themselves or in pairs) can be written down by the same kind of reasoning. They will be illustrated by examples as we proceed. In a succeeding paper they will all be deduced from conventional quantum electrodynamics.

Calculation, from (4), of the transition element between positive energy free electron states gives the Möller scattering of two electrons, when account is taken of the Pauli principle.

The exclusion principle for interacting charges is handled in exactly the same way as for non-interacting charges (I). For example, for two charges it requires only that one calculate $K(3, 4; 1, 2) - K(4, 3; 1, 2)$ to get the net amplitude for arrival of charges at 3 and 4. It is disregarded in intermediate states. The interference effects for scattering of electrons by positrons discussed by Bhabha will be seen to result directly in this formulation. The formulas are interpreted to apply to positrons in the manner discussed in I.

As our primary concern will be for processes in which the quanta are virtual we shall not include here the detailed analysis of processes involving real quanta in initial or final state, and shall content ourselves by only stating the rules applying to them.[8] The result of the analysis is, as expected, that they can be included by the same line of reasoning as is used in discussing the virtual processes, provided the quantities are normalized in the usual manner to represent single quanta. For example, the amplitude that an electron in going from 1 to 2 absorbs a quantum whose vector potential, suitably normalized, is $c_\mu \exp(-ik\cdot x) = C_\mu(x)$ is just the expression (I, Eq. (13)) for scattering in a potential with $\mathbf{A}(3)$ replaced by $\mathbf{C}(3)$. Each quantum interacts only once (either in emission or in absorption), terms like (I, Eq. (14)) occur only when there is more than one quantum involved. The Bose statistics of the quanta can, in all cases, be disregarded in intermediate states. The only effect of the statistics is to change the weight of initial or final states. If there are among quanta, in the initial state, some n which are identical then the weight of the state is $(1/n!)$ of what it would be if these quanta were considered as different (similarly for the final state).

3. THE SELF-ENERGY PROBLEM

Having a term representing the mutual interaction of a pair of charges, we must include similar terms to represent the interaction of a charge with itself. For under some circumstances what appears to be two distinct electrons may, according to I, be viewed also as a single electron (namely in case one electron was created in a pair with a positron destined to annihilate the other electron). Thus to the interaction between such electrons must correspond the possibility of the action of an electron on itself.[9]

This interaction is the heart of the self energy problem. Consider to first order in e^2 the action of an electron on itself in an otherwise force free region. The amplitude $K(2, 1)$ for a single particle to get from 1 to 2 differs from $K_+(2, 1)$ to first order in e^2 by a term

$$K^{(1)}(2, 1) = -ie^2 \int\int K_+(2, 4)\gamma_\mu K_+(4, 3)\gamma_\mu \times K_+(3, 1) d\tau_3 d\tau_4 \delta_+(s_{43}^2). \quad (6)$$

It arises because the electron instead of going from 1 directly to 2, may go (Fig. 2) first to 3, ($K_+(3, 1)$), emit a quantum (γ_μ), proceed to 4, ($K_+(4, 3)$), absorb it (γ_μ), and finally arrive at 2 ($K_+(2, 4)$). The quantum must go from 3 to 4 ($\delta_+(s_{43}^2)$).

This is related to the self-energy of a free electron in the following manner. Suppose initially, time t_1, we have an electron in state $f(1)$ which we imagine to be a positive energy solution of Dirac's equation for a free particle. After a long time $t_2 - t_1$ the perturbation will alter

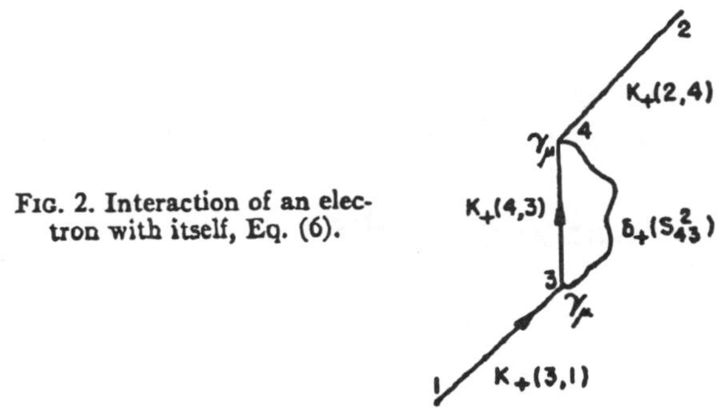

FIG. 2. Interaction of an electron with itself, Eq. (6).

[8] Although in the expressions stemming from (4) the quanta are virtual, this is not actually a theoretical limitation. One way to deduce the correct rules for real quanta from (4) is to note that in a closed system all quanta can be considered as virtual (i.e., they have a known source and are eventually absorbed) so that in such a system the present description is complete and equivalent to the conventional one. In particular, the relation of the Einstein A and B coefficients can be deduced. A more practical direct deduction of the expressions for real quanta will be given in the subsequent paper. It might be noted that (4) can be rewritten as describing the action on a, $K^{(1)}(3, 1) = i\int K_+(3, 5) \times A(5) K_+(5, 1) d\tau_5$ of the potential $A_\mu(5) = e^2 \int K_+(4, 6)\delta_+(s_{56}^2)\gamma_\mu \times K_+(6, 2) d\tau_6$ arising from Maxwell's equations $-\Box^2 A_\mu = 4\pi j_\mu$ from a "current" $j_\mu(6) = e^2 K_+(4, 6)\gamma_\mu K_+(6, 2)$ produced by particle b in going from 2 to 4. This is virtue of the fact that δ_+ satisfies

$$-\Box_2^2 \delta_+(s_{21}^2) = 4\pi\delta(2, 1). \quad (5)$$

[9] These considerations make it appear unlikely that the contention of J. A. Wheeler and R. P. Feynman, Rev. Mod. Phys. 17, 157 (1945), that electrons do not act on themselves, will be a successful concept in quantum electrodynamics.

the wave function, which can then be looked upon as a superposition of free particle solutions (actually it only contains f). The amplitude that $g(2)$ is contained is calculated as in (I, Eq. (21)). The diagonal element ($g=f$) is therefore

$$\int\int \bar{f}(2)\beta K^{(1)}(2,1)\beta f(1)d^3\mathbf{x}_1 d^3\mathbf{x}_2. \tag{7}$$

The time interval $T=t_2-t_1$ (and the spatial volume V over which one integrates) must be taken very large, for the expressions are only approximate (analogous to the situation for two interacting charges).[10] This is because, for example, we are dealing incorrectly with quanta emitted just before t_2 which would normally be reabsorbed at times after t_2.

If $K^{(1)}(2,1)$ from (6) is actually substituted into (7) the surface integrals can be performed as was done in obtaining I, Eq. (22) resulting in

$$-ie^2\int\int \bar{f}(4)\gamma_\mu K_+(4,3)\gamma_\mu f(3)\delta_+(s_{43}^2)d\tau_3 d\tau_4. \tag{8}$$

Putting for $f(1)$ the plane wave $u\exp(-ip\cdot x_1)$ where p_μ is the energy (p_4) and momentum of the electron ($p^2=m^2$), and u is a constant 4-index symbol, (8) becomes

$$-ie^2\int\int(\bar{u}\gamma_\mu K_+(4,3)\gamma_\mu u)$$
$$\times\exp(ip\cdot(x_4-x_3))\delta_+(s_{43}^2)d\tau_3 d\tau_4,$$

the integrals extending over the volume V and time interval T. Since $K_+(4,3)$ depends only on the difference of the coordinates of 4 and 3, $x_{43\mu}$, the integral on 4 gives a result (except near the surfaces of the region) independent of 3. When integrated on 3, therefore, the result is of order VT. The effect is proportional to V, for the wave functions have been normalized to unit volume. If normalized to volume V, the result would simply be proportional to T. This is expected, for if the effect were equivalent to a change in energy ΔE, the amplitude for arrival in f at t_2 is altered by a factor $\exp(-i\Delta E(t_2-t_1))$, or to first order by the difference $-i(\Delta E)T$. Hence, we have

$$\Delta E=e^2\int(\bar{u}\gamma_\mu K_+(4,3)\gamma_\mu u)\exp(ip\cdot x_{43})\delta_+(s_{43}^2)d\tau_4, \tag{9}$$

integrated over all space-time $d\tau_4$. This expression will be simplified presently. In interpreting (9) we have tacitly assumed that the wave functions are normalized so that $(u^*u)=(\bar{u}\gamma_4 u)=1$. The equation may therefore be made independent of the normalization by writing the left side as $(\Delta E)(\bar{u}\gamma_4 u)$, or since $(\bar{u}\gamma_4 u)=(E/m)(\bar{u}u)$ and $m\Delta m=E\Delta E$, as $\Delta m(\bar{u}u)$ where Δm is an equivalent change in mass of the electron. In this form invariance is obvious.

One can likewise obtain an expression for the energy shift for an electron in a hydrogen atom. Simply replace K_+ in (8), by $K_+^{(V)}$, the exact kernel for an electron in the potential, $V=\beta e^2/r$, of the atom, and f by a wave function (of space and time) for an atomic state. In general the ΔE which results is not real. The imaginary part is negative and in $\exp(-i\Delta ET)$ produces an exponentially decreasing amplitude with time. This is because we are asking for the amplitude that an atom initially with no photon in the field, will still appear after time T with no photon. If the atom is in a state which can radiate, this amplitude must decay with time. The imaginary part of ΔE when calculated does indeed give the correct rate of radiation from atomic states. It is zero for the ground state and for a free electron.

In the non-relativistic region the expression for ΔE can be worked out as has been done by Bethe.[11] In the relativistic region (points 4 and 3 as close together as a Compton wave-length) the $K_+^{(V)}$ which should appear in (8) can be replaced to first order in V by K_+ plus $K_+^{(1)}(2,1)$ given in I, Eq. (13). The problem is then very similar to the radiationless scattering problem discussed below.

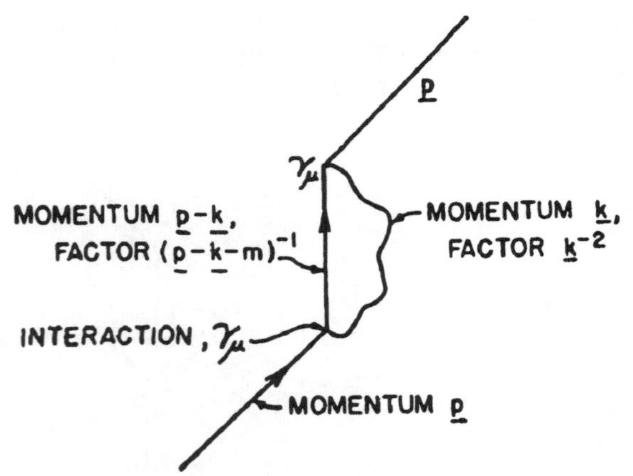

FIG. 3. Interaction of an electron with itself. Momentum space, Eq. (11).

4. EXPRESSION IN MOMENTUM AND ENERGY SPACE

The evaluation of (9), as well as all the other more complicated expressions arising in these problems, is very much simplified by working in the momentum and energy variables, rather than space and time. For this we shall need the Fourier Transform of $\delta_+(s_{21}^2)$ which is

$$-\delta_+(s_{21}^2)=\pi^{-1}\int\exp(-ik\cdot x_{21})k^{-2}d^4k, \tag{10}$$

which can be obtained from (3) and (5) or from I Eq. (32) noting that $I_+(2,1)$ for $m^2=0$ is $\delta_+(s_{21}^2)$ from

[10] This is discussed in reference 5 in which it is pointed out that the concept of a wave function loses accuracy if there are delayed self-actions.

[11] H. A. Bethe, Phys. Rev. 72, 339 (1947).

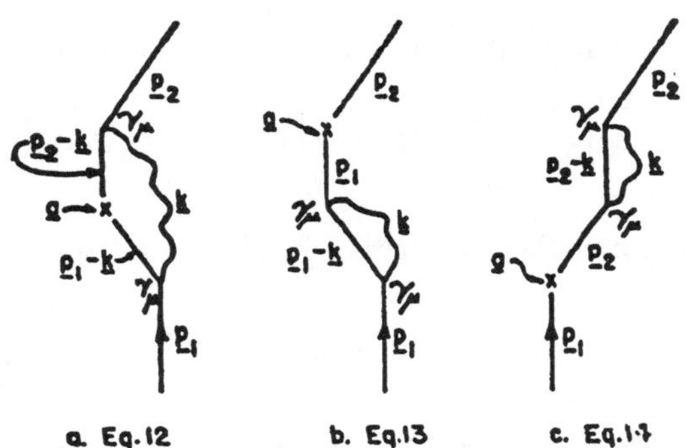

FIG. 4. Radiative correction to scattering, momentum space.

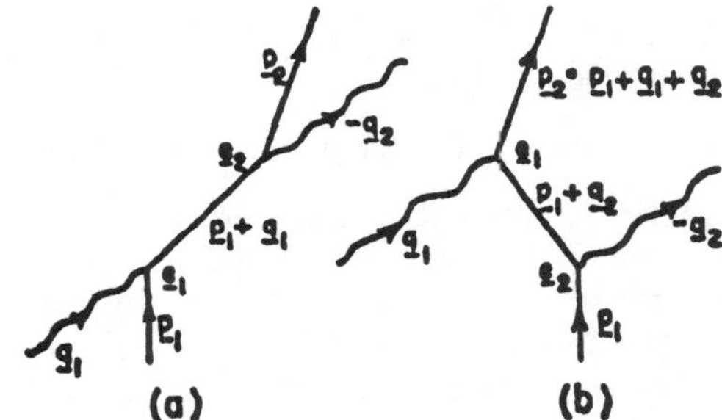

FIG. 5. Compton scattering, Eq. (15).

I, Eq. (34). The k^{-2} means $(k\cdot k)^{-1}$ or more precisely the limit as $\delta\rightarrow 0$ of $(k\cdot k+i\delta)^{-1}$. Further d^4k means $(2\pi)^{-2}dk_1dk_2dk_3dk_4$. If we imagine that quanta are particles of zero mass, then we can make the general rule that all poles are to be resolved by considering the masses of the particles and quanta to have infinitesimal negative imaginary parts.

Using these results we see that the self-energy (9) is the matrix element between $\bar{u}$ and u of the matrix

$$(e^2/\pi i)\int \gamma_\mu(p-k-m)^{-1}\gamma_\mu k^{-2}d^4k, \qquad (11)$$

where we have used the expression (I, Eq. (31)) for the Fourier transform of K_+. This form for the self-energy is easier to work with than is (9).

The equation can be understood by imagining (Fig. 3) that the electron of momentum p emits (γ_μ) a quantum of momentum k, and makes its way now with momentum $p-k$ to the next event (factor $(p-k-m)^{-1}$) which is to absorb the quantum (another γ_μ). The amplitude of propagation of quanta is k^{-2}. (There is a factor $e^2/\pi i$ for each virtual quantum). One integrates over all quanta. The reason an electron of momentum p propagates as $1/(p-m)$ is that this operator is the reciprocal of the Dirac equation operator, and we are simply solving this equation. Likewise light goes as $1/k^2$, for this is the reciprocal D'Alembertian operator of the wave equation of light. The first γ_μ represents the current which generates the vector potential, while the second is the velocity operator by which this potential is multiplied in the Dirac equation when an external field acts on an electron.

Using the same line of reasoning, other problems may be set up directly in momentum space. For example, consider the scattering in a potential $A=A_\mu\gamma_\mu$ varying in space and time as $a\exp(-iq\cdot x)$. An electron initially in state of momentum $p_1=p_{1\mu}\gamma_\mu$ will be deflected to state p_2 where $p_2=p_1+q$. The zero-order answer is simply the matrix element of a between states 1 and 2. We next ask for the first order (in e^2) radiative correction due to virtual radiation of one quantum. There are several ways this can happen. First for the case illustrated in Fig. 4(a), find the matrix:

$$(e^2/\pi i)\int \gamma_\mu(p_2-k-m)^{-1}a(p_1-k-m)^{-1}\gamma_\mu k^{-2}d^4k. \qquad (12)$$

For in this case, first[12] a quantum of momentum k is emitted (γ_μ), the electron then having momentum p_1-k and hence propagating with factor $(p_1-k-m)^{-1}$. Next it is scattered by the potential (matrix a) receiving additional momentum q, propagating on then (factor $(p_2-k-m)^{-1}$) with the new momentum until the quantum is reabsorbed (γ_μ). The quantum propagates from emission to absorption (k^{-2}) and we integrate over all quanta (d^4k), and sum on polarization μ. When this is integrated on k_4, the result can be shown to be exactly equal to the expressions (16) and (17) given in B for the same process, the various terms coming from residues of the poles of the integrand (12).

Or again if the quantum is both emitted and reabsorbed before the scattering takes place one finds (Fig. 4(b))

$$(e^2/\pi i)\int a(p_1-m)^{-1}\gamma_\mu(p_1-k-m)^{-1}\gamma_\mu k^{-2}d^4k, \qquad (13)$$

or if both emission and absorption occur after the scattering, (Fig. 4(c))

$$(e^2/\pi i)\int \gamma_\mu(p_2-k-m)^{-1}\gamma_\mu(p_2-m)^{-1}ak^{-2}d^4k. \qquad (14)$$

These terms are discussed in detail below.

We have now achieved our simplification of the form of writing matrix elements arising from virtual processes. Processes in which a number of real quanta is given initially and finally offer no problem (assuming correct normalization). For example, consider the Compton effect (Fig. 5(a)) in which an electron in state p_1 absorbs a quantum of momentum q_1, polarization vector $e_{1\mu}$ so that its interaction is $e_{1\mu}\gamma_\mu=e_1$, and emits a second quantum of momentum $-q_2$, polarization e_2 to arrive in final state of momentum p_2. The matrix for

[12] First, next, etc., here refer not to the order in true time but to the succession of events along the trajectory of the electron. That is, more precisely, to the order of appearance of the matrices in the expressions.

this process is $e_2(p_1+q_1-m)^{-1}e_1$. The total matrix for the Compton effect is, then,

$$e_2(p_1+q_1-m)^{-1}e_1+e_1(p_1+q_2-m)^{-1}e_2, \qquad (15)$$

the second term arising because the emission of e_2 may also precede the absorption of e_1 (Fig. 5(b)). One takes matrix elements of this between initial and final electron states ($p_1+q_1=p_2-q_2$), to obtain the Klein Nishina formula. Pair annihilation with emission of two quanta, etc., are given by the same matrix, positron states being those with negative time component of p. Whether quanta are absorbed or emitted depends on whether the time component of q is positive or negative.

5. THE CONVERGENCE OF PROCESSES WITH VIRTUAL QUANTA

These expressions are, as has been indicated, no more than a re-expression of conventional quantum electrodynamics. As a consequence, many of them are meaningless. For example, the self-energy expression (9) or (11) gives an infinite result when evaluated. The infinity arises, apparently, from the coincidence of the δ-function singularities in $K_+(4,3)$ and $\delta_+(s_{43}^2)$. Only at this point is it necessary to make a real departure from conventional electrodynamics, a departure other than simply rewriting expressions in a simpler form.

We desire to make a modification of quantum electrodynamics analogous to the modification of classical electrodynamics described in a previous article, A. There the $\delta(s_{12}^2)$ appearing in the action of interaction was replaced by $f(s_{12}^2)$ where $f(x)$ is a function of small width and great height.

The obvious corresponding modification in the quantum theory is to replace the $\delta_+(s^2)$ appearing the quantum mechanical interaction by a new function $f_+(s^2)$. We can postulate that if the Fourier transform of the classical $f(s_{12}^2)$ is the integral over all k of $F(k^2)\exp(-ik\cdot x_{12})d^4k$, then the Fourier transform of $f_+(s^2)$ is the same integral taken over only positive frequencies k_4 for $t_2>t_1$ and over only negative ones for $t_2<t_1$ in analogy to the relation of $\delta_+(s^2)$ to $\delta(s^2)$. The function $f(s^2)=f(x\cdot x)$ can be written* as

$$f(x\cdot x)=(2\pi)^{-2}\int_{k_4=0}^{\infty}\int \sin(k_4|x_4|)$$
$$\times\cos(\mathbf{K}\cdot\mathbf{x})dk_4d^3\mathbf{K}g(k\cdot k),$$

where $g(k\cdot k)$ is k_4^{-1} times the density of oscillators and may be expressed for positive k_4 as (A, Eq. (16))

$$g(k^2)=\int_0^{\infty}(\delta(k^2)-\delta(k^2-\lambda^2))G(\lambda)d\lambda,$$

where $\int_0^\infty G(\lambda)d\lambda=1$ and G involves values of λ large compared to m. This simply means that the amplitude for propagation of quanta of momentum k is

$$-F_+(k^2)=\pi^{-1}\int_0^{\infty}(k^{-2}-(k^2-\lambda^2)^{-1})G(\lambda)d\lambda,$$

rather than k^{-2}. That is, writing $F_+(k^2)=-\pi^{-1}k^{-2}C(k^2)$

$$-f_+(s_{12}^2)=\pi^{-1}\int\exp(-ik\cdot x_{12})k^{-2}C(k^2)d^4k. \qquad (16)$$

Every integral over an intermediate quantum which previously involved a factor d^4k/k^2 is now supplied with a convergence factor $C(k^2)$ where

$$C(k^2)=\int_0^{\infty}-\lambda^2(k^2-\lambda^2)^{-1}G(\lambda)d\lambda. \qquad (17)$$

The poles are defined by replacing k^2 by $k^2+i\delta$ in the limit $\delta\rightarrow 0$. That is λ^2 may be assumed to have an infinitesimal negative imaginary part.

The function $f_+(s_{12}^2)$ may still have a discontinuity in value on the light cone. This is of no influence for the Dirac electron. For a particle satisfying the Klein Gordon equation, however, the interaction involves gradients of the potential which reinstates the δ function if f has discontinuities. The condition that f is to have no discontinuity in value on the light cone implies $k^2C(k^2)$ approaches zero as k^2 approaches infinity. In terms of $G(\lambda)$ the condition is

$$\int_0^{\infty}\lambda^2G(\lambda)d\lambda=0. \qquad (18)$$

This condition will also be used in discussing the convergence of vacuum polarization integrals.

The expression for the self-energy matrix is now

$$(e^2/\pi i)\int\gamma_\mu(p-k-m)^{-1}\gamma_\mu k^{-2}d^4kC(k^2), \qquad (19)$$

which, since $C(k^2)$ falls off at least as rapidly as $1/k^2$, converges. For practical purposes we shall suppose hereafter that $C(k^2)$ is simply $-\lambda^2/(k^2-\lambda^2)$ implying that some average (with weight $G(\lambda)d\lambda$) over values of λ may be taken afterwards. Since in all processes the quantum momentum will be contained in at least one extra factor of the form $(p-k-m)^{-1}$ representing propagation of an electron while that quantum is in the field, we can expect all such integrals with their convergence factors to converge and that the result of all such processes will now be finite and definite (excepting the processes with closed loops, discussed below, in which the diverging integrals are over the momenta of the electrons rather than the quanta).

The integral of (19) with $C(k^2)=-\lambda^2(k^2-\lambda^2)^{-1}$ noting that $p^2=m^2$, $\lambda\gg m$ and dropping terms of order m/λ, is (see Appendix A)

$$(e^2/2\pi)[4m(\ln(\lambda/m)+\tfrac{1}{2})-p(\ln(\lambda/m)+5/4)]. \qquad (20)$$

* This relation is given incorrectly in A, equation just preceding 16.

When applied to a state of an electron of momentum p satisfying $pu=mu$, it gives for the change in mass (as in B, Eq. (9))

$$\Delta m = m(e^2/2\pi)(3\ln(\lambda/m)+\tfrac{3}{4}). \qquad (21)$$

6. RADIATIVE CORRECTIONS TO SCATTERING

We can now complete the discussion of the radiative corrections to scattering. In the integrals we include the convergence factor $C(k^2)$, so that they converge for large k. Integral (12) is also not convergent because of the well-known infra-red catastrophy. For this reason we calculate (as discussed in B) the value of the integral assuming the photons to have a small mass $\lambda_{\min} \ll m \ll \lambda$. The integral (12) becomes

$$(e^2/\pi i)\int \gamma_\mu(p_2-k-m)^{-1}a(p_1-k-m)^{-1} \times \gamma_\mu(k^2-\lambda_{\min}^2)^{-1}d^4kC(k^2-\lambda_{\min}^2),$$

which when integrated (see Appendix B) gives $(e^2/2\pi)$ times

$$\left[2\left(\ln\frac{m}{\lambda_{\min}}-1\right)\left(1-\frac{2\theta}{\tan 2\theta}\right)+\theta\tan\theta + \frac{4}{\tan 2\theta}\int_0^\theta \alpha\tan\alpha d\alpha\right]a + \frac{1}{4m}(qa-aq)\frac{2\theta}{\sin 2\theta}+ra, \qquad (22)$$

where $(q^2)^{\frac{1}{2}}=2m\sin\theta$ and we have assumed the matrix to operate between states of momentum p_1 and $p_2=p_1+q$ and have neglected terms of order $\lambda_{\min}/m$, m/λ, and q^2/λ^2. Here the only dependence on the convergence factor is in the term ra, where

$$r=\ln(\lambda/m)+9/4-2\ln(m/\lambda_{\min}). \qquad (23)$$

As we shall see in a moment, the other terms (13), (14) give contributions which just cancel the ra term. The remaining terms give for small q,

$$(e^2/4\pi)\left(\frac{1}{2m}(qa-aq)+\frac{4q^2}{3m^2}a\left(\ln\frac{m}{\lambda_{\min}}-\frac{3}{8}\right)\right), \qquad (24)$$

which shows the change in magnetic moment and the Lamb shift as interpreted in more detail in B.[13]

We must now study the remaining terms (13) and (14). The integral on k in (13) can be performed (after multiplication by $C(k^2)$) since it involves nothing but the integral (19) for the self-energy and the result is allowed to operate on the initial state u_1, (so that $p_1u_1=mu_1$). Hence the factor following $a(p_1-m)^{-1}$ will be just Δm. But, if one now tries to expand $1/(p_1-m) = (p_1+m)/(p_1^2-m^2)$ one obtains an infinite result, since $p_1^2=m^2$. This is, however, just what is expected physically. For the quantum can be emitted and absorbed at any time previous to the scattering. Such a process has the effect of a change in mass of the electron in the state 1. It therefore changes the energy by ΔE and the amplitude to first order in ΔE by $-i\Delta E\cdot t$ where t is the time it is acting, which is infinite. That is, the major effect of this term would be canceled by the effect of change of mass Δm.

The situation can be analyzed in the following manner. We suppose that the electron approaching the scattering potential a has not been free for an infinite time, but at some time far past suffered a scattering by a potential b. If we limit our discussion to the effects of Δm and of the virtual radiation of one quantum between two such scatterings each of the effects will be finite, though large, and their difference is determinate. The propagation from b to a is represented by a matrix

$$a(p'-m)^{-1}b, \qquad (25)$$

in which one is to integrate possibly over p' (depending on details of the situation). (If the time is long between b and a, the energy is very nearly determined so that p'^2 is very nearly m^2.)

We shall compare the effect on the matrix (25) of the virtual quanta and of the change of mass Δm. The effect of a virtual quantum is

$$(e^2/\pi i)\int a(p'-m)^{-1}\gamma_\mu(p'-k-m)^{-1} \times \gamma_\mu(p'-m)^{-1}bk^{-2}d^4kC(k^2), \qquad (26)$$

while that of a change of mass can be written

$$a(p'-m)^{-1}\Delta m(p'-m)^{-1}b, \qquad (27)$$

and we are interested in the difference (26)–(27). A simple and direct method of making this comparison is just to evaluate the integral on k in (26) and subtract from the result the expression (27) where Δm is given in (21). The remainder can be expressed as a multiple $-r(p'^2)$ of the unperturbed amplitude (25);

$$-r(p'^2)a(p'-m)^{-1}b. \qquad (28)$$

This has the same result (to this order) as replacing the potentials a and b in (25) by $(1-\frac{1}{2}r(p'^2))a$ and

[13] That the result given in B in Eq. (19) was in error was repeatedly pointed out to the author, in private communication, by V. F. Weisskopf and J. B. French, as their calculation, completed simultaneously with the author's early in 1948, gave a different result. French has finally shown that although the expression for the radiationless scattering B, Eq. (18) or (24) above is correct, it was incorrectly joined onto Bethe's non-relativistic result. He shows that the relation $\ln 2k_{\max}-1=\ln\lambda_{\min}$ used by the author should have been $\ln 2k_{\max}-5/6=\ln\lambda_{\min}$. This results in adding a term $-(1/6)$ to the logarithm in B, Eq. (19) so that the result now agrees with that of J. B. French and V. F. Weisskopf, Phys. Rev. 75, 1240 (1949) and N. H. Kroll and W. E. Lamb, Phys. Rev. 75, 388 (1949). The author feels unhappily responsible for the very considerable delay in the publication of French's result occasioned by this error. This footnote is appropriately numbered.

$(1-\frac{1}{2}r(p'^2))b$. In the limit, then, as $p'^2 \rightarrow m^2$ the net effect on the scattering is $-\frac{1}{2}ra$ where r, the limit of $r(p'^2)$ as $p'^2 \rightarrow m^2$ (assuming the integrals have an infra-red cut-off), turns out to be just equal to that given in (23). An equal term $-\frac{1}{2}ra$ arises from virtual transitions after the scattering (14) so that the entire ra term in (22) is canceled.

The reason that r is just the value of (12) when $q^2=0$ can also be seen without a direct calculation as follows: Let us call p the vector of length m in the direction of p' so that if $p'^2=m(1+\epsilon)^2$ we have $p'=(1+\epsilon)p$ and we take ϵ as very small, being of order T^{-1} where T is the time between the scatterings b and a. Since $(p'-m)^{-1} = (p'+m)/(p'^2-m^2) \approx (p+m)/2m^2\epsilon$, the quantity (25) is of order ϵ^{-1} or T. We shall compute corrections to it only to its own order (ϵ^{-1}) in the limit $\epsilon \rightarrow 0$. The term (27) can be written approximately[14] as

$$(e^2/\pi i)\int a(p'-m)^{-1}\gamma_\mu(p-k-m)^{-1} \times\gamma_\mu(p'-m)^{-1}bk^{-2}d^4kC(k^2),$$

using the expression (19) for Δm. The net of the two effects is therefore approximately[15]

$$-(e^2/\pi i)\int a(p'-m)^{-1}\gamma_\mu(p-k-m)^{-1}\epsilon p(p-k-m)^{-1} \times\gamma_\mu(p'-m)^{-1}bk^{-2}d^4kC(k^2),$$

a term now of order $1/\epsilon$ (since $(p'-m)^{-1}\approx(p+m) \times(2m^2\epsilon)^{-1}$) and therefore the one desired in the limit. Comparison to (28) gives for r the expression

$$(p_1+m/2m)\int \gamma_\mu(p_1-k-m)^{-1}(p_1m^{-1})(p_1-k-m)^{-1} \times\gamma_\mu k^{-2}d^4kC(k^2). \quad (29)$$

The integral can be immediately evaluated, since it is the same as the integral (12), but with $q=0$, for a replaced by p_1/m. The result is therefore $r\cdot(p_1/m)$ which when acting on the state u_1 is just r, as $p_1u_1=mu_1$. For the same reason the term $(p_1+m)/2m$ in (29) is effectively 1 and we are left with $-r$ of (23).[16]

In more complex problems starting with a free electron the same type of term arises from the effects of a virtual emission and absorption both previous to the other processes. They, therefore, simply lead to the same factor r so that the expression (23) may be used directly and these renormalization integrals need not be computed afresh for each problem.

In this problem of the radiative corrections to scattering the net result is insensitive to the cut-off. This means, of course, that by a simple rearrangement of terms previous to the integration we could have avoided the use of the convergence factors completely (see for example Lewis[17]). The problem was solved in the manner here in order to illustrate how the use of such convergence factors, even when they are actually unnecessary, may facilitate analysis somewhat by removing the effort and ambiguities that may be involved in trying to rearrange the otherwise divergent terms.

The replacement of δ_+ by f_+ given in (16), (17) is not determined by the analogy with the classical problem. In the classical limit only the real part of δ_+ (i.e., just δ) is easy to interpret. But by what should the imaginary part, $1/(\pi is^2)$, of δ_+ be replaced? The choice we have made here (in defining, as we have, the location of the poles of (17)) is arbitrary and almost certainly incorrect. If the radiation resistance is calculated for an atom, as the imaginary part of (8), the result depends slightly on the function f_+. On the other hand the light radiated at very large distances from a source is independent of f_+. The total energy absorbed by distant absorbers will not check with the energy loss of the source. We are in a situation analogous to that in the classical theory if the entire f function is made to contain only retarded contributions (see A, Appendix). One desires instead the analogue of $\langle F\rangle_{ret}$ of A. This problem is being studied.

One can say therefore, that this attempt to find a consistent modification of quantum electrodynamics is incomplete (see also the question of closed loops, below). For it could turn out that any correct form of f_+ which will guarantee energy conservation may at the same time not be able to make the self-energy integral finite. The desire to make the methods of simplifying the calculation of quantum electrodynamic processes more widely available has prompted this publication before an analysis of the correct form for f_+ is complete. One might try to take the position that, since the energy discrepancies discussed vanish in the limit $\lambda\rightarrow\infty$, the correct physics might be considered to be that obtained by letting $\lambda\rightarrow\infty$ after mass renormalization. I have no proof of the mathematical consistency of this procedure, but the presumption is very strong that it is satisfactory. (It is also strong that a satisfactory form for f_+ can be found.)

7. THE PROBLEM OF VACUUM POLARIZATION

In the analysis of the radiative corrections to scattering one type of term was not considered. The potential

[14] The expression is not exact because the substitution of Δm by the integral in (19) is valid only if p operates on a state such that p can be replaced by m. The error, however, is of order $a(p'-m)^{-1}(p-m)(p'-m)^{-1}b$ which is $a((1+\epsilon)p+m)(p-m) \times((1+\epsilon)p+m)p(2\epsilon+\epsilon^2)^{-2}m^{-4}$. But since $p^2=m^2$, we have $p(p-m) = -m(p-m)=(p-m)p$ so the net result is approximately $a(p-m)b/4m^2$ and is not of order $1/\epsilon$ but smaller, so that its effect drops out in the limit.

[15] We have used, to first order, the general expansion (valid for any operators A, B)

$$(A+B)^{-1}=A^{-1}-A^{-1}BA^{-1}+A^{-1}BA^{-1}BA^{-1}-\cdots$$

with $A=p-k-m$ and $B=p'-p=\epsilon p$ to expand the difference of $(p'-k-m)^{-1}$ and $(p-k-m)^{-1}$.

[16] The renormalization terms appearing B, Eqs. (14), (15) when translated directly into the present notation do not give twice (29) but give this expression with the central p_1m^{-1} factor replaced by $m\gamma_4/E_1$ where $E_1=p_{1\mu}$ for $\mu=4$. When integrated it therefore gives $ra((p_1+m)/2m)(m\gamma_4/E_1)$ or $ra-ra(m\gamma_4/E_1)(p_1-m)/2m$. (Since $p_1\gamma_4+\gamma_4p_1=2E_1$) which gives just ra, since $p_1u_1=mu_1$.

[17] H. W. Lewis, Phys. Rev. 73, 173 (1948).

which we can assume to vary as $a_\mu \exp(-iq\cdot x)$ creates a pair of electrons (see Fig. 6), momenta p_a, $-p_b$. This pair then reannihilates, emitting a quantum $q=p_b-p_a$, which quantum scatters the original electron from state 1 to state 2. The matrix element for this process (and the others which can be obtained by rearranging the order in time of the various events) is

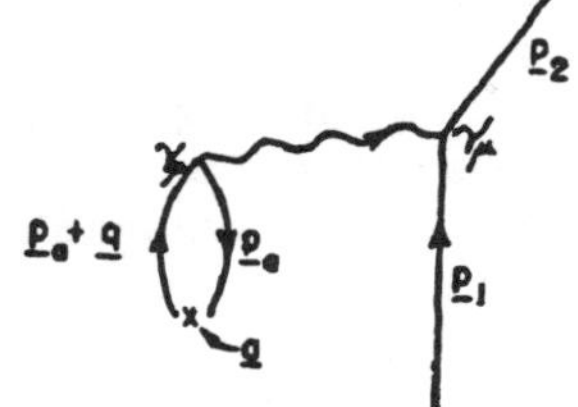

FIG. 6. Vacuum polarization effect on scattering, Eq. (30).

$$-(e^2/\pi i)(\bar{u}_2\gamma_\mu u_1)\int Sp[(p_a+q-m)^{-1} \times\gamma_\nu(p_a-m)^{-1}\gamma_\mu]d^4p_a q^{-2}C(q^2)a_\nu. \quad (30)$$

This is because the potential produces the pair with amplitude proportional to $a_\nu\gamma_\nu$, the electrons of momenta p_a and $-(p_a+q)$ proceed from there to annihilate, producing a quantum (factor γ_μ) which propagates (factor $q^{-2}C(q^2)$) over to the other electron, by which it is absorbed (matrix element of γ_μ between states 1 and 2 of the original electron $(\bar{u}_2\gamma_\mu u_1)$). All momenta p_a and spin states of the virtual electron are admitted, which means the spur and the integral on d^4p_a are calculated.

One can imagine that the closed loop path of the positron-electron produces a current

$$4\pi j_\mu = J_{\mu\nu}a_\nu, \quad (31)$$

which is the source of the quanta which act on the second electron. The quantity

$$J_{\mu\nu}=-(e^2/\pi i)\int Sp[(p+q-m)^{-1} \times\gamma_\nu(p-m)^{-1}\gamma_\mu]d^4p, \quad (32)$$

is then characteristic for this problem of polarization of the vacuum.

One sees at once that $J_{\mu\nu}$ diverges badly. The modification of δ to f alters the amplitude with which the current j_μ will affect the scattered electron, but it can do nothing to prevent the divergence of the integral (32) and of its effects.

One way to avoid such difficulties is apparent. From one point of view we are considering all routes by which a given electron can get from one region of space-time to another, i.e., from the source of electrons to the apparatus which measures them. From this point of view the closed loop path leading to (32) is unnatural. It might be assumed that the only paths of meaning are those which start from the source and work their way in a continuous path (possibly containing many time reversals) to the detector. Closed loops would be excluded. We have already found that this may be done for electrons moving in a fixed potential.

Such a suggestion must meet several questions, however. The closed loops are a consequence of the usual hole theory in electrodynamics. Among other things, they are required to keep probability conserved. The probability that no pair is produced by a potential is not unity and its deviation from unity arises from the imaginary part of $J_{\mu\nu}$. Again, with closed loops excluded, a pair of electrons once created cannot annihilate one another again, the scattering of light by light would be zero, etc. Although we are not experimentally sure of these phenomena, this does seem to indicate that the closed loops are necessary. To be sure, it is always possible that these matters of probability conservation, etc., will work themselves out as simply in the case of interacting particles as for those in a fixed potential. Lacking such a demonstration the presumption is that the difficulties of vacuum polarization are not so easily circumvented.[18]

An alternative procedure discussed in *B* is to assume that the function $K_+(2,1)$ used above is incorrect and is to be replaced by a modified function K_+' having no singularity on the light cone. The effect of this is to provide a convergence factor $C(p^2-m^2)$ for *every* integral over electron momenta.[19] This will multiply the integrand of (32) by $C(p^2-m^2)C((p+q)^2-m^2)$, since the integral was originally $\delta(p_a-p_b+q)d^4p_ad^4p_b$ and both p_a and p_b get convergence factors. The integral now converges but the result is unsatisfactory.[20]

One expects the current (31) to be conserved, that is $q_\mu j_\mu=0$ or $q_\mu J_{\mu\nu}=0$. Also one expects no current if a_ν is a gradient, or $a_\nu=q_\nu$ times a constant. This leads to the condition $J_{\mu\nu}q_\nu=0$ which is equivalent to $q_\mu J_{\mu\nu}=0$ since $J_{\mu\nu}$ is symmetrical. But when the expression (32) is integrated with such convergence factors it does not satisfy this condition. By altering the kernel from K to another, K', which does not satisfy the Dirac equation we have lost the gauge invariance, its consequent current conservation and the general consistency of the theory.

One can see this best by calculating $J_{\mu\nu}q_\nu$ directly from (32). The expression within the spur becomes $(p+q-m)^{-1}q(p-m)^{-1}\gamma_\mu$ which can be written as the difference of two terms: $(p-m)^{-1}\gamma_\mu-(p+q-m)^{-1}\gamma_\mu$. Each of these terms would give the same result if the integration d^4p were without a convergence factor, for

[18] It would be very interesting to calculate the Lamb shift accurately enough to be sure that the 20 megacycles expected from vacuum polarization are actually present.

[19] This technique also makes self-energy and radiationless scattering integrals finite even without the modification of δ_+ to f_+ for the radiation (and the consequent convergence factor $C(k^2)$ for the quanta). See B.

[20] Added to the terms given below (33) there is a term $\frac{1}{4}(\lambda^2-2\mu^2+\frac{1}{3}q^2)\delta_{\mu\nu}$ for $C(k^2)=-\lambda^2(k^2-\lambda^2)^{-1}$, which is not gauge invariant. (In addition the charge renormalization has $-7/6$ added to the logarithm.)

the first can be converted into the second by a shift of the origin of p, namely $p'=p+q$. This does not result in cancelation in (32) however, for the convergence factor is altered by the substitution.

A method of making (32) convergent without spoiling the gauge invariance has been found by Bethe and by Pauli. The convergence factor for light can be looked upon as the result of superposition of the effects of quanta of various masses (some contributing negatively). Likewise if we take the factor $C(p^2-m^2) = -\lambda^2(p^2-m^2-\lambda^2)^{-1}$ so that $(p^2-m^2)^{-1}C(p^2-m^2) = (p^2-m^2)^{-1}-(p^2-m^2-\lambda^2)^{-1}$ we are taking the difference of the result for electrons of mass m and mass $(\lambda^2+m^2)^{\frac{1}{2}}$. But we have taken this difference for *each* propagation between interactions with photons. They suggest instead that once created with a certain mass the electron should continue to propagate with this mass through all the potential interactions until it closes its loop. That is if the quantity (32), integrated over some finite range of p, is called $J_{\mu\nu}(m^2)$ and the corresponding quantity over the same range of p, but with m replaced by $(m^2+\lambda^2)^{\frac{1}{2}}$ is $J_{\mu\nu}(m^2+\lambda^2)$ we should calculate

$$J_{\mu\nu}{}^P=\int_0^\infty [J_{\mu\nu}(m^2)-J_{\mu\nu}(m^2+\lambda^2)]G(\lambda)d\lambda, \quad (32')$$

the function $G(\lambda)$ satisfying $\int_0^\infty G(\lambda)d\lambda=1$ and $\int_0^\infty G(\lambda)\lambda^2 d\lambda=0$. Then in the expression for $J_{\mu\nu}{}^P$ the range of p integration can be extended to infinity as the integral now converges. The result of the integration using this method is the integral on $d\lambda$ over $G(\lambda)$ of (see Appendix C)

$$J_{\mu\nu}{}^P=-\frac{e^2}{\pi}(q_\mu q_\nu-\delta_{\mu\nu}q^2)\left(-\frac{1}{3}\ln\frac{\lambda^2}{m^2}-\left[\frac{4m^2+2q^2}{3q^2}\left(1-\frac{\theta}{\tan\theta}\right)-\frac{1}{9}\right]\right), \quad (33)$$

with $q^2=4m^2\sin^2\theta$.

The gauge invariance is clear, since $q_\mu(q_\mu q_\nu-q^2\delta_{\mu\nu})=0$. Operating (as it always will) on a potential of zero divergence the $(q_\mu q_\nu-\delta_{\mu\nu}q^2)a_\nu$ is simply $-q^2a_\mu$, the D'Alembertian of the potential, that is, the current producing the potential. The term $-\frac{1}{3}(\ln(\lambda^2/m^2))(q_\mu q_\nu-q^2\delta_{\mu\nu})$ therefore gives a current proportional to the current producing the potential. This would have the same effect as a change in charge, so that we would have a difference $\Delta(e^2)$ between e^2 and the experimentally observed charge, $e^2+\Delta(e^2)$, analogous to the difference between m and the observed mass. This charge depends logarithmically on the cut-off, $\Delta(e^2)/e^2 = -(2e^2/3\pi)\ln(\lambda/m)$. After this renormalization of charge is made, no effects will be sensitive to the cut-off.

After this is done the final term remaining in (33), contains the usual effects[21] of polarization of the vacuum. It is zero for a free light quantum ($q^2=0$). For small q^2 it behaves as $(2/15)q^2$ (adding $-\frac{1}{5}$ to the logarithm in the Lamb effect). For $q^2>(2m)^2$ it is complex, the imaginary part representing the loss in amplitude required by the fact that the probability that no quanta are produced by a potential able to produce pairs $((q^2)^{\frac{1}{2}}>2m)$ decreases with time. (To make the necessary analytic continuation, imagine m to have a small negative imaginary part, so that $(1-q^2/4m^2)^{\frac{1}{2}}$ becomes $-i(q^2/4m^2-1)^{\frac{1}{2}}$ as q^2 goes from below to above $4m^2$. Then $\theta=\pi/2+iu$ where $\sinh u=+(q^2/4m^2-1)^{\frac{1}{2}}$, and $-1/\tan\theta=i\tanh u=+i(q^2-4m^2)^{\frac{1}{2}}(q^2)^{-\frac{1}{2}}$.)

Closed loops containing a number of quanta or potential interactions larger than two produce no trouble. Any loop with an odd number of interactions gives zero (I, reference 9). Four or more potential interactions give integrals which are convergent even without a convergence factor as is well known. The situation is analogous to that for self-energy. Once the simple problem of a single closed loop is solved there are no further divergence difficulties for more complex processes.[22]

8. LONGITUDINAL WAVES

In the usual form of quantum electrodynamics the longitudinal and transverse waves are given separate treatment. Alternately the condition $(\partial A_\mu/\partial x_\mu)\Psi=0$ is carried along as a supplementary condition. In the present form no such special considerations are necessary for we are dealing with the solutions of the equation $-\square^2 A_\mu=4\pi j_\mu$ with a current j_μ which is conserved $\partial j_\mu/\partial x_\mu=0$. That means at least $\square^2(\partial A_\mu/\partial x_\mu)=0$ and in fact our solution also satisfies $\partial A_\mu/\partial x_\mu=0$.

To show that this is the case we consider the amplitude for emission (real or virtual) of a photon and show that the divergence of this amplitude vanishes. The amplitude for emission for photons polarized in the μ direction involves matrix elements of γ_μ. Therefore what we have to show is that the corresponding matrix elements of $q_\mu\gamma_\mu=q$ vanish. For example, for a first order effect we would require the matrix element of q between two states p_1 and $p_2=p_1+q$. But since $q=p_2-p_1$ and $(\bar{u}_2p_1u_1)=m(\bar{u}_2u_1)=(\bar{u}_2p_2u_1)$ the matrix element vanishes, which proves the contention in this case. It also vanishes in more complex situations (essentially because of relation (34), below) (for example, try putting $e_2=q_2$ in the matrix (15) for the Compton Effect).

To prove this in general, suppose a_i, $i=1$ to N are a set of plane wave disturbing potentials carrying momenta q_i (e.g., some may be emissions or absorptions of the same or different quanta) and consider a matrix for the transition from a state of momentum p_0 to p_N such

[21] E. A. Uehling, Phys. Rev. 48, 55 (1935), R. Serber, Phys. Rev. 48, 49 (1935).

[22] There are loops completely without external interactions. For example, a pair is created virtually along with a photon. Next they annihilate, absorbing this photon. Such loops are disregarded on the grounds that they do not interact with anything and are thereby completely unobservable. Any indirect effects they may have via the exclusion principle have already been included.

as $a_N\prod_{i=1}^{N-1}(p_i-m)^{-1}a_i$ where $p_i=p_{i-1}+q_i$ (and in the product, terms with larger i are written to the left). The most general matrix element is simply a linear combination of these. Next consider the matrix between states p_0 and p_N+q in a situation in which not only are the a_i acting but also another potential $a\exp(-iq\cdot x)$ where $a=q$. This may act previous to all a_i, in which case it gives $a_N\prod(p_i+q-m)^{-1}a_i(p_0+q-m)^{-1}q$ which is equivalent to $+a_N\prod(p_i+q-m)^{-1}a_i$ since $+(p_0+q-m)^{-1}q$ is equivalent to $(p_0+q-m)^{-1}\times(p_0+q-m)$ as p_0 is equivalent to m acting on the initial state. Likewise if it acts after all the potentials it gives $q(p_N-m)^{-1}a_N\prod(p_i-m)^{-1}a_i$ which is equivalent to $-a_N\prod(p_i-m)^{-1}a_i$ since p_N+q-m gives zero on the final state. Or again it may act between the potential a_k and a_{k+1} for each k. This gives

$$\sum_{k=1}^{N-1}a_N\prod_{i=k+1}^{N-1}(p_i+q-m)^{-1}a_i(p_k+q-m)^{-1}\times q(p_k-m)^{-1}a_k\prod_{j=1}^{k-1}(p_j-m)^{-1}a_j.$$

However,

$$(p_k+q-m)^{-1}q(p_k-m)^{-1}=(p_k-m)^{-1}-(p_k+q-m)^{-1}, \quad (34)$$

so that the sum breaks into the difference of two sums, the first of which may be converted to the other by the replacement of k by $k-1$. There remain only the terms from the ends of the range of summation,

$$+a_N\prod_{i=1}^{N-1}(p_i-m)^{-1}a_i-a_N\prod_{i=1}^{N-1}(p_i+q-m)^{-1}a_i.$$

These cancel the two terms originally discussed so that the entire effect is zero. Hence any wave emitted will satisfy $\partial A_\mu/\partial x_\mu=0$. Likewise longitudinal waves (that is, waves for which $A_\mu=\partial\phi/\partial x_\mu$ or $a=q$) cannot be absorbed and will have no effect, for the matrix elements for emission and absorption are similar. (We have said little more than that a potential $A_\mu=\partial\varphi/\partial x_\mu$ has no effect on a Dirac electron since a transformation $\psi'=\exp(-i\phi)\psi$ removes it. It is also easy to see in coordinate representation using integrations by parts.)

This has a useful practical consequence in that in computing probabilities for transition for unpolarized light one can sum the squared matrix over all four directions rather than just the two special polarization vectors. Thus suppose the matrix element for some process for light polarized in direction e_μ is $e_\mu M_\mu$. If the light has wave vector q_μ we know from the argument above that $q_\mu M_\mu=0$. For unpolarized light progressing in the z direction we would ordinarily calculate $M_x^2+M_y^2$. But we can as well sum $M_x^2+M_y^2+M_z^2-M_t^2$ for $q_\mu M_\mu$ implies $M_t=M_z$ since $q_t=q_z$ for free quanta. This shows that unpolarized light is a relativistically invariant concept, and permits some simplification in computing cross sections for such light.

Incidentally, the virtual quanta interact through terms like $\gamma_\mu\cdots\gamma_\mu k^{-2}d^4k$. Real processes correspond to poles in the formulae for virtual processes. The pole occurs when $k^2=0$, but it looks at first as though in the sum on all four values of μ, of $\gamma_\mu\cdots\gamma_\mu$ we would have four kinds of polarization instead of two. Now it is clear that only two perpendicular to k are effective.

The usual elimination of longitudinal and scalar virtual photons (leading to an instantaneous Coulomb potential) can of course be performed here too (although it is not particularly useful). A typical term in a virtual transition is $\gamma_\mu\cdots\gamma_\mu k^{-2}d^4k$ where the $\cdots$ represent some intervening matrices. Let us choose for the values of μ, the time t, the direction of vector part $\mathbf{K}$, of k, and two perpendicular directions 1, 2. We shall not change the expression for these two 1, 2 for these are represented by transverse quanta. But we must find $(\gamma_t\cdots\gamma_t)-(\gamma_{\mathbf{K}}\cdots\gamma_{\mathbf{K}})$. Now $k=k_4\gamma_t-K\gamma_{\mathbf{K}}$, where $K=(\mathbf{K}\cdot\mathbf{K})^{\frac{1}{2}}$, and we have shown above that k replacing the γ_μ gives zero.[23] Hence $K\gamma_{\mathbf{K}}$ is equivalent to $k_4\gamma_t$ and

$$(\gamma_t\cdots\gamma_t)-(\gamma_{\mathbf{K}}\cdots\gamma_{\mathbf{K}})=((K^2-k_4^2)/K^2)(\gamma_t\cdots\gamma_t),$$

so that on multiplying by $k^{-2}d^4k=d^4k(k_4^2-K^2)^{-1}$ the net effect is $-(\gamma_t\cdots\gamma_t)d^4k/K^2$. The γ_t means just scalar waves, that is, potentials produced by charge density. The fact that $1/K^2$ does not contain k_4 means that k_4 can be integrated first, resulting in an instantaneous interaction, and the $d^3\mathbf{K}/K^2$ is just the momentum representation of the Coulomb potential, $1/r$.

9. KLEIN GORDON EQUATION

The methods may be readily extended to particles of spin zero satisfying the Klein Gordon equation,[24]

$$\Box^2\psi-m^2\psi=i\partial(A_\mu\psi)/\partial x_\mu+iA_\mu\partial\psi/\partial x_\mu-A_\mu A_\mu\psi. \quad (35)$$

[23] A little more care is required when both γ_μ's act on the same particle. Define $x=k_4\gamma_t+K\gamma_{\mathbf{K}}$, and consider $(k\cdots x)+(x\cdots k)$. Exactly this term would arise if a system, acted on by potential x carrying momentum $-k$, is disturbed by an added potential k of momentum $+k$ (the reversed sign of the momenta in the intermediate factors in the second term $x\cdots k$ has no effect since we will later integrate over all k). Hence as shown above the result is zero, but since $(k\cdots x)+(x\cdots k)=k_4^2(\gamma_t\cdots\gamma_t)-K^2(\gamma_{\mathbf{K}}\cdots\gamma_{\mathbf{K}})$ we can still conclude $(\gamma_{\mathbf{K}}\cdots\gamma_{\mathbf{K}})=k_4^2K^{-2}(\gamma_t\cdots\gamma_t)$.

[24] The equations discussed in this section were deduced from the formulation of the Klein Gordon equation given in reference 5, Section 14. The function ψ in this section has only one component and is not a spinor. An alternative formal method of making the equations valid for spin zero and also for spin 1 is (presumably) by use of the Kemmer-Duffin matrices β_μ, satisfying the commutation relation

$$\beta_\mu\beta_\nu\beta_\sigma+\beta_\sigma\beta_\nu\beta_\mu=\delta_{\mu\nu}\beta_\sigma+\delta_{\sigma\nu}\beta_\mu.$$

If we interpret a to mean $a_\mu\beta_\mu$, rather than $a_\mu\gamma_\mu$, for any a_μ, all of the equations in momentum space will remain formally identical to those for the spin 1/2; with the exception of those in which a denominator $(p-m)^{-1}$ has been rationalized to $(p+m)(p^2-m^2)^{-1}$ since p^2 is no longer equal to a number, $p\cdot p$. But p^3 does equal $(p\cdot p)p$ so that $(p-m)^{-1}$ may now be interpreted as $(mp+m^2+p^2-p\cdot p)(p\cdot p-m^2)^{-1}m^{-1}$. This implies that equations in coordinate space will be valid of the function $K_+(2,1)$ is given as $K_+(2,1)=[(i\nabla_2+m)-m^{-1}(\nabla_2^2+\Box_2^2)]iI_+(2,1)$ with $\nabla_2=\beta_\mu\partial/\partial x_{2\mu}$. This is all in virtue of the fact that the many component wave function ψ (5 components for spin 0, 10 for spin 1) satisfies $(i\nabla-m)\psi=A\psi$ which is formally identical to the Dirac Equation. See W. Pauli, Rev. Mod. Phys. 13, 203 (1940).

The important kernel is now $I_+(2, 1)$ defined in (I, Eq. (32)). For a free particle, the wave function $\psi(2)$ satisfies $+\Box^2\psi - m^2\psi = 0$. At a point, 2, inside a space time region it is given by

$$\psi(2) = \int [\psi(1)\partial I_+(2, 1)/\partial x_{1\mu} - (\partial\psi/\partial x_{1\mu})I_+(2, 1)]N_\mu(1)d^3V_1,$$

(as is readily shown by the usual method of demonstrating Green's theorem) the integral being over an entire 3-surface boundary of the region (with normal vector N_μ). Only the positive frequency components of ψ contribute from the surface preceding the time corresponding to 2, and only negative frequencies from the surface future to 2. These can be interpreted as electrons and positrons in direct analogy to the Dirac case.

The right-hand side of (35) can be considered as a source of new waves and a series of terms written down to represent matrix elements for processes of increasing order. There is only one new point here, the term in $A_\mu A_\mu$ by which two quanta can act at the same time. As an example, suppose three quanta or potentials, $a_\mu \exp(-iq_a\cdot x)$, $b_\mu \exp(-iq_b\cdot x)$, and $c_\mu \exp(-iq_c\cdot x)$ are to act in that order on a particle of original momentum $p_{0\mu}$ so that $p_a = p_0 + q_a$ and $p_b = p_a + q_b$; the final momentum being $p_c = p_b + q_c$. The matrix element is the sum of three terms ($p^2 = p_\mu p_\mu$) (illustrated in Fig. 7)

$$\begin{aligned}(p_c\cdot c + p_b\cdot c)&(p_b^2 - m^2)^{-1}(p_b\cdot b + p_a\cdot b)\\ &\times(p_a^2 - m^2)^{-1}(p_a\cdot a + p_0\cdot a)\\ -(p_c\cdot c + p_b\cdot c)&(p_b^2 - m^2)^{-1}(b\cdot a)\\ -(c\cdot b)&(p_a^2 - m^2)^{-1}(p_a\cdot a + p_0\cdot a).\end{aligned} \tag{36}$$

The first comes when each potential acts through the perturbation $i\partial(A_\mu\psi)/\partial x_\mu + iA_\mu\partial\psi/\partial x_\mu$. These gradient operators in momentum space mean respectively the momentum after and before the potential A_μ operates. The second term comes from b_μ and a_μ acting at the same instant and arises from the $A_\mu A_\mu$ term in (a). Together b_μ and a_μ carry momentum $q_{b\mu} + q_{a\mu}$ so that after $b\cdot a$ operates the momentum is $p_0 + q_a + q_b$ or p_b. The final term comes from c_μ and b_μ operating together in a similar manner. The term $A_\mu A_\mu$ thus permits a new type of process in which two quanta can be emitted (or absorbed, or one absorbed, one emitted) at the same time. There is no $a\cdot c$ term for the order a, b, c we have assumed. In an actual problem there would be other terms like (36) but with alterations in the order in which the quanta a, b, c act. In these terms $a\cdot c$ would appear.

As a further example the self-energy of a particle of momentum p_μ is

$$(e^2/2\pi im)\int [(2p-k)_\mu((p-k)^2 - m^2)^{-1} \times (2p-k)_\mu - \delta_{\mu\mu}]d^4k k^{-2}C(k^2),$$

where the $\delta_{\mu\mu} = 4$ comes from the $A_\mu A_\mu$ term and represents the possibility of the simultaneous emission and absorption of the same virtual quantum. This integral without the $C(k^2)$ diverges quadratically and would not converge if $C(k^2) = -\lambda^2/(k^2 - \lambda^2)$. Since the interaction occurs through the gradients of the potential, we must use a stronger convergence factor, for example $C(k^2) = \lambda^4(k^2 - \lambda^2)^{-2}$, or in general (17) with $\int_0^\infty \lambda^2 G(\lambda)d\lambda = 0$. In this case the self-energy converges but depends quadratically on the cut-off λ and is not necessarily small compared to m. The radiative corrections to scattering after mass renormalization are insensitive to the cut-off just as for the Dirac equation.

When there are several particles one can obtain Bose statistics by the rule that if two processes lead to the same state but with two electrons exchanged, their amplitudes are to be added (rather than subtracted as for Fermi statistics). In this case equivalence to the second quantization treatment of Pauli and Weisskopf should be demonstrable in a way very much like that given in I (appendix) for Dirac electrons. The Bose statistics mean that the sign of contribution of a closed loop to the vacuum polarization is the opposite of what it is for the Fermi case (see I). It is ($p_b = p_a + q$)

$$J_{\mu\nu} = \frac{e^2}{2\pi im}\int [(p_{b\mu} + p_{a\mu})(p_{b\nu} + p_{a\nu})(p_a^2 - m^2)^{-1} \times (p_b^2 - m^2)^{-1} - \delta_{\mu\nu}(p_a^2 - m^2)^{-1} - \delta_{\mu\nu}(p_b^2 - m^2)^{-1}]d^4p_a$$

giving,

$$J_{\mu\nu}{}^P = \frac{e^2}{\pi}(q_\mu q_\nu - \delta_{\mu\nu}q^2)\left[\frac{1}{6}\ln\frac{\lambda^2}{m^2} + \frac{1}{9} - \frac{4m^2 - q^2}{3q^2}\left(1 - \frac{\theta}{\tan\theta}\right)\right],$$

the notation as in (33). The imaginary part for $(q^2)^{\frac{1}{2}} > 2m$ is again positive representing the loss in the probability of finding the final state to be a vacuum, associated with the possibilities of pair production. Fermi statistics would give a gain in probability (and also a charge renormalization of opposite sign to that expected).

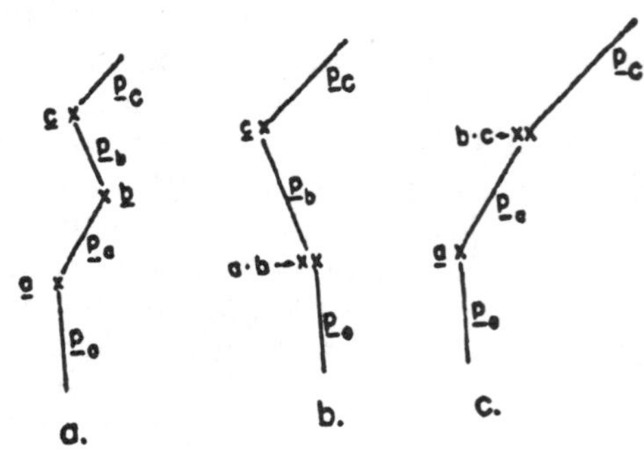

FIG. 7. Klein-Gordon particle in three potentials, Eq. (36). The coupling to the electromagnetic field is now, for example, $p_0\cdot a + p_a\cdot a$, and a new possibility arises, (b), of simultaneous interaction with two quanta $a\cdot b$. The propagation factor is now $(p\cdot p - m^2)^{-1}$ for a particle of momentum p_μ.

10. APPLICATION TO MESON THEORIES

The theories which have been developed to describe mesons and the interaction of nucleons can be easily expressed in the language used here. Calculations, to lowest order in the interactions can be made very easily for the various theories, but agreement with experimental results is not obtained. Most likely all of our present formulations are quantitatively unsatisfactory. We shall content ourselves therefore with a brief summary of the methods which can be used.

The nucleons are usually assumed to satisfy Dirac's equation so that the factor for propagation of a nucleon of momentum p is $(p-M)^{-1}$ where M is the mass of the nucleon (which implies that nucleons can be created in pairs). The nucleon is then assumed to interact with mesons, the various theories differing in the form assumed for this interaction.

First, we consider the case of neutral mesons. The theory closest to electrodynamics is the theory of vector mesons with vector coupling. Here the factor for emission or absorption of a meson is $g\gamma_\mu$ when this meson is "polarized" in the μ direction. The factor g, the "mesonic charge," replaces the electric charge e. The amplitude for propagation of a meson of momentum q in intermediate states is $(q^2-\mu^2)^{-1}$ (rather than q^{-2} as it is for light) where μ is the mass of the meson. The necessary integrals are made finite by convergence factors $C(q^2-\mu^2)$ as in electrodynamics. For scalar mesons with scalar coupling the only change is that one replaces the γ_μ by 1 in emission and absorption. There is no longer a direction of polarization, μ, to sum upon. For pseudoscalar mesons, pseudoscalar coupling replace γ_μ by $\gamma_5=i\gamma_x\gamma_y\gamma_z\gamma_t$. For example, the self-energy matrix of a nucleon of momentum p in this theory is

$$(g^2/\pi i)\int \gamma_5(p-k-M)^{-1}\gamma_5 d^4k(k^2-\mu^2)^{-1}C(k^2-\mu^2).$$

Other types of meson theory result from the replacement of γ_μ by other expressions (for example by $\frac{1}{2}(\gamma_\mu\gamma_\nu-\gamma_\nu\gamma_\mu)$ with a subsequent sum over all μ and ν for virtual mesons). Scalar mesons with vector coupling result from the replacement of γ_μ by $\mu^{-1}q$ where q is the final momentum of the nucleon minus its initial momentum, that is, it is the momentum of the meson if absorbed, or the negative of the momentum of a meson emitted. As is well known, this theory with neutral mesons gives zero for all processes, as is proved by our discussion on longitudinal waves in electrodynamics. Pseudoscalar mesons with pseudo-vector coupling corresponds to γ_μ being replaced by $\mu^{-1}\gamma_5 q$ while vector mesons with tensor coupling correspond to using $(2\mu)^{-1}(\gamma_\mu q-q\gamma_\mu)$. These extra gradients involve the danger of producing higher divergencies for real processes. For example, $\gamma_5 q$ gives a logarithmically divergent interaction of neutron and electron.[25] Although these divergencies can be held by strong enough convergence factors, the results then are sensitive to the method used for convergence and the size of the cut-off values of λ. For low order processes $\mu^{-1}\gamma_5 q$ is equivalent to the pseudoscalar interaction $2M\mu^{-1}\gamma_5$ because if taken between free particle wave functions of the nucleon of momenta p_1 and $p_2=p_1+q$, we have

$$(\bar{u}_2\gamma_5 q u_1)=(\bar{u}_2\gamma_5(p_2-p_1)u_1)=-(\bar{u}_2 p_2\gamma_5 u_1)\\ -(\bar{u}_2\gamma_5 p_1 u_1)=-2M(\bar{u}_2\gamma_5 u_1)$$

since γ_5 anticommutes with p_2 and p_2 operating on the state 2 equivalent to M as is p_1 on the state 1. This shows that the γ_5 interaction is unusually weak in the non-relativistic limit (for example the expected value of γ_5 for a free nucleon is zero), but since $\gamma_5^2=1$ is not small, pseudoscalar theory gives a more important interaction in second order than it does in first. Thus the pseudoscalar coupling constant should be chosen to fit nuclear forces including these important second order processes.[26] The equivalence of pseudoscalar and pseudovector coupling which holds for low order processes therefore does not hold when the pseudoscalar theory is giving its most important effects. These theories will therefore give quite different results in the majority of practical problems.

In calculating the corrections to scattering of a nucleon by a neutral vector meson field (γ_μ) due to the effects of virtual mesons, the situation is just as in electrodynamics, in that the result converges without need for a cut-off and depends only on gradients of the meson potential. With scalar (1) or pseudoscalar (γ_5) neutral mesons the result diverges logarithmically and so must be cut off. The part sensitive to the cut-off, however, is directly proportional to the meson potential. It may thereby be removed by a renormalization of mesonic charge g. After this renormalization the results depend only on gradients of the meson potential and are essentially independent of cut-off. This is in addition to the mesonic charge renormalization coming from the production of virtual nucleon pairs by a meson, analogous to the vacuum polarization in electrodynamics. But here there is a further difference from electrodynamics for scalar or pseudoscalar mesons in that the polarization also gives a term in the induced current proportional to the meson potential representing therefore an additional renormalization of the *mass of the meson* which usually depends quadratically on the cut-off.

Next consider charged mesons in the absence of an electromagnetic field. One can introduce isotopic spin operators in an obvious way. (Specifically replace the neutral γ_5, say, by $\tau_i\gamma_5$ and sum over $i=1$, 2 where $\tau_1=\tau_++\tau_-$, $\tau_2=i(\tau_+-\tau_-)$ and τ_+ changes neutron to proton (τ_+ on proton$=0$) and τ_- changes proton to neutron.) It is just as easy for practical problems simply to keep track of whether the particle is a proton or a neutron on a diagram drawn to help write down the

[25] M. Slotnick and W. Heitler, Phys. Rev. 75, 1645 (1949).

[26] H. A. Bethe, Bull. Am. Phys. Soc. 24, 3, Z3 (Washington, 1949).

matrix element. This excludes certain processes. For example in the scattering of a negative meson from q_1 to q_2 by a neutron, the meson q_2 must be emitted first (in order of operators, not time) for the neutron cannot absorb the negative meson q_1 until it becomes a proton. That is, in comparison to the Klein Nishina formula (15), only the analogue of second term (see Fig. 5(b)) would appear in the scattering of negative mesons by neutrons, and only the first term (Fig. 5(a)) in the neutron scattering of positive mesons.

The source of mesons of a given charge is not conserved, for a neutron capable of emitting negative mesons may (on emitting one, say) become a proton no longer able to do so. The proof that a perturbation q gives zero, discussed for longitudinal electromagnetic waves, fails. This has the consequence that vector mesons, if represented by the interaction γ_μ would not satisfy the condition that the divergence of the potential is zero. The interaction is to be taken[27] as $\gamma_\mu - \mu^{-2} q_\mu q$ in emission and as γ_μ in absorption if the real emission of mesons with a non-zero divergence of potential is to be avoided. (The correction term $\mu^{-2} q_\mu q$ gives zero in the neutral case.) The asymmetry in emission and absorption is only apparent, as this is clearly the same thing as subtracting from the original $\gamma_\mu \cdots \gamma_\mu$, a term $\mu^{-2} q \cdots q$. That is, if the term $-\mu^{-2} q_\mu q$ is omitted the resulting theory describes a combination of mesons of spin one and spin zero. The spin zero mesons, coupled by vector coupling q, are removed by subtracting the term $\mu^{-2} q \cdots q$.

The two extra gradients $q \cdots q$ make the problem of diverging integrals still more serious (for example the interaction between two protons corresponding to the exchange of two charged vector mesons depends quadratically on the cut-off if calculated in a straightforward way). One is tempted in this formulation to choose simply $\gamma_\mu \cdots \gamma_\mu$ and accept the admixture of spin zero mesons. But it appears that this leads in the conventional formalism to negative energies for the spin zero component. This shows one of the advantages of the method of second quantization of meson fields over the present formulation. There such errors of sign are obvious while here we seem to be able to write seemingly innocent expressions which can give absurd results. Pseudovector mesons with pseudovector coupling correspond to using $\gamma_5(\gamma_\mu - \mu^{-2} q_\mu q)$ for absorption and $\gamma_5 \gamma_\mu$ for emission for both charged and neutral mesons.

[27] The vector meson field potentials φ_μ satisfy

$$-\partial/\partial x_\nu(\partial\varphi_\mu/\partial x_\nu - \partial\varphi_\nu/\partial x_\mu) - \mu^2\varphi_\mu = -4\pi s_\mu,$$

where s_μ, the source for such mesons, is the matrix element of γ_μ between states of neutron and proton. By taking the divergence $\partial/\partial x_\mu$ of both sides, conclude that $\partial\varphi_\nu/\partial x_\nu = 4\pi\mu^{-2}\partial s_\nu/\partial x_\nu$ so that the original equation can be rewritten as

$$\Box^2\varphi_\mu - \mu^2\varphi_\mu = -4\pi(s_\mu + \mu^{-2}\partial/\partial x_\mu(\partial s_\nu/\partial x_\nu)).$$

The right hand side gives in momentum representation $\gamma_\mu - \mu^{-2} q_\mu q_\nu \gamma_\nu$ the left yields the $(q^2-\mu^2)^{-1}$ and finally the interaction $s_\mu \varphi_\mu$ in the Lagrangian gives the γ_μ on absorption.

Proceeding in this way find generally that particles of spin one can be represented by a four-vector u_μ (which, for a free particle of momentum q satisfies $q \cdot u = 0$). The propagation of virtual particles of momentum q from state ν to μ is represented by multiplication by the 4–4 matrix (or tensor) $P_{\mu\nu} = (\delta_{\mu\nu} - \mu^{-2} q_\mu q_\nu) \times (q^2-\mu^2)^{-1}$. The first-order interaction (from the Proca equation) with an electromagnetic potential $a \exp(-ik \cdot x)$ corresponds to multiplication by the matrix $E_{\mu\nu} = (q_2 \cdot a + q_1 \cdot a)\delta_{\mu\nu} - q_{2\nu} a_\mu - q_{1\mu} a_\nu$ where q_1 and $q_2 = q_1 + k$ are the momenta before and after the interaction. Finally, two potentials a, b may act simultaneously, with matrix $E'_{\mu\nu} = -(a \cdot b)\delta_{\mu\nu} + b_\mu a_\nu$.

In the presence of an electromagnetic field, whenever the nucleon is a proton it interacts with the field in the way described for electrons. The meson interacts in the scalar or pseudoscalar case as a particle obeying the Klein-Gordon equation. It is important here to use the method of calculation of Bethe and Pauli, that is, a virtual meson is assumed to have the same "mass" during all its interactions with the electromagnetic field. The result for mass μ and for $(\mu^2+\lambda^2)^{\frac{1}{2}}$ are subtracted and the difference integrated over the function $G(\lambda)d\lambda$. A separate convergence factor is not provided for each meson propagation between electromagnetic interactions, otherwise gauge invariance is not insured. When the coupling involves a gradient, such as $\gamma_5 q$ where q is the final minus the initial momentum of the nucleon, the vector potential A must be subtracted from the momentum of the proton. That is, there is an additional coupling $\pm\gamma_5 A$ (plus when going from proton to neutron, minus for the reverse) representing the new possibility of a simultaneous emission (or absorption) of meson and photon.

Emission of positive or absorption of negative virtual mesons are represented in the same term, the sign of the charge being determined by temporal relations as for electrons and positrons.

Calculations are very easily carried out in this way to lowest order in g^2 for the various theories for nucleon interaction, scattering of mesons by nucleons, meson production by nuclear collisions and by gamma-rays, nuclear magnetic moments, neutron electron scattering, etc., However, no good agreement with experiment results, when these are available, is obtained. Probably all of the formulations are incorrect. An uncertainty arises since the calculations are only to first order in g^2, and are not valid if $g^2/\hbar c$ is large.

The author is particularly indebted to Professor H. A. Bethe for his explanation of a method of obtaining finite and gauge invariant results for the problem of vacuum polarization. He is also grateful for Professor Bethe's criticisms of the manuscript, and for innumerable discussions during the development of this work. He wishes to thank Professor J. Ashkin for his careful reading of the manuscript.

APPENDIX

In this appendix a method will be illustrated by which the simpler integrals appearing in problems in electrodynamics can be directly evaluated. The integrals arising in more complex processes lead to rather complicated functions, but the study of the relations of one integral to another and their expression in terms of simpler integrals may be facilitated by the methods given here.

As a typical problem consider the integral (12) appearing in the first order radiationless scattering problem:

$$\int \gamma_\mu(p_2-k-m)^{-1}a(p_1-k-m)^{-1}\gamma_\mu k^{-2}d^4kC(k^2), \quad (1a)$$

where we shall take $C(k^2)$ to be typically $-\lambda^2(k^2-\lambda^2)^{-1}$ and d^4k means $(2\pi)^{-2}dk_1dk_2dk_3dk_4$. We first rationalize the factors $(p-k-m)^{-1}=(p-k+m)((p-k)^2-m^2)^{-1}$ obtaining,

$$\int \gamma_\mu(p_2-k+m)a(p_1-k+m)\gamma_\mu k^{-2}d^4kC(k^2) \times((p_1-k)^2-m^2)^{-1}((p_2-k)^2-m^2)^{-1}. \quad (2a)$$

The matrix expression may be simplified. It appears to be best to do so *after* the integrations are performed. Since $AB=2A\cdot B-BA$ where $A\cdot B=A_\mu B_\mu$ is a number commuting with all matrices, find, if R is any expression, and A a vector, since $\gamma_\mu A=-A\gamma_\mu+2A_\mu$,

$$\gamma_\mu AR\gamma_\mu=-A\gamma_\mu R\gamma_\mu+2RA. \quad (3a)$$

Expressions between two γ_μ's can be thereby reduced by induction. Particularly useful are

$$\begin{aligned}\gamma_\mu\gamma_\mu&=4\\ \gamma_\mu A\gamma_\mu&=-2A\\ \gamma_\mu AB\gamma_\mu&=2(AB+BA)=4A\cdot B\\ \gamma_\mu ABC\gamma_\mu&=-2CBA\end{aligned} \quad (4a)$$

where A, B, C are any three vector-matrices (i.e., linear combinations of the four γ's).

In order to calculate the integral in (2a) the integral may be written as the sum of three terms (since $k=k_\sigma\gamma_\sigma$),

$$\gamma_\mu(p_2+m)a(p_1+m)\gamma_\mu J_1-[\gamma_\mu\gamma_\sigma a(p_1+m)\gamma_\mu +\gamma_\mu(p_2+m)a\gamma_\sigma\gamma_\mu]J_2+\gamma_\mu\gamma_\sigma a\gamma_\tau\gamma_\mu J_3, \quad (5a)$$

where

$$J_{(1;2;3)}=\int(1;k_\sigma;k_\sigma k_\tau)k^{-2}d^4kC(k^2) \times((p_2-k)^2-m^2)^{-1}((p_1-k)^2-m^2)^{-1}. \quad (6a)$$

That is for J_1 the $(1;k_\sigma;k_\sigma k_\tau)$ is replaced by 1, for J_2 by k_σ, and for J_3 by $k_\sigma k_\tau$.

More complex processes of the first order involve more factors like $((p_3-k)^2-m^2)^{-1}$ and a corresponding increase in the number of k's which may appear in the numerator, as $k_\sigma k_\tau k_\nu\cdots$. Higher order processes involving two or more virtual quanta involve similar integrals but with factors possibly involving $k+k'$ instead of just k, and the integral extending on $k^{-2}d^4kC(k^2)k'^{-2}d^4k'C(k'^2)$. They can be simplified by methods analogous to those used on the first order integrals.

The factors $(p-k)^2-m^2$ may be written

$$(p-k)^2-m^2=k^2-2p\cdot k-\Delta, \quad (7a)$$

where $\Delta=m^2-p^2$, $\Delta_1=m_1^2-p_1^2$, etc., and we can consider dealing with cases of greater generality in that the different denominators need not have the same value of the mass m. In our specific problem (6a), $p_1^2=m^2$ so that $\Delta_1=0$, but we desire to work with greater generality.

Now for the factor $C(k^2)/k^2$ we shall use $-\lambda^2(k^2-\lambda^2)^{-1}k^{-2}$. This can be written as

$$-\lambda^2/(k^2-\lambda^2)k^2=k^{-2}C(k^2)=-\int_0^{\lambda^2}dL(k^2-L)^{-2}. \quad (8a)$$

Thus we can replace $k^{-2}C(k^2)$ by $(k^2-L)^{-2}$ and at the end integrate the result with respect to L from zero to λ^2. We can for many practical purposes consider λ^2 very large relative to m^2 or p^2. When the original integral converges even without the convergence factor, it will be obvious since the L integration will then be convergent to infinity. If an infra-red catastrophe exists in the integral one can simply assume quanta have a small mass $\lambda_{\min}$ and extend the integral on L from $\lambda^2_{\min}$ to λ^2, rather than from zero to λ^2.

We then have to do integrals of the form

$$\int(1;k_\sigma;k_\sigma k_\tau)d^4k(k^2-L)^{-2}(k^2-2p_1\cdot k-\Delta_1)^{-1} \times(k^2-2p_2\cdot k-\Delta_2)^{-1}, \quad (9a)$$

where by $(1;k_\sigma;k_\sigma k_\tau)$ we mean that in the place of this symbol either 1, or k_σ, or $k_\sigma k_\tau$ may stand in different cases. In more complicated problems there may be more factors $(k^2-2p_i\cdot k-\Delta_i)^{-1}$ or other powers of these factors (the $(k^2-L)^{-2}$ may be considered as a special case of such a factor with $p_i=0$, $\Delta_i=L$) and further factors like $k_\sigma k_\tau k_\rho\cdots$ in the numerator. The poles in all the factors are made definite by the assumption that L, and the Δ's have infinitesimal negative imaginary parts.

We shall do the integrals of successive complexity by induction. We start with the simplest convergent one, and show

$$\int d^4k(k^2-L)^{-3}=(8iL)^{-1}. \quad (10a)$$

For this integral is $\int(2\pi)^{-2}dk_4d^3K(k_4^2-K\cdot K-L)^{-3}$ where the vector K, of magnitude $K=(K\cdot K)^{\frac{1}{2}}$ is k_1, k_2, k_3. The integral on k_4 shows third order poles at $k_4=+(K^2+L)^{\frac{1}{2}}$ and $k_4=-(K^2+L)^{\frac{1}{2}}$. Imagining, in accordance with our definitions, that L has a small negative imaginary part only the first is below the real axis. The contour can be closed by an infinite semi-circle below this axis, without change of the value of the integral since the contribution from the semi-circle vanishes in the limit. Thus the contour can be shrunk about the pole $k_4=+(K^2+L)^{\frac{1}{2}}$ and the resulting k_4 integral is $-2\pi i$ times the residue at this pole. Writing $k_4=(K^2+L)^{\frac{1}{2}}+\epsilon$ and expanding $(k_4^2-K^2-L)^{-3}=\epsilon^{-3}(\epsilon+2(K^2+L)^{\frac{1}{2}})^{-3}$ in powers of ϵ, the residue, being the coefficient of the term ϵ^{-1}, is seen to be $6(2(K^2+L)^{\frac{1}{2}})^{-5}$ so our integral is

$$-(3i/32\pi)\int_0^\infty 4\pi K^2dK(K^2+L)^{-5/2}=(3/8i)(1/3L)$$

establishing (10a).

We also have $\int k_\sigma d^4k(k^2-L)^{-3}=0$ from the symmetry in the k space. We write these results as

$$(8i)\int(1;k_\sigma)d^4k(k^2-L)^{-3}=(1;0)L^{-1}, \quad (11a)$$

where in the brackets $(1;k_\sigma)$ and $(1;0)$ corresponding entries are to be used.

Substituting $k=k'-p$ in (11a), and calling $L-p^2=\Delta$ shows that

$$(8i)\int(1;k_\sigma)d^4k(k^2-2p\cdot k-\Delta)^{-3}=(1;p_\sigma)(p^2+\Delta)^{-1}. \quad (12a)$$

By differentiating both sides of (12a) with respect to Δ, or with respect to p_τ there follows directly

$$(24i)\int(1;k_\sigma;k_\sigma k_\tau)d^4k(k^2-2p\cdot k-\Delta)^{-4} =-(1;p_\sigma;p_\sigma p_\tau-\tfrac{1}{2}\delta_{\sigma\tau}(p^2+\Delta))(p^2+\Delta)^{-2}. \quad (13a)$$

Further differentiations give directly successive integrals including more k factors in the numerator and higher powers of $(k^2-2p\cdot k-\Delta)$ in the denominator.

The integrals so far only contain one factor in the denominator. To obtain results for two factors we make use of the identity

$$a^{-1}b^{-1}=\int_0^1 dx(ax+b(1-x))^{-2}, \quad (14a)$$

(suggested by some work of Schwinger's involving Gaussian integrals). This represents the product of two reciprocals as a parametric integral over one and will therefore permit integrals with two factors to be expressed in terms of one. For other powers of a, b, we make use of all of the identities, such as

$$a^{-2}b^{-1}=\int_0^1 2xdx(ax+b(1-x))^{-3}, \quad (15a)$$

deducible from (14a) by successive differentiations with respect to a or b.

To perform an integral, such as

$$(8i)\int(1;k_\sigma)d^4k(k^2-2p_1\cdot k-\Delta_1)^{-2}(k^2-2p_2\cdot k-\Delta_2)^{-1}, \quad (16a)$$

write, using (15a),

$$(k^2-2p_1\cdot k-\Delta_1)^{-2}(k^2-2p_2\cdot k-\Delta_2)^{-1}=\int_0^1 2xdx(k^2-2p_x\cdot k-\Delta_x)^{-3},$$

where

$$p_x=xp_1+(1-x)p_2 \quad\text{and}\quad \Delta_x=x\Delta_1+(1-x)\Delta_2, \tag{17a}$$

(note that Δ_x is *not* equal to $m^2-p_x^2$) so that the expression (16a) is $(8i)\int_0^1 2xdx\int(1;k_\sigma)d^4k(k^2-2p_x\cdot k-\Delta_x)^{-3}$ which may now be evaluated by (12a) and is

$$(16a)=\int_0^1(1;p_{x\sigma})2xdx(p_x^2+\Delta_x)^{-1}, \tag{18a}$$

where p_x, Δ_x are given in (17a). The integral in (18a) is elementary, being the integral of ratio of polynomials, the denominator of second degree in x. The general expression although readily obtained is a rather complicated combination of roots and logarithms.

Other integrals can be obtained again by parametric differentiation. For example differentiation of (16a), (18a) with respect to Δ_2 or $p_{2\tau}$ gives

$$(8i)\int(1;k_\sigma;k_\sigma k_\tau)d^4k(k^2-2p_1\cdot k-\Delta_1)^{-2}(k^2-2p_2\cdot k-\Delta_2)^{-2}$$
$$=-\int_0^1(1;p_{x\sigma};p_{x\sigma}p_{x\tau}-\tfrac{1}{2}\delta_{\sigma\tau}(p_x^2+\Delta_x))\times 2x(1-x)dx(p_x^2+\Delta_x)^{-2}, \tag{19a}$$

again leading to elementary integrals.

As an example, consider the case that the second factor is just $(k^2-L)^{-2}$ and in the first put $p_1=p$, $\Delta_1=\Delta$. Then $p_x=xp$, $\Delta_x=x\Delta+(1-x)L$. There results

$$(8i)\int(1;k_\sigma;k_\sigma k_\tau)d^4k(k^2-L)^{-2}(k^2-2p\cdot k-\Delta)^{-2}$$
$$=-\int_0^1(1;xp_\sigma;x^2p_\sigma p_\tau-\tfrac{1}{2}\delta_{\sigma\tau}(x^2p^2+\Delta_x))\times 2x(1-x)dx(x^2p^2+\Delta_x)^{-2}. \tag{20a}$$

Integrals with three factors can be reduced to those involving two by using (14a) again. They, therefore, lead to integrals with two parameters (e.g., see application to radiative correction to scattering below).

The methods of calculation given in this paper are deceptively simple when applied to the lower order processes. For processes of increasingly higher orders the complexity and difficulty increases rapidly, and these methods soon become impractical in their present form.

A. Self-Energy

The self-energy integral (19) is

$$(e^2/\pi i)\int\gamma_\mu(p-k-m)^{-1}\gamma_\mu k^{-2}d^4kC(k^2), \tag{19}$$

so that it requires that we find (using the principle of (8a)) the integral on L from 0 to λ^2 of

$$\int\gamma_\mu(p-k+m)\gamma_\mu d^4k(k^2-L)^{-2}(k^2-2p\cdot k)^{-1},$$

since $(p-k)^2-m^2=k^2-2p\cdot k$, as $p^2=m^2$. This is of the form (16a) with $\Delta_1=L$, $p_1=0$, $\Delta_2=0$, $p_2=p$ so that (18a) gives, since $p_x=(1-x)p$, $\Delta_x=xL$,

$$(8i)\int(1;k_\sigma)d^4k(k^2-L)^{-2}(k^2-2p\cdot k)^{-1}$$
$$=\int_0^1(1;(1-x)p_\sigma)2xdx((1-x)^2m^2+xL)^{-1},$$

or performing the integral on L, as in (8),

$$(8i)\int(1;k_\sigma)d^4kk^{-2}C(k^2)(k^2-2p\cdot k)^{-1}$$
$$=\int_0^1(1;(1-x)p_\sigma)2dx\ln\frac{x\lambda^2+(1-x)^2m^2}{(1-x)^2m^2}.$$

Assuming now that $\lambda^2\gg m^2$ we neglect $(1-x)^2m^2$ relative to $x\lambda^2$ in the argument of the logarithm, which then becomes $(\lambda^2/m^2)(x/(1-x)^2)$. Then since $\int_0^1dx\ln(x(1-x)^{-2})=1$ and $\int_0^1(1-x)dx\ln(x(1-x)^{-2})=-(1/4)$ find

$$(8i)\int(1;k_\sigma)k^{-2}C(k^2)d^4k(k^2-2p\cdot k)^{-1}$$
$$=\left(2\ln\frac{\lambda^2}{m^2}+2;p_\sigma\left(\ln\frac{\lambda^2}{m^2}-\frac{1}{2}\right)\right)$$

so that substitution into (19) (after the $(p-k-m)^{-1}$ in (19) is replaced by $(p-k+m)(k^2-2p\cdot k)^{-1}$) gives

$$(19)=(e^2/8\pi)\gamma_\mu[(p+m)(2\ln(\lambda^2/m^2)+2)-p(\ln(\lambda^2/m^2)-\tfrac{1}{2})]\gamma_\mu$$
$$=(e^2/8\pi)[8m(\ln(\lambda^2/m^2)+1)-p(2\ln(\lambda^2/m^2)+5)], \tag{20}$$

using (4a) to remove the γ_μ's. This agrees with Eq. (20) of the text, and gives the self-energy (21) when p is replaced by m.

B. Corrections to Scattering

The term (12) in the radiationless scattering, after rationalizing the matrix denominators and using $p_1^2=p_2^2=m^2$ requires the integrals (9a), as we have discussed. This is an integral with three denominators which we do in two stages. First the factors $(k^2-2p_1\cdot k)$ and $(k^2-2p_2\cdot k)$ are combined by a parameter y;

$$(k^2-2p_1\cdot k)^{-1}(k^2-2p_2\cdot k)^{-1}=\int_0^1dy(k^2-2p_y\cdot k)^{-2},$$

from (14a) where

$$p_y=yp_1+(1-y)p_2. \tag{21a}$$

We therefore need the integrals

$$(8i)\int(1;k_\sigma;k_\sigma k_\tau)d^4k(k^2-L)^{-2}(k^2-2p_y\cdot k)^{-2}, \tag{22a}$$

which we will then integrate with respect to y from 0 to 1. Next we do the integrals (22a) immediately from (20a) with $p=p_y$, $\Delta=0$:

$$(22a)=-\int_0^1\int_0^1(1;xp_{y\sigma};x^2p_{y\sigma}p_{y\tau}$$
$$-\tfrac{1}{2}\delta_{\sigma\tau}(x^2p_y^2+(1-x)L))2x(1-x)dx(x^2p_y^2+L(1-x))^{-2}dy.$$

We now turn to the integrals on L as required in (8a). The first term, (1), in $(1;k_\sigma;k_\sigma k_\tau)$ gives no trouble for large L, but if L is put equal to zero there results $x^{-2}p_y^{-2}$ which leads to a diverging integral on x as $x\rightarrow0$. This infra-red catastrophe is analyzed by using λ_{min}^2 for the lower limit of the L integral. For the last term the upper limit of L must be kept as λ^2. Assuming $\lambda_{min}^2\ll p_y^2\ll\lambda^2$ the x integrals which remain are trivial, as in the self-energy case. One finds

$$-(8i)\int(k^2-\lambda_{min}^2)^{-1}d^4kC(k^2-\lambda_{min}^2)(k^2-2p_1\cdot k)^{-1}(k^2-2p_2\cdot k)^{-1}$$
$$=\int_0^1p_y^{-2}dy\ln(p_y^2/\lambda_{min}^2) \tag{23a}$$

$$-(8i)\int k_\sigma k^{-2}d^4kC(k^2)(k^2-2p_1\cdot k)^{-1}(k^2-2p_2\cdot k)^{-1}$$
$$=2\int_0^1p_{y\sigma}p_y^{-2}dy, \tag{24a}$$

$$-(8i)\int k_\sigma k_\tau k^{-2}d^4kC(k^2)(k^2-2p_1\cdot k)^{-1}(k^2-2p_2\cdot k)^{-1}$$
$$=\int_0^1p_{y\sigma}p_{y\tau}p_y^{-2}dy-\tfrac{1}{2}\delta_{\sigma\tau}\int_0^1dy\ln(\lambda^2p_y^{-2})+\tfrac{1}{4}\delta_{\sigma\tau}. \tag{25a}$$

The integrals on y give,

$$\int_0^1p_y^{-2}dy\ln(p_y^2\lambda_{min}^{-2})=4(m^2\sin2\theta)^{-1}\left[\theta\ln(m\lambda_{min}^{-1})-\int_0^\theta\alpha\tan\alpha d\alpha\right], \tag{26a}$$

$$\int_0^1p_{y\sigma}p_y^{-2}dy=\theta(m^2\sin2\theta)^{-1}(p_{1\sigma}+p_{2\sigma}), \tag{27a}$$

$$\int_0^1p_{y\sigma}p_{y\tau}p_y^{-2}dy=\theta(2m^2\sin2\theta)^{-1}(p_{1\sigma}+p_{1\tau})(p_{2\sigma}+p_{2\tau})+q^{-2}q_\sigma q_\tau(1-\theta\,\mathrm{ctn}\theta), \tag{28a}$$

$$\int_0^1dy\ln(\lambda^2p_y^{-2})=\ln(\lambda^2/m^2)+2(1-\theta\,\mathrm{ctn}\theta). \tag{29a}$$

These integrals on y were performed as follows. Since $p_2=p_1+q$ where q is the momentum carried by the potential, it follows from $p_2^2=p_1^2=m^2$ that $2p_1\cdot q=-q^2$ so that since $p_y=p_1+q(1-y)$, $p_y^2=m^2-q^2y(1-y)$. The substitution $2y-1=\tan\alpha/\tan\theta$ where θ is defined by $4m^2\sin^2\theta=q^2$ is useful for it means $p_y^2=m^2\sec^2\alpha/\sec^2\theta$ and $p_y^{-2}dy=(m^2\sin 2\theta)^{-1}d\alpha$ where α goes from $-\theta$ to $+\theta$.

These results are substituted into the original scattering formula (2a), giving (22). It has been simplified by frequent use of the fact that p_1 operating on the initial state is m, and likewise p_2 when it appears at the left is replacable by m. (Thus, to simplify:

$$\gamma_\mu p_2 a p_1 \gamma_\mu=-2p_1 a p_2 \text{ by (4a)},$$
$$=-2(p_2-q)a(p_1+q)=-2(m-q)a(m+q).$$

A term like $qaq=-q^2a+2(a\cdot q)q$ is equivalent to just $-q^2a$ since $q=p_2-p_1=m-m$ has zero matrix element.) The renormalization term requires the corresponding integrals for the special case $q=0$.

C. Vacuum Polarization

The expressions (32) and (32') for $J_{\mu\nu}$ in the vacuum polarization problem require the calculation of the integral

$$J_{\mu\nu}(m^2)=-\frac{e^2}{\pi i}\int Sp[\gamma_\mu(p-\tfrac{1}{2}q+m)\gamma_\nu(p+\tfrac{1}{2}q+m)]d^4p \times((p-\tfrac{1}{2}q)^2-m^2)^{-1}((p+\tfrac{1}{2}q)^2-m^2)^{-1}, \quad (32)$$

where we have replaced p by $p-\frac{1}{2}q$ to simplify the calculation somewhat. We shall indicate the method of calculation by studying the integral,

$$I(m^2)=\int p_\sigma p_\tau d^4p((p-\tfrac{1}{2}q)^2-m^2)^{-1}((p+\tfrac{1}{2}q)^2-m^2)^{-1}.$$

The factors in the denominator, $p^2-p\cdot q-m^2+\frac{1}{4}q^2$ and $p^2+p\cdot q-m^2+\frac{1}{4}q^2$ are combined as usual by (8a) but for symmetry we substitute $x=\frac{1}{2}(1+\eta)$, $(1-x)=\frac{1}{2}(1-\eta)$ and integrate η from -1 to $+1$:

$$I(m^2)=\int_{-1}^{+1} p_\sigma p_\tau d^4p(p^2-\eta p\cdot q-m^2+\tfrac{1}{4}q^2)^{-2}d\eta/2. \quad (30a)$$

But the integral on p will not be found in our list for it is badly divergent. However, as discussed in Section 7, Eq. (32') we do not wish $I(m^2)$ but rather $\int_0^\infty[I(m^2)-I(m^2+\lambda^2)]G(\lambda)d\lambda$. We can calculate the difference $I(m^2)-I(m^2+\lambda^2)$ by first calculating the derivative $I'(m^2+L)$ of I with respect to m^2 at m^2+L and later integrating L from zero to λ^2. By differentiating (30a), with respect to m^2 find,

$$I'(m^2+L)=\int_{-1}^{+1} p_\sigma p_\tau d^4p(p^2-\eta p\cdot q-m^2-L+\tfrac{1}{4}q^2)^{-3}d\eta.$$

This still diverges, but we can differentiate again to get

$$I''(m^2+L)=3\int_{-1}^{+1} p_\sigma p_\tau d^4p(p^2-\eta p\cdot q-m^2-L+\tfrac{1}{4}q^2)^{-4}d\eta$$
$$=-(8i)^{-1}\int_{-1}^{+1}(\tfrac{1}{4}\eta^2 q_\sigma q_\tau D^{-2}-\tfrac{1}{2}\delta_{\sigma\tau}D^{-1})d\eta \quad (31a)$$

(where $D=\frac{1}{4}(\eta^2-1)q^2+m^2+L$), which now converges and has been evaluated by (13a) with $p=\frac{1}{2}\eta q$ and $\Delta=m^2+L-\frac{1}{4}q^2$. Now to get I' we may integrate I'' with respect to L as an indefinite integral and *we may choose any convenient arbitrary constant*. This is because a constant C in I' will mean a term $-C\lambda^2$ in $I(m^2)-I(m^2+\lambda^2)$ which vanishes since we will integrate the results times $G(\lambda)d\lambda$ and $\int_0^\infty\lambda^2G(\lambda)d\lambda=0$. This means that the logarithm appearing on integrating L in (31a) presents no problem. We may take

$$I'(m^2+L)=(8i)^{-1}\int_{-1}^{+1}[\tfrac{1}{4}\eta^2q_\sigma q_\tau D^{-1}+\tfrac{1}{2}\delta_{\sigma\tau}\ln D]d\eta+C\delta_{\sigma\tau},$$

a subsequent integral on L and finally on η presents no new problems. There results

$$-(8i)\int p_\sigma p_\tau d^4p((p-\tfrac{1}{2}q)^2-m^2)^{-1}((p+\tfrac{1}{2}q)^2-m^2)^{-1}$$
$$=(q_\sigma q_\tau-\delta_{\sigma\tau}q^2)\left[\frac{1}{9}-\frac{4m^2-q^2}{3q^2}\left(1-\frac{\theta}{\tan\theta}\right)+\frac{1}{6}\ln\frac{\lambda^2}{m^2}\right]$$
$$+\delta_{\sigma\tau}[(\lambda^2+m^2)\ln(\lambda^2m^{-2}+1)-C'\lambda^2], \quad (32a)$$

where we assume $\lambda^2\gg m^2$ and have put some terms into the arbitrary constant C' which is independent of λ^2 (but in principle could depend on q^2) and which drops out in the integral on $G(\lambda)d\lambda$. We have set $q^2=4m^2\sin^2\theta$.

In a very similar way the integral with m^2 in the numerator can be worked out. It is, of course, necessary to differentiate this m^2 also when calculating I' and I''. There results

$$-(8i)\int m^2d^4p((p-\tfrac{1}{2}q)^2-m^2)^{-1}((p+\tfrac{1}{2}q)^2-m^2)^{-1}$$
$$=4m^2(1-\theta\operatorname{ctn}\theta)-q^2/3+2(\lambda^2+m^2)\ln(\lambda^2m^{-2}+1)-C''\lambda^2), \quad (33a)$$

with another unimportant constant C''. The complete problem requires the further integral,

$$-(8i)\int(1;p_\sigma)d^4p((p-\tfrac{1}{2}q)^2-m^2)^{-1}((p+\tfrac{1}{2}q)^2-m^2)^{-1}$$
$$=(1,0)(4(1-\theta\operatorname{ctn}\theta)+2\ln(\lambda^2m^{-2})). \quad (34a)$$

The value of the integral (34a) times m^2 differs from (33a), of course, because the results on the right are not actually the integrals on the left, but rather equal their actual value minus their value for $m^2=m^2+\lambda^2$.

Combining these quantities, as required by (32), dropping the constants C', C'' and evaluating the spur gives (33). The spurs are evaluated in the usual way, noting that the spur of any odd number of γ matrices vanishes and $Sp(AB)=Sp(BA)$ for arbitrary A, B. The $Sp(1)=4$ and we also have

$$\tfrac{1}{4}Sp[(p_1+m_1)(p_2-m_2)]=p_1\cdot p_2-m_1m_2, \quad (35a)$$

$$\tfrac{1}{4}Sp[(p_1+m_1)(p_2-m_2)(p_3+m_3)(p_4-m_4)]$$
$$=(p_1\cdot p_2-m_1m_2)(p_3\cdot p_4-m_3m_4)$$
$$-(p_1\cdot p_3-m_1m_3)(p_2\cdot p_4-m_2m_4)$$
$$+(p_1\cdot p_4-m_1m_4)(p_2\cdot p_3-m_2m_3), \quad (36a)$$

where p_i, m_i are arbitrary four-vectors and constants.

It is interesting that the terms of order $\lambda^2\ln\lambda^2$ go out, so that the charge renormalization depends only logarithmically on λ^2. This is not true for some of the meson theories. Electrodynamics is suspiciously unique in the mildness of its divergence.

D. More Complex Problems

Matrix elements for complex problems can be set up in a manner analogous to that used for the simpler cases. We give three illustrations; higher order corrections to the Møller scatter-

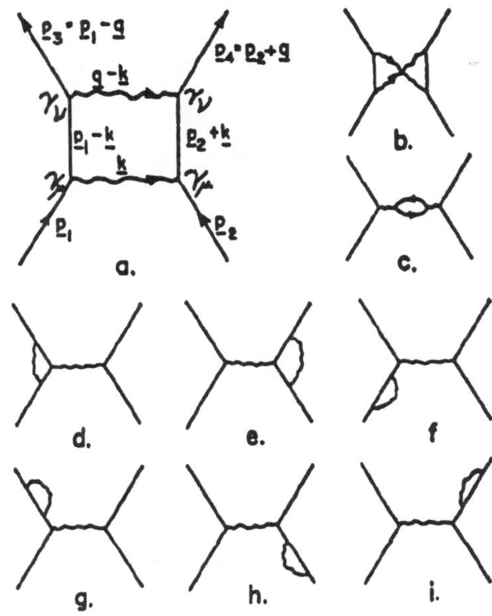

FIG. 8. The interaction between two electrons to order $(e^2/\hbar c)^2$. One adds the contribution of every figure involving two virtual quanta, Appendix D.

ing, to the Compton scattering, and the interaction of a neutron with an electromagnetic field.

For the Møller scattering, consider two electrons, one in state u_1 of momentum p_1 and the other in state u_2 of momentum p_2. Later they are found in states u_3, p_3 and u_4, p_4. This may happen (first order in $e^2/\hbar c$) because they exchange a quantum of momentum $q=p_1-p_3=p_4-p_2$ in the manner of Eq. (4) and Fig. 1. The matrix element for this process is proportional to (translating (4) to momentum space)

$$(\bar{u}_4\gamma_\mu u_2)(\bar{u}_3\gamma_\mu u_1)q^{-2}. \tag{37a}$$

We shall discuss corrections to (37a) to the next order in $e^2/\hbar c$. (There is also the possibility that it is the electron at 2 which finally arrives at 3, the electron at 1 going to 4 through the exchange of quantum of momentum p_3-p_2. The amplitude for this process, $(\bar{u}_4\gamma_\mu u_1)(\bar{u}_3\gamma_\mu u_2)(p_3-p_2)^{-2}$, must be subtracted from (37a) in accordance with the exclusion principle. A similar situation exists to each order so that we need consider in detail only the corrections to (37a), reserving to the last the subtraction of the same terms with 3, 4 exchanged.)

One reason that (37a) is modified is that two quanta may be exchanged, in the manner of Fig. 8a. The total matrix element for all exchanges of this type is

$$(e^2/\pi i)\int(\bar{u}_3\gamma_\nu(p_1-k-m)^{-1}\gamma_\mu u_1)(\bar{u}_4\gamma_\nu(p_2+k-m)^{-1}\gamma_\mu u_2)\cdot k^{-2}(q-k)^{-2}d^4k, \tag{38a}$$

as is clear from the figure and the general rule that electrons of momentum p contribute in amplitude $(p-m)^{-1}$ between interactions γ_μ, and that quanta of momentum k contribute k^{-2}. In integrating on d^4k and summing over μ and ν, we add all alternatives of the type of Fig. 8a. If the time of absorption, γ_μ, of the quantum k by electron 2 is later than the absorption, γ_ν, of $q-k$, this corresponds to the virtual state p_2+k being a positron (so that (38a) contains over thirty terms of the conventional method of analysis).

In integrating over all these alternatives we have considered all possible distortions of Fig. 8a which preserve the order of events along the trajectories. We have not included the possibilities corresponding to Fig. 8b, however. Their contribution is

$$(e^2/\pi i)\int(\bar{u}_3\gamma_\nu(p_1-k-m)^{-1}\gamma_\mu u_1)\times(\bar{u}_4\gamma_\mu(p_2+q-k-m)^{-1}\gamma_\nu u_2)k^{-2}(q-k)^{-2}d^4k, \tag{39a}$$

as is readily verified by labeling the diagram. The contributions of all possible ways that an event can occur are to be added. This means that one adds with equal weight the integrals corresponding to each topologically distinct figure.

To this same order there are also the possibilities of Fig. 8d which give

$$(e^2/\pi i)\int(\bar{u}_3\gamma_\nu(p_3-k-m)^{-1}\gamma_\mu(p_1-k-m)^{-1}\gamma_\nu u_1)\times(\bar{u}_4\gamma_\mu u_2)k^{-2}q^{-2}d^4k.$$

This integral on k will be seen to be precisely the integral (12) for the radiative corrections to scattering, which we have worked out. The term may be combined with the renormalization terms resulting from the difference of the effects of mass change and the terms, Figs. 8f and 8g. Figures 8e, 8h, and 8i are similarly analyzed.

Finally the term Fig. 8c is clearly related to our vacuum polarization problem, and when integrated gives a term proportional to $(\bar{u}_4\gamma_\mu u_2)(\bar{u}_3\gamma_\nu u_1)J_{\mu\nu}q^{-4}$. If the charge is renormalized the term $\ln(\lambda/m)$ in $J_{\mu\nu}$ in (33) is omitted so there is no remaining dependence on the cut-off.

The only new integrals we require are the convergent integrals (38a) and (39a). They can be simplified by rationalizing the denominators and combining them by (14a). For example (38a) involves the factors $(k^2-2p_1\cdot k)^{-1}(k^2+2p_2\cdot k)^{-1}k^{-2}(q^2+k^2-2q\cdot k)^{-2}$. The first two may be combined by (14a) with a parameter x, and the second pair by an expression obtained by differentiation (15a) with respect to b and calling the parameter y. There results a factor $(k^2-2p_x\cdot k)^{-2}(k^2+yq^2-2yq\cdot k)^{-4}$ so that the integrals on d^4k now involve two factors and can be performed by the methods given earlier in the appendix. The subsequent integrals on the parameters x and y are complicated and have not been worked out in detail.

Working with charged mesons there is often a considerable reduction of the number of terms. For example, for the interaction between protons resulting from the exchange of two mesons only the term corresponding to Fig. 8b remains. Term 8a, for example, is impossible, for if the first proton emits a positive meson the second cannot absorb it directly for only neutrons can absorb positive mesons.

As a second example, consider the radiative correction to the Compton scattering. As seen from Eq. (15) and Fig. 5 this scattering is represented by two terms, so that we can consider the corrections to each one separately. Figure 9 shows the types of terms arising from corrections to the term of Fig. 5a. Calling k the momentum of the virtual quantum, Fig. 9a gives an integral

$$\int\gamma_\mu(p_2-k-m)^{-1}e_2(p_1+q_1-k-m)^{-1}e_1(p_1-k-m)^{-1}\gamma_\mu k^{-2}d^4k,$$

convergent without cut-off and reducible by the methods outlined in this appendix.

The other terms are relatively easy to evaluate. Terms b and c of Fig. 9 are closely related to radiative corrections (although somewhat more difficult to evaluate, for one of the states is not that of a free electron, $(p_1+q)^2\neq m^2$). Terms e, f are renormalization terms. From term d must be subtracted explicitly the effect of mass Δm, as analyzed in Eqs. (26) and (27) leading to (28) with $p'=p_1+q$, $a=e_2$, $b=e_1$. Terms g, h give zero since the vacuum polarization has zero effect on free light quanta, $q_1^2=0$, $q_2^2=0$. The total is insensitive to the cut-off λ.

The result shows an infra-red catastrophe, the largest part of the effect. When cut-off at $\lambda_{\min}$, the effect proportional to $\ln(m/\lambda_{\min})$ goes as

$$(e^2/\pi)\ln(m/\lambda_{\min})(1-2\theta\,\mathrm{ctn}\,2\theta), \tag{40a}$$

times the uncorrected amplitude, where $(p_2-p_1)^2=4m^2\sin^2\theta$. This is the same as for the radiative correction to scattering for a deflection p_2-p_1. This is physically clear since the long wave quanta are not effected by short-lived intermediate states. The infra-red effects arise[28] from a final adjustment of the field from the asymptotic coulomb field characteristic of the electron of

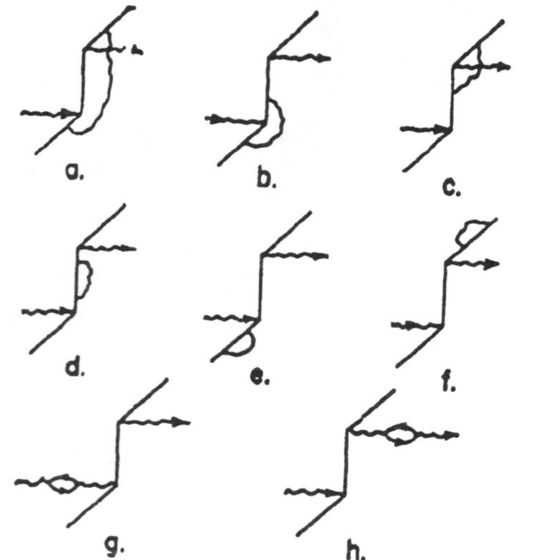

FIG. 9. Radiative correction to the Compton scattering term (a) of Fig. 5. Appendix D.

[28] F. Bloch and A. Nordsieck, Phys. Rev. 52, 54 (1937).

momentum p_1 before the collision to that characteristic of an electron moving in a new direction p_2 after the collision.

The complete expression for the correction is a very complicated expression involving transcendental integrals.

As a final example we consider the interaction of a neutron with an electromagnetic field in virtue of the fact that the neutron may emit a virtual negative meson. We choose the example of pseudoscalar mesons with pseudovector coupling. The change in amplitude due to an electromagnetic field $A=a\exp(-iq\cdot x)$ determines the scattering of a neutron by such a field. In the limit of small q it will vary as $qa-aq$ which represents the interaction of a particle possessing a magnetic moment. The first-order interaction between an electron and a neutron is given by the same calculation by considering the exchange of a quantum between the electron and the nucleon. In this case a_μ is q^{-2} times the matrix element of γ_μ between the initial and final states of the electron, the states differing in momentum by q.

The interaction may occur because the neutron of momentum p_1 emits a negative meson becoming a proton which proton interacts with the field and then reabsorbs the meson (Fig. 10a). The matrix for this process is ($p_2=p_1+q$),

$$\int(\gamma_5 k)(p_2-k-M)^{-1}a(p_1-k-M)^{-1}(\gamma_5 k)(k^2-\mu^2)^{-1}d^4k. \quad (41a)$$

Alternatively it may be the meson which interacts with the field. We assume that it does this in the manner of a scalar potential satisfying the Klein Gordon Eq. (35), (Fig. 10b)

$$-\int(\gamma_5 k_2)(p_1-k_1-M)^{-1}(\gamma_5 k_1)(k_2^2-\mu^2)^{-1} \times(k_2\cdot a+k_1\cdot a)(k_1^2-\mu^2)^{-1}d^4k_1, \quad (42a)$$

where we have put $k_2=k_1+q$. The change in sign arises because the virtual meson is negative. Finally there are two terms arising from the $\gamma_5 a$ part of the pseudovector coupling (Figs. 10c, 10d)

$$\int(\gamma_5 k)(p_2-k-M)^{-1}(\gamma_5 a)(k^2-\mu^2)^{-1}d^4k, \quad (43a)$$

and

$$\int(\gamma_5 a)(p_1-k-M)^{-1}(\gamma_5 k)(k^2-\mu^2)^{-1}d^4k. \quad (44a)$$

Using convergence factors in the manner discussed in the section on meson theories each integral can be evaluated and the results combined. Expanded in powers of q the first term gives the magnetic moment of the neutron and is insensitive to the cut-off, the next gives the scattering amplitude of slow electrons on neutrons, and depends logarithmically on the cut-off.

The expressions may be simplified and combined somewhat before integration. This makes the integrals a little easier and also shows the relation to the case of pseudoscalar coupling. For example in (41a) the final $\gamma_5 k$ can be written as $\gamma_5(k-p_1+M)$ since $p_1=M$ when operating on the initial neutron state. This is

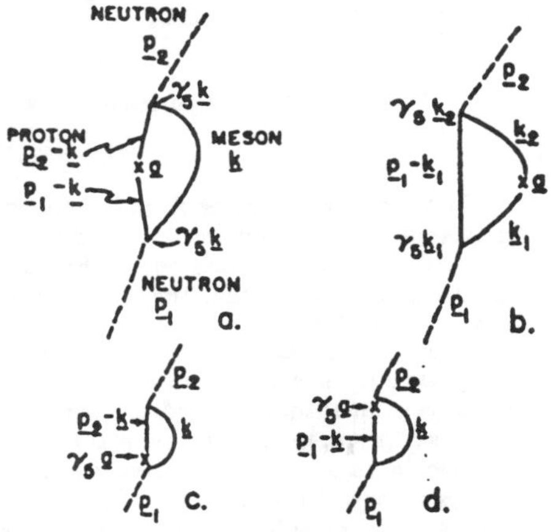

FIG. 10. According to the meson theory a neutron interacts with an electromagnetic potential a by first emitting a virtual charged meson. The figure illustrates the case for a pseudoscalar meson with pseudovector coupling. Appendix D.

$(p_1-k-M)\gamma_5+2M\gamma_5$ since γ_5 anticommutes with p_1 and k. The first term cancels the $(p_1-k-M)^{-1}$ and gives a term which just cancels (43a). In a like manner the leading factor $\gamma_5 k$ in (41a) is written as $-2M\gamma_5-\gamma_5(p_2-k-M)$, the second term leading to a simpler term containing no $(p_2-k-M)^{-1}$ factor and combining with a similar one from (44a). One simplifies the $\gamma_5 k_1$ and $\gamma_5 k_2$ in (42a) in an analogous way. There finally results terms like (41a), (42a) but with pseudoscalar coupling $2M\gamma_5$ instead of $\gamma_5 k$, no terms like (43a) or (44a) and a remainder, representing the difference in effects of pseudovector and pseudoscalar coupling. The pseudoscalar terms do not depend sensitively on the cut-off, but the difference term depends on it logarithmically. The difference term affects the electron-neutron interaction but not the magnetic moment of the neutron.

Interaction of a proton with an electromagnetic potential can be similarly analyzed. There is an effect of virtual mesons on the electromagnetic properties of the proton even in the case that the mesons are neutral. It is analogous to the radiative corrections to the scattering of electrons due to virtual photons. The sum of the magnetic moments of neutron and proton for charged mesons is the same as the proton moment calculated for the corresponding neutral mesons. In fact it is readily seen by comparing diagrams, that for arbitrary q, the scattering matrix to *first order in the electromagnetic potential* for a proton according to neutral meson theory is equal, if the mesons were charged, to the sum of the matrix for a neutron and the matrix for a proton. This is true, for any type or mixtures of meson coupling, to all orders in the coupling (neglecting the mass difference of neutron and proton).

PHYSICAL REVIEW VOLUME 80, NUMBER 3 NOVEMBER 1, 1950

Mathematical Formulation of the Quantum Theory of Electromagnetic Interaction

R. P. FEYNMAN*
Department of Physics, Cornell University, Ithaca, New York

(Received June 8, 1950)

The validity of the rules given in previous papers for the solution of problems in quantum electrodynamics is established. Starting with Fermi's formulation of the field as a set of harmonic oscillators, the effect of the oscillators is integrated out in the Lagrangian form of quantum mechanics. There results an expression for the effect of all virtual photons valid to all orders in $e^2/\hbar c$. It is shown that evaluation of this expression as a power series in $e^2/\hbar c$ gives just the terms expected by the aforementioned rules.

In addition, a relation is established between the amplitude for a given process in an arbitrary unquantized potential and in a quantum electrodynamical field. This relation permits a simple general statement of the laws of quantum electrodynamics.

A description, in Lagrangian quantum-mechanical form, of particles satisfying the Klein-Gordon equation is given in an Appendix. It involves the use of an extra parameter analogous to proper time to describe the trajectory of the particle in four dimensions.

A second Appendix discusses, in the special case of photons, the problem of finding what real processes are implied by the formula for virtual processes.

Problems of the divergences of electrodynamics are not discussed.

1. INTRODUCTION

IN two previous papers[1] rules were given for the calculation of the matrix element for any process in electrodynamics, to each order in $e^2/\hbar c$. No complete proof of the equivalence of these rules to the conventional electrodynamics was given in these papers. Secondly, no closed expression was given valid to all orders in $e^2/\hbar c$. In this paper these formal omissions will be remedied.[2]

In paper II it was pointed out that for many problems in electrodynamics the Hamiltonian method is not advantageous, and might be replaced by the over-all space-time point of view of a direct particle interaction. It was also mentioned that the Lagrangian form of quantum mechanics[3] was useful in this connection. The rules given in paper II were, in fact, first deduced in this form of quantum mechanics. We shall give this derivation here.

The advantage of a Lagrangian form of quantum mechanics is that in a system with interacting parts it permits a separation of the problem such that the motion of any part can be analyzed or solved first, and the results of this solution may then be used in the solution of the motion of the other parts. This separation is especially useful in quantum electrodynamics which represents the interaction of matter with the electromagnetic field. The electromagnetic field is an especially simple system and its behavior can be analyzed completely. What we shall show is that the net effect of the field is a delayed interaction of the particles. It is possible to do this easily only if it is not necessary at the same time to analyze completely the motion of the particles. The only advantage in our problems of the form of quantum mechanics in C is to permit one to separate these aspects of the problem. There are a number of disadvantages, however, such as a lack of familiarity, the apparent (but not real) necessity for dealing with matter in non-relativistic approximation, and at times a cumbersome mathematical notation and method, as well as the fact that a great deal of useful information that is known about operators cannot be directly applied.

It is also possible to separate the field and particle aspects of a problem in a manner which uses operators and Hamiltonians in a way that is much more familiar. One abandons the notation that the order of action of operators depends on their written position on the paper and substitutes some other convention (such that the order of operators is that of the time to which they refer). The increase in manipulative facility which accompanies this change in notation makes it easier to represent and to analyze the formal problems in electrodynamics. The method requires some discussion, however, and will be described in a succeeding paper. In this paper we shall give the derivations of the formulas of II by means of the form of quantum mechanics given in C.

The problem of interaction of matter and field will be analyzed by first solving for the behavior of the field in terms of the coordinates of the matter, and finally discussing the behavior of the matter (by matter is actually meant the electrons and positrons). That is to say, we shall first eliminate the field variables from the equations of motion of the electrons and then discuss the behavior of the electrons. In this way all of the rules given in the paper II will be derived.

Actually, the straightforward elimination of the field

* Now at the California Institute of Technology, Pasadena, California.

[1] R. P. Feynman, Phys. Rev. 76, 749 (1949), hereafter called I, and Phys. Rev. 76, 769 (1949), hereafter called II.

[2] See in this connection also the papers of S. Tomonaga, Phys. Rev. 74, 224 (1948); S. Kanesawa and S. Tomonaga, Prog. Theoret. Phys. 3, 101 (1948); J. Schwinger, Phys. Rev. 76, 790 (1949); F. Dyson, Phys. Rev. 75, 1736 (1949); W. Pauli and F. Villars, Rev. Mod. Phys. 21, 434 (1949). The papers cited give references to previous work.

[3] R. P. Feynman, Rev. Mod. Phys. 20, 367 (1948), hereafter called C.

variables will lead at first to an expression for the behavior of an arbitrary number of Dirac electrons. Since the number of electrons might be infinite, this can be used directly to find the behavior of the electrons according to hole theory by imagining that nearly all the negative energy states are occupied by electrons. But, at least in the case of motion in a fixed potential, it has been shown that this hole theory picture is equivalent to one in which a positron is represented as an electron whose space-time trajectory has had its time direction reversed. To show that this same picture may be used in quantum electrodynamics when the potentials are not fixed, a special argument is made based on a study of the relationship of quantum electrodynamics to motion in a fixed potential. Finally, it is pointed out that this relationship is quite general and might be used for a general statement of the laws of quantum electrodynamics.

Charges obeying the Klein-Gordon equation can be analyzed by a special formalism given in Appendix A. A fifth parameter is used to specify the four-dimensional trajectory so that the Lagrangian form of quantum mechanics can be used. Appendix B discusses in more detail the relation of real and virtual photon emission. An equation for the propagation of a self-interacting electron is given in Appendix C.

In the demonstration which follows we shall restrict ourselves temporarily to cases in which the particle's motion is non-relativistic, but the transition of the final formulas to the relativistic case is direct, and the proof could have been kept relativistic throughout.

The transverse part of the electromagnetic field will be represented as an assemblage of independent harmonic oscillators each interacting with the particles, as suggested by Fermi.[4] We use the notation of Heitler.[5]

2. QUANTUM ELECTRODYNAMICS IN LAGRANGIAN FORM

The Hamiltonian for a set of non-relativistic particles interacting with radiation is, classically, $H=H_p+H_I+H_c+H_{tr}$, where $H_p+H_I=\sum_n \frac{1}{2}m_n^{-1}(\mathbf{p}_n-e_n\mathbf{A}^{tr}(\mathbf{x}_n))^2$ is the Hamiltonian of the particles of mass m_n, charge e_n, coordinate $\mathbf{x}_n$ and momentum $\mathbf{p}_n$ and their interaction with the transverse part of the electromagnetic field. This field can be expanded into plane waves

$$\mathbf{A}^{tr}(\mathbf{x})=(8\pi)^{\frac{1}{2}}\sum_{\mathbf{K}}[\mathbf{e}_1(q_{\mathbf{K}}^{(1)}\cos(\mathbf{K}\cdot\mathbf{x})+q_{\mathbf{K}}^{(2)}\sin(\mathbf{K}\cdot\mathbf{x}))+\mathbf{e}_2(q_{\mathbf{K}}^{(3)}\cos(\mathbf{K}\cdot\mathbf{x})+q_{\mathbf{K}}^{(4)}\sin(\mathbf{K}\cdot\mathbf{x}))] \quad (1)$$

where $\mathbf{e}_1$ and $\mathbf{e}_2$ are two orthogonal polarization vectors at right angles to the propagation vector $\mathbf{K}$, magnitude k. The sum over $\mathbf{K}$ means, if normalized to unit volume, $\frac{1}{2}\int d^3\mathbf{K}/8\pi^3$, and each $q_{\mathbf{K}}^{(r)}$ can be considered as the coordinate of a harmonic oscillator. (The factor $\frac{1}{2}$ arises for the mode corresponding to $\mathbf{K}$ and to $-\mathbf{K}$ is the same.) The Hamiltonian of the transverse field represented as oscillators is

$$H_{tr}=\frac{1}{2}\sum_{\mathbf{K}}\sum_{r=1}^{4}((p_{\mathbf{K}}^{(r)})^2+k^2(q_{\mathbf{K}}^{(r)})^2)$$

where $p_{\mathbf{K}}^{(r)}$ is the momentum conjugate to $q_{\mathbf{K}}^{(r)}$. The longitudinal part of the field has been replaced by the Coulomb interaction,[6]

$$H_c=\tfrac{1}{2}\sum_n\sum_m e_ne_m/r_{nm}$$

where $r_{nm}^2=(\mathbf{x}_n-\mathbf{x}_m)^2$. As is well known,[4] when this Hamiltonian is quantized one arrives at the usual theory of quantum electrodynamics. To express these laws of quantum electrodynamics one can equally well use the Lagrangian form of quantum mechanics to describe this set of oscillators and particles. The classical Lagrangian equivalent to this Hamiltonian is $L=L_p+L_I+L_c+L_{tr}$ where

$$L_p=\tfrac{1}{2}\sum_n m_n\dot{\mathbf{x}}_n^2 \quad (2a)$$

$$L_I=\sum_n e_n\dot{\mathbf{x}}_n\cdot\mathbf{A}^{tr}(\mathbf{x}_n) \quad (2b)$$

$$L_{tr}=\tfrac{1}{2}\sum_{\mathbf{K}}\sum_r((\dot{q}_{\mathbf{K}}^{(r)})^2-k^2(q_{\mathbf{K}}^{(r)})^2) \quad (2c)$$

$$L_c=-\tfrac{1}{2}\sum_n\sum_m e_ne_m/r_{mn}. \quad (2d)$$

When this Lagrangian is used in the Lagrangian forms of quantum mechanics of C, what it leads to is, of course, mathematically equivalent to the result of using the Hamiltonian H in the ordinary way, and is therefore equivalent to the more usual forms of quantum electrodynamics (at least for non-relativistic particles). We may, therefore, proceed by using this Lagrangian form of quantum electrodynamics, with the assurance that the results obtained must agree with those obtained from the more usual Hamiltonian form.

The Lagrangian enters through the statement that the functional which carries the system from one state to another is $\exp(iS)$ where

$$S=\int Ldt=S_p+S_I+S_c+S_{tr}. \quad (3)$$

The time integrals must be written as Riemann sums with some care; for example,

$$S_I=\sum_n\int e_n\dot{\mathbf{x}}_n(t)\cdot\mathbf{A}^{tr}(\mathbf{x}_n(t))dt \quad (4)$$

becomes according to C, Eq. (19)

$$S_I=\sum_n\sum_i\tfrac{1}{2}e_n(\mathbf{x}_{n,i+1}-\mathbf{x}_{n,i})\cdot(\mathbf{A}^{tr}(\mathbf{x}_{n,i+1})+\mathbf{A}^{tr}(\mathbf{x}_{n,i})) \quad (5)$$

so that the velocity $\dot{\mathbf{x}}_{n,i}$ which multiplies $\mathbf{A}^{tr}(\mathbf{x}_{n,i})$ is

$$\dot{\mathbf{x}}_{n,i}=\tfrac{1}{2}\epsilon^{-1}(\mathbf{x}_{n,i+1}-\mathbf{x}_{n,i})+\tfrac{1}{2}\epsilon^{-1}(\mathbf{x}_{n,i}-\mathbf{x}_{n,i-1}). \quad (6)$$

---

[4] E. Fermi, Rev. Mod. Phys. 4, 87 (1932).

[5] W. Heitler, *The Quantum Theory of Radiation*, second edition (Oxford University Press, London, 1944).

[6] The term in the sum for $n=m$ is obviously infinite but must be included for relativistic invariance. Our problem here is to re-express the usual (and divergent) form of electrodynamics in the form given in II. Modifications for dealing with the divergences are discussed in II and we shall not discuss them further here.

In the Lagrangian form it is possible to eliminate the transverse oscillators as is discussed in C, Section 13. One must specify, however, the initial and final state of all oscillators. We shall first choose the special, simple case that all oscillators are in their ground states initially and finally, so that all photons are virtual. Later we do the more general case in which real quanta are present initially or finally. We ask, then, for the amplitude for finding no quanta present and the particles in state $\chi_{t''}$ at time t'', if at time t' the particles were in state $\psi_{t'}$ and no quanta were present.

The method of eliminating field oscillators is described in Section 13 of C. We shall simply carry out the elimination here using the notation and equations of C. To do this, for simplicity, we first consider in the next section the case of a particle or a system of particles interacting with a single oscillator, rather than the entire assemblage of the electromagnetic field.

3. FORCED HARMONIC OSCILLATOR

We consider a harmonic oscillator, coordinate q, Lagrangian $L=\frac{1}{2}(\dot{q}^2-\omega^2q^2)$ interacting with a particle or system of particles, action S_p, through a term in the Lagrangian $q(t)\gamma(t)$ where $\gamma(t)$ is a function of the coordinates (symbolized as x) of the particle. The precise form of $\gamma(t)$ for each oscillator of the electromagnetic field is given in the next section. We ask for the amplitude that at some time t'' the particles are in state $\chi_{t''}$ and the oscillator is in, say, an eigenstate m of energy $\omega(m+\frac{1}{2})$ (units are chosen such that $\hbar=c=1$) when it is given that at a previous time t' the particles were in state $\psi_{t'}$ and the oscillator in n. The amplitude for this is the transition amplitude [see C, Eq. (61)]

$$\langle\chi_{t''}\varphi_m|1|\psi_{t'}\varphi_n\rangle_{S_p+S_0+S_I}$$

$$=\int\int\chi_{t''}^*(x_{t''})\varphi_m^*(q_{t''})\exp i(S_p+S_0+S_I)\cdot\varphi_n(q_{t'})\psi_{t'}(x_{t'})dx_{t''}dx_{t'}dq_{t''}dq_{t'}\mathfrak{D}x(t)\mathfrak{D}q(t)\quad(7)$$

where x represents the variables describing the particle, S_p is the action calculated classically for the particles for a given path going from coordinate $x_{t'}$ at t' to $x_{t''}$ at t'', S_0 is the action $\int\frac{1}{2}(\dot{q}^2-\omega^2q^2)dt$ for any path of the oscillator going from $q_{t'}$ at t' to $q_{t''}$ at t'', while

$$S_I=\int q(t)\gamma(t)dt,\quad(8)$$

the action of interaction, is a functional of both $q(t)$ and $x(t)$, the paths of oscillator and particles. The symbols $\mathfrak{D}x(t)$ and $\mathfrak{D}q(t)$ represent a summation over all possible paths of particles and oscillator which go between the given end points in the sense defined in C, Eq. (9). (That is, assuming time to proceed in infinitesimal steps, ϵ, an integral over all values of the coordinates x and q corresponding to each instant in time, suitably normalized.)

The problem may be broken in two. The result can be written as an integral over all paths of the particles only, of $(\exp iS_p)\cdot G_{mn}$:

$$\langle\chi_{t''}\varphi_m|1|\psi_{t'}\varphi_n\rangle_{S_p+S_0+S_I}=\langle\chi_{t''}|G_{mn}|\psi_{t'}\rangle_{S_p}\quad(9)$$

where G_{mn} is a functional of the path of the particles alone (since it depends on $\gamma(t)$) given by

$$G_{mn}=\left\langle\varphi_m\left|\exp i\int q(t)\gamma(t)dt\right|\varphi_n\right\rangle_{S_0}$$

$$=\int\varphi_m^*(q_{t''})\exp i(S_0+S_I)\varphi_n(q_{t'})dq_{t'}dq_{t''}\mathfrak{D}q(t)$$

$$=\int\varphi_m^*(q_j)\exp i\epsilon\sum_{i=0}^{j-1}\left[\tfrac{1}{2}\epsilon^{-2}(q_{i+1}-q_i)^2-\tfrac{1}{2}\omega^2q_i^2+q_i\gamma_i\right]\cdot\varphi_n(q_0)dq_0a^{-1}dq_1a^{-1}dq_2\cdots a^{-1}dq_j\quad(10)$$

where we have written the $\mathfrak{D}q(t)$ out explicitly (and have set $a=(2\pi i\epsilon)^{\frac{1}{2}}$, $t''-t'=j\epsilon$, $q_{t'}=q_0$, $q_{t''}=q_j$). The last form can be written as

$$G_{mn}=\int\varphi_m^*(q_j)k(q_j,t'';q_0,t')\varphi_n(q_0)dq_0dq_j\quad(11)$$

where $k(q_j,t'';q_0,t')$ is the kernel [as in I, Eq. (2)] for a forced harmonic oscillator giving the amplitude for arrival at q_j at time t'' if at time t' it was known to be at q_0. According to C it is given by

$$k(q_j,t'';q_0,t')=(2\pi i\omega^{-1}\sin\omega(t''-t'))^{-\frac{1}{2}}\times\exp iQ(q_j,t'';q_0,t')\quad(12)$$

where $Q(q_j,t'';q_0,t')$ is the action calculated along the classical path between the end points $q_j,t'';q_0,t'$, and is given explicitly in C.[7] It is

[7] That (12) is correct, at least insofar as it depends on q_0, can be seen directly as follows. Let $\bar{q}(t)$ be the classical path which satisfies the boundary condition $\bar{q}(t')=q_0$, $\bar{q}(t'')=q_j$. Then in the integral defining k replace each of the variables q_i by $q_i=\bar{q}_i+y_i$, $(\bar{q}_i=\bar{q}(t_i))$, that is, use the displacement y_i from the classical path $\bar{q}_i$ as the coordinate rather than the absolute position. With the substitution $q_i=\bar{q}_i+y_i$ in the action

$$S_0+S_I=\int(\tfrac{1}{2}\dot{q}^2-\tfrac{1}{2}\omega^2q^2+\gamma q)dt$$

$$=\int(\tfrac{1}{2}\dot{\bar{q}}^2-\tfrac{1}{2}\omega^2\bar{q}^2+\gamma\bar{q})dt+\int(\tfrac{1}{2}\dot{y}^2-\tfrac{1}{2}\omega^2y^2)dt$$

the terms linear in y drop out by integrations by parts using the equation of motion $\ddot{\bar{q}}=-\omega^2\bar{q}+\gamma(t)$ for the classical path, and the boundary conditions $y(t')=y(t'')=0$. That this should occur should occasion no surprise, for the action functional is an extremum at $q(t)=\bar{q}(t)$ so that it will only depend to second order in the displacements y from this extremal orbit $\bar{q}(t)$. Further, since the action functional is quadratic to begin with, it cannot depend on y more than quadratically. Hence

$$S_0+S_I=Q+\int(\tfrac{1}{2}\dot{y}^2-\tfrac{1}{2}\omega^2y^2)dt$$

so that since $dq_i=dy_i$,

$$k(q_j,t'';q_0,t')=\exp(iQ)\int\exp\left(i\int\frac{1}{2}(\dot{y}^2-\omega^2y^2)dt\right)\mathfrak{D}y(t).$$

The factor following the $\exp iQ$ is the amplitude for a free oscillator to proceed from $y=0$ at $t=t'$ to $y=0$ at $t=t''$ and does not there-

$$Q=\frac{\omega}{2\sin\omega(t''-t')}\Big[(q_f^2+q_0^2)\cos\omega(t''-t')-2q_fq_0$$
$$+\frac{2q_f}{\omega}\int_{t'}^{t''}\gamma(t)\sin\omega(t-t')dt$$
$$+\frac{2q_0}{\omega}\int_{t'}^{t''}\gamma(t)\sin\omega(t''-t)dt$$
$$-\frac{2}{\omega^2}\int_{t'}^{t''}\int_{t'}^{t}\gamma(t)\gamma(s)\sin\omega(t''-t)$$
$$\times\sin\omega(s-t')dsdt\Big]. \quad (13)$$

The solution of the motion of the oscillator can now be completed by substituting (12) and (13) into (11) and performing the integrals. The simplest case is for $m, n=0$ for which case[8]

$$\varphi_0(q_0)=(\omega/\pi)^{\frac{1}{4}}\exp(-\tfrac{1}{2}\omega q_0^2)\exp(-\tfrac{1}{2}i\omega t')$$

so that the integrals on q_0, q_f are just Gaussian integrals. There results

$$G_{00}=\exp\left(-\frac{1}{2}\omega^{-1}\int_{t'}^{t''}\int_{t'}^{t}\exp(-i\omega(t-s))\gamma(t)\gamma(s)dtds\right)$$

a result of fundamental importance in the succeeding developments. By replacing $t-s$ by its absolute value $|t-s|$ we may integrate both variables over the entire range and divide by 2. We will henceforth make the results more general by extending the limits on the integrals from $-\infty$ to $+\infty$. Thus if one wishes to study the effect on a particle of interaction with an oscillator for just the period t' to t'' one may use

$$G_{00}=\exp\bigg(-\frac{1}{4\omega}\int_{-\infty}^{\infty}\int_{-\infty}^{\infty}$$
$$\times\exp(-i\omega|t-s|)\gamma(t)\gamma(s)dtds\bigg) \quad (14)$$

imagining in this case that the interaction $\gamma(t)$ is zero outside these limits. We defer to a later section the discussion of other values of m, n.

Since G_{00} is simply an exponential, we can write it as $\exp(iI)$, consider that the complete "action" for the system of particles is $S=S_p+I$ and that one computes transition elements with this "action" instead of S_p (see C, Sec. 12). The functional I, which is given by

$$I=\tfrac{1}{2}i\omega^{-1}\int\int\exp(-i\omega|t-s|)\gamma(s)\gamma(t)dsdt \quad (15)$$

is complex, however; we shall speak of it as the complex action. It describes the fact that the system at one time can affect itself at a different time by means of a temporary storage of energy in the oscillator. When there are several independent oscillators with different interactions, the effect, if they are all in the lowest state at t' and t'', is the product of their separate G_{00} contributions. Thus the complex action is additive, being the sum of contributions like (15) for each of the several oscillators.

4. VIRTUAL TRANSITIONS IN THE ELECTROMAGNETIC FIELD

We can now apply these results to eliminate the transverse field oscillators of the Lagrangian (2). At first we can limit ourselves to the case of purely virtual transitions in the electromagnetic field, so that there is no photon in the field at t' and t''. That is, all of the field oscillators are making transitions from ground state to ground state.

The $\gamma_{\mathbf{K}}^{(r)}$ corresponding to each oscillator $q_{\mathbf{K}}^{(r)}$ is found from the interaction term L_I [Eq. (2b)], substituting the value of $\mathbf{A}^{tr}(\mathbf{x})$ given in (1). There results, for example,

$$\gamma_{\mathbf{K}}^{(1)}=(8\pi)^{\frac{1}{2}}\textstyle\sum_n e_n(\mathbf{e}_1\cdot\dot{\mathbf{x}}_n)\cos(\mathbf{K}\cdot\mathbf{x}_n)$$
$$\gamma_{\mathbf{K}}^{(3)}=(8\pi)^{\frac{1}{2}}\textstyle\sum_n e_n(\mathbf{e}_1\cdot\dot{\mathbf{x}}_n)\sin(\mathbf{K}\cdot\mathbf{x}_n) \quad (16)$$

the corresponding results for $\gamma_{\mathbf{K}}^{(2)}$, $\gamma_{\mathbf{K}}^{(4)}$ replace $\mathbf{e}_1$ by $\mathbf{e}_2$.

The complex action resulting from oscillator of coordinate $q_{\mathbf{K}}^{(1)}$ is therefore

$$I_{\mathbf{K}}^{(1)}=\frac{8\pi i}{4k}\sum_n\sum_m\int\int e_ne_m\exp(-ik|t-s|)(\mathbf{e}_1\cdot\dot{\mathbf{x}}_n(t))$$
$$\times(\mathbf{e}_1\cdot\dot{\mathbf{x}}_m(s))\cdot\cos(\mathbf{K}\cdot\mathbf{x}_n(t))\cos(\mathbf{K}\cdot\mathbf{x}_m(s))dsdt.$$

The term $I_{\mathbf{K}}^{(3)}$ exchanges the cosines for sines, so in the sum $I_{\mathbf{K}}^{(1)}+I_{\mathbf{K}}^{(3)}$ the product of the two cosines, $\cos A\cdot\cos B$ is replaced by $(\cos A\cos B+\sin A\sin B)$ or $\cos(A-B)$. The terms $I_{\mathbf{K}}^{(2)}+I_{\mathbf{K}}^{(4)}$ give the same result with $\mathbf{e}_2$ replacing $\mathbf{e}_1$. The sum $(\mathbf{e}_1\cdot\mathbf{V})(\mathbf{e}_1\cdot\mathbf{V}')+(\mathbf{e}_2\cdot\mathbf{V})(\mathbf{e}_2\cdot\mathbf{V}')$ is $(\mathbf{V}\cdot\mathbf{V}')-k^{-2}(\mathbf{K}\cdot\mathbf{V})(\mathbf{K}\cdot\mathbf{V}')$ since it is the sum of the products of vector components in two orthogonal directions, so that if we add the product in the third direction (that of $\mathbf{K}$) we construct the complete scalar product. Summing over all $\mathbf{K}$ then, since $\sum_{\mathbf{K}}=\tfrac{1}{2}\int d^3\mathbf{K}/8\pi^3$ we find for the total complex action of all of the transverse oscillators,

$$I_{tr}=i\sum_n\sum_m\int_{t'}^{t''}dt\int_{t'}^{t''}ds\int e_ne_m\exp(-ik|t-s|)$$
$$\times[\dot{\mathbf{x}}_n(t)\cdot\dot{\mathbf{x}}_m(s)-k^{-2}(\mathbf{K}\cdot\dot{\mathbf{x}}_n(t))(\mathbf{K}\cdot\dot{\mathbf{x}}_m(s))]$$
$$\cdot\cos(\mathbf{K}\cdot(\mathbf{x}_n(t)-\mathbf{x}_m(s)))d^3\mathbf{K}/8\pi^2k. \quad (17)$$

fore depend on q_0, q_f, or $\gamma(t)$, being a function only of $t''-t'$. [That it is actually $(2\pi i\omega^{-1}\sin\omega(t''-t'))^{-\frac{1}{2}}$ can be demonstrated either by direct integration of the y variables or by using some normalizing property of the kernels k, for example that G_{00} for the case $\gamma=0$ must equal unity.] The expression for Q given in C on page 386 is in error, the quantities q_0 and q_f should be interchanged.

[8] It is most convenient to define the state φ_n with the phase factor $\exp[-i\omega(n+\frac{1}{2})t']$ and the final state with the factor $\exp[-i\omega(m+\frac{1}{2})t'']$ so that the results will not depend on the particular times t', t'' chosen.

This is to be added to S_p+S_e to obtain the complete action of the system with the oscillators removed.

The term in $(\mathbf{K}\cdot\mathbf{x}'_n(t))(\mathbf{K}\cdot\mathbf{x}'_m(s))$ can be simplified by integration by parts with respect to t and with respect to s [note that $\exp(-ik|t-s|)$ has a discontinuous slope at $t=s$, or break the integration up into two regions]. One finds

$$I_{tr}=R-I_e+I_{\text{transient}} \tag{18}$$

where

$$R=-i\sum_n\sum_m\int_{t'}^{t''}dt\int_{t'}^{t''}ds\int e_ne_m \times\exp(-ik|t-s|)(1-\mathbf{x}'_n(t)\cdot\mathbf{x}'_m(s)) \cdot\cos\mathbf{K}\cdot(\mathbf{x}_n(t)-\mathbf{x}_m(s))d^3\mathbf{K}/8\pi^2k \tag{19}$$

and

$$I_e=-\sum_n\sum_m\int_{t'}^{t''}dt\int e_ne_m \times\cos\mathbf{K}\cdot(\mathbf{x}_n(t)-\mathbf{x}_m(t))d^3\mathbf{K}/4\pi^2k^2 \tag{20}$$

comes from the discontinuity in slope of $\exp(-ik|t-s|)$ at $t=s$. Since

$$\int\cos(\mathbf{K}\cdot\mathbf{R})d^3\mathbf{K}/4\pi^2k^2=\int_0^\infty(kr)^{-1}\sin(kr)dk/\pi=(2r)^{-1}$$

this term I_e just cancels the Coulomb interaction term $S_c=\int L_cdt$. The term

$$I_{\text{transient}}=-\sum_n\sum_m e_ne_m\int\frac{d^3\mathbf{K}}{4\pi^2k^2} \times\Big\{\int_{t'}^{t''}[\exp(-ik(t''-t))\cos\mathbf{K}\cdot(\mathbf{x}_n(t'')-\mathbf{x}_m(t)) +\exp(-ik(t-t'))\cos\mathbf{K}\cdot(\mathbf{x}_n(t)-\mathbf{x}_m(t'))]dt +(2k)^{-1}i[\cos\mathbf{K}\cdot(\mathbf{x}_n(t'')-\mathbf{x}_m(t'')) +\cos\mathbf{K}\cdot(\mathbf{x}_n(t')-\mathbf{x}_m(t')) -2\exp(-ik(t''-t'))\cos\mathbf{K}\cdot(\mathbf{x}_n(t')-\mathbf{x}_m(t''))]\Big\}. \tag{21}$$

is one which comes from the limits of integration at t' and t'', and involves the coordinates of the particle at either one of these times or the other. If t' and t'' are considered to be exceedingly far in the past and future, there is no correlation to be expected between these temporally distant coordinates and the present ones, so the effects of $I_{\text{transient}}$ will cancel out quantum mechanically by interference. This transient was produced by the sudden turning on of the interaction of field and particles at t' and its sudden removal at t''. Alternatively we can imagine the charges to be turned on after t' adiabatically and turned off slowly before t'' (in this case, in the term L_c, the charges should also be considered as varying with time). In this case, in the limit, $I_{\text{transient}}$ is zero.[9] Hereafter we shall drop the transient term and consider the range of integration of t to be from $-\infty$ to $+\infty$, imagining, if one needs a definition, that the charges vary with time and vanish in the direction of either limit.

To simplify R we need the integral

$$J=\int\exp(-ik|t|)\cos(\mathbf{K}\cdot\mathbf{R})d^3\mathbf{K}/8\pi^2k =\int_0^\infty\exp(-ik|t|)\sin(kr)dk/2\pi r \tag{22}$$

where r is the length of the vector $\mathbf{R}$. Now

$$\int_0^\infty\exp(-ikx)dk=\lim_{\epsilon\to0}(-i(x-i\epsilon)^{-1}) =-ix^{-1}+\pi\delta(x)=\pi\delta_+(x)$$

where the equation serves to define $\delta_+(x)$ [as in II, Eq. (3)]. Hence, expanding $\sin(kr)$ in exponentials find

$$J=-(4\pi r)^{-1}((|t|-r)^{-1}-(|t|+r)^{-1}) +(4ir)^{-1}(\delta(|t|-r)-\delta(|t|+r)) =-(2\pi)^{-1}(t^2-r^2)^{-1}+(2i)^{-1}\delta(t^2-r^2) =-\tfrac{1}{2}i\delta_+(t^2-r^2) \tag{23}$$

where we have used the fact that

$$\delta(t^2-r^2)=(2r)^{-1}(\delta(|t|-r)+\delta(|t|+r))$$

and that $\delta(|t|+r)=0$ since both $|t|$ and r are necessarily positive.

Substitution of these results into (19) gives finally,

$$R=-\frac{1}{2}\sum_n\sum_m\int_{-\infty}^{+\infty}\int_{-\infty}^{+\infty}e_ne_m(1-\mathbf{x}'_n(t)\cdot\mathbf{x}'_m(s)) \times\delta_+((t-s)^2-(\mathbf{x}_n(t)-\mathbf{x}_m(s))^2)dtds. \tag{24}$$

The total complex action of the system is then[10] S_p+R. Or, what amounts to the same thing; to obtain

[9] One can obtain the final result, that the total interaction is just R, in a formal manner starting from the Hamiltonian from which the longitudinal oscillators have not yet been eliminated. There are for each $\mathbf{K}$ and cos or sin, four oscillators $q_{\mu\mathbf{K}}$ corresponding to the three components of the vector potential ($\mu=1, 2, 3$) and the scalar potential ($\mu=4$). It must then be assumed that the wave functions of the initial and final state of the $\mathbf{K}$ oscillators is the function $(k/\pi)\exp[-\tfrac{1}{2}k(q_{1\mathbf{K}}^2+q_{2\mathbf{K}}^2+q_{3\mathbf{K}}^2-q_{4\mathbf{K}}^2)]$. The wave function suggested here has only formal significance, of course, because the dependence on $q_{4\mathbf{K}}$ is not square integrable, and cannot be normalized. If each oscillator were assumed actually in the ground state, the sign of the $q_{4\mathbf{K}}$ term would be changed to positive, and the sign of the frequency in the contribution of these oscillators would be reversed (they would have negative energy).

[10] The classical action for this problem is just $Sp+R'$ where R' is the real part of the expression (24). In view of the generalization of the Lagrangian formulation of quantum mechanics suggested in Section 12 of C, one might have anticipated that R would have been simply R'. This corresponds, however, to boundary conditions other than no quanta present in past and future. It is harder to interpret physically. For a system enclosed in a light tight box, however, it appears likely that both R and R' lead to the same results.

transition amplitudes including the effects of the field we must calculate the transition element of $\exp(iR)$:

$$\langle \chi_{t''} | \exp iR | \psi_{t'} \rangle_{S_p} \tag{25}$$

under the action S_p of the particles, excluding interaction. Expression (24) for R must be considered to be written in the usual manner as a Riemann sum and the expression (25) interpreted as defined in C [Eq. (39)]. Expression (6) must be used for $\dot{\mathbf{x}}_n$ at time t.

Expression (25), with (24), then contains all the effects of virtual quanta on a (at least non-relativistic) system according to quantum electrodynamics. It contains the effects to all orders in $e^2/\hbar c$ in a single expression. If expanded in a power series in $e^2/\hbar c$, the various terms give the expressions to the corresponding order obtained by the diagrams and methods of II. We illustrate this by an example in the next section.

5. EXAMPLE OF APPLICATION OF EXPRESSION (25)

We shall not be much concerned with the non-relativistic case here, as the relativistic case given below is as simple and more interesting. It is, however, very similar and at this stage it is worth giving an example to show how expressions resulting from (25) are to be interpreted according to the rules of C. For example, consider the case of a single electron, coordinate $\mathbf{x}$, either free or in an external given potential (contained for simplicity in S_p, not in[11] R). Its interaction with the field produces a reaction back on itself given by R as in (24) but in which we keep only a single term corresponding to $m=n$. Assume the effect of R to be small and expand $\exp(iR)$ as $1+iR$. Let us find the amplitude at time t'' of finding the electron in a state ψ with no quanta emitted, if at time t' it was in the same state. It is

$$\langle \psi_{t''} | 1+iR | \psi_{t'} \rangle_{S_p} = \langle \psi_{t''} | 1 | \psi_{t'} \rangle_{S_p} + i \langle \psi_{t''} | R | \psi_{t'} \rangle_{S_p}$$

where $\langle \psi_{t''} | 1 | \psi_{t'} \rangle_{S_p} = \exp[-iE(t''-t')]$ if E is the energy of the state, and

$$\langle \psi_{t''} | R | \psi_{t'} \rangle_{S_p} = -\tfrac{1}{2}e^2 \int_{t'}^{t''} dt \int_{t'}^{t''} ds \langle \psi_{t''} | (1-\dot{\mathbf{x}}_t \cdot \dot{\mathbf{x}}_s) \times \delta_+((t-s)^2-(\mathbf{x}_t-\mathbf{x}_s)^2) | \psi_{t'} \rangle_{S_p}. \tag{26}$$

Here $\mathbf{x}_s = \mathbf{x}(s)$, etc. In (26) we shall limit the range of integrations by assuming $s<t$, and double the result.

The expression within the brackets $\langle\ \rangle_{S_p}$ on the right-hand side of (26) can be evaluated by the methods described in C [Eq. (29)]. An expression such as (26) can also be evaluated directly in terms of the propagation kernel $K(2,1)$ [see I, Eq. (2)] for an electron moving in the given potential.

The term $\dot{\mathbf{x}}_s \cdot \dot{\mathbf{x}}_t$ in the non-relativistic case produces an interesting complication which does not have an analog for the relativistic case with the Dirac equation. We discuss it below, but for a moment consider in further detail expression (26) but with the factor $(1-\dot{\mathbf{x}}_s \cdot \dot{\mathbf{x}}_t)$ replaced simply by unity.

The kernel $K(2,1)$ is defined and discussed in I. From its definition as the amplitude that the electron be found at $\mathbf{x}_2$ at time t_2, if at t_1 it was at $\mathbf{x}_1$, we have

$$K(\mathbf{x}_2, t_2; \mathbf{x}_1, t_1) = \langle \delta(\mathbf{x}-\mathbf{x}_2)_{t_2} | 1 | \delta(\mathbf{x}-\mathbf{x}_1)_{t_1} \rangle_{S_p} \tag{27}$$

that is, more simply $K(2,1)$ is the sum of $\exp(iS_p)$ over all paths which go from space time point 1 to 2.

In the integrations over all paths implied by the symbol in (26) we can first integrate over all the $\mathbf{x}_i$ variables corresponding to times t_i from t' to s, not inclusive, the result being a factor $K(\mathbf{x}_s, s; \mathbf{x}_{t'}, t')$ according to (27). Next we integrate on the variables between s and t not inclusive, giving a factor $K(\mathbf{x}_t, t; \mathbf{x}_s, s)$ and finally on those between t and t'' giving $K(\mathbf{x}_{t''}, t''; \mathbf{x}_t, t)$. Hence the left-hand term in (26) excluding the $\dot{\mathbf{x}}_t \cdot \dot{\mathbf{x}}_s$ factor is

$$-e^2 \int dt \int ds \int \psi^*(\mathbf{x}_{t''}, t'') K(\mathbf{x}_{t''}, t''; \mathbf{x}_t, t) \delta_+((t-s)^2 - (\mathbf{x}_t-\mathbf{x}_s)^2) \cdot K(\mathbf{x}_t, t; \mathbf{x}_s, s) K(\mathbf{x}_s, s; \mathbf{x}_{t'}, t') \times \psi(\mathbf{x}_{t'}, t') d^3\mathbf{x}_{t''} d^3\mathbf{x}_t d^3\mathbf{x}_s d^3\mathbf{x}_{t'} \tag{28}.$$

which in improved notation and in the relativistic case is essentially the result given in II.

We have made use of a special case of a principle which may be stated more generally as

$$\langle \chi_{t''} | F(\mathbf{x}_1, t_1; \mathbf{x}_2, t_2; \cdots \mathbf{x}_k, t_k) | \psi_{t'} \rangle_{S_p} = \int \chi^*(\mathbf{x}_{t''}) K(\mathbf{x}_{t''}, t''; \mathbf{x}_1, t_1) \cdot K(\mathbf{x}_1, t_1; \mathbf{x}_2, t_2) \cdots \times K(\mathbf{x}_{k-1}, t_{k-1}; \mathbf{x}_k, t_k) K(\mathbf{x}_k, t_k; \mathbf{x}_{t'}, t') \cdot F(\mathbf{x}_1, t_1; \mathbf{x}_2, t_2; \cdots \mathbf{x}_k, t_k) \psi(\mathbf{x}_{t'}) \times d^3\mathbf{x}_{t''} d^3\mathbf{x}_1 d^3\mathbf{x}_2 \cdots d^3\mathbf{x}_k d^3\mathbf{x}_{t'} \tag{29}$$

where F is any function of the coordinate $\mathbf{x}_1$ at time t_1, $\mathbf{x}_2$ at t_2 up to $\mathbf{x}_k$, t_k, and, it is important to notice, we have assumed $t''>t_1>t_2>\cdots t_k>t'$.

Expressions of higher order arising for example from R^2 are more complicated as there are quantities referring to several different times mixed up, but they all can be interpreted readily. One simply breaks up the ranges of integrations of the time variables into parts such that in each the order of time of each variable is definite. One then interprets each part by formula (29).

[11] One can show from (25) how the correlated effect of many atoms at a distance produces on a given system the effects of an external potential. Formula (24) yields the result that this potential is that obtained from Liénard and Wiechert by retarded waves arising from the charges and currents resulting from the distant atoms making transitions. Assume the wave functions χ and ψ can be split into products of wave functions for system and distant atoms and expand $\exp(iR)$ assuming the effect of any individual distant atom is small. Coulomb potentials arise even from nearby particles if they are moving slowly.

As a simple example we may refer to the problem of the transition element

$$\left\langle \chi_{t''} \middle| \int U(\mathbf{x}(t), t)dt \int V(\mathbf{x}(s), s)ds \middle| \psi_{t'} \right\rangle$$

arising, say, in the cross term in U and V in an ordinary second order perturbation problem (disregarding radiation) with perturbation potential $U(\mathbf{x}, t)+V(\mathbf{x}, t)$. In the integration on s and t which should include the entire range of time for each, we can split the range of s into two parts, $s<t$ and $s>t$. In the first case, $s<t$, the potential V acts earlier than U, and in the other range, vice versa, so that

$$\left\langle \chi_{t''} \middle| \int U(\mathbf{x}_t, t)dt \int V(\mathbf{x}_s, s)ds \middle| \psi_{t'} \right\rangle$$

$$= \int_{t'}^{t''} dt \int_{t'}^{t} ds \int \chi^*(\mathbf{x}_{t''}) K(\mathbf{x}_{t''}, t''; \mathbf{x}_t, t)$$

$$\times U(\mathbf{x}_t, t) K(\mathbf{x}_t, t; \mathbf{x}_s, s) V(\mathbf{x}_s, s)$$

$$\cdot K(\mathbf{x}_s, s; \mathbf{x}_{t'}, t')\psi(\mathbf{x}_{t'}) d^3\mathbf{x}_{t''} d^3\mathbf{x}_t d^3\mathbf{x}_s d^3\mathbf{x}_{t'}$$

$$+ \int_{t'}^{t''} dt \int_{t}^{t''} ds \int \chi^*(\mathbf{x}_{t''}) K(\mathbf{x}_{t''}, t''; \mathbf{x}_s, s)$$

$$\times V(\mathbf{x}_s, s) K(\mathbf{x}_s, s; \mathbf{x}_t, t) U(\mathbf{x}_t, t)$$

$$\cdot K(\mathbf{x}_t, t; \mathbf{x}_{t'}, t')\psi(\mathbf{x}_{t'}) d^3\mathbf{x}_{t''} d^3\mathbf{x}_s d^3\mathbf{x}_t d^3\mathbf{x}_{t'} \quad (30)$$

so that the single expression on the left is represented by two terms analogous to the two terms required in analyzing the Compton effect. It is in this way that the several terms and their corresponding diagrams corresponding to each process arise when an attempt is made to represent the transition elements of single expressions involving time integrals in terms of the propagation kernels K.

It remains to study in more detail the term in (26) arising from $\dot{\mathbf{x}}(t)\cdot\dot{\mathbf{x}}(s)$ in the interaction. The interpretation of such expressions is considered in detail in C, and we must refer to Eqs. (39) through (50) of that paper for a more thorough analysis. A similar type of term also arises in the Lagrangian formulation in simpler problems, for example the transition element

$$\left\langle \chi_{t''} \middle| \int \dot{\mathbf{x}}(t)\cdot\mathbf{A}(\mathbf{x}(t), t)dt \int \dot{\mathbf{x}}(s)\cdot\mathbf{B}(\mathbf{x}(s), s)ds \middle| \psi_{t'} \right\rangle$$

arising say, in the cross term in $\mathbf{A}$ and $\mathbf{B}$ in a second-order perturbation problem for a particle in a perturbing vector potential $\mathbf{A}(\mathbf{x}, t)+\mathbf{B}(\mathbf{x}, t)$. The time integrals must first be written as Riemannian sums, the velocity (see (6)) being replaced by $\dot{\mathbf{x}}=\frac{1}{2}\epsilon^{-1}(\mathbf{x}_{i+1}-\mathbf{x}_i)+\frac{1}{2}\epsilon^{-1}(\mathbf{x}_i-\mathbf{x}_{i-1})$ so that we ask for the transition element of

$$\sum_i\sum_j [\tfrac{1}{2}(\mathbf{x}_{i+1}-\mathbf{x}_i)+\tfrac{1}{2}(\mathbf{x}_i-\mathbf{x}_{i-1})]\cdot\mathbf{A}(\mathbf{x}_i, t_i)$$

$$\times[\tfrac{1}{2}(\mathbf{x}_{j+1}-\mathbf{x}_j)+\tfrac{1}{2}(\mathbf{x}_j-\mathbf{x}_{j-1})]\cdot\mathbf{B}(\mathbf{x}_j, t_j). \quad (31)$$

In C it is shown that when converted to operator notation the quantity $(x_{i-1}-x_i)/\epsilon$ is equivalent (nearly, see below) to an operator,

$$(x_{i+1}-x_i)/\epsilon \rightarrow i(Hx-xH) \quad (32)$$

operating in order indicated by the time index i (that is after x_l's for $l\leq i$ and before all x_l's for $l>i$). In nonrelativistic mechanics $i(Hx-xH)$ is the momentum operator p_x divided by the mass m. Thus in (31) the expression $[\frac{1}{2}(\mathbf{x}_{i+1}-\mathbf{x}_i)+\frac{1}{2}(\mathbf{x}_i-\mathbf{x}_{i-1})]\cdot\mathbf{A}(\mathbf{x}_i, t_i)$ becomes $\epsilon(\mathbf{p}\cdot\mathbf{A}+\mathbf{A}\cdot\mathbf{p})/2m$. Here again we must split the sum into two regions $j<i$ and $j>i$ so the quantities in the usual notation will operate in the right order such that eventually (31) becomes identical with the right-hand side of Eq. (30) but with $U(\mathbf{x}_t, t)$ replaced by the operator

$$\frac{1}{2m}\left(\frac{1}{i}\frac{\partial}{\partial\mathbf{x}_t}\cdot\mathbf{A}(\mathbf{x}_t, t)+\mathbf{A}(\mathbf{x}_t, t)\cdot\frac{1}{i}\frac{\partial}{\partial\mathbf{x}_t}\right)$$

standing in the same place, and with the operator

$$\frac{1}{2m}\left(\frac{1}{i}\frac{\partial}{\partial\mathbf{x}_s}\cdot\mathbf{B}(\mathbf{x}_s, s)+\frac{1}{i}\mathbf{B}(\mathbf{x}_s, s)\cdot\frac{\partial}{\partial\mathbf{x}_s}\right)$$

standing in the place of $V(\mathbf{x}_s, s)$. The sums and factors ϵ have now become $\int dt \int ds$.

This is nearly but not quite correct, however, as there is an additional term coming from the terms in the sum corresponding to the special values, $j=i$, $j=i+1$ and $j=i-1$. We have tacitly assumed from the appearance of the expression (31) that, for a given i, the contribution from just three such special terms is of order ϵ^2. But this is not true. Although the expected contribution of a term like $(x_{i+1}-x_i)(x_{j+1}-x_j)$ for $j\neq i$ is indeed of order ϵ^2, the expected contribution of $(x_{i+1}-x_i)^2$ is $+i\epsilon m^{-1}$ [C, Eq. (50)], that is, of order ϵ. In nonrelativistic mechanics the velocities are unlimited and in very short times ϵ the amplitude diffuses a distance proportional to the square root of the time. Making use of this equation then we see that the additional contribution from these terms is essentially

$$im^{-1}\epsilon\sum_i \mathbf{A}(\mathbf{x}_i, t_i)\cdot\mathbf{B}(\mathbf{x}_i, t_i)=im^{-1}\int \mathbf{A}(\mathbf{x}(t), t)\cdot\mathbf{B}(\mathbf{x}(t), t)dt$$

when summed on all i. This has the same effect as a first-order perturbation due to a potential $\mathbf{A}\cdot\mathbf{B}/m$. Added to the term involving the momentum operators we therefore have an additional term[12].

$$\frac{i}{m}\int_{t'}^{t''} dt \int \chi^*(\mathbf{x}_{t''})K(\mathbf{x}_{t''}, t''; \mathbf{x}_t, t)\mathbf{A}(\mathbf{x}_t, t)\cdot\mathbf{B}(\mathbf{x}_t, t) \cdot K(\mathbf{x}_t, t; \mathbf{x}_{t'}, t')\psi(\mathbf{x}_{t'})d^3\mathbf{x}_{t''}d^3\mathbf{x}_t d^3\mathbf{x}_{t'}. \quad (33)$$

In the usual Hamiltonian theory this term arises, of course, from the term $\mathbf{A}^2/2m$ in the expansion of the Hamiltonian

$$H=(2m)^{-1}(\mathbf{p}-\mathbf{A})^2=(2m)^{-1}(\mathbf{p}^2-\mathbf{p}\cdot\mathbf{A}-\mathbf{A}\cdot\mathbf{p}+\mathbf{A}^2)$$

while the other term arises from the second-order action of $\mathbf{p}\cdot\mathbf{A}+\mathbf{A}\cdot\mathbf{p}$. We shall not be interested in non-relativistic quantum electrodynamics in detail. The situation is simpler for Dirac electrons. For particles satisfying the Klein-Gordon equation (discussed in Appendix A) the situation is very similar to a four-dimensional analog of the non-relativistic case given here.

6. EXTENSION TO DIRAC PARTICLES

Expressions (24) and (25) and their proof can be readily generalized to the relativistic case according to the one electron theory of Dirac. We shall discuss the hole theory later. In the non-relativistic case we began with the proposition that the amplitude for a particle to proceed from one point to another is the sum over paths of $\exp(iS_p)$, that is, we have for example for a transition element

$$\langle\chi|1|\psi\rangle=\lim_{\epsilon\to 0}\int\cdots\int\chi^*(\mathbf{x}_N)\Phi_p(\mathbf{x}_N, \mathbf{x}_{N-1}, \cdots\mathbf{x}_0) \cdot\psi(\mathbf{x}_0)d^3\mathbf{x}_0d^3\mathbf{x}_1\cdots d^3\mathbf{x}_N \quad (34)$$

where for $\exp(iS_p)$ we have written Φ_p, that is more precisely,

$$\Phi_p=\Pi_i A^{-1}\exp iS(\mathbf{x}_{i+1}, \mathbf{x}_i).$$

As discussed in C this form is related to the usual form of quantum mechanics through the observation that

$$(\mathbf{x}_{i+1}|\mathbf{x}_i)_\epsilon=A^{-1}\exp[iS(\mathbf{x}_{i+1}, \mathbf{x}_i)] \quad (35)$$

where $(\mathbf{x}_{i+1}|\mathbf{x}_i)_\epsilon$ is the transformation matrix from a representation in which $\mathbf{x}$ is diagonal at time t_i to one in which $\mathbf{x}$ is diagonal at time $t_{i+1}=t_i+\epsilon$ (so that it is identical to $K_0(\mathbf{x}_{i+1}, t_{i+1}; \mathbf{x}_i, t_i)$ for the small time interval ϵ). Hence the amplitude for a given path can also be written

$$\Phi_p=\Pi_i(\mathbf{x}_{i+1}|\mathbf{x}_i)_\epsilon \quad (36)$$

for which form, of course, (34) is exact irrespective of whether $(\mathbf{x}_{i+1}|\mathbf{x}_i)_\epsilon$ can be expressed in the simple form (35).

For a Dirac electron the $(\mathbf{x}_{i+1}|\mathbf{x}_i)_\epsilon$ is a 4×4 matrix (or $4^N\times 4^N$ if we deal with N electrons) but the expression (34) with (36) is still correct (as it is in fact for any quantum-mechanical system with a sufficiently general definition of the coordinate $\mathbf{x}_i$). The product (36) now involves operators, the order in which the factors are to be taken is the order in which the terms appear in time.

For a Dirac particle in a vector and scalar potential (times the electron charge e) $\mathbf{A}(\mathbf{x}, t)$, $A_4(\mathbf{x}, t)$, the quantity $(\mathbf{x}_{i+1}|\mathbf{x}_i)_\epsilon^{(A)}$ is related to that of a free particle to the first order in ϵ as

$$(\mathbf{x}_{i+1}|\mathbf{x}_i)_\epsilon^{(A)}=(\mathbf{x}_{i+1}|\mathbf{x}_i)_\epsilon^{(0)}\exp[-i(\epsilon A_4(\mathbf{x}_i, t_i) -(\mathbf{x}_{i+1}-\mathbf{x}_i)\cdot\mathbf{A}(\mathbf{x}_i, t_i))]. \quad (37)$$

This can be verified directly by substitution into the Dirac equation.[13] It neglects the variation of $\mathbf{A}$ and A_4 with time and space during the short interval ϵ. This produces errors only of order ϵ^2 in the Dirac case for the expected square velocity $(\mathbf{x}_{i+1}-\mathbf{x}_i)^2/\epsilon^2$ during the interval ϵ is finite (equaling the square of the velocity of light) rather than being of order $1/\epsilon$ as in the non-relativistic case. [This makes the relativistic case somewhat simpler in that it is not necessary to define the velocity as carefully as in (6); $(\mathbf{x}_{i+1}-\mathbf{x}_i)/\epsilon$ is sufficiently exact, and no term analogous to (33) arises.]

Thus $\Phi_p^{(A)}$ differs from that for a free particle, $\Phi_p^{(0)}$, by a factor $\Pi_i\exp-i(\epsilon A_4(\mathbf{x}_i, t_i)-(\mathbf{x}_{i+1}-\mathbf{x}_i)\cdot\mathbf{A}(\mathbf{x}_i, t_i))$ which in the limit can be written as

$$\exp\left\{-i\int[A_4(\mathbf{x}(t), t)-\mathbf{x}'(t)\cdot\mathbf{A}(\mathbf{x}(t), t)]dt\right\} \quad (38)$$

exactly as in the non-relativistic case.

The case of a Dirac particle interacting with the quantum-mechanical oscillators representing the field may now be studied. Since the dependence of $\Phi_p^{(A)}$ on $\mathbf{A}$, A_4 is through the same factor as in the non-relativistic case, when $\mathbf{A}$, A_4 are expressed in terms of the oscillator coordinates q, the dependence of Φ on the oscillator coordinates q is unchanged. Hence the entire analysis of the preceding sections which concern the results of the integration over oscillator coordinates can be carried through unchanged and the results will be expression (25) with formula (24) for R. Expression (25) is now interpreted as

$$\langle\chi_{t''}|\exp iR|\psi_{t'}\rangle=\lim_{\epsilon\to 0}\int\chi^*(\mathbf{x}_{t''}^{(1)}, \mathbf{x}_{t''}^{(2)}\cdots) \times\prod_n(\Phi_{p,n}^{(0)}d^3\mathbf{x}_{t''}^{(n)}d^3\mathbf{x}_{t''-\epsilon}^{(n)}\cdots d^3\mathbf{x}_{t'}^{(n)}) \cdot\exp(iR)\psi(\mathbf{x}_{t'}^{(1)}, \mathbf{x}_{t'}^{(2)}\cdots) \quad (39)$$

[12] The term corresponding to this for the self-energy expression (26) would give an integral over $\delta_+((t-t)^2-(\mathbf{x}_t-\mathbf{x}_t)^2)$ which is evidently infinite and leads to the quadratically divergent self-energy. There is no such term for the Dirac electron, but there is for Klein-Gordon particles. We shall not discuss the infinities in this paper as they have already been discussed in II.

[13] Alternatively, note that Eq. (37) is exact for arbitrarily large ϵ if the potential A_μ is constant. For if the potential in the Dirac equation is the gradient of a scalar function $A_\mu=\partial\chi/\partial x_\mu$ the potential may be removed by replacing the wave function by $\psi=\exp(-i\chi)\psi'$ (gauge transformation). This alters the kernel by a factor $\exp[-i(\chi(2)-\chi(1))]$ owing to the change in the initial and final wave functions. A constant potential A_μ is the gradient of $\chi=A_\mu x_\mu$ and can be completely removed by this gauge transformation, so that the kernel differs from that of a free particle by the factor $\exp[-i(A_\mu x_{\mu 2}-A_\mu x_{\mu 1})]$ as in (37).

where $\Phi_{p,n}^{(0)}$, the amplitude for a particular path for particle n is simply the expression (36) where $(\mathbf{x}_{i+1}|\mathbf{x}_i)_\epsilon$ is the kernel $K_{0,n}(\mathbf{x}_{i+1}^{(n)}, t_{i+1}; \mathbf{x}_i^{(n)}, t_i)$ for a free electron according to the one electron Dirac theory, with the matrices which appear operating on the spinor indices corresponding to particle (n) and the order of all operations being determined by the time indices.

For calculational purposes we can, as before, expand R as a power series and evaluate the various terms in the same manner as for the non-relativistic case. In such an expansion the quantity $\mathbf{x}^\cdot(t)$ is replaced, as we have seen in (32), by the operator $i(H\mathbf{x}-\mathbf{x}H)$, that is, in this case by α operating at the corresponding time. There is no further complicated term analogous to (33) arising in this case, for the expected value of $(x_{i+1}-x_i)^2$ is now of order ϵ^2 rather than ϵ.

For example, for self-energy one sees that expression (28) will be (with other terms coming from those with $\mathbf{x}^\cdot(t)$ replaced by α and with the usual β in back of each K_0 because of the definition of K_0 in relativity theory)

$$\langle\psi_{t''}|R|\psi_{t'}\rangle_{S_p}=-e^2\int\psi^*(\mathbf{x}_{t''})K_0(\mathbf{x}_{t''},t'';\mathbf{x}_t,t)\beta\alpha_\mu$$
$$\cdot\delta_+((t-s)^2-(\mathbf{x}_t-\mathbf{x}_s)^2)K_0(\mathbf{x}_t,t;\mathbf{x}_s,s)\beta\alpha_\mu$$
$$\cdot K_0(\mathbf{x}_s,s;\mathbf{x}_{t'},t')\beta\psi(\mathbf{x}_{t'})d^3\mathbf{x}_{t''}d^3\mathbf{x}_td^3\mathbf{x}_sd^3\mathbf{x}_{t'}dtds, \quad (40)$$

where $\alpha_4=1$, $\alpha_{1,2,3}=\alpha_{x,y,z}$ and a sum on the repeated index μ is implied in the usual way; $a_\mu b_\mu=a_4b_4-a_1b_1-a_2b_2-a_3b_3$. One can change $\beta\alpha_\mu$ to γ_μ and ψ^* to $\bar\psi\beta$. In this manner all of the rules referring to virtual photons discussed in II are deduced; but with the difference that K_0 is used instead of K_+ and we have the Dirac one electron theory with negative energy states (although we may have any number of such electrons).

7. EXTENSION TO POSITRON THEORY

Since in (39) we have an arbitrary number of electrons, we can deal with the hole theory in the usual manner by imagining that we have an infinite number of electrons in negative energy states.

On the other hand, in paper I on the theory of positrons, it was shown that the results of the hole theory in a system with a given external potential A_μ were equivalent to those of the Dirac one electron theory if one replaced the propagation kernel, K_0, by a different one, K_+, and multiplied the resultant amplitude by factor C_v involving A_μ. We must now see how this relation, derived in the case of external potentials, can also be carried over in electrodynamics to be useful in simplifying expressions involving the infinite sea of electrons.

To do this we study in greater detail the relation between a problem involving virtual photons and one involving purely external potentials. In using (25) we shall assume in accordance with the hole theory that the number of electrons is infinite, but that they all have the same charge, e. Let the states $\psi_{t'}$, $\chi_{t''}$, represent the vacuum plus perhaps a number of real electrons in positive energy states and perhaps also some empty negative energy states. Let us call the amplitude for the transition in an external potential B_μ, but *excluding virtual photons*, $T_0[B]$, a functional of $B_\mu(1)$. We have seen (38)

$$T_0[B]=\langle\chi_{t''}|\exp iP|\psi_{t'}\rangle \quad (41)$$

where

$$P=-\sum_n\int[B_4(\mathbf{x}^{(n)}(t),t)-\mathbf{x}^{\cdot(n)}(t)\cdot\mathbf{B}(\mathbf{x}^{(n)}(t),t)]dt$$

by (38). We can write this as

$$P=-\sum_n\int B_\mu(x_\nu^{(n)}(t))\dot{x}_\mu^{(n)}(t)dt$$

where $x_4(t)=t$ and $\dot{x}_4=1$, the other values of μ corresponding to space variables. The corresponding amplitude for the same process in the same potential, but *including* all the virtual photons we may call,

$$T_{e^2}[B]=\langle\chi_{t''}|\exp(iR)\exp(iP)|\psi_{t'}\rangle. \quad (42)$$

Now let us consider the effect on $T_{e^2}[B]$ of changing the coupling e^2 of the virtual photons. Differentiating (42) with respect to e^2 which appears only[14] in R we find

$$dT_{e^2}[B]/d(e^2)=\left\langle\chi_{t''}\left|-\frac{i}{2}\sum_n\sum_m\int\int dtds\dot{x}_\mu^{(n)}(t)\dot{x}_\mu^{(m)}(s)\right.\right.$$
$$\left.\left.\cdot\delta_+((x_\nu^{(n)}(t)-x_\nu^{(m)}(s))^2)\exp i(R+P)\right|\psi_{t'}\right\rangle. \quad (43)$$

We can also study the first-order effect of a change of B_μ:

$$\delta T_{e^2}[B]/\delta B_\mu(1)=-i\left\langle\chi_{t''}\left|\sum_n\int dt\dot{x}_\mu^{(n)}\delta^4(x_\alpha^{(n)}(t)-x_{\alpha,1})\right.\right.$$
$$\left.\left.\cdot\exp i(R+P)\right|\psi_{t'}\right\rangle \quad (44)$$

where $x_{\alpha,1}$ is the field point at which the derivative with respect to B_μ is taken[15] and the term (current density) $-\sum_n\int dt\dot{x}_\mu^{(n)}(t)\delta^4(x_\alpha^{(n)}(t)-x_{\alpha,1})$ is just $\delta P/\delta B_\mu(1)$. The function $\delta^4(x_\alpha^{(n)}-x_{\alpha,1})$ means $\delta(x_4^{(n)}-x_{4,1})$

[14] In changing the charge e^2 we mean to vary only the degree to which virtual photons are important. We do not contemplate changes in the influence of the external potentials. If one wishes, as e is raised the strength of the potential is decreased proportionally so that B_μ, the potential times the charge e, is held constant.

[15] The functional derivative is defined such that if $T[B]$ is a number depending on the functions $B_\mu(1)$, the first order variation in T produced by a change from B_μ to $B_\mu+\Delta B_\mu$ is given by

$$T[B+\Delta B]-T[B]=\int(\delta T[B]/\delta B_\mu(1))\Delta B_\mu(1)d\tau_1$$

the integral extending over all four-space $x_{\alpha,1}$.

$\times\delta(x_2^{(n)}-x_{2,1})\delta(x_3^{(n)}-x_{3,1})\delta(x_1^{(n)}-x_{1,1})$ that is, $\delta(2,1)$ with $x_{\alpha,2}=x_\alpha^{(n)}(t)$. A second variation of T gives, by differentiation of (44) with respect to $B_\nu(2)$,

$$\delta^2 T_{e^2}[B]/\delta B_\mu(1)\delta B_\nu(2)$$

$$=-\left\langle \chi_{t''}\left|\sum_n\sum_m\int dt\int ds\dot{x}_\mu^{(n)}(t)\dot{x}_\nu^{(m)}(s)\cdot\delta^4(x_\alpha^{(n)}(t)-x_{\alpha,1})\delta^4(x_\beta^{(n)}(s)-x_{\beta,2})\times\exp i(R+P)\right|\psi_{t'}\right\rangle.$$

Comparison of this with (43) shows that

$$dT_{e^2}[B]/d(e^2)=\tfrac{1}{2}i\int\int(\delta^2 T_{e^2}[B]/\delta B_\mu(1)\delta B_\mu(2))\times\delta_+(s_{12}^2)d\tau_1 d\tau_2 \quad (45)$$

where $s_{12}^2=(x_{\mu,1}-x_{\mu,2})(x_{\mu,1}-x_{\mu,2})$.

We now proceed to use this equation to prove the validity of the rules given in II for electrodynamics. This we do by the following argument. The equation can be looked upon as a differential equation for $T_{e^2}[B]$. It determines $T_{e^2}[B]$ uniquely if $T_0[B]$ is known. We have shown it is valid for the hole theory of positrons. But in I we have given formulas for calculating $T_0[B]$ whose correctness relative to the hole theory we have there demonstrated. Hence we have shown that the $T_{e^2}[B]$ obtained by solving (45) with the initial condition $T_0[B]$ as given by the rules in I will be equal to that given for the same problem by the second quantization theory of the Dirac matter field coupled with the quantized electromagnetic field. But it is evident (the argument is given in the next paragraph) that the rules[16] given in II constitute a solution in power series in e^2 of the Eq. (45) [which for $e^2=0$ reduce to the $T_0[B]$ given in I]. Hence the rules in II must give, to each order in e^2, the matrix element for any process that would be calculated by the usual theory of second quantization of the matter and electromagnetic fields. This is what we aimed to prove.

That the rules of II represent, in a power series expansion, a solution of (45) is clear. For the rules there given may be stated as follows: Suppose that we have a process to order k in e^2 (i.e., having k virtual photons) and order n in the external potential B_μ. Then, *the matrix element for the process with one more virtual photon and two less potentials is that obtained from the previous matrix by choosing from the n potentials a pair, say $B_\mu(1)$ acting at 1 and $B_\nu(2)$ acting at 2, replacing them by $ie^2\delta_{\mu\nu}\delta_+(s_{12}^2)$, adding the results for each way of choosing the pair, and dividing by $k+1$, the present number of photons.* The matrix with no virtual photons ($k=0$) being given to any n by the rules of I, this permits terms to all orders in e^2 to be derived by recursion. It is evident that the rule in italics is that of II, and equally evident that it is a word expression of Eq. (45). [The factor $\frac{1}{2}$ in (45) arises since in integrating over all $d\tau_1$ and $d\tau_2$ we count each pair twice. The division by $k+1$ is required by the rules of II for, there, each diagram is to be taken only once, while in the rule given above we say what to do to add one extra virtual photon to k others. But which one of the $k+1$ is to be identified at the last photon added is irrelevant. It agrees with (45) of course for it is canceled on differentiating with respect to e^2 the factor $(e^2)^{k+1}$ for the $(k+1)$ photons.]

[16] That is, of course, those rules of II which apply to the unmodified electrodynamics of Dirac electrons. (The limitation excluding real photons in the initial and final states is removed in Sec. 8.) The same arguments clearly apply to nucleons interacting via neutral vector mesons, vector coupling. Other couplings require a minor extension of the argument. The modification to the $(x_{i+1}|x_i)_\epsilon$, as in (37), produced by some couplings cannot very easily be written without using operators in the exponents. These operators can be treated as numbers if their order of operation is maintained to be always their order in time. This idea will be discussed and applied more generally in a succeeding paper.

8. GENERALIZED FORMULATION OF QUANTUM ELECTRODYNAMICS

The relation implied by (45) between the formal solution for the amplitude for a process in an arbitrary unquantized external potential to that in a quantized field appears to be of much wider generality. We shall discuss the relation from a more general point of view here (still limiting ourselves to the case of no photons in initial or final state).

In earlier sections we pointed out that as a consequence of the Lagrangian form of quantum mechanics the aspects of the particles' motions and the behavior of the field could be analyzed separately. What we did was to integrate over the field oscillator coordinates first. We could, in principle, have integrated over the particle variables first. That is, we first solve the problem with the action of the particles and their interaction with the field and then multiply by the exponential of the action of the field and integrate over all the field oscillator coordinates. (For simplicity of discussion let us put aside from detailed special consideration the questions involving the separation of the longitudinal and transverse parts of the field.[9]) Now the integral over the particle coordinates for a given process is precisely the integral required for the analysis of the motion of the particles in an unquantized potential. With this observation we may suggest a generalization to all types of systems.

Let us suppose the formal solution for the amplitude for some given process with matter in an external potential $B_\mu(1)$ is some numerical quantity T_0. We mean matter in a more general sense now, for the motion of the matter may be described by the Dirac equation, or by the Klein-Gordon equation, or may involve charged or neutral particles other than electrons and positrons in any manner whatsoever. The quantity T_0 depends of course on the potential function $B_\mu(1)$; that is, it is a functional $T_0[B]$ of this potential. We assume we have some expression for it in terms of B_μ (exact, or to some desired degree of approximation in the strength of the potential).

Then the answer $T_{e2}[B]$ to the corresponding problem in quantum electrodynamics is $T_0[A_\mu(1)+B_\mu(1)]\times\exp(iS_0)$ summed over all possible distributions of field $A_\mu(1)$, wherein S_0 is the action for the field $S_0=-(8\pi e^2)^{-1}\sum_\mu\int((\partial A_\mu/\partial t)^2-(\nabla A_\mu)^2)d^3xdt$ the sum on μ carrying the usual minus sign for space components.

If $F[A]$ is any functional of $A_\mu(1)$ we shall represent by ${}_0|F[A]|_0$ this superposition of $F[A]\exp(iS_0)$ over distributions of A_μ for the case in which there are no photons in initial or final state. That is, we have

$$T_{e2}[B]={}_0|T_0[A+B]|_0. \qquad (46)$$

The evaluation of ${}_0|F[A]|_0$ directly from the definition of the operation ${}_0|\quad|_0$ is not necessary. We can give the result in another way. We first note that the operation is linear,

$${}_0|F_1[A]+F_2[A]|_0={}_0|F_1[A]|_0+{}_0|F_2[A]|_0 \qquad (47)$$

so that if F is represented as a sum of terms each term can be analyzed separately. We have studied essentially the case in which $F[A]$ is an exponential function. In fact, what we have done in Section 4 may be repeated with slight modification to show that

$${}_0\left|\exp\left(-i\int j_\mu(1)A_\mu(1)d\tau_1\right)\right|_0$$

$$=\exp\left(-\tfrac{1}{2}ie^2\int\int j_\mu(1)j_\mu(2)\delta_+(s_{12}^2)d\tau_1d\tau_2\right) \qquad (48)$$

where $j_\mu(1)$ is an arbitrary function of position and time for each value of μ.

Although this gives the evaluation of ${}_0|\quad|_0$ for only a particular functional of A_μ the appearance of the arbitrary function $j_\mu(1)$ makes it sufficiently general to permit the evaluation for any other functional. For it is to be expected that any functional can be represented as a superposition of exponentials with different functions $j_\mu(1)$ (by analogy with the principle of Fourier integrals for ordinary functions). Then, by (47), the result of the operation is the corresponding superposition of expressions equal to the right-hand side of (48) with the various j's substituted for j_μ.

In many applications $F[A]$ can be given as a power series in A_μ:

$$F[A]=f_0+\int f_\mu(1)A_\mu(1)d\tau_1$$

$$+\int\int f_{\mu\nu}(1,2)A_\mu(1)A_\nu(2)d\tau_1d\tau_2+\cdots \qquad (49)$$

where f_0, $f_\mu(1)$, $f_{\mu\nu}(1,2)\cdots$ are known numerical functions independent of A_μ. Then by (47)

$${}_0|F[A]|_0=f_0+\int f_\mu(1){}_0|A_\mu(1)|_0d\tau_1$$

$$+\int\int f_{\mu\nu}(1,2){}_0|A_\mu(1)A_\nu(2)|_0d\tau_1d\tau_2+\cdots \qquad (50)$$

where we set ${}_0|1|_0=1$ (from (48) with $j_\mu=0$). We can work out expressions for the successive powers of A_μ by differentiating both sides of (48) successively with respect to j_μ and setting $j_\mu=0$ in each derivative. For example, the first variation (derivative) of (48) with respect to $j_\mu(3)$ gives

$${}_0\left|-iA_\mu(3)\exp\left(-i\int j_\nu(1)A_\nu(1)d\tau_1\right)\right|_0$$

$$=-ie^2\int\delta_+(s_{34}^2)j_\mu(4)d\tau_4$$

$$\times\exp\left(-\tfrac{1}{2}ie^2\int\int j_\nu(1)j_\nu(2)\delta_+(s_{12}^2)d\tau_1d\tau_2\right). \qquad (51)$$

Setting $j_\mu=0$ gives

$${}_0|A_\mu(3)|_0=0.$$

Differentiating (51) again with respect to $j_\nu(4)$ and setting $j_\nu=0$ shows

$${}_0|A_\mu(3)A_\nu(4)|_0=ie^2\delta_{\mu\nu}\delta_+(s_{34}^2) \qquad (52)$$

and so on for higher powers. These results may be substituted into (50). Clearly therefore when $T_0[B+A]$ in (46) is expanded in a power series and the successive terms are computed in this way, we obtain the results given in II.

It is evident that (46), (47), (48) imply that $T_{e2}[B]$ satisfies the differential equation (45) and conversely (45) with the definition (46) implies (47) and (48). For if $T_0[B]$ is an exponential

$$T_0[B]=\exp\left(-i\int j_\mu(1)B_\mu(1)d\tau_1\right) \qquad (53)$$

we have from (46), (48) that

$$T_{e2}[B]=\exp\left[-\tfrac{1}{2}ie^2\int\int j_\mu(1)j_\mu(2)\delta_+(s_{12}^2)d\tau_1d\tau_2\right]$$

$$\cdot\exp\left[-i\int j_\nu(1)B_\nu(1)d\tau_1\right]. \qquad (54)$$

Direct substitution of this into Eq. (45) shows it to be a solution satisfying the boundary condition (53). Since the differential equation (45) is linear, if $T_0[B]$ is a superposition of exponentials, the corresponding superposition of solutions (54) is also a solution.

Many of the formal representations of the matter system (such as that of second quantization of Dirac electrons) represent the interaction with a fixed potential in a formal exponential form such as the left-hand side of (48), except that $j_\mu(1)$ is an operator instead of a numerical function. Equation (48) may still be used if care is exercised in defining the order of the operators on the right-hand side. The succeeding paper will discuss this in more detail.

Equation (45) or its solution (46), (47), (48) constitutes a very general and convenient formulation of the laws of quantum electrodynamics for virtual processes. Its relativistic invariance is evident if it is assumed that the unquantized theory giving $T_0[B]$ is invariant. It has been proved to be equivalent to the usual formulation for Dirac electrons and positrons (for Klein-Gordon particles see Appendix A). It is suggested that it is of wide generality. It is expressed in a form which has meaning even if it is impossible to express the matter system in Hamiltonian form; in fact, it only requires the existence of an amplitude for fixed potentials which obeys the principle of superposition of amplitudes. If $T_0[B]$ is known in power series in B, calculations of $T_{e^2}[B]$ in a power series of e^2 can be made directly using the italicized rule of Sec. 7. The limitation to virtual quanta is removed in the next section.

On the other hand, the formulation is unsatisfactory because for situations of importance it gives divergent results, even if $T_0[B]$ is finite. The modification proposed in II of replacing $\delta_+(s_{12}^2)$ in (45), (48) by $f_+(s_{12}^2)$ is not satisfactory owing to the loss of the theorems of conservation of energy or probability discussed in II at the end of Sec. 6. There is the additional difficulty in positron theory that even $T_0[B]$ is infinite to begin with (vacuum polarization). Computational ways of avoiding these troubles are given in II and in the references of footnote 2.

9. CASE OF REAL PHOTONS

The case in which there are real photons in the initial or the final state can be worked out from the beginning in the same manner.[17] We first consider the case of a system interacting with a single oscillator. From this result the generalization will be evident. This time we shall calculate the transition element between an initial state in which the particle is in state $\psi_{t'}$ and the oscillator is in its nth eigenstate (i.e., there are n photons in the field) to a final state with particle in $\chi_{t''}$, oscillator in mth level. As we have already discussed, when the coordinates of the oscillator are eliminated the result is the transition element $\langle\chi_{t''}|G_{mn}|\psi_{t'}\rangle$ where

$$G_{mn}=\int\varphi_m^*(q_j)k(q_j,t'';q_0,t')\varphi_n(q_0)dq_0dq_j \qquad (11)$$

where φ_m, φ_n are the wave functions[8] for the oscillator in state m, n and k is given in (12). The G_{mn} can be evaluated most easily by calculating the generating function

$$g(X,Y)=\sum_m\sum_n G_{mn}X^mY^n(m!n!)^{-\frac{1}{2}} \qquad (55)$$

for arbitrary X, Y. If expression (11) is substituted in the left-hand side of (55), the expression can be simplified by use of the generating function relation for the eigenfunctions[8] of the harmonic oscillator

$$\sum_n\varphi_n(q_0)Y^n(n!)^{-\frac{1}{2}}=(\omega/\pi)^{\frac{1}{4}}\exp(-\tfrac{1}{2}i\omega t')$$
$$\times\exp\tfrac{1}{2}[\omega q_0^2-(Y\exp[-i\omega t']-(2\omega)^{\frac{1}{2}}q_0)^2]$$

Using a similar expansion for the φ_m^* one is left with the exponential of a quadratic function of q_0 and q_j. The integration on q_0 and q_j is then easily performed to give

$$g(X,Y)=G_{00}\exp(XY+i\beta^*X+i\beta Y) \qquad (56)$$

from which expansion in powers of X and Y and comparison to (11) gives the final result

$$G_{mn}=G_{00}(m!n!)^{-\frac{1}{2}}\sum_r\frac{m!}{(m-r)!r!}\frac{n!}{(n-r)!r!}$$
$$\times r!(i\beta^*)^{m-r}(i\beta)^{n-r} \qquad (57)$$

where G_{00} is given in (14) and

$$\beta=(2\omega)^{-\frac{1}{2}}\int_{t'}^{t''}\gamma(t)\exp(-i\omega t)dt,$$
$$\beta^*=(2\omega)^{-\frac{1}{2}}\int_{t'}^{t''}\gamma(t)\exp(+i\omega t)dt, \qquad (58)$$

and the sum on r is to go from 0 to m or to n whichever is the smaller. (The sum can be expressed as a Laguerre polynomial but there is no advantage in this.)

Formula (57) is readily understandable. Consider first a simple case of absorption of one photon. Initially we have one photon and finally none. The amplitude for this is the transition element of $G_{01}=i\beta G_{00}$ or $\langle\chi_{t''}|i\beta G_{00}|\psi_{t'}\rangle$. This is the same as would result if we asked for the transition element for a problem in which all photons are virtual but there was present a perturbing potential $-(2\omega)^{-\frac{1}{2}}\gamma(t)\exp(-i\omega t)$ and we required the first-order effect of this potential. Hence photon absorption is like the first order action of a potential varying in time as $\gamma(t)\exp(-i\omega t)$ that is with a positive frequency (i.e., the sign of the coefficient of t in the exponential corresponds to positive energy). The amplitude for emission of one photon involves $G_{10}=i\beta^*G_{00}$, which is the same result except that the potential has negative frequency. Thus we begin by interpreting $i\beta^*$ as the amplitude for emission of one photon $i\beta$ as the amplitude for absorption of one.

Next for the general case of n photons initially and m finally we may understand (57) as follows. We first

[17] For an alternative method starting directly from the formula (24) for virtual photons, see Appendix B.

neglect Bose statistics and imagine the photons as individual distinct particles. If we start with n and end with m this process may occur in several different ways. The particle may absorb in total $n-r$ of the photons and the final m photons will represent r of the photons which were present originally plus $m-r$ new photons emitted by the particle. In this case the $n-r$ which are to be absorbed may be chosen from among the original n in $n!/(n-r)!r!$ different ways, and each contributes a factor $i\beta$, the amplitude for absorption of a photon. Which of the $m-r$ photons from among the m are emitted can be chosen in $m!/(m-r)!r!$ different ways and each photon contributes a factor $i\beta^*$ in amplitude. The initial r photons which do not interact with the particle can be re-arranged among the final r in $r!$ ways. We must sum over the alternatives corresponding to different values of r. Thus the form of G_{mn} can be understood. The remaining factor $(m!)^{-\frac{1}{2}}(n!)^{-\frac{1}{2}}$ may be interpreted as saying that in computing probabilities (which therefore involves the square of G_{mn}) the photons may be considered as independent but that if m are actually equal the statistical weight of each of the states which can be made by rearranging the m equal photons is only $1/m!$. This is the content of Bose statistics; that m equal particles in a given state represents just one state, i.e., has statistical weight unity, rather than the $m!$ statistical weight which would result if it is imagined that the particles and states can be identified and rearranged in $m!$ different ways. This holds for both the initial and final states of course. From this rule about the statistical weights of states the derivation of the blackbody distribution law follows.

The actual electromagnetic field is represented as a host of oscillators each of which behaves independently and produces its own factor such as G_{mn}. Initial or final states may also be linear combinations of states in which one or another oscillator is excited. The results for this case are of course the corresponding linear combination of transition elements.

For photons of a given direction of polarization and for sin or cos waves the explicit expression for β can be obtained directly from (58) by substituting the formulas (16) for the γ's for the corresponding oscillator. It is more convenient to use the linear combination corresponding to running waves. Thus we find the amplitude for absorption of a photon of momentum $\mathbf{K}$, frequency $k=(\mathbf{K}\cdot\mathbf{K})^{\frac{1}{2}}$ polarized in direction $\mathbf{e}$ is given by including a factor i times

$$\beta_{\mathbf{K},\mathbf{e}}=(4\pi)^{\frac{1}{2}}(2k)^{-\frac{1}{2}}\sum_n e_n\int_{t'}^{t''}\exp(-ikt)\times\exp(i\mathbf{K}\cdot\mathbf{x}_n(t))\mathbf{e}\cdot\dot{\mathbf{x}}_n(t)dt \quad (59)$$

in the transition element (25). The density of states in momentum space is now $(2\pi)^{-3}d^3\mathbf{K}$. The amplitude for emission is just i times the complex conjugate of this expression, or what amounts to the same thing, the same expression with the sign of the four vector k_μ reversed. Since the factor (59) is exactly the first-order effect of a vector potential

$$\mathbf{A}^{PH}=(2\pi/k)^{\frac{1}{2}}\mathbf{e}\exp(-i(kt-\mathbf{K}\cdot\mathbf{x}))$$

of the corresponding classical wave, we have derived the rules for handling real photons discussed in II.

We can express this directly in terms of the quantity $T_{e^2}[B]$, the amplitude for a given transition without emission of a photon. What we have said is that the amplitude for absorption of just one photon whose classical wave form is $A_\mu^{PH}(1)$ (time variation $\exp(-ikt_1)$ corresponding to positive energy k) is proportional to the first order (in ϵ) change produced in $T_{e^2}[B]$ on changing B to $B+\epsilon A^{PH}$. That is, more exactly,

$$\int(\delta T_{e^2}[B]/\delta B_\mu(1))A_\mu^{PH}(1)d\tau_1 \quad (60)$$

is the amplitude for absorption by the particle system of one photon, A^{PH}. (A superposition argument shows the expression to be valid not only for plane waves, but for spherical waves, etc., as given by the form of A^{PH}.) The amplitude for emission is the same expression but with the sign of the frequency reversed in A^{PH}. The amplitude that the system absorbs two photons with waves $A_\mu^{PH_1}$ and $A_\nu^{PH_2}$ is obtained from the next derivative,

$$\int\int(\delta^2 T_{e^2}[B]/\delta B_\mu(1)\delta B_\nu(2))A_\mu^{PH_1}(1)A_\nu^{PH_2}(2)d\tau_1 d\tau_2,$$

the same expression holding for the absorption of one and emission of the other, or emission of both depending on the sign of the time dependence of A^{PH_1} and A^{PH_2}. Larger photon numbers correspond to higher derivatives, absorption of l_1 emission of l_2 requiring the (l_1+l_2) derivaties. When two or more of the photons are exactly the same (e.g., $A^{PH_1}=A^{PH_2}$) the same expression holds for the amplitude that l_1 are absorbed by the system while l_2 are emitted. However, the statement that initially n of a kind are present and m of this kind are present finally, does not imply $l_1=n$ and $l_2=m$. It is possible that only $n-r=l_1$ were absorbed by the system and $m-r=l_2$ emitted, and that r remained from initial to final state without interaction. This term is weighed by the combinatorial coefficient $(m!n!)^{-\frac{1}{2}}\binom{m}{r}\binom{n}{r}r!$ and summed over the possibilities for r as explained in connection with (57). Thus once the amplitude for virtual processes is known, that for real photon processes can be obtained by differentiation.

It is possible, of course, to deal with situations in which the electromagnetic field is not in a definite state after the interaction. For example, we might ask for the total probability of a given process, such as a scattering, without regard for the number of photons emitted. This is done of course by squaring the amplitude for the emission of m photons of a given kind and summing on all m. Actually the sums and integrations over the oscillator momenta can usually easily be performed analytically. For example, the amplitude, starting from vacuum and ending with m photons of a given kind, is by (56) just

$$G_{m0}=(m!)^{-\frac{1}{2}}G_{00}(i\beta^*)^m. \qquad (61)$$

The square of the amplitude summed on m requires the product of two such expressions (the $\gamma(t)$ in the β of one and in the other will have to be kept separately) summed on m:

$$\begin{aligned}\textstyle\sum_m G_{m0}{}^*G_{m0}{}'=\sum_m G_{00}{}^*G_{00}{}'(m!)^{-1}\beta^m(\beta'^*)^m \\ =G_{00}{}^*G_{00}{}'\exp(\beta\beta'^*).\end{aligned}$$

In the resulting expression the sum over all oscillators is easily done. Such expressions can be of use in the analysis in a direct manner of problems of line width, of the Bloch-Nordsieck infra-red problem, and of statistical mechanical problems, but no such applications will be made here.

The author appreciates his opportunities to discuss these matters with Professor H. A. Bethe and Professer J. Ashkin, and the help of Mr. M. Baranger with the manuscript.

APPENDIX A. THE KLEIN-GORDON EQUATION

In this Appendix we describe a formulation of the equations for a particle of spin zero which was first used to obtain the rules given in II for such particles. The complete physical significance of the equations has not been analyzed thoroughly so that it may be preferable to derive the rules directly from the second quantization formulation of Pauli and Weisskopf. This can be done in a manner analogous to the derivation of the rules for the Dirac equation given in I or from the Schwinger-Tomonaga formulation[2] in a manner described, for example, by Rohrlich.[18] The formulation given here is therefore not necessary for a description of spin zero particles but is given only for its own interest as an alternative to the formulation of second quantization.

We start with the Klein-Gordon equation

$$(i\partial/\partial x_\mu - A_\mu)^2\psi = m^2\psi \qquad (1A)$$

for the wave function ψ of a particle of mass m in a given external potential A_μ. We shall try to represent this in a manner analogous to the formulation of quantum mechanics in C. That is, we try to represent the amplitude for a particle to get from one point to another as a sum over all trajectories of an amplitude $\exp(iS)$ where S is the classical action for a given trajectory. To maintain the relativistic invariance in evidence the idea suggests itself of describing a trajectory in space-time by giving the four variables $x_\mu(u)$ as functions of some fifth parameter u (rather than expressing x_1, x_2, x_3 in terms of x_4). As we expect to represent paths which may reverse themselves in time (to represent pair production, etc., as in I) this is certainly a more convenient representation, for all four functions $x_\mu(u)$ may be considered as functions of a parameter u (somewhat analogous to proper time) which increase as we go along the trajectory, whether the trajectory is proceeding forward ($dx_4/du>0$) or backward ($dx_4/du<0$) in time.[19] We shall then have a new type of wave function $\varphi(x, u)$ a function of five variables, x standing for the four x_μ. It gives the amplitude for arrival at point x_μ with a certain value of the parameter u. We shall suppose that this wave function satisfies the equation

$$i\partial\varphi/\partial u = -\tfrac{1}{2}(i\partial/\partial x_\mu - A_\mu)^2\varphi \qquad (2A)$$

which is seen to be analogous to the time-dependent Schrödinger equation, u replacing the time and the four coordinates of space-time x_μ replacing the usual three coordinates of space.

Since the potentials $A_\mu(x)$ are functions only of coordinates x_μ and are independent of u, the equation is separable in u and we can write a special solution in the form $\varphi=\exp(\frac{1}{2}im^2u)\psi(x)$ where $\psi(x)$, a function of the coordinates x_μ only, satisfies (1A) and the eigenvalue $\frac{1}{2}m^2$ conjugate to u is related to the mass m of the particle. Equation (2A) is therefore equivalent to the Klein-Gordon Eq. (1A) provided we ask in the end only for the solution of (1A) corresponding to the eigenvalue $\frac{1}{2}m^2$ for the quantity conjugate to u.

We may now proceed to represent Eq. (2A) in Lagrangian form in general and without regard to this eigenvalue condition. Only in the final solutions need we apply the eigenvalue condition. That is, if we have some special solution $\varphi(x, u)$ of (2A) we can select that part corresponding to the eigenvalue $\frac{1}{2}m^2$ by calculating

$$\psi(x)=\int_{-\infty}^{\infty}\exp(-\tfrac{1}{2}im^2u)\,\varphi(x, u)du$$

and thereby obtain a solution ψ of Eq. (1A).

Since (2A) is so closely analogous to the Schrödinger equation, it is easily written in the Lagrangian form described in C, simply by working by analogy. For example if $\varphi(x, u)$ is known at one value of u its value at a slightly larger value $u+\epsilon$ is given by

$$\begin{aligned}\varphi(x, u+\epsilon)=\int \exp i\epsilon\Big[-\frac{(x_\mu-x_\mu')^2}{2\epsilon^2}-\frac{1}{2}\Big(\frac{x_\mu-x_\mu'}{\epsilon}\Big)(A_\mu(x)+A_\mu(x'))\Big] \\ \cdot\varphi(x', u)d^4\tau_{x'}(2\pi i\epsilon)^{-\frac{3}{2}}(-2\pi i\epsilon)^{-\frac{1}{2}} \qquad (3A)\end{aligned}$$

where $(x_\mu-x_\mu')^2$ means $(x_\mu-x_\mu')(x_\mu-x_\mu')$, $d^4\tau_{x'}=dx_1'dx_2'dx_3'dx_4'$ and the sign of the normalizing factor is changed for the x_4 component since the component has the reversed sign in its quadratic coefficient in the exponential, in accordance with our summation convention $a_\mu b_\mu=a_4b_4-a_1b_1-a_2b_2-a_3b_3$. Equation (3A), as can be verified readily as described in C, Sec. 6, is equivalent to first order in ϵ, to Eq. (2A). Hence, by repeated use of this equation the wave function at $u_0=n\epsilon$ can be represented in terms of that at $u=0$ by:

$$\begin{aligned}\varphi(x_{\nu,n}, u_0)=\int \exp-\frac{i\epsilon}{2}\sum_{i=1}^{n}\Big[\Big(\frac{x_{\mu,i}-x_{\mu,i-1}}{\epsilon}\Big)^2 \\ +\epsilon^{-1}(x_{\mu,i}-x_{\mu,i-1})(A_\mu(x_i)+A_\mu(x_{i-1}))\Big] \\ \cdot\varphi(x_{\nu,0}, 0)\prod_{i=0}^{n-1}(d^4\tau_i/4\pi^2\epsilon^2 i). \qquad (4A)\end{aligned}$$

That is, roughly, the amplitude for getting from one point to another with a given value of u_0 is the sum over all trajectories of $\exp(iS)$ where

$$S=-\int_0^{u_0}[\tfrac{1}{2}(dx_\mu/du)^2+(dx_\mu/du)A_\mu(x)]du, \qquad (5A)$$

when sufficient care is taken to define the quantities, as in C. This completes the formulation for particles in a fixed potential but a few words of description may be in order.

In the first place in the special case of a free particle we can define a kernal $k^{(0)}(x, u_0; x', 0)$ for arrival from x_μ', 0 to x_μ at u_0 as the sum over all trajectories between these points of $\exp-i\int_0^{u_0}\frac{1}{2}(dx_\mu/du)^2du$. Then for this case we have

$$\varphi(x, u_0)=\int k^{(0)}(x, u_0; x', 0)\varphi(x', 0)d^4\tau_{x'} \qquad (6A)$$

and it is easily verified that k_0 is given by

$$k^{(0)}(x, u_0; x', 0)=(4\pi^2u_0^2i)^{-1}\exp-i(x_\mu-x_\mu')^2/2u_0 \qquad (7A)$$

for $u_0>0$ and by 0, by definition, for $u_0<0$. The corresponding

[18] F. Rohrlich (to be published).

[19] The physical ideas involved in such a description are discussed in detail by Y. Nambu, Prog. Theor. Phys. 5, 82 (1950). An equation of type (2A) extended to the case of Dirac electrons has been studied by V. Fock, Physik Zeits. Sowjetunion 12, 404 (1937).

kernel of importance when we select the eigenvalue $\frac{1}{2}m^2$ is[20]

$$2iI_+(x,x')=\int_{-\infty}^{\infty}k^{(0)}(x,u_0;x',0)\exp(-\tfrac{1}{2}im^2u_0)du_0$$
$$=\int_0^{\infty}du_0(4\pi^2u_0^2i)^{-1}\exp-\tfrac{1}{2}i(m^2u_0+u_0^{-1}(x_\mu-x_\mu')^2)\quad \text{(8A)}$$

(the last extends only from $u_0=0$ since k_0 is zero for negative u_0) which is identical to the I_+ defined in II.[21] This may be seen readily by studying the Fourier transform, for the transform of the integrand on the right-hand side is

$$\int(4\pi^2u_0^2i)^{-1}\exp(ip\cdot x)\exp-\tfrac{1}{2}i(m^2u_0+x_\mu^2/u_0)d^4\tau_x$$
$$=\exp-\tfrac{1}{2}iu_0(m^2-p_\mu^2)$$

so that the u_0 integration gives for the transform of I_+ just $1/(p_\mu^2-m^2)$ with the pole defined exactly as in II. Thus we are automatically representing the positrons as trajectories with the time sense reversed.

If $\Phi^{(0)}[x(u)]=\exp-i\int_0^{u_0}\frac{1}{2}(dx_\mu/du)^2du$ is the amplitude for a given trajectory $x_\nu(u)$ for a free particle, then the amplitude in a potential is

$$\Phi^{(A)}[x(u)]=\Phi^{(0)}[x(u)]\exp-i\int_0^{u_0}(dx_\mu/du)A_\mu(x)du.\quad \text{(9A)}$$

If desired this may be studied by perturbation methods by expanding the exponential in powers of A_μ.

For interpretation, the integral in (9A) must be written as a Riemann sum, and if a perturbation expansion is made, care must be taken with the terms quadratic in the velocity, for the effect of $(x_{\mu,i+1}-x_{\mu,i})(x_{\nu,i+1}-x_{\nu,i})$ is not of order ϵ^2 but is $-i\delta_{\mu\nu}\epsilon$. The "velocity" dx_μ/du becomes the momentum operator $p_\mu=+i\partial/\partial x_\mu$ operating half before and half after A_μ, just as in the non-relativistic Schrödinger equation discussed in Sec. 5. Furthermore, in exactly the same manner as in that case, but here in four dimensions, a term quadratic in A_μ arises in the second-order perturbation terms from the coincidence of two velocities for the same value of u.

As an example, the kernal $k^{(A)}(x,u_0;x',0)$ for proceeding from $x_\mu',0$ to x_μ,u_0 in a potential A_μ differs from $k^{(0)}$ to first order in A_μ by a term

$$-i\int_0^{u_0}duk^{(0)}(x,u_0;y,u)\tfrac{1}{2}(p_\mu A_\mu(y)+A_\mu(y)p_\mu)k^{(0)}(y,u;x',0)d\tau_y$$

the p_μ here meaning $+i\partial/\partial y_\mu$. The kernel of importance on selecting the eigenvalue $\frac{1}{2}m^2$ is obtained by multiplying this by $\exp(-\frac{1}{2}im^2u_0)$ and integrating u_0 from 0 to ∞. The kernel $k^{(0)}(x,u_0;y,u)$ depends only on $u'=u_0-u$ and in the integrals on u and u_0; $\int_0^{\infty}du_0\int_0^{u_0}du\exp(-\frac{1}{2}im^2u_0)\cdots$, can be written, on interchanging the order of integration and changing variables to u and u', $\int_0^{\infty}du\int_0^{\infty}du'\exp(-\frac{1}{2}im^2(u+u'))\cdots$. Now the integral on u' converts $k^{(0)}(x,u_0;y,u)$ to $2iI_+(x,y)$ by (8A), while that on u converts $k^{(0)}(y,u;x',0)$ to $2iI_+(y,x')$, so the result becomes

$$\int 2iI_+(x,y)(p_\mu A_\mu+A_\mu p_\mu)I_+(y,x')d^4\tau_y$$

as expected. The same principle works to any order so that the rules for a single Klein-Gordon particle in external potentials given in II, Section 9, are deduced.

The transition to quantum electrodynamics is simple for in (5A) we already have a transition amplitude represented as a sum (over trajectories, and eventually u_0) of terms, in each of which the potential appears in exponential form. We may make use of the general relation (54). Hence, for example, one finds for the case of no photons in the initial and final states, in the presence of an external potential B_μ, the amplitude that a particle proceeds from $(x_\mu',0)$ to (x_μ,u_0) is the sum over all trajectories of the quantity

$$\exp-i\left[\frac{1}{2}\int_0^{u_0}\left(\frac{dx_\mu}{du}\right)^2du+\int_0^{u_0}\frac{dx_\mu}{du}B_\mu(x(u))du\right.$$
$$\left.+\frac{e^2}{2}\int_0^{u_0}\int_0^{u_0}\frac{dx_\mu(u)}{du}\frac{dx_\nu(u')}{du'}\delta_+((x_\mu(u)-x_\mu(u'))^2)dudu'\right].\quad \text{(10A)}$$

This result must be multiplied by $\exp(-\frac{1}{2}im^2u_0)$ and integrated on u_0 from zero to infinity to express the action of a Klein-Gordon particle acting on itself through virtual photons. The integrals are interpreted as Riemann sums, and if perturbation expansions are made, the necessary care is taken with the terms quadratic in velocity. When there are several particles (other than the virtual pairs already included) one use a separate u for each, and writes the amplitude for each set of trajectories as the exponental of $-i$ times

$$\frac{1}{2}\sum_n\int_0^{u_0^{(n)}}\left(\frac{dx_\mu^{(n)}}{du}\right)^2du+\sum_n\int_0^{u_0^{(n)}}\frac{dx_\mu^{(n)}}{du}B_\mu(x_\mu^{(n)}(u))du$$
$$+\frac{e^2}{2}\sum_n\sum_m\int_0^{u_0^{(n)}}\int_0^{u_0^{(m)}}\frac{dx_\nu^{(n)}(u)}{du}\frac{dx_\nu^{(m)}(u')}{du'}$$
$$\times\delta_+((x_\mu^{(n)}(u)-x_\mu^{(m)}(u'))^2)dudu',\quad \text{(11A)}$$

where $x_\mu^{(n)}(u)$ are the coordinates of the trajectory of the nth particle.[22] The solution should depend on the $u_0^{(n)}$ as $\exp(-\frac{1}{2}im^2\sum_n u_0^{(n)})$.

Actually, knowledge of the motion of a single charge implies a great deal about the behavior of several charges. For a pair which eventually may turn out to be a virtual pair may appear in the short run as two "other particles." As a virtual pair, that is, as the reverse section of a very long and complicated single track we know its behavior by (10A). We can assume that such a section can be looked at equally well, for a limited duration at least, as being due to other unconnected particles. This then implies a definite law of interaction of particles if the self-action (10A) of a single particle is known. (This is similar to the relation of real and virtual photon processes discussed in detail in Appendix B.) It is possible that a detailed analysis of this could show that (10A) implied that (11A) was correct for many particles. There is even reason to believe that the law of Bose-Einstein statistics and the expression for contributions from closed loops could be deduced by following this argument. This has not yet been analyzed completely, however, so we must leave this formulation in an incomplete form. The expression for closed loops should come out to be $C_v=\exp+L$ where L, the contribution from a

[20] The factor $2i$ in front of I_+ is simply to make the definition of I_+ here agree with that in I and II. In II it operates with $\mathbf{p}\cdot\mathbf{A}+\mathbf{A}\cdot\mathbf{p}$ as a perturbation. But the perturbation coming from (3A) in a natural way by expansion of the exponential is $-\frac{1}{2}i(\mathbf{p}\cdot\mathbf{A}+\mathbf{A}\cdot\mathbf{p})$.

[21] Expression (8A) is closely related to Schwinger's parametric integral representation of these functions. For example, (8A) becomes formula (45) of F. Dyson, Phys. Rev. 75, 486 (1949) for $\Delta_F=\Delta^{(1)}-2i\bar{\Delta}=2iI_+$ if $(2\alpha)^{-1}$ is substituted for u_0.

[22] The form (10A) suggests another interesting possibility for avoiding the divergences of quantum electrodynamics in this case. The divergences arise from the δ_+ function when $u=u'$. We might restrict the integration in the double integral such that $|u-u'|>\delta$ where δ is some finite quantity, very small compared with m^{-2}. More generally, we could keep the region $u=u'$ from contributing by including in the integrand a factor $F(u-u')$ where $F(x)\rightarrow1$ for x large compared to some δ, and $F(0)=0$ (e.g., $F(x)$ acts qualitatively like $1-\exp(-x^2\delta^{-2})$. (Another way might be to replace u by a discontinuous variable, that is, we do not use the limit in (4A) as $\epsilon\rightarrow0$ but set $\epsilon=\delta$.) The idea is that two interactions would contribute very little in amplitude if they followed one another too rapidly in u. It is easily verified that this makes the otherwise divergent integrals finite. But whether the resulting formulas make good physical sense is hard to see. The action of a potential would now depend on the value of u so that Eq. (2A), or its equivalent, would not be separable in u so that $\frac{1}{2}m^2$ would no longer be a strict eigenvalue for all disturbances. High energy potentials could excite states corresponding to other eigenvalues, possibly thereby corresponding to other masses. This note is meant only as a speculation, for not enough work has been done in this direction to make sure that a reasonable physical theory can be developed along these lines. (What little work has been done was not promising.) Analogous modifications can also be made for Dirac electrons.

single loop, is

$$L=2\int_0^\infty l(u_0)\exp(-\tfrac{1}{2}im^2u_0)du_0/u_0$$

where $l(u_0)$ is the sum over all trajectories which close on themselves $(x_\mu(u_0)=x_\mu(0))$ of $\exp(iS)$ with S given in (5A), and a final integration $d\tau_{x(0)}$ on $x_\mu(0)$ is made. This is equivalent to putting

$$l(u_0)=\int(k^{(A)}(x,u_0;x,0)-k^{(0)}(x,u_0;x,0))d\tau_x.$$

The term $k^{(0)}$ is subtracted only to simplify convergence problems (as adding a constant independent of A_μ to L has no effect).

APPENDIX B. THE RELATION OF REAL AND VIRTUAL PROCESSES

If one has a general formula for all virtual processes he should be able to find the formulas and states involved in real processes. That is to say, we should be able to deduce the formulas of Section 9 directly from the formulation (24), (25) (or its generalized equivalent such as (46), (48)) without having to go all the way back to the more usual formulation. We discuss this problem here.

That this possibility exists can be seen from the consideration that what looks like a real process from one point of view may appear as a virtual process occurring over a more extended time.

For example, if we wish to study a given real process, such as the scattering of light, we can, if we wish, include in principle the source, scatterer, and eventual absorber of the scattered light in our analysis. We may imagine that no photon is present initially, and that the source then emits light (the energy coming say from kinetic energy in the source). The light is then scattered and eventually absorbed (becoming kinetic energy in the absorber). From this point of view the process is virtual; that is, we start with no photons and end with none. Thus we can analyze the process by means of our formula for virtual processes, and obtain the formulas for real processes by attempting to break the analysis into parts corresponding to emission, scattering, and absorption.[23]

To put the problem in a more general way, consider the amplitude for some transition from a state empty of photons far in the past (time t') to a similar one far in the future $(t=t'')$. Suppose the time interval to be split into three regions a, b, c in some convenient manner, so that region b is an interval $t_2>t>t_1$ around the present time that we wish to study. Region a, $(t_1>t>t')$, precedes b, and c, $(t''>t>t_2)$, follows b. We want to see how it comes about that the phenomena during b can be analyzed by a study of transitions $g_{ji}(b)$ between some initial state i at time t_1 (which no longer need be photon-free), to some other final state j at time t_2. The states i and j are members of a large class which we will have to find out how to specify. (The single index i is used to represent a large number of quantum numbers, so that different values of i will correspond to having various numbers of various kinds of photons in the field, etc.) Our problem is to represent the over-all transition amplitude, $g(a,b,c)$, as a sum over various values of i, j of a product of three amplitudes,

$$g(a,b,c)=\Sigma_i\Sigma_j\, g_{0j}(c)g_{ji}(b)g_{i0}(a);\qquad(1B)$$

first the amplitude that during the interval a the vacuum state makes transition to some state i, then the amplitude that during b the transition to j is made, and finally in c the amplitude that the transition from j to some photon-free state 0 is completed. The mathematical problem of splitting $g(a,b,c)$ is made definite by the further condition that $g_{ji}(b)$ for given i, j must not involve the coordinates of the particles for times corresponding to regions a or c, $g_{i0}(a)$ must involve those only in region a, and $g_{0j}(c)$ only in c.

To become acquainted with what is involved, suppose first that we do not have a problem involving virtual photons, but just the transition of a one-dimensional Schrödinger particle going in a long time interval from, say, the origin o to the origin o, and ask what states i we shall need for intermediary time intervals. We must solve the problem (1B) where $g(a,b,c)$ is the sum over all trajectories going from o at t' to o at t'' of $\exp iS$ where $S=\int Ldt$. The integral may be split into three parts $S=S_a+S_b+S_c$ corresponding to the three ranges of time. Then $\exp(iS)=\exp(iS_a)\cdot\exp(iS_b)\cdot\exp(iS_c)$ and the separation (1B) is accomplished by taking for $g_{i0}(a)$ the sum over all trajectories lying in a from o to some end point x_{t_1} of $\exp(iS_a)$, for $g_{ji}(b)$ the sum over trajectories in b of $\exp(iS_b)$ between end points x_{t_1} and x_{t_2}, and for $g_{0j}(c)$ the sum of $\exp(iS_c)$ over the section of the trajectory lying in c and going from x_{t_2} to o. Then the sum on i and j can be taken to be the integrals on x_{t_1}, x_{t_2} respectively. Hence the various states i can be taken to correspond to particles being at various coordinates x. (Of course any other representation of the states in the sense of Dirac's transformation theory could be used equally well. Which one, whether coordinate, momentum, or energy level representation, is of course just a matter of convenience and we cannot determine that simply from (1B).)

We can consider next the problem including virtual photons. That is, $g(a,b,c)$ now contains an additional factor $\exp(iR)$ where R involves a double integral $\int\int$ over all time. Those parts of the index i which correspond to the particle states can be taken in the same way as though R were absent. We study now the extra complexities in the states produced by splitting the R. Let us first (solely for simplicity of the argument) take the case that there are only two regions a, c separated by time t_0 and try to expand

$$g(a,c)=\Sigma_i\, g_{0i}(c)g_{i0}(a).$$

The factor $\exp(iR)$ involves R as a double integral which can be split into three parts $\int_a\int_a+\int_c\int_c+\int_a\int_c$ for the first of which both t, s are in a, for the second both are in c, for the third one is in a the other in c. Writing $\exp(iR)$ as $\exp(iR_{cc})\cdot\exp(iR_{aa})\cdot\exp(iR_{ac})$ shows that the factors R_{cc} and R_{aa} produce no new problems for they can be taken bodily into $g_{0i}(c)$ and $g_{i0}(a)$ respectively. However, we must disentangle the variables which are mixed up in $\exp(iR_{ac})$.

The expression for R_{ac} is just twice (24) but with the integral on s extending over the range a and that for t extending over c. Thus $\exp(iR_{ac})$ contains the variables for times in a and in c in a quite complicated mixture. Our problem is to write $\exp(iR_{ac})$ as a sum over possibly a vast class of states i of the product of two parts, like $h_i'(c)h_i(a)$, each of which involves the coordinates in one interval alone.

This separation may be made in many different ways, corresponding to various possible representations of the state of the electromagnetic field. We choose a particular one. First we can expand the exponential, $\exp(iR_{ac})$, in a power series, as $\Sigma_n i^n(n!)^{-1}(R_{ac})^n$. The states i can therefore be subdivided into subclasses corresponding to an integer n which we can interpret as the number of quanta in the field at time t_0. The amplitude for the case $n=0$ clearly just involves $\exp(iR_{aa})$ and $\exp(iR_{cc})$ in the way that it should if we interpret these as the amplitudes for regions a and c, respectively, of making a transition between a state of zero photons and another state of zero photons.

Next consider the case $n=1$. This implies an additional factor in the transitional element; the factor R_{ac}. The variables are still mixed up. But an easy way to perform the separation suggests itself. Namely, expand the $\delta_+((t-s)^2-(\mathbf{x}_n(t)-\mathbf{x}_m(s))^2)$ in R_{ac} as a Fourier integral as

$$i\int\exp(-ik|t-s|)\exp(-i\mathbf{K}\cdot(\mathbf{x}_n(t)-\mathbf{x}_m(s))d^3\mathbf{K}/4\pi^2k.$$

[23] The formulas for real processes deduced in this way are strictly limited to the case in which the light comes from sources which are originally dark, and that eventually all light emitted is absorbed again. We can only extend it to the case for which these restrictions do not hold by hypothesis, namely, that the details of the scattering process are independent of these characteristics of the light source and of the eventual disposition of the scattered light. The argument of the text gives a method for discovering formulas for real processes when no more than the formula for virtual processes is at hand. But with this method belief in the general validity of the resulting formulas must rest on the physical reasonableness of the above-mentioned hypothesis.

For the exponential can be written immediately as a product of $\exp+i(\mathbf{K}\cdot\mathbf{x}_m(s))$, a function only of coordinates for times s in a (suppose $s<t$), and $\exp-i\mathbf{K}\cdot\mathbf{x}_n(t)$ (a function only of coordinates during interval c). The integral on $d^3\mathbf{K}$ can be symbolized as a sum over states i characterized by the value of $\mathbf{K}$. Thus the state with $n=1$ must be further characterized by specifying a vector $\mathbf{K}$, interpreted as the momentum of the photon. Finally the factor $(1-\mathbf{x}'_n(t)\cdot\mathbf{x}'_m(s))$ in R_{ac} is simply the sum of four parts each of which is already split (namely 1, and each of the three components in the vector scalar product). Hence each photon of momentum $\mathbf{K}$ must still be characterized by specifying it as one of four varieties; that is, there are four polarizations.[24] Thus in trying to represent the effect of the past a on the future c we are lead to invent photons of four polarizations and characterized by a propagation vector $\mathbf{K}$.

The term for a given polarization and value of $\mathbf{K}$ (for $n=1$) is clearly just $-\beta_c\beta_a^*$ where the β_a is defined in (59) but with the time integral extending just over region a, while β_c is the same expression with the integration over region c. Hence the amplitude for transition during interval a from a state with no quanta to a state with one in a given state of polarization and momentum is calculated by inclusion of an extra factor $i\beta_a^*$ in the transition element. Absorption in region c corresponds to a factor $i\beta_c$.

We next turn to the case $n=2$. This requires analysis of R_{ac}^2. The δ_+ can be expanded again as a Fourier integral, but for each of the two δ_+ in $\frac{1}{2}R_{ac}^2$ we have a value of $\mathbf{K}$ which may be different. Thus we say, we have two photons, one of momentum $\mathbf{K}$ and one momentum $\mathbf{K}'$ and we sum over all values of $\mathbf{K}$ and $\mathbf{K}'$. (Similarly each photon is characterized by its own independent polarization index.) The factor $\frac{1}{2}$ can be taken into account neatly by asserting that we count each possible pair of photons as constituting just one state at time t_0. Then the $\frac{1}{2}$ arises for the sum over all $\mathbf{K}$, $\mathbf{K}'$ (and polarizations) counts each pair twice. On the other hand, for the terms representing two identical photons ($\mathbf{K}=\mathbf{K}'$) of like polarization, the $\frac{1}{2}$ cannot be so interpreted. Instead we invent the rule that a state of two like photons has statistical weight $\frac{1}{2}$ as great as that calculated as though the photons were different. This, generalized to n identical photons, is the rule of Bose statistics.

The higher values of n offer no problem. The $1/n!$ is interpreted combinatorially for different photons, and as a statistical factor when some are identical. For example, for all n identical one obtains a factor $(n!)^{-1}(-\beta_c\beta_a^*)^n$ so that $(n!)^{-\frac{1}{2}}(i\beta_a^*)^n$ can be interpreted as the amplitude for emission (from no initial photons) of n identical photons, in complete agreement with (61) for G_{m0}.

To obtain the amplitude for transitions in which neither the initial nor the final state is empty of photons we must consider the more general case of the division into three time regions (1B). This time we see that the factor which involves the coordinates in an entangled manner is $\exp i(R_{ab}+R_{bc}+R_{ac})$. It is to be expanded in the form $\sum_i\sum_j h_i''(c)h_{ij}'(b)h_j(a)$. Again the expansion in power series and development in Fourier series with a polarization sum will solve the problem. Thus the exponential is $\sum_r\sum_{l_1}\sum_{l_2}(iR_{ac})^r(iR_{ab})^{l_1}(iR_{bc})^{l_2}(l_1!)^{-1}(l_2!)^{-1}(r!)^{-1}$. Now the R are written as Fourier series, one of the terms containing l_1+l_2+r variables $\mathbf{K}$. Since l_1+r involve a, l_2+r involve c and l_1+l_2 involve b, this term will give the amplitude that l_1+r photons are emitted during the interval a, of those l_1 are absorbed during b but the remaining r, along with l_2 new ones emitted during b go on to be absorbed during the interval c. We have therefore $n=l_1+r$ photons in the state at time t_1 when b begins, and $m=l_2+r$ at t_2 when b is over. They each are characterized by momentum vectors and polarizations. When these are different the factors $(l_1!)^{-1}(l_2!)^{-1}(r!)^{-1}$ are absorbed combinatorially. When some are equal we must invoke the rule of the statistical weights. For example, suppose all l_1+l_2+r photons are identical. Then $R_{ab}=i\beta_b\beta_a^*$, $R_{bc}=i\beta_c\beta_b^*$, $R_{ac}=i\beta_c\beta_a^*$ so that our sum is

$$\sum_{l_1}\sum_{l_2}\sum_r(l_1!l_2!r!)^{-1}(i\beta_c)^{l_2+r}(i\beta_b)^{l_1}(i\beta_b^*)^{l_2}(i\beta_a^*)^{l_1+r}.$$

Putting $m=l_2+r$, $n=l_1+r$, this is the sum on n and m of

$$(i\beta_c)^m(m!)^{-\frac{1}{2}}\left[\sum_r(m!n!)^{\frac{1}{2}}((m-r)!(n-r)!r!)^{-1}\times(i\beta_b^*)^{m-r}(i\beta_b)^{n-r}\right](n!)^{-\frac{1}{2}}(i\beta_a^*)^n.$$

The last factor we have seen is the amplitude for emission of n photons during interval a, while the first factor is the amplitude for absorption of m during c. The sum is therefore the factor for transition from n to m identical photons, in accordance with (57). We see the significance of the simple generating function (56).

We have therefore found rules for real photons in terms of those for virtual. The real photons are a way of representing and keeping track of those aspects of the past behavior which may influence the future.

If one starts from a theory involving an arbitrary modification of the direct interaction δ_+ (or in more general situations) it is possible in this way to discover what kinds of states and physical entities will be involved if one tries to represent in the present all the information needed to predict the future. With the Hamiltonian method, which begins by assuming such a representation, it is difficult to suggest modifications of a general kind, for one cannot formulate the problem without having a complete representation of the characteristics of the intermediate states, the particles involved in interaction, etc. It is quite possible (in the author's opinion, it is very likely) that we may discover that in nature the relation of past and future is so intimate for short durations that no simple representation of a present may exist. In such a case a theory could not find expression in Hamiltonian form.

An exactly similar analysis can be made just as easily starting with the general forms (46), (48). Also a coordinate representation of the photons could have been used instead of the familiar momentum one. One can deduce the rules (60), (61). Nothing essentially different is involved physically, however, so we shall not pursue the subject further here. Since they imply[25] all the rules for real photons, Eqs. (46), (47), (48) constitute a compact statement of all the laws of quantum electrodynamics. But they give divergent results. Can the result *after* charge and mass renormalization also be expressed to all orders in $e^2/\hbar c$ in a simple way?

APPENDIX C. DIFFERENTIAL EQUATION FOR ELECTRON PROPAGATION

An attempt has been made to find a differential wave equation for the propagation of an electron interacting with itself, analogous to the Dirac equation, but containing terms representing the self-action. Neglecting all effects of closed loops, one such equation has been found, but not much has been done with it. It is reported here for whatever value it may have.

An electron acting upon itself is, from one point of view, a complex system of a particle and a field of an indefinite number of photons. To find a differential law of propagation of such a system we must ask first what quantities known at one instant will permit the calculation of these same quantities an instant later. Clearly, a knowledge of the position of the particle is not enough. We should need to specify: (1) the amplitude that the electron is at x and there are no photons in the field, (2) the amplitude the electron is at x and there is one photon of such and such a kind in the field, (3) the amplitude there are two photons, etc. That is, a series of functions of ever increasing numbers of variables. Following this view, we shall be led to the wave equation of the theory of second quantization.

We may also take a different view. Suppose we know a quantity $\Phi_{e^2}[B, x]$, a spinor function of x_μ, and functional of $B_\mu(1)$, defined as the amplitude that an electron arrives at x_μ with no photon in the field when it moves in an arbitrary external unquantized potential $B_\mu(1)$. We allow the electron also to interact with itself,

[24] Usually only two polarizations transverse to the propagation vector $\mathbf{K}$ are used. This can be accomplished by a further rearrangement of terms corresponding to the reverse of the steps leading from (17) to (19). We omit the details here as it is well-known that either formulation gives the same results. See II, Section 8.

but $\Phi_e 2$ is the amplitude at a given instant that there happens to be no photons present. As we have seen, a complete knowledge of this functional will also tell us the amplitude that the electron arrives at x and there is just one photon, of form $A_\mu^{PH}(1)$ present. It is, from (60), $\int(\delta\Phi_e 2[B, x]/\delta B_\mu(1))A_\mu^{PH}(1)d\tau_1$.

Higher numbers of photons correspond to higher functional derivatives of $\Phi_e 2$. Therefore, $\Phi_e 2[B, x]$ contains all the information requisite for describing the state of the electron-photon system, and we may expect to find a differential equation for it. Actually it satisfies ($\nabla=\gamma_\mu\partial/\partial x_\mu$, $B=\gamma_\mu B_\mu$),

$$(i\nabla-m)\Phi_e 2[B, x]-B(x)\Phi_e 2[B, x] +ie^2\gamma_\mu\int\delta_+(s_{x1}^2)(\delta\Phi_e 2[B, x]/\delta B_\mu(1))d\tau_1 \quad (1C)$$

as may be seen from a physical argument.[25] The operator $(i\nabla-m)$ operating on the x coordinate of $\Phi_e 2$ should equal, from Dirac's equation, the changes in $\Phi_e 2$ as we go from one position x to a neighboring position due to the action of vector potentials. The term $B(x)\Phi_e 2$ is the effect of the external potential. But $\Phi_e 2$ may also change for at the first position x we may have had a photon present (amplitude that it was emitted at another point 1 is $\delta\Phi_e 2/\delta B_\mu(1)$) which was absorbed at x (amplitude photon released at 1 gets to x is $\delta_+(s_{x1}^2)$ where s_{x1}^2 is the squared invariant distance from 1 to x) acting as a vector potential there (factor γ_μ). Effects of vacuum polarization are left out.

Expansion of the solution of (1C) in a power series in B and e^2 starting from a free particle solution for a single electron, produces a series of terms which agree with the rules of II for action of potentials and virtual photons to various orders. It is another matter to use such an equation for the practical solution of a problem to all orders in e^2. It might be possible to represent the self-energy problem as the variational problem for m, stemming from (1C). The δ_+ will first have to be modified to obtain a convergent result.

We are not in need of the general solution of (1C). (In fact, we have it in (46), (48) in terms of the solution $T_0[B]=\Phi_0[B, x]$ of the ordinary Dirac equation $(i\nabla-m)\Phi_0[B, x]=B\Phi_0[B, x]$. The general solution is too complicated, for complete knowledge of the motion of a self-acting electron in an arbitrary potential is essentially all of electrodynamics (because of the kind of relation of real and virtual processes discussed for photons in Appendix B, extended also to real and virtual pairs). Furthermore, it is easy to see that other quantities also satisfy (1C). Consider a system of many electrons, and single out some one for consideration, supposing all the others go from some definite initial state i to some definite final state f. Let $\Phi_e 2[B, x]$ be the amplitude that the special electron arrives at x, there are no photons present, and the other electrons go from i to f when there is an external potential B_μ present (which B_μ also acts on the other electrons). Then $\Phi_e 2$ also satisfies (1C). Likewise the amplitude with closed loops (all other electrons go vacuum to vacuum) also satisfies (1C) including all vacuum polarization effects. The various problems correspond to different assumptions as to the dependence of $\Phi_e 2[B, x]$ on B_μ in the limit of zero e^2. The Eq. (1C) without further boundary conditions is probably too general to be useful.

[25] Its general validity can also be demonstrated mathematically from (45). The amplitude for arriving at x with no photons in the field with virtual photon coupling e^2 is a transition amplitude. It must, therefore, satisfy (45) with $T_e 2[B]=\Phi_e 2[B, x]$ for any x. Hence show that the quantity

$$C_e 2[B, x]=(i\nabla-m-B(x))\Phi_e 2[B, x] -ie^2\gamma_\mu\int\delta_+(s_{x1}^2)(\delta\Phi_e 2[B, x]/\delta B_\mu(1))d\tau_1$$

also satisfies Eq. (45) by substituting $C_e 2[B, x]$ for $T_e 2[B]$ in (45) and using the fact that $\Phi_e 2[B, x]$ satisfies (45). Hence if $C_0[B, x]=0$ then $C_e 2[B, x]=0$ for all e^2. But $C_e 2[B, x]=0$ means that $\Phi_e 2[B, x]$ satisfies (1C). Therefore, that solution $\Phi_e 2[B, x]$ of (45) which also satisfies $(i\nabla-m-B(x))\Phi_0[B, x]=0$ (the propagation of a free electron without virtual photons) is a solution of (1C) as we wished to show. Equation (1C) may be more convenient than (45) for some purposes for it does not involve differentiation with respect to the coupling constant, and is more analogous to a wave equation.

PHYSICAL REVIEW VOLUME 96, NUMBER 1 OCTOBER 1, 1954

Conservation of Isotopic Spin and Isotopic Gauge Invariance*

C. N. YANG † AND R. L. MILLS
Brookhaven National Laboratory, Upton, New York

(Received June 28, 1954)

It is pointed out that the usual principle of invariance under isotopic spin rotation is not consistant with the concept of localized fields. The possibility is explored of having invariance under local isotopic spin rotations. This leads to formulating a principle of isotopic gauge invariance and the existence of a **b** field which has the same relation to the isotopic spin that the electromagnetic field has to the electric charge. The **b** field satisfies nonlinear differential equations. The quanta of the **b** field are particles with spin unity, isotopic spin unity, and electric charge $\pm e$ or zero.

INTRODUCTION

THE conservation of isotopic spin is a much discussed concept in recent years. Historically an isotopic spin parameter was first introduced by Heisenberg[1] in 1932 to describe the two charge states (namely neutron and proton) of a nucleon. The idea that the neutron and proton correspond to two states of the same particle was suggested at that time by the fact that their masses are nearly equal, and that the light stable even nuclei contain equal numbers of them. Then in 1937 Breit, Condon, and Present pointed out the approximate equality of $p-p$ and $n-p$ interactions in the ^1S state.[2] It seemed natural to assume that this equality holds also in the other states available to both the $n-p$ and $p-p$ systems. Under such an assumption one arrives at the concept of a total isotopic spin[3] which is conserved in nucleon-nucleon interactions. Experi-

* Work performed under the auspices of the U. S. Atomic Energy Commission.

† On leave of absence from the Institute for Advanced Study, Princeton, New Jersey.

[1] W. Heisenberg, Z. Physik **77**, 1 (1932).

[2] Breit, Condon, and Present, Phys. Rev. **50**, 825 (1936). J. Schwinger pointed out that the small difference may be attributed to magnetic interactions [Phys. Rev. **78**, 135 (1950)].

[3] The total isotopic spin **T** was first introduced by E. Wigner, Phys. Rev. **51**, 106 (1937); B. Cassen and E. U. Condon, Phys. Rev. **50**, 846 (1936).

ments in recent years[4] on the energy levels of light nuclei strongly suggest that this assumption is indeed correct, An implication of this is that all strong interactions such as the pion-nucleon interaction, must also satisfy the same conservation law. This and the knowledge that there are three charge states of the pion, and that pions can be coupled to the nucleon field *singly*, lead to the conclusion that pions have isotopic spin unity. A direct verification of this conclusion was found in the experiment of Hildebrand[5] which compares the differential cross section of the process $n+p\rightarrow\pi^0+d$ with that of the previously measured process $p+p\rightarrow\pi^++d$.

The conservation of isotopic spin is identical with the requirement of invariance of all interactions under isotopic spin rotation. This means that when electromagnetic interactions can be neglected, as we shall hereafter assume to be the case, the orientation of the isotopic spin is of no physical significance. The differentiation between a neutron and a proton is then a purely arbitrary process. As usually conceived, however, this arbitrariness is subject to the following limitation: once one chooses what to call a proton, what a neutron, at one space-time point, one is then not free to make any choices at other space-time points.

It seems that this is not consistent with the localized field concept that underlies the usual physical theories. In the present paper we wish to explore the possibility of requiring all interactions to be invariant under *independent* rotations of the isotopic spin at all space-time points, so that the relative orientation of the isotopic spin at two space-time points becomes a physically meaningless quantity (the electromagnetic field being neglected).

We wish to point out that an entirely similar situation arises with respect to the ordinary gauge invariance of a charged field which is described by a complex wave function ψ. A change of gauge[6] means a change of phase factor $\psi\rightarrow\psi'$, $\psi'=(\exp i\alpha)\psi$, a change that is devoid of any physical consequences. Since ψ may depend on x, y, z, and t, the relative phase factor of ψ at two different space-time points is therefore completely arbitrary. In other words, the arbitrariness in choosing the phase factor is local in character.

We define *isotopic gauge* as an arbitrary way of choosing the orientation of the isotopic spin axes at all space-time points, in analogy with the electromagnetic gauge which represents an arbitrary way of choosing the complex phase factor of a charged field at all space-time points. We then propose that all physical processes (not involving the electromagnetic field) be invariant under an isotopic gauge transformation, $\psi\rightarrow\psi'$, $\psi'=S^{-1}\psi$, where S represents a space-time dependent isotopic spin rotation.

To preserve invariance one notices that in electrodynamics it is necessary to counteract the variation of α with x, y, z, and t by introducing the electromagnetic field A_μ which changes under a gauge transformation as

$$A_\mu'=A_\mu+\frac{1}{e}\frac{\partial\alpha}{\partial x_\mu}.$$

In an entirely similar manner we introduce a B field in the case of the isotopic gauge transformation to counteract the dependence of S on x, y, z, and t. It will be seen that this natural generalization allows for very little arbitrariness. The field equations satisfied by the twelve independent components of the B field, which we shall call the **b** field, and their interaction with any field having an isotopic spin are essentially fixed, in much the same way that the free electromagnetic field and its interaction with charged fields are essentially determined by the requirement of gauge invariance.

In the following two sections we put down the mathematical formulation of the idea of isotopic gauge invariance discussed above. We then proceed to the quantization of the field equations for the **b** field. In the last section the properties of the quanta of the **b** field are discussed.

ISOTOPIC GAUGE TRANSFORMATION

Let ψ be a two-component wave function describing a field with isotopic spin $\frac{1}{2}$. Under an isotopic gauge transformation it transforms by

$$\psi=S\psi', \qquad (1)$$

where S is a 2×2 unitary matrix with determinant unity. In accordance with the discussion in the previous section, we require, in analogy with the electromagnetic case, that all derivatives of ψ appear in the following combination:

$$(\partial_\mu-i\epsilon B_\mu)\psi.$$

B_μ are 2×2 matrices such that[7] for $\mu=1$, 2, and 3, B_μ is Hermitian and B_4 is anti-Hermitian. Invariance requires that

$$S(\partial_\mu-i\epsilon B_\mu')\psi'=(\partial_\mu-i\epsilon B_\mu)\psi. \qquad (2)$$

Combining (1) and (2), we obtain the isotopic gauge transformation on B_μ:

$$B_\mu'=S^{-1}B_\mu S+\frac{i}{\epsilon}S^{-1}\frac{\partial S}{\partial x_\mu}. \qquad (3)$$

The last term is similar to the gradiant term in the gauge transformation of electromagnetic potentials. In analogy to the procedure of obtaining gauge invariant field strengths in the electromagnetic case, we

[4] T. Lauritsen, Ann. Rev. Nuclear Sci. 1, 67 (1952); D. R. Inglis, Revs. Modern Phys. 25, 390 (1953).
[5] R. H. Hildebrand, Phys. Rev. 89, 1090 (1953).
[6] W. Pauli, Revs. Modern Phys. 13, 203 (1941).

[7] We use the conventions $\hbar=c=1$, and $x_4=it$. Bold-face type refers to vectors in isotopic space, not in space-time.

define now

$$F_{\mu\nu}=\frac{\partial B_\mu}{\partial x_\nu}-\frac{\partial B_\nu}{\partial x_\mu}+i\epsilon(B_\mu B_\nu-B_\nu B_\mu). \tag{4}$$

One easily shows from (3) that

$$F_{\mu\nu}'=S^{-1}F_{\mu\nu}S \tag{5}$$

under an isotopic gauge transformation.‡ Other simple functions of B than (4) do not lead to such a simple transformation property.

The above lines of thought can be applied to any field ψ with arbitrary isotopic spin. One need only use other representations S of rotations in three-dimensional space. It is reasonable to assume that different fields with the same total isotopic spin, hence belonging to the same representation S, interact with the same matrix field B_μ. (This is analogous to the fact that the electromagnetic field interacts in the same way with any charged particle, regardless of the nature of the particle. If different fields interact with different and independent B fields, there would be more conservation laws than simply the conservation of total isotopic spin.) To find a more explicit form for the B fields and to relate the B_μ's corresponding to different representations S, we proceed as follows.

Equation (3) is valid for any S and its corresponding B_μ. Now the matrix $S^{-1}\partial S/\partial x_\mu$ appearing in (3) is a linear combination of the isotopic spin "angular momentum" matrices T^i ($i=1, 2, 3$) corresponding to the isotopic spin of the ψ field we are considering. So B_μ itself must also contain a linear combination of the matrices T^i. But any part of B_μ in addition to this, $\bar{B}_\mu$, say, is a scalar or tensor combination of the T's, and must transform by the homogeneous part of (3), $\bar{B}_\mu'=S^{-1}\bar{B}_\mu S$. Such a field is extraneous; it was allowed by the very general form we assumed for the B field, but is irrelevant to the question of isotopic gauge. Thus the relevant part of the B field is of the form

$$B_\mu=2\mathbf{b}_\mu\cdot\mathbf{T}. \tag{6}$$

(Bold-face letters denote three-component vectors in isotopic space.) To relate the $\mathbf{b}_\mu$'s corresponding to different representations S we now consider the product representation $S=S^{(a)}S^{(b)}$. The B field for the combination transforms, according to (3), by

$$B_\mu'=[S^{(b)}]^{-1}[S^{(a)}]^{-1}BS^{(a)}S^{(b)}$$
$$+\frac{i}{\epsilon}[S^{(a)}]^{-1}\frac{\partial S^{(a)}}{\partial x_\mu}+\frac{i}{\epsilon}[S^{(b)}]^{-1}\frac{\partial S^{(b)}}{\partial x_\mu}.$$

But the sum of $B_\mu^{(a)}$ and $B_\mu^{(b)}$, the B fields corresponding to $S^{(a)}$ and $S^{(b)}$, transforms in exactly the same way, so that

$$B_\mu=B_\mu^{(a)}+B_\mu^{(b)}$$

(plus possible terms which transform homogeneously, and hence are irrelevant and will not be included). Decomposing $S^{(a)}S^{(b)}$ into irreducible representations, we see that the twelve-component field $\mathbf{b}_\mu$ in Eq. (6) is the same for all representations.

To obtain the interaction between any field ψ of arbitrary isotopic spin with the $\mathbf{b}$ field one therefore simply replaces the gradiant of ψ by

$$(\partial_\mu-2i\epsilon\mathbf{b}_\mu\cdot\mathbf{T})\psi, \tag{7}$$

where T^i ($i=1, 2, 3$), as defined above, are the isotopic spin "angular momentum" matrices for the field ψ.

We remark that the nine components of $\mathbf{b}_\mu$, $\mu=1, 2, 3$ are real and the three of $\mathbf{b}_4$ are pure imaginary. The isotopic-gauge covariant field quantities $F_{\mu\nu}$ are expressible in terms of $\mathbf{b}_\mu$:

$$F_{\mu\nu}=2\mathbf{f}_{\mu\nu}\cdot\mathbf{T}, \tag{8}$$

where

$$\mathbf{f}_{\mu\nu}=\frac{\partial\mathbf{b}_\mu}{\partial x_\nu}-\frac{\partial\mathbf{b}_\nu}{\partial x_\mu}-2\epsilon\mathbf{b}_\mu\times\mathbf{b}_\nu. \tag{9}$$

$\mathbf{f}_{\mu\nu}$ transforms like a vector under an isotopic gauge transformation. Obviously the same $\mathbf{f}_{\mu\nu}$ interact with all fields ψ irrespective of the representation S that ψ belongs to.

The corresponding transformation of $\mathbf{b}_\mu$ is cumbersome. One need, however, study only the infinitesimal isotopic gauge transformations,

$$S=1-2i\mathbf{T}\cdot\delta\boldsymbol{\omega}.$$

Then

$$\mathbf{b}_\mu'=\mathbf{b}_\mu+2\mathbf{b}_\mu\times\delta\boldsymbol{\omega}+\frac{1}{\epsilon}\frac{\partial}{\partial x_\mu}\delta\boldsymbol{\omega}. \tag{10}$$

FIELD EQUATIONS

To write down the field equations for the $\mathbf{b}$ field we clearly only want to use isotopic gauge invariant quantities. In analogy with the electromagnetic case we therefore write down the following Lagrangian density:[8]

$$-\tfrac{1}{4}\mathbf{f}_{\mu\nu}\cdot\mathbf{f}_{\mu\nu}.$$

Since the inclusion of a field with isotopic spin $\frac{1}{2}$ is illustrative, and does not complicate matters very much, we shall use the following total Lagrangian density:

$$\mathcal{L}=-\tfrac{1}{4}\mathbf{f}_{\mu\nu}\cdot\mathbf{f}_{\mu\nu}-\bar{\psi}\gamma_\mu(\partial_\mu-i\epsilon\boldsymbol{\tau}\cdot\mathbf{b}_\mu)\psi-m\bar{\psi}\psi. \tag{11}$$

One obtains from this the following equations of motion:

$$\begin{gathered}\partial\mathbf{f}_{\mu\nu}/\partial x_\nu+2\epsilon(\mathbf{b}_\nu\times\mathbf{f}_{\mu\nu})+\mathbf{J}_\mu=0,\\ \gamma_\mu(\partial_\mu-i\epsilon\boldsymbol{\tau}\cdot\mathbf{b}_\mu)\psi+m\psi=0,\end{gathered} \tag{12}$$

‡ *Note added in proof.*—It may appear that B_μ could be introduced as an auxiliary quantity to accomplish invariance, but need not be regarded as a field variable by itself. It is to be emphasized that such a procedure violates the principle of invariance. Every quantity that is not a pure numeral (like 2, or M, or any definite representation of the γ matrices) should be regarded as a dynamical variable, and should be varied in the Lagrangian to yield an equation of motion. Thus the quantities B_μ must be regarded as independent fields.

[8] Repeated indices are summed over, except where explicitly stated otherwise. Latin indices are summed from 1 to 3, Greek ones from 1 to 4.

where

$$\mathbf{J}_\mu = i\epsilon\bar{\psi}\gamma_\mu\boldsymbol{\tau}\psi. \tag{13}$$

The divergence of $\mathbf{J}_\mu$ does not vanish. Instead it can easily be shown from (13) that

$$\partial\mathbf{J}_\mu/\partial x_\mu = -2\epsilon\mathbf{b}_\mu\times\mathbf{J}_\mu. \tag{14}$$

If we define, however,

$$\mathfrak{J}_\mu = \mathbf{J}_\mu + 2\epsilon\mathbf{b}_\nu\times\mathbf{f}_{\mu\nu}, \tag{15}$$

then (12) leads to the equation of continuity,

$$\partial\mathfrak{J}_\mu/\partial x_\mu = 0. \tag{16}$$

$\mathfrak{J}_{1,2,3}$ and $\mathfrak{J}_4$ are respectively the isotopic spin current density and isotopic spin density of the system. The equation of continuity guarantees that the total isotopic spin

$$\mathbf{T} = \int \mathfrak{J}_4 d^3x$$

is independent of time and independent of a Lorentz transformation. It is important to notice that $\mathfrak{J}_\mu$, like $\mathbf{b}_\mu$, does not transform exactly like vectors under isotopic space rotations. But the total isotopic spin,

$$\mathbf{T} = -\int \frac{\partial\mathbf{f}_{4i}}{\partial x_i} d^3x,$$

is the integral of the divergence of $\mathbf{f}_{4i}$, which transforms like a true vector under isotopic spin space rotations. Hence, under a general isotopic gauge transformation, if $S\to S_0$ on an infinitely large sphere, $\mathbf{T}$ would transform like an isotopic spin vector.

Equation (15) shows that the isotopic spin arises both from the spin-$\frac{1}{2}$ field ($\mathbf{J}_\mu$) and from the $\mathbf{b}_\mu$ field itself. Inasmuch as the isotopic spin is the source of the $\mathbf{b}$ field, this fact makes the field equations for the $\mathbf{b}$ field nonlinear, even in the absence of the spin-$\frac{1}{2}$ field. This is different from the case of the electromagnetic field, which is itself chargeless, and consequently satisfies linear equations in the absence of a charged field.

The Hamiltonian derived from (11) is easily demonstrated to be positive definite in the absence of the field of isotopic spin $\frac{1}{2}$. The demonstration is completely identical with the similar one in electrodynamics.

We must complete the set of equations of motion (12) and (13) by the supplementary condition,

$$\partial\mathbf{b}_\mu/\partial x_\mu = 0, \tag{17}$$

which serves to eliminate the scalar part of the field in $\mathbf{b}_\mu$. This clearly imposes a condition on the possible isotopic gauge transformations. That is, the infinitesimal isotopic gauge transformation $S=1-i\boldsymbol{\tau}\cdot\delta\boldsymbol{\omega}$ must satisfy the following condition:

$$2\mathbf{b}_\mu\times\frac{\partial}{\partial x_\mu}\delta\boldsymbol{\omega}+\frac{1}{\epsilon}\frac{\partial^2}{\partial x_\mu^2}\delta\boldsymbol{\omega}=0. \tag{18}$$

This is the analog of the equation $\partial^2\alpha/\partial x_\mu^2=0$ that must be satisfied by the gauge transformation $A_\mu' = A_\mu + e^{-1}(\partial\alpha/\partial x_\mu)$ of the electromagnetic field.

QUANTIZATION

To quantize, it is not convenient to use the isotopic gauge invariant Lagrangian density (11). This is quite similar to the corresponding situation in electrodynamics and we adopt the customary procedure of using a Lagrangian density which is not obviously gauge invariant:

$$\mathfrak{L} = -\frac{1}{2}\frac{\partial\mathbf{b}_\mu}{\partial x_\nu}\cdot\frac{\partial\mathbf{b}_\mu}{\partial x_\nu}+2\epsilon(\mathbf{b}_\mu\times\mathbf{b}_\nu)\cdot\frac{\partial\mathbf{b}_\mu}{\partial x_\nu} - \epsilon^2(\mathbf{b}_\mu\times\mathbf{b}_\nu)^2+\mathbf{J}_\mu\cdot\mathbf{b}_\mu-\bar{\psi}(\gamma_\mu\partial_\mu+m)\psi. \tag{19}$$

The equations of motion that result from this Lagrangian density can be easily shown to imply that

$$\frac{\partial^2}{\partial x_\nu^2}\mathbf{a}+2\epsilon\mathbf{b}_\nu\times\frac{\partial}{\partial x_\nu}\mathbf{a}=0,$$

where

$$\mathbf{a}=\partial\mathbf{b}_\mu/\partial x_\mu.$$

Thus if, consistent with (17), we put on one space-like surface $\mathbf{a}=0$ together with $\partial\mathbf{a}/\partial t=0$, it follows that $\mathbf{a}=0$ at all times. Using this supplementary condition one can easily prove that the field equations resulting from the Lagrangian densities (19) and (11) are identical.

One can follow the canonical method of quantization with the Lagrangian density (19). Defining

$$\boldsymbol{\Pi}_\mu = -\partial\mathbf{b}_\mu/\partial x_4+2\epsilon(\mathbf{b}_\mu\times\mathbf{b}_4),$$

one obtains the equal-time commutation rule

$$[b_\mu^i(x), \Pi_\nu^j(x')]_{t=t'} = -\delta_{ij}\delta_{\mu\nu}\delta^3(x-x'), \tag{20}$$

where b_μ^i, $i=1, 2, 3$, are the three components of $\mathbf{b}_\mu$. The relativistic invariance of these commutation rules follows from the general proof for canonical methods of quantization given by Heisenberg and Pauli.[9]

The Hamiltonian derived from (19) is identical with the one from (11), in virtue of the supplementary condition. Its density is

$$\begin{aligned} H&=H_0+H_{\text{int}},\\ H_0&=-\tfrac{1}{2}\boldsymbol{\Pi}_\mu\cdot\boldsymbol{\Pi}_\mu+\frac{1}{2}\frac{\partial\mathbf{b}_\mu}{\partial x_j}\cdot\frac{\partial\mathbf{b}_\mu}{\partial x_j}+\bar{\psi}(\gamma_j\partial_j+m)\psi,\\ H_{\text{int}}&=2\epsilon(\mathbf{b}_i\times\mathbf{b}_4)\cdot\boldsymbol{\Pi}_i-2\epsilon(\mathbf{b}_\mu\times\mathbf{b}_j)\cdot(\partial\mathbf{b}_\mu/\partial x_j)\\ &\quad+\epsilon^2(\mathbf{b}_i\times\mathbf{b}_j)^2-\mathbf{J}_\mu\cdot\mathbf{b}_\mu. \end{aligned} \tag{21}$$

The quantized form of the supplementary condition is the same as in quantum electrodynamics.

[9] W. Heisenberg and W. Pauli, Z. Physik 56, 1 (1929).

PROPERTIES OF THE b QUANTA

The quanta of the **b** field clearly have spin unity and isotopic spin unity. We know their electric charge too because all the interactions that we proposed must satisfy the law of conservation of electric charge, which is exact. The two states of the nucleon, namely proton and neutron, differ by charge unity. Since they can transform into each other through the emission or absorption of a **b** quantum, the latter must have three charge states with charges $\pm e$ and 0. Any measurement of electric charges of course involves the electromagnetic field, which necessarily introduces a preferential direction in isotopic space at all space-time points. Choosing the isotopic gauge such that this preferential direction is along the z axis in isotopic space, one sees that for the nucleons

$$Q=\text{electric charge}=e(\tfrac{1}{2}+\epsilon^{-1}T^z),$$

and for the **b** quanta

$$Q=(e/\epsilon)T^z.$$

The interaction (7) then fixes the electric charge up to an additive constant for all fields with any isotopic spin:

$$Q=e(\epsilon^{-1}T^z+R). \qquad (22)$$

The constants R for two charge conjugate fields must be equal but have opposite signs.[10]

FIG. 1. Elementary vertices for **b** fields and nucleon fields. Dotted lines refer to **b** field, solid lines with arrow refer to nucleon field.

We next come to the question of the mass of the **b** quantum, to which we do not have a satisfactory answer. One may argue that without a nucleon field the Lagrangian would contain no quantity of the dimension of a mass, and that therefore the mass of the **b** quantum in such a case is zero. This argument is however subject to the criticism that, like all field theories, the **b** field is beset with divergences, and dimensional arguments are not satisfactory.

One may of course try to apply to the **b** field the methods for handling infinities developed for quantum electrodynamics. Dyson's approach[11] is best suited for the present case. One first transforms into the interaction representation in which the state vector Ψ satisfies

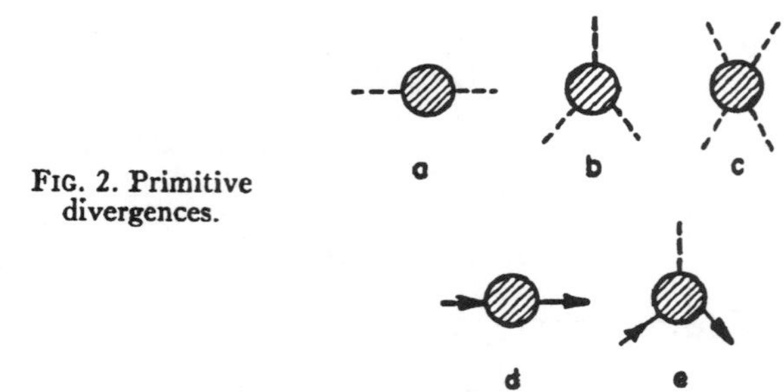

FIG. 2. Primitive divergences.

$$i\partial\Psi/\partial t=H_{\text{int}}\Psi,$$

where H_{int} was defined in Eq. (21). The matrix elements of the scattering matrix are then formulated in terms of contributions from Feynman diagrams. These diagrams have three elementary types of vertices illustrated in Fig. 1, instead of only one type as in quantum electrodynamics. The "primitive divergences" are still finite in number and are listed in Fig. 2. Of these, the one labeled a is the one that effects the propagation function of the **b** quantum, and whose singularity determines the mass of the **b** quantum. In electrodynamics, by the requirement of electric charge conservation,[12] it is argued that the mass of the photon vanishes. Corresponding arguments in the **b** field case do not exist[13] even though the conservation of isotopic spin still holds. We have therefore not been able to conclude anything about the mass of the **b** quantum.

A conclusion about the mass of the **b** quantum is of course very important in deciding whether the proposal of the existence of the **b** field is consistent with experimental information. For example, it is inconsistent with present experiments to have their mass less than that of the pions, because among other reasons they would then be created abundantly at high energies and the charged ones should live long enough to be seen. If they have a mass greater than that of the pions, on the other hand, they would have a short lifetime (say, less than 10^{-20} sec) for decay into pions and photons and would so far have escaped detection.

[10] See M. Gell-Mann, Phys. Rev. **92**, 833 (1953).

[11] F. J. Dyson, Phys. Rev. **75**, 486, 1736 (1949).

[12] J. Schwinger, Phys. Rev. **76**, 790 (1949).

[13] In electrodynamics one can formally prove that $G_{\mu\nu}k_\nu=0$, where $G_{\mu\nu}$ is defined by Schwinger's Eq. (A12). ($G_{\mu\nu}A_\nu$ is the current generated through virtual processes by the arbitrary external field A_ν.) No corresponding proof has been found for the present case. This is due to the fact that in electrodynamics the conservation of charge is a consequence of the equation of motion of the electron field alone, quite independently of the electromagnetic field itself. In the present case the **b** field carries an isotopic spin and destroys such general conservation laws.

Behavior of Neutral Particles under Charge Conjugation

M. GELL-MANN,* *Department of Physics, Columbia University, New York, New York*

AND

A. PAIS, *Institute for Advanced Study, Princeton, New Jersey*

(Received November 1, 1954)

Some properties are discussed of the θ^0, a heavy boson that is known to decay by the process $\theta^0 \rightarrow \pi^+ + \pi^-$. According to certain schemes proposed for the interpretation of hyperons and K particles, the θ^0 possesses an antiparticle $\bar{\theta}^0$ distinct from itself. Some theoretical implications of this situation are discussed with special reference to charge conjugation invariance. The application of such invariance in familiar instances is surveyed in Sec. I. It is then shown in Sec. II that, within the framework of the tentative schemes under consideration, the θ^0 must be considered as a "particle mixture" exhibiting two distinct lifetimes, that each lifetime is associated with a different set of decay modes, and that no more than half of all θ^0's undergo the familiar decay into two pions. Some experimental consequences of this picture are mentioned.

I

IT is generally accepted that the microscopic laws of physics are invariant to the operation of charge conjugation (CC); we shall take the rigorous validity of this postulate for granted. Under CC, every particle is carried into what we shall call its "antiparticle". The principle of invariance under CC implies, among other things, that a particle and its antiparticle must have exactly the same mass and intrinsic spin and must have equal and opposite electric and magnetic moments.

A charged particle is thus carried into one of opposite charge. For example, the electron and positron are each other's antiparticles; the π^+ and π^- and the μ^+ and μ^- mesons are supposed to be pairs of antiparticles; and the proton must possess an antiparticle, the "antiproton".

Neutral particles fall into two classes, according to their behavior under CC:

(a) Particles that transform into themselves, and which are thus their own antiparticles. For instance the photon and the π^0 meson are bosons that behave in this fashion. It is conceivable that fermions, too, may belong to this class. An example is provided by the Majorana theory of the neutrino.

In a field theory, particles of class (a) are represented by "real" fields, i.e., Hermitian field operators. There is an important distinction to be made within this class, according to whether the field takes on a plus or a minus sign under CC. The operation of CC is performed by a unitary operator $\mathcal{C}$. The photon field operator $A_\mu(x)$ satisfies the relation

$$\mathcal{C}A_\mu(x)\mathcal{C}^{-1} = -A_\mu(x), \tag{1}$$

while for the π^0 field operator $\phi(x)$ we have

$$\mathcal{C}\phi(x)\mathcal{C}^{-1} = \phi(x). \tag{2}$$

Equation (1) expresses the obvious fact that the electromagnetic field changes sign when positive and negative charges are interchanged; that the π^0 field must not change sign can be inferred from the observed two-photon decay of the π^0.

We are effectively dealing here with the "charge conjugation quantum number" C, which is the eigenvalue of the operator $\mathcal{C}$, and which is rigorously conserved in the absence of external fields. If only an odd (even) number of photons is present, we have $C = -1(+1)$; if only π^0's are present, $C = +1$; etc. As a trivial example of the conservation of C, we may mention that the decay of the π^0 into an odd number of photons is forbidden.[1]

We may recall that a state of a neutral system composed of charged particles may be one with a definite value of C. For example, the ${}^1S_0$ state of positronium has $C = +1$; a state of a π^+ and a π^- meson with relative orbital angular momentum l has $C = (-1)^l$; etc.

For fermions, as for bosons, a distinction may be made between "odd" and "even" behavior of neutral fields of class (a) under CC. However, the distinction is then necessarily a relative rather than an absolute one.[2] In other words, it makes no sense to say that a single such fermion field is "odd" or "even", but it does make sense to say that two such fermion fields have the same behavior under CC or that they have opposite behavior.

(b) Neutral particles that behave like charged ones in that: (1) they have antiparticles distinct from themselves; (2) there exists a rigorous conservation law that prohibits virtual transitions between particle and antiparticle states.

A well-known member of this class is the neutron N, which can obviously be distinguished from the antineutron $\bar{N}$ by the sign of its magnetic moment. The law that forbids the virtual processes $N \rightleftarrows \bar{N}$ is the law

* On leave from Department of Physics and Institute for Nuclear Studies, University of Chicago, Chicago, Illinois.

[1] For other consequences of invariance under charge conjugation see A. Pais and R. Jost, Phys. Rev. 87, 871 (1952); L. Wolfenstein and D. G. Ravenhall, Phys. Rev. 88, 279 (1952); L. Michel, Nuovo cimento 10, 319 (1953).

[2] This is due to the fact that fermion fields can interact only bilinearly. For example, one easily sees that the interactions responsible for $P \rightarrow N + e^+ + \nu$ would not lead to physically distinguishable results if ν were either an even or an odd Majorana neutrino.

of conservation of baryons,[3] which is, so far as we know, exact, and which states that n, the number of baryons minus the number of antibaryons, must remain unchanged. Clearly all neutral hyperons likewise belong to this class. Although we know of no "elementary" bosons in the same category, we have no *a priori* reason for excluding their existence. [Note that the H atom is an example of a "non-elementary" boson of class (b).]

Particles in this class are represented by "complex" fields, and the operation of charge conjugation transforms the field operators into their Hermitian conjugates.

It is the purpose of this note to discuss the possible existence of neutral particles that seem, at first sight, to belong neither to class (a) nor to class (b).

II

Recently, attempts have been made to interpret hyperon and K-particle phenomena by distinguishing sharply between strong interactions, to which essentially all production of these particles is attributed, and weak interactions, which are supposed to induce their decay. It is necessary to assume that the strong interactions give rise to "associated production "exclusively.[4]

Certain detailed schemes[5] which meet this requirement lead to further specific properties of particles and interactions. In particular, a suggestion has been made about the θ^0 particle, a heavy boson that is known to decay according to the scheme:

$$\theta^0\rightarrow\pi^+ + \pi^- + (\sim 215\text{ Mev}). \tag{3}$$

It has been proposed that the θ^0 possesses an antiparticle $\bar{\theta}^0$ distinct from itself, and that in the absence of the weak decay interactions, there is a conservation law that prohibits the virtual transitions $\theta^0\rightleftarrows\bar{\theta}^0$. [In our present language, we would say that the θ^0 belongs to class (b) if the weak interactions are turned off.] This conservation law also leads to stability of the θ^0 and $\bar{\theta}^0$; moreover, while it permits the reaction $\pi^- + P\rightarrow\Lambda^0+\theta^0$ it forbids the analogous process $\pi^- + P\rightarrow\Lambda^0+\bar{\theta}^0$. In the schemes under consideration this is the same law that forbids the reaction: 2 neutrons$\rightarrow 2\Lambda^0$.

The weak interactions that must be invoked to account for the observed decay (3) evidently cause the conservation law to break down (a fact that is, of course, of little importance for production). This breakdown makes the forbiddenness of the processes $\theta^0\rightleftarrows\bar{\theta}^0$ no longer absolute, as can be seen from the following argument: In the decay (3) the pions are left in a state with a definite relative angular momentum and therefore with a definite value of the charge-conjugation quantum number C. The charge-conjugate process,

$$\bar{\theta}^0\rightarrow\pi^+ + \pi^-, \tag{4}$$

must also occur and must leave the pions in the same state; moreover the reverse of (4) must also be possible, at least as a virtual process. Therefore the virtual transition $\theta^0\rightleftarrows\pi^+ + \pi^-\rightleftarrows\bar{\theta}^0$ is induced by the weak interactions, and we are no longer dealing exactly with case (b).

In order to treat this novel situation, we shall find it convenient to introduce a change of representation. Since the θ^0 and $\bar{\theta}^0$ are distinct, they are associated, in a field theory, with a "complex" field ψ (a non-Hermitian field operator), just as in case (b). Under charge conjugation ψ must transform according to the law:

$$\mathfrak{C}\psi\mathfrak{C}^{-1}=\psi^+,\quad \mathfrak{C}\psi^+\mathfrak{C}^{-1}=\psi, \tag{5}$$

where ψ^+ is the Hermitian conjugate of ψ. Let us now define

$$\psi_1\equiv(\psi+\psi^+)/\sqrt{2} \tag{6}$$

and

$$\psi_2\equiv(\psi-\psi^+)/\sqrt{2}i, \tag{7}$$

so that ψ_1 and ψ_2 are Hermitian field operators satisfying

$$\mathfrak{C}\psi_1\mathfrak{C}^{-1}=\psi_1, \tag{8}$$

and

$$\mathfrak{C}\psi_2\mathfrak{C}^{-1}=-\psi_2. \tag{9}$$

The fields ψ_1 and ψ_2 evidently correspond to class (a); in fact ψ_1 is "even" like the π^0 field and ψ_2 is "odd" like the photon field. Corresponding to these real fields there are quanta, which we shall call θ_1^0 and θ_2^0 quanta. The relationship that these have to the quanta of the complex ψ field, which we have called θ^0 and $\bar{\theta}^0$, may be seen from an example: Let Ψ_1 be the wave-functional representing a single θ_1 quantum in a given state, while Ψ_0 and Ψ_0' describe a θ^0 and a $\bar{\theta}^0$, respectively, in the same state. Then we have

$$\Psi_1=(\Psi_0+\Psi_0')/\sqrt{2}.$$

Thus the creation of a θ_1 (or, for that matter, of a θ_2) corresponds physically to the creation, with equal probability and with prescribed relative phase, of either a θ^0 or a $\bar{\theta}^0$. Conversely, the creation of a θ^0 (or of a $\bar{\theta}^0$) corresponds to the creation, with equal probability and prescribed relative phase, of either a θ_1^0 or a θ_2^0.

The transformation (6), (7) to two real fields could equally well have been applied to a complex field of class (b), such as that associated with the neutron. However, this would not be particularly enlightening. It would lead us, for instance, to describe phenomena involving neutrons and antineutrons in terms of "N_1 and N_2 quanta". Now a state with an N_1 (or N_2) quantum is a mixture of states with different values of the quantum number n, the number of baryons minus the number of antibaryons. But the law of conservation of baryons requires this quantity to be a constant of the motion, and so a mixed state can never arise from a pure one. Since in our experience we deal exclusively

[3] Nucleons and hyperons are collectively referred to as baryons.
[4] A. Pais, Phys. Rev. 86, 663 (1952).
[5] M. Gell-Mann, Phys. Rev. 92, 833 (1953); A. Pais, Proc. Nat, Acad. Sci. U. S. 40, 484, 835 (1954); M. Gell-Mann and A. Pais. *Proceedings of the International Conference Glasgow* (Pergamon Press, London, to be published).

with states that are pure with respect to n, the introduction of N_1 and N_2 quanta can only be a mathematical device that distracts our attention from the truly physical particles N and $\bar{N}$.

On the other hand, it can obviously not be argued in a similar way that the θ_1^0 and θ_2^0 quanta are completely unphysical, for the corresponding conservation law in that case is not a rigorous one. Always assuming the correctness of our model of the θ^0, we still have the θ^0 and $\bar{\theta}^0$ as the primary objects in production phenomena. But we shall now show that the decay process is best described in terms of θ_1^0 and θ_2^0.

The weak interactions, in fact, must lead to very different patterns of decay for the θ_1^0 and θ_2^0 into pions and (perhaps) γ rays; any state of pions and/or γ rays that is a possible decay mode for the θ_1^0 is not a possible one for the θ_2^0, and *vice versa*. This is because, according to the postulate of rigorous CC invariance, the quantum number C is conserved in the decay; the θ_1^0 must go into a state that is even under charge conjugation, while the θ_2^0 must go into one that is odd. Since the decay modes are different and even mutually exclusive for the θ_1^0 and θ_2^0, their rates of decay must be quite unrelated. There are thus two independent lifetimes, one for the θ_1^0, and one for the θ_2^0.

An important illustration of the difference in decay modes of the θ_1^0 and θ_2^0 is provided by the two-pion disintegration. We know that reaction (3) occurs; therefore at least one of the two quanta θ_1^0 and θ_2^0, say θ_1^0, must be capable of decay into two charged pions. The final state of the two pions in the θ_1^0 decay is then even under charge conjugation like the θ_1^0 state itself. These two pions are thus in a state of even relative angular momentum and therefore of even parity. So the θ_1^0 must have even spin and even parity. Now we assume that the θ^0 has a definite intrinsic parity, and therefore the parity and spin of the θ_2^0 must be the same as those of the θ_1^0, both even. If the θ_2^0 were to decay into two pions, these would again be in a state of even relative angular momentum and thus even with respect to charge conjugation. However, the θ_2^0 is itself odd under charge conjugation; its decay into two pions is therefore forbidden.

Alternatively, if the θ_2^0 is the one that actually goes into two pions, then the spin and parity of θ_1^0 and the θ_2^0 are both odd, and so the θ_1^0 cannot decay into two pions.

Of the θ_1^0 and the θ_2^0, that one for which the two-pion decay is forbidden may go instead into $\pi^+ + \pi^- + \gamma$ or possibly into three pions (unless the spin and parity of the θ^0 are 0^+), etc.

While we have seen that the θ_1^0 and θ_2^0 may each be assigned a lifetime, this is evidently not true of the θ^0 or $\bar{\theta}^0$. Since we should properly reserve the word "particle" for an object with a unique lifetime, it is the θ_1^0 and θ_2^0 quanta that are the true "particles". The θ^0 and the $\bar{\theta}^0$ must, strictly speaking, be considered as "particle mixtures."

It should be remarked that the θ_1^0 and the θ_2^0 differ not only in lifetime but also in mass, though the mass difference is surely tiny. The weak interactions responsible for decay cause the θ_1^0 and the θ_2^0 to have their respective small level widths and correspondingly must produce small level shifts which are different for the two particles.

To sum up, our picture of the θ^0 implies that it is a particle mixture exhibiting two distinct lifetimes, that each lifetime is associated with a different set of decay modes, and that *not more than half of all* θ^0*'s* can undergo the familiar decay into two pions.[6]

We know experimentally that the lifetime τ for the decay mode (3) (and hence for all decay modes that may compete with this one) is about 1.5×10^{-10} sec. The present qualitative considerations, even if at all correct in their underlying assumptions, do not enable us to predict the value of the "second lifetime" τ' of the θ^0. Nevertheless, the examples given above of decays responsible for the second lifetime lead one to suspect that[7] $\tau' \gg \tau$. As an illustration of the experimental implications of this situation consider the study of the reaction $\pi^- + P \rightarrow \Lambda^0 + \theta^0$ in a cloud chamber. If the reaction occurs and subsequently $\Lambda^0 \rightarrow P + \pi^-$, $\theta^0 \rightarrow \pi^+ + \pi^-$, there should be a reasonable chance to observe this whole course of events in the chamber, as the lifetime for the Λ^0 decay ($\sim 3.5 \times 10^{-10}$ sec) is comparable to τ. However, if it is true that $\tau' \gg \tau$, it would be very difficult to detect the decay with the second lifetime in the cloud chamber with its characteristic bias for a limited region of lifetime values.[8] Clearly this also means an additional complication in the determination from cloud chamber data as to whether or not production always occurs in an associated fashion. In some such cases the analysis of the reaction $\pi^- + P \rightarrow \Lambda^0 + ?$ may still be pushed further, however, if one assumes that besides the Λ^0 only one other neutral object is formed.[9]

At any rate, the point to be emphasized is this: a neutral boson may exist which has the characteristic θ^0 mass but a lifetime $\neq \tau$ and which may find its natural place in the present picture as the second component of the θ^0 mixture.

One of us, (M.'G.-M.), wishes to thank Professor E. Fermi for a stimulating discussion.

[6] Note that if the spin and parity of the θ^0 are even, then the θ_1^0 may decay into $2\pi^0$'s as well as into $\pi^+ + \pi^-$.

[7] The process $\theta^0 \rightarrow \pi^+ + \pi^- + \gamma$ may occur as a radiative correction to the allowed decay into $\pi^+ + \pi^-$ connected with the lifetime τ; see S. B. Treiman, Phys. Rev. 95, 1360 (1954). The process may also occur as one of the principal decay modes associated with the second lifetime τ'. The latter case may be distinguished from the former not only by the distinct lifetime but also by a different energy spectrum which probably favors higher γ-ray energies; such a spectrum is to be expected in a case where the emission of the γ ray is not just part of the "infrared catastrophe", but is an integral part of the decay process.

[8] See, e.g., Leighton, Wanlass, and Anderson, Phys. Rev. 89, 148 (1953), Sec. III.

[9] See Fowler, Shutt, Thorndike, and Whittemore, Phys. Rev. 91, 1287 (1953).

Question of Parity Conservation in Weak Interactions*

T. D. LEE, *Columbia University, New York, New York*

AND

C. N. YANG,† *Brookhaven National Laboratory, Upton, New York*

(Received June 22, 1956)

The question of parity conservation in β decays and in hyperon and meson decays is examined. Possible experiments are suggested which might test parity conservation in these interactions.

RECENT experimental data indicate closely identical masses[1] and lifetimes[2] of the $\theta^+(\equiv K_{\pi 2}^+)$ and the $\tau^+(\equiv K_{\pi 3}^+)$ mesons. On the other hand, analyses[3] of the decay products of τ^+ strongly suggest on the grounds of angular momentum and parity conservation that the τ^+ and θ^+ are not the same particle. This poses a rather puzzling situation that has been extensively discussed.[4]

One way out of the difficulty is to assume that parity is not strictly conserved, so that θ^+ and τ^+ are two different decay modes of the same particle, which necessarily has a single mass value and a single lifetime. We wish to analyze this possibility in the present paper against the background of the existing experimental evidence of parity conservation. It will become clear that existing experiments do indicate parity conservation in strong and electromagnetic interactions to a high degree of accuracy, but that for the weak interactions (i.e., decay interactions for the mesons and hyperons, and various Fermi interactions) parity conservation is so far only an extrapolated hypothesis unsupported by experimental evidence. (One might even say that the present $\theta-\tau$ puzzle may be taken as an indication that parity conservation is violated in weak interactions. This argument is, however, not to be taken seriously because of the paucity of our present knowledge concerning the nature of the strange particles. It supplies rather an incentive for an examination of the question of parity conservation.) To decide unequivocally whether parity is conserved in weak interactions, one must perform an experiment to determine whether weak interactions differentiate the right from the left. Some such possible experiments will be discussed.

* Work supported in part by the U. S. Atomic Energy Commission.

† Permanent address: Institute for Advanced Study, Princeton, New Jersey.

[1] Whitehead, Stork, Perkins, Peterson, and Birge, Bull. Am. Phys. Soc. Ser. II, 1, 184 (1956); Barkas, Heckman, and Smith, Bull. Am. Phys. Soc. Ser. II, 1, 184 (1956).

[2] Harris, Orear, and Taylor, Phys. Rev. **100**, 932 (1955); V. Fitch and K. Motley, Phys. Rev. **101**, 496 (1956); Alvarez, Crawford, Good, and Stevenson, Phys. Rev. **101**, 503 (1956).

[3] R. Dalitz, Phil. Mag. **44**, 1068 (1953); E. Fabri, Nuovo cimento **11**, 479 (1954). See Orear, Harris, and Taylor [Phys. Rev. **102**, 1676 (1956)] for recent experimental results.

[4] See, e.g., *Report of the Sixth Annual Rochester Conference on High Energy Physics* (Interscience Publishers, Inc., New York, to be published).

PRESENT EXPERIMENTAL LIMIT ON PARITY NONCONSERVATION

If parity is not strictly conserved, all atomic and nuclear states become mixtures consisting mainly of the state they are usually assigned, together with small percentages of states possessing the opposite parity. The fractional weight of the latter will be called $\mathfrak{F}^2$. It is a quantity that characterizes the degree of violation of parity conservation.

The existence of parity selection rules which work well in atomic and nuclear physics is a clear indication that the degree of mixing, $\mathfrak{F}^2$, cannot be large. From such considerations one can impose the limit $\mathfrak{F}^2 \lesssim (r/\lambda)^2$, which for atomic spectroscopy is, in most cases, $\sim 10^{-6}$. In general a less accurate limit obtains for nuclear spectroscopy.

Parity nonconservation implies the existence of interactions which mix parities. The strength of such interactions compared to the usual interactions will in general be characterized by $\mathfrak{F}$, so that the mixing will be of the order $\mathfrak{F}^2$. The presence of such interactions would affect angular distributions in nuclear reactions. As we shall see, however, the accuracy of these experiments is not good. The limit on $\mathfrak{F}^2$ obtained is not better than $\mathfrak{F}^2 < 10^{-4}$.

To give an illustration, let us examine the polarization experiments, since they are closely analogous to some experiments to be discussed later. A proton beam polarized in a direction z perpendicular to its momentum was scattered by nuclei. The scattered intensities were compared[5] in two directions A and B related to each other by a reflection in the x–y plane, and were found to be identical to within $\sim 1\%$. If the scattering originates from an ordinary parity-conserving interaction plus a parity-nonconserving interaction (e.g., $\boldsymbol{\sigma}\cdot\mathbf{r}$), then the scattering amplitudes in the directions A and B are in the proportion $(1+\mathfrak{F})/(1-\mathfrak{F})$, where $\mathfrak{F}$ represents the ratio of the strengths of the two kinds of interactions in the scattering. The experimental result therefore requires $\mathfrak{F} < 10^{-2}$, or $\mathfrak{F}^2 < 10^{-4}$.

The violation of parity conservation would lead to an electric dipole moment for all systems. The magnitude of the moment is

$$\text{moment} \sim e\mathfrak{F} \times (\text{dimension of system}). \qquad (1)$$

[5] See, e.g., Chamberlain, Segrè, Tripp, and Ypsilantis, Phys. Rev. **93**, 1430 (1954).

The presence of such electric dipole moments would have interesting consequences. For example, if the proton has an electric dipole moment $\cong e\times(10^{-16}$ cm), the perturbation caused by the presence of the neighboring $2p$ state of the hydrogen atom would shift the energy of the $2s$ state by about 1 Mc/sec. This would be inconsistent with the present theoretical interpretations of the Lamb shift. Another example is found in the electron-neutron interaction. An electric dipole moment for the neutron $\cong e\times(10^{-18}$ cm) is the upper limit allowable by the present experiments.

By far the most accurate measurement of the electric dipole moment was made by Purcell, Ramsey, and Smith. They gave[6] an upper limit for the electric dipole moment of the neutron of $e\times(5\times10^{-20}$ cm). This value sets the upper limit for $\mathfrak{F}^2$ as $\mathfrak{F}^2<3\times10^{-13}$, which is also the most accurate verification of the conservation of parity in strong and electromagnetic interactions. We shall see, however, that even this high degree of accuracy is not sufficient to supply an experimental proof of parity conservation in the weak interactions. For such a proof an accuracy of $\mathfrak{F}^2<10^{-24}$ is necessary.

QUESTION OF PARITY CONSERVATION IN β DECAY

At first sight it might appear that the numerous experiments related to β decay would provide a verification that the weak β interaction does conserve parity. We have examined this question in detail and found this to be not so. (See Appendix.) We start by writing down the five usual types of couplings. In addition to these we introduce the five types of couplings that conserve angular momentum but do not conserve parity. It is then apparent that the classification of β decays into allowed transitions, first forbidden, etc., proceeds exactly as usual. (The mixing of parity of the *nuclear states* would not measurably affect these selection rules. This phenomenon belongs to the discussions of the last section.) The following phenomena are then examined: allowed spectra, unique forbidden spectra, forbidden spectra with allowed shape, β-neutrino correlation, and $\beta-\gamma$ correlation. It is found that these experiments have no bearing on the question of parity conservation of the β-decay interactions. This comes about because in all of these phenomena no interference terms exist between the parity-conserving and parity-nonconserving interactions. In other words, the calculations always result in terms proportional to $|C|^2$ plus terms proportional to $|C'|^2$. Here C and C' are, respectively, the coupling constants for the usual parity-conserving interactions (a sum of five terms) and the parity-nonconserving interactions (also a sum of five terms.) Furthermore, it is well known[7] that without measuring the spin of the neutrino we cannot distinguish the couplings C from the couplings C' (provided the mass of the neutrino is zero). The experimental results concerning the above-named phenomena, which constitute the bulk of our present knowledge about β decay, therefore cannot decide the degree of mixing of the C' type interactions with the usual type.

The reason for the absence of interference terms CC' is actually quite obvious. Such terms can only occur as a pseudoscalar formed out of the experimentally measured quantities. For example, if three momenta $\mathbf{p}_1$, $\mathbf{p}_2$, $\mathbf{p}_3$ are measured, the term $CC'\mathbf{p}_1\cdot(\mathbf{p}_2\times\mathbf{p}_3)$ may occur. Or if a momentum $\mathbf{p}$ and a spin $\boldsymbol{\sigma}$ are measured, the term $CC'\mathbf{p}\cdot\boldsymbol{\sigma}$ may occur. In all the β-decay phenomena mentioned above, no such pseudoscalars can be formed out of the measured quantities.

POSSIBLE EXPERIMENTAL TESTS OF PARITY CONSERVATION IN β DECAYS

The above discussion also suggests the kind of experiments that could detect the possible interference between C and C' and consequently could establish whether parity conservation is violated in β decay. A relatively simple possibility is to measure the angular distribution of the electrons coming from β decays of oriented nuclei. If θ is the angle between the orientation of the parent nucleus and the momentum of the electron, an asymmetry of distribution between θ and $180°-\theta$ constitutes an unequivocal proof that parity is not conserved in β decay.

To be more specific, let us consider the allowed β transition of any oriented nucleus, say Co^{60}. The angular distribution of the β radiation is of the form (see Appendix):

$$I(\theta)d\theta=(\text{constant})(1+\alpha\cos\theta)\sin\theta d\theta, \qquad (2)$$

where α is proportional to the interference term CC'. If $\alpha\neq0$, one would then have a positive proof of parity nonconservation in β decay. The quantity α can be obtained by measuring the fractional asymmetry between $\theta<90°$ and $\theta>90°$; i.e.,

$$\alpha=2\left[\int_0^{\pi/2}I(\theta)d\theta-\int_{\pi/2}^{\pi}I(\theta)d\theta\right]\bigg/\int_0^{\pi}I(\theta)d\theta.$$

It is noteworthy that in this case the presence of the magnetic field used for orienting the nuclei would automatically cause a spatial separation between the electrons emitted with $\theta<90°$ and those with $\theta>90°$. Thus, this experiment may prove to be quite feasible.

It appears at first sight that in the study of γ-radiation distribution from β-decay products of oriented nuclei one can form a pseudoscalar from the spin of the oriented nucleus and the γ-ray momentum $\mathbf{p}_\gamma$. Thus it may seem to offer another possible experimental test of parity conservation. Unfortunately, the nuclear levels have definite parities, and electromagnetic inter-

[6] E. M. Purcell and N. F. Ramsey, Phys. Rev. 78, 807 (1950); Smith *et al.* as quoted in N. F. Ramsey, *Molecular Beams* (Oxford University Press, London, 1956).

[7] C. N. Yang and J. Tiomno, Phys. Rev. 79, 495 (1950).

actions conserve parity. (Any small mixing of parities characterized by $\mathfrak{F}^2<3\times10^{-15}$ would not affect the arguments here.) Consequently the γ rays carry away definite parities. Thus the observed probability function must be an even function of $\mathbf{p}_\gamma$. This property eliminates the possibility of forming a pseudoscalar quantity. It is therefore not possible to use such experiments as a test of parity conservation.

In β-γ-γ' triple correlation experiments one can, by some rather similar but more complicated reasoning, prove that a measurement of the three momenta cannot supply any information on the question of parity conservation in β decay.

In β-γ correlation experiments the nature of the polarization of the γ ray could provide a test. To be more specific, let us consider the polarization state of γ rays emitted parallel to the β ray. If parity conservation holds for β decay, the γ ray will be unpolarized. On the other hand, if parity conservation is violated in β decay, the γ ray will in general be polarized. However, this polarization must be circular in nature and therefore may not lend itself to easy experimental detection. (The usual ways of measuring polarization through Compton effect, photoelectric effect, and photodisintegration of the deuteron are all incapable of detecting circular polarization. This is because circular polarization is specified by an axial vector parallel to the direction of propagation. From the observed momenta in these detection techniques such an axial vector cannot be formed.) For other directions of γ-ray propagation, elliptical polarization will result if parity is not conserved. This effect will thus be more difficult to detect.

QUESTION OF PARITY CONSERVATION IN MESON AND HYPERON DECAYS

If the weak interactions, such as the β-decay interactions or the decay interactions of mesons and hyperons, do not conserve parity, parity mixing will occur in all interactions by means of second-order processes. To examine this effect let us consider, for example, the decay of the Λ^0:

$$\Lambda^0\rightarrow p+\pi^-.$$

The assumption that parity is not conserved in this decay implies that the Λ^0 exists virtually in states of opposite parities. It could therefore possess an electric dipole moment of a magnitude

$$\text{moment}\sim e\mathcal{G}^2\times(\text{dimension of }\Lambda^0), \qquad (3)$$

where $\mathcal{G}$ is the coupling strength of the decay interaction of the Λ^0. ($\mathcal{G}^2\lesssim10^{-12}$.) The electric dipole moment of the Λ^0 is therefore $\lesssim e\times(10^{-25}\text{ cm})$.

Clearly the proton would have an electric dipole moment of the same order of magnitude. The existence of such a small electric dipole moment is, as we have seen, completely consistent with the present experimental information. Another way of putting this is to observe that by comparing Eq. (3) with Eq. (1), one has

$$\mathfrak{F}\sim\mathcal{G}^2.$$

Since all the weak interaction including β interactions are characterized by coupling strengths $\mathcal{G}^2<10^{-12}$, a violation of parity in weak interactions would introduce a parity mixing characterized by an $\mathfrak{F}^2<10^{-24}$. This is outside the present limit of experimental knowledge, as we have discussed before.

If the weak interactions violate parity conservation, parity would be defined and measured in strong and electromagnetic interactions only, just as strangeness is. Furthermore it is important to notice that with the conservation of strangeness, as with every conservation law, there is an element of arbitrariness introduced into the parity of all systems. The parity of all strange particles would be defined only up to a factor of $(-1)^S$, where S is the strangeness. The parity of the Λ^0 (relative to the nucleons) is therefore a matter of definition. But once this is defined, the parity of other strange particles would be measurable from the strong interactions.

POSSIBLE EXPERIMENTAL TESTS OF PARITY CONSERVATION IN MESON AND HYPERON DECAYS

To have a sensitive unequivocal test of whether parity is conserved in weak interactions, one must decide whether the weak interactions differentiate between the right and the left. This is possible only if one produces interference between states of opposite parities. The mere observation of two decay products of opposite parities originating from a "particle" cannot provide conclusive evidence that parity is not conserved. Such indeed is the state of affairs of the present $\theta-\tau$ puzzle.

As we have discussed before, these interference terms are possible only if the observed quantities can form a pseudoscalar such as $\mathbf{p}_1\cdot(\mathbf{p}_2\times\mathbf{p}_3)$. The observation of Λ^0 decays in association with their production does provide such a possible pseudoscalar and hence a possible test of whether parity is conserved in the Λ^0 decay interaction. Let us consider the experiment

$$\pi^-+p\rightarrow\Lambda^0+\theta^0,\quad \Lambda^0\rightarrow p+\pi^-. \qquad (4)$$

Let $\mathbf{p}_{\text{in}}$, $\mathbf{p}_\Lambda$, and $\mathbf{p}_{\text{out}}$ be, respectively, the momenta in the laboratory system of the incoming pion, the Λ^0, and the decay pion. We define a parameter R as the projection of $\mathbf{p}_{\text{out}}$ in the direction of $\mathbf{p}_{\text{in}}\times\mathbf{p}_\Lambda$. The value of R ranges from approximately -100 Mev/c to approximately $+100$ Mev/c. Switching from a right-handed convention for vector products (which we use) to a left-handed convention means a switch of the sign of R. Parity conservation in the weak decay interaction of Λ^0 can therefore be experimentally checked by investigating whether $+R$ and $-R$ have equal probabilities of occurrence.

To see more clearly the meaning of the parameter R, one transforms $\mathbf{p}_{\text{out}}(\rightarrow\mathbf{p}')$ into the center-of-mass system of Λ^0. The new vector $\mathbf{p}'$ has a constant magnitude $\cong 100$ Mev/c. The frequency distribution of this vector $\mathbf{p}'$ can then be plotted on a spherical surface. Taking the z axis for this sphere to be in the direction of $\mathbf{p}_{\text{in}}\times\mathbf{p}_\Lambda$, one can prove the following two symmetries:

(*a*) The frequency distribution on the sphere remains unchanged under a rotation through 180° around the z axis. This symmetry follows from parity conservation in the strong reaction producing the Λ^0. It does not depend on the nature of the weak interaction.

(*b*) If parity is conserved in the decay interaction of Λ^0, the frequency distribution on the sphere is unchanged under a reflection with respect to the production plane of Λ^0.

To prove statement (*a*), one need only consider the invariance of the production process under a reflection with respect to the production plane defined by $\mathbf{p}_{\text{in}}$ and $\mathbf{p}_\Lambda$. This reflection is the resultant of an inversion and a rotation through 180° around the z axis (which is normal to the production plane). The state of polarization of Λ^0 is thus invariant under a 180° rotation around the z axis, leading to the stated symmetry.[8]

Statement (*b*) follows[8] directly from the assumption that the weak interaction as well as the strong interaction conserves parity. A reflection with respect to the production plane must then leave the whole process invariant.

The frequency distribution of R is just the projection of the distribution on the sphere onto the z axis. An asymmetry between $+R$ and $-R$ therefore implies parity nonconservation in Λ^0 decay. However, if the spin of Λ^0 is unpolarized, no asymmetry[9] can obtain even if parity is not conserved in Λ^0 decay. To obtain a polarized Λ^0 beam, the experiment is therefore best done at a definite nonforward angle of production of Λ^0 and at a definite incoming energy.

The above discussions apply also to any other strange particle decay if (1) the particle has a nonvanishing spin and (2) it decays into two particles at least one of which has a nonvanishing spin, or it decays into three or more particles. Thus the above considerations can be applied also to the decays of $\Sigma^\pm$ and *maybe* also to $K_{\mu 2}{}^\pm$, $K_{\mu 3}{}^\pm$ and $K_{\pi 3}{}^\pm$ ($\equiv\tau^\pm$).

In the decay processes

$$\pi\rightarrow\mu+\nu, \quad (5)$$

$$\mu\rightarrow e+\nu+\nu, \quad (6)$$

starting from a π meson at rest, one could study the distribution of the angle θ between the μ-meson momentum and the electron momentum, the latter being in the center-of-mass system of the μ meson. If parity is conserved in neither (5) nor (6), the distribution will not in general be identical for θ and $\pi-\theta$. To understand this, consider first the orientation of the muon spin. If (5) violates parity conservation, the muon would be in general polarized in its direction of motion. In the subsequent decay (6), the angular distribution problem with respect to θ is therefore closely similar to the angular distribution problem of β rays from oriented nuclei, which we have discussed before. (Entirely similar considerations can be applied to $\Xi^-\rightarrow\Lambda^0+\pi^-$ and $\Lambda^0\rightarrow p+\pi^-$.)

REMARKS

If parity conservation is violated in hyperon decay, the decay products will have mixed parities. This, however, does not affect the arguments of Adair[10] and of Treiman[11] concerning the relationship between the spin of the hyperons and the angular distribution of their decay products in certain special cases.[12]

One may question whether the other conservation laws of physics could also be violated in the weak interactions. Upon examining this question, one finds that the conservations of the number of heavy particles, of electric charge, of energy, and of momentum all appear to be inviolate in the weak interactions. The same cannot be said of the conservation of angular momentum, and of parity. Nor can it be said of the invariance under time reversal. It might appear at first sight that the equality of the life times of $\pi^\pm$ and of those of $\mu^\pm$ furnish proofs of the invariance under charge conjugation of the weak interactions. A closer examination of this problem reveals, however, that this is not so. In fact, the equality of the life times of a charged particle and its charge conjugate against decay through a weak interaction (to the lowest order of the strength of the weak interaction) can be shown to follow from the invariance under proper Lorentz transformations (i.e., Lorentz transformation with neither space nor time inversion). One has therefore at present no experimental proof of the invariance under charge conjugation of the weak interactions. In the present paper, only the question of parity nonconservation is discussed.

[8] This proof for statement (*a*) is correct only if Λ^0 exists as a single particle with a definite parity in the strong interactions, (as discussed in the last section); i.e. if Λ^0 does not exist as two degenerate states $\Lambda_1{}^0$ and $\Lambda_2{}^0$ of opposite parity, as has been suggested [T. D. Lee and C. N. Yang, Phys. Rev. **102**, 290 (1956)]. [It is to be emphasized, that if parity is indeed not conserved in the weak interactions, there would be (at present) no necessity to introduce the complication of two degenerate states of opposite parity at all.] On the other hand, statement (*b*) is correct even if Λ^0 exists as two degenerate states $\Lambda_1{}^0$ and $\Lambda_2{}^0$ of opposite parity. *To summarize, violation of the symmetry stated in (a) implies the existence of the parity doublets $\Lambda_1{}^0$ and $\Lambda_2{}^0$ with a mass difference less than their widths. Violation of the symmetry stated in (b) implies the nonconservation of parity in Λ decay.* See also footnote 12 and T. D. Lee and C. N. Yang, Phys. Rev. (to be published).

[9] Also the interference may accidentally be absent if the relative phase between the two parities in the decay product is 90°. This, however, cannot be the case if time-reversal invariance is preserved in the decay process.

[10] R. K. Adair, Phys. Rev. **100**, 1540 (1955).

[11] S. B. Treiman, Phys. Rev. **101**, 1216 (1956).

[12] The existence of $\Lambda_1{}^0$ and $\Lambda_2{}^0$ of opposite parity may affect these relationships. This is similar to the violation of symmetry (*a*) discussed in footnote 8. See T. D. Lee and C. N. Yang, Phys. Rev. (to be published).

The conservation of parity is usually accepted without questions concerning its possible limit of validity being asked. There is actually no *a priori* reason why its violation is undesirable. As is well known, its violation implies the existence of a right-left asymmetry. We have seen in the above some possible experimental tests of this asymmetry. These experiments test whether the present elementary particles exhibit asymmetrical behavior with respect to the right and the left. If such asymmetry is indeed found, the question could still be raised whether there could not exist corresponding elementary particles exhibiting opposite asymmetry such that in the broader sense there will still be over-all right-left symmetry. If this is the case, it should be pointed out, there must exist two kinds of protons p_R and p_L, the right-handed one and the left-handed one. Furthermore, at the present time the protons in the laboratory must be predominantly of one kind in order to produce the supposedly observed asymmetry, and also to give rise to the observed Fermi-Dirac statistical character of the proton. This means that the free oscillation period between them must be longer than the age of the universe. They could therefore both be regarded as stable particles. Furthermore, the numbers of p_R and p_L must be separately conserved. However, the interaction between them is not necessarily weak. For example, p_R and p_L could interact with the same electromagnetic field and perhaps the same pion field. They could then be separately pair-produced, giving rise to interesting observational possibilities.

In such a picture the supposedly observed right-and-left asymmetry is therefore ascribed not to a basic non-invariance under inversion, but to a cosmologically local preponderance of, say, p_R over p_L, a situation not unlike that of the preponderance of the positive proton over the negative. Speculations along these lines are extremely interesting, but are quite beyond the scope of this note.

The authors wish to thank M. Goldhaber, J. R. Oppenheimer, J. Steinberger, and C. S. Wu for interesting discussions and comments. They also wish to thank R. Oehme for an interesting communication.

APPENDIX

If parity is not conserved in β decay, the most general form of Hamiltonian can be written as

$$\begin{aligned} H_{\rm int} = {} & (\psi_p{}^\dagger\gamma_4\psi_n)(C_S\psi_e{}^\dagger\gamma_4\psi_\nu + C_S'\psi_e{}^\dagger\gamma_4\gamma_5\psi_\nu) \\ & + (\psi_p{}^\dagger\gamma_4\gamma_\mu\psi_n)(C_V\psi_e{}^\dagger\gamma_4\gamma_\mu\psi_\nu + C_V'\psi_e{}^\dagger\gamma_4\gamma_\mu\gamma_5\psi_\nu) \\ & + \tfrac{1}{2}(\psi_p{}^\dagger\gamma_4\sigma_{\lambda\mu}\psi_n)(C_T\psi_e{}^\dagger\gamma_4\sigma_{\lambda\mu}\psi_\nu \\ & + C_T'\psi_e{}^\dagger\gamma_4\sigma_{\lambda\mu}\gamma_5\psi_\nu) + (\psi_p{}^\dagger\gamma_4\gamma_\mu\gamma_5\psi_n) \\ & \times(-C_A\psi_e{}^\dagger\gamma_4\gamma_\mu\gamma_5\psi_\nu - C_A'\psi_e{}^\dagger\gamma_4\gamma_\mu\psi_\nu) \\ & + (\psi_p{}^\dagger\gamma_4\gamma_5\psi_n)(C_P\psi_e{}^\dagger\gamma_4\gamma_5\psi_\nu + C_P'\psi_e{}^\dagger\gamma_4\psi_\nu), \end{aligned} \quad (A.1)$$

where $\sigma_{\lambda\mu} = -\frac{1}{2}i(\gamma_\lambda\gamma_\mu - \gamma_\mu\gamma_\lambda)$ and $\gamma_5 = \gamma_1\gamma_2\gamma_3\gamma_4$. The ten constants C and C' are all real if time-reversal invariance is preserved in β decay. This however, will not be assumed in the following.

Calculation with this interaction proceeds exactly as usual. One obtains, e.g., for the energy and angle distribution of the electron in an allowed transition

$$N(W,\theta)dW\sin\theta d\theta = \frac{\xi}{4\pi^3}F(Z,W)pW(W_0-W)^2 \times\left(1+\frac{ap}{W}\cos\theta+\frac{b}{W}\right)dW\sin\theta d\theta, \quad (A.2)$$

where

$$\xi = (|C_S|^2+|C_V|^2+|C_S'|^2+|C_V'|^2)|M_{\rm F.}|^2 + (|C_T|^2+|C_A|^2+|C_T'|^2+|C_A'|^2)|M_{\rm G.T.}|^2, \quad (A.3)$$

$$a\xi = \tfrac{1}{3}(|C_T|^2-|C_A|^2+|C_T'|^2-|C_A'|^2)|M_{\rm G.T.}|^2 - (|C_S|^2-|C_V|^2+|C_S'|^2-|C_V'|^2)|M_{\rm F.}|^2, \quad (A.4)$$

$$\begin{aligned} b\xi = {} & \gamma[(C_S{}^*C_V+C_SC_V{}^*)+(C_S'{}^*C_V'+C_S'C_V'{}^*)]|M_{\rm F.}|^2 \\ & +\gamma[(C_T{}^*C_A+C_A{}^*C_T)+(C_T'{}^*C_A'+C_A'{}^*C_T')] \\ & \times|M_{\rm G.T.}|^2. \end{aligned} \quad (A.5)$$

In the above expression all unexplained notations are identical with the standard notations. (See, e.g., the article by Rose.[13])

The above expression does not contain any interference terms between the parity-conserving part of the interactions and the parity-nonconserving ones. It is in fact directly obtainable by replacing in the usual expression the quantity $|C_S|^2$ by $|C_S|^2+|C_S'|^2$, and $C_SC_V{}^*$ by $C_SC_V{}^*+C_S'C_V'{}^*$, etc. This rule also holds in general, except for the cases where a pseudoscalar can be formed out of the measured quantities, as discussed in the text.

When a pseudoscalar can be formed, for example, in the β decay of oriented nuclei, interference terms would be present, as explicitly displayed in Eq. (2). In an allowed transition $J\to J-1$ (no), the quantity α is given by

$$\alpha = \beta\langle J_z\rangle/J,$$

$$\beta = \mathrm{Re}\left[C_TC_T'^* - C_AC_A'^* + i\frac{Ze^2}{\hbar cp}(C_AC_T'^*+C_A'C_T^*)\right] \times|M_{\rm G.T.}|^2\frac{v_e}{c}\frac{2}{\xi+(\xi b/W)}, \quad (A.6)$$

where $M_{\rm G.T.}$, ξ, and b are defined in Eqs. (A.3)–(A.5), v_e is the velocity of the electron, and $\langle J_z\rangle$ is the average spin component of the initial nucleus. For an allowed transition $J\to J+1$ (no), α is given by

$$\alpha = -\beta\langle J_z\rangle/(J+1). \quad (A.7)$$

The effect of the Coulomb field is included in all the above considerations.

[13] M. E. Rose, in *Beta- and Gamma-Ray Spectroscopy* (Interscience Publishers, Inc., New York, 1955), pp. 271–291.

PHYSICAL REVIEW VOLUME 105, NUMBER 5 MARCH 1, 1957

Parity Nonconservation and a Two-Component Theory of the Neutrino

T. D. LEE, *Columbia University, New York, New York*

AND

C. N. YANG, *Institute for Advanced Study, Princeton, New Jersey*

(Received January 10, 1957; revised manuscript received January 17, 1957)

A two-component theory of the neutrino is discussed. The theory is possible only if parity is not conserved in interactions involving the neutrino. Various experimental implications are analyzed. Some general remarks concerning nonconservation are made.

RECENTLY the question has been raised[1,2] as to whether the weak interactions are invariant under space inversion, charge conjugation, and time reversal. It was pointed out that although these invariances are generally held to be valid for all interactions, experimental proof has so far only extended to cover the strong interactions. (We group here the electromagnetic interactions with the strong interactions.) To test the possible violation of these invariance laws in the weak interactions, a number of experiments were proposed. One of these is to study the angular distribution of the β ray coming from the decay of oriented nuclei. We have been informed by Wu[3] that such an experiment is in progress. The preliminary results indicate a large asymmetry with respect to the spin direction of the oriented nuclei. Since the spin is an axial vector, its observed correlation with the β-ray momentum (which is a polar vector) can be understood only in terms of a violation of the law of space inversion invariance in β decay.

In view of this information and especially in view of the large asymmetry found, we wish to examine here a possible theory of the neutrino different from the conventionally accepted one. In this theory for a given momentum p the neutrino has only *one* spin state, the spin being always parallel to p. The spin and momentum of the neutrino together therefore automatically define the sense of the screw.

In this theory the mass of the neutrino must be zero, and its wave function need only have two components instead of the usual four. That such a relativistic theory is possible is well known.[4] It was, however, always rejected because of its intrinsic violation of space inversion invariance, a reason which is now no longer valid. (In fact, as we shall see later, in such a theory the violation of space inversion invariance attains a maximum.)

In Sec. 1 we describe this two-component theory of the neutrino. It is then shown in Sec. 2 that this theory is mathematically equivalent to a familiar four-component neutrino formalism for which all parity-conserving and parity-nonconserving Fermi couplings C and C' (as defined in the appendix of reference 1) are always related in the following manner: $C_S=C_S'$, $C_V=C_V'$, etc. or $C_S=-C_S'$, $C_V=-C_V'$, etc. Sections 3 to 8 are devoted to the physical consequences of the theory that can be put to experimental test. In the last section some general remarks about nonconservation are made.

I. NEUTRINO FIELD

1. Consider first the Dirac equation for a free spin-$\frac{1}{2}$ particle with zero mass. Because of the absence of the mass term, one needs only three anticommuting Hermitian matrices. Thus the neutrino can be represented by a spinor function φ_ν which has only two components.[4] The Dirac equation for φ_ν can be written as ($\hbar=c=1$)

$$\sigma\cdot\mathbf{p}\varphi_\nu=i\partial\varphi_\nu/\partial t, \tag{1}$$

where σ_1, σ_2, σ_3 are the usual 2×2 Pauli matrices. The relativistic invariance of this equation for proper Lorentz transformations (i.e., Lorentz transformations without space inversion and time inversion) is well known. In particular, for the space rotations through an angle θ around, say, the z axis, the wave function transforms in the following way:

$$\varphi\rightarrow\exp(-i\sigma_3\theta/2)\phi. \tag{2}$$

The σ matrices are therefore the spin matrices for the neutrino. For a state with a definite momentum $\mathbf{p}$, the energy and the spin along $\mathbf{p}$ are given, respectively, by

$$H=(\sigma\cdot\mathbf{p}),$$
$$\sigma_p=(\sigma\cdot\mathbf{p})/|\mathbf{p}|.$$

They are therefore related by

$$H=|\mathbf{p}|\sigma_p. \tag{3}$$

In the c-number theory, for a given momentum, the particle has therefore two states: a state with positive energy, and with $\frac{1}{2}$ as the spin component along $\mathbf{p}$, and a state with negative energy and with $-\frac{1}{2}$ as the spin component along $\mathbf{p}$.

It is easy to see that in a hole theory of such particles, *the spin of a neutrino* (defined to be a particle in the

---

[1] T. D. Lee and C. N. Yang, Phys. Rev. 104, 254 (1956).

[2] Lee, Oehme, and Yang, Phys. Rev. (to be published).

[3] Wu, Ambler, Hayward, Hoppes, and Hudson. We wish to thank Professor C. S. Wu for informing us of the progress of the experiment.

[4] See, e.g., W. Pauli, *Handbuch der Physik* (Verlag Julius Springer, Berlin, 1933), Vol. 24, 226–227.

positive-energy state) *is always parallel to its momentum while the spin of an antineutrino* (defined to be a hole in the negative-energy state) *is always antiparallel to its momentum* (i.e., the momentum of the antineutrino). Many of the experimental implications discussed in later sections are direct consequences of this correlation between the spin and the momentum of a neutrino. We have remarked in the introduction that such a correlation defines automatically the sense of a screw. With the usual (right-handed) conventions which we adopt throughout this paper, the spin and the velocity of the neutrino represent the spiral motion of a right-handed screw while the spin and the velocity of the antineutrino represent the spiral motion of a left-handed screw.

We shall now discuss some general properties[5] of this neutrino field:

(A) In this theory it is clear that the neutrino state and the antineutrino state cannot be the same. A Majorana theory for such a neutrino is therefore impossible.

(B) The mass of the neutrino and the antineutrino in this theory is necessarily zero. This is true for the physical mass even with the inclusion of all interactions. To see this, one need only observe that all the one-particle *physical* states consisting of one neutrino (or one antineutrino) must belong to a representation of the inhomogeneous proper Lorentz group identical with the representation to which the free neutrino states discussed above belong. For such a representation to exist at all, the mass must be zero.

(C) That the theory does not conserve parity is well known. We see it also in the following way: Under a space inversion P, one inverts the momentum of a neutrino but not its spin direction. Since in this theory the two are always parallel, the operator P applied to a neutrino state leads to a nonexisting state. Consequently the theory is not invariant under space inversion.

(D) By the same reasoning one concludes that the theory is also not invariant under charge conjugation C which changes a particle into its antiparticle but does not change its spin direction or momentum.

(E) It is possible, however, for the theory to be invariant under the operation CP, as this operation changes a neutrino into an antineutrino and simultaneously reverses its momentum while keeping the spin direction fixed. By the Lüders-Pauli theorem[6] it follows that the theory can be invariant under time reversal T.

For the free neutrino field, as described by (1), one can prove that the theory is indeed invariant under time reversal and under CP.

2. We shall in this section indicate how one can use the conventional four-component formalism of the neutrino (with violation of parity conservation) and obtain the same results as the present theory.

We start from Eq. (1) and enlarge the matrices by the following definitions (1 represents a 2×2 unit matrix):

$$\alpha\equiv\begin{pmatrix}\sigma & 0\\ 0 & -\sigma\end{pmatrix},\quad \beta\equiv\begin{pmatrix}0 & 1\\ 1 & 0\end{pmatrix}, \tag{4}$$

$$\psi_\nu\equiv\begin{pmatrix}\phi\\ 0\end{pmatrix}, \tag{5a}$$

$$\gamma\equiv-i\beta\alpha,\quad \gamma_4\equiv\beta,\quad \gamma_5\equiv\gamma_1\gamma_2\gamma_3\gamma_4=\begin{pmatrix}-1 & 0\\ 0 & 1\end{pmatrix}. \tag{6}$$

An immediate consequence of these definitions is

$$\gamma_5\psi_\nu=-\psi_\nu. \tag{7a}$$

The free neutrino part of the Lagrangian is, as usual,

$$L_\nu=\psi_\nu^\dagger\gamma_4\left(\gamma_\mu\frac{\partial}{\partial x_\mu}\right)\psi_\nu, \tag{8}$$

where $\psi_\nu^\dagger$ = Hermitian conjugate of ψ_ν. The most general interaction Lagrangian not containing derivatives for the process

$$n\rightarrow p+e+\bar{\nu} \tag{9a}$$

is exactly as usual; namely, it is the sum of the usual S, V, T, A, and P couplings:

$$+L_{\rm int}=-H_{\rm int}=\sum[-2C_i(\psi_p^\dagger O_i\psi_n)(\psi_e^\dagger O_i\psi_\nu)], \tag{10a}$$

where i runs over S, V, T, A, and P and

$$\begin{aligned}O_S&=\gamma_4,\\ O_V&=\gamma_4\gamma_\mu,\\ O_T&=-\frac{1}{2\sqrt{2}}i\gamma_4(\gamma_\lambda\gamma_\mu-\gamma_\mu\gamma_\lambda),\\ O_A&=-i\gamma_4\gamma_\mu\gamma_5,\\ O_P&=\gamma_4\gamma_5.\end{aligned} \tag{11}$$

It is not difficult to prove that Eqs. (5a) and (7a) are consistent with a relativistic theory even in the presence of the interaction (10a). Another way of proving this is to start from the conventional theory of the neutrino with the interaction Hamiltonian given in (A.1) of reference 1 and observe that when

$$C_S=-C_S',\quad C_V=-C_V',\ \text{etc.} \tag{12a}$$

the neutrino field ψ_ν there always appears in interactions in the combination $(1-\gamma_5)\psi_\nu$. In the explicit representa-

[5] We have received a manuscript from Professor A. Salam on a theory of the neutrino similar to the one discussed in the present paper. He specifically discussed points (A) and (B) that we discuss here. He also gave the Michel parameter for the μ decay that agrees with the ones obtained below in Sec. 6.

[6] G. Lüders, Kgl. Dansk Videnskab. Selskab, Mat.-fys. Medd. 28, No. 5 (1954); W. Pauli, *Niels Bohr and the Development of Physics* (Pergamon Press, London, 1955).

tion that we have adopted above, this means that only the first two components of ψ_ν contribute to the interaction. *All calculations using the conventional theory of the neutrino with the Hamiltonian* (A.1) *of reference* 1 *concerning β decay therefore gives the same result as the present theory if we take the choice of constants* (12a). There exists, however, the possibility that in the decay of the neutron a neutrino[7] is emitted:

$$n \rightarrow p + e + \nu. \tag{9b}$$

The corresponding general form (not including derivatives of the fields) of the Hamiltonian is

$$H_{\text{int}} = \sum [2C_i(\psi_p^\dagger O_i \psi_n)(\psi_e^\dagger O_i \psi_\nu')], \tag{10b}$$

where O_i has been defined in Eq. (11). The field ψ_ν' is a four-component spinor defined in terms of the two-component neutrino field ϕ by

$$\psi_\nu' = \begin{pmatrix} 0 \\ \sigma_2\varphi^\dagger \end{pmatrix}. \tag{5b}$$

From Eq. (6), we see that

$$\gamma_5\psi_\nu' = +\psi_\nu'. \tag{7b}$$

It can be shown that (5b) and (7b) are consistent with a relativistic theory even in the presence of interaction (10b). *It can also be proved that one can use again the Hamiltonian* (A.1) *of reference* 1 *for the conventional theory of the neutrino with the choice of the coupling constants*

$$C_S = C_S', \quad C_V = C_V', \text{ etc.} \tag{12b}$$

and obtain the same result as the present theory.

The two possible choices (12a) and (12b) depend on whether, in the β decay of the neutron, process (9a) or (9b) prevails, i.e., whether a neutrino[7] or an antineutrino is emitted. We shall see in Sec. 3 that experimentally it will be easy to decide which of the two choices is appropriate (if the theory is correct). [We do not consider the possibility here of the simultaneous presence of (9a) and (9b), since the double beta decay process does not seem to be observed experimentally.]

II. EXPERIMENTAL IMPLICATIONS

3. We consider in this section the experiment of the β decay of oriented nuclei already discussed in reference 1, and currently being carried out.[3] For the present theory, according to Eqs. (12a) or (12b), Eq. (A.6) reduces to

$$\beta = \mp\left(\frac{v_e}{c}\right)\frac{|C_T|^2 - |C_A|^2 - (2Ze^2/\hbar cp)\,\text{Im}(C_A C_T^*)}{|C_T|^2 + |C_A|^2}. \tag{13}$$

[7] The neutrino as defined in Sec. 1 is a particle with spin parallel to its momentum representing a right-hand screw. Similarly, the antineutrino as defined there is a particle with spin antiparallel to its momentum representing a left-handed screw. We use this definition throughout the present paper.

The choice of the $\mp$ sign depends on whether

$$n \rightarrow p + e + \bar{\nu} \quad (\bar{\nu} = \text{left-handed screw}),$$
or
$$n \rightarrow p + e + \nu \quad (\nu = \text{right-handed screw}). \tag{14}$$

In writing down (13) the Fierz interference terms has been set equal to zero, which is in conformity with the experimental results,[8] and which implies [see Eq. (A.5) of reference 1]:

$$\text{Real part of } C_A C_T^* = 0. \tag{15}$$

By measuring the momentum dependence of the asymmetry parameter β, one can test whether the present theory is correct.

It is interesting to notice that for a positron emitter the asymmetry parameter has the opposite sign. This is a direct consequence of the fact that in positron and electron emission, the neutrino and antineutrino emitted have opposite spirality.

4. An experiment such as the one being carried out by Cowan and collaborators[9] measures the cross section for neutrino absorption, which can be calculated in both the present theory and the usual theory. Now one determines the magnitude of the β-coupling constants to give the observed lifetimes of nuclei against β decay. The calculated value of the cross section turns out then to be *twice as great* in the present theory as in the usual theory. This follows from the following simple reasoning: The neutrino flux is an experimental quantity independent of the theory. If the neutrinos in a given direction have only one spin state instead of the usual two, by a detailed balancing argument they must have twice the cross section for absorption as the usual ones.

5. In the decay of $\pi^\mp$ mesons at rest, let us consider the component of angular momentum along the direction of $\mathbf{p}_\mu$, the momentum of the μ meson. The orbital angular momentum contributes nothing to this component. The μ spin component is therefore completely determined (irrespective of its total spin) by the spin component of the ν or $\bar{\nu}$. There are then two possibilities:

$$\text{(A)} \quad \begin{aligned} &\pi^+ \rightarrow \mu^+ + \nu, \quad (\mu^+ \text{ spin along } \mathbf{p}_\mu) = +\tfrac{1}{2}, \\ &\pi^- \rightarrow \mu^- + \bar{\nu}, \quad (\mu^- \text{ spin along } \mathbf{p}_\mu) = -\tfrac{1}{2}; \end{aligned} \tag{16}$$

or

$$\text{(B)} \quad \begin{aligned} &\pi^+ \rightarrow \mu^+ + \bar{\nu}, \quad (\mu^+ \text{ spin along } \mathbf{p}_\mu) = -\tfrac{1}{2}, \\ &\pi^- \rightarrow \mu^- + \nu, \quad (\mu^- \text{ spin along } \mathbf{p}_\mu) = +\tfrac{1}{2}. \end{aligned} \tag{17}$$

In each case the μ mesons with fixed $\mathbf{p}_\mu$ form a polarized beam. (It was pointed out in reference 1 that if parity is not conserved in the decay of π mesons, the μ mesons would in general be polarized.) Furthermore, the polarization is now complete (i.e., in a pure state). If this theory of the neutrino is correct, then the $\pi-\mu$ decay is a perfect polarizer of the μ meson, offering a

[8] See, e.g., R. Sherr and R. H. Miller, Phys. Rev. 93, 1076 (1954).

[9] See C. L. Cowan, Jr. *et al.*, Science **124**, 103 (1956).

natural way to measure the spin and the magnetic moment of the μ meson. (It turns out that the $\mu-e$ decay may serve as a good analyzer, as we shall discuss in the next section.)

The choice of the two possibilities (16) and (17) will be further discussed in Sec. 7.

6. For the $\mu^- - e^-$ decay the process can be

$$\mu^- \rightarrow e^- + \nu + \bar{\nu}, \tag{18}$$

or

$$\mu^- \rightarrow e^- + 2\nu, \tag{19}$$

or

$$\mu^- \rightarrow e^- + 2\bar{\nu}. \tag{20}$$

Consider process (18) first. The decay coupling can be written with the notations defined in Eq. (11). (We assume no derivitive coupling.)

$$H_{\text{int}} = \sum_{i=V,A} f_i(\psi_e^\dagger O_i \psi_\mu)(\psi_\nu^\dagger O_i \psi_\nu). \tag{21}$$

It is easy to see that in the present theory, where ψ_ν satisfies (7a), the S-, T-, and P-type couplings do not exist. We have assumed in writing down (21) that the spin of the μ meson is $\frac{1}{2}$. For a μ^- at rest with spin completely polarized, the normalized electron distribution is given by

$$dN = 2x^2[(3-2x) + \xi \cos\theta(1-2x)]dxd\Omega_e(4\pi)^{-1}, \tag{22}$$

where p = electron momentum, $x = p$/maximum electron momentum, θ = angle between electron momentum and the spin direction of the μ, Ω_e = solid angle of electron momentum, and

$$\xi = [|f_V|^2 + |f_A|^2]^{-1}[f_V f_A^* + f_A f_V^*]. \tag{23}$$

The mass of the electron is neglected in this calculation. The decay probability per unit time is ($\hbar = c = 1$):

$$\lambda = M^5[|f_A|^2 + |f_V|^2]/(3\times 2^8 \pi^3), \tag{24}$$

where M is the mass of the μ meson. The spectrum (22) for a nonpolarized μ meson,

$$dN = 2x^2[3-2x]dxd\Omega_e(4\pi)^{-1}, \tag{25}$$

is characterized[8] by a Michel[10] parameter $\rho = \frac{3}{4}$, which is consistent with known[11] experimental results.

One sees that for not too small values of ξ, the spectrum (22) is sensitive to $\cos\theta$, especially in the region of large momentum for the electrons. Therefore the $\mu-e$ decay may turn out to be a very good analyzer of the μ-meson spin.

An analysis of the so-called universality of the Fermi couplings is easier in this theory because there are fewer coupling constants, and also because $\pi-\mu-e$ decay measurements would supply information concerning the parameter ξ of (23).

If process (19) or (20) prevails, the spectrum becomes

$$dN = 12x^2(1-x)dx[1+\eta\cos\theta]d\Omega_e(4\pi)^{-1}. \tag{26}$$

This is characterized[8] by a Michel parameter[10] $\rho = 0$ which is not consistent with experiments.[11] One therefore concludes that (18) is the correct process.

A general theorem concerning the relationship between μ^+ and μ^- decays will be stated in Sec. 9.

7. If experiments should show that in the decay of the π meson, process (16) prevails, and in the β-decay process (9a) prevails, then one would say that the ν (the right-handed screw), the μ^-, and the e^- are light particles, and there is a conservation of light particles. If processes (17) and (9b) prevail, one would say that the $\bar{\nu}$ (the left-handed screw), the μ^-, and the e^- are light particles, and there is a conservation of light particles. Similar concepts have been discussed before.[12]

We have already seen in Sec. 3 that the sign of β in Eq. (13) determines whether

$$n \rightarrow p + e + \bar{\nu} \tag{9a}$$

or

$$n \rightarrow p + e + \nu \tag{9b}$$

is the process for β decay. To decide whether

$$\pi^+ \rightarrow \mu^+ + \nu, \quad (\mu^+ \text{ spin along } \mathbf{p}_\mu) = \tfrac{1}{2} \tag{16}$$

or

$$\pi^+ \rightarrow \mu^+ + \bar{\nu}, \quad (\mu^+ \text{ spin along } \mathbf{p}_\mu) = -\tfrac{1}{2} \tag{17}$$

one will have to determine the spin of μ^+ along its direction of motion.

8. The $\pi-\mu-e$ type experiment discussed in Secs. 6 and 7 can be done with the $K_{\mu 2}-\mu-e$ decays. The analysis is dependent on the spin of $K_{\mu 2}$. If this spin is not zero, the polarization of the μ meson is not necessarily complete. The degree of polarization can be experimentally found by a comparison of the angular distribution of the electrons in $\pi-\mu-e$ decay and in $K-\mu-e$ decay.

Another interesting experiment is to measure the momentum and polarization of the electron emitted in a β decay. A polarization of the electron results only if parity is not conserved; a measurement of this polarization is a measurement of a quantity similar to the parameter β in Eq. (13). The polarization in such a case will be along the direction of the momentum of the electron. Polarization along other directions can result if the momentum of the recoil nucleus is also determined. Theoretical considerations of such possibilities are being made by Dr. R. R. Lewis.

GENERAL REMARKS

9. Some general remarks concerning the conservation and nonconservation of the parity P, the charge conjugation C, and the time reversal T will be made in this section. Except for the last paragraph, no assump-

[10] L. Michel, Proc. Phys. Soc. (London) A63, 514 (1950).
[11] See, e.g., Sargent *et al.*, Phys. Rev. 99, 885 (1955).
[12] E. J. Konopinski and H. M. Mahmoud, Phys. Rev. 92, 1045 (1953).

tion that the neutrino is a two-component wave is made.

Since the preliminary result of the oriented nucleus experiment that there is a strong asymmetry, Eq. (A.6) of reference 1 shows that not only parity, but also charge conjugation is not conserved[2] in β decay. A measurement of the velocity dependence of the asymmetry parameter could supply[2] some information concerning time reversal invariance or noninvariance. If the $\pi-\mu-e$ decay should show any forward-backward asymmetry (as discussed in reference 1, and further analyzed above in Sec. 6 for the two-component neutrino theory), it can be shown from theorem 2 of reference 2 that charge conjugation invariance must be violated in both the $\pi-\mu$ and $\mu-e$ decays.

It is, however, easy to show from the Lüders-Pauli theorem[6] that even if C, T, and P are all not conserved, a stable particle ($e^\pm$ or $p^\pm$, or a deuteron, etc.) must have *exactly* the same mass as its antiparticle.

One can also prove that even if C, T, and P are all not conserved, the e^+ angular distribution in $\pi^+-\mu^+-e^+$ decay is exactly the same as the e^- angular distribution in $\pi^--\mu^--e^-$ decay. The only difference in the two cases is that the average spin of μ^+ along $\mathbf{p}_\mu$ is the opposite of that of μ^- along $\mathbf{p}_\mu$. (The decays are here assumed to occur in free space from $\pi^\pm$ at rest.)

It is further obvious from the Lüders-Pauli theorem[6] that if time reversal invariance is not violated, the operation CP is conserved. This means that the left-right asymmetry that is found in a laboratory is always exactly opposite to that found in the antilaboratory.

Should it further turn out that the two-component theory of the neutrino described above is correct, one would have a natural understanding of the violation of parity conservation in processes involving the neutrino. An understanding of the $\theta-\tau$ puzzle presents now a problem on a new level because no neutrinos are involved in the decay of $K_{\pi 2}$ and $K_{\pi 3}$. Perhaps this means that a more fundamental theoretical question should be investigated: the origin of all weak interactions. Perhaps the strange particles belong to strange representations of the Lorentz group. (Nature seems to make use of simple but odd representations.) It is also interesting to note that the massless electromagnetic field is the cause of the breakdown of the conservation of isotopic spin. The similarity to the massless two-component neutrino field that introduces the nonconservation of parity may not be accidental.

PHYSICAL REVIEW VOLUME 125, NUMBER 3 FEBRUARY 1, 1962

Symmetries of Baryons and Mesons*

MURRAY GELL-MANN
California Institute of Technology, Pasadena, California
(Received March 27, 1961; revised manuscript received September 20, 1961)

The system of strongly interacting particles is discussed, with electromagnetism, weak interactions, and gravitation considered as perturbations. The electric current j_α, the weak current J_α, and the gravitational tensor $\theta_{\alpha\beta}$ are all well-defined operators, with finite matrix elements obeying dispersion relations. To the extent that the dispersion relations for matrix elements of these operators between the vacuum and other states are highly convergent and dominated by contributions from intermediate one-meson states, we have relations like the Goldberger-Treiman formula and universality principles like that of Sakurai according to which the ρ meson is coupled approximately to the isotopic spin. Homogeneous linear dispersion relations, even without subtractions, do not suffice to fix the scale of these matrix elements; in particular, for the nonconserved currents, the renormalization factors cannot be calculated, and the universality of strength of the weak interactions is undefined. More information than just the dispersion relations must be supplied, for example, by field-theoretic models; we consider, in fact, the equal-time commutation relations of the various parts of j_4 and J_4. These nonlinear relations define an algebraic system (or a group) that underlies the structure of baryons and mesons. It is suggested that the group is in fact $U(3)\times U(3)$, exemplified by the symmetrical Sakata model. The Hamiltonian density θ_{44} is not completely invariant under the group; the noninvariant part transforms according to a particular representation of the group; it is possible that this information also is given correctly by the symmetrical Sakata model. Various exact relations among form factors follow from the algebraic structure. In addition, it may be worthwhile to consider the approximate situation in which the strangeness-changing vector currents are conserved and the Hamiltonian is invariant under $U(3)$; we refer to this limiting case as "unitary symmetry." In the limit, the baryons and mesons form degenerate supermultiplets, which break up into isotopic multiplets when the symmetry-breaking term in the Hamiltonian is "turned on." The mesons are expected to form unitary singlets and octets; each octet breaks up into a triplet, a singlet, and a pair of strange doublets. The known pseudoscalar and vector mesons fit this pattern if there exists also an isotopic singlet pseudoscalar meson χ^0. If we consider unitary symmetry in the abstract rather than in connection with a field theory, then we find, as an attractive alternative to the Sakata model, the scheme of Ne'eman and Gell-Mann, which we call the "eightfold way"; the baryons N, Λ, Σ, and Ξ form an octet, like the vector and pseudoscalar meson octets, in the limit of unitary symmetry. Although the violations of unitary symmetry must be quite large, there is some hope of relating certain violations to others. As an example of the methods advocated, we present a rough calculation of the rate of $K^+\rightarrow\mu^++\nu$ in terms of that of $\pi^+\rightarrow\mu^++\nu$.

I. INTRODUCTION

IN connection with the system of strongly interacting particles, there has been a great deal of discussion of possible approximate symmetries,[1] which would be violated by large effects but still have some physical consequences, such as approximate universality of meson couplings, approximate degeneracy of baryon or meson supermultiplets, and "partial conservation" of currents for the weak interactions.

In this article we shall try to clarify the meaning of such possible symmetries, for both strong and weak interactions. We shall show that a broken symmetry, even though it is badly violated, may give rise to certain exact relations among measurable quantities. Furthermore, we shall suggest a particular symmetry group as the one most likely to underlie the structure of the system of baryons and mesons.

We shall treat the strong interactions without approximation, but consider the electromagnetic, weak, and gravitational interactions only in first order.

The electromagnetic coupling is described by the matrix elements of the electromagnetic current operator $ej_\alpha(x)$. Likewise, the gravitational coupling is specified by the matrix elements of the stress-energy-momentum tensor $\theta_{\alpha\beta}(x)$, particularly the component $\theta_{44}=H$, the Hamiltonian density.

The weak interactions of baryons and mesons with leptons are assumed to be given (ignoring possible nonlocality) by the interaction term[2]

$$GJ_\alpha^\dagger J_\alpha^{(l)}/\sqrt{2}+\text{H.c.}, \tag{1.1}$$

where the leptonic weak current $J_\alpha^{(l)}$ has the form

$$J_\alpha^{(l)}=i\bar{\nu}\gamma_\alpha(1+\gamma_5)e+i\bar{\nu}\gamma_\alpha(1+\gamma_5)\mu. \tag{1.2}$$

We shall refer to $J_\alpha(x)$ as the weak current of baryons and mesons. Its matrix elements specify completely the weak interactions with leptons.

It is possible that the full weak interaction may be given simply by the term

$$G(J_\alpha+J_\alpha^{(l)})^\dagger(J_\alpha+J_\alpha^{(l)})/\sqrt{2}, \tag{1.3}$$

although this form provides no explanation of the approximate rule $|\Delta I|=\frac{1}{2}$ in the nonleptonic decays of strange particles. If we can find no *dynamical* explanation of the predominance of the $|\Delta I|=\frac{1}{2}$ amplitude in these decays, we may be forced to assume that in addition to (1.3) there is a weak interaction involving the product

$$GL_\alpha^\dagger L_\alpha/\sqrt{2}, \tag{1.4}$$

of charge-retention currents (presumably not involving leptons); or else we may be compelled to abandon (1.3)

* Research supported in part by U. S. Atomic Energy Commission and Alfred P. Sloan Foundation. A report of this work was presented at the La Jolla Conference on Strong and Weak Interactions, June, 1961.

[1] For example, see the "global symmetry" scheme of M. Gell-Mann, Phys. Rev. 106, 1296 (1957) and J. Schwinger, Ann. Phys. 2, 407 (1957).

[2] We use $\hbar=c=1$. The Lorentz index α takes on the values 1, 2, 3, 4. For each value of α, the Dirac matrix γ_α is Hermitian; so is the matrix γ_5.

altogether. In any case, we shall define the weak current J_α by the coupling to leptons.

We shall assume microcausality and hence the validity of dispersion relations for the matrix elements of the various currents and densities. In addition, we shall sometimes require the special assumption of highly convergent dispersion relations.

Our description of the symmetry group for baryons and mesons is most conveniently given in the framework of standard field theory, where the Lagrangian density L of the strong interactions is expressed as a simple function of a certain number of local fields $\psi(x)$, which are supposed to correspond to the "elementary" baryons and mesons. Recently this type of formalism has come under criticism[3]; it is argued that perhaps none of the strongly interacting particles is specially distinguished as "elementary," that the strong interactions can be adequately described by the analyticity properties of the S matrix, and that the apparatus of field theory may be a misleading encumbrance.

Even if the criticism is justified, the field operators $j_\alpha(x)$, $\theta_{\alpha\beta}(x)$, and $J_\alpha(x)$ may still be well defined (by all their matrix elements, including analytic continuations thereof) and measurable in principle by interactions with external electromagnetic or gravitational fields or with lepton pairs. Since the Hamiltonian density H is a component of $\theta_{\alpha\beta}$, it can be a physically sensible quantity.

In order to make our description of the symmetry group independent of the possibly doubtful details of field theory, we shall phrase it ultimately in terms of the properties of the operators H, j_α, and J_α. In introducing the description, however, we shall make use of field-theoretic models. Moreover, in describing the behavior of a particular group, we shall refer extensively to a special example, the symmetrical Sakata model of Ohnuki *et al.*,[4] Yamaguchi,[5] and Wess.[6]

The order of presentation is as follows: We treat first the hypothesis of highly convergent dispersion relations for the matrix elements of currents; and we show that the notion of a meson being coupled "universally" or coupled to a particular current or density means simply that the meson state dominates the dispersion relations for that current or density at low momenta. Next we discuss the universality of strength of the currents themselves; evidently it cannot be derived from homogeneous linear dispersion relations for the matrix elements of the currents. We show that equal-time commutation relations for the currents fulfill this need (or most of it), and that, in a wide class of model field theories, these commutation rules are simple and reflect the existence of a symmetry group, which underlies the structure of the baryon-meson system even though some of the symmetries are badly violated. We present the group properties in an abstract way that does not involve the details of field theory.

Next, it is asked what group is actually involved. The simplest one consistent with known phenomena is the one suggested. It is introduced, for clarity, in connection with a particular field theory, the symmetrical Sakata model, in which baryons and mesons are built up of fundamental objects with the properties of n, p, and Λ. For still greater simplicity, we discuss first the case in which Λ is absent.

We then return to the question of broken symmetry in the strong interactions and show how some of the symmetries in the group, if they are not too badly violated, would reveal themselves in approximately degenerate supermultiplets. In particular, there should be "octets" of mesons, each consisting of an isotopic triplet with $S=0$, a pair of doublets with $S=\pm1$, and a singlet with $S=0$. In the case of pseudoscalar mesons, we know of π, K, and $\bar{K}$; these should be accompanied by a singlet pseudoscalar meson χ^0, which would decay into 2γ, $\pi^++\pi^-+\gamma$, or 4π, depending on its mass.

In Sec. VIII, we propose, as an alternative to the symmetrical Sakata model, another scheme with the same group, which we call the "eightfold way." Here the baryons, as well as mesons, can form octets and singlets, and the baryons N, Λ, Σ, and Ξ are supposed to constitute an approximately degenerate octet.

In Sec. IX, some topics are suggested for further investigation, including the possibility of high energy limits in which non-conserved quantities become conserved, and we give, as an example of methods suggested here, an approximate calculation of the rate of $K^+\rightarrow\mu^++\nu$ decay from that of $\pi^+\rightarrow\mu^++\nu$ decay.

II. MESONS AND CURRENTS

To introduce the connection between meson states and currents or densities, let us review the derivation[7] of the Goldberger-Treiman relation[8] among the charged pion decay amplitude, the strength of the axial vector weak interaction in the β decay of the nucleon, and the pion-nucleon coupling constant.

The axial vector term in J_α with $\Delta S=0$, $|\Delta\mathbf{I}|=1$, $GP=-1$, can be written as $P_{1\alpha}+iP_{2\alpha}$, where $\mathbf{P}_\alpha$ is an axial vector current that transforms like an isotopic vector. We have, for nucleon β decay,

$$\langle N|\mathbf{P}_\alpha|N\rangle=\bar{u}_f[i\gamma_\alpha F_{ax}(s)+k_\alpha\beta(s)]\gamma_5(\boldsymbol{\tau}/2)u_i, \quad (2.1)$$

where u_i and u_f are the initial and final spinors, k_α is the four-momentum transfer, and $s=-k^2=-k_\alpha k_\alpha$. At

[3] G. F. Chew, Talk at La Jolla Conference on Strong and Weak Interactions, June, 1961 (unpublished).

[4] M. Ikeda, S. Ogawa, and Y. Ohnuki, Progr. Theoret. Phys. (Kyoto) 22, 715 (1959); Y. Ohnuki, *Proceedings of the 1960 Annual International Conference on High-Energy Physics at Rochester* (Interscience Publishers, Inc., New York, 1960).

[5] Y. Yamaguchi, Progr. Theoret. Phys. (Kyoto) Suppl. No. 11, 1 (1959).

[6] J. Wess, Nuovo cimento 10, 15 (1960).

[7] J. Bernstein, S. Fubini, M. Gell-Mann, and W. Thirring, Nuovo cimento 17, 757 (1960). See also Y. Nambu, Phys. Rev. Letters 4, 380 (1960); and Chou Kuang-Chao, Soviet Phys.—JETP 12, 492 (1961).

[8] M. Goldberger and S. Treiman, Phys. Rev. 110, 1478 (1958).

$s=0$ we have just

$$F_{ax}(0)=-G_A/G, \tag{2.2}$$

the axial vector renormalization constant.

The axial vector current is not conserved; its divergence $\partial_\alpha \mathbf{P}_\alpha$ has the same quantum numbers as the pion ($J=0^-$, $I=1$). Between nucleon states we have

$$\langle N|\partial_\alpha \mathbf{P}_\alpha|N\rangle=\bar{u}_f i\gamma_5(\boldsymbol{\tau}/2)u_i[2m_N F_{ax}(s)+s\beta(s)]. \tag{2.3}$$

We may compare this matrix element with that between the vacuum and a one-pion state

$$\langle 0|\partial_\alpha \mathbf{P}_\alpha|\pi\rangle=m_\pi^2(2f_\pi)^{-1}\boldsymbol{\phi}, \tag{2.4}$$

where $\boldsymbol{\phi}$ is the pion wave function and the constant f_π (or at least its square) may be measured by the rate of $\pi^+\rightarrow\mu^++\nu$:

$$\Gamma_\pi=G^2m_\pi m_\mu^2(1-m_\mu^2/m_\pi^2)^2(f_\pi^2/4\pi)^{-1}(64\pi^2)^{-1}. \tag{2.5}$$

It is known that the matrix element (2.3) has a pole at $s=m_\pi^2$ corresponding to the virtual emission of a pion that undergoes leptonic decay. The strength of the pole is given by the product of m_π^2/f_π and the pion-nucleon coupling constant $g_{NN\pi}$. If we assume that the expression in brackets vanishes at large s, we have an unsubtracted dispersion relation for it consisting of the pole term and a branch line beginning at $(3m_\pi)^2$, the next lowest mass that can be virtually emitted:

$$2m_NF_{ax}(s)+s\beta(s)=(g_{NN\pi}/f_\pi)m_\pi^2(m_\pi^2-s)^{-1}$$
$$+\int\sigma_{ax}(M^2)M^2dM^2\,(M^2-s-i\epsilon)^{-1}. \tag{2.6}$$

At $s=0$, we have, using (2.2), the sum rule

$$2m_N(-G_A/G)=g_{NN\pi}/f_\pi+\int\sigma_{ax}(M^2)dM^2. \tag{2.7}$$

Now if the dispersion relation (2.6) is not only convergent but dominated at low s by the term with the lowest mass, then we have the approximate Goldberger-Treiman relation

$$2m_N(-G_A/G)\approx g_{NN\pi}/f_\pi, \tag{2.8}$$

which agrees with experiment to within a few percent.

The success of the relation suggests that other matrix elements of $\partial_\alpha\mathbf{P}_\alpha$ may also obey unsubtracted dispersion relations dominated at low s by the one-pion term. For example, if we consider the matrix element between Λ and Σ, we should arrive at the relation

$$(m_\Lambda+m_\Sigma)(-G_A^{\Lambda\Sigma}/G)\approx g_{\Lambda\Sigma\pi}/f_\pi, \tag{2.9}$$

if Λ and Σ have the same parity, or an analogous relation if they have opposite parity.

If such a situation actually obtains, then it may be said that the pion is, to a good approximation, coupled "universally" to the divergence of the axial vector current. To calculate any g approximately, we multiply the universal constant f_π, the sum of the initial and final masses, and the renormalization factor for the axial vector current.

Now let us turn to the case of a current that is conserved, say the isotopic spin current $\mathfrak{J}_\alpha$ with quantum numbers $J=1^-$, $I=1$. Acting on the vacuum, the operator $\mathfrak{J}_\alpha$ does not lead to any stable one-meson state, but it does lead to the unstable vector meson state ρ at around 750 Mev, which decays into 2π or 4π. For simplicity, let us ignore the rather large width ($\Gamma_\rho\sim 100$ Mev) of the ρ state and treat it as stable. The mathematical complications resulting from the instability are not severe and have been discussed elsewhere.[9,10]

In place of (2.4), then, we have the definition

$$\langle 0|\mathfrak{J}_\alpha|\rho\rangle=m_\rho^2(2\gamma_\rho)^{-1}\boldsymbol{\phi}_\alpha, \tag{2.10}$$

of the constant γ_ρ, where $\boldsymbol{\phi}_\alpha$ is the wave function of the ρ meson. In place of (2.1) or (2.3), we consider the matrix element between nucleon states of the isotopic spin current:

$$\langle N|\mathfrak{J}_\alpha|N\rangle=\bar{u}_f i\gamma_\alpha(\boldsymbol{\tau}/2)u_iF_1{}^V(s)+\text{magnetic term}, \tag{2.11}$$

where $F_1{}^V(s)$ is the familiar isovector form factor of the electric charge of the nucleon, since the electromagnetic current has the form

$$j_\alpha=\mathfrak{J}_{3\alpha}+\text{isoscalar term}. \tag{2.12}$$

If we continue to ignore the width of ρ, we get a dispersion relation like (2.6) with a pole term at m_ρ^2:

$$F_1{}^V(s)=(\gamma_{NN\rho}/\gamma_\rho)m_\rho^2(m_\rho^2-s)^{-1}$$
$$+\int\sigma_1{}^V(M^2)dM^2\,M^2(M^2-s-i\epsilon)^{-1}. \tag{2.13}$$

Here $\gamma_{NN\rho}$ is the coupling constant of ρ to $\bar{u}_f i\boldsymbol{\tau}\gamma_\alpha u_i$, just as $g_{NN\pi}$ is the coupling constant of π to $\bar{u}_f i\boldsymbol{\tau}\gamma_5 u_i$. In this case, we have used an unsubtracted dispersion relation just for convenience.

Since the current is conserved, there is no renormalization and we have

$$F_1{}^V(0)=1, \tag{2.14}$$

giving, in place of (2.7), the sum rule

$$1=\gamma_{\rho NN}/\gamma_\rho+\int\sigma_1{}^V(M^2)dM^2. \tag{2.15}$$

If the dispersion relation is dominated at low s by the ρ term, then we obtain the analog of the Goldberger-Treiman formula:

$$1\approx\gamma_{\rho NN}/\gamma_\rho. \tag{2.16}$$

[9] G. F. Chew, University of California Radiation Laboratory Report No. UCRL-9289, 1960 (unpublished).

[10] M. Gell-Mann and F. Zachariasen, Phys. Rev. 124, 953 (1961).

Now the same reasoning may be applied to the isovector electric form factor of another particle, for example the pion:

$$\langle\pi|\mathfrak{J}_\alpha|\pi\rangle=[i\phi_f^*\times\partial_\alpha\phi_i-i\partial_\alpha\phi_f^*\times\phi_i]F_\pi(s), \quad (2.17)$$

$$F_\pi(s)=(\gamma_{\rho\pi\pi}/\gamma_\rho)m_\rho^2(m_\rho^2-s)^{-1}+\int\sigma_\pi(M^2)dM^2\,M^2(M^2-s-i\epsilon)^{-1}, \quad (2.18)$$

and

$$1=\gamma_{\rho\pi\pi}/\gamma_\rho+\int\sigma_\pi(M^2)dM^2. \quad (2.19)$$

If this dispersion relation, too, is dominated by the ρ pole at low s, then we find

$$1\approx\gamma_{\rho\pi\pi}/\gamma_\rho. \quad (2.20)$$

To the extent that the ρ pole gives most of the sum rule in each case, we have ρ coupled *universally* to the isotopic spins of nucleon, pion, etc., with coupling parameter $2\gamma_\rho$. Such universality was postulated by Sakurai,[11] within the framework of a special theory, in which ρ is treated as an elementary vector meson described by a Yang-Mills field. It can be seen that whether or not such a field description is correct, the *effective* universality ($\gamma_{\rho\pi\pi}\approx\gamma_{\rho NN}\approx\gamma_{\rho KK}$, etc.) is an approximate rule the validity of which depends on the domination of (2.15), (2.19), etc., by the ρ term.

The various coupling parameters $\gamma_{\rho\pi\pi}$, $\gamma_{\rho NN}$, etc., can be determined from the contribution of the ρ "pole" to various scattering processes, for example $\pi+N\rightarrow\pi+N$. But the factors $\gamma_{\rho\pi\pi}/\gamma_\rho$, $\gamma_{\rho NN}/\gamma_\rho$, etc., can also be measured, using electromagnetic interactions.[10]

An approximate determination of $\gamma_{\rho NN}/\gamma_\rho$ was made by Hofstadter and Herman[12] as follows The masses M^2 in the integral in Eq. (2.13) are taken to be effectively vary large, so that (2.13) becomes approximately

$$F_1^V(s)\approx(\gamma_{NN\rho}/\gamma_\rho)m_\rho^2(m_\rho^2-s)^{-1}+1-(\gamma_{\rho NN}/\gamma_\rho). \quad (2.21)$$

Fitting the experimental data on $F_1^V(s)$ with such a formula and using $m_\rho\approx750$ Mev, we obtain $\gamma_{\rho NN}/\gamma_\rho\approx1.4$. (Hofstadter and Herman, with a smaller value of m_ρ, found 1.2.)

III. EQUAL-TIME COMMUTATION RELATIONS

The dispersion relations for the matrix elements of weak or electromagnetic currents are linear and homogeneous. For example, Eq. (2.6) may be thought of as an expression for the matrix element of $\mathbf{P}_\alpha$ between the vacuum and a nucleon-antinucleon pair state. On the right-hand side, the pole term contains the product of the matrix element of $\mathbf{P}_\alpha$ between the vacuum and a one-pion state multiplied by the transition amplitude for the transition from π to $N\bar{N}$ by means of the strong interactions. The weight function $\sigma_{\alpha x}(M^2)$ is just the sum of such products over many intermediate states (such as 3π, 5π, etc.) with total mass M.

Now such linear, homogeneous equations may determine the dependence of the current matrix elements on variables such as s, but they cannot fix the scale of these matrix elements; constants like $-G_A/G$ cannot be calculated without further information. A field theory of the strong interactions, with explicit expressions for the currents, somehow contains more than these dispersion relations. In what follows, we shall extract some of this additional information in the form of equal-time commutation relations between components of the currents. Since these are nonlinear relations, they can help to fix the scale of each matrix element. Moreover, these relations may be the same for the lepton system and for the baryon-meson system, so that universality of strength of the weak interactions, for example, becomes meaningful.[13]

Let us begin our discussion of equal-time commutation relations with a familiar case—that of the isotopic spin **I**. Its components I_i obey the well-known commutation relations

$$[I_i,I_j]=ie_{ijk}I_k. \quad (3.1)$$

In terms of the components $\mathfrak{J}_{i\alpha}$ of the isotopic spin current, we have

$$I_i=-i\int\mathfrak{J}_{i4}d^3x, \quad (3.2)$$

and the conservation law

$$\partial_\alpha\mathfrak{J}_{i\alpha}=0 \quad (3.3)$$

tells us that

$$\dot{I}_i=\int\partial_\alpha\mathfrak{J}_{i\alpha}d^3x=0, \quad (3.4)$$

at all times.

Now the commutator of $\mathfrak{J}_{i4}(\mathbf{x},t)$ and $\mathfrak{J}_{j4}(\mathbf{x}',t)$ must vanish for $\mathbf{x}\neq\mathbf{x}'$, in accorance with microcausality. (Note we have taken the times equal.) If the commutator is not more singular than a delta function, then (3.1) and (3.2) give us the relation

$$[\mathfrak{J}_{i4}(\mathbf{x},t),\mathfrak{J}_{j4}(\mathbf{x}',t)]=-ie_{ijk}\mathfrak{J}_{k4}(\mathbf{x},t)\delta(\mathbf{x}-\mathbf{x}'), \quad (3.5)$$

which can also be obtained in any simple field theory by explicit commutation.[14]

In discussing the various parts of the weak current J_α, we shall have to deal with currents like $\mathbf{P}_\alpha$ that are not

---

[11] J. J. Sakurai, Ann. Phys. 11, 1 (1960).

[12] R. Hofstadter and R. Herman, Phys. Rev. Letters 6, 293 (1961). See also S. Bergia, A. Stanghellini, S. Fubini, and C. Villi, Phys. Rev. Letters 6, 367 (1961).

[13] M. Gell-Mann, *Proceedings of the 1960 Annual International Conference on High-Energy Physics at Rochester* (Interscience Publishers, Inc., New York, 1960).

[14] In some cases explicit commutation may be ambiguous and misleading. For example, a superficial consideration of $[j_i(\mathbf{x},t), j_4(\mathbf{x}',t)]$ for $i=1, 2, 3$ may lead to the conclusion that the expression vanishes. Yet the vacuum expectation value of the commutator can be shown to be a nonzero quantity times $\partial_i\delta(\mathbf{x}-\mathbf{x}')$, and that result is confirmed by more careful calculation. See J. Schwinger, Phys. Rev. Letters 3, 296 (1959).

conserved.[15] Here, too, we may define a quantity analogous to $\mathbf{I}$:

$$D_i = -i\int P_{i4}d^3x, \tag{3.6}$$

but D_i is *not* independent of time:

$$\dot{D}_i = \int \partial_\alpha P_{i\alpha}d^3x \neq 0. \tag{3.7}$$

For the moment, let us restrict our attention to the currents $\mathfrak{J}_\alpha$ and $\mathbf{P}_\alpha$ and the operators $\mathbf{I}$ and $\mathbf{D}(t)$. Since $\mathbf{D}$ is an isovector, we have the relations

$$[I_i,D_j] = [D_i,I_j] = ie_{ijk}D_k, \tag{3.8}$$

but what is the commutator of two components of $\mathbf{D}$? Since $\mathbf{P}_\alpha$ is a physical quantity, so is $\mathbf{D}$ and the question is one with direct physical meaning. We shall give both a general and a specific answer.

In general, we may take the commutators of D's (divided by i), the components of $\mathbf{I}$ and $\mathbf{D}$, the commutators of all of these with one another (divided by i), etc., until we obtain a system of Hermitian operators that is closed under commutation. Any of these operators can be written as a linear combination of N linearly independent Hermitian operators $R_i(t)$, where N might be infinite, and where the commutator of any two R_i is a linear combination of the R_i:

$$[R_i(t),R_j(t)] = ic_{ijk}R_k(t), \tag{3.9}$$

with c_{ijk} real. Such a system is called an algebra by the mathematicians. If we consider the set of infinitesimal unitary operators $1+i\epsilon R_i(t)$ and all possible products of these, we obtain an N-parameter continuous group of unitary transformations. We can refer to (3.9) as the algebra of the group. It is a physically meaningful statement to specify what group or what algebra is generated in this way by the currents $\mathfrak{J}_\alpha$ and $\mathbf{P}_\alpha$. Since a commutation relation like (3.9) is left invariant by a unitary transformation such as $\exp(-it\int Hd^3x)$, the numbers c_{ijk} are independent of time.

A second mathematical statement is also in order, i.e., the specification of the transformation properties of the Hamiltonian density $H(\mathbf{x},t)$ under the group or the algebra. Those R_i for which $[R_i(t),H(\mathbf{x},t)]=0$ are independent of time, but some of them, like D_i, do not commute with H. If all of the R_i commuted with H, then H would belong to the trivial one-dimensional representation of the group. In fact, H behaves in a more complicated way. By commuting all of the $R_i(t)$ with $H(\mathbf{x},t)$, we obtain a linear set of operators, containing H, that form a representation of the group; it may be broken up into the direct sum of irreducible representations. We want to know, then, what group is generated by $\mathbf{I}$ and $\mathbf{D}$ and to what irreducible representations of this group H belongs. Suggested are specific answers to both questions.

Let us look at the vector and axial vector weak currents for the leptons. For the time being, we shall consider only ν and e, ignoring the muon. (In the same way, we shall, in this section, ignore strange particles, and consider only baryons and mesons with $S=0$.) The vector weak current $i\bar{\nu}\gamma_\alpha e$ and the axial current $i\bar{\nu}\gamma_\alpha\gamma_5 e$ can be regarded formally as components of two "isotopic vector" currents for the leptons:

$$\mathfrak{J}_\alpha^{(l)} = i\bar{\xi}\boldsymbol{\tau}\gamma_\alpha\xi/2, \quad \mathbf{P}_\alpha^{(l)} = i\bar{\xi}\boldsymbol{\tau}\gamma_\alpha\gamma_5\xi/2, \tag{3.10}$$

where ξ stands for (ν,e). We can also form the mathematical analogs of $\mathbf{I}$ and $\mathbf{D}$:

$$\mathbf{I}^{(l)} = -i\int \mathfrak{J}_4^{(l)}d^3x, \quad \mathbf{D}^{(l)} = -i\int \mathbf{P}_4^{(l)}d^3x. \tag{3.11}$$

Now in this leptonic case we can easily compute the commutation rules of $\mathbf{I}^{(l)}$ and $\mathbf{D}^{(l)}$:

$$\begin{aligned}&[I_i^{(l)},I_j^{(l)}] = ie_{ijk}I_k^{(l)},\\ &[I_i^{(l)},D_j^{(l)}] = [D_i^{(l)},I_j^{(l)}] = ie_{ijk}D_k^{(l)},\\ &[D_i^{(l)},D_j^{(l)}] = ie_{ijk}I_k^{(l)}.\end{aligned} \tag{3.12}$$

Another way to phrase these commutation rules is to put

$$\begin{aligned}\mathbf{I}^{(l)} &= \mathbf{L}_+^{(l)} + \mathbf{L}_-^{(l)},\\ \mathbf{D}^{(l)} &= \mathbf{L}_+^{(l)} - \mathbf{L}_-^{(l)},\end{aligned} \tag{3.13}$$

and to notice that $\mathbf{L}_+^{(l)}$ and $\mathbf{L}_-^{(l)}$ are two commuting angular momenta [essentially $\boldsymbol{\tau}(1+\gamma_5)/4$ and $\boldsymbol{\tau}(1-\gamma_5)/4$]. The weak current $i\bar{\nu}\gamma_\alpha(1+\gamma_5)e$ is just a component of the current of $\mathbf{L}_+^{(l)}$.

We now suggest that the algebraic structure of $\mathbf{I}$ and $\mathbf{D}$ is exactly the same in the case of baryons and mesons. To (3.1) and (3.8), we add the rule[16,17]

$$[D_i,D_j] = ie_{ijk}I_k, \tag{3.14}$$

which closes the system and makes $\mathbf{I}^+ \equiv (\mathbf{I}+\mathbf{D})/2$ and $\mathbf{I}^- \equiv (\mathbf{I}-\mathbf{D})/2$ two commuting angular momenta. Again, we make the weak current a component of the current of $\mathbf{I}^+$. Evidently the statement that $(\mathbf{I}+\mathbf{D})/2$ is an angular momentum and not some factor times an angular momentum, fixes the scale of the weak current. It makes universality of strength between baryons and leptons meaningful, and it specifies, together with the dispersion relations, the value of such constants as $-G_A/G$.

The simplest way to realize the algebraic structure under discussion in a field-theory model of baryons and mesons is to construct the currents $\mathfrak{J}_\alpha$ and $\mathbf{P}_\alpha$ out of p and n fields just as $\mathfrak{J}_\alpha^{(l)}$ and $\mathbf{P}_\alpha^{(l)}$ are made out of ν and e fields:

$$\mathfrak{J}_\alpha = i\bar{N}\boldsymbol{\tau}\gamma_\alpha N/2, \quad \mathbf{P}_\alpha = i\bar{N}\boldsymbol{\tau}\gamma_\alpha\gamma_5 N/2, \tag{3.15}$$

[15] We assume that the vector weak current with $\Delta S=0$ is just a component of the isotopic spin current $\mathfrak{J}_\alpha$ and thus conserved.

[16] F. Gursey, Nuovo cimento 16, 230 (1960).

[17] M. Gell-Mann and M. Lévy, Nuovo cimento 16, 705 (1960).

where N means (p,n). We then obtain not only the commutation rules (3.1), (3.8), and (3.14), but the stronger rule (3.5) and its analogs:

$$\begin{aligned}&[\mathfrak{J}_{i4}(\mathbf{x},t),P_{j4}(\mathbf{x}',t)]=-ie_{ijk}P_{k4}(\mathbf{x},t)\delta(\mathbf{x}-\mathbf{x}'),\\&[P_{i4}(\mathbf{x},t),P_{j4}(\mathbf{x}',t)]=-ie_{ijk}\mathfrak{J}_{k4}(\mathbf{x},t)\delta(\mathbf{x}-\mathbf{x}').\end{aligned}\tag{3.16}$$

Next we want to use a field-theory model to suggest an answer to the second question—how H behaves under the group or, what is the same thing, under the algebra consisting of $\mathbf{I}$ and $\mathbf{D}$ or of $\mathbf{I}^+$ and $\mathbf{I}^-$. Since $\mathbf{I}^+$ and $\mathbf{I}^-$ are two commuting angular momenta, any irreducible representation of the algebra is specified by a pair of total angular momentum quantum numbers: i_+ for $\mathbf{I}^+$ and i_- for $\mathbf{I}^-$. The total isotopic spin quantum number I is associated with $\mathbf{I}^++\mathbf{I}^-=\mathbf{I}$.

Now we want the vector weak current $\mathfrak{J}_\alpha$ to be the isotopic spin current and to be conserved. Thus H must commute with $\mathbf{I}$; it transforms as an isoscalar, with $I=0$. In order to couple to zero, i_+ and i_- must be equal. So H can consist of terms with $(i_+,i_-)=(0,0)$, $(\frac{1}{2},\frac{1}{2})$, $(1,1)$, $(\frac{3}{2},\frac{3}{2})$, etc. Which of these are in fact present?

The simplest model in which the total isotopic current is given by just (3.15) is the Fermi-Yang[18] model, in which the pion is a composite of nucleon and antinucleon. To write an explicit Lagrangian, it must be decided what form the binding interaction takes. Since a direct four-fermion coupling leads to unpleasant singularities, we whall use a massive neutral vector meson field B^0 coupled to the nucleon current, as proposed by Teller[18] and Sakurai[11]; the exchange of a B^0 gives attraction between nucleon and antinucleon, permitting binding, and it also gives repulsion between nucleons, contributing to the "hard core." The model Lagrangian is then[19]

$$\begin{aligned}L=-\bar{N}\gamma_\alpha\partial_\alpha N-(\partial_\alpha B_\beta-\partial_\beta B_\alpha)^2/4\\-\mu_0^2B_\alpha B_\alpha/2-ih_0B_\alpha\bar{N}\gamma_\alpha N-m_0\bar{N}N.\end{aligned}\tag{3.17}$$

If the mass term for the nucleon were absent, then both $\mathfrak{J}_\alpha$ and $\mathbf{P}_\alpha$ would be conserved; $\mathbf{I}$ and $\mathbf{D}$ would both commute with L and with H. Thus,

$$H=H(0,0)-u_0,\tag{3.18}$$

where $H(0,0)$ transforms according to $(i_+,i_-)=(0,0)$ and the noninvariant term u_0 is just $-m_0\bar{N}N$. To what representation does it belong?

It is easy to see that the field B^0 belongs to $(0,0)$, while $N_L\equiv(1+\gamma_5)N/2$ belongs to $(\frac{1}{2},0)$ and $N_R\equiv(1-\gamma_5)N/2$ belongs to $(0,\frac{1}{2})$. One can thus verify that all terms of (3.17) except the last belong to $(0,0)$, since $\bar{N}\gamma_\alpha N$ or $\bar{N}\gamma_\alpha\partial_\alpha N$ couples $\bar{N}_L$ to N_L and $\bar{N}_R$ to N_R. But the Dirac matrix β, unlike $\beta\gamma_\alpha$, anticommutes with γ_5, so that the last term $-m_0\bar{N}N$ couples $\bar{N}_L$ to N_R and $\bar{N}_R$ to N_L. Thus u_0 belongs to $(\frac{1}{2},\frac{1}{2})$. We have $H=H(0,0)+H(\frac{1}{2},\frac{1}{2})$.

There are four components to the representation $(\frac{1}{2},\frac{1}{2})$ to which $u_0=-H(\frac{1}{2},\frac{1}{2})$ belongs. By commuting $\mathbf{D}$ with u_0, we generate the other three easily and see that they are proportional to $-i\bar{N}\boldsymbol{\tau}\gamma_5N$. In fact $\mathbf{D}$ acts like $\boldsymbol{\tau}\gamma_5/2$, $\mathbf{I}$ like $\boldsymbol{\tau}/2$, u_0 like β, and the other three components like $-i\beta\gamma_5\boldsymbol{\tau}$. Denoting the three new components by v_i, we have

$$\begin{aligned}&[I_i,u_0]=0, &&[D_i,u_0]=-iv_i,\\&[I_i,v_j]=ie_{ijk}v_k, &&[D_i,v_j]=i\delta_{ij}u_0.\end{aligned}\tag{3.19}$$

In the model, there are the even stronger relations for the densities

$$\begin{aligned}[\mathfrak{J}_{i4}(\mathbf{x},t),u_0(\mathbf{x}',t)]=0,\quad &[P_{i4}(\mathbf{x},t),u_0(\mathbf{x}',t)]\\&=-iv_i(\mathbf{x},t)\delta(\mathbf{x}-\mathbf{x}'),\text{ etc.}\end{aligned}\tag{3.20}$$

The noninvariant term u_0 is what prevents the axial vector current from being conserved. Thus one can express the divergence $\partial_\alpha\mathbf{P}_\alpha$ of the current in terms of the commutator of $\mathbf{D}$ with u_0. The conditions for this relation to hold are treated in the appendix and are applicable to all models we discuss. We find simply

$$\partial_\alpha\mathbf{P}_\alpha=-i[\mathbf{D},H]=i[\mathbf{D},u_0]=\mathbf{v},\tag{3.21}$$

and, of course,

$$\partial_\alpha\mathfrak{J}_\alpha=-i[\mathbf{I},H]=0.\tag{3.22}$$

It is precisely the operator $\mathbf{v}$, then, that we used in a dispersion relation in order to obtain the Goldberger-Treiman relation in Sec. II. Acting on the vacuum, it leads mostly to the one-pion state, so that the pion is effectively coupled universally to the divergence of the axial vector current. Thus $\mathbf{v}$ is a sort of effective pion field operator for the Fermi-Yang theory, which has no explicit pion field.

If we insist on a model in which there is a field variable $\boldsymbol{\pi}(x,t)$ then we must complicate the discussion. The total isotopic spin current is no longer given by just (3.15); there is a pion isotopic current term as well. In order to preserve the same algebraic structure of $\mathbf{I}$ and $\mathbf{D}$, one must then modify $\mathbf{P}_\alpha$ as well. Such a theory was described by Gell-Mann and Lévy,[17] who called it the "σ-model".[20] Along with the field $\boldsymbol{\pi}$, we must introduce a scalar, isoscalar field σ' in such a way that $\boldsymbol{\pi}$, σ' transform under the group like $\mathbf{v}$, u_0. Then, just as $\mathfrak{J}_\alpha$ has an additional term quadratic in $\boldsymbol{\pi}$, $\mathbf{P}_\alpha$ requires an additional term bilinear in $\boldsymbol{\pi}$ and σ'.

As we shall see in the next section, the introduction of

[18] E. Fermi and C. N. Yang, Phys. Rev. **76**, 1739 (1949); E. Teller, *Proceedings of the Sixth Annual Rochester Conference on High-Energy Nuclear Physics, 1956* (Interscience Publishers, Inc., New York, 1956).

[19] Conceivably a massive B^0 meson can be described by (3.17) even with $\mu_0=0$. [J. Schwinger, lectures at Stanford University, summer, 1961 (unpublished)]. In that case the noninvariant term in (3.17) is just equal to $\theta_{\alpha\alpha}$ and the traceless part of $\theta_{\alpha\beta}$ commutes with the group elements at equal times. In any case, whether μ_0 is zero or not, the off-diagonal terms in $\theta_{\alpha\beta}$ commute with the group.

[20] In the σ model, explicit commutation of u_0 and $\mathbf{v}$ at equal times gives zero, while in the Fermi-Yang model this is not so; if we take these results seriously, they give us definite physical distinctions among models.

strange particles makes the group much larger. The term u_0 is then a member of a much larger representation, with eighteen components. Thus if a pion field is introduced, fifteen more components are needed as well. Such a theory is too complicated to be attractive; we shall therefore ignore it and concentrate on the simplest generalization of the Fermi-Yang model to strange particles, namely the symmetrical Sakata model.

IV. SYMMETRICAL SAKATA MODEL AND UNITARY SYMMETRY

In the previous section, we proceeded inductively. We showed that starting from physical currents like $\mathfrak{J}_\alpha$ and P_α we may construct a group and its algebra and that it is physically meaningful to specify the group and also the transformation properties of H under the group. We chose the algebraic structure by analogy with the case of leptons and we saw that the simplest field theory model embodying the structure is just the Fermi-Yang model, in which p and n fields are treated just like the ν and e fields for the leptons, except that they are given a mass and a strong "gluon" coupling. The transformation properties of H were taken from the model; H consists, then, of an invariant part $H_{0,0}$ plus a term $(-u_0)$, where u_0 and a pseudoscalar isovector quantity $\mathbf{v}$ belong to the representation $(\frac{1}{2},\frac{1}{2})$ of the group. We then have the commutation rules (3.1), (3.8), (3.14), and (3.19). Microcausality with the assumption of commutators that are not too singular, or else direct inspection of the model, gives the stronger commutation rules (3.5), (3.16), and (3.20) for the densities. The model also gives specific equal-time commutation rules for u_0 and $\mathbf{v}$, which we did not list. All of these properties can be abstracted from the model and considered on their own merits as proposed relations among the currents and the Hamiltonian density.

Now, to argue deductively, we want to include the strange particles and all parts of the weak current J_α and the electromagnetic current j_α. We generalize the Fermi-Yang description to obtain the symmetrical Sakata model and abstract from it as many physically meaningful relations as possible.

It has long been recognized that the qualitative properties of baryons and mesons could be understood in terms of the Sakata model,[21] in which all strongly interacting particles are made out of N, Λ, $\bar{N}$, and $\bar{\Lambda}$ (or at least out of basic fields with the same quantum numbers as these particles).

We write the Lagrangian density for the Sakata model as a generalization of (3.17):

$$\begin{aligned} L= & -\bar{p}\gamma_\alpha p-\bar{n}\gamma_\alpha\partial_\alpha\bar{n}-\bar{\Lambda}\gamma_\alpha\partial_\alpha\Lambda-\tfrac{1}{4}(\partial_\alpha B_\beta-\partial_\beta B_\alpha)^2 \\ & -\tfrac{1}{2}\mu_0^2 B_\alpha B_\alpha-ih_0(\bar{p}\gamma_\alpha p+\bar{n}\gamma_\alpha n+\bar{\Lambda}\gamma_\alpha\Lambda)B_\alpha \\ & -m_{0N}(\bar{n}n+\bar{p}p)-m_{0\Lambda}\bar{\Lambda}\Lambda. \end{aligned} \quad (4.1)$$

According to this picture, the baryons present a striking parallel with the leptons,[22] for which we write the Lagrangian density

$$L_l=-\bar{\nu}\gamma_\alpha\nu-\bar{e}\gamma_\alpha\partial_\alpha e-\bar{\mu}\gamma_\alpha\partial_\alpha\mu-0\cdot(\bar{\nu}\nu+\bar{e}e)-m_\mu\bar{\mu}\mu, \quad (4.2)$$

if we turn off the electromagnetic and weak couplings, along with the ν-e mass difference. Here it is assumed there is only one kind of neutrino.

The only real difference between baryons and leptons in (4.1) and (4.2), respectively, is that the baryons are coupled, through the baryon current, to the field B. It is tempting to suppose that the weak current of the strongly interacting particles is just the expression.

$$i\bar{p}\gamma_\alpha(1+\gamma_5)n+i\bar{p}\gamma_\alpha(1+\gamma_5)\Lambda, \quad (4.3)$$

analogous to Eq. (1.2) for the leptonic weak current $J_\alpha^{(l)}$. Now (4.3) is certainly a reasonable expression, qualitatively, for weak currents of baryons and mesons. As Okun has emphasized,[23] the following properties of the weak interactions, often introduced as postulates, are derivable from (1.1), (1.2), (4.1), and (4.3):

(a) The conserved vector current.[24] In the model under discussion, as in that of Fermi and Yang, $i\bar{p}\gamma_\alpha n$ is a component of the total isotopic spin current.

(b) The rules $|\Delta S|=1$, $\Delta S/\Delta Q=+1$, and $|\Delta \mathrm{I}|=\frac{1}{2}$ for the leptonic decays of strange particles.[25]

(c) The invariance under GP of the $\Delta S=0$ weak current.[26]

(d) The rules $|\Delta S|=1$, $|\Delta \mathrm{I}|=\frac{1}{2}$ or $\frac{3}{2}$ in the nonleptonic decays of strange particles; along with $|\Delta S|=1$, we have the absence of a large K_1^0-K_2^0 mass difference.

The quantitative facts that the effective coupling constants for $|\Delta S|=1$ leptonic decays are smaller than those for $|\Delta S|=0$ leptonic decays and that in nonleptonic decays of strange particles the $|\Delta \mathrm{I}|=\frac{1}{2}$ amplitude greatly predominates over the $|\Delta \mathrm{I}|=\frac{3}{2}$ amplitude are not explained in any fundamental way.[27]

[21] S. Sakata, Progr. Theoret. Phys. (Kyoto) 16, 686 (1956).

[22] A. Gamba, R. E. Marshak, and S. Okubo, Proc. Natl. Acad. Sci. U. S. 45, 881 (1959).

[23] L. Okun, Ann. Rev. Nuclear Sci. 9, 61 (1959).

[24] R. P. Feynman and M. Gell-Mann, Phys. Rev. 109, 193 (1958). See also S. S. Gershtein and J. B. Zeldovich, Soviet Phys.—JETP 2, 576 (1957).

[25] M. Gell-Mann, *Proceedings of the Sixth Annual Rochester Conference on High-Energy Nuclear Physics, 1956* (Interscience Publishers, Inc., New York, 1956). These rules were in fact suggested on the basis of the idea that N and Λ are fundamental. Should the rules prove too restrictive (for example should $\Delta S/\Delta Q=+1$ be violated), then we would try a larger group; in the language of the field-theoretic model, we would assume more fundamental fields. For a discussion of possible larger groups, see M. Gell-Mann and S. Glashow, Ann. Phys. 15, 437 (1961) and S. Coleman and S. Glashow (to be published).

[26] S. Weinberg, Phys. Rev. 112, 1375 (1958).

[27] A possible dynamical explanation of the predominance of $|\Delta \mathrm{I}|=\frac{1}{2}$ is being investigated by Nishijima (private communication). For example, consider the decay $\Lambda\to N+\pi$. A dispersion relation without subtractions is written for the matrix element of $J_\alpha^\dagger J_\alpha$ between the vacuum and a state containing $N+\bar{\Lambda}+\pi$. The parity-violating part leads to intermediate pseudoscalar states with $S=+1$ and with $|\Delta \mathrm{I}|=\frac{1}{2}$ or $\frac{3}{2}$. In the case of $|\Delta \mathrm{I}|=\frac{1}{2}$, there is an intermediate K particle, which may give a large contribution, swamping the term with $|\Delta \mathrm{I}|=\frac{3}{2}$, which has no one-meson state. For the same argument to apply to the parity-conserving part, we need the K' meson of Table III.

TABLE I. A set of matrices λ_i.

$$\lambda_1=\begin{pmatrix}0&1&0\\1&0&0\\0&0&0\end{pmatrix}\quad \lambda_2=\begin{pmatrix}0&-i&0\\i&0&0\\0&0&0\end{pmatrix}\quad \lambda_3=\begin{pmatrix}1&0&0\\0&-1&0\\0&0&0\end{pmatrix}$$

$$\lambda_4=\begin{pmatrix}0&0&1\\0&0&0\\1&0&0\end{pmatrix}\quad \lambda_5=\begin{pmatrix}0&0&-i\\0&0&0\\i&0&0\end{pmatrix}\quad \lambda_6=\begin{pmatrix}0&0&0\\0&0&1\\0&1&0\end{pmatrix}$$

$$\lambda_7=\begin{pmatrix}0&0&0\\0&0&-i\\0&i&0\end{pmatrix}\quad \lambda_8=\begin{pmatrix}1/\sqrt{3}&0&0\\0&1/\sqrt{3}&0\\0&0&-2/\sqrt{3}\end{pmatrix}$$

The electromagnetic properties of baryons and leptons are not exactly parallel in the Sakata model. The electric current (divided by e), which are denoted by j_α, is given by

$$i\bar{p}\gamma_\alpha p \tag{4.4}$$

for the baryons and mesons and by

$$-i(\bar{e}\gamma_\alpha e+\bar{\mu}\gamma_\alpha\mu) \tag{4.5}$$

for the leptons.

Now, we return to the Lagrangian (4.1) and separate it into three parts:

$$L=\bar{L}+L'+L'', \tag{4.6}$$

where $\bar{L}$ stands for everything except the baryon mass terms, while L' and L'' are given by the expressions

$$\begin{aligned}L'&=(2m_{0N}+m_{0\Lambda})(\bar{N}N+\bar{\Lambda}\Lambda)/3,\\ L''&=(m_{0N}-m_{0\Lambda})(\bar{N}N-2\bar{\Lambda}\Lambda)/3.\end{aligned} \tag{4.7}$$

If we now consider the Lagrangian with the mass-splitting term L'' omitted, we have a theory that is completely symmetrical in p, n, and Λ. We may perform any unitary linear transformation (with constant coefficients) on these three fields and leave $\bar{L}+L'$ invariant. Thus in the absence of the mass-splitting term L'' the theory is invariant under the three-dimensional unitary group $U(3)$; we shall refer to this situation as "unitary symmetry."

If we now turn on the mass-splitting, the symmetry is reduced. The only allowed unitary transformations are those involving n and p alone or Λ alone. The group becomes $U(2)\times U(1)$, which corresponds, as we shall see, to the conservation if isotopic spin, strangeness, and baryon number.

For simplicity, let us return briefly to the simpler case in which there is no Λ. The symmetry group is then just $U(2)$, the set of unitary transformations on n and p. We can factor each unitary transformation uniquely into one which multiplies both fields by the same phase factor and one (with determinant unity) which leaves invariant the product of the phase factors of p and n. Invariance under the first kind of transformation corresponds to conservation of nucleons n and p; it may be considered separately from invariance under the class of transformations of the second kind [called by mathematicians the unitary unimodular group $SU(2)$ in two dimensions]. In mathematical language, we can factor $U(2)$ into $U(1)\times SU(2)$.

Each transformation of the first kind can be written as a matrix 1 $\exp i\phi$, where 1 is the unit 2×2 matrix. The infinitesimal transformation is $1+i1\delta\phi$, and so the unit matrix is the infinitesimal generator of these transformations. Those of the second kind are generated in the same way by the three independent traceless 2×2 matrices, which may be taken to be the Pauli isotopic spin matrices τ_1, τ_2, and τ_3. We thus have

$$N\rightarrow(1+i\sum_{k=1}^{3}\delta\theta_k\tau_k/2)N, \tag{4.8}$$

as the general infinitesimal transformation of the second kind. Symmetry under all the transformations of the second kind is the same as symmetry under isotopic spin rotations. The whole formalism of isotopic spin theory can then be constructed by considering the transformation properties of the doublet or spinor (p,n) and of more complicated objects that transform like combinations of two or more such nucleons (or antinucleons).

The Pauli matrices τ_k are Hermitian and obey the rules

$$\begin{aligned}\mathrm{Tr}\,\tau_i\tau_j&=2\delta_{ij},\\ [\tau_i,\tau_j]&=2ie_{ijk}\tau_k,\\ \{\tau_i,\tau_j\}&=2\delta_{ij}1.\end{aligned} \tag{4.9}$$

The invariance under the group $SU(2)$ of isotopic spin rotations corresponds to conservation of the isotopic spin current

$$\mathfrak{J}_\alpha=i\bar{N}\tau\gamma_\alpha N/2,$$

while the invariance under transformations of the first kind corresponds to conservation of the nucleon current $i\bar{N}\gamma_\alpha N/2=n_\alpha$.

Defining the total isotopic spin $\mathbf{I}$ as in (3.2), we obtain for I_i the commutation rules (3.1), which are the same as those for $\tau_i/2$. Likewise the nucleon number is defined as $-i\int n_4 d^3x$ and commutes with $\mathbf{I}$.

We now generalize the idea of isotopic spin by including the third field Λ. Again we factor the unitary transformations on baryons into those which are generated by the 3×3 unit matrix 1 (and which correspond to baryon conservation) and those which are generated by the eight independent traceless 3×3 matrices [and which form the unitary unimodular group $SU(3)$ in three dimensions]. We may construct a typical set of eight such matrices by analogy with the 2×2 matrices of Pauli. We call then $\lambda_1\cdots\lambda_8$ and list them in Table I. They are Hermitian and have the properties

$$\begin{aligned}\mathrm{Tr}\,\lambda_i\lambda_j&=2\delta_{ij},\\ [\lambda_i,\lambda_j]&=2if_{ijk}\lambda_k,\\ \{\lambda_i,\lambda_j\}&=2d_{ijk}\lambda_k+\tfrac{4}{3}\delta_{ij}1,\end{aligned} \tag{4.10}$$

where f_{ijk} is real and totally antisymmetric like the

TABLE II. Nonzero elements of f_{ijk} and d_{ijk}. The f_{ijk} are odd under permutations of any two indices while the d_{ijk} are even.

ijk	f_{ijk}	ijk	d_{ijk}
123	1	118	$1/\sqrt{3}$
147	1/2	146	1/2
156	−1/2	157	1/2
246	1/2	228	$1/\sqrt{3}$
257	1/2	247	−1/2
345	1/2	256	1/2
367	−1/2	338	$1/\sqrt{3}$
458	$\sqrt{3}/2$	344	1/2
678	$\sqrt{3}/2$	355	1/2
...	...	366	−1/2
...	...	377	−1/2
...	...	448	$-1/(2\sqrt{3})$
...	...	558	$-1/(2\sqrt{3})$
...	...	668	$-1/(2\sqrt{3})$
...	...	778	$-1/(2\sqrt{3})$
...	...	888	$-1/\sqrt{3}$

Kronecker symbol e_{ijk} of Eq. (4.9), while d_{ijk} is real and totally symmetric. These properties follow from the equations

$$\begin{aligned} \mathrm{Tr}\lambda_k[\lambda_i,\lambda_j] &= 4if_{ijk}, \\ \mathrm{Tr}\lambda_k\{\lambda_i,\lambda_j\} &= 4d_{ijk}, \end{aligned} \tag{4.11}$$

derived from (4.10).

The nonzero elements of f_{ijk} and d_{ijk} are given in Table II for our choice of λ_i. Even and odd permutations of the listed indices correspond to multiplication of f_{ijk} by ± 1, respectively, and of d_{ijk} by $+1$.

The general infinitesimal transformation of the second kind on the three basic baryons b is, of course,

$$b \to (1+i\sum_{i=1}^{8}\delta\theta_i\lambda_i/2)b, \tag{4.12}$$

by analogy with (4.8). Together with conservation of baryons, invariance under these transformations corresponds to complete "unitary symmetry" of the three baryons. We have factored $U(3)$ into $U(1)\times SU(3)$.

The invariance under transformations of the first kind gives us conservation of the baryon current

$$i\bar{b}\gamma_\alpha b = i\bar{n}\gamma_\alpha n + i\bar{p}\gamma_\alpha p + i\bar{\Lambda}\gamma_\alpha\Lambda, \tag{4.13}$$

while invariance under the second class of transformations would give us conservation of the eight-component "unitary spin" current

$$\mathfrak{F}_{i\alpha} = i\bar{b}\lambda_i\gamma_\alpha b/2 \quad (i=1,\cdots,8). \tag{4.14}$$

Now in fact L'' is not zero and so not all the components of $\mathfrak{F}_{i\alpha}$ are actually conserved. This does not prevent us from defining $\mathfrak{F}_{i\alpha}$ as in (4.14), nor does it affect the commutation rules of the unitary spin density. The total unitary spin F_i is defined by the relation

$$F_i = -i\int \mathfrak{F}_{i4}d^3x, \tag{4.15}$$

at any time and at equal times the commutation rules for F_i follow those for $\lambda_i/2$

$$[F_i,F_j] = if_{ijk}F_k. \tag{4.16}$$

The baryon number, of course, commutes with all components F_i.

It will be noticed that λ_1, λ_2, and λ_3 agree with τ_1, τ_2, and τ_3 for p and n and have no matrix elements for Λ. Thus the first three components of the unitary spin are just the components of the isotopic spin. The matrix λ_8 is diagonal in our representation and has one eigenvalue for the nucleon and another for the Λ. Thus F_8 is just a linear combination of strangeness and baryon number. It commutes with the isotopic spin.

The matrices λ_4, λ_5, λ_6, and λ_7 connect the nucleon and Λ. We see that the components F_4, F_5, F_6, and F_7 of the unitary spin change strangeness by one unit and isotopic spin by a half unit. When the mass-splitting term L'' is "turned on," it is these components that are no longer conserved, while the conservation of F_1, F_2, F_3, F_8, and baryon number remains valid.

V. VECTOR AND AXIAL VECTOR CURRENTS

We may unify the mathematical treatment of the baryon current and the unitary spin current if we define a ninth 3×3 matrix

$$\lambda_0 = (\tfrac{2}{3})^{1/2}1, \tag{5.1}$$

so that the *nine* matrices λ_i obey the rules

$$\begin{aligned} [\lambda_i,\lambda_j] &= 2if_{ijk}\lambda_k \quad (i=0,\cdots,8), \\ \{\lambda_i,\lambda_j\} &= 2d_{ijk}\lambda_k \quad (i=0,\cdots,8), \\ \mathrm{Tr}\lambda_i\lambda_j &= 2\delta_{ij} \quad (i=0,\cdots,8). \end{aligned} \tag{5.2}$$

Here, f_{ijk} is defined as before, except that it vanishes when any index is zero; d_{ijk} is also defined as before, except that it has additional nonzero matrix elements equal to $(\frac{2}{3})^{1/2}$ whenever any index is zero and the other two indices are equal. The baryon current is now $(\frac{3}{2})^{1/2}\mathfrak{F}_{0\alpha}$.

The definitions (4.15) and the equal-time commutation relations (4.16) now hold for $i=0,\cdots,8$. Moreover, there are the equal-time commutation relations

$$[\mathfrak{F}_{i4}(\mathbf{x},t),\mathfrak{F}_{j4}(\mathbf{x}',t)] = -if_{ijk}\mathfrak{F}_{k4}(\mathbf{x},t)\delta(\mathbf{x}-\mathbf{x}') \tag{5.3}$$

for the densities.

The electric current j_α is then

$$j_\alpha = (\sqrt{2}\mathfrak{F}_{0\alpha}+\mathfrak{F}_{8\alpha}+\sqrt{3}\mathfrak{F}_{3\alpha})/2\sqrt{3}, \tag{5.4}$$

while the vector weak current is

$$\mathfrak{F}_{1\alpha}+i\mathfrak{F}_{2\alpha}+\mathfrak{F}_{4\alpha}+i\mathfrak{F}_{5\alpha}. \tag{5.5}$$

We now wish to set up the same formalism for the axial vector currents. We recall that the presence of the symmetry-breaking term L'' did not prevent us from defining the $\mathfrak{F}_{i\alpha}$ and obtaining the commutation rules (5.3) characteristic of the unitary symmetry group $U(3)$.

In the same way, we now remark that if both L''

and L' are "turned off," we have invariance under the infinitesimal unitary transformations

$$b \to (1+i\sum_{i=0}^{8} \delta\psi_i\gamma_5\lambda_i/2)b, \quad (5.6)$$

as well as the infinitesimal transformations

$$b \to (1+i\sum_{i=0}^{8} \delta\theta_i\lambda_i/2)b \quad (5.7)$$

we have used before.[28] Thus the axial vector currents

$$\mathfrak{F}_{i\alpha}{}^5 = i\bar{b}\lambda_i\gamma_\alpha\gamma_5 b/2 \quad (5.8)$$

would be conserved if both L' and L'' were absent. Even in the presence of these terms, we have the commutation rules

$$[\mathfrak{F}_{i4}{}^5(\mathbf{x},t),\mathfrak{F}_{j4}(\mathbf{x}',t)] = -if_{ijk}\mathfrak{F}_{k4}{}^5(\mathbf{x},t)\delta(\mathbf{x}-\mathbf{x}') \quad (5.9)$$

and

$$[\mathfrak{F}_{i4}{}^5(\mathbf{x},t),\mathfrak{F}_{j4}{}^5(\mathbf{x}',t)] = -if_{ijk}\mathfrak{F}_{k4}(\mathbf{x},t)\delta(\mathbf{x}-\mathbf{x}') \quad (5.10)$$

at equal times, We may use the definition

$$F_i{}^5(t) = -i\int \mathfrak{F}_{i4}{}^5 d^3x, \quad (5.11)$$

along with (4.15).

Just as we put $\mathbf{I}=\mathbf{L}_+ + \mathbf{L}_-$ and $\mathbf{D}=\mathbf{L}_+ - \mathbf{L}_-$ in the discussion following Eq. (3.16), so we now write

$$F_i(t) = F_i{}^+(t) + F_i{}^-(t),$$
$$F_i{}^5(t) = F_i{}^+(t) - F_i{}^-(t), \quad (5.12)$$

and it is seen that $F_i{}^+$ and $F_i{}^-$ separately obey the commutation rules

$$[F_i{}^\pm, F_j{}^\pm] = if_{ijk}F_k{}^\pm, \quad (5.13)$$

while they commute with each other:

$$[F_i{}^\pm, F_j{}^\pm] = 0. \quad (5.14)$$

Thus we are now dealing with the group $U(3)$ taken twice: $U(3)\times U(3)$. Factoring each $U(3)$ into $U(1)\times SU(3)$, we have[29] $U(1)\times U(1)\times SU(3)\times SU(3)$. Thus we have defined a left- and a right-handed baryon number and a left- and right-handed unitary spin.

The situation is just as in Sec. III, where we defined a left- and a right-handed isotopic spin and we could have defined a left- and a right-handed nucleon number.

The left- and right-handed quantities are connected to each other by the parity operation P:

$$PF_i{}^\pm P^{-1} = F_i{}^\mp. \quad (5.15)$$

Now that we have constructed the mathematical apparatus of the group $U(3)\times U(3)$ and its algebra, we may inquire how the Hamiltonian density H behaves under the group, i.e., under commutation with the algebra.

In the model, there is, corresponding to (4.6), the formula

$$H = \bar{H} - L' - L'', \quad (5.16)$$

where $\bar{H}$ is the Hamiltonian density derived from the Lagrangian density $\bar{L}$ and is completely invariant under the group. Instead of defining u_0 as in Sec. III, let us put

$$u_0 = L' \propto \bar{b}\lambda_0 b. \quad (5.17)$$

We can easily see that by commutation of u_0 with F_i and $F_i{}^5$ ($i=0,\cdots,8$) at equal times we obtain a set of eighteen quantities:

$$u_i \propto \bar{b}\lambda_i b,$$
$$v_i \propto -i\bar{b}\lambda_i\gamma_5 b. \quad (5.18)$$

In fact F_i acts like $\lambda_i/2$, $F_i{}^5$ like $\lambda_i\gamma_5/2$, u_i like $\beta\lambda_i$, and v_i like $-i\beta\gamma_5\lambda_i$. Thus we have at equal times[30]

$$[F_i,u_j] = if_{ijk}u_k,$$
$$[F_i,v_j] = if_{ijk}v_k,$$
$$[F_i{}^5,u_j] = -id_{ijk}v_k,$$
$$[F_i{}^5,v_j] = id_{ijk}u_k, \quad (5.19)$$

and the stronger relations

$$[\mathfrak{F}_{i4}(\mathbf{x},t),u_j(\mathbf{x}',t)] = -f_{ijk}u_k(\mathbf{x},t)\delta(\mathbf{x}-\mathbf{x}'), \text{ etc.} \quad (5.20)$$

for the densities. All indices run from 0 to 8.

Note that we can now express not only L' (which is defined to be u_0) but L'' as well, since by (4.7) it is proportional to u_8. We have, then,

$$H = \bar{H} = u_0 - cu_8, \quad (5.21)$$

where c is of the order $(m_{0N}-m_{0\Lambda})/m_{0N}$ in the model.

We may now make a series of abstractions from the model. First, we suppose that currents $\mathfrak{F}_{i\alpha}$ and $\mathfrak{F}_{i\alpha}{}^5$ are defined, with commutation rules (5.3), (5.9), and (5.10), and with the weak current given by the analog of (5.5)[31]:

$$J_\alpha = \mathfrak{F}_{1\alpha} + \mathfrak{F}_{1\alpha}{}^5 + i\mathfrak{F}_{2\alpha} + i\mathfrak{F}_{2\alpha}{}^5 + \mathfrak{F}_{4\alpha} + \mathfrak{F}_{4\alpha}{}^5 + i\mathfrak{F}_{5\alpha} + i\mathfrak{F}_{5\alpha}{}^5, \quad (5.22)$$

[28] Actually the Lagrangian (4.1) without the nucleon mass terms is invariant under a larger continuous group of transformations than the one $[U(3)\times U(3)]$ that we treat here. For example, there are infinitesimal transformations in which the baryon fields b acquire small terms in $\bar{b}$. Invariance under these is associated with the conservation of currents carrying baryon number 2. The author wishes to thank Professor W. Thirring for a discussion of these additional symmetries and of conformal transformations, which give still more symmetry.

[29] The groups $U(1)$, $SU(3)$, and $SU(2)$ cannot be further factored in this fashion. They are called *simple*.

[30] Note that even if we use just F_i and $F_i{}^5$ for $i=1,\cdots,8$, or $SU(3)\times SU(3)$ only, we still generate all eighteen u's and v's. [In the two-dimensional case described in Sec. III the situation is different. Using $SU(2)\times SU(2)$, we generate from u_0 only itself and v_1, v_2, v_3; if we then bring in $F_0{}^5$ as well, we obtain three more u's and one more v.] This remark is interesting because the group that gives currents known to be physically interesting is just $U(1)\times SU(3)\times SU(3)$; there is no known physical coupling to $\mathfrak{F}_{0\alpha}{}^5$, the axial vector baryon current.

[31] Note that the *total* weak current, whether for baryons and mesons or for leptons, is just a component of the current of an angular momentum. See reference 13.

while the electric current is given by (5.4). Next, we may take the Hamiltonian density to be of the form (5.21), with $\bar{H}$ invariant and u_i and v_i transforming as in (5.20). Then, if the theory is of the type described in Appendix A, we can calculate the divergences of the currents in terms of the equal-time commutators

$$\begin{aligned}\partial_\alpha \mathfrak{F}_{i\alpha} &= i[F_i,u_0]+ic[F_i,u_8],\\ \partial_\alpha \mathfrak{F}_{i\alpha}{}^5 &= i[F_i{}^5,u_0]+ic[F_i{}^5,u_8],\end{aligned} \tag{5.23}$$

or, explicitly,

$$\begin{aligned}&\partial_\alpha \mathfrak{F}_{i\alpha}=0, \quad (i=0,\,1,\,2,\,3,\,8)\\ &\partial_\alpha \mathfrak{F}_{4\alpha}=(\tfrac{3}{2})^{\frac{1}{2}}u_5, \text{ etc.},\\ &\partial_\alpha \mathfrak{F}_{0\alpha}{}^5=(\tfrac{2}{3})^{\frac{1}{2}}v_0+(\tfrac{2}{3})^{\frac{1}{2}}cv_8,\\ &\partial_\alpha \mathfrak{F}_{1\alpha}{}^5=[(\tfrac{2}{3})^{\frac{1}{2}}+(\tfrac{1}{3})^{\frac{1}{2}}c]v_1, \text{ etc.},\\ &\partial_\alpha \mathfrak{F}_{4\alpha}{}^5=[(\tfrac{2}{3})^{\frac{1}{2}}-(\tfrac{1}{12})^{\frac{1}{2}}c]v_4, \text{ etc.},\\ &\partial_\alpha \mathfrak{F}_{8\alpha}{}^5=[(\tfrac{2}{3})^{\frac{1}{2}}-(\tfrac{1}{3})^{\frac{1}{2}}c]v_8+(\tfrac{2}{3})^{\frac{1}{2}}cv_0.\end{aligned} \tag{5.24}$$

Finally, if we taken the model really seriously, we may abstract the equal-time commutation relations of the u_i and v_i as obtained by explicit commutation in the model.

The relations of Sec. III are all included in those of this section, except that what was called u_0 there is now called u_0+cu_8 and what was called v_i is now called $[(\frac{2}{3})^{\frac{1}{2}}+(\frac{1}{3})^{\frac{1}{2}}c]v_i$ for $i=1,\,2,\,3$.

All of the relations used here are supposed to be exact and are not affected by the symmetry-breaking character of the non-invariant term in the Hamiltonian. In the next section, we discuss what happens if c can be regarded as small in any sense. We may then expect to see some trace of the symmetry under $U(3)$ that would obtain if c were 0 and L'' disappeared. In this limit, N and Λ are degenerate, and all the components F_i of the unitary spin are conserved. The higher symmetry would show up particularly through the existence of degenerate baryon and meson supermultiplets, which break up into ordinary isotopic multiplets when L'' is turned on. These supermultiplets have been discussed previously for baryons and pseudoscalar mesons[4,6] and then for vector mesons.[32–34]

We shall not discuss the case in which both L' and L'' are turned off; that is the situation, still more remote from reality, in which all the axial vector currents are conserved as well as the vector ones.

VI. BROKEN SYMMETRY—MESON SUPERMULTIPLETS

We know that because of isotopic spin conservation the baryons and mesons form degenerate isotopic multiplets, each corresponding to an irreducible representation of the isotopic spin algebra (3.1). Each multiplet has $2I+1$ components, where the quantum number I distinguishes one representation from another and gives us the eigenvalue $I(I+1)$ of the operator $\sum_{i=1}^{3} I_i^2$, which commutes with all the elements of the isotopic spin group. The operators I_i are represented, within the multiplet, by Hermitian $(2I+1)\times(2I+1)$ matrices having the commutation rules (3.1) of the algebra.

If we start from the doublet representation, we can build up all the others by considering combinations of particles that transform like the original doublet. Just as (p,n) form a doublet representation for which the I_i are represented by $\tau_i/2$, the antiparticles $(\bar{n},-\bar{p})$ also form a doublet representation that is equivalent. (Notice the minus sign on the antiproton state or field.) Now, if we put together a nucleon and an antinucleon, we can form the combination

$$\bar{N}N=\bar{p}p+\bar{n}n,$$

which transforms like an isotopic singlet, or the combinations

$$\bar{N}\tau_i N, \quad (i=1,\,2,\,3)$$

which form an isotopic triplet. The direct product of nucleon and antinucleon doublets gives us a singlet and a triplet. Any meson that can dissociate virtually into nucleon and antinucleon must be either a singlet or a triplet. For the singlet state, the components I_i are all zero, while for the three triplet states the three 3×3 matrices, $I_i{}^{jk}$ of the components I_i, are given by

$$I_i{}^{jk}=-i\epsilon_{ijk}. \tag{6.1}$$

Now let us generalize these familiar results to the unitary spin and the three basic baryons b (comprising n, p, and Λ). These three fields or particles form a three-dimensional irreducible representation of the unitary spin algebra (4.16) from which all the other representations may be constructed.

For example, consider a meson that can dissociate into b and $\bar{b}$. It must transform either like

$$\bar{b}b=\bar{p}p+\bar{n}n+\bar{\Lambda}\Lambda,$$

a unitary singlet, or else like

$$\bar{b}\lambda_i b, \quad (i=1,\,\cdots,\,8)$$

a unitary octet.

The unitary singlet is evidently neutral, with strangeness $S=0$, and forms an isotopic singlet. But how does the unitary octet behave with respect to isotopic spin? We form the combinations

$$\begin{aligned}&\left.\begin{aligned}\bar{b}(\lambda_1-i\lambda_2)b/2&=\bar{n}p,\\ \bar{b}\lambda_3 b/\sqrt{2}&=(\bar{p}p-\bar{n}n)/\sqrt{2},\\ \bar{b}(\lambda_1+i\lambda_2)b/2&=\bar{p}n,\end{aligned}\right\}\quad I=1,\,S=0\\ &\left.\begin{aligned}\bar{b}(\lambda_4-i\lambda_5)b/2&=\bar{\Lambda}p,\\ \bar{b}(\lambda_6-i\lambda_7)b/2&=\bar{\Lambda}n,\end{aligned}\right\}\quad I=\tfrac{1}{2},\,S=+1\\ &\left.\begin{aligned}\bar{b}(\lambda_4+i\lambda_5)b/2&=\bar{p}\Lambda,\\ \bar{b}(\lambda_6+i\lambda_7)b/2&=\bar{n}\Lambda,\end{aligned}\right\}\quad I=\tfrac{1}{2},\,S=-1\\ &\bar{b}\lambda_8 b/\sqrt{2}=(\bar{p}p+\bar{n}n-2\bar{\Lambda}\Lambda)/\sqrt{6},\quad I=0,\,S=0,\end{aligned} \tag{6.2}$$

[32] M. Gell-Mann, California Institute of Technology Synchrotron Laboratory Report No. CTSL-20, 1961 (unpublished).
[33] Y. Ne'eman, Nuclear Phys. 26, 222 (1961).
[34] A. Salam and J. C. Ward, Nuovo cimento 20, 419 (1961).

and we see immediately that the unitary octet comprises an isotopic triplet with $S=0$, a pair of isotopic doublets with $S=\pm1$, and an isotopic singlet with $S=0$. All these are degenerate only in the limit of unitary symmetry ($L''=0$); when the mass-splitting term is turned on, the singlet, the triplet, and the pair of doublets should have three somewhat different masses.

The known pseudoscalar mesons (π, K, and $\bar{K}$) fit very well into this picture, provided there is an eighth pseudoscalar meson to fill out the octet. Let us call the hypothetical isotopic singlet pseudoscalar meson χ^0. Since it is pseudoscalar, it cannot dissociate (virtually or really) into 2π. It has the value $+1$ for the quantum number G, so that it cannot dissociate into an odd number of pions either. Thus in order to decay by means of the strong interactions, it must have enough energy to yield 4π. It would then appear as a 4π resonance. The decay into 4π is, however, severely hampered by centrifugal barriers.

If the mass of χ^0 is too low to permit it to decay readily into 4π, then it will decay electromagnetically. If there is sufficient energy, the decay mode $\chi^0\rightarrow\pi^++\pi^-+\gamma$ is most favorable; otherwise[34a] it will decay into 2γ like π^0.

Let us now turn to the vector mesons. The best known vector meson is the $I=1$, $J=1^-$ resonance of 2π, which we shall call ρ. It has a mass of about 750 Mev. According to our scheme, it should belong, like the pion, to a unitary octet. Since it occupies the same position as the π ($I=1, S=0$), we denote it by the succeeding letter of the Greek alphabet.

The vector analog of χ^0 we shall call ω^0 (skipping the Greek letter ψ). It must have $I=0$, $J=1^-$, and $G=-1$ and so it is capable of dissociation into $\pi^++\pi^-+\pi^0$. Presumably it is the 3π resonance found experimentally[35] at about 790 Mev.

In order to complete the octet, we need a pair of strange doublets analogous to K and $\bar{K}$. In the vector case, we shall call them M and $\bar{M}$ (skipping the letter L). Now there is a known $K\pi$ resonance with $I=\frac{1}{2}$ at about 884 Mev. If it is a p-wave resonance, then it fits the description of M perfectly.

In the limit of unitary symmetry, we can have, besides the unitary octet of vector mesons, a unitary singlet. The hypothetical B^0 that we discussed in Sec. III would have such a character. If B^0 exists, then the turning-on of the mass-splitting term L'' mixes the states B^0 and ω^0, which are both *isotopic* singlets.

Other mesons may exist besides those discussed, for example, scalar and axial vector mesons. All those that can associate into $b+\bar{b}$ should form unitary octets or singlets or both, with each octet splitting into isotopic multiplets because of the symmetry-breaking term L''.

TABLE III. Possible meson octets and singlets.

Unitary spin	Isotopic spin	Strangeness	Pseudoscalar	Vector	Scalar	Axial vector
Octet	1	0	π	ρ	π'	ρ'
	1/2	+1	K	M	K'	M'
	1/2	−1	$\bar{K}$	$\bar{M}$	$\bar{K}'$	$\bar{M}'$
	0	0	χ	ω	χ'	ω'
Singlet	0	0	A	B	A'	B'

A list of some possible meson states is given in Table III, along with suggested names for the mesons.

It is interesting that we can predict not only the degeneracy of an octet in the limit $L''\rightarrow0$ but also a sum rule[32] that holds in first order in L'':

$$\begin{aligned}(m_K+m_{\bar{K}})/2&=(3m_\chi+m_\pi)/4,\\(m_M+m_{\bar{M}})/2&=(3m_\omega+m_\rho)/4.\end{aligned}\qquad(6.3)$$

If M is at about 884 Mev and ρ at about 750 Mev, then ω should lie at about 930 Mev according to the sum rule; since it is actually at 790 Mev, the sum rule does not seem to give a good description of the splitting. Perhaps an important effect is the repulsion between the ω^0 and B^0 levels, pushing ω^0 down and B^0 up. For what it is worth, (6.3) gives a χ^0 mass of around 610 Mev.

In the limit of unitary symmetry, not only are the supermultiplets degenerate but their effective couplings are symmetrical. For example, the effective coupling of the unitary pseudoscalar octet to N and Λ takes the form

$$ig_1\{\bar{N}\tau\gamma_5N\cdot\pi+\bar{N}\gamma_5\Lambda K+\bar{\Lambda}\gamma_5N\bar{K}+3^{-\frac{1}{2}}\bar{N}\gamma_5N\chi-2\times3^{-\frac{1}{2}}\bar{\Lambda}\gamma_5\Lambda\chi\},\qquad(6.4)$$

in terms of renormalized "fields." Now, as the term L'' is turned on, the various coupling constants become unequal; instead of calling them all g_1, we refer to them as $g_{NN\pi}$, $g_{N\Lambda K}$, $g_{NN\chi}$, and $g_{\Lambda\Lambda\chi}$, respectively, each of these constants being the measurable renormalized coupling parameter at the relevant pole.

We have written the effective coupling (6.4) as if there were renormalized fields for all the particles involved, but that is only a matter of notation; the mesons can perfectly well be composite. We may simplify the notation still further by constructing a traceless 3×3 matrix Π containing the pseudoscalar "fields" in such a way that (6.4) becomes

$$ig_1\bar{b}\Pi\gamma_5b.\qquad(6.5)$$

We may now write, in a trivial way, other effective couplings in the limit of unitary symmetry. We define a traceless 3×3 matrix W_α containing the "fields" for the vector meson octet just as Π^* contains those for the pseudoscalar octet. We then have the invariant

[34a] *Note added in proof*. H. P. Duerr and W. Heisenberg (preprint) have pointed out the importance of the decay mode $\chi^0\rightarrow3\pi$ induced by electromagnetism. For certain χ masses, it may be a prominent mode.

[35] B. C. Maglić, L. W. Alvarez, A. H. Rosenfeld, and M. L. Stevenson, Phys. Rev. Letters 7, 178 (1961).

effective coupling

$$i\gamma_1 \,\mathrm{Tr} W_\alpha(\Pi\partial_\alpha\Pi-\partial_\alpha\Pi\Pi)/2 \qquad (6.6)$$

in the symmetric limit. When the asymmetry is turned on, the single coupling parameter γ_1 is replaced by the set of different parameters $\gamma_{\rho\pi\pi}$, $\gamma_{\rho KK}$, $\gamma_{\omega KK}$, $\gamma_{MK\pi}$, and $\gamma_{MK\chi}$.

In the same way, we have another effective coupling

$$ih_1 \,\mathrm{Tr}\Pi(\partial_\alpha W_\beta-\partial_\beta W_\alpha)(\partial_\gamma W_\delta-\partial_\delta W_\gamma)e_{\alpha\beta\gamma\delta} \qquad (6.7)$$

in the symmetric limit; in the actual asymmetric case, we define the distinct constants $h_{\pi\omega\rho}$, $h_{\pi MM}$, $h_{\chi\omega\omega}$, $h_{\chi\rho\rho}$, $h_{\chi MM}$, $h_{KM\rho}$, and $h_{KM\omega}$. All of these constants can be measured, in principle, in "pole" experiments, except that for the broad resonances like ρ the poles are well off the physical sheet.

We have generalized the definitions of constants like $g_{NN\pi}$ and $\gamma_{\rho\pi\pi}$, as used in Sec. II, to other particles. The constants γ_ρ and f_π of Sec. II also have analogs, of course, and we define γ_ω, f_K, etc., in the obvious way. In the limit of unitary symmetry, of course, we would have $f_\pi=f_K=f_\chi$ and $\gamma_\rho=\gamma_\omega=\gamma_M$. Likewise, the constant $-G_A/G$ for nucleon β decay would equal the corresponding quantity $-G_A{}^{\Lambda N}/G$ for the β decay of Λ.

VII. BROKEN SYMMETRY—BARYON SUPERMULTIPLETS

What has been done in the previous section may be described mathematically as follows. We considered a three-dimensional representation of the unitary spin algebra (4.16) or of the group $SU(3)$ that is generated by the algebra. It is the *representation to which b* belongs (that is, n, p, and Λ) and we may denote it by the symbol **3**.

The antiparticles $\bar{b}$ belong to the conjugate representation **3***, which is inequivalent[36] to **3**. We have then taken the direct product **3**×**3*** and found it to be given by the rule

$$\mathbf{3}\times\mathbf{3}^*=\mathbf{8}+\mathbf{1}, \qquad (7.1)$$

where **8** is the octet representation and **1** the singlet representation of unitary spin. Each of these is its own conjugate; that is a situation that occurs only when the dimension is the cube of an integer.

There are, of course, more complicated representations to which mesons might belong that are incapable (in the limit of unitary symmetry) of dissociation into $b+\bar{b}$ but capable of dissociation into $2b+2\bar{b}$ or higher configurations. But we might guess that at least the mesons of lowest mass would correspond to the lowest configurations.

Now we want to examine the simplest configurations for baryons, apart from just b. Evidently the next simplest is $2b+\bar{b}$, which poses the problem of reducing the direct product **3**×**3**×**3***; the result is the following:

$$\mathbf{3}\times\mathbf{3}\times\mathbf{3}^*=\mathbf{3}\times\mathbf{1}+\mathbf{3}\times\mathbf{8}=\mathbf{3}+\mathbf{3}+\mathbf{6}+\mathbf{15}. \qquad (7.2)$$

*The six-dimensional representation **6** is composed of an* isotopic triplet with $S=-1$, a doublet with $S=0$, and a singlet with $S=+1$; the fifteen-dimensional representation **15** is composed of a doublet with $S=-2$, a singlet and a triplet with $S=-1$, a doublet and a quartet with $S=0$, and a triplet with $S=+1$.

According to the scheme, then, Ξ should belong to **15**. Where are the other members of the supermultiplet? For $S=-1$ and $S=0$, there are many known resonances, some of which might easily have the same spin and parity as Ξ. For $S=+1$, $I=1$, however, no resonance has been found so far (in K^+-p scattering, for example).

*The hyperon Σ should also be placed in a super*multiplet, which may or may not be the same one to which Ξ belongs; we do not know if the spin and parity of Σ and Ξ are the same, with K taken to be pseudoscalar. If Σ belongs to **6** in the limit of unitary symmetry, then there should be a $\bar{K}N$ resonance in the $I=0$ state.

It is difficult to say at the present time if the baryon states can be reconciled with the model. Further knowledge of the baryon resonances is required.

One curious possibility is that the fundamental objects b are hidden and that the physical N and Λ, instead of belonging to **3**, belong, along with Σ and Ξ, to the representation **15** in the limit of unitary symmetry. *That would require the spins and parities of* N, Λ, Σ, and Ξ to be equal, and it would require a πN resonance in the $p_{\frac{1}{2}}$, $I=\frac{3}{2}$ state as well as a KN resonance in the $p_{\frac{1}{2}}$, $I=1$ state to fill out the supermultiplet.

VIII. THE "EIGHTFOLD WAY"

Unitary symmetry may be applied to the baryons in a more appealing way if we abandon the connection with the symmetrical Sakata model and treat unitary symmetry in the abstract. (An abstract approach is, of course, required if there are no "elementary" baryons and mesons.) Of all the groups that could be generated by the vector weak currents, $SU(3)$ is still the smallest and the one that most naturally gives rise to the rules $|\Delta \mathbf{I}|=\frac{1}{2}$ and $\Delta S/\Delta Q=0, +1$.

There is no longer any reason for the baryons to belong to the **3** representation or the other spinor representations of the group $SU(3)$; the various irreducible spinor representations are those obtained by reducing direct products like **3**×**3**×**3***, **3**×**3**×**3**×**3***×**3***, etc.

Instead, the baryons may belong, like the mesons, to representations such as **8** or **1** obtained by reducing the direct products of equal numbers of **3**'s and **3***'s. It is then natural to assign the stable and metastable baryons N, Λ, Σ, and Ξ to an octet, degenerate in the limit of unitary symmetry. We thus obtain the scheme

[36] In other words, no unitary transform can convert the representations **3** and **3*** into each other. That is easy to see, since the eigenvalues of λ_8 are opposite in sign for the two representations, and changing the signs changes the set of eigenvalues. In the case of the group $SU(2)$ of isotopic spin transformations, the basic spinor representation $I=\frac{1}{2}$ *is* equivalent to the corresponding antiparticle representation.

of Gell-Mann[32] and Ne'eman[33] that we call the "eightfold way." The component F_8 of the unitary spin is now $(\sqrt{3}/2)Y$, where Y is the hypercharge (equal to strangeness plus baryon number).

The baryons of the octet must have the same spin and parity (treating K as pseudoscalar). To first order in the violation of unitary symmetry, the masses should obey the sum rule analogous to (6.3):

$$(m_N+m_\Xi)/2=(3m_\Lambda+m_\Sigma)/4, \qquad (8.1)$$

which agrees surprisingly well with observations, the two sides differing by less than 20 Mev.

To form mesons that transform like combinations of these baryons and their antiparticles, we reduce the direct product 8×8 (remembering that $8=8^*$) and obtain

$$8\times8=1+8+8+10+10^*+27, \qquad (8.2)$$

where **1** and **8** are the singlet and octet representations already discussed; **10** consists of an isotopic triplet with $Y=0$, a doublet with $Y=-1$, a quartet with $Y=+1$, and a singlet with $Y=-2$; **10*** has the opposite behavior with respect to Y; and **27** consists of an isotopic singlet, triplet, and quintet with $Y=0$, a pair of doublets with $Y=\pm1$, a pair of quartets with $Y=\pm1$, and a pair of triplets with $Y=\pm2$. Evidently the known mesons are to be assigned to octets and perhaps singlets, as in Sec. VI. The meson-nucleon scattering resonances must then also be assigned representations among those in (8.2); the absence so far of any observed structure in K-N scattering makes it difficult to place the $I=3/2$, $J=3/2$, π-N resonance in a supermultiplet.

The fact that **8** occurs twice in Eq. (8.2) means that there are two possible forms of symmetrical Yukawa coupling of a meson octet to the baryon octet in the limit of unitary symmetry. As in Sec. VI for the mesons, we form a 3×3 traceless matrix out of the formal "fields" of the baryon octet; call it $\mathcal{B}$. The effective symmetrical coupling of pseudoscalar mesons may then be written as

$$ig_1\alpha\,\mathrm{Tr}\,(\bar{\mathcal{B}}\pi\gamma_5\mathcal{B}+\pi\bar{\mathcal{B}}\gamma_5\mathcal{B})/2$$
$$+ig_1(1-\alpha)\,\mathrm{Tr}\,(\bar{\mathcal{B}}\pi\gamma_5\mathcal{B}-\pi\bar{\mathcal{B}}\gamma_5\mathcal{B})/2. \qquad (8.3)$$

The two types of coupling differ in their behavior under the operation R that exchanges N and Ξ, K and $\bar{K}$, M and $\bar{M}$, etc.; the first term is symmetric while the second is antisymmetric under R. The parameter α just specifies how much of each effective coupling is presented in the limit of unitary symmetry. When we take into account violations of the symmetry, we must define separate coupling constants $g_{NN\pi}$, $g_{NK\Lambda}$, etc., in a suitable way.

Likewise the vector mesons have the general symmetrical coupling

$$i\gamma_1\beta\,\mathrm{Tr}\,(\bar{\mathcal{B}}W_\alpha\gamma_\alpha\mathcal{B}+W_\alpha\bar{\mathcal{B}}\gamma_\alpha\mathcal{B})$$
$$+i\gamma_1(1-\beta)\,\mathrm{Tr}\,(\bar{\mathcal{B}}W_\alpha\gamma_\alpha\mathcal{B}-W_\alpha\bar{\mathcal{B}}\gamma_\alpha\mathcal{B}), \qquad (8.4)$$

where we ignore Pauli moment terms for simplicity. To the extent that the vector meson octet W_α dominates the dispersion relations for the unitary spin current $\mathfrak{F}_{i\alpha}$, then the mesons of W_α are coupled effectively to the components of $\mathfrak{F}_{i\alpha}$, and we have $\beta=0$ in (8.5). Then ρ is effectively coupled to the isotopic spin current, ω to the hypercharge current, and M to the strangeness-changing vector current. The first two of these currents are conserved, and so we have the approximate universality of ρ and ω couplings proposed by Sakurai[11] and discussed in Sec. II. In the limit of unitary symmetry, under the assumptions just mentioned, ρ is effectively coupled to the current of $2\gamma_1\mathbf{I}$ and ω to the current of $2\gamma_1F_8=\sqrt{3}\gamma_1Y$.

The electromagnetic current is now given by the formula

$$j_\alpha=\mathfrak{F}_{3\alpha}+3^{-\frac{1}{2}}\mathfrak{F}_{8\alpha} \qquad (8.5)$$

instead of (5.4), while the weak vector current is still described by Eq. (5.5). If we are to treat the vector and axial vector currents by means of $SU(3)\times SU(3)$, as we did earlier, then the entire weak current is given by (5.22) and we have the commutation rules (5.3), (5.9), and (5.10) for the various components of the currents. The question of the behavior of H under the group $SU(3)\times SU(3)$ should, however, be re-examined for the eightfold way; we shall not go into that question here. But let us consider how the baryon octet transforms in the limit of conserved vector *and* axial vector currents [invariance under $SU(3)\times SU(3)$]. In the Sakata model, the left-handed baryons transformed under (F_i^+,F_i^-) like (3, 1), while the right-handed baryons transformed according to (1, 3). For the eightfold way, there are two simple possibilities for these transformation properties. Either we have (8, 1) and (1, 8) or else we adjoin a ninth neutral baryon (which need not be degenerate with the other eight in the limit of conserved *vector* currents and which need not have the same parity) and use the transformation properties (3, 3*) and (3*, 3). In the first case, the baryons transform like the quantities $\mathfrak{F}_{i\alpha}$ and $\mathfrak{F}_{i\alpha}{}^5$ $(i=1,\cdots,8)$ and in the second case they transform like u_i and v_i $(i=0,\cdots,8)$ of Sec. V.

IX. REMARKS AND SUGGESTIONS

Our approach to the problem of baryon and meson couplings leads to a number of suggestions for new investigations, both theoretical and experimental.

First, the equal-time commutation relations for currents and densities lead to exact sum rules for the weak and electromagnetic matrix elements. As an example, take the commutation rules (3.5) for the isotopic spin current. These do not, of course, depend on any higher symmetry, but they can be used to illustrate the results that can be obtained from the more general relations like (5.3).

Consider the electromagnetic form factor $F_\pi(s)$ of the charged pion, which is just the form factor of the isotopic spin current between one-pion states. Let p and p' be the initial and final pion four-momenta, with $s=-(p-p')^2$. Let K be any four-momentum with $K^2=-m_\pi^2$. Then, taking the matrix element of (3.5) between one-pion states, we obtain the result

$$\begin{aligned}2(p_0+p_0')K_0F_\pi(-(p-p')^2)\\ =(p_0+K_0)(p_0'+K_0)F_\pi(-(p-K)^2)F_\pi(-(p'-K)^2)\\ -(p_0-K_0)(p_0'-K_0)F_\pi(-(p+K)^2)F_\pi(-(p'+K)^2)\\ +\text{inelastic terms},\quad(9.1)\end{aligned}$$

where the inelastic terms come from summing over bilinear forms in the inelastic matrix elements of the current. We see that if there is no inelasticity the form factor is unity. Thus the departure from unity of $F_\pi(s)$ is related to the amount of inelasticity.

A similar relation is familiar in nonrelativistic quantum mechanics:

$$\langle e^{i(p-p')\cdot x}\rangle_{00}=\sum_n\langle e^{i(p-k)\cdot x}\rangle_{0n}\langle e^{i(k-p')\cdot x}\rangle_{n0}.\quad(9.2)$$

If we apply relations like (9.1) to the matrix elements of non-conserved currents like $\mathbf{P}_\alpha$, along with the linear homogeneous dispersion relations for these matrix elements, we can in principle determine constants like $-G_A/G$.

A second line of theoretical investigation is suggested by the vanishing at high momentum transfer of matrix elements of divergences of non-conserved currents, like $\partial_\alpha \mathbf{P}_\alpha$. We should try to find limits involving high energies and high momentum transfers in which we can show that the conservation of helicity, unitary spin, etc., becomes valid. A preliminary effort in this direction has been made by Gell-Mann and Zachariasen.[37]

A third topic of study is the testing of broken symmetry at low energy. Do the mesons fall into unitary octets and singlets? An experimental search for χ^0 is required and also a determination of the spin of K^* at 884 Mev to see if it really is our M meson.

Let us discuss briefly the properties of χ^0. An $I=0$ state of 4π can have two types of symmetry: either totally symmetric (partition [4]) in both space and isotopic spin or else the symmetry of the partition [2+2] in space and in isotopic spin. For a pseudoscalar state, the first type of wave function in momentum space is very complicated. If $\mathbf{p}$, $\mathbf{q}$, and $\mathbf{r}$ are the three momentum differences, it must look like

$$\mathbf{p}\cdot\mathbf{q}\times\mathbf{r}(E_1-E_2)(E_2-E_3)(E_3-E_4)\times(E_1-E_3)(E_1-E_4)(E_2-E_4),$$

times a symmetric function of the energies E_1, E_2, E_3, E_4 of the four pions. On the basis of any reasonable dynamical picture of χ^0, such a wave function should have a very small amplitude. In particular, dispersion theory suggests that the wave function of χ^0 should have large contributions from virtual dissociation into 2ρ, which gives a wave function with [2+2] symmetry.

[37] M. Gell-Mann and F. Zachariasen, Phys. Rev. 123, 1065 (1961).

If [2+2] predominates, then the charge ratio in decay is 2:1 in favor of $2\pi^0+\pi^++\pi^-$ over $2\pi^++2\pi^-$, with $4\pi^0$ absent. If virtual dissociation into 2ρ actually predominates, then the matrix element of the 4π configuration is easily written down and the spectrum of the decay $\chi^0\to 4\pi$ can be calculated.

If χ^0 is lighter than 4π, it will, of course, decay electromagnetically. Even if it is above threshold for 4π, however, the matrix element for decay contains so many powers of pion momenta that electromagnetic decay should be appreciable over a large range of masses. The branching ratio $(\pi^++\pi^-+\gamma)/(4\pi)$ is approximately calculable by dispersion methods. In both cases χ^0 first dissociates into 2ρ. Then either both virtual ρ mesons decay into 2π, or else (in the case where both are neutral) one may decay into $\pi^++\pi^-$, while the other turns directly into γ. If we draw a diagram for such a process, then the constant $\gamma_{\rho\pi\pi}$ is inserted whenever we have a $\rho\pi\pi$ vertex and the constant $em_\rho^2/2\gamma_\rho$ at a ρ-γ junction.[10]

If the meson spectrum is consistent with broken unitary symmetry, we should examine the baryons, and see whether the various baryon states fit into the representations **3**, **6**, and **15** (or the representations **1**, **8**, **10**, **10***, and **27** that arise in the alternative form of unitary symmetry).

If some states are lacking in a given supermultiplet, it does not necessarily prove that the broken symmetry is wrong, but only that it is badly violated. We assume that baryon isobars like the πN $\frac{3}{2}$, $\frac{3}{2}$ resonance are dynamical in nature; there may be some attractive and some repulsive forces in this channel, and the attractive ones have won out, producing the resonance. In the KN channel with $I=1$, for example, it is conceivable that the repulsive ones are stronger (because of symmetry violation), and the analogous $p_{\frac{3}{2}}$ resonance disappears. In such a situation, the concept of broken symmetry at low energies is evidently of little value.

Suppose, however, that the idea of broken unitary symmetry is confirmed for both mesons and baryons, say according to the Sakata picture, in which N and Λ belong to the representation **3** in the limit of unitary symmetry. There are, nevertheless, gross violations of unitary symmetry, and the elucidation of these, both theoretical and experimental, is a fourth interesting subject.

If unitary symmetry were exact, then not only would m_K/m_π equal unity, instead of about 3.5, but f_K^2/f_π^2 would be 1 instead of about 6, and $3G_A^2+G_V^2$ for the β decay of Λ would be equal to $3G_A^2+G_V^2$ for the nucleon instead of being 1/15 as large. All these huge departures from unity represent very serious violations of unitary symmetry.

Yet the relatively small mass difference of N and Λ compared to their masses would seem to indicate, if our model is right, that the constant c in Eq. (5.21) is considerably smaller than unity. It is conceivable that the large mass ratio of K to π comes about because the total mass of the system is so small. It is possible that even with a fairly small c (say $\sim -\frac{1}{10}$) we might explain the gross violations of unitary symmetry. We might try to interpret the large values of $g_{NN\pi}^2/g_{NK\Lambda}^2$, f_K^2/f_π^2, etc., in terms of the large value of m_K^2/m_π^2.

An example of such a calculation, and one that illustrates the various methods suggested in this article, is the following. We try to calculate f_K^2/f_π^2 in terms of m_K^2/m_π^2.

Consider the following vacuum expectation value, written in parametric representation:

$$\langle[\mathfrak{F}_{1\alpha}^5(x),\partial_\beta\mathfrak{F}_{1\beta}^5(x')]\rangle_0=i/(2\pi)^3\int d^4K\, e^{iK\cdot(x-x')}$$
$$\times K_\alpha\epsilon(K)\int dM^2/M^2\delta(K^2+M^2)\rho(M^2). \quad (9.3)$$

Here x and x' are arbitrary space-time points. In terms of (9.3), we have

$$\langle[\partial_\alpha\mathfrak{F}_{1\alpha}^5(x),\partial_\beta\mathfrak{F}_{1\beta}^5(x')]\rangle=1/(2\pi)^3\int d^4K\, e^{iK\cdot(x-x')}$$
$$\times\epsilon(K)\int dM^2\,\delta(K^2+M^2)\rho(M^2). \quad (9.4)$$

Now the contribution of the one-pion intermediate state is easily obtained in terms of the constant f_π^2:

$$\rho(M^2)=\delta(M^2-m_\pi^2)m_\pi^4/4f_\pi^2+\text{higher terms}. \quad (9.5)$$

If $\int\rho(M^2)dM^2/M^2$ converges and if the one-pion term dominates, we have

$$\int\rho(M^2)dM^2/M^2\approx m_\pi^2/4f_\pi^2. \quad (9.6)$$

But from (9.3) we can extract the expectation value of the equal-time commutator of the fourth component of $\mathfrak{F}_{1\alpha}^5$ with $\partial_\beta\mathfrak{F}_{1\beta}^5$; making use of (5.20) and (5.24), we can express the result in terms of $\langle u_0\rangle$ and $\langle u_8\rangle$. Thus we find

$$\int\rho(M^2)dM^2/M^2=[(2/3)^{\frac{1}{2}}+(1/3)^{\frac{1}{2}}c]$$
$$\times[(2/3)^{\frac{1}{2}}\langle u_0\rangle_0+(1/3)^{\frac{1}{2}}\langle u_8\rangle_0], \quad (9.7)$$

assuming convergence.

Now we can do exactly the same thing for $\mathfrak{F}_{4\alpha}^5$ and the K meson, obtaining, in place of the formula

$$m_\pi^2/4f_\pi^2\approx[(\tfrac{2}{3})^{\frac{1}{2}}+(\tfrac{1}{3})^{\frac{1}{2}}c][(\tfrac{2}{3})^{\frac{1}{2}}\langle u_0\rangle_0+(\tfrac{1}{3})^{\frac{1}{2}}\langle u_8\rangle_0], \quad (9.8)$$

the analogous result

$$m_K^2/4f_K^2\approx[(\tfrac{2}{3})^{\frac{1}{2}}-(\tfrac{1}{12})^{\frac{1}{2}}c][(\tfrac{2}{3})^{\frac{1}{2}}\langle u_0\rangle_0-(\tfrac{1}{12})^{\frac{1}{2}}\langle u_8\rangle_0]. \quad (9.9)$$

If c is really small, presumably $\langle u_8\rangle_0$ is also small compared to $\langle u_0\rangle_0$. Then we can, roughly, set (9.8) equal to (9.9), obtaining

$$f_K^2/f_\pi^2\approx m_K^2/m_\pi^2. \quad (9.10)$$

The left-hand side is about 6 and the right-hand side about 10. Thus we can, in a crude approximation, calculate the rate of $K^+\to\mu^++\nu$ in terms of that for $\pi^+\to\mu^++\nu$ and explain one large violation of symmetry in terms of another.

The Goldberger-Treiman formula relating f_π, $g_{NN\pi}$, and $(-G_A/G)$ can also be used for the K particle to give a relation among f_K, $g_{N\Lambda K}$, and $(-G_A/G)$ for the β decay of Λ. Of course, the K-particle pole is much closer to the branch line beginning at $(m_K+2m_\pi)^2$ than the pion pole is to the branch line beginning at $9m_\pi^2$; thus the Goldberger-Treiman formula may be quite bad for the K meson. Still, we may try to use it to discuss the coupling of N and Λ to K and to leptons. We have

$$(m_N+m_\Lambda)(-G_A^{N\Lambda}/G)\approx g_{N\Lambda K}/f_K, \quad (9.11)$$

by analogy with (2.8). Comparing the two formulas, we have

$$(-G_A^{N\Lambda}/G)^2(-G_A/G)^{-2}$$
$$\approx g_{N\Lambda K}^2g_{NN\pi}^{-2}(2m_Nf_\pi)^2[(m_\Lambda+m_N)f_K]^{-2}. \quad (9.12)$$

The ratio of g^2 factors is thought to be ~0.1 from photoproduction of K, while the remaining factor on the right is also ~0.1, so that the Goldberger-Treiman relation leads us to expect a very small axial vector β-decay rate for the Λ, much smaller than the observed one. The observed β decay would be nearly all vector; this prediction of the Goldberger-Treiman formula can easily be checked by observing the electron-neutrino angular correlation in the β decay of Λ, using bubble chambers.

We should, of course, try to predict the value of $g_{N\Lambda K}^2g_{NN\pi}^{-2}$ in terms of m_K^2/m_π^2 just as we did above for f_K^2/f_π^2; however, it is a much harder problem.

When we know more about the coupling constants of the vector mesons (strong coupling constants such as $\gamma_{\omega NN}$, $h_{\omega\pi\rho}$, etc., and coupling strengths of currents such as γ_ω, γ_M, etc.) we will be able to make a survey of the pattern of coupling constants as well as the pattern of masses and see whether the higher symmetry has any relevance. Also it should become clear how well the approximation of dominant low-mass states works, in terms of universality of meson couplings and Goldberger-Treiman relations.[38]

In summary, then, we suggest the use of the equal-time commutators to predict sum rules, attempts to derive high-energy conservation laws and to check them

[38] An interesting relation of the Goldberger-Treiman type is one that holds if the trace $\theta_{\alpha\alpha}$ of the stress-energy-momentum tensor has matrix elements obeying highly convergent dispersion relations. Because of the vanishing of the self-stress, the expectation value of $\theta_{\alpha\alpha}$ in the state of a particle at rest gives the mass of the particle. Rewriting the matrix element as one between the vacuum and a one-pair state, we see that the dispersion relation involves intermediate states with $I=0$, $J=0^+$, $G=+1$. If there is a resonance or quasi-resonance in this channel (like the χ' meson of Table III) and if that resonance dominates the dispersion relation at low momentum transfers, then the coupling of the resonant state to different particles is roughly proportional to their masses. That is just the situation discussed by Schwinger in reference 1 and by Gell-Mann and Lévy in reference 17 for the "σ meson."

experimentally, the search for broken symmetry at low energies, attempts to calculate some violations in terms of others, and efforts to check the highly convergent dispersion relations dominated by low-mass states.

Nowhere does our work conflict with the program of Chew *et al.* of dynamical calculation of the S matrix for strong interactions, using dispersion relations. If something like the Sakata model is correct, then most of the mesons are dynamical bound states or resonances, and their properties are calculable according to the program. Those particles for which there are fundamental fields (like n, p, Λ, and B^0 in the specific field-theoretic model) would presumably occur as CDD poles or resonances in the dispersion relations.[39]

If there are no fundamental fields and no CDD poles, all baryons and mesons being bound or resonant states of one another, models like that of Sakata will fail; the symmetry properties that we have abstracted can still be correct, however. This situation would presumably differ in two ways[10] from the one mentioned above. First, all the masses and coupling constants could be calculated from coupled dispersion relations. Second, certain scattering amplitudes at high energies would show different behavior, corresponding to different kinds of subtractions in the dispersion relations. The second point should be investigated further, as it could lead to experimental tests of the "fundamental" character of various particles.[10,40]

ACKNOWLEDGMENTS

It is a pleasure to thank R. P. Feynman, S. L. Glashow, and R. Block for many stimulating discussions of symmetry, and to acknowledge the great value of conversations with G. F. Chew, S. Frautschi, R. Haag, R. Schroer, and F. Zachariasen about the explanation of approximate universality in terms of highly convergent dispersion relations.

APPENDIX

The field theories of the Fermi-Yang and Sakata models, given by Eqs. (3.17) and (4.1), respectively, belong to a general class of theories, which we now describe.

The Lagrangian density L is given as a function of a number of fields ψ_A and their gradients. The kinetic part of the Lagrangian (consisting of those terms containing gradients) is invariant under a set of infinitesimal unitary transformations generated by N independent Hermitian operators R_i, which may depend on the time. Under the transformations, the various fields ψ_A undergo linear recombinations:

$$\begin{aligned}\psi_A(\mathbf{x},t)\rightarrow\psi_A(\mathbf{x},t)&-i\Lambda_i[R_i(t),\psi_A(\mathbf{x},t)]\\&=\psi_A(\mathbf{x},t)+i\Lambda_i\textstyle\sum_B M_i^{AB}\psi_B(\mathbf{x},t),\end{aligned}\quad\text{(A1)}$$

where Λ_i is the infinitesimal gauge constant associated with the ith transformation. The equal-time commutation rules of the R_i are the same as those of the matrices M_i. Moreover, the set of R_i and linear combinations of R_i is algebraically complete under commutation; in other words, we have an algebra. The matrices M_i are the basis of a representation of the algebra (in general, a reducible representation). It is convenient to take the matrices of the basis to be orthonormal,

$$\mathrm{Tr}M_iM_j=(\mathrm{const})\delta_{ij},\quad\text{(A2)}$$

redefining the R_i accordingly. The structure constants c_{ijk} in the commutation rules

$$\begin{aligned}[M_i,M_j]&=ic_{ijk}M_k,\\ [R_i(t),R_j(t)]&=ic_{ijk}R_k(t),\end{aligned}\quad\text{(A3)}$$

are now real and totally antisymmetric in i, j, and k. We may still perform real rotations in the N-dimensional space of the R_i or the M_i. Suppose, after performing such a rotation, that we can split the R_i into two sets that commute with each other. Then our algebra is the direct sum of two commuting algebras. We continue this process until no further splitting is possible, even after performing rotations. The algebra has then been expressed as the direct sum of *simple* algebras. All the simple algebras have been listed by Cartan.[40] Besides the trivial one-dimensional algebra of $U(1)$ (which is not included by the mathematicians), there are the three-dimensional algebra of $SU(2)$, the eight-dimensional algebra of $SU(3)$, and so forth.

Now let us construct the currents of the operators R_i. We consider the gauge transformation of the second kind

$$\psi_A\rightarrow\psi_A(\mathbf{x},t)-i\Lambda_i(\mathbf{x},t)[R_i(t),\psi_A(\mathbf{x},t)],\quad\text{(A4)}$$

and ask what change it induces in the Lagrange density L. There will be a term in Λ_i and a term in $\partial_\alpha\Lambda_i$, so adjusted[17] that the total change is just the divergence of a four-vector:

$$L\rightarrow L(\mathbf{x},t)-\partial_\alpha\Lambda_i(\mathbf{x},t)R_{i\alpha}(\mathbf{x},t)-\Lambda_i(\mathbf{x},t)\partial_\alpha R_{i\alpha}(\mathbf{x},t).\quad\text{(A5)}$$

We define $R_{i\alpha}$ to be the current of R_i. It can be shown that R_i is in fact given by the relation

$$R_i=-i\int R_{i4}d^3x.\quad\text{(A6)}$$

Now if, for constant Λ_i, the whole Lagrangian is invariant under R_i, then the term in Λ_i in (A4) must vanish; we have $\partial_\alpha R_{i\alpha}=0$. In other words, if there is exact symmetry under R_i, the current $R_{i\alpha}$ is conserved.

If there is a noninvariant part of L with respect to the symmetry operation R_i, then the current will not be conserved. By hypothesis, the noninvariant term (call it u) contains no gradients. Therefore, the effect

[39] L. Castillejo, R. H. Dalitz, and F. J. Dyson, Phys. Rev. 101, 453 (1956).

[40] S. C. Frautschi, M. Gell-Mann, and F. Zachariasen (to be published).

[40] E. Cartan, *Sur la Structure des groupes de transformations finis et continus*, thèse (Paris, 1894; 2nd ed., 1933).

of the transformation (A3) for *constant* Λ_i will be simply to add a term $-i\Lambda_i[R_i,u]$ to the Lagrangian density. We have, then, using (A4), the result

$$\partial_\alpha R_{i\alpha}(\mathbf{x},t)=i[R_i(t),u(\mathbf{x},t)]. \quad \text{(A7)}$$

Since u contains no gradients, it is not only the noninvariant term in the Lagrangian density, but also the negative of the noninvariant term in the Hamiltonian density. The invariant part of H evidently commutes with R_i. Thus we have

$$\partial_\alpha R_{i\alpha}(\mathbf{x},t)=-i[R_i(t),H(\mathbf{x},t)]. \quad \text{(A8)}$$

By considering the transformation properties of H under commutation with the algebra, we generate the divergences of all the currents. The formula obtained by integrating (A6) over space is, of course, very familiar:

$$\dot{R}_i=\int\partial_\alpha R_{i\alpha}d^3x=-i\left[R_i,\int Hd^3x\right]. \quad \text{(A9)}$$

PHYSICAL REVIEW VOLUME 145, NUMBER 4 27 MAY 1966

Spontaneous Symmetry Breakdown without Massless Bosons*

PETER W. HIGGS†
Department of Physics, University of North Carolina, Chapel Hill, North Carolina

(Received 27 December 1965)

We examine a simple relativistic theory of two scalar fields, first discussed by Goldstone, in which as a result of spontaneous breakdown of $U(1)$ symmetry one of the scalar bosons is massless, in conformity with the Goldstone theorem. When the symmetry group of the Lagrangian is extended from global to local $U(1)$ transformations by the introduction of coupling with a vector gauge field, the Goldstone boson becomes the longitudinal state of a massive vector boson whose transverse states are the quanta of the transverse gauge field. A perturbative treatment of the model is developed in which the major features of these phenomena are present in zero order. Transition amplitudes for decay and scattering processes are evaluated in lowest order, and it is shown that they may be obtained more directly from an equivalent Lagrangian in which the original symmetry is no longer manifest. When the system is coupled to other systems in a $U(1)$ invariant Lagrangian, the other systems display an induced symmetry breakdown, associated with a partially conserved current which interacts with itself via the massive vector boson.

I. INTRODUCTION

THE idea that the apparently approximate nature of the internal symmetries of elementary-particle physics is the result of asymmetries in the stable solutions of exactly symmetric dynamical equations, rather than an indication of asymmetry in the dynamical equations themselves, is an attractive one. Within the framework of quantum field theory such a "spontaneous" breakdown of symmetry occurs if a Lagrangian, fully invariant under the internal symmetry group, has such a structure that the physical vacuum is a member of a set of (physically equivalent) states which transform according to a nontrivial representation of the group. This degeneracy of the vacuum permits nontrivial multiplets of scalar fields (which may be either fundamental dynamic variables or polynomials constructed from them) to have nonzero vacuum expectation values, whose appearance in Feynman diagrams leads to symmetry-breaking terms in propagators and vertices. That vacuum expectation values of scalar fields, or "vacuons," might play such a role in the breaking of symmetries was first noted by Schwinger[1] and by Salam and Ward.[2] Under the alternative name, "tadpole" diagrams, the graphs in which vacuons appear have been used by Coleman and Glashow[3] to account for the observed pattern of deviations from $SU(3)$ symmetry.

The study of field theoretical models which display spontaneous breakdown of symmetry under an internal Lie group was initiated by Nambu,[4] who had noticed[5] that the BCS theory of superconductivity[6] is of this type, and was continued by Glashow[7] and others.[8] All these authors encountered the difficulty that their theories predicted, *inter alia*, the existence of a number of massless scalar or pseudoscalar bosons, named "zerons" by Freund and Nambu.[9] Since the models which they discussed, being inspired by the BCS theory, used an attractive interaction between massless fermions and antifermions as the mechanism of symmetry breakdown, it was at first unclear whether zerons occurred as a result of the approximations (including the usual cutoff for divergent integrals) involved in handling the models or whether they would still be there in an exact solution. Some authors,

* This work was partially supported by the U. S. Air Force Office of Scientific Research under grant No. AF-AFOSR-153-64.

† On leave from the Tait Institute of Mathematical Physics, University of Edinburgh, Scotland.

[1] J. Schwinger, Phys. Rev. 104, 1164 (1954); Ann. Phys. (N. Y.) 2, 407 (1957).

[2] A. Salam and J. C. Ward, Phys. Rev. Letters 5, 390 (1960); Nuovo Cimento 19, 167 (1961).

[3] S. Coleman and S. L. Glashow, Phys. Rev. 134, B671 (1964).

[4] Y. Nambu and G. Jona-Lasinio, Phys. Rev. 122, 345 (1961); 124, 246 (1961); Y. Nambu and P. Pascual, Nuovo Cimento 30, 354 (1963).

[5] Y. Nambu, Phys. Rev. 117, 648 (1960).

[6] J. Bardeen, L. N. Cooper, and J. R. Schrieffer, Phys. Rev. 106, 162 (1957).

[7] M. Baker and S. L. Glashow, Phys. Rev. 128, 2462 (1962); S. L. Glashow, *ibid.* 130, 2132 (1962).

[8] M. Suzuki, Progr. Theoret. Phys. (Kyoto) 30, 138 (1963); 30, 627 (1963); N. Byrne, C. Iddings, and E. Shrauner, Phys. Rev. 139, B918 (1965); 139, B933 (1965).

[9] P. G. O. Freund and Y. Nambu, Phys. Rev. Letters 13, 221 (1964).

wishing to identify their zerons with known massive scalar or pseudoscalar mesons, were prepared to spoil the elegance of their theories by adding symmetry-breaking terms to the Lagrangian in order to generate masses.

That zerons must indeed be present in Lorentz invariant field theories in which an internal symmetry breaks down spontaneously was first shown by Goldstone.[10] He clarified the nature of the phenomenon considerably by exhibiting it in a model of a self-interacting scalar field, where the vacuon is the vacuum expectation value of the field itself, rather than that of a bilinear combination of fermion operators which occurs in the BCS model and its progeny. In a theory of this type the breakdown of symmetry occurs already at the level of classical field theory, where vacuons are just nontrivial translationally invariant solutions of the field equations, and zerons, whose existence is readily demonstrated, are small-amplitude waves (superimposed on a "vacuon" solution) whose frequency tends to zero as their wavelength tends to infinity. In a later paper[11] the proof of the Goldstone theorem, as it is now known, was generalized to allow for the possibility that vacuons might be formed from polynomials of any degree in the fundamental field variables of a dynamical system.

During the last few years the problem of how to avoid massless Goldstone bosons has received much attention. Attempts in this direction have been encouraged by the observation that the BCS model does not contain such excitations, provided that Coulomb interactions are taken into account.[12] Klein and Lee[13] showed that in a nonrelativistic theory the spectral representations upon which the more sophisticated proofs of Goldstone's theorem are based are not so restricted in form as to allow the proof to go through, and they conjectured that this might remain true in some relativistic theories. But Gilbert[14] pointed out that their extra terms are ruled out in relativistic theories by the requirement of manifest Lorentz covariance. The present writer restored the status quo to a limited extent by remarking[15] that radiation gauge formulations of gauge field theories, of which electrodynamics is the simplest and best known example, can describe Lorentz-invariant dynamical systems despite the lack of manifest covariance of some of the equations. The freedom which Klein and Lee hoped for in the spectral representations is thereby restored sufficiently to invalidate the Goldstone theorem. From a more physical standpoint one may regard this as an effect of Coulomb interactions, treated now as part of a relativistic field theory.

More recently Guralnik, Hagen, and Kibble[16] and Lange[17] have studied how the failure of global (as distinct from local) conservation laws in spontaneous breakdown theories is related to the existence of Goldstone bosons. Meanwhile, proofs of the theorem have reached new levels of sophistication within the language (or languages) of axiomatic field theory.[18] It has been pointed out that theories of the type proposed in Ref. 15 do not contradict the Goldstone theorem, but rather represent a departure from the assumptions, such as *manifest* covariance and *manifest* causality, upon which it is based. Such considerations do not seem relevant to the possible usefulness of such theories in generating zeron-free models of spontaneous symmetry breakdown, a point which seems to have been overlooked by those[19] who proclaim the failure of the Nambu-Glashow program.

In parallel with the development of "superconductor" models a program has emerged for describing the weak,[20] and possibly also the strong,[21] interactions by an extension of the gauge principle[22] which operates in electrodynamics. According to this principle the symmetry group of the Lagrangian is to be enlarged from global to local (i.e., coordinate-dependent) transformations: To maintain the invariance of derivative terms it is necessary to couple the currents of the group generators to a multiplet of vector fields belonging to the adjoint representation.[23] Like the "superconductor" theories, these gauge theories have suffered from a zero-mass difficulty: The gauge principle appears to guarantee that the associated vector field quanta are massless, in

[10] J. Goldstone, Nuovo Cimento 19, 154 (1961).

[11] J. Goldstone, A. Salam, and S. Weinberg, Phys. Rev. 127, 965 (1962).

[12] P. W. Anderson, Phys. Rev. 112, 1900 (1958).

[13] A. Klein and B. W. Lee, Phys. Rev. Letters 12, 266 (1964).

[14] W. Gilbert, Phys. Rev. Letters 12, 713 (1964).

[15] P. W. Higgs, Phys. Letters 12, 132 (1964).

[16] G. S. Guralnik, C. R. Hagen, and T. W. B. Kibble, Phys. Rev. Letters 13, 585 (1964). These authors appear to attribute the failure of a local conservation law to yield a global conservation law to the lack of manifest covariance of a theory. In fact this happens even in manifestly covariant models of spontaneous breakdown.

[17] R. V. Lange, Phys. Rev. Letters 14, 3 (1965).

[18] R. F. Streater, Proc. Roy. Soc. (London) A287, 510 (1965). The proof of the Goldstone theorem given here is based on axioms which include manifest causality. Radiation gauge theories escape *this* version of the theorem by virtue of their (quite innocuous) acausality. The question of the extent to which one may give up manifest covariance and causality in a theory without losing covariance and causality of the physics which it describes deserves further study. If there are contexts in which it is possible other than the gauge theories which we are discussing, then there are probably other escape routes from the Goldstone theorem. See also D. Kastler, D. W. Robinson, and A. Swieca, Commun. Math. Phys. (to be published).

[19] R. F. Streater, Phys. Rev. Letters 15, 475 (1965). The generalized Goldstone theorem proved by this author and extended by N. Fuchs, Phys. Rev. Letters 15, 911 (1965) also relies on manifest causality and is therefore inapplicable to gauge theories.

[20] S. A. Bludman, Phys. Rev. 100, 372 (1955); Nuovo Cimento 9, 433 (1958); S. L. Glashow, Nucl. Phys. 10, 107 (1959); 22, 579 (1961).

[21] A. Salam and J. C. Ward, Nuovo Cimento 11, 568 (1959); Phys. Rev. 136, B763 (1964); J. J. Sakurai, Ann. Phys. (N. Y.) 11, 1 (1960).

[22] C. N. Yang and R. L. Mills, Phys. Rev. 96, 191 (1954); R. Shaw, dissertation, Cambridge University, 1954 (unpublished); R. Utiyama, Phys. Rev. 101, 1597 (1956).

[23] M. Gell-Mann and S. L. Glashow, Ann. Phys. (N. Y.) 15, 437 (1961).

perturbation theory at least. But the only known massless vector boson is the photon; the existing evidence suggests[24] that all other vector bosons must be massive. In particular, the hypothetical intermediate vector bosons of weak interactions, which in a gauge theory belong to the gauge field multiplet, must be much heavier than the known hadrons. For the most part, advocates of gauge theories have met this difficulty either by spoiling the gauge invariance of their theories with explicit mass terms or by taking comfort from the argument of Schwinger[25] that a sufficiently strong gauge-field coupling might generate mass. Recently, however, it was shown by Englert and Brout[26] and by the present writer[27] that gauge vector mesons acquire mass if the symmetry to whose generators they are coupled breaks down spontaneously, however weak their coupling may be. In Ref. 27 this phenomenon was exhibited in a classical field theory, and it was pointed out that in such a theory the longitudinal polarization of the massive vector excitation replaces the massless scalar excitation which would occur in the absence of coupling to the gauge field. Thus it now appears that the spontaneous breakdown program of Nambu *et al.* and the gauge field program of Salam *et al.* stand or fall together. Each saves the other from its zero-mass difficulty.

The purpose of the present paper is to amplify and substantiate the assertions made in Refs. 15 and 27 by displaying the behavior of the simplest possible relativistic field theory which combines spontaneous breakdown of symmetry under a compact Lie group with the gauge principle. That is to say, we take the symmetry group to be the trivial Abelian group $U(1)$, we take as the fundamental dynamic variables a pair of Hermitian scalar fields $\Phi_1(x)$, $\Phi_2(x)$ together with the Hermitian vector gauge field $A_\mu(x)$, and we induce spontaneous breakdown by means of the simplest $U(1)$-invariant self-interaction of $\Phi_a(x)$ which will do the trick, namely, a combination of quadratic and quartic terms. In the absence of the gauge field coupling, the model is just one which Goldstone[10] first discussed.[28]

In Sec. II the behavior of the small-amplitude wave solutions of the classical field equations is used as a guide in formulating a perturbation theory in which the major effects of spontaneous breakdown are already taken into account in zero order. The radiation gauge commutators of the zero-order approximation are obtained and used to provide an explicit realization of the spectral representation which was proposed in Ref. 15. In Sec. III decay and scattering amplitudes are calculated in lowest order and it is verified that they are gauge-invariant, Lorentz-invariant, and causal despite the lack of manifest covariance and causality of the radiation gauge. In Sec. IV it is shown that the same amplitudes may be derived by a manifestly covariant and causal method from an equivalent Lagrangian which lacks the original symmetry. Finally, our conclusions are summarized in Sec. V, and the way in which coupling between a system of the kind described here and other symmetric dynamical systems may lead to partially conserved currents is sketched.

In subsequent papers we propose to elaborate these considerations, both with regard to symmetry groups and with regard to mechanisms of symmetry breakdown, so as to make closer contact with particle physics.

II. THE MODEL

The Lagrangian density from which we shall work is given by[29]

$$\mathcal{L}=-\tfrac{1}{4}g^{\kappa\mu}g^{\lambda\nu}F_{\kappa\lambda}F_{\mu\nu}-\tfrac{1}{2}g^{\mu\nu}\nabla_\mu\Phi_a\nabla_\nu\Phi_a +\tfrac{1}{2}m_0^2\Phi_a\Phi_a-\tfrac{1}{8}f^2(\Phi_a\Phi_a)^2. \quad (1)$$

In Eq. (1) the metric tensor $g^{\mu\nu}=-1$ $(\mu=\nu=0)$, $+1$ $(\mu=\nu\neq 0)$ or 0 $(\mu\neq\nu)$, Greek indices run from 0 to 3 and Latin indices from 1 to 2. The $U(1)$-covariant derivatives $F_{\mu\nu}$ and $\nabla_\mu\Phi_a$ are given by

$$F_{\mu\nu}=\partial_\mu A_\nu-\partial_\nu A_\mu,$$
$$\nabla_\mu\Phi_1=\partial_\mu\Phi_1-eA_\mu\Phi_2,$$
$$\nabla_\mu\Phi_2=\partial_\mu\Phi_2+eA_\mu\Phi_1.$$

At first sight this theory appears to be scalar electrodynamics augmented by a quartic self-interaction. However, what appears to be the bare-mass term has the wrong sign. In conjunction with the quartic term this feature has the consequence that the field equations

$$\partial_\nu F^{\mu\nu}=ej^\mu,\quad j_\mu=\Phi_2\nabla_\mu\Phi_1-\Phi_1\nabla_\mu\Phi_2,$$
$$\nabla_\mu\nabla^\mu\Phi_a+\tfrac{1}{2}(m_0^2-f^2\Phi_b\Phi_b)\Phi_a=0, \quad (2)$$

in the classical theory possess a coordinate-independent solution

$$A_\mu=0,\quad \Phi_b\Phi_b=m_0^2/f^2. \quad (3)$$

The invariance of the Lagrangian (1) under the local $U(1)$ transformations

$$A_\mu(x)\to A_\mu(x)+e^{-1}\partial_\mu\Lambda(x),$$
$$\Phi_1(x)\to\Phi_1(x)\cos\Lambda(x)+\Phi_2(x)\sin\Lambda(x), \quad (4)$$
$$\Phi_2(x)\to-\Phi_1(x)\sin\Lambda(x)+\Phi_2(x)\cos\Lambda(x),$$

is reflected in the existence of a one-parameter family of static solutions defined by Eq. (3).[30] In the classical

[24] See, for example, S. Weinberg, Phys. Rev. Letters 13, 495 (1964).

[25] J. Schwinger, Phys. Rev. 125, 397 (1962); 128, 2425 (1962).

[26] F. Englert and R. Brout, Phys. Rev. Letters 13, 321 (1964).

[27] P. W. Higgs, Phys. Rev. Letters 13, 508 (1964).

[28] I understand from Dr. Goldstone (private communication) that he and W. Gilbert at one time considered adding a gauge field to the model.

[29] We do not explicitly perform the symmetrizations which a correct quantum-mechanical treatment would demand. They are not necessary for the purposes of the present paper and would, in any case, be dealt with more satisfactorily in a first-order formalism.

[30] Strictly speaking, global $U(1)$ invariance suffices to guarantee this result.

theory, any one of these solutions,

$$\Phi_1=\eta\cos\alpha,\quad \Phi_2=\eta\sin\alpha,$$

where $\eta=m_0/f$, defines a possible asymmetric configuration of stable equilibrium, stability being ensured by the sign of the quartic term in Eq. (1). Quantum mechanically each solution, regarded as the "bare" value of the vacuon $\langle\Phi_a\rangle$, corresponds to a different possible vacuum state.[31]

Let us choose $\alpha=\pi/2$ and linearize the classical field equations (2) by treating A_μ, Φ_1, and $\Phi_2-\eta$ as small quantities. We obtain

$$\partial_\nu F^{\mu\nu}=-e^2\eta^2B^\mu,\quad \partial_\mu B^\mu=0,$$
$$(\Box-m_0^2)\chi=0,$$

in which we have introduced the notation

$$B_\mu=A_\mu-(e\eta)^{-1}\partial_\mu\Phi,$$
$$\Phi=\Phi_1,\quad \chi=\Phi_2-\eta. \tag{5}$$

As we remarked in Ref. 27, these are the linear field equations which, after quantization, describe free vector bosons of mass $e\eta$ and free scalar bosons of mass m_0. The longitudinal vector excitation becomes the Goldstone scalar excitation in the limit $e\rightarrow 0$. The Lagrangian to which these field equations belong is given by

$$\mathcal{L}_0=-\tfrac{1}{4}F_{\mu\nu}F^{\mu\nu}-\tfrac{1}{2}g^{\mu\nu}(\partial_\mu\Phi-m_1A_\mu)(\partial_\nu\Phi-m_1A_\nu)$$
$$-\tfrac{1}{2}g^{\mu\nu}\partial_\mu\chi\partial_\nu\chi-\tfrac{1}{2}m_0^2\chi^2, \tag{6}$$

where we have written m_1 for the vector boson mass $e\eta$.

The foregoing analysis of classical small-amplitude wave propagation suggests the following perturbation theoretic treatment of the quantized theory. We rewrite Eq. (1) in the form $\mathcal{L}=\mathcal{L}_0+\mathcal{L}_{\text{int}}$, where $\mathcal{L}_0$ is given by Eq. (6) apart from a trivial additive constant and

$$\mathcal{L}_{\text{int}}=eA^\mu(\chi\partial_\mu\Phi-\Phi\partial_\mu\chi)-em_1\chi A_\mu A^\mu-\tfrac{1}{2}fm_0\chi(\Phi^2+\chi^2)$$
$$-\tfrac{1}{2}e^2A_\mu A^\mu(\Phi^2+\chi^2)-\tfrac{1}{8}f^2(\Phi^2+\chi^2)^2. \tag{7}$$

Note that the ancestry of (6) plus (7) in the symmetric Lagrangian (1) is embodied in the relation $m_1/m_0=e/f$ between the bare masses and the bare coupling constants. Our perturbation theory now consists in developing transition amplitudes in powers of e and f (or, equivalently, in inverse powers of η), the masses m_0 and m_1 being treated as of order zero.[32] Thus in Eq. (7) all five cubic vertices are of the first degree and all five quartic vertices are of the second in the expansion parameter. It will be found that, with few exceptions, gauge-invariant results are obtained only when all Feynman graphs of the same degree are summed.

We first write down the commutators and propagators corresponding to the bare Lagrangian $\mathcal{L}_0$. Apart from the terms in χ, this is just the second-order Lagrangian of a model proposed by Boulware and Gilbert[33] as an illustration of the possibility of a gauge-invariant theory describing a massive vector boson. We shall study it in a radiation gauge; the Lorentz gauge formulation, which even in quantum electrodynamics leads to unnecessary complications such as redundant states, is here inconsistent with the canonical com, mutation rules, as was pointed out by Guralnik, Hagen- and Kibble.[16] In a radiation gauge defined by the condition

$$(\partial A)+(n\partial)(nA)=0,$$

where n^μ is a constant timelike unit vector and (ab) denotes $a_\mu b^\mu$, the variables A_μ and Φ may be expressed in terms of the massive vector field B_μ which was introduced in Eq. (5):

$$A_\mu=B_\mu+m_1^{-1}\partial_\mu\Phi,$$
$$\Phi=-m_1[(\partial^2)+(n\partial)^2]^{-1}[(\partial B)+(n\partial)(nB)]. \tag{8}$$

Since $\mathcal{L}_0$, when expressed in terms of B_μ and χ, is just the usual second order Lagrangian for free vector and scalar bosons, we may immediately write down the covariant commutators:

$$[B_\mu(x),B_\nu(y)]=-i(g_{\mu\nu}-m_1^{-2}\partial_\mu\partial_\nu)\Delta(x-y,m_1^2),$$
$$[\chi(x),\chi(y)]=-i\Delta(x-y,m_0^2), \tag{9}$$

where $\Delta(x,m^2)=i(2\pi)^{-3}\int d^4k e^{i(kx)}\epsilon(k^0)\delta(k^2+m^2)$. Then Eq. (8) enables us to deduce the nonvanishing commutators of A_μ, Φ, and χ:

$$[A_\mu(x),A_\nu(y)]=-i\{g_{\mu\nu}-[(n_\mu\partial_\nu+n_\nu\partial_\mu)(n\partial)+\partial_\mu\partial_\nu]$$
$$\times[(\partial^2)+(n\partial)^2]^{-1}\}\Delta(x-y,m_1^2),$$
$$[A_\mu(x),\Phi(y)]=-im_1n_\mu(n\partial)$$
$$\times[(\partial^2)+(n\partial)^2]^{-1}\Delta(x-y,m_1^2), \tag{10}$$
$$[\Phi(x),\Phi(y)]=-i(n\partial)^2[(\partial^2)+(n\partial)^2]^{-1}\Delta(x-y,m_1^2),$$
$$[\chi(x),\chi(y)]=-i\Delta(x-y,m_0^2).$$

We also note the commutator relation

$$[B_\mu(x),\Phi(y)]=-im_1^{-1}[(\partial^2)n_\mu-(n\partial)\partial_\mu](n\partial)$$
$$\times[(\partial^2)+(n\partial)^2]^{-1}\Delta(x-y,m_1^2). \tag{11}$$

[31] The orthogonality of the worlds built upon different vacua may be understood as a consequence of the impossibility in the classical theory of a displacement of the system from one static configuration to another, on account of the infinite inertia associated with such a motion. To see this, imagine a one-dimensional model consisting of an infinite uniform elastic string subjected to a force field of cylindrical symmetry about the axis from which transverse displacements Φ_1, Φ_2 are measured. (We omit the gauge field from this do-it-yourself model.) If the force is such that stable equilibrium occurs when the whole string is at a distance η from the axis in any direction, then the system exhibits broken rotational symmetry. Displacement of the string from equilibrium at one orientation to equilibrium at another is impossible, since the moment of inertia about the axis is infinite. Waves on the string do not conserve angular momentum about the axis, since the string as a whole can emit or absorb angular momentum without recoiling.

[32] When one comes to consider radiative corrections, it becomes necessary to make these statements about the renormalized rather than the bare masses and coupling constants.

[33] D. C. Boulware and W. Gilbert, Phys. Rev. 126, 1563 (1962).

The generator $Q(t)$ of infinitesimal *global* $U(1)$ transformations (that is, transformations (4) with Λ constant and infinitesimal) on the hypersurface $(nx)+t=0$ is $\int d\sigma_\mu j^\mu$, where $d\sigma_\mu$ is the volume element of the hypersurface and $j_\mu(x)$ is given by Eq. (2). The invariance of the Lagrangian (1) under these transformations leads to the local conservation law, $\partial_\mu j^\mu=0$. However, even in the absence of the gauge field coupling, the four-dimensional integral of this equation fails to yield the usual global conservation law, $Q(t)=\text{constant}$, since the flux of j_μ across the surface of a large sphere bounding the hypersurface does not tend to zero as the radius tends to infinity. That this is so can be seen by noting that the lowest order approximation to j_μ is $-e\eta^2 B_\mu$ (or $\eta\partial_\mu\Phi$ in the absence of the gauge field): Matrix elements of this operator do not decrease sufficiently rapidly at large spatial distances for the flux term to vanish. (In normal theories the lowest order term in j_μ is quadratic, giving a better asymptotic behavior of the matrix elements.) Strictly speaking, the "operator" $Q(t)$ is now not merely time-dependent but nonexistent, since $\int d\sigma_\mu j^\mu$ diverges as a result of the same bad asymptotic behavior of j^μ. However, certain commutators, such as $[Q(t),\Phi_a(y)]$, do still exist.[24]

The zero-order approximation to the commutator vacuum expectation value $\langle i[j_\mu(x),\Phi_1(y)]\rangle$, upon which so much of the discussion of the Goldstone theorem has centered,[25] is found by replacing j_μ by $-e\eta^2 B_\mu$ and using Eq. (11): It is

$$\eta[(n\partial)\partial_\mu-(\partial^2)n_\mu](n\partial)[(\partial^2)+(n\partial)^2]^{-1}\Delta(x-y,\,m_1^2).$$

Its Fourier transform,

$$-2\pi\eta[(nk)k_\mu-(k^2)n_\mu](nk)\times[(k^2)+(nk)^2]^{-1}\epsilon(k^0)\delta(k^2+m_1^2), \quad (12)$$

provides an explicit realization of a spectral representation of the form obtained in Ref. 15, the Lorentz invariance of the spectrum of intermediate states now being made clear. We are led to conjecture that the vacuum expectation value of the exact commutator may be of the form

$$\langle\Phi_2\rangle[(n\partial)\partial_\mu-(\partial^2)n_\mu](n\partial)[(\partial^2)+(n\partial)^2]^{-1}\times\int_0^\infty dm^2\rho(m^2)\Delta(x-y,\,m^2), \quad (13)$$

where $\rho(m^2)$ is a nonnegative spectral function satisfying the sum rule

$$\int_0^\infty dm^2\rho(m^2)=1.$$

It may be noted that when $m_1=0$ in Eq. (12), corresponding to $e=0$, we recover the manifestly covariant spectral representation $-2\pi\eta k_\mu\epsilon(k^0)\delta(k^2)$ and with it the Goldstone theorem.

We define the propagators of the system described by $\mathcal{L}_0$ to be the quantities $\langle T^*A_\mu(x)A_\nu(y)\rangle$, etc., obtained from the corresponding commutators in Eq. (10) by substituting for Δ the scalar propagator Δ_F given by

$$\Delta_F(x,m^2)=(2\pi)^{-4}\int d^4k\, e^{i(kx)}(k^2+m^2-i\epsilon)^{-1}.$$

Then we may calculate S-matrix elements by using the simple Feynman rules based on the Nishijima-Wick expansion of the expression $T^*\exp(i\int d^4x\,\mathcal{L}_{\rm int})$ for the S-operator in the interaction picture.[26] We thereby avoid the n_μ-dependent terms, in addition to those already introduced by the radiation gauge, which the use of simple chronological ordering and the Dyson-Wick expansion would entail.

III. DECAY AND SCATTERING AMPLITUDES

As an illustration of the physical content of the model we now calculate in lowest order the matrix elements for the simplest processes which it describes. We shall verify that, despite the unpromising appearance of the radiation gauge propagators, these matrix elements are gauge invariant and Lorentz invariant.

In applying the Feynman rules we shall need, in addition to the propagators, the wave functions $a_\mu(k,\sigma)$ and $\phi(k,\sigma)$ which correspond to the annihilation by the operators A_μ and Φ, respectively, of a vector meson from an incoming state of momentum k and spin component σ. They are related by the Fourier transform of Eq. (8) to the usual vector meson wave function $b_\mu(k,\sigma)$, which has the explicit form

$$b^\mu(k,0)=(\omega/m_1)(|\mathbf{k}|/\omega,\mathbf{k}/|\mathbf{k}|),$$
$$b^\mu(k,\pm1)=2^{-1/2}(0,\,\mathbf{e}_1\pm i\mathbf{e}_2), \quad (14)$$

where $\omega=(|\mathbf{k}|^2+m_1^2)^{1/2}$ and $\mathbf{e}_1$, $\mathbf{e}_2$, $\mathbf{k}/|\mathbf{k}|$ form a right-handed orthonormal triad. Actually, all that we shall need is the relation

$$a_\mu=b_\mu+(ik_\mu/m_1)\phi, \quad (15)$$

by which matrix elements may be expressed in terms of wave functions b_μ and ϕ, the desired invariance being achieved by the cancellation of all terms containing factors ϕ. Similar considerations apply to out-

[24] In Ref. 18 it is proved that in a manifestly causal theory this commutator (or at least certain of its matrix elements) is independent of t, despite the breakdown of the global conservation law. The gauge field coupling destroys manifest causality and induces a time dependence in this commutator: In the zero-order approximation it oscillates at a frequency m_1.

[25] See Refs. 11, 13, 14, and 15. In Refs. 14 and 15 it is implied erroneously that the commutator $[Q(t),\Phi_a(y)]$ is independent of t. Fortunately, the discussion of the Goldstone theorem in these papers does not depend on this assumption.

[26] P. T. Matthews, Phys. Rev. 76, 684 (1949); K. Nishijima, Progr. Theoret. Phys. (Kyoto) 5, 405 (1950). The most general conditions for the validity of this expression have been stated by C. S. Lam, Nuovo Cimento 38, 1755 (1965).

going states and associated complex conjugate wave functions.

i. Decay of a Scalar Boson into Two Vector Bosons

The process occurs in first order (four of the five cubic vertices contribute), provided that $m_0 > 2m_1$. Let p be the incoming and k_1, k_2 the outgoing momenta. Then

$$M = i\{e[a^{*\mu}(k_1)(-ik_{2\mu})\phi^*(k_2) + a^{*\mu}(k_2)(-ik_{1\mu})\phi^*(k_1)] - e(ip_\mu)[a^{*\mu}(k_1)\phi^*(k_2) + a^{*\mu}(k_2)\phi^*(k_1)] - 2em_1 a_\mu^*(k_1)a^{*\mu}(k_2) - fm_0\phi^*(k_1)\phi^*(k_2)\}.$$

By using Eq. (15), conservation of momentum, and the transversality $(k_\mu b^\mu(k) = 0)$ of the vector wave functions we reduce this to the form

$$M = -2iem_1 b^{*\mu}(k_1)b_\mu^*(k_2) - iem_1^{-1}(p^2 + m_0^2)\phi^*(k_1)\phi^*(k_2). \quad (16)$$

We have retained the last term, which we shall need in calculating scattering amplitudes; when the incident particle is on the mass shell it vanishes and we are left with the invariant expression

$$M = -2iem_1 b^{*\mu}(k_1)b_\mu^*(k_2). \quad (17)$$

Conservation of angular momentum allows three possibilities for the spin states of the decay products: They may be both right-handed, both left-handed, or both longitudinal ($\sigma_1 = \sigma_2 = +1, -1$, or 0). With the help of the explicit vectors (14), we find

$$M(+1,+1) = M(-1,-1) = 2iem_1,$$
$$M(0,0) = ifm_0(1 - 2e^2/f^2).$$

We note that as $e \to 0$ the amplitudes for decay to transverse states tend to zero, but the amplitude $M(0,0)$ tends to the value ifm_0 which we would calculate from the vertex $-\frac{1}{2}fm_0\Phi^2\chi$ for the decay of one massive into two massless scalar bosons in the original Goldstone model. (The sign change arises from the factor i which is associated with the term ϕ in each b_μ.)

ii. Vector Boson-Vector Boson Scattering

Let k_1, k_2 be the incoming and k_1', k_2' the outgoing momenta. The process occurs as a second-order effect of the cubic vertices, by exchange of a scalar boson in the s, t, or u channel, where $s = -(p_1 + p_2)^2$, $t = -(p_1 - p_1')^2$, $u = -(p_1 - p_2')^2$. It also occurs as a direct effect of two of the quartic vertices. Equation (16) enables us to write down

$$M_s = i^2\{-2em_1 b_\mu^*(k_1')b^{*\mu}(k_2') + em_1^{-1}(s - m_0^2)\phi^*(k_1')\phi^*(k_2')\} \times i(s - m_0^2)^{-1}\{-2em_1 b_\nu(k_1)b^\nu(k_2) + em_1^{-1}(s - m_0^2)\phi(k_1)\phi(k_2)\}$$

and similar expressions for M_t and M_u. The quartic vertices yield a contribution given by

$$M_{\text{direct}} = i(-2e^2)\{a_\mu^*(k_1')a^{*\mu}(k_2')\phi(k_1)\phi(k_2) + 5 \text{ similar terms}\} + i(-3f^2)\phi^*(k_1')\phi^*(k_2')\phi(k_1)\phi(k_2)$$
$$= -2ie^2\{b_\mu^*(k_1')b^{*\mu}(k_2')\phi(k_1)\phi(k_2) + 5 \text{ similar terms}\} + i(4e^2 - 3f^2)\phi^*(k_1')\phi^*(k_2')\phi(k_1)\phi(k_2).$$

It is only when we combine these four contributions that we obtain (after some algebra) the invariant expression

$$M_{\text{total}} = M_s + M_t + M_u + M_{\text{direct}}$$
$$= -4ie^2m_1^2\{(s - m_0^2)^{-1}b^{*\mu}(k_1')b^{*}_\mu(k_2')b_\nu(k_1)b^\nu(k_2) + (t - m_0^2)^{-1}b_\mu^*(k_1')b^\mu(k_1)b_\nu^*(k_2')b^\nu(k_2) + (u - m_0^2)^{-1}b_\mu^*(k_1')b^\mu(k_2)b_\nu^*(k_2')b^\nu(k_1)\}. \quad (18)$$

iii. Vector Boson-Scalar Boson Scattering

Let k, p be the momenta of the incoming vector and scalar boson, respectively, and k', p' be their outgoing momenta. Again there are four contributions, M_s, M_t, M_u, and M_{direct}. In the s and u channels a vector boson is exchanged and it turns out that the various propagators, $\langle T^*A_\mu A_\nu\rangle$, $\langle T^*A_\mu\Phi\rangle$, and $\langle T^*\Phi\Phi\rangle$, occur only in the combination $\langle T^*B_\mu B_\nu\rangle$. We obtain the expression

$$M_s = i^2\{-2em_1 b^{*\mu}(k') + ieq^\mu\phi^*(k')\}i(g_{\mu\nu} + m_1^{-2}q_\mu q_\nu) \times (s - m_1^2)^{-1}\{-2em_1 b^\nu(k) - ieq^\nu\phi(k)\},$$

where $q = k + p$ and $s = -q^2$, and a similar expression for M_u. In the t channel a scalar boson is exchanged, and we find that

$$M_t = i^2\{-3fm_0\}i(t - m_0^2)^{-1}\{-2em_1 b_\mu^*(k')b^\mu(k) + em_1^{-1}(t - m_0^2)\phi^*(k')\phi(k)\},$$

where $t = -(k - k')^2$. Finally, the contribution of the quartic vertices is given by

$$M_{\text{direct}} = i\{-2e^2[b_\mu^*(k') - im_1^{-1}k_\mu'\phi^*(k')] \times [b^\mu(k) + im_1^{-1}k^\mu\phi(k)] - f^2\phi^*(k')\phi(k)\}.$$

Again the four contributions sum to the invariant expression

$$M_{\text{total}} = -2im_1^2\{2e^2(s - m_1^2)^{-1}[b_\mu^*(k')b^\mu(k) + m_1^{-2}p_\mu'b^{*\mu}(k')p_\nu b^\nu(k)] + 3f^2(t - m_0^2)^{-1}b_\mu^*(k')b^\mu(k) + 2e^2(u - m_1^2)^{-1}[b_\mu^*(k')b^\mu(k) + m_1^{-2}p_\mu b^{*\mu}(k')p_\nu' b^\nu(k)]\} - 2ie^2 b_\mu^*(k')b^\mu(k). \quad (19)$$

A similar matrix element may be written down for the process, vector pair ↔ scalar pair, by making appropriate interchanges of incoming and outgoing momenta and wave functions.

iv. Scalar Boson-Scalar Boson Scattering

This is the only simple process in which no invariance problems arise in lowest order: The vertices which are involved contain no vector boson factors. We find that

$$M_{\text{total}}=M_s+M_t+M_u+M_{\text{direct}}$$
$$=-9if^2m_0^2\{(s-m_0^2)^{-1}+(t-m_0^2)^{-1}+(u-m_0^2)^{-1}+(3m_0^2)^{-1}\}. \quad (20)$$

IV. EQUIVALENT LAGRANGIAN

In the previous section we have illustrated the Lorentz and gauge invariance of the model by the somewhat unsophisticated device of performing a few lowest order calculations. From a more sophisticated point of view we remark that Lorentz invariance may be proved by constructing the generators of the Lorentz group and verifying their commutation relations. Provided that the Lagrangian (1) is first properly symmetrized, the proof goes through as in quantum electrodynamics[37]; spontaneous breakdown of the internal symmetry is irrelevant to the argument, which depends only on the equal-time commutators of products of field operators.

Concerning gauge invariance we remark that our result, that (in lowest order at least) decay and scattering amplitudes depend only on the gauge-invariant vector wave functions $b_\mu(k,\sigma)$, suggests that it must be possible to rewrite the theory in a form in which only gauge-invariant variables appear. Indeed, if one were shown only the expressions (17)–(20), he would guess that they had been derived from an interaction Lagrangian given by

$$\mathcal{L}_{\text{int}}'=-em_1B_\mu B^\mu\chi-\tfrac{1}{2}fm_0\chi^3-\tfrac{1}{2}e^2B_\mu B^\mu\chi^2-\tfrac{1}{8}f^2\chi^4. \quad (21)$$

We shall now show that the expressions (7) and (21) are equivalent by finding a transformation of the total Lagrangian (1) which takes the one into the other.

We note that the gauge dependence of the classical dynamic variables may be expressed in the form

$$\Phi_1(x)=R(x)\cos\Theta(x),$$
$$\Phi_2(x)=R(x)\sin\Theta(x), \quad (22)$$
$$A_\mu(x)=B_\mu(x)-e^{-1}\partial_\mu\Theta(x),$$

where $R(x)$ and $B_\mu(x)$ are gauge invariant variables and the transformations (4) take the simple form, $\Theta(x)\rightarrow\Theta(x)-\Lambda(x)$. The classical Lagrangian (1), expressed in terms of the new variables, takes the form

$$\mathcal{L}'=-\tfrac{1}{4}F_{\mu\nu}F^{\mu\nu}-\tfrac{1}{2}g^{\mu\nu}\partial_\mu R\partial_\nu R$$
$$-\tfrac{1}{2}e^2B_\mu B^\mu R^2+\tfrac{1}{4}m_0^2R^2-\tfrac{1}{8}f^2R^4. \quad (23)$$

Gauge invariance here has ensured the disappearance of the variable Θ from the scene. [What we have done is to exploit the freedom which *local* $U(1)$ invariance gives us to "rotate" the entire two-component field $\Phi_a(x)$ onto one of the "axes." In a theory with only *global* $U(1)$ invariance this rotation cannot be performed on the entire field but only on the static solution (3).] The existence of the solution $B_\mu=0$, $R^2=m_0^2/f^2$ suggests the substitution $R(x)=\eta+\chi(x)$. In this way, we find immediately that, apart from an additive constant, $\mathcal{L}'=\mathcal{L}_0'+\mathcal{L}_{\text{int}}'$, where

$$\mathcal{L}_0'=-\tfrac{1}{4}F_{\mu\nu}F^{\mu\nu}-\tfrac{1}{2}m_1^2B_\mu B^\mu-\tfrac{1}{2}g^{\mu\nu}\partial_\mu\chi\partial_\nu\chi-\tfrac{1}{2}m_0^2\chi^2. \quad (24)$$

The expression (24) is the same as (6), except that the exactly gauge-invariant variables B_μ and χ which we have just defined replace their interaction picture counterparts.

We conjecture that the equivalence demonstrated here between the classical Lagrangians (1) and (23) may be extended to the corresponding quantum mechanical operators, provided that careful attention is given to the ordering problems which may arise, for example, in the definition of the current j_μ.[38]

V. DISCUSSION

The foregoing considerations illustrate our contention in Refs. 15 and 27 that the extension of a spontaneously broken internal symmetry of a Lorentz-invariant Lagrangian from global to local transformations not only may but actually does change zerons into the longitudinal states of massive vector bosons. Since we believe the value of this simple model to lie in the insight which it may give into this phenomenon when one looks at it as simple-mindedly as we have here, we shall not go into more difficult questions, such as radiative corrections and renormalization, in the present paper.

We note that in this model the original symmetry is almost unrecognizable in the physical states. Even without the gauge field coupling, the invalidation of the usual argument leading to the conservation of $Q(t)$ [39] by the asymptotic behavior of the term $\langle\Phi_2\rangle\partial_\mu\Phi_1$ in j_μ destroys the commutativity of Q with the Hamiltonian: Consequently, the one-particle states are not eigenstates of Q and the masses within the Φ_a multiplet are not degenerate, but at least the multiplet structure remains. The gauge field coupling obscures even the multiplet structure: The scalar doublet is now incomplete, having lost its formerly massless member to form the longitudinal polarization of the vector singlet.

[37] See B. Zumino, J. Math. Phys. 1, 1 (1960).

[38] J. Schwinger, Phys. Rev. Letters 3, 296 (1959); K. Johnson, Nucl. Phys. 25, 431 (1961).

[39] In passing, we remark that this feature of spontaneous breakdown theories seems to call into question the validity of the results obtained on the basis of chirality conservation by Nambu and his collaborators. See Y. Nambu and D. Lurié, Phys. Rev. 125, 1429 (1962); Y. Nambu and E. Shrauner, *ibid.* 128, 862 (1962); E. Shrauner, *ibid.* 131, 1847 (1963).

In view of the rather drastic nature of the symmetry breakdown which we have just summarized, it is of interest to inquire what happens when this system is coupled to a second in a Lagrangian which retains local $U(1)$ invariance and contains no additional mechanisms for spontaneous breakdown. To be specific, let us take the second system to be a pair of "baryons" of "charges" $\pm\frac{1}{2}$, together with their antiparticles, and let us assume that the Φ-baryon interaction is of the Yukawa type. The total Lagrangian is then given by

$$\mathcal{L}_{\text{total}}=\mathcal{L}(A,\Phi)-\bar{\psi}_a(\gamma^\mu\nabla_\mu+M)\psi_a+g[\Phi_1(\bar{\psi}_1\psi_2+\bar{\psi}_2\psi_1)+\Phi_2(\bar{\psi}_2\psi_2-\bar{\psi}_1\psi_1)], \quad (25)$$

in which $\mathcal{L}(A,\Phi)$ is the expression (1), $\nabla_\mu\psi_1=\partial_\mu\psi_1-\frac{1}{2}eA_\mu\psi_2$, $\nabla_\mu\psi_2=\partial_\mu\psi_2+\frac{1}{2}eA_\mu\psi_1$ and we have, without loss of generality, made a choice of a phase angle in writing down the invariant Yukawa term. But for the presence of this last term, the Lagrangian would be invariant under global $U(1)$ transformations on Φ and ψ *independently*; that is, the symmetry would be $U(1)\times U(1)$ and the currents $j_\mu(\Phi)$ and $j_\mu(\psi)$ would be *separately* conserved. Thus, despite the nonconservation of $Q(\Phi)$ brought about by the structure of the first term in (25), there would still be conservation of $Q(\psi)$.

The Yukawa term reduces the symmetry to $U(1)$, the divergences of the individual currents now being given by

$$\begin{aligned}\partial_\mu j^\mu(\psi)&=g[\Phi_1(\bar{\psi}_1\psi_1-\bar{\psi}_2\psi_2)+\Phi_2(\bar{\psi}_1\psi_2+\bar{\psi}_2\psi_1)]\\&=-\partial_\mu j^\mu(\Phi).\end{aligned} \quad (26)$$

We observe that spontaneous breakdown of the symmetry in the Φ system breaks the symmetry of the ψ system to an extent which depends on the coupling constant g. In the spirit of the perturbative approach which we have been using, we may evaluate the major part of the effects on the ψ system by replacing Φ by its vacuum expectation value. Then in Eq. (25) the term $g\eta(\bar{\psi}_2\psi_2-\bar{\psi}_1\psi_1)$ removes the baryon mass degeneracy, and Eq. (26) becomes

$$\partial_\mu j^\mu(\psi)=g\eta(\bar{\psi}_1\psi_2+\bar{\psi}_2\psi_1)+\text{higher order terms}. \quad (27)$$

If we were to modify the Lagrangian (25) by adding to it such $U(1)$ invariant baryon-antibaryon interactions as would produce a doublet of low-mass scalar bound states with wave functions transforming as $\bar{\psi}_1\psi_2+\bar{\psi}_2\psi_1$ and $\bar{\psi}_2\psi_2-\bar{\psi}_1\psi_1$, then Eq. (27) would be approximately a partial conservation law of the type which has proved so successful for the axial currents of the weak interactions.[40] Moreover, the current $j_\mu(\psi)$ interacts with itself via the massive intermediate vector boson which the $\Phi-A_\mu$ coupling produces.

There appears to be some hope that the basic ingredients of our model, namely the combination of spontaneous symmetry breakdown with the gauge principle, may provide the basis for an understanding of the broken symmetries of high-energy physics. In a subsequent paper we shall discuss models in which the breakdown of higher symmetries such as $SU(3)$ is treated in the same fashion.

ACKNOWLEDGMENTS

I have learned much about the Goldstone theorem from discussions with Dr. G. S. Guralnik, Dr. C. R. Hagen, Dr. T. W. B. Kibble, and Dr. R. F. Streater. I wish to thank Professor Bryce and Professor Cécile DeWitt for the hospitality of the Institute of Field Physics at the University of North Carolina, where this work was completed.

[40] It would be the type of partial conservation law proposed by Y. Nambu, Phys. Rev. Letters 4, 380 (1960) and by M. Gell-Mann, Phys. Rev. 125, 1067 (1962), rather than that proposed by M. Gell-Mann and M. Lévy, Nuovo Cimento 16, 705 (1960), in which the low-lying scalar or pseudoscalar states are treated as elementary rather than composite in the context of a field theory.

A MODEL OF LEPTONS*

Steven Weinberg†
Laboratory for Nuclear Science and Physics Department,
Massachusetts Institute of Technology, Cambridge, Massachusetts
(Received 17 October 1967)

Leptons interact only with photons, and with the intermediate bosons that presumably mediate weak interactions. What could be more natural than to unite[1] these spin-one bosons into a multiplet of gauge fields? Standing in the way of this synthesis are the obvious differences in the masses of the photon and intermediate meson, and in their couplings. We might hope to understand these differences by imagining that the symmetries relating the weak and electromagnetic interactions are exact symmetries of the Lagrangian but are broken by the vacuum. However, this raises the specter of unwanted massless Goldstone bosons.[2] This note will describe a model in which the symmetry between the electromagnetic and weak interactions is spontaneously broken, but in which the Goldstone bosons are avoided by introducing the photon and the intermediate-boson fields as gauge fields.[3] The model may be renormalizable.

We will restrict our attention to symmetry groups that connect the observed electron-type leptons only with each other, i.e., not with muon-type leptons or other unobserved leptons or hadrons. The symmetries then act on a left-handed doublet

$$L \equiv [\tfrac{1}{2}(1+\gamma_5)]\binom{\nu_e}{e} \tag{1}$$

and on a right-handed singlet

$$R \equiv [\tfrac{1}{2}(1-\gamma_5)]e. \tag{2}$$

The largest group that leaves invariant the kinematic terms $-\bar{L}\gamma^\mu\partial_\mu L - \bar{R}\gamma^\mu\partial_\mu R$ of the Lagrangian consists of the electronic isospin $\vec{T}$ acting on L, plus the numbers N_L, N_R of left- and right-handed electron-type leptons. As far as we know, two of these symmetries are entirely unbroken: the charge $Q = T_3 - N_R - \frac{1}{2}N_L$, and the electron number $N = N_R + N_L$. But the gauge field corresponding to an unbroken symmetry will have zero mass,[4] and there is no massless particle coupled to N,[5] so we must form our gauge group out of the electronic isospin $\vec{T}$ and the electronic hyperchange $Y \equiv N_R + \frac{1}{2}N_L$.

Therefore, we shall construct our Lagrangian out of L and R, plus gauge fields $\vec{A}_\mu$ and B_μ coupled to $\vec{T}$ and Y, plus a spin-zero doublet

$$\varphi = \binom{\varphi^0}{\varphi^-} \tag{3}$$

whose vacuum expectation value will break $\vec{T}$ and Y and give the electron its mass. The only renormalizable Lagrangian which is invariant under $\vec{T}$ and Y gauge transformations is

$$\mathcal{L} = -\tfrac{1}{4}(\partial_\mu \vec{A}_\nu - \partial_\nu \vec{A}_\mu + g\vec{A}_\mu \times \vec{A}_\nu)^2 - \tfrac{1}{4}(\partial_\mu B_\nu - \partial_\nu B_\mu)^2 - \bar{R}\gamma^\mu(\partial_\mu - ig'B_\mu)R - L\gamma^\mu(\partial_\mu ig\vec{t}\cdot\vec{A}_\mu - i\tfrac{1}{2}g'B_\mu)L$$
$$-\tfrac{1}{2}|\partial_\mu\varphi - ig\vec{A}_\mu\cdot\vec{t}\varphi + i\tfrac{1}{2}g'B_\mu\varphi|^2 - G_e(\bar{L}\varphi R + \bar{R}\varphi^\dagger L) - M_1{}^2\varphi^\dagger\varphi + h(\varphi^\dagger\varphi)^2. \tag{4}$$

We have chosen the phase of the R field to make G_e real, and can also adjust the phase of the L and Q fields to make the vacuum expectation value $\lambda \equiv \langle\varphi^0\rangle$ real. The "physical" φ fields are then φ^-

and

$$\varphi_1 \equiv (\varphi^0+\varphi^{0\dagger}-2\lambda)/\sqrt{2} \quad \varphi_2 \equiv (\varphi^0-\varphi^{0\dagger})/i\sqrt{2}. \tag{5}$$

The condition that φ_1 have zero vacuum expectation value to all orders of perturbation theory tells us that $\lambda^2 \cong M_1^2/2h$, and therefore the field φ_1 has mass M_1 while φ_2 and φ^- have mass zero. But we can easily see that the Goldstone bosons represented by φ_2 and φ^- have no physical coupling. The Lagrangian is gauge invariant, so we can perform a combined isospin and hypercharge gauge transformation which eliminates φ^- and φ_2 everywhere[6] without changing anything else. We will see that G_e is very small, and in any case M_1 might be very large,[7] so the φ_1 couplings will also be disregarded in the following.

The effect of all this is just to replace φ everywhere by its vacuum expectation value

$$\langle\varphi\rangle = \lambda\begin{pmatrix}1\\0\end{pmatrix}. \tag{6}$$

The first four terms in $\mathfrak{L}$ remain intact, while the rest of the Lagrangian becomes

$$-\tfrac{1}{8}\lambda^2 g^2[(A_\mu^{\ 1})^2+(A_\mu^{\ 2})^2] -\tfrac{1}{8}\lambda^2(gA_\mu^{\ 3}+g'B_\mu)^2-\lambda G_e\bar{e}e. \tag{7}$$

We see immediately that the electron mass is λG_e. The charged spin-1 field is

$$W_\mu \equiv 2^{-1/2}(A_\mu^{\ 1}+iA_\mu^{\ 2}) \tag{8}$$

and has mass

$$M_W = \tfrac{1}{2}\lambda g. \tag{9}$$

The neutral spin-1 fields of definite mass are

$$Z_\mu = (g^2+g'^2)^{-1/2}(gA_\mu^{\ 3}+g'B_\mu), \tag{10}$$

$$A_\mu = (g^2+g'^2)^{-1/2}(-g'A_\mu^{\ 3}+gB_\mu). \tag{11}$$

Their masses are

$$M_Z = \tfrac{1}{2}\lambda(g^2+g'^2)^{1/2}, \tag{12}$$

$$M_A = 0, \tag{13}$$

so A_μ is to be identified as the photon field. The interaction between leptons and spin-1 mesons is

$$\frac{ig}{2\sqrt{2}}\bar{e}\gamma^\mu(1+\gamma_5)\nu W_\mu + \text{H.c.} + \frac{igg'}{(g^2+g'^2)^{1/2}}\bar{e}\gamma^\mu e A_\mu + \frac{i(g^2+g'^2)^{1/2}}{4}\left[\left(\frac{3g'^2-g^2}{g'^2+g^2}\right)\bar{e}\gamma^\mu e - \bar{e}\gamma^\mu\gamma_5 e + \bar{\nu}\gamma^\mu(1+\gamma_5)\nu\right]Z_\mu. \tag{14}$$

We see that the rationalized electric charge is

$$e = gg'/(g^2+g'^2)^{1/2} \tag{15}$$

and, assuming that W_μ couples as usual to hadrons and muons, the usual coupling constant of weak interactions is given by

$$G_W/\sqrt{2} = g^2/8M_W^{\ 2} = 1/2\lambda^2. \tag{16}$$

Note that then the e-φ coupling constant is

$$G_e = M_e/\lambda = 2^{1/4}M_e G_W^{\ 1/2} = 2.07\times 10^{-6}.$$

The coupling of φ_1 to muons is stronger by a factor M_μ/M_e, but still very weak. Note also that (14) gives g and g' larger than e, so (16) tells us that $M_W > 40$ BeV, while (12) gives $M_Z > M_W$ and $M_Z > 80$ BeV.

The only unequivocal new predictions made by this model have to do with the couplings of the neutral intermediate meson Z_μ. If Z_μ does not couple to hadrons then the best place to look for effects of Z_μ is in electron-neutron scattering. Applying a Fierz transformation to the W-exchange terms, the total effective e-ν interaction is

$$\frac{G_W}{\sqrt{2}}\bar{\nu}\gamma_\mu(1+\gamma_5)\nu\left\{\frac{(3g^2-g'^2)}{2(g^2+g'^2)}\bar{e}\gamma^\mu e + \tfrac{3}{2}\bar{e}\gamma^\mu\gamma_5 e\right\}.$$

If $g \gg e$ then $g \gg g'$, and this is just the usual e-ν scattering matrix element times an extra factor $\frac{3}{2}$. If $g \simeq e$ then $g \ll g'$, and the vector interaction is multiplied by a factor $-\frac{1}{2}$ rather than $\frac{3}{2}$. Of course our model has too many arbitrary features for these predictions to be taken very seriously, but it is worth keeping in mind that the standard calculation[8] of the electron-neutrino cross section may well be wrong.

Is this model renormalizable? We usually do not expect non-Abelian gauge theories to be renormalizable if the vector-meson mass is not zero, but our Z_μ and W_μ mesons get their mass from the spontaneous breaking of the symmetry, not from a mass term put in at the beginning. Indeed, the model Lagrangian we start from is probably renormalizable, so the question is whether this renormalizability is lost in the reordering of the perturbation theory implied by our redefinition of the fields. And if this model is renormalizable, then what happens when we extend it to include the couplings of $\vec{A}_\mu$ and B_μ to the hadrons?

I am grateful to the Physics Department of MIT for their hospitality, and to K. A. Johnson for a valuable discussion.

---

*This work is supported in part through funds provided by the U. S. Atomic Energy Commission under Contract No. AT(30-1)2098).

†On leave from the University of California, Berkeley, California.

[1]The history of attempts to unify weak and electromagnetic interactions is very long, and will not be reviewed here. Possibly the earliest reference is E. Fermi, Z. Physik 88, 161 (1934). A model similar to ours was discussed by S. Glashow, Nucl. Phys. 22, 579 (1961); the chief difference is that Glashow introduces symmetry-breaking terms into the Lagrangian, and therefore gets less definite predictions.

[2]J. Goldstone, Nuovo Cimento 19, 154 (1961); J. Goldstone, A. Salam, and S. Weinberg, Phys. Rev. 127, 965 (1962).

[3]P. W. Higgs, Phys. Letters 12, 132 (1964), Phys. Rev. Letters 13, 508 (1964), and Phys. Rev. 145, 1156 (1966); F. Englert and R. Brout, Phys. Rev. Letters 13, 321 (1964); G. S. Guralnik, C. R. Hagen, and T. W. B. Kibble, Phys. Rev. Letters 13, 585 (1964).

[4]See particularly T. W. B. Kibble, Phys. Rev. 155, 1554 (1967). A similar phenomenon occurs in the strong interactions; the ρ-meson mass in zeroth-order perturbation theory is just the bare mass, while the A_1 meson picks up an extra contribution from the spontaneous breaking of chiral symmetry. See S. Weinberg, Phys. Rev. Letters 18, 507 (1967), especially footnote 7; J. Schwinger, Phys. Letters 24B, 473 (1967); S. Glashow, H. Schnitzer, and S. Weinberg, Phys. Rev. Letters 19, 139 (1967), Eq. (13) et seq.

[5]T. D. Lee and C. N. Yang, Phys. Rev. 98, 101 (1955).

[6]This is the same sort of transformation as that which eliminates the nonderivative $\vec{\pi}$ couplings in the σ model; see S. Weinberg, Phys. Rev. Letters 18, 188 (1967). The $\vec{\pi}$ reappears with derivative coupling because the strong-interaction Lagrangian is not invariant under chiral gauge transformation.

[7]For a similar argument applied to the σ meson, see Weinberg, Ref. 6.

[8]R. P. Feynman and M. Gell-Mann, Phys. Rev. 109, 193 (1957).

PHYSICAL REVIEW D VOLUME 2, NUMBER 7 1 OCTOBER 1970

Weak Interactions with Lepton-Hadron Symmetry*

S. L. Glashow, J. Iliopoulos, and L. Maiani†
Lyman Laboratory of Physics, Harvard University, Cambridge, Massachusetts 02139
(Received 5 March 1970)

We propose a model of weak interactions in which the currents are constructed out of four basic quark fields and interact with a charged massive vector boson. We show, to all orders in perturbation theory, that the leading divergences do not violate any strong-interaction symmetry and the next to the leading divergences respect all observed weak-interaction selection rules. The model features a remarkable symmetry between leptons and quarks. The extension of our model to a complete Yang-Mills theory is discussed.

INTRODUCTION

Weak-interaction phenomena are well described by a simple phenomenological model involving a single charged vector boson coupled to an appropriate current. Serious difficulties occur only when this model is considered as a quantum field theory, and is examined in other than lowest-order perturbation theory.[1] These troubles are of two kinds. First, the theory is too singular to be conventionally renormalized. Although our attention is not directed at this problem, the model of weak interactions we propose may readily be extended to a massive Yang-Mills model, which may be amenable to renormalization with modern techniques. The second problem concerns the selection rules and the relationships among coupling constants which are carefully and deliberately incorporated into the original phenomenological Lagrangian. Our principal concern is the fact that these properties are not necessarily maintained by higher-order weak interactions.

Weak-interaction processes, and their higher-order weak corrections, may be classified[2] according to their dependence upon a suitably introduced cutoff momentum Λ. Contributions to the S matrix of the form

$$\sum_{n=1}^{\infty} A_n(G\Lambda^2)^n$$

(where G is the usual Fermi coupling constant and A_n are dimensionless parameters) are called zeroth-order

* Work supported in part by the Office of Naval Research, under Contract No. N00014-67-A-0028, and the U. S. Air Force under Contract No. AF49(638)-1380.

† On leave of absence from the Laboratori di Fisica, Istituto Superiore di Santa, Roma, Italy.

[1] B. L. Ioffe and E. P. Shabalin, Yadern. Fiz. 6, 828 (1967) [Soviet J. Nucl. Phys. 6, 603 (1968)]; Z. Eksperim. i Teor. Fiz. Pis'ma v Redaktsiyu 6, 978 (1967) [Soviet Phys. JETP Letters 6, 390 (1967)]; R. N. Mohapatra, J. Subba Rao, and R. E. Marshak, Phys. Rev. Letters 20, 1081 (1968); Phys. Rev. 171, 1502 (1968); F. E. Low, Comments Nucl. Particle Phys. 2, 33 (1968); R. N. Mohapatra and P. Olesen, Phys. Rev. 179, 1917 (1969).

[2] T. D. Lee, Nuovo Cimento 59A, 579 (1969).

weak effects, terms of the form

$$G\sum_{n=0}^{\infty}B_n(G\Lambda^2)^n$$

are called first-order weak effects, and generally, terms of the form

$$G^l\sum_{n=0}^{\infty}C_{ln}(G\Lambda^2)^n$$

are called lth order. (We are disregarding possible logarithmic dependences on the cutoff.) The zeroth-order terms present us with the dangerous possibility of serious violations of parity and hypercharge in strong interactions. First-order terms include the usual lowest-order contributions (order G) to leptonic and semileptonic processes. However, other first-order terms may yield violations of observed selection rules: There can be $\Delta S=2$ amplitudes, yielding a K_1-K_2 mass splitting, beginning at order $G(G\Lambda^2)$, as well as contributions to such unobserved decay modes as $K_2\to\mu^++\mu^-$, $K^+\to\pi^++l+\bar{l}$, etc., involving neutral lepton pairs, or departures from the leptonic $\Delta S=\Delta Q$ law. We shall say of a model that its divergences are properly ordered if it is true that the zeroth-order terms *do not* yield violations of parity or hypercharge, and if the first-order terms *do* satisfy the observed selection rules of weak-interaction phenomena.

In most conventional formulations of a weak-interaction field theory (say, a vector boson coupled to a quark triplet), the divergences are not properly ordered. Defenders of such theories must argue that there is an effective weak-interaction cutoff which guarantees that the induced higher-order effects are as small as experiment indicates. A remarkably small cutoff,[1] not greater than 3 or 4 GeV, seems necessary. Should such a cutoff be justified, the problem of higher-order departures from known selection rules is solved; all such departures are small.

Others feel that such a small cutoff is implausible and unrealistic, and that one must confront the possibility that $G\Lambda^2$ is large—perhaps obtaining sensible results in the limit $G\Lambda^2\to\infty$. In this case, one may regard all the first-order terms as having the same general magnitude, that of observed weak phenomena, and nth-order terms as having the magnitude naively expected of nth-order weak interactions.

An elegant solution to the problem of the zeroth-order terms was recently discovered, removing the specter of strong violations of parity and hypercharge.[3,4] One assumes a particular form for the breakdown of chiral $SU(3)$: The symmetry-breaking term must transform like the $(3,\bar{3})+(\bar{3},3)$ representation[5]; in a quark model, like the quark mass term. In this case, the zeroth-order weak interactions may be identified as an object belonging to the same representation as the symmetry-breaking term. After an appropriate $SU(3)\times SU(3)$ transformation, their only effect is to cause a renormalization of the symmetry-breaking terms, giving renormalized quark masses.[4] There is no violation of hypercharge or parity. Indeed, from a speculative stability requirement of the symmetry-breaking term under weak and electromagnetic corrections, the correct value of the Cabibbo angle may be deduced.[4]

Although the zeroth-order terms are controlled with an appropriate model of strong interactions, the first-order terms remain troublesome. Indeed, with a quark model, we immediately encounter strangeness-violating couplings of neutral lepton currents and contributions to the neutral kaon mass splitting to order $G(G\Lambda^2)$.[6] (In such a model, departures from $\Delta S=\Delta Q$ first appear at second order.) For this reason, it appears necessary to depart from the original phenomenological model of weak interactions. One suggestion[7] involves the introduction of a large number of intermediaries of spins one and zero, so coupled that the leading divergences are associated with only the diagonal symmetry-preserving interactions; in this fashion a proper ordering of divergences is readily obtained. But this model is an awkward one involving many intermediaries with different spins but degenerate coupling strengths. Few would concede so much sacrifice of elegance to expediency.[8]

We wish to propose a simple model in which the divergences are properly ordered. Our model is founded in a quark model, but one involving four, not three, fundamental fermions; the weak interactions are mediated by just one charged vector boson. The weak hadronic current is constructed in precise analogy with the weak lepton current, thereby revealing suggestive lepton-quark symmetry. The extra quark is the simplest modification of the usual model leading to the proper ordering of divergences. Just as importantly, we argue that universality is preserved, in the sense that the

[3] C. Bouchiat, J. Iliopoulos, and J. Prentki, Nuovo Cimento 56A, 1150 (1968); J. Iliopoulos, *ibid.* 62A, 209 (1969); R. Gatto, G. Sartori, and M. Tonin, Phys. Letters 28B, 128 (1968); Nuovo Cimento Letters 1, 1 (1969).

[4] N. Cabibbo and L. Maiani, Phys. Letters 28B, 131 (1968); Phys. Rev. D 1, 707 (1970).

[5] S. L. Glashow and S. Weinberg, Phys. Rev. Letters 20, 224 (1968); M. Gell-Mann, R. J. Oakes, and B. Renner, Phys. Rev. 175, 2195 (1968).

[6] Of course, one cannot exclude *a priori* the possibility of a cancellation in the sum of the relevant perturbation expansion in the limit $\Lambda\to\infty$.

[7] M. Gell-Mann, M. L. Goldberger, N. M. Kroll, and F. E. Low, Phys. Rev. 179, 1518 (1969).

[8] For other departures from the conventional theory, see, for example, C. Fronsdal, Phys. Rev. 136B, 1190 (1964); W. Kummer and G. Segrè, Nucl. Phys. 64, 585 (1965); G. Segrè, Phys. Rev. 181, 1996 (1969); L. F. Li and G. Segrè, *ibid.* 186, 1477 (1969); N. Christ, *ibid* 176, 2086 (1968). It should be understood that the ingenious conjecture of T. D. Lee and G. C. Wick [Nucl. Phys. B9, 209 (1969)] for removing divergences is logically independent of our analysis. If their hypothesis is correct, the role of the cutoff momentum is played by M_W. Only if M_W is small (~3–4 GeV) would the problems associated with ordering of divergences be solved; otherwise, a modification of the coupling scheme, such as ours, is still necessary.

leading divergent corrections (i.e., the first-order terms) yield a ***common*** renormalization to each of the various observed coupling constants.

The new model is discussed in Sec. I. Since Cabibbo's algebraic notion of universality[9] is maintained, that is to say, the entire weak charges generate the algebra of $SU(2)$, we observe in Sec. II that an extension to a three-component Yang-Mills model may be feasible. In contradistinction to the conventional (three-quark) model, the couplings of the neutral intermediary—now hypercharge conserving—cause no embarrassment. The possibility of a synthesis of weak and electromagnetic interactions is also discussed.

In Sec. III we briefly note some of the implications of the existence of a fourth quark, and finally, in Sec. IV we discuss some of the experimental tests of our model of weak interactions.

I. NEW MODEL

We begin by introducing four quark fields.[10] The three quarks $\mathcal{P}$, $\mathcal{N}$, and λ form an $SU(3)$ triplet, and the fourth, $\mathcal{P}'$, has the same electric charge as $\mathcal{P}$ but differs from the triplet by one unit of a new quantum number $\mathcal{C}$ for charm. The strong-interaction Lagrangian is supposed to be invariant under chiral $SU(4)$, except for a symmetry-breaking term transforming, like the quark masses, according to the $(4,\bar{4})+(\bar{4},4)$ representation. This term may always be put in real diagonal form by a transformation of $SU(4)\times SU(4)$, so that B, Q, Y, $\mathcal{C}$, and parity are necessarily conserved by these strong interactions.

The extra quark completes the symmetry between quarks and the four leptons ν, ν', e^-, and μ^-. Both quadruplets possess unexplained unsymmetric mass spectra, and consist of two pairs separated by one in electric charge.

The weak lepton current may be expressed as

$$J_\mu{}^L=\bar{l}C_L\gamma_\mu(1+\gamma_5)l, \tag{1}$$

where l is a column vector consisting of the four lepton fields (ν, ν', e^-, μ^-) and the matrix C_L is given by

$$C_L=\begin{bmatrix}0&0&1&0\\0&0&0&1\\0&0&0&0\\0&0&0&0\end{bmatrix}. \tag{2}$$

This is a convenient way to rewrite the conventional current. In analogy with this expression, we define the weak hadron current to be

$$J_\mu{}^H=\bar{q}C_H\gamma_\mu(1+\gamma_5)q, \tag{3}$$

where q is the quark column vector $(\mathcal{P}',\mathcal{P},\mathcal{N},\lambda)$ and the matrix C_H must be of the form

$$C_H=\left[\begin{array}{cc|c}0&0&\\0&0&\smash{U}\\\hline 0&0&0\quad 0\\0&0&0\quad 0\end{array}\right] \tag{4}$$

in order for $J_\mu{}^H$ to carry unit charge. Pursuing the analogy further, we demand that the 2×2 submatrix U be unitary, so that the matrix C_H is equivalent to C_L under an $SU(4)$ rotation. Of course, it is not convenient to carry out the transformation making C_H and C_L coincide, for this would destroy the diagonalization of the $SU(4)$-breaking term, the quark masses. Nevertheless, suitable redefinitions of the relative phases of the quarks may be performed in order to make U real and orthogonal, so without loss of generality we write

$$U=\begin{bmatrix}-\sin\theta & \cos\theta\\ \cos\theta & \sin\theta\end{bmatrix}. \tag{5}$$

This is just the form of the weak current suggested in an earlier discussion of $SU(4)$ and quark-lepton symmetry.[10] What is new is the observation that this model is consistent with the phenomenological selection rules and with universality even when all divergent first-order terms [i.e., $G(G\Lambda^2)^n$] are considered.

To see this, we proceed diagrammatically in the quark model ignoring the strong $SU(4)$-invariant interactions.[11] Zeroth-order terms occur only in diagrams with only one external quark line, and give contributions to the quark mass operator of the form

$$\delta M(\gamma k)=\sum A_n(G\Lambda^2)^n\bar{q}M_n\gamma\cdot k(1+\gamma_5)q. \tag{6}$$

The A_n are dimensionless parameters, and the matrix M_n is a symmetric homogeneous polynomial of order n in C_H and of order n in $C_H{}^\dagger$. From the definition of C_H, it is seen that M_n must be a multiple of the unit matrix—again in contradistinction to the $SU(3)$ situation. Now, the zeroth-order terms are $SU(4)$ invariant.

There remains an apparent zeroth-order violation of parity, which may be transformed away because of the simple fashion of chiral $SU(4)$ breaking we have assumed. We simply define new quark fields

$$q_i'=(\alpha+\beta\gamma_5)q_i \tag{7}$$

with the real cutoff-dependent parameters α and β chosen so that the entire (bare plus zeroth-order) mass operator, in terms of q_i', is diagonal and parity conserving. The $SU(4)\times SU(4)$-invariant strong interactions are left unchanged. The procedure is analogous

[9] N. Cabibbo, Phys. Rev. Letters 10, 531 (1963).

[10] B. J. Bjorken and S. L. Glashow, Phys. Letters 11, 255 (1964).

[11] All our results about the zero- and first-order selection rules are trivially extended to the case of an $SU(4)$-invariant strong interaction which consists of a neutral vector boson coupled to quark number, the so-called "gluon" model. The only results of this paper which might be affected by such an interaction are the universality conditions in Eq. (9).

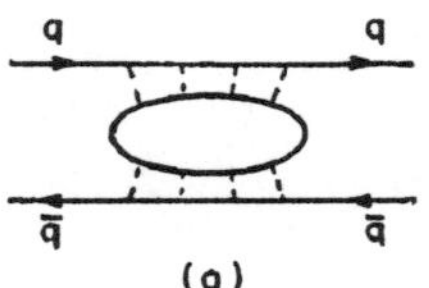

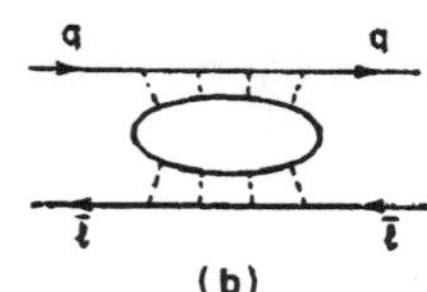

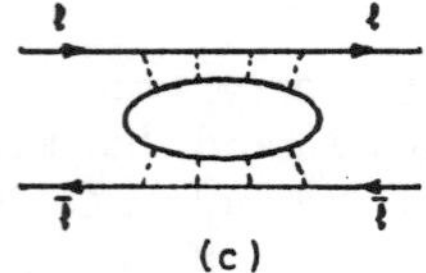

FIG. 1. (a) Connected part of the $q\bar{q} \to q\bar{q}$ amplitude. The crossed (annihilation) channel is also understood. (b) Connected part of the $ql \to ql$ amplitude. (c) Connected part of the $ll \to ll$ amplitude.

to that of Ref. 4, with the difference that the transformation (7) is $SU(4)$ invariant and does not change the definition of strangeness (or charm), or of the Cabibbo angle. An important consequence of the fact that M_n does not depend on the Cabibbo angle is that, unlike the situation in Ref. 4, it is impossible in our case to evaluate the Cabibbo angle by imposing a condition on the leading divergences. We conclude that zeroth-order weak effects are not significant.

We now consider the first-order $G(G\Lambda^2)^n$ terms which are of four types: (i) next-to-the-leading contributions to the quark and lepton mass operators, (ii) leading contributions to quark-quark or quark-antiquark scattering, (iii) leading contributions to quark-lepton scattering, and (iv) leading contributions to lepton-lepton scattering. Graphs with more than two external fermion lines yield no larger than second-order effects. Terms of type (i) are harmless: They contribute to observable nonleptonic $\Delta I = \frac{1}{2}$ processes, but since they cannot give $\Delta Y = 2$, they do not produce a K_1K_2 mass splitting. On the other hand, type-(ii) diagrams could lead to $\mathfrak{N}\bar{\lambda} \to \bar{\mathfrak{N}}\lambda$, possibly giving rise to first-order contributions to the K_1K_2 mass difference, contrary to experiment. Let us show that they do not.

Graphs contributing to type (ii) effects are of the general form shown in Fig. 1(a), where the bubble includes any possible connections among the boson lines, and any number of closed fermion loops. The leading divergent contributions to q-$\bar{q}$ scattering from these graphs have the form

$$T_{HH} = G \sum_{n=2}^{\infty} B_n (G\Lambda^2)^{n-1} [\bar{q}\gamma_\mu (1+\gamma_5) \times B_H^{(n)} q\bar{q}\gamma^\mu (1+\gamma_5) B_H^{(n)\dagger} q], \quad (8)$$

where the B_n are finite dimensionless parameters independent of masses or momenta. It is clear that these first-order terms are independent of all external momenta. The matrix $B_H^{(n)}$ is a polynomial in C_H and $C_H^\dagger$ of order k and l, respectively, with $k+l \leq n$. Furthermore, the charge structure of the quark multiplets allows a change of charge no greater than unity, so that $|k-l|$ must be zero or one, and the matrices $B_H^{(n)}$ are easily computed (see the Appendix) to be

$$B_H^{(n)} = C_H \text{ or } C_H^\dagger \quad (k = l \pm 1) \quad (8')$$

$$= [C_H, C_H^\dagger] \quad (k = l). \quad (8'')$$

Thus, T_{HH} gives rise to contributions with $|\Delta Y| \leq 1$ and, in particular, it does not yield a first-order K_1K_2 mass splitting. Of course, the next-to-the-leading divergences of these graphs will give $\Delta Y = 2$, and do contribute to a second-order K_1K_2 mass difference, agreeing with experiment.

The leading divergences of types (iii) and (iv) give first-order contributions T_{HL} and T_{LL}, to semileptonic and leptonic processes. There will be a 1-to-1 correspondence among the graphs contributing to T_{LL}, T_{HL} [Figs. 1(b) and 1(c)], and T_{HH}. Because the algebraic properties of C_H and C_L are identical, we construct T_{HL} and T_{LL} from T_{HH} by the appropriate substitutions of $q \to L$ and $C_H \to C_L$.

In processes where the lepton charge changes, no violations of observed selection rules occur, but the first-order terms cause a renormalization of observed coupling constants. It is important to note that these renormalizations are common to leptonic and semileptonic processes, so that the relations

$$G_V(\Delta S=0) = G_\mu \cos\theta, \quad G_V(\Delta S=1) = G_\mu \sin\theta \quad (9)$$

remain true when all first-order terms are included. This renormalization is given by the factor $1+\sum B_n (G\Lambda^2)^{n-1}$. A sufficient condition for these renormalizations to be common is the algebraic version of universality—a condition which is satisfied by our model, as well as by the usual three-quark model.

Next, we turn to the induced first-order couplings of hadrons to neutral lepton currents and self-couplings of neutral lepton currents. The neutral lepton currents are generated by the matrix C_L^0 and the neutral hadron currents by the matrix C_H^0, where

$$C_L^0 = [C_L, C_L^\dagger] = \begin{bmatrix} 1 & 0 \\ 0 & -1 \end{bmatrix} = [C_H, C_H^\dagger] = C_H^0. \quad (10)$$

Evidently, there are no induced couplings of neutral lepton currents to strangeness-changing currents. The induced couplings involve the strangeness-conserving current

$$\begin{aligned} J_\mu^0 &= \bar{q}\gamma_\mu C_H^0 (1+\gamma_5) q + \bar{l}\gamma_\mu C_L^0 (1+\gamma_5) l \\ &= \bar{\mathcal{P}}'\gamma_\mu (1+\gamma_5)\mathcal{P}' + \bar{\mathcal{P}}\gamma_\mu (1+\gamma_5)\mathcal{P} - \bar{\mathfrak{N}}\gamma_\mu (1+\gamma_5)\mathfrak{N} \\ &\quad - \bar{\lambda}\gamma_\mu (1+\gamma_5)\lambda + \bar{\nu}'\gamma_\mu (1+\gamma_5)\nu' + \bar{\nu}\gamma_\mu (1+\gamma_5)\nu \\ &\quad - \bar{e}\gamma_\mu (1+\gamma_5) e - \bar{\mu}\gamma_\mu (1+\gamma_5)\mu. \quad (11) \end{aligned}$$

The coupling constant for this new neutral current-current interaction is a first-order expression of the form

$$G\sum_{n=2}^{\infty} C_n(G\Lambda^2)^{n-1}.$$

We anticipate that its strength should be comparable to the strength of the charged leptonic interactions. The new coupling plays no role in observed decay modes, but is should be detectable in accelerator experiments.

In Sec. II we discuss the possible extension of our model to a Yang-Mills model, where the coupling strength of the neutral W to its current is uniquely determined. These neutral lepton couplings constitute the most characteristic and interesting feature of our model. Relevant experimental evidence is discussed in Sec. IV.

II. YANG-MILLS MODEL OF WEAK INTERACTIONS

Divergences appear in our model of weak interactions, but they are properly ordered; observed selection rules are broken only in order $G^2(G\Lambda^2)^n$. But, the model is certainly not renormalizable. There is at least a possibility that a Yang-Mills model of weak interactions may be less singular.[12] In this section, we show how our model can be extended to include a symmetrically coupled triplet of W's. It is possible that W self-couplings can be introduced to give a complete Yang-Mills theory.

The Lagrangian with which we work may be written, in the four-quark model, without electromagnetism,

$$\mathcal{L}=\mathcal{L}_{\rm kin}+\mathcal{L}_s+\mathcal{L}_M+\mathcal{L}_w\,, \qquad (12)$$

where $\mathcal{L}_{\rm kin}$ is the purely kinematic term

$$\mathcal{L}_{\rm kin}=\bar{q}\gamma\cdot pq+\bar{l}\gamma\cdot pl+G_{\mu\nu}G^{\mu\nu}+W_{\mu\nu}{}^\dagger W^{\mu\nu} \qquad (13)$$

describing four free massless quarks, four leptons, and their strong and weak intermediaries ($X_{\mu\nu}$ denotes the antisymmetric curl of X_μ). $\mathcal{L}_s$ denotes the $SU(4)$-invariant strong interaction, most simply

$$\mathcal{L}_s=fG_\mu\bar{q}\gamma^\mu q\,, \qquad (14)$$

and $\mathcal{L}_w$ is the weak interaction

$$\mathcal{L}_w=gW_\mu{}^\dagger[\bar{q}C_H\gamma^\mu(1+\gamma_5)q+\bar{l}C_L\gamma^\mu(1+\gamma_5)l+\text{H.a.}]\,. \qquad (15)$$

The bare-mass term $\mathcal{L}_M$ produces the observed masses of the leptons, the masses of W and G, and gives rise to the observed hierarchy of hadron symmetry,

$$\mathcal{L}_M=\bar{q}M_Hq+\bar{l}M_Ll+m^2G_\mu G^\mu+M^2W_\mu W^\mu\,, \qquad (16)$$

where M_H and M_L are 4×4 matrices. This model gives a complete description of weak-interaction phenomena. The most important new feature is the appearance of neutral currents generated by the most divergent parts of diagrams containing an exchange of W^+, W^- pairs between two fermion lines. The effective coupling strength of these currents is expected to be of order G but, at this stage, we cannot predict its precise numerical value since we are unable to sum the perturbation series. In order to extend this model to a more symmetric one, we introduce an additional weak intermediary W_0 with appropriate couplings.

The couplings of W_0 to hadrons and leptons must be taken to be

$$2^{-1/2}gW_0{}^\mu\{\bar{q}[C_H{}^\dagger,C_H]\gamma_\mu(1+\gamma_5)q+\bar{l}[C_L{}^\dagger,C_L]\gamma_\mu(1+\gamma_5)l\}\,. \qquad (17)$$

We emphasize that the introduction of W_0 is by no means necessary in our model; however, we think that it gives a much more symmetric and aesthetically appealing theory.

In the conventional model of weak interactions, the extension to a three-component vector-meson theory cannot be made without contradicting experiment: The neutral boson leads to strangeness-changing decays involving neutral-lepton currents and to $\Delta S=2$ at order G. This is because the commutator of the conventional weak charge with its adjoint yields a strangeness-violating neutral charge. In our case, the corresponding operator is diagonal, and these difficulties are absent.

It is straightforward to show that the introduction of the neutral current does not spoil the proper ordering of divergences: The observed selection rules are preserved by all terms of order $G(G\Lambda^2)^n$. This is shown in the Appendix.

We note that W_0 is coupled to precisely the same neutral current appearing in the last section as an induced coupling. In the symmetric three-W model, its strength is uniquely predicted. Universality now applies to both charged and neutral couplings. That is to say, the leading divergent corrections to each are the same. The bare relationship

$$G_0=\tfrac{1}{2}G \qquad (18)$$

is preserved by the renormalizations, to first order [i.e., including all terms of order $G(G\Lambda^2)^n$]. This assertion is proved in the Appendix.

The introduction of a neutral W opens the possibility of formulating the weak interactions into a Yang-Mills theory. Self-couplings must be introduced among the W triplet in order to ensure the gauge symmetry. This is accomplished if we choose the Lagrangian in a manifestly gauge-invariant fashion:

$$\mathcal{L}=\bar{q}\gamma\Pi_Hq+\bar{l}\gamma\Pi_Ll+\mathbf{W}_{\mu\nu}\mathbf{W}^{\mu\nu}+G_{\mu\nu}G^{\mu\nu}+\mathcal{L}_M+\mathcal{L}_s\,, \qquad (19)$$

where

$$\Pi_H{}^\mu=\partial^\mu+ig(\mathbf{C}_H\cdot\mathbf{W}^\mu)(1+\gamma_5)\,, \qquad (19')$$

$$\Pi_L{}^\mu=\partial^\mu+ig(\mathbf{C}_L\cdot\mathbf{W}^\mu)(1+\gamma_5)\,, \qquad (19'')$$

[12] See, for example, S. Mandelstam, Phys. Rev. 175, 1580 (1969); M. Veltman, Nucl. Phys. B7, 637 (1968); H. Reiff and M. Veltman, *ibid.* B13, 545 (1969); D. Boulware, Ann. Phys. (N. Y.) 56, 140 (1970); A. A. Slavnov, University of Kiev Report No. ITP 69/20 (unpublished); E. S. Fradkin and I. V. Tyutin, Phys. Letters 30B, 562 (1969). Notice, however, that none of these references consider the far more difficult case of vector mesons coupled to nonconserved currents.

TABLE I. Quark quantum numbers.

	Fractional assignment			Integral assignment		
	Q	Y	$\mathcal{C}$	Q	Y	$\mathcal{C}$
$\mathcal{P}'$	$\frac{2}{3}$	$-\frac{2}{3}$	1	0	0	0
$\mathcal{P}$	$\frac{2}{3}$	$\frac{1}{3}$	0	0	0	-1
$\mathfrak{N}$	$-\frac{1}{3}$	$\frac{1}{3}$	0	-1	0	-1
λ	$-\frac{1}{3}$	$-\frac{2}{3}$	0	-1	-1	-1

and

$$W^{\mu\nu}=\Pi_W{}^{\mu}W^{\nu}-\Pi_W{}^{\nu}W^{\mu}, \qquad (19''')$$

where

$$(\Pi_W{}^{\mu})_{ij}=\delta_{ij}\partial^{\mu}+ig2^{-1/2}(\mathbf{t}\cdot\mathbf{W}^{\mu})_{ij}. \qquad (19'''')$$

The matrix-valued vectors $\mathbf{C}_H$ and $\mathbf{C}_L$ have components $(C,C^{\dagger},2^{-1/2}[C^{\dagger},C])$ in a basis where charge is diagonal, and $\mathbf{t}$ are the usual 3×3 generators of $O(3)$, with t_3 diagonal. The gauge group thus introduced is an exact symmetry of the entire Lagrangian excepting both $\mathcal{L}_M$ and electromagnetism.

The Yang-Mills model is undoubtedly the most attractive way to couple a triplet of vector mesons and the only one for which people have expressed some hope of constructing a renormalizable theory. The massless case has been proved to be renormalizable[12]; however, very little is known about the physically more interesting massive theory. In fact, the naive power counting shows that the highest divergence in a Yang-Mills theory is $g^{2n}\Lambda^N$ with $N=6n$. Notice that in the absence of the self-couplings the corresponding divergences are given, as we have already seen, by $N=2n$. So, at first sight, the Yang-Mills theory seems to be much more divergent than the ordinary coupling of the vector mesons with the currents. However, one can show that the naive limit $N=6n$ can be considerably lowered. We have already been able to show that $N\leq 3n$ and we believe that one can still lower this limit to at least $N=2n$. In other words, we believe that the introduction of the self-couplings does not make the theory more divergent.

Let us briefly consider a more daring speculation. It has long been suspected[13] that there may be a fundamental unity of weak and electromagnetic interactions, reflected phenomenologically by the common vectorial character of their couplings. For this reason, it may have been wrong for us to introduce a gauge symmetry for the weak interactions not shared by electromagnetism. As a more speculative alternative, consider the possibility of a four-parameter gauge group involving $\mathbf{W}$, and an additional Abelian singlet W_S, broken only by the mass term $\mathcal{L}_M$. Suppose, however, that a one-parameter gauge symmetry, corresponding to a linear combination A of W_0 and W_S remains unbroken. Then A must be massless, and may be identified as the photon. The orthogonal neutral combination B is massive, and acts as an intermediary of weak interactions along with $W^{\pm}$. This model could be correct only if the weak bosons are very massive (100 GeV) so that the weak and electromagnetic coupling constants could be comparable. With this model, the relation (18) would not persist, and the weak neutral current would involve $(1-\gamma_5)$ as well as $(1+\gamma_5)$ currents. The precise form of the model would depend on what linear combination of W_0 and W_S is the photon.

[13] J. Schwinger, Ann. Phys. (N. Y.) 2, 407 (1957); S. L. Glashow, Nucl. Phys. 10, 107 (1959); 22, 579 (1961)

III. ANOTHER QUARK MAKES $SU(4)$

Having introduced four quarks, we must consider strong interactions which admit the algebra of chiral $SU(4)$. Does this mean we should expect $SU(4)$ to be an approximate symmetry of nature? Nothing in our argument depends on how much $SU(4)$ is broken; the divergences are necessarily properly ordered. However, for the higher-order nonleading divergences to be as small as they must be, the breaking of $SU(4)$ cannot be too great: The limit on the cutoff Λ is replaced by a limit on Δ, a parameter measuring $SU(4)$ breaking; and from the observed K_1K_2 mass difference we now conclude that Δ must be not larger than 3–4 GeV. Thus, some residue of $SU(4)$ symmetry should persist.

We expect the appearance of charmed hadron states.[10] Meson multiplets, made up of a quark-antiquark pair, must belong to the 15-dimensional adjoint representation of $SU(4)$, consisting of an uncharmed $SU(3)$ singlet and octet, as well as two $SU(3)$ triplets of charm ±1. The structure of baryons depends on the quantum numbers assigned to the quarks. The two simplest possibilities are shown in Table I. For the more conventional fractional charge assignment, the baryons are made up of three quarks, and must belong to one of the representations contained in $4\times4\times4$. The only possibility is a 20-dimensional representation, which contains, besides the baryon octet, a triplet and sextet of charmed states and a doubly charmed triplet. The $j=\frac{3}{2}^{+}$ baryon decuplet belongs to another 20-dimensional representation with a charmed sectet, a doubly charmed triplet, and a triply charmed singlet.

With the integral-charge assignment, the baryon octet must be made of two quarks and an antiquark, the decuplet of three quarks and two antiquarks. The lepton and quark charged spectra now coincide, and the synthesis of weak and electromagnetic interactions appears more plausible. Moreover, there is no difficulty in obtaining the correct value for the π^0 lifetime.

Why have none of these charmed particles been seen? Suppose they are all relatively heavy, say ~2 GeV. Although some of the states must be stable under strong (charm-conserving) interactions, these will decay rapidly ($\sim10^{+13}$ sec^{-1}) by weak interactions into a very wide variety of uncharmed final states (there are about a hundred distinct decay channels). Since the charmed particles are copiously produced only in associated production, such events will necessarily be of very complex topology, involving the plentiful decay products of both charmed states. Charmed particles could easily have escaped notice.

Finally, we briefly comment on the leptonic decay rates of ρ, ω, and ϕ (Γ_ρ, Γ_ω, and Γ_ϕ). Our electric current contains $SU(3)$ singlet as well as octet terms, so that the inequality

$$m_\omega\Gamma_\omega+m_\phi\Gamma_\phi\geq\tfrac{1}{3}m_\rho\Gamma_\rho \tag{20}$$

may be deduced from the Weinberg spectral function sum rules and ω, ϕ, ρ dominance.[14] A stronger result is obtained if we extend Weinberg's Schwinger-term hypothesis to the vector currents of $SU(4)$:

$$m_\omega\Gamma_\omega+m_\phi\Gamma_\phi\geq m_\rho\Gamma_\rho. \tag{21}$$

This result is in poor agreement with experiment, which favors the equality in (20). A resolution of this difficulty that does not abandon the Schwinger-term symmetry requires the introduction of a third $Y=T=0$ vector meson, another partner of ω and ϕ, corresponding to the $SU(4)$ singlet vector current.

IV. EXPERIMENTAL SUGGESTIONS

In this section, we discuss some of the observable effects characteristic of our picture of strong and weak interactions. Firstly, consider the experimental implications of the existence of a new quantum number—charm—broken only by weak interactions. The charmed particles, because they are heavy, are too short lived to give visible tracks. However, they should be copiously produced in hardonic collisions at accelerator energies:

$$(\text{hadron or }\gamma)+(\text{hadron})\rightarrow X^{(+)}+X^{(-)}+\cdots,$$

where $X^{(\pm)}$ are oppositely charmed particles, each rapidly decaying into uncharmed hadrons with or without a charged lepton pair. The purely hadronic decay modes could provide illusory violations of hypercharge conservation in strong interactions. The leptonic decay modes provide a mechanism for the seemingly direct production of one or two charged leptons in hadron-hadron collisions.[15] Conceivably, muons thus produced may be responsible for the anomalous observed angular distribution of cosmic-ray muons in the 10^{12}-eV range,[16] where these directly produced muons may dominate the sea-level muon flux.

Should this last speculation about cosmic rays be correct, we need to revise radically estimates of the flux of ν and $\bar{\nu}$ in this energy range. We expect the charmed particle decays to yield equal numbers of each lepton variety; this gives a flux of electron neutrinos and antineutrinos equal to the muon flux, and 10–100 times greater than other estimates. This fact is of crucial importance to the possible detection of the resonance scattering[17]

$$\bar{\nu}+e^-\rightarrow\bar{\nu}'+\mu^-.$$

Charmed particles may be produced singly by neutrinos in such reactions as

$$\nu'+N\rightarrow\mu^-+X,\quad \bar{\nu}'+N\rightarrow\mu^++X,$$

where the charmed particle X would have a variety of decay modes, including leptonic ones. With the fractional charge assignment, the neutrino processes are suppressed by $\sin^2\theta$ and the antineutrino processes are forbidden. On the other hand, with the integral-charge assignment, the neutrino processes are again proportional to $\sin^2\theta$ while the antineutrino processes are proportional to $\cos^2\theta$.

The second new feature of our model is the appearance of neutral leptonic and semileptonic couplings involving a specified ($Y=0$) current and with a coupling constant comparable with the Fermi constant. Without the introduction of a W_0, we may say only $G_0\sim G$. To be more definite, we shall phrase our arguments in terms of the value $G_0=\frac{1}{2}G$ of Eq. (18).

Let us summarize the presently available data about these interactions.[18] Consider the following three reactions induced by muon neutrinos:

$$\text{(i)}\quad \nu'+e^-\rightarrow\nu'+e^-,$$
$$\text{(ii)}\quad \nu'+p\rightarrow\nu'+p,$$
$$\text{(iii)}\quad \nu'+p\rightarrow\nu'+\pi^++n.$$

None of these neutral couplings have been observed; experimentally, we can only quote limits. From the absence of observed forward energetic electrons in the CERN bubble-chamber experiments, we may conclude

$$G_0\leq G,$$

a limit which is close to but consistent with our prediction.

For reaction (ii), it is found that

$$R=\sigma(\nu'p\rightarrow\nu'p)/\sigma(\bar{\nu}'p\rightarrow\mu^+n)\leq 0.5.$$

Because our neutral current contains both $I=0$ and $I=1$ parts, we cannot unambiguously predict this ratio. In a naive quark model, where the proton consists of only $\mathfrak{N}$ and $\mathcal{P}$ quarks, we find $R=\frac{1}{4}$, again close but consistent.

Finally, we quote the experimental limit on reaction (iii):

$$R'=\sigma(\nu'+p\rightarrow\pi^++n+\nu')/\sigma(\nu'+p\rightarrow\pi^++p+\mu^-)\leq 0.08.$$

[14] S. Weinberg, Phys. Rev. Letters **18**, 507 (1967); T. Das, V. Mathur, and S. Okubo, *ibid.* **19**, 470 (1967).

[15] In a recent experiment, P. J. Wanderer *et al.* [Phys. Rev. Letters **23**, 729 (1969)] have performed a search for W's by measuring the intensity and polarization of prompt energetic muons from the interaction of 28-GeV protons with nuclei. Their results are compatible with the assumption that all 25-GeV prompt muons have electromagnetic origin. There is no indication of the existence of W's. However, the published evidence does not seem to be relevant to the existence of charmed particles, which are produced in pairs, decay into many final states, and are not expected to produce many very energetic muons.

[16] H. E. Bergeson *et al.*, Phys. Rev. Letters **21**, 1089 (1968).

[17] M. G. K. Menon *et al.*, Proc. Roy. Soc. (London) **A301**, 137 (1967).

[18] See D. H. Perkins, in Proceedings of the Topical Conference in Weak Interactions, CERN, 1969 [CERN Report No. 69-7], pp. 1–42 (unpublished).

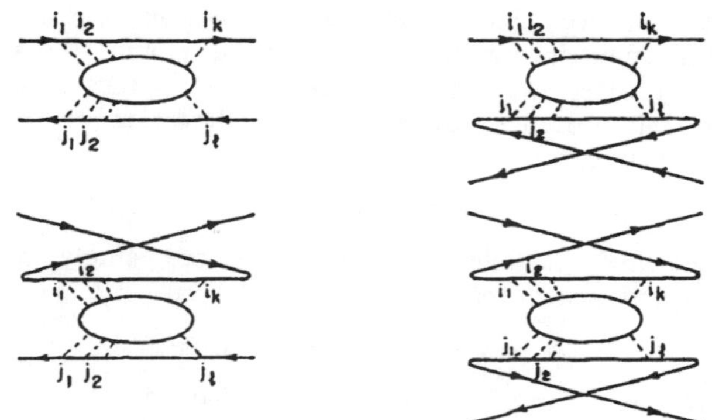

FIG. 2. Decomposition of the $q\bar{q}\to q\bar{q}$ connected amplitude by crossing the external fermion lines.

Because this transition is $\Delta I=1$, we unambiguously predict $R'=\frac{1}{9}$ under the hypothesis of $\Delta(1238)$ dominance. In each of these three reactions, experiment is very close to a decisive test of our model.

In our model, the parity-violating nonleptonic interaction is also changed. In particular, the parity-violating one-pion-exchange nuclear force is no longer suppressed by $\sin^2\theta$.

Next we consider some experiments which could discover the existence of W_0. A simple and attractive possibility is the search for muon tridents in the semiweak reaction[19]

$$\mu^-+Z\to\mu^-+W_0+Z,$$

with the subsequent muonic decay of W_0. Another possibility is the reaction[20]

$$e^+e^-\to\mu^+\mu^-.$$

The interference between the W^0 and photon contributions causes an asymmetry of the μ^+ angular distribution relative to the momentum of the incident e^+ given by

$$\delta=\frac{N_F-N_B}{N_F+N_B}=\frac{3M_W{}^2}{16\sqrt{2}}\frac{G}{\alpha\pi}\frac{s}{s-M_W{}^2},$$

where

$$G=10^{-5}M_p{}^{-2},\quad \alpha=1/137,\quad\text{and}\quad s=4E_e{}^2.$$

Away from the W^0 pole, the effect is rather small (less than 1% for $E_e=3.5$ GeV) and it is masked by a similar effect due to the two-photon contribution. However, the factor $s/(s-M_W{}^2)$ makes the asymmetry increase sharply and change sign near M_W. Therefore, this reaction is an excellent tool to sweep a substantial mass range looking for W's. Another effect, much harder to detect, would be the direct observation of parity violation in $e^+e^-\to\mu^+\mu^-$. This requires the measurement of μ polarization.

Finally, we recall from Sec. III that the $SU(4)$ description of leptonic decays of vector mesons suggested the existence of another strongly coupled neutral $I=0$ vector meson with considerable coupling to lepton pairs. Evidence for its existence could come from colliding beam experiments.

[19] M. Tannenbaum (private communication).

[20] N. Cabibbo and R. Gatto, Phys. Rev. 129, 1577 (1961).

APPENDIX

In this appendix we determine the form of the leading weak corrections to the q-$\bar{q}$, q-l, and l-l amplitudes.

We have already shown that the wave-function renormalization of spinors is the same for both quarks and leptons and contributes a common factor to T_{HH}, T_{HL}, and T_{LL}. Therefore we need consider only the q-$\bar{q}$ amplitude T_{HH}. The other amplitudes T_{HL} and T_{LL} can be obtained from T_{HH} by appropriate substitutions. In the following, we shall omit the common wave-function renormalization factors.

For the sake of clarity, let us first consider our model of weak interactions, where we have three vector bosons symmetrically coupled.

The graphs of Fig. 1(a) can be decomposed into four classes of terms, obtained by keeping the boson lines fixed and reversing the fermion lines, as shown in Fig. 2. We then obtain for the contribution to T_{HH} corresponding to these four classes of diagrams

$$\begin{aligned}T_{HH}{}^{(n,k,l)}&=\bar{q}\gamma_\mu(1+\gamma_5)[C_{i_1}C_{i_2}\cdots C_{i_k}-(-1)^kC_{i_k}C_{i_{k-1}}\cdots C_{i_1}]q\\&\quad\times P_{j_1\cdots j_l;i_1\cdots i_k}\bar{q}^\mu(1+\gamma_5)\\&\quad\times[C_{j_1}C_{j_2}\cdots C_{j_l}-(-1)^lC_{j_l}C_{j_{l-1}}\cdots C_{j_1}]q,\\&\qquad k+l\leq n,\quad k,l\geq 1.\end{aligned}\tag{A1}$$

All the i's and j's go from 1 to 3 and the sum over all indices is understood. $P_{j_1\cdots j_l;i_1\cdots i_k}$ is a tensor made out of the invariant tensors δ_{ij} and ϵ_{ijk}.

It is easy to show that for any k

$$\mathrm{Tr}[C_{i_1}C_{i_2}\cdots C_{i_k}-(-)^kC_{i_k}C_{i_{k-1}}\cdots C_{i_1}]=0.\tag{A2}$$

Therefore, since the interaction is $O(3)$ invariant, the connected part of T_{HH} has the form

$$\begin{aligned}T_{HH}=G\sum_{n=0}^{\infty}b_n(G\Lambda^2)^n\\\times(\bar{q}\gamma_\mu(1+\gamma_5)C_Hq)\cdot(\bar{q}\gamma^\mu(1+\gamma_5)C_Hq).\end{aligned}\tag{A3}$$

In the case where we have only charged bosons, the argument is even simpler. Each of the indices $i_1\cdots i_k$, $j_1\cdots j_l$ appearing in Eq. (A1) takes only two possible values. With the relations

$$\begin{aligned}(C_H)^2=(C_H{}^\dagger)^2=0,\\(C_HC_H{}^\dagger)^2=C_HC_H{}^\dagger,\\(C_H{}^\dagger C_H)^2=C_H{}^\dagger C_H,\end{aligned}\tag{A4}$$

Eq. (A1) explicitly reads

$$\begin{aligned}T_{HH}{}^{(n;k,l)}&=(\bar{q}\gamma_\mu(1+\gamma_5)[C_H,C_H{}^\dagger]q)\\&\quad\times(\bar{q}\gamma^\mu(1+\gamma_5)[C_H,C_H{}^\dagger]q),\quad k=l\\T_{HH}{}^{(n,k,l)}&=(\bar{q}\gamma_\mu(1+\gamma_5)C_Hq)(\bar{q}\gamma^\mu(1+\gamma_5)C_H{}^\dagger q),\\&\qquad k=l+1.\quad\text{Q.E.D.}\end{aligned}$$

PHYSICAL REVIEW D VOLUME 2, NUMBER 8 15 OCTOBER 1970

Broken Scale Invariance in Scalar Field Theory*

CURTIS G. CALLAN, JR.†
California Institute of Technology, Pasadena, California 91109
and
Institute for Advanced Study, Princeton, New Jersey 08540

(Received 4 June 1970)

We use scalar-field perturbation theory as a laboratory to study broken scale invariance. We pay particular attention to scaling laws (Ward identities for the scale current) and find that they have unusual anomalies whose presence might have been guessed from renormalization-group arguments. The scaling laws also appear to provide a relatively simple way of computing the renormalized amplitudes of the theory, which sidesteps the overlapping-divergence problem.

INTRODUCTION

THE perturbation theory of a self-interacting scalar field is about the simplest available model field theory, and a convenient laboratory for testing new ideas in strong-interaction physics. In this paper we shall be concerned with studying the concept of broken scale invariance within such a framework. We shall find that the model calls for some unexpected modifications of our ideas on broken scale invariance. At the same time, the approach suggested by broken scale invariance leads to an interesting, and simple, new approach to renormalization. We hope that this mutual illumination of two interesting questions justifies yet another paper on scalar field theory.

In Sec. I we shall review the general properties of scale invariance as a broken symmetry, leading up to the idea of a scaling law (the analog for scale invariance of PCAC low-energy theorems). In Sec. II we shall see how the general structure of renormalized perturbation theory constrains the allowable form of the scaling law and forces it to differ from naive expectations. In Sec. III we shall show how the existence of the scaling law leads to a simple prescription for computing the renormalized Green's functions of the theory. Finally, in Sec. IV, we shall demonstrate an interesting connection between the scaling law and the predictions of the renormalization group.

I. BROKEN SCALE INVARIANCE

In simple canonical field theories it is possible to introduce an acceptable energy-momentum tensor[1,2] $\Theta_{\mu\nu}$ having the following properties: (a) $\Theta=\Theta_\mu{}^\mu$ is proportional to those terms in the Lagrangian having dimensional coupling constants (such as mass terms); (b) the charge, $D=\int d^3x\, S_0$, formed from the current $S_\mu=\Theta_{\mu\nu}x^\nu$, acts as the generator of scale transformations,

$$[D(x_0),\phi(x)]=-i(d+x\cdot\partial)\phi(x)\,, \tag{1}$$

where d is the dimension of the field; (c) the current S_μ satisfies $\partial^\mu S_\mu=\Theta$ so that it is conserved when there are no dimensional coupling constants in the Lagrangian. With the help of the current S_μ and its equal-time commutation relations with fields, given above, one is able

* Work supported in part by the U. S. Atomic Energy Commission under Contract No. AT(11-1)-68 and by the U. S. Air Force Office of Scientific Research under Contract No. AFOSR 70-1866.
† Alfred P. Sloan Foundation Fellow.

[1] C. G. Callan, Jr., S. Coleman, and R. Jackiw, Ann. Phys. (N. Y.) 59, 42 (1970).
[2] M. Gell-Mann, University of Hawaii Summer School lectures, 1969 (unpublished).

to derive a standard sort of Ward identity[1]

$$\left[n(d-4)+4-\sum p_i\cdot\frac{\partial}{\partial p_i}\right]G(p_1\cdots p_{n-1})=-iF(0,p_1\cdots p_{n-1}),\quad(2)$$

where

$$(2\pi)^4\delta(\sum_{i=1}^{n}p_i)G(p_1\cdots p_{n-1})=\int dx_1\cdots dx_n e^{i\Sigma p_i x_i}\langle 0|T(\phi(x_1)\cdots\phi(x_n))|0\rangle,\quad(3a)$$

$$(2\pi)^4\delta(q+\sum_{i=1}^{n}p_i)F(q,q_1\cdots p_{n-1})=\int dydx_1\cdots dx_n e^{iq\cdot y+i\Sigma p_i x_i}\times\langle 0|T(\Theta(y)\phi(x_1)\cdots\phi(x_n))|0\rangle.\quad(3b)$$

The significance of such an equation is clear: If $\Theta=0$, so that $F=0$, the particle Green's functions satisfy $SG=0$, where S is the linear operator appearing in square brackets in Eq. (2). One can always find the general solution of such an equation, and it turns out to imply that, apart from explicit kinematic factors, G depends only on dimensionless ratios of momentum variables. This is precisely what one expects from naive dimensional reasoning in the event that no dimensional coupling constants are present in the theory. Therefore, the scaling law, Eq. (2), says that the matrix elements F of Θ act as the source of violations of simple dimensional scaling in the particle matrix elements G. It also appears to be of general validity, not depending on the details of the theory, providing a general framework for the study of broken scale invariance. We want to ask whether such a relation, which we shall refer to as a scaling law, actually holds in a simple renormalizable field theory.

It is convenient to define the scaling law for "one-particle-irreducible" Green's functions rather than the full Green's functions defined in Eq. (3). The one-particle-irreducible Green's functions, which we shall denote by $\bar{G}$ and $\bar{F}$, are obtained from the full Green's functions by first throwing away all diagrammatic contributions which fall into disjoint pieces when one internal line is cut, and then dividing out of the remainder one factor of the propagator for each external leg. This simply turns full vertices into proper vertices. The same formal arguments which led to the Ward identity for G allow one to derive a Ward identity for $\bar{G}$:

$$\left[4-nd-\sum p_i\cdot\frac{\partial}{\partial p_i}\right]\bar{G}(p_1\cdots p_{n-1})=-i\bar{F}(0,p_1\cdots p_{n-1}).\quad(4)$$

The difference between the two Ward identities, the factor of $4-nd$ rather than $n(d-4)+4$, arises entirely from the different dimensions of G and $\bar{G}$.

We shall eventually be dealing with a simple theory in which the only dimensional parameter is the particle mass μ. In such a case, ordinary dimensional reasoning requires that $\bar{G}$ have the following dependence on μ:

$$\bar{G}(p_1\cdots p_{n-1})=\mu^{4-n}\Phi(p_1/\mu,\ldots,p_{n-1}/\mu).$$

(It is perhaps worth inspecting a Feynman diagram or two to convince oneself that the dimension of $\bar{G}$ in power of mass is just $4-n$.) Then we have the identity

$$\left[4-n-\sum p_i\cdot\frac{\partial}{\partial p_i}\right]\bar{G}(p_1\cdots p_{n-1})=\mu\frac{\partial}{\partial\mu}\bar{G}(p_1\cdots p_{n-1}),$$

which leads to an equivalent form of Eq. (4):

$$[\mu\partial/\partial\mu+n\delta]\bar{G}(p_1\cdots p_{n-1})=-i\bar{F}(0,p_1\cdots p_{n-1}),\quad(5)$$

where

$$\delta=1-d.$$

From now on, we shall refer to the operator in square brackets as S.

We remarked earlier that d was the dimension of the field. For a scalar field the dimension in powers of mass is unity, so that one expects $d-1=0$. However, as Wilson[3] has pointed out, when there are interactions it is not guaranteed that the naive dimension and the dimension defined by the commutator of the generator of scale transformations with the field [as in Eq. (1)] are the same. Therefore, we shall let the term $(d-1)n$ in Eq. (5) stand. The question now is whether the scaling law of Eq. (5) actually holds in renormalized perturbation theory, and if not, whether there is a simple equation which replaces it.

The question seems interesting since the scaling law is the only obvious direct, and model-independent, expression of how specific forms of scale-invariance breaking affect particle scattering amplitudes. Also it brings in the dimension of the field in a way which may be of phenomenological significance.

II. GENERAL CONSTRAINTS ON SCALING LAW

To settle the questions raised in Sec. I, we shall study perturbation theory for a massive scalar field interacting through $\lambda\phi^4$:

$$\mathfrak{L}=\tfrac{1}{2}(\partial_\mu\phi)(\partial^\mu\phi)-\tfrac{1}{2}\mu^2\phi^2-(\lambda/4!)\phi^4.$$

Since the only term in the Lagrangian with a dimensional coupling constant is the mass term, we expect $\Theta=\mu^2\phi^2$ to be the source for violations of scale invariance.

Since this is a renormalizable theory, there are a finite number of amplitudes which require subtractions.

[3] K. Wilson, Phys. Rev. 179, 1499 (1969).

The standard lore tells us that $\bar{G}^{(2)}$ requires two subtractions, $\bar{G}^{(4)}$ requires one, and $\bar{F}^{(2)}$ requires one, with all other amplitudes requiring no subtractions. That $\bar{F}^{(2)}$ is the only matrix element of Θ requiring a subtraction follows, via the usual power-counting arguments, from our requirement that Θ be proportional to ϕ^2. As a consequence of these subtractions, $\bar{F}^{(2)}$ and $\bar{G}^{(4)}$ are determined up to an arbitrary constant, while $\bar{G}^{(2)}$ is determined up to an arbitrary first-order polynomial in p^2. This arbitrariness means that we can choose $\bar{G}^{(4)}(0)$, $\bar{G}^{(2)}(0)$, $(d/dp^2)\bar{G}^{(2)}(p^2)|_0$, and $\bar{F}^{(2)}(0)$ at will. These parameters, which we call $-i\lambda$, $-i\mu^2$, iZ, and $2m^2$, respectively, play the role of arbitrary parameters in terms of which all the Green's functions of the theory are determined.[4]

The Green's functions of the theory which do not need a subtraction ($\bar{G}^{(n)}$, $n>4$; $\bar{F}^{(n)}$, $n>2$) have the useful property of being directly expressible via the skeleton expansion in terms of $\bar{G}^{(2)}$, $\bar{G}^{(4)}$, and $\bar{F}^{(2)}$. A skeleton diagram for $\bar{G}^{(n)}$ is a Feynman diagram containing no subgraphs identifiable as a contribution to $\bar{G}^{(2)}$ or $\bar{G}^{(4)}$, while a skeleton for $\bar{F}^{(n)}$ is a Feynman diagram containing no subgraphs identifiable as $\bar{G}^{(2)}$, $\bar{G}^{(4)}$, or $\bar{F}^{(2)}$. The skeleton expansion for a given Green's function is obtained by taking all skeleton graphs for that Green's function and replacing all point four-particle vertices by $\bar{G}^{(4)}$, all internal lines by $[-\bar{G}^{(2)}]^{-1}$,[5] and all point insertions of Θ by $\bar{F}^{(2)}$. Since the higher Green's functions are determined once the fundamental Green's functions ($\bar{G}^{(2)}$, $\bar{G}^{(4)}$, and $\bar{F}^{(2)}$) are known, presumably the scaling operator S should be such that $S\bar{G}^{(n)}=-i\bar{F}^{(n)}$ is automatically true for $n>4$ once it is true for $n<4$. The required property of S is that it be "distributive" in the following sense:

Let us consider a particular diagram in the skeleton expansion of $\bar{G}^{(n)}$. If it is a diagram with i vertices and j internal lines, its contribution to $\bar{G}^{(n)}$ is gotten by replacing each vertex by $\bar{G}^{(4)}$ and each internal line by $[-\bar{G}^{(2)}]^{-1}$ and doing the integrations over loop momenta:

$$\bar{G}^{(n)}\sim\int d\,(\text{loop momenta})\,(\bar{G}^{(4)}\cdots\bar{G}^{(4)})_{i\text{ factors}}\times([-\bar{G}^{(2)}]^{-1}\cdots[-\bar{G}^{(2)}]^{-1})_{j\text{ factors}}.$$

What happens when we act on this integral with the scaling operator found in Sec. I, $S=\mu\partial/\partial\mu+n\delta$? By the chain rule for differentiation, $\mu\partial/\partial\mu$ gives a sum of terms in which it acts independently on each factor in the integral for $\bar{G}^{(n)}$:

$$\bar{G}^{(4)}\rightarrow\mu\frac{\partial}{\partial\mu}\bar{G}^{(4)},$$

and

$$[-\bar{G}^{(2)}]^{-1}\rightarrow[-\bar{G}^{(2)}]^{-1}\left[\mu\frac{\partial}{\partial\mu}\bar{G}^{(2)}\right][-\bar{G}^{(2)}]^{-1}.$$

By virtue of the trivial topological relation $n=4i-2j$, the term $n\delta$ has the same behavior. So, if $S=\mu\partial/\partial\mu+n\delta$,

$$\begin{aligned}S\int d\,(\text{loop momenta})\,\bar{G}^{(4)}(1)\cdots[-\bar{G}^{(2)}(j)]^{-1}\\=\int d\,(\text{loop momenta})\,\{\{[S\bar{G}^{(4)}(1)]\bar{G}^{(4)}(2)\cdots\bar{G}^{(4)}(i)+(i-1\text{ similar terms})\}[-\bar{G}^{(2)}(1)]^{-1}\cdots[-\bar{G}^{(2)}(j)]^{-1}\\+\bar{G}^{(4)}(1)\cdots\bar{G}^{(4)}(i)\{[-\bar{G}^{(2)}(1)]^{-1}[S\bar{G}^{(2)}(1)][-\bar{G}^{(2)}(1)]^{-1}[-\bar{G}^{(2)}(2)]^{-1}\cdots[-\bar{G}^{(2)}(j)]^{-1}\\+(j-1\text{ similar terms})\}\},\end{aligned}$$

which is to say that S acts in succession on each vertex and inserts $S\bar{G}^{(2)}$ in succession on each internal line (see Fig. 1). Any S with this property will be called distributive. Obviously, we can add to S any kind of differentiation operation, such as differentiation with respect to coupling constant, without changing its distributive nature.

Now if $S\bar{G}^{(2)}=-i\bar{F}^{(2)}$ and $S\bar{G}^{(4)}=-i\bar{F}^{(4)}$, the above recipe for acting with S on a given skeleton for $\bar{G}^{(n)}$ is as follows: Insert $\bar{F}^{(2)}$ in succession on each internal line and replace each vertex $\bar{G}^{(4)}(p_1p_2p_3)$ in succession by $\bar{F}^{(4)}(0,p_1p_2p_3)$. Inserting $\bar{F}^{(2)}$ on an internal line gives directly a skeleton for $\bar{F}^{(n)}$. Since $\bar{F}^{(4)}$ is a convergent amplitude, it has itself a skeleton expansion. Therefore, the action of replacing a vertex by $\bar{F}^{(4)}$ gives a *sum* of sleketons for $\bar{F}^{(n)}$ in which the vertex in question is replaced in turn by all the skeletons in the expansion for $\bar{F}^{(4)}$. Therefore, S turns a single skeleton for $\bar{G}^{(n)}$ into a sum of skeletons for $\bar{F}^{(n)}$. If, when we sum all skeletons for $\bar{G}^{(n)}$, we get all skeletons for $\bar{F}^{(n)}$, we have the desired result $S\bar{G}^{(n)}\equiv-i\bar{F}^{(n)}$.

Let us show that the above recipe for turning G skeletons into F skeletons exhausts all possibilities. Consider a particular skeleton, S^F for $\bar{F}^{(n)}$. When we

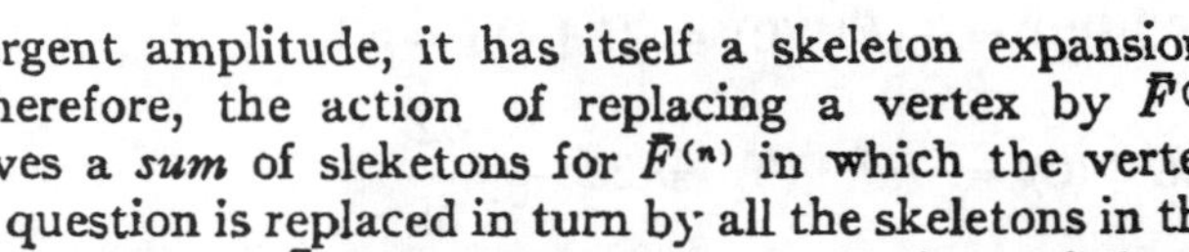

FIG. 1. Action of S on a particular skeleton for $\bar{G}^{(6)}$. The cross stands for insertion of Θ.

[4] The subtraction can be made at any fixed value of momentum. Since we subtract at zero four-momentum, the parameter μ is not identical to the physical mass.

[5] The amplitude which we called $G^{(2)}$ is identical to the propagator. The passage to $\bar{G}^{(2)}$ involves dividing out two factors of the propagator from $G^{(2)}$, so that $\bar{G}^{(2)}\propto[G^{(2)}]^{-1}$.

remove from it the insertion of $\bar{F}^{(2)}$, it becomes a graph $\mathcal{G}$ for $\bar{G}^{(n)}$, which may or may not be a skeleton. Suppose it *is* a skeleton, $S_1{}^G$. Then S^F is created by inserting $\bar{F}^{(2)}$ on some internal line of $S_1{}^G$. Suppose $\mathcal{G}$ is *not* a skeleton. Then it contains a subgraph identifiable as a $\bar{G}^{(2)}$ or a $\bar{G}^{(4)}$ within which is contained the line from which $\bar{F}^{(2)}$ was removed to create $\mathcal{G}$ in the first place. Actually, this subgraph cannot be a $\bar{G}^{(2)}$, since that would mean, on putting back the $\bar{F}^{(2)}$ insertion, that the original graph S^F contained an insertion identifiable as an $\bar{F}^{(2)}$, which is not allowed for an F skeleton. So, $\mathcal{G}$ contains a subgraph identifiable as a $\bar{G}^{(4)}$, which, if shrunk to a point vertex, turns $\mathcal{G}$ into a skeleton $S_2{}^G$ for $\bar{G}^{(n)}$. Also, if we put the $\bar{F}^{(2)}$ insertion back into this subgraph, the subgraph becomes a skeleton for $\bar{F}^{(4)}$. So, in this case, S^F is obtained from $S_2{}^G$, a skeleton for $G^{(n)}$, by replacing one of the vertices by a skeleton for $\bar{F}^{(4)}$. Putting these two cases together, we see that all the sleketons for $\bar{F}^{(n)}$ are obtained by acting on all the skeletons for $\bar{G}^{(n)}$ in precisely the manner described in the previous paragraph. Therefore, a distributive S is guaranteed to satisfy the scaling law $S\bar{G}^{(n)}=-i\bar{F}^{(n)}$ for all n if it is true for $n=2$ and 4.

The scaling operator $S=\mu\partial/\partial\mu+n\delta$ suggested by formal arguments on broken scale invariance is of course distributive. It remains so if we add to it differentiation with respect to any parameter. The only parameter in the theory apart from the mass μ is the coupling constant λ. This suggests a more general form for S:

$$S=\mu\partial/\partial\mu+n\delta(\lambda)+f(\lambda)\partial/\partial\lambda. \qquad (6)$$

The question is whether one has to make use of this freedom. One would be happier with a scaling law *not* involving differentiation with respect to the coupling constant, since that would leave open some hope of direct phenomenological application.

The simplest way to answer this question is to study the scaling-law constraints on the fundamental constants of the theory. One easily finds that

$$S\bar{G}^{(2)}(0)=-i\bar{F}^{(2)}(0) \Rightarrow (1+\delta)\mu^2=m^2,$$

$$S\bar{G}^{(2)}(0)'=-i\bar{F}^{(2)}(0)'\Rightarrow 2\delta Z+f\frac{\partial Z}{\partial\lambda}=\alpha=-\frac{\partial}{\partial p^2}\bar{F}^{(2)}(0,p)\Big|_{p=0},$$

$$S\bar{G}^{(4)}(0)=-i\bar{F}^{(4)}(0) \Rightarrow 4\delta\lambda+f=\beta=\bar{F}^{(4)}(0).$$

The quantities α and β correspond to Green's functions not requiring subtractions, and so can be calculated in terms of the basic parameters of the theory to any desired order. If we make the conventional choice $Z=1$,[6] we can explicitly solve these equations for the parameters δ, f, and m^2:

$$\delta=\tfrac{1}{2}\alpha,$$
$$f=\beta-2\alpha\lambda,$$
$$m^2=\mu^2(1+\tfrac{1}{2}\alpha).$$

If we look at the lowest-order contributions to α and β, we find that they are both $O(\lambda^2)$. This has the immediate consequence that $f\neq0$, except possibly for some specific value of λ.[7] Therefore, we have to live with the general form of the scaling law, Eq. (6). We return to the question of interpretation later on. It should also be noted that the scaling law determines, to any desired order in perturbation, the funny constants δ and f appearing in S.

III. COMPUTING FUNDAMENTAL GREEN'S FUNCTIONS

We have yet to show, of course, that the scaling law is satisfied for the full Green's functions $\bar{G}^{(2)}$ and $\bar{G}^{(4)}$. This is rendered somewhat difficult by the fact that $\bar{G}^{(2)}$ and $\bar{G}^{(4)}$ are beset by overlapping divergences and not easy to calculate by standard techniques. What we can do, however, is to use the scaling law to *compute* $\bar{G}^{(2)}$ and $\bar{G}^{(4)}$ in a systematic way which not only guarantees that they satisfy the scaling law, but automatically solves the overlapping divergence problem. To see how this goes, let us consider the scaling law for $\bar{G}^{(4)}$, slightly rearranged:

$$\mu\frac{\partial}{\partial\mu}\bar{G}^{(4)}(p_1p_2p_3)=-\left(4\delta+f\frac{\partial}{\partial\lambda}\right)\bar{G}^{(4)}(p_1p_2p_3)-i\bar{F}^{(4)}(0;\,p_1p_2p_3)$$
$$=\Phi(p_1p_2p_3;\,\bar{G}^{(2)},\bar{G}^{(4)},\bar{F}^{(2)}). \qquad (7)$$

This is a differential equation for $G^{(4)}$ which can easily be integrated:

$$\bar{G}^{(4)}(p_1p_2p_3)=-i\lambda-\int_0^1\frac{d\alpha}{\alpha}\Phi(\alpha p_1,\alpha p_2,\alpha p_3;\bar{G}^{(2)},\bar{G}^{(4)},\bar{F}^{(2)})$$

[the convergence of the integral is guaranteed by our choice of δ and f, which implies that $\Phi(0,0,0)=0$]. The contribution to this integral of a particular order in λ involves in the integrand the fundamental Green's functions only to *lower* orders in λ [this is because both δ and f are $O(\lambda^2)$ and $\bar{F}^{(4)}$ has a skeleton expansion]. Therefore, we have a systematic scheme for determining $\bar{G}^{(4)}$: If the fundamental Green's functions are known to $O(\lambda^{n-1})$, the requirement that $\bar{G}^{(4)}$ satisfy the scaling law uniquely determines $\bar{G}^{(4)}$ to $O(\lambda^n)$ and no convergence difficulties arise. It should be remembered that the condition $\Phi(0,0,0)=0$ is what is used to determine

[6] This choice of Z does not leave the propagator correctly normalized at the physical particle pole, meaning that a finite rescaling eventually has to be carried out. This does not affect any of our arguments.

[7] It is worth pointing out that we may *impose* $\delta=0$ at the price of fixing Z to be a function of λ such that $\partial Z/\partial\lambda=\alpha/\beta$.

the nth-order contributions to δ and f, thus completing the induction. If we had a scheme for computing $\bar{F}^{(2)}$ in terms of lower-order fundamental Green's functions, we could use the scaling law $S\bar{G}^{(2)}=-i\bar{F}^{(2)}$ to compute $\bar{G}^{(2)}$ systematically in the same way.

$\bar{F}^{(2)}$ is beset with overlapping divergences, as are $\bar{G}^{(2)}$ and $\bar{G}^{(4)}$, and is not easy to handle directly. We can sidestep this problem by studying the scaling law satisfied by $\bar{F}^{(n)}$ rather than $\bar{G}^{(n)}$. The Green's function $\bar{F}^{(n)}(q,p_1\cdots p_n)$ is really quite analogous to $\bar{G}^{(n+1)}$—instead of being a matrix element of $n+1$ identical fields, it is a matrix element of n identical fields plus another field identifiable as the trace of the energy-momentum tensor. Therefore, we expect that when it is operated on by the appropriate scaling operator $\hat{S}$, it yields the matrix element $\bar{H}^{(n)}$ of n identical fields plus *two* traces, one of which carries zero four-momentum:

$$\hat{S}\bar{F}^{(n)}(q,p_1\cdots p_{n-1})=-i\bar{H}^{(n)}(0q;\,p_1\cdots p_{n-1})\,.$$

Such an equation is easy to derive heuristically by standard Ward-identity methods, and one finds

$$\hat{S}=\mu\partial/\partial\mu+n\delta+\bar{\delta}-2\,,$$

where the $\bar{\delta}$ accounts for the difference between the real and naive dimension of Θ in the same way that δ takes care of the possible anomaly in the dimension of the field ϕ. The virtue of this scaling law is that the usual power-counting arguments tell us that all of the matrix elements $\bar{H}^{(n)}$ are primitively convergent, even for $n=2$.[8] Therefore, all $\bar{H}^{(n)}$ have skeleton expansions in terms of the fundamental Green's functions $\bar{G}^{(2)}$, $\bar{G}^{(4)}$, and $\bar{F}^{(2)}$.

The $\bar{F}^{(n)}$ with $n>2$ are all primitively convergent and possess skeleton expansions in terms of $\bar{G}^{(2)}$, $\bar{G}^{(4)}$, and $\bar{F}^{(2)}$. Therefore, we want the corresponding equations $\hat{S}\bar{F}^{(n)}=-i\bar{H}^{(n)}$ to be automatically valid once the fundamental equations $S\bar{G}^{(2)}=-i\bar{F}^{(2)}$, $S\bar{G}^{(4)}=-i\bar{F}^{(4)}$, and $\hat{S}\bar{F}^{(2)}=-i\bar{H}^{(2)}$ are assumed valid. Precisely the same type of argument that led to the requirement that S be distributive then implies that $\hat{S}=S+\bar{\delta}-2$. The quantity $\bar{\delta}$ is then determined by the equation $\hat{S}\bar{F}^{(2)}(0)=-i\bar{H}^{(2)}(0)$ in a perfectly straightforward way, and is found to be $O(\lambda)$.

The scaling law for $F^{(2)}$ can now be rewritten as

$$\begin{aligned}(\mu\partial/\partial\mu-2)\bar{F}^{(2)}(q;\,p)&=(-2\delta-\bar{\delta}-f\partial/\partial\lambda)\bar{F}^{(2)}(q,p)-i\bar{H}^{(2)}(qq;\,p)\\&=X(q,p;\,\bar{G}^{(2)},\bar{G}^{(4)},\bar{F}^{(2)})\,.\end{aligned}$$

Upon integration, this gives

$$F^{(2)}(q,p)f=\bar{F}^{(2)}(0,0)-\int_0^1\frac{d\alpha}{\alpha}X(\alpha q,\alpha p,\bar{G}^{(2)},\bar{G}^{(4)},\bar{F}^{(2)})\,.$$

The convergence of the integral is guaranteed because $\bar{\delta}$ is so chosen that $X(0,0)=0$. Also, this equation determines the nth-order contribution to $\bar{\delta}$ in terms of lower-order quantities. Just as in the analysis of Eq. (7), we see that to calculate the right-hand side of this equation to $O(\lambda^n)$ requires the knowledge of $\bar{G}^{(2)}$, $\bar{G}^{(4)}$, and $\bar{F}^{(2)}$ only up to $O(\lambda^{n-1})$. Therefore, taken together with the corresponding equations for $\bar{G}^{(2)}$ and $\bar{G}^{(4)}$, this equation provides a systematic scheme for computing the fundamental Green's functions of the theory to successively higher orders in perturbation theory in a way automatically consistent with the scaling law. This method also completely avoids divergence difficulties since the calculations always start with amplitudes possessing a skeleton expansion.

The amplitudes $\bar{G}^{(2)}$ and $\bar{G}^{(4)}$ so computed are not obviously identical to those one would determine by the usual methods of renormalized perturbation theory. We therefore are obligated to verify that they possess the usual properties of analyticity and unitarity. If they do, then they must be perfectly acceptable amplitudes, and in fact identical to the amplitudes computed in the usual way.

Let us first consider the question of analyticity. The amplitude $\bar{G}^{(4)}$ is computed from Eq. (7) by successive approximations: The fundamental Green's functions correct to order $n-1$ are inserted on the right-hand side, and the equation is then integrated to get $G^{(4)}$ correct to order n:

$$\bar{G}^{(4)}(p_1p_2p_3)=-i\lambda-\int_0^1\frac{d\alpha}{\alpha}\Phi(\alpha p_1,\alpha p_2,\alpha p_3;\bar{G}^{(4)},\bar{G}^{(2)},\bar{F}^{(2)})\,.$$

If $\Phi(p_1p_2p_3)$ has a cut of the usual form, say, along the line $(2\mu)^2<p_1^2<\infty$, then of course so does

$$\int_0^1\frac{d\alpha}{\alpha}\Phi(\alpha p_1,\alpha p_2,\alpha p_3)\,.$$

In fact, so long as the cuts in Φ are of the usual sort in s, t, u, and masses, then $\bar{G}^{(4)}$ has precisely the same cuts, with different discontinuities across the cuts. But the cuts of Φ to a given order are just the cuts of $\bar{G}^{(4)}$ to lower order and the cuts of $\bar{F}^{(4)}(0,p_1p_2p_3)$ to the same order. Since $\bar{F}^{(4)}(0;\,p_1p_2p_3)$ is constructed via a standard skeleton expansion, it will have the usual cuts which, because the momentum arguments are the same, coincide with those of $\bar{G}^{(4)}$. Since we are proceeding by induction, we assume that $\bar{G}^{(4)}$ to lower order has the proper cuts. Then Φ will have exactly the same singularity structure as we would expect for $\bar{G}^{(4)}$, which means that the next approximation to $\bar{G}^{(4)}$, gotten by integrating Φ, has the right singularity structure. The same sort of arguments, of course, apply to $\bar{G}^{(2)}$ since it is computed in much the same way.

The question of unitarity is less trivial but is attacked in a similar manner. Let us for simplicity consider the

[8] Each insertion of ϕ^2 (remember that $\Theta\propto\phi^2$ in this theory) reduces the degree of divergence by 2. The only divergent matrix element with one insertion of Θ is $\bar{F}^{(2)}$, and its degree of divergence is 0. Therefore, $\bar{H}^{(2)}$ has degree of divergence -2.

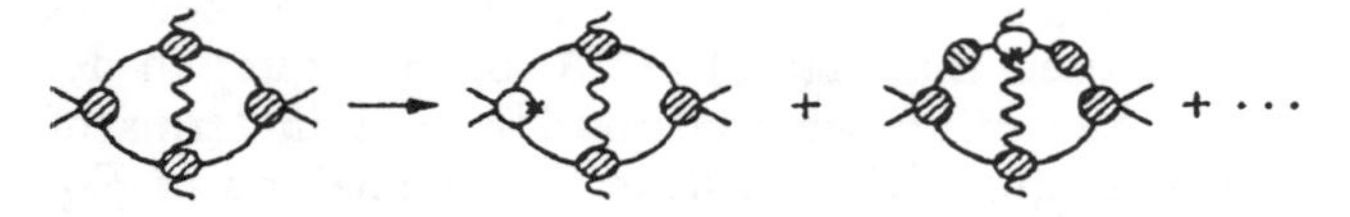

FIG. 2. Action of S on the two-body unitarity integral. The cross stands for insertion of Θ and the wavy line stands for replacing the affected propagators by their discontinuity across the single-particle pole.

amplitude $\bar{G}^{(4)}$ in a kinematic region where only a two-particle cut in the variable $(p_1+p_2)^2$ exists. The normal unitarity prediction for the discontinuity across this cut is

$$D_u\bar{G}^{(4)}(p_1p_2p_3p_4) = \int \frac{d^4l}{(2\pi)^4}\frac{d^4l'}{(2\pi)^4}(2\pi)^4\delta(p_1+p_2+l+l') \times\{\bar{G}^{(4)}(p_1p_2ll')\Delta(l)\Delta(l')\bar{G}^{(4)}(ll'p_3p_4)\},$$

where $\Delta(l)$ is the discontinuity of the propagator across its single-particle pole [if the physical mass is $\bar{\mu}^2$ and the propatagor is conventionally normalized, $\Delta(l) = 2\pi\delta(l^2-\bar{\mu}^2)\theta(l_0)$]. If we act on this equation with the scaling operator S, the distributive property of S means that we get four terms, corresponding to S acting in succession on each of the four terms in curly brackets:

$$SD_u\bar{G}^{(4)}(p_1p_2p_3p_4) = \int \frac{d^4l}{(2\pi)^4}\frac{d^4l'}{(2\pi)^4}(2\pi)^4\delta(p_1+p_2+l+l') \times\{[S\bar{G}^{(4)}(p_1p_2ll')]\Delta(l)\Delta(l')\bar{G}^{(4)}(ll'p_3p_4) +\bar{G}^{(4)}(p_1p_2ll')[S\Delta(l')]\Delta(l)\bar{G}^{(4)}(ll'p_3p_4)+\cdots\}.$$

Since S acting on $\bar{G}^{(4)}$ gives $\bar{F}^{(4)}$ and since S acting on the propagator gives the insertion of $\bar{F}^{(2)}$ on the propagator,[9] these four terms can be represented graphically as in Fig. 2. These terms are immediately recognized as the standard unitarity contributions to the two-particle discontinuity of $-i\bar{F}^{(4)}(0;p_1p_2p_3p_4)$, which, because it is computed from a skeleton expansion, is known to satisfy normal unitarity. Therefore, if we let D stand for the operation of taking the discontinuity, we have, schematically,

$$S[D_u\bar{G}^{(4)}]=D[-\bar{F}^{(4)}]=D[S\bar{G}^{(4)}]=S[D\bar{G}^{(4)}],$$

where the last equation follows because the operations S and taking the discontinuity commute. Therefore, $S[D\bar{G}^{(4)}-D_uG^{(4)}]=0$. We gave arguments for this equation on the two-particle cut only, but it obviously works for any cut and we can take it to be true in general. If we assume that $\bar{G}^{(4)}$ satisfies normal unitarity up to order λ^{n-1}, then arguments of the kind we have often used [see Eq. (7) *et seq.*] imply that to order λ^n, $\bar{G}^{(4)}$ satisfies

$$\mu\frac{\partial}{\partial\mu}[D\bar{G}^{(4)}-D_u\bar{G}^{(4)}]=0.$$

This allows the solution

$$D\bar{G}^{(4)}-D_u\bar{G}^{(4)}=\text{momentum-independent constant}.$$

Since there is a kinematic region where both $D\bar{G}^{(4)}$ and $D_u\bar{G}^{(4)}$ vanish (below the two-particle threshold), the constant on the right-hand side of this equation must in fact vanish. Therefore, we can show inductively that $\bar{G}^{(4)}$ satisfies the usual unitarity relation. Similar arguments, of course, apply to $\bar{G}^{(2)}$.

These arguments for analyticity and unitarity are somewhat sketchy, but presumably could be made rigorous.[10] They seem, however, sufficiently convincing to make us believe the proposed scheme for computing $\bar{G}^{(2)}$ and $\bar{G}^{(4)}$.

To summarize, we have done two things by this rather long argument. First of all, we have shown that the particle amplitudes in this theory satisfy a scaling law, albeit one which differs in a profound way from the one suggested by naive broken-scale-invariance requirements. Secondly, we have shown how this scaling law is used to compute the amplitudes of the theory in a way which automatically avoids all questions of divergence, overlapping or otherwise. Finally, it should be noted that these arguments will generalize in an obvious way to any renormalizable field theory, and might even be of some use in making simpler calculations of higher-order quantities.

IV. INTERPRETATION AND CONNECTION WITH RENORMALIZATION GROUP

At this point we should make some effort to understand why the scaling law takes on the form it does. If we could somehow "turn off" the explicit scale-invariance-breaking terms in the Lagrangian—in this case, the mass term—then the particle amplitudes would satisfy $S\bar{G}^{(n)}=0$. If S were simply $\mu\partial/\partial\mu+n\delta$, this would imply that the functions $\bar{G}^{(n)}$ are homogeneous functions of their momentum arguments of degree $4-nd$, with $d=1-\delta$.[11] This is what one might call naive scaling, appropriately modified for the anomalous dimensions of the fields. "Turning off" the mass terms can actually be achieved in practice by taking appropriate asymptotic limits of momenta, and one would expect the Green's functions to satisfy naive scaling in such limits. In fact, $S=\mu\partial/\partial\mu+n\delta+f\partial/\partial\lambda$, so that even though we can achieve $S\bar{G}^{(n)}=0$ in appropriate asymptotic regions, this does not mean that the $\bar{G}^{(n)}$ satisfy naive scaling in the same limit. It appears that naive scaling is replaced by some restriction on the joint de-

[9] We assume, and can show directly, that S acting on the *discontinuity* Δ of the propagator gives the *discontinuity* of the insertion of $\bar{F}^{(2)}$ on the propagator.

[10] See in this connection T. T. Wu, Phys. Rev. 125, 1436 (1962).

[11] Recall that $\mu\partial/\partial\mu+n\delta=4-nd-\sum p_i\cdot\partial/\partial p_i$.

pendence of $\bar{G}^{(n)}$ on momenta *and* coupling constant. That S contains the term $f(\lambda)\partial/\partial\lambda$ is apparently equivalent to saying that even in the absence of explicit symmetry-breaking terms (mass terms), scale invariance is still broken by some mechanism. The nature of this mechanism is not hard to find: We assumed that in the Lagrangian

$$\mathcal{L}=\tfrac{1}{2}(\partial_\mu\phi)(\partial^\mu\phi)-\tfrac{1}{2}\mu^2\phi^2-\frac{\lambda}{4!}\phi^4$$

only the mass term $\mu^2\phi^2$ contributes to scale-invariance breaking. In fact, any term in the Lagrangian with dimension different from four will contribute.[2] The interaction term $\lambda\phi^4$ has dimension four to lowest order in perturbation theory, but, just as ϕ has a dimension different from one when interactions are considered,[3] so will ϕ^4 have a dimension different from four. Therefore, one can expect the $\lambda\phi^4$ term to contribute to scale-invariance breaking even though it has a dimensionless coupling constant. The interesting thing about the scaling law is that it provides a simple analytic expression of the effect of this implicit sort of symmetry breaking.

We mentioned that whenever the right-hand side of the equation $SG=-iF$ could be neglected, one obtained a constraint on the joint dependence of G on momenta and coupling constant rather than naive scaling. We would like to pursue this somewhat further in order to establish a connection with the renormalization group. Consider for a moment the scaling law for the two-particle amplitude: $S\bar{G}^{(2)}(p^2)=-i\bar{F}^{(2)}(0,p)$. If we take the limit $p^2\to-\infty$, Weinberg's theorem[12] implies $\bar{F}^{(2)}\to(p^0)\times$(powers of $\ln p^2$) while $\bar{G}^{(2)}\to(p^2)\times$(powers of $\ln p^2$). If we collect together all those terms in $\bar{G}^{(2)}$ which are proportional to p^2 and denote them by p^2 $\Phi(\ln(-p^2/\mu^2),\lambda)$, the scaling law clearly implies that $S\Phi=0$, a rather severe restriction on the form of Φ. In fact, since $S=\mu\partial/\partial\mu+f(\lambda)\partial/\partial\lambda+2\delta$ is a linear operator. it is quite easy to get the general solution for Φ:

$$\Phi(z,\lambda)=\hat{\Phi}(z+\eta(\lambda))\exp\left(-\int^{\lambda}\frac{d\lambda'}{2f(\lambda')}\delta(\lambda')\right)$$

where

$$\eta(\lambda)=\int^{\lambda}\frac{d\lambda'}{2f(\mu')},$$

and $\hat{\Phi}$ is an arbitrary function of one variable. This kind of correlation between the asymptotic dependence on momentum and the dependence on coupling constant is typical of renormalization-group arguments,[13] and it is interesting to see how easily it emerges from the scaling law. Similar considerations would apply to other amplitudes than the propagator, and presumably allow one to extract renormalization-group conclusions in an expeditious manner.

[12] S. Weinberg, Phys. Rev. **118**, 838 (1960).
[13] J. D. Bjorken and S. Drell, *Relativistic Quantum Fields* (McGraw-Hill, New York, 1965), Vol. II, p. 368.

CONCLUSIONS

We undertook this investigation in the hope of finding out just how scale invariance is broken in a model field theory. We found that the source (in the sense of the scaling law) for violations of naive dimensional scaling was not simply those terms in the Lagrangian having dimensional coupling constants. The interpretation of this is relatively simple: A term in the Lagrangian will not break scale invariance only if its dimension (as defined by commutation with the dilation generator) is exactly equal to four. But the terms in the Lagrangian with dimensionless coupling constants are guaranteed to have dimension four only to lowest order in perturbation—when the effects of interactions are considered, their dimensions will change and they will contribute to scale-invariance breaking. The surprising thing we found was that these "implicit" breaking terms could be incorporated in the scaling law by a rather simple change in its form. The resulting scaling law probably cannot be used in a direct phenomenological fashion, but by studying a special asymptotic limit we were able to recover the results of the renormalization group. There are other asymptotic limits in which field-theory scattering amplitudes are supposed to have simple forms (the impact-parameter representation, for instance) and one might find useful constraints on such forms by studying their compatibility with the scaling law. Another aspect of this is that the scaling law appears to provide a particularly simple approach to renormalization—one can use it to completely sidestep questions of overlapping divergences and obtain a relatively simple prescription for computing renormalized amplitudes.

Finally, it should be said that we confined ourselves to a scalar-field theory only for the sake of simplicity. It seems quite clear that the general ideas which we have developed are applicable, with simple modifications, to any renormalizable theory. It is not clear that any of this has immediate practical importance. Nonetheless, it is always useful to see an old problem in a new light, and we hope that this, along with whatever clarification of the problems of broken scale invariance we may have achieved, justifies yet another paper on scalar field theory.

Note added in manuscript: For another, not dissimilar, approach to these questions, the reader should consult a recent paper by K. Wilson, this issue, Phys. Rev. D **2**, 1478 (1970).

ACKNOWLEDGMENT

It is a pleasure to acknowledge many discussions with Sidney Coleman, without which this paper could not have been written.

PHYSICAL REVIEW D VOLUME 5, NUMBER 6 15 MARCH 1972

Effects of a Neutral Intermediate Boson in Semileptonic Processes*

Steven Weinberg
Laboratory for Nuclear Science and Department of Physics, Massachusetts Institute of Technology, Cambridge, Massachusetts 02139
(Received 6 December 1971)

The observable effects of a neutral intermediate vector boson in semileptonic processes are considered in the context of a proposed renormalizable theory of the weak and electromagnetic interactions. With strange particles neglected, this theory allows the calculation of neutral-current form factors in terms of charged-current form factors and electromagnetic form factors. The results are neither confirmed nor refuted by present data. One possible method of incorporating the strange particles is briefly discussed.

I. INTRODUCTION

It now appears that a renormalizable theory of the weak and electromagnetic interactions may be constructed from a Yang-Mills theory with spontaneous breaking of gauge invariance.[1] There are many possible theories of this general type, but for the present it seems best to concentrate on one particularly simple model,[2] which requires the smallest possible number of unobserved particles. This proposed model has so far been worked out in detail only as it applies to the area of the weak and electromagnetic interactions of leptons, and within this area involves the usual leptons, photons, a new massive scalar meson, charged massive vector mesons W, and a neutral massive vector meson Z. The neutral vector meson produces striking effects in neutrino-electron scattering, effects which are just on the verge of observability.[3]

This paper will deal with the possible observable effects of the Z boson in semileptonic processes, especially neutrino scattering. The difficult part of this problem has to do with the role of the strange particles. In Sec. II the strange particles are simply ignored altogether, and we find a rather natural extension of the proposed model to semileptonic reactions, which allows the form factors for any neutral-current process to be calculated from the form factors for the corresponding charge-exchange and electron-scattering processes. This theory is applied to neutrino scattering in Sec. III. The chief result, valid for not too large values of the momentum transfer, is that

$$0.15 \leq \frac{\sigma(\nu' + p \to \nu' + p)}{\sigma(\nu' + n \to \mu^- + p)} \leq 0.25 \qquad (1.1)$$

as compared with an experimental value[4] (or upper limit) 0.12 ± 0.06.

The strange particles are considered in Sec. IV. It is found that a four-quark model of Glashow, Maiani, and Iliopoulos[5] naturally explains the absence of neutral strangeness-changing currents, but leads to new terms in the neutral strangeness-conserving current, which could in principle alter predictions like (1.1).

In summary, there is no obvious obstacle to the

extension of this model to the weak and electromagnetic interactions of hadrons, but one feels that a more natural way of incorporating strange particles is needed.

II. NEUTRAL SEMILEPTONIC WEAK INTERACTIONS

The proposed theory is based on an assumed gauge-invariance group of the total Lagrangian,[1,2] consisting of an SU(2) group generated by a "left-handed isospin" $\vec{T}_L$, and a U(1) group generated by $T_{3L}-Q$, where Q is the ordinary electric charge. If we ignore the strange particles altogether, it is natural to identify $\frac{1}{2}(1+\gamma_5)N$ as a $\vec{T}_L$ doublet, where N is the usual nucleon doublet:

$$N=\begin{pmatrix} p \\ n \end{pmatrix}.$$

The hadronic currents generated by $\vec{T}_L$ and Q are then

$$\vec{\mathfrak{J}}^\lambda = i\bar{N}\vec{t}\gamma^\lambda(1+\gamma_5)N + \text{meson terms}, \tag{2.1}$$

$$J^\lambda = i\bar{N}(\tfrac{1}{2}+t_3)\gamma^\lambda N + \text{meson terms}, \tag{2.2}$$

with $\vec{t}$ the usual isospin matrix (half the Pauli matrix). That is, $\vec{\mathfrak{J}}^\lambda$ is just the usual "V minus A" current of β decay, and eJ^λ is the electric current. Gauge invariance dictates that the theory must involve a triplet $\vec{A}_\mu$ of intermediate bosons associated with $\vec{T}_L$, and a singlet B_μ associated with $T_{3L}-Q$. The interaction of these fields with the hadrons is then

$$\mathfrak{L}_h' = \tfrac{1}{2}g\vec{A}_\lambda\cdot\vec{\mathfrak{J}}^\lambda + \tfrac{1}{2}g'B_\lambda(\mathfrak{J}_3^\lambda - 2J^\lambda), \tag{2.3}$$

where g and g' are a pair of independent coupling constants,[1] which must be the same as in the interaction of $\vec{A}_\lambda$ and B_λ with leptons.

The spontaneous breaking[6] of $\vec{T}_L$ conservation generates a nondiagonal mass matrix for the $\vec{A}_\mu$ and B_μ fields. The fields with definite masses are then a charged massive field

$$W_\mu = \frac{1}{\sqrt{2}}(A^1_\mu + iA^2_\mu), \tag{2.4}$$

$$m_W^2 = \tfrac{1}{4}\lambda^2 g^2, \tag{2.5}$$

a neutral massive field

$$Z_\mu = (g^2+g'^2)^{-1/2}(gA^3_\mu + g'B_\mu), \tag{2.6}$$

$$m_Z^2 = \tfrac{1}{4}\lambda^2(g^2+g'^2), \tag{2.7}$$

and the photon field

$$A_\mu = (g^2+g'^2)^{-1/2}(-g'A^3_\mu + gB_\mu), \tag{2.8}$$

$$m_A^2 = 0, \tag{2.9}$$

where λ is a symmetry-breaking parameter.[1] By expressing $\vec{A}_\mu$ and B_μ in terms of W_μ, $W^\dagger_\mu$, Z_μ, and A_μ, we find that (2.3) becomes

$$\mathfrak{L}_h' = \frac{1}{2\sqrt{2}}gW_\mu\mathfrak{J}_W^{\mu\dagger} + \frac{1}{2\sqrt{2}}gW^\dagger_\mu\mathfrak{J}_W^\mu + \tfrac{1}{2}(g^2+g'^2)^{1/2}Z_\mu\mathfrak{J}_Z^\mu - eA_\mu J^\mu, \tag{2.10}$$

where

$$\mathfrak{J}_W^\mu \equiv \mathfrak{J}_1^\mu + i\mathfrak{J}_2^\mu, \tag{2.11}$$

$$\mathfrak{J}_Z^\mu \equiv \mathfrak{J}_3^\mu - \left(\frac{2g'^2}{g^2+g'^2}\right)J^\mu, \tag{2.12}$$

and

$$e = \frac{gg'}{(g^2+g'^2)^{1/2}}.$$

The W, Z, and A fields are also supposed to couple to the leptons through the interaction

$$\begin{aligned}\mathfrak{L}_l' = \frac{ig}{2\sqrt{2}}[\bar{e}\gamma_\lambda(1+\gamma_5)\nu + \bar{\mu}\gamma_\lambda(1+\gamma_5)\nu']W^\lambda + \text{H.c.} + \tfrac{1}{4}i(g^2+g'^2)^{1/2}[\bar{\nu}\gamma_\lambda(1+\gamma_5)\nu + \bar{\nu}'\gamma_\lambda(1+\gamma_5)\nu']Z^\lambda \\ + i(g^2+g'^2)^{-1/2}\{\bar{e}\gamma_\lambda[\tfrac{1}{2}(1-\gamma_5)+\tfrac{1}{4}(1+\gamma_5)(g'^2-g^2)]e + \bar{\mu}\gamma_\lambda[\tfrac{1}{2}(1-\gamma_5)+\tfrac{1}{4}(1+\gamma_5)(g'^2-g^2)]\mu\}Z^\lambda \\ + ie(\bar{e}\gamma_\lambda e + \bar{\mu}\gamma_\lambda\mu)A^\lambda .\end{aligned} \tag{2.13}$$

Hence, at not too high momentum transfers, the charge-exchange semileptonic interactions are described by the effective coupling

$$\mathfrak{L}_x' = \frac{iG_\beta}{\sqrt{2}}[\bar{e}\gamma_\lambda(1+\gamma_5)\nu + \bar{\mu}\gamma_\lambda(1+\gamma_5)\nu']\mathfrak{J}_W^\lambda + \text{H.c.}, \tag{2.14}$$

where

$$G_\beta/\sqrt{2} = g^2/8m_W^2. \tag{2.15}$$

This is just the usual β-decay interaction; the Fermi coupling constant G_β is known from the rate of μ decay or O^{14} decay to have the value

$$G_\beta = 1.0\times10^{-5}m_p^{-2}.$$

In addition, there is an effective neutrino interaction

$$\mathfrak{L}_\nu' = \frac{i(g^2+g'^2)}{8m_Z^2}[\bar{\nu}\gamma_\lambda(1+\gamma_5)\nu + \bar{\nu}'\gamma_\lambda(1+\gamma_5)\nu']\mathfrak{J}_Z^\lambda$$

and an effective P- and C-violating charged-lepton interaction

$$\mathcal{L}_c' = \frac{i}{8m_Z^2}\{\bar{e}\gamma_\lambda[(1-\gamma_5)g'^2 + \tfrac{1}{2}(1+\gamma_5)(g'^2-g^2)]e + \bar{\mu}\gamma_\lambda[(1-\gamma_5)g'^2 + \tfrac{1}{2}(1+\gamma_5)(g'^2-g^2)]\mu\}\mathcal{J}_Z^\lambda .$$

Using (2.15) and our formulas (2.5) and (2.7) for m_Z and m_W gives

$$G_\beta/\sqrt{2} = (g^2+g'^2)/8m_Z^2 , \quad (2.16)$$

so the factors $g^2+g'^2$ cancel, and the effective interactions produced by Z exchange are

$$\mathcal{L}_\nu' = \frac{iG_\beta}{\sqrt{2}}[\bar{\nu}\gamma_\lambda(1+\gamma_5)\nu + \bar{\nu}'\gamma_\lambda(1+\gamma_5)\nu']\mathcal{J}_Z^\lambda , \quad (2.17)$$

$$\mathcal{L}_c' = \frac{-iG_\beta}{\sqrt{2}}[(1-4\sin^2\theta)\bar{e}\gamma_\lambda e + \bar{e}\gamma_\lambda\gamma_5 e + (1-4\sin^2\theta)\bar{\mu}\gamma_\lambda\mu + \bar{\mu}\gamma_\lambda\gamma_5\mu]\mathcal{J}_Z^\lambda , \quad (2.18)$$

where θ is a "mixing angle," defined by

$$\tan\theta \equiv g'/g. \quad (2.19)$$

In terms of this mixing angle, our formula (2.12) for the neutral hadronic current becomes

$$\mathcal{J}_Z^\lambda = \mathcal{J}_3^\lambda - 2\sin^2\theta J^\lambda . \quad (2.20)$$

These results, Eqs. (2.17)–(2.20), are particularly useful, because the form factors for $\mathcal{J}_3^\lambda$ may be determined by an isospin transformation from the form factors for $\mathcal{J}_W^\lambda$, which are measured in charge-exchange neutrino reactions, while the form factors for J^λ are directly measured in electron scattering experiments. Thus the existing data on charge-exchange neutrino reactions and on electron scattering may be used to predict the cross sections for neutral-current weak-interaction processes, without having to worry about the complications introduced by strong interactions.

III. NEUTRINO REACTIONS

Let us consider in detail the simplest and best studied of the neutral-current semileptonic processes, $\nu'+p\to\nu'+p$. Assuming time-reversal invariance and only first-class currents, the form factors for this process may be defined by the formula

$$\langle p'|\mathcal{J}_Z^\lambda|p\rangle = i(2\pi)^{-3}(4E_p'E_p)^{-1/2}\bar{u}_p'[g_V^0\gamma^\lambda + g_A^0\gamma^\lambda\gamma_5 + if_V^0(P_p'+P_p)^\lambda + ih_A^0(P_p-P_p')^\lambda\gamma_5]u_p , \quad (3.1)$$

where g_V^0, g_A^0, f_V^0, and h_A^0 are real dimensionless functions of the invariant squared momentum transfer

$$q^2 = (P_p - P_p')^2 = (\vec{P}_p - \vec{P}_p')^2 - (E_p - E_p')^2 .$$

The corresponding form factors for the charge-exchange semileptonic process $\nu'+n\to\mu^-+p$ are defined by

$$\langle p|\mathcal{J}_W^\lambda|n\rangle = i(2\pi)^{-3}(4E_nE_p)^{-1/2}\bar{u}_p[g_V\gamma^\lambda + g_A\gamma^\lambda\gamma_5 + if_V(P_n+P_p)^\lambda + ih_A(P_n-P_p)^\lambda\gamma_5]u_n , \quad (3.2)$$

while the electromagnetic form factors of the proton and neutron are defined by

$$\langle p'|J^\lambda|p\rangle = i(2\pi)^{-3}(4E_p'E_p)^{-1/2} \times \bar{u}_p'[G^p\gamma^\lambda + iF^p(P_p'+P_p)^\lambda]u_p , \quad (3.3)$$

$$\langle n'|J^\lambda|n\rangle = i(2\pi)^{-3}(4E_n'E_n)^{-1/2} \times \bar{u}_n'[G^n\gamma^\lambda + iF^n(P_n'+P_n)^\lambda]u_n . \quad (3.4)$$

According to Eq. (2.20), the form factors in Eq. (3.1) are given by

$$g_V^0 = \tfrac{1}{2}g_V - 2\sin^2\theta G^p , \quad (3.5)$$

$$f_V^0 = \tfrac{1}{2}f_V - 2\sin^2\theta F^p , \quad (3.6)$$

$$g_A^0 = \tfrac{1}{2}g_A , \quad (3.7)$$

$$h_A^0 = \tfrac{1}{2}h_A . \quad (3.8)$$

In using these relations, it is useful to keep in mind the conserved-vector-current (CVC) relations[7] between the vector form factors

$$g_V = G^p - G^n , \quad f_V = F^p - F^n . \quad (3.9)$$

In particular, the nucleon charges and magnetic moments give at zero momentum transfer

$$G^p(0) = 2.79, \quad G^p(0) - 2m_NF^p(0) = 1 , \quad (3.10)$$

$$G^n(0) = -1.91, \quad G^n(0) - 2m_NF^n(0) = 0 , \quad (3.11)$$

so that

$$g_V(0) = 4.70 , \quad g_V(0) - 2m_Nf_V(0) = 1 , \quad (3.12)$$

and (3.5) and (3.6) give

$$g_V^0(0) = 2.35 - 5.58\sin^2\theta , \quad (3.13)$$

$$g_V^0(0) - 2m_Nf_V^0(0) = \tfrac{1}{2} - 2\sin^2\theta . \quad (3.14)$$

Also, it is known from β-decay experiments that

$$g_A(0) \simeq 1.2 , \quad (3.15)$$

so (3.7) gives

$$g_A^0(0) \simeq 0.6 . \quad (3.16)$$

It is commonly assumed that the q^2 dependence

of G^p, G^n, F^p, F^n is given by the same function $\mathfrak{F}_V(q^2)$:

$$\frac{G^p(q^2)}{G^p(0)}=\frac{G^n(q^2)}{G^n(0)}=\frac{F^p(q^2)}{F^p(0)}=\frac{F^n(q^2)}{F^n(0)}=\mathfrak{F}_V(q^2), \quad (3.17)$$

with

$$\mathfrak{F}_V(0)\equiv 1 .$$

Also, the study of the reaction $\nu'+n\rightarrow p+\mu^-$ has provided a fairly good idea of the q^2 dependence of the axial-vector form factor $g_A(q^2)$, which we may write as

$$g_A(q^2)/g_A(0)=\mathfrak{F}_A(q^2), \quad (3.18)$$

with

$$\mathfrak{F}_A(0)=1 .$$

If we regard $\mathfrak{F}_V(q^2)$ and $\mathfrak{F}_A(q^2)$ as *known*, then the above results provide a complete model of the relevant form factors:

$$g_V^0(q^2)=(2.35-5.58\sin^2\theta)\mathfrak{F}_V(q^2), \quad (3.19)$$

$$g_V^0(q^2)-2m_N f_V^0(q^2)=(\tfrac{1}{2}-2\sin^2\theta)\mathfrak{F}_V(q^2), \quad (3.20)$$

$$g_A^0(q^2)\simeq 0.6\ \mathfrak{F}_A(q^2). \quad (3.21)$$

The differential cross section for the reaction $\nu'+p\rightarrow\nu'+p$ is[8]

$$d\sigma(\nu'+p\rightarrow\nu'+p)/dq^2=\frac{G_\beta^2}{8\pi}\bigg[(g_A^{02}-g_V^{02})\frac{q^2}{2\omega^2}+(g_A^0+g_V^0)^2+(g_A^0-g_V^0)^2\left(1-\frac{q^2}{2m_N\omega}\right)^2 +[(4m_N^2+q^2)f_V^{02}-4m_N f_V^0 g_V^0]\left(2-\frac{q^2(2\omega+m_N)}{2m_N\omega^2}\right)\bigg], \quad (3.22)$$

where ω is the lab-frame incident neutrino energy, and the invariant squared momentum transfer is here given in terms of the kinetic energy T_p' of the recoiling proton by

$$q^2=2m_pT_p'. \quad (3.23)$$

Thus we have an unambiguous prediction of the neutrino scattering cross section, with the mixing angle θ as the single unknown.

In order to compare this prediction with experiment it would be necessary to integrate over the assumed ω spectrum of the incident neutrinos, and then compare the result with the observed distribution of neutrino scattering "candidates" in T_p'. This detailed analysis is beyond the scope of the present work. However, in order to get some idea of the probable results, it is useful to resort to a crude approximation, and neglect all terms in (3.22) which depend explicitly on q^2. In the CERN neutrino experiment of Cundy *et al.*,[4] ω was in the range 1–4 GeV and T_p' was in the range 150–500 MeV, so for most events the parameters $q^2/2m_N\omega$, $q^2/2\omega^2$, and $q^2/4m_N^2$ were indeed rather small. With this approximation, (3.22) becomes simply

$$d\sigma(\nu'+p\rightarrow\nu'+p)/dq^2\simeq\frac{G_\beta^2}{4\pi}[g_A^{02}+(g_V^0-2m_Nf_V^0)^2]. \quad (3.24)$$

(We are not necessarily neglecting the q^2 dependence of the form factors here.) Also, the differential cross section for $\nu'+n\rightarrow p+\mu^-$ is given (for $E_\mu\gg m_\mu$) by a formula[8] identical with (3.22), except that g_V^0, f_V^0, and g_A^0 are replaced with g_V, f_V, and g_A. Applying the same approximations to this formula, we find the cross-section ratio

$$\frac{d\sigma(\nu'+p\rightarrow\nu'+p)/dq^2}{d\sigma(\nu'+n\rightarrow\mu^-+p)/dq^2}\simeq\frac{g_A^{02}+(g_V^0-2m_Nf_V^0)^2}{g_A^2+(g_V-2m_Nf_V)^2}. \quad (3.25)$$

If we now assume that over the limited q^2 range of this experiment the vector and axial-vector form factors exhibit the same q^2 dependence

$$\mathfrak{F}_A(q^2)\simeq\mathfrak{F}_V(q^2),$$

then the form factors in (3.25) may be replaced with their values at $q^2=0$, so that

$$\frac{d\sigma(\nu'+p\rightarrow\nu'+p)/dq^2}{d\sigma(\nu'+n\rightarrow\mu^-+p)/dq^2}\simeq\frac{1}{4}\left[\frac{(1.2)^2+(1-4\sin^2\theta)^2}{(1.2)^2+1}\right]. \quad (3.26)$$

For $\theta=0$, this ratio is just $\frac{1}{4}$, as expected[5] since in this case $\mathfrak{J}_Z^\lambda$ and $\mathfrak{J}_W^\lambda$ are parts of the same isovector current. The experimental evidence in the process $\bar\nu_e+e^-\rightarrow\bar\nu_e+e^-$ sets a bound[3] on $|\theta|$:

$$|\theta|<35^\circ .$$

For $|\theta|\leqslant 45^\circ$, the ratio (3.26) stays within the bounds

$$0.15\leqslant\frac{d\sigma(\nu'+p\rightarrow\nu'+p)/dq^2}{d\sigma(\nu'+n\rightarrow\mu^-+p)/dq^2}\leqslant 0.25, \quad (3.27)$$

with the lower limit attained for $\theta=30^\circ$. This result may be compared with the experimental value of Cundy *et al.*[4]:

$$\frac{\sigma(\nu'+p\rightarrow\nu'+p)}{\sigma(\nu'+n\rightarrow\mu^-+p)}=0.12\pm0.06 .$$

No clear discrepancy has yet emerged, but as already mentioned, a more detailed analysis of the data is needed.

The one other semileptonic process which has been used as a probe of possible neutral-current interactions is the pion-production reaction, $\nu' + p \to \nu' + \pi^+ + n$. In principle, the cross section for this process may be determined from the cross sections for the charge-exchange processes $\nu' + N \to \mu^- + \pi + N$ and from the form factors for the one-photon-exchange processes $e^- + N \to e^- + \pi + N$. However, this analysis is now very complicated, and only the crudest version will be attempted here. First, it is usual to regard the processes $\nu' + N \to \nu' + \pi + N$ and $\nu' + N \to \mu^- + \pi + N$ as being dominated by the 3-3 resonance Δ. Second, the $\sin^2\theta J^\lambda$ contribution to the processes $\nu' + N \to \nu' + \Delta$ and $\nu' + N \to \mu^- + \Delta$ may be somewhat suppressed, because at zero momentum transfer it is only the space components of the *axial-vector* current that can produce the transition $N \to \Delta$, and/or because $\sin^2\theta$ is not too large. If we neglect this term in (2.20) altogether, then the neutral current $\mathcal{J}_Z^\lambda$ is related by an isospin transformation to the β-decay current $\mathcal{J}_W^\lambda$, so that

$$\frac{\sigma(\nu' + p \to \nu' + \Delta^+)}{\sigma(\nu' + p \to \mu^- + \Delta^{++})} \simeq \frac{1}{3}\,.$$

However, all Δ^{++} decay into $\pi^+ + p$, while only $\frac{1}{3}$ of the Δ^+ decay into $\pi^+ + n$, so with these approximations, we predict that

$$\frac{\sigma(\nu' + p \to \nu' + \pi^+ + n)}{\sigma(\nu' + p \to \mu^- + \pi^+ + p)} \simeq \frac{1}{9}\,. \tag{3.28}$$

This prediction has already been obtained, on similar grounds, by Glashow, Maiani, and Iliopoulos.[5] The experimental value given by Cundy *et al.*[4] is

$$\frac{\sigma(\nu' + p \to \nu' + \pi^+ + n)}{\sigma(\nu' + p \to \mu^- + \pi^+ + p)} = 0.08 \pm 0.04\,.$$

Again, the proposed theory is neither confirmed nor refuted.

IV. STRANGE PARTICLES

We have seen that the data on strangeness-conserving neutral-current interactions are not yet good enough to rule out the existence of such currents. In contrast, the upper limits on strangeness-nonconserving processes such as $K^0 \to \mu^+ + \mu^-$, $K^+ \to \pi^+ + \nu + \bar{\nu}$, $\Sigma^+ \to p + e^+ + e^-$, etc., are very stringent, and show that no strangeness-nonconserving neutral currents can have any appreciable coupling to leptons.[9] Thus some way must be found to introduce the strangeness-changing charged currents, without introducing any strangeness-changing neutral currents.

One solution to this problem was suggested by Glashow, Maiani, and Iliopoulos.[5] A new quantum number, the "charm" $\mathfrak{C}$, is introduced, and the $\vec{\mathrm{T}}_L$ baryon doublets are taken as

$$\mathfrak{B}_L = \tfrac{1}{2}(1+\gamma_5)\begin{pmatrix} \mathcal{P} \\ \mathfrak{N}\cos\theta_C + \lambda\sin\theta_C \end{pmatrix}, \tag{4.1}$$

$$\mathfrak{B}_L' = \tfrac{1}{2}(1+\gamma_5)\begin{pmatrix} \mathcal{P}' \\ -\mathfrak{N}\sin\theta_C + \lambda\cos\theta_C \end{pmatrix}, \tag{4.2}$$

where θ_C is Cabibbo's angle; $\mathcal{P}$, $\mathfrak{N}$, and λ are the usual SU(3) quarks with $\mathfrak{C}=0$; and $\mathcal{P}'$ is a fourth quark with $\mathfrak{C}=+1$. The current to which $\vec{\mathrm{A}}_\mu$ couples is then

$$\vec{\mathcal{J}}^\mu = 2i\,\overline{\mathfrak{B}}_L\vec{\mathrm{t}}\gamma^\mu\mathfrak{B}_L + 2i\,\overline{\mathfrak{B}}_L'\vec{\mathrm{t}}\gamma^\mu\mathfrak{B}_L'\,, \tag{4.3}$$

so the W^μ and Z^μ couple to the currents

$$\begin{aligned}\mathcal{J}_W^\mu &= \mathcal{J}_1^\mu + i\,\mathcal{J}_2^\mu \\ &= i\cos\theta_C\bar{\mathcal{P}}\gamma^\mu(1+\gamma_5)\mathfrak{N} + i\sin\theta_C\bar{\mathcal{P}}\gamma^\mu(1+\gamma_5)\lambda \\ &\quad - i\sin\theta_C\bar{\mathcal{P}}'\gamma^\mu(1+\gamma_5)\lambda + i\cos\theta_C\bar{\mathcal{P}}'\gamma^\mu(1+\gamma_5)\mathfrak{N}\,,\end{aligned} \tag{4.4}$$

$$\begin{aligned}\mathcal{J}_Z^\mu &= \mathcal{J}_3^\mu - 2\sin^2\theta J^\mu \\ &= \tfrac{1}{2}i\bar{\mathcal{P}}\gamma^\mu(1+\gamma_5)\mathcal{P} - \tfrac{1}{2}i\,\bar{\mathfrak{N}}\gamma^\mu(1+\gamma_5)\mathfrak{N} \\ &\quad + \tfrac{1}{2}i\,\bar{\mathcal{P}}'\gamma^\mu(1+\gamma_5)\mathcal{P}' - \tfrac{1}{2}i\bar{\lambda}\gamma^\mu(1+\gamma_5)\lambda - 2\sin^2\theta J^\mu\,,\end{aligned} \tag{4.5}$$

where J^μ is the usual electric current. (Note that it is quite unnecessary here to specify any particular charge assignment for the four quarks.) To put this in a more model-independent and useful way, we may write

$$\mathcal{J}_Z^\mu = \mathcal{J}_{N3}^\mu + \tfrac{1}{2}\mathcal{J}_{\mathfrak{C}}^\mu - \tfrac{1}{2}\mathcal{J}_S^\mu - 2\sin^2\theta\, J^\mu\,, \tag{4.6}$$

where $\mathcal{J}_{N3}^\mu$ is the third component of the "V minus A" isospin current

$$\vec{\mathcal{J}}_N^\mu = (\bar{\mathcal{P}}\ \ \bar{\mathfrak{N}})\gamma^\mu(1+\gamma_5)\vec{\mathrm{t}}\begin{pmatrix}\mathcal{P}\\ \mathfrak{N}\end{pmatrix} \tag{4.7}$$

and $\mathcal{J}_{\mathfrak{C}}^\mu$ and $\mathcal{J}_S^\mu$ are the "V minus A" currents of charm and strangeness

$$\mathcal{J}_{\mathfrak{C}}^\mu = i\,\bar{\mathcal{P}}'\gamma^\mu(1+\gamma_5)\mathcal{P}'\,, \tag{4.8}$$

$$\mathcal{J}_S^\mu = i\,\bar{\lambda}\gamma^\mu(1+\gamma_5)\lambda\,. \tag{4.9}$$

Equation (4.4) then gives the charm- and strangeness-conserving part of the charged current as

$$\mathcal{J}_W^\mu(\Delta S = \Delta\,\mathfrak{C} = 0) = \cos\theta_C(\mathcal{J}_{N1}^\mu + i\,\mathcal{J}_{N2}^\mu)\,. \tag{4.10}$$

The differences between this theory and that which was used in Secs. II and III are, first, that a factor $\cos\theta_C$ appears in the β-decay current, and, second that the neutral semileptonic current contains additional terms $\mathcal{J}_{\mathfrak{C}}^\mu$ and $\mathcal{J}_S^\mu$, which persist even in the limit of zero Cabibbo angle. The factor $\cos\theta_C$ merely forces us to use in (2.17) and (2.18) the value of the weak coupling constant obtained from μ decay rather than β decay, so that an additional factor of $1/\cos\theta_C$ should appear on the right-hand sides of Eqs. (3.25) and (3.26). The terms $\mathcal{J}_c^\mu$ and $\mathcal{J}_\xi^\mu$ in (4.6) could have a much more serious effect. In order to preserve the predictions of Sec. III for processes like $\nu' + p \to \nu' + p$, it would be necessary to argue that the nucleon-nucleon matrix elements of $\mathcal{J}_c^\lambda$ and $\mathcal{J}_\xi^\lambda$ are small, perhaps because the nucleon consists primarily of $\mathcal{P}$ and $\mathfrak{N}$ quarks. However, this problem does not arise for $\Delta I = 1$ processes like $\nu' + N \to \nu' + \Delta$, because the new terms in $\mathcal{J}_Z^\lambda$ are purely isoscalar.

Unfortunately, the neutral currents in this theory do not help us to understand the $\Delta I = \frac{1}{2}$ rule.

ACKNOWLEDGMENTS

I am grateful for conversations with Sheldon Glashow, Francis Low, Carlo Rubbia, and Sam Treiman.

---

*This work is supported in part through funds provided by the U. S. Atomic Energy Commission under Contract No. AT(11-1)-3069.

[1]S. Weinberg, Phys. Rev. Letters 19, 1264 (1967); A. Salam, *Elementary Particle Theory*, edited by N. Svartholm (Almquist and Forlag, Stockholm, 1968), p. 367; G. 't Hooft, Nucl. Phys. B35, 167 (1971); B. W. Lee, Phys. Rev. D 5, 823 (1972); S. Weinberg, Phys. Rev. Letters 27, 1688 (1971).

[2]This model is in most respects the same as that proposed by J. Schwinger, Ann. Phys. (N.Y.) 2, 407 (1957); S. L. Glashow, Nucl. Phys. 22, 579 (1961); A. Salam and J. Ward, Phys. Letters 13, 168 (1964). The new element added by the works cited in Ref. 1 is the use of spontaneous symmetry breaking to generate masses without losing renormalizability.

[3]H. H. Chen and B. W. Lee, Phys. Rev. D (to be published); G. 't Hooft, Phys. Letters 37B, 197 (1971).

[4]D. C. Cundy, G. Myatt, F. A. Nezrick, J. B. M. Pattison, D. H. Perkins, C. A. Ramm, W. Venus, and H. W. Wachsmuth, Phys. Letters 31B, 478 (1970).

[5]S. L. Glashow, J. Iliopoulos, and L. Maiani, Phys. Rev. D 2, 1285 (1970). Also see B. J. Bjorken and S. L. Glashow, Phys. Letters 11, 255 (1964).

[6]The role of the hadrons in spontaneous symmetry breaking is considered by S. Weinberg, Phys. Rev. Letters 27, 1688 (1971). Also see J. Schechter and Y. Ueda (unpublished).

[7]R. P. Feynman and M. Gell-Mann, Phys. Rev. 109, 193 (1958).

[8]See, e.g., T. D. Lee and C. N. Yang, Phys. Rev. 126, 2239 (1962). The weak form factors defined by Lee and Yang are the same as those used here, except that they include the factor G_β.

[9]For a recent summary, see S. B. Treiman, in *Phenomenology in Particle Physics 1971*, edited by C. B. Chiu, G. C. Fox, and A. J. G. Hey (California Institute of Technology, Pasadena, 1971).

Ultraviolet Behavior of Non-Abelian Gauge Theories*

David J. Gross† and Frank Wilczek
Joseph Henry Laboratories, Princeton University, Princeton, New Jersey 08540
(Received 27 April 1973)

It is shown that a wide class of non-Abelian gauge theories have, up to calculable logarithmic corrections, free-field—theory asymptotic behavior. It is suggested that Bjorken scaling may be obtained from strong-interaction dynamics based on non-Abelian gauge symmetry.

Non-Abelian gauge theories have received much attention recently as a means of constructing unified and renormalizable theories of the weak and electromagnetic interactions.[1] In this note we report on an investigation of the ultraviolet (UV) asymptotic behavior of such theories. We have found that they possess the remarkable feature, perhaps unique among renormalizable theories, of asymptotically approaching free-field theory. Such asymptotically free theories will exhibit, for matrix elements of currents between on-mass-shell states, Bjorken scaling. We therefore suggest that one should look to a non-Abelian gauge theory of the strong interactions to provide the explanation for Bjorken scaling, which has so far eluded field-theoretic understanding.

The UV behavior of renormalizable field theories can be discussed using the renormalization-group equations,[2,3] which for a theory involving one field (say $g\varphi^4$) are

$$[m\partial/\partial m+\beta(g)\,\partial/\partial g-n\gamma(g)]\,\Gamma_{\text{asy}}{}^{(n)}(g;P_1,\ldots,P_n)=0. \tag{1}$$

$\Gamma_{\text{asy}}{}^{(n)}$ is the asymptotic part of the one-particle-irreducible renormalized n-particle Green's function, $\beta(g)$ and $\gamma(g)$ are finite functions of the renormalized coupling constant g, and m is either the renormalized mass or, in the case of massless particles, the Euclidean momentum at which the theory is renormalized.[4] If we set $P_i=\lambda q_i{}^0$, where $q_i{}^0$ are (nonexceptional) Euclidean momenta, then (1) determines the λ dependence of $\Gamma^{(n)}$:

$$\Gamma^{(n)}(g;P_i)=\lambda^D\Gamma^{(n)}(\bar{g}(g,t);q_i)\exp\left[-n\int_0^t\gamma(\bar{g}(g,t'))\,dt'\right], \tag{2}$$

where $t=\ln\lambda$, D is the dimension (in mass units) of $\Gamma^{(n)}$, and $\bar{g}$, the invariant coupling constant, is the solution of

$$d\bar{g}/dt=\beta(\bar{g}),\quad \bar{g}(g,0)=g. \tag{3}$$

The UV behavior of $\Gamma^{(n)}$ $(\lambda\to+\infty)$ is determined by the large-t behavior of $\bar{g}$ which in turn is controlled by the zeros of β: $\beta(g_f)=0$. These fixed points of the renormalization-group equations are said to be UV stable [infrared (IR) stable] if $\bar{g}\to g_f$ as $t\to+\infty$ $(-\infty)$ for $\bar{g}(0)$ near g_f. If the physical coupling constant is in the domain of attraction of a UV-stable fixed point, then

$$\Gamma^{(n)}(g;P_i)\underset{\lambda\to\infty}{\approx}\lambda^{D-n\gamma(g_f)}\Gamma^{(n)}(g_f;q_i)\exp\left\{-n\int_0^\infty[\gamma(\bar{g}(g,t))-\gamma(g_f)]\,dt\right\}, \tag{4}$$

so that $\gamma(g_f)$ is the anomalous dimension of the field. As Wilson has stressed, the UV behavior is determined by the theory at the fixed point ($g=g_f$).[5]

In general, the dimensions of operators at a fixed point are not canonical, i.e., $\gamma(g_f)\neq 0$. If we wish to explain Bjorken scaling, we must assume the existence of a tower of operators with canonical dimensions. Recently, it has been argued for all but gauge theories, that this can only occur if the fixed point is at the origin, $g_f=0$, so that the theory is asymptotically free.[6,7] In that case the anomalous dimensions of all operators vanish, one obtains naive scaling up to finite and calculable powers of $\ln\lambda$, and the structure of operator products at short distances is that of free-field theory.[7] Therefore, the existence of such a fixed point, for a theory of the strong interactions, might explain Bjorken scaling and the success of naive light-cone or parton-model relations. Unfortunately, it appears that the fixed point at the origin, which is common to all theories, is not UV stable.[8,9] The only exception would seem to be non-Abelian gauge theories, which hitherto have not been explored in this regard.

Let us consider a Yang-Mills theory given by the Lagrangian

$$\mathcal{L} = -\tfrac{1}{4}F_{\mu\nu}{}^a F_a{}^{\mu\nu},$$
$$F_{\mu\nu}{}^a = \partial_\mu A_\nu{}^a - \partial_\nu A_\mu{}^a - gC_{abc}A_\mu{}^b A_\nu{}^c, \tag{5}$$

where the C_{abc} are the structure constants of some (semisimple) Lie group G. Since the theory is massless, the renormalization is performed at an (arbitrary) Euclidean point. For example, the wave-function renormalization constant $Z_3(g, \Lambda/m)$ will be defined in terms of the unrenormalized vector-meson propagator $D_{\mu\nu}{}^{ab}$ (in the Landau gauge),

$$D_{\mu\nu}{}^{ab}(P)|_{P^2=-m^2} = \left(g_{\mu\nu} + \frac{P_\mu P_\nu}{m^2}\right)\frac{iZ_3}{m^2}\delta_{ab}. \tag{6}$$

(For a thorough discussion of the renormalization see the work of Lee and Zinn-Justin.[10]) The renormalization-group equations for this theory are easily derived.[11] In the Landau gauge they are identical with (1). In order to investigate the stability of the origin, it is sufficient to calculate β to lowest order in perturbation theory. To this order we have

$$\beta(g) = \frac{\partial g}{\partial \ln m}\bigg|_{\Lambda \text{ fixed}} = -g\frac{\partial}{\partial \ln\Lambda}\left(\frac{Z_3{}^{3/2}}{Z_1}\right), \tag{7}$$

where Λ is a UV cutoff, and Z_1 the charge-renormalization constant. In Abelian gauge theories $Z_3 = Z_1 = 1 - g^2C\ln\Lambda$ $(C>0)$, as a consequence of gauge invariance and the Källén-Lehman representation, and thus $\beta(g) \cong g^3$ which leads to IR stability at $g=0$. Non-Abelian theories have no such requirement; Z_3 and Z_1 are gauge dependent and can be greater than 1. Thus $\beta(g)$, which must be gauge independent in lowest order, could have any sign at $g=0$. We have calculated Z_1 and Z_3 for the above Lagrangian, and we find that[12]

$$\beta_V = -(g^3/16\pi^2)\tfrac{11}{3}C_2(G) + O(g^5), \tag{8}$$

where $C_2(G)$ is the quadratic Casimir operator of the adjoint representation of the group G: $\sum_{b,c} c_{abc} \times c_{dbc} = C_2(G)\delta_{ad}$ [e.g., $C_2(\mathrm{SU}(N)) = N$]. The solution of (3) is then $\bar{g}^2(t) = g^2/(1 - 2\beta_V g^{-1}t)$, and $\bar{g} \to 0$ as $t\to\infty$ as long as the physical coupling constant g is in the domain of attraction of the origin.[13]

We have thus established that for all non-Abelian gauge theories based on semisimple Lie groups the origin is UV stable. It is easy to incorporate fermions into such a theory without destroying the UV stability. The fermion interaction is given by $L_F = \bar{\psi}(i\gamma\cdot\partial - g\gamma\cdot B^a M^a)\psi$ + mass terms, where M^a are the matrices of some representation R of the gauge group G. The only effect of the fermions is to change the value of $\beta(g)$ by the amount[11]

$$\beta_F(R) = (g^3/16\pi^2)\tfrac{4}{3}T(R), \tag{9}$$

where $\mathrm{Tr}(M^aM^b) = T(R)\delta_{ab}$, $T(R) = C_2(R)d(R)/r$, $d(R)$ is the dimension of the representation R, and r is the order of the group, i.e., the number of generators, and $C_2(R)$ is the quadratic Casimir operator of the representation. Although the fermions tend to destabilize the origin, there is room to spare. For example, in the case of SU(3): $\beta_V = -11$, whereas $\beta_F(\underline{3}) = \frac{2}{3}$, $\beta_F(\underline{8}) = 4$, etc., so that one could accomodate as many as sixteen triplets. One can therefore construct many asymptotically free theories with fermions. The vector mesons, however, will remain massless until the gauge symmetry is spontaneously broken. One might hope that this would be a consequence of the dynamics,[14] but at the present the only known way of achieving this is to introduce scalar Higgs mesons, whose nonvanishing vacuum expectation values break the symmetry.

The introduction of scalar mesons has a very destabilizing effect on the UV stability of the origin. Their contribution to $\beta(g)$ is small; a scalar meson transforming under a complex (real) representation R of the gauge group adds to β a term equal to $\frac{1}{4}$ $(\frac{1}{8})$ of Eq. (9). The problem with scalar mesons is that they necessarily have their own quartic couplings, and one must deal with a new coupling constant. Consider the Lagrangian for the coupling of scalars belonging to a representation R of G:

$$\mathcal{L} = \tfrac{1}{2}[(\partial_\mu - igB_\mu{}^aM^a)\vec{\varphi}]^2 - \lambda(\vec{\varphi}\cdot\vec{\varphi})^2 + V(\vec{\varphi}). \tag{10}$$

where $V(\varphi)$ contains cubic, quadratic, and linear terms in φ (which have no effect on the UV behavior of the theory) plus, perhaps, additional quartic terms invariant under G. The renormalization-group equations have an additional term, $\beta_\lambda(g,\lambda)\partial/\partial\lambda$, and one must investigate the UV stability of the origin $(g=\lambda=0)$ with respect to both g and λ [if there are other quartic invariants in $V(\varphi)$ there will be additional coupling constants to consider]. The structure of the renormalization-group equation for g is unchanged to lowest order, whereas for the coupling constant $\Gamma \equiv \lambda/g^2$ we have[11]

$$d\bar{\Gamma}(\Gamma, t, g^2)/dt = \bar{g}^2[A\bar{\Gamma}^2 + B\bar{\Gamma} + C] \tag{11}$$

(where we have neglected terms of order g^4, $g^4\Gamma$,

$g^4\Gamma^2$, and $g^4\Gamma^3$). In the absence of vector mesons ($g=0$) this equation is UV unstable at $\lambda=0$, since A is strictly positive and λ must be positive.[15] The vector mesons contribute to B and C and tend to stabilize the origin. If the right-hand side of (11) has positive zeros ($C>0$, $B<0$, and $B^2>4AC$), then for Γ less than the larger zero of (11) we will have that $\lambda \to +0$ as $t\to\infty$. We have investigated the structure of these equations for a large class of gauge theories and representations of the scalar mesons. We have found many examples of theories which contain scalar mesons and are UV stable.[11] These include (a) SU(N) if the scalar mesons belong to the adjoint representation for $N\geq 6$; (b) SU(N)$\otimes$SU(N) if the scalars belong to the $(N,\bar{N})$ representation for $N\geq 5$; (c) SU(N) with the scalars transforming as a symmetric tensor for $N\geq 9$; and many others. In all of these models it is necessary for the theory to contain a large number of fermions in order to make β_g small; otherwise $\bar{g}$ approaches zero too rapidly for the vector mesons to stabilize the scalar couplings.

Unfortunately, in none of these models can the gauge symmetry be totally broken by the Higgs mechanism. The requirement that the interactions of the scalar mesons be renormalizable so severely constrains the form of Lagrangian that the ground state invariably is invariant under some non-Abelian subgroup of the gauge group. If one tries to overcome this by larger representations for the scalar mesons, UV instability inevitably occurs.

It thus appears to be very difficult to retain UV stability and break the gauge symmetry by explicitly introducing Higgs mesons. Since the Higgs mesons are so restrictive, we would prefer to believe that spontaneous symmetry breaking would arise dynamically.[14] This is suggested by the IR instability of the theories, which assures us that perturbation theory is not trustworthy with respect to the stability of the symmetric theory nor to its particle content.

With this hope in mind one can construct many interesting models of the strong interactions. One particularly appealing model is based on three triplets[16] of fermions, with Gell-Mann's SU(3)$\otimes$SU(3) as a global symmetry and an SU(3) "color" gauge group to provide the strong interactions. That is, the generators of the strong-interaction gauge group commute with ordinary SU(3)$\otimes$SU(3) currents and mix quarks with the same isospin and hypercharge but different "color." In such a model the vector mesons are neutral, and the structure of the operator product expansion of electromagnetic or weak currents is (assuming that the strong coupling constant is in the domain of attraction of the origin!) essentially that of the free quark model (up to calculable logarithmic corrections).[11]

Finally, we note that theories of the weak and electromagnetic interactions, built on semisimple Lie groups,[17] will be asymptotically free if we again ignore the complications due to the Higgs particles. This suggests that the program of Baker, Johnson, Willey, and Adler[18] to calculate the fine-structure constant as the value of the UV-stable fixed point in quantum electrodynamics might fail for such theories.

---

*Research supported by the U.S. Air Force Office of Scientific Research under Contract No. F-44620-71-C-0180.

†Alfred P. Sloan Foundation Research Fellow.

[1]S. Weinberg, Phys. Rev. Lett. 19, 1264 (1967). For an extensive review as well as a list of references, see B. W. Lee, in Proceedings of the Sixteenth International Conference on High Energy Physics, National Accelerator Laboratory, Batavia, Illinois, 1972 (to be published).

[2]M. Gell-Mann and F. E. Low, Phys. Rev. 95, 1300 (1954).

[3]C. G. Callan, Phys. Rev. D 2, 1541 (1970); K. Symanzik, Commun. Math. Phys. 18, 227 (1970).

[4]The basic assumption underlying the derivation and utilization of the renormalization group equations is that the large Euclidean momentum behavior of the theory is the same as the sum, to all orders, of the leading powers in perturbation theory.

[5]K. Wilson, Phys. Rev. D 3, 1818 (1971).

[6]G. Parisi, to be published.

[7]C. G. Callan and D. J. Gross, to be published.

[8]A. Zee, to be published.

[9]S. Coleman and D. J. Gross, to be published.

[10]B. W. Lee and J. Zinn-Justin, Phys. Rev. D 5, 3121 (1972).

[11]Full details will be given in a forthcoming publication: D. J. Gross and F. Wilczek, to be published.

[12]After completion of this calculation we were informed of an independent calculation of β for gauge theories coupled to fermions by H. D. Politzer [private communication, and following Letter, Phys. Rev. Lett. 30, 1346 (1973)].

[13]K. Wilson has suggested that the coupling constants of the strong interactions are determined to be IR-stable fixed points. For nongauge theories the IR stability of the origin in four-dimensional field theories implies that theories so constructed are trivial, at least in a domain about the origin. Our results suggest that non-Abelian gauge theories might possess IR-stable fixed points at nonvanishing values of the coupling constants.

[14]Y. Nambu and G. Jona-Lasino, Phys. Rev. 122, 345 (1961); S. Coleman and E. Weinberg, Phys. Rev. D 7, 1888 (1973).

[15]K. Symanzik (to be published) has recently suggested that one consider a $\lambda\varphi^4$ theory with a negative λ to achieve UV stability at $\lambda=0$. However, one can show, using the renormalization-group equations, that in such theory the ground-state energy is unbounded from below (S. Coleman, private communication).

[16]W. A. Bardeen, H. Fritzsch, and M. Gell-Mann, CERN Report No. CERN-TH-1538, 1972 (to be published).

[17]H. Georgi and S. L. Glashow, Phys. Rev. Lett. 28, 1494 (1972); S. Weinberg, Phys. Rev. D 5, 1962 (1972).

[18]For a review of this program, see S. L. Adler, in Proceedings of the Sixteenth International Conference on High Energy Physics, National Accelerator Laboratory, Batavia, Illinois, 1972 (to be published).

Reliable Perturbative Results for Strong Interactions?*

H. David Politzer
Jefferson Physical Laboratories, Harvard University, Cambridge, Massachusetts 02138
(Received 3 May 1973)

An explicit calculation shows perturbation theory to be arbitrarily good for the deep Euclidean Green's functions of any Yang-Mills theory and of many Yang-Mills theories with fermions. Under the hypothesis that spontaneous symmetry breakdown is of dynamical origin, these symmetric Green's functions are the asymptotic forms of the physically significant spontaneously broken solution, whose coupling could be strong.

Renormalization-group techniques hold great promise for studying short-distance and strong-coupling problems in field theory.[1,2] Symanzik[2] has emphasized the role that perturbation theory might play in approximating the otherwise unknown functions that occur in these discussions. But specific models in four dimensions that had been investigated yielded (in this context) disappointing results.[3] This note reports an intriguing contrary finding for any generalized Yang-Mills theory and theories including a wide class of fermion representations. For these one-coupling-constant theories (or generalizations involving product groups) the coefficient function in the Callan-Symanzik equations commonly called $\beta(g)$ is negative near $g=0$.

The constrast with quantum electrodynamics (QED) might be illuminating. Renormalization of QED must be carried out at off-mass-shell points because of infrared divergences. For small e^2, we expect perturbation theory to be good in some neighborhood of the normalization point. But what about the inevitable logarithms of momenta that grow as we approach the mass shell or as some momenta go to infinity? In QED, the mass-shell divergences do not occur in observable predictions, when we take due account of the experimental situation. The renormalization-group technique[4] provides a somewhat opaque analysis of this situation. Loosely speaking,[5] the effective coupling of soft photons goes to zero, compensating for the fact that there are more and more of them. But the large-p^2 divergence represents a real breakdown of perturbation theory. It is commonly said that for momenta such that $e^2\ln(p^2/m^2)\sim 1$, higher orders become comparable, and hence a calculation to any finite order is meaningless in this domain. The renormalization group technique shows that the effective coupling grows with momenta.

The behavior in the two momentum regimes is reversed in a Yang-Mills theory. The effective coupling goes to zero for large momenta, but as p^2's approach zero, higher-order corrections become comparable. Thus perturbation theory tells *nothing* about the mass-shell structure of the symmetric theory. Even for arbitrarily small g^2, there is no sense in which the interacting theory is a small perturbation on a free multiplet of massless vector mesons. The truly catastrophic infrared problem makes a symmetric particle interpretation impossible. Thus, though one can well approximate asymptotic Green's functions, to what particle states do they refer?

Consider theories defined by the Lagrangian

$$\mathcal{L}=-\tfrac{1}{4}F_{\mu\nu}{}^{a}F^{a\mu\nu}+i\bar{\psi}_i\gamma\cdot D_{ij}\psi_j, \qquad (1)$$

where

$$F_{\mu\nu}{}^{a}=\partial_\mu A_\nu{}^{a}-\partial_\nu A_\mu{}^{a}+gf^{abc}A_\mu{}^{b}A_\nu{}^{c},$$

and

$$D_{ij}{}^{\mu} = \partial^{\mu}\delta_{ij} - igA^{a\mu}T_{ij}{}^{a},$$

the f^{abc} are the group structure constants, and the T^a are representation matrices corresponding to the fermion multiplet. (One may be interested in models with massless fermions because of their group structure or because they have the same asymptotic forms[6] as massive theories.) The normalizations of the conventionally defined irreducible vertices for n mesons and n' fermions, $\Gamma^{n,n'}$, must refer to some mass M. The renormalization-group equation reads

$$\left(M\frac{\partial}{\partial M} + \beta(g)\frac{\partial}{\partial g} + n\gamma_A(g) + n'\gamma_\psi(g)\right)\Gamma^{n,n'} = 0. \quad (2)$$

Putting it in this form makes use of the first available simplification, proper choice of gauge. Equation (2) describes how finite renormalizations accompanied by a change in g and a rescaling of the fields leave the $\Gamma^{n,n'}$ unchanged. Consider gauges defined by α in the zeroth-order propagator

$$\Delta_{\mu\nu}(p^2) = \frac{-g_{\mu\nu} + p_\mu p_\nu/p^2}{p^2} + \alpha\frac{p_\mu p_\nu}{p^4}.$$

The generalized Ward identities[7] imply that there are no higher-order corrections to the longitudinal part. But if the fields are rescaled as in Eq. (2), α must be changed to leave Γ^2 invariant. Hence α should occur in Eq. (2) much as g does, and one would have to study the $\Gamma^{n,n'}$ for arbitrary α to determine the coefficient functions perturbatively. But for $\alpha = 0$ initially, it remains zero under finite renormalizations; so it suffices to study the theory in a Landau gauge.

To first order, the meson inverse propagator is

$$\Gamma_{\mu\nu}{}^{2ab}(p,-p) = \delta^{ab}(-g_{\mu\nu}p^2 + p_\mu p_\nu)[1 + (\tfrac{13}{3}c_1 - \tfrac{8}{3}c_2)(g/4\pi)^2\ln(-p^2/M^2)], \quad (3)$$

where

$$f_{acd}f_{bcd} = 2c_1\delta_{ab}, \quad \mathrm{tr}(T^aT^b) = 2c_2\delta_{ab},$$

and $c_1 > 0$ and $c_2 \geq 0$. [For SU(2), $c_1 = 1$, $c_2(\text{isodoublet}) = \frac{1}{4}$, and $c_2(\text{isotriplet}) = 1$.] To first order (only), the fermion self-energy is proportional to the self-energy in massless QED, which vanishes in the Landau gauge. Similarly, the contribution to the fermion-vector three-point vertex correction proportional to the first-order QED correction needs no subtractions and contains no reference to M. Calculation of the remaining correction, involving the meson self-coupling, yields

$$\Gamma_{\mu ij}{}^{1,2a}(0,p,-p) = gT_{ij}{}^a\gamma_\mu[1 - \tfrac{3}{2}c_1(g/4\pi)^2\ln(-p^2/M^2)]. \quad (4)$$

Applying Eq. (2) to these functions at their normalization points yields

$$\gamma_\psi(g) = 0 + O(g^4), \quad \gamma_A(g) = (\tfrac{13}{3}c_1 - \tfrac{8}{3}c_2)(g/4\pi)^2 + O(g^4), \quad \beta(g) = -(\tfrac{22}{3}c_1 - \tfrac{8}{3}c_2)g(g/4\pi)^2 + O(g^5).$$

It is also apparent, by inspecting the graphs, that to this order the coupling constants of product groups do not enter into each other's β functions.

For the case where there are no fermions, the coefficient functions can be obtained by setting $c_2 = 0$. (Even though the fermion-vector vertex, which had been used implicitly to define g, is no longer present, it can be simulated by introducing two multiplets of spinor fields with the same group transformations but opposite statistics. The physical effects of internal fermions are canceled by the ghosts —spinor fields with Bose statistics.) Alternatively, one can study the corrections to the three-meson vertex. Define F by

$$\Gamma_{\lambda\mu\nu}{}^{3abc}(p,-p,0) = f^{abc}(p_\lambda g_{\mu\nu} + p_\mu g_{\nu\lambda} - 2p_\nu g_{\lambda\mu})gF(p^2/M^2, g^2). \quad (5)$$

The normalization condition is $F(-1, g^2) = 1$ (up to a phase convention.) To first order

$$F = 1 + \tfrac{17}{6}c_1(g/4\pi)^2\ln(-p^2/M^2) \quad (6)$$

which yields the same β as described above.

The renormalization-group "improved" perturbation theory[4,5] extends results valid near the normalization point by effectively moving that point. The improved vertex functions are constructed from the straightforward perturbative ones, involving a momentum-scale-dependent effective coupling $g'(g,t)$, where $t = \frac{1}{2}\ln(s/M^2)$ and s sets the scale, e.g., $s = \sum(-p_i^2)$. $g'(g,t)$ is defined by

$$\partial g'/\partial t = \beta(g'), \qquad g'(g,0) = g. \quad (7)$$

For the approximate β's derived above, $\beta = -bg^3$,

$$g'^2 \approx g^2/(1+2bg^2t). \tag{8}$$

Thus for a pure meson theory or for theories including not too many fermions (in the sense that $c_2 < \frac{11}{4}c_1$), g' goes to zero for asymptotic momenta, i.e., $t\to\infty$. The $\Gamma^{n,n'}$ show a well-defined slow approach to quasifree field values.

It is worth remembering that successive orders of perturbation theory give the behavior of β for infinitesimal g and, strictly speaking, say nothing about finite g. Making a polynomial fit to a perturbative result for β is pure conjecture.

Hypothesizing that β stays negative (at least into the domain of strong coupling constant) relates all theories defined by Eq. (1) [with g less than the first zero of $\beta(g)$] to the model with g arbitrarily small by a change in mass scale. They all share the same asymptotic Green's functions, differing only by how large is asymptotic.

To utilize this result, we make the following hypothesis: The gauge symmetry breaks down spontaneously as a result of the dynamics. Consequently, the fields obtain (in general massive) particle interpretation—the Higgs phenomenon. As yet, nothing is known about the particle spectrum, the low-energy dynamics, or particles describable only by composite fields. But the Callan-Symanzik analysis says that the asymptotic Green's functions for the "dressed" fundamental fermion and vector fields are the symmetric functions discussed above.[8]

[An alternative is to introduce fundamental scalar fields, in terms of which the group transformation properties of the vacuum can be studied.[9] But these theories are not in general ultraviolet stable in terms of the additional coupling constants that must be introduced. Particular models which are ultraviolet stable as well as spontaneously asymmetric have been found.[10] But gauge theories of fermions (only) have aesthetic attractions, including the possibility of a dynamical determination of the dimensionless coupling constant.[9]]

Hypotheses of this type go back to the work of Nambu and collaborators.[11] In the renormalizable massless theories including scalars that have been studied,[9] infrared instability is a necessary condition for spontaneous symmetry breakdown.[12] The model of Nambu and Jona-Lasinio can be treated by the methods of Coleman and Weinberg.[9] The model is defined by

$$\mathcal{L} = i\bar\psi\gamma\cdot\partial\psi + g_0[(\bar\psi\psi)^2 - (\bar\psi\gamma_5\psi)^2] \tag{9}$$

and the stipulation that the momentum integrals are cut off at some Euclidean mass squared Λ^2. Define a scalar

$$\varphi(x) = g_0\bar\psi(x)\psi(x)$$

and an analogous pseudoscalar, which one can do because of the cutoff. A study of the Green's functions in the one-loop approximation yields all the original results. But the existence of the vacuum-degenerate solution requires the dimensionless parameter characterizing the theory to satisfy $g_0\Lambda^2 > 2\pi^2$. But this is the condition that the one-loop correction to fermion-fermion scattering be at least as important as the tree approximation.

The situation is similar in the renormalizable models. $\lambda\varphi^4$ is stable for small λ because the one-loop corrections are small. But in massless scalar QED, photon-loop corrections of order e^4 can dominate over the lowest order φ-φ scattering (order λ) for both λ and e arbitrarily small. The requirement is just that $\lambda < e^4$. In this light, the problem with the Nambu–Jona-Lasinio model is not its nonrenormalizability but that in the domain of large $g_0\Lambda^2$, where spontaneous breakdown is alleged to occur, higher loop corrections are likely to dominate. (In the framework of the original solution, more complex infinite chains and self-energy graphs dominate over the ones studied.) In theories defined by Eq. (1), composite scalar densities can also be defined and studied in perturbation theory. But the condition that the one-loop approximation imply vacuum degeneracy requires that the expansion parameter be large, rendering the application of perturbation theory suspect.

The author thanks Sidney Coleman and Erick Weinberg, who have offered insights and advice freely, and the latter especially for his help in the computations.

---

*Work supported in part by the U. S. Air Force Office of Scientific Research under Contract No. F44620-70-C-0030.

[1]Of central importance is the work reviewed in K. Wilson and J. Kogut, "The Renormalization Group and the ϵ Expansion" (to be published); K. Johnson and M. Baker, "Some Speculations on High Energy Quantum Electrodynamics" (to be published); S. Adler, Phys. Rev. D 5, 3021 (1972).

[2]K. Symanzik, DESY Report No. DESY 72/73, 1972

(to be published), and references therein.

[3]A. Zee, "Study of the Renormalization Group for Small Coupling Constants" (to be published).

[4]N. N. Bogoliubov and D. V. Shirkov, *Introduction to the Theory of Quantized Fields* (Interscience, New York, 1959).

[5]Definitions of the relevant quantities will be given, but for the general theory and derivations see Refs. 1, 2, and 4; S. Coleman, in the Proceedings of the 1971 International Summer School "Ettore Majorana" (Academic, New York, to be published); S. Coleman and E. Weinberg, Phys. Rev. D 7, 1888 (1973), whose conventions we follow.

[6]Asymptotic refers to a particular set of Euclidean momenta as they are collectively scaled upward.

[7]E. g., B. W. Lee and J. Zinn-Justin, Phys. Rev. D 5, 3121 (1972); G. 't Hooft, Nucl. Phys. B33, 173 (1971), which also include details of Feynman rules, regularization, etc.

[8]Configurations where the symmetric theory has infrared singularities not present in the massive case are discussed in detail by Symanzik (Ref. 2).

[9]Coleman and Weinberg, Ref. 5.

[10]D. Gross and F. Wilczek, preceding Letter [Phys. Rev. Lett. 30, 1343 (1973)].

[11]Y. Nambu and G. Jona-Lasinio, Phys. Rev. 122, 345 (1961).

[12]$\lambda\varphi^4$ theory with $\lambda < 0$ is ultraviolet stable (Ref. 2) and hence infrared unstable but cannot be physically interpreted in perturbation theory. Using the computations of Ref. 9, for $\lambda < 0$ "improved" perturbation theory is arbitrarily good for large field strengths. In particular, the potential whose minimum determines the vacuum decreases without bound for large field.

Asymptotically free gauge theories. II *

David J. Gross† and Frank Wilczek
Joseph Henry Laboratories, Princeton University, Princeton, New Jersey 08540
(Received 27 August 1973)

Deep-inelastic lepton-hadron scattering is analyzed in asymptotically free gauge theories of the strong interactions. The renormalization-group equations for the coefficients of the twist-two operators in the Wilson expansion are reviewed. A careful treatment of the mixing of operators with identical quantum numbers and dimensions is given. The relevant anomalous dimensions of the twist-two operators are calculated to second order in perturbation theory. These are used to calculate the asymptotic q^2 behavior of the moments of the structure functions. It is shown that the approach to the asymptotic limit is logarithmic, that Bjorken scaling is violated by powers of $\ln(-q^2)$, and that the naive light-cone or parton-model relations for the moments of the structure functions are true asymptotic theorems. A new sum rule for the first moment of F_2, in terms of the energy-momentum tensor, is derived. An example of a function whose moments have roughly the correct asymptotic q^2 behavior is constructed. The q^2 behavior of the structure functions for a given x is discussed.

I. INTRODUCTION

In a recent paper[1] we have constructed a class of gauge theories of the strong interactions, which have the remarkable feature of being "asymptotically free." The primary motivation for this proposal is the evidence that Bjorken scaling requires an asymptotically free theory,[2] that only non-Abelian gauge theories can be asymptotically free,[3] and that indeed many non-Abelian gauge theories are asymptotically free.[4-6] Deep-inelastic scattering is therefore the natural arena in which to test our theories. In this paper we shall discuss in detail the properties of lepton-hadron scattering in asymptotically free gauge theories of the strong interactions.

This paper is a sequel to Ref. 1 and should be read in conjunction with it, although the phenomenological discussion of Sec. III can be understood independently (hereafter Ref. 1 will be referred to as paper I, and the prefix I refers to equations of Ref. 1). The general features of deep-inelastic scattering in asymptotically free theories were already described in paper I. These include the logarithmic approach to scaling, the calculable logarithmic deviations from Bjorken scaling, and the validity of the naive or light-cone parton-model relations.

In Sec. II of this paper, we discuss in some detail the application of renormalization-group techniques to the Wilson expansion. This analysis has appeared in many other places[7-9,2] and is included here for the sake of completeness. In particular we discuss the mixing of operators with the same quantum numbers and dimensions. In gauge theories this mixing is particularly annoying since it would appear that "ghost" operators (i.e., operators involving Feynman-Faddeev-Popov ghost fields[10,11]) mix together with ordinary operators. We argue that this mixing can be ignored. This claim is further substantiated by a calculation, which appears in Appendix A, performed in a gauge which is free of Faddeev-Popov ghosts.

In Sec. III we calculate the anomalous dimensions of the relevant operators in the Wilson expansion.[12] Some of these were already presented in paper I. We derive the asymptotic form for the moments of the structure functions, as well as the various relations and sum rules satisfied by these moments. A sum rule for the first moment of the structure functions, the "energy-momentum-tensor sum rule," is derived. An example of an explicit functional form for the structure functions, with roughly the correct asymptotic behavior, is presented and the general features of this function are discussed.

Section IV contains some concluding remarks.

II. THE RENORMALIZATION-GROUP APPROACH TO THE WILSON EXPANSION

In the deep-inelastic scattering of a lepton off a hadron one measures the Fourier transform of the commutator of electromagnetic or weak currents. We define the standard structure functions as follows:

$$\frac{1}{2\pi}\int dy\, e^{iq\cdot y}\langle p|[J_\mu(a;\tfrac{1}{2}y), J_\nu(b;-\tfrac{1}{2}y)]|p\rangle = \frac{p_\mu p_\nu}{m\nu}F_2^{(a,b)}(\nu,q^2) - \frac{g_{\mu\nu}}{m}F_1^{(a,b)}(\nu,q^2) + i\frac{\epsilon_{\mu\nu\sigma\lambda}p_\sigma q_\lambda}{2m\nu}F_3^{(a,b)}(\nu,q^2) + \cdots , \quad (1)$$

where p is the momentum of the hadron, q is the momentum transfer to the hadrons, $\nu=p\cdot q$, spin averages have been taken, and we have suppressed terms proportional to q_μ and q_ν which give terms proportional to the lepton mass when contracted with the leptonic currents. The label a denotes the SU(3) × SU(3) character of the current. The Bjorken limit corresponds to

$$x=\frac{Q^2}{2\nu}\text{ fixed},\quad Q^2=-q^2\to+\infty \tag{2}$$

and probes the commutator for lightlike values of y.

The structure of the product of local operators at short, and at lightlike, distances is given by Wilson's operator-product expansion. In the case of interest this expansion takes the form

$$\begin{aligned}J_\mu(a;\tfrac{1}{2}y)J_\nu(b;-\tfrac{1}{2}y)=&\tfrac{1}{2}g_{\mu\nu}\left(\frac{\partial}{\partial y}\right)^2\frac{1}{y^2-i\epsilon y_0}\sum_{n=0}^{\infty}\sum_i C_{i,1}^{(n)}(a,b;y^2-i\epsilon y_0)\,O^i_{\mu_1\cdots\mu_n}(0)y^{\mu_1}\cdots y^{\mu_n}\\&+\frac{1}{y^2-i\epsilon y_0}\sum_{n=0}^{\infty}\sum_i C_{i,2}^{(n)}(a,b;y^2-i\epsilon y_0)\,O^i_{\mu\nu\mu_1\cdots\mu_n}(0)y^{\mu_1}\cdots y^{\mu_n}\\&+\tfrac{1}{2}i\epsilon_{\mu\nu\sigma\lambda}\frac{\partial}{\partial y^\lambda}\frac{1}{y^2-i\epsilon y_0}\sum_{n=0}^{\infty}\sum_i C_{i,3}^{(n)}(a,b;y^2-i\epsilon y_0)\,O^i_{\sigma\mu_1\cdots\mu_n}y^{\mu_1}\cdots y^{\mu_n}+\cdots\,.\end{aligned} \tag{3}$$

In this expansion we have neglected, as in Eq. (1), gradient terms (involving $\partial/\partial y^\mu$ or $\partial/\partial y^\nu$). The operator $O^i_{\mu_1\cdots\mu_n}(x)$ has spin k (traceless and symmetric) and dimension $n+2$ (or twist $=$dimension $-$ spin $=2$). The label i denotes the various operators of equal twist which might occur in the expansion. $C_{i,k}^{(n)}$ $(k=1,2,3)$ are c-number functions of x^2 and the coupling constants of the theory. They have been normalized to be dimensionless, so that naively (as in free-field theory) they would be nonsingular as $x^2\to0$. The three dots refer to other operators of higher twist, whose contribution to the structure functions is suppressed, to any finite order of perturbation theory, by powers of q^2. (In an asymptotically free theory this suppression is guaranteed to all orders in perturbation theory by the vanishing of the anomalous dimensions of all operators at the fixed point $g=0$. One has logarithmic, but not power, corrections to naive scaling.)

The structure functions F_k in the deep-inelastic region are determined by the behavior of $C_{i,k}^{(n)}$ for small x, and by the hadronic matrix elements of the operators O^i

$$\langle p|O^i_{\mu_1\cdots\mu_n}(0)|p\rangle_{\text{spin average}}=i^n\frac{1}{m}p_{\mu_1}\cdots p_{\mu_n}M_n^i\cdots\,. \tag{4}$$

One cannot simply take the Fourier transform of Eq. (3), since after Fourier transforming it does not converge in the region of interest $(0\le x\le1)$. Instead the Fourier transforms of $C_{i,k}^{(n)}$ are determined by the *moments* of the structure functions.[13,9]

In fact

$$\begin{aligned}\int_0^1 dx\,x^n F_1^{a,b}(x,q^2)&=\sum_i\bar C_{i,1}^{(n+1)}(a,b;q^2)M_i^{n+1}\,,\\\int_0^1 dx\,x^n F_2^{a,b}(x,q^2)&=\sum_i\bar C_{i,2}^{(n)}(a,b;q^2)M_i^{n+2},\\\int_0^1 dx\,x^n F_3^{a,b}(x,q^2)&=\sum_i\bar C_{i,3}^{(n)}(a,b;q^2)M_i^{n+1}\,,\end{aligned} \tag{5}$$

where

$$\bar C_{i,k}^{(n)}(a,b;q^2)=\tfrac{1}{2}i(q^2)^{n+1}\left(-\frac{\partial}{\partial q^2}\right)^n\times\int d^4y\,e^{iq\cdot y}\,\frac{C_{i,k}^{(n)}(a,b;y^2)}{y^2-i\epsilon y_0}\,. \tag{6}$$

In a free-quark model the operators $O^i_{\mu_1\cdots\mu_n}$ are simply

$$O^a_{\mu_1\cdots\mu_n}(x)=S\bar\psi(x)\gamma_{\mu_1}\overleftrightarrow{\partial}_{\mu_2}\cdots\overleftrightarrow{\partial}_{\mu_n}\lambda^a(1\pm\gamma_5)\psi(x)\,, \tag{7}$$

where S denotes symmetrization of the indices $\mu_1\cdots\mu_n$. In that case the functions $\bar C^{(n)}$ are constants, independent of q^2, which can be determined by the light-cone commutator of the currents (see Ref. 14).

In an interacting theory the functions $\bar C^{(n)}$ will be nontrivial functions of q^2 and the coupling constants of the theory (g). To determine the q^2 dependence of $\bar C^{(n)}$ we employ the renormalization-group equations. Let us outline the derivation of these equations for the Wilson coefficients. Consider the contribution of an operator $O^{(n)}(x)$ to the short-distance expansion of the product of $A(x)$ and $B(x)$ (we suppress all tensor and quantum-number labels):

$$A(x)B(-x) \underset{x^\mu\approx 0}{\approx} C^{(n)}(x^2,g)O^{(n)}_{\mu_1\cdots\mu_n}(0)x^{\mu_1}\cdots x^{\mu_n}. \tag{8}$$

If $O^{(n)}(x)$ is the dominant operator for short distances (i.e., the operator of smallest dimension, which we assume for the moment to be unique) then the short-distance behavior of $C^{(n)}$ can be determined by calculations performed as if all masses and dimensional coupling constants were zero (for details see paper I). In that case the only dimensional parameter in the theory is the subtraction point μ introduced to perform the required renormalization. A change in μ can be reabsorbed by a change in the coupling constant and the scale of all operators in the theory. If we consider the effect on Eq. (8) of a change in μ we derive that

$$\left(\mu\frac{\partial}{\partial\mu}+\beta(g)\frac{\partial}{\partial g}+\gamma_A+\gamma_B-\gamma_O\right)C(x^2,g)=0\,, \tag{9}$$

where β is given by Eq. I(4.9), and γ_A (γ_B,γ_O) is the anomalous dimension of the operator A $(B, O^{(n)})$. Thus, for example,

$$\gamma_A(g^2)=\mu\frac{\partial}{\partial\mu}\ln Z_A\Big|_{g_0,\Lambda\text{ fixed}}\,. \tag{10}$$

Z_A is the renormalization constant of the operator A, where we have assumed that A is multiplicatively renormalizable (this is true for all the operators that control the light-cone behavior of current products).

In the case of interest this implies that the functions $\bar{C}^{(n)}(q^2/\mu^2,g)$ satisfy

$$\left(\mu\frac{\partial}{\partial\mu}+\beta(g)\frac{\partial}{\partial g}-\gamma^n(g)\right)\bar{C}^{(n)}(q^2/\mu^2,g)=0\,, \tag{11}$$

where again we assume a unique twist-two operator appears in the Wilson expansion of currents, and $\bar{C}^{(n)}$ is related, by an equation similar to Eq. (6), to the coefficient of the spin-n twist-two operator. Note that the anomalous dimension of conserved or partially conserved currents (by which we mean that the dimension of the divergence of the current is less than four) vanishes for all g.

The solution of this renormalization group equation is expressed in terms of the effective coupling constant $\bar{g}(t,g)$ defined by

$$\frac{d\bar{g}}{dt}=\beta(\bar{g}),\quad \bar{g}(0,g)=g \tag{12}$$

where

$$t=\tfrac{1}{2}\ln(-q^2/\mu^2)\,. \tag{13}$$

In fact

$$\bar{C}^{(n)}(q^2/\mu^2,g)=\bar{C}^{(n)}(1,\bar{g}(t,g))\times\exp\left(-\int_0^t dx\,\gamma^n(\bar{g}(x,g))\right)\,. \tag{14}$$

The large-q^2 behavior of $\bar{C}^{(n)}$ will thus be determined by the large-t behavior of $\bar{g}(t,g)$, which in turn is determined by the fixed points of the renormalization group Eq. (12). In an asymptotically free gauge theory

$$\bar{g}^2(t,g)\underset{t\to\infty}{\sim} b_0^{-1}t^{-1}+b_1b_0^{-3}t^{-2}\ln t+O(1/t^2)\,, \tag{15}$$

where b_0 is given by Eq. I(4.9). The anomalous dimension γ^n will behave for small g like

$$\gamma^n(g)=\gamma^n g^2+O(g^4)\,. \tag{16}$$

Therefore in an asymptotically free theory

$$\bar{C}^{(n)}(q^2/\mu^2,g)\underset{t\to\infty}{\sim}\text{const}[\ln(-q^2)]^{-a_n}\times[\bar{C}^{(n)}(1,0)+O(1/\ln(-q^2))], \tag{17}$$

where

$$a_n=\frac{\gamma^n}{2b_0}\,. \tag{18}$$

From this expression we see that the approach to the asymptotic region is logarithmic, i.e., the corrections to the asymptotic form are suppressed by $\ln(-q^2)$. The asymptotic value of $\bar{C}^{(n)}$ does not exhibit naive (Bjorken) scaling. Instead there are logarithmic deviations, whose magnitude is calculable in second-order perturbation theory. Furthermore the tensor and quantum number *structure* of the operator-product expansion will be that of free-field theory, up to logarithmic corrections, since $\bar{C}^{(n)}$ is evaluated at $g=0$ on the right-hand side of Eq. (17).

In general there will appear in the Wilson expansion many operators $O_i^{(n)}$ with the same quantum numbers and twist. In that case a given $O_i^{(n)}$ is *not* multiplicatively renormalizable, rather one must take linear combinations of these to obtain operators which are multiplicatively renormalizable and have definite dimensions.

Consider the Wilson expansion

$$J_a(x)J_b(-x)=\sum_{i,k}C^{(n)}_{i,k}(x^2-i\epsilon x_0,g)\times O_i^{(n)}(0)_{\mu_1\cdots\mu_n}x^{\mu_1}\cdots x^{\mu_n}+\cdots, \tag{19}$$

where again we suppress the tensor indices of the currents. The label i runs over the set of operators with spin n and dimension $(n+2)$, and the label k denotes the tensorial and SU(3)×SU(3) structure of the expansion. [In other words k stands for the labels $a,b,1,2,3$ in Eq. (3).] Since the operators $O_i^{(n)}$ can mix, a change in μ will, in general, entail a mixing of the various operators. The Wilson coefficients will obey the renormalization group equation

$$\left(\mu\frac{\partial}{\partial\mu}+\beta(g)\frac{\partial}{\partial g}\right)C_{i,k}^{(n)}=\sum_j \gamma_{ij}^{(n)}(g)\,C_{j,k}^{(n)} . \tag{20}$$

The anomalous dimension is now a matrix

$$\begin{aligned}(\underline{\gamma}^{(n)}(g))_{ij}&=\gamma_{ij}^{(n)}(g)\\&=\left(\underline{Z}_0^{-1}\,\mu\frac{\partial}{\partial\mu}\underline{Z}_0\right)_{ij}\Bigg|_{g_0,\Lambda\ \text{fixed}}\end{aligned} \tag{21}$$

and $\underline{Z}_0$ is the matrix of wave-function renormalization constants of the operators $O_i^{(n)}$, i.e.,

$$O_{i\ \text{renormalized}}^{(n)}=\sum_j O_{j\ \text{bare}}^{(n)}(\underline{Z}_0^{-1})_{ji} . \tag{22}$$

The solution of Eq. (20) for the Fourier transform of $C_{i,k}^{(n)}$ is again most easily expressed in terms of $\bar{g}(t,g)$

$$\bar{C}_{i,k}^{(n)}\left(\frac{q^2}{\mu^2},g\right)=\sum_j\left\{T\exp\left[-\int_0^t\underline{\gamma}^{(n)}(\bar{g}(x,g))\,dx\right]\right\}_{ij}\times\bar{C}_{j,k}^{(n)}(1,\bar{g}(t,g)) , \tag{23}$$

where T refers to the fact that the exponential is to be t ordered. In an asymptotically free theory $\bar{g}^2$ vanishes for large t like $b_0^{-1}t^{-1}$. The anomalous-dimension matrix $\underline{\gamma}^{(n)}(\bar{g})$ similarly vanishes like $1/t$. We define

$$\underline{\gamma}^{(n)}(g)=\underline{\gamma}^{(n)}g^2+\underline{r}^{(n)}(g) , \tag{24}$$

where $\underline{r}^{(n)}(g)$ is of order g^4, for small g. Then the large-t behavior of $\bar{C}_{i,k}^{(n)}$ is given by

$$\bar{C}_{i,k}^{(n)}(q^2/\mu^2,g)\underset{t\to\infty}{\sim}\sum_j\left[\exp\left(\frac{-\ln t}{b_0}\underline{\gamma}^{(n)}\right)\underline{M}^{(n)}\right]_{ij}\times\bar{C}_{j,k}^{(n)}(1,0) , \tag{25}$$

where

$$\underline{M}^{(n)}=T\exp\left[-\int_0^\infty dt\exp\left(\frac{\ln t}{b_0}\underline{\gamma}^{(n)}\right)\underline{r}^{(n)}(\bar{g}(t,g))\exp\left(-\frac{\ln t}{b_0}\underline{\gamma}^{(n)}\right)\right] \tag{26}$$

Thus we see that the large-q^2 behavior of the Wilson coefficients is determined by the eigenvalues of $\underline{\gamma}^{(n)}$. These can be calculated by evaluating the wave-function renormalization matrix $\underline{Z}$ to second order in perturbation theory. The structure of Eq. (25) is more transparent if we write the matrix $\underline{\gamma}^{(n)}$ in terms of its eigenvalues $\gamma_i^{(n)}$

$$\underline{\gamma}^{(n)}=\sum_i\gamma_i^{(n)}\underline{P}^i , \tag{27}$$

where $\underline{P}^i$ are projection matrices

$$\underline{P}^i\underline{P}^j=\delta_{ij}\underline{I} ,\qquad \sum_i\underline{P}^i=\underline{I} . \tag{28}$$

Then we have

$$\bar{C}_{i,k}^{(n)}(q^2/\mu^2,g)\underset{-q^2\to\infty}{\sim}\sum_l[\ln(-q^2)]^{-a_l^{(n)}}\times\sum_j(\underline{P}^l\underline{M}^{(n)})_{ij}\,\bar{C}_{j,k}^{(n)}(1,0), \tag{29}$$

where

$$a_l^{(n)}=\frac{1}{2b_0}\gamma_l^{(n)} . \tag{30}$$

The operator-product expansion has the form (in momentum space)

$$J_aJ_b\approx\sum_l[\ln(-q^2)]^{-a_l^{(n)}}\sum_{i,j}O_i^{(n)}[\underline{P}^l\underline{M}^{(n)}]_{ij}\times\bar{C}_{j,k}^{(n)}(1,0) . \tag{31}$$

The dominant operator for large q^2 will thus be picked out by the smallest eigenvalue of $\underline{\gamma}^{(n)}$. The coefficient of this leading operator is not, in general, determined. Although the functions $C_{i,k}^{(n)}(1,0)$ are known (equal to their free-field-theory values) the matrix $\underline{M}^{(n)}$ is not. It depends on the (unknown) behavior of $\underline{\gamma}$ for large coupling constant. The only case in which the coefficient of the dominant operator is known is when the matrix $\underline{P}^l\underline{M}^{(n)}$ vanishes identically for all g. This will be the case if we are considering the coefficient of a conserved or partially conserved current or the energy-momentum tensor. In that case the (appropriately projected) matrix M equals unity and the coefficient of the current, or energy-momentum tensor is determined. In addition $a_l^{(n)}$ vanishes for these operators. These are the only operators for which one is interested in the numerical value of the coefficients since only these operators have known hadronic matrix elements. Also if a particular tensorial or SU(3) ×SU(3) structure holds for the operator-product expansion when $g=0$, it will also be valid when

$-q^2 \to \infty$. This is because the tensorial and SU(3) $\times$SU(3) structure is totally contained in the labels k of $\tilde{C}^{(n)}_{i,k}$ evaluated at $g=0$.

Consequently all current-algebra sum rules (Adler, Gross–Llewellyn Smith, etc.) and relations between moments of the structure functions derived in the parton or naive light-cone models will be true asymptotic theorems in our theories.

III. DEEP-INELASTIC SCATTERING IN ASYMPTOTICALLY FREE GAUGE THEORIES

We now supply the explicit results one obtains by applying the techniques of Sec. II to an asymptotically free gauge theory. In order to ensure sufficient generality we will work with the Lagrangian

$$L_0 = -\tfrac{1}{2}\,\mathrm{Tr}\,F_{\mu\nu}F^{\mu\nu} + \bar{\psi}(i\not{\nabla} - m)\psi\,, \tag{32}$$

where $\nabla_\mu = \partial_\mu + ig\sigma^a B^a_\mu$ is the covariant derivative, the σ^a being the matrices of the representation of the Lie algebra of the gauge group G in the space of fermions. We retain the normalization of Eqs. I(2.3)–(2.4) and omit for brevity the gauge-fixing and ghost terms needed to define (32) properly. For simplicity we assume that G is simple, but allow the fermions to transform according to an arbitrary representation of G.

The class of theories described by (32) includes the case where $G = \mathrm{SU}(3)$ and ψ transforms as a direct sum of three triplets. In this case we may interpret G as the gauge group of "color" and the group which mixes the three G-triplets as ordinary SU(3), so the fermions are just the usual "colored" quarks. As was discussed in paper I, there is some reason to hope that this theory provides a model of hadrons. We will adopt it in the following for illustrative purposes. If $m=0$ we also have ordinary chiral SU(3). Mass terms, and in general symmetry-breaking terms with operator dimension less than four, have no influence on the leading-order effects we discuss in most of this paper. In particular, all symmetries broken by such terms will be reinstated asymptotically, so that for instance all the operators in a given chiral SU(3)$\times$SU(3) multiplet have the same anomalous dimensions.

More complicated models, with charm quarks or a different gauge group (so long as it is non-Abelian), are clearly allowed. It might be interesting to see what the effects of Higgs scalars on the light-cone behavior of asymptotically free theories are (asymptotically free theories with scalar particles were constructed in paper I). One would be faced with some complications due to the presence of at least two dimensionless coupling constants and more involved mixings of lowest-twist operators. We have not carried out this study.

Throughout this paper we shall assume that the gauge group G commutes with the ordinary symmetry group H of the hadronic currents, e.g., chiral SU(3)$\times$SU(3). This excludes models of the Bars-Halpern-Yoshimura type.[15] There are good reasons for this assumption. The spin content of the fundamental charged constituents of the nucleon as measured by the Callan-Gross[16] sum rule points to spin-$\frac{1}{2}$ constituents. A "shielding" mechanism (like that contemplated in paper I) is impossible if one wishes to identify these particles with the observed vector or axial-vector octets, since the shielding mechanism yields only singlets of the gauge group for physical states. The alternative complicated system of Higgs scalars required in Ref. 15 almost certainly destroys asymptotic freedom.

We are interested in the light-cone behavior of the product of weak (by this we mean electromagnetic or truly weak) currents. The operators of twist two appearing in the operator-product expansion of two weak currents near the light cone will be

$$^nO^V_{\mu_1\cdots\mu_n} = i^{n-2}\,S\,\mathrm{Tr}\,F_{\mu_1\alpha}\nabla_{\mu_2}\cdots\nabla_{\mu_{n-1}}F^{\alpha}{}_{\mu_n} - \text{trace terms}, \tag{33}$$

$$^nO^{F\pm,0}_{\mu_1\cdots\mu_n} = \tfrac{1}{2}i^{n-1}\,S\,\bar{\psi}\gamma_{\mu_1}\nabla_{\mu_2}\cdots\nabla_{\mu_n}(1\pm\gamma_5)\psi - \text{trace terms}, \tag{34}$$

$$^nO^{F\pm,a}_{\mu_1\cdots\mu_n} = \tfrac{1}{2}i^{n-1}\,S\,\bar{\psi}\gamma_{\mu_1}\nabla_{\mu_2}\cdots\nabla_{\mu_n}(1\pm\gamma_5)\tfrac{1}{2}\lambda^a\psi - \text{trace terms}, \tag{35}$$

where recall that ∇_μ is the covariant derivative, which is $\partial_\mu + ig\sigma^a B^a_\mu$ acting on fermions and $\partial_\mu + igT^a B^a_\mu$ acting on vectors, while (34) and (35) represent the SU(3)$\times$SU(3) singlet and octet pieces of the fermion operators. The λ^a, $a=1,\ldots,8$, are the standard λ matrices of Gell-Mann. The product of two octet currents will in general also have decimet and 27-plet components; these correspond to higher-twist operators and are therefore suppressed. [The vertices associated with the operators (33)–(35) are given in Fig. 1.]

In addition to the operators $^nO^V$, $^nO^F$ (we often suppress in the following tensor and H indices) there are composite operators formed from Faddeev-Popov ghosts which may have the same twist and therefore are expected to mix with H singlet operators. Arguments to exclude the ghosts on the basis of gauge invariance are unconvincing because the ghosts do in fact mix with the above operators in off-shell matrix elements,

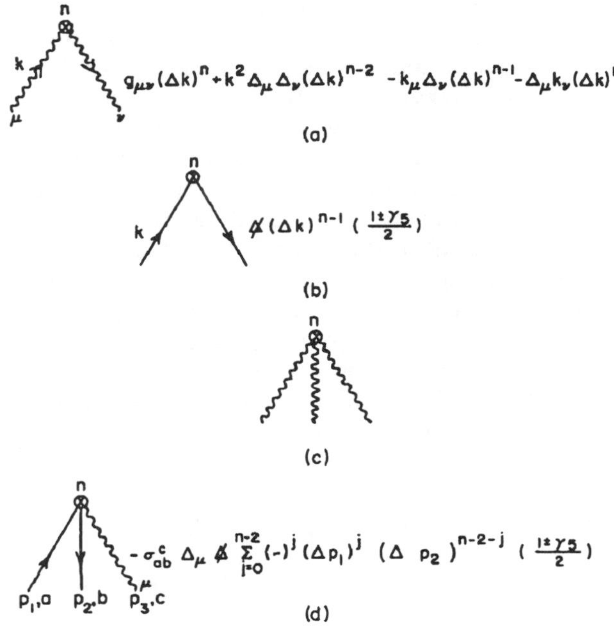

FIG. 1. (a) The vertex for ${}^nO^V$ (order 0). (b) The vertex for ${}^nO^F$ (order 0). (c) The vertex for ${}^nO^V$ (order g). (d) The vertex for ${}^nO^F$ (order g).

as is shown by explicit calculation.

It is argued in detail in Appendix A that one can solve this problem by going to a gauge in which no ghosts are present, so that the mixing does not appear. Moreover, the anomalous dimensions of the gauge-invariant operators ${}^nO^V$, ${}^nO^F$ and their mixing are gauge-independent, so they may be calculated in the standard Fermi-type gauges. In the remainder of the text we shall take the results of Appendix A for granted and ignore the ghost mixings.

According to the prescriptions of Sec. II the light-cone behavior of the coefficients in the Wilson expansion, and thereby the high-q^2 behavior of the moments of the structure functions, will be controlled by the renormalization group Eqs. (20). The only missing ingredient in these equations is the matrix $\underline{\gamma}$.

As was briefly mentioned in Sec. II, operators of the same structure (tensor and symmetry properties) and twist will mix, and only certain linear combinations of them will be multiplicatively renormalizable in the usual sense. This phenomenon appears already in Fig. 2(a), where we see that ${}^nO^V$, which in lowest order vanishes between fermion states, acquires (logarithmically divergent) contributions in higher order. In order to compute the $\underline{\gamma}$ matrix of Eq. (21) to second order in g we must according to Eq. (22) express the renormalized operators in terms of the bare ones.

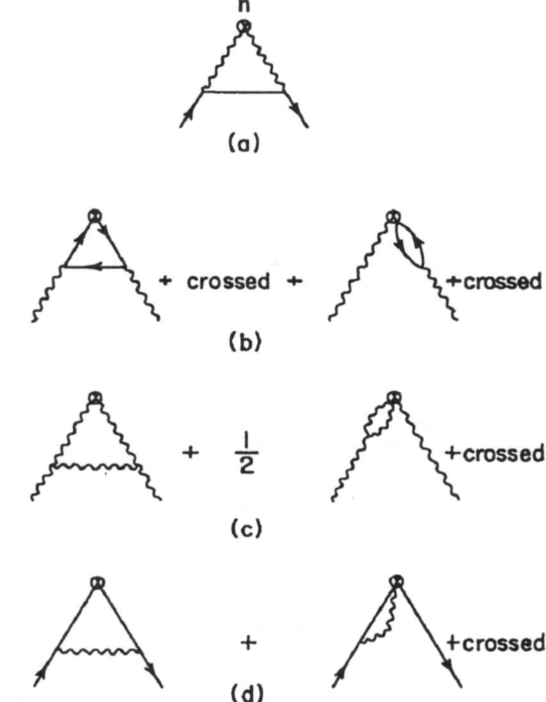

FIG. 2. Graphs for computing $\underline{\gamma}$ (radiative corrections to matrix elements). (a) The matrix element of ${}^nO^V$ between fermion states. (b) The matrix element of ${}^nO^F$ between vector states. (c) The matrix element of ${}^nO^V$ between vector states. (d) The matrix element of ${}^nO^F$ between fermion states.

This is most easily done by taking matrix elements between vector and fermion states, since the bare operators have trivial matrix elements. We are thus lead to computing the logarithmically divergent pieces of the matrix elements of ${}^nO^V$ and ${}^nO^F$ with the same tensor structure (twist) as ${}^nO^V_{\rm bare}$ and ${}^nO^F_{\rm bare}$, between these states. From this we determine the wave-function renormalization matrix and thereby $\underline{\gamma}$. The computations involve evaluating the graphs of Figs. 2(a)–2(d). Because we want to amputate the external propagators there will also be an order-g^2 contribution to the diagonal elements of $\underline{\gamma}$ due to the possibility of self-energy insertions on the propagators, as in Eq. I(5.18).

To summarize the results we introduce the matrix

$$ {}^n\gamma = \begin{pmatrix} {}^n\gamma^F_{FF} & {}^n\gamma^V_{FF} \\ {}^n\gamma^F_{VV} & {}^n\gamma^V_{VV} \end{pmatrix}, \tag{36}$$

where ${}^n\gamma^F_{VV}$ is given by $\mu(\partial/\partial\mu)\ln(Z)^F_{VV}$ and so forth.

The results of the computations are (for n = even, $n \geq 2$)

$$^n\gamma^V_{VV} = \frac{g^2}{8\pi^2}\left[C_2(G)\left(\frac{1}{3} - \frac{4}{n(n-1)} - \frac{4}{(n+1)(n+2)} + 4\sum_2^n \frac{1}{j}\right) + \frac{4}{3}\, T(R)\right], \quad (37)$$

$$^n\gamma^F_{FF} = \frac{g^2}{8\pi^2}\left[C_2(R)\left(1 - \frac{2}{n(n+1)} + 4\sum_2^n \frac{1}{j}\right)\right], \quad (38)$$

$$^n\gamma^F_{VV} = \frac{-g^2}{8\pi^2}\,\frac{4(n^2+n+2)}{n(n+1)(n+2)}\, T(R)\,, \quad (39)$$

$$^n\gamma^V_{FF} = \frac{-g^2}{8\pi^2}\,\frac{2(n^2+n+2)}{n(n^2-1)}\, C_2(R)\,, \quad (40)$$

where, as in paper I, $C_2(G)$ is the quadratic Casimir operator of G evaluated on the adjoint representation $C_2(R)$ is the evaluation of the quadratic Casimir operator of G on the irreducible representation R of G to which the fermions belong,[17] and $T(R)$ is the trace of the square of a matrix in the Lie algebra of the representation R. It should be remarked that the mixing terms (39) and (40) vanish for fermion operators which are not singlets under H (we have been suppressing H labels on fermion operators, they should always be understood). The remarkable feature of Eqs. (37)–(38), characteristic of gauge theories,[2] is the $\sum_2^n 1/j \sim \ln(n)$ term. It arises from the graphs of 2(c) and 2(d) with three lines coming out of the vertex, which in turn corresponds to the fact that A_μ occurs with ∂_μ in the covariant derivative.

These values of $^n\underline{\gamma}$ satisfy the constraints of positivity (the smallest eigenvalue of $^n\underline{\gamma}$ must increase with n), gauge-invariance, and nonrenormalization of the energy-momentum tensor, which provides a check on the calculation (and on the argument of Appendix A).

Having calculated the $\underline{\gamma}$ matrix we can now infer the properties of deep-inelastic scattering according to the methods outlined in Sec. II. First let us review the general features.

1. The approach to the asymptotic region is logarithmic. In other words the leading corrections to the asymptotic forms for the moments of the structure functions will be suppressed by powers of $\ln(-q^2)$. These corrections arise from two sources. First, the fact that the effective coupling $\bar{g}^2(t,g)$ vanishes logarithmically for large t ($\bar{g}^2 \sim 1/t$) means that the order-g^2 corrections to the Wilson coefficients will be suppressed by $1/\ln(-q^2)$ for large $-q^2$. It is hard to estimate how rapidly these corrections vanish in a realistic model, since this will depend on the unknown scale (μ), the actual value of the physical coupling constant, and the large-g^2 behavior of $\beta(g^2)$. The large-g^2 behavior of β is, of course, totally unknown. If, for example, $\beta(\bar{g}^2) = d\bar{g}^2/dt$ were to be linear in $\bar{g}^2$ for large $\bar{g}^2$ then the effective coupling would decrease rapidly (like a power of $-q^2$) from its physical value to small values of $\bar{g}^2$ and only then approach zero logarithmically. In that case one might understand the rapid onset of scaling.

Additional logarithmic corrections occur when more than one operator can contribute to a given moment. This occurs in our theories for the H singlet component of the structure functions. In that case, as explained in Sec. II, the leading asymptotic behavior of the moments will be given by the lowest eigenvalue of the $\underline{\gamma}$ matrix. However, the other eigenvectors of $\underline{\gamma}$ will also contribute terms, which will be suppressed by some (in general noninteger) power of $\ln(-q^2)$. Thus a generic structure function $F(x,q^2)$ will satisfy

$$\int_0^1 dx\, x^n F(x,q^2) \underset{q^2\to-\infty}{\sim} \sum_i (\ln - q^2)^{-a_i^n} \times F_i^n(\ln(-q^2)) \cdots\,, \quad (41)$$

where the a_i^n are proportional to the eigenvalues of $\underline{\gamma}$, and the F's approach constants (at an unknown rate) as $q^2 \to -\infty$.

2. Bjorken scaling is violated by finite powers of logarithms. These logarithmic violations are readily calculated in terms of the $\underline{\gamma}$ matrix evaluated in second-order perturbation theory. Here we must distinguish between the strong symmetry group singlet and nonsinglet pieces of the structure functions. The latter are easy to analyze since there is only one operator, that given in Eq. (35), that contributes. The relevant anomalous dimension is given in Eq. (38). It is, of course, common to all the SU(3)×SU(3) operators that appear in the operator-product expansion. Therefore if $F^{\rm NS}(x,q^2)$ stands for the nonsinglet piece of F_2, xF_1, or xF_3 (independent of the quantum numbers of the currents) we have that

$$\int_0^1 dx\, x^n F^{\rm NS}(x,Q^2) \underset{Q^2\to\infty}{\sim} C^{(n)}_{\rm NS}(\ln Q^2)^{-A^{\rm NS}_{n+2}}\,, \quad (42)$$

where

$$A_n^{\rm NS} = \frac{3C_2(R)}{22C_2(G) - 8T(R)}\left(1 - \frac{2}{n(n+1)} + 4\sum_{k=2}^n \frac{1}{k}\right) \quad (43)$$

and

$$Q_2 = -q^2\,.$$

In the "red, white, and blue" quark model [$H = {\rm SU}(3)$], we have

$$C_2(G) = 3,\quad C_2(R) = \tfrac{4}{3},\quad T(R) = \tfrac{3}{2}\,. \quad (44)$$

The numerical value of these coefficients is small, i.e., $A_2^{NS}=\frac{16}{81}$, $A_4^{NS}=0.36$, etc. An excellent interpolation formula, accurate to 1% already for $n=2$, is

$$A_n^{NS} \simeq 0.296 \ln(n) - 0.051 \ . \tag{45}$$

In treating the singlet piece of the structure functions we must take into account the mixing of the vector-meson and fermionic operators. This mixing occurs for H-singlet operators of both normal and abnormal parity. In the case of abnormal parity the fermion operator mixes with a vector-meson operator given by Eq. (33) with one $F_{\mu\nu}$ replaced by its dual. These operators contribute to the parity-violating structure function F_3. The appropriate $\underline{\gamma}$ matrix in this case will be treated in a subsequent publication. Here we will only deal with H-singlet normal parity operators. Consequently we must diagonalize the γ matrix evaluated above. For $n=2$ this matrix has the form

$$\underline{\gamma}^{(2)} = \frac{g^2}{8\pi^2}\begin{pmatrix} \frac{8}{3}C_2(R) & -\frac{8}{3}C_2(R) \\ \frac{4}{3}T(R) & \frac{4}{3}T(R)\end{pmatrix} . \tag{46}$$

Thus the smallest eigenvalue is zero, so that the corresponding moments will in fact scale. If $F^S(x,q^2)$ stands for the singlet pieces of F_2 or xF_1, then

$$\int_0^1 dx\, x^n F^S(x,Q^2) \underset{Q^2\to\infty}{\sim} C_S^{(n)} (\ln Q^2)^{-A_{n+2}^S} , \tag{47}$$

where

$$A_n^S = \frac{1}{2g^2 b_0}\{{}^n\gamma_{FF}^F + {}^n\gamma_{VV}^V - [({}^n\gamma_{FF}^F - {}^n\gamma_{VV}^V)^2 + 4\,{}^n\gamma_{VV}^{Fn}\gamma_{FF}^V]^{1/2}\} . \tag{48}$$

Since the off-diagonal matrix elements of $\underline{\gamma}$ vanish rapidly (like $1/n$), we have to a very good approximation (1% for $n=4$) that

$$A_2^S = 0 ,$$
$$A_n^S = A_n^{NS} - O\left(\frac{1}{n^2 \ln n}\right), \quad n>2 . \tag{49}$$

The fact that $A_n^S < A_n^{NS}$ is a simple consequence of positivity; however, we note that the difference between these coefficients vanishes rapidly as n increases. Already for $n=4$ it is less than 1%, so that one can hardly differentiate between the single and nonsinglet parts of the structure functions (except of course in the behavior of the first moment).

In asymptotically free gauge theories there are then two stages in the approach to the asymptotic region. In the first stage, when we have reached what might be called the approximate scaling region, the effective coupling constant becomes small and the one-loop approximation to the renormalization-group equation becomes valid. In this region we have scaling up to finite powers of logarithms as previously computed. The rate of approach to this region is not determined by the methods of this paper. In the second stage, which might be called the true asymptotic region, quantities suppressed by powers of logarithms become effectively zero, and the sum rules (energy-momentum and parton-model sum rules) mentioned below become valid.

3. Sum rules and relations between moments of the structure functions which follow from the tensorial and SU(3)×SU(3) structure of the free-quark model are true asymptotic theorems in our theories [if H = SU(3)]. This transpires because we have chosen the strong gauge group G to commute with H [say SU(3)]. Therefore the vector mesons are neutral with respect to the SU(3)×SU(3) charges, and the coefficients $C_V^{(n)}$ vanish when $g=0$. Thus the tensorial and SU(3)×SU(3) structure of the Wilson expansion for large $-q^2$ will be identical to that of free-field theory.

The Adler sum rule[18] is of course valid for all q^2. The Gross–Llewellyn Smith sum rule[19] holds, since for this moment the appropriate anomalous dimension vanishes for all g. It, however, is approached logarithmically, i.e.,

$$\int_0^1 dx[F_3^{\nu p}(x,q^2) + F_3^{\nu n}(x,q^2)] = -6F(q^2) ,$$
$$F(q^2) \underset{q^2\to-\infty}{\sim} 1 + O\left(\frac{1}{\ln(-q^2)}\right) . \tag{50}$$

The Llewellyn Smith relation[20] holds for individual moments

$$\int_0^1 dx\, x^n \{6[F_2^{ep}(x,q^2) - F_2^{en}(x,q^2)] - x[F_3^{\nu p}(x,q^2) - F_3^{\nu n}(x,q^2)]\} \approx O\left(\frac{1}{\ln(-q^2)}\right), \tag{51}$$

as well as the various inequalities between *moments* of the structure functions discussed by Llewellyn Smith[20] and Nachtmann.[21]

All these are relations or sum rules for the *moments* of the structure functions. As will be shown below it does not necessarily follow that the relations are valid for the structure functions themselves. Thus the quantity in parentheses in Eq. (51) does not necessarily vanish for a given x like $1/\ln(-q^2)$ as $q^2 \to -\infty$.

Similarly the Callan-Gross relation holds for moments of the structure function. In other words the ratio of the moments of $F_L = F_2 - 2xF_1$ and F_2 vanishes logarithmically for large $-q^2$:

$$\frac{\int_0^1 dx\, x^n F_L(x, q^2)}{\int_0^1 dx\, x^n F_2(x, q^2)} \underset{q^2\to-\infty}{\sim} O\left(\frac{1}{\ln(-q^2)}\right) . \qquad (52)$$

However this does not necessarily imply that $F_L(x, q^2)/F_2(x, q^2)$ vanishes like $1/\ln(-q^2)$ for fixed x.

Finally we can derive, in our models, a new sum rule which is related to the matrix element of the energy-momentum tensor.[22] The singlet piece of the Wilson expansion will contain the energy-momentum tensor $\theta_{\mu\nu}$ with vanishing anomalous dimension. According to the discussion in Sec. II this will appear, for large $-q^2$, in the form

$$(O_F^{(2)}, O_V^{(2)}) \underline{P}^{(2)} \begin{pmatrix} C_F(1,0) \\ 0 \end{pmatrix} , \qquad (53)$$

where we have set $\underline{P}^{(2)} \underline{M}^{(2)} = \underline{P}^{(2)}$ (since the anomalous dimension of the energy-momentum tensor vanishes), C_F is the coefficient of the fermionic part of $\theta_{\mu\nu}$ $(O_F^{(2)})$, and the coefficient of the vector contribution to $\theta_{\mu\nu}$ $(O_V^{(2)})$ vanishes when $\bar{g}=0$. The projection matrix $\underline{P}^{(2)}$ is given from Eq. (46) as

$$\underline{P}^{(2)} = \frac{1}{2C_2(R)+T(R)} \begin{pmatrix} T(R) & 2C_2(R) \\ T(R) & 2C_2(R) \end{pmatrix} \qquad (54)$$

so that the coefficient of $\theta_{\mu\nu} = \theta_F^{(2)} + \theta_V^{(2)}$ is given, for large $-q^2$, by

$$\frac{T(R)}{2C_2(R)+T(R)} C_F(1,0) . \qquad (55)$$

The net effect of the mixing of operators is to multiply the free-field theory value by r, where

$$r = \frac{T(R)}{2C_2(R)+T(R)}$$

$$= \tfrac{9}{25} \text{ for the "red, white, and blue" quark model} . \qquad (56)$$

The contribution of the energy-momentum tensor to the commutator of two vector (or axial-vector) currents is thus (for a quark model)

$$[J_\mu^a(x), J_\nu^b(-x)] = (\tfrac{1}{3}\operatorname{Tr}\lambda^a\lambda^b)\frac{-i}{2\pi}\,\delta'(x^2) r \times [x^\lambda \theta_{\mu\lambda}(0) x_\nu + x^\lambda \theta_{\nu\lambda} x_\mu - g_{\mu\nu} x^\alpha x^\lambda \theta_{\alpha\lambda}] . \qquad (57)$$

As a consequence the singlet piece of F_2 for electroproduction will satisfy

$$\int_0^1 dx\, F^e_{2,\text{singlet}}(x, q^2) \underset{q^2\to-\infty}{\longrightarrow} \langle Q^2 \rangle r , \qquad (58)$$

where $\langle Q^2 \rangle$ is the average quark charge. In the "red, white, and blue" quark model this has the value $\frac{2}{25}$. This sum rule holds for any hadronic target. However, the corrections to it, arising from the nonsinglet operators, might be very large. They vanish for infinite q^2 but at a rate governed by $A_2^{NS} = 0.2$ (in the "red, white, and blue" quark model).

Similarly in the case of neutrino or antineutrino the singlet contribution to $F_2^{(\nu,\bar{\nu})}$ satisfies (setting the Cabibbo angle equal to zero) in the quark model

$$\int_0^1 dx\, F^{(\nu,\bar{\nu})}_{2,\text{singlet}}(x, q^2) \underset{q^2\to-\infty}{\longrightarrow} \tfrac{4}{3} r . \qquad (59)$$

To improve the convergence one can take linear combinations of structure functions to obtain a pure SU(3) singlet:

$$\int_0^1 dx [6(F_2^{ep} + F_2^{en}) - (F_2^{\nu p} + F_2^{\nu n})] \underset{q^2\to-\infty}{\longrightarrow} \tfrac{4}{3} r . \qquad (60)$$

The value of this sum rule for "red, white, and blue" quarks is 0.48, whereas experimentally one has 0.72 ± 0.28.

We have determined the large-$(-q^2)$ behavior of the moments of the structure functions. What can one say about the q^2 behavior of the functions themselves? It is useful to construct an example of a function with roughly the correct anomalous dimensions. Consider the nonsinglet piece of $F_2^{NS}(x, q^2)$ for electroproduction. It satisfies Eq. (42). If the constants $C_{NS}^{(n)}$ were known then one could construct $F_2^{NS}(x, q^2)$. Since these constants are not known there exist many functions which satisfy Eq. (42). Let us approximate A_n^{NS} by its asymptotic form, $A_n^{NS} = \alpha \ln(n) - \beta$. For the "red, white, and blue" quark model $\alpha = 0.296$, $\beta = 0.051$. Then we have to find a function $F^{NS}(x, q^2)$ which satisfies

$$\lim_{q^2\to-\infty} \int_0^1 dx\, x^n F(x, q^2) = C^{(n)} e^{+\beta L} (n+2)^{-\alpha L},$$

$$L = \ln\ln(-q^2) . \qquad (61)$$

In addition we shall impose Regge behavior, in the sense that for fixed q^2 we demand that $F(x, q^2)$ approach a constant as $x \to 0$.

A solution of this is provided by

$$F(x, q^2) = \frac{e^{\beta L}}{\Gamma(\alpha L + 1)} \int_c^{c-\ln x} dy\, e^{-y} y^{\alpha L} , \qquad (62)$$

where c is an arbitrary constant. This solves Eq. (61) with

$$C^{(n)} = \frac{e^{c(n+1)}}{(n+1)(n+2)} .$$

One can easily construct additional solutions by multiplying Eq. (62) by a polynomial in x. This function has Regge behavior, i.e.,

$$F(0, q^2) = \frac{e^{BL}}{\Gamma(\alpha L+1)} \int_c^\infty dy\, e^{-y} y^{\alpha L} \underset{q^2 \to -\infty}{\sim} [\ln(-q^2)]^B\,. \quad (63)$$

At fixed $x \neq 0, 1$ it behaves, for large $-q^2$, like

$$F(x, q^2) \approx \left(\frac{(c-\ln x)}{\alpha L}\right)^{\alpha L} e^{BL}\,. \quad (64)$$

Accordingly $F(x, q^2)$ will (a) for any $x \neq 0, 1$, vanish faster than any power of $\ln(-q^2)$ for large enough $-q^2$ (in fact $F(x, q^2) < [\ln(-q^2)]^{-\ln\ln\ln(-q^2)}$), and will (b) increase for intermediate values of $-q^2$ for sufficiently small x and decrease for x close to unity. The transition point, at which F does not change as $-q^2$ increases, is roughly $x \approx 1/[\ln(-q^2)]^p$, where p is some positive constant. (c) Finally, $F(x, q^2)$ will show largest deviations from scaling in the vicinity of $x=1$.

Such behavior is more general than this particular example. Specifically (a) is a simple consequence of positivity,

$$\begin{aligned} F(x, q^2) &\simeq \frac{1}{2\epsilon}\int_{x-\epsilon}^{x+\epsilon} dy\, F(y, q^2) \\ &\leq \frac{1}{2\epsilon(x-\epsilon)^n}\int_0^1 dy\, y^n F(y, q^2) \\ &\underset{q^2 \to -\infty}{\sim} F_n[\ln(-q^2)]^{-A_n}\,, \end{aligned} \quad (65)$$

and the fact that $A_n \to \ln n$ as $n \to \infty$.

This example also illustrates the rather small variation in q^2 of the asymptotic form of the structure functions. To test for deviations from scaling one will require large variations of q^2 and measurements in the vicinity of $x=1$. One might expect a 50% variation for $x \approx 0.9$ as $-q^2$ increases from 10 to 50 BeV^2.

This example further illustrates that two different solutions of Eq. (61) might have quite different q^2 behavior for a given value of x.[23] Consider the solution F' to Eq. (61) which is given by Eq. (62) with c replaced by $c' < c$. Then, for large $-q^2$,

$$\frac{F(x, q^2)}{F(x, q^2)} \approx [\ln(-q^2)]^{\ln(c/c')} \geq \left(\frac{c-nx}{c'-nx}\right)^L \geq 1\,.$$

By making the ratio c/c' large we can arrange for F/F' to increase like a large power of $\ln(-q^2)$ for x near unity.

As a consequence the parton-model relations for moments of the structure functions, for example Eq. (51), do not imply that these relations are satisfied for the structure functions themselves. Also the fact that the moments of the longitudinal structure functions decrease like $1/\ln(-q^2)$ relative to the moments of the transverse structure function [Eq. (52)] does not imply that $R(q^2, x) = F_L(q^2, x)/F_2(q^2, x)$ decreases logarithmically for all x. Indeed it might very well increase for x close to unity. Without additional input the only reliable predictions one can make are with respect to the q^2 behavior of the moments.

IV. CONCLUSIONS

A crucial test of asymptotically free gauge theories of the strong interactions is the verification of the q^2 behavior of the moments of the structure functions. In the best of all worlds one could confront an infinite number of these moments with the predicted asymptotic forms (which are determined solely by the gauge group and the fermion representation). An additional test is provided by the energy-momentum sum rule derived above.

In reality, of course, it will be very hard to determine the q^2 behavior of the moments. More than likely the most practical place to look for violations of Bjorken scaling is in the vicinity of threshold ($x \approx 1$). There we expect to see the structure functions decreasing like ever-increasing powers of $\ln(-q^2)$.

Whether the picture described in the paper is consistent with experiment is an open question. The fact that scaling appears to have set in already for rather small values of q^2 is neither explained nor contradicted by our theories. The rate of approach to asymptopia must be determined by nonperturbative methods. This problem, as well as the understanding of the low-energy and on-mass-shell behavior of the theory, requires the development of new theoretical techniques.

APPENDIX A: PROBLEM OF GHOST MIXING

One is tempted to argue that only manifestly gauge-invariant operators appear in the Wilson expansion of the product of two gauge-invariant operators. However, in the case of composite operators constructed from Faddeev-Popov ghost fields it is not clear how to formulate the criterion of gauge invariance. It does not seem that the ghost fields have any simple transformation properties under gauge transformations (in particular, one can choose gauges in which the ghosts are entirely absent). Moreover we will show in a simple example that operators constructed from ghost fields "mix" in a highly nontrivial way with the manifestly gauge-invariant operators of lowest twist which control the Bjorken limit, in the sense that the γ matrix appearing in the renormalization-group equations for these operators has nonvanishing off-diagonal entries. This appendix is devoted first to showing that a nontrivial problem is involved here, and second to describing that argument we used to convince ourselves that the calculations given in the main test do indeed give the correct asymptotic behavior for products of currents in the Bjorken limit. We argue that the anomalous dimensions (renormalization constants) of manifestly gauge-invariant operators are independent of the gauge in which they are computed, so since there exist gauges in which no Faddeev-Popov ghosts are present, the correct values are obtained in gauges with ghosts by ignoring the ghost mixing.

An instructive example of a gauge field theory is quantum electrodynamics with the unusual gauge choice[24]

$$\partial_\mu A^\mu - \tfrac{1}{2}\alpha g A_\mu A^\mu = 0\,, \tag{A1}$$

where g is the coupling constant and α is a gauge parameter. The Feynman rules for this theory are given in Ref. 24. Notice the presence of Faddeev-Popov ghosts.

For simplicity we consider the theory without fermions. In this case we are of course just dealing with free-field theory. The manifestly gauge-invariant operators of lowest twist are

$${}^V O^{(n)}_{\mu_1\cdots\mu_n} = S F_{\mu_1\alpha}\bar\partial_{\mu_2}\cdots\bar\partial_{\mu_{n-1}} F_{\mu_n}{}^{\alpha}\,, \tag{A2}$$

where S denotes symmetrization with respect to the μ's. To compute γ_V, the anomalous dimension of the vector field, to order g^2 we need only consider the logarithmically divergent parts of the two self-energy graphs with vector and ghost loops. The result is[25]

$$\gamma_V = 0\,. \tag{A3}$$

We can now compute ${}^n\gamma^V_{VV}$, the diagonal term in the γ matrix for the operator ${}^V O^{(n)}$, from the radiative corrections to the vertex shown in Fig. 3(a). The result is

$$\gamma^V_{VV} = 0\,. \tag{A4}$$

To see whether these operators mix with the ghost operators we need only check that the anomalous dimension for ${}^V O^{(n)}$, sandwiched between ghost states γ^V_{GG}, does not vanish. A calculation exactly like the one just described for the vectors involving the diagrams of Fig. 4(a) gives

$$\gamma_G = \frac{-\alpha^2 g^2}{16\pi^2}\,,\quad \gamma^V_{GG} = \frac{-1}{n(n-1)}\,\frac{\alpha^2 g^2}{16\pi^2}\,. \tag{A5}$$

Notice for $\alpha = 0$ we are in ordinary Landau gauge, and of course no ghost mixing can occur (there are no ghosts).

This shows there is an operator that mixes with ${}^V O^{(n)}$ which to lowest order (g^0) is given by (ρ is the ghost field)

$${}^G O^{(n)}_{\mu_1\cdots\mu_n}(x) = S\rho^*(x)\bar\partial_{\mu_1}\cdots\bar\partial_{\mu_n}\rho(x) + O(g)\,. \tag{A6}$$

In addition there might be order-g terms in which one of the derivatives is replaced by gA_μ, arising from the diagrams in Fig. 4(b).

To complete the picture we notice that γ^G_{VV}, the anomalous dimension of the ghost operator sandwiched between vector states, does not vanish. This is readily seen by noticing that only Fig. 5(a) can give a term of the structure $g_{\mu\nu}\ln\Lambda^2$ (notations are as in Sec. III). So we know that if we failed to consider other operators the γ matrix

$${}^n\gamma = \begin{pmatrix} {}^n\gamma^V_{VV} & {}^n\gamma^G_{VV} \\ {}^n\gamma^V_{GG} & {}^n\gamma^G_{GG} \end{pmatrix} \tag{A7}$$

could not have zero eigenvalues (${}^n\gamma^V_{VV}=0$, ${}^n\gamma^G_{VV}\neq 0$, ${}^n\gamma^V_{GG}\neq 0$, so $\det\gamma\neq 0$). However, we know that we are dealing with free-field theory, so there must be an operator with zero anomalous dimensions. Evidently there are even more operators which mix, and there is a complicated cancellation mechanism among many operators which are not individually gauge-invariant.

To verify that the same phenomenon occurs in

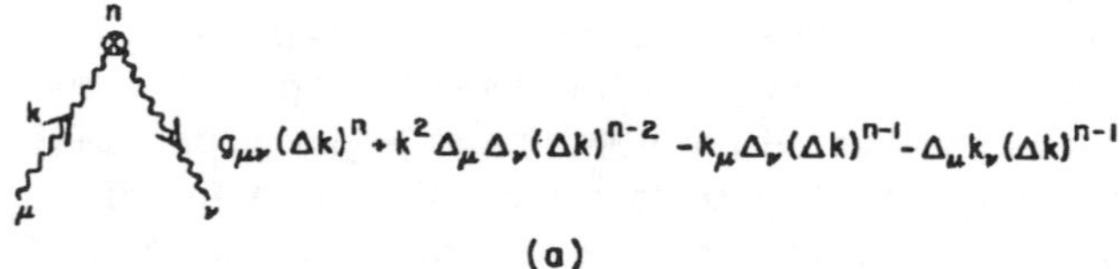

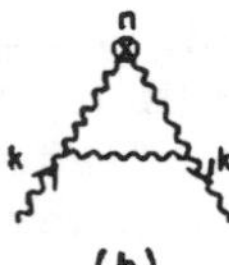

FIG. 3. Graphs for computing γ^V_{VV} in quantum electrodynamics in the gauge of Eq. (A1). (a) The vertex for ${}^V O^{(n)}$. (b) The radiative correction to the matrix element of ${}^V O^{(n)}$ between vector states.

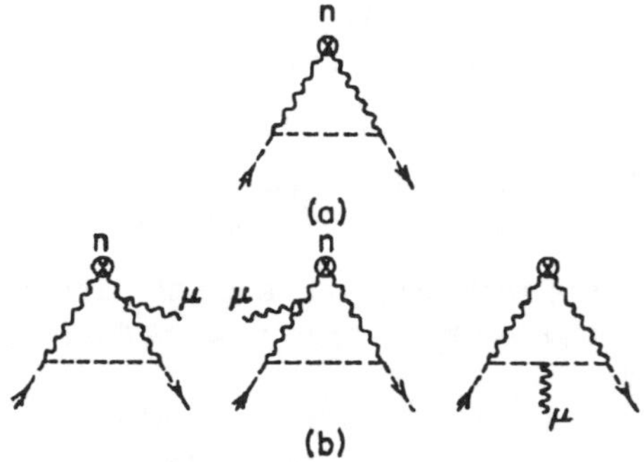

FIG. 4. Mixing of vector and ghost operators. (a) The zero-order contributions to ${}^G O^{(n)}$. (b) The order-g contributions to ${}^G O^{(n)}$.

Yang-Mills theory we will work in a gauge where there are no Faddeev-Popov ghosts.[26] Our gauge is specified by

$$n^\mu A_\mu^a = 0\,, \tag{A8}$$

where n is a fixed vector and a is the group index. (This gauge condition is not Lorentz-invariant.) The absence of Faddeev-Popov ghosts is a consequence of the fact that, under a gauge transformation

$$\delta A_\mu(x) = \partial_\mu \chi(x) - g[\chi(x), A_\mu(x)],$$

$\delta(n^\mu A_\mu^a)$ is independent of A for fields satisfying (A8).

In order to implement gauge conditions like (A8) we add a term $-(1/2\alpha)(n^\mu A_\mu)^2$ to the Lagrangian. This leads to the standard Feynman rules for Yang-Mills theories but without the ghosts and with the propagator modified to be

$$D_{ab}^{\mu\nu}(p) = \frac{i\delta_{ab}}{p^2}\left(-g^{\mu\nu} + \frac{\alpha p^2 - n^2}{(n\cdot p)^2}p^\mu p^\nu + \frac{n^\mu p^\nu + p^\mu n^\nu}{n\cdot p}\right). \tag{A9}$$

The theory in this gauge is not renormalizable by power counting if $\alpha \neq 0$. From now on we will always assume that $\alpha = 0$ and simplify further by taking n^μ lightlike, $n^2 = 0$. Then the propagator is

$$D_{ab}^{\mu\nu}(p) = \frac{i\delta_{ab}}{p^2}\left(-g_{\mu\nu} + \frac{n_\mu p_\nu + p_\mu n_\nu}{n\cdot p}\right). \tag{A10}$$

The calculation of γ_V and ${}^n\gamma_{VV}^V$ is straightforward, following the pattern of Sec. III. The same diagrams [Fig. 2(c)] are involved, except that there is no ghost-loop contribution to the self-energy. The results are

$$\gamma_V = -(g^2/8\pi^2)C_2(G)\tfrac{11}{6}, \tag{A11}$$

$${}^n\gamma_{VV}^V = (g^2/8\pi^2)C_2(G)\times\left(\frac{1}{3} - \frac{4}{n(n-1)} - \frac{4}{(n+1)(n+2)} + 4\sum_2^n \frac{1}{j}\right). \tag{A12}$$

In the calculations we have taken $nk = n\Delta = 0$ (k is the *external* momentum). It turns out that the dangerous-looking denominators in (A10) all cancel from internal propagators, at least in our calculation of logarithmically divergent pieces.

Although γ_V is certainly gauge-dependent, the ${}^n\gamma_{VV}^V$ are not.[27] In fact the ${}^n\gamma_{VV}^V$ have the same values in the ghost-free gauges as in the Fermi-type gauges. In the ghost-free gauges the operator-product expansion for gauge-invariant operators takes a simple form. In particular there appear no ghost operators. The anomalous dimensions of

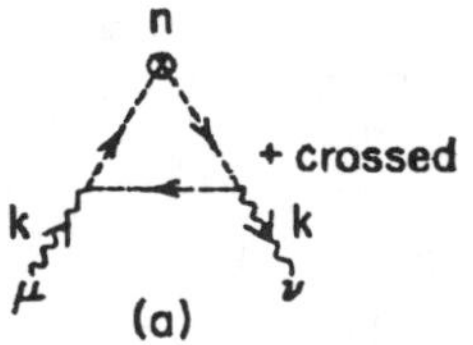

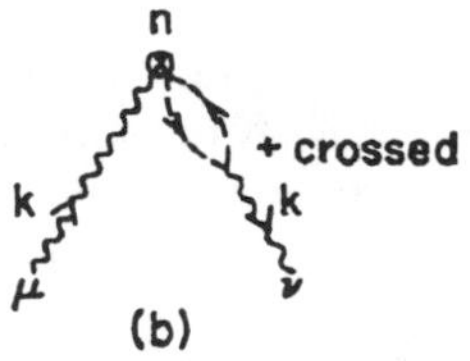

FIG. 5. Contributions to ${}^n\gamma_{VV}^G$. The coefficients of $g_{\mu\nu}$ arise from diagram (a).

the operators appearing in it can, however, be computed in any convenient gauge. This demonstrates why the calculations given in the text are relevant for determining asymptotic behavior in the Bjorken limit, despite the apparent problem of ghost mixing. The same apparent problem arises even in quantum electrodynamics in the unusual gauge (A1), but here, of course, we *know* the correct answer, and it is correctly given by the arguments of this appendix.

APPENDIX B: SAMPLE CALCULATION, ${}^n\gamma_{FF}^F$

In this appendix we will give details of the calculation of ${}^n\gamma_{FF}^F$. The techniques used here should enable the reader to duplicate without too much difficulty most of the numerical results in this paper. The calculation will also clarify the origin of the $\sum_2^n 1/j$ terms in the anomalous dimensions which are characteristic of gauge theories and give pronounced violations of canonical scaling as $n\to\infty$.

The key tool in these calculations is the angular average integral

$$\frac{1}{2\pi^2}\int d\Omega\, k_{\mu_1}\cdots k_{\mu_n} = \frac{2}{(n+2)!!}\sum_{\text{pairs}} g_{\mu_{i(1)}\mu_{i(2)}}\cdots g_{\mu_{i(n-1)}\mu_{i(n)}}, \tag{B1}$$

where on the left-hand side we have the average over the unit sphere in Euclidean 4-space, while on the right-hand side the sum runs over all possible ways of grouping the μ's into pairs. This formula holds for even n; for odd n the left-hand side vanishes. The formula is readily proved by induction.

Let us evaluate the first graph of Fig. 2(d). (Δ is an arbitrary vector which is contracted into all free indices of the fermion tensor.) In the Feynman gauge it is, after some simple Dirac algebra,

$$ig^2C_2(R)\int\frac{d^4p}{(2\pi)^4}\frac{1}{(p^2)^2(p-k)^2}$$
$$\times[2p^2\not\Delta-4(p\cdot\Delta)\not\Delta](\Delta\cdot p)^{n-1}. \qquad \text{(B2)}$$

We are interested in the (logarithmically divergent) coefficient of $(\Delta\cdot k)^{n-1}\not\Delta$ in the expansion of this integral. Throwing away other terms, we write symbolically

$$c_1(\Delta\cdot k)^{n-1}\not\Delta\ln\Lambda^2=ig^2C_2(R)$$
$$\times\int\frac{d^4p}{(2\pi)^4}\frac{2p^2\not\Delta-4(p\cdot\Delta)\not p}{(p^2)^2(p-k)^2}(\Delta\cdot p)^{n-1}, \qquad \text{(B3)}$$

where Λ is an ultraviolet cutoff. Differentiating (B3) on both sides $n-1$ times with respect to k and keeping only logarithmically divergent parts gives

$$c_1(n-1)!\Delta_{\epsilon_1}\cdots\Delta_{\epsilon_{n-1}}\not\Delta\ln\Lambda^2=ig^2C_2(R)\int\frac{d^4p}{(2\pi)^4}\frac{2^{n-1}(n-1)!}{(p^2)^{n+2}}p_{\epsilon_1}\cdots p_{\epsilon_{n-1}}(\Delta\cdot p)^{n-1}[2p^2\not\Delta-4(p\cdot\Delta)\not p]. \qquad \text{(B4)}$$

Now rotating into Euclidean space and using (B1) gives

$$c_1=\frac{-g^2}{16\pi^2}C_2(R)\left(\frac{2}{n(n+1)}\right). \qquad \text{(B5)}$$

The second graph of Fig. 2(d) gives

$$ig^2C_2(R)\int\frac{d^4p}{(2\pi)^4}\frac{(2p\cdot\Delta)\not\Delta}{p^2(p-k)^2}\sum_{j=0}^{n-2}(\Delta\cdot p)^j(\Delta\cdot k)^{n-2-j}. \qquad \text{(B6)}$$

The reader who has worked through the previous example carefully should be able to do this integral "by inspection"; the answer is

$$c_2=\frac{g^2}{16\pi^2}C_2(R)2\sum_{j=2}^{n}\frac{1}{j} \qquad \text{(B7)}$$

with the obvious definition of c_2.

The third graph gives the same number as the second.

Putting these results together gives

$$^n\gamma^F_{FF}-2\gamma_F=\frac{g^2}{8\pi^2}\left(-\frac{2}{n(n+1)}+4\sum_2^n\frac{1}{j}\right)$$

and with our old result γ_F [Eq. I (4.19)] we get the announced result for $^n\gamma^F_{FF}$.

In this calculation it was clear that the origin of the terms in $^n\gamma$ growing logarithmically with n arose from the replacement of ordinary by covariant derivitives. In terms of the "hand-waving argument" of Ref. 2, it appears that covariant derivatives do not "separate" in space the fields they are sandwiched between.

The method outlined here is a very powerful one for calculations of anomalous dimensions at the one-loop level. Other methods have also been proposed for this purpose.[28] It is doubtful that any method of comparable simplicity exists for higher-order calculations.

*Research supported by the U. S. Air Force Office of Scientific Research under Contract No. F-44620-71-C-0108 and by the National Science Foundation under Grant No. GP-30738X.

†Alfred P. Sloan Research Fellow.

[1]D. J. Gross and F. Wilczek, Phys. Rev. D 8, 3633 (1973).

[2]C. G. Callan and D. J. Gross, Phys. Rev. D 8, 4383 (1973).

[3]S. Coleman and D. J. Gross, Phys. Rev. Lett. 31, 851 (1973).

[4]D. J. Gross and F. Wilczek, Phys. Rev. Lett. 30, 1343 (1973).

[5]H. D. Politzer, Phys. Rev. Lett. 30, 1346 (1973).

[6]We have been informed by K. Symanzik that the fact that β is negative for a pure Yang-Mills theory was known to G. 't Hooft, but not published by him.

[7]K. Symanzik, Commun. Math. Phys. 23, 49 (1971).

[8]C. G. Callan, Phys. Rev. D 5, 3202 (1972).

[9]N. Christ, B. Hasslacher, and A. Mueller, Phys. Rev. D 6, 3543 (1972).

[10]R. P. Feynman, Acta. Phys. Polon. 26, 697 (1963).

[11]L. D. Faddeev and V. N. Popov, Phys. Lett. 25B, 29 (1967); V. N. Popov and L. D. Faddeev, Kiev ITP Report 67-36, translation NAL-THY-57, 1973 (unpublished).

[12]After completion of this work we received a preprint by H. Georgi and H. D. Politzer [Phys. Rev. D 9, 416 (1974)] in which electroproduction is discussed in an asymptotically free theory.

[13]R. Jackiw, R. Van Royen, and G. B. West, Phys. Rev. D 2, 2473 (1970); H. Leutwyler and J. Stern, Nucl. Phys. B20, 77 (1970); D. J. Gross, Phys. Rev. D 4, 1059 (1971).

[14]D. J. Gross and S. B. Treiman, Phys. Rev. D 4, 1059 (1971).

[15]I. Bars, M. B. Halpern, and M. Yoshimura, Phys. Rev.

D 7, 1233 (1973).

[16]C. G. Callan and D. J. Gross, Phys. Rev. Lett. 22, 156 (1969).

[17]If the representation R_F if G for the fermion fields is reducible, the situation is a little more complicated. Let us decompose R_F into irreducible representations R_i, $i=1,\ldots,N$ which act on the G multiplets ψ_i (for the moment we assume the ψ_i are H singlets). The fermion operators appearing in the operator-product expansion will be of the form $S\bar\psi_i\gamma_{\mu_1}\nabla_{\mu_2}\cdots\nabla_{\mu_n}\psi_i$ (i *not* summed). The matrix of anomalous dimensions $^n\gamma$ then becomes an $(N+1)\times(N+1)$ matrix. This matrix has a simple structure. There is no mixing among the fermion fields for different i, so there is a diagonal $N\times N$ submatrix of $^n\gamma$. The entries of this matrix for each i are given by (3.6) with $R=R_i$. The mixing terms between fermions and vectors, (3.7) and (3.8), are also to be taken with $R=R_i$. On the other hand in the $^n\gamma_{VV}^F$ term (3.5) we must take $R=R_F$ [note $T(R_F)=\sum_i T(R_i)$]. Finally the H-nonsinglet operators in the operator-product expansion do not mix with the vector operator.

[18]S. Adler, Phys. Rev. 143, 144 (1966).

[19]D. J. Gross and C. Llewellyn Smith, Nucl. Phys. B14, 377 (1969).

[20]C. H. Llewellyn Smith, Nucl. Phys. B17, 277 (1970).

[21]O. Nachtmann, J. Phys. (Paris) 32, 97 (1971); Nucl. Phys. B38, 397 (1972); Phys. Rev. D 5, 686 (1972).

[22]Various forms of an energy-momentum–tensor sum rule have appeared in the literature. They include a sum rule valid in the Sugawara model [C. G. Callan and D. J. Gross, Phys. Rev. Lett. 21, 3081 (1968)], a sum rule derived by G. Mack [Nucl. Phys. B35, 592 (1971)] in which the coefficient of $\theta_{\mu\nu}$ is not determined, and a sum rule derived in parton or light-cone models in which the fraction of energy carried by neutral gluons is unknown [C. H. Llewellyn Smith, Phys. Rep. 3C, No. 5 (1972); H. Fritzsch and M. Gell-Mann, in *Broken Scale Invariance and the Light Cone*, 1971 Coral Gables Conference on Fundamental Interactions at High Energy, edited by M. Dal Cin, G. J. Iverson, and A. Perlmutter (Gordon and Breach, New York, 1971), Vol. 2, p. 1].

[23]We would like to thank Sam Treiman for raising this point, and providing an example which illustrates this phenomena.

[24]G. 't Hooft and M. Veltman, Nucl. Phys. B50, 318 (1973).

[25]In calculating γ_V we consider only the coefficients of $g_{\mu\nu}$ in the leading logarithms for the two-point function. There is a nonvanishing logarithmic term multiplying $P_\mu P_\nu$, which should be interpreted as a gauge renormalization.

[26]These gauges were brought to our attention by S. Coleman.

[27]See however the Appendix of I.

[28]G. 't Hooft, Nucl. Phys. B61, 455 (1973).

Chapter 13

SCIENCE AND TECHNOLOGY

CHARLES H. TOWNES

Introduction

Papers Reprinted in Book

Clement D. Child (1868-1933). (Courtesy of AIP Emilio Segrè Visual Archives.)

Charles Townes is a Professor in the Physics Department at the University of California, Berkeley. He has held positions at the Bell Telephone Laboratory, Columbia University, the Institute for Defense Analysis in Washington, and Massachusetts Institute of Technology. He has also served on the boards of General Motors and the Perkin Elmer Corporation. His research has involved molecular and nuclear structure, quantum electronics, and microwave and infrared astronomy. Awards and honors include the National Medal of Science, the Mees and Thomas Young medal in optics, the Niels Bohr International Gold Medal, NASA's Distinguished Public Service Medal, membership in the Inventors Hall of Fame, and the Nobel Prize awarded in 1964 for fundamental work which has led to the construction of oscillators and amplifiers based on the maser-laser principle.

Owen W. Richardson (1879-1959). (Courtesy of AIP Emilio Segrè Visual Archives.)

It is both fascinating and inspiring to review the many papers in *The Physical Review* over the last 100 years that have contributed strongly to the growth of technology. It is also frustrating not to be able to call more attention to so many of those that must be omitted from this brief list. Such a review of history makes obvious the changes in both physics and technology over the last 100 years. It maps the growth of physics publications in the United States from the late 19th century to the late 1930s, the sharp decrease in the number of publications during World War II as physicists turned towards war work and engineering, and the very rapid increase thereafter with a resulting birth of a number of new technologies.

Some of the earliest contributions of physics to technology were of course in the fields of mechanics, thermodynamics, and light. Only the last was still actively being published when *The Physical Review* came into being and it continued to add many new aspects to technology. For the first few decades of *The Physical Review*, perhaps the most important area of technology to which new physics very directly contributed was communication and electronics; much of the latter was initiated during that period. Notable publications that contributed to electronics of the first few decades of this century include work on electron emission, nonlinear detectors, noise, crystal oscillators, plasmas and discharges, vacuum tubes, and even radar, which was presaged by ionospheric reflection measurements using pulsed radio waves. But there were also other devices outside of electronics that were

Harry Nyquist (1889-1976). (Courtesy of AIP Emilio Segrè Visual Archives.)

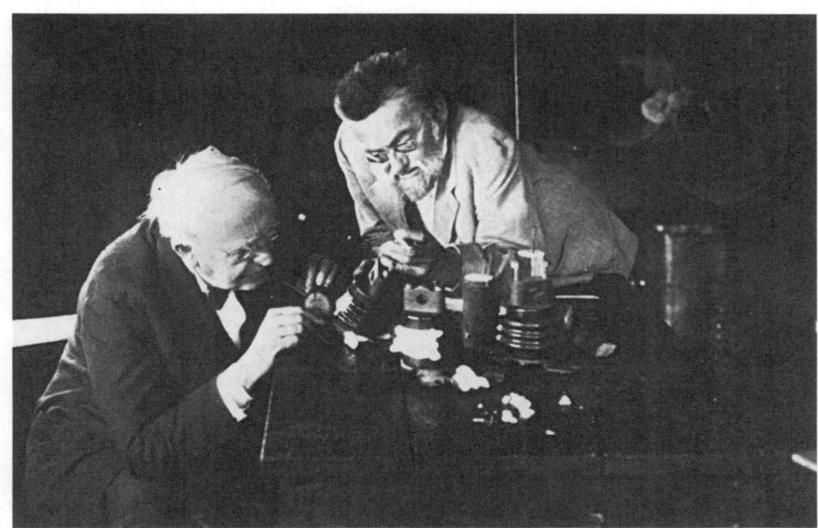

Thomas A. Edison (1847-1931) and **Charles P. Steinmetz (1865-1923).** (Courtesy of General Electric Research Laboratory, AIP Emilio Segrè Visual Archives.)

of great importance to technology, such as vacuum pumps, spectrometers, x-ray tubes, mass spectrometers, electron microscopes, and various other measuring devices.

At a slightly later stage, electronics, communication, and then digital computing received another tremendous boost from solid state physics. Modern solid state electronics was perhaps born in *The Physical Review* with the transistor paper of Bardeen and Brattain. Hence it will seem strange that this and many other solid state papers are not included in this selection. The reason is that there is a separate selection devoted to solid state physics, where this and many other papers that can be considered both basic contributions to the science and initiations of new technological devices are included.

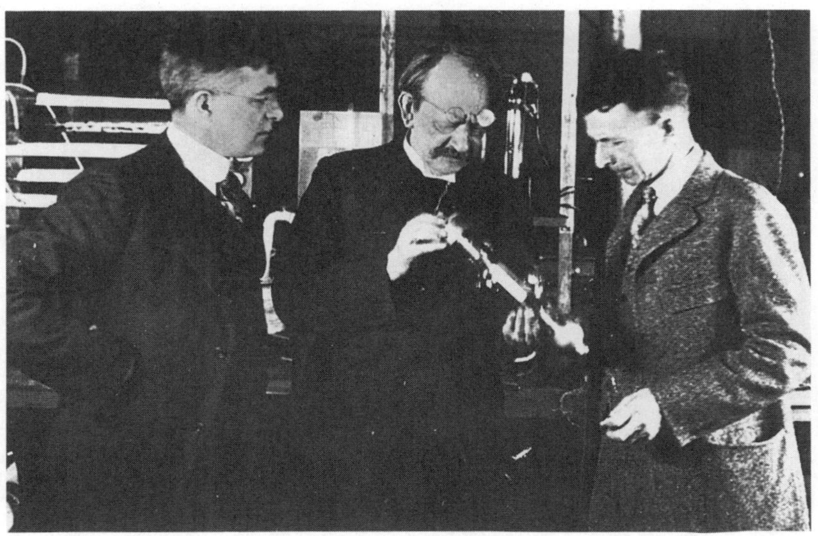

Irving Langmuir (1881-1957), J. J. Thomson (1856-1940), and **William D. Coolidge (1873-1975)** inspecting a pliotron, a high-vacuum tube, 1923 at the General Electric Research Laboratory, Schenectady, New York (Courtesy of General Electric Research & Development Center, AIP Emilio Segrè Visual Archives.)

G. W. Pierce (1872-1956). (Courtesy of AIP Emilio Segrè Visual Archives.)

Van de Graaff accelerator, built in 1931 by Merle A. Tuve and associates at Carnegie Institution's Division of Terrestrial Magnetism, Washington, D.C.. This is the first machine to attain energies of 1 million electron volts. (Courtesy of Carnegie Institution, AIP Emilio Segrè Visual Archives.)

J. Robert Oppenheimer (1904-1967) and **John von Neumann (1903-1957)** in front of Princeton computer. (Courtesy of Mrs. Alan W. Richards.)

Nuclear physics and nuclear energy, initiated in the 1930s, have contributed broadly to technology in a number of other directions. These include the use of nuclear tracers, detection of trace components, and measurements of dense materials. It also includes invention of many types of accelerators, and discovery of the deuteron, fission, and fusion and the resulting nuclear energy. The field has of course had enormous technical impact. However, as in the case of solid state physics, papers covering nuclear physics and its applications are omitted from the present selection because they are largely covered in the chapter on nuclear physics.

Shortly after World War II there were a number of important inventions and discoveries in radio and microwave spectroscopy. Nuclear magnetic resonance, including chemical effects and imaging, and molecular beam resonances, which provide our best long-term timing devices, illustrate these contributions. Many of these pioneering papers are published, as in the case of solid state and nuclear physics, in chapters such as atomic physics or plasma physics. A field that has already contributed much, but where applications are still being intensively developed, is quantum electronics and the various new applications of visible and infrared radiation in intense or highly controlled forms. Since there is no separate section on optics, these developments are represented here more strongly than might otherwise be the case. As did other forms of electronics, light-wave technology now contributes strongly to the "information age," but the entire quantum electronics field extends from the radio range down to wavelengths as short as x-rays. Even with rather liberal representation here, many aspects of it, such as free-electron lasers, x-ray lasers, and a number of solid state masers and lasers, could not be included.

Edward M. Purcell (1912-). (Courtesy of AIP Meggers Gallery of Nobel Laureates, AIP Emilio Segrè Visual Archives.)

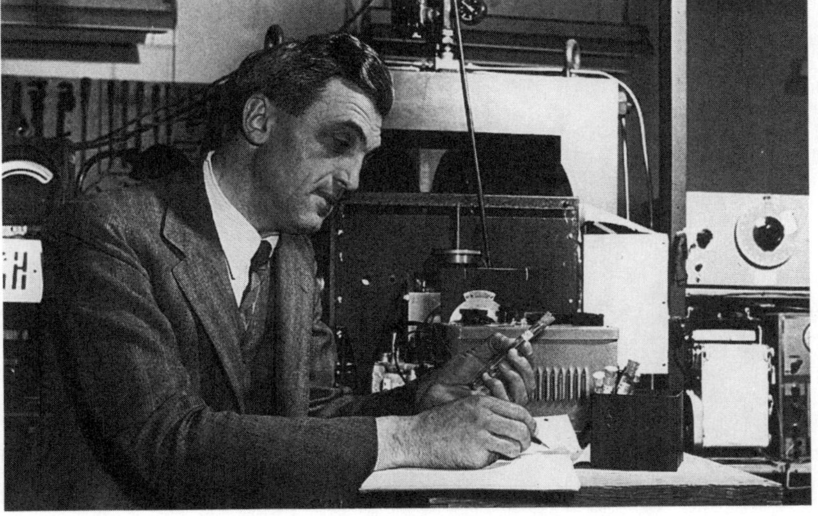

Felix Bloch (1905-1983). (Courtesy of Stanford University, AIP Emilio Segrè Visual Archives.)

Theodore H. Maiman (1927-) and **Irnee J. D'Haenens (1934-)** with replica of laser, on Maiman's induction into Hall of Fame, 1984. (Courtesy of Hughes Research Laboratories, AIP Emilio Segrè Visual Archives.)

Many early papers in *The Physical Review* were written in an inclusive and somewhat more lengthy style than are those today, where frequently an important discovery is first published as a short letter, and even lengthier papers are very compact. Five of the eight earliest papers in this chapter are, for example, longer than 20 pages, and these five would have used up 138 pages, when only 110 pages are allotted to this section on technology. For this reason, it was essential to reproduce only representative and particularly revealing parts of these wonderful papers, while others could be reproduced fully.

In many cases, authors first published a brief note of an important discovery in *The Physical Review* and then a fuller paper before substantial other work elucidated the field. In such cases, if space is available, the fuller paper is published here since it

Charles H. Townes (1915-) on left, **Alfred Kastler (1902-1984)**, **Mary E. Warga (1904-1991)**, **Luis W. Alvarez (1911-1988)**, **Gerhard Herzberg (1904-)**, **Dennis Gabor (1900-)**, and **Arthur Schawlow (1921-)** at Optical Society of America ceremonial session October 1972 in San Francisco. (Courtesy of AIP Emilio Segrè Visual Archives.)

can fairly be described as the one on which much subsequent work was based. To my regret, the important first laser demonstration, Maiman's ruby laser system, is not represented. This is because Maiman first published briefly in *Nature* and then sometime later published a fuller account in *The Physical Review*. However, by the time this latter paper appeared, there were already important publications on the same subject in *The Physical Review*, so that this paper was not itself the crucial exposé it might have been.

There are undoubtedly other new fields of physics that will permeate the technology of the future. What many of these are can only be guessed at this point, and largely for this reason no very recent publications appear here. However, the substantial basic developments in superconductivity, micropositioning of atoms, or nonlinear optics are some of those that are very likely to become increasingly important to technology. Very precise measurements and new developments in the production of special materials or special structures will probably also play increasingly important technological roles. We can only have glimpses and guesses of what the really important papers for technology will be during the coming century, but if history can be projected at all, we can surely look forward to further revolutionary developments emanating from work published in *The Physical Reviews.*

Editor's note:

There are few fields of physics that have not been seminal in the development of technology, as described. While many of the relevant articles were not grouped in this section of the book, they can be found in the chapters The Early Years, Atomic, Nuclear, Condensed Matter and Plasma Physics. They include the important topics of nuclear fission, plasma fusion devices, radiative processes, thermal agitation of electrons in conductors, superconductivity, nuclear induction, transistors, and atom trapping. H.H.S.

Nicolaas Bloembergen (1920-). (Courtesy of Harvard University, AIP Emilio Segrè Visual Archives.)

Francis Bitter (1902-1967). (Courtesy of MIT, AIP Emilio Segrè Visual Archives.)

Papers Reproduced on CD-ROM

C. P. Steinmetz. Notes on the theory of oscillating currents, *Phys. Rev. 1st series* **III**, 335 (1896)

G. W. Pierce. Crystal rectifiers for electric currents and electric oscillations, *Phys. Rev. 1st series* **XXV**, 31–60 (1907)

O. W. Richardson and F. C. Brown. The kinetic energy of the negative ions from hot metals, *Phys. Rev. 1st series* **XXVI**, 409–410(A) (1908)

W. D. Coolidge. A powerful Röntgen ray tube with a pure electron discharge, *Phys. Rev.* **2**, 409–430 (1913)

I. Langmuir. The effect of space charge and residual gases on thermionic currents in high vacuum, *Phys. Rev.* **2**, 450–486 (1913)

I. Langmuir. A high vacuum mercury vapor pump of extreme speed, *Phys. Rev.* **8**, 48–51 (1916)

W. G. Cady. The piezo electric resonator, *Phys. Rev.* **17**, 531(A) (1921)

E. F. Nichols and J. D. Tear. Short electric waves, *Phys. Rev.* **21**, 587–610 (1923)

A. H. Taylor and E. O. Hulburt. The propagation of radio waves over the Earth, *Phys. Rev.* **27**, 189–215 (1926)

W. Schottky. Small-shot effect and flicker effect, *Phys. Rev.* **28**, 74–103 (1926)

G. Breit and M. A. Tuve. A test of the existence of the conducting layer, *Phys. Rev.* **28**, 554–575 (1926)

J. B. Johnson. Thermal agitation of electricity in conductors, *Phys. Rev.* **32**, 97–109 (1928)

L. Tonks and I. Langmuir. A general theory of the plasma of an arc, *Phys. Rev.* **34**, 876–922 (1929)

G. F. Metcalf and B. J. Thompson. A low grid-current vacuum tube, *Phys. Rev.* **36**, 1489–1494 (1930)

L. A. DuBridge. The amplification of small direct currents, *Phys. Rev.* **37**, 392–400 (1931)

W. F. Giauque and D. P. MacDougall. Attainment of temperatures below 1° absolute by demagnetization of $Gd_2(SO_4)_3 \cdot 8H_2O$, *Phys. Rev.* **43**, 768(L) (1933)

J. B. Taylor and I. Langmuir. The evaporation of atoms, ions, and electrons from caesium films on tungsten, *Phys. Rev.* **44**, 423–458 (1933)

C. E. Cleeton and N. H. Williams. Electromagnetic waves of 1.1 cm wave-length and the absorption spectrum of ammonia, *Phys. Rev.* **45**, 234–237 (1934)

L. Marton. Electron microscopy of biological objects, *Phys. Rev.* **46**, 527–528(L) (1934)

K. T. Bainbridge and E. B. Jordan. Mass spectrum analysis. 1. The mass spectrograph. 2. The existence of isobars of adjacent elements, *Phys. Rev.* **50**, 282–296 (1936)

P. A. Čerenkov. Visible radiation produced by electrons moving in a medium with velocities exceeding that of light, *Phys. Rev.* **52**, 378–379(L) (1937)

H. L. Anderson, E. T. Booth, J. R. Dunning, E. Fermi, G. N. Glasoe, and F. G. Slack. The fission of uranium, *Phys. Rev.* **55**, 511–512(L) (1939)

I. I. Rabi, S. Millman, P. Kusch, and J. R. Zacharias. The molecular beam resonance method for measuring nuclear magnetic moments. The magnetic moments of $_3Li^6$, $_3Li^7$ and $_9F^{19}$, *Phys. Rev.* **55**, 526–535 (1939)

H. L. Anderson, E. Fermi, and H. B. Hanstein. Production of neutrons in uranium bombarded by neutrons, *Phys. Rev.* **55**, 797–798(L) (1939)

H. L. Anderson and E. Fermi. Simple capture of neutrons by uranium, *Phys. Rev.* **55**, 1106–1107(L) (1939)

H. L. Anderson, E. Fermi, and L. Szilard. Neutron production and absorption in uranium, *Phys. Rev.* **56**, 284–286 (1939)

J. M. B. Kellogg, I. I. Rabi, N. F. Ramsey, Jr., and J. R. Zacharias. The magnetic moments of the proton and the deuteron. The radiofrequency spectrum of H_2 in various magnetic fields, *Phys. Rev.* **56**, 728–743 (1939)

S. Millman and P. Kusch. On the radiofrequency spectra of sodium, rubidium, and caesium, *Phys. Rev.* **58**, 438–445 (1940)

E. M. McMillan. The synchrotron—A proposed high energy particle accelerator, *Phys. Rev.* **68**, 143–144(L) (1945)

E. M. Purcell, H. C. Torrey, and R. V. Pound. Resonance absorption by nuclear magnetic moments in a solid, *Phys. Rev.* **69**, 37–38(L) (1946)

F. Bloch, W. W. Hansen, and M. Packard. Nuclear induction, *Phys. Rev.* **69**, 127(L) (1946)

F. Bloch. Nuclear induction, *Phys. Rev.* **70**, 460–474 (1946)

F. Bloch, W. W. Hansen, and M. Packard. The nuclear induction experiment, *Phys. Rev.* **70**, 474–485 (1946)

C. H. Townes. The ammonia spectrum and line shapes near 1.25-cm wave-length, *Phys. Rev.* **70**, 665–671 (1946)

N. Bloembergen, E. M. Purcell, and R. V. Pound. Relaxation effects in nuclear magnetic resonance absorption, *Phys. Rev.* **73**, 679–712 (1948)

F. R. Elder, R. V. Langmuir, and H. C. Pollock. Radiation from electrons accelerated in a synchrotron, *Phys. Rev.* **74**, 52–56 (1948)

R. Hofstadter. Alkali halide scintillation counters, *Phys. Rev.* **74**, 100–101(L) (1948)

J. Schwinger. On the classical radiation of accelerated electrons, *Phys. Rev.* **75**, 1912–1925 (1949)

N. F. Ramsey. A new molecular beam resonance method, *Phys. Rev.* **76**, 996(L) (1949)

N. F. Ramsey. A molecular beam resonance method with separated oscillating fields, *Phys. Rev.* **78**, 695–699 (1950)

N. F. Ramsey. Magnetic shielding of nuclei in molecules, *Phys. Rev.* **78**, 699–703 (1950)

SCIENCE AND TECHNOLOGY

E. L. Hahn. Spin echoes, *Phys. Rev.* **80**, 580–594 (1950)

F. Bloch. Nuclear relaxation in gases by surface catalysis, *Phys. Rev.* **83**, 1062–1063(L) (1951)

T. Holstein. Imprisonment of resonance radiation in gases. II., *Phys. Rev.* **83**, 1159–1168 (1951)

N. F. Ramsey. Chemical effects in nuclear magnetic resonance and in diamagnetic susceptibility, *Phys. Rev.* **86**, 243–246 (1952)

R. H. Dicke. The effect of collisions upon the Doppler width of spectral lines, *Phys. Rev.* **89**, 472–473 (1953)

A. Abragam and R. V. Pound. Influence of electric and magnetic fields on angular correlations, *Phys. Rev.* **92**, 943–962 (1953)

H. D. Hagstrum. Auger ejection of electrons from tungsten by noble gas ions, *Phys. Rev.* **96**, 325–335 (1954)

H. D. Hagstrum. Theory of Auger ejection of electrons from metals by ions, *Phys. Rev.* **96**, 336–365 (1954)

J. P. Gordon, H. J. Zeiger, and C. H. Townes. The maser—New type of microwave amplifier, frequency standard, and spectrometer, *Phys. Rev.* **99**, 1264–1274 (1955)

A. T. Forrester, R. A. Gudmundsen, and P. O. Johnson. Photoelectric mixing of incoherent light, *Phys. Rev.* **99**, 1691–1700 (1955)

E. W. Müller and K. Bahadur. Field ionization of gases at a metal surface and the resolution of the field ion microscope, *Phys. Rev.* **102**, 624–631 (1956)

K. Shimoda, T. C. Wang, and C. H. Townes. Further aspects of the theory of the maser, *Phys. Rev.* **102**, 1308–1321 (1956)

N. F. Ramsey. Thermodynamics and statistical mechanics at negative absolute temperatures, *Phys. Rev.* **103**, 20–28 (1956)

N. Bloembergen. Proposal for a new type solid state maser, *Phys. Rev.* **104**, 324–327 (1956)

M. T. Weiss. A solid-state microwave amplifier and oscillator using ferrites, *Phys. Rev.* **107**, 317(L) (1957)

A. L. Schawlow and C. H. Townes. Infrared and optical masers, *Phys. Rev.* **112**, 1940–1949 (1958)

R. J. Collins, D. F. Nelson, A. L. Schawlow, W. Bond, C. G. B. Garrett, and W. Kaiser. Coherence, narrowing, directionality, and relaxation oscillations in the light emission from ruby, *Phys. Rev. Lett.* **5**, 303–305 (1960)

H. M. Goldenberg, D. Kleppner, and N. F. Ramsey. Atomic hydrogen maser, *Phys. Rev. Lett.* **5**, 361–362 (1960)

T. H. Maiman. Stimulated optical emission in fluorescent solids. I. Theoretical considerations, *Phys. Rev.* **123**, 1145–1150 (1961)

T. H. Maiman, R. H. Hoskins, I. J. D'Haenens, C. K. Asawa, and V. Evtuhov. Stimulated optical emission in fluorescent solids. II. Spectroscopy and stimulated emission in ruby, *Phys. Rev.* **123**, 1151–1157 (1961)

A. Javan, W. R. Bennett, Jr., and D. R. Herriott. Population inversion and continuous optical maser oscillation in a gas discharge containing a He–Ne mixture, *Phys. Rev. Lett.* **6**, 106–110 (1961)

P. A. Franken, A. E. Hill, C. W. Peters, and G. Weinreich. Generation of optical harmonics, *Phys. Rev. Lett.* **7**, 118–119 (1961)

D. Kleppner, H. M. Goldenberg, and N. F. Ramsey. Theory of the hydrogen maser, *Phys. Rev.* **126**, 603–615 (1962)

J. A. Armstrong, N. Bloembergen, J. Ducuing, and P. S. Pershan. Interactions between light waves in a nonlinear dielectric, *Phys. Rev.* **127**, 1918–1939 (1962)

H. London, G. R. Clarke, and E. Mendoza. Osmotic pressure of He^3 in liquid He^4, with proposals for a refrigerator to work below 1°K, *Phys. Rev.* **128**, 1992–2005 (1962)

M. Bass, P. A. Franken, A. E. Hill, C. W. Peters, and G. Weinreich. Optical mixing, *Phys. Rev. Lett.* **8**, 18 (1962)

R. N. Hall, G. E. Fenner, J. D. Kingsley, T. J. Soltys, and R. O. Carlson. Coherent light emission from GaAs junctions, *Phys. Rev. Lett.* **9**, 366–368 (1962)

G. Eckhardt, R. W. Hellwarth, F. J. McClung, S. E. Schwarz, D. Weiner, and E. J. Woodbury. Stimulated Raman scattering from organic liquids, *Phys. Rev. Lett.* **9**, 455–457 (1962)

W. E. Lamb, Jr. Theory of an optical maser, *Phys. Rev.* **134**, A1429–A1450 (1964)

C. K. N. Patel. Interpretation of CO_2 optical maser experiments, *Phys. Rev. Lett.* **12**, 588–590 (1964)

R. Y. Chiao, C. H. Townes, and B. P. Stoicheff. Stimulated Brillouin scattering and coherent generation of intense hypersonic waves, *Phys. Rev. Lett.* **12**, 592–595 (1964)

D. Kleppner, H. C. Berg, S. B. Crampton, N. F. Ramsey, R. F. C. Vessot, H. E. Peters, and J. Vanier. Hydrogen-maser principles and techniques, *Phys. Rev.* **138**, A972–A983 (1965)

J. A. Giordmaine and R. C. Miller. Tunable coherent parametric oscillation in $LiNbO_3$ at optical frequencies, *Phys. Rev. Lett.* **14**, 973–976 (1965)

C. K. N. Patel. Efficient phase-matched harmonic generation in tellurium with a CO_2 laser at 10.6 μ, *Phys. Rev. Lett.* **15**, 1027–1030 (1965)

B. D. Josephson. Macroscopic field equations for metals in equilibrium, *Phys. Rev.* **152**, 211–217 (1966)

M. O. Scully and W. E. Lamb, Jr. Quantum theory of an optical maser. I. General theory, *Phys. Rev.* **159**, 208–226 (1967)

M. Sargent III, W. E. Lamb, Jr., and R. L. Fork. Theory of a Zeeman laser. I., *Phys. Rev.* **164**, 436–449 (1967)

C. K. N. Patel and E. D. Shaw. Tunable stimulated Raman scattering from conduction electrons in InSb, *Phys. Rev. Lett.* **24**, 451–455 (1970)

P. W. Smith and T. Hänsch. Cross-relaxation effects in the saturation of the 6328-Å neon-laser line, *Phys. Rev. Lett.* **26**, 740–743 (1971)

T. W. Hänsch, M. D. Levenson, and A. L. Schawlow. Complete hyperfine structure of a molecular iodine line, *Phys. Rev. Lett.* **26**, 946–949 (1971)

T. W. Hänsch, I. S. Shahin, and A. L. Schawlow. High-resolution saturation spectroscopy of the sodium *D* lines with a pulsed tunable dye laser, *Phys. Rev. Lett.* **27**, 707–710 (1971)

K. M. Evenson, J. S. Wells, F. R. Petersen, B. L. Danielson, G. W. Day, R. L. Barger, and J. L. Hall. Speed of light from direct frequency and wavelength measurements of the methane-stabilized laser, *Phys. Rev. Lett.* **29**, 1346–1349 (1972)

L. R. Elias, W. M. Fairbank, J. M. J. Madey, H. A. Schwettman, and T. I. Smith. Observation of stimulated emission of radiation by relativistic electrons in a spatially periodic transverse magnetic field, *Phys. Rev. Lett.* **36**, 717–720 (1976)

W. Neuhauser, M. Hohenstatt, P. Toschek, and H. Dehmelt. Optical-sideband cooling of visible atom cloud confined in parabolic well, *Phys. Rev. Lett.* **41**, 233–236 (1978)

D. J. Wineland and W. M. Itano. Laser cooling of atoms, *Phys. Rev. A* **20**, 1521–1540 (1979)

H. F. Hess, G. P. Kochanski, J. M. Doyle, T. J. Greytak, and D. Kleppner. Spin-polarized hydrogen maser, *Phys. Rev. A* **34**, 1602–1604 (1986)

P. D. Lett, R. N. Watts, C. I. Westbrook, W. D. Phillips, P. L. Gould, and H. J. Metcalf. Observation of atoms laser cooled below the Doppler limit, *Phys. Rev. Lett.* **61**, 169–172 (1988)

Ernest O. Lawrence (1901-1958), Milton S. Livingston (1905-1986), and **C. U. Foulds,** President P.W.W.W. with 85-ton cyclotron at Pelton Water Wheel Works. (Courtesy of AIP Emilio Segrè Visual Archives.)

CRYSTAL RECTIFIERS FOR ELECTRIC CURRENTS AND ELECTRIC OSCILLATIONS.

PART I. CARBORUNDUM.

BY GEORGE W. PIERCE.

TABLE OF CONTENTS.

Introductian. — General H. H. C. Dunwoody of the United States Army has discovered[1] that a crystalline mass of carborundum when supplied with electrodes acts as a receiver for electric waves. In his patents pecification General Dunwoody shows several ways of attaching the electrodes to the crystal. One method is to wind the wires around the two ends of the specimen. Another method is to hold the crystal in a clamp of which the two jaws, insulated from each other, serve as electrodes. The crystal with its electrodes is put into a receiving circuit of a wireless telegraph system with a telephone and battery about the carborundum. The sounds heard in the telephone when a message is being re-

[1] U. S. Patent, No. 837,616, issued December 4, 1906; application filed March 23, 1906.

ceived are like those heard with the electrolytic detector. General Dunwoody also found that the battery could be omitted and the leads of the telephone connected directly about the terminals of the carborundum, and with this arrangement, without the battery, he says that he has read messages from a sending station several hundred miles away.

The present investigation was undertaken in the effort to obtain further knowledge of this interesting property of carborundum. The experiments were extended to include many other crystalline substances, but the present discussion is chiefly confined to carborundum.

Apparatus for Current-voltage Measurements. — Fig. 1 shows a sketch of a form of circuit employed in studying the conductivity of carborundum under various conditions. The crystal of carbo-

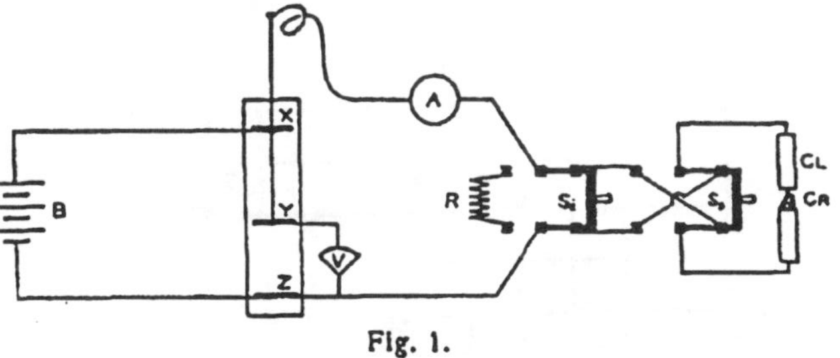

Fig. 1.

rundum, held in a clamp, is shown at *Cr*; *B* is a storage battery; *XYZ* is a potentiometer consisting of two fixed plates of zinc *X* and *Z*, and one movable plate *Y*, immersed in a zinc sulphate solution. By means of the voltmeter *V* the difference of potential between the plates *Y* and *Z* could be read, and the resulting current through the carborundum was given by a galvanometer or milliammeter at *A*. The resistance of the galvanometer was so small in comparison with the resistance of the carborundum that the reading of the voltmeter was practically the drop of voltage in the carborundum.

The switch S_3 enables the observer to reverse the current in the crystal under examination without reversing the galvanometer. A known resistance at *R*, could be thrown into circuit with the galvanometer for the purpose of calibrating it.

This method of experimenting has previously been employed by Eccles,[1] Guthe and Trowbridge[2] and others in the study of the coherer; by Rothmund and Lessing[3] and by Austin[4] and Armagnat[5] in the study of the electrolytic detector.

Current-voltage Curve for Carborundum.—A curve obtained by plotting the current against voltage in an experiment with carborundum is shown in Fig. 2. It is seen that the current through the carborundum is not a linear function of the voltage about it; the apparent resistance of the substance diminishes with increasing current. Curves of approximately this form were obtained by Greenleaf W. Picard[6] in a study of carborundum. This curve also resembles closely the curves obtained by Rothmund and Lessing, by Austin and by Armagnat for the relation of current to voltage in the electrolytic receiver. It resembles also the building up portion of the current-voltage curve in the coherer as obtained by Eccles.

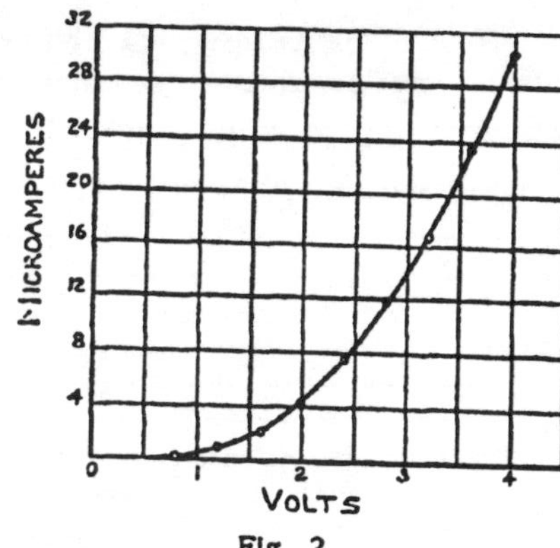

Fig. 2.

Experiments were made by the writer on a great many specimens of carborundum, and curves of approximately the shape shown in Fig. 2 were obtained in all the cases. The value of the current for a given voltage was found to depend on the temperature, and pressure, and on the method of leading the current to the crystal, and was different for different specimens. Before discussing these effects of temperature, pressure, etc., attention is called to a more interesting property of carborundum, the property of unilateral conductivity.

Unilateral Conductivity of Carborundum.—The current through the crystal in one direction under a given electromotive force was

[1] Electrician, 47, pp. 682 and 715, 1901.
[2] Phys. Rev., 11, p. 22, 1900.
[3] Ann. d. Phys., 15, p. 193, 1904.
[4] Bureau of Standards, 2, p. 261, 1906.
[5] Bull. soc. francaise, Session of Apl., 1906, p. 205.
[6] El. World, 48, p. 994, 1906.

found to be different from the current in the opposite direction under the same electromotive force; that is to say, carborundum is unilaterally conductive. This effect may be seen by a reference to

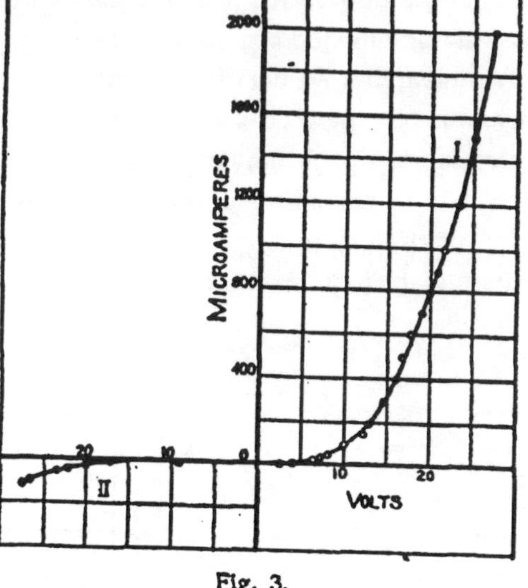

Fig. 3.

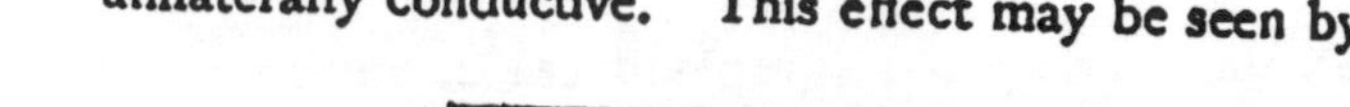

Fig. 3. The branch *I* of the curve shows the current, plotted against voltage, when the current is in one direction; branch *II* the corresponding values of the current obtained when the voltage is reversed. The accompanying table, Table I., contains the numerical values from which these curves were plotted. This property of unilateral conductivity of carborundum has been overlooked by previous experimenters on this substance. The property of unilateral conductivity has, however, been previously found in some of the crystalline metallic oxides and sulphides, by Ferdinand Braun. Reference to Braun's work is given in the historical note on page 58. None of the substances investigated by Braun showed such striking asymmetry as that obtained in the present experiments with carborundum.

In the experiment whose result is shown in Fig. 3 and Table I., the specimen of carborundum was held in a clamp under a pressure of about 500 grams, and it is seen that the current in one direction is 100 times as great as the current in the opposite direction when an electromotive force of 10 volts is applied in the two cases. With increase of current through the specimen, the ratio of the current in the two opposite directions diminishes. At 27.5 volts C_1 is only 17 times C_2.

TABLE I.

Relation of Current to Voltage, Showing unilateral Conductivity of Carborundum.

Volts.	Current in Microamperes.		C_1/C_2.
	C_1, Commutator Left.	C_2, Commutator Right.	
2.2	1		
2.8	2		
4.0	5		
4.7	10		
5.9	20		
6.5	30		
7.3	40		
8.0	50		
10.0	100	1	100
12.1	150		
12.8	200		
14.5	300	5	60
16.0	400		
16.8	500	10	50
17.7	600		
19.4	700		
20.0	800	20	40
21.0	900		
21.9	1,000	30	33
23.2	1,200	50	24
25.0	1,500		
27.5	2,000	120	17

In this particular experiment the piece of carborundum was submerged in an oil bath designed to keep the temperature of the specimen constant. The piece of carborundum was held in a clamp, the jaws of which served to lead the current to the specimen. The oil of which the temperature was 64°C., came freely into contact with the crystal.

Similar effects were obtained at various temperatures between − 10°C. and 100°C., both with and without the use of oil as a bath. A like result was had with different specimens and under different pressures. The relative values of the positive and negative currents, however, varied from piece to piece, and also was different under different conditions of temperature and pressure. The effects of temperature and pressure are investigated below.

An interesting property of some of the specimens is presented in Fig. 4. The curve of current against voltage obtained when the

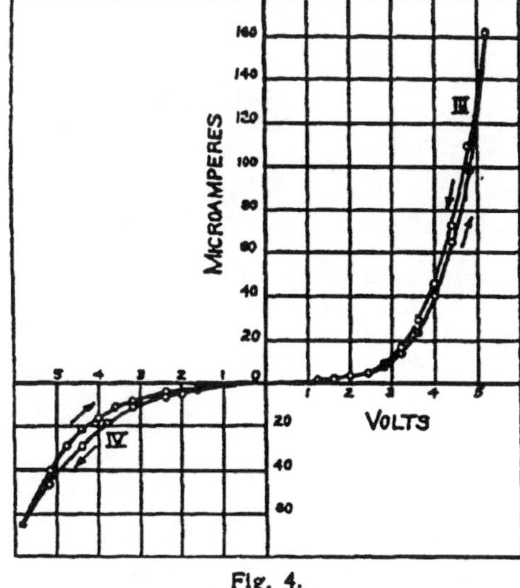

Fig. 4.

voltage was *increased* step by step did not exactly agree with the values of the current obtained with the same voltage when the voltage was decreasing step by step. The difference is indicated by the non-coincidence of the two curves marked *III* in Fig. 4. The arrows indicate the order of succession of the observations. Curve *IV* shows the corresponding effects when the current is in the opposite direction. This effect was apparently due to a slow building up of the current, and after several reversals of the current usually disappeared. The specimen that gave the curves of Fig. 4 was under a pressure of 2 kilograms.

Effects of Pressure.—Current-voltage curves were taken with the same specimen under various pressures. A series of results are shown in the curves of Fig. 5. In taking these observations the specimen was held in a clamp with jaws insulated from one another. One of the jaws was capable of being moved forward without rotation under the action of a screw and spring. The spring was calibrated, so that the pressure on the carborundum could be read off on a scale. Care was taken that the specimen was subjected to a steady compression without twisting. The pressure was in the direction of the current.

In taking the set of observations shown in Fig. 5 the pressure was first made three kilograms, giving the top curve of the series. The pressure was then reduced successively to two kilograms and one kilogram. From the first quadrant of the figure it is seen that

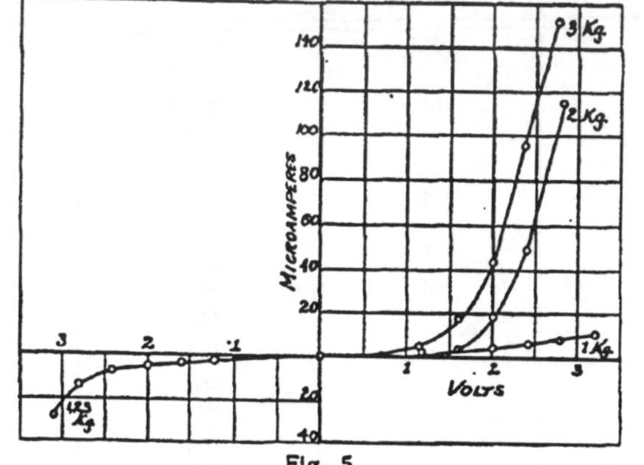

Fig. 5.

the conductivity of the specimen diminishes with diminishing pressure. This is the case when the current is in one direction. With the current in the opposite direction the current-voltage curves for the three different pressures coincide, and for the case of the pressure equal to one kilogram the prevailing conductivity is actually reversed with respect to its direction in the case of the higher pressure.

Several experiments were made with other specimens of carborundum with considerable disparity in the results, and the curves of Fig. 5 cannot be taken to represent a general occurrence.

Many of the results of the experiments here described are apparently confused by difficulties arising at the contact of the metallic electrodes with the crystal. On account of the irregularities of the surfaces of the specimens actual contact with the electrodes in general occurs at small areas. Although this contact is apparently not loose when the pressure on the clamp is several kilograms, it is

evident that the carborundum, which is an exceedingly hard substance, will imbed itself in the electrodes to a depth depending on the pressure, and one cannot be certain that a specimen will return to its original condition when the pressure is put on and taken off or when the specimen is removed from the clamp and again replaced in it. On this account the experiments are sometimes incapable of repetition unless the clamp is left undisturbed.

SUMMARY OF RESULTS.

1. Current voltage curves for carborundum are shown.

2. Carborundum is unilaterally conductive. With one specimen under 10 volts the current in one direction is 100 times the current in the opposite direction. With another specimen platinized on one side the current at 34.5 volts is 527 times as great as the current in the opposite direction under the same voltage. In another case at 30 volts the current in one direction is 4,000 times the current in the opposite direction under the same voltage.

3. As the current increases the efficiency of rectification decreases.

4. A specimen platinized on both sides has a smaller efficiency of rectification and a much lower resistance than a piece not platinized. Though the efficiency of rectification is less with the platinized specimen, the *excess* of one current over the other for a given voltage is much greater with the platinized specimen on account of its low resistance.

5. An oscillogram is given showing the distortion of an alternating current wave by carborundum.

6. A method is shown of employing crystal rectifiers in the construction of alternating current measuring instruments.

7. These instruments are applicable to the measurement of telephonic currents, and may be used in experiments on resonance in telephone circuits.

8. A determination of the temperature coefficient of conductivity of carborundum is given. This coefficient is in the neighborhood of the temperature coefficient of weak electrolytes.

9. A discussion is given of the action of carborundum in General Dunwoody's detector for electric waves.

10. No theory is given as to the cause of the unilateral conductivity of carborundum. A number of other crystals have the same property, and it is proposed to accumulate data in regard to these other substances before attempting an explanation.

JEFFERSON PHYSICAL LABORATORY,
HARVARD UNIVERSITY, CAMBRIDGE, MASS.,
May 6, 1907.

The Kinetic Energy of the Negative Ions from Hot Metals.

By O. W. Richardson and F. C. Brown.

THE kinetic energy of the ions has been measured by observing the rate at which the potential of one of a pair of parallel plates charged up when a small area (about one square millimeter) of the opposite plate was occupied by heated platinum foil. The hot platinum and the plate surrounding it were maintained at zero potential. If the particles are shot off from the foil with a velocity component u perpendicular to the plates they will only reach the upper plate provided $\frac{1}{2}mu^2 > Ve$ where m is their mass and e their electric charge, V being the difference of potential between the plates. The method thus measures the number of particles whose component of energy in the direction perpendicular to the plates (if we may be allowed to use this expression) is greater than a given value. It is evident that the results give the average value of this part of the energy and also the way in which it is distributed among the different particles emitted.

On the assumption that the ions shot off from the hot metal behave like the molecules of a gas it can be shown that the number n which reach the upper plate against a potential V is given by $n = n_0 e^{-\nu Ve/R\theta}$ where n_0 is the number which reach the upper plate when $V = 0$, ν is the number of molecules in 1 c.c. of a gas at normal temperature and pressure, θ is the absolute temperature and R is the constant in the gas equation $pv = R\theta$. This formula is equivalent to $\log i - \log i_0 = -\nu Ve/R\theta$ where i and i_0 are the currents corresponding to n and n_0. This has been tested by plotting the experimental values of $\log i$ and V and the linear relation demanded by theory has been accurately verified. The inclination of this line is equal to $\nu e/R\theta$. Now νe is the quantity of electricity required to set free half a cubic centimeter of hydrogen at 0° C. and 760 m.m. pressure by electrolysis. Substituting the value of this constant and that of θ the temperature of the wire we find the value of R, the gas constant for the negative electrons. Different experiments have given values ranging from 3.5×10^3 to 4.05×10^3 as compared with the standard value 3.7×10^3.

We are thus led to the conclusion that the kinetic energy of the negative electrons emitted by hot metals is the same as that of the translational kinetic energy of the molecules of a gas at the same temperature as that of the metal.

If we admit the truth of the atomic theory and the correctness of the accepted values for the mass of an atom this investigation furnishes what

[1] Abstract of a paper read before the American Physical Society, in New York, February 29, 1908.

we believe to be the first direct test of certain theorems in the kinetic theory of gases relating to the way in which the velocities are distributed among the different atoms.

It is also suggested that a method based on these principles will ultimately be of value, on account of its practical simplicity and theoretical directness, as a standard of thermometry for very high temperatures.

Princeton, N. J.,
February 28, 1908.

Second Series December, 1913 Vol. II, No. 6

THE PHYSICAL REVIEW.

A POWERFUL RÖNTGEN RAY TUBE WITH A PURE ELECTRON DISCHARGE.

By W. D. Coolidge.

§ 1. Introduction.

In an earlier publication attention has been called[1] to the use of wrought tungsten for the anticathode, or target, of a Röntgen tube of the ordinary type. In the development of this target many different designs were made and mounted in tubes, and these tubes were operated on what was then the most powerful Röntgen apparatus on the market, a 10 K.W. transformer coupled to a mechanical rectifying device. The operation of tubes in this manner, to see how much energy it took to ruin the target, gave perhaps an unusual viewpoint. When, as a result of these experiments, a satisfactory form of target had been developed, the writer became interested in studying the remaining limitations in the tube. Some of these limitations are the following:

1. With low discharge currents the vacuum gradually improves, with a consequent increase in the penetrating power of the rays produced.

2. With high discharge currents there are very rapid vacuum changes, sometimes in one direction and sometimes in the other.

3. If a heavy discharge current is continued for more than a few seconds the target is heated to redness and then gives off so much gas that the tube may have to be reëxhausted.

4. If the temperature of the standard copper-backed target is allowed to get up to bright redness, a rapid deposition of metallic copper begins to take place on the bulb, continuing for some time after the cutting off of the current, owing to the very slow rate of cooling from such temperatures in the evacuated space.

5. Of the tubes tested, very many have failed from cracking of the

[1] Coolidge, Trans. Am. Inst. E. E., June, 1912, pp. 870–872.

glass, and this, with one exception, always at the same point; that is, in the zone around the cathode. In many cases there has first been chipping-out of the glass from the inner surface of the tube at this point.

6. The focal spot on the target in many tubes wanders about very rapidly.[1] In many cases where it does not show a tendency to wander, it will be found after a heavy discharge to have permanently changed its location.

7. While it is relatively easy to lower the tube resistance by means of the various gas regulators, it is a relatively slow matter to raise it much.

8. With very heavy discharges, the central portion of the usual massive aluminum cathode melts, and the molten globules so formed are shot right across the tube, flattening themselves out on the glass and sticking to it. When the melted area is small, no harm is done except that the curvature of the cathode at this point may be changed, and the focal spot may, in consequence, be moved.

9. No two tubes are exactly alike in their electrical characteristics.

10. The characteristics are in general far from ideal, in that the penetrating power of the Röntgen rays produced, changes with the magnitude of the discharge current.

11. When operated on a periodically intermittent current, even though it be of constant potential, the tube, of necessity, gives a very heterogeneous bundle of primary Röntgen rays, for the reason that the break-down voltage of the tube is much higher than the running voltage.[2]

[1] In radiographic work, movement of the focal spot during an exposure is of course detrimental to good definition.

[2] See F. Dessauer, Phys. Zeitschr., 14, pp. 246–247 (1913).

§ 4. Detailed Description of Tube No. 147.

This description relates to tube No. 147, which was used in getting the data for the following tables. Fig. 1 shows a complete assembly, while

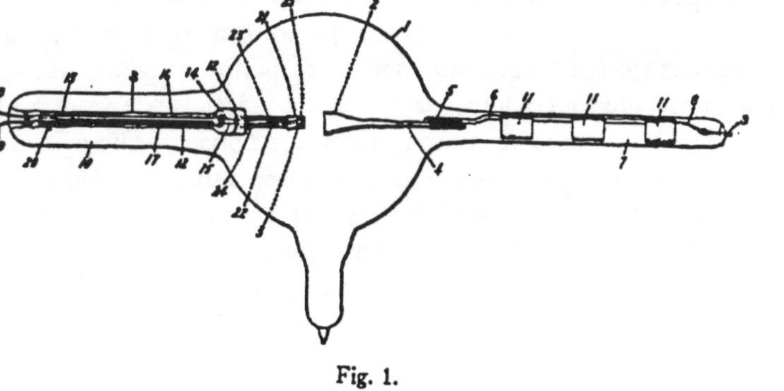

Fig. 1.

Fig. 2 shows an enlarged detail of the cathode and of the front end of the target.

The Cathode.

In the diagrams, 25 is a tungsten filament in the shape of a flat, closely wound spiral. It consists of a wire 0.216 mm. in diameter and 33.4 mm. long with 5½ convolutions, the outermost of which has a diameter of 3.5 mm. It is electrically welded to the ends of two heavy molybdenum wires 14 and 15, to the other extremities of which are welded the two copper wires 16 and 17. These in turn are welded to the platinum wires 18 and 19. The molybdenum wires are sealed directly into a piece of special glass, 12, which has essentially the same temperature coefficient of expansion as molybdenum. This first seal is simply to insure a rigid support for the hot filament, the outer seal being the one relied upon for vacuum tightness. The outer end, 13, of the support tube is of Gernan glass like the bulb itself, and it is therefore necessary to interpose at S a series of intermediate glasses to take care of the difference in expansion coefficients between 12 and 13. The small glass tube, 20, prevents short-circuiting of the copper wires, 16 and 17.

Fig. 2.

The filament is heated by current from a small storage battery which is, electrically, well insulated from the ground.

In the circuit are placed an ammeter and an adjustable rheostat and, by means of the latter, the filament current can be regulated, by very

fine steps, from 3 to 5 amperes. Over this current range, the potential drop through the filament varies from 1.8 to 4.6 volts and the filament temperature from 1890 to 2540 degrees absolute.

The Focusing Device.

This consists of a cylindrical tube of molybdenum, 21. It is 6.3 mm. inside diameter and is mounted so as to be concentric with the tungsten filament, and so that its inner end projects 1.0 mm. beyond the plane of the latter. It is supported by the two stout molybdenum wires, 22 and 23, which are sealed into the end of the glass tube, 12. It is metallically connected to one of the filament leads, at 24.

Besides acting as a focusing device, it also prevents any discharge from the back of the heated portion of the cathode.

The Anticathode or Target.

The anticathode or target, 2, which also serves as anode, consists of a single piece of wrought tungsten, having at the end facing the cathode a diameter of 1.9 cm. (Its weight is about 100 gm.) By means of a molybdenum wire, 5, it is firmly bound to the molybdenum support, 6. This support is made up of a rectangular strip and, riveted to this, three split rings, 11, 11, 11, all of molybdenum. The split rings fit snugly in the glass anode arm, 7. They serve the double purpose of properly supporting the anode and of conducting heat away from the rectangular strip and so preventing too much heat flow to the seal of the lead-in-wire, 9.

The Bulb.

This is of German glass and about 18 cm. in diameter.

§ 7. SUMMARY.

In the foregoing, a new and powerful Röntgen ray tube has been described. It differs in principle from the ordinary type in that the discharge current is purely thermionic in character. Both the tube and the electrodes are as thoroughly freed from gas as possible, and all of the characteristics seem to indicate that positive ions play no appreciable rôle.

The tube allows current to pass in only one direction and can therefore be operated from either direct or alternating current.

The intensity and the penetrating power of the Röntgen rays produced are both under the complete control of the operator, and each can be instantly increased or decreased independently of the other.

The tube can be operated continuously for hours, with either high or low discharge currents, without showing an appreciable change in either the intensity or the penetrating power of the resulting radiations.

The tube in operation shows no fluorescence of the glass and no local heating of the anterior hemisphere.

The starting and running voltage are the same.

The tube permits of the realization of intense homogeneous primary Röntgen rays of any desired penetrating power.

An article bearing especially upon the application of the new tube to radiographic and diagnostic and to therapeutic purposes will appear shortly in one of the Röntgen ray journals.

It is a pleasure to me, in closing, to express my appreciation of the services of Mr. Leonard Dempster, who has assisted me throughout this work.

RESEARCH LABORATORY OF THE GENERAL ELECTRIC CO.,
SCHENECTADY, N. Y.

THE EFFECT OF SPACE CHARGE AND RESIDUAL GASES ON THERMIONIC CURRENTS IN HIGH VACUUM.

BY IRVING LANGMUIR.

WHEN a carbon or metal filament is heated in a vacuum and surrounded by a positively charged metal cylinder, it is well known that electrons are given off by the hot solid. This effect in lamps has been commonly known as the Edison effect and has been rather fully described in the case of carbon lamps by Fleming.[1]

Richardson and others have studied quantitatively the ionization produced by hot solids, especially from heated platinum, and have collected a large amount of data. It has generally been found that the saturation current is independent of the pressure of the gas and increases rapidly with increasing temperature of the filament. However, certain gases were found to have very marked effects; for example, traces of hydrogen were found to enormously increase the saturation current obtained from hot platinum.[2] Recent investigations have shown[3] that at least in some cases the current is due to secondary chemical effects.

Pring and Parker[4] showed that the current obtained from incandescent carbon could be cut down to very small values by progressive purification of the carbon and improvement of the vacuum. They conclude that "the large currents hitherto obtained with heated carbon cannot be ascribed to the emission of electrons from carbon itself, but that they are probably due to some reaction at high temperatures between the carbon, or contained impurities, and the surrounding gases, which involves the emission of electrons." Pring and Parker observed also that the ionization (or rather thermionic current) "increased only very slightly with the temperature above 1800°."

The effect of these publications, together with that of Soddy,[5] who noticed similar effects with a Wehnelt cathode, has been to cast doubt on the existence of a thermionic current in a perfect vacuum and from pure metals. The opinion seems to be gaining ground, especially in Germany,

[1] Phil. Mag., 42, p. 52 (1896).
[2] H. A. Wilson, Phil. Trans., 202, 243 (1903).
[3] Fredenhagen, Ver. d. phys. Ges., 14, 384 (1912).
[4] Phil. Mag., 23, 192 (1912).
[5] Nature, 77, 53 (1907).

that the emission of electrons from incandescent solids is a secondary effect produced by chemical reactions, or at least is caused by the presence of gas.

With the above-mentioned exceptions, it has generally been found that the thermionic current increased with the temperature at a very high rate. The relation between current and temperature was usually accurately represented by Richardson's equation

$$i = a\sqrt{T}e^{-\frac{b}{T}}, \qquad (1)$$

where a and b are constants and i is the saturation current at the absolute temperature T.

If the older values of a and b as found, for example, for carbon, are substituted in the above equation and the currents for very high temperatures (above 2500°) are calculated, values of many amperes or even thousands of amperes per square cm. are usually obtained. This raises the question why in ordinary incandescent lamps very large thermionic currents do not occur.

There is every reason to think that the thermionic current from tungsten should be fairly large. When we run a tungsten lamp up to 2 or even 2.5 times its normal voltage (filament temperature 2900–3400° K.) we should therefore expect to get thermionic currents of several amperes between the two ends of the filament. Simple observation of a lamp run under such conditions indicates that this is not the case. For example, consider a lamp which takes 110 volts and 0.3 ampere when running at normal specific consumption (1.25 watts per candle). By raising the voltage to 250, the temperature of the filament will be brought to about 3000° K. and the current is then about 0.45 ampere. The resistance of the filament has thus increased from 366 ohms up to 555. The total surface of the filament is nearly half a square cm., yet it is evident that if there is any thermionic current between the two ends of the filament, it cannot exceed a few hundredths of an ampere. This apparent discrepancy between the results of calculation by Richardson's equation and the facts observed with a tungsten lamp seemed at first to confirm the growing opinion that in a very high vacuum the thermionic current is very small, if not entirely absent.

Experiments on the Edision effect in tungsten lamps, made some time ago by the writer, throw a great deal of light on the cause of the apparent failure of Richardson's equation at high temperatures. The observations therefore seem of sufficient interest to warrant their publication.

EXPERIMENTS ON EDISON EFFECT IN TUNGSTEN LAMPS.

Some lamps were made containing two single loop (hairpin) tungsten filaments with separate leading-in wires. Each loop could thus be run separately. The lamps were given a specially good lamp exhaust, which involved heating them to 360° for an hour while being exhausted with a mercury pump. A trap immersed in liquid air was placed between the pump and the lamp to condense out water vapor, carbon dioxide and mercury vapor. The filaments were then connected in series and the lamps run at a specific consumption of about 1 watt per candle for fifteen minutes, to drive the gas from the filaments. The lamps were then sealed off from the pump and the filaments were again heated, this time being run at 0.4 watt per candle for a few minutes, to age the filaments and improve the vaccuum (clean-up effect).

Experiments were then undertaken to measure the thermionic currents that flowed across the space between the two filaments when one was heated to various temperatures while the other was connected to a constant source of positive potential of about 125 volts. A milliammeter was connected in series with the cold filament.

When the temperature of the cathode filament was raised to about 2000° K. a current of about 0.0001 ampere was observed to flow between the two filaments. As this temperature was raised the thermionic current rose very rapidly, until at about 2200° K. it was about .0006 ampere. As the temperature was raised above 2200° K., *no further increase in the thermionic current occurred*, even when the filament was heated nearly to the melting-point (3540° K.). By raising the voltage on the anode to about 250 volts, the thermionic current increased to about .0015 ampere. It required, however, a temperature about 200° higher to reach this current than had been found necessary to reach the maximum current at the lower voltage. At temperatures below 2200° K., the current was practically the same with 125 as with 250 volts.

The results of a later and more accurate experiment are given in Fig. 1. The filament used for these measurements consisted of a single loop of drawn tungsten wire, of diameter .0069 cm. and total length of 10.84 cm. and area of 0.234 sq. cm. A similar filament, at a distance of about 1.2 cm., served as anode. Both filaments had been aged a couple of hours at high temperature and the vacuum thus obtained was certainly better than 10^{-6} mm.[1]

The temperatures of the filament were determined from the relation

$$T = \frac{11,230}{7.029 - \log_{10} H},$$

[1] Judging from measurements on similar lamps made by means of the "molecular" gage described by the writer, PHYS. REV., I (2), 337 (1913).

where H is the intrinsic brilliancy of the filament in international candles per sq. cm. of projected area.[1]

The points indicated by small circles (Fig. 1) are experimentally determined. Two different anode voltages were used: 240 and 120 volts, with respect to the negative terminal of the filament which served as hot cathode. The voltage used to heat the cathode varied from about 7 to 15 volts, so the *average* potential difference between anode and cathode

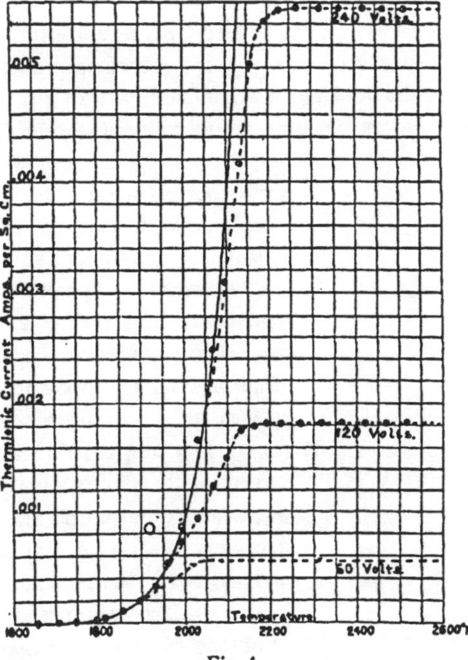

Fig. 1.

was somewhat less than 240 and 120 volts. The curve given for 60 volts was determined from other experiments devised especially to determine the effect of voltage variations.

It is seen from these curves that at low temperatures the current for all three voltages is the same, but that as the temperature is raised the currents at the lower voltages fall below those for higher voltages and finally each in turn reaches a constant value.

By plotting ($\log. i - \frac{1}{2} \log T$) against $1/T$ it was found that all the points on the 240-volt curve up to a temperature of about 2150° lay very close to a straight line. This indicates that these results can be expressed

[1] The derivation of this formula will soon be published, probably in the PHYSICAL REVIEW.

by Richardson's equation. From the slope and position of the line, the values of a and b of Richardson's equation were found to be

$$a = 27 \times 10^6 \text{ (amperes per sq. cm.)},$$
$$b = 55{,}600 \quad \text{(degrees)}.$$

The heavy black curve of Fig. 1 was calculated by plotting Richardson's equation, using these values of a and b. The agreement between this curve and the experiments at low temperature is nearly perfect; in fact, much better than can be seen from Fig. 1.

It is seen that each experimentally determined curve can be divided into three parts:

1. A part which follows Richardson's equation accurately.
2. A part which consists of a horizontal straight line; that is, a part in which the current is independent of the temperature of the filament.
3. A transition curve between these.

The horizontal part of the curve was of particular interest. The current being independent of the temperature of the filament is probably independent of the nature of the cathode. It seemed possible, however, that it might be dependent on the anode. Several experiments were undertaken to determine the factors which governed the value of this new kind of "temperature" saturation current. It was found that it was very largely affected by any one of the four factors:

1. Voltage of anode.
2. Presence of magnetic field.
3. Area of anode.
4. Distance from anode to cathode.

It is especially noteworthy that none of these factors had any influence on the thermionic current over the first part of the curve; *i. e.*, that part which follows Richardson's equation. That is, the constants a and b were not affected by voltage, or magnetic field or distance or area of anode.

After trying out several hypotheses which suggested themselves, it finally occurred to the writer that this temperature saturation might be due to a space charge produced by the electrons between the cathode and anode. The theory of electronic conduction in a space devoid of all positive charges or gas molecules seems to have been strangely neglected. It has apparently always been taken for granted that positive ions are present, or at least a sufficient amount of gas, so that the motion of the electrons follows the laws of diffusion. J. J. Thomson[1] gives the differential equations that apply to the calculation of electron conduction

[1] Conduction of Electricity through Gases, 2d edition, p. 223.

through space, and suggests that a method for the determination of e/m could be worked out in this way. He apparently does not fully integrate the equations or realize their application to ordinary thermionic currents.

Theory of Electronic Conduction in a Space Devoid of Molecules or Positive Ions.

In order to form a clear conception of the problem before us, let us consider (see Fig. 2) two infinite parallel planes, A and B, one of which, A, has the properties of an incandescent solid; that is, we assume that it emits low velocity electrons spontaneously. The other, plane B, we consider to be positively charged.

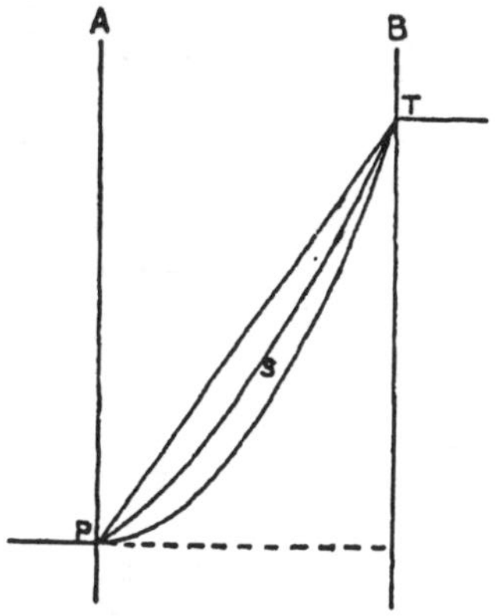

Now if the temperature of the plate A is so low that few or no electrons are emitted, then the potential between the two plates will vary linearly between the two, as indicated by the line PT.

As the temperature of A is raised, electrons are emitted. Under the influence of the field these pass across the space from A to B and thus constitute a current of magnitude i (per sq. cm.).

Fig. 2.

These electrons move with a velocity which depends on the potential drop through which they have passed. Let us assume, as a first rough approximation, that they move with constant velocity across the space. Then there will be in the unit volume a space charge ρ equal to i/v, where v is the velocity of the electrons. If the velocities are uniform, the space charge will be uniform and it follows from Laplace's equation

$$\Delta V = \frac{\partial^2 V}{\partial x^2} + \frac{\partial^2 V}{\partial y^2} + \frac{\partial^2 V}{\partial z^2} = -4\pi\rho \tag{2}$$

that

$$\frac{d^2 V}{dx^2} = -4\pi\rho. \tag{3}$$

If we consider ρ constant and negative (for electrons), we see from this equation that the potential distribution between the two plates takes the form of a parabola, as indicated by the curve PST.

If the temperature of the plate A be increased still further, the electron current increases so that the potential curve finally becomes a parabola with a horizontal tangent at P.

If we assume that the electrons are given off from the plate A with practically no initial velocity, we see that the current cannot increase beyond the point where the potential curve becomes horizontal at P, for any further increase of current would make the potential curve at P slope downwards and the electrons would be unable to move against this unfavorable potential gradient.

In other words, we see that the effect of the space charge is to limit the current. A further increase in the temperature of the plate A would then not cause an increase of current.

Electron Current between Parallel Planes.—Let us now attempt a more rigorous solution of the problem.

It has been shown by Richardson and others that the mean kinetic energy of the thermions is closely equal to that of gas molecules at the same temperature. This indicates that they have velocities so low that very few of them are capable of moving against a negative potential of more than a couple of volts. Since the voltages applied to the anode are much larger than this, we may assume, for convenience, that the electrons are given off by the plate A without initial velocity.

Now let V be the potential at a distance x from the plate A. The kinetic energy of an electron when it has traveled the distance x from the plate will thus be

$$\tfrac{1}{2}mv^2 = Ve. \tag{4}$$

The current (per unit area) carried by the electrons at any place will be

$$i = \rho v. \tag{5}$$

For convenience, we take e and ρ positive even for electrons. Equation (3) thus becomes (in electrostatic units)

$$\frac{d^2V}{dx^2} = 4\pi\rho. \tag{6}$$

These three equations enable us to express V as a function of x and i. By eliminating ρ and v from (4), (5) and (6), we obtain

$$\frac{d^2V}{dx^2} = 2\pi\sqrt{2}\sqrt{\frac{m}{e}}\frac{i}{\sqrt{V}}. \tag{7}$$

Multiply this by $2\cdot dV/dx$ and integrate

$$\left(\frac{dV}{dx}\right)^2 - \left(\frac{dV}{dx}\right)_0^2 = 8\pi i\sqrt{\frac{2mV}{e}}. \tag{8}$$

Now if there is an opposing (negative) potential gradient at the surface of the plate A, *no current* will flow. If there is a positive potential

gradient all the electrons that are given off from the plate A will reach B, so that the current that flows will be determined by Richardson's equation. The case that we are interested in is that in which the current is less than the saturation current and is determined by the voltage of the anode. Evidently for this case the potential gradient at the plate A is zero; that is,

$$\left(\frac{dV}{dx}\right)_0 = 0,$$

whence from (8) by extracting the square root:

$$\frac{dV}{dx} = \sqrt{8\pi i}\sqrt[4]{\frac{2mV}{e}}. \tag{9}$$

Integrating and solving for i, we obtain:

$$i = \frac{\sqrt{2}}{9\pi}\sqrt{\frac{e}{m}}\frac{V^{\frac{3}{2}}}{x^2}. \tag{10}$$

This equation[1] gives the maximum electron current density between two infinite parallel plates with the distance x between them and with a potential difference V. This equation holds only where the initial velocity of the electrons at the plate A is negligible compared to that produced by the potential V. It does not hold at such high voltages that the electrons move with velocities approaching that of light.

Taking $e/m = 1.77 \times 10^7$ E.M. units, reducing to E.S. units and substituting in (10) and then reducing to volt, ampere units, we obtain from equation (10):

$$i = 2.33 \times 10^{-6}\frac{V^{\frac{3}{2}}}{x^2}, \tag{11}$$

where i is the maximum current density in amperes per sq. cm., x is the distance between the plates in centimeters, and V is the potential difference in volts.

[1] Since submitting this paper for publication the attention of the writer has been called to the fact that C. D. Child (PHYS. REV., 32, 492, 1911), has already derived this equation. He has, however, applied it only to the case where the conduction takes place solely by positive ions.

Summary.

It is shown both theoretically and experimentally that the mutual repulsion of electrons (space charge) in a space devoid of positive ions, limits the current that flows from a hot cathode to a cold anode. For parallel plane electrodes of infinite extent, separated by the distance x, and with a potential difference V between them, the maximum current (per unit area) that can flow if no positive ions are present is

$$i = \frac{\sqrt{2}}{9\pi}\sqrt{\frac{e}{m}}\frac{V^{3/2}}{x^2}.$$

For the analogous case of an infinitely long, hot wire, placed concentrically within a cylindrical anode, of radius r, the maximum current per unit length is

$$i = \frac{2\sqrt{2}}{9}\sqrt{\frac{e}{m}}\frac{V^{3/2}}{r\beta^2},$$

where β varies from 0 to 1, according to the diameter of the wire, but for all wires less than 1/20 the diameter of the anode, β is a quantity extremely close to unity.

2. In the presence of gas at pressures above .001 mm., and at voltages above 40 volts, there is usually sufficient production of positive ions to greatly reduce the space charge and thus allow more current to flow than indicated by the above equations.

3. It is shown, contrary to the ordinary opinion, that the general effect of very low pressures of gas is to greatly *reduce* the electron emission from an incandescent metal.

4. This effect is especially marked at low temperatures. In most cases it probably disappears at very high temperatures.

5. The constant b of Richardson's equation

$$i = a\sqrt{T}e^{-\frac{b}{T}}$$

is always increased, in the case of tungsten, by the introduction of oxygen, nitrogen, water vapor, carbon monoxide or dioxide. Argon, however, has no effect on either constant.

[1]Leipziger Berichte. math. phys. Kl., 65, 42, 1913.

6. The normal thermionic current from tungsten in a "*perfect*" vacuum follows Richardson's equation accurately. The constants approximately:

$$a = 34 \times 10^6 \text{ amps. per sq. cm.},$$

$$b = 55{,}500.$$

7. Preliminary data are given for the electron emission from tantalum, molybdenum, platinum and carbon. With these substances also the effect of gases is to greatly decrease the electron emission.

8. The effect of nitrogen in decreasing the thermionic current from tungsten depends on the voltage of the anode. In many cases less current is obtained with 240 volts than with 120 volts. With oxygen, the effect seems independent of the anode voltage.

9. The following theory seems to account for most of the observed phenomena and is apparently not inconsistent with any:

The effect of gases in changing the saturation current is due to the formation of unstable compounds on the surface of the wire. In the cases observed the presence of the compound decreases the electron emission. It is possible, however, that in some cases it might cause an increase. The extent to which the surface is covered by the compound depends on the rate of formation of the compound and on its rate of removal from the surface. The compound may be *formed* on the surface directly by reaction with the gas (for example, oxygen), or by reacting principally with positive ions which strike the surface (nitrogen). The compound may be *removed* from the surface by decomposition, evaporation, or cathodic sputtering (*i. e.*, being driven off by bombardment of positive ions).

10. The experimental conditions which should be met, in order to most easily study the thermionic currents in high vacuum, are discussed. It is pointed out that failure to observe these conditions is probably the cause of other investigators having found that the thermionic currents tend to decrease with increasing purity of the cathode and progressive improvement of the vacuum.

11. It is concluded that with proper precautions the emission of electrons from an incandescent solid in a very high vacuum (pressures below $.10^{-6}$ mm.) is an important specific property of the substance and is not due to secondary causes.

In conclusion the writer wishes to express his appreciation of the valuable assistance of Mr. S. P. Sweetser, and Mr. William Rogers who have carried out most of the experimental part of this investigation.

Research Laboratory,
General Electric Company,
Schenectady, N. Y.

A HIGH VACUUM MERCURY VAPOR PUMP OF EXTREME SPEED.

BY IRVING LANGMUIR.

ASPIRATORS or ejectors in which a blast of steam or air is used to produce a partial vacuum have been in use many years. For example the vacuum in the automatic vacuum brake system is obtained by such a device. The Parsons' vacuum augmenter used in the condensers of steam turbines produces a pressure as low as a few centimeters of mercury.

In these devices the high velocity of the jet of steam causes, according to hydrodynamical principles, a lowering of pressure, so that the air to be exhausted is sucked directly into the jet. If, however, the jet were surrounded by a perfect vacuum it can be readily seen from the principles of the kinetic theory that there would be a blast of gas molecules escaping from the jet into this vacuum. It is therefore not possible to obtain a very high vacuum by means of an instrument based on the principle of the ordinary ejector or aspirator.

The action of the aspirator or ejector, however, really consists of two processes.

1. The process by which the air is drawn into the jet.

2. The action of the jet in carrying the admixed air along into the condensing chamber.

The aspirators cease operating at low pressures because of the failure of the first of these processes. If air at low pressures could be made to enter the jet, and if gas escaping from the jet could be prevented from passing back into the vessel to be exhausted, then it should be possible to construct a jet pump which would operate even at the lowest pressures.

Gaede[1] has recently described a pump (called the diffusion pump) in which a blast of mercury vapor carries along the gas to be exhausted into the condenser in much the same way as the steam aspirator does (Process 2). In order to introduce the gas into the blast of mercury vapor (process 1), Gaede has used diffusion through a porous diaphragm, or through what amounts to the same thing, a slit of a width comparable with the mean free path of the mercury atoms in the blast. A portion of the mercury blast escapes through the slit, and the gas to be exhausted

[1] Ann. Phys., *46*, 357 (1915).

diffuses in against this blast of mercury vapor. This renders it necessary to make the slit very narrow (about 0.1 mm.) and for this reason the speed of the pump is necessarily relatively low.

Gaede defines the speed of a pump S by the equation

$$S = \frac{V}{t} \ln \frac{p_0}{p},$$

where V is the volume of the vessel being exhausted and t is the time required for the pressure to fall from p_0 to p.

Gaede gives some interesting data on the speed of his various pumps when exhausting air at about .001 mm. pressure.

Rotary mercury pump S = 120 cc. per sec.
Molecular pump S = 1,300 c.c. per sec.
Diffusion pump S = 80 c.c. per sec.

The great advantage of the diffusion pump is that the value of S remains constant down to the lowest pressures, whereas with all other forms of pump the value of S rapidly decreases when the pressure becomes much less than .001 mm.

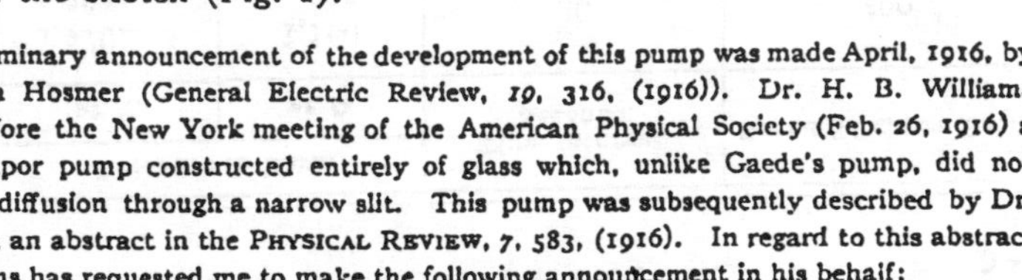

Fig. 1.

Some time ago it occurred to the writer that there should be other methods by which the gas to be exhausted may be introduced (Process 2) into the blast of mercury vapor. The serious limitation of speed imposed by the slowness of diffusion through narrow slits may thus be overcome.

Several methods have been found by which this may be accomplished. One of the types of pump which has given satisfactory results is that shown in the sketch (Fig. 1).[1]

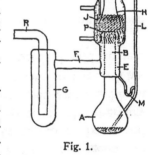

[1] A preliminary announcement of the development of this pump was made April, 1916, by Miss Helen Hosmer (General Electric Review, *19*, 316, (1916)). Dr. H. B. Williams showed before the New York meeting of the American Physical Society (Feb. 26, 1916) a mercury vapor pump constructed entirely of glass which, unlike Gaede's pump, did not depend on diffusion through a narrow slit. This pump was subsequently described by Dr. Williams in an abstract in the PHYSICAL REVIEW, 7, 583, (1916). In regard to this abstract Dr. Williams has requested me to make the following announcement in his behalf:

"In describing recently before the Physical Society a mercury vapor pump, I stated that before carrying out the experiments I had in mind with a view to improving the Gaede type, I learned through Professor Pegram of Columbia that Dr. Irving Langmuir was already working with a pump in which the slit had been discarded and diffusion took place through a wide opening. I therefore constructed my own pump with a wide opening instead of experimenting with various widths as had been my original intention. In writing an abstract I omitted to make this acknowledgment owing to a desire to condense

In this device a blast of mercury vapor passes upward from the heated flask A through the tubes B and C into the condenser D. Surrounding B is an annular space E connecting through F and the trap G with the vessel to be exhausted. The tube C is enlarged into a bulb H just above the upper end of the tube B. This enlargement is surrounded by a water condenser J from which the water is removed at any desired height by means of the tube K which is connected to a water aspirator. The mercury condensing in D and H returns to the flask A by means of the tubes L and M. The tube N connects to the "rough" or "backing" pump which should maintain a pressure considerably lower than the vapor pressure of the mercury in A.

In this pump, the mercury atoms escaping from the upper end of the tube B, radiate out in all directions. A part of them passes up into C, but the larger part strikes the walls of the enlargement H.

If there is no water in the condenser J the mercury which condenses on the walls reëvaporates nearly as fast as it condenses. The molecules passing from the end of the tube B towards the walls H collide with the molecules which reëvaporate and may then be deflected downward into the annular space E. This blast of mercury vapor down through E prevents the gas from F from passing up into H so that under these conditions the gas from F may pass through the pump much more slowly than if no mercury vapor were produced in A.

On the other hand, when cold water circulates through the condenser J, *all* the mercury atoms striking the walls of H are condensed, so that no mercury passes down through E. The gas from F thus passes freely up through E and when it meets the mercury vapor blast at P is blown outward and upward along the walls of the condenser H, and finally forced into the main stream of mercury vapor passing up through C into the condenser D.

The main advantages of this pump are:

1. Simplicity and reliability.
2. Extremely high speed.
3. Absence of lower limits to which the pressure may be reduced.

The speed of a pump like that shown in the sketch has been determined for air and hydrogen.

A vessel of eleven liter capacity was connected to R. Air was admitted to the vessel, and the pressure was found to decrease as follows:

the abstract and because I expected to make such an acknowledgment later in print. The publication of a description of Dr. Langmuir's pump makes it seem desirable that this statement be made at the same time. H. B. Williams."

Time, Seconds.	Pressure in Bars.		Time Interval, Seconds.	Speed S Cc. per Sec.
	at R.	at D.		
0	1,470.	1,160		
30	294	720	> 30	590
60	12.8	218	> 30	1,150
80	0.015	18	> 20	3,700

The rough pump used had a speed of about 200 c.c. per second at pressures of 400 bars, but this speed fell off to 60 c.c. at 40 bars, and became zero at about 10 bars' pressure. The speed of the mercury vapor pump increased rapidly as the pressure decreased and reached a limit of about 4,000 c.c. per second at pressures below 10 bars. Theoretical considerations would indicate that this speed should remain constant down to the very lowest pressures.

Several experiments with hydrogen showed that the maximum speed with this gas was about 7,000 c.c. per second.

The pump shown in the sketch is made of glass. Many pumps operating on these principles have also been constructed of electrically welded sheet iron.

In a subsequent paper the writer will describe in more detail other modifications of mercury vapor pumps, some of which have marked advantages in simplicity of construction and in reliability of operation over that shown here. One particularly efficient type of pump is made wholly of metal.

The writer is greatly indebted to Dr. S. Dushman for assistance in the development of the new pump.

Research Laboratory,
General Electric Company,
Schenectady, N. Y.

The Piezo-Electric Resonator.

By W. G. Cady.

A PLATE or rod suitably prepared from a piezo-electric crystal, and provided with metallic coatings, can be brought into a state of vigorous resonant longitudinal vibration when the coatings are connected to a source of alternating E.M.F. of the right frequency. Under these conditions, the plate reacts upon the electric circuit in a remarkable manner.

The vibrations are a consequence of the so-called "converse" piezo-electric effect, *i.e.*, the deformation resulting from an electric stress; while the periodically strained condition of the plate, through the action of the "direct" effect, sets up a periodic component of polarization which causes the reaction referred to.

On the theoretical side, the phase and magnitude of the counter-polarization in the plate are derived, as well as expressions for the current flowing to the plate and for the total current in the oscillatory circuit. The plate is assumed to be in parallel with the circuit capacity. It is shown that, owing to the piezo-electric polarization, and to the absorption of energy in the plate, the apparent electrostatic capacity and resistance of the plate are not constant, but depend upon the frequency somewhat as does the motional impedance of a telephone receiver. Over a certain range in frequency the capacity may even become negative.

A graphical method is developed for presenting the results of the theory. By an application of this method, it is possible, after making a series of purely electrical observations, to deduce the coefficient of viscosity of the material, even though the absolute value of the piezo-electric constant is not known.

Methods are described for mounting a small plate of piezo-electric crystal upon a rod of any solid elastic substance in such a manner as to excite longitudinal vibrations in the entire rod. The high-frequency viscosity of the rod can then be found, subject, however, to more or less error due to losses in the cement which attaches the crystal plate to the rod.

Experiments on longitudinal vibrations in rods or plates of steel, quartz, and Rochelle salt are described.

The possibility is discussed of using the piezo-electric resonator for a standard of high frequency; for excluding from a circuit oscillations of a given frequency; and as a coupling device to transfer small amounts of power from one circuit to another at a particular frequency.

Wesleyan University, Middletown, Conn.,
February 3, 1921.

THE PROPAGATION OF RADIO WAVES OVER THE EARTH

By A. Hoyt Taylor and E. O. Hulburt

Abstract

Theory of radio wave propagation over the earth.—Larmor's theory of refraction due to the electrons of the Kennelly-Heaviside layer does not explain the "skip distances" for short radio waves (regions of silence around the transmitter which Taylor's measurements showed to be 175, 400, 700 and 1300 miles in radius in the daytime for waves of 40, 32, 21 and 16 meters, respectively, and which are surrounded by zones of strong signals). The range as a function of wave-length shows a minimum for about 200 meters which suggests the introduction of a critical frequency term. If the effect of the magnetic field of the earth on the motion of the electrons is taken into account, as suggested by Appleton and by Nichols and Schelleng, the modification of the Larmor theory necessary to fit it to the experimental facts is secured. A quantitative theory is here developed. The upper atmosphere is assumed to contain N free electrons per cc., and neglecting absorption the dispersion equations are worked out for various modes of polarization of the radio waves. Then the skip distances are computed, making various assumptions as to the electron density distribution. (a) Reflection theory. As a first approximation the layer is taken to be sharply separated from the un-ionized lower atmosphere. At this layer total reflection occurs in accordance with Snell's law. (b) Refraction theory. The following distributions are considered: (1) Density increasing linearly with the height h, beginning at a certain height h_0; (2) Density proportional to h^2; (3) Density proportional to e^h; (4) Density proportional to $h^{\frac{1}{2}}$. Comparison with the experimental skip distances shows good agreement, and indicates that the radio waves which just reach the edge of the zone beyond are refracted around a curved path, reaching in the daytime a maximum height of from 97 miles (case 1, $h_0=21$ miles, and case 2) to 149 miles (case 3). At this height the electron density comes out close to 10^5 electrons per cc. At night the electron density gradient is less and the height is greater. These conclusions agree with physical conceptions from other evidence. From the dispersion equations it follows that for waves of 60 to 200 meters, total reflection may occur from the electron layers at all angles of incidence. From this result, combined with interference between various modes of polarization of the radio rays, a detailed qualitative explanation of many fading phenomena is presented. Further conclusions are: That the ions in the atmosphere have little effect in comparison with the electrons; that for longer waves, the Larmor theory is correct; that short waves are propagated long distances by refraction in the upper atmosphere and reflection at the surface of the earth, not by earth-bound waves; that waves below 14 meters cannot be efficiently used for long distance transmission.

Long Distance Transmission

27. In the light of the foregoing conclusions it is of interest to consider the character of the reception at various distances from the transmitter. No attempt to cover all details will be entered upon, but a few cases deserve remark. It is convenient to make the diagrams from the reflection theory; the conclusions from these will be the same as from a refraction theory. In Fig. 7 the curved line AL is the surface

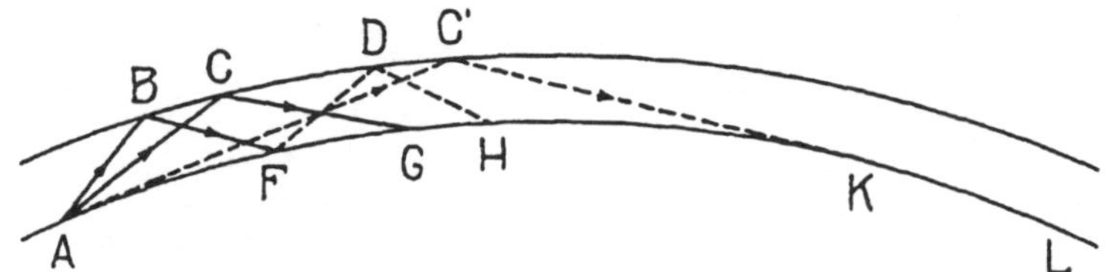

Fig. 7. Paths taken by radio rays in traveling around the earth.

of the earth and BC' the Kennelly-Heaviside layer at a height of 150 miles. Suppose that the radio rays from the transmitter A are confined to the space BAC. The upper limiting ray AB descends to F, where AF is the first skip region, is reflected back to D, down again to the earth at H, etc., continuing around by successive reflections. The lower limiting ray may be tangent to the earth as shown by AC' or may be inclined at an angle to the horizontal in the manner of AC. In the latter case if $AG<AH$, where $AH=2AF$, there exists a second smaller skip region GH. From a continuation of the drawing further skip zones are found at greater distances. These are possible, but perhaps not probable because they become successively smaller and more ill-defined.

If the bundle of transmitted rays is limited on its lower side by the tangent ray AC', which is reflected to the earth again at K, the region KL can be reached by a ray from A only after at least one ground reflection and two layer reflections (refractions). With a layer height of 150 miles AK is 2000 miles. The region FK, on the other hand, can be reached by a ray which has experienced only one layer reflection and no ground reflection. Therefore, because the ground reflecting power may vary with the locality, one might expect possibilities of poorer signals in the region *KL* than in the region *FK*, quite apart from the difference in remoteness; or, more graphically, the poor reception at a station 5,000 miles away may be due to a forest 2500 miles away. This distinction between the regions less and greater than 2,000 miles would seem to be generally valid at all wave-lengths for which the direct ground wave is inoperative and for all the electronic distributions permissible in §§15 to 20.

The observational data relevant to these questions are none too certain, but a recent program of tests with the 25.6 meter transmitter from this station (NKF) has permitted a few conclusions. A portion of the program involved the reception of the signals at every hour of a 24-hour day by some forty stations scattered to 7000 miles distance. In the first place, the first daylight skip zone for 25.6 meters was found to be between 500 and 600 miles, which is in excellent agreement with Fig. 1. Secondly, no secondary skip zone, as *GH*, Fig. 7, appeared; the region from 700 to 1200 miles in daylight being unmistakably one of good signals. This meant merely that the lower transmitted rays were probably near to tangency with the earth, as *AC'*, Fig. 7. In this connection it might be possible to elevate the ray *AC'* until the second skip zone was produced, by properly loading and exciting the transmitting antenna.[20] Thirdly, the signals in the region from 2000 to 3000 miles appeared more uncertain than in the region within the 2000 mile mark. The observations referred mainly to over-land waves and it would be of interest to repeat them with over-sea waves.

In general, for the shorter waves observations by many receiving stations in the United States indicate better signal reception in the region extending from the first skip zone to 2000 miles than in the region between 2000 and 4000 miles where at least one earth reflection is involved. At greater distances, however, from 5000 to 10,000 miles the short wave signals are very reliable, much more so than in the 2000 to 4000 mile zone. This is to be expected because, with increase of distance, there are a greater number of possible ray-paths connecting the transmitter and receiver, so that a local disturbance, such as a poor earth reflection, etc., of any one ray will have small influence on the signal. One might expect the inverse distance law of signal intensity to hold approximately in this region for the short waves.

[20] Van der Pol, Proc. Phys. Soc. London 29, 269 (1916-1917).

At the antipodes of the earth there is a concentration of ray paths and a corresponding increase in signal strength. This has been frequently observed.

For radio communication over longer distances, of the interplanetary order, waves shorter than 40 meters would appear to be best able to pierce our own electron atmosphere as well as that of another planet. The circularly polarized long wave of Eq. (2) is of speculative utility, depending as it does on the intensity of magnetization of the electron atmosphere. Venturing even further into the realm of conjecture, the course of a ray proceeding in interplanetary space would be influenced by the electrons distributed from the sun. If this distribution were impartial in all directions, the electron density increases toward the sun, and the radio ray would be diverted towards the sun. It might pass beyond the influence of the sun after a small deflection, or it might spiral towards the sun until it reached that electron density requisite for total reflection, whereupon it would pursue an enlarging spiral until free again.

Radio Division (A.H.T.),
Heat and Light Division (E.O.H.).
Naval Research Laboratory,
Washington, D.C.
October 17, 1925.

SEPTEMBER, 1926 PHYSICAL REVIEW VOLUME 28

A TEST OF THE EXISTENCE OF THE CONDUCTING LAYER

By G. Breit and M. A. Tuve

Abstract

A method previously proposed for a test of the existence of ionization in the upper atmosphere has been developed, and a definite proof of the *existence of echoes from the upper regions* has been obtained. The echoes are present for 70-meter waves with an 8-mile base near Washington, D. C. The effective height of the layer is between 50 and 130 miles. At times multiple reflections are present. *Radio fading* is shown to be not only an effect of interference between the ground and the reflected waves, but also to a large extent an effect of the presence or absence of reflected waves. A *seasonal variation* in the effective height between summer and fall seems to exist. A smaller *diurnal effect* is also suspected. The height seems greater in the fall than in the summer and greater in the afternoon than in the morning. *Effects of wave-length and of location* have been studied. A quantitative discussion of the results enables one to eliminate too gradual distributions of electron density. The measured retardation is shown to correspond to a height greater than the actual by amounts differing for various polarizations of the refracted waves.

With the development of the sciences of radio and of terrestrial magnetism it has become probable that the upper atmosphere is ionized sufficiently to reflect electromagnetic waves. However, the evidence obtained has not been quite direct until recently, when Appleton and Barnett have completed their work. For this reason we have undertaken an oscillographic study of radio signals with the purpose of observing the echo from the layer. We have described the method previously,[1] but for purposes of completeness it is here described again. A transmitting station is operated so as to send out what is commonly called I.C.W.; it is a set of interrupted trains of waves. The duration of each train is made to be about 1/1000 second. At the receiving end the signal is detected, amplified, and oscillographed. The oscillogram may either show a single set of humps corresponding to a single path transmission, or else it may show two or more sets of humps corresponding to the existence of echoes. From the displacement of the echo with respect to its origin, the retardation between the reflected and the directly received wave is ascertained.

Experimental Arrangements

(a) *Transmission from Bellevue.* Arrangements for transmission have been made with the Naval Research Laboratory (station *NKF*, Bellevue,

[1] M. A. Tuve and G. Breit, Terr. Mag. **30**, 15–16 (1925); G. Breit and M. A. Tuve, Nature **116**, 357 (1925).

Anacostia, D.C.), the Westinghouse Electric and Manufacturing Company (station *KDKA*, Pittsburgh, Pa.), the Radio Corporation of America (station *WSC*, Tuckerton, N. J.), the Bureau of Standards (station *WWV*, Washington, D. C.), and several of the enthusiastic amateurs residing in Washington. The most definite results have been obtained from the Naval Research Laboratory owing to the fortunate relative location of the two laboratories and to the high constancy of the frequency emitted by the *NKF* transmitter. This is achieved by the use of crystal control and makes it superior to any of the other stations we tried for the purpose in question. The interruptions in the wave trains were obtained by supplying the amplifier tubes of the transmitter with alternating current while the master oscillator was fed on direct current. This gave a constant frequency and the required type of modulation. The frequency of alternating current used was 500, and the "on" time of the wave trains has been varied to some extent (1/3 to 1/5) of a cycle) by changing the biasing grid voltages. These arrangements have been kindly made for us by Dr. A. H. Taylor and his assistants, Messrs. L. A. Gebhardt and L. C. Young. After the first definite indications of the reflections have been obtained on 70 meters, the wave form of the transmitter was investigated by means of an oscillograph quite similar to the one used in the reception experiments. A single detector tube was first of all coupled to the antenna of the transmitter by means of a coil with connecting leads shielded in a lead tube, the shield being grounded. The output of the detector was applied to the input of a power amplifier containing four tubes of type 202 (5 watts each) used in parallel. The connections of the amplifier will be described in detail in connection with the receiving apparatus. The output of the amplifier was fed through the oscillograph. Visual observations made by means of a rotating mirror showed that the wave form consisted of a single set of humps. The observations taken by coupling directly to the antenna were not very reliable because they had to be taken directly in the gallery containing the 5 k.w. transmitter, and consequently had to be carried through in a hurry under unfavorable conditions of vibration. The apparatus was consequently transferred into another building of the Naval Research Laboratory, and there the signals were received by means of a small antenna as well as a loop on the single-tube detector as before. The wave form appeared to correspond closely to a single set of humps. However, a slight presence of radiation in the "off" time was detected. The effect could not be due to a reflection, the directly received wave being too strong in comparison. The difficulty was eliminated by using a little additional grid bias in the transmitter. The exact cause of this radiation in the "off" time is unknown, though according to Dr. A. H. Taylor it may be due to certain transient effects in the transmitter circuit. Since care was taken to eliminate the transient effect in all of the transmitters operated by Bellevue, it is safe to assume that the wave form sent out was accurately single-humped with the exception of the first two tests. However, even in these the transient effect was about 1/10 of the reflections observed.

Summary

(1) Groups of radio waves arrive at the receiving station separated from their echoes. This shows that the hypothesis of an ionized upper layer of the atmosphere is correct.

(2) The retardation is such as though the layer were at a height of from 50 to 130 miles. The apparent height is greater in the fall than in the summer for 70-meter waves with an 8-mile base.

(3) Radio fading is present for reflections alone quite independently of interference between the ground and reflected waves.

(4) A quantitative discussion of the possibilities of refraction shows that in most cases the increase of electronic density must be more sudden and discontinuous than that given by a density proportional to the square of the height or else that not all of the possible states of polarization of the waves in the upper atmosphere are present.

The writers are greatly indebted to the various institutions which cooperated in the transmission of signals, particularly so to the Naval Research Laboratory and the Bureau of Standards which has been very generous in loaning apparatus. They also desire to express their sincere thanks to Mr. J. A. Fleming of the Department of Terrestrial Magnetism for his help in making the necessary arrangements and to Mr. C. Huff for the great interest and thoroughness with which he constructed the necessary apparatus.

Department of Terrestrial Magnetism,
Carnegie Institution of Washington,
May 19, 1926.

JULY, 1928 PHYSICAL REVIEW VOLUME 32

THERMAL AGITATION OF ELECTRICITY IN CONDUCTORS

By J. B. Johnson

Abstract

Statistical fluctuation of electric charge exists in all conductors, producing random variation of potential between the ends of the conductor. The effect of these fluctuations has been measured by a vacuum tube amplifier and thermocouple, and can be expressed by the formula $\overline{I^2}=(2kT/\pi)\int_0^\infty R(\omega)|Y(\omega)|^2d\omega$. I is the observed current in the thermocouple, k is Boltzmann's gas constant, T is the absolute temperature of the conductor, $R(\omega)$ is the *real* component of impedance of the conductor, $Y(\omega)$ is the transfer impedance of the amplifier, and $\omega/2\pi=f$ represents frequency. *The value of Boltzmann's constant* obtained from the measurements lie near the accepted value of this constant. *The technical aspects of the disturbance* are discussed. In an amplifier having a range of 5000 cycles and the input resistance R the power equivalent of the effect is $\overline{V^2}/R=0.8\times10^{-16}$ watt, with corresponding power for other ranges of frequency. The least contribution of *tube noise* is equivalent to that of a resistance $R_e=1.5\times10^5 i_p/\mu$, where i_p is the space current in milliamperes and μ is the effective amplification of the tube.

In two short notes[1] a phenomenon has been described which is the result of spontaneous motion of the electricity in a conducting body. The electric charges in a conductor are found to be in a state of thermal agitation, in thermodynamic equilibrium with the heat motion of the atoms of the conductor. The manifestation of the phenomenon is a fluctuation of potential difference between the terminals of the conductor which can be measured by suitable instruments.

The effect is one of the causes of that disturbance which is called "tube noise" in vacuum tube amplifiers.[2] Indeed, it is often by far the larger part of the "noise" of a good amplifier. When such an amplifier terminates in a telephone receiver, and has a high resistance connected between the grid and filament of the first tube on the input side, the effect is perceived as a steady rustling noise in the receiver, like that produced by the small-shot (Schrot) effect under similar circumstances. The use of a thermocouple or rectifier in place of the telephone receiver allows reasonably accurate measurements to be made on the effective amplitude of the disturbance.

It had been known for some time among amplifier technicians that the "noise" increases as the input resistance is made larger. A closer study of this phenomenon revealed the fact that a part of the noise depends on the resistance alone and not on the vacuum tube to which it is connected. The true nature of the effect being then suspected, the temperature of the re-

[1] Johnson, Nature 119, p. 50, Jan. 8, 1927; Phys. Rev. 29, p. 367 (Feb. 1927).

[2] The possibility that under certain conditions the heat motion of electricity could create a measurable disturbance in amplifiers has been recognized on theoretical grounds by W. Schottky (Ann. d. Phys. 57, 541 (1918)). Schottky considered the special case of a resonant circuit connected to the input of a vacuum tube, and concluded that there the effect would be so small as to be masked by the small-shot effect in the tube.

sistance was varied and the result of that test left little doubt that the thermal agitation of electricity in the resistance element was being observed. Further experiments led to the expression of the effect by a formula which, except for a small difference in the numerical constant, was the same as that later developed by Dr. H. Nyquist[3] on a wholly theoretical basis.

The chief results of the measurements may be summarized as follows. The mean-square potential fluctuation over the conductor is proportional to the electrical resistance and the absolute temperature of the conductor. It is independent of the size, shape or material of the conductor. Its apparent magnitude depends on the electrical characteristics of the measuring system as well as on those of the conductor itself. The quantitative data yield a value for Boltzmann's gas constant which agrees well with that obtained by other methods. It is the purpose of this article to describe in more detail these results and the methods by which they were derived, and to discuss the limit which the phenomenon imposes on amplification by vacuum tubes.

Experimental Method and Apparatus

The significance of the mathematical expression for the effect will be developed with the aid of the generalized circuit diagram of Fig. 1. Z is the conductor under investigation, A the amplifier to which it is connected, J the thermocouple ammeter. The amplifier A is characterized by a complex

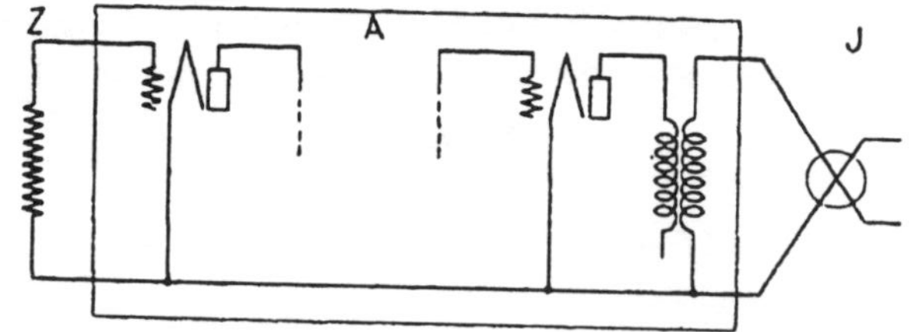

Fig. 1. Simplified diagram of the circuit.

transfer admittance $Y(\omega)$, defined as the ratio of output current to applied input voltage at any frequency $\omega/2\pi=f$. The complex impedance of the element Z has a real resistance component $R(\omega)$ which is also a function of the frequency. The random fluctuation of potential across the input element, arising from the thermal motion of its electric charges, should give rise to a mean-square current $\overline{I^2}$ in the thermocouple according to the equation[4]

$$\overline{I^2}=(2kT/\pi)\int_0^\infty R(\omega)\,|\,Y(\omega)\,|\,^2d\omega. \tag{1}$$

T is here the absolute temperature of the input element and k is Boltzmann's gas constant.

[3] Nyquist, Phys. Rev. 29, 614 (1927); 32, 110 (1928).

[4] Nyquist. Phys. Rev. 32, 110 (1928), Eq. (6).

1130

The value of the integral in this expression may be found by graphic ntegration of the curve formed by the experimentally determined values of $R(\omega)|Y(\omega)|^2$ plotted against ω. In most of the present work the input element Z was a high resistance in parallel with its own shunt capacity and that of its leads and of the input of the amplifier. In such a combination the real resistance component $R(\omega)$ is related to the pure resistance R_0 and the capacity C according to

$$R(\omega)=R_0/(1+\omega^2C^2R_0^2) \qquad (2)$$

Two cases now arise. If the input element and the amplifier are so chosen that $R(\omega)$ does not change much over the frequency range of the amplifier, then it is permissible to use the mean value of $R(\omega)$, obtained to a sufficient degree of approximation by replacing ω by ω_0, the frequency of maximum amplification. The factor $R(\omega)$ can then be placed outside the sign of integration and Eq. (1) becomes

$$\bar{I^2}=\frac{2kTR_0}{\pi(1+\omega_0^2C^2R_0^2)}\int_0^\infty |Y(\omega)|^2 d\omega. \qquad (3)$$

If, on the other hand, $R(\omega)$ cannot be considered constant over the frequency range then the integral must be used as it stands in Eq. (1). The method of determining the various quantities involved in these expressions 1, 2 and 3 will be taken up as the apparatus is described in greater detail.

The amplifier which was used consisted of six stages of audion tubes, suitably coupled, and chosen according to the voltage and power requirements of their various positions. The coupling between tubes was done, with the exception of one interstage, either by transformers or by choke coils and condensers. The exception was a coupling consisting of either a resonant circuit, as shown in Fig. 2, or a band-pass filter, used for the purpose of limiting the amplification to a selected band of frequencies. A comparatively narrow frequency band was admitted by the resonant circuit, the width of the band depending on the relative magnitude chosen for the various components of this system. The band-pass filter, on the other hand, had the frequency range of 500 to 1000 periods per second, with a much sharper cutoff at the limits than the simple resonant circuit.

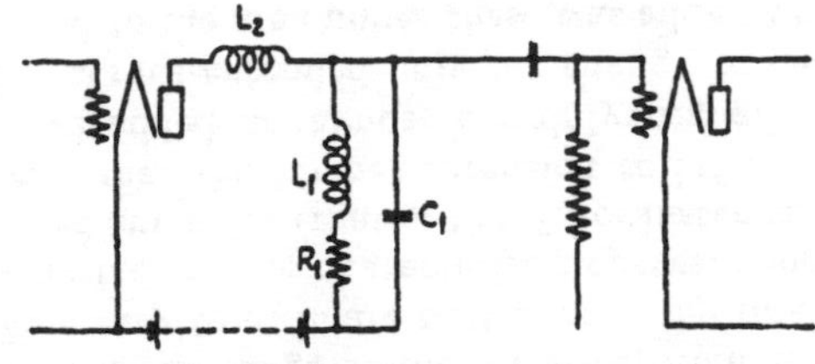

Fig. 2. Diagram of the resonant coupling.

The amplifier was enclosed in a steel cabinet. It was elaborately shielded against electric, magnetic, acoustic and mechanical shocks, but it was not

always entirely free from these disturbances. Regeneration was apparently negligibly small.

The last tube of the amplifier was coupled by a transformer to a vacuum thermocouple, used with a microammeter as the indicating instrument. The couple was calibrated against a direct current meter of established accuracy. Conveniently, the deflection of the microammeter was closely proportional to the square of the current in the thermocouple.

For the calibration of the amplifier, current sufficiently free from harmonics was obtained from a vacuum tube oscillator. The current was measured by a thermocouple similar to that used in the output. An attenuator of the type described by A. G. Jensen[5] was used so that a known fraction of the current was passed through a resistance of one ohm connected across the input of the amplifier, producing a known input voltage. The corresponding output current was observed. Here, however, a certain correction had to be applied. The amplifier itself, and chiefly the first vacuum tube, produced a disturbance which caused the output meter to deflect even without any external source of input voltage. This "zero deflection" was observed frequently and was subtracted from the total deflection. This could be done, since the mean-square currents added linearly in the deflection, and the value of the mean-square current was desired. The quantity $|Y(\omega)|^2$ was then given by the ratio of the mean-square output current to the mean-square input voltage, which in turn was the ratio of the corrected output deflection to the input deflection, times a multiplier derived from the attenuator setting and the calibrations of the couples. Depending on whether absolute or only comparative measurements were to be made, the determination of $|Y(\omega)|^2$ was done at a series of frequencies or only at the single frequency of maximum amplification. In the latter case the factor $|Y(\omega)|^2$ was only a sort of amplif-

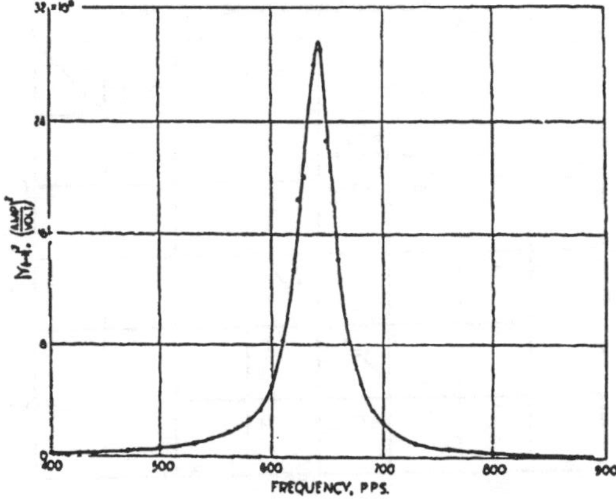

Fig. 3. Typical resonance curve.

[5] Jensen, Phys. Rev. 26, 118 (1925).

cation factor which served as a check on the constancy of the amplifier, while in the former case it was used in the determination of the integral of Eqs. (1) and (3).

A typical curve of amplification factor versus frequency is shown in Fig. 3, obtained from the data for one particular condition of the resonant circuit. The area under curves of this type, establishing the value of the integral, was measured by a planimeter. Since, however, the curves could not be plotted for the full range of the integrals, from zero to infinite frequency, it was necessary to estimate that part of the total area which was limited by the extremes of the characteristic curves. Some of the curves were actually carried out so far (that of Fig. 3, for instance, from 200 to 2200 periods per second) that the further extension to infinite and to zero frequency would have added an almost certainly negligible amount to the total area. Other curves extended only far enough to delimit the greater part of the area, and to the area under these was added a correction arrived at by comparison with the more extensive curves.

An alternating current bridge was used for measuring the resistances R_0 and the capacities C at the frequencies involved. The capacities were of the order 50 mmf. for the input to the amplifier, 10 mmf. for the resistance and from 5 to 150 mmf. for the leads. The effective resistance of the input (grid to filament of the first amplifier tube) was of the order 15 megohms. The measurements of input resistance and capacity were made while the resistance unit was connected in the input of the amplifier, and the amplifier was in the normal operating condition.

Temperature baths were prepared in Dewar flasks when it was desired to keep the resistances at a temperature other than that of the room. The heating or cooling agents were boiling water, melting ice, solid carbon dioxide in acetone, and old liquid air. The temperatures were measured by a platinum resistance thermometer.

There remain to be described the resistance units upon which the experiments were done. These were chosen so as to include materials of different properties, such as high or low resistivity, positive or negative temperature coefficient, metallic or electrolytic conduction, light ions or heavy ions. In value the resistances ranged from a few thousand ohms to a few megohms.

For values of more than one-half megohm commercial "grid leak" resistances were used, made with India ink on paper. Another type of commercial product, made of carbon filament wound on lavite, covered the range of resistance below one megohm. Platinum and copper resistances were used in the form of thin films deposited on glass by evaporation. A resistance of nearly a half megohm was made of Advance wire, wound non-inductively. This was provided with a tap so that one-third or two-thirds of it could be used.

Electrolytic resistances were made of aqueous solutions of the salts NaCl, $CuSO_4$, K_2CrO_4, $Ca(NO_3)_2$, and of a solution of sulphuric acid in ethyl alcohol. Glass tubes about 15 cm long, some of capillary size and some larger, were filled with the different solutions of such strength as to give them

all about the same resistance, and this was repeated with another resistance value.

Measurements and Results

A considerable part of the work consisted of comparative measurements in which the characteristics of the amplifier did not need to be known. In these circumstances only the maximum amplification was determined. It was convenient in such cases to think of the resistance as impressing on the amplifier a mean-square potential $\overline{V^2}$. By this method of comparison was determined the fact that the phenomenon is independent of the material and shape of the resistance unit and of the mechanism of the conduction,[6] but does depend on the electrical resistance. A few of the results are reproduced in Fig. 4. They are expressed in terms of $\overline{V^2}$, the apparent mean-square potential fluctuation, plotted against the resistance component $R(\omega)$. The points lie close to a straight line. The quantity $W=\overline{V^2}/R(\omega)$, which may be

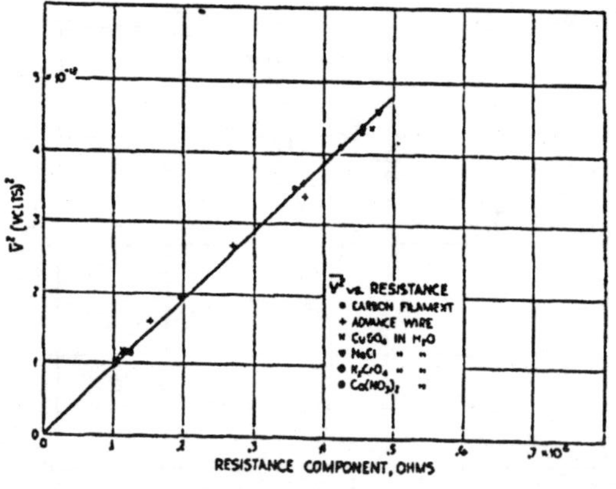

Fig. 4. Voltage-squared vs. resistance component for various kinds of conductors.

called the power equivalent of the effect, is independent of all the variables so far considered, including the electrical resistance itself.

The effect of the shunt capacity C across the conductor is shown in Fig. 5. In this case the abscissae are the measured values of resistance R_0, the circles marking the observed values of the apparent $\overline{V^2}$. These values of $\overline{V^2}$ reach a maximum and then actually decrease as the resistance is indefinitely increased. Obtaining a factor of proportionality K from the initial slope of the curve and using the measured value of C and ω for the calculation of $\overline{V^2}=KR(\omega)=KR_0/(1+\omega^2C^2R_0^2)$, the expected values of $\overline{V^2}$ were calculated. These are represented by the curve in Fig. 5. The course of the calculated curve agrees well with that for the observed points. The agreement was also verified by using a fixed resistance and adding known shunt capacities up to as high as 60,000 mmf.

[6] Resistances such as thermionic tubes and photoelectric cells, perhaps all resistances not obeying Ohm's law, are exceptions to this rule. In these the statistical conditions are different.

The effect of the temperature of the resistance element was studied chiefly by the same comparative method that was used for the varied resistances. The experiments were done on the Advance wire resistance and on carbon filament resistances over the temperature range from −180°C (liquid air) to 100°C (boiling water). They were also done on two liquid resistances made of alcohol and sulphuric acid, covering a range of tempera-

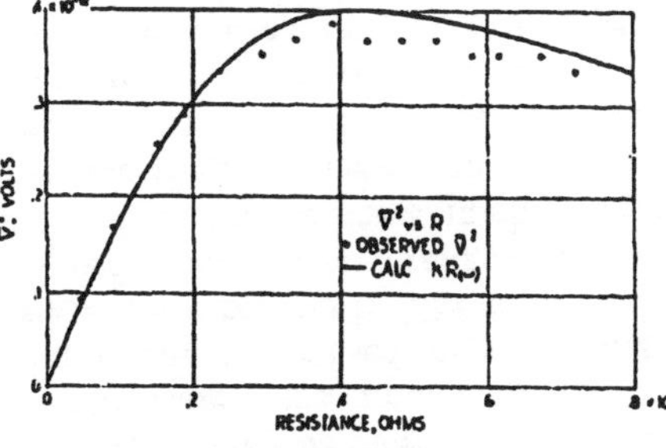

Fig. 5. Voltage-squared vs. resistance with fixed shunt capacity; frequency 635 p.p.s., capacity 577 mmf.

tures from −72°C to 90°C. The resistance values used in the computations were those measured at the various temperatures. Advance wire and carbon filaments changed very little in resistance over the temperature range used for them, while that of the liquid elements changed tremendously, increasing for lower temperatures. In all cases, however, the virtual power $\overline{V^2}/R(\omega)$

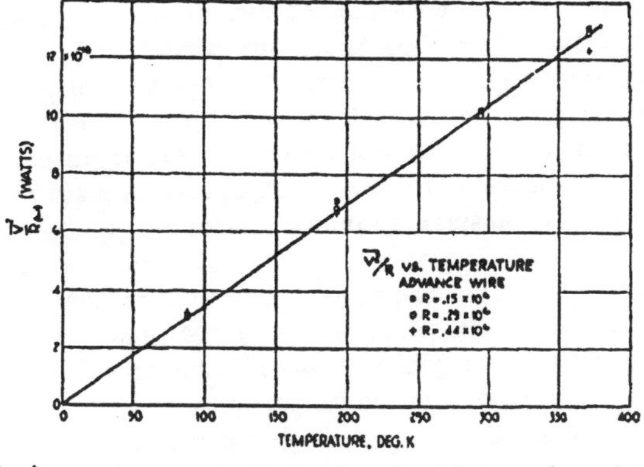

Fig. 6. Apparent power vs. temperature, for Advance wire resistances.

was proportional to the absolute temperature of the resistance element. Fig. 6 shows graphically the results for the three Advance wire resistances. The other resistances gave values falling closely on the same straight line as these.

There is finally to be recounted the verification of Eqs. (1) and (3) as they involve the electrical properties of the measuring system. The method

of obtaining the frequency characteristic of the amplifier, the graphic integration of this curve and the corrections that were applied, have already been described. For these determinations the amplifier was altered in various ways, but usually by changing the resonant circuit forming one of the interstage couplings, or by replacing this circuit by the band-pass filter. The natural frequency of the resonant circuit was changed, by means of the inductance or condenser, over the range of 300 to 2000 periods per second. The sharpness of resonance was varied by changing either the resistance in the resonant circuit or the inductance in series with this circuit and the tube. The area under the characteristic curve could thus be made larger or smaller over a considerable range.

The input resistance element, in all of this work, was kept at only slightly above room temperature. It was of the "grid leak" type for most of those measurements in which the resonant circuit was employed, while the Advance

TABLE I. *Determination of Boltzmann's Constant.*

No.	f	T	$R(\omega)$	$\int_0^\infty \lvert Y(\omega)\rvert^2 d\omega$	$\int_0^\infty \lvert Y(\omega)\rvert^2 R(\omega) d\omega$	ΔS	$\overline{V^2}$	k
	p.p.s.	°K	$\times 10^{-6}$	$\times 10^{-10}$	$\times 10^{-16}$	%	$\times 10^{6}$	$\times 10^{16}$
1	1010	298	.526	.213		12.0	2.7	1.27
2	2023	"	.470	.272		26.2	2.8	1.15
3	1418	"	.508	.361		15.6	3.8	1.09
4	"	"	"	.188		42.3	1.8	.99
5	295	"	.548	.252		3.3	3.1	1.18
6	"	"	"	.202		15.8	2.0	.95
7	302	"	"	.221		12.4	2.7	1.18
8	"	"	"	.195		15.3	2.3	1.13
9	653	"	.541	.747		6.8	10.4	1.14
10	"	"	"	.645		12.9	8.6	1.30
11	1418	"	.508	.286		18.5	3.5	1.26
12	"	"	"	.161		41.3	1.7	1.09
13	1465	"	.505	1.93		18.7	21.2	1.14
14	"	"	"	1.75		20.3	19.1	1.14
15	635	"	.541	.594		8.8	8.9	1.46
16	"	"	"	.139		35.0	2.1	1.47
17	"	"	"	.597		10.9	7.8	1.28
18*	643	295	.44 ±		.439	0	11.0	1.38
19	645	297	"		.396	0	11.1	1.49
20	1830	301	"		.913	0	19.8	1.13
21	500–1000	299	"		.831	0	25.9	1.64
22	"	300	"		.662	0	21.5	1.70
23	"	300	"		.832	0	26.0	1.63

* The resonance curve for this determination is that reproduced in Fig. 3.

wire resistance was used as input element in connection with the band-pass filter. The resistance and capacity of each was measured accurately *in situ*, the resistance being of the order of one-half megohm.

The results will be presented in terms of the value of Boltzmann's constant k, as calculated from the data according to Eqs. (1) or (3). The last column of Table I contains these calculated values, while in the other columns of the table are indicated some of the experimental conditions involved in each determination. The columns of the table indicate, in order, the resonance frequency or frequency range of the selective circuit, the temperature and the ohmic value of the input resistance component, and the value of one or the other infinite integral. The column under ΔS contains the part, in percent, which the estimated area contributes to the total integral. I^2 is the current-squared observed in the thermocouple.

The average of the values of k in the last column of the table is 1.27×10^{-16} ergs per degree, with a mean deviation of 13 percent. The average is 7.5 percent lower than the accepted value[7] of k, 1.372×10^{-16} ergs per degree. The series of measurements, therefore, yields a value of k which is correct within the mean deviation. I believe, however, that the method is capable of a much higher accuracy than that obtained here. This leads to a discussion of the possible sources of error.

An inspection of the tabulated data reveals no systematic relation of the value of k to the numbers in any one column, except that the results of the last three measurements, made with the band-pass filter, are higher than the rest. The deviations are apparently distributed at random. Among the quantities which enter Eq. (1) there can be little question of the accuracy of the temperature T and the resistance component $R(\omega)$. The error is therefore to be sought in the current-squared, I^2, and in the integral of $|Y(\omega)|^2 d\omega$. The latter is indeed open to the criticism that the correction term for the area was often large and therefore uncertain, but this cannot be held against the last six determinations. It is possible, because of feedback through the internal capacities of the tubes, that the amplification of the system was not the same when the resistance alone was connected across the input as when this was shunted by the low resistance of the calibration circuit. It is not clear, however, how such a change in amplification could make the values of k sometimes too large and at other times too small. The other debatable factor, I^2, may be questioned in two respects. There was a "zero reading," due to tube noise, amounting usually to about ten percent of the total reading of the meter used for measuring I^2. This zero reading, however, was always quite constant. The total reading, on the other hand, was not constant but fluctuated over a range comprising about five percent of the total reading. This unsteadiness might be caused by disturbances generated either within the amplifier or coming from the outside, or it might represent a natural fluctuation in the phenomenon itself. Two different procedures were used in obtaining the reading. At first a single datum was used, taken after watching the needle for perhaps thirty seconds. Judgment placed this value nearer the lower limit of the excursion of the needle, since disturbances would tend to make the readings high. For each of the last six sets of measurements, Nos. 18 to 23, the average of a series of readings was used. The meter was read at about one hundred equally spaced intervals of time, before and after obtaining the frequency characteristic of the system. The average of these readings was slightly greater than that arrived at by judging the undisturbed position of the needle.

[7] Int. Crit. Tables, v. 1, p. 18.

Whatever the cause of the deviation, the fact remains that there is essential agreement in all respects between the theory and the experimental results. It is remarkable that the same apparatus by which it is possible to determine the charge on the electron by means of the small-shot effect, can by a slight change in procedure be made to yield an independent measurement of Boltzmann's gas constant.

Technical Aspects of the Phenomenon

Since the thermal agitation places a limit on what can be done by amplifiers, it will be of interest to discuss this limit in terms more commonly used than those of Eq. (1). It is more convenient to speak of a disturbance to an amplifier as a voltage fluctuation at the input than as a current fluctuation at the output which has been used here. A voltage fluctuation across the input of the amplifier cannot, in general, logically be derived from Eq. (1). Certain simplifying conditions may be assumed, however, which make such a derivation plausible. Let us take a circuit having constant amplification over the frequency range f_1 to f_2, zero amplification outside this range. The input resistance component may be assumed to have the constant value R within the frequency range of the circuit. Eq. (1) may then be written

$$I^2=(2kTRY^2/\pi)(\omega_2-\omega_1)=4kTRY^2(f_2-f_1)=4kTR(f_2-f_1)I^2/V^2, \tag{4}$$

where V is the r.m.s. voltage, having any frequency within the pertinent range, applied at the input of the amplifier, and I is the current produced at the output of the circuit. Now the voltage V may be given such a magnitude that the corresponding output current-squared equals that produced by the thermal agitation in the resistance R. The value of V^2 can then be considered equivalent to the voltage-squared generated by the thermal agitation and may be denoted by $\overline{V^2}$. Eq. (4) becomes

$$I^2=4kTR(f_2-f_1)\overline{I^2}/\overline{V^2}\ ;$$

$$\overline{V^2}=4kTR(f_2-f_1)=WR\ ; \tag{5}$$

$$W=\overline{V^2}/R=4kT(f_2-f_1). \tag{6}$$

W is the virtual power, which in its effect on the output meter is equivalent to an actual power of this value dissipated in the input resistance. For an amplifier operated at room temperature and covering the approximate voice frequency range of 5000 cycles this power is

$$W=4\times1.37\times10^{-16}\times300\times5000\times10^{-7}=0.82\times10^{-16}\text{ watt.}$$

If the input resistance is one-half megohm the mean-square voltage fluctuation is $\overline{V^2}=WR=0.82\times10^{-16}\times0.5\times10^{6}=0.41\times10^{-10}$ (volts)^2, to which corresponds an apparent input voltage of 6.4 microvolts. A value very close

[8] The value of 10^{-16} watt given in the cited notes was for the narrower frequency band of a resonant circuit.

to this was actually observed when the experimental amplifier was given approximately the characteristics assumed here.

For input elements other than pure resistance the problem cannot be handled in this simple way, but Eq. (1) should still apply as it has been seen to do in the case of the resistance-capacity combination. When a resonant circuit was used in the input, however, the observed value of $\overline{I^2}$ was somewhat greater than that computed from the characteristics of the system. The difficulty was, no doubt, that it was impossible by means conveniently at hand to shield the inductance coil well enough against magnetic disturbances of external origin.

Towards the problem of reducing the noise in amplifiers caused by thermal agitation the theory makes three suggestions. The first is the use of a low input resistance. This factor, however, is not usually entirely at our disposal, being influenced by the apparatus which supplies the small voltage that is to be amplified. Secondly, the input resistance may be kept at a low temperature. This expedient, too, has practical limitations since the elements which make up the input resistance cannot always conveniently be confined in a small space. The third possible way to reduce the noise consists in making the frequency range of the system no greater than is essential for the proper transmission of the applied input voltage.[9] For the purpose of detecting or measuring a voltage of constant frequency and amplitude one may go to extremes in making the system selective and thereby proportionately reducing the noise. One may, for instance, make use of the great sharpness of tuning obtained by means of mechanical rather than electrical resonance. When, however, the applied voltage varies in frequency or amplitude, the system must have a frequency range large enough to take care of these variations, and the presence of a certain level of noise must be accepted. Beyond this, the chief value of the knowledge gained here lies in the ability it gives to predict the lower limit of noise in any case, so that the impossible will not be expected of an amplifier system.

The noise of thermal agitation is usually the predominant source of disturbance in a well constructed amplifier. Thermionic tubes, however, produce a noise of the same nature, which may exceed the thermal noise when the input resistance component is small. This tube noise is a result of fluctuations in the space current of the tube.[10] When tubes are operated at too low a filament temperature the greater part of these fluctuations are caused by the phenomena named by Schottky "small-shot effect" and "flicker effect." In tubes operated with full space charge limitation of the current, the small-shot effect and flicker effect are largely or perhaps entirely suppressed by the space charge. Under these conditions, and with the grid connected directly to the filament, there is still a remanent fluctuation of current in the tube. In a rough way the amount of this disturbance may be described in terms of the resistance R_e which when connected between the grid and filament of the tube would cause an equal disturbance in the amplifier, so that the total noise is given by the relation

[9] The principle holds for disturbances of other kinds, cf. J. R. Carson, Bell System Tech. J. 4, 265 (1925).

[10] Johnson, Phys. Rev. 26, 71 (1925); A. W. Hull, Phys. Rev. 25, 147 (1925); Phys. Rev. 27, 439 (1926); W. Schottky, Phys. Rev. 28, 74 (1926).

$$\overline{V^2}=W(R+R_e). \tag{7}$$

The order of magnitude of this resistance R_e, for a tube connected for operation in the range of voice frequencies, may be estimated from the working formula

$$R_e=1.5\times10^6 i_p/\mu, \tag{8}$$

the symbol μ standing for the effective amplification of the tube in combination with its external output impedance, i_p for the space current in the tube measured in milliamperes. Consider, as an example of this rule, a tube having a space current $i_p=0.5$ milliampere and an amplification factor $\mu_0=30$, working into an external impedance equal to its own internal plate resistance. The effective amplification is then $\mu=15$, and the minimum noise of the tube should be equal to that of a resistance $R_e=5{,}000$ ohms in the grid circuit of the tube. If the tube works into an amplifier such as was previously considered this tube noise is therefore equivalent to an effective voltage of .9 microvolt impressed on the grid of the tube. It is not suggested that the predicted minimum noise can be attained with every tube. Many tubes, defective in some way or other, have noise levels much higher than the minimum. It is seen, however, that with the best tubes, properly operated, the tube noise is important only when the resistance component of the input circuit is smaller than the equivalent resistance R_e. With resistances greater than this the thermal agitation should contribute the preponderant part of the noise.[11]

It is interesting to note that in another field of measurements the effect of the thermal agitation of charge in conductors has made itself felt. A group of workers[12] in the Netherlands observed that the deflection of a highly sensitive string galvanometer executed random deviations from the zero position. They ascribed this phenomenon, evidently correctly, in part to "Brownian motion of current" in the galvanometer system. Since then the phenomenon has been further investigated,[13] both theoretically and experimentally, with the conclusion that with a galvanometer the measurement of

[11] In my earlier paper (l.c. 10) the relation of noise to amplification for a number of tubes is shown in Fig. 13. When these data were obtained *a resistance of one-half megohm was connected between the grid and filament of the tube*, a connection which was thought more normal than having no resistance in the grid circuit. The lower limit of noise appears therefore to have been "resistance noise" rather than "tube noise" as this term is now used. Since the noise was measured at the output of the amplifier it is natural that the observed noise should have increased with increasing amplification constant of the experimental tube.

[12] W. Einthoven, F. W. Einthoven, W. van der Holst & H. Hirschfeld, Physica 5, 358 (1925).

[13] J. Tinbergen, Physica 5, 361 (1925); G. Ising, Phil. Mag. 1, 827 (1926); F. Zernike, Zeits. f. Physik 40, 628 (1926); A. V. Hill, J. Sc. Instr. 4, 72 (1926).

direct current of less than 10^{-12} ampere becomes unreliable, just as the alternating potential of 10^{-6} volt marks a critical region for amplifiers.

In conclusion, I wish to acknowledge my indebtedness to Dr. O. E. Buckley for the helpful criticism he has given during the course of this work; and to Mr. J. H. Rohrbaugh for assistance in the measurements.

Bell Telephone Laboratories, Incorporated,
December 20, 1927.*

* Received April 19, 1928. Ed.

The Molecular Beam Resonance Method for Measuring Nuclear Magnetic Moments

The Magnetic Moments of $_3Li^6$, $_3Li^7$ and $_9F^{19}$*

I. I. RABI, S. MILLMAN, P. KUSCH, *Columbia University, New York, New York*

AND

J. R. ZACHARIAS, *Hunter College, New York, New York*

(Received January 20, 1939)

A new method of measuring nuclear or other magnetic moment is described. The method, which consists essentially in the measurement of a Larmor frequency in known magnetic fields, is of very general application and capable of the highest precision in absolute and relative measurements. The apparatus consists of two magnets in succession which produce inhomogeneous magnetic fields of oppositely directed gradients. A molecular beam of the substance to be studied possesses a sigmoid path in these magnets and is focused on a suitable detector. A third magnet which produces a homogeneous field is placed in the region between the two deflecting magnets. In this strong homogeneous field the nuclear moments are decoupled from other nuclear moments and from rotational moments of a molecule in a $^1\Sigma$ state, and precess with their Larmor frequency $\nu = \mu H/hI$. An oscillating field perpendicular to the homogeneous field produces transitions to other states of space quantization when the frequency of this field is close to ν. If such transitions take place the molecule is no longer focused on to the detector by the subsequent inhomogeneous field and the observed intensity diminishes. The application of the method to the molecules LiCl, LiF, NaF and Li_2 is described. The nuclear moments of Li^7, Li^6 and F^{19} were found to be 3.250, 0.820 and 2.622 nuclear magnetons, respectively.

THE magnetic moment of the atomic nucleus is one of the few of its important properties which concern both phases of the nuclear problem, the nature of the nuclear forces and the appropriate nuclear model. According to current theories the anomalous moment of the proton is directly connected with the processes from which nuclear forces arise. The question whether the intrinsic moments of the proton and neutron are maintained within the nucleus is part of the problem of two and multiparticle forces between nuclear constituents. With regard to the atomic model it is clear that the nuclear angular momentum does not alone suffice to fix the nature of the wave functions which specify the state of the nucleus. The magnetic moment, on the other hand, is sensitive to the relative contributions of spin and orbital moment and, with the advance of mathematical technique, suffices to decide between the different proposed configurations.

In the light of these considerations it is particularly desirable that nuclear moments be known to high precision because small effects may be of great importance. A case in point is that of $_3Li^7$; according to the calculations of Rose and Bethe[1] the contribution of the orbital motions to the moment of this nucleus is about 10 percent of the total moment. The rest is contributed by the intrinsic proton moment. If the nuclear moment were known to only 10 percent, the importance of this datum would be greatly diminished.

In two letters to this journal,[2, 3] we reported briefly on a new precision method of measuring nuclear moment, and on some results. In this paper we shall give a more detailed account of the method, apparatus and results.

METHOD

The principle on which the method is based applies not only to nuclear magnetic moments but rather to any system which possesses angular momentum and a magnetic moment. We consider a system with angular momentum, J, in units of $h/2\pi$, and magnetic moment μ. In an external magnetic field H_0 the angular mo-

* Publication assisted by the Ernest Kempton Adams Fund for Physical Research of Columbia University.

[1] M. E. Rose and H. A. Bethe, Phys. Rev. **51**, 205 (1937).

[2] I. I. Rabi, J. R. Zacharias, S. Millman and P. Kusch, Phys. Rev. **53**, 318 (1938).

[3] I. I. Rabi, S. Millman, P. Kusch and J. R. Zacharias, Phys. Rev. **53**, 495 (1938).

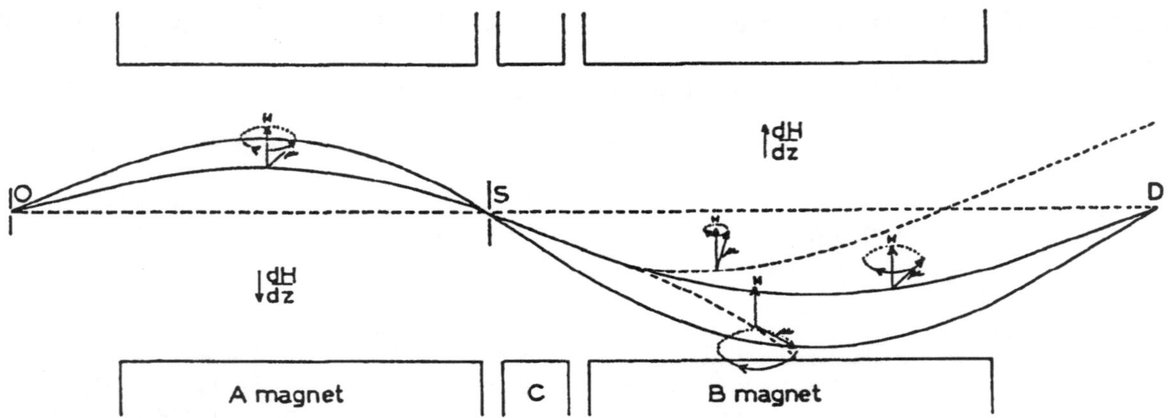

FIG. 1. Paths of molecules. The two solid curves indicate the paths of two molecules having different moments and velocities and whose moments are not changed during passage through the apparatus. This is indicated by the small gyroscopes drawn on one of these paths, in which the projection of the magnetic moment along the field remains fixed. The two dotted curves in the region of the B magnet indicate the paths of two molecules the projection of whose nuclear magnetic moments along the field has been changed in the region of the C magnet. This is indicated by means of the two gyroscopes drawn on the dotted curves, for one of which the projection of the magnetic moment along the field has been increased, and for the other of which the projection has been decreased.

mentum will precess with the Larmor frequency, ν, (in revolutions per sec.) given by,

$$\nu=\mu H_0/Jh. \tag{1}$$

Our method consists in the measurement of ν in a known field H_0. The measurement of ν is the essential step in this method, since H_0 may be measured by conventional procedures. Using Eq. (1) we obtain the gyromagnetic ratio. If, in addition, the angular momentum, J, of the system is known, we can evaluate the magnetic moment μ. In its present state of development our method is not suitable for the measurement of J.

The process by which the precession frequency ν is measured has a rather close analog in classical mechanics. To the system described in the previous paragraph, we apply an additional magnetic field H_1, which is much smaller than H_0 and perpendicular to it in direction. If we consider the initial condition such that H_1 is perpendicular to both the angular momentum and H_0, the additional precession caused by H_1 will be such as to increase or decrease the angle between the angular momentum, J, and H_0, depending on the relative directions. If H_1 rotates with the frequency ν this effect is cumulative and the change in angle between H_0 and J can be made large. It is apparent that if the frequency of revolution, f, of H_1 about H_0 is markedly different from ν, the net effect will be small. Furthermore, if the sense of rotation of H_1 is opposite to that of the precession, the effect will also be small. The smaller the ratio H_1/H_0 the sharper this effect will be in its dependence on the exact agreement between the frequency of precession, ν, and the frequency f.

Any method which enables one to detect this change in orientation of the angular momentum with respect to H_0 can therefore utilize this process to measure the precession frequency and therefore the magnetic moment. The general method here outlined includes not only the magnitude but also the sign of the magnetic moment since the direction of precession depends on whether the magnetic moment vector is parallel or antiparallel to J.

The precise form of the initial conditions previously described is not important and we may consider H_1 initially at any angle ϕ with the plane determined by H_0 and J but still perpendicular to H_0. In fact, according to quantum mechanics, we must consider the initial conditions of an ensemble of systems with a definite projection of J on H_0 as uniformly distributed over ϕ. This only means that some systems will increase and other systems will decrease their projections in the direction of H_0.

In practice it is frequently more convenient to use an oscillating field H_1 rather than a rotating field. Although the situation is not quite as clear as for the rotating field, it is reasonable to expect that the effects will be similar if the oscillating field is sufficiently small. A simple calculation shows that no change in the magnitude of the projection of J on H_0 will occur unless the frequency of oscillation is close to the frequency of precession. The sign of the moment does not affect any processes when a pure oscillating field is used, since it may be considered as the superposition of two oppositely rotating fields. Hence no information as to the sign of the moment can be obtained when this type of field is substituted for a rotating field.

Although the reorientations of the system under the combined influence of H_0 and H_1 may be detected in a number of ways,[4] the most delicate and precise is that of molecular beams.

The arrangement used in our experiment is shown schematically in Fig. 1. A stream of molecules coming from the source, O, in a high vacuum apparatus is defined by a collimating slit, S, and detected by some suitable device at D. The magnets, A and B, produce inhomogeneous magnetic fields, the gradients of which, $d|H|/dz$, are indicated by arrows. When these magnets are turned on, molecules having magnetic moment will be deflected in the direction of the gradient if the projection of the moment, μ_z, along the field is positive, and in the opposite direction if μ_z is negative. A molecule starting from O along the direction OS will be deflected in the z direction by the inhomogeneous A field and will not pass through the collimating slit unless its projected moment is very small or it is moving with very high speed. In general, for a molecule having any moment, μ_z, and any energy, $\frac{1}{2}mv^2$, it is possible to find an initial direction for the velocity of the molecule at the source such that the molecule will pass through the collimating slit. This is indicated by the solid lines in the diagram. If d_A denotes the deflection at the detector from the line OSD suffered by the molecule due to the A field alone, it may be expressed by:

$$d_A=(\mu_z/2mv^2)(d|H|/dz)_A G_A.$$

The deflection in the B field will be in a direction opposite to that in the A field and is given by:

$$d_B=(\mu_z/2mv^2)(d|H|/dz)_B G_B.$$

[4] C. J. Gorter, Physica 9, 995 (1936).

The factors, $(d|H|/dz)_A G_A$ and $(d|H|/dz)_B G_B$, depend only on the geometry of the apparatus and can be adjusted to have the same value. Thus if a molecule of any velocity has the same μ_z in both deflecting fields it will be brought back to the detector by the B field. A simple consideration shows that when the fields A and B are properly adjusted the number of molecules which reaches the detector is the same whether the magnets A and B are on or off. The molecular velocity distribution is also the same.

Magnet C produces the homogeneous field H_0. In addition, there is a device, not pictured in Fig. 1, which produces an oscillating field perpendicular to H_0. If the reorientation which we have described takes place in this region the conditions for deflecting the molecules back to D by means of the B magnet no longer obtain. The molecule will follow one dotted line or the other depending on whether μ_z has become more positive or has changed sign. In fact, if any change in orientation occurs, the molecule will miss the detector and cause a diminution in its reading. We thus have a means of knowing when the reorientation effect occurs.

Since most of the systems in which one is interested have small angular momenta ($<10h/2\pi$) the classical considerations given above have to be reconsidered from the point of view of quantum mechanics. The reorientation process is more accurately described as one in which the system, originally in some state with magnetic quantum number, m, makes a transition to another magnetic level, m'. An exact solution for the transition probability for the case where H_1 rotates and is arbitrary in magnitude was given by Rabi.[5] For the particular case of $J=\frac{1}{2}$ we have,

$$P_{(\frac{1}{2},-\frac{1}{2})}=\frac{\sin^2\theta}{1+q^2-2q\cos\theta}\sin^2\pi tf(1+q^2-2q\cos\theta)^{\frac{1}{2}},$$

where $P_{(\frac{1}{2},-\frac{1}{2})}$ is the probability that the system, originally in the state $m=\frac{1}{2}$ is found in the state $m=-\frac{1}{2}$ after a time t, q the ratio of the frequency of revolution f to the frequency of precession $\nu[\nu=\mu(H_0^2+H_1^2)^{\frac{1}{2}}/Jh]$, and $\tan\theta=H_1/H_0$.

For an oscillating field, in the limit where $H_1/H_0\ll 1$, and in the neighborhood of $f=\nu$ this

[5] I. I. Rabi, Phys. Rev. 51, 652 (1937).

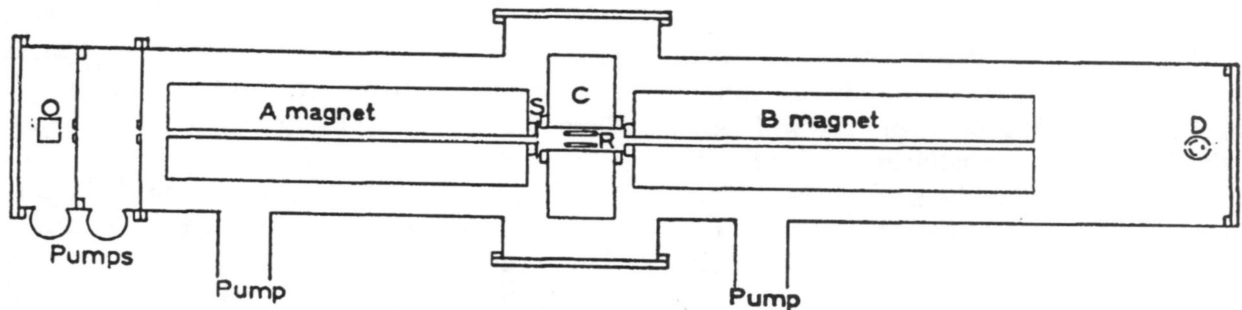

FIG. 2. Schematic diagram of apparatus.

formula becomes:

$$P_{(\frac{1}{2}-\frac{1}{2})}=\frac{\Delta^2}{(1-q)^2+\Delta^2}\sin^2 \pi tf[(1-q)^2+\Delta^2]^{\frac{1}{2}}, \quad (2)$$

where $\Delta=H_1/2H_0$, one-half the ratio of the amplitude of the oscillating field to the static field, and the other symbols retain their meaning. For spins higher than $\frac{1}{2}$ the general formula given by Majorana[6] applies, and

$$P_{(\alpha,\,m,\,m')}=(\cos\tfrac{1}{2}\alpha)^{4J}(J+m)!$$
$$\times(J+m')!(J-m)!(J-m')!$$
$$\times\left[\sum_{\nu=0}^{2J}\frac{(-1)^\nu(\tan\frac{1}{2}\alpha)^{2\nu-m+m'}}{\nu!(\nu-m+m')(J+m-\nu)!(J-m'-\nu)!}\right]^2, \quad (3)$$

where α is defined through $P_{(\frac{1}{2};\,-\frac{1}{2})}=\sin^2(\alpha/2)$. That is, we calculate α for a system which has the same μ/i but with a spin of $\frac{1}{2}$ and subject to the same field, and use it in Eq. (3).

The orders of magnitude involved can be seen from a simple example: consider a system with spin $\frac{1}{2}$ and a moment of 1 nuclear magneton in a field of 1000 gauss and an oscillating field of 10 gauss amplitude. We assume that the system is moving at a speed of 10^5 cm per second which is of the order of thermal velocities, and set $t=l/v=10^{-5}l$. The resonance frequency is

$$\frac{uH}{hi}=\frac{(0.5\times10^{-23})(10^3)}{(6.55\times10^{-27})(\frac{1}{2})}\sim1.5\times10^6 \text{ cycles per sec.,}$$

which fortunately is in a very convenient range of radiofrequencies. To make the $\sin^2$ terms a maximum at $q=1$ we set

$$\pi\times10^{-5}l\times1.5\times10^6\times0.5\times10^{-2}=\pi/2.$$

Solving for l, we obtain $l=6.6$ cm, which is a very convenient length for the oscillating field.

The theoretically simplest systems to which these ideas may be applied in the study of nuclear moment are atoms which are normally in a state with electronic angular momentum equal to zero. If the electronic J is not zero, the interaction of the nuclear spin with the electronic angular momentum is of the order of magnitude of its interaction with the applied field H_0. Moreover, the electronic magnetic moment is so much larger than nuclear moment that the deflections in the A and B fields will be almost entirely due to this electronic moment and the apparatus will accordingly be insensitive to changes in nuclear orientation. Resort must therefore be had to atoms in a state $J=0$ or to molecules in a $^1\Sigma$ state in which all electronic angular momentum is neutralized to the first order. These considerations do not preclude the study, with these methods, of atoms with electronic angular momentum, as such, but rather point out that they are not the most suitable systems for the investigation of nuclear magnetic moment.

As elementary calculations show, the interactions between the nuclear moments of the nuclei in a molecule in a $^1\Sigma$ state and the other angular momentum vectors, such as molecular rotation, are of the order of magnitude of 100 gauss or less. External fields of a few thousand gauss will therefore decouple all the nuclear spins from each other and from the molecular rotation to such a degree that they may be regarded as free. The other interactions will result in a fine structure of constant or decreasing width as H_0 is increased. These effects on the precision can, therefore, be reduced to any assigned value merely by working at suitably high field.

[6] E. Majorana, Nuovo Cimento 9, 43 (1932).

Apparatus

The apparatus (Fig. 2) is contained in a long brass-walled tube divided into three distinct chambers, each with its own high vacuum pumping system. The source chamber contains the oven which is mounted on tungsten pegs. By means of a screw the mount may be moved, under vacuum, in a direction perpendicular to the beam axis. Stopcock grease and Apiezon Q on the screw preserve vacuum even when the screw is turned. The interchamber contains no essential parts of the apparatus, but provides adequate vacuum isolation of the receiving chamber from the gassing of the heated oven, by means of a narrow slit on each end of the chamber. These slits may be moved under vacuum in a manner similar to the oven mount. The receiving chamber contains most of the essential parts of the apparatus: the two deflecting magnets, A and B, the magnet, C, which produces the constant field, the radio-frequency oscillating field, R, the collimating slit, S, and the 1-mil tungsten filament detector, D.

The A and B fields are electromagnets of the type described by Millman, Rabi and Zacharias[7] and are 52 cm and 58 cm long, respectively. The gap is bounded by two cylindrical surfaces, one convex of radius 1.25 mm, and the other concave of radius 1.47 mm. The gap width in the plane of symmetry, defined by the axes of the two cylindrical surfaces, is 1.0 mm. The nature of the field obtained is approximately the same as that produced by two parallel wires with centers 2.5 mm apart and carrying current in opposite directions. Each magnet has four turns of copper windings; current is supplied by a 3000-ampere-hour, 2-volt storage cell. A current of 300 amperes in the windings yields a field of over 12,000 gauss and a gradient of about 100,000 gauss/cm in the gap.

The C magnet, which produces the homogeneous field, is made of annealed Armco iron and is of conventional design. It is wound with 12 turns of $\frac{3}{16}''$ square copper rod to which $\frac{3}{16}''$ copper tubing has been soldered for cooling purposes. Insulation between turns, and between the windings and the magnet, is provided by mica. The pole faces, separated by a gap of $\frac{1}{4}''$, are 10 cm long and 4 cm high. A field of about 23 gauss is realized in the gap per ampere of current in the exciting coils.

In mounting the magnets in the apparatus, care must be taken to avoid regions of weak, rapidly changing fields between magnets. Such regions cause transitions between quantum states of the various magnetic moments associated with the molecule and prevent good refocusing of the beam by the B field. Although the gradient in the B field is necessarily in a direction opposite to that in the A field, the magnetic fields in the planes of symmetry of the two magnets are in the same direction and parallel to that in the C magnet. The magnets are placed as close to each other as the windings will permit. Moreover these windings are completely hidden from the "view" of the molecular beam by mounting slabs of iron as extensions on both ends of the C magnet and on the ends of the A and B magnets facing the C magnet. This arrangement insures a fairly strong field along the entire path of the molecular beam where changes in the over-all magnetic moment of the molecule affects its position at D, i.e., from the beginning of the A field to the end of the B field, and thus limits transitions between the quantum states to the region of the R field, where they may be controlled and studied.

The oscillating field, R, consists of two $\frac{1}{8}''$ copper tubes, 4 cm long, carrying current in opposite directions. These tubes are flattened to permit their insertion between the pole faces of the C magnet when a space of about 1 mm between the tubes is left for the passage of the beam. The plane defined by the centers of these tubes is horizontal and is adjusted to be closely the same as the planes of symmetry of the A and B magnets. These tubes are supported by heavy copper tubing through which electrical and water connections may be made outside the apparatus.

The magnetic field, H_1, produced by a current in the tubes is about 2 gauss/amp. and is approximately vertical and therefore at right angles to the field H_0 produced by the C magnet. The high frequency currents in the tubes are obtained by coupling a loop in series with them

[7] S. Millman, I. I. Rabi and J. R. Zacharias, Phys. Rev. 53, 384 (1938).

to the tank coil of a conventional Hartley oscillator in which an Eimac 250 TL tube is used. The frequencies used for these experiments range from 0.6 to 8 megacycles. The currents producing the oscillating field may be varied from 0 to 40 amperes; the higher currents are more easily obtained at low frequencies.

Procedure

A preliminary line-up of magnets, slits and detector is made by optical means while the apparatus is assembled. If this line-up is sufficiently good, a beam may be sent through the apparatus and a more precise line-up made by means of a triangulation process utilizing the property of rectilinear motion of the molecules in the beam.

The A and B magnets have knife edges at both ends which overlap the gap on the side of the convex pole face by known amounts and extend above the gap into the region above the magnet by a known amount. Since it is impossible to sight through the gaps with a telescope, the preliminary optical line-up is made by sighting on the extensions of the edges in the region above the magnets. In this way it is possible to adjust the plane of symmetry of magnet A to coincide with that of magnet B. The optical line-up is sufficient for this purpose, since no very great precision is needed for this adjustment. It is also possible to adjust optically the lateral position of the magnets as well as the slits and detector to permit a beam to pass through the magnet gaps. The C magnet is lined up so that the median plane of its gap coincides with the centers of the gaps of the A and B magnets. The two wires which produce the radiofrequency field, R, are suspended from a brass plate which is mounted on top of the vacuum chamber, and are so constructed that the width of the assemblage is only very slightly less than the width of the gap in the C magnet. The field R is then arranged in the gap so that it does not short to the poles of the C magnet. Since the width of the gap between the two wires is greater than that of the available working gap in the A and B magnets, this line-up is sufficient for the field R.

A sample of the molecular compound, the magnetic moments of whose constituent nuclei are to be determined, is placed in an oven. The oven is completely closed except for a slit about 0.03 mm wide. It is heated by means of spiral tungsten heaters passed through the oven block and electrically insulated from it by means of quartz tubing. When the temperature of the oven is sufficiently high so that the sample has a vapor pressure of the order of 1 mm of Hg a beam may be observed at the detector and a more precise line-up may be initiated.

In the present apparatus the B magnet is permanently fixed inside the vacuum chamber and all other line-up operations are made with respect to it. By suitable movements of the oven, the collimating slit, and the detector, the beam is shifted until it is cut by each of the two fiduciary knife edges on the B magnet in turn. From a knowledge of the distances separating the various elements involved in a cut-off, it is possible to set the beam parallel to the plane defined by the two edges and to ascertain the distance of the beam from that plane. The only measurements that must be made during this line-up process are the readings of detector positions by means of a calibrated tele-microscope. By successive movements of the oven, collimating slit and detector the beam can be translated parallel to itself by any desired amount. It is thus brought into a position at which one would like to have the plane of the edges of the A field. This field is then moved, in a manner similar to that described for the motion of the oven mount, until its fiduciary edges cut the beam. This operation sets the plane defined by the edges on the A field parallel to the corresponding plane of the B field and at a predetermined distance from it. The beam is then translated to a position approximately midway between these planes.

The experimental criterion which determines the exact position of the beam is that the weakening of a molecular beam at the detector by the A and B fields taken separately must be equal. This may easily be accomplished by a lateral displacement of the beam, since for any such displacement the gradient increases in one of the fields and decreases in the other. When this criterion is satisfied the intensity of the refocused beam with a current of about 300 amp. in the windings of each of the two inhomogeneous

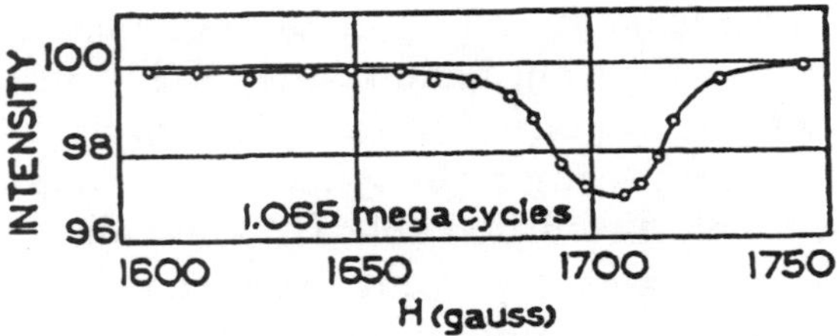

FIG. 3. Resonance curve of the Li^6 nucleus observed in LiCl.

fields is about 95 percent of the beam observed in the absence of fields.

As has been pointed out, the refocusing condition obtains only when there is no change in the space quantization of any of the moment vectors associated with the molecule. If weak fields occur in the region between the A and B fields, transitions may occur. The refocusing becomes good only when the C field itself is fairly large, and the intensity of the refocused beam is an increasing function of the C field up to a value of about 500 gauss. The resonance minima to be described subsequently are usually observed at fields larger than 1000 gauss.

Because the amplifier is not completely shielded from the oscillator and because the steady deflection of the galvanometer associated with the amplifier due to the oscillator is a function of the frequency, observations are made of the beam intensity as a function of the magnetic field, H_0, when the frequency is held fixed. Curves relating the beam intensity to the field H_0, taken for $_3Li^7$, $_3Li^6$ and $_9F^{19}$ are shown in Figs. 3, 4 and 5.

Magnetic and Frequency Measurements

Since the value of the magnetic moment of any nucleus is calculated from an observed magnetic field and an observed frequency it is essential that these quantities be known to a high degree of precision. The frequency of the oscillating magnetic field is determined to better than 0.03 percent by measuring the frequency of the oscillator with a General Radio Type 620A heterodyne frequency meter. It was found that the frequency of the oscillator varied by no more than 0.01 percent during the time required to obtain data on one resonance curve (~15 minutes).

A calibration of the magnetic field of the homogeneous C magnet in terms of the current through the exciting coils was made in the usual way by measuring the ballistic deflection of a galvanometer when a flip coil was pulled from the magnetic field. The galvanometer was calibrated by the use of a 50-millihenry mutual inductance, good to $\frac{1}{5}$ percent. Several flip coils were constructed in this laboratory by winding various types and sizes of insulated wire on carefully measured brass spools. Errors in the magnetic field due to uncertainties in flip coil areas are probably not greater than 0.2 percent.

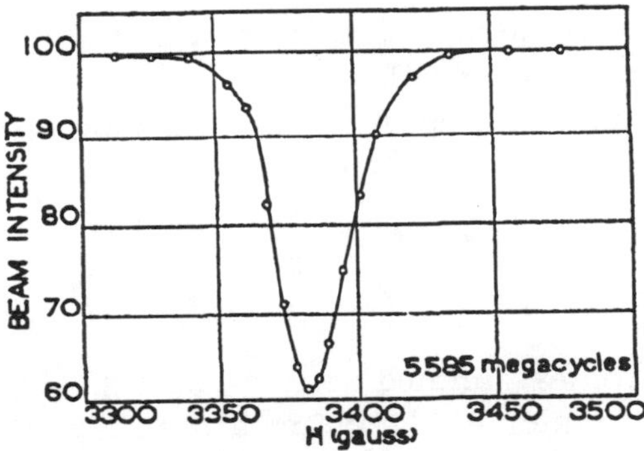

FIG. 4. Resonance curve of the Li^7 nucleus observed in LiCl.

A type K potentiometer was used to measure the potential drop across a shunt in series with the C magnet windings. The same shunt was used both in the calibration and in subsequent work, thereby eliminating the necessity of knowing its resistance accurately.

It is important that the magnetic field always return to the same value for a given magnetizing current. It was found that when a definite, reproducible procedure was used for demagnetizing the homogeneous field and for bringing it up to any state of magnetization, this condition was fulfilled to better than 0.1 percent.

A considerable variation in the value of the mutual inductance was observed, apparently depending on the humidity. The absolute value of the magnetic field is indeterminate to about 0.5 percent due to the uncertainty in the value of the mutual inductance and uncertainty in the areas of the flip coils. This, of course, introduces a corresponding uncertainty in the absolute values of the magnetic moments.

Results

The first nuclei to be studied by this method were $_3Li^6$, $_3Li^7$ and $_9Li^{19}$ in the LiCl, LiF, NaF and

Li_2 molecules. The resonance minima which are obtained are shown in Figs. 3, 4 and 5. For each nucleus the f/H values corresponding to the resonance minima are constant to a very high degree for wide variations of frequency. This shows that we are dealing with a change of nuclear orientation and not with some molecular transition, since such a transition would not possess a frequency proportional to H. A representative sample of the results is shown in Table I. The constancy of f/H also shows that our method of calibration of the C magnet yields accurate results, at least for relative values of the homogeneous field.

The nuclear g is obtained from the observed f/H values by use of the formula

$$g=\frac{4\pi}{e/Mc}\cdot\frac{f}{H}=1.3122\times10^{-3}\frac{f}{H},$$

which follows immediately from Eq. (1) if the magnetic moment μ is measured in units of $eh/4\pi Mc$, the nuclear magneton, and $f=\nu$. The specific charge of the proton in electromagnetic

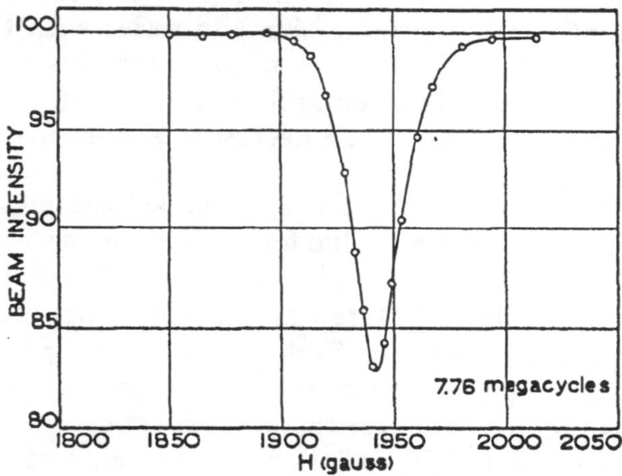

FIG. 5. Resonance curve of the F^{19} nucleus observed in NaF.

units, e/Mc, is obtained directly from the value of the Faraday[8] (9648.9 e.m.u.) and the atomic weight[9] of hydrogen (1.0081). Expressed in this form our experimental results do not depend upon any inaccuracies in e, h or m/M, the ratio of the electronic mass to the mass of the proton. The nuclear spins of Li^6 and Li^7 are known from atomic beam measurements,[10, 11] and that of F^{19} from band spectra.[12] The nuclear moments are obtained directly by multiplying g by i. The nuclear g's, the spins and the magnetic moments are listed in Table II. The values here given are about 0.5 percent lower than, and are to supersede, those published in the preliminary report.[3] The differences are due to the use of a more trustworthy mutual inductance in the calibration of the magnetic field and to an error in the value of the constant $4\pi/(e/Mc)$ previously used. The identification of the resonance minimum with a particular nucleus is made by using the same element in more than one molecule. For example, two of the resonance minima, observed for each of the molecules,[13] LiCl, LiF and Li_2 have f/H values which are the same in all three cases. These must be attributed to the nuclei of the two isotopes of lithium. Since Li^7 is about 12 times as abundant as Li^6 and since the intensity drop at resonance for one of these minima is as much as 60 percent of the refocused beam, this minimum can only be assigned to Li^7. No minimum is definitely assigned to a nucleus unless it has been observed in at least two different molecules.

The accuracy of the nuclear moment values depends solely on a knowledge of the magnetic field, H, at which the Larmor frequency associated with the nuclear magnetic moment is equal to the frequency of the oscillating field. The absolute moment values depend upon the absolute calibration of magnetic standards and cannot at present be taken to be better than 0.5 percent. The relative moment values, on the other hand, do not depend on such standards but merely on the accuracy of the shape of the, magnetization curve for the homogeneous field on the reproducibility of a definite field with the same current in the exciting coils of the homoge-

[8] R. T. Birge, Rev. Mod. Phys. 1, 1 (1929).

[9] Eighth Report of the Committee on Atomic Weights of the International Union of Chemistry, J. Am. Chem. Soc. 60, 737 (1938).

[10] J. H. Manley and S. Millman, Phys. Rev. 51, 19 (1936).

[11] M. Fox and I. I. Rabi, Phys. Rev. 48, 746 (1935).

[12] H. G. Gale and G. S. Monk, Astrophys. J. 69, 77 (1929).

[13] The Li_2 molecules are obtained by heating lithium metal to about 1000°K. At vapor pressures of one mm the beam contains about 0.5 percent molecules. The lithium atoms have magnetic moments of the order of one Bohr magneton and suffer such large deflections in the A field that they do not reach the detector. The problem of working with alkali molecules of the type of Li_2 is solely one of obtaining enough intensity for the molecular beam. The atoms do not interfere with the experiment in any way.

neous field, on the location of a minimum in the resonance curve and on the assumption that any form of interaction tending to broaden the resonance curve and not considered in the simple theory will introduce no asymmetry into the curve. The criterion for the first three points mentioned is the internal consistency of the f/H values obtained under varied conditions. This leads to a precision of about 0.1 percent for the relative moment values of Li^6, Li^7 and F^{19}. From a consideration of the small half-widths observed for the resonance curves (~1 percent) and their symmetrical character it seems unlikely that any interactions are present which will tend to shift the minimum by more than 0.2 or 0.3 percent, if at all.

The simple model which we have used to discuss the principles of the method is, no doubt, insufficient to describe the finer details of the results, such as the width and shape of the resonance curves. For this purpose one must consider the various interactions between the nuclear spins of the different nuclei and their interactions with the rest of the molecular structure. The nature of other perturbations and the physical information which can be obtained from detailed observation of resonance minima will be discussed in another paper.

TABLE I. *Representative values of f/H for Li^6, Li^7 and F^{19}.*

Nucleus	Molecule	f megacycles per second	H gauss	$\frac{f}{H}$
Li^6	LiCl	2.127	3405	624.6
		2.127	3400	625.6
		2.155	3455	623.8
		2.155	3446	625.3
	Li_2	1.714	2742	625.0
		1.714	2744	624.7
	LiF	2.193	3506	625.5
		2.193	3501	626.5
Li^7	LiCl	5.611	3399	1651
		5.610	3400	1650
		6.587	3992	1650
		2.113	1278	1654
		5.552	3383	1651
	LiF	5.621	3401	1653
		6.580	3981	1653
		3.517	2133	1649
	Li_2	3.056	1862	1651
		3.084	1879	1652
		3.129	1907	1651
F^{19}	NaF	5.634	1407	4001
		5.634	1409	3998
		7.799	1949	4001
		7.799	1953	3992
		7.799	1952	3995
	LiF	4.204	1053	3994
		4.204	1055	3986

TABLE II. *Nuclear g's and magnetic moments.*

Nucleus	g	Spin	Moment
${}_3Li^6$	0.820	1	0.820
${}_3Li^7$	2.167	3/2	3.250
${}_9F^{19}$	5.243	1/2	2.622

Discussion

One of the important objects of nuclear moment investigations is to ascertain whether the hyperfine structure of atomic energy levels can be accounted for entirely by the assumption that the nucleus interacts with the external electrons as a small magnet. The effects arising from the finite size of the nucleus and its charge distribution (isotope effect and electric quadrupole moment effect). modify slightly the h.f.s. predicted from this simple assumption but are still within the range of electromagnetic interactions. There may possibly be some other interactions with the electron which are not electromagnetic in nature but more like spin dependent nuclear forces. To this end a comparison of the ratio of the magnetic moments of two isotopes measured by our direct methods with that obtained from the results of h.f.s. measurements on the same isotopes is of interest. Since the electronic wave functions are the same for two isotopes, the ratio of the h.f.s. separations $(\Delta\nu)_1/(\Delta\nu)_2$ of a given atomic energy state should yield the ratio of the moments, μ_1/μ_2, very accurately if no other effect enters. It is to be expected that a discrepancy between these two moment ratios will be very small because of the short time which an electron spends in the region very close to the nucleus.

The ratio of the moments of the lithium isotopes, μ_7/μ_6, found by Manley and Millman[10] by the atomic beam zero moment method of measuring h.f.s. separations is 3.89. Our value is 3.963, which is about two percent higher. It is difficult to be certain, at the present time, that this difference represents a real physical effect rather than an experimental error. Although our value can hardly be off by as much as 0.3 percent, the value given by Manley and Millman may possibly be in error because the Li^6 zero moment peak was not completely resolved from the Li^7 background. Further work along this line is clearly desirable. Another method of studying this question is through very accurate calculations of atomic wave functions (particularly in the case of Li) from which the nuclear moment can be calculated from h.f.s. data to a precision comparable with that of our direct measurements. The present status of this side of the problem is that our value of 3.250 for Li^7 is to be compared with 3.29 obtained from the measurements of Granath[14] on the h.f.s. of Li II, and the calculations of Breit and Doerman.[15] Fox and Rabi[11] find 3.14[16] from atomic beam experiments on Li I and the theory of Goudsmit[17] and that of Fermi and Segrè,[18] while Bartlett, Gibbons and Watson[19] calculate 3.33 from the same data. These differences, though small, may be significant; however, they are, as yet, within the range of accuracy claimed by the calculations.

For $_9F^{19}$, Brown and Bartlett[20] calculate values ranging from 1.9 to 3.8 from the h.f.s. data of Campbell[21] which are to be compared with our value of 2.622. For a discussion of the accuracy and reliability of these calculations see the conclusions of the papers by Bartlett and his co-workers.

From the standpoint of current nuclear theory our results for the ratio of the moments μ_{Li^7}/μ_{Li^6} diverge even more widely from the calculations of Rose and Bethe,[1] than did the previous results of Manley and Millman. In a recent paper, Bethe[22] has sought to improve the previous calculation by the use of an α-particle model. The agreement with experiment is more satisfactory than for the previous theory of Rose and Bethe. Whether this result is accidental remains to be seen from future calculations of other moments with a similar model.

This research has been aided by a grant from the Research Corporation.

[14] L. P. Granath, Phys. Rev. 36, 1018 (1930).
[15] G. Breit and F. W. Doerman, Phys. Rev. 36, 1732 (1930).
[16] The value 3.20 cited by Fox and Rabi for the moment of Li^7 is in error due to a mistake in sign of ds/dn in the factor $(1-ds/dn)$ of the formula of Fermi and Segrè. The effect of the correction factor $(1-ds/dn)$ is to decrease the moment and not to increase it.
[17] S. Goudsmit, Phys. Rev. 43, 636 (1933).
[18] E. Fermi and E. Segrè, Zeits. f. Physik 82, 729 (1933).
[19] J. H. Bartlett, J. J. Gibbons and R. E. Watson, Phys. Rev. 50, 315 (1936).
[20] F. W. Brown and J. H. Bartlett, Phys. Rev. 45, 527 (1934).
[21] J. S. Campbell, Zeits. f. Physik 84, 393 (1933).
[22] H. A. Bethe, Phys. Rev. 53, 842 (1938).

Resonance Absorption by Nuclear Magnetic Moments in a Solid

E. M. Purcell, H. C. Torrey, and R. V. Pound*
Radiation Laboratory, Massachusetts Institute of Technology, Cambridge, Massachusetts
December 24, 1945

IN the well-known magnetic resonance method for the determination of nuclear magnetic moments by molecular beams,[1] transitions are induced between energy levels which correspond to different orientations of the nuclear spin in a strong, constant, applied magnetic field. We have observed the absorption of radiofrequency energy, due to such transitions, in a *solid* material (paraffin) containing protons. In this case there are two levels, the separation of which corresponds to a frequency, ν, near 30 megacycles/sec., at the magnetic field strength, H, used in our experiment, according to the relation $h\nu = 2\mu H$. Although the difference in population of the two levels is very slight at room temperature ($h\nu/kT \sim 10^{-5}$), the number of nuclei taking part is so large that a measurable effect is to be expected providing thermal equilibrium can be established. If one assumes that the only local fields of importance are caused by the moments of neighboring nuclei, one can show that the imaginary part of the magnetic permeability, at resonance, should be of the order $h\nu/kT$. The absence from this expression of the nuclear moment and the internuclear distance is explained by the fact that the influence of these factors upon absorption cross section per nucleus and density of nuclei is just cancelled by their influence on the width of the observed resonance.

A crucial question concerns the time required for the establishment of thermal equilibrium between spins and lattice. A difference in the populations of the two levels is a prerequisite for the observed absorption, because of the relation between absorption and stimulated emission. Moreover, unless the relaxation time is very short the absorption of energy from the radiofrequency field will equalize the population of the levels, more or less rapidly, depending on the strength of this r-f field. In the expectation of a long relaxation time (several hours), we chose to use so weak an oscillating field that the absorption would persist for hours regardless of the relaxation time, once thermal equilibrium had been established.

A resonant cavity was made in the form of a short section of coaxial line loaded heavily by the capacity of an end plate. It was adjusted to resonate at about 30 mc/sec. Input and output coupling loops were provided. The inductive part of the cavity was filled with 850 cm^3 of paraffin, which remained at room temperature throughout the experiment. The resonator was placed in the gap of the large cosmic-ray magnet in the Research Laboratory of Physics, at Harvard. Radiofrequency power was introduced into the cavity at a level of about 10^{-11} watts. The radiofrequency magnetic field in the cavity was everywhere perpendicular to the steady field. The cavity output was balanced in phase and amplitude against another portion of the signal generator output. Any residual signal, after amplification and detection, was indicated by a microammeter.

With the r-f circuit balanced the strong magnetic field was slowly varied. An extremely sharp resonance absorption was observed. At the peak of the absorption the deflection of the output meter was roughly 20 times the magnitude of fluctuations due to noise, frequency, instability, etc. The absorption reduced the cavity output by 0.4 percent, and as the loaded Q of the cavity was 670, the imaginary part of the permeability of paraffin, at resonance, was about $3\cdot 10^{-6}$, as predicted.

Resonance occurred at a field of 7100 oersteds, and a frequency of 29.8 mc/sec., according to our rather rough calibration. We did not attempt a precise calibration of the field and frequency, and the value of the proton magnetic moment inferred from the above numbers, 2.75 nuclear magnetons, agrees satisfactorily with the accepted value, 2.7896, established by the molecular beam method.

The full width of the resonance, at half value, is about 10 oersteds, which may be caused in part by inhomogeneities in the magnetic field which were known to be of this order. The width due to local fields from neighboring nuclei had been estimated at about 4 oersteds.

The relaxation time was apparently shorter than the time ($\sim$ one minute) required to bring the field up to the resonance value. The types of spin-lattice coupling suggested by I. Waller[2] fail by a factor of several hundred to account for a time so short.

The method can be refined in both sensitivity and precision. In particular, it appears feasible to increase the sensitivity by a factor of several hundred through a change in detection technique. The method seems applicable to the precise measurement of magnetic moments (strictly, gyromagnetic ratios) of most moderately abundant nuclei.

It provides a way to investigate the interesting question of spin-lattice coupling. Incidentally, as the apparatus required is rather simple, the method should be useful for standardization of magnetic fields. An extension of the method in which the r-f field has a rotating component should make possible the determination of the sign of the moment.

The effect here described was sought previously by Gorter and Broer, whose experiments are described in a paper[3] which came to our attention during the course of this work. Actually, they looked for dispersion, rather than absorption, in LiCl and KF. Their negative result is perhaps to be attributed to one of the following circumstances: (a) the applied oscillating field may have been so strong, and the relaxation time so long, that thermal equilibrium was destroyed before the effect could be observed—(b) at the low temperatures required to make the change in permeability easily detectable by their procedure, the relaxation time may have been so long that thermal equilibrium was never established.

* Harvard University, Society of Fellows (on leave).
[1] Rabi, Zacharias, Millmann, and Kusch, Phys. Rev. 53, 318 (1938).
[2] I. Waller, Zeits. f. Physik 79, 370 (1932).
[3] Gorter and Broer, *Physica* 9, 591 (1942).

Nuclear Induction

F. Bloch, W. W. Hansen, and Martin Packard
Stanford University, Stanford University, California
January 29, 1946

THE nuclear magnetic moments of a substance in a constant magnetic field would be expected to give rise to a small paramagnetic polarization, provided thermal equilibrium be established, or at least approached. By superposing on the constant field (z direction) an oscillating magnetic field in the x direction, the polarization, originally parallel to the constant field, will be forced to precess about that field with a latitude which decreases as the frequency of the oscillating field approaches the Larmor frequency. For frequencies near this magnetic resonance frequency one can, therefore, expect an oscillating induced voltage in a pick-up coil with axis parallel to the y direction. Simple calculation shows that with reasonable apparatus dimensions the signal power from the pick-up coil will be substantially larger than the thermal noise power in a practicable frequency band.

We have established this new effect using water at room temperature and observing the signal induced in a coil by the rotation of the proton moments. In some of the experiments paramagnetic catalysts were used to accelerate the establishment of thermal equilibrium.

By use of conventional radio techniques the induced voltage was observed to produce the expected pattern on an oscillograph screen. Measurements at two frequencies ν showed the effect to occur at values H of the z field such that the ratio H/ν had the same value. Within our experimental error this ratio agreed with the g value for protons, as determined by Kellogg, Rabi, Ramsey, and Zacharias.[1]

We have thought of various investigations in which this effect can be used fruitfully. A detailed account will be published in the near future.

[1] J. M. B. Kellogg, I. I. Rabi, N. F. Ramsey, and J. R. Zacharias, Phys. Rev. 56, 738 (1939).

Spin Echoes*†

E. L. HAHN‡
Physics Department, University of Illinois, Urbana, Illinois
(Received May 22, 1950)

Intense radiofrequency power in the form of pulses is applied to an ensemble of spins in a liquid placed in a large static magnetic field H_0. The frequency of the pulsed r-f power satisfies the condition for nuclear magnetic resonance, and the pulses last for times which are short compared with the time in which the nutating macroscopic magnetic moment of the entire spin ensemble can decay. After removal of the pulses a non-equilibrium configuration of isochromatic macroscopic moments remains in which the moment vectors precess freely. Each moment vector has a magnitude at a given precession frequency which is determined by the distribution of Larmor frequencies imposed upon the ensemble by inhomogeneities in H_0. At times determined by pulse sequences applied in the past the constructive interference of these moment vectors gives rise to observable spontaneous nuclear induction signals. The properties and underlying principles of these spin echo signals are discussed with use of the Bloch theory. Relaxation times are measured directly and accurately from the measurement of echo amplitudes. An analysis includes the effect on relaxation measurements of the self-diffusion of liquid molecules which contain resonant nuclei. Preliminary studies are made of several effects associated with spin echoes, including the observed shifts in magnetic resonance frequency of spins due to magnetic shielding of nuclei contained in molecules.

I. INTRODUCTION

IN nuclear magnetic resonance phenomena the nuclear spin systems have relaxation times varying from a few microseconds to times greater than this by several orders of magnitude. Any continuous Larmor precession of the spin ensemble which takes place in a static magnetic field is finally interrupted by field perturbations due to neighbors in the lattice. The time for which this precession maintains phase memory has been called the spin-spin or total relaxation time, and is denoted by T_2. Since T_2 is in general large compared with the short response time of radiofrequency and pulse techniques, a new method for obtaining nuclear induction becomes possible. If, at the resonance condition, the ensemble at thermal equilibrium is subjected to an intense r-f pulse which is short compared to T_2, the macroscopic magnetic moment due to the ensemble acquires a non-equilibrium orientation after the driving pulse is removed. On this basis Bloch[1] has pointed out that a transient nuclear induction signal should be observed immediately following the pulse as the macroscopic magnetic moment precesses freely in the applied static magnetic field. This effect has already been reported[2] and is closely related to another effect, given the name of "spin echoes," which is under consideration in this investigation. These echoes refer to spontaneous nuclear induction signals which are observed to appear due to the constructive interference of precessing macroscopic moment vectors after more than one r-f pulse has been applied. It is the purpose of this paper to describe and analyze these effects due to free Larmor precession in order to show that they can be applied for the measurement of nuclear magnetic resonance phenomena, particularly relaxation times, in a manner which is simple and direct.

II. FEATURES OF NUCLEAR INDUCTION METHODS

(A) Previous Resonance Techniques (Forced Motion)

The chief method for obtaining nuclear magnetic resonance has been one whereby nuclear induction signals are observed while an ensemble of nuclear spins is perturbed by a small radiofrequency magnetic field. A large d.c. magnetic field $\bar{H}_0$ establishes a net spin population at thermal equilibrium which provides a macroscopic magnetic moment $\bar{M}_0$ oriented parallel to $\bar{H}_0$. The forced motion of $\bar{M}_0$ is brought about by subjecting the spin ensemble to a rotating radiofrequency field $\bar{H}_1$ normal to $\bar{H}_0$ at the resonance condition $\omega=\omega_0=\gamma H_0$, where γ is the gyromagnetic ratio, ω is the angular radiofrequency, and ω_0 is the Larmor frequency. The techniques which obtain resonance under this condition provide for the application of a driving r-f voltage to an LC circuit tuned to the Larmor frequency. The sample containing the nuclear spins is placed in a coil which is the inductance of the tuned circuit. At resonance a small nuclear signal is induced in the coil and is superimposed upon an existing r-f carrier signal of relatively high intensity. In order to detect this small nuclear signal the r-f carrier voltage is reduced to a low level by a balancing method if the LC circuit is driven by an external oscillator,[1,3] or the LC circuit may be the tuned circuit of an oscillator which is designed to change its level of operation when nuclear resonance absorption occurs.[4,5] In general, a condition exists whereby transitions induced by H_1, which tend

* This research was supported in part by the ONR.
† Reported at the Chicago Meeting of the American Physical Society, November, 1949; Phys. Rev. 77, 746 (1950).
‡ Present address: Physics Dept. Stanford University, Stanford, California.
[1] F. Bloch, Phys. Rev. 70, 460 (1946).
[2] E. L. Hahn, Phys. Rev. 77, 297 (1950).
[3] Bloembergen, Purcell, and Pound, Phys. Rev. 73, 679 (1948).
[4] R. V. Pound, Phys. Rev. 72, 527 (1947); R. V. Pound and W. D. Knight, Rev. Sci. Inst. 21, 219 (1950).
[5] A. Roberts, Rev. Sci. Inst. 18, 845 (1947).

to upset the thermal equilibrium of the spins, are in competition with processes of emission due to lattice perturbations which tend to restore equilibrium. Spin relaxation phenomena, which are measured in terms of the relaxation times T_2 and T_1 (spin-lattice), must be distinguished simultaneously from effects due to the influence of r-f absorption. Consequently the study of resonance absorption line shapes, intensities, and transients must carefully take into account the intensity of H_1 and the manner in which resonance is obtained.

In practice, resonance takes place over a range of frequencies determined by the inhomogeneity of H_0 throughout the sample. For resonances concerning nuclei in liquids it is generally found that the natural line width given by $1/T_2$ on a frequency scale is much narrower than the spread in Larmor frequencies caused by external field inhomogeneities, whereas the converse is true in solids. Therefore steady state resonance lines due to nuclei in liquids are artificially broadened; transient signals are modified in shape and have decay times which are shorter than would otherwise be determined by T_1 and T_2.

(B) Nuclear Induction Due to Free Larmor Precession

The observation of transient nuclear induction signals due to free Larmor precession becomes possible at the resonance condition described above if the r-f power is now applied in the form of intense, short pulses. The r-f inductive coil which surrounds the sample is first excited by the applied pulses and thereafter receives spontaneous r-f signals at the Larmor frequency due to the precessing nuclei. In particular, the echo effect is brought about by subjecting the sample to two r-f pulses in succession (the simplest case) at pulse width $t_w < \tau < T_1$, T_2, where τ is the time interval between pulses. At time τ after the leading edge of the second pulse the echo signal appears. Since H_1 is absent while these signals are observed, no particular attention need be given to elaborate procedures for eliminating receiver saturation effects (as must be done in the forced motion technique) providing that T_2 is large enough to permit observation of echoes at times when the receiver has recovered from saturation due to the pulses. Because the T_2 of nuclei in liquids is generally large enough to favor this condition, the technique for obtaining echoes in this work has been largely confined to the magnetic resonance of nuclei in liquid compounds. Preliminary observations of free induction signals in solids, where T_2 becomes of the order of microseconds, indicate again, however, that procedures must be undertaken for preventing receiver saturation due to intense pulses.

For spin ensembles in liquids it will be shown that the analysis of observed echo signals yields direct information about T_1 and T_2 without requiring consideration of the effect of H_1 on the measured decay of the signal. Because of the inhomogeneity in H_0, the self diffusion of "spin-containing liquid molecules" brings about an attenuation of observed transient signals in addition to the decay due to T_1 and T_2. However, this is only serious for liquids of rather low viscosity which also have a large T_2 for the resonant nuclei concerned, whereas in conventional resonance methods (forced motion), field inhomogeneities obscure a direct measurement of T_2 in liquids over a much wider range of viscosities. The free motion technique, which will hereafter be denoted by the method of spin echoes or free nuclear induction, also reveals in a unique manner differences in resonant frequency between nuclear spins of the same species located in different parts of a single molecule or in different molecules. Such differences have been observed by previous resonance methods,[6,7] and the echo technique gives at least as good a resolution in the measurement of small shifts.

In this investigation the in-phase condition of a precessing spin ensemble is considered to be eventually destroyed because of lattice perturbations which limit the phase memory time of Larmor precession. The precession of an individual spin may be interrupted either because its energy of precession is transferred to neighboring spins in a time $\sim T_2''$ (mutual spin-spin flipping), or because this energy is transferred to the lattice as thermal energy in a time $\sim T_1$. The spread in Larmor frequencies, due to local "z magnetic field" fluctuations at the position of the nucleus caused by neighboring spins and paramagnetic ions, also serves to disturb phase memory (H_0 is in the z direction). In a formal treatment[3] this effect is considered in conjunction with the interaction giving rise to T_2'', and a general relaxation time T_2' is formulated. The inverse of the total relaxation time, $1/T_2 \sim 1/T_2' + 1/T_1$, therefore becomes the uncertainty in frequency of a precessing spin, which can then acquire an uncertainty in phase of the order of one radian in time T_2.

It will be convenient to describe the formation of free induction signals by considering the free precession of individual macroscopic moment vectors $\mathbf{M}_0(\omega_0)$. Each of these vectors has a magnitude at a given ω_0 which is determined by a z magnetic field distribution imposed upon the ensemble by inhomogeneities in H_0. In this spectral distribution $\mathbf{M}_0(\omega_0)$ can be defined as an isochromatic macroscopic moment which consists of an ensemble of nuclear moments precessing in phase at the assigned frequency ω_0. The precessional motion of any $\mathbf{M}_0(\omega_0)$ vector about the total magnetic field (with or without r-f pulses) can be followed regardless of what phases the individual isochromatic moments have with respect to one another throughout the entire spectrum. At the time a short r-f pulse initiates the free precession of $\mathbf{M}_0(\omega_0)$ from a classically non-precessing initial condition at thermal equilibrium, relaxation and possibly diffusion processes begin to diminish the magnitude of

[6] W. G. Proctor and F. C. Yu, Phys. Rev. 77, 717 (1950).
[7] W. C. Dickinson, Phys. Rev. 77, 736 (1950).

the precessing vector $\bar{M}_0(\omega_0)$ as the individual nuclear spins get out of phase with one another or return to thermal equilibrium.

The actual H_0 field which persists at the position of a precessing spin accounts for a given ω_0 and hence for a given $\bar{M}_0(\omega_0)$. In liquids this persisting field and the way it is distributed over the sample is taken to be entirely due to the magnet; any contributions to the local field at the nucleus by neighbors in the lattice average out in a time short compared to a Larmor period. Free induction signals from nuclei in solids, however, indicate that a broad distribution in H_0 exists (compared to a relatively homogeneous external field) which is determined by fixed lattice neighbors, and now this local field distribution does not average out.

The description of free induction effects is simplified by transforming to a coordinate system in which the $x'y'$ plane (Fig. 1) is rotating at some convenient reference angular frequency ω'. This frequency is usually chosen to be the center frequency of a given distribution of isochromatic moments, where the distribution is typically described by a Gaussian or Lorentz (damped oscillator) function. In the next section definite properties of the rotating coordinate representation are presented. The precessional motion as viewed in the rotating system is conveniently followed when (1) $\bar{M}_0(\Delta\omega)$ undergoes forced transient motion during the driving pulse, and (2) when the $\bar{M}_0(\Delta\omega)$ vectors precess freely, where $\Delta\omega=\omega_0-\omega'$ and $\bar{M}_0(\omega_0)=\bar{M}_0(\Delta\omega)$. The condition in (1) has already been analyzed theoretically and experimentally.[8,9,10] Although it is strictly a condition in which $\bar{M}_0(\Delta\omega)$ precesses about the total field $\bar{H}_0+\bar{H}_1$, as viewed in the laboratory system, it has been characterized by the fact that not only does $\bar{M}_0(\Delta\omega)$ appear to precess about the z axis at a high Larmor frequency, but also it appears to nutate with respect to the z axis at a much lower frequency.[11]

III. THEORY AND APPLICATIONS

(A) The Moving Coordinate Representation

Consider the torque equation, with no damping, which describes the precession of $\bar{M}$ as seen in the laboratory system:

$$d\bar{M}/dt=\gamma(\bar{M}\times\bar{H}), \tag{1}$$

where $\bar{H}$ is the total magnetic field. During the application of r-f pulses, $\bar{H}=\bar{H}_0+\bar{H}_1$; and during the free precession of $\bar{M}$ in the absence of pulses, $\bar{H}=\bar{H}_0$. During a pulse it is convenient to transform to a moving coordinate system in which $\omega'=\omega$, and $\bar{H}_1$ is chosen to lie along the x' axis. It will be pointed out, however, that, regardless of the choice of direction of $\bar{H}_1$ in the $x'y'$ plane, the description of the spin echo model presented later is not affected, except under a very special condition. If $D\bar{M}/dt$ is the observed torque in the moving coordinate system, then by a well-known transformation,

$$d\bar{M}/dt=D\bar{M}/dt+\bar{\omega}\times\bar{M}, \tag{2}$$

where $\bar{M}=M(u, v, M_z)$ and $\bar{H}=H(H_1, 0, H_0)$. Combining (1) and (2) we obtain

$$D\bar{M}/dt=\bar{M}\times(\Delta\bar{\omega}+\bar{\omega}_1) \tag{3}$$

as the torque in the moving system during a pulse. The vector $\bar{M}$ is identified with the isochromatic moment $\bar{M}_0(\Delta\omega)$ which appears to precess about the effective field vector $(\Delta\bar{\omega}+\bar{\omega}_1)/\gamma$. Let $(\Delta\omega)_{\frac{1}{2}}$ be the width at half-maximum of an assumed function which describes the distribution of $\bar{M}_0(\Delta\omega)$ over the inhomogeneous external field, and let ω' be the center frequency of this distribution. If, during a pulse, the inequality $1/t_w$, $\omega_1\gg(\Delta\omega)_{\frac{1}{2}}$ applies at resonance $(\omega\approx\omega')$, then the precession of any $\bar{M}_0(\Delta\omega)$ vector will appear to take place practically about the $\bar{\omega}_1$ vector in the moving system. This precessional frequency is given by $\omega_1=\gamma H_1$ (of the order of kilocycles) which appears in the laboratory system as a frequency of nutation superimposed upon a high Larmor precession frequency (~30 Mc). In the rotating system any $\bar{M}_0(\Delta\omega)$ vector will precess in a cone whose axis is in the direction of $\bar{H}_1$ and whose angle is determined by the angle between $\bar{M}_0(\Delta\omega)$ and $\bar{H}_1$ at the time $\bar{H}_1$ is suddenly applied. When $\bar{H}_1$ is suddenly removed, the vector $\bar{M}_0(\Delta\omega)$ is oriented at a fixed angle θ with respect to the z axis, and precesses freely at angular frequency $\Delta\omega$ about the effective magnetic field $\Delta\bar{\omega}/\gamma$ along the z axis. The angle θ will be determined by $\omega_1 t_w$ and the initial conditions established by successive pulses applied in the past.

(B) Simple Vector Model of the Spin Echo

For spin ensembles in liquids a simple vector model will account for the manner in which two applied r-f pulses establish a given spectral distribution of moment components in the $x'y'$ plane, where the axis of the inductive coil is oriented. This distribution then freely precesses to form, by constructive interference, a resultant "echo" in the $x'y'$ plane. This is formulated by integrating a general expression for the $x'y'$ component of the isochromatic moment over all frequencies $\Delta\omega$ imposed by H_0 field inhomogeneities. Purcell[12] has suggested a three-dimensional model of the echo, Fig. 1, which arises in a special case. At $t=0$, when $\bar{H}_1$ is suddenly applied, $\bar{M}_0(\Delta\omega)$ is at thermal equilibrium, aligned parallel to $\bar{H}_0$ along the z axis. During time t_w

[8] N. Bloembergen, *Nuclear Magnetic Relaxation* (Martinus Nijhoff, The Hague, 1948).

[9] H. C. Torrey, Phys. Rev. 76, 1059 (1949).

[10] E. L. Hahn, Phys. Rev. 76, 461 (1949).

[11] This is observed to come about in the laboratory system as the resonance absorption mode becomes modulated at the low nutation frequency. Classically speaking, the term nutation is applied only to the physical top, in which the presence of angular momentum about an axis other than the spin axis is responsible for the nutation. Although a nuclear spin possesses extremely negligible angular momentum about any axis other than its spin axis, the term nutation is convenient to retain here in order to refer to the tipping motion of $\bar{M}_0(\Delta\omega)$ with respect to the z axis.

[12] Private communication.

of the first pulse, let $\bar{M}_0(\Delta\omega)$ precess an angle $\omega_1 t_w = \pi/2$ about $\bar{H}_1$, so that all moment vectors in the spectrum will have nutated into the $x'y'$ plane. Let $\tau \gg 1/(\Delta\omega)_{\frac{1}{2}}$, $T_1 = T_2 = \infty$, and assume a rectangular spectrum over $\Delta\omega$, i.e., $g(\Delta\omega) = \text{const.}$, where $g[(\Delta\omega)_{\frac{1}{2}}] = 0$. During time $t_w \leqslant t \leqslant \tau$, the various isochromatic moment pairs $M_0(+|\Delta\omega|)$, $M_0(-|\Delta\omega|)$, will precess at frequency $\Delta\omega$, maintaining a symmetry about the y' axis but rotating in opposite directions. These precessing moments will attain an isotropic distribution in a time $\gtrsim 2\pi/(\Delta\omega)_{\frac{1}{2}}$ prior to which a free induction decay is observed.[2] At time τ the second r-f pulse, identical with the first one, will rotate the moment pairs from angular positions $\varphi = 3\pi/2 \pm |\Delta\omega|\tau$, $\theta = \pi/2$ to $\varphi = (0, \pi)$, $\theta = \pi - |\Delta\omega|\tau$ in spherical coordinates. During the time interval $\tau + t_w \leqslant t < 2\tau$ all moment vectors interfere destructively with one another and distribute themselves isotropically over a unit sphere until the time $t \approx 2\tau$ when they interfere constructively. At time 2τ all of the moment vectors will have again precessed angles $\Delta\omega\tau$ respectively from their positions at $t = \tau + t_w$ so that they terminate in a figure eight pattern whose equation is $\theta = \varphi$. Free induction for $t \geqslant \tau + t_w$ will be obtained from the linearly polarized component of magnetization

$$v(\Delta\omega, t) = M_0 \sin\Delta\omega\tau \sin\Delta\omega(t-\tau) \tag{4}$$

along the y' axis. The observed induction voltage will be due to the integrated precessing moment

$$V(t) = \int_{-(\Delta\omega)_{\frac{1}{2}}}^{(\Delta\omega)_{\frac{1}{2}}} g(\Delta\omega) v(\Delta\omega, t) d(\Delta\omega) \tag{5}$$

where

$$\int_{-(\Delta\omega)_{\frac{1}{2}}}^{(\Delta\omega)_{\frac{1}{2}}} g(\Delta\omega) d(\Delta\omega) = 1.$$

Therefore, from (4) and (5) we obtain

$$V(t) = \frac{M_0}{2}\left[\frac{\sin(\Delta\omega)_{\frac{1}{2}}(t-2\tau)}{(\Delta\omega)_{\frac{1}{2}}(t-2\tau)} - \frac{\sin(\Delta\omega)_{\frac{1}{2}} t}{(\Delta\omega)_{\frac{1}{2}} t}\right]. \tag{6}$$

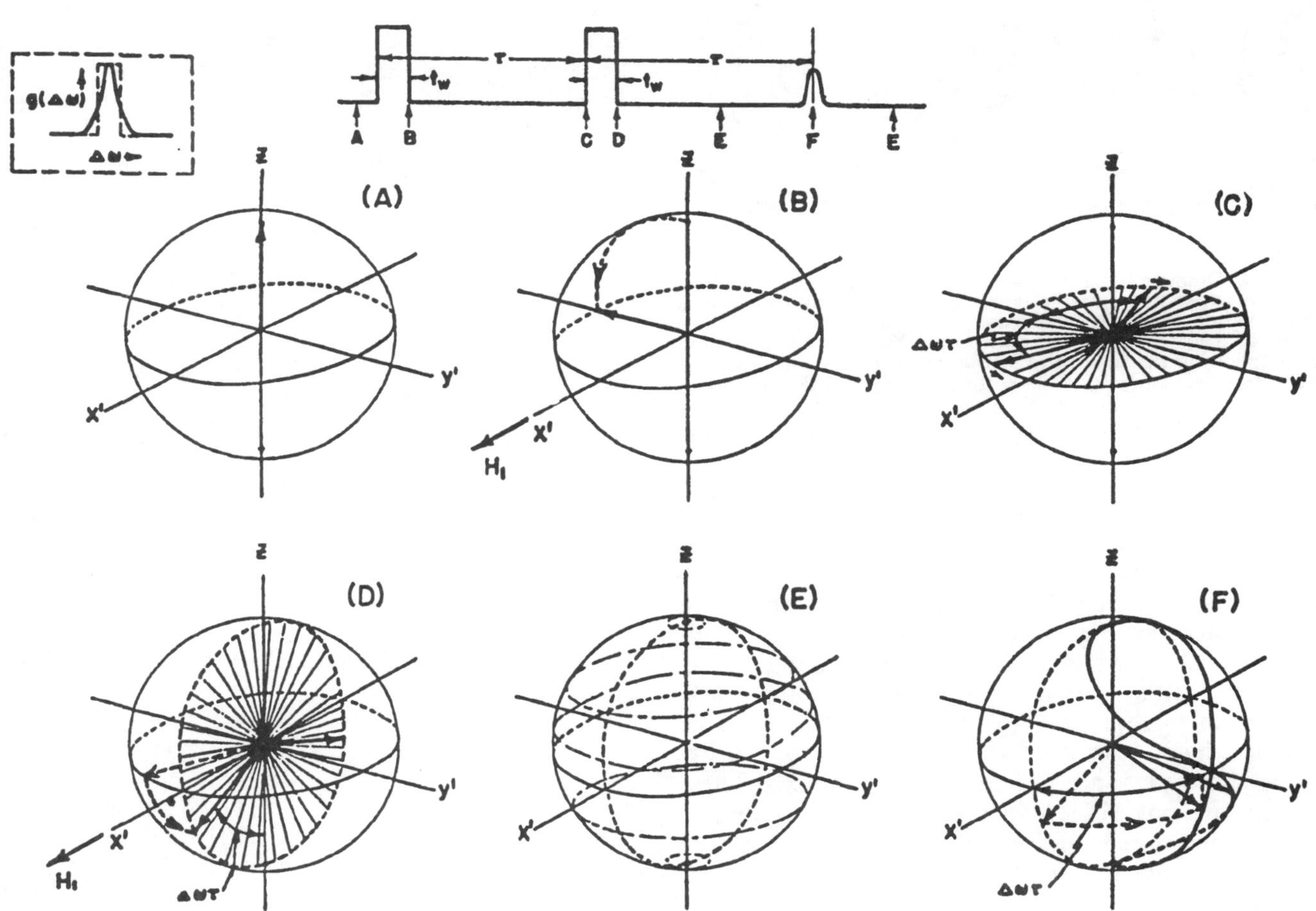

FIG. 1. For the pulse condition $\omega_1 t_w = \pi/2$, the formation of the eight-ball echo pattern is shown in the coordinate system rotating at angular frequency ω. The moment vector monochromats are allowed to travel completely in a time $\tau \gg 1/(\Delta\omega)_{\frac{1}{2}}$ before the second pulse is applied. The echo gives maximum available amplitude at $\omega_1 t_w = 2\pi/3$.

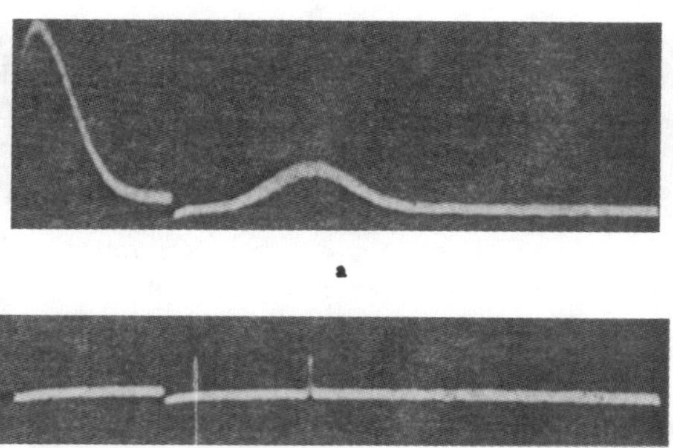

a

b

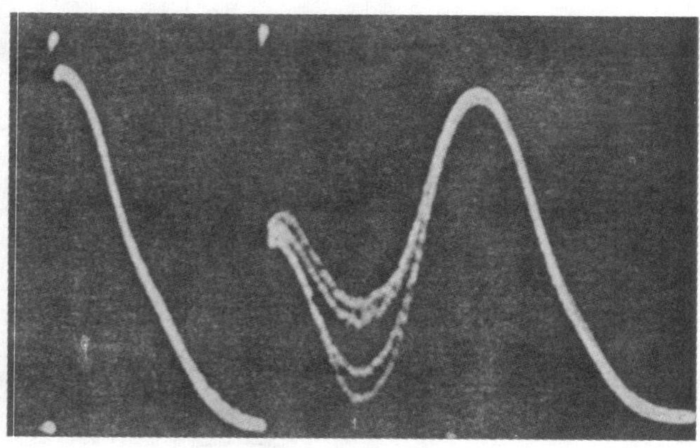

c

FIG. 2. Oscillographic traces for proton echoes in glycerine. The two upper photographs indicate broad and narrow signals corresponding to H_0 fields of good and poor homogeneity. The pulses, scarcely visible, are separated by 0.0005 sec. The induction decay following the first pulse in the top trace has an initial dip due to receiver saturation. The bottom photograph shows random interference of the induction decay with the echo for several exposures. The two r-f pulses are phase incoherent relative to one another.

According to the first term on the right side of (6) the echo maximum occurs at $t=2\tau$, and the signal lasts for $\sim 4\pi/(\Delta\omega)_{\frac{1}{2}}$ seconds. No free induction is predicted after the second pulse for this particular case, which is illustrated in Fig. 2 (top). For extremely large $(\Delta\omega)_{\frac{1}{2}}$ the echo becomes very sharp and the free induction decay after the first pulse becomes practically unobservable. Equation (6) indicates that periodic maxima should occur at times $2\pi/(\Delta\omega)_{\frac{1}{2}}$ sec. apart during the appearance of free induction signals. These maxima are not observed in general for this reason, because the choice of $g(\Delta\omega)$ here does not correspond to experimental conditions. A modulation is observed in particular cases because of an entirely different effect which will be discussed later.

(C) General Analysis

The echo effect will now be treated in a general way, after which some of the simplifying assumptions outlined for the very simple case just described will be applied. By making use of Bloch's equations[1] and choosing $g(\Delta\omega)$ to approximate the actual distribution of spins over H_0, the decay of echo signals due to T_1, T_2, and self-diffusion (in the case of some liquids) can be accounted for. As in the case illustrated above, $\bar{H}_1$ will be chosen to lie along the x' axis in the rotating system for both pulses. Actually $\bar{H}_1$ may appear in any possible position in the $x'y'$ plane during the second pulse since the r-f is not necessarily coherent for both pulses. However, free induction signals will be independent of this random condition as long as $\tau \gg 1/(\Delta\omega)_{\frac{1}{2}}$.[13] This signifies that free induction decay following a single pulse does not interfere with the echo (see Fig. 2, bottom, where this interference effect is shown). Ordinarily the scalar differential equations obtained from (3) are written to include additional torque terms due to relaxation according to Bloch. In the case of echo phenomena it is found that nuclear signals due to precessing nuclear moments contained in liquid molecules (particularly of low viscosity) are not only attenuated by the influence of T_1 and T_2, but also suffer a decay due to self-diffusion of the molecules into differing local fields established by external field inhomogeneities. Consequently, the phase memory of Larmor precession can be destroyed artificially to an appreciable extent. The effect of self-diffusion will be qualitatively accounted for by using Bloch's equations with a diffusion term added:

$$du/dt+[\Delta\omega+\delta(t)]v=-u/T_2 \qquad \text{(7-A)}$$

$$dv/dt-[\Delta\omega+\delta(t)]u+\omega_1 M_z=-v/T_2 \qquad \text{(7-B)}$$

$$dM_z/dt-\omega_1 v=-(M_z-M_0)/T_1. \qquad \text{(7-C)}$$

u and v are the components of magnetization parallel and normal to $\bar{H}_1$ respectively. As time increases from the point where the first pulse is applied, $\delta(t)$ is taken to represent, due to diffusion, the shift in Larmor frequency of the u and v components away from the initial value of $\Delta\omega$. If the decay terms during a pulse are neglected, since t_w is very short compared to all decay time constants, the motion will be simply described by the following solutions of (7):

$$u(t)=(\Delta\omega/\beta)AQ+u(t_i) \qquad \text{(8-A)}$$

$$v(t)=A\sin(\beta t+\xi) \qquad \text{(8-B)}$$

$$M_z(t)=-(\omega_1 A/\beta)Q+M_z(t_i), \qquad \text{(8-C)}$$

where $Q=\cos(\beta t+\xi)-\cos(\beta t_i+\xi)$ and $\beta=[(\Delta\omega)^2+\omega_1^2]^{\frac{1}{2}}$. The constants A, ξ, $u(t_i)$, and $M_z(t_i)$ are determined by initial conditions at the beginning of the pulse ($t=t_i$) and the assumption that $M_z(t_i)^2+u(t_i)^2+v(t_i)^2=M_z(t)^2+u(t)^2+v(t)^2$ during the pulse. When the r-f pulse is removed at $t=t_i'$, then $\omega_1=0$, and Eqs. (8-A) and

[13] In the calculation which follows, $\bar{H}_1$ during the second pulse could be assumed to have an arbitrary angle α with respect to $\bar{H}_1$ which existed during the first pulse. It can easily be shown that all nuclear signals are independent of α and that the direction of the echo resultant will be at an angle $\alpha+\pi/2$ with respect to the direction of $\bar{H}_1$ which was applied during the second pulse.

(8-B) combine to give a solution

$$F(t)=F(t_i')\exp\{-(t-t_i')/T_2+i[\Delta\omega(t-t_i')+\int_{t_i'}^{t}\delta(t'')dt'']\}, \quad (9)$$

where $F=u+iv$ and $t\geqslant t_i'$.

A constant field gradient, $(dH_0/gl)_{av}=G$, shall be assumed to exist throughout the sample, where l is any direction in which the field gradient has the given average value G. The actual direction of $\bar{H}_0$ must vary in the sample. Any precessing moment which experiences a change in the magnitude of $\bar{H}_0$ due to diffusion will adiabatically follow a corresponding change in Larmor frequency of precession which will take place about the new direction of $\bar{H}_0$. Therefore $\bar{H}_1$ will not have the same magnitude during both pulses for a particular spin because the component of the applied r-f field perpendicular to the different directions of $\bar{H}_0$ will differ. Free induction signals will suffer negligible distortion because of this as compared to the distortion caused by variation in direction and magnitude of $\bar{H}_1$ throughout the sample due to coil geometry. For purposes of simplicity, the analysis will not attempt to take into account any sort of inhomogeneity of the $\bar{H}_1$ field.

In (9) let $\delta(t'')=\gamma Gl(t'')$ and

$$\int_{t_i'}^{t}\delta(t'')dt''=\Phi(t)-\Phi(t_i'), \quad (10)$$

where $\Phi(t)-\Phi(t_i')$ is the total phase shift accumulated in a time $t-t_i'$ by a precessing spin due to diffusion. The solution (9) must be averaged over all Φ, using a phase probability function $P(\Phi, t)$, by considering in particular the integral

$$\int_{-\infty}^{\infty}\{\exp i[\Phi(t)-\Phi(t_i')]\}P(\Phi, t)d\Phi=\exp\left[-\frac{kt^3}{3}-i\Phi(t_i')\right], \quad (11\text{-A})$$

where $k=(\gamma G)^2D$, and D is the self-diffusion coefficient of the spin-containing molecule. It can be shown[14] that

$$P(\Phi, t)=(4\pi kt^3/3)^{-\frac{1}{2}}\exp[-\Phi^2/(4\pi kt^3/3)]. \quad (11\text{-B})$$

[14] First one must take into account all possible paths (essentially all possible areas expressed by the integral in (10)) which the diffusing molecule may take in the l, t plane so that the total phase shift accumulated by the precessing spin which the molecule carries with it has a given value which is the same regardless of path length and final position of the molecule. The ordinary diffusion law is assumed to apply in expressing the probability of a given path under the constraint that a certain $\Phi(t)$ be accumulated. The distribution function (11-B) over all phases then follows by applying a standard deviation theorem (see James V. Uspensky, *Introduction to Mathematical Probability* (McGraw-Hill Book Company Inc., New York, 1937), p. 270). The author is indebted to Dr. C. P. Slichter for this derivation.

From Eq. (7-C) the solution for $M_z[\Delta\omega+\delta(t)]$ must be averaged over the probability that the moment vector corresponding to it is precessing at frequency $\Delta\omega+\delta(t)$ at time t. The ordinary diffusion law will be assumed to apply as regards the distance of diffusion l which corresponds to frequency shift δ. General solutions of (7) representing free motion can therefore be written as follows:

$$F(t)=F(t_i')\exp\{-(t-t_i')/T_2-\tfrac{1}{3}kt^3+i[\Delta\omega(t-t_i')-\Phi(t_i')]\} \quad (12)$$

$$M_z(t)=M_0\left[1+\frac{M_i-M_0}{M_0}\exp-(t-t_i')/T_1\right] \quad (13)$$

where

$$M_i=\int_{-\infty}^{\infty}M_z(\Delta\omega+\delta, t_i')P(\delta, t)d\delta \quad (14)$$

and[15]

$$P(\delta, t)=\frac{1}{[4\pi k(t-t_i')]^{\frac{1}{2}}}\exp[-\delta^2/4k(t-t_i')].$$

For the case in which twin pulses are applied, we have at $t=0$, $M_z=M_0$ and $u=v=0$. At $t=t_w$ the moments in the rotating system are obtained from (8). At time τ the r-f pulse is again applied and removed at $t=\tau+t_w$. After the second pulse the initial values of the magnetization components which undergo free motion are as follows:

$$u(\tau+t_w)=\frac{\Delta\omega}{\omega_1M_0}[v(\tau)v(t_w)-u(\tau)u(t_w)]+\frac{u(t_w)M_z(\tau)}{M_0}+u(\tau) \quad (15\text{-A})$$

$$v(\tau+t_w)=\frac{v(t_w)}{M_0}\left[M_z(\tau)-\frac{\Delta\omega}{\omega_1}u(\tau)\right]+v(\tau)\cos\beta t_w \quad (15\text{-B})$$

$$M_z(\tau+t_w)=\frac{1}{M_0}[M_z(\tau)M_z(t_w)+u(\tau)u(t_w)-v(\tau)v(t_w)]. \quad (15\text{-C})$$

The v component, which is an even function in $\Delta\omega$, provides the free induction voltage, whereas the u component is an odd function in $\Delta\omega$ and does not contribute to the integral which will be applied in (18). Imposing the condition $\omega_1\gg\Delta\omega$ and $\tau\gg t_w$ we obtain:

(a) $(t\leqslant\tau)$:

$$v(t, \Delta\omega)\approx-M_0\sin\omega_1t_w\cos\Delta\omega t\exp(-t/T_2-\tfrac{1}{3}kt^3) \quad (16)$$

(b) $(t\geqslant\tau)$:

$$v(t, \Delta\omega)\approx M_0\sin\omega_1t_w[\sin^2\tfrac{1}{2}\omega_1t_w\cos\Delta\omega(t-2\tau)-\cos^2\tfrac{1}{2}\omega_1t_w\cos\Delta\omega t]\exp(-t/T_2-\tfrac{1}{3}kt^3)-M_z(\tau)\sin\omega_1t_w\cos\Delta\omega(t-\tau)\exp[-(t-\tau)/T_2-\tfrac{1}{3}k(t-\tau)^3]. \quad (17)$$

[15] E. H. Kennard, *Kinetic Theory of Gases* (McGraw-Hill Book Company, Inc., New York, 1938), p. 283.

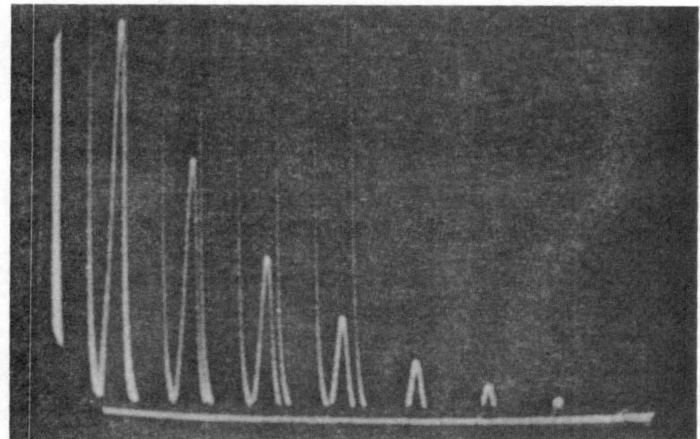

FIG. 3. Multiple exposures of proton echoes in a water solution of $Fe(NO_3)_3$ (2.5×10^{18} Fe^{+++} ions/cc). The faint vertical traces indicate paired pulses which are applied at time intervals $\gg T_2$, with the first pulse of each pair occurring at the same initial position on the sweep. For each pulse pair the interval τ is increased by 1/300 sec. The echoes are spaced 2/300 sec. apart and the measured decay time constant of the echo envelope gives $T_2=0.014$ sec.

The measured signal will be due to the integral

$$V(t)=\int_{-\infty}^{\infty} g(\Delta\omega)v(t,\Delta\omega)d(\Delta\omega). \tag{18}$$

For convenience $g(\Delta\omega)$ is chosen to be a Gaussian distribution:

$$g(\Delta\omega)=(2\pi)^{-\frac{1}{2}}T_2^*\exp[-(\Delta\omega T_2^*)^2/2],\quad T_2^*=(2\ln 2)^{\frac{1}{2}}/(\Delta\omega)_{\frac{1}{2}}, \tag{19}$$

where the integral of $g(\Delta\omega)$ over all $\Delta\omega$ is equal to unity. Integration of (16) and (17) according to (18) gives the following:

(a) $(t\leq\tau)$:

$$V(t)\approx -M_0\sin\omega_1 t_w\exp\left[-\left(\frac{t^2}{2T_2^{*2}}+t/T_2+\frac{kt^3}{3}\right)\right], \tag{20}$$

(b) $(t\geq\tau)$:

$$V(t)\approx M_0\sin\omega_1 t_w\left\{\left(\sin^2\frac{\omega_1 t_w}{2}\right)\exp\left[-\frac{(t-2\tau)^2}{2T_2^{*2}}\right]-\left(\cos^2\frac{\omega_1 t_w}{2}\right)\exp\left(-\frac{t^2}{2T_2^{*2}}\right)\right\}$$
$$\times\exp\left(-\frac{kt^3}{3}-t/T_2\right)-M_z(\tau)\sin\omega_1 t_w$$
$$\times\exp-\left[\frac{1}{2}\left(\frac{t-\tau}{T_2^*}\right)^2+\frac{t-\tau}{T_2}+\frac{k}{3}(t-\tau)^3\right] \tag{21}$$

The echo at $t=2\tau$ is accounted for by the first term in (21) and has a width of $\sim T_2^*$ seconds. The remaining terms in (20) and (21) predict the occurrence of free induction decay signals immediately following the removal of the pulses. Actual shapes of all induction signals are determined mainly by what shape $g(\Delta\omega)$ happens to have due to external field inhomogeneities over the magnet. T_2 will play a significant role in affecting the shape only if $T_2\gtrsim T_2^*$. Signal amplitudes are independent of T_2^* as long as $\omega_1\gg 1/T_2^*$. In practice $g(\Delta\omega)$ is roughly a function which is some compromise between the Gaussian distribution given above and the Lorentz damped oscillator function given by

$$g(\Delta\omega)=\frac{2T_2^*}{1+(\Delta\omega T_2^*)^2}\quad\text{where}\quad T_2^*=\frac{1}{(\Delta\omega)_{\frac{1}{2}}}.$$

(D) Measurement of T_2

If $\frac{1}{3}kt^3\ll t/T_2$ and $T_2^*\ll\tau<T_2$, T_1, then T_2 can be measured directly by plotting the logarithm of the maximum echo amplitude at $t=2\tau$ *versus* arbitrary values of 2τ. Figure 3 illustrates photographs of echoes on the oscilloscope for protons in a water solution of Fe^{+++} ions under these conditions. Figure 4 indicates how the measured T_2 for various concentrations of Fe^{+++} ions agrees with results obtained by Bloembergen, *et al.*[3] using the line width method. The law $C\propto 1/T_2$ is obeyed where C is the number of Fe^{+++} ions/cc for a given sample.

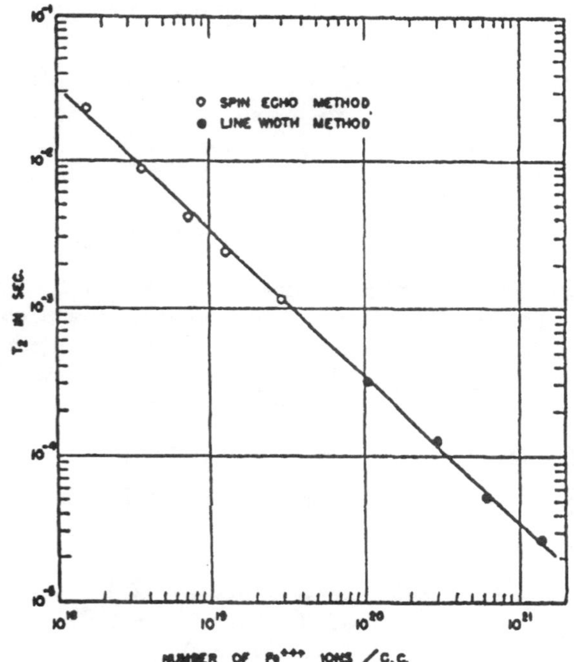

FIG. 4. T_2 measurements from the envelope decay of proton echoes are obtained for given concentrations of $Fe(NO_3)_3$ in H_2O. The plot compares with measurements made by the line width method (see reference 3).

(E) Secondary Spin Echoes

If a third r-f pulse (identical to pulses producing the primary echo) is applied to the sample at a time T with respect to $t=0$, where $2\tau<T<T_2$, additional echoes occur at the following times: $T+\tau$, $2T-2\tau$, $2T-\tau$, $2T$. For $\tau<T<2\tau$ the signal at $2T-2\tau$ is absent but the others remain (see Fig. 5). These additional echoes can be readily predicted by rewriting Eq. (15) such that $\tau+t_w\to T+t_w$, $\tau\to T$, $t_w\to\tau+t_w$ ($\cos\beta t_w$ remains un-

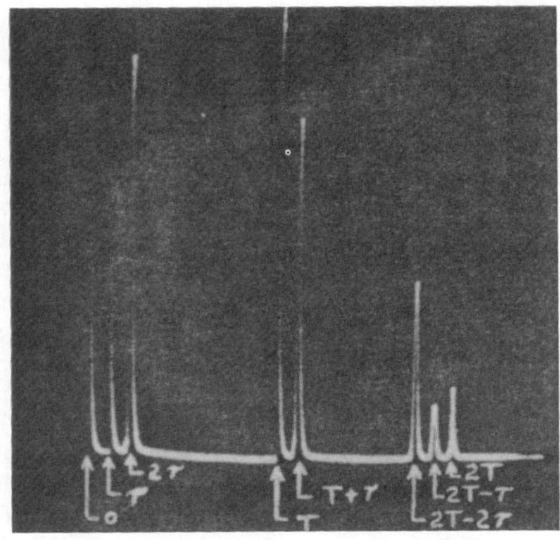

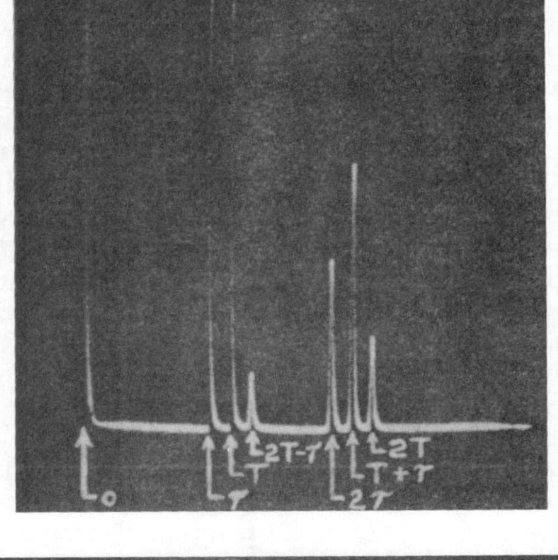

changed in (15-B)) and applying the resulting expressions as initial conditions in (12), (13), and (14). In this manner, by successive application of accumulating initial conditions, the echo pattern resulting from any number and sequence of r-f pulses can be predicted. After integrating $v(t)$ over $\Delta\omega$ for $t \geqslant T > 2\tau$, using $g(\Delta\omega)$ according to (19), the following expression for $V(t)$ is obtained (terms due to induction decay directly following the pulse are omitted and assumed not to interfere with the echoes since $\tau \gg T_2^*$):

$$V(t) \approx \frac{M_0}{2}(\sin^3\omega_1 t_w) \times \exp\left\{-(T-\tau)\left(\frac{1}{T_1}-\frac{1}{T_2}\right)-t/T_2 -\frac{k}{3}[\tau^3+(t-T)^3]-k\tau^2(T-\tau) -\frac{[t-(T+\tau)]^2}{2T_2^{*2}}\right\} \quad \text{(22-A)}$$

$$-M_0\left(\sin\omega_1 t_w \sin^4\frac{\omega_1 t_w}{2}\right) \times \exp\left\{-\frac{[t-(2T-2\tau)]^2}{2T_2^{*2}}-t/T_2-\frac{kt^3}{3}\right\} \quad \text{(22-B)}$$

$$+M_s(\tau)\left(\sin\omega_1 t_w \sin^2\frac{\omega_1 t_w}{2}\right) \times \exp\left\{-\frac{[t-(2T-\tau)]^2}{2T_2^{*2}}-(t-\tau)/T_2 -k(t-\tau)^3/3\right\} \quad \text{(22-C)}$$

$$+\frac{M_0}{4}(\sin^3\omega_1 t_w) \times \exp\left\{-\frac{(t-2T)^2}{2T_2^{*2}}-t/T_2-\frac{kt^3}{3}\right\}. \quad \text{(22-D)}$$

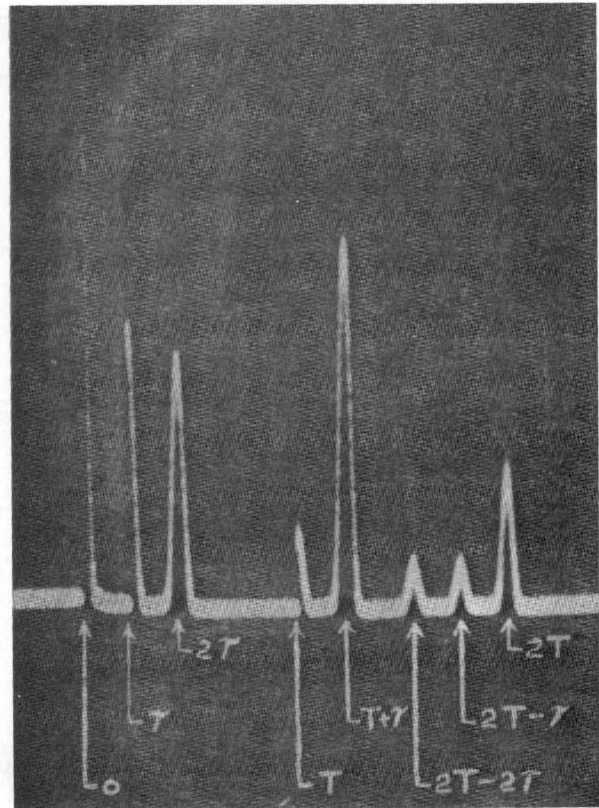

c

FIG. 5. Proton echo patterns in H_2O resulting from three applied r-f pulses. The pulses are visible in the upper two traces, and have a width $t_w \sim 0.5$ msec. In the upper trace $\tau = 0.008$ sec., $T = 0.067$ sec., and for the second trace $\tau = 0.046$ sec. and $T = 0.054$ sec. The bottom photograph shows a similar pattern for the case $T > 2\tau$ where induction decay signals can be seen following very short invisible r-f pulses. Saturation of a narrow band communications receiver, used in the case of the upper two traces, prevents the observation of these signals, whereas a wide band i.f. amplifier makes this observation possible in the bottom photograph.

Term (22-A) provides a "stimulated echo" signal at $T+\tau$. The signal at $2T-2\tau$ (22-B) can be expected qualitatively by considering the "eight ball" alignment in Fig. 1 as equivalent to an initial orientation of moments in a given direction by an imaginary r-f pulse at 2τ. Therefore, it follows that the stimulating pulse at T causes an "image echo" to occur at $2T-2\tau$. The signals at $2T-\tau$ (22-C) and $2T$ (22-D) are essentially primary echoes corresponding to twin pulses at τ, T and 0, T respectively. The signal at $2T$ is modified by the presence of the second r-f pulse at τ so that it does not have the same trigonometric dependence on $\omega_1 t_w$ as do the primary echoes at τ and $2T-\tau$.[16] Experimentally the various echo signals are observed to go through maxima and minima in general agreement with their respective trigonometric dependences on $\omega_1 t_w$ as this quantity is varied. The stimulated echo at $T+\tau$ is particularly interesting and useful in view of the fact that if τ is sufficiently small so that all terms in the exponent of (22-A) are negligible except T/T_1, the signal survives as long as T_1 permits. The remaining echo signals in liquids of low viscosity have maxima which attenuate in a time much shorter than T_1 as T is arbitrarily increased for a given τ. This is due to the diffusion factor $\frac{1}{3}kt^3$ which occurs in the exponents of (22-B), (22-C), and (22-D), but occurs only as $k\tau^2 T$ in (22-A). This property of the stimulated echo is schematically indicated in Fig. 6. The constructive interference at $T+\tau$ is due to moment vectors which previously existed as $M_z(\Delta\omega)$ components distributed in a spectrum approximately as $\cos\Delta\omega\tau$ during the time interval $\tau+t_w\rightarrow T$. This can be seen by noting that $M_z(\tau+t_w)$ has a term $v(\tau)$ proportional to $\cos(\Delta\omega+\delta)\tau$ from (15-C). This cosine distribution becomes smeared due to diffusion and must be averaged over all δ by applying the integral in (14). However, the self-diffusion of spin-containing molecules will not seriously upset this frequency pattern providing $1/\tau \gg \gamma(dH/dl)l(T)$ (let $T\gg\tau$), where $l(T)$ is the effective distance of diffusion in time T over which a shift in Larmor frequency can occur. The attenuation effect of diffusion upon echoes

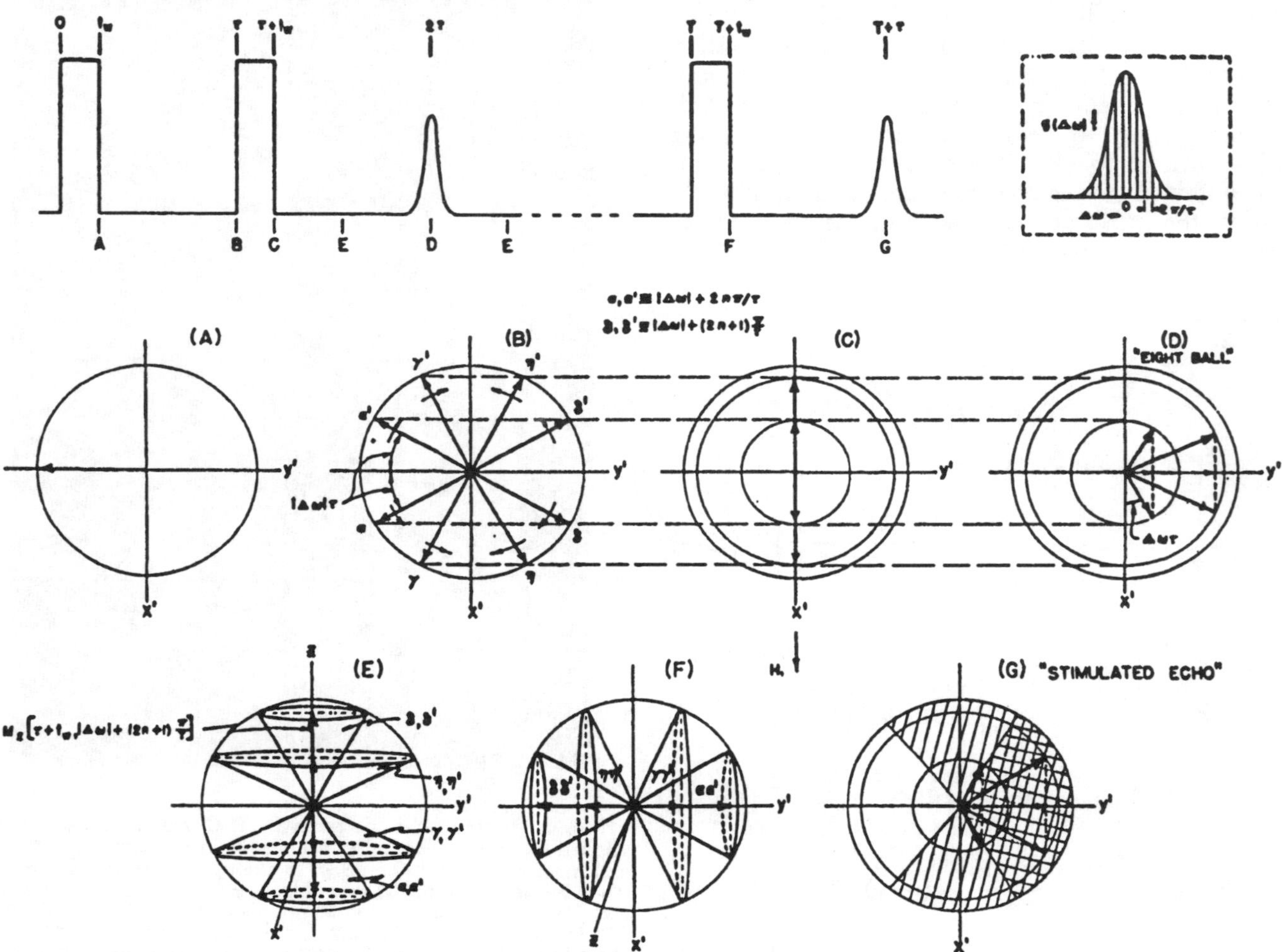

Fig. 6. A vector representation which accounts for the stimulated echo at $t=T+\tau$ is shown under conditions of the special case for the primary echo model in Fig. 1. For a given $|\Delta\omega|$, the symbols α, α' and δ, δ' denote those moments which have Larmor frequencies such that they precess angles $|\Delta\omega|\tau+2n\pi$ and $|\Delta\omega|\tau+(2n+1)\pi$ respectively in time $t=\tau$. n is any integer which applies to frequencies within the spectrum which will lie in a pair of cones corresponding to a specific $|\Delta\omega|$. These cones provide M_z components (after the pulse at τ) which are available for stimulated echo formation after the pulse at T. The shaded area in G indicates the density of moment vectors. The absence of vectors on the $-y'$ side leaves a dimple on the unit sphere.

[16] For $\tau<T<2\tau$ the echo at 2τ is modified and has the coefficient $M_0/4\sin^2\omega_1 t_w$ instead of the one given by (21).

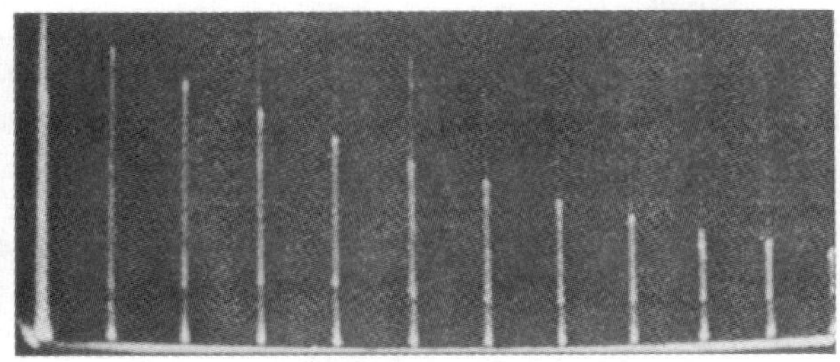

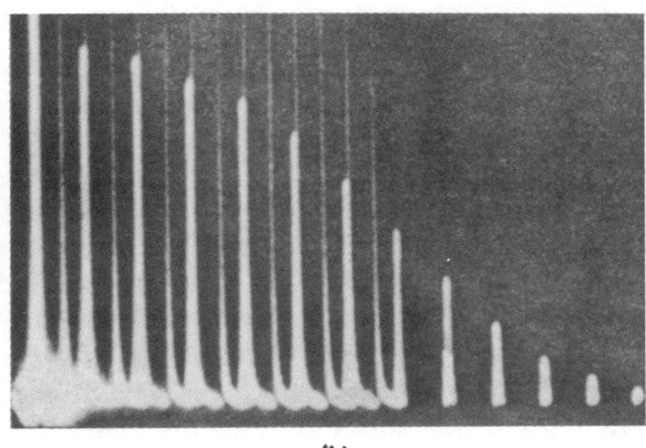

(b)

FIG. 7. A typical exponential plot of stimulated echo amplitudes is shown in the top photograph for protons in H_2O. This is obtained in a manner described for Fig. 3, except that T for the third pulse here is increased by 16/60 sec. intervals while τ is fixed at 0.0039 sec. The measured decay of the envelope is 1.89 sec. which serves as a point on the graph in Fig. 8. The apparent break in intensity in each of the stimulated echoes (seen as vertical traces because of the slow sweep speed) is due to a condition where the echo follows so soon after the stimulating pulse that it superimposes upon the voltage recovery of the receiver detector RC filter.

The bottom photograph indicates approximately an $\exp(-kt^3/3)$ decay law for the primary echo envelope in H_2O. The separation between echoes is 1/60 sec.

whose configuration depends purely upon phase and not frequency is much greater due to the exponential factor $\frac{1}{3}kt^3$ rather than $k\tau^2T$ which occurs only for the stimulated echo.

(F) Measurement of T_1; Qualitative Confirmation of the Diffusion Effect

If the condition $k\tau^2T \ll T/T_1$ is maintained by choosing τ very small, a plot of the logarithm of the stimulated echo maximum amplitude *versus* arbitrary values of T gives a straight line whose slope provides an approximate measure of T_1. In this manner glycerine is found to have a $T_1=0.034$ sec. The self-diffusion coefficient of glycerine is apparently sufficiently small so that T_1 can be measured directly as well as T_2, according to the discussion in III-D. A measured value of $T_2=0.023$ sec. is obtained, which is in substantial agreement with previous measurements.[3,9] The data for T_1 is obtained from oscillographic traces, an example of which is shown in Fig. 7 (top) for protons in distilled water. All relaxation measurements are made at room temperature, at $\omega=30$ Mc. A better value of T_1 is obtainable by plotting $1/T_m$ against τ^2 where T_m is the measured envelope decay time for the stimulated echo which decays as e^{-T/T_m}. A straight line is obtained which has the equation $1/T_m=1/T_1+k\tau^2$, which is seen from the exponential factor in (22-A) where $t=T+\tau$ and $\tau \ll T$. Such a plot is given in Fig. 8 for protons in distilled water (not in vacuum) where the reciprocal of the ordinate intercept gives $T_1=2.3\pm0.1$ sec., in agreement within experimental error with previous measurements.[3,17] Using the value of $D=2\times10^{-5}$ cm²/sec. for the water molecule,[18] the field gradient, G, calculated from the measured slope is 0.9 gauss/cm, which correlates roughly with the actual gradient over the sample. The gradient is expressed as $G\approx(\Delta H)_{\frac{1}{2}}/d$ where $(\Delta H)_{\frac{1}{2}}\sim0.2$ gauss is measured directly from the resonance absorption line width (or echo width) and $d\sim3$ to 4 mm is the average thickness of the cylindrical sample. In Fig. 7 (bottom) the echo envelope for protons in distilled H_2O is reduced to $1/e$ of its maximum amplitude at $t=(3/k)^{\frac{1}{3}}$, since we neglect the decay due to T_2, which is negligible compared to diffusion. The calculated $(\Delta H)_{\frac{1}{2}}$ here is also in rough agreement with the actual field inhomogeneity present. This agreement with the predicted diffusion law confirms the existence at least of a smooth gradient in H_0 over the sample. If the sample is slightly rotated while r-f pulses are applied to obtain echoes, the echo amplitude is markedly reduced as the spin ensemble rotates into varying field inhomogeneity patterns.

(G) The Echo Beat and Envelope Modulation Effects

It has been found that the exact magnetic resonance frequency of nuclear moments of a given species depends upon the type of molecule in which it is contained. It is apparent that the local magnetic field at the position of the nucleus is shifted from the value of the applied external field by an amount which is too large to be accounted for by the normal diamagnetic correc-

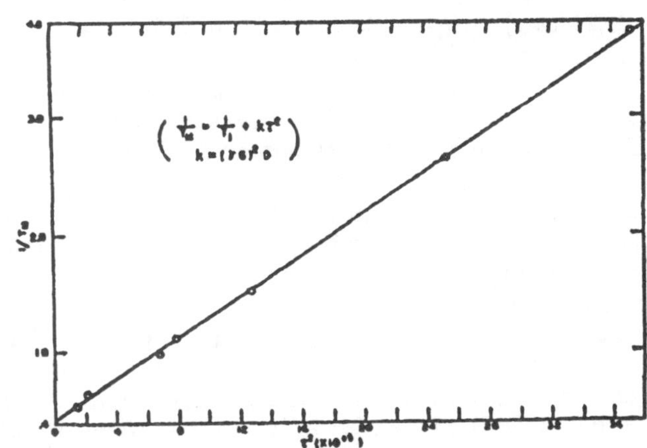

FIG. 8. Stimulated echoe measurement of spin-lattice relaxation time (T_1) of protons in H_2O.

[17] E. L. Hahn, Phys. Rev. 76, 145 (1949).
[18] W. J. C. Orr and J. A. V. Butler, J. Chem. Soc. 1273 (1935).

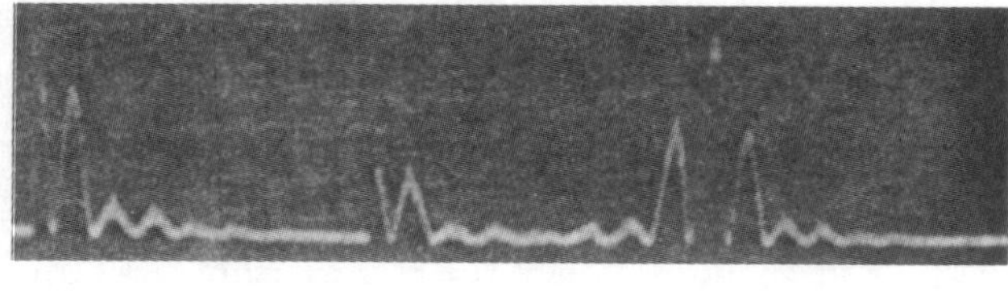

A

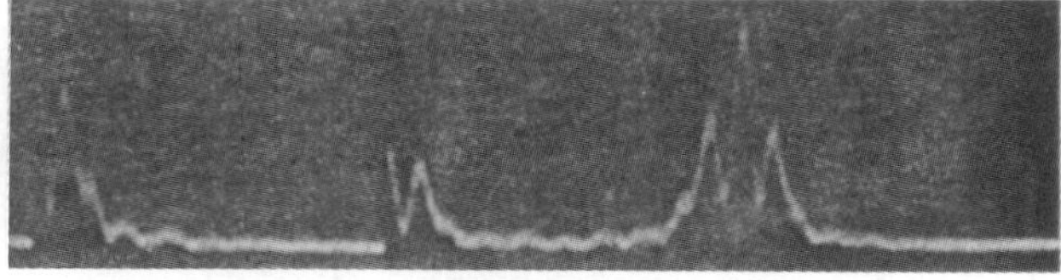

B

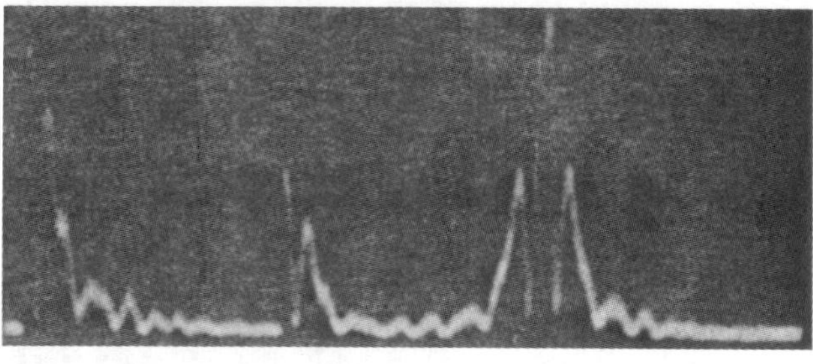

C

FIG. 9–A–B–C. Heterodyne beat signals for different F^{19} resonance frequencies due to the chemical Larmor shift effect.
(A) $CF_2CCl=CCl_2$ and 1,4 difluoro-benzene ($C_6H_4F_2$) mixture
(B) $CF_2CCl=CCl_2$ and 1,2,4 trifluoro-benzene ($C_6H_3F_3$) mixture
(C) 1, trifluoro-methyl 2,3,6 trifluoro-benzene ($C_6H_2F_3CF_3$)

tion.[19] Ramsey has shown that there exists the possibility of a much stronger field shift[20] due to second-order paramagnetism arising from the type of molecule which contains the nuclear spin. The echo technique reveals simultaneously the presence of two or more groups of resonant nuclei having slightly different Larmor frequencies due to such possible shifts in the local field at the nucleus, providing they are of the order of $(\Delta\omega)_{\frac{1}{2}}/\gamma$ gauss in magnitude. Echoes and free induction decay signals are modulated by beat patterns (Fig. 9) due to the fact that two or more spin groups of one species are contained in the same molecule or different molecules and have non-equivalent molecular environments in the same sample. For example, let ω' and ω'' denote respectively the Larmor frequencies at which the rotating coordinate systems of two spin ensembles may precess, and allow symmetric distribution functions $g'(\Delta\omega')$ and $g''(\Delta\omega'')$ to be a maximum for $\Delta\omega'=\omega_0-\omega'=0$, $\Delta\omega''=\omega_0-\omega''=0$. Therefore, identical echo configurations will result in two frames of reference, each rotating with frequencies ω' and ω'' respectively. The r-f induction is due to the magnetization component $v(\Delta\omega, t)\sin\omega t$ for an individual spin group where $v(\Delta\omega, t)$ is described as in (16) and (17). Integration over all frequencies leading to (20) and (21) provides the following total induction:

$$V(t)=V'(t)\sin\omega'(t-t_1')+V''(t)\sin\omega''(t-t_1') \quad (23)$$

$V'(t)$ and $V''(t)$ signify the free induction signals due to each of the spin groups alone. The envelope of the echo signal (Fig. 9-A) is given by

$$V(t)_{\max}=[V'(t)^2+V''(t)^2+2V'(t)V''(t)\times\cos(\omega''-\omega')(t-2\tau)]^{\frac{1}{2}}. \quad (24)$$

As typical examples of this effect it has been found that the signals due to F^{19} nuclei in certain organic compounds yield modulation patterns which obey the heterodyne law expressed by (24). In order to observe this effect the condition $2\pi/(\omega''-\omega')\lesssim T_2^*$ must be attained in order to observe at least one period of the modulation within the lifetime T_2^* of an echo or induction decay signal following a pulse. Consequently, a high degree of homogeneity in the magnetic field must be attained in order to get very good resolution; i.e., to resolve very small shifts in Larmor frequency. It appears that this approach to the determination of very small Larmor shifts has a resolution no better than ordinary magnetic resonance absorption methods[1,3] in which the limitation is also due to external field inhomogeneities. However, the echo method is fast and lends itself more conveniently to search purposes in finding these shifts.[21] Somewhat higher resolution than that available by the normal method can be attained beyond the limitation imposed by field inhomogeneities by introducing into the receiver an r-f signal at a frequency somewhere near the Larmor frequencies present. An audio beat modulation appears having an envelope which is modulated in turn by the Larmor shift beat note. These beats can then be more easily distinguished from noise for the condition $2\pi/(\omega''-\omega')\gtrsim T_2^*$ in favorable cases in which the induction signals are sufficiently intense. Periods of the order of $3, 4T_2^*$ may possibly be observed, in which case Larmor shifts as small as 0.01 gauss, of the order of normal diamagnetic shifts, may be detected, assuming a $(\Delta H)_{\frac{1}{2}}\sim 0.05$ gauss is available out of a total field of 7000 gauss. It can be seen from Fig. 9 that the modulation on the echo and decay signals (following r-f pulses) correlate in pattern. It is significant that the pattern on the echo is always symmetric regardless of the spacing τ between the two r-f pulses. This is understandable in view of the fact that two rotating frames of reference, for example, increase in phase difference by $(\omega''-\omega')\tau$ radians between the pulses. The second pulse produces an initial condition such that the two frames

[19] W. E. Lamb, Phys. Rev. 60, 817 (1941).
[20] This is treated theoretically in a paper in Phys. Rev. 78, 699 (1950), kindly forwarded to the author in advance of publication by Professor N. F. Ramsey.

[21] One must be careful that the observed modulation is not due instead to a condition where the H_0 magnetic field inhomogeneity pattern over the sample has two or more discrete bumps in it. The modulation will again be symmetric on the echo and can only be distinguished from a true beat effect by moving the sample to a different part of the field in the magnet gap and noting whether or not the modulation disappears or varies in frequency.

of reference now rotate into one another by the same amount and coincide at the time 2τ of the echo maximum. This principle is inherent in the echo effect itself: the phase differences of all moment vectors (with respect to the initial orientation established by the first pulse) are effectively cancelled at the time of the echo maximum. This cancellation is made possible by the second pulse. If no further pulses are applied, the echo at 2τ can never repeat itself, as might be expected, because the "eight-ball" configuration is essentially only a single recurrence of the initial in-phase condition of the moment vectors at $t=t_w$, though not quite the same due to a spread in Larmor frequencies.[22]

Fluorine nuclei in the compounds[23] $CF_2CCl=CCl_2$ and $C_6H_4F_2$ (1,4 difluoro-benzene) give induction signals in separate samples in which no significant beat patterns appear. Weak beats may appear due to other fluorine compound impurities used in the synthesis of these compounds. Figure 9-A indicates the beat which results when these two molecules are mixed in liquid form in a single sample such that two fluorine spin groups are in a one-to-one ratio in concentration. The separate molecules contain fluorine atoms located in equivalent positions and therefore cannot give rise to a beat among themselves. A mixture of the two molecules, having fluorine nuclei which are relatively nonequivalent in molecular environment, now reveals a separation of 1.9 kc in Larmor frequency for the two groups in a field of 7500 gauss. According to (24) the modulation pattern goes to a complete null at this frequency since the mixture is adjusted so that $V'(t) = V''(t)$. By observing the normal resonance absorption signal of this mixture on the oscilloscope, using 30 cycle field modulation, two distinct absorption lines are observed, separated by 1.9 kc on a frequency scale. By using a mixture in which the concentration of one molecule exceeds that of the other, the relative difference in intensity of the absorption lines indicates that the fluorine resonance frequency in $C_6H_4F_2$ lies on the high side relative to that in $CF_2CCl=CCl_2$. It is reasonable to expect this if the charge density of the electronic configuration about the fluorine nuclei in $C_6H_4F_2$, being less than that in $CF_2CCl=CCl_2$, can be correlated with a correspondingly smaller negative magnetic shielding correction. This property appears to exist in all mixtures and single molecules so far investigated in which a distinction between spin groups has been made. Within experimental error, the Larmor

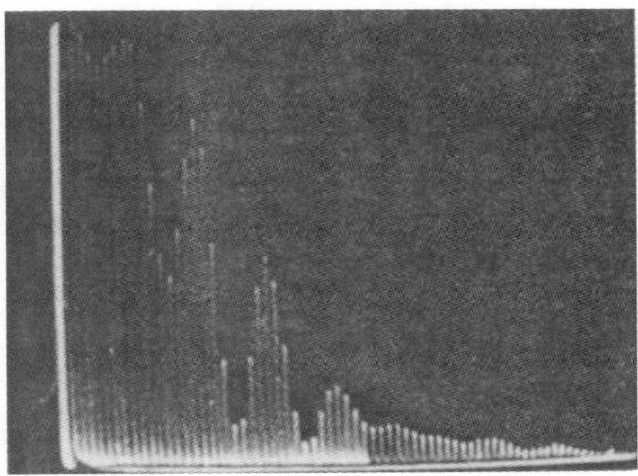

FIG. 10. The echo envelope modulation effect for protons in C_2H_5OH. Paired pulses are applied in the usual manner for obtaining multiple exposures. The echo separation is 1/300 sec. The first echo at the left follows so closely after the r-f pulses that it is not at normal amplitude because of receiver saturation.

frequency shifts observed here appear to be proportional to the applied field, based on measurements made at 7070 and 3760 gauss. Figure 9-B shows the beat pattern due to approximately a one-to-one mixture (in terms of fluorine nuclei) of the compounds 1, 2, 4 trifluoro-benzene and $CF_2CCl=CCl_2$. More than one beat modulation frequency is evident, due obviously to the presence of more than two fundamental spin groups. Figure 9-C shows how a similar complexity in beat pattern arises from a sample of 1-trifluoro-methyl 2, 3, 6 trifluoro-benzene. All observable beat frequencies are of the order of a few kilocycles.

Preliminary studies have been made of another effect which is shown in Fig. 10. The envelope of the normal echo maximum envelope plot is modulated by a beat pattern which is in violation of the normal decay due to self-diffusion and T_2. The envelope shown for C_2H_5OH (period$\approx$0.027 sec.) is an example which is typical for protons in various organic compounds. If the effect is present in the particular substance investigated it is readily observable only if the period of the modulation is shorter than the normal decay time of the echo envelope upon which it is superimposed. Several organic compounds studied so far have been observed to have characteristic periods of the order of 0.1 to 0.01 sec. Modulation patterns in many cases do not contain a single frequency but perhaps several as it appears in C_2H_5OH. The period of the modulation is found in general to be greater than T_2^*, the echo lifetime. This modulation effect cannot be attributed to an interference between several spin groups because the observed echo maximum is always due to the sum of the echo maxima contributed by each of the spin groups alone. This is true regardless of the number of different spin groups present, and therefore the beat frequencies due to such Larmor shifts cannot show up in the envelope of the echo maxima. Within experimental error

[22] It is interesting to note that the configuration at $t=t_w$, namely, $M_{xy}=M_0$, can in principle be exactly repeated at $t=2\tau$ by doubling the second r-f pulse width with respect to the first one which is at the pulse condition $\omega_1 t_w=\pi/2$ (see Fig. 1). Actual experiment indicates that the inhomogeneity in H_1 throughout the sample prevents this from exactly taking place, but shows an increase in the available echo amplitude beyond the optimum amplitude at $\omega_1 t_w=2\pi/3$ (Eq. 21). The stimulated echo at $t=T+\tau$ then nearly disappears.

[23] The fluorine compounds used were kindly provided by Dr. G. C. Finger of the Fluorspar Research Section of the Illinois State Geological Survey, where they were synthesized.

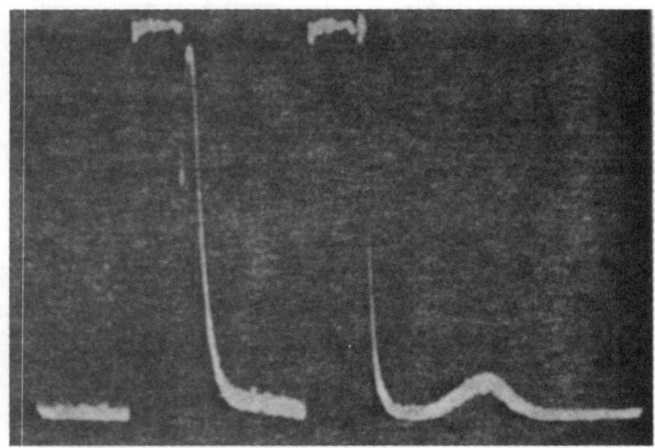

FIG. 11. Free induction signals for protons in paraffin. The echo lasts for $\sim 1.4\times 10^{-5}$ sec. The r-f pulses, about 25 μsec. wide, cause some blocking of the i.f. amplifier. The echo envelope decay time is also of the order of the single echo lifetime.

(five to ten percent) the period of the envelope modulation is found to be inversely proportional to the applied magnetic field. It is possible that an interaction between the nuclear spin and the molecule which contains it causes a periodic reduction of the echo amplitude by a modification of the echo constructive interference pattern. This effect will be treated in a later paper in greater detail.

(H) Free Nuclear Induction in Solids

It has been established in the case of liquids that, if one excludes the effect of diffusion, the lifetime of the nuclear induction decay following a pulse is given by $T_m = T_2T_2^*/(T_2+T_2^*)$, and the single echo lifetime is given by T_2^*. Qualitative observations of free nuclear induction signals due to nuclei in solids, however, indicate that the role played by T_2^* is no longer significant. The ensemble instead precesses in a magnetic field distribution described by a function $G(\delta\omega)$, where $\delta\omega=\omega_0'-\omega'$ and $\omega_0'=\gamma(H_{\rm local}+H_{\rm magnet})$. The local field $H_{\rm local}$, due to lattice neighbors, is superimposed on the externally applied field at the positions of the precessing nuclei. This local field is spread over a width much greater than the width due to the magnet. In one case echoes have been observed for protons in paraffin (Fig. 11) where it appears that the echo and induction decay lifetimes are now given by $\sim 1/(\delta\omega)_{\frac{1}{2}}$ seconds. Extremely intense r-f power is required in order to excite all of the spins over a broad spectrum of Larmor frequencies in a pulse time $t_w \gtrsim 1/(\delta\omega)_{\frac{1}{2}}$ seconds, and therefore the condition $1/t_w$, $\omega_1 \gg (\delta\omega)_{\frac{1}{2}}$ must apply. A striking indication of the predominance of either T_2^* or $1/(\delta\omega)_{\frac{1}{2}}$ is shown by observing how the broad free induction signals from protons in liquid paraffin become very narrow as the paraffin cools and solidifies. It appears that echoes in solids can be observed in principle in a time $\lesssim T_2 \sim 1/(\delta\omega)_{\frac{1}{2}}$, because a given distribution in $H_{\rm local}$ determined by $G(\delta\omega)$ (which now plays the role of $g(\Delta\omega)$ in the case of liquids) is able to last roughly for a time T_2. The local z magnetic field due to neighboring magnetic moments (nuclear spins, paramagnetic ions and impurities) therefore not only depends upon the particular location of these moments with respect to the precessing nucleus, but also upon a time T_2. This time determines how long a given parallel and antiparallel configuration of these neighboring moments can exist with respect to the externally applied field. It follows, therefore, in the case of paraffin, that the stimulated echo, which depends upon frequency memory of the spin distribution, cannot be observed out to times $T+\tau\sim T_1=0.01$ sec., where $T_1 \gg T_2$.

Although the Bloch theory is highly successful in accounting for the echo effect in liquids where T_1 and T_2 are introduced in a phenomenological way, it must be understood in this theory that the predicted natural resonance line shapes will always be described by a damped oscillator resonance function (Lorentz) in the steady state. This corresponds to the observed exponential decay of free induction signal amplitudes in the transient case. This concept does not necessarily apply in general, especially as regards magnetic resonance line shapes in solids in the steady state. It remains to be shown that the properties of free nuclear induction signals in solids are explained by a transient analysis which gives results equivalent to the general steady state treatment formulated by Van Vleck[24] and others.[2,25]

With further refinements in technique for obtaining a sufficiently fast response of the r-f circuits to very short r-f pulses of large intensity (t_w of the order of microseconds at $H_1\sim 20$ to 100 gauss), it may be possible to obtain informative data from free nuclear induction signals in solids which have a T_2 of the order of 10^{-5} to 10^{-6} seconds. It will be of profit to investigate the induction decay which follows single pulses (already found for protons in powdered crystals of NH_4Cl, $(NH_4)_2SO_4$, $MgSO_4\cdot 7H_2O$ and for F^{19} in CaF_2) without the attempt to observe echoes.

IV. EXPERIMENTAL TECHNIQUE

The block diagram in Fig. 12 indicates the necessary components for obtaining the echo effect. All features of the sample, inductive coil, and methods of coupling from the oscillator and to the receiver are typical of nuclear induction techniques and have been discussed in detail elsewhere. A great simplification is introduced here because only a single *LC* tuned circuit is necessary. However, in the case where a narrow band receiver is used ($\sim$8 kc) in place of a very broad band i.f. strip ($\sim$5 Mc) and where r-f pulses are particularly intense, it is convenient to use a bridge balance in order to minimize overloading of the receiver. Two methods have been used to provide r-f power in the form of pulses. A method best suited for r-f amplitude and frequency stability employs a 7.5 Mc crystal-controlled

[24] J. H. Van Vleck, Phys. Rev. 74, 1168 (1948).
[25] G. E. Pake, J. Chem. Phys. 16, 327 (1948).

oscillator whose frequency is quadrupled to 30 Mc and amplified r-f power is then gated through stages whose grids are biased by square wave pulses from a one shot multivibrator controlled in turn by timer pulses. This method is essential for studies of the phases of various echo signals and other effects. The crystal maintains a source of coherent r-f oscillations which can be used to heterodyne weakly with the nuclear signal that has a phase determined initially by these oscillations in the form of intense pulses. The phases of the resulting audio beat frequency oscillations seen superimposed on the echoes then yield certain interesting proofs.

(a) The phase of the audio modulation on all echoes is invariant to any time variation in the spacing between r-f pulses. With respect to a fixed reference in a rotating coordinate system, all echoes form a resultant which is constant in direction due to the fact that the accumulated phase differences before the r-f pulse (at $t=\tau$) are exactly neutralized after it (referring to discussion in III-G).

(b) The negative sign of the "image echo" term (Eq. 22-B) signifies that the resultant of this signal is 180° out of phase relative to the resultants of all other echoes formed before or after it. This is borne out by the fact that the phase of the observed audio modulation on the image echo is exactly 180° out of phase with respect to the modulation pattern on all other echoes. Otherwise, the modulation patterns appear to be identical.

(c) Echoes at $t=2\tau$ are observed not to fluctuate in amplitude when $\tau<T_2^*$ due to the fact that the r-f is coherent for successive pulses, and the phase of the moment configuration prior to the r-f pulse at τ has a definite relationship with respect to the phase of the echo which follows.

With the above method, precautions must be taken to prevent r-f power leakage to the sample during the absence of pulses. The necessity for this precaution is eliminated by turning on and off a high power oscillator (by gating the oscillator grid bias, Fig. 13) which drives the *LC* circuit directly. This method, although not as stable, makes available higher r-f power, and the ability to vary the driving frequency ω is convenient. Although r-f pulses are produced now in random phase, experimental results are the same as long as $\tau\gg T_2^*$. Both

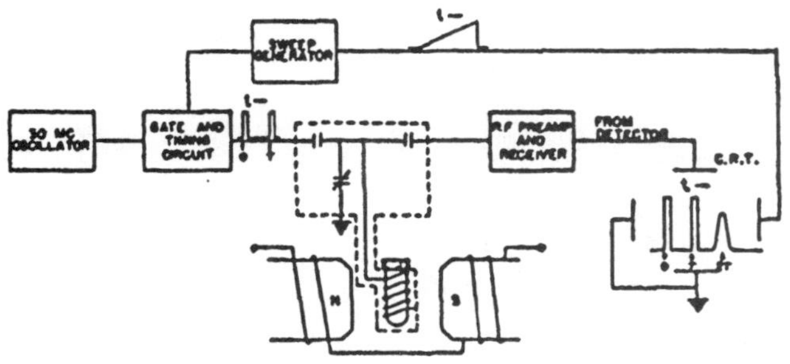

Fig. 12. Arrangement for obtaining spin echoes.

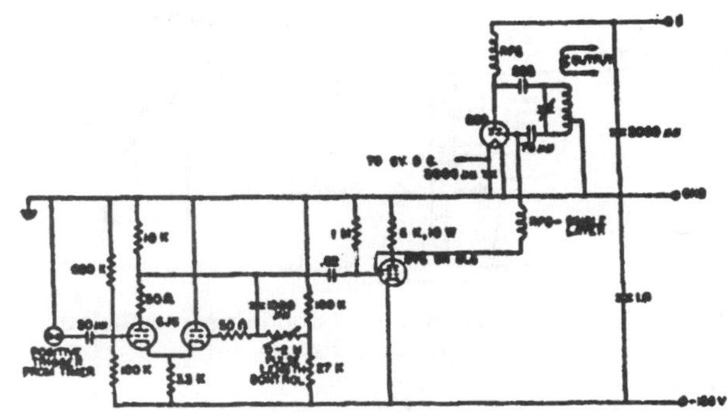

Fig. 13. Gated oscillator.

pulse methods combined provide pulses of $t_w\sim 20$ μsec. to a few milliseconds and $H_1\sim 0.01$ to 50 gauss.

In order to obtain accurate and reproducible data with echoes it is necessary that the dc magnetic field be held constant to at least one part in 10^5 over the length of time during which a set of echo data is being photographed. One might say that some field drift is tolerable within the limits set by the condition $\omega_1\gg(\Delta\omega)_{\frac{1}{2}}$. However, it is advisable even to guard against field drifts less than H_1 gauss because the Fourier amplitudes of all r-f frequency components which resonate with the given Larmor spin frequencies will vary to some degree as the dc field varies. In cases where the decay of echoes is plotted for nuclei in liquids having a long T_1 (several seconds for protons in most organic liquids) and where maximum available signal is desired, the spins must be allowed to return to complete thermal equilibrium between applications of paired pulses. Therefore, if one waits at least five half-lives to obtain a plot such as is shown in Fig. 10, a total time of 17 minutes is required during which the magnetic field must not drift appreciably. In order to minimize the effect of slow field drifts it is convenient to apply paired pulses at a repetition rate whose period is some constant fraction of T_1. During this period the operator has sufficient time to adjust timer switches (reset switches on a conventional scalar unit[17]) in order to provide increasing integers of time between the two pulses. The sample is therefore partially saturated at a level which is practically constant when the pulses are applied, although a small but negligible variation in the level of saturation is introduced as the time τ is systematically increased. The over-all signal to noise ratio of the pattern is reduced but data can be recorded at a convenient speed.

For these experiments the magnetic field is stabilized by means of a separate proton resonance regulator[26] which monitors practically the same magnetic field which is present at the echo sample. The regulator resonance sample is located in the same magnet gap and is subjected to 30 Mc r-f power which is well shielded from the experiment sample. The regulator sample is placed in the inductance of the tuned circuit

[26] To be discussed in a later paper.

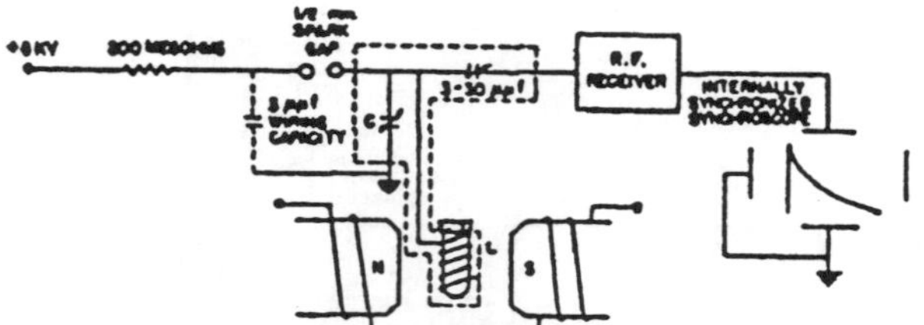

Fig. 14. Simple spark method for obtaining free nuclear induction decay signals.

of a transitron oscillator. A sinusoidally vibrating reed capacitively modulates the frequency of this oscillator within the line width of the regulator proton resonance. When the magnetic field is brought into the resonance value it is locked in and controlled by the regulator. The transitron oscillator r-f voltage level decreases due to resonance absorption and is modulated at the frequency of the vibrating reed. A discriminator circuit utilizes this signal to control a correction current to the magnet in a conventional manner.[27]

In Fig. 14 a sparking technique is noted purely for its novel features of simplicity in demonstrating, qualitatively, free nuclear induction decay signals directly following single random r-f pulses. The spark generated across the gap contains essentially all frequencies and excites for a very short time the tuned circuit in which the sample is located. After the spark extinguishes, the sample in the inductive coil transmits a decaying r-f induction signal to the receiver at the Larmor frequency determined by H_0. Capacitor C needs adjustment such that the tank circuit will resonate approximately in the region of the Larmor frequency (with no spark). Signals can be obtained over a broad range of H_0 without requiring a retuning of C. The observed signal, of course, has random amplitudes since the r-f energy transferred by the sparks is random within a certain range.

[27] M. Packard, Rev. Sci. Inst. 19, 435 (1948).

V. CONCLUDING REMARKS

Simple principles of the free nuclear induction technique have been described and tested, principally with proton and fluorine (F^{19}) signals in liquids. Data which is made available by this technique is to be presented later in more systematic detail. The echo technique appears to be highly suitable as a fast and stable method in searching for unknown resonances. Intense pulses of H_1 provide a broad spectrum of frequencies. This makes possible the observation of free induction signals far from exact resonance. Echo signals have proved useful for the measurement of relaxation times under conditions where interference effects (microphonics, thermal drifts, oscillator noise) encountered in conventional resonance methods are avoided. The self-diffusion effect in liquids of low viscosity offers a means of measuring relative values of the self-diffusion coefficient D, a quantity which is very difficult to measure by ordinary methods. It is of technical interest to consider the possibility of applying echo patterns as a type of memory device.

The formal analysis of the signal-to-noise ratio of the echo method is nearly identical to the treatment already given by Torrey with regard to transient nutations.[9] However, a great practical improvement in eliminating noise is available with the echo technique which cannot be assessed from formal analysis; namely, that H_1 is absent during the observation of nuclear signals, and noise or hum that may be introduced by the oscillator and associated bridge components is avoided.

The author wishes to thank Professor J. H. Bartlett for his counsel in carrying out this research, and is grateful to Dr. C. P. Slichter for the benefit derived from many clarifying discussions with him regarding this work. The author is indebted to H. W. Knoebel for his excellent design and construction work on the electronic apparatus.

The Maser—New Type of Microwave Amplifier, Frequency Standard, and Spectrometer*†

J. P. GORDON,‡ H. J. ZEIGER,§ AND C. H. TOWNES
Columbia University, New York, New York

(Received May 4, 1955)

A type of device is described which can be used as a microwave amplifier, spectrometer, or oscillator. Experimental results are given. When operated as a spectrometer, the device has good sensitivity, and, by eliminating the usual Doppler broadening, a resolution of 7 kc/sec has been achieved. Operated as an oscillator, the device produced a frequency stable to at least 4 parts in 10^{12} in times of the order of a second, and stable over periods of an hour or more to at least a part in 10^{10}. The device is examined theoretically, and results are given for the expected sensitivity of the spectrometer, the stability and purity of the oscillation, and the noise figure of the amplifier. Under certain conditions a noise figure approaching the theoretical limit of unity, along with reasonably high gain, should be attainable.

INTRODUCTION

A TYPE of device is described below can be used as a microwave spectrometer, a microwave amplifier, or as an oscillator. As a spectrometer, it has good sensitivity and very high resolution since it can virtually eliminate the Doppler effect. As an amplifier of microwaves, it should have a narrow band width, a very low noise figure and the general properties of a feedback amplifier which can produce sustained oscillations. Power output of the amplifier or oscillator is small, but sufficiently large for many purposes.

The device utilizes a molecular beam in which molecules in the excited state of a microwave transition are selected. Interaction between these excited molecules and a microwave field produces additional radiation and hence amplification by stimulated emission. We call an apparatus utilizing this technique a "maser," which is an acronym for "microwave amplification by stimulated emission of radiation."

Some results obtained with this device have already been briefly reported.[1] An independent proposal for a system of this general type has also been published.[2] We shall here examine in some detail the general behavior and characteristics of the maser and compare experimental results with theoretical expectations. Particular attention is given to its operation with ammonia molecules. The preceding paper,[3] which will hereafter be referred to as (I), discusses an investigation of the hyperfine structure of the microwave spectrum of $N^{14}H_3$ with this apparatus. Certain of its properties which are necessary for an understanding of the relative intensities of the hyperfine structure components are also discussed there.

BRIEF DESCRIPTION OF OPERATION

A molecular beam of ammonia is produced by allowing ammonia molecules to diffuse out a directional source consisting of many fine tubes. The beam then transverses a region in which a highly nonuniform electrostatic field forms a selective lens, focusing those molecules which are in upper inversion states while defocusing those in lower inversion states. The upper inversion state molecules emerge from the focusing field and enter a resonant cavity in which downward transitions to the lower inversion states are induced. A simplified block diagram of this apparatus is given in Fig. 1. The source, focuser, and resonant cavity are all enclosed in a vacuum chamber.

For operation of the maser as a spectrometer, power of varying frequency is introduced into the cavity from an external source. The molecular resonances are then observed as sharp increases in the power level in the cavity when the external oscillator frequency passes the molecular resonance frequencies.

At the frequencies of the molecular transitions, the beam amplifies the power input to the cavity. Thus the maser may be used as a narrow-band amplifier. Since the molecules are uncharged, the usual shot noise existing in an electronic amplifier is missing, and essentially no noise in addition to fundamental thermal noise is present in the amplifier.

If the number of molecules in the beam is increased beyond a certain critical value the maser oscillates. At the critical beam strength a high microwave energy density can be maintained in the cavity by the beam alone since the power emitted from the beam compensates for the power lost to the cavity walls and coupled wave guides. This oscillation is shown both experimentally and theoretically to be extremely monochromatic.

* Work supported jointly by the Signal Corps, the Office of Naval Research, and the Air Research and Development Command.

† Submitted by J. P. Gordon in partial fulfillment of the requirements of the degree of Doctor of Philosophy at Columbia University.

‡ Now at the Bell Telephone Laboratories, Inc., Murray Hill, New Jersey.

§ Carbide and Carbon Postdoctoral Fellow in Physics, now at Project Lincoln, Massachusetts Institute of Technology, Cambridge, Massachusetts.

[1] Gordon, Zeiger, and Townes, Phys. Rev. 95, 282 (1954).

[2] N. G. Bassov and A. M. Prokhorov, J. Exptl. Theoret. Phys. (U.S.S.R.) 27, 431 (1954). Also N. G. Bassov and A. M. Prokhorov, Proc. Acad. of Sciences (U.S.S.R.) 101, 47 (1945).

[3] J. P. Gordon, preceding paper [Phys. Rev. 99, 1253 (1955)].

APPARATUS

The geometrical details of the apparatus are not at all critical, and so only a brief description of them will be made. Two ammonia masers have been constructed with somewhat different focusers. Both have operated satisfactorily.

A source designed to create a directional beam of the ammonia molecules was used. An array of fine tubes is produced in accordance with a technique described by Zacharias,[4] which is as follows. A $\frac{1}{4}$ in. wide strip of 0.001-in. metal foil (stainless steel or nickel, for example) is corregated by rolling it between two fine-toothed gears. This strip is laid beside a similar uncorregated strip. The corregations then form channels leading from one edge of the pair of strips to the other. Many such pairs can then be stacked together to create a two-dimensional array of channels, or, as was done in this work, one pair of strips can be rolled up on a thin spindle. The channels so produced were about 0.002 in. by 0.006 in. in cross section. The area covered by the array of channels was a circle of radius about 0.2 in., which was about equal to the opening into the focuser. Gas from a tank of anhydrous ammonia was maintained behind this source at a pressure of a few millimeters of mercury.

This type of source should produce a strong but directed beam of molecules flowing in the direction of the channels. It proved experimentally to be several times more effective than a source consisting of one annular ring a few mils wide at a radius of 0.12 in., which was also tried.

The electrodes of the focuser were arranged as shown in Fig. 1. High voltage is applied to the two electrodes marked V, while the other two are kept at ground. Paul *et al.*[5,6] have used similar magnetic pole arrangements for the focusing of atomic beams.

In the first maser which was constructed the inner faces of the electrodes were shaped to form hyperbolas with 0.4-in. separating opposing electrodes. The distance of closest approach between adjacent electrodes was 0.08 in., and the focuser was about 22 in. long. Voltages up to 15 kv could be applied to these electrodes before sparking occurred. In the second maser the electrodes were shaped in the same way, but were separated from each other by 0.16 in. This allowed voltages up to almost 30 kv to be applied, and somewhat more satisfactory operation was obtained since higher field gradients could be achieved in the region between the electrodes. This second focuser was only 8 in. long. Teflon spacers were used to keep the electrodes in place. To provide more adequate pumping of the large amount of ammonia released into the vacuum system from the source the focuser electrodes were hollow and were filled with liquid nitrogen.

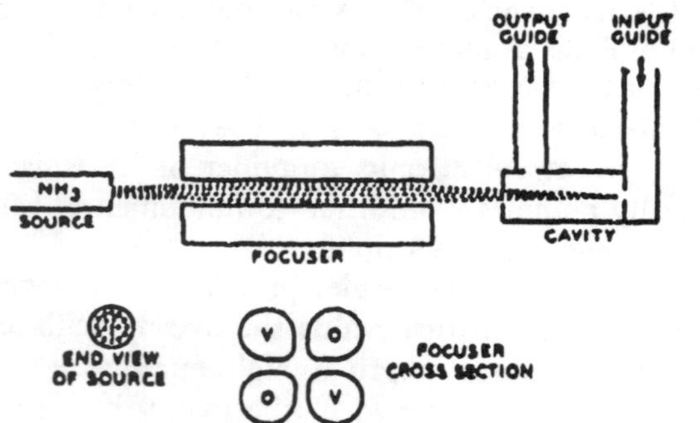

FIG. 1. Simplified diagram of the essential parts of the maser.

[4] J. R. Zacharias and R. D. Haun Jr., Quarterly Progress Report, Massachusetts Institute of Technology Research Laboratory of Electronics, 34, October, 1954 (unpublished).

[5] H. Friedberg and W. Paul, Naturwiss. 38, 159 (1951).

[6] H. G. Bennewitz and W. Paul, Z. Physik 139, 489 (1954).

The resonant cavities used in most of this work were circular in cross section, about 0.6 in. in diameter by 4.5 in. long, and were resonant in the TE_{011} mode at the frequency of interest (about 24 kMc/sec). Each cavity could be turned over a range of about 50 Mc/sec by means of a short section of enlarged diameter and variable length at one end. A hole 0.4 in. in diameter in the other end allowed the beam to enter. The beam traversed the length of the cavity. The cavities were made long to provide a considerable time for the molecules to interact with the microwave field. Only one-half wavelength of the microwave field in the cavity in the axial direction was allowed for reasons which will appear later in the paper. Since the free space wavelength of 24-kMc/sec microwaves is only about 0.5 in., and an axial wavelength of about 9 in. was required in the cavity, the diameter of the cavity had to be very close to the cut-off diameter for the TE_{01} mode in circular wave guide. The diameter of the beam entrance hole was well beyond cutoff for this mode and so very little loss of microwave power from it was encountered. The cavities were machined and mechanically polished. They were made of copper or silver-plated Invar, and had values of Q near 12 000. Some work was also done with cavities in the TM_{01} mode which has some advantages over the TE_{01} mode. However, the measurements described here all apply to the TE_{011} cavities.

Microwave power was coupled into and out from the cavities in several ways. Some cavities had separate input and output wave guides, power being coupled into the cavity through a two-hole input in the end of the cavity furthest from the source and coupled out through a hole in the sidewall of the cavity. In other cavities the sidewall hole served as both input and output, and the end-wall coupling was eliminated. About the same spectroscopic sensitivity was obtained with both types of cavities.

Three MCF 300 diffusion pumps (Consolidated Vacuum Company, Inc.) were used to maintain the necessary vacuum of less than 10^{-6} mm Hg. Nevertheless, due to the large volume of gas released into the system through the source, satisfactory operation has not yet been attained without cooling the focuser electrodes with liquid nitrogen. At 78°K the vapor pressure of ammonia is considerably less than 10^{-6} mm Hg and so the cold electrode surfaces provide a large trapping area which helps maintain a sufficiently low pressure in the vacuum chamber. The pumping could undoubtedly be accomplished by liquid air traps alone; however the diffusion pumps alone have so far proven insufficient. The solidified ammonia which builds up on the focuser electrodes is somewhat of a nuisance as electrostatic charges which distort the focusing field tend to build up on it, and crystals form which can eventually impede the flow of gas. For the relatively short runs, however, which are required for spectroscopic work, this arrangement has been fairly satisfactory.

EXPERIMENTAL RESULTS

Experimental results have been obtained with the maser as a spectrometer and as an oscillator. Although it has been operated as an amplifier, there has as yet been no measurement of its characteristics in this role. Its properties as an amplifier are examined theoretically below.

The reader is referred to (I) for the results obtained from an examination of the hyperfine structure of the $N^{14}H_3$ inversion spectrum with the maser. Resolution of about seven kc/sec was obtained, which is a considerable improvement over the limit of about[7] 65 kc/sec imposed by Doppler broadening in the usual absorption-cell type of microwave spectrometer. This resolution can be improved still further by appropriate cavity design. The sensitivity of the maser was considerably better than that of other spectrometers which have had comparably high resolution.[8–10]

The factors which determine the sensitivity and resolution of the maser spectrometer are discussed in detail below, but we may make a general comment here. The sensitivity of the maser depends in part on the physical separation of quantum states by the focuser and thus on the forces exerted by the focuser on molecules in the various quantum states. For this reason its sensitivity is not simply related to the gas absorption coefficient for a given molecular transition. Each individual case must be examined in detail. Due to the focuser, for example, the sensitivity of the maser varies more rapidly with the dipole moment of the molecule to be studied than does that of the ordinary absorption spectrometer.

The experimental results obtained with the maser in its role as an oscillator agree with the theory given below and show that its oscillation is indeed extremely monochromatic, in fact more monochromatic than any other known source of waves. Oscillations have been produced at the frequencies of the 3–3 and 2–2 inversion lines of the ammonia spectrum, those for the 3–3 line being the stronger. Tests of the oscillator stability were made using the 3–3 line, so we shall limit the discussion to oscillation at this frequency. Other ammonia transitions, or transitions of other molecules could, of course, be used to operate a maser oscillator.

The frequency of the $N^{14}H_3$ 3–3 inversion transition is 23 870 mc/sec. The maser oscillation at this frequency was sufficiently stable in an experimental test so that a clean audio-frequency beat note between the two masers could be obtained. This beat note, which was typically at about 30 cycles per second, appeared on an oscilloscope as a perfect sine wave, with no random phase variations observable above the noise in the detecting system. The power emitted from the beams during this test was not measured directly, but is estimated to be about 5×10^{-10} watt.

The test of the oscillators was made by combining signals from the two maser oscillators together in a 1N26 crystal detector. A heterodyne detection scheme was used, with a 2K50 klystron as a local oscillator and a 30-Mc/sec intermediate-frequency (IF) amplifier. The amplified intermediate frequency signals from the two maser oscillators were then beat together in a diode detector, and their difference, which was then a direct beat between the two maser oscillator frequencies, displayed on an oscilloscope. The over-all band width of this detecting system was about 2×10^4 cps, and the beat note appeared on the oscilloscope with a signal to noise ratio of about 20 to 1.

It was found that the frequency of oscillation of each maser could be varied one or two kc/sec on either side of the molecular transition frequency by varying the cavity resonance frequency about the transition frequency. If the cavity was detuned too far, the oscillation ceased. The ratio of the frequency shift of the oscillation to the frequency shift of the cavity was almost exactly equal to the ratio of the frequency width of the molecular response (that is, the line width of the molecular transition as seen by the maser spectrometer) to the frequency width of the cavity mode. This behavior is to be expected theoretically as will be shown below. The two maser oscillators were well enough isolated from one another so that the beat note could be lowered to about 20 cps before they began to lock together. The appearance of this beat note has been noted above. As perhaps $\frac{1}{10}$-cycle phase variation could have been easily detected in a time of a second (which is about the time the eye normally averages what it observes), the appearance of the beat indicates a spectral purity of each oscillator of at least 0.1 part in 2.4×10^{10}, or 4 parts in 10^{12} in a time of the order of a second.

By using Invar cavities maintained in contact with ice water to control thermal shifts in their resonant frequencies, the oscillators were kept in operation for

[7] Gunther-Mohr, White, Schawlow, Good, and Coles, Phys. Rev. 94, 1184 (1954).
[8] G. Newell and R. H. Dicke, Phys. Rev. 83, 1064 (1951).
[9] R. H. Romer and R. H. Dicke, Phys. Rev. 98, 1160(A) (1955).
[10] M. W. P. Strandberg and H. Dreicer, Phys. Rev. 94, 1393 (1954).

periods of an hour or so with maximum variations in the beat frequency of about 5 cps or 2 parts in 10^{10} and an average variation of about one part in 10^{10}. Even these small variations seemed to be connected with temperature changes such as those associated with replenishing the liquid nitrogen supply in the focusers. Theory indicates that variations of about 0.1°C in temperature, which was about the accuracy of the temperature control, would cause frequency deviations of just this amount.

It was found that the oscillation frequency was slightly dependent on the source pressure and the focuser voltage, both of which affect the strength of the beam. These often produced frequency changes of the order of 20 cycles per second when either voltage or pressure was changed by about 25%. As the cavity was tuned, however, both these effects changed direction, and the null points for the two masers coincide to within about 30 cps. The frequency at which these effects disappear is probably very near the center frequency of the molecular response, so this may provide a very convenient way of resetting the frequency of a maser oscillator without reference to any other external standard of frequency.

THE FOCUSER

In (I) it was shown that forces are exerted by the nonuniform electric field of the focuser on the ammonia molecules, the force being radially inward toward the focuser axis for molecules in upper inversion states and radially outward for molecules in lower inversion states. Molecules in upper inversion states are therefore focused by the field, and only these molecules reach the cavity. Moreover, the quadrupole hyperfine splitting of the upper inversion state was shown to affect the focusing since the flight of the molecules through the focuser is adiabatic with respect to transitions between the different quadrupole levels. As a result the higher energy quadrupole levels are focused considerably more strongly than the lower energy ones. The further slight splitting of the various quadrupole states by the magnetic hyperfine interactions of the hydrogen nuclei has little effect since the molecules make many transitions between these closely spaced levels as they enter and leave the focuser. In regions of high field strength where hyperfine effects are unimportant and can be neglected the energy of the molecules in an electric field may be written as

$$W=W_{\text{rotation}}(J,K)\pm\left\{\left(\frac{h\nu_0}{2}\right)^2+\left(\frac{M_JK}{J(J+1)}\mu\mathcal{E}\right)^2\right\}^{\frac{1}{2}}, \quad (1)$$

where ν_0 is the zero-field inversion frequency, J, K, and M_J specify the rotational state of the molecule relative to the direction of the field, μ is the molecular dipole moment, and $\mathcal{E}$ is the magnitude of the electric field.

With these considerations in mind, an approximate calculation of the total number of molecules in the upper inversion state which are trapped by the potential well of the focuser and delivered to the cavity is fairly straightforward. It involves some computation, since the line used for the oscillation (the main line of the $J=K=3$ inversion transition) is composed of three different but unresolved component transitions between quadrupole sublevels of the inversion states, and therefore the number of molecules trapped by the focuser must be calculated for each of these three sublevels and the results added. This calculation is outlined below. We shall consider in detail the properties of the first maser oscillator, with which the work reported in (I) was done.

The focuser electrodes form approximate equipotentials of the potential $V=V_0r^2\cos2\theta$, where r and θ are cylindrical coordinates of a system whose z axis coincides with the axis of the focuser. 15 kv applied to the high-voltage focuser electrodes establishes an electric field whose magnitude is given by

$$\mathcal{E}=200r, \quad (2)$$

where $\mathcal{E}$ is measured in esu and r is in cm. For simplicity we shall assume that the source is small in area and is located on the axis of the focuser. We shall also assume that all molecules which can travel farther than 0.5 cm from the focuser axis collide with the focuser electrodes and are removed from the beam. From (1) and (2) it is seen that the force ($\mathbf{f}=-\text{grad}W$) on the molecules is radial, and for small field strength is proportional to r. Furthermore it can be seen from energy considerations that all molecules which emerge from the source with radial velocity v_r less than v_{max}, where $\frac{1}{2}mv_{max}^2=W(r=0.5\text{ cm})-W(r=0)$ are held within the focuser by the electric field, while all molecules whose radial velocity is greater than v_{max} collide with the electrodes. Since v_{max} is a function of M_J (M_J is the projection of J on the direction of the electric field of the focuser) the number of molecules focused from a given zero-field quadrupole level depends on the high field distribution of these molecules among the various possible M states.

From kinetic theory, the number of molecules per second emerging from a thin-walled source of area S with radial velocity less than v_{max} is given by

$$N=PSv_0\Omega/(2\pi)^{\frac{1}{2}}kT, \quad (3)$$

where P is the source pressure, $v_0=(kT/m)^{\frac{1}{2}}$ is the most probable velocity of molecules in the beam, T is the absolute temperature, and Ω is a solid angle defined by $\Omega=\pi(v_{max}/v_0)^2$. The number of molecules per second in a given quadrupole level which are focused is therefore

$$N(F_1)=\frac{PSv_0}{(2\pi)^{\frac{1}{2}}kT}f(JKF_1)\sum_{M_J}\varphi(F_1M_J)\Omega(M_J), \quad (4)$$

where $f(JKF_1)$ is the fraction of molecules emerging from the source in the quadrupole state characterized by J, K, and F_1 ($\mathbf{F}_1=\mathbf{J}+\mathbf{I}_N$, where I_N is the spin of the nitrogen nucleus), and $\varphi(F_1, M_J)$ is that fraction of these molecules which, according to the discussion in (I), go adiabatically into the state characterized by the quantum number M_J as they enter the high electric field of the focuser. The total number of molecules per second in the upper inversion state which are delivered to the cavity by the focuser is then just the sum of the $N(F_1)$ for the three quadrupole levels, and so is

$$N(J,K)=\frac{PSv_0}{(2\pi)^{\frac{1}{2}}kT}f(JK)\Omega(JK), \tag{5}$$

where

$$\Omega(JK)=\sum_{F_1}\frac{f(JKF_1)}{f(JK)}\sum_{M_J}\varphi(F_1M_J)\Omega(M_J)$$

is an average solid angle for the upper inversion state, and $f(JK)$ is the fraction of molecules emerging from the source in the upper inversion state of the JK rotational level.

If each of these N molecules could be induced to make a transition to the lower inversion state while in the resonant cavity the total power delivered by the beam would be just $N(JK)h\nu_0$. Actually only about 50 to 75% average transition probability for the molecules in the beam can be obtained due to the variation of their velocities and spatial orientations. Assuming 50% transition probability, a source temperature of 300°K, and the geometry and voltage of the focuser given above, a calculation of the solid angle for the 3–3 line gives $\Omega(3\text{-}3)=4\times10^{-3}$ steradian, and available power of 1.5×10^{-9} watt per square millimeter of source area at 1 mm Hg source pressure.

It is estimated that the total number of molecules emerging from the source in the solid angle from which the upper inversion state molecules are selected is about 10^{16} per second. This estimate comes from knowledge of the number of molecules necessary to induce oscillation. This indicates that the present source is operating fairly inefficiently.

RESONANT CAVITY AND LINE WIDTH

The beam of molecules which enters the resonant cavity is almost completely composed of molecules in the upper inversion state. During their flight through the cavity the molecules are induced to make downward transitions by the rf electric field existing in the cavity. The transition probability for any particular molecule at low field strengths is given from first-order perturbation theory by

$$P_{ab}=\hbar^{-2}|\mu_{ab}|^2\left|\int_0^{L/v}\mathcal{E}(t)e^{-i2\pi\nu_0t}dt\right|^2, \tag{6}$$

where μ_{ab} is the dipole matrix element for the transition, L is the length of the cavity, v is the velocity of the molecule, and $\mathcal{E}(t)$ is the rf electric field at the position of the molecule.

An average transition probability $\bar{P}_{ab}$ can be obtained for all molecules in the beam by averaging over the various velocities, trajectories, and values of $|\mu_{ab}|$ for the molecules in the several states which contribute to each spectral line. The power emitted from the beam is then just

$$P=Nh\nu\bar{P}_{ab}, \tag{7}$$

so $\bar{P}_{ab}$ as a function of the frequency of the applied field determines the line width of the molecular response.

Under the simplifying assumptions that the molecules all travel axially down the length of the cavity, that their velocity is uniform and equal to v_0, and that the cavity is a perfect cylinder with only one-half wavelength of rf field in the axial direction, we find that the emitted power has a maximum at the natural transition frequency ν_B, and a total width at half-maximum of $1.2v_0/L$. If the field is assumed to be uniform along the axis rather than one-half of a sine wave, the corresponding total width at half-maximum is $0.9v_0/L$. This line width of about v_0/L can alternatively be obtained from the uncertainty principle and the finite time of interaction of the molecules with the rf field. Thus $\Delta\nu\approx1/\Delta t$, where Δt is the time of flight of the molecule in the cavity, or L/v_0. The identity of the "Doppler broadening" of the spectral line has essentially disappeared. The sharpness of the molecular response as opposed to that obtained in the usual spectrometer may alternatively be attributed to the long wavelength of the rf field in the cavity in the direction of travel of the beam. If the cavity is excited in a mode in which there is more than one-half wavelength in the direction of travel of the beam, then the molecular emission line, as given by Eq. (6), has two peaks symmetrically spaced about the transition frequency. The frequency separation of these peaks can be associated with the Doppler shift.

Equations (6) and (7) show that for small rf field strengths the emitted power P is proportional to $\mathcal{E}_{max}^2$ and thus to the energy stored in the cavity. For larger field strengths, of course, the molecular transitions begin to saturate and Eq. (6) is no longer sufficient for the calculation of P_{ab}. The effects of this saturation will be considered in detail in a later paper; however, we can say that P_{ab} must certainly be less than 1, and that if a high field strength is maintained in the cavity then the average transition probability $\bar{P}_{ab}(\nu_B)$ will be about 0.5. The total power available from the beam is therefore about $Nh\nu_B/2$. Power saturation in this case is rather similar to that in the usual molecular beam experiment, for which it has been considered by Torrey.[11]

Associated with the power emitted from the beam is an anomalous dispersion, that is, a sharp variation

[11] H. C. Torrey, Phys. Rev. 59, 293 (1941).

in the dielectric constant of the cavity medium due to the beam. These two effects can be considered at the same time by thinking of the beam as a polarizable medium introduced into the cavity, whose average electric susceptibility is given by $\chi=\chi'+i\chi''$. The power emitted from the beam can then be shown directly from Maxwell's equations to be[12]

$$P=8\pi^2\nu_B W\chi'', \tag{8}$$

where W is the energy stored in the cavity. Thus, from Eqs. (7) and (8), χ'' is related to $\bar{P}_{ab}$ by

$$\chi''=Nh\bar{P}_{ab}/8\pi^2 W. \tag{9}$$

The value of χ' is given from χ'' by Kramer's relation,[13] which for a sharp resonance line can be approximated by[14]

$$\chi'(\nu)=\frac{1}{\pi}\int_0^\infty \frac{\chi''(\nu')d\nu'}{\nu'-\nu}. \tag{10}$$

Figure 2 shows the form of χ' and χ'', calculated with the assumptions that all molecules are traveling parallel to the axis of the cavity with uniform velocity, that the cavity is excited in the TE_{011} mode, and that there is a small field strength in the cavity so that $P_{ab}\ll 1$ and Eq. (6) is valid. χ' and χ'' can also be found directly by calculation of the induced dipole moments of the molecules as they traverse the cavity.

If the Q of the cavity is defined in terms of net power loss (i.e., $dW/dt=-2\pi\nu W/Q_C$) then the presence of the beam can be considered as causing a change in the effective Q_C given by $1/Q_{CB}=1/Q_C-4\pi\chi''$, where Q_{CB} and Q_C are respectively the cavity Q's with and without the beam, along with a shift in the resonant frequency of the cavity given by $\nu_{CB}=\nu_C(1-2\pi\chi')$ if $\chi'\ll 1$. These relations can also be easily derived directly from Maxwell's equations, and they will prove important in determining the properties of the maser.

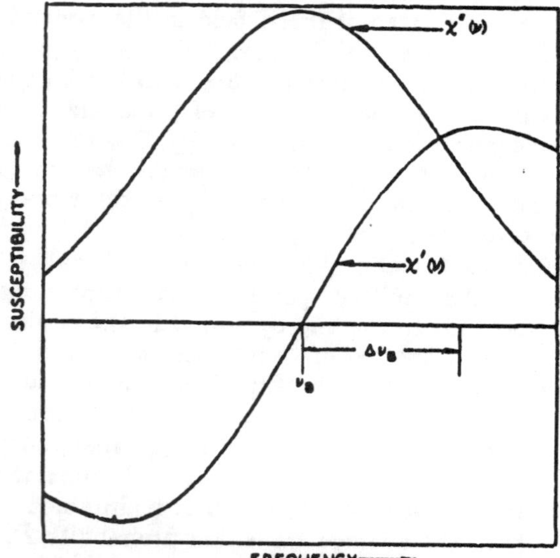

FIG. 2. Real and imaginary parts of the electric susceptibility χ of the molecular beam in the cavity. $\chi=\chi'+i\chi''$.

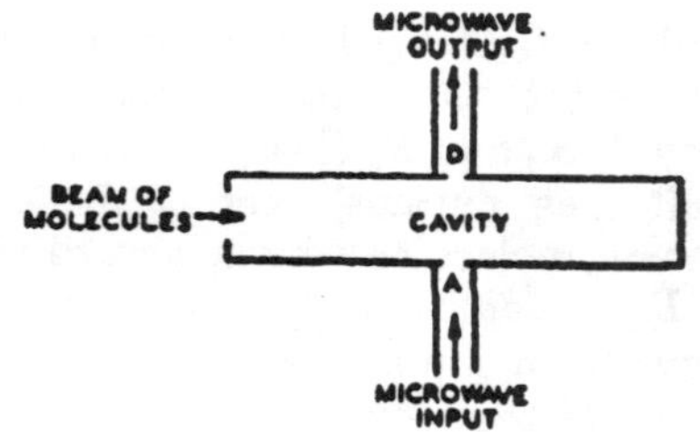

FIG. 3. Schematic diagram of the resonant cavity and molecular beam.

THE MASER SPECTROMETER

Observed Line Shape as a Function of the Cavity Resonant Frequency

Consider the situation shown in Fig. 3. Power P_0 is incident on the cavity from wave guide A, and the power transmitted on out through wave guide D is detected as a function of the frequency of the input power. The power transmitted through the cavity in the absence of the beam is given by[12]

$$P_D(\nu)=\frac{P_0}{Q_A Q_D}\Big/\left[\left(\frac{1}{2Q_L}\right)^2+\left(\frac{\nu-\nu_C}{\nu_C}\right)^2\right], \tag{11}$$

where Q_A and Q_D are defined in terms of the power losses from the cavity to wave guides A and D respectively, and Q_L is the loaded Q of the cavity, given by $1/Q_L=1/Q_C+1/Q_A+1/Q_D$. As was shown in the last section, the change in $P_D(\nu)$ caused by the presence of the beam can be described through variations in Q_C and ν_C near the transition frequency ν_B. Thus in the presence of the beam we find P_D modified to

$$P_{DB}(\nu)=\frac{P_0}{Q_A Q_D}\Big/\left[\left(\frac{1}{2Q_L}-2\pi\chi''(\nu)\right)^2+\left(\frac{\nu-\nu_C+2\pi\nu_C\chi'(\nu)}{\nu_C}\right)^2\right]. \tag{12}$$

As long as the power output P_{DB} is not so high that nonlinearities in the molecular response are important, (12) gives the output power as a function of frequency in the presence of the beam and represents the spectrum which may be observed. For most spectroscopic applications we are interested in the case for which $\chi''(\nu)\ll 1/4\pi Q_L$ for all ν. For this case, an appropriate expansion of Eq. (12) shows that if $\nu_C=\nu_B$, where ν_B is the center frequency of the molecular response, then the presence of the beam shows up as a pip of the shape

[12] J. C. Slater, *Microwave Electronics* (D. Van Nostrand Company, Inc., New York, 1950).

[13] J. H. Van Vleck, *Massachusetts Institute of Technology Radiation Laboratory Report*, 735, see also *Radiation Laboratory Series* (McGraw-Hill Book Company, Inc., New York, 1948), Vol. 13, Chap. 8.

[14] G. E. Pake and E. M. Purcell, Phys. Rev. 74, 1184 (1948).

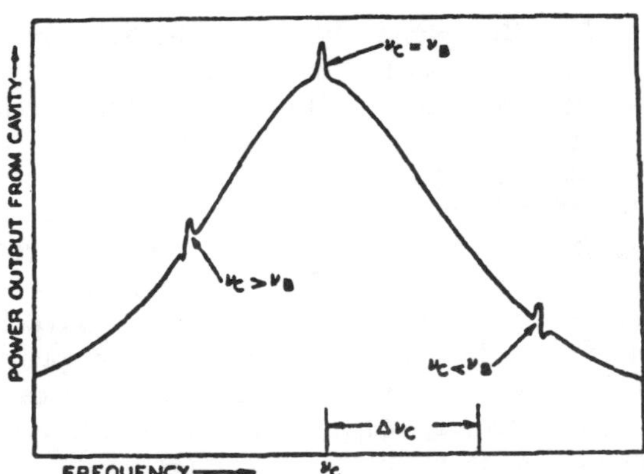

FIG. 4. Spectral line shapes as observed in the maser.

of χ'' superimposed on the cavity mode. The half-width $\Delta\nu_B$ of the molecular response is therefore defined as half the frequency separation of the half-maximum points of $\chi''(\nu)$. If the cavity frequency is altered so that the molecular line appears on the wings of the cavity mode, then the pip due to the beam assumes the shape of $\pm\chi'(\nu)$, the $+$ or $-$ sign depending on whether the line appears on the low or high frequency side of the cavity mode. These three situations are illustrated in Fig. 4.

Sensitivity of the Maser Spectrometer

We have here the usual problem occurring in microwave spectroscopy of detecting a small change in output power P_0 caused by the presence of the molecules. Assume that $\nu_C=\nu_B=\nu$ in Eq. (12). Then

$$P_{DB}=\frac{P_0}{Q_AQ_D}\bigg/\left(\frac{1}{2Q_L}-2\pi\chi''(\nu_B)\right)^2. \tag{13}$$

The change in output power which must be detected is then

$$\delta P_D=P_{DB}-P_D\cong P_0(4Q_L^2/Q_AQ_D)\times[8\pi Q_L\chi''(\nu_B)]. \tag{14}$$

Consider now the change in output voltage δV_D, which is proportional to $(P_D+\delta P_D)^{\frac{1}{2}}-P_D^{\frac{1}{2}}$ or to $\delta P_D/2P_D^{\frac{1}{2}}$ when $\delta P_D\ll P_D$. The noise voltage at the output of a linear detector is just $(FkT\Delta\nu)^{\frac{1}{2}}$, where $\Delta\nu$ is the band width of the detector and F is its over-all noise figure. Thus the voltage signal-to-noise ratio is given by

$$\delta V_D/V_N=[(\delta P_D)^2/4FkTP_D\Delta\nu]^{\frac{1}{2}}. \tag{15}$$

As long as the change in power δP_D due to the beam increases linearly with the power level P_D, the signal to noise ratio from (15) continues to increase. Hence the power input to the cavity should be increased until the transition begins to saturate. For saturation and a given Q_C, it can be shown that maximum sensitivity is achieved by using a small input coupling ($Q_A\gg Q_C$) and a matched output coupling ($Q_D=Q_C$). If we approximate the saturation condition by setting W equal to the level at which the power emitted by the beam is just $\frac{1}{2}Nh\nu_B$, then with a little algebra we find, from Eqs. (8), (14), and the known relationship of W to P_0,[13] that the change in output power δP_D is just $\frac{1}{2}Nh\nu_B$ (the beam emits $\frac{1}{2}Nh\nu_B$ of power, and due to the change in the input match caused by the beam, $\frac{1}{2}Nh\nu_B$ more power enters the cavity through the input coupling hole A. Thus the increase in power input to the cavity is twice the power emitted from the beam. Half of this increase emerges into wave guide D since it was assumed to be matched to the cavity. The power level P_D is now determined by the required energy W from the relation

$$P_D=2\pi\nu_BW/Q_C. \tag{16}$$

This, with (8), gives

$$P_D=P/4\pi\chi''(\nu_B)Q_C=Nh\nu_B/8\pi\chi''(\nu_B)Q_C. \tag{17}$$

Inserting these values for P_D and δP_D in (15) yields

$$\delta V_D/V_N=[\pi Q_CNh\nu_B\chi''(\nu_B)/2FkT\Delta\nu]^{\frac{1}{2}}. \tag{18}$$

This relation gives the sensitivity of the maser, once the value of $\chi''(\nu_B)$ for a given molecule is calculated. χ'' is, of course, related to the average transition probability $\bar{P}_{ab}$ by Eq. (9), so that Eq. (18) can easily be rewritten in terms of the transition probability.

For the ammonia 3–3 line, a calculation of the number of molecules in the 3–3 state necessary to make $\delta V_D/V_N=1$ was done, assuming $F=100$, $\Delta\nu=1$ cps, $Q_C=12\,000$, $T=300°$ and using an approximate calculation of $\chi''(\nu_B)$ based on the considerations of the previous section. The result was 10^9 molecules per second. It is estimated from the value of $\chi''(\nu_B)$ which is necessary to cause oscillation (see next section) that the number of upper inversion state molecules in the 3–3 rotational state in the beam when oscillations occur is at least 10^{13} molecules per second, and experimentally a number about four times this great was achieved. Thus, for ammonia, the maser should have good sensitivity, and the results described in paper (I) show that this is indeed the case.

In the case of the ammonia inversion spectrum, the focuser can effect an almost complete separation of the upper states from the lower states of the transitions. For some other transitions; this ideal state of affairs may not be attainable, but yet the focuser may preferentially focus one of the two states of the transition. In such a case all of the above considerations apply so long as one uses for N just the excess number of molecules in one of the two states. It is, of course, unimportant for spectroscopic purposes whether the more highly focused state is the upper or lower state of the transition. The high sensitivity attained in the observation of the ammonia spectrum with the maser gives promise that it may be generally useful as a microwave spectrometer of very high resolution.

THE MASER OSCILLATOR AND AMPLIFIER

By extending the considerations of the previous section to include amplification of the thermal noise which exists in the cavity, we can discuss the properties of the maser as an oscillator or amplifier. The results of this analysis, which is made below, are as follows:

(1) The center frequency ν_0 of the oscillation is given to a good approximation by the equation

$$\nu_0=\nu_B+\frac{\Delta\nu_B}{\Delta\nu_C}(\nu_C-\nu_B), \qquad (19)$$

where $\Delta\nu_C$ and $\Delta\nu_B$ are respectively the half-widths of the cavity mode and of the molecular emission line; and $\nu_C-\nu_B$ is the difference between the cavity resonant frequency ν_C and the line frequency ν_B.

(2) The total width at half-power of the spectral distribution of the oscillation is approximately

$$2\delta\nu=8\pi kT(\Delta\nu_B)^2/P_B \qquad (20)$$

where T is the temperature and P_B is the power emitted from the beam. Inserting in (20) values which approximate the experimental conditions, $T=300°K$, $\Delta\nu_B=3\times10^3$ cps, $P_B=10^{-10}$ watt, we find $2\delta\nu\approx10^{-2}$ cps, or $\nu_B/\delta\nu=5\times10^{12}$.

(3) If the beam is sufficiently strong, the maser may be used as an amplifier with a gain greater than unity and a noise figure very close to unity.

The argument goes along the following lines. Consider the situation of Fig. 3, the cavity with two wave guides. The whole system will be assumed to be in thermal equilibrium in the absence of the beam. Noise power of amount kT per unit band width is incident on the cavity from each wave guide, and the cavity walls emit noise power within the cavity. Of the noise power incident on the cavity from wave guide A, a certain amount within the frequency range of the cavity mode enters the cavity; part of this power is then absorbed by the cavity walls and part is transmitted on out through wave guide D. A similar situation holds for noise power incident on the cavity from wave guide D; some is absorbed in the cavity and some is transmitted through to wave guide A. The cavity walls emit noise power in the region of the cavity mode and some of this power goes out through each wave guide. When the beam is not present we have assumed the system to be in thermal equilibrium, so there must be kT per unit band width of noise power flowing away from the cavity down each wave guide and there must be kT of noise energy in the cavity mode, as required by the equipartition theorem.

In the presence of the beam thermal equilibrium is upset. The beam, since it is composed solely of upper inversion state molecules, and since the probability for spontaneous decay of these molecules to the lower states is negligible during the time they take to traverse the cavity, contributes no random noise of its own to the rf field of the cavity. What it does is merely to amplify, in a way described by its effect on the loaded Q and resonant frequency of the cavity, all of the noise signals which exist in the cavity. Thus to the noise sources in the wave guides the intrinsic Q of the cavity seems to have been altered; whereas to the noise source within the cavity, the loading on the cavity seems to have changed. In fact, the presence of the beam can be duplicated in the imagination by attaching to the cavity a third wave guide, with a negative Q equal to $\frac{1}{2}\pi\chi''(\nu)$ describing its coupling to the cavity, and by simultaneously shifting the resonant frequency of the cavity by an amount $-2\pi\nu_C\chi'(\nu)$.

From these considerations we will show that in the presence of the beam more than kT of power per unit bandwidth travels down each wave guide away from the cavity. The extra power, of course, comes from the beam. At a certain critical beam intensity this power suddenly becomes large, corresponding to sustained oscillations.

Let $\Delta\nu$ be some arbitrarily small element of the frequency spectrum at frequency ν. Within this range noise power of magnitude $kT\Delta\nu$ is incident on the cavity from each wave guide, independent of ν. Let $P_A\Delta\nu$ be the amount of noise power which enters the cavity from the incident power in wave guide A, and let $P_A'\Delta\nu$ be the total noise power re-emitted into wave guide A from inside the cavity. The presence of the beam will be indicated by an added subscript; i.e., P_{AB} will represent the value of P_A when the beam is present, etc. Similar definitions apply to the output guide D. Since the noise powers generated in the wave guides and in the cavity are completely incoherent with one another, we can simply add power coming from various sources to obtain the total power in any element of the system. Thus the total energy $W\Delta\nu$ stored in the cavity is merely the sum $\sum_i W_i\Delta\nu$ of all the energies due to power coming from the various different power sources. (W is energy per unit band width.)

Consider now the flow of power when no beam is present. The noise power entering the cavity from wave guide A is given by[12]

$$P_A=\frac{kT}{Q_AQ_L^A}\bigg/\left[\left(\frac{1}{2Q_L^A}\right)^2+\left(\frac{\nu-\nu_C}{\nu_C}\right)^2\right], \qquad (21)$$

where $1/Q_L^x=\sum_{m\neq x}1/Q_m$ is just proportional to all the losses from the cavity except that due to Q_x. Of this power P_A, some is absorbed in the cavity walls, and the rest is transmitted on out through wave guide D. The energy per unit band width stored in the cavity due to this power input is

$$W_A=P_AQ_L^A/2\pi\nu$$
$$=\frac{kT}{2\pi\nu Q_A}\bigg/\left[\left(\frac{1}{2Q_L}\right)^2+\left(\frac{\nu-\nu_C}{\nu_C}\right)^2\right], \qquad (22)$$

and the power transmitted on to the output wave guide D is $2\pi\nu W_A/Q_D$, or $P_A Q_L^A/Q_D$. Similar expressions hold for the noise power incident on the cavity from wave guide D. Furthermore, the cavity loss associated with Q_C may be assumed to be due to a third wave guide with coupling characterized by Q_C to a perfectly conducting cavity, so that the energy W_C emitted into the actual cavity from its wall has the same form as (22). The total energy stored in the cavity per unit frequency interval is hence

$$W=W_A+W_D+W_C$$
$$=\frac{kT}{2\pi\nu Q_L}\Big/\left[\left(\frac{1}{2Q_L}\right)^2+\left(\frac{\nu-\nu_C}{\nu_C}\right)^2\right], \quad (23)$$

where

$$1/Q_L=1/Q_A+1/Q_D+1/Q_C.$$

The total energy stored in the cavity, given by $\int_0^\infty W d\nu$, is easily shown to be equal to kT (we make the assumption that $Q_L \gg 1$, so that in the integration the approximation $\nu\approx\nu_C$ may be made) as required by the equipartition theorem. The net noise power flowing in the wave guides A or D is also easily shown to be zero, so that the system is indeed seen to be in thermal equilibrium.

Consider now the case when the beam is present. The noise power incident from each wave guide sees a cavity whose rates of internal loss has been reduced by an amount $4\pi\chi''(\nu)$ by the energy emitted from the beam and whose resonant frequency has been shifted by an amount $-2\pi\nu_C\chi'(\nu)$. Corresponding to Eq. (21), the power entering the cavity from wave guide A in the presence of the beam is

$$P_{AB}=\frac{kT}{Q_{LB}^A Q_A}\Big/\left[\left(\frac{1}{2Q_{LB}}\right)^2+\left(\frac{\nu-\nu_{CB}}{\nu_{CB}}\right)^2\right], \quad (24)$$

where $1/Q_{xB}=1/Q_x-4\pi\chi''(\nu)$ for any x and $\nu_{CB}=\nu_C[1-2\pi\chi'(\nu)]$, while that entering from wave guide D is similarly

$$P_{DB}=\frac{kT}{Q_{LB}^D Q_D}\Big/\left[\left(\frac{1}{2Q_{LB}}\right)^2+\left(\frac{\nu-\nu_{CB}}{\nu_{CB}}\right)^2\right]. \quad (25)$$

The noise energy stored in the cavity due to these two sources is

$$W_{AB}=\frac{P_{AB}Q_{LB}^A}{2\pi\nu}, \quad\text{and}\quad W_{DB}=P_{DB}Q_{LB}^D/2\pi\nu. \quad (26)$$

At the same time, the energy stored in the cavity due to its own internal noise source is changed as though the loading on the cavity has been altered while its internal loss was unaffected. This energy is therefore given by

$$W_{CB}=\frac{kT}{2\pi\nu Q_C}\Big/\left[\left(\frac{1}{2Q_{LB}}\right)^2+\left(\frac{\nu-\nu_{CB}}{\nu_{CB}}\right)^2\right]. \quad (27)$$

Due to the presence of the beam the net noise power emitted from the cavity into the output wave guide D is now no longer zero. The power emerging from the cavity is now

$$P_{DB}'=\frac{2\pi\nu}{Q_D}[W_{CB}+W_{AB}]$$
$$=\frac{kT}{Q_D Q_L^D}\Big/\left[\left(\frac{1}{2Q_{LB}}\right)^2+\left(\frac{\nu-\nu_{CB}}{\nu_{CB}}\right)^2\right]. \quad (28)$$

Thus the additional noise output in wave guide D due to the beam is, from (25) and (28),

$$P_{DN}\equiv P_{DB}'-P_{DB}$$
$$=\frac{kT}{Q_D}4\pi\chi''(\nu)\Big/\left[\left(\frac{1}{2Q_{LB}}\right)^2+\left(\frac{\nu-\nu_{CB}}{\nu_{CB}}\right)^2\right]. \quad (29)$$

The power which must be emitted from the beam to give this amount of power in wave guide D is just

$$P_N\equiv P_{DN}Q_D/Q_L=\frac{kT}{Q_L}4\pi\chi''(\nu)\Big/$$
$$\left[\left(\frac{1}{2Q_L}-2\pi\chi''(\nu)\right)^2+\left(\frac{\nu-\nu_C+2\pi\nu_C\chi'(\nu)}{\nu_C}\right)^2\right], \quad (30)$$

where ν_C has replaced ν_{CB} in the denominator of the last term in the denominator of this expression since $\nu_{CB}\approx\nu_C$. Note that (30) is just equivalent to (24) if the beam is thought of as a wave guide coupled to the cavity with a Q of $-1/4\pi\chi''$.

Expression (30) gives the complete spectrum of the power emitted from the beam due to amplification of the noise signals which are always present in the cavity. The necessary condition for the existence of oscillations as some cavity frequency is evidently that $\chi''(\nu_B)\approx\frac{1}{4}\pi Q_L$.

Assume that the cavity is tuned so that $|\nu_C-\nu_B|\ll\Delta\nu_C$, and then let the beam strength slowly increase so that χ'' increases. Then at the critical beam strength where $\chi''(\nu_B)\to 1/4\pi Q_L$, the total power $\int P_N d\nu$ emitted from the beam approaches infinity accordingly to (30). Obviously, the total power emitted from the beam cannot go to infinity, but is limited to about $\frac{1}{2}Nh\nu_B$. When the power level in the cavity reaches the point at which the molecular transition begins to saturate, χ'' and χ' become functions of the power level, and, of course, vary in such a way that $\int P_N d\nu$ is always less than $\frac{1}{2}Nh\nu_B$. We can, for simplicity, avoid the problem of dealing with this saturation merely by increasing χ'' until $\int P_N d\nu=\frac{1}{2}Nh\nu_B$ and examining the frequency spectrum of the power emitted from the beam at this level of output. Although χ'' is not independent of the electric field strength when saturation occurs, it varies much more slowly with time than does the oscillation, so that it may be considered constant in treating the short-term behavior of the microwave field.

As the critical number of molecules is reached, P_N becomes very large at frequencies very close to ν_B. Hence it is appropriate to expand χ' and χ'' about the center frequency ν_B. This gives approximately

$$\chi'=\chi_0\left(\frac{\nu-\nu_B}{\Delta\nu_B}\right)+\cdots,$$
$$\chi''=\chi_0\left[1-\frac{1}{2}\left(\frac{\nu-\nu_B}{\Delta\nu_B}\right)^2+\cdots\right]. \tag{31}$$

Writing Eq. (30) in terms of (31), and setting $\int P_N d\nu$ equal to P_B, where $P_B=\frac{1}{2}Nh\nu_B$, one obtains

$$P_N\approx 4kT(\Delta\nu_B)^2\Big/\left[(\nu-\nu_0)^2+\left(\frac{4\pi kT}{P_B}(\Delta\nu_B)^2\right)^2\right], \tag{32}$$

where ν_0, the oscillation frequency, is given by the equation

$$\frac{\nu_0-\nu_B}{\nu_0-\nu_C}=-\frac{\Delta\nu_B}{\Delta\nu_C}, \tag{33}$$

or, as in (19),

$$\nu_0=\nu_B+(\nu_B-\nu_C)\Delta\nu_B/\Delta\nu_C \quad \text{if} \quad \Delta\nu_B/\Delta\nu_C\ll 1.$$

The total width $2\delta\nu$ at half-maximum power of this "noise" output is, from (32),

$$2\delta\nu=(8\pi kT/P_B)(\Delta\nu_B)^2 \tag{34}$$

as already stated in (20).

It should be remembered that (32) involves the assumption that the maser is a linear noise amplifier of very high gain. Actually, the noise properties of an oscillator depend to a considerable extent on the non-linearities in its response, or the overload, and (32) does not accurately represent the precise noise spectrum of the maser as an oscillator. However, the approximate width of its noise spectrum is properly given by (34). As in the more usual types of oscillators, this oscillator actually maintains a nearly fixed amplitude of oscillation, but its phase slowly varies with time in a random way, corresponding to a noise spectrum of a width given approximately by (34). A more detailed discussion of noise will be given in a later publication.

The half-width of this oscillation signal is not to be confused with the half-width of the molecular response $\Delta\nu_B$. The latter represents the band width of the maser amplifier at low gain, whereas the former gives the band width of the oscillation signal. The oscillation frequency ν_0 can be varied throughout the range over which the molecules will amplify in accordance with (33) or (19). Hence care must be taken to keep the cavity frequency ν_C constant if it is desired to keep the oscillation frequency constant for any extended period of time.

Noise Figure and Band Width of the Amplifier

The noise figure of the maser amplifier may be easily found from the results of the foregoing sections. Assume that $\nu_C=\nu_B=\nu$, where ν is the frequency of the signal to be amplified. Also assume that the detector has a band width $\Delta\nu_{\text{det}}$ such that $\Delta\nu_{\text{det}}\ll\Delta\nu_B$. Equation (13) gives the signal power at the cavity output, while Eq. (29) gives the noise at the output in excess of kT. Thus we see that the signal-to-noise ratio at the output is just

$$\frac{P_0\Big/\left[Q_AQ_D\left(\frac{1}{2Q_L}-2\pi\chi_0\right)^2\right]}{kT\Delta\nu_{\text{det}}\left[1+\frac{4\pi\chi_0}{Q_D}\Big/\left(\frac{1}{2Q_L}-2\pi\chi_0\right)^2\right]}, \tag{35}$$

where $\chi''(\nu_B)=\chi_0$. At the input to the cavity, the signal to noise ratio is $P_0/kT\Delta\nu_{\text{det}}$. Therefore the noise figure F, which is just the ratio of these two quantities, is

$$F=Q_AQ_D\left[\frac{4\pi\chi_0}{Q_D}+\left(\frac{1}{2Q_L}-2\pi\chi_0\right)^2\right]. \tag{36}$$

At the same time, the power amplification available is, from (13), given simply by

$$\mu=P_{DB}/P_0=\left[Q_AQ_D\left(\frac{1}{2Q_L}-2\pi\chi_0\right)^2\right]^{-1}. \tag{37}$$

It can be shown from (37) that $\mu<1$ if $4\pi\chi_0<1/Q_C$, i.e., if there is a net loss of power within the cavity itself. Thus unless it is possible to produce oscillation by putting lossless reflections in all the wave guides so that $Q_L\to Q_C$, it is also impossible to create an amplifier with a gain greater than unity. In order to obtain a large gain μ, one must have $1/Q_L\approx 4\pi\chi_0$. If the gain is large, then a noise figure approaching unity is attainable by making $1/Q_A\approx +4\pi\chi_0\approx 1/Q_L$. This shows that for high amplification and at the same time a low noise figure, a fairly large input coupling to the cavity and a small output coupling is needed. Furthermore a sufficiently strong beam is required so that the maser is not too far from oscillation.

The maser acts as a regenerative amplifier, as can be seen from (12). Thus under conditions such that $4\pi\chi_0\approx 1/Q_L$ so that the midband gain is high, the band width becomes substantially smaller than $2\Delta\nu_B$.

It might also be noted that a certain amount of modulation of the amplified output is to be expected due to random variations of the number of molecules in the cavity at any time. These effects, however, are proportional to the input signal strength, and so are quite different from thermal noise signals which have no dependence on input power. Furthermore, they represent a modulation of only about one part in 10^6 since there are 10^{12} or more molecules in the cavity at any time. This type of modulation can be neglected when small input signals are considered and is not important under most circumstances. This shot effect and also the effect of power flow through the cavity on the frequency dependence of the amplification will be discussed in more detail in a subsequent paper.

Amplification may also be accomplished using one wave guide as both input and output, and the noise figure of such an amplifier can also approach unity. The amplified output signal might be coupled out and detected through a directional coupler, which would have to have a fairly small coupling so that little of the input power was lost to it. Then so long as the amplified input noise appearing at the detector was large compared to kT, the noise figure of this amplifier would be small.

The maser amplifier may be useful in a restricted range of applications in spite of its narrow band width because of its potentially low noise figure. For example, suppose that the signal to be amplified came from outer space, where the temperature is only a few degrees absolute. Then by making the coupling through the cavity fairly large so that little noise is contributed by the cavity itself, amplification should be attainable while keeping the noise figure, *based on the temperature of the signal source*, fairly low. This might prove to have a considerable advantage over electronic amplifiers. It might also be possible to tune the frequency of a maser amplifier through the use of the Stark or Zeeman effects on the molecular transition frequencies.

ACKNOWLEDGMENTS

The authors would like to express their gratitude to the personnel of the Columbia Radiation Laboratory who assisted in the construction of the experimental apparatus. They would also like to thank Dr. T. C. Wang and Dr. Koichi Shimoda for their assistance and suggestions during the later stages of this work.

PHYSICAL REVIEW VOLUME 104, NUMBER 2 OCTOBER 15, 1956

Proposal for a New Type Solid State Maser*

N. BLOEMBERGEN
Cruft Laboratory, Harvard University, Cambridge, Massachusetts

(Received July 6, 1956)

The Overhauser effect may be used in the spin multiplet of certain paramagnetic ions to obtain a negative absorption or stimulated emission at microwave frequencies. The use of nickel fluosilicate or gadolinium ethyl sulfate at liquid helium temperature is suggested to obtain a low noise microwave amplifier or frequency converter. The operation of a solid state maser based on this principle is discussed.

TOWNES and co-workers[1,2] have shown that microwave amplification can be obtained by stimulated emission of radiation from systems in which a higher energy level is more densely populated than a lower one. In paramagnetic systems an inversion of the population of the spin levels may be obtained in a variety of ways. The "180° pulse" and the "adiabatic rapid passage" have been extensively applied in nuclear magnetic resonance. Combrisson and Honig[2] applied the fast passage technique to the two electron spin levels of a P donor in silicon, and obtained a noticeable power amplification.

Attention is called to the usefulness of power saturation of one transition in a multiple energy level system to obtain a change of sign of the population difference between another pair of levels. A variation in level populations obtained in this manner has been demonstrated by Pound.[3] Such effects have since acquired wide recognition through the work of Overhauser.[4]

Consider for example a system with three unequally spaced energy levels, $E_3>E_2>E_1$. Introduce the notation,

$$h\nu_{31}=E_3-E_1 \quad h\nu_{32}=E_3-E_2 \quad h\nu_{21}=E_2-E_1.$$

Denote the transition probabilities between these spin levels under the influence of the thermal motion of the heat reservoir (lattice) by

$$w_{12}=w_{21}\exp(-h\nu_{21}/kT), \quad w_{13}=w_{31}\exp(-h\nu_{31}/kT),$$
$$w_{23}=w_{32}\exp(-h\nu_{32}/kT).$$

The w's correspond to the inverse of spin lattice relaxation times. Denote the transition probability caused by a large saturating field $H(\nu_{31})$ of frequency

* Supported by the Joint Services.
[1] Gordon, Zeiger, and Townes, Phys. Rev. 99, 1264 (1955).
[2] Combrisson, Honig, and Townes, Compt. rend. 242, 2451 (1956).
[3] R. V. Pound, Phys. Rev. 79, 685 (1950).
[4] A. W. Overhauser, Phys. Rev. 92, 411 (1953).

ν_{31} by W_{13}. Let a relatively small signal of frequency ν_{32} cause transitions between levels two and three at a rate W_{32}. The numbers of spins occupying the three levels n_1, n_2, and n_3, satisfy the conservation law

$$n_1+n_2+n_3=N.$$

For $h\nu_{32}/kT\ll 1$ the populations obey the equations[5]:

$$\frac{dn_3}{dt}=w_{13}\left(n_1-n_3-\frac{N}{3}\frac{h\nu_{31}}{kT}\right)+w_{23}\left(n_2-n_3-\frac{N}{3}\frac{h\nu_{32}}{kT}\right)+W_{31}(n_1-n_3)+W_{32}(n_2-n_3),$$

$$\frac{dn_2}{dt}=w_{23}\left(n_3-n_2+\frac{N}{3}\frac{h\nu_{32}}{kT}\right)+w_{21}\left(n_1-n_2-\frac{N}{3}\frac{h\nu_{21}}{kT}\right)+W_{32}(n_3-n_2), \quad (1)$$

$$\frac{dn_1}{dt}=w_{13}\left(n_3-n_1+\frac{N}{3}\frac{h\nu_{31}}{kT}\right)+w_{21}\left(n_2-n_1+\frac{N}{3}\frac{h\nu_{21}}{kT}\right)-W_{31}(n_1-n_3).$$

In the steady state the left-hand sides are zero. If the saturating field at frequency ν_{31} is very large, $W_{31}\gg W_{32}$ and w's, the solution is obtained

$$n_1-n_2=n_3-n_2=-\frac{1}{3}\frac{hN}{kT}\frac{-w_{23}\nu_{32}+w_{21}\nu_{21}}{w_{23}+w_{12}+W_{32}}. \quad (2)$$

This population difference will be positive, corresponding to negative absorption or stimulated emission at the frequency ν_{32}, if

$$w_{21}\nu_{21}>w_{32}\nu_{32}. \quad (3)$$

If the opposite is true, stimulated emission will occur at the frequency ν_{21}. The following discussion could easily be adapted to this situation. The power emitted by the magnetic specimen is

$$P_{\text{magn}}=\frac{Nh^2\nu_{32}}{3kT}\frac{(w_{21}\nu_{21}-w_{32}\nu_{32})W_{32}}{w_{23}+w_{12}+W_{32}}. \quad (4)$$

For a magnetic resonance line with a normalized response curve $g(\nu)$ and $g(\nu_{\max})=T_2$, the transition probability at resonance is given by

$$W_{32}=\hbar^{-2}|(2|M_x|3)|^2H_x^2(\nu_{32})T_2. \quad (5)$$

For simplicity it has been assumed that the signal field $H(\nu_{32})$ is uniform in the x direction over the volume of the sample. A similar expression holds for W_{31}.

For the moment we shall restrict ourselves to the important case that the signal excitation at frequency ν_{32} is small, $W_{32}\ll w_{23}+w_{31}$. No saturation effects at this transition occur and a magnetic quality factor can be defined by

$$-1/Q_{\text{magn}}=\frac{4P_{\text{magn}}}{\nu_{23}\langle H^2(\nu_{32})\rangle_{\text{Av}}V_c}. \quad (6)$$

Q_{magn} is negative for stimulated emission, $P_{\text{magn}}>0$. V_c is the volume of the cavity, and $\langle H^2\rangle_{\text{Av}}$ represents a volume average over the cavity. The losses in the cavity, exclusive of the magnetic losses or gains in the sample, are described by the unloaded quality factor Q_0. The external losses from the coupling to a wave guide or coaxial line are described by Q_e. Introduce the voltage standing wave ratio β for the cavity tuned to resonance,

$$\beta=(Q_e/Q_0)+(Q_e/Q_{\text{magn}}).$$

The ratio of reflected to incident power is

$$\frac{P_r}{P_i}=\frac{(1-\beta)^2}{(1+\beta)^2}.$$

There is a power gain or amplification, when β is negative or, $-Q_{\text{magn}}^{-1}>Q_0^{-1}$. Oscillation will occur when

$$-Q_{\text{magn}}^{-1}>Q_0^{-1}+Q_e^{-1}=Q_L^{-1},$$

where Q_L is the "loaded Q." The amplitude of the oscillation will be limited by the saturation effect, embodied by the W_{32} in the denominator of Eq. (4). The absolute value of $1/Q_{\text{magn}}$ decreases as the power level increases. In the oscillating region the device will act as a microwave frequency converter. Power input is at the frequency ν_{13}, a smaller power output at the frequency ν_{23}. The balance of power is dissipated in the form of heat through the spin-lattice relaxation and through conductiôn losses in the cavity walls. For $-Q_{\text{magn}}=Q_L$, $\beta=-1$, and the amplification factor would be infinite. The device will act as a stable c.w. amplifier at frequency ν_{23}, if

$$Q_0^{-1}+Q_e^{-1}>-Q_{\text{magn}}^{-1}>Q_0^{-1}. \quad (7)$$

The choice of paramagnetic substance is largely dependent on the existance of suitable energy levels and the existence of matrix elements of the magnet moment operator between the various spin levels. The absorption and stimulated emission process depend directly on this operator, but the relaxation terms (w) also depend on the spin angular momentum operator via spin-orbit coupling terms. It is essential that all off-diagonal elements between the three spin levels under consideration be nonvanishing. This can be achieved by putting a paramagnetic salt with a crystalline field splitting δ in a magnetic field, which makes an angle with the crystalline field axis. The magnitude of the field is such that the Zeeman energy is comparable to the crystalline field splitting. In this case the states with magnetic quantum numbers m_s are all scrambled. This situation is usually avoided to unravel paramagnetic resonance spectra, but occasionally "forbidden lines" have been observed, indicating mixing of the m_s states. For our

[5] In case $h\nu_{31}\sim kT$, the Boltzmann exponential factors cannot be approximated by the linear terms. The algebra becomes more involved without changing the character of the effect.

purposes the mixing up of the spin states by Zeeman and crystalline field interactions of comparable magnitude is essential. The energy levels and matrix elements of the spin angular momentum operator can be obtained by a numerical solution of the determinantal problem of the spin Hamiltonian.[6] The number of electron spin levels may be larger than three. One may choose the three levels between which the operation will take place. The analysis will be similar, but algebraically more complicated. One has a considerable amount of freedom by the choice of the external dc magnetic field, to adjust the frequencies ν_{23} and ν_{13} and to vary the values of the inverse relaxation times w. It is advisable—although perhaps not absolutely necessary—to operate at liquid helium temperature. This will give relatively long relaxation times (between 10^{-2} and 10^{-4} sec), and thus keep the power requirements for saturation down. The factor T in the denominator of Eq. (4) will also increase the emission at low temperature. Although the order of magnitude of the w's is known through the work of Leiden school,[7] there is only one instance where w's have been measured for some individual transitions.[8] Van Vleck's[9] theory of paramagnetic relaxation should be extended to the geometries envisioned in this paper. If a Debye spectrum of the lattice vibrations is assumed, the relaxation times will increase with decreasing frequency at liquid helium temperature, where Raman processes are negligible. This implies that the condition (3) should be easily realizable when $\nu_{32}<\nu_{21}$.

Important applications as a microwave amplifier could, e.g., be obtained for $\nu_{32}=1420$ Mc/sec, corresponding to the interstellar hydrogen line, or to another relatively low microwave frequency used in radar systems. The frequency ν_{31} could be chosen in the X band, $\nu_{31}=10^{10}$ cps. To obtain well scrambled states with these frequency splittings one should have crystalline field splittings between 0.03 cm^{-1} and 0.3 cm^{-1}. Paramagnetic crystals which are suggested by these considerations are nickel fluosilicate[10] and gadolinium ethyl sulfate.[11] These crystals have the additional advantage that all magnetic ions have the same crystalline field and nuclear hyperfine splitting is absent, thus keeping the total number of possible transitions down. The use of magnetically dilute salts is indicated to reduce the line width, increase the value of T_2 in Eq. (5) and to separate the individual resonance transitions.

A single crystal 5% Ni 95% Zn Si $F_6 \cdot 6H_2O$ has a line width of 50 oersted ($T_2=1.2\times10^{-9}$ sec) and an average crystalline field splitting $\delta=0.12$ cm^{-1} for the Ni^{++} ions. With an effective spin value $S=1$ there are indeed three energy levels of importance. The spin lattice relaxation time is about 10^{-4} sec at 2°K as measured in a saturation experiment by Meyer.[12] Further dilution does not decrease the line width, as there is a distribution of crystalline fields in the diluted salt.

A single crystal of 1% Gd 99% La $(C_2H_5SO_4)_3\cdot 9H_2O$ has an effective spin $S=7/2$. In zero field there are four doublets separated respectively by $\delta=0.113$ cm^{-1}, 0.083 cm^{-1}, and 0.046 cm^{-1} as measured at 20°K. These splittings are practically independent of temperature. The line width is 7 oersteds due to the distribution of local fields arising from the proton magnetic moments. This width could be reduced by a factor three by using the deuterated salt. The relaxation time is not known, but should be about the same as in other Gd salts,[7] which give $T_1\sim10^{-2}$ sec at 2°K.

In the absence of detailed calculations for the relaxation mechanism, we shall take $w_{12}=w_{13}=w_{32}=10^4$ sec^{-1} for the nickel salt and equal to 10^2 sec^{-1} for the gadolinium salt. The matrix elements $(2|M_x|3)$, etc., can be calculated exactly by solving the spin determinant. For the purpose of judging the operation of the maser using these salts, we shall take the off-diagonal elements of magnetic moment operator simply equal to $g\beta_0$, where $g=2$ is the Landé spin factor and β_0 is the Bohr magneton. For the higher spin value of the Gd^{+++} some elements will be larger but this effect is offset by the distribution of the ions over eight rather than three spin levels. Take $T=2$°K and $Q_0=10^4$, which is readily obtained in a cavity of pure metal at this temperature. A coaxial cavity may be used which has a fundamental mode resonating at the frequency $\nu_{32}=1.42\times10^9$ cps and a higher mode resonating at $\nu_{31}\approx10^{10}$ cps. Take the volume of the cavity $V=60$ cm^3 and $H_s^2=6\langle H^2\rangle_{Av}$. If these values are substituted in Eqs. (4)–(6), the condition (7) for amplification is satisfied if $N>3\times10^{18}$ for nickel fluosilicate and $N>3\times10^{17}$ for gadolinium ethyl sulfate ($N>10^{17}$ for the deuterated salt). The minimum required number of Ni^{++} ions are contained in 0.02 cm^3 of the diluted nickel salt. The gadolinium salt, diluted to 1% Gd, contains the required number in about the same volume. The critical volume is only 0.006 cm^3 for the deuterated salt. Crystals of appreciably larger size can still be fitted conveniently in the cavity. A c.w. amplifier or frequency converter should therefore be realizable with these substances. A larger amount of power can be handled by these crystals than by the P impurities in silicon which have a very long relaxation time, and require an intermittant operation, and where it is harder to get the required number of spins in the cavity.

So far we have assumed that the width corresponds to the inverse of a true transverse relaxation time T_2.

[6] See, e.g., Bleaney and K. H. W. Stevens, Repts. Progr. in Phys. **16**, 108 (1953).

[7] See, e.g., C. J. Gorter, *Paramagnetic Relaxation* (Elsevier Publishing Company, Amsterdam, 1948).

[8] A. H. Eschenfelder and R. T. Weidner 92, 869 (1953).

[9] J. H. Van Vleck, Phys. Rev. **57**, 426 (1940).

[10] R. P. Penrose and K. H. W. Stevens, Proc. Phys. Soc. (London) **A63**, 29 (1949).

[11] Bleaney, Scovil, and Trenam, Proc. Roy. Soc. (London) **A223**, 15 (1954).

[12] J. W. Meyer, Lincoln Laboratory Report 1955 (unpublished).

Actually the width $1/T_2^*$ is due to an internal inhomogeneity broadening with normalized distribution $h(\nu)$ and $h(\nu_{max})\approx T_2^*$ in both cases. The response curve for a single magnetic ion is probably very narrow indeed, $g(\nu_{max})=T_2\approx T_1$, and $T_1=10^{-4}$ should be used in Eq. (5) rather than $T_2^*=1.2\times10^{-9}$ sec. The response to a weak threshold signal at ν_{32} now originates, however, from a small fraction of the magnetic ions. If $\gamma H(\nu_{32})<1/T_1\approx10^4$ cps, then only T_2^*/T_1 of the ions contribute to the stimulated emission and the net result is the same as calculated above. In most applications the incoming signal will be so weak that this situation will apply, even with a power amplification of 30 or 40 db.

For use as an oscillator or high level amplifier with a field $H(\nu_{32})$ in the cavity larger than $1/\gamma T_1$, one has essentially complete saturation ($W_{32}\gg w_{23}+w_{13}$) in Eq. (4) for those magnetic ions lying in a width $2\pi\Delta\nu=\gamma H(\nu_{32})$ in the distribution $h(\nu)$. One has then for the power emitted instead of Eqs. (4) and (5)

$$P_{magn}=\frac{h^2\nu_{32}}{3kT}N(-w_{32}\nu_{32}+w_{21}\nu_{21})\gamma H(\nu_{31})T_2^*. \quad (8)$$

The power is proportional to the amplitude of the radio frequency field rather than its square. This effect has been discussed in more detail by Portis.[13] It will limit the oscillation or amplification to an amplitude which can be calculated by using Eq. (8) in conjunction with Eqs. (6) and (7).

The driving field $H(\nu_{31})$ will necessarily have to satisfy the condition $\gamma H(\nu_{31})>w_{31}=T_1^{-1}$ to obtain saturation between levels 1 and 3. The power absorbed in the crystal will be proportional to the amplitude $H(\nu_{31})$, and is in order of magnitude given by

$$P_{abs}\sim N\frac{h^2\nu_{13}^2}{3kT}w_{13}\gamma H(\nu_{31})T_2^*. \quad (9)$$

This equation looses its validity if $\gamma H(\nu_{31})>T_2^{*-1}$. In this case the whole line would be saturated, but such excessive power levels will not be used. For $T_1^{-1}<\gamma H(\nu_{31})<T_2^{*-1}$, the effective band width of the amplifier is determined by $H(\nu_{31})$. It is about 0.5 Mc/sec for $H(\nu_{31})=0.2$ oersted. The power dissipated in a specimen of fluosilicate ten times the critical size is 0.5 milliwatt under these circumstances. For the gadolinium salt, also ten times the critical size, either deuterated or not, the dissipation is only 0.005 milliwatt. There should be no difficulty in carrying this amount out of the paramagnetic crystal without excessive heating. The power dissipation in the walls under these conditions will be 5 milliwatts. Liquid helium will boil off at the rate of only 0.01 cc/min due to heating in the cavity. Since helium is superfluid at 2°K, troublesome vapor bubbles in the cavity are eliminated.

The noise power generated in this type of amplifier should be very low. The cavity with the paramagnetic salt can be represented by two resonant coupled circuits as discussed by Bloembergen and Pound.[14] Noise generators are associated with the losses in the cavity walls, kept at 2°K, and with the paramagnetic spin abosrption which is described by an effective spin temperature, associated with the distribution of the spin population. The absolute value of this effective temperature also has the order of magnitude of 1°K. The input is from an antenna, which sees essentailly the radiation temperature of interstellar space. Reflected power is channeled by a circulating nonreciprocal element[15] into a heterodyne receiver, or, if necessary, into a second stage Maser cavity. The circulator makes the connection: antenna→maser cavity→heterodyne receiver→dummy load→antenna. If the antenna is not well matched, the dummy load may be a matched termination kept at liquid helium temperature to prevent extra power from entering the cavity. The input arm of the cavity at frequency ν_{31} will be beyond cutoff for the frequency ν_{32}. The coaxial line passing the signal at ν_{32} between cavity and circulator will contain a rejection filter at frequency ν_{31} to prevent overloading and noise mixing at the mixer crystal of the super heterodyne receiver.

It may be concluded that the realization of a low-noise c.w. microwave amplifier by saturation of a spin level system at a higher frequency seems promising. The device should be particularly suited for detection of weak signals at relatively long wavelength, e.g., the 21-cm interstellar hydrogen radiation. It may also be operated as a microwave frequency converter, capable of handling milliwatt power. More detailed calculations and design of the cavity are in progress.

[13] A. M. Portis, Phys. Rev. 91, 1071 (1953).

[14] N. Bloembergen and R. V. Pound, Phys. Rev. 95, 8 (1954).

[15] C. L. Hogan, Bell System Tech. J. 31, 1 (1952).

PHYSICAL REVIEW VOLUME 112, NUMBER 6 DECEMBER 15, 1958

Infrared and Optical Masers

A. L. SCHAWLOW AND C. H. TOWNES*
Bell Telephone Laboratories, Murray Hill, New Jersey

(Received August 26, 1958)

The extension of maser techniques to the infrared and optical region is considered. It is shown that by using a resonant cavity of centimeter dimensions, having many resonant modes, maser oscillation at these wavelengths can be achieved by pumping with reasonable amounts of incoherent light. For wavelengths much shorter than those of the ultraviolet region, maser-type amplification appears to be quite impractical. Although use of a multimode cavity is suggested, a single mode may be selected by making only the end walls highly reflecting, and defining a suitably small angular aperture. Then extremely monochromatic and coherent light is produced. The design principles are illustrated by reference to a system using potassium vapor.

INTRODUCTION

AMPLIFIERS and oscillators using atomic and molecular processes, as do the various varieties of masers,[1-4] may in principle be extended far beyond the range of frequencies which have been generated electronically, and into the infrared, the optical region, or beyond. Such techniques give the attractive promise of coherent amplification at these high frequencies and of generation of very monochromatic radiation. In the infrared region in particular, the generation of reasonably intense and monochromatic radiation would allow the possibility of spectroscopy at very much higher resolution than is now possible. As one attempts to extend maser operation towards very short wavelengths, a number of new aspects and problems arise, which require a quantitative reorientation of theoretical discussions and considerable modification of the experimental techniques used. Our purpose is to discuss theoretical aspects of maser-like devices for wavelengths considerably shorter than one centimeter, to examine the short-wavelength limit for practical devices of this type, and to outline design considerations for an example of a maser oscillator for producing radiation in the infrared region. In the general discussion, roughly reasonable values of design parameters will be used. They will be justified later by more detailed examination of one particular atomic system.

CHARACTERISTICS OF MASERS FOR MICROWAVE FREQUENCIES

For comparison, we shall consider first the characteristics of masers operating in the normal microwave range. Here an unstable ensemble of atomic or molecular systems is introduced into a cavity which would normally have one resonant mode near the frequency which corresponds to radiative transitions of these systems. In some cases, such an ensemble may be located in a wave guide rather than in a cavity but again there would be characteristically one or a very few modes of propagation allowed by the wave guide in the frequency range of interest. The condition of oscillation for n atomic systems excited with random phase and located in a cavity of appropriate frequency may be written (see references 1 and 2)

$$n \geq hV\Delta\nu/(4\pi\mu^2 Q_c), \qquad (1)$$

where n is more precisely the difference n_1-n_2 in number of systems in the upper and lower states, V is the volume of the cavity, $\Delta\nu$ is the half-width of the atomic resonance at half-maximum intensity, assuming a Lorentzian line shape, μ is the matrix element involved in the transition, and Q_c is the quality factor of the cavity.

The energy emitted by such a maser oscillator is usually in an extremely monochromatic wave, since the energy produced by stimulated emission is very much larger than that due to spontaneous emission or to the normal background of thermal radiation. The frequency range over which appreciable energy is distributed is given approximately by[1]

$$\delta\nu = 4\pi kT(\Delta\nu)^2/P, \qquad (2)$$

* Permanent address: Columbia University, New York, New York.

[1] Gordon, Zeiger, and Townes, Phys. Rev. 99, 1264 (1955).

[2] Combrisson, Honig, and Townes, Compt. rend. 242, 2451 (1956).

[3] N. Bloembergen, Phys. Rev. 104, 329 (1956).

[4] E. Allais, Compt. rend. 245, 157 (1957).

where $\Delta\nu$ is the half-width at half-maximum of the resonant response of a single atomic system, P is the total power emitted, k is Boltzmann's constant, and T the absolute temperature of the cavity walls and wave guide. Since in all maser oscillators at microwave frequencies which have so far been considered, $P \gg kT\Delta\nu$, the radiation is largely emitted over a region very much smaller than $\Delta\nu$, or $\delta\nu \ll \Delta\nu$.

As amplifiers of microwave or radio-frequency energy, masers have the capability of very high sensitivity, approaching in the limit the possibility of detecting one or a few quanta. This corresponds to a noise temperature of $h\nu/k$, which for microwave frequencies is of the order of 1°K.

USE OF MULTIMODE CAVITIES AT HIGH FREQUENCIES

Consider now some of the modifications necessary to operate a maser at frequencies as high as that of infrared radiation. To maintain a single isolated mode in a cavity at infrared frequencies, the linear dimension of the cavity would need to be of the order of one wavelength which, at least in the higher frequency part of the infrared spectrum, would be too small to be practical. Hence, one is led to consider cavities which are large compared to a wavelength, and which may support a large number of modes within the frequency range of interest. For very short wavelengths, it is perhaps more usual to consider a plane wave reflected many times from the walls of such a cavity, rather than the field of a standing wave which would correspond to a mode.

The condition for oscillation may be obtained by requiring that the power produced by stimulated emission is as great as that lost to the cavity walls or other types of absorption. That is,

$$\left(\frac{\mu' E}{\hbar}\right)^2 \frac{h\nu n}{4\pi\Delta\nu} \geq \frac{E^2}{8\pi}\frac{V}{t}, \tag{3}$$

where μ' is the matrix element for the emissive transition, E^2 is the mean square of the electric field. (For a multiresonant cavity, E^2 may be considered identical in all parts of the cavity.) n is the excess number of atoms in the upper state over those in the lower state, V is the volume of the cavity, t is the time constant for the rate of decay of the energy, $\Delta\nu$ is the half-width of the resonance at half maximum intensity, if a Lorentzian shape is assumed. The decay time t may be written as $2\pi\nu/Q$, but is perhaps more naturally expressed in terms of the reflection coefficient α of the cavity walls.

$$t = 6V/(1-\alpha)Ac, \tag{4}$$

where A is the wall area and c the velocity of light. For a cube of dimension L, $t = L/(1-\alpha)c$. The condition for oscillation from (3) is then

$$n \geq \frac{3hV}{8\pi^2\mu^2 t}\frac{\Delta\nu}{\nu}, \tag{5}$$

or

$$n \geq \frac{\Delta\nu}{\nu}\frac{h(1-\alpha)Ac}{16\pi^2\mu^2}. \tag{6}$$

Here μ'^2 has been replaced by $\mu^2/3$, since μ'^2 is the square of the matrix element for the transition which, when averaged over all orientations of the system, is just one-third of the quantity μ^2 which is usually taken as the square of the matrix elment.

In a gas at low pressure, most infrared or optical transitions will have a width $\Delta\nu$ determined by Doppler effects. Then the resonance half-width at half-maximum intensity is

$$\Delta\nu = \frac{\nu}{c}\left(\frac{2kT}{m}\ln 2\right)^{\frac{1}{2}}, \tag{7}$$

where m is the molecular mass, k is Boltzmann's constant, and T the temperature. Because of the Gaussian line shape in this case, expression (6) becomes

$$n \geq \frac{\Delta\nu}{\nu}\frac{h(1-\alpha)Ac}{16\pi^2\mu^2(\pi \ln 2)^{\frac{1}{2}}}, \tag{8}$$

or

$$n \geq \frac{h(1-\alpha)A}{16\pi^2\mu^2}\left(\frac{2kT}{\pi m}\right)^{\frac{1}{2}}. \tag{9}$$

It may be noted that expression (9) for the number of excited systems required for oscillation is independent of the frequency. Furthermore, this number n is not impractically large. Assuming the cavity is a cube of 1 cm dimension and that $\alpha = 0.98$, $\mu = 5\times10^{-18}$ esu, $T = 400$°K, and $m = 100$ amu, one obtains $n = 5\times10^{8}$.

The condition for oscillation, indicated in (5), may be conveniently related to the lifetime τ of the state due to spontaneous emission of radiation by a transition between the two levels in question. This lifetime is given, by well-known theory, as

$$\tau = 3hc^3/(64\pi^4\nu^3\mu^2). \tag{10}$$

Now the rate of stimulated emission due to a single quantum in a single mode is just equal to the rate of spontaneous emission into the same single mode. Hence, $1/\tau$ is this rate multiplied by the number of modes p which are effective in producing spontaneous emission. Assuming a single quantum present in a mode at the resonant frequency, the condition for instability can then be written

$$nh\nu/p\tau \geq h\nu/t_1,$$

or

$$n \geq p\tau/t. \tag{11}$$

This gives a simple expression which may sometimes

be useful, and which is equivalent to (5), since

$$p=\int p(\nu)\frac{(\Delta\nu)^2 d\nu}{(\nu-\nu_0)^2+(\Delta\nu)^2}, \tag{12}$$

where $p(\nu)d\nu$ is the number of modes between ν and $\nu+d\nu$, which is well known to be

$$p(\nu)d\nu=8\pi\nu^2 V d\nu/c^3. \tag{13}$$

From (12) and (13), one obtains for a Lorentzian line shape,

$$p=8\pi^2\nu^2 V\Delta\nu/c^3. \tag{14}$$

Or, for a line broadened by Doppler effects, the corresponding number of effective modes is

$$p=8\pi^2\nu^2 V\Delta\nu/(\pi \ln 2)^{\frac{1}{2}}c^3. \tag{15}$$

If τ and p are inserted into (11) from expressions (10) and (14), respectively, it becomes identical with (5), as one must expect.

The minimum power which must be supplied in order to maintain n systems in excited states is

$$P=nh\nu/\tau=ph\nu/t. \tag{16}$$

This expression is independent of the lifetime or matrix element. However, if there are alternate modes of decay of each system, as by collisions or other transitions, the necessary power may be larger than that given by (16) and dependent on details of the system involved. Furthermore, some quantum of higher frequency than that emitted will normally be required to excite the system, which will increase the power somewhat above the value given by (16). Assuming the case considered above, i.e., a cube of 1-cm dimension with $\alpha=0.98$, $\lambda=10^4$ A, and broadening due to Doppler effect, (16) gives $P=0.8\times10^{-3}$ watt. Supply of this much power in a spectral line does not seem to be extremely difficult.

The power generated in the coherent oscillation of the maser may be extremely small, if the condition of instability is fulfilled in a very marginal way, and hence can be much less than the total power, which would be of the order of 10^{-3} watt, radiated spontaneously. However, if the number of excited systems exceeds the critical number appreciably, e.g., by a factor of two, then the power of stimulated radiation is given roughly by $h\nu$ times the rate at which excited systems are supplied, assuming the excitation is not lost by some process not yet considered, such as by collisions. The electromagnetic field then builds up so that the stimulated emission may be appreciably greater than the total spontaneous emission. For values even slightly above the critical number, the stimulated power is of the order of the power $nh\nu/\tau$ supplied, or hence of the order of one milliwatt under the conditions assumed above.

The most obvious and apparently most convenient method for supplying excited atoms is excitation at a higher frequency, as in optical pumping or a three-level maser system. The power supplied must, of course, be appreciably greater than the emitted power in expression (9). There is no requirement that the pumping frequency be much higher than the frequency emitted, as long as the difference in frequency is much greater than kT/h, which can assure the possibility of negative temperatures. Since, for the high frequencies required, an incoherent source of pumping power must be used, a desirable operating frequency would be near the point where the maximum number of quanta are emitted by a given transition from a discharge or some other source of high effective temperature. This maximum will occur somewhere near the maximum of the blackbody radiation at the effective temperature of such a source, or hence in the visible or ultraviolet region. The number of quanta required per second would probably be about one order of magnitude greater than the number emitted at the oscillating frequency, so that the input power required would be about ten times the output given by (16), or 10 milliwatts. This amount of energy in an individual spectroscopic line is, fortunately, obtainable in electrical discharges.

Very desirable features of a maser oscillator at infrared or optical frequencies would be a high order of monochromaticity and tunability. In the microwave range, a maser oscillator is almost inherently a very monochromatic device. However, a solid state maser can also normally be tuned over a rather large fractional variation in frequency. Both of these features are much more difficult to obtain in the infrared or optical regions. Frequencies of atomic or molecular resonances can in principle be tuned by Stark or Zeeman effects, as they would be in the radio-frequency or microwave range. However, such tuning is usually limited to a few wave numbers (or a few times 10 000 Mc/sec), which represents a large fractional change in the microwave range and only a small fractional change in the optical region. Certain optical and infrared transitions of atoms in solids are strongly affected by neighboring atoms. This may be the result of Stark effects due to internal electric fields or, as in the case of antiferromagnetic resonances, internal magnetic fields may vary enough with temperature to provide tuning over a few tens of wave numbers. Hence variation of temperature or pressure can produce some tuning. However, it appears unreasonable to expect more than a small fractional amount of tuning in an infrared or optical maser using discrete levels.

SPECTRUM OF A MASER OSCILLATOR

Monochromaticity of a maser oscillator is very closely connected with noise properties of the device as an amplifier. Consider first a maser cavity for optical or infrared frequencies which supports a single isolated mode. As in the microwave case, it is capable of detecting, in the limit, one or a few quanta, corresponding to a noise temperature of $h\nu/k$. However, at a wavelength of 10 000 A, this noise temperature is about 14 000°K, and hence not remarkably low. Furthermore, other well-known photon detectors, such as a photoelectric tube, are capable of detecting a single quantum. At such frequencies, a maser has no great advantage over well-known techniques in detecting small numbers of quanta. It does offer the new possibility of coherent amplification. However, if many modes rather than a single one are present in the cavity, a rather large background of noise can occur, the noise temperature being proportional to the number of modes which are confused within the resonance width of the atomic or molecular system. A method for isolation of an individual mode which avoids this severe difficulty will be discussed below.

Let us examine now the extent to which the normal line width of the emission spectrum of an atomic system will be narrowed by maser action, or hence how monochromatic the emission from an infrared or optical maser would be. Considerations were given above concerning the number of excited systems required to produce stimulated power which would be as large as spontaneous emission due to all modes of a multimode cavity which lie within the resonance width of the system. Assume for the moment that a single mode can be isolated. Spontaneous emission into this mode adds waves of random phase to the electromagnetic oscillations, and hence produces a finite frequency width which may be obtained by analogy with expression (2) as

$$\Delta\nu_{osc}=(4\pi h\nu/P)(\Delta\nu)^2, \qquad (17)$$

where $\Delta\nu$ is the half-width of the resonance at half-maximum intensity, and P the power in the oscillating field. Note that kT, the energy due to thermal agitation, has been replaced in expression (15) by $h\nu$, the energy in one quantum. Usually at these high frequencies, $h\nu\gg kT$, and there is essentially no "thermal" noise. There remains, however, "zero-point fluctuations" which produce random noise through spontaneous emission, or an effective temperature of $h\nu/k$.

For the case considered numerically above, $4\pi h\nu\Delta\nu/P$ is near 10^{-6} when P is given by expression (16), so that $\Delta\nu_{osc}\sim10^{-6}\Delta\nu$. This corresponds to a remarkably monochromatic emission. However, for a multimode cavity, this very monochromatic emission is superimposed on a background of stimulated emission which has width $\Delta\nu$, and which, for the power P assumed, is of intensity equal to that of the stimulated emission. Only if the power is increased by some additional factor of about ten, or if the desired mode is separated from the large number of undesired ones, would the rather monochromatic radiation stand out clearly against the much wider frequency distribution of spontaneous emission.

Another problem of masers using multimode cavities which is perhaps not fundamental, but may involve considerable practical difficulty, is the possibility of oscillations being set up first in one mode, then in another—or perhaps of continual change of modes which would represent many sudden jumps in frequency. If the cavity dimensions, density distribution of gas and distribution of excited states remains precisely constant, it seems unlikely that oscillations will build up on more than one mode because of the usual nonlinearities which would allow the most favored mode to suppress oscillations in those which are less favored. However, if many nearby modes are present, a very small change in cavity dimensions or other characteristics may produce a shift of the oscillations from one mode to another, with a concomitant variation in frequency.

SELECTION OF MODES FOR AMPLIFICATION

We shall consider now methods which deviate from those which are obvious extensions of the microwave or radio-frequency techniques for obtaining maser action. The large number of modes at infrared or optical frequencies which are present in any cavity of reasonable size poses problems because of the large amount of spontaneous emission which they imply. The number of modes per frequency interval per unit volume cannot very well be reduced for a cavity with dimensions which are very large compared to a wavelength. However, radiation from these various modes can be almost completely isolated by using the directional properties of wave propagation when the wavelength is short compared with important dimensions of the region in which the wave is propagated.

Consider first a rectangular cavity of length D and with two square end walls of dimension L which are slightly transparent, its other surfaces being perfectly reflecting. Transparency of the end walls provides coupling to external space by a continuously distributed excitation which corresponds to the distribution of field strength at these walls. The resulting radiation produces a diffraction pattern which can be easily calculated at a large distance from the cavity, and which is effectively separated from the diffraction pattern due to any other mode of the cavity at essentially the same frequency.

The field distribution along the end wall, taken as the xy plane, may be proportional, for example, to $\sin(\pi rx/L)\cos(\pi sy/L)$. The resonant wavelength is of the form

$$\lambda=\frac{2}{[(q/D)^2+(r/L)^2+(s/L)^2]^{\frac{1}{2}}}, \qquad (18)$$

where q is the number of half-wavelengths along the z direction. If L is not much smaller than D, and if $q\gg r$ or s, the resonant wavelength is approximately

$$\lambda=\frac{2D}{q}\left[1-\tfrac{1}{2}\left(\frac{Dr}{Lq}\right)^2-\tfrac{1}{2}\left(\frac{Ds}{Lq}\right)^2\right], \qquad (19)$$

which is primarily dependent on q and insensitive to r or s. The direction of radiation from the end walls, however, is critically dependent on r and s. The Fraunhofer diffraction pattern of the radiation has an intensity variation in the x direction given by

$$I \propto (2\pi r)^2 \sin^2\left(\frac{\pi L \sin\theta}{\lambda}+\frac{\pi r}{2}\right) \Big/ \left(\pi r+\frac{2\pi L \sin\theta}{\lambda}\right)^2\left(\pi r-\frac{2\pi L \sin\theta}{\lambda}\right)^2, \quad (20)$$

where θ is the angle between the direction of observation and the perpendicular to the end walls. For a given value of r, the strongest diffraction maxima occur at

$$\sin\theta = \pm r\lambda/2L,$$

and the first minima on either side of the maxima at

$$\sin\theta = \pm r\lambda/2L \pm \lambda/L.$$

Thus the maximum of the radiation from a mode designated by $r+1$ falls approximately at the half-intensity point of the diffraction pattern from the mode designated by r, which is just sufficient for significant resolution of their individual beams of radiation. This provides a method for separately coupling into or out of one or a few individual modes in the multimode cavity. A practical experimental technique for selecting one or a few modes is to focus radiation from the end walls by means of a lens onto a black screen in the focal plane. A suitable small hole in the screen will accept only radiation from the desired mode or modes.

There may, of course, be more than one mode which has similar values of r and s but different values of q, and which radiate in essentially identical directions. However, the frequencies of such modes are appreciably different, and may be sufficiently separated from each other by an appropriate choice of the distance D between plates. Thus if only one mode with a particular value of r and s is wanted within the range of response $2\Delta\nu$ of the material used to produce oscillations, D should be less than $c/4\Delta\nu$. Or, if it is undesirable to adjust D precisely for a particular mode, and approximately one mode of this type is wanted in the range $2\Delta\nu$, one may choose

$$D \approx c/4\Delta\nu. \quad (21)$$

For the conditions assumed above, the value of D given by (21) has the very practical magnitude of about 10 cm.

It is desirable not only to be able to select radiation from a single mode, but also to make all but one or a few modes of the multimode cavity lossy in order to suppress oscillations in unwanted modes. This again can be done by making use of directional properties. Loss may be introduced perhaps most simply by removing the perfectly reflecting walls of the cavity. The "cavity" is then reduced to partially transparent end plates and nonexistent (or lossy) perfectly-matched side walls.

Suppose now that one of the modes of such a cavity is excited by suddenly introducing the appropriate field distribution on one of the end walls. This will radiate a wave into the cavity having directional properties such as those indicated by the diffraction pattern (20). If r and s have their minimum values, the maximum energy occurs near $\theta=0$, and the wave travels more or less straight back and forth between the two plates, except for a gradual spreading due to diffraction. If r or s are larger, the maximum energy occurs at an appreciable angle θ, and the wave packet will wander off the reflecting plates and be lost, perhaps after a number of reflections. Those modes for which θ is large are highly damped and merge into a continuum, since energy radiated into them travels immediately to the walls and is lost from the cavity. However, modes for which θ is quite small may have relatively high Q and hence be essentially discrete.

For estimates of damping, consider first two end plates of infinite extent, but excited only over a square area of dimension L by a distribution which corresponds to one of our original modes. The radiated wave will be reflected back and forth many times, gradually spreading out in the diffraction pattern indicated by (20). If a mode with small values of r and s is used, the wave undergoes reflection every time it travels a distance D, and the rate of loss of energy W is given by the equations

$$dW/dt = -c(1-\alpha)W/D, \qquad W = W_0 e^{-c(1-\alpha)t/D}. \quad (22)$$

The decay time t is then $D/c(1-\alpha)$ rather than that given by (4) for the multimode case, or the effective distance traveled is $D/(1-\alpha)$.

Since the wavelength for modes with small r and s is given by (19), the frequency separation between modes with successive values of q is given by the usual Fabry-Perot condition

$$\delta\nu = c/2D. \quad (23)$$

Thus $\delta\nu \gg 1/t$ and the modes with successive values of q are discrete if $1-\alpha$, the loss on reflection, is much less than unity. On the other hand, the various modes given by small values of r or s and the same value of q are nearly degenerate, since according to (19) their frequency difference is less than $\delta\nu$ given in (23) by the factor r/q, which is of the order 10^{-4} for a typical case. These modes must be separated purely by their directional properties, rather than by their differences in frequency.

After traveling a distance $D/(1-\alpha)$, the radiation resulting from the excitation discussed above will have moved sideways in the x direction along the infinite parallel plates a distance of approximately $D\theta/(1-\alpha)$, where θ is the angle of one of the two large diffraction maxima given by (20). This distance is then

$$x = D\lambda r/[2(1-\alpha)L]. \tag{24}$$

Consider now the case of finite end plates of dimension L without their infinite extension which was assumed immediately above. After a number of reflections, the diffraction pattern would no longer be precisely that given by (20). However, expression (24) would still give a reasonable approximation to the distance of sideways motion, and if this distance is larger than the end-wall dimension L, the radiation will have been lost to the cavity, and the decay time for the mode in question is appreciably shorter than that indicated by (22). This condition occurs when

$$D\lambda r/[2(1-\alpha)L] \gtrsim L, \quad \text{or} \quad r \gtrsim 2(1-\alpha)L^2/D\lambda. \tag{25}$$

Thus to damp out modes with $r \gtrsim 10$, when $L = \frac{1}{2}$ cm, $\alpha = 0.98$, and $\lambda = 10^{-4}$ cm, the separation D between plates needs to be as large as a few centimeters. By choosing L sufficiently small, it is possible to discriminate by such losses between the lowest mode ($r = 1$), and any higher modes. Too small a ratio $L^2/D\lambda$ will, however, begin to appreciably add to the losses from the lowest mode, and hence is undesirable if the longest possible delay times are needed.

The precise distribution of radiation intensity in the plane of the end walls which will give minimum loss, or which will occur during maser oscillation, cannot be very easily evaluated. It must, however, be somewhat like the lowest mode, $r = 1$, $s = 0$. A normal and straightforward method for exciting a Fabry-Perot interferometer is to use a plane wave moving perpendicular to the reflecting plates, and screened so that it illuminates uniformly all but the edge of the plates. Such a distribution may be expressed in terms of the nearly degenerate modes of the "cavity" with various r and s, and the considerable majority of its energy will be found in the lowest mode $r = 1$, $s = 0$, if it is polarized in the y direction. There is, of course, an exactly degenerate mode of the same type which is polarized in the x direction. Any much more complicated distribution than some approximation to uniform illumination or to the lowest mode $r = 1$, $s = 0$ of our rectangular cavity will produce a wider diffraction pattern which would be lost to a detector arranged to accept a very small angle θ near zero, and which would also be subject to greater losses when $L^2/D\lambda$ is small. However, nonuniform distribution of excited atoms between the reflecting plates could compensate for the larger diffraction losses, and in some cases induce oscillations with rather complex distributions of energy.

The above discussion in terms of modes of a rectangular cavity illustrates relations between the arrangement using a Fabry-Perot interferometer and the usual microwave resonant cavity.† An alternative approach which uses the approximation of geometrical optics more directly may also be helpful and clarifying. An atom radiating spontaneously in any direction has a decay time τ given by expression (10). The probability per unit time of emission of a quantum within a given solid angle $\Delta\Omega$ is then

$$\frac{1}{t'} = \frac{16\pi^3\nu^3}{3hc^3}\mu^2\Delta\Omega. \tag{26}$$

Hence if a sufficiently small solid angle is selected from the radiation, the amount of spontaneous emission can be made arbitrarily small. However, if essentially all the stimulated emission emitted from the end-wall of the interferometer is to be collected in a receiver or detector, allowance must be made for diffraction and a solid angle as large as about $(\lambda/L)^2$ must be used, so that the rate of spontaneous emission into the detector is

$$\frac{1}{t'} = \frac{16\pi^3\mu^2\nu}{3hcL^2}. \tag{27}$$

The rate of spontaneous emission (27) within the diffraction angle may be compared with the rate of induced transitions produced by one photon reflected back and forth in the volume L^2D. This rate is, as in (3), $(\mu'E/\hbar)^2(4\pi\Delta\nu)^{-1}$, where $E^2L^2D/8\pi = h\nu$. That is, since $\mu'^2 = \mu^2/3$,

$$\frac{1}{t''} = \frac{8\pi^2\mu^2\nu}{3\Delta\nu L^2Dh}. \tag{28}$$

If D is $c/4\Delta\nu$ as in expression (21), so that there is approximately one and only one interference maximum of the interferometer in a particular direction within the range $2\Delta\nu$ of emission, then (28) becomes

$$\frac{1}{t''} = \frac{32\pi^2\mu^2\nu}{3hcL^2}. \tag{29}$$

Except for a small numerical factor of the order of the accuracy of the approximations used here, $1/t''$ given by (29) may be seen to equal $1/t'$. That is, use of the limiting amount of directional selection reduces the background of spontaneous emission to the same rate as that of stimulated emission due to a single photon. This is similar to the situation of a single mode in a cavity at microwave frequencies. It affords the limit of sensitivity which can be obtained by the usual maser amplifier, and the smallest possible noise for such a system as an oscillator.

† *Note added in proof.*—Use of two parallel plates for a maser operating at short wavelengths has also recently been suggested by A. M. Prokhorov [JETP 34, 1658 (1958)] and by R. H. Dicke [U. S. Patent 2,851,652 (September 9, 1958)]. These sources do not, however, discuss the reduction of excess modes or spontaneous emission.

Consider now the rate of loss of energy from a beam being reflected back and forth between the two end plates in the approximation of geometric optics. If the angle of deviation from the direction perpendicular to the plates is θ, then the additional rate of energy loss from a plane wave due to its spilling off the edges of the reflecting surfaces is

$$dW'/dt=-c\theta W/L. \quad (30)$$

Expression (30) assumes, to be precise, that the deviation is parallel to one edge of the end plate. Thus, when $\theta=(1-\alpha)L/D$, the decay time is one-half that for $\theta=0$. Because of nonlinearities when oscillations set in, it may be seen from expression (3) that only those modes with the largest decay times will fulfill the condition for oscillation. The fraction ϵ of all modes of the "cavity" which have decay times greater than one-half that of the maximum decay time is approximately $(2\theta)^2/2\pi$, or

$$\epsilon=2(1-\alpha)^2L^2/\pi D^2. \quad (31)$$

Letting $(1-\alpha)=1/20$, $L=1$ cm, and $D=10$ cm, one obtains $\epsilon=1.6\times10^{-5}$. This enormously reduces the number of modes which are likely to produce oscillations. Since the total number of modes is, from (14), $(8\pi^2\nu^2\Delta\nu/c^3)L^2D$, this number which may produce oscillations is

$$p'=\frac{16\pi(1-\alpha)^2\nu^2\Delta\nu L^4}{c^3D}. \quad (32)$$

Under the assumptions used above, p' may be found to be approximately 10^5, which is very much smaller than the total number of modes in the multimode cavity, but still may be an inconveniently large number. By using limiting values of the solid angle $(2\theta)^2$ set by diffraction, the number of modes can be further reduced to approximately unity, as was seen above.

FURTHER DISCUSSION OF PROPERTIES OF MASERS USING LARGE DIMENSIONS

It is important to notice that in the parallel plate case a very large amount of spontaneous emission may be radiating in all directions, even though only the very small amount indicated above is accepted in the detector, or is confused with the amplified wave. This property is quite different from the normal case in the microwave or radio-frequency range, and requires a rather rapid rate of supply of excited systems in order to maintain enough for maser action. Furthermore, great care must be taken to avoid scattering of light from undesired modes into the one which is desired. The fraction of spontaneous light which is scattered into the detector must be typically as low as about 10^{-6} or 10^{-7} in order to approach genuine isolation of a single mode.

Admission of a signal into the region between the two parallel plates is very similar to the process involved in a microwave cavity. The partially reflecting surfaces are analogous to coupling holes. If a monochromatic plane wave strikes the outside surface of one of the partially reflecting planes, energy will build up with the region between the planes, and the relations between input wave, energy in the "cavity", and output waves are just analogous to those for a microwave impinging on an appropriate cavity with input and output coupling holes.

Another interesting property of optical or infrared maser action which is associated with directional selection is that a beam of light may be passed through an ensemble of excited states with resulting amplification, but no important change in the wave front or phase. This amplification is just the inverse of an absorption, where it is well known that the wave front and phase are not distorted. Suppose, for example, that parallel light is focused by a lens. If an amplifying medium of excited gas is interposed between the lens and its focal point, the image will be intensified, but not otherwise changed except for some more or less normal effects which may be attributed to the dielectric constant of the gas. The same situation can, in principle, occur for maser amplification of microwaves. However, at these lower frequencies the amplification per unit length is usually so small that an impractically large volume of excited material would be required for amplification of a wave in free space to be evident. There may be a considerable amount of spontaneous emission in all directions, but only a very small fraction of the total spontaneous emission will fall at the focal point of the lens and be superimposed as noise on the intensified image. Noise from spontaneous emission decreases, for example, with the inverse square of distance from the emitting material, whereas the intensity of the focused beam increases as one approaches the focal point.

A SPECIFIC EXAMPLE

As an example of a particular system for an infrared maser, let us consider potassium. Atomic potassium is easily vaporized and has a simple spectrum as indicated by the energy levels shown in Fig. 1. Absorption transitions can occur from the $4s\ ^2S_{\frac{1}{2}}$ ground state only to the various p levels. In particular, the atoms can be excited to the $5p\ ^2P_{\frac{1}{2},\frac{3}{2}}$ by radiation of wavelength 4047 A. Just the right exciting frequency can be obtained from another potassium lamp, whose light is filtered to remove the red radiation at 7700 A. These excited atoms will decay to the $5s$ or $3d$ states in about 2×10^{-7} sec, or more slowly to the $4s$ ground state. However, if excited atoms are supplied fast enough, a sizable population can be maintained in the $5p$ state.

The minimum number of excited atoms required for maser-type oscillation may be found from (6), if the dipole matrix element were known, or from (11) if the lifetime were known. Although the wave functions necessary for obtaining the matrix elements have been calculated,[5,6] only estimates of the matrix element or lifetime can be made at present. The rate at which atoms must be supplied may, however, be obtained without detailed knowledge of the matrix elements by a small modification of expression (11). If τ is the mean life for spontaneous radiation of the desired wavelength, and φ is the fraction of the all decay processes from the upper level which occur in this manner, then the actual mean life in the excited state is $\tau'=\varphi\tau$. The number of atoms needed per second can be obtained from (11) as

$$\frac{n}{\tau'}=\frac{p}{\varphi t}=\frac{8\pi^2 V}{(\pi \ln 2)^{\frac{1}{2}}\varphi t\lambda^3}\frac{\Delta\nu}{\nu}=8\pi(2\pi)^{\frac{1}{2}}\frac{V}{\varphi\lambda^3 ct}\left(\frac{kT}{m}\right)^{\frac{1}{2}}, \quad (33)$$

where λ is the wavelength. Thus, if the fraction φ is known, no other detailed properties of the atomic transition are required to evaluate the rate at which excited atoms must be supplied.

For the particular case of a gas (such as potassium vapor) at sufficiently low pressure that collisions are not too frequent, we can obtain φ from the relative intensities of the various radiative transitions out of the excited state. Observed relative intensities[7] show that for potassium the $5p\rightarrow 3d$ transitions are about 4 times more intense than the $5p\rightarrow 5s$ transitions. Within the $5p\rightarrow 3d$ transitions it follows from elementary angular momentum theory that the $5p_{\frac{1}{2}}\rightarrow 3d_{\frac{1}{2}}$ is the most intense, accounting for 9/15 of the radiation

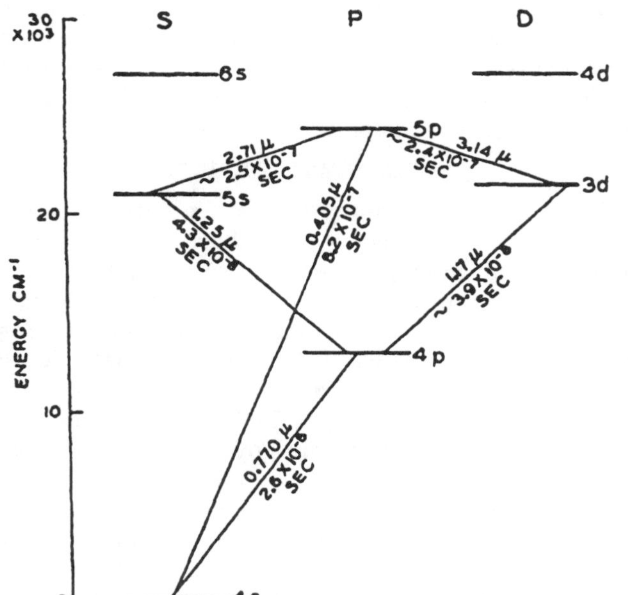

FIG. 1. Low-lying energy levels and transitions of atomic potassium.

emitted. Using the observed intensity ratio to allow for transitions to the $5s$ level, we conclude that about $9/18=\frac{1}{2}$ of those atoms excited to the $5p_{\frac{1}{2}}$ level decay to the $3d_{\frac{1}{2}}$ level. Decay to the $4s$ ground state is almost certainly less likely since we do know that this matrix element is not very large ($f=0.010$,[6] so that $\mu=0.65\times 10^{-18}$ esu). Thus $\varphi=\frac{1}{2}$ for the transition $5p_{\frac{1}{2}}-3d_{\frac{1}{2}}$ at 31 391 A.

Assume now two parallel plates of area 1 cm² and 10 cm apart, having a reflectivity α of 0.98. The decay time t for radiation in the space between the plates is $(10/3\times 10^{10})\times 50$ sec and $V=10$ cm³. For potassium vapor of suitable pressure, $T=435°$K and $m=39$ amu. Hence, from (29), the number of excited atoms needed per second is $dn/dt\geq 2.5\times 10^{15}$.

The energy needed per second is $d/dt(nh\nu)$, where ν is the frequency of the exciting radiation. Its value is 1.2×10^{-3} watt. This energy requirement is quite attainable. Incomplete absorption of the exciting radiation, reflection losses and multiplicity of the atomic states might raise this requirement somewhat. The absorption of the existing radiation is easily calculable and can be adjusted by controlling the density of the vapor:

$$k_0=\frac{1}{\Delta\nu_D}\left(\frac{\ln 2}{\pi}\right)^{\frac{1}{2}}\frac{\pi e^2}{mc}Nf, \quad (34)$$

where k_0 is the absorption coefficient at the peak of the line $\Delta\nu_D$ is the (Doppler) line half-width, e is the electron charge, m is the electron mass, c is the velocity of light, N is the number of initial state atoms per cc, and f is the oscillator strength of the transition; i.e.,

$$k_0=1.25\times 10^{-2}(Nf/\Delta\nu_D).$$

For the exciting transition, 4046 A, in potassium, $\nu_0=7.42\times 10^{14}$ cycles/sec and at 435°K, $\Delta\nu_D=0.84\times 10^9$ cycles/sec. At this temperature the vapor pressure is 10^{-3} mm of mercury, so that in saturated vapor $N=2.5\times 10^{13}$/cc. Since $f=0.10$ for the $4s_{\frac{1}{2}}\rightarrow 5p_{\frac{1}{2}}$ transition,[6] $k_0=3.72$. This is high enough that the exciting radiation would be absorbed in a thin layer; if necessary it can be reduced by changing the pressure or temperature.

The light power for excitation is proportional to

$$\frac{V}{t}=\frac{AD}{(1-\alpha D/c)}=\frac{Ac}{1-\alpha}, \quad (35)$$

where V is the volume of the cavity, t is the decay time for light in the cavity, D is the length of the cavity, A is the cross-section area of the cavity, α is the reflectivity of the end plates, and c is the velocity of light. This is independent of length, so that for a given cross-sectional area the light density needed can be reduced by increasing the length.

[5] L. Biermann and K. Lubeck, Z. Astrophys. 25, 325 (1948).
[6] D. S. Villars, J. Opt. Soc. Am. 42, 552 (1952).
[7] *Viz.*, Tabulation in the *Handbook of Chemistry and Physics*, edited by D. Hodgman (Chemical Rubber Publishing Company, Cleveland, 1957), thirty-ninth edition.

LIGHT SOURCES FOR EXCITATION

A small commercial potassium lamp (Osram) was operated with an input of 15 watts, 60 cycles, and its output was measured. In the red lines (7664–7699 A), the total light output was 28 mw from about 5 cc volume. At the same time, the total output in the violet lines (4044–4047 A) was 0.12 mw,[8] so that the output in $4s-5p_{\frac{1}{2}}$ was 0.08 mw. By increasing the current from 1.5 to 6 amp, with forced air cooling (the outer jacket being removed), the total violet output was increased to 0.6 mw. These outputs are somewhat short of the power level needed, but they may be considerably increased by adjusting discharge conditions to favor production of the violet line, and by using microwave excitation. With a long maser cell, the lamp area can be greatly increased. If necessary, very high peak light powers could be obtained in pulsed operation, although one would have to be careful not to broaden the line excessively.

Another possibility for excitation is to find an accidental coincidence with a strong line of some other element. The $8p$ level of cesium is an example of this type, since it can be excited very well by a helium line. The 4047 A line of mercury is 5 cm^{-1} from the potassium line, and is probably too far away to be useful even when pressure broadened and shifted.

Different modes correspond to different directions of propagation, and we only want to produce one or a few modes. Thus the cavity need only have two good reflecting walls opposite each other. The side walls need not reflect at all, nor do they need to transmit infrared radiation.

Unfortunately, most elements which have simple spectra, are quite reactive. Sapphire has good chemical inertness and excellent infrared transmission, being almost completely transparent as far as about 4 microns wavelength.[9] With such good transmission, the principal reflecting surfaces can be put outside the cell, and hence chosen for good reflectivity without regard to chemical inertness. Thus, one could use gold which has less than 2% absorption in this region, and attain a reflectivity of ~97% with 1% transmission. Even better reflectivity might be obtained with multiple dielectric layers of alternately high- and low-dielectric constant. The inner walls of the sapphire cell would reflect about 5% of the infrared light, and the thickness should be chosen so that the reflections from the two surfaces are in phase. The phase angle between reflections from the two surfaces depends on the thickness and the refractive index. Since sapphire is crystalline and the index is different for ordinary and extraordinary rays, the thickness could be chosen to give constructive interference for one polarization, and destructive interference for the perpendicular polarization. Thus, one could discriminate, if desired, between modes traveling in the same direction with different polarization.

To select just one from among the very many modes possible within the line width, the stimulated emission of radiation with one chosen direction of propagation must be favored. Thus the cell should be made long in the desired direction and fitted with highly reflecting end plates. The desired wave then has a long path as it travels back and forth, and so has a good chance to pick up energy from the excited atoms. A large width decreases the angular discrimination, and increases the pumping power needed.

For the potassium radiation at 3.14×10^{-4} cm wavelength, and $\Delta\nu$ being the Doppler width at 435°K, i.e., $\Delta\nu/\nu_0=1.2\times10^{-6}$, the number of modes is $2.0\times10^6\,V$ from expression (15). If we consider a cavity 1 cm square by 10 cm long, this number is 2.0×10^7, or 3.2×10^6 modes per steradian (forward and backward directions are taken as equivalent for standing waves). The angular separation between modes is then $(32\times10^6/2)^{-\frac{1}{2}}=2.5\times10^{-4}$ radian, where the 2 in the denominator removes the polarization degeneracy. The angular aperture accepted by this cavity is 1/10, but if the end plates had 98% reflectivity, the effective length would be increased by a factor of 50, and the angular aperture reduced to 2×10^{-3} radian. Thus there would be only 8 modes of each polarization within the effective aperture of the cell. Obviously this type of mode selection could be pushed further by making the cavity longer or narrower or more reflecting but this should not be necessary. Furthermore, the emission line does not have constant intensity over the width $\Delta\nu$, and the mode nearest the center frequency would be the first to oscillate at the threshold of emission.

SOLID-STATE DEVICES

There are a good many crystals, notably rare earth salts, which have spectra with sharp absorption lines, some of them having appeared also in fluorescence. In a solid, a concentration of atoms as large as 10^{19} per cc may be obtained. The oscillator strengths of the sharp lines are characteristically low, perhaps 10^{-6}. If the f value is low, radiative lifetimes are long, and in some cases lifetimes are as long as 10^{-3} sec or even more.

If the lifetime is primarily governed by radiation in the desired line, the pumping power required for the onset of stimulated oscillation is independent of the f value, as was shown above. For the atomic potassium level considered earlier, there are several alternative radiative decay paths (to the $4s$ and $3d$ states). In a solid there may also be rapid decay by nonradiative processes. If the storage time is long, because of a small f value, there is more time for competing processes to occur. Even lines which are sharp for solids are likely to be broader than those obtainable in gases. This larger width makes the attainment of maser oscillation more difficult, and it adds greatly to the difficulty of

[8] We are indebted to R. J. Collins for making these measurements.

[9] R. W. Kebler, *Optical Properties of Synthetic Sapphire* (Linde Company, New York).

selecting a single mode. However, there may very well be suitable transitions among the very many compounds.

The problem of populating the upper state does not have as obvious a solution in the solid case as in the gas. Lamps do not exist which give just the right radiation for pumping. However, there may be even more elegant solutions. Thus it may be feasible to pump to a state above one which is metastable. Atoms will then decay to the metastable state (possibly by nonradiative processes involving the crystal lattice) and accumulate until there are enough for maser action. This kind of accumulation is most likely to occur when there is a substantial empty gap below the excited level.

SUMMARY AND HIGH-FREQUENCY LIMITS

The prospect is favorable for masers which produce oscillations in the infrared or optical regions. However, operation of this type of device at frequencies which are still very much higher seems difficult. It does not appear practical to surround an atomic system with cavity walls which would very much affect its rate of spontaneous emission at very short wavelengths. Hence any ensemble of excited systems which is capable of producing coherent amplification at very high frequencies must also be expected to emit the usual amount of spontaneous emission. The power in this spontaneous emission, from expressions (14) and (16), increases very rapidly with frequency—as ν^4 if the width $\Delta\nu$ is due to Doppler effects, or as ν^6 if the width is produced by spontaneous emission. By choice of small matrix elements, $\Delta\nu$ can, in principle, be limited to that associated with Doppler effects, but the increase in spontaneously emitted power as fast as ν^4 is unavoidable.

For a wavelength $\lambda = 10^4$ A, it was seen above that spontaneous emission produced a few milliwatts of power in a maser system of dimensions near one centimeter, assuming reflectivities which seem attainable at this wavelength. Thus in the ultraviolet region at $\lambda = 1000$ A, one may expect spontaneous emissions of intensities near ten watts. This is so large that supply of this much power by excitation in some other spectral line becomes very difficult. Another decrease of a factor of 10 in λ would bring the spontaneous emission to the clearly prohibitive value of 100 kilowatts. These figures show that maser systems can be expected to operate successfully in the infrared, optical, and perhaps in the ultraviolet regions, but that, unless some radically new approach is found, they cannot be pushed to wavelengths much shorter than those in the ultraviolet region.

For reasonably favorable maser design in the short wavelength regions, highly reflecting surfaces and means of efficient focusing of radiation must be used. If good reflecting surfaces are not available, the number of excited systems used must, from (6), be very much increased with a resulting increase in spontaneous emission and difficulty in supply of excited systems. If focusing is not possible, the directional selection of radiation can in principle be achieved by detection at a sufficiently large distance from the parallel plates. However, without focusing the directional selection is much more difficult, and the background of spontaneous emission may give serious interference as noise superimposed on the desired radiation.

Finally, it must be emphasized that, as masers are pushed to higher frequencies, the fractional range of tunability must be expected to decrease more or less inversely as the frequency. The absolute range of variation can be at least as large as the width of an individual spectral line, or as the few wave numbers shift which can be obtained by Zeeman effects. However, continuous tuning over larger ranges of frequency will require materials with very special properties.

ACKNOWLEDGMENTS

The authors wish to thank W. S. Boyle, M. Peter, A. M. Clogston, and R. J. Collins for several stimulating discussions.

PHYSICAL REVIEW
LETTERS

VOLUME 5 OCTOBER 15, 1960 NUMBER 8

ATOMIC HYDROGEN MASER*

H. M. Goldenberg,† D. Kleppner, and N. F. Ramsey
Harvard University, Cambridge, Massachusetts
(Received September 16, 1960)

Up to the present it has not been possible to produce maser oscillations with gaseous atoms due to the weakness of the magnetic dipole radiation matrix elements. However, with sufficiently long interaction times with the radiation field, oscillation can be achieved. Such increased interaction time and correspondingly narrowed resonance widths have been obtained in the present experiment by retaining the atoms in a storage box[1,2] with suitable walls. In this fashion self-sustained emission at the atomic hydrogen hyperfine transition frequency has been observed in an atomic beam maser. Mean interaction times up to 0.3 sec have so far been measured.

A diagram of the apparatus is given in Fig. 1. Atomic hydrogen from a Wood's discharge source passes through a six-pole state selecting magnet which focuses atoms in the $F=1$, $m=0$ and $F=1$, $m=1$ states onto an aperture in a paraffin-coated quartz bulb. The bulb is located in the center of a cylindrical rf cavity, operating in the TE_{011} mode, which is tuned to the ($F=1$, $m=0 \rightarrow F=0$, $m=0$) hyperfine transition frequency, approximately 1420.405 Mc/sec. The conditions are such that the atoms spontaneously radiate to the lower hyperfine level while in the bulb. The atoms make random collisions with the paraffin-covered bulb wall and eventually leave the bulb through the entrance aperture. Due to the small interaction with the paraffin surface they are not seriously perturbed for at least 10^4 collisions.

The criterion for self-sustained oscillation is that the power delivered to the cavity by the atomic beam must equal that lost by the cavity. In our case, with an average interaction time of 0.3 sec and a loaded cavity $Q=60\,000$, the minimum beam flux needed for oscillation is 4×10^{12} particles per seco..d. The maximum power delivered to the cavity by this beam is approximately 10^{-12} watt. However, in the present case the detecting probe is weakly coupled to the cavity so that the detected power is considerably less. The width of the resonance $\Delta\nu_\gamma$ without maser operation is about 1 cps as compared to several kc/sec in a conventional ammonia maser. The bulb used is approximately 15 cm in diameter, and has an entrance aperture of 1-mm radius. It is loaded with paraffin[3] in air, and is then baked while under vacuum.

The signal is detected by mixing it with a 1450-

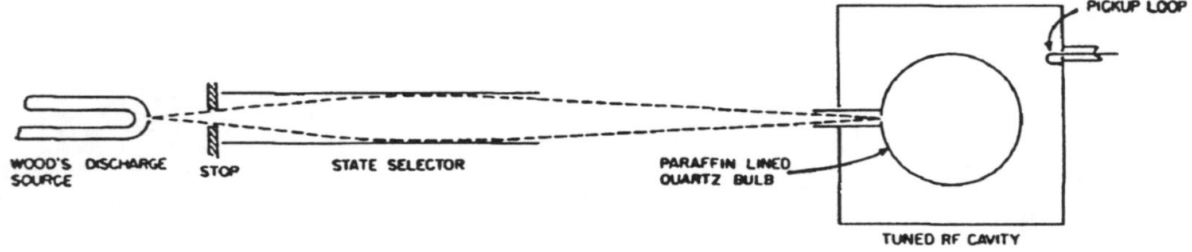

FIG. 1. Schematic diagram of atomic hydrogen maser.

Mc/sec signal obtained from the 5-Mc/sec output from an Atomichron by multiplication in a Gertsch FM4A frequency multiplier. The 29.595-Mc/sec i.f. signal is amplified and mixed in a phase discriminator with a similar signal from a Gertsch AM-1 stabilized oscillator. The resultant low-frequency signal passes through an *RC* filter, which limits the detector's over-all bandwidth, and is displayed on an oscilloscope.

In addition to oscillation, stimulated emission has also been observed. In this case oscillation is completely suppressed by shortening the mean time in the bulb (by means of a larger entrance aperture) or by loading the cavity. A pulse of rf at the transition frequency is fed into the cavity by a second coupling loop (not shown in Fig. 1). If this pulse is of the correct intensity the electron spin of the atom is precessed 90°, and the atom will continue to radiate after the rf has been turned off. A picture of the oscilloscope trace of this radiation is shown in Fig. 2. The amplifier is gated off during the stimulating pulse at the beginning of the sweep. The signal envelope decays with a time constant equal to the mean time in the bulb. The oscillating component of the signal is due to a slight offset in the local oscillator which is introduced for purposes of display. An interesting feature of this method of inducing emission is that those atoms which happen to be in the cavity for greater than the mean time are subject to more than one rf pulse. The mean interaction time of these atoms is then longer than the mean time in the cavity, and is characterized by the total time between the pulses they experience. If the pulse intensity is reduced so that it takes two pulses to put an atom in a state of maximum radiation, then the resonance line has a width characteristic of the pulse repetition frequency, which can be made several times smaller than the natural linewidth in the bulb.

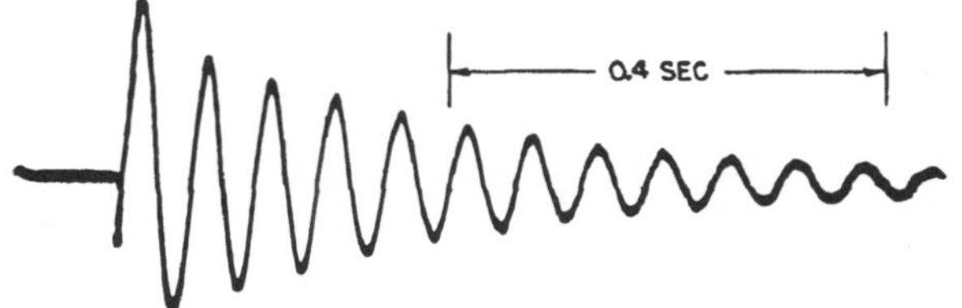

FIG. 2. Pulsed stimulated emission.

The magnetic-field-independent hyperfine transition has been used to observe the field-dependent Zeeman transitions, $F=1,\ m=0 \rightarrow F=1,\ m=1$) and ($F=1,\ m=0,\ \rightarrow F=1,\ m=-1$) by means of a double resonance method. A signal at the Zeeman frequency was coupled into the cavity at the same time that the hyperfine transition was occurring. The intensity of the hyperfine signal then decreased because the mean lifetime of atoms in the $F=1$, $m=0$ state was reduce due to their making transitions to the other $F=1$ states. This technique offers a method of measuring the static magnetic field to high accuracy.

This experiment represents the first successful use of the storage box principle. We believe that the atomic hydrogen maser will allow determination of the hyperfine splitting of the hydrogen isotopes to considerably higher precision than is now known. In addition, as a time standard, the maser may have a stability greatly exceeding that of any other standard yet proposed.

*Research sponsored by the Joint Program of the Office of Naval Research and the U. S. Atomic Energy Commission, and by the National Science Foundation. Some of the equipment was provided by the U. S. Army Signal Corps.

†National Science Foundation Predoctoral Fellow.

[1]N. F. Ramsey, Rev. Sci. Instr. 28, 58 (1956).

[2]D. Kleppner, N. F. Ramsey, and P. Fjelstad, Phys. Rev. Letters 1, 232 (1958).

[3]The paraffin used is a high melting point variety supplied under the trade name "Paraflint" by Moore and Munger, New York, New York.

POPULATION INVERSION AND CONTINUOUS OPTICAL MASER OSCILLATION IN A GAS DISCHARGE CONTAINING A He-Ne MIXTURE

A. Javan, W. R. Bennett, Jr., and D. R. Herriott
Bell Telephone Laboratories, Murray Hill, New Jersey
(Received December 30, 1960)

This Letter is intended to give a summary of the results of experimental determinations of several physical properties of a gaseous discharge consisting of a He-Ne mixture which have led to the successful operation of a continuous-wave maser at five different wavelengths in the near-infrared. Population inversions are achieved between several Ne levels by means of excitation transfer from the metastable He($2\,^3S$) to the $2s$ levels of Ne.[1,2] The maser oscillation takes place in a narrow beam with a diameter of 0.45 inch. The power in our strongest beam at the wavelength 11 530 A is 15 milliwatts. The measured linewidth is in the range of 10 to 80 kc/sec. The angular spread of the beam is less than one minute of arc.

In the present system, the He($2\,^3S$) metastables are used as carriers of energy to excite the levels of Ne. Due to the resonant nature of this process, the excitation cross section is large only to those levels of Ne which fall in energy within a few kT of that of the He($2\,^3S$). Figure 1 gives the pertinent energy levels of He and Ne. As may be seen from Fig. 1,[3] the energy separations between the He($2\,^3S$) and Ne($2s$) states are small enough to insure an appreciable degree of transfer to the $2s$ states; however, the energy gap between the $2\,^3S$ and $2p$ levels is much too large to permit direct population of the $2p$ states in the collision process. Although the $2s_2$ and $2s_4$ levels

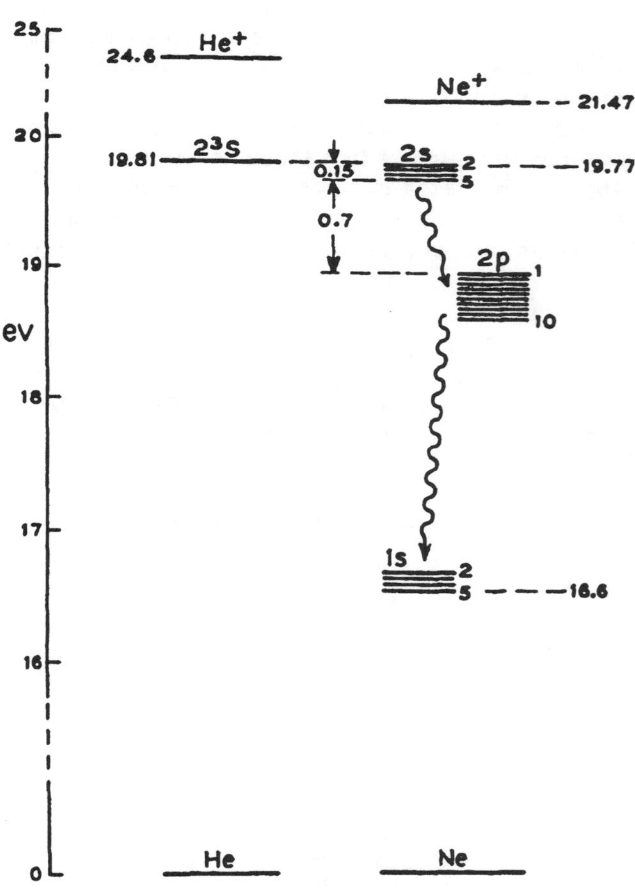

FIG. 1. Energy level diagram of He and Ne atoms.

of Ne may radiate to the Ne ground state, in the limit of complete resonance trapping their lifetimes are determined primarily by radiative decay to the $2p$ levels. Under this condition, the lifetimes of all of the $2s$ levels are about one order of magnitude longer than those of the corresponding $2p$ states. (The decay of the $2p$ levels is due to their radiative decay to the $1s$ levels.) Thus population inversions may be obtained on each of the thirty allowed $2s \rightarrow 2p$ transitions (see below).

In order to determine the transfer cross section and spurious sources of excitation of Ne levels, the afterglow of a pulsed rf discharge containing varying amounts of He and Ne was studied in detail. The time variation of the densities of the $2s$ and $2p$ levels was observed by the fluorescent decay of these levels using appropriate isolated transitions. The time variation of the $2\,^3S$ density was determined through the absorption of the $2\,^3P \rightarrow 2\,^3S$ He transition.

The conditions were found under which the dominant mode of excitation of Ne levels arose from the He($2\,^3S$) transfer. Among other sources of excitation of the Ne levels it was found that inelastic collisions of energetic electrons with Ne metastables[2] (i.e., the $1s$ levels[4]) were by far the most important. It should be noted that certain amounts of $2p$ excitation, arising from sources other than $2s$ - $2p$ cascade, can be tolerated without upsetting the presence of a population inversion. For instance, let us consider the population inversion between the $2s_5$ and $2p_9$ levels. The $2p_9$ is optically connected to only one of the $2s$ levels, namely the $2s_5$. The Einstein A coefficient for the $2s_5 - 2p_9$ transition was measured and found to be at least 25 times smaller than the rate of decay of the $2p_9$ level to within an accuracy of 20%. (This measurement was done by means of two independent techniques. The first was a spectroscopic technique to be described in a later publication. The second is discussed below.) This means that even an order of magnitude added excitation of $2p_9$ over that due to the cascade from the $2s_5$ still yields a population of the $2s_5$ levels larger than twice the population of the $2p_9$. In view of this effect, the adjustment of the electron density was found to be most critical.

In the afterglow, the energetic electrons thermalize within 10 to 20 μsec. The excitation transfer from He($2\,^3S$), however, continues until the time when all of the $2\,^3S$ states are quenched. This effect gives rise to a decay of the $2s$ levels identical to the time variation of the He($2\,^3S$). The afterglow of the $2p$ levels also shows an identical decay due to the cascade transition from the $2s$ levels. Our observations indicate that the time variations of the $2\,^3S$, $2s$, $2p$ levels in the afterglow are indeed identical to a high degree of accuracy. The possibility of the direct transfer from the $2\,^3S$ to the $2p$ levels was ruled out by a process of relative intensity measurement at a fixed time in the afterglow to be described elsewhere.

Study of the fluorescent decay rate of the $2s$ and $2p$ levels as a function of several Ne pressures gave a total velocity-averaged inelastic cross section of $\langle\sigma\rangle = (3.7 \pm 0.5)\times 10^{-17}$ cm^2 for destroying He($2\,^3S$) metastables in two-body collisions with Ne ground-state atoms at 300°K. This cross section can safely be assigned to the excitation transfer to the $2s$ levels of Ne since we have been able to show that other sources of destruction of the He($2\,^3S$) by Ne ground-state atoms such as the one leading[5] to a formation of $HeNe^+$ are negligible. Our cross-section measurement at the same time furnishes the value of the diffusion coefficient of $2\,^3S$ He metastables. Our result is in excellent agreement with the value of (470±25 cm^2/sec) determined by Phelps.

The radiative lifetimes of the $2s$ and $2p$ levels were determined using techniques described previously.[7] Our tentative values of the lifetimes of the $2s_2$, $2s_3$, $2s_4$, and $2s_5$ fall in the range from about 100 to 200 mμsec. Our results for the $2p$ levels are in general agreement with Ladenberg's[8] results. These measurements are still in progress and will be reported more fully at a later date. Each of the $2s$ levels decays into several of the ten $2p$ levels. The measured and estimated branching ratios for the transitions originating from each of the $2s$ levels, together with their total rate of radiative decay obtained from the above lifetimes, give the values of the individual Einstein A coefficients connecting the various $2s$ and $2p$ levels.

From the early stages of this experiment we were able to observe single-pass amplification of a transmitted signal at an infrared frequency where the inverted population was expected. Several electronic detection schemes were devised for this purpose. One of these schemes responded only during a small portion of the time and was in synchronism with the rf pulsed discharge. This technique enabled us to observe a larger value of amplification in the afterglow than during the glow; the latter is to be expected since the added excitation of the $2p$ levels by electron impact is absent in the afterglow. However, in addition to extreme difficulties encountered in the interpretation of the results, the gain measurement alone was not sufficient to shed insight into the nature of the physical processes involved in the discharge.

In view of the above considerations, we were able to determine optimum condition for the maser action prior to the observation of oscillation as described below.

The continuous-wave maser oscillation is observed in a discharge containing 0.1 mm Hg of Ne and 1 mm Hg of He. The optical feedback[9] is obtained through high-reflectance Fabry-Perot plates placed within the gas chamber. This chamber consists of a long quartz tube with an inside diameter of 1.5 cm and length of 80 cm. The quartz tube is terminated at each end with a larger metal chamber containing the high-reflectance plates. Flexible bellows are used in the end chambers to allow external mechanical tuning of the Fabry-Perot plates. Two optically flat windows at the extremes of this system are provided to allow the maser beam to leave the instrument undistorted. The separation of the plates is one meter. The discharge is excited by means of external electrodes using a 28-Mc/sec generator. The estimated power dissipation is around 50 watts.

The high reflectance is achieved by means of 13-layer evaporated dielectric films on the hundredth wavelength flat surface of the fused silica plates. The reflectance is 98.9% with 0.3% transmission in the wavelength range 11 000 A to 12 000 A. Outside of this range of wavelength the reflectance decreases rapidly.

The initial adjustment of the plates prior to starting of the discharge is done using an autocollimator. We have observed oscillations at the following wavelengths: 11 180, 11 530, 11 600, 11 990, and 12 070 A. The strongest oscillation occurs at 11 530 A with an output power of 15 mw.

An image converter tube is used visually and with a camera to observe and record structure of the beam. Two types of beam patterns have been studied.

A cross section of the emergent beam can be intercepted by the photocathode of the image converter and the intensity distribution recorded. This near-field pattern is shown in the two upper photographs of Fig. 2 for two Fabry-Perot plate angles.

The intensity pattern that the beam would form

PLATES PARALLEL

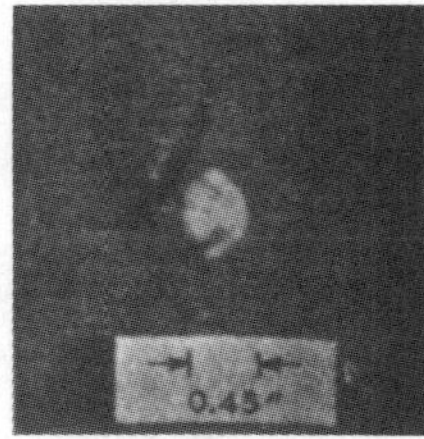

6 SECOND ANGLE

PATTERN IN MASER CAVITY

PLATES PARALLEL

6 SECOND ANGLE

ANGULAR DISTRIBUTION IN BEAM

FIG. 2. Far-field and near-field patterns from maser beam at two plate angles.

at an infinite distance can be observed by imaging the maser beam to a point and observing the distribution of light in this image with an image converter microscope. The intensity distribution in this image represents the angular distribution in the maser beam. The two lower pictures in Fig. 2 show this far-field pattern at the indicated plate positions. The structures seen in the near- and far-field patterns would indicate the existence of several modes at a finite plate angle. The patterns shown in Fig. 2 were made at constant exposure and are therefore greatly overexposed by the increased intensity with parallel plates. The half-intensity width of the parallel plate output is less than one minute of arc.

The phase variation over the area of the beam was studied by placing a double or multiple slit in the beam and recording the diffraction pattern. A nine-slit mask of 0.050-in. spacing and 0.002-in. slit width covering the full 0.45-in. aperture of the beam results in a diffraction pattern shown in Fig. 3. The minima are very low, indicating small phase variation over the aperture. A double-slit pattern at 0.0175-in. is also shown.

Since the linewidth of the maser oscillation is many orders of magnitude narrower than the resolution of the best available spectrometers or

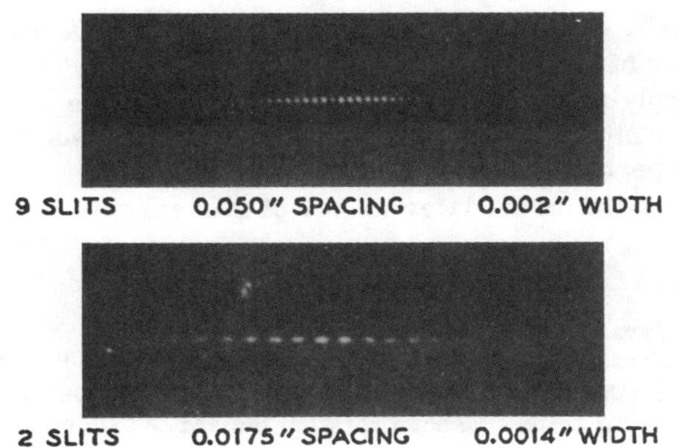

FIG. 3. Far-field diffraction patterns obtained by inserting multiple slits in the maser beam.

interferometers, standard optical techniques could not be used to measure the linewidth. Of the several possible methods examined, the most practical was found to be one based on analysis of the Fourier spectrum observed through a photomultiplier tube of the maser output. The photomultiplier is essentially a perfect square-law detector and should respond to beat frequencies. Hence, one can observe the spectral distribution obtained by beating all of the components of the line. The exact width of the original line determined from this distribution is independent of detailed assumptions on the line shape to within a factor of about two, providing the original line is not multiply peaked. In practice, this spectral distribution will be superimposed on top of the flat spectrum obtained from shot noise in the photomultiplier tube, and the latter may be used to check the bandwidth of the system.

The system was first checked by observing the spectrum from an intense line emitted in an ordinary discharge tube using a spectrum analyzer. The latter was flat within the frequency range of our analyzer, covering from 100 cps to 13 Mc/sec. Using a grating spectrometer to isolate the strong line at 11 530 A, the maser output was next examined. Our initial observations indicated a strong signal starting at zero frequency and extending to a half-width of about 100 kc/sec. Further examination showed the presence of an additional series of anharmonically related beats centered about frequencies other than zero. The location of these beats depended critically on the Fabry-Perot plate alignment. As the angular alignment of the plates was changed from 0 to 3 seconds, the beats became more widely spaced and weaker in intensity. Several possible spurious sources of beats were carefully examined and found not to be responsible for the observed signal. The possibility of "blocking oscillation" of the maser giving rise to an amplitude modulation of the output signal can be ruled out by the following observation: The peak frequency of a beat note is only a function of the plate adjustment and is independent of the factor by which the maser gain exceeds the oscillation threshold. The most plausible explanation found was that the beats arose from simultaneous oscillation in several different modes separated by frequencies in the Mc/sec range. Such a conclusion is not in disagreement with detailed calculations of the oscillatory modes by Fox and Li.[10] A distinct

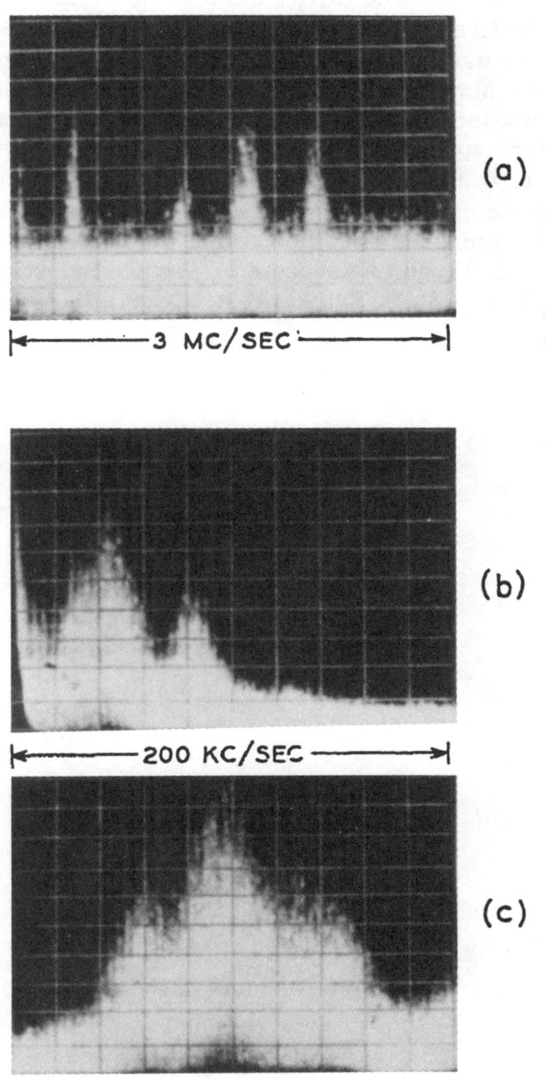

FIG. 4. Frequency spectra of the beat notes between various maser beam components.

correlation was found between the visual structure of the beam observed through the infrared image converter, and the presence of the beats as the angular alignment of the plates was varied. Figure 4(a) shows several beat notes ranging from 50 kc/sec to 2 Mc/sec. Figures 4(b) and 4(c) show the structure of these beats on an expanded scale. It should be noted that in our present system the angular adjustment of the plates is also accompanied by a slight change in plate separation. The important conclusion to be drawn here regards the linewidth of the maser oscillation. As is evident from Figs. 4(b) and 4(c), the beat notes imply linewidths in the range from 10 to 80 kc/sec.

The use of excitation transfer for production of inverted population was suggested[1,2] by one of the authors. The He-Ne mixture described above is the first gaseous system which has led to maser oscillations at optical frequencies. Recently some evidence for the presence of inverted populations in a Hg-Zn mixture has been reported[11] which supports the wide applicability of this principle.[1,2]

In the course of this research, a number of co-workers have contributed in the design, construction, and measurement. We are indebted to F. Muller, P. S. Kubik, A. R. Strnad, D. MacNair, E. Koch, G. J. Wolff, and G. E. Reitter. Dr. A. F. Turner of Bausch & Lomb has been very helpful in supplying the special evaporated reflectance films. In particular we would like to acknowledge the expert and patient assistance given by Mr. E. A. Ballik in the later stages of this experiment.

---

[1]A. Javan, Phys. Rev. Letters 3, 518 (1959).

[2]A. Javan, in Quantum Electronics, edited by C. H. Townes (Columbia University Press, New York, 1960).

[3]For simplicity, Paschen notation is used to describe the Ne energy levels.

[4]Not all of the 1*s* levels are truly metastables: the $1s_2$ and $1s_4$ undergo transitions to the ground state; however, due to the trapping of the resonant ultraviolet photons these levels are effectively long-lived and quasi-metastable.

[5]M. Pahl and U. Weimer, Z. Naturforsch. 12a, 926 (1957).

[6]A. V. Phelps, Phys. Rev. 99, 4, 1307 (1955).

[7]W. R. Bennett, Jr., A. Javan, and E. A. Ballik, Bull. Am. Phys. Soc. 5, 496 (1960).

[8]R. Ladenberg, Revs. Modern Phys. 5, 243 (1933).

[9]A. L. Schawlow and C. H. Townes, Phys. Rev. 112, 1940 (1958).

[10]A. G. Fox and P. Li, Proc. Inst. Radio Engrs. 48, 1904 (1960).

[11]V. K. Ablekov, M. S. Pesiu, and I. L. Fabelinski, Zhur. Eksp. i Teoret. Fiz. 39, 892 (1960).

GENERATION OF OPTICAL HARMONICS*

P. A. Franken, A. E. Hill, C. W. Peters, and G. Weinreich
The Harrison M. Randall Laboratory of Physics, The University of Michigan, Ann Arbor, Michigan
(Received July 21, 1961)

The development of pulsed ruby optical masers[1,2] has made possible the production of monochromatic (6943 A) light beams which, when focussed, exhibit electric fields of the order of 10^5 volts/cm. The possibility of exploiting this extraordinary intensity for the production of optical harmonics from suitable nonlinear materials is most appealing. In this Letter we present a brief discussion of the requisite analysis and a description of experiments in which we have observed the second harmonic (at ~ 3472 A) produced upon projection of an intense beam of 6943A light through crystalline quartz.

A suitable material for the production of optical harmonics must have a nonlinear dielectric coefficient and be transparent to both the fundamental optical frequency and the desired overtones. Since all dielectrics are nonlinear in high enough fields, this suggests the feasibility of utilizing materials such as quartz and glass. The dependence of polarization of a dielectric upon electric field E may be expressed schematically by

$$P = \chi E\left(1 + \frac{E}{E_1} + \frac{E^2}{E_2^{\,2}} + \cdots\right), \tag{1}$$

where E_1, E_2 ... are of the order of magnitude of atomic electric fields (~10^8 esu). If E is sinusoidal in time, the presence in Eq. (1) of terms of quadratic or higher degree will result in P containing harmonics of the fundamental frequency. Direct-current polarizations should accompany the even harmonics.

Let $\vec{p}$ be that part of $\vec{P}$ which is quadratic in $\vec{E}$; that is, $\vec{p}$ is a linear function of the components of the symmetric tensor $\vec{E}\vec{E}$. The eighteen coefficients which occur in this function are subject to restrictions due to the point symmetry of the medium. These restrictions are, in fact, identical with those governing the piezoelectric coefficients. In particular, $\vec{p}$ necessarily vanishes in a material such as glass which is isotropic or contains a center of inversion. For crystalline quartz, however, there are two independent coefficients α and β in terms of which

$$p_x = \alpha(E_x^{\,2} - E_y^{\,2}) + \beta E_y E_z,$$
$$p_y = -\beta E_x E_z - 2\alpha E_x E_y,$$
$$p_z = 0 \tag{2}$$

Table I. The square of the total p perpendicular to the direction of propagation of light through crystalline quartz.

Direction of incident beam	The square of the total p perpendicular to direction of propagation
x ($E_x = 0$)	$p_y^{\,2} + p_z^{\,2} = 0$
y ($E_y = 0$)	$p_z^{\,2} + p_x^{\,2} = \alpha^2 E_x^{\,4}$
z ($E_z = 0$)	$p_x^{\,2} + p_y^{\,2} = \alpha^2(E_x^{\,2} + E_y^{\,2})^2$

(z is the threefold, or optic, axis; x a twofold axis). If a light beam traverses quartz in one of the three principal directions, Eqs. (2) predict the results summarized in Table I. The second-harmonic light should be absent in the first case, dependent upon incident polarization in the second case, and independent of this polarization in the third.

If an intense beam of monochromatic light is focussed into a region of volume V, there should occur an intensity I of second harmonic given (in Gaussian units) by

$$I \sim (\omega^4/c^3)(pv)^2(V/v), \tag{3}$$

where ω is the angular frequency of the second harmonic, c the velocity of light, and v an effective "volume of coherence"; that is, the size of a region within the sample in which there is phase coherence of the p excitation. (This volume may in practice be much smaller than V.) An estimate of v is governed by several considerations. For example, it is probably of no greater extent in the propagation direction than ~ $[n_2 \times (n_2 - n_1)^{-1}]\lambda_2$, where n_1 and n_2 are the indices of refraction for the fundamental and second harmonic frequencies, respectively, and λ_2 is the wavelength of the second harmonic. The lateral extent of this volume is determined in large part by the coherence characteristics of the optical maser. The situation for a maser of the gas discharge[3] type is clearly more favorable in this respect than that for the ruby device.[1,2] For a coherence volume of 10^{-11} cm^3, which we think may be realistic in our case, Eq. (1) indicates

FIG. 1. A direct reproduction of the first plate in which there was an indication of second harmonic. The wavelength scale is in units of 100 A. The arrow at 3472 A indicates the small but dense image produced by the second harmonic. The image of the primary beam at 6943 A is very large due to halation.

that second harmonic intensities as high as a fraction of a percent of the fundamental could be achieved.

In the experiments we have used a commercially available ruby optical maser[4] which produces approximately 3 joules of 6943A light in a one-millisecond pulse. This light is passed through a red filter for the elimination of the xenon flash background and is then brought to a focus inside a crystalline quartz sample. The emergent beam is analyzed by a quartz prism spectrometer equipped with red insensitive Eastman Type 103 spectrographic plates. A reproduction of the first plate in which there was an unambiguous indication of second harmonic (3472 A) is shown in Fig. 1. This plate was exposed to only one "shot" from the optical maser. We believe the following two facts, among others, rule out the possibility of artifact:

(1) The light at 3472 A disappears when the quartz is removed or is replaced by glass.

(2) The light at 3472 A exhibits the expected dependence on polarization and orientation summarized in Table I.

Considerations of the photographic image density and the efficiency of the optical system lead us to believe that the order of 10^{11} second harmonic photons were generated within the quartz sample per pulse.

The production of a second harmonic should be observable in isotropic materials such as glass if a strong bias field were applied to the sample. This bias could be oscillatory, thus producing sidebands on the fundamental frequency and the harmonics.

We would like to thank the staff of Trion Instruments, Inc., for their valuable and sustained cooperation in this work.

---

*This work was supported in part by the U. S. Atomic Energy Commission.

[1]T. H. Maiman, Nature 187, 493 (1960).

[2]R. J. Collins et al., Phys. Rev. Letters 5, 303 (1960).

[3]A. Javan, W. R. Bennet, and D. R. Herriott, Phys. Rev. Letters 6, 106 (1961). Even though the intensity of the gas device is very low compared with ruby masers, the gain in coherence volume and the potential improvement of focussing suggest that the gas maser may be comparable or even superior as a source for optical harmonics.

[4]Trion Instruments, Inc., Model No. TO-3000.

COHERENT LIGHT EMISSION FROM GaAs JUNCTIONS

R. N. Hall, G. E. Fenner, J. D. Kingsley, T. J. Soltys, and R. O. Carlson
General Electric Research Laboratory, Schenectady, New York
(Received September 24, 1962)

Coherent infrared radiation has been observed from forward biased GaAs p-n junctions. Evidence for this behavior is based upon the sharply beamed radiation pattern of the emitted light, upon the observation of a threshold current beyond which the intensity of the beam increases abruptly, and upon the pronounced narrowing of the spectral distribution of this beam beyond threshold. The stimulated emission is believed to occur as the result of transitions between states of equal wave number in the conduction and valence bands.

Several requirements must be fulfilled[1] in order that such stimulated emission can be observed: (a) The electron and hole populations within the active region must be large enough that their quasi-Fermi levels are separated by an energy greater than that of the radiation; (b) losses due to absorption by other processes must be small relative to the gain produced by stimulated emission; and (c) the active region must be contained within a cavity having a resonance which falls in the spectral range within which stimulated emission is possible.

In our structure, the necessary population inversion is produced by injection of carriers from the degenerate n- and p-type end regions of the junction into the transition region. Condition (b) is the most difficult to fulfill, since the stimulated radiation propagates in the plane of the junction where the active region may be only a fraction of a wavelength in thickness and has a wave front which laterally extends many wavelengths into the passive n- and p-type regions bounding the junction plane. Laser action is favored by the large matrix element for band-to-band radiative recombination compared with that for free carrier absorption in GaAs and by the fact that the energy of the emitted radiation is below the absorption threshold of the degenerate material bounding the junction.

The diodes can be described approximately as cubes 0.4 mm on an edge, the junction lying in a horizontal plane through the center. Current is passed through the junction by means of Ohmic contacts attached to the top and bottom faces. The front and back faces are polished parallel to each other and perpendicular to the plane of the junction, in order to satisfy condition (c) above. Current is applied in the form of pulses of 5 to 20 μsec duration with the diode immersed in liquid nitrogen.

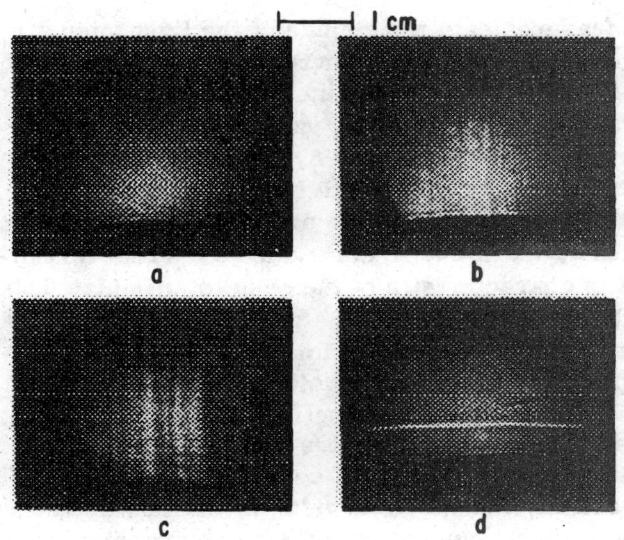

FIG. 1. Radiation patterns observed with image tube a distance d from junction. (a) and (b), diode L-69 below and above threshold, d = 6 cm. (c) Diode L-69 above threshold, d = 15 cm. (d) Diode L-75, d = 5 cm.

We have studied the radiation pattern by means of an infrared converter tube which provides an image that can be examined visually or photographically. Examples of such patterns are given in Fig. 1. The first two photographs show the pattern produced by a diode with the current below and above threshold, with the image converter 6 cm from the diode. Figure 1(c) shows the same diode above threshold with the screen 15 cm away, so that the radiation pattern is correspondingly enlarged. In these photographs the horizontal bands result from interference between the light produced at the junction and its image in the metal disc on which the diode was mounted, and their presence does not imply coherence of the source. On the other hand, the vertical bands which appear only in Figs. 1(b) and 1(c) are due to interference between the waves emitted at various points along the edge of the junction at the front face of the cube, and their appearance is evidence that a definite phase relation exists between the light emitted at various points along the junction, i.e., that the light is coherent. The angular separation between the interference maxima is consistent with the dimensions of the junction and the wavelength of the radiation. However, the relative intensities of the maxima produced by this and other similarly constructed diodes indicate that the radiation is not produced by the most elementary mode of this type of cavity.

Measurements were made of the light intensity from diode L-70 as a function of junction current at 8420 Å on the axis of the beam. Below 5000 A/cm² the light intensity varied linearly with current density. Near 8500 A/cm² the intensity increased very rapidly with current, reaching a value about ten times the extrapolated low-current intensity at 20 000 A/cm². Such a current threshold is characteristic of the onset of stimulated light emission, and it is significant that the azimuthal interference maxima of Fig. 1 make their appearance at this threshold.

Figure 2 shows the spectral distribution measured with the spectrometer located on the beam axis at several values of current density. Below threshold, the spectral width at half maximum is 125 Å, in agreement with the measurements of Keyes and Quist.[2] As the current is increased through threshold, the spectral width decreases suddenly to 15 Å, in a manner which again is characteristic of the onset of stimulated emission. If this 15Å width is due to a single resonant mode so that the emission is homogeneously broadened by the lifetime of this mode, then the oscillations persist only for a period of 1.6×10^{-12} sec or for a time long enough for the radiation to travel 0.1 mm, which is less than the length of the crystal. However, the broadening may instead be due to oscillations occurring at several unresolved modes during the pulse or to a variation in the wavelength of the resonant mode during a pulse. The instrumental resolution was about 3 Å.

Additional modes appear at longer wavelengths when the current is further increased beyond threshold, as illustrated in Fig. 3. The fact that the additional modes appear at lower energy suggests that the principal dissipative process becomes more effective at longer wavelengths, as would be the case for free carrier absorption. The separation of these additional modes is about three times that calculated on the assumption that each corresponds to one additional half wavelength between the plane surfaces, which were 0.32 mm apart in this diode.

Other junctions have been constructed which ex-

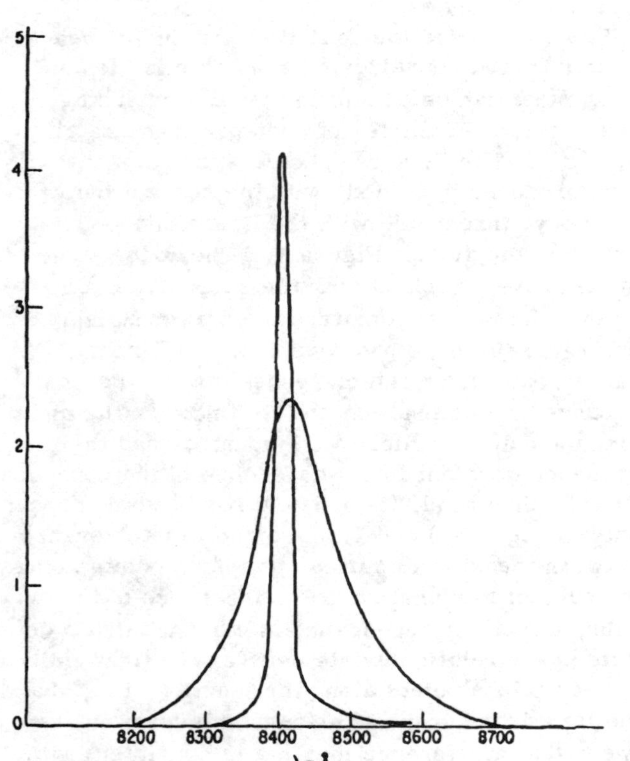

FIG. 2. Spectral distributions from diode L-69 below and above threshold. Different vertical scales.

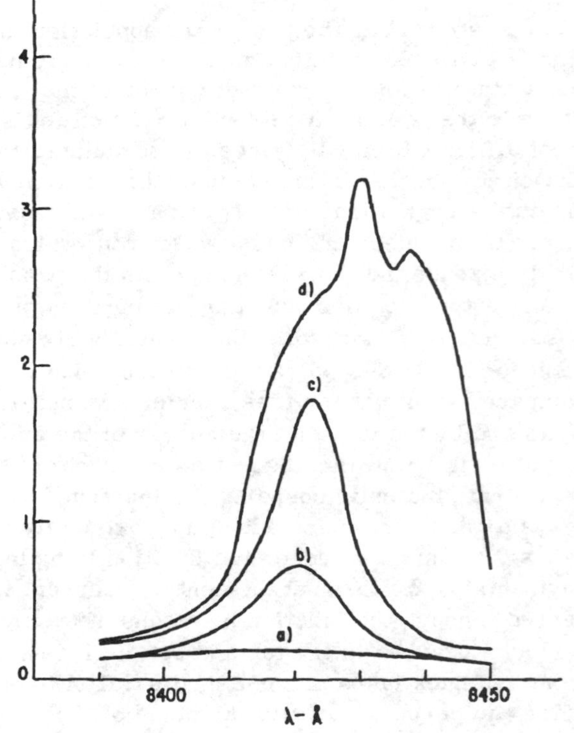

FIG. 3. Spectral distributions of diode L-70 at various currents, same vertical scale for all curves, arbitrary units. (a) 6000 A/cm², (b) 8600 A/cm², (c) 10 400 A/cm², (d) 20 000 A/cm².

hibit quite different characteristics. These produce a radiation pattern which is almost uniform in the azimuthal plane, but which is only a few tenths of a degree wide in the vertical plane, as shown in Fig. 1(d). This pattern implies that there is coherence over a distance of the order of 100 μ in the vertical direction but virtually no spatial coherence in the horizontal direction.

While stimulated emission has been observed in many systems, this is the first time that direct conversion of electrical energy to coherent infrared radiation has been achieved in a solid state device. It is also the first example of a laser involving transitions between energy bands rather than localized atomic levels.

---

[1]M. G. A. Bernard and G. Duraffourg, Physica Status Solidi 1, 699 (1961).

[2]R. J. Keyes and T. M. Quist, Proc. Inst. Radio Engrs. 50, 1822 (1962).

INTERPRETATION OF CO_2 OPTICAL MASER EXPERIMENTS

C. K. N. Patel
Bell Telephone Laboratories, Murray Hill, New Jersey
(Received 27 April 1964)

Optical maser action on a number of rotational transitions of the $\Sigma_u^+ - \Sigma_g^+$ vibrational band of CO_2 has been recently reported.[1] The maser lines were identified as the rotational transitions from $P(12)$ to $P(38)$ of the 0 0^01 - 1 0^00 band and from $P(22)$ to $P(34)$ of the 0 0^01 - 0 2^00 band. We wish to give here a simple theoretical treatment which allows us to interpret the results and especially the fact that no R-branch transitions were seen in maser oscillation. The treatment satisfactorily explains the results and leads to an interesting conclusion that for the vibrational-rotational transitions, optical maser action can be obtained on the P-branch transitions even when no inversion exists between the total population densities in the two vibrational states.

Figure 1 shows pertinent parts of the energy level diagram of CO_2 (Herzberg[2]). The rotational levels belonging to each of the vibrational states are not shown for the sake of simplicity. The upper maser level (for both the bands) $\Sigma_u^+(0\ 0^01)$ is optically connected to the ground state $\Sigma_g^+(0\ 0^00)$ of CO_2 through strongly allowed transitions at 2349.3 cm^{-1}. The lower laser levels $\Sigma_g^+(1\ 0^00$ and $0\ 2^00)$ both decay to the $\Pi_u(0\ 1^10)$ levels through radiative transitions at 720.5 and 618.1 cm^{-1}, respectively, and these transitions are reported to be of medium strength (reference 2). The molecules in the $\Pi_u(0\ 1^10)$ levels decay through strongly allowed transitions at 667.3 cm^{-1} to the ground state of CO_2. Thus the maser scheme looks like a four-level system. The probable excitation and decay processes are shown in Fig. 1 with their appropriate strengths as obtained from reference 2. Alternate lines in the rotational spectrum of $\Sigma_u^+ - \Sigma_g^+$ bands of CO_2 are missing because of symmetry considerations for the linear and symmetric molecule CO_2. Also, the Q-branch – i.e., $\Delta J = 0$ – transitions are forbidden since both the upper and the lower levels have $l = 0$.

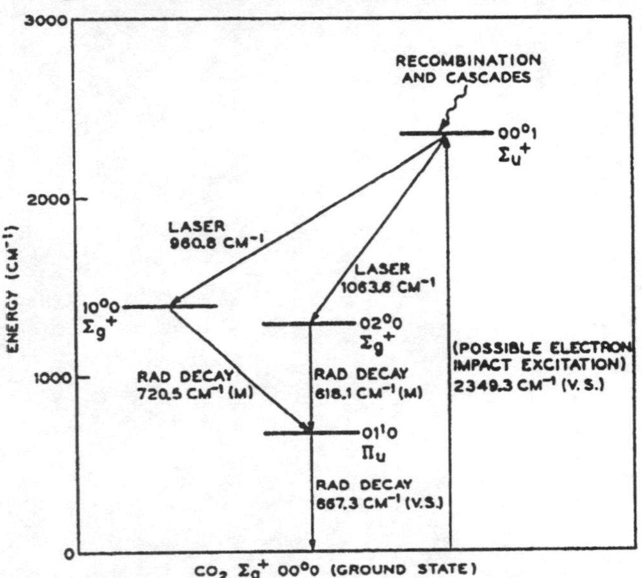

FIG. 1. Pertinent part of energy level diagram of CO_2 showing the maser transitions and other optical transitions with their respective strengths (reference 2).

Now consider a simplified model of a vibrational level in which the rotational level populations are described by a Boltzmann distribution at a temperature T. It can be shown[2] that for a linear and symmetric molecule like CO_2,

$$N_J \approx N(hcB/kT)g_J e^{-F(J)hc/kT} \quad \text{for } hcB/kT \ll 1, \quad (1)$$

where N_J is the population density of the Jth rotational level, $N = \sum_J N_J$, h = Planck's constant, c = velocity of light, B = rotational constant for the particular vibrational level, k = Boltzmann's constant, g_J = statistical weight for the Jth rotational level, and $F(J)$ = energy of Jth rotational level from the 0th rotational level. $F(J)$ is given by

$$F(J) = BJ(J+1) - DJ^2(J+1)^2,$$

with $D \ll B$.

Then the net optical gain coefficient for a rotational transition between vibrational levels 1 and 2 can be shown to be[3]

$$\alpha_{1_J 2_{J\pm 1}} = \left(\frac{\ln 2}{\pi}\right)^{1/2} \frac{16\pi^3 c^3}{3h\Delta\nu_D \lambda_{1_J 2_{J\pm 1}}} \left|\sum R^{1_J 2_{J\pm 1}}\right|^2 \times \left(\frac{N_{1_J}}{g_J} - \frac{N_{2_{J\pm 1}}}{g_{J\pm 1}}\right), \quad (2)$$

assuming that the transitions $1_J - 2_{J\pm 1}$ are primarily Doppler broadened, and where

$$\left|\sum R^{1_J 2_{J\pm 1}}\right|^2$$

is the matrix element[2] for the transition.

According to reference 2, $\left|\sum R^{1_J 2_{J\pm 1}}\right|^2$ can be split up into two parts, one which is dependent on J and the other which is independent of J; i.e.,

$$\left|\sum R^{1_J 2_{J\pm 1}}\right|^2 = K_{12} S_J, \quad (3)$$

where S_J = J-dependent part of the matrix element, and K_{12} is that part of the matrix element which does not depend on J. $S_J = J+1$ for the P branch, and $S_J = J$ for the R branch, where J is the rotational quantum number of the upper level.

Then substituting Eq. (1) in (2), with

$$\Delta\nu_D = \lambda_{1_J 2_{J\pm 1}}^{-2}[(2kT/M)\ln 2]^{1/2},$$

where M = molecular mass, we obtain (a) for P-branch transitions, i.e., for $P(J+1)$,

$$\alpha_{1_J 2_{J+1}} = \frac{8\pi^3 c^4 K_{12}}{3kT}\left(\frac{2\pi kT}{M}\right)^{1/2}(J+1) \times \left\{N_1 B_1 \exp\left[-F_1(J)\frac{hc}{kT}\right] - N_2 B_2 \exp\left[-F_2(J+1)\frac{hc}{kT}\right]\right\}, \quad (4)$$

and (b), for the R-branch, i.e., $R(J-1)$ transitions,

$$\alpha_{1_J 2_{J-1}} = \frac{8\pi^3 c^4 K_{12}}{3kT}\left(\frac{2\pi kT}{M}\right)^{1/2} \times J\left\{N_1 B_1 \exp\left[-F_1(J)\frac{hc}{kT}\right] - N_2 B_2 \exp\left[-F_2(J-1)\frac{hc}{kT}\right]\right\} \quad (5)$$

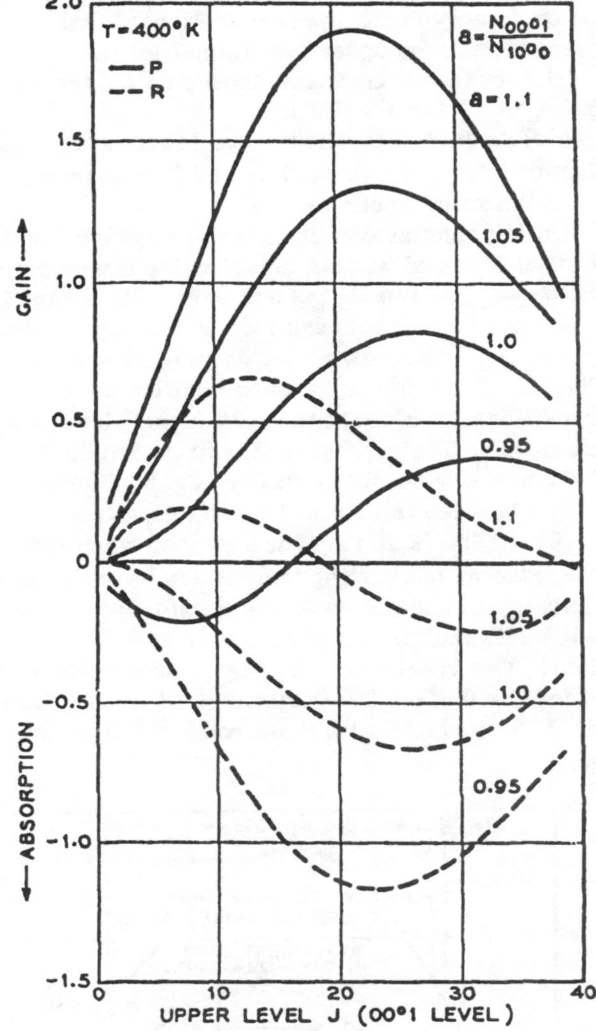

FIG. 2. Normalized gain as a function of upper-level J for P and R branches (for T = 400°K and $N_{0\,0^01}/N_{1\,0^00}$ = 0.95, 1, 1.05, and 1.1).

The gain coefficients calculated from Eqs. (4) and (5) for $0\,0^01$ as the upper level and $1\,0^00$ as the lower level are plotted in Fig. 2 as a function of upper level J. T = 400°K has been assumed and should be reasonable for a gas discharge. The curves are given for various values of $N_{0\,0^01}/N_{1\,0^00}$ and are arbitrarily normalized to

$$\frac{8\pi^3 c^4 K_{12}}{3kT(2\pi kT/M)^{1/2}} N_2.$$

($B_{0\,0^01}$ = 0.3866 cm^{-1}, $B_{1\,0^00}$ = 0.3897 cm^{-1}, and $B_{0\,2^00}$ = 0.3899 cm^{-1}).

From Fig. 2 the following conclusions can be reached immediately:

(a) P-branch transitions show optical gain even

when $N_{0\ 0^01}/N_{1\ 0^00} < 1$, i.e., when the total population density in the lower vibrational level exceeds that in the upper vibrational level.

(b) R-branch transitions show gain only when $N_1/N_2 > 1.02$ for $T = 400°K$.

(c) R-branch transitions have lower optical gain than that for the P-branch transitions starting from the same upper level J.

Similar conclusions can also be reached for the $0\ 0^01$-$0\ 0^02$ band without actual calculations since the $B_{1\ 0^00}$ and the $B_{0\ 2^00}$ are very nearly equal.

The agreement between theory and experiment on the CO_2 maser experiments may be seen in Fig. 3. The $0\ 0^01$-$1\ 0^00$ band oscillates on P transitions from $J = 11$ to $J = 37$ (upper-level J's are used). The strongest transition is that for $J = 23$ and is also shown in Fig. 3. The best fit as can be seen is obtained for $N_{0\ 0^01}/N_{1\ 0^00} = 1.05$. (The fit is regarded as good when the two extreme oscillating transitions have the same optical gain and the peak of the gain curve coincides with the strongest optical-maser transition.) The lower curve in Fig. 3 shows the best fit for the $0\ 0^01$-$0\ 2^00$ transitions which oscillate for $J = 21$ to $J = 33$ with the strongest transition occurring for $J = 27$. Here for best fit, $N_{0\ 0^01}/N_{0\ 2^00} = 1.0$ is required. Thus for the best fit, we have $T = 400°K$, $N_{0\ 0^01} = N_{0\ 2^00} = 1.05 N_{1\ 0^00}$. It should be noted here that absolute value of gain depends upon K_{12} which is not the same for the $0\ 0^01$-$1\ 0^00$ and $0\ 0^01$-$0\ 2^00$ bands. And hence the apparent difference between the oscillation threshold gain for $0\ 0^01$-$1\ 0^00$ and that for $0\ 0^01$-$0\ 2^00$ bands, as seen in Fig. 3, is not significant.

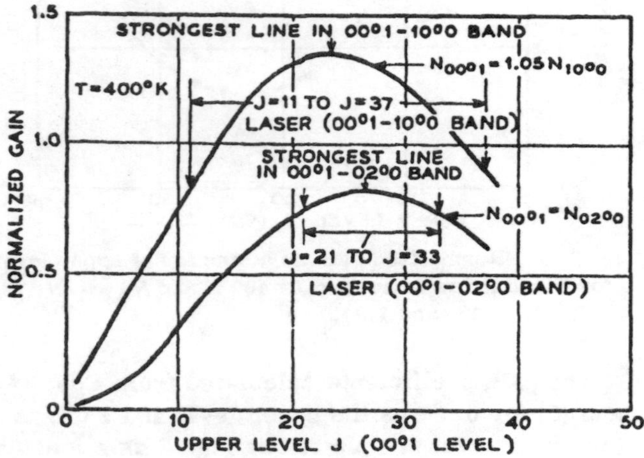

FIG. 3. Normalized gain as a function of upper level J for P branch (for $T = 400°K$ and $N_{0\ 0^01}/N_{1\ 0^00} = 1.05$, $N_{0\ 0^01}/N_{0\ 2^00} = 1$), together with observed laser transitions and strongest laser lines.

Going back to Fig. 2 it can be seen that for $N_{0\ 0^01}/N_{1\ 0^00} = 1$ (or also for $N_{0\ 0^01}/N_{0\ 2^00} = 1$), the R branch does not show optical gain and hence it is quite easy to understand why the R transitions in the $0\ 0^01$-$0\ 2^00$ band do not oscillate. For $N_{0\ 0^01}/N_{1\ 0^00} = 1.05$ we see that the R-branch transitions do show optical gain for low J values, but in all cases, the gain on R transition is lower than that for a P-branch transition starting from the same upper J level. Thus due to competition effects, the P transition will oscillate preferentially. Consequently the populations in 1_J and 2_{J+1} levels will equalize and this will cause a further reduction in the gain on the R transition (i.e., population inversion between 1_J and 2_{J-1} levels). Hence, it is not too surprising to find that R transitions have not been seen in maser oscillation for $0\ 0^01$-$1\ 0^00$ band also. (The last argument holds only in the case when there is no wavelength discriminating device present in the optical-maser cavity to differentiate between λ_R and λ_P. This was the case for the maser experiments reported in reference 1.)

We would like to thank Dr. C. G. B. Garrett, Dr. J. P. Gordon, and Dr. P. K. Tien for critical comments and suggestions on the manuscript.

---

[1]C. K. N. Patel, W. L. Faust, and R. A. McFarlane, Bull. Am. Phys. Soc. 9, 500 (1964).

[2]G. Herzberg, Molecular Spectra and Molecular Structure II, Infrared and Raman Spectra of Polyatomic Molecules (D. Van Nostrand and Company, Inc., Princeton, New Jersey, 1945).

[3]A. C. G. Mitchell and M. W. Zemansky, Resonance Radiation and Excited Atoms (Cambridge University Press, New York, 1961).

Chapter 14

QUANTUM MECHANICS

SHELDON GOLDSTEIN

JOEL L. LEBOWITZ

Introduction

Papers Reprinted in Book

David Bohm (1917-1992). (Courtesy of Still Pictures, London.)

Quantum mechanics is undoubtedly the most successful theory yet devised by the human mind. Not one of the multitude of its calculated predictions has ever been found wanting, even in the last measured decimal place—nor is there any reason to believe that this will change in the foreseeable future. All the same, it is a bizarre theory. Let us quote Feynman,[1] one of the deepest scientist–thinkers of our century and one not known for his intellectual (or any other) modesty, on the subject: "There was a time when the newspapers said that only twelve men understood the theory of relativity. I do not believe there ever was such a time. There might have been a time when only one man did, because he was the only guy who caught on, before he wrote his

Sheldon Goldstein is a Professor of Mathematics at Rutgers University in New Brunswick, New Jersey. He worked for many years on probability theory and the rigorous foundations of statistical mechanics. In particular, he has investigated the ergodic properties of large systems, the existence of steady-state nonequilibrium ensembles, derivations of Brownian motion for interacting particles and of diffusion and subdiffusion limits for random motions in random environments. In recent years he has been concerned with the foundations of quantum mechanics.

Joel L. Lebowitz—see biographical note in Chapter 6.

paper. But after people read the paper a lot of people understood the theory of relativity in some way or other, certainly more than twelve. On the other hand, I think I can safely say that nobody understands quantum mechanics. ... I am going to tell you what nature behaves like. If you will simply admit that maybe she does behave like this, you will find her a delightful, entrancing thing. Do not keep saying to yourself, if you can possibly avoid it, 'but how can it be like that?' because you will get 'down the drain,' into a blind alley from which nobody has yet escaped. Nobody knows how it can be like that."

Feynman's point of view, expressed as usual with great vigor and clarity, characterizes the attitude of most physicists towards the foundations of quantum mechanics, a subject concerned with the meaning and interpretation of quantum theory—at least it did so before the work of Bell[2] and the experiments of Aspect *et al.*[3] (which came after the cited Feynman lecture). Even today, the subject is treated like a poor stepchild of the physics family. It is pretty much ignored in most standard graduate texts, and what is conveyed there is often so mired in misconception and confusion that it usually does more harm than good. To many physicists, it appears that not only does foundational research not lead to genuine scientific progress, but that it is in fact dangerous, with the potential for getting people "down the drain." Even to the more tolerant ones, it often seems that what is achieved merely supports what every good physicist should have known already.

While these perceptions are partly true, they are also in part misapprehensions arising from the, to our taste, much too practical approach taken by many physicists. Basic questions concerning the physical meaning of quantities such as the wave function which we manipulate in our computations are too important to be left to philosophers. One such question, whether the description of a physical system provided by its wave function is complete, is central to the articles reprinted in this chapter.

Albert Einstein (1879-1955) and **Niels Bohr (1885-1962).** (Courtesy of Ehrenfest Collection, AIP Emilio Segrè Visual Archives.)

Einstein, Podolsky, and Rosen (EPR)[4] argue that quantum mechanics provides at best an incomplete description of physical reality. Indeed, they claim that there are situations in which the very predictions of quantum theory demand that there be elements of physical reality—i.e., predetermined, preexisting values for physical quantities, which are *revealed* rather than *created* if and when we measure those quantities—that are not incorporated within the orthodox quantum framework. In the original version of the argument, these elements of reality are the (simultaneous) values of the position and momentum of a particle belonging to an EPR pair—a pair of particles whose quantum state, given by the EPR wave function, involves such strong quantum pair correlations that the position or momentum of one of the particles can be inferred from the measurement of that of the other. By the uncertainty principle, however, the position and momentum of one particle cannot simultaneously be part of the

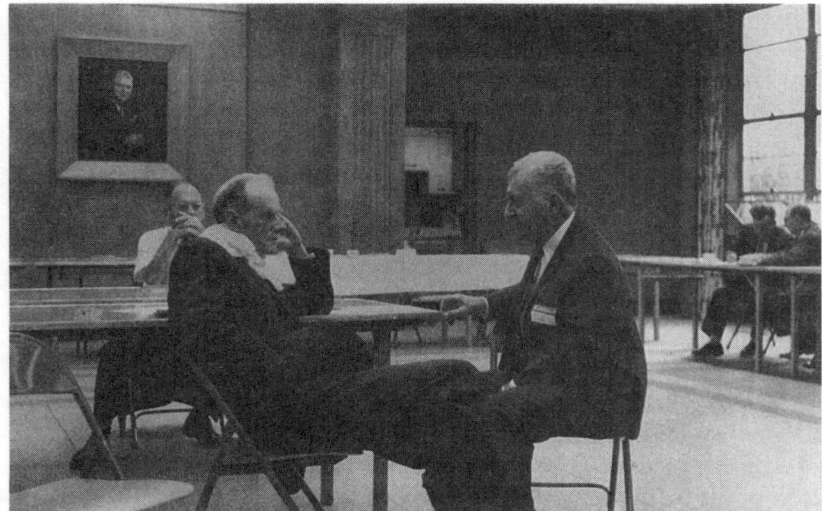

P. A. M. Dirac (1902-1984) and **Boris Podolsky (1896-1966)** in a corridor session, January 1964 [in background: **Donald A. Glaser (1926-), Oldweg Von Roos (1925-),** and **E. P. Wigner (1902-1995)**]. (Courtesy of *Physics Today* Collection, AIP Emilio Segrè Visual Archives.)

quantum description. In the later version of the EPR analysis due to Bohm,[5] which provides the framework for most of the experimental tests of quantum theory that were stimulated by the celebrated Bell's inequality paper,[2] these elements of reality are the values of the (simultaneous) components, in all possible directions, of the spins of the particles belonging to a Bohm-EPR pair—a pair of spin 1/2 particles prepared in the singlet $S = 0$ state—or, in another version, the simultaneous components of photon polarization in a suitable photon pair. We shall call these the Bohm-EPR elements of reality. (They again cannot simultaneously be part of the quantum description, because spin components in different directions do not commute.)

The EPR analysis begins with a criterion of reality: *"If, without in any way disturbing a system, we can predict with certainty ... the value of a physical quantity, then there exists an element of physical reality corresponding to this physical quantity."* EPR continue, "It seems to us that this criterion, while far from exhausting all possible ways of recognizing a physical reality, at least provides us with one such way ... . Regarded not as a necessary, but merely as a sufficient, condition of reality, this criterion is in agreement with classical as well as quantum-mechanical ideas of reality." They then deduce the existence of the relevant elements of reality for an EPR pair from the predictions of quantum theory for the pair. In so doing, however, they crucially require a locality assumption that "the process of measurement carried out on the first system ... does not disturb the second system in any way." EPR conclude as follows: "While we have thus shown that the wave function does not provide a complete description of the physical reality, we left open the question of whether or not such a description exists. We believe, however, that such a theory is possible."

We wish to emphasize that in arguing here for the incompleteness of the quantum description, EPR were not questioning the

Max Born (1882–1970). (Courtesy of AIP Emilio Segrè Visual Archives.)

Erwin Schrödinger (1887–1961). (Courtesy of Weber Collection, AIP Emilio Segrè Visual Archives.)

Eugene P. Wigner (1902–1995). Photograph of charcoal drawing by Peter Geoffrey Cook. (Courtesy of Weber Collection, AIP Emilio Segrè Visual Archives.)

validity of the experimental predictions of quantum theory. On the contrary, they were claiming that these predictions were not only compatible with a more complete description—in particular, one involving their elements of reality—but also demanded one. Elsewhere, Einstein[6] asserts that in "a complete physical description, the statistical quantum theory would...take an approximately analogous position to the statistical mechanics within the framework of classical mechanics."

Niels Bohr,[7] in what is perhaps the definitive statement of his principle of complementarity, disagreed with the EPR conclusion, though he did not take the EPR analysis lightly. The central objection in Bohr's reply is that the EPR reality criterion "contains an ambiguity as regards the meaning of the expression 'without in any way disturbing a system.' Of course, there is...no question of a mechanical disturbance... . But...there is essentially the question of an influence on the very conditions which...constitute an inherent element of the description of any phenomenon to which the term 'physical reality' can be properly attached... ." While, with Bell,[8] we "have very little idea what this means," it does perhaps suggest "the feature of wholeness typical of proper quantum phenomena" elsewhere stressed by Bohr.[9]

Bohm,[10, 11] on the other hand, not only agreed with EPR that the quantum description is incomplete, but showed explicitly how to extend the incomplete quantum description—by the introduction of "hidden variables"—into a complete one, in such a way that the indeterminism of quantum theory is completely eliminated. We shall call Bohm's deterministic completion of nonrelativistic quantum theory Bohmian mechanics. In Bohmian mechanics the hidden variables are simply the positions of the particles, which move, under an evolution governed by the wave function, in what is in effect the simplest possible manner.[12] We should emphasize that Bohmian mechanics is indeed an extension of quantum theory, in the sense that in this theory, as in quantum theory, the wave function evolves autonomously according to Schrödinger's equation. Moreover, it can be shown[12] that the statistical description in quantum theory, given by $\rho = |\psi|^2$, indeed takes, as Einstein wanted, "an approximately analogous position to the statistical mechanics within the framework of classical mechanics."

Bohmian mechanics was ignored by most physicists, but it was taken very seriously by Bell, who declared[13] that "in 1952 I saw the impossible done." Bell quite naturally asked how Bohm had managed to do what von Neumann[14] had proclaimed to be—and almost all authorities agreed was—impossible. (It is perhaps worth noting that despite the almost universal acceptance among physicists of the soundness of von Neumann's proof of the impossibility of hidden variables, undoubtedly based in part on von Neumann's well-deserved reputation as one of the greatest mathematicians of the 20th century, Bell[15] felt that the assumptions made by von Neumann about the requirements for a hidden-variable theory are so unreasonable that "the proof of von Neumann is not merely false but *foolish!*" See also Ref. 16.) His ensuing hidden-variables analysis led to Bell's inequality, which must be satisfied by certain correlations between Bohm-EPR elements of reality—and, of course, by correlations between their measured values. He observed also that quantum theory predicts a sharp violation of the inequality when the quantities in question are measured.

Thus the specific elements of reality to which the EPR analysis would lead (if applied to the Bohm-EPR version) must satisfy correlations that are incompatible with those given by quantum theory. That is, these elements of reality, whatever else they may be, are demonstrably incompatible with the predictions of quantum theory and hence are certainly not part of any completion of it. It follows that there is definitely something wrong with the EPR analysis, since quantum mechanics cannot be (even partially) completed in the manner demanded by this analysis. In other words, had EPR been aware of the work of Bell, they might well have predicted that quantum theory is wrong and proposed an experimental test of Bell's inequality to settle the issue once and for all.

Of course, EPR were not aware of Bell's analysis, but Clauser, Horne, Shimony, and Holt were.[17] Their proposal for an experimental test has led to an enormous proliferation of experiments, the most conclusive of which was perhaps that of Aspect *et al.*[3] included here. The result: Quantum mechanics is right.

We note, however, that the predictions of (nonrelativistic) quantum mechanics—in particular, those for the experimental tests of Bell's inequality—are in complete agreement with the predictions of Bohmian mechanics. Thus the Bohm-EPR elements of reality are not part of Bohmian mechanics! This is because in Bohmian mechanics the result of what we speak of as measuring

Abner E. Shimony (1928–). (Courtesy of Abner E. Shimony, Boston University.)

Yves A. Rocard (1903–1992), Louis duc de Broglie (1892–1987), Maurice duc de Broglie (1875–1960), and Francis Perrin (1901–). (Courtesy of AIP Emilio Segrè Visual Archives.)

John Bell (1928–1990). (Courtesy of Tatiana Fabergé and M. N. Fontaine, CERN.)

Alain Aspect (1947–). (Courtesy of Philip Pearle, Hamilton College.)

a spin component depends as much upon the detailed experimental arrangement for performing the measurement as it does upon anything existing prior to and independent of the measurement. This dependence is an example of what the experts in the hidden-variables field call contextuality[18] (see also Ref. 15), i.e., of the critical importance of not overlooking "the interaction with the measuring instruments which serve to define the conditions under which the phenomena appear."[19]

In fact, just before he arrived at his inequality, Bell noticed that "in this (Bohm's) theory an explicit causal mechanism exists whereby the disposition of one piece of apparatus affects the results obtained with a distant piece. ... Bohm of course was well aware of these features of his scheme, and has given them much attention. However, it must be stressed that, to the present writer's knowledge, there is no *proof* that *any* hidden variable account of quantum mechanics *must* have this extraordinary character. It would therefore be interesting, perhaps, to pursue some further 'impossibility proofs,' replacing the arbitrary axioms objected to above by some condition of locality, or of separability of distant systems."[18] Almost immediately, Bell found his inequality. Thus did Bohmian mechanics lead to Bell's refutation of the EPR claim to have "shown that the wave function does not provide a complete description." At the same time it showed, by explicit example, the correctness of the EPR belief "that such a theory is possible"!

While Bell's analysis, together with the results of experiments such as Aspect's, implies that the EPR analysis was faulty, where in fact did EPR go wrong? Since their only genuine assumption was that of locality quoted above, and since their subsequent reasoning is valid, it is this assumption that must fail, both for quantum theory and for nature herself. Aspect's experiment thus establishes perhaps the most striking implication of quantum theory: Nature is nonlocal! This conclusion is of course implicit in the very structure of quantum theory itself, based as it is on a field—the wave function—which for a many-body system lives not on physical space but on a $3n$-dimensional configuration space, a structure that allows for the entanglement of states of distant systems—as most dramatically realized in the EPR state itself. But while quantum mechanics may someday be replaced by a theory of an entirely different character, we may nonetheless conclude—though there are some who disagree[15]—from Bell and Aspect that the nonlocality it implies is here to stay.

One of the great foundational mysteries that remains very much unsolved is how nonlocality can be rendered compatible with special relativity, i.e., with Lorentz invariance. Here Bohmian mechanics is of no direct help, since it manifestly and fundamentally is not Lorentz invariant. But there is no reason to believe that a more appropriate completion of quantum theory, one that is Lorentz invariant and perhaps even generally covariant, cannot be found. However, one should not expect finding it to be easy.

One lesson of this story is perhaps that we would be wise to place greater trust in the mathematical structure of quantum theory, and less in the philosophy with which quantum theory is so often encumbered. For the EPR problem, the mathematical structure correctly suggests nonlocality, while the philosophy makes the questionable demand that the wave function provide a complete description, at least on the microscopic level. The paper by Aharonov and Bohm[20] included here supports this lesson. Aharonov and Bohm dramatically demonstrate that the electromagnetic vector potential has a reality in quantum theory far beyond what it has classically: A nonvanishing vector potential may generate a shift in an interference pattern for an electron confined to a region in which the magnetic field itself vanishes. The Aharonov-Bohm effect, while rather clear from the role played by the vector potential in Schrödinger's equation, is rather surprising from the perspective of the usual quantum philosophy, which, in attempting to explain quantum deviations from classical behavior, appeals to limitations on what can be measured or known arising from disturbances occurring during the act of measurement that are due to the finiteness of the quantum of action.

Philip Pearle (1936–) and **Yakir Aharonov (1932-).** (Courtesy of Philip Pearle, Hamilton College.)

It is appropriate to mention at this time—even though it is not the focus of any of the five papers included in this chapter—one of the strongest arguments for the conclusion that the quantum mechanical description is incomplete: the notorious measurement problem—or, what amounts to the same thing, the paradox of Schrödinger's cat. The problem is that the after-measurement wave function for system and apparatus arising from Schrödinger's equation for the composite system typically involves a superposition over terms corresponding to what we would like to regard as the various possible results of the measurement—e.g., different pointer orientations. Since it seems rather important that the actual result of the measurement be a part of the description of the after-measurement situation, it is difficult to see how this wave function could be the complete description of this situation. By contrast, with a theory or interpretation in which the description of the after-measurement situation includes, in addition to the wave function, at least the values of the variables that register the result, the measurement problem vanishes. (The remaining problem of then justifying the use of the "collapsed" wave function—corresponding to the actual result—in place of the original one is often confused with the measurement problem. The justification for this replacement is nowadays frequently expressed in terms of decoherence. One of the best descriptions of the mechanisms of decoherence, though not the word itself, can be found in the Bohm article reprinted here; see also Ref. 5. We wish to emphasize, however, as did Bell in his article "Against Measure-

Murray Gell-Mann (1929–) and Richard P. Feynman (1918–1988). (Courtesy of Marshak Collection, AIP Emilio Segrè Visual Archives.)

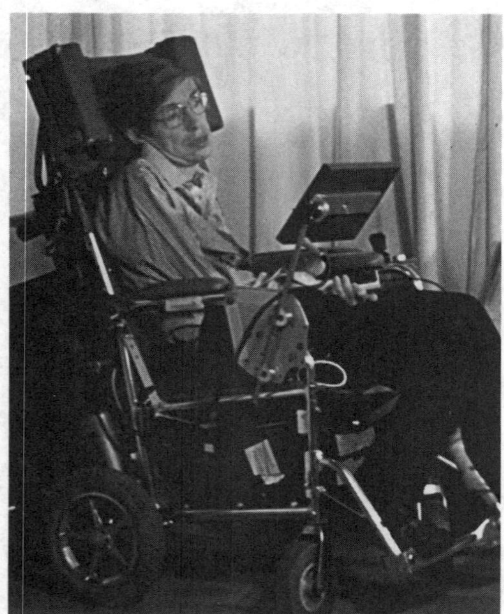

Stephen Hawking (1942–). (Courtesy of Ann Benedeuce.)

ment,"[21] that decoherence *per se* in no way comes to grips with the measurement problem itself.)

The orthodox response to the measurement problem is that we must distinguish between closed systems and open systems—those upon which an external "observer" intervenes. While we do not want to delve into the merits of this response here—nor is this the place to discuss the sundry proposals for alternate interpretations of quantum theory, such as those of Schulman,[22] Pearle,[23, 24] and of Ghirardi, Rimini, and Weber[25, 26] —we do wish to note one particular difficulty, much emphasized of late. This concerns the now-popular subject of quantum cosmology, concerned with the physics of the universe as a whole, certainly a closed system! A formulation of quantum mechanics that makes sense for closed systems seems to be demanded. Bohmian mechanics is one such formulation. Others also now generating a good deal of excitement are due to Griffiths,[27] Omnès,[28] and Gell-Mann and Hartle.[29] All of these exemplify the EPR conclusion "that the wave function does not provide a complete description of the physical reality."

REFERENCES

1. R. Feynman, *The Character of Physical Law* (MIT, Cambridge, 1992), p. 129.
2. J. S. Bell, *On the Einstein Podolsky Rosen Paradox*, Physics **1**, 195–200 (1964) (reprinted in Ref. 8).
3. A. Aspect, J. Dalibard, and G. Roger, *Experimental Test of Bell's Inequalities Using Time-Varying Analyzers*, Phys. Rev. Lett. **49**, 1804–1807 (1982).
4. A. Einstein, B. Podolsky, and N. Rosen, *Can Quantum-Mechanical Description of Physical Reality be Considered Complete?*, Phys. Rev. **47**, 777–780 (1935).
5. D. Bohm, *Quantum Theory* (Prentice-Hall, Englewood Cliffs, NJ, 1951).
6. *Albert Einstein, Philosopher–Scientist*, edited by P.A. Schilpp (Library of Living Philosophers, Evanston, IL, 1949), p. 672.
7. N. Bohr, *Can Quantum-Mechanical Description of Physical Reality be Considered Complete?*, Phys. Rev. **48**, 696–702 (1935).
8. J.S. Bell, *Speakable and Unspeakable in Quantum Mechanics* (Cambridge University, Cambridge, 1987), p. 155.
9. N. Bohr, *Quantum Physics and Philosophy*, Essays on Atomic Physics and Human Knowledge (Wiley, New York, 1963).
10. D. Bohm, *A Suggested Interpretation of the Quantum Theory in Terms of "Hidden" Variables, I*, Phys. Rev. **85**, 166–179 (1952).
11. D. Bohm, *A Suggested Interpretation of the Quantum Theory in Terms of "Hidden" Variables, II*, Phys. Rev. **85**, 180–193 (1952).
12. D. Dürr, S. Goldstein, and N. Zanghi, *Quantum Equilibrium and the Origin of Absolute Uncertainty*, J. Stat. Phys. **67**, 843–907 (1992); *Quantum Mechanics, Randomness, and Deterministic Reality*, Phys. Lett. A **172**, 6–12 (1992).
13. J.S. Bell, *On the Impossible Pilot Wave*, Found. Phys. **12**, 989–999 (1982).
14. J. von Neumann, *Mathematische Grundlagen der Quantenmechanik* (Springer, Berlin, 1932); English translation by R.T. Beyer, *Mathematical Foundations of Quantum Mechanics* (Princeton University, Princeton, NJ, 1955), 324–325.
15. N.D. Mermin, *Hidden Variables and the Two Theorems of John Bell*, Rev. Mod. Phys. **65**, 803–815 (1993).
16. T.J. Pinch, *What Does a Proof Do if it Does Not Prove?, A Study of the Social Conditions and Metaphysical Divisions Leading to David Bohm and John von*

Neumann Failing to Communicate in Quantum Physics, The Social Production of Scientific Knowledge, edited by E. Mendelsohn, P. Weingart, and R. Whitly (Reidel, Boston, 1977), pp. 171–215.

17. J.F. Clauser, M.A. Horne, A. Shimony, and R.A. Holt, *Proposed Experiment to Test Local Hidden-Variable Theories*, Phys. Rev. Lett. **23**, 880–884 (1969).
18. J.S. Bell, *On the Problem of Hidden Variables in Quantum Mechanics*, Rev. Mod. Phys. **38**, 447–452 (1966).
19. N. Bohr, *Discussion with Einstein on Epistemological Problems in Atomic Physics*, in Ref. 6, pp. 199–244.
20. Y. Aharonov and D. Bohm, *Significance of Electromagnetic Potentials in the Quantum Theory*, Phys. Rev. **115**, 485–491 (1959).
21. J.S. Bell, *Against "Measurement"*, Phys. World **3**, 33–40 (1990).
22. L.S. Schulman, *Deterministic Quantum Evolution Through Modification of the Hypotheses of Statistical Mechanics*, J. Stat. Phys. **42**, 689 (1986).
23. P. Pearle, *Reduction of the State-Vector by a Nonlinear Schrödinger Equation*, Phys. Rev. D **13**, 857–868 (1976).
24. P. Pearle, *Combining Stochastic Dynamical Statevector Reduction with Spontaneous Localisation*, Phys. Rev. A **39**, 2277–2289 (1989).
25. G.C. Ghirardi, A. Rimini, and T. Weber, *Unified Dynamics for Microscopic and Macroscopic Systems*, Phys. Rev. D **34**, 470–491 (1986).
26. J.S. Bell, *Are There Quantum Jumps?*, in Ref. 8.
27. R.B. Griffiths, *Consistent Histories and the Interpretation of Quantum Mechanics*, J. Stat. Phys. **36**, 219–272 (1984); *A Consistent Interpretation of Quantum Mechanics Using Quantum Trajectories*, Phys. Rev. Lett. **70**, 2201 (1993).
28. R. Omnès, *Logical Reformulation of Quantum Mechanics I*, J. Stat. Phys. **53**, 893–932 (1988).
29. M. Gell-Mann and J.B. Hartle, *Quantum Mechanics in the Light of Quantum Cosmology*, Complexity, Entropy, and the Physics of Information, edited by W. Zurek (Addison-Wesley, Reading, 1990), pp. 425–458; *Alternative Decohering Histories in Quantum Mechanics*, Proceedings of the 25th International Conference on High Energy Physics, Singapore, 1990, edited by K.K. Phua and Y. Yamaguchi (World Scientific, Singapore, 1991); *Classical Equations for Quantum Systems*, Phys. Rev. D **47**, 3345–3382 (1993).

N. David Mermin (1935–) and **Victor F. Weisskopf (1908–).** (Courtesy of Abner E. Shimony, Boston University.)

Niels Bohr (1885–1962) and **Max Planck (1858–1947).** (Courtesy of Margrethe Bohr Collection, AIP Emilio Segrè Visual Archives.)

QUANTUM MECHANICS

Papers Reproduced on CD-ROM

E. Schrödinger. An undulatory theory of the mechanics of atoms and molecules, *Phys. Rev.* **28**, 1049–1070 (1926)

H. P. Robertson. The uncertainty principle, *Phys. Rev.* **34**, 163–164(L) (1929)

A. Einstein, R. C. Tolman, and B. Podolsky. Knowledge of past and future in quantum mechanics, *Phys. Rev.* **37**, 780–781(L) (1931)

E. Wigner. On the quantum correction for thermodynamic equilibrium, *Phys. Rev.* **40**, 749–759 (1932)

A. Einstein, B. Podolsky, and N. Rosen. Can quantum-mechanical description of physical reality be considered complete?, *Phys. Rev.* **47**, 777–780 (1935)

N. Bohr. Can quantum-mechanical description of physical reality be considered complete?, *Phys. Rev.* **48**, 696–702 (1935)

W. H. Furry. Note on the quantum-mechanical theory of measurement, *Phys. Rev.* **49**, 393–399 (1936)

W. Pauli. The connection between spin and statistics, *Phys. Rev.* **58**, 716–722 (1940)

N. Bohr and L. Rosenfeld. Field and charge measurements in quantum electrodynamics, *Phys. Rev.* **78**, 794–798 (1950)

D. Bohm. A suggested interpretation of the quantum theory in terms of "hidden" variables. I, *Phys. Rev.* **85**, 166–179 (1952)

D. Bohm. A suggested interpretation of the quantum theory in terms of "hidden" variables. II, *Phys. Rev.* **85**, 180–193 (1952)

G. C. Wick, A. S. Wightman, and E. P. Wigner. The intrinsic parity of elementary particles, *Phys. Rev.* **88**, 101–105 (1952)

D. Bohm. Proof that probability density approaches $|\psi|^2$ in causal interpretation of the quantum theory, *Phys. Rev.* **89**, 458–466 (1953)

A. Siegel and N. Wiener. "Theory of measurement" in differential-space quantum theory, *Phys. Rev.* **101**, 429–432 (1956)

D. Bohm and Y. Aharonov. Discussion of experimental proof for the paradox of Einstein, Rosen, and Podolsky, *Phys. Rev.* **108**, 1070–1076 (1957)

H. Salecker and E. P. Wigner. Quantum limitations of the measurement of space-time distances, *Phys. Rev.* **109**, 571–577 (1958)

Y. Aharonov and D. Bohm. Significance of electromagnetic potentials in the quantum theory, *Phys. Rev.* **115**, 485–491 (1959)

H. Araki and M. M. Yanase. Measurement of quantum mechanical operators, *Phys. Rev.* **120**, 622–626 (1960)

Y. Aharonov and D. Bohm. Time in the quantum theory and the uncertainty relation for time and energy, *Phys. Rev.* **122**, 1649–1658 (1961)

Y. Aharonov, P. G. Bergmann, and J. L. Lebowitz. Time symmetry in the quantum process of measurement, *Phys. Rev.* **134**, B1410–B1416 (1964)

E. Nelson. Derivation of the Schrödinger equation from Newtonian mechanics, *Phys. Rev.* **150**, 1079–1085 (1966)

J. F. Clauser, M. A. Horne, A. Shimony, and R. A. Holt. Proposed experiment to test local hidden-variable theories, *Phys. Rev. Lett.* **23**, 880–884 (1969)

S. J. Freedman and J. F. Clauser. Experimental test of local hidden-variable theories, *Phys. Rev. Lett.* **28**, 938–941 (1972)

S. W. Hawking. Black holes and thermodynamics, *Phys. Rev. D* **13**, 191–197 (1976)

P. Pearle. Reduction of the state vector by a nonlinear Schrödinger equation, *Phys. Rev. D* **13**, 857–868 (1976)

M. Lamehi-Rachti and W. Mittig. Quantum mechanics and hidden variables: A test of Bell's inequality by the measurement of the spin correlation in low-energy proton–proton scattering, *Phys. Rev. D* **14**, 2543–2555 (1976)

Y. Aharonov and D. Z. Albert. States and observables in relativistic quantum field theories, *Phys. Rev. D* **21**, 3316–3324 (1980)

A. Peres. Can we undo quantum measurements?, *Phys. Rev. D* **22**, 879–883 (1980)

Y. Aharonov and D. Z. Albert. Can we make sense out of the measurement process in relativistic quantum mechanics?, *Phys. Rev. D* **24**, 359–370 (1981)

W. H. Zurek. Pointer basis of quantum apparatus: Into what mixture does the wave packet collapse?, *Phys. Rev. D* **24**, 1516–1525 (1981)

A. O. Caldeira and A. J. Leggett. Influence of dissipation on quantum tunneling in macroscopic systems, *Phys. Rev. Lett.* **46**, 211–214 (1981)

W. G. Unruh and R. M. Wald. Acceleration radiation and the generalized second law of thermodynamics, *Phys. Rev. D* **25**, 942–958 (1982)

A. Aspect, P. Grangier, and G. Roger. Experimental realization of Einstein-Podolsky-Rosen-Bohm *Gedankenexperiment:* A new violation of Bell's inequalities, *Phys. Rev. Lett.* **49**, 91–94 (1982)

A. Aspect, J. Dalibard, and G. Roger. Experimental test of Bell's inequalities using time-varying analyzers, *Phys. Rev. Lett.* **49**, 1804–1807 (1982)

MAY 15, 1935 PHYSICAL REVIEW VOLUME 47

Can Quantum-Mechanical Description of Physical Reality Be Considered Complete?

A. EINSTEIN, B. PODOLSKY AND N. ROSEN, *Institute for Advanced Study, Princeton, New Jersey*

(Received March 25, 1935)

In a complete theory there is an element corresponding to each element of reality. A sufficient condition for the reality of a physical quantity is the possibility of predicting it with certainty, without disturbing the system. In quantum mechanics in the case of two physical quantities described by non-commuting operators, the knowledge of one precludes the knowledge of the other. Then either (1) the description of reality given by the wave function in quantum mechanics is not complete or (2) these two quantities cannot have simultaneous reality. Consideration of the problem of making predictions concerning a system on the basis of measurements made on another system that had previously interacted with it leads to the result that if (1) is false then (2) is also false. One is thus led to conclude that the description of reality as given by a wave function is not complete.

1.

ANY serious consideration of a physical theory must take into account the distinction between the objective reality, which is independent of any theory, and the physical concepts with which the theory operates. These concepts are intended to correspond with the objective reality, and by means of these concepts we picture this reality to ourselves.

In attempting to judge the success of a physical theory, we may ask ourselves two questions: (1) "Is the theory correct?" and (2) "Is the description given by the theory complete?" It is only in the case in which positive answers may be given to both of these questions, that the concepts of the theory may be said to be satisfactory. The correctness of the theory is judged by the degree of agreement between the conclusions of the theory and human experience. This experience, which alone enables us to make inferences about reality, in physics takes the form of experiment and measurement. It is the second question that we wish to consider here, as applied to quantum mechanics.

Whatever the meaning assigned to the term *complete*, the following requirement for a complete theory seems to be a necessary one: *every element of the physical reality must have a counterpart in the physical theory.* We shall call this the condition of completeness. The second question is thus easily answered, as soon as we are able to decide what are the elements of the physical reality.

The elements of the physical reality cannot be determined by *a priori* philosophical considerations, but must be found by an appeal to results of experiments and measurements. A comprehensive definition of reality is, however, unnecessary for our purpose. We shall be satisfied with the following criterion, which we regard as reasonable. *If, without in any way disturbing a system, we can predict with certainty (i.e., with probability equal to unity) the value of a physical quantity, then there exists an element of physical reality corresponding to this physical quantity.* It seems to us that this criterion, while far from exhausting all possible ways of recognizing a physical reality, at least provides us with one such way, whenever the conditions set down in it occur. Regarded not as a necessary, but merely as a sufficient, condition of reality, this criterion is in agreement with classical as well as quantum-mechanical ideas of reality.

To illustrate the ideas involved let us consider the quantum-mechanical description of the behavior of a particle having a single degree of freedom. The fundamental concept of the theory is the concept of *state*, which is supposed to be completely characterized by the wave function ψ, which is a function of the variables chosen to describe the particle's behavior. Corresponding to each physically observable quantity A there is an operator, which may be designated by the same letter.

If ψ is an eigenfunction of the operator A, that is, if

$$\psi' \equiv A\psi = a\psi, \tag{1}$$

where a is a number, then the physical quantity A has with certainty the value a whenever the particle is in the state given by ψ. In accordance with our criterion of reality, for a particle in the state given by ψ for which Eq. (1) holds, there is an element of physical reality corresponding to the physical quantity A. Let, for example,

$$\psi = e^{(2\pi i/h)p_0 x}, \tag{2}$$

where h is Planck's constant, p_0 is some constant number, and x the independent variable. Since the operator corresponding to the momentum of the particle is

$$p = (h/2\pi i)\partial/\partial x, \tag{3}$$

we obtain

$$\psi' = p\psi = (h/2\pi i)\partial\psi/\partial x = p_0\psi. \tag{4}$$

Thus, in the state given by Eq. (2), the momentum has certainly the value p_0. It thus has meaning to say that the momentum of the particle in the state given by Eq. (2) is real.

On the other hand if Eq. (1) does not hold, we can no longer speak of the physical quantity A having a particular value. This is the case, for example, with the coordinate of the particle. The operator corresponding to it, say q, is the operator of multiplication by the independent variable. Thus,

$$q\psi = x\psi \neq a\psi. \tag{5}$$

In accordance with quantum mechanics we can only say that the relative probability that a measurement of the coordinate will give a result lying between a and b is

$$P(a, b) = \int_a^b \bar{\psi}\psi dx = \int_a^b dx = b - a. \tag{6}$$

Since this probability is independent of a, but depends only upon the difference $b-a$, we see that all values of the coordinate are equally probable.

A definite value of the coordinate, for a particle in the state given by Eq. (2), is thus not predictable, but may be obtained only by a direct measurement. Such a measurement however disturbs the particle and thus alters its state. After the coordinate is determined, the particle will no longer be in the state given by Eq. (2). The usual conclusion from this in quantum mechanics is that *when the momentum of a particle is known, its coordinate has no physical reality.*

More generally, it is shown in quantum mechanics that, if the operators corresponding to two physical quantities, say A and B, do not commute, that is, if $AB \neq BA$, then the precise knowledge of one of them precludes such a knowledge of the other. Furthermore, any attempt to determine the latter experimentally will alter the state of the system in such a way as to destroy the knowledge of the first.

From this follows that either (1) *the quantum-mechanical description of reality given by the wave function is not complete* or (2) *when the operators corresponding to two physical quantities do not commute the two quantities cannot have simultaneous reality.* For if both of them had simultaneous reality—and thus definite values—these values would enter into the complete description, according to the condition of completeness. If then the wave function provided such a complete description of reality, it would contain these values; these would then be predictable. This not being the case, we are left with the alternatives stated.

In quantum mechanics it is usually assumed that the wave function *does* contain a complete description of the physical reality of the system in the state to which it corresponds. At first sight this assumption is entirely reasonable, for the information obtainable from a wave function seems to correspond exactly to what can be measured without altering the state of the system. We shall show, however, that this assumption, together with the criterion of reality given above, leads to a contradiction.

2.

For this purpose let us suppose that we have two systems, I and II, which we permit to interact from the time $t=0$ to $t=T$, after which time we suppose that there is no longer any interaction between the two parts. We suppose further that the states of the two systems before $t=0$ were known. We can then calculate with the help of Schrödinger's equation the state of the combined system I+II at any subsequent time; in particular, for any $t>T$. Let us designate the corresponding wave function by Ψ. We cannot, however, calculate the state in which either one of the two systems is left after the interaction. This, according to quantum mechanics, can be done only with the help of further measurements, by a process known as the *reduction of the wave packet*. Let us consider the essentials of this process.

Let $a_1, a_2, a_3, \cdots$ be the eigenvalues of some physical quantity A pertaining to system I and $u_1(x_1), u_2(x_1), u_3(x_1), \cdots$ the corresponding eigenfunctions, where x_1 stands for the variables used to describe the first system. Then Ψ, considered as a function of x_1, can be expressed as

$$\Psi(x_1, x_2)=\sum_{n=1}^{\infty} \psi_n(x_2)u_n(x_1), \tag{7}$$

where x_2 stands for the variables used to describe the second system. Here $\psi_n(x_2)$ are to be regarded merely as the coefficients of the expansion of Ψ into a series of orthogonal functions $u_n(x_1)$. Suppose now that the quantity A is measured and it is found that it has the value a_k. It is then concluded that after the measurement the first system is left in the state given by the wave function $u_k(x_1)$, and that the second system is left in the state given by the wave function $\psi_k(x_2)$. This is the process of reduction of the wave packet; the wave packet given by the infinite series (7) is reduced to a single term $\psi_k(x_2)u_k(x_1)$.

The set of functions $u_n(x_1)$ is determined by the choice of the physical quantity A. If, instead of this, we had chosen another quantity, say B, having the eigenvalues $b_1, b_2, b_3, \cdots$ and eigenfunctions $v_1(x_1), v_2(x_1), v_3(x_1), \cdots$ we should have obtained, instead of Eq. (7), the expansion

$$\Psi(x_1, x_2)=\sum_{s=1}^{\infty} \varphi_s(x_2)v_s(x_1), \tag{8}$$

where φ_s's are the new coefficients. If now the quantity B is measured and is found to have the value b_r, we conclude that after the measurement the first system is left in the state given by $v_r(x_1)$ and the second system is left in the state given by $\varphi_r(x_2)$.

We see therefore that, as a consequence of two different measurements performed upon the first system, the second system may be left in states with two different wave functions. On the other hand, since at the time of measurement the two systems no longer interact, no real change can take place in the second system in consequence of anything that may be done to the first system. This is, of course, merely a statement of what is meant by the absence of an interaction between the two systems. Thus, *it is possible to assign two different wave functions* (in our example ψ_k and φ_r) *to the same reality* (the second system after the interaction with the first).

Now, it may happen that the two wave functions, ψ_k and φ_r, are eigenfunctions of two noncommuting operators corresponding to some physical quantities P and Q, respectively. That this may actually be the case can best be shown by an example. Let us suppose that the two systems are two particles, and that

$$\Psi(x_1, x_2)=\int_{-\infty}^{\infty} e^{(2\pi i/h)(x_1-x_2+x_0)p}dp, \tag{9}$$

where x_0 is some constant. Let A be the momentum of the first particle; then, as we have seen in Eq. (4), its eigenfunctions will be

$$u_p(x_1)=e^{(2\pi i/h)px_1} \tag{10}$$

corresponding to the eigenvalue p. Since we have here the case of a continuous spectrum, Eq. (7) will now be written

$$\Psi(x_1, x_2) = \int_{-\infty}^{\infty} \psi_p(x_2)u_p(x_1)dp, \quad (11)$$

where

$$\psi_p(x_2) = e^{-(2\pi i/h)(x_2 - x_0)p}. \quad (12)$$

This ψ_p however is the eigenfunction of the operator

$$P = (h/2\pi i)\partial/\partial x_2, \quad (13)$$

corresponding to the eigenvalue $-p$ of the momentum of the second particle. On the other hand, if B is the coordinate of the first particle, it has for eigenfunctions

$$v_x(x_1) = \delta(x_1 - x), \quad (14)$$

corresponding to the eigenvalue x, where $\delta(x_1 - x)$ is the well-known Dirac delta-function. Eq. (8) in this case becomes

$$\Psi(x_1, x_2) = \int_{-\infty}^{\infty} \varphi_x(x_2)v_x(x_1)dx, \quad (15)$$

where

$$\varphi_x(x_2) = \int_{-\infty}^{\infty} e^{(2\pi i/h)(x - x_2 + x_0)p}dp$$
$$= h\delta(x - x_2 + x_0). \quad (16)$$

This φ_x, however, is the eigenfunction of the operator

$$Q = x_2 \quad (17)$$

corresponding to the eigenvalue $x + x_0$ of the coordinate of the second particle. Since

$$PQ - QP = h/2\pi i, \quad (18)$$

we have shown that it is in general possible for ψ_k and φ_r to be eigenfunctions of two noncommuting operators, corresponding to physical quantities.

Returning now to the general case contemplated in Eqs. (7) and (8), we assume that ψ_k and φ_r are indeed eigenfunctions of some noncommuting operators P and Q, corresponding to the eigenvalues p_k and q_r, respectively. Thus, by measuring either A or B we are in a position to predict with certainty, and without in any way disturbing the second system, either the value of the quantity P (that is p_k) or the value of the quantity Q (that is q_r). In accordance with our criterion of reality, in the first case we must consider the quantity P as being an element of reality, in the second case the quantity Q is an element of reality. But, as we have seen, both wave functions ψ_k and φ_r belong to the same reality.

Previously we proved that either (1) the quantum-mechanical description of reality given by the wave function is not complete or (2) when the operators corresponding to two physical quantities do not commute the two quantities cannot have simultaneous reality. Starting then with the assumption that the wave function does give a complete description of the physical reality, we arrived at the conclusion that two physical quantities, with noncommuting operators, can have simultaneous reality. Thus the negation of (1) leads to the negation of the only other alternative (2). We are thus forced to conclude that the quantum-mechanical description of physical reality given by wave functions is not complete.

One could object to this conclusion on the grounds that our criterion of reality is not sufficiently restrictive. Indeed, one would not arrive at our conclusion if one insisted that two or more physical quantities can be regarded as simultaneous elements of reality *only when they can be simultaneously measured or predicted.* On this point of view, since either one or the other, but not both simultaneously, of the quantities P and Q can be predicted, they are not simultaneously real. This makes the reality of P and Q depend upon the process of measurement carried out on the first system, which does not disturb the second system in any way. No reasonable definition of reality could be expected to permit this.

While we have thus shown that the wave function does not provide a complete description of the physical reality, we left open the question of whether or not such a description exists. We believe, however, that such a theory is possible.

Can Quantum-Mechanical Description of Physical Reality be Considered Complete?

N. BOHR, *Institute for Theoretical Physics, University, Copenhagen*

(Received July 13, 1935)

It is shown that a certain "criterion of physical reality" formulated in a recent article with the above title by A. Einstein, B. Podolsky and N. Rosen contains an essential ambiguity when it is applied to quantum phenomena. In this connection a viewpoint termed "complementarity" is explained from which quantum-mechanical description of physical phenomena would seem to fulfill, within its scope, all rational demands of completeness.

IN a recent article[1] under the above title A. Einstein, B. Podolsky and N. Rosen have presented arguments which lead them to answer the question at issue in the negative. The trend of their argumentation, however, does not seem to me adequately to meet the actual situation with which we are faced in atomic physics. I shall therefore be glad to use this opportunity to explain in somewhat greater detail a general viewpoint, conveniently termed "complementarity," which I have indicated on various previous occasions,[2] and from which quantum mechanics within its scope would appear as a completely rational description of physical phenomena, such as we meet in atomic processes.

The extent to which an unambiguous meaning can be attributed to such an expression as "physical reality" cannot of course be deduced from *a priori* philosophical conceptions, but—as the authors of the article cited themselves emphasize—must be founded on a direct appeal to experiments and measurements. For this purpose they propose a "criterion of reality" formulated as follows: "If, without in any way disturbing a system, we can predict with certainty the value of a physical quantity, then there exists an element of physical reality corresponding to this physical quantity." By means of an interesting example, to which we shall return below, they next proceed to show that in quantum mechanics, just as in classical mechanics, it is possible under suitable conditions to predict the value of any given variable pertaining to the description of a mechanical system from measurements performed entirely on other systems which previously have been in interaction with the system under investigation. According to their criterion the authors therefore want to ascribe an element of reality to each of the quantities represented by such variables. Since, moreover, it is a well-known feature of the present formalism of quantum mechanics that it is never possible, in the description of the state of a mechanical system, to attach definite values to both of two canonically conjugate variables, they consequently deem this formalism to be incomplete, and express the belief that a more satisfactory theory can be developed.

Such an argumentation, however, would hardly seem suited to affect the soundness of quantum-mechanical description, which is based on a coherent mathematical formalism covering automatically any procedure of measurement like that indicated.* The apparent contradiction in

---

[1] A. Einstein, B. Podolsky and N. Rosen, Phys. Rev. **47**, 777 (1935).

[2] Cf. N. Bohr, *Atomic Theory and Description of Nature*, I (Cambridge, 1934).

* The deductions contained in the article cited may in this respect be considered as an immediate consequence of the transformation theorems of quantum mechanics, which perhaps more than any other feature of the formalism contribute to secure its mathematical completeness and its rational correspondence with classical mechanics. In fact, it is always possible in the description of a mechanical system, consisting of two partial systems (1) and (2), interacting or not, to replace any two pairs of canonically conjugate variables (q_1p_1), (q_2p_2) pertaining to systems (1) and (2), respectively, and satisfying the usual commutation rules

$$[q_1p_1]=[q_2p_2]=ih/2\pi,$$
$$[q_1q_2]=[p_1p_2]=[q_1p_2]=[q_2p_1]=0,$$

by two pairs of new conjugate variables (Q_1P_1), (Q_2P_2) related to the first variables by a simple orthogonal transformation, corresponding to a rotation of angle θ in the planes (q_1q_2), (p_1p_2)

$$q_1=Q_1\cos\theta-Q_2\sin\theta \qquad p_1=P_1\cos\theta-P_2\sin\theta$$
$$q_2=Q_1\sin\theta+Q_2\cos\theta \qquad p_2=P_1\sin\theta+P_2\cos\theta.$$

Since these variables will satisfy analogous commutation rules, in particular

$$[Q_1P_1]=ih/2\pi, \qquad [Q_1P_2]=0,$$

it follows that in the description of the state of the combined system definite numerical values may not be assigned to both Q_1 and P_1, but that we may clearly assign

fact discloses only an essential inadequacy of the customary viewpoint of natural philosophy for a rational account of physical phenomena of the type with which we are concerned in quantum mechanics. Indeed the *finite interaction between object and measuring agencies* conditioned by the very existence of the quantum of action entails —because of the impossibility of controlling the reaction of the object on the measuring instruments if these are to serve their purpose—the necessity of a final renunciation of the classical ideal of causality and a radical revision of our attitude towards the problem of physical reality. In fact, as we shall see, a criterion of reality like that proposed by the named authors contains—however cautious its formulation may appear—an essential ambiguity when it is applied to the actual problems with which we are here concerned. In order to make the argument to this end as clear as possible, I shall first consider in some detail a few simple examples of measuring arrangements.

Let us begin with the simple case of a particle passing through a slit in a diaphragm, which may form part of some more or less complicated experimental arrangement. Even if the momentum of this particle is completely known before it impinges on the diaphragm, the diffraction by the slit of the plane wave giving the symbolic representation of its state will imply an uncertainty in the momentum of the particle, after it has passed the diaphragm, which is the greater the narrower the slit. Now the width of the slit, at any rate if it is still large compared with the wave-length, may be taken as the uncertainty Δq of the position of the particle relative to the diaphragm, in a direction perpendicular to the slit. Moreover, it is simply seen from de Broglie's relation between momentum and wave-length that the uncertainty Δp of the momentum of the particle in this direction is correlated to Δq by means of Heisenberg's general principle

$$\Delta p \Delta q \sim h,$$

which in the quantum-mechanical formalism is a direct consequence of the commutation relation for any pair of conjugate variables. Obviously the uncertainty Δp is inseparably connected with the possibility of an exchange of momentum between the particle and the diaphragm; and the question of principal interest for our discussion is now to what extent the momentum thus exchanged can be taken into account in the description of the phenomenon to be studied by the experimental arrangement concerned, of which the passing of the particle through the slit may be considered as the initial stage.

Let us first assume that, corresponding to usual experiments on the remarkable phenomena of electron diffraction, the diaphragm, like the other parts of the apparatus,—say a second diaphragm with several slits parallel to the first and a photographic plate,—is rigidly fixed to a support which defines the space frame of reference. Then the momentum exchanged between the particle and the diaphragm will, together with the reaction of the particle on the other bodies, pass into this common support, and we have thus voluntarily cut ourselves off from any possibility of taking these reactions separately into account in predictions regarding the final result of the experiment,—say the position of the spot produced by the particle on the photographic plate. The impossibility of a closer analysis of the reactions between the particle and the measuring instrument is indeed no peculiarity of the experimental procedure described, but is rather an essential property of any arrangement suited to the study of the phenomena of the type concerned, where we have to do with a feature of *individuality* completely foreign to classical physics. In fact, any possibility of taking into account the momentum exchanged between the particle and the separate parts of the apparatus would at once permit us to draw conclusions regarding the "course" of such phenomena,—say through what particular slit of the second diaphragm the particle passes on its way to the photographic plate—which would be quite incompatible with the fact that the probability of the particle reaching a given element of area on this plate is determined not by the presence of any particular slit, but by the positions of all the slits of the second diaphragm within reach

---

such values to both Q_1 and P_2. In that case it further results from the expressions of these variables in terms of (q_1p_1) and (q_2p_2), namely

$$Q_1 = q_1 \cos\theta + q_2 \sin\theta, \qquad P_2 = -p_1 \sin\theta + p_2 \cos\theta,$$

that a subsequent measurement of either q_2 or p_2 will allow us to predict the value of q_1 or p_1 respectively.

of the associated wave diffracted from the slit of the first diaphragm.

By another experimental arrangement, where the first diaphragm is not rigidly connected with the other parts of the apparatus, it would at least in principle* be possible to measure its momentum with any desired accuracy before and after the passage of the particle, and thus to predict the momentum of the latter after it has passed through the slit. In fact, such measurements of momentum require only an unambiguous application of the classical law of conservation of momentum, applied for instance to a collision process between the diaphragm and some test body, the momentum of which is suitably controlled before and after the collision. It is true that such a control will essentially depend on an examination of the space-time course of some process to which the ideas of classical mechanics can be applied; if, however, all spatial dimensions and time intervals are taken sufficiently large, this involves clearly no limitation as regards the accurate control of the momentum of the test bodies, but only a renunciation as regards the accuracy of the control of their space-time coordination. This last circumstance is in fact quite analogous to the renunciation of the control of the momentum of the fixed diaphragm in the experimental arrangement discussed above, and depends in the last resort on the claim of a purely classical account of the measuring apparatus, which implies the necessity of allowing a latitude corresponding to the quantum-mechanical uncertainty relations in our description of their behavior.

The principal difference between the two experimental arrangements under consideration is, however, that in the arrangement suited for the control of the momentum of the first diaphragm, this body can no longer be used as a measuring instrument for the same purpose as in the previous case, but must, as regards its position relative to the rest of the apparatus, be treated, like the particle traversing the slit, as an object of investigation, in the sense that the quantum-mechanical uncertainty relations regarding its position and momentum must be taken explicitly into account. In fact, even if we knew the position of the diaphragm relative to the space frame before the first measurement of its momentum, and even though its position after the last measurement can be accurately fixed, we lose, on account of the uncontrollable displacement of the diaphragm during each collision process with the test bodies, the knowledge of its position when the particle passed through the slit. The whole arrangement is therefore obviously unsuited to study the same kind of phenomena as in the previous case. In particular it may be shown that, if the momentum of the diaphragm is measured with an accuracy sufficient for allowing definite conclusions regarding the passage of the particle through some selected slit of the second diaphragm, then even the minimum uncertainty of the position of the first diaphragm compatible with such a knowledge will imply the total wiping out of any interference effect—regarding the zones of permitted impact of the particle on the photographic plate—to which the presence of more than one slit in the second diaphragm would give rise in case the positions of all apparatus are fixed relative to each other.

In an arrangement suited for measurements of the momentum of the first diaphragm, it is further clear that even if we have measured this momentum before the passage of the particle through the slit, we are after this passage still left with a *free choice* whether we wish to know the momentum of the particle or its initial position relative to the rest of the apparatus. In the first eventuality we need only to make a second determination of the momentum of the diaphragm, leaving unknown forever its exact position when the particle passed. In the second eventuality we need only to determine its position relative to the space frame with the inevitable loss of the knowledge of the momentum exchanged between the diaphragm and the particle. If the diaphragm is sufficiently massive in comparison with the particle, we may even arrange the procedure of measurements in such a way that the diaphragm after the first determination of its momentum will remain at rest in some unknown position relative to the

* The obvious impossibility of actually carrying out, with the experimental technique at our disposal, such measuring procedures as are discussed here and in the following does clearly not affect the theoretical argument, since the procedures in question are essentially equivalent with atomic processes, like the Compton effect, where a corresponding application of the conservation theorem of momentum is well established.

other parts of the apparatus, and the subsequent fixation of this position may therefore simply consist in establishing a rigid connection between the diaphragm and the common support.

My main purpose in repeating these simple, and in substance well-known considerations, is to emphasize that in the phenomena concerned we are not dealing with an incomplete description characterized by the arbitrary picking out of different elements of physical reality at the cost of sacrifying other such elements, but with a rational discrimination between essentially different experimental arrangements and procedures which are suited either for an unambiguous use of the idea of space location, or for a legitimate application of the conservation theorem of momentum. Any remaining appearance of arbitrariness concerns merely our freedom of handling the measuring instruments, characteristic of the very idea of experiment. In fact, the renunciation in each experimental arrangement of the one or the other of two aspects of the description of physical phenomena,—the combination of which characterizes the method of classical physics, and which therefore in this sense may be considered as *complementary* to one another,—depends essentially on the impossibility, in the field of quantum theory, of accurately controlling the reaction of the object on the measuring instruments, i.e., the transfer of momentum in case of position measurements, and the displacement in case of momentum measurements. Just in this last respect any comparison between quantum mechanics and ordinary statistical mechanics,—however useful it may be for the formal presentation of the theory,—is essentially irrelevant. Indeed we have in each experimental arrangement suited for the study of proper quantum phenomena not merely to do with an ignorance of the value of certain physical quantities, but with the impossibility of defining these quantities in an unambiguous way.

The last remarks apply equally well to the special problem treated by Einstein, Podolsky and Rosen, which has been referred to above, and which does not actually involve any greater intricacies than the simple examples discussed above. The particular quantum-mechanical state of two free particles, for which they give an explicit mathematical expression, may be reproduced, at least in principle, by a simple experimental arrangement, comprising a rigid diaphragm with two parallel slits, which are very narrow compared with their separation, and through each of which one particle with given initial momentum passes independently of the other. If the momentum of this diaphragm is measured accurately before as well as after the passing of the particles, we shall in fact know the sum of the components perpendicular to the slits of the momenta of the two escaping particles, as well as the difference of their initial positional coordinates in the same direction; while of course the conjugate quantities, i.e., the difference of the components of their momenta, and the sum of their positional coordinates, are entirely unknown.* In this arrangement, it is therefore clear that a subsequent single measurement either of the position or of the momentum of one of the particles will automatically determine the position or momentum, respectively, of the other particle with any desired accuracy; at least if the wave-length corresponding to the free motion of each particle is sufficiently short compared with the width of the slits. As pointed out by the named authors, we are therefore faced at this stage with a completely free choice whether we want to determine the one or the other of the latter quantities by a process which does not directly interfere with the particle concerned.

Like the above simple case of the choice between the experimental procedures suited for the prediction of the position or the momentum of a single particle which has passed through a slit in a diaphragm, we are, in the "freedom of choice" offered by the last arrangement, just concerned with a *discrimination between different experimental procedures which allow of the unambiguous use of complementary classical concepts*. In fact to measure the position of one of the particles can mean nothing else than to establish a correlation between its behavior and some

* As will be seen, this description, apart from a trivial normalizing factor, corresponds exactly to the transformation of variables described in the preceding footnote if (q_1p_1), (q_2p_2) represent the positional coordinates and components of momenta of the two particles and if $\theta=-\pi/4$. It may also be remarked that the wave function given by formula (9) of the article cited corresponds to the special choice of $P_2=0$ and the limiting case of two infinitely narrow slits.

instrument rigidly fixed to the support which defines the space frame of reference. Under the experimental conditions described such a measurement will therefore also provide us with the knowledge of the location, otherwise completely unknown, of the diaphragm with respect to this space frame when the particles passed through the slits. Indeed, only in this way we obtain a basis for conclusions about the initial position of the other particle relative to the rest of the apparatus. By allowing an essentially uncontrollable momentum to pass from the first particle into the mentioned support, however, we have by this procedure cut ourselves off from any future possibility of applying the law of conservation of momentum to the system consisting of the diaphragm and the two particles and therefore have lost our only basis for an unambiguous application of the idea of momentum in predictions regarding the behavior of the second particle. Conversely, if we choose to measure the momentum of one of the particles, we lose through the uncontrollable displacement inevitable in such a measurement any possibility of deducing from the behavior of this particle the position of the diaphragm relative to the rest of the apparatus, and have thus no basis whatever for predictions regarding the location of the other particle.

From our point of view we now see that the wording of the above-mentioned criterion of physical reality proposed by Einstein, Podolsky and Rosen contains an ambiguity as regards the meaning of the expression "without in any way disturbing a system." Of course there is in a case like that just considered no question of a mechanical disturbance of the system under investigation during the last critical stage of the measuring procedure. But even at this stage there is essentially the question of *an influence on the very conditions which define the possible types of predictions regarding the future behavior of the system.* Since these conditions constitute an inherent element of the description of any phenomenon to which the term "physical reality" can be properly attached, we see that the argumentation of the mentioned authors does not justify their conclusion that quantum-mechanical description is essentially incomplete. On the contrary this description, as appears from the preceding discussion, may be characterized as a rational utilization of all possibilities of unambiguous interpretation of measurements, compatible with the finite and uncontrollable interaction between the objects and the measuring instruments in the field of quantum theory. In fact, it is only the mutual exclusion of any two experimental procedures, permitting the unambiguous definition of complementary physical quantities, which provides room for new physical laws, the coexistence of which might at first sight appear irreconcilable with the basic principles of science. It is just this entirely new situation as regards the description of physical phenomena, that the notion of *complementarity* aims at characterizing.

The experimental arrangements hitherto discussed present a special simplicity on account of the secondary role which the idea of time plays in the description of the phenomena in question. It is true that we have freely made use of such words as "before" and "after" implying time-relationships; but in each case allowance must be made for a certain inaccuracy, which is of no importance, however, so long as the time intervals concerned are sufficiently large compared with the proper periods entering in the closer analysis of the phenomenon under investigation. As soon as we attempt a more accurate time description of quantum phenomena, we meet with well-known new paradoxes, for the elucidation of which further features of the interaction between the objects and the measuring instruments must be taken into account. In fact, in such phenomena we have no longer to do with experimental arrangements consisting of apparatus essentially at rest relative to one another, but with arrangements containing moving parts,—like shutters before the slits of the diaphragms,—controlled by mechanisms serving as clocks. Besides the transfer of momentum, discussed above, between the object and the bodies defining the space frame, we shall therefore, in such arrangements, have to consider an eventual exchange of energy between the object and these clock-like mechanisms.

The decisive point as regards time measurements in quantum theory is now completely analogous to the argument concerning measurements of positions outlined above. Just as the transfer of momentum to the separate parts of the apparatus,—the knowledge of the relative positions of which is required for the description of the phenomenon,—has been seen to be entirely uncontrollable, so the exchange of energy between the object and the various bodies, whose relative motion must be known for the intended use of the apparatus, will defy any closer analysis. Indeed, it is *excluded in principle to control the energy which goes into the clocks without interfering essentially with their use as time indicators.* This use in fact entirely relies on the assumed possibility of accounting for the functioning of each clock as well as for its eventual comparison with other clocks on the basis of the methods of classical physics. In this account we must therefore obviously allow for a latitude in the energy balance, corresponding to the quantum-mechanical uncertainty relation for the conjugate time and energy variables. Just as in the question discussed above of the mutually exclusive character of any unambiguous use in quantum theory of the concepts of position and momentum, it is in the last resort this circumstance which entails the complementary relationship between any detailed time account of atomic phenomena on the one hand and the unclassical features of intrinsic stability of atoms, disclosed by the study of energy transfers in atomic reactions on the other hand.

This necessity of discriminating in each experimental arrangement between those parts of the physical system considered which are to be treated as measuring instruments and those which constitute the objects under investigation may indeed be said to form a *principal distinction between classical and quantum-mechanical description of physical phenomena.* It is true that the place within each measuring procedure where this discrimination is made is in both cases largely a matter of convenience. While, however, in classical physics the distinction between object and measuring agencies does not entail any difference in the character of the description of the phenomena concerned, its fundamental importance in quantum theory, as we have seen, has its root in the indispensable use of classical concepts in the interpretation of all proper measurements, even though the classical theories do not suffice in accounting for the new types of regularities with which we are concerned in atomic physics. In accordance with this situation there can be no question of any unambiguous interpretation of the symbols of quantum mechanics other than that embodied in the well-known rules which allow to predict the results to be obtained by a given experimental arrangement described in a totally classical way, and which have found their general expression through the transformation theorems, already referred to. By securing its proper correspondence with the classical theory, these theorems exclude in particular any imaginable inconsistency in the quantum-mechanical description, connected with a change of the place where the discrimination is made between object and measuring agencies. In fact it is an obvious consequence of the above argumentation that in each experimental arrangement and measuring procedure we have only a free choice of this place within a region where the quantum-mechanical description of the process concerned is effectively equivalent with the classical description.

Before concluding I should still like to emphasize the bearing of the great lesson derived from general relativity theory upon the question of physical reality in the field of quantum theory. In fact, notwithstanding all characteristic differences, the situations we are concerned with in these generalizations of classical theory present striking analogies which have often been noted. Especially, the singular position of measuring instruments in the account of quantum phenomena, just discussed, appears closely analogous to the well-known necessity in relativity theory of upholding an ordinary description of all measuring processes, including a sharp distinction between space and time coordinates, although the very essence of this theory is the establishment of new physical laws, in the comprehension of which we must renounce the customary separation of space and time ideas.*

* Just this circumstance, together with the relativistic invariance of the uncertainty relations of quantum mechanics, ensures the compatibility between the argumentation outlined in the present article and all exigencies of relativity theory. This question will be treated in greater detail in a paper under preparation, where the writer will in particular discuss a very interesting paradox suggested by Einstein concerning the application of gravitation theory to energy measurements, and the solution of which offers an especially instructive illustration of the generality of the argument of complementarity. On the same occasion a more thorough discussion of space-time measurements in quantum theory will be given with all necessary mathematical developments and diagrams of experimental

The dependence on the reference system, in relativity theory, of all readings of scales and clocks may even be compared with the essentially uncontrollable exchange of momentum or energy between the objects of measurements and all instruments defining the space-time system of reference, which in quantum theory confronts us with the situation characterized by the notion of complementarity. In fact this new feature of natural philosophy means a radical revision of our attitude as regards physical reality, which may be paralleled with the fundamental modification of all ideas regarding the absolute character of physical phenomena, brought about by the general theory of relativity.

---

arrangements, which had to be left out of this article, where the main stress is laid on the dialectic aspect of the question at issue.

PHYSICAL REVIEW VOLUME 85, NUMBER 2 JANUARY 15, 1952

A Suggested Interpretation of the Quantum Theory in Terms of "Hidden" Variables. I

DAVID BOHM*
Palmer Physical Laboratory, Princeton University, Princeton, New Jersey
(Received July 5, 1951)

The usual interpretation of the quantum theory is self-consistent, but it involves an assumption that cannot be tested experimentally, *viz.*, that the most complete possible specification of an individual system is in terms of a wave function that determines only probable results of actual measurement processes. The only way of investigating the truth of this assumption is by trying to find some other interpretation of the quantum theory in terms of at present "hidden" variables, which in principle determine the precise behavior of an individual system, but which are in practice averaged over in measurements of the types that can now be carried out. In this paper and in a subsequent paper, an interpretation of the quantum theory in terms of just such "hidden" variables is suggested. It is shown that as long as the mathematical theory retains its present general form, this suggested interpretation leads to precisely the same results for all physical processes as does the usual interpretation. Nevertheless, the suggested interpretation provides a broader conceptual framework than the usual interpretation, because it makes possible a precise and continuous description of all processes, even at the quantum level. This broader conceptual framework allows more general mathematical formulations of the theory than those allowed by the usual interpretation. Now, the usual mathematical formulation seems to lead to insoluble difficulties when it is extrapolated into the domain of distances of the order of 10^{-13} cm or less. It is therefore entirely possible that the interpretation suggested here may be needed for the resolution of these difficulties. In any case, the mere possibility of such an interpretation proves that it is not necessary for us to give up a precise, rational, and objective description of individual systems at a quantum level of accuracy.

1. INTRODUCTION

THE usual interpretation of the quantum theory is based on an assumption having very far-reaching implications, *viz.*, that the physical state of an individual system is completely specified by a wave function that determines only the probabilities of actual results that can be obtained in a statistical ensemble of similar experiments. This assumption has been the object of severe criticisms, notably on the part of Einstein, who has always believed that, even at the quantum level, there must exist precisely definable elements or dynamical variables determining (as in classical physics) the actual behavior of each individual system, and not merely its probable behavior. Since these elements or variables are not now included in the quantum theory and have not yet been detected experimentally, Einstein has always regarded the present form of the quantum theory as incomplete, although he admits its internal consistency.[1–5]

Most physicists have felt that objections such as those raised by Einstein are not relevant, first, because the present form of the quantum theory with its usual probability interpretation is in excellent agreement with an extremely wide range of experiments, at least in the domain of distances[6] larger than 10^{-13} cm, and, secondly, because no consistent alternative interpretations have as yet been suggested. The purpose of this paper (and of a subsequent paper hereafter denoted by II) is, however, to suggest just such an alternative interpretation. In contrast to the usual interpretation, this alternative interpretation permits us to conceive of each individual system as being in a precisely definable state, whose changes with time are determined by definite laws, analogous to (but not identical with) the classical equations of motion. Quantum-mechanical probabilities are regarded (like their counterparts in classical statistical mechanics) as only a practical necessity and not as a manifestation of an inherent lack of complete determination in the properties of matter at the quantum level. As long as the present general form of Schroedinger's equation is retained, the physical results obtained with our suggested alternative interpretation are precisely the same as those obtained with the usual interpretation. We shall see, however, that our alternative interpretation permits modifications of the mathematical formulation which could not even be described in terms of the usual interpretation. Moreover, the modifications can quite easily be formulated in such a way that their effects are insignificant in the atomic domain, where the present quantum theory is in such good agreement with experiment, but of crucial importance in the domain of dimensions of the order of 10^{-13} cm, where, as we have seen, the present theory is totally inadequate. It is thus entirely possible that some of the modifications describable in terms of our suggested alternative interpretation, but

* Now at Universidade de São Paulo, Faculdade de Filosofia, Ciencias, e Letras, São Paulo, Brasil.

[1] Einstein, Podolsky, and Rosen, Phys. Rev. 47, 777 (1933).

[2] D. Bohm, *Quantum Theory* (Prentice-Hall, Inc., New York, 1951), see p. 611.

[3] N. Bohr, Phys. Rev. 48, 696 (1935).

[4] W. Furry, Phys. Rev. 49, 393, 476 (1936).

[5] Paul Arthur Schilp, editor, *Albert Einstein, Philosopher-Scientist* (Library of Living Philosophers, Evanston, Illinois, 1949). This book contains a thorough summary of the entire controversy.

[6] At distances of the order of 10^{-13} cm or smaller and for times of the order of this distance divided by the velocity of light or smaller, present theories become so inadequate that it is generally believed that they are probably not applicable, except perhaps in a very crude sense. Thus, it is generally expected that in connection with phenomena associated with this so-called "fundamental length," a totally new theory will probably be needed. It is hoped that this theory could not only deal precisely with such processes as meson production and scattering of elementary particles, but that it would also systematically predict the masses, charges, spins, etc., of the large number of so-called "elementary" particles that have already been found, as well as those of new particles which might be found in the future.

not in terms of the usual interpretation, may be needed for a more thorough understanding of phenomena associated with very small distances. We shall not, however, actually develop such modifications in any detail in these papers.

After this article was completed, the author's attention was called to similar proposals for an alternative interpretation of the quantum theory made by de Broglie[7] in 1926, but later given up by him partly as a result of certain criticisms made by Pauli[8] and partly because of additional objections raised by de Broglie[7] himself.† As we shall show in Appendix B of Paper II, however, all of the objections of de Broglie and Pauli could have been met if only de Broglie had carried his ideas to their logical conclusion. The essential new step in doing this is to apply our interpretation in the theory of the measurement process itself as well as in the description of the observed system. Such a development of the theory of measurements is given in Paper II,[9] where it will be shown in detail that our interpretation leads to precisely the same results for all experiments as are obtained with the usual interpretation. The foundation for doing this is laid in Paper I, where we develop the basis of our interpretation, contrast it with the usual interpretation, and apply it to a few simple examples, in order to illustrate the principles involved.

2. THE USUAL PHYSICAL INTERPRETATION OF THE QUANTUM THEORY

The usual physical interpretation of the quantum theory centers around the uncertainty principle. Now, the uncertainty principle can be derived in two different ways. First, we may start with the assumption already criticized by Einstein,[1] namely, that a wave function that determines only probabilities of actual experimental results nevertheless provides the most complete possible specification of the so-called "quantum state" of an individual system. With the aid of this assumption and with the aid of the de Broglie relation, $\mathbf{p}=\hbar\mathbf{k}$, where $\mathbf{k}$ is the wave number associated with a particular fourier component of the wave function, the uncertainty principle is readily deduced.[10] From this derivation, we are led to interpret the uncertainty principle as an inherent and irreducible limitation on the precision with which it is correct for us even to conceive of momentum and position as simultaneously defined quantities. For if, as is done in the usual interpretation of the quantum theory, the wave intensity is assumed to determine only the probability of a given position, and if the $\mathbf{k}$th Fourier component of the wave function is assumed to determine only the probability of a corresponding momentum, $\mathbf{p}=\hbar\mathbf{k}$, then it becomes a contradiction in terms to ask for a state in which momentum and position are simultaneously and precisely defined.

A second possible derivation of the uncertainty principle is based on a theoretical analysis of the processes with the aid of which physically significant quantities such as momentum and position can be measured. In such an analysis, one finds that because the measuring apparatus interacts with the observed system by means of indivisible quanta, there will always be an irreducible disturbance of some observed property of the system. If the precise effects of this disturbance could be predicted or controlled, then one could correct for these effects, and thus one could still in principle obtain simultaneous measurements of momentum and position, having unlimited precision. But if one could do this, then the uncertainty principle would be violated. The uncertainty principle is, as we have seen, however, a necessary consequence of the assumption that the wave function and its probability interpretation provide the most complete possible specification of the state of an individual system. In order to avoid the possibility of a contradiction with this assumption, Bohr[3,5,10,11] and others have suggested an additional assumption, namely, that the process of transfer of a single quantum from observed system to measuring apparatus is inherently unpredictable, uncontrollable, and not subject to a detailed rational analysis or description. With the aid of this assumption, one can show[10] that the same uncertainty principle that is deduced from the wave function and its probability interpretation is also obtained as an inherent and unavoidable limitation on the precision of all possible measurements. Thus, one is able to obtain a set of assumptions, which permit a self-consistent formulation of the usual interpretation of the quantum theory.

The above point of view has been given its most consistent and systematic expression by Bohr,[3,5,10] in terms of the "principle of complementarity." In formulating this principle, Bohr suggests that at the atomic level we must renounce our hitherto successful practice of conceiving of an individual system as a unified and precisely definable whole, all of whose aspects are, in a manner of speaking, simultaneously and

[7] L. de Broglie, *An Introduction to the Study of Wave Mechanics* (E. P. Dutton and Company, Inc., New York, 1930), see Chapters 6, 9, and 10. See also Compt. rend. 183, 447 (1926); 184, 273 (1927); 185, 380 (1927).

[8] *Reports on the Solvay Congress* (Gauthiers-Villars et Cie., Paris, 1928), see p. 280.

† *Note added in proof.*—Madelung has also proposed a similar interpretation of the quantum theory, but like de Broglie he did not carry this interpretation to a logical conclusion. See E. Madelung, Z. f. Physik 40, 332 (1926), also G. Temple, *Introduction to Quantum Theory* (London, 1931).

[9] In Paper II, Sec. 9, we also discuss von Neumann's proof [see J. von Neumann, *Mathematische Grundlagen der Quantenmechanik* (Verlag, Julius Springer, Berlin, 1932)] that quantum theory cannot be understood in terms of a statistical distribution of "hidden" causal parameters. We shall show that his conclusions do not apply to our interpretation, because he implicitly assumes that the hidden parameters must be associated only with the observed system, whereas, as will become evident in these papers, our interpretation requires that the hidden parameters shall also be associated with the measuring apparatus.

[10] See reference 2, Chapter 5.

[11] N. Bohr, *Atomic Theory and the Description of Nature* (Cambridge University Press, London, 1934).

unambiguously accessible to our conceptual gaze. Such a system of concepts, which is sometimes called a "model," need not be restricted to pictures, but may also include, for example, mathematical concepts, as long as these are supposed to be in a precise (i.e., one-to-one) correspondence with the objects that are being described. The principle of complementarity requires us, however, to renounce even mathematical models. Thus, in Bohr's point of view, the wave function is in no sense a conceptual model of an individual system, since it is not in a precise (one-to-one) correspondence with the behavior of this system, but only in a statistical correspondence.

In place of a precisely defined conceptual model, the principle of complementarity states that we are restricted to complementarity pairs of inherently imprecisely defined concepts, such as position and momentum, particle and wave, etc. The maximum degree of precision of definition of either member of such a pair is reciprocally related to that of the opposite member. This need for an inherent lack of complete precision can be understood in two ways. First, it can be regarded as a consequence of the fact that the experimental apparatus needed for a precise measurement of one member of a complementary pair of variables must always be such as to preclude the possibility of a simultaneous and precise measurement of the other member. Secondly, the assumption that an individual system is completely specified by the wave function and its probability interpretation implies a corresponding unavoidable lack of precision in the very conceptual structure, with the aid of which we can think about and describe the behavior of the system.

It is only at the classical level that we can correctly neglect the inherent lack of precision in all of our conceptual models; for here, the incomplete determination of physical properties implied by the uncertainty principle produces effects that are too small to be of practical significance. Our ability to describe classical systems in terms of precisely definable models is, however, an integral part of the usual interpretation of the theory. For without such models, we would have no way to describe, or even to think of, the result of an observation, which is of course always finally carried out at a classical level of accuracy. If the relationships of a given set of classically describable phenomena depend significantly on the essentially quantum-mechanical properties of matter, however, then the principle of complementarity states that no single model is possible which could provide a precise and rational analysis of the connections between these phenomena. In such a case, we are not supposed, for example, to attempt to describe in detail how future phenomena arise out of past phenomena. Instead, we should simply accept without further analysis the fact that future phenomena do in fact somehow manage to be produced, in a way that is, however, necessarily beyond the possibility of a detailed description. The only aim of a mathematical theory is then to predict the statistical relations, if any, connecting these phenomena.

3. CRITICISM OF THE USUAL INTERPRETATION OF THE QUANTUM THEORY

The usual interpretation of the quantum theory can be criticized on many grounds.[5] In this paper, however, we shall stress only the fact that it requires us to give up the possibility of even conceiving precisely what might determine the behavior of an individual system at the quantum level, without providing adequate proof that such a renunciation is necessary.[6] The usual interpretation is admittedly consistent; but the mere demonstration of such consistency does not exclude the possibility of other equally consistent interpretations, which would involve additional elements or parameters permitting a detailed causal and continuous description of all processes, and not requiring us to forego the possibility of conceiving the quantum level in precise terms. From the point of view of the usual interpretation, these additional elements or parameters could be called "hidden" variables. As a matter of fact, whenever we have previously had recourse to statistical theories, we have always ultimately found that the laws governing the individual members of a statistical ensemble could be expressed in terms of just such hidden variables. For example, from the point of view of macroscopic physics, the coordinates and momenta of individual atoms are hidden variables, which in a large scale system manifest themselves only as statistical averages. Perhaps then, our present quantum-mechanical averages are similarly a manifestation of hidden variables, which have not, however, yet been detected directly.

Now it may be asked why these hidden variables should have so long remained undetected. To answer this question, it is helpful to consider as an analogy the early forms of the atomic theory, in which the existence of atoms was postulated in order to explain certain large-scale effects, such as the laws of chemical combination, the gas laws, etc. On the other hand, these same effects could also be described directly in terms of existing macrophysical concepts (such as pressure, volume, temperature, mass, etc.); and a correct description in these terms did not require any reference to atoms. Ultimately, however, effects were found which contradicted the predictions obtained by extrapolating certain purely macrophysical theories to the domain of the very small, and which could be understood correctly in terms of the assumption that matter is composed of atoms. Similarly, we suggest that if there are hidden variables underlying the present quantum theory, it is quite likely that in the atomic domain, they will lead to effects that can also be described adequately in the terms of the usual quantum-mechanical concepts; while in a domain associated with much smaller dimensions, such as the level associated with the "fundamental length" of the order of 10^{-13} cm, the hidden variables may lead to completely new effects not consistent with the extrapolation of the present quantum theory down to this level.

If, as is certainly entirely possible, these hidden variables are actually needed for a correct description at small distances, we could easily be kept on the wrong track for a long time by restricting ourselves to the usual interpretation of the quantum theory, which excludes such hidden variables as a matter of principle. It is therefore very important for us to investigate our reasons for supposing that the usual physical interpretation is likely to be the correct one. To this end, we shall begin by repeating the two mutually consistent assumptions on which the usual interpretation is based (see Sec. 2):

(1) The wave function with its probability interpretation determines the most complete possible specification of the state of an individual system.

(2) The process of transfer of a single quantum from observed system to measuring apparatus is inherently unpredictable, uncontrollable, and unanalyzable.

Let us now inquire into the question of whether there are any experiments that could conceivably provide a test for these assumptions. It is often stated in connection with this problem that the mathematical apparatus of the quantum theory and its physical interpretation form a consistent whole and that this combined system of mathematical apparatus and physical interpretation is tested adequately by the extremely wide range of experiments that are in agreement with predictions obtained by using this system. If assumptions (1) and (2) implied a unique mathematical formulation, then such a conclusion would be valid, because experimental predictions could then be found which, if contradicted, would clearly indicate that these assumptions were wrong. Although assumptions (1) and (2) do limit the possible forms of the mathematical theory, they do not limit these forms sufficiently to make possible a unique set of predictions that could in principle permit such an experimental test. Thus, one can contemplate practically arbitrary changes in the Hamiltonian operator, including, for example, the postulation of an unlimited range of new kinds of meson fields each having almost any conceivable rest mass, charge, spin, magnetic moment, etc. And if such postulates should prove to be inadequate, it is conceivable that we may have to introduce nonlocal operators, nonlinear fields, S-matrices, etc. This means that when the theory is found to be inadequate (as now happens, for example, at distances of the order of 10^{-13} cm), it is always possible, and, in fact, usually quite natural, to assume that the theory can be made to agree with experiment by some as yet unknown change in the mathematical formulation alone, not requiring any fundamental changes in the physical interpretation. This means that as long as we accept the usual physical interpretation of the quantum theory, we cannot be led by any conceivable experiment to give up this interpretation, even if it should happen to be wrong. The usual physical interpretation therefore presents us with a considerable danger of falling into a trap, consisting of a self-closing chain of circular hypotheses, which are in principle unverifiable if true. The only way of avoiding the possibility of such a trap is to study the consequences of postulates that contradict assumptions (1) and (2) at the outset. Thus, we could, for example, postulate that the precise outcome of each individual measurement process is in principle determined by some at present "hidden" elements or variables; and we could then try to find experiments that depended in a unique and reproducible way on the assumed state of these hidden elements or variables. If such predictions are verified, we should then obtain experimental evidence favoring the hypothesis that hidden variables exist. If they are not verified, however, the correctness of the usual interpretation of the quantum theory is not necessarily proved, since it may be necessary instead to alter the specific character of the theory that is supposed to describe the behavior of the assumed hidden variables.

We conclude then that a choice of the present interpretation of the quantum theory involves a real physical limitation on the kinds of theories that we wish to take into consideration. From the arguments given here, however, it would seem that there are no secure experimental or theoretical grounds on which we can base such a choice because this choice follows from hypotheses that cannot conceivably be subjected to an experimental test and because we now have an alternative interpretation.

4. NEW PHYSICAL INTERPRETATION OF SCHROEDINGER'S EQUATION

We shall now give a general description of our suggested physical interpretation of the present mathematical formulation of the quantum theory. We shall carry out a more detailed description in subsequent sections of this paper.

We begin with the one-particle Schroedinger equation, and shall later generalize to an arbitrary number of particles. This wave equation is

$$i\hbar\partial\psi/\partial t=-(\hbar^2/2m)\nabla^2\psi+V(\mathbf{x})\psi. \quad (1)$$

Now ψ is a complex function, which can be expressed as

$$\psi=R\exp(iS/\hbar), \quad (2)$$

where R and S are real. We readily verify that the equations for R and S are

$$\frac{\partial R}{\partial t}=-\frac{1}{2m}[R\nabla^2S+2\nabla R\cdot\nabla S], \quad (3)$$

$$\frac{\partial S}{\partial t}=-\left[\frac{(\nabla S)^2}{2m}+V(\mathbf{x})-\frac{\hbar^2}{2m}\frac{\nabla^2R}{R}\right]. \quad (4)$$

It is convenient to write $P(\mathbf{x})=R^2(\mathbf{x})$, or $R=P^{\frac{1}{2}}$ where $P(\mathbf{x})$ is the probability density. We then obtain

$$\frac{\partial P}{\partial t}+\nabla\cdot\left(P\frac{\nabla S}{m}\right)=0, \tag{5}$$

$$\frac{\partial S}{\partial t}+\frac{(\nabla S)^2}{2m}+V(\mathbf{x})-\frac{\hbar^2}{4m}\left[\frac{\nabla^2 P}{P}-\frac{1}{2}\frac{(\nabla P)^2}{P^2}\right]=0. \tag{6}$$

Now, in the classical limit ($\hbar\rightarrow 0$) the above equations are subject to a very simple interpretation. The function $S(\mathbf{x})$ is a solution of the Hamilton-Jacobi equation. If we consider an ensemble of particle trajectories which are solutions of the equations of motion, then it is a well-known theorem of mechanics that if all of these trajectories are normal to any given surface of constant S, then they are normal to all surfaces of constant S, and $\nabla S(\mathbf{x})/m$ will be equal to the velocity vector, $\mathbf{v}(\mathbf{x})$, for any particle passing the point $\mathbf{x}$. Equation (5) can therefore be re-expressed as

$$\partial P/\partial t+\nabla\cdot(P\mathbf{v})=0. \tag{7}$$

This equation indicates that it is consistent to regard $P(\mathbf{x})$ as the probability density for particles in our ensemble. For in that case, we can regard $P\mathbf{v}$ as the mean current of particles in this ensemble, and Eq. (7) then simply expresses the conservation of probability.

Let us now see to what extent this interpretation can be given a meaning even when $\hbar\neq 0$. To do this, let us assume that each particle is acted on, not only by a "classical" potential, $V(\mathbf{x})$ but also by a "quantum-mechanical" potential,

$$U(\mathbf{x})=\frac{-\hbar^2}{4m}\left[\frac{\nabla^2 P}{P}-\frac{1}{2}\frac{(\nabla P)^2}{P^2}\right]=\frac{-\hbar^2}{2m}\frac{\nabla^2 R}{R}. \tag{8}$$

Then Eq. (6) can still be regarded as the Hamilton-Jacobi equation for our ensemble of particles, $\nabla S(\mathbf{x})/m$ can still be regarded as the particle velocity, and Eq. (5) can still be regarded as describing conservation of probability in our ensemble. Thus, it would seem that we have here the nucleus of an alternative interpretation for Schroedinger's equation.

The first step in developing this interpretation in a more explicit way is to associate with each electron a particle having precisely definable and continuously varying values of position and momentum. The solution of the modified Hamilton-Jacobi equation (4) defines an ensemble of possible trajectories for this particle, which can be obtained from the Hamilton-Jacobi function, $S(\mathbf{x})$, by integrating the velocity, $\mathbf{v}(\mathbf{x})=\nabla S(\mathbf{x})/m$. The equation for S implies, however, that the particles moves under the action of a force which is not entirely derivable from the classical potential, $V(\mathbf{x})$, but which also obtains a contribution from the "quantum-mechanical" potential, $U(\mathbf{x})=(-\hbar^2/2m)\times\nabla^2 R/R$. The function, $R(\mathbf{x})$, is not completely arbitrary, but is partially determined in terms of $S(\mathbf{x})$ by the differential Eq. (3). Thus R and S can be said to codetermine each other. The most convenient way of obtaining R and S is, in fact, usually to solve Eq. (1) for the Schroedinger wave function, ψ, and then to use the relations,

$$\psi=U+iW=R[\cos(S/\hbar)+i\sin(S/\hbar)],$$
$$R^2=U^2+V^2;\quad S=\hbar\tan^{-1}(W/U).$$

Since the force on a particle now depends on a function of the absolute value, $R(\mathbf{x})$, of the wave function, $\psi(\mathbf{x})$, evaluated at the actual location of the particle, we have effectively been led to regard the wave function of an individual electron as a mathematical representation of an objectively real field. This field exerts a force on the particle in a way that is analogous to, but not identical with, the way in which an electromagnetic field exerts a force on a charge, and a meson field exerts a force on a nucleon. In the last analysis, there is, of course, no reason why a particle should not be acted on by a ψ-field, as well as by an electromagnetic field, a gravitational field, a set of meson fields, and perhaps by still other fields that have not yet been discovered.

The analogy with the electromagnetic (and other) field goes quite far. For just as the electromagnetic field obeys Maxwell's equations, the ψ-field obeys Schroedinger's equation. In both cases, a complete specification of the fields at a given instant over every point in space determines the values of the fields for all times. In both cases, once we know the field functions, we can calculate force on a particle, so that, if we also know the initial position and momentum of the particle, we can calculate its entire trajectory.

In this connection, it is worth while to recall that the use of the Hamilton-Jacobi equation in solving for the motion of a particle is only a matter of convenience and that, in principle, we can always solve directly by using Newton's laws of motion and the correct boundary conditions. The equation of motion of a particle acted on by the classical potential, $V(\mathbf{x})$, and the "quantum-mechanical" potential, Eq. (8), is

$$m\,d^2\mathbf{x}/dt^2=-\nabla\{V(\mathbf{x})-(\hbar^2/2m)\nabla^2 R/R\}. \tag{8a}$$

It is in connection with the boundary conditions appearing in the equations of motion that we find the only fundamental difference between the ψ-field and other fields, such as the electromagnetic field. For in order to obtain results that are equivalent to those of the usual interpretation of the quantum theory, we are required to restrict the value of the initial particle momentum to $\mathbf{p}=\nabla S(\mathbf{x})$. From the application of Hamilton-Jacobi theory to Eq. (6), it follows that this restriction is consistent, in the sense that if it holds initially, it will hold for all time. Our suggested new interpretation of the quantum theory implies, however, that this restriction is not inherent in the conceptual structure. We shall see in Sec. 9, for example, that it is quite consistent in our interpretation to contemplate modifications in the theory, which permit an arbitrary relation between $\mathbf{p}$ and $\nabla S(\mathbf{x})$. The law of force on the particle can, however, be so chosen that in the atomic domain, $\mathbf{p}$ turns out to be very nearly equal to $\nabla S(\mathbf{x})/m$, while in processes involving very small distances, these two quantities may be very different. In this way, we can improve the analogy between the ψ-field and the electromagnetic field (as well as between quantum mechanics and classical mechanics).

Another important difference between the ψ-field and the electromagnetic field is that, whereas Schroedinger's equation is homogeneous in ψ, Maxwell's equations are inhomogeneous in the electric and magnetic fields. Since inhomogeneities are needed to give rise to radiation, this means that our present equations imply that the ψ-field is not radiated or absorbed, but simply changes its form while its integrated intensity remains constant. This restriction to a homogeneous equation is, however, like the restriction to a homogeneous equation is, however, like the restriction to $\mathbf{p}=\nabla S(\mathbf{x})$, not inherent in the conceptual structure of our new interpretation. Thus, in Sec. 9, we shall show that one can consistently postulate inhomogeneities in the equation governing ψ, which produce important effects only at very small distances, and negligible effects in the atomic domain. If such inhomogeneities are actually present, then the ψ-field will be subject to being emitted and absorbed, but only in connection with processes associated with very small distances. Once the ψ-field has been emitted, however, it will in all atomic processes simply obey Schroedinger's equation as a very good approximation. Nevertheless, at very small distances, the value of the ψ-field would, as in the case of the electromagnetic field, depend to some extent on the actual location of the particle.

Let us now consider the meaning of the assumption of a statistical ensemble of particles with a probability density equal to $P(\mathbf{x})=R^2(\mathbf{x})=|\psi(\mathbf{x})|^2$. From Eq. (5), it follows that this assumption is consistent, provided that ψ satisfies Schroedinger's equation, and $\mathbf{v}=\nabla S(\mathbf{x})/m$. This probability density is numerically equal to the probability density of particles obtained in the usual interpretation. In the usual interpretation, however, the need for a probability description is regarded as inherent in the very structure of matter (see Sec. 2), whereas in our interpretation, it arises, as we shall see in Paper II, because from one measurement to the next, we cannot in practice predict or control the precise location of a particle, as a result of corresponding unpredictable and uncontrollable disturbances introduced by the measuring apparatus. Thus, in our interpretation, the use of a statistical ensemble is (as in the case of classical statistical mechanics) only a practical necessity, and not a reflection of an inherent limitation on the precision with which it is correct for us to conceive of the variables defining the state of the system. Moreover, it is clear that if in connection with very small distances we are ultimately required to give up the special assumptions that ψ satisfies Schroedinger's equation and that $\mathbf{v}=\nabla S(\mathbf{x})/m$, then $|\psi|^2$ will cease to satisfy a conservation equation and will therefore also cease to be able to represent the probability density of particles. Nevertheless, there would still be a true probability density of particles which is conserved. Thus, it would become possible in principle to find experiments in which $|\psi|^2$ could be distinguished from the probability density, and therefore to prove that the usual interpretation, which gives $|\psi|^2$ only a probability interpretation must be inadequate. Moreover, we shall see in Paper II that with the aid of such modifications in the theory, we could in principle measure the particle positions and momenta precisely, and thus violate the uncertainty principle. As long as we restrict ourselves to conditions in which Schroedinger's equation is satisfied, and in which $\mathbf{v}=\nabla S(\mathbf{x})/m$, however, the uncertainty principle will remain an effective practical limitation on the possible precision of measurements. This means that at present, the particle positions and momenta should be regarded as "hidden" variables, since as we shall see in Paper II, we are not now able to obtain experiments that localize them to a region smaller than that in which the intensity of the ψ-field is appreciable. Thus, we cannot yet find clear-cut experimental proof that the assumption of these variables is necessary, although it is entirely possible that, in the domain of very small distances, new modifications in the theory may have to be introduced, which would permit a proof of the existence of the definite particle position and momentum to be obtained.

We conclude that our suggested interpretation of the quantum theory provides a much broader conceptual framework than that provided by the usual interpretation, for all of the results of the usual interpretation are obtained from our interpretation if we make the following three special assumptions which are mutually consistent:

(1) That the ψ-field satisfies Schroedinger's equation.

(2) That the particle momentum is restricted to $\mathbf{p}=\nabla S(\mathbf{x})$.

(3) That we do not predict or control the precise location of the particle, but have, in practice, a statistical ensemble with probability density $P(\mathbf{x})=|\psi(\mathbf{x})|^2$. The use of statistics is, however, not inherent in the conceptual structure, but merely a consequence of our ignorance of the precise initial conditions of the particle.

As we shall see in Sec. 9, it is entirely possible that a better theory of phenomena involving distances of the order of 10^{-13} cm or less would require us to go beyond the limitations of these special assumptions. Our principal purpose in this paper (and in Paper II) is to show, however, that if one makes these special assumptions, our interpretation leads in all possible experiments to the same predictions as are obtained from the usual interpretation.[9]

It is now easy to understand why the adoption of the

usual interpretation of the quantum theory would tend to lead us away from the direction of our suggested alternative interpretation. For in a theory involving hidden variables, one would normally expect that the behavior of an individual system should not depend on the statistical ensemble of which it is a member, because this ensemble refers to a series of similar but disconnected experiments carried out under equivalent initial conditions. In our interpretation, however, the "quantum-mechanical" potential, $U(\mathbf{x})$, acting on an individual particle depends on a wave intensity, $P(\mathbf{x})$, that is also numerically equal to a probability density in our ensemble. In the terminology of the usual interpretation of the quantum theory, in which one tacitly assumes that the wave function has only one interpretation; namely, in terms of a probability, our suggested new interpretation would look like a mysterious dependence of the individual on the statistical ensemble of which it is a member. In our interpretation, such a dependence is perfectly rational, because the wave function can consistently be interpreted both as a force and as a probability density.[12]

It is instructive to carry our analogy between the Schroedinger field and other kinds of fields a bit further. To do this, we can derive the wave Eqs. (5) and (6) from a Hamiltonian functional. We begin by writing down the expression for the mean energy as it is expressed in the usual quantum theory:

$$\bar{H}=\int\psi^*\left(-\frac{\hbar^2}{2m}\nabla^2+V(\mathbf{x})\right)\psi d\mathbf{x}$$

$$=\int\left\{\frac{\hbar^2}{2m}|\nabla\psi|^2+V(\mathbf{x})|\psi|^2\right\}d\mathbf{x}.$$

Writing $\psi=P^{\frac{1}{2}}\exp(iS/\hbar)$, we obtain

$$\bar{H}=\int P(\mathbf{x})\left\{\frac{(\nabla S)^2}{2m}+V(\mathbf{x})+\frac{\hbar^2}{8m}\frac{(\nabla P)^2}{P^2}\right\}d\mathbf{x}. \quad (9)$$

We shall now reinterpret $P(\mathbf{x})$ as a field coordinate, defined at each point, $\mathbf{x}$, and we shall tentatively assume that $S(\mathbf{x})$ is the momentum, canonically conjugate to $P(\mathbf{x})$. That such an assumption is appropriate can be verified by finding the Hamiltonian equations of motion for $P(\mathbf{x})$ and $S(\mathbf{x})$, under the assumption that the Hamiltonian functional is equal to $\bar{H}$ (See Eq. (9)). These equations of motion are

$$\dot{P}=\frac{\delta\bar{H}}{\delta S}=-\frac{1}{m}\nabla\cdot(P\nabla S),$$

$$\dot{S}=-\frac{\delta\bar{H}}{\delta P}=-\left[\frac{(\nabla S)^2}{2m}+V(\mathbf{x})-\frac{\hbar^2}{4m}\left(\frac{\nabla^2P}{P}-\frac{1}{2}\frac{(\nabla P)^2}{P^2}\right)\right].$$

[12] This consistency is guaranteed by the conservation Eq. (7). The questions of why an arbitrary statistical ensemble tends to decay into an ensemble with a probability density equal to $\psi^*\psi$ will be discussed in Paper II, Sec. 7.

These are, however, the same as the correct wave Eqs. (5) and (6).

We can now show that the mean particle energy averaged over our ensemble is equal to the usual quantum mechanical mean value of the Hamiltonian, $\bar{H}$. To do this, we note that according to Eqs. (3) and (6), the energy of a particle is

$$E(\mathbf{x})=-\frac{\partial S(\mathbf{x})}{\partial t}=\left[\frac{(\nabla S)^2}{2m}+V(\mathbf{x})-\frac{\hbar^2}{2m}\frac{\nabla^2R}{R}\right]. \quad (10)$$

The mean particle energy is found by averaging $E(\mathbf{x})$ with the weighting function, $P(\mathbf{x})$. We obtain

$$\langle E\rangle_{\substack{\text{ensemble}\\\text{average}}}=\int P(\mathbf{x})E(\mathbf{x})d\mathbf{x}$$

$$=\int P(\mathbf{x})\left[\frac{(\nabla S)^2}{2m}+V(\mathbf{x})\right]d\mathbf{x}-\frac{\hbar^2}{2m}\int R\nabla^2Rd\mathbf{x}.$$

A little integration by parts yields

$$\langle E\rangle_{\substack{\text{ensemble}\\\text{average}}}=\int P(\mathbf{x})\left[\frac{(\nabla S)^2}{2m}+V(\mathbf{x})+\frac{\hbar^2}{8m}\frac{(\nabla P)^2}{P^2}\right]d\mathbf{x}=\bar{H}. \quad (11)$$

5. THE STATIONARY STATE

We shall now show how the problem of stationary states is to be treated in our interpretation of the quantum theory.

The following seem to be reasonable requirements in our interpretation for a stationary state:

(1) The particle energy should be a constant of the motion.

(2) The quantum-mechanical potential should be independent of time.

(3) The probability density in our statistical ensemble should be independent of time.

It is easily verified that these requirements can be satisfied with the assumption that

$$\psi(\mathbf{x},t)=\psi_0(\mathbf{x})\exp(-iEt/\hbar)$$
$$=R_0(\mathbf{x})\exp[i(\Phi(\mathbf{x})-Et)/\hbar]. \quad (12)$$

From the above, we obtain $S=\Phi(\mathbf{x})-Et$. According to the generalized Hamilton-Jacobi Eq. (4), the particle energy is given by

$$\partial S/\partial t=-E.$$

Thus, we verify that the particle energy is a constant of the motion. Moreover, since $P=R^2=|\psi|^2$, it follows that P (and R) are independent of time. This means that both the probability density in our ensemble and the quantum-mechanical potential are also time independent.

The reader will readily verify that no other form of solution of Schroedinger's equation will satisfy all three of our criteria for a stationary state.

Since ψ is now being regarded as a mathematical representation of an objectively real force field, it follows that (like the electromagnetic field) it should be everywhere finite, continuous, and single valued. These requirements will guarantee in all cases that occur in practice that the allowed values of the energy in a stationary state, and the corresponding eigenfunctions are the same as are obtained from the usual interpretation of the theory.

In order to show in more detail what a stationary state means in our interpretation, we shall now consider three examples of stationary states.

Case 1: "s" State

The first case that we shall consider is an "s" state. In an "s" state, the wave function is

$$\psi = f(r)\exp[i(\alpha - Et)/\hbar], \tag{13}$$

where α is an arbitrary constant and r is the radius taken from the center of the atom. We conclude that the Hamilton-Jacobi function is

$$S = \alpha - Et.$$

The particle velocity is

$$\mathbf{v} = \nabla S = 0.$$

The particle is therefore simply standing still, wherever it may happen to be. How can it do this? The absence of motion is possible because the applied force, $-\nabla V(\mathbf{x})$, is balanced by the "quantum-mechanical" force, $(\hbar^2/2m)\nabla(\nabla^2 R/R)$, produced by the Schroedinger ψ-field acting on its own particle. There is, however, a statistical ensemble of possible positions of the particle, with a probability density, $P(\mathbf{x}) = (f(r))^2$.

Case 2: State with Nonzero Angular Momentum

In a typical state of nonzero angular momentum, we have

$$\psi = f_n^{\,l}(r)P_l^{\,m}(\cos\theta)\exp[i(\beta - Et + \hbar m\phi)/\hbar], \tag{14}$$

where θ and ϕ are the colatitude and azimuthal polar angles, respectively, $P_l^{\,m}$ is the associated Legendre polynomial, and β is a constant. The Hamilton-Jacobi function is $S = \beta - Et + \hbar m\phi$. From this result it follows that the z component of the angular momentum is equal to $\hbar m$. To prove this, we write

$$L_z = xp_y - yp_x = x\partial S/\partial y - y\partial S/\partial x = \partial S/\partial\phi = \hbar m. \tag{15}$$

Thus, we obtain a statistical ensemble of trajectories which can have different forms, but all have the same "quantized" value of the z component of the angular momentum.

Case 3: A Scattering Problem

Let us now consider a scattering problem. Because it is comparatively easy to analyze, we shall discuss a hypothetical experiment, in which an electron is incident in the z direction with an initial momentum, p_0, on a system consisting of two slits.[11] After the electron passes through the slit system, its position is measured and recorded, for example, on a photographic plate.

Now, in the usual interpretation of the quantum theory, the electron is described by a wave function. The incident part of the wave function is $\psi_0 \sim \exp(ip_0 z/\hbar)$; but when the wave passes through the slit system, it is modified by interference and diffraction effects, so that it will develop a characteristic intensity pattern by the time it reaches the position measuring instrument. The probability that the electron will be detected between $\mathbf{x}$ and $\mathbf{x}+d\mathbf{x}$ is $|\psi(\mathbf{x})|^2 d\mathbf{x}$. If the experiment is repeated many times under equivalent initial conditions, one eventually obtains a pattern of hits on the photographic plate that is very reminiscent of the interference patterns of optics.

In the usual interpretation of the quantum theory, the origin of this interference pattern is very difficult to understand. For there may be certain points where the wave function is zero when both slits are open, but not zero when only one slit is open. How can the opening of a second slit prevent the electron from reaching certain points that it could reach if this slit were closed? If the electron acted completely like a classical particle, this phenomenon could not be explained at all. Clearly, then the wave aspects of the electron must have something to do with the production of the interference pattern. Yet, the electron cannot be identical with its associated wave, because the latter spreads out over a wide region. On the other hand, when the electron's position is measured, it always appears at the detector as if it were a localized particle.

The usual interpretation of the quantum theory not only makes no attempt to provide a single precisely defined conceptual model for the production of the phenomena described above, but it asserts that no such model is even conceivable. Instead of a single precisely defined conceptual model, it provides, as pointed out in Sec. 2, a pair of complementary models, *viz.*, particle and wave, each of which can be made more precise only under conditions which necessitate a reciprocal decrease in the degree of precision of the other. Thus, while the electron goes through the slit system, its position is said to be inherently ambiguous, so that if we wish to obtain an interference pattern, it is meaningless to ask through which slit an individual electron actually passed. Within the domain of space within which the position of the electron has no meaning we can use the wave model and thus describe the subsequent production of interference. If, however, we

[11] This experiment is discussed in some detail in reference 2, Chapter 6, Sec. 2.

tried to define the position of the electron as it passed the slit system more accurately by means of a measurement, the resulting disturbance of its motion produced by the measuring apparatus would destroy the interference pattern. Thus, conditions would be created in which the particle model becomes more precisely defined at the expense of a corresponding decrease in the degree of definition of the wave model. When the position of the electron is measured at the photographic plate, a similar sharpening of the degree of definition of the particle model occurs at the expense of that of the wave model.

In our interpretation of the quantum theory, this experiment is described causally and continuously in terms of a single precisely definable conceptual model. As we have already shown, we must use the same wave function as is used in the usual interpretation; but instead we regard it as a mathematical representation of an objectively real field that determines part of the force acting on the particle. The initial momentum of the particle is obtained from the incident wave function, $\exp(ip_0z/\hbar)$, as $p=\partial s/\partial z=p_0$. We do not in practice, however, control the initial location of the particle, so that although it goes through a definite slit, we cannot predict which slit this will be. The particle is at all times acted on by the "quantum-mechanical" potential, $U=(-\hbar^2/2m)\nabla^2R/R$. While the particle is incident, this potential vanishes because R is then a constant; but after it passes through the slit system, the particle encounters a quantum-mechanical potential that changes rapidly with position. The subsequent motion of the particle may therefore become quite complicated. Nevertheless, the probability that a particle shall enter a given region, $d\mathbf{x}$, is as in the usual interpretation, equal to $|\psi(\mathbf{x})|^2d\mathbf{x}$. We therefore deduce that the particle can never reach a point where the wave function vanishes. The reason is that the "quantum-mechanical" potential, U, becomes infinite when R becomes zero. If the approach to infinity happens to be through positive values of U, there will be an infinite force repelling the particle away from the origin. If the approach is through negative values of U, the particle will go through this point with infinite speed, and thus spend no time there. In either case, we obtain a simple and precisely definable conceptual model explaining why particles can never be found at points where the wave function vanishes.

If one of the slits is closed, the "quantum-mechanical" potential is correspondingly altered, because the ψ-field is changed, and the particle may then be able to reach certain points which it was unable to reach when both slits were open. The slit is therefore able to affect the motion of the particle only indirectly, through its effect on the Schroedinger ψ-field. Moreover, as we shall see in Paper II, if the position of the electron is measured while it is passing through the slit system, the measuring apparatus will, as in the usual interpretation, create a disturbance that destroys the interference pattern. In our interpretation, however, the necessity for this destruction is not inherent in the conceptual structure; and as we shall see, the destruction of the interference pattern could in principle be avoided by means of other ways of making measurements, ways which are conceivable but not now actually possible.

6. THE MANY-BODY PROBLEM

We shall now extend our interpretation of the quantum theory to the problem of many bodies. We begin with the Schroedinger equation for two particles. (For simplicity, we assume that they have equal masses, but the extension of our treatment to arbitrary masses will be obvious.)

$$i\hbar\frac{\partial\psi}{\partial t}=-\frac{\hbar^2}{2m}(\nabla_1^2\psi+\nabla_2^2\psi)+V(\mathbf{x}_1,\mathbf{x}_2)\psi.$$

Writing $\psi=R(\mathbf{x}_1,\mathbf{x}_2)\exp[iS(\mathbf{x}_1,\mathbf{x}_2)/\hbar]$ and $R^2=P$, we obtain

$$\frac{\partial P}{\partial t}+\frac{1}{m}[\nabla_1\cdot P\nabla_1S+\nabla_2\cdot P\nabla_2S]=0, \tag{16}$$

$$\frac{\partial S}{\partial t}+\frac{(\nabla_1S)^2+(\nabla_2S)^2}{2m}+V(\mathbf{x}_1,\mathbf{x}_2)-\frac{\hbar^2}{2mR}[\nabla_1^2R+\nabla_2^2R]=0. \tag{17}$$

The above equations are simply a six-dimensional generalization of the similar three-dimensional Eqs. (5) and (6) associated with the one-body problem. In the two-body problem, the system is described therefore by a six-dimensional Schroedinger wave and by a six-dimensional trajectory, specifying the actual location of each of the two particles. The velocity of this trajectory has components, ∇_1S/m and ∇_2S/m, respectively, in each of the three-dimensional surfaces associated with a given particle. $P(\mathbf{x}_1,\mathbf{x}_2)$ then has a dual interpretation. First, it defines a "quantum-mechanical" potential, acting on each particle

$$U(\mathbf{x}_1,\mathbf{x}_2)=-(\hbar^2/2mR)[\nabla_1^2R+\nabla_2^2R].$$

This potential introduces an additional effective interaction between particles over and above that due to the classically inferrable potential $V(\mathbf{x})$. Secondly, the function $P(\mathbf{x}_1,\mathbf{x}_2)$ can consistently be regarded as the probability density of representative points $(\mathbf{x}_1,\mathbf{x}_2)$ in our six-dimensional ensemble.

The extension to an arbitrary number of particles is straightforward, and we shall quote only the results here. We introduce the wave function, $\psi=R(\mathbf{x}_1,\mathbf{x}_2,\cdots\mathbf{x}_n)\exp[iS(\mathbf{x}_1,\mathbf{x}_2\cdots\mathbf{x}_n)/\hbar]$ and define a $3n$-dimensional trajectory, where n is the number of particles, which describes the behavior of every particle in the system. The velocity of the ith particle is $\mathbf{v}_i=\nabla_iS(\mathbf{x}_1,\mathbf{x}_2\cdots\mathbf{x}_n)/m$. The function $P(\mathbf{x}_1,\mathbf{x}_2\cdots\mathbf{x}_n)=R^2$ has two interpretations. First, it defines a "quantum-mechanical" potential

$$U(\mathbf{x}_1, \mathbf{x}_2 \cdots \mathbf{x}_n) = -\frac{\hbar^2}{2mR} \sum_{s=1}^{n} \nabla_s^2 R(\mathbf{x}_1, \mathbf{x}_2 \cdots \mathbf{x}_n). \quad (18)$$

Secondly, $P(\mathbf{x}_1, \mathbf{x}_2 \cdots \mathbf{x}_n)$ is equal to the density of representative points $(\mathbf{x}_1, \mathbf{x}_2 \cdots \mathbf{x}_n)$ in our $3n$-dimensional ensemble.

We see here that the "effective potential," $U(\mathbf{x}_1, \mathbf{x}_2, \cdots \mathbf{x}_n)$, acting on a particle is equivalent to that produced by a "many-body" force, since the force between any two particles may depend significantly on the location of every other particle in the system. An example of the effects of such a force is given by the exclusion principle. Thus, if the wave function is antisymmetric, we deduce that the "quantum-mechanical" forces will be such as to prevent two particles from ever reaching the same point in space, for in this case, we must have $P=0$.

7. TRANSITIONS BETWEEN STATIONARY STATES—THE FRANCK-HERTZ EXPERIMENT

Our interpretation of the quantum theory describes all processes as basically causal and continuous. How then can it lead to a correct description of processes such as the Franck-Hertz experiment, the photoelectric effect, and the Compton effect, which seem to call most strikingly for an interpretation in terms of discontinuous and incompletely determined transfers of energy and momentum? In this section, we shall answer this question by applying our suggested interpretation of the quantum theory in the analysis of the Franck-Hertz experiment. Here, we shall see that the apparently discontinuous nature of the process of transfer of energy from the bombarding particle to the atomic electron is brought about by the "quantum-mechanical" potential, $U=(-\hbar^2/2m)\nabla^2 R/R$, which does not necessarily become small when the wave intensity becomes small. Thus, even if the force of interaction between two particles is very weak, so that a correspondingly small disturbance of the Schroedinger wave function is produced by the interaction of these particles, this disturbance is capable of bringing about very large transfers of energy and momentum between the particles in a very short time. This means that if we view only the end results, this process presents the aspect of being discontinuous. Moveover, we shall see that the precise value of the energy transfer is in principle determined by the initial position of each particle and by the initial form of the wave function. Since we cannot in practice predict or control the initial particle positions with complete precision, we are also unable to predict or control the final outcome of such an experiment, and can, in practice, predict only the probability of a given outcome. Because the probability that the particles will enter a region with coordinates, $\mathbf{x}_1, \mathbf{x}_2$, is proportional to $R^2(\mathbf{x}_1, \mathbf{x}_2)$, we conclude that although a Schroedinger wave of low intensity can bring about large transfers of energy, such a process is (as in the usual interpretation) highly improbable.

In Appendix A of Paper II, we shall see that similar possibilities arise in connection with the interaction of the electromagnetic field with charged matter, so that electromagnetic waves can very rapidly transfer a full quantum of energy (and momentum) to an electron, even after they have spread out and fallen to a very low intensity. In this way, we shall explain the photoelectric effect and the Compton effect. Thus, we are able in our interpretation to understand by means of a causal and continuous model just those properties of matter and light which seem most convincingly to require the assumption of discontinuity and incomplete determinism.

Before we discuss the process of interaction between two particles, we shall find it convenient to analyze the problem of an isolated single particle that happens to be in a nonstationary state. Because the field function ψ is a solution of Schroedinger's equation, we can linearly suppose stationary-state solutions of this equation and in this way obtain new solutions. As an illustration, let us consider a superposition of two solutions

$$\psi = C_1\psi_1(\mathbf{x})\exp(-iE_1t/\hbar) + C_2\psi_2(\mathbf{x})\exp(-iE_2t/\hbar),$$

where C_1, C_2, ψ_1, and ψ_2 are real. Thus we write $\psi_1 = R_1$, $\psi_2 = R_2$, and

$$\psi = \exp[-i(E_1+E_2)t/2\hbar]\{C_1R_1\exp[-i(E_1-E_2)t/2\hbar] + C_2R_2\exp[i(E_1-E_2)t/2\hbar]\}.$$

Writing $\psi = R\exp(iS/\hbar)$, we obtain

$$R^2 = C_1^2R_1^2(\mathbf{x}) + C_2^2R_2^2(\mathbf{x}) + 2C_1C_2R_1(\mathbf{x})R_2(\mathbf{x})\cos[(E_1-E_2)t/2\hbar], \quad (19)$$

$$\tan\left\{\frac{S+(E_1-E_2)t/2}{\hbar}\right\} = \frac{C_2R_2(\mathbf{x})-C_1R_1(\mathbf{x})}{C_2R_2(\mathbf{x})+C_1R_1(\mathbf{x})}\tan\left\{\frac{(E_1-E_2)t}{2\hbar}\right\}. \quad (20)$$

We see immediately that the particle experiences a "quantum-mechanical" potential, $U(\mathbf{x}) = (-\hbar/2m)\nabla^2 R/R$, which fluctuates with angular frequency, $\omega = (E_1 - E_2)/\hbar$, and that the energy of this particle, $E = -\partial S/\partial t$, and its momentum $\mathbf{p} = \nabla S$, fluctuate with the same angular frequency. If the particle happens to enter a region of space where R is small, these fluctuations can become quite violent. We see then that, in general, the orbit of a particle in a nonstationary state is very irregular and complicated, resembling Brownian motion more closely than it resembles the smooth track of a planet around the sun.

If the system is isolated, these fluctuations will continue forever. The result is quite reasonable, since as is well known, a system can make a transition from one stationary state to another only if it can exchange energy with some other system. In order to treat the problem of transition between stationary states, we must therefore introduce another system capable of exchanging energy with the system of interest. In this section, we shall discuss the Franck-Hertz experiment, in which this other system consists of a bombarding particle. For the sake of illustration, let us suppose that we have hydrogen atoms of energy E_0 and wave function, $\psi_0(\mathbf{x})$, which are bombarded by particles that can be scattered inelastically, leaving the atom with energy E_n and wave function, $\psi_n(\mathbf{x})$.

We begin by writing down the initial wave function, $\Psi_i(\mathbf{x}, \mathbf{y}, t)$. The incident particle, whose coordinates are represented by $\mathbf{y}$ must be associated with a wave packet, which can be written as

$$f_0(\mathbf{y}, t)=\int e^{i\mathbf{k}\cdot\mathbf{y}}f(\mathbf{k}-\mathbf{k}_0)\exp(-i\hbar k^2t/2m)d\mathbf{k}. \quad (21)$$

The center of this packet occurs where the phase has an extremum as a function of $\mathbf{k}$, or where $\mathbf{y}=\hbar\mathbf{k}_0t/m$.

Now, as in the usual interpretation, we begin by writing the incident wave function for the combined system as a product

$$\Psi_i=\psi_0(\mathbf{x})\exp(-iE_0t/\hbar)f_0(\mathbf{y}, t). \quad (22)$$

Let us now see how this wave function is to be understood in our interpretation of the theory. As pointed out in Sec. 6, the wave function is to be regarded as a mathematical representation of a six-dimensional but objectively real field, capable of producing forces that act on the particles. We also assume a six-dimensional representative point, described by the coordinates of the two particles, $\mathbf{x}$ and $\mathbf{y}$. We shall now see that when the combined wave function takes the form (22) involving a product of a function of $\mathbf{x}$ and a function of $\mathbf{y}$, the six-dimensional system can correctly be regarded as being made up of two independent three-dimentional subsystems. To prove this, we write

$$\psi_0(\mathbf{x})=R_0(\mathbf{x})\exp[iS_0(\mathbf{x})/\hbar] \quad \text{and}$$
$$f_0(\mathbf{y}, t)=M_0(\mathbf{y}, t)\exp[iN_0(\mathbf{y}, t)/\hbar].$$

We then obtain for the particle velocities

$$d\mathbf{x}/dt=(1/m)\nabla S_0(\mathbf{x}); \quad d\mathbf{y}/dt=(1/m)\nabla N_0(\mathbf{y}, t), \quad (23)$$

and for the "quantum-mechanical" potential

$$U=-\frac{\hbar^2\{(\nabla_x^2+\nabla_y^2)R(\mathbf{x}, \mathbf{y})\}}{2mR(\mathbf{x}, \mathbf{y})}$$
$$=\frac{-\hbar^2}{2m}\left\{\frac{\nabla^2R_0(\mathbf{x})}{R_0(\mathbf{x})}+\frac{\nabla^2M_0(\mathbf{y}, t)}{M_0(\mathbf{y}, t)}\right\}. \quad (24)$$

Thus, the particle velocities are independent and the "quantum-mechanical" potential reduces to a sum of terms, one involving only $\mathbf{x}$ and the other involving only $\mathbf{y}$. This means that the particles move independently. Moreover, the probability density, $P=R_0^2(\mathbf{x})\times M_0^2(\mathbf{y}, t)$, is a product of a function of $\mathbf{x}$ and a function of $\mathbf{y}$, indicating that the distribution in $\mathbf{x}$ is statistically independent of that in $\mathbf{y}$. We conclude, then, that whenever the wave function can be expressed as a product of two factors, each involving only the coordinates of a single system, then the two systems are completely independent of each other.

As soon as the wave packet in $\mathbf{y}$ space reaches the neighborhood of the atom, the two systems begin to interact. If we solve Schroedinger's equation for the combined system, we obtain a wave function that can be expressed in terms of the following series:

$$\Psi=\Psi_i+\sum_n\psi_n(\mathbf{x})\exp(-iE_nt/\hbar)f_n(\mathbf{y}, t), \quad (25)$$

where the $f_n(\mathbf{y}, t)$ are the expansion coefficients of the complete set of functions, $\psi_n(\mathbf{x})$. The asymptotic form of the wave function is[14]

$$\Psi=\Psi_i(\mathbf{x}, \mathbf{y})+\sum_n\psi_n(\mathbf{x})\exp\left(-\frac{iE_nt}{\hbar}\right)\int f(\mathbf{k}-\mathbf{k}_0)$$
$$\times\frac{\exp[ik_n\cdot\mathbf{r}-(\hbar k_n^2/2n)t]}{r}g_n(\theta, \phi, \mathbf{k})d\mathbf{k}, \quad (26)$$

where

$$\hbar^2k_n^2/2m=(\hbar^2k_0^2/2m)+E_0-E_n$$
$$\text{(conservation of energy).} \quad (27)$$

The additional terms in the above equation represent outgoing wave packets, in which the particle speed, $\hbar k_n/m$, is correlated with the wave function, $\psi_n(\mathbf{x})$, representing the state in which the hydrogen atom is left. The center of the nth packet occurs at

$$r_n=(\hbar k_n/m)t. \quad (28)$$

It is clear that because the speed depends on the hydrogen atom quantum number, n, every one of these packets will eventually be separated by distances which are so large that this separation is classically describable.

When the wave function takes the form (25), the two particles system must be described as a single six-dimensional system and not as a sum of two independent three-dimensional subsystems, for at this time, if we try to express the wave function as $\psi(\mathbf{x}, \mathbf{y})=R(\mathbf{x}, \mathbf{y})\times\exp[iS(\mathbf{x}, \mathbf{y})/\hbar]$, we find that the resulting expressions for R and S depend on $\mathbf{x}$ and $\mathbf{y}$ in a very complicated way. The particle momenta, $\mathbf{p}_1=\nabla_xS(\mathbf{x}, \mathbf{y})$ and $\mathbf{p}_2=\nabla_yS(\mathbf{x}, \mathbf{y})$, therefore become inextricably interdependent. The "quantum-mechanical" potential,

$$U=-\frac{\hbar^2}{2mR(\mathbf{x}, \mathbf{y})}(\nabla_x^2R+\nabla_y^2R)$$

ceases to be expressible as the sum of a term involving $\mathbf{x}$ and a term involving $\mathbf{y}$. The probability density,

[14] N. F. Mott and H. S. W. Massey, *The Theory of Atomic Collisions* (Clarendon Press, Oxford, 1933).

$R^2(\mathbf{x}, \mathbf{y})$ can no longer be written as a product of a function of $\mathbf{x}$ and a function of $\mathbf{y}$, from which we conclude that the probability distributions of the two particles are no longer statistically independent. Moreover, the motion of the particle is exceedingly complicated, because the expressions for R and S are somewhat analogous to those obtained in the simpler problem of a nonstationary state of a single particle [see Eqs. (19) and (20)]. In the region where the scattered waves $\psi_n(\mathbf{x})f_n(\mathbf{y}, t)$ have an amplitude comparable with that of the incident wave, $\psi_0(\mathbf{x})f_0(\mathbf{y}, t)$, the functions R and S, and therefore the "quantum-mechanical" potential and the particle momenta, undergo rapid and violent fluctuations, both as functions of position and of time. Because the quantum-mechanical potential has $R(\mathbf{x}, \mathbf{y}, t)$ in the denominator, these fluctuations may become very large in this region where R is small. If the particles happen to enter such a region, they may exchange very large quantities of energy and momentum in a very short time, even if the classical potential, $V(\mathbf{x}, \mathbf{y})$ is very small. A small value of $V(\mathbf{x}, \mathbf{y})$ implies, however, a correspondingly small value of the scattered wave amplitudes, $f_n(\mathbf{y}, t)$. Since the fluctuations become large only in the region where the scattered wave amplitude is comparable with the incident wave amplitude and since the probability that the particles shall enter a given region of $\mathbf{x}$, $\mathbf{y}$ space is proportional to $R^2(\mathbf{x}, \mathbf{y})$, it is clear that a large transfer of energy is improbable (although still always possible) when $V(\mathbf{x}, \mathbf{y})$ is small.

While interaction between the two particles takes place then, their orbits are subject to wild fluctuations. Eventually, however, the behavior of the system quiets down and becomes simple again. For after the wave function takes its asymptotic form (26), and the packets corresponding to different values of n have obtained classically describable separations, we can deduce that because the probability density is $|\psi|^2$, the outgoing particle must enter one of these packets and stay with that packet thereafter (since it does not enter the space between packets in which the probability density is negligibly different from zero). In the calculation of the particle velocities, $\mathbf{V}_1=\nabla_x S/m$, $\mathbf{V}_2=\nabla_y S/m$, and of the quantum-mechanical potential, $U=(-\hbar^2/2mR)(\nabla_x^2 R+\nabla_y^2 R)$, we can therefore ignore all parts of the wave function other than the one actually containing the outgoing particle. It follows that the system acts as if it had the wave function

$$\Psi_n=\psi_n(\mathbf{x})\exp\left(\frac{iE_n t}{\hbar}\right)\int f(\mathbf{k}-\mathbf{k}_0) \times\frac{\exp\{i[\mathbf{k}_n\cdot\mathbf{r}-(\hbar k_n^2 t/2m)t]\}}{r}g_n(\theta, \phi, \mathbf{k})d\mathbf{k}, \quad (29)$$

where n denotes the packet actually containing the outgoing particle. This means that for all practical purposes the complete wave function (26) of the system may be replaced by Eq. (29), which corresponds to an atomic electron in its nth quantum state, and to an outgoing particle with a correlated energy, $E_n'=\hbar^2 k_n^2/2m$. Because the wave function is a product of a function of $\mathbf{x}$ and a function of $\mathbf{y}$, each system once again acts independently of the other. The wave function can now be renormalized because the multiplication of Ψ_n by a constant changes no physically significant quantity, such as the particle velocity or the "quantum-mechanical" potential. As shown in Sec. 5, when the electronic wave function is $\psi_n(\mathbf{x})\exp(-iE_n t/\hbar)$, its energy must be E_n. Thus, we have obtained a description of how it comes about that the energy is always transferred in quanta of size E_n-E_0.

It should be noted that while the wave packets are still separating, the electron energy is not quantized, but has a continuous range of values, which fluctuate rapidly. It is only the final value of the energy, appearing after the interaction is over that must be quantized. A similar result is obtained in the usual interpretation if one notes that because of the uncertainty principle, the energy of either system can become definite only after enough time has elapsed to complete the scattering process.[15]

In principle, the actual packet entered by the outgoing particle could be predicted if we knew the initial position of both particles and, of course, the initial form of the wave function of the combined system.[16] In practice, however, the particle orbits are very complicated and very sensitively dependent on the precise values of these initial positions. Since we do not at present know how to measure these initial positions precisely, we cannot actually predict the outcome of such an interaction process. The best that we can do is to predict the probability that an outgoing particle enters the nth packet within a given range of solid angle, $d\Omega$, leaving the hydrogen atom in its nth quantum state. In doing this, we use the fact that the probability density in $\mathbf{x}$, $\mathbf{y}$ space is $|\psi(\mathbf{x}, \mathbf{y})|^2$ and that as long as we are restricted to the nth packet, we can replace the complete wave function (26) by the wave function (29), corresponding to the packet that actually contains the particle. Now, by definition, we have $\int|\psi_n(\mathbf{x})|^2 d\mathbf{x}=1$. The remaining integration of

$$\left|\int f(\mathbf{k}-\mathbf{k}_0)\frac{\exp\{i[k_n r-(\hbar k_n^2/2m)t]\}}{r}g_n(\theta, \phi, k)d\mathbf{k}\right|^2$$

over the region of space corresponding to the nth outgoing packet leads, however, to precisely the same probability of scattering as would have been obtained by applying the usual interpretation. We conclude, then, that if ψ satisfies Schroedinger's equation, that if $\mathbf{v}=\nabla S/m$, and that if the probability density of particles is $P(\mathbf{x}, \mathbf{y})=R^2(\mathbf{x}, \mathbf{y})$, we obtain in every respect

[15] See reference 2, Chapter 18, Sec. 19.

[16] Note that in the usual interpretation one assumes that *nothing* determines the precise outcome of an individual scattering process. Instead, one assumes that all descriptions are inherently and unavoidably statistical (see Sec. 2).

exactly the same physical predictions for this problem as are obtained when we use the usual interpretation.

There remains only one more problem; namely, to show that if the outgoing packets are subsequently brought together by some arrangement of matter that does not act on the atomic electron, the atomic electron and the scattered particle will continue to act independently.[17] To show that these two particles will continue to act independently, we note that in all practical applications, the outgoing particle soon interacts with some classically describable system. Such a system might consist, for example, of the host of atoms of the gas with which it collides or of the walls of a container. In any case, if the scattering process is ever to be observed, the outgoing particle must interact with a classically describable measuring apparatus. Now all classically describable systems have the property that they contain an enormous number of internal "thermodynamic" degrees of freedom that are inevitably excited when the outgoing particle interacts with the system. The wave function of the outgoing particle is then coupled to that of these internal thermodynamic degrees of freedom, which we represent as $y_1, y_2, \cdots y_s$. To denote this coupling, we write the wave function for the entire system as

$$\Psi=\sum_n \psi_n(\mathbf{x})\exp(-iE_nt/\hbar)f_n(\mathbf{y}, y_1, y_2\cdots y_s). \quad (30)$$

Now, when the wave function takes this form, the overlapping of different packets in $\mathbf{y}$ space is not enough to produce interference between the different $\psi_n(\mathbf{x})$. To obtain such interference, it is necessary that the packets $f_n(\mathbf{y}, y_1, y_2, \cdots y_s)$ overlap in every one of the $S+3$ dimensions, $\mathbf{y}, y_1, y_2\cdots y_s$. The reader will readily convince himself, by considering a typical case such as a collision of the outgoing particle with a metal wall, that it is overwhelmingly improbable that two of the packets $f_n(y_1, y_1, y_2\cdots y_s)$ will overlap with regard to every one of the internal thermodynamic coordinates, $y_1, y_2, \cdots y_s$, even if they are successfully made to overlap in $\mathbf{y}$ space. This is because each packet corresponds to a different particle velocity and to a different time of collision with the metal wall. Because the myriads of internal thermodynamic degrees of freedom are so chaotically complicated, it is very likely that as each of the n packets interacts with them, it will encounter different conditions, which will make the combined wave packet $f_n(\mathbf{y}, y_1, \cdots y_s)$ enter very different regions of $y_1, y_2\cdots y_s$ space. Thus, for all practical purposes, we can ignore the possibility that if two of the packets are made to cross in $\mathbf{y}$ space, the motion either of the atomic electron or of the outgoing particle will be affected.[18]

[17] See reference 2, Chapter 22, Sec. 11, for a treatment of a similar problem.

[18] It should be noted that exactly the same problem arises in the usual interpretation of the quantum theory for (reference 16), for whenever two packets overlap, then even in the usual interpretation, the system must be regarded as, in some sense, covering the states corresponding to both packets simultaneously. See reference 2, Chapter 6 and Chapter 16, Sec. 25. Once two packets have obtained classically describable separations, then, both in the usual interpretation and in our interpretation the probability that there will be significant interference between them is so overwhelmingly small that it may be compared to the probability that a tea kettle placed on a fire will happen to freeze instead of boil. Thus, we may for all practical purposes neglect the possibility of interference between packets corresponding to the different possible energy states in which the hydrogen atom may be left.

8. PENETRATION OF A BARRIER

According to classical physics, a particle can never penetrate a potential barrier having a height greater than the particle kinetic energy. In the usual interpretation of the quantum theory, it is said to be able, with a small probability, to "leak" through the barrier. In our interpretation of the quantum theory, however, the potential provided by the Schroedinger ψ-field enables it to "ride" over the barrier, but only a few particles are likely to have trajectories that carry them all the way across without being turned around.

We shall merely sketch in general terms how the above results can be obtained. Since the motion of the particle is strongly affected by its ψ-field, we must first solve for this field with the aid of "Schroedinger's equation." Initially, we have a wave packet incident on the potential barrier; and because the probability density is equal to $|\psi(\mathbf{x})|^2$, the particle is certain to be somewhere within this wave packet. When the wave packet strikes the repulsive barrier, the ψ-field undergoes rapid changes which can be calculated[19] if desired, but whose precise form does not interest us here. At this time, the "quantum-mechanical" potential, $U=(-\hbar^2/2m)\nabla^2R/R$, undergoes rapid and violent fluctuations, analogous to those described in Sec. 7 in connection with Eqs. (19), (20), and (25). The particle orbit then becomes very complicated and, because the potential is time dependent, very sensitive to the precise initial relationship between the particle position and the center of the wave packet. Ultimately, however, the incident wave packet disappears and is replaced by two packets, one of them a reflected packet and the other a transmitted packet having a much smaller intensity. Because the probability density is $|\psi|^2$, the particle must end up in one of these packets. The other packet can, as shown in Sec. 7, subsequently be ignored. Since the reflected packet is usually so much stronger than the transmitted packet, we conclude that during the time when the packet is inside the barrier, most of the particle orbits must be turned around, as a result of the violent fluctuations in the "quantum-mechanical" potential.

9. POSSIBLE MODIFICATIONS IN MATHEMATICAL FORMULATION LEADING TO EXPERIMENTAL PROOF THAT NEW INTERPRETATION IS NEEDED

We have already seen in a number of cases and in Paper II we shall prove in general, that as long as we

[19] See, for example, reference 2, Chapter 11, Sec. 17, and Chapter 12, Sec. 18.

assume that ψ satisfies Schroedinger's equation, that $\mathbf{v}=\nabla S(\mathbf{x})/m$, and that we have a statistical ensemble with a probability density equal to $|\psi(\mathbf{x})|^2$, our interpretation of the quantum theory leads to physical results that are identical with those obtained from the usual interpretation. Evidence indicating the need for adopting our interpretation instead of the usual one could therefore come only from experiments, such as those involving phenomena associated with distances of the order of 10^{-13} cm or less, which are not now adequately understood in terms of the existing theory. In this paper we shall not, however, actually suggest any specific experimental methods of distinguishing between our interpretation and the usual one, but shall confine ourselves to demonstrating that such experiments are conceivable.

Now, there are an infinite number of ways of modifying the mathematical form of the theory that are consistent with our interpretation and not with the usual interpretation. We shall confine ourselves here, however, to suggesting two such modifications, which have already been indicated in Sec. 4, namely, to give up the assumption that $\mathbf{v}$ is necessarily equal to $\nabla S(\mathbf{x})/m$, and to give up the assumption that ψ must necessarily satisfy a homogeneous linear equation of the general type suggested by Schroedinger. As we shall see, giving up either of those first two assumptions will in general also require us to give up the assumption of a statistical ensemble of particles, with a probability density equal to $|\psi(\mathbf{x})|^2$.

We begin by noting that it is consistent with our interpretation to modify the equations of motion of a particle (8a) by adding any conceivable force term to the right-hand side. Let us, for the sake of illustration, consider a force that tends to make the difference, $\mathbf{p}-\nabla S(\mathbf{x})$, decay rapidly with time, with a mean decay time of the order of $\tau=10^{-13}/c$ seconds, where c is the velocity of light. To achieve this result, we write

$$m\frac{d^2\mathbf{x}}{dt^2}=-\nabla\left\{V(\mathbf{x})-\frac{\hbar^2}{2m}\frac{\nabla^2 R}{R}\right\}+\mathbf{f}(\mathbf{p}-\nabla S(\mathbf{x})), \quad (31)$$

where $\mathbf{f}(\mathbf{p}-\nabla S(\mathbf{x}))$ is assumed to be a function which vanishes when $\mathbf{p}=\nabla S(\mathbf{x})$ and more generally takes such a form that it implies a force tending to make $\mathbf{p}-\nabla S(\mathbf{x})$ decrease rapidly with the passage of time. It is clear, moreover, that f can be so chosen that it is large only in processes involving very short distances (where $\nabla S(\mathbf{x})$ should be large).

If the correct equations of motion resembled Eq. (31), then the usual interpretation would be applicable only over times much longer than τ, for only after such times have elapsed will the relation $\mathbf{p}=\nabla S(\mathbf{x})$ be a good approximation. Moreover, it is clear that such modifications of the theory cannot even be described in the usual interpretation, because they involve the precisely definable particle variables which are not postulated in the usual interpretation.

Let us now consider a modification that makes the equation governing ψ inhomogeneous. Such a modification is

$$i\hbar\partial\psi/\partial t=H\psi+\xi(\mathbf{p}-\nabla S(\mathbf{x}_i)). \quad (32)$$

Here, H is the usual Hamiltonian operator, $\mathbf{x}_i$, represents the actual location of the particle, and ξ is a function that vanishes when $\mathbf{p}=\nabla S(\mathbf{x}_i)$. Now, if the particle equations are chosen, as in Eq. (31), to make $\mathbf{p}-\nabla S(\mathbf{x}_i)$ decay rapidly with time, it follows that in atomic processes, the inhomogeneous term in Eq. (32) will become negligibly small, so that Schroedinger's equation is a good approximation. Nevertheless, in processes involving very short distances and very short times, the inhomogeneities would be important, and the ψ-field would, as in the case of the electromagnetic field, depend to some extent on the actual location of the particle.

It is clear that Eq. (32) is inconsistent with the usual interpretation of the theory. Moreover, we can contemplate further generalizations of Eq. (32), in the direction of introducing nonlinear terms that are large only for processes involving small distances. Since the usual interpretation is based on the hypothesis of linear superposition of "state vectors" in a Hilbert space, it follows that the usual interpretation could not be made consistent with such a nonlinear equation for a one-particle theory. In a many-particle theory, operators can be introduced, satisfying a nonlinear generalization of Schroedinger's equation; but these must ultimately operate on wave functions that satisfy a linear homogeneous Schroedinger equation.

Finally, we repeat a point already made in Sec. 4, namely, that if the theory is generalized in any of the ways indicated here, the probability density of particles will cease to equal $|\psi(\mathbf{x})|^2$. Thus, experiments would become conceivable that distinguish between $|\psi(\mathbf{x})|^2$ and this probability; and in this way we could obtain an experimental proof that the usual interpretation, which gives $|\psi(\mathbf{x})|^2$ *only* a probability interpretation, must be inadequate. Moreover, we shall show in Paper II that modifications like those suggested here would permit the particle position and momentum to be measured simultaneously, so that the uncertainty principle could be violated.

ACKNOWLEDGMENT

The author wishes to thank Dr. Einstein for several interesting and stimulating discussions.

THE PHYSICAL REVIEW

A journal of experimental and theoretical physics established by E. L. Nichols in 1893

SECOND SERIES, VOL. 115, NO. 3 AUGUST 1, 1959

Significance of Electromagnetic Potentials in the Quantum Theory

Y. AHARONOV AND D. BOHM
H. H. Wills Physics Laboratory, University of Bristol, Bristol, England

(Received May 28, 1959; revised manuscript received June 16, 1959)

In this paper, we discuss some interesting properties of the electromagnetic potentials in the quantum domain. We shall show that, contrary to the conclusions of classical mechanics, there exist effects of potentials on charged particles, even in the region where all the fields (and therefore the forces on the particles) vanish. We shall then discuss possible experiments to test these conclusions; and, finally, we shall suggest further possible developments in the interpretation of the potentials.

1. INTRODUCTION

IN classical electrodynamics, the vector and scalar potentials were first introduced as a convenient mathematical aid for calculating the fields. It is true that in order to obtain a classical canonical formalism, the potentials are needed. Nevertheless, the fundamental equations of motion can always be expressed directly in terms of the fields alone.

In the quantum mechanics, however, the canonical formalism is necessary, and as a result, the potentials cannot be eliminated from the basic equations. Nevertheless, these equations, as well as the physical quantities, are all gauge invariant; so that it may seem that even in quantum mechanics, the potentials themselves have no independent significance.

In this paper, we shall show that the above conclusions are not correct and that a further interpretation of the potentials is needed in the quantum mechanics.

2. POSSIBLE EXPERIMENTS DEMONSTRATING THE ROLE OF POTENTIALS IN THE QUANTUM THEORY

In this section, we shall discuss several possible experiments which demonstrate the significance of potentials in the quantum theory. We shall begin with a simple example.

Suppose we have a charged particle inside a "Faraday cage" connected to an external generator which causes the potential on the cage to alternate in time. This will add to the Hamiltonian of the particle a term $V(x,t)$ which is, for the region inside the cage, a function of time only. In the nonrelativistic limit (and we shall assume this almost everywhere in the following discussions) we have, for the region inside the cage, $H=H_0+V(t)$ where H_0 is the Hamiltonian when the generator is not functioning, and $V(t)=e\phi(t)$. If $\psi_0(x,t)$ is a solution of the Hamiltonian H_0, then the solution for H will be

$$\psi=\psi_0 e^{-iS/\hbar},\quad S=\int V(t)dt,$$

which follows from

$$i\hbar\frac{\partial\psi}{\partial t}=\left(i\hbar\frac{\partial\psi_0}{\partial t}+\psi_0\frac{\partial S}{\partial t}\right)e^{-iS/\hbar}=[H_0+V(t)]\psi=H\psi.$$

The new solution differs from the old one just by a phase factor and this corresponds, of course, to no change in any physical result.

Now consider a more complex experiment in which a single coherent electron beam is split into two parts and each part is then allowed to enter a long cylindrical metal tube, as shown in Fig. 1.

After the beams pass through the tubes, they are combined to interfere coherently at F. By means of time-determining electrical "shutters" the beam is chopped into wave packets that are long compared with the wavelength λ, but short compared with the length of the tubes. The potential in each tube is determined by a time delay mechanism in such a way that the potential is zero in region I (until each packet is well inside its tube). The potential then grows as a function of time, but differently in each tube. Finally, it falls back to zero, before the electron comes near the

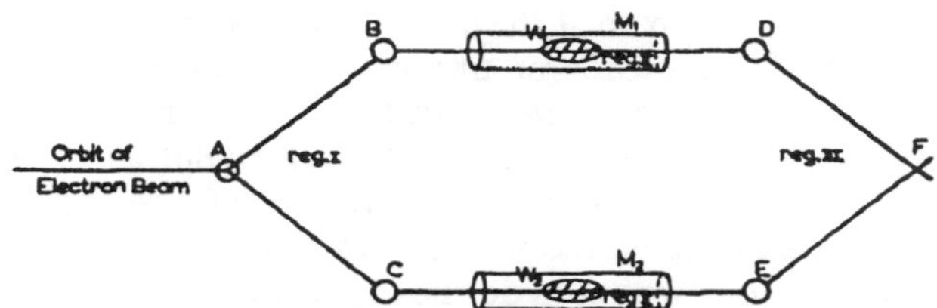

FIG. 1. Schematic experiment to demonstrate interference with time-dependent scalar potential. A, B, C, D, E: suitable devices to separate and divert beams. W_1, W_2: wave packets. M_1, M_2: cylindrical metal tubes. F: interference region.

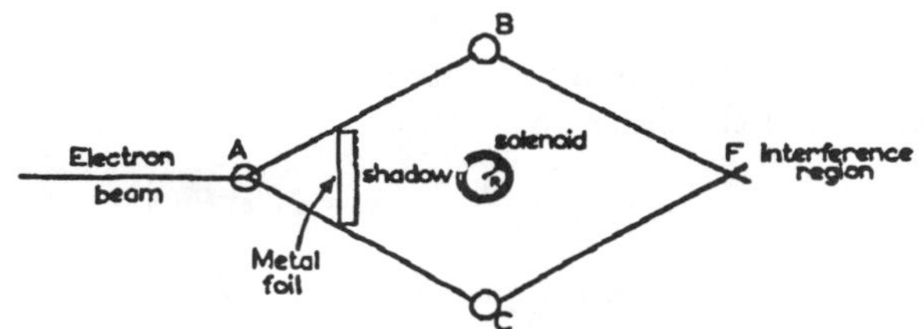

FIG. 2. Schematic experiment to demonstrate interference with time-independent vector potential.

other edge of the tube. Thus the potential is nonzero only while the electrons are well inside the tube (region II). When the electron is in region III, there is again no potential. The purpose of this arrangement is to ensure that the electron is in a time-varying potential without ever being in a field (because the field does not penetrate far from the edges of the tubes, and is nonzero only at times when the electron is far from these edges).

Now let $\psi(x,t)=\psi_1^0(x,t)+\psi_2^0(x,t)$ be the wave function when the potential is absent (ψ_1^0 and ψ_2^0 representing the parts that pass through tubes 1 and 2, respectively). But since V is a function only of t wherever ψ is appreciable, the problem for each tube is essentially the same as that of the Faraday cage. The solution is then

$$\psi=\psi_1^0e^{-iS_1/\hbar}+\psi_2^0e^{-iS_2/\hbar},$$

where

$$S_1=e\int\varphi_1dt,\quad S_2=e\int\varphi_2dt.$$

It is evident that the interference of the two parts at F will depend on the phase difference $(S_1-S_2)/\hbar$. Thus, there is a physical effect of the potentials even though no force is ever actually exerted on the electron. The effect is evidently essentially quantum-mechanical in nature because it comes in the phenomenon of interference. We are therefore not surprised that it does not appear in classical mechanics.

From relativistic considerations, it is easily seen that the covariance of the above conclusion demands that there should be similar results involving the vector potential, $\mathbf{A}$.

The phase difference, $(S_1-S_2)/\hbar$, can also be expressed as the integral $(e/\hbar)\oint\varphi dt$ around a closed circuit in space-time, where φ is evaluated at the place of the center of the wave packet. The relativistic generalization of the above integral is

$$\frac{e}{\hbar}\oint\left(\varphi dt-\frac{\mathbf{A}}{c}\cdot d\mathbf{x}\right),$$

where the path of integration now goes over any closed circuit in space-time.

As another special case, let us now consider a path in space only (t=constant). The above argument suggests that the associated phase shift of the electron wave function ought to be

$$\Delta S/\hbar=-\frac{e}{c\hbar}\oint\mathbf{A}\cdot d\mathbf{x},$$

where $\oint\mathbf{A}\cdot d\mathbf{x}=\int\mathbf{H}\cdot d\mathbf{s}=\phi$ (the total magnetic flux inside the circuit).

This corresponds to another experimental situation. By means of a current flowing through a very closely wound cylindrical solenoid of radius R, center at the origin and axis in the z direction, we create a magnetic field, $\mathbf{H}$, which is essentially confined within the solenoid. However, the vector potential, $\mathbf{A}$, evidently, cannot be zero everywhere outside the solenoid, because the total flux through every circuit containing the origin is equal to a constant

$$\phi_0=\int\mathbf{H}\cdot d\mathbf{s}=\int\mathbf{A}\cdot d\mathbf{x}.$$

To demonstrate the effects of the total flux, we begin, as before, with a coherent beam of electrons. (But now there is no need to make wave packets.) The beam is split into two parts, each going on opposite sides of the solenoid, but avoiding it. (The solenoid can be shielded from the electron beam by a thin plate which casts a shadow.) As in the former example, the beams are brought together at F (Fig. 2).

The Hamiltonian for this case is

$$H=\frac{[\mathbf{P}-(e/c)\mathbf{A}]^2}{2m}.$$

In singly connected regions, where $\mathbf{H}=\nabla\times\mathbf{A}=0$, we can always obtain a solution for the above Hamiltonian by taking $\psi=\psi_0e^{-iS/\hbar}$, where ψ_0 is the solution when $\mathbf{A}=0$ and where $\nabla S/\hbar=(e/c)\mathbf{A}$. But, in the experiment discussed above, in which we have a multiply connected region (the region outside the solenoid), $\psi_0e^{-iS/\hbar}$ is a non-single-valued function[1] and therefore, in general, not a permissible solution of Schrödinger's equation. Nevertheless, in our problem it is still possible to use such solutions because the wave function splits into two parts $\psi=\psi_1+\psi_2$, where ψ_1 represents the beam on

[1] Unless $\phi_0=nhc/e$, where n is an integer.

one side of the solenoid and ψ_2 the beam on the opposite side. Each of these beams stays in a simply connected region. We therefore can write

$$\psi_1=\psi_1^0 e^{-iS_1/\hbar},\quad \psi_2=\psi_2^0 e^{-iS_2/\hbar},$$

where S_1 and S_2 are equal to $(e/c)\int \mathbf{A}\cdot d\mathbf{x}$ along the paths of the first and second beams, respectively. (In Sec. 4, an exact solution for this Hamiltonian will be given, and it will confirm the above results.)

The interference between the two beams will evidently depend on the phase difference,

$$(S_1-S_2)/\hbar=(e/\hbar c)\oint \mathbf{A}\cdot d\mathbf{x}=(e/\hbar c)\phi_0.$$

This effect will exist, even though there are no magnetic forces acting in the places where the electron beam passes.

In order to avoid fully any possible question of contact of the electron with the magnetic field we note that our result would not be changed if we surrounded the solenoid by a potential barrier that reflects the electrons perfectly. (This, too, is confirmed in Sec. 4.)

It is easy to devise hypothetical experiments in which the vector potential may influence not only the interference pattern but also the momentum. To see this, consider a periodic array of solenoids, each of which is shielded from direct contact with the beam by a small plate. This will be essentially a grating. Consider first the diffraction pattern without the magnetic field, which will have a discrete set of directions of strong constructive interference. The effect of the vector potential will be to produce a shift of the relative phase of the wave function in different elements of the gratings. A corresponding shift will take place in the directions, and therefore the momentum of the diffracted beam.

3. A PRACTICABLE EXPERIMENT TO TEST FOR THE EFFECTS OF A POTENTIAL WHERE THERE ARE NO FIELDS

As yet no direct experiments have been carried out which confirm the effect of potentials where there is no field. It would be interesting therefore to test whether such effects actually exist. Such a test is, in fact, within the range of present possibilities.[2] Recent experiments[3,4] have succeeded in obtaining interference from electron beams that have been separated in one case by as much as 0.8 mm.[3] It is quite possible to wind solenoids which are smaller than this, and therefore to place them between the separate beams. Alternatively, we may obtain localized lines of flux of the right magnitude (the magnitude has to be of the order of $\phi_0=2\pi c\hbar/e\sim 4\times 10^{-7}$ gauss cm^2) by means of fine permanently magnetized "whiskers".[5] The solenoid can be used in Marton's device,[3] while the whisker is suitable for another experimental setup[4] where the separation is of the order of microns and the whiskers are even smaller than this.

[2] Dr. Chambers is now making a preliminary experimental study of this question at Bristol.

[3] L. Marton, Phys. Rev. 85, 1057 (1952); 90, 490 (1953). Marton, Simpson, and Suddeth, Rev. Sci. Instr. 25, 1099 (1954).

[4] G. Mollenstedt, Naturwissenschaften 42, 41 (1955); G. Mollenstedt and H. Düker, Z. Physik 145, 377 (1956).

In principle, we could do the experiment by observing the interference pattern with and without the magnetic flux. But since the main effect of the flux is only to displace the line pattern without changing the interval structure, this would not be a convenient experiment to do. Instead, it would be easier to vary the magnetic flux within the same exposure for the detection of the interference patterns. Such a variation would, according to our previous discussion, alter the sharpness and the general form of the interference bands. This alteration would then constitute a verification of the predicted phenomena.

When the magnetic flux is altered, there will, of course, be an induced electric field outside the solenoid, but the effects of this field can be made negligible. For example, suppose the magnetic flux were suddenly altered in the middle of an exposure. The electric field would then exist only for a very short time, so that only a small part of the beam would be affected by it.

4. EXACT SOLUTION FOR SCATTERING PROBLEMS

We shall now obtain an exact solution for the problem of the scattering of an electron beam by a magnetic field in the limit where the magnetic field region tends to a zero radius, while the total flux remains fixed. This corresponds to the setup described in Sec. 2 and shown in Fig. 2. Only this time we do not split the plane wave into two parts. The wave equation outside the magnetic field region is, in cylindrical coordinates,

$$\left[\frac{\partial^2}{\partial r^2}+\frac{1}{r}\frac{\partial}{\partial r}+\frac{1}{r^2}\left(\frac{\partial}{\partial\theta}+i\alpha\right)^2+k^2\right]\psi=0, \qquad (1)$$

where $\mathbf{k}$ is the wave vector of the incident particle and $\alpha=-e\phi/ch$. We have again chosen the gauge in which $A_r=0$ and $A_\theta=\phi/2\pi r$.

The general solution of the above equation is

$$\psi=\sum_{m=-\infty}^{\infty} e^{im\theta}[a_m J_{m+\alpha}(kr)+b_m J_{-(m+\alpha)}(kr)], \qquad (2)$$

where a_m and b_m are arbitrary constants and $J_{m+\alpha}(kr)$ is a Bessel function, in general of fractional order (dependent on ϕ). The above solution holds only for $r>R$. For $r<R$ (inside the magnetic field) the solution has been worked out.[6] By matching the solutions at $r=R$ it is easily shown that only Bessel functions of positive order will remain, when R approaches zero.

[5] See, for example, Sidney S. Brenner, Acta Met. 4, 62 (1956).

[6] L. Page, Phys. Rev. 36, 444 (1930).

This means that the probability of finding the particle inside the magnetic field region approaches zero with R. It follows that the wave function would not be changed if the electron were kept away from the field by a barrier whose radius also went to zero with R.

The general solution in the limit of R tending to zero is therefore

$$\psi=\sum_{m=-\infty}^{\infty} a_m J_{|m+\alpha|}e^{im\theta}. \quad (3)$$

We must then choose a_m so that ψ represents a beam of electrons that is incident from the right ($\theta=0$). It is important, however, to satisfy the initial condition that the current density,

$$\mathbf{j}=\frac{\hbar(\psi^*\nabla\psi-\psi\nabla\psi^*)}{2im}-\frac{e}{mc}\mathbf{A}\psi^*\psi, \quad (4)$$

shall be constant and in the x direction. In the gauge that we are using, we easily see that the correct incident wave is $\psi_{\text{inc}}=e^{-ikx}e^{-i\alpha\theta}$. Of course, this wave function holds only to the right of the origin, so that no problem of multiple-valuedness arises.

We shall show in the course of this calculation that the above conditions will be satisfied by choosing $a_m=(-i)^{|m+\alpha|}$, in which case, we shall have

$$\psi=\sum_{m=-\infty}^{\infty}(-i)^{|m+\alpha|}J_{|m+\alpha|}e^{im\theta}.$$

It is convenient to split ψ into the following three parts: $\psi=\psi_1+\psi_2+\psi_3$, where

$$\psi_1=\sum_{m=1}^{\infty}(-i)^{m+\alpha}J_{m+\alpha}e^{im\theta},$$

$$\psi_2=\sum_{m=-\infty}^{-1}(-i)^{m+\alpha}J_{m+\alpha}e^{im\theta},$$

$$=\sum_{m=1}^{\infty}(-i)^{m-\alpha}J_{m-\alpha}e^{-im\theta}, \quad (5)$$

$$\psi_3=(-i)^{|\alpha|}J_{|\alpha|}.$$

Now ψ_1 satisfies the simple differential equation

$$\frac{\partial\psi_1}{\partial r'}=\sum_{m=1}^{\infty}(-i)^{m+\alpha}J_{m+\alpha}'e^{im\theta}$$

$$=\sum_{m=1}^{\infty}(-i)^{m+\alpha}\frac{J_{m+\alpha-1}-J_{m+\alpha+1}}{2}e^{im\theta}, \quad r'=kr \quad (6)$$

where we have used the well-known formula for Bessel functions:

$$dJ_\gamma(r)/dr=\tfrac{1}{2}(J_{\gamma-1}-J_{\gamma+1}).$$

As a result, we obtain

$$\frac{\partial\psi_1}{\partial r'}=\frac{1}{2}\sum_{m'=0}^{\infty}(-i)^{m'+\alpha+1}J_{m'+\alpha}e^{i(m'+1)\theta}$$
$$-\frac{1}{2}\sum_{m'=2}^{\infty}(-i)^{m'+\alpha-1}J_{m'+\alpha}e^{i(m'-1)\theta} \quad (7)$$
$$=\frac{1}{2}\sum_{m'=1}^{\infty}(-i)^{m'+\alpha}J_{m'+\alpha}e^{im'\theta}(-ie^{i\theta}+i^{-1}e^{-i\theta})$$
$$+\tfrac{1}{2}(-i)^{\alpha}[J_{\alpha+1}-ie^{i\theta}J_\alpha].$$

So

$$\partial\psi_1/\partial r'=-i\cos\theta\psi_1+\tfrac{1}{2}(-i)^{\alpha}(J_{\alpha+1}-iJ_\alpha e^{i\theta}).$$

This differential equation can be easily integrated to give

$$\psi_1=A\int_0^{r'}e^{ir'\cos\theta}[J_{\alpha+1}-iJ_\alpha e^{i\theta}]dr', \quad (8)$$

where

$$A=\tfrac{1}{2}(-i)^{\alpha}e^{-ir'\cos\theta}.$$

The lower limit of the integration is determined by the requirement that when r' goes to zero, ψ_1 also goes to to zero because, as we have seen, ψ_1 includes Bessel functions of positive order only.

In order to discuss the asymptotic behavior of ψ_1, let us write it as $\psi_1=A[I_1-I_2]$, where

$$I_1=\int_0^{\infty}e^{ir'\cos\theta}[J_{\alpha+1}-ie^{i\theta}J_\alpha]dr',$$
$$I_2=\int_r^{\infty}e^{ir'\cos\theta}[J_{\alpha+1}-ie^{i\theta}J_\alpha]dr'. \quad (9)$$

The first of these integrals is known[7]:

$$\int_0^{\infty}e^{i\beta r}J_\alpha(kr)=\frac{e^{i[\alpha\arcsin(\beta/k)]}}{(k^2-\beta^2)^{\frac{1}{2}}}, \quad 0<\beta<k, \quad -2<\alpha.$$

In our cases, $\beta=\cos\theta$, $k=1$, so that

$$I_1=\left[\frac{e^{i\alpha(\frac{1}{2}\pi-|\theta|)}}{|\sin\theta|}-ie^{i\theta}\frac{e^{i(\alpha+1)(\frac{1}{2}\pi-|\theta|)}}{|\sin\theta|}\right]. \quad (10)$$

Because the integrand is even in θ, we have written the final expression for the above integral as a function of $|\theta|$ and of $|\sin\theta|$. Hence

$$I_1=e^{i\alpha(\frac{1}{2}\pi-|\theta|)}\left[\frac{ie^{-i|\theta|}-ie^{i\theta}}{|\sin\theta|}\right]$$
$$=0 \quad \text{for } \theta<0,$$
$$=e^{-i\alpha\theta}2i^{\alpha} \quad \text{for } \theta>0, \quad (11)$$

where we have taken θ as going from $-\pi$ to π.

[7] See, for example, W. Gröbner and N. Hofreiter, *Integraltafel* (Springer-Verlag, Berlin, 1949).

We shall see presently that I_1 represents the largest term in the asymptotic expansion of ψ_1. The fact that it is zero for $\theta<0$ shows that this part of ψ_1 passes (asymptotically) only on the upper side of the singularity. To explain this, we note that ψ_1 contains only positive values of m, and therefore of the angular momentum. It is quite natural then that this part of ψ_1 goes on the upper side of the singularity. Similarly, since according to (5)

$$\psi_2(r',\theta,\alpha)=\psi_1(r',-\theta,-\alpha),$$

it follows that ψ_2 will behave oppositely to ψ_1 in this regard, so that together they will make up the correct incident wave.

Now, in the limit of $r'\to\infty$ we are allowed to take in the integrand of I_2 the first asymptotic term of J_α,[8] namely $J_\alpha\to(2/\pi r)^{\frac{1}{2}}\cos(r'-\frac{1}{2}\alpha-\frac{1}{4}\pi)$. We obtain

$$I_2=\int_r^\infty e^{ir'\cos\theta}(J_{\alpha+1}-ie^{i\theta}J_\alpha)dr'\to C+D, \tag{12}$$

where

$$C=\int_r^\infty e^{ir'\cos\theta}[\cos(r'-\tfrac{1}{2}(\alpha+1)\pi-\tfrac{1}{4}\pi)]\frac{dr'}{(r')^{\frac{1}{2}}}\left(\frac{2}{\pi}\right)^{\frac{1}{2}},$$
$$D=\int_r^\infty e^{ir'\cos\theta}[\cos(r'-\tfrac{1}{2}\alpha-\tfrac{1}{4}\pi)]\frac{dr'}{(r')^{\frac{1}{2}}}\left(\frac{2}{\pi}\right)^{\frac{1}{2}}(-i)e^{i\theta}. \tag{13}$$

Then

$$C=\int_r^\infty e^{ir'\cos\theta}[e^{i[r'-\frac{1}{2}(\alpha+1)\pi-\frac{1}{4}\pi]}+e^{-i[r'-\frac{1}{2}(\alpha+1)\pi-\frac{1}{4}\pi]}]\frac{dr'}{(2\pi r')^{\frac{1}{2}}}$$
$$=\left(\frac{2}{\pi}\right)^{\frac{1}{2}}\frac{(-i)^{\alpha+\frac{3}{2}}}{(1+\cos\theta)^{\frac{1}{2}}}\int_{[r'(1+\cos\theta)]^{\frac{1}{2}}}^\infty\exp(+iz^2)dz$$
$$+\left(\frac{2}{\pi}\right)^{\frac{1}{2}}\frac{i^{\alpha+\frac{3}{2}}}{(1-\cos\theta)^{\frac{1}{2}}}\int_{[r'(1-\cos\theta)]^{\frac{1}{2}}}^\infty\exp(-iz^2)dz, \tag{14}$$

where we have put

$$z=[r'(1+\cos\theta)]^{\frac{1}{2}}\quad\text{and}\quad z=[r'(1-\cos\theta)]^{\frac{1}{2}},$$

respectively.

Using now the well-known asymptotic behavior of the error function,[9]

$$\int_a^\infty\exp(iz^2)dz\to\frac{i}{2}\frac{\exp(ia^2)}{a},$$
$$\int_a^\infty\exp(-iz^2)dz\to\frac{-i}{2}\frac{\exp(-ia^2)}{a}, \tag{15}$$

we finally obtain

$$C=\left[\frac{(-i)^{\alpha+\frac{1}{2}}}{(2\pi)^{\frac{1}{2}}}\frac{e^{ir'}}{[r'(1+\cos\theta)^2]^{\frac{1}{2}}}+\frac{i^{\alpha+\frac{1}{2}}}{(2\pi)^{\frac{1}{2}}}\frac{e^{-ir'}}{[r'(1-\cos\theta)^2]^{\frac{1}{2}}}\right]e^{ir'\cos\theta}, \tag{16}$$

$$D=\left[\frac{(-i)^{\alpha-\frac{1}{2}}}{(2\pi)^{\frac{1}{2}}}\frac{e^{ir'}}{[r'(1+\cos\theta)^2]^{\frac{1}{2}}}+\frac{i^{\alpha-\frac{1}{2}}}{(2\pi)^{\frac{1}{2}}}\frac{e^{-ir'}}{[r'(1-\cos\theta)^2]^{\frac{1}{2}}}\right]e^{ir'\cos\theta}(-i)e^{i\theta}. \tag{17}$$

Now adding (16) and (17) together and using (13) and (9), we find that the term of $1/(r')^{\frac{1}{2}}$ in the asymptotic expansion of ψ_1 is

$$\frac{(-i)^{\frac{1}{2}}}{2(2\pi)^{\frac{1}{2}}}\left[(-1)^\alpha\frac{e^{ir'}}{(r')^{\frac{1}{2}}}\frac{1+e^{i\theta}}{1+\cos\theta}+i\frac{e^{-ir'}}{(r')^{\frac{1}{2}}}\frac{1-e^{i\theta}}{1-\cos\theta}\right]. \tag{18}$$

Using again the relation between ψ_1 and ψ_2 we obtain for the corresponding term in ψ_2

$$\frac{(-i)^{\frac{1}{2}}}{2(2\pi)^{\frac{1}{2}}}\left[(-1)^{-\alpha}\frac{e^{ir'}}{(r')^{\frac{1}{2}}}\frac{1+e^{-i\theta}}{1+\cos\theta}+i\frac{e^{-ir'}}{(r')^{\frac{1}{2}}}\frac{1-e^{-i\theta}}{1-\cos\theta}\right]. \tag{19}$$

Adding (18) and (19) and using (11), we finally get

$$\psi_1+\psi_2\to\frac{(-i)^{\frac{1}{2}}}{(2\pi)^{\frac{1}{2}}}\left[\frac{ie^{-ir'}}{(r')^{\frac{1}{2}}}+\frac{e^{ir'}}{(r')^{\frac{1}{2}}}\frac{\cos(\pi\alpha-\frac{1}{2}\theta)}{\cos(\frac{1}{2}\theta)}\right]+e^{-i(r'\cos\theta+\alpha\theta)}. \tag{20}$$

There remains the contribution of ψ_3, whose asymptotic behavior is [see Eq. (12)]

$$(-i)^{|\alpha|}J_{|\alpha|}(r')\to(-i)^{|\alpha|}\left(\frac{2}{\pi r'}\right)^{\frac{1}{2}}\cos(r'-\tfrac{1}{4}\pi-\tfrac{1}{2}|\alpha|\pi).$$

Collecting all terms, we find

$$\psi=\psi_1+\psi_2+\psi_3\to e^{-i(\alpha\theta+r'\cos\theta)}\pm\frac{e^{ir'}}{(2\pi ir')^{\frac{1}{2}}}\sin\pi\alpha\frac{e^{-i\theta/2}}{\cos(\theta/2)}, \tag{21}$$

where the $\pm$ sign is chosen according to the sign of α.

The first term in equation (21) represents the incident wave, and the second the scattered wave.[10] The scattering cross section is therefore

$$\sigma=\frac{\sin^2\pi\alpha}{2\pi}\frac{1}{\cos^2(\theta/2)}. \tag{22}$$

[8] E. Jahnke and F. Emde, *Tables of Functions* (Dover Publications, Inc., New York, 1943), fourth edition, p. 138.

[9] Reference 8, p. 24.

[10] In this way, we verify, of course, that our choice of the a_m for Eq. (3) satisfies the correct boundary conditions.

When $\alpha = n$, where n is an integer, then σ vanishes. This is analogous to the Ramsauer effect.[11] σ has a maximum when $\alpha = n + \frac{1}{2}$.

The asymptotic formula (21) holds only when we are not on the line $\theta = \pi$. The exact solution, which is needed on this line, would show that the second term will combine with the first to make a single-valued wave function, despite the non-single-valued character of the two parts, in the neighborhood of $\theta = \pi$. We shall see this in more detail presently for the special case $\alpha = n + \frac{1}{2}$.

In the interference experiment discussed in Sec. 2, diffraction effects, represented in Eq. (21) by the scattered wave, have been neglected. Therefore, in this problem, it is adequate to use the first term of Eq. (21). Here, we see that the phase of the wave function has a different value depending on whether we approach the line $\theta = \pm\pi$ from positive or negative angles, i.e., from the upper or lower side. This confirms the conclusions obtained in the approximate treatment of Sec. 2.

We shall discuss now the two special cases that can be solved exactly. The first is the case where $\alpha = n$. Here, the wave function is $\psi = e^{-ikx} e^{-i\alpha\theta}$, which is evidently single-valued when α is an integer. (It can be seen by direct differentiation that this is a solution.)

The second case is that of $\alpha = n + \frac{1}{2}$. Because $J_{(n+\frac{1}{2})}(r)$ is a closed trigonometric function, the integrals for ψ can be carried out exactly.

The result is

$$\psi = \frac{i^{\frac{1}{2}}}{\sqrt{2}} e^{-i(\frac{1}{2}\theta + r' \cos\theta)} \int_0^{[r'(1+\cos\theta)]^{\frac{1}{2}}} \exp(iz^2) dz. \quad (23)$$

This function vanishes on the line $\theta = \pi$. It can be seen that its asymptotic behavior is the same as that of Eq. (2) with α set equal to $n + \frac{1}{2}$. In this case, the single-valuedness of ψ is evident. In general, however, the behavior of ψ is not so simple, since ψ does not become zero on the line $\theta = \pi$.

5. DISCUSSION OF SIGNIFICANCE OF RESULTS

The essential result of the previous discussion is that in quantum theory, an electron (for example) can be influenced by the potentials even if all the field regions are excluded from it. In other words, in a field-free multiply-connected region of space, the physical properties of the system still depend on the potentials.

It is true that all these effects of the potentials depend only on the gauge-invariant quantity $\oint \mathbf{A} \cdot d\mathbf{x} = \int \mathbf{H} \cdot d\mathbf{s}$, so that in reality they can be expressed in terms of the fields inside the circuit. However, according to current relativistic notions, all fields must interact only locally. And since the electrons cannot reach the regions where the fields are, we cannot interpret such effects as due to the fields themselves.

[11] See, for example, D. Bohm, *Quantum Theory* (Prentice-Hall, Inc., Englewood Cliffs, New Jersey, 1951).

In classical mechanics, we recall that potentials cannot have such significance because the equation of motion involves only the field quantities themselves. For this reason, the potentials have been regarded as purely mathematical auxiliaries, while only the field quantities were thought to have a direct physical meaning.

In quantum mechanics, the essential difference is that the equations of motion of a particle are replaced by the Schrödinger equation for a wave. This Schrödinger equation is obtained from a canonical formalism, which cannot be expressed in terms of the fields alone, but which also requires the potentials. Indeed, the potentials play a role, in Schrödinger's equation, which is analogous to that of the index of refration in optics, The Lorentz force $[e\mathbf{E} + (e/c)\mathbf{v} \times \mathbf{H}]$ does not appear anywhere in the fundamental theory, but appears only as an approximation holding in the classical limit. It would therefore seem natural at this point to propose that, in quantum mechanics, the fundamental physical entities are the potentials, while the fields are derived from them by differentiations.

The main objection that could be raised against the above suggestion is grounded in the gauge invariance of the theory. In other words, if the potentials are subject to the transformation $A_\mu \rightarrow A_\mu' = A_\mu + \partial\psi/\partial x_\mu$, where ψ is a continuous scalar function, then all the known physical quantities are left unchanged. As a result, the same physical behavior is obtained from any two potentials, $A_\mu(x)$ and $A_\mu'(x)$, related by the above transformation. This means that insofar as the potentials are richer in properties than the fields, there is no way to reveal this additional richness. It was therefore concluded that the potentials cannot have any meaning, except insofar as they are used mathematically, to calculate the fields.

We have seen from the examples described in this paper that the above point of view cannot be maintained for the general case. Of course, our discussion does not bring into question the gauge invariance of the theory. But it does show that in a theory involving only local interactions (e.g., Schrödinger's or Dirac's equation, and current quantum-mechanical field theories), the potentials must, in certain cases, be considered as physically effective, even when there are no fields acting on the charged particles.

The above discussion suggests that some further development of the theory is needed. Two possible directions are clear. First, we may try to formulate a nonlocal theory in which, for example, the electron could interact with a field that was a finite distance away. Then there would be no trouble in interpreting these results, but, as is well known, there are severe difficulties in the way of doing this. Secondly, we may retain the present local theory and, instead, we may try to give a further new interpretation to the potentials. In other words, we are led to regard $A_\mu(x)$ as a physical variable. This means that we must be able to define the physical difference between two quantum states which differ only by gauge transformation. It will be shown in a future paper that in a system containing an undefined number of charged particles (i.e., a superposition of states of different total charge), a new Hermitian operator, essentially an angle variable, can be introduced, which is conjugate to the charge density and which may give a meaning to the gauge. Such states have actually been used in connection with recent theories of superconductivity and superfluidity[12] and we shall show their relation to this problem in more detail.

ACKNOWLEDGMENTS

We are indebted to Professor M. H. L. Pryce for many helpful discussions. We wish to thank Dr. Chambers for many discussions connected with the experimental side of the problem.

[12] See, for example, C. G. Kuper, *Advances in Physics*, edited by N. F. Mott (Taylor and Francis, Ltd., London, 1959), Vol. 8, p. 25, Sec. 3, Par. 3.

Experimental Test of Bell's Inequalities Using Time-Varying Analyzers

Alain Aspect, Jean Dalibard,[(a)] and Gérard Roger
Institut d'Optique Théorique et Appliquée, F-91406 Orsay Cédex, France
(Received 27 September 1982)

Correlations of linear polarizations of pairs of photons have been measured with time-varying analyzers. The analyzer in each leg of the apparatus is an acousto-optical switch followed by two linear polarizers. The switches operate at incommensurate frequencies near 50 MHz. Each analyzer amounts to a polarizer which jumps between two orientations in a time short compared with the photon transit time. The results are in good agreement with quantum mechanical predictions but violate Bell's inequalities by 5 standard deviations.

PACS numbers: 03.65.Bz, 35.80.+s

Bell's inequalities apply to any correlated measurement on two correlated systems. For instance, in the optical version of the Einstein-Podolsky-Rosen-Bohm *Gedankenexperiment*,[1] a source emits pairs of photons (Fig. 1). Measurements of the correlations of linear polarizations are performed on two photons belonging to the same pair. For pairs emitted in suitable states, the correlations are strong. To account for these correlations, Bell[2] considered theories which invoke common properties of both members of the

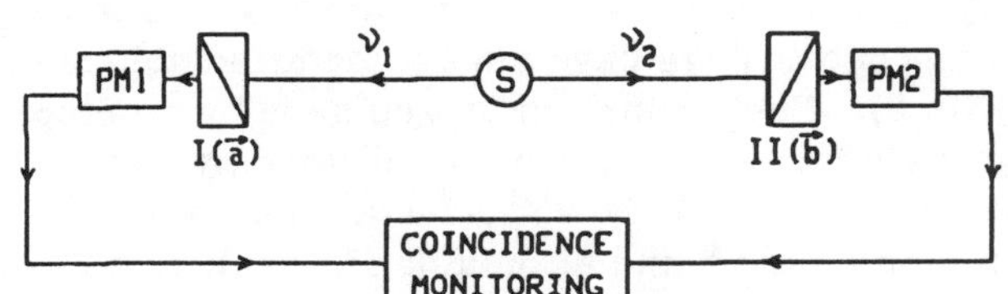

FIG. 1. Optical version of the Einstein-Podolsky-Rosen-Bohm *Gedankenexperiment*. The pair of photons ν_1 and ν_2 is analyzed by linear polarizers I and II (in orientations $\vec{a}$ and $\vec{b}$) and photomultipliers. The coincidence rate is monitored.

pair. Such properties are referred to as supplementary parameters. This is very different from the quantum mechanical formalism, which does not involve such properties. With the addition of a reasonable locality assumption, Bell showed that such classical-looking theories are constrained by certain inequalities that are not always obeyed by quantum mechanical predictions.

Several experiments of increasing accuracy[3-5] have been performed and clearly favor quantum mechanics. Experiments using pairs of visible photons emitted in atomic radiative cascades seem to achieve a good realization of the ideal *Gedankenexperiment*.[5] However, all these experiments have been performed with static setups, in which polarizers are held fixed for the whole duration of a run. Then, one might question Bell's locality assumption, that states that the results of the measurement by polarizer II does not depend on the orientation $\vec{a}$ of polarizer I (and vice versa), nor does the way in which pairs are emitted depend on $\vec{a}$ or $\vec{b}$. Although highly reasonable, such a locality condition is not prescribed by any fundamental physical law. As pointed out by Bell,[2] it is possible, in such experiments, to reconcile supplementary-parameter theories and the experimentally verified predictions of quantum mechanics: "The settings of the instruments are made sufficiently in advance to allow them to reach some mutual rapport by exchange of signals with velocity less than or equal to that of light." If such interactions existed, Bell's locality condition would no longer hold for static experiments, nor would Bell's inequalities.

Bell thus insisted upon the importance of "experiments of the type proposed by Bohm and Aharonov,[6] in which the settings are changed during the flight of the particles." In such a "timing experiment," the locality condition would then become a consequence of Einstein's causality, preventing any faster-than-light influence.

In this Letter, we report the results of the first experiment using variable polarizers. Following our proposal,[7] we have used a modified scheme (Fig. 2). Each polarizer is replaced by a setup involving a switching device followed by two polarizers in two different orientations: $\vec{a}$ and $\vec{a}'$ on side I, and $\vec{b}$ and $\vec{b}'$ on side II. Such an optical switch is able to rapidly redirect the incident light from one polarizer to the other one. If the two switches work at random and are uncorrelated, it is possible to write generalized Bell's inequalities in a form similar to Clauser-Horne-Shimony-Holt inequalities[8]:

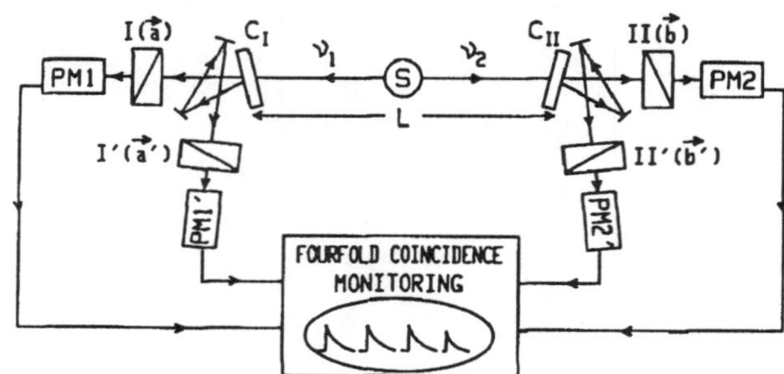

FIG. 2. Timing experiment with optical switches. Each switching device ($C_{\mathrm{I}}, C_{\mathrm{II}}$) is followed by two polarizers in two different orientations. Each combination is equivalent to a polarizer switched fast between two orientations.

$$-1 \leq S \leq 0,$$

with

$$S = \frac{N(\vec{a},\vec{b})}{N(\infty,\infty)} - \frac{N(\vec{a},\vec{b}')}{N(\infty,\infty')} + \frac{N(\vec{a}',\vec{b})}{N(\infty',\infty)} + \frac{N(\vec{a}',\vec{b}')}{N(\infty',\infty')} - \frac{N(\vec{a}',\infty)}{N(\infty',\infty)} - \frac{N(\infty,\vec{b})}{N(\infty,\infty)}.$$

The quantity S involves (i) the four coincidence counting rates [$N(\vec{a},\vec{b})$, $N(\vec{a}',\vec{b})$, etc.] measured in a *single run;* (ii) the four corresponding coincidence rates [$N(\infty,\infty)$, $N(\infty',\infty)$, etc.] with all polarizers removed; and (iii) two coincidence rates [$N(\vec{a}',\infty)$, $N(\infty,\vec{b})$] with a polarizer removed on each side. The measurements (ii) and (iii) are performed in auxiliary runs.

In this experiment, switching between the two channels occurs about each 10 ns. Since this delay, as well as the lifetime of the intermediate level of the cascade (5 ns), is small compared to L/c (40 ns), a detection event on one side and the corresponding change of orientation on the other side are separated by a spacelike interval.

The switching of the light is effected by acousto-optical interaction with an ultrasonic standing wave in water.[9] As sketched in Fig. 3 the incidence angle is equal to the Bragg angle, $\theta_B = 5 \times 10^{-3}$ rad. It follows that light is either transmitted straight ahead or deflected at an angle $2\theta_B$. The light is completely transmitted when the amplitude of the standing wave is null, which occurs twice during an acoustical period. A quarter of a period later, the amplitude of the standing wave is maximum and, for a suitable value of the acoustical power, light is then fully

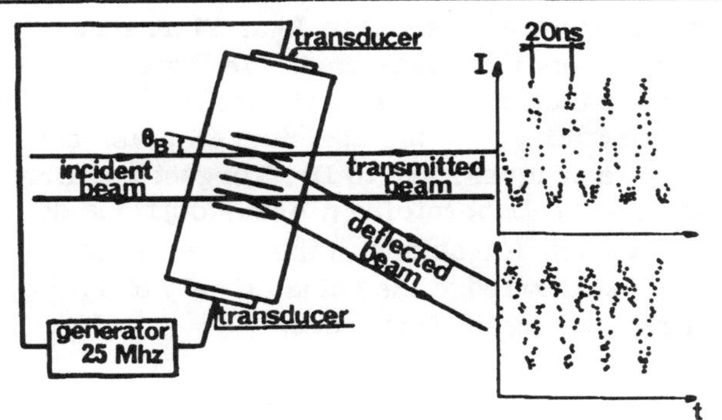

FIG. 3. Optical switch. The incident light is switched at a frequency around 50 MHz by diffraction at the Bragg angle on an ultrasonic standing wave. The intensities of the transmitted and deflected beams as a function of time have been measured with the actual source. The fraction of light directed towards other diffraction orders is negligible.

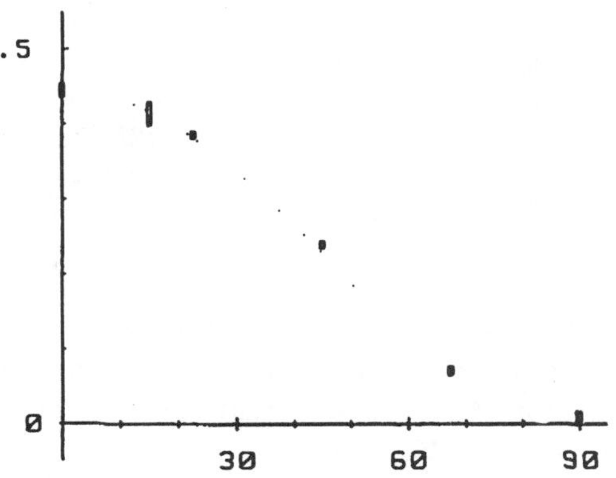

FIG. 4. Average normalized coincidence rate as a function of the relative orientation of the polarizers. Indicated errors are ± 1 standard deviation. The dashed curve is not a fit to the data but the predictions by quantum mechanics for the actual experiment.

deflected. This optical switch thus works at twice the acoustical frequency.

The ultrasonic standing-wave results from interference between counterpropagating acoustic waves produced by two electroacoustical transducers driven in phase at about 25 MHz. In auxiliary tests with a laser beam, the switching has been found complete for an acoustical power about 1 W. In the actual experiment, the light beam has a finite divergence, and the switching is not complete (Fig. 3).

The other parts of the experiment have already been described in previous publications.[4,5] The high-efficiency well-stabilized source of pairs of correlated photons, at wavelengths $\lambda_1 = 422.7$ nm and $\lambda_2 = 551.3$ nm, is obtained by two-photon excitation of a $(J=0)\rightarrow(J=1)\rightarrow(J=0)$ cascade in calcium.

Since each switch is 6 m from the source, rather complicated optics are required to match the beams with the switches and the polarizers. We have carefully checked each channel for no depolarization, by looking for a cosine Malus law when a supplementary polarizer is inserted in front of the source. These auxiliary tests are particularly important for the channels which involve two mirrors inclined at 11°. They also yield the efficiencies of the polarizers, required for the quantum mechanical calculations.

The coincidence counting electronics involve four double-coincidence-counting circuits with coincidence windows of 18 ns. For each relevant pair of photomultipliers, we monitor nondelayed and delayed coincidences. The true coincidence rate (i.e., coincidences due to photons emitted by the same atom) are obtained by subtraction. Simultaneously, a time-to-amplitude converter, followed by a fourfold multichannel analyzer, yields four time-delay spectrums. Here, the true coincidence rate is taken as the signal in the peak of the time-delay spectrum.[4]

We have not been able to achieve collection efficiencies as large as in previous experiments,[4,5] since we had to reduce the divergence of the beams in order to get good switching. Coincidence rates with the polarizers removed were only a few per second, with accidental coincidence rates about one per second.

A typical run lasts 12 000 s, involving totals of 4000 s with polarizers in place at a given set of orientations, 4000 s with all polarizers removed, and 4000 s with one polarizer removed on each side. In order to compensate the effects of systematic drifts, data accumulation was alternated between these three configurations about every 400 s. At the end of each 400-s period, the raw data were stored for subsequent processing with the help of a computer.

At the end of the run, we average the true coincidence rates corresponding to the same configurations for the polarizers. We then compute the relevant ratios for the quantity S. The statistical accuracy is evaluated according to standard statistical methods for photon counting. The processing is performed on both sets of data: that obtained with coincidence circuits, and that obtained with the time-to-amplitude converter. The two methods have always been found to be consistent.

Two runs have been performed in order to test Bell's inequalities. In each run, we have chosen a set of orientations leading to the greatest predicted conflict between quantum mechanics and Bell's inequalities $[(\vec{a},\vec{b})=(\vec{b},\vec{a}')=(\vec{a}',\vec{b}')=22.5°; (\vec{a},\vec{b}')=67.5°]$. The average of the two runs yields

$$S_{\rm expt}=0.101\pm 0.020,$$

violating the inequality $S\leq 0$ by 5 standard deviations. On the other hand, for our solid angles and polarizer efficiencies, quantum mechanics predicts $S_{\rm QM}=0.112$.

We have carried out another run with different orientations, for a direct comparison with quantum mechanics. Figure 4 shows that the agreement is excellent.

The new feature of this experiment is that we change the settings of the polarizers, at a rate greater than c/L. The ideal scheme has not been completed since the change is not truly random, but rather quasiperiodic. Nevertheless, the two switches on the two sides are driven by different generators at different frequencies. It is then very natural to assume that they function in an uncorrelated way.

A more ideal experiment with random and complete switching would be necessary for a fully conclusive argument against the whole class of supplementary-parameter theories obeying Einstein's causality. However, our observed violation of Bell's inequalities indicates that the experimental accuracy was good enough for pointing out a hypothetical discrepancy with the predictions of quantum mechanics. No such effect was observed.[10]

This work has been carried out in the group of Professor Imbert, who we thank for his support. We thank all the technical staff of the Institut d'Optique, especially André Villing for the electronics. We are indebted to Jean-Pierre Passérieux, from the Département de Physique Nucléaire et Basse Energie at Centre d'Etudes National de Saclay, for his assistance in fast-coincidence techniques, and to Dr. Torguet and Dr. Gazalé, from Laboratoire d'Acoustooptique de Valenciennes, for help with the optical switches. We also acknowledge the valuable help of Philippe Grangier during the final runs.

---

[a]Permanent address: Laboratoire de Physique de l'Ecole Normale Supérieure, F-75231 Paris Cédex 05, France.

[1]A. Einstein, B. Podolsky, and N. Rosen, Phys. Rev. 47, 777 (1935); D. Bohm, *Quantum Theory* (Prentice-Hall, Englewood Cliffs, N.J., 1951).

[2]J. S. Bell, Physics (N.Y.) 1, 195 (1965).

[3]J. F. Clauser and A. Shimony, Rep. Prog. Phys. 41, 1981 (1978). (This paper is an exhaustive review of the subject); F. M. Pipkin, in *Advances in Atomic and Molecular Physics*, edited by D. R. Bates and B. Bederson (Academic, New York, 1978) (a comprehensive review).

[4]A. Aspect, P. Grangier, and G. Roger, Phys. Rev. Lett. 47, 460 (1981).

[5]A. Aspect, P. Grangier, and G. Roger, Phys. Rev. Lett. 49, 91 (1982).

[6]D. Bohm and Y. Aharonov, Phys. Rev. 108, 1070 (1957). The suggestion of this thought-experiment was already given by D. Bohm, Ref. 1.

[7]A. Aspect, Phys. Lett. 54A, 117 (1975), and Phys. Rev. D 14, 1944 (1976).

[8]J. F. Clauser, M. A. Horne, A. Shimony, and R. A. Holt, Phys. Rev. Lett. 23, 880 (1969).

[9]A. Yariv, *Quantum Electronics* (Wiley, New York, 1975).

[10]Let us emphasize that such results cannot be taken as providing the possiblity of faster-than-light communication. See, for instance, A. Aspect, J. Phys. (Paris), Colloq. 42, C263 (1981).

AUTHOR INDEX

This index contains all of the articles in this volume and on the CD-ROM. Articles that can be found on the CD-ROM only are indicated in italics by the letters "CD" followed by the year(s) of publication in parentheses; articles that appear in the book are indicated by page numbers followed by the year(s) of publication in parentheses. All articles printed in the book are also present on the CD-ROM.

AUTHOR INDEX

C

AUTHOR INDEX

AUTHOR INDEX

H

N

O

P

Q

R

S

AUTHOR INDEX

NAME LIST

A

Elihu Abrahams
Yakir Aharonov
Guenter Ahlers
Berni Julian Alder
Hannes Alfvén
Ralph Asher Alpher
Luis Walter Alvarez
Carl David Anderson
John D. Anderson
Philip Warren Anderson
Vladimir I. Arnold
Taro Asano
Arthur Ashkin
Alain Aspect

B

Robert Fox Bacher
Kenneth Tompkins Bainbridge
George Allen Baker, Jr.
George C. Baldwin
Elizabeth Urey Baranger
Michael N. Barber
John Bardeen
Samuel J. Barnett
Henry Herman Barschall
Carl Barus
Rodney J. Baxter
Jesse Beams
Antoine Henri Becquerel
Frederick Bedell
John S. Bell
Giuseppe Benfatto
William Ralph Bennett
Peter Bergmann
Ted H. Berlin
Hans Albrecht Bethe
Homi Bhabha
Gerd Binnig
Raymond Thayer Birge
Garrett Birkhoff
James D. Bjorken
Patrick Maynard Blackett
Frederic Columbus Blake
John Markus Blatt
Ilan Blech
Felix Bloch
Katharine Blodgett
Nicolaas Bloembergen
Jacques Blons
Nikolai N. Bogolyubov
David Bohm
Aage Bohr
Niels Bohr
Ludwig Boltzmann
Roberto Bonetti
Walther Wilhelm Bothe
Louis P. Bouckaert
Richard Theodore Brackmann
William Henry Bragg
William Lawrence Bragg
Carl H. Brans
Walter Houser Brattain
Gregory Breit
Ferdinand G. Brickwedde
Percy Williams Bridgman
Léon Nicolas Brillouin
David Allan Bromley
Gerald E. Brown
Laurie Brown
Keith Brueckner
Clifford C. Butler
Stuart Thomas Butler

C

Walter Guyton Cady
John Werner Cahn
Curtis G. Callan
Herbert Bernard Callen
John Franklin Carlson
Hendrik Brugt Gerhard Casimir
B. Cassen
Pavel Aleksejevič Čerenkov
James Chadwick
Owen Chamberlain
Leroy Chang
Katherine L. Chen
Clement Dexter Child
Boris Chirikov
James H. Christenson
John Francis Clauser
Jacob Clay
William Weber Coblentz
John D. Cockcroft
Keith Codling
Stirling Auchincloss Colgate
Arthur Holly Compton
Karl Talylor Compton
Edward Uhler Condon
Marcello Conversi
William E. Cooke
William David Coolidge
Leon N. Cooper
Bruce Cork
André Frédéric Cournand
Clyde L. Cowan
Stuart J. B. Crampton
Horace Richard Crane
Charles Louis Critchfield
James Watson Cronin

D

Thibaut Damour
Karl Kelchner Darrow
Raymond Davis
Clinton Joseph Davisson
John Myrick Dawson
Petrus Debye
Hans Georg Dehmelt
Arthur Jeffrey Dempster
David Mathias Dennison
Martin Deutsch
Carlo Di Castro
Robert Henry Dicke
Paul Adrien Maurice Dirac
Cyril Domb
Gerhard Dorda
Lee A. Du Bridge
Freeman John Dyson

E

Carl Eckart
Arthur Stanley Eddington
Thomas Alva Edison
Paul Ehrenfest
Albert Einstein
Leonard Eisenbud
Philip A. Ekstrom
Guy T. Emery
Torleif Ericson
Leo Esaki
Immanuel Estermann
Hans Euler

F

Ugo Fano
Eugene Feenberg
Mitchell Feigenbaum
Enrico Fermi
Herman Feshbach
Richard Phillips Feynman
Michael Ellis Fisher
Lennard A. Fisk
Val Logsdon Fitch
Wade Lanford Fite
Georgii Nikolaevich Flerov
Vladimir I. Fock
Leslie L. Foldy
Henry Michael Foley
Scott E. Forbush
Joseph Ford
A. Theodore Forrester
Claude Russell Fountain
James Franck
Il'ya Mikhailovich Frank
Daniel Friedan
Herbert Friedman
Otto Robert Frisch
Alessandro Carlo Fubini

G

Thomas Francis Gallagher
Giovanni Gallavotti
George Anthony Gamow
Eugene Gardner
Jagadesh Bahari Garg
Richard Laurence Garwin
Murray Gell-Mann
Christoph Gerber
Edward Gerjuoy
Lester Halbert Germer
Gian Carlo Ghirardi
Ivar Giaever
Josiah Willard Gibbs
Julian Howard Gibbs
Donald Arthur Glaser
Sheldon Lee Glashow
Roy Jay Glauber
Marvin Leonard Goldberger
Gerson Goldhaber
Maurice Goldhaber
Sheldon Goldstein
Jeffrey Goldstone
Jerry Paul Gollub
James Power Gordon
Arthur Charles Gossard
Kurt Gottfried
Samuel Abraham Goudsmit
Denis Gratias
Melville S. Green
Robert Budington Griffiths
T. H. Gronwall
David Jonathan Gross
Richard A. Gudmundsen
Ronald W. Gurney

H

Lawrence R. Hafstad
Bertrand Halperin
Theodor Wolfgang Hänsch
William W. Hansen
James Burkett Hartle
Paul Hartman
Douglas Hartree
Walter Hauser
Carl Heiles
Werner Heisenberg
Walter Heitler
Michel Hénon
Robert Herman
Donald R. Herriott
Victor Franz Hess
Georg Hevesy
Norman Paulson Heydenburg
Peter W. Higgs
Lillian Hoddeson
Robert Hofstadter
Pierre Claude Hohenberg
Theodore T. Holstein
Richard Arnold Holt
Michael A. Horne
Vernon Willard Hughes
Albert Wallace Hull
Gordon Ferrie Hull
Friedrich Hund

I

John Iliopoulos
Herbert Eugene Ives
Dimitry D. Iwanenko

J

John David Jackson
Robert C. Jaklevic
Robert Jastrow
Ali Javan
James Hopwood Jeans
J. Hans Daniel Jensen
Montgomery Hunt Johnson
Philip O. Johnson
Thomas Hope Johnson
Giovanni Jona-Lasinio
Brian Josephson
Einar Juliusson

K

Mark Kac
Leo P. Kadanoff
Hartmut Kallmann
Richard Kaufmann
Jerome Meryl Blake Kellogg
Edwin C. Kemble
E. F. Kingsbury
John Gamble Kirkwood
G. Stanley Klaiber
Daniel Kleppner
Walter Kohn
Werner Kolhörster
Andrei M. Kolmogorov
Steven Elliot Koonin
Benzion Kozlovsky
Robert Kraichnan
Hendrik A. Kramers
Ralph de Laer Kronig
Martin David Kruskal
Ryugo Kubo
Franz Newell Devereux Kurie
Polykarp Kusch

L

Willis Eugene Lamb
John Lambe
Lev Davydovich Landau
Donald Newton Langenberg
Irving Langmuir
Cesare M. G. Lattes
Robert Betts Laughlin
Ernest Orlando Lawrence
Joel L. Lebowitz
Leon Lederman
Tsung Dao Lee
Andrew Lenard
Louis Leprince-Ringuet
Joseph S. Levinger
Harold Warren Lewis
Willard Frank Libby
Albert Joseph Libchaber
Donald C. Licciardello
Elliot Lieb
Bernard Abram Lippmann
Milton Stanley Livingston
Edward Norton Lorenz
Francis Low
Aleksandr M. Lyapunov
Theodore Lyman

M

Shang-keng Ma
Stefan Machlup
Robert Phyfe Madden
Luciano Maiani
Theodore Harold Maiman
Stanley Mandelstam
Paul Manneville
Robert Eugene Marshak
Paul Cecil Martin
Ladislaus Laszlo Marton
Daniel Charles Mattis
Jean Maurer
Maria Goeppert Mayer

NAME LIST

James Edgar Mercereau
N. David Mermin
Ernest George Merritt
Nicolas Metropolis
Peter Meyer
André Michaudon
Albert Abraham Michelson
Marcel Migeotte
Robert Andrews Millikan
Sidney Millman
Robert Laurence Mills
Philip McCord Morse
Jürgen Kurt Moser
Ben R. Mottelson
Duane O. Muhleman
Dietrich Müller
Robert Sanderson Mulliken
George M. Murphy

N

Seth H. Neddermeyer
John William Negele
David Robert Nelson
Werner Neuhauser
Edward Leamington Nichols
Ernest Fox Nichols
Alfred Otto Carl Nier
Sven Gösta Nilsson
Kenneth L. Nordvedt

O

Thomas F. O'Malley
Roland Omnès
Lars Onsager
Jacob Robert Oppenheimer

P

Martin E. Packard
Henri E. Padé
Abraham Pais
Ettore Pancini
Wolfgang K. H. Panofsky
Eugene Newman Parker
William Henry Parker
William E. Parkins
John Pasta
Simon Pasternack
Wolfgang Pauli
Linus Carl Pauling
Daniel Paya
Philip Pearle
P. James E. Peebles
Rudolf Ernst Peirels
Chaim L. Pekeris
Oliver Penrose
Michael Pepper
Donald B. Perkins
Martin Lewis Perl
Konstantin Antonovich Petrjak
Arthur Van Rensselaer Phelps
Oreste Piccioni
Robert W. Pidd
George Washington Pierce
David Pines
Max Planck
Boris Podolsky
Henri Poincaré
Dirk Polder
H. David Politzer
Alexander M. Polyakov
Yves Pomeau
Isaac Yakovlevich Pomeranchuk
Charles E. Porter
Robert Vivian Pound
Cecil Frank Powell
Richard D. Present
Ilya Prigogine
Edward Mills Purcell

Q

Zongan Qiu

R

Isidore Isaac Rabi
Giulio Racah
T. V. Ramakrishnan
Reuven Ramaty
Norman Foster Ramsey
William Rarita
Glen A. Rebka
Erich Rudolph Alexander Regener
Tullio Regge
Frederick Reines
Robert C. Retherford
Arthur Rich
Lewis Richardson
Owen Williams Richardson
Burton Richter
Alberto Rimini
George D. Rochester
Wilhelm Conrad Roentgen
Heinrich Rohrer
Vladimir A. Rokhlin
Nathan Rosen
Leonard Rosenberg
Arianna W. Rosenbluth
Marshall N. Rosenbluth
Bruno Rossi
John M. Rowell
John Rowlinson
David Ruelle
Henry Norris Russell
Ernest Rutherford

S

Abdus Salam
Arthur Leonard Schawlow
Leonard Isaac Schiff
John Robert Schrieffer
Erwin Schrödinger
Lawrence S. Schulman
George Schulz
Arthur A. Schupp
Melvin Schwartz
Silvan Samuel Schweber
Paul B. Schwinberg
Julian Seymour Schwinger
Marlan Orvil Scully
Glenn Theodore Seaborg
George A. Seielstad
Frederick Seitz
Irwin I. Shapiro
Dan Shechtman
Stephen Shenker
Abner Eliezer Shimony
William Shockley
Clifford Glenwood Shull
Carl Ludwig Siegel
Arnold Herbert Silver
John Alexander Simpson
Oliver Cecil Simpson
Yakov G. Sinai
Dmitry V. Skobeltzyn
John Clarke Slater
Stephen Smale
J. Samuel Smart
Phillip T. Smith
Roman Smoluchowski
Henry DeWolf Smyth
Larry Spruch
Jack Steinberger
Charles Proteus Steinmetz
Otto Stern
Edward C. Stevenson
Horst Ludwig Stormer
Jabez Curry Street
Hinko Henry Stroke
Andrew W. Sunyar
Harry Leonard Swinney
Kurt Symanzik

T

Floris Takens
Igor' Evgen'evich Tamm
John Torrence Tate
Barry Norman Taylor
John Bradshaw Taylor
Joseph Hooton Taylor
J. D. Tear
Augusta H. Teller
Edward Teller
Robert G. Thomas
Joseph John Thomson
Gerardus 't Hooft
Samuel Chao Chung Ting
Michael Tinkham

NAME LIST

William Tobocman
Richard Chase Tolman
Sin-itiro Tomonaga
Henry Cutler Torrey
Charles Hard Townes
Sam Bard Treiman
George Henry Trilling
Daniel Chee Tsui
René Turlay
Merle Anthony Tuve

U

Edwin A. Uehling
George Eugene Uhlenbeck
Stanislaw Ulam
Harold Clayton Urey

V

Joseph Valasek
James Alfred Van Allen
Robert Jemison Van de Graaff
Robert S. Van Dyck
Léon Charles Prudent van Hove
Nico van Kampen
John Hasbrouck Van Vleck
Robert F. C. Vessot
Klaus von Klitzing
John von Neumann
Carl F. von Weizsäcker

W

Herbert Wagner
Thomas E. Wainright
Ernest Thomas Sinton Walton
Gregory H. Wannier
Gleb Wataghin
Tullio Weber
Arthur Gordon Webster
Franz J. Wegner
Edmund Weibel
Steven Weinberg
Victor Frederick Weisskopf
Theodore E. Welton
John Archibald Wheeler
Milton Grandison White
Benjamin Widom
Eugene Paul Wigner
Frank Wilczek
Clifford Will
James Gerard Williams
Lawrence A. Wills
Kenneth Wilson
David Jeffrey Wineland
Richard Wolfgang
Calvin Wong
Robert Williams Wood
Louis F. Wouters
Siegfried A. Wouthuysen
Chien-Shiung Wu

Y

Chen-Ning Yang
Lloyd A. Young
Horace P. Yuen
Hideki Yukawa

Z

Norman Julius Zabusky
Jerrold Reinach Zacharias
Herbert J. Zeiger
George Zweig